The chemistry of

the cyclopropyl group

Volume 2

THE CHEMISTRY OF FUNCTIONAL GROUPS

A series of advanced treatises under the general editorship of
Professors Saul Patai and Zvi Rappoport

The chemistry of alkenes (2 volumes)
The chemistry of the carbonyl group (2 volumes)
The chemistry of the ether linkage
The chemistry of the amino group
The chemistry of the nitro and nitroso groups (2 parts)
The chemistry of carboxylic acids and esters
The chemistry of the carbon–nitrogen double bond
The chemistry of amides
The chemistry of the cyano group
The chemistry of the hydroxyl group (2 parts)
The chemistry of the azido group
The chemistry of acyl halides
The chemistry of the carbon–halogen bond (2 parts)
The chemistry of the quinonoid compounds (2 volumes, 4 parts)
The chemistry of the thiol group (2 parts)
The chemistry of the hydrazo, azo and azoxy groups (2 parts)
The chemistry of amidines and imidates (2 volumes)
The chemistry of cyanates and their thio derivatives (2 parts)
The chemistry of diazonium and diazo groups (2 parts)
The chemistry of the carbon–carbon triple bond (2 parts)
The chemistry of ketenes, allenes and related compounds (2 parts)
The chemistry of the sulphonium group (2 parts)
Supplement A: The chemistry of double-bonded functional groups (2 volumes, 4 parts)
Supplement B: The chemistry of acid derivatives (2 volumes, 4 parts)
Supplement C: The chemistry of triple-bonded functional groups (2 volumes, 3 parts)
Supplement D: The chemistry of halides, pseudo-halides and azides (2 volumes, 4 parts)
Supplement E: The chemistry of ethers, crown ethers, hydroxyl groups
and their sulphur analogues (2 volumes, 3 parts)
Supplement F: The chemistry of amino, nitroso and nitro compounds and their derivatives (2 parts)
The chemistry of the metal–carbon bond (5 volumes)
The chemistry of peroxides
The chemistry of organic selenium and tellurium compounds (2 volumes)
The chemistry of the cyclopropyl group (2 volumes, 3 parts)
The chemistry of sulphones and sulphoxides
The chemistry of organic silicon compounds (2 parts)
The chemistry of enones (2 parts)
The chemistry of sulphinic acids, esters and their derivatives
The chemistry of sulphenic acids and their derivatives
The chemistry of enols
The chemistry of organophosphorus compounds (3 volumes)
The chemistry of sulphonic acids, esters and their derivatives
The chemistry of alkanes and cycloalkanes
Supplement S: The chemistry of sulphur-containing functional groups
The chemistry of organic arsenic, antimony and bismuth compounds
The chemistry of enamines (2 parts)
The chemistry of organic germanium, tin and lead compounds

UPDATES

The chemistry of α-haloketones, α-haloaldehydes and α-haloimines
Nitrones, nitronates and nitroxides
Crown ethers and analogs
Cyclopropane derived reactive intermediates
Synthesis of carboxylic acids, esters and their derivatives
The silicon–heteroatom bond
Syntheses of lactones and lactams
The syntheses of sulphones, sulphoxides and cyclic sulphides

Patai's 1992 guide to the chemistry of functional groups—*Saul Patai*

The chemistry of
the cyclopropyl group

Volume 2

Edited by
ZVI RAPPOPORT
The Hebrew University, Jerusalem

1995

JOHN WILEY & SONS
CHICHESTER–NEW YORK– BRISBANE–TORONTO–SINGAPORE
An Interscience® Publication

Telephone: National 01243 779777
International (+44) 1243 779777

Other Wiley Editorial Offices

John Wiley & Sons, Inc., 605 Third Avenue,
New York, NY 10158-0012, USA

Jacaranda Wiley Ltd, 33 Park Road, Milton,
Queensland 4064, Australia

John Wiley & Sons (Canada) Ltd, 22 Worcester Road,
Rexdale, Ontario M9W 1L1, Canada

John Wiley & Sons (SEA) Pte Ltd, 37 Jalan Pemimpin #05-04,
Block B, Union Industrial Building, Singapore 2057

Library of Congress Cataloging-in-Publication Data

The Chemistry of the cyclopropyl group.

(The Chemistry of functional groups)
'An Interscience publication.'
1. Cyclopropane. I. Rappoport, Zvi. II. Title:
Cyclopropyl group. III. Series.
QD305.H9 C48 1987 547'.511 87-10440
ISBN 0 471 90658 1 (Part 1)
ISBN 0 471 91737 0 (Part 2)
ISBN 0 471 91738 9 (Set)

British Library Cataloguing in Publication Data

A catalogue record for this book is available from the British Library

ISBN 0 471 94074 7

Typeset in 9/10 pt Times by Thomson Press (India) Ltd, New Delhi, India.
Printed and bound in Great Britain by Biddles Ltd, Guildford, Surrey
This book is printed on acid-free paper responsibly manufactured from sustainable forestation, for which at least two trees are planted for each one used for paper production.

To
Shinji Kobayashi

My eyes, ears and mouth in Japan

Contributing authors

John E. Baldwin	Department of Chemistry, Syracuse University, Room 1–014, Center for Science and Technology, Syracuse, New York 13244–4100, USA
Ronald F. Childs	Department of Chemistry, McMaster University, 1280 Main Street West, Hamilton, Ontario L8S 4M1, Canada
Dieter Cremer	Department of Theoretical Chemistry, University of Göteborg, Kemigarden 3, S-41296 Göteborg, Sweden
George Elia	Department of Chemistry, McMaster University, 1280 Main Street West, Hamilton, Ontario L8S 4M1, Canada
Zeev Goldschmidt	Department of Chemistry, Bar–Ilan University, Ramat-Gan 52900, Israel
Brian Halton	Department of Chemistry, Victoria University of Wellington, P.O. Box 600, Wellington, New Zealand
Piotr Kaszynski	Department of Chemistry, Vanderbilt University, Nashville, Tennessee 37235, USA
Elfi Kraka	Department of Theoretical Chemistry, University of Göteborg, Kemigarden 3, S-41296 Göteborg, Sweden
Joel F. Liebman	Department of Chemistry and Biochemistry, University of Maryland, Baltimore County Campus, 5401 Wilkens Avenue, Baltimore, Maryland 21228–5398, USA
Kirill A. Lukin	Department of Chemistry, The University of Chicago, 5735 South Ellis Avenue, Chicago, Illinois 60637, USA
M. Anthony McKervey	School of Chemistry, The Queen's University of Belfast, David Keir Building, Belfast, BT9 5AG, Northern Ireland, UK
Josef Michl	Department of Chemistry and Biochemistry, University of Colorado at Boulder, Campus Box 215, Boulder, Colorado 80309–0215, USA
Shinya Nishida	Department of Chemistry, Faculty of Science, Hokkaido University, Sapporo 060, Japan
Masakazu Ohkita	Department of Chemistry, Faculty of Science, Hokkaido University, Sapporo 060, Japan
George A. Olah	Loker Hydrocarbon Research Institute and Department of Chemistry, University of Southern California, Los Angeles, California 90089–1661, USA
G. K. Surya Prakash	Loker Hydrocarbon Research Institute and Department of Chemistry, University of Southern California, Los Angeles, California 90089–1661, USA

Prakash V. Reddy	Loker Hydrocarbon Research Institute and Department of Chemistry, University of Southern California, Los Angeles, California 90089–1661, USA
Béla Rozsondai	Eötvös University Budapest, Structural Chemistry Research Group, Hungarian Academy of Sciences, Puskin u. 11–13, P.O. Box 117, H–1088 Budapest, Hungary
Kalman Josef Szabo	Department of Theoretical Chemistry, University of Göteborg, Kemigarden 3, S-41296 Göteborg, Sweden
Takashi Tsuji	Department of Chemistry, Faculty of Science, Hokkaido University, Sapporo 060, Japan
Tao Ye	School of Chemistry, The Queen's University of Belfast, David Keir Building, Belfast, BT9 5AG, Northern Ireland, UK
Nikolai S. Zefirov	Department of Chemistry, Moscow State University, Moscow, Russia
Howard E. Zimmerman	Department of Chemistry, University of Wisconsin Madison, 1101 University Avenue, Madison, Wisconsin 53706, USA

Foreword

The two earlier parts of *The Chemistry of the Cyclopropyl Group* appeared in 1987, and an offspring volume *Cyclopropane Derived Reactive Intermediates* by G. Boche and H. M. Walborsky appeared in the Updates series in 1990. The subject has advanced considerably in the last eight years and the present volume contains new material on recent advances in topics covered in the previous volume such as the synthesis of cyclopropanes, their preparation by metal catalysed reactions, their stereomutations and thermochemistry, and [1.1.1]propellanes. The previous chapter dealing with theory and structural chemistry was widely extended to two chapters and description of the long-lived observable cyclopropylcarbinyl cations supplement the earlier treatment of their solvolytic generation. The book includes several chapters on topics which were not, or were only marginally, covered earlier such as cycloproparenes, photochemistry of cyclopropanes, spiroannulated cyclopropanes and organometallic derivatives.

There are three differences between this and previously published volumes in the series. First, two extensive chapters deal with cyclopropyl homoconjugation from both the experimental and the theoretical points of view. Whereas homoconjugation does not necessarily involve cyclopropyl rings and the non-cyclic structure can be of major importance, we believe that an in-depth discussion of this topic fits a volume devoted to the cyclopropyl group. Second, the chapter on photochemistry mainly gives an account of the contributions of its author, a pioneer in the field, rather than a comprehensive review. Third, the book opens with a chapter related to the history of cyclopropanes before 1916, for which we reproduce here the 'historical' chapter from the 1916 PhD Thesis of J. B. Conant, 'A study of certain cyclopropane derivatives'. We hope that presenting the facsimile of the original thesis which retains the original 'trimethylene' nomenclature will add the historical flavour and at the same time will emphasize the progress made in the field during this century. Inclusion of this chapter was suggested by Professor J. D. Roberts, and Professor F. H. Westheimer was of great help in obtaining the permission of the Harvard Archives and the Conant family to print this chapter. My sincere thanks are due to all of them.

The literature coverage is mostly up to 1994.

Editing of the book started in Jerusalem and continued in a large part in Fukuoka, Japan, where I spent the last year as a visitor in the Institute for Fundamental Research of Organic Chemistry, Kyushu University. Thanks are due to members of the Institute for showing me the well-known Japanese hospitality during the period. It is also a pleasure to congratulate one of the authors, Prof. G. A. Olah, for receiving the 1994 Nobel prize for Chemistry.

I will be grateful to readers who will call my attention to mistakes or omissions in the present volume, as well as in other volumes of the Functional Groups Series.

Fukuoka
May 1995

ZVI RAPPOPORT

The Chemistry of Functional Groups
Preface to the series

The series 'The Chemistry of Functional Groups' was originally planned to cover in each volume all aspects of the chemistry of one of the important functional groups in organic chemistry. The emphasis is laid on the preparation, properties and reactions of the functional group treated and on the effects which it exerts both in the immediate vicinity of the group in question and in the whole molecule.

A voluntary restriction on the treatment of the various functional groups in these volumes is that material included in easily and generally available secondary or tertiary sources, such as Chemical Reviews, Quarterly Reviews, Organic Reactions, various 'Advances' and 'Progress' series and in textbooks (i.e. in books which are usually found in the chemical libraries of most universities and research institutes), should not, as a rule, be repeated in detail, unless it is necessary for the balanced treatment of the topic. Therefore each of the authors is asked not to give an encyclopaedic coverage of his subject, but to concentrate on the most important recent developments and mainly on material that has not been adequately covered by reviews or other secondary sources by the time of writing of the chapter, and to address himself to a reader who is assumed to be at a fairly advanced postgraduate level.

It is realized that no plan can be devised for a volume that would give a complete coverage of the field with no overlap between chapters, while at the same time preserving the readability of the text. The Editors set themselves the goal of attaining reasonable coverage with moderate overlap, with a minimum of cross-references between the chapters. In this manner, sufficient freedom is given to the authors to produce readable quasi-monographic chapters.

The general plan of each volume includes the following main sections:

(a) An introductory chapter deals with the general and theoretical aspects of the group.

(b) Chapters discuss the characterization and characteristics of the functional groups. i.e. qualitative and quantitative methods of determination including chemical and physical methods, MS, UV, IR, NMR, ESR and PES—as well as activating and directive effects exerted by the group, and its basicity, acidity and complex-forming ability.

(c) One or more chapters deal with the formation of the functional group in question, either from other groups already present in the molecule or by introducing the new group directly or indirectly. This is usually followed by a description of the synthetic uses of the group, including its reactions, transformations and rearrangements.

(d) Additional chapters deal with special topics such as electrochemistry, photochemistry, radiation chemistry, thermochemistry, syntheses and uses of isotopically labelled compounds, as well as with biochemistry, pharmacology and toxicology. Whenever applicable, unique chapters relevant only to single functional groups are also included (e.g. 'Polyethers', 'Tetraaminoethylenes' or 'Siloxanes').

This plan entails that the breadth, depth and thought-provoking nature of each chapter will differ with the views and inclinations of the authors and the presentation will necessarily be somewhat uneven. Moreover, a serious problem is caused by authors who deliver their manuscript late or not at all. In order to overcome this problem at least to some extent, some volumes may be published without giving consideration to the originally planned logical order of the chapters.

Since the beginning of the Series in 1964, two main developments have occurred. The first of these is the publication of supplementary volumes which contain material relating to several kindred functional groups (Supplements A, B, C, D, E, F and S). The second ramification is the publication of a series of 'Updates', which contain in each volume selected and related chapters, reprinted in the original form in which they were published, together with an extensive updating of the subjects, if possible, by the authors of the original chapters. A complete list of all mentioned volumes published to date will be found on the page opposite the main title page of this book. Unfortunately the publication of the 'Updates' had to be discontinued for economic reasons.

Advice or criticism regarding the plan and execution of this series will be welcomed by the Editors.

The publication of this series would never have been started, let alone continued, without the support of many persons in Israel and overseas, including colleagues, friends and family. The efficient and patient co-operation of staff-members of the publisher also rendered us invaluable aid. Our sincere thanks are due to all of them.

The Hebrew University
Jerusalem, Israel

SAUL PATAI
ZVI RAPPOPORT

Contents

1. Prologue: History of cyclopropane derivatives before 1916 — 1
   James Bryant Conant
2. General and theoretical aspects of the cyclopropyl group — 43
   Dieter Cremer, Elfi Kraka and Kalman Josef Szabo
3. Structural chemistry of cyclopropane derivatives — 139
   Béla Rozsondai
4. Interrelations in the thermochemistry of cyclopropanes — 223
   Joel F. Liebman
5. Recent advances in synthesis of cyclopropanes — 261
   Masakazu Ohkita, Shinya Nishida and Takashi Tsuji
6. Cyclopropane photochemistry — 319
   Howard E. Zimmerman
7. Cyclopropyl homoconjugation, homoaromaticity and homoantiaromaticity: Theoretical aspects and analysis — 339
   Dieter Cremer, Ronald F. Childs and Elfi Kraka
8. Cyclopropyl homoconjugation: Experimental facts and interpretations — 411
   Ronald F. Childs, Dieter Cremer and George Elia
9. Thermal stereomutations of cyclopropanes and vinylcyclopropanes — 469
   John E. Baldwin
10. Organometallic derivatives of cyclopropanes and their reactions — 495
    Zeev Goldschmidt
11. Metal catalysed cyclopropanation — 657
    Tao Ye and M. Anthony McKervey
12. Cycloproparenes — 707
    Brian Halton
13. [1.1.1.]Propellanes — 773
    Piotr Kaszynski and Josef Michl
14. Long-lived cyclopropylcarbinyl cations — 813
    George A. Olah, Prakash V. Reddy and G. K. Surya Prakash

15. Spiroannulated cyclopropanes 861
Kirill A. Lukin and Nikolai S. Zefirov

Author index 887

Subject index 947

List of abbreviations used

Ac	acetyl (MeCO)
acac	acetylacetone
Ad	adamantyl
AIBN	azoisobutyronitrile
Alk	alkyl
All	allyl
An	anisyl
Ar	aryl
Bz	benzoyl (C_6H_5CO)
Bu	butyl (also *t*-Bu or Bu^t)
CD	circular dichroism
CI	chemical ionization
CIDNP	chemically induced dynamic nuclear polarization
CNDO	complete neglect of differential overlap
Cp	η^5-cyclopentadienyl
Cp*	η^5-pentamethylcyclopentadienyl
DABCO	1,4-diazabicyclo[2.2.2]octane
DBN	1,5-diazabicyclo[4.3.0]non-5-ene
DBU	1,8-diazabicyclo[5.4.0]undec-7-ene
DIBAH	diisobutylaluminium hydride
DME	1,2-dimethoxyethane
DMF	*N*,*N*-dimethylformamide
DMSO	dimethyl sulphoxide
ee	enantiomeric excess
EI	electron impact
ESCA	electron spectroscopy for chemical analysis
ESR	electron spin resonance
Et	ethyl
eV	electron volt
Fc	ferrocenyl
FD	field desorption
FI	field ionization

FT	Fourier transform
Fu	furyl (OC_4H_3)
GLC	gas liquid chromatography
Hex	hexyl (C_6H_{13})
c-Hex	cyclohexyl (C_6H_{11})
HMPA	hexamethylphosphortriamide
HOMO	highest occupied molecular orbital
HPLC	high performance liquid chromatography
i-	iso
Ip	ionization potential
IR	infrared
ICR	ion cyclotron resonance
LAH	lithium aluminium hydride
LCAO	linear combination of atomic orbitals
LDA	lithium diisopropylamide
LUMO	lowest unoccupied molecular orbital
M	metal
M	parent molecule
MCPBA	*m*-chloroperbenzoic acid
Me	methyl
MNDO	modified neglect of diatomic overlap
MS	mass spectrum
n	normal
Naph	naphthyl
NBS	*N*-bromosuccinimide
NCS	*N*-chlorosuccinimide
NMR	nuclear magnetic resonance
Pc	phthalocyanine
Pen	pentyl (C_5H_{11})
Pip	piperidyl ($C_5H_{10}N$)
Ph	phenyl
ppm	parts per million
Pr	propyl (also *i*-Pr or Pr^i)
PTC	phase transfer catalysis or phase transfer conditions
Pyr	pyridyl (C_5H_4N)
R	any radical
RT	room temperature
s-	secondary
SET	single electron transfer
SOMO	singly occupied molecular orbital

t-	tertiary
TCNE	tetracyanoethylene
TFA	trifluoroacetic acid
THF	tetrahydrofuran
Thi	thienyl (SC_4H_3)
TLC	thin layer chromatography
TMEDA	tetramethylethylene diamine
TMS	trimethylsilyl or tetramethylsilane
Tol	tolyl (MeC_6H_4)
Tos or Ts	tosyl (*p*-toluenesulphonyl)
Trityl	triphenylmethyl (Ph_3C)
Xyl	xylyl ($Me_2C_6H_3$)

In addition, entries in the 'List of Radical Names' in *IUPAC Nomenclature of Organic Chemistry*, 1979 Edition, Pergamon Press, Oxford, 1979, p. 305–322, will also be used in their unabbreviated forms, both in the text and in formulae instead of explicitly drawn structures.

CHAPTER 1

Prologue: History of cyclopropane derivatives before 1916

JAMES BRYANT CONANT

Department of Chemistry, Harvard University, Cambridge, Massachusetts, USA

This is the first chapter from the PhD Thesis of James Bryant Conant 'A Study of Certain Cyclopropane Derivatives' which was presented to the Faculty of Arts and Sciences, Harvard University on May 1st, 1916, and reprinted here with the permission of Harvard Archives and the Conant family (see also Foreword).

PART I.

In 1881 Freund[1] prepared the hydrocarbon cyclopropane by the action of sodium on trimethylene bromide, $CH_2Br-CH_2-CH_2Br$. This was the first synthesis of a three-membered carbon ring, although Reboul[2] had tried previously to prepare this hydrocarbon by the same method, but had failed. Gustavson[3], shortly after Freund's discovery, found that zinc in a medium of alcohol gave much better results than sodium. The discoverer of this new gas, at that time called trimethylene, showed it to be an isomer of the known hydrocarbon propylene. Subsequent investigators confirmed this and pointed out that while many of its reactions were remarkably like propylene, it was not, however, oxidized by potassium permanganate solution[4]. The ease with which the new isomer reacted with dilute sulphuric acid and fuming hydriodic acid was quite remarkable; and it slowly combined with bromine in the presence of sunlight. The products of these reactions were normal propyl alcohol, $CH_3CH_2CH_2OH$, (by boiling the sulphuric acid solution with water any sulphate, $CH_3CH_2CH_2SO_4H$, was changed to alcohol), normal propyl iodide and trimethylene bromide, respectively. With the exception of the action with potassium permanganate all these reactions are very similar to those of ethylene and its homologues. In fact, this great similarity in reactions between the isomers propylene and trimethylene, together with some thermochemical data, led Berthelot[1] to consider them as "dynamic isomers", differing only in their content of chemical energy.

In spite of the analogy in the reactions of propylene and trimethylene there seems to have been no doubt in the minds of these early investigators (with perhaps the exception of Berthelot) of the cyclic structure of the compound prepared from trimethylene bromide. This was probably because it is impossible, if one accepts the ordinary views of structural organic chemistry, to write any formula for trimethylene other than that of a three-membered ring. Its addition reactions must then involve ring cleavage and consequently the ring in this, the simplest of cyclopropane compounds, must be considered

[1] Monat., 2, 642; 3, 625. J. Pr. C., 26, 367.
[2] C. R., 76, 1775.
[3] J. Pr. C., 36, 300; 50, 382.
[4] B., 21, 1236, 1282.

[1] Ann. Chim. & Phys., 4, (7) 107.

as quite unstable.

More recent researches have brought to light more facts concerning trimethylene. At 400°C, the gas is changed largely into its isomer propylene.[1] This process of isomerization, though at first disputed,[2] has since been proved beyond doubt,[3] and moreover it has been shown that very much lower temperatures will cause the rearrangement if a catalyst is used. By the use of finely divided platinum[4] a temperature of only 100° need be employed, while zinc chloride also causes the change to take place at moderate temperatures.[5] Recently it has been found that certain gaseous reactions such as the oxidation of nitric oxide to nitrogen peroxide cause trimethylene to change into propylene even at ordinary temperatures.[6] This case of so-called resonance is very similar to that discovered by Skraup[7] in the transformation of maleic into fumaric acid; whatever the true explanation of this may be it will apply equally to these two very different cases. In spite of this apparent ease of

1. B., 29, 1297.
2. B., 31, 3067.
3. B., 32, 702, 1965.
4. B., 36, 2014; Z.P..C., 41, 735.
5. Ann. chim. & phys., 20, (7) 27.
6. Z., '08, II, 1424.
7. Monat., 12, 107.

rearrangement neither long standing (eight years) at ordinary temperatures[1] nor the action of light[2] causes any isomerization of cyclopropane. When subjected to modern catalytic reduction trimethylene reduces[3] at a temperature of 80° - 120° (using nickel as the catalyst); cyclobutane was similarly reduced at a somewhat higher temperature, ethylene at a lower one. A peculiar reaction, noted by Gustavson[4], may be mentioned; in the presence of aluminum chloride trimethylene reacts with benzene to give propyl and isopropyl benzene.

The reactions of the halogens on trimethylene lead to a large number of products which depend on the conditions of the experiment. Chlorine acts in the presence of sunlight to give principally a monochlor substitution product,[5] and a little dichlor compound which probably has both chlorine atoms attached to the same carbon atom. It was later found that besides these ring compounds several straight chain chlorides could be obtained[6] in small amounts, the principal one being trimethylene chloride, $CH_2Cl-CH_2-CH_2Cl$. Chlorine acts thus in

1. Ann. chim. & phys., 20, (7) 27.
2. Z.P..C., 41, 735.
3. B., 40, 4457.
4. J.Pr.C., 50, 382 (note).
5. J.Pr.C., 42, 495; 43, 396.
6. J.Pr.C., 50, 380.

two ways; principally substituting a hydrogen on the ring and partially adding and breaking the ring to give a straight chain compound.

Bromine unlike chlorine has almost no substituting action and though the speed of its reaction and the products obtained vary greatly with the conditions, in no case has a cyclopropane bromide been found[1]. A complete study of the reaction of bromine and trimethylene has been carried out by Gustavson[2] and the essential points of his investigations will be briefly summarized. In the dark the reaction is very slow, but the addition of a little hydrobromic acid or water (which reacts with the bromine to give hydrobromic acid) increases the speed greatly, while the addition of aluminum bromide is even more effective. The following products are formed in all three cases in amounts varying with the conditions: $CH_2Br-CH_2-CH_2Br$, $CH_2Br-CHBr-CH_3$, $CH_2Br-CH_2-CH_3$. If the reaction is carried out in the presence of sunlight it proceeds rapidly and the only product is trimethylene bromide. If the reaction were simply one of ring cleavage and the addition of bromine it would take place as follows and the sole product

1. J.Pr.C., 59,302.
2. J.Pr.C., 62,273.

would be trimethylene bromide:

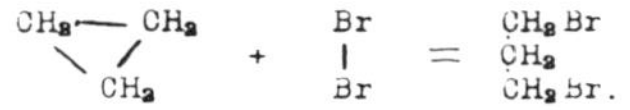

This is the actual result in the presence of sunlight, but the formation of the other bromides in the dark and the function of the catalysts needs further explanation. In this connection a study of the bromination of 1,1 dimethyl cyclopropane is of great value. This compound on treatment with bromine gives two products: $(CH_3)_2>CBr-CHBr-CH_3$ and $(CH_3)_2>CBr-CH_2-CH_3$ [1] but no $(CH_3)_2>CBr-CH_2-CH_2Br$ which would be expected from the reaction,

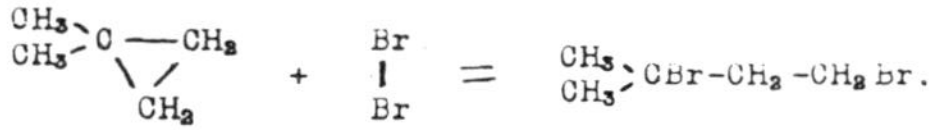

The explanation of this result and of the peculiar points in the bromination of trimethylene is as follows: a small amount of hydrobromic acid is always at first formed (by the action of moisture or by a little bromine substituting in the ring), this hydrobromic acid adds to the ring to give a normal bromide of the formula $(CH_3)_2>\overset{1}{C}Br-\overset{2}{C}H_2-\overset{3}{C}H_3$ in the one case and $\overset{1}{C}H_2Br-\overset{2}{C}H_2-\overset{3}{C}H_3$ in the other; a hydrogen on

1. J.Pr.C., 62,270.

carbon atom number 2 of these bromides is then easily substituted by bromine with the evolution of hydrobromic acid which causes more of the cyclic compound to react and thus the final bromides are $\frac{CH_3}{CH_3}>CBr-CHBr-CH_3$ or $CH_2Br-CHBr-CH_3$ as the case may be. This view finds additional support in the fact that dimethylcyclopropane and trimethylene both react easily with hydrobromic acid and the resulting straight chain bromides easily react with a molecule of bromine. In the case of trimethylene besides this reaction with hydrobromic acid, there is also the reaction with a molecule of bromine to give trimethylene bromide; by increasing the concentration of hydrobromic acid the speed of the reaction with this reagent is increased and the amounts of the bromides, $CH_2Br-CHBr-CH_3$ and $CH_2Br-CH_2-CH_3$ are found to increase. The complete reaction can then be expressed by the two following equations:

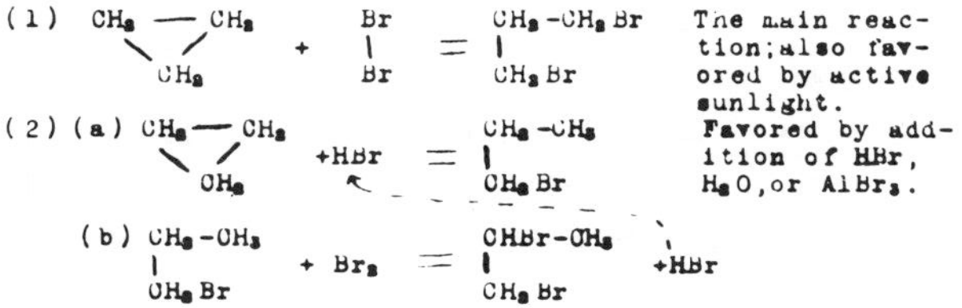
The main reaction; also favored by active sunlight.
Favored by addition of HBr, H_2O, or $AlBr_3$.

The only derivatives of trimethylene that can be directly prepared from the gas itself are the mono and dichlor compounds. In spite of the fact that the reactions of these two compounds have been but little studied, several facts of importance are known. The halogen atom is apparently unreactive[1] and with alkaline reagents can be neither replaced nor removed in the form of hydrochloric acid. This is significant if further investigation should show it to be a general property of cyclopropane halides, as it is quite parallel to the unreactivity of the halogen atoms attached to an unsaturated carbon atom, the relation between three-membered rings and double bonds again being noticeable. The substitution of chlorine for hydrogen on the ring decreases the reactivity of the ring towards bromine[2]; the ease of reaction of the dichlorcyclopropane being less than that of the monochlor and the monochlor less than that of trimethylene itself.[3] An attempt to prepare chlorcyclopropane by the action of zinc on $CH_2Br-CHCl-CH_2Br$ gave only allyl chloride[4] $CH_2=CH-CH_2Cl$. It seems probable that the first product of the reaction might have been chlor-

1. J.Pr.C., 42,495; 43,396.
2. Or perhaps hydrobromic acid; cf. bromination of trimethylene.
3. J.Pr.C., 42,495; 43,396.
4. J.Pr.C., 43,396.

cyclopropane but that it was either rearranged by the action of the zinc bromide or split by some halogen acids liberated during the action to give $CH_2Br-CH_2-CH_2Cl$ which then lost hydrobromic acid. In support of this view it is significant that monochlorcyclopropane readily combines with sulphuric acid[1], and that it also reacts with acetic acid and potassium acetate to give allyl acetate[1], probably thus:

$$\underset{CH_2}{CH_2-CHCl} + C_2H_4O_2 = \begin{array}{l}CH_2-CH_2Cl\\ |\\ CH_2-OC_2H_3O\end{array} + KC_2H_3O_2 = \begin{array}{l}CH=CH_2\\ |\\ CH_2-OC_2H_3O\end{array} + KCl + C_2H_4$$

By methods* quite similar to that used by Freund

* The preparation of 1,1,2 trimethylcyclopropane is perhaps an exception; this compound was prepared from a tertiary bromide whereas in all other cases a secondary bromide has been used. The well known ease with which a teriary bromide loses hydrobromic acid to give an unsaturated compound has led Östling[2] to question the structure of 1,1,2 trimethylcyclopropane. Zelinsky[3] who prepared it states that it slowly decolorizes potassium permanganate. Certain products of the action of nitric acid[4], however, support the cyclopropane structure very strongly and the compound can best be considered to have a cyclic structure.

1. J.Pr.C.,46,157.
2. J.C.S.,101,457.
3. B.,34,2856.
4. Z.,'14,I,1497.

in the preparation of trimethylene there can be obtained a number of methyl substitution products of cyclopropane. Thus by the action of zinc on the following bromides the corresponding cyclopropane derivatives result.

Bromide.	Cyclopropane resulting by the action of zinc.	
$CH_3CH_2Br-CH_2-CH_2Br$	CH_3CH-CH_2 / CH_2	Methylcyclopropane[1]
$CH_2Br-C(CH_3)_2-CH_2Br$	$(CH_3)_2C-CH_2$ / CH_2	1,1 dimethylcyclopropane[2]
$CH_3CHBr-CH_2-CHBr-CH_3$	$CH_3CH-CHCH_3$ / CH_2	1,2 dimethylcyclopropane[3]
$CH_3CHBr-CH(CH_3)CHBr-CH_3$	$CH_3CH-CHCH_3$ / CH_2	1,2,3 trimethylcyclopropane[4]
$(CH_3)_2CBr-CH_2-CHBr-CH_3$	$(CH_3)_2C-CH_2$ / $CHCH_3$	1,1,2 trimethylcyclopropane[5]

The reactions of these cyclic hydrocarbons show that the introduction of a methyl group has the opposite effect from the introduction of a halogen atom. The halogenated cyclopropanes are less reactive towards bromine than trimethylene but towards the same reagent methylcycloprpane is more reactive than the simple hydrocarbon and 1,1 dimethyl-

1. B.,28,21.
2. J.Pr.C.,58,458.
3. J.C.S.,101,457.
4. B.,34,2856.
5. B.,34,2856.

cyclopropane even still more reactive[1]; the 1,2 dimethyl and the 1,2,3 trimethyl[2] and the 1,1,2 trimethyl compounds[3] also combine more readily with bromine than trimethylene itself. It has also been found[4] that 1,1 dimethylcyclopropane is catalytically reduced much faster at 200° than is trimethylene.

The other reagents which react with trimethylene also combine with the methyl derivatives at least as readily and generally with greater ease. The ring cleavage always takes place between the carbon atom having the greatest number of substituent groups and the one having the least number; in the case of reagents having the general formula HX (where X is an acid radical) the X goes to the carbon atom holding the most groups. For example:

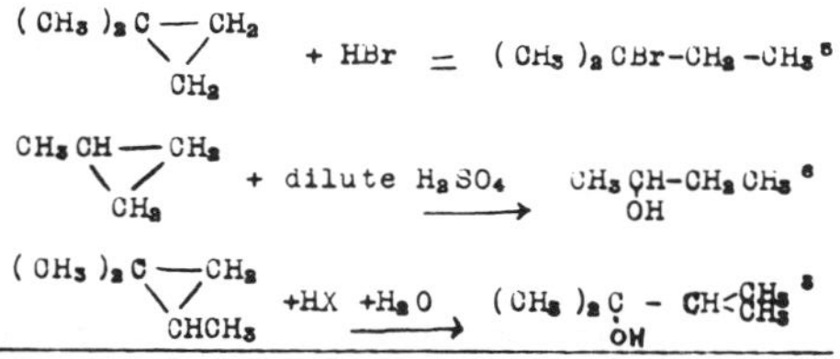

1. J.C.S., 101, 457; J.Pr.C., 62, 271.
2. J.C.S., 101, 457; B., 34, 2867.
3. Z., '14, I, 1997.
4. B., 36, 2015.
5. J.Pr.C., 62, 271.
6. B., 28, 21.

A complete analogy to the behavior of these cyclic compounds can be found in the ethylenic series of hydrocarbons. Ethylene and propylene do not react with hydrobromic acid in glacial acetic acid, but $(CH_3)_2C=CH_2$ and $(CH_3)_2C=C(CH_3)_2$ do[1]; thus the introduction of a methyl group has increased the reactivity of the unsaturated compound, as it did the reactivity of cyclopropane. The addition of hydrobromic acid gives mainly $(CH_3)_2CBr-CH_3$, the X of the HX going to the most substituted carbon atom just as it was found to do in the case of the methylated cyclopropanes.

Two other hydrocarbons were for many years considered to be cyclopropane derivatives. These are the so-called vinyl-trimethylene and ethylidene trimethylene. Gustavson[2] by the action of zinc on the tetrabromide of pentaerythrite, $C(CH_2Br)_4$, prepared a hydrocarbon of the composition C_5H_8. This compound formed a dibromide when treated with bromine and on oxidation with potassium permanganate gave a glycol, $C_5H_{10}O_2$, which on further oxidation yielded α-oxyglutaric acid. From the method of preparation and these reactions Gustavson proposed the formula

1. Z., '03, II, 334; '04, II, 691.
2. J.Pr.C., 54, 97.

13

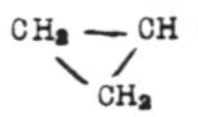 ,and called the compound vinyl-trimethylene. Its formation was supposed to take place in this manner:

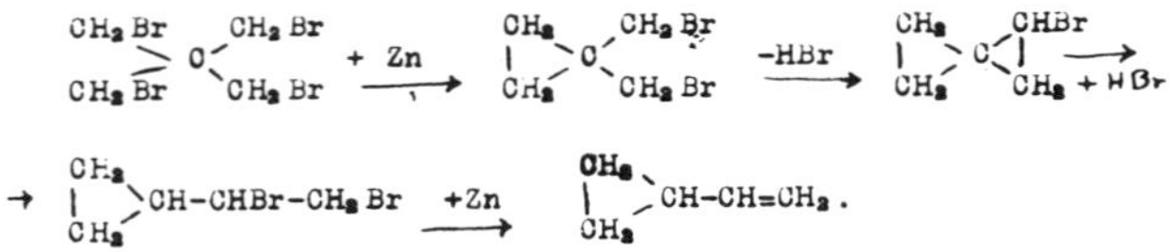

Hydrobromic and hydrochloric acids apparently reacted with the compound with ring cleavage. Hydriodic acid, however, gave an iodide from which an isomeric hydrocarbon could be obtained by the action of potassium hydroxide. This hydrocarbon was assigned the formula of ethylidene-trimethylene:

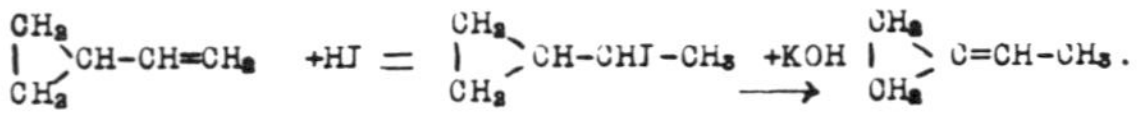

The new substance readily combined with bromine and hydriodic acid, showing its unsaturated character.

The evidence for the structure of the two hydrocarbons was considered somewhat dubious even by Gustavson, particularly in as much as oxidation with

14

lead dioxide gave ketopentamethylene, 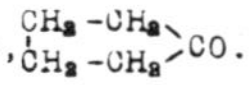

The same author pointed out the possibility of a cyclobutane formula for vinyl-trimethylene which would be as consistent with the facts as the other structure. The cyclopropane formulae and names have been copied in the literature, however.

Demjanow[1] in 1908 prepared a crystalline nitrosite of vinyl-trimethylene in the usual way. By reduction of this nitrosite cyclobutanone resulted; this pointed toward the cyclobutane formula for the hydrocarbon, as the following scheme shows:

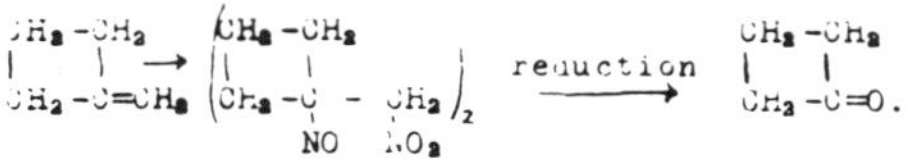

Faworsky[2] has very recently settled the question by the oxidation of ethylidene-trimethylene. If vinyl-trimethylene is represented as above then ethylidene-trimethylene must be 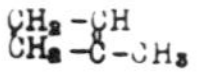 and on oxidation should give laevulinic acid.

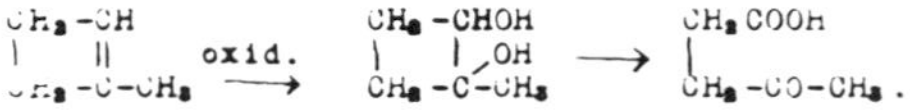

This was found to be the case and there can thus be

1. B., 41, 915.
2. B., 47, 1648.

but little doubt as to the structure of the substance.* Its formation from the tetrabromide must thus involve a considerable intramolecular rearrangement.

Leaving the halogen and methyl substitution products of trimethylene, there can be seen the possibility of countless cyclopropane compounds of different types obtained by substitution of the hydrogen atoms of CH_2—CH_2 (bridged by CH_2) with the great variety of groups of organic chemistry. Since only a relatively small number of these numerous possibilities are known, the task of reviewing the subject becomes a more hopeful undertaking. The number of cyclopropane derivatives described is comparatively small because each new type of compound generally requires a new or modified method of preparation. The reason for this can readily be understood when it is remembered on the one hand that, unlike the benzene ring, the trimethylene ring is so unstable that no atom or group, with the exception of chlorine, can be directly introduced into it, and on the other hand that chlorcyclopropane, unlike the alkyl halides, is extremely unreactive and the ordinary methods of synthesis cannot be used with it.

* Another paper has appeared giving an alternate formula (B., 40, 3883). The experimental facts have, however, been recently shown (Z., '14, II, 1266) to be erroneous.

Therefore to synthesize most cyclopropanes a process must be used which starts from a straight chain compound and ends with a cyclic one; this formation of a three-membered ring may be accomplished by several methods and each method will be dealt with as it occurs in the discussion of the various substances.*

It will simplify the review of the published experimental work on these cyclopropane compounds to divide them into classes according to the number and position of the substituent groups. First the mono and disubstitution products having the group or groups attached to the same carbon atom may well be studied. Of these compounds the simplest would be, perhaps, the alcohol cyclopropanol, CH_2—$CHOH$ (bridged by CH_2). It was hoped that by the action of zinc on CH_2Br-$CHOH$-CH_2Br this substance might be obtained, the only product, however, was allyl alcohol, CH_2=CH-CH_2OH[1]. In fact neither this nor any cyclopropane compound having a hydroxyl group on a ring carbon atom has ever been prepared. It seems very probable that this is not because of lack of proper experimental methods but because this cyclopropanol is extremely unstable; for example,

1. J. Pr. C., 46, 158.

* Substances whose reactions have not been sufficiently studied to throw light on the problems of cyclopropane chemistry will generally be mentioned only in foot-notes. They are added for the sake of completeness.

to explain the formation of allyl alcohol in the case just cited, it seems quite rational to assume as an intermediary unstable compound the hydroxy cyclopropane, which subsequently rearranged with ring cleavage to give allyl alcohol. Aminocyclopropane, $CH_2—CHNH_2$ / CH_2 (ring) on treatment with nitrous acid would be expected to give the desired alcohol but again allyl alcohol is obtained[1], probably by the same rearrangement. Here once more the similarity between these compounds and unsaturated ones is manifest. An ethylene alcohol having the hydroxyl group on an unsaturated carbon atom is very unstable and generally exists only in equilibrium with its rearrangement product, a ketone. (Cf. keto-enol tautomerism.) The fact that the groupings $-CH{=}C(OH)-$ and $CH_2—CHOH$ / CH_2 (ring) should both be unstable is a very striking proof of the relationship of the ring and double bond and is not in the least vitiated by the fact that the former grouping rearranges to a saturated ketone, the latter to an α,β-unsaturated alcohol.

Cyancyclopropane, or the nitrile of cyclopropane-carboxylic acid, is easily obtained from γ-chlor-

1. Z., '05, I, 1709.

butyronitrile by the elimination of hydrochloric acid by means of potassium hydroxide.[1]

$$CH_2CN{-}CH_2{-}CH_2Cl + KOH = \underset{CH_2}{CH_2—CHCN} + KCl + H_2O.$$

No reactions have been tried with this compound which would throw light on the question of ring stability. Its chief importance is that on boiling with water it is transformed into the corresponding acid; under these conditions, at least, there is no tendency for the ring to break.

The monocarboxylic acid just mentioned can also be obtained by heating the cyclopropane 1,1 diacid which loses carbon dioxide at about 180° and gives about seventy per cent of the cyclopropane monoacid. The dibasic acid was for many years the only source of the monobasic acid, but the newer method of preparation from the nitrile is much more advantageous. This cyclopropane acid is much less reactive towards addition reagents than any of the ring compounds thus far dealt with. It is not attacked by cold saturated aqueous hydrobromic acid[2] (although it dissolves in it) and is likewise unreactive towards bromine in the cold[3]. In fact its anilide can be converted into

1. Z., '99, I, 975.
2. J.C.S., 67, 118.
3. Ann., 227, 24.

the parabrom anilide by the action of bromine[1] without ring cleavage. Towards potassium permanganate it shows the characteristics of the cyclopropane series, attacking the reagent only very slowly on standing.

Although the ring is not as easily opened in this acid as it is in many of the other cyclopropane derivatives, it is by no means entirely unreactive. When heated with hydrobromic acid to 175° ring cleavage takes place, hydrobromic acid adds, and γ-brombutyric acid results[2]. This addition of hydrobromic acid is similar to the reaction between dimethylcyclopropane and hydrobromic acid at low temperatures. There is one very important difference, however, namely that while in the case of the hydrocarbon the bromine goes to the carbon atom having the greatest number of groups, in the acid it goes to the one having the least, for the product is <u>not</u> $CH_3-CH_2-CHBr-COOH$ but is $CH_2Br-CH_2-CH_2-COOH$. This is a general property of cyclopropane compounds having the group $-\overset{O}{C}-$ connected to a carbon atom of the ring; when such compounds react with reagents of the type HX the X goes to the β-position in the ring and the resulting

1. B., <u>38</u>, 2534.
2. J.C.S., <u>67</u>, 118.

straight chain compound has the X in the γ-position.

$$\underbrace{CH_2 - CH-COOH}_{\diagdown\ CH_2\ \diagup} + \overset{H}{\underset{X}{|}} \longrightarrow \begin{array}{l} CH_2-COOH \\ CH_2 \\ CH_2X. \end{array}$$

The same sort of phenomenon is found in ethylenic compounds having similar groups. Acrylic acid, $CH_2=CH-COOH$, for example, reacts with hydrobromic acid to give CH_2Br-CH_2-COOH, the bromine going to the carbon atom farthest from the substituting group. This peculiar mode of addition of hydrobromic acid to cyclopropane acids thus finds complete analogy in the unsaturated series and adds still another connecting link between these two classes of compounds. Besides the reaction with hydrobromic acid another reaction in known involving the cleavage of the ring in the monocarboxylic acid. Although without action in the cold, bromine at 60° for sixty days[1] with the chloride of the cyclopropane acid reacts to give a dibrom addition product which is α,γ-dibrombutyril chloride; the corresponding acid bromide is obtained by starting with the acid, red phosphorus and bromine.

From the chloride of the monobasic acid the amide can be easily obtained and from this by the usual Hofmann reaction the cyclopropane amine[2],

1. Z., '09, II, 1130.
2. Z., '01, II, 579.

$CH_2 \genfrac{}{}{0pt}{}{}{\diagdown CH_2} CH\text{-}NH_2$ with CH_2. This compound is only of passing interest, but it is worth noting that it is not attacked by fuming hydrobromic acid. On treating with nitrous acid as has been mentioned, allyl alcohol results, and on attempting to oxidize the amine to a ketone the ring likewise appears to break[1]. The reactions of this compound as well as those of cyancyclopropane have been too little studied to enable one to make any statement as to the effects of these groups on the stability of the ring. In as much as the monocarboxylic acid was comparatively stable, it will be of great interest to investigate the effect of substituting the other hydrogen atom on the same carbon atom by carboxyl. A compound having such a grouping is the 1,1 dibasic acid which is of great theoretical and historical importance, as it was one of the first cyclopropane derivatives synthesized.

Almost simultaneously in 1883 Fittig and Roder in Germany and Perkin in England prepared the ester of cyclopropane 1,1 diacid by the action of ethylene dibromide on sodium malonic ester; the former published

1. Z.,'05,J,1703.

two preliminary notes on his research in 1883[1] and a summary in 1885[2], the latter published his work in 1884[3]. Both investigators obtained the same product and reported similar properties but differed radically in the interpretation of their work. Fittig believed the reaction took place in the following way to give an unsaturated acid which he called vinaconic acid:

$$CH_2Br\text{-}CH_2Br + NaC\genfrac{}{}{0pt}{}{COOC_2H_5}{COOC_2H_5}[4] \longrightarrow CH_2Br\text{-}CH_2\text{-}C\genfrac{}{}{0pt}{}{COOC_2H_5}{H}\bigg|_{COOC_2H_5}$$
$$CH_2 = CH\text{-}C\genfrac{}{}{0pt}{}{COOC_2H_5}{H}\bigg|_{COOC_2H_5}$$

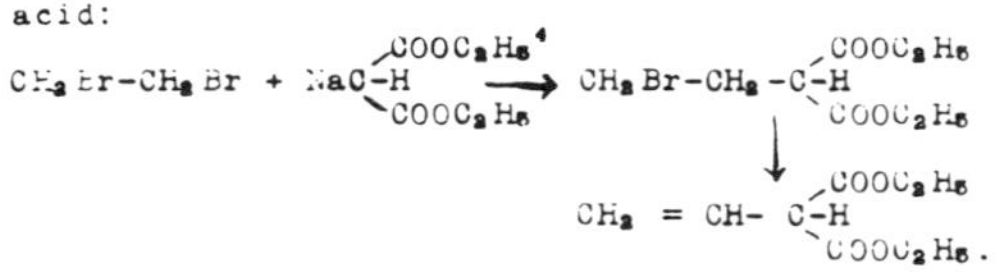

Perkin represented the reactions by the following mechanism:

$$\genfrac{}{}{0pt}{}{CH_2Br}{\big|\ CH_2Br} + Na_2C\genfrac{}{}{0pt}{}{COOC_2H_5}{COOC_2H_5} = \genfrac{}{}{0pt}{}{CH_2}{CH_2}\diagdown C \diagup \genfrac{}{}{0pt}{}{COOC_2H_5}{COOC_2H_5} + 2NaBr.$$

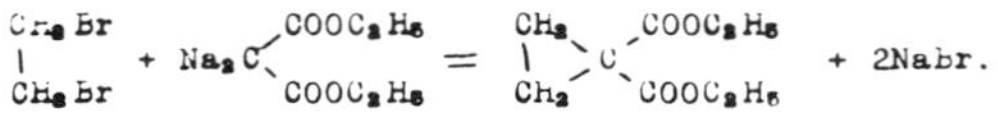

This latter reaction involves the use of sufficient sodium to correspond to two atoms, but Fittig showed that the same product resulted if only one equivalent of sodium was used. (A more detailed discussion of the mechanism of this and similar reactions will be found on page 42.)

1. B.,16,372,2592.
2. A.n.,227,13.
3. B.,17,54.
4. The sodium of sodium malonic ester and similar compounds will be written for convenience in combination with the carbon atoms.

23

The controversy centered on the structure of the free acids which were obtained from the esters by hydrolysis, and the question was not easily settled for apparently the reactions of the compound could be interpreted in two ways. In perfect accord with the unsaturated formula the acid added hydrobromic acid easily[1] to give a product which was shown to be $CH_2Br-CH_2-CH<(COOH)_2$; on heating the "vinaconic acid", carbon dioxide was evolved and a monobasic acid and a lactone $CH_2 - CH_2 - CH_2$ (O—CO) were obtained. The formation of the lactone and the addition of hydrobromic acid were evidence for the formula $CH_2 = CH - C(H)(COOH)_2$

On the other hand, if this were the correct formula the compound should readily add bromine and reduce potassium permangante; the first reaction it gave only in the sunlight, the other not at all[2]. These facts, its stability towards reduction with sodium amalgam[3], the value of its magnetic rotation and the fact that it behaved like a disubstituted malonic acid[4] * settled the dispute in favor of Perkin.

1. Ann., 227, 19; J.C.S., 47, 812.
2. Ann., 294, 125; J.C.S., 47, 812; B., 18, 1734.
3. B., 23, 704; J.C.S., 47, 812.
4. Cf. B., 18, 1734; B., 19, 1049; J.C.S., 47, 820; Ann., 294, 129.
* Perkin showed that the ester would not react with sodium alcoholate and benzoyl chloride, thus proving the absence of a replaceable hydrogen atom.

24

From this short summary of a rather long controversy it can be seen that it is often not an easy matter to decide between the possibilities of a cyclopropane or an unsaturated formula. It often happens that the reactions of these two classes are so much alike that only a great deal of skilful work can differentiate between the two alternatives.

Besides the reactions just mentioned which show the great reactivity of the ring in this dibasic acid, several others should be noted. On boiling with dilute sulphuric acid[1] and then treating with water γ-oxy-ethylmalonic acid, $CH_2OH-CH_2-C(H)(COOH)_2$, is found as the product. When the ester is treated with sodium malonic ester the ring breaks, sodium malonic ester adds and the product obtained on acidifying is butane tetracarboxylic ester[2].

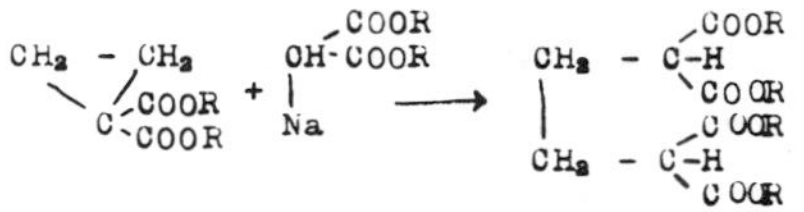

This reaction with sodium malonic ester is entirely in accord with what would be expected from the rela-

1. Ann., 227, 18.
2. J.C.S., 67, 108.

tion of this dibasic acid (and ester) to unsaturated esters, for as is well known the latter add sodium malonic ester with great ease. (Michael's reaction. See page 25.)

Although the monocarboxylic acid was found to be considerably less reactive than trimethylene itself, the 1,1 dibasic acid appears to be fully as reactive as the hydrocarbon towards all reagents except bromine. While the substitution of one hydrogen atom of cyclopropane with carboxyl somewhat reduces the reactivity of the ring, the introduction of a second carboxyl on the same atom causes the compound to be more reactive towards addition reagents than the parent hydrocarbon.

Two compounds closely allied to the 1,1 diacid are the benzoyl- and acetyl-cyclopropane-1-acids. They are prepared by the action of ethylene dibromide on the sodium derivatives of acetacetic and benzoyl--acetic esters respectively[1]. The reaction may be expressed as follows: (Cf. page 42.)

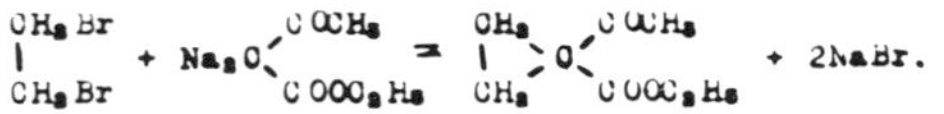

1. B., 16, 21136.

The ester so obtained, unlike most acetoacetic ester derivatives, may be hydrolyzed to the free acid without decomposition. Their general behavior is like that of the 1,1 diacid, the same reactivity of the ring being characteristic of both the benzoyl and acetyl compounds. The ester when treated with cold hydrobromic acid gives γ-bromethylacetacetic ester[1]; if the acid is used, carbon dioxide is evolved and γ-brompropyl ketone results[2]:

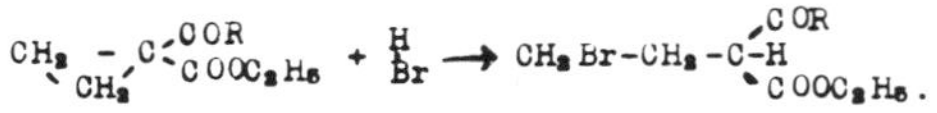

On boiling with water the acid loses carbon dioxide besides adding the elements of water and forms acetopropyl alcohol.[5] On reduction with sodium amalgam a hydroxy cyclic compound* is at first formed which is then further reduced with ring cleavage to a hydroxy butyric acid derivative, $RCHOH\text{-}CH(C_2H_5)COOH$, (R being CH_3 or C_6H_5.)[3] On treatment with phosphorus pentachloride ring cleavage also takes place, a compound of uncertain structure being the product[4]. The

1. J.C.S., 51, 833.   5. J.C.S. 61, 67; 55, 358; 51, 829
3. J.C.S., 51, 824.
2. J.C.S., 47, 842.
4. J.C.S., 51, 841.

* As there is no proof of the formula assigned to this first product, it may be $R\text{-}CO\text{-}CH(C_2H_5)\text{-}COOH$, which is then further reduced to the alcohol.

general analogy of the reactions of the compounds to those of the 1,1 diacid is abundant proof of their cyclic structure. However, at the time of the preparation of these two ketonic substances little was known about the chemistry of cyclopropane and a formula representing a five-ring containing oxygen was urged by some[1] instead of the cyclopropane structure. Such a possibility was removed by the preparation of crystalline oximes of the acids in question[2].

On heating, both the acetyl and benzoyl acids lose carbon dioxide and the only product is the acetyl or benzoyl cyclopropane[3]. The ketones thus obtained readily form oximes[4], thus definitely proving that they are not oxygen ring compounds. Furthermore they have been connected with other cyclopropanes by the oxidation of the acetyl compound to cyclopropane monocarboxylic acid[5] and by the formation of the benzoylcyclopropane from the chloride of this same acid and benzene[6]. Acetyl cyclopropane has also been obtained by the following reaction, potas-

1. B., 22, 210; J.Pr.C., 37, 492.
2. J.C.S., 59, 867.
3. J.C.S., 59, 858; 51, 831.
4. J.C.S., 59, 815; 47, 844; B., 36, 1374, 1795.
5. Z., '98, II, 474.
6. Z., '12, I, 1458.

sium hydroxide being used to eliminate the hydrobromic acid.[1] *

$$CH_3COCH_2\text{-}CH_2\text{-}CH_2Br + KOH = CH_3COCH\genfrac{}{}{0pt}{}{\diagup CH_2}{\diagdown CH_2} + KBr.$$

The ring in these monosubstituted cyclopropanes is reactive; they combine with cold hydrobromic acid[2], and are readily reduced by sodium amalgam to alcohols of the type $RCHOH\text{-}CH_2\text{-}CH_2\text{-}CH_3$; on boiling with dilute sulfuric acid, the acetyl compound gives acetopropyl alcohol[3], although when heated alone with water to 200° it is not altered. The reduction of

* Lipp, who performed this experiment, prepared the acetopropyl bromide by hydrolyzing the product obtained by the action of ethylene dibromide and monosodium acetacetic ester. He believed the reaction to run thus:

$$\begin{matrix} COOC_2H_5 \\ | \\ CHNa \\ | \\ COCH_3 \end{matrix} + CH_2BrCH_2Br = \begin{matrix} COOC_2H_5 \\ | \\ CH\text{-}CH_2\text{-}CH_2Br \\ | \\ COCH_3 \end{matrix} = \begin{matrix} CH_2\text{-}CH_2\text{-}CH_2Br \\ | \\ CO \\ | \\ CH_3 \end{matrix}.$$

The product of the action of potassium hydroxide on the bromide he assumed to be the anhydride of acetopropyl alcohol, but it is in reality acetylcyclopropane, as Perkin showed[4]. The formation of Lipp's bromide is not to be represented by the above reaction but is rather formed by the splitting of his acetylcyclopropane-1-acid by the concentrated acid he used in acidifying and also from methyldehydropentone-carboxylic acid (a by-product).

1. B., 22, 1210.
2. Z., '12, I, 1458.
3. J.C.S., 59, 8??;
4. J.C.S., 59, 867.

these compounds and the acids from which they are derived is of interest. Most cyclopropane compounds (the 1,1 acid for example) are not attacked by sodium amalgam, but in the cases of these ketonic substances, probably because of the carbonyl group, ring cleavage and addition of hydrogen takes place.

Ethylene dibromide reacts with the sodium derivative of cyanacetic ester in the same way as it does with acetacetic ester, and the cyclopropane ester thus obtained gives on careful hydrolysis the cyancyclopropane-1-acid[1], $CH_2 - C(CN)(COOH)$ (ring closed by CH_2). On boiling with alkalies the nitrile group is also hydrolyzed and the 1,1 diacid results[1]. In fact this is the best method of making the dibasic acid for the yields are much greater than by the ordinary process. Few reactions of the cyan-acid have been studied but in all probability the reactivity of the ring is comparable with that of the 1,1 diacid. Sodium cyanacetic ester easily reacts with the cyclic ester to form ethyl α,δ dicyanadipate, $(CN)(COOH)CH-(CH_2)_2-CH(CN)(COOH)$, ring cleavage, of course, having taken place[2]. Attempts to prepare cyancyclopropane by heating the acid were unsuccessful[3].

1. Z., '06, I, 822; J.C.S., 75, 922.
2. J.C.S., 95, 692.
3. J.C.S., 75, 927.

Phenylcyclopropane was recently prepared from cinnamic ester and hydrazine hydrate[1]. These two substances react to give a compound having a five membered ring containing nitrogen which probably has this formula: $C_6H_5CH - CH_2 - CH_2$ (ring closed by $N = N$). The pyrazolene, as it is called, on heating with platinized clay loses nitrogen and phenylcyclopropane is the result. This monosubstitution product is a reactive substance, polymerizing with concentrated sulphuric acid and adding hydrobromic acid. The polymerization is quite explicable, for styrene, $C_6H_5CH{=}CH_2$, the analagous ethylenic compound, also easily undergoes polymerization.

This concludes the consideration of the first class of cyclopropane derivatives with the exception of certain monosubstitution products which may be considered as side chain derivatives of cyclopropane and whose reactions are not of much interest as regards the properties of the ring, and hence will be discussed later. The next class to be examined will be the 1,2 disubstitution products, that is compounds of the general formula $CH_2 - CHX$ (ring closed by CHY).

1. Z., '13, II, 2129.

If the formula of the cyclopropane 1,2 diacid, CH_2 - CH-COOH / CH-COOH, be examined it will be seen that two spatial isomers are possible, one, the cis, having the two carboxyl groups on the same side of the plane of the ring, the other, the trans, having them on opposite sides:

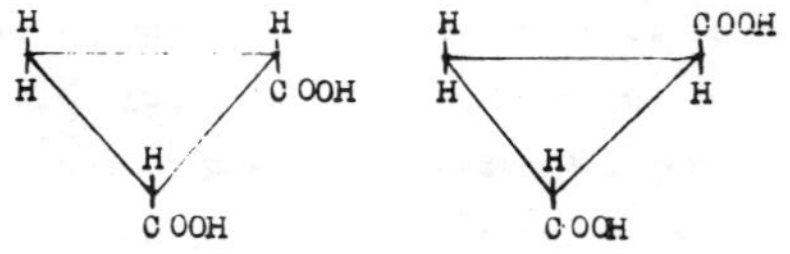

(I) cis — (II) trans

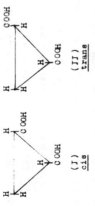

Two acids are actually known which can be readily converted into each other and one of which gives an anhydride and is therefore designated as the cis acid. Cyclopropane 1,1,2 tribasic acid (loc.cit.) on heating loses carbon dioxide and gives the anhydride of the cis acid[1], which on treatment with water gives the dibasic acid. The anhydride also results[2] from the heating of the 1,1,2,2 tetrabasic acid (loc.cit.). The trans acid can be made by eliminating hydrobromic acid from monobromglutaric ester with diethylaniline

1. B., 17, 1186.
2. Ann., 256, 197.

or potassium hydroxide,[1]

$CHBr-COOC_2H_5$ / CH_2 / $CH_2COOC_2H_5$ → CH_2 - $CH-COOC_2H_5$ / $CH-COOC_2H_5$,

no unsaturated compound being formed. The best method of preparing the 1,2 acids and one that introduces a new method was discovered by Buchner and has been applied to the preparation of a great number of cyclopropane compounds. Diazoacetic ester can be added to many unsaturated compounds (in general those having a "negative" group) to give a five-membered nitrogen ring compound. The pyrazolene thus formed loses nitrogen on heating and a three-membered carbon ring closes. (Cf. phenylcyclopropane.) In this particular case[2] acrylic ester was the starting point and the ester of the trans acid was the product, together with some unsaturated ester.

$CH_2 = CH-COOC_2H_5 + N_2CH-COOC_2H_5$ ⟶

CH_2 - CH - $COOC_2H_5$ / N / CH - N= / $COOC_2H_5$ ⟶ N_2 + CH_2 - $CH-COOC_2H_5$ / $CH-COOC_2H_5$

Another reaction similar to this has been used to synthesize the same cyclic acid from fumaric ester. Diazomethane, N_2CH_2, forms with this latter compound

1. J.C.S., 87, 365; Z., '00, I, 284.
2. B., 23, 701.

a pyrazolene which can be decomposed to give the 1,2 diacid[1].

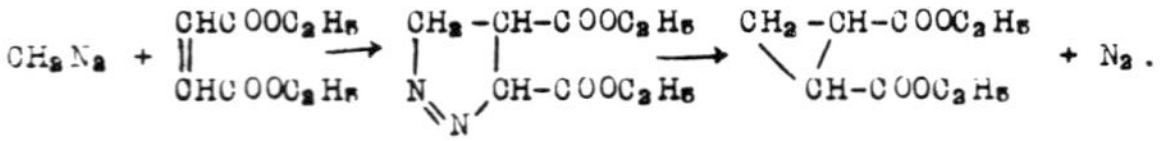

The mechanism of this change from a five-membered to a three-membered ring by loss of nitrogen is not understood and a discussion of it must therefore be omitted in this consideration of the reactions.

As regards the structure of the 1,2 diacids, they are undoubtedly ring compounds for all other possible acids of the formula $C_6H_8O_4$ (glutaconic, mesaconic, itaconic and citraconic) are known, and moreover they fail to oxidize potassium permanganate[2], reduce with sodium amalgam[2], or decolorize bromine[3]. The only difference in the chemical properties of the two acids is anhydride formation; the cis acid on boiling with acetyl chloride readily gives an anhydride, the trans remains unchanged[4]. The two isomers can be transformed into each other with considerable ease[5]; the silver salt of the trans acid when heated gives the cis anhydride, the cis acid

1. B., 27, 1888.
2. B., 23, 705.
3. Ann., 284, 200.
4. Ann., 284, 208.
5. Ann., 284, 216, 218.

rearranges to the trans acid when heated to 150 - 245° under conditions which render anhydride formation impossible, that is, in the presence of fifty percent acid or fused alkali.

Up to this point the relation of cyclopropane compounds to unsaturated ones has often been emphasized and a large amount of evidence proving the analogy has been collected. From this one might conclude that all three-membered carbon rings were similarly reactive, and would add reagents as easily as an ethylenic linkage. A study of the reactions of the 1,2 acids shows, however, that this generalization is not wholly true, for these compounds give none of the addition reactions which have hitherto been characteristic of the three-carbon ring. Hydrobromic acid, a reagent which has not failed up to this point, does not react with these cyclic compounds even after they have stood dissolved in it for five days[1]; hydrochloric acid at 180° for six hours likewise is without effect[2] and bromine at a high temperature in the presence of red phosphorus does not cause ring cleavage but substitutes a hydrogen of the ring with bromine[3]. Besides

2. Ann., 284, 200.
3. B., 38, 1600.

this conclusive proof of the great stability of the ring, the acids are unchanged on heating with dilute sulphuric acid at 150° [1], or with fused alkali to 245°. In strong contrast with the 1,1 diacid the trans 1,2 diacid distills unchanged at 200° (30mm.) and the cis acid at the same temperature goes over into its anhydride[2], in neither case does any ring cleavage occur. The high temperature reaction with bromine is the first case that has been thus far met in which bromine substitutes without the ring breaking; it is comparable with the action of chlorine on trimethylene itself. Both the 1,2 acids give as a product of this reaction two isomeric dibromsubstitution products which on reduction (zinc and acid) regenerate the cyclopropane trans acid[3]; this is good proof of their cyclic structure as dibromglutaric acid (the other possible formula) under the same conditions gives only glutaric acid.

Optical isomerism is possible in the cyclopropane series and if the spatial formulas of the three dibasic acids, 1,1 and two 1,2 acids, are examined it will be seen that only one, the trans 1,2 acid, should

1. Ann., 284, 200. Buchner speaks of a trace of oxy-acid being formed, but gives no formula.
2. Ann., 284, 200.
3. B., 38, 1601.

show this property. In beautiful agreement with this, Buchner and Heide[1] succeeded in resolving the trans acid into two optical antipodes, but could not resolve the other two. These experiments afford additional evidence for the cyclic structure of the 1,2 acids.

The apparent anomaly in the behavior of the cyclopropane 1,2 diacid towards addition reagents leads at once to the question, is this due primarily to the fact that they are 1,2 disubstituted cyclopropanes? An answer to this can be found by examining other 1,2 compounds. The behavior of 1,2 dimethylcyclopropane toward bromine it will be remembered was more energetic than that of trimethylene itself. Two methyl groups in the 1,2 positions, then, seem to increase the reactivity of the ring.

Methylcyclopropane-2-phenyl, methylcyclopropane-2-isopropyl and 1,2 diphenylcyclopropane[1] have been prepared[2] by the same sort of reaction as was used for making phenylcyclopropane. The starting point was benzalacetone, isobutylideneacetone and benzalacetophenone respectively and the resulting pyrazolenes had the following formulas:

1. B., 38, 3112.
2. Z., '12, II, 1927; '13, II, 2133.
3. C.A. 9, 3051.

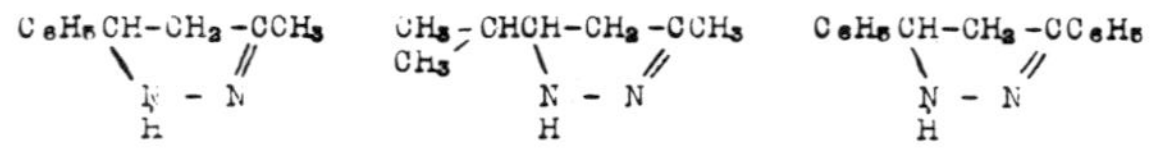

The cyclopropane compounds which result by the elimination of nitrogen have a reactive ring. With hydrobromic acid they form the straight chain bromides, $C_6H_5CHBr-CH(CH_3)_2$, $(CH_3)_2CH-CHBr-CH(CH_3)_2$, and $C_6H_5-C(CH_3)=CHC_6H_5$ (by loss of hydrobromic acid from the addition product), the ring breaking and addition taking place according to the usual rule of hydrocarbons.

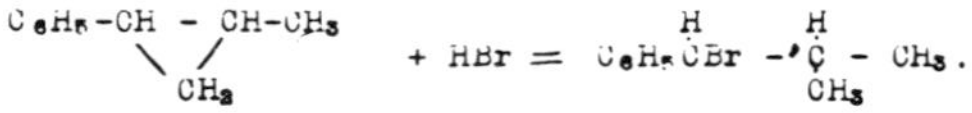

Several 1,2,cyclopropanes containing a ketone group have been prepared by the elimination of hydrochloric acid from certain γ-chlorketones[1]. Thus methylcyclopropane-2-acetyl, methylcyclopropane-2-benzoyl and cyclopropane tolyl ketone have been made but their reactions have not been studied.

Methylcyclopropane-2-acid is obtained by loss of carbon dioxide from the corresponding dibasic acid (page 43). Its properties have been little investigated but the fact that it easily reacts with bromine

1. C.A., 9, 2057.

in the cold to give a dibrom addition product[1] is evidence for its reactivity. No other disubstitution products have been sufficiently studied to throw light on the problem of the reactivity of the ring, but it may be well to consider the others for the sake of completeness.

Phenylcyclopropane-2-acid was prepared[2] by the action of diazoacetic ester on styrene, nitrogen being evolved during the reaction; its relations as regards the reactivity of the ring have not been studied. Attempts to prepare phenylcyclopropane by heating the calcium salt failed, phenylpropene and styrene resulting. If diazoacetic ester is heated with phenylbutadiene nitrogen is slowly evolved and a cyclopropane compound of the following formula is the product[3]:

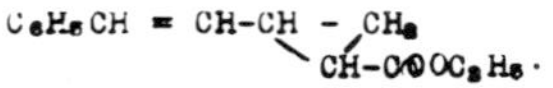

This styrenecyclopropane-2-acid adds bromine readily and is oxidized easily to benzoic acid and cyclopropane 1,2 diacid, thus proving its structure. Its further reactions are too uncertain to be worth considering. It seems to reduce readily to a straight chain saturated compound but to add only one molecule of hydrobromic acid while retain-

1. Ann., 294, 132.
2. B., 36, 3782.
3. B., 37, 2104.

ing its cyclc structure.

It seems entirely probable from what has been stated concerning 1,2 compounds in general that the great stability of the cyclopropane 1,2 acid is not because it is a 1,2 disubstitution product, since the other disubstitution products of this type having two alkyl groups, an alkyl and a phenyl, or an alkyl and a carboxyl are reactive substances. The introduction of a second carboxyl cannot be a sufficient explanation as the 1,1 diacid was much more reactive than the corresponding monobasic acid; this peculiar effect on the reactivity of the ring is in some way connected with the fact that there are two carboxyl groups on adjoining carbon atoms. What other combinations, if any, will also greatly diminish the reactivity of the cyclopropane ring? A new problem has thus arisen, namely the effect of groups on the reactivity of the ring. That there might be such an effect has been foreshadowed from the very first, as it was seen that the introduction of chlorine, methyl and carboxyl groups altered to some extent the properties of the substance. The present difficulty is, however, much more marked, for there almost seem to be two distinct

classes of cyclopropane compounds, one with a reactive ring, the other with an unreactive one. A consideration of the tri-substituted cyclopropanes will bring out this difference even more clearly.

Cyclopropane derivatives having three groups are of two types, the 1,1,2 and the 1,2,3 trisubstitution products. While a number of such compounds have been prepared, in only a few cases has enough work been done to allow of a generalization as regards the reactivity of the ring. The following table, however, correlates as much evidence as can be brought to bear on the question. It will be seen from this table that the reactivity of the compounds varies greatly according to the different groups; the stable compounds are those containing carboxyl and phenyl groups while the most reactive are the hydrocarbons. (See following page.)

The 1,1,2 triacid is prepared by the action of sodium malonic ester on either dibrompropionic ester[1] or bromacrylic ester[2].* The latter reaction, as

1. B., 17, 1186; Ann., 284, 216.
2. Am. Chem. J., 9, 122; J. Pr. C., 35, 132.
* Another method of preparing this acid is by the action of diethylaniline on

Summary of the reactions of
the trisubstituted cyclopropanes.

Substance	HBr or dilute H_2SO_4	Reaction with Bromine	Reaction on heating alone
1,1,2 triacid	-----	-----	loss of CO_2; no ring cleavage.
1-methyl 2,2 acid	adds cold	adds in sunlight	75 p.c. ring cleavage.
1-methyl-2-acid 2-acetyl	On boiling with water gives acetoisobutyl alcohol.		
1,2,3 triacid	Stable even on heating	-----	Forms anhydride.
1 methyl 2,3 acid (trans)	-----	not attacked when cold	Ring cleavage.
1 ethyl 2,3 acid (trans)	-----	-----	Ring cleavage.
1 phenyl 2,3 acid	Stable even on heating	not attacked	Forms anhydride.
1,2,3 tri-methyl	-----	Reacts easily	--------
1,1,2 tri-methyl	-----	Reacts easily	--------
1,1 dimethyl 2 isobutenyl	Reacts	-----	---------
1,1 dimethyl 2 isobutyl	Reacts	-----	--------

*(cont...) $(COOC_2H_5)_2CBr-CH_2-CH_2-COOC_2H_5$.[1] This reaction gave about eighty per cent of glutaconic ester and about twenty per cent of the cyclopropane compound. This 1,1,2 acid should exist in

1. J.C.S., 101, 249.

subsequent investigations have shown[1], takes place by the addition of malonic ester to the unsaturated ester, the substituted malonic ester thus formed is converted into its sodium derivative and sodium bromide is finally eliminated, closing the cyclopropane ring.

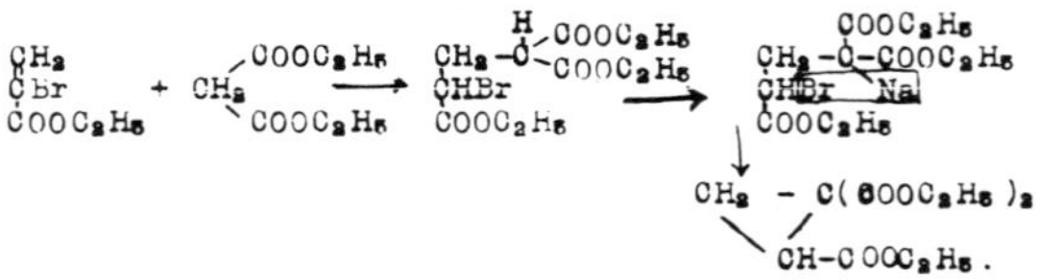

These researches, which are due to Michael, show that in all probability the reactions between many α,β-dibromides and sodium malonic ester should be interpreted as follows: the dibromide first loses hydrobromic acid under the influence of the alkaline reagent (sodium alcoholate, as the reaction is usually carried out in the presence of alcohol), an α-brom unsaturated compound is thus formed which reacts with the malonic ester in the manner outlined above for bromacrylic ester. In the case of dibromides which

(cont.) two optically active forms, the 1,2,3 acid should not; experiments have confirmed this prediction.

1. J.Pr.C., 35, 349.
2. B., 38, 3118.

do not easily lose hydrobromic acid, the first reaction is probably the replacement of one of the bromine atoms by the malonic ester residue, the sodium derivative of this new ester is then formed and the ring closed by splitting out of sodium bromide. This process is illustrated by the preparation of another 1,1,2 compound, methylcyclopropane 1,1 acid, from propylene bromide[1].

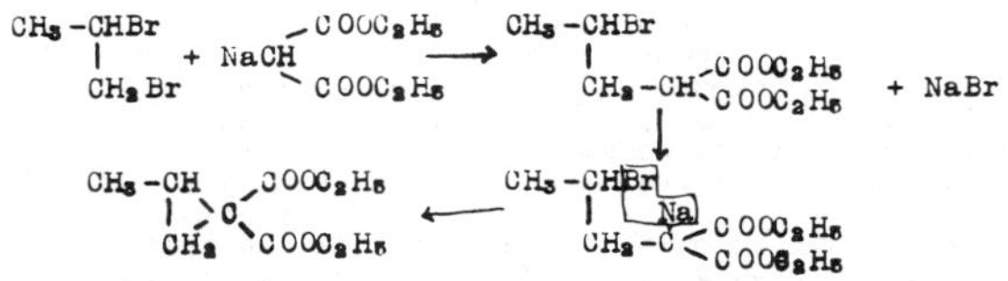

These two compounds, the 1,1,2 acid and the methylcyclopropane 2,2 acid are interesting to contrast as they differ only in that a carboxyl group in one is replaced by a methyl group in the other. On heating to 184 - 190° the first loses carbon dioxide, but no ring cleavage products result[2]; the second at a temperature of 190° also loses carbon dioxide but about 75 per cent of the product is valerolactone[3],

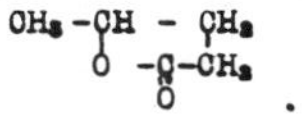

1. Ann., 294, 111. (Considerable evidence is given in the article in support of this mechanism.)
2. Ann., 284, 203; B., 17, 1187.
3. Ann., 294, 131.

The difference of stability of these two compounds is thus very marked*. Moreover, the methyl compound shows its reactivity by combining with cold hydrobromic acid, and with bromine in the presence of sunlight[1].

The action of propylene bromide and sodium acetacetic or benzoylacetic esters proceeds as would be expected and methylcyclopropane-2-benzoyl-2 acid and methylcyclopropane-2-acetyl-2-acid are the products. The latter on boiling with water loses carbon dioxide with ring cleavage and acetoisobutylalcohol results.**[2]

* Michael found that on treating the ester of the 1,1,2 acid with sodium ethylate a yellow color, was produced. On treating this solution with ethyl iodide or benzoyl chloride some inorganic salt was precipitated but no products of the reaction could be isolated[3]. It is possible that sodium ethylate splits the ring of this ester and the resulting sodium derivative reacts with a halide. (Cf. page 91)

** Ruhemann prepared phenylcarboxylic-aconitic ester by the action of phenylmalonic ester on chlorfumaric ester in the presence of sodium ethylate. This ester, $C_6H_5C(COOC_2H_5)_2-C(COOC_2H_5)=CH-COOC_2H_5$, he claims[4] gave on hydrolysis the anhydride of (over)

1. Ann., 294, 121, 125.
2. B., 17, 1443; J.C.S., 61, 72, 84.
3. J.Pr.C., 35, 135.
4. J.C.S., 81, 1212.

A representative of the second type of trisubstituted cyclopropane is the 1,2,3 triacid. It is prepared by Buchner's method, either fumaric ester or maleic ester being used to combine with the diazo ester[1]. Although two isomers of this acid are possible, only one, the trans (that is, one having two carboxyl groups on one side of the plane of the ring) is known;* it readily gives an anhydride on

** (cont.) phenyl-cyclopropane-1,2-dicarboxylic acid:

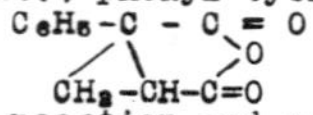

. This seems to be an entirely reaction and as the proof of the constitution of the cyclopropane compound is very scanty, its structure must be considered to be still problematical.

unlikely

By heating the pyrazolene formed by the addition of diazoacetic ester to itaconic ester a tribasic cyclopropane acid of the following formula results:[2]

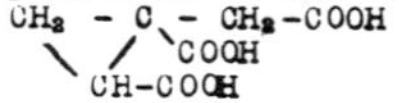

. No important reactions of this compound are known.

* This acid is considered to be the trans form largely because of its physical properties.[3] It is possible that the other form has been prepared by Perkin[4] by heating a 1,1,2,3 acid which he prepared from dibromsuccinic ester and sodium malonic ester. The acid differed in melting point from the one prepared by Buchner and formed no anhydride. A repetition of Perkin's work by Buchner[5] gave negative results. (over)

1. B., 21, 2640.
2. B., 27, 879.
3. Ann., 284, 210.
4. J.C.S., 47, 826.
5. loc.cit.

heating. The acid may also be obtained by the hydrolysis of the 1,1,2,2,3,3 hexaester, three molecules of carbon dioxide being lost in the saponification[1], and in still another way by heating the 1,1,2,3 acid.[2] The tribasic acid is quite as unreactive as the 1,2 dibasic acid; heated with hydrobromic or hydrochloric acid for three days it is unchanged,[3] and boiling with dilute sulphuric acid produces no alteration though continued heating with this reagent at 150° apparently causes ring cleavage though no product could be isolated.[4] It is unattacked by bromine in chloroform solution[5] even when heated for five hours.

Thus the introduction of a carboxyl group on another carbon atom increases the stability of the 1,1 acid and leaves the already stable 1,2 acid unchanged, while the introduction of a methyl group in the 1,1 acid increases a little the reactivity of

*(cont.) Kötz[6] states that he prepared a 1,2,3 acid identical with Perkin's but refers to Buchner's paper and gives no melting point or properties for the compound.

1. J.Pr.C., 68, 166.
2. Ann., 229, 89; B., 23, 2584.
3. B., 23, 2583; Ann., 284, 200.
4. Ibid.
5. B., 21, 2642.
6. J.Pr.C., 68, 166.

this reactive compound. Quite similarly the introduction of an alkyl group in the 1,2 acids changes these very unreactive substances to fairly reactive ones. Methylcyclopropane 2,3 acid, which is one of these alkyl derivatives of the 1,2 acid has been obtained in both a cis and a trans form by the elimination of carbon dioxide from methyl 2,2,3,3 tetracarboxylic ester[1] (hydrolysis and loss of carbon dioxide at the same time), and from methyl 2,3,3 triacid[2] respectively. The reactions of the latter form are unknown but the former on heating for some time rearranges into the unsaturated acid[2], methyl citraconic acid: $CH_3-CH_2-\underset{\|}{C}-COOH$ / $HC-COOH$. The trans ethylcyclopropane 2,3 acid prepared in an entirely similar manner, on heating also undergoes ring cleavage and a lactonic acid, $CH_3CH-CH - CH-COOH$ / $O - C-CH_3$ / O, is the product[3]. In both these cases it should be noticed that the ring splitting took place between a carbon atom carrying an alkyl group and one carrying a carboxyl, not between two carboxyl groups, $CH_3-CH-CH-COOH$ / $CH-COOH$ $\longrightarrow$ $CH_3-CH_2-C-COOH$ / $\|$ / $CH-COOH$.

1. [redacted] J. Pr. C. 75, 443
2. B., 36, 1087.
3. J. Pr. C., 75, 481.

If instead of an alkyl group a phenyl be substituted in the 1,2 acid a quite different result is produced; no decrease in the stability of the ring can be observed. This phenylcyclopropane 2,3 acid is unreactive towards bromine, reduction with sodium amalgam, concentrated hydrobromic acid or fused alkali[1]; on heating to 282 - 340° it gives its anhydride without any ring cleavage taking place[2]. The spatial configuration of this acid is worthy of attention as considerable care has been taken in working it out. The acid, whether prepared by Buchner's method from diazoacetic ester[3] or by hydrolyzing the corresponding tetraester[4], is obtained in only one modification, although the three following are possible.

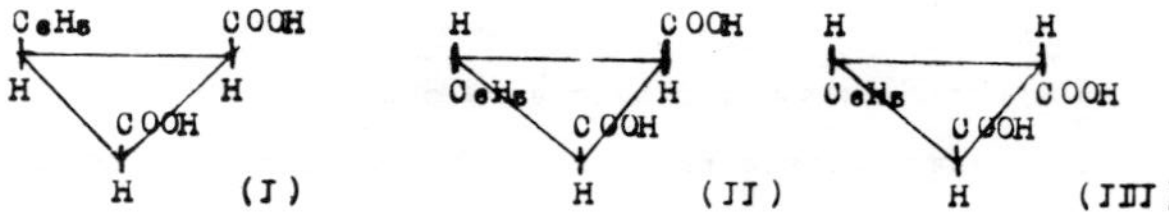

(I) (II) (III)

Formula III cannot represent the known acid since this forms an anhydride easily which it would not do if the two carboxyl groups were on opposite

1. B., 21, 2646; 25, 1148.
2. B., 21, 2646; 25, 1153.
3. B., 21, 2637.
4. J. Pr. C., 68, 163; 75, 490.

sides of the plane of the ring. To decide between I and II Buchner[1] oxidized the phenyl group to carboxyl by first nitrating it, then reducing the nitro group to an amino group, and finally oxidizing the whole phenyl group to carboxyl; the tribasic acid obtained was the known 1,2,3 acid which has the cis-trans configuration. Hence the correct configuration for the phenyl acid, if no rearrangement took place, is that shown by formula II.*

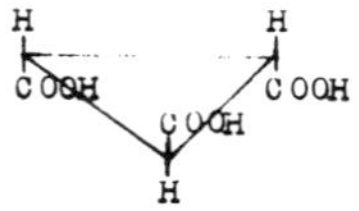

The cyclopropane hydrocarbons which have three substituent groups are all reactive substances. The 1,1,2 and 1,2,3 trimethyl compounds have been discussed. Besides these, two 1,1,2 compounds are known which have been prepared by the same method as was used for making phenylcyclopropane, that is, from an

* Attempts to rearrange the phenyl acid into the trans form by heating to 290° with fused alkali yielded a small amount of an isomeric acid[2]. This acid was unattacked by potassium permanganate and was probably the cis-phenyl cis-trans dicarboxylic acid; not enough was at hand to investigate its properties.

1. B., 36, 3780.
2. B., 36, 3782.

unsaturated ketone, hydrazine hydrate, and decomposition of the pyrazolene formed from these two compounds. Thus from phorone, $(CH_3)_2C=CH-CO-CH=C(CH_3)_2$, 1,1 dimethyl-2-isobutenylcyclopropane is formed; and from this by reduction some 1,1 dimethyl-2-isobutyl compound results[1]. Both substances react with hydrobromic acid and give straight chain bromides of the following formulas:

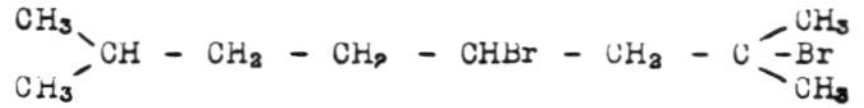

from:

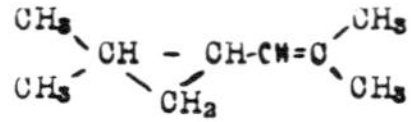

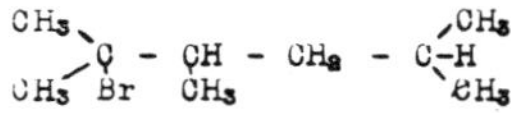

from

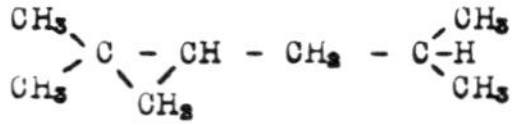

The first reaction is the only known exception to the rule previously stated, that in the case of cyclopropane hydrocarbons the ring breaks between the carbon atoms having the most and the fewest substituents.

Paal[2] supposed that he obtained two isomeric

1. Z., '13, II, 2130.
2. B., 36, 2425.

tribenzoylcyclopropanes among the products of the action of sodium ethylate or metallic sodium and ω-iodacetophenone. The compounds are of no interest as regards this discussion as their structure is uncertain and their reactions are few and indefinite.

Another compound whose formula has about the same degree of probability as tribenzoylcyclopropane is tricyancyclopropane. It was prepared by the action of barium hydroxide on tricyantricarboxylic ester[1] (page 63). No reactions of the compound have been recorded.

Several new methods of preparation are illustrated by the 1,1,2,2 cyclopropanes. The tetraester of the 1,1,2,2 acid was first prepared by Perkin[2] and later by Dressel[3] by the action of bromine on the disodium derivative of methylenebismalonic ester, two molecules of sodium bromide being eliminated.

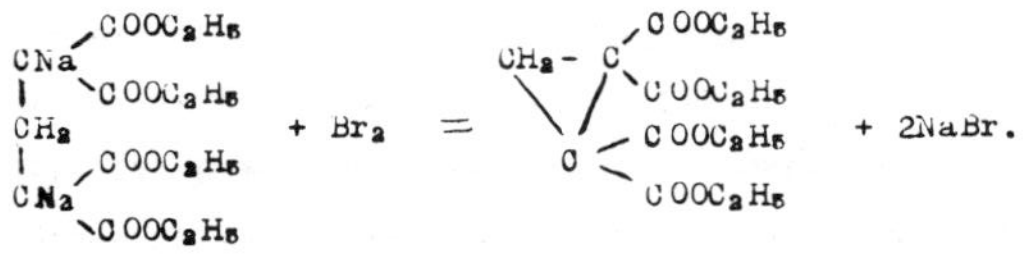

This reaction, which involves the elimination of two

1. B., 34, 3704.
2. B., 19, 1056.
3. Ann., 256, 196; Cf. also Stohmann, J.Pr.C., 45, 476.

atoms of a metal from the 1,3 position by means of halogen is the converse of the familiar elimination of bromine from a 1,3 dibromide by means of a metal. This same compound can be prepared by the converse method for if 1,3 dibrommethylenebismalonic ester is treated with metallic sodium in benzol, or with sodium methylate or ammonia in alcohol the cyclopropane ring is closed[1]. Still another way of forming this same cyclopropane compound involves the principle of coupling two molecules together to close the ring. On treating the sodium derivative of ethane tetracarboxylic ester with methylene iodide two molecules of sodium iodide are split out, and a cyclic ester is the product.[2] *

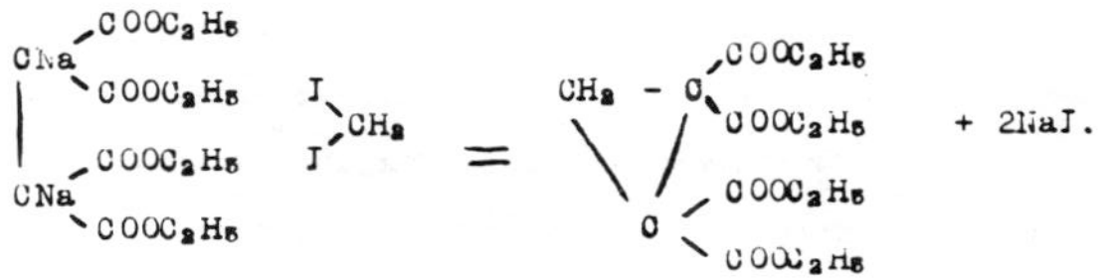

This reaction does not possess great scope as attempts

1. J.Pr.C., 77, 43, 55, 56.
2. J.Pr.C., 68, 167, 169.
* This ester is a solid and on hydrolysis yielded the cis form of the 1,2 diacid. An isomeric liquid ester was also obtained which when hydrolyzed gave the trans 1,2 diacid; no explanation is given.

to use other dihalogen compounds such as ethylidene iodide, CH_3CHI_2, failed. On attempting to prepare a hexamethylene derivative by the action of disodium methylenebismalonic ester and the corresponding 1,3 dibromide, the tetraester of cyclopropane resulted instead of a six-membered ring[1]. The exact mechanism of this reaction is not known.

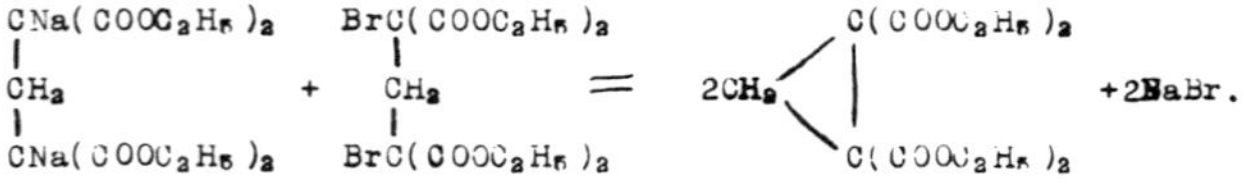

Another cyclopropane of the same type is 1,2 dimethylcyclopropane 1,2 diacid. This compound is formed by the elimination of hydrochloric acid from chlortrimethylsuccinic ester.[2] *

Both this acid and the 1,1,2,2 acid are unreactive as might be expected from their relation to the 1,2

* Another isomeric acid is obtained from bromtrimethylsuccinic anhydride[3]; it is probably an unsaturated acid, the formula for the bromanhydride perhaps being 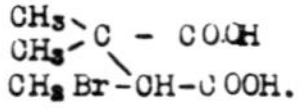

1. J.C.S., 83,780; 87,358.
2. J.C.S., 91,1954.
3. G.I.C., (2), 30,503.

diacid. The former is not attacked by bromine except very slowly and then with the evolution of hydrobromic acid, the latter on heating to 200° gives off carbon dioxide forming the anhydride of the cis 1,2 acid but no ring cleavage takes place[1]. Although the evidence is somewhat slight for both these compounds it is entirely probable that they do not contain a very reactive ring.

Besides these two tetrasubstitution products just mentioned, which apparently have an unreactive ring, another type of tetrasubstituted cyclopropane is known, whose reactions vary greatly with the substituent groups. These are the 1,1,2,3 compounds and a summation of their typical reactions is given in the table which heads the following page; they are treated in somewhat more detail in the next few pages.

In these substances, as in the 1,2,3 derivatives, it is interesting to compare two cyclopropanes differing only by the substitution of a carboxyl group by methyl; two such compounds are the 1,1,2,3 tetra acid and methylcyclopropane 2,2,3 triacid:

1. Ann., 256,196; J.Pr.C., 45,477; 77,49.

Summary of the reactions of

the tetrasubstituted cyclopropanes.

Substance	Reaction with HBr or dil. H_2SO_4	Reaction with Bromine	Reaction when heated alone	Remarks.
1,1,2,2 tetra-acid	-----	-----	Ring stable loses CO_2.	1,1,2,2 compounds.
1,2 dimethyl diacid	-----	not attacked.	-----	
1,1,2,3 tetra-acid	-----	-----	Ring stable loses CO_2.	
1 methyl 2,2,3 triacid.	-----	-----	Ring stable. loses CO_2; heated with H_2O at 200°, ring cleavage.	
1 acetic acid 1, 2,3, triacid	-----	-----	Ring breaks on hydrolysis.	
Phenyl 2,2,3 triacid	-----	-----	Ring breaks on hydrolysis.	
1,1 dimethyl 2,3 acid	No action cold; at 100° ring cleavage.	not attacked.	Stable; cis form gives anhydride.	

The esters of both these acids are obtained by similar reactions, the first by the action of sodium malonic ester on dibromsuccinic ester[1] or brommaleic ester,[2 3] the second from sodium malonic ester and either dibrombutyric or brom (or chlor) crotonic

1. B., 17, 1652; J.C.S., 47, 823.
2. Ann., 229, 91.
3. Ann., 284, 224.

ester[1,2] *. These reactions are of the usual type discussed on page 42.

The ring of the tetrabasic acids, like that of the 1,2, the 1,1,2,2 and the 1,2,3 acids is stable on heating, that is, the substance loses carbon dioxide at 190°-200°, but no ring cleavage takes place, the 1,2,3 acid being the product[3]; the methyl 2,2,3 triacid loses carbon dioxide without cleavage and the anhydride of the corresponding 1,2 acid is the product.[1] These reactions would indicate ring stability on the part of both acids; in the case of the 1,1,2,3 acid no further evidence is known, but the methyl compound has been further investigated and found to be somewhat reactive. On heating with water to 210° the ring splits and a mixture of methylitaconic and methylparaconic acids results.[4]

* The properties of the tetrabasic acid prepared from dibromsuccinic ester by Perkin differed from the acids obtained in other ways. Buchner repeated this work, however, and got only the known acid. (Cf. note on page 45.) The 1,1,2,3 acid is also obtained by the action of barium hydroxide on dicyancyclopropane 1,2,3 tricarboxylate.[5]

1. B., 36, 1085, 1087.
2. B., 17, 2833.
3. Ann., 284, 224; 229, 89
4. B., 36, 1087.
5. B., 33, 2980; 34, 370.

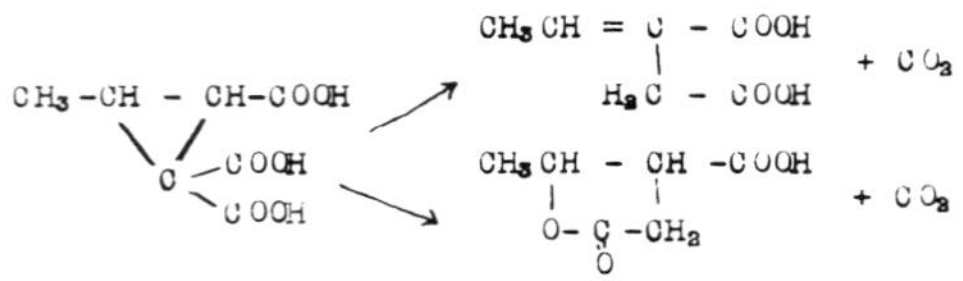

A methylcyclopropane tribasic acid isomeric with the above is methylcyclopropane 1,2,3 triacid. The ester of this acid is made by heating the pyrazolene obtained from diazoacetic ester and citraconic ester[1]. The acid resulting from the ester by hydrolysis does not attack permanganate and on distilling gives an anhydride. No further reactions are known.

The methods of preparation of most cyclopropane acids involve first the preparation of the cyclic ester and subsequent hydrolysis to the free acid. That such hydrolysis did not change the cyclic structure has been seen to be true in all the cases thus far examined; two exceptions to this rule will now be studied, for they are both 1,1,2,3 compounds. Aconitic ester and diazoacetic ester give by the usual reactions a cyclopropane ester as follows[2]:

1. B., 27, 877.
2. B., 27, 871.

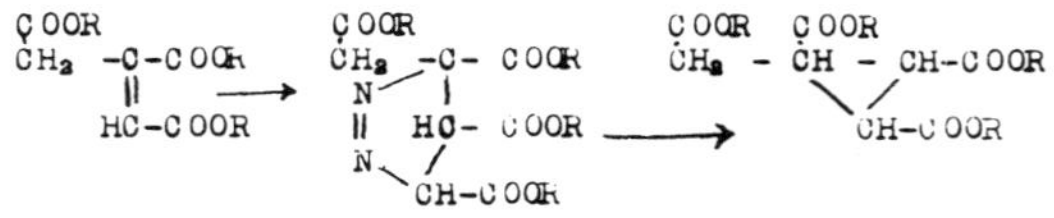

This ester of cyclopropaneacetic acid, 1,2,3 triacid could not be successfully hydrolyzed to a solid acid by sodium or barium hydroxide, but heated with sodium carbonate at 65° for thirty hours some solid acid could be obtained. This acid was not the expected cyclic tetrabasic acid but a lactonic acid of the formula COOH-CH_2-CH - CH(COOH) - CH_2 (O —— C=O). The ring must therefore have split during hydrolysis, probably as follows:

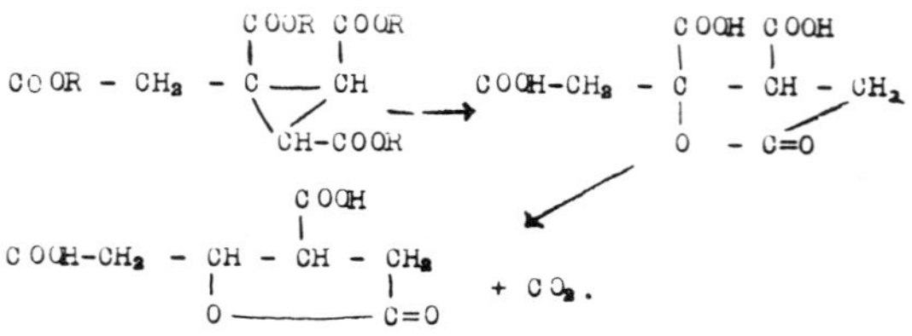

Sodium malonic ester and dibromcinnamic ester react and a compound results which analyses for phenylcyclopropane 2,2,3 triester and is not oxidized by potassium permanganate[1]. On attempting to prepare the tribasic acid by saponification with aqueous sodium hydroxide phenylparaconic acid is the product[2].

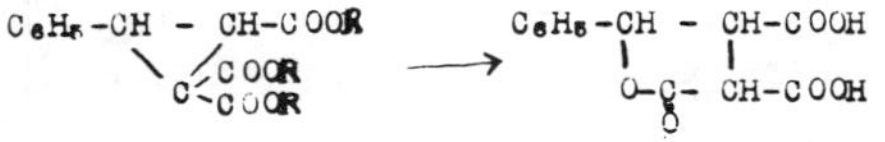

The splitting of the ring in these two cases may or may not be significant. If the hydrolysis was carried out in the same manner as was usual with most of the other cyclopropane esters, then these two substances undoubtedly behave differently from all other esters, and this must probably be due to some peculiar reactivity of the ring. It may be that in one or both of these cases conditions were sufficiently diverse to vitiate any comparison with other compounds. That this might be true in the first instance is quite likely as heating with sodium carbonate at 65° for thirty hours is somewhat different from ordinary methods of hydrolysis.

Caronic acid, or 1,1 dimethylcyclopropane 2,3 diacid is a compound of some interest because of

1. B., 25, 1154.
2. B., 36, 3776.

its relation to caron which in turn in related to carvon and other dicyclic terpenes. Caron on oxidation with potassium permanganate gives two isomeric acids which are named cis and trans caronic acids. They are the geometric isomers of 1,1 dimethylcyclopropane, 2,3 acid[1].

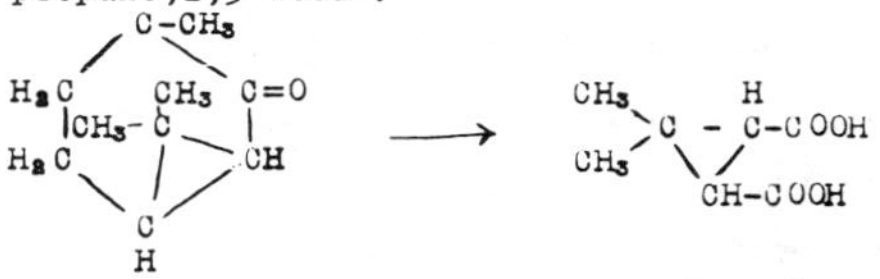

Caron

Caronic acid.

The constitution of the acids thus obtained was well established by their subsequent preparation from α- brom β,β-dimethylglutaric acid, alcoholic potash being used to eliminate the hydrobromic acid.[2] * Both

* Some lactone of hydroxydimethylglutaric acid is also formed. An interesting point in connection with this closing of the ring is that the hydrogen ethyl salt of the bromine compound, $COOH-CHBr-C(CH_3)_2-CH_2-COOC_2H_5$, if treated with sodium carbonate gave a very good yield of a lactone, $COOH-CH-C(CH_3)_2-CH_2$ (O——C=O), but if the same hydrogen ethyl salt were boiled with alcoholic potash an almost quantitative yield of trans-caronic acid resulted[3]. The caronic acids have also been obtained[4] by the hydrolysis of the ester of dimethylcyclopropane 1,1,2,2 tetracarboxylic acid, two molecules of carbon dioxide being eliminated during the saponification.

1. B., 29, 2797.
2. J.C.S., 75, 50.
3. J.C.S., 75, 56-59.
4. J.Pr.C., 75, 502.

acids are unattacked by cold bromine, hot dilute sulphuric acid or cold hydrobromic acid; on heating with the latter reagent to 100°, however, ring cleavage occurs and a lactonic acid, terebic acid, results.

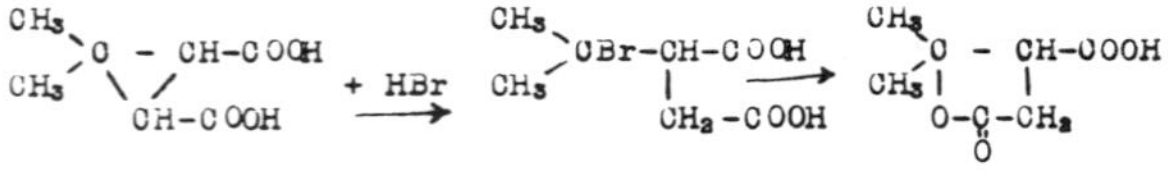

terebic acid.

The ring in this disubstituted 1,2 acid is thus less stable than in the parent acid, but still quite unreactive.

The penta- and hexa-substitution derivatives need be considered but briefly as their only interest lies in the fact that they are useful in the synthesis of some of the other types whose reactions have been more studied. By the action of bromine on disodium propane tetracarboxylic ester, the ester of the 1,1,2,2,3,3 acid was obtained[1].

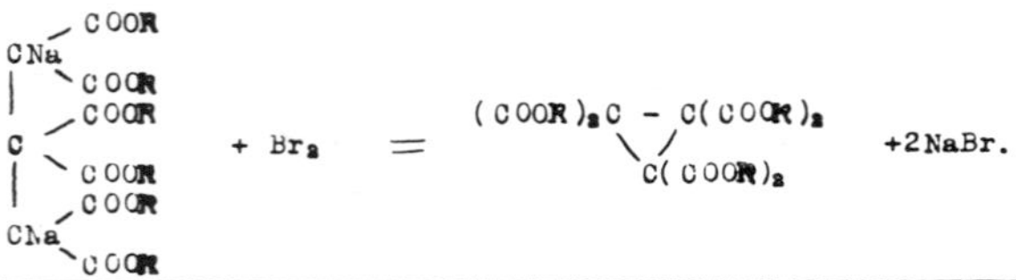

1. J.Pr.C., 68, 166.

This ester on hydrolysis gives the 1,2,3 acid by losing three molecules of carbon dioxide. A similar loss of carbon dioxide has resulted when most other esters of these classes (penta-and hexa-substitution products) have been hydrolyzed and therefore only the esters of the compounds are known and no reactions have been tried with them. Methylcyclopropane 2,2,3,3 tetraester is formed by a reaction similar to the one just mentioned[1], the disodium derivative of ethylidenebismalonic ester and bromine being used. The dimethylcyclopropane 2,2,3,3 tetraester was prepared by the elimination of hydrobromic acid from monobromisopropylidenebismalonic ester.[2]

By one or both of these two typical reactions the esters of the 2,2,3,3 acids of the following substituted cyclopropanes can be prepared: ethylcyclopropane[3], phenylcyclopropane[4], and trichlormethylcyclopropane[5]. These tetraesters have never been hydrolyzed to the corresponding tetraacids; the only products obtained have been the 1,2 acids which are dealt with elsewhere. No conclusions can there-

1. J.Pr.C., 68, 157.
2. J.Pr.C., 75, 494.
3. J.Pr.C., 75, 479.
4. J.Pr.C., 68, 162; 75, 489.
5. J.Pr.C., 75, 486.

fore be stated as regards the reactivity of the ring in the penta- and hexa-substituted compounds.

Two pentasubstituted cyclopropanes prepared by Perkin[1] (cf.page 72) are of some interest. As seen from formulas I and II they are substitution products

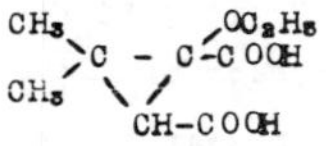

I

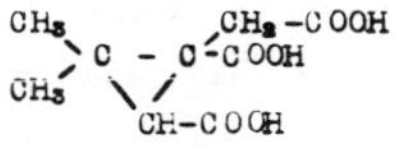

II

of caronic acid. No reactions of II are known but I reacts with both sulphuric acid and hydrobromic acid, the ring splitting between the carbon atoms carrying the carboxyl groups and a lactone resulting. This is an unusual way for the ring to cleave with an acid reagent and is the one exception noted to the general way in which substituted 1,2 acids react.

If the sodium derivative of cyanacetic ester is treated with halogen or with bromcyanacetic ester a number of products can be isolated[2]. Among these are two cyclopropane derivatives, 1,2,3 tricyan, 1,2,3 tricarboxylate, and dicyantricarboxylate. The formation of these compounds involves the coupling together of three moleculws of the cyanacetic ester by some

1. J.C.S., 79, 732.
2. B., 33, 2976; 34, 3704.

mechanism not at present evident. As has been previously mentioned (page 51.), the 1,2,3 tricyancyclopropane-1,2,3 tricarboxylate gives on hydrolysis[1] tricyancyclopropane; cyclopropane 1,1,2,3 tetraacid and an acid containing one cyanogen group are probably also produced. The dicyan compound gives as one of the products of hydrolysis the 1,1,2,3 cyclopropane tetraacid[1]. (Cf. note, page 56.)*

The next class of cyclopropane derivatives to be considered is that comprising the group of side-chain compounds. These compounds are monosubstituted

* By the interaction of cyanacetic ester, ammonia and ketones of the formula $CH_3-CO-C_nH_{2n+1}$ compounds of the following formula[2] result:

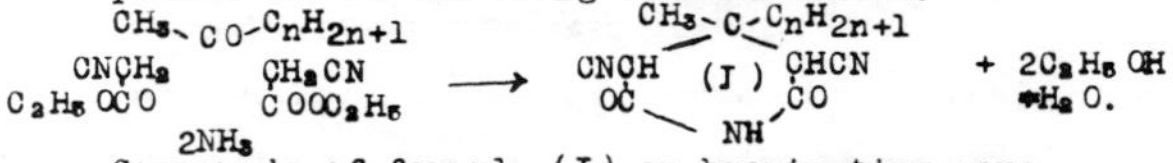

Compounds of formula (I) on brominating give dibromides (II) which on heating spontaneously lose bromine and a cyclopropane ring is closed (III)

(II) (III)

No interesting reactions of these compounds have been recorded and their structure is somewhat uncertain.

1. B., 33, 2976; 34, 3704.
2. Z., '98, II, 544; '99, II, 439.

derivatives whose reactions and methods of preparation are of minor interest in a study of the properties of the cyclopropane ring. There are, however, certain interesting points in connection with their study and they will be briefly considered.

Cyclopropylcarbinol and its derivatives make a convenient starting point for this discussion. Unfortuneately the subject is somewhat confused because the first workers in the field, Henry and Dall, misinterpreted some of their results. Potassium hydroxide eliminates hydrochloric acid from γ-chlorbutyrilnitrile forming cyclopropane cyanide. (Cf. page 18 .) By reduction an amine is obtained[1] which can be converted into an alcohol with nitrous acid[1], and a number of halides and esters can be formed from this alcohol. This substance was considered to be cyclopropylcarbinol. Demjanow, however, showed that the alcohol was in reality a mixture of cyclobutanol and cyclopropylcarbinol[2]. He succeeded in preparing the pure cyclopropylcarbinol by the reduction of ethyl cyclopropanecarboxylate. The formation of the cyclobutanol from the cyclopropane amine involves a

1. Z., '01, I, 1357; '02, I, 913.
2. B., 40, 4393; Z., '03, II, 489.

peculiar shifting of linkages but the same investigator has shown this to be a quite general reaction of cyclic amines and explains the phenomena by assuming the existence of an intermediate dicyclic compound[1]. The pure cyclopropylcarbinol so produced was oxidized to cyclopropyl aldehyde which in turn could be oxidized to the well-known cyclopropane monobasic acid.[2] By this series of interesting reactions the proof of the structure of Demjanow's carbinol seems to be made certain.

The bromide was prepared from the cyclopropyl carbinol obtained by reduction and from the halide a Grignard compound was formed. This compound treated with formaldehyde gave among other products cyclobutylcarbinol.[3] Whether this is to be attributed to isomerization or to the presence of cyclobutanol in the cyclopropyl alcohol is uncertain.

Demjanow states that methylcyclopropylcarbinol results from the reduction of acetylcyclopropane.[3] This seems very unlikely in as much as Perkin[4] found that acetylcyclopropane readily reduced to a straight chain compound (propylmethyl-

1. B., 40, 4393; Z., '03, II, 489.
2. B., 40, 4397.
3. Z., '14, I, 1998.
4. J.C.S., 59, 874, 886. (Demjanow seems to have overlooked this research.)

carbinol). Demjanow's carbinol differs in boiling point from the saturated carbinol by only three degrees and it is highly probable that this is the compound he had, in which case his subsequent transformations are of no significance. The same chemist reduced the oxime of acetylcyclopropane to an amine[1] and this amine on treatment with nitrous acid yielded the alcohol just considered, which was written as cyclopropylmethylcarbinol. In view of the probable structure of this latter compound it seems as if ring cleavage had taken place either during the reduction of the oxime or on subsequent treatment with nitrous acid.

Acetylcyclopropane treated with methylmagnesium iodide is reported to give[2] a cyclopropyldimethyl carbinol. On attempting to dehydrate this compound with oxalic acid a hexylene oxide probably results, $(CH_3)_2$-CH-CH_2-CH_2-CH_2 (O bridge), which the investigator explains by ring cleavage. It would be expected from general theoretical considerations (page 229) that the Grignard reagent would add in the 1,4 position to the ring and ketone group and an unsaturated alcohol or rather its

1. Z., '14, 1998. (Oxime of benzoylcyclopropane, Z., '12, I, 1459.
2. B., 34, 2884, 3887.

tautomer, a ketone, would be the product. This ketone

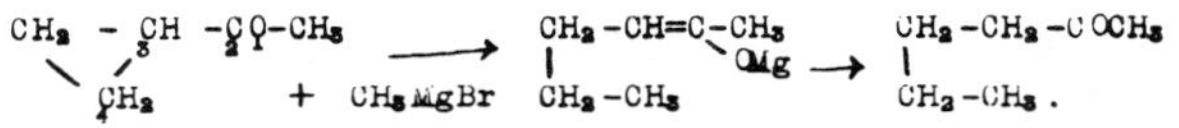

might lose water when treated with oxalic acid forming an oxide; evidence is lacking to decide between these possibilities. An entirely similar set of transformations have been tried with benzoylcyclopropane and methylmagnesium iodide[1]. If ring cleavage took place in this case by 1,4 addition phenylbutylketone would result. This compound[2] has almost the same boiling point as that given for the product of the reaction; the refractive index of the ketone, however, is n_D^{20} = 1.5350 while that found for the supposed cyclopropylphenylmethylcarbinol is n_D^{20} = 1.515; this difference is probably significant and argues for the cyclic structure of the product.

Bruylants[3], following in the steps of Henry and Dall, prepared a number of compounds by the action of the Grignard reagent on cyclopropyl cyanide; Michiels[4] worked along similar lines and like the other investigators, prepared a series of alcohols,

1. Z., '12, I, 1459.
2. Z., '01, I, 1613.
3. Z., '09, I, 1859.
4. Z., '11, I, 67.

halides and hydrocarbons which he considered to be side chain derivatives of cyclopropane. These compounds are of no particular interest and their structure cannot be looked upon as being definitely established, as no proof of constitution is given for any of them.

Kishner[1] studied the reaction of the Grignard reagent on the ester of cyclopropane monocarboxylic acid. He obtained a compound which he considered to be cyclopropyldimethylcarbinol, and from this he obtained an unsaturated hydrocarbon, cyclopropylmethylethylene, $CH_2-CH-C=CH_2$ (ring closed by CH_2; CH_3 on C), The cyclic structure of this compound seems to have been established, as acetylcyclopropane is one of the products of oxidation. The hydrocarbon heated with dilute sulphuric acid gives hexyleneoxide and the carbinol combines with cold hydrobromic acid forming a straight chain bromide; thus the ring in these compounds is apparently quite reactive.

An interesting set of reactions has been carried out with sodamide, benzoylcyclopropane and alkyl halides.[2]

1. Z.,'11,II,363.
2. Z.,'12,II,497.

In this way methylbenzoylcyclopropane, allylbenzoylcyclopropane and benzylbenzoylcyclopropane have been prepared. The last-named compound on further treatment with sodamide splits off benzene and gives benzylcyclopropanecarboxylic acid. Unfortunately no further reactions have been tried with these compounds.

Very little is known about cyclopropane derivatives which contain a double bond in the cyclic system. Only two series of such substances have been prepared and an exact proof of their structure is lacking.

Mereshkowski[1] has recently claimed to have prepared methylcyclopropene (II) by the action of zinc on the tribromide (I):

(I) $(CH_3)(CH_2Br)CBr-CH_2Br$ (II) $CH_3-C=CH$ (ring closed by CH_2)

So few reactions have been tried with the product that its structure cannot be considered as even probable. Feist[2] has studied the acid obtained by the action of alkali on bromisodehydracetic

1. Z.,'14,I,2160.
2. B.,26,750; J.C.S.,87,1062.

ester (I). He considered the acid to be a methylcyclopropenedicarboxylic acid (II).

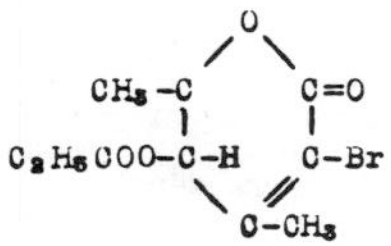
(I)

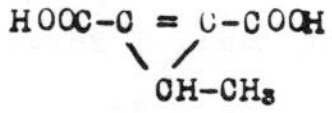
(II)

This acid readily formed a dibromide which on reduction yielded an isomeric acid, to which Feist assigned the formula (III):

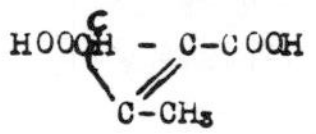
(III)

The dibromide formed from the ester of acid (II), however, when treated with zinc regenerated the original substance, and did not give acid (III). With an aqueous solution of bromine ring cleavage took place between the atoms carrying the carboxyl groups and a lactonic acid was apparently formed. While no direct evidence is at hand for either formula III or II it is extremely difficult to find any other formula for a dibasic acid of this composition. However, Buchner[1] apparently had a dibromide (IV) very similar to that obtained from acid (II) by

1. B., 38, 1001.

Feist (V), which on treatment with zinc and acid or

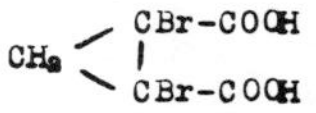
(IV)

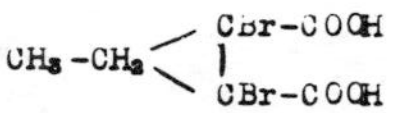
(V)

sodium amalgam gave the saturated cyclopropane acid and not an unsaturated cyclic acid as claimed by Feist; it is possible that the difference in conditions might account for this apparent inconsistency but this type of reaction must be further studied before the structure of Feist's acids can be considered as definitely established.

Perkin[1] attempted to prepare a cyclopropene acid by eliminating two molecules of hydrobromic acid from α,γ,dimethyldibromglutaric ester. All the reagents which he tried failed to give him the desired product, an ethoxy derivative (I) being formed instead.

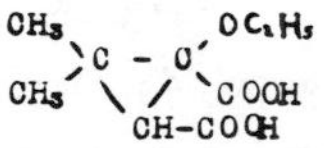
(I)

Perkin believed that a cyclopropene compound was first formed but that it subsequently added alcohol. This view was supported by the observation that on carrying out the reaction in the presence of sodium malonate, malonic ester apparently added instead of

1. J.C.S., 79, 732.

alcohol, giving compound (II). If this explanation

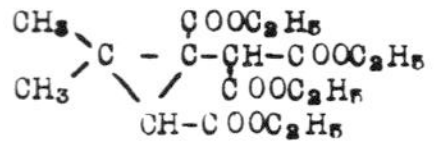

(II)

were true it would be contradictory to Feist's observations as his cyclopropene compound (of almost identical structure) did not add alkaline or acid reagents to form a saturated cyclic compound. Perkin's assumption seems unnecessary as there is the possibility of direct replacement of a bromine atom by the ethoxy or malonic ester group.

This general historical survey of cyclopropane chemistry can be concluded with a consideration of some of the more important physico-chemical data which have been obtained in this field.

A study of the thermo-chemical data obtained by Berthelot, Stohman and others shows that the heat of combustion of a cyclopropane compound is slightly greater than that of the corresponding unsaturated compound, the heat of formation being correspondingly greater. That trimethylene itself is richer in energy than propylene was shown by Berthelot[1] who found that the heat given off when bromine combined with propylene

1. Ann.de Chim.Phys.,(6) 30,560; (7) 4,110.

was 9.4 kilogram calories less than the heat given off in the reaction between bromine and trimethylene. Very nearly the same difference was also noted when sulphuric acid was used in the place of bromine.

In the following table are given all the heats of combustion of cyclopropane compounds that could be found in the literature; the analogous unsaturated compounds are given where possible for comparison. It will be noticed that the total energy of the 1,1 dibasic acid and the 1,2 acid are practically the same. This is very interesting when it is remembered that their reactivities were very different and that two isomers so closely related as fumaric and maleic acids have an energy content differing by 6.2 kg.Kal. An interesting article by Stohman[1] correlates a great deal of thermochemical data concerning straight chain and cyclic compounds but as the discussion only slightly concerns cyclopropane chemistry the paper need not be reviewed here. The table is entered on the following page.

1. J.Pr.C.,48,490.

Substance.	Heat of combustion at constant pressure in kg.-Kal. per g.m.w.	Reference.
Trimethylene	499.4 507.0	T B
Propylene	492.74 499.3	L L
Dichlorcyclopropane	426.0 426.5	B B
Acetylcyclopropane	698.0	Z
Cyclopropanecarboxylic acid	479.7	R.L.
Crotonic acid	472.3	L.
Cyclopropane 1,1 diacid dimethyl ester	827.7	R.L.
Cyclopropane 1,1 diacid	483.4	S
Cyclopropane 1,2 diacid	484.4	S
Itaconic acid	477.7	L
Mesaconic acid	499.[illegible]	L
Citraconic acid	487.4	L
Cyclopropane 1,1,2,2 tetracarboxylic acid	483.2	S
Cyclopropane 1,1,2,2 tetramethyl ester	1170.9	S

References.

T Thomsen's Thermochemistry, page 369.
B Berthelot, Ann.de Chim.& Phys.,(6) 28,126,565.
L Landolt and Bornstein Tables.
R.L. Roth, observer: listed in Landolt's Tables.
S Stohman, J.Pr.C.,40,202; 48,490.
Z Zubow, Z.,'02,I,161.

The dissociation constants for a few cyclopropane acids have been measured and are given below. (From Landolt and Bornstein, 4th edition, page 1170; a search through the literature has disclosed no other determinations.) A few unsaturated and saturated acids are given for comparison. No generalizations can be drawn from these results but is is important to note that the 1,1 acid is a very strong acid compared with most organic acids, being of about the same strength as oxalic acid. The 1,1,2 triacid, however, is weaker, the introduction of another carboxyl group decreasing the strength of the acid. Unfortunately no data for unsaturated malonic acid derivatives is obtainable, so that no comparison can be made. For the other cyclopropane acids the value of K is in most cases of the same order of magnitude as found for the isomeric unsaturated acid, but larger than for the saturated acids. The table follows on page 77.

Substance	K
cyclopropanecarboxylic acid	1.7×10^{-5}
crotonic acid	$2. \times 10^{-5}$
isocrotonic acid	3.6×10^{-5}
vinylacetic acid	3.8×10^{-5}
cyclopropane 1,1,diacid	2.0×10^{-2}
cyclopropane 1,2 diacid (cis)	4.0×10^{-4}
cyclopropane 1,2 diacid (trans)	2.06×10^{-4}
glutaconic acid	1.83×10^{-4}
mesaconic acid	$8. \times 10^{-4}$
citraconic acid	3.4×10^{-3}
itaconic acid	1.5×10^{-4}
cyclopropane 1,1,2 acid	9.1×10^{-3}
malonic acid	1.6×10^{-3}
glutaric acid	4.7×10^{-5}

Only a few words need be said about the refractive index and other optical measurements. Perkin (senior) determined the magnetic rotation[1] of a few cyclopropane derivatives and the values so obtained were used as an argument for the cyclic formula in the pioneer days of cyclopropane chemistry.

1. J.C.S., 67, 117.

Smiles in his book on "Chemical Constitution and Physical Properties", page 270, gives a summary of the more important cyclopropane compounds whose refractive index has been measured together with the observed and calculated values for the molecular refraction. The increment due to the presence of the cyclopropane ring varies from 0.66 to 1.3. This great variability shows that it would be very uncertain to dra any conclusions as to structure from the molecular refraction of these compounds. Östling[1] has collected the data for some three and four membered rings and claims that a fairly consistent value of 0.7 can be taken for the average increment due to the cyclopropane ring. He shows, however, that in some dicyclic systems this increment is much larger than 0.7 and also in some very simple cyclopropane derivatives the value may be nearly twice the average. Haller[2] has measured the refractive indices of numerous cyclopropane derivatives of rather questionable structure and claims to have obtained evidence for the conjugation of the ring with a carbonyl group. The change in the

1. J.C.S., 101, 457.
2. Z., '12, II, 497, 1926.

increment is small, however, and in view of the great uncertainty of this quantity, the results can not be regarded as very significant.

An extremely interesting and important branch of cyclopropane chemistry lies in the field of dicyclic compounds. A survey of this subject would involve too much af a digression from the point at issue, the chemical properties of the cyclopropane ring, - and hence must be left aside. With this exception a complete review has been made of the literature which concerns cyclopropane derivatives and while it is not safe to claim that all the published work has been covered, it is hoped that the omissions are few and unimportant.

* * * * * *

CHAPTER 2

General and theoretical aspects of the cyclopropyl group

DIETER CREMER, ELFI KRAKA and KALMAN JOSEF SZABO

Department of Theoretical Chemistry, University of Göteborg, Kemigården 3, S-41296 Göteborg, Sweden

I. INTRODUCTION 44
II. STRUCTURE AND TOPOLOGY 45
III. ELECTRONIC STRUCTURE—ORBITAL DESCRIPTION 48
A. Walsh Orbitals and Refined Walsh Orbitals 49
B. Förster–Coulson–Moffitt Orbitals and Non-orthogonal Valence Bond Hybrid Orbitals 55
C. SCF Orbitals, Orbital Energies and Ionization Potentials 60
IV. ELECTRON DENSITY DISTRIBUTION AND CHEMICAL BONDING 61
A. Analysis of the Difference Electron Density Distribution 62
B. Analysis of the Electron Density Distribution 64
C. Analysis of the Laplacian of the Electron Density Distribution 68
D. Surface Delocalization, π-Complexes and Bonding 68
V. ENERGY AND STABILITY 73
A. Analysis of Ring Strain in Terms of Atomic Energies 74
B. Analysis of Ring Strain in Terms of Bond Energies 75
C. Dissection of the Molecular Energy into Strain Contributions 77
D. Dissection of the Molecular Wave Function 79
E. Strain Energy, σ-Aromaticity and Surface Delocalization 82
VI. GEOMETRY 83
A. Geometry of Cyclopropane 85
B. Substituent Effects on the Geometry of the Cyclopropyl Group 86
1. The molecular orbital description 87
2. The electron density description 90
3. Substituted cyclopropanes 93
C. Rationalization of the Geometry of Three-membered Rings 96
VII. VIBRATIONAL SPECTRA 98
VIII. ONE-ELECTRON PROPERTIES 106
IX. NMR SPECTRA 109

The chemistry of the cyclopropyl group, Vol. 2
Edited by Z. Rappoport

X. EXCITED STATES AND ULTRAVIOLET ABSORPTION SPECTRA . . . 111
XI. CHARGED CYCLOPROPYL GROUPS . . . 113
A. Cyclopropyl Anion . . . 113
B. Protonated Cyclopropane . . . 114
C. Cyclopropyl Radical Cation . . . 116
D. Cyclopropyl Cation . . . 117
XII. THE CYCLOPROPYL RADICAL . . . 120
XIII. FUSED CYCLOPROPANES . . . 123
A. Bicyclo[1.1.0]butane . . . 123
B. [1.1.1]Propellane . . . 124
XIV. FORMATION AND REACTIVITY . . . 125
A. Formation Reactions . . . 125
B. Thermal Ring-opening and Stereomutation Reactions . . . 126
C. Electrophilic Ring-opening Reactions and Insertion Reactions . . . 126
XV. ACKNOWLEDGEMENTS . . . 129
XVI. REFERENCES . . . 129

I. INTRODUCTION

The chemical and physical properties of cyclopropane (**1**) and the cyclopropyl group significantly differ from those of other alkanes or cycloalkanes. Best known is the double-bond-like behaviour of a cyclopropane ring in conjugation with a π-system. Also unusual are the length of its CC and CH bonds, upfield shifts of its proton and ^{13}C NMR signals, its UV spectrum and its (for a three-membered ring) relatively large stability. Contrary to other cycloalkanes, **1** undergoes ring opening upon attack by an electrophile, but resists substitution.

Many of the peculiarities of cyclopropane and the cyclopropyl group have been described in detail in previous review articles[1–5] and in many theoretical investigations that focused on the electronic structure of **1**[6–13]. A considerable part of the *ab initio* work on **1** up to the year 1987 has been reviewed by Wiberg in the pervious volume on the *Chemistry of the Cyclopropyl Group*[14]. Some other theoretical work has also been covered in review articles by Runge (on the chiroptical properties of the cyclopropyl group)[15], by Morris (on the NMR and infrared spectra of cyclopropanes)[16], by Ballard (on the photoelectron spectra of cyclopropanes)[17], by Battiste and Coxon (on the acidity and basicity of cyclopropanes)[18] and by Tidwell (on conjugation and substituent properties of cyclopropanes)[19], which appeared also in the previous volume on the *Chemistry of the Cyclopropyl Group*. However, since these articles were written, a considerable amount of new research on **1** has been published that will be covered in the present article. Also, in each of the previous articles a particular aspect of the theory of the cyclopropyl group was presented since a general account on the theoretical work of **1** was not intended. For example, Wiberg[14] concentrated in particular on his own work on **1** and its derivatives to get a consistent presentation of the theory of the cyclopropyl group.

In the present article, *ab initio* investigations on **1** and the cyclopropyl group published in the last seven years are summarized. But apart from this, we will also include some of the theoretical work that appeared before 1987, but were not mentioned in Wiberg's review. We will also shortly repeat some of the theoretical aspects of the cyclopropyl group already included in Wiberg's work, because this is necessary for a readable presentation. Finally, we will stress electronic features of the cyclopropyl group that have an impact on the general understanding of phenomena such as the delocalization of electrons or chemical bonding. For example, investigation of the electronic structure of **1** is essential for a

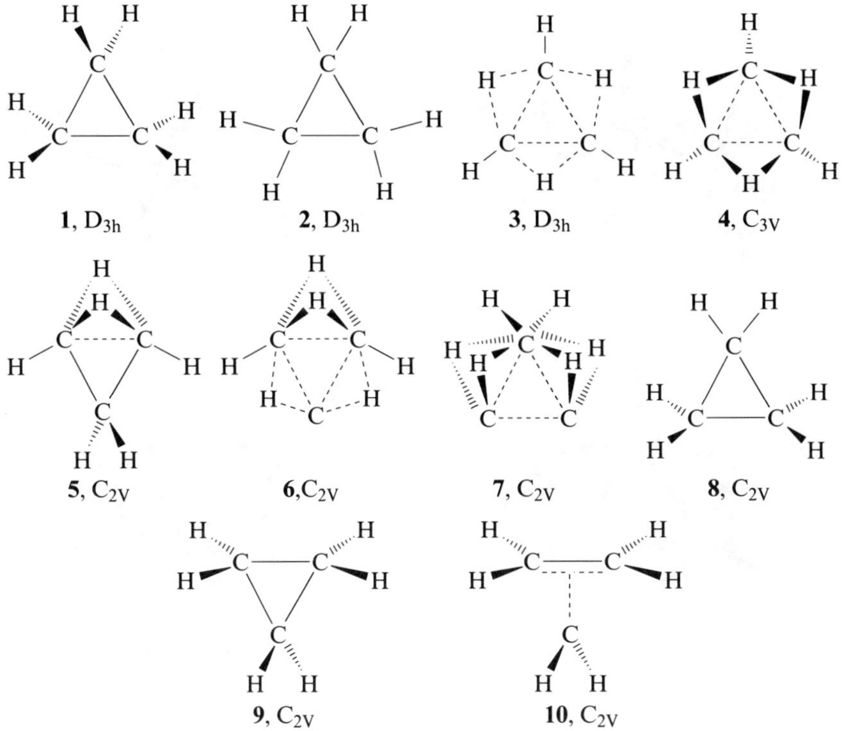

FIGURE 1. Possible structural formulas for a molecule with a C_3-ring and the stoichiometric formula C_3H_6

critical evaluation of new concepts such as surface delocalizaton of electrons (as opposed to ribbon or volume delocalization) or σ-aromaticity (as opposed to π-aromaticity)[8–13, 20].

In this way, we attempt to present a state-of-the-art account of the theoretical work on the cyclopropyl group that also leaves room for controversial descriptions even if we have to sacrifice the quality of consistency in the presentation. The inclusion of conflicting results will show an important point: although the cyclopropyl group is one of the best investigated functional groups (both experimentally and theoretically), there are still open questions with regard to its electronic structure, molecular properties and reactivity. It is the purpose of this chapter to give a fair presentation of all theoretical aspects of the cyclopropyl group, not only those that are no longer subject to theoretical discussions. To fulfill this objective, we have covered the literature up to July 1994.

II. STRUCTURE AND TOPOLOGY

Cyclopropane, C_3H_6, is the smallest cycloalkane. If one considers that the three C atoms have to form a ring, then only few chemically reasonable possibilities will remain to attach the H atoms to the carbon ring. In structures **1** and **2** shown in Figure 1, all C atoms are tetravalent while in structures **3**, **4** and **5** some or all C atoms are penta-coordinated because H atoms take a bridge position between the C atoms thus leading to non-classical cycloalkane structures. Structures **6** and **7** lead to hexa- and octa-coordinated C with the H atoms clustering in one region of the molecule.

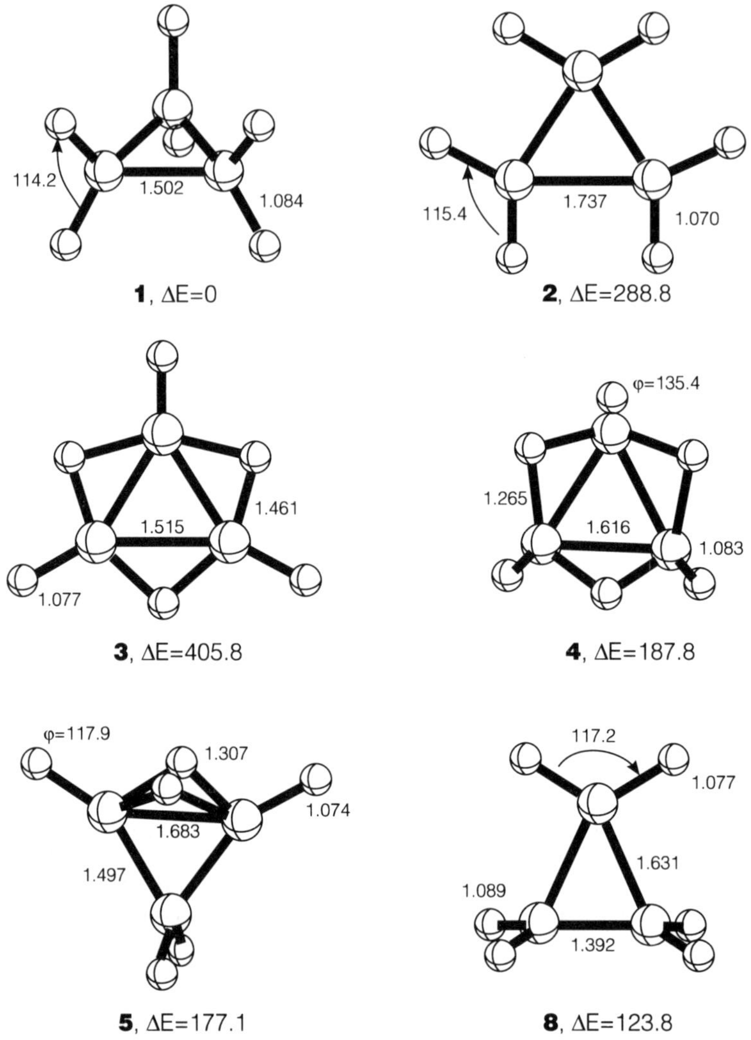

FIGURE 2. Geometries (bond lengths in Å, angles in deg) and relative energies (in kcal mol^{-1}) of possible structures of C_3H_6 [MP2/6-31G(d) calculations]. Structure **3** has a triplet ground state while all other structures possess singlet ground states. Angle φ denotes the folding of the plane $CH_{bridge}C$ out of the ring plane

Considering that cyclopropane possesses D_{3h} symmetry, structures **4**, **5**, **6** and **7** can be excluded. Considering further that all C atoms in cyclopropane are tetravalent, only **1** and **2** remain as possible cyclopropane structures. Finally, structure **1** is left as the only possibility if one considers that each CH_2 group in the molecule has to be perpendicular to the ring plane.

Although structures **2**–**7** are rather unusual alternatives for **1**, it is interesting to compare their properties with those of **1**. As shown in Figure 2, structure **2** with all CH_2 groups in the plane of the carbon ring possesses a relatively high energy [289 kcal mol^{-1}, MP2/6-

31G(d) calculations] that would be sufficient to break not just one CC (CH) but several CC (CH) bonds. This is because structure **2** contains three times a 'planar C atom' (i.e. all four bonds of tetravalent C lying in a plane) which is a configuration that leads to two rather than four bonding electrons and two non-bonding pπ-electrons. In the case of **2**, there would be just six electrons for 12 CC and CH bonds, thus leading to a highly electron-deficient hydrocarbon with rather long CC bonds (1.74 Å, Figure 2). These are also weakened by electron repulsion between six π-type lone-pair electrons located at the C atoms. Hence, **2** is not an electronically feasible structure. Even if only one CH_2 group in cyclopropane is rotated into the ring plane as in structure **8** (Figure 1), the energy increases to 124 kcal mol^{-1} (Figure 2) as has been shown by Schleyer and coworkers[21]. A planar C atom in a three-membered ring can only be stabilized if hetero atoms with (a) π-electron acceptor and (b) σ-electron donor ability are incorporated into the ring. This is accomplished for diboracyclopropane, which is more stable by 13 kcal mol^{-1} in the planar structure than in the non-planar structure[22].

Structure **3** may be thought of as containing sp^2-hydridized C atoms connected by three H bridges. There would be six electrons from the C atoms and three from the bridging H atoms for cyclic bonding, which means that one σ-electron would be unpaired. Similarly, one of the three π-electrons would be also unpaired, thus leading to a triplet biradical if the two unpaired electrons would have parallel spin or an open-shell singlet biradical if the σ- and the π-electron would obtain parallel spins. In any case, structure **3** would be highly unstable [ΔE = 406 kcal mol^{-1} for the triplet ground state, UMP2/6-31G(d) calculations, Figure 2]. Calculations suggest CC distances of 1.52 Å and CH_{bridge} distances of 1.46 Å.

Structure **3** is considerably stabilized if the H bridges and the normal CH bonds can bend out of the C plane thus leading to the C_{3V}-symmetrical structure **4**. In **4**, σ- and π-orbitals mix and, as a consequence, there is pairing of all electrons and an involvement of all electrons into bonding. This decreases the relative energy to 188 kcal mol^{-1}, which is still too high to compete with structure **1** since CC and CH bond dissociation energies are much smaller than 188 kcal mol^{-1}. The three CC distances increase to 1.62 Å while CH_{bridge} distances decrease to 1.27 Å (Figure 2). The CH_{bridge} C plane is folded by the angle φ =135° out of the ring plane.

In structure **5**, one normal CH_2 group is included into the ring with the HCH plane standing orthogonal to the ring plane. The adjacent CC bonds are normal, which is confirmed by bond lengths of 1.50 Å close to the values found for **1** (Figure 2). The third CC bond is bridged by two H atoms, which leads to some lengthening of the interaction distances (CC: 1.68 Å, CH_{bridge}: 1.31 Å). Structure **5** is somewhat more stable than **4** (ΔE = 177 kcal mol^{-1}), but still too unstable to be of any relevance.

Since both structures **6** and **7** are highly unstable, we can conclude that the cyclopropane structure **1** is by far the most stable cyclic C_3H_6 system possible. It complies with the rules of classical carbon chemistry [tetravalent C, (distorted) tetragonal geometries] and, therefore, it is relatively stable.

In Figure 3, the chemical relationship of **1** to other compounds is indicated by simple changes in its structure. Ionization will lead to the cyclopropyl radical cation (reaction 1, Section XI. C), homolytic or heterolytic CH bond dissociation (reactions 2,3,4) to cyclopropyl radical (Section XII), cation (Section XI. D) and anion (Section XI. A) and CC bond rupture to the trimethylene biradical (reaction 5, Section XIV.B) and propene (reaction 6). Reaction 7 is actually more interesting in the reverse mode as the ethene–carbene cycloaddition reaction leading to **1** (Section XIV. A). Reaction 8 is just formal (Section V. A) and reactions 9,10 and 11 bring **1** into relation with its carbene, with cyclopropene and cyclopropyne, which will not be discussed in this chapter. The protonation reaction 12 (Section XI. B) can lead to edge or corner protonated **1** while the CH_2 insertion reaction 13 sets **1** into relation with its higher homologue cyclobutane (Section XIV. C). Most of these reactions will be discussed in this review article, but a comment with regard to reaction 7

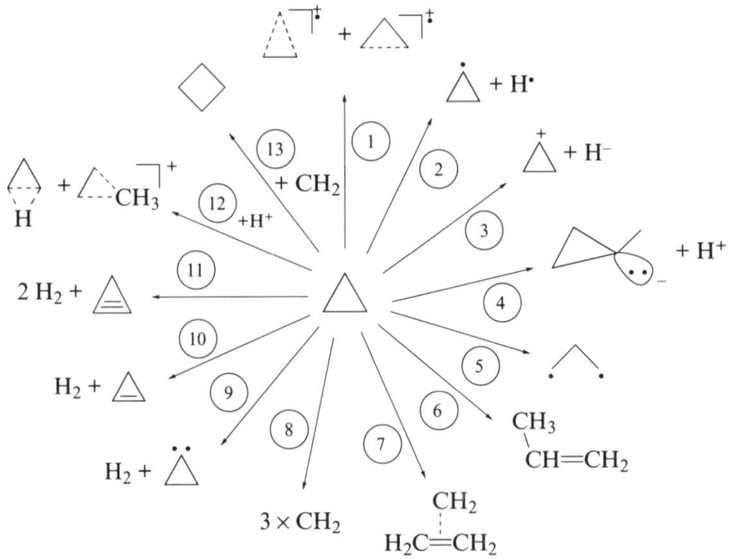

FIGURE 3. Possible reactions of cyclopropane and its relationship to isomeric hydrocarbons

is appropriate here. If this reaction would proceed via a C_{2V}-symmetrical structure **9** (Figure 1), further distortion of **1** would lead from the three-membered ring **9** to the π-complex **10**. There is an important relationship between π-complexes and three-membered rings that helps to understand the electronic structure and CC bonding in **1** (Section IV. D). This will be discussed in more detail in the following sections.

III. ELECTRONIC STRUCTURE—ORBITAL DESCRIPTION

The key to the understanding of physical properties and chemical reactivity of **1** is found in the electronic structure of the molecule, which can be described in terms of molecular orbitals (MOs), valence bond (VB) orbitals or its electron density distribution $\rho(\mathbf{r})$. Numerous investigations have considered the MOs of **1**[23–33] and, therefore, one could expect that a review article on cyclopropane appearing in the year 1995 can skip this part by just referring to one of the previous review articles[1–4, 14–20]. However, there is considerable confusion among chemists with regard to the appropriate MOs of cyclopropane, which needs clarification.

It is a wide-spread assumption that the electronic structure of **1** can be rationalized by using either Förster–Coulson–Mofitt orbitals[23, 24] or Walsh orbitals[25], which are considered to represent equivalent orbital sets and to lead to similar descriptions of bonding in the C_3 ring of **1**. The Walsh orbitals have attracted special attention, in particular among organic chemists, since they seem to be close to the canonical (delocalized) SCF MOs of **1** and seem to explain

(a) the π character of its CC bonds,
(b) substituent effects on ring geometry and stability,
(c) the ability of the cyclopropyl group to conjugate with other groups,
(d) the relationship between cyclopropane and π-complexes[9, 11–13, 34–44] and
(e) delocalization of electrons in the plane of the ring[23, 9–13].

The Förster–Coulson–Mofitt orbitals[23, 24], on the other hand, are considered to represent localized (bond) orbitals that help to visualize the bent bond character of the CC bonds of **1** and to explain its ring strain. They seem to be less suited to analyse substituent–ring interactions or the conjugative properties of **1**.

However, these views are largely incorrect. Walsh orbitals suffer from a number of deficiencies that one has to know before using them.

1. Heilbronner and coworkers[31] have shown that Walsh orbitals and Förster–Coulson–Moffitt orbitals are not equivalent. For example, only the Förster–Coulson–Moffitt orbitals but not the Walsh orbitals are part of the manifold of bonding cyclopropane orbitals. Therefore, a correct description of bonding in **1** in terms of Walsh orbitals is not possible.

2. Because of this, interactions between CH_2 apex group and CH_2CH_2 basal fragment cannot correctly be described using Walsh orbitals. For example, the 'σ-bridged-π-character'[32] of HOMO 12 (see Section VI. C) is lost and, by this, also the possibility of describing the exact relationship between three-membered rings and π-complexes[9, 11–13, 34–44].

3. It is not possible to describe the properties of the cyclopropyl radical cation correctly on the basis of the Walsh orbitals (see Section XI. C and, there, Reference 233).

4. The hybridization scheme used by Walsh (sp^2, p for C) disguises the pseudo-π-character of some of the CH_2 orbitals of **1**[29, 30].

5. It is not possible to describe the interactions of **1** with π-donor substituents (let alone σ-donor/σ-acceptor substituents) on the basis of Walsh orbitals. For this purpose one has to retreat to the full set of SCF orbitals[33].

6. The conjugative propensity of the cyclopropyl group is wrongly explained to depend only on the overlap between Walsh orbitals and adjacent π-orbitals while in reality it is also a consequence of the HOMO energies of **1** that facilitate delocalization of electrons into low-lying π^* orbitals[31].

If one wants to use Walsh orbitals for discussing the properties of the cyclopropyl group, one has to refine them.

A. Walsh Orbitals and Refined Walsh Orbitals

Walsh[25] derived the CC MOs for **1** by considering that the molecule is made up of CH_2 units, each of which has a set of sp^2 hybrid orbitals and a p-orbital. However, a somewhat more realistic picture of the MOs is obtained by assuming a set of sp hybrid orbitals complemented by two p orbitals for each C atom. In Figure 4, these orbitals are classified[11, 12] as radially oriented (with regard to the centre of the ring) sp_{in} and sp_{out} orbitals and tangentially oriented (with regard to the perimeter of the ring) p_{ip} (in-plane) and p_{op} (out-of-plane) orbitals. Orbitals sp_{out} and p_{op} form together with the in-phase and the out-of-phase combination of the two 1s orbitals of the H···H unit the $\sigma(CH_2)$ and $\pi(CH_2)$ orbitals of **1** while sp_{in} and p_{ip} lead to CC σ- and π-orbitals of the ring. In Figure 4 it is shown that the orbital set $\{sp_{in}, sp_{out}, p_{ip}, p_{op}\}$ can be used for the construction of the MOs of any cycloalkane, where ethene is included as a 'two-membered cycloalkane'. The radially oriented CC orbitals are σ-type for ethene and **1** (formally, they enclose angles of 30° with the internuclear connection lines in the case of **1**), while the tangentially oriented CC orbitals have π-character (they enclose angles of 60° with the internuclear connection line for **1**). Cremer[11,12] pointed out that the nature of the CC orbitals changes when going from small to larger rings: The σ(CC) orbitals of ethene and **1** become the π-type ribbon orbitals of a larger ring and the π-orbitals become σ-type ribbon orbitals. In this case, there is a close relationship between ethene and **1** since for these molecules orbitals differ from those of a normal cycloalkane.

The tangential p_{ip} orbitals form a Hückel system for even-membered rings but a Möbius system for odd-membered rings. However, this seems to be of little consequence because it has been shown that both Hückel and Möbius orbital systems have always an aromatic

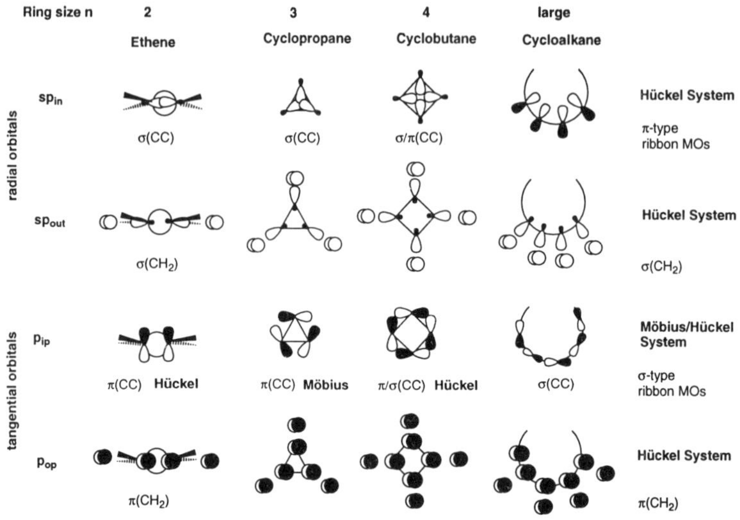

FIGURE 4. Basis orbitals for (planar) cycloalkanes with ring size *n*. Basis orbitals comprise the radially oriented sp_{in} and sp_{out} hybrid orbitals at C, the tangentially oriented p_{ip} (in plane) and p_{op} (out of plane) orbitals at C as well as in-phase and out-of-phase combinations of the two 1s(H) orbitals, which combine with sp_{out} and p_{op} orbitals, respectively. Note that ethene is included as a 'two-membered ring' where the ring is symbolized by two bent bonds and the CH bonds are shown to define the orientation of the ring plane (parallel to the drawing plane). The nature of each orbital set (σ or π, CC or CH_2 bonding, Hückel or Möbius type) is given

electron configuration ($n = 2$: 2; $n = 3$: 4; $n = 4$: 6 electrons, etc.) Molecule **1** as any other cycloalkane possesses only subshells with aromatic electron ensembles (see Figure 4) and, therefore, an aromatic/antiaromatic classification of cycloalkanes is not possible on a topological basis[11, 12].

In Figure 5, an orbital interaction diagram is shown, in which radial and tangential orbitals are combined to the MOs of **1**. The orbital interaction diagram is based on a SCF calculation with a large basis set and differs with regard to the order of the unoccupied MOs considerably from previously published diagrams and orbital sets that were based on Extended Hückel type or minimal basis set SCF calculations[29–31]. It is well known that the virtual (unoccupied) MOs of a SCF calculation have little chemical significance because virtual MOs are just a by-product of the mathematical procedure to calculate the occupied MOs. Nevertheless, one uses the virtual MOs of minimal basis set calculations to illustrate energy ordering, shape and nodal properties of unoccupied MOs. In this spirit, all previous presentations of the MOs of **1** have been made.

However, it seems to us more interesting to present unoccupied MOs in the way they are suggested by large basis set HF calculations. This, of course, leads to the inclusion of diffuse virtual orbitals (Rydberg orbitals) with orbital energies close to zero into the set of unoccupied MOs, but diffuse virtual MOs are needed to reproduce the measured ultraviolet absorption spectrum of **1** in a CI calculation (see Section X). Therefore, the ordering of virtual orbitals given in Figures 5 and 6 (below) has some justification, although it may be generally criticized on the basis that virtual MOs have no physical meaning.

The six Walsh orbitals of **1** are formed from the (radial) sp_{in} and (tangential) p_{ip} starting orbitals. They may be denoted as ω_O, ω_S, ω_A (CC bonding) and ω_O*, ω_S*, ω_A* orbital (CC antibonding). Two important improvements of the original Walsh orbitals ω are indicated in Figure 5:

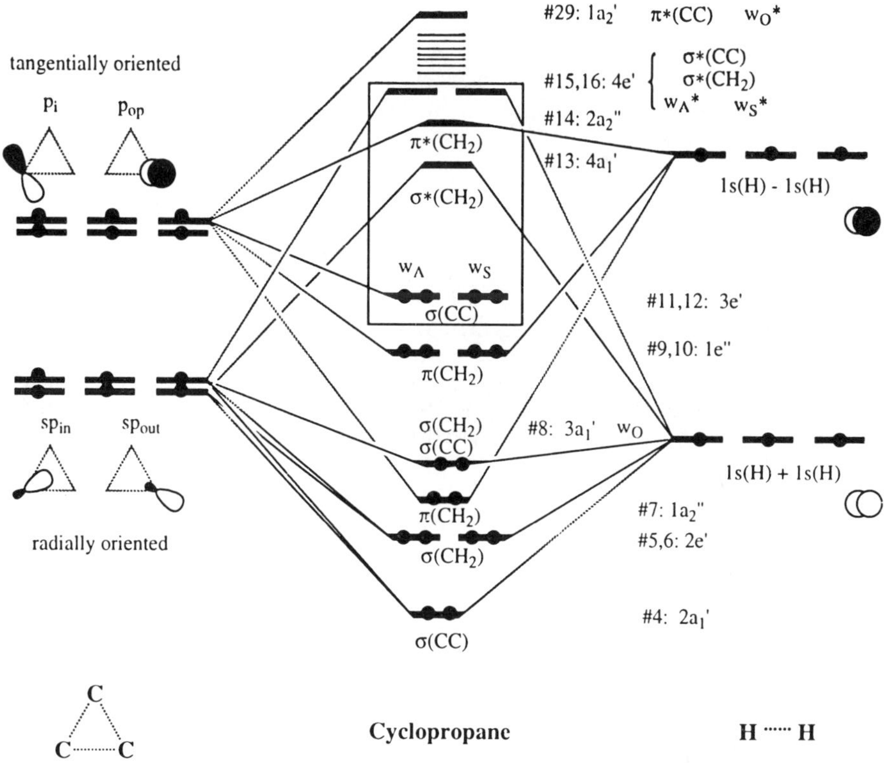

FIGURE 5. Orbital interaction diagram for cyclopropane according to SCF calculations with a large basis set. Basis orbitals for the three C atoms and for the three H···H units are given on the left and the right side, respectively. Resulting MOs are characterized with regard to their symmetry, σ/π character and their bonding nature (stars indicate antibonding nature). They are numbered including core orbitals $1a_1'$ and $1e'$. Orbital mixing between $3e'$ and $4e'$ MOs is indicated by a frame around the corresponding orbital levels. CC bonding and antibonding MOs that correspond to one of the refined Walsh orbitals are denoted by the appropriate symbol, namely w_O, w_A, w_S, w_O*, w_A* or w_S*

1. The σ(CC) and $\sigma(CH_2)$ orbitals of a_1' symmetry can mix. As a result, the $2a_1'$ (MO 4) orbital obtains more s-character and becomes almost pure CC bonding while the $3a_1'$ (MO 8, w_o) orbital gets more p-character, mixes with the in-phase combination of the HH orbitals and adopts both CC and CH_2 bonding character.

2. The refined Walsh orbitals $3e'$ (MOs 11, 12; w_A and w_S) and $4e'$ (MOs 15, 16; w_A* and w_S*) result from mixing between the original Walsh orbitals ω (indicated in Figure 5 by a frame around these orbitals) as shown in equations 1–4:

MO 11: $$w_A = N_A(\omega_A + \lambda\omega_A^*) \quad (1)$$

MO 15: $$w_A^* = N_A^*(\omega_A - \lambda\omega_A^*) \quad (2)$$

MO 12: $$w_S = N_S(\omega_S + \lambda\omega_A^*) \quad (3)$$

MO 16: $$w_S^* = N_S^*(\omega_S - \lambda\omega_S^*) \quad (4)$$

where N is a normalization factor and λ a mixing coefficient. As a result of the mixing, HOMOs $3e'$ (w_A and w_S) can be classified as 'π-bridged-π orbital' (π-orbital bridge at C1) and 'σ-bridged-π-orbital' (σ-orbital bridge at C1, see Scheme 1)[32].

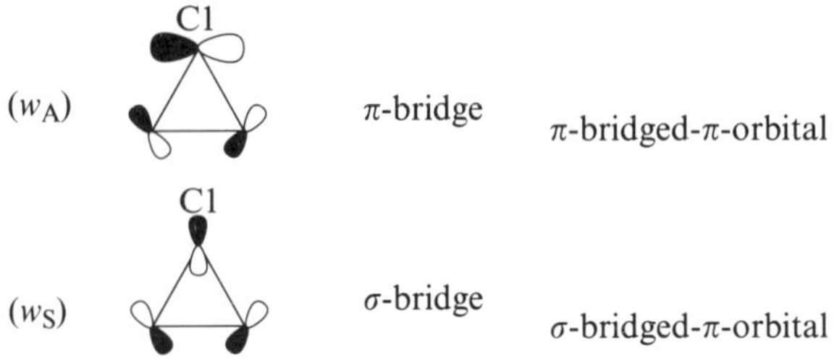

SCHEME 1

Mixing of 3e' and 4e' orbitals increases (a) the π-character (in the C2C3 bond) of the original ω_A and the ω_S* orbitals and adds (b) the σ-orbital bridge to orbitals ω_S and ω_A*. This is shown more clearly in Figure 6, where both the refined Walsh orbitals w (in the form of simple orbital pictures) and the final canonical SCF MOs (in the form of perspective three-dimensional drawings) of **1** are given. The electronic structure of **1** can only be rationalized if the full set of MOs derived from $\{sp_{in}, sp_{out}, p_{ip}, p_{op}\}$ by appropriate orbital mixing (i.e. including refined Walsh orbitals w) is used.

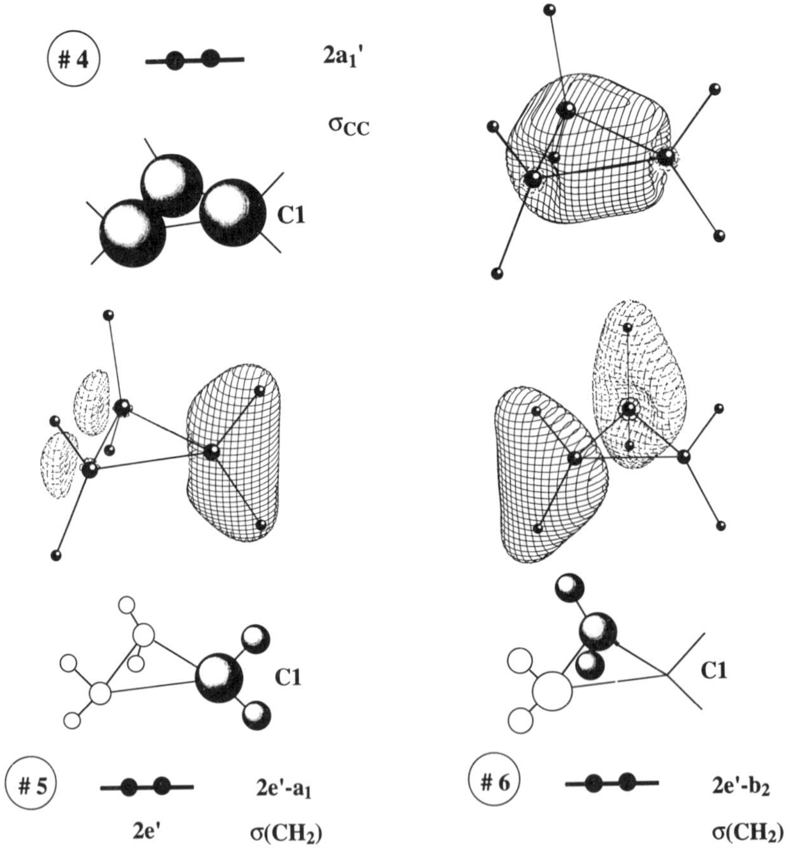

FIGURE 6. (Caption on page 55)

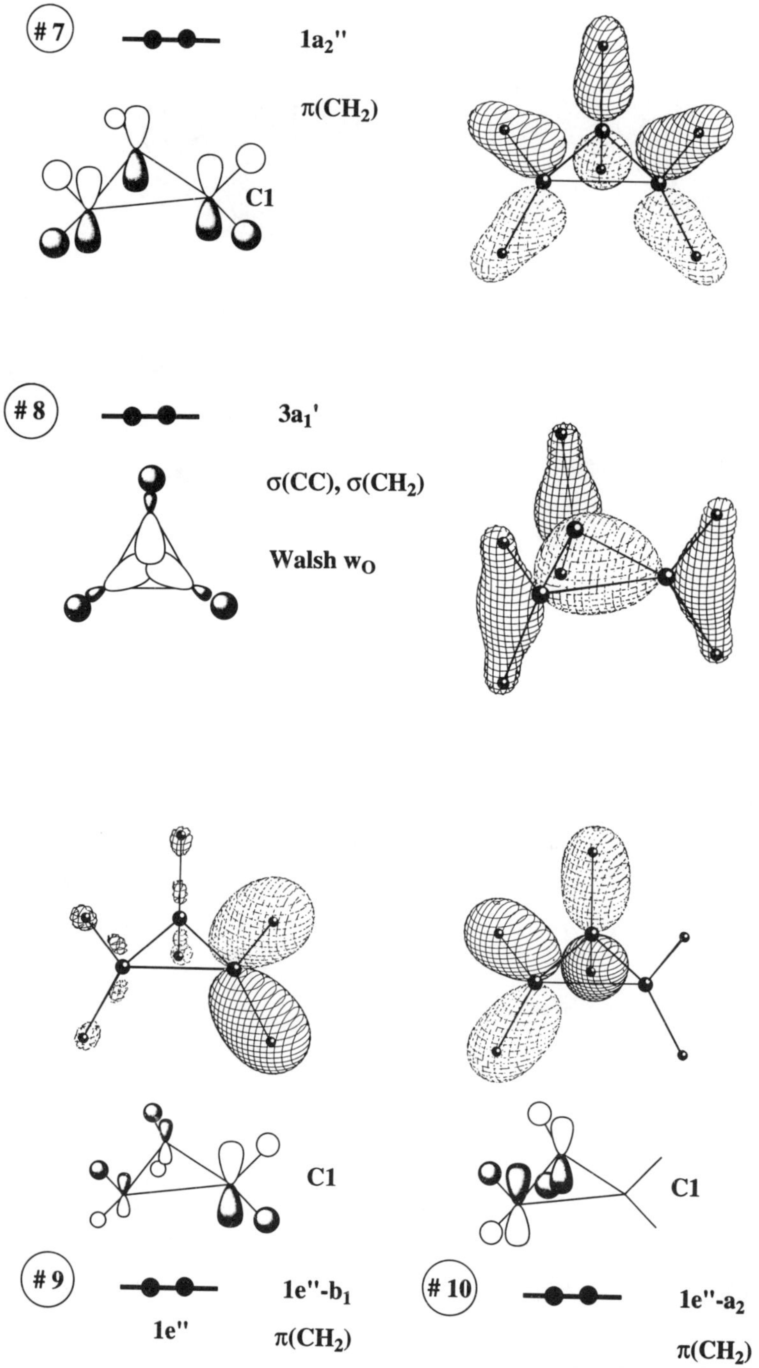

FIGURE 6. (*continued*)

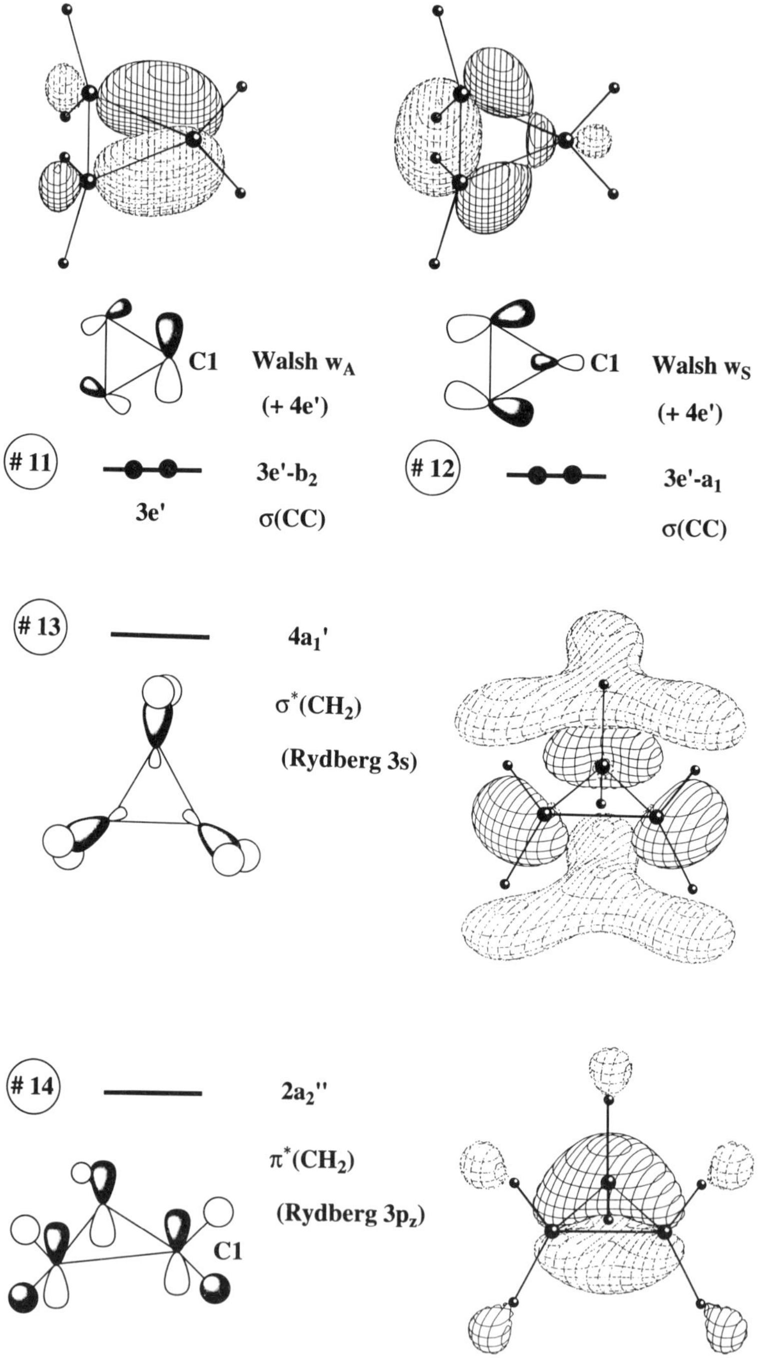

FIGURE 6. (*continued*)

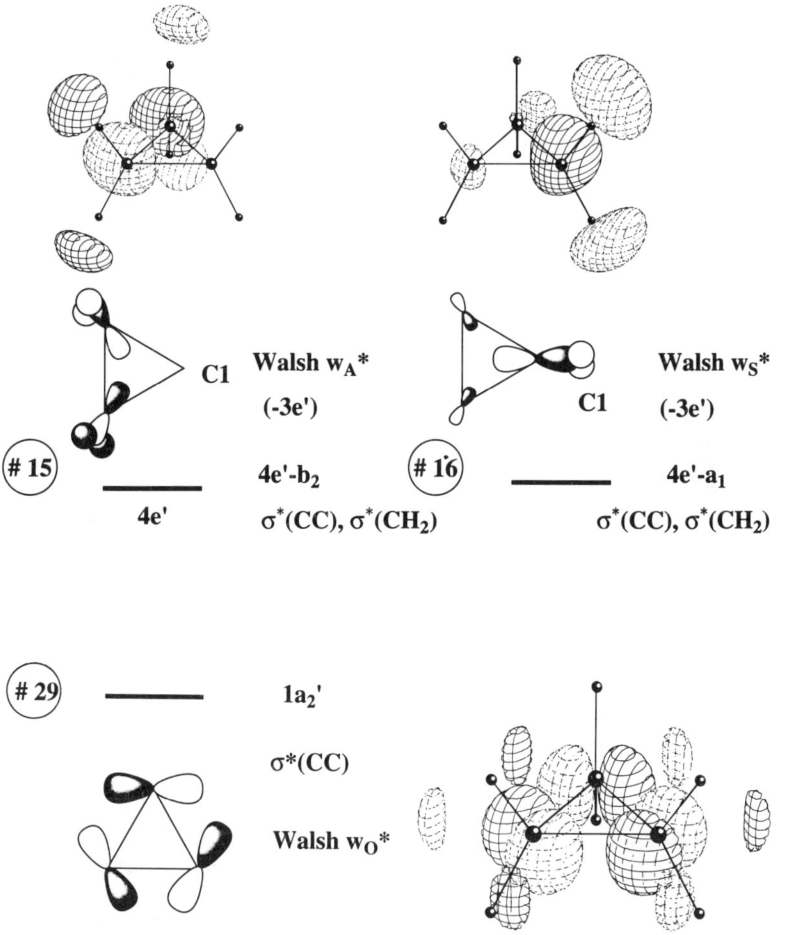

FIGURE 6. Hartree–Fock SCF orbitals of cyclopropane as obtained with a DZ + P basis set. The corresponding orbitals of the orbital interaction diagram shown in Figure 5 are given below or on the left side of each SCF MO. MOs are characterized with regard to their symmetry, σ/π character, their anti/bonding nature and possible Rydberg character (unoccupied MOs). For degenerate MOs, additional symmetry designations are given, which the MOs would obtain in case of a C_{2v} distortion along the C_2 axis that passes through C1. MOs are numbered including core orbitals $1a_1'$ and $1e'$. Orbital mixing between $3e'$ and $4e'$ MOs is indicated. CC bonding and antibonding MOs that correspond to one of the refined Walsh orbitals are denoted by the appropriate symbol, namely w_O, w_A, w_S, w_O*, w_A* or w_S*

B. Förster–Coulson–Moffitt Orbitals and Non-orthogonal Valence Bond Hybrid Orbitals

Coulson and Moffitt[23] established a bent bond model of **1** by elaborating ideas first proposed by Förster[24]. They determined sp^n (CC) and sp^m (CH) hybrid orbitals with optimal hybridization ratios n and m to describe bonding in **1**. Calculations showed that, for **1**, the p-character of the CC hybrid orbitals has to be increased from sp^3 to sp^4 while the s-

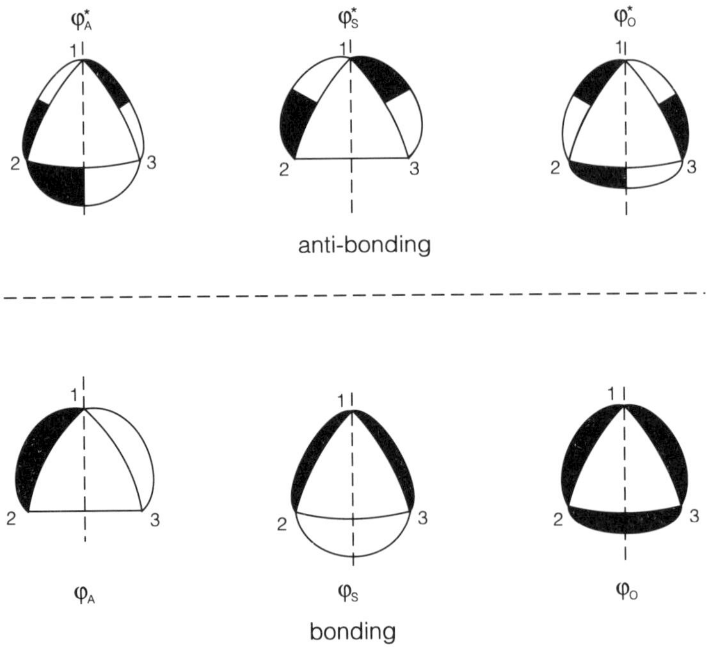

FIGURE 7. Förster–Coulson–Moffitt bent bond orbitals of cyclopropane. Reprinted from E. Honegger, E. Heilbronner and A. Schmelzer, *Nouv. J. Chim.*, **6**, 519 (1982) by permission of Gauthier-Villars Publishers

character of the CH hybrid orbitals increases from sp^3 to sp^2. Bonding and antibonding CC bond orbitals obtained from these hybrid orbitals are shown in Figure 7.

The Förster–Coulson–Moffitt orbitals (φ_0, φ_A, φ_S, φ_0*, φ_A*, φ_S*) reveal that the CC hybrid orbitals are considerably bent in an attempt to avoid the geometrical angle $\alpha = 60°$ and to come close to the strain-free tetrahedral angle. Bending of the hybrid orbitals can only be achieved by an increase of their p character. The corresponding orbital energies are increased and the CC bond is weakened. Hence, the Förster–Coulson–Moffitt orbitals suggest weakening of the CC bonds and a strained three-membered ring as a result of bond weakening.

The results of the Coulson–Moffitt model have been verified many times using different energy minimization criteria for obtaining the best hybridization ratios (*n* and *m*) and determining angles between bent bond orbitals. Some results are summarized in Table 1[7,23,31,45–47].

According to SCF calculations followed by a Boys localization[48], the CC bonding hybrid orbitals deviate by 28° from the CC internuclear connection line, which leads to a interorbital angle of 115°, i.e. 55° larger than the geometrical angle $\alpha = 60°$ [7]. Wardeiner and coworkers[46] have used the ^1J(CH) coupling constant of **1** and the Müller–Pritchard equation[49] to derive an interorbital angle of 103°. Hence, the original description of CC bonding in **1** by Coulson and Moffitt is essentially confirmed and adds support to the usefulness of the Förster–Coulson–Moffitt bent bond orbitals.

Heilbronner and coworkers[31] have shown that for all practical purposes, Förster–Coulson–Moffitt orbitals can be used as representatives for the final SCF orbitals of **1** while this is only true for the refined Walsh orbitals *w* given in equations 1–4 rather than the original Walsh

TABLE 1. Hybridization and interorbital angles in cyclopropane using orthogonal hybrid orbitals

Author(s)	Year	n in sp^n	m in sp^m	Interorbital angle (deg)		Ref.
		CC	CH	CCC	HCH	
Coulson and Moffitt	1949	4.12	2.28	104	116	23
Randic and Maksic	1965	4.91	2.02	102	120	45
Newton	1977	3.38	1.86	115	117	7
Honegger and coworkers	1982	3.21	1.94	117	116	31a
Wardeiner and coworkers	1982	4.58	2.12	103	118	46a
Inagaki and coworkers	1994	4.0	2.0			47

orbitals ω. The use of Förster–Coulson–Moffitt orbitals also helps to correct another common belief based on Walsh orbital descriptions, namely that the ability of the cyclopropyl group to conjugate with unsaturated groups results from the presence of tangential p_{ip}-orbitals. However, calculations by Heilbronner and coworkers[31] reveal that it is the orbital energy of the $3e'$-MOs well above those of other alkanes or cycloalkanes and close to that of the π-orbital of an alkene that leads to delocalization of cyclopropyl CC bonding electrons and the observed conjugative propensity.

Recently, two interesting VB investigations of **1** have been published, which provide insight into the nature of bond orbitals and bonding of **1** and some reference molecules[50,51]. Hamilton and Palke[50] applied a steepest-descent technique to obtain optimized hybrid orbitals and optimized overlap within a perfect-pairing VB approach. They used a DZ + P basis set of Slater-type orbitals (!) and the experimental geometry of **1**. Contrary to previous descriptions of **1** that used orthogonal hybrid orbitals and constraints on the composition (amount of s or p character on one centre is constant) and number of hybrids, the VB description leads to non-orthogonal hybrid orbitals and, accordingly, the hybridization ratios of CC and CH hybrid orbitals are not directly related. Eliminating the orthogonality constraint between CC and CH hybrid orbitals increases their s-character, which can only be analysed by comparing hybridization ratios for suitable reference molecules calculated with the same type of VB wave function.

TABLE 2. Hybridization and overlap values obtained with non-orthogonal hybrid orbitals within a valence bond calculation of cyclopropane[a]

Molecule	CC Bond		CH Bond	
	sp^n	overlap	sp^m	overlap
CH_4			$sp^{1.514}$	0.831
$H_3C{-}CH_3$	$sp^{1.381}$	0.832	$sp^{1.524}$	0.829
$H_2C{=}CH_2$[b]	$sp^{1.370}$	0.823	$sp^{1.561}$	0.830
Cyclopropane	$sp^{1.706}$	0.827	$sp^{1.348}$	0.833

[a]From Reference 50.
[b]The ethene double bond is composed of two banana bonds, one above and one below the plane of the nuclei. Therefore, the CC overlap value has to be multiplied by a factor 2 to get the total CC bond overlap.

As can be seen from Table 2, the amount of s-character of the CH hybrid orbital of **1** is indeed increased relative to that of the CH hybrid orbitals of ethene, ethane and methane while at the same time the s-character of the CC hybrid orbitals is decreased. This seems to confirm predictions based on model calculations with orthogonal hybrid orbitals. However, closer inspection of the data in Table 2 reveals that s-character of the CH hybrid orbitals does not parallel expected trends in CH bond strengths. For example, the s-character of the CH hybrid orbitals of ethene is lower than that of methane and ethane, while overlap values suggest similar values for the three molecules with **1** possessing a clearly larger CH overlap value (Table 2). This contradicts the well-documented fact that the CH bond in ethene is significantly stronger than the CH bonds of alkanes. Obviously, the general belief that the bond strength increases as the overlap of the two orbitals in a VB pair is not always confirmed by fully optimized VB calculations.

In this connection, a recent spin-coupled VB calculation of **1** by Karadakov, Gerratt, Cooper and Raimondi[51] is interesting. This approach removes the strong orthogonality constraint of GVB calculations and makes use of the whole spin space and not just of the perfect pairing spin function of other less general one-configuration approaches. In this way, the energy obtained with the single-configuration spin-coupled VB wave function comes in most cases close to that of its CASSCF counterpart. Calculations were done with a DZ+P basis of Gaussian functions for **1**, cyclobutane and propane at experimental geometries. The active space was limited to six, eight and four singly-occupied non-orthogonal CC hybrid orbitals corresponding to the fact that the three molecules contain three, four and two symmetry-equivalent CC bonds.

TABLE 3. Overlap values calculated using non-orthogonal hybrid orbitals within valence bond calculations of propane, cyclopropane and cyclobutane[a]

Overlap	Propane	Cyclopropane	Cyclobutane
Bonding overlap	0.817	0.803	0.798
Geminal overlap	0.290	0.340	0.309
Vicinal through-ring overlap	0.087	0.207	0.144
Vicinal π-type overlap	—	0.049	0.124
1,3-Overlap	−0.092	—	−0.022
			0.081

[a]From Reference 51. In cyclopropane, bonding overlap corresponds to overlap between hybrid orbitals 1 and 2, geminal overlap to overlap between 2 and 3, vicinal through-ring overlap to overlap between 1 and 3, vicinal π-type oveerlap between 1 and 4. Compare with Figure 17.

Karadakov and coworkers[51] have published overlap matrices for their optimized non-orthogonal CC hybrid orbitals that show some interesting trends in bonding and non-bonding hydrid orbital overlap (see Table 3). Although the authors claim that CC bonding is similar in **1** and cyclobutane as documented by calculated bonding overlap values (this argument is in itself very problematic, since on the same basis the Hamilton–Palke results would suggest the same CC bonding for ethane and **1**!), comparison of their overlap values reveals a striking difference between **1** on the one side and propane and cyclobutane on the other:

(1) Geminal overlap is 10–20% larger in **1** than in the two other compounds.

(2) Vicinal through-ring overlap is more than 40% larger for **1** than in the case of cyclobutane.

(3) Antibonding 1,3 overlap does not exist for **1**, but takes considerable values in propane and cyclobutane.

Nonbonding overlap is significantly larger for **1** and leads to a shift of electron density into the three-membered ring. This is one important aspect of the VB calculations of Karadakov and coworkers (not mentioned by the authors)[51], which becomes obvious when looking at the VB hybrid orbitals depicted in Figure 8.

No matter whether calculated within the perfect pairing VB approach or by the spin-coupled VB approach, in both cases the CC hybrid orbital extends outside the three-membered ring as expected by the schematic representations in Figure 7, *but also inside*

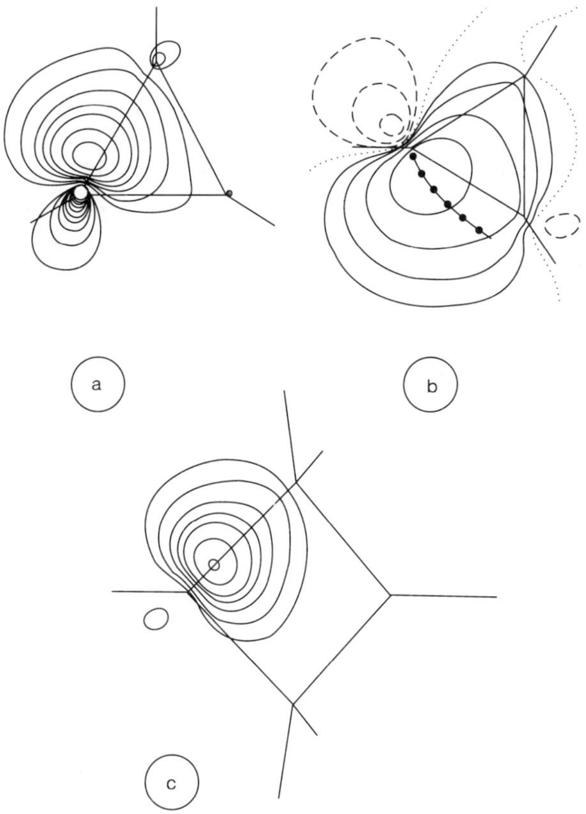

FIGURE 8. Calculated valence bond orbitals of cyclopropane and cyclobutane in the form of contour line diagrams plotted in the plane of the carbon rings. Because of the chosen topside view, only CC and one of the CH bonds of each CH_2 group can be seen in the form of solid lines connecting atom positions. (a) Orbital density plot of one of the six symmetry-equivalent valence orbitals of cyclopropane obtained from spin-coupled valence bond calculations. (b) Orbital amplitude plot of one of the six symmetry-equivalent valence orbitals of cyclopropane obtained from perfect-pairing valence bond calculations. Solid lines denote positive, dashed lines negative and the dotted line the zero amplitude contour line. Orbital bending is indicated by •–•–• where each point • is given by the maximum orbital amplitude. (c) Orbital density plot of one of the eight symmetry-equivalent valence orbitals of cyclobutane obtained from spin-coupled valence bond calculations. Diagrams have been reconstructed according to calculations by (a,c) P. B. Karadakov, J. Gerratt, D. L. Cooper and M. Raimondi, *J. Am. Chem. Soc.*, **116**, 7714 (1994) and (b) J. G. Hamilton and W. E. Palke, *J. Am. Chem. Soc.*, **115**, 4159 (1994)

the ring covering the whole ring surface, which cannot be predicted from the pictures in Figure 7.

Hamilton and Palke[50] determined the interorbital angle between the CC hybrid orbitals to be 123° at the position of the C nucleus given by the direction of the maximum orbital amplitude (see –•–•– line in Figure 8). With increasing distance from the nucleus the orbital bends toward the internuclear connection line as reflected by a decrease of the interorbital angle to 100°. Bending of the orbitals will lead to strain and, since the bending of the CC hybrid orbitals for **1** is clearly larger than for the CC orbitals of cyclobutane (Figure 8), there should be higher strain in the case of **1** in line with common chemical thinking.

C. SCF Orbitals, Orbital Energies and Ionization Potentials

In Figure 6 perspective drawings of Hartree–Fock SCF orbitals are given. They reveal that the construction of MOs from the basis orbitals shown in Figure 4 leads to a reasonable description of SCF MOs, but their exact form can only be obtained from appropriate mixing of Walsh orbitals of the same symmetry. Worthy of note is the difference in shape of the $3e'$ HOMOs (#11 and #12) and the a_1' σ-MOs ($2a_1'$, $3a_1'$, #4 and #8) that is of relevance for a detailed description of the electronic structure and chemical bonding in **1**.

A complete set of orbital energies, including core orbitals and a large number of unoccupied orbitals, has been published by Segal and coworkers, who used a DZ basis augmented by ring-centred diffuse functions[52]. Valence shell orbital energies or ionization potentials, which can be related to orbital energies via the Koopmans theorem, have been published over the years by various authors[39, 52–59]. Some of these data can be found in Ballard's review on the photoelectron spectroscopy of **1** and related three-membered rings[17]. Therefore, only a short summary of orbital energies, calculated ionization potentials and their comparison with experimental vertical ionization potentials of **1** is give in Table 4.

TABLE 4. Orbital energies and ionization potentials of cyclopropane[a]

Orbital		Ionization potential (eV)				
number	sym.	LCAO-Xα Ref. 59	HAM/3 Ref. 56	*ab initio* Ref. 55	Green Ref. 55, 58	Exp Ref. 60
# 11,12	$3e'$	11.9	11.0	11.5	10.7	10.6
# 9, 10	$1e''$	13.1	13.3	13.9	13.0	13.0
# 8	$3a_1'$	16.0	15.4	17.0	15.7	15.7
# 7	$1a_2''$	17.1	16.7	18.3	16.8	16.7
# 5,6	$2e'$	19.1	20.1	22.9	19.9	19.5
# 4	$2a_1'$	27.3	26.1			26.3

[a]Direct calculation of the ionization potential by LCAO-Xα, HAM/3 and Green's function techniques or via the Koopmans theorem by *ab initio* techniques.

All calculations confirm the sequence of orbitals given in Figure 6. In particular, they underline the magnitude of the lowest ionization potential (10.6 eV[60]), which according to the Koopmans theorem[61] gives a $3e'$ orbitals energy similar to that of the ethene π-orbital (10.5 eV[60]). Probably, the most reliable calculations of the ionization of **1** are those by von Niessen and coworkers using Green's functions[55,58]. But also HAM/3 calculations by Fridh[56] and LCAO-Xα calculations of ionization potentials by Alti and coworkers[59] are quite satisfactory (apart from the value of the first ionization potential). Hartree–Fock calculations of ionization potentials on the basis of the Koopmans theorem[61] depend on the basis set used, but are on the average 8% too large.

Wiberg and coworkers[62] have calculated the vertical and adiabatic ionization potential of **1** in a comparative study including fused cyclopropanes. Using MP4/6-31G(d) and spin contamination corrected UMP4/6-31G(d) energies calculated at MP2/6-31G(d) geometries of the parent compound and radical cation, they obtained an adiabatic ionization potential (relaxation of geometry of the radical cation) of 9.48 eV and a vertical ionization

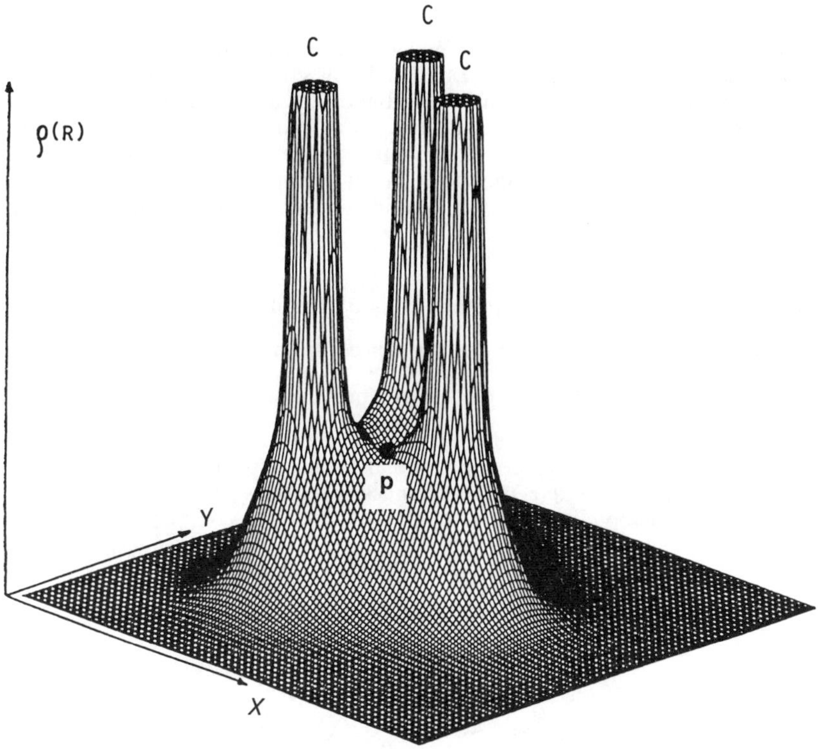

FIGURE 9. Perspective drawing of the calculated electron density distribution $\rho(\mathbf{r})$ in the plane of the cyclopropane ring [HF/6–31 G(d,p) calculations]. Point **p** denotes the position of the bond critical point between two neighbouring C atoms. For better presentation, density values above 14 e Å^{-3} are cut off

potential (geometry of parent molecule used for radical cation) of 10.43 eV, where the latter ionization potential has to be compared with an experimental value of 10.6 eV. Although there seems to be reasonable agreement between theory and experiment, calculational results are flawed by the fact that a transition state rather than the ground state of the cyclopropyl radical cation (see Section XI. C) was calculated[62].

IV. ELECTRON DENSITY DISTRIBUTION AND CHEMICAL BONDING

Analysis of the total electron density distribution $\rho(\mathbf{r})$ of a molecule is useful since $\rho(\mathbf{r})$, contrary to MOs and wave functions, is an observable quantity that can be determined both experimentally and theoretically[63]. As shown by Hohenberg and Kohn[64], the energy of a molecule in a non-degenerate ground state is a function of $\rho(\mathbf{r})$. All physical and chemical properties of a molecule depend in some way on the electron density distribution. Accordingly, it is plausible that analysis of $\rho(\mathbf{r})$ should lead to primary information of electronic structure and chemical bonding of **1**.

In Figure 9, a perspective drawing of the calculated HF/6-31G(d,p) electron density distribution $\rho(\mathbf{r})$ in the plane of the C atoms of **1** is shown[11]. The electron density takes maximal values at the positions of the three nuclei and decreases exponentially in off-nucleus directions. This is typical of all molecular $\rho(\mathbf{r})$-distributions and obscures many details of the density distribution that relate to bonding, accumulation of density in the region of lone-pair electrons, holes in the valence shell in the direction of unoccupied orbitals or to other anisotropies of the electron density distribution at a bonded atom. There exist three major ways to unravel details of electronic structure and bonding from $\rho(\mathbf{r})$, namely (a) the analysis of difference electron density distributions $\Delta\rho(\mathbf{r})$, (b) the analysis of $\rho(\mathbf{r})$ by the virial partitioning method and (c) analysis of $\rho(\mathbf{r})$ via its Laplacian. All three methods have been used to investigate **1** and results of these investigations are summarized in Sections IV. A, IV. B and IV. C.

A. Analysis of the Difference Electron Density Distribution

The difference density distribution is defined in equation 5[65]:

$$\Delta\rho(\mathbf{r}) = \rho[\text{molecule}]\text{-}\rho[\text{promolecule}] \tag{5}$$

where the promolecular density distribution is conventionally constructed by summing over spherically averaged atomic densities, with the atoms kept in the positions they adopt in the molecule. A positive $\Delta\rho(\mathbf{r})$ in the internuclear region is generally considered to be indicative of covalent bonding.

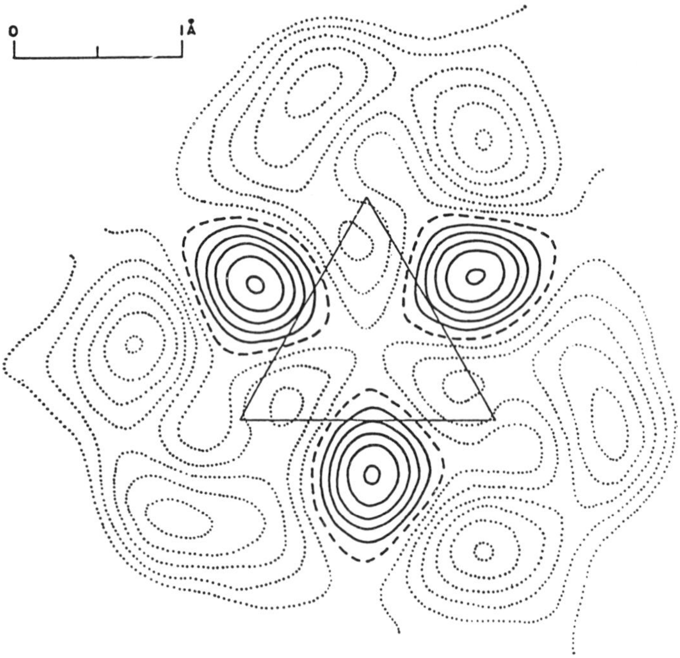

FIGURE 10. Contour line diagram of the difference electron density distribution in the ring plane of *cis,cis*-1,2,3-tricyanocyclopropane as obtained by X-ray diffractometric measurements. Solid lines are in the regions with positive difference densities, dotted lines in regions with negative difference densities. Dashed lines correspond to zero values. Reprinted from A. Hartman and F. L. Hirshfeld, *Acta Crystallogr.*, **20**, 80 (1966) by permission of the International Union of Crystallography

TABLE 5. Description of atoms in molecules and chemical bonds in terms of the properties of $\rho(\mathbf{r})$[a]

Chemical term	Term used in density analysis	Comment
Atom	Nucleus + basin Ω	Basin filled by trajectories that terminate at the nucleus as the attractor of these trajectories
Atomic volume	Volume V of basin	$\int_\Omega d\mathbf{r} = V$
Atomic charge	Atomic charge	$\int_\Omega \rho(\mathbf{r})\, d\mathbf{r} = Q$
Atomic dipole moment	Atomic dipole moment	$\int_\Omega \rho(\mathbf{r})\, \mathbf{r}^n d\mathbf{r} = \mu$ with $n = 1$
Atomic energy	Atomic energy	$\int_\Omega H(\mathbf{r})\, d\mathbf{r} = E$
Interatomic surface	Zero-flux surface S	Internuclear surface through which the flux of $\nabla\rho(\mathbf{r})$ is zero (see equation 6)
Covalent bond	Bond path	Necessary condition: Existence of a MED path linking the bonded atoms Sufficient condition: Negative energy density $H(\mathbf{p})$ at the bond critical point $\mathbf{p}$
	Bond critical point $\mathbf{p}$	Saddle point of $\rho(\mathbf{r})$
Bond energy	Bond energy BE	$\mathrm{BE(A,B)} = \alpha(\mathrm{A,B})\, N(\mathrm{A,B})/R(\mathrm{A,B})^2$ with $\mathrm{N(A,B)} = \mathbf{R}(\mathrm{A,B}) \int_{AB} dS(\mathbf{r})\, \rho(\mathbf{r})\, \mathbf{n}_A(\mathbf{r})$
Bond length r_e	Bond path length r_b	r_b is larger than the geometrical distance r_e for bent bonds
Bond angle α_e	Interpath angle β	Angle between geminal bond paths; β is larger than the geometrical angle α in strained rings
Bent bond character	Distance d	Deviation d of bond path from internuclear connection line
Bond order	Bond order n (local property)	$n(\mathrm{AB}) = \exp\{a[\rho(\mathbf{p}) - b]\}$ where a and b are determined from two suitable reference bonds AB with fixed bond order n
π Character	Bond ellipticity Anisotropy of $\rho(\mathbf{p})$ (local property)	$\varepsilon = (\lambda_1/\lambda_2) - 1$ where λ_i ($\lambda_1 \le \lambda_2 \le \lambda_3$) are the curvatures of $\rho(\mathbf{p})$ along the principal axes calculated from the Hessian of $\rho(\mathbf{p})$; λ_1 is called the hard curvature, λ_2 the soft curvature
Molecular geometry	Molecular graph	Network of bond paths plus bond critical points
Molecular structure	Molecular structure	Equivalent class of molecular graphs

[a]For an explanation of geometrical parameters and bond parameters in the case of cyclopropane, see below.

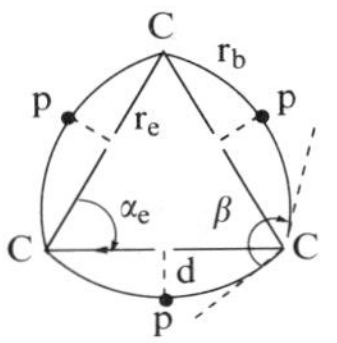

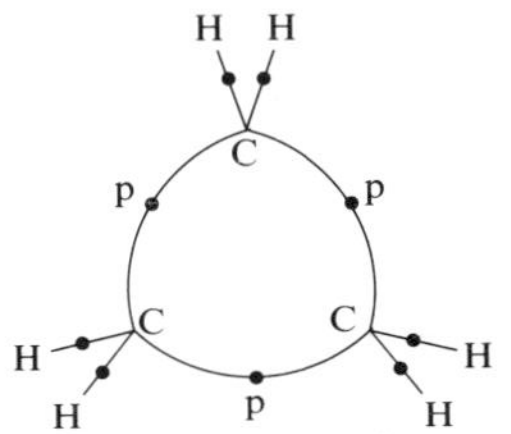

Molecular Graph

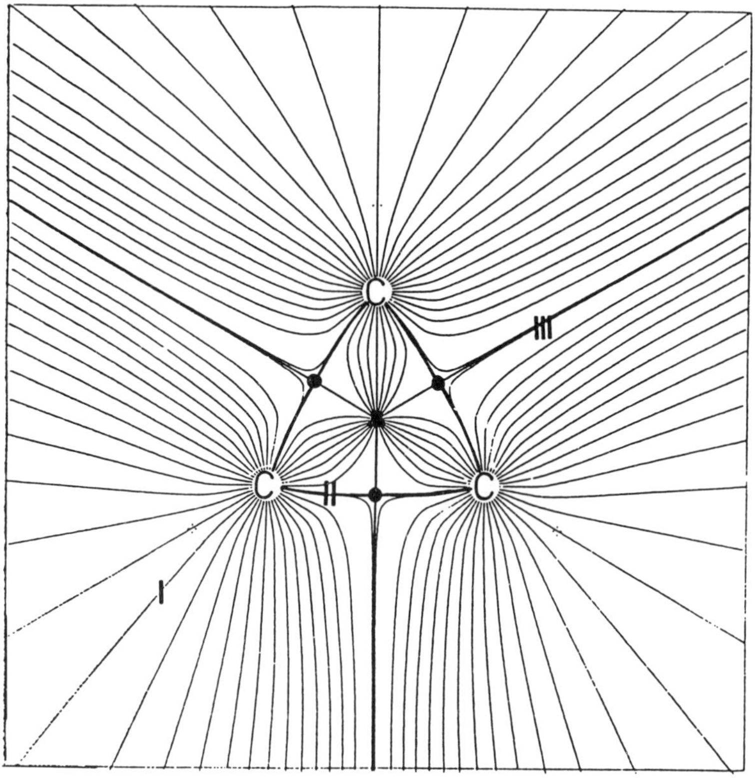

FIGURE 11. Gradient vector field of the HF/6–31 G(d,p) electron density distribution ρ (**r**) calculated for the plane of the cyclopropane ring. Bond critical points **p** are denoted by dots. There are three different types of trajectories: type 1 trajectories start at infinity or the centre of the ring and end at a carbon nucleus; type II trajectories (heavy lines) define the bond path linking two neighbouring carbon atoms; type III trajectories form the three zero-flux surfaces between the C atoms (in the two-dimensional display only their traces can be seen). They terminate at the bond critical points

Distributions $\Delta\rho(\mathbf{r})$ have been determined for various derivatives of **1** by both *ab initio* and X-ray diffraction studies[66]. In Figure 10, a contour line diagram of $\Delta\rho(\mathbf{r})$ in the ring plane of *cis*, *cis*-1,2,3-tricyanocyclopropane is shown[66a]. Positive difference densities are found between the three C atoms, but the $\Delta\rho(\mathbf{r})$ maxima are displayed up to 0.3 Å from the internuclear axis[66]. The displacement of the maxima is considered to indicate the bent bond character of the CC bonds of **1**.

B. Analysis of the Electron Density Distribution

Since difference electron densities, deformation densities or valence electron densities are not observable quantities, and since the Hohenberg–Kohn theorem[64] applies only to the total electron density, much work has concentrated on the analysis of $\rho(\mathbf{r})$. The accepted analysis method today is the virial partitioning method by Bader and coworkers[67], which is based on a quantum mechanically well-founded partitioning of the molecular space into subspaces using $\rho(\mathbf{r})$. Since Bader could show (a) that these subspaces normally contain just one nucleus and (b) that the virial theorem applies to the subspaces, the subspaces are considered to be atomic subspaces and the partitioning of the molecular space into atomic subspaces is called the virial partitioning method. This method has been described in several review articles[13,76] and, therefore, only some essential terms of the method are summarized in Table 5.

The virial partitioning method is based on the calculation and analysis of the gradient vector field $\nabla\rho(\mathbf{r})$ corresponding to the distribution $\rho(\mathbf{r})$[67]. Figure 11 shows the vector field $\nabla\rho(\mathbf{r})$ calculated in the ring plane of **1**. One can distinguish three types of trajectories, indicated in Figure 11 by I, II and III. Type I trajectories start at infinity or at the ring centre and terminate at one of the three C nuclei of **1**, which are the (three-dimensional) attractors of type I trajectories. All type I trajectories that terminate at a particular C nucleus fill a subspace (basin) associated with this nucleus.

Type II trajectories start at a point **p** in the internuclear region between two bonded atoms and end at one of the two nuclei in question. There are just two trajectories per bond, which together define a path of maximum electron density (MED path) that is visible in the perspective drawing of $\rho(\mathbf{r})$ shown in Figure 9. Each lateral displacement from the MED path leads to a decrease of $\rho(\mathbf{r})$. The point **p** corresponds to the minimum of $\rho(\mathbf{r})$ along the path and to a saddle point of $\rho(\mathbf{r})$ in three dimensions.

Saddle point **p** is the sink of type III trajectories, i.e. it is the attractor of all trajectories in directions perpendicular to the MED path. Type III trajectories form a surface, which reaches from infinity (source of type III trajectories) to a line **L** through the center of the ring and perpendicular to the ring plane. The flux of $\nabla\rho(\mathbf{r})$ vanishes for all surface points (equation 6):

$$\nabla\rho(\mathbf{r})\,\mathbf{n}(\mathbf{r}) = 0 \qquad \mathbf{r} \in S \tag{6}$$

where **n** is the unit vector normal to the surface S, which has been called the zero-flux surface[67]. There are three zero-flux surfaces in Figure 11, which meet at line **l** and separate the three basins of the three C atoms of **1**. Other zero-flux surfaces are between C and H atoms of **1**. Hence, the zero-flux surfaces partition the molecular space into subspaces, each containing one atomic nucleus. Since there are just few exceptions to this observation, one has called the subspaces atomic basins and has considered them to represent the space of an atom in a molecule. Molecule **1** has nine of these atomic subspaces, three of which, namely those of the C atoms, can be recognized in Figure 11.

The three MED paths (type II trajectories)between the C atoms of **1** correspond to the three CC bonds. Since MED paths can also be found in the case of van der Waals interactions, Cremer and Kraka[13,68,69] suggested using the local energy density $H(\mathbf{r})$ (equation 7):

$$H(\mathbf{r}) = G(\mathbf{r}) + V(\mathbf{r}) \tag{7}$$

[where $G(\mathbf{r})$ is the local kinetic energy density and $V(\mathbf{r})$ the local potential energy density] at the saddle point **p** to distinguish between covalent bonds and closed-shell (van der Waals) interactions. It is of general understanding that the formation of a covalent bond is accompanied by delocalization of electrons, decrease in kinetic energy, orbital contraction and lowering of the potential energy[70]. If $H(\mathbf{p}) < 0$, then this will suggest a reduction of the kinetic energy density and a dominance of the potential energy density at the saddle point **p** [$V(\mathbf{p})$ is always < 0 and $G(\mathbf{p})$ is always > 0]. Hence, accumulation of electron density in the bonding region at the point **p** is stabilizing. On the other hand, if $H(\mathbf{p}) \geq 0$, then electron density at point **p** will be destabilizing and indicates closed-shell interactions as for van der Waals interactions, ionic bonding or H-bonding. Therefore, Cremer and Kraka suggested that the existence of a MED path between two nuclei can be considered as the necessary condition and $H(\mathbf{p}) < 0$ as the sufficient condition for covalent bonding[13,68,69].

A clear definition of (covalent) bonding is essential for describing the strained CC bonds of **1**. Cremer and Kraka have shown in an electron density investigation of various three-membered rings that the CC bond paths are significantly bent (Table 6)[9]. The bond saddle point **p** is shifted by 0.06 Å from the midpoint of the internuclear connection line and the bond path length r_b(CC) (1.506 Å, Table 6) is almost 0.01 Å longer than the internuclear distance r_e(CC) (1.497 Å, Table 6). In addition, the interpath angle β(CCC) is 79° due to the bending of the bonds. The deviation of β from a tetrahedral angle in **1** is less dramatic as suggested by the geometrical angle $\alpha = 60°$ and, therefore, it is physically more reasonable to use the interpath angle β(CCC) rather than the geometrical angle α(CCC) to assess the ring strain of **1** (or other cycloalkanes)[9–12].

TABLE 6. Description of cyclopropane and some related compounds in terms of the properties of the electron density distribution $\rho(\mathbf{r})$[a]

Property	Structural element	Cyclopropane	Aziridine	Oxirane	Cyclobutane
Distance r_e (Å)	CC	1.497	1.470	1.453	1.544
Distance d (Å)		0.060	0.080	0.094	0.038
Bond length r_b (Å)		1.506	1.486	1.476	1.547
ρ (**p**) (e Å^{-3})		1.681	1.763	1.819	1.680
Bond order[b]		1.00	1.06	1.08	1.00
Ellipticity ε (**p**)		0.49	0.39	0.31	0.02
Distance r_e (Å)	CX	1.497	1.449	1.401	1.544
Distance d (Å)		0.060	0.043	0.004	0.038
Bond length r_b (Å)		1.506	1.455	1.404	1.547
ρ (**p**) (e Å^{-3})		1.681	1.823	1.771	1.680
Bond order[b]		1.00	0.97	0.96	1.00
Ellipticity ε (**p**)		0.49	0.50	0.88	0.02
Geometrical angle α	CXC	60	60.9	62.4	88.6
Bond angle β (deg)		78.8	76.4	75.8	95.6
Geometrical angle α	CCX	60	59.5	58.8	88.6
Bond angle β (deg)		78.8	77.3	72.8	95.6
ρ (**c**) (e Å^{-3})	Ring	1.379	1.485	1.533	0.554
ξ (%)[c]		82.0	83.1	85.8	32.9
Ellipticity ε (**c**)[d]		0	0.445	0.094	0

[a] HF/6–31G(d) calculations from Reference 9. For an explanation of property terms, see Table 5 and text.
[b] All bond orders are normalized.
[c] Ratio [ρ (**c**)/ρ (**p**)$_{av}$] 100 where ρ (**p**)$_{av}$ is the average of all ρ (**p**) values.
[d] Ellipticity at the ring critical point **c**.

Bond order n and π-character ε can be extracted from the properties of $\rho(\mathbf{r})$ at the bond critical point **p** (see Table 5). The calculated CC bond order of **1** ($n = 1.00$),which is obtained from the electron density at the local point **p**, is not sensitive to the bending of the CC bond as has been pointed out by Cremer and Kraka[11]. This is in line with the observation made in VB calculations that the overlap between CC hybrid orbitals in **1** is similar to that of other cycloalkanes (see Table 3 and Section V. B)[51].

More interesting is the calculated ellipticity ε of the CC bonds (Table 6), which is as high as that of the π-bond in ethene. However, contrary to ethene, the soft curvature of $\rho(\mathbf{r})$ (Table 5) is in the plane of the carbon ring, i.e. the electron density extends from the bond critical point toward the centre of the ring as well as outside of the ring. This is in line with the orbital description of **1** and the expected π-character of its CC bonds substantiated in many experimental investigations. Investigation of $\rho(\mathbf{r})$ of **1** reveals that *bending of a formal CC σ bond leads to an admixture of π character*.

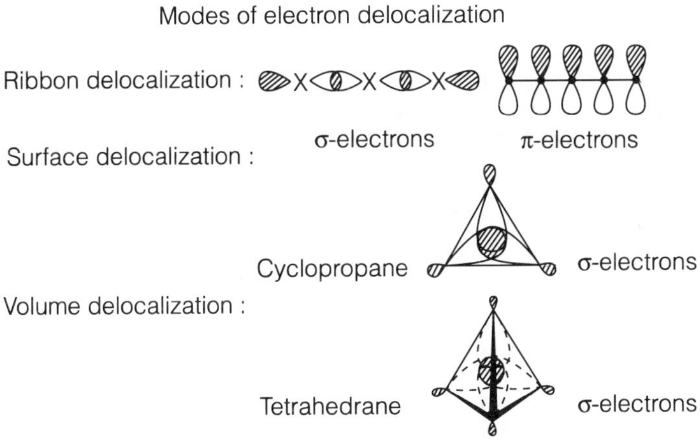

FIGURE 12. Possible modes of electron delocalization. Reprinted from D.Cremer *Tetrahedron*, **44**, 7427 (1988) with kind permission from Elsevier Science Ltd., The Boulevard, Langford Lane, Kidlington OX5 1GB, UK

Another characteristic property of the electron density of **1** is its relatively high value at the centre **c** of the ring (more than 80% of that at the CC bond critical point). Density is smeared out over the ring surface and concentrated at its centre because of the occupation of the w_O -orbital (MO #8, $3a_1'$, Figure 6), which has the character of a 'surface orbital'. Cremer and Kraka[9, 11, 13] have termed this phenomenon 'surface delocalization' of electrons, to be distinguished from ribbon delocalization and volume delocalization of electrons (Figure 12)[12].

Surface delocalization is characterized by ξ , which is the percentage of negative charge at the ring critical point relative to that at the bond critical point. The value of ξ is 82% for **1**, but below 35% for ring molecules such as cyclobutane (Table 6) or benzene. For substituted **1**, interactions between ring and substituents influence the π-character of the CC bonds and surface delocalization. Depending on the nature of the substituent (see Section VI) the bond ellipticity of the vicinal CC bonds exceeds that of the distal CC bond or vice versa. For the same reason, $\rho(\mathbf{c})$ at the ring critical point becomes anisotropical, i.e. the ellipticity at the ring critical point, $\varepsilon(\mathbf{c})$, becomes > 0 because $\lambda_1 \neq \lambda_2$ (for **1**, $\lambda_1 = \lambda_2$, Table 6). The soft curvature of the ring ellipticity $\varepsilon(\mathbf{c})$ always points to the CC bond(s) with the largest π-character. Hence, a direction can be assigned to surface delocalization (indicated by a double-headed arrow) in substituted three-membered rings, which is the direction of the soft curvature of $\varepsilon(\mathbf{c})$ (see Scheme 2)[9].

Surface delocalization is not found for cyclobutane or larger cycloalkanes[10]. Furthermore, it does not appear for cyclotrisilane since in this case the overlap within the surface orbital is not sufficient to bring enough electron density into the centre of the ring[71].

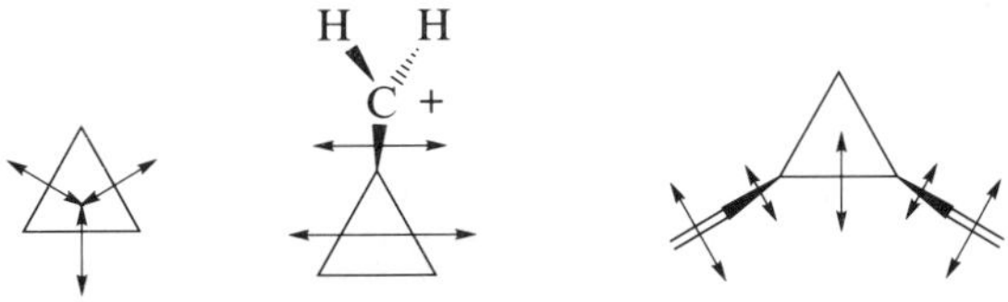

SCHEME 2

C. Analysis of the Laplacian of the Electron Density Distribution

It is a disadvantage of the analysis of $\rho(\mathbf{r})$ that many features of the orbital description of a molecule are not reflected by the properties of the electron density. However, this gap between orbital and $\rho(\mathbf{r})$ description can be closed by investigating the Laplacian of $\rho(\mathbf{r})$, $\nabla^2\rho(\mathbf{r})$[72]. The Laplacian of any scalar function is negative where the scalar function concentrates, and it is positive where the scalar function is depleted. This becomes obvious, when considering the second derivative of a general function $f(x)$:

$$\lim_{\Delta x \to 0}\{f(x) - \tfrac{1}{2}[f(x-\Delta x) + f(x+\Delta x)]\} = -\tfrac{1}{2}\lim_{\Delta x \to 0}\{[f(x+\Delta x) - f(x) - [f(x) - f(x-\Delta x)]\}$$
$$= -\tfrac{1}{2}(d^2f/dx^2)\,dx^2$$

If the second derivative, and hence the curvature of f, is negative at x, then f at x will be larger than the average of f at all neighbouring points, i.e. f concentrates at point x[73]. Therefore $-\nabla^2\rho(\mathbf{r})$, which is the second derivative of a function depending on three coordinates x, y and z, has been called the Laplace concentration of the electron density distribution. Furthermore, the Laplacian of $\rho(\mathbf{r})$ provides the link between electron density $\rho(\mathbf{r})$ and energy density $H(\mathbf{r})$ via a local virial theorem (equation 8)[67],

$$(\hbar^2/4m)\,\nabla^2\rho(\mathbf{r}) = 2G(\mathbf{r}) + V(\mathbf{r}) = G(\mathbf{r}) + H(\mathbf{r}) \qquad (8)$$

where $G(\mathbf{r})$, $V(\mathbf{r})$ and $H(\mathbf{r})$ are kinetic, potential and total energy density distribution[68, 69]. Integration of the Laplacian over an atomic basin or the total molecular space leads to zero, i.e. the fluctuations in $\nabla^2\rho(\mathbf{r})$ are such that local depletion of negative charge $[\nabla^2\rho(\mathbf{r}) > 0]$ and local concentration of negative charge $[\nabla^2\rho(\mathbf{r}) < 0]$ cancel each other, both for the atoms in a molecule as well for the molecule itself. Investigation of the Laplace concentration in the valence shell of an atom reveals maxima (lumps) and minima (holes) that can be associated with the amplitudes and the form of the HOMO and LUMO of this atom[13, 72]. This applies also to molecules and leads to a visualization of the frontier orbitals via the Laplace concentration of the electron density. Hence, the Laplacian $\nabla^2\rho(\mathbf{r})$ bridges the gap between the orbital and the density description of the electronic structure of a molecule.

In Figure 13, the calculated Laplace concentration, $-\nabla^2\rho(\mathbf{r})$, of **1** is shown in the plane of the ring[9–12]. The position of the C nuclei can be easily recognized by the 1s concentration peaks. Inner shell and valence region are separated by a sphere of charge depletion. Laplace concentration in the valence shell of the C atoms is distorted in a way such that there are concentration lumps in the direction of each bond (two of which are not visible, since they are in the direction of the CH bonds and therefore outside the reference plane). The concentration lumps in the ring plane can be associated with the $3e'$ HOMOs (MOs #11, #12) of **1** while the arrangement of the holes at the C atoms resembles the shape of the $1a_2'$ LUMO of the C_3 ring (lower-lying unoccupied MOs are of either $\sigma^*(CH_2)$ or $\pi^*(CH_2)$ character).

The contour line diagram of $-\nabla^2\rho(\mathbf{r})$ reveals concentration of electronic charge (dashed contour lines) not only in the CC bonding regions, but also inside the ring, thus confirming surface delocalization of electrons. Surface delocalization of σ-electrons inside the ring implies that there is a relatively low kinetic energy density $G(\mathbf{r})$ and a relatively large potential energy density $|V(\mathbf{r})|$. *Electrons stay longer inside the ring since they experience there the stabilizing attraction of three C nuclei.* The electrostatic potential of the three C nuclei is homomorphic with $\rho(\mathbf{r})$, which indicates that electron–nucleus attraction is supporting σ-electron delocalization inside the ring of **1**.

D. Surface Delocalization, π-Complexes and Bonding

Surface delocalization has been confirmed by various other authors. Coulson and Moffitt[23] were the first to note that there is a plateau of relatively high electron density

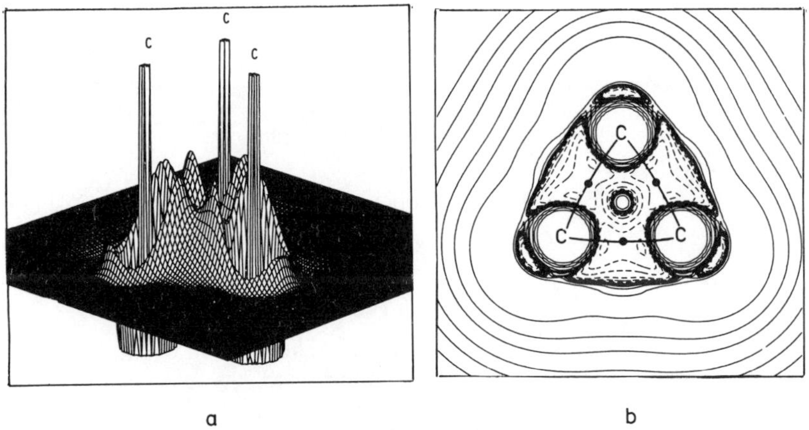

FIGURE 13. (a) Perspective drawing of the HF/6–31G(d,p) Laplace concentration $-\nabla^2\rho(\mathbf{r})$ of cyclopropane depicted in the ring plane. Inner-shell concentrations are indicated by the atomic symbol C. For a better presentation values above and below a threshold are cut off. (b) contour line diagram of the Laplace concentration shown in (a). Bond paths are indicated by heavy solid lines and bond critical points by dots. Dashed lines are in regions where electronic charge is concentrated and solid lines in regions where charge is depleted. Inner-shell concentrations are not shown. Reprinted from D. Cremer *Tetrahedron*, **44**, 7427 (1988) with kind permission from Elsevier Science Ltd., The Boulevard, Langford Lane, Kidlington OX5 1GB, UK

inside the ring of **1**. Schwarz and coworkers[74] calculated that $\rho(\mathbf{r})$ is increased by 0.16 e Å^{-3} in the centre of the ring relative to a promolecular density formed by three spherical isolated atoms. These authors attributed the increase in density to surface delocalization of electrons, which leads to a decrease in their kinetic energy and a subsequent contraction of the carbon AOs. This in turn lowers the potential energy as well as the total energy, restores the virial relation and leads to CC bond shortening.

Ahlrichs and Ehrhardt[75] calculated shared electron numbers for **1**. While bonding is normally reflected by two-centre contributions and negligible contributions from three and four-centre terms, a CCC shared electron number of 0.3 was calculated for **1**, which is indicative of significant three-centre bonding.

Recent VB calculations by Hamilton and Palke[50] reveal that optimized (non-orthogonal) CC hybrid orbitals for **1** cover the whole ring surface (see Section III. B, Figure 8) and take the character of surface orbitals that bring via six-fold overlap electron density into the surface of the ring, thus leading to surface delocalization of electrons and increased stability of the ring. For cyclobutane, hybrid orbitals are much more confined to the bonding region and therefore overlap inside the ring is much smaller[51]. In this case, it is not justified to speak about surface orbitals and surface delocalization of electrons (Figure 8).

Inagaki and coworkers[47] carried out a configuration analysis of the HF wave function of **1** (see also Section V. D). Using the coefficients for ground and excited configurations as a measure for electron delocalization, they found no indication of σ-electron delocalization in **1**. For ring molecules such as cyclobutane, they observed antibonding geminal electron delocalization that reduces the electron density inside the ring. But in the case of **1**, antibonding geminal electron delocalization is significantly lower than in other cycloalkanes. As a consequence, the electron density inside the C_3 ring of **1** is higher than inside a C_4 ring of cyclobutane in line with the calculated $\rho(\mathbf{r})$.

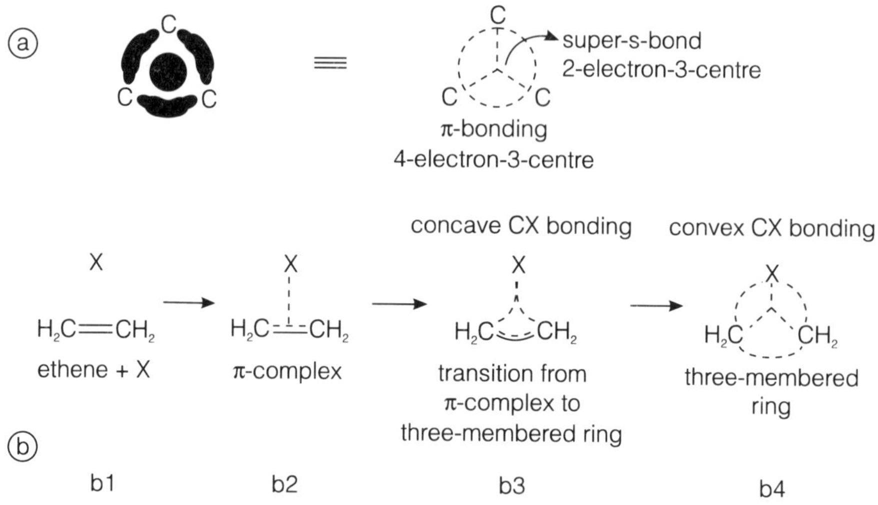

FIGURE 14. (a) Bonding in cyclopropane. On the left side the regions of relatively large internuclear electron density are indicated schematically. On the right side, 2-electron 3-centre bonding ('super-σ bond') and peripheral 4-electron 3-centre bonding ('π-bonds') are given by dashed lines. (b) Transition from ethene + X (b1) to a π-complex (b2, b3) and a three-membered ring (b3, b4)

Cremer[12] has pointed out that the term *delocalization* can be used in a quantum mechanical sense (delocalization of electrons in the space of two or more bonded atoms) and a heuristic sense (delocalization implies non-additivity of bond properties). The use of the term delocalization in the language of configuration (orbital) interactions (delocalization of electrons from occupied bond orbitals of the ground state configuration into antibonding orbitals of an excited configuration) does not coincide with these meanings. Clearly, the term surface delocalization has been based on the quantum mechanical meaning of electron delocalization, which can be translated to the language of configuration (orbital) interactions by describing **1** as a resonance hybrid of three π-complexes [delocalization of CH_2 electrons into antibonding (π^*) orbital of CH_2CH_2], which implies that the reference function is that of the π-complex and not that of **1** (reference for benzene is cyclohexatriene and not benzene itself).

Surface delocalization can be considered as the result of occupying the $3a_1'$ surface orbital (MO #8, w_O, Figure 6) in **1** and the π-character of the CC bonds as the result of occupying the two $3e'$ HOMOs (MOs #11 and 12; w_A and w_S, Figure 6). Hence, bonding in **1** is exceptional since the C atoms are held together by (see Figure 14):

1. a central 2-electron 3-centre bond ('super-σ bond') and
2. two peripheral 2-electron 3-centre bonds ('π-bonds')[12].

Hamilton and Palke[50] have critically considered the question of a 2-electron 3-centre bond. They compare CC hybrid orbitals calculated at the VB level (see Section III. B) with the orbitals representing the 2-electron 3-centre bonds in diborane, B_2H_6. Since the latter fully envelope the bridging H atoms and since this is not the case for the CC hybrid orbitals of **1**, Hamilton and Palke conclude that 2-electron 3-centre bonding does not play any role in **1**.

Since orbitals are not observable objects, it is clear that orbital sets related to each other by a unitary transformation, all will lead to valid descriptions of the electronic structure of the molecule although they may suggest very different orbital interactions leading to bond-

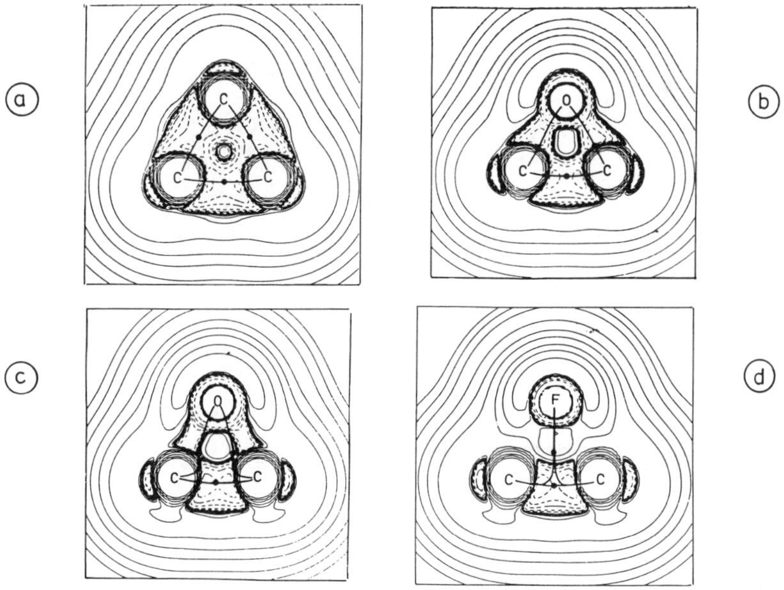

FIGURE 15. Molecular graphs and Laplace concentrations $-\nabla^2\rho(\mathbf{r})$ of (a) cyclopropane, (b) oxirane, (c) protonated oxirane and (d) halogen-bridged fluororethyl cation. The reference plane contains the nuclei of the heavy atoms. Bond paths are indicated by heavy solid lines and bond critical points by dots. Dashed lines are in regions where electronic charge is concentrated $[-\nabla^2\rho(\mathbf{r})>0]$ and solid lines in regions where charge is depleted $(-\nabla^2\rho(\mathbf{r})<0)$. Inner-shell concentrations are not shown. [HF/6–31G(d) calculations from Reference 9]. Reprinted with permission from D. Cremer and E. Kraka, *J. Am. Chem. Soc.*, **107**, 3800 (1985). Copyright (1985) American Chemical Society

ing. Using delocalized MOs, surface delocalization seems to be best explained by a 2-electron 3-centre bond caused by the occupation of the $3a_1'$ MO (MO#8, Figure 6). However, a description of CC bonding in **1** by hybrid orbitals optimized at the VB level of theory[51] suggests that the extension of the CC hybrid orbitals over the ring surface and their mutual overlap inside the ring (Section III. B, Table 3) leads to a transfer of electron density into the ring and to surface delocalization of electrons. Hence, both orbital descriptions lead to the same result. It is a matter of taste and suitability which orbital description is preferred and which bond description is considered to be more useful.

Dewar[34, 36, 43] was the first to point out a relationship between three-membered rings and π-complexes. His idea was later ventilated by a number of other authors[35, 37–45]. Cremer and Kraka[9] demonstrated, on the basis of electron density studies, that there is a continuous transition from three-membered rings to π-complexes depending on the electronegativity of the constituent atoms (see Figures 14 and 15). Molecule **1** is a three-membered ring with convex (outwardly) bent ring bonds. But if a CH_2 group in **1** is replaced by a more electronegative group X such as, e.g., O, OH^+ or F^+ (Figure 15b, c, d), then the CX bonds will be gradually bent inwardly toward the centre of the ring (concave bent bonds, Figure 15). For X = F^+, concave bending of the CX bond paths is so strong that the two paths coincide largely and form the T-structure of a π-complex (Figures 14 and 15d). Cremer and Kraka[9, 13] explained the relationship between three-membered rings and π-complexes by an orbital model, in which the orbitals of the apex group X interact in two ways (interactions 1 and 2) with the orbitals of a basal group A=A (Figure 16).

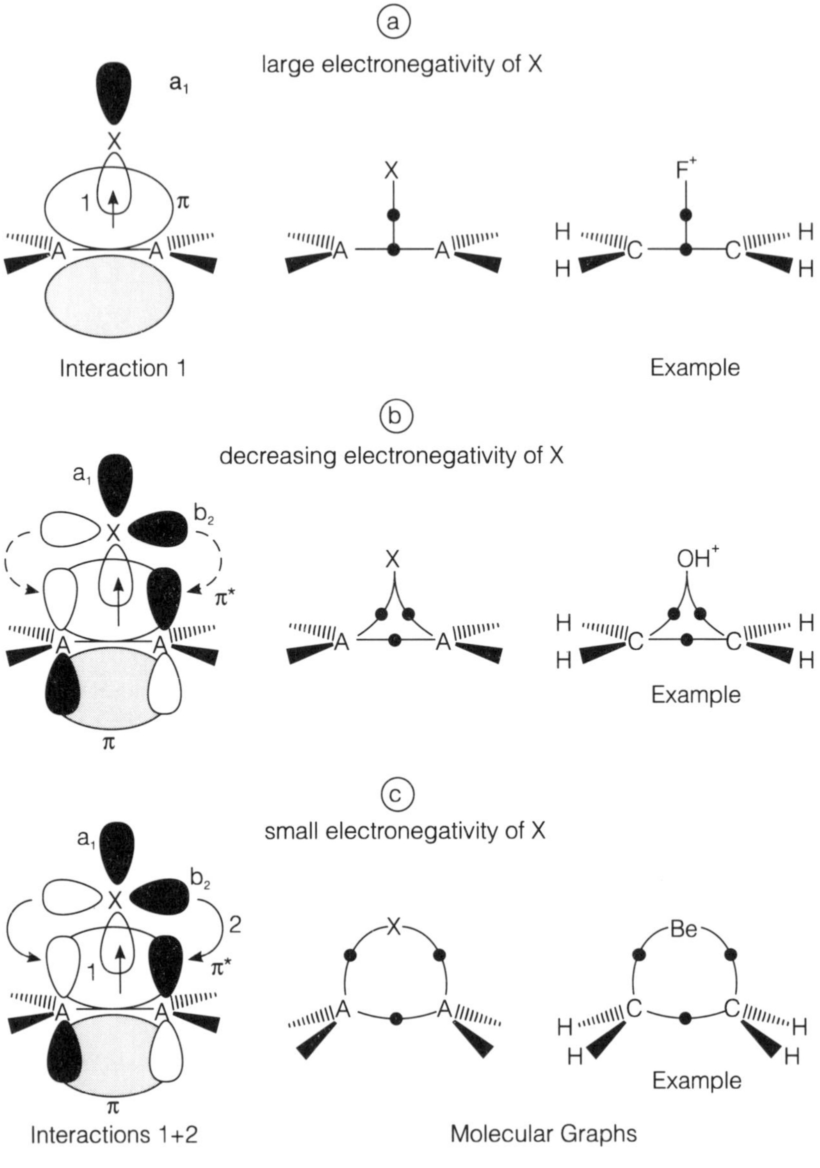

FIGURE 16. MO description of donor–acceptor interactions between the basal group A_2H_4 (A = C, Si etc.) and the apex group X. The relevant orbitals (a_1 and b_2 symmetry) are shown on the left side and the corresponding molecular graphs together with an appropriate example on the right side of each diagram. The direction of charge transfer caused by orbital interactions 1 and 2 is indicated by arrows (dashed arrows indicate reduced charge transfer). Top: π-Complex. Middle: Three-membered ring with concave bent bonds. Bottom: Three-membered ring with convex bent bonds. The electronegativity of X decreases from top to bottom

Interaction 1: The basal group A=A donates electrons from a π MO into a suitable low lying a_1 MO (C_{2v} symmetry assumed) thus establishing a build-up of electron density in the centre of the ring.

Interaction 2: Back donation from a relatively high-lying b_2 MO into the π*MO of A=A leads to an accumulation of electron density in the periphery of the ring, which determines the bending of bond paths.

Three different cases can be considered[9, 13, 71]:

(a) If interactions 1 and 2 are of comparable magnitude, then a three-membered ring with convex bent bonds will be formed (Figure 16c, example: **1** in Figure 15a).

(b) If interaction 2 is reduced because of the electronegativity of X (reduced donor capacity of X), a three-membered ring with concave bent bonds results (Figure 16b, example: X = OH^+, Figure 15c).

(c) If back donation is totally suppressed because of the large electronegativity difference between X and A, then there is just interaction 1, which leads to the T-structure of a π-complex, i.e. X is bound to the bond critical point of A=A (Figure 16a, example: X = F^+, Figure 15d).

According to the relationship between three-membered rings and π-complexes, cyclopropane can be considered as a resonance hybrid of three equivalent methylene–ethene π-complexes[12, 13]. Of course, such π-complexes do not exist but this is also true in the case of the two cyclohexatriene resonance structures normally used to present benzene. Spin-coupled valence bond calculations of Karadakov and coworkers[51] reveal that there is a small but significant contribution of 3.7% to the electronic structure of **1** resulting from π-complex structures (see Section V. E). This indicates that the π-complex description of **1** is not totally unreasonable and, although seldom used, helps to unravel some of the peculiarities of bonding in **1**:

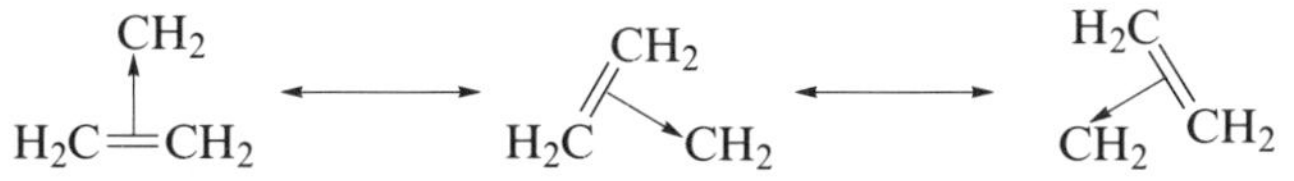

(a) There are different types of CC interactions in **1**. The orbital interaction 1 leads to a three-centre two-electron bond and surface delocalization. Orbital interaction 2 is responsible for the two peripheral 2-electron 3-centre bonds ('π-bonds').

(b) The π-complex description explains the π-character of the CC bonds.

(c) It further suggests increased s-character in the CH bonds and a concomitant strengthening of the CH bonds, which is experimentally confirmed.

(d) In a methylene, ethene π-complex typical CC distances are 2 Å and 1.34 Å, respectively. The CC bonds of **1** should be between these two values (because of resonance) and not necessarily identical with those of normal CC bonds.

Resonance between three π-complex structures might lead to stabilization of **1** in the sense of π-aromatic stabilization involving the six CC bond electrons. Therefore, Dewar[8] has discussed the stability of **1** in terms of a σ-aromatic stabilization (Section V). However, spin-coupled valence bond theory clearly shows that **1** cannot be considered as the σ-aromatic analogue to π-aromatic benzene[51]. The π-complex description of **1** is a (very formal) model description, which should be discarded as soon as it leads to conflicting descriptions of the properties of **1**. This will be discussed in Section V.

V. ENERGY AND STABILITY

Several authors have pointed out that the conventional strain energy (CSE) of **1** (27.5 kcal mol^{-1}) is about of the same magnitude as that of cyclobutane (26.5 kcal mol^{-1}) and therefore **1** must be stabilized in some way relative to cyclobutane[8, 9, 76]. In the last 15 years a number of investigations by Dewar[8], Cremer[9–13], Schleyer[76], Allen[32], Wiberg[14, 77–79], Inagaki[47, 80] and other authors[81–83] have been published to pin down the electronic causes for the unusual strain energy of **1**. These studies have also helped to clarify the concept of ring strain in general and the strain of three-membered rings in particular[5, 11–13]. Since little of this work is covered in the previous review by Wiberg[14], we will summarize here the most important results of investigations concerning stability and strain of **1**.

The CSE of **1** is determined by comparison with the energy of a suitable reference compound. Theoretically, this has been done by defining homodesmotic reactions such as those in equations 9 and 10[84]:

$$\mathbf{1} + 3CH_3CH_3 \longrightarrow 3CH_3CH_2CH_3 \tag{9}$$

$$C_nH_{2n} + nCH_3CH_3 \longrightarrow nCH_3CH_2CH_3 \tag{10}$$

which are formal reactions and for which the reaction energy can be easily determined by *ab initio* calculations. Alternatively, CSE values can be obtained by the traditional group equivalent method that has been transferred to *ab initio* theory by Wiberg[85] and by Ibrahim and Schleyer[86]. Applications of these approaches to **1** and other cycloalkanes in order to calculate CSE values have been amply discussed in previous reviews on strained hydrocarbons[5, 11, 14, 77] and need not to be described here.

Four different ways have been pursued to rationalize the CSE of **1** with the help of electronic structure calculations.

(1) Dissection of the energy into atomic energies with the help of the virial partitioning method.

(2) Dissection of the energy into bond energies.

(3) Dissection of the energy into strain contributions in a molecular mechanics related fashion.

(4) Dissection of the molecular wave function by a configuration analysis.

Each of these approaches leads to a different description of strain and stability of **1** and therefore the question is whether the various rationalizations of strain and stability are compatible.

A. Analysis of Ring Strain in Terms of Atomic Energies

Cremer and Gauss[10] were the first to show that the strain energy of **1** can be derived from atomic energies obtained by the virial partitioning method. These authors calculated charges and energy of the CH_2 group in **1**, cyclobutane and propane at the HF/6-31G(d,p) level of theory. The 6-31G(d,p) basis set was chosen because it contains polarization functions for both C and H atoms and, accordingly, guarantees a balanced description of geometry, CH bond polarity and charge distribution. For all molecules, the H atoms were found to be negatively charged, suggesting a larger electronegativity for H than for C (see also Section VIII). However, the difference in electronegativities between H and C was reduced when going from propane to cyclobutane and **1**, i.e. with increasing strain the electronegativity of the C atom increases (Table 7). This leads to two opposing effects:

(1) Increasing strain diminishes atomic volume V_Ω (Table 5) and atomic charge Q_Ω of the H atoms and, as a consequence, the H atoms are destabilized, i.e. their energy becomes more positive.

(2) At the same time, the C atoms accumulate more electronic charge in their atomic basin and therefore their energy becomes more negative, indicating stabilization of the atoms.

The total effect on the energy of a CH_2 group in case of increased strain, namely the destabilization of two H atoms and the stabilization of a C atom, is destabilizing, as can be seen when using an appropriate CH_2 reference group such as, e.g., the CH_2 group of cyclo-

TABLE 7. Atomic energies and atomic charges of cyclopropane, cyclobutane and propane as calculated by the virial partitioning method[a]

Property	Atom/group	Cyclopropane	Cyclobutane	Propane	Reference CH_2 (cyclohexane)
Energy	C	−37.7126	−37.6896	−37.6378	
	H	−0.6552	−0.6687 (eq)	−0.6738	
			−0.6691 (ax)		
	CH_2	−39.0230	−39.0274	−38.9854	−39.0379
Rel. energy	CH_2	9.3	6.6	32.9	0
Strain energy		27.9	26.4	–	0
Charge	C	105.6	171.9	222.9	
	H	−52.8	−87.0 (eq)	−92.2	
			−84.9 (ax)		
	CH_2	0	0	38.5	0

[a] From Reference 10. Absolute energies in hartree, relative energies and strain energies in kcal mol^{-1}, charges in melectron according to HF/6-31G(d,p) calculations. The reference group —CH_2— has properties almost identical with those of the CH_2 group of cyclohexane.

hexane. Cremer and Gauss[10] pointed out that the CH_2 group of propane is not suitable as a reference group since it is positively charged (Table 7), contrary to the CH_2 groups in cycloalkanes such as **1** or cyclohexane. The positive charge results from the larger group electronegativity of CH_3 (three electronegative H atoms) compared to CH_2 (two H atoms) and can only be balanced by filling up the atomic basins of the CH_2 group of propane by negative charge until electroneutrality is reached. This corresponds to an adjustment of the charge of the two CH_3 groups in propane to that of the CH_3 groups in ethane. A new hypothetical CH_2 reference group is derived that comprises the total subspace of the CH_2 group and a part of the subspaces of the two adjacent C atoms, which is indicated by the notation —CH_2—. Its energy is equal to $E(CH_2, \text{propane}) + 2\{E(CH_3,\text{propane}) - E(CH_3, \text{ethane})\}$, i.e. ethane is used to get the appropriate reference group. This is identical to the way of getting homodesmotic strain energies from the formal reaction 9 (*vide infra*) or to directly using a 'diagonal reference state'[5c] such as the CH_2 group in cyclohexane, which has almost the same properties as the group —CH_2—. (Using cyclohexane as a diagonal reference state would have been easier, but was not feasible because of computational reasons.)

Cremer and Gauss[10] could obtain in this way the CSEs of **1**, cyclobutane and other small cycloalkanes. Later, their work was extended to other strained hydrocarbons by using the same approach but smaller basis sets[78, 79]. A puzzling result of these investigations is the fact that ring strain seems to be a result of destabilization of the H atoms, which in turn is a result of a charge transfer from H to C caused by the increased electronegativity of C. This, of course, can be connected with hybridization models of Coulson and Moffitt[23] and others[7, 45–47, 50]. An increase in angle strain leads to increased p-character of the hybrid orbitals that constitute the CC bonds. The hybrid orbitals for the CH bonds obtain more s-character, i.e. the electronegativity of the C atom increases and C attracts negative charge from the H atoms with increasing angle strain. This is exactly reflected by the atomic charges and energies calculated with the virial partitioning method. Hence, the dissection of the molecular energy into atomic energies leads to a description of ring strain, which is consistent with other interpretations, but it does not explain the similarity in the CSEs of **1** and cyclobutane.

B. Analysis of Ring Strain in Terms of Bond Energies

Bond energies are not observable quantities and therefore they can only be calculated on the basis of a suitable model. Several attempts have been made on the basis of the maximum overlap model[45] or semiempirical calculations[87] to rationalize the stability of **1** in terms of CC bond weakening (because of angle strain) slightly compensated by simultaneous CH bond strengthening because of the higher s-character of the hybrid orbitals forming the CH bonds. Cremer and Gauss[10] have used the virial partitioning method to calculate CC and CH bond energies from first principles. Virial partitioning of the electron density distribution $\rho(\mathbf{r})$ is based on the zero-flux surfaces S(A,B) (see equation 6 and Table 5) that separate the atomic basins of bonded atoms A and B in a molecule. The bond energy BE(A,B) should be proportional to the electron density N(A,B) in the surface S(A,B) and the forces exerted on this density. For non-polar and weakly polar bonds, the second factor can be covered by a proportionality constant α(A,B) so that the bond energy is given by equation 11[88]:

$$\mathrm{BE(A,B)} = \alpha(\mathrm{A,B})\, N(\mathrm{A,B})/R(\mathrm{A,B})^2 \tag{11}$$

where N(A,B) is determined by equation 12:

$$N(\mathrm{A,B}) = \mathbf{R}(\mathrm{A,B}) \oint_{\mathrm{AB}} dS(\mathbf{r})\, \rho(\mathbf{r})\, \mathbf{n}_{\mathrm{A}}(\mathbf{r}) \tag{12}$$

where **R**(A,B) is the vector from the nucleus of A to the nucleus of B, and $\mathbf{n}_A$ is a unit vector normal to the surface, outwardly directed from A.

TABLE 8. Bond energies and strain energies of cyclopropane, cyclobutane and propane as calculated by the virial partitioning method[a]

Property	Bond group	Cyclopropane	Cyclobutane	Propane
Bond energy	C—C	71.0	73.9	81.9
	C—H	106.6	105.9	105.5
Strain energy	C—C	32.7	32.0	0
Stabilization energy[b]	C—H	−6.6	−3.2	0
Error in atomization energy		1.4	−3.1	0.6
Total strain energy		27.5	25.7	0

[a] From Reference 10. Bond and strain energies in kcal mol^{-1} according to HF/6–31G(d,p) calculations. The strain energy is derived from the difference in CC bond energies and CH stabilization energies corrected for errors in the theoretical atomization energies.
[b] Stabilization due to hybridization effects in CH bonding.

In Table 8, HF/6-31G(d,p) bond energies for **1**, cyclobutane and propane are given, which indicate a decrease in BE(CC) from 82 to 71 kcal mol^{-1} when going from propane to **1**, thus leading to a strain energy of 34 kcal mol^{-1} (including small corrections in calculated atomization energies[10]). The strain energy of the ring is slightly reduced by an increase of the CH bond energy from 105.5 in propane (*sec*-CH bond) to 106.6 kcal mol^{-1} in **1** (increased s-character of CH hybrid orbitals). Hence, the final strain energy (27.5 kcal mol^{-1}, Table 8) is in good agreement with the accepted CSE of **1**. For cyclobutane, the corresponding values are 73.9 kcal mol^{-1} (BE of CC), 105.9 (BE of CH) and 26 kcal mol^{-1} (final SE, Table 8). Hence, the SE of **1** results from CC bond destabilization caused by angle bending and the concomitant increase of p-character in the hybrid orbitals forming the CC bonds[10].

The difference in the values of the CC bond energies for **1** and cyclobutane does not contradict observations made for bond orders and overlap values, namely that

(a) bond orders n for **1** and cyclobutane are identical (n = 1.00, Table 6[9,10]), and

(b) CC bonding overlap for **1** and cyclobutane as calculated at the VB level of theory is almost identical (Table 3)[51].

Both bond order n and CC bonding overlap S are local quantities[11]. The bond order is derived from the value of $\rho(\mathbf{r})$ at position **p** and S contains just the CC hybrid orbital overlap. Bond energies by definition are global rather than local quantities because they are derived by appropriate dissection of atomization energies and, accordingly, contain all bonding and non-bonding interactions between two atoms. If one wants to estimate bond energies from overlap values, one has to consider both bonding and non-bonding overlap (see Table 3). In addition, one has to include effects such as bond bending and bond polarity, where the former determines hybridization and electronegativity of the atoms forming the bond and the latter is a result of the electronegativity difference between the atoms of a bond. In the case of **1**, the bond energy should be given by a relationship such as equation 13:

$$\text{BE (CC)} = a\, I_C\, S(\text{CC}) + b \qquad (13)$$

where a and b are constants, S(CC) is the CC overlap and I_C the ionization potential of a C atom in the C_3 ring. Equation 11 is exactly of the form of equation 13 with α(A,B) taking the part of I and N(A,B) the part of S.

Neither bond order or CC overlap values alone can lead to an estimate of the CC bond energy and, in this respect, *the observation that both* **1** *and cyclobutane possess similar CC bonding overlap*[51] *does not say anything with regard to the CC bond strength.*

As in the case of the calculation of *in situ* atomic energies, it is possible to give a reasonable explanation for the strain of **1** by comparing calculated bond energies, but it is not possible to explain the similar stability of **1** and cyclobutane on this basis. For example, it is not clear why the CC bonds in **1** and cyclobutane have comparable strengths. It has been suggested by Schleyer[76] that CH bond strengthening in **1** is much larger than that of the *sec*-CH bond in propane. For example, the CH bond dissociation enthalpy DH of **1** (106.3 kcal mol^{-1}) is 11.2 kcal mol^{-1} higher than the one for the secondary CH bond of propane (95.1 kcal mol^{-1})[89]. However, DH values do not necessarily reflect the magnitude of bond energies since they depend on the stability of both reactant and product. In this way, the large DH(CH) value of **1** may just reflect the increase in ring strain when the cyclopropyl radical is formed.

C. Dissection of the Molecular Energy into Strain Contributions

According to the classical definition of ring strain introduced by Baeyer at the end of the last century[90], a three-membered ring should be much more strained than a four-membered ring. Its bond angles α deviate from the standard, strain-free CCC angle (109.5°) by $\Delta\alpha$ = 49.5° while those of the planar four-membered ring deviate by just 19.5°. According to Hooke's law, the Baeyer strain energy (angle bending strain energy) as given in equation 14:

$$\Delta E(\text{Baeyer}) = n(k_\alpha/2)(\Delta\alpha)^2 \qquad (14)$$

(where n is the size of the ring and k_α is the CCC bending force constant) should be 173 kcal mol^{-1} for **1** and 36 kcal mol^{-1} for cyclobutane if k_α(CCC) = 1.071 mdyn Å rad^{-2} of propane[91] is used as an appropriate bending force constant.

Cremer and Gauss[10] have pointed out that it is unrealistic to use geometrical angles α and the bending force constant of propane to calculate the Baeyer strain energy of **1**. These authors chose the CC bond path length r_b (1.506 Å compared to r_e =1.497 Å, Table 6) and the interpath angles β(CCC) (79° compared to α_e = 60°, Table 6) rather than the geometrical parameters to get realistic stretching and angle strain energies. Furthermore, they pointed out that the bending force constant of propane depends on 1,3-non-bonded interactions, which are totally missing in **1**. In molecular mechanics, this problem is solved by considering the bending force constant k as an adjustable parameter that takes values between 0.45 and 0.8 mdyn Å rad^{-2} to reproduce measured molecular properties[92]. Cremer and Gauss[10] determined k(CCC) = 0.583 mdyn Å rad^{-2} using the strain energy of cyclobutane in the absence of 1,3-CC non-bonded repulsion (Dunitz–Schomaker strain energy)[93].

Table 9. Dissection of the strain energy of cyclopropane[a]

Strain	Cyclopropane	Cyclobutane	Reference	Comment	Method[Ref.]
Stretching	0.5	1.0	ethane	k_r (CC) = 4.57 mdyn Å^{-1} k_r (CH) = 4.88 mdyn Å^{-1} r (CC) = 1.5268 Å r (CC) = 1.0858 Å	exp[91] exp[91] HF/6–31G(d,p)[10] HF/6–31G(d,p)[10]
			Note: Bond path lengths r_b are used for bent bonds		
Baeyer (angle bending)	46.3	13.0	propane	k_b (CCC) = 0.583 mdyn Å rad^{-2} k_α (CCH) = 0.656 mdyn Å rad^{-2} k_α (HCH) = 0.550 mdyn Å rad^{-2} $\alpha = 109.5°$	 exp[91] exp[91]
			Note: Bond path angles β are used for bent bonds. According to Hooke's law, the Baeyer strain energy is 41.3 kcal mol^{-1}; anharmonicity effects lead to another 5 kcal mol^{-1}		
Pitzer (bond eclipsing)	4.0	3.9	ethane	$V_3 = 3.0$ $\tau = 60°$	HF/6-31G(d,p)[10]
Dunitz–Schomaker (non-bonded interactions)	0	12.0	cyclobutane *Note*: Results scaled as a function of k_β (CCC) to reproduce CSE and inversion barrier of cyclobutane		CNDO/2[93]
Total	50.8	29.9			
CH strengthening	– 6.4	– 2.8	from *ab initio* bond energies *Note*: Rehybridization leads to higher s-character		HF/6-31G(d,p)[10]
σ Delocalization	– 16.4	0	Difference between CSE and ΔE = 50.8 –6.4 = 44.4 kcal mol^{-1}		
CSE	28.0	27.1	Conventional Strain Energy		

[a] All energy values in kcal mol^{-1}

Table 9 shows the various strain energies calculated for **1** and cyclobutane by Cremer and Gauss[10]. The Baeyer strain energy of **1** is 46 kcal mol^{-1}, including an estimated 5 kcal mol^{-1} from anharmonicity effects, while the Baeyer strain energy of cyclobutane is just 13 kcal mol^{-1}. However, the four-membered ring is destabilized by 12 kcal mol^{-1} because of 1,3-CC repulsion (Dunitz–Schomaker strain). Stretching strain and Pitzer strain add together just 4–5 kcal mol^{-1} to the total strain energy, which is 51 kcal mol^{-1} in the case of **1** and 30 kcal mol^{-1} for cyclobutane. These values are in line with an expected increase of the strain energy with decreasing ring size. However, they must be corrected for stabilizing effects, namely

(a) CH bond strengthening as a result of increased s-character of the hybrid orbitals forming the CH bond orbitals, and

(b) surface delocalization of σ-electrons[9].

If one takes the bond strengthening effects calculated with equations 11 and 12 (see Table 8), then surface delocalization of σ-electrons must add about 16 kcal mol^{-1} to the stability of **1** to lead to a CSE of 28 kcal mol^{-1} [10].

D. Dissection of the Molecular Wave Function

While the three approaches discussed above lead to a quantitative reproduction of the SE of **1**, a dissection of the HF wave function suggested by Inagaki and coworkers[47, 80] provides qualitative arguments for a rationalization of strain and stability of **1**. The method is based on a configuration analysis of the HF wave function, Ψ_{HF} according to equation 15:

$$\Psi_{HF} = C_G \Phi_G + \Sigma C_T \Phi_T \qquad (15)$$

where Φ_G is the ground state configuration and Φ_T are singly excited configurations. The ground configuration is made up from localized bond orbitals each of which is a linear combination of hybrid atomic orbitals located at bonded atoms. Optimal hybrid atomic orbitals (with regard to the s/p ratio) are determined by maximizing the coefficient C_G of the ground state configuration[47, 80].

The admixture of singly-excited configurations Φ_T to Φ_G to represent the HF wave function indicates electron transfer from bonding to antibonding orbitals and, accordingly, can be interpreted in terms of electron delocalization, which can be measured by the ratio C_T/C_G. Alternatively, one can define interbond populations by equation 16[47, 80]:

$$\mathrm{IBP}_{ij} = 2\Sigma_p n_p c_{pi} c_{pj} s_{ij} \qquad (16)$$

where n_p is the occupation number of the pth MO, c_{pi} the expansion coefficient of the ith bond orbital for the pth MO and s_{ij} the overlap between bond orbitals i and j. The method of configuration analysis leads to

(a) the percentage of s-character of the hybrid orbitals to describe the electronegativity of atoms forming the ring;

(b) atomic bond populations and overlap of hybrid orbitals to describe the strength of bonds;

(c) overlap repulsion indices $\mathrm{IBP}_{\sigma\sigma}$ between geminal bonding orbitals to describe angle strain upon bond angle deformation and

(d) interbond population $\mathrm{IBP}_{\sigma\sigma^*}$ between bonding and antibonding orbitals to describe delocalization effects.

In Table 10, results of the configuration analysis of HF/6-31G(d) wave functions of several three-membered rings are shown. Data indicate that[47]:

(1) The s-character (p-character) of the hybrid orbitals forming the ring bonds decreases (increases) when going from the acyclic reference compound to the four- and then to the three-membered ring or when going to a system with more electronegative heavy atoms.

TABLE 10. Conventional strain energies (CSE), hybridizations, s-character, overlap values, overlap repulsions and geminal delocalizations of propane, cyclobutane, cyclopropane and their heterologues with X = NH, O, SiH_2, PH, S from Reference 47[a]

Parameter	Acyclic compound	Four-membered ring	Three-membered ring
	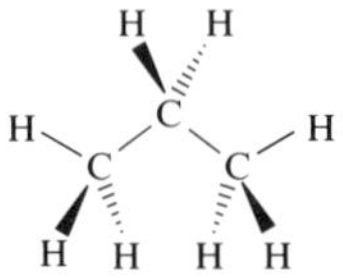	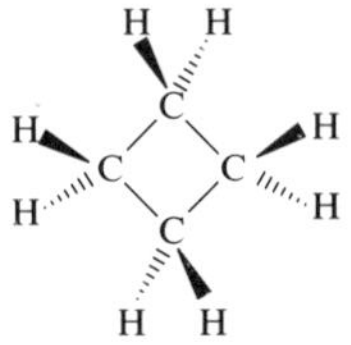	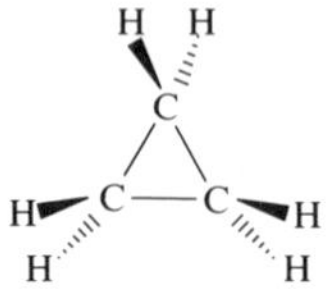
	(11)	**(12)**	**(1)**
CSE		26.6	28.7
sp^n(CC) [s-character in%]	sp$^{2.9}$ [26]	sp$^{3.2}$ [24]	sp$^{4.0}$ [20]
sp^m(CH) [s-character in%]	sp$^{3.1}$ [24]	sp$^{2.8}$ [26]	sp$^{2.3}$ [30]
S(CC)	0.61	0.58	0.56
S(CH)	0.65	0.66	0.67
IBP$\sigma\sigma$	−0.021	−0.046	−0.132
IBP$\sigma\sigma$*	−0.011	−0.017	−0.003
	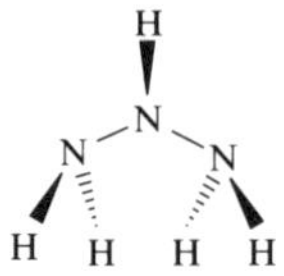	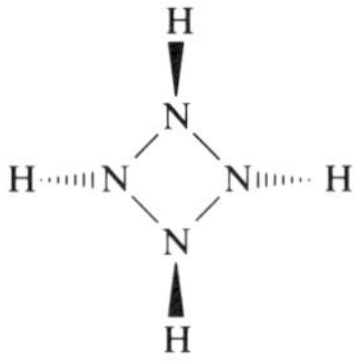	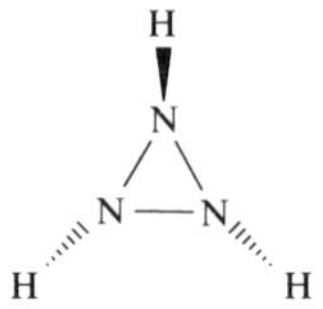
	(13)	**(14)**	**(15)**
CSE		33.4	31.8
sp^n(NN) [s-character in%]	sp$^{3.3}$ [23]	sp$^{4.8}$ [17]	sp$^{7.2}$ [12]
S(NN)	0.57	0.52	0.48
S(NH)	0.64; 0.64	0.63	0.60; 0.63
IBP$\sigma\sigma$	−0.010	−0.033	−0.095
IBP$\sigma\sigma$*	−0.014	−0.022	−0.004
	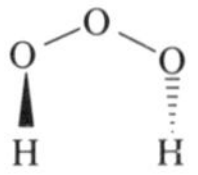	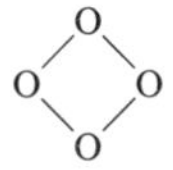	
	(16)	**(17)**	**(18)**
CSE		57.0	38.7
sp^n(OO) [s-character in%]	sp$^{6.7}$ [13]	sp$^{8.6}$ [10]	sp$^{11.9}$ [8]
sp^m(lp) [s-character in%]	sp$^{1.7}$ [37]	sp$^{1.5}$ [40]	sp$^{1.4}$ [42]
S(OO)	0.42	0.39	0.37
IBP$\sigma\sigma$	−0.006	−0.018	−0.061
IBP$\sigma\sigma$*	−0.007	−0.011	0.004

Parameter	Acyclic compound	Four-membered ring	Three-membered ring
	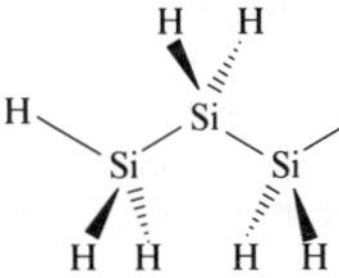	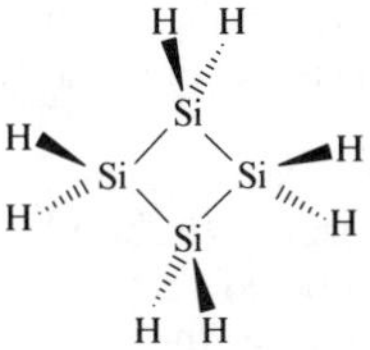	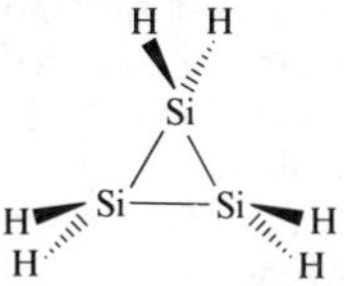
	(19)	**(20)**	**(21)**
CSE		16.6	38.8
sp^n(SiSi) [s-character in%]	$sp^{2.9}$ [26]	$sp^{3.4}$ [23]	$sp^{3.6}$ [22]
sp^m(SiH) [s-character in%]	$sp^{3.1}$ [24]	$sp^{2.7}$ [27]	$sp^{2.5}$ [28]
S(SiSi)	0.48	0.45	0.44
S(SiH)	0.37	0.36	0.35
IBP$\sigma\sigma$	−0.004	−0.024	−0.087
IBP$\sigma\sigma$*	−0.005	−0.012	−0.021
	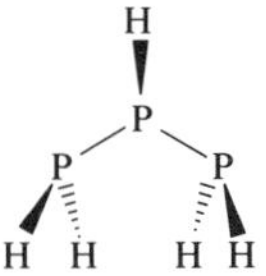	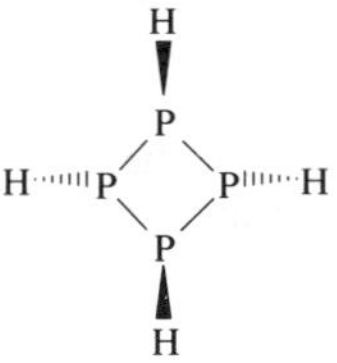	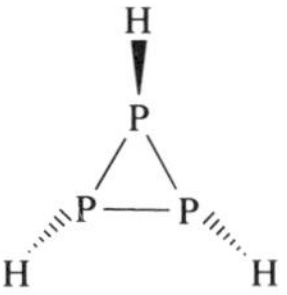
	(22)	**(23)**	**(24)**
CSE		9.5	11.2
sp^n(PP) [s-character in%]	$sp^{7.2}$ [12]	$sp^{9.6}$ [9]	$sp^{14.3}$ [6]
S(PP)	0.41	0.39	0.36
S(PH)	0.56	0.56	0.56
IBP$\sigma\sigma$	−0.002	−0.013	−0.058
IBP$\sigma\sigma$*	−0.004	−0.007	−0.002
	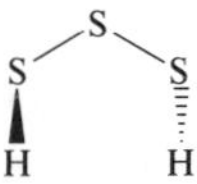	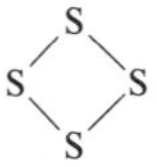	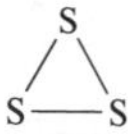
	(25)	**(26)**	**(27)**
CSE		39.6	28.9
sp^n(SS) [s-character in%]	$sp^{9.6}$ [9]	$sp^{12.6}$ [7]	$sp^{16.3}$ [6]
sp^m(lp) [s-character in%]	$sp^{1.5}$ [40]	$sp^{1.3}$ [44]	$sp^{1.3}$ [44]
S(SS)	0.40	0.37	0.36
IBP$\sigma\sigma$	−0.0021	−0.013	−0.055
IBP$\sigma\sigma$*	−0.003	−0.006	−0.002

[a] Conventional strain energies (CSE) in kcal mol^{-1} from HF/6-31G(d) calculations. Hybridization ratios *n* and *m*, overlap *S*, and interbond population values from a configuration analysis[47].

(2) With increasing p-character, bond populations and the overlap between hybrid orbitals forming the bond orbital decrease. (We note that calculations based on non-

orthogonal hybrid orbitals might lead to different trends in orbital overlap[51].) This indicates bond weakening in the order: open-chain compound < four-membered ring < three-membered ring.

(3) Destabilizing (antibonding) overlap repulsions between geminal bonds increase (i.e. $IBP_{\sigma\sigma}$ becomes more negative) in the order: open-chain compound < four-membered ring < three-membered ring.

(4) Bonds XH become stronger as indicated by overlap and bond populations (hybrid orbitals of ring atoms forming the bonds possess more s-character) in the order: open-chain compound < four-membered ring < three-membered ring. However, in the case of ring atoms with lone-pair orbitals, the increase in s-character is observed only for the lone-pair orbitals and the peripheral ring bonds become actually weaker rather than stronger.

(5) Geminal delocalization measured by the interbond populations $IBP_{\sigma\sigma^*}$ is destabilizing (antibonding) in most cases. For Si compounds, the destabilizing character of geminal delocalization increases from the open-chain compound to the four-membered and the three-membered ring. This means that geminal delocalization and overlap repulsion both lead to a decrease of the electron density between geminal bonds and that depression of electron density increases with increasing angle strain. However, if the p-character of the hybrid orbitals forming the ring increases because of an increase in electronegativity of the ring atoms X, then geminal delocalization in three-membered rings will become less antibonding [e.g. X= C (**1**), N (**15**)] or even bonding [e.g. X= O (**18**)].

While observations (1)–(4) are in line with other descriptions of the strain in **1** or three-membered rings, geminal delocalization seems to play a decisive role with regard to the relative stability of three-membered versus four-membered ring. Bonding geminal delocalization leads to a build-up of electron density inside the ring and surface delocalization. An increase of geminal delocalization (surface delocalization) reduces angle strain and leads to higher stability of the ring. This can be found in the series **1** (X = C), **15** (X = N), **18** (X = O) as well as in the series **21** (X = Si), **24** (X = P), **27** (X = S) where, for the hetero atom rings, the CSE is smaller in the three- than in the four-membered rings. Hence, the strain energy of **18** is 39 kcal mol^{-1} compared to 57 kcal mol^{-1} in the case of **17**. In this way, **1** is not so peculiar with regard to its stability as one might think.

In summary, the configuration analysis explains the peculiar stability of **1** and confirms the existence of surface delocalization in **1**. This leads to extra-stabilization of the ring and a reduction in the ring strain, so that it becomes comparable to that of cyclobutane. The configuration analysis can provide only trends but no energy data that lead to a dissection of the total strain energy. Accordingly neither interbond populations nor delocalization indices correlate with calculated strain energies.

E. Strain Energy, σ-Aromaticity and Surface Delocalization

Dewar[8] has pointed out that there is an analogy between the =HC—CH= group of a conjugated cyclopolyene and the —CH_2— group of a cycloalkane. He has argued that **1** and benzene are isoconjugate. While benzene is stabilized by a system of six delocalized π-electrons leading to π-aromaticity, **1** is stabilized by a sextet of delocalized σ-electrons leading to σ-aromaticity. Dewar[8] has related known properties of **1**, such as

(1) its surprisingly low strain energy,

(2) its relatively short CC bond lengths,

(3) its relatively high CC bond strengths,

(4) upfield shifts of its ^{1}H (1 ppm) and ^{13}C (20 ppm) NMR signals compared to those of other alkanes,

(5) its electronic interactions with substituents and

(6) its ability to enhance conjugation in homoaromatic systems

to a possible σ-aromatic stabilization. He estimated that the σ-aromatic stabilization energy might be as large as 55 kcal mol^{-1}. Schleyer[76] and, independently, Grev and

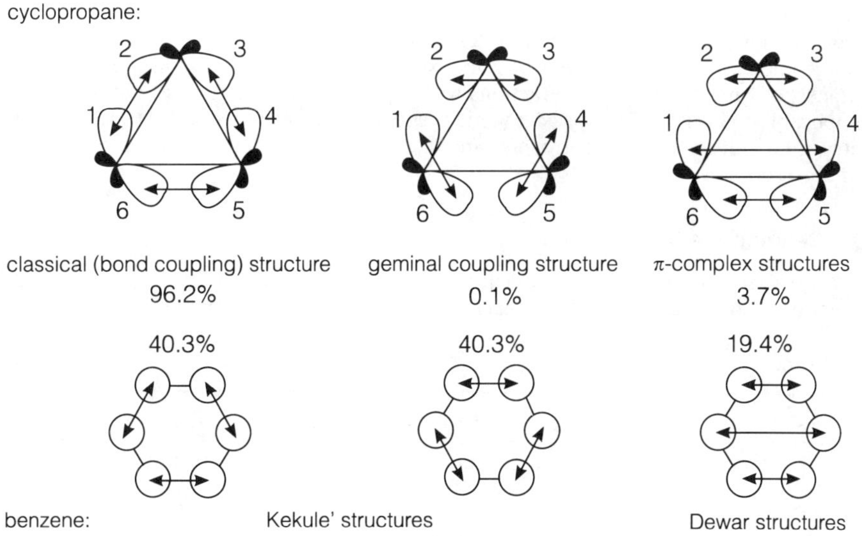

FIGURE 17. Schematic representation of the symmetry-unique spin-coupling patterns in cyclopropane (above) and benzene (below). In the case of cyclopropane, carbon hybrid orbitals and, in the case of benzene, carbon pπ orbitals are shown. For each structure, Gallup–Norbeck occupation numbers as determined by spin-coupled valence bond theory are given. All data from Reference 51

Schaefer[94] have criticized the concept of σ-aromaticity. Schleyer in particular noted that CH bond strengthening in **1** and 1,3-CC non-bonded repulsion in cyclobutane could easily explain the similarity in strain energies of the three- and four-membered rings. Karadakov and coworkers[51], who carried out a spin-coupled VB calculation of **1**, found no indication that more than one spin-coupling pattern plays any significant role in the description of the electronic structure of **1** (see Figure 17). Two resonance structures with equally large weight as in the case of benzene (40.3%, Figure 17) do not exist in the case of **1** and therefore the authors reject the possibility of σ-aromaticity.

It is interesting to note that the spin coupling patterns that correspond to a π-complex (1–4, 2–3, 5–6 and two equivalent patterns obtained by cyclic permutation, see Figure 17) contribute to the wave function by 3.7%, which is not large but suggests that the resonance description of **1** in terms of ethene–methylene π-complexes is not totally wrong and may become more important for reacting **1** or other three-membered rings.

Cremer[12] has summarized the pros and cons for invoking the term σ-aromaticity. His major conclusion was that all the peculiar properties of **1** can be rationalized in terms of surface delocalization of σ-electrons *without invoking σ-aromaticity*. This is of advantage since surface delocalization is based on an experimental fact, namely the observed increase in electron density inside the ring. Surface delocalization adds to the stability of **1** and lowers the strain energy. It is, however, a question as to how the energy of **1** is dissected and whether the stabilization energy resulting from surface delocalization is 16 kcal mol^{-1} or some other value[10].

VI. GEOMETRY

Investigation of the geometry of **1**, related three-membered rings and substituted **1** has been a major research goal of structural chemists for decades[5, 26, 27, 32, 33, 71, 83, 94–125] since these studies provided primary information on the electronic structure of **1**, its ability to conjugate with unsaturated groups and its interactions with substituents in general. In addition, the geometry of three-membered rings has been the testing ground for various theoretical models of ring–substituent interactions with regard to their usefulness, limitations and applicability. In view of this massive work, it is surprising that even today the exact r_e geometry of **1** is still a matter of controversy and that not all geometries of substituted **1** are well understood.

A. Geometry of Cyclopropane

Calculated and experimental geometries of **1** are summarized in Table 11[10, 62, 126–133]. There are two experimental investigations that derived the r_e (equilibrium) geometry from measured

TABLE 11. Geometry of cyclopropane[a]

Method	r (CC) (Å)	r (CH) (Å)	α (HCH) (deg)	Ref.
HF/4-21G	1.515	1.071	114.5	131
HF/4-31G	1.503	1.072	113.7	128
HF/[4s2p/2s]	1.519	1.073	113.8	129
HF/6-31G (d)	1.497	1.076	114.0	128
HF/6-31G (d,p)	1.497	1.076	114.1	10
HF/[4s2p1d/2s1p]	1.503	1.078	114.4	132
HF/[5s3p2d/3s2p]	1.499	1.072	114.4	132
HF/[5s3p2d/3s2p]	1.497	1.073	114.4	133
MP2/6-31G (d)	1.502	1.084	114.2	62
MP2/[4s2p1d/2s1p]	1.513	1.083	115.1	130
MP2/[5s3p2d/3s2p]	1.507	1.076	115.2	133
CISD(FC)/[4s2p1d/2s1p]	1.510	1.080	114.7	130
MP4(SDTQ)/[5s3p2d/3s2p]	1.514	1.079	115.1	133
CCSD(T)/[5s3p2d/3s2p]	1.514	1.079	115.0	133
r_z (Electron diffraction)	1.5127 (12)	1.0840 (20)	114.5 (9)	126
r_z (Microwave)	1.5157 (23)	1.0797 (34)	115.47 (38)	127
r_e (exp)	1.501 (5)	1.084 (5)	114.5 (9)	126
r_e (exp)	1.5101 (23)	1.0742 (29)	115.85 (33)	127

[a] For the experimentally based geometries, uncertainties are given in parentheses.

r_0 and r_z geometries of **1**[126, 127]. Endo, Chang and Hirota[127] measured the microwave spectrum of **1**-1,1-d_2 and determined its rotational and centrifugal distortion constants. Using these data together with published rotational constants for **1** (B_0, C_0) and **1**-d_6 (B_0), these authors obtained r_0 and r_z structures in good agreement with previously published r_0 and r_z structures[126]. In addition, Endo, Chang and Hirota derived an r_e geometry of **1** utilizing reported vibration–rotation constants and considering just the CC and CH third-order anharmonicity constants as adjustable parameters. This r_e geometry (Table 11) differs considerable from an r_e geometry of **1** published earlier by Yamamoto, Nakata, Fukuyama and Kuchitsu[126], who investigated **1** by a joint analysis of electron diffraction intensities and spectroscopic data including rotational constants for vibrationally excited states. Kuchitsu and co-workers assumed that α_e(HCH) is equal to α_z(HCH) and used effective CC and CH stretching anharmonicity constants derived from the rotational constants for the v_{11} vibrational state[126].

The two r_e geometries differ by 0.01 Å and 1.3° with regard to C—C (C—H) bond lengths and HCH angles, respectively, which is well outside the error bars given in the two investigations (see Table 11)[126, 127]. Considering the fact that r_e geometrical parameters of small molecules can be given with a precision of 0.001 Å (0.5°) or better[134], the deviation of the two r_e geometries suggests that at least one of the published r_e geometries is seriously in error.

Ab initio geometries of **1** (Table 11) vary over a relatively large region[10, 33, 62,128–133]: CC bond lengths from 1.497 to 1.519 Å, CH bond lengths from 1.072 to 1.084 Å and HCH angles from 113.5 to 115.2°. Considering, however, the known trends of calculated equilibrium geometries as dependent on method and basis set, it is rather straightforward to suggest the most likely r_e geometry of **1**[135].

At the HF level of theory, the largest basis set leads to the shortest CC and CH bond lengths and the largest HCH angle. In general, bond lengths decrease with increasing basis set, which has to do with the redistribution of electron density in a molecule upon basis set enlargement. Since an electron 'sees' only the average field of all other electrons at the HF level, electronic charge can accumulate close to the nuclei to increase stabilizing electron–nucleus attractions. The more basis functions are available to describe the area around the nuclei, the more electrons are packed close to the nuclei, thus effectively shielding the nuclei with regard to each other. As a consequence, nuclear repulsion is reduced and internuclear distances are decreased. Short bond distances will lead to an increase in non-bonded repulsion of partially positive H atoms and thereby cause an enlargement of the HCH angles. Since HF/large basis set calculations will underestimate bond distances, HF is not the appropriate method for high-accuracy determinations of r_e geometries. Of course, it may be the case that HF/small or medium basis set calculations accidentally lead to accurate r_e geometries because of a fortuitous cancellation of basis set and correlation errors.

As can be seen from Table 11, inclusion of electron correlation leads to an increase of both CC and CH bond lengths. This is due to the fact that correlated movements of the electrons exclude clustering of electrons in the vicinity of the nuclei. Instead, electrons have to spread out in the molecule to avoid close contacts and destabilizing Coulomb repulsion. The nuclei become deshielded, nuclear repulsion is increased and, hence, bond distances become longer for correlation corrected *ab initio* methods. These effects are the larger the more correlation effects are included into the *ab initio* method. For example, MP2 covers just pair correlation effects, which is sufficient to get useful descriptions of r_e geometries for singly bonded molecules. However, in the case of molecules with multiple bonds, triple and quadruple excitations have to be included to obtain reliable r_e geometries. This suggests the use of methods such as MP4[136] or CCSD(T)[137], where the latter method is correct to fourth-order perturbation theory but contains in addition infinite-order effects resulting from S, D and T excitations[138].

Calculated CC bond distances of **1** obtained with various basis sets at the MP2 level[136] range from 1.502 to 1.513 Å. Both MP4/TZ + 2P and CCSD(T)/TZ + 2P calculations[133] suggest an even longer CC bond length of 1.514 Å, which is close to the experimentally based value of Endo and coworkers[127]. In view of the fact that the CC bonds of **1** possess considerable π-character, the inclusion of T excitations seems to be necessary and suggests higher reliability to MP4 and CCSD(T) (rather than MP2) results, which clearly support the r_e geometry predicted by Endo and coworkers[127].

The geometrical parameters of **1** are characteristically different from those of other alkanes (Table 12)[10, 62, 139–143]. Compared to propane, both CC and CH bonds are shorter (0.02 and 0.01 Å, respectively) and HCH angles are larger (8°). Differences between the geometrical parameters of **1** and cyclobutane are similarly large or, as in the case of the CC bond (1.510 vs 1.548 Å, Tables 11 and 12), even larger. These characteristic differences are a result of the peculiar electronic structure of **1** (see Section III–V), which lend the molecule similarity to an alkene. This is confirmed by comparing the geometrical parameters of **1** and ethene: CH bond lengths and HCH angles have indeed similar values (Tables 11 and 12) while the CC bond of **1** is closer to that of an alkane than to that of an alkene.

If the cyclopropane ring is fused with another cyclopropane ring, thus leading to bicyclo[1.1.0]butane, all CC bonds become shorter but, in particular, the bridge bond C1C3, which decreases to 1.47 Å according to theory (Table 12)[62, 141]. The experimental value is longer (1.50 Å) and also suggests r(C1C3) > r(C1C2)[142], which may be doubted in view of the increase in π-character of the bridge bond (see Section XII. A). If three cyclopropane

TABLE 12. Geometries of propane, ethene, cyclobutane and some fused cyclopropanes[a]

Molecule	Parameter	Method				Reference	
		HF 4-31G	HF 6-31G(d)[b]	MP2 6-31G(d)	Exp.	Theor	Exp
Propane	CC	1.530	1.528	1.526	1.526	10	139
	sec-CH	1.085	1.088	1.096	1.092		
	CH_i	1.083	1.086	1.094	1.096		
	CH_o	1.084	1.088	1.095	1.096		
	CCC	112.6	112.8	112.4	112.7		
	CCH_i	110.9	111.1	111.5			
	CCH_o	111.3	111.1	110.8			
	H_iCH_o	107.4	107.8	106.3			
	H_oCH_o	107.8	107.6	107.9			
	H_sCH_s	106.4	106.3	106.3	106.3		
Ethene	CC	1.315	1.317	1.335	1.339	10,	139
	CH	1.073	1.076	1.085	1.085	115	
	HCH	116.1	116.4	116.5	117.8		
Cyclobutane	CC	1.554	1.547	1.543	1.548	10	140
	CH	1.081	1.085	1.094	1.092		
	CCC	89.3	89.0	87.9	87.2		
	HCH	108.5	108.3	108.8			
	Puckering angle		15.0	30.8	25		
Bicyclo[1.1.0]butane	C1C2	1.502	1.489	1.492	1.489	62,	142
	C1C3	1.478	1.466	1.469	1.497	128	
	C1H	1.062	1.070	1.080	1.071		
	$C2H_{eq}$	1.074	1.078	1.088	1.093		
	$C2H_{ax}$	1.076	1.083	1.092	1.093		
	C1C2C3	59.0	58.9	60.2	60.0		
	C2C1C4	97.6	97.9		98.3		
	C3C1H	133.4	132.1	128.1	128.4		
	$C1C2H_{eq}$	117.3	117.5	117.0			
	$C1C2H_{ax}$	119.4	119.5	119.2			
	HC2H	113.9	114.0	114.1	115.6		
	Puckering angle		120.9	122.4	122.7		
[1.1.1]Propellane	C1C2	1.528	1.502	1.514	1.525	62,	143
	C1C3	1.600	1.543	1.592	1.596	128	
	C2H	1.070	1.075	1.088	1.106		
	HCH	114.7	114.5	114.9	116.0		

[a] Bond distances in Å, angles in deg. Subscripts s, i, o, ax, and eq denote secondary, in-plane and out-of-plane H atoms in case of propane and axial and equatorial H atoms in case of cyclobutane and bicyclobutane.
[b] Some geometries have been calculated at the HF/6-31G(d,p) level.

rings are fused to lead to [1.1.1]propellane, the bond C1C3 will become longer (1.60 Å, Table 12), which has been traced to an unusual type of bonding (see Section XII. B)[32, 62, 141, 143]

B. Substituent Effects on the Geometry of the Cyclopropyl Group

The influence of substituents on the bond lengths of **1** has received considerable attention in the past decades[26, 27, 33, 95–123]. Because of the peculiar electronic structure of **1**, substituent

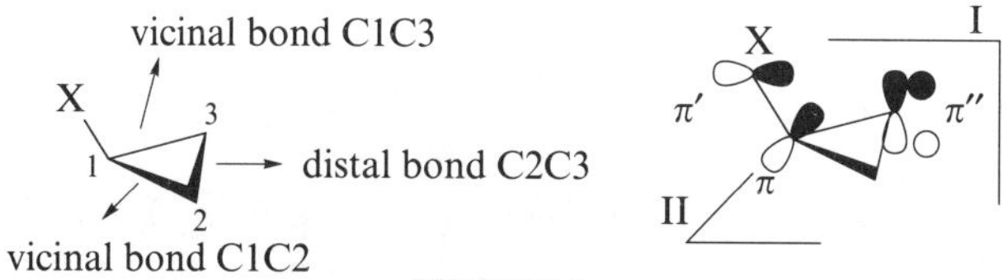

SCHEME 3

effects on its geometry are more pronounced than in the case of other alkanes. Depending on the nature of substituents, both C1C2 (vicinal) and C2C3 (distal) bond lengths (Scheme 3) deviate from the CC bond lengths in **1**. Changes in vicinal and distal bond lengths usually have opposite sign, i.e. if C1C2 is elongated, C2C3 is shortened and *vice versa*. Allen and coworkers[99] have collected experimental geometries on substituted **1** and have analysed changes in geometry in a systematic way. These data are complemented by a considerable number of calculated equilibrium geometries of substituted cyclopropanes (Table 13), which are discussed in the following. Two alternative models have been developed to rationalize substituent effects on the geometry of **1**: (a) an MO model[33] and (b) an electron density model[97].

1. The molecular orbital description

Clark and coworkers[33] have distinguished between four different classes of substituents, namely π-acceptor, π-donor, σ-acceptor and σ-donor substituents. This classification is based on possible 2-electron–2-orbital interactions involving substituent and cyclopropane MOs. Prerequisites for these interactions are comparable orbital energies for substituent and ring and a sufficiently large primary overlap between the orbitals involved. The latter requirement implies a large amplitude of the interacting cyclopropane orbital at C1, which is the location of the substituent X (compare with Scheme 3). This excludes all MOs of **1**, but those in Scheme 4, where degenerate MOs are characterized by symmetry notations they would obtain in (distorted) C_{2v}-symmetrical **1** (distortion along the C_2 axis that passes through C1) and the MO ordering obtained with a large basis set (Figure 6) is used rather

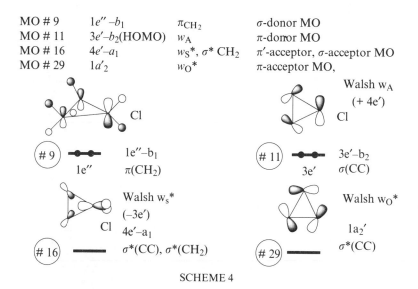

MO # 9	$1e''-b_1$	π_{CH_2}	σ-donor MO
MO # 11	$3e'-b_2$(HOMO)	w_A	π-donor MO
MO # 16	$4e'-a_1$	w_S^*, σ^* CH_2	π'-acceptor, σ-acceptor MO
MO # 29	$1a'_2$	w_O^*	π-acceptor MO,

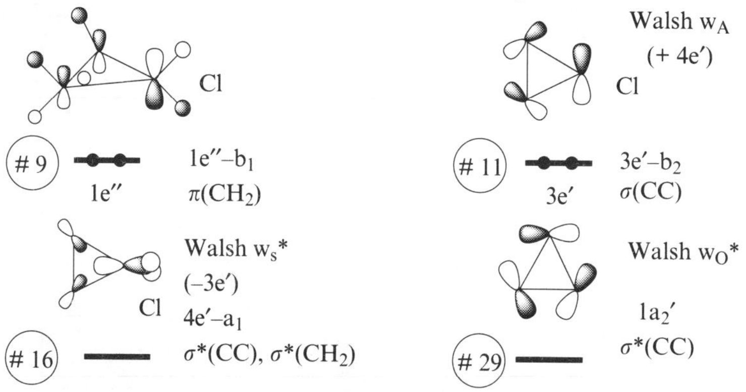

SCHEME 4

TABLE 13. Distortion of the cyclopropane ring upon substitution at C1 by substituent X. Predictions by the molecular orbital (MO) and electron density (ED) model[a,b,33,97]

X	Conformation	Method/basis	C1C2	C2C3	CX	Change of		Prediction by						Reliability	
			vicinal	distal		C1C2	C2C3	MO model[33]			ED model[97]			MO	ED
								Type	C1C2	C2C3	Type	C1C2	C2C3		
Li		HF/6-31G(d)[98]	1.521	1.491		l	s	σ-donor	l	s	σ-repeller	l	s	yes	yes
		HF/4-31G[33]	1.533	1.500	1.963										
BeH		HF/6-31G(d)[98]	1.522	1.484		l	s	σ-donor	l	s	σ-repeller	l	s	yes	yes
		HF/4-31G[33]	1.531	1.490	1.678										
BH_2	bisected	HF/6-31G(d)[98]	1.527	1.473		l	s	π-acceptor	l	s	π-attractor	l	s	yes	yes
	perpend.		1.503	1.499		l	s	σ-donor	l	s	σ-repeller	l	s	yes	yes
	bisected	HF/4-31G[33]	1.536	1.478	1.534										
	perpend.		1.510	1.505	1.561										
CH_2^+	bisected	MP2/6-31G(d)[111]	1.645	1.414	1.356	l	s	π-acceptor	l	s	π-attractor	l	s	yes	yes
		HF/4-31G	1.647	1.415	1.351										
CH_2	bisected	HF/DZ[96]	1.524	1.530		s	l	π-donor	l	l	π-repeller	s	l	no	yes
CH_3	staggered	HF/6-31G(d)[98]	1.497	1.501		s	l	σ-donor	l	s	σ-repeller	l	s	no	no
								π′-donor	l	s	π-repeller	s	l	no	yes
								π-acceptor	l	s				no	
								π-donor	l	l				no	
		HF/4-31G[33]	1.503	1.506	1.510										
	eclipsed		1.503	1.501	1.519	l	s	σ-donor	l	s	σ-repeller	l	s	yes	yes
		MW[116]	1.514	1.514	1.513										
		ED[117]	1.509	1.509	1.517										
CN		HF/4-21G[102]	1.525	1.505	1.435	l	s	π-acceptor	l	s	π-attractor	l	s	yes	yes
		MW[101]	1.528	1.500											
COF		HF/6-31G(d)[115]	1.511	1.497	1.476	l	s	π-acceptor	l	s	π-attractor	l	s	yes	yes
CH_2=CH	*s-trans*	HF/4-21G[102a]	1.522	1.510	1.482	l	s	π-acceptor	l	s	π-attractor	l	s	yes	yes
								π-donor	l	l					
	s-cis		1.522	1.508	1.491	l	s	π-acceptor	l	s	π-attractor	l	s	yes	yes
		ED[118]	1.522	1.522	1.475										
NH_2	*anti*	HF/6-31G(d)[98]	1.494	1.500		s	l	σ-acceptor	s	l	σ-attractor	s	l	yes	yes

								π′-donor	s	l	π-repeller	s	l	no	yes
		HF/4-31G[33]	1.500	1.503	1.428										
		MW[106]	1.486	1.513	1.462										
NC		HF/6-31G(d)[114]	1.495	1.500	1.405	s	l	σ-acceptor	s	l	σ-attractor	s	l	yes	yes
		exp.[122]	1.523	1.513	1.380	l	s								
NO_2	bisected	HF/4-21G[113]	1.514	1.503	1.458	l	s	π-acceptor	l	s	π-attractor	l	s	yes	yes
	perpend.		1.497	1.522	1.475	s	l	σ-acceptor	s	l	σ-attractor	s	l	yes	yes
O^-		HF/DZ[96]	1.525	1.532		s	l	σ-acceptor	s	l	σ-attractor	s	l	yes	yes
								π,π′-donor	l	l	π-repeller	s	l	yes	yes
OH	*anti*	HF/6-31G(d)[98]	1.489	1.513		s	l	σ-acceptor	s	l	σ-attractor	s	l	yes	yes
								π′-donor	l	s	π-donor	s	l	no	yes
								π-donor	l	l				no	
		HF/4-31G[33]	1.492	1.517	1.408										
		HF/DZ[96]	1.498	1.525											
F		HF/6-31G(d)[98]	1.481	1.512	1.355	s	l	σ-acceptor	s	l	σ-attractor	s	l	yes	yes
								π-donor	l	l	π-repeller	s	l	no	yes
								π′-donor	l	s				no	
		HF/4-31G[33]	1.480	1.505	1.395										
SiH_3	staggered	HF/6-31G(d)[98]	1.511	1.489		l	s	σ-donor	l	s	σ-repeller	l	s	yes	yes
								π-acceptor	l	s	π-attractor	l	s	yes	yes
		MW[119]	1.520	1.508	1.853										
PH_2	*anti*	HF/6-31G(d)[98]	1.506	1.490		l	s	σ-donor	l	s	σ-repeller	l	s	yes	yes
											π-repeller	s	l		no
SH	*anti*	HF/6-31G(d)[98]	1.493	1.504		s	l	σ-acceptor	s	l	σ-attractor	s	l	yes	yes
Cl		HF/6-31G(d)[98]	1.488	1.503	1.758	s	l	σ-acceptor	s	l	σ-attractor	s	l	yes	yes
		MW[120]	1.513	1.515	1.740										
Br		HF/6-31G(d)[112]	1.489	1.502	1.939	s	l	σ-acceptor	s	l	σ-attractor	s	l	yes	yes
		exp.[121]	1.512	1.521	1.905										
GeH_3		HF/3-21G(d)[123]	1.506	1.491	1.895	l	s	σ-donor	l	s	σ-repeller	l	s	yes	yes
		ED[123]	1.521	1.502	1.924										
I		HF/6-31G(d)[112]	1.492	1.500	2.125	s	l	σ-acceptor	s	l	σ-attractor	s	l	yes	yes

[a] Distances in Å. For X = NH_2, OH, PH_2, SH, the conformation is given by a dihedral angle H1—C1—Y—lp of 180°.
[b] The abbreviations l, s indicate whether vicinal (distal) bonds are longer or shorter than distal (vicinal) bonds as predicted by the various models. In the last column the reliability of the various predictions is checked on the basis of the data given in the 4th and 5th columns.

than that resulting from minimal basis set calculations (see Reference 33). However, this does not affect the discussion of substituent–ring interactions. The π MOs of substituent X and ring are classified as shown in Scheme 3[33, 96]. The symmetry plane of a C_s symmetrical X-**1** (plane I in Scheme 3) is the nodal plane for π MOs while π' MOs are lying in this plane and π'' MOs are perpendicular to the ring plane (plane II in Scheme 3).

Depopulation or population of MOs #9, 11 and 16, but not the $1a'_2$ MO 29, leads to opposing changes in vicinal and distal bond lengths. Accordingly, one can expect characteristic changes in the geometry of **1** upon substitution by a π/σ- acceptor/donor substituent (Table 14).

TABLE 14. Substituent effects on the geometry of cyclopropane as a function of interactions between substituent and cyclopropane MOs[a, 33]

Substituent type	MO of cyclopropane involved	Change in vicinal bonds C1C2	Change in distal bond C2C3	Favoured substituent conformation
π-Acceptor	$3e'$–b_2 (w_A), #11 π-type	longer	shorter	bisected
π-Donor	$4e'$–a_1 (w_S*), #16 π'-type	longer	shorter	perpendicular
σ-Acceptor	$1e''$–b_1 (π_{CH_2}), #9	shorter	longer	
σ-Donor	$4e'$–a_1 (w_S*), #16	longer	shorter	

[a] Compare with Schemes 3 and 4. See also Figure 6.

The effect of *π-acceptors*, first treated by Hoffmann[26] and, independently, by Günther[27], involves an interaction between HOMO $3e'$–b_2 (#11, refined Walsh orbital w_A), which is C1C2 bonding and C2C3 antibonding, with a vacant p or π* orbital of substituent X and a subsequent depopulation of the C2C3 antibonding Walsh orbital w_A (Figure 6), thus leading to lengthening of the vicinal and shortening of the distal bond. This implies for substituents such as CH_2^+ or BH_2 a bisected rather than a perpendicular conformation in order to guarantee sufficient π, π overlap.

π-Donor substituents will transfer charge via an occupied π orbital into the $1a'_2$ (MO #29, w_O*) orbital, but this will lead to lengthening of all CC bonds of **1**[27]. The interaction of a π'-donor MO of the substituent with the $4e'$–a_1 MO (#16, w_S*) of **1** should be equally strong or even stronger because of the lower orbital energy of the latter. Charge transfer into $4e'$–a_1 leads to lengthening of the vicinal bonds and shortening of the distal bond.

σ-Acceptor substituents withdraw electron density from the $1e'$–b_1 of **1** (MO #9). This results in lengthening of the distal and shortening of vicinal CC bonds, which is just opposite to the effects of π-acceptor substituents. Since the $1e''$–b_1 MO is a π'' MO with π_{CH_2} bonding character, the effect on the CC bonds can be considered to be indirect.

The $4e'$–a_1 MO#16 of **1** can interact both with a π- and a σ-donor substituent. Hence, *σ-donor substituents* have the same effects as π-donor substituents, namely a lengthening of vicinal and a shortening of the distal CC bond of **1**.

2. *The electron density description*

Cremer and Kraka[97] based their model on a 'principal of avoidance of geminal and vicinal charge concentrations', which they derived from the analysis of the Laplace concentration, $-\nabla^2\rho(\mathbf{r})$, in the valence shell of bonded atoms. As noted in Section IV. C, the Laplacian of $\rho(\mathbf{r})$ reflects the shell structure of an atom. Upon bond formation, the 'valence

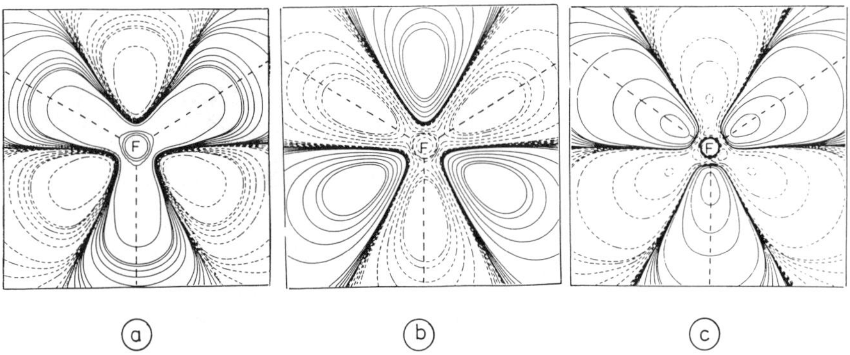

FIGURE 18. Concentration of electron density in the lone-pair regions of F in CH_3F. Contour-line diagrams have been drawn with regard to the plane that contains the F nucleus and is perpendicular to the CF bond axis. Positions of the methyl CH bonds are given by (heavy) dashed lines. (a) Energy density $H(\mathbf{r})$. (b) Electron density $\rho(\mathbf{r})$. (c) Laplace concentration $-\nabla^2\rho(\mathbf{r})$. In order to amplify effects, difference maps are used, i.e. $H(\mathbf{r})$, $\rho(\mathbf{r})$ and $\nabla^2\rho(\mathbf{r})$ are plotted with regard to CH_3F with the methyl group rotated by 60° as reference. Dashed lines indicate areas with larger stabilizing energy density (lower electron density, larger concentration) and solid lines areas with lower stabilizing energy density (larger electron density, smaller charge concentration). (HF/6-31 G(d) calculations from Reference 97.) Reprinted with permission from D. Cremer and E. Kraka, *J. Am. Chem. Soc.*, **107**, 3811 (1985). Copyright (1985) American Chemical Society

shell' is distorted in the way that maxima (lumps) appear in the direction of the bonds with adjacent atoms. Since all changes in the Laplacian of $\rho(\mathbf{r})$ have to cancel within the boundaries of an atom (no matter whether isolated or bonded), the formation of concentration lumps implies the formation of concentration minima (holes) of $-\nabla^2\rho(\mathbf{r})$ in other regions of the valence sphere. Cremer and Kraka observed that the pattern of lumps and holes in the valence sphere of bonded atoms shows some regularities that are best described as the result of an avoidance of geminal and vicinal concentration lumps. For example, the electron concentration of an F atom of F_2 will be cylindrical if viewed along the bond axis. However, if one F atom is replaced by a methyl group as in CH_3F, then the Laplace concentration $-\nabla^2\rho(\mathbf{r})$ will become larger (smaller) in the regions that are staggered (eclipsed) with regard to the CH bonds. There is a preference for staggering of non-bonded charge concentrations (see Figure 18), which has also been founded for other molecules[97].

Substituents distort the pattern of concentration lumps and holes in the valence shell of an atom by either pulling bonded (σ-) or non-bonded (π-) lumps in the direction of the substituent (σ/π-attractors) or pushing them closer to the atom in question (σ/π-repellers). Accordingly, the substituents of **1** can be classified by their σ-attractor, σ-repeller, π-attractor or π-repeller ability. Typical distortions caused by these substituents in the valence shell of the three C atoms are shown in Figure19.

σ-Attractors (Figure 19a) distort the valence sphere of the adjacent C1 atom by extending it on the frontside and compressing it on the backside. The C1 nucleus is better shielded in the direction of C2 and C3 and, as a consequence, the vicinal bonds become shorter, which, in turn, leads to a lengthening of the distal bond. σ-Repellers have the reverse effect (Figure 19b). Holes or charge concentrations in the π-region of a π-attractor/repeller substituent draw charge concentration at the C1 atom out of or into the direction of the vicinal bonds, which results in bond lengthening (Figure 19c) or bond shortening (Figure 19d). On this basis, the substituent effects summarized in Table 15 can be predicted (Figure 19).

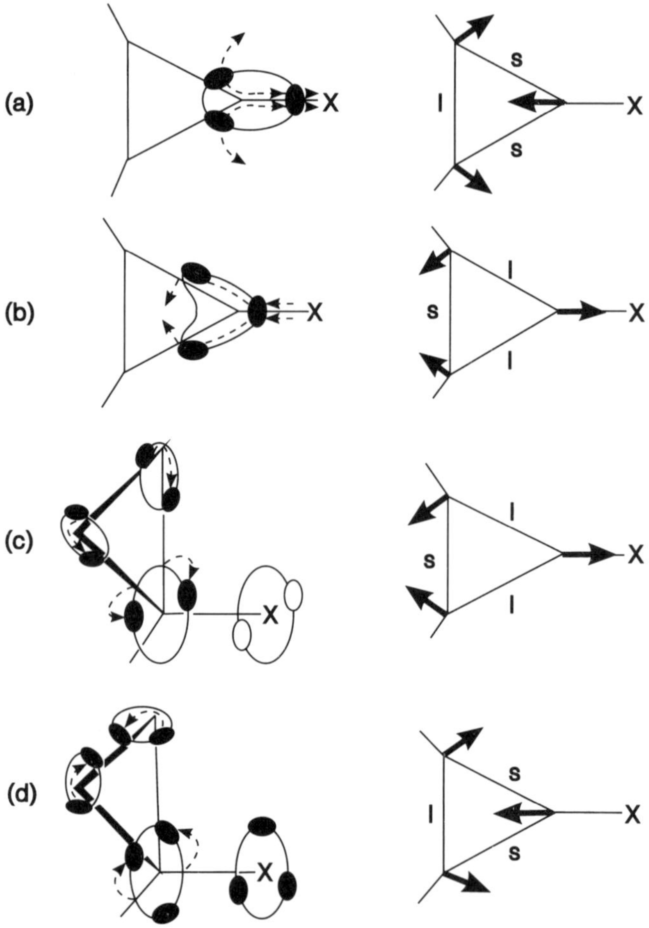

FIGURE 19. Schematic representation of distortions of the valence sphere of three-membered ring atoms upon C1 substitution by a substituent X with (a) σ-attractor, (b) σ-repeller, (c) π-attractor and (d) π-repeller ability. Distorted valence spheres are indicated by (large) ellipses or circles. Small ellipses depict locations in the valence sphere or the bond region with charge compression (solid) or charge expansion (open). Dashed arrows indicate the direction of distortions of the valence spheres. The nuclei of the three-membered ring move into the areas with charge compression as indicated by the heavy arrows in the diagrams on the right side. The corresponding bond length changes are denoted by s (short) and l (long). Reprinted with permission from D. Cremer and E. Kraka, *J. Am. Chem. Soc.*, **107**, 3811 (1985). Copyright (1985) American Chemical Society

In Table 13, calculated and measured geometry data of substituted cyclopropanes are compared with predictions of the MO model of Clark and coworkers (Table 14)[33] and those of the electron density model by Cremer and Kraka (Table 15)[97]. From the comparison, it becomes clear that both models lead to similar predictions, but differ with regard to some

TABLE 15. Substituent effects on the geometry of cyclopropane according to the electron density (ED) model of Cremer and Kraka[a,97]

Type of substituent	Change in vicinal bonds C1C2 and C1C3	Change in distal bond C2C3
σ-Attractor	shorter	longer
σ-Repeller	longer	shorter
π-Attractor		
bisected	longer	shorter
perpendicular	(longer)	(shorter)
π-Repeller	shorter	longer

[a] Compare with Figure 19.

important details. It seems that the MO model cannot be applied in some cases without additional assumptions (see next Section).

3. Substituted cyclopropanes

Cyclopropyl carbinyl cation[111]. The most dramatic example for a π-acceptor/π-attractor substituent effect is provided by the cyclopropyl carbinyl cation. In its bisected conformation, the vicinal CC bonds are considerably longer (1.65 Å) and the distal CC bond considerably shorter (1.41 Å) than the bonds in **1** (Table 13). The CH_2^+ group stabilizes **1** considerably more in the bisected form than in the perpendicular form (13 kcal mol^{-1}). Since the CC^+ bond possesses partial double bond character, it is shorter (1.35–1.36 Å, Table 13) than a normal CC^+ bond (1.51 Å).

Cyclopropylborane[33,98]. A strong π-acceptor (π-attractor) effect is also found for the BH_2 substituent that, according to HF calculations, prefers by 7 kcal mol^{-1} the bisected over the perpendicular conformation (Table 16, NMR measurements suggest a similar energy difference[119]). In the former conformation, the vicinal bonds are elongated to 1.53 Å while the distal bond is shortened to 1.47 Å (Table 13). Compared to the CH_2^+ group, substituent–ring interactions are weaker because of a longer C1—X distance (smaller overlap) and a larger energy difference between donor and acceptor orbital. In the perpendicular conformation, the π-acceptor effect is turned off thus leading to changes in the geometry of **1** that identify BH_2 as a σ-donor (σ-repeller) substituent.

Cyanocyclopropane[97, 101, 102] *and isocyanocyclopropane*[114, 122]. According to Skancke and Boggs[102a], the CN group is a π-acceptor that leads to lengthening of vicinal and shortening of distal CC bond. A similar effect could also be expected for the isocyano group, and CC bond lengths (lengthening of vicinal, shortening of distal bond) in accordance with this have been measured spectroscopically[122]. On the other hand, HF/6-31G(d) calculations by Reynders and Schrumpf[114] suggest the reverse CC bond length pattern for isocyanocyclopropane. Calculation of Davidson–Roby populations[144] indicates a σ-acceptor nature of the NC substituent in agreement with calculated ring bond lengths. In view of the difficulties of getting multiple bonds correctly described[135], further calculations have to clarify the true nature of the isocyano substituent.

Nitrocyclopropane. Skancke[113] reported that in the bisected conformer of nitrocyclopropane, the vicinal bonds are longer and the distal bond is shorter than the CC bonds of **1**. This is a result of charge transfer from the w_A orbital of **1** to the empty pπ orbital of the nitro group. Rotation of the nitro group leads to an energy increase by 3.2–4.7 kcal mol^{-1} [145,146]. At the same time, the vicinal bond lengths are decreased while the distal bond length is increased. Skancke explained these changes by interactions between the symmet-

TABLE 16. Energy differences ΔE (kcal mol^{-1}) between the different rotamers of mono-substituted cyclopropanes calculated at the Hartree–Fock level with various basis sets

Substituent	Conformation[a]	Basis set	ΔE	Reference
BH_2	bisected	6-31G(d,p)	0.0	98 (see also 33)
	perpendicular		7.2	
CH_3	staggered	6-31G(d,p)	0.0	98
	eclipsed		2.9	
NH_2	180°	6-31G(d,p)	0.0	98
	43°		2.6	
	0°		3.0	
	106°		4.9	
OH	72°	6-31G(d,p)	0.0	98
	180°		2.5	
	163°		2.5	
	0°		2.7	
SiH_3	staggered	6-31G(d,p)	0.0	98
	eclipsed		1.7	
GeH_3	staggered	3-21G(d)	0.0	123
	eclipsed		1.4	
PH_2	180°	6-31G(d,p)	0.0	98
	48°		1.5	
	0°		2.7	
	108°		3.6	
SH	75°	6-31G(d,p)	0.0	98
	180°		2.8	
	156°		2.8	
	0°		3.6	
NO_2	staggered	4-21G	0.0	113
	eclipsed		6.6	
CH_2SH	*gauche*	3-21G(d)	0.0	148
	anti		3.3	
COF	0°	6-31G(d)	0.0	115
	90°		6.1	
	180°		0.1	
	270°		6.0	

[a] H—C—X—H torsional angles for OH and SH are given. For NH_2 or PH_2, H—C—X—lp torsional angles are given, where lp is the lone pair, which is assumed to be *anti* to the bisector N(HH) and P(HH). For CH_2SH and COF, the conformation is determined by the torsion angles H—S—C—C and H—C—C—F, respectively.

ric Walsh MO w_S and the empty pπ orbital of the nitro group. However, this interpretation does not consider the fact that the amplitude of w_S is rather low at C1 and therefore overlap between the interacting orbitals is relatively small. It is more reasonable to consider the NO_2 group in the perpendicular conformation as a σ-acceptor or σ-attractor that leads to the observed changes in the CC bonds by interaction with the $1e''$–b_1 MO #9 of **1**.

Halocyclopropanes[33, 96–98, 112]. According to Clark and coworkers,[33] F acts predominantly as σ-acceptor. If π-donor ability is invoked for F, then controversial geometry effects are predicted by the MO model. Predictions by the electron density model of Cremer and Kraka[97] are consistent, no matter whether σ-attractor or π-repeller ability of F is considered (Table 13). The other halogens are also σ-attractors/π-repellers but their effects on the geometry of **1** decrease in the order

$$F > Cl \approx Br > I$$

Substitution by two F atoms[97, 105, 131] at C1 reduces vicinal bond lengths by 0.02 Å (1.465 Å, HF/4-31G) and increases the distal bond length by 0.04 Å (1.532 Å), thus indicating additivity for the two F effects.

Cyclopropanol and cyclopropanethiol[33, 97, 98]. Calculated geometry changes of **1** upon substitution by an OH or SH group (*anti* conformation) can only be explained by the MO model by assuming σ-acceptor character for the substituent. Again, the electron density model leads to prediction of the correct geometry changes, no matter whether a σ-attractor or π-repeller nature of the OH (SH) substituent is assumed. In the most stable conformation, OH and SH groups are in a *gauche* position (dihedral angle 72° and 75°, respectively; Table 16). The rotational barriers will be 2.7 kcal mol^{-1} (OH) and 3.6 kcal mol^{-1} (SH group) if rotation proceeds via the *syn* form, but 2.5 and 2.8 kcal mol^{-1}, respectively, if rotation proceeds via the *anti* form.

Cyclopropylamine[33, 98, 106] *and cyclopropylphosphine*[98]. The NH_2 and PH_2 groups both prefer an *anti* conformation. Calculated changes in vicinal and distal bond length can only be explained on the basis of the MO model of Clark and coworkers[33] if one assumes a dominant σ-acceptor nature for NH_2 and σ-donor nature for PH_2. Again, the electron density model does also allow π-repeller ability for the NH_2 group. The calculated rotational barriers of NH_2 and PH_2 are 4.9 and 3.6 kcal mol^{-1} (Table 16)[98].

Methyl-,[33, 97, 98] *silyl-*[98, 119] *and germylcyclopropane*[123]. Although changes in the geometry of **1** caused by a methyl group are small, they represent a critical test for the applicability of MO[33] and the ED model[97]. Changes obtained for the staggered conformation of the methyl group suggest σ-acceptor ability for the substituent. However, this prediction is in conflict with the known large electronegativity of the C atoms in the three-membered ring, which forces a methyl group to act as a σ-donor rather than a σ-acceptor. Hyperconjugative interactions between substituent and ring could also lead to a π-donor/acceptor character of the methyl group. However, none of these possibilities leads to the observed pattern of vicinal and distal bond length changes (see Table 13). The electron density model suggests both σ-repeller and π-repeller character for methyl with a preponderance of the latter. In the eclipsed conformation, the σ-repeller (σ-donor) nature of the substituent prevails.

The silyl and germyl groups are σ-donors and π-acceptors[123]. Calculated CC distances in silyl-[98, 119] and germyl cyclopropane[123] suggest that the π-acceptor character is significantly more pronounced for a silyl than a germyl substituent.

Vinylcyclopropane[102a]. The vinyl group can act both as π-donor and π-acceptor. This is indicated by lengthening of all ring bonds, where the vicinal bonds become longer than the distal bond. The same bond length pattern is obtained by considering the π-attractor propensity of the vinyl group on the basis of the electron density model[97].

Lithiocyclopropane[33, 98]. Lithium is a σ-donor according to the MO interpretation[33] and a σ-repeller according to the electron density description[97]. Both models predict a lengthening of the vicinal and a shortening of the distal CC bond, which is confirmed by *ab initio* calculations (Table 13).

Cyclopropylmethylene anion and cyclopropyl oxide anion[96]. Clearly, CH_2^- (in its bisected conformation) and O^- are π-donor substituents that would lead to a lengthening of all ring bonds (π-donor effect) with the vicinal bonds becoming longer than the distal bond (π'-donor effect, see Scheme 3) according to the MO model. However, the reverse geometry effect (vicinal CC bonds shorter than distal CC bond) is calculated. This geometry change is correctly predicted by the electron density model of Cremer and Kraka[97] on the basis of the π-repeller nature of CH_2^- and O^-. In the latter case, the resulting geometry effect is enhanced by the σ-attractor nature of the substituent.

Summary. The electron density model of substituent–ring interactions functions better than the MO model, which is not surprising since the electron density covers all MO effects while any MO model will simplify orbital interactions by selecting just a few important ones. This, of course, leads to incorrect descriptions in cases where many orbital interactions contribute to observed properties.

With the aid of virial partitioning of the electron density it has been shown that the cyclopropyl group is more electronegative than the 2-propyl group[97]. This is in line with hybridization models and the analysis of isodesmic reactions. Clark and coworkers[33] characterize the cyclopropyl group in the following way:

σ-acceptor ability: methyl > cyclopropyl > 2-propyl
σ-donor ability: methyl > 2-propyl > cyclopropyl

If the cyclopropyl group is compared with a 2-propyl group, then electropositive substituents (σ-donor substituents such as Li) stabilize the ring while strongly electronegative groups such as F, OH, etc. destabilize the ring.

Cyclopropyl is a fairly strong π-donor and therefore it is stabilized by π-acceptor substituents, which implies that the interacting π orbitals overlap sufficiently. Cyclopropyl can also act as a π'-acceptor provided it interacts with a strong donor that possesses high-lying occupied π'-donor orbitals (Scheme 3). In exceptional cases such as $X = CH_2^-$, cyclopropyl turns out as a π-acceptor that accepts charge in its $1a_2'$ MO.

C. Rationalization of the Geometry of Three-membered Rings

The analysis of the geometry and electronic structure of the cyclopropyl group has led to a better understanding of three-membered ring geometries in general. Several comparisons of the geometry of **1** with that of heterocyclopropanes have been made utilizing *ab initio* and experimental data[6,9,27,32,43,47,71,76,94,124,125,130,147–153]. It turns out that there are some simple trends in the geometry of three-membered rings as shown in Table 17[9,32,147].

TABLE 17. Effect of a hetero atom on the geometry of cyclopropane. Transition from a three-membered ring to a π-complex[a]

Parameter		Basal group $H_2C{=}CH_2$ + Apex group X			
	X =	BH[151]	CH_2[97]	NH[97]	O[97]
CC		1.544	1.498	1.470	1.453
XC		1.534	1.498	1.449	1.401
γ		145.5	150	162.5	158.5
	X =	AlH[149]	SiH_2[124]	PH[124]	S[124]
CC		1.600	1.553	1.492	1.473
XC		1.904	1.855	1.853	1.811
γ		135.2	141.2	148.3	151.6
	X =	NH_2^+[97]	OH^+[97]	F^+[97]	H^+[97]
CC		1.488	1.446	1.445	1.371
XC		1.460	1.498	1.533	1.306
γ		160.0	165.5	171.3	178.3

[a] Distances in Å, CC(HH) angle γ in deg. HF/6-31 G(d) or HF/DZ + P calculations as given in References 97, 124, 149 and 151.

With increasing electronegativity of group (atom) X, both the CC and the CX bond length decrease and the (HH)CC angle γ increases until it becomes finally close to 180°. Changes in the CX bond lengths just reflect the decrease in the covalent radius of the heavy

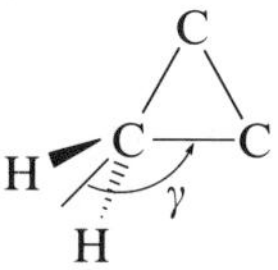

atom of X as this becomes more electronegative. With increasing electronegativity of X, there is a transition from three-membered rings to π-complexes that leads to a gradual conversion of the basal CH_2CH_2 group of the ring into an ethene fragment only loosely bound to the apex group X. This transition has been made visible and rationalized by electron density studies[9, 13] (see Section IV. D, Figures 14 and 15) and orbital considerations[13, 32, 43, 71] (Figure 16 in Section IV. D). Hence for X = BH, CH_2, NH and O, the increase in electronegativity reduces back donation via a b_2-symmetrical orbital of X to the π^* MO of the ethene unit. Since π-back donation is primarily responsible for distortions of the ethene unit (CC bond lengthening; pyramidalization of the CH_2 groups and deviation from planarity), a reduction in π-back donation has the CC bond length and the (HH)CC angle approaching the values of ethene.

Similar trends can also be observed for X = AlH[149], SiH_2[32b, 47, 124], PH[47, 124] and S[47, 124] where, however, CC bond shortening and the increase in γ are somewhat smaller because of the decrease in electronegativity when going up in the columns of the periodic table.

Since π-back donation requires CX bonding overlap in the π-bridged π-orbital (MO #11, w_A, Figure 6) of the three-membered ring and since this will be reduced with increasing electronegativity of X, one should expect that a decrease in back donation might lead to a weakening of bonds CX and, hence, to a weakening of the stability of the three-membered ring (see Scheme 5). Accordingly, one could expect that heterocyclopropanes of the second row will be more stable than their analogues from the first row of the periodic table. Allen[32] has pointed out that this is not the case because three-membered ring bonding is also supported by CX bonding overlap in the σ-bridged π-orbital. This orbital, however, becomes more bonding in the way the CH_2CH_2 unit approaches ethene. Accordingly, the σ-bridged π-orbital can partially compensate a loss in bonding overlap and retain the stability of three-membered rings, such as aziridine or oxirane similar to that of **1**[32]. The σ-bridged π-orbital is also responsible for transferring electron density into the surface of the ring and adding in this way to surface delocalization, which is a stabilizing factor of the three-membered ring[9, 13, 32, 71].

The situation changes, however, if the electronegativity of the group X increases drastically as in the series X = NH_2^+, OH^+, F^+[9]. π-Back donation is stepwise decreased to zero and π-complexes are formed. This has been demonstrated via the electron density analysis,which reveals that convex bent bonds change into concave (inwardly curved) bent

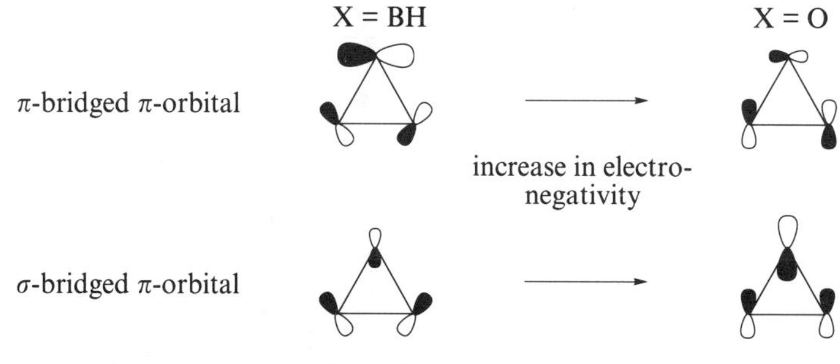

SCHEME 5

bonds ($X = OH^+$) and, finally, collapse to one bond path that connects X and the midpoint of the CH_2CH_2 unit, thus yielding the T structure of a π-complex[9]. In these cases, interactions between X and CH_2CH_2 become so weak that the CX distance increases rather than decreases with the electronegativity of the heavy atom of X.

The σ-donation π-back donation model has been used to explain bonding in a large variety of three-membered rings with different apex group X (atoms from the first, second and third rows in the periodic table) and different basal groups (H_2CCH_2, BHBH, OO, HCCH, NN, H_2SiSiH_2, H_2GeGeH_2, etc.)[9, 32, 43, 47, 71, 94, 124, 148–152].

VII. VIBRATIONAL SPECTRA

High-resolution infrared spectra[154, 155], reliable experimentally based force fields and a detailed normal coordinate analysis[156, 157] have been available for **1** for a relatively long time. The high symmetry (D_{3h}) or **1** facilitates an analysis of its 14 fundamental vibrations (derived from 21 non-redundant coordinates), which transform as three a_1', one a_2', one a_1'', two a_2'', four e' and three e'' and which can be described by using symmetry coordinates such as bond stretching, CH_2 angle deformation, CH_2 wagging, CH_2 rocking or CH_2 twisting motion (see Table 18 and Figure 20).

Ab initio investigations of the vibrational spectra of **1**[129, 130, 158, 159] have been exclusively based on the harmonic approximation, and therefore they were primarily concerned to find

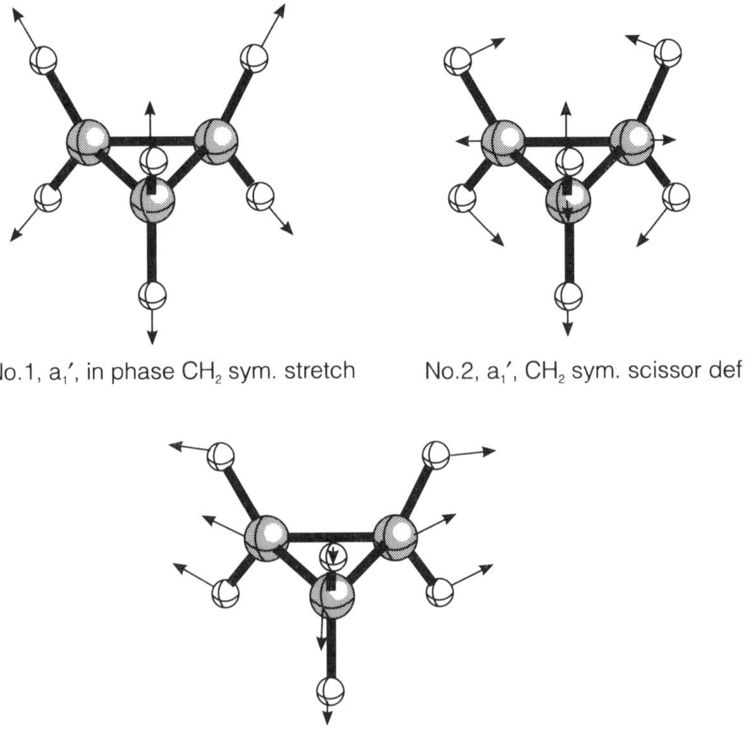

No.1, a_1', in phase CH_2 sym. stretch

No.2, a_1', CH_2 sym. scissor def

No.3, a_1', C—C ring breath

FIGURE 20. (Caption on page 100)

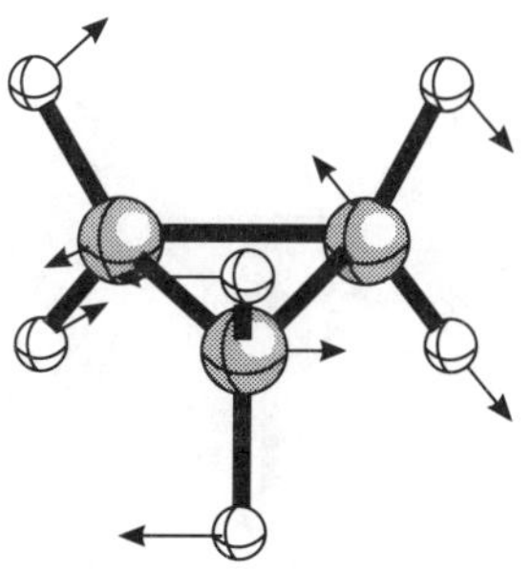

No.4, a_2', CH_2 wag

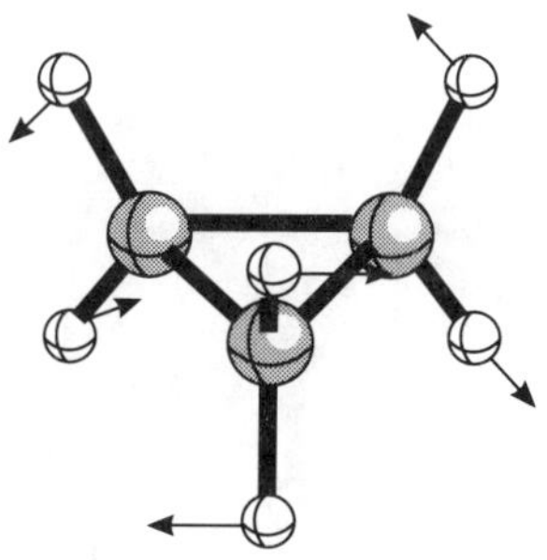

No.5, a_1'', CH_2 twist

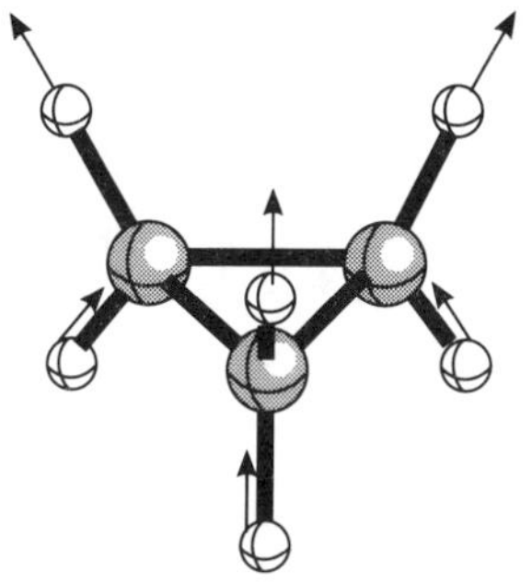

No. 6, a_2'', in phase CH_2 asym. stretch

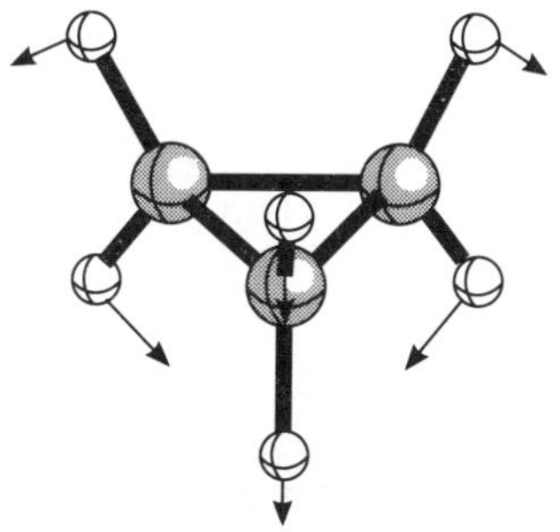

No. 7, a_2'', CH_2 rock

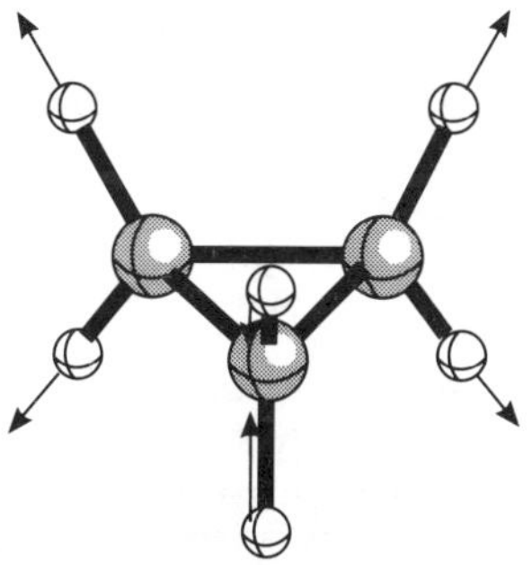

No. 8, e', out of phase CH_2 sym. stretch

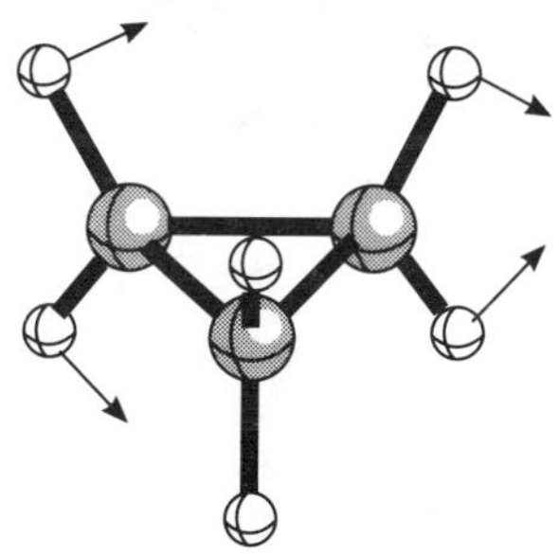

No. 9, e', CH_2 asym scissor def

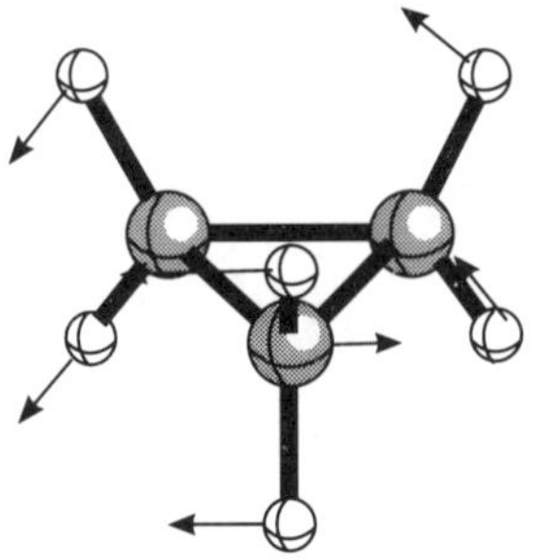

No. 10, e′, CH_2 wag

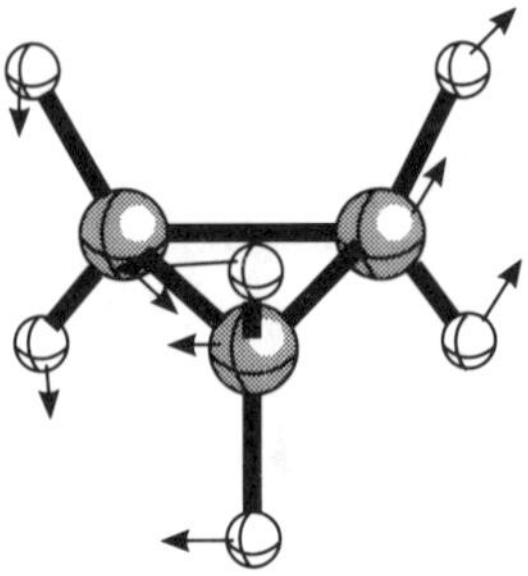

No. 11, e′, C—C ring def

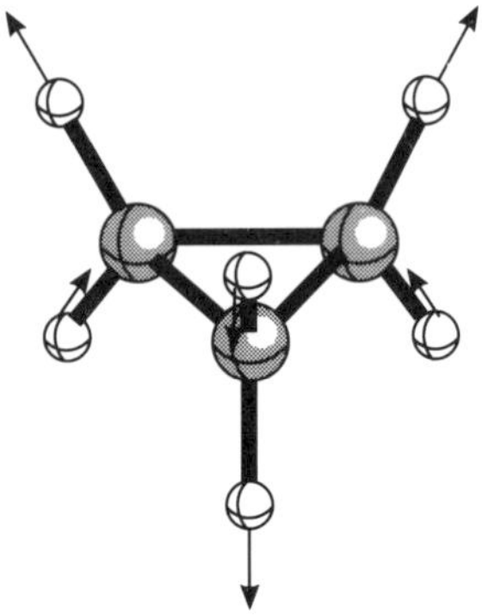

No. 12, e″, out of phase CH_2 asym. stretch

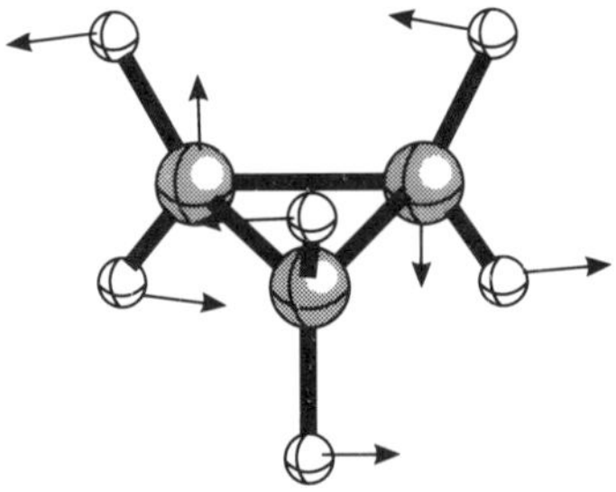

No. 13, e″, CH_2 rock + CH_2 twist

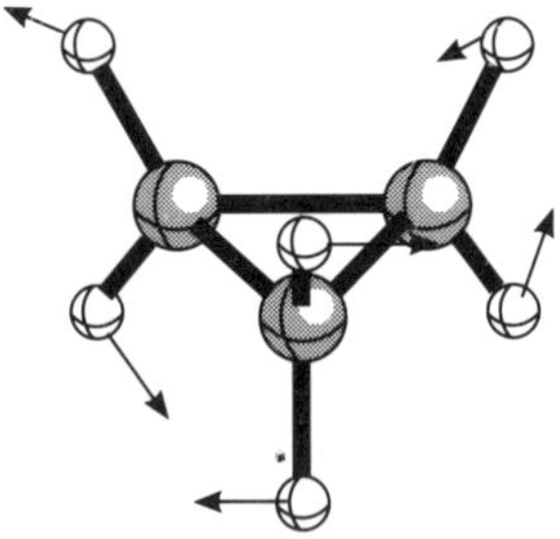

No. 14, e″, CH_2 twist + CH_2 rock

FIGURE 20. Vibrational modes of cyclopropane as obtained at the HF/6-31G(d,p) level of theory. Arrows indicate the direction and amplitude of each atomic motion. Symmetry assignments and a characterization of each mode is also given in line with the notations used in Table 18

TABLE 18. Comparison of observed fundamental vibrational frequencies (ν_i) and calculated harmonic frequencies (ω_i) (in cm^{-1}) for cyclopropane

#	Sym	Characterization	Exp.[a]	*Ab initio*					
				HF/DZ[b]	HF/6-31 G(d)[c]		HF/TZ + 2P[d]	MP2/DZ + P[e]	
					unscaled	scaled		unscaled	scaled
1	a_1'	in-phase CH_2 symmetric stretch	3038	3340	3300	3130	3219	3239	3077
2		CH_2 scissoring deformation	1479	1660	1667	1570	1657	1575	1496
3		C—C ring breathing	1188	1288	1299	1238	1269	1244	1182
4	a_2'	CH_2 wag	1070	1266	1220	1158	1212	1098	1043
5	a_1''	CH_2 twist	1126	1260	1261	1197	1256	1188	1129
6	a_2''	in-phase CH_2 asymmetric stretch	3102	3438	3385	3211	3377	3352	3184
7		CH_2 rock	854	930	920	873	917	890	846
8	e'	out-of-phase CH_2 symmetric stretch	3024	3320	3286	3117	3278	3229	3067
9		CH_2 scissoring deformation	1438	1618	1602	1515	1599	1515	1439
10		CH_2 wag	1028	1204	1187	1113	1171	1098	1043
11		C—C ring deformation	868	931	959	911	931	918	872
12	e''	out-of-phase CH_2 asymmetric stretch	3082	3416	3363	3189	3082	3335	3168
13		CH_2 rock + CH_2 twist	1188	1328	1323	1271	1320	1239	1177
14		CH_2 twist + CH_2 rock	739	832	802	744	739	773	734

[a] From Reference 157.
[b] From Reference 129, calculated with a (9s5p/4s) [4s2p/2s] basis set.
[c] From Reference 158. In this work, frequencies were calculated by numerical differences, which leads to small inaccuracies. Values have been corrected by using analytical second derivatives. Scaled values were obtained by scaling the diagonal force constants.
[d] From Reference 132, calculated with a (9s5p2d/4s2p) [4s2p2d/4s2p] basis set.
[e] From Reference 130, calculated with a (9s5p1d/4s1p) [4s2p1d/2s1p] basis set. The calculated frequencies were scaled by 0.95 in this work.

out about the differences between calculated and experimental vibrational frequencies (force constants). Force constants and vibrational frequencies of **1** have been calculated at the HF[129, 130, 158, 159] and MP2[130] level of theory using minimal (STO-3G)[158], DZ[129, 158], DZ + P[130, 158] and TZ + 2P basis sets[159]. HF/STO-3G frequencies are too large by 20–30%[158], HF/DZ frequencies by 10–20%[129] while inclusion of polarization functions into a split valence or DZ basis set reduces the deviation to 10–15%[158]. The MP2/DZ + P frequencies agree with the measured ones within 5–10%[130]. The largest discrepancies between calculated harmonic (ω_i) and experimentally observed frequencies (ν_i) are found for the stretching motions, for which anharmonic effects are considerable.

These discrepancies result (a) from the harmonic approximation used in all calculations [ω_i (theory) > ν_i (exp)], (b) the known deficiencies of minimal and DZ basis sets to describe three-membered rings [polarization functions are needed to describe small CCC bond angles: ω_i(DZ + P) > ω_i(DZ) > ω_i(minimal basis)] and (c) the need of electron correlated wave functions to correctly describe the curvature of the potential energy surface at a minimum energy point [ω_i(SCF) > ω_i(MP2)]. Because of these relationships, one can expect that

$$\omega_i(\text{SCF}) > \omega_i(\text{MP2}) > \nu_i(\text{exp})$$

holds at least for the stretching frequencies, but in the case of **1** this is also true for all other frequencies (Table 18)[130]. By appropriate scaling of calculated frequencies for force constants [scaling factors[160]: 0.87 (HF) to 0.93 (MP2); the best scaling factor for the MP2 frequencies of Table 18 is 0.95], predicted frequencies of **1** are within 1–3% of experimental values (Table18)[158].

For CH stretching modes and CH_2 deformation modes, all calculations lead to the ordering of frequencies observed experimentally (Table 18). The correct assignment of vibrational frequencies in the mid-frequency region (1100–1300 cm^{-1}) is ring stretch (ν_3, 1188 cm^{-1}), CH_2 twist (ν_5, 1126 cm^{-1}) and CH_2 wag (ν_4, 1070 cm^{-1}), which is correctly reproduced by using polarization functions but not by DZ basis sets, thus underlining the necessity of augmented basis sets.

Recently, attempts have been made to attain electronic structure information on **1** by analysing its normal modes in terms of 'internal modes' that are largely localized in molecular fragments which, in turn, can be characterized by a single internal parameter such as a bond length (diatomic fragments), bond angles (triatomic fragment) or dihedral angles (tetraatomic fragments)[161–163]. In this way, each 'internal mode' corresponds to an internal geometry parameter, which can be used for the normal-mode analysis. Normally, the description of molecular vibrations is carried out in terms of normal coordinates, which are in most cases linear combinations of several internal parameters. This has to do with the fact that vibrational motions described by normal coordinates are symmetry-adapted, i.e. each normal mode of the molecule transforms as one of the irreducible representations of the molecular point group[164]. Accordingly, a specific internal coordinate will contribute to several normal modes and each normal mode is delocalized in a similar way as canonical MOs are delocalized.

Boatz and Gordon[161] and, recently, Konkoli, Larsson and Cremer[162, 163] have worked out methods for assigning vibrational frequencies to individual internal coordinate motions. These procedures can be considered in many ways as being similar to the calculation of localized MOs from canonical SCF MOs. They yield, in the case of the Boatz–Gordon method, 'intrinsic frequencies'[161], which in their nature are averaged frequencies without an associated mode and which have the disadvantage of depending on a careful construction of the molecular geometry from non-redundant and (in case of symmetry) redundant internal coordinates. The method by Konkoli and coworkers leads to 'adiabatic motions'and 'adiabatic frequencies' that do not depend on redundant parameters[162, 163].

Analysis of the vibrational normal modes obtained at the HF/6-31G(d,p) level of theory in terms of adiabatic modes provides the basis for a quantitative dissection of the

TABLE 19. Analysis of the normal modes of cyclopropane using, as 'internal parameter' modes, adiabatic modes from Reference 163[a]

#	Sym	Freq.	Characterization	Number of internal parameters
1	a_1'	3300	CH stretch (96%)	CH: (6 × 16%)
2		1667	CH_2 scissoring def (81%) + CC stretch (18%)	CH_2: (3 × 27%) + CC (3 × 6%)
3		1299	CC stretch (90%)	CC stretch: (3 × 30%)
4	a_2'	1220	CH_2 wag (99%)	CH_2 wag (3 × 33%)
5	a_1''	1261	CH_2 twist (99%)	CH_2 twist (3 × 33%)
6	a_2''	3385	CH_2 stretch (100%)	CH stretch (6 × 16.7%)
7		920	CH_2 rock (99%)	CH_2 rock (3 × 33%)
8	e'	3286	CH stretch (100%)	CH stretch (4 × 25%)
		3286	CH stretch (66%)	CH stretch (2 × 33%)
9		1602	CH_2 def (98%)	CH_2 def (66% + 17% + 16%)
		1602	CH_2 def (98%)	CH_2 def (2 × 49%)
10		1187	CH_2 wag (86%) + CC stretch (9%)	CH_2 wag (48% + 38%) + CC (9%)
		1187	CH_2 wag (86%) + CC stretch (14%)	CH_2 wag (57 + 9 + 20%) + CC stretch (8 + 6%)
11		959	CC stretch (88%)	CC stretch (64% + 24%)
		959	CC stretch (96%)	CC stretch (55 + 41%)
12	e''	3363	CH stretch (98%)	CH stretch (2 × 25% + 2 × 24%)
		3363	CH stretch (66%)	CH stretch (2 × 33%)
13		1323	CH_2 rock (34%) + CH_2 twist (49%)	CH_2 rock (34%) + CH_2 twist (25 + 24%)
		1323	CH_2 rock (51%) + CH_2 twist (33%)	CH_2 rock (26 + 25%) + CH_2 twist (33%)
14		802	CH_2 twist (56%) + CH_2 rock (30%)	CH_2 twist (2 × 28%) + CH_2 rock (30%)
		802	CH_2 rock (44%) + CH_2 twist (37%)	CH_2 rock (2 × 22%) + CH_2 twist (37%)

[a] All frequencies in cm^{-1}. HF/6–31G(d,p) calculations. Each normal mode is dissected into internal parameter associated vibrations according to Reference 162. Compare with Figure 20.

former as shown in Table 19[163]. The dissection is based on a comparison of normal-mode vectors and adiabatic vectors through force constants, and therefore it is similar to an analysis in terms of potential energy contributions[164] but has several advantages with regard to the latter analysis[162].

The normal modes of **1** (Figure 20) are easy to identify because most of them involve motions associated with the same type of internal parameter, e.g. all six CH bond lengths (mode 1) or all three CH_2 twisting parameters (mode 5). Strong coupling between different types of internal parameters can only be found for modes 2,10,13 and 14 (Table 19). In the first two cases, CC stretching motions are mixed in, which is obvious from the pictorial representations in Figure 20. However, these representations are sometimes misleading as can be seen from mode 3. According to the pictorial representation, one might expect that the ring breathing motion is connected with a CH scissoring or CH stretching motion, but the adiabatic analysis shows that mode 3 does not involve CH_2 scissoring or CH stretching. (The arrows at the H atoms result from the movement of the C atoms.) Modes 13 and 14 are a result of strong coupling between adiabatic CH_2 rocking and CH_2 twisting motions that is described quantitatively in the adiabatic mode analysis (Table 19).

Adiabatic frequencies of **1** are compared in Table 20 with those of some other hydrocarbons[163]. The adiabatic CC frequency is about 80 and the adiabatic CH stretching frequency about 100 cm^{-1} larger than the corresponding values for cyclohexane. Compared to ethene, adiabatic CH stretching frequencies are almost identical, which is in line with the high dissociation energy of the CH bond of **1** (see Section V. E)[89]. The same observation has been made by McKean using isolated CH frequencies obtained by appropriate deuteration of **1**[165].

TABLE 20. Adiabatic internal frequencies of cyclopropane and some simple hydrocarbons from Reference 163[a]

Molecule	CC stretch	CH stretch	HCH def	CH_2 twist	CH_2 rock	CH_2 wag
Ethene	1798	3344	1626		1121	
Cyclopropane	1169	3328	1614	1072	1017	1196
Cyclobutane	1114	3222 (ax)	1621	1201	927	1368
		3233 (eq)				
Cyclohexane	1132	3172 (ax)	1621	1286	1012	1416
		3200 (eq)				
Propane	1143	3192	1623	1293	984	1427

[a] All frequencies in cm^{-1}. HF/6-31G(d,p) calculations.

Since adiabatic frequencies provide direct information on the curvature of the potential energy surface and since the curvature is related to the dissociation energy (a large D_e should lead to a large positive curvature), adiabatic frequencies can be correlated with dissociation energies (but not bond energies, which are just averages over dissociation energies). Therefore, a suggested increase in the CC dissociation energy (the activation energy of CC bond breakage for **1** is 61.0 compared to 62.5 kcal mol^{-1} for cyclobutane[166]) is not necessarily a contradiction with regard to the relatively small CC bond energies calculated for **1** (Section V. B and Table 8).

With the development of analytical energy derivative methods[135, 167], the calculation of vibrational frequencies (second derivatives of the energy with regard to atomic coordinates) and infrared absorption intensities (derivatives of the energy with regard to components of electronic field and atomic coordinates, i.e. dipole moment derivatives) both at the HF and correlation corrected levels has become routine[168]. There are six (two a_2'' + four e')

infrared active vibrational modes, which is confirmed by theory (Table 21)[163]. The intensity pattern agrees well with experimental infrared data, but a direct comparison between calculated and measured absolute infrared intensities[157] has not been carried out so far.

TABLE 21. Integrated infrared intensities and calculated infrared intensities of cyclopropane

		Experiment		HF/6–31G(d,p)		
#	Sym	Freq (cm^{-1})	IR intensity[157] ($cm^2\,mmol^{-1}$)	Freq (cm^{-1})	IR intensity[173] ($km\,mol^{-1}$)	Characterization according to Exp.
6	a_2''	3102	0.974	3385	54.6	strong
7		854	0.058	919	0.03	very weak
8	e'	3024	1.274	3286	73.2	strong
9		1438	0.126	1602	0.4	weak
10		1028	1.976	1187	8.0	strong
11		868	3.584	959	49.5	very strong

There is considerable interest by experimentalists in infrared intensities, because these can be used to describe electronic charge reorganizations in vibrating molecules[169]. For this purpose, atomic polar tensors[170] and various effective charges[157, 169] have been derived, which are related directly to dipole moment derivatives. Effective charges can be compared to calculated atomic charges (see Section VIII).

Several features of the vibrational spectrum of **1** are characteristic also for its heterocyclic analogues, such as oxirane and aziridine. Komornicki and coworkers[158] showed that the three ring molecules have six types of internal coordinates in common (CH stretch, CH_2 deformation, CH_2 twist, CH_2 rock, CH_2 wag and ring motions; see Figure 20). The experimental order for the CH stretching motions of **1** is ν_6 (a_2'', 3102 cm^{-1}), ν_{12} (e'', 3082 cm^{-1}), ν_1 (a_1', 3038 cm^{-1}) and ν_8 (e', 3024 cm^{-1}); see Table 18. This order is maintained for all three molecules. In addition, the mode with the highest symmetry leads to the highest frequency.

Kaupert, Heydtmann and Thiel[112] calculated the vibrational spectrum of monohalogenated **1** at the HF level using the 6-31G(d) basis set and effective core potentials with DZ + P basis sets for Cl, Br and I. Reduction from D_{3h} to C_s symmetry leads to considerable coupling between modes (exceptions: C—H stretching and CH_2-deformation modes) of **1**. Vibrational frequencies that are influenced by the halogen substituent are shifted to lower values with increasing mass of the halogen.

Marstokk and Møllendal[171] investigated the equilibrium conformation and the rotational potential of cyclopropanemethanethiol using microwave and vibrational spectroscopy in connection with *ab initio* calculations. They found a heavy-atom *gauche* conformation with the thiol H atom residing over a vicinal CC bond of the ring to be most stable. Comparison of experimental and theoretical vibrational spectra suggested the existence of a weak hydrogen bond involving the SH group and the 'quasi-π' bond of the cyclopropane ring.

An investigation of the vibrational spectrum of cyclopropylcarbonyl fluoride was carried out by Durig and coworkers using HF/3-21G theory[115]. The authors could assign all frequencies of *cis* and *trans* conformations and analyse normal modes in terms of potential energy contributions using appropriate symmetry coordinates. The calculated conformational stability and rotational barriers [HF/6-31G(d) and HF/3-21G] were compared with results obtained from the far-infrared spectrum.

Vibrational circular dichroism (VCD) reflects the stereochemistry of a chiral molecule[172]. According to Stephens[173], analysis and prediction of VCD spectra can be carried out on the basis of calculated vibrational rotational strengths R_i, which depend on dipole strengths (determined by atomic polar tensors $\mathbf{P}_{\alpha\beta}^{\lambda}$; see equation 17) and tensors $\mathbf{I}_{\alpha\beta}^{\lambda}$ (equation 18) that describe changes of the molecular wave function upon changes in atomic coordinates and magnetic field components:

$$\mathbf{P}_{\alpha\beta}^{\lambda} = [(\partial\mu_{el})\beta/\partial x_{\lambda\alpha}]\mathbf{R}_e \qquad (\alpha,\beta = x, y, z) \qquad (17)$$

$$\mathbf{I}_{\alpha\beta}^{\lambda} = \langle(\partial\psi_G(\mathbf{Rz})/\partial x_{\lambda\alpha})_{\mathbf{R}_e} | (\partial\psi_G(\mathbf{R}_e, H_\beta)/\partial H_\beta)\rangle_{H_\beta = 0} \qquad (18)$$

where μ_{el} is the electric dipole moment of the ground state, $\partial x_{\lambda\alpha}$ the Cartesian displacement coordinate from $\mathbf{R}_e$ of nucleus λ, $\mathbf{R}_e$ the equilibrium geometry, $\psi_G(\mathbf{R})$ the electronic wave function of the ground state, and $\psi_G(\mathbf{R}_e, H_\beta)$ is the wave function in the presence of a magnetic field perturbation. Using these definitions, Jalkanen and coworkers[174] investigated the VCD spectrum of *trans*-1(*S*),2(*S*)-dicyanocyclopropane at the HF/4-31G level. Frequencies, relative absorption intensities and VCD intensities were in reasonable agreement with experimental values.

Lazzeretti and coworkers[175] calculated nuclear electric and electromagnetic shielding tensors for **1** and oxirane. These properties are related to atomic polar tensors and atomic axial tensors used by infrared and VCD spectroscopists. The authors demonstrated that they could obtain fairly accurate sum rules for atomic polar tensors and atomic axial tensors with relatively little computational effort.

VIII. ONE-ELECTRON PROPERTIES

Although **1** is one of the best investigated molecules, there is, apart from data concerning its electron density distribution, very little information available on its one-electron properties. In principle, accurate data could be obtained by correlation-corrected *ab initio* methods, but almost nothing has been done in this direction, which of course has to do with the fact that experimental data on one-electron properties of **1** are also rare, and therefore, it is difficult to assess the accuracy and usefulness of calculated one-electron properties such as higher multipole moments, electric field gradients, etc.

The most important one-electron property of **1** is its electron density distribution $\rho(\mathbf{r})$, which has been discussed in Section IV. Apart from X-ray and neutron diffraction studies, information on $\rho(\mathbf{r})$ is also obtained from experimentally based atomic charges and measured multipole moments of a molecule. Zerbi and coworkers[169] use integrated infrared-absorption intensities (Section VII) to derive atomic densities. Since infrared intensities result from changes in the molecular dipole moment upon activating vibrational modes, one can express the total dipole moment as a sum of bond dipole moments, and hence dipole moment derivatives in terms of bond dipole moment derivatives. In this way, it is possible to evaluate from infrared intensities bond dipole moments and, with known bond lengths r, effective atomic charges q from bond dipole moments $\mu(\text{Bond}) = qr$. In Table 22, experimentally based atomic charges are compared with calculated atomic charges for small hydrocarbons with different hybridizations. It is well known that the electronegativity of a C atom increases with increasing s-character, which is nicely reflected by the virial charges. In addition, virial charges suggest similar hybridizations for **1** and ethene as far as the CH hybrid orbitals are concerned.

The H charges derived from infrared intensities seem to confirm the increase in the electronegativity of the C atom with increasing s-character. However, the corresponding C charges reveal that the electronegativity change from ethene to acetylene is not correctly described and that a large electronegativity difference between **1** and ethene is predicted. Mulliken charges also fail to reproduce the increase in the C electronegativity when going from ethene to acetylene. They suggest similarity between ethene and **1**, flawed however by the fact that Mulliken charges suggest larger electronegativity (s-character) for C in **1**.

TABLE 22. Experimentally derived and calculated atomic densities [electron]

Molecule	Infrared intensities[a]		Mulliken[b]		Virial partitioning[c]	
	C	H	C	H	C	H
CH_4	−0.260	0.065	−0.472	0.118	0.244	−0.061
$H_3C—CH_3$	−0.135	0.045	−0.335	0.112	0.237	−0.079
Cyclopropane	−0.170	0.085	−0.261	0.130	0.104	−0.052
$H_2C{=}CH_2$	−0.268	0.134	−0.254	0.127	0.082	−0.041
$HC{\equiv}CH$	−0.208	0.208	−0.233	0.233	−0.121	0.121

[a] Derived from infrared intensities (Reference 169). Carbon charges were determined from H charges by symmetry arguments.
[b] HF/6-31G(d,p) calculations. This work.
[c] HF/6-31G(d,p)//HF/6-31G(d) calculations (Reference 79).

Hence, neither experimentally based charges nor Mulliken charges provide a consistent picture and, in addition, they show little resemblance (apart from the sign of q) contrary to what has been claimed previously[169].

Virial charges[67] have been criticized because (a) they show little similarity to Mulliken charges, (b) they are often very large and (c) they lead to bond polarities in contradiction to established chemical thinking. For example, virial charges suggest a $C^+—H^-$ bond polarity while Mulliken charges and intensity-based charges predict $C^-—H^+$ bond polarity. This seems to result from the basically different definition of atomic charges used by various authors. For example, experimental atomic charges represent effective quantities derived to fit both molecular dipole moment and infrared intensities; i.e. *they absorb the effects of (true) atomic charges and atomic dipole moments, where the latter result from the anisotropy of the electron density at an atom.* In the virial partitioning method, atomic charges and atomic dipole moments (multipole moments) are calculated separately and their values may cancel largely in the expression for the molecular dipole moment[67]. Hence, effective atomic charges and true atomic charges can differ considerably where, of course, it should be more difficult to discuss effective charges since they contain the cumulative effect of at least two quantities. The equal splitting of overlap populations to get Mulliken charges mixes in higher multipole moments (as, e.g., becomes obvious from Stone's distributed multipole analysis[176]) and, accordingly, lends them the character of an effective rather than a pure atomic charge.

The multipole moments of a molecule indicate anisotropies in the molecular charge distribution. Since **1** is an uncharged molecule with D_{3h} symmetry, the first non-vanishing molecular moment of **1** is its quadrupole moment, which has been determined experimentally by measuring the birefringence induced by an electric field gradient and is $5.3 \pm 0.7 \times 10^{-40}$ C m^2(1.6 ± 0.2 Buckingham)[177]. At the HF/[5s3p2d/3s1p] level, a value of 8.4×10^{-40} C m^2 (2.5 Buckingham) was obtained by Amos and Williams[178]. Since possible hyperpolarizability effects would decrease rather than increase the experimental quadrupole moment, the relatively large difference between experimental and calculated quadrupole moments could not be explained. In the same investigation, calculated values of octupole and hexadecapole moments of **1** are also given (see Table 23).

The experimental values of the polarizability $\alpha = \frac{1}{3}(2\alpha_{xx} + \alpha_{zz})$ of **1** and its anisotropy $\Delta\alpha = \alpha_{zz} - \alpha_{xx}$ are 5.50 and −0.74 Å^3, which have been obtained by extrapolating polarizabilities measured at optical frequencies to their static limits[179]. Amos and Williams calculated for α and $\Delta\alpha$ 5.03 and −0.67 Å^3 at the HF/[5s3p2d/3s1p] level, which are about 10% too small, typical of HF/large basis set calculations. As can be seen from Table 23, $\alpha_{xx} = \alpha_{yy}$ (components in the ring plane) is larger than α_{zz} (perpendicular to the ring plane). Since an

TABLE 23. Multipole moments and polarizability α of cyclopropane[a]

Method	Θ (10^{-40} C m^2)	Ω (10^{-50} C m^3)	Φ (10^{-60} C m^4)	
HF/[5s3p2d/3s1p][178]	8.40	19.05	−48.91	
exp.[177]	5.3 ± 0.7			
	α_{xx} (Å^3)	α_{zz} (Å^3)	$\Delta\alpha = \alpha_{zz} - \alpha_{xx}$ (Å^3)	$\alpha = \frac{1}{3}(2\alpha_{xx} + \alpha_{zz})$ (Å^3)
HF/[5s3p2d/3s1p][178]	5.26	4.59	−0.67	5.03
exp.[179]	5.74	5.00	−0.74	5.50

[a] z is the threefold axis of cyclopropane and x is one of its C_2 axes. With this definition, $\Theta = \Theta_{zz}$, $\Omega = \Omega_{xxx}$, and $\Phi = \Phi_{zzzz}$. Other non-zero elements of the multipole tensor are related to these by symmetry.

important part of the polarizability is the flow of electrons between bonds, and since it is much easier to displace charge along a bond than across it, the component of α parallel to a bond chain is always larger than those perpendicular to it. In the case of a small ring such as **1**, it is easier to move charge in the ring plane than perpendicular to it as reflected by α_{xx} and α_{zz}. However, compared to the polarizability component perpendicular to the C_2 axis of propane in the plane of the three C atoms (and by this almost 'parallel' to the C bond chain), the α_{zz} value is significantly smaller, which indicates that it is more difficult to polarize the electronic charge in the plane of **1** than along the CCC chain of propane. On the other hand, one has to consider that the molecular volume of propane is larger than that of **1**.

Assuming that the molecule can be approximated as a perfectly conducting sphere (radius r) with volume V, the dipole moment μ induced in the sphere upon its placement in an electric field of magnitude F is given by equation 19:

$$\mu = r^3 F \tag{19}$$

which means that the molecular polarizability is equated with a volume measure. Gough[180] has shown that calculated molecular volumes defined by the contour for which the electron density is 0.001 a.u. indeed correlate with the molecular polarizability calculated at the HF/(9s5p1d/4s1p)[4s2p1d/2s1p] level (see Figure 21). According to this correlation, the actual molecular polarizability of **1** is about 10% larger than that predicted from the molecular volume. This is confirmed when predicting molecular polarizabilities from CH_3 (1.69 Å^3) and CH_2 group increments (1.47 Å^3) derived from calculated polarizabilities of ethane and cyclohexane[180]. While α values of alkanes can be accurately reproduced, the α value of **1** (4.42 Å^3) is smaller than the theoretical value. This is similar to the case of ethene (see also Figure 21), which suggests that the polarization of negative charge in particular in the plane of the ring is larger than for a normal alkane and that **1** and ethene are related in this respect.

Gough[180] has investigated polarizability derivatives associated with CH and CC bond stretching, which can be compared with parameters derived from Raman trace scattering cross sections. The mean of the CH polarizability derivative of **1** is slightly larger than that for CH (CH_2) in propane. Components along the bond are larger than those perpendicular to the bond. From all investigated alkanes the CC polarizability derivative of **1** is the smallest one, which has to do with the fact that a CC stretching mode in **1** involves three rather than two C atoms.

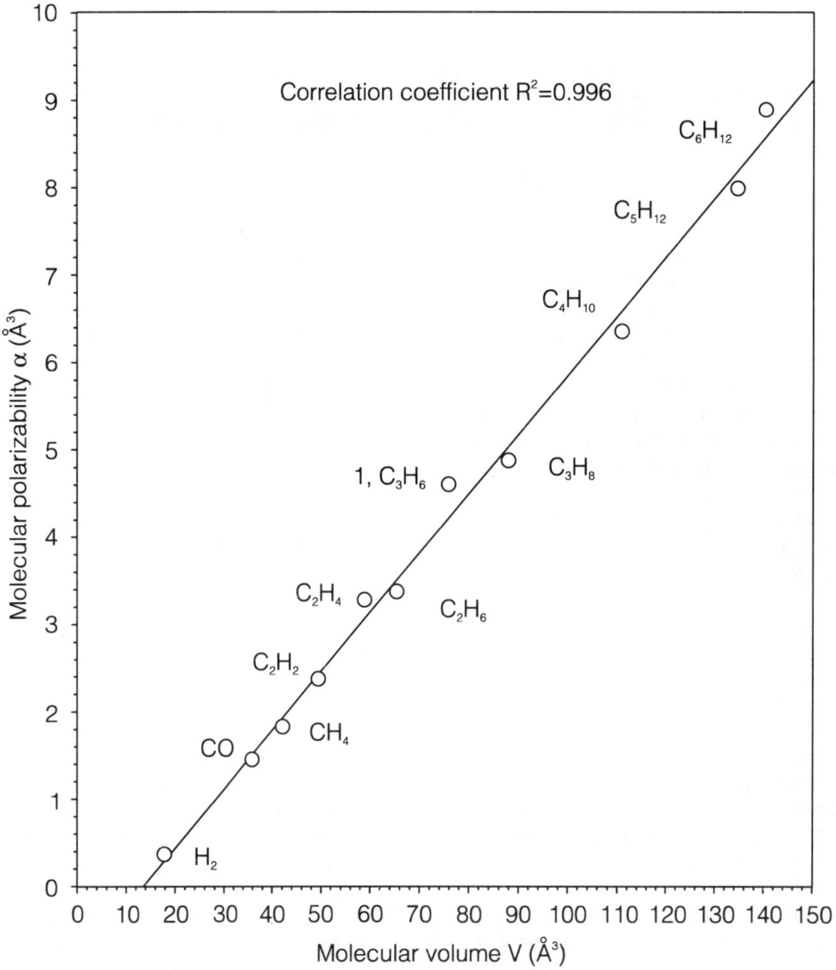

FIGURE 21. Correlation of molecular polarizabilities α with molecular volumes *V* for a number of hydrocarbons including cyclopropane. Data from Reference 180

IX. NMR SPECTRA

The rapid development of analytical energy gradient methods[135, 167] in the 70s and 80s has made accessible a large number of molecular properties to routine *ab initio* calculations. Certainly, one of the most important steps (after calculating molecular geometries and molecular vibrations) was the calculation of NMR chemical shifts by *ab initio* methods. For a long time, this information could not be provided by *ab initio* theory, since a routine calculation of NMR parameters with sufficient accuracy was not possible. There were several approaches to calculate NMR chemical shift data, of which the best known is probably the method based on *gauge independent atomic orbitals* (GIAOs)[181], originally suggested by London[182] and later used by Pople[183], Ditchfield[184] and others[185]. However, the computer-time consuming integral evaluation over GIAOs prevented applications to larger molecules. This changed when Pulay and coworkers[186] improved the GIAO method by implementing modern techniques for integral and integral derivative evaluation. However, this development took place years after the routine calculation of NMR chemical shifts at the *ab initio* level was solved by Kutzelnigg and Schindler[187]. These authors solved the gauge problem inherent in all calculations of magnetic properties with the help of localized MOs rather than GIAOs. Accordingly, they coined their method *individual gauge for localized orbitals* (IGLO)[187, 188].

The work by Kutzelnigg and Schindler triggered further developments in the field of NMR chemical shift calculations. Beside the IGLO program, several other *ab initio* methods are today available for routine calculations of magnetic properties of molecules: (1) The LORG (localized orbital/local origin) method by Hansen and Bouman[189]; (2) GIAO-SCF in the version of Pulay and coworkers[186]; (3) GIAO-MBPT2 by Gauss to calculate correlation corrected NMR chemical shifts at the second-order many-body perturbation theory level[190]; (4) GIAO-MBPT3 and GIAO-MBPT (SDQ)-4 by Gauss to get third-order and fourth-order corrections to GIAO values[191]; (5) MC-IGLO by Kutzelnigg and coworkers for problems that require a MCSCF wave function[192]. In addition, other methods to obtain correlation-corrected NMR chemical shift values have been described in the literature[193].

Since *ab initio* calculations lead to the determination of the full shielding tensor of each nucleus of **1**, one should have expected that these calculations would have helped to rationalize ^{1}H and ^{13}C NMR spectra of **1**, which are unusual with regard to measured shift values: The proton shift (δ 0.12 ppm relative to TMS[194]) is upfield by more than 1 ppm compared to the shift values of suitable reference compounds (cyclohexane: δ 1.44 ppm; cyclobutane: δ 1.96 ppm[194]) while the ^{13}C shift (isotropic shift: δ –4.0 ppm[195]; shift in liquid: δ –2.8 ppm relative to TMS[196]) is more than 20 ppm upfield (cyclohexane: δ 27 ppm; cyclobutane: δ 23 ppm)[196]. Wiberg[14] has summarized in his review article on the cyclopropyl group the work that has been carried out by both theoreticians and experimentalists to rationalize the ^{13}C NMR chemical shift of **1**. Today, it is clear that the observed upfield shift of more than 20 ppm is largely due to the tensor component perpendicular to the ring plane[197]. This is also reflected by the anisotropy and asymmetry of the ^{13}C shielding tensor of **1** calculated by Hansen and Bouman with their LORG method[198]. These authors suggested an analysis of shielding tensors in terms of the shielding response vector **T**, which can be displayed pictorially in the same way as one displays MOs (see Figure 6). Three-dimensional contour line diagrams of **T** were used to rationalize ^{13}C shielding tensors in cyclopropene and other three-membered rings[198]. However, the ^{13}C shielding tensor of **1** was not discussed in this work.

So far, speculations which attribute the observed upfield shifts of the NMR signals of **1** to a ring current of the σ-electrons[199] (in line with the idea of σ-aromaticity[8]) have not been refuted. However, a less spectacular rationalization of the NMR chemical shifts in terms of local anisotropy contributions caused by the unique electron distribution of **1** (see Section IV) may also be possible[200].

The determination of NMR chemical shifts by either IGLO, LORG or GIAO turns out to be very sensitive with regard to the geometry used[201–204]. Experimental geometries are not that useful in this connection since very often they are not accurate enough, represent different geometries (r_z, r_s, r_o, r_a, r_g, r_v, etc.) or suffer from intermolecular interactions in condensed phases. *Ab initio* geometries provide a consistent description of molecules that does not suffer from the ambiguities of experimental geometries. Many calculations have shown that reasonable NMR chemical shifts are obtained if the geometry of the molecule in question has been optimized at a correlation-corrected level of theory such as second-order perturbation theory (MP2) using DZ, DZ + P or better basis sets. Since the calculated NMR chemical shifts clearly depend on the geometry, an agreement between experimental and theoretical shifts not only means a clear identification but also a geometry determination of the molecule in question. On the other hand, if theoretical and experimental shifts differ considerably, other possible geometries or structures have to be tested[205].

Schleyer was the first to fully realize the sensitivity of calculated NMR chemical shifts with regard to molecular geometry and he used this for *ab initio*/IGLO/NMR-based structural determinations in many cases including carbocations, boron and organolithium compounds[201, 202]. A recent assessment of this approach suggests that 'structural assignments based on the *ab initio*/IGLO/NMR method are quickly approaching a confidence level that rivals modern day X-ray diffraction determinations of molecular structures'[206].

The *ab initio*/IGLO/NMR method has been used to determine the relative distribution and stability difference of the cyclopropylcarbinyl cation and cyclobutyl cation in solution[207]. Agreement between ^{13}C IGLO chemical shifts and experimental shifts could only be obtained when assuming a rapid equilibrium between the two cations. Over the range of temperatures considered (–61 to –132° C), a cyclobutyl cation structure with an axial H atom and short 1,3-distances of 1.65 Å (bicyclobutonium ion structure) was found to be more stable by 0.5 kcal mol^{-1} [207]. For the gas phase, however, the cyclopropylcarbinyl cation was calculated to be 0.26 kcal mol^{-1} more stable [MP4/6-31G(d)//MP2/6-31G(d) calculations including vibrational corrections][207].

Cremer and coworkers investigated a number of potentially homoconjugated cyclopropyl compounds such as the monohomotropylium cation[203, 205], the 1,4- and 1,3-bishomotropylium cation[208], the trishomotropylium cation[209], the barbaralyl cation[210] and the cyclobutenyl cation[211]. All these cations have the choice between a closed cyclopropyl structure (**Ia**), an open cyclopolyenyl structure (**Ic**) and an intermediate structure (**Ib**) as demonstrated in the case of the monohomotropylium cation.

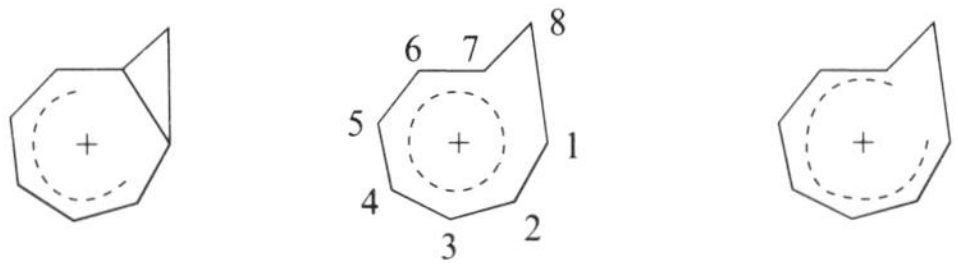

(**Ia**) (**Ib**) (**Ic**)

For all cations considered, ^{13}C NMR spectra have been measured in solution while direct structural information on the question as to whether structure **Ia**, **Ib** or **Ic** corresponds to a minimum energy form was completely missing. Therefore, the structural problem was solved by utilizing the *ab initio*/IGLO/NMR method. Since results of this work will be reviewed in another chapter of this volume (see Chapter 7), we refrain from discussing calculated NMR data for potentially homoconjugated cyclopropyl compounds at this point.

There is little computational work on NMR coupling constants since the *ab initio* methods for calculating these quantities are still at an infant stage. A discussion of the experimental work on NMR coupling constants of **1** and related ring compounds can be found in the review article of Wiberg[14].

X. EXCITED STATES AND ULTRAVIOLET ABSORPTION SPECTRA

The observed ultraviolet absorption spectrum of **1**[212] (for a display of the recorded spectrum see Section I.G of Wiberg's review article on the cyclopropyl group[14]) consists of three broad bands, of which the first (60,000 to 66,000 cm^{-1}; 7.44–8.18 eV) possesses a maximum at 63,000 cm^{-1} (7.8 eV; oscillator strength f 0.12), far to the red of most saturated absorbers and in the region of the $\pi \rightarrow \pi^*$ transitions of unsaturated molecules. This observation has been one of the reasons to bring **1** into relation with ethene and to associate π-character with its CC bonds. The second ultraviolet absorption band extends from 67,000 to 72,000 cm^{-1} (8.3–8.9 eV) with a maximum at 70,000 cm^{-1} (8.7 eV, $f = 0.04$) while the third, which is the most intense band ($f = 0.7$), is located at 83,000 cm^{-1} (10.3 eV) extending over the range from 74,000 to 85,000 cm^{-1} (9.2–10.5 eV). The first and the third band possess a discrete structure superimposed on a broad continuous background, which has been brought into connection with the Rydberg character of the first and partial Rydberg character of the third band. The second band may have either valence or Rydberg character according to experiment[212].

The classic *ab initio* work on excited states of **1** is the SCF plus limited CI study by Buenker and Peyerimhoff from 1969, which gave a first basis for analysing its ultraviolet absorption spectrum and investigating ring opening to trimethylene[213]. A decade later, Goldstein, Vijaya and Segal carried out an extensive *ab initio* CI investigation of absorption and magnetic circular dichroism (MCD) spectra of **1** using the improved computational possibilities of the eighties[52]. Segal and coworkers used a (9s5p/5s)[4s2p/1s] basis augmented by diffuse s, p and d functions (exponent 0.02) located at the ring centre to describe in particular the lower Rydberg states of **1**. The CI calculations (based on 32,542 spin eigenfunctions) were carried out by using perturbational CI techniques[52]. Some of the results obtained in this work are summarized in Table 24.

TABLE 24. Comparison of calculated excitation energies[52] and experimental absorptions of cyclopropane[a]

State	Primary excitation	Excitation energy (eV)	f^r	f^{∇}	$f^{r\cdot\nabla}$	Exp. UV spectrum[212]
$1^3E'$	$3e' \rightarrow 4a_1'$ (0.85)	7.47	not electric-dipole-allowed			
$1^1E'$	$3e' \rightarrow 4a_1'$ (0.90)	7.61	0.009	0.013	0.011	7.4–8.2, $f = 0.12$
$1^1A_2'$	$3e' \rightarrow 4e'$ (0.68)	8.07	not electric-dipole-allowed			
$1^1A_1'$	$3e' \rightarrow 4e'$ (0.68)	8.08	not electric-dipole-allowed			
$2E'$	$3e' \rightarrow 4e'$ (0.66)	8.11	0.127	0.101	0.113	
$1A_2''$	$3e' \rightarrow 2e''$ (0.68)	8.72	0.007	0.001	0.005	
$1A_1''$	$3e' \rightarrow 2e''$ (0.67)	8.79	not electric-dipole-allowed			
$3E'$	$3e' \rightarrow 5a_1'$ (0.95)	8.85	0.002	0.001	0.002	8.3–8.9, $f = 0.04$
$1E''$	$3e' \rightarrow 2e''$ (0.68)	8.96	not electric-dipole-allowed			
$4E'$	$3e' \rightarrow 5e'$ (0.69)	9.08	0.002	0.001	0.001	
$2A_2'$	$3e' \rightarrow 5e'$ (0.69)	9.09	not electric-dipole-allowed			
$2A_1'$	$3e' \rightarrow 5e'$ (0.69)	9.19	not electric-dipole-allowed			
$5E'$	$3e' \rightarrow 6a_1'$ (0.88)	9.92	0.001	0.004	0.002	weak Rydberg at 9.9 eV,
$6E'$	$1e'' \rightarrow 2a_2''$ (0.97)	10.56	0.135	0.076	0.102	9.2–10.5, maximum 10.3,
$2E''$	$1e'' \rightarrow 4e'$ (0.70)	10.62	not electric-dipole-allowed			
$2A_2''$	$1e'' \rightarrow 4e'$ (0.63)	10.79	0.072	0.053	0.062	valence with Rydberg
$7E'$	$3e' \rightarrow 1a_2'$ (0.64)	11.94	0.376	0.244	0.303	superimposed, $f = 0.7$

[a] All calculated excited states are listed together with the main contribution (corresponding CI coefficient given in parentheses).

Molecule **1** possesses a large number of degenerate electronic energy levels and an unusually high density of electronic states in the energy region between the onset of optical absorption and the first ionization potential. Because of its D_{3h} symmetry, transitions from the ground state to the A_2'' and E' excited states are electric dipole allowed. CI calculations suggest that eight states (six E' + two A_2'') are involved in the first three absorptions (Table 24)[50].

The first optical absorption band is made up of two states, $1E'$ ($3e' \rightarrow 4a_1'$) and $2E'$ ($3e' \rightarrow 4e'$), which can be viewed as excitations to 3s and $3p_{x,y}$ Rydberg orbitals[52].

There are three dipole allowed singlet states in the region of the second absorption band, $1A_2''$, $3E'$ and $4E'$, all of which possess approximate 3d Rydberg character (Table 24). Experimental and calculated oscillator strength ($f_{exp} = 0.04, f_{theor} = 0.01$ for all three states) differ, which could not be resolved[52]. The principal transition of the third absorption has frequently been assigned to the transition $3e' \rightarrow 1a_2'$, but theoretical values for the energy of the state, for which the transition $3e' \rightarrow 1a_2'$ is the primary contribution, are 2–3 eV higher than experimental values. Instead, CI calculations suggest that the third band is the result of a transition to a $3p_z$ Rydberg state ($6E'$, $1e'' \rightarrow 2a_2''$; 10.56 eV; $f = 0.135$), a transition to a state with partial Rydberg character ($2A_2''$, $1e'' \rightarrow 4e'$; 10.79 eV; $f = 0.072$) and a transition to a state with valence character ($7E'$, $3e' \rightarrow 1a_2'$, $1e'' \rightarrow 2e''$; 11.94 eV; $f = 0.376$). Again, the calculated total oscillator strength (0.2) is smaller than the experimental one (0.7, Table 24), which in this case seems to be a deficiency of the basis set used to describe a possible mixing in of higher Rydberg states[52].

Although the investigation of Segal and coworkers gives a first basis for the understanding of the ultraviolet absorption spectrum of **1**, it leaves a number of open questions, which can only be answered by more extended calculations. Such calculations are possible and have been done for derivatives of **1** (see, e.g., the MR-CI calculations for fluoro- and methyl-cyclopropanone with more than 50,000 spin adapted configurations[214]) but not for **1** itself. Therefore, further calculations are needed to get additional information on the excited states of **1**.

XI. CHARGED CYCLOPROPYL GROUPS

A. Cyclopropyl Anion

Rapid and significant progress in gas-phase carbanion chemistry in the last decade has promoted gas-phase acidity measurements even for such weakly acidic hydrocarbons such as **1**[215]. In addition, theoretical investigations of negative ions using *ab initio* MO methods have also produced new thermochemical and structural data for isolated carbanions which exhibit impressive accuracy when compared with experimental results (Table 25)[216].

TABLE 25. Proton affinity (PA) and relative stability with regard to the CH_3^- anion (ΔE of equation 20) given for cyclopropane and some other carbanions[a]

Carbanion	Method	ΔE	Proton Affinity calc.	Proton Affinity exp.	Reference
Cyclopropyl	HF/4-31+G	−2.1	414.5	412	216 217
	MP2/6-31+G(d)//HF/6-31G(d)	−4.0			216
Cyclopropylmethyl	HF/4-31+G	−2.7	413.9		216
Vinyl	HF/4-31+G	−9.7	406.9	408.0	216
	HF/3-21+G	−11.2	420.5		215
Allyl	HF/4-31+G	−28.0	388.6	391.0	216
	HF/3-21+G	−27.8	403.9		215
Ethynyl	HF/4-31+G	−47.9	368.7	375.4	216
2-Propyl	HF/4-31+G	5.9	422.5	419.0	216

[a] Energies in kcal mol^{-1}.

Basis sets used for carbanions have to include diffuse functions because anions generally have low ionization potentials, i.e. there is a pair of (or a single) electrons in the form of a diffuse charge cloud that extends relatively far from the nuclei and therefore is easily lost. Without diffuse functions, even larger basis sets such as DZ + P are not entirely successful either in the calculation of absolute acidities or in the ordering of acidities.

Froelicher, Freiser and Squires[215] calculated the gas-phase geometry of the cyclopropyl anion at the HF/3-21+G level of theory. In case of C1 deprotonation, vicinal (C1C2: 1.562 Å) and distal (C2C3: 1.531 Å) bond lengths become longer than the CC bond lengths of **1** (Section VI. A). The cyclopropyl anion favours pyramidal geometry at the anionic centre to avoid additional ring strain. The calculated activation barrier for carbanion inversion is 18.4 kcal mol^{-1}. The changes in the geometry of the cyclopropyl ring upon deprotonation can be rationalized in terms of changes in electronegativity and hybridization of atom C1. The hybrid orbital that accommodates the electron lone pair tries to adopt as much s-character as possible to stabilize the non-bonded charge distribution (see discussion of heterocyclopropanes in Section V. D). This will increase the p-character of the CH hybrid orbital which, according to a calculated HC1C2 angle of 111°, should be close to sp^3 hybridization.

Froelicher and coworkers[215] and Schleyer and coworkers[216] calculated the proton affinity of several carbanions (Table 25). Especially, the HF/4-31+G results are in good agreement with experimental values. Calculated proton affinities are compared to that of CH_3^- with the help of the isodesmic reaction given in equation 20.

$$CH_3^- + RH \longrightarrow CH_4 + R^- \tag{20}$$

The ethynyl anion is calculated to be the most stable carbanion followed by the allyl, vinyl and cyclopropyl carbanion. The latter is only 2–4 kcal mol^{-1} more stable than CH_3^- but significantly more stable than the 2-propyl carbanion (Table 25), which is destabilized by the two methyl groups (due to 2-orbital–4-electron destabilization between the filled $p\pi$-orbital of the carbanion and the occupied pseudo-π orbitals of the methyl groups). This is a general observation for alkyl anions with the exception of the cyclopropyl carbanion, which is stabilized by a methyl group when compared to CH_3^- (Table 25). The effect of methyl substituents at the anionic centre depends on the electronegativity of the atom to which the CH_3 group is attached. The large electronegativity of C1 of the cyclopropyl anion relative to that of the anionic C of 2-propyl or methyl anion explains the observed difference in proton affinities (Table 25).

Schleyer and coworkers[216] calculated the electron affinity of ethyl, 2-propyl, cyclobutyl and cyclopropyl radicals. Apart from the cyclopropyl radical, these radicals have negative electron affinities suggesting that the corresponding anions cannot be observed as long-lived species in the gas phase. For the cyclopropyl radical, an electron affinity of 5.1 kcal mol^{-1} was predicted[216], in reasonable agreement with the experimental value of 8 kcal mol^{-1} [217]. Accordingly, it is probable that the cyclopropyl anion is the only saturated carbanion that can be observed experimentally in the gas phase.

B. Protonated Cyclopropane

The reactivity of **1** in substitution reactions is markedly different from that of other cycloalkanes. An electrophilic substitution of **1** is followed by opening of the three-membered ring to a 2-propyl cation. Therefore, protonation of **1** as the simplest electrophilic attack has been extensively investigated, both experimentally[218–220] and computationally[221–223].

Koch, Liu and Schleyer[223] explored the $C_3H_7^+$ potential energy surface at the MP2 and MP4//MP2 level of theory using a TZ + P basis [6-311G(d,p)]. Their results are summarized in Figure 22 (structures **28**–**33**). The global minimum of the $C_3H_7^+$ surface is occupied by the 2-propyl cation (**28**) that possesses C_2 symmetry rather than C_{2v} symmetry as is com-

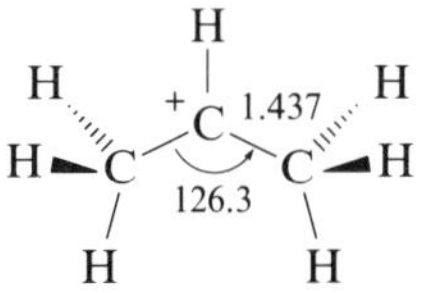

28, $\Delta E = 0$, C_2 (0)

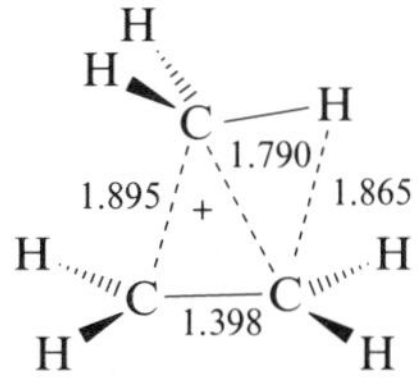

29, $\Delta E = 7.2$, C_s (0)

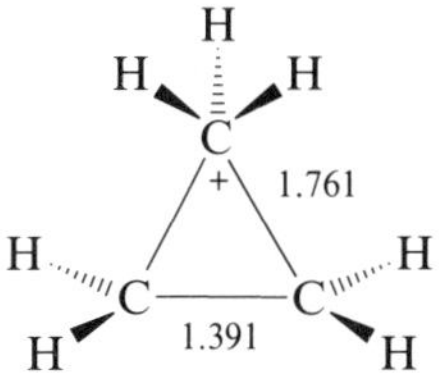

30, $\Delta E = 7.3$, C_s (1)

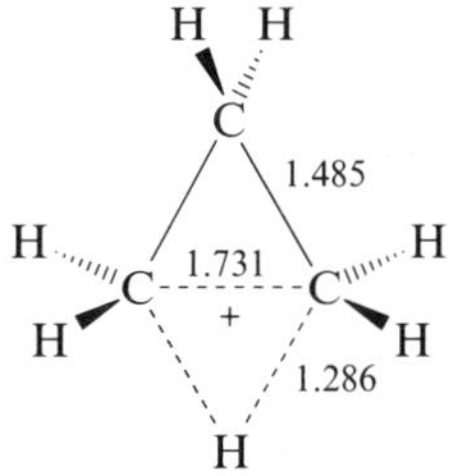

31, $\Delta E = 8.6$, C_{2V} (0)

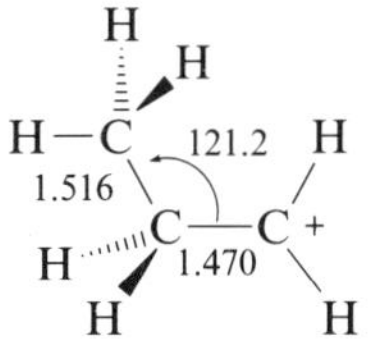

32, $\Delta E = 19.3$, C_s (1)

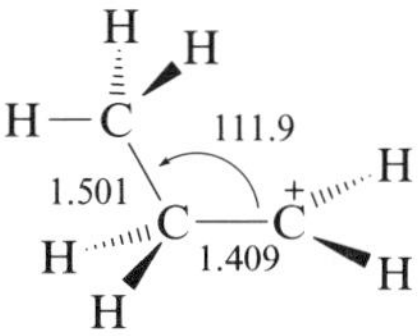

33, $\Delta E = 20.5$, C_s (1)

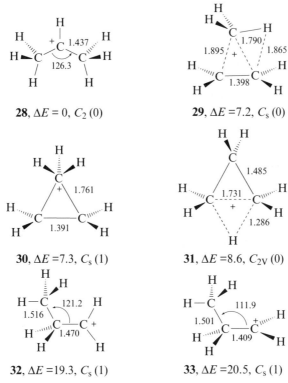

FIGURE 22. MP2/6-311G(d,p) geometries and MP4(SDTQ, frozen core)/6-311G((d,p) relative energies at MP2 geometries for protonated cyclopropane and related compounds. For each geometry, the number of imaginary frequencies (0: energy minimum; 1: first-order transition state) is given in parentheses. All data from Reference 223

monly believed. Protonation of **1** may lead in the first step to the corner (**29**) or edge protonated isomers (**31**), which are 7.2 and 8.6 kcal mol^{-1} above the global minimum (Figure 22). The C1C2 bond length of the corner potonated isomer is longer [1.895 and 1.790 Å at MP2/6–311G(d,p), Figure 22], while its C2C3 bond is shorter (1.398 Å) than the CC bonds in **1** (see Section VI. A). According to Dewar and coworkers[222], **29** is best characterized as a π-complex between methyl cation and ethene.

Structure **30** in Figure 22 corresponds to the transition state of methyl rotation (CH_3 rotated by 30° with regard to **29**), which is just 0.1 kcal mol^{-1} above cation **29** according to MP4(FC)/6-311G(d,p)//MP2/6-311G(d,p) calculations[223] [0.7 kcal mol^{-1} according to MP4SDQ/6-31G(d)//MP2/6-31G(d) calculations[222]]. This means that **29** is essentially a symmetric species with a rapidly rotating methyl group.

Edge protonated **1** (isomer **31**) is positioned in a very flat minimum which, if zero-point energy corrections are considered, may also be a transition state[223]. In any case, hydrogen scrambling in **29** (via ion **31**) will be also rapid considering a barrier of just 1.4 kcal mol^{-1}. The 1-propyl cation structures **32** and **33** are both transition states (there exists no minimum energy structure corresponding to 1-propyl cation), which are passed on the way to ring opening of **29** to give **28** (H migration takes place without any barrier). The calculated barrier for the process **28** $\rightarrow$ **29** (19.3 kcal mol^{-1} [223], Figure 22) is somewhat higher than the activation energy (16.3 ± 0.4 kcal mol^{-1} [224]) determined in non-nucleophilic (super acid) media. The calculated proton affinity of **1** is 180.0 kcal mol^{-1} [223], which is in excellent agreement with the experimental value of 179.8 kcal mol^{-1} [218].

C. Cyclopropyl Radical Cation

Investigation of **1** upon removal of an electron has intrigued both theoreticians[225–233] and experimentalists[234, 235] because the properties of the formed cyclopropyl radical cation provide a basis to test MO models of three-membered rings (the Walsh MO model cannot explain the properties of the cyclopropyl radical cation[233]). Ejection of an electron from the $3e'$ MOs of **1** (MOs #11,12 in Figure 6) leads to the Jahn–Teller unstable electron configuration $^2E'$ (see Section X), which can be stabilized by distortion to the C_{2v}-symmetrical states $^2A_1$ and $^2B_2$ (first-order Jahn–Teller effect, see Figure 23).

If an electron is ejected from the $3e'$–a_1 MO, which is C2C3 bonding, elongation of the distal bond can be expected accompanied by a lowering of the $3e'$–b_2 MO because of reduction of its C2C3 antibonding overlap (Figure 23). The reverse effect should occur for electron removal from the $3e'$–b_2 MO, namely (a) shortening of the distal bond, (b) lengthening of the vicinal bonds and (c) lowering of the energy of the $3e'$–a_1 MO. Accordingly, the $^2A_1$ state possesses a structure that is trimethylene-like and characterized by a short vicinal [1.474 Å at UMP2/6-31 G(d)] and a long distal [1.826 Å at UMP2/6-31G(d)] bond according to calculations by Krogh-Jespersen and Roth (see Figure 21)[231]. The $^2B_2$ state, on the other hand, is best described as π complex formed between a methylene cation and ethene because of its long vicinal [1.665 Å at UMP2/6-31G(d)] and short distal [1.410 Å at UMP2/6-31G(d) bond[231].

Krogh-Jespersen and Roth[231] identified the $^2B_2$ state as a transition state with one imaginary frequency and the $^2A_1$ state as the ground state of the cyclopropyl radical cation. At

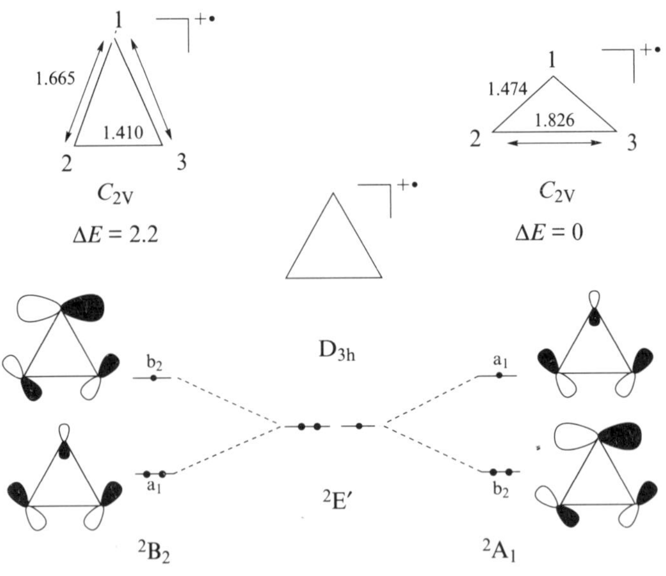

FIGURE 23. First-order Jahn–Teller distortion of D_{3h} symmetrical cyclopropyl radical cation. Geometries [UMP2/6-31G(d) calculations] and relative energy [CISD/6-31G(d)] from Reference 231

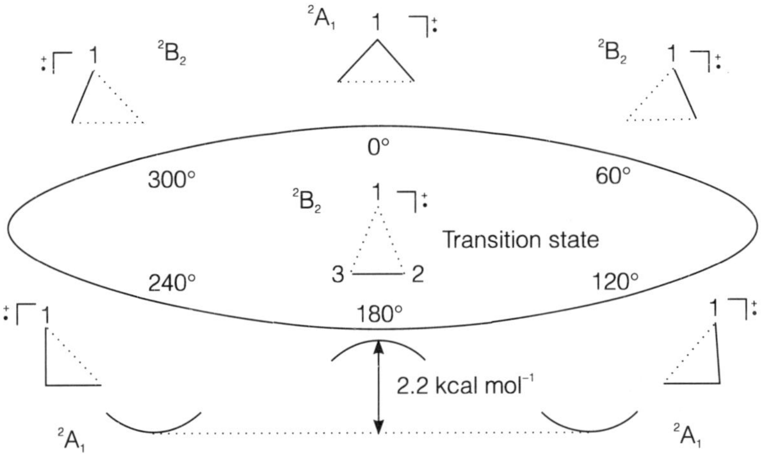

FIGURE 24. Pseudorotation cycle of the cyclopropyl radical cation. Relative energy [CISD/6–31G(d)] of the $^2A_1$ ground state and the $^2B_2$ transition state from Reference 231

the UMP2/6-31G(d) level the energy difference between the two states is 2.5 kcal mol^{-1}, while CI with all single and double substitutions CISD/6-31G(d)//UHF/6-31G(d) + zero point energy correction] predicts the $^2B_2$–$^2A_1$ energy difference to be 2.2 kcal mol^{-1}. MP4(SDTQ)/6-31G(d,p)//ROHF/6-31G(d) calculations by Hudson and coworkers[233] lead to an energy difference of 2 kcal mol^{-1}. Hence, the radical cation can undergo rapid pseudorotation (Figure 24), which is in line with ESR coupling constant measurements at 77 K[234, 235]. Lunell and coworkers determined isotropic hyperfine coupling constants of the cyclopropyl radical cation at the CI level of theory, which turned out to be in reasonable agreement with measured values[230].

Krogh-Jespersen and Roth[231] also investigated Jahn–Teller distortions in mono-, di- and tetramethyl substituted cyclopropyl radical cations. In all cases, the ground state of the corresponding radical cation can be considered to be derived from a $^2A_1$-type state and (the) transition state(s) of pseudorotation from (a) $^2B_2$-type state(s). Several minima and transition states were found along the pseudorotation path where the most stable form corresponds to an asymmetrical $^2A_1$-type state.

The possible existence of a trimethylene radical cation was investigated as a reaction to claims by experimentalists that this cation had been observed in the ring-opening reaction of the cyclopropyl radical cation[236, 237]. Borden and coworkers[227] calculated the energy difference between these two radical cations to be as large as 19 kcal mol^{-1} with little chance to observe the trimethylene radical cation on the $C_3H_6^{\bullet+}$ potential energy surface. UMP2 calculations suggest that the cyclopropane radical cation can rearrange to the propane radical cation in two stages [barrier: 19 kcal mol^{-1}, reaction energy: –10.3 kcal mol^{-1} at UMP2/6-31G(d) + zero point energy corrections]: (a) ring opening via a conrotatory transition state and (b) migration of a H atom (with just a 0.2 kcal mol^{-1} barrier).

D. Cyclopropyl Cation

The cyclopropyl cation (**34**) corresponds to a stationary point on the $C_3H_5^+$ potential energy surface[238, 239], the global minimum of which is occupied by the allyl cation (**35**). Table 26 summarizes relative energies of various $C_3H_5^+$ isomers (**34**–**39**) that were investigated at the HF, MP2 and MP4(SDQ) levels of theory[238]. Cation **34** is about 36 kcal mol^{-1} less

TABLE 26. Relative energies (kcal mol^{-1}) of cyclopropyl cation and other $C_3H_5^+$ cations[238]

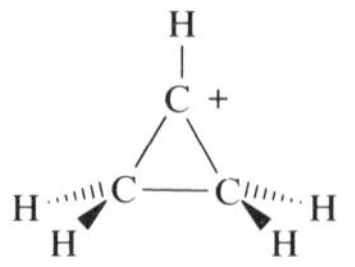

(34)

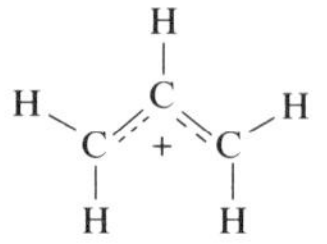

(35)

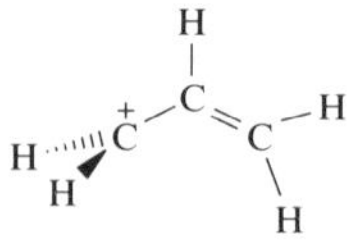

(36)

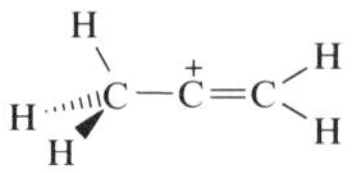

(37)

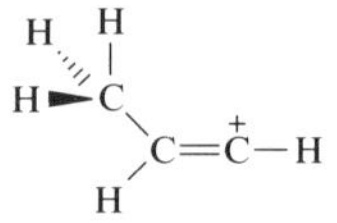

(38)

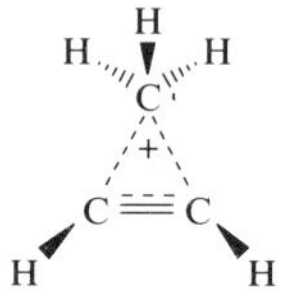

(39)

Cation	Sym	HF/6–31G(d,p) //HF/6–31G(d)	MP2/6–31G(d,p) //HF/6–31G(d)	MP4(SDQ)/6–31G(d,p)^a //HF/6–31G(d)
Cyclopropyl (**34**)	C_{2v}	37.8	37.0	35.0
Allyl (**35**)	C_{2v}	0	0	0
Perpendicular allyl (**36**)	C_s	33.7	37.7	34.9
2-Propenyl (**37**)	C_s	16.1	14.2	11.6
1-Propenyl (**38**)	C_s	32.3	33.1	30.2
Corner protonated cyclopropene (**39**)	C_s	42.6	30.9	33.6

^{a} Estimated from MP4(SDQ)/6-31G(d) values.

stable than cation **35**, 2 kcal mol^{-1} less stable than corner protonated cyclopropene (**39**) and 1 kcal mol^{-1} less stable than the perpendicular allyl cation (**36**). It has been speculated that **34** is the transition state for the stereomutation of the planar allyl cation[238]. This is confirmed by MP2/6-31G(d,p) calculations[240], which show that the cyclopropyl cation is located at a first-order transition state possessing one imaginary frequency of b_1 symmetry that describes a disrotatory movement of the CH_2 groups in line with a stereomutation process of the allyl cation. Alternatively, stereomutation can also follow a stepwise route via perpendicular allyl cations **36** that are somewhat more stable than **34** (Table 26) and therefore lead to lower stereomutation barriers[238].

In Figure 25, geometries and relative stabilities of some 1-substituted cyclopropyl cations are compared with the corresponding 2-allyl cations according to HF/6-31G(d)//HF/3-21G results of Lien and Hopkinson[239]. In all cases, the vicinal bond C^+C bonds (1.43–1.46 Å) are much shorter than the distal CC bond (1.52–1.58 Å). Increase of the positive charge at C1 caused by electron-withdrawing substituents (e.g. F or CN) leads to shortening of vicinal and lengthening of the distal bond, which is in line with an increase in ring strain as a result of increased s-character at C1 and a stretching of the distal bond. Electron-donating substituents have the opposite effect and stabilize the ring.

Of the substituents considered in Figure 25, The amino group is the strongest π-donor and, not surprisingly, the α-aminocyclopropyl cation is more stable than the 2-aminoallyl cation by 23.4 kcal mol^{-1} at the HF/6-31G(d)//HF/3-21G level of theory. This is in line with

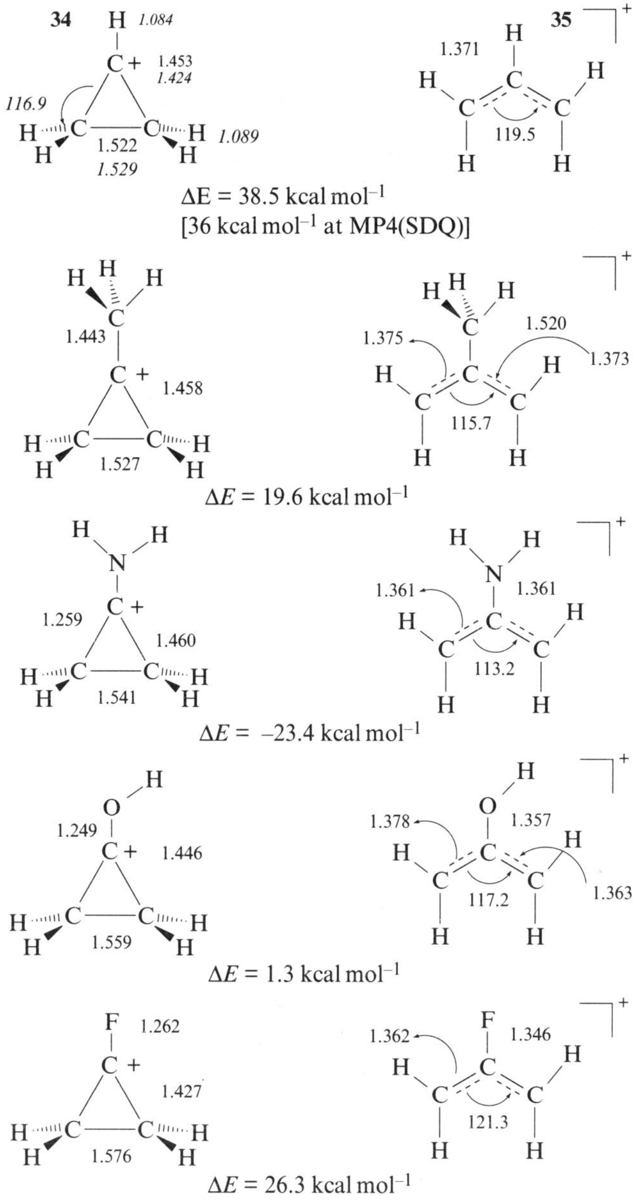

FIGURE 25. Comparison of geometries and relative energies of 1-substituted cyclopropyl cations and the corresponding 2-allyl cations according to HF/6-31G(d)//HF/3-21G calculations of Lien and Hopkinson[239]. MP4(SDQ) value from Reference 238. Geometrical parameters in italics from MP2/6-31G(d,p) calculations (K. J. Szabo and D. Cremer, unpublished results). Bond lengths in Å, bond angles in degrees

the observation and characterization of α-dimethylamino- and α-methylaminocyclopropyl cations by NMR spectroscopy[241–243].

Stabilization of **34** by a hydroxy group is smaller, as is suggested by the small energy difference of 1.3 kcal mol^{-1} between α-hydroxycyclopropyl and 2-hydroxyallyl cation in favour of the latter cation (see Figure 25). Experimentally, the α-methoxycyclopropyl cation appears to be a stable intermediate in substitution reactions in solution[244], and there is evidence for the independent existence of both the α-methoxycyclopropyl and 2-methoxyallyl cations in the gas phase[245].

XII. THE CYCLOPROPYL RADICAL

The CH bond dissociation enthalpy DH of **1** is 106.3 kcal mol^{-1}, which is 11.2 kcal mol^{-1} larger than that of the secondary CH bond of propane (95.1 kcal mol^{-1})[89a]. At the UMP2/6-31G(d) level, a dissociation energy DE of 107.7 kcal mol^{-1} is calculated, which can be improved by using isogyric reactions (the number of unpaired electron spins is preserved) such as equation 21:

$$RH + R'\bullet \longrightarrow R\bullet + R'H \quad (21)$$

provided the dissociation energy (enthalpy) of the reaction given in equation 22:

$$R'H \longrightarrow R'\bullet + H\bullet \quad (22)$$

is known exactly (e.g. R′H = H_3CH, Table 27) and can be used to derive DE values for RH from the calculated reaction energy of reaction 21.

TABLE 27. CH dissociation enthalpies DH(298) and changes in CH dissociation enthalpies (ΔDH) and energies (ΔDE) of cyclopropane and various small hydrocarbons according to the reaction[a]:

$$R{-}H + CH_3\bullet \longrightarrow R\bullet + CH_4$$

Radical–H	DH[b]	ΔDH[b]	ΔDH[c]	ΔDH[d]	ΔDE[e]
Methyl–H	105.1 ± 0.2	0	0	0	0
Ethyl–H	98.2 ± 1	−6.9	−4.6	−3.6	−3.6
n-Propyl–H	97.9 ± 1	−7.2	−4.0	−3.6	−3.5
i-Propyl–H	95.1 ± 1	−10.0	−5.8	−6.2	−6.8
Cyclopropyl–H	106.3 ± 0.3	1.2		5.2	3.2
Cyclobutyl–H	96.5 ± 1	−8.6		−3.1	
Cyclopentyl–H	94.5 ± 1	−10.6			
Cyclohexyl–H	95.5 ± 1	−9.6			
Vinyl–H	110 ± 2	4.9		5.2	10.4
Phenyl–H	110.9 ± 2	5.8		6.3	

[a] All energies in kcal mol^{-1}.
[b] Reference 89a.
[c] Reference 89c.
[d] Reference 165.
[e] Reference 76a.

In Table 27, CH dissociation enthalpies and differences of DH or DE values are compared[89, 165, 76a]. They clearly confirm that CH dissociation for **1** requires a significantly larger energy (up to 11 kcal mol^{-1}) than for other cycloalkanes or propane. It is just 4 kcal mol^{-1} smaller than the CH dissociation enthalpy for ethene or benzene, as can be seen from the

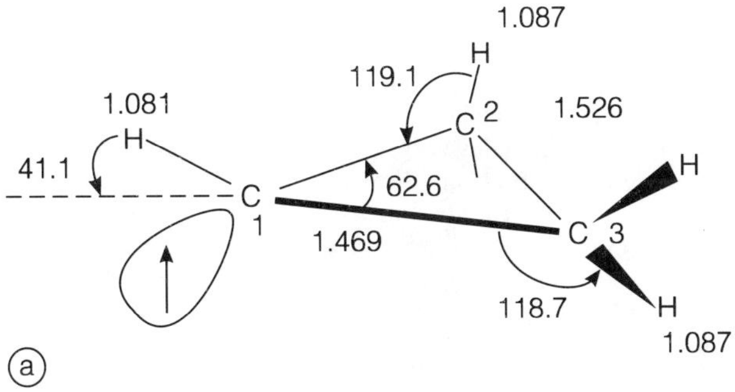

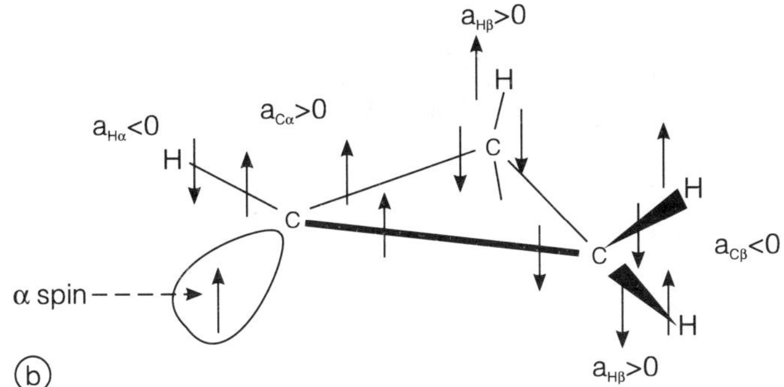

(**40**)

FIGURE 26 (a) MP2/6-31G(d) geometry of the cyclopropyl radical **40** according to Reference 246. (b) Spin-coupling scheme of the cyclopropyl radical according to the 'intra-atomic Hund rule' and spin coupling within bonds. Hyperfine splitting a_N at nucleus N will be > 0 (< 0) if valence electrons at N possess α (β) spin

data in Table 27. The relatively large value of the CH dissociation energy of **1** may partially reflect the strengthening of the CH bond, but as to the major part, it simply reflects the increase in strain energy when going to the cyclopropyl radical (**40**). In general, an alkyl radical prefers a planar trigonal geometry with bonding angles of 120° at the radical centre. Reduction of these angles to 60° in the cyclopropyl radical costs much more energy than reduction from a tetrahedral angle (109.5°) to a CCC angle to 60° in **1**.

This is also the reason why the cyclopropyl radical prefers a non-planar rather than a planar geometry at the radical centre. A pyramidalization angle of 41° has been calculated for **40** (Figure 26)[246–250], which indicates a 17° decrease compared to the corresponding angle for **1** [57.1° = ½ α_e(HCH); see the MP2/6-31G(d) value in Table 11]. Hence, the s-character of the CC hybrid orbitals increases at the radical centre, thus causing a widening of the C2C1C3 angle (62.6°), shortening of the vicinal bonds (1.469 Å), lengthening of the distal bond C2C3 (1.526 Å) and increased ring strain. One can also view **40** as a derivative of **1**, which has lost an electron from the highest CH bonding MO. This is the $\pi(CH_2)$ MO #9, which has C1C2 antibonding and C2C3 bonding character (see Figure 6, Section III.A). Accordingly, removal of an electron (as caused by an extremely strong σ-electron acceptor; see Section VI. B, Table 14) leads to shortening of the vicinal and lengthening of the distal CC bond.

Barone and coworkers[250] have investigated the inversion potential of **40** using UHF in connection with MP2, MP3, MP4, QCISD(T) correlation methods and a Huzinaga–Dunning DZ + P basis. At the highest level of theory [QCISD (T)/DZ + P], they calculated an inversion barrier of 3.66 kcal mol^{-1}, which is somewhat larger than the inversion barrier measured for the methylcyclopropyl radical (3.11 kcal mol^{-1} [251]). Barone and coworkers also calculated the vibrational levels of the inverting cyclopropyl radical by treating the C1—H movement with a one-dimensional Schrödinger equation. In this way, they determined splitting of vibrational levels below the barrier top (H tunneling) (v = 0: 1.1 cm^{-1}; v = 1: 35.9 cm^{-1}), rate constant for inversion at 344 K (4 × 1011 s^{-1}; estimate derived from trapping experiments: 1012 s^{-1} [252]) and the true inversion barrier that measures from the ground vibrational level (v = 0) to the first vibrational level above the barrier top (v = 2). The latter value was calculated to be 3.06 kcal mol^{-1}, in much better agreement with the experimental value of 3.11 kcal mol^{-1} for the methylcyclopropyl radical[251].

Barone and coworkers[250] also determined EPR hyperfine splittings a_N of the radical **40** at the UMP2/DZ + P level of theory using the Fermi contact operator and a finite field method with an increment size of 0.001 a.u. Expectation values of a_N, $<a_N>$, at higher temperatures T were calculated by assuming a Boltzmann population of vibrational levels according to equation 23:

$$<a>_T = \sum_{j=0} <a>_j \exp[(\varepsilon_0 - \varepsilon_j)/k_B T] / \sum_{j=0} \exp[(\varepsilon_0 - \varepsilon_j)/k_B T] \qquad (23)$$

with k_B being the Boltzmann constant and ε_j corresponding to the energies of the vibrational levels. In this way, the data in Table 28 were obtained, and clearly show that vibrational averaging generally improves the agreement between computed experimental hyperfine splittings a_N[250].

TABLE 28. Isotropic hyperfine splittings (Gauss) of the cyclopropyl radical computed at the UMP2/DZ + P level[250]

Atom	a_N (min)	$\langle a_N \rangle_{0+}$	$\langle a_N \rangle_{0-}$	$\langle a_N \rangle_{1+}$	$\langle a_N \rangle_{1-}$	$\langle a_N \rangle_2$	$\langle a_N \rangle_{T=302}$	a_N (exp)^b
H^α	−5.7	−7.6	−7.5	−13.9	−11.1	−18.5	−7.8	−6.7
H^β	18.2^a	18.8	18.8	21.0	19.9	22.6	18.9	23.5
C^α	116.9	111.1	111.5	91.5	101.0	77.5	77.5	95.9
C^β	−7.3	−7.6	−7.6	−8.8	−8.3	−9.8	−9.8	—

^{a} Mean between *syn* and *anti* values.
^{b} Experimental values obtained at T = 77K (H^α and H^β) and T = 203 K (C^α and C^β). From Reference 252.

Considering the spin coupling scheme shown in Figure 26, the signs of the hyperfine splittings result from (a) spin alignment according to an 'intra-atomic Hund rule'[253] and (b) spin coupling within a bond. The low negative value of $a_{H\alpha}$ (−6.7 G[252]; for methyl radical a_H = −23 G[253]) and the relatively large positive value of $a_{C\alpha}$ (95.9 G[252]; for methyl radical a_C = 38 G[252]) confirm the pyramidal geometry at C^α and provide a rough estimate of the inversion barrier[252].

Experimental a_N values have been obtained at T = 77 K[252b, c] (H^α and H^β) and T = 203 K[252a] (C^α and C^β). They do not show any significant T dependence up to 220 K, which is in line with expectation value calculations between 0 and 200 K (Table 28). For temperatures considerably larger than 200 K, higher vibrational levels are occupied that lead to considerable H tunneling (v = 1+, 1−; Table 28) or a large amplitude vibration involving pyramidal and inverted radical form (v = 2), thus decreasing the hyperfine coupling constant $a_{H\alpha}$ and $a_{C\alpha}$ to values typical of a planar alkyl radical.

Cometta-Morini, Ha and Oth[246] have investigated the vibrational spectra of cyclopropyl and allyl radicals using DZ + P basis sets at the UHF and UMP2 level. Calculated harmonic frequencies were scaled according to different procedures involving measured and calculated vibrational frequencies of propene and **1**. The authors improved previous mode assignments[247] and related all calculated harmonic frequencies to frequencies of the experimental infrared spectrum that had been recorded by Holtzhauer and coworkers at 18 K in the argon matrix after photochemically induced ring closure of the allyl radical to radical **40**[254].

The heats of formation for allyl and cyclopropyl radical are 66.5 ± 2.7 and 43.7 ± 2.2 kcal mol^{-1}, which suggests an isomerization enthalpy of −22.8 ± 4.9 kcal mol^{-1} for radical **40**[255]. The activation energy of the gas-phase thermolysis of **40** leading to the allyl radical was measured to be 22 ± 2 kcal mol^{-1} [256, 257]. The reaction can proceed with the CH_2 groups moving in a disrotatory or conrotatory mode. Contrary to the ring opening of cyclopropyl cation or anion, neither Woodward–Hoffmann rules, orbital and state correlation diagrams nor PMO arguments make any valid prediction with regard to the preferred ring-opening mode. Therefore, Olivella, Sole and Bofill[249] have studied thermal ring opening of radical **40** into the allyl radical at the UHF and the CASSCF level of theory employing a 3-21G basis. The authors demonstrate that the reaction proceeds via a highly non-symmetric transition state with unequal vicinal bond lengths (1.424 and 1.485 Å at CASSCF/3-21G[249]) and one CH_2 group having rotated by about 24° while the other is still orthogonal to the ring plane. The CC bond, being broken, has a length of 2.066 Å in the transition state[249]. The rotation of the second CH_2 group takes place in the last phase of the ring-opening process after the distal bond is fully broken and the C1C2 π-bond is formed.

The non-synchronous rotation of the two methylene groups implies a common transition state for both disrotatory and conrotatory ring opening, which may split into different transition states under the impact of a substituent. CASSCF/3-21G and CASSCF/6-31G(d) calculations lead to reaction energies which are far too negative (< −30 kcal mol^{-1}) while UMP2/6-31G(d) predicts for this process a reaction energy of −20.2 kcal mol^{-1} [246] in good agreement with experiment[255]. This reflects the importance of dynamic correlation corrections and the fact that the UHF wave function provides a reasonable description of radicals. Furthermore, Olivella and coworkers[249] show that quartet contamination of the UHF doublet wave function leads to some useful electron correlation that helps to obtain a reasonable transition state description at the UHF level. However, an accurate prediction of the transition state energy is only obtained at the CASSCF/6-31G(d)//CASSCF/3-21G level (3 electrons in 3 active orbitals leading to 8 doublet spin-adapted configuration state functions) after zero point energy corrections (21.9 kcal mol^{-1})[249].

XIII. FUSED CYCLOPROPANES

A. Bicyclo[1.1.0]butane

This is the most strained of all bicyclic alkanes[79, 258, 259]. Calculated geometries of bicyclobutane (Table 12, Section VI) indicate that the bridgehead (C1C3) bond length (1.47 Å) is significantly shorter than the CC bonds in **1** while other geometrical parameters are similar to the corresponding parameters in **1**. Dependence of calculated bond lengths on the applied theoretical method and basis set agree with trends described in Section VI.A for **1**. HF/4-31G leads to the longest CC bond lengths, while inclusion of polarization functions at the HF level results in considerable shortening of the CC bonds. MP2/6-31G(d) predicts somewhat longer bond lengths than those obtained at the HF/6-31(d) level. Apart from the C1C3 bond length, the experimental value of which should be questioned, the MP2/6-31G(d) geometry is in good agreement with experimental data.

The C1C3 bridgehead bond is the most strained part of the molecule (CSE = 68.6 kcal mol^{-1} [258]). According to Newton and Schulman[260], it is formed from hybrid orbitals of nearly pure p character inclined at an angle of *ca* 30° with respect to the bond vector; a π character of 26% has been calculated for the central bond from localized bond hybrids. This early work is largely confirmed by electron density studies of Wiberg and coworkers[79], who found that C1 (C3) carries a strong negative charge. This suggests strongly increased s-character and a large electronegativity of the bridgehead carbon[79]. Calculation of the atomic energies with the help of the virial partitioning method predicts the methylene groups rather than the CH groups to the more destabilized, which can be easily traced to a shift of electronic charge from the former to the latter [$q(CH_2) = 0.068$, $q(CH) = -0.067$[79]]. This description, however, is not very helpful since it disguises the fact that the strain of the central bond dominates the chemical behaviour of bicyclobutane.

Jackson and Allen[32] calculated the difference electron density (see Section IV.A) of bicyclo[1.1.0]butane. They found that the maximum of positive difference density is displaced by about 0.35 Å from the C1C3 internuclear axis, which is significantly larger than the corresponding displacement calculated for cyclopropane (0.2 Å) and suggests increased bending for the central CC bond. These authors also noted that there is accumulation of charge in the non-bonded regions of the C1C3 bond, which supports the prediction that the C1C3 bond is formed by overlap of nearly pure p orbitals.

B. [1.1.1]Propellane

[1.1.1]propellane possesses two bridgehead C atoms (C1, C3), for which all four bonds are on one side of a plane perpendicular to the C1C3 axis and containing the bridgehead C atom. Such C atoms have been termed inverted C atoms, which should lead to large molecular strain, high reactivity, peculiar bonding features and exceptional geometries. Nevertheless, it was possible to synthesize [1.1.1]propellane and to determine many of its properties[14]. The strain energy of [1.1.1]propellane (CSE = 98 kcal mol^{-1} [14]) is indeed much larger than the sum of the strain energies (81 kcal mol^{-1}) of the three cyclopropane rings that constitute the molecule. Contrary to bicyclo[1.1.0]butane, the bridgehead distance C1C3 [1.592 Å at MP2/6-31G(d), Table 12] is longer in [1.1.1]propellane than the CC bond in **1** while the other CC bonds (C1C2) are similar to the CC bond in **1**. Comparing published *ab initio* geometries of [1.1.1]propellane listed in Table 12, it becomes obvious that the MP2/6-31G(d) geometry agrees best with the experimental geometry.

The most intriguing aspect of [1.1.1]propellane is the nature of the interactions between its bridgehead atoms C1 and C3[32, 258, 261]. Newton and Schulman[261] were the first to point out that the localized MO associated with the central CC bond (C1C3) is non-bonding or even antibonding in the centre of the molecule. Jackson and Allen[32a] compared the highest occupied σ(CC) MO in eclipsed ethane, **1**, bicyclo[1.1.0]butane and [1.1.1]propellane and demonstrated that in each of these cases the CC MO has bonding and non-bonding components, but that the latter increase and become larger than the bonding components when going to [1.1.1]propellane. While a normal bonding σ(CC) MO increases in energy upon CC stretching, the σ(CC) MO of [1.1.1]propellane decreases in energy. The calculated difference electron density of its central CC bond is negative (loss of electron density) while there is a relatively large positive electron density in the non-bonding region[32]. This is in line with other calculations[258] and an X-ray study of Chakrabarti, Dunitz and coworkers[262] that also suggested charge loss in the region between formally bonded inverted C atoms. Inspection of the calculated difference electron density reminds one of two (almost parallel) p-type charge distributions in the non-bonding region of C1C3, which are connected by three filaments of positive difference electron density resulting from contributions of the three bridging CH_2 groups.

Although there seems to be no direct σ-type (along the C1C3 axis) and π-type bonding between the bridgehead C atoms, σ-components of the CH_2 groups lead to a (three fold) σ-bridged π-bond[32] as was already discussed in Section VI.C in connection with the bonding in three-membered rings. The σ-bridged π-bond is responsible for a close-to-normal CC bond length and the relative low reactivity of [1.1.1]propellane with an estimated stability of 65 kcal mol^{-1} with regard to its singlet biradical[258].

XIV. FORMATION AND REACTIVITY

An impressive amount of theoretical work has been carried out in recent years with regard to a computational description of reactions of **1**. We will consider here just some of the more important *ab initio* investigations and leave a detailed discussion of this work to the more specialized chapters in this volume.

A. Formation Reactions

According to Woodward–Hoffmann rules, the addition of singlet methylene ($^1A_1$) to alkene is symmetry-allowed in a C_s but not a C_{2v} symmetrical mode (see Scheme 6)[263], The stereochemistry of the reacting alkene will be retained, thus leading to a stereospecific addition reaction. This has been confirmed by Zurawski and Kutzelnigg[264] on the basis of CEPA/DZ + P calculations. At large distances (> 1.8 Å) the angle of approach is close to 90°. Electrons are transferred from ethene to methylene, which can be considered as an electrophile. At smaller distances, methylene becomes a nucleophile, which transfers electrons to the π^* MO of ethene. Zurawski and Kutzelnigg found no barrier for the reaction, i.e. ethene and singlet methylene form **1** in a strongly exothermic reaction [ΔE(CEPA/DZ + P) = –109.2 kcal mol^{-1} compared to an experimental value of $\Delta_R H(298)$ = –105 kcal mol^{-1} [264]]. An investigation on the addition of substituted carbenes to ethene was carried out by Rondan, Houk and Moss[265], but the level of theory applied in this work was moderate (HF/STO-3G, HF/4-31G) and therefore results were of just qualitative value.

Moreno and coworkers[266] published a study on the triplet carbene–ethene addition reaction. This process should involve two steps, namely the formation of a triplet trimethylene 1,3-diradical as an intermediate followed by intersystem crossing and formation of **1**. The intermediate may live long enough to permit rotation at the CC bonds. In this way, the stereochemistry of the alkene will be lost and a non-stereospecific addition takes place. Moreno and coworkers[266] calculated for the first step of the reaction $CH_2(^3B_2) + H_2C{=}CH_2$ a barrier of 11 kcal mol^{-1} and a reaction energy of –26 kcal mol^{-1} at the MP2/3-21G level. (It has to be mentioned in this connection that the 3-21G basis is far too small to lead to reliable energies in correlation calculations and therefore results are just of a qualitative nature[267].) Activation energies varied from 5 to 17 kcal mol^{-1} if CH_2 was replaced by the triplet state of CH(CN), CH(BeH) and CHLi[266].

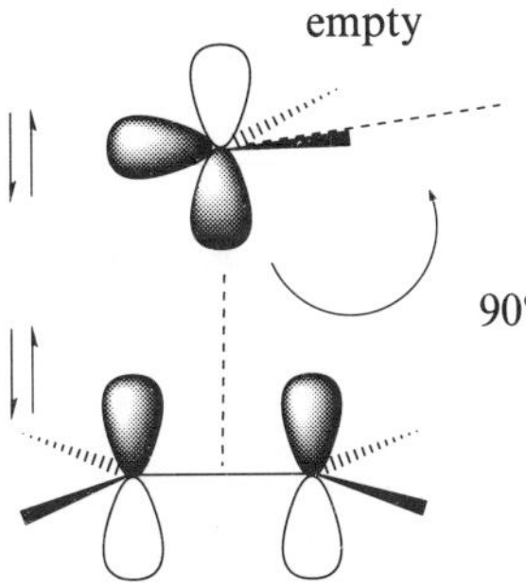

SCHEME 6

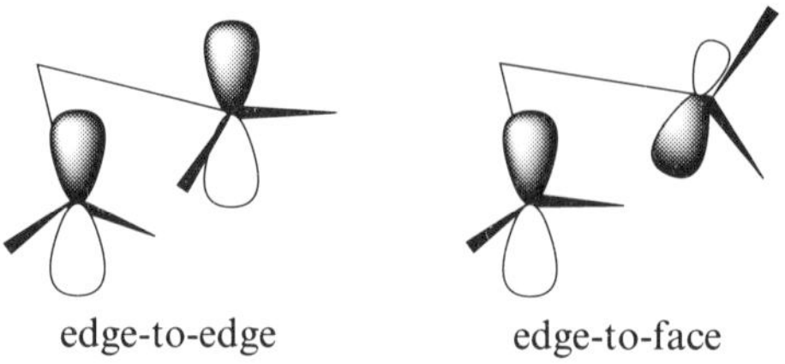

SCHEME 7

B. Thermal Ring-opening and Stereomutation Reactions

Both ring opening and isomerization to propene or stereomutation of **1** require activation energies of about 65 and 61 kcal mol^{-1}, respectively[268]. They lead to an intermediate trimethylene singlet diradical that has attracted a lot of attention by quantum chemists because it represents a challenging test ground for *ab initio* methods involving multiple-configuration electronic states. Use of these methods has clarified that the trimethylene singlet diradical is not a stable intermediate, but that it can occur in various forms located in shallow minima along the reaction path of the stereomutation of **1**[129, 132, 269–275].

The thermal stereomutation of **1** may involve one or more of the following three mechanistic possibilities: (a) rotation of a single methylene group (Smith mechanism[276]), (b) cleavage of a CC bond to give a trimethylene diradical intermediate in which random loss of stereochemistry is competitive with ring closure (Benson mechanism[277]), (c) coupled, simultaneous rotation of two methylene groups (Hoffmann mechanism[263]). Independent investigations by Yamaguchi, Schaefer and Baldwin[132] (TCSCF/TZ + 2P and CISD+Q/TZ + 2P calculations) as well as Getty, Davidson and Borden[275] [GVB/6-31G(d) and CISD/6-31G(d) calculations] revealed that the potential energy surface for methylene group rotation is actually much more complicated than previously believed. Both the edge-to-edge and edge-to-face conformers (Scheme 7) of the trimethylene radical are second-order transition states (hilltops in a two-dimensional space spanned by the methylene torsion angles), which are surrounded by eight (in the case of the edge-to-edge form) and two (in the case of the edge-to-face form) stationary points of lower order (minima and first-order transition states) as is shown in Figure 27[275]. However, differences between the energies of the stationary points are small so that one can expect that better methods and basis sets may lead to somewhat different pictures.

The barrier for stereomutation is calculated to be close to 61 kcal mol^{-1} (CISD+Q/TZ + 2P)[132, 275]. Conrotatory double rotation of the methylene groups is about 1 kcal mol^{-1} more favourable than single rotation of a methylene group. Calculations further reveal that disrotatory rotation is also possible with a barrier 0.5 kcal mol^{-1} larger than conrotatory rotation (Figure 27).

C. Electrophilic Ring-opening Reactions and Insertion Reactions

Electrophilic ring opening of **1** has been studied extensively experimentally and theoretically[18]. The simplest electrophilic ring-opening reaction is protonation of **1**, which has already been discussed in Section XI.B. The first step in this process leads to corner protonated **1**, which decomposes via a transition state with 1-propyl cation structure and subsequent H migration to form the 2-propyl cation (calculated barrier: 12.1 kcal mol^{-1} [223]). Edge protonated **1** is not necessarily involved in this process since its relative energy (with regard to corner protonated **1**) is 1.4 kcal mol^{-1} and since it may correspond to a transition state (of H scrambling) rather than an energy minimum[223].

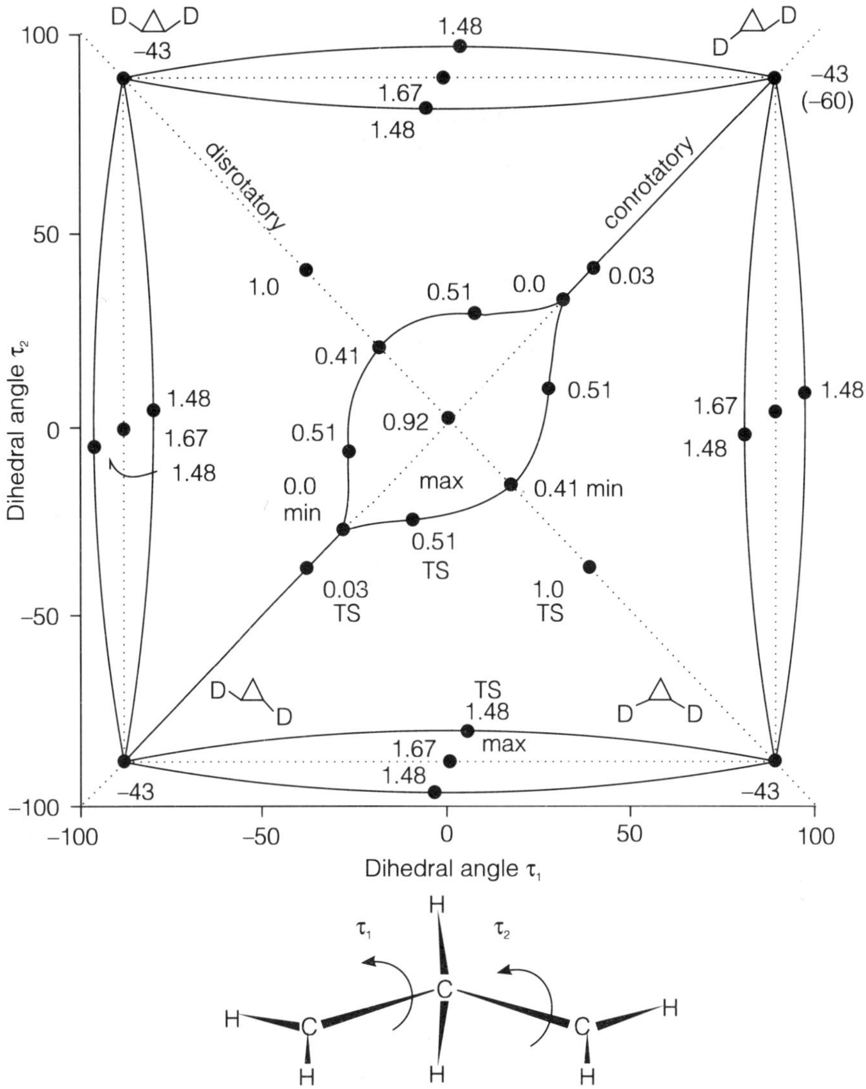

FIGURE 27. Schematic description of the potential energy surface (PES) of cyclopropane-trimethylene in terms of the two dihedral angles τ_1 and τ_2 according to GVB/6-31G(d) calculations by S. J. Getty, E. R. Davidson and W. T. Borden (Reference 275). Deuterated cyclopropane forms are shown to indicate the stereochemistry of the possible stereomutation routes (conrotatory and disrotatory) connecting the four cyclopropane forms in the τ_1, τ_2 space shown. For each calculated stationary point (•) on the τ_1, τ_2 PES, relative energy (in kcal mol^{-1}) and character of the stationary point (min: minimum; max: maximum; TS: transition state) are given. Reference energy is the energy of the most stable trimethylene form with $\tau_1 = \tau_2 = 30.6°$. Note that at the GVB level the stability of cyclopropane is underestimated by about 18 kcal mol^{-1}. The most likely reaction paths in τ_1, τ_2 space for cyclopropane stereomutation are given by solid lines. Note that both the edge-to-edge ($\tau_1 = \tau_2 = 0°$) and the edge-to-face [$\tau_1 = 0°$ ($\pm 90°$), $\tau_2 = \pm 90°$ ($0°$)] conformers correspond to small local maxima on the τ_1, τ_2 PES

Electrophilic attack by halonium ions X^+ may occur in a completely different manner according to HF and MP2 investigations of Yamabe and coworkers[278], who found edge rather than corner attack the preferred reaction mode in the case of X^+. For $X^+ = Cl^+$ or Br^+, a four-membered ring is formed which is more stable by about 20 kcal mol^{-1} than corner complexed $C_3H_6X^+$. The authors explained this preference with possible charge transfer interactions between the p orbitals of X^+ and the $3e'$ HOMOs (Figure 6, MOs #11, 12, w_A and w_S) either in a σ-bridge or π-bridge fashion (establishing a two-fold σ-bridged and π-bridged orbital, respectively; see Sections VI.C and XII.B).

Upon attack by a nucleophile, the four-membered ring collapses in a zig-zag manner into a 3-halopropyl derivative. Alternatively, the four-membered ring can open by rotation and then react with an approaching nucleophile. Calculated energy barriers should be comparable according to modelling the reaction in the absence of nucleophile. Since the authors also investigated protonation of **1**, an assessment of the accuracy of their calculations can be made by comparison with the MP4/TZ + P calculations by Koch, Liu and Schleyer[223]. The latter authors obtained for edge protonated **1**, a relative energy of 1.4 kcal mol^{-1} while Yamabe and coworkers get 12.3 kcal mol^{-1} [278].

Ring opening by metals or metal complexes seems to proceed by a similar mechanism. Alex and Clark[279] studied ring opening and isomerization of **1** catalysed by complexation with the Be radical cation. They used the 6-31G(d) basis and various methods ranging from projected UHF to projected UMP4 and to QCISD(T). According to their computational results, the reaction leads in the first step to an edge oriented $Be^{\bullet+}$–cyclopropane complex ($\Delta E = -57$ kcal mol^{-1}), which rearranges in an endothermic step ($\Delta E = 12.3$ kcal mol^{-1}, $\Delta E = 29$ kcal mol^{-1}) to a metallacyclobutane radical cation intermediate (in line with results obtained for X^+ attack[278]). From this intermediate, isomerization to a propene–$Be^{\bullet+}$ complex or CC bond cleavage forming a carbene–ethene–$Be^{\bullet+}$ can occur. Parallels to transition metal catalysis of the ring opening of **1** were observed[279].

Ring opening by palladium(II) compounds ($PdCl_2$, $PdCl_4^{2-}$, $PdCl^+$) were directly investigated by Blomberg, Siegbahn and Bäckvall[280] using CASSCF and contracted CI calculations. These authors found that the Pd compound prefers a corner attack, especially if it has the possibility of ionizing into a cationic complex. Hence, for $PdCl^+$ an activation energy of just 5 kcal mol^{-1} was estimated while edge activation should require about 25 kcal mol^{-1}. In the case of Pd(0), edge activation (activation energy 17 kcal mol^{-1}) leading to palladacyclobutane was calculated to be more favourable than corner attack (activation energy 30 kcal mol^{-1})[281]. For $PdCl_2$ and $PdCl_4^{2-}$, very high barriers (25–45 kcal mol^{-1}) were found for both corner and edge activation[280]. It was argued that the energy differences between the different modes of attack are directly related to excitation energies of the atomic states of Pd involved.

Gano, Gordon and Boatz[282] investigated singlet methylene and silylene insertion into **1** using the 3-21G* basis at the HF, MP2, MP3, and MP4 level of theory. At the highest level of theory, activation barriers of 2.3 and 20 kcalmol^{-1} were obtained for $CH_2(^1A_1)$ and $SiH_2(^1A_1)$ insertion. These values were about 40 kcalmol^{-1} lower than for the corresponding ethane insertion reactions. Clearly, it is much easier to insert methylene or silylene in a strained rather than an unstrained bond. A similar observation was made by Siegbahn and Blomberg[283], who studied a large number of transition metal insertion reactions including as reagents ethane, cyclopropane and cyclobutane. The authors used a variety of methods [CASSCF, coupled pair functional methods, CCSD(T)] and a (17s13p9d3f/9s5p1d/5s1p)[7s6p4d1f/3s2p1d/3s1p] basis set. Palladium was found to insert into **1** without any barrier while calculated barriers for ethane and cyclobutane were 23 and 7 kcal mol^{-1} [283].

The work of recent years clearly indicates that many puzzles of the chemistry connected with **1** have already been solved with the aid of modern *ab initio* calculations. Many more mechanistic questions will be solved in the near future provided a quantum chemist spends enough time and computer sources to carry out the necessary *ab initio* calculations. In particular, the routine calculation of equilibrium geometries combined with the determination of vibrational frequencies at relatively high levels of theory helps to avoid previous confusion that was caused by the discussion of first- or even second-order transition states as stable reaction intermediates. Important steps have already been made in the area of reactions with transition metal complexes and further *ab initio* work in this area will open new synthetic routes for experimentalists.

XV. ACKNOWLEDGEMENTS

This work was supported by the Swedish Natural Science Research Council (NFR). K.J. Szabo thanks the NFR for a postdoctoral stipend. All calculations needed to complement data from the literature were done on the CRAY YMP/416 of the Nationellt Superdatorcentrum (NSC), Linköping, Sweden. The authors thank Z. Konkoli, L. Olsson and A. Larsson for technical assistance with the calculations and the NSC for a generous allotment of computer time.

XVI. REFERENCES

1. M. Charton, in *The Chemistry of Alkenes: Olefinic Properties of Cyclopropanes* (Ed. J. Zabicky), Vol.2, Wiley, New York, 1970.
2. D. Wendisch, in *Methoden der Organischen Chemie*, Vol. IV, 3, E. Houben-Weyl-Müller, Thieme Verlag, Stuttgart, 1971, p. 17.
3. A. De Meijere, *Angew. Chem.*, **91**, 867 (1979); *Angew. Chem., Int. Ed. Engl.*, **18**, 809 (1979).
4. R. Gleiter, *Top. Curr. Chem.*, **86**, 197 (1979).
5. (a) J. F. Liebman and A. Greenberg, *Chem. Rev.*, **76**, 311 (1976).
   (b) J. L. Franklin, *Ind. Eng. Chem.*, **41**, 1070 (1949).
   (c) D. Van Vechten and J. F. Liebman, *Isr. J. Chem.*, **21**, 105 (1981).
6. W. A. Lathan, L. Radom, P. C. Hariharan, W. J. Hehre and J. A. Pople, *Top. Curr. Chem.*, **40**, 1 (1973).
7. M. D. Newton, in *Modern Theoretical Chemistry* (Ed. H.F. Schaefer III), Vol. 4, Plenum Press, New York, 1977, p. 223.
8. (a) M. J. S. Dewar, *J. Am. Chem. Soc.*, **106**, 669 (1984).
   (b) M. J. S. Dewar, *Bull. Soc. Chim. Belg.*, **88**, 957 (1979).
   (c) M. J. S. Dewar and M. L. McKee, *Pure Appl. Chem.*, **52**, 1431 (1980).
9. D. Cremer and E. Kraka, *J. Am. Chem. Soc.*, **107**, 3800 (1985).
10. D. Cremer and J. Gauss, *J. Am. Chem. Soc.*, **108**, 7467 (1986).
11. D. Cremer and E. Kraka, in *Molecular Structure and Energetics* (Eds. J. F. Liebman and A. Greenberg), Vol. 7, VCH Publishers, Deerfield Beach, 1988, p. 65.
12. D. Cremer, *Tetrahedron*, **44**, 7427 (1988).
13. E. Kraka and D. Cremer, in *Theoretical Models of Chemical Bonding, Part 2: The Concept of the Chemical Bond* (Ed. Z. B. Maksic), Springer-Verlag, Heidelberg, 1990, p. 453.
14. K. B. Wiberg, in *The Chemistry of the Cyclopropyl Group* (Ed. Z. Rappoport), Wiley, New York, 1987, p. 1.
15. W. Runge, in *The Chemistry of the Cyclopropyl Group* (Ed. Z. Rappoport), Wiley, New York, 1987, p. 28.
16. D. G. Morris, in *The Chemistry of the Cyclopropyl Group* (Ed. Z. Rappoport), Wiley, New York, 1987, p. 101.
17. R. E. Ballard, in *The Chemistry of the Cyclopropyl Group* (Ed. Z. Rappoport), Wiley, New York, 1987, p.213.
18. M. A. Battiste and J. M. Coxon, in *The Chemistry of the Cyclopropyl Group* (Ed. Z. Rappoport), Wiley, New York, 1987, p. 255.
19. T. T. Tidwell, in *The Chemistry of the Cyclopropyl Group* (Ed. Z. Rappoport), Wiley, New York, 1987, p. 565.
20. For a balanced account of these concepts, see V. I. Minkin, M.N. Glukhovtsev and B. Ya. Simkin, *Aromaticity and Antiaromaticity, Electronic and Structural Aspects*, Wiley, New York, 1994.

21. (a) J. B. Collins, J. D. Dill, E. D. Jemmis, Y. Apeloig, P. v. R. Schleyer, R. Seeger and J. A. Pople, *J. Am. Chem. Soc.*, **98**, 5419 (1976).
(b) For a summary see: W. J. Hehre, L. Radom, P. v. R. Schleyer and J. A. Pople, *Ab initio Molecular Orbital Theory*, Wiley, New York, 1986.
22. K. Krogh-Jespersen, D. Cremer, D. Poppinger, J. A. Pople, P. v. R. Schleyer and J. Chandrasekhar, *J. Am. Chem. Soc.*, **101**, 4843 (1979).
23. C. A. Coulson and W. E. Moffitt, *Philos. Mag.*, **40**,1 (1949).
24. T. Förster, *Z. Phys. Chem. B*, **43**, 58 (1939).
25. A. D. Walsh, *Trans. Faraday Soc.*, **45**, 179 (1949).
26. (a) R. Hoffmann, *Tetrahedron Lett.*, 2907 (1970).
(b) R. Hoffmann and R. B. Davidson, *J. Am. Chem. Soc.*, **93**, 5699 (1971).
(c) R. Hoffmann, *Special Lectures of the XXIIIrd International Congress of Pure and Applied Chemistry*, Vol. 2, Butterworths, London, 1971, p. 233.
(d) R. Hoffmann, H. Fujimoto, J. R. Swenson and C.-C. Wan, *J. Am. Chem. Soc.*, **95**, 7644 (1973).
27. H. Günther, *Tetrahedron Lett.*, 5173 (1970).
28. R. Hoffmann and W.-D. Stohrer, *J. Am. Chem. Soc.*, **93**, 6941 (1971).
29. C. W. Jorgensen and L. Salem, *The Organic Chemist's Book of Orbitals*, Academic Press, New York, 1973.
30. R. F. Hout, Jr., W. J. Pietro and W. J. Hehre, in *A Pictorial Approach to Molecular Structure and Reactivity*, Wiley, New York, 1984.
31. (a) E. Honegger, E. Heilbronner, A. Schmelzer and W. Jian-Qi, *Isr. J. Chem.*, **22**, 3 (1982).
(b) E. Honegger, E. Heilbronner and A. Schmelzer, *Nouv. J. Chim.*, **6**, 519 (1982).
32. (a) J. E. Jackson and L. C. Allen, *J. Am. Chem. Soc.*, **106**, 591 (1984).
(b) D. B. Kitchen, J. E. Jackson and L. C. Allen, *J. Am. Chem. Soc.*, **112**, 3408 (1990).
(c) C. Liang and L. C. Allen, *J. Am. Chem. Soc.*, **113**, 1878 (1991).
33. T. Clark, G. W. Spitznagel, R. Klose and P. v. R Schleyer, *J. Am. Chem. Soc.*, **106**, 4412 (1984).
34. (a) M. J. S. Dewar, *Nature*, **156**, 748 (1945).
(b) M. J. S. Dewar, *J. Chem. Soc.*, **406**, 777 (1946).
35. A. D. Walsh, *Nature*, **159**, 167, 712 (1947).
36. M. J. S. Dewar, *Bull. Soc. Chim. Fr.*, C71 (1951).
37. M. J. S. Dewar and A. P. Marchand, *Annu. Rev. Phys. Chem.*, **16**, 321 (1965).
38. J. Chatt and L. A. Duncanson, *J. Am. Chem. Soc.*, **75**, 2939 (1953).
39. E. Kochanski and J. M. Lehn, *Theor. Chim. Acta*, **14**, 281 (1969).
40. R. Hoffmann, H. Fujimoto, J. R. Swenson and C.-C. Wan, *J. Am. Chem. Soc.*, **95**, 7644 (1973).
41. M.-M. Rohmer and B. Roos, *J. Am. Chem. Soc.*, **97**, 2025 (1975).
42. G. L. Delker, Y. Wang, G. D. Stucky, L. R. Lambert, Jr., C. K. Haas and D. Seyferth, *J. Am. Chem. Soc.*, **98**, 1779 (1976).
43. M. J .S. Dewar and G. P. Ford, *J. Am. Chem. Soc.*, **101**, 183 (1979).
44. H. B. Yokelson, A. J. Millevolte, G. R. Gillette and R. West, *J. Am. Chem. Soc.*, **109**, 6865 (1987).
45. M. Randic and Z. B. Maksic, *Theor. Chim. Acta*, **3**, 59 (1965).
46. (a) J. Wardeiner, W. Luettke, R. Bergholz and R. Machinek, *Angew. Chem., Int. Ed. Engl.*, **94**, 872 (1982).
(b) For a discussion of hybrid orbitals see: W. A. Bingel and W. Luettke, *Angew. Chem., Int. Ed. Engl.*, **20**, 899 (1981).
47. S. Inagaki, Y. Ishitani and T. Kakefu, *J. Am. Chem. Soc.*, **116**, 5954 (1994).
48. (a) J. M. Foster and S. F. Boys, *Rev. Mod. Phys.*, **32**, 300 (1960).
(b) S. F. Boys, in *Localized Orbitals and Localized Adjustment Functions in Quantum Theory of Atoms, Molecules and the Solid State* (Ed. P. O. Löwdin), Academic Press, New York, 1966, p. 253.
49. N. Mueller and D.E. Pritchard, *Chem. Phys.*, **31**, 1471 (1959).
50. J. G. Hamilton and W. E. Palke, *J. Am. Chem. Soc.*, **116**, 4159 (1994).
51. P. B. Karadakov, J. Gerratt, D. L. Cooper and M. Raimondi, *J. Am. Chem. Soc.*, **116**, 7714 (1994).
52. E. Goldstein, S. Vijaya and G. A. Segal, *J. Am. Chem. Soc.*, **102**, 6198 (1980).
53. H. Basch, M. B. Robin, N. A. Kuebler, C. Baker and D. W. Turner, *J. Chem. Phys.*, **51**, 52 (1969).
54. (a) A. Skancke, *J. Mol. Struct.*, **30**, 95 (1976).
(b) For ionization potential calculations based on the Koopmans theorem, see also J. Kao and L. Radom, *J. Am. Chem. Soc.*, **100**, 379 (1978) and
(c) H. L. Hase, C. Mühler and A. Schweig, *Tetrahedron*, **34**, 2983 (1978).

55. W. von Niessen, L. S. Cederbaum and W. P. Kraemer, *Theor. Chim. Acta*, **44**, 85 (1977).
56. C. Fridh, *J. Chem. Soc., Faraday Trans. 2*, **75**, 993 (1979).
57. J. R. Collins and G. A. Gallup, *J. Am. Chem. Soc.*, **104**, 1530 (1982).
58. G. Bieri, L. Åsbrink and W. von Niessen, *J. Electron Spectrosc.*, **27**, 129 (1982).
59. G. de Alti, P. Decleva and A. Lisini, *J. Mol. Struct. (Theochem)*, **108**, 129 (1984).
60. K. Kimura, S. Katsumata, Y. Achiba, T. Yamazaki and S. Iwata, in *Handbook of He (I) Photoelectron Spectra of Fundamental Organic Molecules*, Halsted Press, New York, 1981.
61. T. Koopmans, *Physica*, **1**, 280 (1931).
62. K. B. Wiberg, C. M. Hadad, S. Sieber and P. v. R. Schleyer, *J. Am. Chem. Soc.*, **114**, 5820 (1992).
63. (a) P. Becker, in *Electron and Magnetization Densities in Molecules and Crystals*, Vol. 48, NATO Advanced Study Institutes, Series B: Phys., Plenum Press, New York, 1980.
(b) P. Coppens and M. B. Hall, in *Electron Distributions and the Chemical Bond*, Plenum Press, New York, 1981.
64. P. Hohenberg and W. Kohn, *Phys. Rev.*, **136 B**, 864 (1964).
65. P. Coppens and E. D. Stevens, *Adv. Quantum Chem.*, **10**, 1 (1977).
66. (a) A. Hartman and F. L. Hirshfeld, *Acta Crystallogr.*, **20**, 80 (1966).
(b) C. J. Fritchie, Jr., *Acta Crystallogr.*, **20**, 27 (1966).
(c) T. Ito and T. Sakurai, *Acta Crystallogr., Sect B*, **B29**, 1594 (1973).
(d) D. A. Matthews and G. D. Stucky, *J. Am. Chem. Soc.*, **93**, 5954 (1971).
67. (a) R. F. W. Bader and T. T. Nguyen-Dang, *Adv. Quantum Chem.*, **63** (1981).
(b) R. F. W. Bader, T. T. Nguyen-Dang and Y. Tal, *Rep. Prog. Phys.*, **44**, 893 (1981).
(c) R. F. W. Bader, in *Atoms in Molecules—A Quantum Theory*, Oxford University Press, Oxford, 1990.
(d) R. F. W Bader, P. L. A. Popelier and T. A. Keith, *Angew. Chem.*, **106**, 647 (1994).
68. D. Cremer and E. Kraka, *Croat. Chem. Acta*, **57**, 1265 (1984).
69. D. Cremer and E. Kraka, *Angew. Chem., Int. Ed. Engl.*, **23**, 627 (1984).
70. (a) K. Ruedenberg, in *Localization and Delocalization in Quantum Chemistry* (Eds. O. Chelvet, R. Daudel, S. Diner and J. P. Malrieu), Vol. 1, Reidel, Dordrecht, 1975, p. 223.
(b) K. Ruedenberg, *Rev. Mod. Phys.*, **34**, 326 (1962).
(c) C. Edmiston and K. Ruedenberg, *J. Phys. Chem.*, **68**, 1628 (1963).
(d) E. M. Layton Jr. and K. Ruedenberg, *J. Phys. Chem.*, **68**, 1654 (1963).
(e) R. R. Rue and K. Ruedenberg, *J. Phys. Chem.*, **68**, 1676 (1963).
71. D. Cremer, J. Gauss and E. Kraka, *J. Mol. Struct. (Theochem)*, **169**, 531 (1988).
72. (a) R. F. W. Bader and H. Essén, *J. Chem. Phys.*, **80**, 1943 (1984).
(b) R. F. W. Bader, P. L. MacDougall and C. D. H. Lau, *J. Am. Chem. Soc.*, **106**, 1594 (1984).
73. See, e.g., P.M. Morse and H. Feshbach, in *Methods of Theoretical Physics*, Vol. 1, McGraw-Hill, New York, 1953, p. 6.
74. D. K. Pan, J.-N. Gao, H.-L. Liu, M. -B. Huang and W. H. E. Schwarz, *Int. J. Quantum Chem.*, **29**, 1147 (1986).
75. R. Ahlrichs and C. Ehrhardt, *Chemie in unserer Zeit*, **19**, 120 (1985).
76. (a) P. v. R. Schleyer, in *Substituent Effects in Radical Chemistry*, Nato ASI Series (Eds. H.G. Viehe, R. Janoschek and R. Merenyi), Reidel, Dordrecht, 1986, p. 61.
(b) P. v. R. Schleyer, A. F. Sax, J. Kalcher and R. Janoschek, *Angew. Chem.*, **99**, 374 (1987).
77. K. B. Wiberg, *Angew. Chem., Int. Ed Engl.*, **25**, 312 (1986).
78. K. B. Wiberg, R. F. W. Bader and C. D. H. Lau, *J. Am. Chem. Soc.*, **109**, 985 (1987).
79. K. B. Wiberg, R. F. W. Bader and C. D. H. Lau, *J. Am. Chem. Soc.*, **109**, 1001 (1987).
80. S. Inagaki, N. Goto and K. Yoshikawa, *J. Am. Chem. Soc.*, **113**, 7144 (1991).
81. P. Politzer and J.-S. Murray, in *Structure and Reactivity* (Eds. J. F. Liebman and A. Greenberg), VCH Publishers, New York, 1988, p. 1.
82. T.S. Slee, in *Modern Models of Bonding and Delocalization* (Eds. J.F. Liebman and A. Greenberg), VCH Publishers, New York, 1988, p. 63.
83. B. Liu and D. Kang, *J. Chem. Inf. Comput. Sci.*, **34**, 418 (1994).
84. P. George, M. Trachtman, C.W. Bock and A.M. Brett, *Tetrahedron*, **32**, 317 (1976).
85. K. B. Wiberg, *J. Comput. Chem.*, **5**, 197 (1984).
86. M. R. Ibrahim and P. v. R. Schleyer, *J. Comput. Chem.*, **6**, 157 (1985).
87. C. Zhixing, *Theor. Chim. Acta*, **68**, 365 (1985).
88. (a) R. F. W. Bader, T. Tang, Y. Tal and F. W. Biegler-König, *J. Am. Chem. Soc.*, **104**, 946 (1982).
(b) R. F .W. Bader, *J. Chem. Phys.*, **73**, 2871 (1980).

(c) F. W. Biegler-König, R. F. W. Bader and T. Tal, *J. Comput. Chem.*, **3**, 317 (1982).
89. (a) D. F. McMillen and D. M. Golden, *Am. Rev. Phys. Chem.*, **33**, 493 (1982).
(b) M. H. Baghal-Vayjooee and S. Benson, *J Am. Chem. Soc.*, **101**, 2840 (1979).
(c) W. Tsang, *J. Am. Chem. Soc.*, **107**, 2872 (1985).
90. (a) A. von Baeyer, *Chem. Ber.*, **18**, 2269 (1885).
(b) For an evaluation of von Baeyer's work see: R. Huisgen, *Angew. Chem., Int. Ed. Engl.*, **25**, 297 (1986).
91. (a) R. G. Snyder and J. M. Schachtschneider, *Spectrochim. Acta*, **21**, 169 (1965).
(b) R. G. Snyder and G. Zerbi, *Spectrochim. Acta*, **23 A**, 39 (1967).
92. (a) U. Buckert and N. L. Allinger, in *Molecular Mechanics*, ACS Monograph 177, Washington D.C., 1980.
(b) S.-J. Chang, D. McNally, S. Shary-Tehrany, M. J. Hickey and R.H. Boyd, *J. Am. Chem. Soc.*, **92**, 3109 (1970).
(c) N. L. Allinger, M. T. Tribble, M. A. Millerand and D. H. Wertz, *J. Am. Chem. Soc.*, **93**, 1637 (1971).
(d) See also discussion in: F. H. Westheimer, in *Steric Effects in Organic Chemistry* (Ed. M. S. Newman), Wiley, New York, 1956, p. 523.
(e) E. M. Engler, J. D. Andose and P. v. R. Schleyer, *J. Am. Chem. Soc.*, **95**, 8005 (1973).
93. N. L. Bauld, J. Cesak and R. L. Holloway, *J. Am. Chem. Soc.*, **99**, 8140 (1977).
94. R. S. Grev and H. Schaefer III, *J. Am. Chem. Soc.*, **109**, 6577 (1987).
95. J. D. Dill, A. Greenberg and J. F. Liebman, *J. Am. Chem. Soc.*, **101**, 6814 (1979).
96. S. Durmaz and H. Kollmar, *J. Am. Chem. Soc.*, **102**, 6942 (1980).
97. D. Cremer and E. Kraka, *J. Am. Chem. Soc.*, **107**, 3811 (1985).
98. K. B. Wiberg and K. E. Laidig, *J. Org. Chem.*, **57**, 5092 (1992).
99. For experimental geometries see:
(a) F. H. Allen, *Acta Crystallogr., Sect. B*, **B36**, 81 (1980).
(b) F. H. Allen, *Acta Crystallogr., Sect. B*, **B37**, 890 (1981).
(c) F. H. Allen, *Tetrahedron*, **38**, 645 (1982).
(d) F. H. Allen, O. Kennard and R. Taylor, *Acc. Chem. Res.*, **16**, 146 (1983).
100. R. E. Penn and J. E. Boggs, *J. Chem. Soc., Chem. Commun.*, 667 (1972).
101. R. Pearson, Jr., A. Choplin and V. W. Laurie, *J. Chem. Phys.*, **62**, 4859 (1975).
102. (a) A. Skancke and J. E. Boggs, *J. Mol. Struct.*, **51**, 267 (1979).
(b) G. R. DeMaré and M.R. Peterson, *J. Mol. Struct.*, **89**, 213 (1982).
(c) Also see: S.W. Staley, A. E. Howard, M. D. Harmony, S. N. Mathur, M. Kattija-Ari, J.-I. Choe and G. J. Lind, *J. Am. Chem. Soc.*, **102**, 3639, (1980).
(d) W. J. Hehre, *J. Am. Chem. Soc.*, **94**, 6592 (1972).
103. A. Skancke, *Acta Chem. Scand., Ser. A*, **A36**, 637 (1982).
104. A. Greenberg, J. F. Liebman, W. R. Dolbier, Jr., K. S. Medinger and A. Skancke, *Tetrahedron*, **39**,1533 (1983).
105. (a) A. T. Peretta and V. W. Laurie, *J. Chem. Phys.*, **62**, 2469 (1975).
(b) C. A. Deakyne, L. C. Allen and N. C. Craig, *J. Am. Chem. Soc.*, **99**, 3895 (1977).
(c) C. A. Deakyne, L. C. Allen and V. W. Laurie, *J. Am. Chem. Soc.*, **99**, 1343 (1977).
(d) A. Skancke, E. Flood and J. E. Boggs, *J. Mol. Struct.*, **40**, 263 (1977).
106. (a) D. K. Hendricksen and M. D. Harmony, *J. Chem. Phys.*, **51**, 700 (1969).
(b) M. D. Harmony, R. E. Bostrom and D. J. Hendricksen, *J. Chem. Phys.*, **62**, 1599 (1975).
(c) S. N. Mathur and M. D. Harmony, *J. Chem. Phys.*, **69**, 4316 (1978).
107. A. Skancke and J. E. Boggs, *J. Mol. Struct.*, **50**, 173 (1978).
108. (a) A. Skancke, *J. Mol. Struct.*, **42**, 235 (1977).
(b) L. Hedberg, K. Hedberg and J. E. Boggs, *J. Chem. Phys.*, **77**, 2996 (1982).
109. (a) J.-M. André, M.-C. André and G. Leroy, *Bull. Soc. Chim. Belg.*, **80**, 265 (1971).
(b) M. Eckert-Maksic and Z. B. Maksic, *J. Mol. Struct.*, **86**, 325 (1982).
110. H. Oberhammer and J. E. Boggs, *J. Mol. Struct.*, **57**, 175 (1979).
111. M. Saunders, K. E. Laidig, K. B. Wiberg and P. v. R. Schleyer, *J. Am. Chem. Soc.*, **110**, 7652 (1988).
112. C. Kaupert, H. Heydtmann and W. Thiel, *Chem. Phys.*, **156**, 85 (1991).
113. A. Skancke, *Acta Chem. Scand.*, **A36**, 637 (1982).
114. P. Reynders and G. Schrumpf, *J. Mol. Struct. (Theochem)*, **150**, 297 (1987).
115. J. R. Durig Jr., A.-Y. Wang and T. S. Little, *J. Mol. Struct. (Theochem)*, **244**, 117 (1991).

116. R. G. Ford and R. A. Beaudet, *J. Chem. Phys.*, **48**, 4671 (1968)
117. A. W. Klein and G. Schrumpf, *Acta Chem. Scand., Ser. A*, **A35**, 425 (1981).
118. A. de Meijere and W. Lüttke, *Tetrahedron*, **25**, 2047 (1969).
119. V. Typke, *J. Mol. Spectrosc*, **77**, 117 (1979).
120. R. H. Schwendman, G. D. Jacobs and T. M. Krigas, *J. Chem.Phys.*, **40**, 1022 (1964).
121. M. L. Lamm and B. P. Dailey, *J. Chem. Phys.*, **49**, 1588 (1968).
122. W. H. Taylor, M. D. Harmony, D. A. Cassada and S. W. Staley, *J. Chem, Phys.*, **81**, 5379 (1984).
123. M. Dakkouri, *J. Am. Chem. Soc.*, **113**, 7109 (1991).
124. J. A. Boatz and M. S. Gordon, *J. Phys. Chem.*, **93**, 3025 (1989).
125. H. Grützmacher and H. Pritzkow, *Angew. Chem., Int. Ed. Engl.*, **30**, 1017 (1991).
126. S. Yamamoto, M. Nakata, T. Fukayama and K. Kuchitsu, *J. Phys. Chem.*, **89**, 3298 (1985).
127. Y. Endo, M. C. Chang and E. Hirota, *J. Mol. Spectrosc.*, **126**, 63 (1987).
128. K. B. Wiberg and J. J. Wendeloski, *J. Am. Chem. Soc.*, **104**, 5679 (1982).
129. Y. Yamaguchi, Y. Osamura and H. F. Schaefer, *J. Am. Chem. Soc.*, **105**, 7506 (1983).
130. E. D. Simandiras, R. D. Amos, N. C. Handy, T. J. Lee, J. E. Rice, R. B. Remington and H. F. Schaefer, *J. Am. Chem. Soc.*, **110**, 1388 (1988).
131. J. E. Boggs and K. Fan, *Acta Chem. Scand.*, **A42**, 595 (1988).
132. Y. Yamaguchi, H. F. Schaefer and J. E. Baldwin, *Chem. Phys. Lett.*, **185**, 143 (1991).
133. D. Cremer and K. J. Szabo, Unpublished, 1994.
134. M. D. Harmony, *Acc. Chem. Res.*, **25**, 321 (1992).
135. J. Gauss and D. Cremer, *Adv. Quantum Chem.*, **27**, 101 (1990).
136. MP2: (a) C. Møller and M. S. Plesset, *Phys Rev.*, **46**, 618 (1934).
(b) J. A. Pople, J. S. Binkley and R. Seeger, *Int. J. Quantum Chem., Symp.*, **10**, 1 (1976).
(c) MP3, MP4: R. Krishnan and J. A. Pople, *Int. J. Quantum Chem.*, **14**, 91 (1978).
(d) R. Krishnan, M. J. Frisch and J. A. Pople, *J. Chem. Phys.*, **72**, 4244 (1980).
137. K. Raghavachari, G. W. Trucks, J. A. Pople and M. Head-Gordon, *Chem. Phys. Lett.*, **157**, 479 (1989).
138. (a) Z. He and D. Cremer, *Theor. Chim. Acta*, **84**, 305 (1993).
(b) Z. He and D. Cremer, *Int. J. Quantum Chem., Symp.*, **25**, 43 (1991).
139. (a) M. D. Harmony, V. W. Laurie, R. L. Kuczkowski, R. H. Schwendeman, D. A. Ramsay, F. J. Lovas, W. A. Lafferty and A. G. Maki, *J. Phys. Chem. Ref. Data*, **8**, 619 (1979).
(b) For C_2H_4 see: K. Kuchitsu, *J. Chem. Phys.*, **44**, 906 (1966).
140. A. Almenningen, O. Bastiansen and P. N. Skancke, *Acta Chem. Scand.*, **15**, 711 (1961).
141. K. B. Wiberg, *J. Am. Chem. Soc.*, **105**, 1227 (1983).
142. K. W. Cox, M. D. Harmony, G. Nelson and K. B. Wiberg, *J. Am. Chem. Soc.*, **90**, 3395 (1968).
143. L. Hedberg and K. Hedberg, *J. Am. Chem. Soc.*, **107**, 7257 (1985).
144. (a) E. R. Davidson, *J. Chem. Phys.*, **46**, 3320 (1967).
(b) K. R. Roby, *Mol. Phys.*, **27**, 81 (1974).
145. A. R.Mochel, C. O. Britt and J. E. Boggs, *J. Chem. Phys.*, **58**, 3221 (1973).
146. J. R. Holtzclaw, W. C. Harris and S. F. Bush, *J. Raman Spectrosc.*, **9**, 257 (1980).
147. B. K. Stalick and J. A. Ibers, *J. Am. Chem. Soc.*, **93**, 3779 (1971).
148. D. A. Horner, R. S. Grev and H. F. Schaefer III, *J. Am. Chem. Soc.*, **114**, 2094 (1992).
149. Y. Xie and H. F. Schaefer, *J. Am. Chem. Soc.*, **112**, 5393 (1990).
150. R. H. Nobes, W. R. Rodwell, W. J. Bouma and L. Radom, *J. Am. Chem. Soc.*, **103**, 1913 (1981).
151. P. H. M. Buldzelaar, A. J. Kos, T. Clark and P. v. R. Schleyer, *Organometallics*, **4**, 429 (1985).
152. M. Bühl, P. v. R. Schleyer, M. A. Ibrahim and T. Clark, *J. Am. Chem. Soc.*, **113**, 2466 (1991).
153. J.-S. Murray, J. M. Seminario, P. Lane and P. Politzer, *J. Mol. Struct. (Theochem)*, **207**, 193 (1990).
154. J. L. Duncan and D. C. McKean, *J. Mol. Spectrosc.*, **27**, 117 (1968).
155. J. L. Duncan and D. Ellis, *J. Mol. Spectrosc.*, **28**, 540 (1968).
156. J. L. Duncan and G. R. Burns, *J. Mol. Spectrose.*, **30**, 253 (1969).
157. I. W. Levin and R. A. R. Pearce, *J. Chem. Phys.*, **69**, 2196 (1978).
158. A. Komornicki, F. Pauzat and Y. Ellinger, *J. Phys. Chem.*, **87**, 3847 (1983).
159. Y. Yamaguchi, H. F. Schaefer III and J. E. Baldwin, *Chem. Phys. Lett.*, **185**, 143 (1991).
160. For a discussion of scaling factors, see Reference 21.
161. J. A. Boatz and M. S. Gordon, *J. Phys. Chem.*, **93**, 1819 (1989).
162. Z. Konkoli, A. Larsson and D. Cremer, *J. Chem. Phys.*, to be published.
163. Z. Konkoli, A. Larsson and D. Cremer, *J. Phys. Chem.*, to be published.

164. (a) E. B. Wilson, Jr., J. C. Decius and P. C. Cross, in *Molecular Vibrations*, Dover, New York, 1980.
(b)W. D. Gwinn, *J. Chem. Phys.*, **55**, 477 (1971).
165. D. C. McKean, *Chem. Soc. Rev.*, **7**, 399 (1978).
166. (a) J. A. Berson, L. D. Pedersen and B. K. Carpenter, *J. Am. Chem. Soc.*, **98**, 122 (1976).
(b) S. W. Benson, in *Thermochemical Kinetics*, Wiley, London, 1976.
167. For reviews on analytical derivative methods, See:
(a) P. Jørgensen and J. Simons (Eds.), *Geometrical Derivatives of Energy Surfaces and Molecular Properties*, Reidel, Dordrecht, 1986.
(b) P. Pulay, *Adv. Chem. Phys.*, **67**, 241 (1987).
(c) H. B. Schlegel, *Adv. Chem. Phys.*, **67**, 249 (1987).
(d) R. D. Amos, *Adv. Chem. Phys.*, **67**, 99 (1987).
168. HF second derivatives:
(a) J. A. Pople, R. Krishnan, H. B. Schlegel and J. S. Binkley, *Int. J. Quantum Chem., Symp.*, **13**, 325 (1979).
MP2 second derivatives:
(b) N. C. Handy, R. D. Amos, J. F. Gaw, J. E. Rice and E. D. Simandiras, *Chem. Phys. Lett.*, **120**, 151 (1985).
(c) R. J. Harrison, G. B. Fitzgerald, W. D. Laidig and R. J. Bartlett, *Chem. Phys. Lett.*, **124**, 291 (1986).
169. For a review see:
(a) C. Castiglioni, M. Gussoni and G. Zerbi, *J. Mol. Struct.*, **141**, 341 (1986).
(b) M. Gussoni, C. Castiglioni and G. Zerbi, *J. Mol. Struct. (Theochem)*, **138**, 203 (1986).
(c) M. Gussoni, C. Castiglioni, M. N. Ramos, M. Rui and G. Zerbi, *J. Mol. Struct.*, **224**, 445 (1990) and references cited therein.
170. (a) W. B. Person and J. H. Newton, *J. Chem. Phys.*, **61**, 1040 (1974).
(b) J. H. Newton and W. B. Person, *J. Chem. Phys.*, **64**, 3036 (1976).
(c) J. H. Newton, R. A. Levine and W. R.Person, *J. Chem. Phys.*, **67**, 3282 (1977).
(d) W. B. Person and J. Overend, *J. Chem. Phys.*, **66**, 1442 (1977).
171. K.-M. Marstokk and H. Møllendal, *Acta Chem. Scand.*, **45**, 354 (1991).
172. (a) G. Holzawrth, E. C. Hsu, H. S. Mosher, T. R. Faulkner and A. Moscowitz, *J. Am. Chem. Soc.*, **96**, 251 (1974).
(b) L. A. Nafie, J. C. Cheng and P. J. Stephens, *J. Am. Chem. Soc.*, **97**, 3842 (1975).
(c) L. A. Nafie, T. A. Keiderling and P. J. Stephens, *J. Am. Chem. Soc.*, **98**, 2715 (1976).
173. P. J. Stephens, *J. Phys. Chem.*, **89**, 748 (1985).
174. K. J. Jalkanen, P. J. Stephens, R. D. Amos and N. C. Handy, *J. Am. Chem. Soc.*, **109**, 7193 (1987).
175. P. Lazzeretti, R. Zanasi, T. Prosperi and A. Lapiccirella, *Chem. Phys. Lett.*, **150**, 515 (1988).
176. A. J. Stone, *Chem. Phys. Lett.*, **83**, 233 (1981).
177. J. H. Williams, Ph.D. Thesis, University of Cambridge (1980).
178. R. D. Amos and J. H. Williams, *Chem. Phys. Lett.*, **84**, 104 (1981).
179. M. P. Bogaard, A. D. Buckingham, R. K. Pierens and A. H. White, *J. Chem. Soc., Faraday Trans. 1*, **74**, 3008 (1978).
180. K. M. Gough, *J. Chem. Phys.*, **91**, 2424 (1989).
181. (a) H. Hameka, *Mol. Phys.*, **1**, 203 (1958).
(b) H. Hameka, *Rev. Mod. Phys.*, **34**, 87 (1962).
(c) D. Zeroka and H. F. Hameka, *J. Chem. Phys.*, **45**, 300 (1966).
182. (a) F. London, *Naturwissemschaften*, **15**, 187 (1937).
(b) F. London, *J. Phys. Rad.*, **8**, 397 (1937).
183. (a) J.A. Pople, *J. Chem. Phys.*, **37**, 53 (1962).
(b) J.A. Pople, *J. Chem. Phys.*, **37**, 60 (1962).
184. (a) R. Ditchfield, *J. Chem. Phys.*, **56**, 5688 (1972).
(b) R. Ditchfiled, *J. Chem. Phys.*, **65**, 3123 (1976).
(c) R. Ditchfield, *Mol. Phys.*, **27**, 789 (1974).
(d) R. Ditchfield, in *Topics in Carbon-13 NMR Spectroscopy*, Vol. I, Wiley, New York, 1974.
185. (a) V. Galasso, *Theor. Chim. Acta*, **63**, 35 (1983).
(b) M. Jaszunski and L. Adamowicz, *Chem. Phys. Lett.*, **79**, 133 (1981).
(c) P. Lazzeretti and R. Zanasi, *J. Chem. Phys.*, **105**, 12 (1983).
186. K. Wolinski, J. F. Hinton and P. Pulay, *J. Am. Chem. Soc.*, **112**, 8251 (1990).

187. (a) W. Kutzelnigg, *Isr. J. Chem.*, **19**, 193 (1980).
(b) M. Schindler and W. Kutzelnigg, *J. Chem. Phys.*, **76**, 1919 (1982).
188. (a) M. Schindler and W. Kutzelnigg, *J. Am. Chem. Soc.*, **105**, 1360 (1983).
(b) M. Schindler and W. Kutzelnigg, *Mol. Phys.*, **48**, 781 (1983).
(c) M. Schindler and W. Kutzelnigg, *J. Am. Chem. Soc.*, **109**, 1021 (1987).
(d) M. Schindler, *J. Am. Chem. Soc.*, **109**, 5950 (1987).
(e) M. Schindler, *Magn. Res. Chem.*, **26**, 394 (1988).
(f) M. Schindler, *J. Am. Chem. Soc.*, **110**, 6623 (1988).
(g) M. Schindler, *J. Chem. Phys.*, **88**, 7638 (1988).
(h) W. Kutzelnigg, M. Schindler and U. Fleischer, in *NMR, Basic Principles and Progress* (Eds. P. Diehl, E. Fluck, H. Günther, R. Kosfeld and J. Seelig), Vol. 23, Springer, Berlin, 1991, p. 165.
189. A. E. Hansen and T. D. Bouman, *J. Chem. Phys.*, **82**, 5035 (1985).
190. (a) J. Gauss, *Chem. Phys. Lett.*, **191**, 614 (1992).
(b) J. Gauss, *J. Chem. Phys.*, **99**, 3629 (1993).
191. J. Gauss, *Chem. Phys. Lett.*, in press.
192. (a) W. Kutzelnigg, C. van Wüllen, U. Fleischer and R. Franke, *Proceedings of the NATO Advanced Workshop on the Calculation of NMR Shielding Constants and their Use in the Determination of the Geometric and Electronic Structures of Molecules and Solids*, 1992.
(b) C. van Wüllen and W. Kutzelnigg, *Chem. Phys. Lett.*, **205**, 563 (1993).
193. (a) T. D. Bouman and A. E. Hansen, *Chem. Phys. Lett.*, **175**, 292 (1990).
(b) J. Geertsen and J. Oddershedde, *J. Chem. Phys.*, **90**, 301 (1984).
(c) J. Geertsen, P. Jørgensen and D. L. Yaeger, *Comp. Phys. Rep.*, **2**, 33 (1984).
(d) G. T. Darborn and N. C. Handy, *Mol. Phys.*, **49**, 1277 (1983).
(e) M. Jaszunski and A. Sadlej, *Theor. Chim. Acta*, **40**, 157 (1975).
194. J .W. Emsley, J. Feeney and L. H. Sutcliffe, *High Resolution Magnetic Resonance*, Vol. 2, Pergamon, Oxford, 1966.
195. K. W. Zilm, A. J. Beller, D. M. Grant, J. Michl, T. C. Chou and E. L. Allred, *J. Am. Chem. Soc.*, **103**, 2119 (1981).
196. G. C. Levy, R. L. Lichter and G. L. Nelson, in *Carbon-13 Nuclear Magnetic Resonance Spectroscopy*, 2nd ed., Wiley-interscience, New York, 1980.
197. A. M. Orendt, J. C. Facelli, M. D. Grant, J. Michl, F. H. Walker, W. P. Daily, S. T. Waddell, K. B. Wiberg, M. Schindler and W. Kutzelnigg, *Theor. Chim. Acta*, **68**, 421 (1985).
198. A. E. Hansen and T. D. Bouman, *J. Chem. Phys.*, **91**, 3552 (1989).
199. (a) See Reference 8a
(b) C. D. Poulter, R. S. Boikess, J. I. Brauman and S. Winstein, *J. Am. Chem. Soc.*, **94**, 2291 (1972).
(c) W. R. Bley, *Mol. Phys.*, **20**, 491 (1971).
(d) R. C. Benson and W. H. Flygare, *J. Chem. Phys.*, **58**, 2651 (1973).
200. R. F. Childs, M. J. McGlinchey and A. Varadarajan, *J. Am. Chem. Soc.*, **106**, 5974 (1984).
201. See, e.g.:
(a) P. Buzek, P.v.R. Schleyer and S. Sieber, *Chemie in unserer Zeit*, **26**, 116 (1992).
(b) M. Bühl and P.v.R. Schleyer, in *Electron Deficient Boron and Carbon Clusters* (Eds. G.A. Olah, K. Wade and R.E. Williams), Wiley, New York, 1991.
(c) M. Bühl, N. J. R. v. E. Hommes, P. v. R. Schleyer, U. Fleischer and W. J. Kutzelnigg, *J. Am. Chem. Soc.*, **113**, 2459 (1991).
202. See e.g.:
(a) D. Hnyk, E. Vajda, M. Buehl and P. v. R. Schleyer, *Inorg. Chem.*, **31**, 2464 (1992).
(b) M. Buehl and P.v.R. Schleyer, *J. Am. Chem. Soc.*, **114**, 477 (1992).
(c) M. Buehl, P.v.R. Schleyer and M. L. McKee, *Heteroat. Chem.*, **2**, 499 (1991).
(d) M. Buehl and P.v.R. Schleyer, *Angew. Chem.*, **102**, 962 (1990).
(e) P.v.R. Schleyer, M. Büehl, U. Fleischer and W. Koch, *Inorg. Chem.*, **29**, 153 (1990).
(f) P.v.R. Schleyer, W. Koch, B. Liu and U. Fleischer, *J. Chem. Soc., Chem. Commun.*, 1098 (1989).
(g) M. Bremer, K. Schoetz, P.v.R. Schleyer, U. Fleischer, M. Schindler, W. Kutzelnigg, W. Koch and P. Pulay, *Angew. Chem.*, **101**, 1063 (1989).
203. D. Cremer, F. Reichel and E. Kraka, *J. Am. Chem. Soc.*, **113**, 9459 (1991).
204. F. Reichel, Ph.D. Dissertation, University of Köln, 1991.
205. D. Cremer, L. Olsson, F. Reichel and E. Kraka, *Isr. J. Chem.*, **33**, 369 (1993).

206. T. Onak, J. Tseng, M. Diaz, D. Tran, J. Arias, S. Herrera and D. Brown, *Inorg. Chem.*, **32**, 487 (1993).
207. M. Saunders, K. E. Laidig, K. B. Wiberg and P.v.R. Schleyer, *J. Am. Chem. Soc.*, **110**, 7652 (1988).
208. D. Cremer, P. Svensson, E. Kraka, Z. Konkoli and P. Ahlberg, *J. Am. Chem. Soc.*, **115**, 7457 (1993).
209. D. Cremer, P. Svensson and K.J. Szabo, to be published.
210. D. Cremer, P. Svensson, E. Kraka and P. Ahlberg, *J. Am. Chem. Soc.*, **115**, 7445 (1993).
211. S. Sieber, P.v.R. Schleyer, A. H. Otto, J. Gauss, F. Reichel and D. Cremer, *J. Phys. Org. Chem.*, **6**, 445 (1993).
212. (a) P. Wagner and A. B. F. Duncan, *J. Chem. Phys.*, **21**, 516 (1953).
(b) M. B. Robin, *Higher Excited States of Polyatomic Molecules*, Vol. I, Academic Press, New York, 1975.
213. R. J. Buenker and S. D. Peyerimhoff, *J. Phys. Chem.*, **73**, 1299 (1969).
214. T.-K. Ha and W. Cencek, *Chem. Phys. Lett.*, **177**, 463 (1991).
215. S. W. Froelicher, B. S. Freiser and R. R. Squires, *J. Am. Chem. Soc.*, **108**, 2853 (1986).
216. P.v.R. Schleyer, G. W. Spitznagel and J. Chandrasekhar, *Tetrahedron Lett.*, **27**, 4411 (1986).
217. C. H. DePuy, V. M. Bierbaum and M. Damrauer, *J. Am. Chem. Soc.*, **106**, 4051 (1984).
218. S. G. Lias, J. L. Liebman and R. D. Levin, *J. Phys. Chem. Ref. Data*, **13**, 695 (1984).
219. S.-L. Chon and J. L. Franklin, *J. Am. Chem. Soc.*, **94**, 6347 (1972).
220. J. C. Schulz, F. A. Houle and J. L. Beauchamp, *J. Am. Chem. Soc.*, **106**, 3917 (1984).
221. K. Raghavachari, R. A. Whiteside, J. A. Pople and P.v.R. Schleyer, *J. Am. Chem. Soc.*, **103**, 5649 (1981).
222. M. J .S. Dewar, E. F. Healy and J. M. Ruiz, *J. Chem. Soc., Chem. Commun.*, 943 (1987).
223. W. Koch, B. Liu and P. v. R. Schleyer, *J. Am. Chem. Soc.*, **111**, 3479 (1989).
224. M. Saunders, P. Vogel, E. L. Hagen and J. Rosenfeld, *Acc. Chem. Res.*, **6**, 53 (1973).
225. W. J. Bouma, D. Poppinger and L. Radom, *Isr. J. Chem.*, **23**, 21 (1983).
226. D. D. M. Wayner, R. J. Boyd and D. R. Arnold, *Can. J. Chem.*, **63**, 3283 (1985).
227. P. Du, D. A. Hrovat and W. T. Borden, *J. Am. Chem. Soc.*, **110**, 3405 (1988).
228. C. E. Hudson, M. S. Ahmed, J. C. Traeger, C. S. Giam and D. J. McAdoo, *Int. J. Mass Spectrom. Ion Processes*, **113**, 117 (1992).
229. J. A. Booze and T. Baer, *J. Phys. Chem.*, **96**, 5710 (1992).
230. S. Lunell, L. Yin and M.-B. Huang, *Chem. Phys.*, **139**, 293 (1989).
231. K. Krogh-Jespersen and H. D. Roth, *J. Am. Chem. Soc.*, **114**, 8388 (1992).
232. L. A. Eriksson and S. Lunell, *J. Am. Chem Soc.*, **114**, 4532 (1992).
233. C. E. Hudson, C. S. Giam and D. J. McAdoo, *J. Org. Chem.*, **58**, 2017 (1993).
234. (a) M. Iwasaki, K. Toriyama and K. Nunome, *J. Chem. Soc., Chem. Commun.*, 202 (1983).
(b) K. Ohta, H. Nakatsuji, H. Kubodera and T. Shida, *Chem. Phys.*, **76**, 271 (1983).
235. S. Lunell, M.-B. Huang and A. Lund, *Faraday Discuss. Chem. Soc.*, **78**, 35 (1984).
236. (a) X.-Z. Qin and F. Williams, *Chem. Phys. Lett.*, **112**, 79 (1984).
(b) X.-Z. Qin and F. Williams, *Tetrahedron*, **42**, 6301 (1986).
237. T. M. Sack, D. L. Miller and M. L. Gross, *J. Am. Chem. Soc.*, **107**, 6795 (1985).
238. (a) K. Raghavachari, R. A. Whiteside, J. A. Pople and P.v.R. Schleyer, *J. Am. Chem. Soc.*, **103**, 5649 (1981).
(b) L. Radom J. A. Pople and P.v.R. Schleyer, *J. Am. Chem. Soc.*, **95**, 8193 (1973).
(c) For a summary, see Reference 21b, Chapter 7.
239. M. H. Lien and A. C. Hopkinson, *J. Phys. Chem.*, **88**, 1513 (1984).
240. D. Cremer and K. J. Szabo, unpublished results.
241. E. Jongejan, W. J. M. van Tilborg, Ch. H. V. Dusseau, H. Steinberg and Th. J. deBoer, *Tetrahedron Lett.*, 2359 (1972).
242. E. Jongejan, H. Steinberg and Th. J. deBoer, *Tetrahedron Lett.*, 397 (1972).
243. E. Jongejan, H. Steinberg and Th. J. DeBoer, *Recl. Trav. Chim. Pays-Bas*, **98**, 66 (1979).
244. J. R. van der Vecht, H. Steinberg and Th. J. deBoer, *Recl. Trav. Chim. Pays-Bas.*, **96**, 313 (1977).
245. M. W. M. van Tilborg, R. van Doorn and N. M. M. Nibbering, *J. Am. Chem. Soc.*, **101**, 7617 (1979).
246. C. Cometta-Morini, T.-K. Ha and J. F. M. Oth, *J. Mol. Struct. (Theochem)*, **188**, 79 (1989).
247. M. Dupuis and J. Pacansky, *J. Chem. Phys.*, **76**, 2511 (1982).
248. M. M. H. Lien and A. C. J. Hopkinson, *J. Comput. Chem.*, **6**, 2764 (1985).
249. S. Olivella, A. Solé and J. M. Bofill, *J. Am. Chem. Soc.*, **112**, 2160 (1990).

250. V. Barone, C. Minichino, H. Faucher, R. Subra and A. Grand, *Chem. Phys. Lett.*, **205**, 324 (1993).
251. (a) S. Deycars, L. Hughes, J. Lustzyk and K. U. Ingold, *J. Am. Chem. Soc.*, **109**, 4954 (1987).
(b) See also: R. W. Fessenden and R. H. Schuler, *J. Chem., Phys.*, **39**, 2147 (1963).
c) K. S. Chen, D. G. Edge and G. K. Kochi, *J. Am. Chem. Soc.*, **95**, 7036 (1973).
252. L. J. Johnston and K. U. Ingold, *J. Am. Chem. Soc.*, **108**, 2343 (1986).
253. J. E. Wertz and J .R. Bolton, *Electron Spin Resonance*, McGraw-Hill, New York, 1972.
254. (a) K. Holtzhauer, Ph.D. Thesis, ETH, Zürich, 1987.
(b) K. Holtzhauer, C. Cometta-Morini and J. F. M. Oth, *J. Phys. Org. Chem.*, **3**, 219 (1990).
255. D. J. DeFrees, R. T. Mclver and W. J. Hehre, *J. Am. Chem. Soc.*, **102**, 3334 (1980).
256. (a) G. Greig and J. C. C. Thynne, *Trans. Faraday Soc.*, **62**, 3338 (1966).
(b) G. Greig and J. C. C. Thynne, *Trans. Faraday Soc.*, **63**, 1369 (1967).
(c) J. A. Kerr, A. Smith and A. F. Trotman-Dickenson, *J. Chem. Soc. (A)*, 1400 (1969).
257. R. Walsh, *Int. J. Chem. Kinet.*, **2**, 71 (1970).
258. K. B. Wiberg, *J. Am. Chem. Soc.*, **105**, 1227 (1983).
259. P. H. M. Budzelaar, E. Kraka, D. Cremer and P.v.R. Schleyer, *J. Am. Chem. Soc.*, **108**, 561 (1986).
260. M. D. Newton and J. M. Schulman, *J. Am. Chem. Soc.*, **94**, 767 (1972).
261. M. D. Newton and J. M. Schulman, *J. Am. Chem. Soc.*, **94**, 773 (1972).
262. P. Chakrabarti, P. Seiler, J. D. Dunitz, A.-D. Schluter and G. J. Szeimies, *J. Am. Chem. Soc.*, **103**, 7378 (1981).
263. R. Hoffmann, *J. Am. Chem. Soc.*, **90**, 1475 (1968).
264. B. Zurawski and W. Kutzelnigg, *J. Am. Chem. Soc.*, **100**, 2654 (1978).
265. N. G. Rondan, K. N. Houk and R. A. Moss, *J. Am. Chem. Soc.*, **102**, 1770 (1980).
266. M. Moreno, J. M. Lluch, A. Oliva and J. Bertrán. *J. Mol. Struct. (Theochem)*, **164**, 17 (1988).
267. See, e.g., A. Szabo and N. S. Ostlund, *Modern Quantum Chemistry, Introduction to Advanced Electronic Structure Theory*, MacMillan, New York, 1982 and references cited therein.
268. (a) Activation energy for the reaction to give propene: I. E. Klein and B. S. Rabinovitch, *Chem. Phys.*, **35**, 439 (1978).
(b) B. S. Rabinovitch, *Chem. Phys.*, **67**, 201 (1982).
(c) The difference between the activation energies was determined to be 3.7 kcal mol^{-1}: E.V. Waag and B. S. Rabinovitch, *J. Phys. Chem.*, **76**, 1965 (1972).
(d) See also: W. Doering, *Proc. Natl. Acad. Sci. U.S.A.*, **78**, 5279 (1982).
269. C. Doubleday, J.W. Mclver and M. Page, *J. Am. Chem. Soc.*, **104**, 6533 (1982).
270. Y. Yamaguchi and H. F. Schaefer III, *J. Am. Chem. Soc.*, **106**, 5115 (1984).
271. T. R. Furlani and H. F. King, *J. Chem. Phys.*, **82**, 5577 (1985).
272. L. Carlacci, C. Doubleday, T. R. Furlani, H. F. King and J. W. Mclver, *J. Am. Chem. Soc.*, **109**, 5323 (1987).
273. C. Doubleday, J. W. Mclver and M. Page, *J. Phys. Chem.*, **92**, 4367 (1988).
274. S. Olivella and J. Salvadori, *Int. J. Quantum Chem.*, **37**, 713 (1990).
275. S. J. Getty, E. R. Davidson and W. T. Borden, *J. Am. Chem. Soc.*, **114**, 2085 (1992).
276. (a) F. T. Smith, *J. Chem. Phys.*, **29**, 235 (1958).
(b) H. Kollmar, *J. Am. Chem. Soc.*, **95**, 966 (1973).
277. (a) S. W. Benson, *J. Chem. Phys.*, **34**, 521 (1961)
(b) S. W. Benson and P. S. Nangia, *J. Chem. Phys.*, **38**, 18 (1963).
(c) H. E. O'Neil and S. W. Benson, *J. Chem. Phys.*, **72**, 1866 (1968).
(d) H. E. O'Neil and S. W. Benson, *Int. J. Chem. Kinet.*, **2**, 42 (1970).
(e) S. W. Benson, *Thermochemical Kinetics*, Wiley, New York, 1971, p. 83.
278. S. Yamabe, T. Minato, M. Seki and S. Inagaki, *J. Am. Chem. Soc.*, **110**, 6047 (1988).
279. A. Alex and T. Clark, *J. Am. Chem. Soc.*, **114**, 10897 (1992).
280. M. R. A. Blomberg, P. E. M. Siegbahn and J.-E. Bäckvall, *J. Am. Chem. Soc.*, **109**, 4450 (1987).
281. J.-E. Bäckvall, E. E Björkman, L. Pettersson, P. E. M. Siegbahn and A. Strich, *J. Am. Chem. Soc.*, **107**, 7408 (1985).
282. D. R. Gano, M. S. Gordon and J. A. Boatz, *J. Am. Chem. Soc.*, **113**, 6711 (1991).
283. P. E. M. Siegbahn and M. R. A. Blomberg, *J. Am. Chem. Soc.*, **114**, 10548 (1992).

CHAPTER 3

Structural chemistry of cyclopropane derivatives

BÉLA ROZSONDAI

Structural Chemistry Research Group of the Hungarian Academy of Sciences, Eötvös University, P.O. Box 117, H-1431 Budapest, Hungary
Fax: +36 1 266 3899; e-mail: H2378roz@ELLA.HU

I. INTRODUCTION 140
II. CYCLOPROPANE 143
A. Structure and Electron-density Distribution 143
B. Gas-phase Complexes 146
III. SUBSTITUTED CYCLOPROPANES 146
A. Alkyl and Cycloalkyl Cyclopropanes 147
1. Alkyl cyclopropanes 148
2. Bicyclopropyl derivatives 149
3. Alkyl derivatives with functional groups 150
B. Vinyl and Aryl Cyclopropanes 153
1. Vinyl derivatives 153
2. Aryl derivatives 156
C. Carbonyl Derivatives 160
1. Carboxylic acids 161
2. 1-Aminocyclopropanecarboxylic acid and derivatives 170
3. Carboxylic acid derivatives 172
D. Ethynyl and Cyano Cyclopropanes 175
E. Halogenides 176
F. Ethers and Sulfur Derivatives 179
G. Nitrogen and Phosphorus Derivatives 181
1. Aminocyclopropanes 181
2. Cyclopropyl isocyanate and isothiocyanate 182
3. Nitrocyclopropanes 183
4. Phosphorus derivatives 185
H. Silyl and Germyl Cyclopropanes 185
IV. CYCLOPROPYLIDENES 187
V. CYCLOPROPENES 190
VI. CYCLOPROPENYLIDENES AND RELATED SYSTEMS 194

The chemistry of the cyclopropyl group, Vol. 2
Edited by Z. Rappoport

A. Methylenecyclopropenes and Cyclopropenones 194
B. Ionic Species 197
VII. CYCLOPROPANE RINGS IN POLYCYCLIC SYSTEMS 198
A. Bicyclo[n.1.0]alkanes 198
B. Cycloproparenes 201
C. Bridged Ring Systems, Cage Molecules 202
1. Bridged bicyclo[1.1.0]butanes 202
2. [1.1.1]Propellanes 203
3. Tetrahedranes 205
4. Semibullvalenes and barbaralanes 206
5. Nortricyclenes 208
6. Prismanes 208
7. Quadricyclanes 209
8. Triasteranes and octahedranes 210
9. Other structures 211
VIII. ACKNOWLEDGMENTS 212
IX. REFERENCES 212

I. INTRODUCTION

The cyclic triad of methylene groups has fascinated many chemists, and has inspired experimental and theoretical studies on cyclopropane and derivatives.

Cyclopropanes resemble olefins in some respects, or are intermediate between alkenes and saturated hydrocarbons. The cyclopropane ring is involved in a great variety of types of chemical reactions. It participates in hydrogen bonds, forms complexes, ionic and radical species, or charge-transfer arrays. As a constituent in a molecular framework, it provides, as does a double bond, a relatively rigid core, a firm support and rotational axes for attached bulky groups. Although cyclopropane itself has a lower strain than expected, structures containing this cyclic unit rank among the strained organic molecules. The three-membered carbon ring is synthesized in living organisms, often as part of a ring system, and is also introduced in laboratory products in order to modify the structure and biological activity of a molecule. Some of these points will be touched on later in this chapter in the discussion of molecular structure.

Professor Dunitz remembers an episode from the time (1959) when he completed the X-ray-crystallographic study of cyclododecane at ETH Zürich[1]. He met the famous Professor Leopold Ruzicka on the steps of the chemistry building and mentioned to him that he had just determined the *structure* of cyclododecane. "Ahah," Ruzicka replied, "*die* kenne ich schon lange".* He meant constitution as expressed by a structural formula, Dunitz meant the *geometrical aspect* of molecular structure, which will be given in terms of bond lengths, bond angles and torsional angles in the present review.

More than two decades ago in this series, Charton reviewed the olefinic properties, bonding models and molecular structures of cyclopropanes[2]. In the previous volume on the cyclopropyl group, Wiberg discussed, based mainly on theoretical calculations, the structures, spectra and energies of cyclopropanes[3]. Sparse structural data and references to experimental studies occur in some other chapters. Recent reviews have focused on the geometrical structure[4,5] or on the strain and energetics[6,7] of cyclopropanes. Allen analyzed the geometry and conformation of cyclopropane derivatives, based on all relevant experimental data available, mainly from crystallographic studies[8–12].

*I thank Professor Jack D. Dunitz and Verlag Helvetica Chimica Acta for their kind permission to quote these words here.

This chapter surveys structural information derived from *experimental* studies, and covers the period from about 1985 (the closing date of the previous cyclopropyl volume) through 1993. The structures of free molecules deserve special attention, since their geometry is governed solely by intramolecular forces. Crystal structure studies, on the other hand, give valuable information on possible interactions with the environment and its structural effects in solid and liquid phases, in real chemical and biological systems. Results of theoretical molecular orbital (MO) or molecular mechanics (MM) calculations are often utilized in a complex analysis of experimental data, and will be cited here occasionally for comparison. Only brief mention will be made of bonding models of cyclopropane and electronic interaction of the ring with substituents. Detailed depictions of orbitals and interactions and reference to earlier work are found in many of the reviews and papers cited and in other chapters of this volume.

The basic experimental techniques of structure determination, i.e. microwave spectroscopy (MW) and electron diffraction (ED) for the gas phase and X-ray diffraction (XD) for crystals, as well as the physical meaning of parameters obtained (r_s, r_z, r_a, r_g, r_α etc.) are summarized, and further references are given in a monograph edited by Domenicano and Hargittai[13] and in Landolt–Börnstein[14]. Theoretical calculations yield the equilibrium structure, the parameters (r_e) of the hypothetically motionless molecule. r_0 parameters are obtained directly from a set of zero-point rotational constants. The r_s substitution structure is more consistent and is calculated from shifts of rotational constants upon a systematic isotopic substitution. The effective distance parameter derived directly from ED is r_a. Vibrational corrections yield then r_g, the average internuclear distance, and r_α, the distance between average nuclear positions, both at thermal equilibrium. r_α^0 is the zero-point value of r_α, and is essentially the same as r_z, derived from rotational constants. Error estimates of experimental structural data are quoted from the original papers, and will be given in parentheses in units of the last digit of the parameter. SI units are used with a few exceptions; 1 Å = 100 pm = 10^{-10} m, 1D = 1 debye = 1.336×10^{-30} Cm; energy units have been converted by 1 cal = 4.184 J or are sometimes given as the wave number of the associated radiation, 1 cm^{-1} for 11.96 J mol^{-1}. I have calculated some structural parameters of interest from crystallographic unit cell data and fractional atomic coordinates.

The conformation of a substituted cyclopropane may be defined in different ways. Jason and Ibers characterized the relative orientation of orbitals in phenylcyclopropanes by the acute angle (θ) between the vector of the distal bond in the three-membered ring and the vector normal to the (mean) plane of the benzene ring[15]. Torsion about the bond linking the substituent is more commonly described by dihedral angle τ(M—C1—C4—X), where M is the midpoint of the distal bond (Figure 1). The two angles were found[15] to have similar values within 5°. The forms with $\tau = 0°$ and 180° are termed *bisected* conformation, those with $\tau = \pm 90°$ *perpendicular* conformation. Besides I will use *syn* or *synperiplanar* (*sp*) and *synclinal* (*sc*), *anti* or *ap* and *ac*, according to IUPAC recommendations[16], and also *gauche* for *sc*.

Factors like volatility of the substance, size and complexity of the molecules, and parameter correlation affect the limits of applicability of ED and MW. X-ray crystallography, on the other hand, has become a routine tool for structure elucidation (structure now in Ruzicka's interpretation), and geometrical data are often not printed in the journal but are deposited as supplementary material. Structural data of free molecules are available in printed form[14] and in a database at the University of Ulm[17], and crystal-phase data are found in the Cambridge Structural Database[18] (CSD).

About a hundred new structures containing a three-membered carbocyclic ring are added to the CSD in a year. The wealth and diversity of the relevant material render a thorough and systematic treatment of all classes of cyclopropane derivatives impossible. I have selected mainly simple molecules and systems of some structural interest, and have resorted to earlier studies in some cases in order to include some basic molecules. Fused

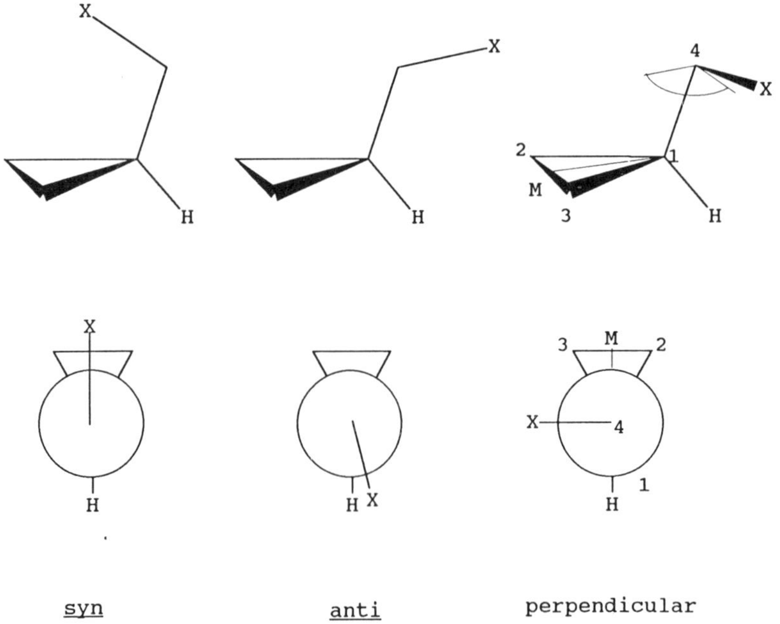

FIGURE 1. Conformations of cyclopropyl derivatives, characterized by dihedral angle τ(M—C1—C4—X): perspective views (above) and Newman projections down the C4—C1 bond (below). M is the midpoint of the C2—C3 bond

ring systems and cage molecules will be treated only briefly; some interesting classes of compounds and natural products have been completely neglected. Spiroannulated systems are discussed by Zefirov and Lukin in this volume.

(1)

In the following sections, cyclopropane (**1**) and its molecular complexes are discussed first. Molecules are then classified, as far as possible, according to the 'bonding scheme' of the three-membered ring: the number (and type) of bonds it has internally and to its neighbors, and the topology of polycyclic ring systems. Substituted derivatives of **1** have six bonds (**2**), linking atoms or groups to the C_3 ring. Classes of substituents (functional groups) form the basis of further subdivision. Methylenecyclopropane and cyclopropanone are characterized by five connections (**3**), cyclopropene by four (**4**) and molecules **5**–**7** by three connections. A 'bidentate' substituent can occupy geminal or *cis* or,

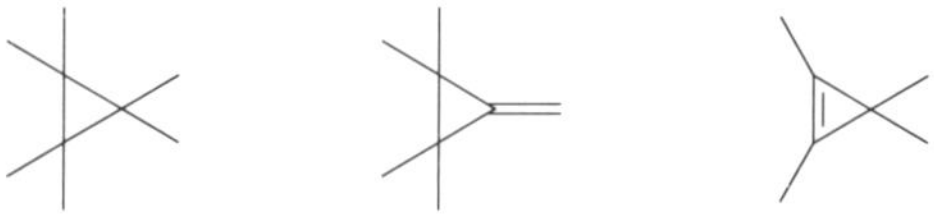

(2) **(3)** **(4)**

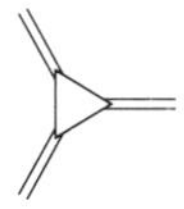
(5)

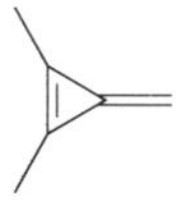
(6)

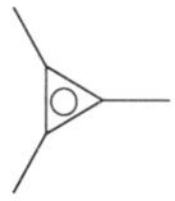
(7)

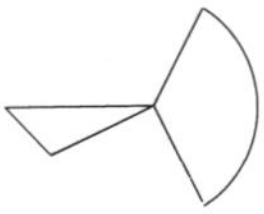
(8)

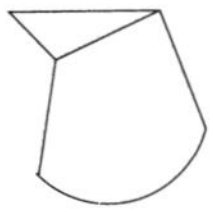
(9)

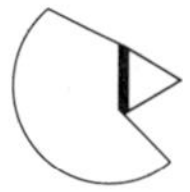
(10)

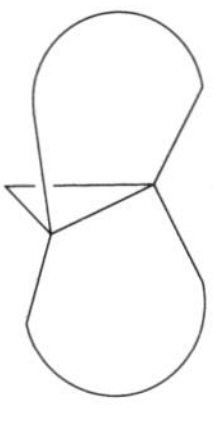
(11)

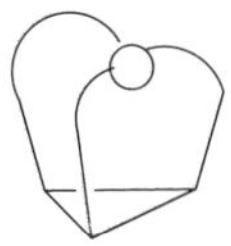
(12)

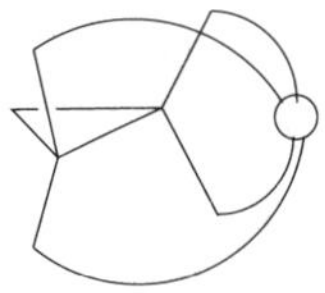
(13)

rather rarely, *trans* positions (**10**), formally generating a spiro derivative (**8**), a bicycloalkane or a cycloproparene (**9**), etc. Combinations thereof are found in polycyclic and cage molecules (**11–13**). (The small circle in **12** and **13** represents any atom, group or ring system having three and four connections to the C_3 ring.)

II. CYCLOPROPANE

A. Structure and Electron-density Distribution

The geometric and electronic structure of the parent cyclopropane molecule **1** has continued to attract the interest of investigators since its structure was first determined in an early gas-phase electron-diffraction study[19]. The goal of theoretical studies has been to find better bonding models for **1** and derivatives, to interpret and predict stabilities, chemical rearrangements, spectroscopic and other properties, geometrical changes, and interactions between the ring and substituents.

The structure of **1** has been redetermined recently by different methods (Table 1). The agreement with the results of an earlier ED study[20] is quite good.

As **1** is a nonpolar symmetric top with D_{3h} symmetry, it should have no pure rotational spectrum, but it acquires a small dipole moment by partial isotopic substitution or through centrifugal distortion. In recent analyses of gas-phase data, rotational constants from earlier IR and Raman spectroscopic studies, and those for cyclopropane-1,1-d_2 and for an excited state of the ν_{11} C—C stretching vibration were utilized[21,22]. Anharmonicity constants for the C—C and C—H bonds were determined in both works. It is the r_z parameters, then from r_z the equilibrium r_e structure, that can be derived and compared from both the ED and the MW data by appropriate vibrational corrections. Variations due to different representations of molecular geometry are of the same magnitude as stated uncertainties. The r_e parameters from experiment agree satisfactorily with the results of high-level theoretical calculations (Table 1).

TABLE 1. Structural parameters of cyclopropane (**1**) from experiment[a] and *ab initio* calculations (r_e)

		C—C (Å)	C—H (Å)	H—C—H (deg)	Reference
ED	r_a	1.510 (2)	1.089 (3)	115.1 (10)	20
ED+SP[b]	r_g	1.5139 (12)	1.099 (2)		21
	r_z	1.5127 (12)	1.084 (2)	114.5 (9)	
	r_e	1.501 (4)	1.083 (5)	114.5 (9)[c]	
MW+SP[b]	r_z	1.516 (2)	1.080 (3)	115.5(4)	22
	r_e	1.510 (2)	1.074 (3)	$115.8_5(3_3)$	
XD		1.4991 (7)[d]	1.029 (5)[d]	114.5[e]	23b
4–21G[f]		1.515	1.071	114. 5	24
4–31G		1.502	1.071	113. 8	25
6–31G*		1.4974		114.0[g]	26
MP2/6–31G*		1.502	1.084	114.2	27
MP2/6–311G (MC)*		1.507	1.083	114.7	28

[a] Some parameter values and uncertainties have been rounded from the original data.
[b] Combined analysis with the inclusion of rotational constants from infrared or Raman spectroscopic studies.
[c] Assumed to be equal to $\angle_z$(H—C—H).
[d] Average value.
[e] Average calculated from the original data[23c].
[f] 4–21G split-valence basis set.
[g] Calculated from ∠C—C—H 118.14°.

Nijveldt and Vos determined[23], by a careful analysis of XD data, the structure and electron-density distribution of **1** (at 94 K), bicyclopropyl and vinylcyclopropane. Figure 2 shows sections of the electron-density map of **1**. The molecule lies at a mirror plane in the orthorhombic $Cmc2_1$ crystal; the plane bisects the C2—C1—C2′ angle. Some deviation from the ideal D_{3h} symmetry is apparent in the density map and in differences between independent geometrical parameters. The largest distortion is found in the H—C—H angle at C1, which is wider by 3.9 (14)° than the other two H—C—H angles[23a,c]. The systematic shift between bond lengths in the gas phase and in the crystal may arise from libration in the crystal or from other sources. Correction for rigid-body motion could not be done for **1**, and is expected to be larger than in vinylcyclopropane where correction resulted[23c] in an average lengthening of C—C bonds by 0.0076 Å.

The electron-density plots (Figure 2) show marked electron concentrations outside the C_3 triangle, in complete agreement with the notion of bent σ bonds. The position of a density maximum may be characterized by the angle between the C—C bond and the line connecting a carbon nucleus and the maximum. These angles are 13 (1)° and 15 (1)° in **1** (see Table 3 in Reference 23c). The bond path[29], the line connecting points of maximum charge densities between a pair of bonded atoms, is a curved line for a C—C bond of **1**. The bond path angle is 78.88° from 6–31G* *ab initio* calculations[26,29].

A new valence-bond (VB) approach[30] gives qualitatively the same picture of electron distribution and bent C—C bonds in **1** as the MO calculations. The starting bond directions at the carbon nucleus form an angle of 123°, and the weighted average 'interorbital bond angle' is 110°. Calculated average hybridizations of the carbon orbitals (which cannot be directly compared with orthogonal hybrids used in MO theory), viz. $sp^{1.706}$ for the C—C bonds and $sp^{1.348}$ for the C—H bonds, indicate that the C—H bonds in **1** have more s character and are stronger than in other alkanes. This fact should be accounted for when calculating strain energies[30]. Different bonding models have yielded estimates of $sp^{3.69}$ to sp^5 hybrids for the C—C bonds within the cyclopropane ring, and sp^2 to $sp^{2.49}$ for bonds to substituents (see the review[9] by Allen for references). The analysis of a large body of bond length data has lead to $sp^{4.26}$ hybrids for the intraring bonds[11], and $sp^{2.22}$ for the bonds to substituents[9].

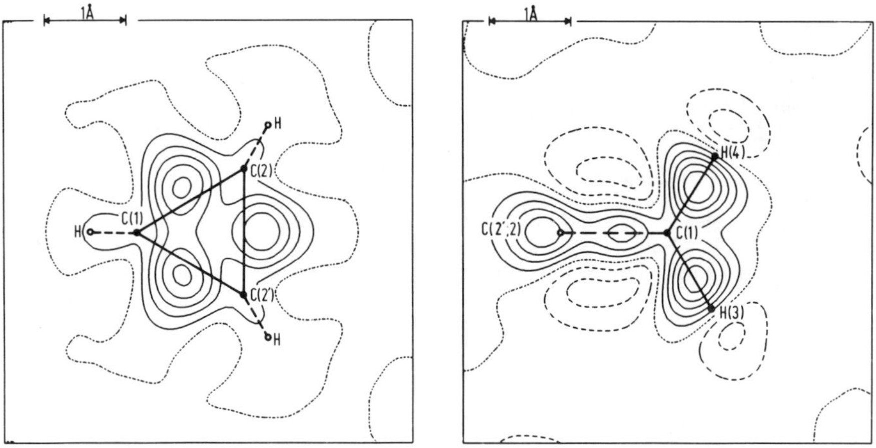

(a) (b)

FIGURE 2. Dynamic filtered deformation electron-density distribution of cyclopropane (**1**). Sections (a) in the ring plane and (b) in the crystallographic mirror plane, perpendicular to the ring plane. Contours are at 0.05 e $Å^{-3}$ intervals, dashed lines represent negative areas. Reproduced by permission of the International Union of Crystallography from Reference 23b

Unexpectedly, the ring strain[30,31] in **1**, 116 kJ mol^{-1}, is about the same as in cyclobutane, 110 kJ mol^{-1}. So it is interesting to compare their geometric parameters.

The carbon–carbon bond is shorter in **1** than in the other cycloalkanes: the bond length has a maximum in cyclobutane with increasing ring size[32]. A similar variation of C—O and C—S bond lengths is observed in the analogous heterocyclic molecules[33,34].

C—C	c-C_3H_6	c-C_4H_8	c-C_5H_{10}	c-C_6H_{12}
r_g (Å)	1.5139 (12)	1.554 (1)	1.546_0 (1_2)	1.535 (2)
Reference	21	35	36	37

The remaining C—C bond will be shorter if one of the methylene groups in **1** is replaced by an NH group, then by O; the change is the same from the NH to the O analog as from PH to S:

C—C	$(CH_2)_2NH$	$(CH_2)_2O$	$(CH_2)_2PH$	$(CH_2)_2S$
r_s (Å)	1.481 (1)	1.466 (2)	1.502 (5)	1.484 (5)
Reference	38	39	40	41

The C—H bond is shorter in **1**, in line with its higher s character, than in cyclobutane and in the CH_2 group of propane, and is comparable to C(sp^2)—H distances in benzene and ethene:

C—H	c-C_3H_6	c-C_4H_8	C_3H_8	C_6H_6	CH_2=CH_2
r_z (Å)	1.084 (2)	1.093 (3)	1.096 (2)	1.083 (6)	1.0868 (13)
Reference	21	35	42	43	44

The narrow C—C—C bond angle in **1** is accompanied by a wide H—C—H angle, $\angle_z$ 114.5 (9)° (Table 1). The methylene bond angles in propane are[42] $\angle_z$ C—C—C 112.0 and H—C—H 107.8°. The values for **1** stand somewhere near one of the extremes in empirical relationships between bond lenghts and bond angles in C—CH_2—C fragments[45].

B. Gas-phase Complexes

The bent quasi-π C—C bond of **1** is capable of forming different types of intra- and intermolecular bonding, e.g. it may serve as proton acceptor in hydrogen-bonded complexes. Structural characteristics of the complexes have been obtained from their MW spectra.

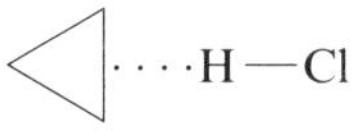

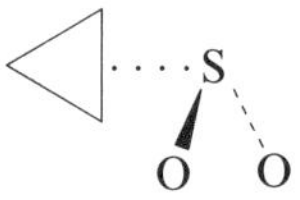

(14) **(15)**

The complex with HCl (**14**) has C_{2v} symmetry. The HCl molecule lies in the C_2 axis in the ring plane, and the H atom points to the middle of a C—C edge[46]. The distance of the Cl atom from that midpoint (M) is 3.567 Å. Similar is the orientation of HF[47] to the ring with M···F 3.021 (3) Å and of HCN[48] with M···C 3.475 (2) Å. The distance increases and the stretching force constant of the complex bond decreases in analogous π complexes from **1** to acetylene and to ethylene[48]. The hydrogen bond is directed to the center of the carbon–carbon bond in these cases. There is nearly linear hydrogen bond in the complex of **1** with water[49]. The bonding H atom and the O atom are in the ring plane, and the O—H line points to one of the C—C centers (M), with an M···H distance of 2.34 Å. The complex with SO_2 (**15**) has a different geometry and C_s symmetry[50]. The S atom lies in the ring plane at 3.203 Å from a C—C bond center, and the O—S—O plane is nearly parallel to this bond. The distance is larger in the acetylene and ethylene complexes, but the orientation of SO_2 is also different (see Reference 50 for references). The complex molecules perform intramolecular tunneling motion that leads to line splittings in the spectrum[49–51].

Oxirane and thiirane also form complexes with proton donors, but the hydrogen bond is oriented towards one of the lone pairs of the heteroatom[52].

III. SUBSTITUTED CYCLOPROPANES

Structures are arranged in this section according to the kind of atom attached to the cyclopropane ring. Substituted derivatives with an sp^3, sp^2 and sp carbon atom are followed by those with halogens, oxygen, nitrogen etc. Mixed derivatives are discussed under the substituent group of our primary interest.

Numerous systematic studies on substituted cyclopropanes, also on simple molecules, have been performed recently. The interpretation of geometric effects and the nature of electronic interactions between the cyclopropane ring and substituents have been a matter of much controversy. π-acceptor substituents as a carbonyl, vinyl, phenyl, cyano etc. give rise to a shortening of the distal ring bond and a lengthening of the adjacent bonds, and prefer the bisected conformation[8,9]. Geminal difluoro and dichloro substitution causes the opposite geometrical change, but other factors than π conjugation seem to be important (cf. Section III. E).

Allen described[8] the asymmetry of the three-membered ring (**16**) in terms of δ_n, the deviation of bond length r_n (n = 1, 2, 3) from the average in the same molecule: $\delta_n = r_n -$

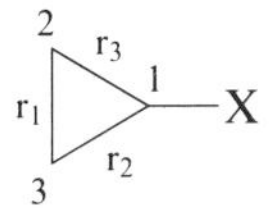

(16)

$(r_1 + r_2 + r_3)/3$. The geometrical effect of a single substituent X or of a pair of geminal substituents is then characterized by $\delta(X) = \delta_1$, the lengthening (a negative value when a shortening occurs) of the opposite bond compared to the average. In multiply substituted cyclopropanes, additivity of the substituent effects is assumed and has been demonstrated[8,53]. It is also assumed, as a rule, that the effect of a substituent on the vicinal bonds is symmetric, i.e. $\delta_2 = \delta_3 = -\delta_1/2$.

Asymmetry parameters $\delta(X)$ have been obtained for a number of substituents[8]. Several factors limit the reliability of such derived parameters, e.g. the validity of additivity and other assumptions, the influence of conformation, electronic and steric interactions of substituents, additional strain in polycyclic systems, crystal packing effects and size and quality of the data sample.

A. Alkyl and Cycloalkyl Cyclopropanes

Alkyl and cycloalkyl substituents seem to have little effects on the geometry of the cyclopropane ring. There are hardly any reliable data on ring distortion in simple molecules. Recent studies have concentrated on highly strained structures (Table 2).

TABLE 2. Structural parameters of alkyl cyclopropanes and bicyclopropyl derivatives[a]

Compound		$(C—C)_{ring}$	$C1—C_{me}$	C1—C4	M—C1—C4	$C4—C1—C_{me}$	Reference
17	ED, r_a	1.509 (1)	1.517 (2)		123.3 (1)[b]		54
18	MW, r_0	1.520	1.514		122.2[b]		55
19	ED, r_a	1.502 (4)[c]	1.519 (18)[c]	1.535 (14)	121 (5)	117 (2)	56
20	ED, r_a	1.506 (3)	1.516 (10)	1.540 (5)	126.6[d]	114.7 (13)	56
21	ED, r_a	1.508 (3)	1.519 (4)		123.2 (2)[b]		54
22[e]	XD	1.501		1.532	130.8[f]		57
23	ED, r_a	1.507 (3)		1.499 (16)	126.4[g]		58
23	XD	1.5050 (3)[h]		1.4924 (4)	124.70 (5)	115.0 (4)[i]	23c
24	ED, r_a	1.510 (5)	1.530 (11)	1.508[d]	124.7 (6)	115.1 (9)	59
25[j]	XD	1.510	1.461	1.508[k]	126.6[l]	112.7 (3)[k]	60

[a]Distances are in Å, angles in degrees. The average C—C distance in the ring is given. C_{me} is the methyl carbon, M is the midpoint of the C2—C3 bond.
[b]Angle $M—C1—C_{me}$.
[c]From the refinement with fixed M—C1—C4 and $C4—C1—C_{me}$ angles.
[d]Fixed in the final refinements.
[e]Mean values.
[f]Calculated from angles C3—C1—C2 59.4° and C2—C1—C4 124.6°.
[g]Calculated from angles C3—C1—C2 60° (assumed) and C2—C1—C4 120.9(10)°.
[h]Reference 23b.
[i]Angle C4—C1—H.
[j]Mean values, corrected for libration.
[k]The C1—C(N) bond and the C4—C1—C(N) angle, respectively.
[l]Calculated from atomic coordinates.

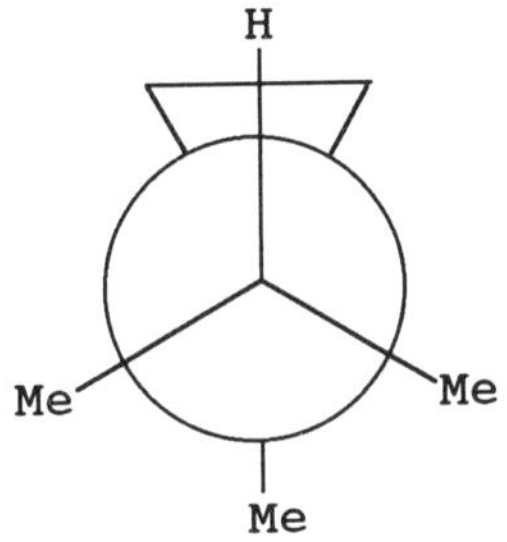

FIGURE 3. The conformation of 1-isopropyl-1-methylcyclopropane (**19**). Newman projection down the C4—C1 bond

1. Alkyl cyclopropanes

The structures and conformational behavior of 1-isopropyl-1-methylcyclopropane (**19**) and 1-*tert*-butyl-1-methylcyclopropane (**20**) have been studied by ED and theoretical calculations[56]. The distortion of the ring could not be determined from the experiment. The mean C—C lengths in the ring are similar to those in the Me derivatives (Table 2). The central bonds C1—C4 are elongated compared to C—Me bonds; the M—C1—C4 angle in **20** is opened for steric reasons. The strain at a C—C bond relaxes and the bond shortens when a pair of attached geminal Me groups is replaced by a —CH_2—CH_2— moiety, forming a smaller angle in the three-membered ring than the Me—C—Me angle[56]. Compare, e.g., the central C—C bond distances[61] in Me_3C—CMe_3 1.583 (10), **20** 1.540 (5) and **24** 1.508 Å (Table 2). **20** exists in the bisected staggered conformation. Energy calculations indicated a second minimum for **19**, but only the lower energy form with a bisecting position of the (Me_2)C—H bond is present according to ED[56] (Figure 3).

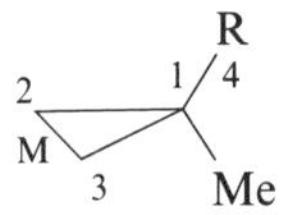

(**17**) R = H
(**18**) R = Me
(**19**) R = *i*-Pr
(**20**) R = *t*-Bu

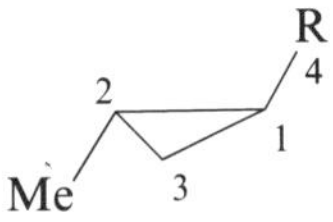

(**21**)

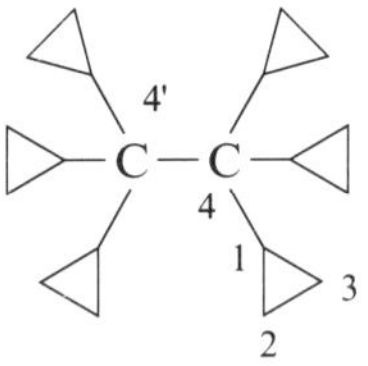

(**22**)

The changes in the length of a C—C single bond may be interpreted as an effect of hybridization. An analysis of XD data indicates that the bond from the cyclopropyl ring to a substituent can be described approximately as an $sp^{2.22}$ hybrid[9], and the mean C—C bond length between the C_3 ring and a C(sp^3) atom is 1.519(2) Å.

Hexacyclopropylethane (**22**) has an effective S_6 point group symmetry in the triclinic $P\bar{1}$ crystal[57]. The asymmetric unit contains four half molecules. The high steric strain is clearly seen in the geometric structure of the molecule. The rings adopt a propeller-like arrangement (Figure 4) with an average dihedral angle[57] C2—C1—C4—C4′ of 90.7°, which means a rotation of the rings by about 120° from the *syn* bisecting position of the C4—C4′ bond. The bond angle C2—C1—C4 124.6° is wide, and the central C—C bond is extremely long, 1.636 Å! The C1—C4 bonds are also longer than C1—C_{me} in **17** (Table 2). The distal bonds in the rings, 1.490 Å on the average, are slightly shorter than the vicinal bonds, 1.507 Å.

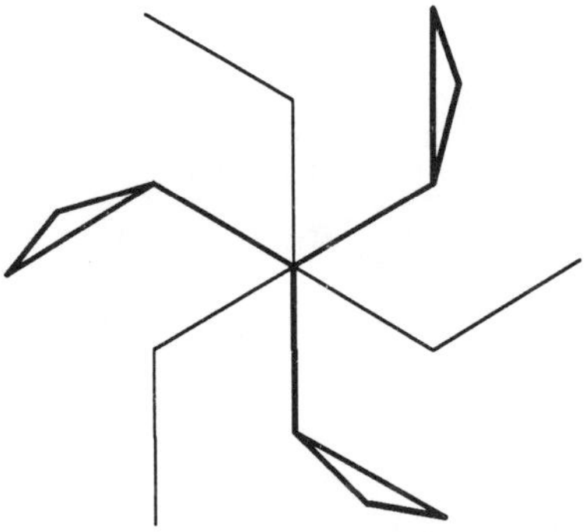

FIGURE 4. Hexacyclopropylethane (**22**) molecule projected down the central C—C bond. Drawn after Reference 57

2. Bicyclopropyl derivatives

Bicyclopropyl (**23**) is a mixture of about equal amounts of the *anti* and the *sc* (*gauche*) conformer in the gas phase[58]. 1,1′-dimethylbicyclopropyl[59] (**24**), on the other hand, exists in the *sc* form with a dihedral angle of 58.0(15)°. Equal ring bond lengths had to be assumed in the ED analysis (Table 2). Parameters obtained are consistent with the MM3 calculated values for **23**[62].

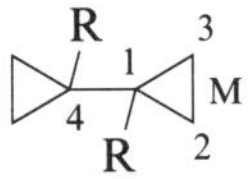

(**23**) R = H
(**24**) R = Me
(**25**) R = CN

The crystal of **23** is orthorhombic[23a], space group *Cmca*. The molecules possess crystallographic $2/m$ (C_{2h}) point group symmetry, which involves the exact *anti* bisected form. This is the less stable form in the gas phase, as deduced from their relative abundance[58]. The deformation electron-density maps have been determined at 100 K by XD[23b]. The density distribution is slightly noncylindric in the section perpendicular to the central C—C bond, less than in the case of vinylcyclopropane (see Figure 7b below), indicating a smaller conjugative interaction than in vinylcyclopropane[23c]. The ring asymmetry is not significant in **23**. The M—C1—C4 angle (Table 2) is somewhat larger than in **17** and in vinylcyclopropane (Table 3).

The molecules of **25** have the same *sc* conformation with τ 57.9° as gas-phase **24**, and C_2 symmetry in the orthorhombic *Pbcn* crystal[60]. Mean bond lengths are the same in the two molecules (Table 2). The cyano substituent causes a shortening of C2—C3 to 1.489 Å in **25**. Steric repulsion of *gauche* CH_2 groups explains the widening of C3—C1—C(N) 122.3(3)° relative to C2—C1—C(N) 120.4(3)°, but C1—C3 1.515 Å is shorter than C1—C2 1.526 Å (corrected bond lengths)[60].

26 is the addition product of dichloroketene and an olefin[63]. The C4—Me bond is in *syn* bisecting position to the three-membered ring, and angles C4—C1—C2 are slightly widened to 122.5°. The ring is nearly symmetric, the C—C bond (mean) 1.503 and the

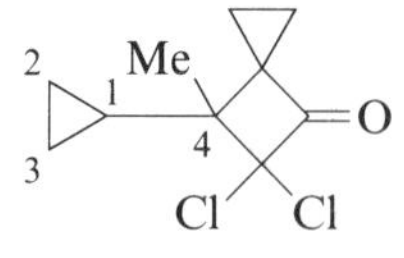

(26)

C1—C4 bond 1.507(3) Å are as usual (Table 2), and indicate no strain in this part of the molecule.

3. Alkyl derivatives with functional groups

The conformational behavior of alkyl cyclopropanes substituted in the alkyl chain is controlled either by steric repulsion between substituent groups and the three-membered ring, or by special interactions, e.g. hydrogen bonding with the bent quasi-π C—C bonds of the cyclopropane ring. The staggered conformation about the bond to the substituent seems to be preferred, and bulky groups turn away from the C_3H_5 ring.

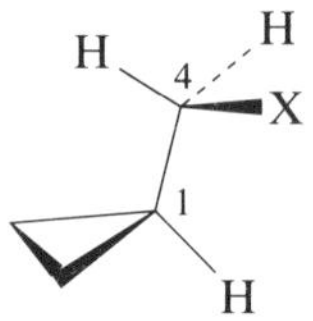

(27) X = Br
(28) X = SiH_3
(29) X = SiF_3

The C—Br bond is *anticlinal* (*ac*) to the ring bisector in the prevailing conformer of **27**[64]; the *syn* form has higher energy by 30 kJ mol^{-1}. The distal C—C bond in the ring is shorter by about 0.03 Å than the adjacent bonds from a joint ED and *ab initio* analysis. The *ac* forms have been found to be predominant in **28** and **29** from vibrational spectra and *ab initio* calculations[65].

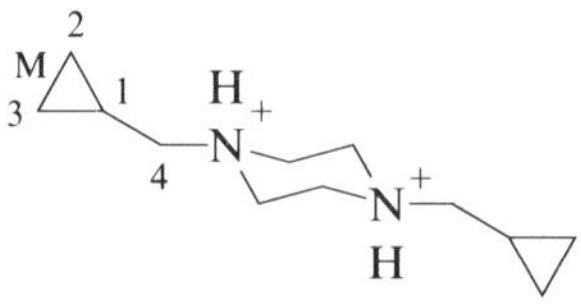

(30)

A piperazine derivative (**30**) of **17** exhibits inotropic activity[66]. The dication in the crystalline dihydrochloride has a center of symmetry and a chair-form ring[66]. The $C_3H_5CH_2$ units occupy *trans* equatorial positions, and take themselves *ac* conformations, thus giving an extended shape to the cationic species. The M—C1—C4—N dihedral angle is (as calculated from coordinates) 125.2°, and some other parameters are: C1—C2 (mean) 1.498, C2—C3 1.490 (3), C1—C4 1.489 (3) Å, C2—C1—C4 (mean) 118.4°.

The crystallographic study of some alkyl cyclopropane derivatives was motivated by the observation of an unusual long-range coupling (^4J) in their ^1H NMR spectra[67] between a H on the ring and a methylene hydrogen of the *trans* substituent. This coupling through four bonds is one of the analogies between cyclopropanes and allylic systems. The required zigzag form of the H31—C3—C2—C4—H chain is achieved by a suitable conformation and partially hindered rotation of the ring substituents[68]. The chemical shift of the proton (H42) which is coupled with H31 in **31a–d** appears at a lower field than that of H41, while the order of shifts is reversed in **31e**. This may be related to the different conformations of **31a–d** in the crystal from that of **31e**[68]. The Ph substituent is close to the perpendicular conformation in these molecules, and the C—C bond opposite to it is the shortest in the ring.

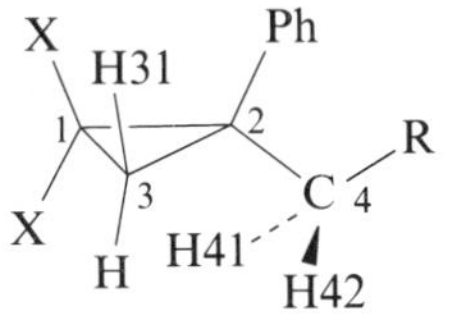

(31)

(**a**) X = Br, R = CH=CH_2
(**b**) X = Br, R = COOH
(**c**) X = Cl, R = COOH
(**d**) X = Cl, R = Br
(**e**) X = Cl, R = $O_2CC_6H_4NO_2$-*p*

The bent σ bonds of the cyclopropane ring, like the bonds of an olefin or acetylene, can form weak intramolecular hydrogen bonds with proton-donating substituent groups. Marstokk and Møllendal have studied such derivatives by MW and other spectroscopic methods, accompanied by *ab initio* calculations, which helped to find possible low-energy forms[69–74]. Observed MW transitions were enough to identify prevailing conformers, and to fit a few of their parameters, assuming the rest of them.

Two hydrogen-bonded forms of (cyclopropylmethyl)amine (**32**) have been found[70] in the gas phase, with different conformations about the C—NH_2 bond and *gauche* form of the H—C1—C4—N sequence having dihedral angles of about 60°. One *gauche* H—C1—C4—O form of 1-cyclopropylethanol (**33**) could be assigned in the MW spectra[71]. The ED intensities of 2-cyclopropylethanol (**34**) were fitted by a mixture of three conformers, in accord with relative energies from 6-31G** *ab initio* calculations[72]. The average of the ring C—C bond lengths, 1.519 (8) Å, was obtained from the ED analysis. The most abundant form (50 percent) is stabilized by a hydrogen bond, has a (+*sc*, +*sc*, –*sc*) H—C1—C4—C5—O—H chain (Figure 5) and is the only one that was identified by MW. Both forms of *trans*-2-methylcyclopropylmethanol (**36**) occur in the gas phase: it seems that the *trans* 2-methyl substituent only slightly biases the choice of the OH group, whether it turns towards the C—C bond vicinal or distal to Me[73]. The thiol group also interacts with the C_3 ring. The only conformer of cyclopropylmethanethiol[74] (**35**) (Figure 6) is similar to the H-bonded *gauche* form of its oxygen analog[75].

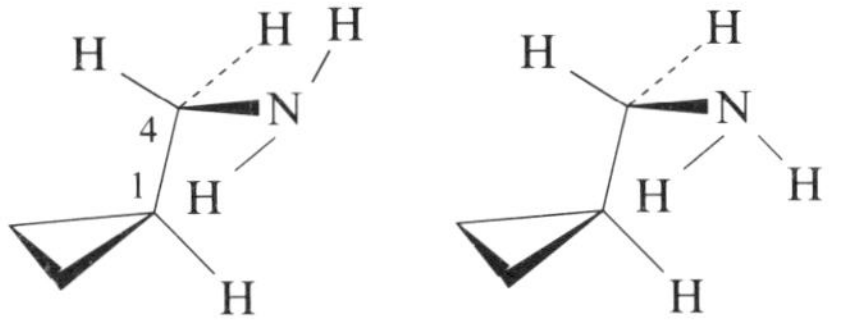

(32)

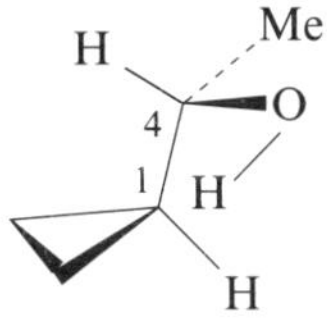

(33)

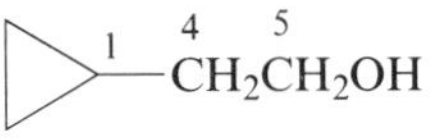

(34)

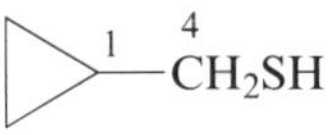

(35)

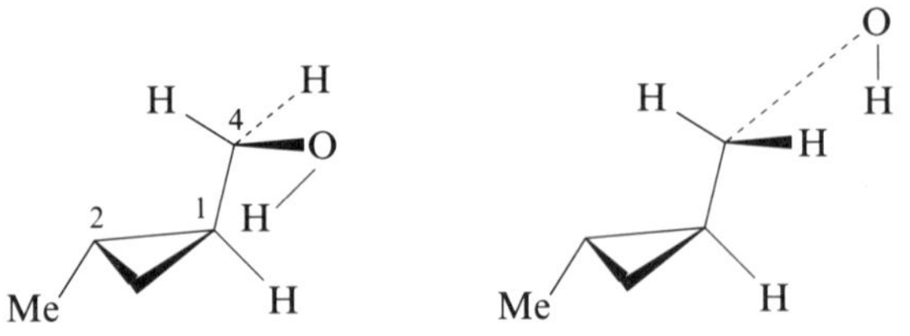

(**36**)

The common feature of the conformers described here is, regarding the torsion about the C1—C4 bond, that the substituent's C—C, C—N, C—O or C—S bond occupies the *ac* position to the ring bisector. The (N)H···C, (O)H···C and (S)H···C distances in these molecules (to one of the carbon atoms anchoring the bent σ bond) are about 2.6 to 3.1 Å, and the N—H···C etc. angles 70° to 110°, indicating rather weak linkages.

The three-membered ring of analogous oxirane derivatives offers three acceptor sites for a H-bonding substituent, viz. the ring C—C bond, the vicinal C—O bond and the oxygen atom, of which the latter seems to be preferred (see References 69, 76 and works cited therein).

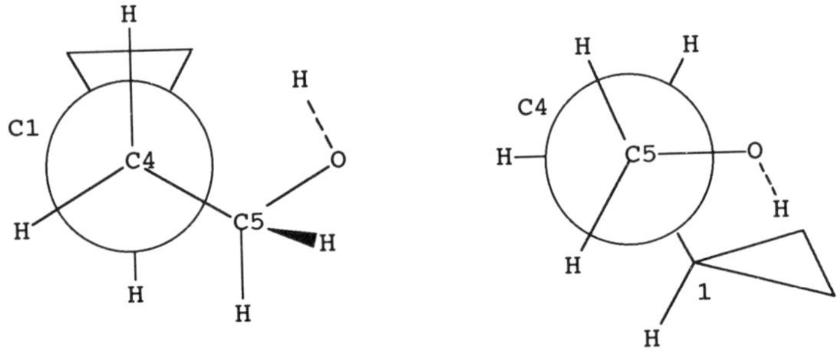

FIGURE 5. The (+*sc*, +*sc*, –*sc*) form of 2-cyclopropylethanol (**34**). Newman projections down the C4—C1 bond (left) and the C5—C4 bond (right)

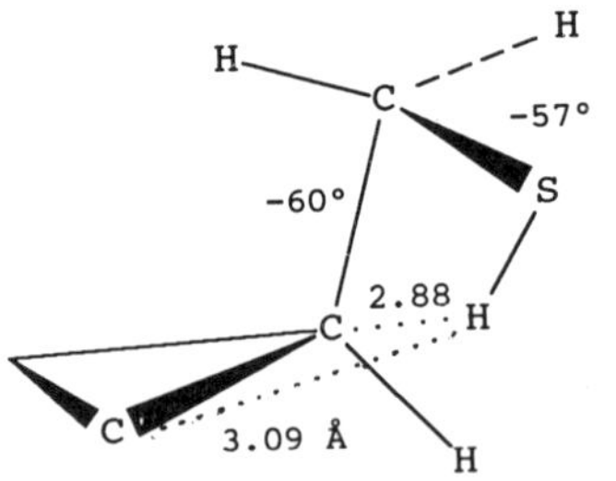

FIGURE 6. The conformation of cyclopropylmethanethiol (**35**). Dihedral angles of the H—C—C—S—H chain and the short H···C distances (Å) are indicated

B. Vinyl and Aryl Cyclopropanes

1. *Vinyl derivatives*

Main structural parameters of vinylcyclopropane (**37**) from recent ED (at 273 K) and low-temperature (94 K) XD studies are collected in Table 3. The large discrepancy in the C—C=C angles might be due to correlation between similar internuclear distances and to assumptions in the ED analysis[77]. The agreement with the earlier ED results[79] may be regarded as satisfactory. The gas consists of the bisected *anti* conformer in 77 (3) percent and of the less stable *gauche* (*sc*) form[77]. The energy difference of 4.3 kJ mol^{-1} calculated from their ratio is in line with the *ab initio* result[80], 5.18 kJ mol^{-1}. The torsional angle is 56 (6)° in the *sc* form.

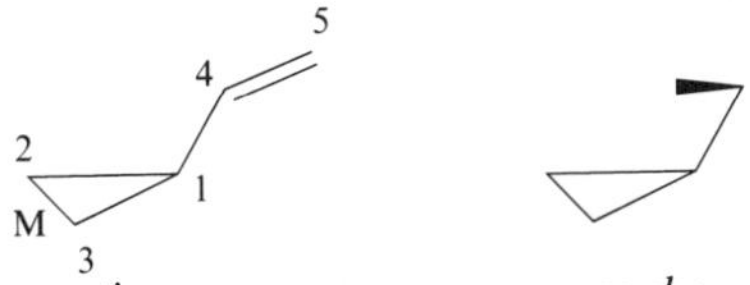

anti *gauche*

(**37**)

(*ap*, *ap*) (*ap*, +*sc*)

(**38**)

TABLE 3. Structural parameters of vinylcyclopropane (**37**) and *trans*-1,2-divinylcyclopropane (**38**)[a]

Bond or angle	**37** ED[b], r_a	**37** XD[c]	**38** ED[f], r_a
C1—C2	1.525 (1)	1.520[d]	1.523 (16)
C2—C3	1.500	1.5096	1.513 (9)
C1—C4	1.470 (2)	1.4783	1.474 (4)
C=C	1.336 (1)	1.3383	1.335 (2)
M—C1—C4	123.9 (3)	123.62[e]	124.7 (9)
C—C=C	127.3 (3)	124.51	125.9 (5)
Reference	77	23c	78

[a]Bond lengths are in Å, angles in degrees. M is the midpoint of the distal bond.
[b]Data for the bisected *anti* form. Differences of some parameters between the *anti* and the *sc* conformer were assumed. Uncertainties are given as one standard deviation.
[c]Parameters corrected for librational motion. Estimated standard deviations are 0.0004 and 0.0005 Å for the uncorrected bond lengths, 0.03 and 0.04° for the angles.
[d]Mean value.
[e]The angle between the C1—C4 bond and the ring plane.
[f]Data for the *ap* vinyl groups. The same *ap* to *sc* bond angle increments were assumed as for **37**. Uncertainties are twice the standard deviations.

The molecules of **37** are located in general positions in the monoclinic $P2_1$ crystal[23a], and deviate slightly from the bisected *anti* form, the only one present in the solid, having a torsional angle of 177.80 (5)°. The average ring bond length in **37**, 1.5090 (3) Å (uncorrected[23b]), is larger than in **1** (Table 1). The relative shortening of the distal bond in **37** is accompanied by an outward rotation of the HCH groups by 1.8 (6)°, each about an axis through the C nucleus and normal to the ring plane[23c]. The shortening, 0.025 (15) Å, was obtained with a large estimated uncertainty from the ED study[77].

The ring asymmetry and conformation of **37** indicate some conjugation between the cyclopropyl and the vinyl groups. A convincing experimental evidence for conjugation is that the dynamic electron-density distribution about the central C1—C4 bond has non-cylindric character, of course to a smaller extent than about the C=C double bond (Figures 7b and c). The bent bonds in the cyclopropyl ring are again clearly seen in the section across the C2—C3 line (Figure 7a). The *syn* form of **37**, which is also favored by conjugation, is not realized because of nonbonded H···H interactions[77,78].

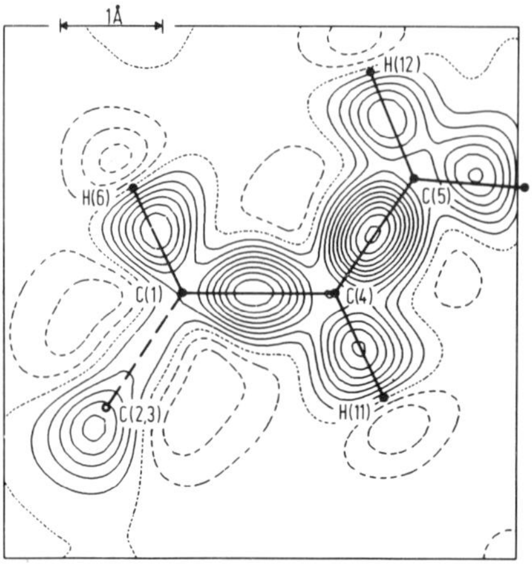

(a)

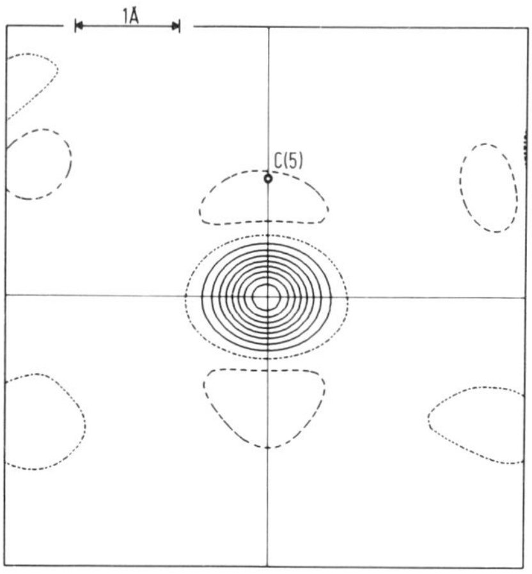

(b)

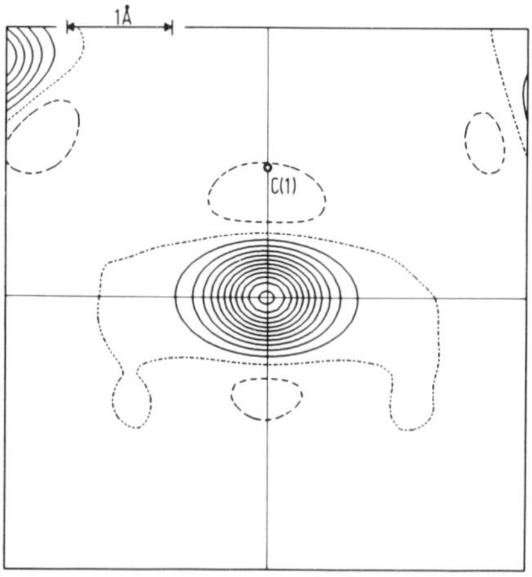

(c)

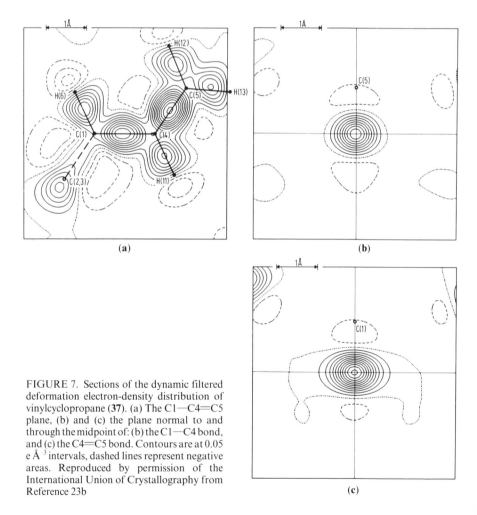

FIGURE 7. Sections of the dynamic filtered deformation electron-density distribution of vinylcyclopropane (**37**). (a) The C1—C4=C5 plane, (b) and (c) the plane normal to and through the midpoint of: (b) the C1—C4 bond, and (c) the C4=C5 bond. Contours are at 0.05 e Å^{-3} intervals, dashed lines represent negative areas. Reproduced by permission of the International Union of Crystallography from Reference 23b

The ED analysis of *trans*-1,2-divinylcyclopropane (**38**) encounters even larger difficulties[78]. Combinations of *antiplanar* and *synclinal* vinyl groups should be considered (see formulae **38** for examples). The conformational composition, (*ap*, *ap*) 58 (30) percent, (*ap*, +*sc*) and (*ap*, −*sc*) 29 (56) percent, (*sc*, *sc*) (including all combinations of + and − orientations) 13 (28) percent, which was obtained with some simplifications, merging contributions that are hardly distinguishable by ED, corresponds roughly to a distribution for two independent vinyl groups, as estimated from the percentages for **37**[78]. The ring bond between the substituents is longer, though not significantly, than the two other bonds (Table 3). The junction and geometry of the vinyl groups are similar to those in **37**; the C—C=C angle seems to be more realistic. Structural data do not indicate any extended conjugation between the two vinyl groups[78]. Compared to the parameters in propene[81], r_g C—C(=) 1.506 (3), C=C 1.324 (2) Å and C—C=C 124.3 (4)°, the double bonds are slightly longer, the bonds between the ring and the vinyl group are shorter in **37** and **38**.

The vinyl substituent on the cyclopropane ring prefers the bisected *syn* or *anti* conformation, and the mean distal bond shortening in the C_3 ring is $\delta_1 = -0.022$ (4) Å, smaller in magnitude than that for the C=O group[8]. The marginal ring asymmetry observed in other forms indicates[8] that π donation from the cyclopropyl to the vinyl group is limited to narrow ranges of the dihedral angle τ about 0° and 180°. The mean C—C bond length from the ring to a vinyl group is 1.480 (4) Å if only hybridization has to be considered, and 1.470 (6) Å with an additional effect of conjugation[9].

The chloro derivative **39**, which has been studied by ED and molecular mechanics (MM) calculations[82], exhibits the perpendicular conformation. The torsional angle is 91 (3)° from ED; the barrier to rotation at the *ap* position is 21(16) kJ mol^{-1}. Structural data were obtained with large uncertainties; the mean ring bond length is 1.518 (11) Å.

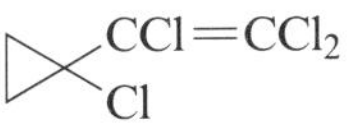

(39)

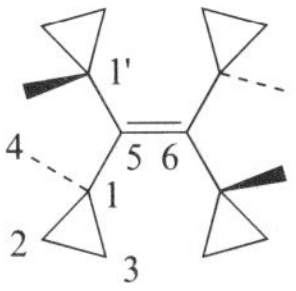

(40)

The structure of tetrakis(1-methylcyclopropyl)ethene (**40**) has been determined at −100 °C by XD[83]. The highly strained molecule possesses a nearly perfect (noncrystallographic) D_2 symmetry in the triclinic $P\bar{1}$ crystal (Figure 8), with the most twisted double bond known at that date. The mean torsional angle about the double bond is 19.7°. Selected mean bond lengths (Å) and bond angles (deg) are listed here[83]:

C1—C2	1.524	C2—C1—C4	113.8
C1—C3	1.506	C3—C1—C4	115.9
C2—C3	1.494	C2—C1—C5	115.7
C1—C4	1.519	C3—C1—C5	120.6
C1—C5	1.514	C1—C5—C1′	111.9
C5=C6	1.353	C1—C5=C6	124.1

The double bond is longer than in ethene, 1.313 Å from XD[84], and r_g 1.337 (2) Å from ED[85], and in **37**, and the C—C(=) single bonds are also longer than in **37** (Table 3). The cyclopropyl groups are rotated by 66.5° from the *syn* bisected form, which would be

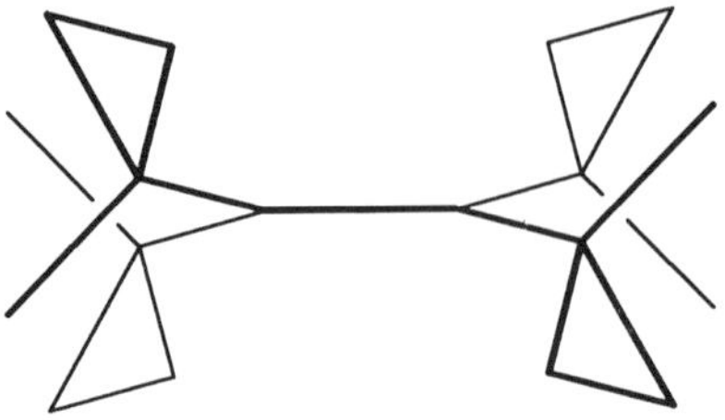

FIGURE 8. Tetrakis(1-methylcyclopropyl)-ethene (**40**) molecule viewed along an axis of approximate two-fold symmetry. Drawn after Reference 83

favorable for conjugation with the C=C bond. Thus a small shortening of the distal bond is observed; the difference between vicinal bonds is caused by steric strain. As seen from the bond angles, the $C_2C{=}C$ fragments are planar; no pyramidalization of the sp^2 carbon atoms occurs[83].

2. *Aryl derivatives*

Numerous phenyl derivatives of cyclopropane have been investigated by XD, and the distortion of the cyclopropane ring and conformation have been assessed[8,15,86,87]. A shortening of the distal bond in the C_3 ring has been found in different rotational forms, δ_1 = −0.018 (2) Å in the sample studied[8]. The perpendicular conformation dominates, due probably to steric interactions with other substituents. Electron donation from a cyclopropyl orbital to the Ph ring is effective in the bisected conformation, and causes distal bond shortening. Another orbital may *accept* electrons from the Ph ring in perpendicular position, resulting in a similar distortion of the C_3 ring[8]. It must be remembered, however, that the cyclopropyl group is a good electron donor but a weak electron acceptor[88].

Two simple phenyl derivatives have been studied by low-temperature XD: phenylcyclopropane (**41**) at −100 °C and *cis*-1,2-diphenylcyclopropane (**42**) at −40 °C (Table 4). Structural data of **43** are also listed in the Table.

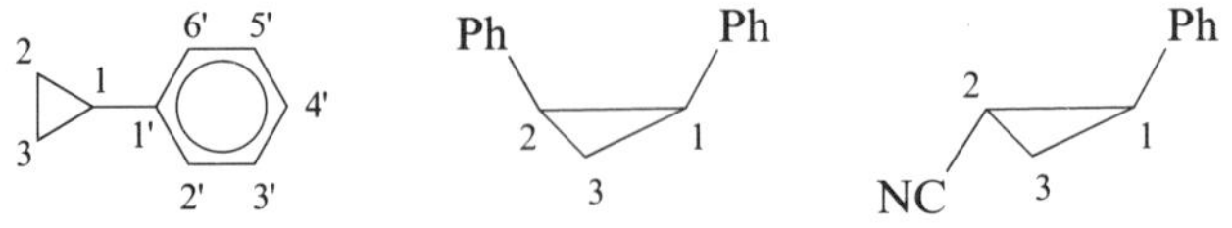

(**41**) (**42**) (**43**)

Two independent molecules are found in the monoclinic *Pa* crystal of **41**[89], both in the bisected form. The distal bond in the cyclopropane ring is shorter by 0.018 Å than the vicinal bonds. The phenyl groups of **42**[90] are rotated (as calculated from atomic coordinates) by 19° and −85° from the bisecting position, away from each other. Such conformationally asymmetric molecules make up the chiral orthorhombic $P2_12_12_1$ crystals. The phenyl rings are suitably oriented for coordinating a metal (ion). Indeed, there is an indication in the ^{71}Ga NMR spectra that **42** forms a complex with Ga(I) in solution like alkyl aromatic molecules do[90]. The mean C—C length in the three-membered ring, 1.506 Å, is the same as in **41**, 1.507 Å (uncorrected). Noteworthy is that the C2—C3 bond, which is distal to the bisecting Ph group, has markedly shortened, and the bond between the *cis* substituents has

TABLE 4. Bond lengths (Å) and bond angles (deg) of phenylcyclopropane (**41**), *cis*-1,2-diphenylcyclopropane (**42**) and *trans*-1-cyano-2-phenylcyclopropane (**43**) from XD[a]

Bond or angle	**41**[b]	**41**[c]	**42**[d]	**43**
C1—C2	1.520 (5)	1.513	1.525 (4)	1.509 (4)
C2—C3	1.502 (7)	1.494	1.489 (4)	1.501 (4)
C3—C1			1.504 (3)[e]	1.492 (4)
C1—C1′	1.476	1.471	1.488	1.478 (3)
C3—C1—C1′		121.3	122.7	121.1
C1′—C2′	1.407	1.400	1.387	1.393
C2′—C3′	1.384	1.381	1.388	1.380
C3′—C4′	1.395	1.388	1.376	1.375
C6′—C1′—C2′		117.4	117.8	117.4 (3)
C1′—C2′—C3′		121.4	121.3	121.4
C2′—C3′—C4′		120.5	119.8	119.8
C3′—C4′—C5′		119.0	120.0	120.3 (3)
Reference	89	89	90	91

[a]Average values of equivalent parameters are listed, except for those parameters of **42** and **43** where uncertainties are given. The numbering of atoms in **43** does not follow conventions (see formulae).
[b]Bond lengths corrected for libration.
[c]Uncorrected parameters.
[d]Parameters of the phenyl ring have been calculated from atomic coordinates and unit cell parameters given in the paper.
[e]Opposite to the perpendicular Ph group.

lengthened, while the C3—C1 bond, which is opposite to the perpendicular Ph group, lies in between (Table 4). Steric and conjugation effects appear superimposed here, and additivity of substituent effects[8] might not be valid even in such a simple molecule as **42**.

The effect of the cyano substituent seems to dominate in **43**; the bond opposite to it is the shortest[91]. The C—C (N) bond has the same length, 1.447 (4) Å, as in other cyanocyclopropanes (Section III.D). The phenyl group adopts the bisected conformation. Crystallization leads to a spontaneous resolution, which is rather rare among phenylcyclopropanes, and a conglomerate of chiral $P2_12_12_1$ crystals is formed[91].

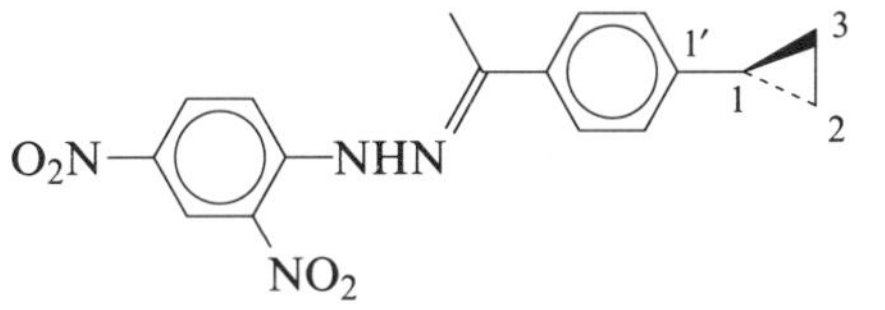

(**44**)

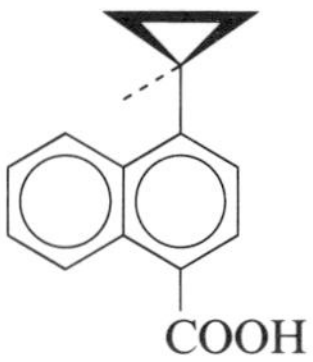

(**45**)

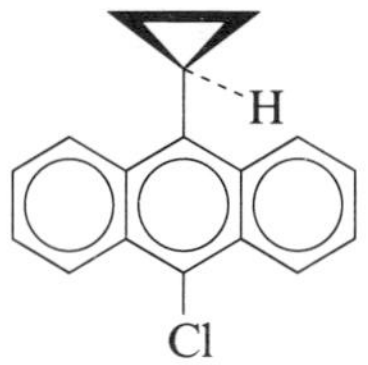

(**46**)

That conjugation effects depend on the torsional angles has been demonstrated by the structural changes in aryl cyclopropanes **44**–**46**, studied by MM calculations and XD[92]. The torsional barrier of the cyclopropyl group and its propensity to perpendicular orientation increase on going from the phenyl derivative **44** to **45** and **46**, because of interactions with hydrogens of the aromatic system[92]. Conjugation and shortening of the distal bonds and of the C1— C1′ bonds should diminish as the angle of torsion (τ) from the bisected form increases (XD, Å, deg):

	C1—C2 (mean)	C2—C3	C1—C1′	τ
44	1.497	1.457 (5)	1.471 (4)	0.6
45	1.459	1.445 (7)	1.498 (5)	54
46	1.499	1.491 (5)	1.503 (4)	88

(Trends are probably blurred by libration of molecules in the crystal.)

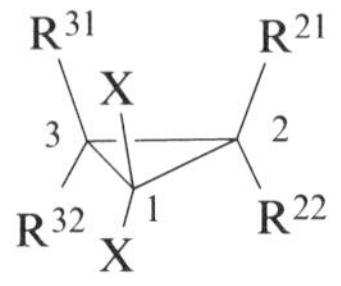

(**47**) X = Cl
(**48**) X = H

	R^{21}	R^{22}	R^{31}	R^{32}
47a	Ph	H	Ph	H
47b	p-C_6H_4OMe	H	p-C_6H_4OMe	H
47c	p-$C_6H_4OCH_2Ph$	H	Ph	H
47d	H	Ph	p-C_6H_4OOCMe	H
47e	Ph	H	p-C_6H_4OMe	Ph
48a	p-C_6H_4OMe	p-C_6H_4CN	p-C_6H_4OMe	p-C_6H_4CN
48b	p-C_6H_4CN	p-C_6H_4CN	Ph	Ph
48c	Ph	p-C_6H_4CN	p-C_6H_4OMe	Ph

The antiestrogenic activity of **47a** motivated structural studies of *gem*-dichloro compounds **47**[93,94]. The orientations of the *cis*-diaryl rings in **47** and in the well-known

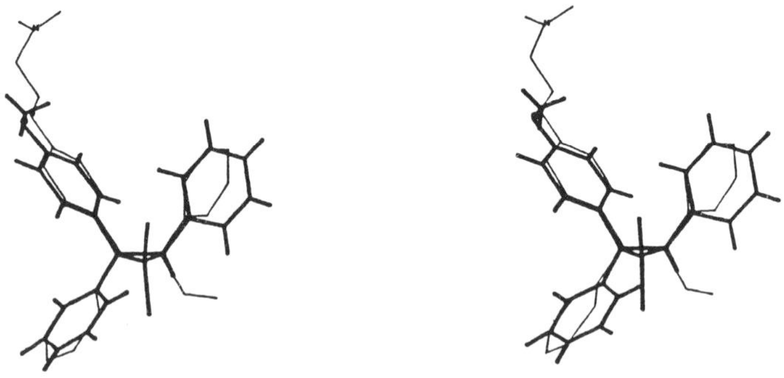

FIGURE 9. Stereoview of molecule A in the crystal of **47e** (thick lines) with a superimposed tamoxifen (**49**) molecule (thin lines). Reproduced by permission of the International Union of Crystallography from Reference 94

ethylenic antiestrogen, tamoxifen (**49**), are remarkably similar (Figure 9). A considerable deviation from **49** is seen only in one of the *trans* aryl substituents of **47d** and **e**. One of the aryl rings in **47** (at the C2 atom) takes a nearly bisecting conformation, the other ring and both of the geminal rings in **47e** are in perpendicular position or not far from it.

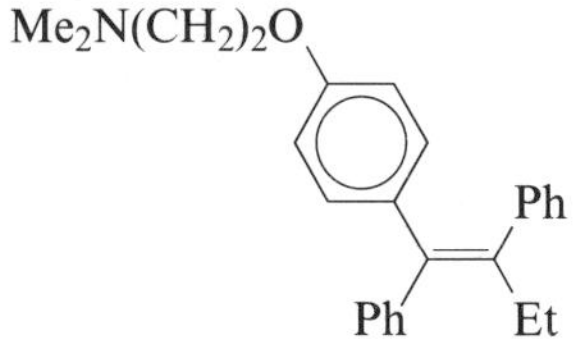

(49)

In each of the tetraaryl systems[87] (**48**), the members of a pair of *trans* aryl rings are very close to the perpendicular form (θ 80° to 90°, see the Introduction and Reference 15), the other rings have θ angles of 16° to 61°. This conformation results partly from conjugation with the cyclopropane ring, and partly from steric strain of the substituents, which is apparent in some bonds to the substituents and some bond angles formed by them. For instance, the bonds to the geminal aryl groups in **47e** are slightly longer[94], C3—C_{ph} 1.507 (3) and 1.511 (3) Å, than C2—C_{ph} 1.496 (3) Å and the bonds to the aryl substituents[93] in **47a–d**, 1.491 (2) to 1.495 (2) Å. The average bond lengths in the C_3 rings of *cis* **47a** 1.516 (15), **47b** 1.521 (13), and **47c** 1.514 (16) Å are also larger than in *trans* **47d**, 1.508 (7) Å[93], while those in the tetraaryl molecules (**48**) seem to be smaller, 1.388 (10), 1.384 (11) and 1.38 (1) Å[87], all with large standard deviations. Molecular mechanics (MM2) calculations have shown that the conformations found in the crystals of **48a–c** are very close to the forms of minimum energy[93]. A somewhat larger deviation is observed in the *trans* molecule **47d**, which is characterized by a broad energy minimum.

The cyclopropane ring shows the expected asymmetry in **47** and **48**: the bond between the aryl substituents and opposite to the *gem*-dichloro group is the longest in all molecules, and C1—C3 which lies opposite the bisecting aryl group is the shortest in **47a–d**. The C2—C3 bond is extremely long[87] in **48a**, 1.575 (7) and **48b**, 1.573 (6) Å. A modified scheme has been proposed to predict the ring asymmetry in phenyl cyclopropanes[93]. The changing effect of a phenyl substituent is taken into account in its additive term by the factor $\cos\theta$, which describes its orientation.

The length of the bond connecting the cyclopropyl group and the Ph substituent can be compared with a mean of 1.490 (15) Å over a sample of $n = 90$ observations (sample standard deviation in parentheses) for aryl cyclopropanes[95].

If the phenyl substituent acts on the geometry of the cyclopropyl ring, the reverse action is also expected. Mean parameters of the Ph group are also shown in Table 4. Substituent effects on the geometry of the benzene ring have been extensively studied[96,97]. In accord with the predictions of the VSEPR model[98], an electron-releasing (σ-donor) substituent would produce a closing of the endocyclic *ipso* angle (at the substituent) and a lengthening of the adjacent bonds in the phenyl ring[96,97]. The *ipso* angle has decreased from the regular 120° in **41**, **42** and **43**, and the *ortho* angle has increased half that value. A lengthening of the (average) C1′—C2′ bond is apparent in **41**, although deviations between crystallographically nonequivalent bonds are of the same magnitude. Similar distortions of the benzene ring are found in toluene[99], *p*-xylene[99] and ethylbenzene[100], where the *ipso* angles are 118.7 (4), 117.1 (3) and 118.3 (10)°, respectively (see also Reference 96). A decrease of the *ipso* angle may be induced by π conjugation with the substituent, either accepting or donating[96], and this notion is in harmony with the pronounced π-donor ability of the cyclopropyl ring[88].

C. Carbonyl Derivatives

The carbonyl group as a π-acceptor substituent causes a shortening of the distal bond in the cyclopropyl ring; the mean asymmetry parameter[8] is $\delta_1 = -0.026$ (5) Å. The *syn* form is preferred, but conjugative interaction seems to be retained for a larger range (±30°) of the torsional angle, so *sc* and *ac* conformations are also common[8]. The vicinal bonds are usually elongated, and often differently, compared to cyclopropane.

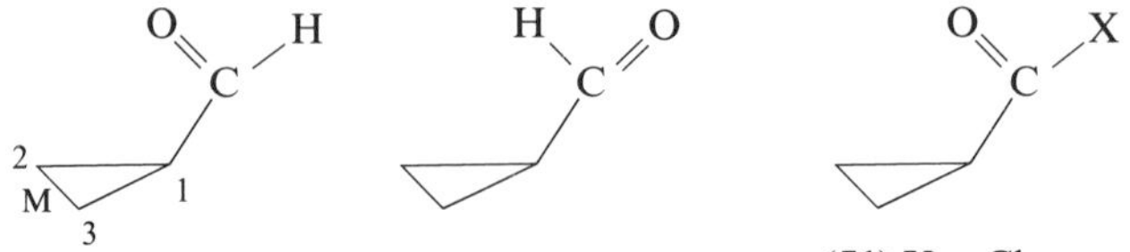

(50)

(51) X = Cl
(52) X = Me

Very few recent structural data on simple aldehydes or ketones are available. Cyclopropanecarboxaldehyde (**50**) is, according to an ED study, a mixture of 55 (10) percent *syn* with the *anti* form in the vapor at room temperature[101]. Raman spectra[102] have shown the *syn* form to be more stable by 60 cm^{-1} in the gas phase, but the *anti* conformer, which has the higher dipole moment[103], is preferred in the liquid and solid[102]. A recent *ab initio* study gave, with 6–31G* basis and full geometry optimization, 114 cm^{-1} energy difference, and barriers of 2034 and 1920 cm^{-1} for the *syn* to *anti* and *anti* to *syn* transition, respectively[104]. Durig and coworkers[104] calculated structural parameters of **50** from rotational constants, using the planar moments[103] as suggested by Penn and Boggs[105], and some constraints from the *ab initio* calculation[104]. The mean values for the two conformers (calculated from the original data), C1—C2 1.521, C2—C3 1.494, C1—C(O) 1.487, C=O 1.210 Å, C2—C1—C(O) 116.7° and C—C=O 123.2°, agree fairly well with the results of the earlier ED study[101], r_g C—C (mean) 1.507 (2), C=O 1.216 Å, M—C1—C(O) 121.0 (12)° and C—C=O 122.0 (18)°. The *syn* bisected form has been found predominant, besides a smaller amount of the form with an *anti* C=O group, in the ED investigation[106] of cyclopropylcarbonyl chloride (**51**) and cyclopropyl methyl ketone (**52**).

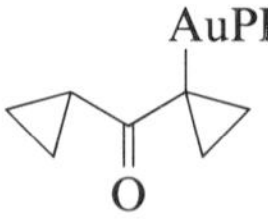

(53)

An aurated derivative (**53**) of dicyclopropyl ketone has been synthesized and studied[107]. Distal ring bonds are shortened, 1.488 (9) and 1.468 (9) Å, the one opposite to the metal substituent being marginally shorter; the average length of the vicinal bonds is 1.518 Å. The difference may be due either to the effect of the $AuPPh_3$ group, or to the deviation from C_{2h} symmetry in the organic moiety[107]. The C_3 ring at Au adopts an *sc* conformation, the other is nearly *syn* bisecting to the C=O group.

Protonated ketones have been studied recently by XD[108]. The cyclopropylcarbinyl cations **54** and **55** have different bisected conformations in their crystals with SbF_6^-. The distal bond is considerably shortened, more than in neutral ketones. Parameters C1—C2 (mean), C2—C3, C1—C4, C4—O and C1—C4—O are 1.516 (8), 1.418 (12), 1.405 (10), 1.256 (8) Å and 116.5 (6)° for **54**, and 1.537, 1.448 (9), 1.461 (9), 1.268 (8) Å and 115.0 (6)° for **55**, respectively.

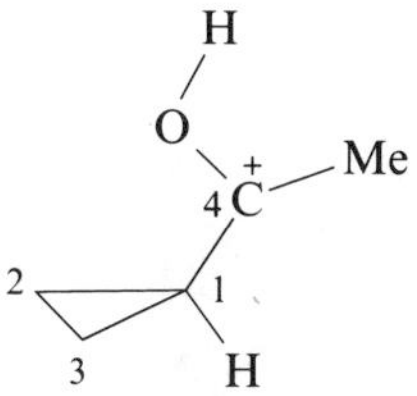

(54)

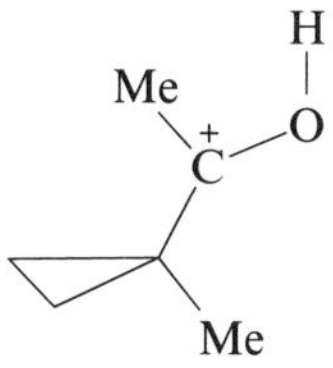

(55)

1. Carboxylic acids

Structural parameters of some cyclopropanecarboxylic acids are compiled in Table 5.

Bond length changes in the carboxylic acids are in accord with Allen's analysis[8], and with a conjugation between the C=O group and ring orbitals. The C—C bond opposite to the *sp* carboxyl group is the shortest in the ring (except for **66t**), or even shorter than in **1**. The C—C(O) bond connecting the *sp* COOH group is the shorter one in comparable cases, and makes a larger acute angle to the ring plane, as can be seen from the narrower C—C—C(O) angles. In the *cis* 1,2-diacids forming an intramolecular H bond (**65c**, **66c**), the COOH groups are bent apart: the internal C2—C1—C(O) and C1—C2—C(O) bond angles are larger than the C3—C1—C(O) and C3—C2—C(O) angles (Table 5).

TABLE 5. Structural parameters of cyclopropanecarboxylic acids[a]

Compound	C1—C2	C1—C3 C2—C3	C1—C(O) C2—C(O)	C2—C1—C(O) C1—C2—C(O)	C3—C1—C(O) C3—C2—C(O)	Reference
56[b]	1.524 (3)	1.493 (7)[c]	1.478 (5)	116.6 (3)		109
59	1.535[d]	1.488 (3)[c]	1.467 (3)	123.9[d]		110
60[d]	1.534	1.462[c]	1.485[e]	115.3[e]		111
			1.483	118.2		
63c	1.514 (4)	1.505 (4)	1.465 (4)[e]	120.9 (3)[e]	118.4 (3)[e]	112
		1.471 (4)[f]	1.480 (4)	122.6 (3)	122.0 (3)	
63t	1.530 (3)	1.488 (3)	1.480 (2)	116.1 (2)	119.7 (1)	112
64c	1.502 (3)	1.539 (3)	1.469 (3)[e]	120.1 (2)[e]	121.4 (2)[e]	113
		1.491 (3)[f]	1.496 (3)	121.2 (2)	125.8 (2)	
64t	1.528 (4)	1.514 (3)	1.474 (2)	117.2 (2)	124.2 (2)	113
65c[g]	1.518	1.528	1.487[e]	124.0[e]	119.5[e]	114
		1.519[f]	1.500	129.0	123.7	
66c[h]	1.520 (5)	1.531 (4)	1.469 (5)[e]	124.9 (3)[e]	121.6 (3)[e]	115
		1.519 (4)[f]	1.493 (5)	129.6 (3)	122.9 (3)	
66t[g]	1.496	1.540	1.488	119.0	118.1	116
		1.536	1.483	121.7	115.8	

[a]Distances in Å, angles in degrees. XD studies unless noted.
[b]r_g and $\angle_\alpha$ parameters from a joint ED, MW and *ab initio* analysis for the so-called dynamic model. Mean values for the two conformers present in the vapor.
[c]r (C2—C3), opposite to the COOH group(s).
[d]Mean values.
[e]The substituent with the *syn* bisecting (*sp*) C=O group.
[f]Opposite to the *sp* C=O group.
[g]Calculated from the atomic coordinates, given in the original paper for **66t**, and kindly provided by Dr. M. Czugler for **65c**.
[h]In the inclusion compound **66c**·EtOH (1/1).

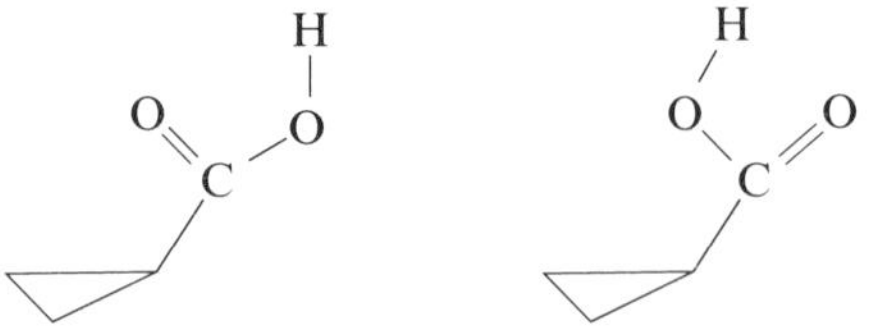

syn *anti*

(56)

Cyclopropanecarboxylic acid (**56**) has been studied recently by a combination of ED, MW and *ab initio* calculations[109] (Table 5). The ED data are consistent either with a mixture of a *syn* bisecting, 65 (5) percent at 323 K, and of an *anti* conformer, or with a 'dynamic model' that treats the equilibrium of conformers in terms of the torsional potential function. The *anti* conformer could not be detected in the MW spectrum, so it should have higher energy by at least 3 kJ mol^{-1}. The barrier to rotation is about 30 kJ mol^{-1}. The H—O—C=O group itself takes the *syn* conformation in both forms. The geometry of the COOH group, C=O, C—O, C—C=O and C—C—O, 1.214 (3), 1.349 (3) Å, 124.1 (6)° and 112.1 (4)°, is similar to that in acetic acid[117] (MW study, r_s), 1.205 (4), 1.352 (4) Å, 125.4 (4)° and 111.7 (3)°, respectively. The C1—C(O) bond, 1.478 (5) Å, is shorter in **56** than in acetic acid, 1.503 (5) Å.

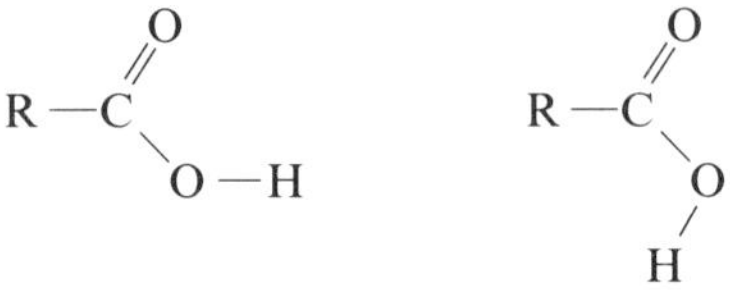

syn *anti*

(57)

Crystal-phase molecular structures and packings of carboxylic acids have been classified and discussed in detail[118]. The *syn* form of the carboxyl group (**57**), which occurs also in **56**, is more stable; *anti* H—O—C=O has been found in 1,2-dicarboxylic acids with an intramolecular hydrogen bond[118]. Both the *syn* and the *anti* carboxyl group may participate in two hydrogen bondings: the OH group acts as a proton donor, and the carbonyl oxygen as an acceptor. The prevalent form of association between carboxylic acid molecules in the crystal is the cyclic dimer, known in the liquid and gas phases, too. Vapor-phase complexes with F_3CCOOH have been identified by low-resolution MW spectroscopy[119], e.g. **58**. The specific chain form in which "each carboxyl group is linked to two neighbors via single O—H···O(=C) bonds", the *catemer* motif as Leiserowitz termed it, is rarely observed[118]. Cyclopropane derivatives present interesting combinations of these structural elements.

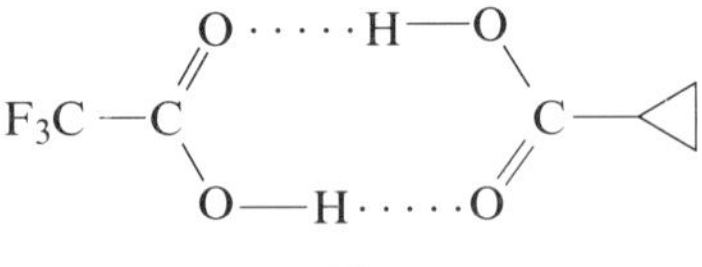

(58)

Cyclic dimers are formed by pairs of hydrogen bonds in the monoclinic $P2_1/n$ crystal of **59**[110], a typical case of monocarboxylic acids (Figure 10). The partners in the dimer are related by a center of symmetry. The carbonyl group adopts the usual nearly bisecting *syn*

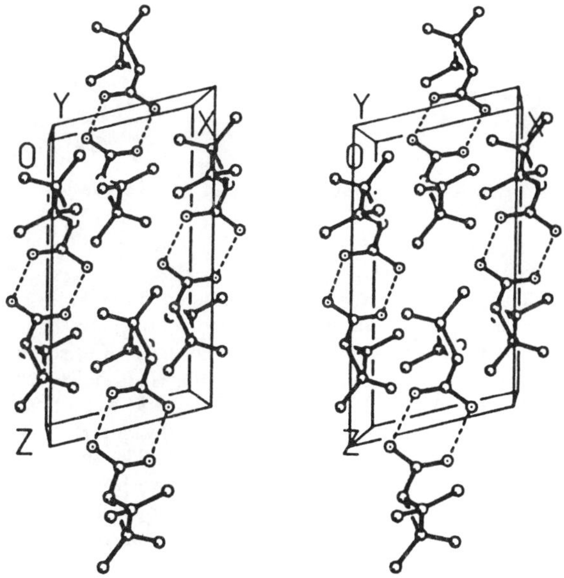

FIGURE 10. Hydrogen-bonded cyclic dimers in the crystal of 2,2,3,3-tetramethylcyclopropanecarboxylic acid (**59**), stereoview. Reproduced by permission of the International Union of Crystallography from Reference 110

conformation. The interesting pattern of bond angles in the C_2CMe_2 moieties reveals considerable steric strain (Figure 11). The COOH group is also pushed away from the Me groups, as compared to its position in free **56** (see bond angles in Table 5), but its dimensions change only slightly: C=O, C—O, C—C=O and C—C—O are (distances corrected for libration[110]) 1.234, 1.327 Å, 125.8 (2)° and 111.9 (2)°.

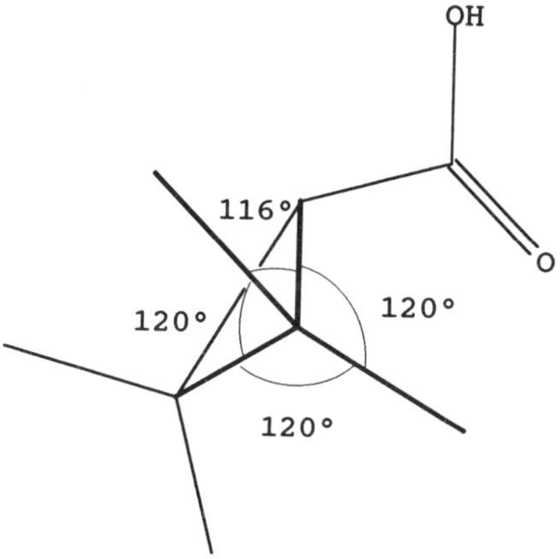

FIGURE 11. Mean bond angles at the methyl substituents in **59**

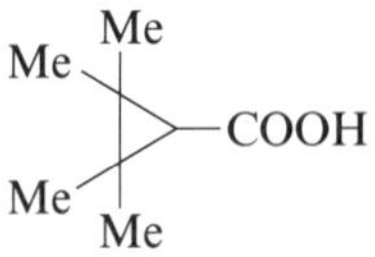

(59)

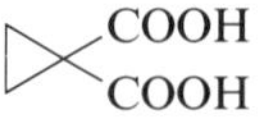

(60)

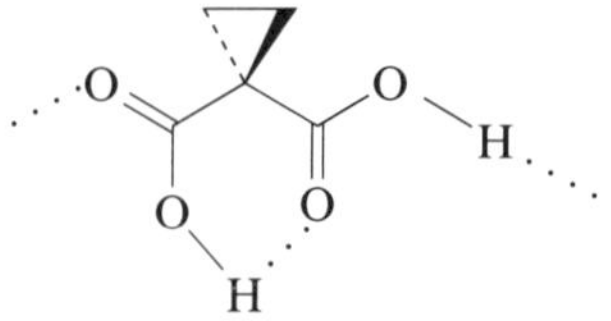

(60a)

Cyclopropane-1,1-dicarboxylic acid (**60**) forms an intramolecular H bond in solution, evidenced by the large value of $K_1/K_2 = 4.06 \times 10^4$, the ratio of its dissociation constants[111]. The conformation in the crystal is fixed by the intramolecular H bond. The COOH groups deviate somewhat from the *syn* and *anti* bisecting positions, and have their O—H *anti* and *syn*, respectively, to the C═O bond[111]. The bridgehead O···O distance is 2.536 Å, angle O—H···O 153° in a nearly planar six-membered ring. The molecules are linked by H bonds (**60a**), O···O 2.641 Å, O—H···O 175°. In the acid salt of **60**, *c*-$C_3H_4(COOH)$ COOK·0.5H_2O (**61**), there is an even shorter intramolecular H bond, O···O 2.41 Å, O—H···O 161°, between the two bisecting acid groups[120]. The anion of the diammonium salt, *c*-$C_3H_4(COONH_4)_2$ (**62**), has C_2 symmetry in the orthorhombic *Pcnb* crystal, it lacks the intramolecular H bond and the COO^- groups are rotated by 24° (calculated from atomic coordinates) from the bisecting orientations[121]. Nevertheless, mean bond parameters (Å, deg) are very similar in the two anions:

	C1—C2	C2—C3	C1—C(O)	C2—C1—C(O)
61	1.517	1.471	1.501	116.6
62	1.517 (3)	1.474 (4)	1.505 (2)	118.1 (1), 113.4 (1)

The *cis* isomer of cyclopropane-1,2-dicarboxylic acid (**63c**), on the other hand, has no intramolecular H bond in the crystal, but both of its COOH groups are engaged in a cyclic pair of intermolecular H bonds, giving rise to chains[112] (Figure 12a). One of the C═O groups is rotated by 55° from its *syn* position, so it is *synclinal* (*sc*). That the shortening of the opposite C—C bond is hardly significant may be an indication of a weaker conjugative interaction (Table 5). Molecules of the *trans* diacid (**63t**) have crystallographic twofold symmetry, and are linked in the same way across centers of inversion as in **63c** but in a different geometry[112] (Figure 12b). The C═O groups are in nearly *syn* bisecting position.

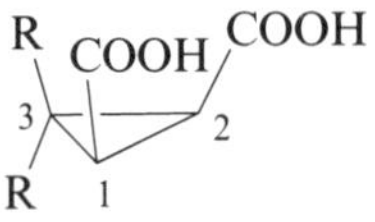

cis (**c**)

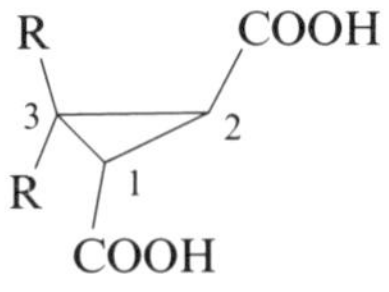

trans (**t**)

(**63c,t**) R = H
(**64c,t**) R = Me
(**65c**) R = Ph
(**66c,t**) R = *p*-C_6H_4Me

The intramolecular situation is quite similar in the 3,3-dimethyl derivatives, *cis*- and *trans*-caronic acid[113]. The *cis* isomer (**64c**) has a *syn* (*sp*) and a *sc* COOH group rotated by 75°, and the C—C bond opposite to *sp* is the shortest in the ring (Table 5). The molecules form an exotic type of packing in the crystal: a pair of *sp* COOH groups combine across an inversion center, and these cyclic dimers are linked by the *sc* COOH groups in a catemer fashion to give a two-dimensional network (Figure 13). Parameters O···O and O—H···O are 2.667 (2) Å and 174 (2)° in the cyclic linkages, and 2.656 (2) Å and 161 (2)° in the

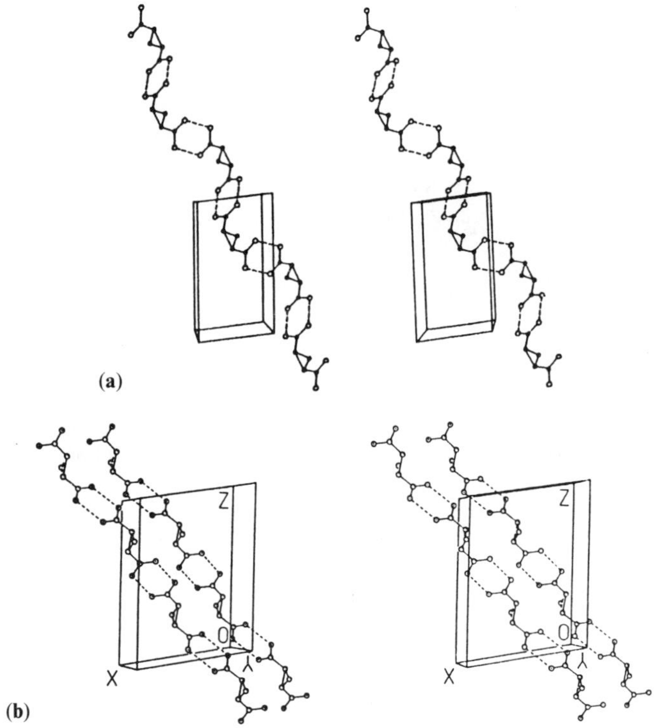

FIGURE 12. Chains of H-bonded molecules of (a) *cis*- (**63c**) and (b) *trans*-cyclopropane-1,2-dicarboxylic acid (**63t**). Stereo packing plots. Reproduced by permission of the International Union of Crystallography from Reference 112

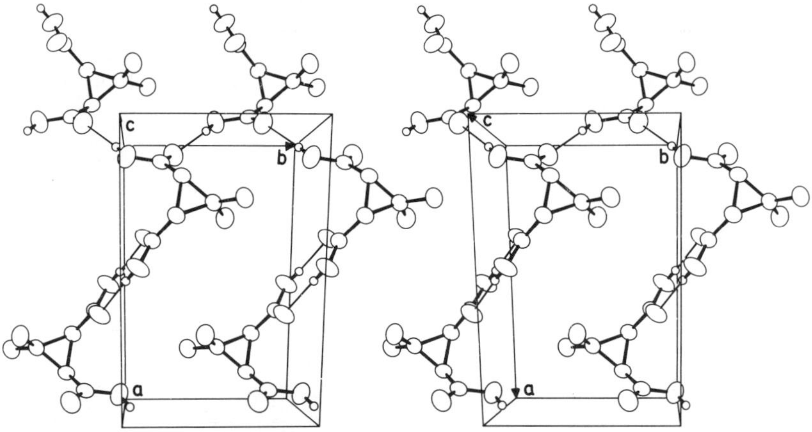

FIGURE 13. Combination of cyclic and catemer-type H bonding in the crystal of *cis*-caronic acid (**64c**). Stereo packing diagram. Reproduced by permission of the International Union of Crystallography from Reference 113

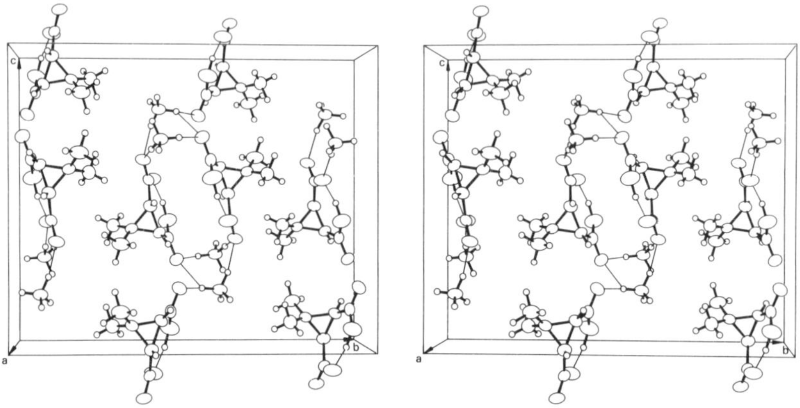

FIGURE 14. The unit cell of ammonium hydrogen *cis*-caronate (**67**). Stereoview, projected along the *a* axis. Reproduced by permission of R. Oldenbourg Verlag GmbH from Reference 122

catemer. The molecules possess C_2 symmetry in the *C*2/*c* crystal of the *trans* isomer (**64t**), and show the usual pattern of dicarboxylic acids with cyclic pairs, linked in chains by the skeletons of molecules[113]. The wider C3—C1—C(O) and C3—C2—C(O) angles in **64c** and **64t**, compared to those in **63c** and **63t**, indicate a steric effect of the Me groups.

Contrary to the parent diacid, acid salts of **64c** form intramolecular H bonds in solution and in the crystal[122] (Figure 14). In ammonium hydrogen *cis*-caronate, *c*-$Me_2C_3H_2(COOH)COONH_4$ (**67**), N—H···O bonds with N···O 2.809 (6) to 2.957 (6) Å connect anions and cations to build layers corresponding to the plate form of crystals (Figure 14).

The effects of the COOH and the cyano groups accumulate in 1-cyanocyclopropanecarboxylic acid[123] (**68**), causing a substantial shortening of the opposite bond and lengthening of the vicinal bonds (Table 6). Bond lengths corrected for libration are systematically longer. Parameters can be compared with those of the acids (Table 5) and of the cyano derivatives (Section III.D; earlier work is cited in Reference 123). The COOH group adopts the *syn* bisected conformation. The molecules lie on crystallographic mirror planes and are linked by intermolecular O—H···N hydrogen bonds in zigzag chains along the *a* axis in the orthorhombic *Pnma* crystal[123].

TABLE 6. Parameters of 1-cyanocyclopropanecarboxylic acid (**68**) from an XD study[123]

	Bond lengths (Å)		Bond angles (deg)	
	uncorrected	corrected		
C1—C2	1.529 (3)	1.541	C2—C1—C(O)	117.5 (2)
C2—C3	1.449 (5)	1.467	C1—C=O	123.2 (3)
C1—C(O)	1.490 (4)	1.504	C1—C—O	112.1 (3)
C=O	1.193 (3)	1.201	(O)C—C1—C(N)	117.1 (3)
C—O	1.322 (4)	1.331	C2—C1—C(N)	117.2 (2)
C1—C(N)	1.445 (4)	1.455	C1—C≡N	180.0 (5)
C≡N	1.132 (4)	1.140		

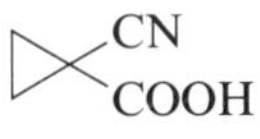

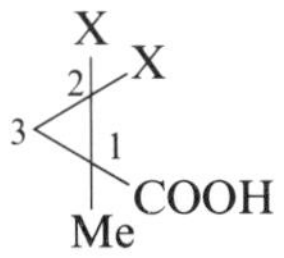

(**68**) (**69a**) X = Cl
(**69b**) X = Br

Noncentrosymmetric cyclic dimers are formed by two molecules of the same chirality in **69a**[124]. Enantiomeric pairs of dimers then build the centrosymmetric $P2_1/n$ crystal. The COOH groups in the pair of molecules are rotated by 37° (*sc*) and –146° (*ac*) from *syn*, thus the C=O or the C—O bond eclipses one of the ring C—C bonds. The same conformers build the centrosymmetric dimers of **69b**, but the COOH groups occupy the two positions probably in a disordered distribution[124]. Bonds C1—C2, C1—C3 and C2—C3 are 1.523 (3), 1.522 (3) and 1.483 (3) Å (mean values) in **69a**, and 1.518 (6), 1.518 (6) and 1.497 (6) Å in **69b**; the shortest bond is opposite to COOH.

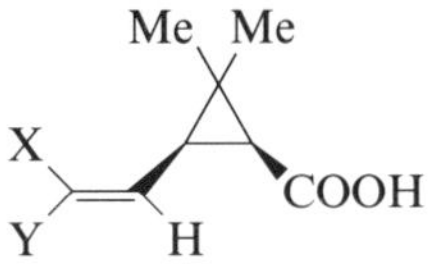

(**70**) X = Cl, Y = CF_3 (**71**) X, Y = Cl

Some derivatives of **56** have been investigated because they are related to the widely used pyrethroid insecticides. Pure cyhalothric acid (**70**) forms H-bonded centrosymmetric dimers in the crystal[125]. Newly discovered clathrates with aromatic molecules may facilitate the separation of isomers[126]. The structures of the acid **71** and of its 2/1 inclusion compound with benzene have been determined[126].

Small-ring molecules, most often hydrogen-bonding carboxylic acid hosts, form the basis of a new family of inclusion compounds[114–116,127].

The pure host *cis*-3,3-diphenylcyclopropane-1,2-dicarboxylic acid[114] (**65c**) has a short intramolecular H bond with O···O 2.513 (3) Å and O—H···O 162°. The molecules are linked by strong linear H bonds, O···O 2.590 (3) Å, O—H···O 180°. The unique catemer motif formed this way (**72**) consists of alternating intra- and intermolecular H bonds and corresponding *anti* and *syn* H—O—C=O groups, translated along the crystallographic *a* axis (Figure 15). These chiral chains are related by twofold screw axes 2_1 parallel to *b*, thus building the enantiomorphous crystal of the monoclinic $P2_1$ space group. The molecular structure of **66c** has been determined[115] in its crystalline inclusion compound with ethanol.

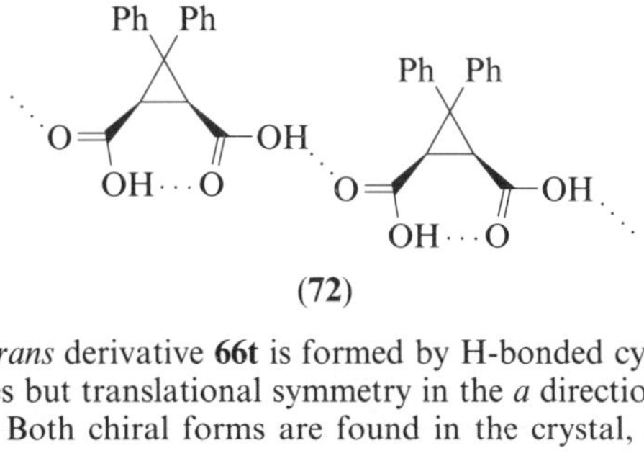

(**72**)

The chain of the *trans* derivative **66t** is formed by H-bonded cyclic pairs of carboxyl groups, and possesses but translational symmetry in the *a* direction[116] (Figure 16). This chain is thus chiral. Both chiral forms are found in the crystal, related by centers of

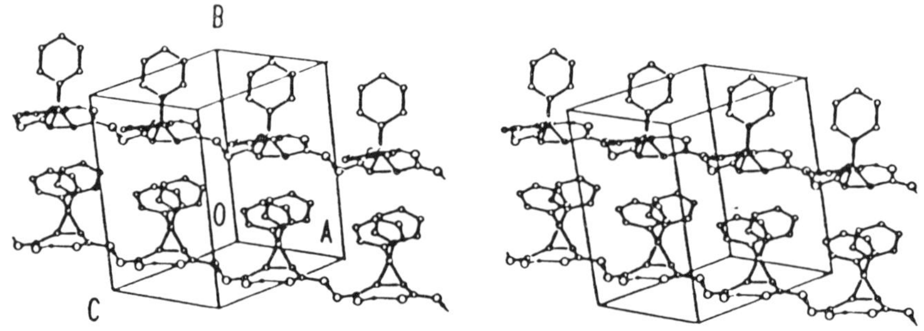

FIGURE 15. Stereoscopic packing diagram of *cis*-3,3-diphenylcyclopropane-1,2-dicarboxylic acid (**65c**). Reproduced with permission from E. Weber *et al.*, *J. Am. Chem. Soc.*, **111**, 7866–7872. Copyright (1989) American Chemical Society

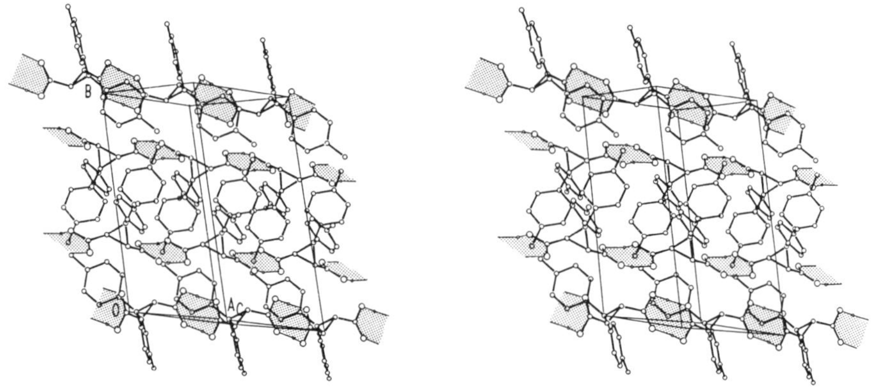

FIGURE 16. Stereoscopic packing diagram of *trans*-3,3-bis(4-methylphenyl)cyclopropane-1,2-dicarboxylic acid (**66t**). Reproduced by permission of Kluwer Academic Publishers from Reference 116

inversion, giving rise to the centrosymmetric triclinic $P\bar{1}$ structure. The H bonds are characterized by O···O 2.6 to 2.7 Å, O—H···O about 170°, and *syn* H—O—C=O groups.

The conformations of 3,3-diaryl derivatives **65c**, **66c** and **66t** indicate a compromise between conjugative, steric and H-bonding effects. One of the COOH groups is in the *ap* position to the ring in the *cis* diacids. The COOH groups are rotated by 18° to 41° from bisecting the ring towards eclipsing one of the C—C bonds, so that they form the intramolecular hydrogen bond in **65c** and **66c**, and avoid the bulky Ph groups in **66t**. (Dihedral angles have been calculated from atomic coordinates). The Ph planes are close to the perpendicular position in the *cis* 1,2-diacids **65c** and **66c**, but are rotated from perpendicular by about 28° and 42° away from the COOH groups in *trans* **66t**.

Hydrogen bonding is one of the factors that control the character of inclusion crystals. If the COOH groups of a *cis*-1,2-dicarboxylic acid are engaged in an intramolecular H bond, only one donor and one acceptor function remain for binding kindred neighbors (as in **72**) or guests. *Trans* diacids have both COOH groups free. Thus, **66t** builds an 1/2 clathrate[116] with Me_2SO through O(H)···O(S) linkages of 2.6 Å, angles O—H···O(S) 170°.

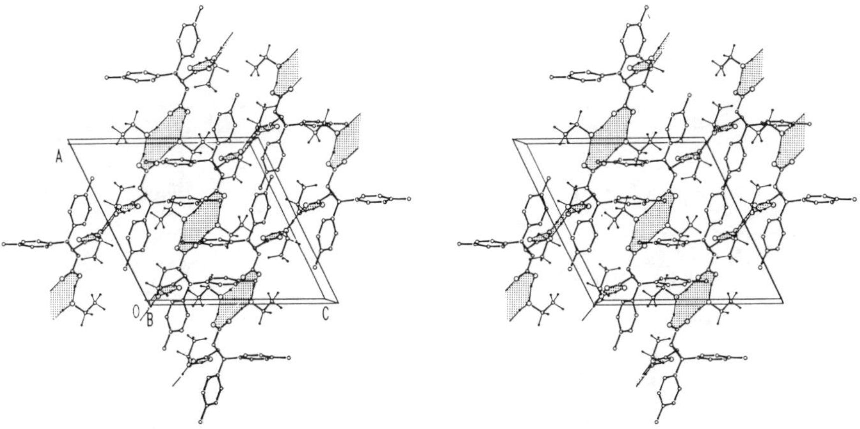

FIGURE 17. Stereo packing illustration of the inclusion compound of **66t** with EtOH (1/2). Reproduced by permission of The Royal Society of Chemistry from Reference 115

Alcohol guests are inserted in the H-bonded system of the diacids, usually giving 1/1 host/guest ratio for *cis*, and 1/2 for *trans* derivatives[127]. The crystal of **65c**·*t*-BuOH (1/1) with a *cis* host contains helices of alternating host and guest molecules, and these identical helices make up the enantiomorphous orthorhombic $P2_12_12_1$ crystals[114]. This bonding scheme is quite different from that in the pure host substance. The related *cis* diacid **66c** forms analogous helices with EtOH (1/1) but inverse pairs of helices combine to give the centrosymmetric monoclinic *C*2/*c* crystals[115]. The *trans* **66t** essentially retains its bonding topology in its clathrate with EtOH (1/2); just the closed loops of cyclic dimers are widened by insertion of two EtOH molecules between the pair of COOH groups[115] (Figure 17). Opposite molecules in the ring are related by an inversion center, thus the chains and the monoclinic $P2_1/c$ crystal are centrosymmetric. The H-bond O···O distances and O—H···O angles in this twelve-membered ring are: (O=C)O—H···O(Et) 2.565 (4), 2.560 (4) Å and 162°, 163°; (C=)O···H—O(Et) 2.698 (3), 2.708 (4) Å, and 156°, 153°. Here and in the other two clathrates, the alcoholic OH group takes a double role, and it seems that the span is wider and the bridging angle narrower when it acts as a proton donor, than in the case when its O atom is an acceptor.

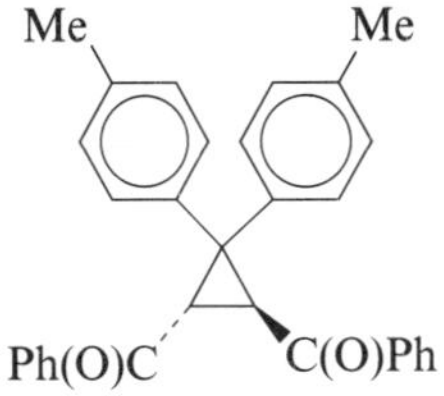

(73)

The structure of another type of inclusion compound, **73**·MeCN (1/1), suggests that the interaction between the ketone host and MeCN is purely of steric character[114,127] (Figure 18). The packing is highly sensitive, and clathrate forming will disappear with any changes in the size of either the host or the guest molecules[127].

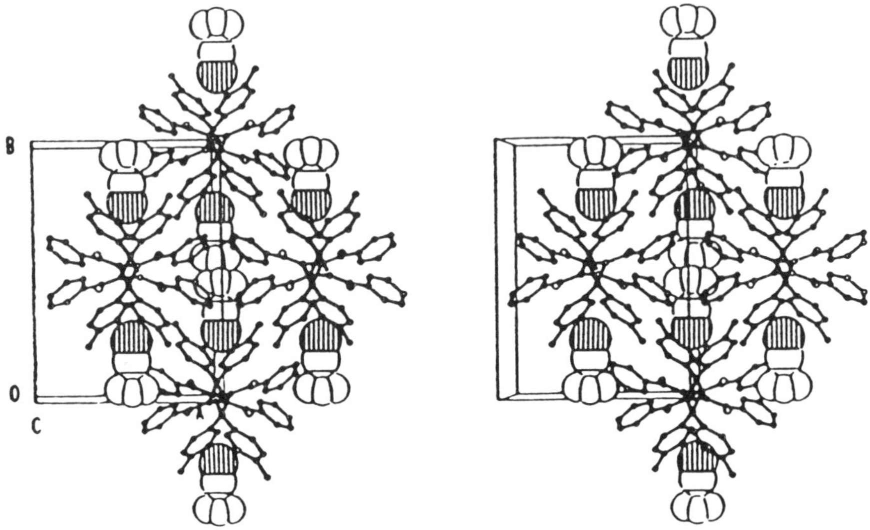

FIGURE 18. Packing in the crystal clathrate of **73** with MeCN (1/1). Stereoscopic view. The guest molecules are shown in van der Waals representation. Reproduced with permission from E. Weber *et al.*, *J. Am. Chem. Soc.*, **111**, 7866–7872. Copyright (1989) American Chemical Society

2. 1-Aminocyclopropanecarboxylic acid and derivatives

1-Aminocyclopropanecarboxylic acid (**74**) has attracted special interest (a) as a modified model of natural α-amino acids, in the synthesis of bioactive peptide analogs, potential enzyme inhibitors[128–130], or e.g. a derivative with sweetness receptor activity[131] and (b) together with its Schiff-base metal complexes[132], as the biosynthetic precursor of ethene, the plant ripening hormone[133,134]. (See these sources and the review in Reference 135 for further references.)

(74)

The crystal molecular structure of **74** has been reported[133] from an XD study of the hemihydrate **74**·$0.5H_2O$. Two independent zwitterionic molecules exist in the crystal. One of the oxygen atoms in one molecule of **74** and the water molecule which is linked to it through a H bond are disordered. The COO^- and NH_3^+ groups take *sp* positions to the ring. Some geometrical parameters are shown in Figure 19. Bond lengths and angles compare well with those of cyclopropylamine (see compound **121** in Table 10, Section III.G.1). The C—C bond to the COO^- group is of the same length as in **67**[122], 1.500 (6) Å. The N—C—C(O) angle formed by the geminal substituents on the cyclopropane ring is larger than the tetrahedral value, the same as this angle in glycine[136], 113.0 (3)°, but larger than in alanine[137], 110.1(8)°, both in the gas phase. The C—O bond lengths in **74** are in the narrow range of 1.239 (1) to 1.257 (1) Å[133], or 1.248 (5) to 1.266 (4) Å from another study[128], as expected for the COO^- group in the zwitterionic form. Amino acid and water molecules are linked in each combination by N—H···O and O—H···O bonds of 2.66 to 2.89 Å in the crystal[128].

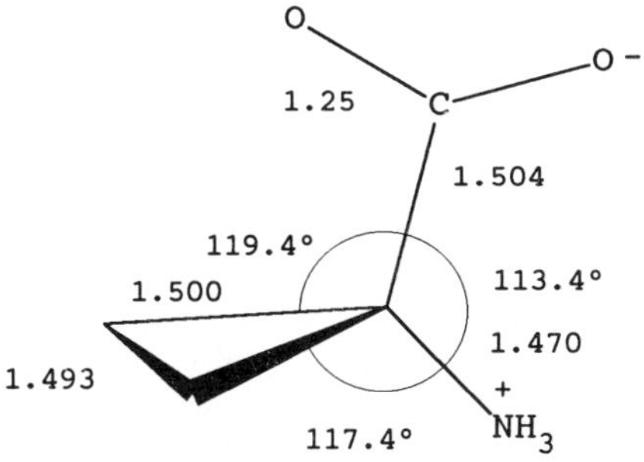

FIGURE 19. Bond lengths (Å) and bond angles (deg) in 1-aminocyclopropanecarboxylic acid (**74**). Mean values are shown as calculated from the reported data[133]

The centrosymmetric triclinic space group was chosen in the first crystallographic analysis[133] (although reported as $P1$), with $Z = 4$ formula units in the unit cell. In the other XD study of the same crystal[128], the centrosymmetric triclinic space group $P\bar{1}$ was used as well, but with a rather different assignment and a double volume of the unit cell, and accordingly with $Z = 8$ and with four independent molecules in the asymmetric unit. (By the way, the formula weight given includes a surplus of a half water molecule above $C_4H_7NO_2{\cdot}0.5H_2O$.) This discrepancy warrants a revision of unit cell parameters.

(75)

(76–82)

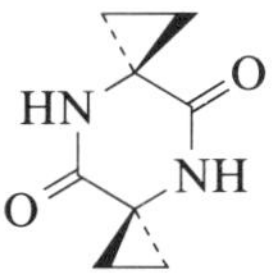

(83)

Peptide analogs containing the aminocyclopropanecarboxylic acid unit (**75**), abbreviated Ac_3c, have special conformational properties, first of all restricted flexibility. A number of oligopeptides up to *t*-BuOOC—$(Ac_3c)_4$—OMe·$2H_2O$ and other derivatives have been investigated recently by X-ray crystallography[128–130]. The N—C—C(O) angles at the cyclopropane rings are really large; the mean is 116.7° in the five linear peptides studied[129]. The benzyloxycarbonyl and hydroxamic acid derivatives are potential enzyme inhibitors[130]. The distal bond in the C_3 ring is shorter than the vicinal bonds, e.g. 1.484 (4) and 1.504 Å, respectively, in **77**[130]. The Ac_3c residue prefers the uncommon 'bridge' region of the conformational map[128,130], i.e. dihedral angles φ and ψ are in ranges $60° < |\varphi| < 120°$ and $0° < |\psi| < 25°$ (Table 7). Acids **76** and **82** take the unusual semiextended form with a *trans* (*ap*) N—C—C(O)—O (H) group[128]. An intramolecular N···H—N bond in the dipeptide ester H—$(Ac_3c)_2$—OMe (**81**) leads to a novel type *cis* (*sp*) H_2N—C—C(O)—N conformation (Figure 20). The C—C(O)—N—C peptide group is *trans* in **81** (Figure 20), but it is *cis* in the nearly planar six-membered ring of the cyclic dipeptide *cyclo*-$(Ac_3c)_2$ (**83**).

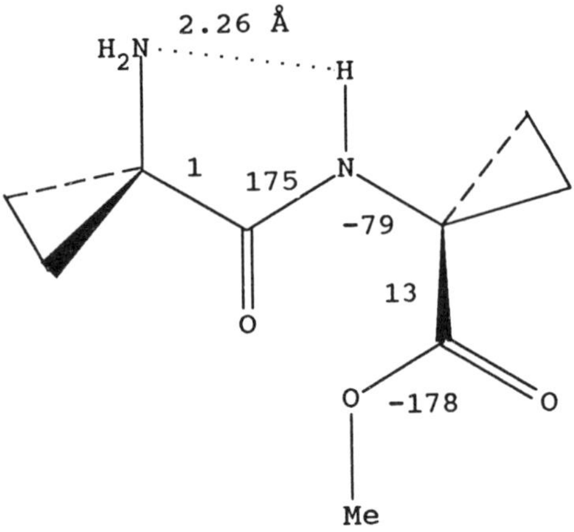

FIGURE 20. The conformation of and intramolecular hydrogen bond in the dipeptide ester H—$(Ac_3c)_2$—OMe (**81**). Dihedral angles (deg) of the H_2N—C—C(O)—N—C—C(O)—O—Me chain are indicated[128,129]

TABLE 7. Dihedral angles (deg) in Ac_3c (**75**) derivatives[a]

Compound	R	X	φ	ψ
76	$PhCH_2OOC$	OH	−67.6 (9)	165.6 (7)
77[b]	$PhCH_2OOC$	NHOH	71.9 (3)	9.0 (3)
78	$PhCH_2OOC$	OMe	−73.2 (7)	1.6 (7)
79	*t*-BuCO	OH	−72.0 (6)	−8.7 (7)
80, α	*t*-BuOOC	OH	−66.5 (13)	−15.7 (16)
80, β	*t*-BuOOC	OH	−81.2 (6)	1.8 (6)
81[c]	H-Ac_3c	OMe	−79.0 (4)	−13.4 (5)
82	*p*-BrC_6H_4CO	OH	74.1 (28)	165.4 (20)

[a] φ is the dihedral angle R—N—C—C(O), ψ is N—C—C(O)—X. Data are from Reference 128 unless noted otherwise.
[b] Reference 130.
[c] References 128 and 129.

3. Carboxylic acid derivatives

Ring formation in *cis*-cyclopropane-1,2-dicarboxylic anhydride[138] (**84**) leads to *ac* conformation of the C=O groups, rotated by 155° from *syn*, and to strikingly narrower (and unequal) bond angles C2—C1—C(O) 105.6 (2)° and C3—C1—C(O) 113.4 (2)° than in the parent diacid **63c** (Table 5). Bond lengths of **84**, C1—C2 1.505 (3), C1—C3 1.497 (3), C—C(O) 1.470 (2) Å, have somewhat changed relative to **63c**. The molecule lies on a crystallographic mirror plane.

There are two independent molecules in the monoclinic $P2_1/c$ crystal of hexamethyl cyclopropanehexacarboxylate[139] (**85**), studied at −120 °C. Molecule A possesses an approximate D_3 symmetry (Figure 21). The COOMe groups apparently avoid each other, and no intramolecular contact within van der Waals limits has been found. The *sc* C=O groups

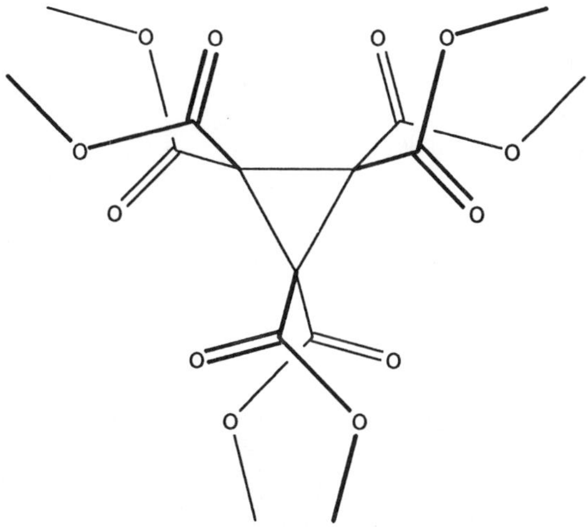

FIGURE 21. Sketch of molecule A in the crystal of hexamethyl cyclopropanehexacarboxylate (**85**)

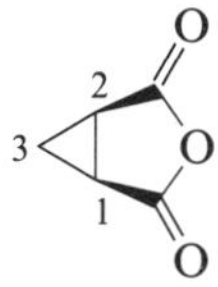

(84)

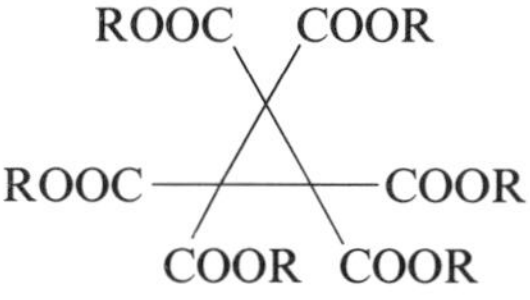

(**85**) R = Me
(**86**) R = Et

are rotated by 12° to 23° from eclipsing a ring C—C bond towards the perpendicular position, except for one COOMe group in molecule B, where it is the C—OMe bond that takes a similar position. This group is also unique in its bond angles, and is asymmetrically bonded to the ring, with C(ring)—C—C(O) 118.0 (3) and 124.4 (3)°, C—C=O 120.2 (3)° and C—C—O 112.8 (3)°, while the ranges of the corresponding angles in the other groups are 117.7 to 120.4°, 124.0 to 125.5° and 107.8 to 109.9°. Each Me—O—C=O group has the *syn* form. Mean bond lengths show no effect of conjugation between C=O groups and ring: C—C(ring) 1.511, C—C(O) 1.523, C=O 1.196, (O=)C—O 1.329, and (H_3)C—O 1.463 Å.

The even more congested hexaethyl ester (**86**) presents similar variations in two crystal modifications[140]. In the orthorhombic crystal, *Pccn*, the molecules have twofold symmetry, close to molecular D_3 symmetry. This chiral conformer and its mirror image make up the achiral crystal. The *sc* C=O groups are rotated by 53, 48 and 53° from the *syn* position, i.e. about halfway to the perpendicular form, quite similarly to **85**. The C—C(O) bonds are longer[140], mean 1.515 Å, than in noncrowded molecules (Table 5), just as in **85**. The mean ring C—C bond length is 1.516 Å. The C(ring)—C—C(O) angles range from 117.9 (3) to 119.9(2)°, thus the angles between the geminal bonds are smaller than in 1,1-disubstituted derivatives **60** and **68** (Tables 5 and 6). The O=C—O—C fragments are *syn* and nearly planar; the dihedral angles about the O—Et bonds are 146, 151 and 162°. The Et groups perform large-amplitude librational motion. The molecular structure is similar in the $P2_1$ monoclinic form[141]. The largest deviation is found in one of the O—Et dihedral angles, which is –104°. The mean ring C—C length is 1.507 Å.

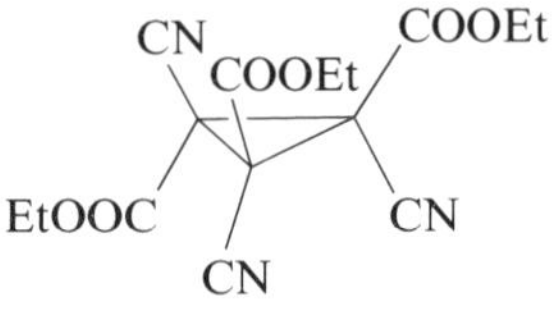

(87)

The interesting asymmetry in the crystal molecular structure of **87** is attributed to conjugative and steric effects[142]. Two C=O groups are nearly *syn* to the ring, but one of the *cis* ester groups is rotated by 76°, near to the perpendicular position, away from the vicinal C=O group. This substituent forms a markedly small bond angle of 109.9(4)° to the geminal CN, while the analogous angles at the other ring atoms are 115.6° (mean) in **87** and 117.1(3)° in **68** (Table 6). The ring C—C bond opposite to the perpendicular ester group is very long at 1.552 (6) Å. The other two ring bonds, which have two distal substituents in favorable position for π donation, are 1.522 (6) and 1.513 (6) Å. The C—C(O) bonds are relatively long, 1.530 Å on the average[142].

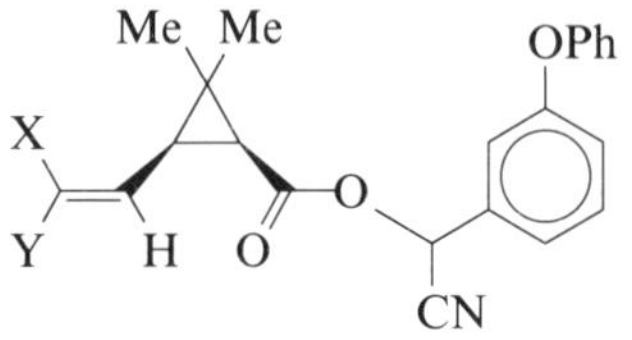

(88) X, Y = Cl
(89) X = COOBu-*t*, Y = H
(90) X = $COOCH(CF_3)_2$, Y = H

Several pyrethroid insecticides have been studied by XD. Among the ester derivatives[143–145] are **88**, the α-cyano-3-phenoxybenzyl ester[143] of **71** (Section III.C.1) and related compounds[144] (**89**, **90**). Relations of biological activity with conformation and absolute configuration have been sought. Uncertainties of structural parameters are large in some cases. Mean C—C bond lengths in the three-membered ring are 1.525, 1.531 (4) and 1.523 (5) Å in **88–90**, respectively. The cyclopropanecarboxylic moiety builds a rigid connection between more flexible parts of the molecules.

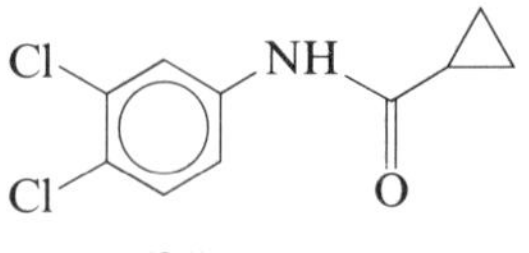

(91)

The carboxamide substituent causes a shortening of the distal bond, e.g. in **91**[146] the vicinal bonds are 1.503 (4) and 1.505 (4) Å; the distal bond is 1.454 (4) Å.

D. Ethynyl and Cyano Cyclopropanes

The cyano group as a π-acceptor substituent induces a shortening of the distal bond in the cyclopropyl ring. A mean asymmetry parameter $\delta_1 = -0.017$ (2) Å has been found[8].

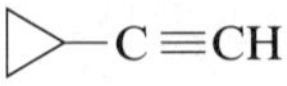

(92)

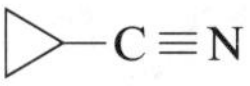

(93)

The structures of cyclopropylacetylene (**92**) and cyanocyclopropane (**93**) were studied in several laboratories within a few years. Parameters from the more recent experimental investigations are shown in Table 8.

Ring bond lenghts of **92** could not be distinguished in the earlier ED study[150], but their average, r_a 1.510 (1) Å, as well as the other parameters, compare well with those from the joint ED+MW analysis[147], which gave a mean ring r_g(C—C) of 1.514 Å. The C—C(≡) single bond to the substituent is 1.440 (3) and 1.445 (8) Å from the two studies; the MW result[148] is lower by 0.02 Å. A similarly short bond has been found in **93** by MW (Table 8). The shortening of the distal bond in **92** and **93** relative to the vicinal bonds and to cyclopropane indicates that both substituents accept π electrons from the ring. Geometries and other features in cyclopropyl and vinyl derivatives have been compared and possible σ-bond and conjugation effects considered[148]. Present geometrical data, however, do not allow a definite conclusion concerning the relative π-acceptor abilities of the ethynyl and cyano groups, nor on the relative conjugative or hybridization effects in cyclopropyl and vinyl derivatives.

Bond lengths in **92** and **93** can be compared with those in MeC≡CH[151], r_s C—C 1.4586 (2), C≡C 1.2066 (2) Å and in MeC≡N[152] r_s C—C 1.458 (9), C≡N 1.157 (9) Å.

The molecular structure of **93** has been determined[149] also by low-temperature XD at −85 °C (Table 8). The molecules are bisected by mirror planes in the orthorhombic *Pnma* crystal, and have no special intermolecular contacts. Uncorrected bond lengths are systematically smaller than the MW data, but curiously not for the C—C(≡) single bond.

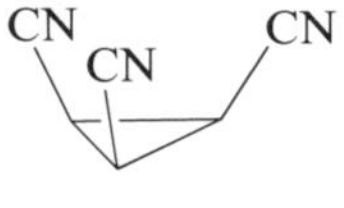

(94c)

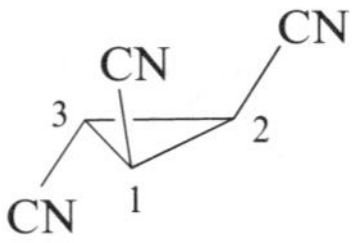

(94t)

TABLE 8. Bond lengths (Å) and bond angles (deg) of cyclopropylacetylene (**92**) and cyanocyclopropane (**93**)

Bond or angle	**92** ED+MW, r_g	**92** MW, r_s	**93** MW, r_s	**93** XD
C1—C2	1.526 (7)	1.527 (6)[a]	1.529 (5)	1.509 (1)
C2—C3	1.490 (14)	1.503 (7)[a]	1.500 (3)	1.476 (1)
C—C (≡)	1.445 (8)	1.422 (6)	1.420 (6)	1.434 (2)
C≡C	1.213 (2)	1.211 (4)		
C≡N			1.161 (4)	1.142 (2)
C—C—C(≡)	118.8 (4)[b]	119.3 (4)[a]	118.7 (4)	119.21 (7)
Reference	147	148	148	149

[a] 'Near-r_s' parameter, obtained with some assumptions.
[b] $\angle_\alpha$ parameter.

The molecules of *trans*-1,2,3-tricyanocyclopropane (**94t**) also lie on mirror planes in the *Pnma* crystal[153]. The C—C(N) and C≡N bond lengths in solid **93**, **94t** and in its *cis* isomer[154] (**94c**) are similar. The ring C—C bond flanked with the *cis* CN groups in **94t**, 1.524 Å, is longer than the other C—C bonds, 1.511 Å, and their mean is close to that in **94c**, 1.519 Å (values corrected for libration). Hexacyanocyclopropane[155] (**95**) crystallizes with three molecules of dioxane in the rhombohedral space group *R*3*c*. One of the two independent molecules of **95** lies on a crystallographic C_3 axis. Its ring C—C distance of 1.530 (6) Å is longer than in the less substituted cyano derivatives above, but markedly shorter than the very long C—C single bonds in 1,1,2,2-tetracyanoethane[156], 1.562 (1) Å, and in *cis*-2,5-diphenyl-3,3,4,4-tetracyanopyrrolidine[157] (**96**), 1.598 (3) Å! A similar eclipsed arrangement of cyano groups occurs in the cyclohexadienyl derivative **97**[158]. The ring bond C1—C2 between the cyano groups is elongated to 1.565 Å, while, interestingly, the cyano groups are tilted towards each other as shown by the bond angles C2—C1—C(N) 115.8° and C3—C1—C(N) 119.3°. Other bond lengths are C1—C3 1.509, C—C(N) 1.462 and C≡N 1.135 Å (calculated from atomic coordinates given, mean values). A slight deviation from linearity of the C—C≡N fragment has been found in these molecules, the largest being about 5° in **95**[155].

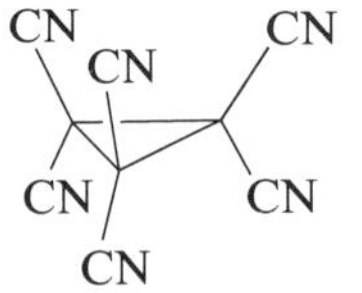

(**95**)

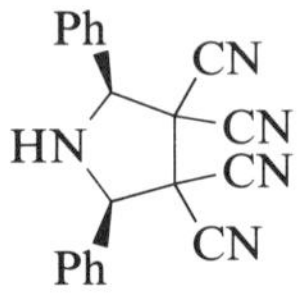

(**96**)

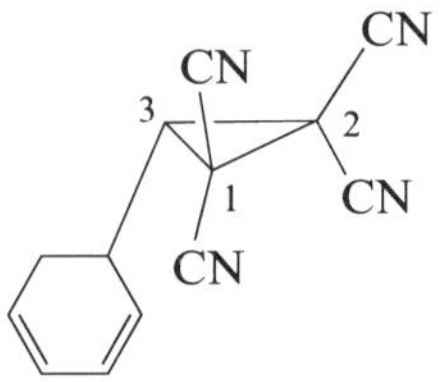

(**97**)

E. Halogenides

Extensive gas-phase structural studies have been performed on fluoro-substituted cyclopropanes. Important structural parameters are collected in Table 9.

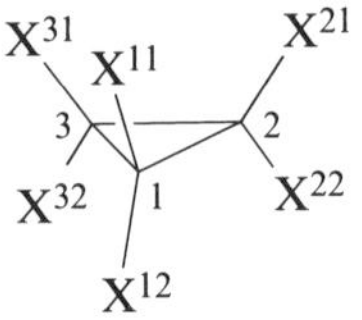

(**98**–**112**) for X^{11}–X^{32}, see Table 9

A fluorine substituent shortens the adjacent C—C bonds and lengthens the opposite bond in the cyclopropane ring. The lengthening of the C—C bond opposite to a CF_2 group is remarkable (cf **98** and **104** in Table 9), but the mean C—C distance is smaller even in these molecules than in cyclopropane (**1**). In the other fluoro derivatives, all C—C bonds are shorter than in **1**. The C—C bonds appear to be longer, the C—F bonds shorter in a *cis* isomer than in the *trans* isomer[160], and the same applies to C—C and C—F bonds in *cis* compared to *trans* CHF—CHF moieties within the same molecule[163]. *Ab initio* calculations with a 4–21G split valence basis set reflect the above trends quite well, except for the distinction of effects in the *cis* and *trans* forms[24]. The additivity of substituent effects, which

TABLE 9. Bond lengths (Å) and bond angles (deg) in halogen-substituted cyclopropanes[a]

Compound	X^{11}	X^{12}	X^{21}	X^{22}	X^{31}	X^{32}
(98)	F	F				
(99)	F		F			
(100)	F			F		
(101)	F		F		F	
(102)	F		F			F
(103)	F	F	F	F		
(104)	F	F	F		F	
(105)	F	F	F	F	F	F
(106)	Cl					
(107)	Cl	Cl				
(108)	Cl		Cl		Cl	
(109)	Cl	Cl	Cl	Cl	Cl	Cl
(110)	Br					
(111)	Br		Br		Br	
(112)	Br	Br	Br	Br	Br	Br

Compound	Method	C1—C2	C2—C3	C—X	X—C—X	Reference
98	MW, r_0	1.464 (2)	1.553 (1)	1.355 (2)	108.4 (2)	159
99	MW	1.488 (3)	1.503 (4)	1.368 (6)	111.3 (4)[b]	160
100	MW	1.466 (4)	1.488 (5)	1.383 (3)	111.3 (4)[b]	161
101	MW	1.507 (1)		1.354 (1)	112.3 (2)[b]	162
102	MW	1.500 (3)	1.478 (10)	1.367 (8)[c]	109.4 (8)[b,c]	163
				1.387 (8)[d]	114.7 (15)[b,d]	
103	MW	1.471 (3)	1.497 (10)	1.344 (4)	109.9 (4)	164
104	MW	1.481 (20)	1.533 (3)	1.355 (4)[e]	110.1 (20)	165
105	ED, r_a	1.505 (3)		1.314 (1)	112.2 (10)	166
106	MW, r_s	1.513 (4)	1.515 (1)	1.740 (3)	118.7 (3)[f]	167
107	ED+MW, r_{av}	1.494 (3)	1.535 (9)	1.756 (2)	112.6 (2)	168
108[g]	XD	1.504[h]		1.750[h]	120.8[f]	169
109[g]	XD	1.537[h]		1.747[h]	112.2 118.9[f]	170
110	ED, r_g, $\angle_\alpha$	1.501 (6)	1.534 (12)	1.920 (4)	118.7 (6)[f]	171
111[g]	XD	1.465		1.910	122.2[f]	172
112[g]	XD	1.544		1.891	110.7 119.5[f]	172

[a]Only the halogen substituents are indicated, and H atoms are omitted in the first half of the table. Partial r_s parameters are given for the fluorides, except for **98** and **105**.
[b]Angle H—C—X.
[c]Substituents (*cis*) at C1 and C2.
[c]Substituents (*trans*) at C3.
[e]Bonds C2—F21 and C3—F31, with r(C1—F11) = 1.327 Å and r(C1—F12) = 1.357 Å assumed.
[f]Angle C—C—X.
[g]Mean values.
[h]Corrected for libration.

was anticipated on theoretical[53] and experimental[8] grounds, fails for heavily fluorinated cyclopropanes. Empirical relationships have been proposed to predict C—C and C—F bond distances in these molecules[165].

Structural changes induced by a halogen substituent in the cyclopropane ring have been interpreted in different ways, by local rehybridization, by a σ-electron shift to the electronegative substituent, or by π donation from the substituent's lone pairs, explaining only a part of the experimental findings and theoretical results. (See Reference 88 for references.) The cyclopropyl group is a good π donor, but functions only exceptionally as a π acceptor, and fluorine substituents act as σ acceptors and not as π donors[88]. Besides, the trends in some geometric and energetic effects induced by different substituents are very similar in cyclopropyl and 2-propyl systems[88]. On the basis of an electron-density model, the π-repeller character of fluorine has been emphasized, and substituent effects on ring strain and geometry have been explained by changes of σ aromaticity[25]. A simultaneous σ-accepting and π-donating action of fluorine was not indicated[25].

The orientation of a bond in the CH_2, CHF and CF_2 groups can be described by two angles, e.g. by the angle of the bond to the ring plane, and the angle of its projection to the bisector of the C—C—C angle[160,161,163]. While the C—F bonds in the fluoro derivatives take approximately the C—H bond orientations of the parent molecule **1**, C—H bond positions are changeable[160,161,163]. This may also be shown by the bond angles, e.g., in **102**[163] (deg): H1—C1—C2 119.9 (3), H1—C1—C3 124.8 (12), H3—C3—C1 117.0 (11), F1—C1—C2 118.9 (5), F1—C1—C3 116.7 (6), F3—C3—C1 118.5 (9), and in **1**[20]: H—C—C 117.7 (4). That is, the C—F bonds and C3—H3 are oriented as the C—H bonds in **1**, but the *cis* C—H bonds have tilted closer to each other and towards the ring plane (cf angles H—C3—F and H—C1—F for **102** in Table 9). Other H—C—F angles and the F—C—F angles are smaller than H—C—H in **1** (Table 1). The geminal F···F distance is fairly constant[164] at 2.2 Å, in accord with the intramolecular 1,3 nonbonded radius[173] of fluorine, $r_{1,3}$(F) 1.08 Å. The *cis* F···F distances are[164] about 2.7 Å, twice the van der Waals radius. The longer vicinal F···F distance of 2.81 Å in **103** is achieved, in spite of the shorter C—C and C—F bonds, by a displacement of the F atoms towards the ring plane, which leads to a closure of the F—C—F angles[164] (Table 9).

Ring geometry and substituent positions change analogously in oxirane upon fluorine substitution (see References 160, 163, 174 and references cited therein).

Up-to-date gas-phase data are available only for a few purely chloro- or bromo-substituted cyclopropanes (Table 9). The distal bond is lengthened in **107** and **110**, but no ring asymmetry was detected in **106**. The mean C—C length is close to that in **1**. Two possible models of **110** have been obtained from a joint analysis of ED and MW data and *ab initio* calculations[171], but unfortunately, neither of them reproduces the A_0 rotational constants from the earlier and from a more recent MW study[175].

The molecules of **108** and **111** have imposed mirror symmetry in the crystal; they are close to the ideal C_{3v}, and **109** to D_{3h} symmetry. Deviations from ideal symmetry (and uncertainties of parameters) are larger in **112**. The *cis* Cl···Cl distance of 3.30 Å in the trichloro derivative **108** is strikingly shorter[169] than the van der Waals diameter[176], 3.60 Å. This strain is only partly released by a small displacement of the Cl atoms towards the ring plane compared to **106** (see the C—C—Cl angles in Table 9). At the same time, the corrected mean C—C length in **108** is slightly smaller (!) than the values for free **1** (Table 1). Also, the uncorrected ring C—C distances in **108** are 1.492 (5) and 1.489 (6) Å, just smaller than the mean uncorrected bond length, 1.4991 (7) Å, in crystalline **1** (Table 1). The hard-sphere model of nonbonded interactions is certainly not valid here[169]. The mere concept of intramolecular 1,3 nonbonded radii is indeed incompatible with the hard-sphere model. The geminal Cl···Cl distance in the hexachloro derivative (**109**), 2.87 Å (calculated from the coordinates), is the double of the radius[173] $r_{1,3}$(Cl) 1.44 Å.

The *cis* Cl···Cl contact decreases further to 3.22 Å in **109**, accompanied by a closing of the C—C—Cl angles and a lengthening of the ring C—C bonds[170] (Table 9). Steric effects play similarly an important role in the structures of **111** and **112**[172]. Mean intramolecular *cis* Br···Br distances are 3.50 Å[172] and 3.40 Å (calculated from the original data), respectively, contrasted with the van der Waals distance[176] 3.90 Å. The geminal distance in **112**, 3.11 Å, is even shorter than twice the 1,3 radius[177] $r_{1,3}$(Br) 1.59 Å. The astonishingly short C—C bond in **111** might be paralleled[172] to an easier opening of the C—C—Br than of the C—C—Cl angle (Table 9).

F. Ethers and Sulfur Derivatives

Up-to-date structural data on simple alkoxy and alkylthio cyclopropanes are scarcely available.

The ring bonds in a cyclopropanol derivative[178] (**113**) are equal within uncertainties, and the C—O bond is 1.407 (8) Å. The OH hydrogen was not located. According to *ab initio* calculations, the OH substituent prefers the *ac* orientation[88], and this has been found in cyclopropanol, $\tau = 106\,(5)°$, from MW spectra[179].

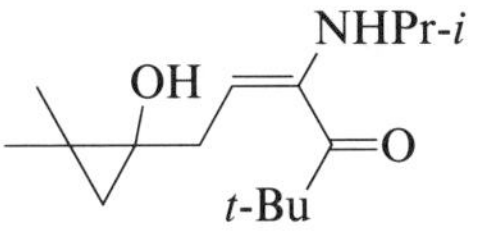

(**113**)

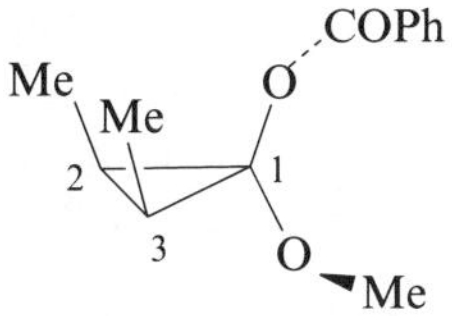

(**114**)

In a hemiacetal derivative (**114**) of cyclopropanone, the methoxy and the benzoyloxy substituents adopt *ac* conformations on different sides of the O—Cl—O plane[180]. The dihedral angles in the Me—O—Cl—O—C(Ph) chain are −71.9 (2)° and −69.9 (3)°. The distal bond in the ring, C2—C3 1.543 (4) Å, is substantially longer than the vicinal bonds, 1.479 (4) and 1.486 (3) Å. The C—O(Me) bond is short, 1.378 (3) Å, like in **117a**, while the length of the C1—O(benzoyl) bond, 1.422 (2) Å, is similar to the C—O bond lengths in dialkyl ethers (see below). Interesting is the bond pattern in **115**, studied[181] by XD at −130°C: C1—C2 1.542 (8), C1—C3 1.542 (7) and C2—C3 1.471 (8) Å. The C—OEt bonds are 1.408 (6) and 1.410 (6) Å.

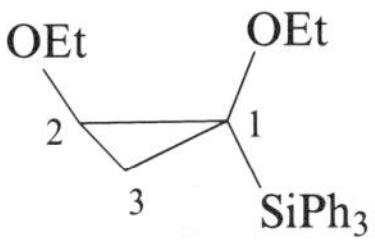

(**115**)

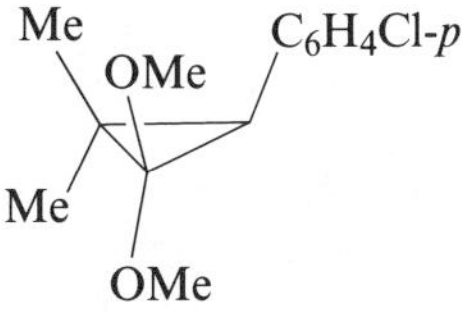

(**116**)

Derivatives with methoxy (**116**[182], **117a**[183]), alkylthio and arylthio groups (**117b–j**)[183–188] and with other, mostly π-accepting substituents in addition, have been studied by XD. Effects of these groups on the ring geometry have been assessed. Assuming additivity, the asymmetry parameters[8] (δ, see the beginning of Section III) were fitted to a set of experimental bond lengths. Tinant, Viehe and coworkers[183] obtained δ(OMe) = −0.016 (8) Å, δ(SR) = −0.020 (6) Å (taking it the same for R = Me, *t*-Bu and Ph) and $\delta(NMe_2)$ = −0.005 (8) Å, among others. These electron-releasing (donor) substituents should thus produce a shortening of the distal bond. Søtofte and Crossland[182], however, calculated $\delta(OMe)_2$ = 0.036 Å for geminal dimethoxy substituents. Here only one and two molecules included bear MeO groups, accompanied by substituents which have (presumably) larger effects. It would be desirable to cover a larger set of consistent experimental data on simple derivatives. An unexpectedly long ring bond, C1—C2 of 1.580 (4) Å in **117e**, is noteworthy[186].

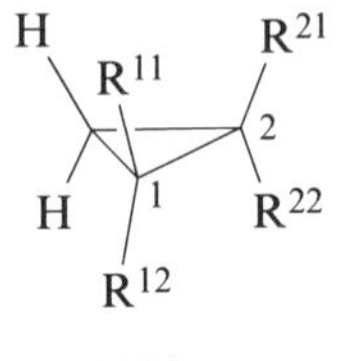

(117)

	R^{11}	R^{12}	R^{21}	R^{22}
(a)	CN	OMe	CN	OMe
(b)	CN	SMe	SMe	CN
(c)	CN	SPh	SMe	CN
(d)	Cl	COOMe	SBu-*t*	CN
(e)	CN	COOMe	SBu-*t*	CN
(f)	CN	Ph	SBu-*t*	CN
(g)	CN	SPh	SBu-*t*	CN
(h)	CN	SPh	CN	SPh
(i)	CN	SBu-*t*	CN	SBu-*t*
(j)	COOMe	SPh	COOMe	SPh

The O—Me and the S—R bonds take *ac* positions to the three-membered ring. The mean C(ring)—OMe bond is 1.390 (2) Å in **116**[182] and 1.371 (3) Å in **117a**[183]. The bonds from the C_3 ring to S have mean lengths of 1.796 (3) in **117b**[183]; C—SMe 1.762 (8), C—SPh 1.793 (8) in **117c**[184]; 1.783 (3) in **117d**[185]; 1.772 (3) in **117e**; 1.784 (4) in **117f**; C—SBu-*t* 1.782 (4), C—SPh 1.798 (4) in **117g**[186]; 1.785 in **117h**[187]; and 1.782 Å in **117j**[188]. These values are between mean values over large samples of *n* observations[95] (standard deviations in parentheses) for C(sp^2)—O (in enol ethers) 1.354 (16), $n = 40$ and C(sp^3)—O (in dialkyl ethers) 1.426 (19), $n = 236$, and for C(sp^2)—S 1.751 (17), $n = 61$ and C(sp^3)—S 1.819 (19), $n = 242$.

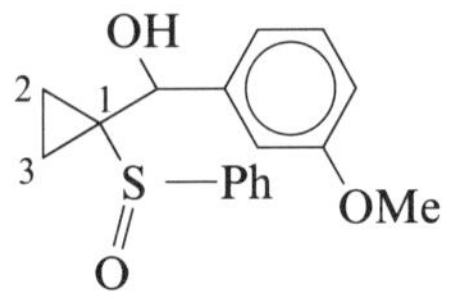

(118)

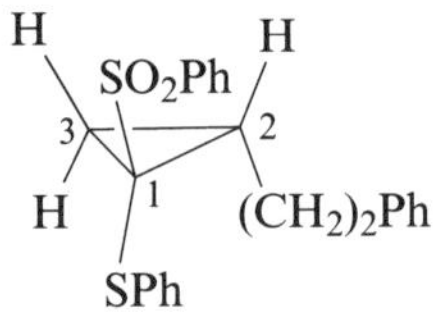

(119)

Cyclopropanes with sulfur substituents have received much interest in synthetic chemistry. Two recent studies of sulfinyl and sulfonyl derivatives will be mentioned here. Relative configurations of two diastereomers of **118** have been determined by XD[189]. The O—S—C1—C2 dihedral angle of 36.6 (4)° means a *syn* bisecting S—O bond (rotated away by 3°) in the S*R* diastereomer. Bond length and angles S—C1, S—C1—C2 and S—C1—C3 are 1.817 (4) Å, 115.6 (3) and 116.0(3)°. The S—O bond adopts the *ac* orientation in R*R*, O—S—C1—C2 –172.5 (4)° (rotation from *syn* –141°, calculated from coordinates), and the sulfur lone pair is above the cyclopropane ring. Corresponding parameters are 1.785 (6) Å, 112.2 (4) and 113.6 (3)°. This substituent in R*R* has a shorter bond to the C_3 ring, and is inclined closer to it, due probably to different conjugation in the two diastereomers.

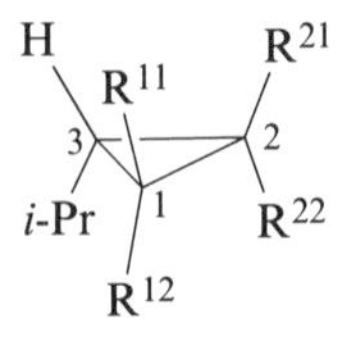

	R^{11}	R^{12}	R^{21}	R^{22}
(120a)	SO_2Ph	SPh	H	SPh
(120b)	SO_2Ph	SO_2Ph	H	SO_2Ph
(120c)	SO_2Ph	SO_2Ph	SO_2Ph	H
(120d)	S(O)(NTs)Ph	SPh	H	SPh

The phenylsulfonyl substituents in **119** and **120a–d** show the effect expected from π-accepting groups: the vicinal bonds are lengthened, the distal bond is shortened[190]. Bond C2—C3, which is opposite to the geminal substituents, is the shortest in these molecules. When going from **120a** to **120b**, i.e. replacing SPh by SO_2Ph, ring bond lengths remain practically the same, while C2—C3 is a little shorter in **120c**, which is the isomer of **120b**. Bond lengths C1—C2, C2—C3 and C3—C1 are: in **120a** 1.526 (3), 1.505 (3) and 1.532 (3); in **120b** 1.524 (3), 1.504 (3) and 1.539 (3); in **120c** 1.526 (3), 1.489 (3) and 1.537 (3) Å. The *N*-(*p*-tolylsulfonyl)sulfoximide group seems to have an even larger effect than SO_2Ph: the C1—C2 bond is 1.546 (3) and 1.537 (3) Å in the two independent molecules of **120d**, longer than in **120a**[190].

G. Nitrogen and Phosphorus Derivatives

1. Aminocyclopropanes

More recent studies of cyclopropylamine (**121**) have resolved discrepancies concerning its structure (see References 191–193 and references cited therein). The previous MW study[194] indicated a considerable shortening of the C1—C2 bond, but this was not supported by theoretical calculations[88,191]. The improved, "hybrid" r_s structure[191] retains the merits of the r_s representation but eliminates the ambiguity caused by small substitution coordinates, such as for atom C1. The new MW study of **121** (Table 10) has established only a small shortening of the vicinal C1—C2 bond, and no change in the C2—C3 bond compared to **1**. The NH_2 group takes a "perpendicular-like" conformation, in which the bisectors of the C2—C1—C3 and the H—N—H angles are in *antiplanar* position (see structure), in accord with *ab initio* results[26,88]. The nitrogen bond geometry is more pyramidal in **121**[191] than in NH_3, H—N—H 108.2 (11)°[197], and in $MeNH_2$, H—N—H 106.0 (6)° and C—N—H 111.5 (7)°[198]. The C—N bond is shorter than in $MeNH_2$, r_z 1.471 (3) Å[198], and makes a larger acute angle, 59.4 (5)°[191], to the ring plane than, e.g., the C—H bonds in **1**, 57.8 (2)° (cf Table 1).

TABLE 10. Structural parameters of cyclopropylamine (**121**), 1-amino-1-methyl- (**122**) and 1-amino-1-phenylcyclopropane (**123**)[a]

Bond or angle	**121** MW, r_s[b]	**121** MP2/6–31G*	**121** XD[c]	**122** XD[c]	**123** XD[c,d]
C1—C2	1.499 (8)	1.499	1.499 (3)[d]	1.510[d]	1.520 (3)
C2—C3	1.512 (3)	1.504	1.506 (3)	1.518	1.503 (4)
C1—C4				1.503 (2)	1.507
C—N	1.452 (7)	1.442	1.457 (2)	1.455	1.451
N—H	1.026 (7)	1.019			
C2—C1—N	116.1(4)	116.2	117.9[d]	115.1[d]	113.7
C2—C1—C4				118.4[d]	117.9
C4—C1—N				117.3 (1)	119.8
C—N—H	108.3(8)	110.0	108.7[d]		
H—N—H	105.4 (7)	106.9	110.3 (15)		
Reference	191	191	192	195	196

[a]Bond lengths are in Å, bong angles in degrees.
[b]Modified r_s structure.
[c]Bond lengths corrected for libration.
[d]Mean values.

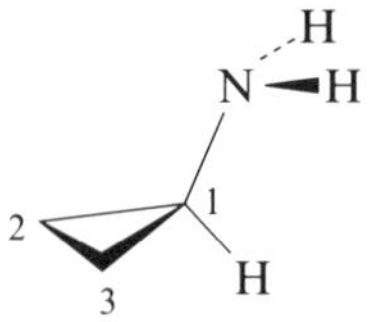

(121)

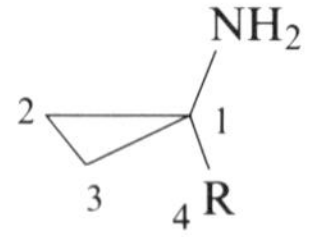

(122) R = Me
(123) R = Ph

A simultaneous crystallographic study (submitted for publication just a week later) gave essentially the same structure[192] (Table 10). There are weak N—H···N hydrogen bonds of N···N 3.22 Å in the crystal.

The structural features of **121** are primarily attributed to rehybridization at the C1 atom, induced by the relatively electronegative NH_2 group, and to four-electron repulsion between the nitrogen lone pair and filled cyclopropyl orbitals[191]. Perpendicular conformation is preferred by the π-donor NH_2 group, although the π-acceptor ability of cyclopropyl is less pronounced[88]. Far-infrared and Raman spectra of **121** reveal the presence of a less stable *gauche* form in the vapor and the liquid, in about 6 percent at room temperature[193]. Only the *anti* form exists in the crystal[192,193]. The relatively high enthalpy difference[193] of 8.4 $kJ\,mol^{-1}$ is consistent with the repulsive interactions mentioned above[191]. A similar number was obtained by 6-31G** calculations[26], with a barrier to rotation (above the lower energy state) of 20.5 $kJ\,mol^{-1}$.

The amino group has the same perpendicular *anti* conformation in derivatives of **121**, and the Ph group takes the bisecting position in **123**. The C(ring)—N bond forms larger acute angles to the ring plane as the size of the geminal substituent increases from H, Me through Ph, but its length changes only slightly (Table 10). It is difficult to tell whether small substituent effects are additive or not in **122** and **123**; the vicinal C—C bond lengths (uncorrected) in **123**, e.g., scatter as much as 0.028 Å in the four independent molecules in the crystal[196].

2. *Cyclopropyl isocyanate and isothiocyanate*

The *syn* and the *anti* conformers of cyclopropyl isocyanate[199] (**124**) and cyclopropyl isothiocyanate[200] (**125**) have been assigned in the MW spectra. The energy difference of the two forms should be very small. In liquid **124**, the *anti* form is more stable by 39 (5) cm^{-1}, determined from the temperature dependence of the Raman spectra, but the *syn* is preferred in the crystal[199]. Some information on the molecular geometry could be obtained from the MW data (Table 11). Structural parameters of **125** were derived from a joint analysis of rotational constants and ED data, using spectroscopic data for vibrational corrections[201]. The *anti* form has an abundance of 72 (5) percent at 35 °C. The ring asymmetry is indiscernible. If the C1—N bond is elongated in the *syn* form, why is the C2—C1—N angle narrower, and why do we get the opposite change in **124**? Large uncertainties and the assumptions made in the latter case may give an explanation. These molecules perform low-frequency large-amplitude motions, torsion about the C1—N bond and bending of the C1—N=C angle, which are not adequately represented by the quasi-rigid models with superimposed small-amplitude vibrations as used in structure analysis[201]. More recent MW data indicate a complicated conformational behavior[203]. The N=C=O and N=C=S groups deviate from linearity, as also observed in the free acids HNCO[204] and HNCS[205]. The structural features of these molecules are reproduced only by sufficiently high-level *ab initio* calculations[201].

TABLE 11. Structural parameters of cyclopropyl isocyanate (**124**) and isothiocyanate (**125**) and isocyanocyclopropane (**126**)[a]

Bond or angle	**124**[b] E = O		**125**[c] E = S		**126**[b]
	anti	*syn*	*anti*	*syn*	
C1—C2	1.520[d]		1.520 (3)		1.521 (7)
C2—C3	1.515[d]		1.515 (3)		1.513 (5)
C1—N	1.417 (10)	1.407 (14)	1.387 (5)	1.413 (5)	1.377 (8)
N=C	1.210[d]		1.193 (3)		1.176 (5)
C=E	1.170[d]		1.574 (3)		
C2—C1—N	117.9 (11)	120.3 (24)	118.9 (6)	116.2 (15)	123.4 (6)[e]
C1—N=C	136.9 (38)	138.6 (13)	149.1 (15)	150.8 (17)	180
N=C=E	172.6[d]		177.7 (20)		
Reference	199		201		202

[a]Bond lengths are in Å, angles in degrees. Parameters which are not shown separately have been assumed to be equal for the *anti* and the *syn* conformer of **124** and **125**.
[b]r_0 parameters fitted to the MW spectra.
[c]r^*_{av} parameters. Effects of zero-point vibration of the large-amplitude torsional motion have been removed by extrapolation.
[d]Assumed values.
[e]Angle M—C1—N, M is the midpoint of the C2—C3 bond.

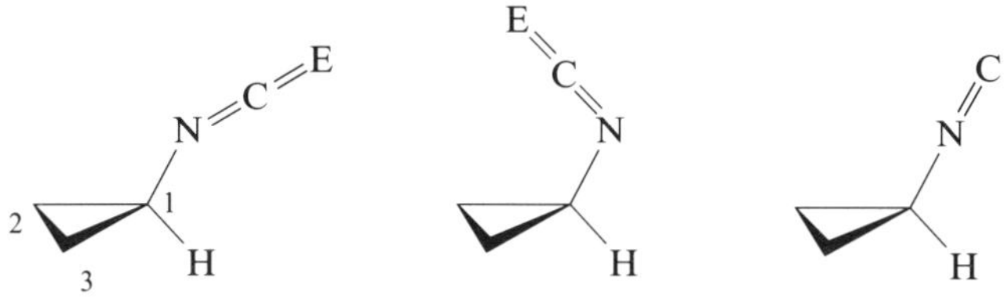

anti *syn*

(**124**) E = O
(**125**) E = S

(**126**)

The isocyano group has been found to be linear within ±1° at the nitrogen atom in isocyanocyclopropane[202] (**126**). The structure has been determined from the MW spectra with some assumptions (Table 11). There is little difference between distal and vicinal bonds; the latter are slightly elongated compared to **1**.

3. Nitrocyclopropanes

1,1′-Dinitrobicyclopropyl (**127**) adopts the *sc* form with 69° dihedral angle and C_2 symmetry in the crystal[206]. The NO_2 groups are in bisecting positions to the rings as in free nitrocyclopropane[207]. The intramolecular O···O contact between the two nitro groups is 2.95 Å, nearly the same as in the corresponding acyclic molecule, 2,3-dimethyl-2,3-dinitrobutane[206] (**129**). The same conformation about the central bond is found in bicyclopropyl derivatives **24** and **25** (Section III.A.2). The central C—C bond of 1.479 Å and the distal bonds of 1.489 Å are shortened in **127**, the vicinal bonds, 1.512 Å (all corrected for libration), are unchanged as compared to bicyclopropyl (**23**) (Table 2). The structure of **128**[208] is similar to that of **127**.

The C—C bond adjacent to both substituents is the longest in *trans*-1,2-dinitrocyclopropane[209] (**130**): C1—C2 1.506 (6), C1—C3 1.473 (7) and C2—C3 1.477 (6) Å.

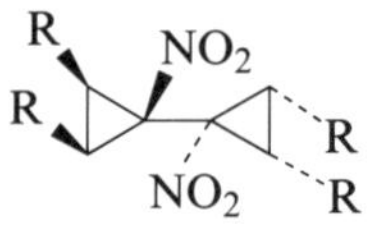

(**127**) R = H
(**128**) R = Me

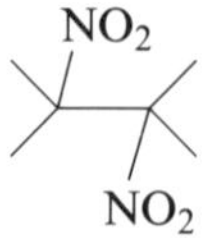

(**129**)

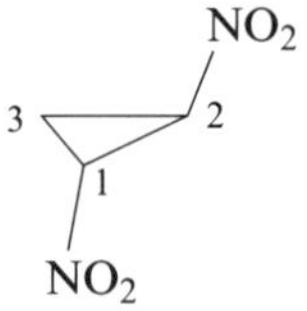

(**130**)

Both NO_2 groups take the bisected conformation in the crystal; the C—NO_2 bond lengths are different, 1.474 (7) and 1.452 (7) Å.

The recently prepared nitrocyclopropanes **131–134** present an interesting conformational behavior of geminal π-accepting substituents[208]. In the 1-nitrocyclopropanecarboxylic acids (**131** and **132**) the group that is *trans* to the Me substituents bisects the ring, C=O being *syn* in **132**, while the *cis* planar group is rotated by about 95°. The nitro group is bisecting in **133**; both groups take the perpendicular orientation in **134**. Conformational choice seems to be affected by steric repulsion of the Me groups as well as of the geminal substituent[208]. Conjugation, which is effective in the bisected form, lengthens ring bonds adjacent to the substituent(s), and shortens the opposite bond, and the bond to the substituent tends to be shorter (Table 12). The changes with conformation are larger for the carboxyl group, indicating that it is a better π acceptor[208]. Noteworthy is the short C—NO_2 bond in **133** and the longer distal bond and longer bonding to the substituents in **134**.

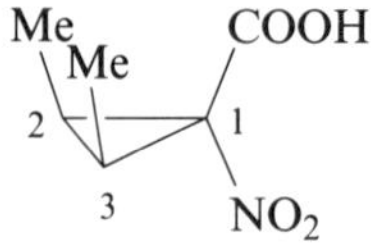

(**131**)

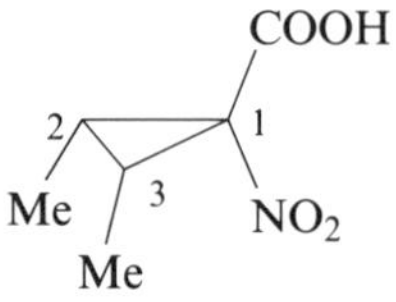

(**132**)

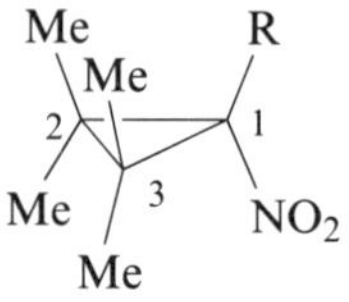

(**133**) R = CN
(**134**) R = $CONH_2$

TABLE 12. Bond lengths (Å) in nitrocyclopropanes[208]

Compound	C1—C2[a]	C2—C3	C1—NO_2	C1—C(O)
131	1.515	1.497 (2)	1.471 (2)	1.503 (3)
132	1.517	1.487 (3)	1.476 (3)	1.476 (3)
133	1.558	1.517 (4)	1.466 (4)	
134	1.513	1.525 (4)	1.494 (3)	1.521 (3)

[a]Mean values of C1—C2 and C1—C3.

4. Phosphorus derivatives

Phosphorus derivatives of **1** have attracted interest as a new type of chelating ligands. The phosphorus lone pairs of the *cis* substituents in **135** are oriented similarly as in $(Ph_2P)_2C{=}C(PPh_2)_2$, just the framework geometries are somewhat different[210]. The conformation of the geminal PPh_2 groups in **135** and **136**[211] is less suitable for coordination. The C—C bond distal to $(PPh_2)_2$ is the shortest in **135**[210]: C1—C2 1.548 (3), C1—C3 1.512 (3) and C2—C3 1.495 (4) Å, and the P—C bonds to the C_3 ring are 1.827 (3) to 1.858 (2) Å. Similar P—C bond lengths have been found in a sulfide[212], **137**, 1.840 (2) Å, but a shorter bond in the oxide[213] **138**, 1.793 (6) Å. Mean bond lengths in the C_3 ring of **137** are: adjacent 1.518, opposite 1.481 Å.

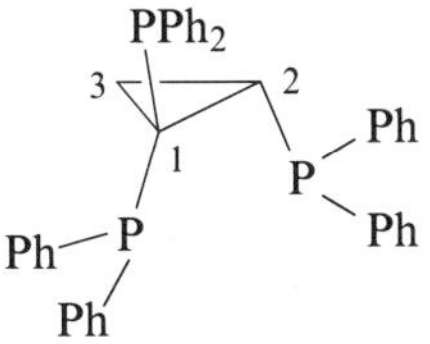

(135)

(136)

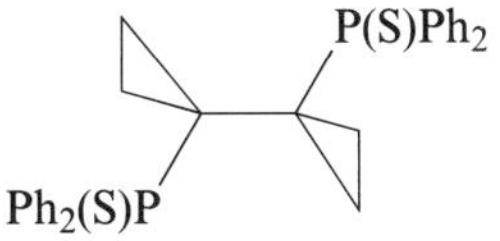

(137)

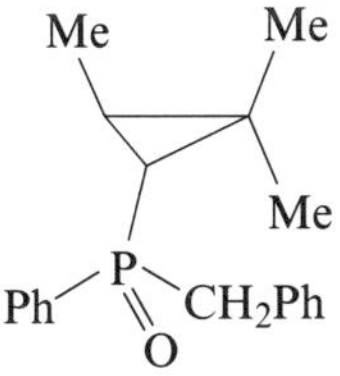

(138)

In triphenylphosphonium cyclopropylide[214] (**139**), the carbanion is pyramidal with an angle M—C1—P of 122°. One of the P—C_{ph} bonds is in *syn* bisecting orientation, the corresponding C1—P—C_{ph} angle, 117.0 (3)° is larger than tetrahedral and this Ph ring is perpendicular to the C1—P—C_{ph} plane. Bonds in the C_3 ring are: adjacent 1.526 (9), distal 1.499 (10) Å. The P^+—C^- bond of 1.696 (6) Å is very short, compared to the mean 1.880 (15) Å of a sample of $n = 35$ for P^+—$C(sp^3)$, where the other bonds of $C(sp^3)$ are to C or H only[95]. The P—C bond lengths in the cyclopropylphosphanes **135** and **136** are comparable to the mean P—$C(sp^3)$, 1.855 (19) Å, $n = 23$, while P—C in **138** is somewhat shorter than the mean P(=O)—$C(sp^3)$, 1.813 (17) Å, $n = 84$[95].

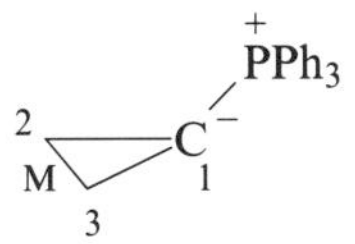

(139)

H. Silyl and Germyl Cyclopropanes

Structures and conformations of silyl and germyl cyclopropanes have been studied by spectroscopic methods, ED and *ab initio* calculations. A recent review is available[215]. Ring geometries are similar in the c-$C_3H_5AX_3$ molecules (**140**) (Table 13): the distal ring bonds are shorter than, the vicinal bonds longer than, and the average C—C bond lengths are about the same as, in **1**. Such changes are characteristic of π-accepting substituents. One of the A—X bonds is in *syn* position to the ring bisector C1—M, i.e. the AX_3 group staggers the C—H and C—C bonds at C1. A small tilt (about 3° or less) of the SiX_3 and GeH_3 groups towards the ring is indicated by the ED data, in accord with theoretical calculations[215].

The rotational barrier of the SiH_3 group is definitely larger in c-$C_3H_5SiH_3$ than in $MeSiH_3$, while there is only a little difference between the corresponding barriers for CH_3

TABLE 13. Structural parameters of cyclopropylsilanes and -germanes c-$C_3H_5AX_3$ (**140**)[a]

AX_3		C1—C2	C2—C3	C—A	M—C1—A	Reference
CH_3	ED, r_a	1.509 (1)[b]		1.517 (2)	123.3 (1)	54
SiH_3	MW, r_0	1.520	1.508	1.853[c]	124.2	216
SiH_3	ED, r_a	1.528 (2)	1.490 (4)	1.840 (2)	124.1 (3)	217
SiF_3	ED, r_a, $\angle_\alpha$	1.522 (8)	1.490[d]	1.807 (2)	124.5 (9)	218
$SiCl_3$	ED, r_a	1.528 (3)	1.492[d]	1.814 (5)	125.7 (7)	215, 219
GeH_3	ED, r_a	1.521 (7)	1.502 (9)	1.924 (2)	124.5 (16)	220
GeH_3	MW, r_s	1.520 (2)	1.505 (2)	1.9170 (9)	124.6 (2)	221

[a]Bond lengths are in Å, angles in degrees. M is the midpoint of the C2—C3 bond.
[b]Mean ring C—C bond length.
[c]Assumed value.
[d]r(C1—C2) – r(C2—C3) was assumed from the *ab initio* calculations.

(Table 14). Effects of the silicon d orbitals, first of all their conjugative interaction with the ring orbitals, have been invoked to explain this observation and geometrical changes[215,217], although such effects might as well not influence rotational barriers[235]. The SiH_3 rotational barriers are, indeed, equal in $EtSiH_3$ and c-$C_3H_5SiH_3$, as are the CH_3 barriers in the analogous molecules. These barriers are 6.7 and 7.1, 14.2 and 12.1 kJ$\,mol^{-1}$, respectively, from 6–31G** calculations[26] (with 6-31G* geometries). Interestingly, the barrier is lower[235], 5.0 kJ$\,mol^{-1}$, in c-$C_3H_5SiF_3$. The trend of decreasing barriers is clear as bond lengths increase in the series CH_3, SiH_3, GeH_3 (Table 14).

The long Si—C(ring) bond of 1.905 (3) Å in **141** is probably due to steric effects[236]. The C—C bonds adjacent to the $SiMe_3$ and Ph substituents are 1.512 (3) and 1.513 (3); the distal bond is 1.484 (4) Å.

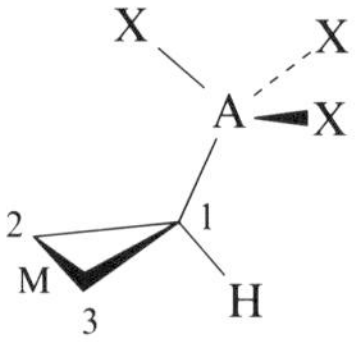

(**140**)

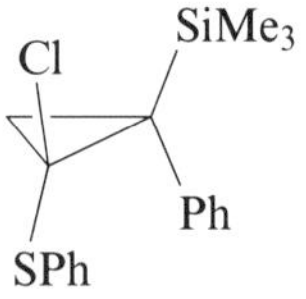

(**141**)

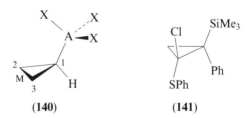

TABLE 14. Rotational barrier (kJ$\,mol^{-1}$) of the AH_3 group in R—AH_3 molecules

R	AH_3		
	CH_3	SiH_3	GeH_3
Me	12.1	7.1	5.2
Reference	222	223	224
Et	13.2	8.3	5.9
Reference	225	226	227
CH_2=CH	8.4	6.2	5.2
Reference	228	229	230
c-C_3H_5	12.7	8.3, 8.0	5.7
Reference	231	232, 233	234

IV. CYCLOPROPYLIDENES

This class of cyclopropane derivatives is characterized by five bonds linking substituents. The molecules can formally be derived by replacing one pair of geminal bonds in substituted cyclopropanes (**2**) by an exocyclic double bond (**3**). (It would have been better to name these bonds[237] 'semicyclic', but the term 'exocyclic' has been accepted.) The analogy is more than formal. Just as the geminal F_2 substituent (and, to a smaller extent, Cl_2), also the $=CH_2$ and the $=O$ substituents induce a relative lengthening (δ_1) of the distal bond by about 0.06 Å, and a shortening of the vicinal bonds[8] (see also References 5 and 238).

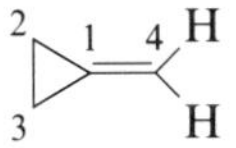

(**142**)

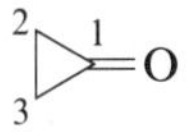

(**143**)

The structures of the parent molecules methylenecyclopropane (**142**) and cyclopropanone (**143**) have been determined by MW. Parameters are in good agreement with results of theoretical calculations (Table 15). The C=C bond in **142** has the same length as in ethene[85], r_g 1.337 (2) Å. The 2-methyl and 2-phenyl derivatives of **142** have been investigated by low-temperature XD[5]. The bond opposite to $=CH_2$ is not influenced; the ring bond between $=CH_2$ and the bisecting Ph group is elongated to 1.487 Å.

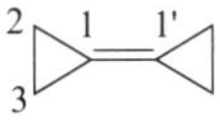

(**144**)

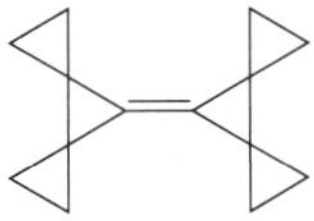

(**145**)

The double bond in bicyclopropylidene (**144**) is shorter than in **142** and in ethene. Bond distances (Å) from XD (–40 °C) have large uncertainties because of poor crystal quality,

TABLE 15. Structural parameters of methylenecyclopropane (**142**) and cyclopropanone (**143**)[a]

Bond or angle	**142**		**143**	
	r_s	r_e	r_s	r_e
C1—C2	1.4570 (14)	1.464	1.475 (17)	1.469
C2—C3	1.5415 (3)	1.536	1.575 (12)	1.567
C=C	1.3317 (14)	1.325		
C=O			1.191 (20)	1.212
H—C2—H	113.5 (10)[c]	114.2	114.1 (20)	114.6
θ[b]	150.8 (10)[c]	148.6	150.9 (20)	153.7
H—C4—H	114.3 (10)[c]	117.2		
Reference	239	27	240	241

[a]Bond lengths are in Å, angles in degrees. r_e parameters are from *ab initio* MP2/6-31G* calculations.
[b]Angle between the H—C2—H plane and bond C2—C3.
[c]r_0 parameter.

and are smaller than those from ED (20 °C), probably due to librational motion in the crystal[242]:

144	C1=C1′	C1—C2 (mean)	C2—C3
XD	1.304(8)	1.440	1.504(7)
ED	1.314(1)	1.468(1)	1.554(2)

The central C=C bond in the spiro derivative[243] **145** is 1.305 (4) Å (XD).

The ring in allenylidenecyclopropane (**146**) is less distorted than in **142**, while the C=C bond adjacent to the ring is shorter. Bond lengths from an XD study[244] are: C1—C2 1.480, C2—C3 1.518, C1=C4 1.281 and C4=C5 1.308 Å; the latter is about the same as in allene[245], r_g 1.3129 (9) Å. In the allene-type molecule **147**, the central C1=C1′ bond, 1.293 (4) Å, is shorter than in allene. Mean ring bond lengths are C1—C2 1.482 (3) and C2—C3 1.526 (5) Å from an XD study[246].

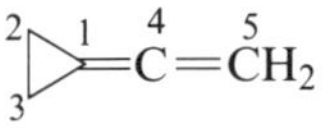

(**146**)

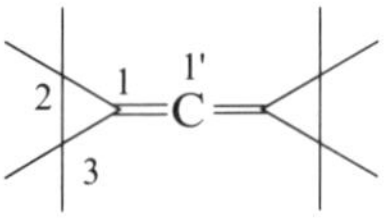

(**147**)

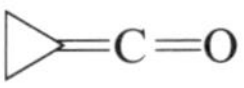

(**148**)

The dynamic properties of cyclopropylidenemethanone (**148**) have been derived from its MW spectra[247]. The potential function for the vibration of the =C=O group out of the ring plane has two minima at bending angles of 17.0 (1)°. The barrier at the planar position is 38.1 (8) cm^{-1}. Since the ground vibrational state lies only 5 cm^{-1} below the barrier, the molecule can be regarded essentially as having a planar heavy-atom skeleton[248].

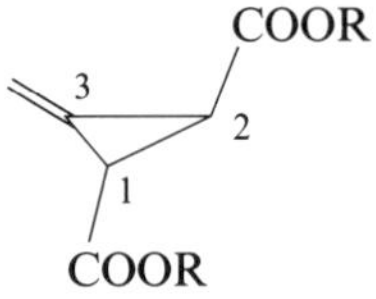

(**149**) R = Me
(**151**) R = H

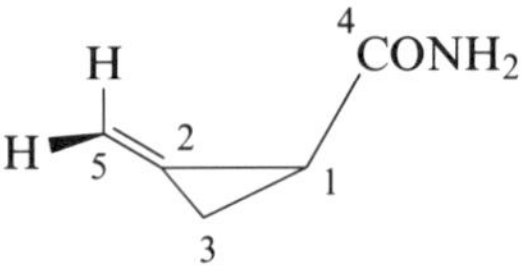

(**150**)

The isomerization of Feist's acid dimethyl ester (**149**) is a well-studied example of the methylenecyclopropane rearrangement[249]. A pre-indication of this [1,3] sigmatropic carbon shift has been demonstrated in the structure of 2-methylenecyclopropanecarboxamide (**150**), determined by XD[250], and augmented recently by a neutron-diffraction (ND) study[251]. Racemic **150** builds a monoclinic $P2_1/c$ crystal with N—H···O intermolecular H bonding. The unit cell volume is reduced by 7.3 percent on cooling from room temperature (XD) to 20 K (ND), but bond lengths and angles are practically unchanged in the two very similar independent molecules[251]. The largest shift occurs in the C1—C3 distance, (mean) 1.548 (XD) to 1.523 Å (ND). Since thermal librations cause an apparent shortening of bonds (cf Tables 4 and 6), an opposite effect would be expected[251]. Ring geometry and the position of the attached CH_2 group are distorted so as to prefigure a breaking of the C1—C3 bond (the longest), altering C2—C3 (the shortest) to a double bond, and forming a new bond C1—C5. Bond angles at C2 are accordingly unequal. In addition, a tilt of the CH_2 group out of the ring plane, towards the $CONH_2$ group is observed[251]. The dihedral angles between the H—C5—H and C1—C2—C3 planes are 1.0 (4)° (molecule A) and 4.9 (5)° (molecule B). Mean bond lengths (Å) and angles (deg) are:

	C1—C2	C2—C3	C1—C3	C1—C2=C5	C3—C2=C5
150	1.473	1.437	1.523	146.0	150.8

Bond angles C2—C1—C4 and C3—C1—C4 are 116.9 (3) to 117.3 (3)°, i.e. equal within uncertainties. It is worth noting that the C1···C5 and C3···C5 distances are uniformly 2.67 Å (in both molecules A and B, calculated from atomic coordinates). The structures of **150** and related molecules have been calculated with 6-31G* basis sets[252]. Crystallographic results are well reproduced, except for the out-of-plane tilt of the (exocyclic) CH_2 group, which is attributed to packing effects. Similar distortions as in the ring and at the CH_2 junction of methylenecyclopropanes are expected also in analogous cyclopropanones and cyclopropanimines having π-accepting substituents in favorable (bisecting) conformation[252].

The ring and methylene geometries[253] in Feist's acid (**151**), C1—C2 1.546 (4), C1—C3 (mean) 1.463 and C=C 1.313 (4) Å, are similar to those in **142** (Table 15, note the different numbering). In both molecules, the bond opposite to C=CH_2 is elongated, and the adjacent bonds are shortened. The COOH groups in **151** adopt *sc* conformations, and are disordered as concerns the interchange of their C=O and C—OH groups[253].

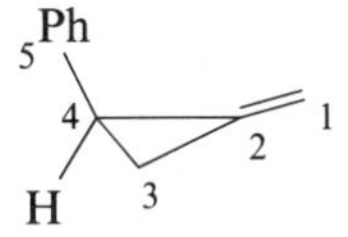

(152)

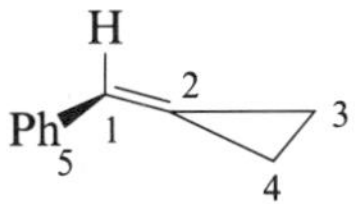

(153)

The rearrangement of 1-methylene-2-phenylcyclopropane (**152**) to **153** leads to an equilibrium[254] at about 160 °C. The crystal molecular structures of both substances have been determined recently[254]. Bond lengths (Å) in the two molecules are:

	C1=C2	C2—C3	C2—C4	C3—C4	C4—C5	C1—C5
152	1.308 (5)	1.461 (5)	1.481 (5)	1.546 (5)	1.499 (5)	
153	1.324 (2)	1.465 (2)	1.464 (2)	1.540 (2)		1.471 (2)

The Ph group is not far from the bisecting conformation in **152**; the two rings are nearly coplanar in **153**.

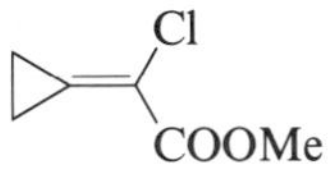

(154)

2-Chlorocyclopropylidenacetates are versatile starting materials in the synthesis of cyclic terpenoid systems[255]. In the methyl ester (**154**), the two independent molecules lie in mirror planes in the orthorhombic *Pnma* crystal, thus the heavy-atom skeletons are completely planar[255]. The distal bond, 1.548 (8) Å, and the vicinal bonds, 1.453 (7), 1.457 (7) Å, have the same lengths as in **142**; the C=C bond is shortened to 1.311 (7) Å. Reactivity in Diels–Alder and Michael additions cannot be directly deduced from the geometrical features.

V. CYCLOPROPENES

As cyclopropanes have olefin-like properties, the presence of a double bond in the three-membered ring adds some acetylenic character to the methine protons. X-ray crystallographic and gas-phase structural studies on cyclopropenes have been reviewed by Allen[11] and recently by Boese[5].

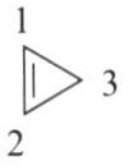

(155)

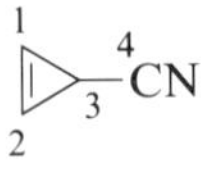

(156)

New r_s parameters of cyclopropene (**155**) have been calculated[256] from existing MW data[257]. A near-equilibrium structure has also been derived from scaled moments of inertia[256] (Table 16). The lengths of the C—C single bond and the methylene C—H bond and H—C—H angle are similar to those in **1** (Table 1). The C=C bond is, however, considerably shorter than in ethene[85], r_g 1.337 (2) Å, and (=)C—H is between C—H in ethene (Section II.A) and in acetylene[256], r_s 1.0586 and r_e 1.0547 Å. Bond-length relations indicate that the methylene carbon in **155** uses approximately the same hybrid orbitals as **1**, $sp^{4.26}$ and $sp^{2.22}$ (Section II.A), to form bonds within the ring and to substituents, while the —CH= carbon in **155** is characterized by $sp^{2.68}$ and $sp^{1.19}$ hybrids, respectively[11].

The structural data of 3-cyanocyclopropene (**156**) shown in Table 16 should be close to r_s parameters, although the complete substitution structure could not be obtained from the available rotational constants[259]. Compared to **155**, the cyano substituent has only a small effect on the ring bonds in **156**, while it creates a large difference of 0.029 Å between vicinal and distal bonds in cyanocyclopropane (**93**) (Table 8). Besides, the C—CN bond is longer, 1.453 (6) Å, in **156** than in **93**, 1.420 (6) Å. *Ab initio* 6-31G* calculations reproduce these structural differences, and lead to the conclusion that the cyano group is more stabilizing in **156** than in **93**, but two- and four-electron interactions with the ring double bond are responsible for the smaller geometrical effects in **156**[259].

The structure of 3-vinylcyclopropene (**157**) has been determined at 103 K by XD[5]. The vinyl group is in *ap* position favorable for conjugation with the ring orbitals. The distal

TABLE 16. Structural parameters of cyclopropene (**155**) and 3-cyanocyclopropene (**156**)[a]

Bond or angle	**155** r_s	**155** $r_e^{\,b}$	**155** $r_e^{\,c}$	**156** r_s
C=C	1.2968	1.2934 (3)	1.289	1.292 (3)
C1—C3	1.5100	1.5051 (3)	1.500	1.511 (5)
C3—C4				1.453 (6)
C≡N				1.162 (5)
C1—H	1.0715	1.0719 (6)	1.075	1.071 (4)
C3—H	1.0900	1.0853 (8)	1.088	1.092 (3)
H—C=C	149.86	149.95 (6)	150.0	152.3 (3)
H—C3—H	114.61	114.32 (12)	113.2	
H—C3—C4				114.2 (3)
M—C3—C4				122.9 (6)
Reference	256	256	258	259

[a]Bond lengths are in Å, angles in degrees. M is the midpoint of the C1=C2 bond.
[b]Near-r_e structure from MW data.
[c]From CSID/6–31G* *ab initio* calculations.

bond is shortened compared to **155**. Bond lengths are C1=C2 1.279 (1), C1—C3 1.517, C3—C4 1.476 (1) and C4=C5 1.329 (1) Å, the latter about the same as in ethene.

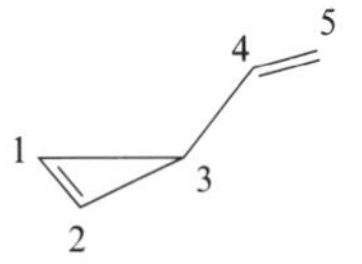

(157)

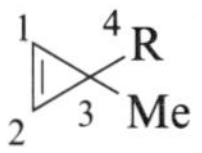

(**158**) R = CH=CH_2
(**159**) R = Ph

The conformation and geometries of **158** and **159** have been studied by ED. Because of similar internuclear distances, severe correlations between parameters are encountered in the analysis. The *syn* form of **158** was identified; preliminary bond lengths and angles were reported[260]. Vibrational spectra indicate the presence of *anti* and *sc* conformers, the former being more stable by 6.5 kJ mol^{-1} according to *ab initio* calculations[261]. Two models of **159** were found to fit the experimental ED data: either an *sc* form with a torsional angle of about 38°, or a mixture of 64 (2) percent of the *syn* bisected form and the perpendicular form[262]. Ring bond lengths are C=C 1.308 (4), 1.312 (4) and C—C 1.510 (3), 1.517 (6) Å for the two models, respectively.

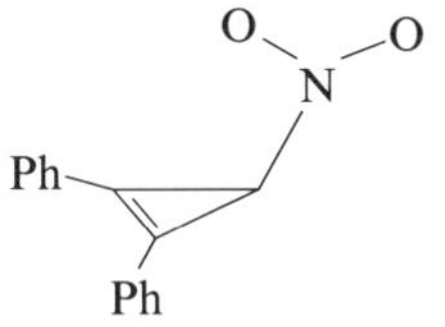

(160)

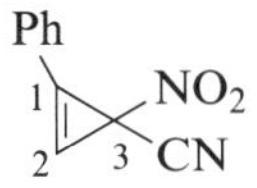

(161)

The nitro group in 1,2-diphenyl-3-nitrocyclopropene (**160**) bisects the cyclopropene ring; the two phenyl groups lie nearly in the C_3 ring plane[263]. A lengthening of the bond distal to the π-accepting NO_2 group would be expected, even in the presence of the Ph substituents. However, the distal C=C bond has about the same length, 1.300 (6) Å, as in **155**, and the vicinal bonds, 1.483 (6) and 1.482 (2) Å, are somewhat shortened. The C—NO_2 bond, 1.523 (6) Å, is longer than in nitro derivatives of **1** (Section III.G.3), just like the C—CN bond is longer in **156** than in **93** (see above). These features are attributed to some ionic character of the C—N bond in **160**, in accord with its facile ionization in solution[263]. The distal C=C bond and the C—NO_2 bond are slightly shorter in **161**, 1.283 (2) and 1.508 (2) Å, and accordingly no solvolysis of **161** is observed in methanol solution[264]. The C—CN bond length, 1.447 (2) Å, is the same as in **156**. The NO_2 group bisects the C_3 ring, the Ph group is nearly coplanar with it.

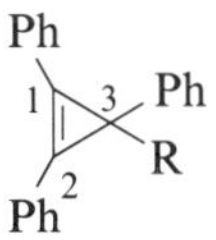

(**162a**) R = H
(**162b**) R = *i*-Pr

The Ph—C=C—Ph arrangement is nearly coplanar in **162a** and **162b**; the 3-Ph group takes the bisecting conformation[265]. The substituents seem to have little effect on the ring bond lengths. The steric repulsion of the *i*-Pr group influences the bonding geometry of the geminal Ph substituent. Some relevant parameters are (Å, deg, average where appropriate):

	C1=C2	C1—C3	C1—C_{ph}	C3—C_{ph}	C1—C3—C_{ph}
162a	1.293 (4)	1.516	1.433	1.470 (4)	121.7 (2)
162b	1.298 (5)	1.516	1.443	1.509 (7)	116.1 (3)

Similarly, in the PF_6^- salt of the pyridinium derivative (**163**), the two rings bonded to the C=C carbons are nearly coplanar with the C_3 ring[266]. The single bonds in the ring, C—C (mean) 1.462 Å, are shorter than in **155**; other bonds have their usual lengths, C1=C2 1.295 (14), C1—N 1.374 (8), C3—N (mean) 1.475 Å.

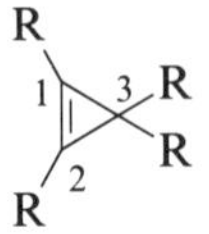

(163)

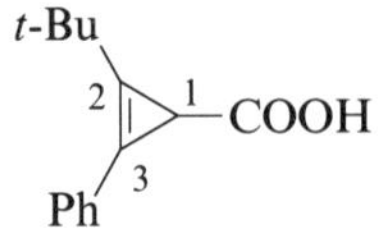

(164)

Coplanarity of the aromatic substituent with the cyclopropene ring is favored by conjugation[267]. This is also found in **164**[268]. Bond lengths are comparable to those listed above, C2=C3 1.290 (2), C1—C2 1.530 (2), C1—C3 1.523 (2), C1—C(O) 1.478 (2), C2—C(Me_3) 1.489 (2) and C3—C_{ph} 1.453 (2) Å. The COOH group takes the bisected form, and interestingly, it is the OH group (*syn*) that is directed to the ring. The usual H-bonded cyclic dimers are formed across crystallographic inversion centers[268].

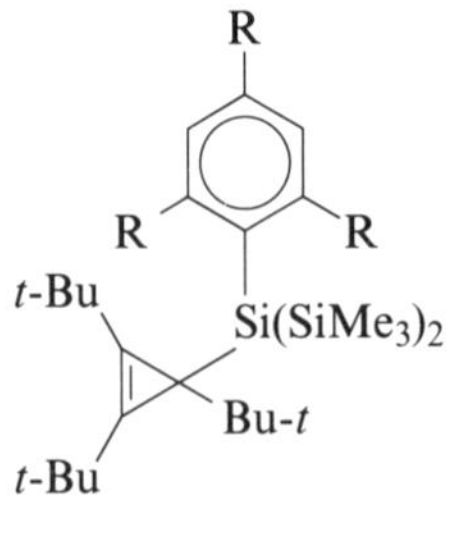

(165a) R = H
(165b) R = Me
(165c) R = *i*-Pr

The ring double bonds are as usual, the single bonds are elongated in the crowded silyl derivatives **165a**, **165b**[269] and **165c**[270]: C=C 1.284 (6), 1.281 (6) and 1.279 (6) Å, and C—C 1.533 (6) to 1.548 (6) Å. The C(ring)—Si bonds, 1.940 (4), 1.959 (4) and 1.988 (4) Å, respectively, are extremely long, and become longer as the size of the aromatic group increases.

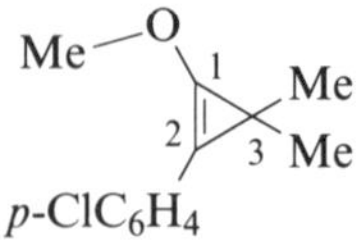

(166)

Substantial ring bond asymmetry is observed in **166**[271]. The bond opposite to the MeO group is relatively long, C2—C3 1.542 (3), C1—C3 1. 475 (3) Å. The aromatic substituent deviates by 12.6° from coplanarity with the C_3 ring; the O—Me bond is *syn* to the double bond. Bonds C1=C2 of 1.294 (3) Å, C1—O 1.336 (2), C2—C_{ar} 1.437 (3), C3—C_{me} 1.511 (3) and 1.518 (3) Å are as expected.

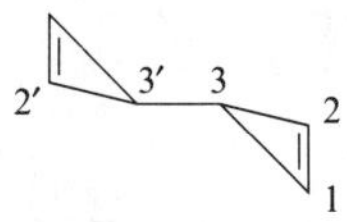

(167)

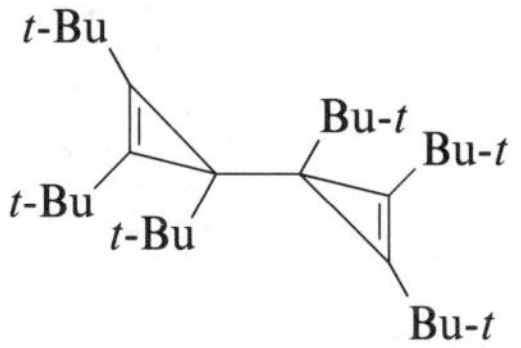

(168)

3,3′-Bicyclopropenyl (**167**), one of the valence isomers of benzene, was prepared a few years ago (1989). Its crystal structure has been determined very recently[272] at 103 K. The hexa-*tert*-butyl derivative (**168**) has been studied at 120 K by XD[273]. The molecule of **167** possesses an inversion center and *anti* conformation; **168** has C_2 symmetry and *sc* conformation in the crystal (Figure 22). Some parameters are shown in Table 17. The ring geometry in **167** is the same as in cyclopropene (**155**) (Table 16); bond length C3—C3′ and angle M—C3—C3′ are similar to those in bicyclopropyl (**23**) and derivatives (Table 2). Photoelectron spectra indicate a 2:1 *ap*:*sc* mixture in the gas phase; the *sc* form with τ = 45–50° has higher energy by 2.0 kJ mol^{-1}, according to 3-21G *ab initio* calculations[272]. The *sc* form in **168**[273] is fixed by the bulky *t*-Bu groups; the dihedral angle τ C1—C3—C3′—C2′

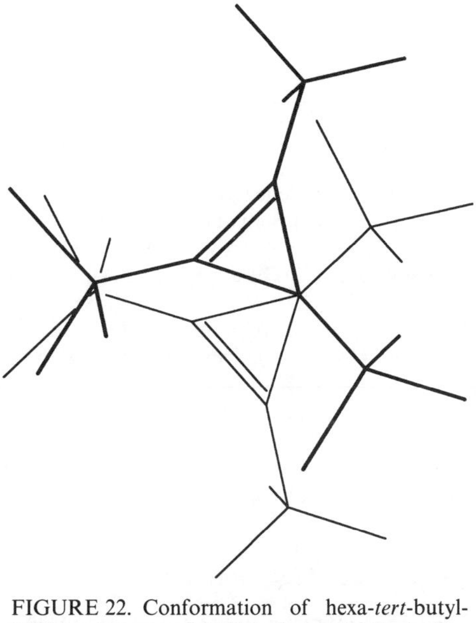

FIGURE 22. Conformation of hexa-*tert*-butyl-3,3′-bicyclopropenyl (**168**) in the crystal. Projection down the central C—C bond, drawn after Reference 273

TABLE 17. Structural parameters of 3,3′-bicyclopropenyl[272] (**167**) and of its hexa-*tert*-butyl derivative[273] (**168**) from XD[a]

Bond	**167**	**168**	Angle	**167**	**168**
C1=C2	1.290 (1)	1.294 (2)	C1—C3—C2	50.5 (1)	49.6 (1)
C1—C3	1.510 (1)	1.536 (2)	C1—C3—C3′	120.9 (1)	120.0 (1)
C2—C3	1.513 (1)	1.550 (2)	C2—C3—C3′	121.1 (1)	120.6 (2)
C1—C11		1.519 (3)	M—C3—C3′	124.7	
C2—C21		1.526 (2)			
C3—C31		1.578 (3)			
C3—C3′	1.503 (1)	1.570 (3)			

[a]Bond lengths in Å, angles in degrees. M is the center of the C1=C2 bond; C11, C21 and C31 are the central atoms of the *t*-Bu groups.

of 87.3° is larger than it would be in the minimum for **167**. The C—C single bonds in the rings and to the *t*-Bu substituents are substantially and asymmetrically elongated (Table 17).

VI. CYCLOPROPENYLIDENES AND RELATED SYSTEMS

This section is a brief survey of systems belonging formally to bonding patterns **5**, **6** and **7**, where three bonds radiate from the three-membered ring.

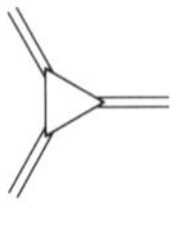

(169)

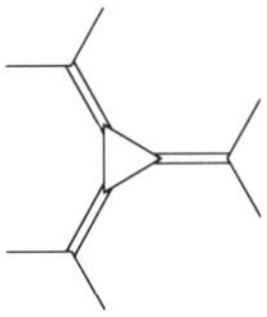

(170)

Trimethylenecyclopropane, [3]radialene (**169**), which is one of the benzene isomers and the smallest member of the radialene family, is characterized by a cyclic arrangement of cross-conjugated π-electron systems. The synthesis and properties of radialenes have been reviewed recently by Hopf and Maas[237]. The C—C and C=C bond lengths in **169**, 1.453 (3) and 1.343 (3) Å, determined by ED[14,274], are similar to those in its hexamethyl derivative (**170**), 1.451 (11) and 1.333 (1) Å, respectively, from an XD study[275]. The double bonds are of the same length as in ethene. The ring bond lengths are equal to the vicinal C1—C2 bond length in methylenecyclopropane (**142**) (Table 15). It seems that the effects of the additional opposite and adjacent =CH_2 groups are cancelled when further two methylene groups are attached to the ring of **142**. The C—C bonds are shorter in the three-membered ring of [3]radialenes than in the larger rings of higher radialenes[237]. As we have seen (Section II.A), also a lengthening of C—C bonds occurs from cyclopropane to cyclobutane.

A. Methylenecyclopropenes and Cyclopropenones

The structure and electronic character of methylenecyclopropene or triafulvene (**171**) have been studied by *ab initio* calculations and MW spectroscopy[27]. Experimental and calculated structural parameters are compared in Table 18. The r_s C3=C4 bond length is

TABLE 18. Bond lengths (Å) and bond angles (deg) in methylenecyclopropene (**171**) and cyclopropenylidene (**172**)

Bond or angle	**171**		**172**	
	r_s	r_e [a]	r_s	r_e [b]
C1=C2	1.323 (3)	1.326	1.3242	1.3240 (6)
C1—C3	1.441 (6)	1.445	1.4195	1.4157 (5)
C3=C4	1.332 (6)	1.330		
C1—H		1.080	1.0754	1.0755 (9)
C4—H		1.083		
C1—C3—C2	54.7		55.61	
H—C1=C2		148.1	149.83	148.89 (22)
H—C4—H		118.0		
Reference	27	27	276	256

[a]From *ab initio* MP2/6-31G* calculations.
[b]Near-equilibrium structure from scaled experimental moments of inertia.

the same as in **142** (Table 15). The C1=C2 bond is longer than in cyclopropene (**155**) (Table 16), and shorter than in cyclopropenylium cations (Section VI.B). It is concluded from the geometrical data and calculated electron distributions that polar resonance form **171a** has a contribution of about 1/5 to **171**, resulting from π delocalization and σ polarization[27]. **171** has a remarkably large dipole moment, 1.90 (2) D. Resonance apparently does not affect the C3=C4 bond length. The molecule is nonaromatic[27].

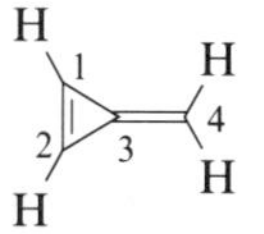
(171)

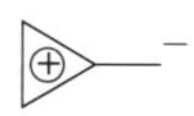
(171a)

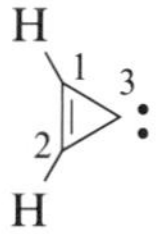
(172)

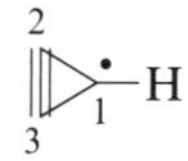
(173)

The discovery of cyclopropenylidene radical (**172**) in laboratory and in space was reported[277] in 1985. Its ring geometry is similar to that of **171** (Table 18). The cyclic $^{\bullet}C_3H$ radical (**173**) is another interstellar species, observed also in laboratory. It has C_{2v} symmetry and a nearly equilateral triangular shape in the $^2B_2$ state. The bond lengths are from MW spectra[278], r_s (H)C—C 1.3750, C≡C 1.3745 and C—H 1.0747 Å.

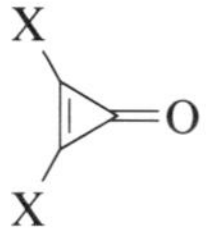
(174) X = H
(175) X = F

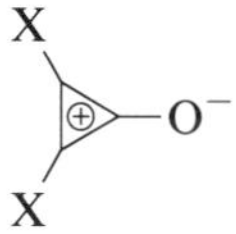
(174a)
(175a)

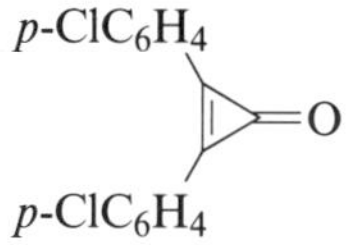
(176)

The geometry and the electronic structure of cyclopropenone (**174**) have been reported from a new MW study and from theoretical calculations[241]. The revised r_s parameters are in good agreement with MP2/6-31G* *ab initio* r_e values[241] and with results of more recent calculations[258] (Table 19). It is instructive to compare bond lengths in related molecules as

TABLE 19. Structural parameters (Å, deg) of cyclopropenone (**174**), difluorocyclopropenone (**175**) and bis(*p*-chlorophenyl)cyclopropenone (**176**)

Bond or angle	**174**		**175**		**176**
	MW, r_s	r_e[a]	MW, r_0	r_e[a]	XD
C=C	1.349 (3)	1.337	1.324	1.324	1.368 (3)
C—C	1.423 (5)	1.422	1.453	1.439	1.418 (3)
C=O	1.212 (6)	1.202	1.192	1.190	1.217 (4)
C—X	1.079 (2)	1.078	1.314	1.301	1.447 (2)
X—C=C	144.3 (1)	144.8	145.7	146.3	149.1
Reference	241	258	258	258	279

[a]From *ab initio* CSID/6-31G* calculations.

the authors did, e.g. the C—C single bonds in three-membered rings (Å, some values rounded, see the relevant sections for sources):

1 1.513

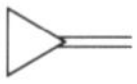
142 1.457

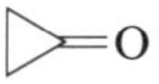

143 1.475

155 1.510

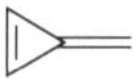
171 1.441

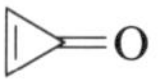

174 1.423

When a double bond is introduced in the C_3 ring, the largest change (shortening) in the adjacent single bond occurs from **143** to **174**. The dipole moment[241] of **174** is large, 4.39 D, larger than that of **143**[240], 2.67 D. The structure, electron distribution and stability of **174** allow the conclusion that this molecule is moderately aromatic[241] (see **174a**).

The structure, charge distribution and stability of the recently synthesized difluorocyclopropenone (**175**) have been studied by NMR and MW spectroscopy and by *ab initio* calculations[258]. Experimental moments of inertia and some differences of bond lengths from the theoretical calculations were used in the combined analysis. The C=C bond is shorter, but the C—C bonds are longer than the corresponding bonds in the nonfluorinated molecule (**174**) (Table 19). π-electron distributions etc. from the quantum chemical calculations suggest that π delocalization and aromaticity are very similar in **174** and **175** (see **175a**), but the geometrical effects may not be evident because of the large σ effects of fluorine[258].

The structure of bis(*p*-chlorophenyl)cyclopropenone (**176**) has been redetermined by XD[279]. The molecule is nearly planar in the monoclinic *C*2/*c* crystal. Main bond parameters are also given in Table 19.

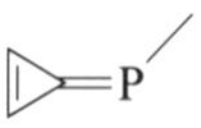

(177)

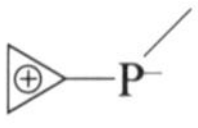

(177a)

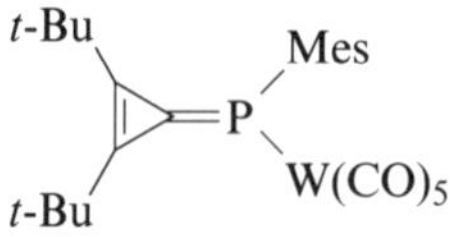

Mes = Mesityl
(178)

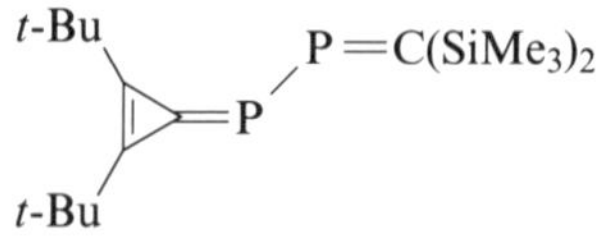

(179)

Phospha-analogs of **171** have recently been prepared[280]. *Ab initio* calculations have confirmed that the phosphorus atom, unlike in phosphaalkenes, is negatively charged in these phosphatriafulvenes (**177**), and the cationic center is stabilized in the aromatic three-membered ring[280] (see **177a**). The short bond opposite to the phosphorus substituent in **178**[280], C=C 1.321 (7), C—C (mean) 1.425 and in **179**[281], C=C 1.331 (7), C—C 1.412 Å, indicates double-bond character. The P atom lies nearly in the plane of the C_3 ring. The P=C bond to the C_3 ring in **178**, 1.679 (5) Å, is shorter than in **179**, 1.731 (5) Å. They are considerably shorter than the P—C single bonds in cyclopropylphosphanes, but are similar to the bond of ionic character in cyclopropylide **139** (Section III. G.4).

B. Ionic Species

The Hückel aromatic cyclopropenyl cation and its derivatives have been studied by theoretical calculations and XD. The tri-*tert*-butylcyclopropenylium ion (**180**) possesses crystallographic C_s symmetry in its hydrogen dichloride[282], $(t\text{-}Bu_3C_3)^+ [HCl_2]^-$. The ring is nearly equilateral with C—C of 1.371 (1) and 1.373 (1) Å; the *t*-Bu—C(ring) bonds are 1.485 (2) and 1.489 (1) Å. The unusual $[HCl_2]^-$ anion lies in the mirror plane in the crystal, Cl···Cl 3.141 (1) Å, Cl—H···Cl 175 (2)°. A similar average C—C distance, 1.377 (12) Å, has been found in the three-membered ring of the 1,2,3-triphenylcyclopropenylium cation[283] (**181**) in its hexabromotellurate(IV), $(Ph_3C_3)_2[TeBr_6]$.

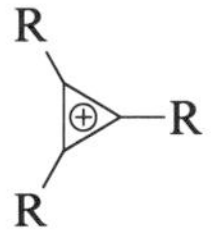

(**180**) R = *t*-Bu
(**181**) R = Ph

The hexacyanotrimethylenecyclopropanide anion (**182**), a derivative of [3]radialene (**169**), forms electron-transfer salts with interesting stacking patterns and magnetic behavior[284]. In the paramagnetic and the antiferromagnetic phases of the Me_4N^+ salt[284a], the average C—C bond length in the ring is 1.391 (5) and 1.402 (3) Å, respectively, longer than in the cyclopropenylium ions. The C(ring)—$C(CN)_2$ distances of 1.363 (17) and 1.367 (2) Å are longer than the C=C bonds in **169** and in **142** (Table 15). Rather large variations in the geometry of the radical anion **182** are found in different environments[284].

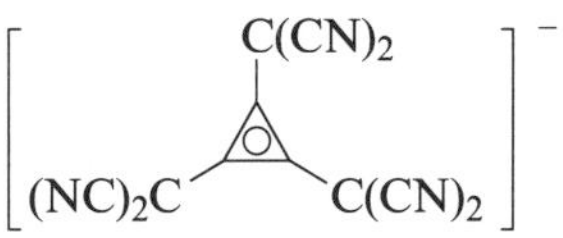

(**182**)

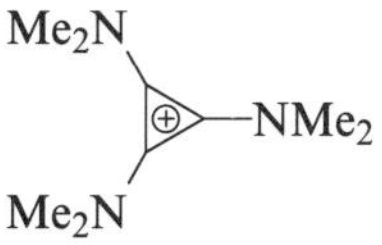

(**183**)

The 1,2,3-tris(dimethylamino)cyclopropenylium ion, $\{(Me_2N)_3C_3\}^+$ (**183**), forms intensely colored charge-transfer solids with certain anions, while the ionic components themselves are colorless[285]. The crystal with $[NbOCl_4(OH_2)]^-$ is orange, with $SbCl_6^-$ violet, with $NbCl_6^-$ dark blue and with $TaCl_6^-$ dark red. The latter two salts form rhombohedral $R\bar{3}$ crystals[285b]. Their ring C—C bonds, 1.380 and 1.371 Å, as well as the C(ring)—N bonds, 1.322 and 1.323 Å, respectively, are very short. The N atoms have planar bond configuration, and the N—CH_3 bonds are 1.446 to 1.467 Å.

The structure of 2,3-dihydroxycyclopropen-1-one, deltic acid (**184**), which has been determined by XD at 135 K, reflects significant contributions from cyclopropenylium or dipolar resonance forms[286]. The difference between formally double bonds and single

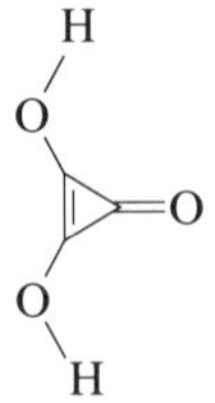

(**184**)

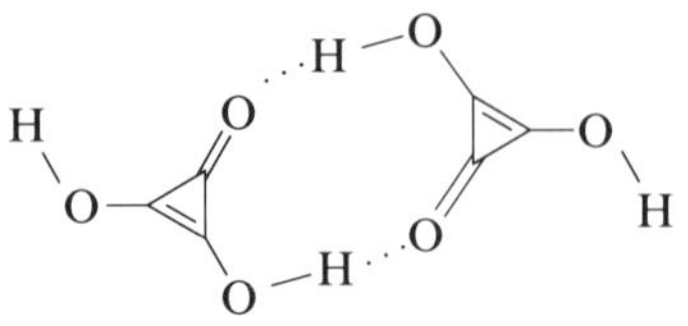

(**184a**)

bonds is small, C=C 1.373 (1), C—C 1.397 (1), C=O 1.265 (1), C—O 1.301 (1) Å and C—C—O 152.4 (1)°. The molecules are planar, and are bisected by mirror planes in the orthorhombic *Pnam* crystal. Cyclic pairs of H bonds link the molecules in chains (see **184a**), similarly as in carboxylic acids. It is remarkable that each carbonyl group accepts two strong H bonds, O···O 2.555(1) Å and O—H···O 178.4°.

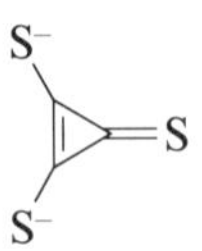

(**185**)

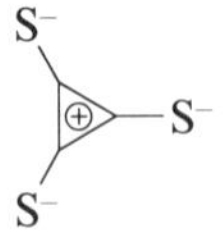

(**185a**)

The trithiodeltate dianion (**185**) in its phosphonium salt, $(Ph_3MeP)_2$ $(c\text{-}C_3S_3)\cdot 3H_2O$, is very close to D_{3h} symmetry and is aromatic (see **185a**)[287]. The mean C—C bond length of 1.405 Å is longer than in cyclopropenylium ions; the C—S bond length, 1.676 Å, indicates partial double-bond character[287].

VII. CYCLOPROPANE RINGS IN POLYCYCLIC SYSTEMS

Sections VII.A and VII.B cover cyclic derivatives which contain the cyclopropane or cyclopropene unit in a fused ring system according to scheme **9** or **10**. Structures of many simple molecules of this class were determined long ago. Some of them will be mentioned here; further structural data and references are given in previous reviews. Bridged derivatives of bicyclo[1.1.0]butane are discussed in Sections VII. C.1 and VII. C.2.

A. Bicyclo[*n*.1.0]alkanes

The bonding geometry at the bridgehead carbon atoms in bicyclo[*n*.1.0]alkanes and in propellanes is highly distorted from the tetrahedral. This gives rise to high strain in these molecules[12]. Allen has surveyed variations of geometry and strain in derivatives of bicyclo[1.1.0]butane (**186**) and homologues[12], and has examined linear correlations between geometric parameters in 1,3-bridged derivatives (propellanes) and also in 2,4-bridged derivatives of **186**[288].

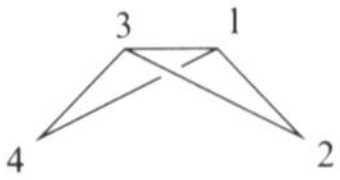

(**186**)

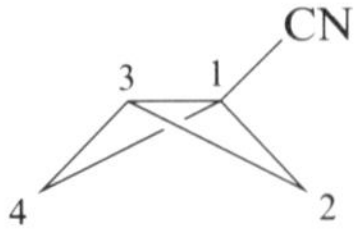

(**187**)

The dihedral angle between the three-membered rings in **186** is 121.7 (5)° from the MW spectra[289]. The bridge bond C1—C3, r_s 1.497 (5) Å, is of the same length as the side bonds C1—C2, 1.498(5) Å, shorter than in **1**. The shape of this four-membered ring is very different from that of cyclobutane[35], where the dihedral angle is 152.1 (16)°, and the C—C bonds are r_z 1.552 (1) Å. Geometric parameters of the 1-cyano derivative **187** have been obtained recently from MW with some assumptions[290]. The bonds opposite to the CN group are shortened: r_0 C1—C3 1.518, C1—C2 1.514 and C2—C3 1.498 Å. The dihedral angle of the two ring planes is 122.5°, and bond angle C3—C1—C(N) 129.1°. Calculations at the RHF/6-31G* and MP2/6-31G* levels have been performed[290]. It is concluded from the electronic structures that the bicyclobutyl group is a better electron donor than cyclopropyl.

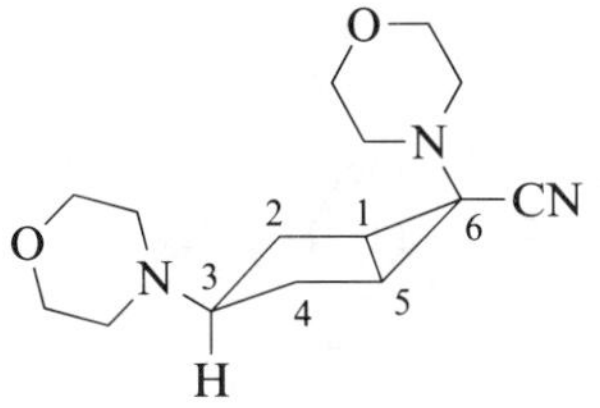

(188)

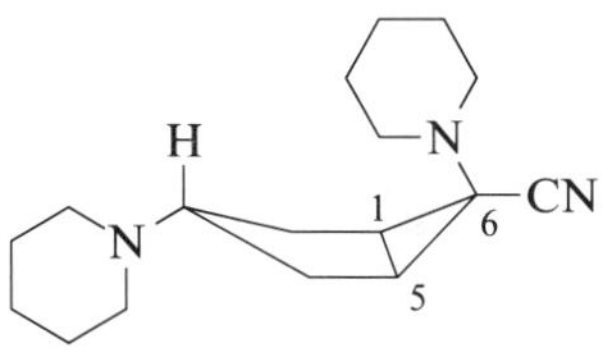

(189)

Bicyclo[3.1.0]hexane[291] and derivatives usually prefer the boat conformation[292]. The chair form has been found in **188**[293], with angles between planes 1–5–6/1–2–4–5 of 67.5° and 2–3–4/1–2–4–5 of −32.4°. A similar molecule, **189**, however, adopts the boat form, with interplanar angles of 67.8 and 21.6°, respectively[293,294]. The substituent at C3 is equatorial in both molecules, thus the resulting N···N distance between the two morpholino groups, 4.28 Å, is nearly the same as between the piperidinyl groups[293], 4.34 Å. The flap angle and bonds at the cyclopropane part are not influenced by the conformation of the five-membered ring. The common bond of the rings is the shortest: C1—C5, C1—C6 and C5—C6 are 1.488 (3), 1.503 (3) and 1.502 (3) Å for **188**; 1.481 (2), 1.508 (2) and 1.509 (2) Å for **189**. The chair conformation has been found[295] at −68 °C in the aza analogue **190** with an equatorial N—Me bond, and dihedral angles of 65 and −29.5°. The cation **191** has again a boat form, albeit flat at the nitrogen end, characterized by 65.8 and 8.1°. Though the N···N distance is only 2.996 Å, hydrogen bonding is formed to the bromide anion instead of within the cation[296]. The interplanar angles in the chair-form boranamine **192**[297] are large, 67.3 and −32.6°, and here the bridge bond is slightly longer, 1.500 (8) Å, than the other bonds in the C_3 ring, 1.487 (5) Å.

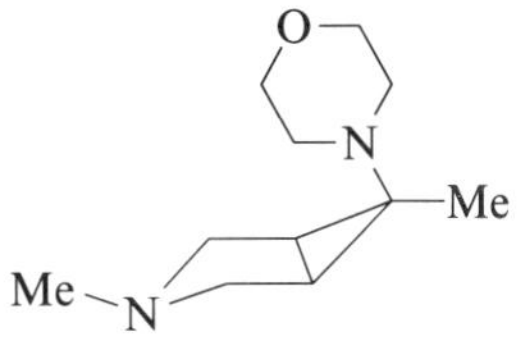

(190)

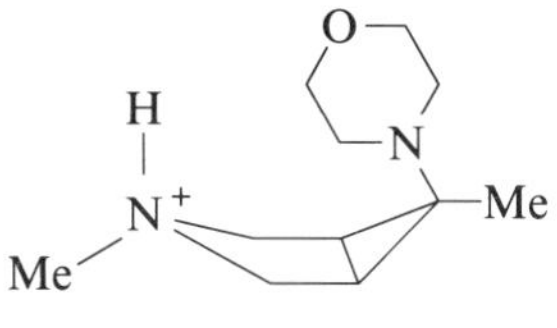

(191)

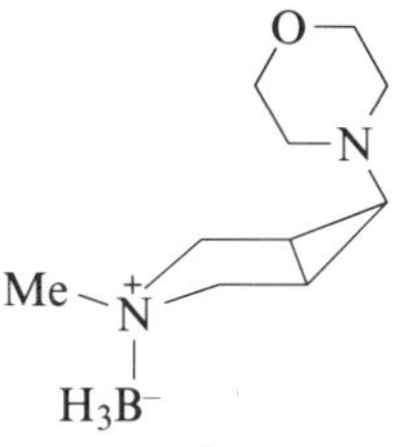

(192)

The two isomers of **193** have a similar boat and *exo* combination of the two rings[298]. This unusual conformation of a norcarane derivative is attributed to the presence of bulky substituents[298]. The different orientation of the CN group in the isomers induces different conformations of the equatorial 2,2-dichloro-3,3-dimethylcyclopropyl substituent. The longest bonds in the C_3 rings, both in the substituent and in the bicyclo[4.1.0]heptane system, are those opposite to the *gem*-Cl_2 group, 1.516 (8) to 1.532 (4) Å, compared to the vicinal C—C bonds, 1.469 (8) to 1.507 (5) Å.

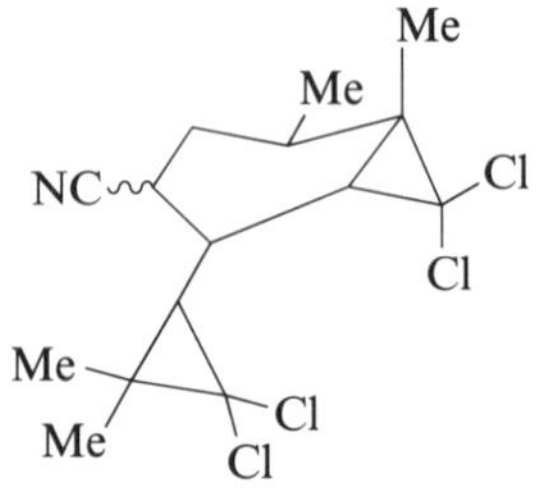

(193)

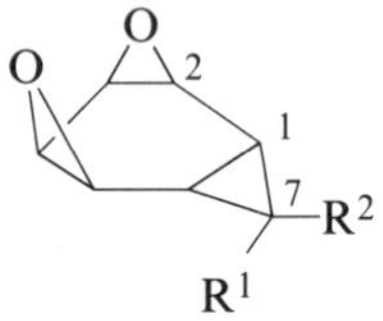

(**194a**) R^1 = Me, R^2 = COOMe

(**194b**) R^1 = Ph, R^2 = Me

The six-membered rings are closely planar in the norcaradiene diepoxides **194a** and **194b**[299]. The **194a** molecule has a mirror plane in the orthorhombic *Pnma* crystal. The *syn* bisecting C=O group causes a shortening of the distal bond, 1.496 (7) Å, and is bent towards the cyclopropane ring, angle C1—C7—C(O) of 112.5 (3)° being small. The vicinal bonds, 1.524 (5) Å, are of the same length as in the equilateral C_3 ring of **194b**, where the Ph group adopts the perpendicular conformation. Angles of the type C7—C1—C2 at the junction of the two rings are close to 120°.

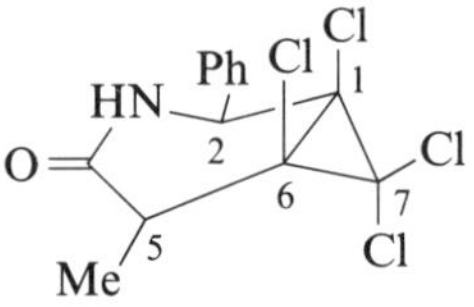

(195)

The three-membered ring in **195**[300] is equilateral, C—C 1.511(3) to 1.521(3) Å; the *gem*-Cl_2 substitution seems to have no significant effect. The bond angles at the shared atoms of the rings are similar to those in simple substituted cyclopropanes, C7—C1—C2 121.3 (2), C7—C6—C5 121.8 (2)°. The C_3 ring takes the *exo* position to the boat-form piperidinone ring.

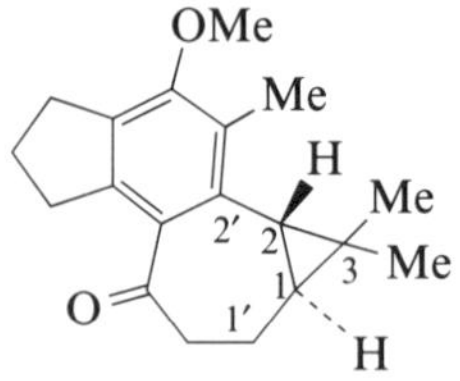

(196)

Synthetic and structural studies have been performed on jatropholones and other natural products containing the bicyclo[5.1.0]octane unit[301]. The uncommon *trans* anellation to the cyclopropane ring (see **10**) has been found in **196**[302]. The bond common with the twisted seven-membered ring is the shortest, C1—C2 1.506, C1—C3 1.514 and C2—C3 1.530 Å. Bond angles demonstrate the strain in this ring system, e.g. C3—C1—C1′ 134.7, C2—C1—C1′ 114.4, C3—C2—C2′ 132.9 and C1—C2—C2′ 114.1° (calculated from atomic coordinates).

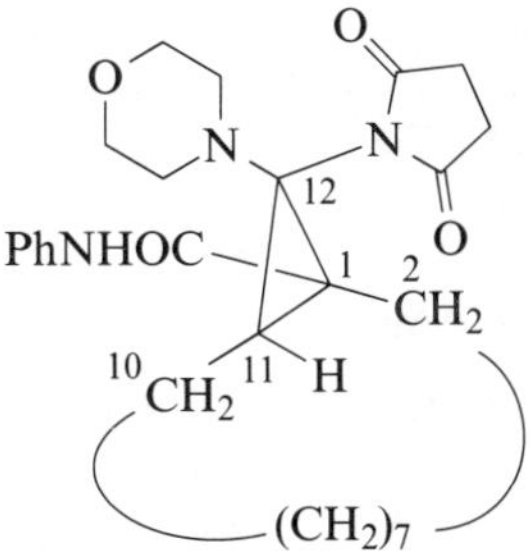

(197)

The reaction of a cyclododecene derivative with succinimide leads to the interesting *trans*-bicyclic system[303] **197**. The relation of ring bonds is as expected, the bond opposite to the carboxamide substituent being the shortest, C1—C11 1.507 (4), C1—C12 1.529 (4) and C11—C12 1.496 (4) Å. The ring fused to cyclopropane is larger here, but some widening of bond angles due to *trans* anellation is still seen, C12—C1—C2 122.7(3) and C12—C11—C10 124.0 (3)°.

B. Cycloproparenes

The structures of cycloproparenes[5,10] and the chemistry and properties of alkylidenecycloproparenes[304] have been reviewed. Much experimental and theoretical work has been devoted to strain and stability and to 'bond fixation' in such fused aromatic systems[305], resulting in much controversy. The anellation of small rings to a benzene ring gives rise to substantial angular deformation and bond distance change.

Just a few structures will be described here. Boese in his recent review gives more examples and discusses some hitherto unpublished results[5].

The structure of 1*H*-cyclopropabenzene (**198a**) has been studied at 120 K by XD[306]. Angles in the benzene ring indicate high strain due to anellation of the three-membered ring (average values): C6—C1—C2 124.5 (2), C1—C2—C3 113.2 (2) and C2—C3—C4 122.4 (2)°. Aromatic π delocalization in the benzene ring seems to have been retained, only the bond common to the rings is shorter, C1—C6 1.334 (4) Å, compared to the average of the other five bond lengths, 1.378 Å. The bonds to the methylene group are longer, C1—C7 1.498 (3) Å. Bond C1—C6 is longer than C=C in cyclopropene (**155**) (Table 16). The derivative **198b** is characterized by similar parameters[306]; just the C1—C7 bonds next to the bulky substituents are elongated to 1.541 (1) Å.

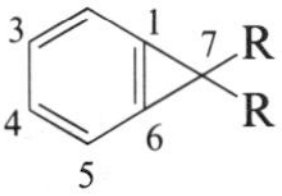

(198a) R = H
(198b) R = *i*-Pr$_3$Si

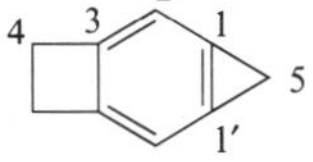

(199)

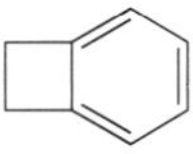

(200)

The structure and electron-density distribution of the highly strained molecule 3,4-dihydro-1*H*-cyclobuta[*a*]cyclopropa[*d*]benzene (**199**) have been determined at 100 K by XD[307]. The bonds, C1—C5 1.508, C1—C1′ 1.349, C1—C2 1.385 Å, etc., are longer than the corresponding bonds in **198a** and **200**, and the bond angle at C2 in the central ring is very small, 109.2°. Electron-density maxima reside outside the three- and four-membered rings, e.g. 0.27 Å from the C1—C1′ line within the central ring. The lines connecting C5 with the two adjacent maxima make an angle of 98°, while angle C1—C5—C1′ is 53.2°. The expansion of the rings is consistent with photoelectron spectra, and has a stabilizing effect (see references in Reference 307).

C. Bridged Ring Systems, Cage Molecules

The systematic treatment of such structures is beyond the scope of the present review. The majority of examples selected belongs to bonding type **12**, constructed of two or more three-membered rings and other structural elements. Propellanes are of the **11** type. A monograph on cage molecules has appeared recently[308].

1. Bridged bicyclo[1.1.0]butanes

The electron-density distribution of a bicyclo[1.1.0]butane (**186**) derivative (**201**) has been determined at 98 K by XD[309]. This molecule represents a transition from **186** to [1.1.1]propellane (**205**) systems with inverted carbon atom configuration, since the C1—H1 and the C1—C3 bonds lie on the same side of the C2—C1—C4 plane, and C1—H1 makes an angle of 167.2° to this plane. It has been demonstrated[309] that as this angle[12] (the 'inversion parameter') decreases, the maximum difference electron density along the bridge bond diminishes and turns to negative as in the case of bridged propellanes **207** and **208** (Section VII.C.2). This peak is 0.12 e Å^{-3} along the C1—C3 bond in **201**, lower than along the other sides of the three-membered rings (Figure 23). The bridge bond is 1.511 Å (calculated from atomic coordinates), slightly longer than in **186**.

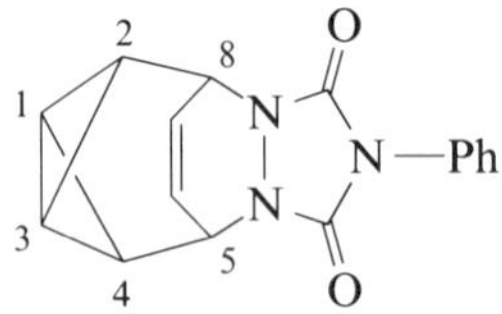

(201)

In related tricyclo[2.1.0.0^{2,5}]pentane (**202**) systems[310,311] substituted at C1, C5 and C3, the dihedral angle between the three-membered rings is about 95 to 99°. The bridge bond C1—C5 shortens as the bond angles C5—C1—R and C1—C5—R to the R substituents open, following a linear correlation[310]. Irngartinger and Lukas have also found a linear relationship with the fold angle of the bicyclobutane unit[288].

The structure of **203** has been redetermined[312] by refinements of the XD data in the centrosymmetric orthorhombic space group *Pnca*. The molecule has C_2 symmetry in the crystal. The side bonds in the bicyclobutane unit are longer, C1—C2 1.540 (6) Å, the bridge bond is shorter, C1—C6 1.442 (5) Å, and the fold angle of 110.7° (calculated from the coordinates) is narrower than in free **186**.

Parameters of the octabisvalene derivative[313] **204** also fit the correlations for bicyclobutanes. The bridge bonds are 1.505 (3) and 1.501 (3) Å, the dihedral angles 118.0 (1) and

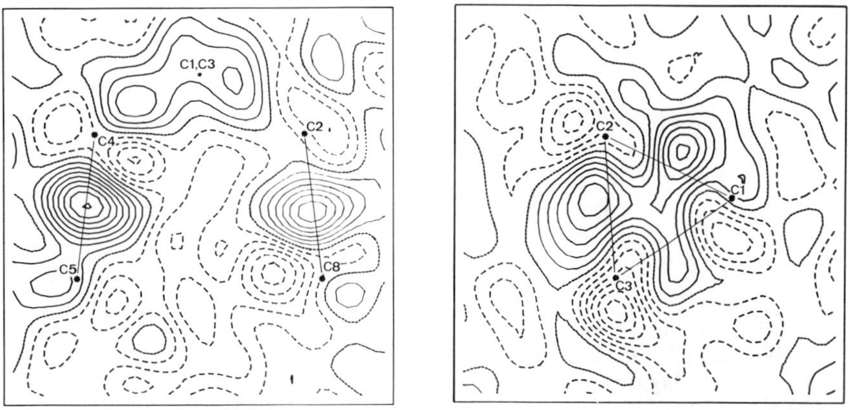

(a) (b)

FIGURE 23. Difference electron-density maps of **201**. (a) Section in the plane through the mid-point of and perpendicular to the bridging C1—C3 bond of the bicyclobutane part. (b) Section in the plane of the three-membered ring C1, C2, C3. Contour lines are at 0.05 e $Å^{-3}$ intervals. Full lines indicate positive, dashed lines negative regions. Reproduced by permission of the International Union of Crystallography from Reference 309

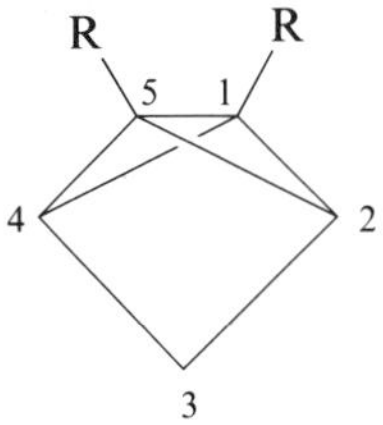

(202)

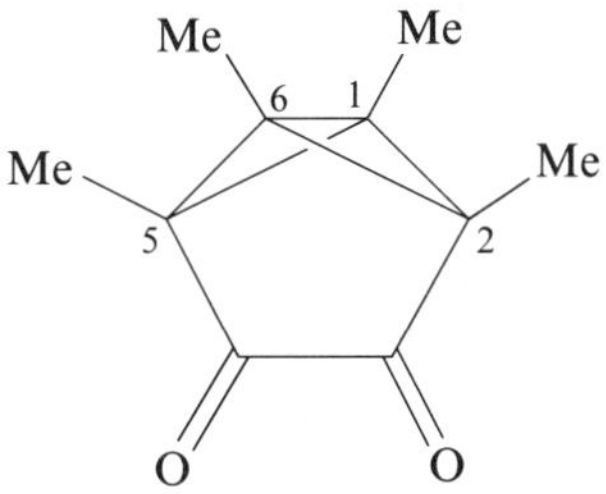

(203)

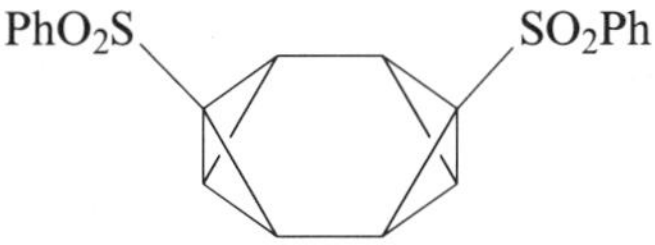

(204)

117.9 (2)°. The side bonds in the C_3 rings which are opposite to the electron-withdrawing $PhSO_2$ substituents are shorter (on the average), 1.495 Å, than the adjacent bonds, 1.512 Å.

2. [1.1.1]Propellanes

The parent molecule, tricyclo[1.1.1.$0^{1,3}$]pentane, [1.1.1]propellane (**205**), may be imagined as a 1,3-bridged **186**. Its remarkable geometric and electronic structure, the inverted geometry at the axial carbon atoms, have attracted the interest of theoreticians and experimentalists[29,314–316].

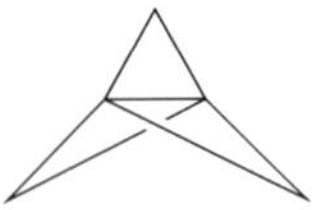

(205)

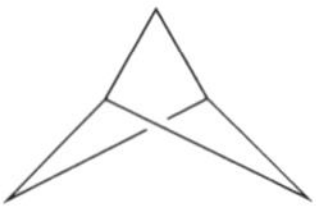

(206)

The structure of the free molecule **205** has been determined from ED data[316] with a joint use of rotational constants. The central bond is very long, r_g 1.596 (5) Å; the side bonds of 1.525 (2) Å are also longer than in **1** (Table 1). The C—H bond is 1.106 (5) Å, angle H—C—H 116.0 (19)°. These are in excellent agreement with results of MP2/6-311G(MC)* calculations[28], the highest-level structure yet reported for **205**, i.e., C—C central 1.602, C—C side 1.521, C—H 1.087 Å and H—C—H 115.3°.

The C—C bond in the related free bicyclo[1.1.1]pentane[317] (**206**) is r_a 1.557 (2) Å. The C···C distance, 1.874 (4) Å, which corresponds to the axial bond in **205**, is longer than the bond but is very short for a nonbonded distance. This parameter is rather sensitive to substituents in the 1 and 3 positions[314b].

There are four independent molecules in the crystal of **205** at 138 K; two of them show rotational disorder about their central bonds[318]. Another phase formed on cooling consists of twinned crystals, which cannot be used for a structure determination[318]. Approximate D_{3h} sysmmetry and bond lengths of about 1.60 and 1.53 Å (corrected for the large thermal motion) in the ordered molecules at 138 K correspond to those in the gas phase.

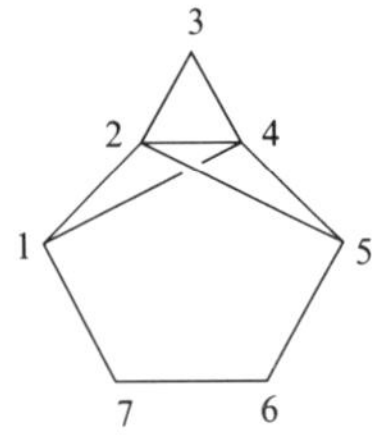

(207)

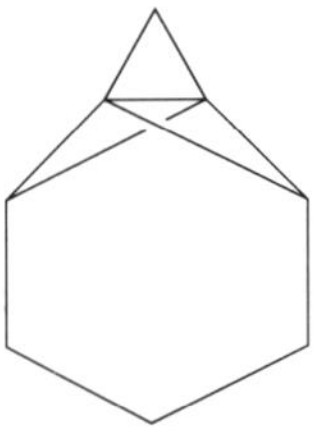

(208)

The structure and electron-density distribution of two derivatives **207** and **208** have been determined at 81 K by XD[319]. The symmetry of the propellane part is close to D_{3h} in both molecules. The bridgehead distances of 1.587 and 1.585, and the mean side bond lengths 1.525 and 1.528 Å (corrected for libration), respectively, are essentially the same as in free **205**. The deformation electron-density maps clearly show the expected maxima for the bent side bonds, and also the electron density from their overlap in the ring centers (Figure 24). No charge accumulation is observed between the bridgehead nuclei; in fact, there is a negative region of difference density between them (Figure 24), which may come from experimental or numerical sources or from the choice of the spherical-atom reference model[319]. The most remarkable feature of these maps is the diffuse positive region, which is contiguous to the bent-bond density peaks at the extensions of the line connecting the bridgehead atoms[319] (Figure 24b), and which was also obtained for **205** in a theoretical analysis of charge concentrations[29].

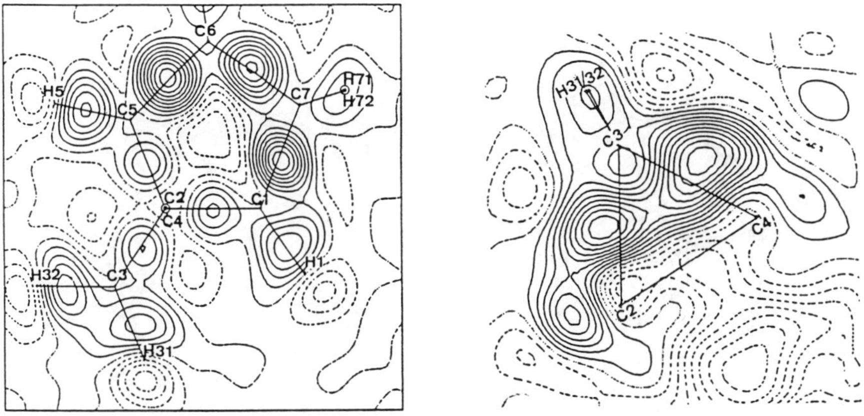

(a) **(b)**

FIGURE 24. Electron-density difference maps of **207**. (a) Section in the plane of atoms C1, C3 and C5, approximately perpendicular to the C2—C4 axis of the propellane unit. Note the density peaks in the centers of the three-membered rings, around the projection of C2 and C4. Contour lines are at 0.05 e Å^{-3} intervals. (b) Section in the plane of the three-membered ring C2, C3, C4. Contour lines are at 0.025 e Å^{-3} intervals. Full lines mark positive, dashed lines negative regions. Reproduced by permission of Verlag Helvetica Chimica Acta from Reference 319

3. *Tetrahedranes*

Tetrahedranes belong to the most strained molecules[320]. The molecular structure of tetra-*tert*-butyltetrahedrane (**209**) has been determined[321] in its low-temperature modification at 213 K. The molecule lies on a mirror plane in the hexagonal $P6_3/m$ crystal, and is close to T_d symmetry. The C—C bonds in the tetrahedron, 1.485 (4) Å, are shorter than in **186**, but longer than expected from geometric correlations (cf Section VII. C.1) for a bicyclo[1.1.0]butane, which has a 2,4 bridge through one bond and an accordingly small dihedral angle between the three-membered ring planes[321]. The bonds to the *t*-Bu groups are 1.490 (5) Å. The methyl carbon atoms are characterized by large and anisotropic vibrational parameters in the crystal, which may be caused either by large-amplitude torsional vibrations or by torsional disorder (Figure 25).

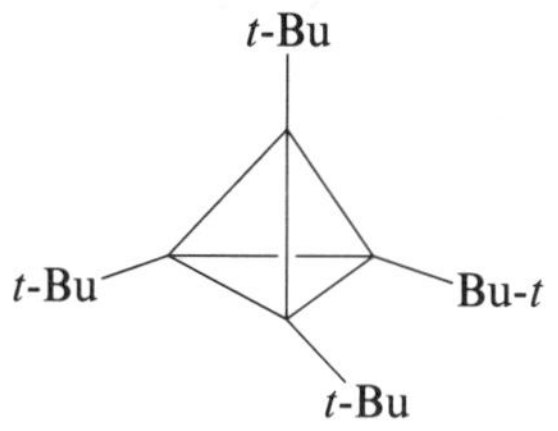

(209)

If **209** is crystallized at 213 K from ethyl acetate saturated with argon, inclusion crystals are formed which contain Ar atoms in one of the two types of octahedral holes in the close-

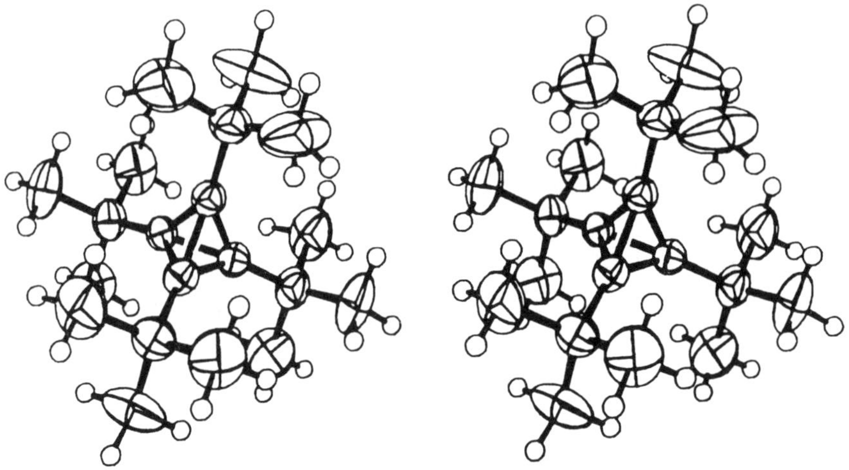

FIGURE 25. Stereoview of thermal ellipsoids (50 percent probability) for the carbon atoms in tetra-*tert*-butyltetrahedrane (**209**) at 213 K. Reproduced by permission of VCH Verlagsgesellschaft mbH from Reference 321

packed-spheres arrangement of molecules of **209**[322] (Figure 26). The occupancy of these sites was 26 percent in the crystal studied at 103 K. The two positions of the disordered *t*-Bu groups could be determined. The C—C bond length in the tetrahedrane skeleton was found to be 1.497 Å. The deformation electron-density maps show the bent bonds: the maximum of electron density is shifted by 0.37 Å from the C—C line[322], which corresponds to a bending of the bond by 26°.

4. Semibullvalenes and barbaralanes

The Cope rearrangement of semibullvalenes (**210**), the fast interchange of the bridged cyclopropane and the open allylic ends has been much studied in solution and in crystal (see e.g. the works cited in Reference 323). The C2—C8 bond lengths and the C4···C6 nonbonded distances span unusually wide ranges in various substituted derivatives of **210**, viz. 1.578 (3) to 1.990 (4) Å for C2—C8 and 1.865 (10) to 2.352 (3) Å for C4···C6 in the molecules listed[323]. The distances found in phenylsulfonyl[323] (**211a** and **211b**) and other derivatives have been interpreted as average values arising from the coexisting valence tautomers and have been used, along with estimated limiting values (the components) for these distances, to calculate the ratio of the two nonequivalent tautomers and the equilibrium constant of the interconversion[323]. Thus, degeneracy, which is observed in solution, may be lifted in the solid, due to intermolecular forces[323]. One of the tautomers predominates in the crystal in many cases.

(**210**)

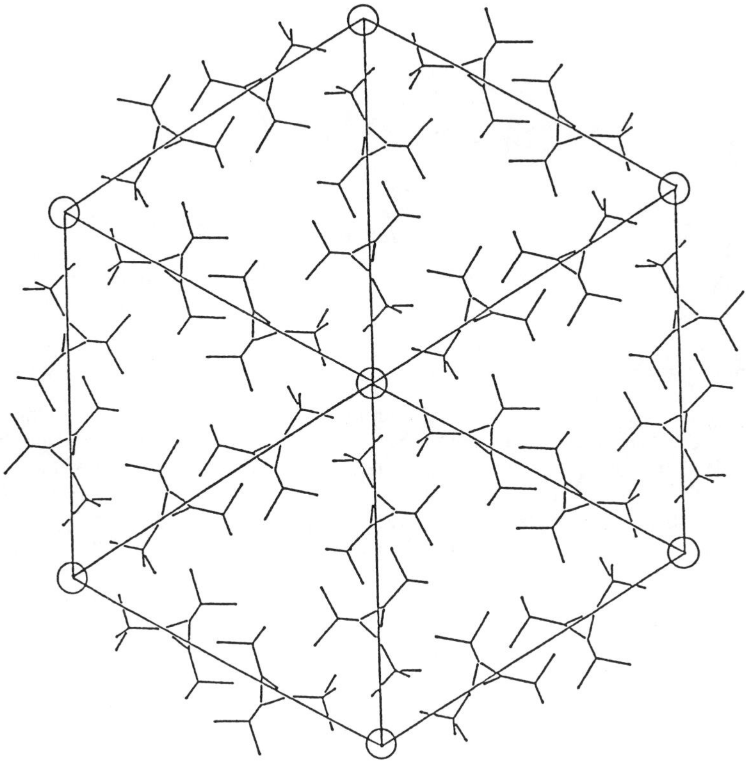

FIGURE 26. Molecular packing of the inclusion crystal of **209** and argon at 103 K. Positions of Ar atoms are marked by circles. Projection down the *c* axis. Reproduced by permission of VCH Verlagsgesellschaft mbH from Reference 322

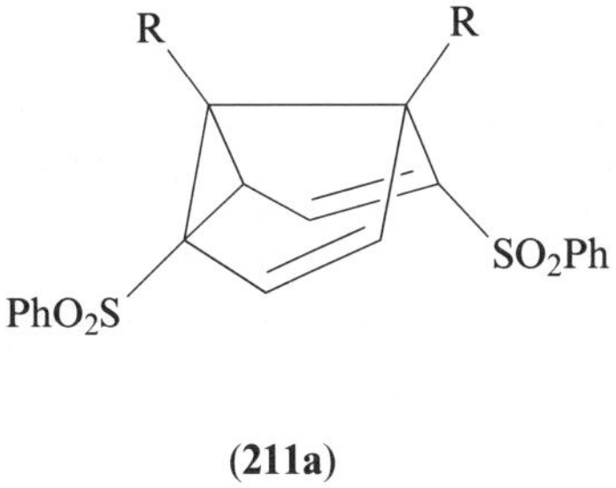

(211a)

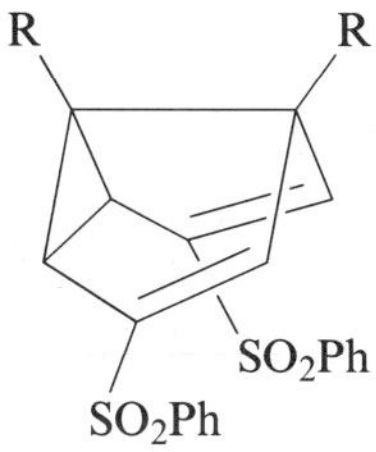

(211b)

Cyano and phenyl derivatives **212a**[324] and **212b**[325] of barbaralane exhibit the same phenomena. They are orange-red in the crystal, form orange solutions and the color intensity strongly depends on the temperature. The degeneracy of the rearrangement disappears in the crystal, and the equilibrium is shifted towards one of the tautomeric forms. Their ratio is estimated from the relevant average distances to be 9:1 in both **212a** and **212b**. The C2—C8 bonds are 1.665 (4) and 1.677 (3) Å, the C4$\cdots$C6 distances 2.412 (5) and

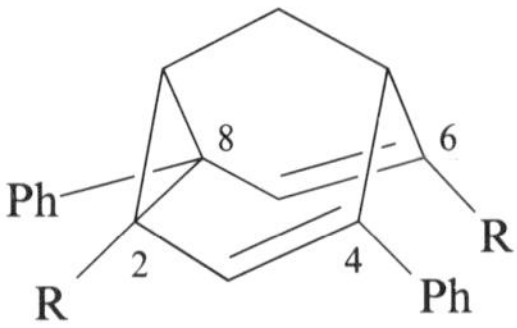

(**212a**) R = Ph
(**212b**) R = CN

2.37 (1) Å, respectively, and the other C—C bonds in the C_3 rings range from 1.493 (4) to 1.507 (3) Å.

5. Nortricyclenes

One example among tricyclo[2.2.1.0^{2,6}]heptane (nortricyclene) derivatives and hetero analogs studied will be mentioned here. The nortricyclene skeleton is distorted in the cyclic sulfate (**213**) of the *endo,endo*-3,5-diol, but the molecule retains approximate mirror symmetry in the crystal[326]. The chair-form six-membered ring C3—C4—C5—O—S—O is flattened at the O—S—O corner compared to the C3—C4—C5 corner. The C_3 ring is equilateral with a mean C—C bond length of 1.497 Å. Bonds are pulled towards the three-fold axis of the cyclopropane ring, as bond angles (taking the average of nearly symmetric pairs) C1—C6—C5 105.6, C2—C6—C5 106.1 and C6—C1—C7 107.6° demonstrate.

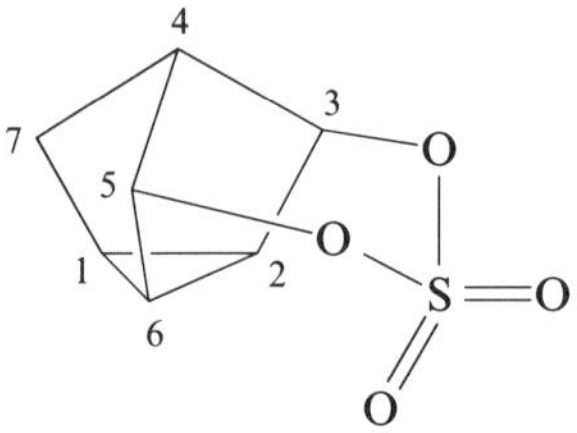

(**213**)

6. Prismanes

The crystal structure of the prismane derivative **214** was determined simultaneously in two laboratories. The XD measurement was performed at room temperature in Cleveland, Ohio[327] and at 89 (2) K in Heidelberg[328], which readily explains the nearly systematic deviations between cell parameters and between reported bond lengths. Geometrical parameters of **214** do not really reflect the high strain in this molecule. The bonds connecting the C_3 rings have nearly equal lengths[328] of (Ph)C—C(CO) 1.551(3), (Me)C—C(Me) 1.561 (3) and 1.562 (3) Å, similar to r_g (C—C) 1.554 (1) Å in free cyclobutane[35]. In the three-membered rings, the bonds distal to the Ph and to the COOMe groups are the shortest, 1.521 (2), 1.527 (2), 1.538 (2) and 1.499 (2), 1.541 (2), 1.564 (2) Å, respectively. The Ph ring takes the perpendicular conformation, and has a small effect; the C=O group is *syn* bisecting. The bond lengths are nearly the same in hexamethylprismane (**215**), which was studied by ED[329]: the bonds connecting the C_3 rings are r_g 1.551(5) Å, those at the fusion of a three- and a four-membered ring are 1.540 (9) Å. In the parent prismane, the respective bond lengths have been calculated at the MP2/6-31G(d) level[330] to be 1.549 and 1.518 Å. The prismene isomer having the double bond at a C_3-ring/C_4-ring edge is predicted to be more stable than the other isomer[330].

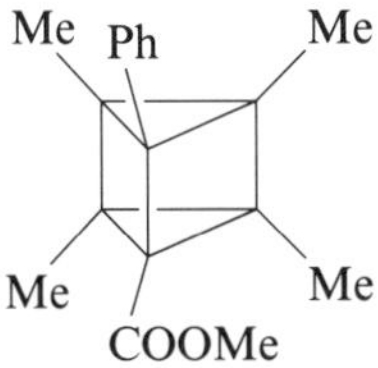

(214)

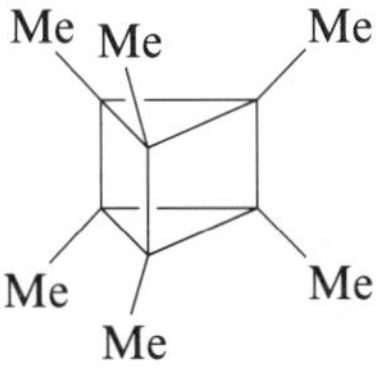

(215)

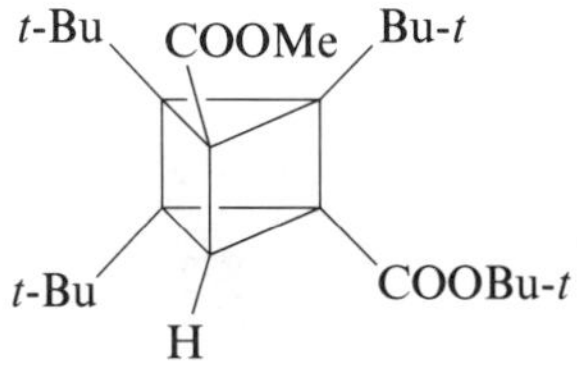

(216)

In derivative **216**[331], the bonds vicinal to the bisecting COOMe group in the C_3 ring are elongated, 1.579 (3) and 1.567 (3) Å; the distal bond is short, 1.528 (3) Å. At the same time, the vicinal bond in the C_4 ring is shorter, 1.506 (4) Å, than the other two parallel bonds, 1.553 (3) and 1.594 (3) Å. The latter very long bond connects the two *t*-Bu substitution sites. The C=O group of the COOBu-*t* substituent takes the *ac* conformation with τ 120.4 (3)°, which is unfavorable for conjugation with the prismane skeleton[328,331].

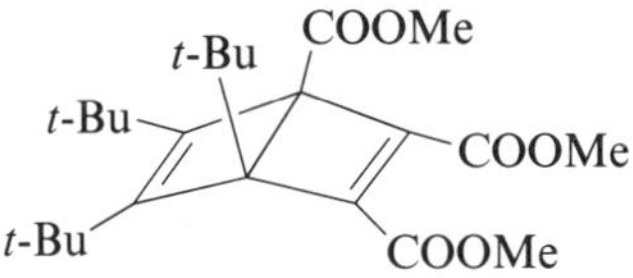

(217)

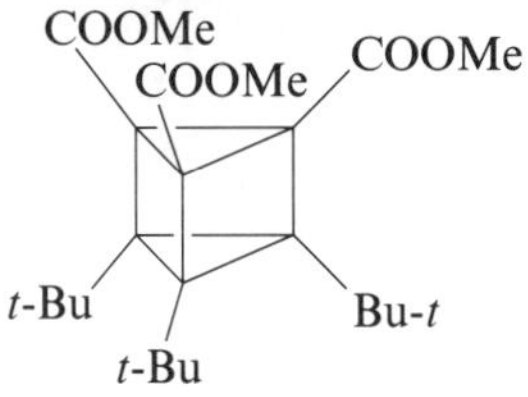

(218)

The photoisomerization of Dewar benzene derivative **217** leads to the prismane derivative **218** with an unexpected and nicely symmetric substitution pattern[332]. Preliminary room-temperature XD observations helped minimize measurement time at –174 °C and possibly avoid decomposition due to X-rays[332]. The molecule has a threefold symmetry axis in the cubic $P2_13$ crystal. The disordered ester groups take two *ac* positions in a ratio of 7:3. The more abundant form has a torsional angle τ of 136.0 (3)°; the other form is its approximate mirror image. Distortions of the prismane skeleton may be attributed to steric rather than to conjugative effects[332]. The bond lengths on the *t*-Bu side, 1.555 (3) Å, are longer than in the C_3 ring with the ester groups, 1.521 (3) Å. The connecting bonds, 1.560 (3) Å, have similar lengths as in **215**.

7. Quadricyclanes

Tetracyclo[3.2.0.0^{2,7}.0^{4,6}]heptane, quadricyclane (**219**), may be imagined as derived from prismane by the insertion of one methylene group. The ED experimental data, even if combined with *ab initio* results, are consistent with several models[333]. Individual bond lengths have been obtained with large uncertainties; the mean C—C bond length is 1.526 (2) Å. Bond angles at the C_3 rings are widened from 90° in prismane to C1—C2—C3 104.3 and C2—C1—C7 110.6° in **219**; the angle C1—C7—C4 of 98.7 (6)° in the methylene bridge remains very small even so. Rotational constants from a more recent MW study[334] would favor a model between those obtained in the two ED studies[333].

The crystal structure of the hexachloro derivative **220** has been redetermined[335]. The bonds in the three-membered rings are relatively long, the shortest being opposite to the COOMe groups: 1.522 (6) and 1.518 (6) Å; the adjacent bonds are 1.530 (6) to 1.547 (6) Å.

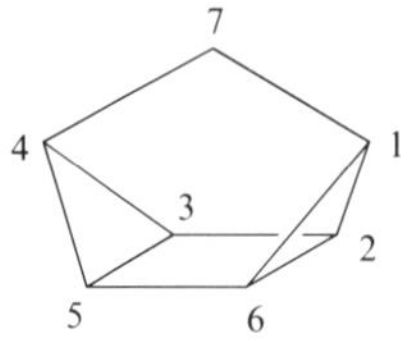

(**219**)

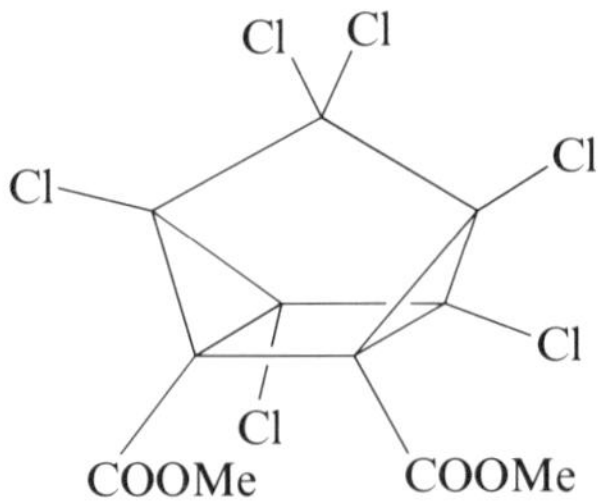

(**220**)

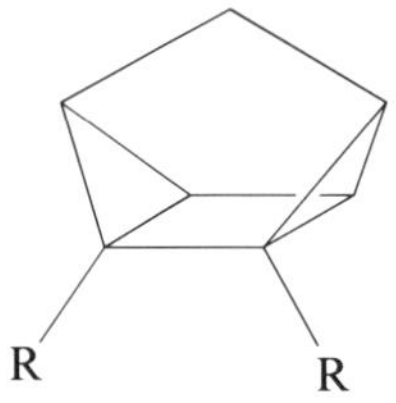

(**221**) R = CN
(**222**) R = CH_2OH
(**223**) R = CMe_2OH

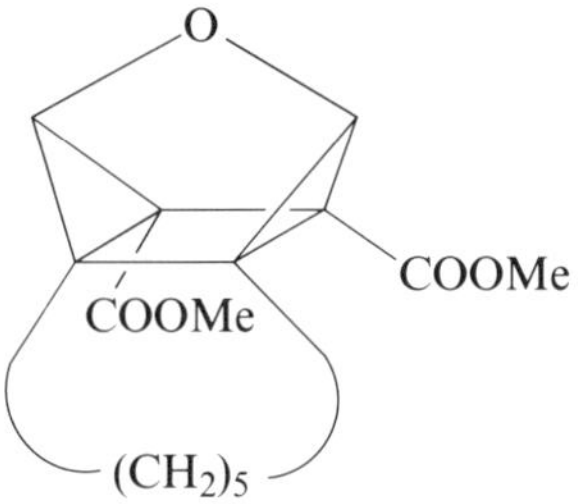

(**224**)

The two independent molecules in the crystal of **221** have similar geometries[336] as **219**. The distortion of the three-membered rings is comparable to that in cyanocyclopropane (**93**) (Table 8): the bonds distal to the CN groups are shorter, 1.484 (3) to 1.496 (3) Å, the vicinal bonds are somewhat longer, 1.520 (3) to 1.534 (3) Å, than in cyclopropane (**1**). The average of the C—CN bond lengths is[336] 1.414 (5) Å. Intramolecular hydrogen bonds have been found in **222** and **223**[337].

Oxygen analogs of **219** have been studied by XD, e.g. **224**[338]. Interaction between the substituents creates large deviations from the bisected forms. Nevertheless, the shortest bonds in the three-membered rings are those opposite to the ester groups[338].

8. Triasteranes and octahedranes

Triasterane (**226**) and the octahedrane **227** discussed below are related geometrically to prismane (**225**): two three-membered rings in parallel planes are bonded directly in **225**, through three methylene bridges in **226**, and through a crown (or, avoiding this royal term, through a chair-form ring) of six CH groups in **227**. The C_3 rings stagger in **227**, and are eclipsed in **225** and **226**. Three rectangles form the side faces of **225**, three six-membered boat rings are joined in **226** and six five-membered rings in **227**.

The fascinating story of asteranes, the synthesis of the first members by Hans Musso, has been told by Hopf[339] and in the fundamental paper by Trætteberg, Lüttke and coworkers[340]. The structure of triasterane, tetracyclo[$3.3.1.0^{2,8}.0^{4,6}$]nonane (**226**), has been determined only recently by gas-phase ED[340], based on experimental data from an earlier unpublished study. The C—C bond length is r_a 1.508 (5) Å in the three-membered rings. The geometry of the methylene bridges, C—CH_2 1.520 (4) Å and C—CH_2—C 112.2 (3)°, is similar to that in propane[42], r_a 1.531 (3) Å and $\angle_\alpha$ 112.4 (12)°. The distance between the

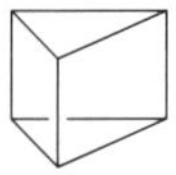

(225)

(226)

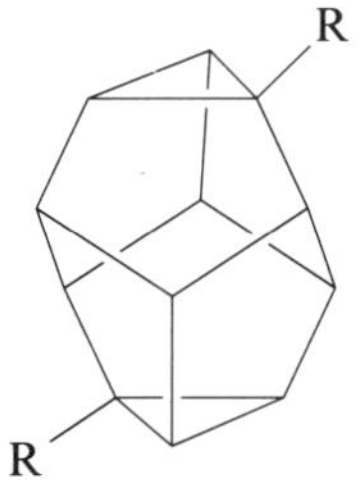

(227) R = H
(228) R = COOMe

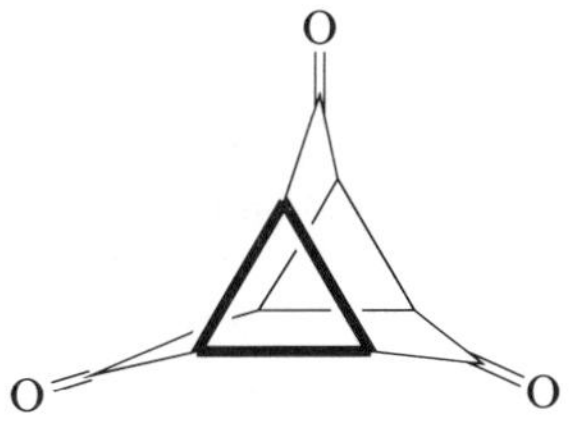

(229)

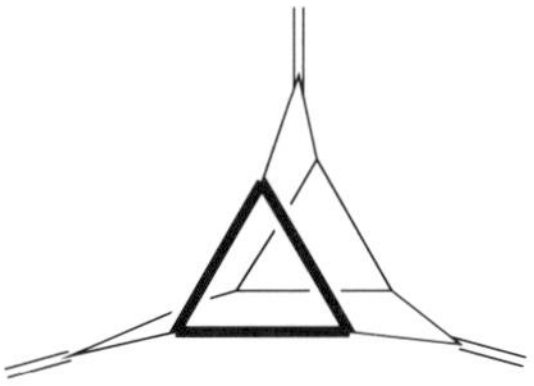

(230)

(231)

C_3 planes, 2.52 Å, is about the same as in the derivatives of **226** with C=O (**229**), C=CH_2 (**230**) and spirocyclopropane bridges (**231**) instead of CH_2, although the latter have shorter C—C bonds but wider C—C—C angles[341] (mean, from XD), 1.481 (2), 1.470 (4) and 1.487 (3) Å, and 115.4 (3), 114.9 (2) and 113.6 (3)°, respectively. The bond lengths in the C_3 rings, 1.517 (4), 1.517 (2) and 1.506 (3), are only slightly different from that in **226**. The geometry and energy of **226** have been compared to those of related molecules and subunits using molecular mechanics and quantum chemical calculations[340]. The larger strain in tetra- and pentaasterane compared to **226** arises mainly from the larger torsional strain in the planar four- and five-membered rings, which are puckered in the free cycloalkanes. **226** has also the smallest van der Waals energy in this series, because of the largest H···H distances between CH_2 groups of the lateral boat-form cyclohexane units[340].

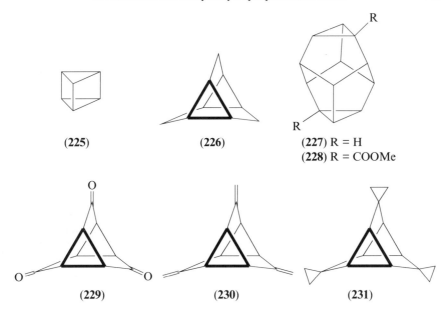

The molecule of *p*-[$3^2.5^6$]octahedrane (**227**) has the rare D_{3d} point group symmetry, and forms tetragonal crystals in space group *R*-3*c*. The structure of its 2,10-COOMe derivative (**228**) has also been determined[342]. The bond in the three-membered ring in **227** is 1.507 (4) Å, while the bonds adjacent to the substituent in **228** are longer, 1.529 (1) Å, the distal bonds shorter, 1.495 (1) Å.

9. Other structures

This survey will be closed by structures of the type **13**, where two pairs of geminal bonds of the cyclopropane ring are joined by a 'four-dentate' substituent. Bridged propellanes (Section VII.C.2) are also members of this class.

Tetracyclic isomers **232a** and **232b** are formed in a thermal depolymerization process[343]. The same $O_2SiOSiO_2$ unit is linked to the cyclopropane frame in two different ways: the SiOSi plane is parallel with the C_3 ring in **232a**, and is perpendicular to it in **232b**. Substantial differences between parameters of **232a** and **232b** are found in bond angles Si—O—Si of 138.9 (9) and 132.9 (5)°, and Si—O—C (mean) of 119.8 (11) and 126.4 (8)°, respectively, due to the different sizes of the rings containing these angles[343].

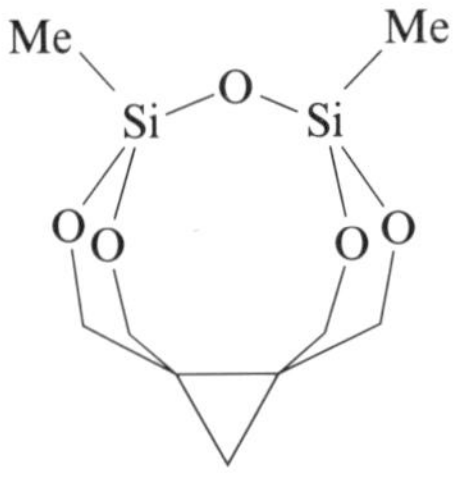

(232a)

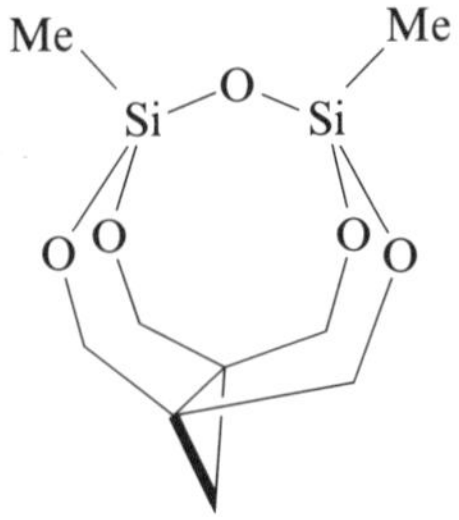

(232b)

The parent fullerene cyclopropane, 1,2-methanobuckminsterfullerene, $C_{61}H_2$ (**233**, a fragment is shown), has been isolated as a dark reddish-brown solid, which dissolves in toluene giving a pink solution[344]. The structure of **233** was ascertained by the use of NMR coupling constants. Accordingly, the CH_2 group is linked to C_{60} at a "6,6 short bond", and forms a closed fullerene[344], instead of forming an open annulene (fulleroid) isomer at a "6,5 long bond".

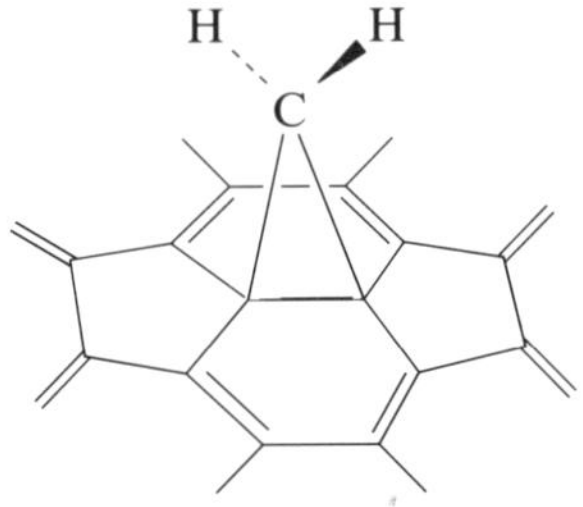

(233)

VIII. ACKNOWLEDGMENTS

My thanks are due to Professor István Hargittai for encouragement and support. It is a pleasure to thank Dr. Gustavo Portalone (Rome) for a complete list of references to crystallographic studies. I am grateful to Dr. Mátyás Czugler (Budapest) for valuable help and advice, to Dr. Marwan Dakkouri (Ulm) for unpublished data and to the many colleagues who have sent me reprints and original copies of figures. Authors and publishers kindly granted permission to reproduce copyrighted material. Financial support from the Hungarian Scientific Research Foundation is acknowledged (OTKA, No. 2103). The lasting solicitude of my wife has helped me to accomplish my task.

IX. REFERENCES

1. H.-B. Bürgi and J. D. Dunitz, *Helv. Chim. Acta*, **76**, 1115 (1993).
2. M. Charton, in *The Chemistry of Alkenes* (Ed. J. Zabicky), Vol. 2, Chap. 10, Wiley/Interscience, London, 1970, pp. 511–610.
3. K. B. Wiberg, in *The Chemistry of the Cyclopropyl Group* (Ed. Z. Rappoport), Part 1, Chap. 1, Wiley, Chichester, 1987, pp. 1–26.
4. A. I. Ioffe, V. A. Svyatkin and O. M. Nefedov, *Stroenie proizvodnykh tsiklopropana* (Structure of the Derivatives of Cyclopropane), Nauka, Moskva, 1986.

5. R. Boese, in *Advances in Strain in Organic Chemistry* (Ed. B. Halton), Vol. 2, Chap. 5, JAI Press, Greenwich, Connecticut, 1992, pp. 191–254.
6. A. Greenberg and T. A. Stevenson, in *Molecular Structure and Energetics* (Eds. J. F. Liebman and A. Greenberg), Vol. 3, Chap. 5, VCH Publishers, Deerfield Beech, Florida, 1986, pp. 193–266.
7. J. F. Liebman and A. Greenberg, in *The Chemistry of the Cyclopropyl Group* (Ed. Z. Rappoport), Part 2, Chap. 18, Wiley, Chichester, 1987, pp. 1083–1119.
8. F. H. Allen, *Acta Crystallogr.*, **B36**, 81 (1980).
9. F. H. Allen, *Acta Crystallogr.*, **B37**, 890 (1981).
10. F. H. Allen, *Acta Crystallogr.*, **B37**, 900 (1981).
11. F. H. Allen, *Tetrahedron*, **38**, 645 (1982).
12. F. H. Allen, *Acta Crystallogr.*, **B40**, 306 (1984).
13. A. Domenicano and I. Hargittai (Eds.), *Accurate Molecular Structures*, Oxford University Press, Oxford, 1992.
14. *Landolt–Börnstein Numerical Data and Functional Relationships in Science and Technology*, New Series, Vol. II/7: *Structure Data of Free Polyatomic Molecules* (Eds. K.-H. Hellwege and A. M. Hellwege); Vol. II/15: *Supplement* (Eds. K.-H. Hellwege and A. M. Hellwege); Vol. II/21: *Supplement* (Ed. K. Kuchitsu), Springer-Verlag, Berlin, Heidelberg, 1976, 1987, 1992.
15. M. E. Jason and J. A. Ibers, *J. Am. Chem. Soc.*, **99**, 6012 (1977).
16. *IUPAC Rules for the Nomenclature of Organic Chemistry, Section E: Stereochemistry, Recommendations*, Pergamon Press, Oxford, 1974.
17. J. Vogt, *Struct. Chem.*, **3**, 147 (1992); *Acta Chim. Hung.*, **129**, 333 (1992); *J. Mol. Spectrosc.*, **155**, 413 (1992).
18. F. H. Allen, in *Accurate Molecular Structures* (Eds. A. Domenicano and I. Hargittai), Chap. 15, Oxford University Press, Oxford, 1992, pp. 355–378.
19. O. Bastiansen and O. Hassel, *Tidsskr, Kjemi, Bergves. Metallurgi*, **6**, 71 (1946); *Chem. Abstr.*, **40**, 6059[9] (1946).
20. O. Bastiansen, F. N. Fritsch and K. Hedberg, *Acta Crystallogr.*, **17**, 538 (1964).
21. S. Yamamoto, M. Nakata, T. Fukuyama and K. Kuchitsu, *J. Phys. Chem.*, **89**, 3298 (1985).
22. Y. Endo, M. C. Chang and E. Hirota, *J. Mol. Spectrosc.*, **126**, 63 (1987).
23. (a) D. Nijveldt and A. Vos, *Acta Crystallogr.*, **B44**, 281 (1988).
(b) D. Nijveldt and A. Vos, *Acta Crystallogr.*, **B44**, 289 (1988).
(c) D. Nijveldt and A. Vos, *Acta Crystallogr.*, **B44**, 296 (1988).
24. J. E. Boggs and K. Fan, *Acta Chem. Scand.*, **A42**, 595 (1988).
25. D. Cremer and E. Kraka, *J. Am. Chem. Soc.*, **107**, 3811 (1985).
26. K. B. Wiberg and K. E. Laidig, *J. Org. Chem.*, **57**, 5092 (1992).
27. T. D. Norden, S. W. Staley, W. H. Taylor and M. D. Harmony, *J. Am. Chem. Soc.*, **108**, 7912 (1986).
28. N. V. Riggs, U. Zoller, M. T. Nguyen and L. Radom, *J. Am. Chem. Soc.*, **114**, 4354 (1992).
29. K. B. Wiberg, R. F. W. Bader and C. D. H. Lau, *J. Am. Chem. Soc.*, **109**, 985 (1987).
30. J. G. Hamilton and W. E. Palke, *J. Am. Chem. Soc.*, **115**, 4159 (1993).
31. S. W. Benson, *Thermochemical Kinetics*, 2nd ed., Wiley, New York, 1976.
32. K. Kuchitsu, in *Molecular Structure and Properties* (Ed. G. Allen), Chap. 6, MTP International Review of Science, Physical Chemistry, Series 1, Vol. 2, Butterworths, University Park Press, London, Baltimore, 1972, pp. 203–240.
33. V. S. Mastryukov, E. L. Osina and L. V. Vilkov, *Zh. Strukt. Khim.*, **16**, 850 (1975); *Chem. Abstr.*, **84**, 73176f (1976).
34. I. Hargittai, *The Structure of Volatile Sulphur Compounds*, Akadémiai Kiadó, Budapest and D. Reidel, Dordrecht, 1985.
35. T. Egawa, T. Fukuyama, S. Yamamoto, F. Takabayashi, H. Kambara, T. Ueda and K. Kuchitsu, *J. Chem. Phys.*, **86**, 6018 (1987).
36. W. J. Adams, H. J. Geise and L. S. Bartell, *J. Am. Chem. Soc.*, **92**, 5013 (1970).
37. J. D. Ewbank, G. Kirsch and L. Schäfer, *J. Mol. Struct.*, **31**, 39 (1976).
38. B. Bak and S. Skaarup, *J. Mol. Struct.*, **10**, 385 (1971).
39. C. Hirose, *Bull. Chem. Soc. Jpn.*, **47**, 1311 (1974).
40. M. T. Bowers, R. A. Beaudet, H. Goldwhite and R. Tang, *J. Am. Chem. Soc.*, **91**, 17 (1969).
41. K. Okiye, C. Hirose, D. G. Lister and J. Sheridan, *Chem. Phys. Lett.*, **24**, 111 (1974).
42. T. Iijima, *Bull. Chem. Soc. Jpn.*, **45**, 1291 (1972).

43. K. Tamagawa, T. Iijima and M. Kimura, *J. Mol. Struct.*, **30**, 243 (1976).
44. E. Hirota, Y. Endo, S. Saito, K. Yoshida, I. Yamaguchi and K. Machida, *J. Mol. Spectrosc.*, **89**, 223 (1981).
45. V. S. Mastryukov and E. L. Osina, *J. Mol. Struct.*, **36**, 127 (1977); I. Yu. Schapin and V. S. Mastryukov, *J. Mol. Struct.*, **268**, 307 (1992).
46. A. C. Legon, P. D. Aldrich and W. H. Flygare, *J. Am. Chem. Soc.*, **104**, 1486 (1982).
47. L. W. Buxton, P. D. Aldrich, J. A. Shea, A. C. Legon and W. H. Flygare, *J. Chem. Phys.*, **75**, 2681 (1981).
48. S. G. Kukolich, *J. Chem. Phys.*, **78**, 4832 (1983).
49. A. M. Andrews, K. W. Hillig, II and R. L. Kuczkowski, *J. Am. Chem. Soc.*, **114**, 6765 (1992).
50. A. M. Andrews, K. W. Hillig, II and R. L. Kuczkowski, *J. Chem. Phys.*, **96**, 1784 (1992).
51. R. L. Kuczkowski and A. Taleb-Bendiab, in *Structures and Conformations of Non-Rigid Molecules* (Eds. J. Laane, M. Dakkouri, B. van der Veken and H. Oberhammer), NATO ASI Series C, Vol. 410, Kluwer, Dordrecht, 1993, pp. 257–276.
52. A. C. Legon and A. L. Wallwork, *J. Chem. Soc., Faraday Trans.*, **86**, 3975 (1990); A. C. Legon, A. L. Wallwork and H. E. Warner, *J. Chem. Soc., Faraday Trans.*, **87**, 3327 (1991); A. C. Legon, C. A. Rego and A. L. Wallwork, *J. Chem. Phys.*, **97**, 3050 (1992); A. C. Legon and D. J. Millen, *Chem. Soc. Rev.*, **16**, 467 (1987) and references cited therein.
53. C. A. Deakyne, L. C. Allen and N. C. Craig, *J. Am. Chem. Soc.*, **99**, 3895 (1977).
54. A. W. Klein and G. Schrumpf, *Acta Chem. Scand.*, **A35**, 425 (1981).
55. S. N. Mathur and M. D. Harmony, *J. Mol. Struct.*, **57**, 63 (1979).
56. M. Trætteberg and W. Lüttke, *Croat. Chem. Acta*, **64**, 295 (1991).
57. W. Bernlöhr, H.-D. Beckhaus, K. Peters, H. G. von Schnering and C. Rüchardt, *Chem. Ber.*, **117**, 1013 (1984).
58. K. Hagen, G. Hagen and M. Trætteberg, *Acta Chem. Scand.*, **26**, 3649 (1972).
59. A. de Meijere and M. Trætteberg, *J. Mol. Struct.*, **161**, 97 (1987).
60. G. Schrumpf and P. G. Jones, *Acta Crystallogr.*, **C43**, 1594 (1987).
61. L. S. Bartell and T. L. Boates, *J. Mol. Struct.*, **32**, 379 (1976).
62. V. S. Mastryukov, K. Chen, L. R. Yang and N. L. Allinger, *J. Mol. Struct.*, **280**, 199 (1993).
63. N. A. Donskaya, A. G. Bessmertnykh, O. E. Grushina, S. V. Lindeman, V. I. Petrov, Yu. T. Struchkov and Yu. S. Shabarov, *Zh. Org. Khim.*, **23**, 2369 (1987); *Chem. Abstr.*, **109**, 109872f (1988).
64. M. Dakkouri and S. Zeeb, to be published; S. Zeeb, *Thesis*, University of Ulm, Ulm (1993).
65. J. R. Durig, T. S. Little, Xiang Zhu and M. Dakkouri, *J. Mol. Struct.*, **293**, 15 (1993).
66. N. H. Dung, B. Viossat, M. C. de Sévricourt and M. Robba, *Chem. Pharm. Bull.*, **35**, 1093 (1987).
67. A. K. Kocharian, A. V. Nilssen, A. Pettersen, C. Rømming and L. K. Sydnes, *Acta Chem. Scand.*, **A42**, 463 (1988).
68. L. K. Sydnes, A. Pettersen, F. Drabløs and C. Rømming, *Acta Chem. Scand.*, **45**, 902 (1991).
69. H. Møllendal, in *Structures and Conformations of Non-Rigid Molecules* (Eds. J. Laane, M. Dakkouri, B. van der Veken and H. Oberhammer), NATO ASI Series C, Vol. 410, Kluwer, Dordrecht, 1993, pp. 277–301.
70. K.-M. Marstokk and H. Møllendal, *Acta Chem. Scand.*, **A38**, 387 (1984).
71. K.-M. Marstokk and H. Møllendal, *Acta Chem. Scand.*, **A39**, 429 (1985).
72. H. Hopf, K.-M. Marstokk, A. de Meijere, C. Mlynek, H. Møllendal, A. Sveiczer, Y. Stenstrøm and M. Trætteberg, *Acta Chem. Scand.*, **47**, 739 (1993).
73. K.-M. Marstokk and H. Møllendal, *Acta Chem. Scand.*, **46**, 861 (1992).
74. K.-M. Marstokk and H. Møllendal, *Acta Chem. Scand.*, **45**, 354 (1991).
75. A. Bhaumik, W. V. F. Brooks, S. C. Dass and K. V. L. N. Sastry, *Can. J. Chem.*, **48**, 2949 (1970); W. V. F. Brooks and C. K. Sastry, *Can. J. Chem.*, **56**, 530 (1978).
76. K.-M. Marstokk and H. Møllendal, *Acta Chem. Scand.*, **47**, 444 (1993).
77. M. Trætteberg, P. Bakken, A. Almenningen and W. Lüttke, *J. Mol. Struct.*, **189**, 357 (1988).
78. M. Trætteberg, A. Almenningen, G. Schrumpf and S. Martin, *J. Mol. Struct.*, **189**, 345 (1988).
79. A. de Meijere and W. Lüttke, *Tetrahedron*, **25**, 2047 (1969).
80. B. Klahn and V. Dyczmons, *J. Mol. Struct.*, **122**, 75 (1985).
81. I. Tokue, T. Fukuyama and K. Kuchitsu, *J. Mol. Struct.*, **17**, 207 (1973).
82. S. H. Schei and A. de Meijere, *Tetrahedron*, **41**, 1973 (1985).
83. J. Deuter, H. Rodewald, H. Irngartinger, T. Loerzer and W. Lüttke, *Tetrahedron Lett.*, **26**, 1031 (1985).

84. G. J. H. van Nes and A. Vos, *Acta Crystallogr.*, **B35**, 2593 (1979).
85. L. S. Bartell, E. A. Roth, C. D. Hollowell, K. Kuchitsu and J. E. Young, Jr., *J. Chem. Phys.*, **42**, 2683 (1965).
86. J. W. Lauher and J. A. Ibers, *J. Am. Chem. Soc.*, **97**, 561 (1975).
87. T. S. Cameron, A. Linden and K. Jochem, *Acta Crystallogr.*, **C46**, 2110 (1990).
88. T. Clark, G. W. Spitznagel, R. Klose and P. v. R. Schleyer, *J. Am. Chem. Soc.*, **106**, 4412 (1984).
89. J. S. A. M. de Boer, B. O. Loopstra and C. H. Stam, *Recl. Trav. Chim. Pays-Bas*, **106**, 537 (1987).
90. H. Schmidbaur, W. Bublak, A. Schier, G. Reber and G. Müller, *Chem. Ber.*, **121**, 1373 (1988).
91. M. Draux, I. Bernal and R. Fuchs, *Struct. Chem.*, **2**, 127 (1991).
92. R. E. Drumright, R. H. Mas, J. S. Merola and J. M. Tanko, *J. Org. Chem.*, **55**, 4098 (1990).
93. M. B. Hossain, J. L. Wang, D. van der Helm, R. A. Magarian, M. T. Griffin and B. W. Day, *Acta Crystallogr.*, **B47**, 511 (1991).
94. Li Du, M. B. Hossain, Xinhua Ji, D. van der Helm, R. A. Magarian and B. W. Day, *Acta Crystallogr.*, **C48**, 887 (1992).
95. F. H. Allen, O. Kennard, D. G. Watson, L. Brammer, A. G. Orpen and R. Taylor, *J. Chem. Soc., Perkin Trans. 2, Suppl.*, pp. S1–S19 (1987).
96. A. Domenicano, in *Stereochemical Applications of Gas-Phase Electron Diffraction* (Eds. I. Hargittai and M. Hargittai), Part B: *Structural Information for Selected Classes of Compounds*, Chap. 7, VCH Publishers, New York, 1988, pp. 281–324.
97. A. Domenicano, in *Accurate Molecular Structures* (Eds. A. Domenicano and I. Hargittai), Chap. 18, Oxford University Press, Oxford, 1992, pp. 437–468.
98. R. J. Gillespie and I. Hargittai, *The VSEPR Model of Molecular Geometry*, Allyn and Bacon, Boston, London, 1991.
99. A. Domenicano, G. Schultz, M. Kolonits and I. Hargittai, *J. Mol. Struct.*, **53**, 197 (1979).
100. P. Scharfenberg, I. Hargittai and B. Rozsondai, *9th Austin Symposium on Molecular Structure*, Austin, Texas, 1982, Abstracts A16, pp. 83–85.
101. L. S. Bartell and J. P. Guillory, *J. Chem. Phys.*, **43**, 647 (1965).
102. J. R. Durig and T. S. Little, *Croat. Chem. Acta*, **61**, 529 (1988).
103. H. N. Volltrauer and R. H. Schwendeman, *J. Chem. Phys.*, **54**, 260 (1971).
104. J. R. Durig, Fusheng Feng, T. S. Little and A.-Y. Wang, *Struct. Chem.*, **3**, 417 (1992).
105. R. E. Penn and J. E. Boggs, *J. Chem. Soc., Chem. Commun.*, 666 (1972).
106. L. S. Bartell, J. P. Guillory and A. T. Parks, *J. Phys. Chem.*, **69**, 3043 (1965).
107. E. G. Perevalova, I. G. Bolesov, Yu. T. Struchkov, I. F. Leschova, E. S. Kalyuzhnaya, T. I. Voyevodskaya, Yu. L. Slovokhotov and K. I. Grandberg, *J. Organomet. Chem.*, **286**, 129 (1985).
108. R. F. Childs, M. D. Kostyk, C. J. L. Lock and M. Mahendran, *J. Am. Chem. Soc.*, **112**, 8912 (1990) and references cited therein.
109. K.-M. Marstokk, H. Møllendal and S. Samdal, *Acta Chem. Scand.*, **45**, 37 (1991).
110. P. G. Jones and G. Schrumpf, *Acta Crystallogr.*, **C43**, 1752 (1987).
111. M. A. M. Meester, H. Schenk and C. H. MacGillavry, *Acta Crystallogr.*, **B27**, 630 (1971).
112. G. Schrumpf and P. G. Jones, *Acta Crystallogr.*, **C43**, 1748 (1987).
113. S. M. Jessen, *Acta Crystallogr.*, **C48**, 106 (1992).
114. E. Weber, M. Hecker, I. Csöregh and M. Czugler, *J. Am. Chem. Soc.*, **111**, 7866 (1989).
115. I. Csöregh, O. Gallardo, E. Weber, M. Hecker and A. Wierig, *J. Chem. Soc., Perkin Trans. 2*, 1939 (1992).
116. I. Csöregh, O. Gallardo, E. Weber, M. Hecker and A. Wierig, *J. Incl. Phenom.*, **14**, 131 (1992).
117. B. P. van Eijck and E. van Zoeren, *J. Mol. Spectrosc.*, **111**, 138 (1985).
118. L. Leiserowitz, *Acta Crystallogr.*, **B32**, 775 (1976).
119. E. M. Bellott, Jr. and E. B. Wilson, *Tetrahedron*, **31**, 2896 (1975).
120. A. Dubourg, E. Fabregue, L. Maury and J.-P. Declercq, *Acta Crystallogr.*, **C46**, 1394 (1990).
121. P. Bréhin, J. Kozelka and C. Bois, *Acta Crystallogr.*, **C48**, 2094 (1992).
122. H. Küppers and S. M. Jessen, *Z. Kristallogr.*, **203**, 167 (1993).
123. P. G. Jones and G. Schrumpf, *Acta Crystallogr.*, **C43**, 1576 (1987).
124. C. Rømming and L. K. Sydnes, *Acta Chem. Scand.*, **B41**, 717 (1987).
125. E. Horn, M. R. Snow and E. R. T. Tiekink, *Acta Crystallogr.*, **C43**, 2459 (1987).
126. D. Dvořák, J. Závada, M. C. Etter and J. H. Loehlin, *J. Org. Chem.*, **57**, 4839 (1992).
127. E. Weber, M. Hecker, I. Csöregh and M. Czugler, *Mol. Cryst. Liq. Cryst.*, **187**, 165 (1990).
128. G. Valle, M. Crisma, C. Toniolo, E. M. Holt, M. Tamura, J. Bland and C. H. Stammer, *Int. J. Pept. Protein Res.*, **34**, 56 (1989).

129. E. Benedetti, B. Di Blasio, V. Pavone, C. Pedone, A. Santini, M. Crisma, G. Valle and C. Toniolo, *Biopolymers*, **28**, 175 (1989).
130. V. Busetti, M. Crisma and C. Toniolo, *Acta Crystallogr.*, **C48**, 1464 (1992).
131. C. Mapelli, M. G. Newton, C. E. Ringold and C. H. Stammer, *Int. J. Pept. Protein Res.*, **30**, 498 (1987).
132. K. Aoki, Ninghai Hu and H. Yamazaki, *Inorg. Chim. Acta*, **186**, 253 (1991).
133. M. C. Pirrung, *J. Org. Chem.*, **52**, 4179 (1987).
134. J. E. Baldwin, R. M. Adlington, G. A. Lajoie, C. Lowe, P. D. Baird and K. Prout, *J. Chem. Soc., Chem. Commun.*, 775 (1988).
135. H.-W. Liu and C. T. Walsh, in *The Chemistry of the Cyclopropyl Group* (Ed. Z. Rappoport), Part 2, Chap. 16, Wiley, Chichester, 1987, pp. 959–1025.
136. K. Iijima, K. Tanaka and S. Onuma, *J. Mol. Struct.*, **246**, 257 (1991).
137. K. Iijima and B. Beagley, *J. Mol. Struct.*, **248**, 133 (1991).
138. P. G. Jones and G. Schrumpf, *Acta Crystallogr.*, **C43**, 1755 (1987).
139. S. V. Lindeman, Yu. T. Struchkov, M. N. Elinson, S. K. Fedukovich and G. I. Nikishin, *Zh. Strukt. Khim.*, **30**, No.3, 171 (1989); *Chem. Abstr.*, **111**, 124314k (1989).
140. G. Schrumpf, P. G. Jones and G. M. Sheldrick, *Acta Crystallogr.*, **C43**, 1758 (1987).
141. P. G. Jones and G. Schrumpf, *Acta Crystallogr.*, **C43**, 2015 (1987).
142. P. G. Jones and G. Schrumpf, *Acta Crystallogr.*, **C43**, 1579 (1987).
143. F. Baert and A. Guelzim, *Acta Crystallogr.*, **C47**, 606 (1991).
144. F. Baert, A. Guelzim and G. Germain, *Acta Crystallogr.*, **C47**, 768 (1991).
145. F. Hamzaoui, J. Lamiot and F. Baert, *Acta Crystallogr.*, **C49**, 818 (1993).
146. P. G. Jones, A. J. Kirby and R. J. Lewis, *Acta Crystallogr.*, **C46**, 78 (1990).
147. K. Tamagawa and R. L. Hilderbrandt, *J. Phys. Chem.*, **87**, 3839 (1983).
148. M. D. Harmony, R. N. Nandi, J. V. Tietz, J.-I. Choe, S. J. Getty and S. W. Staley, *J. Am. Chem. Soc.*, **105**, 3947 (1983).
149. C. T. Kiers, J. S. A. M. de Boer, D. Heijdenrijk, C. H. Stam and H. Schenk, *Recl. Trav. Chim. Pays-Bas*, **104**, 7 (1985).
150. A. W. Klein and G. Schrumpf, *Acta Chem. Scand.*, **A35**, 431 (1981).
151. A. Dubrulle, D. Boucher, J. Burie and J. Demaison, *J. Mol. Spectrosc.*, **72**, 158 (1978).
152. J. Demaison, A. Dubrulle, D. Boucher, J. Burie and V. Typke, *J. Mol. Spectrosc.*, **76**, 1 (1979).
153. P. G. Jones and G. Schrumpf, *Acta Crystallogr.*, **C43**, 1179 (1987).
154. A. Hartman and F. L. Hirshfeld, *Acta Crystallogr.*, **20**, 80 (1966).
155. V. M. Anisimov, A. B. Zolotoi, M. Y. Antipin, P. M. Lukin, O. E. Nasakin and Yu. T. Struchkov, *Mendeleev Commun.*, 24 (1992).
156. J.-P. Declercq, B. Tinant, A. Parfonry, M. Van Meerssche, E. Legrand and M. S. Lehmann, *Acta Crystallogr.*, **C39**, 1401 (1983).
157. A. B. Zolotoi, A. N. Lyshchikov, P. M. Lukin, A. I. Prokhorov, O. E. Nasakin, A. Kh. Bulai and L. O. Atovmyan, *Dokl. Akad. Nauk SSSR*, **313**, 110 (1990); *Chem. Abstr.*, **114**, 23740c (1991).
158. R. Lang, C. Herzog, R. Stangl, E. Brunn, M. Braun, M. Christl, E.-M. Peters, K. Peters and H. G. von Schnering, *Chem. Ber.*, **123**, 1193 (1990).
159. A. T. Perretta and V. W. Laurie, *J. Chem. Phys.*, **62**, 2469 (1975).
160. H. Justnes, J. Zozom, C. W. Gillies, S. K. Sengupta and N. C. Craig, *J. Am. Chem. Soc.*, **108**, 881 (1986).
161. S. K. Sengupta, H. Justnes, C. W. Gillies and N. C. Craig, *J. Am. Chem. Soc.*, **108**, 876 (1986).
162. C. W. Gillies, *J. Mol. Spectrosc.*, **59**, 482 (1976).
163. R. N. Beauchamp, J. W. Agopovich and C. W. Gillies, *J. Am. Chem. Soc.*, **108**, 2552 (1986).
164. R. N. Beauchamp, C. W. Gillies and N. C. Craig, *J. Am. Chem. Soc.*, **109**, 1696 (1987).
165. R. N. Beauchamp, C. W. Gillies and J. Z. Gillies, *J. Mol. Spectrosc.*, **144**, 269 (1990).
166. J. F. Chiang and W. A. Bernett, *Tetrahedron*, **27**, 975 (1971).
167. R. H. Schwendeman, G. D. Jacobs and T. M. Krigas, *J. Chem. Phys.*, **40**, 1022 (1964).
168. L. Hedberg, K. Hedberg and J. E. Boggs, *J. Chem. Phys.*, **77**, 2996 (1982).
169. G. Schrumpf and P. G. Jones, *Acta Crystallogr.*, **C43**, 1182 (1987).
170. G. Schrumpf and P. G. Jones, *Acta Crystallogr.*, **C43**, 1185 (1987).
171. C. J. Marsden, L. Hedberg and K. Hedberg, *J. Phys. Chem.*, **92**, 1766 (1988).
172. G. Schrumpf and P. G. Jones, *Acta Crystallogr.*, **C43**, 1188 (1987).
173. (a) C. Glidewell, *Inorg. Chim. Acta*, **12**, 219 (1975).
(b) L. S. Bartell, *J. Chem. Phys.*, **32**, 827 (1960).

174. T. T. Raw and C. W. Gillies, *J. Mol. Spectrosc.*, **128**, 195 (1988).
175. H. Li, M. C. L. Gerry and W. Lewis-Bevan, *J. Mol. Spectrosc.*, **144**, 51 (1990).
176. L. Pauling, *The Nature of the Chemical Bond*, 3rd ed., Cornell University Press, Ithaca, New York, 1960.
177. (a) C. Glidewell, *Inorg. Chim. Acta*, **20**, 113 (1976).
(b) C. Glidewell, *Inorg. Chim. Acta*, **36**, 135 (1979).
178. J.-P. Declercq, N. de Kimpe, M. Palamareva and N. Schamp, *Bull. Soc. Chim. Belg.*, **94**, 301 (1985).
179. J. N. Macdonald, D. Norbury and J. Sheridan, *J. Chem. Soc., Faraday Trans. 2*, **74**, 1365 (1978).
180. B. Föhlisch, E. Gehrlach, J. J. Stezowski, P. Kollat, E. Martin and W. Gottstein, *Chem. Ber.*, **119**, 1661 (1986).
181. J. Kron and U. Schubert, *J. Organomet. Chem.*, **373**, 203 (1989).
182. I. Søtofte and I. Crossland, *Acta Chem. Scand.*, **46**, 131 (1992).
183. B. Tinant, S. Wu, J.-P. Declercq, M. Van Meerssche, W. Masamba, A. De Mesmaeker and H. G. Viehe, *J. Chem. Soc., Perkin Trans. 2*, 1045 (1988).
184. B. Tinant, J.-P. Declercq and M. Van Meerssche, *Bull. Soc. Chim. Belg.*, **93**, 921 (1984).
185. B. Tinant, J.-P. Declercq and M. Van Meerssche, *Acta Crystallogr.*, **C41**, 597 (1985).
186. B. Tinant, S. Wu, J.-P. Declercq, M. Van Meerssche, A. De Mesmaeker, W. Masamba, R. Merényi and H. G. Viehe, *J. Chem. Soc., Perkin Trans. 2*, 535 (1985).
187. B. Tinant, J.-P. Declercq and M. Van Meerssche, *J. Chem. Soc., Perkin Trans. 2*, 541 (1985).
188. B. Tinant, J.-P. Declercq and M. Van Meerssche, *Acta Crystallogr.*, **C43**, 2343 (1987).
189. P. Thinapong, M. Pohmakotr, B. Skelton and A. H. White, *Acta Crystallogr.*, **C49**, 54 (1993).
190. P. L. Bailey, C. T. Hewkin, W. Clegg and R. F. W. Jackson, *J. Chem. Soc., Perkin Trans. 1*, 577 (1993).
191. M. Rall, M. D. Harmony, D. A. Cassada and S. W. Staley, *J. Am. Chem. Soc.*, **108**, 6184 (1986).
192. J. S. A. M. de Boer, H. Schenk and C. H. Stam, *Recl. Trav. Chim. Pays-Bas*, **105**, 434 (1986).
193. V. F. Kalasinsky and J. R. Durig, *Spectrochim. Acta*, **44A**, 1331 (1988).
194. S. N. Mathur and M. D. Harmony, *J. Chem. Phys.*, **69**, 4316 (1978).
195. J. S. A. M. de Boer and C. H. Stam, *Recl. Trav. Chim. Pays-Bas*, **110**, 317 (1991).
196. J. S. A. M. de Boer and C. H. Stam, *Recl. Trav. Chim. Pays-Bas*, **109**, 375 (1990).
197. K. Kuchitsu, J. P. Guillory and L. S. Bartell, *J. Chem. Phys.*, **49**, 2488 (1968).
198. T. Iijima, *Bull. Chem. Soc. Jpn.*, **59**, 853 (1986); T. Iijima, H. Jimbo and M. Taguchi, *J. Mol. Struct.*, **144**, 381 (1986).
199. J. R. Durig, R. J. Berry and C. J. Wurrey, *J. Am. Chem. Soc.*, **110**, 718 (1988).
200. J. R. Durig, A. B. Nease, R. J. Berry, J. F. Sullivan, Y. S. Li and C. J. Wurrey, *J. Chem. Phys.*, **84**, 3663 (1986).
201. J. R. Durig, J. F. Sullivan, R. J. Berry and S. Cradock, *J. Chem. Phys.*, **86**, 4313 (1987).
202. W. H. Taylor, M. D. Harmony, D. A. Cassada and S. W. Staley, *J. Chem. Phys.*, **81**, 5379 (1984).
203. C. Heldmann and H. Dreizler, *Z. Naturforsch.*, **45a**, 1175 (1990).
204. K. Yamada, *J. Mol. Spectrosc.*, **79**, 323 (1980).
205. K. Yamada, M. Winnewisser, G. Winnewisser, L. B. Szalanski and M. C. L. Gerry, *J. Mol. Spectrosc.*, **79**, 295 (1980).
206. Y. Kai, P. Knochel, S. Kwiatkowski, J. D. Dunitz, J. F. M. Oth, D. Seebach and H.-O. Kalinowski, *Helv. Chim. Acta*, **65**, 137 (1982).
207. A. R. Mochel, C. O. Britt and J. E. Boggs, *J. Chem. Phys.*, **58**, 3221 (1973).
208. P. E. O'Bannon, P. J. Carroll and W. P. Dailey, *Struct. Chem.*, **1**, 491 (1990).
209. P. A. Wade, W. P. Dailey and P. J. Carroll, *J. Am. Chem. Soc.*, **109**, 5452 (1987).
210. H. Schmidbaur, S. Manhart and A. Schier, *Chem. Ber.*, **126**, 2259 (1993).
211. H. Schmidbaur, T. Pollok, R. Herr, F. E. Wagner, R. Bau, J. Riede and G. Müller, *Organometallics*, **5**, 566 (1986).
212. K. Dziwok, J. Lachmann, D. L. Wilkinson, G. Müller and H. Schmidbaur, *Chem. Ber.*, **123**, 423 (1990).
213. Mazhar-Ul-Haque, J. Ahmed, W. Horne, S. E. Cremer, P. K. Kafarski and P. W. Kremer, *J. Cryst. Spectr. Res.*, **19**, 1009 (1989).
214. H. Schmidbaur, A. Schier, B. Milewski-Mahrla and U. Schubert, *Chem. Ber.*, **115**, 722 (1982).
215. M. Dakkouri, in *Structures and Conformations of Non-Rigid Molecules* (Eds. J. Laane, M. Dakkouri, B. van der Veken and H. Oberhammer), NATO ASI Series C, Vol. 410, Kluwer, Dordrecht, 1993, pp. 491–517.

216. V. Typke, *J. Mol. Spectrosc.*, **77**, 117 (1979).
217. M. Dakkouri and V. Typke, *J. Mol. Struct.*, **158**, 323 (1987).
218. M. Dakkouri and V. Typke, *J. Mol. Sruct.*, **320**, 13 (1994).
219. M. Dakkouri and A. Seyfarth-Jacob, to be published; A. Seyfarth-Jacob, *Thesis*, University of Ulm, Ulm (1992).
220. M. Dakkouri, *J. Am. Chem. Soc.*, **113**, 7109 (1991).
221. K. J. Epple, *Thesis*, University of Ulm, Ulm (1990).
222. E. Hirota, Y. Endo, S. Saito and J. L. Duncan, *J. Mol. Spectrosc.*, **89**, 285 (1981).
223. M. Wong, I. Ozier and W. L. Meerts, *J. Mol. Spectrosc.*, **102**, 89 (1983).
224. V. W. Laurie, *J. Chem. Phys.*, **30**, 1210 (1959).
225. G. Bestmann, W. Lalowski and H. Dreizler, *Z. Naturforsch.*, **40a**, 271 (1985).
226. J. R. Durig, P. Groner and A. D. Lopata, *Chem. Phys.*, **21**, 401 (1977).
227. J. R. Durig, A. D. Lopata and P. Groner, *J. Chem. Phys.*, **66**, 1888 (1977).
228. E. Hirota, *J. Mol. Spectrosc.*, **34**, 516 (1970).
229. Y. Shiki, A. Hasegawa and M. Hayashi, *J. Mol. Struct.*, **78**, 185 (1982).
230. J. R. Durig, K. L. Kizer and Y. S. Li, *J. Am. Chem. Soc.*, **96**, 7400 (1974).
231. J. R. Villarreal and J. Laane, *J. Chem. Phys.*, **62**, 303 (1975).
232. J. Laane, E. M. Nour and M. Dakkouri, *J. Mol. Spectrosc.*, **102**, 368 (1983).
233. V. Typke, I. Botskor and K.-H. Wiedenmann, *J. Mol. Spectrosc.*, **120**, 435 (1986).
234. M. B. Kelly, J. Laane and M. Dakkouri, *J. Mol. Spectrosc.*, **137**, 82 (1989).
235. T. S. Little, M. Qtaitat, J. R. Durig, M. Dakkouri and A. Dakkouri, *J. Raman Spectrosc.*, **21**, 591 (1990).
236. E. Schaumann, C. Friese and G. Adiwidjaja, *Tetrahedron*, **45**, 3163 (1989).
237. H. Hopf and G. Maas, *Angew. Chem., Int. Ed. Engl.*, **31**, 931 (1992).
238. C. A. Deakyne, L. C. Allen and V. W. Laurie, *J. Am. Chem. Soc.*, **99**, 1343 (1977).
239. V. W. Laurie and W. M. Stigliani, *J. Am. Chem. Soc.*, **92**, 1485 (1970).
240. J. M. Pochan, J. E. Baldwin and W. H. Flygare, *J. Am. Chem. Soc.*, **91**, 1896 (1969).
241. S. W. Staley, T. D. Norden, W. H. Taylor and M. D. Harmony, *J. Am. Chem. Soc.*, **109**, 7641 (1987).
242. M. Trætteberg, A. Simon, E.-M. Peters and A. de Meijere, *J. Mol. Struct.*, **118**, 333 (1984).
243. S. Zöllner, H. Buchholz, R. Boese, R. Gleiter and A. de Meijere, *Angew. Chem., Int. Ed. Engl.*, **30**, 1518 (1991).
244. R. Boese, W. E. Billups and M. M. Haley, unpublished results, cited in Reference 5.
245. Y. Ohshima, S. Yamamoto, M. Nakata and K. Kuchitsu, *J. Phys. Chem.*, **91**, 4696 (1987).
246. M. Eckert-Maksić, S. Zöllner, W. Göthling, R. Boese, L. Maksimović, R. Machinek and A. de Meijere, *Chem. Ber.*, **124**, 1591 (1991).
247. R. D. Brown, P. D. Godfrey, B. Kleibömer, R. Champion and P. S. Elmes, *J. Am. Chem. Soc.*, **106**, 7715 (1984); R. D. Brown, P. D. Godfrey and B. Kleibömer, *J. Mol. Spectrosc.*, **118**, 317 (1986).
248. R. D. Brown, in *Structures and Conformations of Non-Rigid Molecules* (Eds. J. Laane, M. Dakkouri, B. van der Veken and H. Oberhammer), NATO ASI Series C, Vol. 410, Kluwer, Dordrecht, 1993, pp. 99–112.
249. W. v. E. Doering and H. D. Roth, *Tetrahedron*, **26**, 2825 (1970).
250. D. G. Van Derveer, J. E. Baldwin and D. W. Parker, *J. Org. Chem.*, **52**, 1173 (1987).
251. A. J. Schultz, D. G. Van Derveer, D. W. Parker and J. E. Baldwin, *Acta Crystallogr.*, **C46**, 276 (1990).
252. B. E. Thomas, IV and K. N. Houk, *J. Org. Chem.*, **57**, 4437 (1992).
253. N. Ramasubbu and K. Venkatesan, *Acta Crystallogr.*, **B38**, 976 (1982).
254. W. R. Roth, M. Winzer, H.-W. Lennartz and R. Boese, *Chem. Ber.*, **126**, 2717 (1993).
255. D. Spitzner, A. Engler, P. Wagner, A. de Meijere, G. Bengtson, A. Simon, K. Peters and E.-M. Peters, *Tetrahedron*, **43**, 3213 (1987).
256. R. J. Berry and M. D. Harmony, *Struct. Chem.*, **1**, 49 (1990).
257. W. M. Stigliani, V. W. Laurie and J. C. Li, *J. Chem. Phys.*, **62**, 1890 (1975).
258. C. A. Jacobs, J. C. Brahms, W. P. Dailey, K. Beran and M. D. Harmony, *J. Am. Chem. Soc.*, **114**, 115 (1992).
259. S. W. Staley, T. N. Norden, C.-F. Su, M. Rall and M. D. Harmony, *J. Am. Chem. Soc.*, **109**, 2880 (1987).

260. E. G. Atavin, I. G. Bolesov, V. S. Mastryukov and M. Trætteberg, *14th Austin Symposium on Molecular Structure*, Austin, Texas, 1992, Abstracts S2, p. 85.
261. G. Baranović, M. Eckert-Maksić, M. Golić and J. R. Durig, *J. Raman Spectrosc.*, **24**, 31 (1993).
262. M. Trætteberg, A. de Meijere and I. N. Domnin, *J. Mol. Struct.*, **128**, 207 (1985).
263. C. J. Cheer, D. Bernstein, A. Greenberg and P.-C. Lyu, *J. Am. Chem. Soc.*, **110**, 226 (1988).
264. P. E. O'Bannon, P. J. Carroll and W. P. Dailey, *Struct. Chem.*, **2**, 133 (1991).
265. I. N. Domnin, J. Kopf, S. Keyaniyan and A. de Meijere, *Tetrahedron*, **41**, 5377 (1985).
266. A. S. Feng, D. V. Speer, S. G. DiMagno, M. S. Konings and A. Streitwieser, *J. Org. Chem.*, **57**, 2902 (1992).
267. J. D. Korp, I. Bernal and R. Fuchs, *Can. J. Chem.*, **61**, 50 (1983).
268. J. Kopf, G. Starova, V. Plotkin and I. N. Domnin, *Acta Crystallogr.*, **C46**, 1707 (1990).
269. D. B. Puranik, M. P. Johnson and M. J. Fink, *Organometallics*, **8**, 770 (1989); M. J. Fink and D. B. Puranik, *Organometallics*, **6**, 1809 (1987).
270. M. P. Johnson, D. B. Puranik and M. J. Fink, *Acta Crystallogr.*, **C47**, 126 (1991).
271. I. Søtofte and I. Crossland, *Acta Chem. Scand.*, **43**, 168 (1989).
272. R. Boese, D. Bläser, R. Gleiter, K.-H. Pfeifer, W. E. Billups and M. M. Haley, *J. Am. Chem. Soc.*, **115**, 743 (1993).
273. G. Maier, A. Schick, I. Bauer, R. Boese and M. Nussbaumer, *Chem. Ber.*, **125**, 2111 (1992).
274. E. A. Dorko, J. L. Hencher and S. H. Bauer, *Tetrahedron*, **24**, 2425 (1968).
275. H. Dietrich, *Acta Crystallogr.*, **B26**, 44 (1970).
276. M. Bogey, C. Demuynck, J. L. Destombes and H. Dubus, *J. Mol. Spectrosc.*, **122**, 313 (1987).
277. P. Thaddeus, J. M. Vrtilek and C. A. Gottlieb, *Astrophys. J.*, **299**, L63 (1985).
278. S. Yamamoto, S. Saito, M. Ohishi, H. Suzuki, S. Ishikawa, N. Kaifu and A. Murakami, *Astrophys. J.*, **322**, L55 (1987); S. Yamamoto and S. Saito, *Astrophys. J.*, **363**, L13 (1990); S. Yamamoto and S. Saito, *12th International Symposium on Free Radicals*, Susono, Japan, 1990, III P17.
279. K. Peters and H. G. von Schnering, *Chem. Ber.*, **118**, 2147 (1985).
280. E. Fuchs, B. Breit, H. Heydt, W. Schoeller, T. Busch, C. Krüger, P. Betz and M. Regitz, *Chem. Ber.*, **124**, 2843 (1991).
281. B. Breit, H. Memmesheimer, R. Boese and M. Regitz, *Chem. Ber.*, **125**, 729 (1992).
282. R. Boese and N. Augart, *Z. Kristallogr.*, **182**, 32 (1988).
283. B. A. Borgias, R. C. Scarrow, M. D. Seidler and W. P. Weiner, *Acta Crystallogr.*, **C41**, 476 (1985).
284. (a) S. C. Abrahams, P. Marsh and L. A. Deuring, *Acta Crystallogr.*, **B44**, 263 (1988).
(b) M. D. Ward, P. J. Fagan, J. C. Calabrese and D. C. Johnson, *J. Am. Chem. Soc.*, **111**, 1719 (1989).
(c) J. S. Miller, M. D. Ward, J. H. Zhang and W. M. Reiff, *Inorg. Chem.*, **29**, 4063 (1990).
285. (a) H. N. Schäfer, H. Burzlaff, A. M. H. Grimmeis and R. Weiss, *Acta Crystallogr.*, **C47**, 1808 (1991).
(b) H. N. Schäfer, H. Burzlaff, A. M. H. Grimmeis and R. Weiss, *Acta Crystallogr.*, **C48**, 795 (1992).
(c) H. N. Schäfer, H. Burzlaff, A. M. H. Grimmeis and R. Weiss, *Acta Crystallogr.*, **C48**, 912 (1992).
286. D. Semmingsen and P. Groth, *J. Am. Chem. Soc.*, **109**, 7238 (1987).
287. G. Baum, F.-J. Kaiser, W. Massa and G. Seitz, *Angew. Chem., Int. Ed. Engl.*, **26**, 1163 (1987).
288. H. Irngartinger and K. L. Lukas, *Angew. Chem., Int. Ed. Engl.*, **18**, 694 (1979).
289. K. W. Cox, M. D. Harmony, G. Nelson and K. B. Wiberg, *J. Chem. Phys.*, **50**, 1976 (1969); *J. Chem. Phys.*, **53**, 858 (1972).
290. W. H. Taylor, M. D. Harmony and S. W. Staley, *Struct. Chem.*, **2**, 167 (1991).
291. V. S. Mastryukov, E. L. Osina, L. V. Vilkov and R. L. Hilderbrandt, *J. Am. Chem. Soc.*, **99**, 6855 (1977).
292. V. S. Mastryukov, *J. Mol. Struct.*, **244**, 291 (1991).
293. E. Vilsmaier, J. Fath, C. Tetzlaff and G. Maas, *J. Chem. Soc., Perkin Trans. 2*, 1895 (1993).
294. E. Vilsmaier, J. Fath and G. Maas, *Synthesis*, 1142 (1991).
295. C. Tetzlaff, V. Butz, E. Vilsmaier, R. Wagemann, G. Maas, A. Ritter von Onciul and T. Clark, *J. Chem. Soc., Perkin Trans. 2*, 1901 (1993).
296. V. Butz, E. Vilsmaier and G. Maas, *J. Chem. Soc., Perkin Trans. 2*, 1907 (1993).

297. E. Vilsmaier, C. Tetzlaff, V. Butz and G. Maas, *Tetrahedron*, **47**, 8133 (1991).
298. O. N. Kataeva, I. A. Litvinov, V. A. Naumov, V. V. Ratner and B. A. Arbuzov, *Zh. Strukt. Khim.*, **30**, No. 3, 122 (1989); *Chem. Abstr.*, **111**, 164641w (1989).
299. W. Adam, F. Adamsky, F.-G. Klärner, E.-M. Peters, K. Peters, H. Rebollo, W. Rüngeler and H. G. von Schnering, *Chem. Ber.*, **116**, 1848 (1983).
300. B. Tinant, J.-P. Declercq, P. Rouchy and N. Guillot, *Acta Crystallogr.*, **C47**, 2002 (1991).
301. M. D. Taylor, G. Minaskanian, K. N. Winzenberg, P. Santone and A. B. Smith, III, *J. Org. Chem.*, **47**, 3960 (1982); A. B. Smith, III, N. J. Liverton, N. J. Hrib, H. Sivaramakrishnan and K. N. Winzenberg, *J. Org. Chem.*, **50**, 3239 (1985).
302. P. J. Carroll, N. J. Liverton and A. B. Smith, III, *Acta Crystallogr.*, **C42**, 1594 (1986).
303. P. Altmeier, E. Vilsmaier and G. Wolmershäuser, *Tetrahedron*, **45**, 3189 (1989).
304. B. Halton and P. J. Stang, *Acc. Chem. Res.*, **20**, 443 (1987); B. Halton, *Chem. Rev.*, **89**, 1161 (1989).
305. W. Koch, M. Eckert-Maksić and Z. B. Maksić, *J. Chem. Soc., Perkin Trans. 2*, 2195 (1993) and references cited therein.
306. R. Neidlein, D. Christen, V. Poignée, R. Boese, D. Bläser, A. Gieren, C. Ruiz-Pérez and T. Hübner, *Angew. Chem., Int. Ed. Engl.*, **27**, 294 (1988).
307. D. Bläser, R. Boese, W. A. Brett, P. Rademacher, H. Schwager, A. Stanger and K. P. C. Vollhardt, *Angew. Chem., Int. Ed. Engl.*, **28**, 206 (1989).
308. G. A. Olah (Ed.), *Cage Hydrocarbons*, Wiley, New York, 1990.
309. H. Irngartinger, W. Reimann, R. Lang and M. Christl, *Acta Crystallogr.*, **B46**, 234 (1990).
310. H. Irngartinger, R. Jahn, H. Rodewald, Y. H. Paik and P. Dowd, *J. Am. Chem. Soc.*, **109**, 6547 (1987).
311. H. Irngartinger, A. Goldmann, U. Huber-Patz, P. Garner, Y. H. Paik and P. Dowd, *Acta Crystallogr.*, **C44**, 1472 (1988).
312. A. L. Spek and P. van der Sluis, *Acta Crystallogr.*, **C46**, 1357 (1990).
313. C. Rücker, H. Prinzbach, H. Irngartinger, R. Jahn and H. Rodewald, *Tetrahedron Lett.*, **27**, 1565 (1986).
314. (a) K. B. Wiberg, *Acc. Chem. Res.*, **17**, 379 (1984).
(b) K. B. Wiberg, *Tetrahedron Lett.*, **26**, 599 (1985).
315. K. B. Wiberg, W. P. Dailey, F. H. Walker, S. T. Waddell, L. S. Crocker and M. Newton, *J. Am. Chem. Soc.*, **107**, 7247 (1985).
316. L. Hedberg and K. Hedberg, *J. Am. Chem. Soc.*, **107**, 7257 (1985).
317. A. Almenningen, B. Andersen and B. A. Nyhus, *Acta Chem. Scand.*, **25**, 1217 (1971); see also J. F. Chiang and S. H. Bauer, *J. Am. Chem. Soc.*, **92**, 1614 (1970).
318. P. Seiler, *Helv. Chim. Acta*, **73**, 1574 (1990).
319. P. Seiler, J. Belzner, U. Bunz and G. Szeimies, *Helv. Chim. Acta*, **71**, 2100 (1988).
320. G. Maier, *Angew. Chem., Int. Ed. Engl.*, **27**, 309 (1988).
321. H. Irngartinger, A. Goldmann, R. Jahn, M. Nixdorf, H. Rodewald, G. Maier, K.-D. Malsch and R. Emrich, *Angew. Chem., Int. Ed. Engl.*, **23**, 993 (1984).
322. H. Irngartinger, R. Jahn, G. Maier and R. Emrich, *Angew. Chem., Int. Ed. Engl.*, **26**, 356 (1987).
323. H. Quast, J. Carlsen, R. Janiak, E.-M. Peters, K. Peters and H. G. von Schnering, *Chem. Ber.*, **125**, 955 (1992).
324. H. Quast, K. Knoll, E.-M. Peters, K. Peters and H. G. von Schnering, *Chem. Ber.*, **126**, 1047 (1993).
325. H. Quast, E. Geißler, T. Herkert, K. Knoll, E.-M. Peters, K. Peters and H. G. von Schnering, *Chem. Ber.*, **126**, 1465 (1993).
326. K. A. Potekhin, A. I. Yanovskii, Yu. T. Struchkov, V. D. Sorokin, V. V. Zhdankin, A. S. Koz'min and N. S. Zefirov, *Dokl. Akad. Nauk SSSR*, **301**, 119 (1988); *Chem. Abstr.*, **110**, 192779v (1989).
327. R. Srinivasan, Y. Hu, M. F. Farona, E. A. Zarate and W. J. Youngs, *J. Org. Chem.*, **52**, 1167 (1987).
328. H. Irngartinger, D. Kallfaß, E. Litterst and R. Gleiter, *Acta Crystallogr.*, **C43**, 266 (1987).
329. R. R. Karl, Jr., Y. C. Wang and S. H. Bauer, *J. Mol. Struct.*, **25**, 17 (1975).
330. V. Jonas and G. Frenking, *J. Org. Chem.*, **57**, 6085 (1992).
331. H. Wingert, H. Irngartinger, D. Kallfaß and M. Regitz, *Chem. Ber.*, **120**, 825 (1987).

332. G. Maier, I. Bauer, U. Huber-Patz, R. Jahn, D. Kallfaß, H. Rodewald and H. Irngartinger, *Chem. Ber.*, **119**, 1111 (1986).
333. L. Doms, H. J. Geise, C. van Alsenoy, L. van den Enden and L. Schäfer, *J. Mol. Struct.*, **129**, 299 (1985); K. Mizuno, T. Fukuyama and K. Kuchitsu, *Chem. Lett.*, 249 (1972).
334. B. Vogelsanger and A. Bauder, *J. Mol. Spectrosc.*, **136**, 62 (1989).
335. W. H. Watson, I. Tavanaiepour, A. P. Marchand and P. R. Dave, *Acta Crystallogr.*, **C43**, 1356 (1987).
336. H. Irngartinger, R. Jahn, H. Rodewald, C. T. Kiers and H. Schenk, *Acta Crystallogr.*, **C42**, 847 (1986).
337. H. Irngartinger and R. Jahn, *Croat. Chem. Acta*, **64**, 289 (1991).
338. W. Tochtermann, C. Vogt, E.-M. Peters, K. Peters, H. G. von Schnering and E.-U. Würthwein, *Chem. Ber.*, **124**, 2577 (1991).
339. H. Hopf, *Chem. Ber.*, **125**, No. 2, pp. I–XXIV (1992).
340. B. Ahlquist, A. Almenningen, B. Benterud, M. Trætteberg, P. Bakken and W. Lüttke, *Chem. Ber.*, **125**, 1217 (1992).
341. H. Irngartinger, J. Hauck, A. de Meijere, K. Michelsen and R. Machinek, *Isr. J. Chem.*, **29**, 147 (1989).
342. C.-H. Lee, S. Liang, T. Haumann, R. Boese and A. de Meijere, *Angew. Chem., Int. Ed. Engl.*, **32**, 559 (1993).
343. B. de Ruiter, J. E. Benson, R. A. Jacobson and J. G. Verkade, *Inorg. Chem.*, **29**, 1065 (1990).
344. A. B. Smith, III, R. M. Strongin, L. Brard, G. T. Furst, W. J. Romanow, K. G. Owens and R. C. King, *J. Am. Chem. Soc.*, **115**, 5829 (1993).

CHAPTER **4**

Interrelations in the thermochemistry of cyclopropanes

JOEL F. LIEBMAN

Department of Chemistry and Biochemistry, University of Maryland, Baltimore County Campus, 5401 Wilkens Avenue, Baltimore, Maryland 21228–5398, USA
Fax: (+1)410-455-2608; e-mail: jliebman@umbc2.umbc.edu

I. INTRODUCTION 224
A. Is Another Review Necessary? 224
B. Scope, Definitions and Limitations 224
C. Chapter Contents 225
II. CYCLOPROPANES AND OLEFINS 227
A. Cyclopropane and its Methylated and Alkylated Analogs 227
B. Spiropentane, Methylenecyclopropane and Allene 228
C. Cyclopropanation of Cycloalkenes 229
D. 'Homologous' Spiro-joined Cyclopropanes 229
E. Conjugated Double Bonds and Cyclopropanes 231
F. The Conjugation of Double Bonds with Cyclopropanes 233
G. Acetylenes, Diacetylenes and Cyclopropanes 235
H. Acetylenes and Cyclopropenes 235
I. A Brief Interlude on Homoaromaticity and Homoantiaromaticity 235
J. Dicarbon and Its Cyclopropanated Analogs 236
III. PHENYL VERSUS CYCLOPROPYL 237
A. Alkyl Substituted Derivatives 237
B. Unsaturated 'Hydrocarbyl' Derivatives 239
C. Hetero-atom Containing Substituted Derivatives 239
D. Multiply Substituted Derivatives 240
E. 'Strings' of Cyclopropanes and Benzenes 241
IV. STRAIN ENERGIES OF SPECIES WITH THREE- AND FOUR-MEMBERED RINGS 242
A. Cyclopropanes and Cyclobutanes 242
B. Introduction of sp^2 Carbons into Cyclopropanes and Cyclobutanes 243
C. Ring Fusion of Cyclopropanes and Cyclobutanes 244
D. Another Interlude on Homoaromaticity and Homoantiaromaticity 245

The chemistry of the cyclopropyl group, Vol. 2
Edited by Z. Rappoport

E. Benzoannelation of Cyclopropene and Cyclobutene 245
F. Do Many Rings Mean Many Problems? 246
G. Some Thoughts on the Aromaticity of Cyclopropenones 248
V. ACKNOWLEDGMENTS 249
VI. REFERENCES AND COMMENTARY 249

I. INTRODUCTION

A. Is Another Review Necessary?

As evidenced by the current volume on cyclopropanes appearing so soon after an earlier set[1] and an update volume as well[2] in the 'Patai Series', species with three-membered rings (3MR) remain of considerable interest to the organic chemistry community. The interplay of structure, energetics and reactivity of these species have figured prominently in these volumes and in numerous other reviews written both before and after[3]. As such, it can honestly be asked if another review on the thermochemistry of cyclopropanes is really desirable. Most assuredly, there are relevant new thermochemical data including a 'respectable' number of directly measured enthalpies of formation. Yet if the author of the current chapter, as well as several other cyclopropane reviews[3–6], were *merely* to update, correct and 'fill in the holes of' these earlier studies[7], it is doubtful that such a contribution would be particularly welcome in the current volume. Editors we have dealt with have often emphasized that 'each of the authors is asked *not* to give an encyclopedic coverage[8]. It must be admitted that for the current case of the thermochemistry of cyclopropane and its derivatives, not *that* much new data are available to merely add to the earlier text. What now is to be written? After all, self-plagiarism is tolerated in only small doses.

B Scope, Definitions and Limitations

Even before any decision was made regarding which compounds to include in the text, it was imperative for the author to decide on the scope of the chapter, and on terminology, units and the choice of 'primary' vs 'secondary' bibliographic citations. To begin with, it was necessary to answer the obvious question 'what is thermochemistry?'. For this chapter, the relatively restricted scope of 'enthalpy of formation' (written variously ΔH_f, ΔH_f° and $\Delta_f H_m^\circ$) was chosen: no discussion will be conducted on other thermochemical properties such as entropy, heat capacity and excess enthalpy. Additionally (following thermochemical convention[9]), much as the temperature and pressure were tacitly assumed to be 25°C ('298 K') and 1 atmosphere (taken as either 101,325 or 100,000 Pa) respectively[9], the units were immediately chosen to be kJ instead of kcal (where 4.184 kJ ≡ 1 kcal, 1 kJ = 0.2390 kcal).

In this chapter, intermolecular forces are viewed as 'complications' and 'nuisances': it is the molecule *per se* that is of interest. Therefore, unless explicitly noted to the contrary, any species of interest in this chapter is to be assumed in the (ideal) gas phase. Most organic compounds are 'naturally' liquids or solids under the thermochemically desired conditions, much less as found by the synthetically or mechanistically inclined chemist. 'Corrections' are naturally made by using enthalpies of vaporization (v) and of sublimation(s), defined by equations 1a and 1b:

$$\Delta H_v \equiv \Delta H_f(\mathrm{g}) - \Delta H_f(\mathrm{l}) \quad (1a)$$

$$\Delta H_s \equiv \Delta H_f(\mathrm{g}) - \Delta H_f(\mathrm{s}) \quad (1b)$$

where g, l and s refer to gas, liquid and solid. Phase change enthalpies were obtained from whatever source available: the choice to maximize the use of gas-phase data and minimize that from the liquid or solid forces expediencies on the author. In the absence of data from experimental measurements, enthalpies of vaporization for hydrocarbons will usually be estimated using the generally accurate (± 2 kJ mol^{-1}) 2-parameter equation[10], CHLL2 (equation 2a),

$$\Delta H_v = 4.7\tilde{n}_C + 1.3n_Q + 3.0 \quad (2a)$$

where $\tilde{n}_c$ and n_Q are the numbers of nonquaternary and quaternary carbons, respectively, for the compound of interest. Enthalpies of vaporization for species containing heteroatom (non C, H) containing substituents X will likewise be estimated using the related CHLP equation[11] (equation 2b),

$$\Delta H_v = 4.7\tilde{n}_C + 1.3n_Q + b(X) + 3.0 \quad (2b)$$

where b(X) is the substituent-dependent enthalpy of vaporization parameter[12]. No effort will be made to estimate enthalpies of sublimation—to date, success is more limited and the procedures are more embryonic[13].

The simple choice was made to cite secondary sources over primary. This simplifies the text and so benefits the current reader. It may offend the original author. Having to decide between reader and author, the former wins. Unless explicitly asserted to the contrary, enthalpies of formation of all cyclopropane derivatives, and other three-membered ring (3MR) species, will be taken from our earlier review[3] (where admittedly a disproportionate number of values were immediately accepted and so incorporated from the earlier review of Kolesov and Kozina[14]). Relatedly, all unreferenced enthalpies of formation of noncyclopropanes and other non-3MRs come from the archive of Pedley and his coworkers[15]. (Two additional useful sources of information are the earlier, but incompletely 'incorporated', thermochemical archives, by Kharasch[16] and by Stull, Westrum and Sinke[17].)

By choosing to compare the enthalpies of formation of cyclopropanes with those of other classes of compounds, a relatively simple and abbreviated text results. Rewardingly, numerous data are amenable to coherent comparison, and so both prediction and explanation are often realizable goals. Patterns are presented to the reader for his/her education, enjoyment, entertainment. But frustratingly, very often seemingly related compounds have numerically disparate and conceptually incompatible data. At these more frustrating occasions, the problems have been acknowledged here with the explicit hope that some reader will provide some needed clarifying experiment. The various sections of the text are now outlined: it is in these sections that seeming successes and seeming failures are chronicled. It is hoped that the reader shares the author's conclusion that cyclopropane and its derivatives continue to educate, entertain, encourage.

C. Chapter Contents

The first nonintroductory section of the text starts with the observation made early in this century that cyclopropanes have significant olefinic character. That is, corresponding ethylene and cyclopropane derivatives have significant similarities. There are literature comparisons of the thermochemistry of direct counterparts such as the parent species (**1** and **2**, X = H), propene and methylcyclopropane with X = Me, and of methyl acrylate and methyl cyclopropanecarboxylate[18] with X = COOMe. But the chemistry of substituted ethylenes is far richer than just that of vinyl compounds. One can retrieve enthalpy of formation data for cumulenes ('cumulenated' olefins) such as allene (**3**) and both *cis*- and *trans*-2,3,4-hexatriene (**4a** and **4b**), and for conjugated olefins such as 1,3-butadiene (**5**) and both (Z)- and (E)-1,3,5-hexatriene (**6a** and **6b**). For the cyclopropane chemist it is natural to ask: are spiropentane (**7**), bicyclopropyl [cyclopropylcyclopropane, **8**] and *cis*- and *trans*-1,1′;2′,1″-tercyclopropyl [1,2-dicyclopropylcyclopropane, **9a** and **9b**] the natural counterparts of **3**, **5**, **6a** and **6b**? These and related issues are discussed in Section II

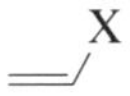

(1)

(2)

(3)

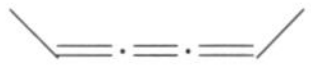

(4a)

(4b)

(5)

(6a)

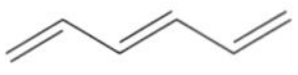

(6b)

(7)

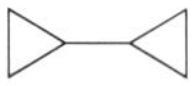

(8)

(9a)

(9b)

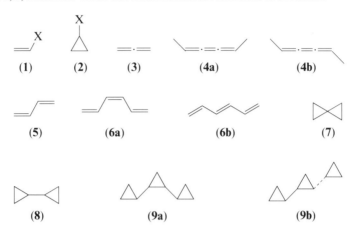

The next section makes use of the much more recent observation[19] that there is a nearly constant difference of the enthalpies of formation of corresponding vinyl and phenyl derivatives. If vinyl relates to cyclopropyl, and vinyl also relates to phenyl, then how do corresponding cyclopropyl and phenyl derivatives relate? Conceptually, vinylcyclopropane (**10**), also identified as **1**, X = Cypr and **2**, X = Vi) and styrene (**11**, X = Vi, also identified as **1**, X = Ph) are thus relatable. Likewise, relatable are cyclopropylamine (**2**, X = NH_2) and aniline (**11**, X = NH_2)[18]. This thermochemical comparison of benzene and cyclopropane derivatives is not merely a check of two purported identities in terms of a third, arithmetically derivable, identity. Benzene is the archetypical π-delocalized aromatic species from which understanding of this widespread phenomenon evolves. Cyclopropane is the paradigm of σ-aromatic species from which understanding of this more exotic phenomenon evolves[20]. Benzene and cyclopropane are thus naturally paired as conceptual models for delocalization and aromaticity. Section III discusses these and related issues.

It is well-recognized that the hydrocarbons cyclopropane and cyclobutane have nearly identical strain energies, and so these 'microcycles' have been quite naturally paired in numerous treatments of molecular strain. How similar are cyclobutylamine (**12**, X = NH_2, **13**, $n = 4$, X = NH_2) and cyclopropylamine (**2**, X = NH_2, **13**, $n = 3$, X = NH_2) and other correspondingly monosubstituted cyclobutanes and cyclopropanes[18]? What about

(10)

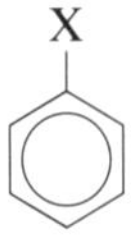

(11)

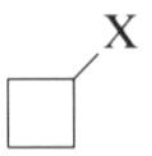

(12)

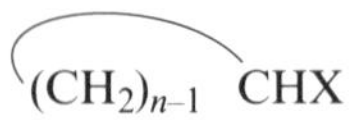

(13)

(14)

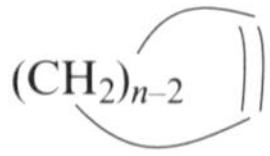

(15)

(16)

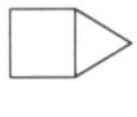

(17)

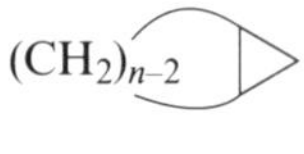

(18)

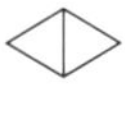

(19)

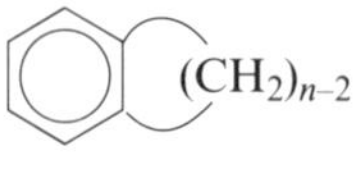

(20)

cyclobutene (**14**, i.e. **15**, $n = 4$) and cyclopropene (**16**, i.e. **15**, $n = 3$) and their derivatives? What about ring fused species such as bicyclo[2.1.0]pentane (**17**, i.e. **18**, $n = 4$) and bicyclobutane (**19**, i.e. **18**, $n = 3$); benzocyclopropene and benzocyclobutene (**20**, $n = 3$ and 4, respectively)? Section IV discusses these and related issues.

We admit now that we have almost completely ignored three-membered ring-containing ions and other reactive intermediates, as well as all hetero- and metallo-derivatives.

II. CYCLOPROPANES AND OLEFINS

The conceptual interrelations of species with three-membered rings and those with double bonds are among the oldest and best-established in the study of strained species[18,21]. These interrelations have ramifications in the understanding of reactivity and stability patterns, spectra, geometry and electronic delocalization and have been expressed in terms of the Walsh (Walsh–Sugden), σ and π orbitals, bent and banana bonds, and cyclopropane and 'cycloethane'[22]. For this chapter we will but consider some of the thermochemical aspects.

A. Cyclopropane and its Methylated and Alkylated Analogs

If cyclopropane may be thermochemically compared with ethylene, then methylcyclopropane may likewise be compared with propene. In the same way, the isomeric 1,1-, *cis*-1,2- and *trans*-1,2-dimethylcyclopropane **21a**, **21b**, **21c** with X = Me) are relatable in turn to the isomeric dimethylethylenes: isobutylene, (*Z*)- and (*E*)-2-butene. But what is meant by 'thermochemically compared'? We start with the observation that cyclopropane and ethylene have fortuitously nearly identical gas-phase enthalpies of formation, (53.3 ± 0.6) and (52.5 ± 0.4) kJ mol^{-1}. Can we then conclude that gas-phase methylcyclopropane and propene (methylethylene) have nearly the same enthalpy of formation, and that the former is higher than the latter by (0.8 ± 0.7) kJ mol^{-1}? In fact, methylcyclopropane has a higher enthalpy of formation[23] than propene by 24.3–[20.0 (± 0.8)] = [4.3 (± >0.8)] kJ mol^{-1}. The two differences, [0.8 (± 0.7)] and [4.3 (± >0.8)] kJ mol^{-1}, are hardly the same.

What about the remaining comparisons mentioned above? Let us recast the numbers to read that methylation decreases the enthalpy of formation of ethylene by (52.5 ± 0.4)–(20.0 ± 0.8) = (32.5 ± 0.9) kJ mol^{-1} and of cyclopropane by (53.3 ± 0.6) – 24.3 = [29.0 (± > 0.6)] kJ mol^{-1}. Let us calibrate our thinking by looking at other hydrocarbon enthalpy of formation changes upon methylation. These additional numbers for comparison are the methylation enthalpy changes of

(a) acetylene to form methylacetylene (propyne),
(b) ethane to form propane,
(c) the methylene carbon of propane to form isobutane.

The difference of acetylene and propyne is (43.3 ± 1.1) kJ mol^{-1} suggesting there are special stabilizing π-effects of methyl with a triple bond that are of greater magnitude than with a double bond. The difference of ethane and propane is (20.9 ± 0.6) kJ mol^{-1}. This suggests an ever smaller methylation enthalpy change as one proceeds from 'sp' to 'sp^2' to 'sp^3' hybridized carbon. It also legitimizes the description of cyclopropane as rather olefinic since the enthalpy changes for cyclopropane are closer to ethylene than for the terminal carbon of propane. However, the difference of propane and isobutane is found to be (30.5 ± 0.9) kJ mol^{-1}, nearly identical to the enthalpy changes accompanying methylation of ethylene and cyclopropane. In that both the central carbon of propane and the carbons of cyclopropane are all bonded to two other carbons and two hydrogens, this nearly identical methylation enthalpy would suggest that cyclopropane is alicyclic and also saturated, much as propane is aliphatic and also saturated. But saturated and olefinic are antonyms. There is an apparent contradiction unless we assert the near-equality of methylation enthalpies is fortuitous. Is it? Maybe we should simply, empirically, operationally say that methylation of ethylene is [3.5 (± >1.1)] $kJ\,mol^{-1}$ more stabilizing than methylation of cyclopropane, and not concern ourselves with the nature of hybridization and of bonding in any of the above species.

The above methylation enthalpies of ethylene and propene can be recast in terms of the formal 'cyclopropanation' reaction given in equation 3.

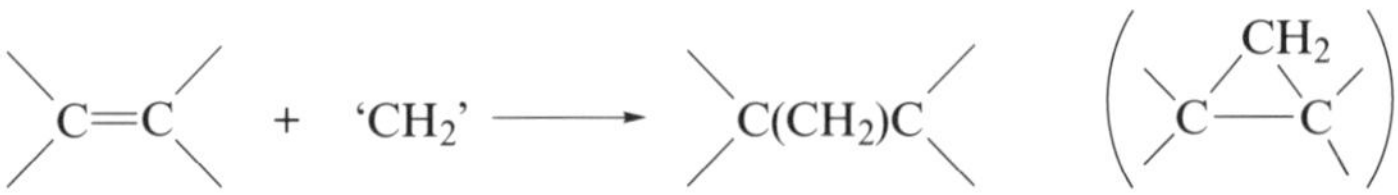

(3)

From the above-cited enthalpies of formation and construction of a suitable Hess cycle, we deduce that the cyclopropanation of propene is more endothermic than that of ethylene by [3.5 (± >1.1)] $kJ\,mol^{-1}$. This *ca* 3 $kJ\,mol^{-1}$ effect is corroborated by cyclopropanation of the monosubstituted ethyl species; the gas-phase enthalpy of formation difference of 1-butene and ethylcyclopropane is [3.3 (± >1.0)] $kJ\,mol^{-1}$. Consider now disubstituted derivatives. For the 1,1-dimethyl case and the *cis*- and *trans*-1,2 cases, the corresponding differences are (8.7 ± 1.5), [8.8 (± >1.0)] and [7.6 (± >1.0)] $kJ\,mol^{-1}$, respectively, favoring the olefin. For the *cis*- and *trans*-1,2-diethyl species[24] (**21b** and **21c**, X = Et) the cyclopropanation reactions are essentially thermoneutral; we lack thermochemical data for 1,1-diethylcyclopropane (**21a**, X = Et) and so lack the remaining diethyl comparison. For the trisubstituted 1,1,2-trimethyl species (**22**, X = Me) a difference of [–24.9 (± >1.1)] $kJ\,mol^{-1}$ 'favoring' the olefin is found while for the l,l-dimethyl-2-ethyl species[25] (**22**, X = Et) the diffrence is 6.5 (± >1.2) $kJ\,mol^{-1}$. For the 1,1,2,2-tetramethyl species[26] (**23**), the difference is [–17.4 (± >1.2)] $kJ\,mol^{-1}$.

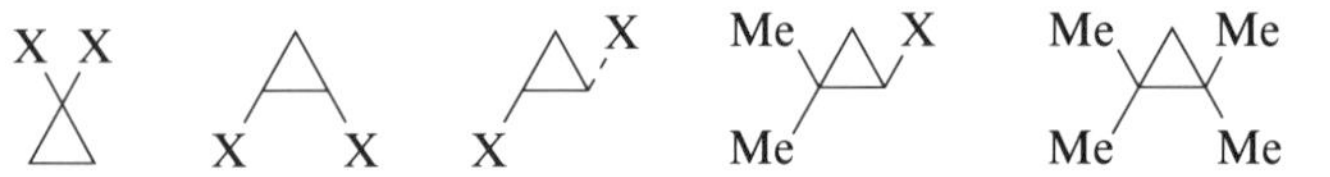

(**21a**) (**21b**) (**21c**) (**22**) (**23**)

These cyclopropane vs olefin enthalpy of formation differences do not increase monotonically with the number of substituents. Worse yet, differences are found to be of differing signs. That is, there is no obvious pattern for *all* of the enthalpies of the cyclopropanation reaction 3. Neglecting any enthalpic contribution from the 'CH_2' or 'cyclopropanation reagent'[27] and considering only un-, mono- and di-substituted olefins and cyclopropanes, we find the enthalpy of reaction 3 is rather coarsely equal to (3 ± 2) $kJ\,mol^{-1}$ per alkyl substituent[28]. We will now accept this 3 $kJ\,mol^{-1}$ per alkyl substituent for reaction 3, where we admit the absence of justification and motivation other than arithmetic and convenience. We are forced to tolerate discrepancies of a few $kJ\,mol^{-1}$ per substituent: differences between our correlations/models and experiment of several $kJ\,mol^{-1}$ must be considered as acceptable for our analysis[29].

B. Spiropentane, Methylenecyclopropane and Allene

If cyclopropane may be thermochemically compared with ethylene, then spiropentane (**7**) may be thermochemically compared with methylenecyclopropane (**24**), and methylenecyclopropane (**24**) with allene (**3**). Said differently, if there were a constant change of enthalpy of formation on 'cyclopropanation' of a double bond into a three-membered ring, then the difference of the enthalpies of formation of cyclopropane and ethylene would equal the differences found between methylenecyclopropane and allene, and between spiropentane and methylenecyclopropane. The previous section suggests we try to keep

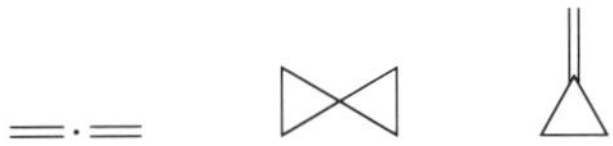

4. Interrelations in the thermochemistry of cyclopropanes

$$=\cdot= \qquad \bowtie \qquad \triangle$$

$$\quad(3) \qquad\quad (7) \qquad (24)$$

substitution patterns intact. We thus consider the transformation of isobutene into 1,1-dimethylcyclopropane as our paradigm. Reaction 3 is now rewritten as the formal reaction 4 to reflect the *gem*-disubstitution:

$$\begin{array}{c}\diagdown\\ \diagup\end{array}\!\!C\!=\!CH_2 + \text{'CH}_2\text{'} \longrightarrow \begin{array}{c}\diagdown\\ \diagup\end{array}\!\!C(CH_2)_2 \qquad (4)$$

Again neglecting any enthalpic contribution from the 'CH$_2$' reagent, we find for the isobutene → 1,1-dimethylcyclopropane and allene → methylenecyclopropane (3 → 24) transformations, the comparable enthalpy differences, (8.7 ± 1.5) and (10.0 ± 2.2) kJ mol^{-1}, respectively[30]. However, for the methylenecyclopropane → spiropentane (24 → 7) transformation, the difference is (−15.3 ± 2.0) kJ mol^{-1}. The last cyclopropanation difference hardly equals the others, and it is admittedly distressing that the cyclopropanation reaction enthalpy differences have nonuniform signs[31].

C. Cyclopropanation of Cycloalkenes

A significant lack of uniformity is seen for the formal cyclopropanation of *n*-carbon cycloalkenes to form bicyclo[(*n*−2).1.0]alkanes via the formal reaction shown in equation 5.

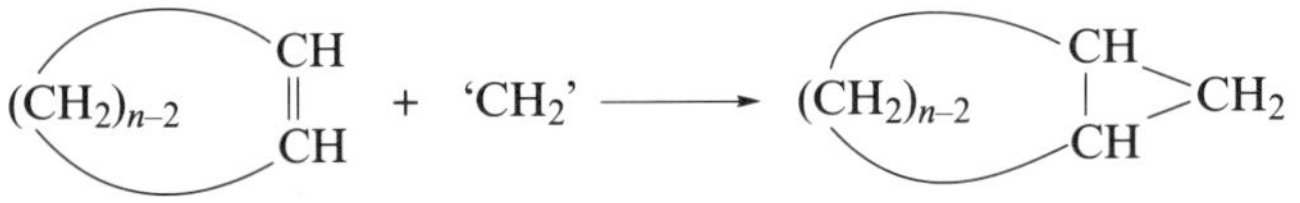

$$(CH_2)_{n-2}\!\!\begin{array}{c}\diagup CH\\ \| \\ \diagdown CH\end{array} + \text{'CH}_2\text{'} \longrightarrow (CH_2)_{n-2}\!\!\begin{array}{c}\diagup CH\\ | \quad\diagdown CH_2\\ \diagdown CH\diagup\end{array} \qquad (5)$$

$$\qquad(15) \qquad\qquad\qquad\qquad (18)$$

Again neglecting the contribution of the 'CH$_2$' reagent, the reaction enthalpies for $n = 3$–7, *cis*-8- and *trans*-8- are $n = 3$, (−59.9 ± 2.6)[32]; $n = 4$, (+ 2.0 ± 2.0); $n = 5$, (+ 4.7 ± 2.5); $n = 6$, (+ 6.3 ± 2.8)[33]; $n = 7$, (+ 3.1 ± 2.6)[34]; $n = 8$, *cis*-, (+ 6.4 ± 2.2) and *trans*-, (−8.4 ± 3.1)[35] kJ mol^{-1}. Because the bicycloalkanes are disubstituted cyclopropanes, the analysis in the previous section would have suggested a change of some 6 kJ mol^{-1}. The $n = 4, 5, 6, 7$ and *cis*-8- cases are all in agreement. However, the $n = 3$ and *trans*-8- cases are significantly deviant[36].

D. 'Homologous' Spiro-joined Cyclopropanes

We recognize that the conceptual relation of ethylene with cyclopropane, and allene with spiropentane, proceeds naturally through to species with increasing numbers of cumulenated double bonds and increasing numbers of spiro linkages. While no enthalpy of formation data for any unsubstituted species containing three or more cumulenated double bonds is seemingly available from experiment, that of both 1,4-dimethyl derivatives, *cis*- and *trans*-2,3,4-hexatriene, **4a** and **4b**, are known from hydrogenation enthalpy measurements[37]. Since the two values are indistinguishable, there is presumably negligible interaction between the two methyl groups. We may thus estimate the desired butatriene enthalpy of formation by assuming thermoneutrality for equation 6:

$$MeCH{=}C{=}C{=}CHMe + 2RCH{=}CH_2 \longrightarrow CH_2{=}C{=}C{=}CH_2 + 2\,(E)\text{-}RCH{=}CHMe \qquad (6)$$

For R = H, Me, Vi, Et and (E)-CH=CHMe, the derived values are [329.9 (± >1.0)], [327.7 (± >1.8)], [332.7 (± >2.0)], [328.5 (± >2.1)] and [328.3 (± > 2.0)][38] kJ mol^{-1}. Alternatively, we may assume thermoneutrality for the related equatlon 7:

$$MeCH{=}C{=}C{=}CHMe + 2RCH{=}C{=}CH_2 \longrightarrow CH_2{=}C{=}C{=}CH_2 + 2RCH{=}C{=}CHMe \qquad (7)$$

For R = H and Me, the derived values are [321.3 (± >1.9)] and [323.3 (± >1.4)] kJ mol^{-1}, respectively[39]. A value of [326 (± > 4)] kJ mol^{-1} for ΔH_f(g, 1,2,3-butatriene) is thus credible. What is found for the cyclopropanation enthalpies of butatriene? There are seemingly no relevant data for either of its monocyclopropanation products, dimethylenecyclopropane (**25a**) or vinylidenecyclopropane (**25b**). The two dicyclopropanation products have comparable enthalpies of formation: dicyclopropylidene (**26**), 286.6 (l) and 324.3 (g), and methylenespiropentane (**27**), 287.0 (l) and 320.9 (g), respectively[40]. The 2–6 kJ mol^{-1} decrease in enthalpies of formation for gaseous dicyclopropanated products is not particularly in accord with the 3 kJ mol^{-1} increase per alkyl substituent of cyclopropanation of simple olefins. However, in that the allene → methylenecyclopropane → spiropentane (**3** → **23** → **7**) enthalpy of formation changes are still enigmatic, and error bars are absent for the dicyclopropanated products, we do not fret. But we eagerly await more thermochemical data.

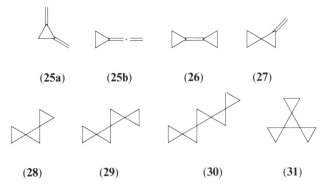

(**25a**)

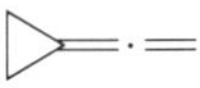

(**25b**)

(**26**)

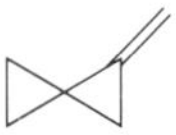

(**27**)

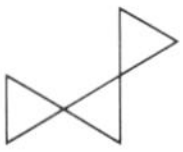

(**28**)

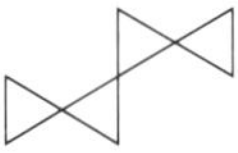

(**29**)

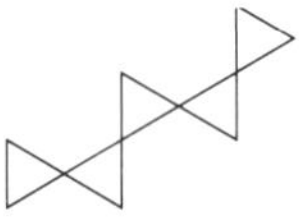

(**30**)

(**31**)

We now briefly discuss the formally homologous series defined by an increasing number of three-membered rings, $n(\Delta)$: cyclopropane, spiropentane (**7**), dispiro[2.0.2.1]heptane (**28**), trispiro[2.0.0.2.1.1]nonane (**29**) and tetraspiro[2.0.0.0.2.1.1.1]undecane (**30**) for which measured enthalpies of formation and measured and estimated values of vaporization are available[41] (Table 1). Do these compounds form a homologous series with regard to their enthalpies of formation? Statistically they do. Using just the experimentally measured enthalpies of formation for the gaseous species (and therefore temporarily omitting the value for **29** because its enthalpy of vaporization was estimated), we derive the linear relation shown in equation 8:

$$\Delta H_f = 122.0n(\Delta) - 64.2 \qquad (n = 4,\ r^2 = 0.9996) \qquad (8)$$

This equation predicts a value for the 'missing' compound of 423.8 kJ mol^{-1}, some 6 kJ mol^{-1} different from that suggested in Reference 42, and fits the other, totally measured,

TABLE 1. Heats of formation of the homologous series cyclopropane (**2**), X = H; spiropentane (**7**), dispiro[2.0.2.1]heptane (**28**), trispiro[2.0.0.2.1.1]nonane (**29**) and tetraspiro[2.0.0.0.2.1.1.1]undecane (**30**) (in kJ mol^{-1})

$n(\Delta)$	Compound	ΔH_f(l)	Lit. ΔH_v	Est. ΔH_v[a]	ΔH_f(g)
1	**2**, X = H			17.0	53.3 ± 0.6[b]
2	**7**	157.7 ± 0.8[b]	27.5 ± 0.0[b]	23.0	185.2 ± 0.8[b]
3	**28**	267.8 ± 3.4[c]	35.0 ± 0.5	29.0	302.8 ± 3.5[c]
4	**29**	387.8 ± 6.8[c]	42.1[d]	35.0	429.9[e]
5	**30**	488.6 ± 6.8[c]	55.4 ± 0.4[c]	41.0	544.0 ± 6.8[c]

[a] Estimates based on use of the CHLL2 estimation approach.
[b] Values taken from our archives.
[c] Experimentally measured values of Reference 41.
[d] This value is an estimate from Reference 41, where it was equated with the experimentally measured value for its isomer trispiro[2.0.2.0.2.0]nonane (**31**) at its melting point of 303 K. (Temperature corrections of the real liquid to a theoretical liquid at 298 K should be small. By comparison, using the enthalpy of sublimation of the solid at 298 K would contain little useful thermochemical information about the compound of interest).
[e] This value is the sum of the experimentally measured enthalpy of formation of liquid compound (at 303 K) and measured enthalpy of vaporization. See footnote *d* above.

numbers to comparable accuracy, although we must admit some incredulity at the value for tetraspiro[2.0.0.0.2.1.1.1]undecane (**30**) because its enthalpy of vaporization is so much larger than expected[42]. This linearity of enthalpies of formation with number of rings is reminiscent of the linearity of enthalpies of formation of conjugated polyolefins with number of double bonds: ethylene, (52.5 ± 0.4); 1,3-butadiene (**5**), (110.0 ± 1.2); (*E*)-1,3.5-hexatriene (**6b**), (167.8 ± 2.5) kJ mol^{-1} [43].

We close this subsection with a brief discussion of the 'branched Y-shaped' trispiro[2.0.2.0.2.0]nonane (**31**). This species has an enthalpy of formation of (381.5 ± 4.5) (l), (423.6 ± 4.9) kJ mol^{-1} (g), some 6 kJ mol^{-1} lower than that of its 'linear N-shaped' isomer, trispiro[2.0.0.2.1.1]nonane (**29**). It is not obvious what comparison between the isomers, if any, is meaningful here, although we note that isobutane has an enthalpy of formation 9 kJ mol^{-1} less than the isomeric *n*-butane, but '1,1-divinylethylene' (**32**) has an enthalpy of formation some 43 kJ mol^{-1} more than either of the isomeric '1,2-divinylethylenes'[43], **6a** and **6b**. However, we close this subsection by returning to reaction 5 where we interrelate trispiro[2.0.2.0.2.0]nonane (**31**) with [3]-radialene (**33**). Since the enthalpy of reaction 5 is taken to be some (3 ± 2) kJ mol^{-1} per substituent on the olefin and there are three *gem*-substituted double bonds, we expect the trispiro compound to have a higher enthalpy of formation than the radialene by *ca* (18 ± 12) kJ mol^{-1}. The experimentally reported values agree within the experimental error [(423.6 ± 4.9) – (396 ± 12)] = (28 ± 13) kJ mol^{-1}. The two numbers overlap: the agreement is astonishing given the number of assumptions made in deriving the enthalpy of formation of the radialene[44].

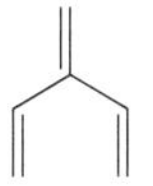

(**32**)

(**33**)

E. Conjugated Double Bonds and Cyclopropanes

The energetics consequences of the conjugation of double bonds has long been known and will not be reviewed here. What about cyclopropanation of conjugated olefins? The simplest processes we can consider are the conceptual sequential conversions of 1,3-butadiene to vinylcyclopropane to bicyclopropyl, **5** → **10** → **8**. Were there no conjugation energy at all in 1,3-butadiene, or if cyclopropanation did not change this stabilization, we would expect these reactions to be *ca* 3 kJ mol^{-1} endothermic since in both cases we are considering monosubstituted olefins and cyclopropanes. The first energy difference is (17.2 ± 1.7), and the second is (2.1 ± 3.8) kJ mol^{-1}. It would appear that vinylcyclopropane is considerably destabilized relative to butadiene, while bicyclopropyl shows no particular stabilization or destabilization relative to vinylcyclopropane: (2.1 ± 3.8) kJ mol^{-1} is consistent with our earlier value of (3 ± 2) kJ mol^{-1} for monosubstituted cyclopropanes. Said differently, vinylcyclopropane and bicyclopropyl lack the conjugative stabilization of butadiene. Is it possible that there is some experimental error? Most assuredly, yes: there always is that possibility[45].

Let us now recast the analysis of the simplest polyene series: ethylene, 1,3-butadiene and (*E*)-1,3,5-hexatriene. The enthalpies of formation increase in the order: (52.5 ± 0.4), (110.0 ± 1.1) and (167.8 ± 2.5) kJ mol^{-1} (see Reference 38 for the last value). The change of enthalpies of formation is essentially constant, (56.5 ± 1.2) and (57.8 + 2.7) kJ mol^{-1}, respectively.

New insights may be gained by saying the above in a different way. Consider the sequential 'oligomerization/dehydrogenation' reactions in equations 9a and 9b[46]:

$$CH_2{=}CH_2 + CH_2{=}CH_2 \longrightarrow \underset{\mathbf{(5)}}{CH_2{=}CH{-}CH{=}CH_2} + H_2 \tag{9a}$$

$$\underset{}{CH_2{=}CH_2} + \underset{\mathbf{(5)}}{CH_2{=}CH{-}CH{=}CH_2} \longrightarrow \underset{\mathbf{(6b)}}{CH_2{=}CH{-}CH{=}CH{-}CH{=}CH_2} + H_2 \tag{9b}$$

These reactions are both endothermic by a nearly constant 5 kJ mol^{-1}, (5.0 ± 1.3) and (5.3 ± 2.8), respectively. By contrast, the sequential 'oligomerization/dehydrogenation' reactions of saturated alkanes (equations 10a and 10b)

$$Me{-}CH_3 + CH_3{-}Me \longrightarrow Me{-}CH_2CH_2{-}Me + H_2 \tag{10a}$$

$$Me{-}CH_3 + CH_3{-}CH_2CH_2Me \longrightarrow Me{-}CH_2CH_2CH_2CH_2{-}Me + H_2 \tag{10b}$$

are endothermic by a nearly constant 42 kJ mol^{-1}, (42.0 ± 0.9) and (42.3 ± 1.3). That reactions 9a and 9b are so much less endothermic than the corresponding reactions 10a and 10b is consonant with butadiene and hexatriene having significant conjugative stabilization.

We now discuss the related series wherein there are three double bonds or cyclopropanes. How does homoconjugation involving cyclopropanes compare to conjugation involving olefins? Paralleling the nearly constant 57 kJ mol^{-1} difference between the enthalpies of formation of ethylene (**1**, X = H) and butadiene (**5**), and butadiene (**5**) and (*E*)-hexatriene (**6b**), the first related difference of cyclopropane (**2**, X = H) and bicyclopropyl (**8**) is (75.8 ± 3.6) kJ mol^{-1}. We lack the enthalpy of formation data of either *cis*- or *trans*-1,1′;2′,1″-tercyclopropyl (**9a** or **9b**) and so the second related difference cannot be found directly from experimental measurements. However, the enthalpies of formation of their 1′-methyl derivatives have been reported[47]: **34a** ≡ *cis*(l), (134.1 ± 1.2), *cis*(g), 181.0; *trans*(l), (131.9 ± 1.2), **34b** ≡ *trans*(g), 179.2 kJ mol^{-1}. The first approximation of the enthalpy of formation assumes the change of enthalpies of formation upon demethylation to form either *cis*- or *trans*-1,1′;2′,1″-tercyclopropyl, i.e. **34** → **9**, is the same as that of 1,1-dimethylcyclopropane on forming methylcyclopropane (**20a** → **2**, X = Me). So doing, we derive an enthalpy of formation of either tercyclopropyl[48] to be *ca* 212 kJ mol^{-1}.

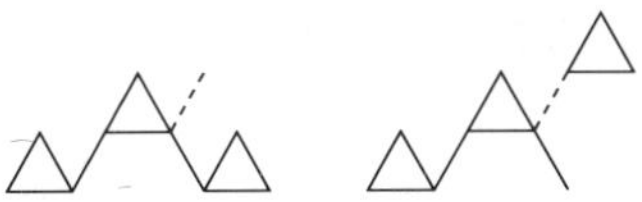

(34a) **(34b)**

The second approximation refines the first by noting there is *cis*-destabilization of the central cyclopropane in either methyltercyclopropyl, either by the two attached cyclopropyls or by the methyl and one of the attached cyclopropyls. Approximating the destabilization by the enthalpy of formation difference of *cis*- and *trans*-1,2-dimethylcyclopropane or of *cis*- and *trans*-1,2-diethylcyclopropane (**20b** vs **20c**, with X = Me or Et) gives us an estimated enthalpy of formation of either tercyclopropyl of *ca* 208 kJ mol^{-1}. The difference of enthalpies of formation of 'tercyclopropyl' (**9**) and bicyclopropyl (**8**) is some 79 kJ mol^{-1}, within the error bars of that earlier given for bicyclopropyl (**8**) and cyclopropane[49] (**2**, X = H).

Let us turn now to the oligomerization/dehydrogenation reactions (equations 11a and 11b):

$$c\text{-}(CH_2)_2CH_2 + c\text{-}CH_2(CH_2)_2 \longrightarrow c\text{-}(CH_2)_2CH{-}CH(CH_2)_2\text{-}c + H_2 \qquad (11a)$$

(8)

and

$$c\text{-}(CH_2)_2CH{-}CH(CH_2)_2\text{-}c + c\text{-}CH_2(CH_2)_2 \longrightarrow$$

(8)

$$c\text{-}(CH_2)_2CH{-}CH(CH_2)CH{-}CH(CH_2)_2\text{-}c + H_2 \qquad (11b)$$

(9)

These reactions are endothermic by (22.3 ± 3.7) and *ca* 26 kJ mol^{-1}, respectively, and seemingly interpolate the energies of the olefinic and acyclic reactions (equations 8a and 8b, and 9a and 9b). But, there is a perhaps better acyclic reaction for comparing cyclopropanes, one involving secondary carbons. The more appropriate corresponding acyclic reaction (equation 12)

$$Me_2CH_2 + CH_2Me_2 \longrightarrow Me_2CH{-}CHMe_2 + H_2 \qquad (12)$$

has the very similar endothermicity, namely (21.1 ± 1.2) kJ mol^{-1}. This documents that there is very little net ground state stabilization arising from the interaction between cyclopropane rings[49].

F. The Conjugation of Double Bonds with Cyclopropanes

We have concluded earlier in this section that the delocalization energy of two double bonds to form a conjugated diene was greater than that of two cyclopropanes to form bicyclopropyl. What is the delocalization energy associated with the interaction of a double bond with a cyclopropane? There are many probes of this interaction[50]. The first probe of this interaction we discuss contrasts the energetics of the oligomerization/dehydrogenation reactions (equations 13a and 13b),

$$c\text{-}(CH_2)_2CH_2 + CH_2{=}CH_2 \longrightarrow c\text{-}(CH_2)_2CH{-}CH{=}CH_2 + H_2 \qquad (13a)$$

(**1**, X = H) (**10**)

$$Me_2CH_2 + CH_2{=}CH_2 \longrightarrow Me_2CH{-}CH{=}CH_2 + H_2 \qquad (13b)$$

The first reaction of the above pair is endothermic by (21.4 ± 1.5) kJ mol^{-1} and the second by (24.6 ± 1.0) kJ mol^{-1}. This suggests that vinylcyclopropane is stabilized relative to 2-vinylpropane (3-methyl-1-butene) by 24.6 – 21.4, or (3.2 ± 1.8) kJ mol^{-1}. Alternatively, one can contrast the hydrogenation enthalpies of vinylcyclopropane and 2-vinylpropane (equations 14a and l4b),

$$c\text{-}(CH_2)_2CH{-}CH{=}CH_2 + H_2 \longrightarrow c\text{-}(CH_2)_2CH{-}Et \qquad (14a)$$

(**10**)

$$Me_2CH{-}CH{=}CH_2 + H_2 \longrightarrow Me_2CH{-}Et \qquad (14b)$$

The two reactions are exothermic by [123.8 (± >1.5)] kJ mol^{-1} and (126.1 ± 1.3) kJ mol^{-1}, respectively. From this, we conclude that vinylcyclopropane is stabilized relative to 2-vinylpropane by 126.1 – 123.8 = 2.3 (± >1.3) kJ mol^{-1}. Finally, we consider reactions 15a and 15b, our last pair, which are again hydrogenations:

$$c\text{-}(CH_2)_2CH{-}CH{=}CH_2 + H_2 \longrightarrow Me_2CH{-}CH{=}CH_2 \qquad (15a)$$

(**10**)

$$c\text{-}(CH_2)_2CH{-}Et + H_2 \longrightarrow Me_2CH{-}Et \qquad (15b)$$

(**2**, X = Et)

Reaction 15a is exothermic by (154.8 ± 1.5) kJ mol^{-1}, and reaction 15b is exothermic by [157.1 (± >1.0)] kJ mol^{-1}. From this, we conclude that vinylcyclopropane is stabilized relative to ethylcyclopropane by 157.1 – 154.8 = [2.3 (± >1.8)] kJ mol^{-1}. In all three cases, we conclude that there is stabilization associated with vinylcyclopropane of some [3 (± >2)] kJ mol^{-1}. Even with optimistic use of error bars (i.e. with the + sign), this stabilization value seems rather small. Yet, an altogether different documentation of the smallness of the double-bond cyclopropane conjugation is shown by the less than 1 kJ mol^{-1} difference in the enthalpies of formation of the isomeric bicyclo[4.1.0]hept-2-ene (**35a**) and bicyclo[4.1.0]hept-3-ene (**35b**)[51].

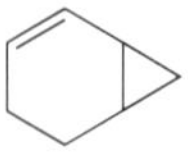

(**35a**)

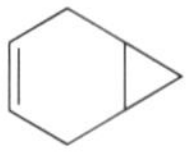

(**35b**)

G. Acetylenes, Diacetylenes and Cyclopropanes

The thermochemistry of acetylenes is surprisingly sparse[52]. Yet it is rich enough to allow some interesting comparisons. Paralleling the 'oligomerization/dehydrogenation' reaction of olefins and alkanes (equations 9 and 10), we now suggest equation 16:

$$X^1{-}C{\equiv}CH + HC{\equiv}C{-}X^2 \longrightarrow X^1{-}C{\equiv}C{-}C{\equiv}C{-}X^2 + H_2 \qquad (16)$$

(36)

For $X^1 = X^2 =$ Me[53], *n*-Bu[54], Cypr[55] and MeC(=CH_2)[56], reactions 16 are, respectively, [7.6 (± >5.1)] kJ mol^{-1} endothermic, [6 (± >2.2)] endothermic, (22.2 ± 2.5) exothermic and 23 kJ mol^{-1} exothermic. It appears that only in the cases of extended conjugation or homoconjugation upon 'oligomerization/dehydrogenation', i.e. with the 'chromophores' (or formally conjugating groups) cyclopropyl or isopropenyl, is reaction 16 exothermic. That is, going from 2 'chromophores' + 2 chromophores → 4 chromophores results in *ca* 30 kJ mol^{-1} stabilization more than is found for 1 chromophore + 1 chromophore → 2 chromophores[57]. This is surprising because the 'mixed' reaction with X^1 = Cypr and X^2 = Me, corresponding to 2 chromophores + 1 chromophore → 3 chromophores, is endothermic by [7.8 (± >8.2)] kJ mol^{-1}. It would appear that the 'not-as' extended (3 chromophore) conjugation has not particularly provided stabilization, since nearly the same endothermicity was found when X^1 and X^2 were both alkyl[58].

H. Acetylenes and Cyclopropenes

Acetylenes and cyclopropenes (**37**) are related to each other in the same formal way as olefins and cyclopropanes are[59]. Recall that cyclopropanation of ethylene is almost slightly endothermic and the endothermicity was asserted to increase by some (3 ± 2) kJ mol^{-1} per alkyl group. Cyclopropanation of acetylene to form cyclopropene[32] (**37**, $X^1 = X^2 =$ H) is endothermic by (48.9 ± 2.6) kJ mol^{-1}. Cyclopropanation of propyne (monomethylacetylene) to form l-methylcyclopropene (**37**, X^1 = Me, X^2 = H) has an increased endothermicity of (58.7 ± 1.4) kJ mol^{-1}. By contrast, the cyclopropanation of 2-butyne (dimethylacetylene) using a derived value for the enthalpy of formation of 1,2-dimethylcyclopropene (**37**, $X^1 = X^2$ = Me) has an accompanying endothermicity of only 41 kJ mol^{-1}. We suspect that the last value is in error[60] and so suggest remeasurement of the enthalpy of formation of dimethylcyclopropene as well as measuring the enthalpy of formation of other cyclopropenes[61].

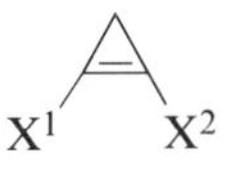

(37)

I. A Brief Interlude on Homoaromaticity and Homoantiaromaticity

If conjugation contributes more stabilization energy for two double bonds than homoconjugation for an olefin and cyclopropane, then it should not surprise anyone that aromaticity provides more stabilization than homoaromaticity[62]. A simple comparison is between the *ca* 9 kJ mol^{-1} increase in enthalpy of formation on going from (*Z*)-2-butene to *cis*-1,2-dimethylcyclopropane and the [202 – (82.6 ± 1.7)] ≈ 119 kJ mol^{-1} on going from

1,3,5-cyclohexatriene (**38**) to bicyclo[4.1.0]hepta-2,4-diene (**39**) (i.e. from benzene to norcaradiene). Likewise, it is reasonable that homoantiaromaticity is less destabilizing than antiaromaticity, even if there are no experimentally measured enthalpies of formation of cyclobutadiene (**40**) (or derived from a measurement on a substituted derivative) to compare with bicyclo[2.1.0]pent-2-ene[63] (**41**): use of high-accuracy computational theory to derive an enthalpy of formation of cyclobutadiene gives an analogous *decrease* of (79 ± 20) kJ mol^{-1} on cyclopropanation[64].

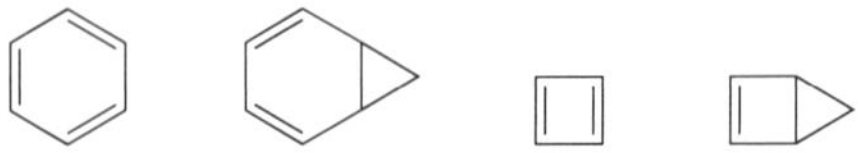

(38) **(39)** **(40)** **(41)**

Alternatively, we can deduce some homoaromatic stabilization of bicyclo[4.1.0]hepta-2,4-diene and homoantiaromatic destabilization of bicyclo[2.1.0]pent-2-ene from enthalpies associated with hydrogenation to form the saturated bicycloalkanes. The bicycloheptadiene hydrogenation enthalpy (i.e. **39** → **18**, $n = 6$) is *ca* 200 kJ mol^{-1}, to be compared with (229.6 ± 1.3) kJ mol^{-1} for the hydrogenation of the corresponding monocyclic 1,3-cyclohexadiene to cyclohexane. The difference of these numbers, some 30 kJ mol^{-1}, may be taken as a definition of the homoaromaticity of bicyclo[4.1.0]hepta-2,4-diene: it is a significant amount, even if it is far smaller than the (87.3 ± 1.8) kJ mol^{-1} aromaticity of benzene as deduced from its corresponding hydrogenation enthalpy to form cyclohexene. The bicyclopentene hydrogenation enthalpy (i.e. **41** → **18**, $n = 4$) is 175 kJ mol^{-1} while the corresponding monocyclic cyclobutene hydrogenation enthalpy is but (128.3 ± 1.6) kJ mol^{-1}. The difference between these numbers, some 57 kJ mol^{-1}, may be taken as a definition of the homoantiaromaticity of bicyclo[2.1.0]pent-2-ene: it is a significant amount, even if it is far smaller than the 166 kJ mol^{-1} for the antiaromaticity of cyclobutadiene as deduced from its corresponding hydrogenation to form cyclobutene[64]. We remind the reader that the degree of aromaticity, antiaromaticity, homoaromaticity and homoantiaromaticity are model-dependent. We are thus not surprised nor disappointed with a literature value of 24.2 kJ mol^{-1} for the homoaromatic stabilization of bicyclo[4.1.0]hepta-2,4-diene[65]. Relatedly, a homoantiaromatic destabilization of *only* 41.4 kJ mol^{-1} for bicyclo[2.1.0]pent-2-ene and homoaromatic stabilization and values of *ca* 14 kJ mol^{-1} have been offerred for the 'propano' and 'benzo' forms of bicyclo[4.1.0]hepta-2,4-diene, i.e. [4.3.1]propella-2,4-diene and 1*a*,7*b*-dihydronaphtho[*a*]cyclopropene[66], species **42** and **43**, respectively.

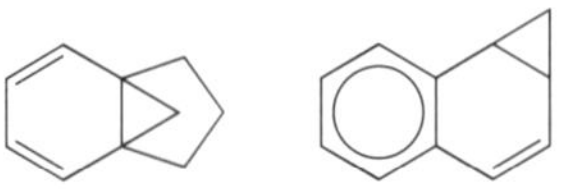

(42) **(43)**

J. Dicarbon and Its Cyclopropanated Analogs

We close this section relating double bonds and cyclopropanes with dicarbon and its formal cyclopropanated analogs: cyclopropyne (**44**), $\Delta^{1,3}$-bicyclo[1.1.0]butene (**45**) and [1.1.1]propellane (**46**). That is, we consider the series of compounds C_2, $C_2(CH_2)$, $C_2(CH_2)_2$ and $C_2(CH_2)_3$. Admittedly, C_2 is not a hydrocarbon and somehow does not really belong in the current series, but analysis making use of its inclusion in its reasoning is conceptually no worse than

(44)

(45)

(46)

(a) the use of graphite to understand polynuclear aromatic hydrocarbons,
(b) the use of diamond to understand alicyclic hydrocarbons,
(c) the use of cycloethane to understand cycloalkanes

and

(d) the use of methane (and *a fortiori* H_2) to understand *n*-alkanes.

From the inorganic compendium[9] we find an enthalpy of formation of C_2 of 831.9 kJ mol^{-1}. Cyclopropyne remains but a chemist's chimera, but only a short time ago, so were $\Delta^{1,3}$-bicyclo[1.1.0]butene and [1.1.1]propellane. From the classical (albeit gas-phase) combustion calorimetric measurement of the enthalpy of formation of bicyclobutane and modern gas-phase ion measurements, the gas-phase enthalpy of formation of $\Delta^{1,3}$-bicyclo-[1.1.0]butene has been determined[67] to be (544 ± 42) kJ mol^{-1}. From reaction calorimetry and some judicious estimates, the enthalpy of formation of gaseous [1.1.1]propellane has been measured[68] and then chronicled[3] among other, mostly 'normal' 3MR species, as (351 ± 4) kJ mol^{-1}. Yet, before we get too complacent, it is clear that anything even approximating a constant cyclopropanation energy is not found in the series of current interest: dicarbon[69] and its formal cyclopropanated analogs **44**–**46**.

III. PHENYL VERSUS CYCLOPROPYL

A. Alkyl Substituted Derivatives

As noted above, it is long-established that cyclopropane has olefinic character. If this cyclopropane–ethylene analogy were strictly thermochemically correct, then the difference of the enthalpies of formation of vinyl-X and cyclopropyl-X would not depend on the affixed group X. Equivalently, the enthalpy of formation difference quantity δ_{17} (Vi, Cypr, X) would be independent of X (equation 17).

$$\begin{aligned}\delta_{17}(\text{Vi, Cypr; X}) &\equiv \Delta H_f(\text{g, ViX}) - \Delta H_f(\text{g, CyprX})\\ &= \Delta H_f(\mathbf{1}) - \Delta H_f(\mathbf{2})\end{aligned} \quad (17)$$

More recently, the similarity of phenyl-X and vinyl-X was noted[19], and were this analogy strictly thermochemically correct as well, then the enthalpy of formation difference quantity δ_{18}(Ph, Vi; X) would also be independent of X as well (equation 18).

$$\begin{aligned}\delta_{18}(\text{Ph, Vi; X}) &\equiv \Delta H_f(\text{g, PhX}) - \Delta H_f(\text{g, ViX})\\ &= \Delta H_f(\mathbf{11}) - \Delta H_f(\mathbf{1})\end{aligned} \quad (18)$$

Were these difference quantities δ_{17}(Vi, Cypr; X) and δ_{18}(Ph, Vi; X) independent of X, so would δ_{19}(Ph, Cypr; X) be (equations 19 and 20) as this last difference quantity is 'merely' the numerical sum of the first two.

$$\begin{aligned}\delta_{19}(\text{Ph, Cypr; X}) &= \Delta H_f(\text{g, PhX}) - \Delta H_f(\text{g, CyprX})\\ &= \Delta H_f(\mathbf{11}) - \Delta H_f(\mathbf{2})\end{aligned} \quad (19)$$

$$\delta_{19}(\text{Ph, Cypr; X}) = \delta_{17}(\text{Vi, Cypr; X}) + \delta_{18}(\text{Ph, Vi; X}) \quad (20)$$

We now explicitly consider several groups X for discussion. It is important to emphasize that this thermochemical comparison of benzene and cyclopropane derivatives is not just a check of two purported identities in terms of a third, arithmetically derived, identity. We recognize benzene as the archetypical π-delocalized aromatic species. It is from benzene that we calibrate our understanding of other aromatic species, whether they be multiring polynuclear hydrocarbons such as naphthalene and coronene, 'microring' aromatics and antiaromatics such as cyclopropenyl cation and anion, respectively, or π-poor or π-rich aromatics such as pyridine and pyrrole, respectively. Cyclopropane is the paradigm of σ-aromatic species from which our understanding of this phenomenon evolves[20]. Benzene and cyclopropane are thus naturally paired as conceptually archetypical models for electronic delocalization.

We start with their simplest systems and we first consider the parent hydrocarbons benzene and cyclopropane themselves. This enthalpy of formation difference δ_{19}(Ph, cypr; H) is (29.3 ± 0.9) kJ mol^{-1}. Turning now to the simple alkyl derivatives, and **1** and **11**, X = Me in particular, we recall that there are no direct experimental data for the enthalpy of formation of gaseous methylcyclopropane but only for the corresponding liquid. Accepting the archival enthalpy of vaporization value from Kolesov and Kozina[14] is equivalent to accepting their value for the enthalpy of formation of methylcyclopropane,.namely 24.3 kJ mol^{-1}. We so deduce a value for δ_{19}(Ph, Cypr; Me) of [26.1 (± >0.9)] kJ mol^{-1}, some 3 kJ mol^{-1} different fom that for the parent hydrocarbons. We are not particularly bothered by this 3 kJ mol^{-1} discrepancy—we recall in footnote 23 a 6 kJ mol^{-1} spread of values suggested for the enthalpy of vaporization of methylcyclopropane. It is conceptually simplest, and procedurally most precise, to use the identical approach to compare ethylbenzene and ethylcyclopropane, for there are no enthalpy of vaporization measurements for the latter 3MR species. Encouragingly, consistency of results is obtained[70]—the value of δ_{19}(Ph, Cypr; Et) equal to 26.5 kJ mol^{-1} is nearly identical to that of δ_{19}(Ph, Cypr; Me).

It is not obvious what to expect when the substituent is a branched alkyl group. After all, steric interactions are expected to be larger with these substituents than for the earlier unbranched and fewer carbon alkyl substituents. Of course, to make the comparison, enthalpies of formation must be available for both the benzene and cyclopropane derivatives. The sole thermochemical representatives of these branched alkyl Xs which qualify are *i*-Pr and *s*-Bu, and here enthalpy of formation data has been reported only for the liquid phase, –46 and –75 kJ mol^{-1}, respectively. Now what? One option is to estimate the desired enthalpies of vaporization for isopropylcyclopropane and *sec*-butylcyclopropane. Kolesov and Kozina's values[14] for the enthalpies of vaporization of methyl and ethylcyclopropane are within 2 kJ mol^{-1} of that derived using the CHLL2 rule. Using this latter approach in the absence of values from the former, we estimate the enthalpies of vaporization of gaseous isopropyl- and *sec*-butylcyclopropane to be 31 and 36 kJ mol^{-1}, respectively. The enthalpies of formation of isopropyl and *sec*-butylcyclopropane are thus deduced to be –15 and –39 kJ mol^{-1}, respectively. From these numbers and the archivally recorded, experimentally measured enthalpies of formation of gaseous isopropyl and *sec*-butyl benzene, we deduce δ_{19}(Ph, Cypr; *i*-Pr) and δ_{19}(Ph, Cypr; *s*-Bu) equal to 19 and 22 kJ mol^{-1}.

Recall the earlier differences for PhX and CyprX of 26 kJ mol^{-1} for X = Me and Et[71], respectively. Let us introduce the second difference quantity $\delta\delta_{21}$(Ph, Cypr; X^1, X^2) in equation 21:

$$\delta\delta_{21}(\mathrm{Ph, Cypr}; X^1, X^2) = \delta_{19}(\mathrm{Ph, Cypr}; X^1) - \delta_{19}(\mathrm{Ph, Cypr}; X^2) \qquad (21)$$

We admit some surprise, if not distress, that $\delta\delta_{21}$(Ph, Cypr; Me or Et, *i*-Pr or *sec*-Bu) is as large as *ca* 6 kJ mol^{-1}.

One can alternatively derive enthalpy of formation difference values explicitly for the liquid phase (equation 22):

$$\delta\delta_{22}(\mathrm{l; Ph, Cypr; X}) = \Delta H_f(\mathrm{l, PhX}) - \Delta H_f(\mathrm{l, CyprX}) \qquad (22)$$

For X = Me, Et, *i*-Pr and *s*-Bu, these values are –10.7, –12.5, –5 and –9 kJ mol^{-1}, respectively. We can now make corrections for the difference in enthalpies of vaporization of the corresponding phenyl and cyclopropyl species. A particularly simple approach uses the CHLL2 protocol, but now 'differentiated' to result in equation 23:

$$\delta\Delta H_v = 4.7\delta\tilde{n}_C + 1.3\delta n_Q \quad (23)$$

Here $\delta\tilde{n}_C$ and δn_Q are the difference of the numbers of nonquaternary and quaternary carbons, respectively, for the two compounds of interest. In the current case of substituted benzenes and cyclopropanes, $\delta\tilde{n}_C$ and δn_Q are respectively 3 and 0, and so the differential enthalpy of vaporization $\delta\Delta H_v$ equals *ca* 14 kJ mol^{-1}. For X = Me, Et, *i*-Pr and *s*-Bu, the newly generated values of δ_{19}(Ph, Cypr; X) are some 25, 26, 19 and 23 kJ mol^{-1}, respectively.

But are the values of 25 and 26 kJ mol^{-1} for the Me and Et cases so disparate from the 19 and 23 kJ mol^{-1} values for *i*-Pr and *s*-Bu? It may be argued that the isopropyl value is in error because the other values are nearly the same. It may also be argued that the *sec*-butyl value is in error because the steric repulsion of a branched alkyl group with phenyl is less than with cyclopropyl. It is clear that the experimental data and/or our understanding of alkylcyclopropanes is somehow defective.

B. Unsaturated 'Hydrocarbyl' Derivatives

Let us now turn to hydrocarbyl, but unsaturated, derivatives of cyclopropane and compare their enthalpies of formation with the corresponding derivatives of benzene. The archetypical comparison is that of X = vinyl, i.e. we will now contrast vinylcyclopropane and styrene. Using the choice of enthalpy of formation of the former as made in Reference 42, we find δ_{19}(Ph, Cypr; CH=CH_2) equals (20.7 ± 2.2) kJ mol^{-1} suggesting that phenyl and vinyl conjugate better with each other than do cyclopropyl and vinyl. This is corroborated by comparison with the difference of the enthalpy of formation of (*E*)-*β*-methylated styrene[72] with the archival value for 1-propenylcyclopropane: δ_{19}(Ph, Cypr; —CH=CHMe) = (17.3 ± 2.1) kJ mol^{-1}. By contrast, the difference for *α*-methylstyrene and isopropenylcyclopropane is increased to 25.7 kJ mol^{-1}, significantly higher than that found for its linear isomer and the unmethylated vinyl case, i.e. δ_{19}(Ph Cypr; —C(Me)=CH_2) is significantly larger than δ_{19}(Ph, Cypr; —CH=CHMe) and δ_{19}(Ph, Cypr; —CH=CH_2). We do not adequately know how to disentangle steric and electronic effects as to explain the isomeric differences or *α*-methylation effects. It may be telling that it was the same research group that reported the enthalpies of formation of isopropylcyclopropane and isopropenylcyclopropane. It is interesting to note that the small and thus sterically innoccous, but rather electron-withdrawing, ethynyl group results in an enthalpy of formation difference δ_{19}(Ph, Cypr; —C≡CH) of only (8.6 ± 4.3) kJ mol^{-1} [73], while with the sterically larger cyclopropyl and phenyl groups, the corresponding differences of δ_{19}(Ph, Cypr; Cypr) and δ_{19}(Ph, Cypr; Ph) are (20.8 ± 3.7) and (30.9 ± 2.3) kJ mol^{-1}, respectively. No explanation is apparent for $\delta\delta_{21}$(Ph, Cypr; CyprX, Ph) thus equaling *ca* 10 kJ mol^{-1}.

C. Hetero-atom Containing Substituted Derivatives

If the thermochemical consequences of substituents that contain only carbon and hydrogen are a source of confusion, those of hetero-atom containing ones are even more so. For the *σ*- and *π*-withdrawing X = CN, the value of δ_{19}(Ph, Cypr; X) is increased to (33.9 ± 2.3) kJ mol^{-1} and for the likewise *σ*- and *π*-withdrawing X = COMe and COOMe[74], the differences are (28.6 ± 1.9) and (16.8 ± 2.8) kJ mol^{-1}, respectively. For the *σ*-withdrawing but *π*-donating NH_2, the difference is decreased to (10.1 ± 1.2) kJ mol^{-1}. We close this discussion of the comparison of phenyl and cyclopropyl derivatives with a mention of the bromides. While experimental data on the enthalpy of formation of cyclopropyl bromide is limited to the liquid, (27.6 ± 4.4) kJ mol^{-1}, we derive an enthalpy of formation for the gas of 59 kJ mol^{-1} using the CHLP vaporization protocol. This gives us a value of some [46 (> ± 6)] kJ mol^{-1} for δ_{19}(Ph, Cypr; Br). Alternatively, we may use the liquid-phase value for cyclopropyl bromide and make use of a simplified version of the 'differentiated' CHLP equation (equation 24)[75].

$$\delta\Delta H_v = 4.7\delta\tilde{n}_C + 1.3\delta n_Q \quad (24)$$

We thus deduce a value of [47 (> ± 6)] kJ mol^{-1} for δ_{19}(Ph, Cypr; Br). How do we explain the diverse values of δ_{19}(Ph, Cypr; X) for the various X groups[76]?

D. Multiply Substituted Derivatives

Having documented that we are perplexed by the differences we found for the enthalpies of formation of simple cyclopropyl and phenyl derivatives, it is premature and perhaps foolish to go from singly to doubly substituted species and expect any better understanding. Steric repulsion between vicinal groups on a cyclopropane may be expected, and hence the *trans*-isomer of 1,2-dimethylcyclopropane is expected to be more stable than the *cis*. This is correct: the enthalpies of formation of *trans*-1,2- and *cis*-1,2-dimethylcyclopropane (**20c** and **20b**, X = Me) respectively, are (–30.7 ± 0.8) and (–26.3 ± 0.6) kJ mol^{-1} for the liquids, and –3.8 and 1.7 kJ mol^{-1} for the gases. The 1,1-dimethyl isomer (**20a**, X = Me) is more stable yet: (–33.3 ± 0.7) and (–8.2 ± 1.2) kJ mol^{-1} for the liquid and gas, respectively. What comparisons with benzene derivatives can we make[77]? With the 1,1-disubstituted cyclopropane isomer, seemingly none. With the *cis*-1,2-isomer, let's try the *o*-disubstituted benzene, *o*-xylene (**47a**, X = Me). With the *trans*-1,2-isomer, let's try the *p*-xylene isomer (**47b**,

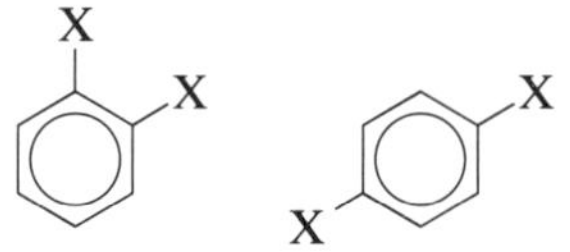

(47a) **(47b)**

X = Me). After all, the groups are further away—and, the *o*- and *p*-positions on a benzene ring are generally more related to each other than either is to the *m*-position[78]. For the diethyl case, there are no enthalpy of formation data for the 1,1-isomer (**20a**, X = Et) but the *trans*-1,2- and *cis*-1,2–species (**20c** and **20b**, X = Et, respectively) have gas-phase enthalpies of formation of (–49.0 ± 2.3) and (–44.5 ± 1.5) kJ mol^{-1}. The enthalpy difference for these two gas-phase isomers, (–4.5 ± 2.7) kJ mol^{-1}, is comparable to that found in the liquid phase, (–3.4 ± 1.8) kJ mol^{-1} and so suggests the same information content is found using either phase. More precisely, by analogy to equation 19, let us now define the δ_{25}(Phv^o, Cyprv^c; X) and δ_{26}(Phv^p, Cyprv^t; X) terms[79] in equations 25 and 26. The new symbols are explained in footnote 79.

$$\delta_{25}(\text{Phv}^o, \text{Cyprv}^c; \text{X}) = \Delta H_f(\text{g}, o\text{-C}_6\text{H}_4\text{X}_2) - \Delta H_f(\text{g}, cis\text{-1,2-}c\text{-C}_3\text{H}_4\text{X}_2) \quad (25)$$

$$\delta_{26}(\text{Phv}^p, \text{Cyprv}^t; \text{X}) = \Delta H_f(\text{g}, p\text{-C}_6\text{H}_4\text{X}_2) - \Delta H_f(\text{g}, trans\text{-1,2-}c\text{-C}_3\text{H}_4\text{X}_2) \quad (26)$$

By analogy to equation 19 we have equations 27 and 28:

$$\delta_{27}(\text{l}; \text{Phv}^o, \text{Cyprv}^c; \text{X}) = \Delta H_f(\text{l}, o\text{-C}_6\text{H}_4\text{X}_2) - \Delta H_f(\text{l}, cis\text{-1,2-}c\text{-C}_3\text{H}_4\text{X}_2) \quad (27)$$

$$\delta_{28}(\text{l}; \text{Phv}^p, \text{Cyprv}^t; \text{X}) = \Delta H_f(\text{l}, p\text{-C}_6\text{H}_4\text{X}_2) - \Delta H_f(\text{l}, trans\text{-1,2-}c\text{-C}_3\text{H}_4\text{X}_2) \quad (28)$$

Indeed, we can make use of either liquid- or gaseous-phase enthalpies of formation despite our prejudice for the latter because we may invoke a 'constant' 14 kJ mol^{-1} $\delta\Delta H_v$ differential vaporization enthalpy correction to the former[80]. That is, should some enthalpy of vaporization be unavailable from experiment, we invoke the simple, approximate, equations 29 and 30:

$$\delta_{29}(\text{Phv}^o, \text{Cyprv}^c; \text{X}) = \delta_{27}(\text{l}; \text{Phv}^o, \text{Cyprv}^c; \text{X}) + 14 \quad (29)$$

$$\delta_{30}(\text{Phv}^p, \text{Cyprv}^t; \text{X}) = \delta_{28}(\text{l}; \text{Phv}^p, \text{Cyprv}^t; \text{X}) + 14 \quad (30)$$

From archival sources we deduce $\delta_{29}(\text{Phv}^o, \text{Cyprv}^c; \text{Me}) = 17.4$ kJ mol^{-1} and $\delta_{30}(\text{Phv}^p, \text{Cyprv}^t; \text{Me}) = 21.8$ kJ mol^{-1}, while the estimated $\delta_{29}(\text{Phv}^o, \text{Cyprv}^c; \text{Me})$ and $\delta_{29}(\text{Phv}^p, \text{Cyprv}^t; \text{X})$ are 16 and 20 kJ mol^{-1}, respectively. In that we lack measured enthalpy of vaporization data for any diethylbenzene, we cannot derive either $\delta_{29}(\text{Phv}^o, \text{Cyprv}^c; \text{Et})$ or $\delta_{30}(\text{Phv}^p, \text{Cyprv}^t; \text{Et})$ from the use of experimentally measured quantities alone. Both $\delta_{29}(\text{Phv}^o, \text{Cyprv}^c; \text{Et})$ and $\delta_{30}(\text{Phv}^p, \text{Cyprv}^t; \text{Et})$ are easy to calculate, and both shown to equal 27 kJ mol^{-1}. These alkyl values of 16, 20, 25 and 25 kJ mol^{-1} are consistent with what was found in the comparison of the corresponding monosubstituted cyclopropanes and benzenes[81].

We turn now to the comparison of *solid cis*-cyclopropane-1,2-dicarboxylic acid[82] (**20b**, X = COOH) and its benzene relative, phthalic acid (**47a**, X = COOH). The difference of enthalpies of formation is 17 kJ mol^{-1}, a number that needs to be corrected by enthalpies of sublimation to the desired gas-phase difference. Neither datum is available from experiment[83]. If we equate differential enthalpies of sublimation and of vaporization—a highly suspect approach, but what can we use that is better[84]—we deduce $\delta_{29}(\text{Phv}^o, \text{Cyprv}^c; \text{COOH})$ is *ca* 31 kJ mol^{-1}. Recall that the value for the rather much related $\delta_{19}(\text{Ph, Cypr; COOMe})$ was *ca* 20 kJ mol^{-1}. Are the results consonant? Given all of our assumptions, we are not *that* wrong.

We close this section with a brief discussion of the sole example of thermochemically comparable hexasubstituted species, hexafluorobenzene (**48**, X = F) and hexafluorocyclopropane (**49**, X = F). Complications of vicinal and geminal fluorine substitution are inescapable[85] although we are spared ambiguities of substitution pattern and of phase. The difference of gas-phase enthalpies of formation is 25 kJ mol^{-1}: we remember the 29 kJ mol^{-1} for the parent, unsubstituted, hydrocarbons . This agreement is *too* good to be true[86].

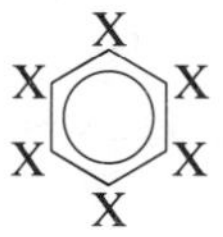

(48)

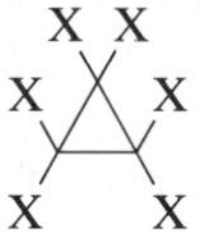

(49)

E. 'Strings' of Cyclopropanes and Benzenes

Earlier in this chapter we talked about the enthalpies of formation of singly and then multiply substituted cyclopropanes and the correspondingly substituted benzenes. It is now time to talk about multiple cyclopropanes and the corresponding benzene derivatives. In Section II.D, the enthalpies of formation of the isomeric (*cis*)- and (*trans*)-1,1′;2′,1″-tercyclopropyl (**9a** and **9b**) were estimated to be 208 kJ mol^{-1}. It was also asserted that the difference of this value and the enthalpy of formation of bicyclopropyl (**8**) was nearly identical to the difference of the enthalpies of formation of bicyclopropyl and cyclopropane. Let us now rewrite the oligomerization/dehydrogenation reactions 11a and 11b as reactions 31a and 31b:

$$2\text{CyprH} \longrightarrow \text{Cypr—Cypr} + H_2 \quad (31a)$$

$$\text{Cypr—Cypr} + \text{CyprH} \longrightarrow \text{Cypr—Cyprv}^{t}\text{—Cypr} + H_2 \quad (31b)$$

and note they are endothermic by (22.3 ± 3) and 25 kJ mol^{-1}, respectively. Consider now the corresponding reactions for benzene, reactions 32a and 32b:

$$2\text{PhH} \longrightarrow \text{Ph—Ph} + H_2 \quad (32a)$$

$$\text{Ph—Ph} + \text{PhH} \longrightarrow \text{Ph—Phv}^{p}\text{—Ph} + H_2 \quad (32b)$$

Reaction 32a is endothermic by (16.2 ± 2.3) kJ mol^{-1}. We would think that equation 32b has very nearly the same endothermicity. In fact, using a derived[87] enthalpy of formation of gaseous *p*-terphenyl (**50**) of 273.8 kJ mol^{-1}, we find that reaction 32b is endothermic by

(**50**)

an acceptable (9.8 ± >4.2) kJ mol^{-1}. We close by looking at the 'co-oligomerization/dehydrogenation reactions' 33a and 33b:

$$\text{PhH} + \text{CyprH} \longrightarrow \text{Ph—Cypr} + H_2 \quad (33a)$$

$$\text{Ph—Cypr} + \text{PhH} \longrightarrow \text{Ph—Cyprv}^{t}\text{—Ph} + H_2 \quad (33b)$$

The first reaction is endothermic by (14.6 ± 1.2) kJ mol^{-1}. Our experience suggests that the second reaction should be endothermic by a comparable amount. In fact, from the numbers in our archives, we derive an inexplicitly small endothermicity of only (0.3 ± 1.1) kJ mol^{-1}.

IV. STRAIN ENERGIES OF SPECIES WITH THREE- AND FOUR-MEMBERED RINGS

Cyclopropane and cyclobutane are the smallest saturated strained rings; it has long been recognized that the strain energies of cyclopropane and cyclobutane are nearly identical. In this section we will not discuss the origin of this phenomenon[88]. Rather, we will see how well this near-equality survives substitution and other functionalization of these rings.

A. Cyclopropanes and Cyclobutanes

The closeness of the numerical values for the strain energies of cyclopropane and cyclobutane can be recast in numerous ways. Among the simplest are the differences of the enthalpies of formation of *real* cyclopropane and cyclobutane with that equated with 3/6 and 4/6 of that of cyclohexane. These two values are 115.2 and 110.7 kJ mol^{-1}. Equivalently, the strain energy of a species c-$(CH_2)_n$ is defined as the enthalpy of the gas-phase reaction 34:

$$c\text{-}(CH_2)_n \longrightarrow n/6[c\text{-}(CH_2)_6] \quad (34)$$

The simplest strain energy difference analysis thus considers the difference in the enthalpies of formation of cyclopropane and cyclobutane: whatever be their strain energies, were the strain energies equal, then gaseous cyclobutane would have an enthalpy of formation lower than that of gaseous cyclopropane by the contribution of one strainless methylene increment[89], namely *ca* –20.6 kJ mol^{-1}. From experiment for parent hydrocarbons, the difference is found to be (–25.1 ± 0.7) kJ mol^{-1}. In the absence of cyclopropane and cyclobutane having their real, and hence distinct, electronic and steric interactions with substituents, then the same *ca* –25 kJ mol^{-1} difference would be found between *any* gaseous, identically substituted pair of cyclobutane and cyclopropane. By the differential enthalpy of vaporization analysis used earlier in comparing cyclopropane and benzene derivatives, we conclude that the difference of the enthalpies of vaporization of liquid, identically substituted cyclopropanes and cyclobutanes would be that of one carbon, 4.7 kJ mol^{-1}, and so the difference of their enthalpies of formation some (–25.1 – 4.7) ≈ –30 kJ mol^{-1}.

For the pair of *liquid* methyl and ethylcycloalkanes (**2** and **12**, X = Me and Et), chosen so that we don't need to estimate phase-change enthalpies, the differences are (–46.2 ± 1.6) and (–34.2 ± 1.1) kJ mol^{-1}. These numbers are significantly different from the above expectations. For cyanocyclobutane and cyanocyclopropane (**12** and **2**, X = CN) with their σ- and π-electron-withdrawing cyano substituent, the liquid-phase difference is (–38.7 ± 1.5) kJ mol^{-1} and that of the gas is (–37.8 ± 1.5) kJ mol^{-1}. For the species with the likewise σ- and π-withdrawing carbomethoxy substituted species (**12** and **2**, X = COOMe), the liquid-phase[90] enthalpy of formation difference is (–49.0 ± 2.0) kJ mol^{-1}. For the σ-withdrawing but π-donating amino substituted species (**12** and **2**, X = NH_2), the liquid-phase difference is –40.2 ± 0.9 and that of the gas is (–35.8 ± 1.0) kJ mol^{-1}. Finally, for the pair of species with the weakly σ-electron-withdrawing and weakly π-electron-donating bromo substituent (**12** and **2**, X = Br), we have the reaction calorimetry derived enthalpy of formation of the liquid bromocyclopropane[91], (27.6 ± 4.4) kJ mol^{-1}, and the appearance potential derived enthalpy of formation of gaseous bromocyclobutane[92], 36.8 kJ mol^{-1}. Using the CHLP protocol to derive the enthalpy of vaporization of the former species, the difference for the gases is –22.1 kJ mol^{-1}. Remembering identity equation 35:

$$\Delta H_{\text{vap}} = -\Delta H_{\text{cond}} \tag{35}$$

and so 'running in reverse' the CHLP protocol to find the enthalpy of condensation (cond), the difference of enthalpies of formation for the liquids is –26.9 kJ mol^{-1}.

It appears that almost regardless of the substituent[93], the substituted cyclobutane is more stabilized relative to the corresponding cyclopropane than we would have thought based on the parent hydrocarbons. One interpretation is that cyclobutane is electronically and sterically much more flexible than cyclopropane: cyclobutane is 'an artful dodger'[94] capable of varying orbital symmetries, stretching bond lengths and nonbonding distances, opening dihedral angle, tilting methylenes. Yet, it has also been more prosaically suggested that the measurement of the enthalpy of formation of the unsubstituted cyclobutane is in error[95] by *ca* 10 kJ mol^{-1}, i.e. gaseous cyclobutane has an enthalpy of formation *ca* 35 kJ mol^{-1} lower than that of cyclopropane. A new measurement for the enthalpy of formation of cyclobutane is in order. Yet, somehow, we will be disappointed if the latter interpretation of substituent effects of cyclobutanes and cyclopropanes is corroborated. Indeed, we would then ask 'why is cyclobutane "so normal"?'.

B. Introduction of sp^2 Carbons into Cyclopropanes and Cyclobutanes

Let us take as a given that gaseous cyclobutane has an enthalpy of formation either *ca* 25 or 35 kJ mol^{-1} lower than gaseous cyclopropane, the two values differing by which value is taken for the enthalpy of formation of cyclobutane. It was seen that regardless of electronic or steric effects, substituents seem to make changes of the order of 10 to 20 kJ mol^{-1} beyond that ring contraction difference[96]. Let us effect a more dramatic change than just replacing a hydrogen by some substituent: let us transform sp^3, tetracoordinate, carbon to sp^2, tricoordinate, carbon. The simplest transformation is that to methylenecyclobutane (**51**) and methylenecyclopropane (**24**) because it corresponds to the change of but one ring carbon (cf species **52** with n = 4 and 3, respectively). We find a gas-phase enthalpy of formation change of (79.0 ± 1.9) kJ mol^{-1}. Is this change, far larger than that for the unsubstituted cycloalkanes, due to destabilization of methylenecyclopropane and/or is it due to stabilization of methylenecyclobutane? One way of appraising this is to look at the 'methylenation' enthalpies (equation 36) of cyclopropane, cyclobutane and propane[97]. These three numbers are (147.2 ± 1.9), (93.1 ± 0.9) and (87.8 ± 1.0) kJ mol^{-1}, respectively. The closeness of the last two numbers shows that cyclobutane is comparatively normal (i.e. more like the unstrained, acyclic propane) while cyclopropane is considerably different. Indeed, the earlier similarity of olefins and cyclopropanes now briefly returns: the methylenation of ethylene to form allene is accompanied by a comparable change of some (138.0 ± 1.3) kJ mol^{-1}.

(51)

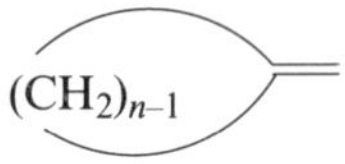

(52)

(36)

We now look at the change associated with two carbons in cyclopropane and cyclobutane being sp^2 instead of sp^3, i.e. we consider cyclopropene (**16**) and cyclobutene (**14**). The difference of their enthalpies of formation is (120.4 ± 2.9) kJ mol^{-1}. Is this still larger change due to destabilization of cyclopropene and/or stabilization of cyclobutene (cf species **15** with n = 3 and 4)? One way of appraising this is to look at the 'olefination' enthalpies (equation 37) of cyclopropane, cyclobutane and butane. These three numbers are (223.8 ± 2.6), (128.3 ± 1.6) and (118.5 ± 1.2) kJ mol^{-1} showing that cyclobutane is comparatively normal (i.e. more like the unstrained, acyclic propane) while cyclopropane is considerably different.

$$-CH_2CH_2- \longrightarrow (Z)\text{-}CH{=}CH- \qquad (37)$$

How do we compare the *ca* 79 and 224 kJ mol^{-1} ring changes affected by transforming one and two carbons from sp^3 to sp^2? Remember there was a 25 to 35 (read 30 ± 5) kJ mol^{-1} ring contraction enthalpy change before any substituents were introduced and hybridization was changed. The hybridization-affected enthalpy changes are (79–30) ≈ 50 and (224 – 128) ≈ 95 kJ mol^{-1}. The introduction of two sp^2 carbons into a three-membered ring is twice as unfavorable as introducing one.

C. Ring Fusion of Cyclopropanes and Cyclobutanes

We will now consider the difference of the enthalpies of formation of saturated bicyclics containing cyclopropane and cyclobutane rings, i.e. the bicyclo[n.1.0]- and bicyclo[n.2.0]-

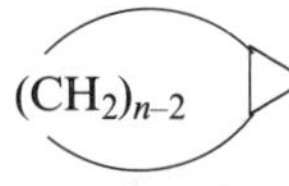

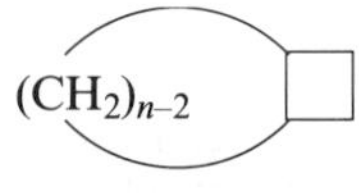

(18) **(53)**

alkanes, species **18** and **53**, respectively. The $n = 0$ case is more commonly recognized in terms of the common names, cyclopropene and cyclobutene, and for which there is an enthalpy of formation difference of (120.4 ± 2.9) kJ mol^{-1}. For the $n = 1$ case, the difference is reduced to (58.5 ± 1.5) kJ mol^{-1} and for the $n = 2$, 3 and 4 cases, the differences approach normalcy with the values of [34.0 (± >1.0)][98], 36.2[99] and (27.5 ± 3.8)[100] kJ mol^{-1}.

D. Another Interlude on Homoaromaticity and Homoantiaromaticity

Equation 37 and the accompanying analysis documents that cyclobutane is essentially normal, i.e. cyclobutene and cyclobutane are not particularly different in terms of their strain energy. One might therefore expect that the enthalpy of formation difference of bicyclo[2.1.0]pentane (**17**) and bicyclo[2.2.0]hexane (**54**) of 34.0 kJ mol^{-1} (the above $n = 2$ case) would approximately equal that of the unsaturated bicyclo[2.1.0]pent-2-ene (**41**) and bicyclo[2.2.0]hex-2-ene (**55**). However, the difference is *ca* 83 kJ mol^{-1} for the latter pair[101]. The

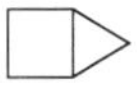
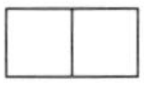
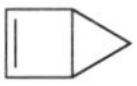
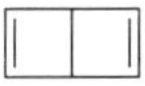

(17) **(54)** **(41)** **(55)**

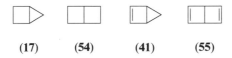

difference of these two numbers, 83 and 34 kJ mol^{-1}, may be ascribed to the homoantiaromaticity of the bicyclopentene: indeed, the new value of 49 kJ mol^{-1} is encouragingly close to our earlier deduced value[102] of 57 kJ mol^{-1}.

What is found for the homoaromatic stabilization for bicyclo[4.1.0]hepta-2,4-diene (**39**)? The natural comparison in the current section is between the enthalpies of formation of this species and that of bicyclo[4.2.0]octa-2,4-diene (**56**, $n = 4$). We find[103] a difference of *ca* 13 kJ mol^{-1}. By contrast, the difference of the enthalpies of formation of bicyclo[4.1.0]heptane (**18**) and bicyclo[4.2.0]octane (**53**, $n = 4$) is 27.5 kJ mol^{-1}. This suggests *ca* 15 kJ mol^{-1} homoaromatic stabilization for bicyclo[4.1.0]hepta-2,4-diene, in qualitative agreement with the earlier presented value[99] of 30 kJ mol^{-1}.

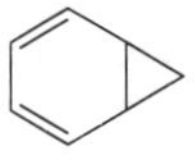
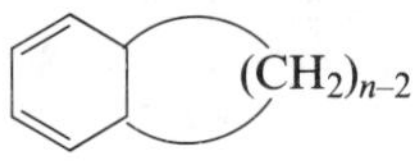

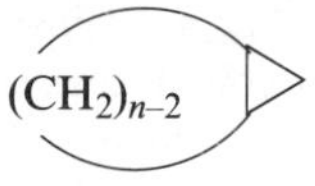

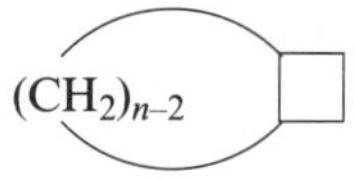

(39) **(56)** **(18)** **(53)**

E. Benzoannelation of Cyclopropene and Cyclobutene

For relatively unstrained medium-size rings, benzoannelation seems to exert a rather small effect on enthalpies of formation[104]. For example, the transformation of cyclopentene, cyclopentadiene and cyclohexene to indane (**20**, $n = 5$), indene and tetralin (**20**, $n = 6$) results in a change of gas-phase enthalpies of formation of (26.8 ± 2.2), (29.1 ± 2.6) and (31.0 ± 2.1) kJ mol^{-1}. For the rather strained cyclobutene, annelation to form benzocyclobutene (**20**, $n = 4$) results in a change of enthalpy of formation[105] of [43.9 (± >1.5)] kJ mol^{-1}. And, for the highly strained cyclopropene, the change upon formation of benzocyclopropene (**20**, $n = 3$) has been increased to some (90 ± 5) kJ mol^{-1}. This phenomenal increase of strain upon benzoannelation of the cyclopropene has been discussed at considerable length[106] in terms of bond fixation: these changes may be related to the relative stability of cycloalkane rings with appended endo- and exocyclic double bonds. But we lack thermochemical information about any exo-methylene or keto derivative of benzocyclopropene or of benzocyclobutene to parallel the available data for the related, but one-ring, species.

F. Do Many Rings Mean Many Problems?

Consider the formal reactions (equations 38 and 39) of polycyclic hydrocarbons prismane (**57**) and quadricyclane (**59**) and '>CH—', the strainless tertiary carbon increment, to form cubane (**58**) and homocubane (**60**), respectively. In both formal reactions, two

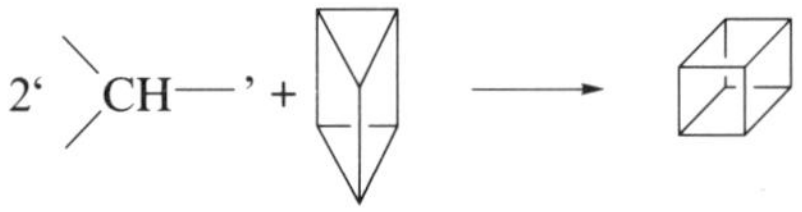

(38)

(**57**) (**58**)

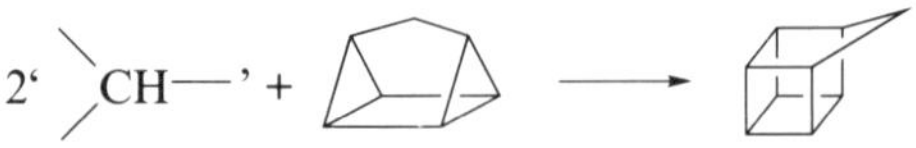

(39)

(**59**) (**60**)

three-membered rings are transformed into two four-membered rings. In that the strain energies of three- and four-membered rings are nearly identical and the enthalpy of formation of '>CH—' is small[107], it appears plausible that both reactions are essentially thermoneutral. Although there is really not that much ambiguity in the numerical assignment of the enthalpy of formation associated with the '>CH—' group[107], let us bypass that problem altogether by considering reaction 40:

$$\mathbf{58} + \mathbf{59} \longrightarrow \mathbf{60} + \mathbf{57} \qquad (40)$$

Counting groups, bonds and rings (and types of rings) and affirming the constancy of each on both sides of the above reaction suggests thermoneutrality is a reasonable expectation—regardless of the exo- or endo-thermicities of the individual reactions 38 and 39. What is found using experimentally measured enthalpies of formation?

Of all the possible multiring hydrocarbons, cubane (**58**) is one of only three that is composed of but one type of group (>CH—), one type of bond (>CH—CH<) and one type of ring. To date, cubane stands alone of these three species[108] for which a measured enthalpy of combustion, and hence of formation, is available for both the solid and gaseous states[109]. Indeed, cubane has been used as a paradigm for the >CH— group and concomitant understanding of strain energy[107]. Yet recently, through analysis of its 1,4-dicarbomethoxy derivative[110], new derived values for the enthalpy of formation of solid and gaseous cubane have been suggested which differ from the earlier measured quantities by *ca* 45 kJ mol^{-1}.

Quadricyclane (**59**) has been a popular compound for thermochemical investigations: Reference 3 cites eight different direct (combustion enthalpy) or indirect (isomerization or other reaction enthalpy) determinations. While approximate consensus for the enthalpy of formation of quadricyclane has seemingly been achieved, it is nonetheless disconcerting that the direct measurement of enthalpy of combustion is the most disparate—this result differs by *ca* 60 kJ mol^{-1} from the others.

Homocubane (**60**), per se, has not been directly studied but its gas-phase enthalpy of formation has been derived as part of calorimetric studies[111] on its 4-carboxylic acid and its 4-carbomethoxy derivatives (species **61**, X = COOH and COOMe). The results were shown

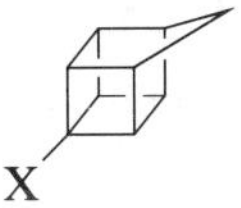

(61)

to be consonant with those from calculational theory, but the same scrutiny from the thermochemical community that was applied to cubane and quadricyclane has not come to bear on this species. And except for the painful precedent of the preceding complications for cubane[112], it would have been natural to ask ‘why *should* there be any doubts?’.

There are also no experimental thermochemical studies[113] of prismane (**57**), per se, but measurements of the enthalpy of isomerization of its hexamethyl derivative to hexamethylbenzene have been reported[114]. These experimentally measured isomerization enthalpies differ by some 40 kJ mol^{-1}. There are also measurements of the isomerization enthalpy for hexakis(trifluoromethyl)prismane to the corresponding benzene[115].

At this stage, we can give up and conclude ‘many rings mean many problems’. But let us proceed using the author's preferences (or should we admit prejudices) for enthalpies of formation. Starting with the enthalpy of formation of cubane, the preferred value is that which was earlier measured, namely (622.1 ± 3.7) kJ mol^{-1}. If the new value for the enthalpy of formation of cubane were correct, the new value of the enthalpy of the group ‘$>$CH—’ derived from it would be seriously dissonant from other suggested values[104]. And that, at least for now, is too stiff a price to pay for choosing the suggested alternative enthalpy of formation of cubane. The enthalpy of formation of quadricyclane is accepted from the analysis of Reference 3. Acknowledging that combustion calorimetry can err, the value of (339.1 ± 2.3) kJ mol^{-1} is adopted here.

Among the conclusion in the study of homocubane and its carboxylic acid and ester derivatives[112] is that the gas-phase reaction shown in equation 41 is essentially thermoneutral. The enthalpy of formation of gaseous homocubane is thus *ca* 390 kJ mol^{-1}.

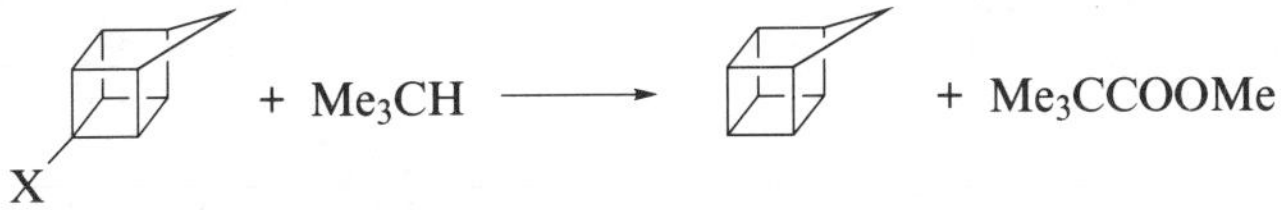

(41)

(**61**), X = COOMe (**60**)

As noted above, there are no measurements on the enthalpy of formation of prismane, per se, but experimental studies of the rearrangement enthalpy of hexamethylprismane to hexamethylbenzene have been reported[114]. But the two relevant numerical values, 382 and (345 ± 3) kJ mol^{-1} respectively, are markedly different from each other, and from the *ca* 570 kJ mol^{-1} suggested by modern calculational theory[113] for the parent species. While the measured rearrangement enthalpy[115] of hexakis(trifluoromethyl)prismane to hexakis(trifluoromethyl)benzene is *ca* 135 kJ mol^{-1} smaller than the above hexamethyl results, so be it: CF_3 substitution is significantly different from that of Me[116]. But it eludes us why there is a nearly 40 kJ mol^{-1} difference between the two measurements for the rearrangement enthalpy of hexamethylprismane[117], and a *ca* 200 kJ mol^{-1} difference between the parent and hexamethyl prismane[118].

G. Some Thoughts on the Aromaticity of Cyclopropenones

Cyclopropenones (**62**) are an important class of nonbenzenoid aromatics[119], and so have an interesting conceptual blend of stabilizing and destabilizing molecular features. Yet the only cyclopropenone derivative for which there is available gas-phase enthalpy of formation data[120] is the additionally conjugated diphenyl species **62**, $X^1 = X^2$ = Ph, namely 316.7 ± 8.2 kJ mol^{-1}. The natural cyclopropane/cyclobutane comparison of a cyclopropenone is with the corresponding cyclobutenone. The only cyclobutenone for which there is enthalpy of formation data[121] is the 3-phenyl derivative **63**, X = Ph, namely 86.4 kJ mol^{-1}. Given the

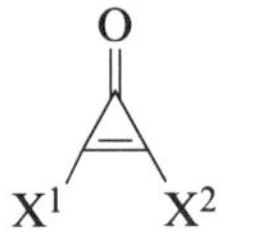

(**62**)

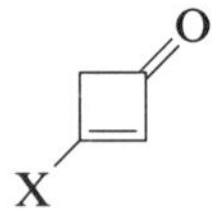

(**63**)

different substitution patterns, how then do we compare cyclopropenones and cyclobutenones? While our prejudice would have suggested that comparison of the parent species with all affixed groups be H, it seems simpler to 'remove' but one phenyl from diphenylcyclopropenone to form the monophenylated species **64**, X = Ph or **62**, X^1 = Ph, X^2 = H. That is, we endeavor to find a thermoneutral phenyl transfer reaction:

(42)

(**62**), $X^1 = X^2$ = Ph (**64**)

But what Y do we choose? It may cogently be argued that the two phenyl groups and the cojoining double bond mimic those of stilbene. But, does this mimicry mean *cis*- or *trans*-stilbene? Although the two phenyls in diphenylcyclopropenone are markedly splayed out compared to *cis*-stilbene, they are still closer than in the *trans*-isomer. Admitting this ambiguity, we let Y = (*Z*)-CH=CHPh. From the archival enthalpy of formation of gaseous diphenylcyclopropenone, we derive the corresponding enthalpy of formation of gaseous monophenyl species to be (212.3 ± 8.6) kJ mol^{-1}. We thus conclude that gaseous phenylcyclopropenone has an enthalpy of formation some 125 kJ mol^{-1} higher than 3-phenylcyclobutenone. This hardly seems suggestive of any stabilization for the cyclopropenone upon remembering that cyclopropane has an enthalpy of formation higher than cyclobutane by but (24.9 ± 0.9) kJ mol^{-1}. Then we remember the effect of introducing trigonal carbons into three- and four-membered rings. While introduction of one trigonal carbon (so that the comparison is between methylenecyclopropane and methylenecyclobutane) increases that difference to (79.0 ± 2.0) kJ mol^{-1}, introduction of two trigonal carbons (by comparing cyclopropene and cyclobutene) increases the difference to (120.4 ± 2.9) kJ mol^{-1}. If three trigonal carbons could be introduced into a three-membered ring without any possibility of additional delocalization, we would have expected the difference to be increased to at least 160 kJ mol^{-1}.

However, since the difference is only 126 kJ mol^{-1} it suggests there is an additional stabilization found in cyclopropenones of some 35 kJ mol^{-1}. This value is small, but it should be remembered that the already conjugated cyclobutenone is the reference species for the discussion of cyclopropenone aromaticity. More precisely, from the data in Reference 121 we derive the difference of the enthalpies of formation of 3-phenyl-2-cyclobutenone and the saturated 3-phenylcyclobutanone to be *ca* 76 kJ mol^{-1}. By contrast, the difference of the enthalpies of formation of cyclobutene and cyclobutane is (128.3 ± 1.6) kJ mol^{-1}. The *ca* 50 kJ mol^{-1} difference between these two numbers can be understood as the resonance energy of cyclobutenones, and is suggestive of homoaromatic stabilization in these species[122]. We wish to emphasize that we are not implying that cyclopropenones have less stabilization than cyclobutenones. Rather, the additional 'aromatic' stabilization relative to its reference species is surprisingly, and perhaps disappointingly, small. Indeed, the total stabilization energy for cyclopropenones is *ca* (30 + 50) ≈ 80 kJ mol^{-1}, a figure rather reminiscent from earlier analyses of this class of species. Consistency is maintained: cyclopropane and its derivatives continue to educate, entertain, encourage.

V. ACKNOWLEDGMENTS

The author wishes to thank James S. Chickos, Eugene S. Domalski, Arthur Greenberg, Sharon G. Lias and Suzanne W. Slayden for numerous discussions of science and of style on the energetics of general organic compounds and, in particular, of cyclopropanes and other compounds with three-membered rings. The author also thanks those individuals, mentioned in footnotes, who shared with him unpublished data and/or insights.

VI. REFERENCES AND COMMENTARY

1. *The Chemistry of the Cyclopropyl Group*, Z. Rappoport (Ed.), Wiley, Chichester, 1987 (2 parts).
2. G. Boche and H. M. Walborsky, *Cyclopropane Derived Reactive Intermediates*, update U4 from *The Chemistry of Functional Groups*, Wiley, Chichester, 1991.
3. J. F. Liebman and A. Greenberg, *Chem. Rev.*, **89**, 1225 (1989). This source alone cites some seventeen other reviews that relate directly to the energetics of cyclopropanes. In addition, pp. 1228–1231 provide four tables that represent our major archives for the enthalpies of formation of compounds of cyclopropanes and other compounds with three-membered rings.
4. J. F. Liebman and A. Greenberg, *Chem. Rev.*, **76**, 311 (1976).
5. A. Greenberg and J. F. Liebman, *Strained Organic Molecules*, Academic Press, New York, 1978.
6. J. F. Liebman and A. Greenberg, in *The Chemistry of the Cyclopropyl Group, Part 1* (Ed. Z. Rappoport), Chap. 18, Wiley, Chichester, 1987, p. 1083.
7. The aphorism about scientific data: 'there's more than you think and less than you need' provides both an excuse and a challenge for the would be archivist of data. Personal experience has shown it perilously easy to miss relevant thermochemical data. Measurements of enthalpies of formation apparently lie fallow in conference abstracts and in patents and are scatterred throughout papers 'officially' on synthesis, structure, mechanisms and reactivity. Even computational theoretical papers have been known to contain new thermochemical data!
8. *Guide for Authors in Preparation of Manuscripts for The Chemistry of Functional Groups series edited by Saul Patai*, Wiley, Chichester, p. 3.
9. For the diligent reader, thermochemical conventions are well-discussed in D. D. Wagman, W. H. Evans, V. B. Parker, R. H. Schumm, I. Halow, S. M. Bailey, K. L. Churney and R. L. Nuttall, *'The NBS Tables of Chemical Thermodynamic Properties: Selected Values for Inorganic and C_1 and C_2 Organic Substances in SI Units', J. Phys. Chem. Ref. Data*, **11** (1982), Supplement 2. However, the various subtleties expressed in this source, such as the above-cited ambiguities in temperature and pressure, have but negligible effect on any of the conclusions about cyclopropane and its derivatives in this chapter: the data are too inexact and the concepts we employ are simply too 'sloppy' to be affected.
10. J. S. Chickos, A. S. Hyman, L. H. Ladon and J. F. Liebman, *J. Org. Chem.*, **46**, 4294 (1981). In

the absence of a literature value suggested for a desired enthalpy of vaporization for a hydrocarbon, this CHLL2 protocol will be used implicitly to estimate the desired quantity.
11. J. S. Chickos, D. G. Hesse, J. F. Liebman and S. Y. Panshin, *J. Org. Chem.*, **53**, 3435 (1988). In the absence of a literature value suggested for a desired enthalpy of vaporization for a substituted hydrocarbon, this CHLP protocol will be used implicitly to estimate the desired quantity.
12. We acknowledge that methylcyclopropane is no less a substituted cyclopropane as is methyl cyclopropanecarboxylate. However, for the former species we use the hydrocarbon equation 3 with $\tilde{n}_C = 4$, and for the latter species we use the substituted compound equation 4 with $b = 10.5$ (for esters) and $\tilde{n}_C = 5$ (remember the carbonyl carbon in calculating the number of nonquaternary carbons).
13. Extensions of equation 4 to the enthalpy of sublimation of hydrocarbons have been made by Chickos, Liebman and their coworkers in two guises. The first combines the CHLL2 equation with the experimentally measured enthalpy of fusion [$\Delta H_{fus} \equiv \Delta H_f(l) - \Delta H_f(s)$], and reported in J. S. Chickos, R. Annunziata, L. H. Ladon, A. S. Hyman and J. F. Liebman, *J. Org. Chem.*, **51**, 4311 (1986). The second approach requires only the generally available melting point of the compound, but this simplification is at the expense of an extensive collection of parameters needed to estimate total phase change entropies: see J. S. Chickos, D. G. Hesse and J. F. Liebman, in *Energetics of Organometallic Species* (Ed. J. A. Martinho Simões), NATO ASI, Series C, Vol. 367, Kluwer, Dordrecht, 1992.
14. V. P. Kolesov and M. P. Kozina, *Russ. Chem. Rev.*, **56**, 912 (1986).
15. J. B. Pedley, R. D. Naylor and S. P. Kirby, *Thermochemical Data of Organic Compounds*, 2nd ed., Chapman & Hall, New York, 1986.
16. M. S. Kharasch, *Bur. Stand. J. Res.*, **2**, 359 (1929).
17. D. R. Stull, E. F. Westrum, Jr. and G. C. Sinke, *The Chemical Thermodynamics of Organic Compounds*, Wiley, New York, 1969. In turn, a rather large number of data 'incorporated' were taken from Kharasch's earlier compendium, our Reference 16.
18. The reader is addressed to A. Greenberg and T. A. Stevenson, in *Molecular Structure and Energetics: Studies of Organic Molecules*, Vol. 3 (Eds. J. F. Liebman and A. Greenberg), VCH, Deerfield Beach, 1986, for a discussion of this regularity in the framework of linear free energy relations involving monosubstituted derivatives of cyclopropane, c-C_3H_5X. The current extension to include multisubstituted and multiring derivatives is not part of this earlier study. It is left to the reader to decide whether the current study presents creative interrelations and useful insights, or suggests careless (or even callous) indifference to the chemistry and even to the chemist.
19. (a) J. F. Liebman, in *Molecular Structure and Energetics: Studies of Organic Molecules* (Eds. J. F. Liebman and A. Greenberg), VCH, Deerfield Beach, 1986.
(b) P. George, C. W. Bock and M. Trachtman, in *Molecular Structure and Energetics: Biophysical Aspects* (Eds J. F. Liebman and A. Greenberg), VCH, New York, 1987.
(c) Y.-R. Luo and J. L. Holmes, *J. Phys. Chem.*, **96**, 568 (1992).
20. See, for example,
(a) N. J. Demjanov, *Ber.* **40**, 4393, 4961 (1907).
(b) M. Charton, in *The Chemistry of Alkenes* (Ed. J. Zabicky), Vol. 2. Chap. 10, Wiley, Chichester, 1970, p. 573.
(c) L. N. Ferguson, *Highlights of Alicyclic Chemistry*, Chap. 3, Franklin Press, Palisade, NJ, 1973.
(d) A. Greenberg and J. F. Liebman, *Strained Organic Molecules*, Chap. 2, Academic Press, New York, 1978.
(e) However, also see A. J. Gordon, *J. Chem. Educ.*, **44**, 461 (1967), for a rare dissenting view.
21. (a) M. J. S. Dewar, *Bull. Soc. Chim. Belg.*, **88**, 957 (1979).
(b) M. J. S. Dewar, in *Modern Models of Bonding and Delocalization* (Eds. J. F. Liebman and A. Greenberg), VCH, New York, 1988.
(c) D. Cremer and E. Kraka, in *Structure and Reactivity* (Eds. J. F. Liebman and A. Greenberg), VCH, New York, 1988.
22. The term 'cycloethane', as used here and in Reference 20d, generally makes more immediate sense to individuals used to molecular models with flexible bonds, whether they be metal springs or plastic tubes, than to those individuals who are used to model sets with 'lobes' or 'fans' to convey orbitals. The term cycloethane is also in interesting historical and linguistic counterpoint to the early chemical literature wherein trimethylene was often used in lieu of cyclopropane. For

example, the archival review by Kharasch[16] uses the name trimethylene instead of cyclopropane and uses both trimethylenecarboxylic acid as the 'first name' for c-$(CH_2)_2CHCOOH$ and cyclopropanecarboxylic acid as its synonym. See also Chapter 1 of this volume.

23. Error bars are not available for the enthalpy of formation of gaseous methylcyclopropane: we find no experimental measurements for the enthalpy of vaporization of methylcyclopropane and so, following Reference 3, we accept Kolesov and Kozina's[14] estimated value. To proceed with our analysis we cannot concern ourselves with small discrepancies in estimates for enthalpies of vaporization. For example, Kolesov and Kozina's suggested value for methylcyclopropane is 22.6 kJ mol^{-1} while Trouton's rule gives a value of 27.5 kJ mol^{-1}. Using the CHLL2 approach, the enthalpy of vaporization of methylcyclopropane is found to be 21.7 kJ mol^{-1}. We do not attempt to give error bars for these estimates.
24. (a) The enthalpies of formation of gaseous (Z)- and (E)-3-hexene were obtained from that of gaseous hexane, and the dilute hydrocarbon solution-phase enthalpies of hydrogenation of the olefins were reported in D. W. Rogers, E. Crooks and K. Dejroongraung, *J. Chem. Thermodyn.*, **19**, 1209 (1987).

    (b) In this analysis, as for all hydrogenation enthalpies reported in this chapter unless otherwise said, we assumed that the reported value is the same as for that of the gas as suggested in D. W. Rogers, O. A. Dagdagan and N. L. Allinger, *J. Am. Chem. Soc.*, **101**, 671 (1979) for hydrogenations in dllute hydrocarbon solution.
25. The enthalpy of formation of 2-methyl-2-pentene was obtained using the same procedure, assumptions and reference as that of its 3-hexene isomers (cf the paper in Reference 24a).
26. The enthalpy of formation of 2,3-dimethyl-2-butene was obtained using the same procedure, assumptions and reference as that of its 3-hexene isomers (cf the paper in Reference 24a).
27. This contribution is neglected because it is not obvious what we should use. The decision is legitimized by noting the contribution is a constant for all formally related reactions with comparably substituted species.
28. We must admit we have really no reason to disregard the truly discrepant differences for the 1,1,2-trimethyl- and 1,1,2,2-tetramethyl- species except that

    (a) invoking 'majority rule', i.e. most of the other compounds seem to form a pattern, albeit tenuous and inexact.

    (b) these data lack error bars that would allow us to more carefully appraise the results.
29. The reader will recall the *ca* $\pm$ 1 kJ mol^{-1} error bars associated with methylating ethylene, acetylene, ethane and propane, even though both the reactant and product hydrocarbons are among the thermochemically best understood of any ever studied. The reader will also recall discrepancies with measured as well as estimated enthalpies of vaporization. Furthermore, error bars associated with measurements of enthalpy of formation are often several kJ mol^{-1} in magnitude, even for unstrained and therefore comparatively simple hydrocarbons. For example, while the error bars for the enthalpies of formation of cyclohexane and its monomethyl derivative are both under 1 kJ mol^{-1}, the error bars for all seven of the dimethylcyclohexanes are between 1.7 and 1.9 kJ mol^{-1}. Perhaps suggestive of considerable sloppiness or uncertainty in the measurements, studies of the enthalpies of many cyclopropanes and other strained species lack error bars.
30. We note that the cyclopropanation of ketene to form cyclopropanone, isoelectronically and otherwise formally related to the cyclopropanation of allene to form methylenecyclopropane, has an altogether different enthalpy of reaction. This value is (63 $\pm$ 4) kJ mol^{-1} favoring ketene and is credible in terms of considerable resonance stabilization of the type $CH_2{=}C{=}O \leftrightarrow {}^{-}CH_2{-}C{\equiv}O^{+}$ and of a 3-atom π system that includes the double bond and the appropriate, antisymmetric (i.e. difference) combination of the oxygen lone pairs. Further documentation of the stabilization of ketene is seen by comparing the energetics of the hydrogenation reaction

    $$CH_2{=}C{=}X + H_2 \rightarrow MeCH{=}X$$

    for X = O and CH_2. For X = O, the reaction is (118.6 $\pm$ 1.7) kJ mol^{-1} exothermic, while for X = CH_2, the reaction is (170.5 $\pm$ 1.4) kJ mol^{-1} exothermic.

    It is now acknowledged that in our 3MR archival tables[3] the formula and data are correctly given for cyclopropan*one* but we wrote the name as cyclopropan*e*. We apologize for not catching this error earlier.
31. That the enthalpy changes are of opposite sign is, in fact, easily remedied. If ethylene is under-

stood as cycloethane, and it and cyclopropane were both strainless, then the cyclopropanation reagent could be taken as a 'strainless methylene group' or 'universal methylene increment'. Accepting the analysis of J. D. Cox and G. Pilcher (*Thermochemistry of Organic and Organometallic Compounds*, Academic Press, New York, 1970), this —CH_2 moiety has an enthalpy of formation of –20.6 $\mathrm{kJ\,mol^{-1}}$. Explicitly including this value as part of the thermochemical arithmetic, we find all three enthalpy changes now have the same sign. However, so doing does not lessen the lack of arithmetic (numerical) uniformity between the difference of allene and methylenecyclopropane, and of methylenecyclopropane and spiropentane, nor is the conceptual problem of widely differing values truly ameliorated.

32. We here acknowledge the fact that in our 3MR archival tables[3], the formula and data are correctly given for cycloprop*e*ne but we wrote the name as cycloprop*a*ne. We apologize for this misprint.
33. We are changing our choice as to the preferred value for the enthalpy of formation of bicyclo[4.1.0]heptane from (11.6 ± 1.3) to (1.3 ± 2.7) $\mathrm{kJ\,mol^{-1}}$. In our earlier thermochemical compendium for the enthalpies of formation of cyclopropane derivatives[3], we had accepted the recommendation of Kolesov and Kozina[14] instead of that of Pedley and his coworkers[15]. More precisely, Kolesov and Kozina cited the gas-phase enthalpy of formation values of (11.5 ± 1.3) $\mathrm{kJ\,mol^{-1}}$ [using the vaporization data of A. I. Druzhinina and R. M. Varushchenko, *Russ. J. Phys. Chem.*, **52**, 1115 (1978)] and the earlier one of (1.3 ± 2.7) $\mathrm{kJ\,mol^{-1}}$ from S. Chang, D. McNally, S. Shary-Tehrany, M. J. Hickey and R. H. Boyd, *J. Am. Chem. Soc.*, **92**, 3109 (1970). Kolesov and Kozina[14] chose the former value. Pedley and his coworkers[15] did not cite the Russian paper at all, but cited the later American paper and an earlier conference proceedings that had nearly the identical value (R. H. Boyd, C. Shieh, S. Chang and D. McNally, 'Thermodynamik-symposium' Paper II 7, Heidelberg, 1967). The change of preferred value arises from this cycloalkene/bicycloalkane comparison. Had we stayed with our earlier choice for the enthalpy of formation of bicyclo[4.1.0]heptane, the enthalpy of formation difference with cyclohexene enigmatically balloons to (16.6 ± 1.5) $\mathrm{kJ\,mol^{-1}}$.
34. To be better in accord with the olefin-cyclopropane thermochemical regularity in the current paper, the preferred value for the enthalpy of formation of *cis*- bicyclo[5.1.0]octane is now taken as (–6.1 ± 2.4) instead of (–16.8 ± 1.5) $\mathrm{kJ\,mol^{-1}}$, see V. A. Luk'yanova, L. P. Timofeeva, M. P. Kozina, I. V. Kazimirchik and T. A. Kotel'nikova, *Russ. J. Phys. Chem.*, **65**, 436 (1991).
35. The enthalpy of formation of (*E*)-cyclooctene and of *trans*-bicyclo[6.1.0]nonane were taken from the semiprimary/semiarchival reference for enthalpies of hydrogenation and of formation, W. R. Roth, O. Adamczak, R. Breuckmann, H.-W. Lennartz and R. Boese, *Chem. Ber.*, **124**, 2499 (1991). As with References 3 and 14–17, we maintain the practice of generally citing archival sources rather than the original reference.
36. One may argue that the $n = 3$ case might be expected to be abnormal in that 'superstrain' (i.e. excess of strain beyond that of component rings, as suggested in Reference 4) might be expected to be maximized for fusion of small, strained rings. Perhaps the *trans*-**8** case is strained enough that 'superstrain' again becomes important, although then the result from $n = 4$ becomes enigmatic because no 'superstrain' is apparent in this case.
37. See the study of W. R. Roth and his coworkers, op. cit., Reference 35.
38. The necessary enthalpy of formation of gaseous (*E*), (*E*)-2,4-hexadiene was taken from W. Fang and D. W. Rogers, *J. Org. Chem.*, **57**, 2294 (1992). Said differently, demethylation of a gaseous strainless olefin XCH=CHMe to form XCH=CH_2 is accompanied by a (33 ± 2) $\mathrm{kJ\,mol^{-1}}$ increase in the enthalpy of formation.
39. Said differently, demethylation of a gaseous strainless allene XCH=C=CHMe to form XCH=C=CH_2 is accompanied by a (29 ± 2) $\mathrm{kJ\,mol^{-1}}$ increase in the enthalpy of formation.
40. These new values, unknown to the current author when preparing Reference 3, are from L. P. Timofeeva, T. S. Kutznetsova, K. A. Lukin and S. I. Kozhushkov, 'Enthalpies of Formation of Hydrocarbons Containing Three-Membered Rings', paper 2.3, p. 62, in 'Problems in Calorimetry and Chemical Thermodynamics: Proceedings of the XIIth All-Union Conference', Poster Vol. 1, Gorky, 1988.
41. V. A. Luk'yanova, S. M. Pimenova, V. P. Kolesov, T. S. Kutznetsova O. V. Kokoreva, S. I. Kozhushkov and N. S. Zefirov, *Russ. J. Phys. Chem.*, **67**, 114 (1993).
42. We note that should we use the CHLL2 approach to estimate the enthalpies of vaporization, all of the experimentally measured phase change values in Reference 41 are 'too' large. Recall the

two parameters in the CHLL2 procedure, the number of nonquaternary and quaternary carbons—the former contribute more extensively to the vaporization enthalpy because the latter are generally 'buried' compared to the former. It would appear that the quaternary carbons in these spiro-joined cyclopropanes are not as buried as they are normally. However, nothing prepares us for the value of (55.4 ± 0.4) kJ mol^{-1} for the vaporization enthalpy of tetraspiroundecane—this value is within a kJ mol^{-1} of that for the totally nonquaternary, unbranched, extended/distended hydrocarbon with the same number of carbons, namely *n*-undecane. No explanation is apparent.

43. The 1,1-isomer of divinylethylene is to be recognized as 3-methylene-1,4-pentadiene (or 2-vinylbutadiene) while the 1,2-isomer is the 1,3,5-hexatriene discussed above. The enthalpy of formation of (*E*)-1,3,5-hexatriene is from the study of Rogers and his coworkers (cited in Reference 38). The enthalpy of formation of 3-methylene-1,4-pentadiene (or 2-vinylbutadiene) is derived by summing the enthalpies of demethylation (some 31 kJ mol^{-1} from notes 38 and 39) and of formation of 3-ethylidene-1,4-pentadiene (2-vinyl-1,3-pentadiene) from Reference 35.
44. T. Bally, H. Baumgartel, U. Buchler, E. Haselbach, W. Lohr, J. P. Maier and J. Vogt, *Helv. Chim. Acta*, **61**, 741 (1978). See also H. M. Rosenstock, J. Dannacher and J. F. Liebman, *Radiat. Phys. Chem.*, **20**, 7 (1982).
45. Liebman and Greenberg[3] derived the value (127.2 ± 1.3) kJ mol^{-1} for the enthalpy of formation of gaseous vinylcyclopropane using the hydrogenation results of W. R. Roth, W. Kirmse, W. Hoffmann and H. W. Lenartz, *Chem. Ber.*, **115**, 2508 (1982), a study 'needless to say' selected by Roth and his coworkers in Reference 35. A consonant value of (131.2 ± 1.5) kJ mol^{-1} was selected by Kolesov and Kozina[14] citing the combustion and vaporization studies of N. D. Lebedeva, N. M. Gutner and L. F. Nazarova, 'Termodinamika Organicheskikh Soedinenii', Izd. Gor'kobsk. Gos. Univ., Gorky, 1977, No. 6, p. 26. Pedley and his coworkers[15] gave a not-that-different value of (122.5 ± 4.2) kJ mol^{-1} from P. J. C. Fierens and J. Nasielski, *Bull. Soc. Chim. Belg.*, **71**, 187 (1962), but this value is for the liquid! Accepting the enthalpy of vaporization of (28.7 ± 1.3) kJ mol^{-1} from Lebedeva and her coworkers, we derive a third enthalpy of formation of gaseous vinylcyclopropane of (151.2 ± 4.4) kJ mol^{-1}. This value is clearly different from the others, and so at least *some* value is seriously in error.
46. This type analysis was discussed by J. F. Liebman, *Struct. Chem.*, **3**, 449 (1992) as part of a thermochemical comparison of dienes and α-diketones.
47. S. M. Pimenova, M. P. Kozina and V. P. Kolesov, *Thermochim. Acta*, **221**, 139 (1993).
48. We admit being a little sloppy with regard to whether we are talking about the *cis*- or *trans*-isomer because Pimenova and her coworkers[47] estimated the enthalpies of vaporization of the two compounds. They suggested values of 46.9 and 47.3 kJ mol^{-1}. While these values are nearly identical to that estimated by the CHLL2 approach, we are somehow not convinced of this equality—we have already seen that discrepancies of several kJ mol^{-1} in the values of enthalpies of vaporization of cyclopropane derivatives arise in the current chapter.
49. This conclusion was derived by Pimenova and her coworkers[47] and by A. E. Deezer, W. Lüttke, A. deMeijere and C. T. Mortimer, *J. Chem. Soc. (B)*, 648 (1966) as part of their calorimetric investigations of the methyltercyclopropyls and bicyclopropyl, respectively.
50. T. T. Tidwell, in Reference 1, Chap. 10, p. 565.
51. This finding, reported in Reference 35, parallels the enthalpy of formation difference of 1,3-cyclohexadiene and 1,4-cyclohexadiene [cf the combustion calorimetric results of V. A. Luk'yanòva, L. P. Timofeeva, M. P. Kozina, V. N. Kirin and A. V. Tarakanova, *Russ. J. Phys. Chem.*, **65**, 439 (1991)]. Interestingly, in that the bicycloheptenes have enthalpies of formation some 14 kJ mol^{-1} higher than the cyclohexadienes while our cyclopropanation analysis gives a difference of (6 ± 4) kJ mol^{-1}, suggests that both bicycloheptenes are somewhat destabilized.
52. T. Kasprzycka-Guttman, in *The Chemistry of Functional Groups, Supplement C: The Chemistry of Triple Bonded Functional Groups, Part 3* (Ed. S. Patai), Wiley, Chichester, 1994.
53. V. A. Luk'yanova, S. M. Pimenova, L. P. Timovfeeva, M. P. Kozina, V. P. Kolesov and A. V. Tarakanova, *Russ. J. Phys. Chem.*, **66**, 2031 (1992). In the current study, we used their experimentally measured enthalpies of formation and estimated enthalpies of vaporization. It is interesting to note that the estimated enthalpies of vaporization for these diynes are *ca* 10 kJ mol^{-1} higher than the CHLL2 estimate, a finding that parallels the *ca* 5 kJ mol^{-1} higher enthalpy of vaporization of the isomeric butynes over what is found using the CHLL2 estimation approach.
54. We derived the enthalpy of formation of l-hexyne from the hydrogenation study of D. W. Rogers, O. A. Dagdagan and N. L. Allinger, op. cit., cited in Reference 24. The enthalpy of formation of

the gaseous 5,7-dodecadiyne was obtained from that of the liquid with an estimated enthalpy of vaporization using the CHLL2 protocol with the above 9 kJ mol^{-1} correction for a diyne.

55. This number used the measured enthalpies of formation and of vaporization of dicyclopropylbutadiyne from Reference 53. Paralleling our analysis of enthalpies of vaporization in that reference, the experimental measurement of the latter resulted in a value some 9 kJ mol^{-1} higher than the CHLL2 estimate.
56. N. D. Lebedeva, V. L. Ryadnenko, N. N. Kiseleva and L. F. Nazarova, 'Enthalpies of Formation of Isopropenylacetylene and Bis(isopropenyl)diacetylene', paper O-2, p. 91 in 'Problems in Calorimetry and Chemical Thermodynamics: Proceedings of the VIIth All-Union Conference', Vol. 1, Chernogolovka, USSR, 1977.
57. For clarification, the stabilization value '*ca* 30' is the difference of {22.2 or 23} and {−7.6 or −6} kJ mol^{-1}.
58. The reader will recall equations 9 and 10 are generally *ca* 5 and 42 kJ mol^{-1} endothermic. However, we generally lack enthalpies of formation for the compounds with the same affixed X as the above with which to compare reactions 9, 10 and 16. Possibly, though very unlikely, the reaction of two conjugated dienes to form a linearly conjugated tetraene will show considerable additional stabilization. We note one comparison of related 4 and 3 chromophore species. Consider the 'de-acetylenation' reaction

$$X^1{-}C{\equiv}C{-}C{\equiv}C{-}X^2(g) \rightarrow X^1{-}C{\equiv}C{-}X^2(g) + 2C(\text{graphite, s})$$

for the four affixed $X^1 = X^2 = X$ of interest. For X = Me, this reaction is exothermic by [231.7 (± >4.0)] kJ mol^{-1} and for X = Bu, using the requisite enthalpy of formation of gaseous 5-decyne as obtained the same way as for 1-hexyne in Reference 54, results in an exothermicity of reaction of essentially identical [231.6 (± >4.3)] kJ mol^{-1}.

For X = Cypr the exothermicity is [214.5 (± >2.4)] kJ mol^{-1}, where ΔH_f(g, Cypr—C≡C—Cypr) was cited by Luk'yanova and her coworkers[53] as taken from S. M. Pimenova's thesis. For X = MeC(=CH_2) case, we lack enthalpy of formation data for MeC(=CH_2)C≡CC(=CH_2)Me and (excepting the admittedly atypical benzyne) for any substance containing the C=C—C≡C—C=C substructure. However, we may estimate the missing number through a rather lengthy path of assumptions and approximations. We begin with ΔH_f(g, PhC≡CPh) which is directly obtained by summing the archival enthalpy of formation of the solid and the measured enthalpy of sublimation from W. V. Steele, B. E. Gammon, N. K. Smith, J. S. Chickos, A. Greenberg and J. F. Liebman, *J. Chem. Thermodyn.*, **17**, 505 (1985). This value of 403 kJ mol^{-1} is then decreased by *ca* 32 kJ mol^{-1} (cf Reference 19) per phenyl group to give us an estimated 339 kJ mol^{-1} for ΔH_f(g, ViC≡CVi). Analogously to References 38 and 39, we find an average value of −36 kJ mol^{-1} for the enthalpy of α-demethylation of a —C(=CH_2)Me. This results in ΔH_f [g, MeC(=CH_2)C≡CC(=CH_2)Me] equal to *ca* 266 kJ mol^{-1}. The de-acetylenation reaction with X = MeC(=CH_2)— is thus 228 kJ mol^{-1} exothermic, very similar to that of X = Me and Bu. It would appear X = Cypr is the outlier.

59. We have once earlier used the term 'cycloethene' for acetylene in the context of defining the strain energy of acetylene in terms of ethylene, i.e. ΔH_f(g, C_2H_2) vs [$2\Delta H_f$(g, C_2H_4) − ΔH_f(g, C_2H_6)], in order to compare that quantity with the strain energies of cyclopropene and other cycloalkenes; cf Reference 4, p. 94.
60. We thank Donald W. Rogers (personal communication) for telling us of his experiences trying to catalytically hydrogenate diphenylcyclopropenone. His hydrogenation calorimetric results were unreproducible and suggestive of considerable, but unsystematic ring cleavage, reactions. We wonder if the same did not occur in the original experiments on dimethylcyclopropene—after all, R. B. Turner, P. Goebel, B. J. Malon, W. von E. Doering, J. F. Coburn, Jr. and M. Pomerantz, *J. Am. Chem. Soc.*, **90**, 4315 (1968) report nonstoichiometric hydrogen uptake.
61. The following compounds are the sole cyclopropenes (in increasing number of carbons) for which there are literature values of enthalpies of formation: methylenecyclopropene, 1,3,3-trimethylcyclopropene, benzocyclopropene, naphtho[*b*]cyclopropene, diphenylcyclopropenone and 3-methyl-1,2,3-triphenylcyclopropene, the second and last species having been missed by the earlier archiving and was reported in S. M. Pimenova, I. N. Domnin, A. M. Lakshin and V. V. Takhistov, 'Enthalpies of Formation of Cyclopropenes and their Ions', paper O-21, p. 41, 'Problems in Calorimetry and Chemical Thermodynamics: Proceedings of the XIth All-Union Conference', Vol. 2, Novosibirsk, USSR, 1986. Of these new cyclopropenes, only the trimethyl-

cyclopropene potentially tells us anything about alkylation effects on the enthalpy of formation of cyclopropenes. More precisely, from the available data, the enthalpy of hydrogenation of 1,3,3-trimethylcyclopropene to 1,l,2-trimethylcyclopropane is deduced to be exothermic by [230.2 (± >1.9)] kJ mol^{-1}. Because two methyls are unavoidably *cis* to each other in the saturated product, the *gem*-dimethylation should result in a decreased enthalpy of hydrogenation relative to that for 1-methylcyclopropene to methylcyclopropane. But the latter exothermicity is only [220.3 (± >1.1)] kJ mol^{-1}, nearly identical to the (223.6 ± 2.6) kJ mol^{-1} for hydrogenation of cyclopropene to cyclopropane. This suggests that the enthalpy of formation of 1,3,3-trimethylcyclopropene and/or 1,l,2-trimethylcyclopropane is in error—we have already argued for the latter possibility but cannot preclude the former as well. For completeness, we note that the enthalpy of hydrogenation of 1,2-dimethylcyclopropene to (*cis*)-1,2-dimethylcyclopropane is 189.8 kJ mol^{-1}, somehow too low to be credulous and suggestive of an error in the enthalpy of formation of the former, a conclusion already enunciated in our study.

62. The energetics consequence of homoaromaticity and homoantiaromaticity are much more readily documented in ions than in neutrals; see, for example,
(a) W. L. Jorgensen, *J. Am. Chem. Soc.*, **97**, 3082 (1975); **98**, 6784 (1976).
(b) L. A. Paquette, *Angew. Chem., Int. Ed. Engl.*, **17**, 106 (1978).
(c) G. R. Stevenson, in *Molecular Structure and Energetics: Studies of Organic Molecules*, Vol. 3 (Eds. J. F. Liebman and A. Greenberg), VCH, Deerfield Beach, 1986.
An example of the inherent complexities in assigning homoaromaticity in neutral hydrocarbons, even when accompanied by thermochemical, molecular mechanical and/or quantum chemical analysis, is shown by the 'competing' studies of the energetics of triquinacene:
(a) J. F. Liebman, L. A. Paquette, J. R. Peterson and D. W. Rogers, *J. Am. Chem. Soc.*, **108**, 4311 (1986).
(b) M. A. Miller, J. M. Schulman and R. L. Disch, *J. Am. Chem. Soc.*, **110**, 7681 (1988).
(c) M. J. S. Dewar and A. J. Holder, *J. Am. Chem. Soc.*, **111**, 5384 (1989).
(d) D. W. Rogers, S. A. Loggins, S. D. Samuel, M. A. Finnerty and J. F. Liebman, *Struct. Chem.*, **1**, 481 (1990).
63. We hesitate to use a 'proof by existence' to document our assertion—that is, bicyclo[2.1.0]pent-2-ene is isolable under conventional chemical conditions while cyclobutadiene is not. As such, our analysis (derived from that of Jorgensen[67a]) must use quantum chemical calculations whenever the thermochemistry of cyclobutadiene is needed.
64. For our analysis in this section, we have made use of Melius' BAC-MP4 quantum chemical calculations from which one finds an enthalpy of formation of cyclobutadiene of (413 ± 20) kJ mol^{-1} (Carl F. Melius, personal communication).
65. W. R. Roth, F.-G. Klärner, G. Soepert and H.-W. Lennartz, *Chem. Ber.*, **125**, 217 (1992).
66. These homoaromaticity values have been taken from the study by Roth and his coworkers[35,65], where they combined hydrogenation enthalpies and their new MM2ERW force field. We use their enthalpies of formation values for [4.3.1]propella-2,4-diene and 1*a*,7*b*-dihydronaphtho-[*a*]cyclopropene, along with the archival values for the much more prosaic indane and naphthalene, to derive the cyclopropanation enthalpies of the latter two substances. The resulting values 132 and 88 kJ mol^{-1} make sense in terms of the earlier enunciated value of 119 kJ mol^{-1} for benzene. After all, the first species is a rather strained propellane (see the chapter by D. Ginsburg in Reference 1) and so indane should be harder to cyclopropanate than benzene. Conversely, because napthalene is less aromatic than benzene, the 1,2-cyclopropanation enthalpy of the former should be less than that of benzene. Indeed, neglecting any (admittedly small) vinyl and/or phenyl cyclopropane conjugation energy, we should think that the two reactions
(i) naphthalene + bicyclo[4.1.0]cyclohepta-2,4-diene → benzene + 1*a*,7*b*-dihydronaphtho-[*a*]cyclopropene
and
(ii) naphthalene + butadiene → benzene + styrene
should have comparable enthalpies. The first exothermicity is the difference of the cyclopropanation enthalpies of naphthalene and benzene, 31 kJ mol^{-1}. The latter, employing well-established archival enthalpies of formation, is (29.8 ± 2.5) kJ mol^{-1}. The agreement is astonishing.
67. P. K. Chou and S. R. Kass, *J. Am. Chem. Soc.*, **113**, 697 (1991).
68. Despite the above text, we still find an aura about [1.1.1]propellane and cite both the original primary paper and a review article describing its diverse chemistry written but a few years later:

K. B. Wiberg, W. P. Dailey, F. H. Walker, S. T. Waddell, L. S. Crocker and M. Newton, *J. Am. Chem. Soc.*, **107**, 7247 (1985) and K. B. Wiberg, *Chem. Rev.*, **89**, 973 (1989).

69. We must admit to being somewhat disingenuous. By including dicarbon in our 'homologous series', we glibly implied it had a quadruple bond between the two carbons. In fact, the dissociation energy of C_2 into atomic carbon is *ca* 602 kJ mol^{-1}, considerably less than that of ethylene into methylene, namely *ca* 728 kJ mol^{-1}. (The relevant enthalpies of formation of C_2, C and CH_2 were taken form Wagman and coworkers[9].) Yet, a triple bond in cyclopropyne also seems wrong, as opposed to 'cyclopropene-1,2-diyl', and the suggested nonplanarity of $\Delta^{1,3}$-bicyclo[1.1.0]butene naively exacerbates the strain. And, if [1.1.1]propellane now appears 'normal', it is only because small propellanes are now well-accepted parts of the chemical landscape (e.g. see the chapters by D. Ginsburg and K. B. Wiberg in Reference 1).

70. For ethylcyclopropane, there is *only* a 4.2 kJ mol^{-1} spread in the values of its enthalpy of vaporization as suggested in Kolesov and Kozina's archive[14], and those derived from Trouton's rule and the CHLL2 protocol.

71. Were we totally consistent, i.e. we used the CHLL2 rule for the enthalpies of vaporization of methyl and ethylcyclopropane instead of accepting Kolesov and Kozina's estimate[14], the differences for the methylated and ethylated benzene vs the corresponding cyclopropane would grow to 27.0 and 28.3 kJ mol^{-1}.

72. The enthalpies of formation of both α- and (E)-β-methylstyrenes. J.-L. M. Abboud, P. Jiménez, M. V. Roux, C. Turrión, C. Lopez-Domingo, A. Pododensin, D. W. Rogers and J. F. Liebman, *J. Phys, Org. Chem.*, in press.

73. The enthalpy of formation of phenylacetylene was obtained by using the enthalpy of hydrogenation reported by D. W. Rogers and F. J. McLafferty, *Tetrahedron*, **27**, 3765 (1971), the archival value for the enthalpy of formation of gas phase ethylbenzene, and the condensed phase–gas phase hydrogenation enthalpy equivalence of D. W. Rogers, O. A. Dagdagan and N. L. Allinger, op. cit., cited in Reference 24. Admittedly, there is an earlier enthalpy of formation measurement, that of T. L. Flitcroft and H. A. Skinner, *Trans. Faraday Soc.*, **54**, 644 (1958) from which we would likewise have derived a δ_{19}(Ph, Cypr; —C≡CH) value of 33.7 ± 4.8 kJ mol^{-1}.

74. We combined the archival enthalpy of formation with the enthalpy of vaporization reported by R. Fuchs and J. H. Hallman, *Can. J. Chem.*, **61**, 503 (1983).

75. The most general 'differentiated' CHLP equation is written

$$\delta\Delta H_v = 4.7\delta\tilde{n}_C + 1.3\delta n_Q + \delta b$$

since this allows for changes in the substituent b. For example, while in this chapter we wanted to compare the enthalpies of vaporization of singly substituted benzenes and cyclopropanes, one can imagine wanting to compare the vaporization enthalpies of a set of nitro compounds and methyl carboxylates, RNO_2 and RCOOMe.

76. Of course, we can always claim our lack of understanding is because there are errors in the experimental measurements. It has been noted that isopropyl-containing compounds often show surprising difficulty in obtaining complete combustion. (See Abboud and his coworkers[72] for citation of 20 calorimetric runs on *i*-PrCOOMe for which all had imcomplete combustion, and of another 20 on *i*-PrCONMe$_2$ for which only two were seemingly complete, albeit numerically discordant.) But no such difficulty has been evinced in the literature for the combustion of the cyclopropyl species.

77. Interestingly, cyclopropane and benzene have the same number of isomers for a given number of identical substituents: ignoring optical isomers, cyclopropane and benzene have one type of monosubstituted derivative, three types of disubstituted, three types of trisubstituted, three types of tetrasubstituted and one type apiece of penta- and hexasubstitution. But does this seemingly accidental counting equivalence have any thermochemical consequences?

78. For example, we recall discussions of *o*- and *p*-directing vs *m*-directing groups for electrophilic aromatic substitution, the importance of *o*- and *p*-quinonoid resonance structures vs the irrelevance of *m*-quinonoid long bonded resonance structures for stabilization of 'push–pull' disubstituted benzenes.

79. We invent in this chapter the generic symbols Phv and Cyprv for —C_6H_4— and —*c*-C_3H_4— where the two prongs of the 'v' are mnemonics for the two bonds used by the phenylene and 'cyclopropylene' (i.e. species **20a**, **20b** and **20c** *sans* the Xs or excess substituent groups). The

superscripts '*o*' and '*p*' accompanying Phv are for *ortho* and *para* respectively, and '*m*' would have been used for *meta*; the superscripts accompanying Cyprv, '*c*' and '*t*', are for *cis* and *trans*, and '*g*' would have been used for *gem*.

80. There are three more carbons in an arbitrary benzene derivative than the corresponding cyclopropane, and so the enthalpy of the former is automatically calculated by the CHLL2 protocol as 3 (4.7) kJ$\,$mol^{-1} higher. The word 'constant' is used in quotes because the CHLL2 predictions are usually only accurate to $\pm$ 2 kJ$\,$mol^{-1}.
81. We admit to presenting no benzene counterpart for *gem*-disubstituted cyclopropanes. Then again, there are almost no thermochemical data with which to make comparisons save for the three dimethylcyclopropanes. For example, the enthalpy of formation of 1,1-dichlorocyclopropane is available as are those of the three dichlorobenzenes, but we lack thermochemical data for the monosubstituted cyclopropyl chloride and for both 1,2-dichlorocyclopropanes.
82. For example, we really don't know what to make of the 100-year-old finding for *cis*-cyclopropane-1,2-dicarboxylic acid. We used the enthalpy of formation of $-$799 kJ$\,$mol^{-1}, reported by D. R. Stull, E. F. Westrum, Jr. and G. C. Sinke[17], which was taken from Kharasch's compendium[16] and originally reported by F. Stohmann and C. Klever, *J. Prakt. Chem.*, **45**, 475 (1892). It is tempting to disregard this finding because it is for the solid state. Besides, the same data evaluation chain involving century-old data for solid *cis*-cyclobutane-1,2-dicarboxylic acid gives an enthalpy of formation of $-$816 kJ$\,$mol^{-1} while a much more recent study gives ($-$838.1 $\pm$ 4.0) kJ mol^{-1} [see N. M. Gutner, V. L. Ryadnenko, N. A. Karpenko, E. F. Mahinya and N. N. Kiseleva, 'Standard Enthalpies of Formation of Several Cyclic Derivatives', paper O-4, p. 196, 'Problems in Calorimetry and Chemical Thermodynamics: Proceedings of the Xth All-Union Conference', Vol. 1, Part 2, Chernogolovka, USSR, 1984; *Chem. Abstr.*, **104**, 148235s (1986)]. The *ca* 20 kJ$\,$mol^{-1} discrepancy is admittedly disconcerting, primarily because we do not know enough to ascribe it to (presumably) improvements in sample purity and calorimetric technology. One should not immediately invalidate old thermochemical results. For example, paralleling the result of the enthalpy of combustion of 2,4-hexadiyne of $-$3553.7 $\pm$ 4.9 kJ$\,$mol^{-1} of Luk'yanova and her coworkers[53] is the over-100-year-old result of W. Louguinine [*Compt. Rend.*, **106**, 1472 (1888)] of $-$3547.2 kJ$\,$mol^{-1}!
83. We really should not be surprised by the absence of these data: on heating, these acids are reasonably expected to dehydrate to form their thermochemically useless anhydrides.
84. In order for the sublimation enthalpy difference between two compounds M^1 and M^2, $\Delta H_S(M^1) - \Delta H_S(M^2)$, to equal the corresponding vaporization enthalpy difference $\Delta H_V(M^1) - \Delta H_V(M^2)$, it is necessary that the enthalpies of fusion, $\Delta H_{fus}(M^1)$ and $\Delta H_{fus}(M^2)$, be equal. From our experiences that generated the studies reported in Reference 13, there is generally no reason for optimism.
85. B. E. Smart, in *Molecular Structure and Energetics: Studies of Organic Molecules* (Eds. J. F. Liebman and A. Greenberg), VCH, Deerfield Beach, 1986.
86. What sensitizes us to the possibility of problems is the following comparison of the enthalpies of formation of tetrafluoro derivatives. The suggested enthalpy of formation of gaseous 1,1,2,2-tetrafluorocyclopropane is ($-$590 $\pm$ 42) kJ mol^{-1}. We would hope that the error bars are large enough to allow us to compare this result with that of *any* tetrafluorobenzene. The sole tetrafluorobenzene for which there is enthalpy of formation data is the liquid 1,2,4,5-species, ($-$683.7 $\pm$ 1.8) kJ mol^{-1}: the necessary enthalpy of vaporization must be estimated. It has been amply documented that fluorination exerts little effect on enthalpies of vaporization [see J. F. Liebman, in *Fluorine-Containing Molecules: Structure, Reactivity, Synthesis, and Applications* (Eds. J. F. Liebman, A Greenberg and W. R. Dolbier, Jr.), VCH, New York, 1988]. For example, benzene, and its mono-, all three di-, penta- and hexa-fluoro derivatives have values spanning a range of under 3 kJ$\,$mol^{-1}. It is safe to assume that 1,2,4,5-tetrafluorobenzene will have a comparable enthalpy of vaporization, namely *ca* (36 $\pm$ 2) kJ$\,$mol^{-1}. The resulting gas-phase enthalpy of formation of 1,2,4,5-tetrafluorobenzene, *ca* $-$648 kJ$\,$mol^{-1}, lies some (60 $\pm$ 40) kJ$\,$mol^{-1} lower than that of 1,1,2,2-tetrafluorocyclopropane. The discrepancy is painful.
87. For this analysis, we derive the desired number by summing the following quantities:
(i) the measured enthalpy of formation of solid *p*-terphenyl [(150.5 $\pm$ 4.2) kJ$\,$mol^{-1}] found on p. 53 of the Research Report of Chemical Thermodynamics Laboratory, Faculty of Science, Osaka University, Vol. 7, for the year 1986 (edited by H. Suga),
(ii) the measured enthalpy of sublimation (120.6 kJ$\,$mol^{-1} at a mean temperature of 363 K), reported by H. Hoyer and W. Peperle, *Z. Elektrochem.*, **62**, 61 (1958) and

(iii) the sublimation temperature correction for 298 K using the heat capacity correction equation 11 of J. S. Chickos, S. Hosseini, D. G. Hesse and J. F. Liebman, *Struct. Chem.*, **4**, 271 (1993) where the heat capacity of solid *p*-terphenyl was assumed to equal that of its *o*-isomer, measured earlier by S. S. Chang and A. B. Bestul, *J. Chem. Phys.*, **56**, 503 (1972).

88. See, for example,
(a) K. B. Wiberg, in Reference 1, Chap.1, p. 1.
(b) D. Cremer and E. Kraka, in *Structure and Reactivity* (Eds. J. F. Liebman and A. Greenberg), VCH, New York, 1988.
(c) M. J. S. Dewar, in *Modern Models of Bonding and Delocalization* (Eds. J. F. Liebman and A. Greenberg), VCH, New York, 1988.

89. To the extent we are studying only hydrocarbons and not particularly concerned about accuracy in deriving enthalpy effects of better than a kJ mol^{-1}, we can simply accept the nearly identical values (though different analyses and vantage points) derived by J. D. Cox and G. Pilcher, *Thermochemistry of Organic and Organometallic Compounds*, Academic Press, New York, 1970 and of J. F. Liebman and D. Van Vechten, in *Molecular Structure and Energetics: Physical Measurements* (Eds. J. F. Liebman and A Greenberg), VCH, Deerfield Beach, 1987. However, complications can arise when considering substituted species—the reader should still note the complementary discussions of S. W. Slayden and J. F. Liebman, in *The Chemistry of Functional Groups, Supplement E2: The Chemistry of Hydroxyl, Ether and Peroxide Groups* (Ed. S. Patai), Chap. 4, Wiley, Chichester, 1993, p. 103 and J. F. Liebman, K. S. K. Crawford and S. W. Slayden, in *The Chemistry of Functional Groups, Supplement S: The Chemistry of Sulphur-containing Functional Groups* (Eds. S. Patai and Z. Rappoport), Chap. 4, Wiley, Chichester, 1993, p. 197.

90. The reader will note our continuing tendency to use thermochemical data from either the gaseous or liquid states. Recall the discussion in Reference 82 on solid *cis*-cyclopropane-1,2-dicarboxylic acid and *cis*-cyclobutane-1,2-dicarboxylic acids: the difference of the 'original' enthalpies of formation is 14 kJ mol^{-1}. For completeness, we note using the same reference chain (from Stohmann and Klever[83] to Kharasch[16] to Stull, Westrum and Sinke[17] to us) we find a 17 kJ mol^{-1} difference for the enthalpies of formation of cyclopropane- and cyclobutane- 1,1-dicarboxylic acids. These two sets of data are consistent. Yet, the corresponding acyclic species (as solids) $(CH_2)_3(COOH)_2$ and $(CH_2)_4(COOH)_2$ have enthalpies of formation that differ by (34.3 ± 1.4) kJ mol^{-1}. This is confusing.

91. T. Holm, *J. Chem. Soc., Perkin Trans. 2*, 464 (1981).

92. J. C. Traeger, *Org. Mass Spectrom.*, **24**, 5591 (1989).

93. It is somehow rather perverse that the cyclopropane/cyclobutane pair that obey our expectations thermochemically best are those employing the least classical and poorest precedented, thermochemical methodologies. Even the computational theorist is thwarted in large part because of comparatively less experience with the 4th row element, Br.

94. See A. Greenberg and T. A. Stevenson, op. cit. (Reference 18), p. 236.

95. See R. Fuchs and J. H. Hallman, op. cit. (Reference 74). We share these authors' conclusion that the sole calorimetric study of the enthalpy of formation of cyclobutane is somehow 'incomplete', and so invites suspicion that it is also inaccurate. [See S. Kaarsemaker and J. Coops, *Recl. Trav. Chim. Pays-Bas*, **71**, 261 (1952).]

96. For simplicity in both writing and reading the chapter, it has been assumed the archival value for the enthalpy of formation of cyclobutane is correct.

97. The reader may feel it was unfair to compare propane in the 'methylenation' reaction and butane in the 'olefination' reaction. The methylenation energy of butane is (90.3 ± 1.2) kJ mol^{-1}, quite indistinguishable from that of propane. Olefination of propane is unfair because it changes a $—CH_3$ into a $=CH_2$, a process not found in any of the other transformations. Finally, we note that the corresponding transformation of ethylene, i.e. its olefination into acetylene, is accompanied by 'merely' a (175.7 ± 0.8) kJ mol^{-1} change, not nearly as large as the cyclopropane to cyclopropene change.

98. The experimentally measured enthalpy of formation of bicyclo[2.2.0]hexane is taken from the study of Roth and his coworkers[35].

99. The enthalpy of formation of bicyclo[3.2.0]heptane is taken as the MM2ERW calculated value of Roth and his coworkers[35]. Despite the considerable success of this force field, we are hesitant to give error bars to any individual value.

100. Recall the revised value for the enthalpy of formation of bicyclo[4.1.0]heptane discussed in Reference 33.
101. The experimentally measured enthalpy of formation of bicyclo[2.2.0]hex-2-ene is taken from the study of Roth and his coworkers[35].
102. This comparison of unsaturated and saturated, three- and four-membered rings containing bicyclic hydrocarbons, used here as a method of ascertaining the homoaromaticity of the unsaturated three-membered ring species, is, in fact, equivalent to the hydrogenation enthalpy analysis of Roth and his coworkers[65] and one of the theoretical models suggested by Jorgensen[62a].
103. Roth and his coworkers[35] give two values for the enthalpy of formation of bicyclo[4.2.0]octa-2,4-diene. We have chosen 189.2 kJ mol^{-1} because it is closer to their MM2ERW calculated value.
104. See J. F. Liebman, in *Cyclophanes*, Vol. 1 (Eds. P. M. Keehn and S. M. Rosenfeld), Academic Press, New York, 1983.
105. Roth and his coworkers[35] give three values for the enthalpy of formation of benzocyclobutene to decide among. We opt for the average of the two reported by Roth and his coworkers that, in fact, differ by less than 2 kJ mol^{-1} and additionally are nearly identical to the MM2ERW calculated value.
106. B. Halton, *Chem. Rev.*, **89**, 1161 (1989).
107. D. Van Vechten and J. F. Liebman, *Isr. J. Chem.*, **21**, 105 (1981).
108. The other two *possible* chemically realizable species are tetrahedrane and dodecahedrane. [We do not consider here the $(CH)_n$ polymer that corresponds to saturated infinite sheets of graphite.]
109. B. D. Kybett, S. Carroll, P. Natalis, D. W. Bonnell, J. L. Margrave and J. L. Franklin, *J. Am. Chem. Soc.*, **88**, 626 (1966).
110. D. R. Kirklin, K. L. Churney and E. S. Domalski, *J. Chem. Thermodyn.*, **21**, 1105 (1989).
111. H. D. Beckhaus, C. Rüchardt and M. Smiser, *Thermochim. Acta*, **79**, 149 (1984).
112. It is indeed unfortunate that the authors of Reference 110 chose to study the dicarbomethoxy derivative and the authors of Reference 111 chose the monocarbomethoxy derivative. Had the degree of substitution been the same, then equation 40 could have been rewritten
(i) 'substituted cubane' + quadricyclane → 'substituted homocubane' + prismane.
113. D. W. Rogers, F. J. McLafferty, W. Fang and Y. Qi, *Struct. Chem.*, **4**, 161 (1993) and the many methods and many references discussed therein.
114. (a) J. F. M. Oth, *Angew. Chem., Int. Ed. Engl.*, **7**, 646 (1968) and *Recl. Trav. Chim. Pays-Bas*, **87**, 1185 (1968).
(b) W. Adam and J. C. Chung, *Int. J. Chem. Kinet.*, **1**, 487 (1969).
115. D. M. Lemal and L. H. Dunlap, Jr., *J. Am. Chem. Soc.*, **94**, 6562 (1972). In this paper, these authors introduce the concept of the 'perfluoroalkyl effect' to explain the relative (kinetic and/or thermodynamic) stabilization of strained species relative to their 'normal' counterparts by CF_3 and other R_f groups.
116. A. Greenberg, J. F. Liebman and D. Van Vechten, *Tetrahedron*, **36**, 1161 (1980).
117. The immediate answer we give is quite simple: one study[114a] refers to measurements at 60 °C while the other[114b] refers to measurements at 178 °C. Intuitively, we would prefer the lower temperature study because
(a) there is presumably less decomposition of either reactant or product to a third, thermochemically irrelevant, and chemically uncharacterized species;
(b) a smaller temperature correction is needed to correct the results to the idealized 25 °C of the thermochemist's standard state.
However, we have no idea how much either measurement is 'contaminated' by the effects of reason (a) and the correction suggested in (b) is expected to be relatively small.
118. The difference of the rearrangement enthalpy of hexamethylprismane and prismane can reasonably be assumed to be the sum of
(i) six times the difference of the rearrangement enthalpy of monomethylprismane into monomethylbenzene (i.e. toluene),
(ii) six times the repulsion energy of the *cis*-methyl groups in 1,2-dimethylcyclopropane,
(iii) three times the repulsion energy of the *cis*-methyl groups in 1,2-dimethylcyclobutane, and
(iv) the negative of six times the repulsion energy of two *o*-groups in 1,2-dimethylbenzene (i.e. *o*-xylene).

However, we lack the first difference, and indeed we lack thermochemical data of any type on any monosubstituted prismane. Let us label this unknown quantity ΔH_{iso}^{prs}. The second difference is expressible as the difference of the enthalpies of formation of the isomeric *cis*- and *trans*-1,2-dimethylcyclopropane: 5.5 kJ mol^{-1} [3]. The third difference is likewise unavailable and, indeed, there are fewer calorimetric data for cyclobutane derivatives than for cyclopropane derivatives. Assuming (admittedly, falsely) substituent effects on cyclopropanes and cyclobutanes are identical, the difference is 5.5 kJ mol^{-1}. Should substituent effects on cyclobutanes be simulated by those of cyclohexanes, the difference is now 7.8 ± 2.6 kJ mol^{-1}. The last thermochemical quantity may be estimated as the difference of the enthalpies of formation of the isomeric *o*- and *p*-xylene, 1.1 ± 1.4 kJ mol^{-1}. For all of the results to be consistent, then $6\Delta H_{iso}^{prs} + 45 \approx 200$ kJ mol^{-1} and so $\Delta H_{iso}^{prs} \approx 25$ kJ mol^{-1}. In that the methylation enthalpies of ethylene, benzene and cyclopropane are so similar (cf Sections II. A and III. A), the methylation of cyclobutane enthalpy cannot be too different. But the resulting *ca* 25 kJ mol^{-1} for ΔH_{iso}^{prs} seems untenable. We are confused.

119. The lack of additional thermochemical data is admittedly frustrating because there are many other documented cyclopropenones [e.g., see W. E. Billups and A. W. Moorehead, in *The Chemistry of the Cyclopropyl Group* (Ed. Z. Rappoport), Chap. 24, Wiley, Chichester, 1987, p. 1533] which have higher stability and alternative amounts of conjugation with the affixed substituents.
120. The enthalpy of formation of gaseous diphenylcyclopropenone is that presented in our major cyclopropane archive[3].
121. R. E. Krall and J. D. Roberts, in *American Chemical Society, Division of Petroleum Chemistry Symposium*, Vol. 3, B63–B68 (1958).
122. We reiterate that both homoaromaticity and aromaticity are more pronounced in ions than in related neutrals. In the tour-de-force of computational theory, S. Sieber, P. v. R. Schleyer, A. H. Otto, J. Gauss, F. Reichel and D. Cremer [*J. Phys. Org. Chem.*, **6**, 445 (1993)] document considerable homoaromatic stabilization of the cyclobutenyl cation. Yet the difference of the enthalpy of formation they calculate from their quantum chemical cations, 1021 kJ mol^{-1}, is only 54 kJ mol^{-1} lower than that archivally recommended for the cyclopropenium ion [S. G. Lias, J. E. Bartmess, J. F. Liebman, J. L. Holmes, R. D. Levin and W. G. Mallard, *J. Phys. Chem. Ref. Data*, **17** (1988), Supplement 1]. This last number allows us to conclude that the aromatic stabilization of cyclopropenium ion exceeds the homoaromatic stabilization of cyclobutenyl cation by at least (160 – 54) = 106 kJ mol^{-1}. Regrettably, inadequate time and space does not allow us to discuss the energetics of ions containing three-membered rings in the current chapter.

CHAPTER **5**

Recent advances in synthesis of cyclopropanes

MASAKAZU OHKITA, SHINYA NISHIDA and TAKASHI TSUJI

Department of Chemistry, Faculty of Science, Hokkaido University, Sapporo 060, Japan
Fax: +81 11 706 4924

I. INTRODUCTION . . . 262
II. 1,3-BOND FORMATION . . . 262
A. 1,3-Elimination of Two Heteroatoms . . . 262
B. 1,3-Elimination of HX; Intramolecular Displacement Reactions . . . 264
C. Cyclization of Allylic Derivatives . . . 270
1. γ-Substituted Michael acceptors and nucleophiles . . . 270
2. Reactions of other allylic derivatives . . . 270
D. Cyclization of 3-Butenyl Derivatives . . . 273
E. Miscellaneous . . . 276
III. COMBINATION OF C_2 AND C_1 BUILDING BLOCKS . . . 279
A. Cyclopropanation of Carbon–Carbon Multiple Bonds with Carbenes and Carbenoids . . . 279
1. Methylene . . . 280
2. Alkyl-, alkenyl- and arylcarbenes . . . 285
3. Alkenylidenes . . . 287
4. Monohalo- and dihalocarbenes . . . 287
5. Acyl- and alkoxycarbonylcarbenes . . . 290
6. Carbenes α-substituted with heteroatoms other than halogen . . . 294
B. Cyclopropanation of Michael Acceptors . . . 296
1. With ylides . . . 296
2. With α-halocarbanions and related species . . . 299
C. Cyclopropanation of Active Methylene Compounds . . . 302
1. With 1,2-dihaloethanes and related compounds . . . 302
2. With Michael acceptors carrying a leaving group at the α-carbon atom . . . 303
D. Cyclopropanation with Diazo Compounds via 1-Pyrazolines . . . 304
E. Miscellaneous . . . 306
IV. REFERENCES . . . 310

The chemistry of the cyclopropyl group, Vol. 2
Edited by Z. Rappoport

I. INTRODUCTION

The purpose of this review is to give a summary of recent developments in the synthesis of three-membered carbocycles. This article is a sequel to Chapter 7 in the previous volume in this series published in 1987[1] and most of the references are taken from the period 1985–1993. Since the sheer volume of applications reported precludes complete coverage here, citations are limited to the reactions suggesting new synthetic aspects. This review is divided into two sections, namely 1,3-bond formation and combination of C_2 and C_1 building blocks. In the former section, intramolecular cyclization reactions to give three-membered carbocycles are covered. In the latter section are treated cyclopropane-producing intermolecular cyclization and cycloaddition reactions including a few examples of combination of three C_1 building blocks. More detailed information on the related, specific subjects may be found in other chapters and in recent reviews, a list of which is given in Reference 2.

II. 1,3-BOND FORMATION

A. 1,3-Elimination of Two Heteroatoms

Reductive 1,3-elimination reaction of alkyl dihalides constitutes one of the classical methods for the preparation of cyclopropyl derivatives and is particularly useful for the synthesis of highly strained polycyclic hydrocarbons. A new preparation method of [1.1.1]propellane, more versatile than the original Wiberg's method, has been devised[3,4]. Thus, treatment of 1,1-dibromo-2,2-bis(chloromethyl)cyclopropane with alkyllithium or lithium powder affords [1.1.1]propellane by two successive 1,3-eliminations of halogens by way of 1-bromo-2-(chloromethyl)bicyclo[1.1.0]butane (equation 1). This method has been

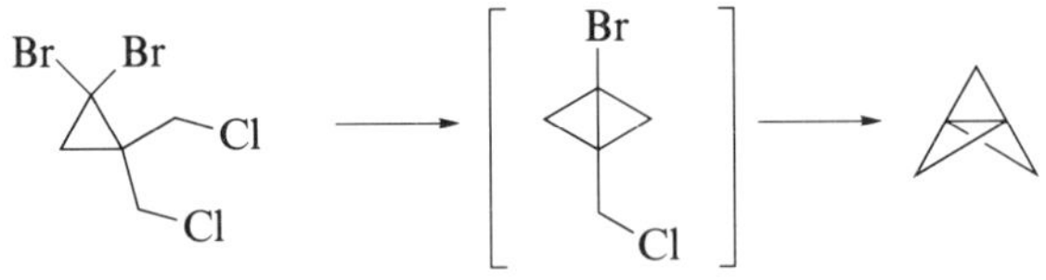

(1)

MeLi/ether
or Li/triglyme-decane (4:1)

70% (ether solution)
25–38% (solvent free)

applied successfully for the preparation of [1.1.1]propellane derivatives including annelated **1**[3, 5] and the functionalized diether **2** (equation 2)[6]. An extremely strained, condensed

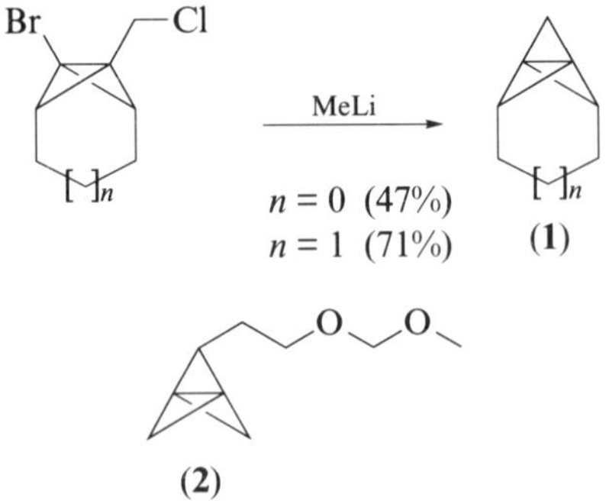

(2)

polycyclopropane, tricyclo[2.1.0.0[1,3]]pentane **3**, has been generated as a thermally labile species using a similar synthetic strategy (equation 3)[7]. The reactions of *cis*-alkenes bear-

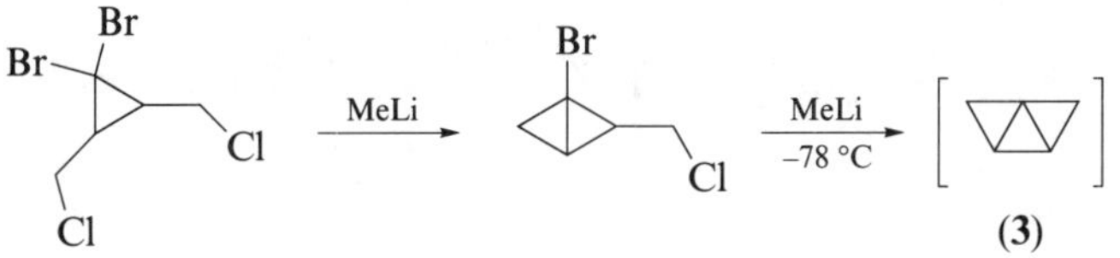

(3) (3)

ing iodine and chloromethyl as the olefinic vicinal substituents with alkyllithium provides 1,2-disubstituted cyclopropenes in good yields (equation 4)[8]. When this method is applied for the preparation of mono-substituted cyclopropenes, however, yields are poor. The precursor mixed dihalides used in these reactions are readily prepared from propargyl alcohols via their carbometallation followed by iodonolysis and chlorination.

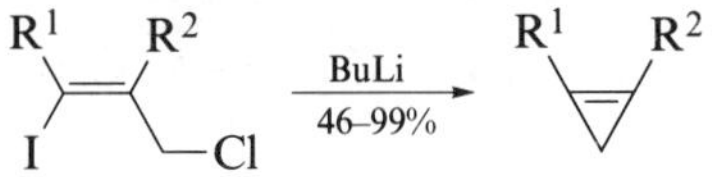

R^1 = $SiMe_3$, Ph, *n*-Hex
R^2 = alkyl, Me, *n*-Pr, *n*-Bu, Ph (4)

Alkyltin compounds carrying a leaving group at the γ-position undergo 1,3-elimination to produce cyclopropanes[9]. γ,δ-Epoxyalkyltin derivatives are converted to cyclopropylmethyl alcohols under catalysis of a Lewis acid (equation 5)[10]. γ-Hydroxyalkyl selenides

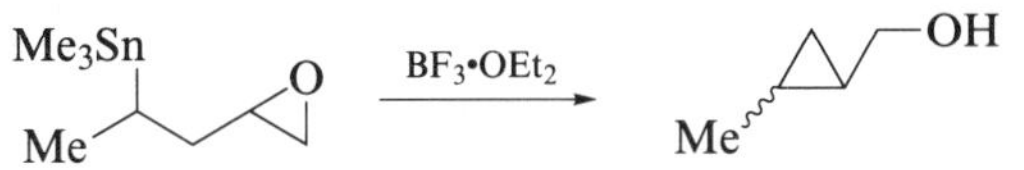

(5)

are transformed into cyclopropanes with reasonably high stereospecificity on sequential reactions with benzenesulfonyl chloride and butyllithium[11], while the reaction of the corresponding tin derivatives is completely stereospecific with inversion of configuration at each of the reaction sites (equation 6)[12]. These reactions proceed via the initial transmeta-

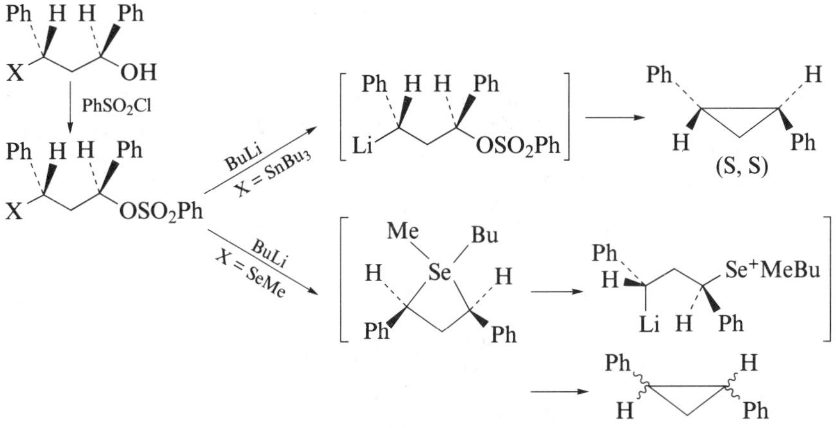

(6)

(S,S) + (R,R)

X = $SnBu_3$ 99.5% ee (67%, *trans*/*cis* 100:0)
SeMe 46% ee (74%, *trans*/*cis* 97:3)

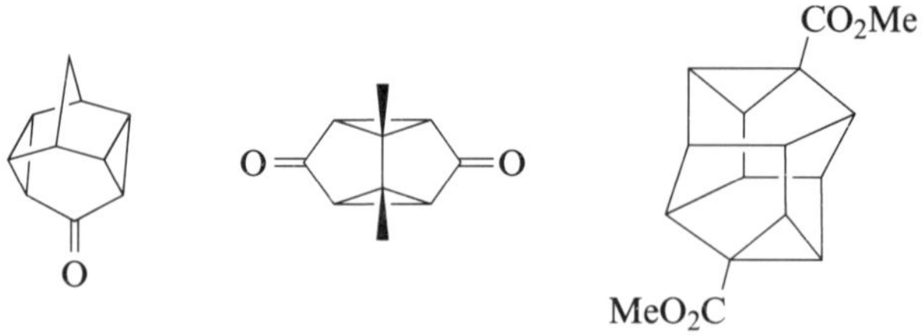

SCHEME 1

lation with butyllithium followed by the intramolecular nucleophilic substitution. It is suggested that the decreased stereospecificity in the former reaction arises from the formation of selenacyclobutane intermediate in competition with the transmetalation[13].

B. 1,3-Elimination of HX; Intramolecular Displacement Reactions

Active methylene and methine compounds bearing a leaving group (X) on the γ-carbon atom can afford cyclopropyl derivatives via 1,3-elimination of HX. 1,2-Elimination to give alkenes and direct nucleophilic substitution by base may compete with the 1,3-elimination, particularly in the preparation of excessively strained cyclopropyl derivatives. The preferred reaction course is, however, highly dependent on reaction conditions, especially on the nature of the base and solvent employed, as exemplified by the reactions of **4** (equation 7)[14].

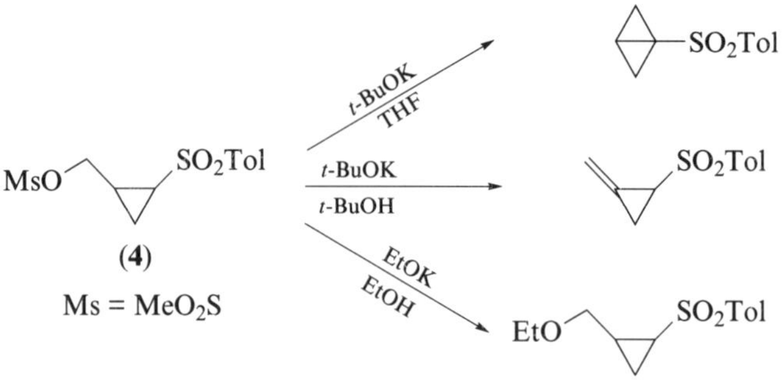

(7)

The use of ethoxide, which is less basic but more nucleophilic than *t*-butoxide, tends to produce substitution products. It has been pointed out that 1,2-elimination becomes predominant in the reactions of γ-sulfonylalkyl methanesulfonates with *t*-BuOK/*t*-BuOH when the increment in enthalpy upon cyclization via 1,3-elimination exceeds 160 kJ mol^{-1}[15].

A variety of three-membered carbocycles including cyclopropylcarbonyl and -sulfonyl derivatives, cyclopropylcarbonitriles and -methanols, nitrocyclopropanes, cyclopropanols and cyclopropylamines have been prepared via the 1,3-elimination of HX. Some representative cyclopropyl derivatives recently prepared by this method are shown in Scheme 1[16–18] and in equations 8–26. Conversion of chelated homoserine, **5**,to chelated 2-amino-4-bromobutyrate and treatment with aqueous base directly affords chelated 1-aminocyclopropane-1-carboxylate (equation 8)[19]. The 1,3-elimination in **6** interestingly leads to the preferential formation of the *cis* isomer, from which **7**, a key structural element of synthetic pyrethroid insecticides, is obtained (equation 9)[20]. A sulfur substituent can serve both as an activating group and as a leaving group in this type of reaction and, thus, 1,3-bis(phenylthio)propane affords cyclopropyl phenyl sulfide upon treatment with butyl-

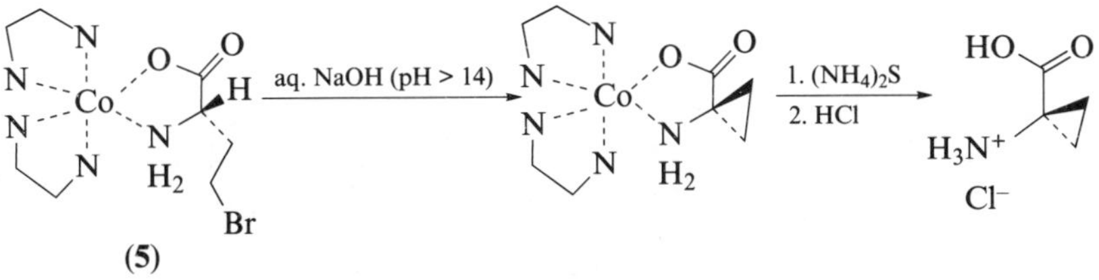

(8)

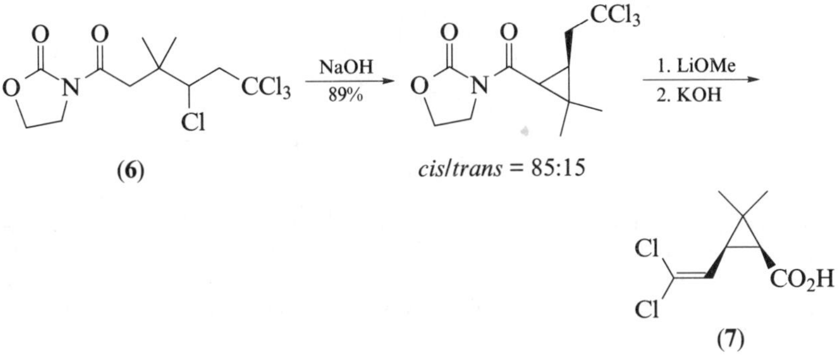

(9)

lithium. The 2-carbamoyl derivative, **8**, undergoes successive cyclization and chelation-controlled lithiation to give **9**, when treated with excess butyllithium, which provides bicyclic γ-butyrolactones, **10**, upon exposure to aldehydes (equation 10)[21]. Sequential

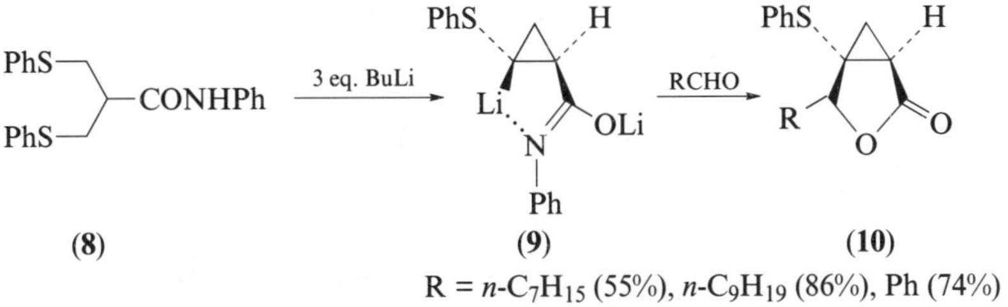

(10)

treatment of the 2-formyl derivative with alkyllithium and DMF stereoselectively furnishes bicyclic lactols, **11**, in an analogous manner (equation 11)[22]. An asymmetric cyclization

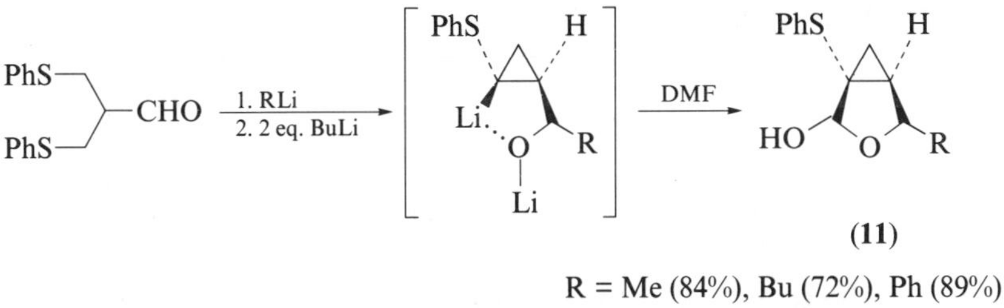

(11)

is realized with the carbamoyl derivatives modified with chiral amines and *trans*-2-(phenylthio)cyclopropanecarboxamides are obtained in diastereomeric excesses (de) up to 83% (equation 12)[23].

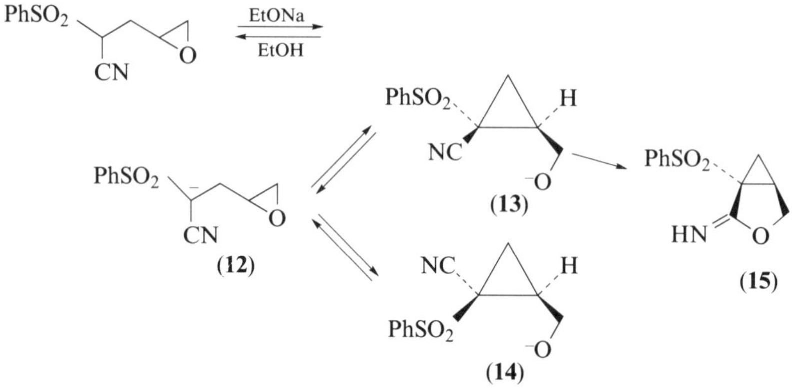

2-Oxiranyl carbanions undergo regioselective intramolecular substitution to give cyclopropylmethoxide ions. Thus, epoxidation of γ,δ-unsaturated active methylene and methine compounds followed by treatment with appropriate base provides substituted cyclopropylmethanols (equation 13)[24]. The use of chiral epoxides, e.g. prepared by

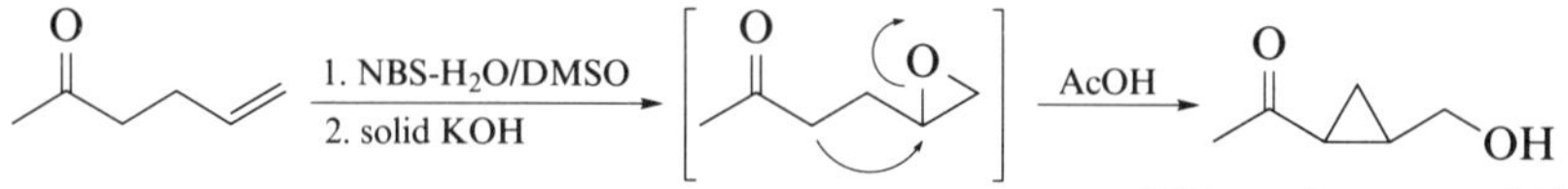

(13)

87%, *cis*/*trans* = > 99/1

Sharpless epoxidation, leads to the formation of optically active cyclopropylmethanols (equation 14)[25,26]. The cyclization/ring opening of α-cyano-α-sulfonyl-γ,δ-epoxycarbanion,

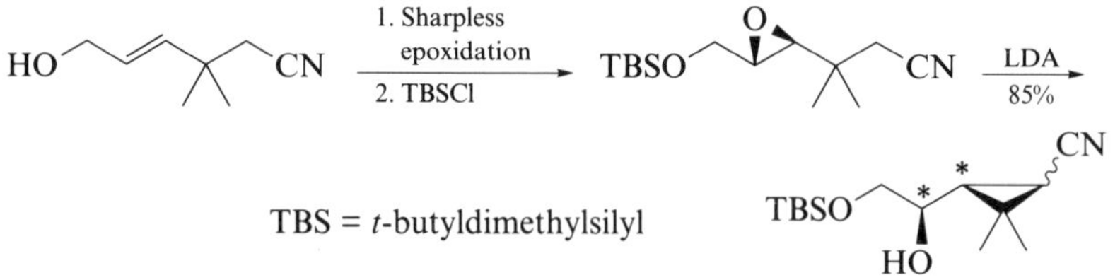

(14)

trans/*cis* = 2.2

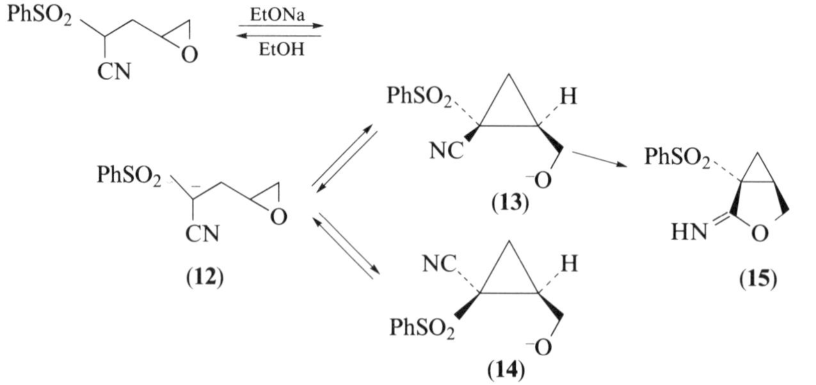

(15)

12, to give a diastereomeric pair of cyclopropylmethoxide ions, **13** and **14**, is reversible and a *cis*-fused bicyclic imine, **15**, is eventually produced stereoselectively in a good yield (equation 15)[27].

Treatment of γ-nitro alcohols with diethyl azodicarboxylate DEAD and triphenylphosphine affords nitrocyclopropanes with inversion of configuration at the α-carbon via the intramolecular Mitsunobu reaction involving carbon nucleophiles stabilized by the nitro group (equation 16)[28]. The reaction works best with nitro compounds ($pK_a < 17$) and is not applicable to the sulfonyl derivatives ($pK_a \sim 25$).

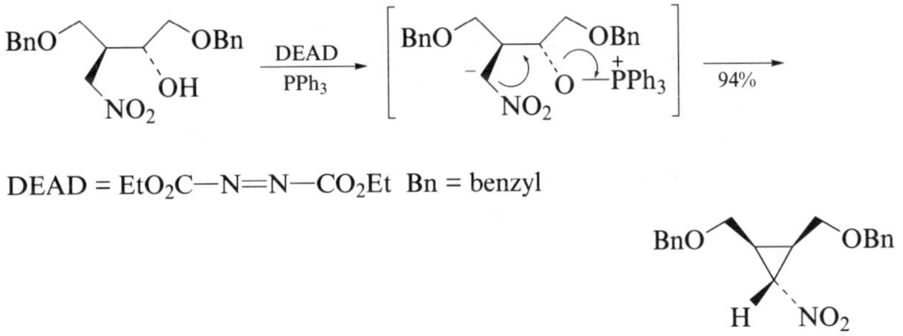

(16)

Stereodivergent synthesis of almost enantiomerically pure and stereospecifically deuterium-labeled 2,2-dimethylcyclopropanols has been achieved by treating 2,2-dimethylpropane-1,3-diol dicarbamate with alkyllithium in the presence of (–)-sparteine (equation 17)[29].

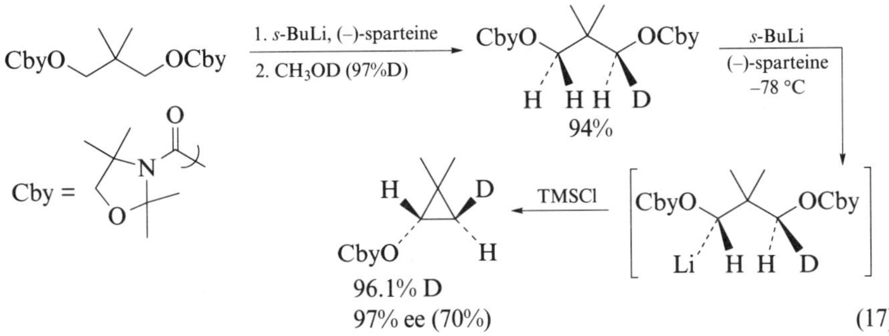

(17)

Stereochemical investigation on the vinylogous 1,5-elimination reactions of **16** to give cyclopropylalkenes has revealed high *syn* stereoselectively of the intramolecular $S_N{}'$ process (equation 18)[30]. The stereoselectivity of the cyclization reaction of **17** to give the (*E*)-isomer is improved by the increase in steric bulkiness of the sulfonyl substituent (equation 19)[31].

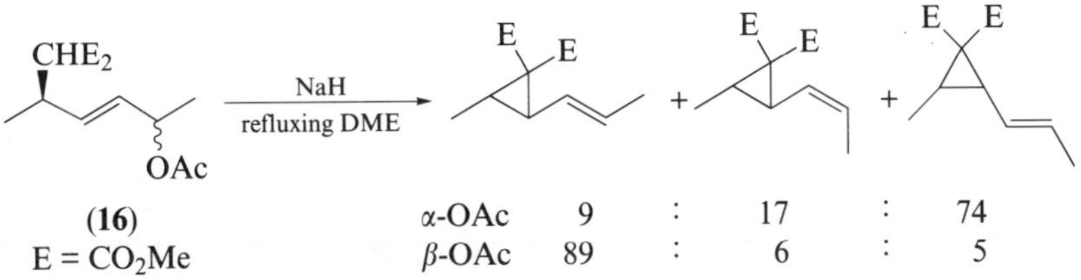

(18)

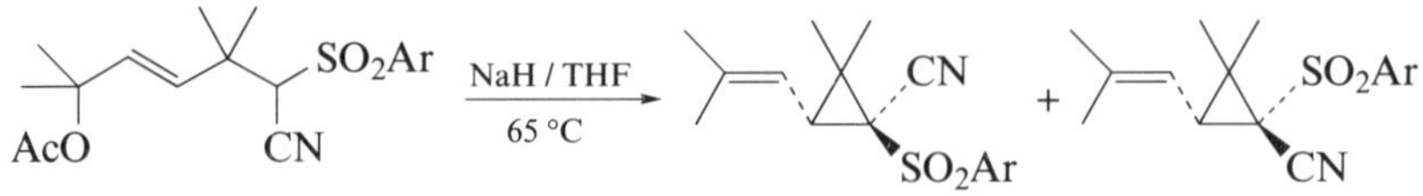

(19)

(**17**)

Ar = *p*-tolyl 80 : 20
mesityl 100 : 0

Preparation of *trans*-2-vinylcyclopropanecarboxylic ester in high diastereomeric excess has been accomplished by the reaction of 6-bromo-4-hexenoic ester carrying a chiral alcohol moiety[32]. Treatment of readily available β-allenic tosylates with LDA or alkyllithium provides an expeditious, although stereorandom, synthesis of alkynylcyclopropanes (equation 20)[33]. Since the cyclization is expected to proceed with inversion of configuration at the carbon bearing the leaving group and the starting β-allenic alcohols are available in high enantiomeric purity, this method provides an easy entry to chiral, non-racemic cyclopropanes possessing an alkyne side-chain.

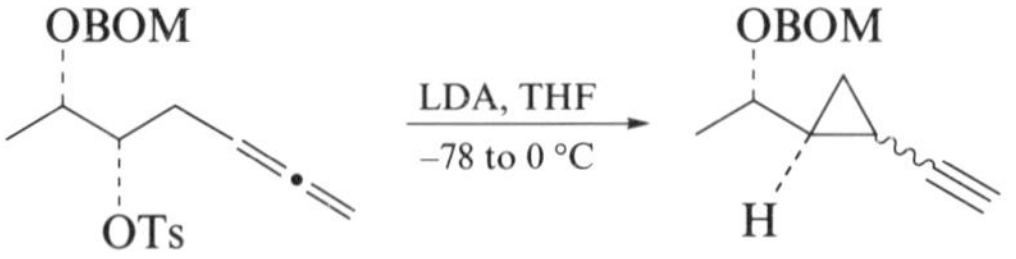

91%, dr = 2 : 1 (20)

BOM = benzyloxycarbonyl

The vinylogous elimination to give alkenylcyclopropanes may also be effected via π-allyl palladium complexes[34,35]. The palladium(0)-catalyzed substitution of allylic esters with stabilized carbon nucleophiles via π-allyl palladium derivatives stereospecifically proceeds with net retention (double inversion) of configuration. Thus, the chirality of an allylic substrate is transferred to resultant alkenylcyclopropanes in the intramolecular $S_{N'}$ reaction via π-allyl palladiums (equation 21)[36,37].

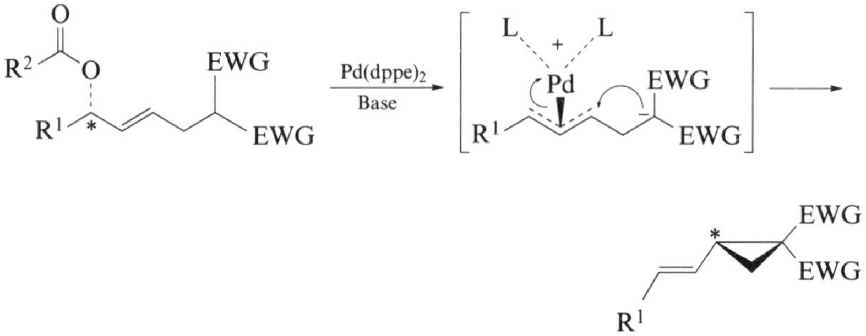

(21)

dppe = 1,2-bis(diphenylphosphino)ethane; EWG = electron-withdrawing group

An asymmetric preparation of alkenylcyclopropanes has also been realized by the use of palladium(0) complexes carrying chiral ferrocenylphosphine ligands (equation 22)[38]. The requisite π-allyl palladium intermediates can also be generated from allene and methylenecyclopropane derivatives, **18**[39] and **19**[40], in the presence of palladium(0) complex and alkenyl or aryl halide (equations 23 and 24). The cobalt complexes, **20**, similarly afford the corresponding alkenylcyclopropanes upon exposure to LDA (equation 25)[41].

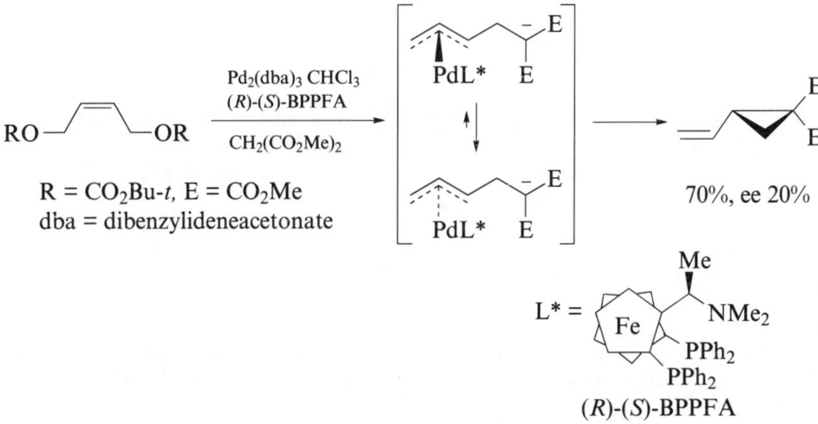
Pd$_2$(dba)$_3$ CHCl$_3$
(R)-(S)-BPPFA
CH$_2$(CO$_2$Me)$_2$
RO
OR
R = CO$_2$Bu-t, E = CO$_2$Me
dba = dibenzylideneacetonate
PdL*
E
70%, ee 20%
L* =
Fe
Me
NMe$_2$
PPh$_2$
PPh$_2$
(R)-(S)-BPPFA

(22)

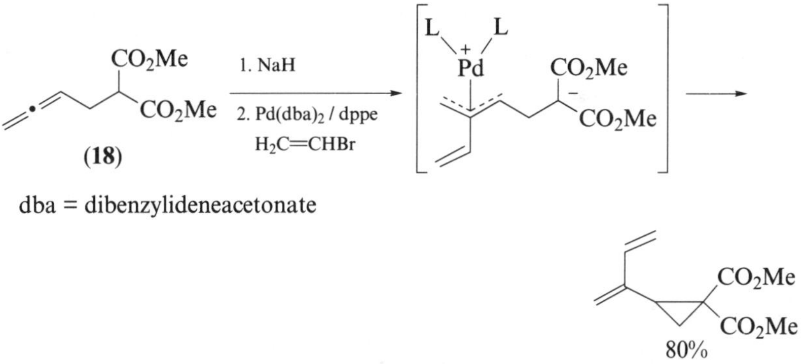
CO$_2$Me
CO$_2$Me
(18)
1. NaH
2. Pd(dba)$_2$ / dppe
H$_2$C=CHBr
L
L
Pd
dba = dibenzylideneacetonate
80%

(23)

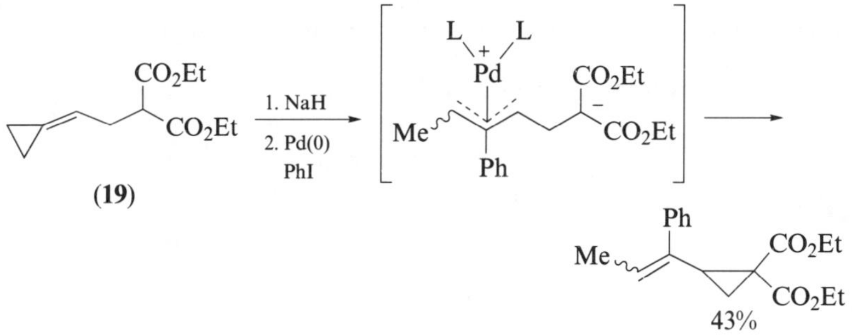
CO$_2$Et
CO$_2$Et
(19)
1. NaH
2. Pd(0)
PhI
L
L
Pd
Me
Ph
43%

(24)

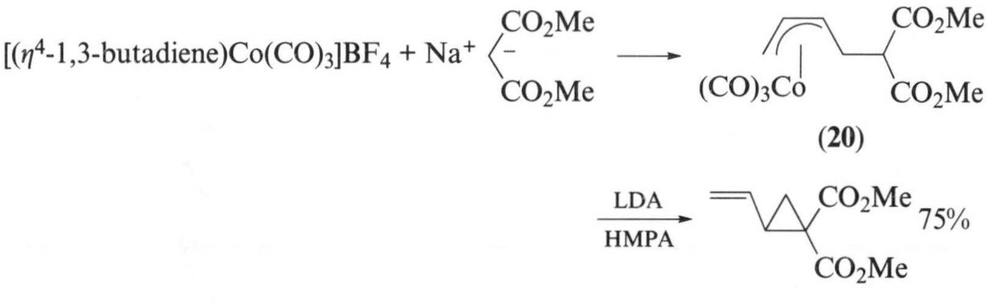
[(η^4-1,3-butadiene)Co(CO)$_3$]BF$_4$ + Na$^+$
CO$_2$Me
CO$_2$Me
(CO)$_3$Co
(20)
LDA
HMPA
75%

(25)

Treatment of β-chloro-*N*-benzylimines, **21**, with *t*-BuOK produces *N*-benzylidenecyclopropylamines via 1,5-dehydrochlorination. Hydrolysis of the imines provides cyclopropylamines (equation 26)[42].

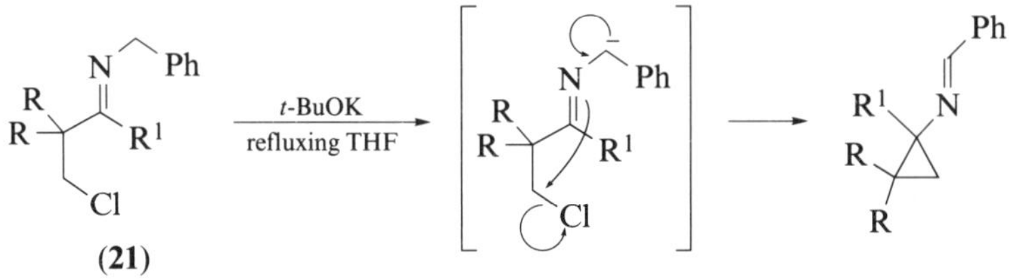

(21) (26)

Phenylhydrazones of β-stannyl ketones undergo oxidative radical cyclization to afford azocyclopropanes upon treatment with an oxidizing agent such as DDQ, $Pb(OAc)_4$ and NBS (equation 27)[43].

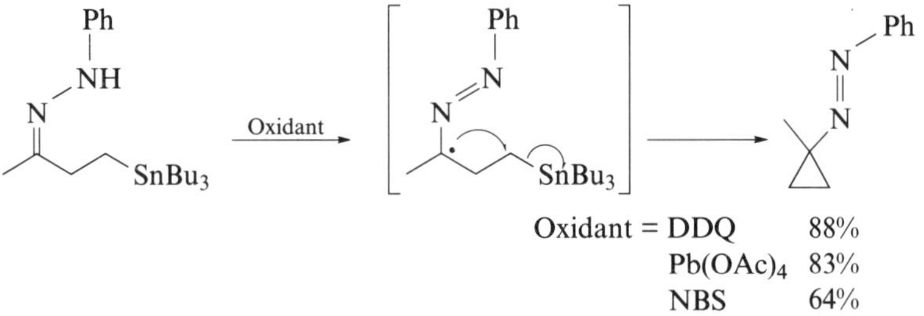

(27)

C. Cyclization of Allylic Derivatives

1. γ-Substituted Michael acceptors and nucleophiles

The Michael addition of nuleophiles to acceptors carrying a leaving group at the γ-position leads to intermediates capable of undergoing an intramolecular displacement to afford three-membered carbocycles (equation 28). A major side-reaction is direct S_N2

(28)

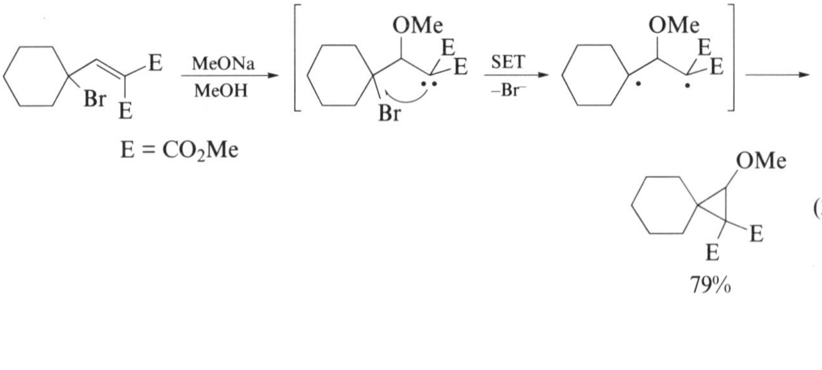

(29)

substitution of the leaving group. This type of transformation is applicable not only to the primary and secondary halides, but also to the sterically congested tertiary derivatives which are generally reluctant to undergo direct nucleophilic substitution. For the reactions of the latter, it is suggested that the second step may proceed via initial single electron transfer to generate 1,3-biradicals prior to cyclization (equation 29)[44].

The preference for conjugate addition exhibited by α,β-unsaturated acylphosphoranes even toward the addition of such hard nucleophiles as alkyllithiums is exploited for the preparation of 2-alkylcyclopropylacyl compounds from γ-halogenated acylphosphoranes and alkyllithiums (equation 30)[45]. 1-Aminocyclopropanecarboxylic acid derivatives have

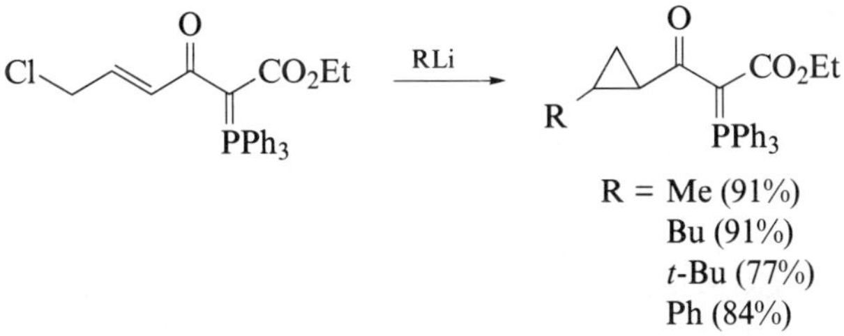

(30)

been synthesized by the related tandem Michael addition–cyclization[46,47]. The reactions of γ-bromonitroalkenes with 2-nitropropyl anion to give nitrocyclopropanes, **22**, are promoted by irradiation of visible light and completely suppressed in the presence of oxygen. Initiation of the reaction by electron transfer from the latter to the former has been proposed (equation 31)[48].

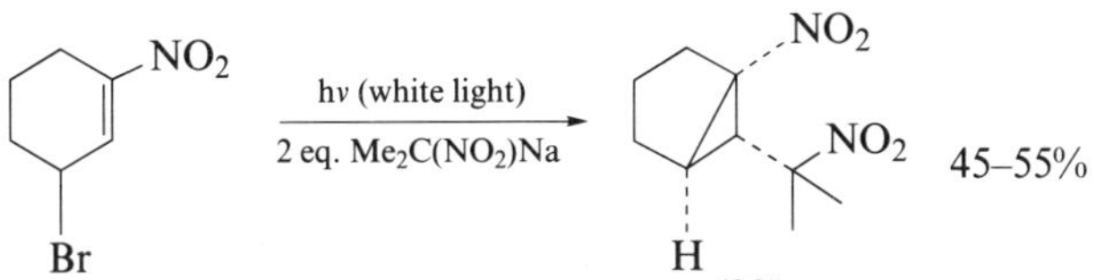

(22)

(31)

An asymmetric cyclopropane synthesis has been achieved by the addition–cyclization reaction of a chiral α,β-unsaturated sulfoxide with a Grignard reagent (equation 32)[49].

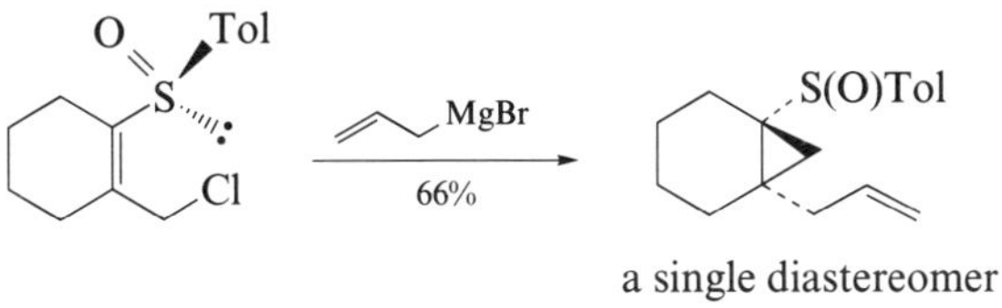

(32)

2. Reactions of other allylic derivatives

Nucleophilic attack on a π-allyl ligand of a metal complex occurs in general at one of the terminal carbons to afford allylated products. The attack, however, may be directed to the central carbon atom of the π-allyl group to produce cyclopropyl derivatives by appropriate choice of nucleophile, metal ligand and reaction conditions (equation 33). A variety of nucleophiles ($pK_a \geq 20$) including ester and ketone enolates and α-sulfonyl carbanions react with

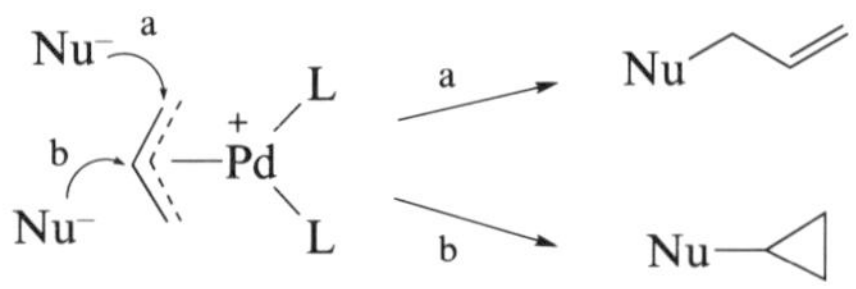

(33)

dimeric (π-allyl)palladium chloride in the presence of TMEDA under carbon monoxide atmosphere to give the corresponding α-cyclopropyl derivatives in good yields (equation 34)[50,51]. The nature of the counterion of the nucleophile plays an important role and the yield

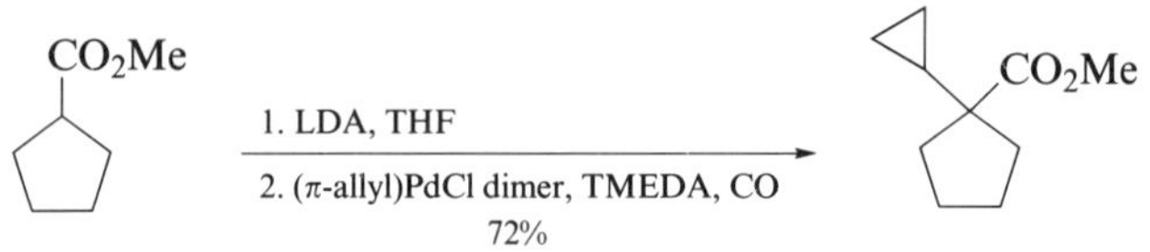

(34)

of cyclopropanes appears to increase when the nucleophile becomes more ionic, i.e. Nu^-K^+ is preferred to Nu^-Li^+. TMEDA is thought to exert a double function, i.e. not only complexing palladium and thus increasing the electron density on it but also activating the nucleophile. The reactions of certain (π-allyl)iridium and -rhodium complexes with relatively hard nucleophiles such as enolate ions proceed in the same manner to produce metallacyclobutane intermediates, which afford cyclopropyl derivatives upon iodonolysis (equations 35 and 36)[52]. The formation of the metallacyclobutanes is a kinetically controlled, reversible process and rearrangement of the intermediates to more conventional allylated products takes place thermally or catalytically in the presence of a Lewis acid.

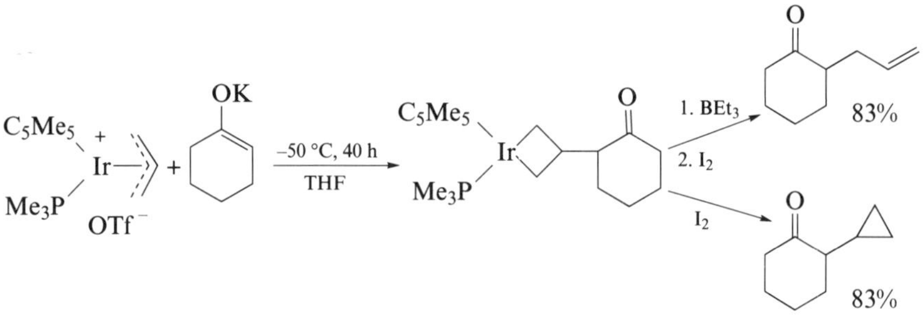

(35)

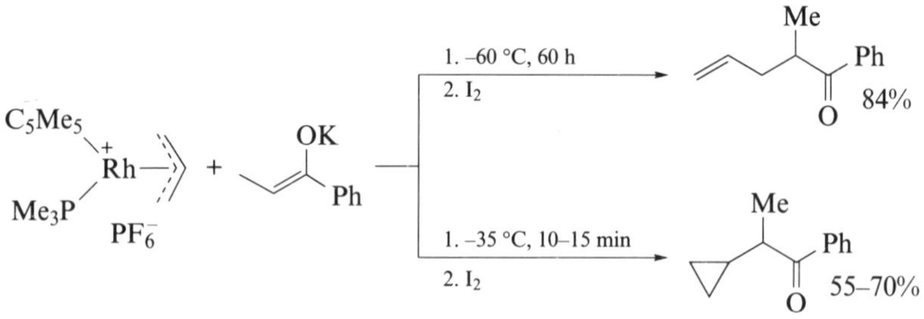

(36)

The reactions of γ-boryl enamines with aldehydes afford (2-aminocyclopropyl)methyl alcohols in modest yields. The use of the chiral boryl derivatives leads to the optically active alcohols (equation 37)[53].

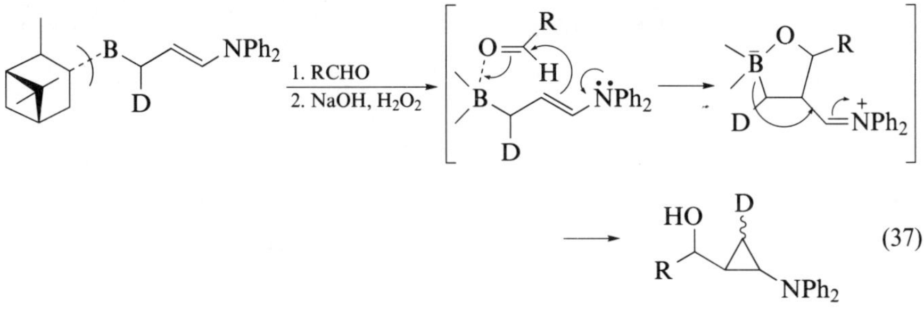

(37)

R = Me, Bu, Hex, Ph 10–26%

D. Cyclization of 3-Butenyl Derivatives

The equilibrium between homoallyl and cyclopropylmethyl radicals favors the former (i.e. $k_f >> k_c$) in the parent as well as in simple, alkylated systems (equation 38). Hence, to

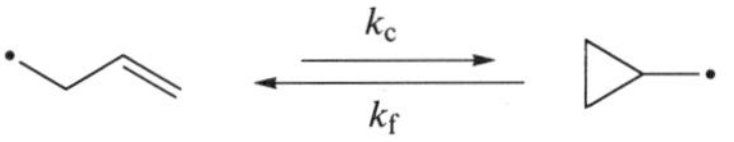

$k_c = 0.8\text{–}1.3 \times 10^4\ s^{-1}$ at 25 °C
$k_f = 1.3 \times 10^8\ s^{-1}$ at 25 °C (38)

exploit the equilibration for the preparation of three-membered carbocycles from the corresponding homoallylic derivatives, effective stabilization of the latter relative to the former or preferential interception of the latter has to be devised. The introduction of unsaturated groups capable of conjugating with the radical center of the latter, but not with that of the former (equations 39 and 40)[54–56], or destabilization of the former relative to the latter by steric constraint (equation 41)[57] represents the first case. The introduction of a radicofugal group at the position vicinal to the radical center of the latter (equation 42)[58]

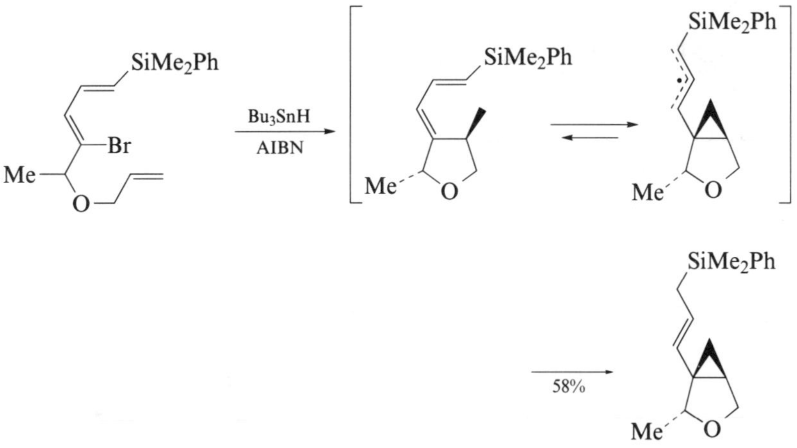

(39)

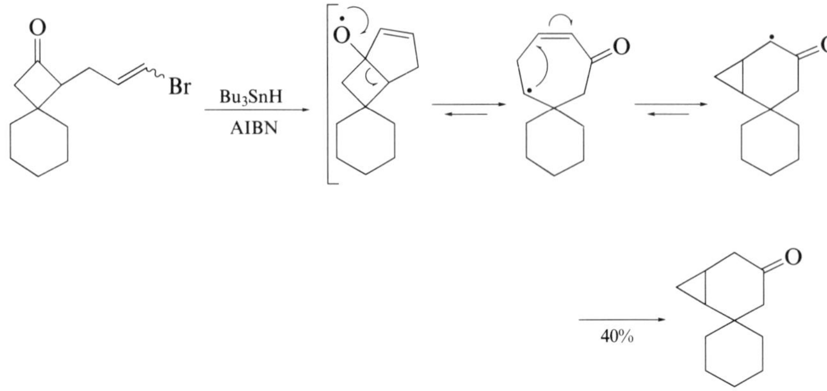

(40)

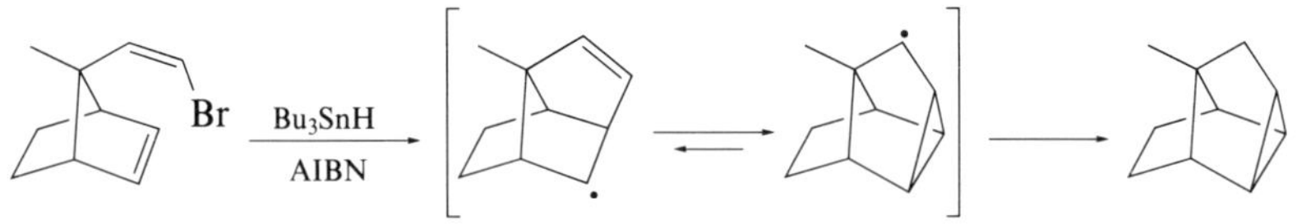

(41)

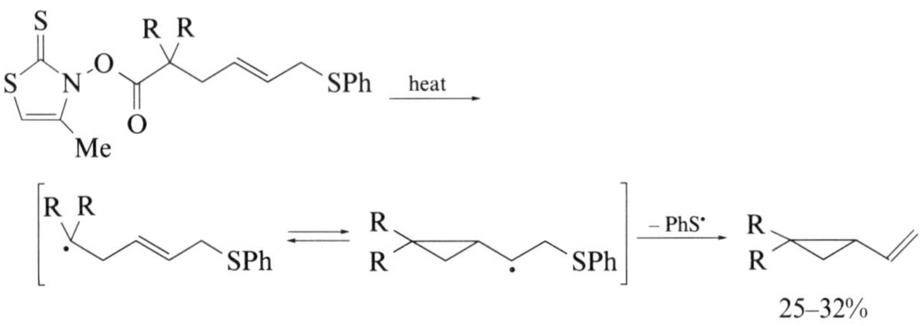

(42)

and specific oxidation or reduction of the latter to produce stabilized carbocations (equation 43)[59] or carbanions (equation 44)[60], respectively, represent the second strategy.

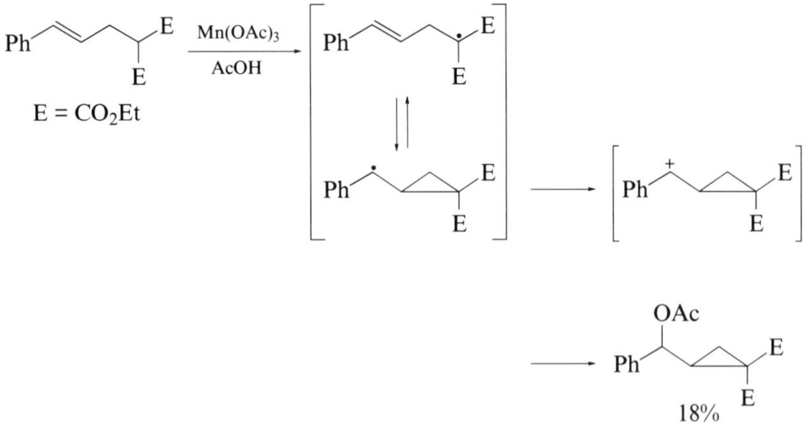

(43)

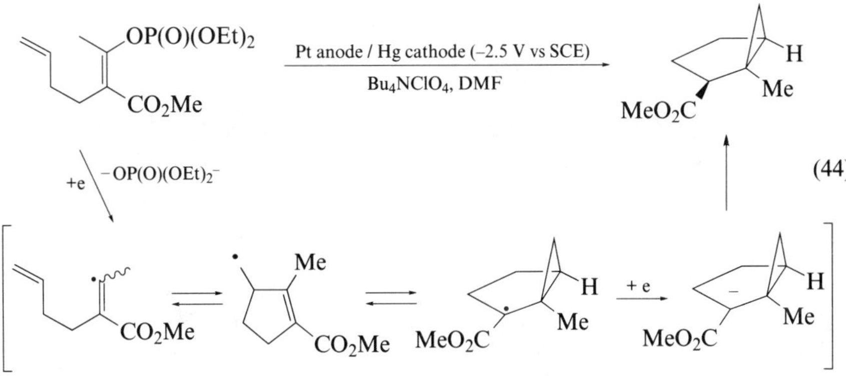

(44)

3-Butenyl derivatives bearing electron-withdrawing groups on the unsaturated carbons may be cyclized to produce cyclopropyl rings via electron transfer from reducing agents (equation 45)[61]. Compound **23** affords **25** upon treatment with (trimethylstannyl)lithium, presumably through the intermediacy of allylstannane, **24** (equation 46)[62]. The sulfonylation of **26** with trifluoromethanesulfonic or *p*-toluenesulfonic anhydride directly affords a cyclopropane-containing product, **27**. The rearrangement has been interpreted as a vinylogous Grob fragmentation (equation 47)[63].

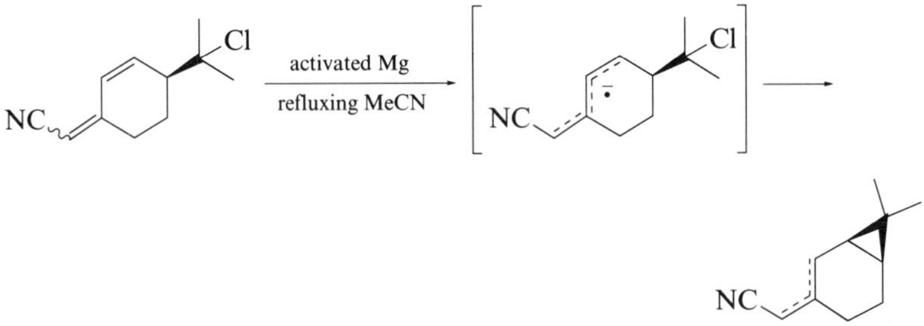

(45)

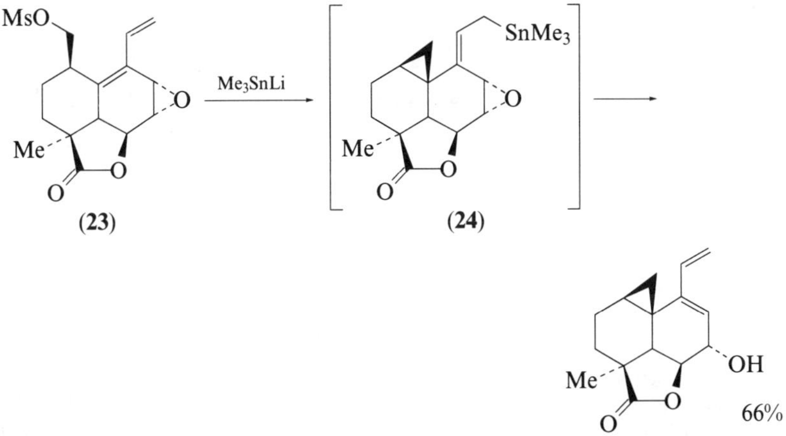

(23) (24) (25)

(46)

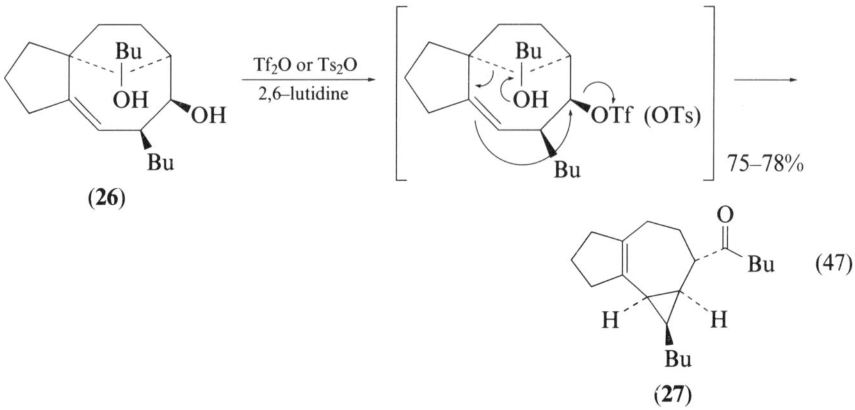

(26) (47) (27)

E. Miscellaneous

Treatment of functionalized vinyllithium, **28**, with allylmagnesium bromide in the presence of zinc bromide leads to a stereoselectively substituted, metallated cyclopropane, **29**, via a tandem metalla–Claisen–cyclization reaction (equation 48)[64]. The produced cyclo-

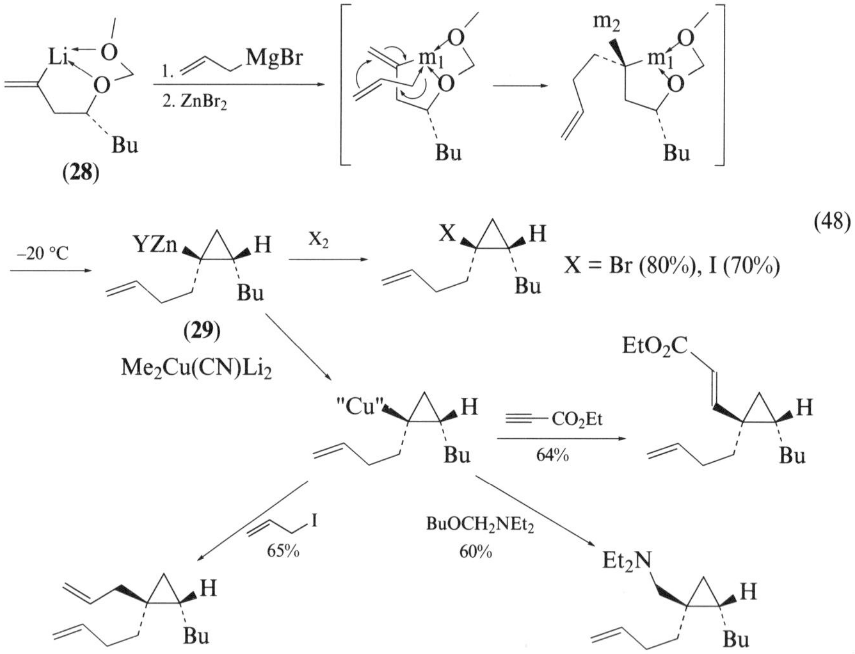

(28) (29) (48)

propylzinc reagent serves as a unique nucleophilic reagent bearing a quaternary chiral carbon, as exemplified by the conversion into the corresponding bromo- and iodocyclopropanes with bromine and iodine, respectively, and into a copper reagent, via metal–metal exchange, from which a variety of stereoselectively 1,1,2-trialkylated cyclopropanes are prepared. The analogous reaction of (*Z*)-lithio allylic ether, **30**, with allylzinc bromide provides a 2,3-disubstituted cyclopropylzinc reagent with a high diastereomeric purity (equation 49)[65].

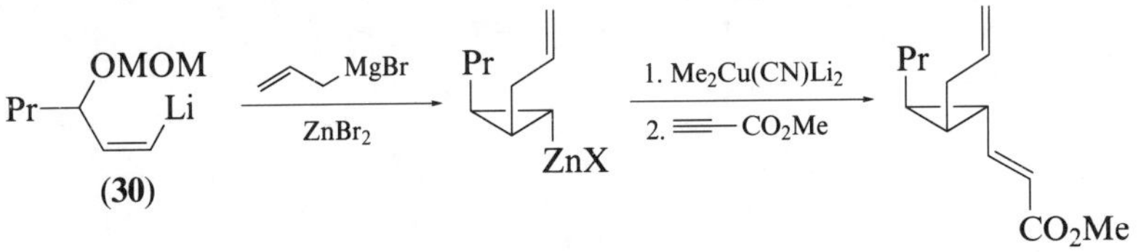

(30) (49)

MOM = methoxymethyl X = OMOM or OMe 68%, dr = > 95 : 5

Tin(IV) chloride-catalyzed biomimetic transformation of allylic oxiranes with a tethered carboxyl group into 2-(hydroxymethyl)cyclopropyl lactones has been reported (equation 50)[66]. Essentially identical results are obtained from the (*E*)- and (*Z*)-isomers.

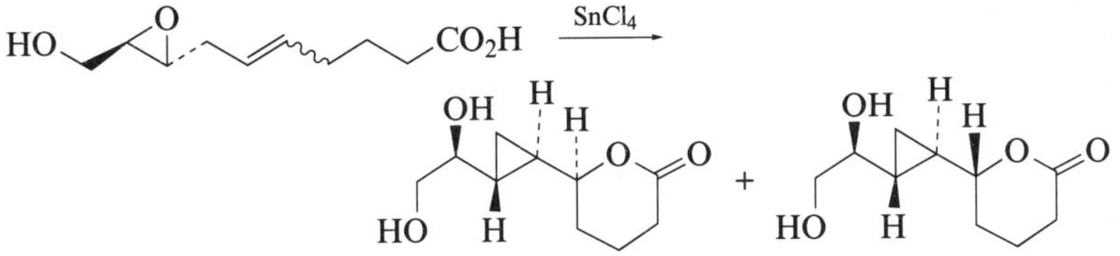

(50)

54% from *E*, 44% from *Z* (*ca* 1.5 : 1)

(Perfluoroalkyl)methyl-substituted electrophilic cyclopropanes are synthesized in excellent yields by a chromium (III) chloride/iron powder-promoted reaction of perfluoroalkyl iodides with allylamalonic ester and its analogues (equation 51)[67]. Oxidative coupling of

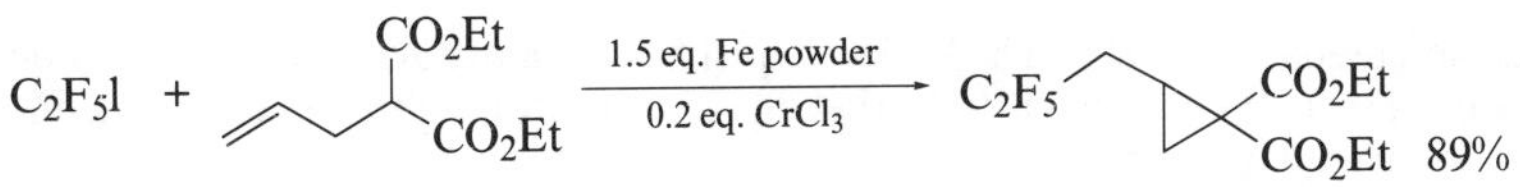

(51)

stabilized 1,3-dicarbanions to furnish functionalized cyclopropanes has been reported for dianions derived from 1,3-dinitroalkanes (equation 52)[68] and a bicyclic diketone (equation 53)[69]. Extrusion of sulfur atom from thietanes to give cyclopropanes is effected with Raney nickel. Spirocyclopropyl lactams, **31** and **32**, have been prepared via photochemical [2 + 2] cycloaddition of alkenes to the corresponding thiocarbonyl derivatives followed by the sul-

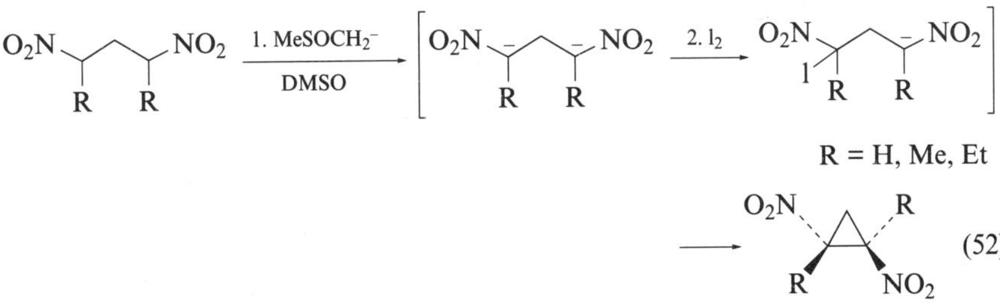

R = H, Me, Et

(52)

23–36%

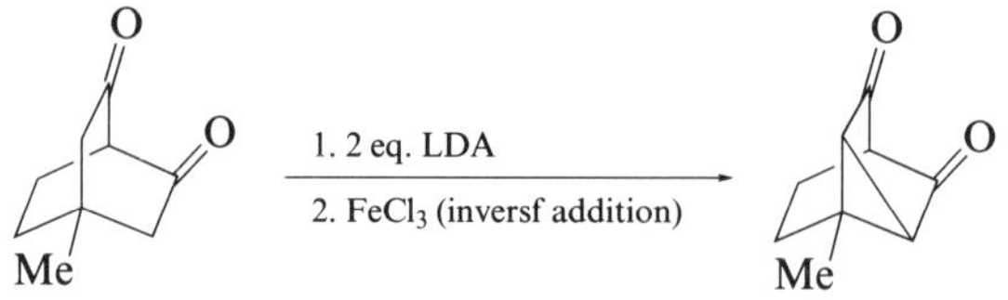

54%

(53)

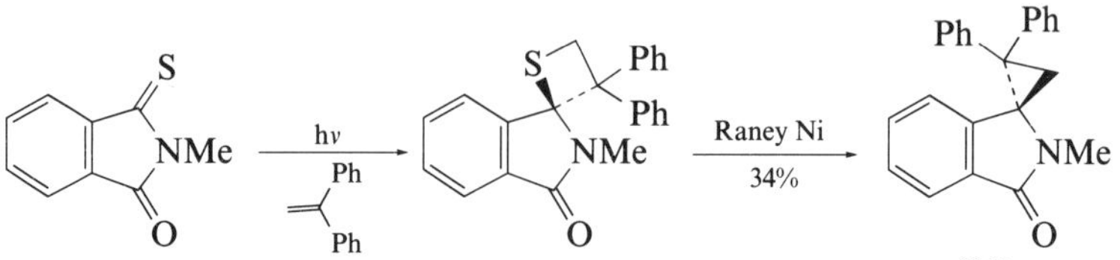

(54)

(31)

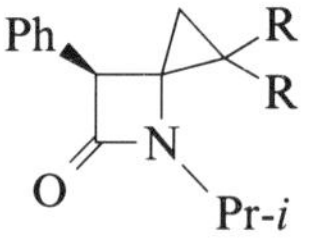

(32) R = Me, Ph

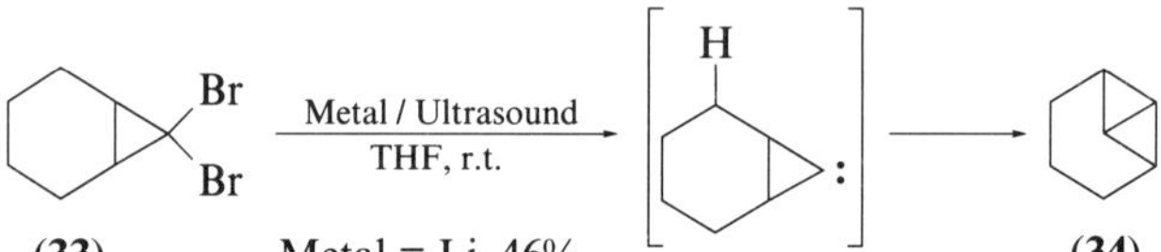

(55)

(33) Metal = Li 46% (34)
Mg 44%

fur extrusion (equation 54)[70,71]. The known transformation of **33** into **34** via intramolecular carbene insertion is conveniently carried out by sonicating a mixture of **33** and lithium or magnesium (equation 55)[72].

A number of new methods for the preparation of cyclopropanols from carbonyl derivatives via 1,3-bond formation between the carbonyl and C_β carbons have been developed. *Trans*-2-alkylcyclopropanols are stereoselectively produced from 2- or 3-substituted acrolein upon exposure to chromium(II) chloride in the presence of a catalytic amount of nickel chloride in DMF (equation 56)[73]. 2,3-Disubstituted acroleins are, in contrast, inert to the chromium reagent. Treatment of β-stannyl carbonyls with titanium(IV) chloride affords cyclopropanols in good yields when the substrates are ketones not bearing β-alkyl

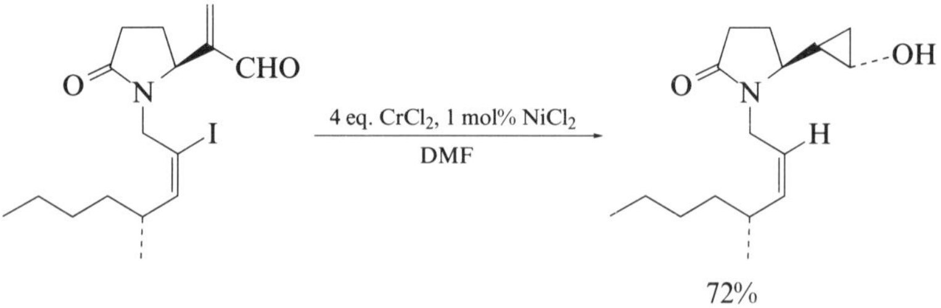

(56)

72%

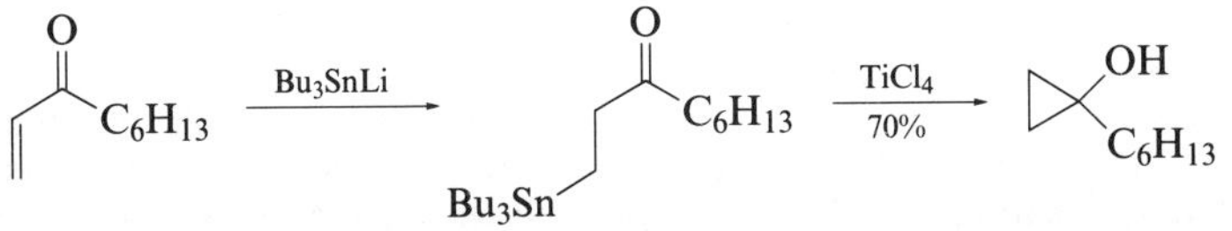
(57)

substituent or aldehydes (equation 57)[74]. Cyclopropanol derivatives are, however, not obtained from the β-alkylated β-stannyl ketones owing to their susceptibility to secondary ring cleavage under the reaction conditions. All the types of β-stannyl carbonyls examined afford cyclopropyl phenyl sulfides in the presence of Me_3SiSPh. The β-stannyl carbonyls used in the above reactions are readily prepared by the conjugate addition of trialkylstannyllithium to α,β-unsaturated carbonyl derivatives. Sequential treatment of ethyl 3bromopropionate with a Grignard reagent in the presence of samarium iodide and chlorotrimethylsilane provides 1-substituted cyclopropyl trimethyl silyl ether in a good yield (equation 58)[75]. Allyl and propargyl β-iodopropionate are converted into 1-allyloxy-

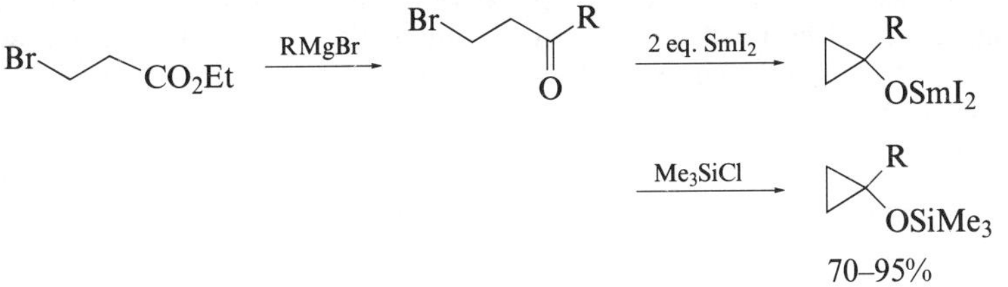
(58)

and 1-propargyloxy-1-siloxycyclopropanes, respectively, upon treatment with zinc–copper couple and silylating agents (equation 59)[76]. Under the conventional reaction

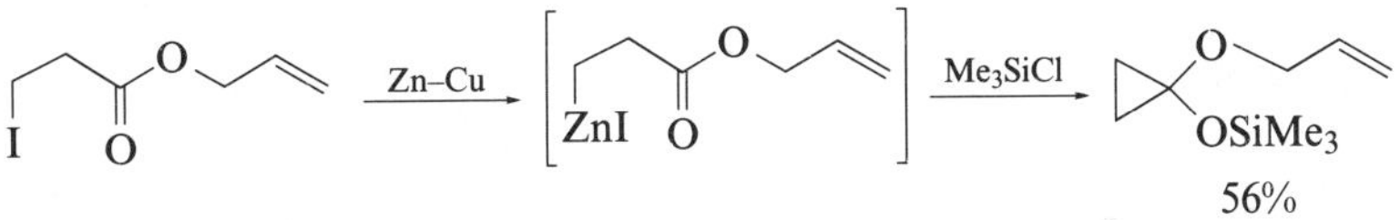
(59)

conditions (Na/chlorotrimethylsilane), those substrates failed to give the cyclopropanone acetals. Di-*tert*-alkyl ketones may be cyclized to give polyalkylated cyclopropanols by treatment with strong bases (equation 60)[77].

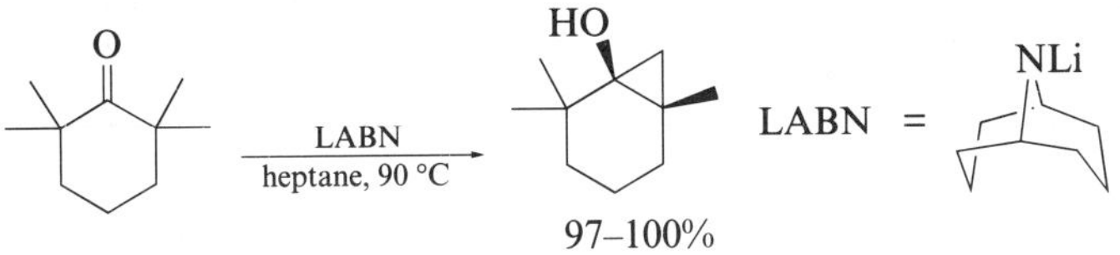
(60)

III. COMBINATION OF C_2 AND C_1 BUILDING BLOCKS

A. Cyclopropanation of Carbon–Carbon Multiple Bonds with Carbenes and Carbenoids

In this section, the classification of carbenes and carbenoids bearing two different substituents is made by giving priority to the functional group which will appear in the later subsection. For example, a cyclopropanation with :C(CH=CHR^1) COOR^2 is described in Section III.A.5 rather than in Section III.A.2. It should be noted that not in all the reactions described in this section was it experimentally confirmed that the reactions proceed via carbenes or carbenoids. Reactions in which involvement of divalent species is only apparent and formation of cyclopropane proceeds via a radical or an ionic intermediate may also be included in this section.

1. Methylene

The Simmons–Smith reaction provides a highly versatile, convenient means of preparing cyclopropyl compounds by methylene transfer to olefinic substrates and is in widespread use[78–86]. Characteristics of the Simmons–Smith reaction may be summarized as follows. It is an efficient, stereospecific transfer of methylene to a variety of alkenes under mild conditions, not hindered by the presence of many types of functional groups including hydroxyl, amine, carbonyl and carboxylic ester, usually free from side reactions such as olefin isomerization and insertion of the carbenoid into C—H bonds, and subject to the directing effects of basic functional groups containing oxygen or nitrogen atom which often enable regio- and stereoselective cyclopropanations of functionalized substrates. While the conventional Simmons–Smith reagent (diiodomethane/zinc–copper couple or diethylzinc) gives satisfactory results for a wide range of olefinic substrates, modifications of the reaction and the development of related new methods have actively been explored. Insufficient reactivity of dibromomethane has largely precluded its use in the reaction, though it is less expensive and easier to purify and store than diiodomethane. By sonication of the reaction mixture[87], or better by the addition of a small amount of titanium tetrachloride[88] or acetyl chloride[89], the cyclopropanation using dibromomethane is markedly promoted to give products in yields as good as or better than the reaction employing diiodomethane (Table 1). The cyclopropanation with dibromomethane can also be achieved in an electrolysis cell fitted with a sacrificial zinc anode[90]. The organozinc reagent derived from diethylzinc and chloroidomethane has been found to be generally more reactive than that from diiodomethane[91].

TABLE 1. Yields (%) of the cyclopropanated reaction of several alkenes in the Simmons–Smith reaction with CH_2X_2 (X = Br, I) and several promoters

Alkene	CH_2I_2 not promoted	CH_2Br_2 not promoted	CH_2Br_2 sonication	CH_2Br_2 $TiCl_4$	CH_2Br_2 MeCOCl
1-Octene	48	18	28	48	61
Cyclohexene	57	38	60	58	61
Cyclooctene	58	28	72	73	88

Treatment of alkenes with trialkylaluminum and diiodomethane produces the corresponding cyclopropanes[92] stereospecifically and usually free from serious side reactions as is the case with the corresponding organozinc reagent[93]. This methodology, however, exhibits a remarkable regioselectivity which is complementary to that observed in the Simmons–Smith reaction and its variants: multiply unsaturated allylic alcohols undergo cyclopropanation selectively at an olefinic site far from the hydroxy group (equations 61 and 62). The diastereoselectivity observed in the cyclopropanations of allylic silanes with this reagent has been rationalized in terms of stereoelectronic factors and allylic strain (equation 63)[94].

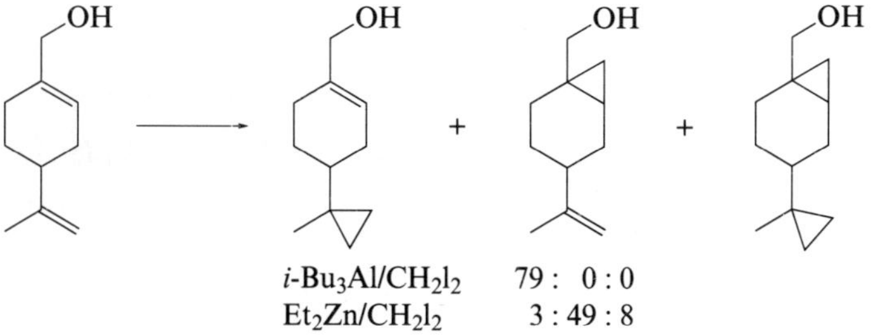

(61)

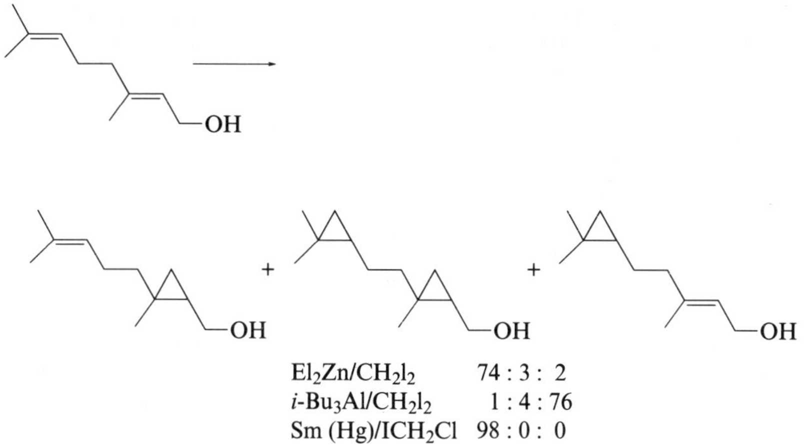

(62)

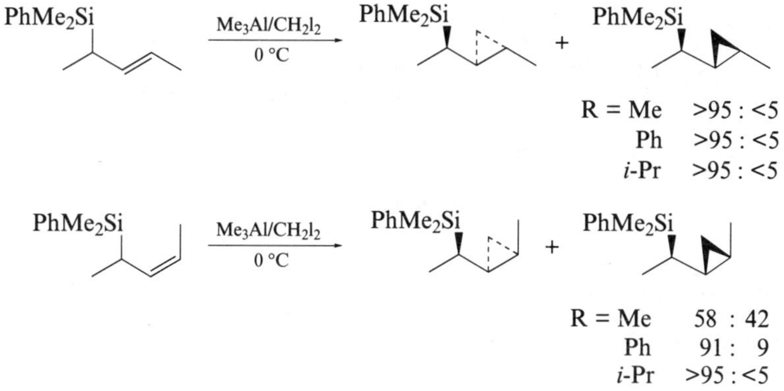

(63)

The directing effect of proximal oxygen substituents on the regio- and stereoselectivity of the Simmons–Smith reaction is well substantiated. Reagent generated from samarium metal and diiodomethane exhibits even higher chemoselectivity and specifically cyclopropanates allylic alcohols (equation 62)[95, 96]. Neither simple alkenes nor double bonds isolated from the hydroxy group by more than two carbons appear to undergo cyclopropanation. The extent of diastereoselectivity observed in the samarium-based cyclopropanations of secondary allylic alcohols is comparable to or higher than that under the traditional Simmons–Smith conditions[96, 97]. Allylic alcohols bearing a geminal alkene substituent and tertiary allylic alcohols, however, afford only poor yields of cyclopropanes[96]. Exploiting the chemoselectivity of this reagent, 2-alkylidenecyclopropanemethanols are prepared from α-allenic alcohols (equation 64)[98]. In addition to

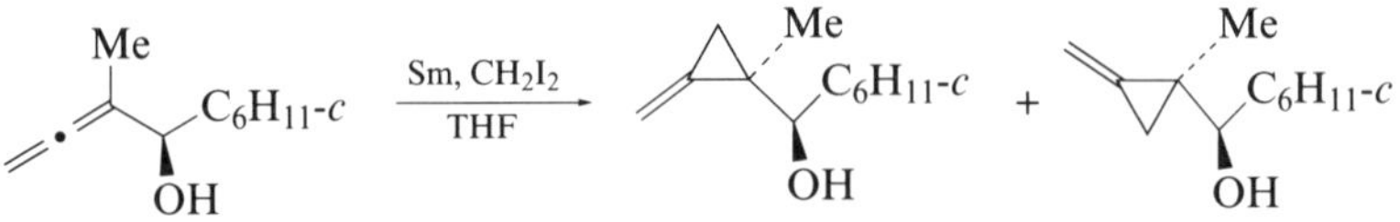

(64)

allylic alcohols, the diiodomethane/samarium system is capable of cyclopropanating lithium enolates to produce cyclopropanols in a one-pot synthesis from ketones (equation 65)[99, 100]. α-Halogenated ketones are directly and regioselectively converted into cyclopropanols via samarium enolates (equation 66)[100].

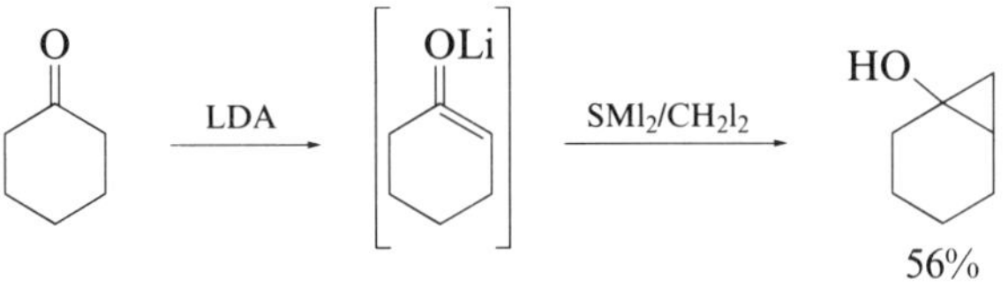

(65)

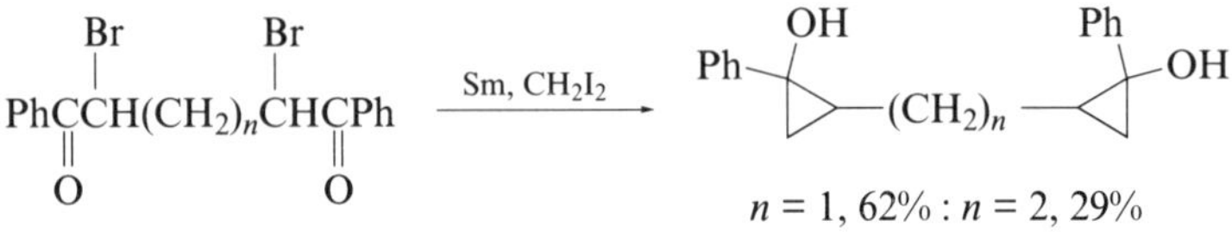

(66)

$n = 1$, 62% : $n = 2$, 29%

Palladium-catalyzed methylene transfer from diazomethane has proved effective for the cyclopropanation of 1-alkenylboronic acid esters[101, 102], allylic alcohols and amines[103], 1-oxy-1,3-butadienes[103] and allenes[104, 105]. Readily accessible iron complex (η^5–C_5H_5)$(CO)_2FeCH_2S^+Me_2{\bullet}BF_4^-$ **35** undergoes direct reaction with a range of alkenes to give cyclopropanes (equation 67)[106, 107]. The salt is sensitive to steric effects and the reaction proceeds

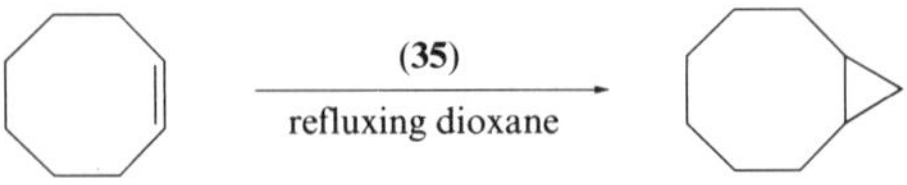

(67)

most reliable for mono- and disubstituted alkenes with high stereospecificity. The reactive intermediate generated from **35** is electrophilic in character and the cyclopropanation of 2-phenylsulfonyl-1,3-butadienes with **35** exclusively takes place at the unsubstituted double bond (equation 68)[108].

(68)

$Me_3S(O)I/NaH$	100 : 0
Zn–Cu/CH_2I_2	83 : 17
η^5-$Cp(CO)_2FeCH_2S^+Me_2 \cdot BF_4^-$	0 : 100

Asymmetric induction in the cyclopropanations of unsaturated substrates with methylene has been extensively investigated. A propensity of the Simmons–Simth and related reagents to make coordination to basic atoms is most frequently exploited. Treatment of α,β-unsaturated aldehyde acetals derived from the aldehydes and chiral dialkyl tartrates or 2,4-pentanediol, with diiodomethane/diethylzinc in hexane, produces cyclopropanecarboxaldehyde acetals with high diastereoselectivity (equation 69)[109,110]. Uniformly good diastereoselectivity has also been realized in the cyclopropanations of chiral acetals

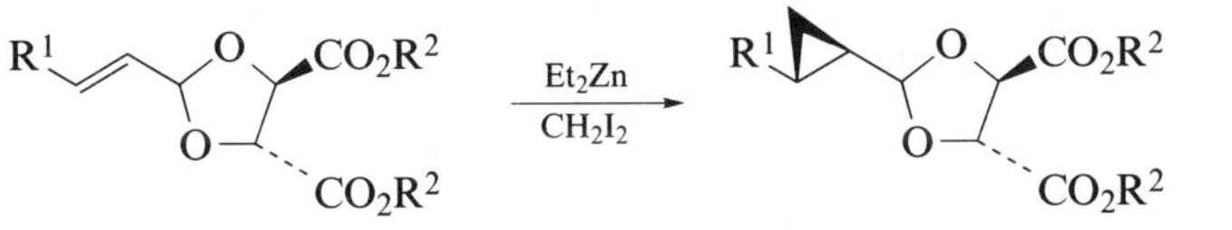

(69)

R^1 = Me, Pr, Ph; R^2 = Et, *i*-Pr

87–94% de (82–95%)

of medium to large ring α,β-unsaturated ketones with diiodomethane/zinc–copper couple in refluxing ether (equation 70)[111–113] and exploited for the synthesis of chiral natural products[114–116]. A number of chiral 1,2-diol auxiliaries for the acetals have been examined in terms of the efficiency of diastereoselection (equation 71)[111, 117]. Diastereoselection under

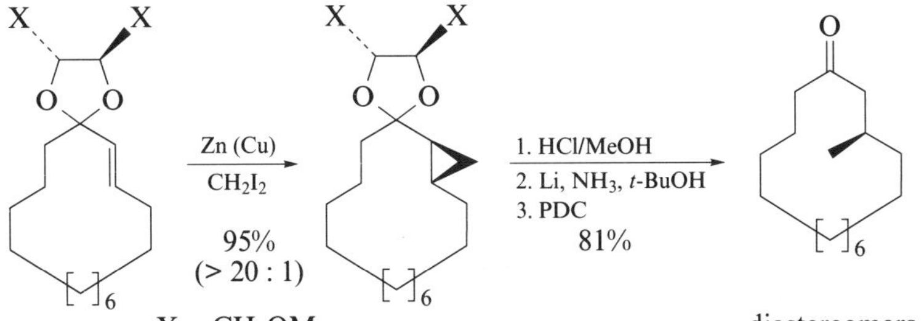

(70)

X = CH_2OMe

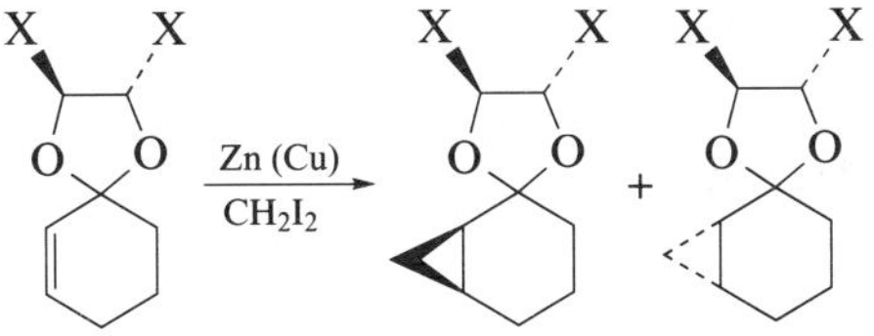

X	diastereomers
X = CH_2OCH_2Ph	9 : 1 (98%)
CH_2OMe	5 : 1 (86%)
CH_2OH	1 : 2 (50%)
CO_2Me	1.5 : 1 (37%)
CMe_2OMe	9 : 1 (86%)
$CH_2CH_2CH_2Ph$	>9 : <1 (92%)
CH_3	4 : 1 (91%)

(71)

the latter conditions, however, is uniformly poor for α,β-unsaturated aldehyde acetals, but better for acetals of the corresponding methyl ketones[112, 113, 118]. The observed stereoselection has been rationalized by assuming preferential coordination of the organozinc reagent to one of the acetal oxygens prior to the delivery of methylene to the double bond. Highly diastereoselective Simmons–Smith reactions have also been achieved for enol ethers derived from ketones and chiral 1,3-diol auxiliaries (equation 72)[119–122] and for 1-alkenylboronic esters modified by chiral diols (equation 73)[123]. Cyclopropanols of high ee are obtained from the products of those reactions. For the asymmetric cyclopropanations of allylic alcohols by the Simmons–Smith reagent, the substrates have not necessarily to be modified with chiral auxiliaries. Thus, the reaction of allylic alcohols in the presence of a

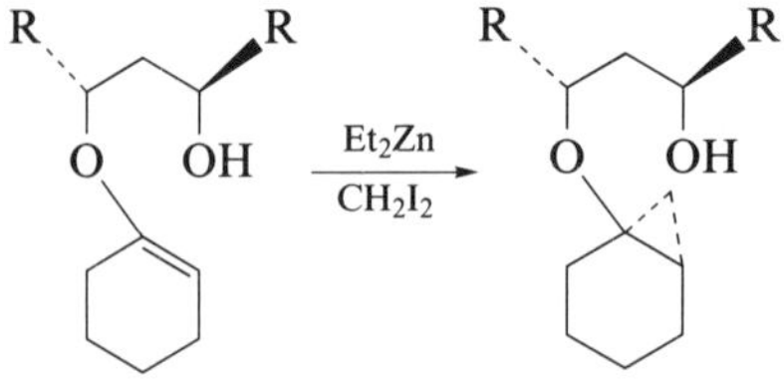

R = Me (DME) 95% de (55%)
R = *i*-Pr (ether) 99.4% de (72%) (72)

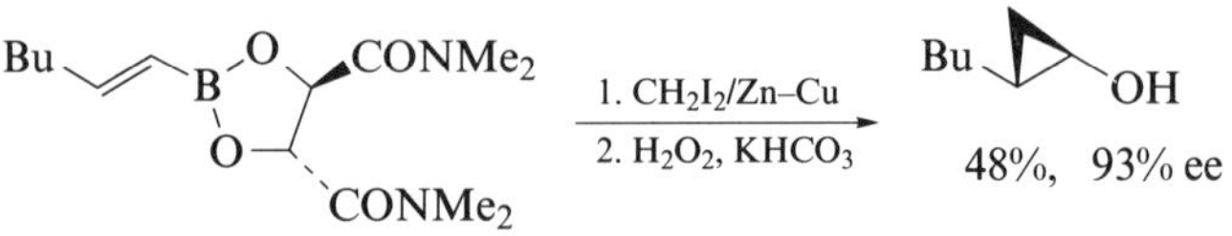

(73)

stoichiometric amount of tartaric ester provides cyclopropylmethyl alcohols in 70–80% ee (equation 74)[124]. The asymmetric Simmons–Smith cyclopropanations of allylic alcohols have also been achieved by using a catalytic amount of chiral disulfonamide **36** (equation 75)[125]. Upon use of a given enantiomer of **36**, the cyclopropanation occurs from the same

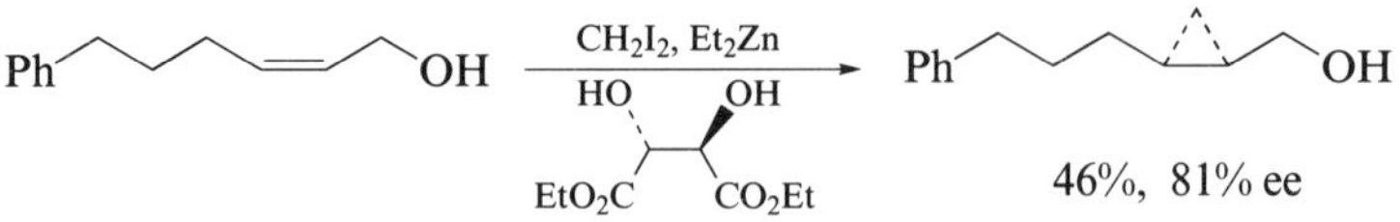

(74)

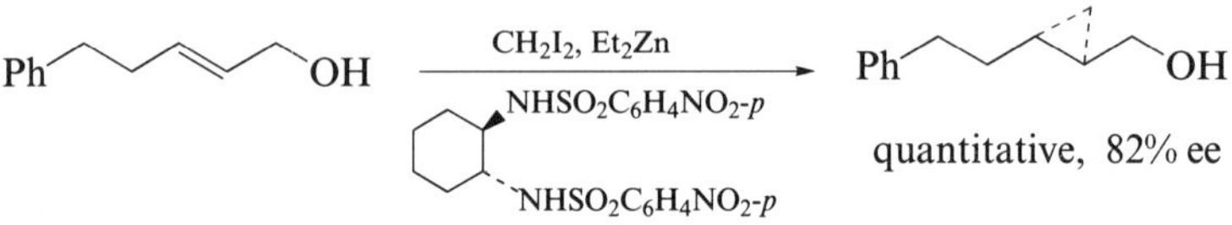

(75)

(36)

enantioface of olefin regardless of its geometry. 2-Substituted cyclopropanecarboxylic acid derivatives of high optical purity are obtained via the methylenation of α,β-unsaturated acyl ligands bound to a chiral iron auxiliary [(η^5–C_5H_5)Fe(CO)(PPh_3)] (equation 76)[126,127]. Asymmetric cyclopropane syntheses with chloroiodomethane/Sm(Hg)[128] and diazomethane/Pd(II)[129,130] have also been reported (equation 77).

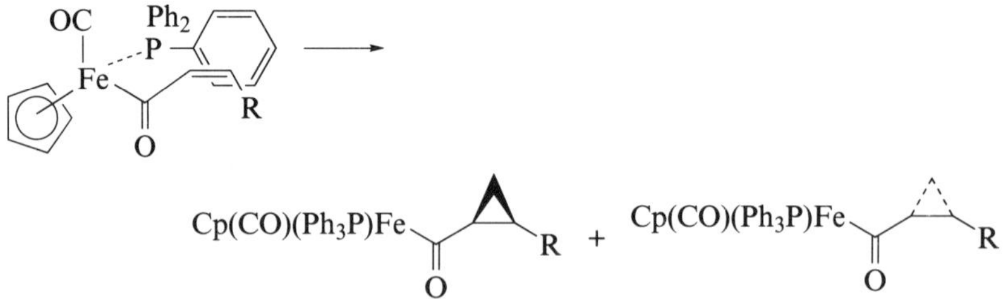

(76)

	$CH_2I_2/Et_2Zn/ZnCl_2$	CH_2I_2/Et_3Al
R = Me	91% (9 : 1)	74% (16 : 1)
Pr	91% (14 : 1)	86% (18 : 1)
i-Pr	93% (24 : 1)	49% (24 : 1)

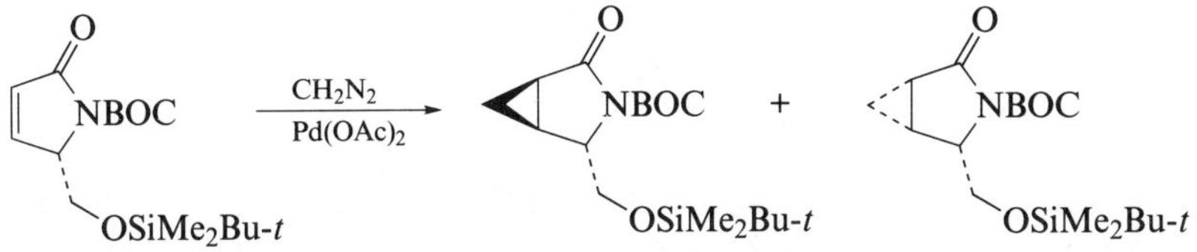

(77)

BOC = *t*-butoxycarbonyl

100% (9 : 1)

2. Alkyl-, alkenyl- and arylcarbenes

Treatment of a variety of α,β-unsaturated carbonyl compounds with zinc and 1,2-bis(chlorodimethylsilyl)ethane leads to the formation of organozinc carbenoids which may be trapped with olefins to give alkenyl- and arylcyclopropanes (equation 78)[131]. With

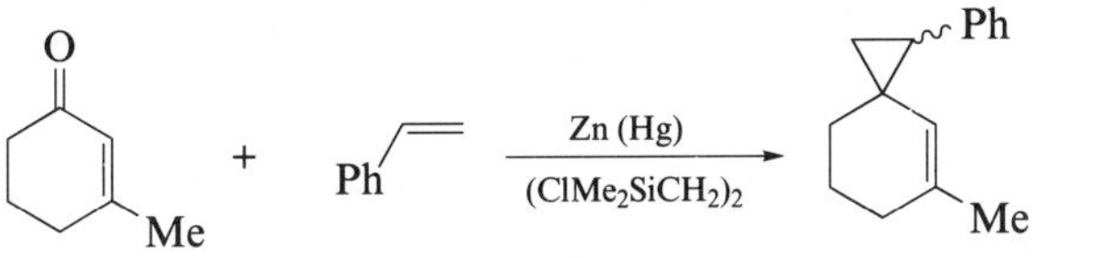

(78)

59%, $E:Z = 1:11$

nonaromatic α,β-unsaturated carbonyl compounds, however, steric shielding around the β-carbon seems essential to a successful alkylidene transfer. An efficient synthesis of *gem*-dimethylcyclopropanes from 2,2-dibromopropane and alkenes is achieved in an electrolysis cell fitted with a sacrificial zinc anode (equation 79)[90]. Samarium-promoted

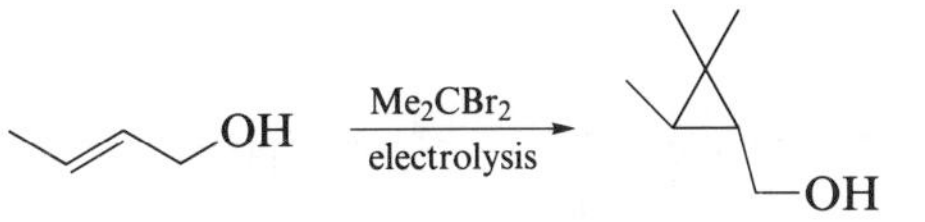

(79)

51%

cyclopropanation of cyclohexen-3-ol with 1,1-diiodoethane provides methylcyclopropanes with high diastereoselectivity[95]. 2-Alkenylcyclopropanes are obtained from the $Rh_2(OAc)_4$-catalyzed reaction of 3-diazopropenes with electron-rich alkenes (equation 80)[132]. Several techniques have been devised for the generation of electrophilic iron

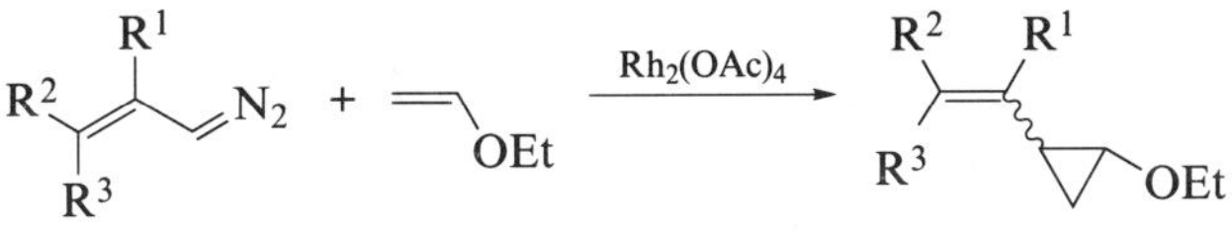

(80)

$R^1 = R^2 = R^3 = Me$	64%	$(Z):(E) = 5.0:1$
$R^1 = H, R^2 = R^3 = Me$	23%	$5.8:1$
$R^1 = Cl, R^2 = R^3 = H$	56%	$5.2:1$
$R^1 = R^2 = R^3 = H$	63%	$2.8:1$

carbenoids $(\eta^5\text{–}C_5H_5)(CO)_2Fe^+{=}CR_2$ capable of transferring alkylidenes and arylidenes to alkenes (equation 81)[2c, 133–135]. When reacted with simple alkenes, the iron carbene

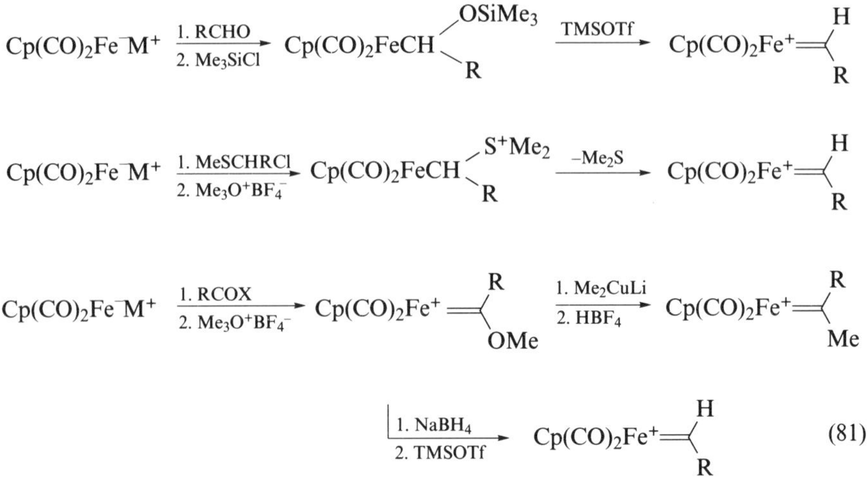

(81)

complexes exhibit high *cis* selectivity (equation 82)[135]. Intramolecular alkylidene transfer reaction of iron carbenoid is also investigated[136]. Alkyne-tethered Fischer carbene complexes of molybdenum react with a variety of electron-deficient alkenes by way of

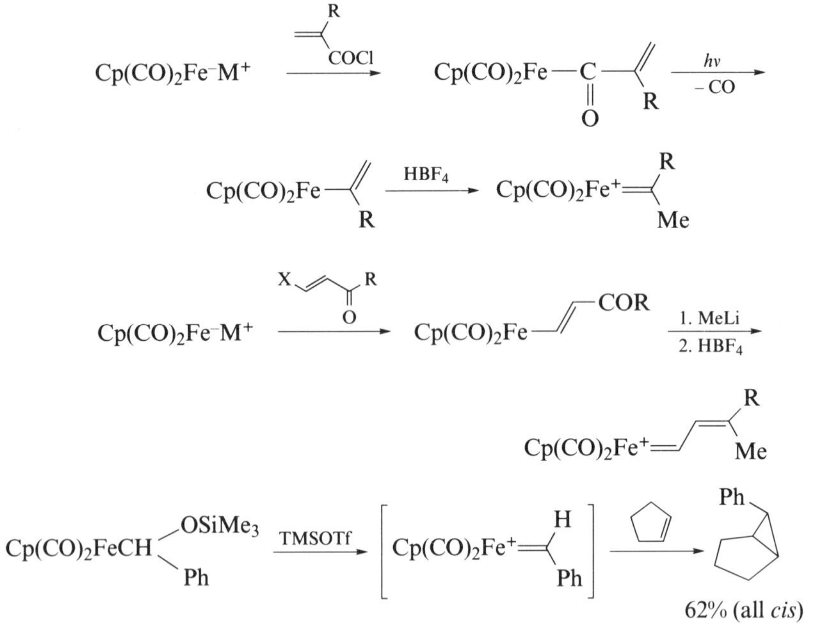

(82)

vinylcarbene complexes generated *in situ* via intramolecular addition of the carbenoid to the triple bond (equation 83)[137, 138]. The success of cyclization to form the vinyl carbenoids is very dependent on the length of the tether. The corresponding chromium and tungsten complexes also produce the adducts, but the yields are significantly lower.

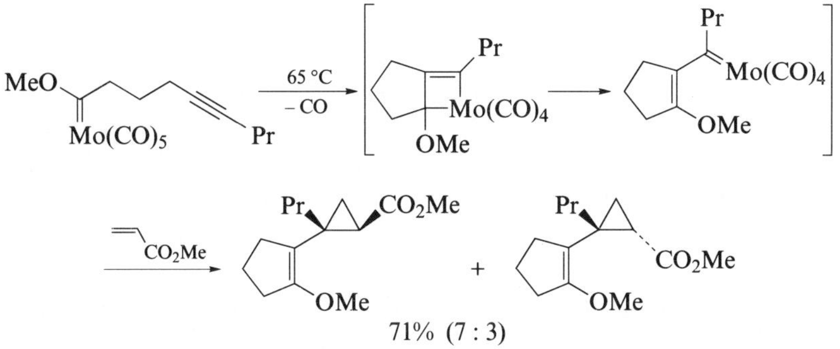

(83)

Asymmetric ethylidene transfer has been achieved in the reactions of 1-cyclohexenyl ethers carrying a chiral auxiliary with 1,1-diodoethane/diethylzinc[139]. Asymmetric induction in the reaction of diazofluorene with fumaric esters bearing chiral alcohol moieties has been investigated (equation 84)[140,141]. Kinetics of intramolecular cyclopropanation in 2-(ethenyloxy)phenylmethylene has been discussed in terms of the methylene's spin multiplicity[142].

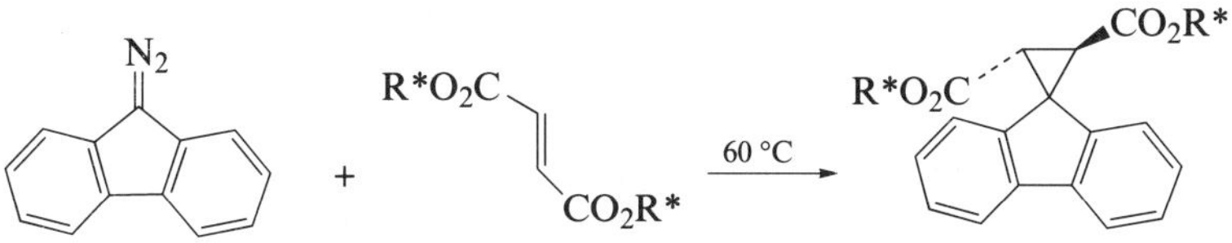

(84)

R* = (–)-menthyl 50% de (76%)
(–)-8-phenylmenthyl 90% de (79%)

3. Alkenylidenes

The development of new procedures in this area in the last decade seems scarce. 3, 3-Disubstituted alkenylidenes are generated from 2,2-disubstituted 1,1-dibromocyclopropanes under phase-transfer-catalysis (PTC) conditions and added to a variety of electron-rich alkenes to give vinylidenecyclopropanes in good yields (equation 85)[143].

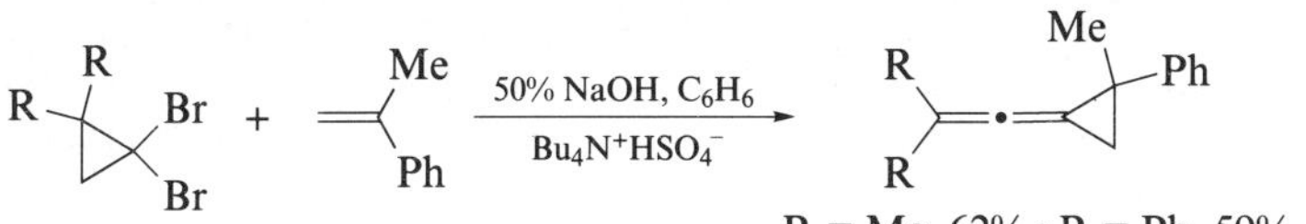

(85)

R = Me, 62% ; R = Ph, 59%

4. Monohalo- and dihalocarbenes

A variety of new reagents capable of transferring halomethylenes to alkenes has been developed. A new zinc difluorocarbenoid generated from dibromodifluoromethane and zinc transfers difluoromethylene to alkenes in excellent yields (equation 86)[144]. 1,1-Difluorocyclopropanes and -cyclopropenes are also obtained by treating alkenes[145, 146] and alkynes[147], respectively, with a reagent system of $CBr_2F_2/PPh_3/KF$/18-crown-6 in DME (equation 87).

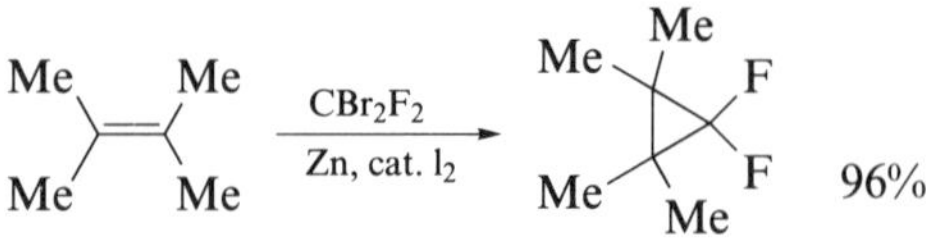

(86)

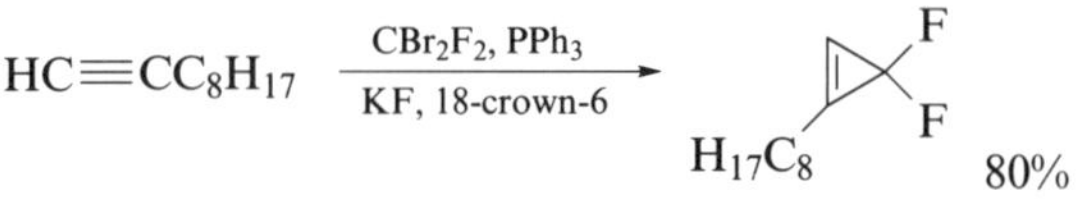

(87)

The preparation of dichlorocarbene from $CCl_3CO_2TMS/KF/Bu_4N^+X^-$ under mild heating[148] and from CH_2X_2 (X = Cl, Br)/CCl_4[149] or $CXCl_3$ (X = Cl, Br)[150] under PTC conditions has been reported. New methods for the generation of mixed dihalocarbenes, :CFCl and :CClBr, have been devised. Thus, treatment of $CFCl_3$ or CF_2Cl_2 with low valent titanium prepared *in situ* from $TiCl_4$ and $LiAlH_4$ produces an organotitanium reagent capable of transferring chlorofluoromethylene to alkenes (equation 88)[151, 152]. Bromochlorocarbene has been generated from several sources including $CH_2Br_2/PhCCl_3$[153], $CHBr_3/CCl_4$ or C_2Cl_6[154], and $CHCl_3/CBr_4$[154] under PTC conditions and added to alkenes.

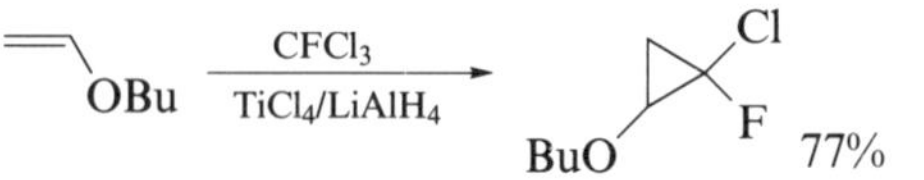

(88)

Improvements and new findings in making dihalocyclopropanes under PTC conditions have been reported. The addition of pinacol as a co-catalyst has a beneficial effect on the yields of dibromo- and dichlorocarbene adducts[155]. Triphenylsulfonium chloride functions as an efficient PT catalyst (equation 89)[156]. The selectivities of PT-catalytically generated

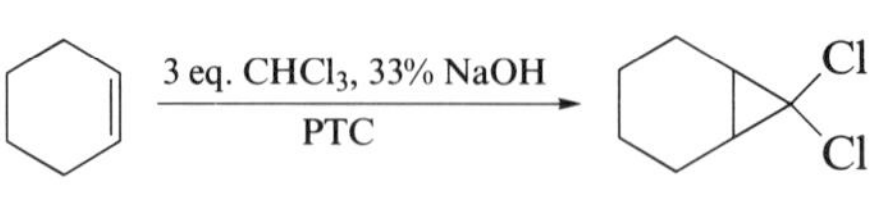

PTC =	
$Ph_3S^+Cl^-$	94%
$Ph_3S^+Br^-$	45%
$Bu_4N^+Br^-$	20%
None	0%

(89)

dichloro- and dibromocarbenes toward simple alkenes are independent of the structure of the catalyst, suggesting the involvement of free carbenes[157]. The dihalocarbene addition reactions of electron-acceptor-substituted alkenes, however, suffer from competing Michael additions of trihalomethyl anions to the electrophilic double bond. Ammonium ions with a small, sterically unhindered head group, e. g. RN^+Me_3, and benzo-crown ethers favor processes involving the carbenes, whereas large delocalized cations such as Ph_4As^+ and $Ph_3P{=}N^+{=}PPh_3$ foster Michael additions (equation 90)[158, 159]. The diastereoselectivity observed in the addition of dichlorocarbene to allylic alcohols has been rationalized

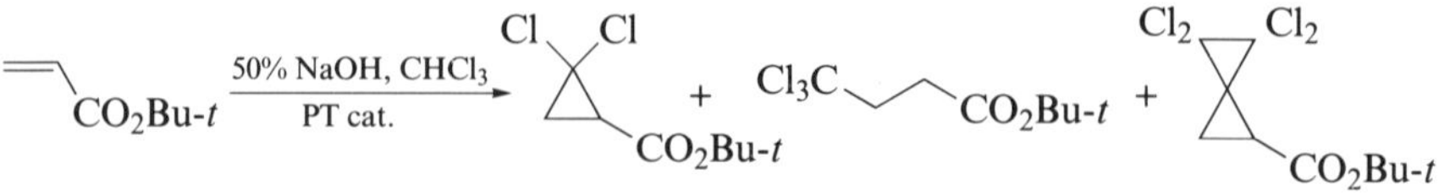

$Me_4N^+Cl^-$	57%	5%	—
$PhCH_2Et_3N^+Cl^-$	25%	9%	—
$Ph_4As^+Cl^-$	17%	19%	12%

(90)

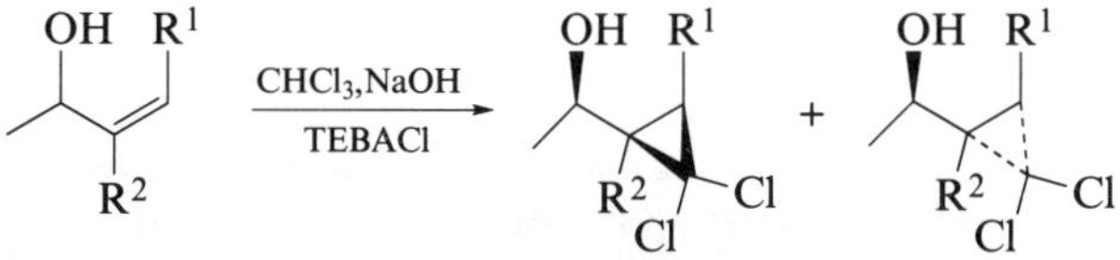

(91)

$R^1 = H, R^2 = Me$ 1 : 1 (75%)
$R^1 = Me, R^2 = H$ 55 : 1 (77%)

in terms of allylic strain in the substrates (equation 91)[160]. The enantioselective addition of dibromocarbene to a chiral alkene has also been investigated[118, 161]. Dihalocarbene adducts of ketene acetals are often thermally too labile to be isolated. The lability may be reduced to a manageable level by using cyclic acetals of ketene such as **37** (equation 92)[162].

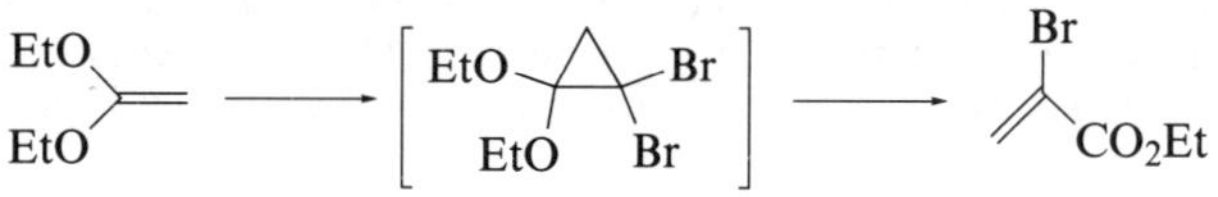

(92)

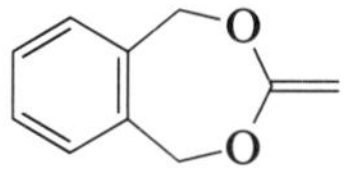

(37)

The generation of arylchlorocarbenes under neutral conditions has been achieved by treating $PhCCl_2SiMe_3$ with KF/18-crown-6 (equation 93)[163] and by sonicating phenylchlorodiazirine[164]. Full papers concerning the preparation of halogenated vinylcarbenes from polyhalopropenes by α-elimination[165] and from polyhalocyclopropenes via

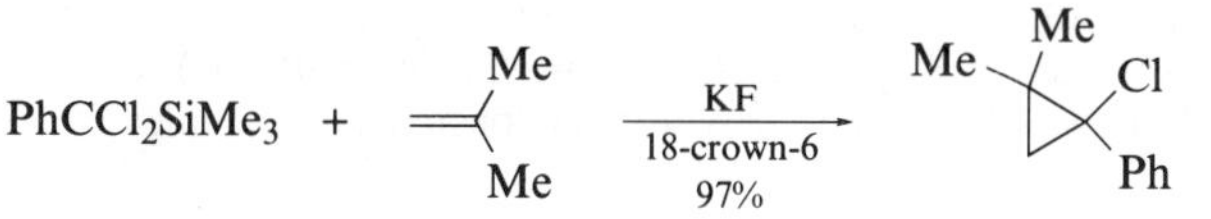

(93)

thermal ring cleavage[166–168] and their additions to alkenes have appeared. The resulting halogenated vinylcyclopropanes have proved to be valuable synthetic intermediates (equation 94)[167, 169, 170].

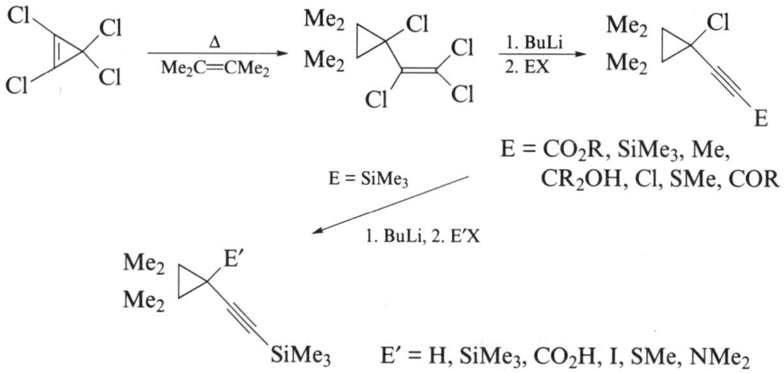

(94)

5. *Acyl- and alkoxycarbonylcarbenes*

The generation and addition of acyl- and alkoxycarbonylcarbenes to alkenes is usually carried out by using the corresponding diazo derivatives under catalysis of metallic compounds. The metal-catalyzed cyclopropanation reactions with diazo compounds are described in detail by McKervey in Chapter 11, hence dealt with rather briefly in this chapter. The readers who are interested in the preparation of acyl- and alkoxycarbonyl-substituted cyclopropanes are requested to refer also to Chapter 9.

The most frequently used metallic catalysts for acyldiazo- and (alkoxycarbonyl)diazomethanes are complexes or salts of rhodium, palladium and copper[2a]. Alkenylboronic esters[101, 102], *N*-silylated allylamines[171] and acetylenes[172, 173] are successfully cyclopropanated with diazocarbonyl compounds under catalysis of one of those metal derivatives. Newly developed metallic catalysts for diazoacetic esters include polymer-bound, quantitatively recoverable Rh(II) carboxylate salts[174], Cu(II) supported on NAFION ion exchange polymer[175], ruthenacarborane clusters[176], $Rh_2(NHCOCH_3)_4$ which produces cyclopropanes with substantially enhanced *trans* (*anti*) selectivity as shown below[177] and $(\eta^5\text{-}C_5H_5)$

trans/*cis* (*anti*/*syn*) ratio

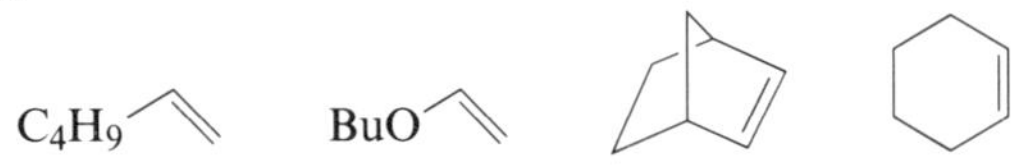

	C_4H_9-alkene	BuO-alkene	bicyclic alkene	cyclohexene
$Rh_2(OAc)_4$	1. 2	1. 3	1. 5	2. 0
$Rh_2(NHCOMe)_4$	1. 7	2. 8	3. 6	10

$(CO)_2Fe^+(THF)\cdot BF_4^-$ [178]. A transition state model for the *syn* stereoselective cyclopropanations of alkenes with diazoacetic ester by Rh–porphyrin catalysts has been proposed[179]. Alkenes[180], conjugated dienes[181, 182] and enol ethers[183] are stereoselectively cyclopropanated with Rh(II)-stabilized 1-(alkoxycarbonyl)vinyl carbenoids derived from the diazo precursors and $Rh_2(OAc)_4$ (equation 95). The $Cu(acac)_2$-catalyzed reactions of $Me_3SiCH_2COCHN_2$ with alkenes provide the expected adducts in good yields[184].

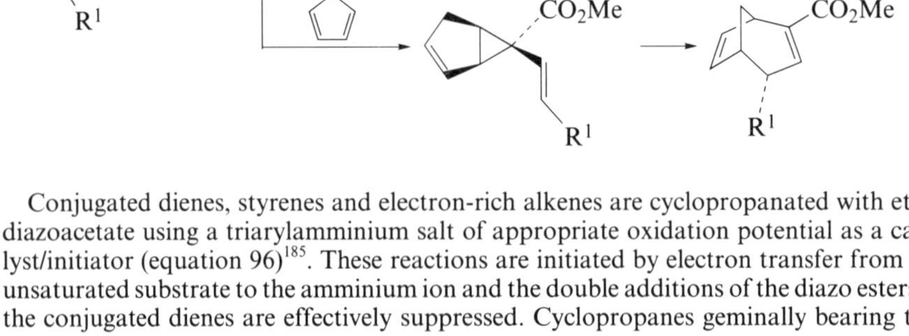

(95)

Conjugated dienes, styrenes and electron-rich alkenes are cyclopropanated with ethyl diazoacetate using a triarylamminium salt of appropriate oxidation potential as a catalyst/initiator (equation 96)[185]. These reactions are initiated by electron transfer from the unsaturated substrate to the amminium ion and the double additions of the diazo esters to the conjugated dienes are effectively suppressed. Cyclopropanes geminally bearing two

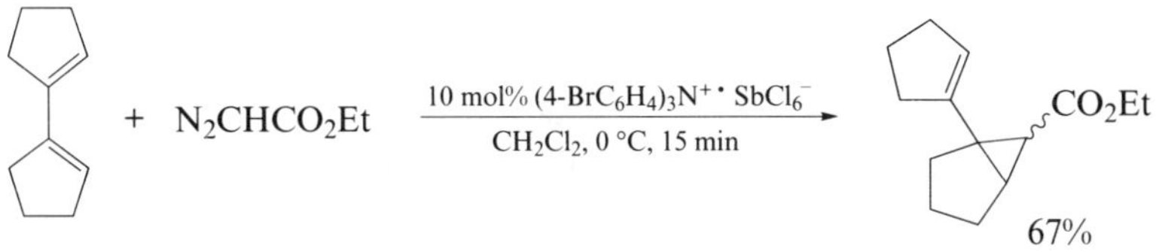

(96)

electron-withdrawing substituents are prepared from the reaction of alkenes with malonic ester or related derivatives in the presence of iodine under solid–liquid PTC conditions (equation 97)[186]. The stereochemical integrity of the alkenes, however, may not be retained.

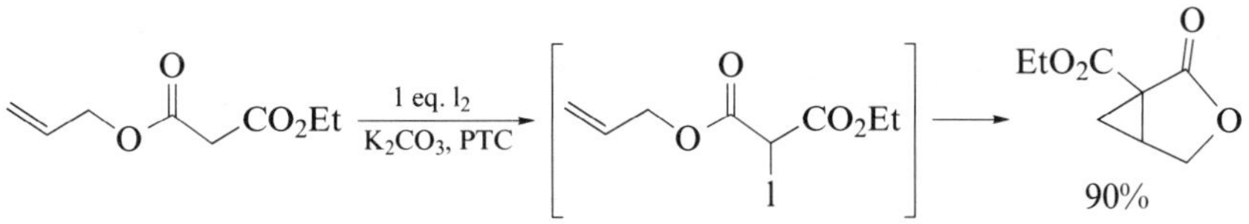

(97)

Intramolecular cyclopropanations of α-diazocarbonyl compounds have been widely utilized as a means of constructing complex polycyclic compounds stereoselectively in short steps (e.g. equation 98)[187–198]. The choice of the ligand of the Rh(II) catalyst, Rh_2L_4,

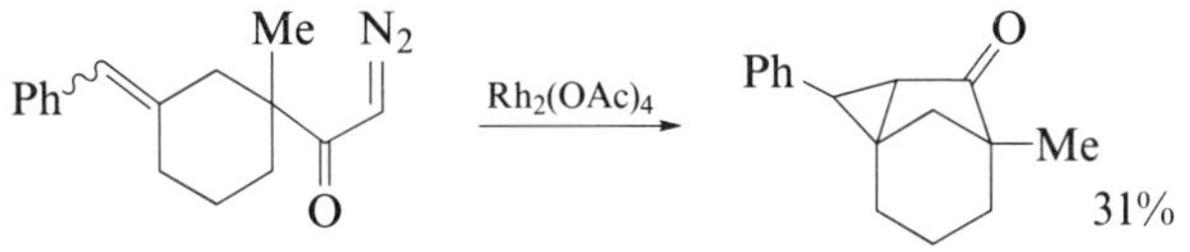

(98)

markedly influences the distribution of products from the substrates in which the intramolecular cyclopropanation process, is in competition with aromatic substitution and/or insertion into adjacent aliphatic C—H bonds. By changing the ligand from carboxylate to carboxamide, the chemoselective cyclopropanation in preference to the latter transformations may be realized (equation 99)[199]. Exposure of α-diazoketones bearing tethered alkyne

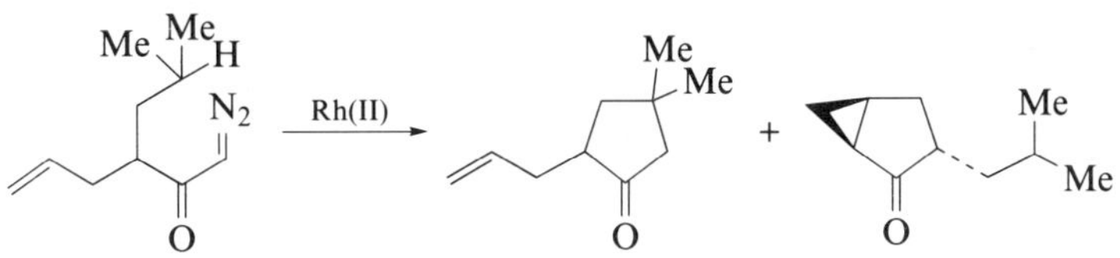

(99)

$Rh_2(OAc)_4$	56 : 44	
$Rh_2(pfb)_4$	100 : 0	pfb = perfluorobutyrate
$Rh_2(cap)_4$	0 : 100	cap = caprolactam

units to a Rh(II) catalyst results in cyclization of a ketocarbenoid to give intermediates in which carbene-like reactivity has been transferred to one of the alkyne carbons. The cyclized intermediate may be trapped by a neighboring unsaturated bond or, externally, by electron-rich alkenes or alkynes (equation 100)[200, 201].

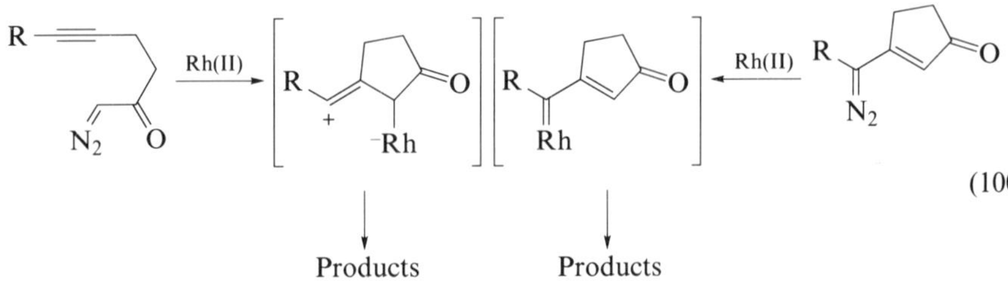

(100)

Asymmetric cyclopropanations of alkenes and alkynes with α-diazocarbonyl compounds have been extensively explored in recent years and a number of very effective chiral catalysts have been developed[2i]. Copper complexes modified with such chiral ligands as salicylaldimines **38**[202, 203], semicorrins **39**[204–208], bis(oxazolines) **40**[209–211] and bipyridines **41**[212] have

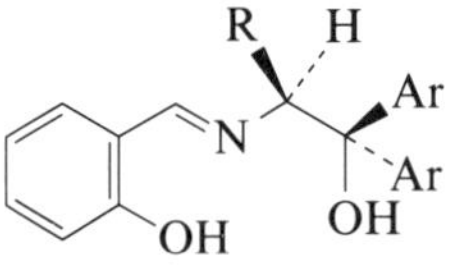

R = Me, *i*-Pr, *i*-Bu, $PhCH_2$

Ar =
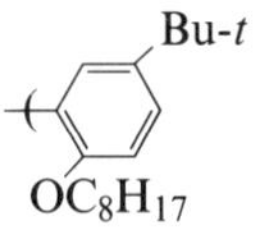

(38)

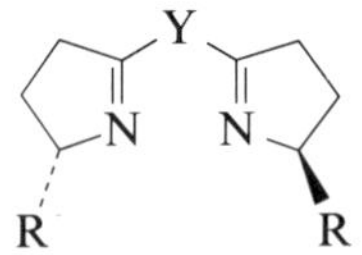

Y = CHCN, NMe
R = CO_2Me, CH_2OSiMe_2Bu-*t*, CMe_2OH

(39)

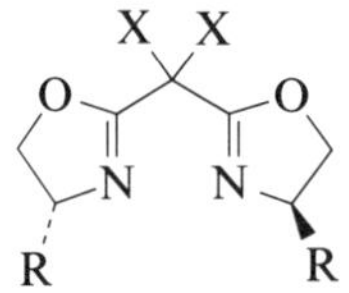

X = H, Me
R = $CHMe_2$, CMe_3

(40)

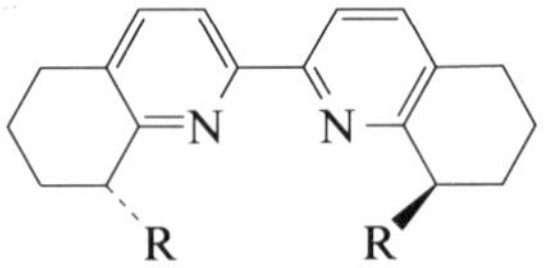

R = *i*-Pr, CMe_2OMe, $SiMe_3$, $SiEt_3$

(41)

been shown to be very effective catalysts for the enantioselective cyclopropanations of alkenes by diazoacetic esters and an extremely high level of selectivity (ee > 95%) has been realized. Rh(II) catalysts bearing chiral carboxamide ligands derived from pyroglutaric acid provide exceptionally high enantioselectivity for the cyclopropanations of certain alkenes[215] and 1-alkynes (equation 101)[216] by diazoacetic esters and for the intramolecular reactions of allylic[215] and homoallylic[216] diazoacetates (equation 102). Highly diastereo- and enantioselective cyclopropanations of alkenes by vinyldiazoacetic esters have been achieved with the chiral Rh(II) catalyst, **42** (equation 103)[217, 218] or using α-hydroxy esters as chiral auxiliaries (equation 104)[219, 220]. Asymmetric induction in the reactions of alkenes with chiral *N*-(diazoacetyl)oxazolidones[221] and of chiral allylic amines with diazoacetic esters[129, 222], and in the intramolecular cyclopropanations of α-diazocarbonyl derivatives bearing chiral auxiliaries[223, 224] has also been explored (equation 105).

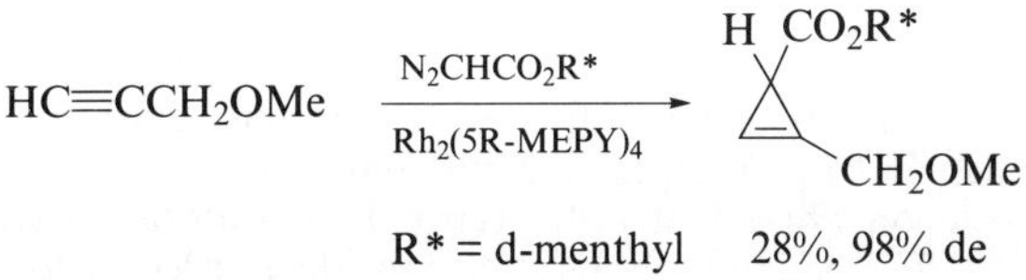

(101)

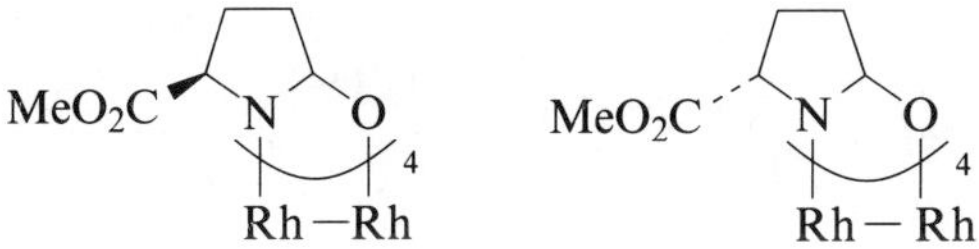

Rh$_2$(5R-MEPY)$_4$ Rh$_2$(5S-MEPY)$_4$

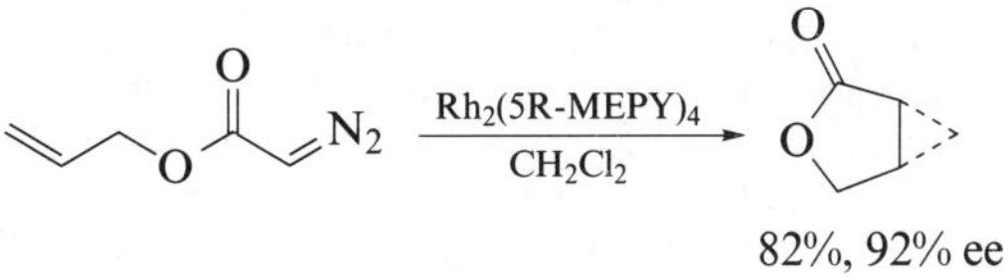

(102)

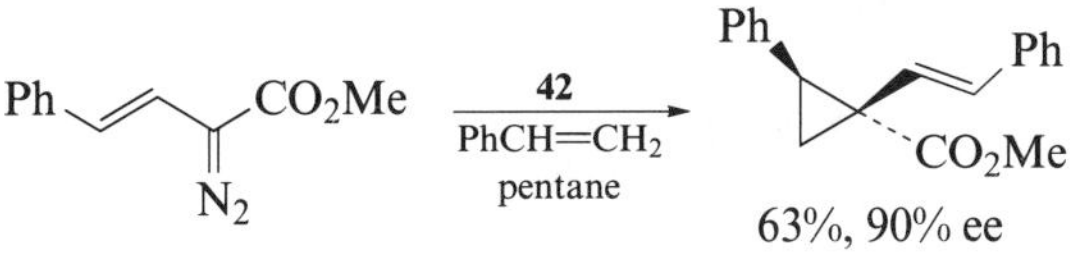

(103)

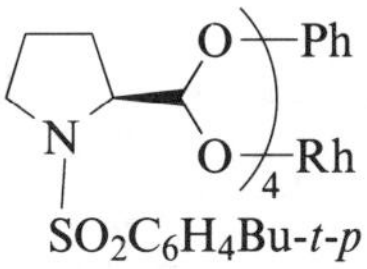

(42)

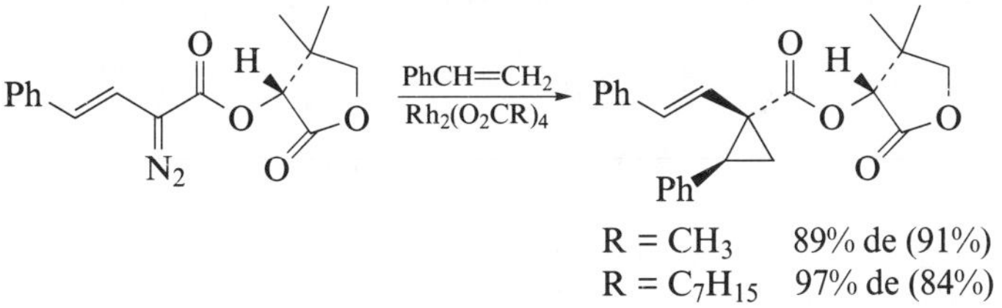

(104)

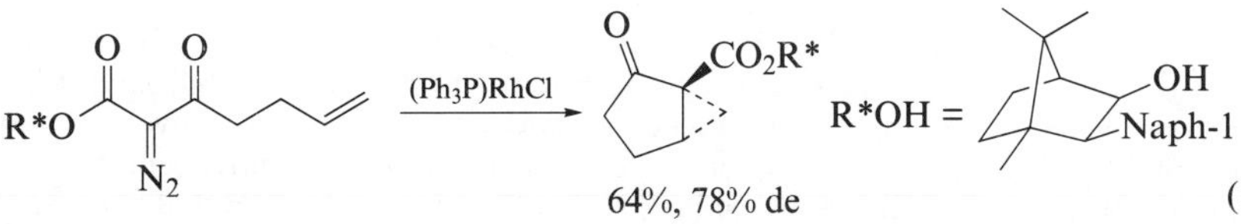

(105)

6. Carbenes α-substituted with heteroatoms other than halogen

Highly electrophilic nitromethylenes [:$CRNO_2$ (R = H, CF_3, COPh, CO_2R, CN, SO_2Ar)] are added to alkenes via $Rh_2(OAc)_4$-catalyzed decomposition of the corresponding diazo precursors. The reaction works best with electron-rich, sterically undemanding alkenes (equation 106)[225–227]. The addition of chiral glycosylidene carbenes to electron-deficient

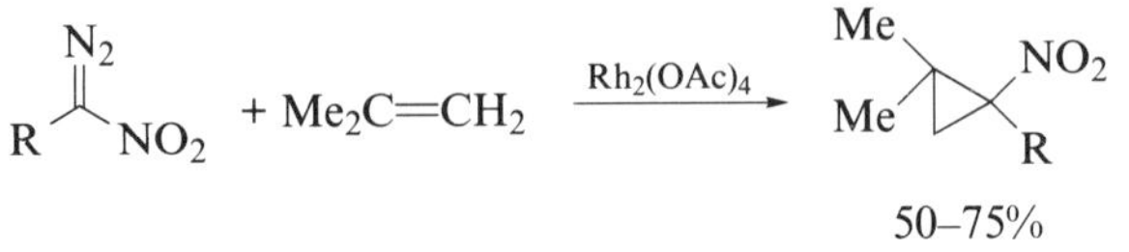

(106)

alkenes is found to be only poorly stereoselective[228–230]. An improved preparation of cyclopropyl phosphonates from phosphoryldiazomethane and alkenes has appeared[231]. Fluoro(phenoxy)carbene[232] and cyano(benzyloxy)carbene[233] are thermally generated from the corresponding diazirines. The former behaves as an ambiphile toward alkenes. It has been demonstrated that the philicity of phenylphosphorylcarbene [:CPhPO (OMe) (OX)] is dramatically changed from slightly electrophilic to highly nucleophilic in going from the ester carbene (OX = OMe) to the anionic carbene (OX = O^-Na^+) (equation 107)[234]. Generation of (alkoxy)sulfonylcarbenes [:$C(OR)SO_2Ar$] by α-elimination and their stereospecific additions to enol ethers are reported[235]. Chloro(phenylthio)methylene is transferred from $PhSCHCl_2$ to vinylsilanes with high (*Z*) -stereoselectivity under PTC conditions[236].

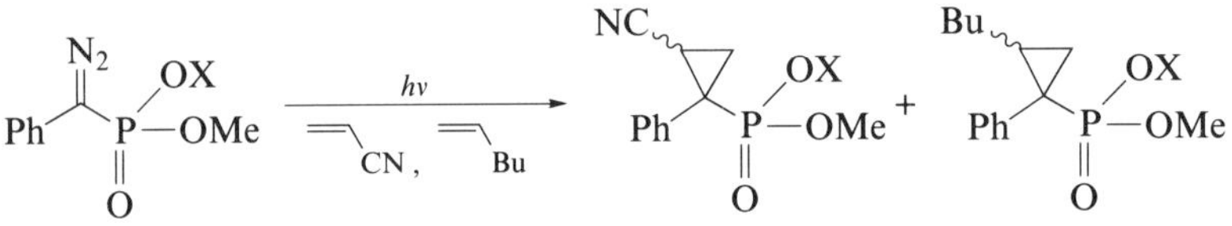

(107)

$k(\mathrm{CN}) / k(\mathrm{Bu}) = 3.34$ (OX = OMe)

> 340 (OX = O^-Na^+)

A variety of geminally substituted cyclopropyl ethers are synthesized employing Fischer carbene complexes [$(CO)_5M{=}CR^1(OR^2)$; M = Cr, Mo, W; R^1 = alkyl, alkenyl and aryl] as alkoxycarbene sources. Electron-deficient alkenes and conjugated dienes are suitable substrates for the reaction (equation 108)[237–245]. Electron-rich enol ethers and enamines are also

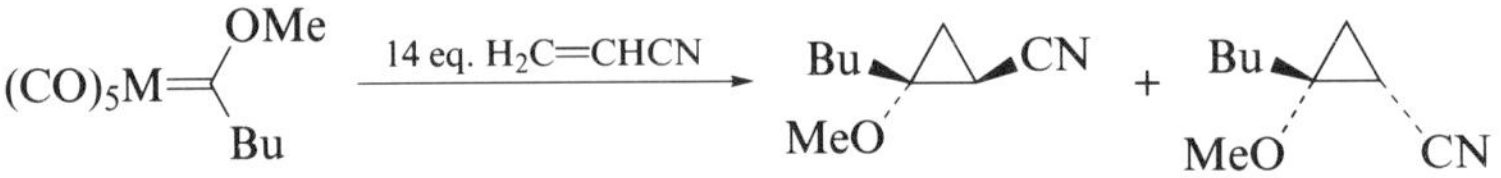

(108)

M = Mo (THF / 65 °C / 1 h)	32%	23%
Cr (PhMe / 80 °C / 2.5 h)	26%	24%
W (PhMe / 110 °C / 5 h)	39%	30%

cyclopropanated with the metal–carbene complexes (equation 109)[243, 246, 247]. The reactions of silyl enol ethers with acyloxycarbene–chromium complexes generated *in situ* from chromium acylate salts and acyl halides provide 1,2-dioxygenated cyclopropanes (equation 110)[243]. Fischer carbene complexes of chromium and molybdenum readily monocyclopropanate 1,3-dienes in good to excellent yields with high level of peri-, regio-

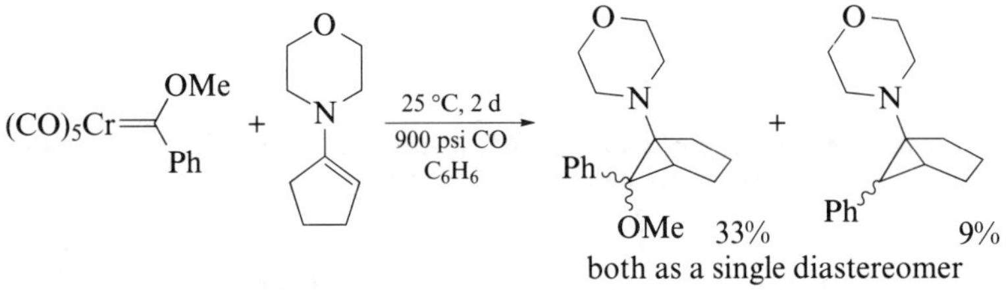

(109)

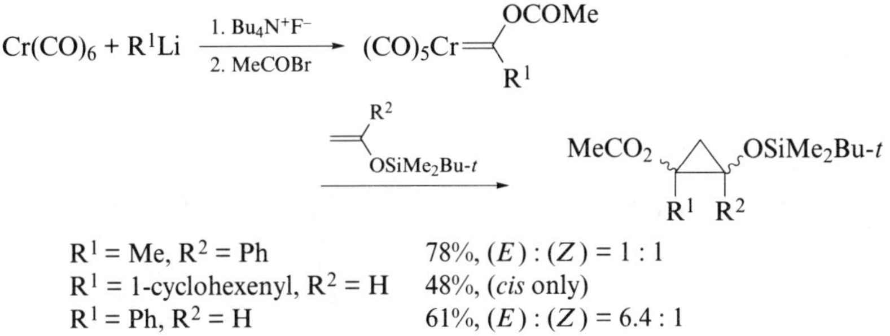

(110)

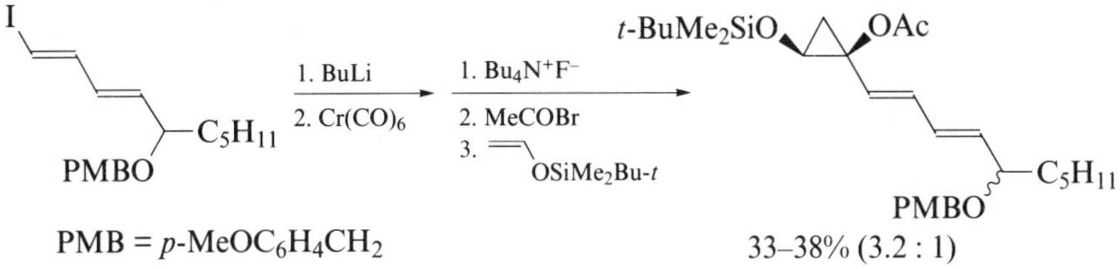

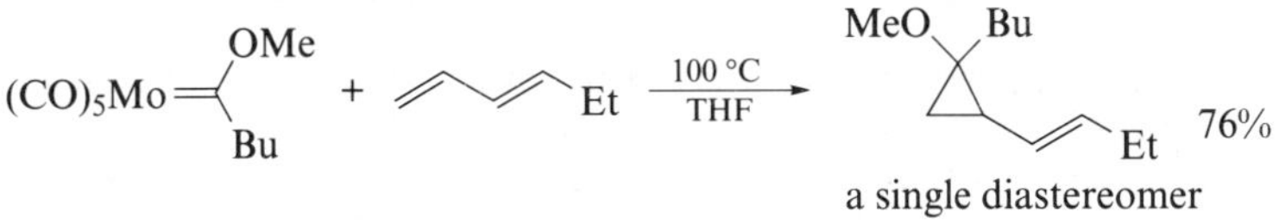

(111)

and stereoselectivity (equation 111)[240, 245–247]. Unactivated alkenes and 1,3-dienes which lack a readily accessible s-*cis* conformer seem, in general, incapable of participating in the cyclopropanation by those complexes for the intramolecular reaction[248, 249]. An improved procedure has been reported for the preparation of molybdenum carbene complexes[244], which react with electron-deficient alkenes to give cyclopropanes under milder conditions than chromium- and tungsten-derived complexes. (Cyclopropyl)methoxycarbene–chromium complex reacts with electron-deficient alkenes and 1,3-dienes without complication arising from opening of the cyclopropyl ring (equation 112)[242].

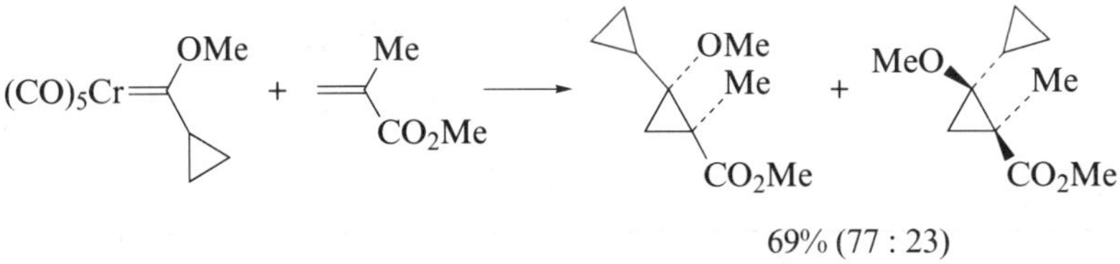

(112)

B. Cyclopropanation of Michael Acceptors

1. With ylides

An unsaturated system capable of serving as Michael acceptor may be cyclopropanated with ylides. The reaction proceeds stepwise via a zwitterionic intermediate. Hence, in general, it is not stereospecific. The degree of stereospecificity actually observed, however, depends on the rate of cyclization relative to that of rotation about the relevant single bond in the zwitterionic intermediate **43** under the reaction conditions (equation 113). The major

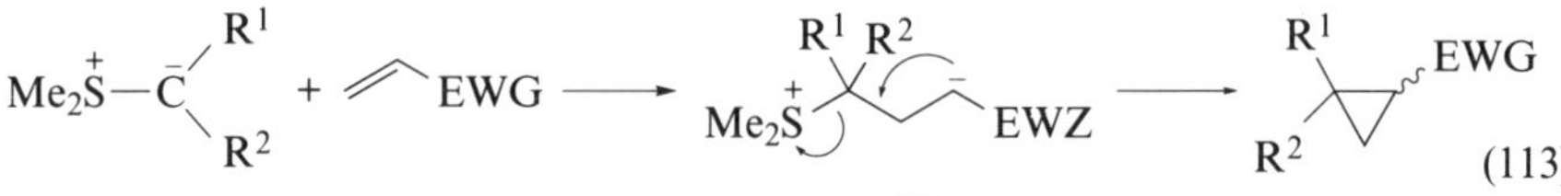

(**43**)

(113)

side reactions encountered in this type of cyclopropanation include oxirane formation (for α,β-unsaturated ketones), acylation of the ylides (for α,β-unsaturated esters) and γ-lactam formation (for α,β-unsaturated amides). The stereochemistry of the cyclopropanations of Michael acceptors by ylides may be understood in terms of the addition of ylide to the less hindered side of the double bond (steric approach control) and subsequent cyclization in the readily accessible stable conformation. It has been suggested that the stereochemistry in the methylene transfer reaction from ylide to Michael acceptor, e.g. in the reaction in equation 114, may be predictable on the basis of conformational analysis of the acceptor

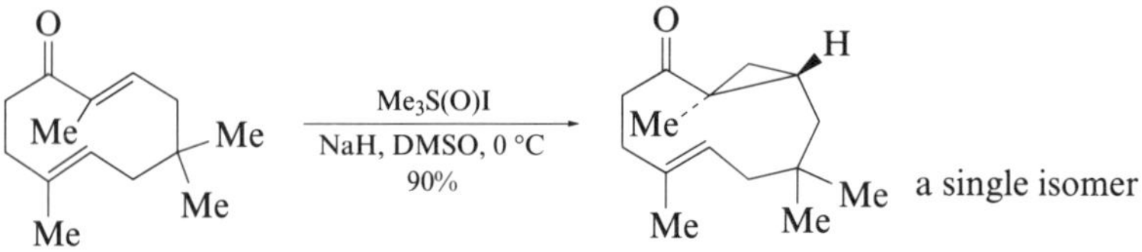

(114)

and the intermediate zwitterion by force-field calculations[250]. The first step of the reaction is generally reversible. Apparently unusual *endo*-selective cyclopropanation of **44** by $Me_2S(O){=}CH_2$[251] may be accounted for by taking the reversibility into consideration (equation 115).

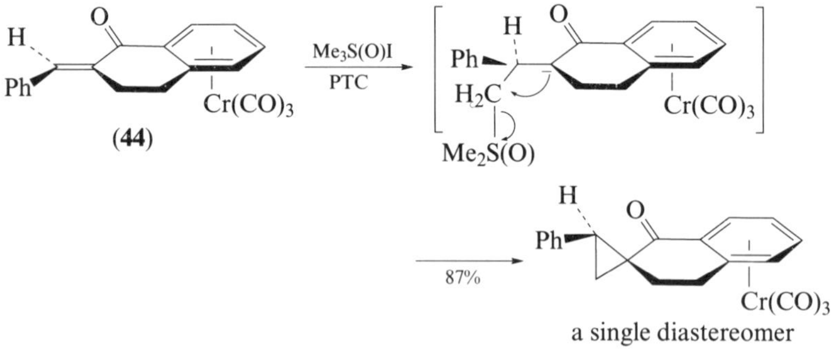

(115)

Sulfur ylides are most frequently employed in this methodology. The cyclopropanations of Michael acceptors with arsenic[252], selenium[253] and tellurium ylides[254] have also been

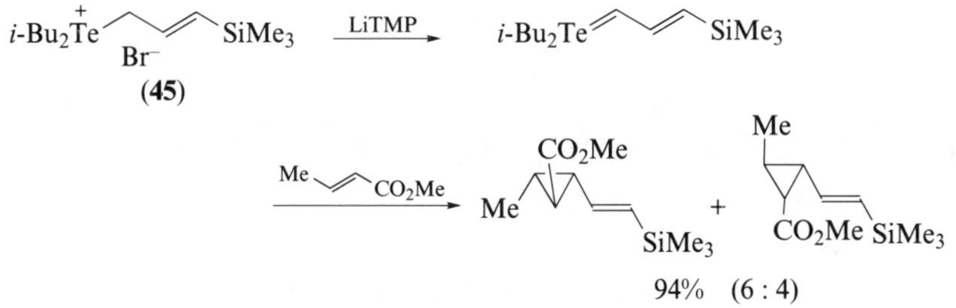

(116)

LiTMP = 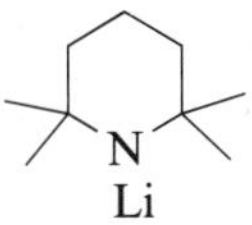

explored. The tellurium ylide generated from **45** reacts with a number of α,β-unsaturated esters to give (trimethylsilylvinyl)cyclopropane derivatives with high stereoselectivity in good yields (equation 116)[254]. Trifluoromethyl-substituted pyrethroids have been synthesized exploiting the reactivity of 1,1-bis(trifluoromethyl)alkenes toward sulfur ylides (equation 117)[255–257]. Cyclopropyl acylsilanes are obtained by the reaction of α,β-unsaturated carbonyl compounds with **46** (equation 118)[258].

(117)

R = CH_2OBz (44%), $CH{=}CCl_2$ (42%), $CH(OEt)_2$ (84%)

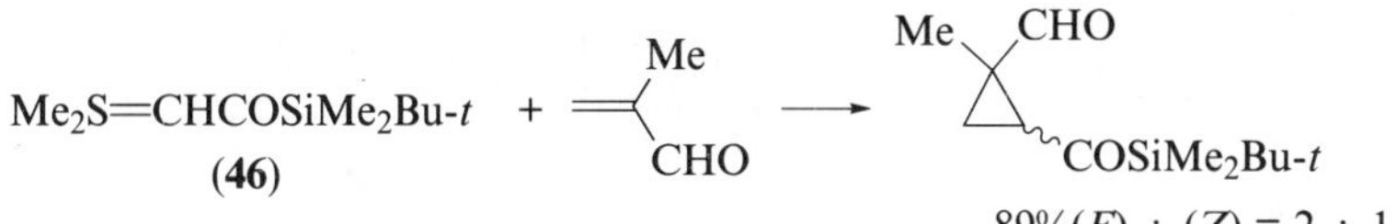

(118)

Asymmetric cyclopropanations of Michael acceptors with ylides have been explored for a number of substrates modified with chiral auxiliaries. The reactions of **47** with $Ph_3P{=}CMe_2$, **48** (equation 119)[259] and of **49** with **50** (equation 120)[260,261] proceed with excel-

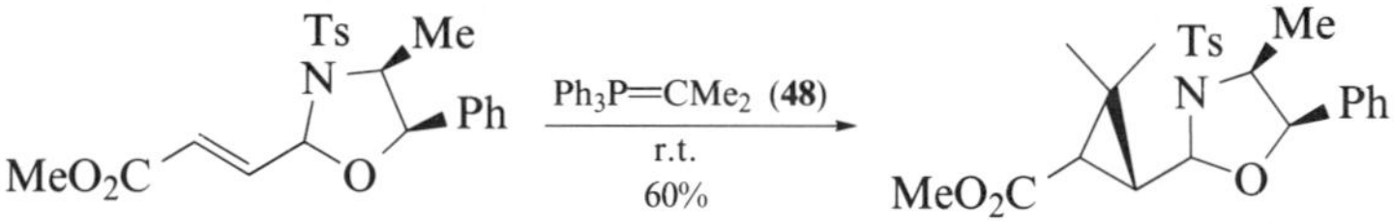

(119)

(**47**) a single diastereomer

lent π-face selectivity to afford a single diastereomer, respectively, whereas that of **51** with **48** is nonstereoselective (equation 121)[262]. The cyclopropanations of chiral bicyclic α,β-unsaturated lactams, **52**, mainly, if not exclusively, provide kinetically controlled *endo-syn*-adducts. The observed stereoselectivity has been rationalized in terms of steric approach control and stereoelectronic factors (equation 122)[263,264]. The stereochemistry of isopropy-

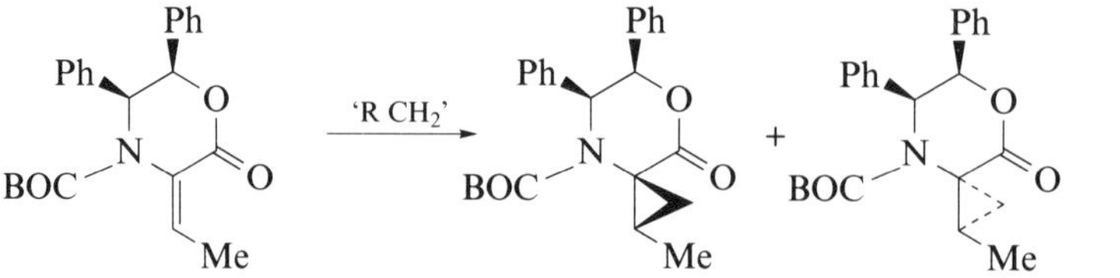

(49) (120)

$Ph(Et_2N)S(O){=}CH_2$ (**50**)	1 : 0	(82%)
$Me_3S(O){=}CH_2$	2–3 : 1	(77%)
CH_2N_2	1 : 1.6	(91%)

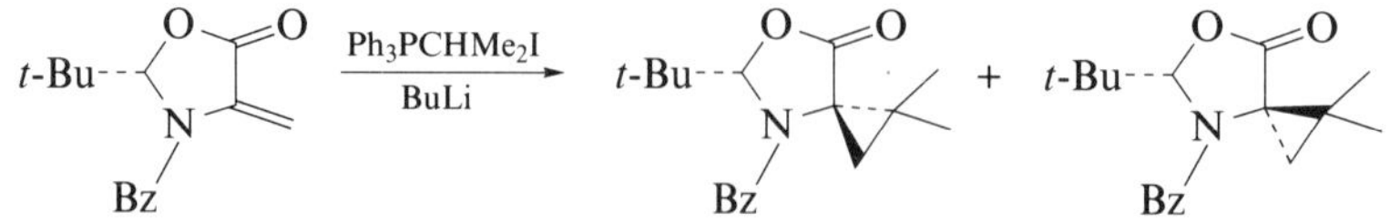

(121)

(51)		
r.t.	1 : 1	
–78 °C/8 h	No reaction	Bz = benzoyl

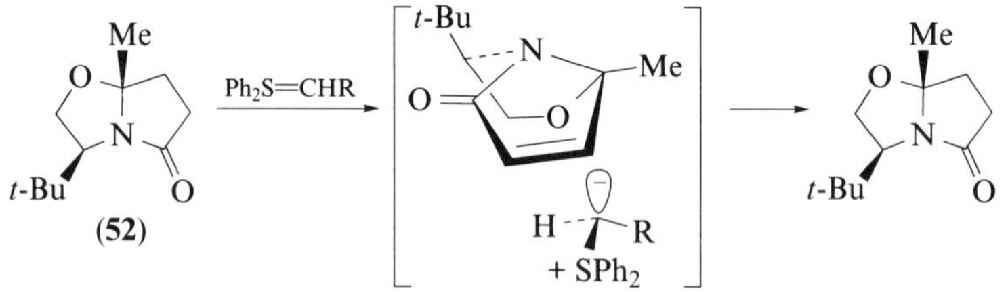

(52) (122)

R = Me	*syn* / *anti* = 4.6 : 1	(64%)
$CH{=}CH_2$	31 : 1	(65%)
Ph	7 : 1	(58%)

lidene transfer from three reagents, **48**, $Ph_2S{=}CMe_2$ **53** and *i*-PrS (O) (NLiTs)$=CMe_2$ (**54**), to chiral γ-alkoxy-α,β-unsaturated carbonyl compounds derived from 2-glyceraldehyde and natural tartaric acid has been investigated in detail (equation 123)[265–268]. Stereochemical control in most of the reactions is very high (de = 88–99%). The alkylidence transfer from **53** to the α,β-unsaturated carbonyls is stereospecific and the geometrical isomerism in the substrates is retained through the reaction, whereas that from **48** or **54** stereoselectively produces the *trans* isomers irrespective of the geometry of the unsaturated carbonyls. Except for the reactions of the (*Z*)-isomers with **54**, the π-facial stereoselection by the reagents is also very high: the reaction occurs from the *Re* face with

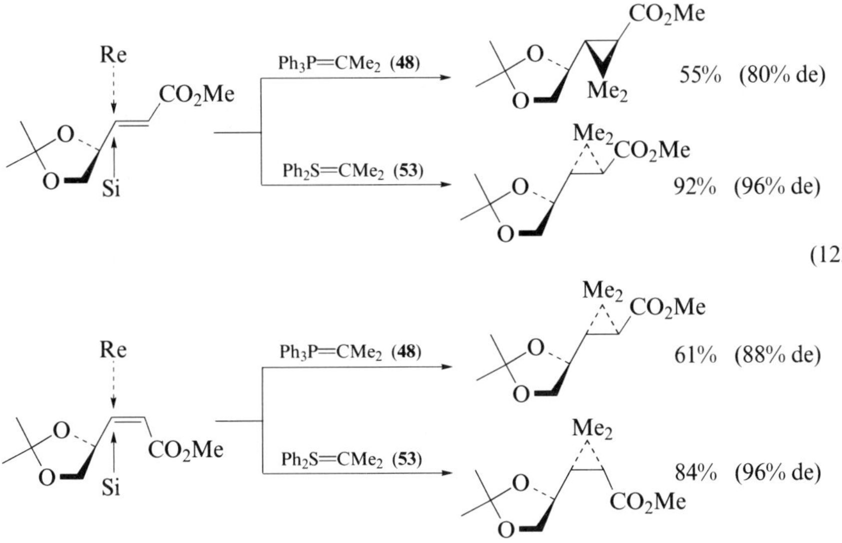

(123)

all the substrates and reagents examined, at the exclusion of (*E*)-unsaturated carbonyls and **48** or **54** which instead react from the *Si* face. Highly stereoselective methylene transfer from $Me_2S(O) = CH_2$ to a chiral vinyl sulfoxide has also been achieved (equation 124)[269].

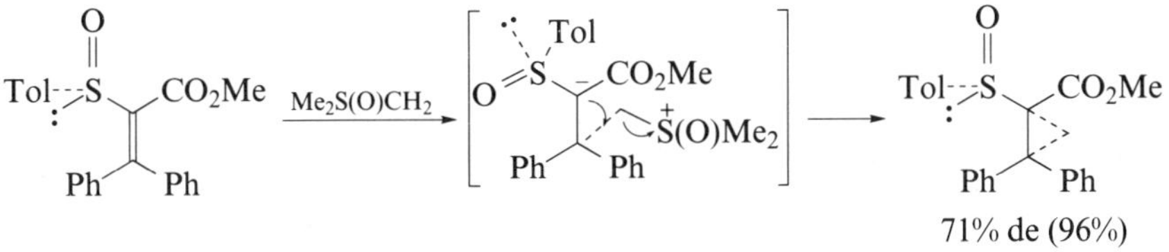

(124)

2. *With α-halocarbanions and related species*

Functionalized cyclopropyl derivatives can be prepared via Michael addition of carbon nucleophiles carrying a leaving group at the α-carbon followed by intramolecular substitution as depicted in equation 125. Anion-stabilizing α-substituents in the nucleophiles

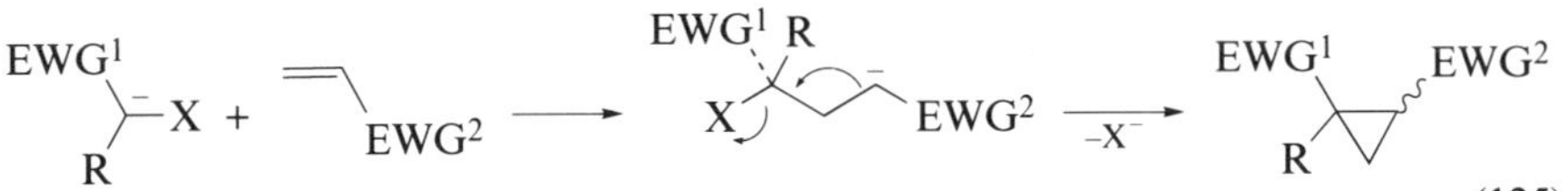

(125)

such as nitro, sulfonyl, phenylthio, and alkylseleno groups can function as the leaving group as well. A variety of stabilized carbanions has been utilized in this type of reactions[270–272]. Lithium enolates derived from α-haloacylsilanes react with a wide variety of electron-deficient alkenes to afford cyclopropyl acylsilanes (equation 126)[258]. The enolates of α-bromocrotonate and its more highly substituted derivatives undergo addition to

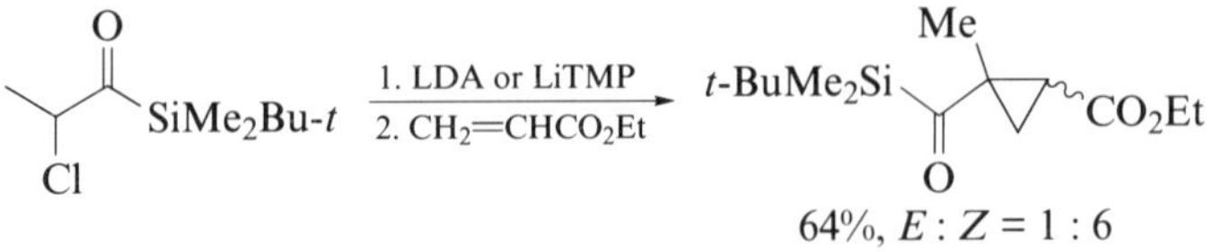

(126)

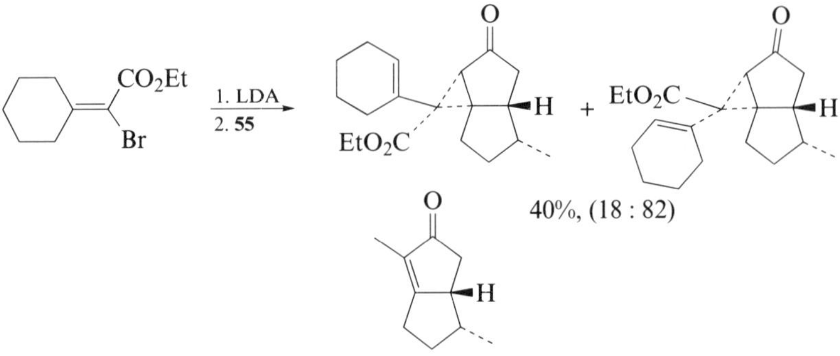

(127)

(55)

enones followed by cyclization to give the corresponding vinylcyclopropanes in excellent yields[273]. The use of a chiral enone such as **55** leads to the adducts with complete diastereofacial selectivity (equation 127). The tandem Michael addition-cyclization involving α-chloro ketones and α,β-unsaturated ketones is promoted by *N*-stannyl carbamate to furnish (*E*)-1,2-diacylcyclopropanes stereoselectively under neutral conditions[274].

Cyclopropanations of α,β-unsaturated ketones with sulfur-substituted methylenes have been achieved in several ways using this methodology. Sequential treatment of conjugated enones with $(PhS)_3CLi$, *s*-BuLi and electrophiles produces phenylthiocyclopropanes (equation 128)[275]. Generation of lithium bicyclo[1.1.0]butane-2-olates as intermediates has

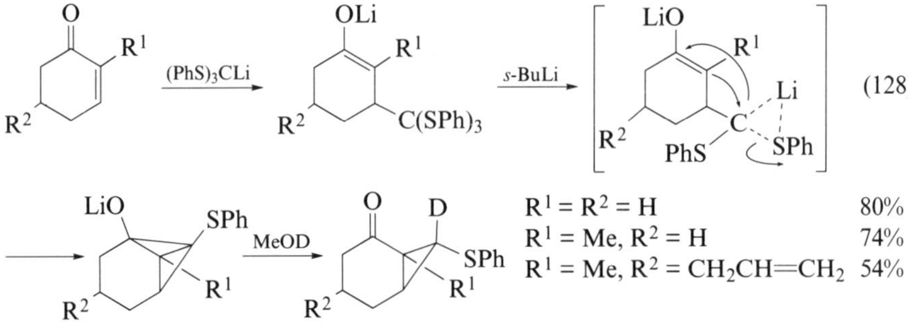

(128)

been suggested for the reactions employing 2-cyclohexenones as starting materials. 1-Chloro-1-(phenylsulfinyl)cyclopropanes are obtained from dichloromethyl phenyl sulfoxide and enones[276]. The cyclopropanations of enones with bis(phenylthio)methylene, phenylthio(trimethylsilyl)methylene or phenylthio-substituted vinylmethylene are brought about by conjugate addition of an appropriate phenylthio-stabilized carbanion followed by treatment of the resulting enolate anion with CuOTf (equation 129)[277]. Cyanonaphthalenes, quinoxalines and naphthyridines are doubly cyclopropanated with chloromethyl aryl sulfones[278]. Moderately activated alkenes such as ketene dithioacetals,

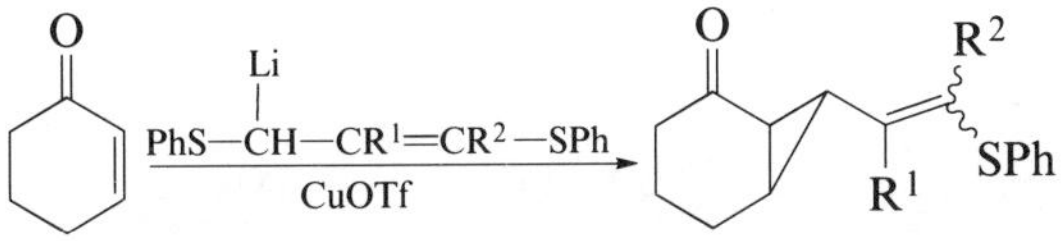

$R^1 = R^2 = H$ 78%
$R^1 = Me, R^2 = H$ 70% (129)
$R^1 = H, R^2 = SPh$ 83%

1-phenylvinylsilanes and 1-phenylthio-1-(trimethylsilyl)ethene are also successfully cyclopropanated with trimethylsilyl-[279], vinyl-[280] and aryl-substituted[281] alkylidenes by using the corresponding phenylthio- or methylseleno-substituted organolithiums (equation 130).

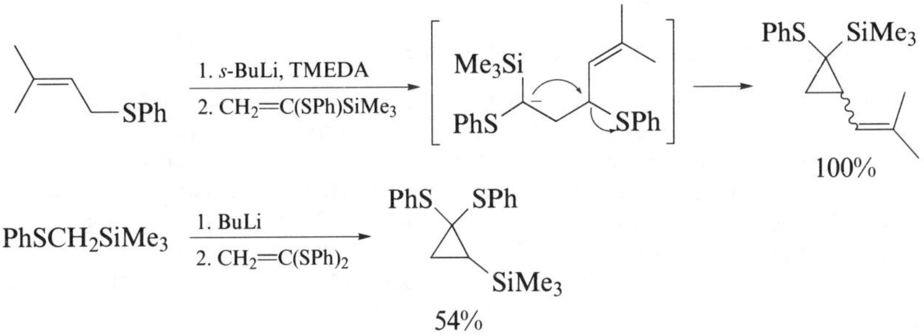

(130)

Nitroalkanes[282–288] as well as alkyl sulfones[283,288] can serve as useful and general alkylidene transfer reagents for highly activated Michael acceptors such as alkylidenemalonic esters (equation 131). Alumina-supported potassium fluoride is an effective reagent for the reaction of nitroalkanes[282]. The alkylidene transfer from sulfones may be promoted by $Ni(acac)_2$[289].

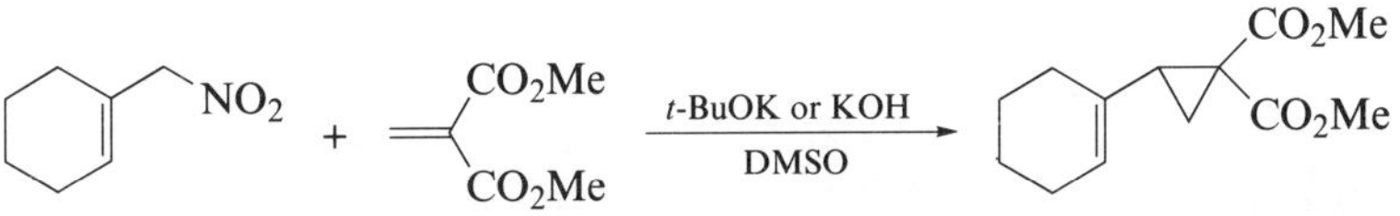

(131)

Cyclopropanes 1,1,2-trisubstituted with electron-withdrawing groups are prepared by treating electron-deficient alkenes with dibromomalonic ester or related active methylene dibromides in the presence of trialkylstibine (equation 132))[290,291], dibutyl telluride[292] or

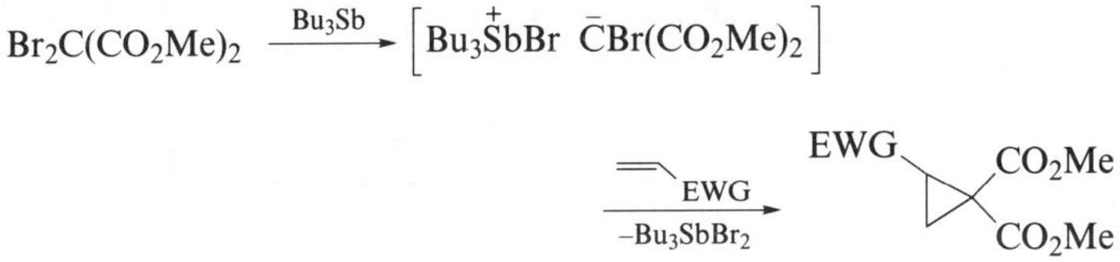

(132)

EWG = CHO (86%), COMe (84%), CO_2Me (43%), CN (59%)

indium metal/LiI[293]. As a modification of this methodology, the requisite α-halocarbanions are generated via the addition of nucleophiles to α-bromoacrylic esters. Subsequent Michael addition–cyclization stereoselectively produces *cis*-1,2-dialkoxycarbonylcyclopropanes (equation 133)[294]. 3,3-Disubstituted allyl sulfones undergo dimerization–cyclization to give 2-alkenylcyclopropyl sulfones, via isomerization to vinyl sulfones prior to the dimerization, upon exposure to base[295].

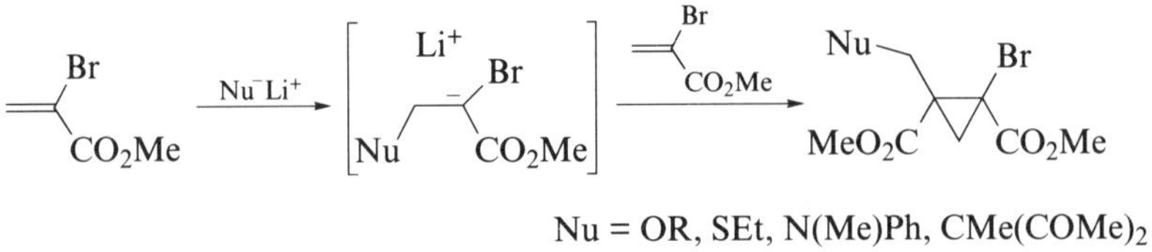

(133)

C. Cyclopropanation of Active Methylene Compounds

1. *With 1,2-dihaloethanes and related compounds*

Treatment of diethyl malonate and related compounds with 1,2-dihaloethane in the presence of base constitutes a classical method of cyclopropane synthesis[296–300]. The reaction can be conveniently carried out under PTC conditions. An improved method utilizing solid–liquid phase transfer catalysis has been reported[298]. The reaction of dimethyl or diethyl malonate with 1,2-dibromoalkanes except for 1,2-dibromethane tends to give only low yields of 2-alkylcyclopropane-1, 1-dicarboxylic esters. By the use of di-*tert*-butyl malonate, their preparations in satisfactory yields are realized (equation 134)[297]. The 2-alkylcyclopropane derivatives are also obtained from the reaction of dimethyl malonate and cyclic sulfates derived from alkane-1, 2-diols (equation 135)[301]. Asymmetric synthesis

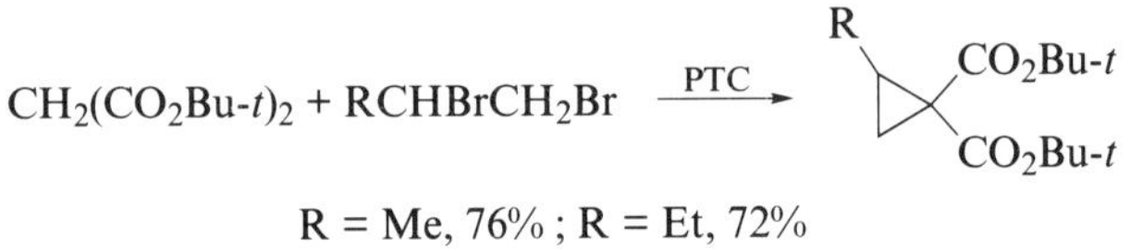

(134)

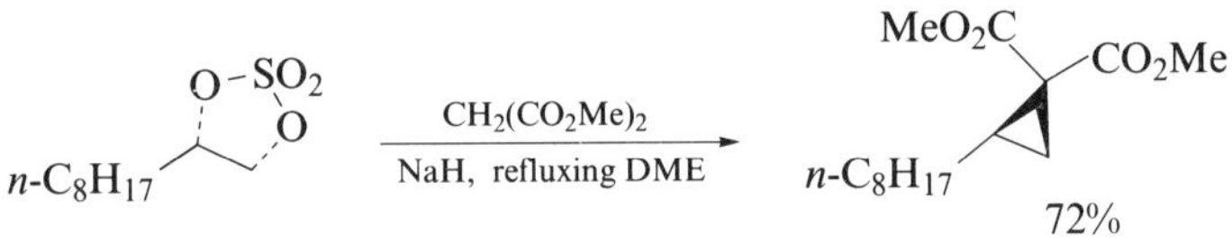

(135)

of stereospecifically deuterium-labeled 1-aminocyclopropanecarboxylic acid has been achieved by using a bis-lactim ether, cyclo[α-methyl-Phe-Gly] as a chiral source (equation 136)[302]. The reactions of active methylene compounds with 1,4-dihalo-2-alkenes usually provide vinylcyclopropane derivatives via allylic rearrangements in preference to cyclopentene derivatives[286–300].

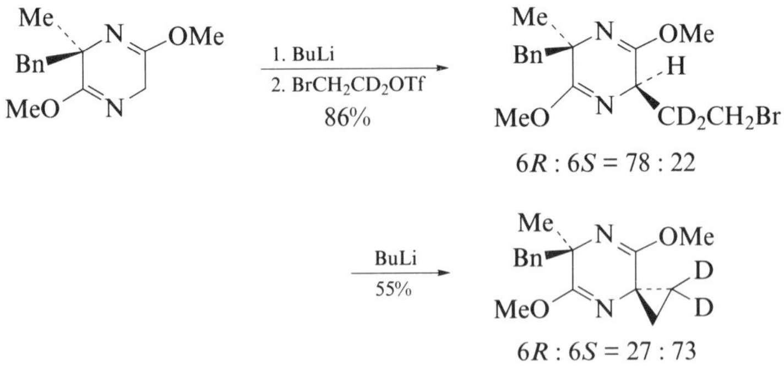

(136)

Oxiranes can also be used as the cyclopropanation reagents, provided that the intermediate alkoxides or alcohols are sulfonated prior to the second cyclization step. Vinyl-[303] and ethynylcyclopropanes[304], cyclopropanes[305] and methylenecyclopropanes[306] have been synthesized in this manner using appropriately substituted oxiranes (equation 137). The

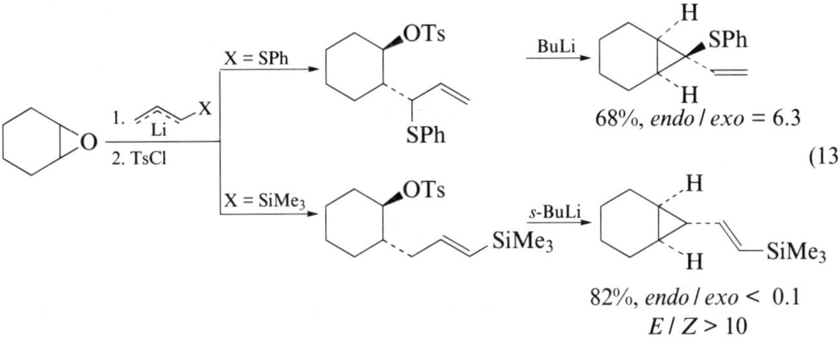

(137)

tandem cyclopropanation-lactonizations of malonic esters with epichlorohydrin provide cyclopropane- annelated γ-butyrolactones (equation 138). The reactions mainly proceed via initial attack at the epoxide ring by malonate anion followed by Payne rearrangement and intramolecular epoxide ring opening. The slight loss of stereochemical fidelity is incurred as a consequence of direct displacement of the chloride competing with the above process[307,308]. In contrast, the reaction of malonate anion with oxiranylmethyl triflate proceeds almost exclusively via the direct displacement of triflate followed by ring opening of the oxirane to give the product with high stereospecificity[307,309].

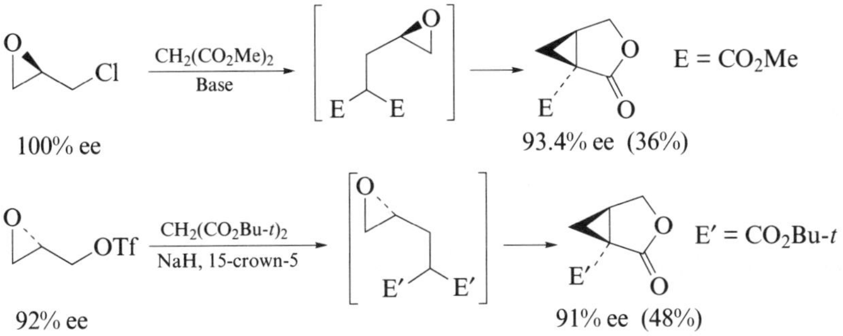

(138)

2. *With Michael acceptors carrying a leaving group at the α-carbon atom*

Michael acceptors which carry a good leaving group at the α-carbon atom or whose electron-withdrawing group itself can serve as the leaving group may be cyclopropanated by active methylene compounds under basic conditions via a prototropic shift subsequent to the Michael addition as outlined in equation 139. Thus, the basicity of the carbanions involved must be balanced to allow the requisite prototropic shift; otherwise, the reaction will be very slow or will not work.

A variety of cyclopropyl derivatives has been prepared utilizing this methodology from malonic ester anion or related stabilized carbanions and Michael acceptors such as **56**, **57**[310] and **58**[311]. The reactions are nonstereospecific in general as expected from the mechanism

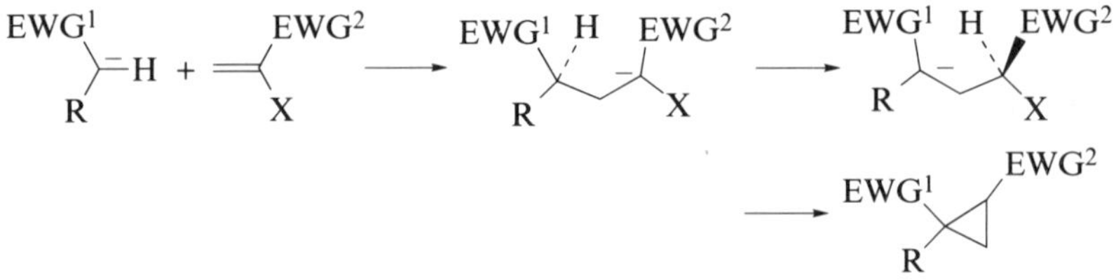

(139)

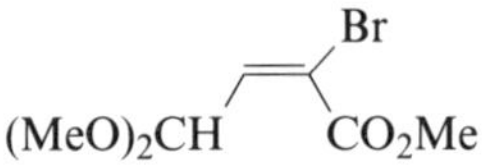

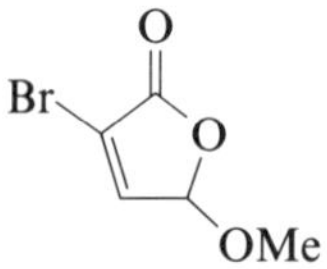

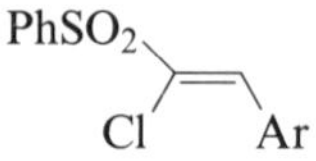

(56) **(57)** **(58)**

involved. A new stereoselective synthesis of 2-alkyl-2-formylcyclopropanecarbonitriles from aliphatic enamines and 2-haloacrylonitriles has been developed (equation 140)[312]. A

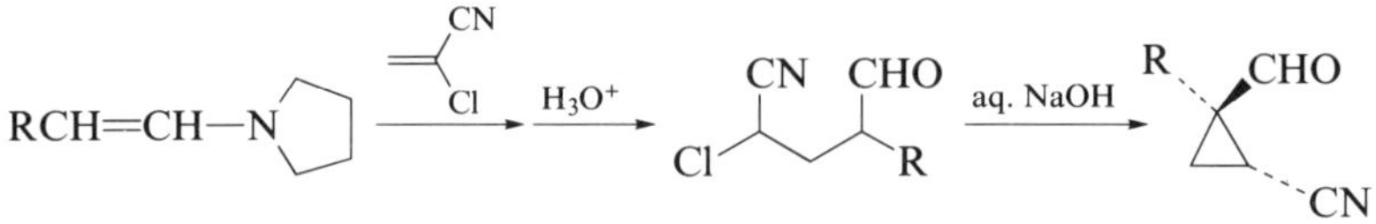

(140)

one-pot reaction of 2-alkenyltriphenylphosphonium salt with the sodium salt of β-ketoester affords 1-acyl-2-alkylcyclopropanecarboxylic ester. Isomerization of the former to 1-alkenylphosphonium salt prior to the addition of the stabilized carbanion is postulated (equation 141)[313].

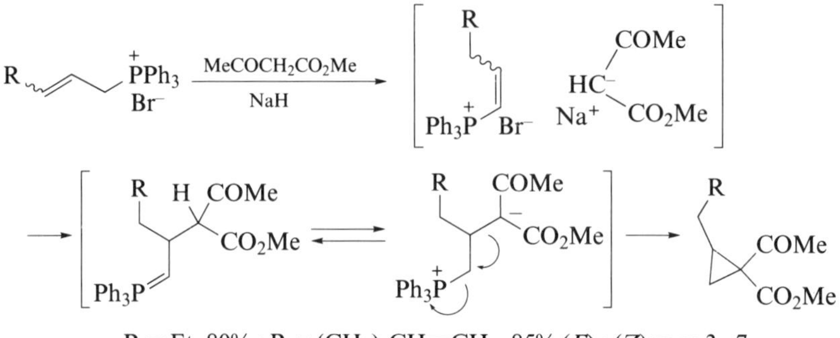

(141)

R = Et, 80% ; R = $(CH_2)_2CH{=}CH_2$, 85% (*E*) : (*Z*) = ca 3 : 7

D. Cyclopropanation with Diazo Compounds via 1-Pyrazolines

Diazo compounds have been extensively used in the preparation of three-membered carbocycles either as carbene sources or as precursors for 1-pyrazolines or 3*H*-pyrazoles. Nitrogen extrusion from pyrazolines is particularly valuable for the synthesis of alkylcyclopropanes, since the direct carbene route is impractical, as a matter of fact, owing to rapid intramolecular processes in alkylcarbenes. The cycloaddition of diazo compounds to unsaturated bonds to give 1-pyrazolines and 3*H*-pyrazoles usually proceed in a concerted manner, and hence is stereospecific. In the subsequent nitrogen extrusion from the adducts,

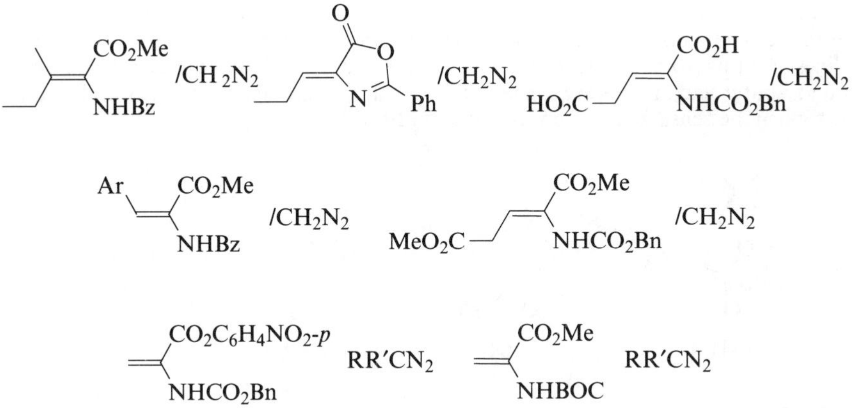

SCHEME 2

however, the stereochemical integrity of the adduct may not be retained. The reactivity of the diazo compound toward alkenes is reduced by substitution of the α- carbon with electron-withdrawing groups, while that of the alkene toward the addition of diazo derivatives is enhanced both by conjugation with an electron-withdrawing group and by substitution with an electron-donating group.

A variety of highly functionalized cyclopropanes has been prepared by this two-step route. Shown in Scheme 2 are the combinations of unsaturated substrate/diazo compound from which various 1-aminocyclopropanecarboxylic acid derivatives have been prepared via the corresponding 1-pyrazolines[314–320]. Asymmetric synthesis of cyclopropyl derivatives has also been achieved through this route by exploiting the diastereoselectivity in the cycloaddition of diazoalkanes to chiral unsaturated substrates. Thus, the addition of diazomethane to a chiral benzylidenediketopiperazine, **59**, derived from (*S*)-proline occurs regio- and stereoselectively to afford a single diastereomer, from which enantiomerically pure **60** has been obtained via photochemical nitrogen extrusion and hydrolysis (equation 142)[321]. The reaction of 5-arylidene Meldrum acid modified with *l*-menthone, **61**, similarly produces a single stereoisomer **62** (equation 143)[322]. The observed high diastereoselectivity

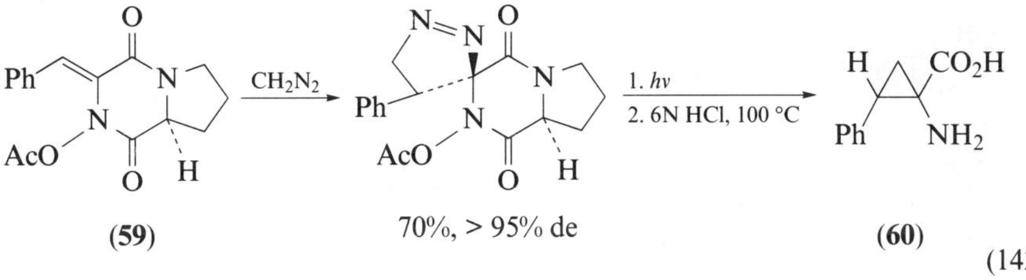

(142)

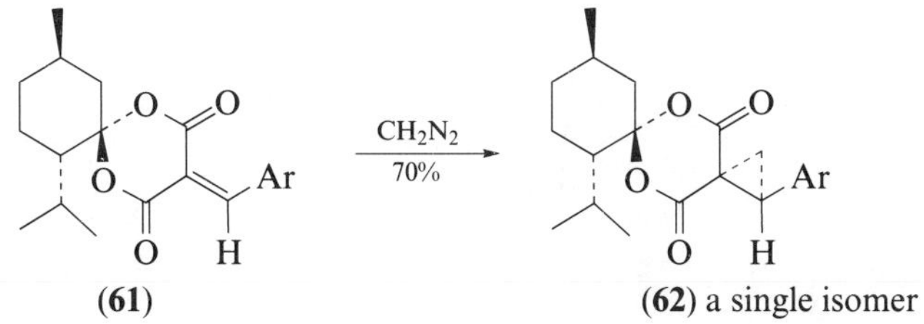

(143)

in the reaction of **61** is accounted for in terms of the π-facial differentiation resulting from the adoption of a boat conformation by the dioxane ring. A pair of regioisomeric 1-pyrazolines resulting from the π-face selective addition of 2-diazopropane to a chiral α,β-unsaturated γ-butyrolactone, **63**, are converted to a single bicyclic lactone, **64**, with high ee upon benzophenone-sensitized photolysis (equation 144)[323].

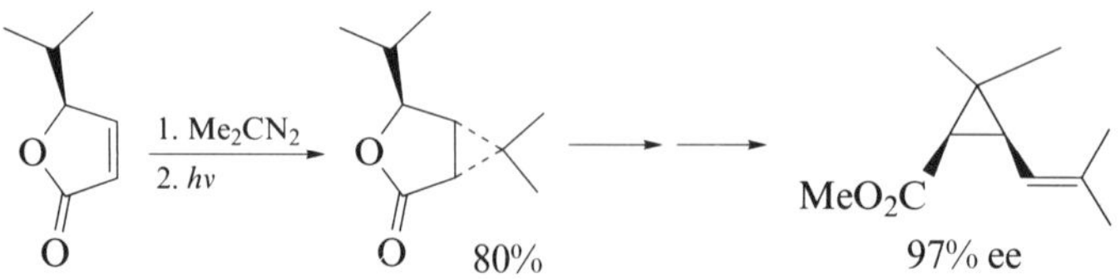

(144)

(63) **(64)**

E. Miscellaneous

Successive treatment of (–)-dimethyl succinate with LiTMP and bromochloromethane provides (*S*, *S*)-*trans*-cyclopropane-1,2-dicarboxylic ester in 99% de (equation 145)[324].

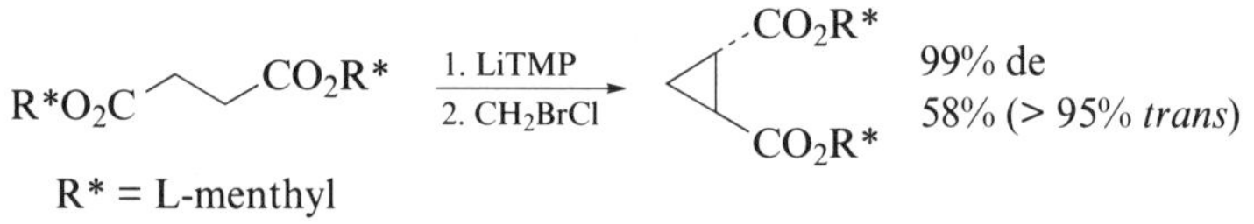

(145)

A convenient, short-step synthesis of enantiomerically pure 2-methylenecyclopropylmethanol has been achieved by treating optically pure epichlorohydrin with methylenephosphorane (equation 146)[325]. Vinyl ketones are cyclopropanated with (*E*)-1-(phenylseleno)-2-(trimethylsilyl)ethene in the presence of $SnCl_4$ to give *trans*-2-substituted cyclopropyl ketones via a formal [2 + 1] cycloaddition accompanied by 1,2-trimethylsilyl migration (equation 147)[326]. The addition of 3-phenylthio- or 3-ethoxyallylzinc bromide to

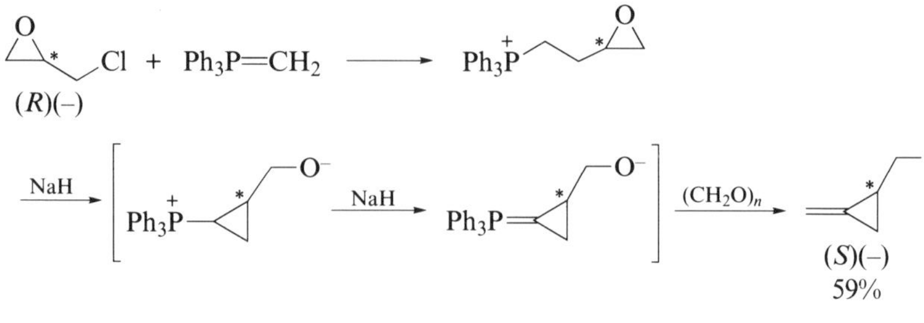

(146)

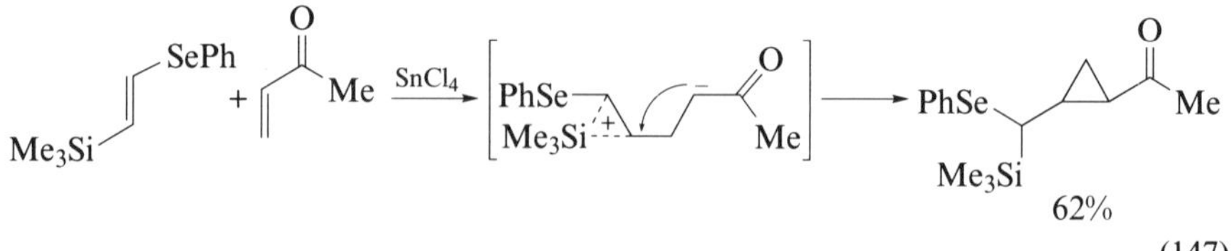

(147)

$PhSCH{=}CHCH_2ZnBr$ + $RCH{=}CHMgBr$

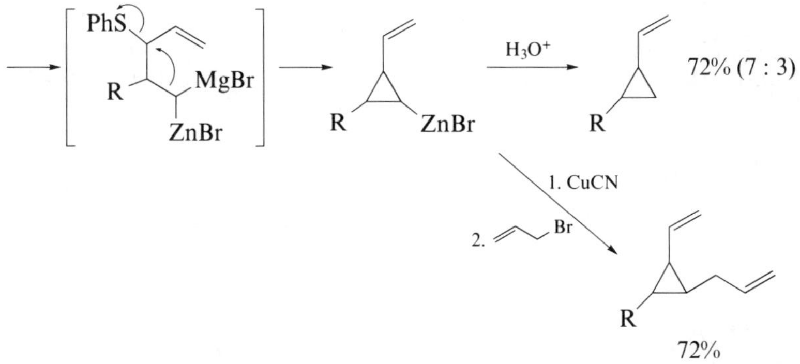

(148)

alkenyllithium or -magnesium derivatives furnishes 2-vinyl-3-alkylcyclopropylzinc bromides with a good diastereoselectivity via intramolecular substitution in *gem*-dimetallic intermediates (equation 148)[327,328]. The produced organozinc reagents serve as nucleophiles for further reactions with a variety of electrophiles. Successive treatment of a mixture of a chloro(methylthio)methylcarbonyl derivative and a conjugated diene with tin(IV) chloride and triethylamine affords 2-alkenyl-1-(methylthio)cyclopropanecarbonyl compound via [2 + 4] polar cycloaddition followed by [2,3] Wittig rearrangement (equation 149)[329]. A

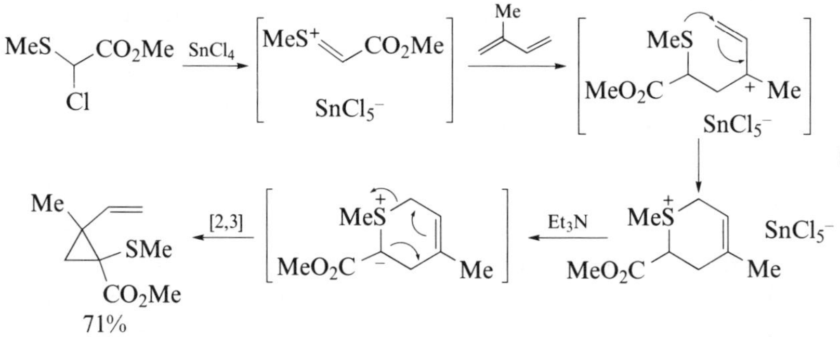

(149)

variety of substituted allylcyclopropanes is obtained from the reactions of alkynes with zirconocene–ethylene complex and homoallylic bromides (equation 150)[330]. An intramolecular cyclopropanation method alternative to the conventional diazoketone route, i.e. a

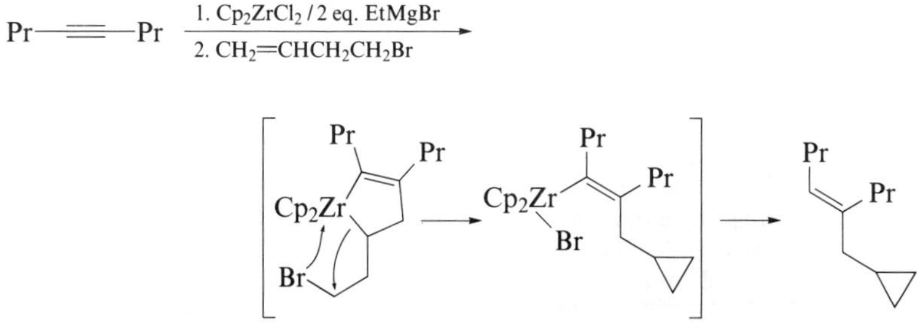

(150)

CuCl-catalyzed decomposition of iodonium ylides prepared from β-keto esters and diacetoxyiodobenzene, has been developed (equation 151)[331]. 1-Methylbenzvalene is obtained in a good yield by treating a mixture of lithium cyclopentadienide and 1,1-dichloroethane with butyllithium[332]. The tandem cyclization substitution in 1-selenyl-5-hexenyllithiums derived from corresponding selenacetals via selenium/lithium exchange produces bicyclo[3.1.0]hexane derivatives[333].

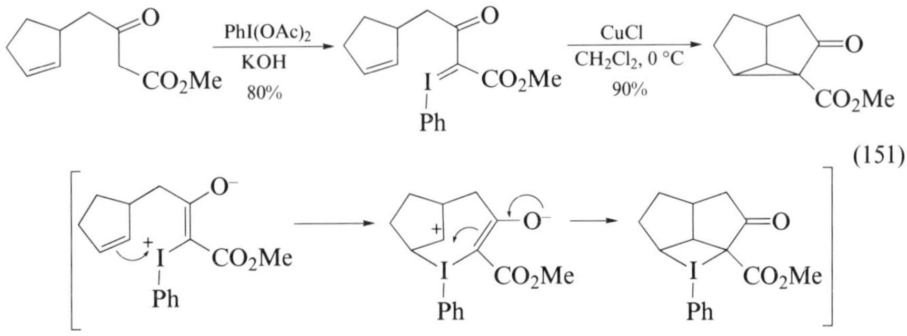

(151)

A number of new methods for the preparation of cyclopropanols has been developed (see Sections II.B and II.E for the preparation of cyclopropanols via 1,3-bond formation; see also Section III. A.1). Treatment of carboxylic esters with ethylmagnesium bromide in the presence of Ti (OPr-*i*)$_4$ affords 1-substituted cyclopropanols in excellent yields (equation 152)[334]. Similar transformation of carboxylic esters to cyclopropanols is realized by

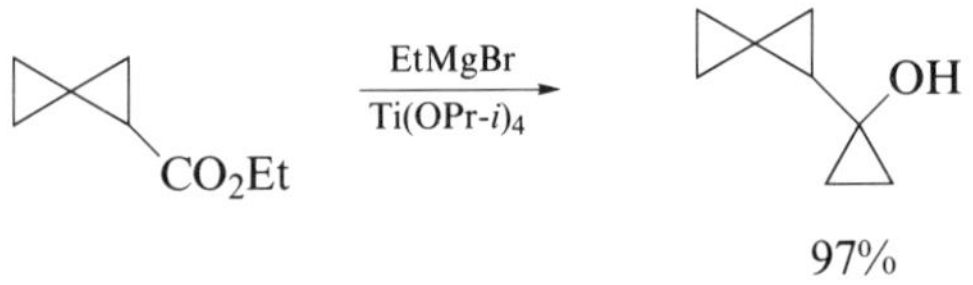

(152)

tandem olefination cyclopropanations of carboxylic esters with diiodomethane/samarium (equation 153)[335]. 1-Substituted cyclopropanols are also obtained via the successive treat-

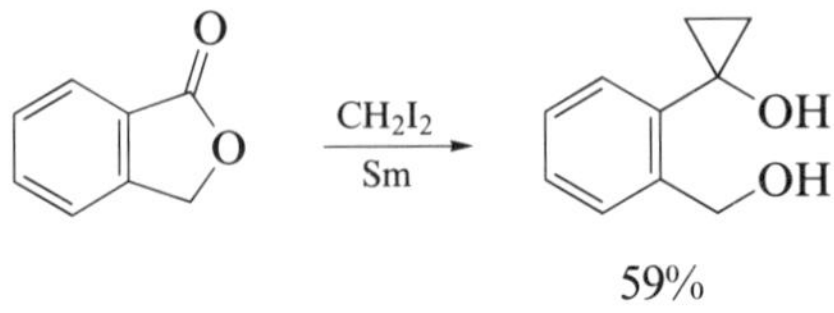

(153)

ment of acyl chlorides with chloromethyllithium and Li powder[336]. The reaction of α,β-epoxyketones with trialkylstannyllithium provides diastereomerically pure cyclopropanols for which the structure **65** is assigned (equation 154)[337]. The cathodic cyclocoupling of styrenes with aliphatic esters produces *trans*-2-arylcyclopropanols[338].

Treatment of 1,6- and 1,7-eneynes with Fischer carbene complexes provides bicyclic cyclopropane-containing skeletons through a coupling of well-established mode of carbene/alkyne reactions to generate presumed vinylcarbene intermediates (equation 155)[339–341]. The same type of transformation is brought about in superior yields by metal acylate salts prepared *in situ* from $Cr(CO)_6$ and organolithiums of Grignard reagents[342].

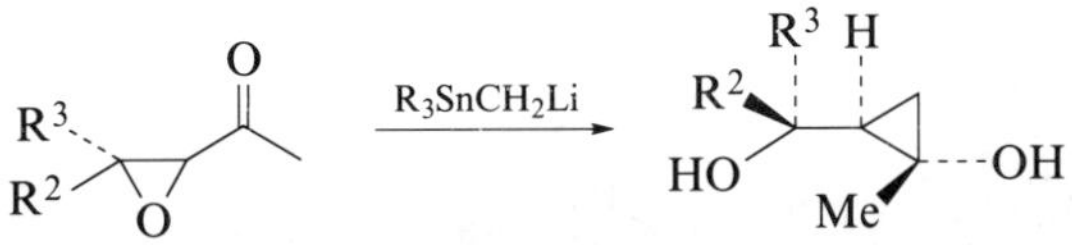

(65)

(154)

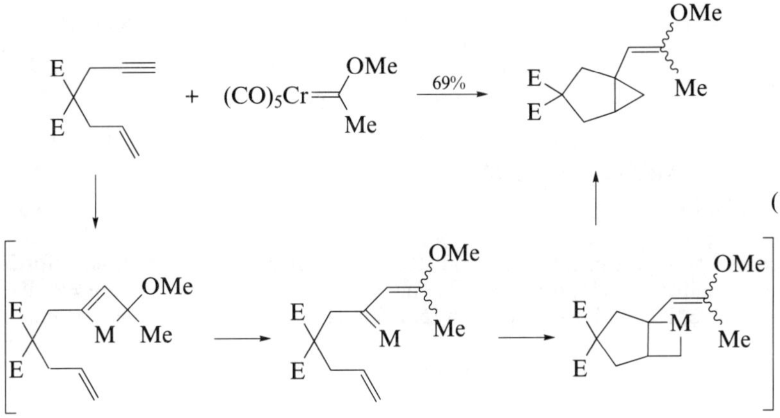

(155)

$E = CO_2Me$ $M = Cr(CO)_4$

Certain substituted, 1,6-enynes and related dienynes undergo palladium-catalyzed cyclization to give polycyclic structures containing a three-membered ring via cascade carbopalladation (equation 156)[343–348]. Transformation of **66** into **67** is representative[346].

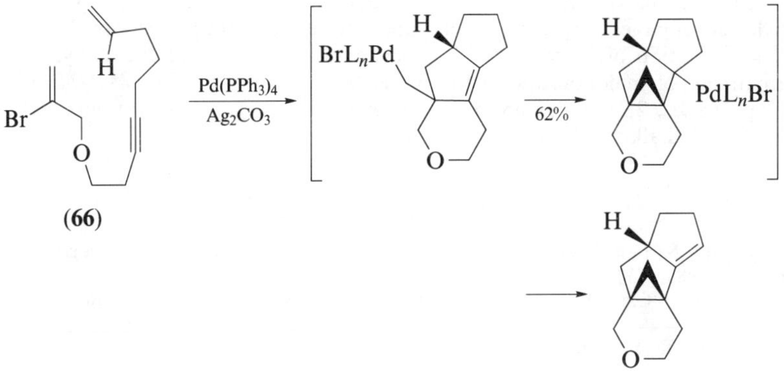

(66)

(67)

(156)

Norbornenes specifically undergo Pd(0)-catalyzed cyclopropanation by 1-acetoxy-3-silyl-2-propanones to afford acetylcyclopropane-annelated products[349]. Methyleneoxazolidinones, ketone α-carbonates and 5-methylene-1,3-dioxolane-2-ones can also be used as the cyclopropanation reagent (equation 157)[350–352].

Cyclopropanes multiply-substituted with electron-withdrawing groups are prepared by a number of methods (see also Sections III. A.5, III.B and III. C). The Cu $(acac)_2$-catalyzed or photochemical reactions of alkenes with bis(phenylsulfonyl)methylide, which is easily

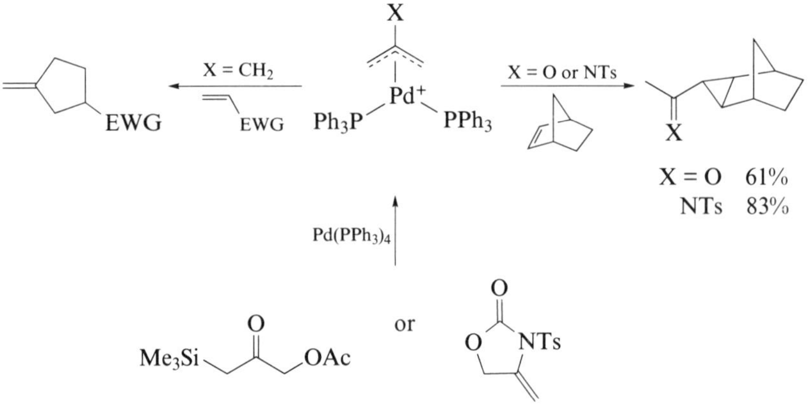

(157)

obtainable from bis(phenylsulfonyl)methane and diacetoxyiodobenzene, affords 1,1-bis(phenylsulfonyl)cyclopropanes[353,354]. The cyclotrimerization of α-bromoacetophenones in a mixture of K_2CO_3/DMF (equation 158)[355] and the Cu(acac)$_2$-catalyzed decomposition

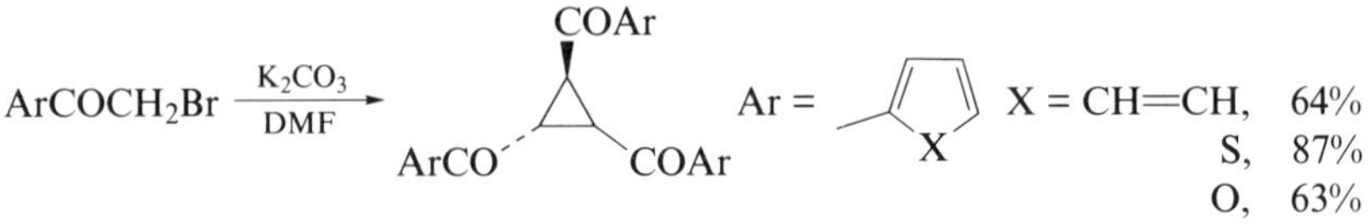

(158)

of α-diazoacetophenones in the presence of bis(*p*-methoxyphenyl)selenide furnish *r*-1, *c*-2, *t*-3-triaroylcyclopropanes[356]. Electrolysis of cyanoacetic esters in the presence of aldehydes leads to the formation of 3-substituted 1,2-dicyanocyclopropane-1,2-dicarboxylates[357]. Similar treatment of malononitrile in the presence of ketones produces 1,1,2,2-tetracyanocyclopropanes. 2,2,3,3-Tetracyanocyclopropanecarboxylic esters are obtained by the reaction of 1,1-dialkoxy-2-bromoalkenes with tetracyanoethylene[358].

IV. REFERENCES

1. T. Tsuji and S. Nishida, in *The Chemistry of the Cyclopropyl Group* (Ed. Z. Rappoport), Wiley, Chichester, 1987, p.307.
2. (a) M. P. Doyle, 'Catalytic methods for metal carbene transformation', *Chem. Rev.*, **86**, 919 (1986).
(b) L. A. Paquette, 'Silyl-substituted cyclopropanes as versatile synthetic reagents', *Chem. Rev.*, **86**, 733 (1986).
(c) M. Brookhart and W. B. Studabaker, 'Cyclopropanes from reactions of transition-metal-carbene complexes with olefins', *Chem. Rev.*, **87**, 411 (1987).
(d) J. R. Y. Salaün, 'Synthesis and synthetic applications of 1-donor substituted cyclopropanes with ethynyl, vinyl, and carbonyl groups', *Top. Curr. Chem.*, **144**, 1 (1988).
(e) H.-U. Reissig, 'Donor–acceptor-substituted cyclopropanes: versatile building blocks in organic synthesis', *Top. Curr. Chem.*, **144**, 73 (1988).
(f) W. E. Billups, W. A. Rodin and M. M. Haley, 'Cycloproparenes', *Tetrahedron*, **44**, 1305 (1988).
(g) K. B. Wiberg, 'Small-ring propellanes', *Chem. Rev.*, **89**, 975 (1989).
(h) J. R. Y. Salaün, 'Optically active cyclopropanes', *Chem. Rev.*, **89**, 1247 (1989).

(i) M. P. Doyle. 'Chiral catalysts for enantioselective carbenoid cyclopropanation reactions', *Rec. Trav. Chim. Pays-Bas.*, **110**, 305 (1991).
3. K. Semmler, G. Szeimies and J. Belzner, *J. Am. Chem. Soc.*, **107**, 6410 (1985).
4. J. Belzner, U. Bunz, K. Semmler, G. Szeimies, K. Opitz and A.-D. Schlüter, *Chem. Ber.*, **122**, 397 (1989).
5. J. Belzner, B. Gareiss, K. Polborn, W. Schmid, K. Semmler and G. Szeimies, *Chem. Ber.*, **122**, 1509 (1989).
6. R. Freudenberger, W. Lamer and A.-D. Schlüter, *J. Org. Chem.*, **58**, 6497 (1993).
7. (a) K. B. Wiberg and J. V. McClusky, *Tetrahedron Lett.*, **28**, 5411 (1987).
(b) K. B. Wiberg, N. Murdie, J. V. McClusky and C. M. Hadad, *J. Am. Chem. Soc.*, **115**, 10653 (1993).
8. A. T. Stoll and E.-I. Negishi, *Tetrahedron Lett.*, **26**, 5671 (1985).
9. (a) I. Fleming and C. J. Urch, *J. Organomet. Chem.*, **285**, 173 (1985).
(b) I. Fleming and M. Rowley, *J. Chem. Soc., Perkin Trans. 1*, 2259 (1987).
10. T. Sato, M. Watanabe and E. Murayama, *Synth. Commun.*, **17**, 781 (1987).
11. A. Krief, M. Hobe, W. Dumont, E. Badoui, E. Guittet and G. Evrard, *Tetrahedron Lett.*, **33**, 3381 (1992).
12. A. Krief and M. Hobe, *Tetrahedron Lett.*, **33**, 6527 (1992).
13. A. Krief and M. Hobe, *Tetrahedron Lett.*, **33**, 6529 (1992).
14. S. W. Roberts and C. J. M. Stirling, *J. Chem. Soc., Chem. Commun.*, 170 (1991).
15. S. M. Jeffery, S. Niedoba and C. J. M. Stirling, *J. Chem. Soc., Chem. Commun.*, 650 (1992).
16. R. J. Israel and R. K. Murray, Jr., *J. Org. Chem.*, **50**, 4703 (1985).
17. R. Gleiter, G. Jähne, G. Müller, M. Nixdorf and H. Irngartinger, *Helv. Chim. Acta*, **69**, 71 (1986).
18. C.-H. Lee, S. Liang, T. Haumann, R. Boese and A. de Meijere, *Angew. Chem., Int. Ed. Engl.*, **32**, 559 (1993).
19. P. M. Angus, B. T. Golding and A. M. Sargeson, *J. Chem. Soc., Chem. Commun.*, 979 (1993).
20. W. A. Kleschick, *J. Org. Chem.*, **51**, 5429 (1986).
21. K. Tanaka, K. Minami and A. Kaji, *Chem. Lett.*, 809 (1987).
22. K. Tanaka, H. Matsuura, I. Funaki and H. Suzuki, *J. Chem. Soc., Chem. Commun.*, 1145 (1991).
23. K. Tanaka, I. Funaki, A. Kaji, K. Minami, M. Sawada and T. Tanaka, *J. Am. Chem. Soc.*, **110**, 7185 (1988).
24. L. Dechoux, M. Ebel, L. Jung and J. F. Stambach, *Tetrahedron Lett.*, **34**, 7405 (1993).
25. M. Ando, K. Wada and K. Takase, *Tetrahedron Lett.*, **26**, 235 (1985).
26. L. Lambs, N. P. Singh and J.-F. Biellmann, *Tetrahedron Lett.*, **32**, 2637 (1991).
27. F. Benedetti, F. Berti and A. Risaliti, *Tetrahedron Lett.*, **34**, 6443 (1993).
28. J. Yu and J. R. Falck, *J. Org. Chem.*, **57**, 3757 (1992).
29. D. Hoppe, M. Paetow and F. Hintze, *Angew. Chem., Int. Ed. Engl.*, **32**, 394 (1993).
30. J. E. Bäckvall, J. O. Vågberg and J. P. Genêt, *J. Chem. Soc., Chem. Commun.*, 159 (1987).
31. J. P. Genet, A. Denis and F. Charbonnier, *Bull. Soc. Chim. Fr.*, 793 (1986).
32. D. Dorsch, E. Kunz and J. G. Helmchen, *Tetrahedron Lett.*, **26**, 3319 (1985).
33. S. Pyo, J. F. Skowron, III and J. K. Cha, *Tetrahedron Lett.*, **33**, 4703 (1992).
34. J. E. Bäckvall, J. O. Vagberg, C. Zercher, J. P. Genet and A. Denis, *J. Org. Chem.*, **52**, 5430 (1987).
35. Y. Hanzawa, S. Ishizawa and Y. Kobayashi, *Chem. Pharm. Bull.*, **36**, 4209 (1988).
36. F. Colobert and J.-P. Genêt, *Tetrahedron Lett.*, **26**, 2779 (1985).
37. J.-P. Genêt and J. M.Gaudin, *Tetrahedron*, **43**, 5315 (1987).
38. T. Hayashi, A. Yamamoto and Y. Itoh, *Tetrahedron Lett.*, **29**, 669 (1988).
39. M. Ahmar, B. Cazes and J. Gore, *Tetrahedron Lett.*, **26**, 3795 (1985).
40. G. Fournet, G. Balme and J. Gore, *Tetrahedron Lett.*, **28**, 4533 (1987).
41. L. S. Barinelli, Z. Li and K. M. Nicholas, *Organometallics*, **8**, 2474 (1989).
42. (a) N. de Kimpe, P. Sulmon and N. Schamp, *Tetrahedron Lett.*, **30**, 5029 (1989).
(b) N. de Kimpe, P. Sulmon and M. Boeykens, *Tetrahedron*, **47**, 3389 (1991).
43. H. Nishiyama, H. Arai, Y. Kanai, H. Kawashima and K. Itoh, *Tetrahedron Lett.*, **27**, 361 (1986).
44. H. M. Walborsky and M. Topolski, *Tetrahedron Lett.*, **34**, 7681 (1993).
45. M. P. Cooke, Jr. and J. Y. Jaw, *J. Org. Chem.*, **51**, 758 (1986).

46. M. Joucla, M. E. Goumzili and B. Fouchet, *Tetrahedron Lett.*, **27**, 1677 (1986).
47. (a) E. Vilsmaier, R. Adam, C. Tetzlaff and R. Cronauer, *Tetrahedron*, **45**, 3683 (1989).
(b) P. Altmeier, E. Vilsmaier and G. Wolmershäuser, *Tetrahedron*, **45**, 3189 (1989).
48. W. R. Bowman, D. S. Brown, C. T. W. Leung and A. P. Stutchbury, *Tetrahedron Lett.*, **26**, 539 (1985).
49. T. Imanishi, T. Ohra, K. Sugiyma, Y. Ueda, Y. Takemoto and C. Iwata, *J. Chem. Soc., Chem. Commun.*, 269 (1992).
50. H. M. R. Hoffmann, A. R. Otte and A. Wilde, *Angew. Chem., Int. Ed. Engl.*, **31**, 234 (1992).
51. A. Wilde, A. R. Otte and H. M. R. Hoffmann, *J. Chem. Soc., Chem. Commun.*, 615 (1993).
52. E. B. Tjaden and J. M. Stryker, *J. Am. Chem. Soc.*, **112**, 6420 (1990).
53. A. G. M. Barrett and M. A. Seefeld, *Tetrahedron*, **49**, 7857 (1993).
54. G. P. Boldrini, F. Mancini, É. Tagliavini, C. Trombini, and A. Umani-Ronchi, *J. Chem. Soc., Chem. Commun.*, 1680 (1990).
55. W. Zhang and P. Dowd, *Tetrahedron Lett.*, **33**, 7307 (1992).
56. W.-W. Weng and T.-Y. Luh, *J. Org. Chem.*, **58**, 5574 (1993).
57. R. C. Denis, J. Rancourt, E. Ghiro, F. Boutonnet and D. Gravel, *Tetrahedron Lett.*, **34**, 2091 (1993).
58. Z. Cekovic and R. Saicic, *Tetrahedron Lett.*, **31**, 6085 (1990).
59. A. Citterio, R. Sebastiano and M. Nicolini, *Tetrahedron*, **49**, 7743 (1993).
60. P. G. Gassmann and C. Lee, *J. Am. Chem. Soc.*, **111**, 739 (1989).
61. T. F. Braish and P. L. Fuchs, *Synth. Commun.*, **15**, 549 (1985).
62. M. Y. Chu-Moyer and S. J. Danishefsky, *J. Am. Chem. Soc.*, **114**, 8333 (1992).
63. M. Harmata and S. Elahmad, *Tetrahedron Lett.*, **34**, 789 (1993).
64. D. Beruben, I. Marek, L. Labaudiniere and J.-F. Normant, *Tetrahedron Lett.*, **34**, 2303 (1993).
65. D. Beruben, I. Marek, L. Labaudiniere, J.-F. Normant and N. Platzer, *Tetrahedron Lett.*, **34**, 7575 (1993).
66. J. D. White and M. S. Jensen, *J. Am. Chem. Soc.*, **115**, 2970 (1993).
67. C.-M. Hu and J. Chen, *Tetrahedron Lett.*, **34**, 5957 (1993).
68. P. A. Wade, W. P. Dailey and P. J. Carroll, *J. Am. Chem. Soc.*, **109**, 5452 (1987).
69. M.-A. Poupart and L. A. Paquette, *Tetrahedron Lett.*, **29**, 269 (1988).
70. M. Machida, K. Oda, E. Yoshida and Y. Kanaoka, *Tetrahedron* , **42**, 4691 (1986).
71. H. Aoyama, H. Sagae and A. Hosomi, *Tetrahedron Lett.*, **34**, 5951 (1993).
72. L. Xu, F. Tao and T. Yu, *Tetrahedron Lett.*, **26**, 4231 (1985).
73. D. Montgomery, K. Reynolds and P. Stevenson, *J. Chem. Soc., Chem. Commun.*, 363 (1993).
74. (a) T. Sato, M. Watanabe and E. Murayama, *Tetrahedron Lett.*, **27**, 1621 (1986).
(b) T. Sato, M. Watanabe, T. Watanabe, Y. Onoda and E. Murayama, *J. Org. Chem.*, **53**, 1894 (1988).
75. S.-I. Fukuzawa, Y. Niimoto and S. Sakai, *Tetrahedron Lett.*, **32**, 7691 (1991).
76. K. Yasui, K. Fugami, S. Tanaka, Y. Tamaru, A. Ii, Z. Yoshida and M. R. Saidi, *Tetrahedron Lett.*, **33**, 785 (1992).
77. C. S. Shiner, A. H. Berks and A. M. Fisher, *J. Am. Chem. Soc.*, **110**, 957 (1988).
78. K. Saigo, T. Yamashita, A. Hongu and M. Hasegawa, *Synth. Commun.*, **15**, 715 (1985).
79. R. F. Cunico and C. -P. Kuan, *J. Org. Chem.*, **50**, 5410 (1985).
80. M. Grignon-Dubois and J. Dunogués, *J. Organometal. Chem.*, **309**, 35 (1986).
81. L. Fitjer, H.-J. Scheuermann, U. Klages, D. Wehle, D. S. Stephenson and G. Binsch, *Chem. Ber.*, **119**, 1144 (1986).
82. M. E. Scheller and B. Frei, *Helv. Chim. Acta*, **69**, 44 (1986).
83. M. Asaoka, K. Takenouchi and H. Takei, *Chem. Lett.*, 921 (1988).
84. R. A. Moss, B. Wilk, K. Krogh-Jespersen and J. D. Westbook, *J. Am. Chem. Soc.*, **111**, 6729 (1989).
85. P. G. Gassman, S. M. Bonser and K. Mlinaric-Majerski, *J. Am. Chem. Soc.*, **111**, 2652 (1989).
86. S. D. Rychnovsky and J. Kim, *Tetrahedron Lett.*, **32**, 7219 (1991).
87. E. C. Friedrich, J. M. Domek and R. Y. Pong, *J. Org. Chem.*, **50**, 4640 (1985).
88. E. C. Friedrich, S. E. Lunetta and E. J. Lewis, *J. Org. Chem.*, **54**, 2388 (1989).
89. E. C. Friedrich and E. J. Lewis, *J. Org. Chem.*, **55**, 2491 (1990).
90. S. Durandetti, S. Sibille and J. Perichon, *J. Org. Chem.*, **56**, 3255 (1991).
91. S. E. Denmark and J. P. Edwards, *J. Org. Chem.*, **56**, 6974 (1991).
92. K. Maruoka, Y. Fukutani and H. Yamamoto, *J. Org. Chem.*, **50**, 4412 (1985).

93. Y. Ukaji and K. Inomata, *Chem. Lett.*, 2353 (1992).
94. I. Fleming, A. K. Sarkar and A. P. Thomas, *J. Chem. Soc., Chem. Commun.*, 157 (1987).
95. G. A. Molander and J. B. Etter, *J. Org. Chem.*, **52**, 3942 (1987).
96. G. A. Molander and L. S. Harring, *J. Org. Chem.*, **54**, 3525 (1989).
97. M. Lautens and P. H. M. Delanghe, *J. Org. Chem.*, **57**, 798 (1992).
98. M. Lautens and P. H. M. Delanghe, *J. Org. Chem.*, **58**, 5037 (1993).
99. T. Imamoto and N. Takiyama, *Tetrahedron Lett.*, **28**, 1307 (1987).
100. T. Inamoto, T. Hatajima, N. Takiyama, T. Takeyama, Y. Kamiya and T. Yoshizawa, *J. Chem. Soc., Perkin Trans. 1*, 3127 (1991).
101. P. Fontani, B. Carboni, M. Vaultier and R. Carrie, *Tetrahedron Lett.*, **30**, 4815 (1989).
102. P. Fontani, B. Carboni, M. Vaultier and G. Maas, *Synthesis*, 605 (1991).
103. Y. V. Tomilov, A. B. Kostitsyn, E. V. Shulishov and O. M. Nefedov, *Synthesis*, 246 (1990).
104. N. S. Zefirov, K. A. Lukin and A. Y. Timofeeva, *J. Org. Chem.(USSR)*, **23**, 2246 (1987).
105. N. S. Zefirov, S. I. Kozhushkov, T. S. Kuznetsova, K. A. Lukin and I. V. Kazimirchik, *J. Org. Chem (USSR)*, **24**, 605 (1988).
106. M. N. Mattson, J. P. Bays, J. Zakutansky, V. Stolarsky and P. Helquist, *J. Org. Chem.*, **54**, 2467 (1989).
107. E. J. O'Connor, S. Brandt and P. Helquist, *J. Am. Chem. Soc.*, **109**, 3739 (1987).
108. J.-E. Bäckvall, C. Löfström, S. K. Juntunen and M. Mattson, *Tetrahedron Lett.*, **34**, 2007 (1993).
109. I. Arai, A. Mori and H. Yamamoto, *J. Am. Chem. Soc.*, **107**, 8254 (1985).
110. A. Mori, I. Arai, H. Yamamoto, H. Nakai and Y. Arai, *Tetrahedron*, **42**, 6447 (1986).
111. E. A. Mash, K. A. Nelson and P. C. Heidt, *Tetrahedron Lett.*, **28**, 1865 (1987).
112. E. A. Mash and K. A. Nelson, *Tetrahedron*, **43**, 679 (1987).
113. E. A. Mash and K. A. Nelson, *J. Am. Chem.Soc.*, **107**, 8256 (1985).
114. E. A. Mash and J. A. Fryling, *J. Org. Chem.*, **52**, 3000 (1987).
115. E. A. Mash, *J. Org. Chem.*, **52**, 4142 (1987).
116. K. A. Nelson and E. A. Mash, *J. Org. Chem.*, **51**, 2721 (1986).
117. E. A. Mash and D. S. Torok, *J. Org. Chem.*, **54**, 250 (1989).
118. M. P. de Frutos, M. D. Fernández, E. Fernández-Alvarez and M. Bernabé, *Tetrahedron Lett.*, **32**, 541 (1991).
119. T. Sugimura, T. Futagawa and A. Tai, *Tetrahedron Lett.*, **29**, 5775 (1988).
120. T. Sugimura, T. Futagawa, M. Yoshikawa and A. Tai, *Tetrahedron Lett.*, **30**, 3807 (1989).
121. T. Sugimura, M. Yoshikawa, T. Futagawa and A. Tai, *Tetrahedron*, **46**, 5955 (1990).
122. T. Sugimura, T. Futagawa and A. Tai, *Chem. Lett.*, 2291 (1990).
123. T. Imai, H. Mineta and S. Nishida, *J. Org. Chem.*, **55**, 4986 (1990).
124. Y. Ukaji, M. Nishimura and T. Fujisawa, *Chem. Lett.*, 61 (1992).
125. H. Takahashi, M. Yoshioka, M. Ohno and S. Kobayashi, *Tetrahedron Lett.*, **33**, 2575 (1992).
126. P. W. Ambler and S. G. Davies, *Tetrahedron Lett.*, **29**, 6979 (1988).
127. P. W. Ambler and S. G. Davies, *Tetrahedron Lett.*, **29**, 6983 (1988).
128. M. Kabat, J. Kiegiel, N. Cohen, K. Toth, P. M. Wovkulich and M. R. Uskokovic, *Tetrahedron Lett.*, **32**, 2343 (1991).
129. N. Kurokawa and Y. Ohfune, *Tetrahedron Lett.*, **26**, 83 (1985).
130. K. Shimamoto and Y. Ohfune, *Tetrahedron Lett.*, **30**, 3803 (1989).
131. W. B. Motherwell and L. R. Roberts, *J. Chem. Soc., Chem. Commun.*, 1582 (1992).
132. A. de Meijere, T.-J. Schulz, R. R. Kostikov, F. Graupner, T. Murr and T. Bielfeldt, *Synthesis*, 547 (1991).
133. C. P. Casey, W. H. Miles and H. Tukada, *J. Am. Chem. Soc.*, **107**, 2924 (1985).
134. M. Brookhart, W. B. Studabaker and G. R. Husk, *Organometallics*, **6**, 1141 (1987).
135. R. M. Vargas, R. D. Theys and M. M. Hossain, *J. Am. Chem. Soc.*, **114**, 777 (1992) and references cited therein.
136. R. S. Iyer, G.-H. Kuo and P. Helquist, *J. Org. Chem.*, **50**, 5898 (1985).
137. D. F. Harvey and M. F. Brown, *Tetrahedron Lett.*, **32**, 5223 (1991).
138. D. F. Harvey and M. F. Brown, *J. Am. Chem. Soc.*, **112**, 7806 (1990).
139. T. Sugimura, T. Katagiri and A. Tai, *Tetrahedron Lett.*, **33**, 367 (1992).
140. L. M. Tolbert and M. B. Ali, *J. Am. Chem.Soc.*, **107**, 4589 (1985).
141. K. Okada, F. Samizio and M. Oda, *Chem. Lett.*, 93 (1987).
142. W. Kirmse and G. Hömberger, *J. Am. Chem. Soc.*, **113**, 3925 (1991).

143. K. Isagawa, K. Mizuno, H. Sugita and Y. Otsuji, *J. Chem. Soc., Perkin Trans. 1*, 2283 (1991).
144. W. R. Dolbier, Jr., H. Wojtowicg and C. R. Burkholder, *J. Org. Chem.*, **55**, 5420 (1990).
145. Y. Bessard, U. Müller and M. Schlosser, *Tetrahedron*, **46**, 5213 (1990).
146. M. Schlosser and Y. Bessard, *Tetrahedron*, **46**, 5222 (1990).
147. Y. Bessard and M. Schlosser, *Tetrahedron*, **47**, 7323 (1991).
148. E. V. Dehmlow and W. Leffers, *J. Organometal. Chem.*, **288**, C41 (1985).
149. A. Jonczyk and P. Balcerzak, *Tetrahedron Lett.*, **30**, 4697 (1989).
150. M. Fedoryñsky, A. Dzikliñsha and A. Joñczyk, *Justus Liebigs Ann. Chem.*, 297 (1990).
151. W. R. Dolbier, Jr. and C. R. Burkholder, *Tetrahedron Lett.*, **29**, 6749 (1988).
152. W. R. Dolbier, Jr. and C. R. Burkholder, *J. Org. Chem.*, **55**, 589 (1990).
153. P. Balcerzak and A. Jonczyk, *Synthesis*, 857 (1991).
154. L. Xu and F. Tao, *Synth. Commun.*, **18**, 2117 (1988).
155. E. V. Dehmlow, H.-C. Raths and J. Soufi, *J. Chem. Res. (S)*, 334 (1988).
156. S. Kondo, Y. Takeda and K. Tsuda, *Synthesis*, 862 (1989).
157. E. V. Dehmlow and U. Fastabend, *J. Chem. Soc., Chem. Commun.*, 1241 (1993).
158. M. Fedorynski, W. Ziólkowska and A. Jonczyk, *J. Org. Chem.*, **58**, 6120 (1993).
159. E. V. Dehmlow and J. Wilkenloh, *Chem. Ber.*, **123**, 583 (1990).
160. F. Mohamadi and W. C. Still, *Tetrahedron Lett.*, **27**, 893 (1986).
161. P. de Frutos, D. Fernández, E. Fernández-Alvarez and M. Bernabé, *Tetrahedron*, **48**, 1123 (1992).
162. P. Dowd, C. Kaufman, P. Kaufman and Y. H. Paik, *Tetrahedron Lett.*, **26**, 2279 (1985).
163. R. F. Cunico and K. S. Chu, *Synth. Commun.*, **17**, 271 (1987).
164. A. K. Bertram and M. T. H. Liu, *J. Chem. Soc., Chem. Commun.*, 467 (1993).
165. S. Keyaniyan, W. Göthling and A. de Meijere, *Chem. Ber.*, **120**, 395 (1987).
166. J. Al-Dulayymi, M. S. Baird and H. H. Hussain, *Tetrahedron Lett.*, **30**, 2009 (1989).
167. T. Liese and A. de Meijere, *Chem. Ber.*, **119**, 2995 (1986).
168. W. Weber and A. de Meijere, *Chem. Ber.*, **118**, 2450 (1985).
169. S. Keyaniyan, M. Apel, J. P. Richmond and A. de Meijere, *Angew. Chem., Int. Ed. Engl.*, **24**, 770 (1985).
170. M. Es-Sayed, C. Gratkowski, N. Krass, A. I. Meyers and A. de Meijere, *Tetrahedron Lett.*, **34**, 289 (1993).
171. K. Paulini and H.-H. Reissig, *Justus Liebigs Ann. Chem.*, 455 (1991).
172. N. Condé-Petiniot, A. J. Hubert, A. F. Noels, R. Warin and P. Teyssié, *Bull. Soc. Chim. Belg.*, **95**, 649 (1986).
173. G. Maier and B. Wolf, *Synthesis*, 871 (1985).
174. D. E. Bergbreiter, M. Morvant and B. Chen, *Tetrahedron Lett.*, **32**, 2731 (1991).
175. W. A. Nugent and F. J. Waller, *Synth. Commun.*, **18**, 61 (1988).
176. A. Demonceau, E. Saive, Y. de Froidmont, A. F. Noels and A. J. Hubert, *Tetrahedron Lett.*, **33**, 2009 (1992).
177. M. P. Doyle, K.-L. Loh, K. M. DeVries and M. S. Chinn, *Tetrahedron Lett.*, **28**, 833 (1987).
178. W. J. Seitz, A. K. Saha, D. Casper and M. M. Hossain, *Tetrahedron Lett.*, **33**, 7755 (1992).
179. K. C. Brown and T. Kodadek, *J. Am. Chem. Soc.*, **114**, 8336 (1992).
180. H. M. L. Davies, T. J. Clark and L. A. Church, *Tetrahedron Lett.*, **30**, 5057 (1989).
181. H. M. L. Davies, H. D. Smith and O. Korkor, *Tetrahedron Lett.*, **28**, 1853 (1987).
182. H. M. L. Davies, T. J. Clark and H. D. Smith, *J. Org. Chem.*, **56**, 3817 (1991).
183. H. M. L. Davies and B. Hu, *J. Org. Chem.*, **57**, 3186 (1992).
184. O. Tsuge, S. Kanemasa, T. Suzuki and K. Matsuda, *Bull. Chem. Soc. Jpn.*, **59**, 2851 (1986).
185. G. Stufflebeme, K. T. Lorenz and N. L. Bauld, *J. Am. Chem. Soc.*, **108**, 4234 (1986).
186. L. Toke, G. T. Szabó, Z. Hell and G. Toth, *Tetrahedron Lett.*, **31**, 7501 (1990).
187. P. Ceccherelli, M. Curini, M. C. Marcotullio and E. Wenkert, *J. Org. Chem.*, **51**, 738 (1986).
188. N. Cagnoli, P. Ceccherelli, M. Curini, M. C. Marcotullio and E. Wenkert, *Synth. Commun.*, **17**, 126 (1987).
189. A. M. P. Koskinen and L. Muñoz, *J. Org. Chem.*, **58**, 879 (1993).
190. A. M. P. Koskinen and L. Muñoz, *J. Chem. Soc., Chem. Commun.*, 1373 (1990).
191. I. D. Reingold and J. Drake, *Tetrahedron Lett.*, **30**, 1921 (1989).
192. A. Padwa, S. F. Hornbuckle, G. E. Fryxell and P. D. Stull, *J. Org. Chem.*, **54**, 817 (1989).
193. U. Burger and D. Zellweger, *Helv. Chim. Acta*, **69**, 676 (1986).
194. T. Imanishi, M. Yamashita, M. Matsui, F. Ninbari, T. Tanaka and C. Iwata, *Chem. Pharm. Bull.*, **36**, 1351 (1988).

195. T. Hudlicky, B. C. Ranu, S. M. Naqvi and A. Srnak, *J. Org. Chem.*, **50**, 123 (1985).
196. J. Adams, C. Lepine-Frenette and D. M. Spero, *J. Org. Chem.*, **56**, 4494 (1991).
197. T. Hudlicky, G. Sinai-Zingde, M. G. Natchus, B. C. Ranu and P. Papadopolous, *Tetrahedron*, **43**, 5685 (1987).
198. S. H. Kang, W. J. Kim and Y. B. Chae, *Tetrahedron Lett.*, **29**, 5169 (1988).
199. A. Padwa, D. J. Austin, S. F. Hornbuckle and M. A. Semones, *J. Am. Chem. Soc.*, **114**, 1874 (1992).
200. (a) A. Padwa, D. J. Austin, Y. Gareau, J. M. Kassir and S. L. Xu, *J. Am. Chem. Soc.*, **115**, 2637 (1993).
(b) A. Padwa, K. E. Krumpe and L. Zhi, *Tetrahedron Lett.*, **30**, 2633 (1989).
(c) A. Padwa, D. J. Austin and S. L. Xu, *Tetrahedron Lett.*, **32**, 4103 (1991).
(d) A. Padwa, U. Chiacchio, D. J. Fairfax, J. M. Kassir, A. Litrico, M. A. Semones and S. L. Xu, *J. Org. Chem.*, **58**, 6429 (1993).
(e) P. H. Mueller, J. M. Kassir, M. A. Semones, M. D. Weingarten and A. Padwa, *Tetrahedron Lett.*, **34**, 4285 (1993).
201. (a) T. R. Hoye and C. J. Dinsmore, *Tetrahedron Lett.*, **32**, 3755 (1991).
(b) T. R. Hoye and C. J. Dinsmore, *J. Am. Chem. Soc.*, **113**, 4343 (1991).
202. T. Aratani, *Pure Appl. Chem.*, **57**, 1839 (1985).
203. W. G. Dauben, R. T. Hendricks, M. J. Luzzio and H. P. Ng, *Tetrahedron Lett.*, **31**, 6969 (1990).
204. H. Fritschi, U. Leutenegger and A. Pfaltz, *Helv. Chim. Acta*, **71**, 1553 (1988).
205. H. Fritschi, U. Leutenegger and A. Pfaltz, *Angew. Chem., Int. Ed. Engl.*, **25**, 1005 (1986).
206. H. Fritschi, U. Leutenegger, K. Siegmann, A. Pfaltz, W. Keller and C. Kratky, *Helv. Chim. Acta*, **71**, 1541 (1988).
207. T. Kunz and H.-U. Reissig, *Tetrahedron Lett.*, **30**, 2079 (1989).
208. U. Leutenegger, G. Umbricht, C. Fahrni, P. von Matt and A. Pfaltz, *Tetrahedron*, **48**, 2143 (1992).
209. D. A. Evans, K. A. Woerpel, M. M. Hinman and M. M. Faul, *J. Am. Chem. Soc.*, **113**, 726 (1991).
210. R. E. Lowenthal, A. Abiko and S. Masamune, *Tetrahedron Lett.*, **31**, 6005 (1990).
211. R. E. Lowenthal and S. Masamune, *Tetrahedron Lett.*, **32**, 7373 (1991).
212. (a) K. Itoh, S. Tabuchi and T. Katsuki, *Synlett*, 575 (1992).
(b) K. Itoh and T. Katsuki, *Tetrahedron Lett.*, **34**, 2661 (1993).
213. M. P. Doyle, B. D. Brandes, A. P. Kazala, R. J. Pieters, M. B. Jarstfer, L. M. Watkins and C. T. Eagle, *Tetrahedron Lett.*, **31**, 6613 (1990).
214. M. N. Protopopova, M. P. Doyle, P. Müller and D. Ene, *J. Am. Chem. Soc.*, **114**, 2755 (1992).
215. M. P. Doyle, R. J. Pieters, S. F. Martin, R. E. Austin, C. J. Oalmann and P. Müller, *J. Am. Chem. Soc.*, **113**, 1423 (1991).
216. S. F. Martin, C. J. Oalmann and S. Liras, *Tetrahedron Lett.*, **33**, 6727 (1992).
217. M. Kennedy, M. A. McKervey, A. R. Maguire and G. H. P. Roos, *J. Chem. Soc., Chem. Commun.*, 361 (1990).
218. H. M. L. Davies and D. K. Hutcheson, *Tetrahedron Lett.*, **34**, 7243 (1993).
219. H. M. L. Davies and W. R. Cantrell, Jr., *Tetrahedron Lett.*, **32**, 6509 (1991).
220. H. M. L. Davies, N. J. S. Huby, W. R. Cantrell, Jr. and J. L. Olive, *J. Am. Chem. Soc.*, **115**, 9468 (1993).
221. M. P. Doyle, R. L. Dorow, J. W. Terpstra and R. A. Rodenhouse, *J. Org. Chem.*, **50**, 1663 (1985).
222. K. Yamanoi and Y. Ohfune, *Tetrahedron Lett.*, **29**, 1181 (1988).
223. S. R. Wilson, A. M. Venkatesan, C. E. Augelli-Szafran and A. Yamin, *Tetrahedron Lett.*, **32**, 2339 (1991).
224. D. F. Taber, J. C. Amedio, Jr. and K. Raman, *J. Org. Chem.*, **53**, 2984 (1988).
225. P. E. O'Bannon and W. P. Dailey, *J. Org. Chem.*, **54**, 3096 (1989).
226. P. E. O'Bannon and W. P. Dailey, *Tetrahedron Lett.*, **29**, 987 (1988).
227. P. E. O'Bannon and W. P. Dailey, *Tetrahedron Lett.*, **30**, 4197 (1989).
228. A. Vasella and C. A. A. Waldraff, *Helv. Chim. Acta*, **74**, 585 (1991).
229. J.-P. Praly, Z. el Kharraf and G. Descotes, *Tetrahedron Lett.*, **31**, 444 (1990).
230. A. Vasella, C. Witzig and R. Husi, *Helv. Chim. Acta*, **74**, 1362 (1991).
231. R. T. Lewis and W. B. Motherwell, *Tetrahedron Lett.*, **29**, 5033 (1988).
232. R. A. Moss, G. Kmiecik-Lawrynowicz and K Krough-Jespersen, *J. Org. Chem.*, **51**, 2168 (1986).

233. R. A. Moss, T. Zdrojewski, K. Krogh-Jespersen, M. Wlostowski and A. Matro, *Tetrahedron Lett.*, **32**, 1925 (1991).
234. H. Tomioka and K. Hirai, *J. Chem. Soc., Chem. Commun.*, 362 (1989).
235. K. Schank, A.-M. A. A. Wahab, P. Eigen and J. Jager, *Tetrahedron*, **45**, 6667 (1989).
236. E. Schaumann, C. Friese and G. Adiwidjaja, *Tetrahedron*, **45**, 3163 (1989).
237. M. Brookhart and W. B. Studabaker, *Chem. Rev.*, **87**, 411 (1987).
238. M. P. Doyle, *Recl. Trav. Chim. Pays-Bas*, **110**, 305 (1991).
239. A. Wienand and H.-U. Reissig, *Tetrahedron Lett.*, **29**, 2315 (1988).
240. M. Buchert and H.-U. Reissig, *Tetrahedron Lett.*, **29**, 2319 (1988).
241. A. Wienand and H.-U. Reissig, *Chem. Ber.*, **124**, 957 (1991).
242. J. W. Herndon and S. U. Tumer, *Tetrahedron Lett.*, **30**, 4771 (1989).
243. C. K. Murray, D. C. Yang and W. D. Wulff, *J. Am. Chem. Soc.*, **112**, 5660 (1990).
244. D. F. Harvey and M. F. Brown, *Tetrahedron Lett.*, **31**, 2529 (1990).
245. D. F. Harvey and K. P. Lund, *J. Am. Chem. Soc.*, **113**, 8916 (1991).
246. W. D. Wulff, D. C. Yang and C. K. Murray, *Pure Appl. Chem.*, **60**, 137 (1988).
247. W. D. Wulff, D. C. Yang and C. K. Murray, *J. Am. Chem. Soc.*, **110**, 2653 (1988).
248. B. C. Söderberg and L. S. Hegedus, *Organometallics*, **9**, 3113 (1990).
249. A. Parlier, H. Rudler, N. Platzer, M. Fontanille and A. Soum, *J. Organometal Chem.*, **287**, C8 (1985).
250. T. Takahashi, Y. Yamashita, T. Doi and J. Tsuji, *J. Org. Chem.*, **54**, 4273 (1989).
251. S. Ganesh, K. M. Sathe, M. Nandi, P. Chakrabarti and A. Sarkar, *J. Chem. Soc., Chem. Commun.*, 224 (1993).
252. (a) Y. Shen and Q. Liao, *Synthesis*, 321 (1988).
(b) Y. Shen and Q. Liao, *J. Organometal. Chem.*, **371**, 31 (1989).
253. N. N. Magdesieva, T. A. Sergeeva and N. V. Averina, *J. Org. Chem.* (*USSR*), **22**, 1894 (1986).
254. Y.-Z. Huang, Y. Tang, Z.-L. Zhou and J.-L. Huang, *J. Chem. Soc., Chem. Commun.*, 7 (1993).
255. T. Taguchi, A. Hosoda, Y. Torisawa, A. Shimazaki, Y. Kobayashi and K. Tsushima, *Chem. Pharm. Bull.*, **33**, 4085 (1985).
256. U. M. Nägele and M. Hanack, *Justus Liebigs Ann. Chem.*, 847 (1989).
257. H. Mack and M. Hanack, *Justus Liebigs Ann. Chem.*, 833 (1989).
258. J. S. Nowick and R. L. Danheiser, *Tetrahedron*, **44**, 4113 (1988).
259. A. Bernardi, C. Scolastico and R. Villa, *Tetrahedron Lett.*, **30**, 3733 (1989).
260. R. M. Williams and G. J. Fegley, *J. Am. Chem. Soc.*, **113**, 8796 (1991).
261. R. M. Williams and G. J. Fegley, *J. Org. Chem.*, **58**, 6933 (1993).
262. R. Chinchilla, C. Nájera, S. Garcia-Grandas and A. Menéndez-Velázquez, *Tetrahedron Lett.*, **34**, 5799 (1993).
263. A. I. Meyers, J. L. Romine and S. A. Fleming, *J. Am. Chem. Soc.*, **110**, 7245 (1988).
264. D. Romo and A. I. Meyers, *J. Org. Chem.*, **57**, 6265 (1992).
265. A. Krief and W. Dumont, *Tetrahedron Lett.*, **29**, 1083 (1988).
266. A. Krief, W. Dumont and P. Pasau, *Tetrahedron Lett.*, **29**, 1079 (1988).
267. A. Krief, W. Dumont, P. Pasau and P. Lecomte, *Tetrahedron*, **45**, 3039 (1989).
268. A. Krief and P. Lecomte, *Tetrahedron Lett.*, **34**, 2695 (1993).
269. C. Hamdouchi, *Tetrahedron Lett.*, **33**, 1701 (1992).
270. W. von der Saal,. R. Reinhardt, H.-M. Seidenspinner, J. Stawitz and H. Quast, *Justus Liebigs Ann. Chem.*, 703 (1989).
271. T. Hudlicky, L. R. Kwart, L.-Q. Li and T. Bryant, *Tetrahedron Lett.*, **29**, 3283 (1988).
272. R. Yoneda, K. Santo, S. Harusawa and T. Kurihara, *Synth. Commun.*, **17**, 921 (1987).
273. T. Hudlicky, A. Fleming and L. Radesca, *J. Am. Chem. Soc.*, **111**, 6691 (1989).
274. A. Bojilova, A. Trendafilova, C. Ivanov and N. A. Rodios, *Tetrahedron*, **49**, 2275 (1993).
275. K. Ramig, M. Bhupathy and T. Cohen, *J. Org. Chem.*, **54**, 4404 (1989).
276. C. Mahidol, V. Reutrakul, C. Panyachotipun, G. Turongsomboon, V. Prapansiri and B. M. R. Bandara, *Chem. Lett.*, 163 (1989).
277. T. Cohen and M. Myers, *J. Org. Chem.*, **53**, 457 (1988).
278. M. Makosza, T. Glinka, S. Ostrowski and A. Rykowski, *Chem. Lett.*, 61 (1987).
279. E. Schaumann, C. Friese and C. Spanka, *Synthesis*, 1035 (1986).
280. T. Cohen, J. P. Scherbine, S. A. Mendelson and M. Myers, *Tetrahedron Lett.*, **26**, 2965 (1985).
281. A. Krief, P. Barbeaux and E. Guittet, *Synlett.*, 509 (1990).
282. J.-M. Mélot, F. Texier-Boullet and A. Foucaud, *Synthesis*, 364 (1987).

283. N. Ono, T. Yanai, I. Hamamoto, A. Kamimura and A. Kaji, *J. Org. Chem.*, **50**, 2806 (1985).
284. A. Krief, L. Hevesi, G. Chaboteaux, P. Mathy, M. Sevrin and M. J. de Vos, *J. Chem. Soc., Chem. Commun.*, 1693 (1985).
285. J. H. Babler and K. P. Spina, *Tetrahedron Lett.*, **26**, 1923 (1985).
286. R. B. Mitra and L. Muthusubramanian, *Synth. Commun.*, **19**, 2515 (1989).
287. G. Chaboteaux and A. Krief, *Bull. Soc. Chim. Belg.*, **94**, 495 (1985).
288. A. Krief, M. J. Devos and M. Sevrin, *Tetrahedron Lett.*, **27**, 2283 (1986).
289. Y. Gai, M. Julia and J.-N. Verpeaux, *Synlett*, 269 (1991).
290. C. Chen, Y.-Z. Huang and Y. Shen, *Tetrahedron Lett.*, **29**, 1033 (1988).
291. C. Chen, Y. Liao and Y.-Z. Huang, *Tetrahedron*, **45**, 3011 (1989).
292. T. Matsuki, N. X. Hu, Y. Aso, T. Otsubo and F. Ogura, *Bull. Chem. Soc. Jpn.*, **62**, 2105 (1989).
293. S. Araki and Y. Butsugan, *J. Chem. Soc., Chem. Commun.*, 1286 (1989).
294. M. Joucla, B. Fouchet, J. le Brun and J. Hamelin, *Tetrahedron Lett.*, **26**, 1221 (1985).
295. Y. Takikawa, K. Osanai, S. Sasaki and K. Shimada, *Chem. Lett.*, 1939 (1987).
296. M. Pirrung and G. M. McGeehan, *Angew. Chem., Int. Ed. Engl.*, **24**, 1044 (1985).
297. J. E. Baldwin, R. M. Adlington and B. J. Rawlings, *Tetrahedron Lett.*, **26**, 481 (1985).
298. J. Heiszman, I. Bitter, K. Harsányi and L. Töke, *Synthesis*, 738 (1987).
299. M. W. Thomsen, B. M. Handwerker, S. A. Katz and S. A. Fisher, *Synth. Commun.*, **18**, 1433 (1988).
300. R. K. Podder, R. K. Sarkar and S. C. Ray, *Ind. J. Chem.*, **27B**, 530 (1988).
301. Y. Gao and K. B. Sharpless, *J. Am. Chem. Soc.*, **110**, 7538 (1988).
302. P. Subramanian and R. W. Woodard, *J. Org. Chem.*, **52**, 15 (1987).
303. (a) E. Schaumann, A. Kirschning and F. Narjes, *J. Org. Chem.*, **56**, 717 (1991).
(b). F. Narjes, O. Bolte, D. Icheln, W. A. König and E. Schaumann, *J. Org. Chem.*, **58**, 626 (1993).
304. F. Narjes and E. Schaumann, *Synthesis*, 1168 (1991).
305. P. Jankowski and J. Wicha, *J. Chem. Soc., Chem. Commun.*, 802 (1992).
306. B. Achmatowicz, M. M. Kabat, J. Krajewski and J. Wicha, *Tetrahedron*, **48**, 10201 (1992).
307. K. Burgess and K.-K. Ho, *J. Org. Chem.*, **57**, 5931 (1992).
308. M. C. Pirrung, S. E. Dunlap and U. P. Trinks, *Helv. Chim. Acta*, **72**, 1301 (1989).
309. F. Zaman, A. Fatima and W. Voelter, *Justus Liebigs Ann. Chem.*, 1101 (1991).
310. F. Fariña, M. C. Maestro, M. V. Martin and M. L. Soria, *Tetrahedron*, **43**, 4007 (1987).
311. I. Yamamoto, T. Sakai, K. Ohta, K. Matsuzaki and K. Fukuyama, *J. Chem. Soc., Perkin Trans. 1*, 2785 (1985).
312. A. Dancsó, M. Kajtar-Peredy, C. Szántay and G. Kalaus, *Synthesis*, 1116 (1985).
313. D. Jacoby, J. P. Celerier, H. Petit and G. Lhommet, *Synthesis*, 301 (1990).
314. J. L. Marco, B. Sánchez, M. D. Fernández and M. Bernabé, *Justus Liebigs Ann. Chem.*, 1099 (1991).
315. I. Arenal, M. Bernabé, E. Fernández-Alvarez and S. Penadés, *Synthesis*, 773 (1985).
316. Y.-F. Zhu, T. Yamazaki, J. W. Tsang, S. Lok and M. Goodman, *J. Org. Chem.*, **57**, 1074 (1992).
317. V. P. Srivastava, M. Roberts, T. Holmes and C. H. Stammer, *J. Org. Chem.*, **54**, 5866 (1989).
318. L. F. Elrod, E. M. Holt, C. Mapelli and C. H. Stammer, *J. Chem. Soc., Chem. Commun.*, 252 (1988).
319. T. Wakamiya, Y. Oda, H. Fujita and T. Shiba, *Tetrahedron Lett.*, **27**, 2143 (1986).
320. J. Blasco, C. Cativiela and M. D. D. de Villegas, *Synth. Commun.*, **17**, 1549 (1987).
321. M. D. Fernández, M. P. de Frutos, J. L. Marco, E. Fernández-Alvarez and M. Bernabé, *Tetrahedron Lett.*, **30**, 3101 (1989).
322. M. Sato, H. Hisamichi and C. Kaneko, *Tetrahedron Lett.*, **30**, 5281 (1989).
323. M. Franck-Neumann, M. Sedrati, J.-P. Vigneron and V. Bloy, *Angew. Chem., Int. Ed. Engl.*, **24**, 996 (1985).
324. A. Misumi, K. Iwanaga, K. Furuta and H. Yamamoto, *J. Am. Chem. Soc.*, **107**, 3343 (1985).
325. K. Okuma, Y. Tanaka, K. Yoshihara, A. Ezaki, G. Koda, H. Ohta, K. Hara and S. Kashimura, *J. Org. Chem.*, **58**, 5915 (1993).
326. S. Yamazaki, S. Katoh and S. Yamabe, *J. Org. Chem.*, **57**, 4 (1992).
327. P. Knochel and J. F. Normant, *Tetrahedron Lett.*, **27**, 5727 (1986).
328. P. Knochel, C. Xiao and M. C. P. Yeh, *Tetrahedron Lett.*, **29**, 6697 (1988).
329. H. Ishibashi, M. Okada, H. Nakatani, M. Ikeda and Y. Tamura, *J. Chem. Soc., Perkin Trans. 1*, 1763 (1986).

330. T. Takahashi, D. Y. Kondakov and N. Suzuki, *Tetrahedron Lett.*, **34**, 6571 (1993).
331. R. M. Moriarty, O. Prakash, R. K. Vaid and L. Zhao, *J. Am. Chem. Soc.*, **111**, 6443 (1989).
332. M. Christl, P. Kemmer and B. Mattauch, *Chem. Ber.*, **119**, 960 (1986).
333. A. Krief and P. Barbeaux, *Tetrahedron Lett.*, **32**, 417 (1991).
334. (a) O. G. Kulinkovich, S. V. Sviridov, D. A. Vasilevskii and T. S. Pritytskaya, *J. Org. Chem. (USSR)*, **25**, 2027 (1989).
(b) O. G. Kulinkovich, S. V. Sviridov and D. A. Vasilevskii, *Synthesis*, 234 (1991).
(c) A. de Meijere, S. I. Kozhushkov, T. Spaeth and N. S. Zefirov, *J. Org. Chem.*, **58**, 502 (1993).
335. T. Imamoto, Y. Kamiya, T. Hatajima and H. Takahashi, *Tetrahedron Lett.*, **30**, 5149 (1989).
336. J. Barluenga, J. L. Fernandez-Simon, J. M. Concellon and M. Yus, *Synthesis*, 584 (1987).
337. T. Sato, T. Kikuchi, N. Sootome and E. Murayama, *Tetrahedron Lett.*, **26**, 2205 (1985).
338. T. Shono, M. Ishifune, H. Kinugasa and S. Kashimura, *J. Org. Chem.*, **57**, 5561 (1992).
339. P. F. Korkowski, T. R. Hoye and D. B. Rydberg, *J. Am. Chem. Soc.*, **110**, 2676 (1988).
340. T. R. Hoye and G. M. Rehberg, *Organometallics*, **9**, 3014 (1990).
341. T. R. Hoye and G. M. Rehberg, *Organometallics*, **8**, 2070 (1989).
342. T. R. Hoye and G. M. Rehberg, *J. Am. Chem. Soc.*, **112**, 2841 (1990).
343. Y. Zhang and E.-I. Negishi, *J. Am. Chem. Soc.*, **111**, 3454 (1989).
344. R. Grigg, M. J. Dorrity and J. F. Malone, *Tetrahedron Lett.*, **31**, 1343 (1990).
345. R. Grigg, V. Sridharan and S. Sukirthalingam, *Tetrahedron Lett.*, **32**, 3855 (1991).
346. F. E. Meyer, P. J. Parsons and A. de Meijere, *J. Org. Chem.*, **56**, 6487 (1991).
347. F. E. Meyer, H. Henniges and A. de Meijere, *Tetrahedron Lett.*, **33**, 8039 (1992).
348. B. M. Trost and A. S. K. Hashmi, *Angew. Chem., Int. Ed. Engl.*, **32**, 1085 (1993).
349. B. M. Trost and S. Schneider, *J. Am. Chem. Soc.*, **111**, 4430 (1989).
350. K. Ohe, T. Ishihara, N. Chatani and S. Murai, *J. Am. Chem. Soc.*, **112**, 9646 (1990).
351. S. Ogoshi, T. Morimoto, K.-I. Nishio, K. Ohe and S. Murai, *J. Org. Chem.*, **58**, 9 (1993).
352. K. Ohe, H. Matsuda, T. Ishihara, S. Ogoshi, N. Chatani and S. Murai, *J. Org. Chem.*, **58**, 1173 (1993).
353. L. Hadjiarapoglou, S. Spyroudis and A. Varvoglis, *J. Am. Chem. Soc.*, **107**, 7178 (1985).
354. L. Hadjiarapoglou and A. Varvoglis, *J. Chem. Soc., Perkin Trans. 1*, 2839 (1988).
355. A. Saba, *J. Chem. Res. (S)*, 288 (1990).
356. T. Ibata and M. Kashiuchi, *Bull. Chem. Soc. Jpn.*, **59**, 929 (1986).
357. M. N. Elinson, T. L. Lizunova, B. I. Ugrak, M. O. Dekaprilevich, G. I. Nikishin and Y. T. Struchkov, *Tetrahedron Lett.*, **34**, 5795 (1993).
358. J.-Y. Lee and H. K. Hall, Jr., *J. Org. Chem.*, **55**, 4963 (1990).

CHAPTER **6**

Cyclopropane photochemistry

HOWARD E. ZIMMERMAN

Chemistry Department, University of Wisconsin, Madison, Wisconsin 53706, USA
Fax: (+1) 608-262-0381; e-mail: Zimmerman@Bert.Chem.Wisc.Edu

I. INTRODUCTION 319
II. REACTIONS GIVING RISE TO THREE-MEMBERED RING PHOTOPRODUCTS 319
A. The Type-A Rearrangement of 2,5-Cyclohexadienones 319
B. The Rearrangement of 4-Arylcyclohexenones 321
C. The Di-π-methane Rearrangement 323
III. PHOTOCHEMICAL REACTIONS OF THREE-MEMBERED RING COMPOUNDS 327
A. Type-B Reactions; Rearrangements of the Bicyclic Photoproducts of Type-A Processes 327
B. The Bicycle Rearrangement 329
C. Photochemical Stereoisomerizations of Aryl, Vinyl and Acyl Cyclopropanes 334
D. Conjugated Cyclopropanes Giving Rearranging Diradicals 336
IV. SUMMARY 337
V. ACKNOWLEDGMENT 337
VI. REFERENCES 337

I. INTRODUCTION

This chapter deals with the author's photochemical research involving cyclopropyl rings, both in reactants and in products. A very large fraction of cyclopropane chemistry is photochemical. Additionally, photochemistry involving three-membered rings is especially intriguing and unusual in its variety. In writing this article, I have divided the presentation into two categories: (i) examples where reactions afford cyclopropyl rings, and (ii) reactions of cyclopropyl compounds.

II. REACTIONS GIVING RISE TO THREE-MEMBERED RING PHOTOPRODUCTS

A. The Type-A Rearrangement of 2,5-Cyclohexadienones

A particularly intriguing rearrangement is the 'Type-A' transformation of 2,5-cyclohexadienones, a reaction typified by the photochemical conversion of Santonin (**1**) to

The chemistry of the cyclopropyl group, Vol. 2
Edited by Z. Rappoport

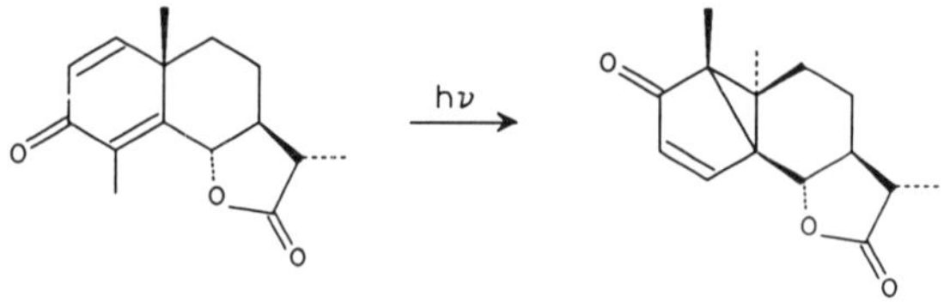

(1)

(1) (2)

Lumisantonin (**2**) (equation 1)[1]. The reaction mechanism derived from the study of a simpler analog, namely 4,4-diphenyl-2,5-cyclohexadienone[2], is outlined in Scheme 1. The mechanism begins with excitation to the singlet n-π^* state and intersystem crossing to the n-π^* triplet, written as 'step 1'. The n-π^* triplet has been shown[3,4] to exhibit a high β,β-bond order. This corresponds to step 2 in Scheme 1, namely β,β-sigma bond formation (i.e. between carbons 3 and 5). Intersystem crossing (step 3) of the resulting triplet diradical is rapid due to spin–orbit coupling so characteristic of oxygen atoms having odd-electron density and also orbitals in two orthogonal planes. This decay of the triplet leads to a ground state species which is an oxyallyl zwitterion. In the final step (i.e. step 4) the zwitterion rearranges to afford 6,6-diphenylbicyclo[3.1.0]hex-3-en-2-one. As has been noted[2], this mechanism accounts for the Santonin to Lumisantonin rearrangement as well as for a very large variety of Type-A processes.

Interestingly, the so-called 'Zimmerman–Schuster' oxyallyl zwitterion may be generated independently by a Favorskii-like route, and leads to the same reaction products as obtained photochemically[5]. In addition, the stereochemistry of the rearrangement of the zwitterion to the bicyclo[3.1.0]hex-3-en-2-one could be established, since it was found that with two different aryl groups, the *endo* aryl group in the zwitterion appeared *endo* in the bicyclic photoproduct and the *exo* group remained *exo*[6,7].

The reaction regioselectivity can be controlled by a β-methoxyl or β-cyano group. In each case, the substituent appears β on the π-bond of the residual enone moiety in the bicyclic photoproduct[8,9]. Furthermore, regioselectivity of the oxyallyl zwitterionic intermediate generated without light using the Favorskii approach cited above is the same.

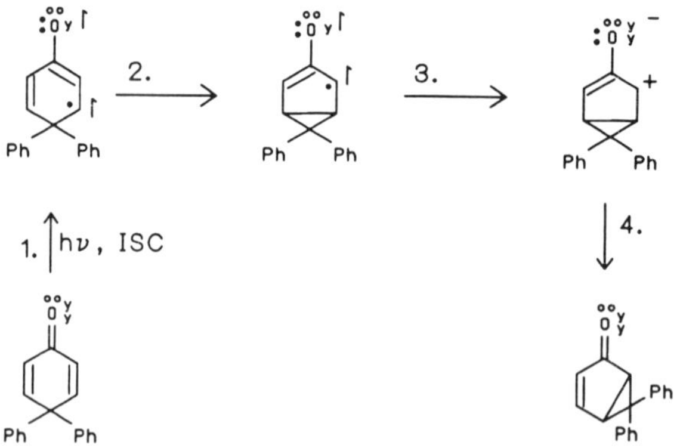

SCHEME 1. Mechanism of the Type-A cyclohexadienone rearrangement

Another type of photochemistry without light involves the actual formation of the triplet reactant excited state by thermolysis of appropriately selected dioxetanes as depicted in equation 2[10,11].

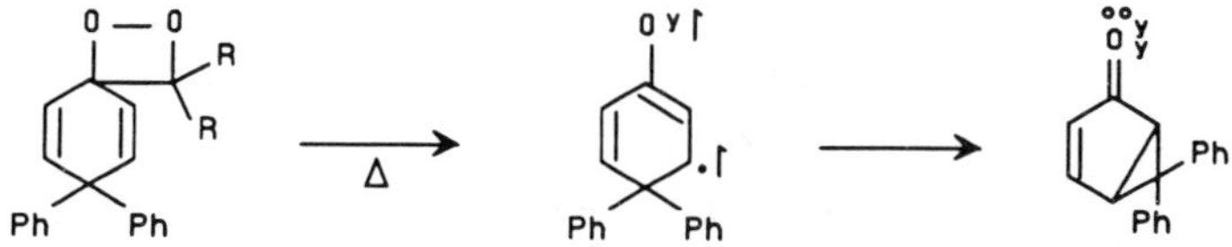

(2)

In the preparation of the Type-A photoproducts, it is critical to generate the n-π* excited states. For example, in the case of 4,4-di-β-naphthyl-2,5-cyclohexadienone[12], the desired bicyclo[3.1.0] photoproduct results from the dioxetane approach, which specifically gives a T_2 (i.e. an n-π*) triplet, but the direct irradiation gives both the desired bicyclic product and also 3,4-dinaphthylphenol. The latter arises from T_1 in which it is the naphthyl group which is excited. Sensitization affords only T_1 which, being naphthyl excited, does not lead to the desired three-ring bicyclic photoproduct and affords only 3,4-dinaphthylphenol. However, in the direct irradiation, it is practical to selectively quench the longer-lived and undesired T_1 using 1,3-cyclohexadiene as a triplet energy acceptor.

B. The Rearrangement of 4-Arylcyclohexenones

4-Arylcyclohexenones are closely related in structure to the dienones just discussed. Also, the lowest energy triplet has an n-π* configuration and bicyclic[3.1.0] ketones are formed. However, both the reaction mechanism and the photoproduct structures differ from that of the Type-A process.

A requirement for the enone rearrangement is the presence of a π-moiety at carbon-4 (i.e. at the γ-carbon). The rearrangement proceeds with migration of this group to C-3 (the β-carbon). The preferred stereochemical course leads preferentially to the 6-*endo* phenyl configuration (i.e. *trans* phenyl groups here). The overall reaction is shown in equation 3

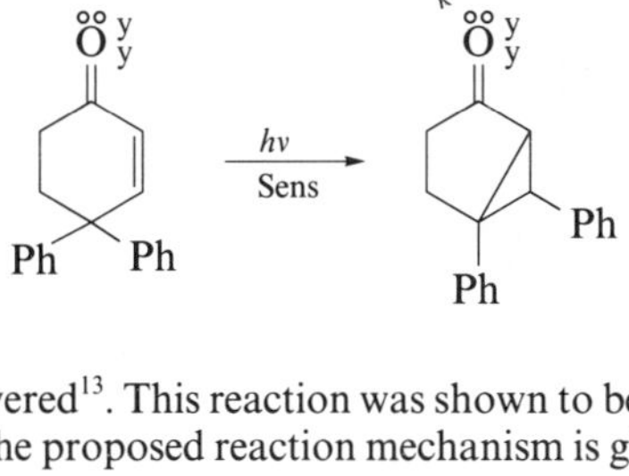

(3)

for the first example uncovered[13]. This reaction was shown to be a triplet process[14] in analogy to the dienone case. The proposed reaction mechanism is given in equation 4[13–15]. The

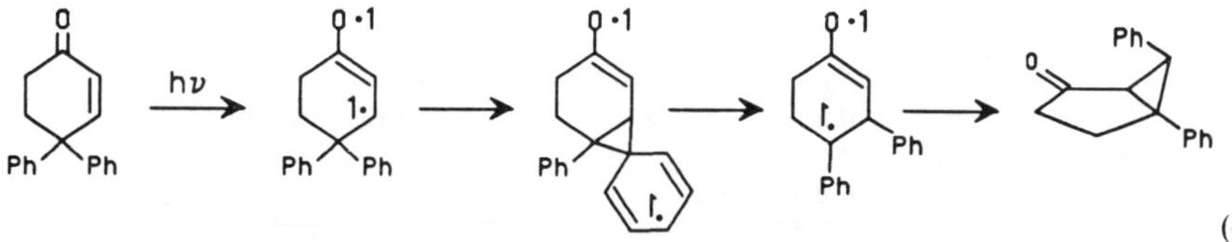

(4)

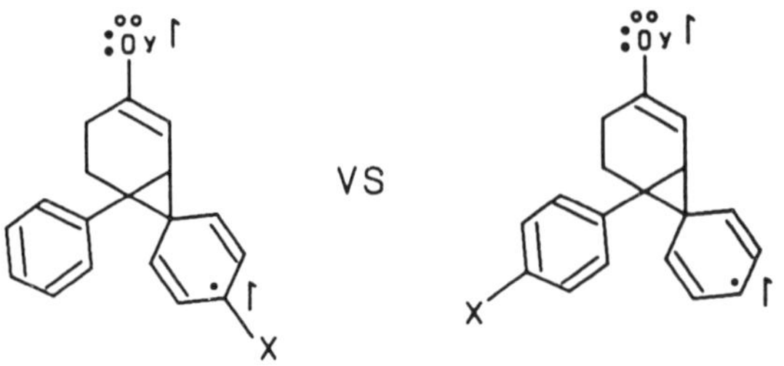

FIGURE 1. Alternative half-migrated species

reaction efficiency was found to be less than that of the corresponding Type-A rearrangement. The rearrangement was shown to have an activation energy of *ca* 10 kcal mol^{-1} on the triplet hypersurface. This can be attributed to the loss of aromaticity of the migrating phenyl group as it bridges to the β-carbon atom[16]. In the same study it was found that the reaction efficiency is independent of wavelength while being dependent on the reaction temperature. This evidence indicates that the reacting enone triplet is thermally equilibrated with its environment prior to rearranging.

A particularly interesting result was found in the competition of 4-phenyl versus 4-*p*-substituted phenyl groups[17]. It was determined that both anisyl and *p*-cyanophenyl groups migrate to the β-carbon in preference to phenyl groups. Figure 1 shows two 'half-migrated', bridged triplet diradical species. If a methoxyl or cyano group is *para* on the bridging phenyl group, extra stabilization results due to the ability of these groups to delocalize odd-electron density. If the β-carbon of the enone triplet were either electron-deficient or electron-rich, both anisyl and cyanophenyl would not show this preferential migratory behavior. We can conclude that the β-carbon primarily has odd-electron character in the n-π* excited state.

The rearrangement is not limited to cyclohexenones. For example, 4,4-diphenylcyclopentenone and 4-phenyl-4-methylcyclopentenone rearranged in modestly analogous fashion[18]. However, here the initially formed 'housones' could be detected only at low temperatures (*ca* –140 °C). At higher temperatures the housone fragments to the unsaturated ketene. Equation 5 shows the reaction course as well as a likely mechanism for reversion of the housone to the ketene. This proceeds via a 1,4-diradical, and such species are known[19–22] to undergo a 1,4- (2,3) fragmentation.

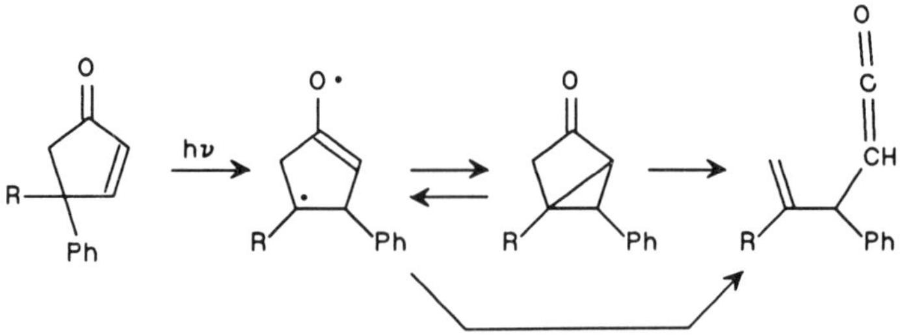

(5)

A further aspect of the photochemistry of cyclic enones to afford bicyclic photoproducts is the observation that the aryl groups at C-4 may be moieties with low triplet energies, such as naphthyl or biphenyl. When light is absorbed by such groups, the singlet excitation is transferred to the remote carbonyl moiety. This is known to act as a 'singlet–triplet switch' by virtue of spin–orbit coupling arising from the two perpendicular orbitals (i.e. py and π). The resulting triplet excitation is then equilibrated with the group at C-4 which migrates with excellent efficiency[23,24]. Hence this variation of the enone rearrangement may well be better described as a di-π-methane rearrangement from a π-π^* excited state in which the excitation is available in the 4-aryl group.

C. The Di-π-methane Rearrangement

In 1966 we observed that barrelene rearranged to an isomer, that we termed 'semibullvalene' [25].* The NMR spectrum of this isomer was enigmatic, suggesting the presence of only three types of hydrogen atoms. We soon recognized that semibullvalene was subject to an exceptionally facile degenerate Cope rearrangement which accounted for only three types of hydrogens on the NMR time scale, and the rearrangement could be written as in equation 6. While the reaction is of intrinsic interest and involves two intriguing com-

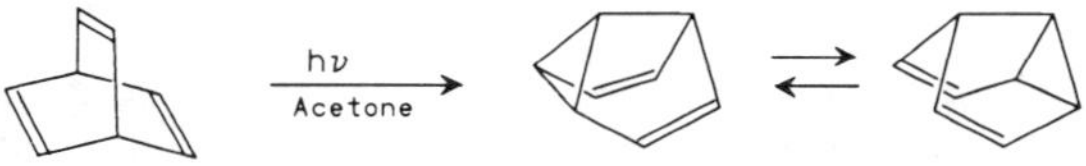

(6)

pounds as reactant and product, the more important consequence of this finding was the reaction mechanism and the realization that the reaction was general[26]. The essential structural feature of the rearrangement is outlined in equation 7. This equation is just schematic. As it implies, the minimal requirement for the reaction is having two π-groups bonded to a single atom, here an sp^3 hybridized carbon. Equation 7 gives vinyl and aryl groups as

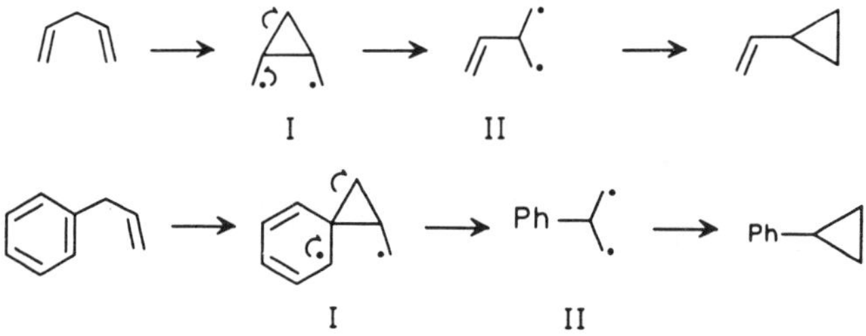

(7)

examples. However, other π-groups which are possible include ethynyl, acyl and acylimino. With an acyl group present, this is the 'oxa-di-π-methane' [27] and with an acylimino group it is the 'aza-di-π-methane' [28] variation. The reaction has been one of the most widely used photochemical reactions for forming three-membered ring compounds.

The reaction has particularly broad application. One typical acyclic example is shown in equation 8[29]. This particular example proceeds only on direct irradiation, which con-

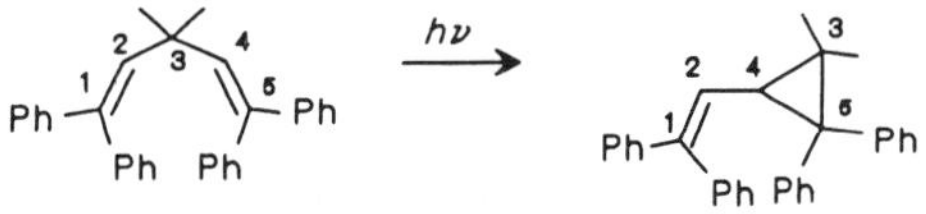

(8)

*Interestingly, the original referees did not like the terms 'barrelene' and 'semibullvalene' with the comment that these were unesthetic and that it was unlikely that any further compounds of these structures would be encountered.

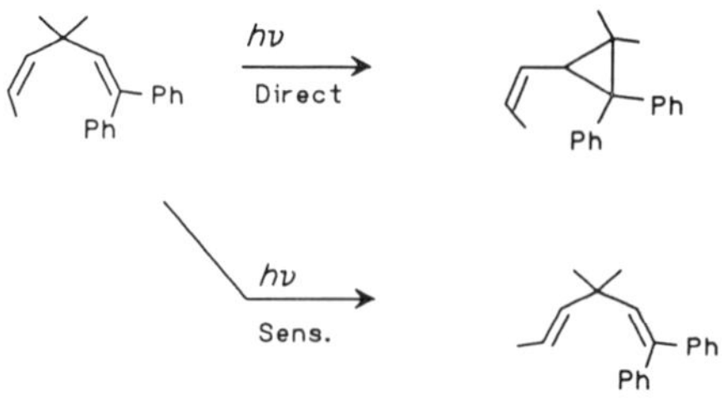

SCHEME 2. Singlet vs triplet reactivity of an acyclic di-π-methane reactant

trasts with the barrelene to semibullvalene case above where a sensitizer is required. Thus, a generalization was derived in which acyclic di-π-methane reactants rearrange efficiently on direct irradiation by way of the singlet excited states and tend to be unreactive on sensitization. Conversely, cyclic di-π-methane reactants having constrained π-bonds, react via the triplet excited state. The main factor controlling the reactivity of triplet reactants is the 'free-rotor effect' [30–32] in which double-bond twisting of the triplet leads to decay to ground state. Note Scheme 2, for example, where the singlet rearranges to the anticipated di-π-methane product while the triplet undergoes stereoisomerization.

Before proceeding further we should note that in equation 7 there are two diradicals. One is 'Diradical I', labeled I, and the other is 'Diradical II', labeled II. In the case of triplets both are reaction intermediates, while in the case of the singlet reactions these appear more likely to be mere points on the excited state hypersurface[32,33,42].

A further point of interest is the reaction regiochemistry. For the reaction in Scheme 2 we might consider two possible vinylcyclopropane products. Thus the intermediate cyclopropyldicarbinyl diradical might open in two ways. However, only one product is formed. In Scheme 3 this is shown more definitively where the two modes of 'unzipping' the cyclopropyl dicarbinyl diradical are depicted. A general rule is that the less stabilized diradical center utilizes its odd-electron density to open the three-membered ring. This leads onwards to the more stabilized 'Diradical II' [32].

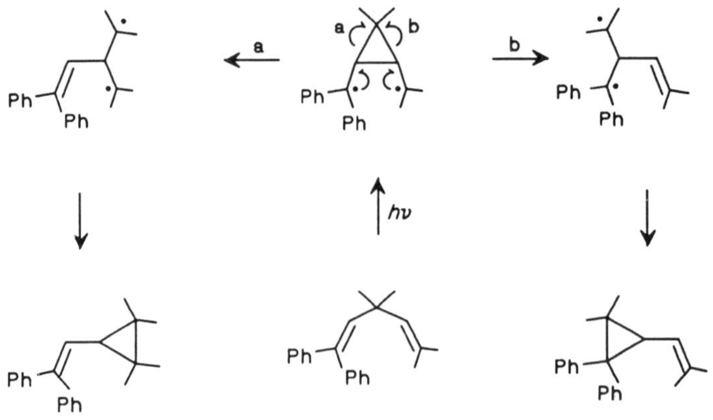

SCHEME 3. Control of the di-π-methane regioselectivity

Another aspect of the reaction is the stereochemistry. There are three points of interest—C-1, C-3 and C-5 in a 1,4-pentadiene type di-π-methane reactant. If C-5 is taken as the carbon of that vinyl group which remains in the photoproduct, that stereochemistry—*cis* or *trans*—is retained[32]. The reaction stereochemistry at C-3 (the 'methane carbon') has been shown to be inversion of configuration[33,34] (equations 9a and 9b). The situation at C-1 is

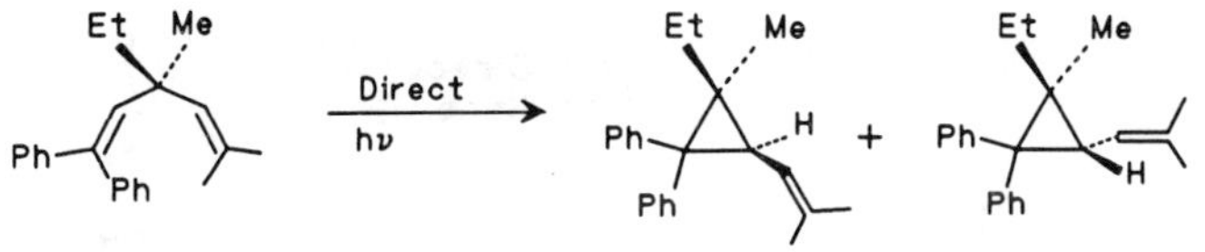

(9a)

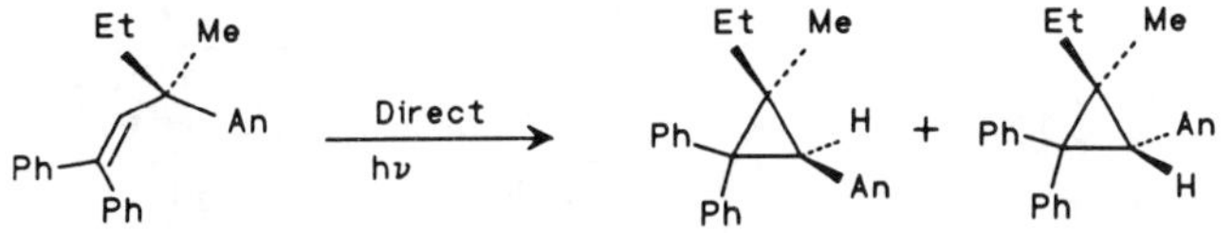

(9b)

equally interesting wherein the *cis* isomer of the vinyl reactant affords the photoproduct with the *cis* substituent being oriented *cis* on the three-membered ring relative to the vinyl side-chain[35] (cf equations 9c and 9d). The C-5 stereochemistry, wherein the *cis* or *trans*

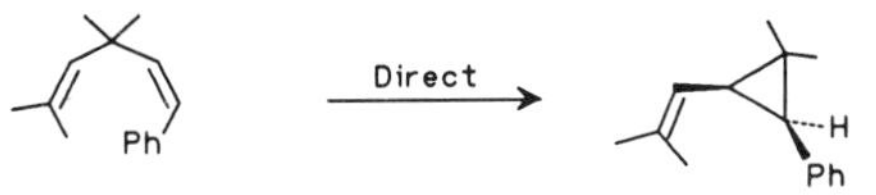

(9c)

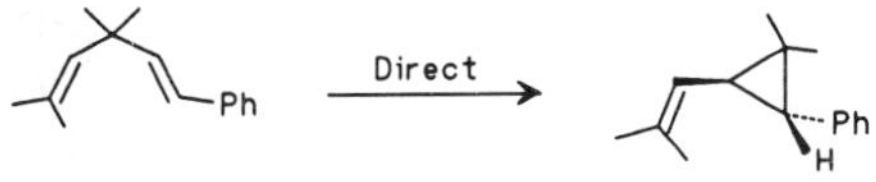

(9d)

double-bond configuration is retained, and the C-3 stereochemistry might be construed as arising from least motion. However, the situation at C-1 is not so obvious; here the double-bond stereochemistry is transferred to the configuration on the product three-membered ring. The overall stereochemistry, including that at C-1, may be understood[30] as a singlet excited state allowed transformation involving a cyclic Möbius[36] array of six orbitals with an odd number of sign inversions (e.g. one plus–minus overlap).

Now we turn to selected examples of interest. One involves a multiplicity dependence of the regioselectivity. This is illustrated in Scheme 4. While one simplistically might write a single intermediate cyclopropyldicarbinyl diradical structure for both singlet and triplet multiplicity, the regioselectivity of 'unzipping' depends on which species is involved[37].

Two other examples of directed regioselectivity of a singlet di-π-methane rearrangement are shown in equations 10a and 10b. These follow a general pattern in which electron-withdrawing groups tend to appear on the three-membered ring of the photoproduct while electron-donating groups are found on the vinyl group. This can be understood from the finding that in S_1 the cyclopropyldicarbinyl carbons are electron-rich relative to the odd-electron centers of diradical II[38,39].

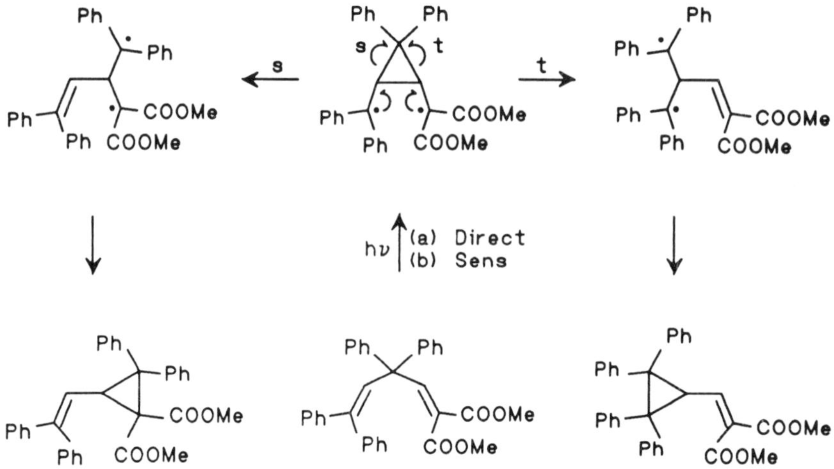

SCHEME 4. Dependence of a di-π-methane regioselectivity on multiplicity

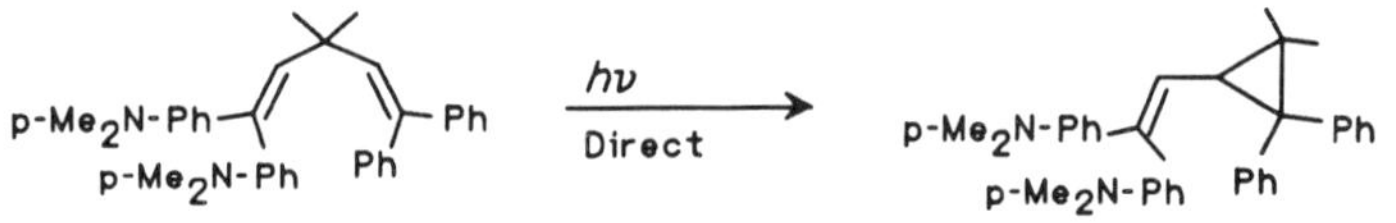

(10a)

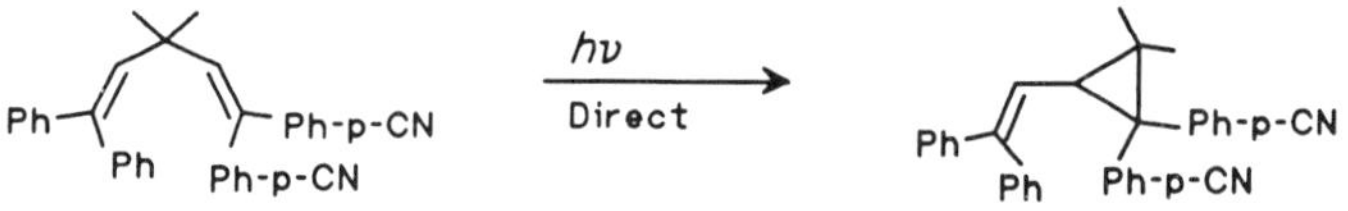

(10b)

The di-π-methane rearrangement of bicyclic systems is particularly extensive and fruitful in leading to products with three-membered rings. One typical example is that of β,β,-naphthobarrelene which affords the corresponding naphthosemibullvalene (equation 11).

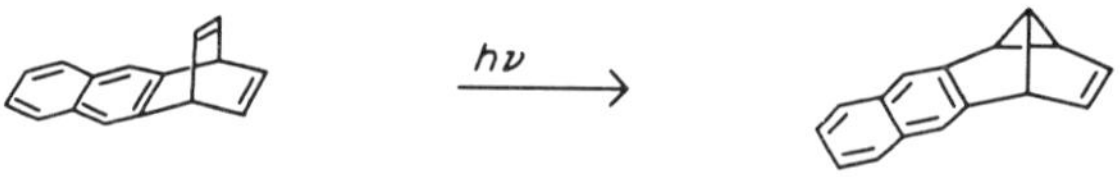

(11)

The reaction proceeds via the triplet and the mechanism involves vinyl–vinyl rather than naphtho–vinyl bridging[40]. In the case of such triplet rearrangements, it has been shown[41] that the cyclopropyldicarbinyl diradicals (i.e. Diradical I) are actual reaction intermediates of finite lifetime and that, when two such alternative diradicals are possible, formation of the lower energy one determines the reaction regioselectivity[42].

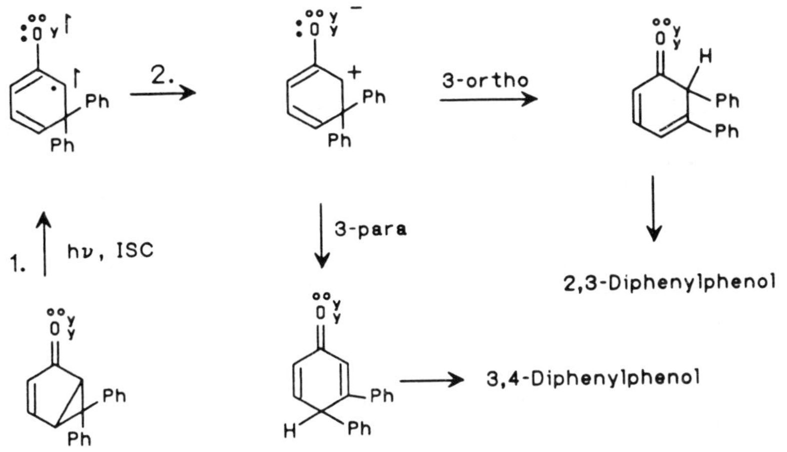

SCHEME 5. The Type-B photochemistry of a bicyclo[3.1.0]hex-3-en-2-one

III. PHOTOCHEMICAL REACTIONS OF THREE-MEMBERED RING COMPOUNDS

A. Type-B Reactions; Rearrangements of the Bicyclic Photoproducts of Type-A Processes

The irradiation of bicyclo[3.1.0]hex-3-en-2-ones leads to photoproducts deriving from fission of the internal three-membered ring bond. For example, the reaction of 6,6-diphenylbicyclo[3.1.0]hex-3-en-2-one affords both 2,3-diphenylphenol and also 3,4-diphenylphenol[43,44]. This is outlined in Scheme 5. In near-neutral reaction media the 2,3-diphenylphenol isomer is preferred, while in acidic medium the 3,4-diphenylphenol predominates[44]. It is clear that the free Type-B zwitterion gives rise to preferential migration to C-2 while the protonated counterpart prefers migration to C-4. The source of the preference for migration to C-2 is not found in the relative positive charges at C-2 versus C-4, since it is C-4 of the zwitterion which is found computationally to be more electron-deficient. Rather control is by enolate delocalization as shown in Scheme 6. It is seen that the 2,3-phenyl-bridged zwitterion has one extra enolate resonance structure compared with the 3,4-bridged counterpart. However, in the case of the protonated bridged species, enolate delocalization is no longer a factor and the calculated greater electron deficiency at C-4 in the zwitterion is controlling.

One might question whether in Scheme 6 we have properly written aryl migration only after intersystem crossing to the Type-B zwitterion. *A priori*, one might write a mechanism in which the initially formed excited state undergoes internal three-membered ring bond fission followed by aryl migration while still a triplet. After aryl migration, intersystem crossing would lead to a phenolic photoproduct. That it is the zwitterion and not the six-ring triplet diradical which rearranges was established in a study of the photochemistry of the stereoisomeric *endo*- and *exo*-6-*p*-cyanophenyl-6-phenylbicyclo[3.1.0]hex-3-en-2-ones[45] (Scheme 7). If the six-membered ring triplet diradical undergoes migration precedent indicates a preferential cyanophenyl migration, while if it is the Type-B zwitterion which rearranges a phenyl migration to a positive center should be preferred. The observation was that phenyl migrates exclusively.

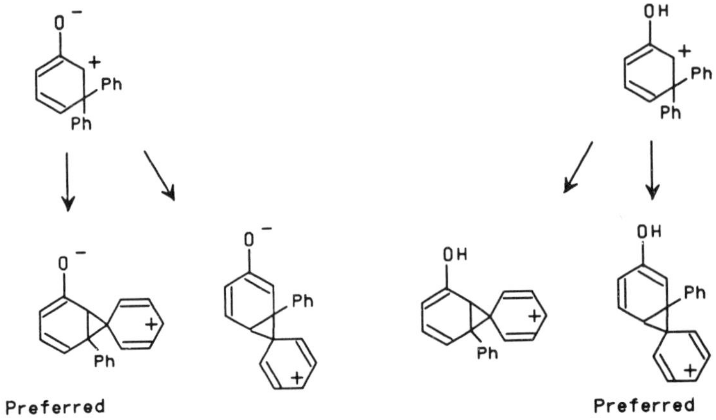

SCHEME 6. 2,3-Phenyl migration versus 3,4-phenyl migration in the Type-B rearrangement

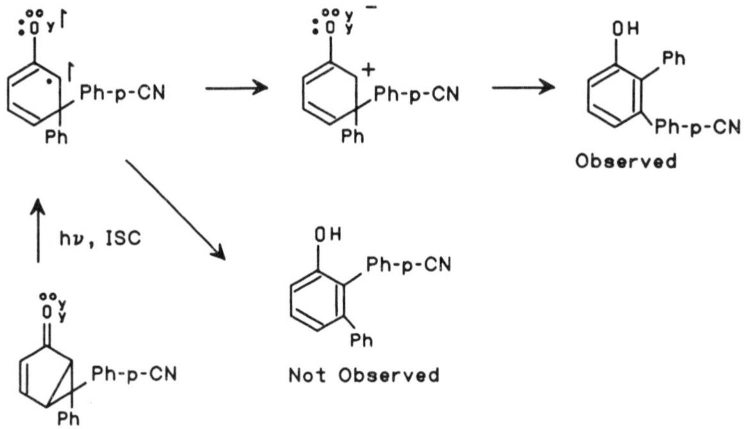

SCHEME 7. Competitive phenyl versus cyanophenyl migration in the Type-B rearrangement

In addition, it was found[45] that both the *endo* and *exo* stereoisomers of the cyanophenyl phenyl bicyclo[3.1.0] ketone afforded the same 2,3- versus 3,4-phenyl as well as the phenyl over cyanophenyl migration preference. This provided further support for the planar zwitterionic intermediate.

A particularly convincing additional result was the dark generation of the Type-B zwitterion as shown in equation 12[46]. Not only were the same two diphenylphenols

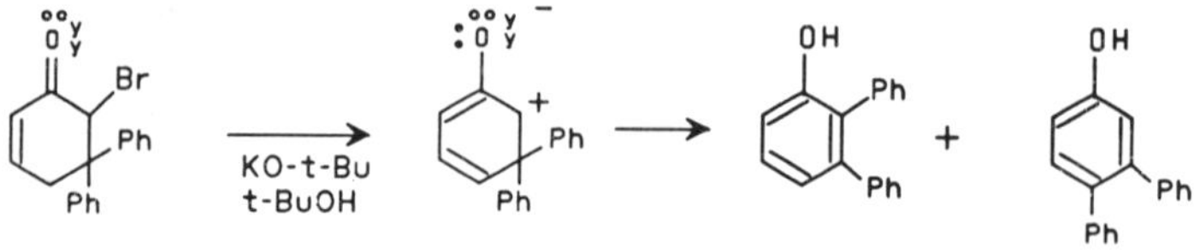

(12)

obtained as in the photochemistry, but also their ratio in a variety of solvents of varying polarity varied identically to that observed for the Type-B photochemistry.

B. The Bicycle Rearrangement

We termed a particularly esoteric and intriguing rearrangement the 'Bicycle Rearrangement' for reasons which become obvious[47]. Equation 13 gives one early example[47a]. Superficially, the reaction appears to involve loss of an isopropylidene carbene and its readdition to the exocyclic bouble bond. However, this is not the reaction mechanism as can be seen from the examples in equations 14a and 14b. Here the *endo* and *exo*

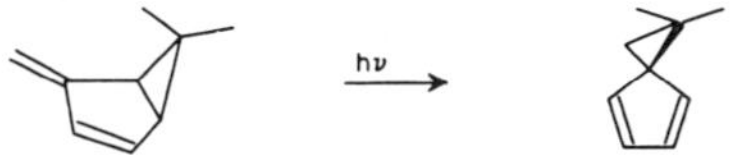

(13)

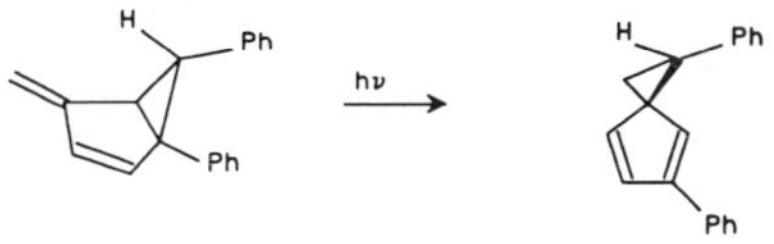

(14a)

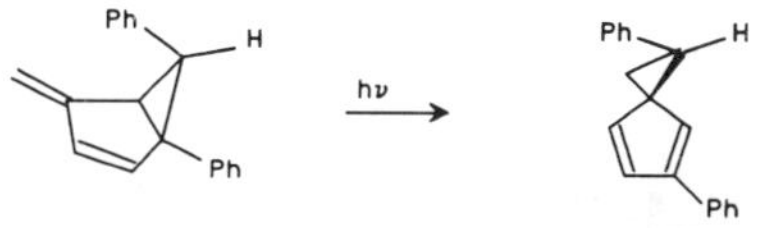

(14b)

stereoisomers, differing in the configurations at what would have to be the carbenoid carbon, are largely retained in the reaction. The *endo* reactant affords the *anti* spiro photoproduct and the *exo* reactant leads to the *syn* spiro photoproduct. The mechanisms available to such reactants are outlined in Scheme 8.

The evidence for such a mechanism results from both the reaction stereochemistry and also from the observation of minor diphenyltoluene by-products (Scheme 8). The major pathway is outlined using heavy arrows. This can be seen to afford the stereospecificity of equations 14a and 14b. Additionally, each of the diradical intermediates and intermediate excited states—**B**, **C**, **E**, **F** and **G**—undergo a minor extent of internal bond fission [i.e. Grob or 2,3- (1,4) fragmentation] to afford a diphenyltoluene with the corresponding ring skeleton. The basis for the choice of a main pathway versus the minor ones comes from the observed stereochemistry.

We can now paraphrase the reaction topology as in Scheme 9, which depicts the divalent carbon as moving along the surface of a five-membered ring to lead to the spiro product with three alternative pathways. One is the direct route, shown in Scheme 8 affording the observed stereochemistry. A second involves the divalent carbon moving around the five-membered ring clockwise; this leads to the reversed stereochemistry. However, an alternative is the 'overshoot-backup' pathway in which the bicycling carbon moves one bond beyond the side-chain and then backs up onto the *exo*-methylene group. This, too, affords the minor stereochemistry.

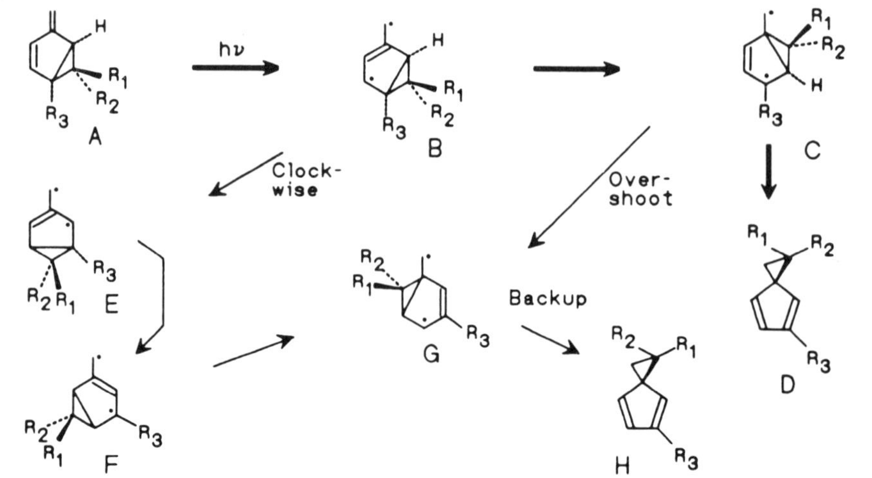

SCHEME 8. Available 'bicycle' reaction mechanisms. Main route: heavy arrows; minor pathways: light arrows; R^1 or R^2 = Ph, R^3 = Ph

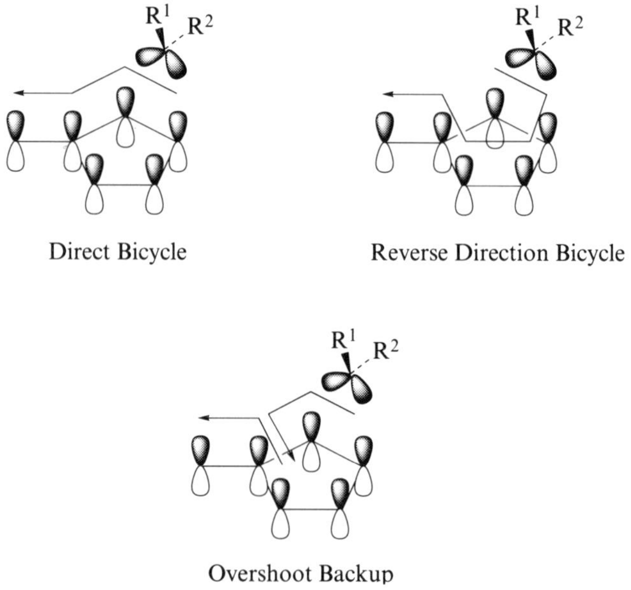

SCHEME 9. The bicycle reaction topology

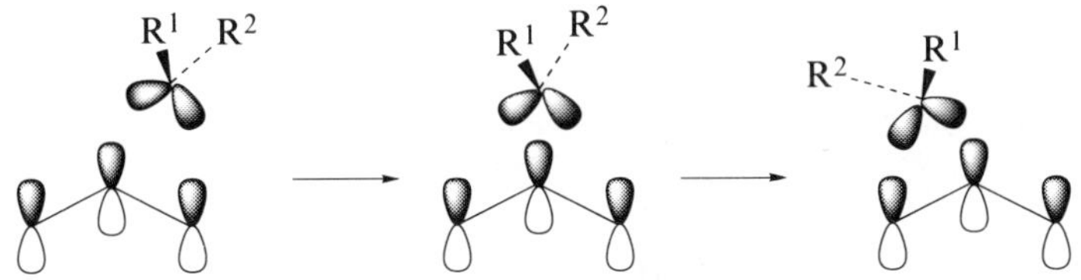

Pivot Mechanism:

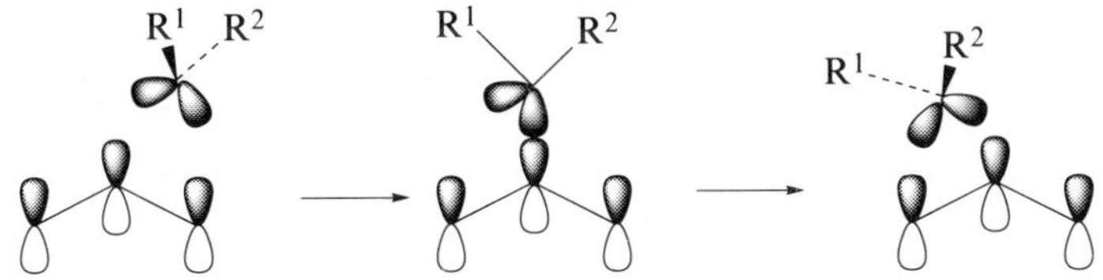

SCHEME 10. Bicycle and pivot alternatives for a walk rearrangement

It should be noted that the observed bicycle motion is one of two *a priori* possibilties as shown in Scheme 10, which includes the alternative, 'pivot' process. Interestingly, an even number of pivot steps would give rise to the same stereochemistry as the bicycle process, although an odd number of pivot steps affords the reverse stereochemistry relative to the bicycle mechanism. The differentiation between these possibilities resulted from the observation that the endo-aryl reactants of the type in Scheme 11 show a marked enhancement of the 1,4- (2,3) fragmentation at the end of an odd number of bicycle steps. The *exo* stereoisomers proceed nicely onward to the spiro bicycle photoproduct. This is understood on the basis of van der Waals repulsions of the *endo*-aryl groups, which leads to enhanced fragmentation due to relief of these repulsions. This, then identifies the aryl groups as *endo* after an odd number of steps.

A particularly intriguing set of reactions involving bicycling steps as well as a di-π-methane rearrangement and a 1,4- (2,3) fragmentation is outlined in Scheme 12[28]. Here there is a superficial paradox. Thus, diradical *D* is formed in two ways. It formed from the benzo di-π reactant (labeled 'Di-π') as well as the benzo bicyclic reactant labeled 'Bicyclic' in Scheme 12. However, diradical *D* reacts differently depending on its mode of formation.

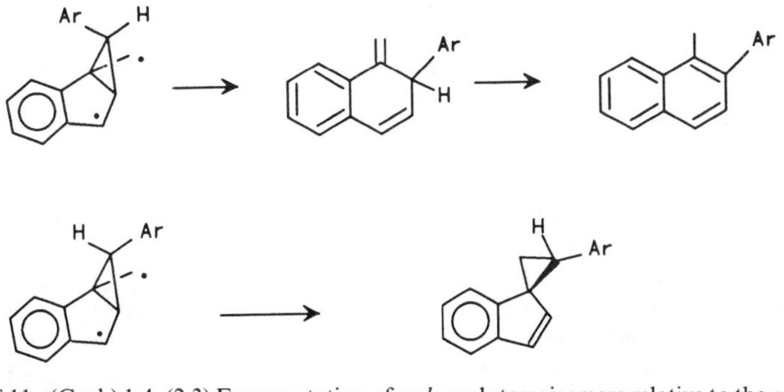

SCHEME 11. (Grob) 1,4- (2,3) Fragmentation of *endo* aryl stereoisomers relative to the *exo* isomers

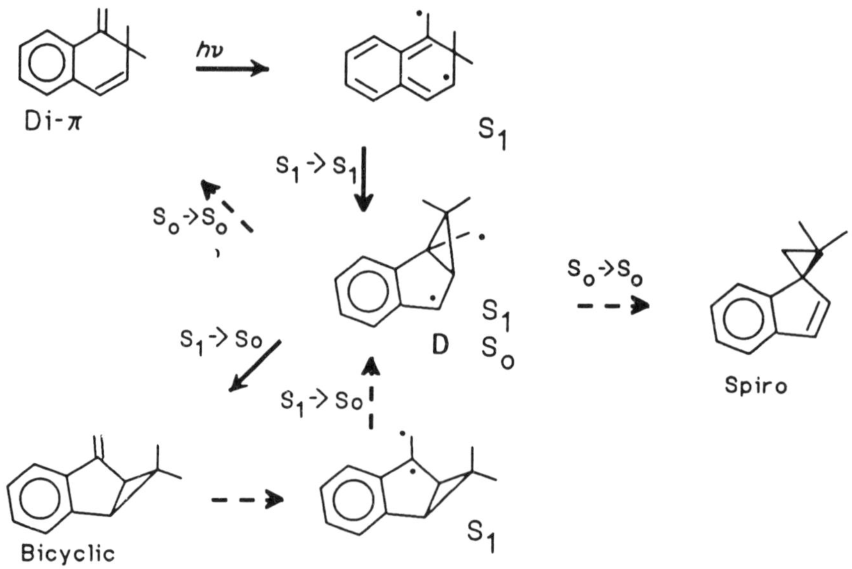

SCHEME 12. Interrelationship between bicycling and the di-π-methane rearrangement. States of one diradical giving different reactivities

From the di-π reactant it affords the benzobicyclic, while on irradiation of the benzobicyclic as the reactant the spiro photoproduct is formed along with lesser amounts of the di-π compound. Clearly, the one basic structure *D* is involved as an intermediate. However, a solution to the paradox was obtained from SCF-CI computations, which revealed that a HOMO–LUMO crossing occurred only between the bicyclic molecule and diradical *D*. The CI computations agreed in revealing the presence of a bifunnel only in this branch of the reaction scheme. Note Scheme 13, which gives the MO correlations in the form of a triptych. At the center of the triptych are those MOs of diradical D which are closest to nonbonding and of most interest. One branch of the triptych corresponds to the di-π reactant and is labeled *B*. A second branch corresponds to the [3.1.0] bicyclic and is labeled *A*. The third branch leads to the spiro photoproduct and is labeled *C*. For the two possible reactants, the di-π and the [3.1.0] bicyclic, the configurations shown are singly excited. It is seen that the di-π reactant has no MO crossing interposed between *B* and *D* and thus leads necessarily to the electronically excited (i.e. S_1) diradical *D*. Only on branch *A* leading to the [3.1.0] bicyclic is there a HOMO–LUMO crossing permitting internal conversion and product formation; hence the formation of only this product starting with the di-π reactant. Differently, starting with the [3.1.0] bicyclic reactant on triptych branch *A*, the HOMO–LUMO crossing affords a 'bifunnel' and decay to ground state S_0 diradical *D*. Inspection of the available triptych branches shows that only branches *B* and *C* lead to ground state configurations. Thus, from the [3.1.0] bicyclic reactant both spiro and di-π photoproducts are possible and observed.

This reasoning was found to be paralleled in configuration interaction computations, and a similar triptych was obtained in which surfaces rather than MOs were drawn. Such a triptych has an avoided crossing and bifunnel only on branch *A* and the reasoning using this variation is parallel to that above. Such computations were used for virtually all of the bicycle-type reactions discussed above.

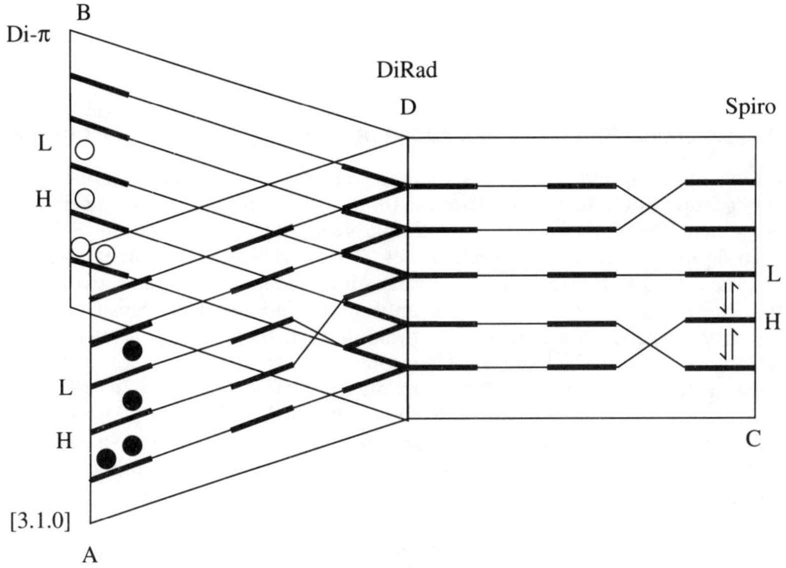

SCHEME 13. Triptych MO correlation diagram for the reaction in Scheme 12

Acyclic systems also are able to undergo the bicycle rearrangement as shown in equations 15a and 15b[21,37,48]. A clue to the reaction mechanism was found in the observation of

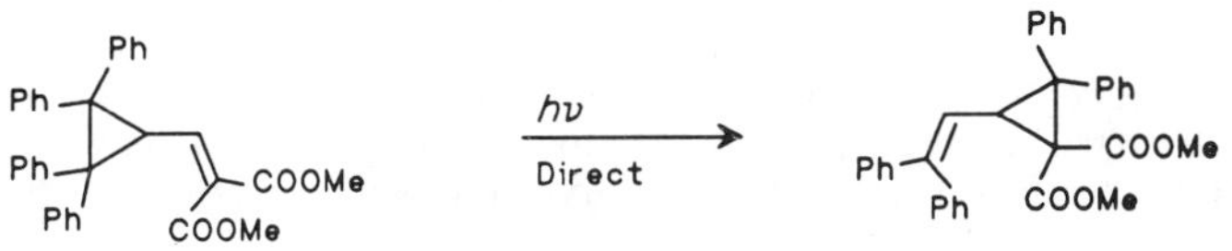

(15a)

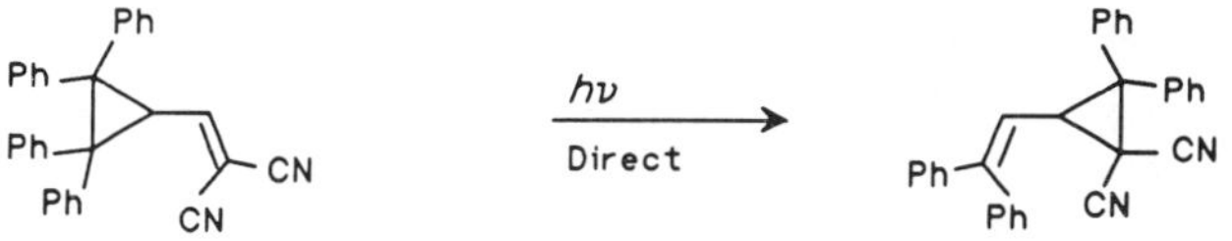

(15b)

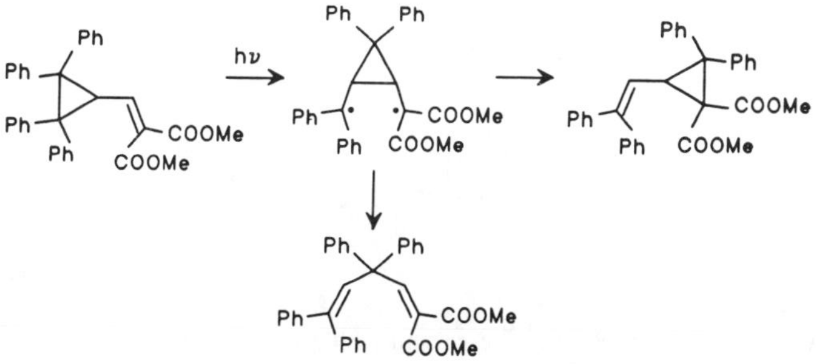

(16)

a product of bicycling of the benzhydryl moiety along the carbon skeleton only half-way, followed by fragmentation of the cyclopropyl–dicarbinyl diradical thus formed (equation 16). This process may be termed a 'reverse di-π-methane rearrangement'.

C. Photochemical Stereoisomerizations of Aryl, Vinyl and Acyl Cyclopropanes

Cyclopropyl compounds with three-membered rings, conjugated to some electron delocalizing group, often undergo σ-bond fission with resultant stereoisomerization or rearrangement. One rather interesting example is that of the epimerization of the *cis*- and *trans*-5,6-diphenylbicyclo[3.1.0]hexan-2-ones. Thus, photolysis of either stereoisomer leads to the other, with formation of a steady state[49]. Scheme 14 outlines two alternative reaction mechanisms. Each utilizes the antibonding odd-electron density of the n-π* excited state to open an α,β σ-bond. The question is whether it is the out-of-plane bond *a* or the internal bond *b* which breaks. The problem was solved when it was noted that the two reaction mechanisms afford different enantiomers. If 'Mechanism A' is operative, then photoproduct Enantiomer A will be formed, while if 'Mechanism B' is utilized, then photoproduct Enantiomer B will be observed. Simple reductive three-membered ring opening of reactant and product affords 3-phenyl-3-benzylcyclopentanones, which then have the same configuration if Mechanism A is operating and the reverse optical rotation if Mechanism B is operating. The experimental observation was that, for the conversion of *trans* to *cis* bicyclic ketones, only Mechanism A operated. But in the *cis* to *trans* conversion, while mainly Mechanism A was used, Mechanism B was found to operate to a significant extent.

Superficially this appears to be a violation of microscopic reversibility. However, since the reaction is not adiabatic, we are not dealing with the interconversion of two species. Each reaction proceeds from excited state of reactant to ground state of product and microscopic reversibility is not relevant.

A second reaction of this type is of interest because it reveals the nature of the diradical produced. Here we have 1-benzoyl-2-aryl-3-phenylcyclopropanes. The aryl groups in the study were *p*-cyanophenyl, anisyl and phenyl[50]. It was observed that the 2,3-*trans,trans* isomer, in which the benzoyl group was *trans* to both aryl groups, on photolysis afforded the *cis–trans* isomers in which the substituted aryl group was now *cis* to benzoyl. One might

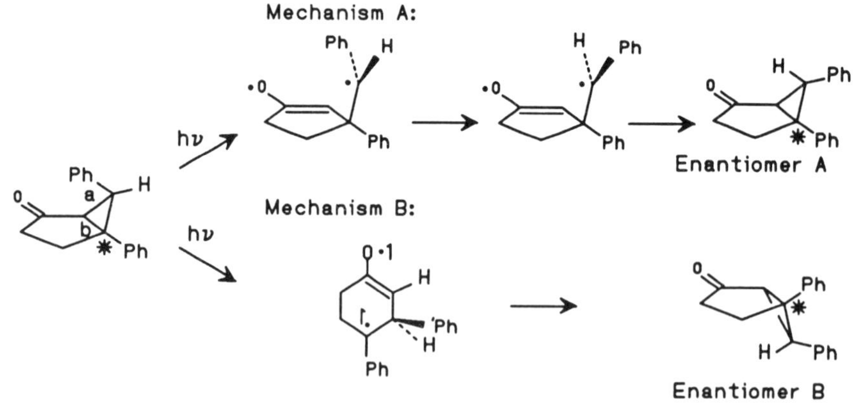

SCHEME 14. Mechanisms for the stereoisomerization of the bicyclo[3.1.0] cyclohexenones

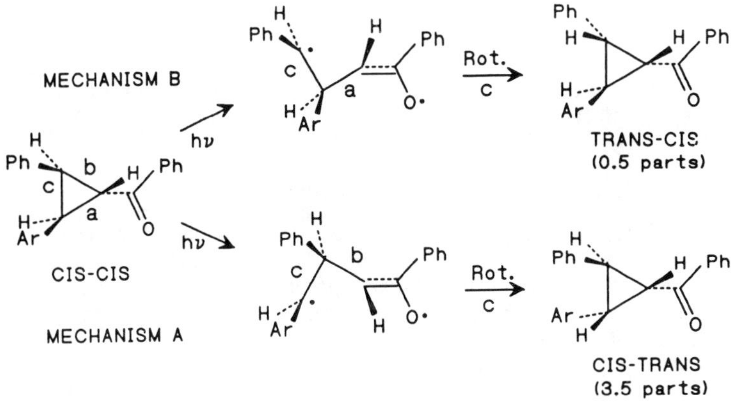

SCHEME 15. Mechanisms of stereoisomerization of keto cyclopropanes

envisage three reaction mechanisms, two of which are depicted in Scheme 15: (a) one in which α,β-sigma bond *a* was severed to afford the diradical having the odd-electron on the *p*-substituted aryl ring, followed by a single rotation about sigma bond *c* and reclosure to form the *cis–trans* benzoyl cyclopropane; (b) a similar mechanism but with differing regioselectivity in which scission of α,β bond *b* to afford an ordinary benzylic odd-electron center, followed by rotation about sigma bond *c*, and then reclosure to afford the *trans–cis* isomer. As noted in Scheme 15, the preference for the *cis–trans* isomer is 3.5:1. However, there is a third mechanistic possibility which could lead to this preference. In this mechanism (c) bond *b* is severed preferentially, followed by a 'double rotation' in which there is rotation about bonds *a* and *c*. While this does leads to the observed *cis–trans* diastereomer, it would afford the opposite enantiomer from that shown in Scheme 15.

In this investigation[50] the resolved reactant was studied and this third mechanistic possibility was found not to occur. However, a minor portion of the *trans–cis* isomer was found to result from fission of bond *a* followed by the same type of double rotation (i.e. about bonds *b* and *c*). Overall, bond *a* to bond *b* fission was found to occur in a 8:1 ratio. The driving force for this regioselectivity is the greater odd-electron stabilization by the *p*-substituted aryl group and the three-membered ring opening occurs as in the preceding bicyclic case by utilization of the antibonding electron density of the π-system of the n-π* excited state.

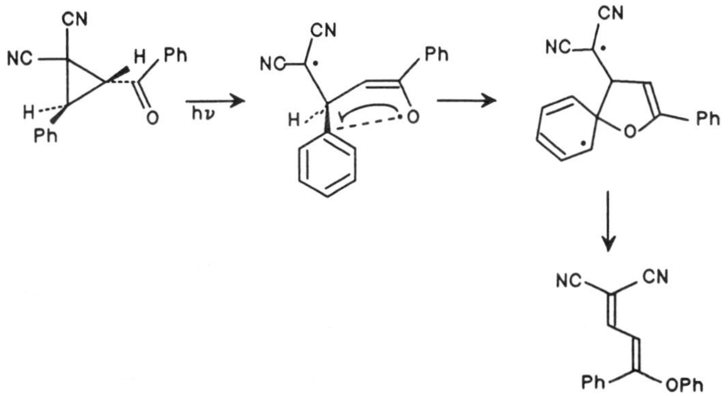

(17)

A related example is given in equation 17[51]. Here, both roles of the n-π* excited state are involved. The antibonding electron density opens that α,β bond which will afford a dicyano-stabilized odd-electron center. Its oxygen p_y orbital attacks the *ipso* carbon of the phenyl group on the cyclopropane ring to give the spiro-diradical species shown. This then opens to photoproduct.

D. Conjugated Cyclopropanes Giving Rearranging Diradicals

One intriguing type of reaction is a π-conjugated cyclopropane so designed that, on three-membered ring opening, the resulting 1,3-diradical has groups substituted at C-2 which can migrate to one of the diradical centers. One example[32] begins with a vinylcyclopropane. This, on direct irradiation, affords a 1,3-singlet diradical with two methyl groups at the central (i.e. C-2) carbon. Such 1,3-singlet diradicals rearrange as shown in equation 18.

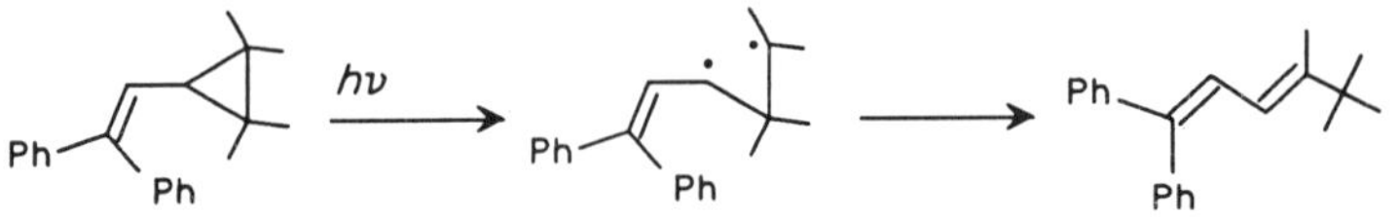

(18)

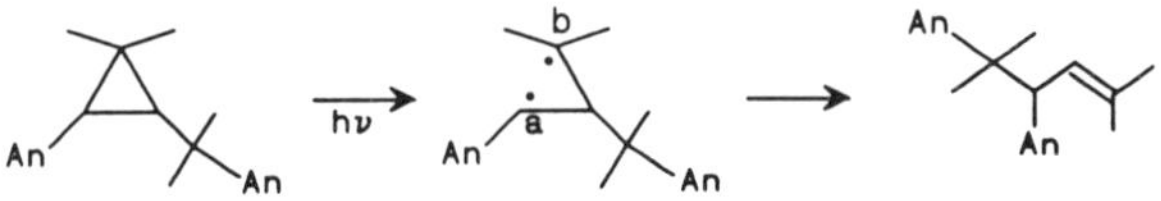

(19)

An = anisyl

A somewhat similar example[52], in equation 19, appears to be partially intermolecular since there is evidence for some extrusion of a *p*-methoxycumenyl free radical. The evidence suggested that *ca* 12% of the reaction is occurring by dissociation–recombination. This example, we note, is not typical since the *p*-methoxycumenyl group is much more subject to extrusion as an odd-electron species than (e.g.) methyl.

The migration of alkyl groups was found to be general and several examples[53] are outlined in equation 20. As noted in equation 20, a competing reaction is the common 1,5-

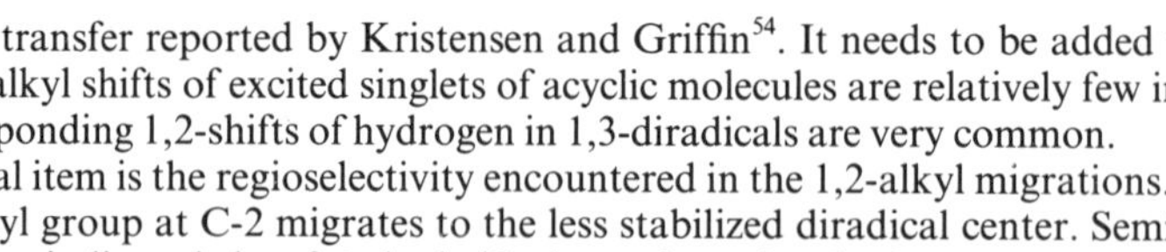

(20)

Ar = Ph, An, *p*-CNC_6H_4, biphenylyl

hydrogen transfer reported by Kristensen and Griffin[54]. It needs to be added that while these 1,2-alkyl shifts of excited singlets of acyclic molecules are relatively few in number, the corresponding 1,2-shifts of hydrogen in 1,3-diradicals are very common.

One final item is the regioselectivity encountered in the 1,2-alkyl migrations. In general, the alkyl group at C-2 migrates to the less stabilized diradical center. Semiempirical calculations indicated that for the half-migrated species the lower energy S_1 is that in which the methyl group migrates to the less delocalized center, while for S_0 the reverse prediction is obtained. Thus, it appears that the regioselectivity is determined while the rearranging molecule is still on the excited singlet hypersurface and has not yet approached ground state.

IV. SUMMARY

Although this is a review which has focused on the author's research, the wealth of varied cyclopropane photochemistry is apparent even from this collection of various representative examples. The author has also limited his selection to examples which fall into discrete categories and where generalizations are possible.

V. ACKNOWLEDGMENT

This work would not have been possible without the support, financial and moral, of these efforts by the National Science Foundation. That so many of the coauthors in my references are now independently professionally active in research attests to the excitement of these research endeavors.

VI. REFERENCES

1. D. H. R. Barton, P. DeMayo and M. Shafiq, *J. Chem. Soc.*, 1215 (1961).
2. H. E. Zimmerman and D. I. Schuster, *J. Am. Chem. Soc.*, **84**, 4527 (1962).
3. (a) H. E. Zimmerman and J. S. Swenton, *J. Am. Chem. Soc.*, **86**, 1436 (1964).
   (b) H. E. Zimmerman and J. S. Swenton, *J. Am. Chem. Soc.*, **89**, 906 (1967).
4. H. E. Zimmerman, R. W. Binkley, J. J. McCullough and G. A. Zimmerman, *J. Amer. Chem. Soc.*, **89**, 6589 (1967).
5. H. E. Zimmerman, D. Döpp and P. S. Huyffer, *J. Am. Chem. Soc.*, **88**, 5352 (1966).
6. H. E. Zimmerman, D. S. Crumrine, D. Döpp and P. S. Huyffer, *J. Am. Chem. Soc.*, **91**, 434 (1969).
7. H. E. Zimmerman and D. S. Crumrine, *J. Am. Chem. Soc.*, **90**, 5612 (1968).
8. H. E. Zimmerman and R. L. Pasteris, *J. Org. Chem.*, **45**, 4868 (1980).
9. H. E. Zimmerman and R. L. Pasteris, *J. Org. Chem.*, **45**, 4876 (1980).
10. H. E. Zimmerman and G. E. Keck, *J. Am. Chem. Soc.*, **97**, 3527 (1975).
11. H. E. Zimmerman, G. E. Keck and J. L. Pflederer, *J. Am. Chem. Soc.*, **98**, 5574 (1976).
12. H. E. Zimmerman and D. C. Lynch, *J. Am. Chem. Soc.*, **107**, 7745 (1985).
13. H. E. Zimmerman, B. R. Cowley, C-Y. Tseng and J. W. Wilson, *J. Am. Chem. Soc.*, **86**, 947 (1964).
14. H. E. Zimmerman and K. G. Hancock, *J. Am. Chem. Soc.*, **90**, 3749 (1968).
15. H. E. Zimmerman and R. L. Morse, *J. Am. Chem. Soc.*, **90**, 954 (1968).
16. H. E. Zimmerman and W. R. Elser, *J. Am. Chem. Soc.*, **91**, 887 (1969).
17. H. E. Zimmerman, R. D. Rieke and J. R. Scheffer, *J. Am. Chem. Soc.*, **89**, 2033 (1967).
18. (a) H. E. Zimmerman and R. D. Little, *J. Chem. Soc., Chem. Commun.*, 698 (1972).
   (b) H. E. Zimmerman and R. D. Little, *J. Am. Chem. Soc.*, **94**, 8256 (1972).
19. H. E. Zimmerman, R. J. Boettcher, N. E. Buehler, G. E. Keck and M. G. Steinmetz, *J. Am. Chem. Soc.*, **98**, 7680 (1976).
20. H. E. Zimmerman, 'The Di-π-Methane Rearrangement', in *Organic Photochemistry* (Ed. A. Padwa), Vol. 11, Mercel Dekker, New York, 1991.
21. H. E. Zimmerman, D. Armesto, M. G. Amezua, T. P. Gannett and R. P. Johnson, *J. Am. Chem. Soc.*, **101**, 6367 (1979).
22. H. E. Zimmerman and R. E. Factor, *J. Am. Chem. Soc.*, **102**, 3538 (1980).
23. H. E. Zimmerman, R. K. King, J.-H. Xu and C. E. Caufield, *J. Am. Chem. Soc.*, **107**, 7724 (1985).
24. H. E. Zimmerman, C. E. Caufield and R. K. King, *J. Am. Chem. Soc.*, **107**, 7732 (1985).
25. (a) H. E. Zimmerman and G. L. Grunewald, *J. Am. Chem. Soc.*, **86**, 1434 (1964).
   (b) H. E. Zimmerman and G. L. Grunewald, *J. Am. Chem. Soc.*, **88**, 183 (1966).
26. H. E. Zimmerman, R. W. Binkley, R. S. Givens and M. A. Sherwin, *J. Am. Chem. Soc.*, **89**, 3932 (1967).
27. (a) J. W. Swenton, *J. Chem. Educ.*, **46**, 217 (1969).
   (b) R. S. Givens and W. F. Oettle, *Chem. Commun.*, 1164 (1969).

(c) W. G. Dauben, M. S. Kellogg, J. I. Seeman and W. A. Spitzer, *J. Am. Chem. Soc.*, **92**, 1786 (1970).
28. (a) D. Armesto, W. M. Horspool, M. Apoita, M. G. Gallego and A. Ramos, *J. Chem. Soc., Perkin Trans. 1*, 2035 (1989).
(b) D. Armesto, M. G. Gallego and W. M. Horspool, *Tetrahedron Lett.*, **31**, 2475 (1990).
29. H. E. Zimmerman and P. S. Mariano, *J. Am. Chem. Soc.*, **91**, 1718 (1969).
30. H. E. Zimmerman and A. C. Pratt, *J. Am. Chem. Soc.*, **92**, 6267 (1970).
31. (a) H. E. Zimmerman, K. S. Kamm and D. P. Werthemann, *J. Am. Chem. Soc.*, **97**, 3718 (1975).
(b) H. E. Zimmerman, F. X. Albrecht and J. J. Haire, *J. Am. Chem. Soc.*, **97**, 3726 (1975).
32. H. E. Zimmerman and A. C. Pratt, *J. Am. Chem. Soc.*, **92**, 6259 (1970).
33. H. E. Zimmerman, J. D. Robbins, R. D. McKelvey, C. J. Samuel and L. R. Sousa, *J. Am. Chem. Soc.*, **96**, 4630 (1974).
34. H. E. Zimmerman, T. P. Gannett and G. E. Keck, *J. Am. Chem. Soc.*, **100**, 323 (1978).
35. H. E. Zimmerman, P. Baeckstrom, T. Johnson and D. W. Kurtz, *J. Am. Chem. Soc.*, **96**, 1459 (1974).
36. (a) H. E. Zimmerman, *J. Am. Chem. Soc.*, **88**, 1564 (1966).
(b) H. E. Zimmerman, *Acc. Chem. Res.*, **4**, 272 (1971).
37. H. E. Zimmerman and R. E. Factor, *Tetrahedron*, **37**, Supplement 1, 125 (1981).
38. H. E. Zimmerman and B. R. Cotter, *J. Am. Chem. Soc.*, **96**, 7445 (1974).
39. H. E. Zimmerman and W. T. Gruenbaum, *J. Org. Chem.*, **43**, 1997 (1978).
40. H. E. Zimmerman and C. O. Bender, *J. Am. Chem. Soc.*, **92**, 4366 (1970).
41. H. E. Zimmerman, A. G. Kutateladze, Y. Meakawa and J. E. Mangette, *J. Am. Chem. Soc.*, **116**, 9795 (1994).
42. H. E. Zimmerman, H. M. Sulzbach and M. B. Tollefson, *J. Am. Chem. Soc.*, **115**, 6548 (1993).
43. H. E. Zimmerman and D. I. Schuster, *J. Amer. Chem. Soc.*, **84**, 4527 (1962).
44. H. E. Zimmerman, J. Nasielski, R. Keese and J. S. Swenton, *J. Am. Chem. Soc.*, **88**, 4895 (1966).
45. H. E. Zimmerman and G. L. Grunewald, *J. Am. Chem. Soc.*, **88**, 183 (1966).
46. H. E. Zimmerman and G. A. Epling, *J. Am. Chem. Soc.*, **94**, 7806 (1972).
47. (a) H. E. Zimmerman, P. Hackett, D. F. Juers, J. M. McCall and B. Schröder, *J. Am. Chem. Soc.*, **93**, 3653 (1971).
(b) H. E. Zimmerman and T. P. Cutler, *J. Org. Chem.*, **43**, 3283 (1978).
(c) See Reference 22.
(d) H. E. Zimmerman *Chimia*, **36**, 423 (1982).
48. H. E. Zimmerman, F. L. Oaks and P. Campos, *J. Am. Chem. Soc.*, **111**, 1007 (1989).
49. H. E. Zimmerman, K. G. Hancock and G. Licke, *J. Am. Chem. Soc.*, **90**, 4892 (1968).
50. (a) H. E. Zimmerman, S. S. Hixson and E. F. McBride, *J. Am. Chem. Soc.*, **92**, 2000 (1970).
(b) H. E. Zimmerman and C. M. Moore, *J. Am. Chem. Soc.*, **92**, 2023 (1970).
(c) H. E. Zimmerman and T. W. Flechtner, *J. Am. Chem. Soc.*, **92**, 6931 (1970).
51. H. E. Zimmerman and R. W. Binkley, *Tetrahedron Lett.*, **26**, 5859 (1985).
52. H. E. Zimmerman and A. P. Kamath, *J. Am. Chem. Soc.*, **110**, 900 (1988).
53. H. E. Zimmerman and J. A. Heydinger, *J. Org. Chem.*, **56**, 1747 (1991).
54. H. Kristensen and G. W. Griffin, *Tetrahedron Lett.*, 3259 (1966).

CHAPTER 7

Cyclopropyl homoconjugation, homoaromaticity and homoantiaromaticity—Theoretical aspects and analysis

DIETER CREMER,

Department of Theoretical Chemistry, University of Göteborg, S-41296 Göteborg, Kemigården 3, Sweden
Fax: +46-31772 2933; e-mail: CREMER@OC.CHALMERS.SE

RONALD F. CHILDS

Department of Chemistry, McMaster University, Hamilton, Ontario, Canada, L8S 4M1.
Fax: +1-905 521 1993; e-mail: RCHILDS@MCMAIL.CIS.MCMASTER.C

and

ELFI KRAKA

Department of Theoretical Chemistry, University of Göteborg, S-41296 Göteborg, Kemigården 3, Sweden

I. INTRODUCTION . . . 340
A. Models in Chemistry . . . 345
B. Organization of the Chapter . . . 345
II. DEFINITION OF HOMOCONJUGATION AND HOMOAROMATICITY—BASIC CONSIDERATIONS . . . 346
A. From Conjugation to Cyclopropyl Homoconjugation . . . 347
1. Conjugation . . . 347
2. Homoconjugation . . . 348
B. From a Topological to a Chemical Definition of Homoconjugation . . . 355
1. The concept of electron or bond delocalization . . . 355

The chemistry of the cyclopropyl group, Vol. 2
Edited by Z. Rappoport

2. A definition of homoconjugation based on the concept of bond (electron) delocalization . . . 356
3. The choice of appropriate reference compounds . . . 357
C. Homoaromaticity and Homoantiaromaticity . . . 360
D. Homoconjugation and the Topology of the Potential Energy Surface: From Homoaromaticity to Frozen Transition States . . . 362
III. THEORETICAL ASPECTS OF DEFINING, DETECTING AND DESCRIBING HOMOCONJUGATION AND HOMOAROMATICITY . . . 364
A. Winstein's Definition of Homoaromaticity . . . 365
B. Description of Homoconjugative Interactions in Terms of Orbital Overlap . . . 366
1. The homotropenylium cation as a test case . . . 368
C. The PMO Description of Homoaromaticity . . . 370
D. Description of a Homoconjugative Bond by Bond Orders and Other Interaction Indices . . . 373
E. The Electron Density Based Definition of a Homoconjugative Bond . . . 375
F. Description of Homoaromaticity in Terms of the Properties of the Electron Density Distribution . . . 378
G. Energy-based Definitions of Homoaromaticity . . . 381
1. Direct calculation of homoaromatic stabilization energies . . . 382
2. Homoaromatic stabilization energies from calculations with a model Hamiltonian . . . 384
3. Evaluation of homoaromatic stabilization energies by using isodesmic reactions . . . 384
4. Evaluation of homoaromatic stabilization energies by using homodesmotic reactions . . . 387
5. Homoconjugative resonance energies from force field calculations . . . 389
IV. *AB INITIO* EXAMINATIONS OF HOMOCONJUGATION . . . 390
A. Basic Requirements . . . 391
B. Investigation of the Homotropenylium Cation . . . 394
1. *Ab initio* calculations of geometry and energy . . . 394
2. Determination of the equilibrium geometry by the *ab initio*/chemical shift/NMR method . . . 396
C. Toward a General Definition of Homoaromaticity . . . 399
1. Bond homo(anti)aromaticity caused by cyclopropyl homoconjugation . . . 399
2. No-bond homoaromaticity . . . 401
3. General remarks . . . 402
V. CONCLUSIONS AND OUTLOOK . . . 403
VI. ACKNOWLEDGEMENTS . . . 406
VII. REFERENCES . . . 406

I. INTRODUCTION

Cyclopropyl homoconjugation describes conjugation of a cyclopropyl group with one or several unsaturated groups corresponding to double or triple bonds, cationic centres with empty pπ orbitals or any other conjugative group. Cyclopropyl homoconjugation is based on the π-character of the cyclopropyl bonds, which is amply documented in the literature[1] and which is discussed extensively in Chapter 2 of this volume by Cremer, Kraka and Szabo[2]. Formally, the single bond of the cyclopropyl group should act as an insulator and lead to an interruption of π-conjugation. As a consequence of the π-character of cyclo-

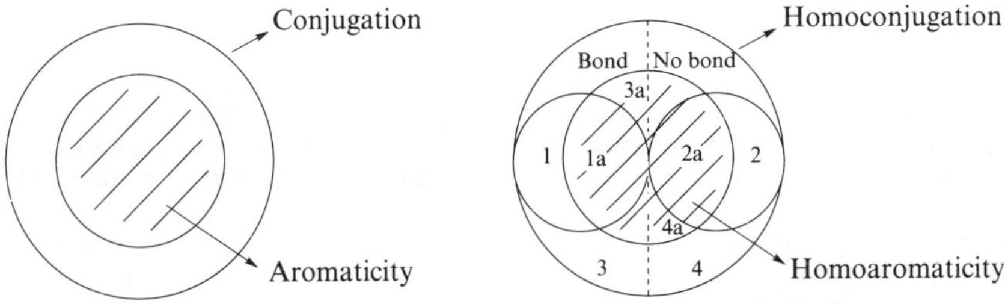

FIGURE 1. Conjugation and aromaticity, homoconjugation and homoaromaticity. Region 1: cyclopropyl homoconjugation ($X = CH_2$); region 2: no-bond homoconjugation ($X = CH_2$); region 3: bond homoconjugation in general ($X \neq CH_2$); region 4: no-bond homoconjugation in general ($X \neq CH_2$). Region 1a: cyclopropyl homoaromaticity ($X = CH_2$); region 2a: no-bond homoaromaticity ($X = CH_2$); region 3a: bond homoaromaticity in general ($X \neq CH_2$); region 4a: no-bond homoaromaticity in general ($X \neq CH_2$). Hence homoconjugation (outer circle) covers regions 1 and 3 (bond homoconjugation) as well as 2 and 4 (no-bond homoconjugation) while homoaromaticity is confined to the inner circle (shaded area) with regions 1a and 3a (bond homoaromaticity) as well as 2a and 4a (no-bond homoaromaticity)

propyl bonds, inclusion of the cyclopropyl ring into a conjugated chain or ring results only in a change in the degree of conjugation but not in its interruption or suppression. In this way the term cyclopropyl homoconjugation expresses homologation of conjugation along a π-chain to systems that include, beside a π-chain or π-cycle, one or several formal σ-bonds of cyclopropyl rings.

Cyclopropyl homoconjugation is a special case of homoconjugation as is indicated in Figure 1. Homoconjugation can be found in molecules **1** or **2**, where $X = CH_2$ and atoms 1 and 3 are connected by a bond (cyclopropyl homoconjugation, region 1 in Figure 1) or just by through-space interactions (no bond homoconjugation, region 2). Molecules **1** and **2** are often connected by a rearrangement process such as the valence tautomeric reaction shown in equation 1. They can also be related as neighbouring forms on the potential energy surface (PES) in the direction of the interaction distance, where either **1** or **2** is located at an energy minimum while the other form is just a transient point on the PES. Cyclopropyl homoconjugation and no bond homoconjugation are strongly related phenomena of one and the same molecular system. Therefore, a review on cyclopropyl homoconjugation must also consider no bond homoconjugation in order to describe, analyse and understand the electronic reasons leading to cyclopropyl homoconjugation.

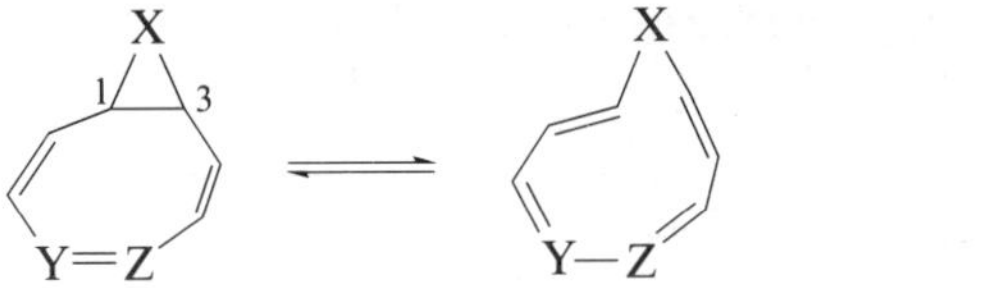

(1)

(1) **(2)** $X = CH_2$

Homoconjugation can also occur when X in **1** and **2** represents groups other than CH_2. Examples of homoconjugation are known for $X = CH$, CH_2CH_2, heteroatoms etc. as will be shown in this and the following review article by Childs, Cremer and Elia[3]. All cases with $X \neq CH_2$ are collected in regions 3 and 4 of Figure 1. In general, it is reasonable to distinguish between *bond homoconjugation*, when conjugative interactions are mediated through a bond (regions 1 and 3, Figure 1), and *no bond homoconjugation*, when conjugative interactions are mediated through space rather than through a bond (regions 2 and 4). The exact borderline between these regions is a matter of theoretically based or experimentally based definitions, which we will discuss in this review.

There is another important aspect that has to be considered in connection with cyclopropyl homoconjugation and that becomes clear when considering the homology between conjugation and homoconjugation. Among the many molecules that show conjugation, there is an interesting subgroup of molecules with cyclic conjugation involving $4q + 2$ π-electrons. This subgroup of molecules is aromatic, which means that they are stabilized by cyclic electron delocalization to an extent that can no longer be explained by the conjugative effects typical of polyenes or normal cyclopolyenes. As indicated in Figure 1, aromaticity is the chemically interesting kernel of conjugation, and therefore more than 40 years of chemical research have been devoted to aromatic molecules as is amply documented in many textbooks and review articles[4–10].

In the same way as aromaticity is the interesting core of conjugation, homoaromaticity is the interesting kernel of homoconjugation (shaded area in Figure 1). One can distinguish between cyclopropyl homoaromaticity (region 1a, Figure 1) and no bond homoaromaticity (region 2a). Since chemical research always focuses more on the exceptional cases of chemical behaviour, the actual topic of a review on cyclopropyl homoconjugation has to include cyclopropyl homoaromaticity, which of course is inseparably linked to no bond homoaromaticity. Homoaromaticity is also not limited to $X = CH_2$, and therefore one has to distinguish between bond homoaromaticity in general (regions 1a and 3a) and no bond homoaromaticity (regions 2a and 4a) in general.

Cyclic electron delocalization does not always lead to stabilization. If $4q$ π-electrons are involved it can lead to destabilization and antiaromaticity. Therefore, an integral part of the concept of aromaticity is the concept of antiaromaticity[4–10]. Antiaromaticity is retained to some extent if cyclic delocalization of $4q$ electrons is mediated through homoconjugative interactions. In this case, one speaks of homoantiaromaticity and, according to the classification given above, one can differentiate between cyclopropyl homoantiaromaticity or, in general, bond homoantiaromaticity and no bond homoantiaromaticity.*

Aromaticity has many facets and the question is whether homoaromaticity has a similar manifold of facets. Beside π-aromaticity, there is σ-aromaticity, radial aromaticity, three-dimensional aromaticity, spherical aromaticity and the extension of aromaticity to heteroatomic molecules (heteroaromaticity, compare with Figure 2), not to speak of the many outdated classifications in this connection[10]. The larger class of homoaromatic molecules can be related to π-aromatic compounds although some more sophisticated differentiation may be appropriate. π-Aromatic compounds may have a planar geometry (e.g. benzene, tropylium cation etc.) or a distorted geometry deviating to some extent from planarity (e.g. bridged annulenes). The majority of π-homoaromatic molecules possesses a distorted, non-planar ring geometry (e.g. homotropenylium cation, homocyclopropenium cation, etc.) while only a few (mostly controversial) examples are known with planar geometry and pure pπ, pπ overlap (see the discussion in Section II).

The term in-plane aromaticity[11] has been used for molecules such as the didehydrophenyl cation (see Section II). However, we stress that the compounds in question are homoaromatic rather than aromatic molecules, which can be directly related to σ-aromaticity[12]. Therefore, the appropriate notation should be homo-σ-aromaticity rather than in-plane aromaticity. The concept of σ-aromaticity is very controversial. One can completely avoid this term by referring to the mode of electron delocalization as was done by Cremer[13]. While π-aromaticity and homo-π-aromaticity are connected with ribbon delocalization of electrons along a conjugative cycle, molecules that have been considered to be either σ-aromatic or homo-σ-aromatic (in-plane aromatic) seem to prefer delocalization of electrons over a surface defined by the participating atoms (see the discussion in

*Some authors have used the term antihomoaromaticity. We think that this term may be misleading since it implies either a system that, despite homoaromaticity, is destabilized or the anti form of a homoaromatic system, which may be considered as the aromatic form itself.

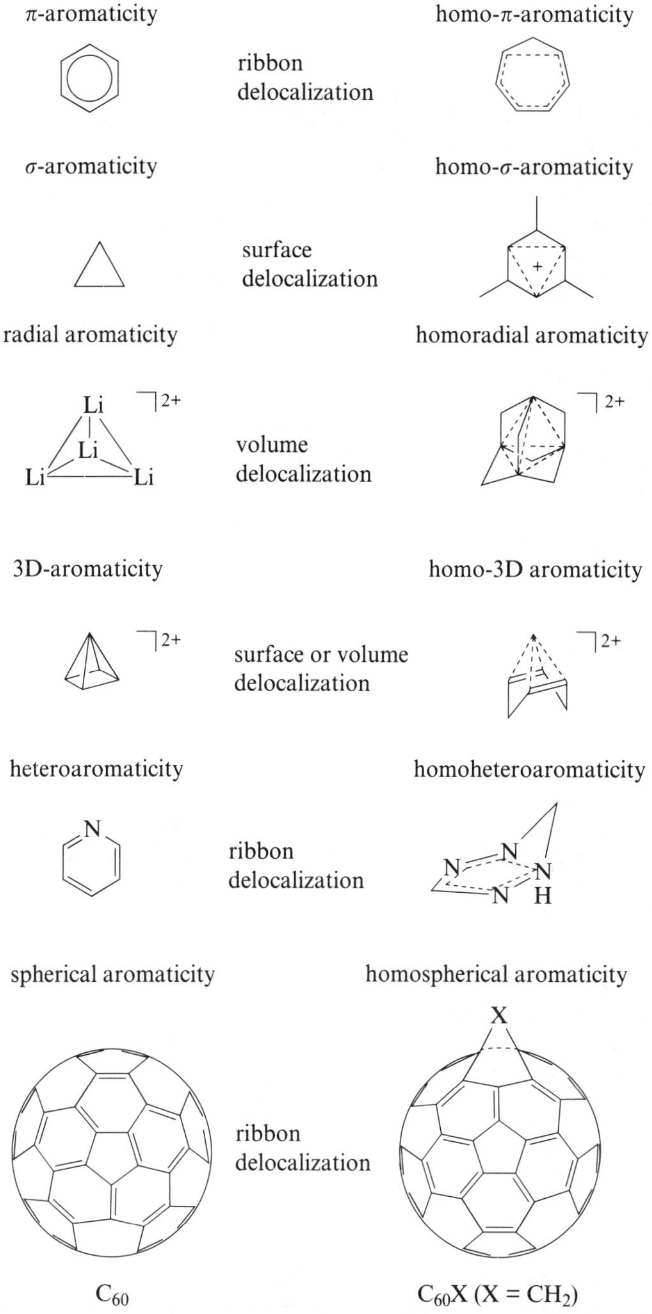

FIGURE 2. Types of aromaticity and homoaromaticity. The preferred electron delocalization mode is given in each case according to Cremer[13]

Chapter 2 of this volume[2]). Therefore, σ-aromatic and homo-σ-aromatic molecules are molecules with surface delocalization of electrons (Figure 2).

Extension of the concepts of ribbon and surface delocalization of electrons to three-dimensions leads to volume delocalization and covers cases of radial aromaticity and three-dimensional (3D) aromaticity[10]. As we will show later the most convicing example of radial aromaticity, namely the 1,3-dehydro-5,7-adamantanediyl cation (Figure 2), is actually an example of homoradial aromaticity. Also, there exist several examples of homo-3D aromaticity that are normally listed under 3D-aromaticity (for an example, see Figure 2). Finally, a number of examples have been investigated that can be classified as homoheteroaromatic systems (Figure 2). It may be only a matter of time until the first molecule with homospherical aromaticity has been synthesized and investigated.

Research on homoconjugative and homoaromatic molecules has been at the centre of organic chemistry for more than 40 years. It was initiated by the epochal investigations of Winstein and his coworkers[14,15] (for a description of Winstein's work, see the following chapter by Childs, Cremer and Elia[3]). Since then it has developed rapidly, as is amply documented in hundreds of publications and research reports. Work on homoconjugation and homoaromaticity was supported by and has influenced important developments in synthesis, kinetics, structure determination, thermochemistry and spectroscopy. The results of this work have improved our understanding of chemical bonding, electronic structure theory, structure and stability and the reactivity behaviour of molecules. Therefore, it has been reviewed in special reports[15–23] and textbooks[6–10] at regular intervals with the latest review appearing in 1994[23].

Any review on cyclopropyl homoconjugation is also a review on cyclopropyl homoaromaticity and has to consider closely related phenomena such as bond and no-bond homoconjugation (homoaromaticity) in general. Furthermore, it will automatically reflect developments, discoveries, problems and ambiguities in the field of aromaticity. While research on aromaticity concentrated in the beginning on planar closed-shell molecules with a clear separation between σ- and π-bonding, studies on homoaromatic molecules led to new dimensions in the realm of aromaticity. First, a clear separation between σ- and π-bonding was no longer possible and one had to realize that partial σ-bonding also leads to conjugation, which, for example, considerably helped an understanding of the electronic structure of bridged annulenes. Also, a formal σ-bond of a cyclopropyl group was accepted as a conjugative element, which could replace a double bond to some extent.

While these new ideas were still understandable against the background of classical bonding theory, it was difficult for a chemist of the fifties and sixties to accept that electron delocalization is not limited to following the framework of bonds but can also occur through space without the transmitting mechanism of any partially or fully developed 2-electron bond. This new mechanism of electron delocalization could only be accepted by inventing the 'homoconjugative (homoaromatic) bond' and by speaking about 'non-classical bonding' and 'non-classical structures'. These terms in a way reflected the difficulties chemists had to preserve classical bonding theory and to comprehend the new type of conjugation through space without bonding[24]. The full acceptance of the new mechanism of electron delocalization required the giving up of dogmas in classical bonding models and therefore it took its time. However, the transition from cyclopropyl homoconjugation to no-bond homoconjugation has led directly to a basic understanding and differentiation of interactions through bond and through space, to a comprehension of the forces acting in transition states, to a distinction between short-range and long-range forces and to improved knowledge about non-bonded interactions in general.

Although our knowledge about the electronic forces that act in homoconjugated molecules has increased considerably during the last 40 years, there is still considerable confusion concerning exactly how to define homoconjugation and homoaromaticity. Several attempts by distinguished reviewers have helped to classify the experimental material, but the synthesis and investigation of new unexpected examples of homoconjugated and homoaromatic molecules have maintained a constant level of confusion. Questions about homoaromaticity for neutral or anionic molecules[25,26], the range of homoaromatic interactions or the energetic consequences have not thus far been settled satisfactorily[14–23]. These questions can only be answered in a clear way for a limited number of homoconjugated systems which has led to the somewhat provocative opinion[25,26] that the concepts of homoconjugation and homoaromaticity apply just to a relatively small number of molecules and therefore their general value may be questioned. Before accepting these claims, one has to remember that both homoconjugation and homoaromaticity are models and as such their basis is simplification.

A. Models in Chemistry

Models are an important part of organic chemistry and they form the working framework within which a large body of factual material can be organized, understood and used in a predictive manner. Some models such as those of the chemical bond, the division of bonds into σ- and π-types, conjugation, molecular orbitals or electronegativity, have become so engrained that frequently it is forgotten that these are still models.

For a model to be useful and effective it should have three characteristics[27], First, it should have some physical basis. Second, it should be simple and readily understood. Lastly, it should have predictive capability. There will always be failures of simple models to account quantitatively for a particular phenomenon. In these instances it is always tempting to develop a further model as, for example, continues to be done with substituent constants, rather than using the discrepancies to provide valuable insights into the reasons for the discrepancy between the model and the system in question.

The concepts of homoconjugation and homoaromaticity build on the ideas of the chemical bond, conjugation and aromaticity. In this chapter, while refining the concepts somewhat, we seek to retain the simplicity of the original models as developed by Winstein and colleagues[14,15]. We recognize that there will be failures of the models to account for all the properties of the molecules discussed here; however, we feel that the simplicity and predictive power of these models justify their retention as part of the working framework of the organic chemist.

B. Organization of the Chapter

Contrary to previous review articles, we will present our account of cyclopropyl homoconjugation, homoconjugation in general and homoaromaticity in two parts organized in two closely connected chapters. In the current chapter, we will thoroughly discuss the theoretical basis and description of homoconjugation and homoaromaticity. In the following Chapter[3] we will review the history and development of the concept of homoconjugation and homoaromaticity. Following this, experimental and theoretical work on specific homoconjugated and homoaromatic systems will be reviewed on a selected basis where ample use is made of basic considerations and definitions worked out in this chapter.

The remainder of the current chapter is organized into four major sections. In Section II, we will focus on the definition of homoconjugation and homoaromaticity starting from a topological angle of perspective and then proceeding to chemically relevant definitions. In Section III, we will examine the theoretical basis for defining, detecting and characterizing homoconjugation and homo(anti)aromaticity. Various theoretical tools, such as orbital overlap, PMO theory, electron density analysis and energy decomposition, will be discussed to obtain useful descriptions and definitions of homoconjugation and homoaromaticity. Ample reference will be made to the theoretical description of the cyclopropyl group, and this section is closely connected with Chapter 2 of this volume in which the theory of cyclopropane and the cyclopropyl group is reviewed by Cremer, Kraka and Szabo[2].

Following Section III, there will be a section (Section IV) in which the basic requirements for an *ab initio* investigation of homoconjugated molecules are sketched. As an illustrative example, the *ab initio* investigation of the homotropenylium cation will be described in detail where special emphasis is laid on an assessment of those molecular properties that are a direct reflection of the homoaromatic character of the molecule. The section will close by deriving detailed definitions and requirements for homoaromaticity and homoantiaromaticity that are adjusted to the more recent developments in the field.

The chapter concludes with a reflective section that provides the link between this more theoretically oriented review on cyclopropyl homoconjugation and the following chapter, which will concentrate on specific examples of homoconjugation and homoaromaticity[3]. In addition, we will point out some directions for future work on cyclopropyl homoconjugation and homoconjugation in general.

II. DEFINITION OF HOMOCONJUGATION AND HOMOAROMATICITY—BASIC CONSIDERATIONS.

In terms of homoconjugation, there are two basic starting points for a particular system as illustrated in equation 1 for cyclic conjugated systems with a single interruption. It is possible to start with a closed form, **1**, and consider its conjugation or one can start with an open form, **2**, and consider through-space interactions. Homoconjugation does not require that the closed form consists of a cyclopropyl ring. However, in practice many known examples formally involve a cyclopropane or three-membered ring form as the ring-closed valence tautomer. It is this that has led the editor to include a discussion of homoconjugation in a volume on the chemistry of the cyclopropyl group. It should be stressed that in many systems to be discussed the starting point is an open form and linkage to a cyclopropane can at times seem tenuous.

In this and the following review[3], we are concerned particularly with cyclopropyl homoconjugation and not simply the conjugation of a cyclopropyl group to an unsaturated centre. This differentiation may on first sight be confusing, since most authors tend to use these terms synonymously. Clearly, one can take the view that cyclopropyl conjugation is also cyclopropyl homoconjugation. But there are fine differences between these terms, which one can use for a better understanding of homoconjugation in general. In the case of cyclopropyl conjugation, the emphasis is on the conjugative ability of the cyclopropyl group (as indicated in Scheme 1) and therefore the term is suited for the description of conjugation in substituted cyclopropanes.

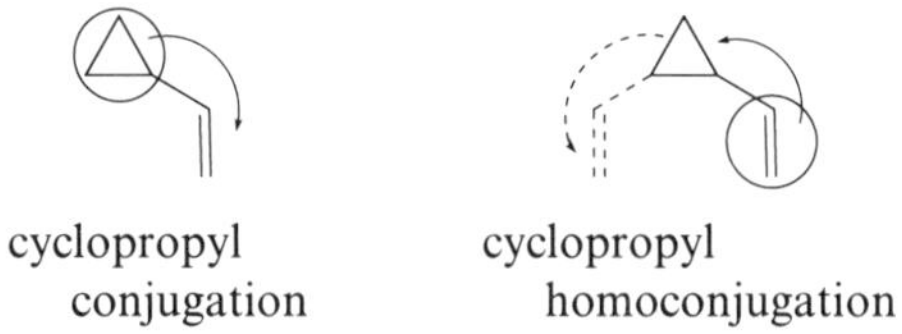

SCHEME 1. Difference between cyclopropyl conjugation and cyclopropyl homoconjugation

In the case of cyclopropyl homoconjugation, the emphasis is more on the mode of conjugation (i.e. homoconjugation) and therefore this term is better suited for homoconjugated systems including a cyclopropyl ring (Scheme 1). Cyclopropyl conjugation can be considered as a normal phenomenon similar to conjugation in polyenes. Cyclopropyl homoconjugation implies some changes in the electronic structure of a molecule over and above what one knows about molecules with cyclopropyl conjugation. The distinction is important, since it reduces the huge number of molecules with a cyclopropyl group in conjugation with some unsaturated substituent to a relatively small number of exceptional molecules with cyclopropyl homoconjugation. The focus of our attention is on delocalization of electrons through space or through a cyclopropane bond (equation 2), thus leading to cyclopropyl groups with unusual bonding. Cyclopropyl conjugation does not necessarily embrace this 'transmission' aspect of homoconjugation and its main focus is normally on the effect of a cyclopropane as a substituent.

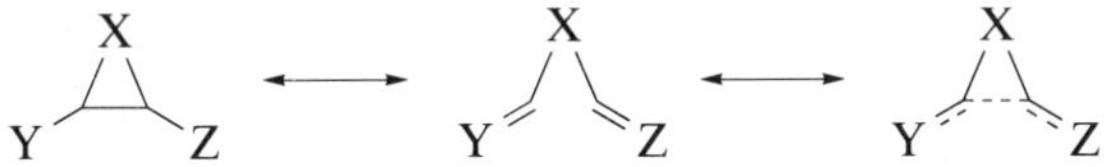

(2)

Homoconjugation can be a linear phenomenon; that is, one can be concerned with conjugation and electron delocalization through space between the ends of two unsaturated fragments. The special and most important case is where the unsaturated fragment or fragments are combined in a cyclic system such that a through-space interaction potentially leads to a cyclically delocalized system **3** (equation 3), which can be stabilized by homoaromaticity. Structure **3** has to be distinguished from **1** and **2** (equation 1) since the latter have different bonding patterns and geometries. Structure **3** can be considered as a bond–no bond resonance hybrid of resonance structures **3a** and **3b** which, of course, possess the same geometry as **3** but are different from **1** and **2**.

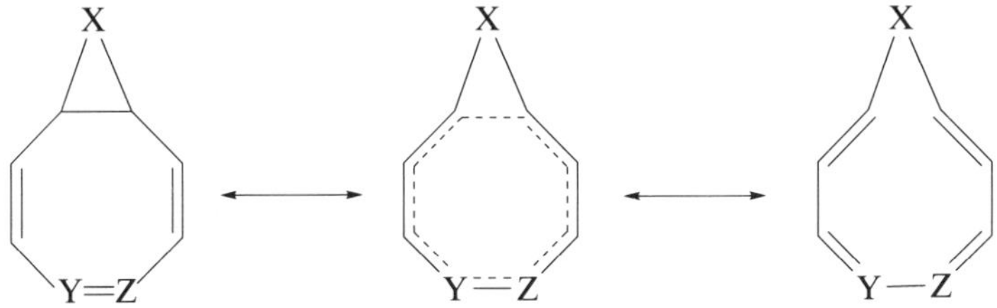

(**3a**) (**3**) (**3b**)

(3)

A. From Conjugation to Cyclopropyl Homoconjugation

Since the terms homoconjugation, cyclopropyl-(homo)conjugation, homoaromaticity, etc. are used by different authors in different ways, a review on cyclopropyl homoconjugation requires some clarification of these basic terms.

1. Conjugation

Originally, the term conjugation was used in a topological sense indicating a particular arrangement of bonds[13]. For example, double bond conjugation implies that each pair of double bonds in a conjugated system be separated by only one single bond. Such a bond arrangement leads to significant interactions of the π-MOs of the double bonds and, as a consequence, to delocalized π-MOs. The term conjugation was extended to orbital language where it describes particular orbital interactions (π-conjugation, σ-conjugation) given by the topology of the molecule. Conjugation implies an alternation between stronger and weaker orbital interactions leading to a corresponding alternation of the resonance integrals β. This is illustrated in Scheme 2 for π- and σ-conjugated systems. (We note that σ-conjugation is superfluous on a topological basis because, topologically, conjugation requires two different bond types.) In the former case, the resonance integrals β^{I} (larger) and β^{II} (smaller) each describe interatomic interactions. In σ-conjugated systems the stronger interactions (β^{I}) are intraatomic and the weaker (β^{II}) interatomic[12].

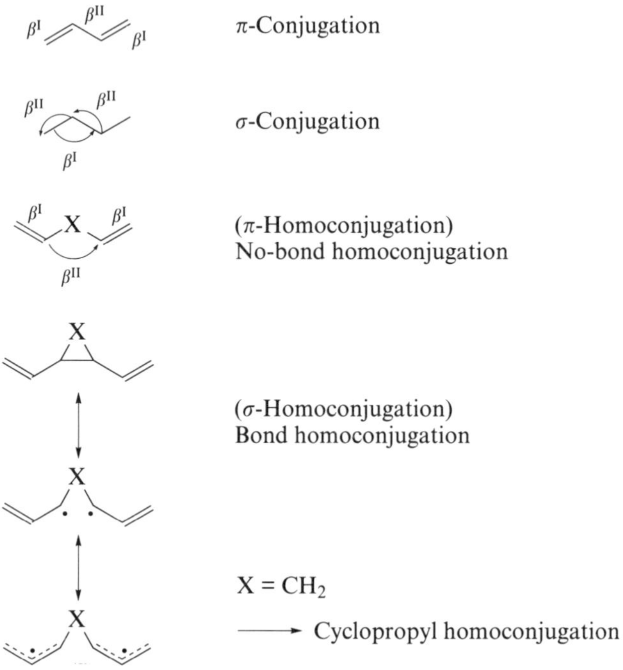

SCHEME 2. Conjugation versus homoconjugation

2. Homoconjugation

Since conjugation was originally based on a topological definition, one should also initially define homoconjugation in a similar manner. Thus when double bond conjugation is interrupted by a saturated group X (e.g. CH_2, Scheme 2), then conjugation can be restored to some extent by through-space interactions between the double bonds and their associated π-orbitals that are separated by the group X. In this way, a single conjugated system is re-established. The interaction bridging the group X was called homoconjugation because it leads to a system that is iso-conjugate with the unperturbed conjugated π-system. Thus, homologation of the unperturbed conjugated π-system by insertion of a saturated group X leads to the homoconjugated π-system.

Cyclopropyl homoconjugation is a special case of homoconjugation. It will occur if $X = CH_2$ and there are sufficiently strong interactions between the double bonds such that a bond is formed.

In order to clarify the role of cyclopropyl homoconjugation within the concept of homoconjugation, it is appropriate to classify the various types of homoconjugative interactions according to the following seven critera:

(a) the number of interruptions in the conjugated chain,

(b) the nature of the orbital interactions,
(c) the nature of the saturated group X,
(d) the type of orbital overlap,
(e) the nature and type of the interacting groups in the conjugated systems,
(f) the charge and multiplicity of the molecule and
(g) the state of the molecule in which homoconjugative interactions become important.

These criteria are discussed in separate sections below.

a. The number of interruptions. In principle, any number of interruptions is possible given a long enough conjugated chain. The number of interruptions is specified by the prefix mono-, bis-, etc. as shown in Scheme 3. For example, homologation of the tropylium cation by insertion of a CH_2 group leads to the monohomotropenylium ion (Scheme 4). Introduction of a second CH_2 group will formally yield either 1,2-, 1,3- or 1,4-bishomotropenylium ion. Formally, there is a possibility of inserting a third CH_2 group, which will lead to a 1,3,5-trishomotropenylium ion (Scheme 4). The cations shown in Scheme 4 are clearly related and the nomenclature used above stresses this point. We will use this type of nomenclature throughout this and the following review[3].

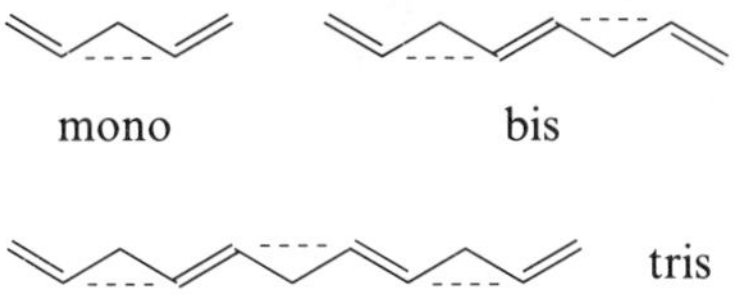

SCHEME 3. Mono-, bis- and tris-homoconjugated molecules

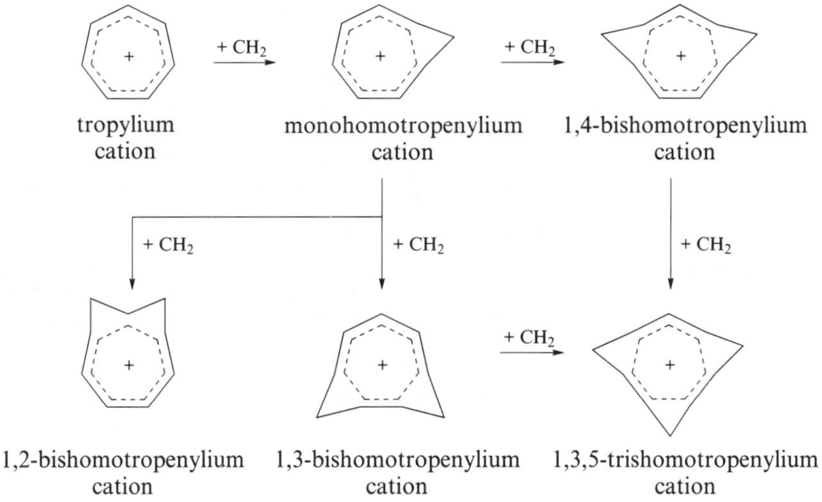

SCHEME 4. Mono-, bis- and trishomoconjugation of the tropenylium cation

b. Nature of orbital interactions. Homoconjugative interactions range from weak through-space interactions to normal bonding interactions. Cyclopropyl homoconjugation implies that a cyclopropyl group has been formed, i.e. that the interacting centres are connected by a bond. Some authors have described this situation by the term 'σ-homoconjugation' to indicate that the conjugative chain is formally closed by a σ-bond (see Scheme 2). We think that such a term is not appropriate since it disguises the fact that it is actually the partial π-character of the connecting cyclopropyl bond[1,2] that leads to conjugation and also incorrectly implies the existence of π-homoconjugation as being the counterpart of cyclopropyl homoconjugation (see Section A.4 below).

It is more appropriate to couple the term 'cyclopropyl homoconjugation' with bond homoconjugation as the conjugative interactions are mediated by a cyclopropyl bond. The term 'bond homoconjugation' is more general than cyclopropyl homoconjugation since it covers all cases in which a homoconjugated system is formed via a bond, irrespective of whether this bond is part of a cyclopropyl, cyclobutyl or any other ring (compare with Figure 1).

It should be stressed that homoconjugation in general does not necessarily require the existence of a bond. Conjugative interactions can be mediated through space by appropriate overlap between the orbitals involved. Through-space interactions between orbitals have been amply documented in the literature in connection with either homoconjugation or other phenomena[28]. They are clearly a function of the distance R between the interacting centres. Indeed, it is convenient to classify the type of interactions as a function of R (Figure 3). For large values of R, overlap and through-space interactions between the

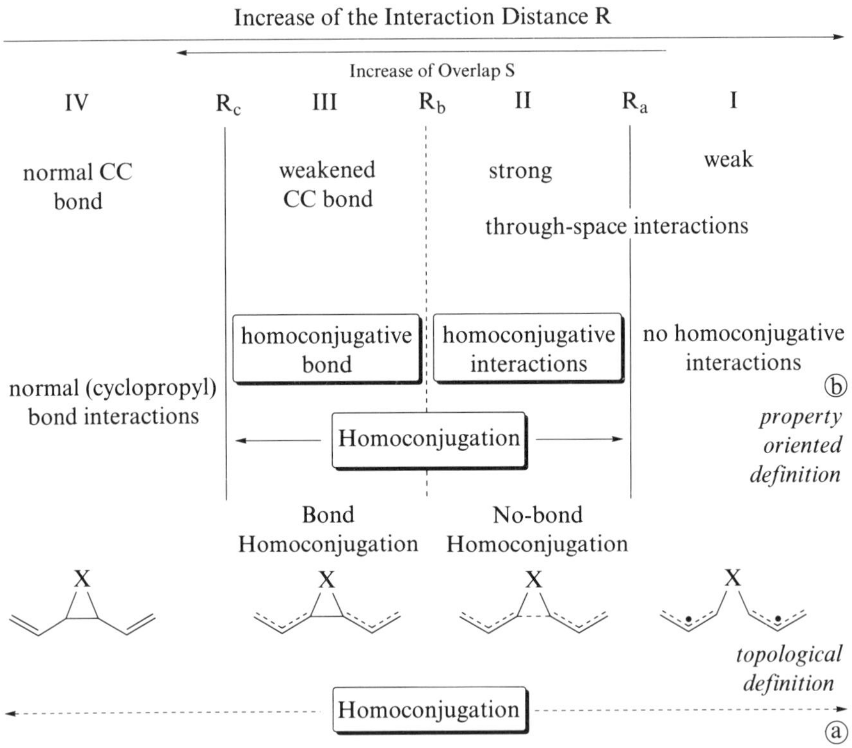

FIGURE 3. Topological (a) and property oriented (b) definition of homoconjugation based on the value R of the distance between interacting atoms in a potentially homoconjugated system. Types of interactions for increasing interaction distance (decreasing overlap S) are given for a hydrocarbon

orbitals involved are weak and probably have little consequence for the properties of a molecule. Although one can speak of homoconjugation in a topological sense, this is not very useful from a chemical point of view as the molecule would not chemically be significantly different from other related molecules with two isolated conjugated systems.

With a decrease of R, a point R_a will be reached at which through-space interactions are sufficiently strong to become chemically significant with the formation of a single homoconjugative system. In this case, we have 'no-bond homoconjugation' (Figure 3).

A further decrease of R will lead to the piont R_b (Figure 3) at which a weak bond is established between the interacting centres. At this point no bond homoconjugation turns into bond homoconjugation or, for the special case X = CH_2, into cyclopropyl homoconjugation.

Further reduction in R leads to the point R_c (Figure 3) at which the bond is fully formed and normal cyclopropyl substituent interactions occur. Homoconjugation ceases to be a relevant chemical factor at point R_c and the molecule can be adequately described in terms of a cyclopropyl substituted system.

It is also interesting to note what formally happens when R is further shortened from the point R_c. In the case of cyclopropyl conjugation, significant shortening between the interacting centres will lead to a conversion of the cyclopropyl ring into an ethylene–methylene complex (see Chapter 2)[2]. Conjugation is established at the cost of losing a methylene group from the system.

As shown in Figure 3, a decrease of the interaction distance R leads to a continuous change from weak through-space interactions to no-bond homoconjugation, bond homoconjugation (cyclopropyl homoconjugation), weak bond interactions between cyclopropyl ring and substituent and, finally, normal conjugation. It is not likely that there is any molecule for which such a transition can be monitored by experimental means. Chemical relevance is only achieved in those situations where a global (local) energy minimum exists on the potential energy surface (PES) with a value of R between R_a and R_c that is deep enough to be detected experimentally (see Section II.D).

c. Nature of the saturated group X. For cyclopropyl homoconjugation, X is equal to CH_2. This is the most common case of homoconjugation and the reason homoconjugation is discussed in a volume on the cyclopropyl group rather than in other volumes of *The Chemistry of the Functional Group* series. However, in principle, homoconjugation is also possible for X = CH_2CH_2 or any other group. Bond homoconjugation would then lead to a cyclobutyl ring (cyclobutyl homoconjugation), a cycloalkyl ring or some other ring. Despite speculation on potential homoconjugation involving cyclobutane or higher rings, no experimental evidence is available to indicate the presence of any conjugative interaction between the two double bonds of **4**, **5** or their derivatives. The absence of homoconjugation in these higher systems points to the special electronic properties of the cyclopropyl ring (see Chapter 1 of this volume[2]). However, it is too early to completely exclude the possibility of cyclobutyl or other types of homoconjugation as their potential depends more on the type of orbital interactions involved (see Section 2.d below) and the steric situation of the molecule than on the nature of X.

(4)

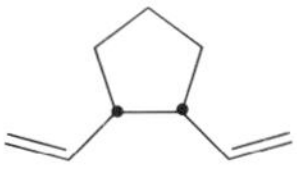

(5)

Considerable homoconjugative interactions could also be retained if X in a homoconjugative system is a heteroatomic group such as NH, O, SiH_2, PH, S, etc. Indeed, it is possible to encounter homoconjugative interactions with almost any heteroatomic group

X provided steric factors are suitable to allow bonding or through-space overlap between the interacting orbitals. The critical factors for homoconjugative interactions are orbital overlap and the difference between the energies of the interacting orbitals. Orbital overlap depends on the geometry (distance R, Figure 3) of a molecule, which in turn is a consequence of topological and steric factors such as bridges, rings, tetrahedral and pyramidal centres, etc. Orbital energies depend on the nature of the conjugated sub-chains and also the nature of the group X. In principle, any bridging group X is possible in terms of homoconjugation as long as orbital overlap and similarity of the energies of the interacting orbitals are guaranteed. In order to clarify further these two requirements we must consider the types of orbitals which can be involved in homoconjugative interactions.

d. Type of orbital overlap. One could classify homoconjugative interactions according to the type of the interacting orbitals. However, this would complicate the description of homoconjugation because there are both a large variety of potential interacting orbitals and, moreover, it is not always clear how to describe the orbitals in question. It is far easier to follow the original suggestion of Winstein[14] and to classify the type of orbital overlap rather than the interacting orbitals themselves. For both bond homoconjugation and no-bond homoconjugation, one can distinguish three different possibilities of orbital overlap. These are π,π-, $\sigma/\pi,\sigma/\pi$- and σ,σ-types of overlap. Examples for the various types of overlap are shown in Figure 4.

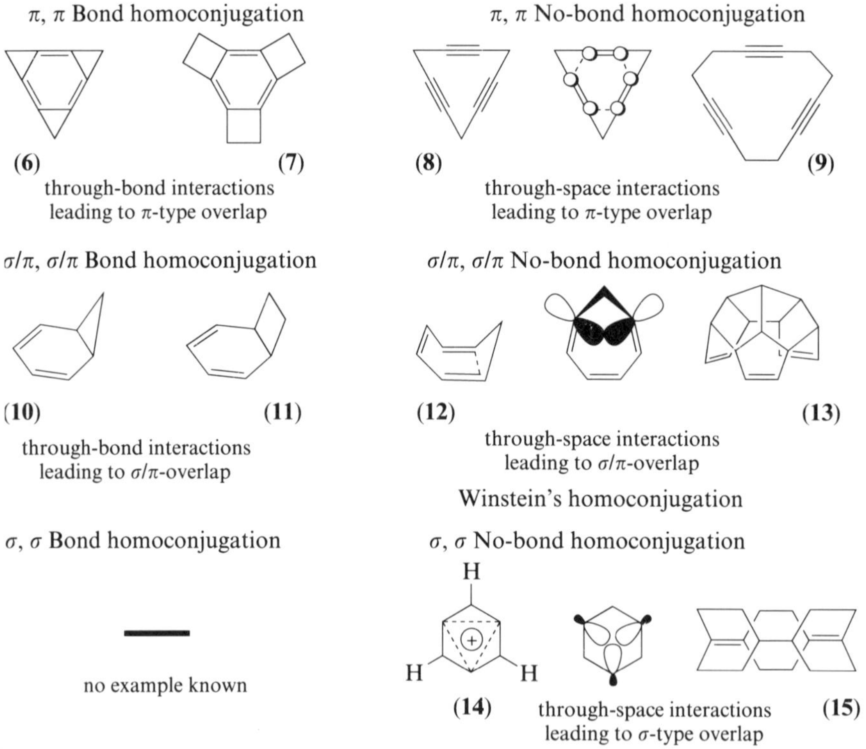

FIGURE 4. Possible types of bond and no-bond homoconjugation classified according to orbital overlap between the interacting centres

Pure π-type overlap will occur, for example, with tricyclopropabenzene or tricyclobutabenzene. There has been a long-lasting debate as to whether these compounds show a Mills–Nixon effect (alternation of bond lengths enforced by cyclopropane and cyclobutane annelation). However, all existing evidence suggests that the benzene ring with its electron delocalization is fully retained and that ring annelation does not 'freeze out' one of the resonance structures of benzene[29–31]. It may be misleading, therefore, to consider tricyclopropabenzene as a trishomoconjugated system from a chemical point of view. However, since at this point we are only considering homoconjugation from a topological perspective, it is possible to consider tricyclopropabenzene and related systems as potential examples of trishomoconjugated systems with π,π-type overlap.

Calculations suggest that tricyclopropabenzene, **6**, and tricyclobutabenzene, **7**, suffer from large strain energies and, as a result, their valence tautomers, [3]pericyclyne (**8**) and 1,5,9-cyclododecatriyne (**9**), are more stable[32]. Both **8** and **9** are examples of potential π,π-type no-bond homoconjugation. Since the overlap between out-of-plane π-orbitals decreases more rapidly than that between in-plane π-orbitals with increase in the distance R, interactions between the former (leading to π-type overlap) are much smaller than those between the latter (leading to σ-type overlap). As a result, **8** and **9** may be better examples of σ,σ-rather than π,π-type no-bond homoconjugation.

Overlap and through-space interactions are increased as soon as pπ-orbitals are tilted toward each other, thus mixing with σ-orbitals and attaining partial σ-character. This is the situation which Winstein described when he spoke of an overlap situation, which is between σ and π[14]. A $\sigma/\pi,\sigma/\pi$-type overlap can occur for both bond and no-bond homoconjugation as shown in Figure 4. Formal examples of the first case are norcaradiene, **10**, or bicyclo[4.2.0]octadiene, **11**, while the classical example of the second case is cycloheptatriene, **12**. In trienes such as *cis*, *cis*, *cis*-1,4,7-cyclononatriene, triquinacene or C_{16}-hexaquinanacene, **13**, the overlap changes gradually from $\sigma/\pi,\sigma/\pi$-type to σ,σ-type, involving in the latter case the in-plane π-orbitals. C_{16}-hexaquinanacene is a potential example of a σ,σ-type of homoconjugative overlap.

There are several potential examples of σ,σ-type no-bond homoconjugation in the literature, of which the 3,5-dehydrophenyl cation, **14**, is probably best known (see also Figure 2)[33]. If the sp^2 hybrid orbitals at position 1, 3 and 5 point toward the centre of the ring, they can overlap and form a homoconjugative 2-electron–3-centre system. In plane overlap between π orbitals, σ-type overlap can also be expected for tetracyclo-[8.2.2.2^{2,5}.2^{6,9}]-1,5,9-octadecatriene, **15**, in which three double bonds are kept face to face by frames made out of cyclohexane rings[34]. Finally, [3]pericyclyne and 1,5,9-cyclododecatriyne could be considered to be better examples for σ,σ-type rather than π,π-type overlap as mentioned above.

Bond homoconjugation via σ,σ-type overlap is identical with σ-conjugation, and if the latter occurs it cannot be distinguished from the former (Scheme 5). As such it is not reasonable to use the term σ,σ-bond homoconjugation.

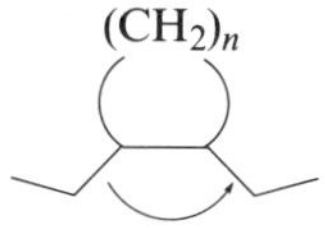

SCHEME 5. σ-Conjugation

e. Nature and type of interacting groups. Usually these are conjugated π-systems with one or more C atoms. The simplest such system would be a carbinyl group, —CH_2^+, joined to a cyclopropyl ring thus leading to the cyclopropylcarbinyl cation, **16**. Homoconjugation involving the CH_2^+ group and the vicinal three-membered ring bonds leads to a homologue

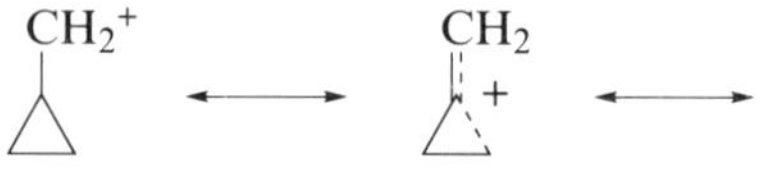

(16)

of the allyl cation. Similarly, vinylcyclopropane is a homologue of butadiene and divinylcyclopropane a homologue of hexatriene. In principle, any charged or uncharged polyenyl group can function as a sub-group in potentially homoconjugated systems.

Little is known about the extent to which heteroatoms can be incorporated into the conjugated chain of a homoconjugative system. In principle it should be possible to include heteroatoms with lone-pair electrons that can contribute their n-type electrons. Alternatively, replacement of the CH_2^+ group by BH_2 or other groups with empty pπ-orbitals should also lead to a retention of homoconjugation (see the next chapter[3]).

Recent investigations by Szabo and Cremer[35] have shown that Si can also participate in homoconjugation and homoaromaticity. This may also be true for atoms such as Ge or Sn. There are many more possibilities of homoconjugative interactions than have been considered thus far in the literature.

f. Charge and multiplicity of the molecule. The total charge of a molecule affects the energies of the interacting orbitals in a homoconjugative system. For example, in the case of the cyclopropylcarbinyl cation, the positive charge at the CH_2 group lowers the energy of the empty pπ-orbital. Interactions with the Walsh orbitals of the cyclopropyl ring are facilitated since they depend on the energy difference between the interacting orbitals (see Chapter 2 of this volume[2]). Charge transfer from the ring to the empty pπ orbital leads to electron delocalization and homoconjugation.

In general, one can expect that positively charged systems are better suited for homoconjugation than neutral or negatively charged molecules. The positive charge is mostly accompanied by relatively low-lying unoccupied orbitals, which can interact with high-lying occupied orbitals. Indeed, homoconjugation and homoaromaticity are best established for cationic systems while they are still controversial for neutral and anionic systems.

Most homoconjugative molecules studied so far represent closed-shell singlet systems with multiplicity $2S + 1 = 1$. Open-shell systems with higher multiplicity are normally very labile and, as a result, can only be studied in detail if electron delocalization leads to significant stabilization of the system in question. In general, homoconjugative electron delocalization cannot guarantee high stabilization energies (see Section III) and therefore homoconjugative effects are too small to be observed in connection with open-shell systems of higher multiplicity.

Nevertheless, there exist a number of free radicals ($2S + 1 = 2$, doublet state), for which ESR measurements suggest non-classical structures as a result of homoconjugative (homoaromatic) interactions. In all cases of radical homoaromaticity considered so far, the radicals in question are actually radical cations. This again emphasizes the important role of a positive charge in connection with homoconjugation[36].

g. The state of the molecule. Homoconjugation has been exclusively detected and investigated for molecules in their ground state. Definitely, it will also play a role for molecules in their excited states, but since excitation energies are normally much larger than homoconjugative stabilization energies, homoconjugation can only be a second-order effect, which will be difficult to detect and to investigate.

The situation will be different if one considers the transition states of chemical reactions. In a bond-forming reaction, the reaction partners will already interact with each other before all bonds are formed, i.e. most of the interactions occur through-space. If the reacting molecules both possess conjugated π-systems as in the case of pericyclic reactions, homoconjugative systems can be formed and homoconjugative interactions will lead to a lowering of the energy of the transition state. Examples are the valence tautomeric rearrangements of cyclopentadiene/bicyclo[2.1.0]pentene, cycloheptatriene/norcaradiene, etc. and related reactions.

One could even go one step further and argue that the important aspect of a pericyclic reaction is the through-space interaction between atoms about to form a bond. The situation is similar to that in no-bond homoconjugative systems and therefore pericyclic transition states resemble homoconjugative systems with or (mostly) without bridging between the conjugation partners. As a matter of fact, pericyclic reactions have been amply investigated with regard to the possibility of through-space interactions and electron delocalization in their transition states. The focus of these investigations has not been homoconjugation, but the translation of the Woodward–Hoffmann orbital symmetry rules[37] into the Evans–Dewar–Zimmermann electron counting rules[38]. In this latter approach, emphasis is placed on the identification of aromatic or antiaromatic electron ensembles participating in the formation of bonds in transition states. Typical interaction distances between C atoms, that are about to form a CC bond, are 1.8 to 2.4 Å in a transition state, and therefore one could speak of homoconjugation and homoaromaticity rather than conjugation and aromaticity in an orbital symmetry-allowed pericyclic transition state.

From the seven classification criteria discussed above, it becomes clear that most homoconjugative systems studied so far belong in the class of monohomoconjugated singlet ground-state cations, in which a π-conjugated electron system is closed by σ/π, σ/π overlap. However, the classification given above also shows that many other homoconjugated systems are possible, the question being only which of these many possibilities is of chemical relevance.

B. From a Topological to a Chemical Definition of Homoconjugation

The topological definition of homoconjugation outlined above, although quite useful, does not necessarily say anything about the possible chemical consequences of such a conjugation. If homoconjugation does not lead to any changes in the properties of a molecule that are interesting enough to be investigated, then the classification of the molecules as being homoconjugated is not very useful.

1. The concept of electron or bond delocalization

When homoconjugation leads to electron or bond delocalization, and thereby to a change in the properties of a molecule, homoconjugation becomes chemically relevant. In fact, electrons are always delocalized over the space of a molecule. However, it has turned out that it is extremely useful to consider bonding, lone-pair and inner-shell electrons to be essentially 'localized' in the bond, lone-pair or core region, respectively. This assumption is the basis of the concept of electron or bond localization and reflects the fact that many properties of a molecule can be reproduced in terms of bond or atom contributions. Of course, neither bond localization nor electron localization refers to any observable molecular property. They simply suggest that most molecules behave as if their bonds were localized and that their properties can be reproduced with the help of bond increments. With the concept of bond localization a large body of experimental data on molecular properties can be rationalized, i.e. bond or electron localization is a heuristic concept[13].

Within the concept of bond or electron localization, the meaning of the term electron (de) localization is changed:

Electrons or bonds will be considered to be localized if the properties of the molecule can be explained in terms of bond contributions. If this is not the case, electrons and bonds are considered to be delocalized.

It is clear from this definition that bond (orbital) conjugation is far more common than bond (electron) delocalization. *Conjugation does not always lead to bond delocalization* and, accordingly, it is not correct to use the two terms indiscriminately[13]. It is well known that polyenes are typical examples of double bond conjugation but, as has been demonstrated by Dewar and coworkers[39], their heats of formation as well as other properties can be reproduced by appropriate bond increments. *Thus polyenes are not examples of bond or electron delocalization.*

It is also misleading to consider delocalized π or σ MOs as an indication of bond (electron) delocalization. As canonical MOs are always delocalized, one could localize the MOs and check whether they are all confined to bond regions or whether certain MOs possess long orbital tails outside the bond region. In the latter case, one could anticipate bond delocalization. However, this classification would be wrong since MOs reflect the properties of single electrons. *The concept of bond or electron delocalization is based on the collective properties of all electrons.* Hence, localized MOs with long tails could just indicate π-orbital conjugation and not bond (electron) delocalization.

2. A definition of homoconjugation based on the concept of bond (electron) delocalization

Since conjugation and homoconjugation are parallel concepts, it is logical to base a chemically relevant definition of homoconjugation on the concept of bond (electron) delocalization:

A molecule will be considered to be a homoconjugative system if

(a) *it fulfils the topological requirements of homoconjugation* (*interruption of a conjugative chain by one or more saturated groups*) *and*

(b) *if its properties cannot be explained in terms of bond or group contributions of the two separated conjugative systems.*

Homoconjugation thus involves electron and bond delocalization (in the heuristic sense) in the homoconjugative system.

This definition helps to clarify which molecular properties will reflect the homoconjugative character of a system in question. For example, many authors cite ESR hyperfine coupling constants, ionization potentials or calculated orbital energies as indicators for homoconjugation. However, these properties are properties of single electrons that just reflect properties of the orbitals (within the Koopmans approximation) such as σ- or π-conjugation that result from the topology of the molecule. For example, it is not appropriate to speak in the case of norbornadiene of no-bond homoconjugation between the two double bonds based on the fact that the photoelectron spectrum of the molecule indicates a splitting between the π-levels as a result of through-space interactions[40]. The splitting simply reflects the tendency of electrons to delocalize (in the quantum mechanical sense of the word) if spatial arrangement and overlap give them this possibility. Therefore, single electron properties such as ionization potentials can confirm homoconjugation only in the topological sense. This may be considered to be a useful confirmation of the topology and (to some extent at least) of the geometry of the molecule. However, they do not say anything about the chemical consequences of homoconjugation as these result from bond (electron) delocalization mediated by through-space interactions or a cyclopropyl bond.

For example, norbornadiene does not possess a stability that is significantly different from that of a diene without any through-space interactions. Its heat of formation can be fully reproduced by appropriate group contributions[41]. Therefore, norbornadiene is not an example for bond (electron) delocalization but rather of through-space interactions between double bonds.

Clearly, only those properties of potentially homoconjugated systems that are the collective ones of all electrons can be used to meaningfully assess the extent of bond (electron) delocalization. In principle, these can be energy, geometry, dipole moment, polarizability, NMR chemical shifts, diamagnetic susceptibility, etc. In most cases, some form of the molecular energy has been used to assess the extent of bond (electron) delocalization but other molecular properties such as bond length alternation parameters, NMR chemical shifts or diamagnetic susceptibility exaltation values have also been used. Utilizing these properties in connection with suitable reference compounds, the two distance values R_a and R_c in Figure 3 can be fixed. The distance R_a is that distance for which through-space interactions turn into chemically relevant homoconjugative interactions. Similarly, the distance R_c is that distance for which homoconjugative interactions are replaced by normal bond interactions between cyclopropyl ring and neighbouring groups.

3. The choice of appropriate reference compounds.

Since energy is the most important molecular property for a chemist, the following discussion will focus on the energies of homoconjugative systems. Comparable arguments are valid for the other properties that depend on all electrons of the molecule.

Homoconjugation can lead to a bond (electron) delocalization energy, which reveals an excess stability of the system when compared to suitable reference compounds. *The selection of appropriate reference compounds is essential for the definition of homoconjugation.*

The problem of selection of a suitable reference is of course also found for other basic concepts in chemistry such as aromaticity, strain or even the covalent bond[27]. By choosing the wrong reference compound, a concept can become so vague that it distracts one from viewing the few exceptional observations one wants to describe. This may be illustrated by considering for a moment the description of conjugated systems. If one were to use the CC bonds in ethane and ethene as reference bonds for the description of polyenes, each of these larger molecules would show an appreciable bond delocalization energy. Even the higher alkenes such as propene, butene, etc. would have a bond delocalization energy because they contain at least one $C(sp^2)$—$C(sp^3)$ single bond that is more stable than the $C(sp^3)$—$C(sp^3)$ reference bond of ethane. Strictly speaking by using this definition there would only be one normal alkene, namely ethene.

Dewar and coworkers[39] have pointed out that an adequate description of polyenes has to be based on a set of reference molecules and reference bonds that comprises not only a $C(sp^3)$—$C(sp^3)$ bond, but also $C(sp^2)$—$C(sp^3)$, $C(sp^2)$—$C(sp^2)$ reference bonds in order to separate normal cases of bond conjugation from exceptional cases which show bond delocalization. When this extended set of reference bonds is used, polyenes as well as radialenes and many cyclopolyenes are described as normal conjugated π-systems without any significant extra stabilization from bond delocalization[39].

A similar point has been made by Roth and coworkers[41,42] who investigated potentially homoconjugated systems using a modified MM2 force field, MM2ERW. The latter contained as reference bonds not only $C(sp^2)$—$C(sp^3)$, $C(sp^2)$—$C(sp^2)$, etc. bonds needed for the description of polyenes, but also the $C(sp^2)$—C(cyclopropyl) single bond taken from vinylcyclopropane in order to adequately describe cyclopropyl-substituted compounds **17–26** listed in Table 1. With this extended set of reference bonds, Roth and coworkers were able to reproduce experimental heats of formation ΔH_f^0 of polyenes and cyclopropyl-substituted molecules with an accuracy of $\pm$ 0.5 kcal mol^{-1}. In particular, calculated and experimental ΔH_f^0 values of cyclopropyl conjugated molecules such as **20**, **21**, **23**, **25** or **26** (Table 1) agree within 0.1 kcal mol^{-1}. If these molecules were to benefit from

TABLE 1. Comparison of experimental and MM2ERW heats of formation ΔH_f^0 (kcal mol^{-1}) for cyclopropyl conjugated molecules[41]

Molecule		ΔH_f^0 (exp)	ΔH_f^0 (MM2ERW)	RE[a]
17		12.7	12.7	0
18	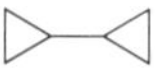	30.9	30.2	0.7
19		44.2	44.4	−0.2
20		30.4	30.4	0
21	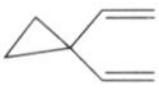	48.2	48.7	−0.5
22		9.1	9.2	−0.1
23		37.8	37.8	0
24	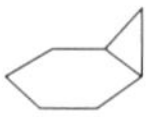	0.3	-0.0_3	0.3
25	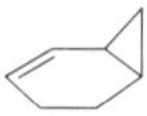	28.9	28.8	0.1
26	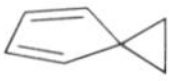	56.8	56.9	−0.1

[a]Resonance energies (RE) are all close to zero.

bond (electron) delocalization, ΔH_f^0 (MM2ERW) values (based on the additivity of bond energies) would turn out to be larger than experimental ΔH_f^0 values. The difference between experimental and calculated enthalpies would be the homoconjugative stabilization energy or resonance energy (RE). The reproducibility of experimental ΔH_f^0 values of all the compounds in Table 1 clearly indicates that the degree of cyclopropyl conjugation is similar to that of the reference molecule, vinylcyclopropane. There is no evidence of any special cyclopropyl homoconjugation.

 (4)

Staley, on the basis of the temperature dependence of the equilibrium in equation 4, has suggested that the stabilization energy resulting from a homoconjugative interaction between a cyclopropyl group and a double bond is 1.1 kcal mol^{-1} [43]. However, this stabilization energy becomes zero when vinylcyclopropane is used as the appropriate reference. This does not mean that π-conjugation vanishes. It simply means that the larger number of homoconjugated systems in a topological point of view is reduced to a smaller number of interesting cases for which homoconjugation leads to exceptional chemical behaviour. Roth and coworkers[41] have underlined this point by noting the similarity of measured heats of hydrogenation of vinylcyclopropane and isopentene (Scheme 6).

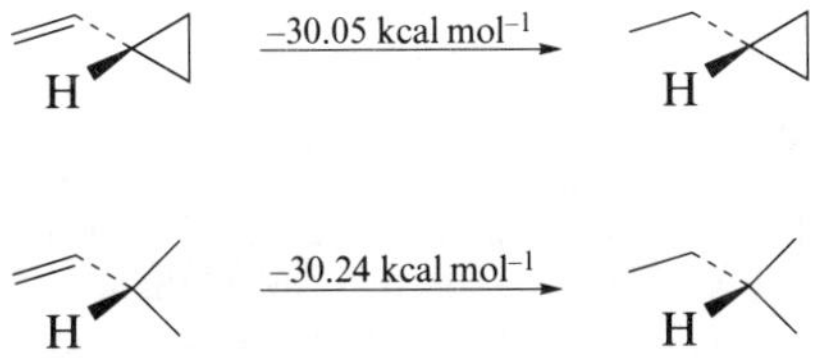

SCHEME 6. Heats of hydrogenation of vinylcyclopropane and isopentene[41]

When vinylcyclopropane is used as the key reference compound, the only examples of significant bond (electron) delocalization energies are cyclic homoconjugative systems with potentially aromatic electron ensembles (compare compounds **27**–**38** in Table 2). For example, norcaradienes **33** and **34** possess small, but significant bond delocalization energies RE of about 3 kcal mol^{-1} (Table 2). Similarly, the RE values of the cycloheptatrienes **30**, **31** and **32** are between 4 and 6 kcal mol^{-1}.

TABLE 2. Resonance energies (RE) obtained from experimental and MM2ERW calculated heats of formation ΔH_f^0 (kcal mol^{-1}) for homoconjugated molecules[41,42]

Molecule		ΔH_f^0 (exp)	ΔH_f^0 (MM2ERW)	RE
27		79.6	69.7	9.9
28		60.0	53.4	6.6
29		50.1	49.7	0.4
30		44.6	48.7	−4.1
31	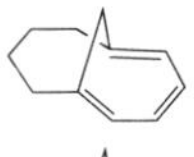	35.0	39.4	−4.4
32	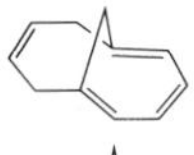	62.3	68.4	−6.1
33	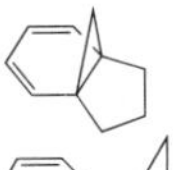	46.0	49.5	−3.5
34	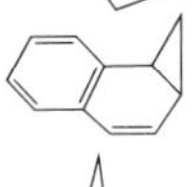	57.0	60.1	−3.1
35	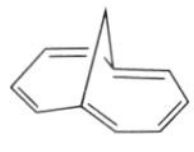	77.1	94.3	−17.2
36	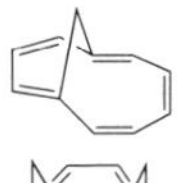	95.5	101.3	−5.8
37	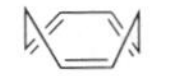	49.9	51.7	−1.8
38		45.2	44.4	0.8

As the topological definition of homoconjugation is combined with the heuristic concept of bond (electron) delocalization, the concept becomes chemically more relevant since it narrows down the number of possible homoconjugative cases. One result of this more rigorous definition is that homoconjugation without the possibility of cyclic electron delocalization ceases to be an interesting phenomenon. There are hardly any examples of significant homoconjugative bond (electron) delocalization in non-homoaromatic compounds (see the following chapter[3]). In view of this it is not surprising that the terms homoconjugation and homoaromaticity are often incorrectly used as synonyms. Such a usage is inaccurate since homoconjugation is a more general term than homoaromaticity in the same way as conjugation is more general than aromatic conjugation (aromaticity; see Figure 1, Section I).

C. Homoaromaticity and Homoantiaromaticity

Predictions as to the chemical relevance of homoconjugation in the sense of aromatic delocalization of electrons can be made if the topological concept of homoconjugation is connected with an electron count. An electron count can help in suggesting whether a homoconjugative interaction could lead to a change in the stability of the molecule. This is particularly useful if the molecule in question represents a cyclic system that by homoconjugative interactions can form an aromatic or antiaromatic ring system. This is indicated in Figure 5. Homoconjugation connected with an electron count leads to the prediction of potential homoaromaticity, which may or may not be verified by experiment.

Homoaromaticity can be characterized in the same way as homoconjugative interactions by:

(1) the number of interruptions in the conjugated chain,
(2) the nature of the orbital interactions,

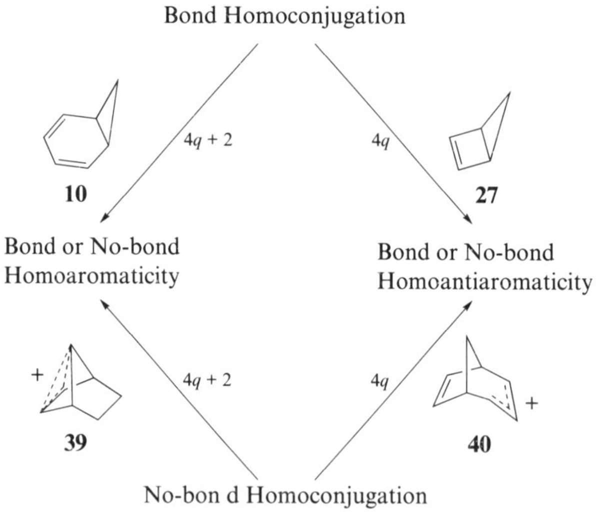

FIGURE 5. From bond and no-bond homoconjugation to bond and no-bond homoaromaticity or homoantiaromaticity. In each case the number of cyclically delocalized electrons ($4q$ or $4q + 2$) is given

(3) the nature of the saturated group X,
(4) the type of orbital overlap,
(5) the nature and type of the interacting groups in the conjugated systems,
(6) the charge and multiplicity of the molecule and
(7) the state of the molecule in which homoconjugative interactions are important.

Hence one can speak of mono-, bis-, tris-homoaromaticity, etc. (1), of bond and no-bond homoaromaticity (2), of cyclopropyl or cyclobutyl homoaromaticity (3), homoaromaticity mediated by π,π-, $\sigma/\pi,\sigma/\pi$- and σ,σ-type overlap (4), homoaromaticity involving heteroatoms, triple bonds etc. (5), neutral, cationic or anionic homoaromaticity as well as radical homoaromaticity (6), homoaromaticity in ground, excited or transition states (7).

In addition, one can classify homoaromaticity with regard to the number of electrons involved, i.e. 2, 6, 10 or, in general, $4q + 2$ electrons.

Apart from this, it would be appropriate, although never done in practice, to define homoaromaticity with regard to a molecular property (such as energy, geometry, chemical shifts, etc.) in comparison to the reference(s) used. Various molecular properties reflect the special electronic features of homoaromatic systems with different sensitivity. Thus, for example, it is possible that NMR chemical shifts could suggest weak bond (electron) delocalization while an analysis of the molecular energy does not provide any indication of homoaromatic character.

In the case of homoantiaromaticity, characterization is more difficult. Antiaromaticity describes a situation in which electron delocalization leads to destabilization. Clearly, if through-space interactions would close a cyclic system to form an antiaromatic electron ensemble, the molecule would adopt another conformation that would help to avoid antiaromatic electron delocalization. Of course, steric factors may enforce through-space interactions as in **40** (Figure 5). However, simple deformations of the molecule can reduce through-space interactions to an insignificant level.

Formally, no-bond homoantiaromaticity cannot be ruled out. However, *de facto* it will not play any major role in determining the chemistry of $4q$-electron systems.

Similarly, bond homoantiaromaticity may be of little importance. For example, in the case of **27** bond homoconjugation would lead to bond (cyclopropyl) homoantiaromaticity (enforced delocalization of 4 π-electrons). However, it is also possible that homoconjugation could involve the peripheral C—C bonds of the cyclopropyl ring and in this way avoid the formation of an antiaromatic π-electron ensemble and instead form a peripheral aromatic electron ensemble. Cremer and coworkers[27,44], following earlier suggestions by Childs, Winstein and coworkers[45], pointed out this possibility in their investigation of the geometry and electron density distribution of various potentially homoantiaromatic molecules. Their observation is in line with the electronic structure of potentially antiaromatic π-electron systems such as bicyclo[6.2.0]decapentaene, **41**, which avoids the cyclobutadiene structure, **41a**, and instead exists as the peripheral 10-π-electron system **41b**[46].

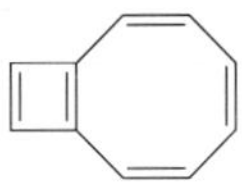

(41a)

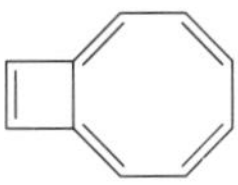

(41b)

There is, however, an important difference between examples **27** and **41**. The later compound forms a Hückel-aromatic orbital system in **41b** while the former compound adopts a Möbius orbital system with $4q + 2$ electrons, i.e. **27** is Möbius antiaromatic although six electrons participate in cyclic delocalization (see Section III. B). This is in line with a destabilizing resonance energy of 9.9 kcal mol^{-1} (Table 2) calculated with the MM2ERW method[41,42].

We conclude that each case of potential bond or, more specifically, cyclopropyl homoantiaromaticity has to be considered separately. Detailed investigations have to clarify whether homoantiaromaticity is of any chemical relevance or whether the molecule has reorganized into a non-homoaromatic electronic structure.

D. Homoconjugation and the Topology of the Potential Energy Surface: From Homoaromaticity to Frozen Transition States

The potentially homoaromatic system **42** can adopt three different structures: **42a**, **42b** and **42c**. The bicyclic structure **42a** corresponds to the situation of bond or cyclopropyl homoconjugation leading to the delocalization of $4q + 2$ electrons and, therefore, to bond or cyclopropyl homoaromaticity. Structure **42c** represents a monocyclic form with weak 1,3-through-space interactions that do not lead to homoconjugation and, accordingly, this structure can be considered to possess an open π-electron system. Finally, structure **42b** corresponds to a no-bond (homoconjugative) homoaromatic system characterized by strong through-space interactions and cyclic electron delocalization.

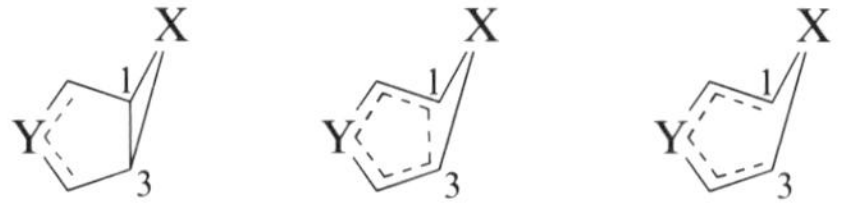

(42a) **(42b)** **(42c)**

$X = CH_2$, $Y = (CH)_n^q$ ($q = +1, 0, -1$)

In Figure 6, one-dimensional cuts through the PES in the direction of the 1,3-interaction distance R in **42** are shown. Either structures **42a**, **42b**, **42c** or all three of them can occupy stationary points on the potential energy surface (PES) and, according to the topology of the PES, different chemical situations can be distinguished.

Situation 1. A minimum exists only for the bicyclic structure **42a**. The PES may in fact be less steep in the direction of a hypothetical form **42b** because of the possibility of slightly stabilizing through-space interactions.

Situation 2. A minimum exists only for the open from **42c**. The PES may again be less steep in the direction of a hypothetical form **42b** because of the possibility of slightly stabilizing through-space interactions.

Situation 3. There is only a minimum for the non-classical form **42b**. The importance of homoaromaticity is reflected by the curvature of the PES at the minimum **42b** (steepness of the PES).

Situation 4. Two minima exist on the PES which correspond to the classic forms **42a** and **42c**. Interconversion of these forms leads to a transition state, which corresponds to the non-classical form **42b**.

There are further possibilities, namely that **42a**,**b** or **b**,**c** or **a**,**b**,**c** occupy two or three minima on the PES. Although these possibilities cannot be fully excluded, they are not likely since the characteristic interaction distances R are very similar and therefore would imply minima on the PES that are very close to each other. Small geometrical changes would lead from one minimum to the other and, since small geometrical changes normally imply small energy changes, one of the minima may be just a shallow energy well that is chemically not detectable. Hence, for all practical purposes, the possibilities 1 to 4 outlined above would seem to be the most likely.

Situation 1 (**42a**) corresponds to cyclopropyl homoconjugation (cyclopropyl homoaromaticity), the subject of this review article. Situation 2 (**42c**) may be only interesting in

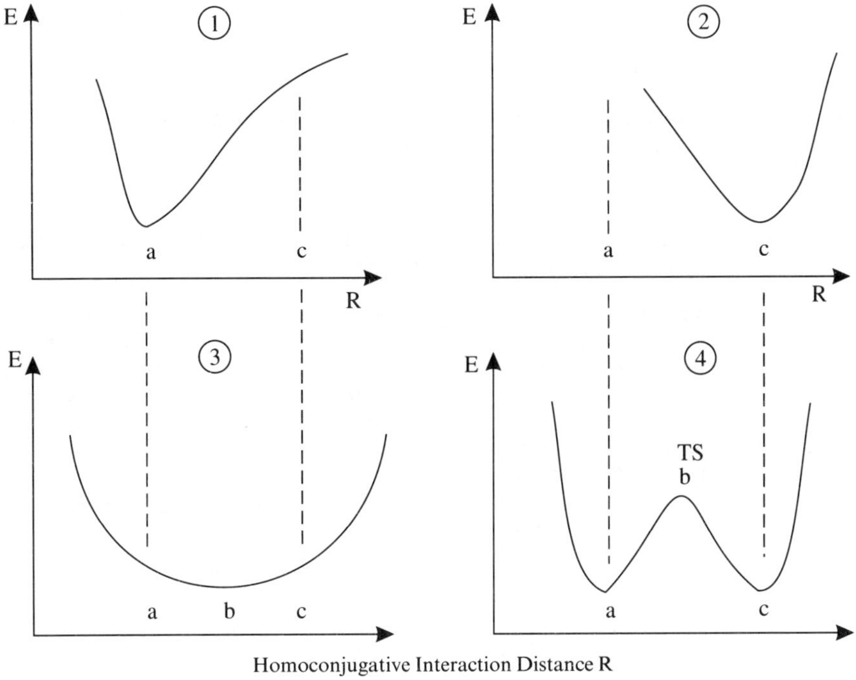

FIGURE 6. One-dimensional representations of the potential energy surface (PES) of molecule **42** shown as a function of the interaction distance *R*. Situation 1 corresponds to the bicyclic molecule **42a**, situation 2 to the open monocyclic molecule **42c**, situation 3 to the no-bond homoaromatic molecule **42b** with non-classical structure and situation 4 to a valence tautomeric equilibrium between **42a** and **42c** with the homoaromatic form **42b** being the transition state. See text

connection with weak through-space interactions while situation 3 (**42b**) represents an example of no-bond homoconjugation and no-bond homoaromaticity.

Situation 4 corresponds to a valence tautomeric rearrangement between the homoconjugative form **42a** and the open monocyclic form **42c** as already discussed in connection with equation 1. The transition state of the rearrangement may be stabilized by no-bond homoconjugation. This has been discussed in various ways using orbital symmetry and electron counting models (see Section II.A)[37,38]. Situation 4 represents an example of an aromatic transition state since, in the terminology of the Dewar–Evans–Zimmermann rules, one cannot distinguish between aromaticity and homoaromaticity (Section II.A). However, as the interaction distances in a transition state are normally outside the range of typical bond lengths, it would be most appropriate to speak in cases corresponding to situation 4 of a homoaromatic or a homoantiaromatic transition state.

It is interesting to consider further the relationship between situations 3 and 4. Situation 3 will be reached if the transition state in 4 is sufficiently stabilized so that its relative energy drops below those of the valence tautomeric forms **42a** and **42c**. In other words, situation 3 corresponds to a *frozen transition state*[47]. *A no-bond homoaromatic compound is simply the realization of a frozen TS.*

This relationship explains some of the fascination that homoaromaticity had, and still has, for chemists. Knowledge about transition states is very important for an understanding of chemical reactions, yet it is the most difficult information to obtain experimentally.

Concepts such as the 'frozen transition state'[47] or the 'frozen reaction path' (Bürgi–Dunitz reaction path)[48] were developed to obtain direct experimental information on transition states. So far, all attempts have failed to realize a frozen transition state experimentally, Cremer and coworkers[49–58] have shown that a no-bond homoaromatic compound (PES situation 3), such as the homotropenylium cation, corresponds to a frozen transition state, and therefore its investigation provides ample information about the properties of transition states.

III. THEORETICAL ASPECTS OF DEFINING, DETECTING AND DESCRIBING HOMOCONJUGATION AND HOMOAROMATICITY

Cyclopropyl homoconjugation can be easily detected and described as long as one retains its topological definition. This also holds to some extent in the case of no-bond homoconjugation. However, as soon as one has to assess the chemical relevance of homoconjugation and to determine a homoconjugative bond (electron) delocalization energy, one needs, as mentioned in Section I, suitable reference compounds for comparison.

The problem of the reference compound is inherent to most chemical concepts[27]. By definition *a suitable reference compound is a compound that possesses the same properties as the target compound with the exception of the electronic and structural features to be investigated.* In most cases, such a compound cannot be found since changes in the (electronic) structure automatically lead to changes in all properties and hinder meaningful comparison. This is the reason why many chemical concepts are discussed at a qualitative rather than a quantitative level. In fact, as has been forcefully described by Binsch, attempts to quantify a concept very often lead to the collapse of the whole concept[59]. This potential collapse-by-quantification problem exists for the concepts of homoconjugation and homoaromaticity just as it does for the concept of aromaticity.

Homoaromaticity as a chemical concept is based on the concepts of aromaticity and homoconjugation. In its simplest form aromaticity is an electron counting concept. Thus if there are $4q + 2$ π-electrons in a planar (or nearly planar) cyclic system, then this is considered to be aromatic*.

At the simplest level (Section I), homoconjugation can be based on a topological footing. At more sophisticated levels, aromaticity as well as homoconjugation are treated as orbital concepts. This means that in the case of homoaromaticity one has to check whether the structure (geometry) and topology of the molecule in question allow through-space overlap to close a cyclic π-system and whether available π-electrons occupy bonding rather than antibonding π-orbitals. If this is the case, a homoaromatic bond (electron) delocalization energy (resonance energy) can result that should reflect the stability of the compound in question. Alternatively, a bond equalization index, diamagnetic susceptibility exaltation or some other property could define the homoaromatic character of the compound. In this way, homoaromaticity is generally easier to describe than homoconjugation as the latter does not necessarily lead to a significant bond (electron) delocalization energy (see above).

The experimental assessment of homoaromaticity is often based on working definitions of homoaromaticity that are influenced by the context of the experimental measurements and available reference data. Such a working definition may be incompatible with other definitions, be limited to a small set of related compounds and frequently be rather vague with regard to a general understanding of homoaromaticity. However, it can be useful to

*This prediction is only correct when the π-electrons fill an aromatic subshell of MOs. As soon as antibonding MOs are filled, aromatic stabilization is no longer guaranteed. Thus, trioxacyclopropane, the cyclic isomer of ozone, possesses 6 π-electrons, but 4 of them occupy antibonding MOs and, therefore, the molecule is relatively unstable.

apply these working definitions of homoaromaticity when research is focused on a limited series of compounds. Difficulties arise as soon as one tries to translate experimental assessments of homoaromatic character so that they comply with a more general quantitative assessment of the phenomenon. Conclusions about homoaromatic character based on some isolated observations can cease to be pertinent within a general concept of homoaromaticity.

In the following, rather than discussing the many definitions of homoconjugation or homoaromaticity that have been expressed by various experimentalists, we compare the few theoretically based attempts to derive a more general definition of homoaromaticity.

A. Winstein's Definition of Homoaromaticity

A first basis for the definition of the term homoaromaticity emerges out of Winstein's impressive work[14,15]. Winstein was careful not to restrict homoaromaticity to just those cases where the carbon framework of an aromatic system is interrupted by a single atom bridge but rather generalized the situation to include other possible bridges such as the —CH_2CH_2— group[14,15]. According to Winstein, the key issue is the presence of an appropriate geometry for orbital overlap through-space rather than the number and type of intervening atoms. As for the nature of the interactions through-space, Winstein speaks of 'electron delocalization across intervening carbon atoms'[14,15] when he defines homoconjugation and homoaromaticity in general terms, thus avoiding any specification of the electronic forces leading to homoconjugative interactions. At other places, e.g. when he discusses the norbornenyl cation and related compounds, Winstein speaks of 2-electron 3-centre bonding and the existence of 'partial (homoaromatic) bonds of bond order between 0 and 1'[14,15]. Formulations such as the latter have led various authors to state that a basic requirement of Winstein's definition of homoaromaticity is the existence of 'a 1,3 bond closing the cyclic conjugation'[44].

However, in order to set Winstein's assessment of a homoaromatic bond into the correct context, one has to consider the understanding of chemical bonding at Winstein's time, which was predominantly based on Hückel MO (HMO) theory. Within HMO theory, the definition of a chemical bond is vauge[60]. A chemical bond between neighbouring atoms is imposed by setting a resonance integral for this particular atom,atom interaction to a preselected value. Solution of the HMO equations leads to a matrix of atom,atom interaction indices which are called 'bond orders' irrespective of whether or not they correspond to an interaction for which a resonance integral has been set. In case of no-bond homoconjugative interactions, small 'bond orders' ≥ 0 were calculated which were considered as an indication of a covalent bond. As a result, even for relatively large homoconjugative interaction distances weak 'homoaromatic bonds' were predicted by HMO theory.

In view of the vague knowledge of bonding in the sixties, we consider it more appropriate to stress Winstein's general understanding of homoaromaticity, which covered both bond and no-bond homoaromaticity[14]. Winstein's requirements for homoaromaticity can be listed as follows:

The potentially aromatic system **43** *with* $(4q+2)$ *π-electrons will be homoaromatic if*:

1. *the system is closed by electron delocalization across the homoconjugatively connected atoms,*
2. *the interaction or bond index of the 1,3 link is between 0 and 1,*
3. *orbital overlap of the participating p-AOs at centres 1 and 3 is neither σ nor π but intermediate between these two (Scheme 7) and*
4. *the* $(4q+2)$ *π-electrons are fully delocalized over the resulting closed cycle, thus leading to net stabilization.*

(4q + 2) π-electrons

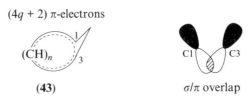

(43) σ/π overlap

SCHEME 7. σ/π-Overlap in the homoaromatic hydrocarbon **43**

Setting out the requirements for homoaromaticity in this manner, it should be easy to distinguish homoaromatic from non-homoaromatic molecules. Clearly, an appropriate geometry or structure of the species in question is required. This pertains not only to the appropriate placement of the AOs at the homoconjugative centres but also to the structural changes associated with the cyclic delocalization of (4q + 2) π-electrons. This cyclic delocalization should also be reflected by the stability of the system and its spectroscopic properties, including in particular its NMR spectrum.

Although Winstein's definition set the basis for an understanding of the phenomenon, neither experimental nor theoretical tools were available at his time to quantitatively assess the chemical consequences of homoaromaticity or to clearly distinguish between homoaromatic molecules and molecules with normal cyclopropyl–substituent interactions or molecules that experience just some weak through-space interactions. Even today, a detailed definition of geometric and electronic requirements for homoaromaticity is still outside the possibilities of a modern theory of homoconjugation. This becomes particularly clear when considering Winstein's requirement of a homoaromatic net stabilization energy (point 4). Throughout Winstein's early work he repeatedly stressed delocalization energy and stability of homoaromatic systems[14,15]. However, the homoaromatic stabilization energy is much smaller in magnitude and even more difficult to define than the aromatic stabilization energy. Hence, even modern theory has its problems when it comes to puffing substance into the basic requirement of homoaromaticity as formulated by Winstein. We will show this in the following by considering the various steps that have been taken in the last 25 years to obtain a more quantitative assessment of homoconjugation and homoaromaticity.

B. Description of Homoconjugative Interactions in Terms of Orbital Overlap

Various authors have tried to define overlap values S at which through-space interactions may lead to homoconjugative interactions[61]. Although S can be related to resonance integrals and bond energies[62], it is in general not possible to give a specific value of S for any atom–atom interaction at which bonding starts. Similarly, it is difficult to define on the basis of overlap values a generally applicable rule that predicts the change from weak through-space interactions to homoconjugative (or homoaromatic) through-space interactions. Nevertheless, a serious attempt has been made to develop at least a working condition for the description of homoaromatic CC interactions. This involves an assessment of π,π overlap in a situation where a molecule is distorted from planarity and considerable mixing between σ- and π-orbitals occurs.

Haddon solved this issue by employing a π-orbital axis vector (POAV) analysis in which the orientation of a pπ-orbital is determined with regard to the σ-bonds and to neighbouring pπ-orbitals in a conjugated system[63]. The POAV is perpendicular in the case of a planar π-system but has to be approximated by POAV1 or POAV2 for a non-planar π-system. In the first case (POAV1) the POAV is assumed to form equal angles to the three σ-bonds at the same atom, while in the second case (POAV2) it is given by the direction of that hybrid orbital which is orthogonal to the σ-orbitals (see Figure 7).

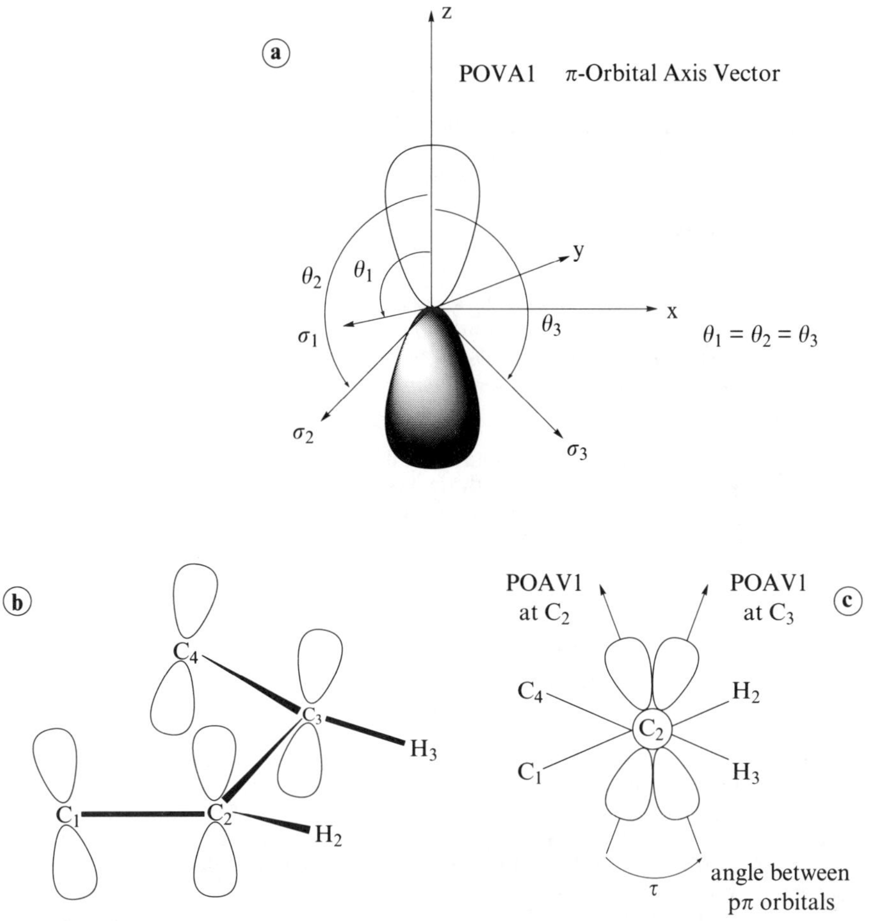

FIGURE 7. (a) Definition of the π-orbital axis vector (POAV1). The angles θ_1, θ_2 and θ_3 between π-orbital and σ-bonds are equal. (b) Part of a distorted π-system. (c) The misalignment between the POAV1 at C2 and that at C3 is measured by the angle τ

On a quantum chemical basis, the latter definition is preferred, although in practice descriptions with either POAV1 or POAV2 are similar. Both definitions re-install the σ–π separability in non-planar systems. However, the π-orbital is now a hybrid orbital rather than a pπ-orbital that is locally orthogonal (POAV2) to the σ-orbitals at the same atom. With the POAV analysis, any misalignment of p-orbitals in non-planar geometries can be determined by the dihedral angle τ between two POAVs (see Figure 7). In addition, the total overlap between the orbitals of two neighbouring atoms in non-planar geometries can be divided into a σ-part (S_σ) and a π-part (S_π).

As a reference value for S_π, Haddon suggested the pπ,pπ overlap integral S^B between nearest neighbours in benzene (R = 1.3964 Å, $S_\pi{}^B$ = 0.246) and to define the fractional overlap $\eta = S_\pi/S_\pi{}^B$. The fractional overlap reflects the degree of pπ,pπ overlap that has developed for a given bond. A π-bond is fully developed for values of η close to 1, while chemically relevant non-bonded overlap can be expected for smaller η values. A value of $\eta = 0.2$ was taken as a suitable threshold value above which $p\pi,p\pi$ overlap becomes significant. This value was based on the second-nearest-neighbour overlap integral in benzene ($\eta = 0.14$) and corresponds to about $S = 0.05$[63].

With these definitions, useful descriptions of non-planar π-systems and potentially homoconjugated systems have been developed. The descriptions are particularly attractive to experimentalists because orbital overlap is accepted as a major contributing factor to bonding and it is easy to visualize. Although Haddon did not formally define homoaromaticity in his work[63] one can use his threshold value of S to define homoaromaticity in the following way:

A potentially aromatic system X with $(4q + 2)$ *π-electrons will be homoaromatic if*

1. *the system is closed by a 1,3 homoconjugative interaction with* $\eta = S_{13}/S^{B} > 0.2$,
2. *the misalignment between the π-AOs at centres* 1 *and* 3 *is larger than* 0° *but lower than* 90° *thus leading to orbital overlap between σ and π and,*
3. *the* $(4q + 2)$ *π-electrons are fully delocalized in the resulting homoconjugative cycle.*

This definition covers both bond and no-bond homoaromaticity with a clear distinction between the possibility of insignificant through-space interactions with $\eta < 0.2$.

It is interesting to apply this definition, which is a clear improvement over Winstein's original definition of homoaromaticity, to a particular case, namely the homotropenylium cation.

1. The homotropenylium cation as a test case

The homotropenylium cation is the *prima facie* example of homoaromaticity, and therefore any useful definition of homoaromaticity has to cover this example. Haddon has calculated [at the HF/6–31G(d) level of theory using 5 d functions] the PES of the homotropenylium cation as a function of the 1,7 interaction distance by optimizing the geometry of the molecule for fixed values of $R(1,7)$[64]. The results of his POAV analysis are summarized in Figure 8.

As can be seen from Figure 8, the fractional overlap η remains significant over the whole range of the PES between $R(1,7) = 1.6$ and 2.6 Å. Using Haddon's definition of homoaromaticity, homoaromatic interactions have already started at 2.6 Å ($\eta = 0.2$) and rapidly develop to a homoaromatic bond with decreasing $R(1,7)$ distance. At the equilibrium distance [R(1,7) = 2 Å, see Section IV.B], $\eta = 0.75$, at $R(1,7) = 1.85$ Å a C(1)C(7) π-bond comparable to those in benzene seems to be fully developed ($\eta = 1.0$) and for further decrease of C(1)C(7) to 1.6 Å the π-bond becomes similar to that of ethylene in terms of overlap ($\eta = 1.5$). Another feature that indicates homoaromaticity is the equilibration of η values of all ring bonds in bond length regions that correspond to the equilibrium geometry. Finally, the ring (bond) current J of the homotropenylium cation was shown to attain a maximum in the region 1.6 to 1.8 Å at a point close to the equilibrium value of $R(1,7)$[64].

The description of the homotropenylium ion on the basis of the POAV analysis confirms the homoaromatic character of the cation. However, the analysis suggests in addition some consequences of homoconjugative delocalization, which are difficult to accept. These are:

1. The overlap parameter η suggests homoaromatic delocalization of electrons for a large range of distances (1.6 to 2.6 Å). This is also found for other molecules and, as a result, molecules such as bridged annulenes with interaction distances of 2.4 Å and more are all described as being homoaromatic[63]. The result is that Haddon's overlap-based definition of homoaromaticity becomes as general as the topology-based definition.

2. The results for the homotropenylium cation suggest that CC interaction distances of up to 2 Å and more lead to a homoaromatic bond. This is in agreement with some of

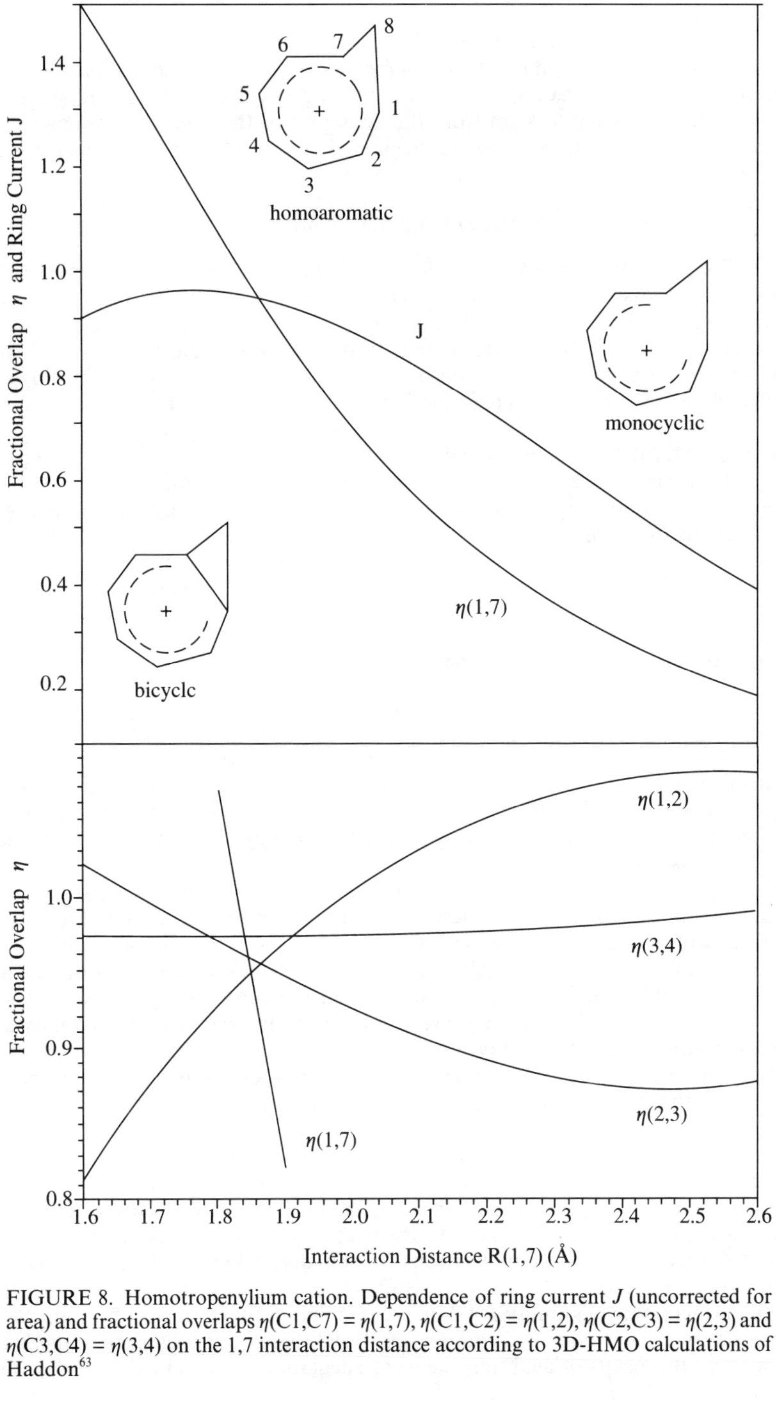

FIGURE 8. Homotropenylium cation. Dependence of ring current J (uncorrected for area) and fractional overlaps $\eta(C1,C7) = \eta(1,7)$, $\eta(C1,C2) = \eta(1,2)$, $\eta(C2,C3) = \eta(2,3)$ and $\eta(C3,C4) = \eta(3,4)$ on the 1,7 interaction distance according to 3D-HMO calculations of Haddon[63]

Winstein's predictions, but in clear contradiction to other observations which suggest that CC bonds are dissolved at much shorter distances.

3. In view of the calculated bond lengths for the homotropenylium cation it is difficult to accept that the 1,7 interaction should be the strongest π-bond in the cyclic system[62].

These difficulties seem to stem from the definition of the overlap parameter η. This parameter is based on the next-neighbour pπ,pπ overlap value in benzene according to equation 5:

$$\eta = \{S(s,s) + S(s,p\sigma) + S(p\sigma,p\sigma) + S(p\pi,p\pi)\}/S^{B}(p\pi,p\pi) \quad (5)$$

where the inclusion of $S(s,s) + S(s,p\sigma) + S(p\sigma,p\sigma)$ indicates that the π-orbitals of the POAV analysis always possess an admixture of σ-character in case of non-planar π-systems. For a given $R > 1.3$ Å, $S(p\sigma,p)$ is considerably larger than $S(p\pi,p\pi)$ and, as a result, the overlap parameter is artificially increased with increasing pyramidalization (non-planarity) of the interacting π-centres. This unwanted effect can only be balanced by calibrating η with $S^{B}(s,s)$, $S^{B}(s,p\sigma)$, $S^{B}(p\sigma,p\sigma)$ of benzene weighted according to the appropriate hybridization ratios.

Apart from the definition of η, the fixing of a threshold value for η is arbitrary because it does not combine this value with any observable property of the molecule (energy, geometry, etc.). In addition, the determination of η does not help to distinguish homoconjugated interactions from homoconjugated bonds, i.e. it does not specify point R_b in Figure 3. In summary, the use of fractional overlap values for the description of homoaromatic character is misleading since it exaggerates the magnitude of these interactions.

C. The PMO Description of Homoaromaticity

Early MO descriptions of homoaromatic compounds were based on Hückel MO (HMO) theory. Through-space interactions between interacting C centres were modelled by assuming a value for the resonance integral β. For example, in the case of the homotropenylium cation, Winstein took β (C1,C7) = 0.5 β_0 and obtained a resonance energy comparable to that of the tropenylium cation[14]. He concluded that, despite the insertion of the CH_2 group into the π-system of the tropenylium cation, delocalization of π-electrons is largely retained.

Inclusion of a saturated group into a π-system leads to a perturbation of π-delocalization and therefore a qualitative MO description of homoaromatic compounds is best done on the basis of perturbational MO (PMO) theory[28,65]. Almost at the same time, several researchers independently formualted the PMO description of homoaromatic compounds[66–73]. Haddon showed that stabilization energies resulting from homoaromatic interactions decrease with increasing ring size, which means that increased stability due to homoaromatization will be readily offset by steric and strain effects. Hence, homoaromaticity can only be observed for relatively small rings. Other qualitative insights were gained as to the influence of substituents, a second homoconjugated linkage and ring annelation[68,71].

Hehre[66,69] and independently Jörgensen[72,73] pointed out that the Möbius and Hückel description of homoconjugated molecules (Figure 9) is consistent with the assumed homoaromtic and homoantiaromatic character of these compounds. However, it was also realized that in the general case such a classification might not be sufficient to describe subtle differences in orbital interactions, which determine the homo(anti)aromatic character of a molecule.

With the increasing possibilities of doing semi-empirical or even small basis set *ab initio* calculations on homoconjugated compounds, PMO theory was used both to predict and to rationalize the results of quantum chemical calculations on potentially homoaromatic

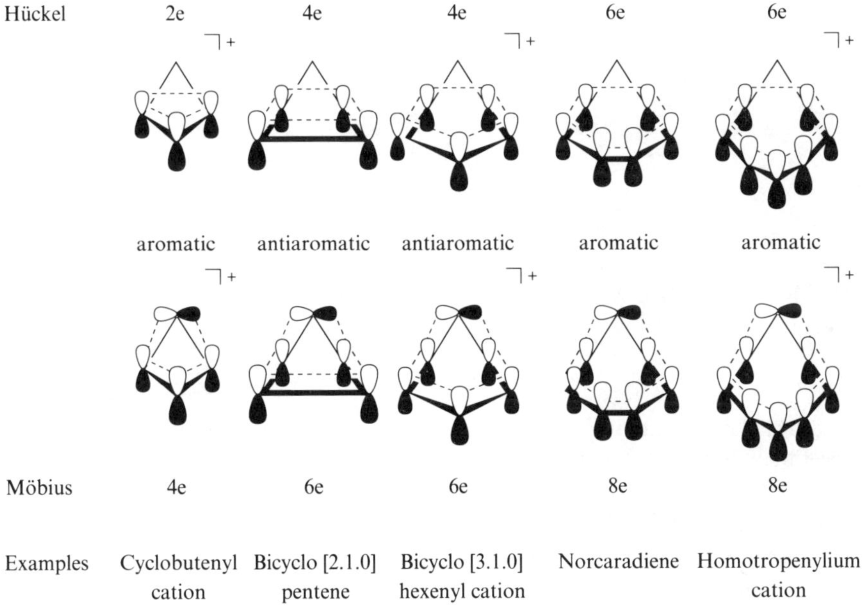

FIGURE 9. Hückel and Möbius orbital systems for homoconjugated molecules. In each case, the number of participating electrons (e) is given and classification according to aromatic or antiaromatic character indicated

molecules[66,69,72,73]. Various ways of dissecting a homoconjugated compound into fragments were considered and the PMO analysis of frontier orbital interactions between the fragments was used to predict the stability of the system in question and to classify the molecule as being homoaromatic or homoantiaromatic. For example, molecules with cyclopropyl homoconjugation were dissected into cyclopropyl and polyene units. For the cyclopropyl unit, bonding and antibonding Walsh MOs (compare with Section III of Chapter 2 of this Volume)[2] were considered, while for the polyene unit π(HOMO) and π(LUMO) were included into the analysis.

As shown in Figure 10 for the cases of bicyclo[2.1.0]pentene and bicyclo[3.1.0]hexenyl cation, four-electron w_S–π(HOMO) or w_A–π(HOMO) interactions are always destabilizing. They are partially or fully balanced by stabilizing two-electron w_A–π(HOMO) or w_S–π(HOMO) interactions, where overlap between interacting orbitals and the difference in orbital energies must be considered to give reliable predictions on whether four-electron destabilizing or two-electron stabilizing effects dominate the relative energy of the compound in question. For example, for bicyclo[2.1.0]pentene the former effect is larger than the latter and, accordingly, the molecule is destabilized and can be considered as homoantiaromatic (see Section II.D). However, for the bicyclo[3.1.0]hexenyl cation, stabilizing two-electron interactions are larger because they involve the much lower-lying π(LUMO) of the allyl cation [compared to π(LUMO) of ethene], and therefore the bicyclo[3.1.0]hexenyl cation might be even slightly stabilized according to PMO theory. Hence the classification of the bicyclo[3.1.0]hexenyl cation as being homoanti- or nonaromatic will be quite problematic if one analyses it just by PMO theory (Figure 10).

Although the PMO analysis becomes increasingly complex for larger systems, some useful predictions can be made from simplified orbital interaction diagrams. For the

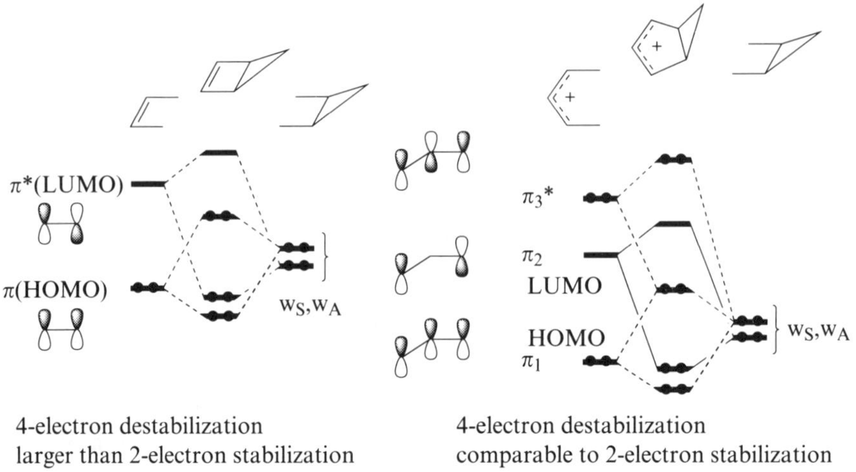

FIGURE 10. PMO interaction diagrams for the frontier orbitals of bicyclo[2.1.0]pentene (left) and bicyclo[3.1.0]hexenyl cation (right) according to Jörgensen[73]. The Walsh orbitals of the cyclopropyl ring are denoted by w_s and w_a, electrons by dots. Compare with Figure 11

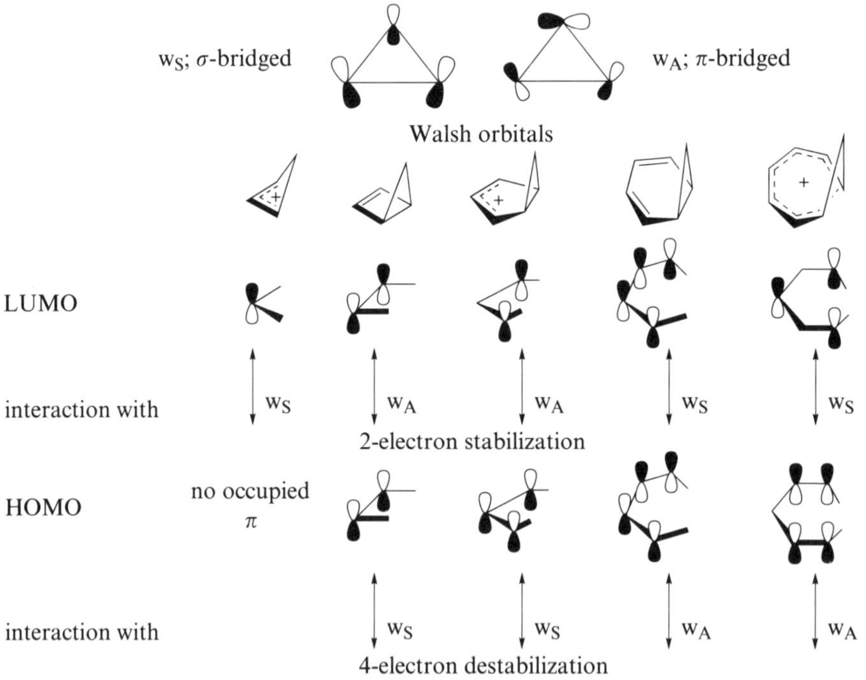

FIGURE 11. Frontier orbital interactions in homoconjugated molecules classified according to stabilizing 2-electron and destabilizing 4-electron interactions (compare with Figure 10)

potentially homoaromatic compounds in Figure 11, Walsh MO w_S is involved in the stabilizing interactions, which means that negative charge is transferred from the fusion bond of cyclopropyl to π(LUMO) of the polyene system. This leads to

(1) a lengthening of the fusion bond,
(2) a delocalization of negative (or positive) charge and
(3) bond equalization in the polyene (note that negative charge is transferred into the C=C antibonding MO of the butadiene unit of norcaradiene; Figure 11).

Also, the bonds between the polyene and the cyclopropyl ring are shortened due to bonding primary overlap between w_S of cyclopropyl and π(LUMO) of the polyene.

If the antibonding Walsh MO is involved in the two-electron stabilizing interactions (bicyclo[2.1.0]pentene and bicyclo[3.1.0]hexenyl cation, Figure 11), then charge transfer to π(LUMO) of the polyene will lead to

(1) lengthening of the cyclopropyl bonds adjacent to the fusion bond,
(2) a shortening of the fusion bond,
(3) delocalization of negative (and positive) charge and
(4) bond equalization in the polyene (note that negative charge is transferred into the C=C antibonding MO of the ethene unit of bicyclo[2.1.0]pentene; Figures 10 and 11).

In summary, PMO theory predicts that two different situations can occur: For the potentially homoaromatic compounds of Figures 10 and 11, electron delocalization takes place in the cyclopolyene part of a formally bicyclic compound. As a result of charge transfer, the fusion bond lengthens and may finally open up so that a formally monocyclic compound with homoaromatic character is formed. The degree of electron delocalization, the exact structure (geometry) and the degree of homoaromaticity depend on the nature of the π(LUMO) of the polyene as well as on strain and steric effects invoked by geometry changes accompanying homoaromatic electron delocalization.

In the potentially homoantiaromatic molecules of Figure 11, electron delocalization occurs along the periphery of a bicyclic system, involving in this way $4q + 2$ rather than $4q$ electrons. Since, however, the corresponding orbital system is of Möbius rather than Hückel type (Figure 9), delocalization of $4q + 2$ electrons leads to overall destabilization rather than stabilization.

Jörgensen[72,73] also applied PMO theory to cyclobutyl-fused analogues of the molecules shown in Figure 11. He observed no significant orbital interactions between the degenerate pair of cyclobutane HOMOs and the π-MOs of the polyene unit in line with observations Haddon had made[68]. Hence, cyclobutyl homoconjugation leading to homoaromaticity was excluded by Jörgensen. In fact, he suggested the use of cyclobutyl-fused molecules as suitable (similarly strained) reference compounds for cyclopropyl homoconjugated molecules with potential homoaromaticity or homoantiaromaticity (see Section III.G)[73].

D. Description of a Homoconjugative Bond by Bond Orders and Other Interaction Indices

It has been always tempting to use bond orders as descriptors for homoconjugative or homoaromatic interactions. For example, PMO theory predicts typical changes in the bond order due to homoconjugative interactions. An increase or decrease of the bond order depends on the number of electrons involved and the dominance of either 2-electron 2-orbital stabilizing or 4-electron 2-orbital destabilizing interactions[28,65,71]. This led Jörgensen suggesting the use of π-bond orders obtained by semi-empirical methods as a gauge for homoaromaticity[72]. A basic problem of this approach is that for both bonding and non-bonding situations, bond orders larger than zero can be obtained. Therefore, it would be appropriate to speak of atom,atom interaction indices and to use the term bond order only for true bonding situations. However, since no generally applicable criteria are known to define a bond via its bond order, the latter term has to be used indiscriminately for both bonding and non-bonding situations, which of course considerably reduces its value.

As typical of many other attempts to describe homoconjugative interactions with the help of bond orders, we mention here recent investigations of Williams, Kurtz and Farley[74]. These authors used various semi-empirical methods (MNDO, AM1, MINDO-CI, AM1-CI) to study cycloheptatriene, 1,6-methano[10]annulene, elassovalene and some other potentially homoaromatic compounds. For the 1,6 interactions in cycloheptatriene and 1,6-methano[10]annulene, small bond orders < 0.1 were calculated suggesting the absence of homoconjugative interactions although homoaromatic character is generally accepted in the case of the 1,6-methano[10]annulene. The authors concluded from this that bond orders seem to be of no use as possible discriminators of homoconjugative interactions[74].

As an alternative to using bond orders Williams, Kurtz and Farley suggested using the two-centre energy terms E (AB) (A and B are interacting atoms) that one obtains upon partitioning of semi-empirical NDO energies into mono- and bicentric contributions. This procedure was originally suggested by Fischer and Kollmar[75] and later used by Dewar and Lo[76], who showed that E (AB) provides a measure of bond strength. A negative value of E (AB) implies strong bonding between atoms A and B while positive values indicate destabilizing interactions between A and B. Also, calculated E (AB) values correlate with the corresponding bond lengths R (AB) as has been shown by various authors[76,77].

Williams and coworkers[74] found that E (AB) values, contrary to bond orders, lead to reasonable predictions with regard to the homoconjugative or homoaromatic character of systems such as 1,6-methano[10]annulene or semibullvalene provided the semi-empirical MNDO or AM1 method is connected with limited Configuration Interaction (CI) of the 2×2 or 4×4 type. However, Dewar and Lo[76], who studied various Cope rearrangements with the help of two-centre energies at the MINDO/2 level, found that E (AB) values are actually too large (four times normal bond energies) to provide a reliable measure of bond strength. In the case of CC bonds their zero-point value can be found in the region stretching from 2.2 to 3 Å depending on the semi-empirical method used. Apart from this, energy partitioning into mono- and bicentric terms cannot be extended to *ab initio* methods because of the occurrence of a large number of 3- and 4-centre energy terms which lead to a sizeable contribution to the energy.

In recent years, one has frequently based the analysis of bond orders on natural bond orbitals (NBO) and natural localized MOs (NLMOs)[78]. Calculations of NBO bond orders for homoaromatic systems such as the cyclobutenyl (**44**) or the homotropenylium cation (**45**)[79] lead to significant bond orders of 0.5–0.7 (Figure 12) for the homoconjugative interactions C1,C3 (R = 1.74 Å) and C1,C7 (R = 1.91 Å), respectively, thus supporting Winstein's expectation of partial bonds[14]. Inspection of the NBO bond orders of Figure 12 also reveals that they are strongly alternating along the closed homoaromatic cycles, predicting bond strengths that do not parallel the corresponding small changes in the bond lengths. It seems that *the NBO analysis pushes the homoaromatic system in the direction of a bicyclic structure with considerable bond alternation* rather than the bond equalization caused by homoaromatic electron delocalization (see Section IV.B).

These deficiencies of the NBO analysis in the case of homoconjugated molecules seems to result from two critical steps of the method[78]: (1) Atomic densities are spherically averaged, which means that anisotropies of the atomic densities caused by neighbouring atoms can only be re-introduced by an orthogonalization process. (2) The occupancy-weighted symmetric orthogonalization procedure used in the NBO analysis enforces the best Lewis structure, i.e. it seeks the next classical structure and, accordingly, may not be suited for describing a homoaromatic system with a non-classical structure.

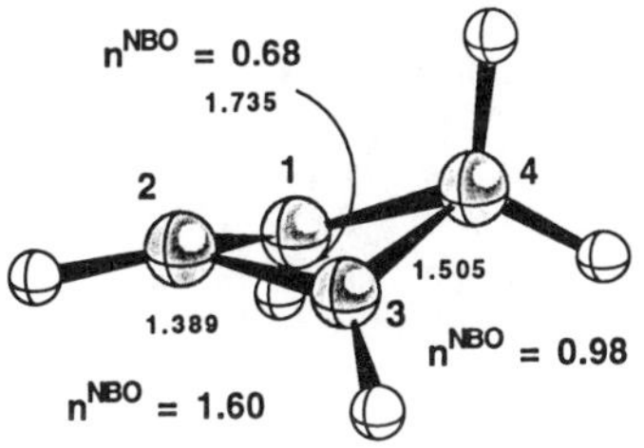

(**44**)

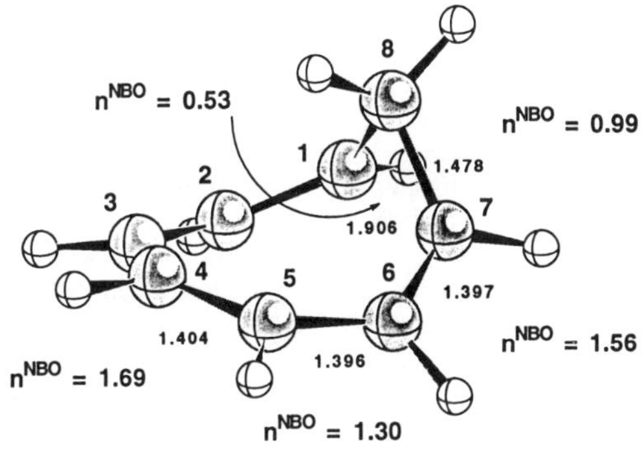

(**45**)

FIGURE 12. NBO bond orders n^{NBO} of homocyclopropenium (**44**) and homotropenylium cation (**45**) obtained from HF/6-31G(d) calculations[79]. MP2 bond lengths (in Å) are also given[49,56]

A better description of homoaromatic systems is provided by bond orders based on the virial partitioning analysis of the total electron density distribution, which we will discuss in the next section.

E. The Electron Density Based Definition of a Homoconjugative Bond

A more promising attempt to define a homoconjugative bond has been made by Cremer and coworkers[27,44,49–58] on the basis of *ab initio* calculations and the topological analysis of the electron density distribution ρ (**r**)[80]. The distribution ρ (**r**) takes a characteristic form in the case of molecules. At the positions of the nuclei, ρ (**r**) attains maximal values. In the off-nucleus direction, ρ (**r**) decreases exponentially and approaches zero for large **r**. This is different if one considers the region between two nuclei belonging to bonded atoms. In this region, ρ (**r**) adopts fairly large values. *The nuclei are connected by a path of maximum electron density* (*MED path*). Any lateral displacement from the MED path leads to a decrease in ρ (**r**). The position **p** of the minimum of ρ (**r**) along the MED path is a point which can be used to characterize the density distribution in the internuclear region. The position **p** corresponds to a maximum of ρ (**r**) in the directions perpendicular to the path, i.e. it is a first-order saddle point of ρ (**r**) in three dimensions.

Bader and coworkers[81] have shown that the saddle point **p** is fully characterized by the first and second derivatives of ρ (**r**) with regard to **r**: The gradient of ρ (**r**), $\nabla\rho$ (**r**), vanishes at **p** and two of the three eigenvalues (curvatures) λ_i ($i = 1,2,3$) of the Hessian matrix of ρ (**r**), i.e. the matrix of second derivatives, are negative. The curvatures λ_1 and λ_2 perpendicular to the MED path are negative while the curvature λ_3 along the MED path is positive due to the minimum of ρ (**r**) in this direction.

If one analyses the gradient of ρ (**r**) not only at the point **p** but also at other points in molecular space, then the gradient vector field of ρ (**r**) will be obtained[81]. The gradient vector ρ (**r**) always points in the direction of a maximum increase in ρ (**r**). Thus, each such vector is directed toward some neighbouring point. By calculating $\nabla\rho$ (**r**) at a continuous succession of points, a trajectory of $\nabla\rho$ (**r**), the path traced out by the gradient vector of ρ (**r**), is obtained.

In the gradient vector field of a diatomic molecule AB (or any general molecule), one can distinguish three types of trajectories: First, there are just two trajectories that connect the nuclei of bonded atoms and the intermediate saddle point and, in this way, define the MED path. Then, there is a class of trajectories which all terminate at the saddle point **p** and form a surface S (AB) perpendicular to the MED path separating the regions of the two bonded atoms. The surface S (AB) is called the zero-flux surface and can be considered as an interatomic surface. Finally, there is a class of trajectories which terminates at the nucleus and forms the basin of the corresponding atom.

Analysis of the electron density distribution ρ (**r**) of numerous molecules has revealed that there exists a one-to-one relation between MED paths, saddle points **p** and interatomic surfaces on the one side and chemical bonds on the other[27,81,82]. However, low-density MED paths can also be found in the case of non-bonding interactions between two molecules in a van der Waals complex[82]. To distinguish covalent bonding fron non-bonded or van der Waals interactions, Cremer and Kraka have given two conditions for the existence of a covalent bond between two atoms A and B[82,83]:

1. *Atoms A and B have to be connected by a MED path.* The existence of a MED path implies a saddle point **p** of the electron density distribution ρ (**r**) as well as a zero-flux surface S (AB) between atoms A and B (necessary condition).
2. *The local energy density H* (**p**) *is stabilizing*, i.e. it must be smaller than zero (sufficient condition).

The local energy density H (**r**) in the bonding area is defined by equation 6[82,83]:

$$H(\mathbf{r}) = G(\mathbf{r}) + V(\mathbf{r}) \tag{6}$$

where G (**r**) is a local kinetic energy density and V (**r**) is the local potential energy density. The distribution V (**r**) is always negative while G (**r**) is always positive. If H (**r**) is negative, then the local potential energy density V (**r**) will dominate and an accumulation of electronic charge in the inter-nuclear region will be stabilizing. In this case, the MED path and saddle point **p** correspond to bond path and bond critical point and can be used to characterize the covalent bond[27,82,83].

If H (**r**) is zero or positive in the inter-nuclear region, then there will be closed-shell interactions between the atoms in question, typical of ionic bonding, hydrogen bonding or van der Waals interactions[81].

If one correlates the calculated electron density at the bond critical point **p** and the CC bond distance R for a variety of hydrocarbons, a linear relationship will be obtained which holds for both single, double, triple, aromatic and homoaromatic CC bonds. On the basis of this relationship, a bond order n(CC) has been defined according to equation 7[80,82]:

$$n = \exp\{a[\rho(\mathbf{p}) - b]\} \tag{7}$$

where the constants a and b have been determined by assigning Lewis bond orders of 1, 2 and 3 to ethane, ethene and acetylene.

Another insight into the nature of a covalent bond is provided by analysing the anisotropy of the electron density distribution ρ (**r**) at the bond critical point **p**[80]. For the CC double bond, the electron density extends more into space in the direction of the π orbitals than perpendicular to them. This is reflected by the eigenvalues λ_1 and λ_2 of the Hessian matrix, which give the curvatures of ρ (**r**) perpendicular to the bond axis. The ratio λ_1 to λ_2 has been used to define the bond ellipticity ε according to equation 8[80]:

$$\varepsilon = \lambda_1/\lambda_2 - 1 \tag{8}$$

The value of ε is a measure of the anisotropy of ρ (**r**) at **p**. A direction has been assigned to ε, namely the direction of the soft curvature given by the eigen vector associated with λ_2. This direction is called the major axis of λ[80]. It is normally indicated by a double-headed arrow.

Although a distinction between σ and π electrons is no longer appropriate when analysing ρ (**r**), it is nevertheless appealing to relate the bond ellipticity ε to the π *character* of a double bond[80,82].

For planar π systems the major axis of ε is always perpendicular to the molecular plane, i.e. all major axes are parallel[27,80]. One can say that the bond ellipticities overlap completely. In the case of benzene the values of ε are all equal, indicating that the π electrons are fully delocalized (see Scheme 8). For conjugated systems such as *trans*-1,3-butadiene and cyclobutadiene, the bond ellipticities of the double bonds propagate to some extent into the formal single bonds, revealing that the latter possess partial π character. The degree of π conjugation can be quantitatively assessed by the calculated n and ε values[27,80,82]. In the same way, the extent of homoconjugation in a molecule is reflected by the calculated values of n and ε[27].

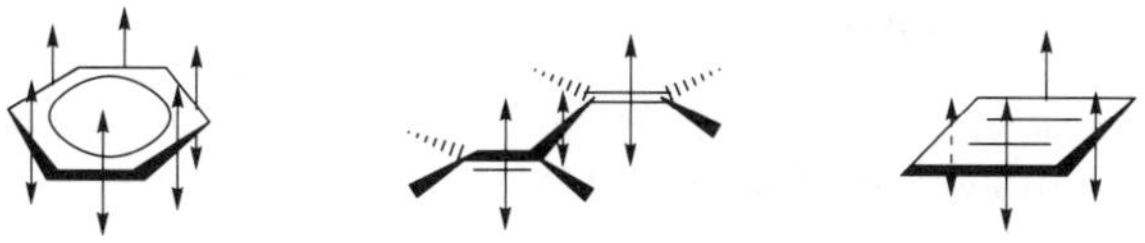

SCHEME 8. Schematic presentation of bond ellipticities in the cases of benzene, 1,3-butadiene and cyclobutadiene. Major axes of bond ellipticities are indicated by double-headed arrows

The Cremer–Kraka criteria for covalent bonding together with calculated bond orders and bond ellipticities have helped in many cases to distinguish covalent bonding from non-covalent, ionic or electrostatic interactions and to characterize covalent bonding in molecules with both classical and non-classical structures[27,84]. They have also been used to distinguish a homoconjugated bond from homoconjugative through-space interactions. In Figure 13, calculated MP2/6-31G(d) bond orders n are given for the homocyclopropenium cation, **44**[56], and the homotropenylium cation, **45**[49], at their equilibrium geometries.

Neither **44** nor **45** possesses a homoaromatic covalent bond between the interacting C atoms and, accordingly, these cations represent examples of no-bond homoconjugation

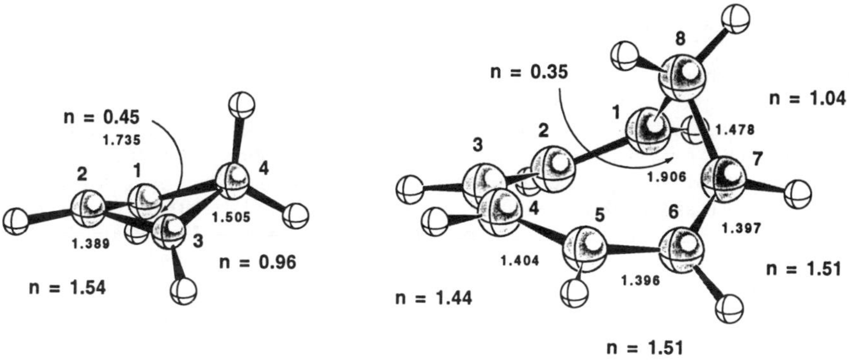

(**44**) (**45**)

FIGURE 13. Topological CC bond orders n of homocyclopropenium (**44**) and homotropenylium cation (**45**) calculated from the MP2 electron density distribution ρ(**r**) at the bond critical points. Note that n values for C1,C3 of **44** and C1,C7 of **45** correspond to interaction indices. MP2 bond lengths (in Å) are also given[49,56]

and no-bond homoaromaticity[49,56]. Utilizing the calculated electron density at the middle points between the interacting atoms C1, C3 (**44**) and C1, C7 (**45**) in connection with the bond-order equation 6, one obtains homoconjugative interaction indices of 0.45 and 0.35 at interaction distances of 1.74 and 1.91 Å. These values reflect a significant amount of electron delocalization thus leading to an equilibration of bond orders and bond lengths in the homoaromatic ring system, which is particularly nicely reflected by the bond orders of the homotropenylium cation[49]. They range between 1.44 and 1.51, i.e. they take values typical of an aromatic molecule such as benzene[27,80].

F. Description of Homoaromaticity in Terms of the Properties of the Electron Density Distribution

Cremer and coworkers have shown that the analysis of $\rho(\mathbf{r})$ provides a basis for a rigorous definition of homoaromaticity[27,44]. Utilizing the definitions of covalent bonding, bond order, π-character and π-delocalization (Section III.E), they translated Winstein's definition of homoaromaticity[14] (Section III.A) into density language[27,44]:

A cyclic system with $(4q+2)$ *π-electrons will be homoaromatic if*

1. *the system is closed by a 1,3-bond path with a bond critical point* **p** *(C1, C3) and* $H(\mathbf{p}) < 0$;
2. *the bond order n of the 1,3-bond is* $0 < n < 1$,
3. *the π-character of the 1,3-bond as measured by the bond ellipticity ε is larger than that of cyclopropane*: ε *(C1, C3)* $>$ ε *(cyclopropane) and*
4. *the major axis of ε (C1, C3) overlaps effectively with those of the neighbouring bonds.*

This is a quantitative definition of homoaromaticity that is generally applicable and helps to specify exactly the point R_b in Figure 3, at which cyclopropyl homoconjugation starts. However, this definition is much more stringent than Winstein's definition because it excludes all those systems with 1,3-interactions that do not lead to a bond path (no-bond homoaromaticity). Hence, it describes homoaromaticity only for the case of cyclopropyl homoconjugation. For example, Kraka and Cremer have used this approach to describe cyclopropyl homoconjugation in norcaradiene (**10**)[27,54].

Calculated bond orders and bond ellipticities of **10** (Figure 14a) reveal that about 6 π-electrons ($\Sigma n = 7.73$; number of π-electrons $= 2(7.73 - 5) = 5.5$; the C1C6 bond is excluded from the summation of σ-bonds because of the known π-character of cyclopropyl ring bonds, see Chapter 1 of this volume[2]) are delocalized in the six-membered ring. The C1C6 bond is relatively weak ($n = 0.85$) and possesses substantial π-character as indicated by a large ellipticity. π-Electron delocalization leads to a π-character of the formal CC single bonds, which possess ellipticities of 0.12 and 0.13. Hence, norcaradiene is described by the density analysis as a weakly homoaromatic system. This is in line with resonance energy calculations of Roth and coworkers[41,42] on annelated norcaradienes **33** and **34** (Table 2, see Sections II.C and III.G, and also the discussion in the following chapter[3]).

In a similar way, the electronic structure of potentially homoantiaromatic molecule can be investigated. Cremer and coworkers[44] have investigated the bicyclo[2.1.0]pent-2-ene **27** and the bicyclo[3.1.0]hexenyl cation **46** (Figure 14b and 14c), which have been described as being homoantiaromatic[73]. An interaction of the ene or allylic 2π system with the cyclopropane unit, as encountered similarly in norcaradiene, would entail destabilizing 4π electron interactions. However, both **27** and **46** avoid Hückel-homoantiaromatic 4π electron delocalization as is clearly revealed by calculated n and ε values (Figure 14b and 14c). In the case of ion **46**, the properties of the bond C1C3, including its length (1.501 Å[44]), are those of a CC bond in an isolated cyclopropane (see Chapter 1 of this Volume[2]). On the other hand, the two external bonds of the three-membered ring, C1C2 and C2C3, are

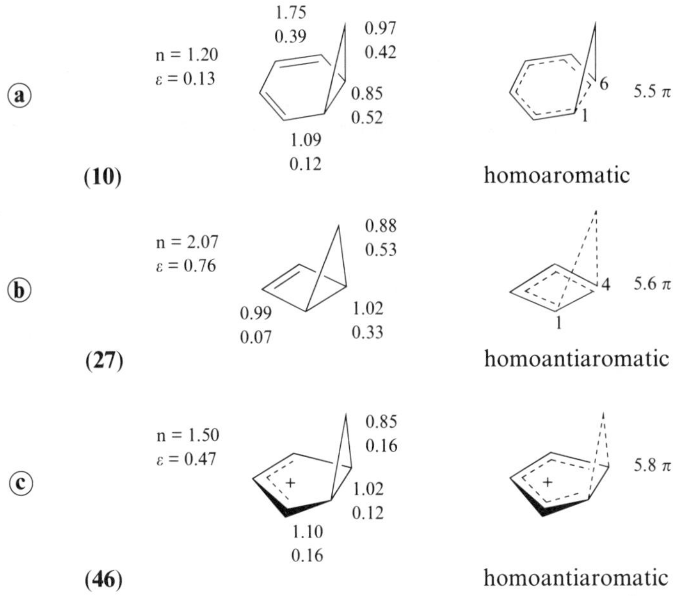

FIGURE 14. CC bond orders n and bond ellipticities ε of cyclopropyl homoconjugated molecules: (a) norcaradiene, (b) bicyclo[2.1.0]pentene, (c) bicyclo[3.1.0]hexenyl cation. On the right, the preferred mode of electron delocalization is indicated by dashed lines. Also given is the number of delocalized electrons as calculated from topological bond orders. See text

lengthened (1.535 Å); they are the weakest bonds in the cation with n being just 0.85. Their ellipticities substantially overlap the ellipticities of the neighbouring bonds in the five-membered ring[44].

The analysis of ρ (**r**) of cation **46** suggests that the ring of the six outer bonds forms a conjugated system. Their bond orders sum to 6.9, equivalent to four single bonds and a π-system of approximately six electrons. The labile character of bonds C1C2 and C2C3 accounts for the perambulatory properties of the cyclopropane ring (see following chapter[3]).

Similar observations can also be made for bicyclo[2.1.0]pent-2-ene (Figure 14b). Again, the external bonds rather than the bridging bond of the three-membered ring are labile as reflected by n = 0.88 and ε = 0.53. Approximately six electrons are delocalized on the perimeter of the five-membered ring. Electronic charge is delocalized over the entire surface of the three-membered ring and this surface is conjugated with the π-system of the adjoining ring. The direction of surface delocalization is parallel rather than perpendicular to the bond C1C3, indicating that the latter is excluded from conjugation[27].

Obviously, the density description suggests that homoantiaromatic molecules prefer Möbius $4q + 2$ electron systems rather than Hückel $4q$ systems. This is in line with the PMO analyses of Hehre[66,69] and Jörgensen[72,73] (see Section III.C).

From the electron density analysis it becomes clear that the cyclopropyl group is an electronic chameleon that can adjust to the different electron delocalization situations. Of course the real reason for this flexibility of the cyclopropyl group stems from the phenomenon of surface delocalisation (see Chapter 2 of this volume)[27,85]. The three CC bonds of the cyclopropyl ring possess considerable π-character as revealed by the bond ellipticities. However, contrary to the ellipticity of an alkene double bond, the directions of the soft curvatures of the ring bonds are not perpendicular but lie in the plane of the carbon nuclei. This means that in the ring plane electron density extends both toward the ring centre and toward the outside of the ring. This is unique for three-membered rings since for cyclobutane and larger rings the bond ellipticities are vanishingly small[85]. The smearing out of electron density in the surface of the three-membered ring has been termed surface delocalization of electrons and stabilization of the cyclopropyl ring has been attributed to this phenomenon[85].

It has also been shown that surface delocalization can adopt a preferential direction if the cyclopropyl group interacts with a π-conjugated system. There are basically two directions of surface delocalization as indicated in Scheme 9 for vinylcyclopropane (**20**) and divinylcyclopropane **47**. In homoaromatic molecules, surface delocalization is perpendicular to the 1,3-bond while in homoantiaromatic molecules it is parallel to the 1,3-bond[27,85].

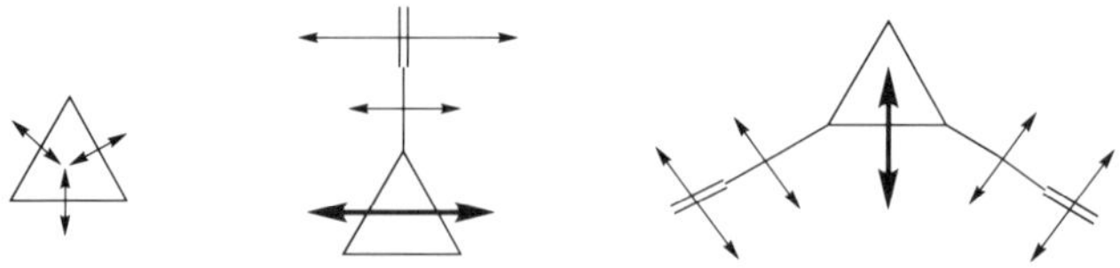

(**17**) (**20**) (**47**)

SCHEME 9. Schematic presentation of surface delocalization in cyclopropane (**17**), vinylcyclopropane (**20**) and 1,2-divinylcyclopropane (**47**). Major axes of bond ellipticities are indicated by arrows; the direction of surface delocalization in **20** and **47** is given by a bold arrow

An extension of the density description to through-space interactions is in principle possible as shown in Section III.E. For example, for the homocyclotropenylium cation (**45**) a 1,7 homoconjugative interaction index n of 0.35 is calculated. But this does not indicate at what n value homoconjugative interactions cease to play a role, i.e. for which n the point R_a in Figure 3 is reached. For example, planar **45**, for which homoconjugative interactions should be marginally small, still possesses an interaction index $n = 0.21$ ($R = 2.675$ Å)[49] suggesting that homoconjugative interactions become small for n values between 0.2 and 0.3. This example shows that no bond homoconjugation can only be described with the help of the electron density analysis if for each compound investigated a suitable reference molecule is found and a comparison of bond orders and interaction indices is carried out.

Alternatively, one could investigate the Laplace concentration of the electron density, $-\nabla^2\rho$ (**r**), rather than ρ (**r**) itself. The Laplace concentration indicates regions in the molecule in which negative charge concentrates and is depleted[27,82,83,86]. Therefore, it is the correct quantity to reveal changes in the electronic structure due to through-space interactions leading to homoaromaticity.

Figure 15 presents a schematic view of how the atomic subspaces C1, C6 and C11 of 1,6-methano[10]annulene (**35**) change upon an approach of C1 to C6. Bond paths (solid lines between atoms), bond critical points (dots) and the traces of the zero-flux surfaces S(A, B) (perpendicular to bond paths) that separate the atomic subspaces are shown in Figure 15a. Clearly, the subspace C11 extends less and less into the region between C1 and C6 until the surfaces of C1 and C6 coincide and a bond path between C1 and C6 is formed. At the same time, the Laplace concentration between C1 and C6 gradually increases and coverges to the one found for a three-membered ring. As shown in Figure 15b, this change corresponds to the valence tautomerism of the 1,6-methano[10]annulene to bisnorcaradiene[27,54].

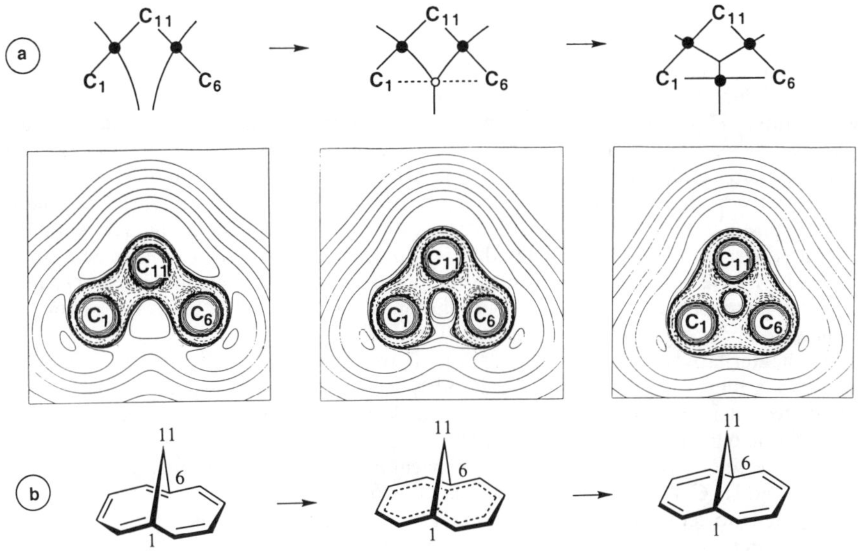

FIGURE 15. Valence tautomerization of 1,6-methano[10]annulene (**35**) to bisnorcaradiene (tricyclo[4.4.1.$0^{1,6}$]undeca-2,4,7,9-tetraene). (a) Schematic representation of atomic subspaces of C1, C6 and C11 for the three molecular forms shown at the bottom. Solid lines between atoms denote bond paths, dots denote bond critical points and a circle denotes the creation of a bond critical point in the moment of a structural catastrophe[27]. The traces of the zero-flux surfaces that separate the atomic basins of C1, C6 and C11 are given as light solid lines. (b) Contour line diagrams of the Laplace concentration $-\nabla^2\rho(\mathbf{r})$ of the molecular forms shown at the bottom of the figure given for the plane of the nuclei C1, C6 and C11. Dashed contour lines denote concentration of negative charge [$\nabla^2\rho(\mathbf{r}) < 0$] and solid contour lines denote depletion of negative charge [$\nabla^2\rho(\mathbf{r}) > 0$]. HF/6-31G(d,p) calculations[27].

An analysis of the Laplace concentration, $-\nabla^2\rho$ (**r**), yields information about the extent of through-space interactions and the concomitant changes in the molecular properties. Hence, a clear distinction between the various modes of intermolecular interactions should be possible. However, a quantification of these changes again needs an appropriate reference, something which in most cases is not present. Therefore, a description of homoconjugative interactions in terms of the Laplace concentrations has only been applied in selected cases[49–58] but has not been worked out to a more general description of no-bond homoaromaticity.

G. Energy-based Definitions of Homoaromaticity

Energy is certainly the most important property of a molecule. Thus, homoconjugative effects leading to changes in the electronic structure of a molecule should be first assessed by investigating changes in the molecular energy. As with conjugation and aromaticity, determination of a homoconjugative delocalization energy or resonance energy leads to a direct measure and description of homoconjugation. If the homoconjugative delocalization (resonance) energy is negative (< -2 kcal mol^{-1}), one can speak of a homoaromatic stabilization (resonance, delocalization) energy; if it is positive (> 2 kcal mol^{-1}) one can speak of a homoantiaromatic destabilization (resonance, delocalization energy). In this connection, ± 2 kcal mol^{-1} are used as threshold values above or below which (de) stabilization energies become significant.

Resonance energies can be directly calculated by semi-empirical methods via appropriate energy partitioning provided that the conjugated π-system is planar and all π-bonds are well defined. However, in the case of a non-planar homoconjugative molecule with non-classical atom, atom interactions (through-space or through-bond), this approach is no longer possible and, as a result, early attempts to estimate resonance energies by assuming arbitrary resonance integrals are of no or just qualitative value. Although *ab initio* theory should provide a better basis for a direct calculation of homoconjugative resonance energies, basic problems also exist with this approach, which are discussed in Section III. G. 1. There have been some interesting attempts to determine homoconjugative resonance energies from calculations with model Hamiltonians (Section III.G.2), but again their application has been limited to semi-empirical methods.

As an alternative to the direct calculation of homoaromatic stabilization energies, there is the possibility of describing the stability of homoconjugated compounds by the calculated energies of formal reactions. Two classes of reactions have been used in this connection, namely isodesmic and homodesmotic reactions, both of which involve the use of suitable reference molecules to obtain meaningful stabilization energies. In Section III.G.3 and III.G.4, we will discuss the use of formal reactions in connection with homoconjugated molecules. Finally, in Section III.G.5, we will report on one approach to determining homoconjugative resonance energies that, although based on force-field rather than quantum-chemical calculations, has been quite successful and seems to provide the most reliable resonance energies at this moment, at least for neutral homoconjugated compounds.

1. Direct calculation of homoaromatic stabilization energies

The stabilization of an aromatic molecule is given by its resonance energy RE, which is the energy (or enthalpy) difference between the aromatic system and the corresponding reference system containing localized non-resonating double bonds[4–10]. Since the latter energy is not a measurable quantity, various ways have been suggested to deduce its value from additivity relationships of bond increments taken from suitable reference compounds. The most successful definition of a resonance energy in this connection is the Dewar resonance energy, which is based on the atomization energies of (linear) polyenes as appropriate reference states, i.e. the resonance energies of polyenes are taken to be zero within this approach[39,65]. The advantage of the Dewar resonance energy as a stability measure for aromatic molecules results from the fact that comparison is made with a real conjugated compound with similar bonding features rather than with a hypothetical model compound with non-resonating multiple bonds. In this way, the number of molecules with significant resonance energies is considerably reduced and chemical research focuses on those few cases with really unusual bonding features.

Since the Dewar resonance energy differs from REs derived for a hypothetical reference system with regard to the bond energy ascribed to a C—C single bond, RE values can be normalized by using the RE value of 1,3-butadiene (or appropriate butadiene derivatives) according to equation 9[87]:

$$\text{RE(normalized)} = \text{RE} - k\,\text{RE(1,3-butadiene)} \quad (9)$$

where k is the number of C—C single bonds in the reference state.

A direct calculation of RE for planar conjugated compounds using *ab initio* theory is possible, as has been demonstrated by Kollmar[87]. Calculations comprise the following steps:

(a) A model wave function, in which the SCF π-MOs are replaced by strictly localized π-MOs representing non-resonating multiple bonds, has to be determined.

(b) Since the geometry of the non-resonating π-system is not the same as that for the resonating π–system, the geometry of the reference state with the model wave function has to be optimized.

(c) Energy comparison between the model reference state and original π-system leads to the determination of vertical (geometry frozen at equilibrium values of the original π-system) and adiabatic (optimized geometries used in both cases) RE values.

(d) RE values are normalized according to equation 9[87].

Using this procedure and correlation-corrected *ab initio* methods, reasonable resonance energies can be obtained for planar aromatic (or antiaromatic) compounds[87]. However, there are basic problems in extending this approach to homoaromatic molecules. As a result of the non-planarity of most homoaromatic compounds there is considerable σ–π mixing. The determination of a model wave function with non-resonating double bonds is not trivial. It will require extensive re-optimization of the σ-MOs because localization of the π-MOs leads to different σ–π mixing. The optimal geometry of the reference state will differ much more from that of the target system as in the case of planar aromatic systems. In view of these difficulties, it is not surprising that a direct calculation of RE of an homoaromatic molecule by *ab initio* methods has not so far been reported.

In addition, there is also the question of how to use suitable reference molecules to obtain normalized RE values which correspond to Dewar resonance energies. In the case of cyclopropyl homoconjugation, butadiene is clearly the wrong reference molecule to consider the two C—C single bonds *a* adjacent to the fusion bond *f* (see Scheme 10). The

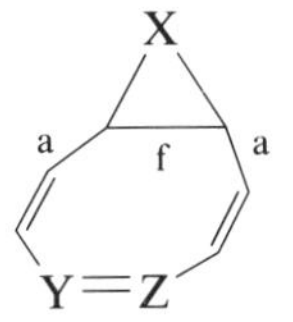

SCHEME 10. Description of fusion bond *f* and single bonds *a*

appropriate reference molecule would be vinylcyclopropane, and therefore normalization can only be achieved by determining and using the RE value of vinylcyclopropane according to equation 10:

$$\text{RE(normalized)} = \text{RE} - (k\text{-}2)\,\text{RE(butadiene)} - 2\,\text{RE(vinylcyclopropane)} \qquad (10)$$

in the case of a monohomoaromatic compound. (For bis-, tris- or multihomoaromatic systems, methylcyclopropylcarbinyl cation, bicyclopropyl and other reference molecules can become important to set up appropriate normalization equations.) Apart from this, one has to consider a problem that is already inherent in the calculation of RE(normalized) for aromatic compounds, but does not become obvious immediately. The appropriate butadiene conformation to be used in the normalization process should be the *cis* form rather than the *trans* form (see discussion in Section II.C). For homoaromatic molecules, one has to use the RE values of distorted butadiene and vinylcyclopropane forms in order to mimic exactly the conformation of the target compound. Although this problem could be solved, it is not clear to which extent steric interactions, e.g. in the *cis* forms, might spoil results since they will be different for target and reference molecules. Much more research is needed in this direction to find out whether RE values can also be obtained for homoconjugated molecules with aromatic or antiaromatic electron ensembles.

2. Homoaromatic stabilization energies from calculations with a model Hamiltonian

An elegant alternative to the direct calculation of RE values is based on the following consideration. If an impenetrable wall is built between the interacting centres of a homoconjugated molecule, a model system will be obtained that should be identical to the original molecule with regard to strain, hyperconjugative, inductive, etc. effects, but should differ in energy because of the absence of homoconjugation. Hence, the difference in molecular energies should be a direct measure of the homoaromatic or homoantiaromatic resonance energy.

This approach is quite suitable for semi-empirical NDO methods. The impenetrable wall can be simulated by defining a model Hamiltonian which does not contain any interactions between atoms separated by the wall. Within NDO theory, this goal is simply achieved by setting all resonance integrals of the Fock matrix that would lead to interactions through the wall equal to zero[88]. The difference between original energy and the energy obtained for the impenetrable wall model leads to the interaction energy in question, e.g. a homoaromatic stabilization energy. In this way, conjugative and hyperconjugative effects can be studied. Schweig and coworkers[89,90] have used this approach to describe conjugation in a number of cyclopolyenes with heteroatoms. Wirth and Bauld[91] have used the same approach to study homoaromaticity in the case of the cyclobutenyl cation and related ions. However, a systematic extension of this approach to homoaromatic systems in general has never been carried out. This may be due to the fact that at the NDO level of theory confusing results were obtained. For example, homoaromatic stabilization energies were predicted to be larger in planar (30 kcal mol^{-1}) than puckered cyclobutenyl cation (6.5 kcal mol^{-1}). The larger stability of the latter form could only be explained by invoking nonclassical σ-delocalization effects[91].

Apart from these confusing predictions, a verification of semi-empirical results by *ab initio* calculations is not possible because the dropping of certain Fock matrix elements should be accompanied by the dropping of the corresponding overlap matrix elements which leads to singularities in the overlap matrix. Weinhold and coworkers[78] have suggested an alternative approach based on localized MOs, but this approach can only be applied for the investigation of hyperconjugative effects.

3. Evaluation of homoaromatic stabilization energies by using isodesmic reactions

An isodesmic reaction[92] is a formal reaction, in which the number of electron pairs as well as formal chemical bond types are conserved while the relationships among the bonds are altered. A subclass of the isodesmic reactions is the class of bond separation energies, in which all formal bonds of a molecule are separated into two-heavy-atom molecules containing the same type of bonds. Stoichiometric balance is achieved for the bond separation energies by adding an appropriate number of one-heavy-atom hydrides to the left side of the reaction[92].

In Scheme 11, isodesmic bond separation reactions for homotropenylium cation (**45**), cycloheptatriene (**30**) and norcaradiene (**10**) are given together with calculated HF/3-21G reaction energies[93]. The latter comprise ring strain, inductive and hyperconjugative effects beside homoconjugative effects. Barzaghi and Gatti[93] have compared the isodesmic bond separation energies with those of suitable reference compounds to estimate homoconjugative stabilization effects (see Scheme 11). From the comparison of **45** with **47** (reactions 12 and 11) and **30** with **48** (reactions 14 and 13), they concluded that **45** and **30** retain 68% and 43%, respectively, of the resonance energy of the parent compounds i.e. tropylium cation and benzene while **10** is already slightly destabilized showing no homoaromaticity[93].

These results are contrary to all other observations and calculations. They reflect the danger of a careless use of bond separation reactions in connection with homoconjugated

Isodesmic bond separation reaction					Calculated ΔE	Reaction
47	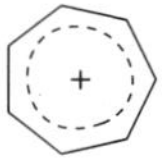	$+ 6\,CH_4 + CH_3^+$	⟶	$2\,CH_3CH_3 + 2\,CH_3CH_2^+ + 3\,CH_2{=}CH_2$	80	(11)
45	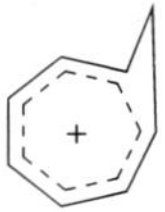	$+ 7\,CH_4 + CH_3^+$	⟶	$3\,CH_3CH_3 + 2\,CH_3CH_2^+ + 3\,CH_2{=}CH_2$	55	(12)
48		$+ 6\,CH_4$	⟶	$3\,CH_3CH_3 + 3\,CH_2{=}CH_2$	60	(13)
30		$+ 7\,CH_4$	⟶	$4\,CH_3CH_3 + 3\,CH_2{=}CH_2$	26	(14)
10	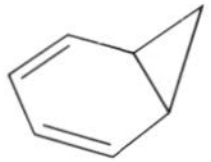	$+ 9\,CH_4$	⟶	$6\,CH_3CH_3 + 2\,CH_2{=}CH_2$	–5	(15)
10	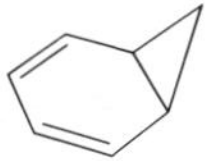	$+ CH_2{=}CH_2 + 2\,CH_4$	⟶	 $+ 2\,CH_3CH_3$	–31	(16)

SCHEME 11. Isodesmic bond separation energies (kcal mol^{-1}) calculated at the HF/3-21G level[93]

molecules. Clearly, the molecules compared in Scheme 11 differ with regard to both strain, inductive and hyperconjugative effects and therefore are far from being suited for a comparison of (homo)conjugative resonance energies. For example, the large difference in the bond separation energies of **30** and **10** simply results from the fact that the conversion of ethene + 2 CH_4 into two ethane molecules (see reaction 16) is exothermic by 21 kcal mol^{-1}. Since the actual energy difference between **30** and **10** is wrongly predicted by HF/3-21 G to be 10 kcal mol^{-1} [93], a misleading reaction energy of – 31 kcal mol^{-1} is obtained for the formal reaction 16 in Scheme 11, thus suggesting a large destabilization of norcaradiene **10**.

The stabilization energies obtained from isodesmic reaction energies become only useful if differences in strain, inductive, hyperconjugative or other effects are known. A possible solution to this problem has been suggested by Jörgensen[73], who investigated the cyclobutyl analogues to cyclopropyl homoconjugated molecules. He found that the cyclobutyl group does not participate in homoconjugation. Since the strain energies of cyclobutane and cyclopropane are similar[94], it is likely that strain energies in cyclopropyl homoconjugated molecules and in their cyclobutyl analogues are also similar. Therefore, the latter are the ideal reference compounds to determine homoaromatic resonance energies in cyclopropyl homoconjugated systems.

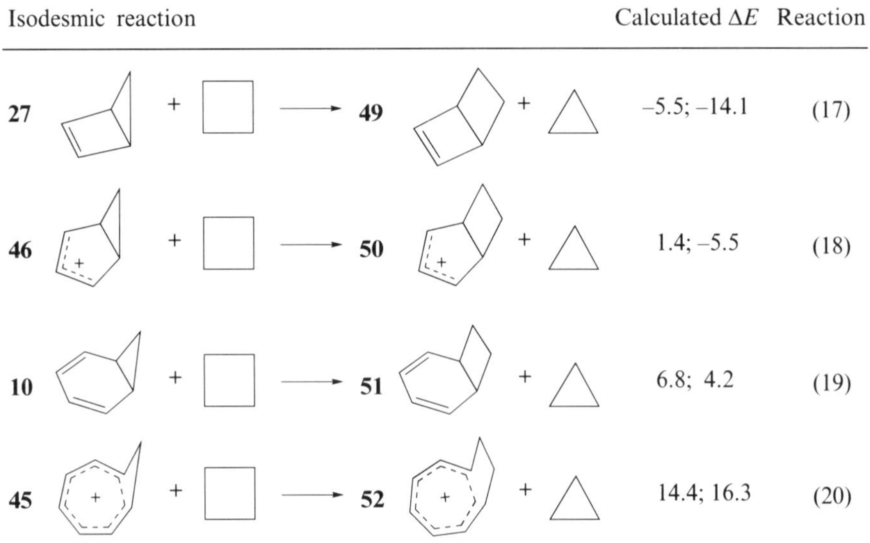

Isodesmic reaction
Calculated ΔE
Reaction
27
+
49
+
–5.5; –14.1
(17)
46
+
50
+
1.4; –5.5
(18)
10
+
51
+
6.8; 4.2
(19)
45
+
52
+
14.4; 16.3
(20)

SCHEME 12. Isodesmic reactions based on cyclobutyl derivatives **49**–**52**. Reaction energies from MINDO/3 (first entry) and EHT calculations (second entry) in kcal mol^{-1} [73]

Energies for reactions 17 to 20 have been calculated by Jörgensen at the MINDO/3 and EHT level of theory[73]. They suggest that the potentially homoantiaromatic electron systems **27** and **46** are destabilized leading to exothermic reaction energies (the positive MINDO/3 value of **46** is probably a consequence of neglect of differential overlap[73]) while the potentially homoaromatic molecules **10** and **45** are stabilized leading to endothermic reaction energies. However, a caveat is also necessary in this case. A slight variation of the reactions in Scheme 12 by using also the saturated analogues of **10**, **27**, **45**, **46** and **49**–**52** as suitable reference compounds led to conflicting results as for the homo(anti)aromatic character of the target compounds. Jörgensen[73] attributed this to deficiencies of the semi-empirical methods used, but also possible are significant differences in the strain energies of the bicyclic molecules used as references.

4. Evaluation of homoaromatic stabilization energies by using homodesmotic reactions

A homodesmotic reaction is a formal reaction, for which extraneous energy contributions arising from changes in hybridization and A—H bonding are minimized by keeping equal numbers of each type of A(sp^m) B(sp^n) bond and each type of A(sp^m) H_k group in reactants and products. The concept of homodesmotic reactions was developed by George and coworkers[95] and used to calculate resonance energies of π-systems and strain energies of cyclic compounds with considerable success[95,96]. In selected cases, it has also been used to calculate homoaromatic stabilization energies.

In Scheme 13, homodesmotic reaction energies are given for compounds **27**, **10** and **44**. Reaction 21 indicates that the homocyclopropenium cation is actually destabilized by 42 kcal mol^{-1} according to MP4 (SDTQ)/6–311G(d,p) calculations of Sieber and coworkers[56]. The major part of this destabilization energy is due to ring strain diminished by homoconjugative and hyperconjugative stabilization energies. Since the latter may be small [the CH_2 group is located in the nodal plane of the π(LUMO) of the allyl system] and the former are reduced in planar cyclobutenyl cation, the ring strain energy can be estimated by using planar **44** in equation 21 rather than the more stable puckered form. This leads to a ring strain energy of 50 kcal mol^{-1} and, accordingly, to an estimate of the homoaromatic delocalization energy of 8 kcal mol^{-1}, which is identical with the barrier of ring inversion[56].

Homodesmotic reactions have to be based on appropriate reference compounds to be suitable for the calculation of homoconjugative resonance energies. They may also be extended to balance strain and other effects on both sides of the formal reaction as is demonstrated in equations 22–24 of Scheme 13. Suitable reference molecules for homoconjugative compounds **27** and **10** are vinylcyclopropane and 1,3-butadiene. However, relating **27** to vinylcyclopropane (equation 22, Scheme 13) leads to a homoconjugative destabilization energy of 44 kcal mol^{-1}, which is contaminated by the strain energy of the cyclobutene ring of **27** (29.2 kcal mol^{-1}). When correcting for ring strain by extending equation 22 to 23, a homoconjugative destabilization energy of 14.8 kcal mol^{-1} results. This value still contains the strain energy caused by annelation of a cyclobutene ring to a cyclopropane ring which can be estimated by the homodesmotic reaction splitting bicyclo[2.1.0]pentane into cyclobutane and cyclopropane. In this way, the corrected homodesmotic reaction 24 is obtained, which provides an improved balance of ring strain energies. According to equation 24, the homoconjugative destabilization energy (resonance energy) of **27** is 11.7 kcal mol^{-1}.

A final correction is needed because the vinylcyclopropane units in **27** or **10** do not adopt the stable *trans* forms but less stable *gauche* forms. This leads to a reduction of the homoconjugative resonance energy by another 2 kcal mol^{-1}. The final resonance energy is 9.7 kcal mol^{-1}, clearly indicating the homoantiaromatic character of **27**.

In the case of **10**, the strain of the six-membered ring is small, as is the strain energy due to ring annelation. Utilizing heats of formations for *cis*-1,3-butadiene, *gauche*-vinylcyclopropane[41,97] and norcaradiene[54], a homoaromatic stabilization energy of 4 kcal mol^{-1} is calculated in line with a description of **10** as a cyclopropyl homoaromatic 6π electron system.

The derivation of the resonance energies for **27** and **10** reveals that (a) homodesmotic reactions are well suited to compensate for the different electronic effects that hinder the calculation of pure homoconjugated resonance energies, (b) use of a homodesmotic reaction such as 24 requires the inclusion of many reference compounds, which of course can lead to considerable error progression in the calculated reaction energy, and (c) the

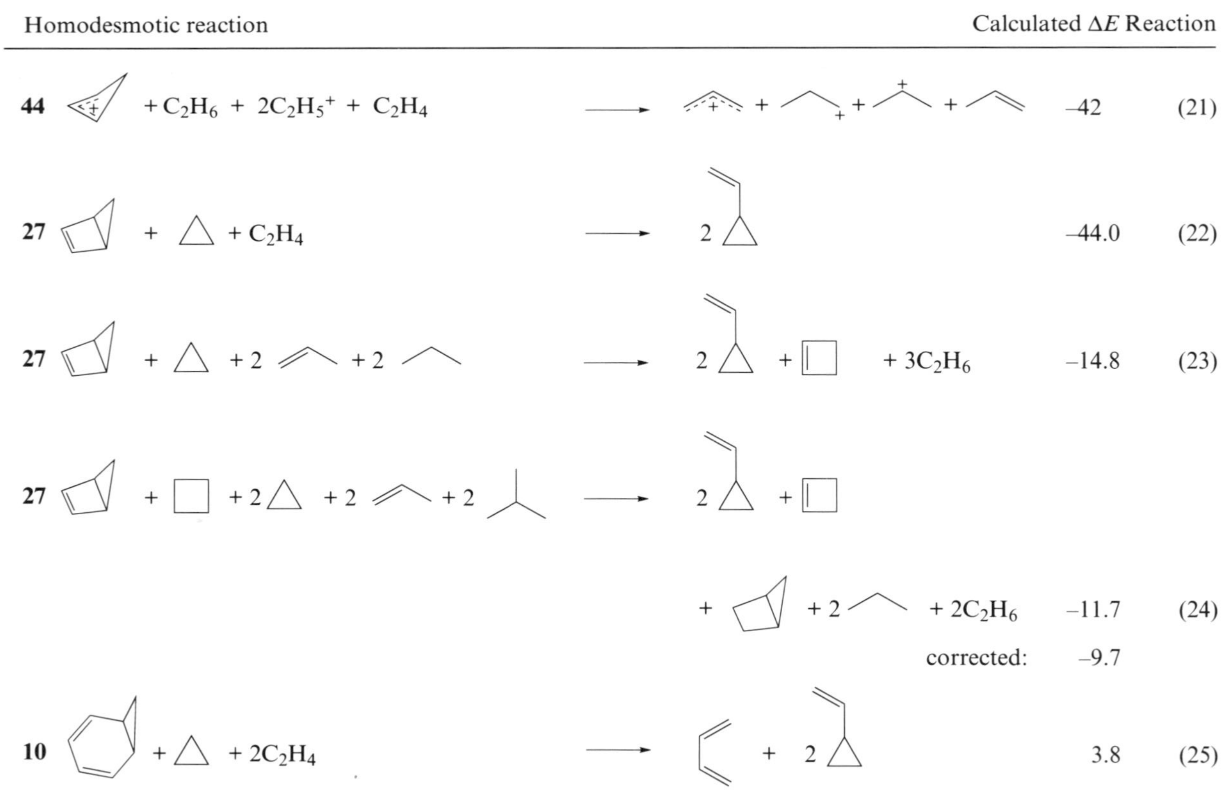

Homodesmotic reaction
Calculated ΔE Reaction
44
+ C2H6 + 2C2H5+ + C2H4
–42
(21)
27
+ C2H4
2
–44.0
(22)
27
+ 2
+ 2
2
+ 3C2H6
–14.8
(23)
27
+ 2
+ 2
+ 2
2
+ 2
+ 2C2H6
–11.7
(24)
corrected:
–9.7
10
+ 2C2H4
2
3.8
(25)

SCHEME 13. Homodesmotic reaction energies according to *ab initio* data for **44**[56] and **10**[54] and experimental heats of formation for **27**[97] in kcal mol^{-1}

calculation is necessarily based on assumptions. For example, if annelation of a cyclobutene ring with a cyclopropane ring leads to a considerably larger strain increase than that calculated for the annelation of a cyclobutane ring with cyclopropane in the case of bicyclo[2.1.0]pentane, then the value of the resonance energy will be overestimated severely.

5. Homoconjugative resonance energies from force field calculations

Roth and coworkers[41,42] have chosen carefully recalibrated force fields to predict reliable heats of formation with errors intended by the authors to be as small as ± 0.5 kcal mol^{-1}. They started with the MM2 force field of Allinger[98] and added to this parameters for the C(sp^2)—C(sp^2) and the C(cyclopropyl)—C(sp^2) single bonds of reference compounds such as substituted 1,3-butadienes and vinylcyclopropanes[41]. Particular care was given to the correct description of the torsion potential of the reference compounds. The modified MM2 force field (MM2ERW) developed by Roth and coworkers describes polyenes and cyclopolyenes in terms of localized bond structures without any reference to quantum chemical methods such as the Pariser–Parr–Pople (PPP) approach (see Section II.C)[41].

Because of the additional calibration, the MM2ERW force field leads to heats of formation of conjugated polyenes or conjugated systems containing the cyclopropyl group in close agreement with experimental heats of formation (see Table 1 in Section II.C). With 1,3-butadiene and vinylcyclopropane as reference compounds, none of these molecules possesses any extra stabilization. This, however, is different for the potentially homoconjugated molecules listed in Table 2 of Section II.C.

MM2ERW force field calculations lead to heats of formations for cycloheptatriene (**30**), the bridged cycloheptatrienes **31** and **32** and the norcaradienes **33** and **34** which are 3–6 kcal mol^{-1} larger than the experimental values, thus suggesting a homoaromatic resonance (electron delocalization) energy (RE) of this magnitude. Although calculated RE values are rather small, they reflect the expected trends depending on the magnitude of overlap between the interacting centres. Thus, planar cycloheptatriene **29** does not benefit from any homoaromatic electron delocalization because of the negligible overlap between parallel pπ orbitals at C1 and C6. Similarly, the large interaction distances in cyclononatriene **38** (R = 2.45 Å) reduces the stabilization energy to a negligible amount. Molecules **27** and **28** with the unfavourable Möbius 6-electron ensembles are destabilized by 9.9 and 6.6 kcal mol^{-1} [41,42] in line with PMO predictions (Section III.C) and RE values based on homodesmotic reaction energies (Section III.G.4).

The homoaromatic RE values in Table 2 cannot directly be related to aromatic REs normally cited in the literature. This becomes obvious when considering MM2ERW REs of aromatic compounds: they are all larger than REs derived from experimental heats of formation with the help of homodesmotic reaction energies. For example, the RE value of benzene is calculated to be 25.9 kcal mol^{-1} [41] while the accepted homodesmotic RE value is 21.6 ± 1.5 kcal mol^{-1} (relative to 1,3-butadiene)[96]. Deviations of up to 20 kcal mol^{-1} and more are obtained for aromatic compounds such as naphthalene (MM2ERW: 40.1; accepted: 30.3 ± 2.6 kcal mol^{-1}), anthracene (MM2ERW: 51.5; accepted: 36.6 ± 5 kcal mol^{-1}), pyrene (MM2ERW: 68.6; accepted: 53.6 ± 5.9 kcal mol^{-1}), phenanthrene (MM2ERW: 58.6; accepted: 43 ± 5 kcal mol^{-1}) or tetracene (MM2ERW: 67.5; accepted: 47.4 ± 6.6 kcal mol^{-1}), where the difference compared to the corresponding homodesmotic REs increases with the number of annelated benzene rings[41,96].

These deviations are the result of the fact that the MM2ERW force field makes explicit use of the rotational potential of 1,3-butadiene. Thus, *cis*-1,3-butadiene is used as the appropriate reference conformation of butadiene for benzene and other aromatic molecules. The *cis* form of butadiene is about 3.5 kcal mol^{-1} higher in energy than the *trans* form, of which about 2 kcal mol^{-1} may be due to steric interactions and about 1.4 kcal mol^{-1} to a decrease in electron delocalization (as a result of approaching the unfavourable cyclobutadiene form)[41]. Since *cis*-butadiene is contained three times in benzene, the resonance energy of benzene taken relative to *cis*-butadiene is 4.2 kcal mol^{-1} higher than the value normally given relative to *trans*-butadiene. Similar considerations apply to naphthalene, anthracene, etc. for which, in addition to *cis*- and *trans*-butadiene, various methylated butadienes have been used as a reference. While it is easy to renormalize the REs of Roth and coworkers to *trans*-butadienes, the REs of Roth are reasonable since they are based on reference systems that agree better with the actual target molecules than the reference systems normally used in the literature.

In the case of homoconjugated molecules, similar considerations have to be made when comparing MM2ERW REs with RE values based on different reference molecules or reference conformations. Since the MM2ERW values correspond to the actual conformation taken by vinylcyclopropane or 1,3-butadiene in the homoconjugative molecule, their magnitude is 3–4 kcal mol^{-1} larger in the case of a potential 6π electron system. Inspection of Table 2 (Section II.C) reveals that this is about the magnitude of the MM2ERW RE values of potentially homoaromatic molecules such as cycloheptatrienes **30**, **31**, **32** and norcaradienes **33**, **34**. Accordingly, descriptions based on the calculation of homodesmotic reaction energies, that use *trans*-vinylcyclopropane and *trans*-butadiene as references, get in these cases no or vanishingly small homoaromatic REs. *This explains some of the confusion, which has accompanied the discussion as to whether molecules such as cycloheptatriene or norcaradiene are homoaromatic*[4–10,27].

It remains to be questioned whether one should not use in general Roth's approach of picking both the right reference molecule and the right reference conformation[41,42]. In principle, this should be possible since, for molecules such as butadiene or vinylcyclopropane, the full rotational potentials have been carefully investigated by both experimental means and *ab initio* methods. On the other hand, using *cis* forms as appropriate reference conformations leads to an artificial increase of homoaromatic REs. The destabilization of a *cis* form results not only from unfavourable π-electron interactions (e.g. by through-space formation of an antiaromatic 4π-system) but also from destabilizing steric interactions not present in the target compound. Because of σ–π mixing in non-planar conformers, it is difficult to separate steric and delocalization effects for any arbitrary conformation of the reference system. However, if one disregards steric effects, calculated homoaromatic stabilization energies will contain, beside the electron delocalization effect, a small but significant steric stabilization energy, which has nothing to do with the concept of homoconjugation.

IV. *AB INITIO* EXAMINATIONS OF HOMOCONJUGATION

The *ab initio* investigation of homoconjugated molecules in general and cyclopropyl homoconjugated molecules in particular is not trivial, and requires a careful choice of method, basis set and level of geometry optimization. In addition methods for calculating other molecular properties such as charge distribution, NMR chemical shifts, magnetic susceptibility and susceptibility exaltations, vibrational spectra, etc. have to be carefully selected, which is beyond the level of routine work in quantum chemistry. Therefore, we will discuss in Section IV.A the basic requirements for a reliable *ab initio* description of homoconjugated molecules. In Section IV.B, we will describe the *ab initio* investigation of the homotropenylium cation to demonstrate practical aspects of *ab initio* calculations on homoaromatic compounds and to show how *ab initio* theory can lead to a more complete picture of homoconjugation and homoaromaticity. Finally, in Section IV.C, we will take steps toward a more general definition of homoaromaticity based on the results of *ab initio* calculations.

A. Basic Requirements

In the seventies and eighties, *ab initio* calculations on potentially homoaromatic molecules were preferentially carried out with the Hartree–Fock (HF) method using minimal or double-zeta (DZ) basis sets. However, neither HF nor small basis sets are appropriate to describe a homoaromatic system. In the case of cyclopropyl homoconjugation, the use of a DZ + P basis set is mandatory since polarization (P) functions are needed to describe the bond arrangements of a three-membered ring.

If one wants to scan the whole region of bond and no-bond homoconjugative structures, even a DZ + P basis set may not be sufficient. Through-space interactions at distances of 2–3 Å are mediated by diffuse density distributions in the tail region of the wave function. Accordingly, the valence region and the tail region of the wave function have to be described in a balanced way. This is not possible by using one of the energy-optimized standard basis sets. The basis set has to include diffuse functions that lead to a correct account of diffuse density distributions. Various recipes are nowadays available to add diffuse functions to standard DZ + P basis sets[99].

The problem of selecting the correct basis set becomes simpler when cationic molecules have to be investigated. The positive charge leads to a contraction of orbitals and wave function, and therefore a correct description of the tail region is no longer that important. In this case, a standard DZ + P basis set may already lead to reasonable results for positively charged homoaromatic molecules. However, such a basis will definitely be too small if no-bond homoaromatic anions are investigated. This has to be considered when evaluating the reliability of the many HF/small basis set calculations from the seventies and eighties.

A major calculational problem is the correct description of bond equalization and bond alternation in conjugated systems. HF theory exaggerated bond alternation by making formal double bonds too short and formal single bonds too long. This trend is enhanced by the use of larger basis sets, which indicates that only a correlation-corrected method can compensate for these deficiencies. Promising results have been obtained with second-order Møller–Plesset (MP2) perturbation theory, which may be considered as one of the simplest correlation-corrected *ab initio* methods nowadays available[100].

If a molecule with no-bond homoaromaticity is investigated, the system in question possesses a non-classical structure with an interaction distance typical of a transition state rather than a closed-shell equilibrium structure. One can consider no-bond homoconjugative interactions as a result of extreme bond stretching and the formation of a singlet biradical, i.e. a low-spin open-shell system. Normally such a situation can only be handled by a multi-determinant description, but in the case of a homoaromatic compound the two single electrons interact with adjacent π-electrons and form together a delocalized electron system, which can be described by a single determinant *ab initio* method provided sufficient dynamic electron correlation is covered by the method.

MP2 theory, which includes all doubly excited configurations (pair correlation effects) but neglects any coupling between these excitations, is the right method to describe non-classical structures and stretched bond situations[101,102]. But it also exaggerates their stability and therefore leads to an imbalance between classical bicyclic structures and non-classical homoaromatic structures. It is a typical experience with HF and MP2 calculations of potentially homoaromatic molecules that the former method predicts the classical bicyclic or open structure while the latter method predicts the non-classical homoaromatic structure to be more stable (see Sections II.D and IV.B). In general, one can say that MP2 results should be closer to reality than HF results, in particular with regard to calculated geometries and the assessment of bond equalization in conjugated molecules. However, to get reliable stabilization energies one certainly has to go beyond MP2 calculations. This is particularly true when the energy difference between bicyclic and a potential no-bond homoaromatic form is relatively small, which is quite often the case[49–58].

Extension to third-order MP (MP3) theory[103] normally leads to a decrease of homoaromatic stabilization energies because at the MP3 level the coupling between double excitations is included and, accordingly, an overestimation of dynamic pair correlation effects is partially reduced. Although MP3 is more accurate than MP2, it is quite unattractive for *ab initio* investigations since it just corrects the correlation effects introduced at the MP2 level without including any new correlation effects. These are introduced at the fourth-order MP (MP4) level[104] in the form of single (S), double (D), triple (T) and quadruple (Q) excitation effects. MP4(SDQ) already provides an important correction with regard to MP2 homoaromatic stabilization energies. S excitations lead to orbital relaxation effects and Q excitations cover pair, pair correlation effects. The overestimation of the stability of structures with stretched bonds is largely corrected by a coupling of D with Q excitations[101,102]. Hence, MP4(SDQ) represents a relatively inexpensive correlation method for homoaromatic systems that may not be described correctly by MP2.

T excitation effects have turned out to be essential for an accurate description of non–classical systems. Although the contribution of a single T excitation to the correlation energy is rather small, the large number of T excitations leads to sizeable effects, which must not be neglected if very accurate homoaromatic stabilization energies are desired. However, with MP4, there is the danger that T effects are overestimated since TT as well as ST, DT and QT coupling effects enter perturbation theory not before fifth-order MP (MP5)[105,106]. In general, it is a disadvantage of any MPn description that calculated molecular properties oscillate between even-order and odd-order results, since the former introduce new correlation effects while the latter just install the coupling between the new correlation effects[101,102,106]. This means that at one order of perturbation theory one fuels the 'perturbation engine', while in the next order one pushes the 'break' thus causing an oscillatory approach to the true value of the property calculated. Relative energies and geometries oscillate very often between MP1 (= HF) and MP2 values and, in critical cases, it is difficult to predict at what level oscillations are dampened out[106].

One can avoid these problems by using Coupled Cluster (CC) theory[107], which contains infinite-order effects and therefore does not lead to the oscillatory behaviour of properties calculated with MPn[108]. Homoaromatic stabilization energies have been calculated for smaller molecules with CCSD(T) or QCISD(T)[54,56]. These are CC methods, which cover S and D excitations and, in addition, include T effects in a perturbational way[109,110]. They represent some of the most accurate single determinant *ab initio* methods available today that can be applied in a routine way.

There have been just a few investigations of homoaromatic molcules with other than HF, MP or CC methods. Therefore, it is justified to concentrate on the latter and refrain from a lengthy discussion as to how GVB, MCSCF, Cl, MRD-Cl, etc. might lead to a reasonable account of homoconjugative interactions.

Apart from the choice of method and basis set, the geometry optimization of a homoaromatic compound is an essential factor. The optimization of all geometrical parameters is a must for all state-of-the-art *ab initio* calculations. Use of experimental, semiempirical or standard geometries will lead to relatively large errors in the calculated energies. Similarly, one has to warn against the use of *ab initio* results based on partial geometry optimizations or HF/small basis set optimized geometries. Reliable are geometries obtained at the HF/DZ + P, MP2/DZ + P or any higher level of theory[49–58].

Another criterion for the reliability of *ab initio* data is the testing of the character of calculated stationary points by vibrational frequencies. These reveal whether the calculated geometry corresponds to a minimum point on the PES (all eigen values of the Hessian matrix of second derivatives are positive), a first-order saddle point (one eigen value is negative, i.e. one gets one imaginary frequency) or any higher-order saddle point with two, three, etc. negative eigen values. In addition, calculated frequencies are needed for calculating zero-point energy and other vibrational corrections to relative energies, which can be quite important.

Apart from energy and geometry, calculation of the charge distribution in a potentially homoaromatic molecule is very informative. This can be done by the Mulliken population or the NBO/NLMO analysis[78]. In this way, gross atomic charges are obtained, which reflect localization or delocalization of electrons in the molecule. Similar information is obtained from bond orders and bond ellipticities, which are results of the topological analysis of the electron density distribution[80–84]. The latter is based on the virial partitioning of the total electron density distribution[81]. Virial partitioning leads to the most complete and certainly most reliable analysis of electron density features. As described in Section III.E and III.F, one obtains bond orders, π-character, etc. at very low computational cost. In addition, atomic charges and other atomic properties can be determined, although this requires expensive numerical integration.

In the last ten years, NMR chemical shift calculations have become a most valuable asset to *ab initio* descriptions of molecules. This development was triggered by the work of Kutzelnigg and Schindler on the IGLO (Individual Gauge for Localized Orbitals) method[111], which made it possible to calculate reliable relative chemical shifts for rather large molecules in an efficient way. Beside the IGLO method, several other *ab initio* methods are available today for routine calculations of magnetic properties of molecules: (1) The LORG (localized orbital/local origin) method by Hansen and Bouman[112]; (2) GIAO-HF in the version of Pulay and coworkers[113]; (3) GIAO-MP2, GIAO-MP3 and GIAO-MP4 (SDQ) by Gauss to calculate correlation-corrected NMR chemical shifts with second-, third- and fourth-order many-body perturbation theory[114]; (4) MC-IGLO by Kutzelnigg and coworkers for problems that require a MCSCF wave function[115].

The use of *ab initio* methods for the calculation of NMR chemical shifts was pushed forward by Schleyer and his coworkers in collaboration with the Kutzelnigg group or other groups[116]. The success of this research tremendously increased the acceptance of *ab initio* results in general and *ab initio* NMR results in particular among experimentally working chemists.

Calculations with IGLO, LORG or GIAO have led to a wealth of NMR chemical shift data and to a new dimension in the cooperation between quantum chemists and experimentalists as is amply documented in the literature[111–116]. Beside energies and geometries, quantum chemists can nowadays offer experimentalists detailed NMR chemical shift data which provide a direct link between theory and experiment, so that calculated energies and geometries become more meaningful for the experimentalist. NMR chemical shift calculations have not only been used to describe the magnetic properties of molecules but also to identify unknown compounds by comparison of experimental and theoretical shift values, to determine equilibrium geometries, to investigate conformational changes, to elucidate the mechanism of molecular rearrangements, to determine solvent effects on NMR data, to identify complexation or coordination of solute molecules by solvent molecules, to detect electronic structure changes caused by the medium and to describe chemical bonding, to mention just some of the many possibilities that have opened to quantum chemists[111–116].

This is the background for using NMR chemical shifts in *ab initio* studies on potentially homoaromatic compounds. Very often these compounds have been generated as labile intermediates in solutions of super acids and therefore no other molecular properties than NMR data are available. In this situation, the calculation of NMR chemical shifts provides the only bridge from theory to experiment. It leads to a determination of relative energy, geometry and other properties of the molecule as will be described in Section IV.B. Reliable values in the case of relative ^{13}C chemical shifts can already be obtained with a DZ + P basis at the IGLO level although more accurate values require basis sets of TZ + P quality[111].

With the available *ab initio* methods, one can also calculate infrared, Raman and ultraviolet spectra as well as many other molecular properties. However, none of these properties has been used extensively in investigations of homoaromatic compounds and therefore we refrain from discussing basic requirements in calculating them by *ab initio* methods.

B. Investigation of the Homotropenylium Cation

In the following, we will discuss *ab initio* descriptions of the prototype of homoaromatic molecules, namely the homotropenylium cation (**45**). Theoretical work on this cation reflects all the problems involved in an *ab initio* investigation of homoaromatic compounds[44,49,53,56,64,69,73,79,93,117–119].

1. *Ab initio calculations of geometry and energy*

Experimental measurements of the geometry of cation **45** have not been possible so far. There is just indirect information on the molecular geometry coming from NMR data, UV measurements or other sources[120]. Direct information on the geometry of **45** is only provided by quantum chemical calculations (see also the discussion in the following chapter[3]).

In line with the discussion given in Section II.E, three different structures are possible (Scheme 14), namely a classical bicyclic structure **45a** that can benefit from cyclopropyl homoconjugation, then a classical monocyclic open structure **45c**, that should possess normal conjugation of a cyclopolyene, and finally a non-classical no-bond homoaromatic structure **45b** with a cyclic 6π electron system formed by 1,7 through-space interactions.

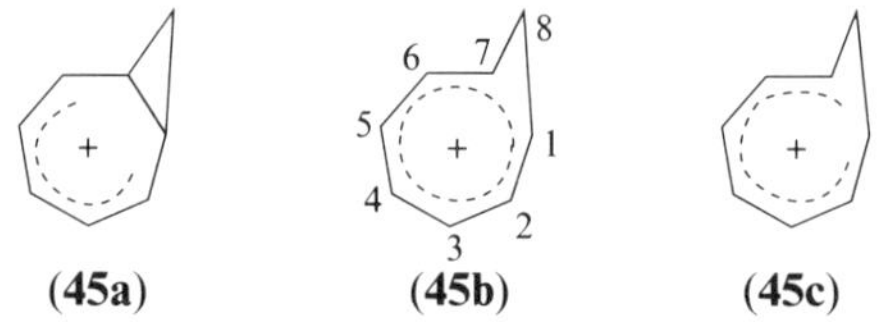

(45a) **(45b)** **(45c)**

SCHEME 14. Possible structures of the homotropenylium cation

HF/STO-3G as well as semi-empirical calculations predict for **45** the bicyclic structure **45a** while HF/6–31 G(d) calculations suggest the open structure **45c**[44,49,64,68,69,93,117]. X-ray structure determinations of substituted homotropenylium cations make the situation even more confusing for deciding the correct geometry for the parent cation[121,122]. For example, 2-hydroxy-**45** was found to possess a relatively short 1,7 distance in line with the bicyclic structure **45a**[121] while 1-ethoxy-**45** has a 1,7 distance of 2.4 Å in line with the open structure **45c** (see also the discussion in the following chapter[3])[122].

To clarify the structural problem of **45**, it is of advantage to calculate the PES in the direction of the 1,7 interaction. This is done by selecting fixed values of $R(1,7)$ between 1.5 and 2.5 Å and optimizing the molecular geometry for each chosen $R(1,7)$ value. Normally, HF/DZ + P or MP2/DZ + P provide a reasonable description of geometrical parameters depending on the interaction distance, although the latter itself may not be described well because of reasons discussed in Section IV.A. As soon as a number of trial geometries is generated, their relative energies can be tested by various methods to find the true shape of the PES in the direction of the interaction distance. This procedure has been used by Cremer and coworkers to investigate homoconjugative interactions in a number of potentially homoaromatic molecules[49–58]. In the following section it is described for cation **45**.

The HF/6–31 G(d) PES in the direction of the 1,7 interaction distance is shown in Figure 16 which, because of the deficiencies of the HF approach, predicts the global minimum of

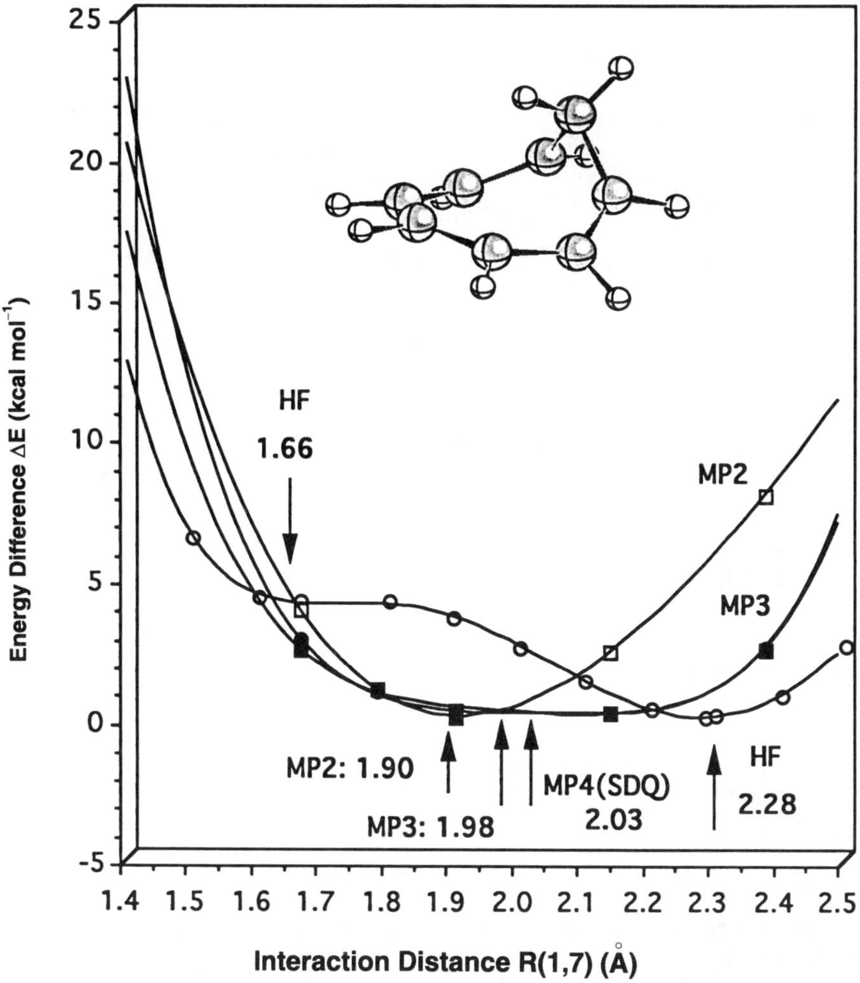

FIGURE 16. Potential energy surface of the homotropenylium cation (**45**) as a function of the C1,C7 interaction distance according to HF, MP2, MP3 and MP4(SDQ) calculations with the 6-31G(d) basis set. Positions of the energy minima are indicated by arrows

45 to be located at $R(1,7) = 2.285$ Å and a local minimum at $R(1,7) = 1.664$ Å[49]. The energy difference between the two minima is 4.1 kcal mol^{-1}, in agreement with similar calculations of Haddon[64].

However, the shape of the PES changes completely when correlation-corrected calculations are carried out (Figure 16)[49]. The MPn/6–31 G(d) (n = 2, 3, 4) PES for **45** possesses only one minimum in the direction of the 1,7 coordinate. This is located between 1.9 and 2.1 Å {1.901 (MP2), 1.985 (MP3) and 2.031 Å [MP4 (SDQ)][49]}, i.e. in a region typical of a non-classical structure **45b**. A 1,7 distance of *ca* 2 Å is accompanied by almost complete bond equalization in the seven-membered ring. The CC bond lengths vary between 1.396 and 1.404 Å with an average CC bond length (without C1—C7) of 1.399 Å (Figure 17).

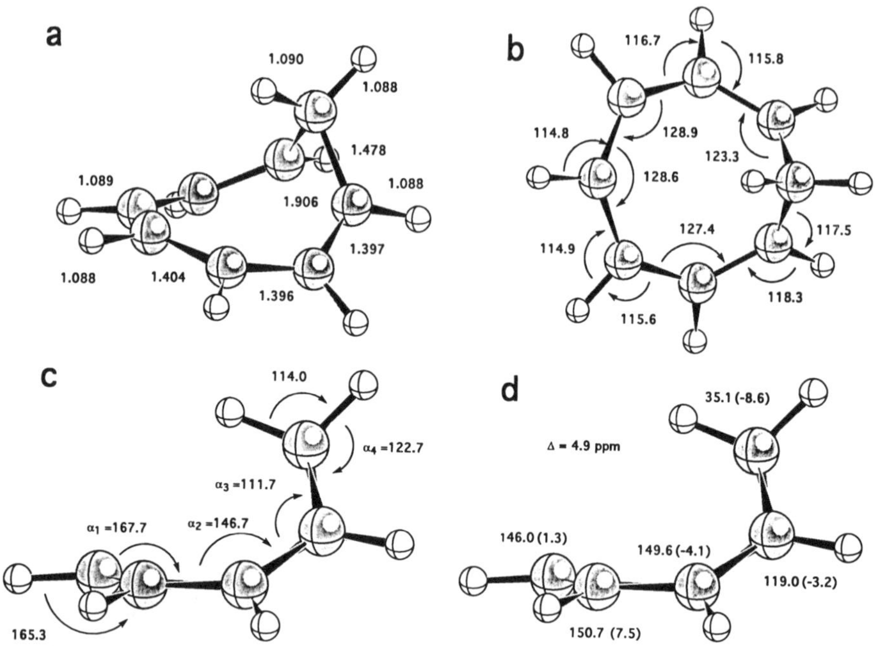

FIGURE 17. MP2/6-31G(d) equilibrium geometry and IGLO/6-31G(d,p) ^{13}C NMR chemical shifts of the homotropenylium cation (**45**). (a) Bond lengths in Å. (b) Bond angles in deg. (c) Folding angles in deg. (d) ^{13}C NMR chemical shifts in ppm relative to TMS calculated at the MP2 geometry with the distance C1,C7 determined at the MP4(SDQ) level of theory. Numbers in parentheses give the deviation of calculated shift values from experimental ^{13}C shifts. The mean deviation between calculated and experimental shifts is denoted by Δ[49]

Noteworthy is the fact that the CH_2 bridge is bent inward by 12° (see Figure 17c). In this way, the *endo* H atom takes a position about 2.2 Å above the centre of the seven-membered ring. The H atoms at the periphery of the ring are slightly bent downward away from the bridge (see Figure 17).

A feature, which becomes apparent from Figure 16, is the flatness of the PES in the region between 1.5 and 2.5 Å[49]. At MP4 (SDQ)/6–31 G(d), a change in the equilibrium value of $R(1,7)$ by ± 0.4 Å leads to an energy change of just 3.5 kcal mol^{-1} corresponding to a (harmonic) force constant of just 0.2 mdyn Å^{-1}[49].

2. Determination of the equilibrium geometry by the ab initio/ chemical shift/NMR method

An alternative approach to test the homoaromatic character of a molecule is based on the calculation and analysis of NMR chemical shift values[49–58]. The determination of NMR chemical shifts by *ab initio* methods such as IGLO, LORG or GIAO turns out to be very sensitive with regard to the geometry used[49,111–116]. *Ab initio* geometries provide a consistent description of molecules that does not suffer from the ambiguities of experimental geometries. Many calculations have shown that reasonable NMR chemical shifts are obtained for HF/DZ + P or MP2/DZ + P optimized geometries[49,111–116,118]. Since the calculated NMR chemical shifts clearly depend on the geometry, *an agreement between experimental and theoretical shifts not only means a clear identification but also a geometry determination of the molecule in question.* On the other hand, if theoretical and experimental shifts differ considerably, other possible geometries or structures have to be tested.

Schleyer was the first to fully realize the sensitivity of calculated NMR chemical shifts with regard to molecular geometry and use this for *ab initio*/IGLO/NMR-based structural determinations in many cases[116]. Cremer and coworkers realized the usefulness of this approach for the determination of geometries of potentially homoaromatic compounds not amenable to experiment[49–58].

In Figure 18 experimental and IGLO/6–31G(d) ^{13}C chemical shifts of **45** are compared[47]. Experimental ^{13}C chemical shifts[120] do not agree with chemical shifts calculated for the two HF/6–31G(d) minima structures or any structure close to **45a** or **45c**. Mean deviations Δ between calculated and experimental ^{13}C shifts are as large as 40 ppm, thus exceeding normal IGLO/6–31G(d,p) errors by a factor of 6 and more. Clearly, IGLO ^{13}C

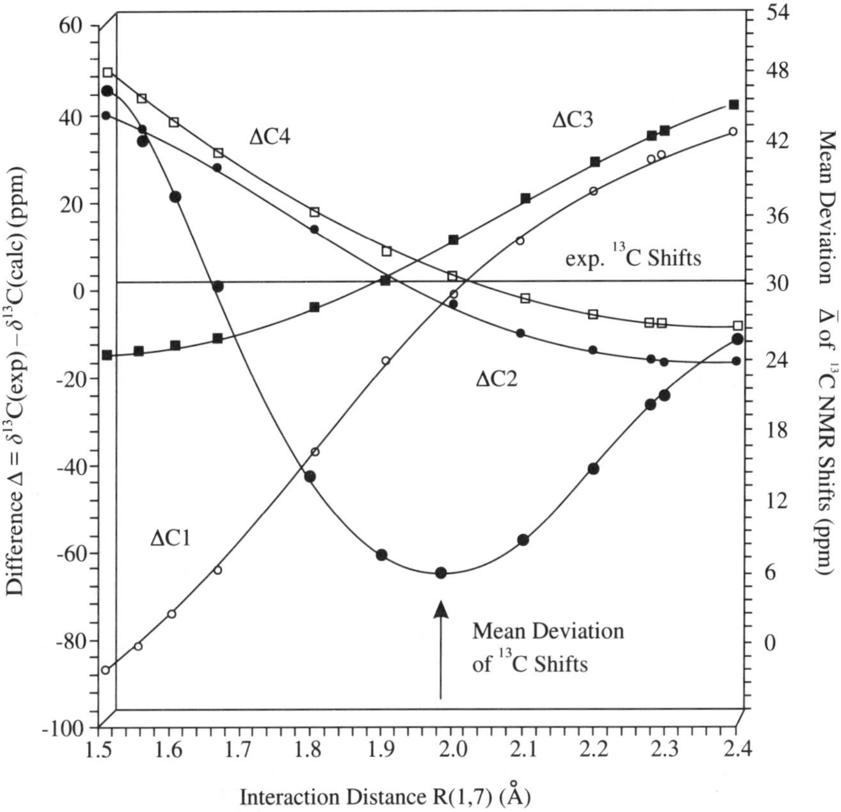

FIGURE 18. Differences between IGLO/6-31G(d,p) ^{13}C NMR chemical shifts and experimental shift values of the homotropenylium cation (**45**) given as a function of the interaction distance C1,C7. The zero line corresponds to the experimental shifts. Also given is the mean deviation $\bar{\Delta}$ of calculated chemical shifts. The minimum of $\bar{\Delta}$ (indicated by an arrow) defines the equilibrium value of the interaction distance C1,C7. All shift values in ppm

chemical shifts suggest that neither **45a** nor **45b** represents the true equilibrium structure of **45**.

If differences Δ for atoms C1–C7 are plotted as a function of the distance $R(1,7)$, all the curves are found to intersect the zero line corresponding to the experimental ^{13}C values at a C1,C7 distance between 1.9 and 2.0 Å (Figure 18). In this region of $R(1,7)$, the best agreement between IGLO and experimental ^{13}C chemical shifts for **45** is found as reflected by a mean deviation $\bar{\Delta}$ of 6 ppm [Figure 18, $R(1,7) = 1.97$ Å][49]. This implies that the PES of **45** in the direction of the $R(1,7)$ coordinate possesses a single minimum rather than the double minimum calculated at the HF level of theory (Figure 16).

In Figure 19, IGLO/6–31G(d) differences $\delta H_a - \delta H_b$ and magnetic susceptibilities $-\chi$ are also plotted as a function of the interaction distance $R(1,7)$. The difference between the shifts of *endo* (H_a) and *exo* proton (H_b) at C8 as well as the magnetic susceptibility are sensitive to electronic delocalization in the potential ring C1—C7. If, for a particular $R(1,7)$ value between 1.5 and 2.5 Å, homoaromatic 6π delocalization becomes a maximum, then this will lead to a large exaltation of $|-\chi|$ as well as large diamagnetic shielding of the *endo* proton H_a and, therefore, to a large difference $\delta H_a - \delta H_b$[49].

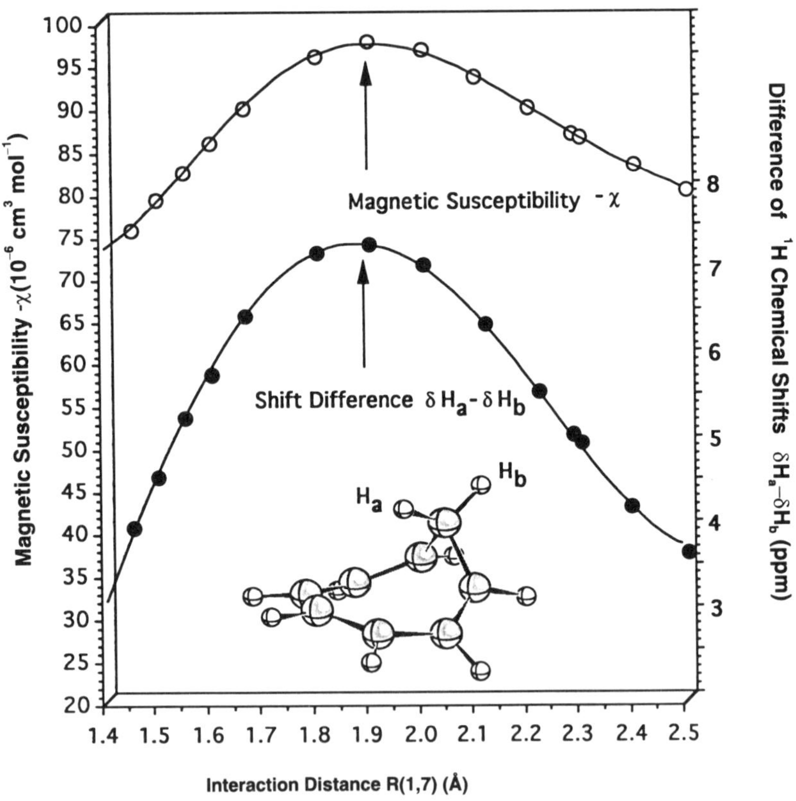

FIGURE 19. IGLO/6-31G(d,p) magnetic susceptibility $|-\chi|$ and shift difference δH_a–δH_b of the homotropenylium cation (**45**) as a function of the interaction distance C1,C7. In each case, the position of the maximum is given by an arrow

Both $\delta H_a - \delta H_b$ and magnetic susceptibility $|-\chi|$ adopt maximal values at $R(1,7)$ distances close to 1.9 Å, suggesting that for this 1,7 distance (homoaromatic) 6π delocalization is strong. Since 6π delocalization will add to the stability of **45**, the location of the maximum of both the difference $\delta H_a - \delta H_b$ and $|-\chi|$ provides further support for the prediction that the equilibrium structure of **45** is characterized by the homoconjugated structure **45b** and a 1,7 distance close to 1.9 Å[49].

In conclusion, the calculated magnetic properties of cation **45** suggest that:

(a) the homotropenylium cation possesses a single minimum PES along the 1,7 coordinate (Figures 16 and 18);

(b) the preferred 1,7 distance is close to 2 Å;

(c) 6π electron delocalization at this $R(1,7)$ value increases the difference $\delta H_a - \delta H_b$ and the magnetic susceptibility $|-\chi|$ to maximum values.

Magnetic susceptibility and chemical shifts are sensitive antenna by which changes in the electronic structure due to geometrical and conformational changes can be measured and analysed. Therefore, they can be used to detect homoaromatic electron delocalization in a compound such as cation **45**. Their sensitivity may be illustrated by the data in Figure 17. If the MP2 equilibrium geometry of **45** is used, the mean deviation between experimental and calculated shift values is 6.2 ppm. However, utilizing the optimal MP4 (SDQ) 1,7 distance (see Figure 16) for the shift calculations, the mean deviation drops to 4.9 ppm caused by an improvement in the ^{13}C shift for atoms C1 and C7[49]. Since this shift value is probably most sensitive to a correct description of 1,7 interactions at the true C1,C7 distance, the improvement of the mean deviation $\bar{\Delta}$ indicates that MP4 (SDQ) provides the best account of homoconjugative interactions and the resulting equilibrium geometry **45b**.

C. Toward a General Definition of Homoaromaticity

Ab initio theory provides exact data on many molecular properties not amenable to experiment. In this way, it leads to a largely complete description of homoconjugated molecules and helps to identify and characterize homoconjugative systems with bond or no-bond homo(anti)aromaticity. A reliable description will be obtained if the PES is systematically scanned in the direction of the interaction distance using correlation-corrected methods with sufficiently large basis sets as described in the case of the homotropenylium cation (Section IV.B). Mere inspection of the PES reveals whether situation 1, 2, 3 or 4 of Figure 6 (see Section II.D) is given. Use of the Cremer–Kraka criteria of covalent bonding[27,82,83] will show whether calculated equilibrium structures belong to the class of bond or no-bond homoaromatic compounds, i.e. the position of point R_b (see Figure 3) can be clearly determined. However, it is not possible to decide by a single criterion whether the compound in question benefits from homoconjugative interactions or exhibits just normal cyclopropyl conjugation or weak through-space interactions, i.e. where points R_a and R_c (see Figure 3) are located. To answer this question a whole series of checks has to be carried out as is listed in the following for the two basic possibilities of bond and no-bond homoconjugation.

1. *Bond homo(anti)aromaticity caused by cyclopropyl homoconjugation*

Bond homoaromaticity is literally identical with cyclopropyl homoaromaticity since no examples involving cyclobutyl or other rings are reported in the literature. Nevertheless, it is advisable to define bond homoaromaticity in a general way that leaves open the question whether there is any bond homoaromaticity beyond cyclopropyl homoaromaticity.

A homoaromatic system is characterized by the following properties:

(1) The interacting centres are connected by a bond path with a bond critical point **p**, at which the energy density distribution $H(\mathbf{r})$ is stabilizing [$H(\mathbf{p}) < 0$].

(2) The bond order n of the closing bond is between 0 and 1, thus indicating a partial bond.

(3) The π-character of the closing bond as measured by the bond ellipticity ε is larger than that of cyclopropane.

(4) Electron delocalization in the cyclic system is characterized by:

(a) a relatively large degree of bond equalization with bond lengths differing from those of normal single or double bonds,

(b) calculated bond orders and bond ellipticities that are approaching those of an aromatic π system,

(c) the major axes of the bond ellipticities of the cyclic system formed by the homoconjugative bond overlap effectively,

(d) in case of charged molecules, positive or negative charge is delocalized throughout the cyclic system.

(5) The number N of π-electrons participating in electron delocalization, $N = 2\Sigma_i n_i - 2N_\sigma$ (where N_σ is the number of formal σ-bonds), is close or identical to $4q + 2$ ($q = 0,1,2,\cdots$).

In a similar way, a homoantiaromatic system formed by bond (cyclopropyl) homoconjugation can be described. There is, however, one major difference between homoaromatic and homoantiaromatic systems (observed in the case of cyclopropyl homoconjugation) that separates homoaromaticity from aromaticity. While aromaticity and antiaromaticity involve different numbers of electrons ($4q + 2$ or $4q$), homoaromaticity and homoantiaromaticity both involve $4q + 2$ electrons but differ with regard to the delocalization modes of these electrons, which are best described by the direction of surface delocalization in a three-membered ring (Scheme 15):

(6) For a homoaromatic system, surface delocalization in the cyclopropyl ring is perpendicular to the bridging bond, thus forming a Hückel aromatic electron ensemble which is delocalized in just one part of the bi(poly)cyclic system.

(7) For a homoantiaromatic system, surface delocalization in the cyclopropyl ring is parallel to the bridging bond, thus forming a Möbius antiaromatic electron ensemble delocalized along the periphery of the bi(poly)cyclic ring system.

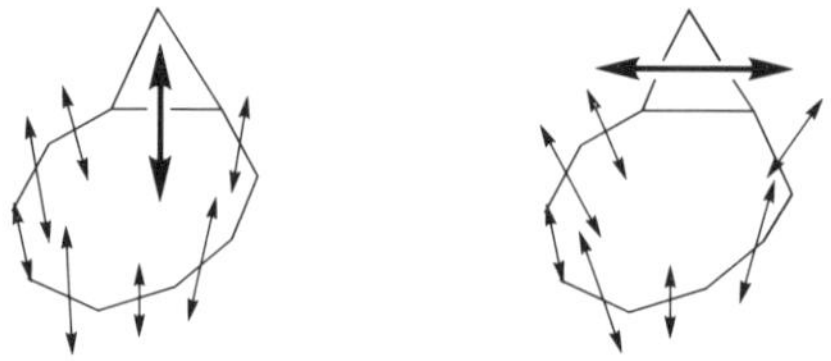

homoaromatic system

homoantiaromatic system

SCHEME 15. Surface delocalization in homoaromatic and homoantiaromatic molecules. Major axes of bond ellipticities are indicated by arrows; the direction of surface delocalization in the three-membered ring is given by a bold arrow

Homoconjugative electron delocalization leads to stabilization or destabilization of the molecule, which can be determined provided correct reference compounds with appropriate reference conformations are chosen. In the case of cyclopropyl homoconjugation, these should be vinylcyclopropane, 1,3-butadiene and their methyl derivatives. The discussion in Section III.G clearly shows that by the use of either correctly chosen homodesmotic reactions or appropriately parametrized force fields, reliable resonance energies for homoconjugated molecules can be calculated.

(8) The resonance energy of a homoaromatic molecule is ≤ –2 kcal mol^{-1} and that of a homoantiaromatic molecule ≥ 2 kcal mol^{-1}. Typical values are between |2| and |10| kcal mol^{-1} indicating that homoconjugative stabilization or destabilization is normally just a matter of a few kcal mol^{-1}.

The many homoaromatic stabilization energies or homoantiaromatic destabilization energies published in the literature very often are contaminated by energies resulting from strain, hyperconjugative or inductive effects, and therefore care must be taken if those values are used to decide the homo(anti)aromatic character of a molecule. Also, one has to warn against comparing homoconjugative resonance energies with resonance energies of aromatic or antiaromatic compounds published in the literature. In most cases, the latter are defined with regard to the *trans* rather than the *cis* form of 1,3-butadiene and therefore they are too low. The argument that norcaradiene (RE = 4 kcal mol^{-1}, Section III.G.4) possesses about 19% of the resonance energy of benzene (21 kcal mol^{-1}, Section III.G.5) is wrong, since the latter value has been obtained using *trans*-1,3-butadiene as a reference. The resonance energy of benzene relative to *cis*-1,3-butadiene is 25.9 kcal mol^{-1} (Section III.G.5) and accordingly norcaradiene covers just 15% of this value.

There are no systematic investigations that clarify how the magnetic properties of a molecule change if cyclopropyl homoconjugation leads to cyclopropyl homoaromaticity. However, in view of the sensitivity of magnetic properties with regard to homoconjugation in the case of no-bond homoaromaticity (see Section IV.B and the following section), it is likely that further research will unravel the dependence of magnetic properties on the extent of cyclopropyl homoconjugation. Such relationships will definitely add to the list of criteria that characterizes cyclopropyl homo(anti)aromaticity.

2. No-bond homoaromaticity

All investigations carried out so far suggest that no-bond homoaromaticity in the case of hydrocarbons manifests itself in the following way:

(1) A cyclic system is formed by strong though-space interactions via interaction distances between 1.8 and 2.2 Å with an optimal value at about 2 Å.

(2) There is no path of maximum electron density between the interacting atoms which, according to Cremer–Kraka[27,82,83], is a necessary condition for covalent bonding. However, interaction indices derived from the electron density distribution are as large as 30% of the bond order of a normal single bond.

(3) Through-space interactions are confirmed by the Laplace concentration $-\nabla^2\rho(\mathbf{r})$ that reveals polarization of the electron density at the interacting atoms.

(4) Electron delocalization in the cyclic system is characterized by:

(a) a relatively large degree of bond equalization with bond lengths differing from those of normal single or double bonds (averaged bond length close to 1.40 ± 0.01 Å),

(b) calculated bond orders and bond ellipticities that are approaching those of an aromatic π system,

(c) the major axes of the bond ellipticities of the cyclic system formed overlap effectively,

(d) in case of charged molecules, positive or negative charge is delocalized throughout the cyclic system.

(5) The number N of π-electrons participating in electron delocalization, $N = 2\Sigma_i n_i - 2N_\sigma$ (where N_σ is the number of formal σ-bonds), is close or identical to $4q + 2$ ($q = 0, 1, 2, \cdots$).

Contrary to bond homoantiaromaticity, very little is known about no-bond homoantiaromaticity[123]. In a potentially no-bond homoantiaromatic molecule, there is often the possibility of avoiding strong destabilizing through-space interactions either by valence tautomeric rearrangements of by conformational changes that lead to larger interaction distances and, accordingly, reduced through-space interactions. Therefore, the energetic consequences of no-bond homoconjugation are considered just for homoaromatic molecules.

(6) The resonance energy of a homoaromatic molecule is ≤ -2 kcal mol^{-1}. Typical values are between -2 and -10 kcal mol^{-1} indicating that homoconjugative stabilization is normally just a matter of a few kcal mol^{-1}.

For the magnetic properties of a no-bond homoaromatic molecule, one can expect typical values.

(7) Because of electron delocalization, there should be a significant equalization of ^{13}C chemical shifts in the cyclic system.

(8) For the magnetic susceptibility $|-\chi|$ determined as a function of the interaction distance, a maximum value should be calculated for the homoaromatic system, i.e. the exaltation of the magnetic susceptibility should indicate homoaromatic electron delocalization.

(9) If the system in question possesses a CH_2 group located above the ring in a similar way to the case of the homotropenylium cation, the shift difference between *endo*- and *exo*-oriented proton should also adopt a maximum value for the homoaromatic system.

3. General remarks

According to the definition given above, both bond (cyclopropyl) and no-bond homoaromaticity can occur for cationic (many examples), neutral (several examples) and anionic systems (few examples) with a frequency that can be explained on the basis of PMO theory (Section III.C). Apart from a few exceptions[124], homoaromaticity has just been observed for hydrocarbons, but recent calculations indicate that homoconjugative interactions can also be expected for Si-containing analogues of homoaromatic systems[35]. In principle, there is no reason to exclude homoaromaticity for heteroatom-containing systems. The only question is how to detect homoconjugative interactions in the presence of strong inductive, anomeric, hyperconjugative or steric effects. Too little work has been done in this direction to clarify whether homoconjugation is an important factor in heteroatomic molecules.

As stressed in Section II.D, homoaromaticity plays an important role in the transition states of certain pericyclic reactions. The valence tautomeric rearrangement of cycloheptatriene to norcaradiene is an example par excellence, as has been demonstrated by calculations of Kraka and Cremer[54]. Other examples have been described by Grimme and coworkers[125]. There seems to be a special relationship between transition states and systems with no-bond homoaromaticity. The latter possess geometries and other properties typical of transition states. By proper substituiton, they can be pushed into the classical bicyclic form **42a** or the classical monocyclic form **42c** as has been demonstrated elegantly by Childs and coworkers[121,122] in the case of the homotropenylium cation. Therefore, *it is appropriate to consider no-bond homoaromatic systems as frozen transition states* (Section II.D), i.e. transition states that by homoconjugative electron delocalization have been energetically lowered below the energies of the two classic forms **42a** and **42c** of a valence tautomeric rearrangement **42a** to **42c**. *No-bond homoaromatic molecules, such as the homotropenylium or the 1,4-bishomotropenylium cation, are the first frozen transition states discovered so far*[49–55].

The description of no-bond homoaromatic systems as frozen transition states is in line with the observation that their PES is rather flat in the direction of the interaction distance. This means that (a) homoaromatic stabilization energies are small (see Table 2 and the discussion presented above) and (b) relatively small energy increases lead to relatively large changes in geometry as well as other properties. Homoaromaticity is mostly accompanied by a delicate balance between (destabilizing) strain and (stabilizing) through-space or through-bond interactions, and therefore *small perturbations due to substituent, counter ion or media effects may disrupt no-bond homoaromatic delocalization.* This has to be kept in mind when trying to confirm homoaromatic character by investigating substituted derivatives of the target molecule.

The possibilities of experiment are often very limited when it comes to the detection, verification and description of homo(anti)aromatic character. These limitations, however, can be compensated by combining experimental with theoretical tools. This holds in particular with regard to the measurement and calculation of the magnetic properties of potentially homoaromatic molecules. Using the NMR/chemical shifts/*ab initio* approach of Cremer and coworkers[49–58] which combines experimental and calculated shift values, it is possible to determine geometry (in particular with regard to the interaction distance), relative energy, electron delocalization and many other properties of the compound in question. *The shift values themselves as well as the magnetic susceptibility exaltation provide sensitive detectors for homoconjugative electron delocalization.* Since one has just started to use these tools to investigate homoconjugated molecules, one can foresee for the future many surprising insights into an electronic phenomenon that, although generally known and accepted for a long time, was only very vaguely described and defined in the textbooks and review articles of the past.

V. CONCLUSIONS AND OUTLOOK

Research on (cyclopropyl) homoconjugation and homoaromaticity has inspired generations of chemists to develop new strategies and techniques for exploring the possibilities of homoaromatic electron delocalization. Quantum Chemistry has accompanied experiment through 40 years of homoaromaticity research and has resulted in many useful descriptions and explanations. Nevertheless, it took until the nineties before theory was able to provide the exact data on molecular properties that are needed in research on homoaromatic molecules. In this respect one must mention the availability of Coupled Cluster methods for high accuracy calculations of energies and geometries as well as the availability of *ab initio* methods for calculating magnetic properties of a molecule (see Section IV.A).

Ab initio research on homoconjugation and homoaromaticity has to fulfil several requirements and tasks to lead to a reliable account of the homoaromatic character of homoconjugated molecules.

(1) *Calculation of reliable energies and accurate geometries.* State-of-the-art *ab initio* methods can fulfil this requirment, if necessary even for medium-sized molecules. It is doubtful whether such a statement can also be made without any reservation for semi-empirical methods. Although improved descriptions of energy and geometry are obtained with methods such as MNDO-Cl or AM1-Cl in selected cases, one cannot rely on this in general. Since investigations with these methods always included configuration interaction in a limited way, one had to decide from case to case how many MOs are considered in the Cl treatment to get reliable results. Thus in the case of investigations based on MNDO or AM1 in connection with limited Cl one cannot speak of a predictive approach, more of an ad hoc adjustment of theory to reproduce experimental facts already known.

(2) *Potential energy surface (PES) scans.* *Ab initio* research on homoaromatic compounds always requires some exploration of the PES rather than just the investigation of equilibrium geometries. As discussed in Sections II.D and IV.B, it is essential to determine the shape of the PES as a function of homoconjugative interaction distances. The number and location of all stationary points have to be determined so that one can distinguish

between (homoconjugated) classical structures, (homoaromatic) non-classical structures or the existence of valence tautomeric equilibria (Figure 6, Section II.D).

(3) *Calculation of magnetic properties.* A most valuable tool in *ab initio* studies on homoconjugated molecules is the calculation of magnetic properties. NMR chemical shifts or shift differences are very sensitive with regard to geometric changes and charge delocalization. In addition, the magnetic susceptibility represents a useful antenna for electron delocalization. Theory has just begun to use the calculation of magnetic properties as a descriptive tool, and therefore one can expect that future *ab initio* investigations will lead to new and probably unexpected insights into the electronic nature of potentially homoaromatic molecules.

(4) *Electron density analysis.* Another important part of *ab initio* research on homoaromatic molecules is the electron density analysis along the lines worked out by Cremer and coworkers[27,82–84] and based on Bader's virial partitioning method[81]. Applying the Cremer–Kraka criterion of covalent bonding[82,83], a differentiation between bond and no-bond homoconjugation (homoaromaticity) (determination of point R_b in Figure 3) becomes possible. In addition, the electron density analysis provides useful molecular parameters (bond orders, interaction indices and π-character indices) that provide an assessment of the degree of electron delocalization. Analysis of the Laplace concentration gives insight into the degree of through-space interactions, although this has not been put on a quantitative basis so far.

(5) *Calculation of homoaromatic resonance energies.* A new element of future *ab initio* work on homoconjugated molecules must be the calculation of homoaromatic resonance energies. In this respect, the analysis of homodesmotic reaction energies can provide a reasonable basis for getting reliable resonance energies (Section III.G). The only method currently available to obtain accurate resonance energies is the molecular mechanics approach of Roth. This approach demonstrates how resonance energies have to be calculated, although it suffers from all the limitations normally encountered by molecular mechanics methods. For example, it depends strongly on the availability of exact structural, conformational and thermochemical data. Since the latter are only available for certain classes of neutral homoaromatic molecules, Roth's method is applicable only to neutral compounds, not to the many interesting cationic and anionic homoconjugated molecules. The future will show whether extension of the force field parametrization can be based on accurate *ab initio* rather than experimental data.

(6) *Investigation of environmental effects.* As has been stressed in this chapter, homoaromaticity is just a matter of a few kcal mol^{-1} stabilization energy in most cases, and therefore environmental effects may have a large impact on structure, stability and other properties of a homoaromatic compound. Future work in theory (as well as in experiment) has to clarify how environmental effects can influence electron delocalization, through-space interactions and bonding in homoaromatic molecules. The theoretical methods are now available to calculate solvent and counter ion effects (for homoaromatic ions in solution) or to study intermolecular and crystal packing forces in the solid state.

By complying with these guidelines, *ab initio* theory should be able to answer some of the pending questions in research on homoconjugation and homoaromaticity.

Although the structural elements supporting cyclopropyl homoaromaticity and no-bond homoaromaticity are now generally understood, it is not clear under what conditions a homoconjugated molecule will prefer to occupy a single minimum or to adopt classical forms connected by a valence tautomeric equilibrium. Of course, one can explain that the norcaradiene/cycloheptatriene system is characterized by a valence tautomeric equilibrium while the homotropenylium cation possesses a single minimum PES. This has simply to do with the fact that in the cyclopropyl carbinyl cation (embedded in the homotropenylium cation) the vicinal bonds are much more labile than the corresponding bonds in vinylcyclopropane (embedded in norcaradiene) (see Chapter 2 in this volume[2]). But this qualitative explanation is not sufficient if one wants to predict on a quantitative basis under which structural and electronic conditions a valence tautomeric equilibrium degenerates to a single minimum situation.

Actually, this question focuses on the generation of a frozen transition state situation. Can each valence tautomeric equilibrium between an unconstrained cyclopropyl homoconjugated compound and its open monocyclic counterpart be manipulated in such a way that a frozen transition state is obtained? There are results pointing in this direction (see the following chapter[3]), however at the moment one is far from being able to generalize any of these observations.

The investigation of homoaromatic moleclues with frozen transition state character is certainly one of the most fascinating goals in chemistry. Transition states are transient points on the PES where the reacting molecule does not stay any longer than at any other non-equilibrium point. Accordingly, there are only a few experimental ways of getting indirect evidence on the nature of transition states. On the other hand, chemists need exact knowledge about transition states in order to steer and manipulate chemical reactions. The freezing of a transition state provides a very attractive way of getting direct evidence on its properties. For example, the investigation of the homotropenylium cation reveals that stabilizing electron delocalization is very effective at distances of 2 Å. This can also be assumed for the transition state of a pericyclic reaction characterized by an aromatic ensemble of electrons. As a consequence, the transition state energy is relatively small compared to the dissociation energies of normal CC bonds. CC bond formation or bond rupture in pericyclic reactions can be manipulated using rather mild steering and regulating devices (temperature, solvents, etc.) compared to the brute forces needed for dissociation of a molecule into its atoms.

Another interesting, but only little investigated aspect concerns electron delocalization and homoaromaticity in three dimensions (homoradial aromaticity, homo-3D aromaticity, homospherical aromaticity (Figure 2, Section I). An increasing number of examples are becoming available suggesting that electron delocalization is not just a one-dimensional phenomenon (along the acyclic or cyclic chain of atoms in the form of ribbon delocalization; Figure 2 and Reference 13) but can also be two- or three-dimensional, i.e. in the form of surface or volume delocalization[13]. Examples of three-dimensional homoaromaticity will be discussed in the following chapter by Childs, Cremer and Elia[3]. From this discussion it will become evident that much more research is needed to fully understand the electronic structure of compounds with 'three-dimensional' homoaromaticity.

Homoconjugation influences the reactivity and internal rearrangements of polycyclic unsaturated hydrocarbons such as semibullvalene, bullvalene, barbaralyl cation, etc. The facile rearrangements of these compounds become possible because of stabilization of their transition states by homoconjugative (homoaromatic) interactions. As described in the case of the barbaralyl cation[57], electron delocalization involves all parts of the molecule and makes all carbon–carbon bonds prone to dissociation and reformation. The rapid rearrangements of the barbaralyl cation (all barriers $\leq 5\ \mathrm{kcal\,mol^{-1}}$ [57]) lead to a complete exchange of all nine carbon positions at temperatures above –150 °C and an equilibration of their properties. Again, a cyclopropylcarbinyl cation group, now in the form of a divinylcyclopropylcarbinyl cation, is responsible for the labile character of the barbaralyl cation. *Ab initio* calculations show that the transition states of the degenerate rearrangements of the barbaralyl cation benefit from no-bond homoaromaticity.

It should be possible to further reduce the energy barriers to internal rearrangements of the barbaralyl cation or related polycyclic compounds by relatively small changes in the structure or by appropriate substitution. In this way, a situation should be reached in which rearrangements are so fast that the molecule in question no longer possesses a fixed structure. Then, the molecule will be best described by a ball or spherical surface of electron density in which the nuclei swim just obeying Coulomb's law but otherwise taking all possible positions on the surface. For such a molecule, the evolutionary path from chaos to the ordered structure of an assembly of nuclei in a molecule would be reversed and a new field of elementary investigations would become possible.

VI. ACKNOWLEDGEMENTS

This work was supported by the Swedish Natural Science Research Council (NFR) and the Natural Science and Engineering Research Council of Canada. All calculations needed to complement data from the literature were done on the CRAY YMP/416 of the Nationellt Superdatorcentrum (NSC), Linköping, Sweden. The authors thank the NSC for a generous allotment of computer time. RFC thanks MacMaster University for a research leave during which this work was conducted.

VII. REFERENCES

1. L. N. Ferguson, *Highlights of Alicyclic Chemistry*, Part 1, Chap. 3, Franklin, Palisades, New York, 1973; M. Charton, in *The Chemistry of Alkenes*, Vol. 2 (Ed. J. Zabicky), Wiley-Interscience, New York, 1970, p. 511; D. Wendisch, in *Methoden der Organischen Chemie*, Vol. 4, Houben-Weyl, Ed. E. Müller, Thieme, Stuttgart, 1971, p.3; M. Y. Lukina, *Russ. Chem. Rev.*, **31**, 419 (1962).
2. D. Cremer, E. Kraka and K. J. Szabo, in *The Chemistry of the Cyclopropyl Group*, Chap. 2, Wiley, Chichester, 1995.
3. R. F. Childs, D. Cremer and G. Elia, in *The Chemistry of the Cyclopropyl Group*, Chap. 8, Wiley, Chichester, 1995.
4. G. M. Badger, *Aromatic Character and Aromaticity*, Cambridge University Press, 1969.
5. D. Lewis and D. Peters, *Facts and Theories of Aromaticity*, Macmillan, London, 1975.
6. D. Lloyd, *Non-benzenoid Conjugated Carbocyclic Compounds*, Elsevier, Amsterdam, 1984.
7. P. J. Garratt, *Aromaticity*, Wiley, New York, 1986.
8. E. D. Bergmann and B. Pullman (Eds.), *Aromaticity, Pseudoaromaticity, Antiaromaticity*, Jerusalem Symposium on Qunatum Chemistry and Biochemistry, Vol. 3, Israel Academy of Science and Humanities, 1971.
9. W. J. le Noble, *Highlights of Organic Chemistry*, Marcel Dekker, New York, 1974.
10. V. I. Minkin, M. N. Glukhovtsev and B. Y. Simkin, *Aromaticity and Antiaromaticity, Electronic and Structural Aspects,* Wiley, New York, 1994.
11. A. B. McEwen and P. v. R. Schleyer, *J. Org. Chem.*, **51**, 4357 (1986).
12. M. J. S. Dewar, *Bull. Soc. Chim. Belg.*, **88**, 957 (1979); M. J. S. Dewar and M. L. Mckee, *Pure. Appl. Chem.*, **52**, 1431 (1980); M. J. S. Dewar, *J. Am. Chem. Soc.*, **106**, 669 (1984).
13. D. Cremer, *Tetrahedron*, **44**, 7427 (1988).
14. S. Winstein, *Spec. Publ. Chem. Soc.*, **21**, 5 (1967); S. Winstein, *Q. Rev. Chem. Soc.*, **23**, 141 (1969).
15. S. Winstein, *Carbonium Ions*, **3**, 965 (1972).
16. P. R. Story and B. C. Clark, Jr., *Carbonium Ions*, **3**, 1007 (1972).
17. J. Haywood-Farmer, *Chem. Rev.*, **74**, 315 (1974).
18. P. M. Warner, *Top. Nonbenzenoid Aromat. Chem.*, 2 (1976).
19. L. A. Paquette, *Angew. Chem., Int. Ed. Engl.*, **17**, 106 (1978).
20. R. F. Childs, *Acc. Chem. Res.*, **17**, 347 (1984).
21. R. F. Childs, M. Mahendran, S. D. Zweep, G. S. Shaw, S. K. Chadda, N. A. D. Burke, D. E. George, R. Faggiani and C. J. L. Lock, *Pure Appl. Chem.*, **58**, 111 (1986).
22. A. T. Balaban, M. Banciu and V. Ciorba, in *Annulenes, Benzo-, Hetero-, Homo- derivatives and their Valence Isomers*, Vol. III, Chap. 9, CRC Press, Boca Raton, Florida, 1987, p. 144.
23. R. V. Williams and H. A. Kurtz, *Adv. Phys. Org. Chem.*, **29**, 273 (1994).
24. N. C. Deno, *Prog. Phys. Org. Chem.*, **2**, 129 (1964).

25. J. B. Grutzner and W. L. Jorgensen, *J. Am. Chem. Soc.*, **103**, 1272 (1981).
26. E. Kaufmann, H. Mayr. J. Chandrasekhar and P. v. R. Schleyer, *J. Am. Chem. Soc.*, **103**, 1375 (1981).
27. E. Kraka and D. Cremer, in *Theoretical Models of Chemical Bonding, The Concept of the Chemical Bond* (Ed. Z. B. Maksic), Springer-Verlag, Berlin, 1990, p. 453.
28. T. A. Albright, J. K. Burdett and M. H. Whangbo, *Orbital Interactions in Chemistry*, Wiley, New York, 1985.
29. R. Neidlein, D. Christen, V. Poignée, R. Boese, D. Bläser, A. Gieren, C. Ruiz-Pérez and T. Hübner, *Angew. Chem.*, **100**, 292 (1988); R. Boese and D. Bläser, *Angew. Chem.*, **100**, 293 (1988).
30. R. Boese, D. Bläser, W. E. Billups, M. M. Haley, A. H. Maulitz, D. L. Mohler and K. P. C. Vollhardt, *Angew. Chem.*, **106**, 321 (1994).
31. J. S. Siegel, *Angew. Chem.*, **106**, 1808 (1994).
32. M. J. S. Dewar and M. K. Holloway, *J. Chem. Soc., Chem. Commun.*, 1188 (1984).
33. P. v. R. Schleyer, H. Jiao, M. N. Glukhovtsev, J. Chandrasekhar and E. Kraka, *J. Am. Chem. Soc.*, **116**, 10129 (1994).
34. J. E. McMurry, G. J. Haley, J. R. Matz, J. C. Clardy, G. van Duyne, R. Gleiter, W. Schäfer and D. H. White, *J. Am. Chem. Soc.*, **106**, 5018 (1984); J. E. McMurry, G. J. Haley, J. R. Matz, J. C. Clardy, G. van Duyne, R. Gleiter, W. Schäfer and D. H. White, *J. Am. Chem. Soc.*, **108**, 2932 (1986). For similar potentially homoantiaromatic compounds, see: J. E. McMurry, G. J. Haley, J. R. Matz, J. C. Clardy and J. Mitchell, *J. Am. Chem. Soc.*, **108**, 515 (1986); K. B. Wiberg, M. G. Matturro, P. J. Okarma and M. E. Jason, *J. Am. Chem. Soc.*, **106**, 2194 (1984).
35. K. J. Szabo and D. Cremer, *J. Am. Chem. Soc.*, to be published.
36. R. V. Williams and H. A. Kurtz, *Adv. Phys. Org. Chem.*, **29**, 273 (1994); H. D. Roth and C. J. Abelt, *J. Am. Chem. Soc.*, **108**, 2013 (1986); H. D. Roth, *Acc. Chem. Res.*, **20**, 343 (1987).
37. R. B. Woodward and R. Hoffmann, *J. Am. Chem. Soc.*, **87**, 395, 2046, 2511 (1965); R. B. Woodward, *Spec. Publ. Chem. Soc.*, **21**, 217 (1967); R. B. Woodward and R. Hoffmann, in *The Conservation of Orbital Symmetry*, Verlag Chemie, GmbH, Weinheim, 1970.
38. M. J. S. Dewar, *Angew. Chem., Int. Ed. Engl.*, **10**, 761 (1971); H. E. Zimmerman, *Acc. Chem. Res.*, **5**, 272 (1971).
39. M. J. S. Dewar and G. J. Gleicher, *J. Am. Chem. Soc.*, **87**, 685 (1965); M. J. S. Dewar and C. de Llano, *J. Am. Chem. Soc.*, **91**, 789 (1969); M. J. S. Dewar and R. C. Dougerthy, *The PMO Theory of Organic Chemistry*, Plenum Press, New York, 1975; A. L. H. Chung and M. J. S. Dewar, *J. Chem. Phys.*, **42**, 756 (1965).
40. M. N. Paddon-Row and K. D. Jordan, in *Modern Models of Bonding and Delocalization* (Eds. J. F. Liebman and A. Greenberg), VCH Publishers, New York, 1988, p. 115; E. Heibronner, *Isr, J. Chem.*, **10**, 143 (1972).
41. W. R. Roth, O. Adamczak, R. Breuckmann, H.-W. Lennartz and R. Boese, *Chem. Ber.*, **124**, 2499 (1991).
42. W. R. Roth, M. Böhm, H.-W. Lennartz and E. Vogel, *Angew. Chem.*, **95**. 1011 (1983); W. R. Roth, F.-G. Klärner, G. Siepert and H.-W. Lennartz, *Chem. Ber.*, **125**, 217 (1992).
43. S. W. Staley, *J. Am. Chem. Soc.*, **89**, 1532 (1967).
44. D. Cremer, E. Kraka, T. S. Slee, R. F. W. Bader, C. D. H. Lau and T. T. Nguyen-Dang, *J. Am. Chem. Soc.*, **105**, 5069 (1983).
45. R. F. Childs, M. Sakai, B. D. Parrington and S. Winstein, *J. Am. Chem. Soc.*, **96**, 6403 (1974).
46. D. Cremer and T. Schmidt, *J. Org. Chem.*, **50**, 2684 (1985).
47. R. Hoffmann and W.-D. Stohrer, *J. Am. Chem. Soc.*, **93**, 6941 (1971); M. J. S. Dewar and C. Jie, *Tetrahedron*, **44**, 1351 (1988); H. Quast, R. Janiak, E.-M. Peters, K. Peters and H. G. v. Schnering, *Chem. Ber.*, **125**, 969 (1992).
48. H. B. Bürgi, *Inorg. Chem.*, **12**, 2321 (1973); H. B. Bürgi, J. D. Dunitz and E. Shefter, *J. Am. Chem. Soc.*, **95**, 5065 (1973).
49. D. Cremer, F. Reichel and E. Kraka, *J. Am. Chem. Soc.*, **113**, 9459 (1991); D. Cremer, L. Olsson, F. Reichel and E. Kraka, *Isr. J. Chem.*, **33**, 369 (1993).
50. D. Cremer, P. Svensson, E. Kraka, Z. Konkoli and P. Ahlberg, *J. Am. Chem. Soc.*, **115**, 7457 (1993); P. Svensson, F. Reichel, P. Ahlberg and D. Cremer, *J. Chem. Soc., Perkin Trans. 2*, 1463 (1991).
51. D. Cremer, P.Svensson and K. J. Szabo, *J. Org. Chem.*, to be published.
52. K. J. Szabo, E. Kraka and D. Cremer, *J. Am. Chem. Soc.*, to be published.

53. D. Cremer, P. Svensson, F. Reichel and K. J. Szabo, to be published.
54. E. Kraka and D. Cremer, *J. Am. Chem. Soc.*, to be published.
55. K. J. Szabo and D. Cremer, to be published.
56. S. Sieber, P. v. R. Schleyer, A. H. Otto, J. Gauss, F. Reichel and D. Cremer, *J. Phys. Org. Chem.*, **6**, 445 (1993).
57. D. Cremer, P. Svensson, E. Kraka and P. Ahlberg, *J. Am. Chem. Soc.*, **115**, 7445 (1993).
58. K. J. Szabo and D. Cremer, *J. Org. Chem.*, **60**, 2257 (1995).
59. G. Binsch, *Naturwissenschaften*, **60**, 369 (1973).
60. L. Salem, *The Molecular Orbital Theory of Conjugated Systems*, W. A. Benjamin, New York, 1966; E. Heilbronner and H. Bock, *Das HMO-Modell und seine Anwendung—Grundlagen und Handhabung*, Verlag Chemie, Weinheim, 1968.
61. L. A. Paquette, T. G. Wallis, T. Kempe, G. G. Christoph, J. P. Springer and J. Clardy, *J. Am. Chem. Soc.*, **99**, 6946 (1977); G. G. Christoph, J. L. Muthard, L. A. Paquette, M. C. Böhm and R. Gleiter, *J. Am. Chem. Soc.*, **100**, 7782 (1978); L. A. Paquette, *Angew. Chem., Int. Ed. Engl.*, **17**, 106 (1978).
62. L. Pauling, *The Nature of the Chemical Bond*, 3rd edn., Cornell University Press, Ithaca, New York, 1960; L. Pauling, *J. Am. Chem. Soc.*, **53**, 1367 (1931); J. E. Kilpatrik and R. Spitzer, *J. Chem. Phys.*, **14**, 46 (1946); M. Randic and Z. Maksic, *Theor. chim. Acta*, **3**, 59 (1965).
63. R. C. Haddon, *Acc. Chem. Res.*, **21**, 243 (1988); R. C. Haddon, *J. Am. Chem. Soc.*, **109**, 1676 (1987); R. C. Haddon, *Chem. Phys. Lett.*, **125**, 231 (1986); R. C. Haddon, *J. Am. Chem. Soc.*, **108**, 2837 (1986); R. C. Haddon and L. T. Scott, *Pure Appl. Chem.*, **58**, 137 (1986).
64. R. C. Haddon, *J. Am. Chem. Soc.*, **110**, 1108 (1988).
65. M. J. S. Dewar and R. Dougherty, *The PMO Theory of Organic Chemistry*, Plenum Press, New York, 1975; M. J. S. Dewar, *The Molecular Orbital Theory of Organic Chemistry*, McGraw-Hill, New York, 1969.
66. W. J Hehre, *J. Am. Chem. Soc.*, **95**, 5807 (1973).
67. W. L. Jörgensen and W. T. Borden, *J. Am. Chem. Soc.*, **95**, 6649 (1973).
68. R. C. Haddon, *Tetrahedron Lett.*, 2797 (1974); R. C. Haddon, *Tetrahedron Lett.*, 4303 (1974); R. C. Haddon, *Tetrahedron Lett.*, 863 (1975).
69. W. J. Hehre, *J. Am. Chem. Soc.*, **96**, 5207 (1974).
70. A. J. P. Devaquet and W. J. Hehre, *J. Am. Chem. Soc.*, **96**, 3644 (1974).
71. R. C. Haddon, *J. Am. Chem. Soc.*, **97**, 3608 (1975).
72. W. L. Jörgensen, *J. Am. Chem. Soc.*, **97**, 3082 (1975).
73. W. L. Jörgensen, *J. Am. Chem. Soc.*, **98**, 6784 (1976).
74. R. V. Williams, H. A. Kurtz and B. Farley, *Tetrahedron*, **44**, 7455 (1988); R. V. Williams and H. A. Kurtz, *J. Org. Chem.*, **53**, 3626 (1988); R. V. Williams and H. A. Kurtz, *J. Chem. Soc., Perkin Trans. 2*, 147 (1994).
75. H. Fischer and H. Kollmar, *Theor. Chim. Acta*, **16**, 163 (1970).
76. M. J. S. Dewar and D. H. Lo, *J. Am. Chem. Soc.*, **93**, 7201 (1971).
77. D. Cremer, Ph.D. Thesis, Köln, 1972.
78. A. E. Reed, L. A. Curtiss and F. Weinhold, *Chem. Rev.*, **88**, 899 (1988); A. E. Reed, R. B. Weinstock and F. Weinhold, *J. Chem. Phys.*, **83**, 735 (1985); J. P. Foster and F. Weinhold, *J. Am. Chem. Soc.*, **102**, 7211 (1980); T. K. Brunck and F. Weinhold, *J. Am. Chem. Soc.*, **101**, 1700 (1978); A. E. Reed and F. Weinhold, *J. Chem. Phys.*, **78**, 4066 (1983).
79. S. Sieber and P. v. R. Schleyer, unpublished results.
80. R. F. W. Bader, T. S. Slee, D. Cremer and E. Kraka, *J. Am. Chem. Soc.*, **105**, 5061 (1983).
81. R. F. W. Bader, T. T. Nguyen-Dang and Y. Tal, *Rep. Prog. Phys.*, **44**, 893 (1981); R. F. Bader and T. T. Nguyen-Dang, *Adv. Quantum Chem.*, **14**, 63 (1981); R. F. W. Bader, in *Atoms in Molecules—A Quantum Theory*, Oxford University Press, Oxford, 1990; R. F. W. Bader, P. L. A. Popelier and T. A. Keith, *Angew. Chem.*, **106**, 647 (1994).
82. D. Cremer and E. Kraka, *Croat. Chem. Acta*, **57**, 1259 (1984).
83. D. Cremer and E. Kraka, *Angew. Chem., Int. Ed. Engl.*, **23**, 627 (1984).
84. D. Cremer, in *Modelling of Structure and Properties of Molecules* (Ed. Z. B. Maksic), Ellis Horwood, Chichester, 1988, p. 125; D. Cremer, J. Gauss, P. v. R. Schleyer and P. H. M. Budzelaar, *Angew. Chem., Int. Ed. Engl.*, **23**, 370 (1984); D. Cremer and C. W. Bock, *J. Am. Chem. Soc.*, **108**, 3375 (1986); W. Koch, G. Frenking, J. Gauss, D. Cremer, A. Sawaryn and P. v. R. Schleyer, *J. Am. Chem. Soc.*, **108**, 5732 (1986); W. Koch, G. Frenking, J. Gauss and D. Cremer, *J. Am. Chem. Soc.*, **108**, 5808 (1986); P. H. M. Budzelaar, D. Cremer, M. Wallasch,

E-U. Würthwein and P. v. R. Schleyer, *J. Am. Chem. Soc.*, **109**, 6290 (1987); W. Koch, G. Frenking, J. Gauss, D. Cremer and J. R. Collins, *J. Am. Chem. Soc.*, **109**, 5917 (1987); D. Cremer, J. Gauss and E. Kraka, *J. Mol. Struct. (Theochem)*, **169**, 531 (1988); G. Frenking, W. Koch, F. Reichel and D. Cremer, *J. Am. Chem. Soc.*, **112**, 4240 (1990).

85. D. Cremer and E. Kraka, *J. Am. Chem. Soc.*, **107**, 3800, 3811 (1985).
86. R. F. W. Bader and H. Essén, *J. Chem. Phys.*, **80**, 1943 (1984); R. F. W. Bader, P. L. Mac Dougall and C. D. H. Lau, *J. Am. Chem. Soc.*, **106**, 1594 (1984).
87. H. Kollmar, *J. Am. Chem. Soc.*, **101**, 4832 (1979).
88. N. C. Baird, *Theor. Chim. Acta*, **16**, 239 (1970).
89. C. Müller, A. Schweig and H. Vermeer, *Angew. Chem.*, **86**, 275 (1974).
90. H. L. Haase and A. Schweig, *Tetrahedron*, **29**, 1759 (1973); W. Schäfer, A. Schweig and F. Mathey, *J. Am. Chem. Soc.*, **98**, 407 (1976).
91. D. Wirth and N. L. Bauld, *J. Comput. Chem.*, **1**, 189 (1980).
92. W. J. Hehre, R. Ditchfield, L. Radom and J. A. Pople, *J. Am. Chem. Soc.*, **92**, 4796 (1970); L. Radom, W. J. Hehre and J. A. Pople, *J. Am. Chem. Soc.*, **93**, 289 (1971); W. J. Hehre, L. Radom, P. v. R. Schleyer and J. A. Pople, *Ab Initio Molecular Orbital Theory*, Wiley, New York, 1986.
93. M. Barzaghi and C. Gatti, *J. Chim. Phys.*, **84**, 783 (1987).
94. E. Kraka and D. Cremer, in *Molecular Structure and Energetics, Structure and Reactivity*, Vol. 7 (Eds. J. F. Liebman and A. Greenberg), VCH Publishers, New York, 1988. p. 65; D. Cremer and J. Gauss, *J. Am. Chem. Soc.*, **108**, 7467 (1986).
95. P. George, M. Trachtman, C. W. Bock and A. M. Brett, *Theor. Chim. Acta*, **38**, 121 (1975).
96. P. George, M. Trachtman, C. W. Bock and A. M. Brett, *J. Chem. Soc., Perkin Trans. 2*, 1222 (1976); P. George, M. Trachtman, C. W. Bock and A. M. Brett, *Tetrahedron,* **32**, 317, 1357 (1976).
97. J. D. Cox and G. Pilcher, *Thermochemistry of Organic and Organometallic Compounds*, Academic Press, New York, 1970.
98. N. L. Allinger, *J. Am. Chem. Soc.*, **99**, 8127 (1977); U. Burkert and N. L. Allinger, *Molecular Mechanics*, ACS Monograph 177, Washington, 1982.
99. E. R. Davidson and D. Feller, *Chem. Rev.*, **86**, 681 (1986).
100. J. S. Binkley and J. A. Pople, *Int. J. Quantum Chem.*, **9**, 229 (1975).
101. E. Kraka, J. Gauss and D. Cremer, *J. Mol. Struct. (Theochem)*, **234**, 95 (1991).
102. J. Gauss and D. Cremer, *Adv. Quantum Chem.*, **23**, 205 (1992).
103. J. A. Pople, J. S. Binkley and R. Seeger, *Int. J. Quantum Chem., Symp.*, **10**, 1 (1976).
104. R. Krishnan and J. A. Pople, *Int. J. Quantum Chem.*, **14**, 91 (1978); R. Krishnan, M. J. Frisch and J. A. Pople, *J. Chem. Phys.*, **72**, 4244 (1980); R. J. Bartlett and I. Shavitt, *Chem. Phys. Lett.*, **50**, 190 (1977); R. J. Bartlett and G. D. Purvis, *J. Chem. Phys.*, **68**, 2114 (1978); R. J. Bartlett, H. Sekino and G. D. Purvis, *Chem. Phys. Lett.*, **98**, 66 (1983).
105. S. Kucharski and R. J. Bartlett, *Adv. Quantum Chem.*, **18**, 281 (1986); S. Kucharski, J. Noga and R. J. Bartlett, *J. Chem. Phys.*, **90**, 7282 (1989); K. Raghavachari, J. A. Pople, E. S. Replogle and M. Head-Gordon, *J. Phys. Chem.*, **94**, 5579 (1990).
106. Z. He and D. Cremer, *Int. J. Quantum Chem., Symp.*, **25**, 43 (1991).
107. R. J. Bartlett, *J. Phys. Chem.*, **93**, 1697 (1989).
108. Z. He and D. Cremer, *Theor. Chim. Acta*, **85**, 305 (1993).
109. K. Raghavachari, G. W. Trucks, J. A. Pople and E. S. Replogle, *Chem. Phys. Lett.*, **158**, 207 (1989).
110. J. A. Pople, M. Head-Gordon and K. Raghavachari, *J. Chem. Phys.*, **87**, 5968 (1987).
111. W. Kutzelnigg, *Isr. J. Chem.*, **19**, 193 (1980); M. Schindler and W. Kutzelnigg, *J. Chem. Phys.*, **76**, 1919 (1982); M. Schindler and W. Kutzelnigg, *J. Am. Chem. Soc.*, **105**, 1360 (1983); M. Schindler and W. Kutzelnigg, *Mol. Phys.*, **48**, 781 (1983); M. Schindler and W. Kutzelnigg, *J. Am. Chem. Soc.*, **109**, 1021 (1987); M. Schindler, *J. Am. Chem. Soc.*, **109**, 5950 (1987); M. Schindler, *Magn. Reson. Chem.*, **26**, 394 (1988); M. Schindler, *J. Am. Chem. Soc.*, **110**, 6623 (1988); M. Schindler, *J. Chem. Phys.*, **88**, 7638 (1988); W. Kutzelnigg, M. Schindler and U. Fleischer, *NMR, Basic Principles and Progress*, Vol. 23, Springer, Berlin, 1989.
112. A. E. Hansen and T. D. Bouman, *J. Chem. Phys.*, **82**, 5035 (1985).
113. K. Wolinski, J. F. Hinton and P. Pulay, *J. Am. Chem. Soc.*, **112**, 8251 (1990).
114. J. Gauss, *Chem. Phys. Lett.*, **191**, 614 (1992); J. Gauss, *J. Chem. Phys.*, **99**, 3629 (1993); J. Gauss, *Chem. Phys. Lett.*, in Press.

115. C. van Wüllen and W. Kutzelnigg, *Chem. Phys. Lett.*, **205**, 563 (1993); W. Kutzelnigg, C. van Wüllen, U. Fleischer and R. Franke, in *Proceedings of the NATO Advanced Workshop on The Calculation of NMR Shielding Constants and Their Use in the Determination of the Geometric and Electronic Structures of Molecules and Solids*, 1992.
116. M. Bühl and P. v. R. Schleyer, in *Electron Deficient Boron and Carbon Clusters* (Eds. G. A. Olah, K. Wade and R. E. Williams), Wiley, New York, 1991; M. Bühl, N. J. R. v. E. Hommes, P. v. R. Schleyer, U. Fleischer and W. Kutzelnigg, *J. Am. Chem. Soc.*, **113**, 2459 (1991). See also: D. Hnyk, E. Vajda, M. Bühl and P. v. R. Schleyer, *Inorg. Chem.*, **31**, 2464 (1992); M. Bühl and P. v. R. Schleyer, *J. Am. Chem. Soc.*, **114**, 477 (1992); M. Bühl, P. v. R. Schleyer and M. L. McKee, *Heteroat. Chem.*, **2**, 499 (1991); M. Bühl and P. v. R. Schleyer, *Angew. Chem.*, **102**, 962 (1990); P. v. R. Schleyer, M. Bühl, U. Fleischer and W. Koch, *Inorg. Chem.*, **29**, 153 (1990); P. v. R. Schleyer, W. Koch, B. Liu and U. Fleischer, *J. Chem. Soc., Chem. Commun.*, 1098 (1989); M. Bremer, K. Schoetz, P. v. R. Schleyer, U. Fleischer, M. Schindler, W. Kutzelnigg, W. Koch and P. Pulay, *Angew. Chem.*, **101**, 1063 (1989).
117. R. C. Haddon. *J. Org. Chem.*, **44**, 3608 (1979).
118. P. Buzek, P. v. R. Schleyer and S. Sieber, *Chem. Unserer Zeit*, **26**, 116 (1992).
119. M. Barzaghi and C. Gatti, *J. Mol. Struct. (Theochem)*, **167**, 275 (1988): M. Barzaghi and C. Gatti, *J. Mol. Struct. (Theochem)*, **166**, 431 (1988).
120. L. A. Paquette, M. J. Broadhurst, P. Warner, G. A. Olah and G. A. Liang, *J. Am. Chem. Soc.*, **105**, 3386 (1973); S. Winstein, H. D. Kaesz, C. G. Kreiter and E. C. Friedrich, *J. Am. Chem. Soc.*, **87**, 3267 (1965); H. J. Dauben, J. D. Wilson and J. L. Laity, in *Nonbenzenoid Aromatics* (Ed. J. P. Snyder), Vol. 2, Academic Press, New York, 1967, p. 167.
121. R. F. Childs, A. Varadarajan, C. J. L. Lock, R. Faggiani, C. A. Fyfe and R. E. Wasylishen, *J. Am. Chem. Soc.*, **104**, 2452 (1982).
122. R. F. Childs. R. Faggiani, C. J. L. Lock and M. Mahendran, *J. Am. Chem. Soc.*, **108**, 3613 (1986).
123. C. F. Wilcox, Jr., D. A. Blain, J. Clardy, G. van Duyne, R. Gleiter and M. Eckert-Maksic, *J. Am. Chem. Soc.*, **108**, 7693 (1986).
124. C. P. R. Jennison, D. Mackay, K. N. Watson and N. J. Taylor, *J. Org. Chem.*, **51**, 3043 (1986).
125. A. Bertsch, W. Grimme, G. Reinhardt, H. Rose and P. M. Warner, *J. Am. Chem. Soc.*, **110**, 5112 (1988).

CHAPTER 8

Cyclopropyl homoconjugation —Experimental facts and interpretations

RONALD F. CHILDS

Department of Chemistry, McMaster University, Hamilton, Ontario, L8S 4M1, Canada.
Fax: +1-905-521-1993; e-mail: RCHILDS@MCMAIL.CIS.MCMASTER.CA

DIETER CREMER

Department of Theoretical Chemistry, University of Göteborg, S-41296 Göteborg, Kemigården 3, Sweden
Fax: +46-31-772-2933; e-mail: CREMER@OC.CHALMERS.SE

and

GEORGE ELIA

Department of Chemistry, McMaster University, Hamilton, Ontario, L8S 4M1, Canada.

I. INTRODUCTION 412
A. Requirements and Criteria for Homoconjugation and Homoaromaticity 413
B. Role of Homoconjugation and Homoaromaticity in Organic Chemistry 415
C. Organization of the Chapter 416
II. HOMOCONJUGATION IN ACYCLIC SYSTEMS 416
III. MONOHOMOAROMATIC AND HOMOANTIAROMATIC CATIONS 418
A. Homotropenylium Ions 418
1. NMR and magnetic properties of homotropenylium ions 419
2. Absorption spectra 420
3. Structural studies 421
4. Thermochemical measurements 423
5. Theoretical calculations 425
B. Homocyclopropenium Ions and Related 2π-Electron Systems 427
1. Boron analogues of the homocyclopropenium ions 430

The chemistry of the cyclopropyl group, Vol. 2
Edited by Z. Rappoport

C. Bicyclo[3.1.0]hexenyl and Cyclohexadienyl Cations 431
1. Structures of the cations 434
2. Rearrangements of bicyclo[3.1.0]hexenyl cations 435
IV. HIGHER HOMOAROMATIC CATIONS 439
A. Bishomotropenylium Ions 440
B. Bishomocyclopropenium Ions 444
C. Trishomocyclopropenium Ions 447
D. Bishomoantiaromatic Cations 449
V. NEUTRAL HOMOAROMATIC SYSTEMS 450
A. Monohomoaromatic Systems 451
B. Bis- and Higher Homoaromatic Systems 455
VI. HOMOAROMATIC ANIONS 457
VII. CONCLUDING REMARKS 459
VIII. ACKNOWLEDGEMENTS 462
IX. REFERENCES 460

I. INTRODUCTION

Homoconjugation has a long and important history in organic chemistry. Its importance goes well beyond the specific phenomenon itself as the concept has led to a large amount of work directed towards understanding conjugation and its role in determining the properties of an organic molecule.

The initial suggestion that a remote double bond could be involved in a displacement reaction was made by Shoppee in 1946 in order to account for the stereochemical results of the transformations of 3-cholesteryl derivatives[1]. Building on the considerable work then underway on neighbouring group participation in ionization reactions[2,3], Shoppee suggested that the double bond in the cholesteryl system, **1**, could function in a similar manner to Lewis base containing groups such as acetate or alkoxy, etc.

Further studies on the cholesteryl/*i*-cholesteryl system were reported by Dodson and Reigel[4] and particularly by Winstein and his coworkers[5,6]. This early work showed that the interconversion of the two cholesteryl derivatives, one of which has an open or 'homoallyl' form **1** and the other a 'cyclopropylcarbinyl' form **3**, could be understood in terms of the intermediate **2** (equation 1). Electron delocalization in **2** was suggested to occur across the intervening carbon atom rather than between adjacent carbon atoms as in normal conjugated systems.

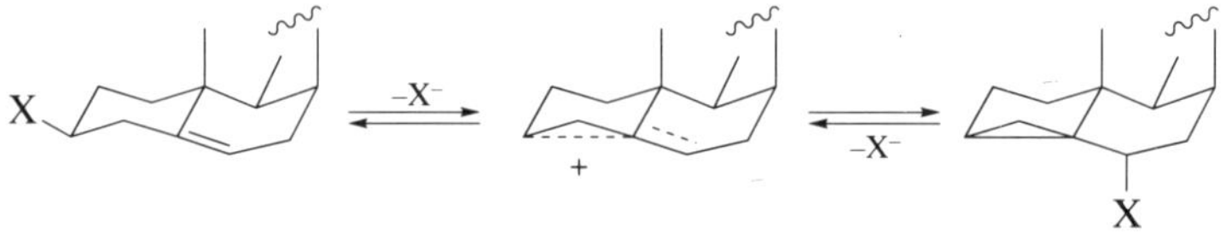

(1)

(1) **(2)** **(3)**

In 1950 both Winstein, Walborsky and Schreiber[7] and Roberts, Bennett and Armstrong[8] independently broadened the concept from the cholesteryl system starting point to a range of other examples including the norbornenyl cations. Winstein and colleagues used the terms homoallyl and homoconjugation to describe the phenomenon, terms which have become widely adopted. On the other hand, Roberts and coworkers suggested the phenomenon be called hyperconjugation. The importance of the correct geometry for homoconjugation was recognized at this early stage of development. Simonetta and Winstein made early use of theory in the form of Hückel calculations to explore the phenomenon[9].

The expansion of the concept to encompass cyclic electron delocalization or homoaromaticity occurred in the late 1950s. In 1956 Applequist and Roberts pointed out that the cyclobutenyl cation resembles the cyclopropenium cation[10]. Doering and colleagues suggested that the cycloheptatriene carboxylic acids could be regarded as planar pseudo-aromatic type structures with a homoconjugative interaction between C(1) and C(6)[11]. Based on the results of solvolytic studies on the bicyclo[3.1.0]hexyl system, Winstein set out the general concept of homoaromaticity in 1959[12,13].

Since the original development of the concepts of homoconjugation and homoaromaticity there has been a very large amount of work carried out to probe, test and find other examples of molecules or ions whose properties can be understood in this context. Several reviews of this work have appeared[14–22].

The concept of homoconjugation is now over 45 years old. Like many models in organic chemistry, while not without some original sceptics[23], it was initially embraced with enthusiasm and used to account for the properties of a wide range of systems. Subsequent to the original spate of claims there has been some detailed questioning of the notion. Some claims have been shown to be unjustified. Indeed, this questioning has continued to the extent that in the case of homoaromaticity, it has been suggested that there are only a limited number of medium ring cations where the phenomenon is important[24].

A. Requirements and Criteria for Homoconjugation and Homoaromaticity

In terms of homoconjugation, there are two basic starting points for a particular system. These are well illustrated in equation 1 for the cholesterol/*i*-cholesterol system. It is possible to start with an open form, **1**, and consider through-space interactions, or one can start from a closed form, in this case a cyclopropyl system, **3**, and consider its conjugation. Homoconjugation does not require that the closed form consists of a cyclopropyl ring. However, in practice most, if not all, known examples formally involve a cyclopropane or three-membered ring form as the ring-closed valence tautomer. It is this that has led the editor to include a discussion of homoconjugation in a volume on the chemistry of the cyclopropyl group. However, in many systems to be discussed in this chapter the starting point is an 'open' structure and linkage to the closed, or cyclopropane, form can at times seem tenuous.

In this review we are concerned particularly with cyclopropyl homoconjugation and not simply the conjugation of a cyclopropyl group to an unsaturated centre. The distinction is important. The focus of our attention is on conjugation and delocalization of electrons through space or through a cyclopropane bond (equation 2). Cyclopropyl conjugation, on the other hand, does not necessarily embrace this 'transmission' aspect of homoconjugation and its main focus is normally on the effect of a cyclopropane as a substituent.

(2)

Homoconjugation can be a linear phenomenon. That is, one can be concerned with conjugation and electron delocalization through space between two unsaturated fragments. The special and most important case is where the unsaturated fragment or fragments are combined in a cyclic system such that a through-space interaction potentially leads to a cyclically delocalized system (equation 3), which can be stabilized by homoaromaticity. In short, cyclopropyl homoconjugation is a special case of homoconjugation; homoconjugation is particularly important when manifested in terms of homoaromaticity (or homoantiaromaticity) in (bi)cyclic systems. As a result, any discussion of cyclopropyl homoconjugation implies also a discussion of homoconjugation and homoaromaticity. These latter terms are of course embedded in the broader concepts of conjugation and aromaticity.

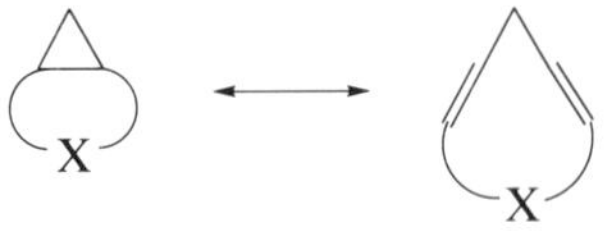

$X = (CH)_n$ $X = (CH)_{n-2}$ (3)

An advanced description of homoconjugation and homoaromaticity has been given by Cremer and coworkers[25–28] who distinguish between *bond* and *no-bond homoconjugation*, the former covering cyclopropyl homoconjugation and the latter all the various possibilities of through-space homoconjugation. Cremer and coworkers base their classification on a clear definition of covalent bonding[29] and a careful determination of molecular properties with the aid of high-level *ab initio* theory. Based on the distinction between bond and no-bond homoconjugation, they derived a detailed definition and set of requirements for homoaromaticity which can be summarized as follows[25–28]:

(1) the system in question should possess one or more homoconjugative interactions (either through bond or through space) closing cyclic conjugation;

(2) the bond or interaction indices of the homoconjugative interactions should be significantly greater than zero, thus indicating either a partial bond (cyclopropyl homoconjugation) or substantial through-space interactions (no-bond homoconjugation);

(3) electron delocalization in the closed cyclic system should be characterized by:

(a) effective overlap between the π-orbitals of the cyclic system,

(b) bond orders and π-character indices that are approaching those of an aromatic π-system,

(c) delocalization of positive or negative charge throughout the cyclic system in case of charged molecules,

(d) a relatively large degree of bond equalization with bond lengths differing from those of normal single or double bonds;

(4) for either cyclopropyl or no-bond homoaromatic systems the number of π-electrons participating in cyclic electron delocalization should be close to $4q + 2$;

(5) homoaromaticity should lead to a stabilizing resonance energy $> 2\ \mathrm{kcal\,mol^{-1}}$;

(6) no-bond homoaromatic systems should possess exceptional magnetic properties[25,27,28] that should lead to:

(a) significant equalization of ^{13}C chemical shifts in the cyclic system,

(b) the magnetic susceptibility, χ, adopting a maximum value for an unconstrained homoaromatic system, i.e. the exaltation of the magnetic susceptibility indicates homoaromatic electron delocalization,

(c) a large chemical shift difference between the *endo*- and *exo*-oriented protons when the system in question possesses a CH_2 group properly located above the ring.

In the case of bond (cyclopropyl) homoconjugation, Cremer and coworkers found that potentially homoantiaromatic $4q$ electron systems prefer to delocalize along the periphery of the bicyclic system, thus increasing the number of electrons involved in cyclic delocalization from $4q$ to $4q + 2$[26,30]. Hence, in line with orbital descriptions suggested by Hehre[31], homoantiaromaticity seems to result from an antiaromatic Möbius $4q + 2$ electron system[32] rather than an antiaromatic Hückel $4q$ electron system[25–28]. In any case, homoantiaromaticity is reflected by a destabilizing resonance energy[25–28].

Setting out the requirements for homoaromaticity and homoantiaromaticity in the manner above, in principle, makes it easy to identify a system as homoaromatic. Clearly, an appropriate geometry or structure of the species in question is required. This pertains not only to the appropriate placement of the AOs at the homoconjugative centres, but also to the structural changes associated with the cyclic delocalization of $(4q + 2)$ π-electrons. The cyclic delocalization should also be reflected in the stability of a system and its spectroscopic properties, including particularly its NMR spectrum. The use of high-level theory in conjugation with experimental observations to examine electron delocalization, bonding, structure, stability and magnetic properties is also important.

B. Role of Homoconjugation and Homoaromaticity in Organic Chemistry

The initial formulation of the concept of homoaromaticity provided a major stimulus to the probing of the boundaries of electron delocalization in organic molecules and ions as is clearly demonstrated by the extensive amount of work reported in this area[13–22]. Some of the basic questions which have arisen in this work are those which test the limits of this type of delocalization. These include, for example, the important issue of just how far a chemical bond can be distorted for it still to be considered a bond and of importance in the description and understanding of the properties of a molecule. Related to this fundamental question is that of the degree to which a cyclopropyl group can conjugate. In most instances the homoconjugative or homoaromatic 'bond' has a bond order less than one and the work performed in this area has led to a much better understanding of the role and importance of these 'fractional' bonds in organic chemistry.

While the initial formulation of homoaromaticity pre-dated the introduction of orbital symmetry by some eight years[33], the two concepts are inextricably linked[34]. This is most evident when pericyclic reactions are considered from the perspective of aromatic or antiaromatic transitions states[35] and the Hückel/Möbius concept[31]. The inter-relationship can be demonstrated by the electrocyclic reaction shown in Scheme 1[36].

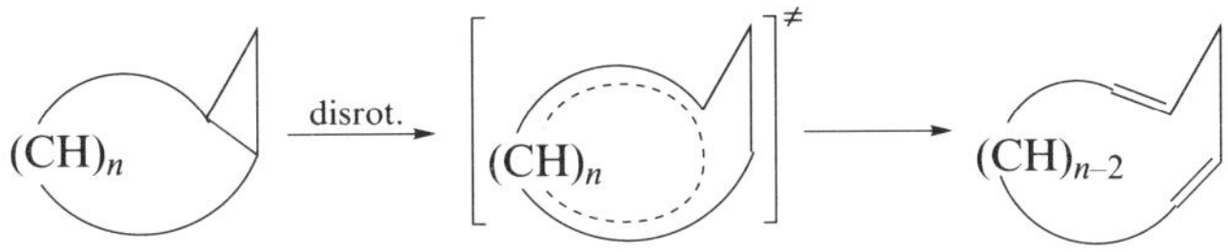

(4) (6) (5)

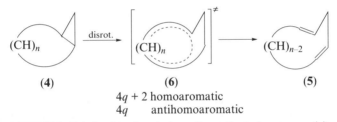

$4q + 2$ homoaromatic
$4q$ antihomoaromatic

SCHEME 1. Relationship of an electrocyclic reaction to homoaromaticity

The closed and open forms, **4** and **5**, respectively, represent the formal starting and end points of an electrocyclic reaction. In terms of this pericyclic reaction, the transition state **6** can be analysed with respect to its configurational and electronic properties as either a stabilized or destabilized Hückel or Möbius transition state. Where **4** and **5** are linked by a thermally allowed disrotatory process, then **6** will have a Hückel-type configuration. Where the process involves $(4q + 2)$ electrons, the electrocyclic reaction is thermally allowed and **6** can be considered to be homoaromatic. In those instances where the **4/5** interconversion is a $4q$ process, then **6** is formally an homoantiaromatic molecule or ion.

A key question in terms of homoaromaticity is the profile of the potential energy surface linking **4** and **5** and, in particular, where the energy minima occur on this surface.

In the $4q$ case, **6** is not an important contributor to the ground state description of the properties of either **4** or **5**. However, with **4** there are alternative modes of homoconjugation possible that involve the external cyclopropane bonds[30,32,37,38]. This is shown in Scheme 2 for the bicyclo[3.1.0]hexenyl cation. This alternative mode of conjugation of a cyclopropane in a $4q$ situation, an option not available to the parent $4q$ antiaromatic unsaturated ring systems[39], leads to a fundamentally different set of properties and reactions of these systems as compared to the potentially homoaromatic $4q + 2$ cases.

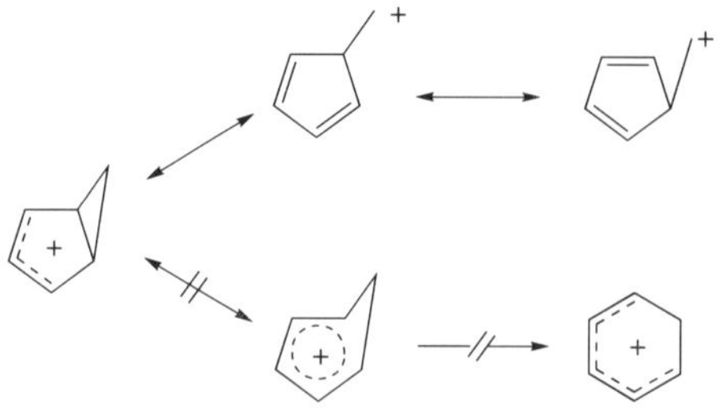

SCHEME 2. Delocalization in the bicyclo[3.1.0]hexenyl cation

C. Organization of the Chapter

The remainder of this chapter is organized into a series of sections which examine the currently available results on a variety of homoconjugative systems. The review is selective in terms of its coverage with examples being chosen that illustrate the issues at hand. Readers are referred to other reviews cited earlier for more comprehensive but, in most cases, less detailed accounts[14–22].

We start with an examination of some examples of acyclic systems in which there is evidence or the possibility of cyclopropyl homoconjugation. We then move on to a broader examination of homoaromatic systems, treating cationic, neutral and anionic systems in separate sections. The results of experimental work and theoretical examinations are integrated so as to provide a cohesive overview of each system. In order to limit the size of the chapter, we refrain from reviewing in detail systems such as the bridged annulenes and radical species. The chapter concludes with a reflective section that seeks to draw together theory with experiment and point out new directions for future work.

II. HOMOCONJUGATION IN ACYCLIC SYSTEMS

As was shown in the initial work involving the cholesteryl/*i*-cholesteryl system **1**–**3**, it is, in principle, possible to approach a homoconjugated or homoaromatic system from two directions. These formally involve either starting from a closed cyclopropane or equivalent ring and allowing this to conjugate with an appropriate π-system or, alternatively, starting with an open-chain π-system, or systems, and allowing a through-space interaction to occur between the end of the system(s).

The conjugative properties of a cyclopropane have been examined extensively. Early work in this area has been reviewed by Charton[40], Story and Clark[17], de Meijere[41] and an extensive overview of more recent work has been prepared by Tidwell[42]. Recent updates on this topic have been provided by Cremer and colleagues[26,43,44].

Studies on the ability of a cyclopropane to conjugate have involved a wide variety of approaches including spectroscopic, thermochemical, structural and theoretical examinations[42]. Two overall thrusts are apparent in the reported work. One approach has been to investigate the impact of a cyclopropyl substituent on the properties of an attached functional group or molecule. Typically, techniques used include measurement of the substituent parameters of a cyclopropyl group, determination of the acidity or basicity of functional groups attached to a cyclopropane, or measurement of the impact of a cyclopropyl substituent on the absorption spectrum of an attached chromophore[42].

The second approach has been to focus attention on the properties of the cyclopropane itself. Thermochemical measurements and particularly structural studies involving both experiment and theory are the principal methods used. It is this second approach to studying cyclopropyl homoconjugation which is discussed in this chapter.

Allen has examined in detail the effect of conjugation on the structure of a cyclopropane[45–47]. The analysis, which was based on the available X-ray structures of cyclopropyl derivatives, showed that there were systematic changes in the geometry of the cyclopropane ring associated with π-electron acceptor groups such as the carbonyl group[45]. Conjugation of cyclopropyl with an acceptor group leads to a lengthening of the two vicinal and a shortening of the distal cyclopropane bonds. As summarized for **7** in Table 1, the contraction of the distal bond was found to be approximately twice as large as the lengthening of the vicinal bonds as compared to an average cyclopropane C—C bond distance (1.504 Å).

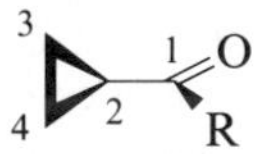

(**7**)

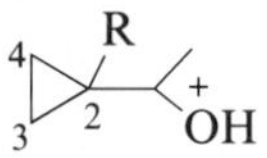

(**8**) R = H
(**9**) R = Me

The structural studies on cyclopropane derivatives have shown that there is a strong conformational preference observed in the structures of the cyclopropyl materials with the plane of the cyclopropane being aligned with that of the π-system of the acceptor group[45,48]. Similar conclusions have also been reached on the basis of UV studies[42,49] and theoretical calculations[50–52,44] (see also the discussion in Chapter 2).

Allen estimated that conjugation of a cyclopropane with an attached substituent was about 70% as effective as conjugation with a double bond[45]. This estimate is consistent with the 60% figure derived by Pete on the basis of an analysis of the UV spectra of cyclopropyl-containing systems[53].

The geometric changes resulting from conjugation of a cyclopropane with a conventional π-acceptor are greatly magnified when the acceptor group becomes positively charged. The cyclopropylcarbinyl cation, the formal archetype of these systems, has been studied extensively. Several reviews of this work exist[54], including one in this volume. In the parent cation the delocalized cyclopropylcarbinyl structure has been shown to be almost of the same energy as a bicyclobutonium ion[55] with a strong cross-ring interaction[56–58]. Both of these ions are significantly more stable than the corresponding homoallyl structure. The barrier to the interconversion of the bicyclobutonium and cyclopropylcarbinyl ions is small and substituents on the cation can profoundly alter the relative energies of the cyclopropylcarbinyl/bicyclobutonium/homoallyl forms of the parent cation.

The crystal structures of several different cyclopropylcarbinyl cations, each with a hydroxy function on the carbinyl carbon, have been reported[59–61]. Of these structures some five represent relatively simple systems and in each of these cases a bisected or close to

TABLE 1. Internuclear distances (Å) of some cyclopropyl compounds and cations

Bond	**7**[a]	**8**[b]	**9**[b]	**10**[c]
C(1)—O	—	1.256(8)	1.268(8)	
C(1)—C(2)	—	1.405(10)	1.461(9)	1.343
C(2)—C(3)	1.517	1.516(8)	1.529(9)	2.159
C(2)—C(4)	—	1.516(8)	1.545(8)	1.543
C(3)—C(4)	1.478	1.418(12)	1.448(9)	1.453

[a]References 43–45.
[b]References 59–61.
[c]Reference 56.

bisected conformation is observed. As can be seen from the data for the methyl-substituted systems **8** and **9**, summarized in Table 1, large bond-distance distortions of the three-membered rings are observed. In the case of **8**, the distal bond distance is remarkably short, being almost the same as is found for the C—C bond length in benzene.

While it is clear that a cyclopropyl group can effectively conjugate with an adjacent π-acceptor, particularly where this acceptor is a positively charged group, and the structure of the cyclopropane ring can be substantially modified, the question of importance in terms of homoconjugation and homoaromaticity in general is whether conjugation can be transmitted through a cyclopropane ring.

Early work based on the solvolysis of substituted cyclopropylcarbinyl systems[62,63], spectroscopic studies[64], calculations[65] and the structural work analysed by Allen[45,46] all point to the cyclopropane ring being poor at the transmission of conjugation. For example, Pews and Ojha estimated that the cyclopropyl ring is about 27% as effective as a vinyl group in its ability to transmit conjugation[66]. Wilcox, Loew and Hoffman have discussed the conjugative properties of a cyclopropane in terms of molecular orbital theory and suggested that the LUMO of the parent cation has relatively small coefficients at all of the cyclopropyl carbons, resulting in comparatively small π-interactions by substituents[67].

In cations the situation would appear to be somewhat different. Wiberg and colleagues have examined the effect of substituents on the cyclobutenium/cyclopropylcarbinyl/homoallyl cation energy surface using *ab initio* calculations at the MP2/6-31G(d) level[56]. In general, π-donor substituents at the 3-position were found to favour a cyclopropylcarbinyl form of the ion with significant homoallyl character (Scheme 3). The optimized geometry of the *trans*-hydroxy substituted system, **10**, is summarized in Table 1. It is clear that there can be large structural changes associated with substitution of a π-electron donor on the cyclopropyl ring of one of these ions. It can be seen that the C(2)—C(3) bond in **10**, is extremely long with a calculated distance of 2.159 Å. The cation has a very asymmetric 'three-membered ring'.

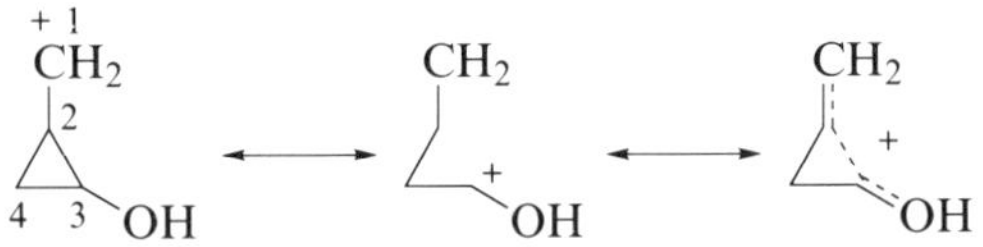

(10)

SCHEME 3. The 3-hydroxycyclopropylcarbinyl cation

Overall it is clear that, while in neutral systems a cyclopropane ring is not particularly effective at transmitting conjugation, this situation can change when very strong π-acceptor groups such as carbenium ions are present or when the cyclopropane is part of a cyclic situation[68].

III. MONOHOMOAROMATIC AND HOMOANTIAROMATIC CATIONS

A. Homotropenylium Ions

In many respects the homotropenylium ion can be considered to be the archetype or 'benzene' of homoaromatic systems. It is not only one of the earliest examples of a homoaromatic system to be described, but well more than forty substituted derivatives of the homotropenylium cation have now been reported[69]. These substituted ions have been examined by a broad range of experimental techniques and theoretical methods. It is interesting to note that unlike the trishomocyclopropenium cation, the initial homoaromatic system to be studied, characterization of the homotropenylium ion did not rely on a solvolytic approach. Rather from the outset, the characterization methods employed were the then relatively new stable ion techniques and, particularly, direct characterization of the ion by NMR spectroscopy.

Access to homotropenylium ions can be achieved by two general routes. The first involves the addition of an electrophile to a cyclooctatetraene or cyclooctatetraene derivative, an approach which can be considered to correspond to a homoallyl route (Scheme 4). In this route the electrophile is generally attached stereoselectively to the *endo* position on C(8)[18,70–74]. The second approach involves the ionization of a bicyclo[5.1.0]octadienyl derivative. This is the cyclopropylcarbinyl approach (Scheme 4). This route has the potential of generating a wide range of differently substituted cations; however, the starting materials can be difficult to access[75–78].

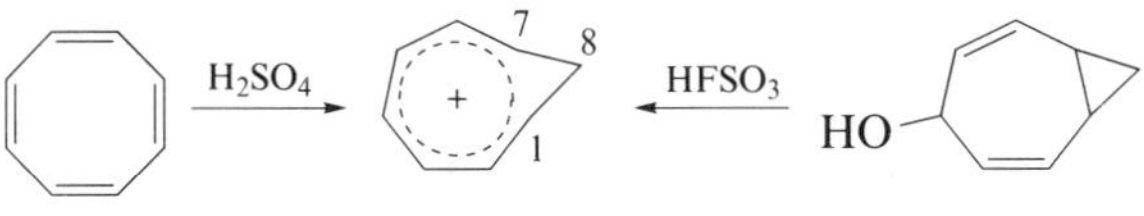

SCHEME 4. Routes to the homotropenylium ion

The original report of the preparation of the homotropenylium ion was by Pettit and coworkers in 1962. They showed that treatment of cyclooctatetraene with strong acids led to the formation of a stable $C_8H_9^+$ cation that exhibited remarkable properties[79]. Apart from its stability, which allowed for the isolation of its salts as solids, the key feature noted was its unusual ^{1}H NMR spectrum in which the resonances attributable to the two protons of a methylene group were non-equivalent and separated by 5.86 ppm[80]. In fact, one of the proton resonances was found at higher field than tetramethylsilane (TMS) at −0.73 ppm, a remarkable position for any hydrocarbon let alone a cation. Pettit and colleagues suggested that the structure of the cation corresponded to bicyclo[5.1.0]octadienyl cation with extensive delocalization of the internal cyclopropane bond. The unusual chemical shifts found for the methylene proton resonances were attributed to an induced ring current in **11**[81,82] although Deno expressed a contrary view[23].

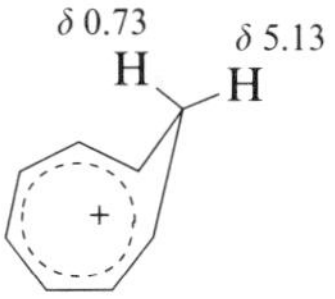

(11)

(12) M = Mo
(13) M = Cr
(14) M = W

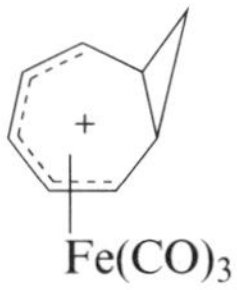

(15)

Since Pettit's original report, the homotropenylium ion has been studied extensively. The resulting large body of work has previously been reviewed[14–22]. The account given here highlights the different lines of evidence for the electronic nature of the cation.

1. NMR and magnetic properties of homotropenylium ions

Subsequent to the initial work outlined above, high field ^{1}H and ^{13}C NMR studies of **11** and model compounds supported the original suggestion for its structure[80,83]. In terms of model compounds one of the key approaches taken was to use the metal carbonyl complexes of the homotropenylium ion system. Thus the molybdenum, **12**, chromium, **13**, and tungsten tricarbonyl, **14**, derivatives were all considered to be homoaromatic while the corresponding iron complex, **15**, in which only four electrons are donated from the π-system to the metal, was suggested to have a localized structure[84].

A distinctive feature of the ^{1}H NMR spectra of homotropenylium ion derivatives is that the magnitude of the chemical shift difference ($\Delta\delta$) between the 8-*exo* and 8-*endo* proton resonances is found to be dependent on the nature and position of substituents on the 'seven-membered' ring[72,78,85,86]. In all cases, a donor substituent on one of the basal ring carbons attenuates the chemical shift difference. For example, the 1- and 2-hydroxy substituted homotropenylium ions, **16** and **17**, have $\Delta\delta$ values only some 50–55% of that found for the parent cation[77,85]. Systematic variation in the donor properties of the oxygen substituent using the Lewis acid scale developed by Childs and colleagues[87] led to a linear change in $\Delta\delta$[69].

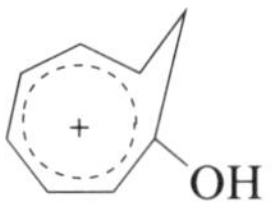

(16)

(17)

It has been suggested that this attenuation of $\Delta\delta$ with substitution is attributable to a reduction in cyclic electron delocalization in the homoaromatic ring and a consequent attenuation in the induced ring current[86]. However, as will be shown later, substitution also results in some fairly major changes in the structure of the homotropenylium ion and the impact of these structural changes on the relative position of the C(8) protons and/or ring current have not been disentangled.

The key underlying assumption of all the NMR studies of the homotropenylium systems is that there is an induced ring current when the ions are in a magnetic field and that this effects the two C(8) protons in a different manner. Dauben, Wilson and Laity measured the diamagnetic susceptibility of **11** and showed that it has a susceptibility exaltation which is similar in magnitude to that of the tropylium ion[88].

Winstein and colleagues carried out a ring current calculation for **11** using the Johnson–Bovey[89] approach and by assuming that atoms C(1)–C(7) adopted a planar configuration[14]. This planar configuration was subsequently shown to be incorrect and a further ring current calculation was undertaken by Childs, McGlinchey and Varadarajan in 1984[90]. This second calculation used as a starting point the known geometry of the 2-hydroxyhomotropenylium ion[91]. It also took into account local anisotropic contributions. Using this approach it was possible to account for the large chemical shift difference of the C(8) protons. However, there were two surprising results of this more recent work. First, both the 8-*exo* and 8-*endo* protons were found to be shielded, albeit the shielding of the former resonance was found to be small compared to the latter. Second, the local anisotropic contribution to the chemical shifts of the C(8) protons was very significant and accounted for more than 40% of the total calculated chemical shift difference. The intrinsic chemical shift of the two protons in the absence of an induced ring current was estimated to be 5.5 ppm, indicating that C(8) cannot be considered to be a cyclopropyl like carbon.

2. *Absorption spectra*

Winstein and coworkers measured the UV spectrum of the homotropenylium ion and showed that its long-wavelength absorption band (λ_{max} 313 nm) was intermediate between that of the tropylium (λ_{max} 273.5 nm) and heptatrienyl (λ_{max} 470 nm) cations[70]. Using a Hückel molecular orbital approach the C(1)—C(7) bond order was estimated to be 0.56.

3. *Structural studies*

The structures of five different homotropenylium or closely related systems determined using X-ray crystallographic techniques have been reported. These ions are **18**[91], **19**[92], **20**[19], **21**[93] and **22**[94].

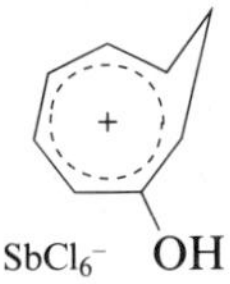

(18)

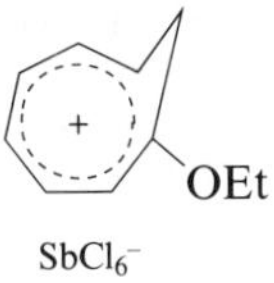

(19)

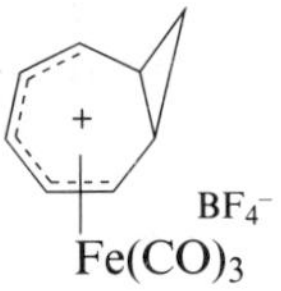

(20)

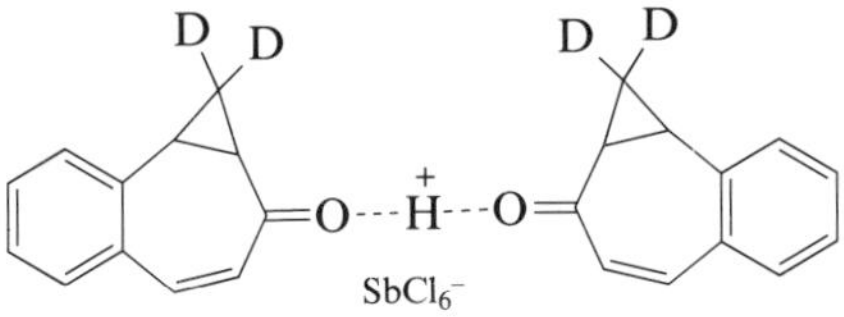

(21)

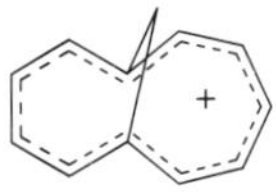

(22)

Up to this time there has been no report of the experimental determination of the structure of the parent homotropenylium ion. The three simplest systems that have been studied are **18**, **19** and the iron complex **20**. Cations **18** and **19** each have an oxygen-containing electron-donor substituent and, as such, appear to have smaller induced ring currents than the parent ion. In fact **18** and **19** have almost identical chemical shift differences ($\Delta\delta = 3.10$ ppm) between the two C(8) protons. In the case of **20**, $\Delta\delta$ is very small and it was considered to be a non-cyclically delocalized model for the bicyclo[5.1.0]heptadienyl cation[69].

Each of the cations **18**, **19** and **20** were found to adopt similar shallow boat-type conformations with C(8) being positioned over the 'seven-membered' ring. Despite their similarity in conformation, the cations were found to be substantially different in terms of their internuclear distances, particularly the C(1)–C(7) distance (Table 2).

TABLE 2. Selected bond distances (Å) of homotropenylium and related ions

Bond	**19**	**18**	**20**	**21**	**22**	**11**
C(1)—C(7)	2.284(5)	1.626(8)	1.474(11)	1.544(8)	2.293(3)	1.906
C(1)—C(8)	1.480(5)	1.488(7)	1.498(11)	1.515(9)	1.487(3)	1.478
C(7)—C(8)	1.506(6)	1.488(7)	1.497(12)	1.471(9)	1.487(3)	1.478
C(1)—C(2)	1.407(5)	1.422(12)	1.488(10)	1.439(10)	1.373(5)	1.397
C(2)—C(3)	1.359(5)	1.37(2)	1.395(10)	1.425(10)	1.416(6)	1.396
C(3)—C(4)	1.420(5)	1.378(13)	1.402(10)	1.341(8)	1.375(5)	1.404
C(4)—C(5)	1.364(6)	1.378(13)	1.400(10)	1.446(7)	1.375(5)	1.404
C(5)—C(6)	1.429(7)	1.37(2)	1.393(11)	1.413(7)	1.416(6)	1.396
C(6)—C(7)	1.337(5)	1.422(12)	1.495(11)	1.484(7)	1.373(5)	1.397
$\Delta\delta^a$ (ppm)	3.10	3.10	0.18		—	5.86
References	92	91	69/19	93	94	27

[a] $\Delta\delta$ is the chemical shift difference between the two resonances of the bridging (C_8) methylene protons.

As can be seen from the data in Table 2, the C(1)–C(7) distance in **18** is 1.626(8) Å while in **19** the same internuclear distance is 2.284(5) Å. In both instances, the bond distance is substantially longer than that normally found for a cyclopropane, 1.504 Å[45] (see also Chapter 2 of this volume)[44], or in the iron complex **20** (Table 2).

It has been suggested that the difference in the C(1)–C(7) distance in the two cations can be understood in terms of the effect of the strongly electron-donating oxygen donating substituents on the relative importance of the various resonance structures in the two cations (cf Scheme 5)[92,93]. A 2-hydroxy group favours those resonance structures which have a closed cyclopropane bond, as in these structures charge can be stabilized by the hydroxy group. Conversely, hydroxy or alkoxy substituents at C(1) will favour resonance structures with an open cyclopropyl bond.

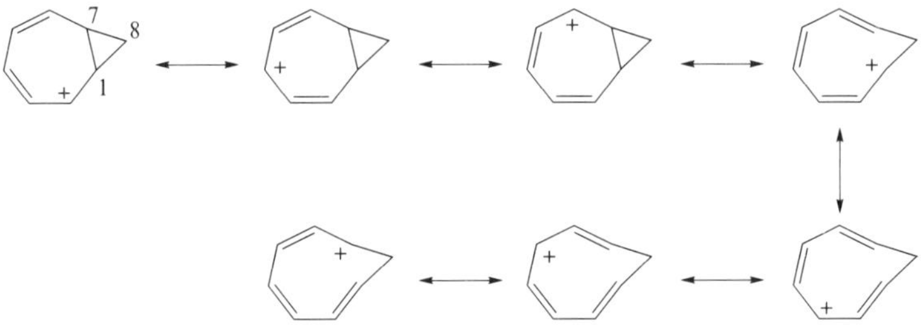

SCHEME 5. Resonance structures of the homotropenylium ion

The long C(1)–C(7) distance in **18** and **19** are consistent with a homoaromatic formulation of each of these cations. However, as was pointed out in the introduction to this review, homoaromaticity requires more than just the presence of a long homoconjugate bond and there should be structural changes throughout the molecule that are consistent with cyclic delocalization. The conformations of **18** and **19** are such that there can be effective overlap of the cyclic π-systems[95,96].

In the case of **18**, the various C—C bond distances in the basal 'seven-membered' ring are consistent with there being a significant degree of cyclic electron delocalization and the ion being classified as homoaromatic. The C(1),C(7) bond order of **18** was estimated to be 0.56 on the basis of the measured internuclear distance[91]. It is interesting to note that the bond distances found for **18** are substantially different from those reported for the protonated cyclopropyl ketones discussed above.

With **19**, Childs and colleagues pointed out that although its conformation is suitable for cyclic electron delocalization, the internuclear distances found for the basal ring carbons are not consistent with such a formulation of its structure[92]. Rather, the C—C bond distances around the ring indicate a progressive increase in bond alternation on proceeding from C(1)—C(2) to C(6)—C(7). Similar patterns of increasing bond length alternation have been observed with several other 1-substituted polyenyl cations[97,98]. It was concluded that the structure of **19** was completely consistent with it being considered to be a 1-ethoxyheptatrienyl cation. As such, **19** would appear to fail the third criterion established at the outset and not be homoaromatic, despite it exhibiting a large chemical shift difference between the C(8) protons.

The concern that the solid state structures do not represent those in the solution phase in which the ^{1}H NMR data were obtained was addressed by comparing the NMR spectra in both phases. Thus it was demonstrated for **17** and **18** through the use of solid state CPMAS and solution ^{13}C NMR spectroscopy that there were no fundamental differences in the structure or charge distribution of the cation in solution or the solid state[91,92].

The structure of **20** was suggested to be fully consistent with a 5C/4π bonding of the iron atom to the cation and the presence of a fully formed cyclopropane ring as proposed on the basis of its NMR spectrum[69,92].

Analysis of the structure of the unusual dimeric cation **21** led Childs and coworkers to conclude that this system could best be regarded as exhibiting cyclopropylcarbinyl-like delocalization[93]. The bridged ion **22** again exhibits a large homoconjugative internuclear distance (Table 2)[94], and the question arises as to whether there is a significant C(1)—C(7) overlap or whether it can best be regarded as a perturbed annulene.

4. Thermochemical measurements

A variety of approaches have been used to assess the importance of homoaromatic delocalization on the thermodynamic stability of homotropenylium ions. The earliest of these involved measurement of the barrier to ring inversion of stereoselectively labelled homotropenyliums ions.

Winstein and colleagues reported that the 8-*endo*-D cation, **11-endo-D**, underwent a slow exchange reaction with the corresponding *exo* derivative (Scheme 6)[70]. The barrier to this process was found to be 22.3 kcal mol^{-1}. Winstein's postulate that the process involved a ring inversion process and the planar cyclooctatrienyl cation **23** was later confirmed by Berson and Jenkins[99]. The barrier to the inversion process can then be regarded as indicative of the difference in energy between **11** and **23** and, as such, a measure of the extra stability of the homotropenylium ion over the planar, linearly conjugated **23**. However, it would be wrong to attribute all of this energy difference to homoaromatic stabilization. There are major differences in strain energy between **11** and **23** and these also have to be taken into account.

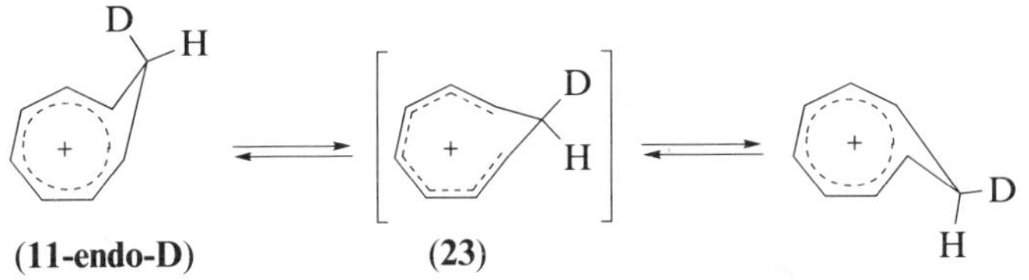

(11-endo-D) **(23)**

SCHEME 6. Inversion of the homotropenylium ring

The reported barriers to the inversion of C(8) of substituted homotropenylium ions vary over a wide range of energies (Table 3)[72,77,100,101]. The highest barrier reported is for **24e**[77] and the smallest for **25**[100]. The differences in these barriers can be understood in terms of the effect of basal ring substituents on the C(1)–C(7) internuclear distance as described above.

TABLE 3. Barriers to inversion of homotropenylium cations

Cation	Barrier to inversion (kcal mol^{-1})	Reference
24a: R8 = D, R1 = R2 = H	22.3	70
24b: R8 = Cl, R1 = R2 = H	22.6	72
24c: R8 = D; R1 = OMe, R2 = H	19.6	101
24d: R8 = Me, R1 = OH, R2 = H	17.1	77
24e: R8 = R1 = H, R2 = OH	>27.0	77
25	9.5	100

(24)

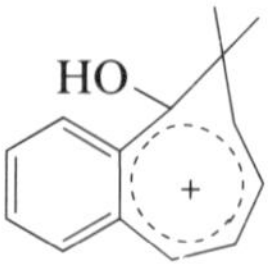

(25)

A more direct approach to measurement of homoaromatic stabilization was taken by Childs and colleagues in which the heats of protonation of the series of ketones **26–31** in FSO_3H were measured[102]. Of particular interest are the differences in heats of protonation for the various ketones and these data are summarized in Scheme 7.

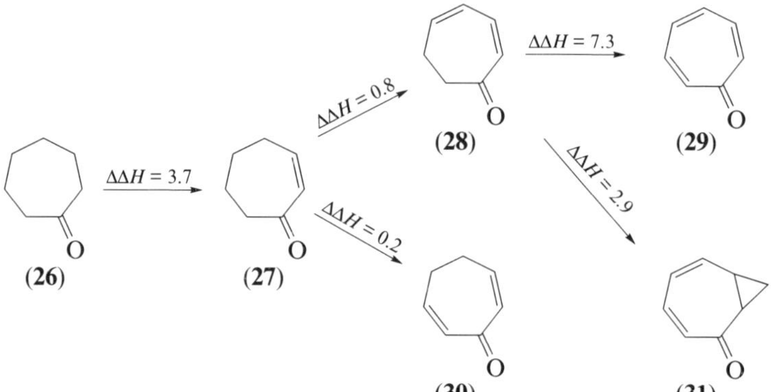

SCHEME 7. Differences in heats of protonation ($kcal\,mol^{-1}$)

As can be seen, the introduction of a conjugated double bond into the seven-membered ring of **26** increases the heat of protonation by 3.7 $kcal\,mol^{-1}$. As excepted, the effect of introducing a second double bond is much smaller (**27** → **28** or **27** → **30**). There is, however, a major effect on introduction of the third double bond, **28** → **29**, where a very large incremental jump in the heat of protonation is observed. This large discontinuity is associated with the well established aromaticity of the hydroxytropylium ion[39]. The effect of the introduction of a cyclopropane, **31**, is smaller (40%) than that of the third double bond, but there is still a substantial discontinuity. It was concluded by Childs and colleagues[102] that the 2-hydroxyhomotropenylium ion, **17**, is homoaromatic and that homoaromatic delocalization is an important factor in determining the overall stability of this cation. The evidence for homoaromatic stabilization of the 1-hydroxyhomotropenylium ion was less clear-cut. This has been confirmed by *ab initio* calculations carried out by Cremer and Colleagues[103].

A further, less direct way in which homoaromatic stabilization of substituted homotropenylium ions has been assessed is by measurement of their rates of interconversion and equilibrium positions between isomeric homotropenylium and other ions. This has been achieved with the 8,8-dimethyl-substituted systems which have been shown to undergo a series of circumambulatory rearrangements, as demonstrated in Scheme 8[77]. It was reported that the order of stability of the isomeric ions was **34** < **32** < **33** < **35** < **36**, with the free-energy differences between the various hydroxyhomotropenylium ions being relatively small (total spread being *ca* 2 $kcal\,mol^{-1}$), while **36** was found to be 15.5 $kcal\,mol^{-1}$ more stable than **33**[77,69]. The surprising feature of this series of rearrangements was the finding

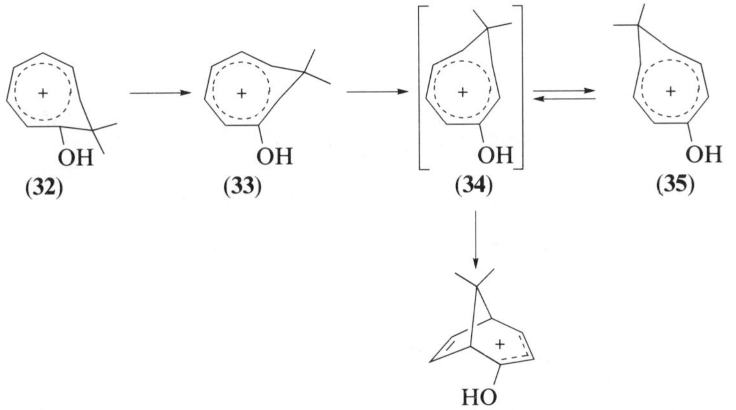

(32) (33) (34) (35) (36)

SCHEME 8. Isomerization of **32**

that **36**, formally a bishomoantiaromatic ion, was more stable than any of the isomeric homotropenylium ions.

The position of the equilibrium between the 8,8-dimethylhomotropenylium ions and the bicyclo[3.2.1]octadienyl system was shown to be very dependent on the nature of any additional substituents[77,104]. Thus the parent system **37** was shown to rearrange to **39** via **38** at low temperatures (Scheme 9). In the case of the ring methylated cations, an equilibrium was found to exist between the bicyclo[3.2.1]octadienyl and homotropenylium ions.

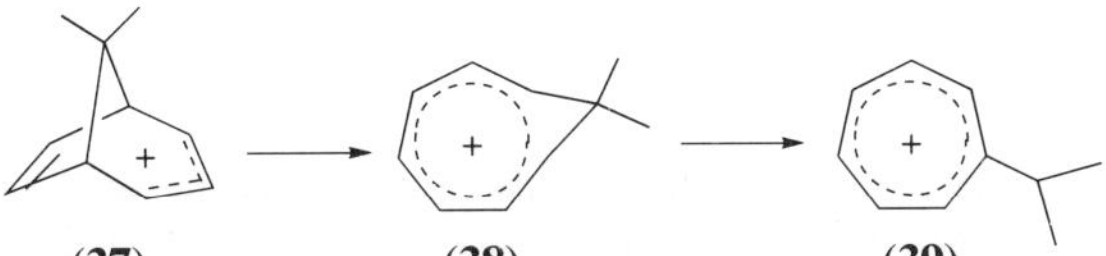

(37) (38) (39)

SCHEME 9. Isomerization of **37**

On the basis of the measurement of the heats of protonation and isomerization and estimates of the heats of formation of the starting ketones, it has been shown that the additional delocalization energy in **33** as compared to **36** is 2.8 kcal mol^{-1}[69].

Scott and colleagues have suggested that there is a qualitative relationship between the magnitude the chemical shift difference of the methylene proton resonances of the bridging carbon $\Delta\delta$ for a series of isomeric hydroxy-substituted systems and their relative thermodynamic stability[86].

5. *Theoretical calculations*

A number of different groups have undertaken theoretical examinations of the structure and properties of the homotropenylium ion. These include Hehre[31,105], Goldstein and Hoffmann[34], Jorgensen[38], Haddon[106–108], Cremer, Bader and coworkers[80], Barzaghi and Gatti[109], Schleyer, Cremer and coworkers[110,111] and Cremer and coworkers (see Section IV. B of Chapter 7[25,27,103].

The underlying reason for the large number of studies is that the calculated ground state structure of the homotropenylium ion is dependent on the level of theory used. In particular, the calculated C(1)–C(7) internuclear distance, and the shape of the potential energy surface as this distance is varied, have been found to be very sensitive to the choice of the theoretical method.

The highest level of calculations have been reported by Haddon[106] in 1988 (HF and MP2 single-point calculations) and Cremer and coworkers[25,27] in 1991 [HF and MP2 geometry optimizations, MP3 and MP4(SDQ)single-point calculations]. In each of these studies it was concluded, in contrast to earlier work, that there is a single energy minimum on the homotropenylium potential energy surface (PES) as the C(1)–C(7) distance is altered. The PES was found to be relatively flat as a function of change in the C(1)–C(7) distance. The minimum energy form had a C(1)–C(7) distance close to 2 Å (Haddon 1.91 and Cremer 2.03 Å).

Cremer and colleagues also calculated the ^1H and ^{13}C NMR chemical shifts as a function of the C(1)–C(7) distance[25,27]. Comparison of these calculated and observed shift values of the homotropenylium ion confirmed a C(1)–C(7) distance of 2 Å. At this distance the chemical shift difference between the *exo* and *endo* C(8) protons became a maximum. In addition the calculated magnetic susceptibility adopted a maximum value, i.e., a distinct susceptibility exaltation was found for the C(1)–C(7) equilibrium distance of 2 Å[25,27].

Analysis of the MP2 electron density distribution of the homotropenylium ion in its equilibrium geometry indicated, according to the Cremer–Kraka criterion of covalent bonding, that atoms C(1) and C(7) are connected by strong through-space interactions (interaction index 0.35) rather than a covalent bond[25,27]. However, Cremer pointed out that the lack of a bond does not exclude electron delocalization in the seven-membered ring closed by the C(1)–C(7) interactions. On the contrary, the 2 Å distance found for the 1,7 distance seems to be sufficient for effective electron delocalization. This was confirmed by: (a) the similarity in the calculated CC bond distances which are all close to 1.4 Å, indicating almost perfect bond equalization, (b) calculated CC bond orders of 1.5, typical of an aromatic system, (c) the π-character of all CC bonds in the 'seven-membered' ring and (d) the high degree of equalization of the positive charge in the 'seven-membered' ring. As a consequence of this equalization of the charge, the ^{13}C chemical shifts were found to be very similar.

In the context of the criteria for homoaromaticity given in Section I. A above, the calculated and measured properties of the homotropenylium cation suggest that it is the prototype of a no-bond homoaromatic molecule[25,27]. However, this conclusion seems to be at odds with the estimates of the homoaromatic resonance energy[27,106]. Using a bicyclic form of the homotropenylium ion with a fixed C(1)–C(7) distance of 1.5 Å as an internal reference, Cremer and colleagues calculated a stabilization energy of just 4 kcal mol^{-1} [MP4(SDQ)/DZ + P] for the homotropenylium ion[27]. However, this value reflects the extra stabilization caused by no-bond homoconjugation compared to normal cyclopropyl homoconjugation. Despite this problem with a reference state, both Cremer[27] and Haddon[106] point out that homoaromaticity is just a matter of a few kcal mol^{-1}. This is consistent with the experimental thermochemical results outlined above and the calculated resonance energies of neutral homoaromatic compounds (Section III. G in Reference 25).

The potential energy surface of the homotropenylium ion is rather flat in the 1,7 direction. This means that external effects, such as any 'seven-membered' ring substituents, will have a profound effect on the C(1)–C(7) distance. Experimentally, this has been found to be the case, as shown for **18** and **19** above. In addition, Cremer and colleagues have confirmed this C(1)–C(7) distance dependence on substitution by high-level *ab initio* calculations on hydroxy-substituted homotropenylium ions[103].

Scott and Hashemi have examined the effect of constraining the C(1)–C(7) distance in the homotropenylium ring by linking these two atoms by a three carbon bridge (termed a 'molecular caliper'), **40**[112]. While the structure of **40** has not been determined, the C(1)–C(7)

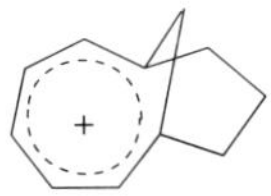

(40)

distance is expected to be small and in the range of 1.5–1.7 Å, and not possibly in the 2 Å range of the parent ion. The chemical shift difference of the methylene protons of **40** was found to be 5.08 ppm, some 87% of that found for the parent system.

In summary, the four major lines of approach to understanding the properties of the homotropenylium ion all lead to the same conclusion, namely that this ion is a cyclically delocalized, homoaromatic system. It should be stressed that this conclusion has been reached by using a combination of a battery of magnetic, spectroscopic, thermochemical, structural and theoretical techniques and these all give a consistent picture of the nature of the electron delocalization in the cation.

B. Homocyclopropenium Ions and Related 2π-Electron Systems

Roberts and coworkers, investigating the ionization reactions of cyclobutene derivatives, found that the resulting cyclobutenyl ions were unusually stable[10,113]. They suggested that rather than regarding these ions as simple allyl cations, their properties were consistent with a C(1),C(3) interaction and cyclic delocalization of the π-electrons. As such, these 2π-electron systems were considered to be the homoaromatic counterparts of the well established, aromatic cyclopropenium ions[39].

Since this original work, a large number of studies of the cyclobutenyl/homocyclopropenium ion, **41**, and its derivatives have been reported. The nature of their electronic structure has been probed using a variety of experimental techniques and theoretical methods[14–21]. The approaches employed parallel those used with the homotropenylium system and include an early examination of the UV spectra of the ions[114]. However, in contrast to its 6π-electron counterpart, fewer thermochemical measurements have been reported for **41** and its derivatives. As for theoretical treatments, the publication of Schleyer, Otto, Cremer and colleagues gives a good summary of previous work[111]. As these latter authors point out, the key question comes down to the nature of the potential energy surface as a function of the C(1)–C(3) internuclear distance (Scheme 10) and, in particular, what are the relative energies of the bicyclobutyl, homocyclopropenium and cyclobutenyl models for the structure of this cation.

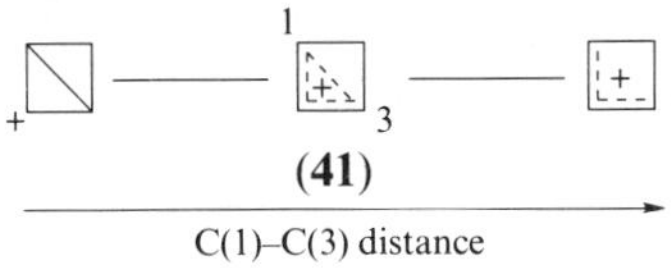

SCHEME 10. Cations on the $C_4H_5^+$ potential energy surface

Olah and collaborators have reported the synthesis and characterization of the parent ion **41** as a stable species (Scheme 11)[115]. The ^1H and ^{13}C NMR spectra of **41** indicated that it exists in a non-planar, envelope-type conformation. Variable-temperature studies demonstrated that **41** undergoes an isomerization that interconverts the *exo* and *endo* protons. It was assumed that this process involved a ring inversion (Scheme 12), rather than a more deep-seated rearrangement such as a circumambulation. Subsequent studies have shown that circumambulation of C(4) around the basal 'three-membered' ring in homocyclopropenium ions is a high-energy process[116].

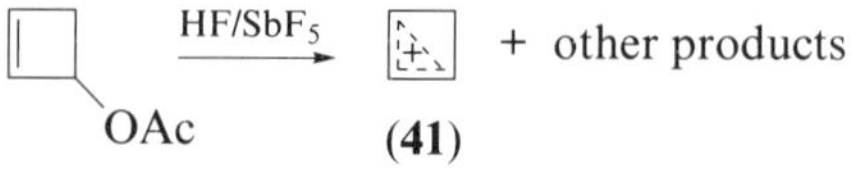

SCHEME 11. Formation of **41**

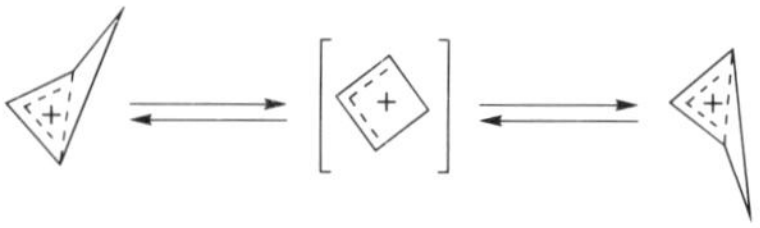

SCHEME 12. Ring inversion of **41**

The barrier to the isomerization in **41** (Scheme 12) was found to be 8.4 kcal mol^{-1}, a value substantially lower than that reported for the homotropenylium ion[70]. Olah suggested that the energy barrier for inversion of **41** was a measure of its homoaromatic stabilization[115]. However, other factors than homoaromatic delocalization contribute to this energy difference.

The finding that **41** has a non-planar conformation does not necessarily mean that there is a significant C(1)–C(3) π-interaction and that the system can be classified as homoaromatic[115]. Olah and colleagues addressed this issue by carefully examining the NMR chemical shifts of the allylic carbons of **41** and related derivatives and comparing these with comparable open-chain allyl cations. The key and important feature to emerge from these comparisons was that there is a fundamental difference in the charge distribution in **41** as compared to a conventional allyl cation such as **42**. In **42**, the ^{13}C resonances of C(1)/C(3) occur at lower field than that of C(2) ($\Delta\delta_{C(1)-C(2)} = +89.0$ ppm). In **41**, the reverse is the case and the C(2) resonance is significantly further downfield as compared to the resonances of C(1)/C(3). ($\Delta\delta_{C(1)-C(2)} = -54.1$ ppm). The NMR spectra of **41** and its charge distribution are entirely consistent with a significant C(1)—C(3) bonding interaction.

(42)

Olah and coworkers showed that substitution on the unsaturated basal ring carbons of the homocyclopropenium ion has a considerable effect on the magnitude of the chemical shift difference between the ^{13}C NMR chemical shifts of C(1)/C(3) and C(2) of the homocyclopropenium cations[115]. With 1,3-diphenyl substituents the system was shown to behave like a typical allyl cation with the C(1)/C(3) resonances being downfield that of C(2) (e.g. $\Delta\delta = +38.6$ ppm for **43**)[117]. With methyl substituents e.g. in **44**, the chemical shifts C(1), C(2) and C(3) are the same ($\Delta\delta = 0$ ppm).

Direct evidence as to the impact of substituents on the structure of the cyclobutenyl/homocyclopropenium ion system comes from X-ray crystal structures of five different

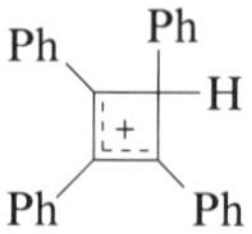

(43)

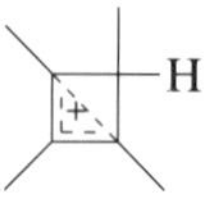

(44)

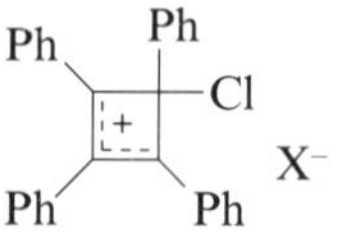

(45) X = $SnCl_5^-$
(46) X = $Nb_2OCl_9^-$

(47)

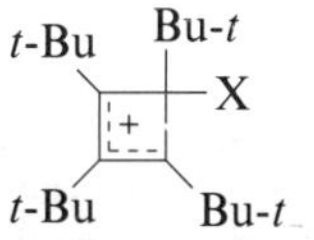

(48) X = H; I_3^-
(49) X = OH; SbF_6^-

salts: **45**[118], **46**[119], **47**[120], **48** and **49**[121]. Key information on each of these cations is summarized in Table 4. The terms used in this Table are defined in Scheme 13.

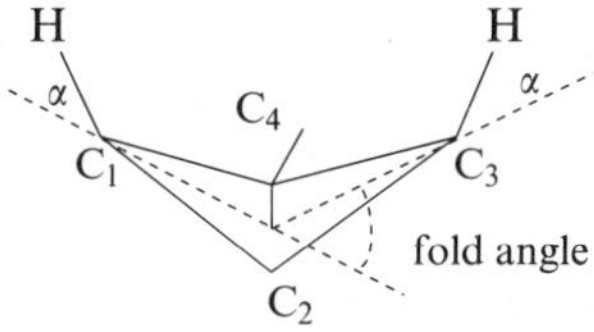

SCHEME 13. Definition of structural parameters used in Table 4

In the phenyl-substituted salts **45** and **46**, the four-membered ring is found to be nearly planar with a relatively large distance between C(1) and C(3). The conformations of the phenyl groups on C(1) and C(3) are such that they can effectively conjugate with the allylic system.

In contrast to the planar structures of **45** and **46**, the four-membered rings in each of the cations **47–49** are non-planar. As shown in Scheme 13, the conformation of the rings can be described in terms of the angle between planes defined by C(1),C(2),C(3) and C(1),C(3),C(4). As can be seen from Table 4, the angles between these planes are large and, as a result, C(1) and C(3) are brought much closer together than is found in the planar conformations of **45** and **46**. Maier and colleagues[121] pointed out that not only are p-orbitals C(1) and C(3) angled towards each other as a result of the bending of the four-membered rings of **47–49**, but p-orbital overlap is enhanced by a distortion of the substituents on these atoms away from their trigonal planes (α in Scheme 13). This type of distortion is consistent with the suggestion of Haddon concerning the importance of the π-orbital axis vector in non-planar systems[107,122].

The structures of ions **47–49** are fully consistent with them being classified as homocyclopropenium or homoaromatic ions. On the other hand, **45** and **46** are clearly cyclobutenyl in character. The observations of Olah and colleagues on the chemical shift

TABLE 4. Selected structural data for cyclopropenium ions and related compounds

Ion or compound	Reference	Method used	C(1)–C(3) distance (Å)	Fold angle (°)	α (°)	δ (ppm)
41 (bent)	111	MP2/6-31G(d)	1.735	35.5	18.2	
41 (planar)	111	MP2/6-31G(d)	1.972	0	0	
46	119	X-ray	2.032(10)	4.4	3.1/2.5	38.6
47	120	X-ray	1.775(4)	31.5	13.8	2.3
48	121	X-ray	1.806(6)	37.3	11.4	–40.0
49	121	X-ray	1.833(4)	36.4	10.1/16.6	–23.4
52	128	X-ray	1.915(4)	31.2	na[a]	
54	131	X-ray	1.792(15)	32.9	na[a]	

[a]Not available.

differences and, hence, charge distribution of the allylic carbons as a function of substituent are fully compatible with the more recent structural information.

Further information on the electron delocalization in **41** was provided by Katz and Gold. These workers pointed out that the UV spectra of alkyl-substituted derivatives of **41** were intermediary in position (λ_{max} *ca* 250 nm) between that of corresponding allyl or cyclopropenium cations (λ_{max} *ca* 300 and 185 nm, respectively)[123].

A large number of theoretical treatments of **41** have been reported[31,38,107,124–126]. Most of these earlier calculations did not yield a non-planar structure for **41** that is in accord with the experimental observations.

The most extensive study of the structure of **41** was reported recently by Schleyer, Otto, Cremer and coworkers[111]. As has been described for the homotropenylium cation, their approach involved high-level calculations [MP4(SDQ)/6-31G(d) and IGLO]. These authors concluded that **41** is homoaromatic with a bent structure, relatively short C(1)–C(3) distance (1.737 Å), a considerable 1–3 interaction index and nearly equal charges on C(1), C(2) and C(3), the basal ring carbons. The calculated chemical shifts and barrier of inversion of **41** agreed well with those observed by Olah and colleagues[115]. Various estimates of the stabilization energy of **41** were made.

It is clear from the consistent results of the various approaches used to probe the structure of **41** that this cation can be properly regarded as a homoaromatic system that meets the requirements set out in Section I.A above (see also Chapter 7[25]). Substitution of the cation has also been demonstrated to lead to large changes in the structure and by no means can all the derivatives of **41** be classified as homoaromatic[127]. This sensitivity of homoaromatic delocalization to substitution parallels that demonstrated with the homotropenylium cations.

1. Boron analogues of the homocyclopropenium ions

As was mentioned in the introduction, cyclopropyl homoconjugation and homoaromaticity need not be restricted to simple carbocyclic systems. In the case of the homocyclopropenium ions, two ring substituted boron analogues have been reported. In order for the systems to retain 2π-electrons they must, in the case of **50**, with a single boron replacement, be neutral or, in the case of **51**, with two boron atoms, be negatively charged. While these systems can rightly be regarded as potentially neutral or anionic examples of homoaromaticity, their isolobal relationship to the homocyclopropenium cations makes their inclusion in this section a logical choice.

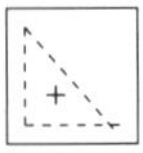

(41)

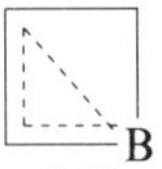

(50)

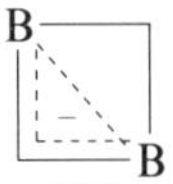

(51)

Berndt and colleagues have reported the preparation and characterization of **52**[128] and **53**[129]. The latter ion was characterized by NMR methods and these indicated the existence of a strong 1,3 interaction. Calculations reported by Cremer and colleagues on the parent system, **50**, indicated that this too would have a folded conformation with a substantial 1,3 interaction[130]. More recent *ab initio* calculations on **51** again indicated the adoption of a folded conformation (fold angle 23°)[131]. A significant B(1)–B(3) interaction was found with a 1,3 interaction index of 0.24. The calculated barrier of inversion in **54** was 4.3 kcal mol^{-1}.

The anion **54** has recently been reported by Berndt and coworkers[131]. Again, this species adopts a folded conformation (Table 4), which brings the two boron atoms into close

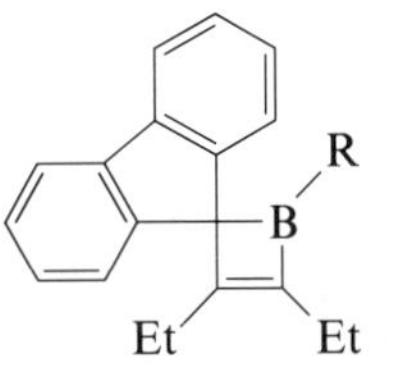

(52)

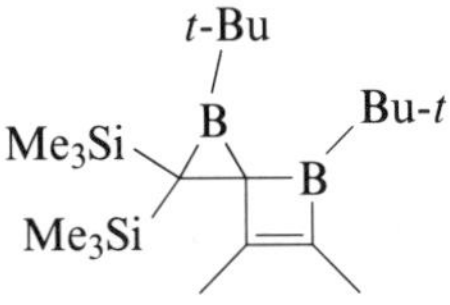

(53)

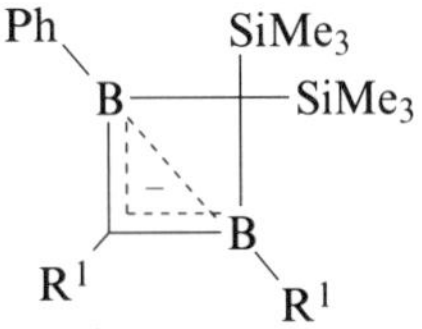

R^1 = Durenyl

(54)

proximity [B–B distance 1.729(15) Å]. Ring inversion was found to occur with a barrier of 7.9 kcal mol^{-1}, a barrier essentially the same as that reported by Olah for **41**. Analysis of the ^{13}C NMR chemical shifts of **54** as compared to model systems indicated that they were consistent with a cyclic homoaromatic 2π-electron delocalization.

Ab initio calculations for the parent anion **51** were also reported[131]. A similar conformation and structure was calculated for **51** as was determined for **54** (fold angle 34°, B–B distance 1.859 Å). The barrier to inversion in **52** was estimated to be 7.4 kcal mol^{-1}.

The structures and electron delocalization in these boron-substituted derivatives of the cyclobutenyl/homocyclopropenium cations are fully consistent with their designation as homoaromatic systems.

C. Bicyclo[3.1.0]hexenyl and Cyclohexadienyl Cations

Just as the unusual stability and reactivity of benzene are placed into their proper context by comparison with cyclobutadiene and cyclooctatetraene[39], the $4n$-electron homologues of benzene, it is instructive to compare the formally homoantiaromatic bicyclo [3.1.0]hexenyl/cyclohexadienyl cation systems with the homocyclopropenium and homotropenylium ions (Scheme 14). Such a comparison not only puts in context the properties of the latter two homoaromatic cations, but also reveals a different mode of cyclopropyl conjugation that occurs in the $4n$-electron systems.

$(4n + 2)$ $(4n)$ $(4n + 2)$

SCHEME 14. The monohomo- and homoantiaromatic cation series

While considerable work has been reported on the bicyclo[3.1.0]hexenyl cation and its derivatives, the results of these studies have not been reviewed as extensively as those of the corresponding homoaromatic systems. The most detailed accounts of these systems are those of Koptyug[132] and Barkhash[133]. Numerous reviews on the cyclohexadienyl cations have appeared[132,134].

The initial work on the bicyclo[3.1.0]hexenyl system was reported by de Vries[135] and Winstein and Battiste in 1960[136]. It was shown that acetolysis of the tosylate **55** occurred with a 10^{10}-fold acceleration over neopentyl tosylate. The ionization of **55** was found to be anchimerically assisted with the predominant kinetic product of the reaction being the homofulvene **56**. Small amounts of the acetate **57** were also present. Pentamethylbenzene, the anticipated product, was notably absent under kinetic control conditions.

Further insight into this system was provided by the stable ion studies reported by Childs, Sakai and Winstein in 1968[137]. These workers generated the cation **58** from **56** in

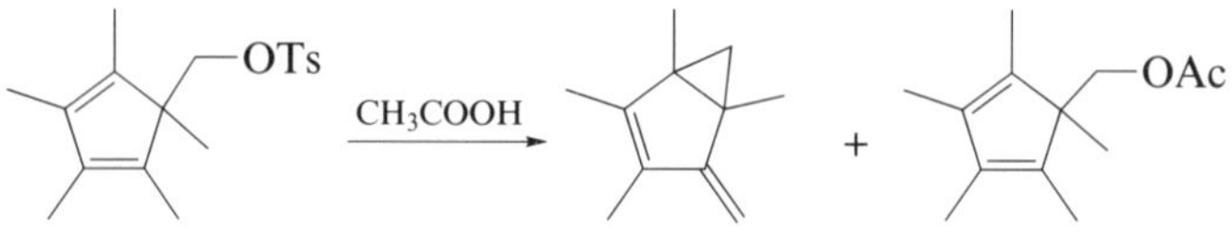

major trace

(55) (56) (57)

super-acid solution and showed that while its conversion to **59** occurred cleanly, the rearrangement was a relatively slow process involving a substantial activation barrier (Scheme 15). Subsequent work showed unequivocally that this isomerization involved a formally symmetry-forbidden electrocyclic ring-opening reaction and not a more deep-seated rearrangement process[138]. It was suggested that charge delocalization in **58** involved the two external cyclopropane bonds rather than the internal one.

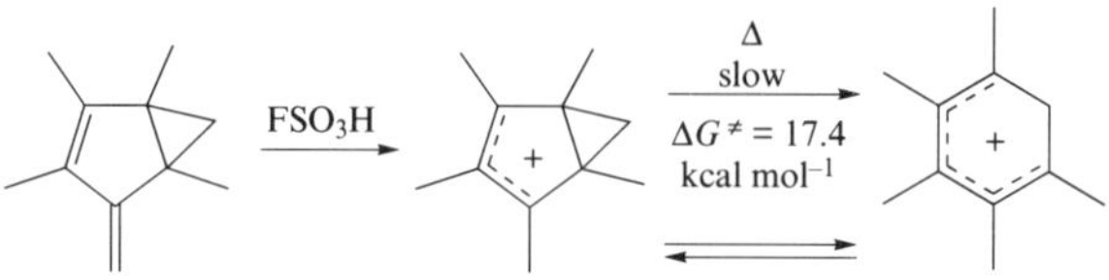

$h\nu$ (–78 °C)

(56) **(58)** **(59)**

SCHEME 15. Interconversion of the bicyclo[3.1.0]hexenyl and cyclohexadienyl cations

Childs and Winstein also showed that irradiation of **59** led to a clean photoisomerization and the formation of a photostationary state consisting of **58** and **59**[139–141]. Thus, in contrast to the homotropylium or homocyclopropenium ion systems, the 'open' and 'closed' forms of these 4*n* systems are interconverted in the first excited rather than ground state.

Berson and colleagues studied the parent cation **61** as a long-lived species in SO_2ClF as well as a transient species under solvolytic conditions[142]. It was again found that ring opening to the benzenium ion involved a substantial activation energy (Scheme 16). Consistent with this finding was the observation that solvolysis of **60** led to the formation of bicyclo[3.1.0]hexenyl derivatives and not benzene[143].

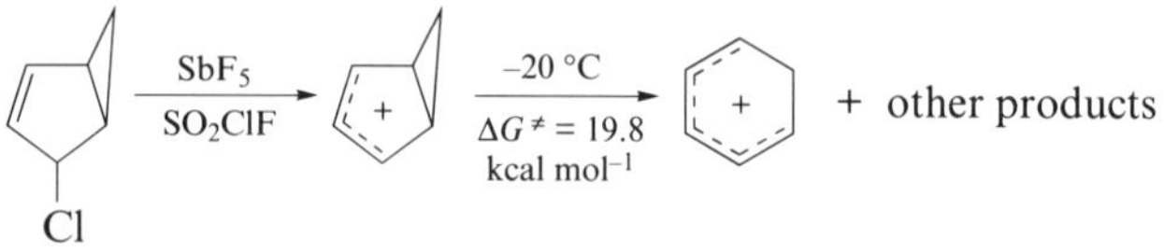

(60) **(61)**

SCHEME 16. Formation of **61**

These early studies on the bicyclo[3.1.0]hexenyl/cyclohexadienyl cation system have been amply reinforced by many additional studies[144,145]. Taken together, these results clearly show that the potential energy surface linking the bicyclo[3.1.0]hexenyl and cyclohexadienyl cations has energy minima corresponding to each of these two structures and that the cyclically delocalized homoconjugate structure is a transition state for their interconversion. This is illustrated in Scheme 17 for the hexamethyl-substituted bicyclohexenyl and cyclohexadienyl cations where a detailed thermochemical investigation has been undertaken[146]. There is a fundamental difference in the nature of the potential energy surface linking cations such as **62** and **63** to that found in the homocyclopropenium and homotropenylium ions. In the latter case, the cation occupies a single minimum on the potential energy surface, whereas ions **62** and **63** correspond to two local minima connected by an intermediate transition state. This transition state is located at about the same interaction distance as the no-bond homoaromatic homotropenylium and homocyclopropenium ions. The consequences of these differences are discussed in Section II. D of the accompanying review[25].

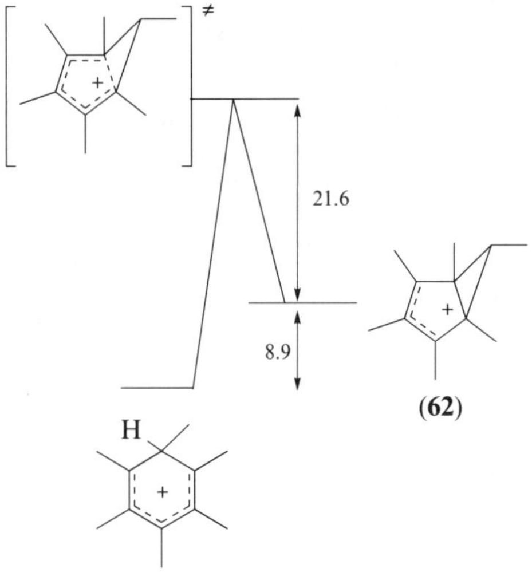

SCHEME 17. Isomerization of **62** (energies in kcal mol^{-1})

Information on electron delocalization in the bicyclo[3.1.0]hexenyl cations is available from their reported NMR spectra[137–145]. Data obtained with a variety of systems point to a completely different charge delocalization pattern to that found with the homotropenylium ions. For example, Olah and colleagues have obtained the ^{13}C NMR spectrum of the parent ion[114], **61**, and compared this with those of **42** and **11**. As can be seen from the data summarized in Scheme 18, the chemical shifts of the five-membered ring carbons of **61** resemble those of the cyclopentenyl cation. There is a considerable difference in chemical shifts, and hence charge distribution, at C(2), C(4) and C(3) of **61**. There is no evidence for the fairly even charge distribution as is found for the homotropenylium and homocyclopropenium ions (see previous Sections III. A and III. B). It was also noted by Olah that the chemical shift of C(6) is consistent with large delocalization to this position, i.e. to conjugation of the allyl system of **61** with the external cyclopropyl bonds.

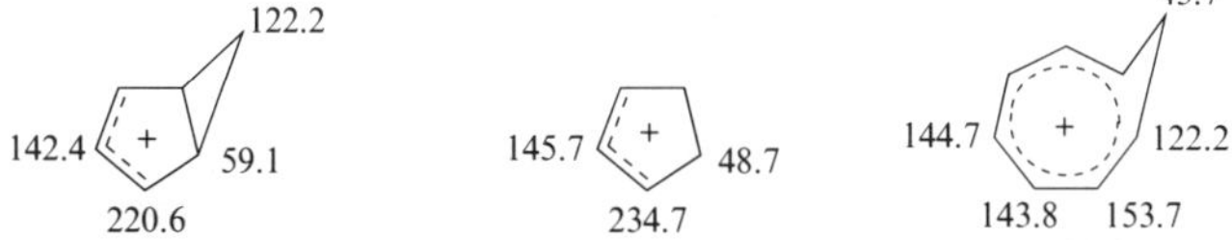

(61) **(42)** **(11)**

SCHEME 18. ^{13}C NMR shifts (ppm)

There have been a large number of studies of the NMR spectra of cyclohexadienyl cations[147,132,133]. These systems behave as open-chain hexadienyl cations and no C(1)–C(5) homoconjugative interaction is evident or needed to account for their spectroscopic properties.

1. *Structures of the cations*

The structures of the bicyclo[3.1.0]hexenyl and cyclohexadienyl cations have been examined using X-ray crystallography as well as theory. Crystal structures of cyclohexadienyl cations such as **62**[148], **63**[149] and **64**[150] have been reported. Unfortunately, the errors associated with these determinations are relatively large and, as a result, no significance can be placed on variations in individual C—C bond distances in the unsaturated fragments. However, it is clear in the cases where the C(6) substituents are the same that the cations adopt a planar or near-planar conformation. Where the C(6) substituents are dissimilar, the cations adopt a shallow envelope-type conformation with C(6) being out of the plane of the pentadienyl unit. However, there is no suggestion in any of these structures for any significant C(1)–C(5) through-space interaction.

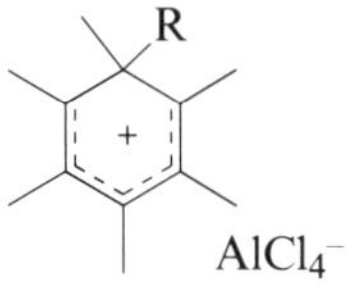

(**62**) R = Me
(**63**) R = Ph

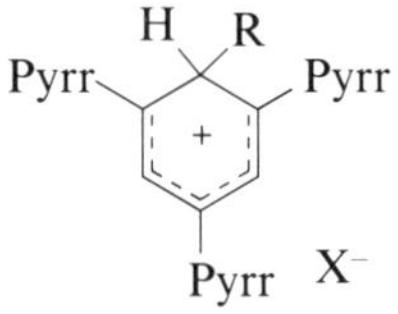

(**64**) R = H; X = ClO_4^-
Pyrr = *N*-Pyrrolyl

In terms of the bicyclo[3.1.0]hexenyl ions, the structure of the protonated ketone **65** was determined by Childs, Lock and colleagues[60]. The bond distances associated with the protonated carbonyl group and unsaturated portion of **65** were completely consistent with those expected for a protonated enone (Table 5). However, comparison of the cyclopropyl portion of structure **65** with that of the 2-hydroxyhomotropenylium ion (*vide supra*) and

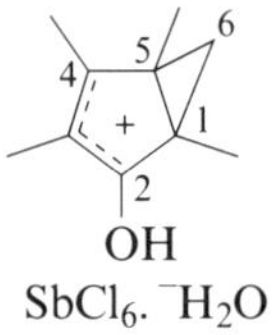

(**65**)

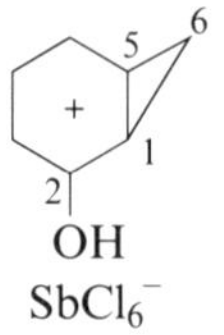

(**66**)

TABLE 5. Selected bond distances (Å) for **65**, **66**[a] and **61**

Compound	C1—C2	C2—C3	C3—C4	C4—C5	C5—C6	C1—C6	C1—C5
65[b]	1.474	1.410	1.351	1.510	1.501	1.547	1.511
66[c]	1.403			1.502	1.433	1.534	1.559
61[d]	1.492	1.392	1.392	1.492	1.535	1.535	1.501

[a]Numbering scheme for **66** is non-standard as shown on the structure.
[b]Average σ is 0.008 Å.
[c]Average σ is 0.011 Å.
[d]Calculated structure STO-3G[30].

other protonated cyclopropyl ketones, e.g. **66**[61], revealed several unusual features. In the first instance, the homoconjugate bond, C(1)—C(5), distance is the same as that of a normal cyclopropyl ring. It is much shorter than encountered with that found for the comparable internuclear distance in the homotropenylium ion or that expected for a comparable bond in a non-cyclically conjugated protonated cyclopropyl ketone. Secondly, it was suggested that the C(1)—C(2) and C(4)—C(5) bond distances were in each case significantly longer than those expected for comparable bonds between a protonated carbonyl carbon atom or vinyl carbon atom and a cyclopropyl carbon, respectively. The authors tentatively suggested that the cation adopts a structure which minimized interaction of the internal cyclopropane bond with the allyl portion of the system. The relatively long C(1)—C(6) bond was taken to be suggestive of conjugative involvement of this external cyclopropane bond.

No other experimental structure determinations of bicyclo[3.1.0]hexenyl cations have been reported. Theoretical calculations have been undertaken by Hehre[105,151] and Cremer and colleagues[30]. While in each of these instances the calculations were at a low level, the results are consistent with the presence of a short C(1)—C(5) and long C(1)—C(6)/C(5)—C(6) internuclear distances for the parent cation.

While more work is required in order to complete our understanding of the structure of **65**, the picture that emerges from these studies is that of a cation in which conjugation involves the allyl portion of the five-membered ring with the two external cyclopropane bonds. Hehre pointed out[105,151] that such a conjugated system is formally Möbius system with 6π electrons and thus formally antiaromatic (see the discussion in Section III. C of the preceding chapter[25]). Cremer and coworkers[26,30] concluded on the basis of an electron density analysis of **65** that close to six electrons are involved in electron delocalization along the periphery of the bicyclic system in a similar mode as was found for bicyclo[2.1.0]pent-2-ene (see the discussion in Section V. A. and the preceding chapter[25]). Applying the definitions given in Section I. A above, the latter system is clearly homoantiaromatic in view of its energetic and structural properties. Similarly, it is likely that the bicyclo[3.1.0]hexenyl cation falls into the same category. However, a detailed analysis of its relative stability has to confirm this characterization.

2. *Rearrangements of bicyclo[3.1.0]hexenyl cations*

The basic differences in electron delocalization between the homoaromatic homotropenylium and homocyclopropenium ions and the bicyclo[3.1.0]hexenyl cations result in fundamentally different reactions of these cations. As was noted earlier, the homotropenylium and homocyclopropenium ions undergo a characteristic ring-inversion process which interconverts the *exo* and *endo* substituents on the methylene bridge. With **61** and its derivatives no such reaction occurs. Rather, two different types of thermal isomerization occur. The first of these is the irreversible rearrangement to the cyclohexadienyl ions mentioned above. The second thermal isomerization involves a circumambulation of the methylene group around the periphery of the five-membered ring[137–143,145,152].

Typical examples of circumambulatory rearrangements of bicyclo[3.1.0]hexenyl cations are shown in Schemes 19 and 20. Swatton and Hart reported the isomerization shown in Scheme 19 in 1967 and proposed that the observed deuterium scrambling could be accounted for on the basis of a cyclopropyl walk reaction[153]. This circumambulation is comparable to that proposed by Zimmerman and Schuster as part of the sequence of reactions involved in the type A photorearrangement of 2,5-cyclohexadienones[154].

Childs and Winstein observed the rapid, five-fold degenerate circumambulation of the hepta- and hexamethylbicyclo[3.1.0]hexenyl cations in 1968[139] (Scheme 20) and subse-

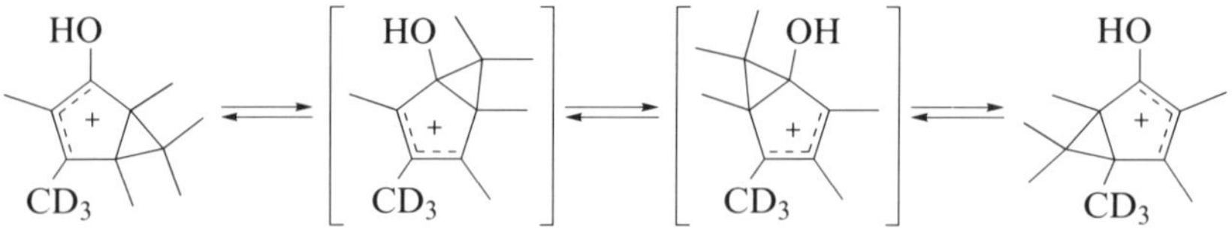

SCHEME 19. Circumambulatory rearrangement of a protonated bicyclohexenone

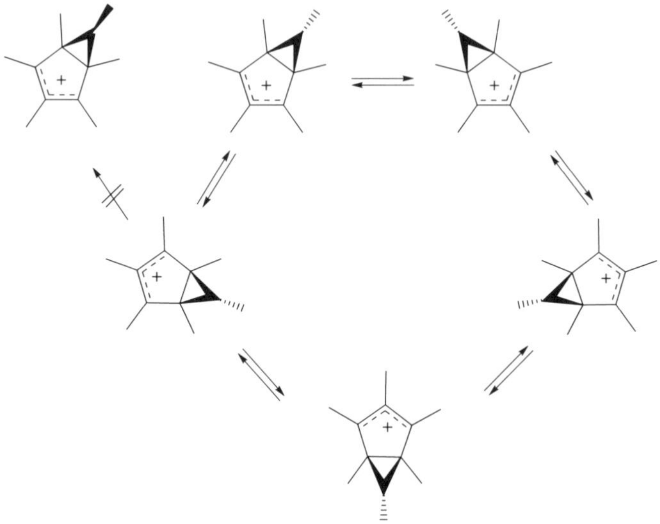

SCHEME 20. Degenerate circumambulatory rearrangement of a hexamethylbicyclo[3.1.0]hexenyl cation

quently went on to show the remarkably high stereoselectivity associated with these reactions[140]. Rearrangements of the parent ion have been examined by Berson and colleagues[142] (Scheme 21).

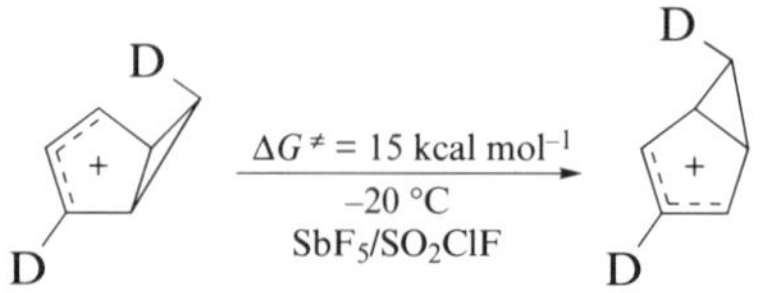

SCHEME 21. Circumambulation of **61D**

Overall, these rearrangements are facile, low activation energy processes which occur with very high stereoselectivity. The energy barriers to the isomerizations are lowered by electron-donating substituents placed on C(6). In fact, Childs and Zeya have shown that with the appropriate choice of substituents it is possible to invert the energies of the ground and transition states for these circumambulations[155]. For example, as is shown in Scheme 22, the Lewis acid complexes of the 5-acylpentamethylcyclopentadienes undergo facile circumambulatory rearrangements in which the corresponding bicyclo[3.1.0]hexenyl cations are now transition states for these degenerate isomerizations.

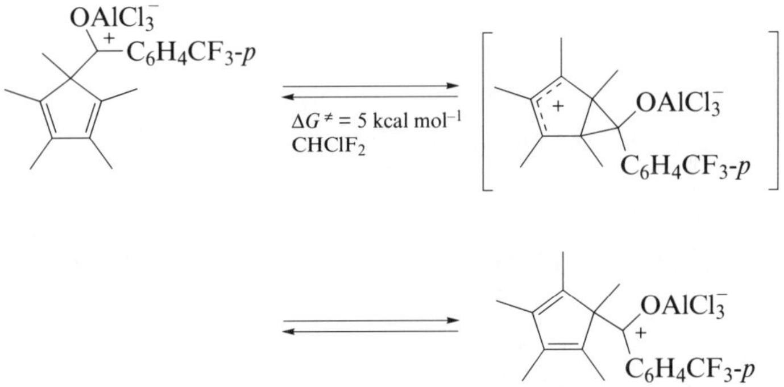

SCHEME 22. Circumambulation of a cyclopentadienylcarbinyl cation

It is interesting to note that the bicyclo[3.1.0]hex-3-en-2-yl radical corresponding to the cation **61** also undergoes a degenerate circumambulatory rearrangement. The barrier to this rearrangement is lower than the ring-opening reaction to give the cyclohexadienyl radical[156].

Berson and Jenkins have looked for a comparable circumambulation in the parent homotropenylium ion **11** using the 4-deuterium labelled ion (Scheme 23)[157]. They were unable to detect the occurrence of any circumambulation prior to decomposition of the ion and, as a result, it was only possible to obtain a lower limit of 27 kcal mol^{-1} for the barrier for circumambulation. Hehre calculated (HF/STO-3G) the barrier to thermally induced circumambulation in **11** as being 43 kcal mol^{-1} [105,151]. Once more it is clear that there is a fundamental difference in the properties of the bicyclo[3.1.0]hexenyl cations and the homotropenylium ions which can be attributed to the difference in electron delocalization of the two systems.

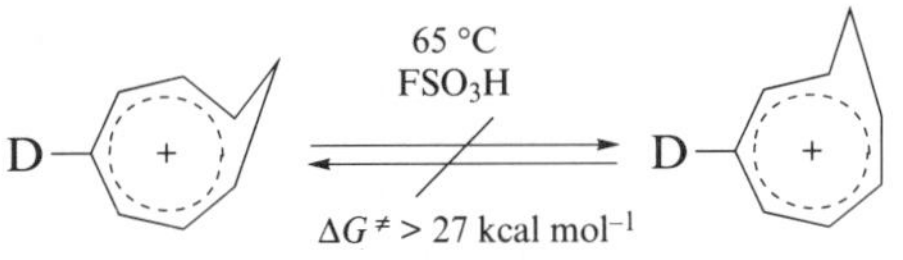

SCHEME 23. Examination of circumambulation in **11D**

It is possible to detect thermally induced circumambulations in certain substituted homotropenylium ions. Hehre in his theoretical study (HF/STO-3G) of **11** suggested that the placement of methyl substituent at C(8) of the homotropenylium ions should reduce the barrier to the thermally induced circumambulation[105,151]. The correctness of this prediction was confirmed by Childs and Varadarajan who reported the generation of the deuterium labelled cation **67** and the measurement of the rate of deuterium scrambling associated with circumambulation of C(8) around the 'seven-membered' ring (Scheme 24)[158]. The barrier for this circumambulation was found to be 14.5 kcal mol^{-1}. Scott and Brunsvold have also reported the occurrence of a circumambulation in a bridged homotropylium ion, **68** (Scheme 25)[159].

Circumambulation of the bridging methylene carbon of the homocyclopropenium ion has not been observed experimentally. Devaquet and Hehre, in examining this reaction using theory (HF/STO-3G), have suggested that circumambulation is a relatively high-energy process that will take place preferentially by the formally symmetry-forbidden pathway[160]. Koptyug and colleagues have reported that a circumambulation of the bridg-

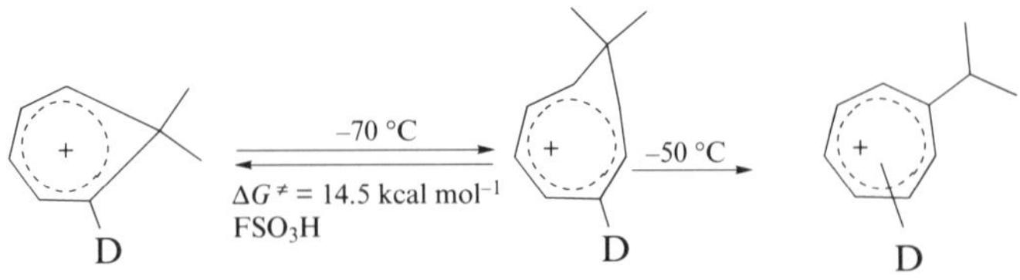

(67)

SCHEME 24. Rearrangement of 8,8-dimethylhomotropenylium ion

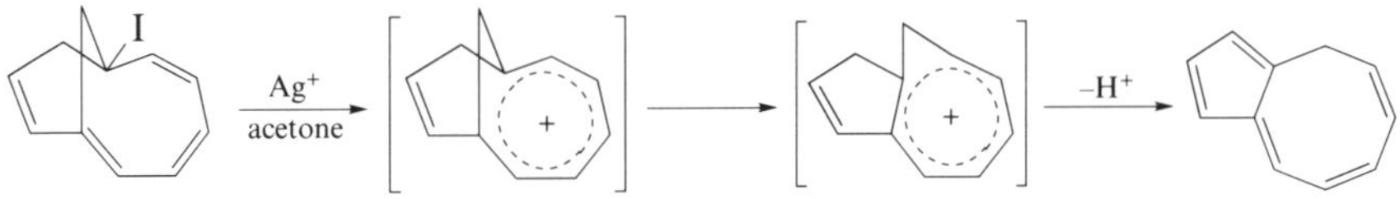

(68)

SCHEME 25. Rearrangement involving the bridged homotropenylium ion **68**

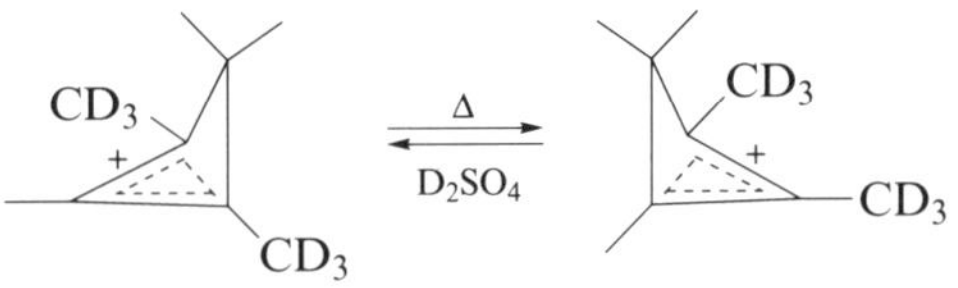

SCHEME 26. Degenerate rearrangement of a homocyclopropenium ion

ing $C(CH_3)_2$ group can be detected with the labelled pentamethyl cation (Scheme 26)[161]. Specific deuterium-labelling experiments ruled out the alternative 1,2-methyl shift process[162].

The different electronic properties of the bicyclo[3.1.0]hexenyl, homotropenylium and homocyclopropenium ions is reflected in the nature of the transition state for the two circumambulatory rearrangements. For example, in the case of **61**, migration of C(6) involves inversion at C(6) leading to an overall retention of stereochemistry. This least motion allowed process occurs with the conservation of orbital symmetry[163]. In the case of **11**, the symmetry-allowed rearrangement formally involves migration with retention of configuration at C(8) which would lead to a net interconversion of the *exo* and *endo* substituents. Such a rearrangement places severe geometry constraints on the transition state for migration.

The geometric limitation for migration with retention of configuration should be lifted in the first excited state of the homotropenylium ions. Childs and Rogerson showed that this was indeed the case and reported a number of examples where photochemically induced circumambulations take place. An example is given in Scheme 27[164].

In concluding this section on the homoaromatic and homoantiaromatic cations, it should be pointed out that in the case of both the homotropenylium and bicyclo[3.1.0]

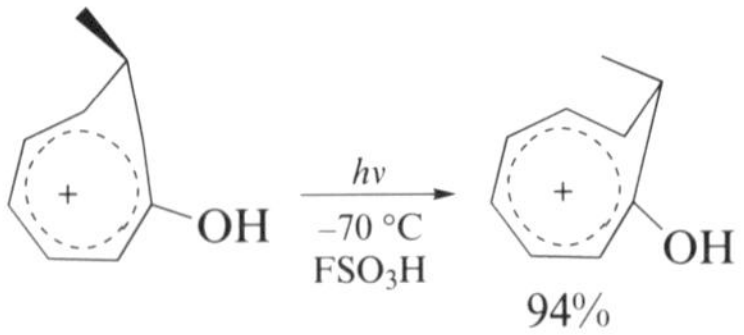

SCHEME 27. Photochemical rearrangement of a homotropenylium ion

hexenyl cations there are other isomeric ions which can be produced by different types of rearrangement of the bridging methylene group. This has already been pointed out for the homotropenylium ions in Scheme 8 where their interconversion with the bicyclo[3.2.1]octadienyl ions was presented. As was pointed out earlier, the bicyclooctadienyl cation is formally a bishomoantiaromatic system. In contrast, in the case of **69**, the isomeric bicyclo[2.1.0]hexenyl ions, **70**, can be obtained by a 1,2 sigmatropic shift of C(6), Scheme 28). The bicyclo[2.1.0]hexenyl ions are formally bishomoaromatic systems and thus might be considered to be more stable than **69**. In fact with the hexamethyl-substituted ions such as **70**, which have been extensively studied by Hogeveen and Volger[165] and Paquette, Olah and colleagues[166], this has been shown not to be the case. Thermochemical measurements by Childs, Mulholland and Nixen[87] have shown that the bicyclo[2.1.1] ions are less stable ($\Delta H = 7.8$ kcal mol^{-1}) than the bicyclo[3.1.0]systems (Scheme 28).

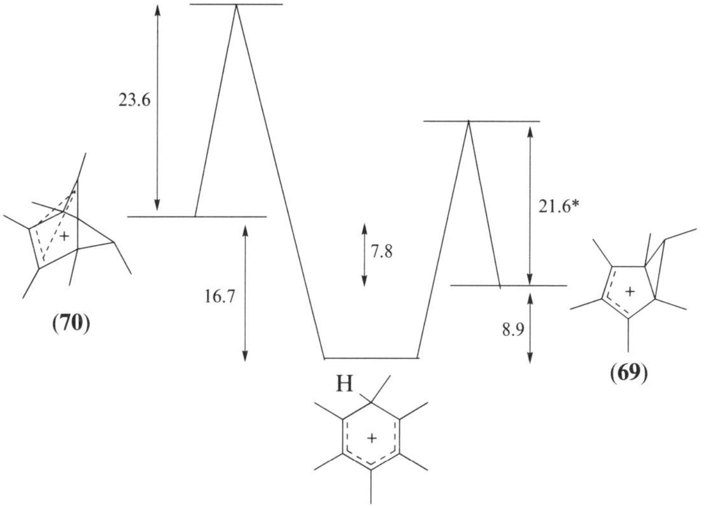

SCHEME 28. Thermochemical relationship between bicyclohexenyl cations (all values are enthalpies except for a $\Delta G^{\neq}$ value marked with* and are in kcal mol^{-1})

In summary, the bicyclo[3.1.0]hexenyl cations clearly show that homoconjugation is an important factor in determining the chemistry and properties of these cationic systems. The properties of the bicyclo[3.1.0]hexenyl cations are sharply different from those of the homotropenylium and cyclopropenium ions, reinforcing the designation of the latter two cations as being examples of homoaromatic systems.

IV. HIGHER HOMOAROMATIC CATIONS

As was mentioned in the introductory sections of this chapter, in principle it is possible to insert more than one homoconjugative interaction in an aromatic system. For example, two homoconjugative linkages would give a potentially bishomoaromatic molecule. In addition to the number and relative positions around the base ring of the bridging groups associated with the homoconjugative linkages, it is also possible for these bridges to adopt a *cis* or *trans* orientation with respect to each other. This possibility is illustrated in Scheme 29 for the 1,4-bishomotropenylium ion.

cis-1,4-bishomotropenylium ion *trans*-1,4-bishomotropenylium ion

SCHEME 29. *Cis* and *trans* bishomotropenylium ions

A further general point needs to be made before specific systems are examined. In the introduction, we laid out the two general approaches to homoconjugative systems. These involved the 'bond' and 'no-bond' starting points. We also pointed out that, in terms of the context of this chapter on cyclopropyl-homoconjugation, the formal starting point is the 'bond' or closed cyclopropane ring. Depending on the relative placement of the homoconjugative linkages, it may not be possible in a given system to have as a formal starting point a resonance structure with all of the linkages being in the closed or cyclopropane form.

A. Bishomotropenylium Ions

Early work on the bishomotropenylium ions has been reviewed by Paquette[18] and a more recent account given by Wiliams and Kurtz[21]. In this section, we will pick up the highlights and particularly emphasize recent work on these cations that helps to define the nature of their electron delocalization and structure.

In 1970 Ahlberg, Harris and Winstein reported the preparation of **71** and **72**, the first examples of bishomotropenylium ions[167]. These 1,4-bishomoaromatic cations were prepared by ionization of the corresponding barbaralyl systems as shown in Scheme **30**. The formation of **71** and **72** proceeds by way of an initial barbaralyl cation, the structure and nature of which has been the subject of a considerable amount of work[168–170]. The initially formed unsubstituted barbaralyl cation rearranges to **71** at –125 °C.

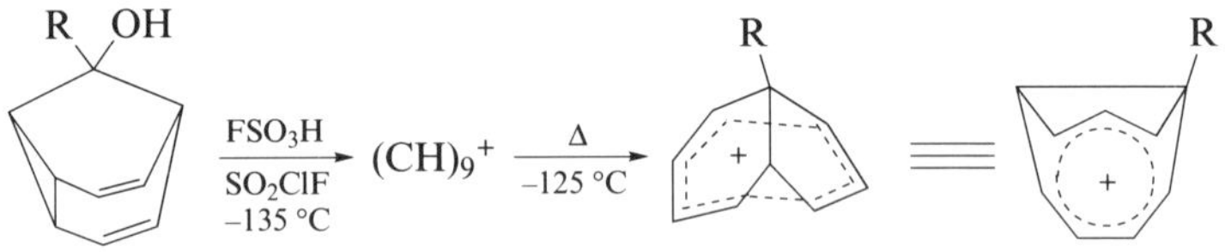

R = H or Me

(**71**) R = H
(**72**) R = Me

SCHEME 30. Formation of bridged bishomotropenylium ions

Since these original reports, a variety of different routes have been used to prepare these ions[171–174]. A related 1,4-bishomotropenylium ion, **73**, was prepared independently by Roberts, Hamberger and Winstein[175] and Schröder and colleagues[176].

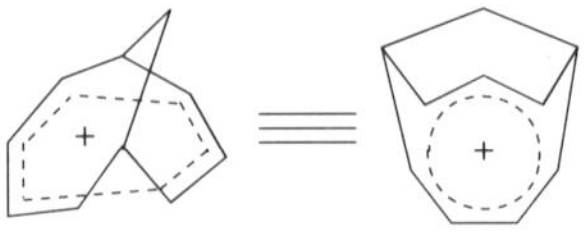

(73)

The ease of formation of **71, 72** and **73** and their relative stability belies the difficulties which have been encountered in defining their structures and mode of charge delocalization. Most of the recent work has been conducted on **71** and it is on this system that most attention will be focussed.

The key issue in adequately understanding the structure and properties of **71** again lies in defining the shape of the potential energy surface linking the open, closed and delocalized structures, **71-O, 71-C** and **71-D**, respectively (Scheme 31). Is **71-D** a transition state linking the two other species or is it the overall energy minimum (see Section I of preceding chapter[25])?

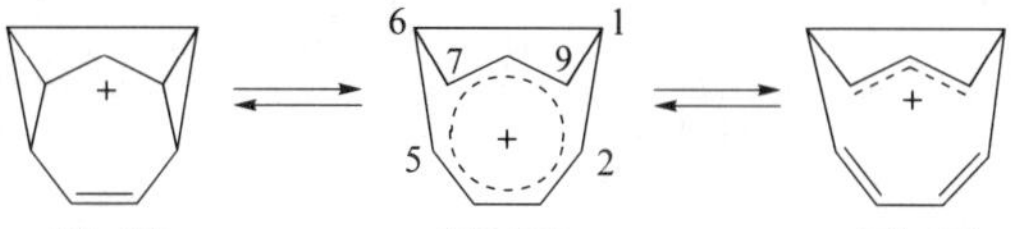

(71-C) **(71-D)** **(71-O)**

SCHEME 31. Bishomotropenylium ion potential energy surface

The closed form **71-C** has not been considered to be the energy minimum on the potential energy surface. Not only are the properties of the observed cation inconsistent with the presence of two cyclopropane rings, but this ion is expected to be highly strained. Distinction between **71-D** and **71-O** has proved to be more difficult.

Examination of the ^{1}H NMR spectrum of **71** provided comparatively little definitive information on its structure and electron delocalization[171]. In particular, the absence of *exo* and *endo* protons on the bridging carbons meant that the difference in their chemical shifts, the traditional but overly simplistic indicator of homoaromatic delocalization, could not be used. Detailed analyses of the ^{1}H NMR spectra of the cations suggested that their structures could be understood in terms of a bishomotropenylium formulation rather than the non-cyclically delocalized cation **71-O**. Further evidence on the nature of the charge delocalization was provided by the ^{13}C NMR spectra of **71** and, in particular, their ^{13}C–^{13}C coupling constants[177]. All of the NMR evidence is consistent with the delocalized structure **71-D** and a fairly even distribution of charge over the basal ring carbons. However, examination of these systems solvolytically provided less than convincing evidence for the importance of a bishomotropenylium formulation of the structure of **71**[171–178].

More recent work on these cations has focussed on the theoretical examination of their structure and stability. Initial theoretical approaches using semi-empirical methods led to no firm conclusions[179]. The most recent calculations of the potential energy surface linking the two ions were reported by Cremer, Ahlberg and colleagues in 1993[28]. These high level calculations [MP2, MP3, MP4(SDQ) employing DZ+P basis set, IGLO calculations and MP2 electron density analysis] showed, in contrast to earlier Hartree–Fock results[180], that **71-D** rather than **71-O** is the energy minimum. However, the homoaromatic stabilization of **71** was found to be small and only of the order of 3 kcal mol^{-1}.

The key features of the calculated structure of **71** were homoconjugative interaction distances close to 2.0 Å and equalization of both the various C—C bond distances and the atomic charges (Table 6). Cremer and colleagues matched the calculated ^{13}C NMR chemical shifts with those observed for **71** as a function of the homoconjugative distance and found a good fit for **71** (mean deviation 8.5 ppm) with a fold angle of 93° and a distance of 2.1 Å (Table 7). As can be seen from the data in Table 7, structure **71-O** gave a very poor fit of the NMR chemical shifts (mean deviation 19 ppm). The authors note that the chemical shifts of the C(7)/C(9) resonances are particularly sensitive to changes in the C(5)/C(7) [C(2)/C(9)] interaction distance and that an exact fit of the observed chemical shift of 155.5 ppm can be obtained with a small reduction in this interaction distance.

TABLE 6. Calculated structures of **71**[28,169,180].

Bond	**71-D**[a]	**71-D**[b]	**71-O**[c]
C(2)—C(3)/C(4)—C(5)	1.380 Å	1.348 (Å)	1.357 (Å)
C(3)—C(4)	1.411	1.427	1.442
C(7)—C(8)/C(8)—C(9)	1.393	1.378	1.390
C(2)—C(9)/C(5)—C(7)	2.147	2.075	2.356
Fold angle (deg)	100.0	93.2	113.8

[a]Geometry at calculated [MP4(SDQ)-MP2/6-31G(d)] energy minimum.
[b]Geometry calculated [MP2/6-31G(d)] for fold angle of 93° obtained from best fit of IGLO calculated chemical shifts.
[c]MP2/6-31G(d) calculated geometry.

TABLE 7. Calculated (IGLO) and observed ^{13}C chemical shifts (ppm) of **71**[28,169,180]

Carbon	**71-D**	**71-O**	Experiment
C(1)/C(6)	45.9	46.4	52.6
C(2)/C(5)	121.5	121.2	117.2
C(3)/C(4)	145.4	141.9	140.2
C(7)/C(9)	176.9	227.4	155.5
C(8)	145.4	148.3	144.0

It should be noted here that it was the exploration of the **71** potential energy surface which led to the establishment of the criteria for homoaromaticity set out at the start of this chapter and the accompanying review[25].

Overall, it is clear that **71** can rightly be considered to be a bishomoaromatic system. However, it was pointed out by Cremer, Ahlberg and colleagues that the relatively small stabilization energy associated with **71** means that substitution could have a major effect on the relative energies of **71-D** and **71-O**. The classification of **71** as bishomoaromatic does not mean that derivatives of **71** will also have a comparable electron delocalization pattern and structure.

Cation **71** has a bridge which links the 'cyclopropyl' carbons and thus maintains an appropriate orientation for homoconjugation. While such a linkage is not formally required for homoaromatic delocalization, it is an open question as to whether examples of bishomotropenylium ions exist which lack such a framework.

Protonation of **74** in FSO_3H at low temperatures was shown by Warner and Winstein to give a cation which was suggested to be the *cis*-1,3-bishomotropenylium cation, **75**[181]. The main evidence for the structure of **75** was its ^1H and ^{13}C NMR spectra[18,182]. However, the ^1H NMR spectrum showed only a small difference in chemical shifts between the *exo* and *endo* bridge protons ($\Delta\delta = 1.91$ ppm). The authors suggested that the pattern of the shifts of the resonances of the vinyl protons and the unsaturated carbons was indicative of a somewhat more even distribution of charge than is found in a pentadienyl cation. They concluded that **75** should be considered as being homoaromatic.

However, comparison of the ^{13}C NMR spectrum of **75** with the model compounds raises in our minds serious questions as to the validity of these claims. Apart from the ^{13}C chemical shifts of the C(1)/C(5) carbon resonances (137.0 ppm) the rest of the resonances are in positions typical of a dienyl cation and an isolated double bond [C(6) and C(7)]. The large spread in the ^{13}C chemical shifts reported for **75** indicated a very uneven charge distribution in the ion.

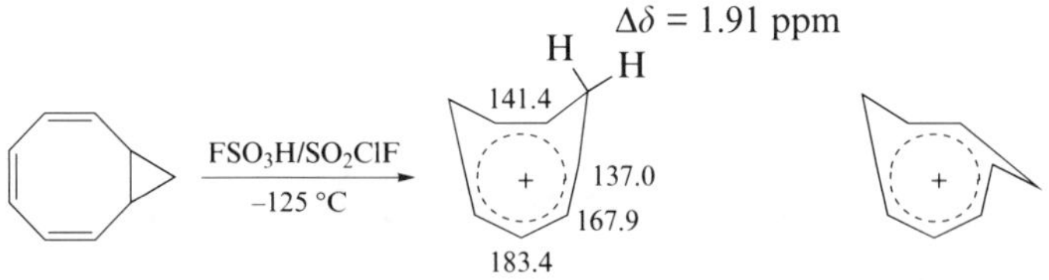

(74) (75) (76)

Recent *ab initio* calculations by Cremer and coworkers carried out for the *cis* isomer **75** and the corresponding *trans* isomer **76** provide convincing evidence that both isomers preferentially exist in 'classical' valence tautomeric forms rather than delocalized homoaromatic systems[183]. The *cis* isomer **75** was shown to prefer the closed form by about 7 kcal mol^{-1} while **76** prefers the open form by 4 kcal mol^{-1}.

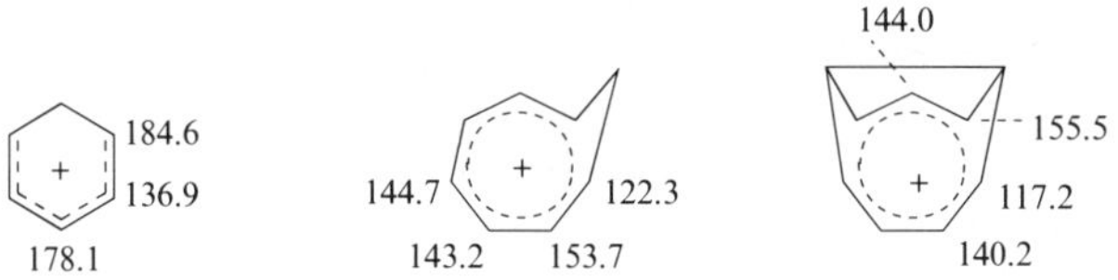

(77) (11) (71)

The *trans* isomer **76** was found to be more stable than the *cis* form **75** by 6 kcal mol^{-1}. Since the experimental ^{13}C NMR shift values exclude the possibility of a closed form, it would appear that Warner and Winstein prepared **76** rather than **75**. Cremer and coworkers also pointed out that agreement between the calculated and experimental ^{13}C NMR spectrum can only be obtained by re-assigning the shifts assigned to C(1) and C(2)[183]. When this is done, the chemical shifts resemble those of **77** rather than the homoaromatic models **11** or **71**. This conclusion is supported by the calculated geometry of **76** which suggests the existence of an almost isolated double bond that does not interact with the pentadienyl system (Scheme 32)[183].

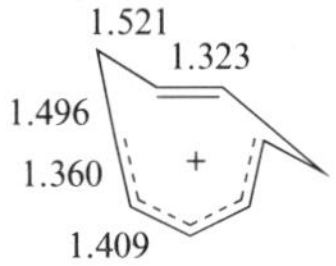

SCHEME 32. Calculated bond distances of **76**

Cremer and colleagues concluded the **76** is non- or only very weakly, bishomoaromatic[183]. However, the transition state for the valence tautomeric interconversions of **75** and **76** was found to possess all the characteristics of homoaromatic electron delocalization.

Other attempts to prepare non-bridged *cis* and *trans* bishomotropenylium ions have also not been successful. Childs and Corver prepared 1,4-bridged cations **78** and **79** by protonation of the corresponding ketones[184]. It was concluded on the basis of ^{1}H NMR studies that neither cation could be regarded as being homoaromatic. It would appear that in the *cis*-isomer, steric interactions between the two methylene groups prevent the cation from attaining an appropriate geometry for cyclic homoconjugation of the two cyclopropane rings. In the case of the *trans*-isomer, the steric problem associated with the two methylene groups is absent. However, it is not possible for the cation to adopt a boat-type conformation, as has been found for the parent homotropenylium ion, that allows both cyclopropanes to be situated for effective homoconjugation. The results with **78** and **79**, both potentially 1,4-bishomotropenylium ions, are consistent with the findings of Cremer and

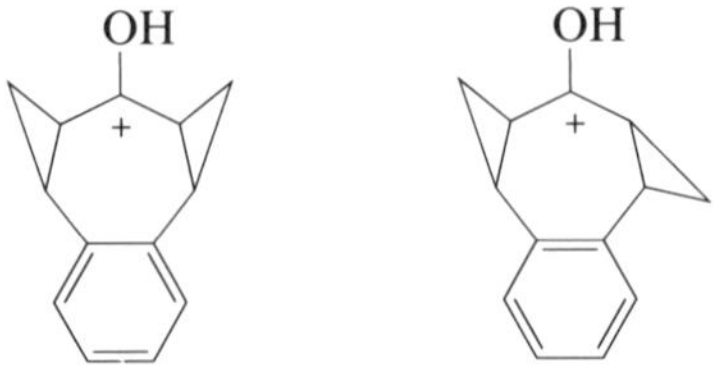

(78) **(79)**

colleagues, discussed above, of the very small homoaromatic stabilization energy associated with the 1,4-bishomotropenylium ions.

There have been several attempts to prepare trishomotropenylium ions. However, these have all been unsuccessful[185].

B. Bishomocyclopropenium Ions

A large amount of work has been reported on potentially bis- or trishomocyclopropenium ions. Many reviews exist including those by Winstein[14,16], Story and Clark[17], Williams and Kurtz[21], Koptyug[132], Hogeveen and Kwant[186], Lenoir and Siehl[187] and a particularly full account by Barkhash[133]. The topic also overlaps that of the pyramidal cations and related boranes[188]. In this present section we will again not give an exhaustive account but concentrate on a few key systems where there has been some recent definitive work reported.

In terms of systems lacking a bridging framework to hold the homoconjugative linkages in an appropriate geometry, the cyclopentenyl cation **80** is the important example. In order for homoconjugative stabilization to be effective in **80**, the cation must exist in a non-planar configuration such that the double bond and C(3), the cationic centre, are placed in a suitable geometry (Scheme 33). In the planar conformation, no homoconjugative stabilization is possible but rather an inductive destabilization of the charge at C(4) by the double bond is expected. Calculations by Schleyer and colleagues at the MP2(FC)/6-31G(d)//6-31G(d) level indicate that the non-planar form **80** is about 19 kcal mol^{-1} lower in energy than the planar form, **81**[189]. More recent high-level calculations of Szabo, Kraka and Cremer revealed that MP2 seriously exaggerates the energy difference[190]. At MP4/DZ+P using MP2 geometries, the non-planar form **80** is 6 kcal mol^{-1} more stable than **81**. The envelope form of **80** is strongly folded (fold angle *ca* 90°), thus reducing the distance between the two interacting atoms C(3) and C(5) from 2.33 Å in **81** to 1.76 Å in **80**. Cremer and colleagues pointed out that both hyperconjugative interactions and strain effects favour the planar form **81**[190]. This means that the energy difference of 6 kcal mol^{-1} provides only a lower limit to the true homoaromatic resonance energy of **80**.

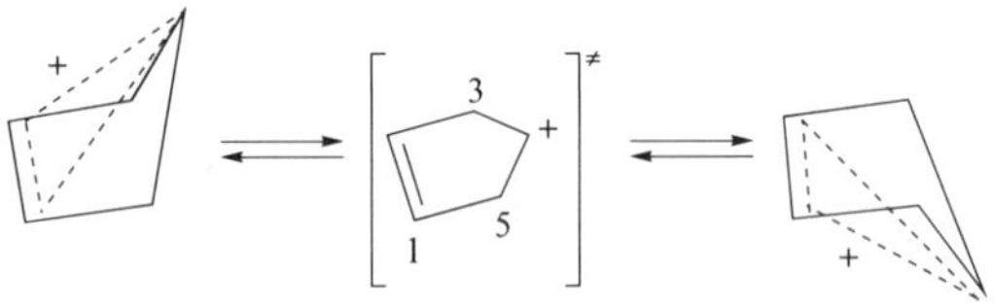

(80) **(81)** **(80′)**

SCHEME 33. Ring inversion of **80**

Despite the seeming stability of **80**, all attempts to prepare the cation as a stable species in super-acid media have been unsuccessful[191,192]. The retention of configuration at C(4) in solvolysis of cyclopentene derivatives suggests that there is some involvement of the double bond in charge delocalization; however, this participation is not clearly reflected in the rates of ionization of these systems[22,193]. It is interesting that the calculations on the borolene **82** corresponding to **80** indicate that the potential energy surface is almost flat with only very shallow minima for the bent forms[22].

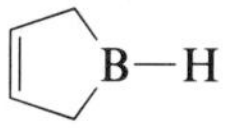

(82)

There have been a larger number of reports on the preparation of bridged bishomocyclopropenium ions in which the homoconjugative bridges are linked together by an appropriate carbon framework. Most of these systems have the general structure **83** and the special case **85** where the bridge is a —CH=CH— unit. The reviews cited above cover much of the work reported on these systems. The focus here will be on recent results with the 7-norbornenyl, **84**, and the 7-norbornadienyl cations, **85**.

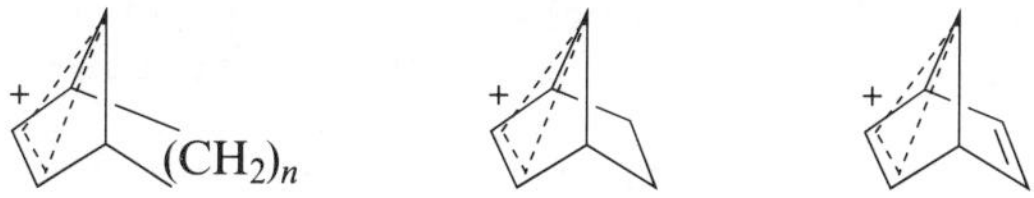

(83) $n = 1$–4 **(84)** **(85)**

The initial report on the formation of **84** was by Winstein and colleagues in 1955[194]. These workers noted the remarkable anchimeric acceleration and high regioselectivity of product formation associated with the ionization of *anti*-7-norbornenyl derivatives. The 7-norbornadienyl derivatives were found to show even larger rate accelerations[195]. These unusually large rate accelerations were attributed to neighbouring group participation or homoconjugation. Later these ions were called bishomocyclopropenium ions[196]. Brown and Bell offered a contrary view and suggested that **84** could be considered to be a rapidly equilibrating pair of cyclopropylcarbinyl cations (Scheme 34)[197].

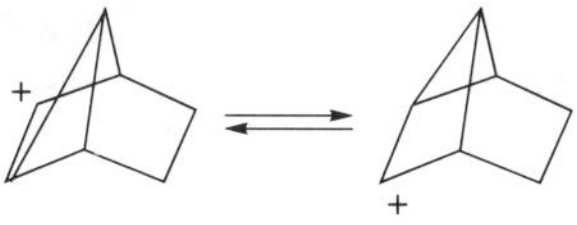

SCHEME 34. Equilibrating cyclopropylcarbinyl cations

Both **84** and **85** have been prepared as stable ions in super-acid media and their NMR properties studied[198–201]. The conclusions reached from the extensive amount of work done with the parent and a variety of substituted systems is that both ions can be considered to be bishomoaromatic. In the case of **85** it should be noted that the C(7) bridge was found to lean towards and interact with one of the double bonds. Cation **85** was found to undergo an inversion process in which there is an interchange of the participating double bond (Scheme 35). Winstein and coworkers were able to place a lower limit of 19.6 kcal mol^{-1} on

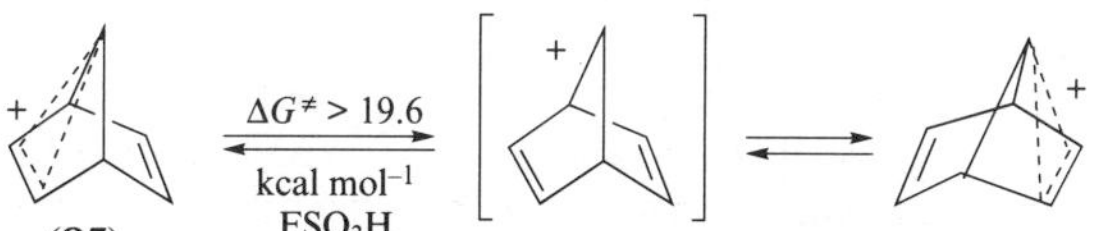

(85)

SCHEME 35. Interconversion of 7-norbornadienyl cations

TABLE 8. Structures of 7-norbornenyl cations and related compounds[a]

Cation	Method	C(2)—C(3)	C(2)—C(7)	Reference
86	X-ray	1.38(1)	1.86(1)	202
84	6-31G(d)	1.380	1.938	203
85	6-31G(d)	1.380	1.719	203
85	MP2(FU)/6-31G(d)	1.400	1.701	203
87	MP2(FU)/6-31G(d)	1.396	1.775	204
87	MP2(FU)/6-31G(d)	1.392	1.759	204
89	X-ray	1.375	1.864	205
90	X-ray		1.814	205

[a]All distances given in Å.

the barrier to this process[198]. The value of this barrier can be thought of as an estimate of the stabilization achieved by homoconjugative participation of the double bond with C(7). Substitution at C(7) was found to lower the barrier to inversion.

Structural information on these systems was recently provided by two groups. Laube reported the X-ray structure determination of the norbornenyl cation **86**[202] while Schleyer and colleagues published the results of high-level calculations on **84** and **85**[203].

While **86** has a phenyl substituent at C(7) and, as has been noted above for the dienyl system, this substitution could substantially reduce the need for interaction of C(7) with the double bond, the structure found by Laube is fully consistent with a bishomoaromatic formulation (Table 8). The C(7) bridge was found to lean towards the double bond bearing the two methyl groups giving C(2)–C(7)/C(3)–C(7) distance of 1.86(1) Å. This distance is well within the range encountered for the homoconjugative distance in homoaromatic systems. The C(2)–C(3) distance of 1.38(1) Å is intermediate in length between that of C,C single and double bonds.

The calculations of Schleyer and colleagues at the MP2(FU)/6-31G(d) level gave very similar results for **85** and an even more distorted structure for **84**. The calculated ^{13}C NMR spectra of these ions were comparable to those observed experimentally; however, no systematic structural changes have yet been reported which optimize the fit of the calculated chemical shifts.

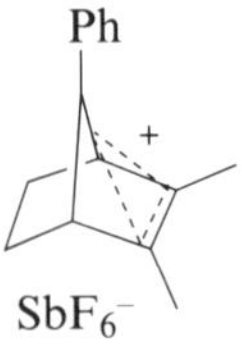

(86)

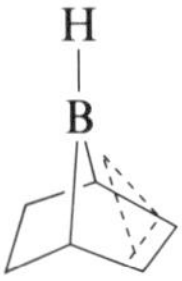

(87)

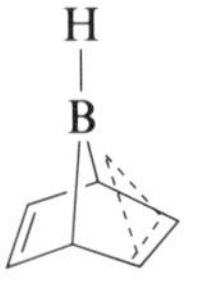

(88)

Recently, work has been reported on the 7-boranorbornene (**89**) and 7-boranorbornadiene (**90**) systems related to **84** and **85**. Schleyer and colleagues have examined the parent molecules **87** and **88** theoretically [MP2(FU)/6-31G(d)] and concluded that they have very similar structures and electron delocalizations to the related cations (Table 8)[203,204]. X-ray structures of substituted derivatives of **89** and **90** have been determined and these have very similar distorted conformations to those of **84**, **85** and **86**[205]. Calculated (IGLO) NMR chemical shifts of **87** and **88** correspond well with those observed experimentally.

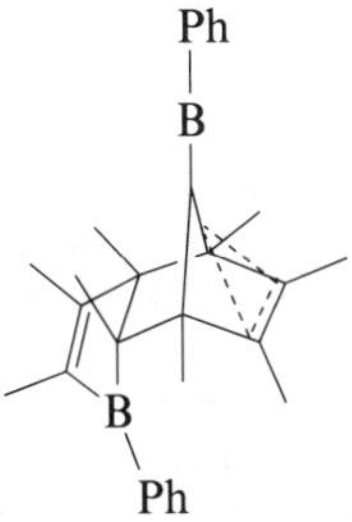

(89)

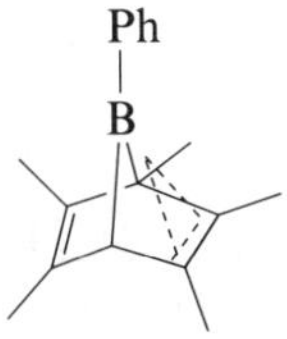

(90)

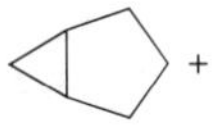

(91)

C. Trishomocyclopropenium Ions

The trishomocyclopropenium species **91** has been an important cation in terms of the development of the concept of homoaromaticity. It also provides an example of a different mode of cyclopropyl homoconjugation to that encountered in the systems discussed thus far. In **91**, the cyclopropane is formally interacting with the remote positive charge in an edge-on manner. All of the previous examples discussed have involved a cyclopropyl carbinyl-type interaction with the conjugating group being joined to the cyclopropane.

Winstein, Sonnenberg and de Vries first proposed the intermediacy of **91** in order to account for the solvolytic properties and products of derivatives of **92**[206]. A major advance in understanding the properties and structure of **91** came from the laboratories of Masamune and colleagues, who reported the preparation of the cation as a stable species in super-acid media[207], a preparation later repeated by Olah and colleagues[192,208] and Kelly and coworkers (Scheme 36)[209].

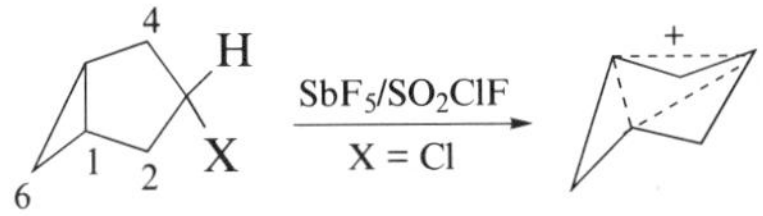

(92) (93)

SCHEME 36. Formation of trishomocyclopropenium ion

The key feature of the NMR spectra of **91** is its simplicity. Thus the ^{13}C NMR spectrum consists of only two resonances at 4.9 and 17.6 ppm, indicating either a symmetrical trishomocyclopropenium cation, **93**, or rapid equilibration between three equivalent structures (Scheme 37). The positions of the ^{13}C NMR resonances of the cation strongly suggested the formulation of its structure as the trishomocyclopropenium ion, **93**[210]. This conclusion was reinforced by the preparation of the deuterated cation and examination of the isotopic perturbation of its ^{13}C chemical shifts[208,211], and measurement of the ^{13}C–^{1}H coupling constants[209].

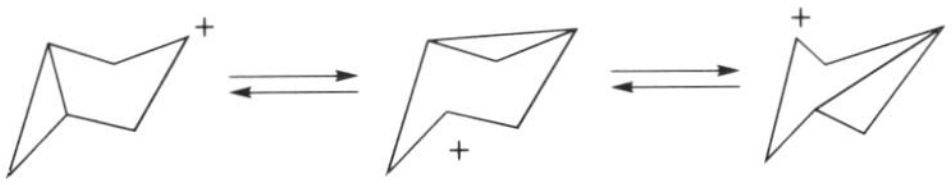

SCHEME 37. Equilibration between classical bicyclohexyl cations

Szabo, Kraka and Cremer have recently carried out an extensive *ab initio* investigation of **93** using MP2, MP3 and MP4 in conjugation with DZ+P basis, calculating chemical shifts and magnetic susceptibility and analyzing orbitals and electron density distributions[190]. In addition, they studied the mechanisms of formation and decomposition of **93** and several hetero analogues. According to MP2 optimizations, **93** possesses a 1,3 interaction distance of 1.824 Å resulting from a folding of the formal five-membered ring by 87 °, similar to that found for ion **80**[190]. The authors concluded on the basis of the calculated energetic, structural, electronic and magnetic properties that **93** fulfils the criteria for homoaromaticity given above in Section I. A and in the preceding chapter[25]. As such **93** is clearly the prototype of a trishomoaromatic cation.

Cremer and colleagues pointed out that an exact determination of the homoaromatic resonance energy of **93** is not possible because of strong hyperconjugative, strain and inductive effects present in the cation[190]. However, they suggested a value of 17.4 kcal mol^{-1} as a lower limit to the true resonance energy. This indicates that homoconjugative electron delocalization is much more developed in **93** than in any other homoaromatic system studied thus far. The high-field shift of the ^{13}C NMR signals by about 350 ppm and the magnetic susceptibility exaltation found by Cremer and coworkers confirm this description.

Cremer and colleagues found that **93** can undergo an inversion (with a barrier of 26 kcal mol^{-1}) to an envelope conformation that is 17 kcal mol^{-1} less stable than **93**. This envelope conformation provides a reasonable reference for estimating the homoaromatic resonance energy of **93**. The envelope form was found to rearrange readily to the slightly more stable (3 kcal mol^{-1}) bicyclo[3.1.0]hex-2-yl cation by a shift of the *trans* hydrogen atom at C(2) (energy barrier 1 kcal mol^{-1}) (Scheme 38)[190].

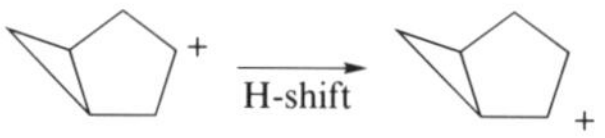

SCHEME 38. Isomerization of the envelope form of **93**

Replacement of C(6) of **93** by various heteroatoms was found by Cremer and coworkers to decrease the homoaromatic resonance energy with increase in the electronegativity of the heteroatom (24 kcal mol^{-1} when X = BH to 4 kcal mol^{-1} when X = O)[190]. Silicon atoms in positions 1,3 and 5 were found to give homoaromatic analogues of **93**; however, it was found difficult to segregate homoaromatic stabilization and Si—C—C^+ hyperconjugation effects.

A variety of other more highly bridged ions would appear to provide further examples of trishomocyclopropenium ions. These include the 7-norbornene and norbornadiene homologues **94**[212,207] and **95**[213]. In general these more complex systems have been studied in less detail than some of the simpler systems reviewed here. The chemistry of these systems is covered in the existing reviews.

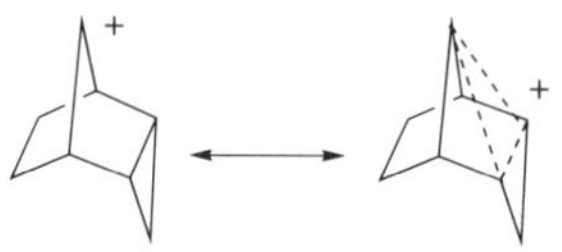

(94)

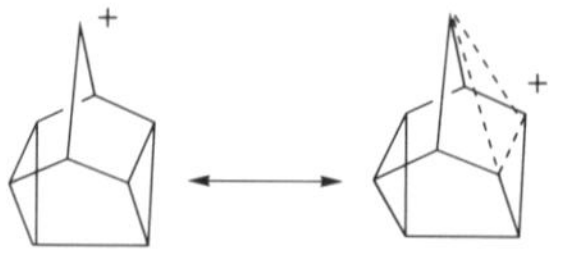

(95)

Szabo and Cremer have also investigated the possibility of cyclobutyl homoconjugation in **96** by determining its structure, conformation and energetics[214]. They found no evidence for a non-classical structure. The only stable form was the classical envelope form of the ion. Upon forced ring inversion (barrier 9 kcal mol^{-1}), **97** rather than a trishomoaromatic species was formed. Ion **97** contains a centre-protonated spirocyclopentane unit with intriguing structural and electronic properties. The authors point out that **97** is the 'missing link' between the bicyclo[3.2.0]hept-3-yl and the 7-norbornyl cations on the $C_7H_{11}^+$ potential energy surface. Cation **97** was found at a local energy minimum surrounded by relatively high barriers which provide kinetic stability for the ion.

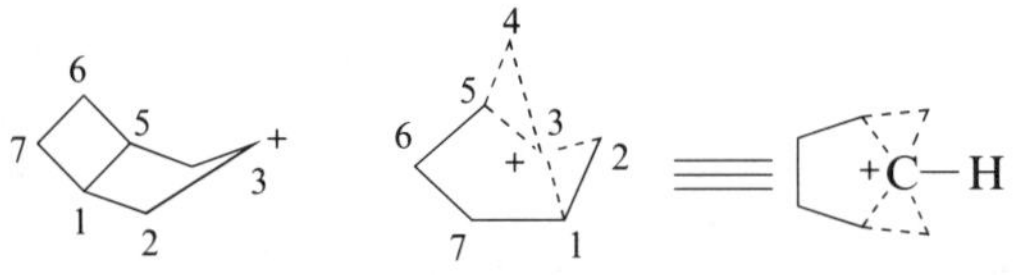

(96) **(97)**

An interesting and somewhat different system worth noting here is one based on the adamantyl framework. Scott and Pincock provided evidence for cyclopropyl participation in the ionization of **98** and suggested that the trishomocyclopropenium ion **99** was formed[215]. Recently Bremer, Schleyer and colleagues have produced and characterized the stable dication **100** from the difluoride **101**[216]. The ^{13}C NMR spectrum of **100** and calculations were suggested to be fully consistent with the formation of the caged pyramidal cation. MP2/6-31G(d) interaction distances were found to be 2.084 Å, somewhat longer than for the trishomocyclopropenium ion discussed above (1.824 Å). Dication **100** is an example of a three-dimensional homoaromatic molecule (homoradial aromaticity[25]) with six homoconjugative linkages, built up from the linkage of four trishomocyclopropenium ions.

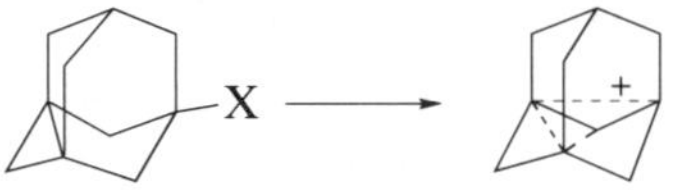

(98) **(99)**

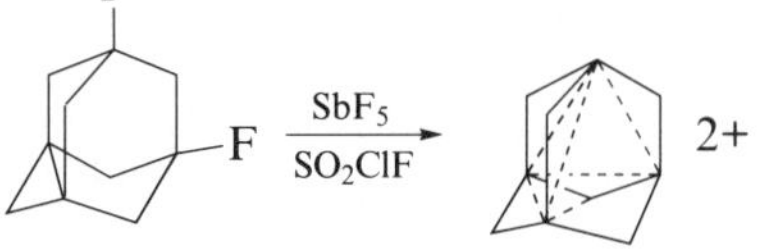

(101) **(100)**

D. Bishomoantiaromatic Cations

This section will be quite short in that we are unaware of any examples of cations which display bishomoantiaromatic character. There are several potential candidates including derivatives of **102** and **103** in which a bishomocyclopentadienyl-type delocalization could occur. Examples of these cations are known; however, there is no evidence for any significant degree of cyclic delocalization[217–222].

The clearest example is that of **104** described by Winstein and colleagues[217]. This cation, which rearranges to the isomeric 1- and 5-methyl-7-norbornadienyl ions at −125 °C, exhibits a ^{1}H NMR spectrum which is typical of a cyclopentenyl cation and an isolated cyclobutene.

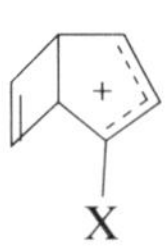

(102) R = H
(104) R = Me

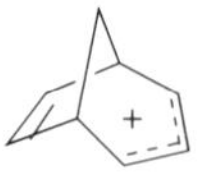

(103)

In concluding this section on the polyhomoaromatic cations, several points stand out:

First, there is only a limited number of examples of systems which can be classified as being cyclically delocalized systems and, thus, termed homoaromatic. These include a number of bridged bishomotropenylium ions and a somewhat larger series of bridged or caged bis- and trishomocyclopropenium ions.

Second, the only non-bridged or non-caged example which has a strong weight of evidence pointing to its homoaromatic delocalization is the parent trishomocyclopropenium ion.

Third, no examples of bishomoaromatic systems with a *trans* orientation of the rings are known.

Fourth, the extra stability associated with homoaromatic delocalization in the bishomotropenylium ions is small and insufficient to overcome any strain associated with conformational changes required for a system to achieve a geometry suitable for homoconjugation. This means that there will likely be few other related 6π-cations described which will be found to be homoaromatic.

Lastly, no examples of bishomoantiaromatic cations are known.

V. NEUTRAL HOMOAROMATIC SYSTEMS

There has been considerable controversy over whether there are any existing examples of neutral homoaromatic systems[223]. Perhaps as a result of the difficulties in this area, a large amount of work has been reported[224,14–22].

We assert in this review that, at this point in time, there are several examples of neutral molecules which have been shown to display either bond or no-bond homoaromaticity. These include, in addition to the boranes mentioned above in Section III. B, cycloheptatriene, norcaradiene, bridged cycloheptatrienes and norcaradienes, semibullvalenes, barbaralanes, bridged annulenes, etc. Confirmation of the homoaromatic character of these systems comes from thermochemical and spectroscopic studies, and force field and *ab initio* calculations. In particular, the work of Roth and coworkers must be mentioned in this connection in that they were the first to provide reliable resonance energies of a large number of these neutral molecules[225,226]. These authors have also demonstrated that systems such as bicyclo[2.1.0]pentene are homoantiaromatic.

The major confusion as to the existence of neutral homoaromatic systems results from the fact that homoaromatic resonance energies, contrary to aromatic resonance energies, are normally less than 10 kcal mol^{-1} [225,226]. In general it is difficult to separate homoaromatic resonance energies from energies that are due to steric strain, hyperconjugative or inductive effects. These difficulties and the possible ways they are overcome are discussed extensively in the preceding chapter[25]. We concentrate here on the results of this analysis of these systems coupled with related experimental work. The focus in this section will be on cycloheptatriene, norcaradiene, semibullvalene and a limited number of other potentially homoaromatic systems.

A. Monohomoaromatic Systems

Cycloheptatriene, **105**, and its valence tautomer norcaradiene, **106**, have been studied extensively[227]. The seven-membered ring of **105** adopts a boat-type conformation with C(7) forming the 'prow' of the boat[228]. The barrier to interconversion of the two boat forms, a process that exchanges the *exo* and *endo* C(7)-protons, is relatively low (6 kcal mol^{-1})[229]. Cycloheptatriene is related to **106** by a thermally allowed disrotatory process[33,36] involving an aromatic transition state[31]. The bicyclic valence tautomer **106**, an important species in terms of the chemistry of the system, is considerably less stable than **105**. Substitution at C(7) can dramatically alter the relative stability of the two valence tautomers and, in certain instances, make **106** the preferred valence tautomer[230–232]. Attempts have been made by Dunitz and coworkers to map the course of the ring closure of bridged derivatives of **105** to **106**[233]. In priniciple, such an approach could reveal intermediary structures, as has been attempted with semibullvalene; however, the nature of the bridges makes segregation of the properties of the **105/106** component difficult.

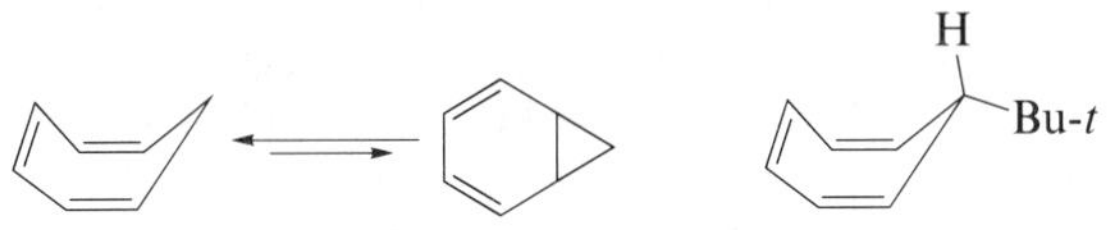

(105) **(106)** **(107)**

While **105** and **106** cannot be viewed as resonance structures of a common delocalized species, the question at issue is whether **105** and **106** can individually be regarded as homoaromatic systems? In terms of thermochemical evidence, it has been recently pointed out that the available experimental data are among the most precise data available in the thermochemical literature[234]. The issue comes down to the interpretation of these data and the proper consideration of contributing factors such as strain, etc.

Liebman and coworkers conclude from three different approaches that **105** is homoaromatically stabilized to the extent of about 6 kcal mol^{-1} [234]. This result is consistent with the results of Roth and his coworkers, who have given a value of 4.5 kcal mol^{-1} as the homoaromatic resonance energy of **105**[225,226]. This latter value is based on the experimental heats of formation coupled with molecular mechanics calculations, as outlined in Section II. G of the preceding chapter[25]. This approach largely separates steric and conjugative effects and can be considered to give a reliable value.

Further evidence supportive of cyclic delocalization in **105** comes from its magnetic properties. Dauben and colleagues measured the diamagnetic susceptibility exaltation, Λ, of **105** and found a value of –8.5 (10^6 cm^3 mol^{-1})[235]. Childs and Pikulik extended these measurements to a series of 7-substituted-cycloheptatrienes and found that the magnitude of Λ increased with the steric size of the substituent[236]. The largest exaltation found was for 7-*t*-butylcycloheptatriene, **107**, Λ = –14.8, a value which, surprisingly, is larger than that found for benzene itself (–13.7)[a]. It was suggested that variation in Λ with substitution at C(7) was related to changes in the conformation of the seven-membered ring and, in particualr, to changes in the relative orientation and positions of C(1) and C(6).

The diamagnetic susceptibility measurements are consistent with the NMR properties of cycloheptatrienes. Pikulik and Childs compared the ^{1}H NMR chemical shifts of the C(7) proton of 7-substituted cycloheptatrienes and the corresponding 1,4-cycloheptadienes and showed that there was a considerable upfield shift of the resonance of the former protons

[a] Comparison of the magnitude of diamagnetic susceptibility exaltations of different systems must take into account the area of the unsaturated cyclic system. In the case of benzene and cycloheptatrienes (homobenzene) the areas of the cyclic π-systems are similar.

when in the pseudo-axial position[237]. The chemical shifts of the methylene protons of **105** and related bridged cycloheptatrienes can be accounted for on the basis of an induced ring current.

The conclusions reached on the basis of magnetic and thermochemical results for **105** are supported by structure investigations[228] and recent high-level *ab initio* calculations [MP2 and CCSD(T)][238]. This recent theoretical work qualifies and corrects older or less reliable *ab initio* or semi-empirical calculations[239–243].

The boat conformation of **105** is appropriate for a C(1)–C(6) interaction. Cremer, Dick and Christen[239] have pointed out that the existing experimentally based structural data (ED and MW) are rather imprecise in terms of the exact measurement of conformation and the 1–6 distance. Kraka and Cremer[238] have found the boat form of **105** to be rather flat in its stern (MP2 fold angle 152°, just 28° from a planar ring form), but strongly folded in its bow (MP2 fold angle 123°, i.e. 57° from a planar form). As a result the pπ orbitals at C(1) and C(6) can orient in a way that small but significant overlap is possible despite a C(1)–C(6) distance (MP2) of 2.39 Å[238].

Comparison of the calculated C—C bond distances (MP2) of **105** [C(1)=C(2) 1.357, C(2)—C(3) 1.439, C(3)=C(4) 1.371, C(1)—C(7) 1.497 Å] with *trans*-butadiene [C(1)=C(2) 1.341, C(2)—C(3) 1.461 Å][244] shows that a considerable degree of bond equalization is present in **105**. Cremer and Kraka suggest that this is the consequence of homoconjugative electron delocalization[238]. This conclusion was confirmed by these authors from the values of the calculated bond orders and π-character indices. These results are contrary to predictions obtained at lower, less reliable levels of theory[239–243].

In conclusion, the energy, geometry and magnetic properties of **105** are fully consistent with it being classified as a neutral homoaromatic molecule with a small, but significant, resonance energy of 4.5 kcal mol^{-1}.

A similar conclusion can be reached in terms of norcaradiene, **106**. Roth and colleagues[225,226] estimated a resonance energy of 5.6 kcal mol^{-1} for **106** based on an assumed heat of formation ΔH_f° of 49.6 kcal mol^{-1} derived from *ab initio* calculations of Cremer and Dick[240] and Schulman, Disch and Sabio[241]. Experimental estimates of the enthalpy difference between **105** and **106** range from 4 to 4.5 kcal mol^{-1} [245]. However, these estimates are too low in view of the recent CSSD(T)/DZ+P calculations (at MP2/DZ+P geometries including MP2 ZPE corrections) of Kraka and Cremer which predict an enthalpy difference of 5.5 kcal mol^{-1} and a heat of formation of 50.1 kcal mol^{-1} [238]. The latter value is in good agreement with Roth's estimate given above. However, Kraka and Cremer's resonance energy of **106** is just 3.8 kcal mol^{-1} (see preceding chapter, Section II. G[25]). The bridged system **108** and **109**, in which the norcaradiene forms are more stable than the open cycloheptatriene forms, were reported to have homoaromatic resonance energies of 3.5 and >5.7 kcal mol^{-1} [225,226], values which are comparable to that calculated by Kraka and Cremer[238].

(108)

(109)

Calculations of the geometry of **106** have been reported by a number of groups[238–243]. The most accurate data are those from the calculations of Kraka and Cremer based on MP2/DZ+P and CCSD(T)/DZ+P calculations[238]. The C(1)–C(6) distance in the latter work was found to be 1.572 Å, which is clearly longer than the value found for cyclopropane. The C(1)—C(6) bond order is significantly smaller than 1, indicating a partial bond between these atoms. π-Electron delocalization, as reflected by the ratio of calculated bond lengths [C(1)—C(2) 1.461, C(2)—C(3) 1.357, C(3)—C(4) 1.446 Å], appeared to be comparable or slightly weaker than that in **105**. However, it should be noted that the six-membered ring in **106** is formed by a partial bond with considerable π-character.

There is limited experimental magnetic information on norcaradienes and most of the currently available information comes from theory. Pikulik and Childs examined the ^1H NMR spectra of the type **110** and pointed out that the chemical shifts of the C(7) and C(1)/C(6) proton resonances were anomalous as compared to complexes of related cyclopropyl compounds such as **111** and **112**[246]. They argued that the chemical shifts of **110** could be accounted for on the basis of an induced ring current.

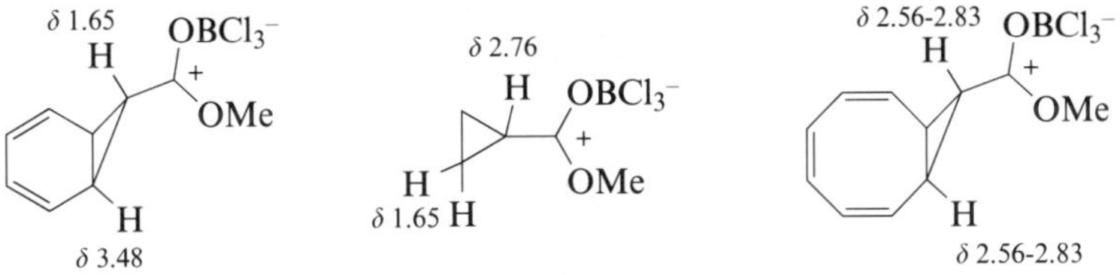

(110) **(111)** **(112)**

Kraka and Cremer have calculated the ^1H and ^{13}C NMR chemical shifts and magnetic susceptibility as a function of interaction distance of both the cycloheptatriene and norcaradiene systems[238]. They point out that both the magnetic susceptibility and the shift difference between the *endo* and *exo* protons at C(7) are at a maximum at the transition state for the valence tautomeric rearrangement between the two systems. The transition state is characterized by a C(1)–C(6) distance of 1.864 Å and an almost complete equalization of C—C bond lengths, bond orders, atomic charges and ^{13}C shifts of C(2)—C(5). Although a resonance energy could not be calculated for the transition state, Kraka and Cremer showed that it could be considered to exhibit no-bond homoaromaticity with a homoconjugative electron delocalization which probably exceeds that of **105** and **106**[238].

Kinetic studies on the decarbonylation of **113** and **114** are of interest in terms of the homoaromatic character of **105/106**. It has been reported that the rate of decarbonylation of **113** was 1×10^5 times greater than that of **114**[247]. It was suggested that this reactivity difference was due to partial opening of the cyclopropane in the transition state and overlap between the Walsh orbitals of the cyclopropane and the rehybridizing s-orbitals of the breaking bonds. Homoaromatic electron delocalization in a norcaradiene/cycloheptatriene-like transition state is only possible with the *anti*-isomer **113**.

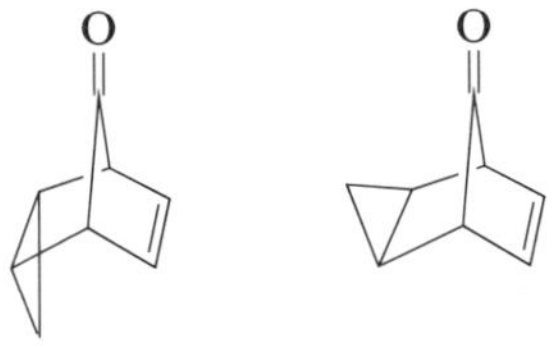

(113) **(114)**

The kinetic measurements carried out by Grimme, Warner and coworkers for the cycloreversion reaction shown in Scheme 39 are consistent with the effects seen with **113** and **114**[248]. The ratio *k*(**115**-*endo*)/*k*(**115**-*exo*) was found to be 8.9×10^3 at 164.5 °C which corresponds to a $\Delta\Delta G^{\neq} = 7.9$ kcal mol^{-1} [248]. These authors also attribute the high reactivity of the *anti*-isomer **115** to the possibility of 6π-electron delocalization in the developing norcaradiene component of the transition state.

In the case of **116** the work of Roth and coworkers suggests that cyclopropyl homoconjugation leads to destabilization of 10 kcal mol^{-1} [225,226]. This result has been confirmed by

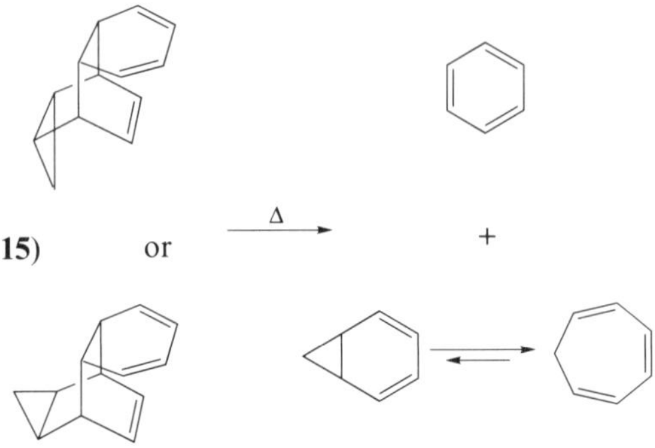

(115)

SCHEME 39. Cycloreversion reactions

ab initio calculations as described in the preceding chapter[25]. The magnitude of the difference between the resonance energies of **106** and **116** is startling. Jörgensen noted the unique properties of **116** on the basis of a MO theory analysis of the molecule[249].

The structure of **116** has been calculated by Karka and Cremer[26] at the HF/6-31G(d) level of theory and by Skancke and coworkers at a lower level[250]. The intriguing feature of the calculated structure is the relatively long C(1)–C(5) and C(4)–C(5) distances as compared to the C(1)–C(4) distance and the corresponding distances in **106**. Kraka and Cremer[26] have reported calculated bond orders of 0.88 for the external cyclopropane bonds of **116**. This magnitude of these bond orders suggests a peripheral delocalization of electrons[25].

Bicyclo[2.1.0]pentene, **116**, can be considered to be the prototype of a neutral homoantiaromatic molecule. The types of structural and bonding effects found for this molecule parallel in many respects those found for the bicyclo[3.1.0]hexenyl cation reported above. Further studies on both of these 4*q* systems will likely be rewarding in terms of fully understanding the nature of cyclopropyl homoconjugation and homoantiaromaticity.

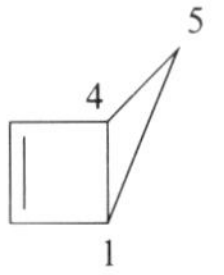

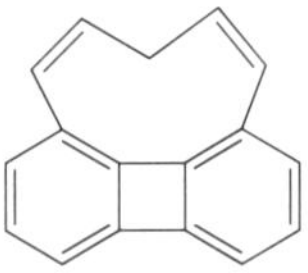

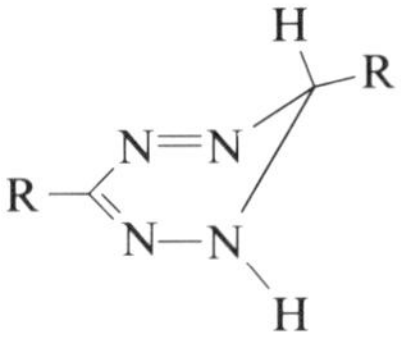

(116) **(117)** **(118)**

Neutral homoantiaromaticity has also been invoked for **117** by Wilcox and coworkers[251]. A shift difference of 2.25 ppm was found for the methylene protons of **117** which was interpreted as arising from a paratropic ring current and local anisotropies. Support for the importance of the homoantiaromaticity in describing **117** was also suggested from a consideration of its UV and PE spectra. However, we note our earlier caution about the use of PE as a criterion for homoaromaticity.

The dihydrotetrazines **118** were suggested by Van der Plas and colleagues to be homoaromatic molecules on the basis of their NMR spectra, particularly the anomalous chemical shifts of the C(6) protons[252]. The more recent work of Mackay and colleagues using acyl derivatives nicely reinforces the induced ring current model for the anomalous chemical shifts[253]. On the other hand, the structures of dihydrotetrazines provide a less compelling case for cyclic delocalization with large N(1)–N(5) internuclear distances.

B. Bis- and Higher Homoaromatic Systems

Homotropylidene (**119**), semibullvalene (**120**) and the related barbaralane (**121**) and bullvalene (**122**) systems undergo degenerate Cope-type rearrangements. The transition states for these rearrangements have been considered to be examples of bishomoaromatic 6-electron systems. The barriers to these degenerate rearrangements are the smallest in semibullvalene (Scheme 40). As such, this system has been studied intensively as it has been considered to be the most likely platform for bishomoaromatic delocalization in a neutral system[254]. Following the suggestion of Doering and colleagues of diradical character in the transition states of analogous acyclic systems, there has been debate about the nature of the transition state in these degenerate Cope rearrangements[255]. With semibullvalene it would appear that a homoaromatic transition state is of lower energy than one possessing diradical character[256].

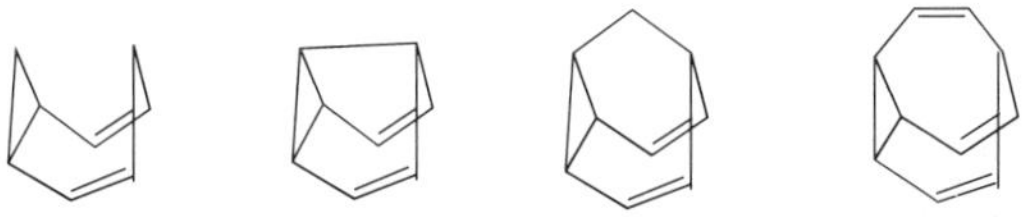

(**119**) (**120**) (**121**) (**122**)

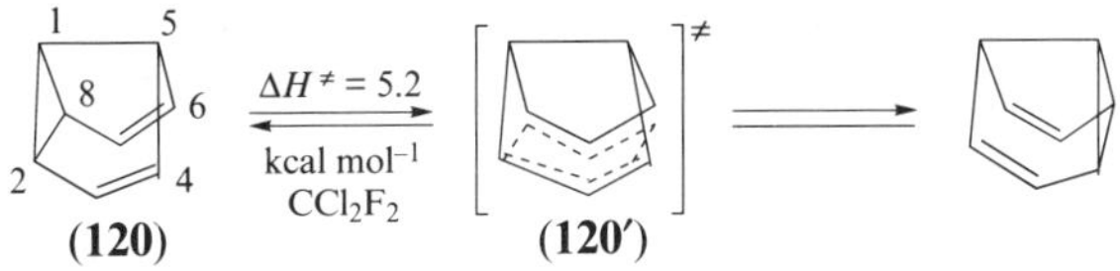

SCHEME 40. Degenerate rearrangement of semibullvalene

As is shown in Scheme 40, the activation enthalpy of the degenerate valence isomerization in **120** is only 5.2 kcal mol^{-1} [257]. Recent *ab initio* calculations by Szabo and Cremer give values of 4.0 (MP2/DZ+P+ZPE) and 6.5 kcal mol^{-1} (MP4/DZ+P+ZPE) for this barrier[258]. Calculated geometrical, electron density and magnetic properties of the transition state clearly indicate it as being bishomoaromatic with a C(2)–C(8) distance of 2.03 Å (MP2). These results confirm the earlier expectations based on MO theory and semi-empirical calculations.

Hoffmann and Stohrer[231], using MO theory, and Dewar and Lo, using MINDO/2[256], predicted that appropriately placed substituents could alter the relative energies of **120** and **120′** and possibly make the intermediary structure lower in energy than **120**. While a wide range of substituted semibullvalenes have been made and systems with remarkably low barriers to degenerate valence tautomerism found, the experimental quest for a symmetrical semibullvalene system has not been fruitful. However, recent semi-empirical calculations by Williams and Kurtz suggest that ethano-bridging in positions 2 and 8 as well as 4 and 6 in **126** (Scheme 41), will lead to the stable homoaromatic form **120′** as being the energy minimum on the potential energy surface[242,259]. If this could indeed be demonstrated experimentally, then **126** could be regarded as a frozen transition state[231,256,260,261].

Convincing evidence for the existence of homoaromatic semibullvalenes corresponding to **120′** has recently been provided by Szabo and Cremer on the basis of high-level *ab initio* calculations (MP2 and MP4)[258]. These authors investigated a series of substituted systems. The results of this study did not support the previous claims based on semi-empirical calculations that 1,5-dimethylsemibullvalene and 3,7-diazasemibullvalene should exist in the 'frozen transition state form' **120′**[262]. However, the *ab initio* investigation revealed that **123**–**127** (Scheme 41) increasingly in the order given, prefer to exist as (homoaromatic) 'frozen transition state' structures. For **123** and **124**, the classical and non-classical forms

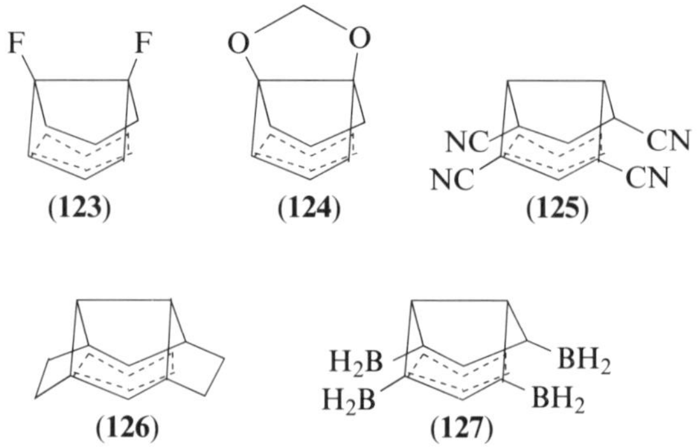

SCHEME 41. Semibullvalenes calculated to possess symmetrical ground state structures

differ by 1 kcal mol^{-1} or less. **125** was shown to prefer the homoaromatic structure by 2 kcal mol^{-1}. In support of the results of Williams, Kurtz and Farely[242,259], **126** was found to be more stable than the classical forms by 6 kcal mol^{-1}. In the case of **127** the energy difference was found to be 8 kcal mol^{-1}. The calculated homo-interaction distances were found to range from 2 to 2.2 Å. Electron density and magnetic properties were indicative of homoaromaticity, meeting the criteria set out in this and the preceding chapter[25].

In view of the clear homoaromatic character of **120′** as demonstrated by the recent work on **123–127**, the question arises as to whether cyclic homoconjugation is important for the ground state of semibullvalene, **120**.

The structures of semibullvalene derivatives have been determined. It was initially suggested that there was a systematic variation in the C(2)–C(8) and C(4)–C(6) distances as the barrier to interconversion of the two valence tautomers varied. Such a variation would be good evidence for cyclic delocalization. However, recent work of Jackman, Quast and coworkers using CP-MAS ^{13}C NMR has shown that the earlier work did not take into account the presence of two valence tautomers[263].

The C(2)–C(8) distances of substituted semibullvalenes are typically found to be *ca* 1.58 Å[264], a distance which is longer than that of a typical cyclopropane bond[45–47]. While AM1 calculations reported by Dewar and Jie[260] do not replicate this long internal cyclopropane bond, the recent calculations of Szabo and Cremer discussed above give a value of 1.58 Å[258]. These latter authors also find evidence of significant π-character in the cyclopropane bond of **120** indicative of electron delocalization and some degree of homoconjugation in the parent molecule.

Baxter, Cowley and coworkers have reported a solid state investigation of the structure of a 9-phosphabarbaralane and suggested this has a symmetrical structure corresponding to a 'frozen transition state'[265]. In solution the molecule exists as a classical structure.

Several systems have been examined in the context of potential tris- and higher homoaromatic systems[21]. These include *cis,cis,cis*-1,4,7-cyclononatriene (**128**), triquinacene (**129**), hexaquinacene and the cyclic polyacetylenes such as **130**. The conformations of some of these systems are such that they could be considered to be examples of 'in plane' homoaromatic systems[266].

In general, there is no strong evidence to support homoaromatic formulations of the structures of any of these systems. There are indications from PE spectroscopy of some degree of interaction between the unsaturated fragments of these molecules. However, as we have pointed out, PE spectroscopy as a technique has limited value in probing homoaromaticity. Magnetic evidence has either not been examined in detail in most systems or, where chemical shifts have been examined, is not definitive. Thermochemical

(128)

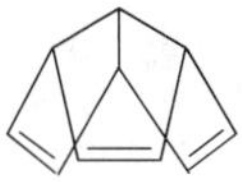

(129)

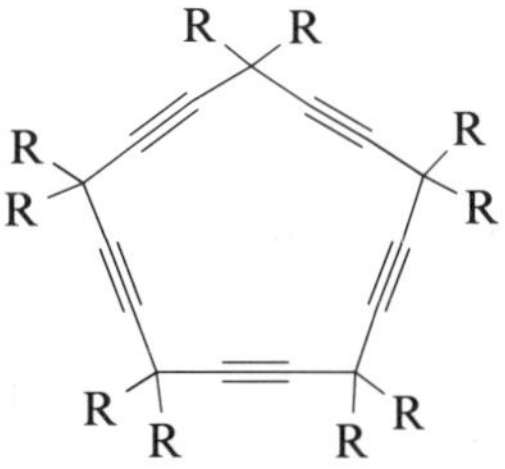

(130)

approaches have been used. For example, Roth and colleagues have found a small destabilization for **128**, a result in line with other evidence for this system[225,226].

Considerable attention has been paid to the thermochemical results obtained by Paquette and coworkers for **129** in which homoaromatic stabilization was suggested[267] and later supported by Rogers and colleagues[234]. However, more recent work has shown that this is not the case[268]. Similarly, Scott and coworkers have shown that any resonance stabilization in **130**, R = Me, if present, is small and cannot be quantified given the accuracy of the thremochemical methods available with these large molecules[269].

In summing up this section on neutral homoaromatic compounds we point out that a considerable number of neutral molecules have been identified as benefiting from homoconjugative electron delocalization. These include cycloheptatriene as well as several bridged derivatives of these molecules. We anticipate that further work on these systems and the related homoantiaromatic bicyclo[2.1.0]pentene will prove rewarding.

The bishomoaromatic neutral systems are of particular interest. Evidence for the importance of neutral homoaromatic delocalization appears to exist solely with certain substituted semibullvalenes. In terms of the latter systems the best candidates for experimental work appear to be **126** and **127**.

There are no neutral molecules with trishomoaromatic character. This is not surprising, given the small size of the resonance energies associated with neutral homoaromatic molecules and the magnitude of the strain effects associated with a potential trishomoaromatic system.

VI. HOMOAROMATIC ANIONS

Anionic systems represent a problematical area with respect to homoaromaticity. Williams and Kurtz in their review summarize the position as there being no anions which are currently recognized as being homoaromatic[21]. In our view, the situation is not as simple as this and there well may be examples of homoaromatic anions. Certainly, this is an area of considerable scrutiny at present and the issues are far from being fully settled.

With cations it was the monohomo systems which showed the clearest evidence for homoaromaticity, *vide supra*. However, with the corresponding anion **131** and its derivatives, there is no experimental or theoretical evidence to suggest that homoconjugation is important[270]. The deliberate attempt of Tolbert and Rajca to bias the system by placing phenyl substituents at C(2)/C(5) (cf **132**), did not provide a sufficient driving force to make homoconjugation a significant factor[271].

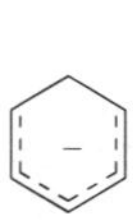

(131)

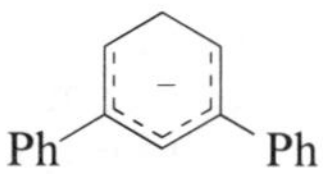

(132)

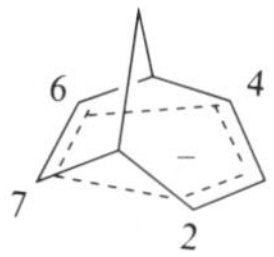

(133)

The bridged bicyclic system **133** represents the earliest example of an anion that was claimed to be homoaromatic[272]. Studies of this system actively continue and the importance of homoconjugation in accounting for the properties of the anion has been a matter of some controversy and debate.

Homoaromatic delocalization in **133** was initially invoked in order to account for its enhanced stability and NMR properties[273]. However, this explanation was challenged in 1981 by two different groups. On the basis of calculations Grutzner and Jorgensen[24] (MINDO/3) as well as Mayr, Schleyer and colleagues[273] (MNDO and STO-3G) concluded independently that the properties of **133** could be accounted without resort to homoaromatic delocalization. Moreover, they also stated more generally that homoaromatic stabilization was not expected to be an important phenomenon in anions.

Matters were not allowed to rest with these conclusions. Brown and colleagues criticized the work of both groups and suggested on the basis of HF/STO-3G calculations that homoconjugation was in fact important in **133** and that the negative charge was delocalized to the C(6)—C(7) ethylene fragment[274]. Christl and coworkers provided further NMR evidence to support the claim for homoaromatic stabilization of **133** and its phenyl-substituted derivatives[275]. NMR evidence supporting some degree of charge delocalization was presented by Köhler and Hertkorn[276] while Trimitsis and Zimmermann cautioned against the use of chemical shifts to probe for homoconjugative interactions in **133** and its derivatives[277]. More direct approaches to understanding the nature of **133** were provided by acidity measurements on the two hydrocarbons **134** and **135**. Solution-phase acidities of **134** were reported by Washburn[278] and measurements in the gas phase were undertaken by Lee and Squires[279]. These latter results showed that **134** possesses a very high gas-phase acidity that is nearly 10 kcal mol^{-1} greater than that of **135**.

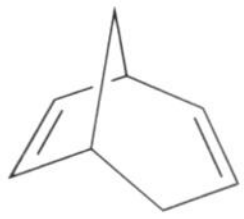
(134)

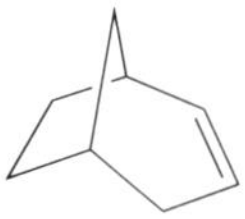
(135)

Structural information on **133** was provided by Köhler and coworkers who, in 1986, reported the isolation of its lithium salt and the determination of its crystal structure[280]. This structure showed the lithium cation to be situated on the *endo* surface of the anion and coordinated with both the allyl portion and the C(6)—C(7) double bond. Key internuclear distances of the anionic portion of the salt are summarized in Scheme 42.

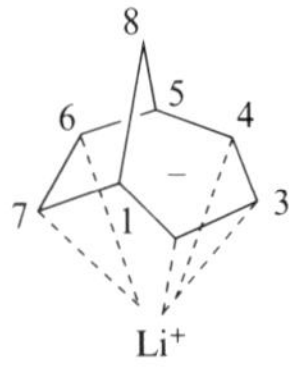

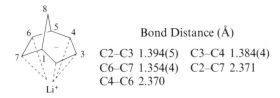

Bond Distance (Å)

C2–C3 1.394(5)	C3–C4 1.384(4)
C6–C7 1.354(4)	C2–C7 2.371
C4–C6 2.370	

SCHEME 42. Structure of the lithium salt of **133**

In 1986 two groups, each using theoretical calculations, again questioned the evidence for significant homoaromatic delocalization in **133**. Schleyer and colleagues, using MNDO, argued that the properties of **133** could be accounted for on the basis of hyperconjugation and gegenion interactions[281]. Lindh and colleagues, on the basis of CASSCF with minimal and split basis sets, suggested that in addition to the gegenion stabilization an electrostatic factor, in which the quadrupole moment in the C(6)—C(7) bridge stabilizes the charge in the allyl portion of the ion, was important[282].

This second major challenge to the need for homoaromatic delocalization to account for the properties of **133** has led to a further series of reports on its properties. These include work of Trimitsis and his group examining the rates of deuterium exchange, which led them to conclude that there was no special stability of derivatives of **133**[283]. A counter-view was expressed by Tuncay and colleagues, also based on exchange experiments[284]. Christl and Müller have continued to examine the NMR spectra of aryl-substituted derivatives of **133** and the impact of counter-ion on the properties of these salts[285]. Hertkorn and Köhler[286] have also addressed the question of cation/anion interactions and reached the same conclusion as Christl, that the properties of **133** cannot be accounted for on the basis of specific ion-pairing. Christl and colleagues have also prepared the anion **136** and, in comparing its properties with those of **137**, reached the conclusion that **136** should be considered to be homoaromatic[287].

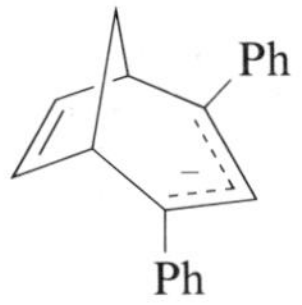

(**136**)

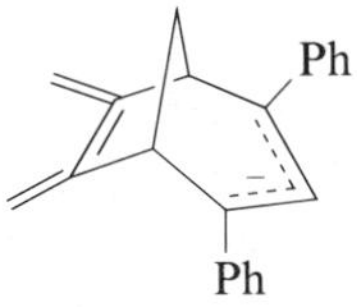

(**137**)

Squires, in a recent review[288], examines and dismisses the explanations put forward by Schleyer and colleagues[281] and Lindh and coworkers[282] as being unable to account for the large difference in acidities of **134** and **135**. He reaches the conclusion that **133** does exhibit homoaromatic delocalization.

In our view the question remains open as to the importance of homoaromatic delocalization in determining the properties of **133**. There is a wealth of experimental evidence available, much of which points to such a delocalization. However, we are troubled by the absence of high-level theoretical calculations of the structure of **133** and its magnetic properties to back up this claim.

There are other anions for which the claim of homoaromatic delocalization has been made. Work on these systems is relatively old and has been reviewed extensively[14–21]. Overall, it is not clear there are any good examples of anions which are homoaromatic. Perhaps, with futher work, **133** will be demonstrated to be an example; however, it is clear that homoaromatic delocalization is not generally going to be an important phenomenon in carbanions.

VII. CONCLUDING REMARKS

At the outset of this chapter we presented a series of criteria for homoaromaticity (Section I. A). The criteria were developed as a result of the detailed theoretical consideration of homoaromaticity given in the previous chapter[25]. The criteria seek to go further than a simple topological definition of homoaromaticity which, coupled with an electron count and NMR spectrum, have frequently been the sole basis for the classification of a system as homoaromatic. In the subsequent sections of this chapter we have presented a detailed consideration of a selected range of potentially homoaromatic molecules and ions in the light of these criteria. It is clear from these analyses that there are a range of systems, including charged and neutral moledules, which can be classified as being homoaromatic.

We would stress that it is important in the consideration of a molecule or ion as a homoaromatic system to use as wide a range of the various criteria we have suggested as is possible. It is clear from the work we have described that it is essential to couple high-level calculations with experimental observations in order to fully understand these systems. In particular, we point to the very important recent advance that uses a geometry optimization technique based on the comparison of calculated ^{13}C NMR chemical shifts with those observed experimentally. This powerful combination of theory and experiment is one which should routinely be used with all potentially homoaromatic or homoantiaromatic systems.

A further key point to make is the desirability of examining related $4q$ as well as $4q + 2$ systems. We believe that the results we have outlined in Sections III and V well demonstrate the additional information that comes in placing a potentially homoantiaromatic system in juxtaposition with its homoaromatic counterparts. Further work with the $4q$ systems is required in many series of systems.

Most of the work reported in this area is limited to carbocyclic systems. The recent developments with the boron analogues of the cyclobutenyl/homocyclopropenium and norbornenyl/norbornadienyl cations point to the potential importance of cyclopropyl homoconjugation and homoaromaticity in a much wider sphere of organic systems. This will likely be an area where there will be considerable further work.

The concepts of cyclopropyl homoconjugation and homoaromaticity have a long history in organic chemistry. Work in this field has passed through various phases. At this point we have largely left the stage where lots of new candidates are being proposed as homoaromatic systems. The last few years have seen a re-emphasis on a detailed examination of the core of basic homoaromatic and homoantiaromatic molecules. As we hope this chapter will show, the results of this 'mature' phase of the investigation of homoaromaticity have and, indeed, are still leading to a deeper understanding of the role and importance of the concept in organic chemistry.

VIII. ACKNOWLEDGEMENTS

The work in Canada was supported by the Natural Science and Engineering Research Council of Canada (NSERC). RFC would like to thank McMaster University for a research leave, which allowed for the writing of this chapter, and the many coworkers whose contributions have helped refine his understanding of the concept of homoaromaticity.

In Sweden, the work was supported by the Swedish Natural Science Research Council (NFR). All calculations needed to complement data from the literature were done on the CRAY YMP/464 of the Nationellt Superdatorcentrum (NSC) Linköping, Sweden. The authors thank the NSC for a generous allotment of computer time and S. Nilsson-Lill for help in proof reading.

IX. REFERENCES

1. C. W. Shoppee, *J. Chem. Soc.*, 1147 (1946).
2. C. K. Ingold, in *Structure and Mechanism in Organic Chemistry*, Cornell University Press, Ithaca, New York, 1953.
3. S. Winstein and R. E. Buckles, *J. Am. Chem. Soc.*, **64**, 2780 (1942); S. Winstein and E. Grunwald, *J. Am. Chem. Soc.*, **70**, 828 (1948).
4. R. M. Dodson and B. Riegel, *J. Org. Chem.*, **13**, 424 (1948).
5. S. Winstein and R. Adams, *J. Am. Chem. Soc.*, **70**, 838 (1948).
6. S. Winstein and A. H. Schlesinger, *J. Am. Chem. Soc.*, **70**, 3528 (1948).
7. S. Winstein, H. M. Walborsky and K. Schreiber, *J. Am. Chem. Soc.*, **72**, 5795 (1950).
8. J. D. Roberts, W. Bennett and R. Armstrong, *J. Am. Chem. Soc.*, **72**, 3329 (1950).
9. M. Simonetta and S. Winstein, *J. Am. Chem. Soc.*, **76**, 18 (1954).

10. D. E. Applequist and J. D. Roberts, *J. Am. Chem. Soc.*, **78**, 4012 (1956).
11. W. von E. Doering, G. Laber, R. Vonderwahl, N. F. Chamberlain and R. B. Williams, *J. Am. Chem. Soc.*, **78**, 5448 (1956).
12. S. Winstein, J. Sonnenberg and L. deVries, *J. Am. Chem. Soc.*, **81**, 6523 (1959).
13. S. Winstein, *J. Am. Chem. Soc.*, **81**, 6524 (1959).
14. S. Winstein, *Spec. Publ. Chem. Soc.*, **21**, 5 (1967); *Quart. Rev. Chem. Soc.*, **23**, 141 (1969).
15. P. M. Warner, *Top. Nonbenzenoid Aromat. Chem.*, 2 (1976).
16. S. Winstein, in *Carbonium Ions*, Vol. III (Ed. G. A. Olah and P. v. R. Schleyer), Wiley, New York, 1972, pp 965–1005.
17. P. R. Story and B. C. Clark, Jr., in *Carbonium Ions*, Vol. III (Ed. G. A. Olah and P. v. R. Schleyer), Wiley, New York, 1972, pp. 1007–1098.
18. L. A. Paquette, *Angew. Chem., Int. Ed. Engl.*, **17**, 106 (1978).
19. R. F. Childs, *Acc. Chem. Res.*, **17**, 347 (1984).
20. A. T. Balaban, M. Banciu and V. Ciorba, in *Annulenes, Benzo-, Hetero-, Homo-derivatives and their Valence Isomers*, Vol. III. Chap. 9, CRC Press Inc., Boca Raton, Florida, 1987, pp. 144–163.
21. R. V. Williams and H. A. Kurtz, *Adv. Phys. Org. Chem.*, **29**, 273 (1994).
22. R. F. Childs, M. Mahendran, S. D. Zweep, G. S. Shaw, S. K. Chadda, N. A. D. Burke, B. E. George, R. Faggiani and C. J. L. Lock, *Pure Appl. Chem.*, **58**, 111 (1986).
23. N. C. Deno, *Prog. Phys. Org. Chem.*, **2**, 129 (1964).
24. J. B. Grutzner and W. L. Jorgensen, *J. Am. Chem. Soc.*, **103**, 1372 (1981); E. Kaufmann, H. Mayr, J. Chandrasekhar and P. v. R. Schleyer, *J. Am. Chem. Soc.*, **103**, 1375 (1981).
25. D. Cremer, R. F. Childs and E. Kraka, see the preceding chapter (Chapter 7) in this volume and references cited therein.
26. E. Kraka and D. Cremer, in *Theoretical Models of Chemical Bonding, The Concept of the Chemical Bond* (Ed. Z. B. Maksić), Springer-Verlag, Berlin, 1990, p. 453; D. Cremer, *Tetrahedron*, **44**, 7427 (1988).
27. D. Cremer, F. Reichel and E. Kraka, *J. Am. Chem. Soc.*, **113**, 9459 (1991); D. Cremer, L. Olsson, F. Reichel and E. Kraka, *Israel J. Chem.*, **33**, 369 (1993).
28. D. Cremer, P. Svensson, E. Kraka, Z. Konkoli and P. Ahlberg, *J. Am. Chem. Soc.*, **115**, 7457 (1993); P. Svensson, F. Reichel, P. Ahlberg and D. Cremer, *J. Chem. Soc., Perkin Trans. 2*, 1463 (1991).
29. D. Cremer and E. Kraka, *Angew. Chem., Int. Ed. Engl.*, **23**, 627 (1984); *Croat. Chem. Acta*, **57**, 1259 (1984); D. Cremer, in *Modelling of Structure and Properties of Molecules* (Ed. Z. B. Maksić), Ellis Horwood, Chichester, 1988, p. 125.
30. D. Cremer, E. Kraka, T. S. Slee, R. F. W. Bader, C. D. H. Lau and T. T. Nguyen-Dang, *J. Am. Chem. Soc.*, **105**, 5069 (1983).
31. W. J. Hehre, *J. Am. Chem. Soc.*, **96**, 5207 (1974).
32. H. E. Zimmerman, *J. Am. Chem. Soc.*, **88**, 1564, 1566 (1966); *Acc. Chem. Res.*, **4**, 272 (1971); E. Heilbronner, *Tetrahedron Lett.*, 1923 (1964).
33. R. B. Woodward and R. Hoffmann, *J. Am. Chem. Soc.*, **87**, 395, 2046, 2511 (1965); R. B. Woodward, *Spec. Publ. Chem. Soc.*, **21**, 217 (1967); R. B. Woodward and R. Hoffmann, in *The Conservation of Orbital Symmetry*, Verlag Chemie, GmbH, Weinheim, 1970.
34. M. J. Goldstein and R. Hoffmann, *J. Am. Chem. Soc.*, **93**, 6193 (1971).
35. M. J. S. Dewar, in *The Molecular Orbital Theory of Organic Chemistry*, McGraw-Hill, New York, 1969; *Tetrahedron Suppl.*, **8**, 75 (1966).
36. For reviews on pericyclic reactions see: T. L. Gilchrist and R. C. Storr, in *Organic Reactions and Orbital Symmetry*, Cambridge University Press, Cambridge, 1972; G. B. Gill and M. R. Willis, in *Pericyclic Reactions*, Chapman-Hall, London, 1974; I. Fleming, in *Frontier Orbitals and Organic Chemical Reactions*, Wiley, London, 1976; A. P. Marchand and R. E. Lehr (Eds), *Pericyclic Reactions*, Vols. 1 and 2, Academic Press, New York, 1977.
37. W. J. Hehre and A. J. P. Devaquet, *J. Am. Chem. Soc.*, **98**, 4370 (1976).
38. W. L. Jorgensen, *J. Am. Chem. Soc.*, **98**, 6784 (1976); **97**, 3082 (1975).
39. P. J. Garratt, in *Aromaticity*, Wiley, New York, 1986; D. Lewis and D. Peters, in *Facts and Theories of Aromaticity*, McMillan, London, 1975.
40. M. Charton, in *The Chemistry of Alkenes*, Vol. 2 (Ed. J. Zabicky), Interscience–Wiley, London, 1970, pp 511–610.
41. A. de Meijere, *Angew. Chem., Int. Ed. Engl.*, **18**, 809 (1979).

42. T. T. Tidwell, in *The Chemistry of the Cyclopropyl Group*(Ed. Z. Rappoport), Wiley, London, 1987, pp.565–632.
43. E. Kraka and D. Cremer, in *Molecular Structure and Energetics, Structure and Reactivity*, Vol. 7 (Eds. J. F. Liebman and A. Greenberg), VCH Publishers, New York, 1988, p. 65; D. Cremer and J. Gauss, *J. Am. Chem. Soc.*, **108**, 7467 (1986).
44. D. Cremer, E. Kraka and K. J. Szabo, in *The Chemistry of the Cyclopropyl Group* (Ed. Z. Rappoport), Chap. 2, Wiley, Chichester, 1995.
45. F. H. Allen, *Acta Crystallogr., Sect. B*, **B36**, 81 (1980).
46. F. H. Allen, *Acta Crystallogr., Sect. B*, **B37**, 890 (1981).
47. F. H. Allen, O. Kennard and R. Taylor, *Acc. Chem. Res.*, **16**, 146 (1983).
48. R. E. Dumright, R. H. Mas, J. S. Merola and J. M. Tanko, *J. Org. Chem.*, **55**, 4098 (1990).
49. F. A. Van-Catledge, D. W. Boerth and J. Kao, *J. Org. Chem.*, **47**, 4096 (1982).
50. D. Cremer and E. Kraka, *J. Am. Chem. Soc.*, **107**, 3811 (1985).
51. K. B. Wiberg and K. E. Laidig, *J. Org. Chem.*, **57**, 5092 (1992).
52. T. Clark, G. W. Spitznagel, R. Close and P. v. R. Schleyer, *J. Am. Chem. Soc.*, **106**, 4412 (1984).
53. J.-P. Pete, *Bull. Soc. Chim. France*, 357 (1967).
54. H. G. Richey, Jr., in *Carbonium Ions*, Vol. III (Eds. G. A. Olah and P. v. R. Schleyer), Wiley, New York, 1972, pp. 1201–1294; K. B. Wiberg and B. A. Hess Jr., in *Carbonium Ions*, Vol. III (Eds. G. A. Olah and P. v. R. Schleyer), Wiley, New York, 1972, pp. 1295–1345; E. C. Friedrich, in *The Chemistry of the Cyclopropyl Group* (Ed. Z. Rappoport), Wiley, Chichester, 1987, pp. 633–700; H. C. Brown (with comments by P. v. R. Schleyer), in *The Non-classical Ion Problem*, Chap. 5, Plenum, New York, 1977; G. A. Olah, V. Prakash-Reddy and G. K. Prakash, *Chem. Rev.*, **92**, 69 (1992).
55. R. H. Mazur, W. N. White, D. A. Semenow, C. C. Lee, M. S. Silver and J. D. Roberts, *J. Am. Chem. Soc.*, **81**, 4390 (1959).
56. K. B. Wiberg, D. Shobe and G. L. Nelson, *J. Am. Chem. Soc.*, **115**, 10645 (1993).
57. M. Saunders, K. E. Laidig, K. B. Wiberg and P. v. R. Schleyer, *J. Am. Chem. Soc.*, **110**, 7652 (1988); W. Koch, B. Liu and D. J. DeFrees, *J. Am. Chem. Soc.*, **110**, 7325 (1988).
58. P. C. Myhre, G. G. Webb and C. S. Yannoni, *J. Am. Chem. Soc.*, **112**, 8992 (1990); H. Vancik, V. Gabelica, D. E. Sunko, P. Buzek and P. v. R. Schleyer, *J. Phys. Org. Chem.*, **6**, 427 (1993).
59. R. F. Childs, R. Faggiani, C. J. L. Lock, M. Mahendran and S. D. Zweep, *J. Am. Chem. Soc.*, **108**, 1692 (1986).
60. S. K. Chadda, R. F. Childs, R. Faggiani and C. J. L. Lock, *J. Am. Chem. Soc.*, **108**, 1694 (1986).
61. R. F. Childs, M. D. Kostyk, C. J. L. Lock and M. Mahendran, *J. Am. Chem. Soc.*, **112**, 8912 (1990).
62. P. v. R. Schleyer and G. W. Van Dine, *J. Am. Chem. Soc.*, **88**, 2321 (1966).
63. D. F. Eaton and T. G. Traylor, *J. Am. Chem. Soc.*, **96**,1226 (1974); J. M. Harris, J. R. Moffatt, M. G. Case, F. W. Clarke, J. S. Polley, T. K. Morgan, Jr., T. M. Ford and R. K. Murray, Jr., *J. Org. Chem.*, **47**, 2740 (1982); V. Buss, R. Gleiter and P. v. R. Schleyer, *J. Am. Chem. Soc.*, **93**, 3927 (1971); J. M. Stewart and G. K. Pagenkopf, *J. Org. Chem.*, **34**, 7 (1969).
64. A. B. Turner, R. E. Lutz, N. S. McFarlane and D. W. Boykin, Jr., *J. Org. Chem.*, **36**, 1107 (1971); G. Montaudo and C. G. Overberger, *J. Org. Chem.*, **38**, 804 (1973); A. L. Goodman and R. H. Eastman, *J. Am. Chem. Soc.*, **86**, 908 (1964).
65. L. D. Kispert, C. Engelman, C. Dyas and C. U. Pittman, Jr., *J. Am. Chem. Soc.*, **93**, 6948 (1971); C. A. Deakyne, L. C. Allen and N. C. Craig, *J. Am. Chem. Soc.*, **99**, 3895 (1977); M. B. Formicheva and V. A. Zubkov, *J. Struct. Chem.*, **20**, 631 (1979).
66. R. G. Pews and N. D. Ojha, *J. Am. Chem. Soc.*, **91**, 5769 (1969).
67. C. F. Wilcox, L. M. Loew and R. Hoffmann, *J. Am. Chem. Soc.*, **95**, 8192 (1973).
68. R. S. Brown and T. G. Traylor, *J. Am. Chem. Soc.*, **95**, 8025 (1973).
69. A. Varadarajan, Ph.D. Thesis, McMaster University, Hamilton, Ontario, Canada, 1983.
70. S. Winstein, C. G. Kreiter and J. I. Brauman, *J. Am. Chem. Soc.*, **88**, 2047 (1966).
71. C. E. Keller and R. Pettit, *J. Am. Chem. Soc.*, **88**, 604 (1966).
72. G. Boche, W. Hechtl, H. Huber and R. Huisgen, *J. Am. Chem. Soc.*, **89**, 3344 (1967); J. Gasteiger and R. Huisgen, *Tetrahedron Lett.*, 3665 (1972); R. Husigen, G. Boche and H. Huber, *J. Am. Chem. Soc.*, **89**, 3345 (1967); R. Huisgen and J. Gasteiger, *Angew. Chem., Int. Ed. Engl.*, **11**, 1104 (1972); *Tetrahedron Lett.*, 3661 (1972).
73. L. A. Paquette, J. R. Malpass and T. J. Barton, *J. Am. Chem. Soc.*, **91**, 4714 (1969).

74. G. I. Fray and R. G. Saxton, in *The Chemistry of Cyclooctatetraene and its Derivatives*, Cambridge University Press, Cambridge, 1978.
75. J. D. Holmes and R. Pettit, *J. Am. Chem. Soc.*, **85**, 2531 (1963).
76. J. A. Berson and J. A. Jenkins, *J. Am. Chem. Soc.*, **94**, 8907 (1972).
77. R. F. Childs and C. V. Rogerson, *J. Am. Chem. Soc.*, **98**, 6391 (1976); **100**, 649 (1978); **102**, 4159 (1980).
78. O. L. Chapman and R. A. Fugiel, *J. Am. Chem. Soc.*, **91**, 215 (1969).
79. J. L. Rosenburg, Jr., J. E. Mahler and R. Pettit, *J. Am. Chem. Soc.*, **84**, 2842 (1962).
80. P. Warner, D. L. Harris, C. H. Bradley and S. Winstein, *Tetrahedron Lett.*, 4013 (1970).
81. S. Winstein, H. D. Kaesz, C. G. Kreiter and E. C. Friedrich, *J. Am. Chem. Soc.*, **87**, 3267 (1965).
82. C. E. Keller and R. Pettit, *J. Am. Chem. Soc.*, **88**, 606 (1966).
83. L. A. Paquette, M. J. Broadhurst, P. Warner, G. A. Olah and G. Liang, *J. Am. Chem. Soc.*, **95**, 3386 (1973); J. F. M. Oth, D. M. Smith, U. Prange and G. Schröder,*Angew. Chem., Int. Ed. Engl.*, **12**, 327 (1973).
84. H. D. Kaesz, S. Winstein and C. G. Kreiter, *J. Am. Chem. Soc.*, **88**, 1319 (1966); R. Aumann and S. Winstein, *Tetrahedron Lett.*, 903 (1970); G. N. Schrauzer, *J. Am. Chem. Soc.*, **83**, 2966 (1961); A. Davison, W. McFarlane, L. Pratt and G. Wilkinson, *J. Chem. Soc.*, 4821 (1962).
85. M. Brookhart, M. Ogliaruso and S. Winstein, *J. Am. Chem. Soc.*, **89**, 1965 (1967).
86. L. T. Scott, M. Oda and M. M. Hashemi, *Chem. Lett.*, 1759 (1986).
87. R. F. Childs, D. L. Mulholland and A. Nixon, *Can. J. Chem.*, **60**, 801, 809 (1982).
88. H. J. Dauben, J. D. Wilson and J. L. Laity, in *Nonbenzenoid Aromatics*, Vol. II (Ed. J. P. Snyder), Academic Press, New York, 1971, p. 167.
89. C. E. Johnson, Jr. and F. A. Bovey, *J. Chem. Phys.*, **29**, 1012 (1958).
90. R. F. Childs, M. J. McGlinchey and A. Varadarajan, *J. Am. Chem. Soc.*, **106**, 5974 (1984).
91. R. F. Childs, A. Varadarajan, C. J. L. Lock, R. Faggiani, C. A. Fyfe and R. E. Wasylishen, *J. Am. Chem. Soc.*, **104**, 2452 (1982).
92. R. F. Childs, R. Faggiani, C. J. L. Lock and M. Mahendran, *J. Am. Chem. Soc.*, **108**, 3613 (1986).
93. R. F. Childs, R. Faggiani, C. J. L. Lock and A. Varadarajan, *Acta Crystallogr., Sect. C*, **C40**, 1291 (1984).
94. R. Destro and M. Simonetta, *Acta Crystallogr., Sect. B.*, **B35**, 1846 (1979).
95. R. Haddon, *J. Am. Chem. Soc.*, **109**, 1676 (1987).
96. R. C. Haddon and L. Scott, *Pure Appl. Chem.*, **58**, 137 (1986); R. C. Haddon, *J. Am. Chem. Soc.*, **108**, 2837 (1986); *J. Phys. Chem.*, **91**, 3719 (1987).
97. R. F Childs, R. M. Orgias, C. J. L. Lock and M. Mahendran, *Can. J. Chem.*, **71**, 836 (1993).
98. B. D. Santarsiero, M. N. G. James, M. Mahendran and R. F. Childs, *J. Am. Chem. Soc.*, **112**, 9416 (1990).
99. J. A. Berson and J. A. Jenkins, *J. Am. Chem. Soc.*, **94**, 8907 (1972).
100. R. F. Childs, M. Mahendran, M. Sivapalan and P. Nguyen, *Chem. Commun.*, 27 (1989).
101. M. S. Brookhart and M. A. M. Atwater, *Tetrahedron Lett.*, 4399 (1972).
102. R. F. Childs, D. L. Mulholland, A. Varadarajan and S. Yeroushalmi, *J. Org. Chem.*, **48**, 1431 (1983).
103. D. Cremer, P. Svensson, F. Reichel and K. J. Szabo, to appear.
104. R. F. Childs and A. Varadarajan, *Can. J. Chem.*, **63**, 418 (1985).
105. W. J. Hehre, *J. Am. Chem. Soc.*, **95**, 5807 (1973).
106. R. C. Haddon, *J. Am. Chem. Soc.*, **110**, 1108 (1988).
107. R. C. Haddon, *J. Org. Chem.*, **44**, 3608 (1979).
108. R. C. Haddon, *Tetrahedron Lett.*, 2797 (1974); 863 (1975); *J. Am. Chem. Soc.*, **97**, 3608 (1975); *Aust. J. Chem.*, **30**, 1 (1977); *Croat. Chem. Acta*, **57**, 1165 (1984).
109. M. Barzaghi and C. Gatti, *J. Mol. Struct. (Theochem.)*, **43**, 431, 275 (1988); *J. Chim. Phys., Phys. Chim. Biol.*, **84**, 783 (1987).
110. P. Buzek, P. v. R. Schleyer and S. Sieber, *Chem. Unserer Zeit*, **26**, 116 (1992).
111. S. Sieber, P. v. R. Schleyer, A. H. Otto, J. Gauss, F. Reichel and D. Cremer, *J. Phys. Org. Chem.*, **6**, 445 (1993).
112. L. T. Scott and M. M. Hashemi, *Tetrahedron*, **42**, 1823 (1986).
113. E. J. Smutny, M. J. Caserio and J. D. Roberts, *J. Am. Chem. Soc.*, **82**, 1793 (1960); E. F. Kiefer and J. D. Roberts, *J. Am. Chem. Soc.*, **84**, 784 (1962); S. L. Manatt, M. Vogel, D. Knutson and J. D. Roberts, *J. Am. Chem. Soc.*, **86**, 2645 (1964).

114. T. J. Katz and E. H. Gold, *J. Am. Chem. Soc.*, **86**, 1600 (1964).
115. G. A. Olah, J. S. Staral, R. J. Spear and G. Liang, *J. Am. Chem. Soc.*, **97**, 5489 (1975); G. A. Olah, J. S. Staral and G. Liang, *J. Am. Chem. Soc.*, **96**, 6233 (1974).
116. R. F. Childs, *Tetrahedron*, **38**, 567 (1982).
117. A. E. Lodder, J. W. Hann, L. J. M. Ven and H. M. Buck, *Recl. Trav. Chim. Pays-Bas*, **92**, 1040 (1973).
118. R. F. Bryan, *J. Am. Chem. Soc.*, **86**, 733 (1964).
119. Von E. Hey, F. Weller and K. Dehnicke, *Z. Anorg. Allg. Chem.*, **502**, 45 (1983).
120. C. Krüger, P. J. Roberts, Y.-H. Tsay and J. B. Koster, *J. Organometal. Chem.*, **78**, 69 (1974).
121. G. Maier, R. Emrich, C.-D. Malsch, K.-A. Schneider, M. Nixdorf and H. Irngartinger, *Chem. Ber.*, **118**, 2798 (1985).
122. R. C. Haddon, *Acc. Chem. Res.*, **21**, 243 (1988).
123. E. H. Gold and T. J. Katz, *J. Org. Chem.*, **31**, 372 (1966).
124. A. J. P. Devaquet and W. J. Hehre, *J. Am. Chem. Soc.*, **96**, 3644 (1974).
125. M. Schindler, *J. Am. Chem. Soc.*, **109**, 1020 (1987); D. R. Kelsey, *J. Chem. Res. (S)*, 44 (1986); P. C. Hariharan and J. A. Pople, *Chem. Phys. Lett.*, **16**, 217 (1972); *Theor. Chim. Acta*, **28**, 213 (1973); J. M. Bofill, J. Castells, S. Olivella and A. Sole, *J. Org. Chem.*, **53**, 5148 (1988); M. J. S. Dewar and W. Thiel, *J. Am. Chem. Soc.*, **99**, 4899, 4907 (1977); M. J. S. Dewar, E. G. Zoebisch, E. F. Healy and J. J. P. Stewart, *J. Am. Chem. Soc.*, **107**, 3902 (1985); R. C. Bingham, M. J. S. Dewar and D. H. Lo, *J. Am. Chem. Soc.*, **97**, 1285, 1294, 1302 (1975); M. J. S. Dewar, D. H. Lo and C. A. Ramsden, *J. Am. Chem. Soc.*, **97**, 1307 (1975); J. J. P. Stewart, *J. Comput. Chem.*, **10**, 221 (1989).
126. R. C. Haddon and R. Raghavachari, *J. Am. Chem. Soc.*, **105**, 118 (1983).
127. G. A. Olah, G. Liang, L. A. Paquette and W. P. Melaga, *J. Am. Chem. Soc.*, **98**, 4327 (1976).
128. C. Pues, G. Baum, W. Massa and A. Berndt, *Z. Naturforsch.*, **B43**, 275 (1988).
129. R. Wehrmann, H. Klusik and A. Berndt, *Angew. Chem., Int. Ed. Engl.*, **23**, 369 (1984).
130. D. Cremer, J. Gauss, P. v. R. Schleyer and P. H. M. Budzelaar, *Angew. Chem., Int. Ed. Engl.*, **23**, 370 (1984).
131. P. Willerhausen, C. Kybart, N. Stamatis, W. Massa, M. Bühl, P. v. R. Schleyer and A. Berndt, *Angew. Chem., Int. Ed. Engl.*, **31**, 1238 (1992).
132. V. A. Koptyug, *Top. Curr. Chem.*, **122**, 1–245 (1984).
133. V. A. Barkhash, *Top. Curr. Chem.*, **116/117**, 1–265 (1984).
134. D. M. Brouwer, E. L. Mackor and C. MacLean, in *Carbonium Ions*, Vol. II (Eds. G. A. Olah and P. v. R. Schleyer), Wiley, New York, 1970, pp. 837–897.
135. L. de Vries, *J. Am. Chem. Soc.*, **82**, 5242 (1960).
136. S. Winstein and M. Battiste, *J. Am. Chem. Soc.*, **82**, 5244 (1960).
137. R. F. Childs, M. Sakai and S. Winstein, *J. Am. Chem. Soc.*, **90**, 7144 (1968).
138. R. F. Childs, M. Sakai, B. D. Parrington and S. Winstein, *J. Am. Chem. Soc.*, **96**, 6403 (1974).
139. R. F. Childs and S. Winstein, *J. Am. Chem. Soc.*, **90**, 7146 (1968).
140. R. F. Childs and S. Winstein, *J. Am. Chem. Soc.*, **96**, 6408 (1974).
141. R. F. Childs and B. D. Parrington, *J. Chem. Soc. (D)*, 1540 (1970).
142. P. Vogel, M. Saunders, N. M. Hasty, Jr. and J. A. Berson, *J. Am. Chem. Soc.*, **93**, 1551 (1971).
143. J. A. Berson and N. M. Hasty, Jr., *J. Am. Chem. Soc.*, **93**, 1549 (1971).
144. G. A. Olah, G. Liang and S. P. Jindal, *J. Org. Chem.*, **40**, 3259 (1975).
145. V. A. Koptyug, L. I. Kuzubova, I. S. Isaev and V. I. Mamatyuk, *J. Chem. Soc. (D)*, 389 (1969); I. S. Isaev, V. I. Mamatyuk, T. G. Egorova, L. I. Kuzubova and V. A. Koptyug, *Bull. Acad. Sci. USSR, Chem. Sci. Div.*, 1954 (1969); I. S. Isaev, V. I. Mamatyuk, L. I. Kuzubova, T. A. Gordymova and V. I. Koptyug, *J. Org. Chem. USSR*, **6**, 2493 (1970); V. A. Koptyug, L. I. Kuzubova, I. S. Isaev and V. I. Mamatyuk, *J. Org. Chem. USSR*, **6**, 1854 (1970); V. A. Koptyug, V. I. Mamatyuk, L. I. Kuzubova and I. S. Isaev, *Bull. Acad. Sci. USSR, Chem. Sci. Div.*, 1524 (1969); V. I. Mamatyuk, A. I. Rezvukhin, I. S. Isaev, V. I. Buraev and V. A. Kotyug, *J. Org. Chem. USSR*, **10**, 662 (1974).
146. R. F. Childs and D. L. Mulholland, *J. Am. Chem. Soc.*, **105**, 96 (1983).
147. G. A. Olah, R. H. Schlosberg, R. D. Porter, Y. K. Mo, D. P. Kelly and Gh. D. Mateescu, *J. Am. Chem. Soc.*, **94**, 2034 (1972); G. A. Olah, R. H. Schlosberg, D. P. Kelly and Gh. D. Mateescu, *J. Am. Chem. Soc.*, **92**, 2546 (1970).
148. G. I. Brodkin, Sh. M. Nagi, I. Yu. Bagryanskaya and Yu. V. Gatilov, *J. Struct. Chem. USSR*, **25**, 440 (1984).

149. N. C. Baenziger and A. D. Nelson, *J. Am. Chem. Soc.*, **90**, 6602 (1968).
150. F. Effenberger, F. Reisinger, K. H. Schönwälder, P. Bäuerle, J. J. Stezowski, K. H. Jogun, K. Schöllkopf and W.-D. Stohrer, *J. Am. Chem. Soc.*, **109**, 882 (1987).
151. W. J. Hehre, *J. Am. Chem. Soc.*, **95**, 8908 (1973).
152. R. F. Childs, *Tetrahedron*, **38**, 567 (1982).
153. D. W. Swatton and H. Hart, *J. Am. Chem. Soc.*, **89**, 5075 (1967).
154. H. E. Zimmerman and D. I. Schuster, *J. Am. Chem. Soc.*, **84**, 4527 (1962); H. E. Zimmerman, *Pure Appl. Chem.*, **9**, 493 (1964).
155. R. F. Childs and M. Zeya, *J. Am. Chem. Soc.*, **96**, 6418 (1974).
156. S. Olivella and A. Solé, *J. Am. Chem. Soc.*, **113**, 8628 (1991); R. Sustmann and F. Lübbe, *J. Am. Chem. Soc.*, **98**, 6037 (1976); *Chem. Ber.*, **112**, 42 (1979).
157. J. A. Berson and J. A. Jenkins, *J. Am. Chem. Soc.*, **94**, 8907 (1972).
158. R. F. Childs and A. Varadarajan, *Can. J. Chem.*, **63**, 418 (1985).
159. L. T. Scott and W. R. Brunsvold, *J. Am. Chem. Soc.*, **100**, 6535 (1978).
160. A. J. P. Devaquet and W. J. Hehre, *J. Am. Chem. Soc.*, **96**, 3644 (1974); W. J. Hehre and A. J. P. Devaquet *J. Am. Chem. Soc.*, **98**, 4370 (1976).
161. I. A. Shleider, I. S. Isaev and V. A. Koptyug, *J. Org. Chem. USSR*, **8**, 1357 (1972).
162. P. B. J. Driessen and H. Hogeveen, *J. Am. Chem. Soc.*, **100**, 1193 (1978).
163. R. B. Woodward and R. Hoffmann, in *The Conservation of Orbital Symmetry*, Academic Press, New York, 1969.
164. R. F. Childs and C. V. Rogerson, *J. Am. Chem. Soc.*, **98**, 6391 (1976); **100**, 649 (1978); **102**, 4159 (1980).
165. H. Hogeveen and H. C. Volger, *Recl. Trav. Chim. Pays-Bas*, **87**, 385, 1042 (1968); **88**, 353 (1969).
166. L. A. Paquette, G. R. Krow, J. M. Bollinger and G. A. Olah, *J. Am. Chem. Soc.*, **90**, 7147 (1968).
167. P. Ahlberg, D. L. Harris and S. Winstein, *J. Am. Chem. Soc.*, **92**, 2146, 4454 (1970).
168. C. Engdahl, G. Jonsäll and P. Ahlberg, *J. Am. Chem. Soc.*, **105**, 891 (1983).
169. D. Cremer, P. Svensson, E. Kraka and P. Ahlberg, *J. Am. Chem. Soc.*, **115**, 7445 (1993)
170. M. B. Huang, O. Goscinski, G. Jonsäll and P. Ahlberg, *J. Chem. Soc., Perkin Trans. 2*, 305 (1983); J. Bella, J. M. Poblet, A. Demoulliens and F. Volatron, *J. Chem. Soc., Perkin Trans. 2*, 37 (1989).
171. P. Ahlberg, J. B. Grutzner, D. L. Harris and S. Winstein, *J. Am. Chem. Soc.*, **92**, 3478 (1970).
172. P. Ahlberg, D. L. Harris, M. Roberts, P. Warner, P. Seidl, M. Sakai, D. Cook, A. Diaz, J. P. Dirlam, H. Hamberger and S. Winstein, *J. Am. Chem. Soc.*, **94**, 7064 (1972).
173. P. Ahlberg, G. Jonsäll and C. Engdahl, *Adv. Phys. Org. Chem.*, **19**, 223 (1983).
174. C. Engdahl and P. Ahlberg, *J. Chem. Res. (S)*, 342 (1977).
175. M. Roberts, H. Hamberger and S. Winstein, *J. Am. Chem. Soc.*, **92**, 6346 (1970).
176. G. Schröder, U. Prange, N. S. Bowman and J. F. M. Oth, *Tetrahedron Lett.*, 3251 (1970); G. Schröder, U. Prange, B. Putzl, J. Thio and J. F. M. Oth, *Chem. Ber.*, **104**, 3406 (1971).
177. G. Jonsäll and P. Ahlberg, *J. Am. Chem. Soc.*, **108**, 3819 (1986).
178. D. Cook, A. Diaz, J. P. Dirlam, D. L. Harris, M. Sakai, S. Winstein, J. C. Barborak and P. v. R. Schleyer, *Tetrahedron Lett.*, 1405 (1971).
179. S. Yaneda, S. Winstein and Z. Yoshida, *Bull. Chem. Soc. Jpn.*, **45**, 2510 (1972); M. B. Huang and G. Jonsäll, *Tetrahedron*, **41**, 6055 (1985).
180. P. Svensson, F. Reichel, P. Ahlberg and D. Cremer, *J. Chem. Soc., Perkin Trans. 2*, 1463 (1991).
181. P. Warner and S. Winstein, *J. Am. Chem. Soc.*, **93**, 1284 (1971).
182. L. A. Paquette, M. J. Broadhurst, P. Warner, G. A. Olah and G. Liang, *J. Am. Chem. Soc.*, **95**, 3386 (1973).
183. K. J. Szabo, E. Kraka and D. Cremer, *J. Org. Chem.*, to appear.
184. R. F. Childs and H. A. Corver, *J. Am. Chem. Soc.*, **94**, 6201 (1972).
185. K. Ohkata and L. A. Paquette, *J. Am. Chem. Soc.*, **102**, 1082 (1980); R. B. Du Vernet, M. Glanzmann and G. Schröder, *Tetrahedron Lett.*, 3071 (1978); L. A. Paquette, P. B. Lavrikand and R. H. Summerville, *J. Org. Chem.*, **42**, 2659 (1977).
186. H. Hogeveen and P. W. Kwant, *Acc. Chem. Res.*, **8**, 413 (1975).
187. D. Lenoir and H.-U. Siehl, in Houben–Weyl's *Method. Org. Chemie.* (Ed. M. Hanack), Vol. 19c, Thieme Verlag, Stuttgart, 1990, p. 1.
188. H. Schwarz, *Angew. Chem., Int. Ed. Engl.*, **20**, 991 (1981); G. A. Olah, G. K. S. Prakash, R. E. Williams, L. D. Field and K. Wade, in *Hypercarbon Chemistry*, Wiley, New York, 1987.

189. P. v. R. Schleyer, T. W. Bentley, W. Koch, A. J. Kos and H. Schwarz, *J. Am. Chem. Soc.*, **109**, 6953 (1987).
190. K. J. Szabo, E. Kraka and D. Cremer, *J. Am. Chem. Soc.*, to appear.
191. M. Saunders and R. J. Berger, *J. Am. Chem. Soc.*, **94**, 4049 (1972).
192. G. A. Olah, G. K. S. Prakash, T. N. Rawdah, D. Wittaker and J. C. Rees, *J. Am. Chem. Soc.*, **101**, 3935 (1979).
193. J. B. Lambert, R. B. Finzel and C. A. Belec, *J. Am. Chem. Soc.*, **102**, 3281 (1980); J. B. Lambert and R. B. Finzel, *J. Am. Chem. Soc.*, **105**, 1954 (1983).
194. S. Winstein, M. Shatavsky, C. Norton and R. B. Woodward, *J. Am. Chem. Soc.*, **77**, 4183 (1955).
195. S. Winstein and C. Ordronneau, *J. Am. Chem. Soc.*, **82**, 2084 (1960).
196. W. G. Woods, R. A. Carboni and J. D. Roberts, *J. Am. Chem. Soc.*, **78**, 5653 (1956).
197. H. C. Brown and H. M. Bell, *J. Am. Chem. Soc.*, **85**, 2324 (1963).
198. M. Brookhart, A. Diaz and S. Winstein, *J. Am. Chem. Soc.*, **88**, 3135 (1966); H. G. Richey and R.K. Lustgarten, *J. Am. Chem. Soc.*, **88**, 3136 (1966).
199. P. R. Story and M. Saunders, *J. Am. Chem. Soc.*, **84**, 4876 (1962).
200. M. Brookhart, R. K. Lustgarten and S. Winstein, *J. Am. Chem. Soc.*, **89**, 6352 (1967); R. K. Lustgarten, M. Brookhart and S. Winstein, *J. Am. Chem. Soc.*, **94**, 2437 (1972).
201. G. Olah and G. Liang, *J. Am. Chem. Soc.*, **97**, 6803 (1975).
202. T. Laube, *J. Am. Chem. Soc.*, **111**, 9224 (1989).
203. J. M. Schulman, R. L. Disch, P. v. R. Schleyer, B. Bühl, M. Bremer and W. Koch, *J. Am. Chem. Soc.*, **114**, 7897 (1992).
204. M. Bremer, K. Schlötz, P. v. R. Schleyer, U. Fleischer, M. Schindler, W. Kutzelnigg, W. Koch and P. Pulay, *Angew. Chem., Int. Ed. Engl.*, **28**, 1042 (1989).
205. P. J. Fagan, E. G. Burns and J. C. Calabrese, *J. Am. Chem. Soc.*, **110**, 2979 (1988); P. J. Fagan, Results cited in Reference 35.
206. S. Winstein, J. Sonnenberg and L. de Vries, *J. Am. Chem. Soc.*, **81**, 6523 (1959).
207. S. Masamune, S. Sakai, A. V. K. Jones and T. Nakashima, *Can. J. Chem.*, **52**, 855, 857 (1974); S. Masamune, M. Sakai and A. V. K. Jones, *Can. J. Chem.*, **52**, 858 (1974)
208. G. K. S. Prakash, M. Arvanghi and G. A. Olah, *J. Am. Chem. Soc.*, **107**, 6017 (1985).
209. D. P. Kelly, J. J. Giansiracusa, D. R. Leslie, I. D. McKern and G. C. Sinclair, *J. Org. Chem.*, **53**, 2497 (1988).
210. P. v. R. Schleyer, D. Lenoir, P. Mison, G. Liang, G. K. S. Prakash and G. A. Olah, *J. Am. Chem. Soc.*, **102**, 683 (1980).
211. H.-U. Siehl, *Adv. Phys. Org. Chem.*, **23**, 63 (1987).
212. H. Tanida, T. Tsuji and T. Irie, *J. Am. Chem. Soc.*, **89**, 1953 (1967); J. S. Haywood-Farmer and R. E. Pincock, *J. Am. Chem. Soc.*, **91**, 3020 (1969); M. Battiste, C. L. Deyrup, R. E. Pincock and J. Haywood-Farmer, *J. Am. Chem. Soc.*, **89**, 1854 (1967).
213. R. M. Coates and J. L. Kirkpatrick, *J. Am. Chem. Soc.*, **92**, 4883 (1970); R. M. Coates and E. R. Fretz, *J. Am. Chem. Soc.*, **97**, 2538 (1975); M. Saunders and M. R. Kates, *J. Am. Chem. Soc.*, **102**, 6867 (1980); W. L. Jorgenson, *Tetrahedron Lett.*, 3033 (1976).
214. K. J. Szabo and D. Cremer, *J. Org. Chem.*, **60**, 2257 (1995).
215. W. R. Scott and R. E. Pincock, *J. Am. Chem. Soc.*, **95**, 2040 (1973).
216. M. Bremer, P. v. R. Schleyer, K. Schötz, M. Kausch and M. Schindler, *Angew. Chem., Int. Ed. Engl.*, **26** 761 (1987).
217. R. K. Lustgarten, M. Brookhart and S. Winstein, *J. Am. Chem. Soc.*, **90**, 7364 (1968).
218. H. Hart and M. Kuzuya, *J. Am. Chem. Soc.*, **97**, 2459 (1975).
219. H. Hogeveen and E. M. G. A. van Kruchten, *Recl. Trav. Chim. Pays-Bas*, **96**, 61 (1977); *J. Org. Chem.*, **42**, 1472 (1977).
220. A. Diaz, M. Sakai and S. Winstein, *J. Am. Chem. Soc.*, **92**, 7477 (1970).
221. E. Kaufmann, H. Mayr, J. Chandrasekhar and P. v. R. Schleyer, *J. Am. Chem. Soc.*, **103**, 1375 (1981).
222. J.-H. Shin, *Bull. Korean Chem. Soc.*, **3**, 66 (1988).
223. K. N. Houk, R. W. Gandour, R. W. Strozier, N. G. Rondan and L. A. Paquette, *J. Am. Chem. Soc.*, **101**, 6797 (1979).
224. L. T. Scott, *Pure Appl. Chem.*, **58**, 105 (1986); R. Gleiter and W. Schafer, *Acc. Chem. Res.*, **23** 369 (1990); M. F. Falcetta, K. I. Jordan, J. E. McMurry and M. N. Paddon-Row, *J. Am. Chem. Soc.*, **112**, 579 (1990).

225. W. R. Roth, O. Adamczak, R. Breuckmann, H.-W. Lennartz and R. Boese, *Chem. Ber.*, **124**, 2499 (1991).
226. W. R. Roth, M. Böhm, H.-W. Lennartz and E. Vogel, *Angew. Chem.*, **95**, 1011 (1983); W. R. Roth, F.-G. Klärner, G. Siepert and H.-W. Lennartz, *Chem. Ber.*, **125**, 217 (1992).
227. G. Maier, *Angew Chem., Int. Ed. Engl.*, **6**, 402 (1967); E. Vogel, *Pure Appl. Chem.*, **20**, 237 (1969).
228. M. Traetteberg, *J. Am. Chem. Soc.*, **86**, 4265 (1964); S. S. Butcher, *J. Chem. Phys.*, **42**, 1833 (1965).
229. F. A. L. Anet, *J. Am. Chem. Soc.*, **86**, 458 (1964); F. R. Jensen and L. A. Smith, *J. Am. Chem. Soc.*, **86**, 956 (1964).
230. E. Ciganek, *J. Am. Chem. Soc.*, **89**, 1454, 1458 (1967).
231. R. Hoffmann, *Tetrahedron Lett.*, 2907 (1970); R. Hoffmann and W.-D. Stohrer, *J. Am. Chem. Soc.*, **93**, 6941 (1971).
232. J. F. Liebman and A. Greenberg, *Chem. Rev.*, **89**, 1225 (1989).
233. H. B. Bürgi, E. Shefter and J. D. Dunitz, *Tetrahedron*, **31**, 3089 (1975).
234. D. W. Rogers, A. Podosenin and J. F. Liebman, *J. Org. Chem.*, **58**, 2589 (1993); D. W. Rogers, S. A. Loggins, S. D. Samuel, M. A. Finnerty and J. F. Liebman, *Struct. Chem.*, **1**, 481 (1990).
235. H. J. Dauben, Jr., J. D. Wilson and J. L. Laity, in *Non-benzenoid Aromatics*, Vol. 2. (Ed. J. P. Snyder), Academic Press, New York, 1971, p. 167.
236. R. F. Childs and I. Pikulik, *Can. J. Chem.*, **55**, 259 (1977).
237. I. Pikulik and R. F. Childs, *Can. J. Chem.*, **55**, 251 (1977).
238. E. Kraka and D. Cremer, *J. Am. Chem. Soc.*, to appear.
239. D. Cremer, B. Dick and D. Christen, *J. Mol. Struct.*, **110**, 277 (1984).
240. D. Cremer and B. Dick, *Angew. Chem., Int. Ed. Engl.*, **21**, 865 (1982).
241. J. M. Schulman, R. L. Disch and M. L. Sabio, *J. Am. Chem. Soc.*, **106**, 7696 (1984).
242. R. V. Williams, H. A. Kurtz and B. Farley, *Tetrahedron*, **44**, 7455 (1988).
243. T.-H. Tang, C. S. Q. Lew, Y.-P. Cui, B. Capon and I. G. Csizmadia, *J. Mol. Struct.*, **305**, 149 (1994).
244. A. Almenningen, O. Bastiansen and M. Traetteberg, *Acta Chem. Scand.*, **12**, 1221 (1958).
245. T. Tsuji, S. Teretake and H. Tanida, *Bull. Chem. Soc. Jpn.*, **42**, 2033 (1969); R. Huisgen, *Angew. Chem., Int. Ed. Engl.*, **9**, 751 (1970); P. M. Warner and S.-H. Lu, *J. Am. Chem. Soc.*, **102**, 331 (1980).
246. I. Pikulik and R. F. Childs, *Can. J. Chem.*, **53**, 1818 (1975).
247. H. Tanida, T. Tsuji and T. Irie, *J. Am. Chem. Soc.*, **89**, 1953 (1967); B. Halton, M. A. Battiste, R. Rehberg, C. L. Deyrup and M. E. Brennan, *J. Am. Chem. Soc.*, **89**, 5964 (1967); M. A. Battiste and J. W. Nebzydosky, *J. Am. Chem. Soc.*, **92**, 4450 (1970); S. C. Clarke and B. L. Johnson, *Tetrahedron Lett.*, 617 (1967); D. M. Birney and J. A. Berson, *Tetrahedron*, **42**, 1561 (1986).
248. A. Bertsch, W. Grimme, G. Reinhardt, H. Rose and P. M. Warner, *J. Am. Chem. Soc.*, **110**, 5112 (1988).
249. W. L. Jörgensen, *J. Am. Chem. Soc.*, **97**, 3082 (1975).
250. P. N. Skancke, K. Yamashita and K. Morokuma, *J. Am. Chem. Soc.*, **109**, 4157 (1987).
251. C. F. Wilcox, Jr., D. A. Blain, J. Clardy, G. Van Duyne, R. Gleiter and M. Eckert Maksic, *J. Am. Chem. Soc.*, **108**, 7693 (1986).
252. A. Counotte-Potman, H. C. van der Plas and B. van Veldhuizen, *J. Org. Chem.*, **46**, 2138 (1981); C. H. Stam, A. Counotte-Potman and H. C. van der Plas, *J. Org. Chem.*, **47**, 2856 (1982).
253. C. P. R. Jennison, D. Mackay, K. N. Watson and N. J. Taylor, *J. Org. Chem.*, **51**, 3043 (1986).
254. G. Schröder, J. F. M. Oth and R. Merényi, *Angew. Chem., Int. Ed. Engl.*, **4**, 752 (1965); L. T. Scott and M. Jones, Jr., *Chem. Rev.*, **72**, 181 (1972).
255. W. v. Doering, V. G. Toscano and G. H. Beasley, *Tetrahedron*, **27**, 5299 (1971).
256. M. J. S. Dewar and D. H. Lo, *J. Am. Chem. Soc.*, **93**, 7201 (1971).
257. A. K. Cheng, F. A. L. Anet, J. Mioduski and J. Meinwald, *J. Am. Chem. Soc.*, **96**, 2887 (1974).
258. K. J. Szabo and D. Cremer, to appear.
259. R. V. Williams and H. A. Kurtz, *J. Org. Chem.*, **53**, 3626 (1988); *J. Chem. Soc., Perkin Trans. 2*, 147 (1994).
260. M. J. S. Dewar and C. Jie, *Tetrahedron*, **44**, 1351 (1988).
261. H. Quast, R. Janiak, E.-M. Peters, K. Peters and H. G. v. Schnering, *Chem. Ber.*, **125**, 969 (1992).
262. L. S . Miller, K. Grohmann and J. J. Dannenberg, *J. Am. Chem. Soc.*, **105**, 6862 (1983).

263. L. M. Jackman, A. Benesi, A. Mayer, H. Quast, E.-M. Peters, K. Peters and H. G. v. Schnering, *J. Am. Chem. Soc.*, **111**, 1512 (1989).
264. Y. C. Wang and S. H. Bauer, *J. Am. Chem. Soc.*, **94**, 5651 (1972); L. A. Paquette, W. E. Volz, M. A. Beno and G. G. Christoph, *J. Am. Chem. Soc.*, **97**, 2562 (1975).
265. S. A. Weisman, S. G. Baxter, A. M. Arif and A. H. Cowley, *J. Am. Chem. Soc.*, **108**, 529 (1986).
266. A. B. McEwan and P. v. R. Schleyer, *J. Org. Chem.*, **51**, 4357 (1986).
267. J. F. Liebman, L. A. Paquette, J. R. Peterson and D. W. Rogers, *J. Am. Chem. Soc.*, **108**, 8267 (1986).
268. M. A. Miller, J. M. Schulman and R. L. Disch, *J. Am. Chem. Soc.*, **110**, 7681 (1988); M. J. S. Dewar and A. J. Holder, *J. Am. Chem. Soc.*, **111**, 5384 (1989); A. J. Holder, *J. Comput. Chem.*, **14**, 251 (1993); J. W. Storer and K. N. Houk, *J. Am. Chem. Soc.*, **114**, 1165 (1992).
269. L. J. Schaad, B. A. Hess, Jr. and L. T. Scott, *J. Phys. Org. Chem.*, **6**, 316 (1993); L. T. Scott, M. J. Cooney, D. W. Rogers and K. Dejroongruang, *J. Am. Chem. Soc.*, **110**, 7244 (1988).
270. G. A. Olah, G. Ascensio, H. Mayr and P. v. R. Schleyer, *J. Am. Chem. Soc.*, **100**,4347 (1978); A. J. Birch, A. L. Hinde and L. Radom, *J. Am. Chem. Soc.*, **102**, 6430 (1980).
271. L. M. Tolbert and A. Rajca, *J. Org. Chem.*, **50**, 4805 (1985).
272. J. M. Brown and J. L. Occolowitz, *J. Chem. Soc., Chem. Commun.*, 376 (1965); *J. Chem. Soc. (B)*, 411 (1968); J. M. Brown, *J. Chem. Soc., Chem. Commun.*, 638 (1967); S. Winstein, M. Ogliaruso, M. Sakai and J. M. Nicholson, *J. Am. Chem. Soc.*, **89**, 3656 (1967).
273. E. Kaufmann, H. Mayr, J. Chandrasekhar and P. v. R. Schleyer, *J. Am. Chem. Soc.*, **103**, 1375 (1981).
274. J. M. Brown, R. J. Elliott and W. G. Richards, *J. Chem. Soc., Perkin Trans. 2*, 485 (1982).
275. M. Christl, H. Leininger and D. Brückner, *J. Am. Chem. Soc.*, **105**, 4843 (1983); M. Christl and D. Brückner, *Chem. Ber.*, **119**, 2025 (1986).
276. F. H. Köhler and N. Hertkorn, *Chem. Ber.*, **116**, 3274 (1983).
277. G. B. Trimitsis and P. Zimmermann, *J. Chem. Soc., Chem. Commun.*, 1506 (1984).
278. W. N. Washburn, *J. Org. Chem.*, **48**, 4287 (1983).
279. R. E. Lee and R. R. Squires, *J. Am. Chem. Soc.*, **108**, 5078 (1986).
280. N. Hertkorn, F. H. Köhler and G. Müller, *Angew. Chem., Int. Ed. Engl.*, **25**, 468 (1986).
281. P. v. R. Schleyer, E. Kaufmann, A. J. Kos, H. Mayr and J. Chandrasekhar, *J. Chem. Soc., Chem. Commun.*, 1583 (1986).
282. R. Lindh, B. O. Roos, G. Jonsäll and P. Ahlberg, *J. Am. Chem. Soc.*, **108**, 6554 (1986).
283. G. Trimitsis, F.-T. Lin, R. Eaton, S. Jones, M. Trimitsis and S. Lane, *J. Chem. Soc., Chem. Commun.*, 1704 (1987); G. Trimitsis, J. Rimoldi, M. Trimitsis, J. Balog, F.-T. Lin, A. Marcus, K. Somayajula, S. Jones, T. Hendrickson and S. Kincaid, *J. Chem. Soc., Chem. Commun.*, 237 (1990).
284. A. Tuncay, M. A. Caroll, L. A. Labeots and J. M. Pawlak, *J. Chem. Soc., Chem. Commun.*, 1590 (1990).
285. M. Christl and H. Müller, *Chem. Ber.*, **126**, 529 (1993).
286. N. Hertkorn and F. H. Köhler, *Z. Naturforsch.*, **45b**, 848 (1990).
287. M. Christl, R. Less and H. Müller, *J. Chem. Soc., Chem. Commun.*, 153 (1994).
288. R. R. Squires, *Acc. Chem. Res.*, **25**, 461 (1992).

CHAPTER 9

Thermal stereomutations of cyclopropanes and vinylcyclopropanes

JOHN E. BALDWIN

Department of Chemistry, Syracuse University, Syracuse, New York 13244, USA

I. INTRODUCTION 469
II. HISTORICAL SYNOPSIS 470
III. EXPERIMENTAL STUDIES 471
IV. CORRELATIONS OF REACTION RATE CONSTANTS 476
V. THEORETICAL MODELS AND CALCULATIONAL RESULTS . . . 479
VI. DISPARATE KINETIC RESULTS 484
VII. CONCLUSIONS 487
VIII. ACKNOWLEDGMENTS 488
IX. REFERENCES 488

I. INTRODUCTION

When a substituted cyclopropane is heated, it may isomerize to one or another geometrical isomer or to its enantiomer. These thermal stereomutation reactions have attracted extensive experimental and theoretical efforts directed toward understanding in detail just how such reactions occur.

The potential energy surfaces (or free energy surfaces) governing the thermal behavior of such relatively small molecules, one may conjecture, might well be amenable to complete and reliable assessments by modern computational methods, and this possibility has been pursued tenaciously for nearly 30 years. On the experimental side there have been equally determined efforts to define as completely as possible the relative importance of conceptually distinct and possibly distinguishable mechanisms for these reactions. Numerous review articles on the thermal chemistry of cyclopropanes[1–16] and of vinylcyclopropanes[17–29] provide convincing testimony to the significance this topic has been accorded by chemists in recent decades.

The chemistry of the cyclopropyl group, Vol. 2
Edited by Z. Rappoport

After a brief historical recapitulation, the substantial body of experimental and theoretical work on these thermal epimerization reactions reported over the past 40 years is summarized. Of primary concern here are examples of stereomutations involving monocyclic, stereochemically unconstricted and minimally substituted molecules. Experimental studies of more heavily substituted cyclopropanes[30–33], attempts to generate trimethylene diradical intermediates from pyrazolines[34,35] and the fascinating and still incompletely understood thermal chemistry of bicyclo[2.1.0]pentanes[36–42], 2-methylenebicyclo[2.1.0]pentanes[43–46], bicyclo[3.1.0]hex-2-enes[47–57] and related reactions such as the pyrolysis of cyclopropane at 1200 °C to give products such as cyclopentadiene and toluene[58] are neglected, in spite of obvious mechanistic interrelationships.

The stereomutations of cyclopropanes and vinylcyclopropanes apparently involve homolytic carbon–carbon bond breaking and bond making and possibly trimethylene diradical intermediates and transition structures. Such reactions of substituted cyclopropanes occur at moderate temperatures in the absence of wall-catalyzed processes thanks to relatively weak C—C bonds. If the mechanistic issues posed by these stereomutations were largely resolved, and the reactions fairly well described or rationalized through theory, one could proceed to address more complex thermal isomerization reactions with greater confidence. One could, for example, pursue work on experimental determinations and mechanistic rationales of reaction stereochemistry for the structural isomerizations of cyclopropanes and vinylcyclopropanes with better prospects for success. While stereochemical information on vinylcyclopropane to cyclopentene rearrangements is beginning to become available[59–64], the stereochemistry of a simple cyclopropane to propene isomerization, a conversion which could take place through four distinct paths, has yet to be reported.

II. HISTORICAL SYNOPSIS

With the initial synthesis of cyclopropane in 1882[65,66], and the report of its thermal structural isomerization to propene in 1896[67], this simplest of cyclic hydrocarbons began its extraordinarily fruitful stimulation of fresh insights on fundamental problems in organic chemistry, ranging from basic concepts of ring strain[68] and structural isomerism to questions of thermochemistry[69] and reactivity and of σ aromaticity[70,71]. And from the beginning there was controversy, extending a few years before suitably authoritative commentators confirmed the fact that cyclopropane is indeed converted thermally to propene[72].

Early kinetic studies on the structural isomerization of cyclopropane to propene provided estimates of activation parameters[73–75] and prompted speculation that the reaction might well involve a trimethylene diradical intermediate. This possibility seemed reinforced when the thermal interconversion of the *cis* and *trans* isomers of 1,2-d_2-cyclopropane at 414 to 474 °C (equation 1) was reported in 1958[76]. This structurally degenerate isomerization was found to be substantially faster than conversion to deuterium-labeled propenes—about 24 times faster at the high pressure limit[76,77].

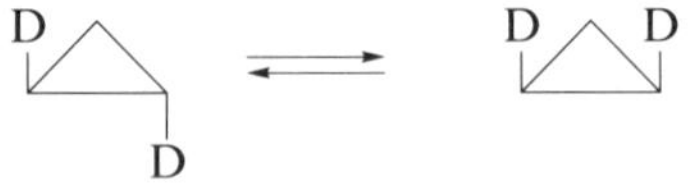

(1)

The history of vinylcyclopropane goes back to Gustavson's reported synthesis of vinylcyclopropane ('vinyltrimethylene') in 1896[78], a publication which occasioned considerable controversy, for the major C_5H_8 compound actually formed in his preparation proved to be spiropentane. Authentic vinylcyclopropane was secured by Demjanow and Dojarenko in 1922[79], but nevertheless both structures remained in dispute. Authorities as eminent as

C. K. Ingold denied the existence of spiropentane, even as an unstable intermediate product[80]; F. C. Whitmore considered both spiropentane and vinylcyclopropane to be '*incapable of existence*'[81]. Whatever the doubts then, they have not survived: the structures of cyclopropane, vinylcyclopropane and spiropentane are now known in considerable detail[82–89].

The thermal rearrangement of vinylcyclopropane to cyclopentene was uncovered in 1960[90,91]. That vinylcyclopropanes, like other cyclopropanes, may undergo *cis,trans* isomerizations was inferred in 1964 when *trans*-1-vinyl-2-methylcyclopropane was thermally converted to mostly (4*Z*)-1,4-hexadiene, a product formed at much lower temperatures from *cis*-1-vinyl-2-methylcyclopropane[92]. The reversible interconversion of the *cis* and *trans* isomers of 1-vinyl-2-d-cyclopropane (equation 2) was reported soon thereafter, in 1967[93–96]. Additional examples, including cases showing both geometrical isomerization and enantiomerization processes, soon followed.

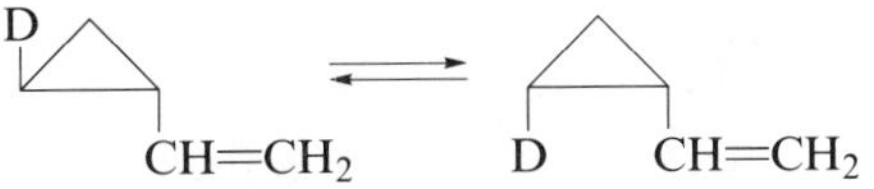

(2)

III. EXPERIMENTAL STUDIES

Though not of primary importance to the topic of this chapter, the cyclopropane to propene structural isomerization, it must be noted, has played a prominent role in the development of theory for unimolecular isomerizations and it continues to attract experimental work. The marked dependence of rate on gas pressure and on isotopic substitution exhibited by this isomerizaton provides fundamental insights on intramolecular and vibrational energy transfers and redistributions[97–114].

The activation parameters at the high pressure limit reported by Chambers and Kistiakowsky in 1934[75] (log A 15.2, E_a 65.0 kcal mol^{-1}) have been confirmed within close limits by numerous subsequent studies[115–135]; two of the most recent investigations report values of 14.7, 62.5 and 14.9, 63.9[134,135].

Activation parameters for the high pressure gas-phase approach of 1,2-d_2-cyclopropanes to *cis, trans* equilibrium (equation 1) have been reported as log A, E_a (kcal mol^{-1}) of 16.0, 64.2 and 16.4, 65.1[76,77]. From pressure-dependent measurements of rate constants and calculations based on RRKM theory, the threshold energy E_0 for the *cis, trans* isomerization has been estimated to be 61.1 kcal mol^{-1} and 61.3 kcal mol^{-1} [136–138].

The determination that the *cis* and *trans* isomers of 1,2-d_2-cyclopropane could be interconverted thermally prompted experimental work on similar reactions exhibited by other substituted cyclopropanes to define activation parameters for representative instances of the isomerization and to begin to discriminate among alternative mechanistic suggestions. The 1,2-dimethyl- and 1-ethyl-2-methylcyclopropanes, for examples, were shown to approach *cis, trans* equilibrium with activation parameters log A, E_a (kcal mol^{-1}) of 15.25, 59.4 and 15.08, 58.9[139,140].

To distinguish between simultaneous or only consecutive epimerizations at CHD centers, kinetic experiments based on the four isomers of 1-methyl-2,3-d_2-cyclopropane were designed[141]. The phenomenological rate constants associated with the various reactions leading from one isomer to another employ subscripts to designate the carbon atom at which an epimerization occurs; k_i indicates a one-center epimerization at C(i), and k_{ij} is used for a two-center epimerization at C(i) and C(j) (Scheme 1). In this case, with all four isomers present in equal concentrations at equilibrium, the time dependence of each is governed by four rate constants: k_1, $k_2 = k_3$, $k_{12} = k_{13}$ and k_{23}. But, at best, only three kinetic parameters may be found experimentally: k_1, k_{23} and $(k_2 + k_{12})$.

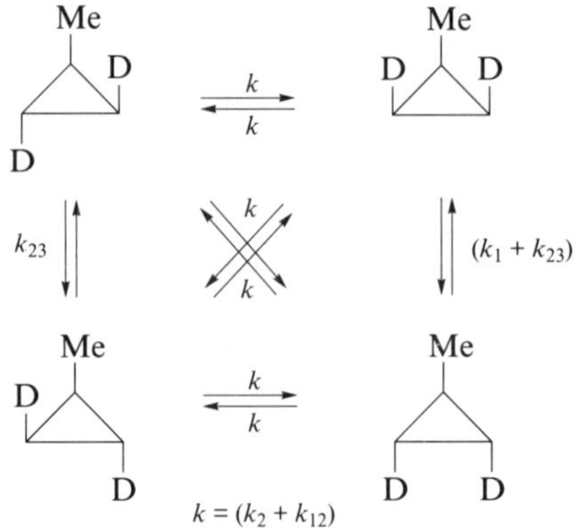

SCHEME 1

Through this work Setser and Rabinovitch[141] were the first to articulate clearly the fundamental agenda for experimental studies: to determine rate constants for various one-center and two-center epimerization events, and use the experimental facts to discriminate among alternative mechanistic models. But in practice, synthetic and analytical limitations restricted their study to the 1-methyl-2,3-d_2-cyclopropane isomers so that only one rate constant could be measured, a rate constant for approach of all isomers to equilibrium. The temperature dependence of this rate constant, $2(k_1 + k_2 + k_{12} + k_{23})$, led to the activation parameters log $A = 15.35$, $E_a = 60.5$ kcal mol^{-1} [141]. Thus, both practical and conceptual challenges were evident: better synthetic and analytical methods would be required to advance the agenda, and it was not at all clear how one might overcome the awkward fact that four rate constants had to be found, but in principle only three kinetic parameters might be measured.

Kinetic work on the isomeric 1,2-diphenylcyclopropanes (Scheme 2) made evident a substantial reduction in E_a and thus implied a stabilization of trimethylene diradical transition structure(s) by phenyl substituents[142]. In further work with 0.2 M (–)-1,2-diphenylcyclopropane in 1-butanol, Crawford and Lynch[143] uncovered a direct route from one *trans* antipode to the other: at 220.7 °C the measured ratio of rate constants k_{rac} (for loss of optical activity) to k_{tc} (for *trans* to *cis* geometrical isomerization) was found to be 1.49 ± 0.05 and since k_{rac} is $(2k_{12} + 2k_1)$, and k_{tc} is $2k_1$ (Scheme 2), the implication is that one-center epimerizations ($2k_1$) are favored over the two-center epimerization process (k_{12}) by

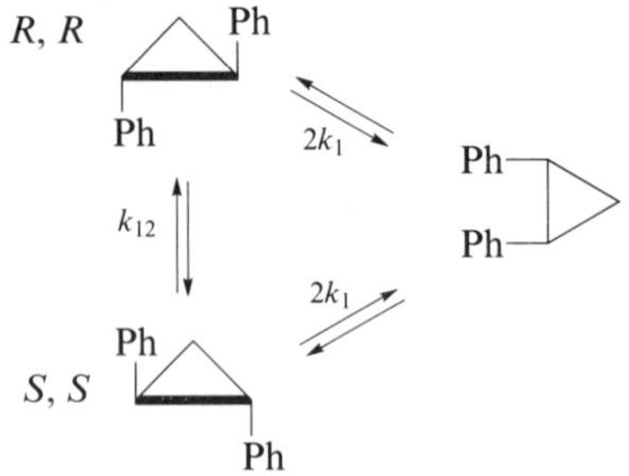

SCHEME 2

a 4:1 ratio. Thus any mechanistic formulation based on only k_1 or only k_{12} being kinetically significant cannot be correct. A recent duplication of this kinetic study under gas-phase conditions at 234 °C gave a very similar outcome, $(2k_1) : (k_{12}) = 2.3:1$[144].

Thermal isomerizations of (–)-1,2-divinylcyclopropane at 170 °C showed that $(2k_1):k_{12}$ was 2.4:1[145]. Similar detailed kinetic work with chiral 1-ethyl-2-methylcyclopropanes provided further evidence that one-center and two-center epimerizations were kinetically competitive. From a chiral *trans* isomer, at 377.2 °C, using a kinetic analysis dependent upon the assumption that only C(1)—C(2) bond cleavage led to isomerizations (the 'most-substituted-bond hypothesis'), the ratio of mechanistic rate constants $(k_1 + k_2):k_{12}$ was reported to be 2.1:1[146–148]. A more detailed investigation of these reactions, using 1,2,3-d_3 analogs of **1** such as **2** and an analysis independent of the 'most-substituted-bond hypothesis', led to an independent evaluation of this ratio, with full confirmation of the result: $(k_1 + k_2):k_{12} = 2.0:1$ at 380 °C[149].

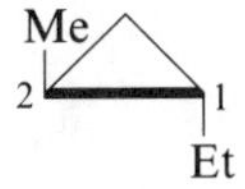

(1)

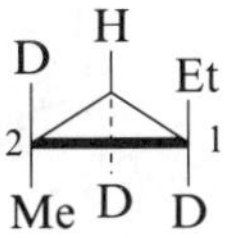

(2)

Kinetic studies by Doering and his collaborators at Harvard[150–154] based on five sets of chiral 1,2-disubstituted cyclopropanes, with 1-cyano, 2-(phenyl or propen-2′-yl or -(*E*)-propenyl or phenylethynyl) (**3**) and 1-phenyl-2-(propen-2′-yl) (**4**) substitution, established the ralative rotational propensities of these substituents and tested the proposition that they might be related to substituent moments of inertia. In all of these cases, the balance between one-center and two-center epimerizations from a *trans* isomer, reflected in $(k_1 + k_2):k_{12}$, was fairly constant, ranging from 1.4:1 to 2.1:1. The kinetic advantages for one-center epimerizations at cyano-substituted carbons for the four cases studied were modest and not especially system-dependent: the $k_1:k_2$ ratios were 2.5, 2.2, 2.4 and 1.8, thus establishing that rotational propensities are not dictated by some simple function of the moments of inertia of substituents.

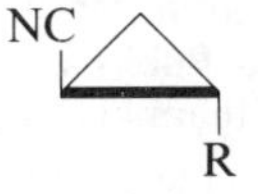

R = Ph, C(Me)=CH$_2$,
(*E*)-CH=CH—Me, C≡C-Ph

(3)

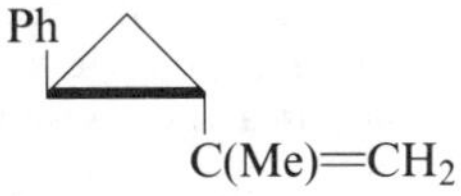

(4)

Concerns over the limits of validity of the 'most-substituted-bond hypothesis' prompted kinetic work on three sets of cyclopropanes having a deuterium as a stereochemical marker at C(3); one thus had the potential of detecting one-center epimerization at C(3), an indication of C(1)—C(3) or C(2)—C(3) bond cleavage. Reasonably enough, these studies demonstrated that when good radical stabilizing groups such as phenyl and cyano are positioned at both C(1) and C(2) (cf **5**), only the isomerizations expected from C(1)—C(2) bond cleavages are seen[155–157]; but the difference in radical stabilizing ability between hydrogen or deuterium and methyl or ethyl is not substantial enough to suppress k_3 epimerization events from being in evidence for 1-ethyl-2-methylcyclopropanes[149] and 1-cyano-2-methylcyclopropanes (**6**)[158,159].

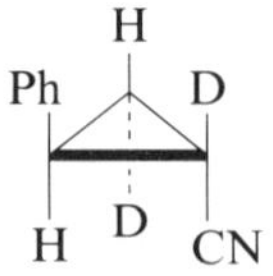

(**5**)

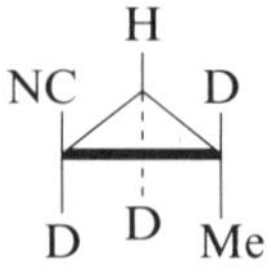

(**6**)

Another exploitation of stereochemically specific deuterium labeling at C(3) was exemplified in an approach to determine the relative one-center epimerization rate constants in a *cis*-1-cyano-2-isobutenylcyclopropane. Thanks to the deuterium labeling (equation 3) the distinction could be made without resort to chiral compounds, and the k_1:k_2 ratio was found to be (3.9 ± 0.5):1 at 206.7 °C[160].

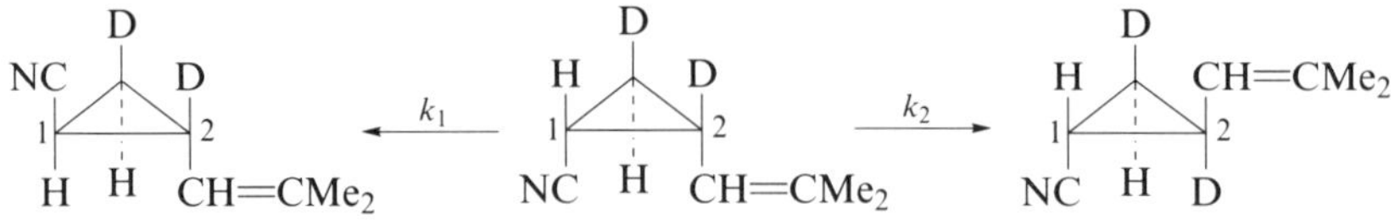

(3)

Specific deuterium labeling at C(3) was applied as well in the chiral spirocyclic system **7**: all three possible one-center epimerization events are kinetically competitive at 198.9 °C[161].

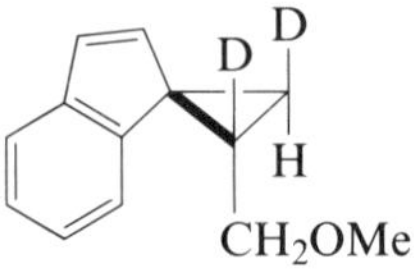

(**7**)

The thermal stereomutations of deuterium-labeled phenylcyclopropanes (Scheme 3) were studied in a progressive manner. First, the racemic and both achiral isomers were synthesized to provide material for kinetic work and to verify analytical methods[162]. The isomerizations among these three isomers at 309.3 °C were followed using either ^2H decoupled ^1H NMR spectroscopy or Raman spectroscopy; the two kinetic parameters $(k_1 + k_{23}) = 0.36 \times 10^{-5}\,s^{-1}$ and $(k_2 + k_{12}) = 1.07 \times 10^{-5}\,s^{-1}$ at 309.3 °C were measured. Published spectra of both sorts for authentic samples of *syn, anti* and *trans* isomers, and of thermal reaction mixtures, provided

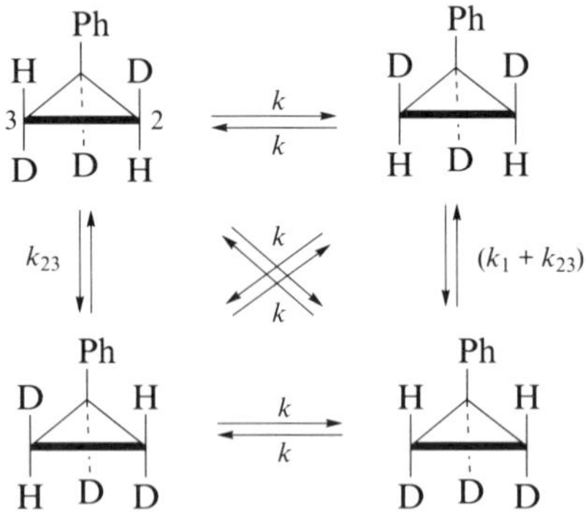

$k = (k_2 + k_{12})$

SCHEME 3

a readily accessible appreciation of the analytical methodology. Analytical results from the two independent spectroscopic methods were indistinguishable[162].

A complete solution to the kinetic problem was attained through further studies of (2*R*, 3*R*)-1,2,3-d$_3$-phenylcyclopropane and (1*R*, 2*S*, 3*R*)-1,2,3-d$_3$-phenylcyclopropane-2-^{13}C[163]. Reaction mixtures from the first were analyzed by NMR and by Raman spectroscopy, and with the aid of the chiral lanthanide shift reagent Eu(hfc)$_3$ on each derived mixture of deuterium-labeled benzoylcyclopropanes. Concentration versus reaction time data for all four isomers led to $k_{23} = 0$ and $k_1 = 0.36 \times 10^{-5}$ s^{-1}. From kinetic work based on the ^{13}C, d$_3$-labelled substrate (equation 4) the final distinction between k_{12} and k_2 reactions was secured: $k_{12} = 0.20 \times 10^{-5}$ s^{-1} and $k_2 = 0.87 \times 10^{-5}$ s^{-1}. Thanks to the ^{13}C,d$_3$ labeling, stereomutations allowed for equilibrations among eight rather than four isomers, and the distinction between k_2 and k_{12} products could be made[163].

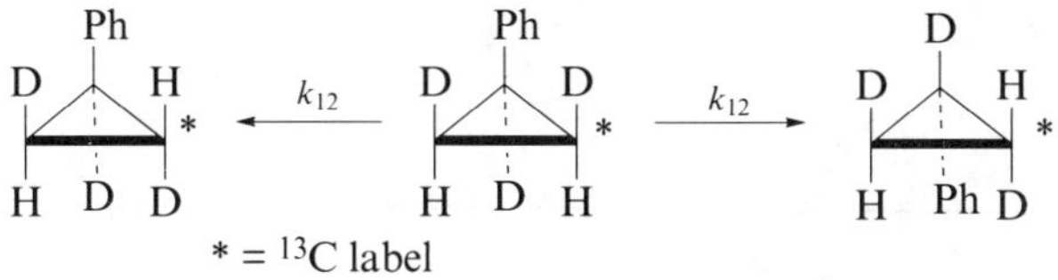

(4)

As an independent test of these rate constants, they were used to predict the time dependence of optical activity for the isomers formed by thermal rearrangement of (–)-*cis*-1-phenyl-2-d-cyclopropane. The predicted and subsequently observed racemization vs. time characteristics of this system was proved to be in quite satisfactory agreement[164].

A different experimental approach to the relative importance of one-center and two-center epimerizations in cyclopropane itself was based on the isomeric 1-^{13}C-1,2,3-d$_3$-cyclopropanes[165–169]. Here each carbon has the same substituents, one hydrogen and one deuterium, and should be equally involved in stereomutation events; secondary carbon-13 kinetic isotope effects or diastereotopically distinct secondary deuterium kinetic isotope effects may be safely presumed to be inconsequential. Unlike the isomeric 1,2,3-d$_3$-cyclopropanes (two isomers, only one phenomenological rate constant, for approach to *syn*, *anti* equilibrium), the 1-^{13}C-1,2,3-d$_3$-cyclopropanes provide four isomers and two distinct observables since there are two chiral forms as well as two *meso* structures (Scheme 4). Both chiral isomers were synthesized, and the phenomenological rate constants at 407 °C were found to be $k_i = (4k_{12} + 8k_1) = (4.63 \pm 0.19) \times 10^{-5}$ s^{-1} and $k_\alpha = (4k_{12} + 4k_1) = (3.10 \pm 0.07) \times 10^{-5}$ s^{-1}. The ratio of rate constants k_1:k_{12} is thus 1.0 ± 0.2: both one-center and two-center

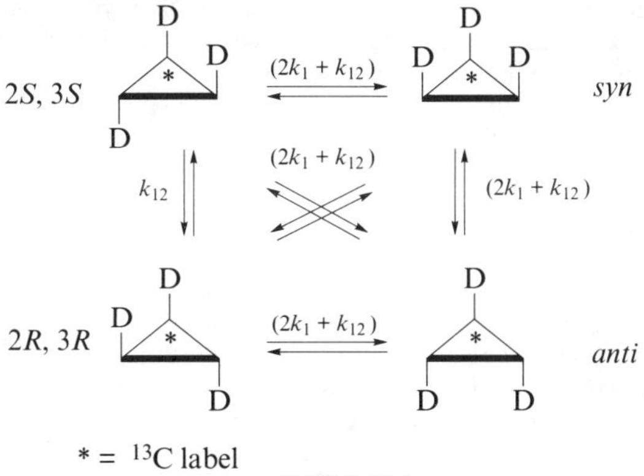

SCHEME 4

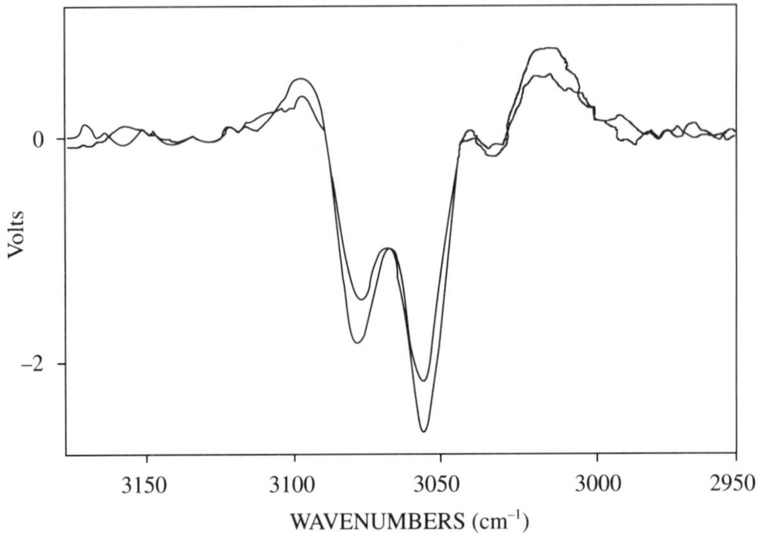

FIGURE 1. Gas-phase vibrational circular dichroism (VCD) spectra for (2*S*, 3*S*)-1-^{13}C-1,2,3,-d_3-cyclopropane (more intense spectrum) and for a 360 min, 407 °C thermal reaction product mixture of ^{13}C, d_3-cyclopropanes (less intense spectrum)[165,166]

thermal epimerizations participate in the stereomutations to approximately equal extents[165–169].

Thermal reaction mixtures in this work were analyzed by FTIR spectroscopy, tunable diode laser spectroscopy[170] and vibrational circular dichroism[171–176]. A pair of representative VCD spectra are shown in Figure 1. The more intense spectrum is an optically pure reference sample of the (2*S*, 3*S*) isomer at 53.59 torr; the other is a thermal reaction product mixture from this isomer after 360 min at 407 °C and gas chromatographic isolation, recorded at 81.30 torr[165,166]. From the measured $\Delta A/\Delta A_{ref}$ absorption intensity ratio and the pressures, one may calculate that the sample retains 51.65% of its original optical activity, which compares well with the value calculated from the k_α value obtained from the least-squares fit of all five VCD experimental points (51.2%). These spectra, obtained in the gas phase with a few mg of chromatographically purified labeled cyclopropanes, demonstrate the promise of VCD for assessing enantiomeric excess in situations where classical polarimetric methods would be of limited utility.

Other experimental work on the stereomutations of labeled phenylcyclopropanes and of 1,2-d_2-cyclopropanes is discussed below.

IV. CORRELATIONS OF REACTION RATE CONSTANTS

An empirical correlation of enantiomerization rate constants (k_{12}) for chiral *trans*-1-R^1-2-R^2-cyclopropanes was communicated in 1988[177]; when calculated values of $\Delta G^\ddagger$ (k_{12}) are plotted against the sum of radical stabilization energies (SE) of the two substituents, a linear correlation is evident. More recent work has provided a few more experimental points; an updated version of this correlation is shown in Figure 2. Kinetic data from 20 kinetic studies reported by six different research groups are accommodated by this simple empirical $\Delta G^\ddagger$ (k_{12}) versus $\{SE(R^1CH_2\cdot) + SE(R^2CH_2\cdot)\}$ correlation.

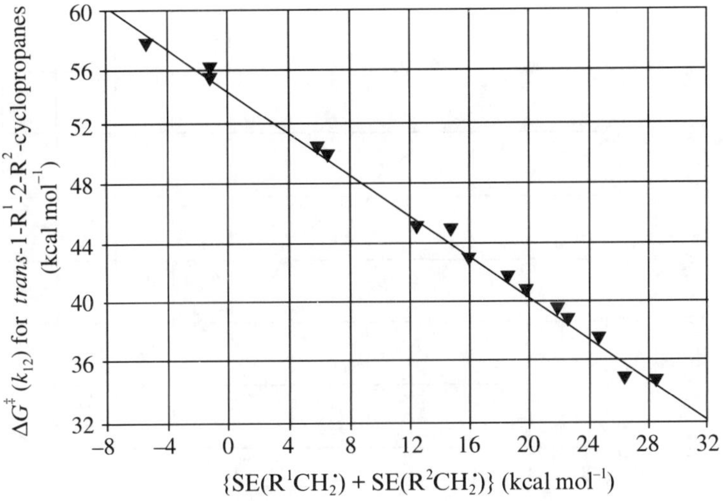

FIGURE 2. Empirical correlation of $\Delta G^{\ddagger}(k_{12})$ for *trans*-1-R^1-2-R^2-cyclopropanes with the sum of radical stabilization energies for $R^1CH_2^{\cdot}$ and $R^2CH_2^{\cdot}$. The 20 examples are taken from References 59, 62, 64, 143–158, 162, 163, 166 and 178–181

Most of the radical stabilization energy parameters utilized in Figure 2 (R = H or D, −2.65; Me or Et, – 0.5; CN, 6.6; vinyl or propen-2-yl, 13.2; (*E*)-propenyl, 15.4) are SE^0 values calculated by Leroy and coworkers[182–184]; those for phenyl (9.4) and cinnamyl (19.1) were set empirically. These particular parameters are serviceable and reasonable, but cannot be considered 'preferred'. Other plausible radical stabilization energy values[185–199] lead to very similar correlations which demonstrate, as Figure 2 demonstrates, that the rate constants for two-center epimerizations of cyclopropanes and vinylcyclopropanes are sensitive to the radical stabilizing capacities of substituent groups in a simple additive fashion. Whatever 1,3-disubstituted trimethylene diradical intermediates or transition structutes may be involved respond kinetically to substituents as though each locus of radical character were thermochemically independent.

A similar plot for $\Delta G^{\ddagger}$ ($k_1 + k_2$) versus {SE($R^1CH_2^{\cdot}$) + SE ($R^2CH_2^{\cdot}$)} may be attempted, based on the 20 kinetic studies included in the correlation of Figure 2 and five additional cases for which k_{tc} rate constants are available and may be taken as fair approximations to ($k_1 + k_2$). To include these *trans*-1,2-disubstituted cyclopropanes [Me,CO_2Me; CN,CN; OMe,vinyl; Ph,vinyl; (*E*)-propenyl, (*E*)-propenyl], two additional stabilization energy terms are required; SE^0 values are used for both ($MeOCH_2$ = 3.4; MeO_2CCH_2 = +0.5). The attempted correlation (Figure 3) based on 25 studies contributed by nine different research groups is very similar to the one displayed in Figure 2; the two least-squares correlation equations give different intercepts (54.6 in Figure 2; 53.6 in Figure 3) and essentially equal slopes (– 0.704; – 0.699) and R^2 values (0.99). The trends are consistent; whatever 1,3-trimethylene diradical transition structures are involved in one- center epimerizations, they respond to substituents as though each radical locus were sensitive to radical stabilizing substituents independently. In both Figures 2 and 3, vinylcyclopropanes correlate right along with other cyclopropanes. The ($k_1 + k_2$):k_{12} ratio remains generally constant as the sum of radical stabilization energies varies over nearly 35 kcal mol^{-1}. Whatever trimethylene diradical transition structures may be involved for k_{12} reactions, they or very similar transition structures are involved for ($k_1 + k_2$) processes as well. The ($k_1 + k_2$):k_{12} ratio for

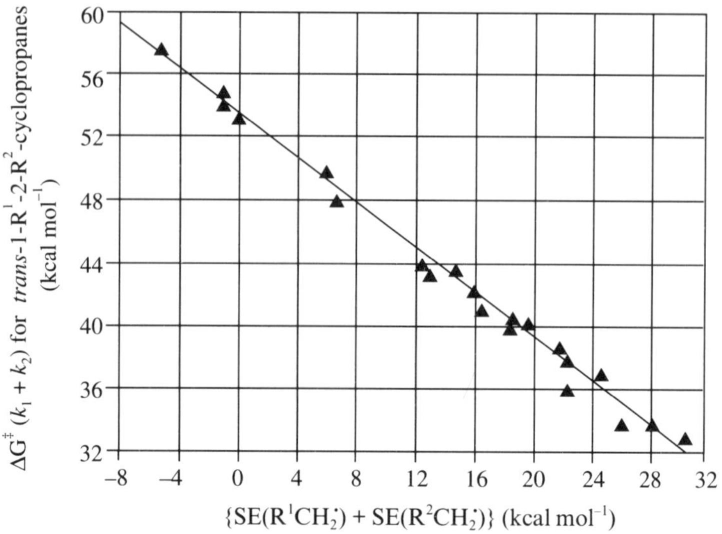

FIGURE 3 Empirical correlation of $\Delta G^{\ddagger}(k_1 + k_2)$ for *trans*-1-R^1-2-R^2-cyclopropanes with the sum of radical stabilization energies for $R^1CH_2^{\bullet}$ and $R^2CH_2^{\bullet}$. The 25 examples are taken from the References cited in the caption to Figure 2 and References 200–205

an 'average system' as defined by the correlations should be somewhat (but only slightly) temperature dependent, with$(k_1 + k_2):k_{12}$ about 2.1 at 400 °C and 3.2 at 160 °C.

An extreme example of the facilitation of stereomutations through radical stabilization of trimethylene structures is provided by the homopentafulvenes shown in equation 5. Interconversion between the two isomers occurs rapidly at 24 °C (half-time about 45 min), a consequence of the unusually large radical stabilization energy of cyclopentadienyl[206].

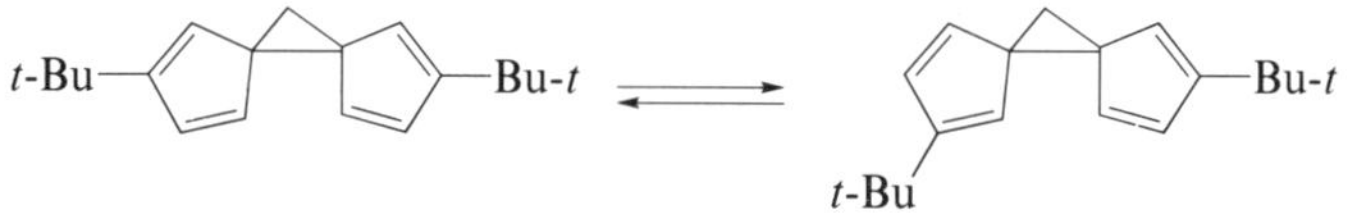

(5)

These empirical correlations provide some basis for estimating radical stabilization energies for groups that have yet to be subjected to experimental or calculational assessments. For *trans*-1-cyano-2-(phenylethynyl)cyclopropane, for example, at 190.7 °C in decalin, $k_{12} = 2.18 \times 10^{-6}\ s^{-1}$ and $(k_1 + k_2) = 4.47 \times 10^{-6}\ s^{-1}$ [152,153]. Hence, according to the two correlations, the radical stabilization energy of Ph—C≡C—$CH_2^{\bullet}$ is estimated to be 14.4 or 14.7 kcal mol^{-1}—more than the SE^0 value for H—C≡C—$CH_2^{\bullet}$ (10.6) but less than the stabilization energy value for cinnamyl (19.1 kcal mol^{-1}).

When one attempts to extend these empirical correlations once again, now considering rate constants for vinylcyclopropane to cyclopentene rearrangements, a fair linear correlation is obtained (Figure 4). The correlation line has an intercept of 56.7 kcal mol^{-1} and a slope of -0.690 ($R^2 = 0.99$). The rate constants utilized were corrected for symmetry (a factor of 1/2 for vinylcyclopropane, and of 1/4 for 1,l-dicyclopropylethene) and the radical stabilization energies of $\cdot CH_2CR{=}CH_2$ for R=Me or cyclopropyl were taken to be identical. The rate constants for vinylcyclopropane to cyclopentene rearrangements respond

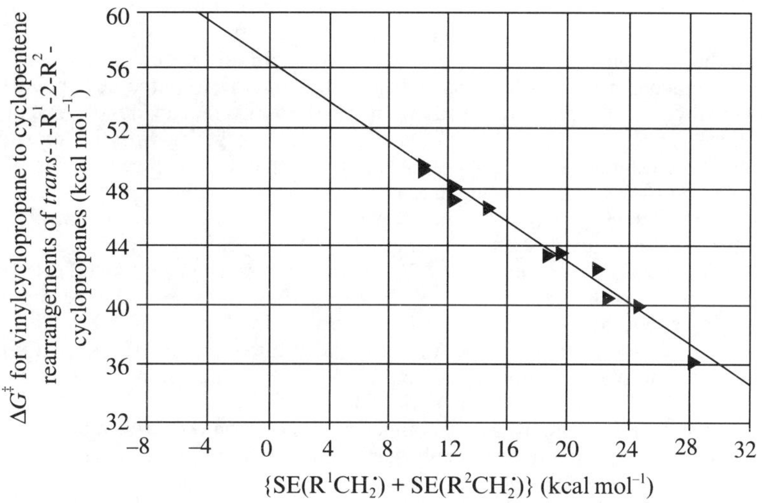

FIGURE 4. Empirical correlation of $\Delta G^{\ddagger}(k_{12})$ for vinylcyclopropane to cyclopentene rearranfements for *trans*-1-R^1-2-R^2-cyclopropanes with the sum of radical stabilization energies for $R^1CH_2^{\cdot}$ and $R^2CH_2^{\cdot}$. The 15 examples are taken from References 59, 62, 64, 144, 150–154, 180, 181 and 207–211

to radical stabilizing substituents just as do k_{12} and $(k_1 + k_2)$ rate constants for thermal stereomutations. Very similar transition structures are implicated.

The relatively constant kinetic disadvantage experienced by an average vinylcyclopropane rearrangement, compared with a $(k_1 + k_2)$ stereomutation, amounts to a $\Delta\Delta G^{\ddagger}$ of about 3 kcal mol^{-1}. This may well be associated with configurationally distinct sets of diradical structures with those of *E* stereochemistry favored thermodynamically over those of *Z* stereochemistry by about 3 kcal mol^{-1}. Only the latter may lead to cyclopentene products[212].

Thus a substantial body of experimental evidence shows that 1,2-disubstituted cyclopropanes, including vinylcyclopropanes, react thermally to give isomeric cyclopropanes through both one-center and two-center epimerizations, with $(k_1 + k_2):k_{12}$ ratios from 1.4 to 4. Rate constants for both $(k_1 + k_2)$ and k_{12} events respond to the capacity of substituents to stabilize adjacent radicals in a regular fashion consistent with trimethylene diradical transition structures. Rate constants for vinylcyclopropane structural isomerizations do as well, thus reinforcing the notion that these reactions are nonconcerted diradical mediated reactions.

V. THEORETICAL MODELS AND CALCULATIONAL RESULTS

The structural isomerization of cyclopropane to propene does not involve radical chain processes[213]; if a trimethylene diradical intermediate is involved, it must be of such a short lifetime that it may not be easily trapped through gas-phase intermolecular reactions[214,215]. Yet the trimethylene diradical hypothesis does account for the thermal interconversion of *cis* and *trans* 1,2-d$_2$-cyclopropanes in a most plausible manner[76]. Homolytic cleavage of one carbon–carbon bond would form a 1,3-diradical intermediate; rotations of terminal methylene groups in this trimethylene diradical followed by reformation of the cyclopropane ring would rationalize the isomerization[216–218]. According to this model, the net outcome of a stereomutation process would be dictated by the relative magnitudes of rate constants for reclosure compared with rate constants for rotations; it featured equilibrating trimethylene diradicals in a substantial energy well, estimated by O'Neal and Benson through thermochemical analogies to be some 9.3 kcal mol^{-1} deep[218]. It soon became a widely accepted formalism for rationalizing stereochemical outcomes for reactions thought to involve trimethylenes, whether generated from cyclopropanes or pyrazolines or other precursors: observed product ratios could be used, together with the 'most-substituted-bond hypothesis' and a steady-state kinetic treatment, to deduce *k*(closure):*k*(rotation) ratios for the diradical intermediate.

Examples of such kinetic treatments were provided by work on chiral 1,1,2,2-tetramethylcyclopropane-d_6[30] and *trans*-l-ethyl-2-methylcyclopropane[146–148]. At 350.2 °C, the first substrate approached *cis, trans* equilibrium with rate constant k_i and suffered loss of optical activity with a rate constant k_α; The k_i:k_α ratio was 1.7:1[30]. The second substituted cyclopropane, at 377.2 °C, exhibited kinetic behavior dictated by k_i:k_α = 2.0:1. Using steady-state kinetic treatments and the most-substituted-bond hypothesis, these rate constant ratios were calculationally transformed into *k*(cyclization):*k*(rotation) ratios of 11:1 and 0.29:1, ratios different by a factor of 38.

The classical diradical intermediate/competitive cyclization versus terminal-methylene-group rotation model did not go unchallenged. Smith was prompt to suggest in 1958 that stereomutations might take place without intervention of an intermediate, through a transition structure accommodating epimerization at one carbon only[219]. His conceptual model was quickly countered with various rhetorical objections and it prompted a clear-headed mechanistic insight: one might distinguish between a diradical-mediated mechanism, which permitted both one-center and two-center epimerizations, from the Smith mechanism, which might govern only one-center epimerizations, by determining experimentally the various k_i and k_{ij} rate constants shown by a substituted cyclopropane. The l-methyl-2,3-d_2-cyclopropanes, the isomers under consideration as this insight was formulated, served to demonstrate how difficult the experimental undertaking might be[141].

The classical trimethylene diradical model for cyclopropane stereomutations featured a deep potential energy well for the intermediate and internal rotations about 10 times faster than ring closure[218]. Improved experimental estimates for the heat of formation for the singlet trimethylene diradical, and for the energy required for the isomerizations of 1,2-d_2-cyclopropanes, place the diradical in an extremely shallow energy well, one on the order of only 1 kcal mol^{-1} deep[220,221]. The classical model can never accommodate $(k_1 + k_2)$:k_{12} ratios less than 2, and several substituted cyclopropanes exhibit such ratios. Thus the model, at least in its original form, seems thermochemically flawed and unable to accommodate experimental data for some systems.

Hoffmann's interests in cyclopropane and new ways of appreciating bonding and chemical reactivity recognized that alternative geometric configurations of trimethylene might be of significance[222–216]. Aided by extended Hückel calculations, he suggested that face-to-face (**FF**), edge-to-edge (**EE**) and edge-to-face (**EF**) versions of trimethylene should be considered[225].

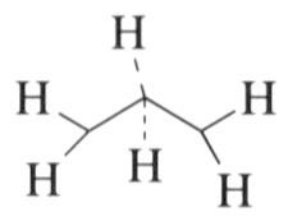

FF

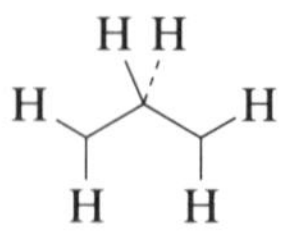

EE

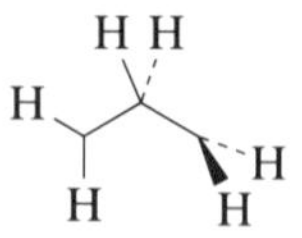

EF(ts)

He found that the **EE** trimethylene species was associated with a shallow minimum on the singlet state energy surface, and that this intermediate structure could progress to the much deeper valley of cyclopropane most easily, surmounting an activation barrier of only about 1 kcal mol^{-1}, through a conrotatory motion of both methylene groups. Conrotatory motion from a *trans*-1,2-disubstituted cyclopropane to an **EE** trimethylene intermediate and on to form a new cyclopropane would give the mirror image version of the starting material. This prophetic analysis, it will be noted, made no pronouncement regarding net one-center versus two-center epimerizations: it concentrated attention on the hypothetical behavoir of an **EE** trimethylene diradical in a very shallow minimum.

Other semiempirical calculations[227–231] which followed Hoffmann's work contributed to discussions of cyclopropane stereomutations chemistry, but have been largely superseded by various *ab initio* efforts[232–270]. The **EE** trimethylene 'π-diradical' according to perturbational theory was found to have no preferred motion for reclosure[237]. The **FF** trimethylene was ascertained to be neither an intermediate nor a transition structure[238]. One cannot break the cyclopropane ring just by elongating one carbon–carbon bond: to get anywhere, bond-breaking must be coordinated with vibrations corresponding to rotations and changes in pyramidalization[150,151]. The **EF(ts)** trimethylene was determined to be a transition structure for one-center or geometrical isomerization[239].

Thorough investigations by Salem and his collaborators[235–242] attempted the extraordinarily ambitious task of probing the full 21-dimensional hyperspace for the cyclopropane–trimethylene system using SCF molecular orbital methods, a minimal basis set of Slater orbitals, and 3×3 configuration interaction in the transition state region. This effort resulted in a detailed and visually accessible representation for the entire (nondynamic) pathway for the isomerization, and the quantitative results were found to agree with the generalization that, 'at least in 1,2-disubstituted cyclopropanes, the overall rate constants for geometrical isomerization and optical isomerization are comparable'[242].

Another, quite independent theoretical assessment, based on *ab initio* calculations and the generalized valence bond method, found that the 'barrier height for *cis,trans* isomerization of cyclopropane is essentially the same (calculated value, 60.5 kcal mol^{-1}) whether one or both of the thermal CH_2 groups are rotated after opening of the CC bond'[243]. Thus, in 1972, there seemed to be general agreement among theoreticians that the stereomutations of cyclopropane should take place with k_1 about equal to k_{12}.

These calculational efforts, agreeing so well in their main conclusions with experimentally defined generalizations, pointed toward a very auspicious future for computational chemistry applied to the reactions of relatively large organic molecules.

These efforts stimulated more profound consideration of diradical structures and states[242–246] and led to further calculations, applying ever higher levels of theory and more rigorous computational methods[247–270]. While most of this work concentrated on the structures and energies of selected stationary points and on singlet and triplet state distinctions, rather than on thorough examinations of the whole singlet hypersurface, there were trajectory calculations[247] directed toward understanding the stereomutations of cyclopropane[248–258]. These calculations, based on the same calculated potential energy surface held to be consistent with the generalization that rate constants for geometrical isomerizations and enantiomerizations are comparable, now seemed consistent with k_{12} being much larger than k_1, a revised judgement apparently responsive to experimental work which concluded that k_{12} is much larger that k_1 (see below).

With the conrotatory path a reasonable model for optical isomerization, and the **EF(ts)** mediated path a serviceable rationale for geometrical isomerization, the possible contribution of a disrotatory ring opening of cyclopropane did not receive full consideration. Yet its potential to contribute, and to complicate the situation, did not pass entirely un-noticed[268].

Recent calculations, using more powerful computational methods, have shown that the conrotatory path does not pass through a C_2-symmetric transition structure[269,270]: the conrotatory route diverges as one methylene rotates in conrotatory fashion slightly faster than

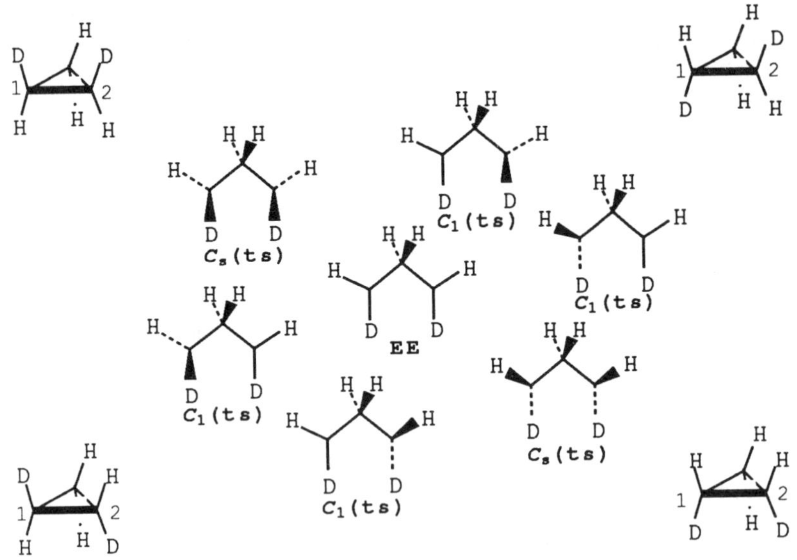

FIGURE 5. Schematic diagram for stereomutation paths interrelating 1,2-d_2-cyclopropanes and a particular **EE** intermediate. The four ***C*₁(ts)** structures link **EE** with chiral isomers; the two ***C*ₛ(ts)** structures separate **EE** and *cis*-1,2-d_2-cyclopropane

the other, resulting in two C_1-symmetric transition structures $\boldsymbol{C_1}$**(ts)** leadinq to the **EE** trimethylene intermediate. This intermediate is viewed as having C_{2V} molecular symmetry[271] and some vibrational flexibility: changes in C—C—C bond angle, H—C(2)—H bond angle, and limited-amplitude in-phase and out-of-phase wagging and twisting distortions of the terminal methylene groups give it C_{2v} molecular symmetry and access to six exit channels (Figure 5)[270].

The schematic representation of Figure 5 illustrates the paths relating one **EE** intermediate with four cyclopropane structures. An alternative schematic, Figure 6, shows the paths relating one cyclopropane with four **EE** intermediates by way of six transition structures. In addition to the paths given explicitly in Figures 5 and 6, there are 4 direct paths by way of **EF** transition structures relating each cyclopropane with one-center epimerization products.

Calculations have also verified that the C_s-symmetric transition structure $\boldsymbol{C_s}$**(ts)** for disrotatory ring opening must play a role in cyclopropane stereomutations[268,270]. At four high-levels of theory, including DZP-CISD + Q calculations[272,273], there is no significant energetic advantage favoring $\boldsymbol{C_s}$**(ts)** over **EF(ts)**, or vice versa; they are each only some 0.9 kcal mol^{-1} above the $\boldsymbol{C_1}$**(ts)** structure[270].

$\boldsymbol{C_1}$**(ts)** $\boldsymbol{C_s}$**(ts)** **EF(ts)**

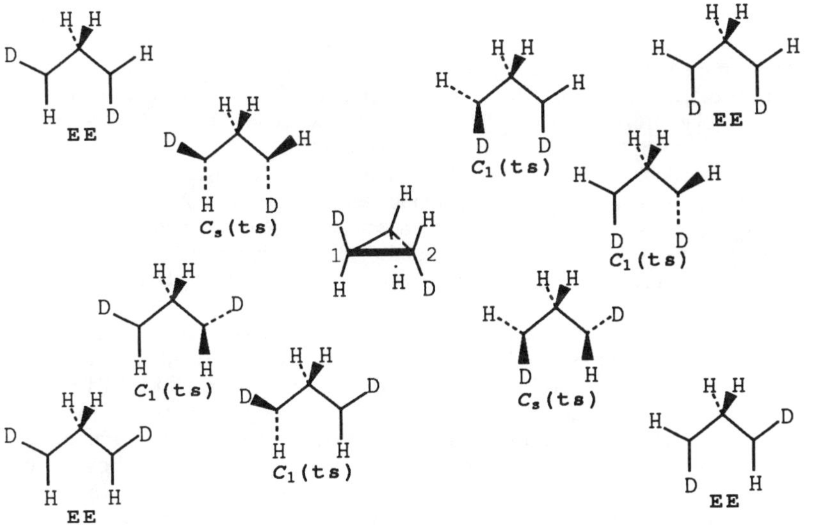

FIGURE 6. Schematic diagram for stereomutation paths interrelating (*R*, *R*)1,2-d$_2$-cyclopropanes with four **EE** intermediates. The four ***C***$_1$**(ts)** transition states lead to C_{2v}-symmetric d$_2$-**EE** structures, while the two ***C***$_s$**(ts)** structures separate the cyclopropane from C_s-symmetric d$_2$-**EE** isomers

This more detailed version of the energy surface near the **EE** trimethylene species implies that this intermediate may be formed from or may be converted to cyclopropane by way of a disrotatory process in addition to energetically slightly favored conrotatory modes. Reactions corresponding to net two-centered epimerizations ('optical isomerizations', k_{12}) will involve formation and further reaction of the trimethylene intermediate in conrotatory, conrotatory or disrotatory, disrotatory fashion. Whenever the intermediate is formed and then reacts further according to stereochemically different modes, in either sequence, it contributes to a net 'geometrical isomerization' (k_1 or k_2) process which occurs as well by way of **EF(ts)** structures. Recent highly correlated single reference based *ab initio* calculations lead to similar conclusions[274].

The conception of paths available for stereomutations is extended so that alternative paths from a starting material to a product are recognized to be kinetically competitive. The overall balance between net one-center and two-center epimerization events thus depends upon the energies of specific transition structures and the number of paths involved. For simplicity, one may talk of one conrotatory path between an **EE** trimethylene intermediate and a cyclopropane (while remembering that it has two branches!) and then enumerate the options: cyclopropane, or *trans*-1,2-d$_2$-cyclopropane, may react to give its mirror image form through two conrotatory,conrotatory paths and two disrotatory, disrotatory paths. It may form *cis*-1,2-d$_2$-cyclopropane through four paths mediated by **EF(ts)** and through eight paths from trimethylene intermediates.

At a temperature of 407 °C (or, very nearly, at 422.5 °C) the assumption that $\Delta\Delta E^{\ddagger}$ may be used as a fair approximation for $\Delta\Delta G^{\ddagger}$ and application of the standard steady-state approximation for the trimethylene intermediate leads to an overall estimation of one-center and two-center epimerizations. With the C_1**(ts)** favored over **EF(ts)** and C_s**(ts)** by 0.9 kcal mol^{-1}, this analysis gives k_1:k_{12} equal to 1.06[270]. According to this conventional model for partitionings among the several paths from trimethylene intermediates, and taking into account the alternative routes to net k_1 and k_{12} products, 56% of overall one-center epimerization takes place through the four **EF(ts)** structures while 44% results from stereochemically dissimilar paths to and from trimethylene intermediates[270].

These recent calculations for cyclopropanes and trimethylene transition structures, and for isotopically labeled analogs, have provided vibrational frequencies from which secondary deuterium isotope effects have been calculated. Getty, Davidson and Borden found that reactions dependent upon C_1**(ts)** at 422.5 °C should be associated with secondary deuterium kinetic isotope effects favoring access to 1,2-d_2-C_1**(ts)** over 1,3-d_2-C_1**(ts)** structures by a factor of 1.13[269].

Calculations based on the steady-state treatment of trimethylene intermediates, and one-center epimerization as well by way of **EF(ts)** structures, provided k_H:k_D values of 1.293 and 1.324 for k_1 and k_{12} for 1,2-d_2-cyclopropane, and of 1.328 and 1.368 for 1,2,3-d_3-cyclopropane: the balance given by the ratio k_1:k_{12}, according to these calculations, is not very sensitive to deuterium labeling, for the ratios are 1.06, 1.08 and 1.09 for unlabeled, 1,2-d_2- and 1,2,3-d_3-cyclopropanes. Thus the theoretically based k_1:k_{12} ratios for cyclopropanes and these deuterium-labeled cyclopropanes agree closely with the experimentally secured ratio k_1:k_{12} = 1.0 ± 0.2 found for the 1-^{13}C-1, 2, 3-d_3-cyclopropanes[166].

The vision of cyclopropane stereomutations stemming from this detailed consideration of the paths available to the parent hydrocarbon–trimethylene diradical is one featuring multiple paths to and from each conceptually distinct version of cyclopropane. That the empirical correlations of Figures 2 and 3 hold up so well implies that substituents play a critical role in determining how readily a particular C—C bond in a cyclopropane will break thermally, but modify only slightly the overall balance dictating net k_1 and k_{ij} products. Not large-scale rotations of substituents at the termini of a trimethylene diradical but relatively small vibrational displacements of carbons and hydrogens of the trimethylene unit dictate just how an intermediate will partition or which alternative **EF(ts)** may be kinetically advantageous.

All of the experimental and theoretical work on the stereomutations of cyclopropanes and vinylcyclopropanes covered above seems consistent with and understandable in terms of kinetically significant involvements of C_1**(ts)**, C_s**(ts)** and **EF(ts)** structures and partitionings of **EE** trimethylene intermediates resulting in the formation of k_1, k_2 and k_{12} products at comparable rates. For *trans*-1,2-disubstituted cyclopropanes, neither the Smith mechanism (one-center stereomutations only) nor any two-center-only formulation can be correct, as demonstrated by Crawford and Lynch in 1968[143] and reinforced by numerous subsequent studies (Figures 2 and 3).

VI. DISPARATE KINETIC RESULTS

The experimental and theoretical work published by the early 1970s viewed the stereomutations of cyclopropanes as kinetically competitive one-center and two-center stereomutations; some details, especially regarding relative rate constants for one-center epimerizations which defined relative rotational propensities, remained unclear, but all agreed that neither the Smith mechanism (one-center only) nor any two-center-only formulation for stereomutations could be sufficient. Thus when kinetic studies[275–278] on the isomerizations shown by chiral samples of l-phenyl-2-d-cyclopropane and 1,2-d_2-cyclopropane purported to show that, actually, two-center stereomutations were kinetically dominant, many were stimulated to fresh speculations and accommodations. Theoretical work at times hinted that the parent hydrocarbon might be an exceptional case and might

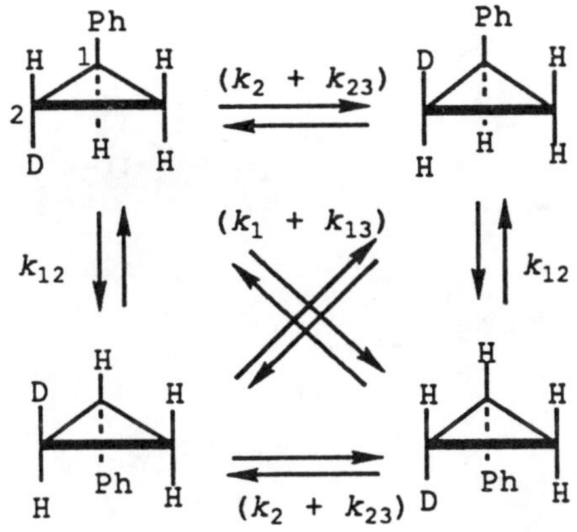

SCHEME 5

well suffer stereomutations with k_{12} much larger than k_1, or turned to aspects of the cyclopropane–trimethylene surface of no direct relevance to this kinetic issue. There were no theory-based suggestions as to why phenylcyclopropane might be exceptional.

The reported stereomutation behavior of these cyclopropanes called for more detailed experimental investigations as well as theoretical scrutiny. Just why they did not follow the general pattern summarized by $(k_1 + k_2)$:k_{12} ranging from 1.4:1 to 4:1 observed for all other cyclopropanes needed to be understood, and the experimental design and execution challenges that the kinetic situations entailed needed to be addressed and surmounted.

The kinetic situation appropriate to the 1975 work on 1-phenyl-2-d-cyclopropanes[275,277,278] involved four isomers, as presented in Scheme 5. There are potentially three kinetic observables and five unknowns. The kinetic work undertaken and the analytical methodology applied could not determine concentrations of all four isomers as functions of time; an estimate of $k_i = 2(k_1 + k_2 + k_{13} + k_{23})$ at 309.5 °C was obtained by infrared spectroscopy. Thermal reaction mixtures starting from a chiral 50:50 *cis*:*trans* mixture, each thought to be of the same optical purity, were monitored by polarimetry and interpreted through the relationship $k_\alpha = 2(k_1 + k_{12} + k_{13})$. When chiral *trans* starting material was employed, the expected non-first-order decay of optical activity was not observed; instead, a simple first-order racemization was seen corresponding to the same value of k_α. The data were analyzed by numerical integrations and the given relationships together with some assumptions, led to the conclusion that the major component of stereomutation at C(2) was a synchronous double rotation: the k_{12}:$(k_2 + k_{23})$ ratio was estimated to be 4:1 and k_1 was deduced to be negligible. Thus every epimerization at C(1) was apparently accompanied by a synchronous rotation at C(2) or C(3).

Further kinetic work with chiral samples of 1-phenyl-2,3-d_2-cyclopropane[279] led to $(k_2 + k_{12}) = 1.11 \times 10^{-5}\ s^{-1}$ at 309.5 °C, a value fully consistent with Baldwin's and Barden's work based on 1-phenyl-1,2,3-d_3-cyclopropanes[162–164]: $(k_2 + k_{12}) = 1.07 \times 10^{-5}\ s^{-1}$ at 309.3 °C.

What is one to make of the 1975 experiments[275,277,278] leading to the conclusions that k_{12}:$(k_2 + k_{23}) = 4$:1 and $k_1 : k_{12} = 0$? They and the 1984 work which reported experimental values for all four mechanistic rate constants and the ratios k_{12}:$(k_2 + k_{23}) = 0.23$ and k_1:$k_{12} = 1.8$ cannot both be correct. One response is to remain 'objective' and quite neutral, and to consider the experimental situation simply unresolved. Another is to try to make an independent assessment of the two sets of experiments by critically evaluating the experimental designs, the synthetic procedures and characterizations of labeled phenylcyclopropanes, the analytical methodologies employed and the number of experimental rate

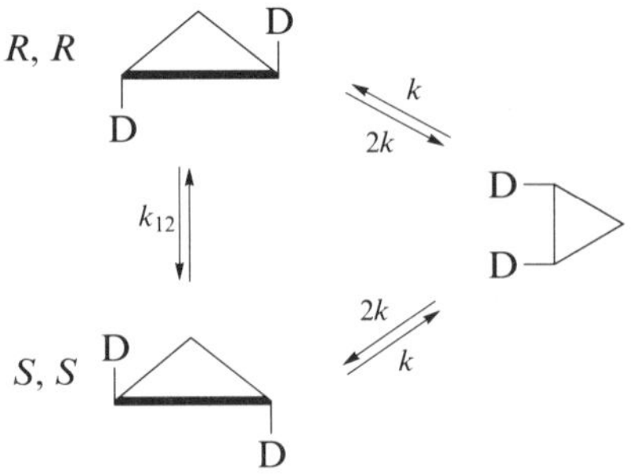

$$k = k_1 + k_1' + k_{13}$$

SCHEME 6

constants measured in each effort. Still another is to secure fresh experimental data relevant to the problem, if it must be considered unresolved, but over the past decade no one has committed to this undertaking.

At any rate, by 1984 the kinetic study of 1,2-d_2-cyclopropanes reported in 1975[276,278] became the only unchallenged claim of kinetically favored two-center epimerization and of a predominance of the 'theoretically predicted double rotation mechanism'.

The kinetic situations posed by the 1,2-d_2-cyclopropanes is summarized in Scheme 6. The experimentally accessible rate constants for overall loss of optical activity, k_α, and for approach to *cis,trans* equilibrium, k_i, are related to the mechanistic rate constants by the equalities $k_\alpha = (2k_{12} + 2k)$ and $k_i = 4k$, with $k = (k_1 + k_1' + k_{13})$. Here k_1 stands for the one-center epimerization rate constant when C(1)–C(2) breaks, and k_1' for one-center turnover at C(l) when C(l)—C(3) breaks. There are thus two observables and four mechanistic rate constants. If $k_1':k_1$ and $k_{13}:k_{12}$ ratios are assigned reasonable values based on assumed kinetic isotope effects, one is left with only k_1 and k_{12} as unknown mechanistic rate constants to be deduced from the observable k_i and k_α kinetic parameters.

The reported experimentally estimated phenomenological rate constants together with reported calculated standard deviations for gas-phase isomerizations conducted at 422.5 °C were $k_\alpha = (6.33 \pm 0.14) \times 10^{-5}\,s^{-1}$ (by polarimetry at 365 nm on neat samples at 3 °C using a pressure cell), $k_i = (6.75 \pm 0.12) \times 10^{-5}\,s^{-1}$ (by infrared spectroscopy monitoring the 846 cm^{-1} absorption shown by *cis*-1,2-d_2-cyclopropane) and $k_i = (7.39 \pm 1.23) \times 10^{-5}\,s^{-1}$ (by infrared spectroscopy based on the 1036 cm^{-1} absorption of the *cis* isomer[275]. These rate constants give $k_i:k_\alpha$ ratios of (1.07 ± 0.03) and (1.17 ± 0.20), or from 1.03 to 1.37, at the one-standard-deviation level of uncertainty. The $k_i:k_\alpha$ values, when combined with the assumptions that both $k_1':k_1$ and $k_{13}:k_{12}$ are equal to 1.10, correspond to $k_1:k_{12}$ ratios as low as 0 and as high as 0.52.

This experimental result is qualitatively in agreement with the kinetic outcome $k_1:k_{12} = 1.0 \pm 0.2$ obtained with $^{13}C,d_3$-labeled cyclopropanes at 407 °C. At the one-standard-deviation level of uncertainty, the k_1: k_{12} ratios from the two studies could be 0.52 and 0.80, corresponding to k_{12} being 49% or 39% of $(k_1 + k_2 + k_{12})$—not a dramatic or mechanistically dichotomous difference.

The 1975 work[275] measured k_i starting with samples of *trans*-1,2-d_2-cyclopropane estimated to be (62.6 ± 3.6)% d_2 and *cis:trans* 2.3: 98.7. Since both cyclopropane-d_1 and *trans*-1, 2-d_2-cyclopropane have absorption bands near the 846 cm^{-1} feature associated with the *cis* isomer, and since no attempt was made to analyze this complex spectral region using curve-fitting or factor-analysis methods, it seems likely that the k_i value used to build the case for k_{12} being much larger than k_1 may have been vulnerable to nontrivial systematic error. Computer generated spectra combining various proportions of authentic spectra for *cis*-1,2-d_2-cyclopropane (strong bands at 845 and 848 cm^{-1}), the *trans*-l, 2-d_2 isomer (852 cm^{-1} medium, and 857 cm^{-1} strong), and for cyclopropane-d_1 (strong band at 852 cm^{-1}) in this region make clear that the apparent concentrations of *cis*-d_2-cyclopropane determined by the method used[275] depend sensitively on the relative concentrations of *cis*-d_2 and d_1 components, and that the concentrations of the *cis* isomer were probably underestimated in the kinetic samples examined in 1975.[280]

The 1036 cm^{-1} absorption band together with the 1041 cm^{-1} absorption associated with the *trans* isomer appear to provide a particularly reliable measure of relative *cis* versus *trans* concentrations[281–182]. Only by disregarding the k_i value based on the 1036 cm^{-1} absorption data and by passing over probable systematic errors in the estimation of k_i from infrared data at 846 cm^{-1} was it possible to conclude that k_{12} is very much greater than k_1, and that the double rotation mechanism predominates by a substantial factor.

A study of the stereomutations of 1,2-d_2-cyclopropanes published in 1990 which claimed, hopefully but unjustifiably, to have measured a k_i:k_α ratio of 1.09 ± 0.05, is now recognized to be of no mechanistic relevance, for the two-parameter, three-data-point least-squares rate constant reported[282] for k_α is associated with intolerable error limits: the 95% confidence interval places k_α between 3.9 and $20.8 \times 10^{-5}\ s^{-1}$.[282]

VII. CONCLUSIONS

Enough substituted cyclopropanes have now been subjected to careful kinetic studies so that a characteristic pattern of reactivity and stereochemical preferences has emerged. Substituents facilitate stereomutations in proportion to their ability to stabilize 1,3-trimethylene diradical structures. The $\Delta G^\ddagger$ values for both k_{12} and $(k_1 + k_2)$ stereomutation rate constants relate linearly with consistent measures of substituent radical stabilization energies with equal sensitivities. Experimentally determined $(k_1 + k_2)$:k_{12} ratios do not vary widely: they range from 1.4 to 2.5 over a fair diversity of substituents. Neither do k_i:k_j ratios vary widely. The majority fall between 1:1 and 2.5:1; the largest yet reported gives k_2(CHD) a symmetry corrected kinetic advantage over k_1(CDPh) in 1-phenyl-1,2,3-d_3-cyclopropanes of 5:1.

The $\Delta G^\ddagger$ values for vinylcyclopropane to cyclopentene isomerizations show the same sensitivity to radical stabilizing substituents, implying that they too involve diradical transition structures.

The latest theoretical treatment of cyclopropane stereomutations predicts[270] that k_1 and k_{12} are about equal for cyclopropane and for 1,2,3-d_3-cyclopropanes, in fair agreement with an experimental value k_1:k_{12} = 1.0 ± 0.2 determined with ^{13}C,d_3-labeled cyclopropanes[166].

While there remain open questions on a few experimental aspects of early work on the stereomutations of 1-phenyl-2-d-cyclopropanes and 1,2-d_2-cyclopropanes, the preponderance of data and theory now provides a consistent understanding of the thermal stereomutations of cyclopropanes: multiple paths and three types of diradical transition structures are involved. Evolving theory relevant to cyclopropane stereomutations and to vinylcyclopropane to cyclopentene isomerizations, and to other 1,3-carbon shifts[283–288], may well provide more detailed insights, rationales and predictions.

The interplay of experiment and theory is rather more complicated here than in, say, the now-resolved controversy over the ground state structure of triplet methylene[289–293]. The two approaches to the problems posed by stereomutation chemistry have been fully complementary. Neither one independently was able to provide a full picture; together they have made sense out of observational data and calculational inferences. The common effort has served to advance both enterprises, and each may be expected to contribute significantly in its own way toward clarifications of other troublesome problems in unimolecular rearrangement chemistry.

VIII. ACKNOWLEDGMENTS

I thank the National Science Foundation for support of our work on cyclopropane stereomutations, now through CHE-9100246, and the coworkers and colleagues who have contributed so invaluably to our studies in this area.

IX. REFERENCES

1. E. Vogel, *Angew. Chem.*, **74**, 829 (1962).
2. H. M. Frey, *Adv. Phys. Org. Chem.*, **4**, 148 (1966).
3. H. M. Frey and R. Walsh, *Chem. Rev.*, **69**, 103 (1969).
4. M. R. Willcott, R. L. Cargill and A. B. Sears, *Prog. Phys. Org. Chem.*, **9**, 25 (1972).
5. R. G. Bergman, in *Free Radicals* (Ed. J. Kochi), Vol. 1, Chap. 5, Wiley, New York, 1973.
6. H.-D. Martin, *Nachr. Chem. Techn.*, **22**, 412 (1974).
7. J. A. Berson, *Annu. Rev. Phys. Chem.*, **28**, 111 (1977).
8. J. A. Berson, in *Rearrangements in Ground and Excited States* (Ed. P. de Mayo), Academic Press, New York, 1980, pp. 311–390.
9. W. T. Borden, in *Reactive Intermediates*, Vol. II (Eds. M. Jones and R. A. Moss), Chap. 5, Wiley, New York, 1981.
10. J. J. Gajewski, *Israel J. Chem.*, **21**, 169 (1981).
11. J. J. Gajewski, *Hydrocarbon Thermal Isomerizations*, Academic Press, New York, 1981, pp. 27–42.
12. P. B. Dervan and D. A. Dougherty, in *Diradicals* (Ed. W. T. Borden), Chap. 3, Wiley, New York, 1982.
13. B. K. Carpenter, *Determination of Organic Reaction Mechanisms*, Wiley, New York, 1984, pp. 62–67.
14. B. K. Carpenter, in *The Chemistry of the Cyclopropyl Group, Part 2* (Ed. Z. Rappoport), Wiley, Chichester; 1987, pp. 1027–1082.
15. K. Hiroi, *Annu. Rep. Tohoku Coll. Pharm.*, 1–60 (1988); *Chem. Abstr.*, **112**, 20695 q (1990).
16. G. Boche and H. M. Walborsky, *Cyclopropane Derived Reactive Intermediates*, Wiley, Chichester, 1990.
17. W. von E. Doering and W. R. Roth, *Angew. Chem., Int. Ed. Engl.*, **2**, 115 (1963).
18. C. D. Gutsche and D. Redmore, *Carbocyclic Ring Expansion Reactions*, Academic Press, New York, 1968, pp. 163–170.
19. S. Sarel, J. Yovell and M. Sarel-Imber, *Angew. Chem., Int. Ed. Engl.*, **7**, 577 (1968).
20. E. M. Mil'vitskaya, A. V. Tarakanova and A. F. Plate, *Russ. Chem. Rev.*, **45**, 469 (1976).
21. J. J. Gajewski, *Hydrocarbon Thermal Isomerizations*, Academic Press, New York, 1981, pp. 81–87.
22. T. Hudlicky, T. M. Kutchan and S. M. Naqvi, *Org. React.*, **33**, 247 (1985).
23. J. Salaün, in *The Chemistry of the Cyclopropyl Group, Part 2* (Ed. Z. Rappoport), Chap. 13, Wiley, Chichester, 1987, pp. 849–857.
24. B. K. Carpenter, in *The Chemistry of the Cyclopropyl Group, Part 2* (Ed. Z. Rappoport), Chap. 17, Wiley, Chichester, 1987, pp. 1045–1054.
25. Z. Goldschmidt and B. Crammer, *Chem. Soc. Rev.*, **17**, 229 (1988).
26. H. N. C. Wong, M.-Y. Hon, C.-W. Tse, Y.-C. Yip, J. Tanko and T. Hudlicky, *Chem. Rev.*, **89**, 165 (1989).
27. J. Salaün, *Chem. Rev.*, **89**, 1247 (1989).
28. T. Hudlicky, F. Rulin, T. C. Lovelace and J. W. Reed, *Stud. Nat. Prod. Chem.*, **3** (*Stereosel. Synth., Pt B*), 3–72 (1989).
29. T. Hudlicky and J. W. Reed, in *Comprehensive Organic Synthesis* (Ed. L. A. Paquette), Vol. 5, Chap. 8.1, Pergamon Press, Oxford, 1991.
30. J. A. Berson and J. M. Balquist, *J. Am. Chem. Soc.*, **90**, 7343 (1968).

31. G. Mass, *Chem. Ber*, **112**, 3241 (1979).
32. D. R. Arnold, D. D. M. Wayner and M. Yoshida, *Can. J. Chem.*, **60**, 2313 (1982).
33. W. von E. Doering, L. R. Robertson and E. E. Ewing, *J. Org. Chem.*, **48**, 4280 (1983).
34. R. J. Crawford and A. Mishra, *J. Am. Chem. Soc.*, **88**, 3963 (1966).
35. R. J. Crawford and D. M. Cameron, *Can. J. Chem.* **45**, 691 (1967).
36. R. Criegee and A. Rimmelin, *Chem. Ber.*, **90**, 414 (1957).
37. J. P. Chesick, *J. Am. Chem. Soc.*, **84**, 3250 (1962).
38. S. L. Buchwalter and G. L. Closs, *J. Am. Chem. Soc.*, **97**, 3857 (1975); **101**, 4688 (1979).
39. J. E. Baldwin and J. Ollerenshaw, *J. Org. Chem.*, **46**, 2116 (1981).
40. F. D. Coms and D. A. Dougherty, *Tetrahedron Lett.*, **29**, 6039 (1988); *J. Am. Chem. Soc.*, **111**, 6894 (1989).
41. M. S. Herman and J. L. Goodman, *J. Am. Chem. Soc.*, **110**, 2681 (1988).
42. C. D. Sherrill, E. T. Seidl and H. F. Schaefer, *J. Phys. Chem.*, **96**, 3712 (1992).
43. U. H. Brinker and W. Ergle, *Angew. Chem., Int. Ed. Engl.*, **26**, 1260 (1987).
44. G. D. Andrews and J. E. Baldwin, *J. Org. Chem.*, **53**, 4624 (1988).
45. S. W. Ham, W. Chang and P. Dowd, *J. Am. Chem. Soc.*, **111**, 4130 (1989).
46. W. R. Roth, F. Bauer and R. Breuckmann, *Chem. Ber.*, **124**, 2041 (1991).
47. R. J. Ellis and H. M. Frey, *J. Chem. Soc., A*, 553 (1966).
48. W. von E. Doering and J. B. Lambert, *Tetrahedron*, **19**, 1989 (1963).
49. E. J. Corey and H. Uda, *J. Am. Chem. Soc.*, **85**, 1788 (1963).
50. R. A. Clark, *Tetrahedron Lett.*, 2279 (1971).
51. D. L. Garin and D. J. Cooke, *J. Chem. Soc., Chem. Commun.*, 33 (1972).
52. R. S. Cooke and U. H. Andrews, *J. Org. Chem.*, **38**, 2725 (1973).
53. T. L. Rose, R. J. Seyse and P. M. Crane, *Int. J. Chem. Kinet.*, **6**, 899 (1974).
54. R. S. Cooke and U. H. Andrews, *J. Am. Chem. Soc.*, **96**, 2974 (1974).
55. J. E. Baldwin and K. E. Gilbert, *J. Am. Chem. Soc.*, **98**, 8283 (1976).
56. D. L. Garin, *Tetrahedron Lett.*, 3035 (1977).
57. W. D. Huntsman, J. P. Chen, K. Yelekci, T.-K. Yin and L. J. Zhang, *J. Org. Chem.*, **53**, 4357 (1988).
58. N. Friedmann, H. H. Bovce and S. L. Miller, *J. Org. Chem.*, **35**, 3230 (1970).
59. G. D. Andrews and J. E. Baldwin, *J. Am. Chem. Soc.*, **98**, 6705 (1976).
60. J. J. Gajewski and J. M. Warner, *J. Am. Chem. Soc.*, **106**, 802 (1984).
61. J. J. Gajewski and M. P. Squicciarini, *J. Am. Chem. Soc.*, **111**, 6717 (1989).
62. J. E. Baldwin and N. D. Ghatlia, *J. Am. Chem. Soc.*, **113**, 6273 (1991).
63. J. J. Gajewski and L. P. Olson, *J. Am. Chem. Soc.*, **113**, 7432 (1991).
64. J. E. Baldwin and S. Bonacorsi, *J. Am. Chem. Soc.*, **115**, 10621 (1993).
65. A. Freund, *Monatsh. Chem.*, **3**, 625 (1882); *J. Prakt. Chem.*, **26**, 367 (1882).
66. I. Z. Siemion, *Wiad. Chem.*, **37**, 509 (1983); *Chem. Abstr.*, **101**, 89783g (1984).
67. S. Tanatar, *Ber. dtsch. Chem. Ges.*, **29**, 1297 (1896); **32**, 702 (1899).
68. K. B. Wiberg, *Angew. Chem., Int. Ed. Engl.*, **25**, 312 (1986).
69. P. E. M. Berthelot, *Thermochimie: Données et Lois Numériques*, Vol. II, Paris, 1897, pp. 406–407.
70. M. J. S. Dewar, *J. Am. Chem Soc.*, **106**, 669 (1984).
71. J. F. Liebman and A. Skanke, *Struct. Chem.*, **2**, 201 (1991).
72. M. Berthelot, *Compt. rend.*, **129**, 483 (1899).
73. M. Trautz and K. Winkler, *J. Prakt. Chem.* **104**, 53 (1922).
74. S. Z. Roginskii and F. H. Rathmann, *J. Am. Chem. Soc.*, **55**, 2800 (1933).
75. T. S. Chambers and G. B. Kistiakowsky, *J. Am. Chem. Soc.*, **56**, 399 (1934).
76. B. S. Rabinovitch, E. W. Schlag and K. B. Wiberg, *J. Chem. Phys.*, **28**, 504 (1958).
77. E. W. Schlag and B. S. Rabinovitch, *J. Am. Chem. Soc.*, **82**, 5996 (1960).
78. G. Gustavson, *J. Prakt. Chem.*, **54**, 97 (1896).
79. N. J. Demjanow and M. Dojarenko, *Ber. dtsch. Chem. Ges.*, **55B**, 2718 (1922).
80. C. K. Ingold, *J. Chem. Soc.*, **123**, 1706 (1923).
81. F. C. Whitmore, *Organic Chemistry*, D. Van Nostrand, New York, 1937, p. 632.
82. M. J. Murray and E. H. Stevenson, *J. Am. Chem. Soc.*, **66**, 812 (1944).
83. R. van Volkenburgh, K. W. Greenlee, J. M. Derfer and C. E. Boord, *J. Am. Chem. Soc.*, **71**, 3595 (1949).
84. G. Dallinga, R. K. van der Draai and L. H. Toneman, *Recl. Trav. Chim. Pays-Bas*, **87**, 897 (1968).

85. W. H. Weber, D. H. Leslie, C. W. Peters and R. W. Terhune, *J. Mol. Spectrosc.*, **81**, 316 (1980).
86. Y. Endo, M. C. Chang and E. Hirota, *J. Mol. Spectrosc.*, **126**, 63 (1987).
87. T. Brupbacher, C. Styger, B. Vogelsanger, I. Ozier and A. Bauder, *J. Mol. Spectrosc.*, **138**, 197 (1989).
88. D. Nijveldt and A. Vos, *Acta Crystallogr. Sect. B: Struct. Sci.*, **B44**, 281, 289, 296 (1988).
89. D. Blaeser, R. Boese, U. H. Brinker and K. Gomann, *J. Am.Chem. Soc.*, **111**, 1501 (1989).
90. E. Vogel, R. Palm and K. H. Ott, Unpublished results; see E. Vogel, *Angew. Chem.*, **72**, 4 (1960), note 162.
91. C. G. Overberger and A. E. Borchert, *J. Am. Chem. Soc.*, **82**,1007 (1960).
92. R. J. Ellis and H. M. Frey, *J. Chem. Soc.*, 5578 (1964).
93. M. R. Willcott and V. H. Cargle, *J. Am. Chem. Soc.*, **89**, 723 (1967); **91**, 4310 (1969).
94. M. R. Willcott and V. H. Cargle, *J. Am. Chem. Soc.*, **91**, 4310 (1969).
95. V. H. Cargle, Ph.D. Dissertation, University of Houston, 1969.
96. M. R. Willcott and V. Cargle, unpublished results; see Reference 4.
97. N. B. Slater, *Proc. R. Soc. London, Ser. A*, **218**, 224 (1953).
98. N. B. Slater, *Theory of Unimolecular Reactlons*, Cornell University Press, Ithaca, 1959.
99. N. B. Slater, *J. Chem. Soc.*, 606 (1961).
100. B. S. Rabinovitch, D. W. Setser and F. W. Schneider, *Can. J. Chem.*, **39**, 2609 (1961).
101. G. M. Wieder and R. A. Marcus, *J. Chem. Phys.*, **37**, 1835 (1962).
102. L. S. Mayants, *Bull. Acad. Sci. USSR, Div. Chem. Sci.*, 937 (1969).
103. B. Dill and H. Heydtmann, *Int. J. Chem. Kinet.*, **9**, 321 (1977).
104. A. W. Yau and H. O. Prichard, *Can. J. Chem.*, **56**, 1389 (1978).
105. C. B. Moore, *J. Chem. Phys.*, **43**, 2979 (1965).
106. R. J. Malins and D. C. Tardy, *Chem. Phys. Lett.*, **57**, 289 (1978).
107. D. C. Tardy and R. J. Malins, *J. Phys. Chem.*, **83**, 93 (1979).
108. B. De Barros Neto and R. E . Bruns, *J. Chem. Phys.*, **71**, 5042 (1979).
109. B. D. Barton, D. F. Kelley and B. S. Rabinovitch, *J. Phys.Chem.*, **84**, 1299 (1980).
110. W. Forst and A. P . Penner, *J. Chem. Phys.*, **72**, 1435 (1980).
111. R. J. McCluskey and R. W. Carr, *J. Phys. Chem.*, **82**, 2637 (1978).
112. E. Kamaratos, J. F. Burkhalter, D. G. Kiel and B. S. Rabinovitch, *J. Phys. Chem.*, **83**, 984 (1979).
113. D. J. Nesbitt and S. R. Leone, *Chem. Phys. Lett.*, **87**, 123 (1982).
114. J. D. Chen, J. Häger and W. Krieger, *Chem. Phys. Lett.*, **90**, 366 (1982).
115. W. E. Falconer, T. H. Hunter and A. F. Trotman-Dickenson, *J. Chem. Soc.*, 609 (1961).
116. G. L. Pratt, *J. Chem. Soc.*, 6050 (1963).
117. B. R. Davis and D. S. Scott, *Ind. Eng. Chem., Fundamentals*, **3**, 20 (1964).
118. H. Miyama and T. Takeyama, *Bull. Chem. Soc. Jpn.*, **38**, 2189 (1965).
119. J. N. Bradley and M. A. Frend, *Trans. Faraday Soc.*, **67**, 72 (1971).
120. E. A. Dorko, D. B. McGhee, C. E. Painter, A. J. Caponecchi and R. W. Crossley, *J. Phys. Chem.*, **75**, 2526 (1971).
121. D. W. Johnson, O. A. Pipkin and C. M. Sliepcevich, *Ind. Eng. Chem., Fundamentals*, **11**, 244 (1972).
122. E. A. Dorko, R. W. Crossley, U. W. Grimm, G. W. Mueller and K. Scheller, *J. Phys. Chem.*, **77**, 143 (1973).
123. D. M. Ruthven, *Ind. Eng. Chem., Fundamentals*, **12**, 262 (1973).
124. D. W. Johnson and C. M. Sliepcevich, *Ind. Eng. Chem., Fundamentals*, **12**, 262 (1973).
125. P. Jeffers, D. Lewis and M. Sarr, *J. Phys. Chem.*, **77**, 3037 (1973).
126. J. A. Barnard, A. T. Cocks and R. K. Y. Lee, *J. Chem. Soc., Faraday Trans. 1*, **70**, 1782 (1974).
127. G. B. Skinner, *Int. J. Chem. Kinet.*, **9**, 863 (1977).
128. D. K. Lewis, S. E. Geisler and M. S. Brown, *Int. J. Chem.Kinet.*, **10**, 277 (1978).
129. P. M. Jeffers and J. Northing, *Int. J. Chem. Kinet.*, **11**, 915 (1979).
130. R. J. Malins and D. C. Tardy, *Int. J. Chem. Kinet.*, **11**, 1007 (1979).
131. B. V. O'Grady, *J. Chem. Educ.*, **62**, 709 (1985).
132. D. C. Tardy and B. S. Rabinovitch, *J. Phys. Chem.*, **89**, 2442 (1985).
133. C. R. Moylan and J. I. Brauman, *Int. J. Chem. Kinet.*, **18**, 379 (1986) and references cited there in.
134. Y. Hidaka and T. Oki, *Chem. Phys. Lett.*, **141**, 212 (1987).
135. U. Holm and K. Kerl, *Ber. Bunsenges. Phys. Chem.*, **94**, 1414 (1990).
136. M. C. Lin and K. J. Laidler, *Trans. Faraday Soc.*, **64**, 94, 927 (1968).

137. E. V. Waage and B. S. Rabinovitch, *J. Phys. Chem.*, **76**, 1695 (1972).
138. M. Menzinger and R. Wolfgang, *Angew. Chem., Int. Ed. Engl.*, **8**, 438 (1969).
139. M. C. Flowers and H. M. Frey, *Proc. R. Soc. London, Ser. A*, **257**, 122 (1960).
140. C. S. Elliott and H. M. Frey, *J. Chem. Soc.*, 900 (1964).
141. D. W. Setser and B. S. Rabinovitch, *J. Am. Chem. Soc.*, **86**, 564 (1964).
142. L. B. Rodewald and C. H. DePuy, *Tetrahedron Lett.*, 2951 (1964).
143. R. J. Crawford and T. R. Lynch, *Can. J. Chem.*, **46**, 1457 (1968).
144. L. A. Asuncion, Syracuse University, unpublished.
145. M. Arai and R. J. Crawford, *Can. J. Chem.*, **50**, 2158 (1972).
146. W. L. Carter and R. G. Bergman, *J. Am. Chem. Soc.*, **90**, 7344 (1968).
147. R. G. Bergman, *J. Am. Chem. Soc.*, **91**, 7405 (1969).
148. R. G. Bergman and W. L. Carter, *J. Am. Chem. Soc.*, **91**, 7411 (1969).
149. J. E. Baldwin and C. B. Selden, *J. Am. Chem. Soc.*, **115**, 2239 (1993).
150. W. von E. Doering and K. Sachdev, *J. Am. Chem. Soc.*, **96**, 1168 (1974).
151. W. von E. Doering and K. Sachdev, *J. Am. Chem. Soc.*, **97**, 5512 (1975).
152. E. A. Barsa, Ph.D. Dissertation, Harvard University, 1977.
153. W. von E. Doering and E. A. Barsa, *Tetrahedron Lett.*, 2495 (1978).
154. W. von E. Doering and E. A. Barsa, *Proc. Natl. Acad. Sci. U.S.A.*, **77**, 2355 (1980).
155. J. E. Baldwin and C. G. Carter, *J. Am. Chem. Soc.*, **100**, 3942 (1978).
156. J. E. Baldwin and C. G. Carter, *J. Am. Chem. Soc.*, **101**, 1325 (1979).
157. J. E. Baldwin and C. G. Carter, *J. Org. Chem.*, **48**, 3912 (1983).
158. J. E. Baldwin and C. G. Carter, *J. Am. Chem. Soc.*, **104**, 1362 (1982).
159. On the applicability of the 'most-substituted-bond hypothesis' to the photochemistry of cyclopropanes, see B. Scholl and H.-J. Hansen, *Helv. Chim. Acta*, **69**, 1936 (1986).
160. W. von E. Doering and Y. Yamashita, *J. Am. Chem. Soc.*, **105**, 5368 (1983).
161. J. E. Baldwin and K. A. Black, *J. Am. Chem. Soc.*, **106**, 1029 (1984).
162. J. E. Baldwin, T. W. Patapoff and T. C. Barden, *J. Am. Chem. Soc.*, **106**, 1421 (1984).
163. J. E. Baldwin and T. C. Barden, *J. Am. Chem. Soc.*, **106**, 5312 (1984).
164. J. E. Baldwin and T. C. Barden, *J. Am. Chem. Soc.*, **106**, 6364 (1984).
165. S. J. Cianciosi, Ph.D. Dissertation, Syracuse University, 1990.
166. S. J. Cianciosi, N. Ragunathan, T. B. Freedman, L. A. Nafie, D. K. Lewis, D. A. Glenar and J. E. Baldwin, *J. Am. Chem. Soc.*, **113**, 1864 (1991).
167. T. B. Freedman, S. J. Cianciosi, N. Ragunathan, J. E. Baldwin and L. A. Nafie, *J. Am. Chem. Soc.*, **113**, 8298 (1991).
168. J. E. Baldwin and S. J. Cianciosi, *J. Am. Chem. Soc.*, **114**, 9401 (1992).
169. J. E. Baldwin, S. J. Cianciosi, D. A. Glenar, G. J. Hoffman, I-W. Wu and D. K. Lewis, *J. Am. Chem. Soc.*, **114**, 9408 (1992).
170. D. A. Glenar and D. K. Lewis, *Appl. Spectrosc.*, **43**, 283 (1989).
171. M. A. Lowe, G. A. Segal and P. J. Stephens, *J. Am. Chem. Soc.*, **108**, 248 (1986).
172. J. S. Chickos, A. Annamalai and T. A. Keiderling, *J. Am. Chem. Soc.*, **108**, 4398 (1986).
173. S. C. Yasui and T. A. Keiderling, *J. Am. Chem. Soc.*, **109**, 2311 (1987).
174. S. J. Cianciosi, K. M. Spencer, T. B. Freedman, L. A. Nafie and J. E. Baldwin, *J. Am. Chem. Soc.*, **111**, 1913 (1989).
175. K. M. Spencer, S. J. Cianciosi, J. E. Baldwin, T. B. Freedman and L. A. Nafie, *Appl. Spectrosc.*, **44**, 235 (1990).
176. T. B. Freedman, K. M. Spencer, N. Ragunathan, L. A. Nafie, J. A. Moore and J. M. Schwab, *Can. J. Chem.*, **69**, 1619 (1991).
177. J. E. Baldwin, *J. Chem. Soc., Chem. Commun.*, 31 (1988).
178. I. G. Bolesov, U. I. Sein, A. S. Koz'min and R. Ya. Levina, *J. Org. Chem. USSR*, **5**, 1655 (1969).
179. T. Schmidt, Ph.D. Dissertation, Ruhr-Universität Bochum, 1972; cited in Reference 152.
180. R. H. Newman–Evans, R. J. Simon and B. K. Carpenter, *J. Org. Chem.*, **55**, 695 (1990).
181. S. Bonacorsi, Syracuse University, unpublished results.
182. G. Leroy, *Adv. Quantum Chem.*, **17**, 1–95 (1985).
183. M. Sana and G. Leroy, *Ann. Soc. Sci. Bruxelles*, **101**, 23 (1987).
184. G. Leroy, M. Sana, C. Wilante and R. M. Nemba, *J. Mol. Struct.*, **198**, 159 (1989).
185. W. von E. Doering and G. H. Beasley, *Tetrahedron*, **29**, 2231 (1973).
186. W. Tsang, *Int. J. Chem. Kinet.*, **10**, 821 (1978).
187. A. M. de P. Nicholas and D. R. Arnold, *Can. J. Chem.*, **62**, 1850 (1984).

188. D. A. Robaugh and S. E. Stein, *J. Am. Chem. Soc.*, **108**, 3224 (1986).
189. J. W. Timberlake, in *Substituent Effects in Radical Chemistry* (Eds. H. G. Viehe, Z. Janousek and R. Merényi), D. Reidel, Dordrecht, 1986, pp. 271–281.
190. W. von E. Doerlng, W. R. Roth, R. Breuckmann, L. Figge, H.-W. Lennartz, W.-D . Fessner and H. Prinzbach, *Chem. Ber.*, **121**, 1 (1988).
191. J. A. Seetula, J. J. Russell, and D. Gutman, *J. Am. Chem. Soc.*, **112**, 1347 (1990).
192. D. Gutman, *Acc. Chem. Res.*, **23**, 375 (1990).
193. H. Hippler and J. Troe, *J. Phys. Chem.*, **94**, 3803 (1990).
194. M. Lehd and F. Jensen, *J. Org. Chem.*, **56**, 884 (1991).
195. W. R. Roth, F. Bauer, A. Beitat, T. Ebbrecht and M. Wüstefeld, *Chem. Ber.*, **124**, 1453 (1991).
196. W. von E. Doering, W. R. Roth, F. Bauer, M. Boenke, R. Breuckmann, J. Ruhkamp and O. Wortmann, *Chem. Ber.*, **124**, 1461 (1991).
197. P. W. Seakins, M. J. Pilling, J. T. Niiranen, D. Gutman and L. N. Krasnoperov, *J. Phys. Chem.*, **96**, 9847 (1992).
198. M. B. Coolidge, D. A. Hrovat and W. T. Borden, *J. Am. Chem. Soc.*, **114**, 2354 (1992).
199. W. von E. Doering, K. D. Belfield and J. He, *J. Am. Chem. Soc.*, **115**, 5414 (1993).
200. C. Ullenius, P. W. Ford and J . E . Baldwin, *J. Am . Chem. Soc.*, **94**, 5910 (1972).
201. J. E . Baldwin and C. Ullenius, *J. Am. Chem. Soc.*, **96**, 1542 (1974).
202. J. M. Simpson and H. G. Richey, *Tetrahedron Lett.*, 2545 (1973).
203. W. von E. Doering, G. Horowitz and K. Sachdev, *Tetrahedron*, **33**, 273 (1977).
204. J. J. Gajewski, R. J. Weber, R. Baum, M. L. Manion and B. Hymen, *J. Am. Chem. Soc.*, **99**, 816 (1977).
205. E. N. Marvell and C. Lin, *J. Am. Chem. Soc.*, **100**, 877 (1978).
206. K. Hafner and G. F. Thiele, *J. Am. Chem. Soc.*, **107**, 5526 (1985).
207. M. C. Flowers and H. M. Frey, *J. Chem. Soc.*, 3547 (1961).
208. C. A. Wellington, *J. Phys. Chem.*, **66**, 1671 (1962).
209. H. M. Frey and D. C. Marshall, *J. Chem. Soc.*, 3981 (1962).
210. G. R. Branton and H. M. Frey, *J. Chem. Soc.*, 1342 (1966).
211. W. R. Roth and J. König, *Justus Liebigs Ann. Chem.*, **688**, 28 (1965).
212. W. von E. Doering and W. R. Roth, *Tetrahedron*, **19**, 715 (1963).
213. R. H. Lindquist and G. K. Rollefson, *J. Chem. Phys.*, **24**, 725 (1956).
214. J. R. McNesby and A. S. Gordon, *J. Chem. Phys.*, **25**, 582 (1956).
215. M. C. Flowers and H. M. Frey, *J. Chem. Soc.*, 2758 (1960).
216. S. W. Benson, *J. Chem. Phys.*, **34**, 521 (1961).
217. S. W. Benson and P. S. Nangia, *J. Chem. Phys.*, **38**, 18 (1963).
218. H. E. O'Neal and S. W. Benson, *J. Phys. Chem.*, **72**, 1866 (1968).
219. F. T. Smith, *J. Chem. Phys.*, **29**, 235 (1958).
220. W. von E. Doering, *Peter A. Leemakers Symposium Lecture*,Wesleyan University, 1981; ACS. Science Symposium Series.
221. W. von E. Doering, *Proc. Natl. Acad. Sci. U.S.A.*, **78**, 5279 (1981).
222. R. Hoffmann, *Tetrahedron Lett.*, 3819 (1965).
223. R. Hoffmann, *Abstracts, 151st ACS National Meeting*, Pittsburgh, PA, March 1966, Paper 109 K.
224. R. Hoffmann, *Trans. N. Y. Acad. Sci.*, 475 (1966).
225. R. Hoffmann, *J. Am. Chem. Soc.*, **90**, 1475 (1968).
226. R. Hoffmann, *Acc. Chem. Res.*, **4**, 1 (1971).
227. A. Gavezzotti and M. Simonetta, *Tetrahedron Lett.*, 4155 (1975).
228. K. Jug, *Theor. Chim. Acta*, **42**, 303 (1976).
229. W. W. Schoeller, *Z. Naturforsch.*, **34a**, 858 (1979).
230. W. W. Schoeller, *J. Chem. Soc., Perkin Trans. 2*, 366 (1979).
231. S. Durmaz and H. Kollmar, *J. Am. Chem. Soc.*, **102**, 6942 (1980).
232. R. J. Buenker and S. D. Peyerimhoff, *J. Phys. Chem.*, **73**, 1299 (1969).
233. A. K. Q. Siu, W. M. St. John and E. F. Hayes, *J. Am. Chem. Soc.*, **92**, 7249 (1970).
234. J. A. Pople, *Int. J. Quantum Chem., Symp.*, **5**, 175 (1971).
235. L. Salem, J. Durup, G. Bergeron, D. Cazes, X. Chapuisat and H. Kagan, *J. Am. Chem. Soc.*, **92**, 4472 (1970).
236. L. Salem, *Acc. Chem. Res.*, **4**, 322 (1971).
237. L . Salem, *Chem. Commun.*, 981 (1970).

238. L. Salem, in *Reaction Transition States* (Ed J.-E. Dubois), 21st Annual Meeting (Sept. 1970) of the Société de Chimie Physique; Gordon and Breach, New York, 1972.
239. L. Salem, *Bull. Soc. Chim. Fr.*, 3161 (1970).
240. Y. Jean and L. Salem, *Chem. Commun.*, 382 (1971).
241. Y. Jean, L. Salem, J. S. Wright, J. A. Horsley, C. Moser and R. M. Stevens, *Pure Appl. Chem., Suppl. (23rd Congr., Boston)*, **1**, 197 (1971).
242. J. A. Horsley, Y. Jean, C. Moser, L. Salem, R. M. Stevens and J. S. Wright, *J. Am. Chem. Soc.*, **94**, 279 (1972).
243. P. J. Hay, W. J. Hunt and W. A. Goddard, *J. Am. Chem. Soc.*, **94**, 638 (1972).
244. L. Salem and C. Rowland, *Angew. Chem., Int. Ed. Engl.*, **11**, 92 (1972).
245. L. Salem, *Pure Appl. Chem.*, **33**, 317 (1973).
246. V. Bonacic–Koutecky, J. Koutecky and J. Michl, *Angew. Chem., Int. Ed. Engl.*, **26**, 170 (1987).
247. I. S. Y. Wang and M. Karplus, *J. Am. Chem. Soc.*, **95**, 8160 (1973).
248. Y. Jean, X. Chapuisat and L. Salem, *J. Am. Chem. Soc.*, **97**, 3646 (1975).
249. X. Chapuisat and Y. Jean, *J. Am. Chem. Soc.*, **97**, 6325 (1975).
250. X. Chapuisat, Y. Jean and L. Salem, *Chem. Phys. Lett.*, **37**, 119 (1976).
251. X. Chapuisat and Y. Jean, *Top. Curr. Chem.*, **68**, 1 (1976).
252. L. Salem, in *The New World of Quantum Chemistry* (Eds. B. Pullman and R. Parr), D. Reidel, Dordrecht, 1976, pp. 241–269.
253. X. Chapuisat, *Bull. Soc. Chim. Belges*, **85**, 937 (1976).
254. X. Chapuisat, *Ber. Bunsenges. Phys. Chem.*, **81**, 203 (1977).
255. X. Chapuisat, *Chem. Phys. Lett.*, **63**, 389 (1979).
256. X. Chapuisat and Y. Jean, *Quantum Theory Chem. React.*, **1**, 25 (1980).
257. Y. Jean, *Quantum Theory Chem. React.*, **1**, 53 (1980).
258. L. Salem, *Electrons in Chemical Reactions: First Principles*, Wiley, New York, 1982.
259. K. Yamaguchi, A. Nishio, S. Yabushita and T. Fueno, *Chem. Phys. lett.*, **53**, 109 (1978).
260. S . Kato and K. Morokuma, ,*Chem. Phys. Lett.*, **65**, 19 (1979).
261. C. Doubleday, J. W. McIver and M. Page, *J. Am. Chem. Soc.*, **104**, 6533 (1982).
262. A. H. Goldberg and D. A. Dougherty, *J. Am. Chem. Soc.*, **105**, 284 (1983).
263. Y. Yamaguchi, Y. Osamura and H. F. Schaefer, *J. Am. Chem. Soc.*, **105**, 7506 (1983).
264. Y. Yamaguchi and H. F. Schaefer, *J. Am. Chem. Soc.*, **106**, 5115 (1984).
265. T. R. Furlani and H. F. King, *J. Chem. Phys.*, **82**, 5577 (1985).
266. L. Carlacci, C. Doubleday, T. R. Furlani, H. F. King and J. W. McIver, *J. Am. Chem. Soc.*, **109**, 5323 (1987).
267. C. Doubleday, J. W. McIver and M. Page, *J. Phys. Chem.*, **92**, 4367 (1988).
268. Y. Yamaguchi, H. F. Schaefer and J. E. Baldwin, *Chem. Phys. Lett.*, **185**, 143 (1991).
269. S. J. Getty, E. R. Davidson and W. T. Borden, *J. Am. Chem. Soc.*, **114**, 2085 (1992). In Scheme I, the first-order rate constant for conversion of *cis*-1,2-d_2-cyclopropane to each antipode of the *trans* isomer is a factor of 2 too large. See also S. J. Getty, D. A. Hrovat and W. T. Borden, *J. Am. Chem. Soc.*, **116**, 1521 (1994).
270. J. E. Baldwin, Y. Yamaguchi and H. F. Schaefer, *J. Phys. Chem.*, **98**, 7513 (1994).
271. P. R. Bunker, *Molecular Symmetry and Spectroscopy*, Academic Press, New York, 1979.
272. E. R. Davidson, in *The World of Quantum Chemistry* (Eds. R. Daudel and B. Pullman), D. Reidel, Dordrecht, 1974, pp. 17–30.
273. S. R. Langhoff and E. R. Davidson, *Int. J. Quantum Chem.*, **8**, 61 (1974).
274. E. S. Replogle and J. A. Pople, *Abstracts, 203rd ACS National Meeting*, San Francisco, April 1992; PHYS 245 E. S. Replogle, Ph.D. Dissertation, Carnegie-Mellon University, 1992.
275. L. D. Pedersen, Ph.D. Dissertation, Yale University, 1975.
276. J . A. Berson and L . D . Pedersen, *J. Am. Chem. Soc.*, **97**, 238 (1975).
277. J. A. Berson, L. D. Pedersen and B. K. Carpenter, *J. Am. Chem. Soc.*, **97**, 240 (1975).
278. J. A. Berson, L. D. Pedersen and B. K. Carpenter, *J. Am. Chem. Soc.*, **98**, 122 (1976); **99**, 7399 (1977).
279. J. T. Wood, J. S. Arney, D. Cortes and J. A. Berson, *J. Am. Chem. Soc.*, **100**, 3855 (1978).
280. T. B. Freedman, F. Long and K. A. Villarica, Syracuse University, unpublished results.
281. C. P. Casey and L. J. Smith, *Organometallics*, **8**, 2288 (1989); C. P. Casey and L. J. Smith-Vosejpka, *Organometallics*, **11**, 738 (1992).
282. S. J. Cianciosi, N. Ragunathan, T. B. Freedman, L. A. Nafie and J. E. Baldwin, *J. Am. Chem. Soc.*, **112**, 8204 (1990).

283. B. K. Carpenter, *J. Am. Chem. Soc.*, **107**, 5730 (1985).
284. N. G. Rondan and K. N. Houk, *J. Am. Chem. Soc.*, **107**, 2099 (1985).
285. B. K. Carpenter, *Adv. Molecular Modeling*, **1**, 41 (1988).
286. B. K. Carpenter, *J. Org. Chem.*, **57**, 4645 (1992).
287. B. K. Carpenter, *Acc. Chem. Res.*, **25**, 520 (1992).
288. K. N. Houk, Y. Li and J. D. Evanseck, *Angew. Chem., Int. Ed. Engl.*, **31**, 682 (1992).
289. C. F. Bender and H. F. Schaefer, *J. Am. Chem. Soc.*, **92**, 4984 (1970).
290. H. F. Schaefer, *Science*, **231**, 1100 (1986).
291. E. Wasserman, *Science*, **232**, 1319 (1986).
292. H. F. Schaefer, *Sclence*, **232**, 1319 (1986).
293. E. Wasserman and H. F. Schaefer, *Science*, **233**, 829 (1986).

CHAPTER **10**

Organometallic derivatives of cyclopropanes and their reactions

ZEEV GOLDSCHMIDT

Department of Chemistry, Bar–Ilan University, Ramat–Gan 52900, Israel

I. INTRODUCTION 498
II. METAL DERIVATIVES OF CYCLOPROPANE 498
A. Cyclopropyl–Metal Compounds 498
1. Direct synthesis 499
2. Exchange reactions 499
a. Metal–halogen exchange 499
b. Metathesis 500
c. Metallation 504
d. Metal exchange 505
3. Addition 506
a. Cyclopropanation 506
b. Carbometallation 506
c. Oxidative coupling 507
d. Oxymercuration 508
4. Insertion 508
a. Into a C—H bond 508
b. Into a C—C bond 509
5. Cyclization 509
6. Decarbonylation and metal migration 510
B. Cyclopropylcarbinyl–Metal Compounds 511
1. Main group elements 511
a. Reactions with alkali metals 511
b. Metal–halogen exchange 512
c. Carbocyclization 513
d. Cyclopropanation 514
2. Transition metal complexes 515
a. Metal–halogen exchange 515
b. Hydrometallation 519

The chemistry of the cyclopropyl group, Vol. 2
Edited by Z. Rappoport

c. Carbocyclization 519
d. Insertion 521
e. Catalysis 521
C. Cyclopropylcarbene Complexes 522
1. Preparation 522
2. Reactions 524
a. Without ring cleavage 524
b. Addition reactions 527
c. Cycloadditions with alkynes 528
d. Rearrangements 533
e. Reaction intermediates 534
D. Cyclopropylcarbyne Complexes 534
1. Preparation 534
2. Reactions without ring cleavage 535
a. Ligand substitution reactions 535
b. Ring substitution reactions 536
c. Reactions with phosphines 536
d. Reactions with strong acids 536
e. Photooxidation 537
3. Reactions involving ring cleavage 538
a. Thermal reactions in chlorinated solvents 538
b. Reactions with HCl 538
c. Reactions with nucleophiles 541
d. Photooxygenation in chlorinated solvents 542
E. Vinylcyclopropane–Metal Derivatives 544
1. Main group elements 544
a. Exchange reactions 544
b. Carbometallations 546
c. Rearrangements 546
2. Transition metal complexes 547
a. σ-Bonded complexes 547
b. π-Complexes of vinylcyclopropanes 549
3. Conjugated vinylcyclopropane complexes 550
a. Acyclic η^4-dienyl complexes. 550
b. Cyclic complexes 551
F. Alkynylcyclopropane–Metal derivatives 557
1. Cyclopropylethynyl–metal derivatives 557
2. 1-Alkynylcyclopropyllithium 560
3. Dimetallation 561
4. Reduction reactions 561
5. Hydration 562
6. η^1-Transition metal complexes. 562
7. Reactions involving metal π-bonding 563
8. Cyclopropylalkynylcarbene complexes 564
III. CYCLOPROPYLIDENE COMPLEXES 566
A. Tungsten Complexes 566
B. Iron Complexes 567
1. μ-Cyclopropylidene complexes 567
2. Cyclopropenylium complexes 568
IV. METAL DERIVATIVES OF CYCLOPROPENE 568
A. σ-Bonded Compounds 569
1. 1-Cyclopropenylmetal compounds 569

a. Main group elements . . . 569
b. Transition metal complexes . . . 571
2. 3-Cyclopropenylmetal compounds . . . 573
a. Preparation . . . 573
b. Reactions . . . 576
B. π-Complexes . . . 578
1. Synthesis and characterization . . . 578
a. Ligand exchange . . . 578
b. Coupling of alkynes with alkylidene complexes . . . 582
c. β-Hydride elimination . . . 583
2. Reactions . . . 583
a. Isomerization to vinyl carbene complexes . . . 583
b. Rearrangement to metallacyclobutenes . . . 585
c. Reaction with alkynes and alkenes . . . 587
d. Oxidative coupling reactions . . . 589
V. η^3-CYCLOPROPENYL COMPLEXES (METALLATETRAHEDRANES) . . . 589
A. Preparation . . . 589
1. Oxidative addition of cyclopropenylium salts . . . 589
2. Metallacyclobutadiene rearrangements . . . 593
3. Reaction of carbyne complexes with alkynes . . . 594
B. Reactions . . . 595
1. Ligand exchange . . . 595
a. Exchange of CO by phosphorus, nitrogen and oxygen ligands . . . 595
b. Exchange of CO by alkenes and alkynes . . . 598
c. Exchange of ligands by cyclopentadienyl and related groups . . . 598
d. Metathetical exchanges of halogen and nitrogen ligands . . . 599
2. Ring enlargement reactions . . . 600
3. Addition reactions . . . 600
C. Structural Aspects . . . 601
1. Solid state structure . . . 601
2. Electronic structure . . . 603
3. Dynamic behavior . . . 603
VI. CYCLOPROPENYLIDENE COMPLEXES . . . 606
A. Cyclopropenylidene Complexes . . . 606
1. Preparation . . . 606
a. From 3,3-dichlorocyclopropenes . . . 606
b. From halocyclopropenylium salts . . . 609
c. From alkoxycyclopropenylium salts . . . 610
d. From lithiocyclopropenylium salts . . . 610
e. By desulfurization and deselenization . . . 610
B. Metallacyclopropenylium Compounds . . . 610
1. Preparation . . . 611
a. Main group metal compounds . . . 611
b. Transition metal complexes . . . 612
2. Reactions . . . 614
3. NMR spectra . . . 617
VII. METAL DERIVATIVES OF METHYLENECYCLOPROPANE . . . 619
A. σ-Bonded Compounds . . . 619
1. Metallation . . . 619
2. Metal–halogen exchange . . . 621
3. Reactions with metals . . . 623

B. π-Complexes 623
1. Preparation 624
a. Ligand exchange reactions 624
b. Cyclopropanation of allenes 627
2. NMR spectra 628
3. Reactions 629
a. Chloropalladation 629
b. Hydroplatination 633
c. Reactions of η^2-methylenecyclopropane complexes 634
d. η^4-Trimethylenemethane complexes 638
e. The Pauson–Khand reaction 638
4. Catalyzed cycloaddition reactions 639
a. Intermolecular cycloadditions 641
b. Intramolecular cycloadditions 642
VIII. METHYLENECYCLOPROPENE COMPLEXES (TRIAFULVENES) . 644
IX. ACKNOWLEDGMENTS 647
X. REFERENCES 647

I. INTRODUCTION

Although the chemistry of cyclopropane as a functional group has been thoroughly reviewed in the past 25 years, particularly in this series[1] in 1987 and in an earlier volume[2] in 1971, the interaction of cyclopropanes with metals has received only occasional attention and has not yet been recognized as a separate topic in its own right.

The last decade has seen a tremendous increase in the number of publications on metal coordinated cyclopropanes which parallels the fast expansion of organometallic chemistry in general. Still, most recent reviews on organometallic aspects of cyclopropanes deal specifically with catalyzed reactions, neglecting the rich stoichiometric chemistry of organometallic cyclopropanes. This chapter is to our knowledge the first attempt to provide a systematic coverage of the stoichiometric interaction of metals with the ubiquitous three-membered ring. Due to the ability of metals to coordinate with unstable organic ligands, the variety of compound types considered here extends beyond those usually encountered in cyclopropane chemistry to include also cyclopropylcarbene, cyclopropylcarbyne, cyclopropylidene, η^3-cyclopropenyl and cyclopropenylidene complexes.

The reader will no doubt notice that catalyzed reactions receive here only sparse attention, usually through citation of previous surveys. We feel that they merit a separate chapter to complement existing reviews, and trust that their exclusion will not detract from the integrity of this chapter.

In accord with the general theme of this series, the chapter is organized principally according to the type of cyclopropyl functional groups as ligands. The literature covered in this chapter has been surveyed up to 1994.

II. METAL DERIVATIVES OF CYCLOPROPANE

A. Cyclopropyl–Metal Compounds

A wide variety of organometallic methods of carbon–metal σ-bond formation[3] have been employed in the synthesis of cyclopropyl–metal compounds. Consequently, with few exceptions the preparation of almost any main group or transition metal cyclopropyl compound has precedence in the literature. In this section only an overview, rather than a comprehensive survey, of the essential synthetic methods of cyclopropyl–metal compounds and their specific reactivity will be given, covering metals from most groups of the periodic table.

1. Direct synthesis

The reaction of metals with alkyl halide to form alkylmetal compounds is typical for lithium and magnesium. This so-called direct synthesis was applied to the preparation of both cyclopropyllithium[4] and derivatives[5], and cyclopropylmagnesium halides[6,7] under the standard preparative conditions of Grignard reagents without ring cleavage (equation 1).

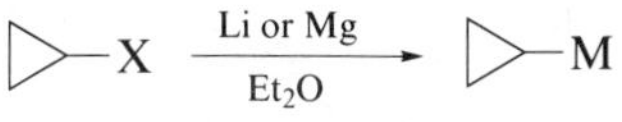

(1)

X = Cl, Br M = Li, MgX

These reagents are key intermediates in synthesis of cyclopropane derivatives. The structure and stereochemistry of lithium and magnesium cyclopropyl derivatives as well as their extensive use in organic synthesis have been thoroughly reviewed in previous[1] and updated[8] volumes of this series and no attempt will be made to elaborate on their chemistry here. Their extreme importance in the organometallic chemistry of cyclopropanes will be further revealed as this chapter progresses.

2. Exchange reactions

a. Metal–halogen exchange. Metal–halogen exchange reactions are most commonly used for preparation of cyclopropyllithium compounds from the corresponding halocyclopropanes. The reaction is particularly useful for selective lithiation of *gem*-dihalocyclopropanes. A recent representative example is the preparation of optically pure (*R*)-1-lithio-1-fluoro-2,2-diphenylcyclopropane from the corresponding (+)-(*S*)-bromide and butyllithium in THF–ether, at –100 °C (equation 2)[9].

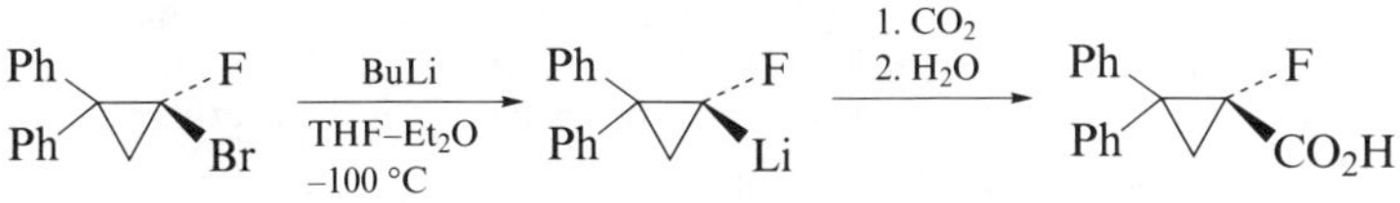

(2)

The retention of configuration at the exchange position was indicated by isolation of the pure (+)-(*R*)-1-fluoro-2,2-diphenylcyclopropanecarboxylic acid, obtained by quenching the reaction mixture with carbon dioxide, and hydrolysis. Extensive use of a variety of cyclopropyllithium and magnesium[10] derivatives, prepared by this method, is made in the synthesis of other metallacyclopropanes by metathesis reactions (*vide infra*). More recently, geminal dilithiocyclopropanes have been detected in the reduction of dihalocyclopropanes by lithium 4,4′-di-*t*-butylbiphenyl[11] and Li(Naph) radical anion[12], at low temperatures (–75 to –95 °C).

Among transition metal complexes, the ubiquitous dicarbonylcyclopentadienyliron (Fp) complexes[13] are the first, and perhaps the best, representatives to demonstrate the utility of metal–halogen exchange reactions in metallacyclopropane synthesis. Thus, reaction of the readily available sodium dicarbonylcyclopentadienyliron $[Cp(CO)_2Fe]Na$ (FpNa) with the parent cyclopropyl bromide[14] and derivatives[14–16] gave in moderate yields the corresponding cyclopropane–Fp complexes (equation 3).

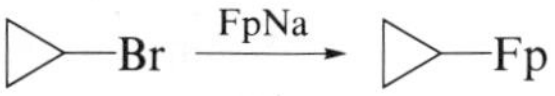

(3)

Fp = $Cp(CO)_2Fe$

The metal–halogen exchange reaction has been extensively used in preparing cyclopropylcarbonyl complexes of manganese[17–19], rhenium[17], iron[17,20–24], rhodium[25], iridium[17,25] and tin[26] from cyclopropylacyl halides and the corresponding carbonylmetal complexes [M] (M=$Mn(CO)_5^-$, $Re(CO)_5^-$, $CpFe(CO)_2^-$, $[RhCl(C_2H_4)_2]_2$, $[IrCl(C_8H_{14})_2]_2$, $Ir(CO)Cl_2(PPh_3)_2$ or Me_6Sn_2; see equation 4). These acylmetal compounds have been the subject of numerous studies of carbonyl migration and decarbonylation reactions (*vide infra*).

(4)

$ML_n = M(CO)_5$, M = Mn, Re
$MCl_2(PMe_2Ph)_2$, M = Rh, Ir
$Ir(CO)Cl_2(PPh_3)_2$, $Cp(CO)_2Fe$, $SnMe_3$

b. Metathesis. The metathesis reactions of cyclopropylmagnesium bromide with mercury dichloride to form dicyclopropylmercury[7], and cyclopropyllithium with tin tetrachloride to give tetracyclopropyltin (equation 5)[4] are perhaps the earliest examples of the most useful general preparative method of cyclopropyl metal compounds.

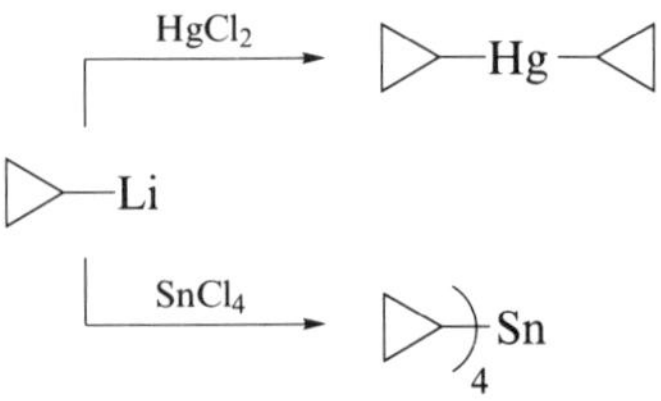

(5)

Analogous reactions of α-bromocyclopropyllithium[27] and magnesium chloride[10] with a variety of silicon, germanium, tin, lead (group 14) and mercury[7] (group 12) mono- and dihalides at temperatures below –70 °C similarly gave the corresponding mono- and dicyclopropyl metal derivatives, selectively with retention of the configuration at the carbon (equation 6). The related α-bromocyclopropylzincates are useful reagents for the stereoselective alkylation of cyclopropanes (*vide infra*)[28].

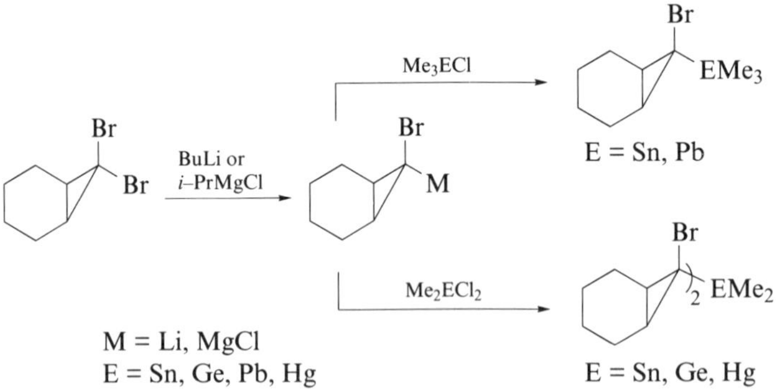

(6)

More recently, the doubly lithiated derivatives of [4.1.1]- and [3.1.1]propellasilanes underwent bridging metathesis at the bicyclobutane ends with dihalides of germanium, tin and transition metal titanium, to give the corresponding highly strained metallacyclic propellanes (equation 7)[29].

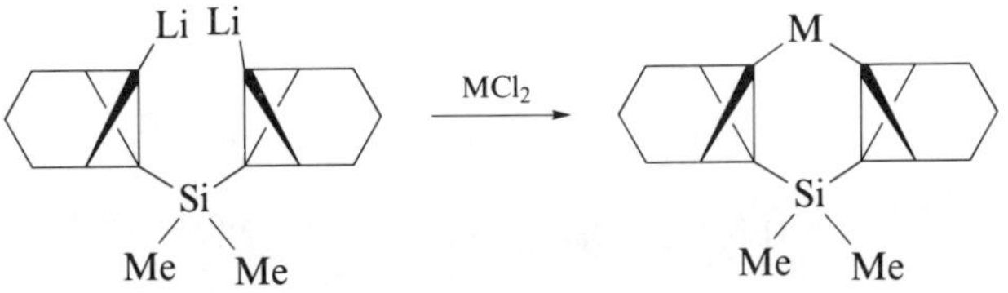

(7)

M = Me_2Ge, Me_2Sn, Cp_2Ti

The relative weakness of the carbon–tin bond provided a useful method for carbon–carbon bond formation by cross-coupling reaction with acyl halides. This reaction has been applied for the two-step synthesis of α-ketocyclopropyl sulfones (equation 8)[30].

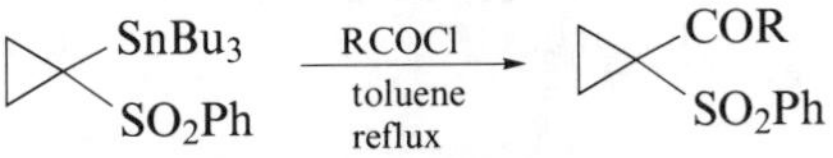

(8)

R = Me, Pr, *i*–Pr, Bu, *t*–Bu, Ph

Many group 11 copper[31] and gold[32] cyclopropyl derivatives have been prepared by the lithium metal metathesis reaction. Chiral (*R*)-1-cuprous-1-isocyano-2,2-diphenylcyclopropane was generated from the corresponding chiral (*S*)-1-lithiocyclopropane derivative by reaction with anhydrous cuprous iodide in THF, at –72 °C (equation 9)[33]. While the lithium compound is configurationally unstable at temperatures above –50 °C, the copper analogue is configurationally stable even at 23 °C. Thus deuterolysis of the latter after 2 h at this temperature revealed a high deuterium content and a high optical purity of the isocyanocyclopropane obtained.

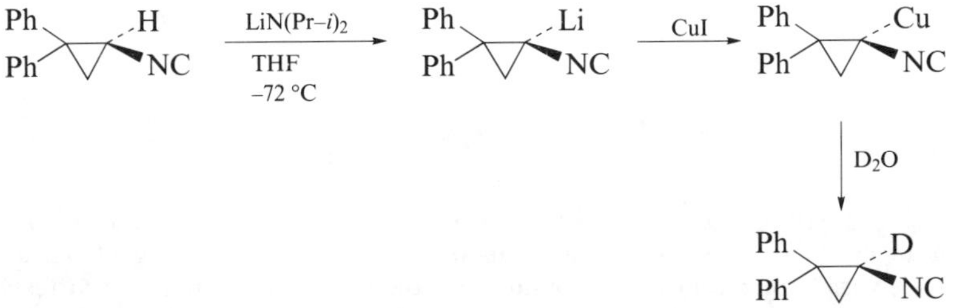

(9)

The related lithium(phenylthio)cyclopropylcuprate, prepared from cyclopropyllithium and PhSCu, has been extensively used in the synthesis of *β*-cyclopropyl enones by coupling with the corresponding *β*-iodo enones (equation 10)[34–36]. Repetitive regioselective coupling of ethynylcyclopropane units by this method led to polyspirocyclopropyl acetylenes as precursors to [N]pericyclynes[37].

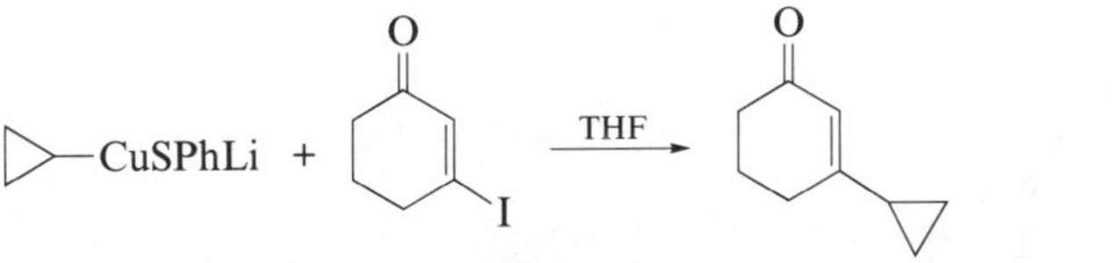

(10)

Aurated cyclopropane complexes are prepared by reaction of 1-lithiodicyclopropyl ketone[38a] and 1-lithio-*N,N*-dialkylcyclopropylamides with triphenylphosphinegold complexes (equation 11)[38b]. The crystalline gold complexes provide a rare opportunity to look at the solid state structure of a linear cyclopropyl metal complex, revealing its most stable bisected conformation.

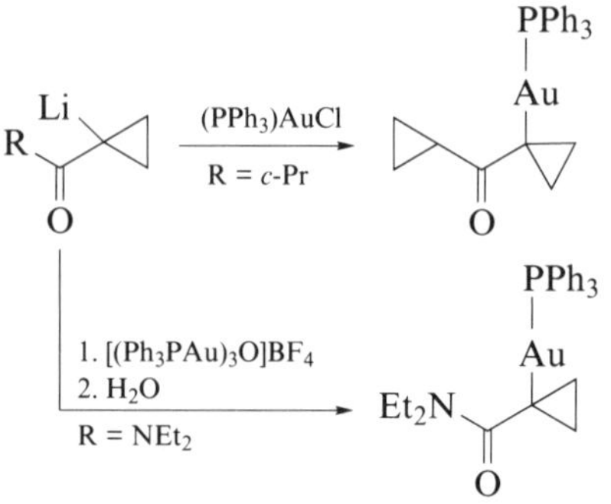

(11)

Metathesis reactions between organyllithium (or magnesium) compounds and transition metal halide complexes are commonly found in groups 4 (Ti, Zr), 6 (W), 8 (Fe, Ru), 9 (Ir) and 10 (Ni, Pt). The metathesis reaction of cyclopropyllithium and $Cp(CO)_2FeBr$ (FpBr) is the first example of a transition metal complex having a σ-bonded cyclopropane ring (equation 12)[14]. This complex undergoes typical[13,39] proximal ring cleavage with ionic electrophiles such as tetrafluoroboric acid and trityl tetrafluoroborate to form η^2-cationic complexes, and formal 3+2 cycloaddition reactions with the electrophilic tetracyanoethylene (TCNE) and SO_2 (equation 12).

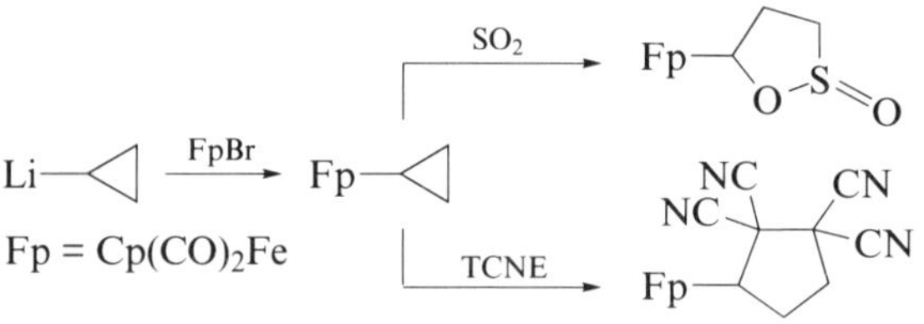

(12)

Bis(cyclopropyl)titanocene is prepared from cyclopropyllithium and dichlorotitanocene in 95% yield. The complex has been utilized as an olefination reagent in reactions with a wide variety of carbonyl compounds, including esters, furnishing the corresponding methylenecyclopropane derivatives (equation 13)[40].

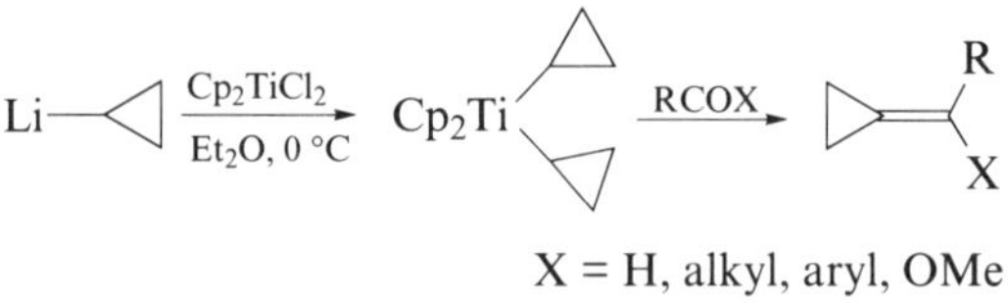

(13)

Dichlorotitanocene, like the tin and germanium dihalides, gives titanocene metallacyclic complexes upon reaction with dilithiated [*n*.1.1]propellanes (n = 3,4) (*vide supra*)[29]. The analogous bis(cyclopropyl)zirconium complexes, prepared by the same method[41], undergo elimination of one cyclopropyl group in the presence of trimethylphosphine to give η^2-cyclopropene complexes (see Section IV.B), and react with diphenylacetylene affording metallabicyclo[3.1.0]hexene complexes (see Section II.E).

The reaction of 7,7-dichlorodibenzonorcaradiene with BuLi gave 7-chloro-7-*exo*-lithiodibenzonorcaradiene which, upon subsequent metal–halogen exchange reaction with $Cp(CO)_3WI$, gave the corresponding *exo*-tungstenio complex (equation 14)[42]. Upon photolysis, this σ-complex undergoes ligand-to-metal rearrangement of the halogen atom and displacement of a carbonyl ligand to give an unstable cyclopropylidene intermediate complex (see Section III), which upon further irradiation undergoes skeletal rearrangement to the stable η^2-dibenzocyclooctatetraene complex.

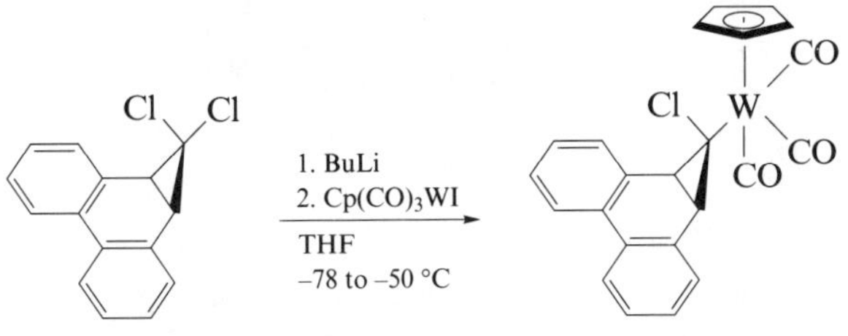

(14)

Chlororuthenium complex $Cp(PPh_3)_2RuCl$ reacts slowly (6 d, 20 °C) with cyclopropylmagnesium bromide to give the normal metathesis product $Cp(PPh_3)_2Ru(c\text{-}Pr)$ in 18% yield[43]. The complex undergoes thermolysis at 120 °C to give $Cp(PPh_3)Ru(\eta^3\text{-allyl})$. However, the analogous reaction with the 2-allylcyclopropyl Grignard reagents led to exchange and a loss of one PPh_3 group, affording the corresponding η^1,η^2-cyclopropyl complexes (equation 15)[44]. These complexes are thermally stable, presumably due to their rigid conformation, which prevents isomerization via β-H elimination.

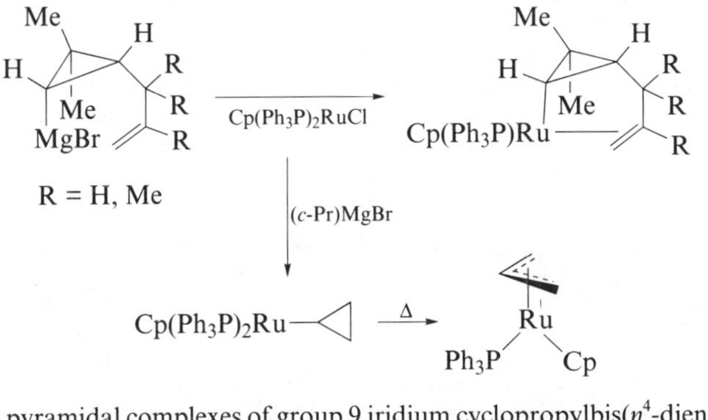

(15)

The square pyramidal complexes of group 9 iridium cyclopropylbis(η^4-diene) complexes were recently synthesized in poor yields (3–5%) by metathesis reaction between (η^4-diene)$_2$ IrCl and cyclopropyllithium or cyclopropylmagnesium bromide (equation 16)[45].

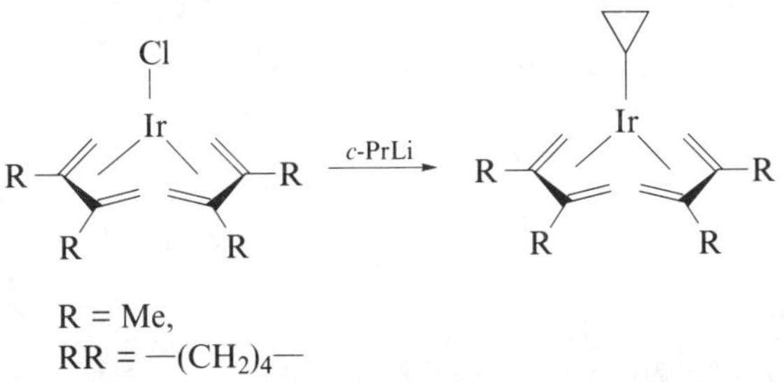

(16)

Similarly, group 10 trigonal nickel complex $Cp(PPh_3)Ni(c\text{-}Pr)$ was prepared from $Cp(PPh_3)NiCl$ and cyclopropylmagnesium bromide[46]. This complex, in analogy with the aforementioned Fp(c-Pr) (*vide supra*), undergoes ring cleavage by protonation with fluorosulfonic acid in CH_2Cl_2 at –80 °C, to give a π-propenyl intermediate complex which by subsequent reaction with methyllithium expels propene, affording $Cp(PPh_3)NiMe$ (equation 17).

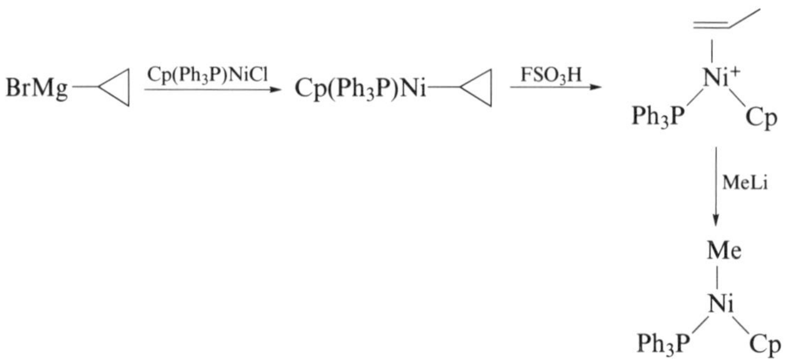

(17)

Finally, platinum complex *cis*-$(PMe_2Ph)_2PtCl_2$ undergoes metathesis with cyclopropyllithium in ether, to give the square planar *cis*-$(PMe_2Ph)_2Pt(Pr\text{-}c)_2$, which is considerably more stable than the analogous open-chain alkyl complexes (equation 18)[47]. Consequently, it undergoes a variety of reactions which retain the cyclopropyl ring. Thus, reaction with HCl afforded *trans*-$(PMe_2Ph)_2PtCl(Pr\text{-}c)$, oxidative addition of MeI gave the octahedral $(PMe_2Ph)_2(c\text{-}Pr)_2PtMeI$ in which the incoming groups are *trans* to each other, and the reaction with allyl or benzyl bromide led to *cis*-addition of bromine to form $(PMe_2Ph)_2(c\text{-}Pr)_2PtBr_2$ with the two phosphine ligands in *trans* position. However, ring opening does occur upon reaction of *trans*-$(PMe_2Ph)_2(c\text{-}Pr)PtCl$ with $AgNO_3$, to yield, after addition of KPF_6, the trigonal $[(PMe_2Ph)_2Pt(\eta^3\text{-allyl})]PF_6$ cation complex[47].

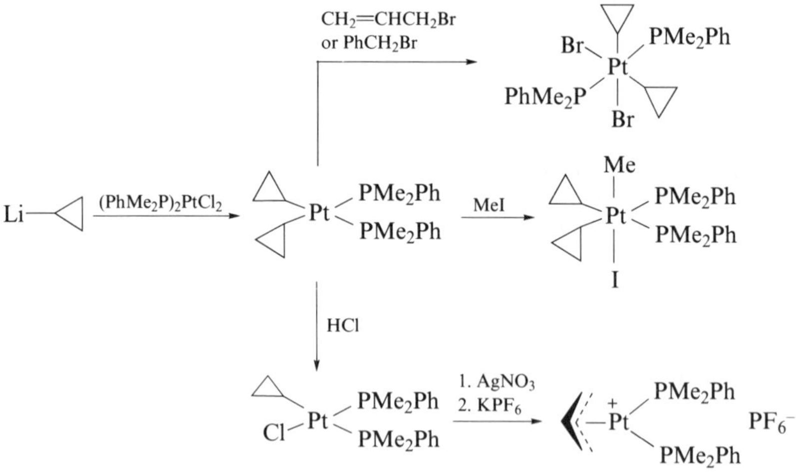

(18)

c. Metallation. Lithiation of cyclopropanes with strong lithium bases, like lithium diisopropylamide (LDA) and butyllithium, occurs when the acidity of the exchanged cyclopropyl hydrogen is sufficiently increased either by adjacent electron-withdrawing groups [COOR, COSR, CN, NO_2, SO_2Ph, $PO(OEt)_2$] (equation 19)[30,48–53] or by ring strain, e.g. the bridgehead protons of bicyclo[1.1.0]butanes[54]. Some bicyclobutanes undergo lithiation at both bridgehead positions[55], whereas others, like the [*n*.1.1]propellane dimers (*vide supra*)[29] which have two bicyclobutane moieties, undergo dilithiation.

(19)

EWG = CO_2R, COSBu-*t*, $CONMe_2$, CN, NC, SO_2Ph, $PO(OEt)_2$, NO_2
E = alkyl, SMe, $SiMe_3$, CO_2H, R_2COH (R = H, alkyl, aryl)
LG = leaving group

Notably, the lithium enolates have the planar methylenecyclopropane-type structure[56], but give C-alkylation products[49–52]. X-ray structure analysis of the lithium enolate[56] and bicyclobutyllithium[57] TMEDA complexes revealed that both crystallize as lithium bridging dimers.

d. Metal exchange. Tricyclopropylaluminum dimer $Al_2(Pr\text{-}c)_6$ was prepared by mercury–aluminum exchange reaction of dicyclopropylmercury and trimethylaluminum dimer, and removal of the more volatile product in vacuum (equation 20)[58]. X-ray crystal analysis of the aluminum dimer revealed an unusual structure in which two cyclopropyl groups are bridging between the two aluminum atoms, forming a puckered bicyclobutane-like structure[59].

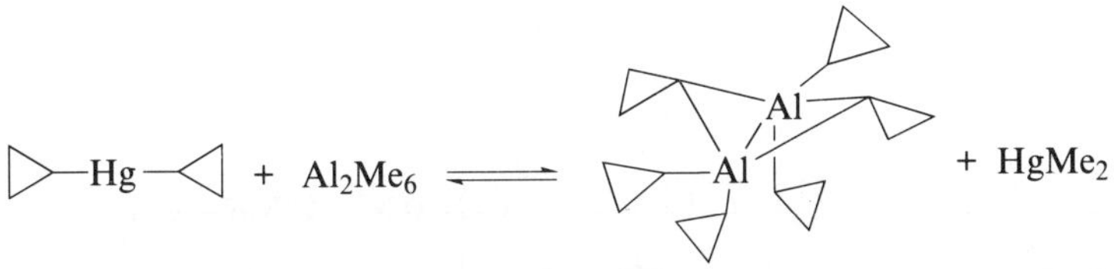

(20)

The classical tin–lithium exchange reaction was first applied to cyclopropanes in the stereospecific transformation of 7-*endo*-bromo-7-*exo*-trimethyltinnorcarane to the corresponding 7-*exo*-lithio compound with BuLi in THF, at –95 °C (*vide supra*)[10]. The retention of configuration in cyclopropyl tin–lithium exchange reactions has also been confirmed in monocyclic 2-metallacyclopropane carboxamides[60]. More recently, this metal exchange reaction was used for the preparation of sensitive methylenecyclopropanes (equation 21)[61].

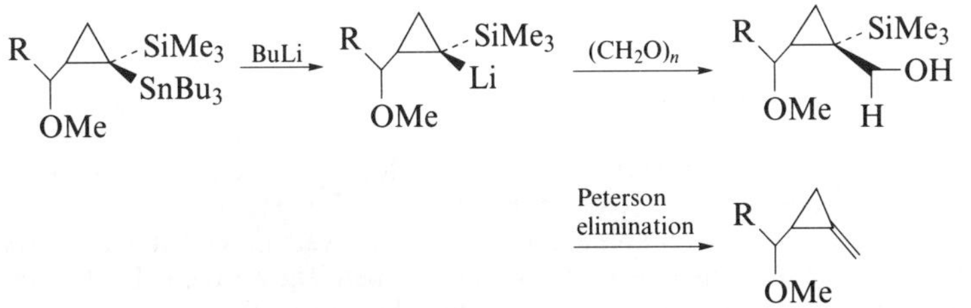

(21)

R = *i*-Pr, *t*-Br, *c*-Hex, *n*-C_7H_{15}

3. Addition

a. Cyclopropanation. Allylic alcohols bearing vinylic tin substituents, undergo highly diastereoselective cyclopropanation in the presence of samarium($HgCl_2$)/CH_2I_2 in THF at –78 °C, to give cyclopropyltin compounds (equation 22)[62]. It is interesting to compare the mild condition used in this synthesis to the early attempts to prepare cyclopropyltin compounds by heating bis(trimethyltin)dibromomethane at 160–180 °C in the presence of cyclohexene in a sealed tube for 4 days. Under these conditions a mixture of fragmentation products was formed containing only trace amounts of the expected 7-bromo-7-trimethyltinnorcarane[63].

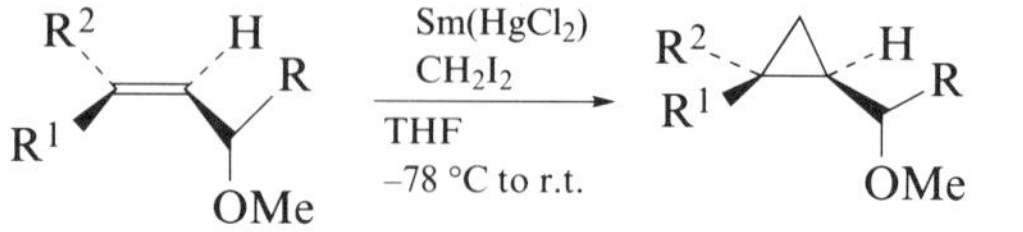

(22)

R = Me, Pr, cyclohexyl
$R^1 = Bu_3Sn$, R^2 = Bu, Me_3Si
R^1 = Bu, $R^2 = Bu_3Sn$

b. Carbometallation. 3,3-Dimethylcyclopropene undergoes *cis*-carbometallation reactions with alkyl and vinyl Grignard reagents, to afford the corresponding 2-alkylated cyclopropylmagnesium halides (equation 23)[64,65].

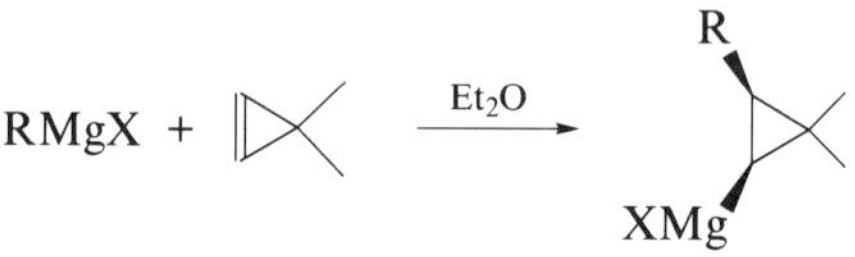

(23)

X = Cl, Br
R = $CH{=}CH_2$, $CH_2CH_2CH{=}CH_2$, $(CH_2)_3C(Me){=}CH_2$

Primary allylmagnesium halides[64,65] and allylzinc[66] reagents react similarly, presumably by an ene-type reaction, to give selectively the cyclopropyl derivative in which the more substituted allylic carbon is attached to the ring (equation 24). Allylzinc compounds have been employed in the same manner in carbozincation of cyclopropenone acetals[67].

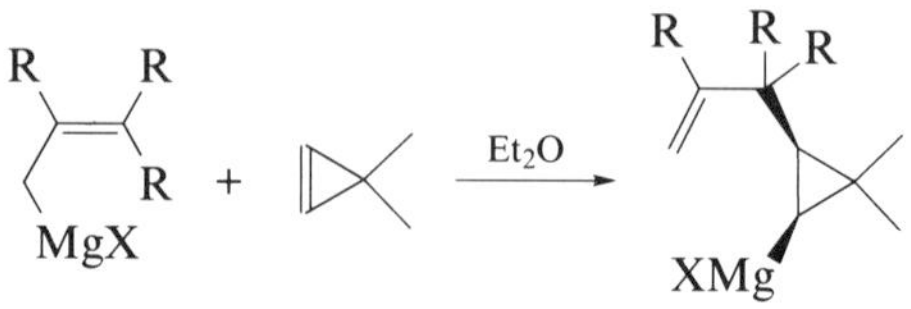

(24)

X = Cl, Br
R = H, Me

The analogous *cis*-carbocupration of cyclic cyclopropenone ketales takes place with R_2CuLi and Gilman reagents $R_2CuMgBr$ (R=Me, Bu, Hex)[68,69]. Carbocupration of chiral cyclopropenone ketals enables the diastereoselective synthesis of quaternary carbon centers (equation 25)[69]. The selectivity is highly dependent on the substituents and the reagents. Best results are obtained with the phenyl-substituted cyclopropenes and Me_2CuLi.

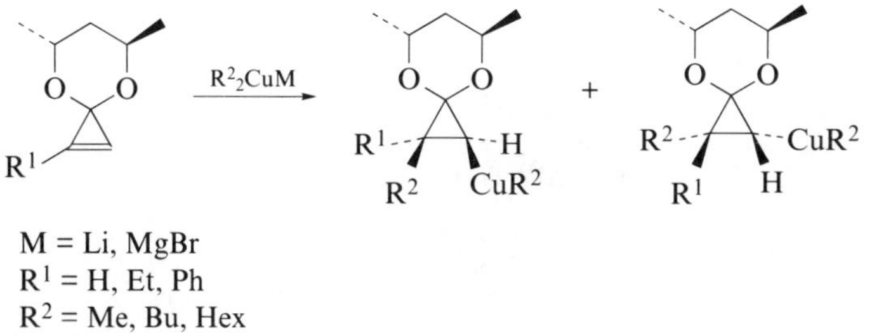

(25)

Reaction of the Tebe reagent [$Cp_2TiClCH_2AlMe_2$] with 3,3-dimethylcyclopropene and 4-(*N*,*N*-dimethylamino)pyridine (DMAP) affords the 2-titanabicyclo[2.1.0]pentane, formally obtained by 2+2 cycloaddition of titanomethylene complex $CpTi{=}CH_2$ and the cyclopropene (equation 26)[70]. Upon reaction with phosphines, the metallacycle undergoes ring opening by the alternative 2+2 cycloreversion, to give a new titanocarbene complex. The reaction with benzophenone affords 3,3-dimethyl-1,1-diphenyl-1,4-pentadiene, suggesting the intermediacy of the same carbene complex. This complex has important applications in ring-opening polymerization of cyclic olefins, e.g. norbornene by the olefin metathesis reaction[71].

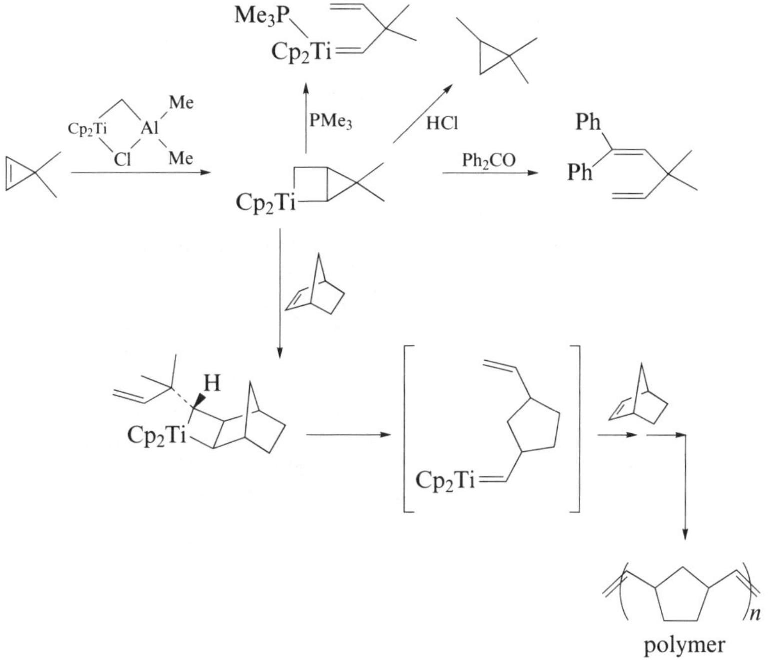

(26)

c. *Oxidative coupling.* Oxidative coupling is the process in which a metal complex and two olefins interact in a formal [1+2+2] cycloaddition reaction, to give a pentametallacycle. This process has been shown to be crucial in the initial stages of transition-metal-catalyzed dimerization, codimerization and cyclooligomerization of cyclopropenes[72,73]. Oxidative coupling reactions are typical for group 10 transition metals (Ni, Pd, Pt), but were also observed with rhodium (group 9), and group 4 titanium and zirconium. Thus, reaction of (bipy)(COD)Ni with two equivalents of 3,3-dimethylcyclopropene in ether at –70 °C gave the corresponding bipyridyl metallacycle complex in 92% yield, whose structure was confirmed by X-ray crystal analysis (equation 27)[74].

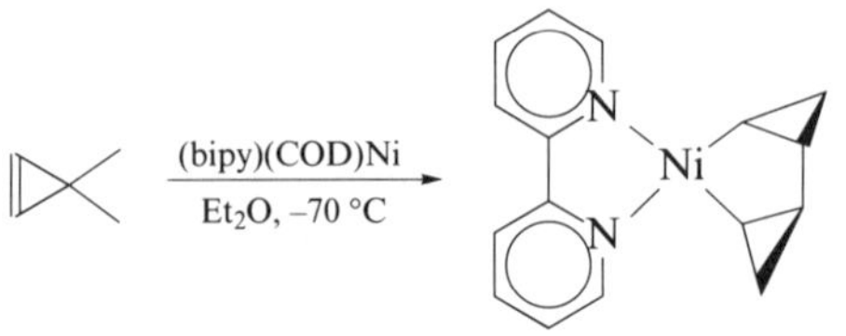

(27)

Similarly, reaction of the cyclopropene with $(Ph_3P)_2Ni$[75], $(Me_3P)_2Pt$[76], $Cp_2Ti(PMe_3)_2$[77], $Cp_2Zr(butene)(PMe_3)$[77] or $CpPd(\eta^3\text{-allyl})$[78] in the presence of phosphines gave the corresponding diphosphine metallacycles. However, the reaction of the palladium complex with excess of cyclopropene, under the same conditions, afforded a metallacyclotetramer complex[78] and the corresponding rhodacyclatrimer was obtained upon reaction of the cyclopropene with $(Ph_3P)RhCl$ (equation 28)[79]. X-ray structure analysis shows that the latter metallacycle has a *syn*, *anti*, *syn* relationship between the three-membered rings.

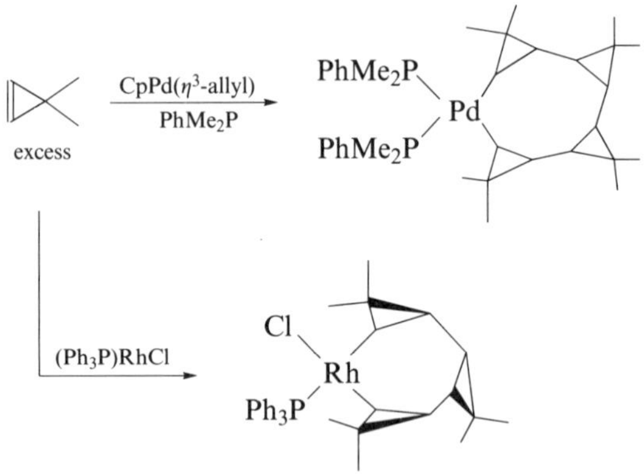

(28)

d. Oxymercuration. Oxymercuration of (diphenylmethylene)cyclopropane with mercuric acetate in $THF–H_2O$ gave regioselectively the acetoxymercurial adduct in which the metal is bonded to the cyclopropane ring (equation 29)[80]. The stereochemistry was confirmed by reductive demetallation with $NaBH_4$ and hydrolysis to diphenylcyclopropylcarbinol.

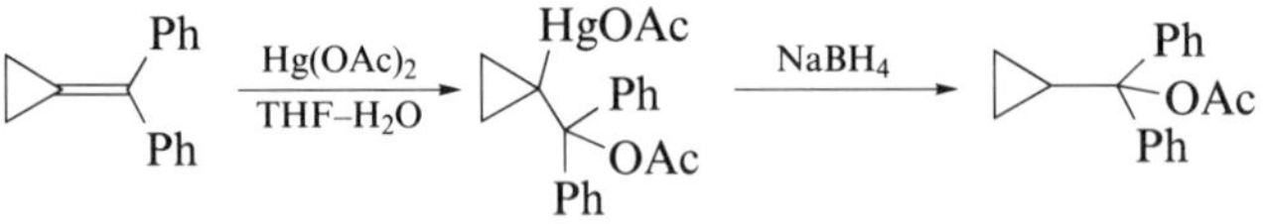

(29)

4. Insertion

a. Into a C—H bond. Generation of the coordinatively unsaturated rhodium fragment $[Cp^*Rh(PMe_3)]$ ($Cp^* = \eta^5\text{-}C_5Me_5$) at –60 °C in cyclopropane by either photolysis of $Cp^*(PMe_3)RhH_2$ or thermal decomposition of $Cp^*(PMe_3)Rh(neopentyl)H$ results in C—H bond insertion to yield the cyclopropylrhodium hydride complex Cp*(PMe$_3$)RhH(Pr-*c*) (equation 30)[81]. The complex rearranges in arene solvents at −20 °C, via intramolecular C—C metal insertion, to the corresponding rhodacyclobutane complex, in 65% yield.

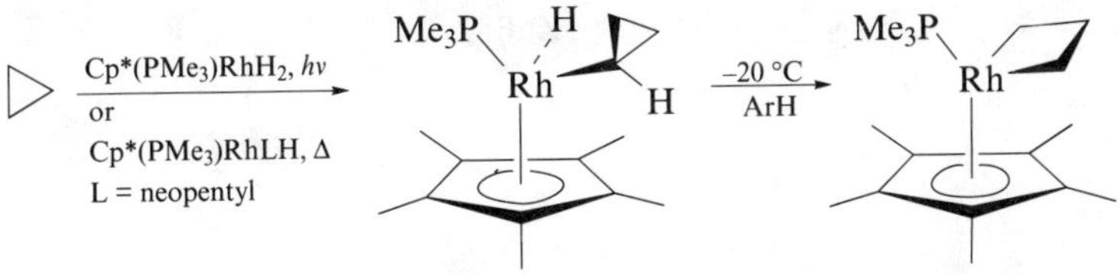

(30)

b. Into a C—C bond. Bridged spiropentanes react with Zeise's dimer [PtCl$_2$(C$_2$H$_4$)]$_2$ in CHCl$_3$ at room temperatures to give, after addition of pyridine, the two spiro-metallacyclobutane isomers which result from metal insertion into the exocyclic C—C bonds of the cyclopropane (equation 31)[82]. The structure of one of the isomers was confirmed by X-ray crystal analysis.

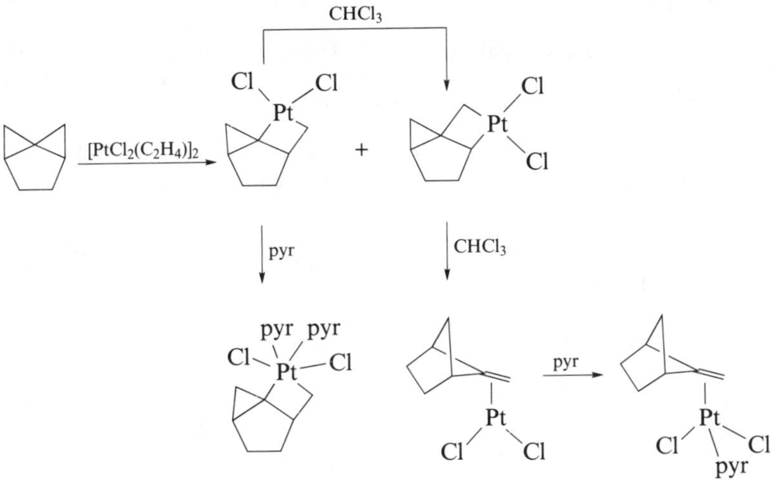

(31)

When the reaction is followed by NMR, the initial 1:1 ratio between the isomers changes as the cyclopropyl isomer rearranges to the cyclopropylcarbinyl counterpart. Further rearrangement of the latter finally afforded ring-opened [η^2-5-methylenebicyclo[2.2.1]-hexane]PtCl$_2$ complex.

5. Cyclization

The addition of 3-(phenylthio)allylzinc bromide to 1-octenylmagnesium bromide leads to the formation of 2-hexyl-3-vinylcyclopropylzinc bromide (equation 32)[83], presumably by intramolecular cyclization of the initially formed *gem*-dimetallic homoallylic intermediate. Hydrolysis gave a mixture of two isomeric 2-hexyl vinylcyclopropanes in a 7:3 ratio.

More recently, an analogous combination of lithium, magnesium, zinc and copper reagents led to the stereocontrolled synthesis of cyclopropanes, illustrated for the synthesis of *trans*-divinylcyclopropane (equation 33)[84].

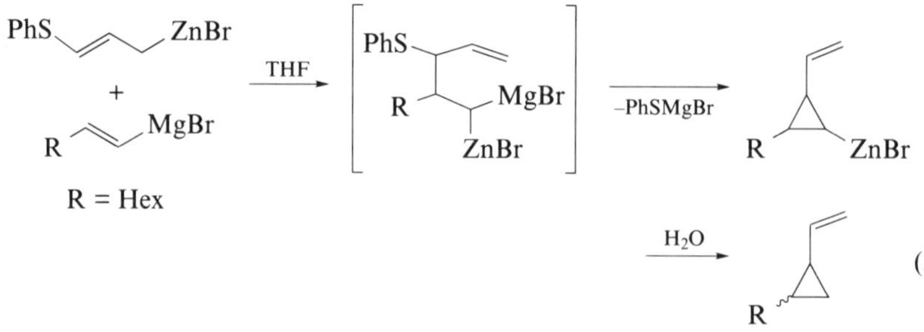

(32)

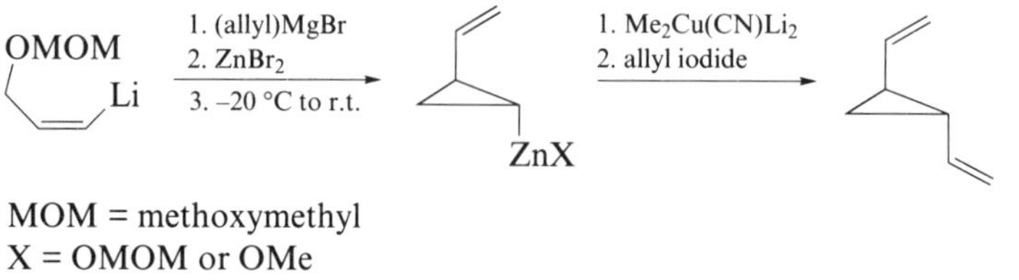

(33)

The photochemical induced cyclization of allyl Grignard reagents by a formal electrocyclic allylic-anion ring closure has been reported to give cyclopropylmagnesium bromides, which were quenched with carbon dioxide to the corresponding cyclopropanecarboxylic acids (equation 34)[85]. Woodward–Hoffmann rules suggest a disrotatory course for such photochemical 4*n*-electron systems. However, this could not be established due to the rapid *E*/*Z* equilibration of the allyl Grignard compounds in solution.

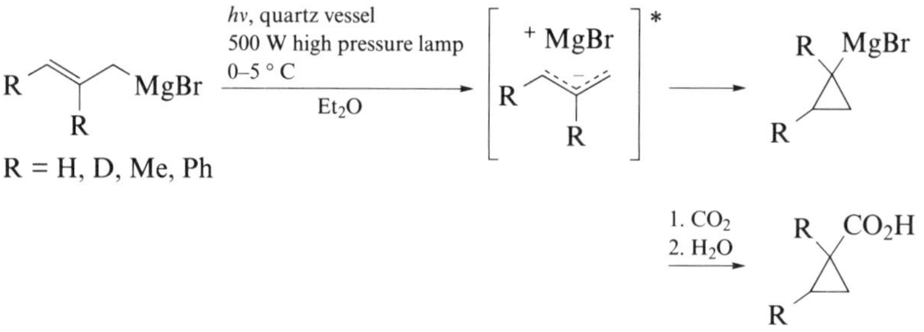

(34)

6. Decarbonylation and metal migration

Although much effort has been devoted to decarbonylation of cyclopropylcarbonyl metal complexes (*vide supra*), only (cyclopentadienyl)dicarbonyliron (Fp) derivatives have been successfully decarbonylated either photochemically[22,24] or using Wilkinson's rhodium catalyst $[(PPh_3)_2RhCl]_2$[21] (equation 35). Further decarbonylation by irradiation led to metallacyclopentane formation, whereas thermal decomposition resulted in the formation of the corresponding Cp(CO)Fe(allyl) complexes.

Finally, an interesting cyclopropyl-to-metal migration occurs upon reaction of cyclopropylcarbonyl chloride with the chlorocycloocteneiridium complex $[IrCl(C_8H_{14})_2]_2$ in the presence of $PMePh_2$ (equation 36)[25]. The initially formed pentacoordinated acylcyclopropane complex $IrCl_2(COC_3H_5)(PMePh_2)_2$ and the octahedral cyclopropyl iridium complexes *cis*-$IrCl_2(CO)(C_3H_5)(PMePh_2)_2$ (2:1 equilibrium mixture) isomerize slowly to the isomeric $IrCl_2(CO)(C_3H_5)(PMePh_2)_2$ where the two phosphine groups are *trans* to each

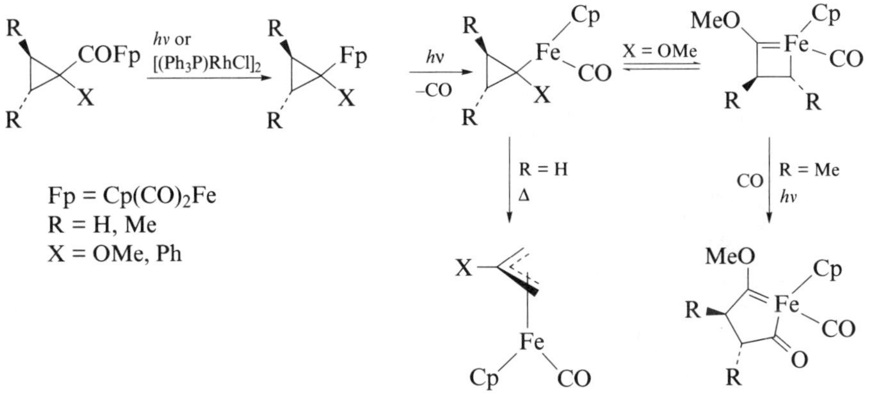

(35)

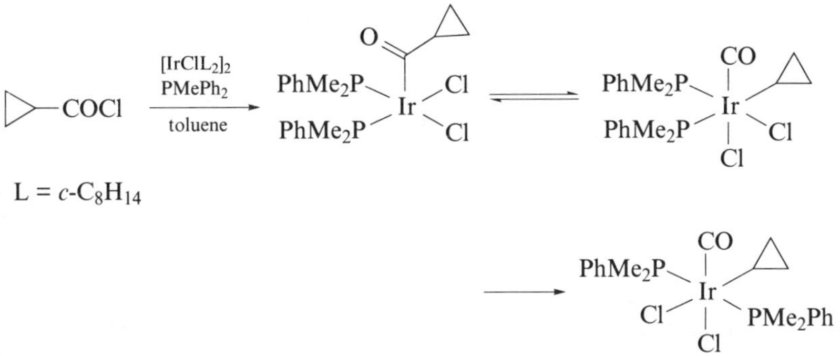

(36)

other. Notably, the analogous reaction with the rhodium complex afforded only the pentacoordinated acylcyclopropane complex, which resisted further rearrangement.

B. Cyclopropylcarbinyl–Metal Compounds

1. Main group elements

a. Reactions with alkali metals. Cyclopropylmethyl bromide readily gives a Grignard reaction with magnesium which easily ring-opens ($t_{1/2}$ = 2h at –24 °C) to form 3-butenylmagnesium bromide[86]. On the other hand, tetramethylcyclopropylmethyl magnesium chloride is stable[87]. It has been suggested that in general a cyclopropylcarbinyl–homoallyl equilibria is set up and the ratio between the cyclic and open-chain forms depends upon alkyl substitution (equation 37). However, the presence of a cyclopropylcarbinyl metal in such equilibrium has rarely been detected, although labeling experiments demonstrated that the α- and β-carbon atoms of the 3-butenylmagnesium halides interchange position, suggesting the intermediacy of the cyclopropylmethylmagnesium halide (equation 38)[88].

The cyclopropylcarbinyl anion is stabilized by phenyl groups attached to the carbinyl carbon toward ring opening, provided that potassium or sodium is the counterion. However, replacement of the potassium by lithium or magnesium leads to immediate

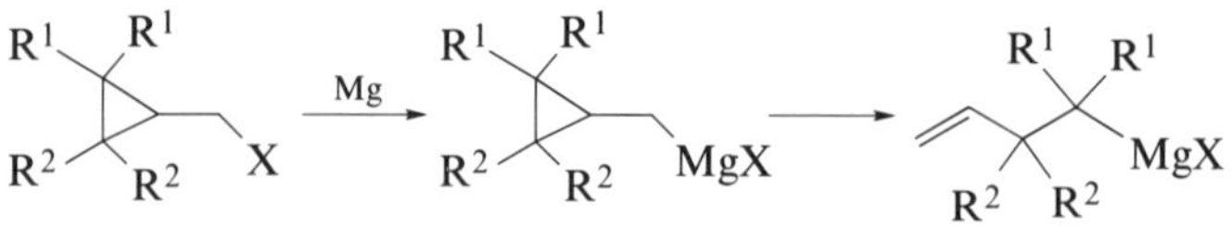

(37)

$R^1 = R^2 = H$
$R^1 = H, R^2 = Me$
$R^1 = R^2 = Me$
$X = Cl, Br$

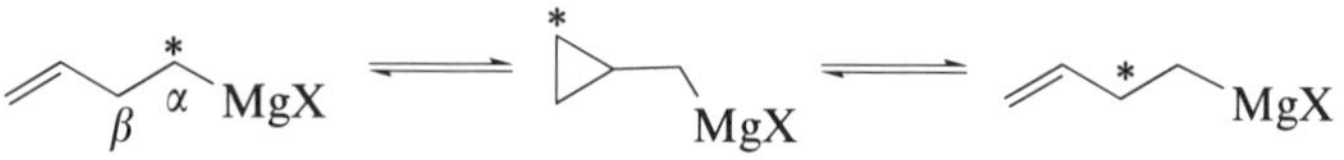

(38)

rearrangement and formation of the corresponding covalent 3-butenyl organometallic compound (equation 39). Replacement of the potassium by mercury or cobalt in ether gave (diphenylmethyl)cyclopropane[89].

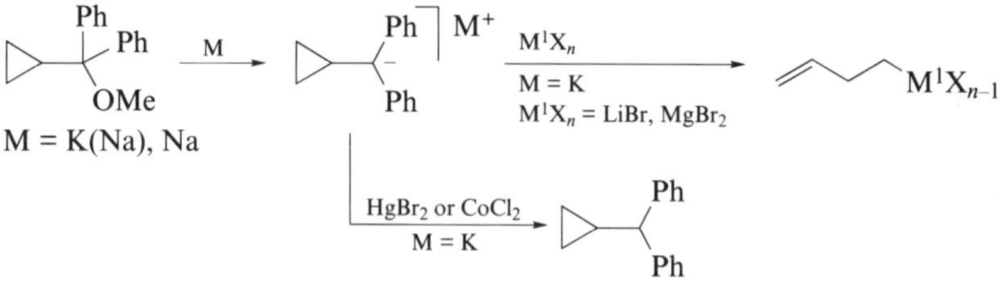

(39)

Interestingly, the related lithium derivative 1-butyl-1-diphenylmethylcyclopropane, obtained by carbometallation of diphenylmethylenecyclopropane with BuLi, is stable and gives the cyclic hydrocarbon upon hydrolysis (equation 40)[90]. On the other hand, α-phenyl styrene reacts similarly with BuLi to give only ring-cleavage products.

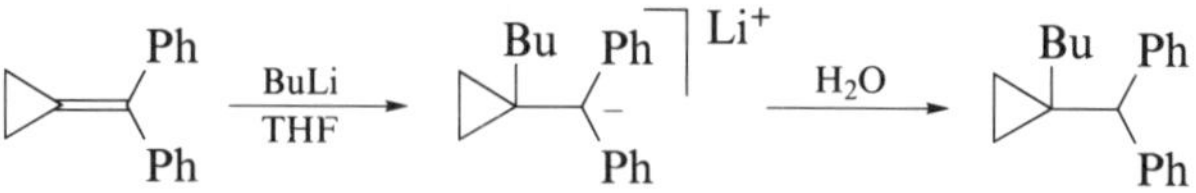

(40)

Scission of the cyclopropane σ-bond is also observed in the intermediate cyclopropylmethyl dilithio derivative obtained from 2-cyclopropyl-1,1-diphenylethylene upon reaction with lithium metal affording the open-chain dilithio-diphenylpentenyl salt (equation 41)[91]. Again, no experimental evidence for the presence of such intermediate was found though the analogous dilithioalkyne derivative was observed (*see below*)[92].

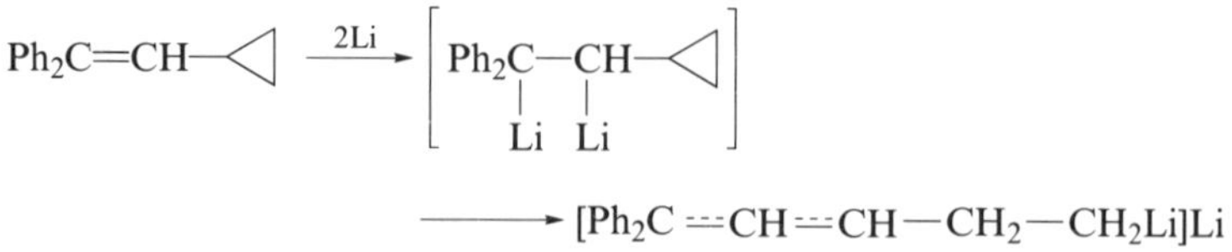

(41)

b. Metal–halogen exchange. Cyclopropylmethyl halides may exchange the halogen by metal essentially by two competing reaction pathways, one which involves the intermediate free cyclopropylcarbinyl radical and the other proceeding *via* nucleophilic addition. Typical examples are the reactions of trialkylstannate R_3Sn^- (R = Me, Bu, Ph)[93] (equation 42) and phenylselenide $PhSe^-$ anions (equation 43) with cyclopropylmethyl halides. In general[94,95], when rearrangement to 3-butenyl metal derivatives is observed, a radical is involved whereas retention of the cyclopropylcarbinyl structure suggests a nucleophilic displacement. However, later studies[96] employing anion trapping agents such as *t*-butylamine and radical scavengers like dicyclohexylphosphine suggest that the cyclopropylcarbinyl anion can also rearrange, making the mechanistic picture quite obscure.

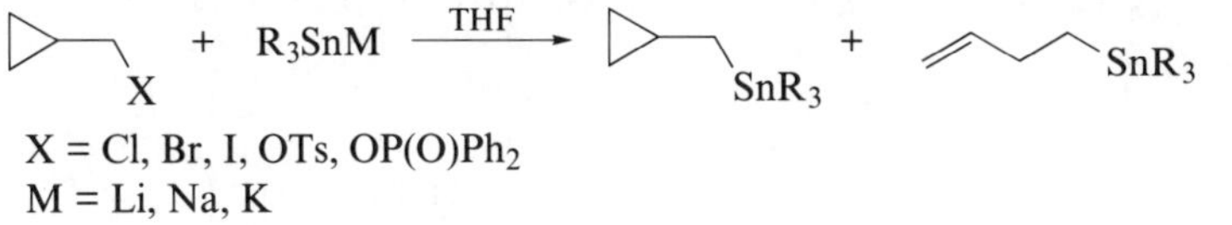

(42)

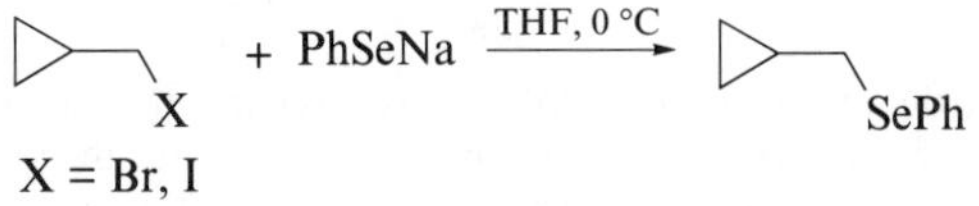

(43)

Cyclopropylcarbinyltrimethylstannanes react with electrophiles either by ring cleavage to 3-butenyl derivatives or alternatively by substituting a methyl group at tin, depending on the electrophile and the solvent[97,98]. Thus, the reaction of (2-methylcyclopropylcarbinyl)trimethylstannane with sulfur dioxide, trifluoroacetic acid (TFA) and iodine in chloroform proceeds with ring cleavage and addition selectively at the unsubstituted cyclopropyl methylene group (equation 44). However, in methanol, iodination and TFA acidolysis occur exclusively at tin, whereas ring cleavage takes place with SO_2.

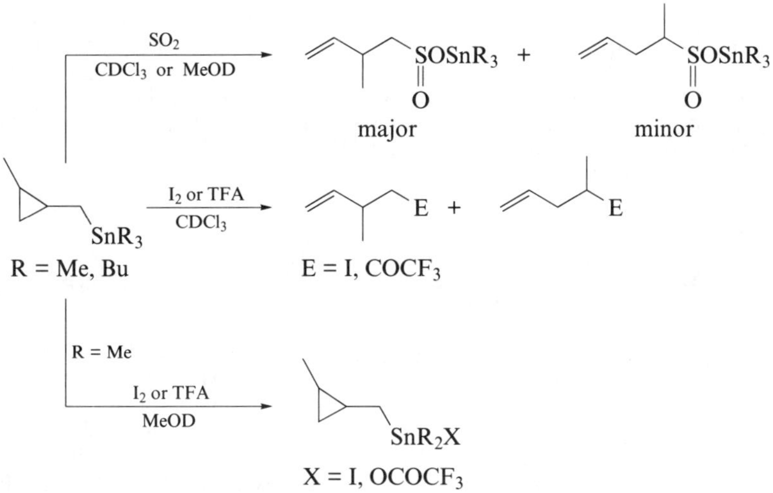

(44)

c. Carbocyclization. Acetoxymercuration of norbornadiene (NBD)[99] under thermodynamic control gave a 3:1 mixture of *exo,exo*- and *exo,endo*-3-acetoxynortricyclyl-5-mercuric chloride (equation 45)[100]. The *exo,endo*-isomer was independently prepared by metal exchange reaction of PhHgOAc with the palladium nortricyclane analogue (*vide infra*), and mercuric chloride catalyzed rearrangement of the bicylic *exo,exo*-2-

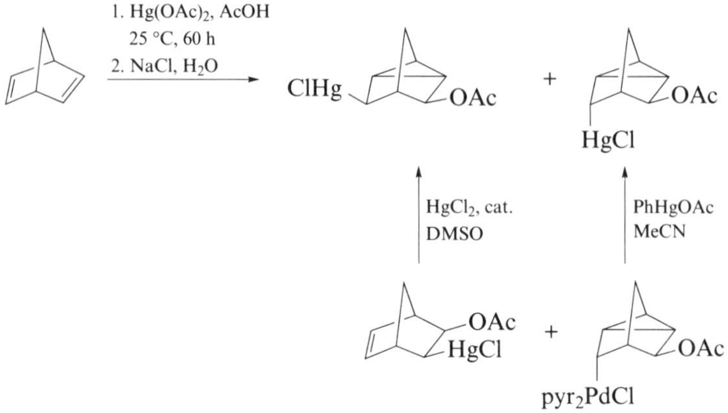

(45)

acetoxynorborn-5-enyl-3-mercuric chloride in DMSO gave selectively the *exo,exo*-nortricyclane isomer. When, however, the acetoxynortricyclane palladium complex is treated instead with Ph_2Hg, acetate-metal exchange occurs together with 5-*endo*-phenylation (equation 46)[101,102].

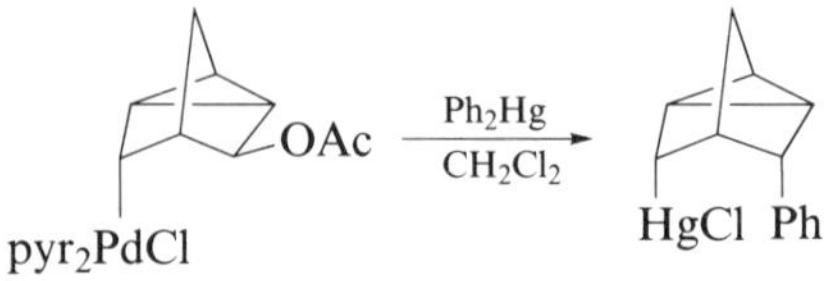

(46)

The versatile selenenylating agent N-(phenylseleno)phthalimide (NPSP) is an effective carbocyclization mediator, which is capable of effecting acid catalyzed cyclization reactions from open-chain olefins, including the formation of cyclopropanes[103]. This is demonstrated by the quantitative cyclization of 3-butenyltrimethyltin to (cyclopropylmethylseleno)benzene upon treatment with 1.1 equivalents of NPSP in CH_2Cl_2 at 25 °C under acid (e.g. p-TsOH) catalysis (equation 47)[104].

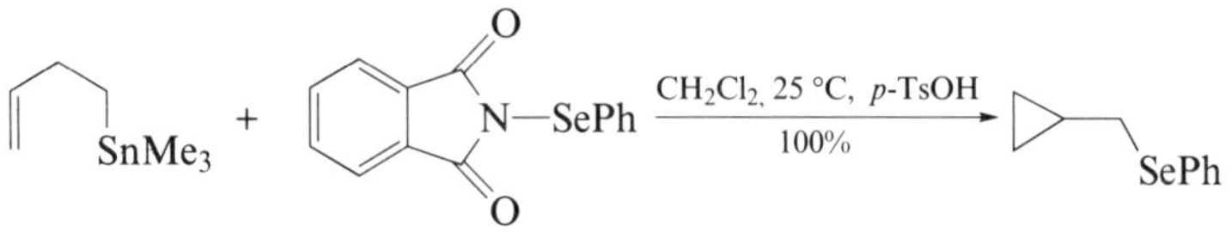

(47)

d. Cyclopropanation. An unusual one-step synthesis of silylated selenocyclopropylcarbinyl compounds by acid catalyzed cyclopropanation-silyl migration of 1,2-selenosilyl olefins with unsaturated carbonyl compounds, has recently been described[105,106]. Thus, reaction of *trans*-1-(phenylseleno)-2-(trialkylsilyl)ethenes with acrolein and vinyl ketones in the presence of $SnCl_4$ afforded *trans*-1-acyl-2-(1-(phenylseleno)-1-(trialkylsilyl)methyl)-cyclopropanes (equation 48). The two geminal hetero groups serve as a protecting group in olefination reactions such as with Zn–CH_2Br_2–$TiCl_4$, and can be later removed by oxidation with $NaIO_4$ to give 2-formyl–vinylcyclopropane derivatives.

The proposed reaction course involves the initial Michael-type addition of the olefin to the Lewis acid activated unsaturated carbonyl compound, forming regioselectively a selenocarbenium ion (*i*). Subsequent 1,2-silyl migration (*ii*), Se-bridging (*iii*) and 1,3-ring

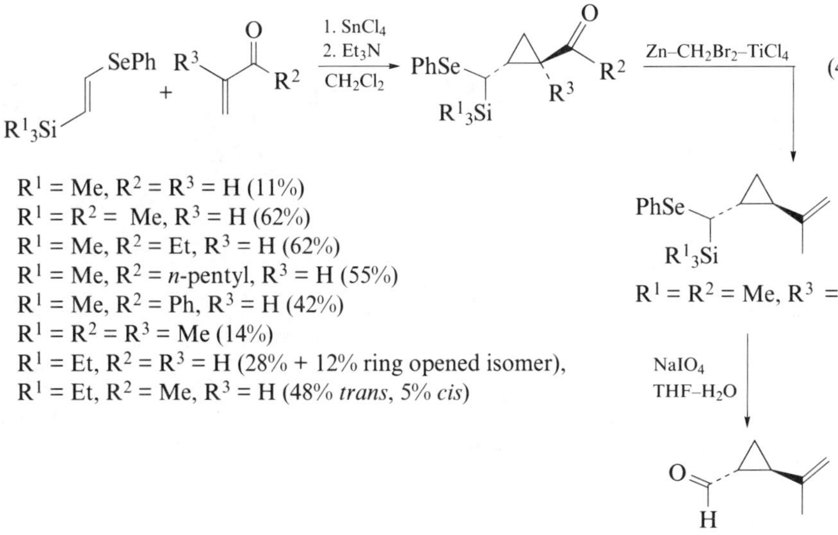

(48)

R^1 = Me, R^2 = R^3 = H (11%)
R^1 = R^2 = Me, R^3 = H (62%)
R^1 = Me, R^2 = Et, R^3 = H (62%)
R^1 = Me, R^2 = *n*-pentyl, R^3 = H (55%)
R^1 = Me, R^2 = Ph, R^3 = H (42%)
R^1 = R^2 = R^3 = Me (14%)
R^1 = Et, R^2 = R^3 = H (28% + 12% ring opened isomer),
R^1 = Et, R^2 = Me, R^3 = H (48% *trans*, 5% *cis*)

closure (*iv*) furnish the product (Scheme 1). This pathway is supported by *ab initio* MO calculations which demonstrate that the seleno-bridged intermediate (step *iii*) is more stable than the initial selenocarbenium ion.

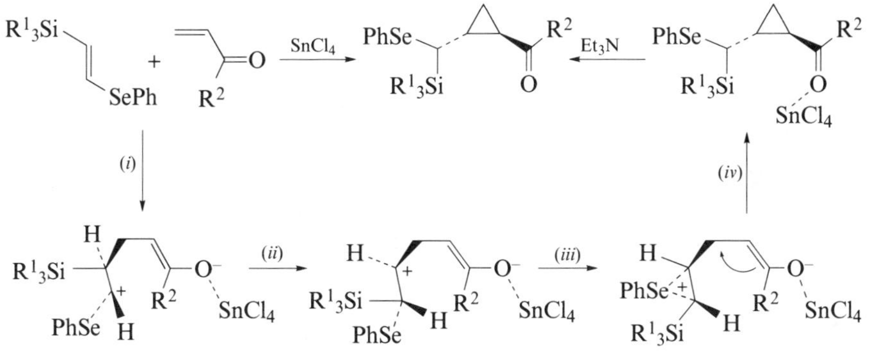

SCHEME 1

2. Transition metal complexes

a. Metal–halogen exchange. Cyclopentadienylmetal carbonyl anion complexes of the type $CpM(CO)_n^-$ (M = Mo, $n = 3$[107]; M = Fe, $n = 2$; M = Ni, $n = 1$) react with cyclopropylmethyl halides to give isolable cyclopropylcarbinyl complexes with moderate stability (equation 49). Ring cleavage to the isomeric 3-butenyl iron complex occurred (30%) during the reaction of $CpFe(CO)_2^-$ (Fp^-) with cyclopropylmethyl iodide but not with the bromide (<3%)[94,108] or benzenesulfonate[109], suggesting that the reaction with iodide proceeds to a significant extent through a cyclopropylcarbinyl radical intermediate.

(Cyclopropylcarbinyl)Fp undergoes a formal 4+1 reaction with sulfur dioxide to give the corresponding sulfone, and a 4+2 reaction with TCNE to give a 6-membered ring

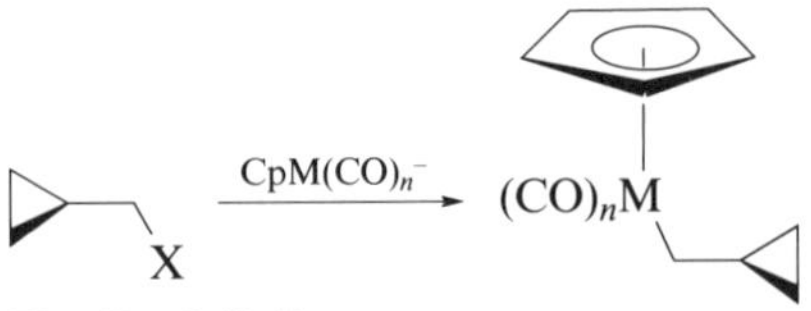

(49)

X = Br, I, BsO
M = Mo, n = 3
M = Fe, n = 2
M = Ni, n = 1

(cyclohexyl)Fp adduct (equation 50)[109]. The reaction of (cyclopropylcarbinyl)Fp with carbenium hydride abstractors proceeds in two distinctive ways, depending on the carbenium reagent. With $Ph_3C^+ X^-$ ($X = BF_4$, PF_6) addition occurs at the γ ring position, leading to the ring-opened (η^2-5,5,5-triphenyl-1-pentene)Fp cation complex. When [(benzocyclobutenium)Fp]PF_6 complex is employed γ-hydride abstraction prevails, affording [(η^2-butadiene)Fp]PF_6 complex salt (equation 51)[110]. Finally, the nickel complex is converted by thermolysis, or preferentially by photolysis to the (η^1,η^2-1-oxopent-4-enyl)NiCp complex (equation 52)[111].

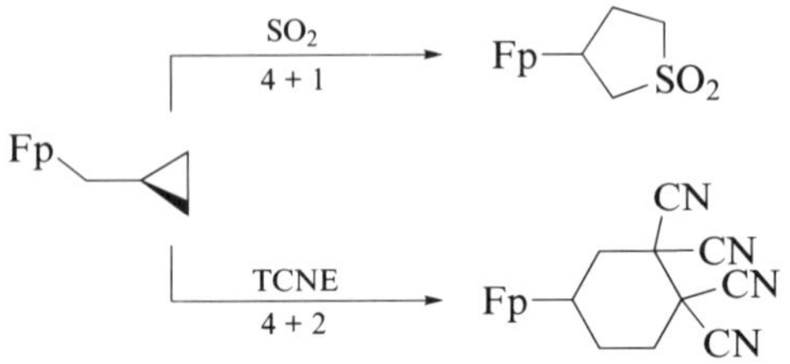

(50)

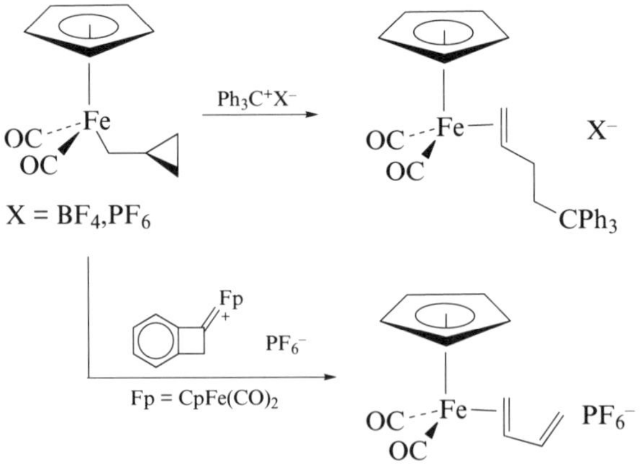

(51)

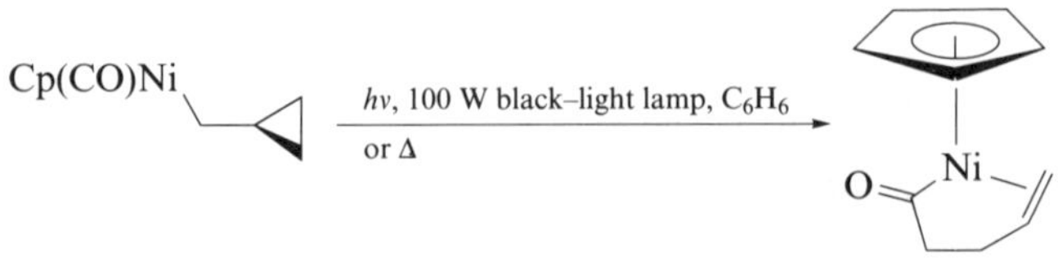

(52)

The sandwich complexes $CoCp_2$ and $CpFe(Me_6C_6)$ were proved to be excellent radical scavengers in oxidative addition reactions with cyclopropylmethyl halides serving as radical clocks[112]. Cyclopropylmethyl iodide reacts slowly (3 days) with $CoCp_2$ in refluxing THF to give a 1:12 mixture of the cyclopropylmethyl and the rearranged 3-butenyl cyclopentenyl cobalt complexes (equation 53). The more efficient (but less selective) radical scavenger $CpFe(Me_6C_6)$ reacts with the corresponding bromide at ambient temperature, furnishing a mixture of 4 isomeric complexes in high yields (equation 54). The cyclopentenyl– and cyclohexadienyl–cyclopropylmethyl regioisomers are obtained in a ratio of 1.7:1, whereas the corresponding pair of 3-butenyl isomers gives a ratio of 1:3.5. The overall ratio of unrearranged to rearranged products is 1:3.5, reflecting the higher efficiency of the iron complex to trap radicals compared to the cobalt counterpart. The reducing ethylenediaminechromium(II) complexes, which are much poorer radical scavengers, give only rearranged products upon reaction with cyclopropylmethyl chloride[113].

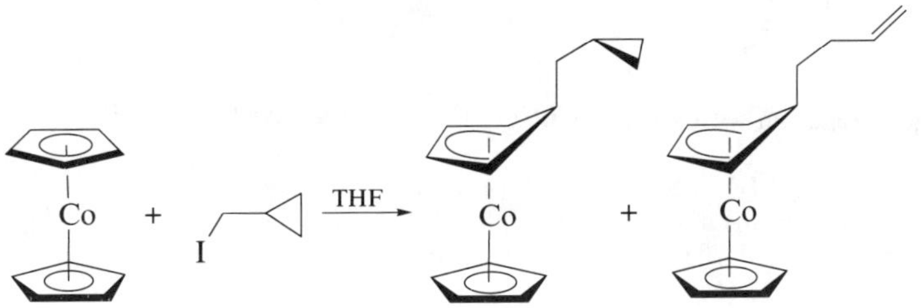

(53)

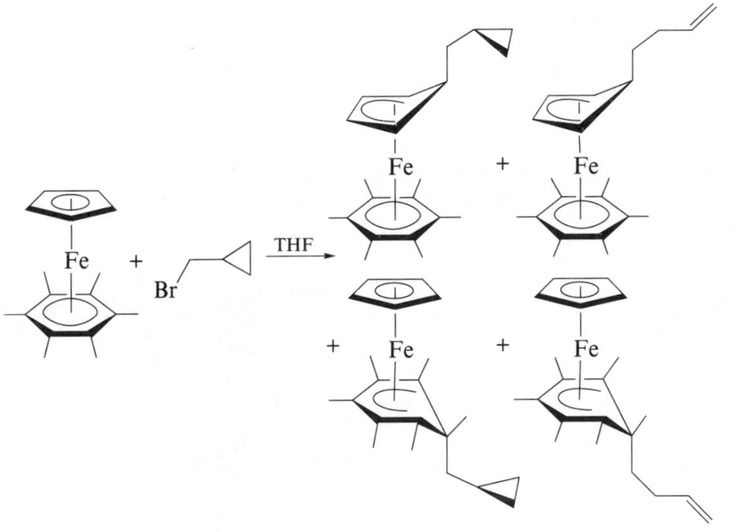

(54)

Transition metal $CpM(CO)_n^-$ anions (M = Mo, W, $n = 3$; M = Fe, $n = 2$; M = Ni, $n = 1$) react with 1,1,1-tris(halogenomethyl)ethane $MeC(CH_2X)_3$ (X = Br, I) in THF, to give mononuclear cyclopropylcarbinyl complexes (equation 55)[114]. The iron and tungsten complexes are stable enough to be isolated but the molybdenum complex decomposes on the alumina chromatographic column during purification attempts. The nickel complex, however, is highly unstable and further rearranges under the reaction conditions, to give the ring-opened (η^1,η^2-4-methyl-1-oxopent-4-enyl)NiCp complex (*vide supra*). Notably,

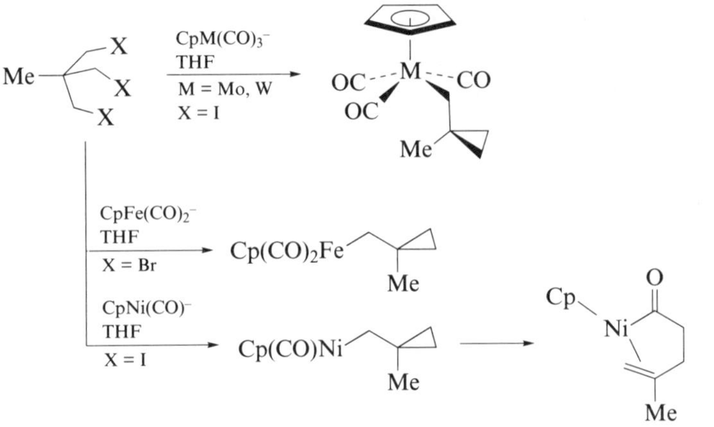

(55)

the analogous reaction with group 9 $CpCo(CO)_4^-$ failed to give the desired carbocyclization reaction.

Proton and ^{13}C NMR spectra show typical high field signals for the cyclopropane protons (0.54–0.62 ppm) and carbons (CH_2 19.8–20.3, C 21.1–22.3 ppm). The MCH_2 protons appear at 1.60–1.70 ppm, and the corresponding carbons resonate at 2.0 (M = W), 14.0 (M = Mo) and 16.1 (M = Fe) ppm.

Further confirmation of the structure comes from the X-ray crystal structure analysis of the cyclopropylcarbinyl tungsten complex $CpW(CO)_3(C_5H_9)$ which reveals a tetragonal bipyramidal geometry at the metal, and a typical bisected conformation of the cyclopropylcarbinyl–M moiety. The sigma C—M bond distance of 234 pm is expectedly much longer than the W—CO bonds (115 pm).

The reaction follows the stoichiometry which requires 3 equivalents of the anion for each tris(halogenomethyl)ethane, furnishing one equivalent of the cyclopropylcarbinyl complex and one dinuclear complex $[CpM(CO)_n]_2$. One possible mechanism is shown in Scheme 2, in which the key step proposed for ring formation is a metathesis involving the metal, the halogen and two carbon atoms.

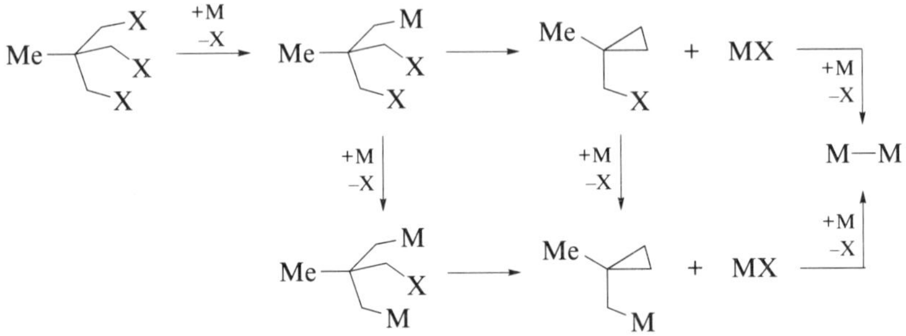

SCHEME 2

Analogously, reaction of group 7 metal carbonyl anions $M(CO)_5^-$ (M = Mn, Re) with $MeC(CH_2I)_3$ gave cyclopropylcarbinyl complexes. However, whereas rhenium produces the normal η^1-metal bonded cyclopropylcarbinyl complex, the manganese complex furnished (cyclopropylacetyl)$Mn(CO)_5$, where carbonyl insertion into the manganese–alkyl bond occurred (equation 56)[114].

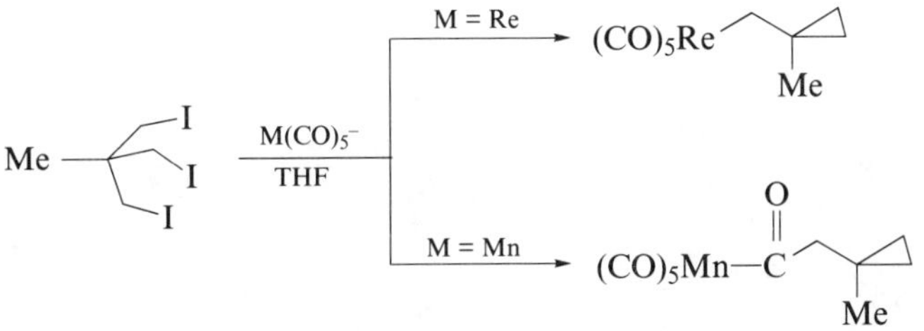

(56)

b. Hydrometallation. Cyclopropylmethyl metal complexes of titanium(III) have been obtained by reaction of Cp_2TiCl_2, Grignard reagent and methylenecyclopropane derivatives. The Grignard reagent presumably affords a titanium hydride intermediate which adds to the double bond (*vide infra*, section VII.B.3.b)[115]. Subsequent demetallation by HCl in ether or in THF gave the corresponding methylcyclopropanes (equation 57). The analogous reaction of group 10 complex Cp_2Ni failed to give an observable intermediate or a cyclopropane. Instead, rearrangement occurs to give (3-butenyl)NiCp derivatives[116]. A related cleavage of methylenecyclopropane by Rh(I) complexes has also been described[117].

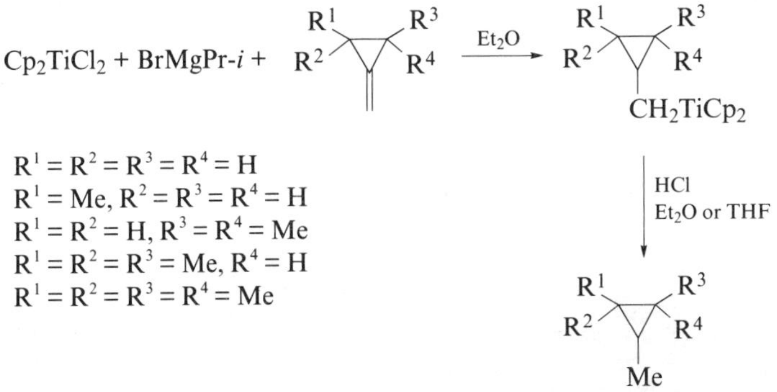

(57)

c. Carbocyclization. The norbornenyl-palladium complexes 3-*exo*-methoxy- and 3-*exo*-acetoxy- di-μ-chlorobis[(η^1,η^2-2,5,6)-norbornen-5-yl-*endo*-palladium][118–120], prepared by reaction of NaOMe and AgOAc with (η^2,η^2-norbornadiene)$PdCl_2$, undergo a formal homoallyl to cyclopropylcarbinyl cyclization and ligand exchange, upon reaction with metal ligands such as pyridine, triphenylphosphine and bis(diphenylphosphino)ethane (DPPE), to give the corresponding η^1-*exo,endo*-3-methoxy- and 3-acetoxynortricyclyl-5-palladium complexes (equation 58)[121,122]. Substitution of the metal is effected by halogens and metanolic CO to give the corresponding halides and esters, respectively. Reduction occurs with $LiAlH_4$, and alkylative mercuration[101,102] by diphenyl- and diisobutenylmercury gives the corresponding *endo,endo*-nortricyclic mercury compounds.

Treatment of (η^2,η^2-norbornadiene)$PdCl_2$ complex with Ph_2Hg or vinyl mercuric chloride derivatives ($R^1R^2C{=}CH$)HgCl (R^1,R^2 = H, Me, *t*-Bu, Cl) in THF or methanol furnished the corresponding *endo,endo*-3-phenyl- and *endo,endo*-3-vinyl-5-nortricyclyl palladium complexes, respectively (equation 59)[102,123–126]. This approach, employing a thiophene mercury compound combined with subsequent carbonylation (*vide supra*), has been used in the synthesis of prostaglandins[127]. Notably, the reaction of the platinum analogue failed to give nortricyclyl metal derivatives. The reaction with Ph_2Hg gave instead a sequential exchange of the chlorine ligands by phenyl groups[102,123].

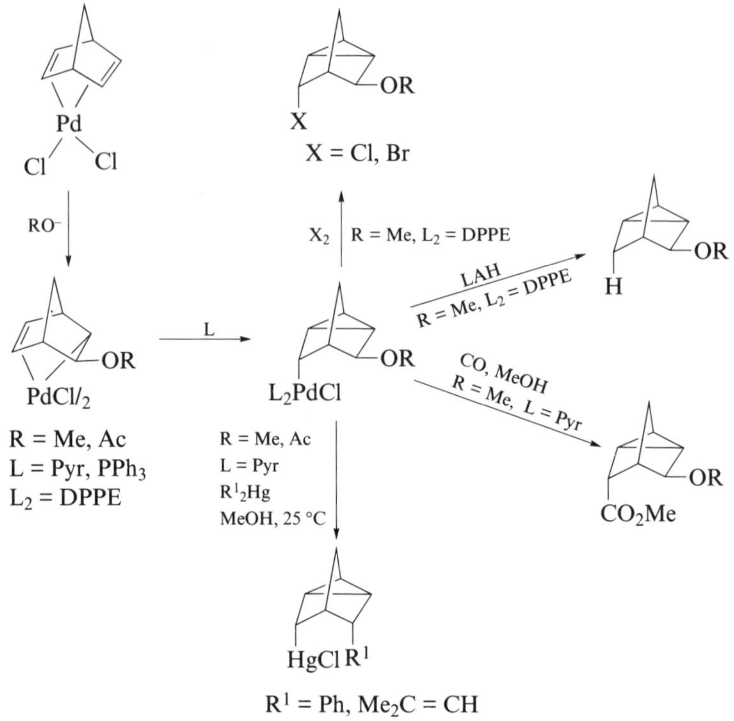
Pd
Cl
Cl
RO⁻
PdCl/2
OR
R = Me, Ac
L = Pyr, PPh3
L2 = DPPE
L
L2PdCl
OR
X2
R = Me, L2 = DPPE
X
OR
X = Cl, Br
LAH
R = Me, L2 = DPPE
H
OR
CO, MeOH
R = Me, L = Pyr
CO2Me
OR
R = Me, Ac
L = Pyr
R¹2Hg
MeOH, 25 °C
HgCl R¹
R¹ = Ph, Me2C = CH

(58)

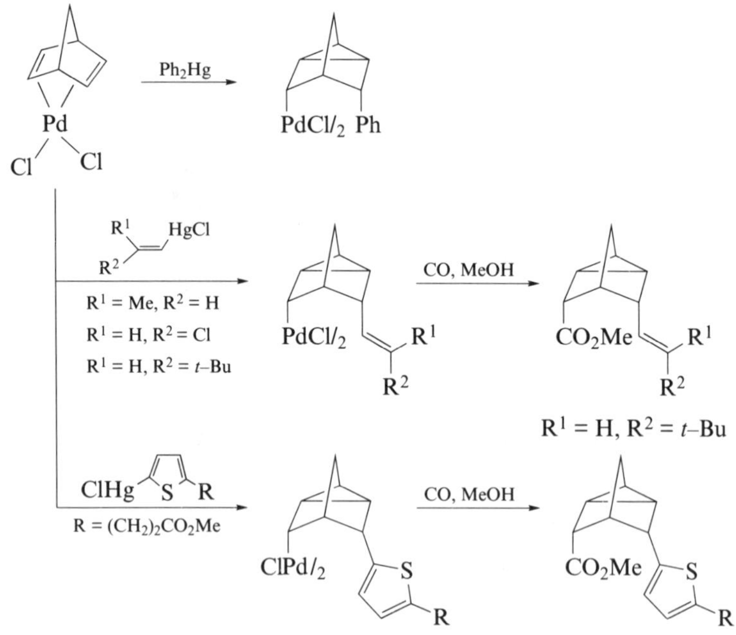
Pd
Cl
Cl
Ph2Hg
PdCl/2 Ph
R¹
HgCl
R²
R¹ = Me, R² = H
R¹ = H, R² = Cl
R¹ = H, R² = t–Bu
PdCl/2
R¹
R²
CO, MeOH
CO2Me
R¹
R²
R¹ = H, R² = t–Bu
ClHg
S
R
R = (CH2)2CO2Me
ClPd/2
S
R
CO, MeOH
CO2Me
S
R

(59)

d. Insertion. Photolysis of quadricyclane and 2 equivalents of $Fe(CO)_5$ in ether afforded an acyliron metallacycle as the main product, together with monocyclic fragmentation products (equation 60a). Quadricyclanone reacts similarly with $Fe(CO)_5$ upon irradiation, affording two isomeric acylmetallacycles, both having the cyclopropylcarbinyl iron moiety (equation 60b)[128]. An earlier work employing $[Rh(CO)_2Cl]_2$ gave similarly the analogous acylrhodium metallacycle having a Cl-bridged polymeric structure[129]. The rhodium complex undergoes a rapid stoichiometric reaction with triphenylphosphine in $CHCl_3$ to give a 6-membered metallacyclic dimer. The insertion reactions presumably arise from initial oxidative addition of quadricyclane to the metal and subsequent carbonyl insertion into the metal–carbon bond.

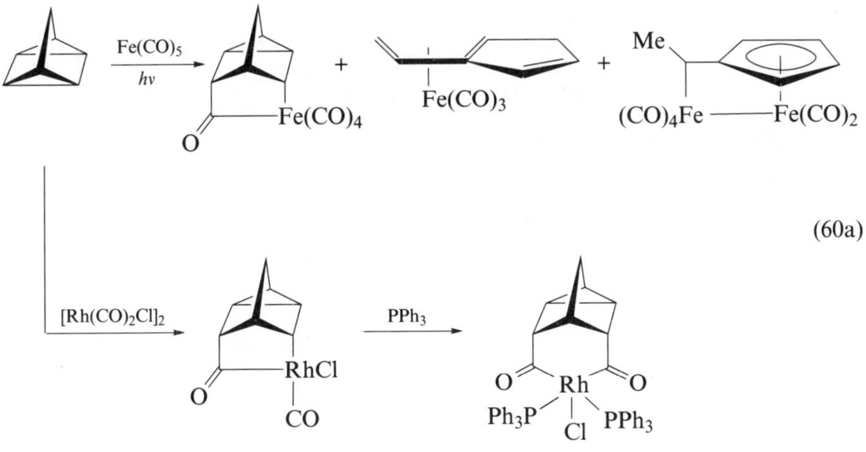

(60a)

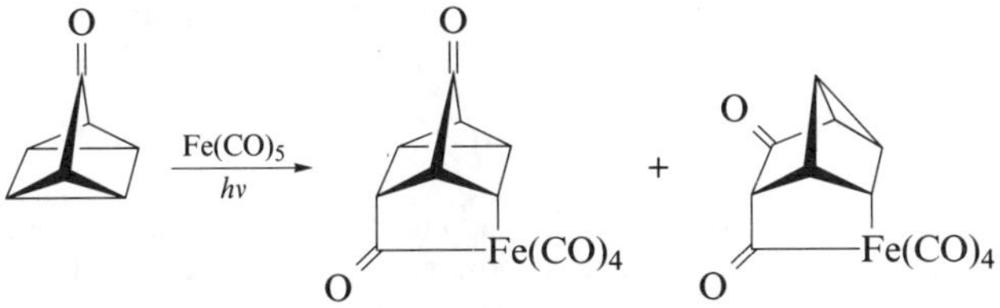

(60b)

e. Catalysis. To conclude this section it is perhaps worthwhile to connect the above stoichiometric reactions with the catalytic processes. (Cyclopropylmethyl)metal complexes are often involved as intermediates in catalytic reactions, particularly those of methylenecyclopropane (mcp) (*see below*). The Ni(0)-catalyzed cyclodimerizations and codimerizations are of special interest since, under careful reaction conditions, these intermediates can be isolated. Thus, 4-nickeladispiro[2.2.2.1]nonane 2,2′-bipyridine (Bipyr) complex is obtained either directly from methylenecyclopropane and (Bipyr)Ni(COD) in THF (–78 to –30 °C) or from (Bipyr)Ni(η^2-methylenecyclopropane) and free methylenecyclopropane in toluene (0 °C) (equation 61)[130,131]. At higher temperatures (32 °C) mixtures are obtained containing monospiro isomers. These metallacyclic cyclopropylmethyl complexes undergo e.g. reductive elimination by methyl acrylate to give the same dimers which are produced under catalytic conditions.

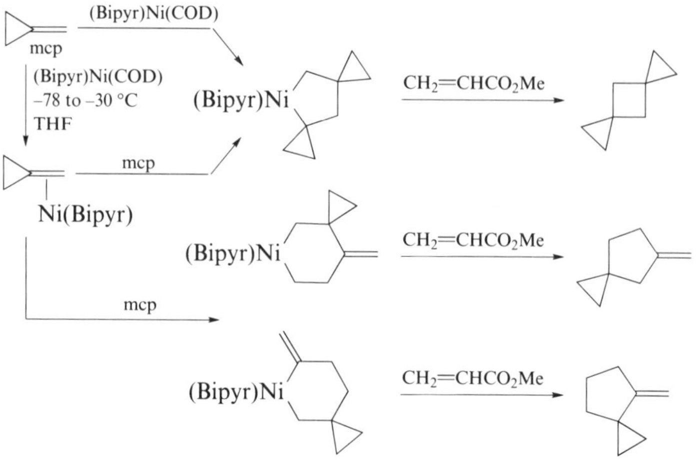

(61)

C. Cyclopropylcarbene Complexes

Cyclopropylcarbene complexes of the type $L_nM{=}C(XR^1)R^2$ ($X = O, S$; R^1 = alkyl, aryl; R^2 = cyclopropyl) having a stabilizing heteroalkyl (XR^1) group on the electrophilic carbene ligand (Scheme 3) have found widespread application in organic synthesis. These so-called Fischer carbene complexes are best known via their group 6 transition metal carbonyl complexes $(CO)_5M{=}(OR^1)R^2$(M = Cr, Mo, W)[132]. Much less abundant are the Schrock-type cyclopropylcarbene complexes $L_nM{=}CR^1R^2$ where no heteroatom is bound to the carbene carbon atom[133].

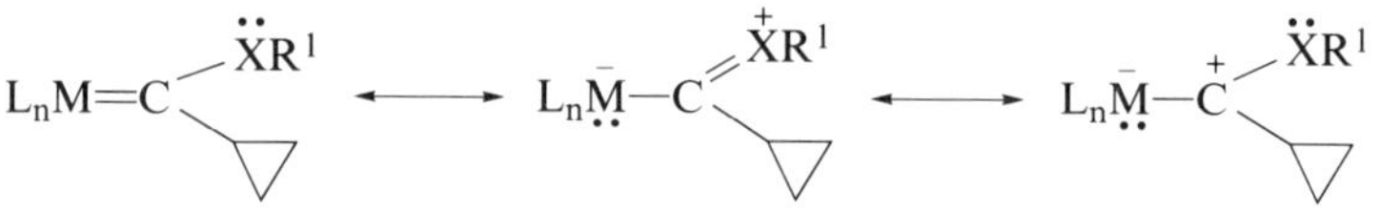

SCHEME 3

Alkoxycyclopropylcarbene complexes display characteristic ^1H and ^{13}C NMR signals for the α-cyclopropyl proton and carbene carbon, respectively. The α-protons resonate at the narrow 3.1–3.5 ppm range. The carbene carbons absorb significantly downfield from the carbonyl ligands (*ca* 215 ppm) over the range of 325–360 ppm.

1. Preparation

While the first transition metal carbene complex was reported in 1964[134], the first *cyclopropyl*carbene complex salt $[(CO)_5Cr{=}C(O^-)(c\text{-}Pr)]NMe_4^+$ was only synthesized about a decade later, by addition of cyclopropyllithium to chromium hexacarbonyl, followed by tetramethylammonium bromide[135]. Subsequent reaction with trimethyloxonium fluoroborate gave methoxycarbene complex $(CO)_5Cr{=}C(OMe)(c\text{-}Pr)$ (equation 62)[136].

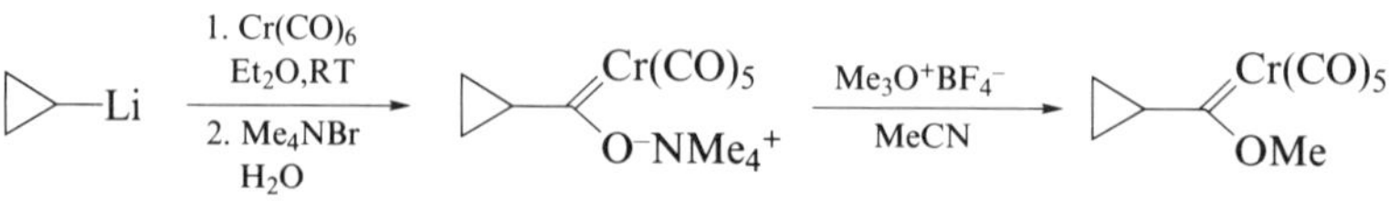

(62)

When the alkoxy group is more complex, the ammonium salt is first treated with acetyl chloride in CH_2Cl_2 to give the intermediate acetate complex $(CO)_5M{=}C(OAc)(c\text{-}Pr)$[136] which, on subsequent addition of the appropriate alcohol, gives the desired alkoxy cyclopropyl carbene complex $(CO)_5M{=}C(OR)(c\text{-}Pr)$ [M = Mo, W; R = $(CH_2)_nCH{=}CH_2$, $(CH_2)_nSMe$ (n = 2,3), $(CH_2)_3C{\equiv}CPh$, Ph] (equation 63)[137,138].

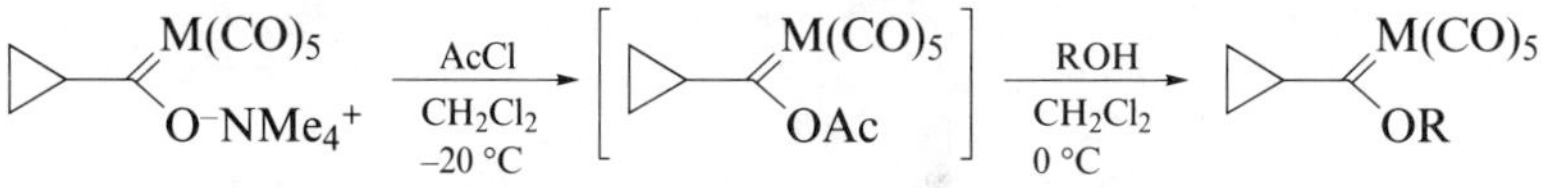

(63)

M = Mo, W
R = $(CH_2)_nCH{=}CH_2$, $(CH_2)_nSMe$, $(CH_2)_3C{\equiv}CPh$, Ph

Preparation of ethylthio-[138] and phenylthio-carbene complexes[139] was similarly achieved by treatment of the corresponding isolated tetrabutylammonium acylate complexes in CH_2Cl_2 at –40 °C with acetyl chloride, then with the corresponding thiols at 0 °C as illustrated in equation 64 for the phenylthio derivatives.

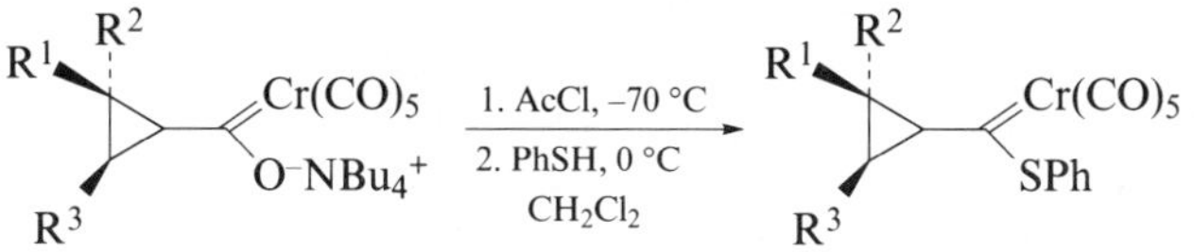

(64)

$R^1 = R^2 = R^3 = H$
$R^1 = R^2 = Me, R^3 = H$
$R^1 = R^2 = H, R^3 = n\text{-}C_6H_{13}$ (*trans*)
$R^1R^3 = {-}(CH_2)_4{-}$; $R^2 = H$ (*exo*)
$R^1R^3 = {-}(CH_2)_5{-}$; $R^2 = H$ (*exo*, *endo*)

For most methoxycarbene complexes a 'one pot' modification of the above method is utilized. This involves direct alkylation of the initially formed lithium acylate carbene complex with trifluoromethanesulfonate or with methyl fluorosulfonate. The method is successfully employed for preparation of chromium[140] as well as molybdenum and tungsten monocyclic (equation 65), bicyclic (equation 66) and tricyclic carbene complexes (equation 67)[137,138,140,141].

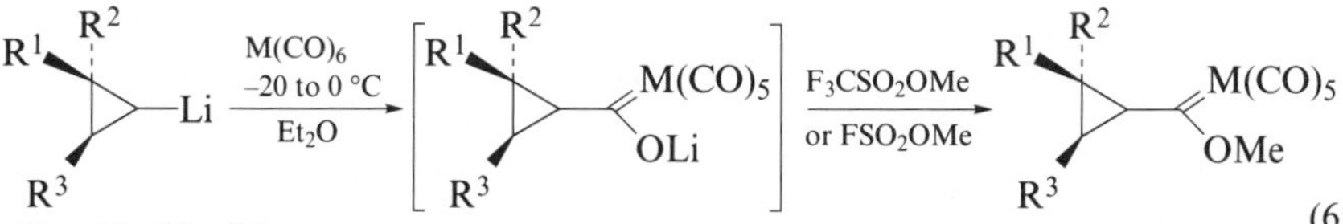

(65)

M = Cr, Mo, W
$R^1 = R^2 = R^3 = H$
$R^1 = R^2 = H, R^3 = Ph$
$R^1 = R^2 = Me, R^3 = H$
$R^1 = R^3 = Me, R^2 = H$ (*cis*)
$R^1 = Me, R^2 = vinyl, R^3 = H$ (isomeric mixture)

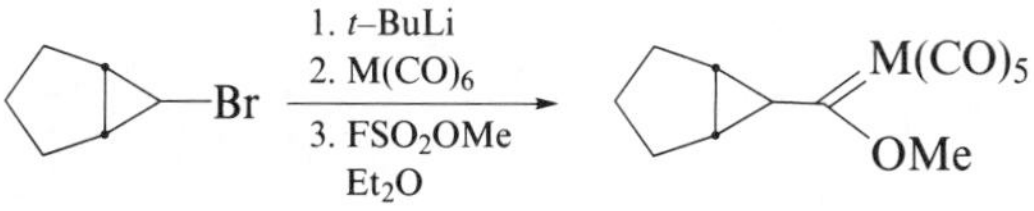

(66)

M = Mo, W

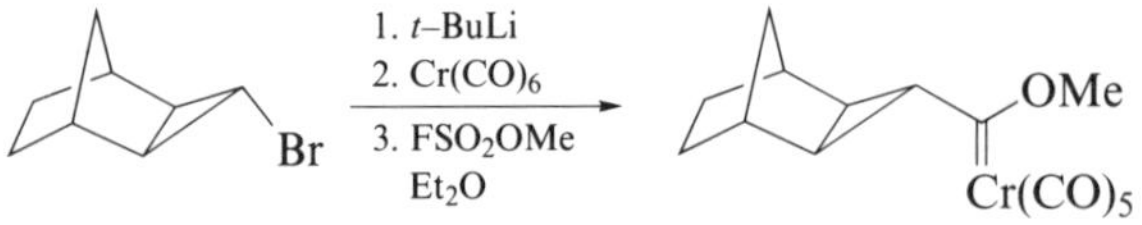

(67)

A different strategy is utilized to prepare Fischer-type cyclopropylmethoxycarbenium complexes. $[LM_nC{=}(OMe)(c\text{-}Pr)]^+$ by alkylation of cyclopropylcarbonyl complexes c-PrC(O)ML_n (*vide supra*). Thus, treating acyl complex FpCO(c-Pr), obtained from the reaction of NaFp [Fp = $Cp(CO)_2$] and c-PrCOCl, with $Me_3O^+BF_4^-$ in CH_2Cl_2 gives cyclopropylmethoxycarbenium fluoroborate complex salt [Fp=C(OMe)(c-Pr)]BF_4 (equation 68) which can be further reduced to the corresponding Schrock-type cyclopropyl carbene complex (*vide infra*, Section II.C.2)[142].

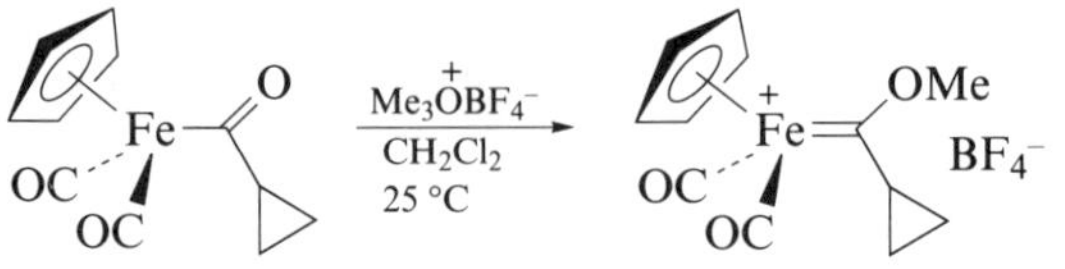

(68)

2. Reactions

a. Without ring cleavage. (i) Interconversion of Fischer complexes. The key precurser to heteroatom-stabilized (Fischer) cyclopropylcarbene complexes is the acylate salt obtained by reaction of group 6 $M(CO)_6$ with cyclopropyllithium, which may be isolated as a tetraalkyl ammonium salt, and further reacts with electrophiles. For example, tetramethylammonium chromium complex $(CO)_5Cr{=}C(O^-)(c\text{-}Pr)]Me_4N^+$ undergoes potonation, methylation and acetylation to form the corresponding hydroxy, methoxy and acetate cyclopropylcarbene chromium complexes (Scheme 4)[136]. The methoxy group can

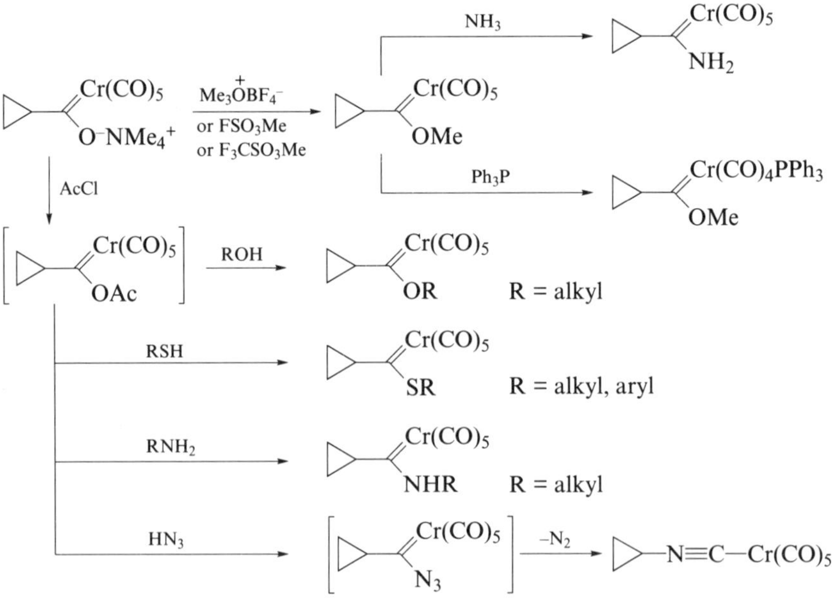

SCHEME 4

be substituted with an amino group by reaction with ammonia, and one ligand carbonyl is substituted by phosphine[140] upon reaction with triphenylphosphine. The acetate, which could not be isolated, can be exchanged with alcohols[143], thiols[138,139] and optically active amines[144] to give alkoxy, thio and amino complexes, respectively. Finally, the unusual reaction with HN_3 leads, via loss of nitrogen and rearrangement[136], to the cyclopropyl isocyanide complex $(CO)_5Cr(CNC_3H_5)$.

(ii) Reactions of carbenium complexes. Sodium borohydride reduction of cyclopropylmethoxycarbenium fluoroborate complex salt [Fp=C(OMe)(*c*-Pr)]BF_4, under basic conditions at –78 °C, gave cyclopropylcarbinyl iron complex FpCH(OMe)(*c*-Pr) (equation 69)[142]. This complex is only stable below 0 °C under nitrogen. It reacts with trimethylsilyl triflate at –78 °C to give another labile complex, identified as the cyclopropylcarbenium fluoroborate complex [Fp=CH(*c*-Pr)]BF_4 by NMR spectroscopy (^{1}H resonance at 16.7 ppm, ^{13}C at 365 ppm) (*vide supra*). The latter readily transfers cyclopropylcarbene to olefins, to give dicyclopropyl derivatives (*vide infra*).

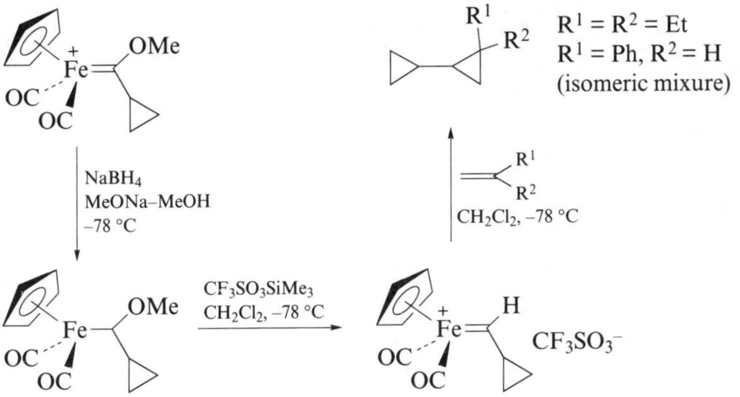

(69)

(iii) Reactions with alkenes (cyclopropanations). The reaction of Fischer cyclopropylcarbene chromium complexes with electron-poor olefins, namely esters and amides but not α,β-unsaturated ketones, leads to bicyclopropyl compounds in good yields, accompanied by minor amounts of ring-opened monocyclopropyl olefin[145,146]. A typical reaction between $(CO)_5Cr$=C(OMe)(*c*-Pr) and methyl acrylate in refluxing THF gave a mixture of *cis*- and *trans*-methoxybicyclopropanecarboxylates (*trans*:*cis* = 46:54) in 64% yields together with 5% of methyl 4-cyclopropyl-4-methoxy-2-butenoate (equation 70). Addition of phosphine to the reaction mixture results in decrease in reaction yields (39%) and

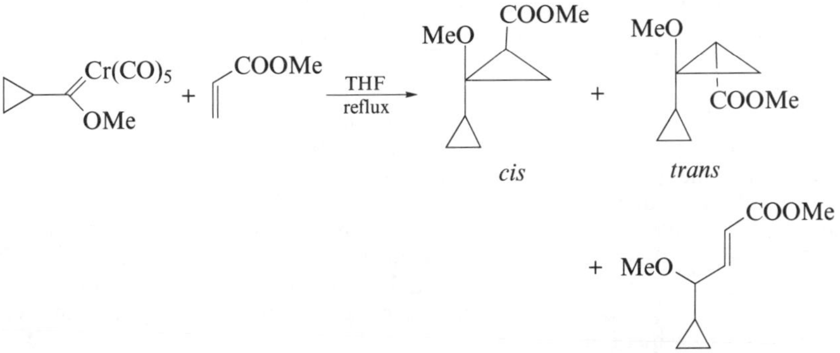

(70)

increase in *cis*-isomer ratio (*trans*:*cis* = 26:74). Thus the carbene complex functions mainly as a cyclopropanating agent affording donor–acceptor-substituted cyclopropanes[147], together with minor amounts of allylic C—H insertion products. Increase in the number of substituents on the alkene, particularly at the β-position, reduces the yields of cyclopropanation products due to steric effects, and improves the yields of the insertion product. The reaction appears to be stereoselective with regard to retention of the stereochemistry of the original olefin substituents, except for the reaction with maleate where both the expected *cis*-diester and the isomerized *trans* counterpart are formed (equation 71).

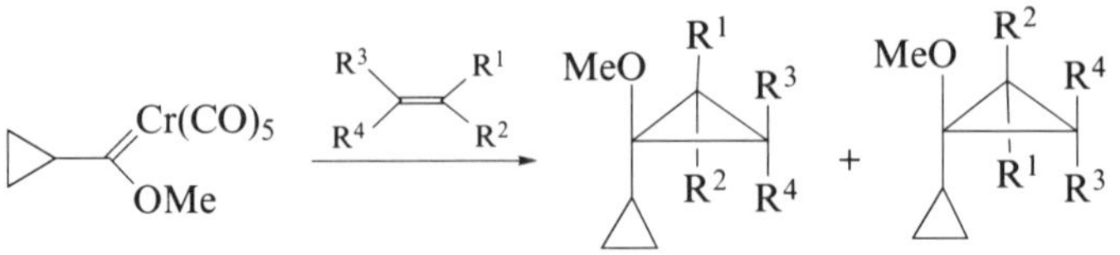

(71)

R[1]	R[2]	R[3]	R[4]	% Cyclopropanation (% insertion)	*trans*:*cis*
COOMe	H	H	H	63 (9)	48:52
COOMe	H	H	COOMe	62(18)	
COOMe	H	COOMe	H	37 (12)	
COOMe	Me	H	H	69	22:78
COOMe	H	H	Me	66	>95% *cis*
$CONMe_2$	H	H	H	81	21:79
COOMe	H	H	Ph	5	0:100
COOMe	H	Me	Me	0 (7)	
COOMe	Me	H	Me	50	14:86

Simple alkenes, norbornene and styrene do not undergo cyclopropanation or insertion reactions with cyclopropyl(methoxy)carbene chromium pentacarbonyl complex. However, the conjugated 1-vinylcyclopentene is cyclopropanated under the reaction conditions at the terminal double bond, affording an isomeric mixture (*trans*:*cis* = 40:60), in 66% yield (equation 72).

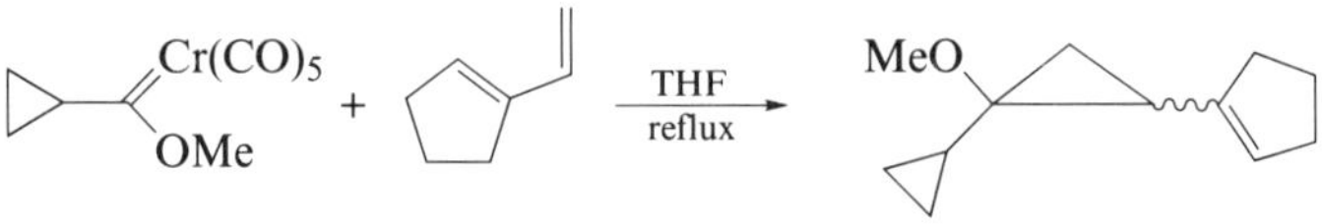

(72)

trans:*cis* = 40:60

(iv) Photocycloadditions. Photolysis of Fischer chromium carbene complexes under CO pressure produces species that react with olefins as if they were ketenes, to give cyclobutanones. Thus, the photocycloaddition of cyclopropyl(methoxy)chromium carbene complex to cyclohexadiene affords stereoselectively the 2+2 adduct 8-cyclopropyl-8-methoxybicyclo[4.2.0]oct-2-en-7-one, in which the carbonyl group is in the γ-position to the double bond, and the major (95%) stereoisomer has the intact cyclopropyl group in the *syn* configuration (equation 73)[143]. The intramolecular photocycloaddition of the corresponding 3-methyl-3-buten-1-oxy complex gave 1-cyclopropyl-5-methyl-2-oxabicyclo-[3.2.0]heptan-7-one (equation 74).

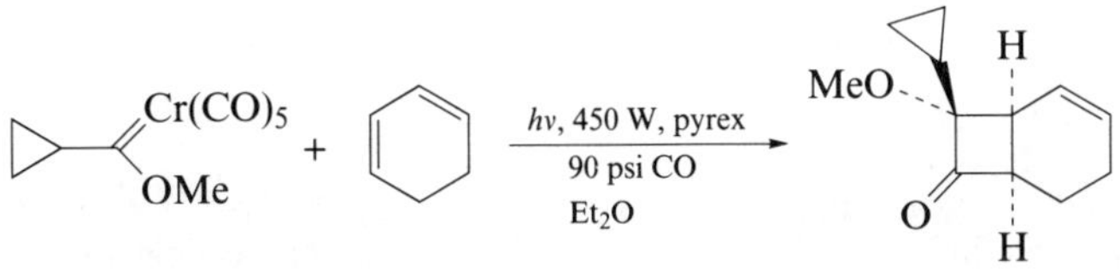

(73)

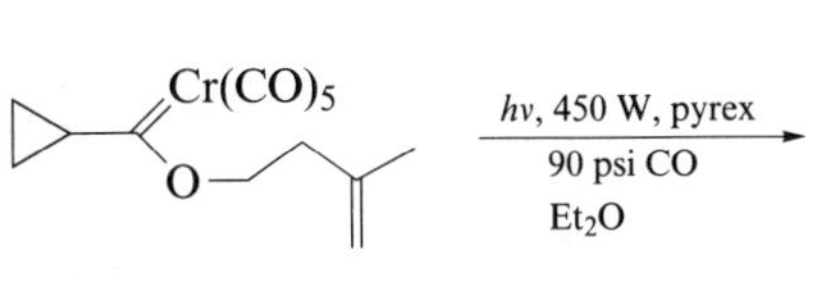

(74)

Similarly, photolysis of chromium complex $(CO)_5Cr{=}C(OMe)(c\text{-}Pr)$ in the presence of two equivalents of the optically active 3-ethenyl-6(*S*)-phenyl-2-oxazolidinone under 90 psi CO pressure produced in high regio- and diastereo-selectively the 3-oxazolidine-substitued (2*R*,3*S*)-cyclobutanone, in optical purity of ≥ 97% de (equation 75)[148].

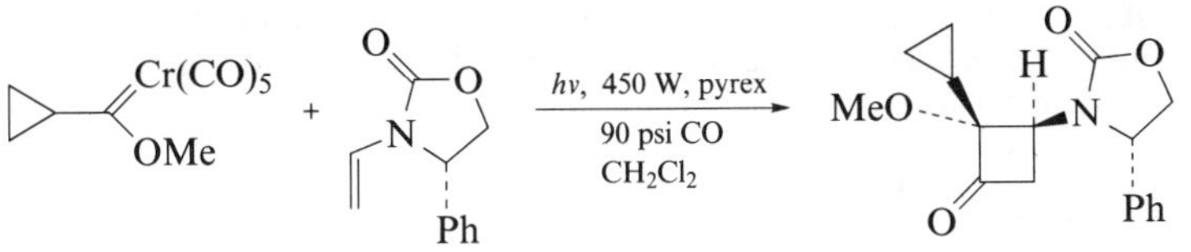

(75)

b. Addition reactions. Cyclopropyl(phenylthio)carbene chromium complexes undergo ring cleavage by stereocontrolled 1,5-addition of halogens (Br_2, I_2) and pseudohalogens (e.g. PhSeCl), which is spontaneously followed by demetallation to give 1,4-dihalo-1-phenylthioalkene derivatives (equation 76)[139]. Reaction occurs predominantly at the more substituted ring position and with inversion of configuration. As with other electron-deficient cyclopropanes, e.g. acyl derivatives, the reaction presumably involves initial activation by the electrophilic halogen (as Lewis acid) on the metal, and a subsequent ring cleavage by nucleophilic halide γ-substitution. This mechanism is supported by the reaction with phenylselenium chloride, where the chlorine ends up at the alkyl position and the phenylseleno group is at the alkenyl position (equation 77). This product is accompanied by the corresponding dichlorothioalkene, which results from further reaction of the selenoolefin with PhSeCl.

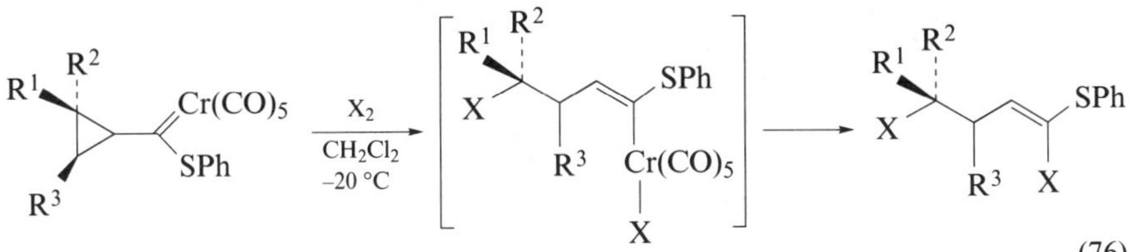

(76)

X = Br, I
$R^1 = R^2 = R^3 = H$
$R^1 = R^2 = Me$, $R^3 = H$
$R^1 = R^2 = H$, $R^3 = n\text{-}C_6H_{13}$ (*trans*)
$R^1R^3 = -(CH_2)_4-$, $R^2 = H$ (*exo*)
$R^1R^3 = -(CH_2)_5-$, $R^2 = H$ (*exo, endo*)

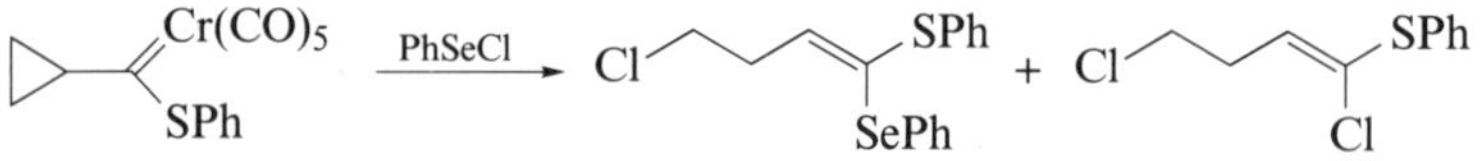

(77)

Cyclopropylalkoxycarbene complexes are much less reactive towards nucleophilic reactions. The methoxy chromium complex does, however, react with iodine to give directly methyl 4-iodobutyrate, derived by hydrolysis of the expected diiodovinyl ether (equation 78)[139].

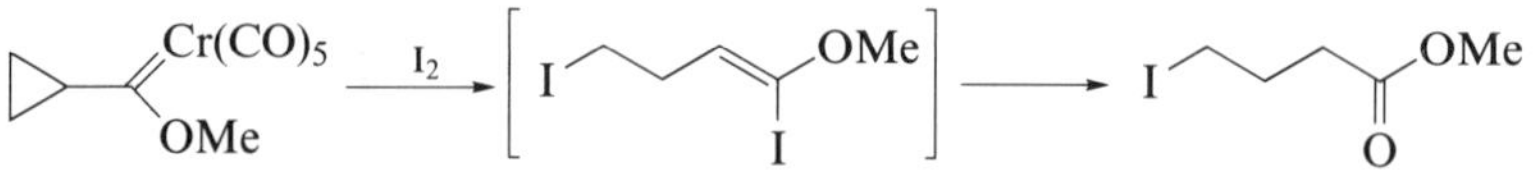

(78)

c. Cycloadditions with alkynes. Cycloaddition reactions of cyclopropylcarbene carbonyl complexes of group 6 (Cr, Mo, W) with alkynes have become an important synthetic route to 5 and 7 membered rings. In general, chromium complexes are preferentially used for construction of the 5-membered rings, whereas molybdenum and tungsten are mainly used to prepare 7-membered ring systems. Molybdenum carbene complexes typically display greater reactivity than either the chromium or tungsten counterparts.

(i) Chromium complexes. Reaction of cyclopropyl(methoxy)carbene chromium pentacarbonyl complexes with terminal and internal alkynes in refluxing wet dioxane (1% water) gives mainly 5-mono- and 4,5-disubstituted 3-methoxy cyclopentenone, respectively, and ethylene (alkenes) derived from the two noncarbenoid cyclopropyl carbons (equation 79)[140,149,150]. The reaction with internal alkynes give mixtures of both *cis* and *trans* isomers. The ratio between the isomers depends on the reflux time, with the *trans*-isomer predominating at longer times. With nonsymmetrically internal alkynes the larger group appears at the 5-position adjacent to carbonyl, and small amounts of isomeric 2,3-disubstituted 4-methoxycyclopentenone are detected in some reactions.

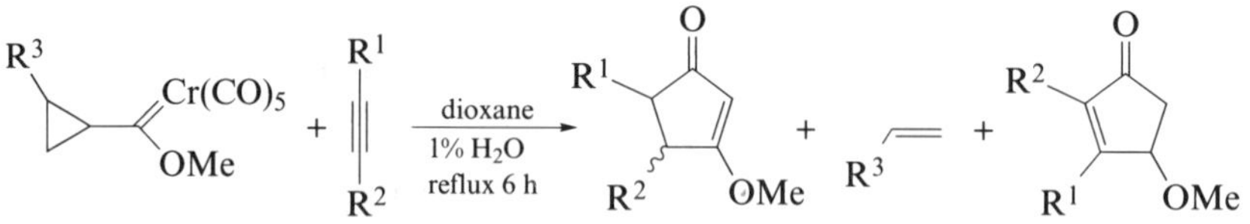

(79)

	trans:cis
$R^1 = Pr, R^2 = R^3 = H$	
$R^1 = c\text{-}Pr, R^2 = R^3 = H$	
$R^1 = Ph, R^2 = R^3 = H$	
$R^1 = i\text{-}Pr, R^2 = Me, R^3 = H$	53:47
$R^1 = R^2 = Pr, R^3 = H$	50:50
$R^1 = Ph, R^2 = Me, R^3 = H$	73:27
$R^1 = R^2 = Ph, R^3 = H$	85:15
$R^1 = R^2 = R^3 = Ph$	
$R^1 = (CH_2)_4OX, R^2 = R^3 = H; X = H, SiMe_2Bu\text{-}t, SO_2Me$	

The effect of the solvent and the temperature on the product composition is crucial. THF and benzene provide primarily the cyclopentanones, but the proportions of the *cis*-isomer are considerably higher than in dioxane. When acetonitrile or DMF is used in the cycloaddition with diphenylacetylene, a complex mixture of products is obtained (equation 80)[140].

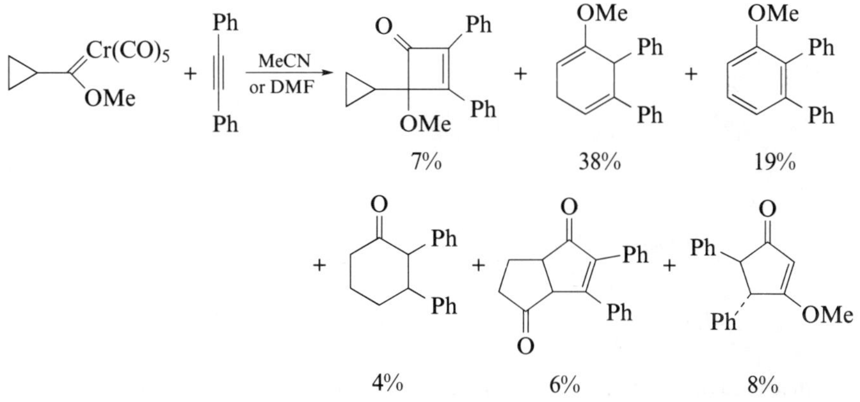

(80)

Extension of the reaction to propargylic cyclobutenols results in a tandem carbene insertion–semipinacol rearrangement of the cyclobutenol ring, leaving the cyclopropyl ring intact (equation 81)[151].

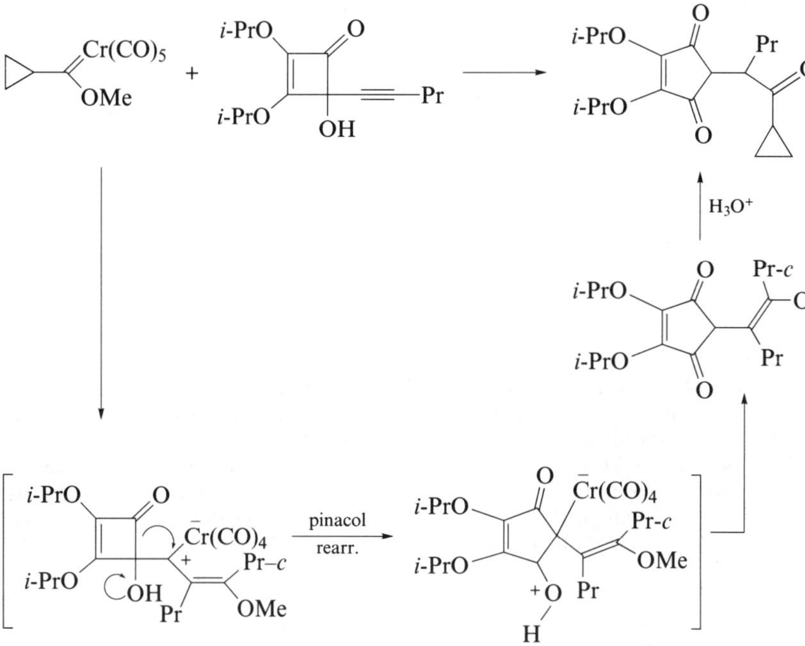

(81)

The reaction with 1,6-heptadiyne gives cyclopropylphenol, presumably via insertion–intramolecular–cycloaddition steps. Again, the cyclopropyl ring is retained (equation 82)[140].

When cyclopropylidenemethyl(ethoxy)carbene chromium pentacarbonyl is treated with diphenylacetylene in THF, a mixture of products is obtained, from which only the tricarbonyl(5-ethoxy-6,7-diphenylbenzofuran)chromium complex could be isolated, in

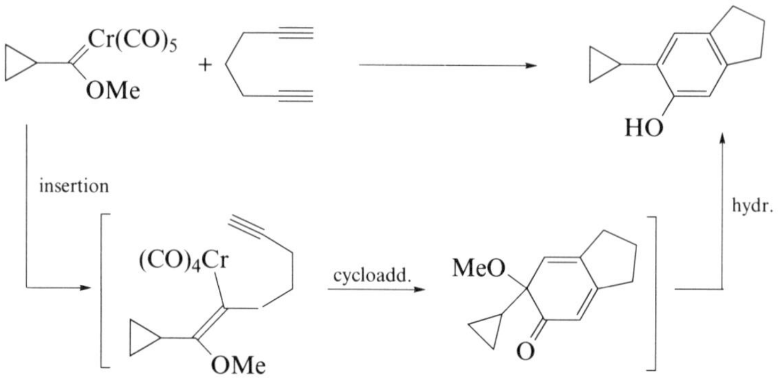

(82)

14% yield (equation 83)[152]. In analogy to related disubstituted vinyl carbene complexes, it is assumed that a spirocyclohexadienone chromium complex is intermediate in the reaction. The related cyclopropylvinyl carbene complex (see Section II.F) reacts with phenylacetylene under similar conditions to give indeed the corresponding cyclopropyl cyclohexadienone (equation 84)[152].

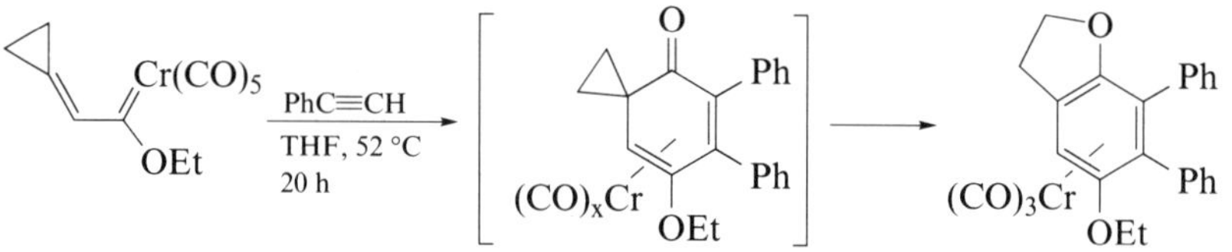

(83)

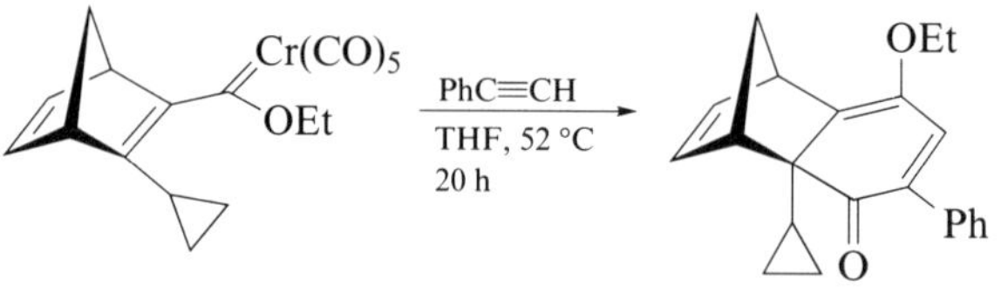

(84)

Interestingly, the (β-dimethylaminovinylcyclopropyl)carbene chromium pentacarbonyl reacts (see Section II.F) with terminal alkynes, presumably via insertion intermediate, to give substituted cyclopropylcyclopentadienes, which readily hydrolyze to the corresponding cyclopropylcyclopentenones (equation 85)[153].

(ii) Molybdenum and tungsten complexes. In contrast to chromium complexes, the general reaction between alkynes and cyclopropylcarbene molybdenum and tungsten complexes leads to cycloheptadienones and furanones (equation 86)[137,138,141]. The cycloaddition is general for internal alkynes, but fails for terminal alkynes, and is less facile when the cyclopropyl ring is substituted. Furanone formation can be suppressed by addition of phosphine. The reaction with nonsymmetrical disubstituted acetylenes is highly regioselective, affording cycloheptadienones with the large substituent (R_L) occupying the position α to carbonyl, whereas in the furanones the reverse is true and the small substituent (R_S) is at the α position.

Similarly, thermolysis of cyclopropyl(5-phenyl-4-pentynoxy)carbene complexes of molybdenum and tungsten gave via intramolecular coupling the corresponding 3,4,5,7-tetrahydro-5-phenylcyclohepta[*b*]pyran-6(2*H*)-one (equation 87)[137,138,149]. As with intermolecular cycloadditions, the tungsten complex is considerably less reactive than the molybdenum complex. Therefore thermolysis of the tungsten complex requires higher

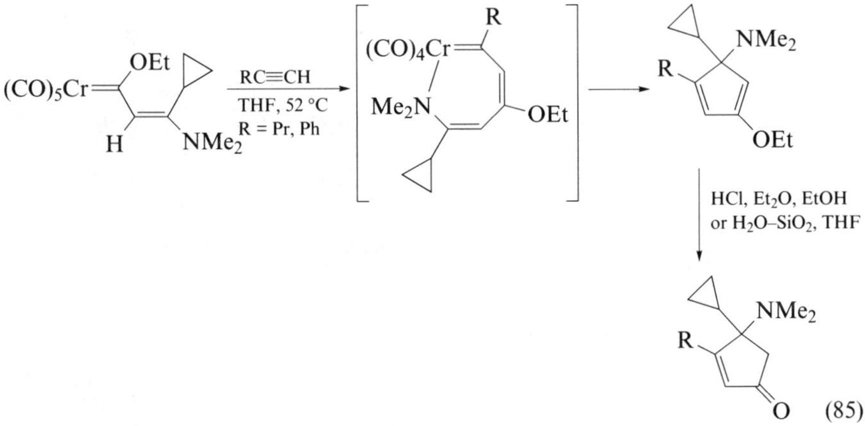

(85)

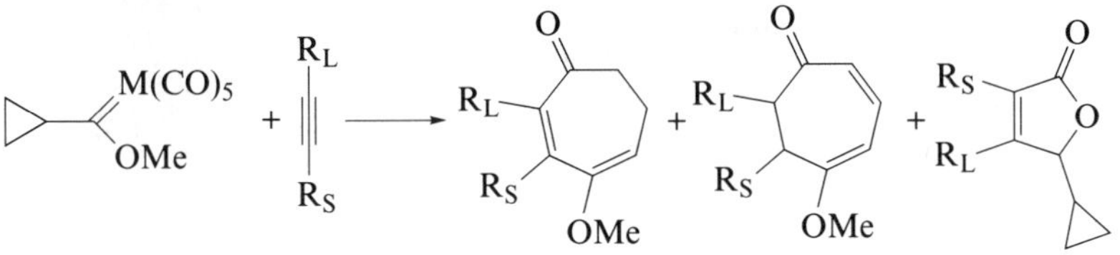

(86)

M = Mo, W
$R_L = R_S = Pr$
$R_L = R_S = Ph$
$R_L = Ph, R_S = Me$
$R_L = Ph, R_S = SiMe_3$

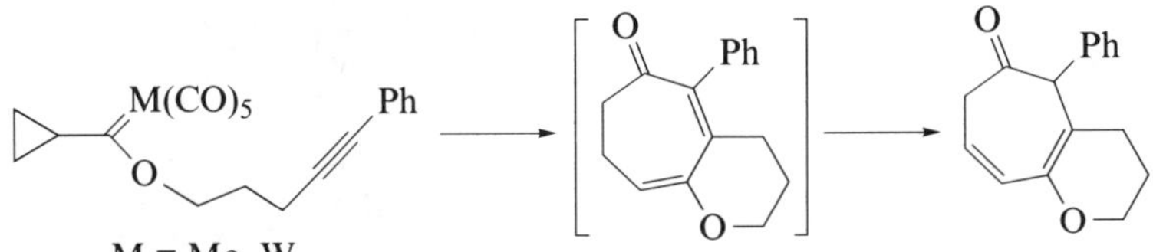

(87)

M = Mo, W

temperatures (100 °C, xylene vs 65 °C in THF for the molybdenum complex) and addition of 1,2-bis(diphenylphosphino)benzene, to obtain optimal yields.

(iii) Mechanistic aspects. Any mechanistic consideration of the reaction of cyclopropylcarbene complexes with alkynes must take into account common pathways leading to (a) cyclopentenones from the chromium complexes (equation 85) and (b) cycloheptadienones and furanones from molybdenum and tungsten complexes (equation 86) (*vide supra*). The proposed coupled mechanism[138,140] involves initial cycloaddition (i), followed by sequential rearrangement to cyclopropyl–vinylcarbene complex (*E*–methoxy isomer) (ii), ring enlargement to 7-membered metallacycle (iii), carbonyl insertion (iv) and π-complexation (v), to give an η^1,η^5 ketene complex intermediate as the point of divergence (Scheme 5)[140].

The *chromium* complex cyclizes to a 5-membered complex (vi), which loses ethylene and the metal (vii), to give methoxycyclopentadienone. Subsequent reduction (viii) under the reaction conditions by Cr(0)–H_2O affords the methoxycyclopentenone. The correspond-

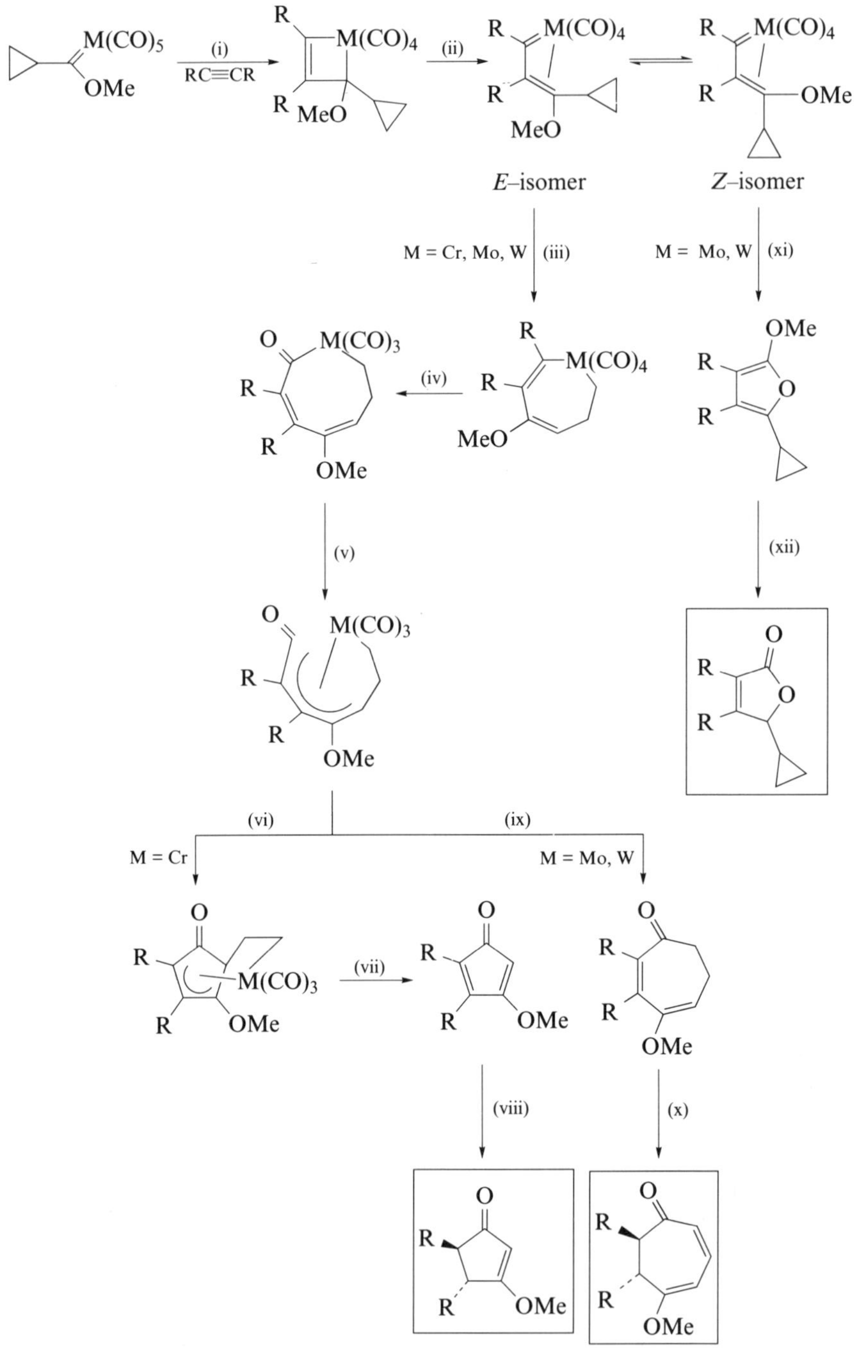
M(CO)5
OMe
(i)
RC≡CR
R
M(CO)4
R
MeO
(ii)
R
M(CO)4
R
MeO
R
M(CO)4
R
OMe
E–isomer
Z–isomer
M = Cr, Mo, W
(iii)
M = Mo, W
(xi)
O
M(CO)3
R
R
OMe
(iv)
R
M(CO)4
R
MeO
OMe
R
O
R
(v)
(xii)
O
M(CO)3
R
R
OMe
O
R
O
R
(vi)
(ix)
M = Cr
M = Mo, W
O
R
M(CO)3
R
OMe
(vii)
O
R
R
OMe
O
R
R
OMe
(viii)
(x)
O
R
R
OMe
O
R
R
OMe

SCHEME 5

ing *molybdenum* and *tungsten* intermediate complexes undergo instead reductive elimination (ix) to a labile cycloheptadienone, which ultimately rearranges (x) to the more stable isolated cycloheptadienone. Finally, *furanone* formation is rationalized[138] at an early stage from the cyclopropyl–vinylcarbene complex *Z*-methoxy isomer (see Scheme 5). This cannot undergo ring enlargement to the metallacycle, but instead ring-closes (xi) by reductive-elimination to 5-cyclopropyl-2-methoxyfuran, and further hydrolyzes (xii) to the furanone.

d. Rearrangements. Fischer cyclopropyl(methoxy)carbene complexes are reluctant to undergo ring-opening reactions even at 100 °C in the absence of alkynes. However, the corresponding vinylcyclopropyl carbene complexes and their phenyl-substituted counterparts are thermally labile. Thermolysis of the vinyl carbene complexes in refluxing THF (65 °C), or at 100 °C in dioxane or in DMF for the less reactive phenyl complexes, led to formation of 5-vinyl- and 5-phenyl-2-methoxycyclopentenones, respectively (equation 88)[150]. Under the reaction conditions some isomerization of the vinylic double bond to the more substituted carbonyl-conjugated position occurs, affording the corresponding 5-alkylidene-2-methoxycyclopentenones.

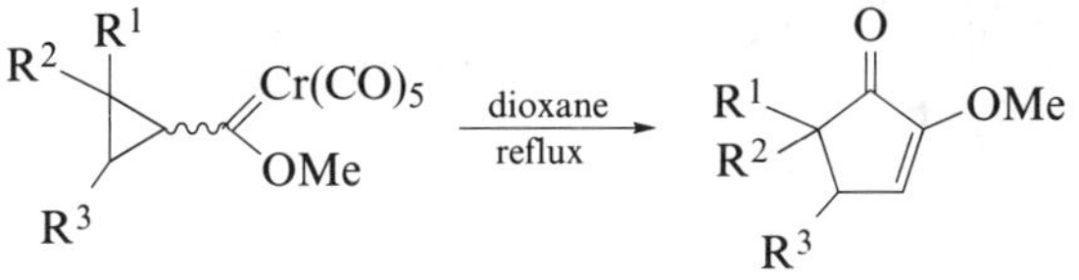

(88)

R^1 = vinyl, $R^2 = R^3$ = H (49%)
R^1 = vinyl, R^2 = Me, R^3 = H (43%)
R^1 = vinyl, R^2R^3 = —$(CH_2)_4$— (67%)
R^1 = vinyl, R^2R^3 = —$(CH_2)_5$— (57%)
R^1 = 1-propenyl, R^2 = H, R^3 = Me (53%)
R^1 = 1-cyclopentenyl, $R^2 = R^3$ = H (69%)
R^1 = Ph, $R^2 = R^3$ = H (22%)
R^1 = Ph, R^2 = Me, R^3 = H (27%)

The rearrangement exhibits some stereochemical preference for *cis*-vinyl carbene complex (with respect to the metal) compared to the *trans*-isomer. Thus, 2-methyl-2-*cis*-vinyl cyclopropyl (methoxy) carbene chromium pentacarbonyl rearranges to 5-methyl-5-vinyl-2-methoxycyclopentenone approximately 4 times faster (THF, 52 °C) than the *trans*-isomer, which in turn rearranges faster than phenyl derivatives. This suggests that vinyl complexes undergo initial Cope-type rearrangement to form metallacycloheptadienes, which then rearrange to π-allyl complexes. Subsequent CO insertion and reductive elimination leads to the vinylcyclopentenones (equation 89)[150].

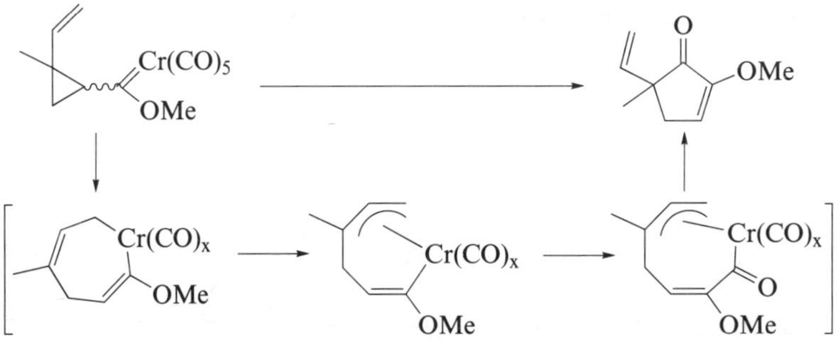

(89)

e. Reaction intermediates. The intermediacy of a cyclopropylpalladacarbene complex in the construction of unusual vinylcyclopropane cyclopentenoids by tetrakis(alkoxycarbonyl)palladacyclopentadiene (TCPC) catalyzed cycloaddition reactions of yndienes and methyl pentadienoate has been recently suggested (equation 90)[154]. The mechanism involves a palladacyclopentene-to-palladacyclopropylcarbene rearrangement followed by a unique participation of a palladavinylcarbene complex as the 4π component in the Diels-Alder reaction.

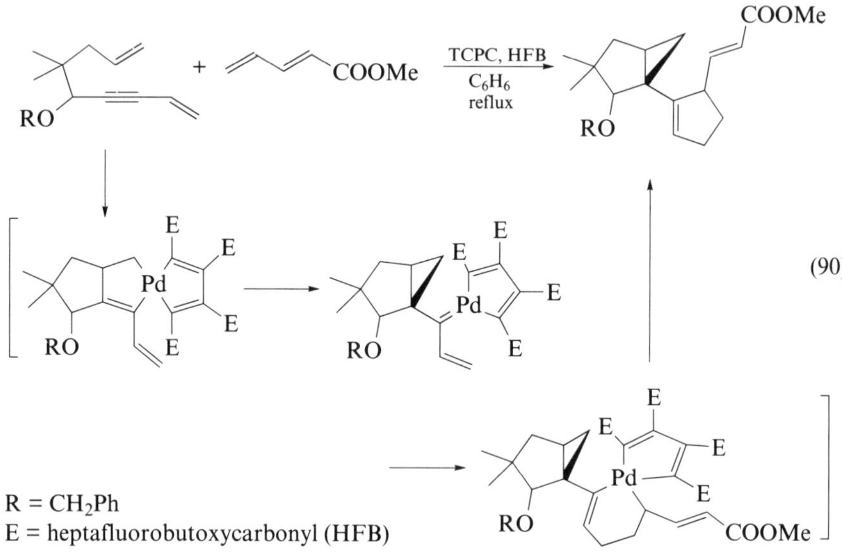

(90)

D. Cyclopropylcarbyne Complexes

1. Preparation

The synthesis of cyclopropyl carbyne complexes follows the general Fischer synthesis of carbyne complexes from alkoxycarbene complexes[155], typical of transition metals of group 6 (Cr, Mo, W). Thus, addition of cyclopropyllithium to chromium and tungsten hexacarbonyl followed by alkylation of the acylmetallate intermediate with triethyloxonium fluoroborate gave cyclopropyl ethoxycarbene complexes which, upon subsequent reaction with boron tribromide at –25 °C, afforded the corresponding *trans*-bromotetracarbonyl cyclopropylcarbyne complexes (equation 91)[156]. However, whereas the monotungsten

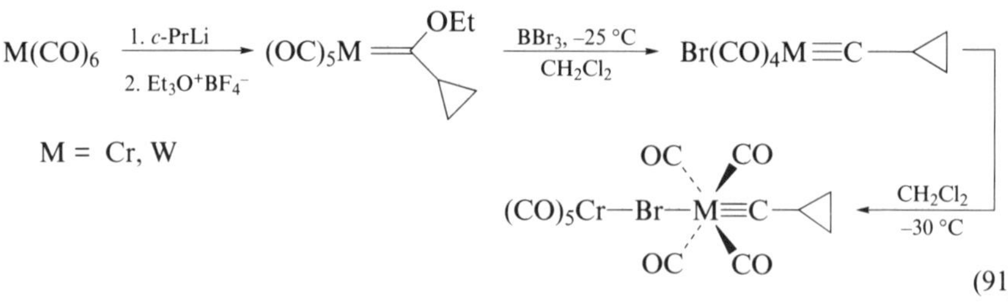

(91)

carbyne complex is the only product obtained in the reaction with BBr_3, a bromo-bridged dinuclear carbyne complex *trans*-(μ-$Br(CO)_5Cr)(CO)_4Cr{\equiv}C(c$-Pr) is observed in the reaction of the corresponding chromium carbene complex. The structure of the dinuclear complex was confirmed by X-ray crystal analysis.

Preparation of the corresponding chlorocarbyne complexes of molybdenum and tungsten was achieved by modification of the classical Fischer synthesis (equation 91), except that the acylmetallate intermediate was isolated as the tetramethylammonium salt, and further reacted with oxalyl chloride in CH_2Cl_2 (equation 92)[157,158]. The carbyne complexes $X(CO)_4M{\equiv}C(c$-Pr) (X = Cl, Br; M = Mo, W) typically react further with cyclopentadienyl anion, leading to the isolobal complexes $Cp(CO)_2M{\equiv}C(c$-Pr) (M = Mo, W) in which the halogen and two carbonyls *fac* ligands are exchanged by a cyclopentadienyl group[158,159].

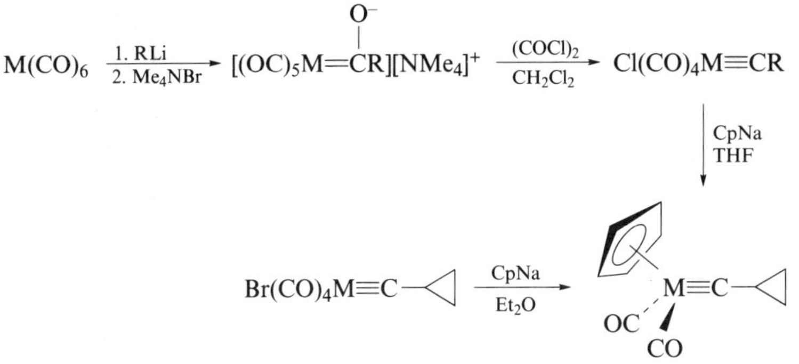

(92)

2. Reactions without ring cleavage

a. Ligand substitution reactions. Trimethyl and triphenyl phosphites efficiently exchange carbonyl ligands to various extents. At room temperature, the smaller trimethyl phosphite group exchanges three of the four carbonyl ligands in a series of ring substituted cyclopropylcarbynyl complexes of Mo and W, affording the corresponding triphosphites (equation 93)[157,158]. The cyclopropyl molybdenum triphosphite undergoes further exchange of the remaining carbonyl by $P(OMe)_3$, to give $Cl[P(OMe)_3]_4Mo{\equiv}C(c$-Pr) (equation 94). The more bulky triphenyl phosphite substitutes only two carbonyl ligands (equation 95). Subsequent treatment of the chlorophosphites with cyclopentadienyl sodium in THF afforded the corresponding cyclopentadienyl phosphite carbyne complexes[158].

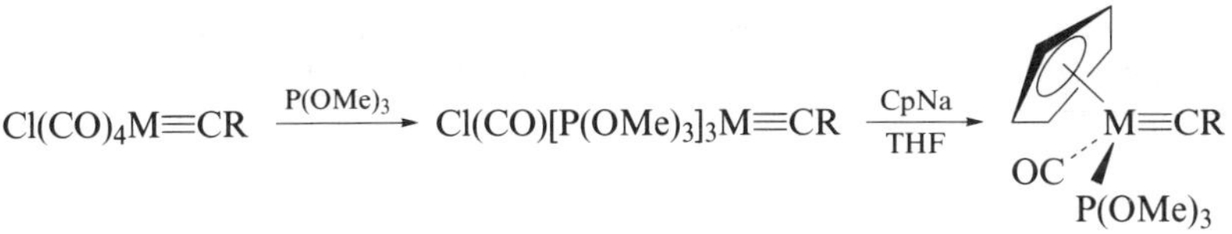

(93)

M = Mo, W	R = cyclopropyl
M = Mo	R = 2-ethylcyclopropyl
M = Mo	R = 2-phenylcyclopropyl
M = Mo	R = 2,2-dimethylcyclopropyl (diastereomeric mixture)
M = Mo, W	R = *cis*- or *trans*-2,3-dimethylcyclopropyl (diastereomeric mixtures)
M = Mo	R = 2,2,3,3-tetramethylcyclopropyl
M = Mo	R = bicyclo[4.1.0]hept-7-yl

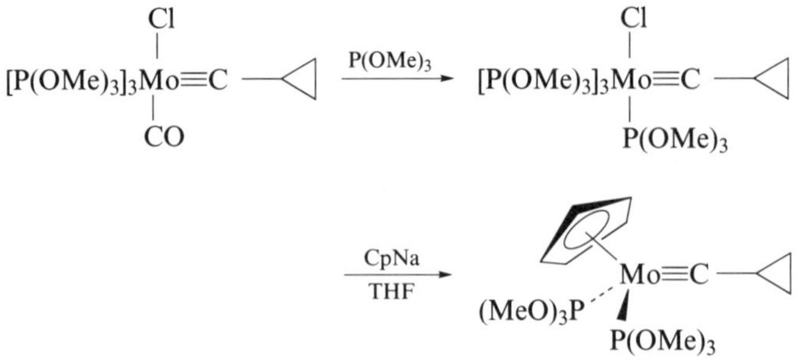

(94)

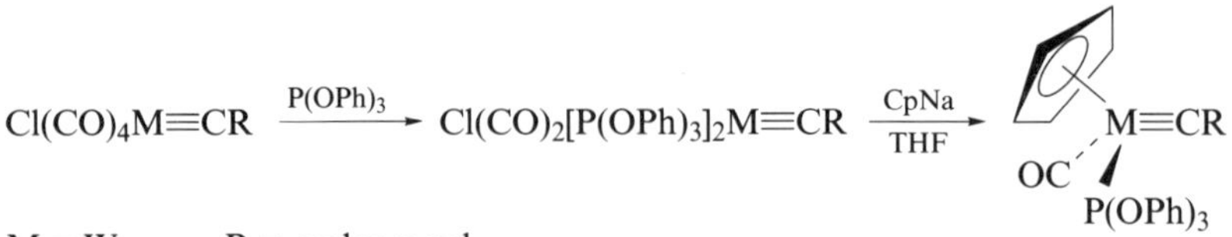

(95)

M = W R = cyclopropyl

M = Mo R = *trans*-2,3-dimethylcyclopropyl (diastereomeric mixture)

b. Ring substitution reactions. Deprotonation of the cyclopropyl proton of carbyne complexes is effected with BuLi in THF at –78 °C, yielding a cyclopropylidenecarbene anionic complex which was quenched with D_2O to give the labelled cyclopropyl carbynyl complex (equation 96)[160]. Alternatively, alkylation of the deprotonated carbyne complex anion with allyl bromide leads to the corresponding allylcyclopropyl carbyne derivative. Similarly, reaction with other electrophiles, such as acyl chlorides and chloroformates, afforded carbyne complexes of cyclopropyl ketones and esters, respectively (equation 97).

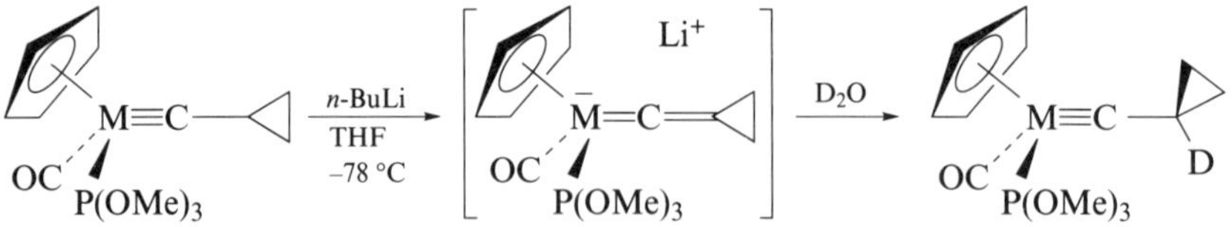

(96)

c. Reactions with phosphines. Trimethylphosphine, unlike the phosphites, did not give simple ligand exchange products upon reaction with cyclopropylcarbynyl complexes $Cp(CO)_2W{\equiv}C(c\text{-Pr})$. Instead, a nucleophilic addition of the phosphine occurred which concurrently induced a typical alkylidyne–carbonyl coupling reaction[161] to give a η^2-cyclopropylketenyl complex (equation 98)[159]. Although no kinetic studies have been reported on nucleophile-induced alkylidyne–carbonyl coupling reactions, it is assumed that the nucleophilic attack occurs at the $M{\equiv}C$ π^* orbital in the coupling plane[161].

d. Reactions with strong acids. The alkylidyne–carbonyl coupling reaction is not only induced by nucleophiles but also by electrophiles. Thus, treatment of the tungsten complex $Cp(CO)_2W{\equiv}C(c\text{-Pr})$ with trifluoroacetic acid afforded the bistrifluoroacetato-η^2-cyclopropylacetyl complex $Cp(CF_3COO)_2WCOCH_2\text{-}(c\text{-Pr})$ in 93% yield, without ring cleavage (equation 99)[162]. Analogously, treating $Cp(CO)[P(OMe)_3]W{\equiv}C(c\text{-Pr})$ with ethereal solution of HCl resulted in the formation of the dichloro-η^2-acyl complex

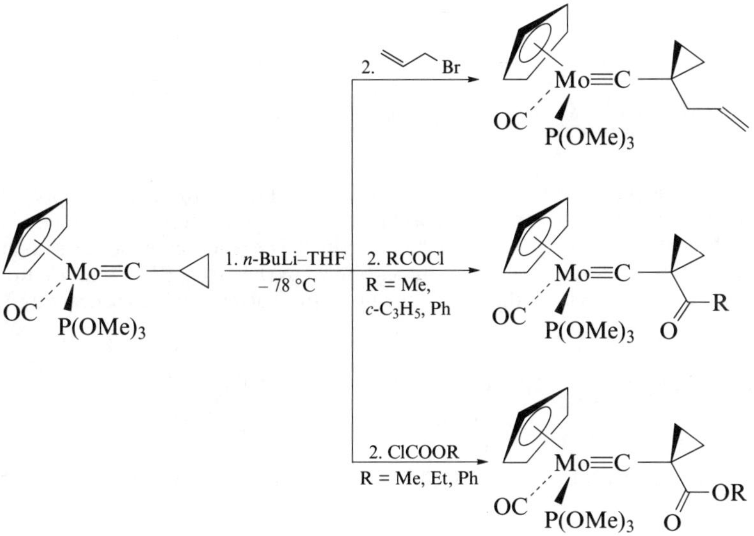

(97)

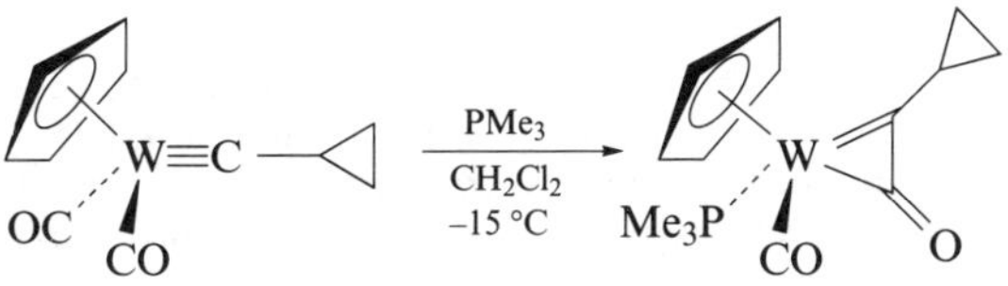

(98)

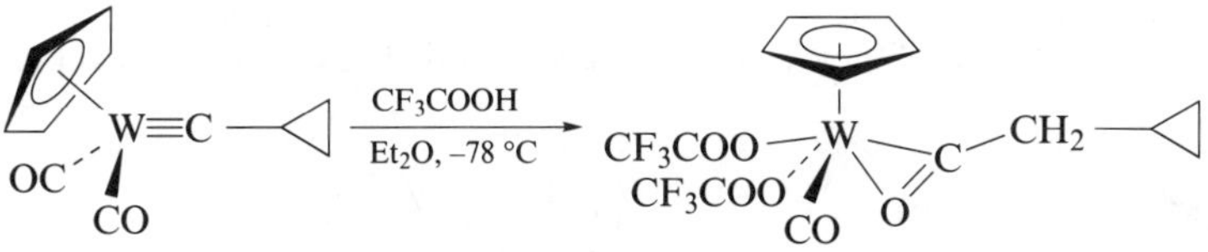

(99)

$CpCl_2[P(OMe)_3]WCOCH_2(c\text{-Pr})$ in 90% yield (equation 100)[158]. Molybdenum complexes are more vulnerable to ring-opening reactions with HCl, particularly the acyl-substituted complexes (*see below*).

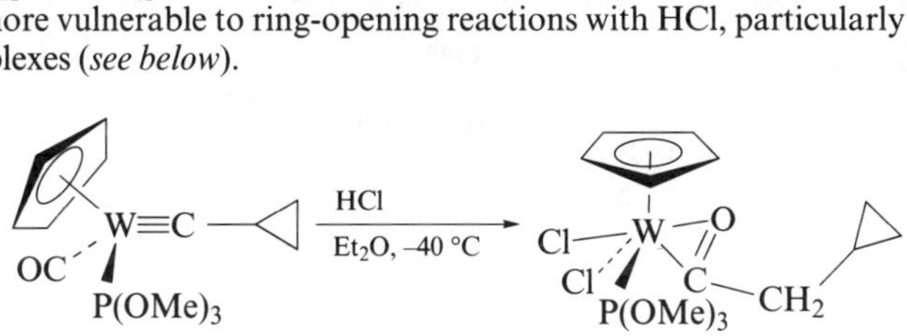

(100)

e. Photooxidation. Photolysis of $Cp[P(OMe)_3](CO)W{\equiv}C(c\text{-Pr})$ in chlorinated solvents like $CHCl_3$ or CH_2Cl_2 containing trimethylphosphine leads to formation of the cationic complex $[Cp(PMe_3)_2(Cl)W{\equiv}C(c\text{-Pr})]Cl$ in about 60% yield (equation 101)[157]. The presence of phosphine is crucial for keeping the cyclopropyl group intact. In the absence of phosphine, ring-opening reactions occur (*see below*).

The reaction mechanism was proposed to involve electron transfer from the d–π^* metal to ligand charge transfer excited state (MLCT transition) to the chlorinated solvent,

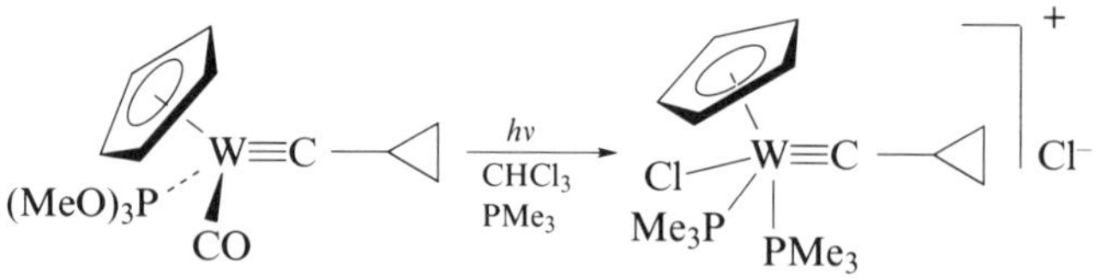

(101)

followed by a rapid ligand exchange in the resulting $17e^-$ carbyne species, probably via associated substitution. Abstraction of a chlorine atom from a second molecule of solvent by $[Cp(PMe_3)_2W{\equiv}C(c\text{-}Pr)]Cl$ gave the final product. The reaction sequence generates two equivalents of $\cdot CHCl_2$ radicals, which are scavenged by the excess PMe_3. No reaction takes place when the complex is irradiated in nonchlorinated solvent, such as benzene or tetrahydrofuran (equation 102).

$$CpL_1L_2M{\equiv}CR \xrightarrow[CHCl_3]{h\nu} [CpL_1L_2M{\equiv}CR]^+ + Cl^- + \cdot CHCl_2$$
$$\downarrow PMe_3 \qquad (102)$$
$$[Cp(PMe_3)_2M{\equiv}CR]^+Cl^- \xrightarrow{CHCl_3} [Cp(PMe_3)_2ClM{\equiv}CR]^+Cl^- + \cdot CHCl_2$$

3. *Reactions involving ring cleavage*

a. Thermal reactions in chlorinated solvents. Thermal decomposition of cyclopropyl carbyne complexes bearing electron-withdrawing groups, in $CHCl_3$, afforded mainly oxymetallacycles accompanied by various amounts of 3-substituted cyclopentenones (equation 103)[160]. The acylcyclopropyl carbyne complexes convert cleanly to the corresponding metallacycles with only traces of cyclopentenones. However, cyclopropyl ester carbyne complexes gave significant amounts of cyclopentenones (equation 103). Interestingly, photolysis of these complexes gave essentially a reversed product ratio.

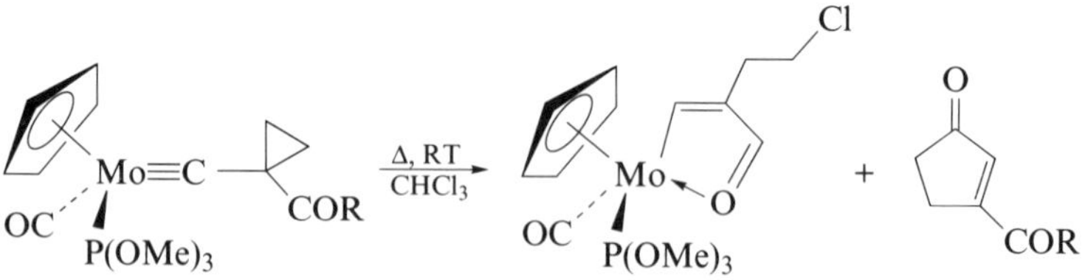

(103)

R = Me	product ratios	50:3
R = *c*-Pr		54:3
R = Ph		81:0
R = OMe		27:23
R = OEt		38:26
R = OPh		44:17

Thermolysis of an allyl-cyclopropyl carbyne complex under the mild conditions used for the acyl carbynes did not result in an oxymetallacycle or a cyclopentenone. Instead, 2-cyclohexenone-5-spirocyclopropane was obtained in low yields (5%) together with moderate yields (43%) of the open-chain (3-allyl-η^4-pentadienal) molybdenum complex (equation 104)[163].

b. Reactions with HCl. Acyl-substituted cyclopropylcarbynyl molybdenum complexes undergo similar carbonyl–carbyne coupling to form ketenyl complexes. However, only

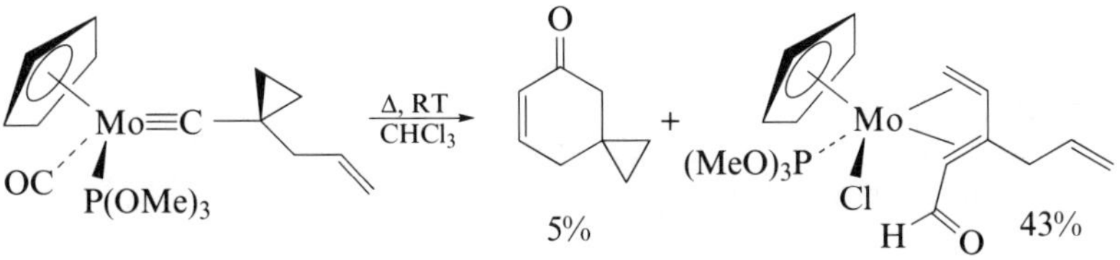

(104)

the esters gave ketenyl complexes as main products, together with small amounts of cyclopentenone esters (equation 105). The ketones gave essentially ring-opened oxymetallacyclic complexes (equation 106) which result from nucleophilic ring cleavage by chloride (*see below*)[158,160]. The *trans*-2,3-dimethylcyclopropyl carbyne complex gave, upon reaction with HCl in ether, a 1:1 mixture of ketenyl and cyclopentenyl complexes (equation 107)[158].

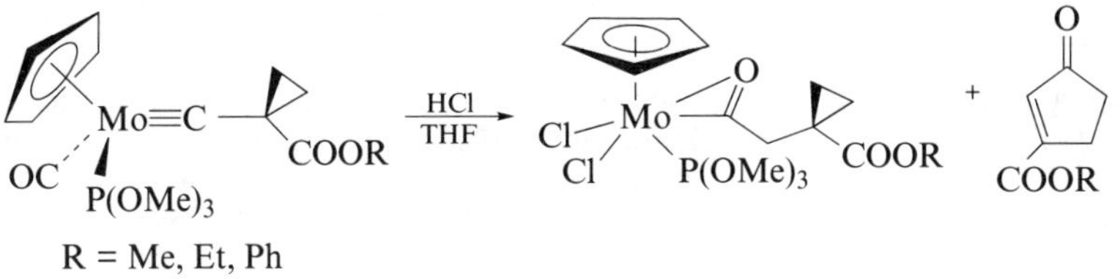

(105)

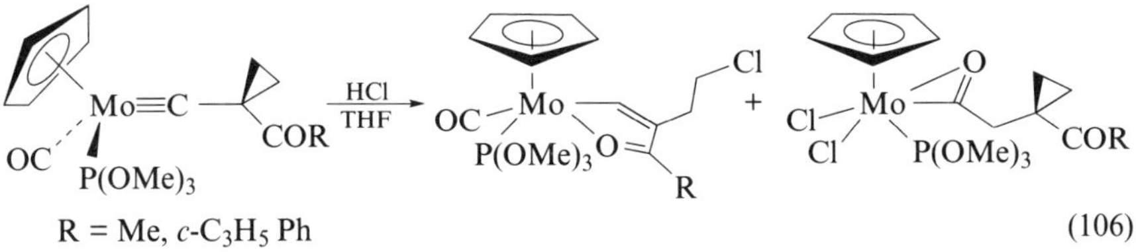

(106)

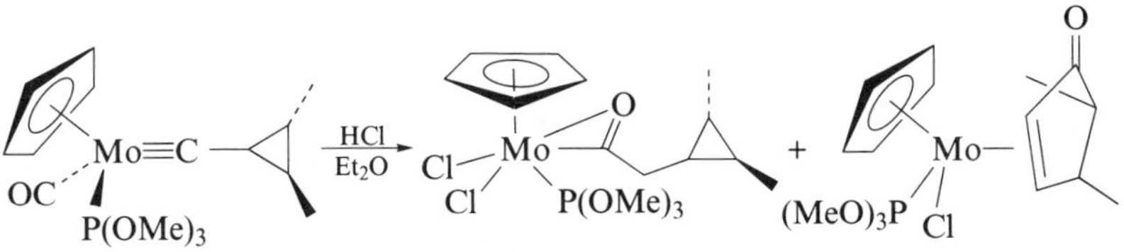

(107)

Although bis(phosphite) carbyne complex $Cp[P(OMe)_3]_2Mo{\equiv}C(c\text{-}Pr)$ is incapable of undergoing carbonyl insertion reactions, it adds 1 equivalent of HCl in ether forming the ring-opened η^4-butadiene complex $Cp[P(OMe)_3](Cl)Mo(\eta^4\text{-butadiene})$ in 15% yield, and $P(OMe)_3$ in equal amounts (equation 108)[158,164]. Careful analysis of the reaction using two equivalents of HCl reveals the presence of the metal hydride complex $Cp[P(OMe)_3]_2Cl_2MoH$ as the main products (70%), and free butadiene. It was furthermore shown that the two molybdenum complexes are not interconvertible under the reaction conditions and both the yields and products ratio are invariant with temperature in the range of –40 °C to room temperature and the amount of added HCl (1 or 2 equivalents).

These results suggest the presence of two competing pathways to products, which depend upon the location of protonation at the M≡C carbyne bond. Charged controlled protonation at the carbyne carbon followed by nucleophilic attack of the Cl^- leads to the butadiene complex. Frontier control of protonation results in attack at the metal center, leading ultimately to the hydride complex[164]. This has been verified by reaction of the

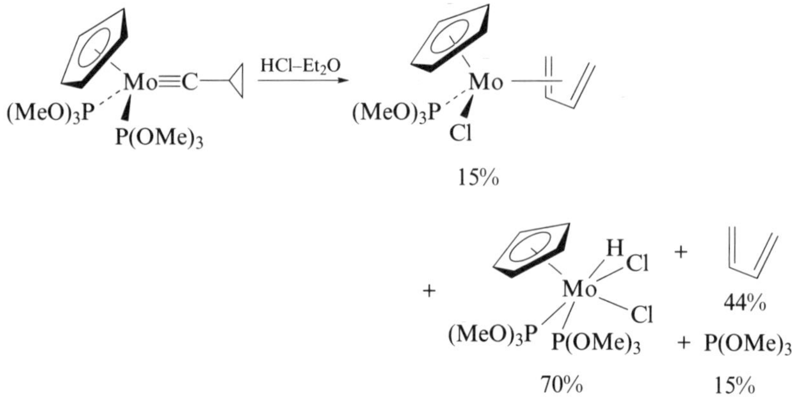

(108)

bis(phosphite) complex with the noncoordinating fluoroboric acid, which gave the separable carbyne hydride complex fluoroborate $[Cp\{P(OMe)_3\}_2(H)Mo{\equiv}C(c\text{-}Pr)]BF_4$. Addition of Cl^- converts the hydride, via a reversible hydrogen migration, to the diene complex $Cp[P(OMe)_3](Cl)Mo(\eta^4\text{-butadiene})$. In the absence of a nucleophile a slow rearrangement takes place at room temperature in $CHCl_3$, affording diene fluoroborate complex $[Cp\{P(OMe)_3\}_2Mo(\eta^4\text{-butadiene})]BF_4$, which is stable towards further substitution by Cl^- (equation 109).

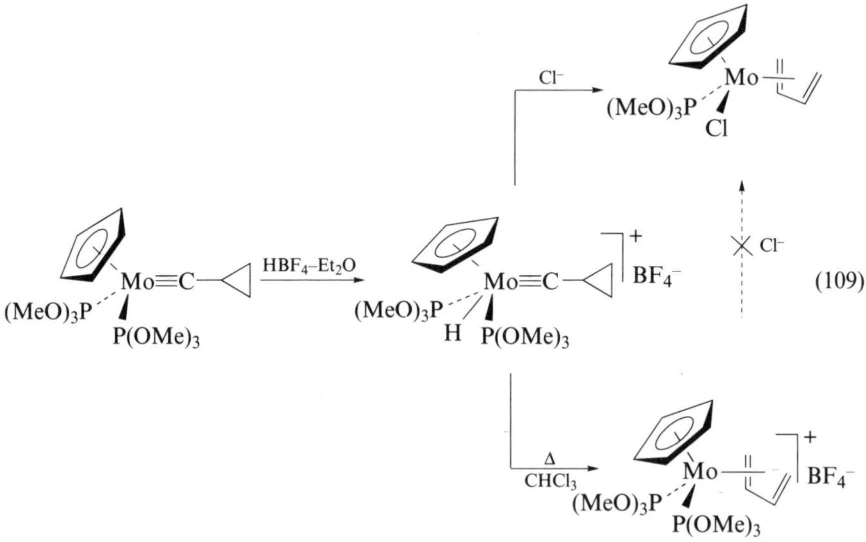

(109)

Addition of HCl to the cyclopropyl–deuterated carbyne complex gave a single stable *syn*-isomer of the butadiene complex, labeled exclusively at the inner butadiene carbon facing the phosphite ligand (equation 110). However, the methyl substituted complex gave a mixture of *syn* and *anti* isoprene regioisomers, which only slowly isomerizes to the more stable *anti* regioisomer[164]. The structure of the *anti* regioisomer was secured by X-ray crystallography.

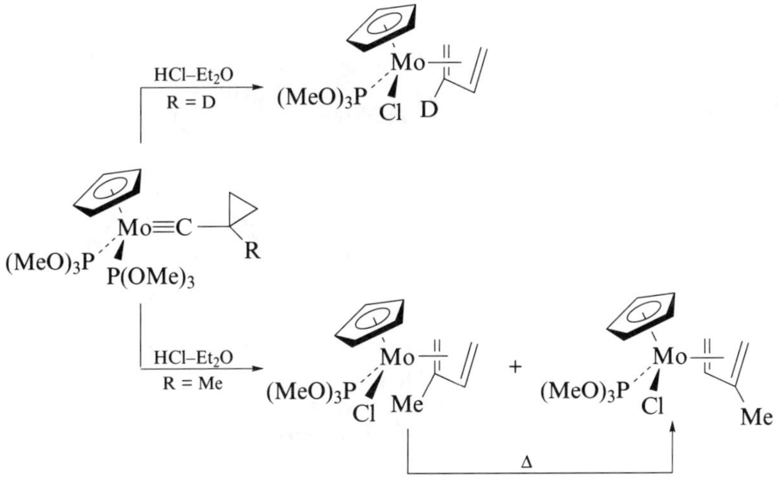

(110)

The proposed mechanism of the ring cleavage reaction of HCl (and other protic acids) with cyclopropyl carbynyl complexes involves addition of HCl across the carbyne triple bond to give a carbene complex as key intermediate. In the absence of a carbonyl ligand this is followed by ring expansion to a metallacyclopentene complex, β-hydrogen elimination and reductive elimination to the diene complex (equation 111)[164].

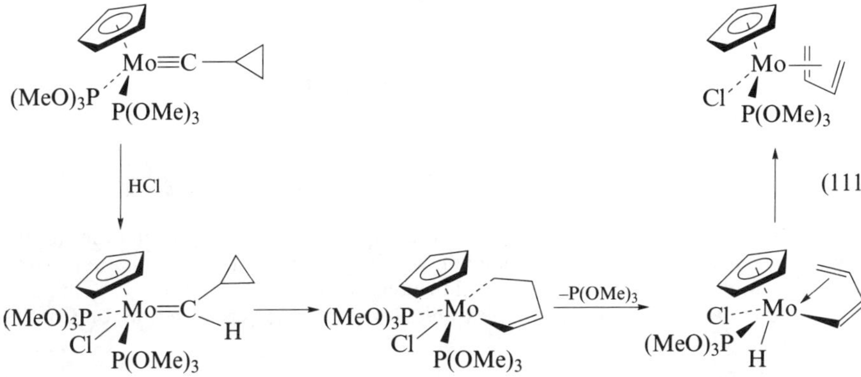

(111)

c. Reactions with nucleophiles. Like uncoordinated electron-deficient cyclopropanes which are known to undergo nucleophilic ring opening, the analogous cyclopropyl carbyne complexes bearing acyl and ester groups react with a variety of nucleophilic reagents (equation 112)[160].

Typically, acetyl bromide gives (bromoethyl)oxymetallacyclopentene. Consistent with the increased nucleophilicity of bromide relative to chloride anion, the rate of ring opening was significantly accelerated. Ring opening was also observed with nucleophiles other than halides, like aniline, thiophenol and water (on neutral silica gel column), though the reactions are notably slower. No reaction was observed with either acetate or hydroxide anions even after prolonged exposure. Formation of the oxymetallacycles is believed to involve the initial homoconjugate nucleophilic addition to form a metal–vinylidene anionic intermediate, followed by protonation and ring closure.

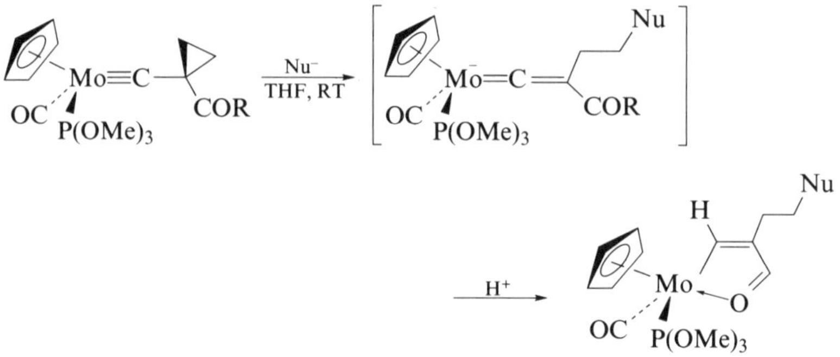

(112)

R = Me, Nu = Br, PhS, PhNH, OH
R = Ph, Nu = OH
R = OMe, Nu = PhNH
R = *c*-Pr, Nu = OH

d. Photooxygenation in chlorinated solvents. When photooxidation of cyclopropyl carbyne complexes containing a carbonyl ligand is carried out in the absence of added phosphine ligands (*vide supra*), the site of reactivity is switched from the metal atom to the carbyne moiety. Photolysis of the carbyne complex $Cp(CO)[P(OMe)_3M{\equiv}C$(*c*-Pr) (M = Mo, W) in $CHCl_3$ results in conversion of the carbyne ligand to cyclopentenone, and formation of the octahedral inorganic complex $Cp\{P(OMe)_3\}(CO)MCl_3$ (equation 113)[158].

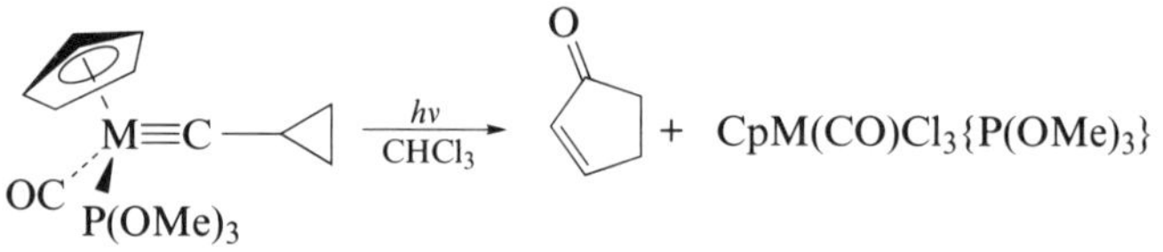

(113)

M = Mo, W

A series of cyclopropyl substituted carbyne complexes was photolyzed in order to follow the regio- and stereochemical outcome of the photooxidation. Irradiation of the deuterated molybdenum complex gave cyclopentenone exclusively labeled at C3 (equation 114). In the analogous tungsten complex only 90% of the deuterium resides on C3; the remaining 10% labeled the 2-position[158].

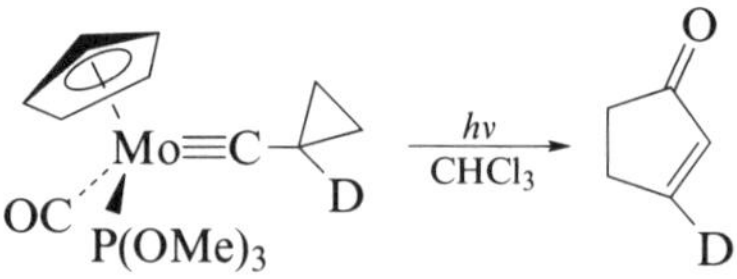

(114)

Photolysis of 2-substituted cyclopropyl carbyne complexes afforded 4-substituted cyclopentenones (equation 115). Irradiation of the diastereomeric mixture of either *cis*- or *trans*-2,3-dimethylcyclopropyl carbyne complexes gave selectively the *trans*-3,4-dimethylcyclopentenone, indicating a rapid cyclopropane photoisomerization prior to photooxidation (equation 116). The *cis* configuration is, however, retained in fused ring systems. Thus photooxidation of bicyclo[4.1.0]hept-7-yl carbyne complex results in formation of the *cis*-fused hexahydroindenone (equation 117). Finally, photolysis of 2,2,3,3-tetramethylcyclopropyl carbene complex afforded only tetramethylethylene[158].

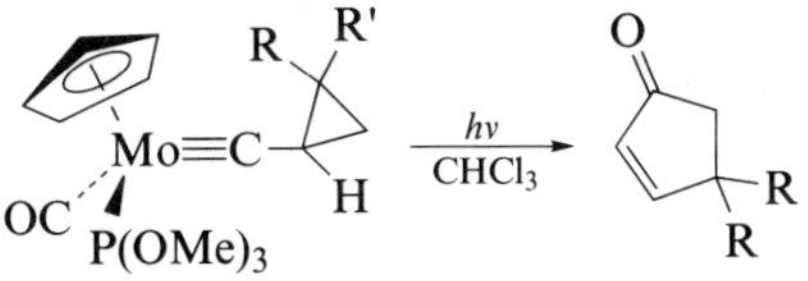

(115)

R = H, R′ = Et
R = H, R′ = Ph
R = R′ = Me

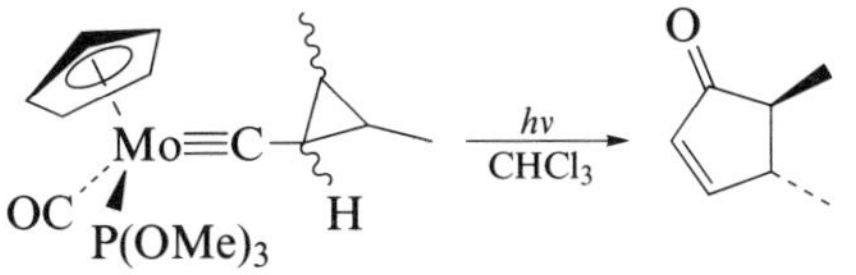

(116)

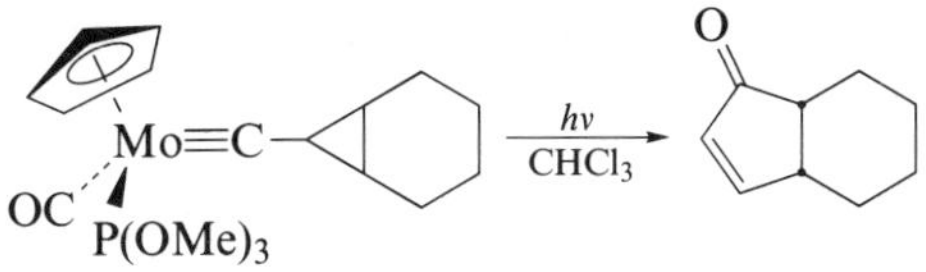

(117)

Acyl and ester derivatives of cyclopropyl carbyne complexes undergo similar photooxidation in chloroform to give the corresponding 3-substituted cyclopentenones. However, this is accompanied by oxymetallacycles, derived from nucleophilic cleavage of the cyclopropane ring by chloride (equation 118)[160]. Oxymetallacycles are typical reaction products of nucleophiles with cyclopropyl carbyne complexes bearing electron-withdrawing groups (*vide supra*).

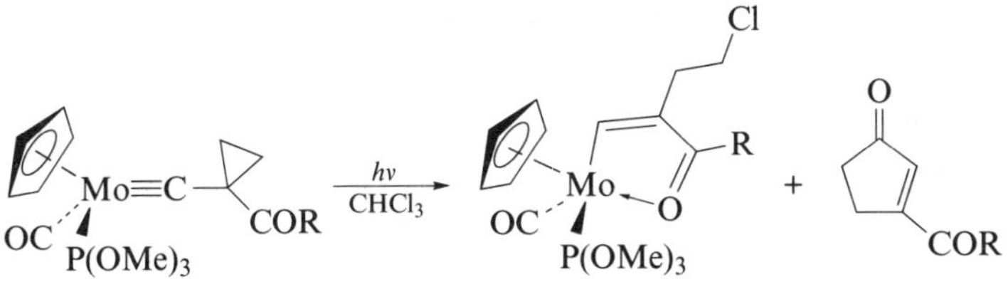

(118)

R = Me	products ratio	10:15
R = *c*-Pr		6:13
R = Ph		10:21
R = OMe		0:42
R = OEt		0:38
R = OPh		5:12

Mechanistic studies are consistent with photochemical electron transfer from the carbyne complex to chloroform followed by H atom abstraction. Ring expansion then occurs to give a metallacyclopentene, which undergoes carbonyl insertion. Finally, reductive elimination yields the cyclopentenone complex that slowly releases the free enone (equation 119)[158].

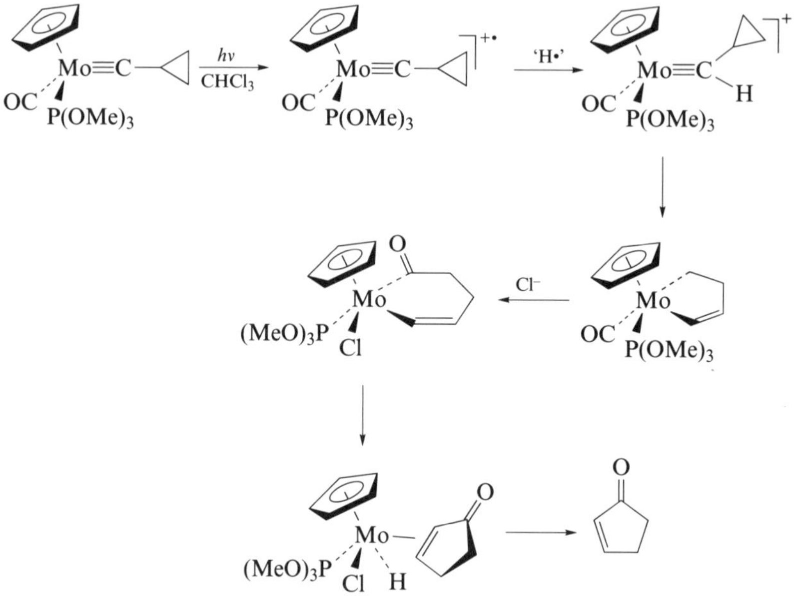

(119)

E. Vinylcyclopropane–Metal Derivatives

The chemistry of alkenylcyclopropanes is dominated by the vinylcyclopropane (vcp) to cyclopentene rearrangement via a formal 1,3-sigmatropic shift (equation 120). This transformation is greatly facilitated by metal catalysis, which not only reduces the activation energy to rearrangement but also significantly improves the control of stereochemistry. Much of the literature concerning the interaction of vinylcyclopropanes with metals with strong emphasis on catalysis has been continuously reviewed over the past 25 years and need not be repeated here[165–171]. We will rather overview in this section the stoichiometric interactions of vinylcyclopropanes with metals and their consequences.

(120)

1. Main group elements

a. Exchange reactions. *Cis*-(2-bromo)-1-vinylcyclopropane undergoes metal halogen interchange with *i*-PrLi in pentane containing 6% of Et_2O at 0 °C, to give the 2-lithio-1-vinylcyclopropane with complete retention of configuration. This was established by obtaining the pure *cis*-ester derivative after quenching the reaction mixture with CO_2 followed by diazomethane esterification (equation 121)[171].

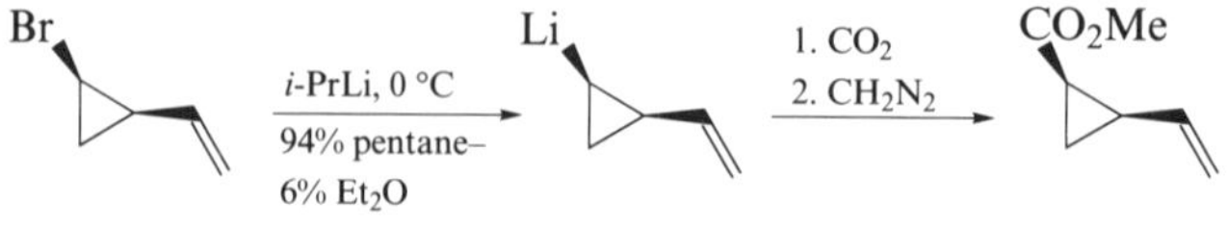

(121)

Similarly, lithiation of 2,2-dibromo-1-vinylcyclopropane at lower temperatures provides the 2-lithio compound, which can be alkylated (e.g. with MeI) or protonated (by ethereal HBr) to the corresponding monobromo hydrocarbons, respectively (equation 122)[172].

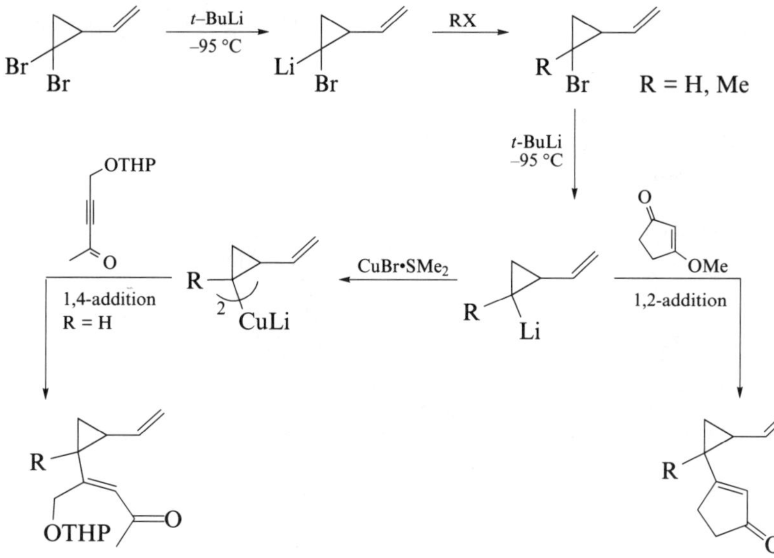

(122)

2-Lithio-1-vinylcyclopropane undergoes an efficient metal exchange reaction with [$CuBr.SMe_2$] to give the cuprate complex. The two metal reagents are important building blocks in the synthesis of divinylcyclopropanes, each one exhibiting the opposite regioselectivity upon reaction with unsaturated carbonyl compounds. This is demonstrated by the 1,2-addition of the lithio reagent to 3-methoxy-2-cyclopentenone, to give after hydrolysis and dehydration 3-(1-*R*-2-vinylcyclopropyl)-2-cyclopentenone (R = H, Me), and the selective 1,4-addition of the cuprate complex to acetylenic ketones to form the analogous divinyl cyclopropyl ketones. A related coupling reaction of 3-iodo-2-cyclohexenone with another vinylcyclopropyl copper variant also leads to the corresponding divinylcyclopropane derivative (equation 123)[173].

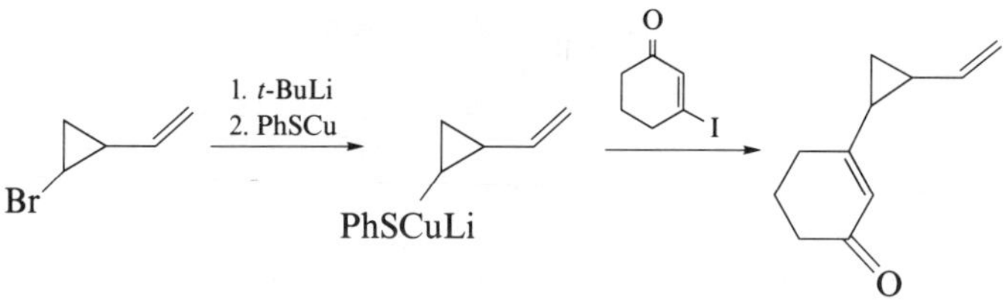

(123)

1-Lithio-1-vinylcyclopropanes are prepared either by metal–halogen exchange (equation 124) or by selenium–lithium metathesis reactions (equation 125)[174]. The greater stability and ready availability of the 1-vinylselenocyclopropanes compared to their bromo counterparts make the seleno compounds preferable precursors for the lithio derivatives. 1-Lithio-1-vinylcyclopropanes like their 2-lithio analogues readily undergo alkylation and carboxylation reaction to give the corresponding 1-substituted alkyl- and carboxylic acid vinylcyclopropane derivatives, respectively.

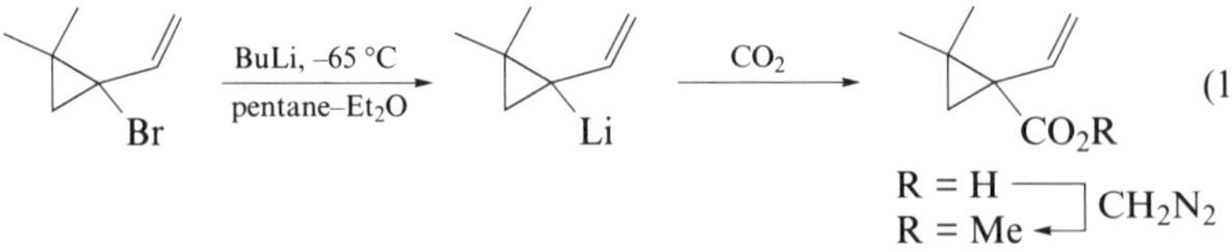

(124)

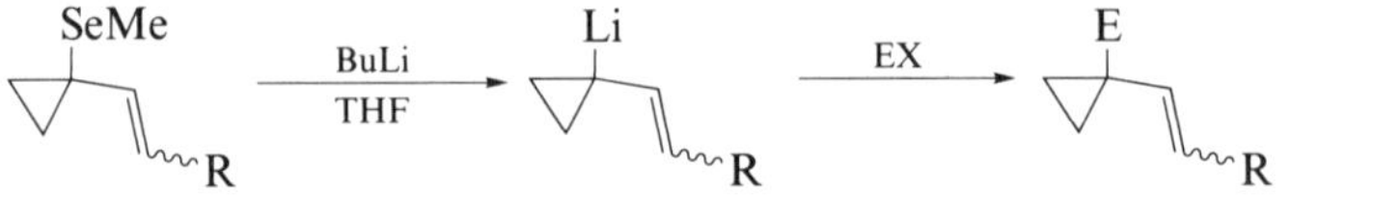

(125)

R = C_6H_{13}, Ph (*Z*/*E* mixtures)

EX = H_2O, E = H
EX = MeI, E = Me
EX = $C_{10}H_{21}Br$, E = $C_{10}H_{21}$
EX = CO_2, E = COOH

b. Carbometallations. Vinylmagnesium chloride readily adds to 3,3-dimethylcyclopropene in a *cis* fashion, to give 3,3-dimethyl-2-vinylcyclopropylmagnesium chloride (equation 126)[64]. This Grignard reagent undergoes normal carboxylation reactions with carbon dioxide and ethyl chloroformate to form the corresponding carboxylic acid derivatives. However, rearrangement occurs upon metathesis with nickelocene to give (η^3-4,4-dimethylcyclopentenyl)NiCp in 73% yield, presumably via initial vinylcyclopropane–cyclopentene rearrangement followed by a further metal hydride rearrangement[65].

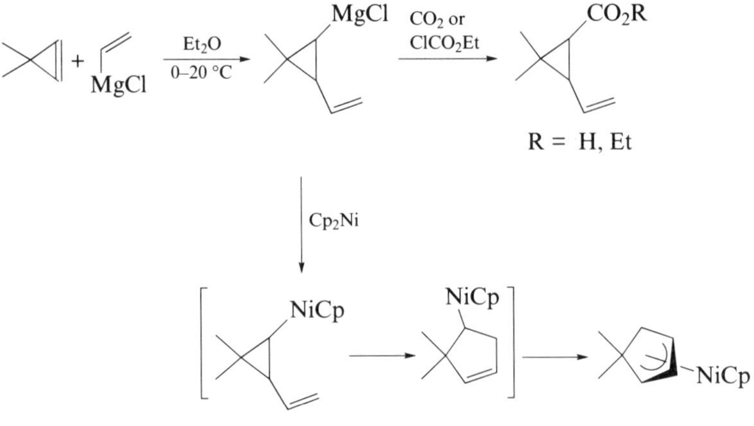

(126)

c. Rearrangements. 1-Seleno-1-vinylcyclopropanes undergo [2,3]-sigmatropic rearrangements to alkylidene cyclopropanes via either the phenylselenoxides or the selenonium ylides (equation 127)[174,175]. The oxides which are obtained in situ by reaction of the alkenyl selenides with H_2O_2 or HIO_4 rearrange to the corresponding carbinols in the presence of piperidine as the base. The selenonium salts, obtained by methylation of the selenides using $MeSO_3F$, or $MeI/AgBF_4$, rearrange to the terminal alkylidenecyclopropane selenides by *t*-BuOK in DMSO. Finally, treatment of 1-seleno-1-vinylcyclopropanes with equimolecular amounts of *p*-TsOH in benzene results in deselenylation and rearrangement to substituted cyclobutanones.

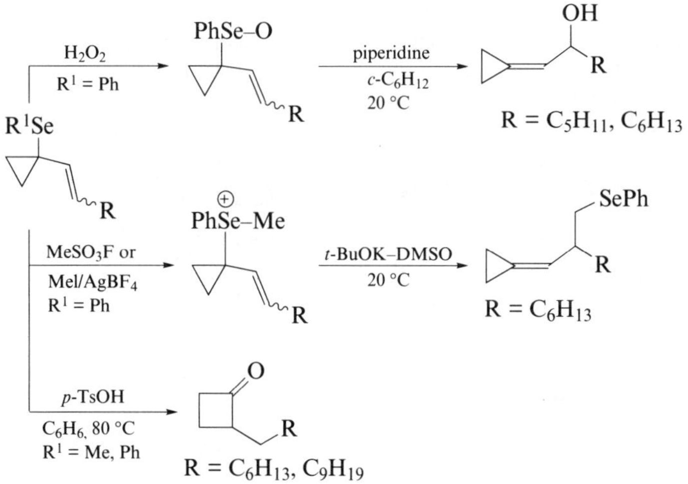

(127)

2. Transition metal complexes

a. σ-Bonded complexes. Group 4 titanium and zirconium alkyne- and alkyne-(trimethylphosphine)–metallocene complexes react readily with 3,3-dimethyl- and 3,3-diphenyl-cyclopropene to give the corresponding bis(η^5-cyclopentadienyl)-2-metallabicyclo[3.1.0]hex-3-ene derivatives in high yields (equation 128)[176].

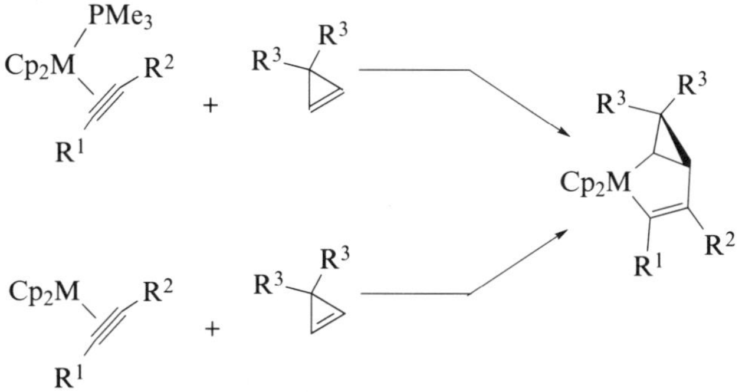

(128)

$R^1 = R^2 = Ph$, $R^3 = Me$; M = Ti, Zr
$R^1 = SiMe_3$, $R^2 = Ph$, $R^3 = Me$; M = Ti, Zr
$R^1 = R^2 = SiMe_3$, $R^3 = Me$; M = Ti, Zr
$R^1 = R^2 = R^3 = Me$; M = Ti
$R^1 = R^2 = Me$, $R^3 = Ph$; M = Zr
$R^1 = SiMe_3$, $R^2 = R^3 = Ph$; M = Zr
$R^1 = R^2 = R^3 = Ph$; M = Zr
$R^1 = R^2 = SiMe_3$, $R^3 = Ph$; M = Ti, Zr

The codimerization reaction with unsymmetrically substituted silylalkynes is highly regioselective, specifically giving the isomer with the silylated carbon attached to the metal. In an analogous manner, metallabicyclo[3.1.0]hexene derivatives of both titanium and zirconium can be prepared by the codimerization of 1,2-diphenylcyclopropene-(trimethylphosphine)metallocenes with 2-butyne (equation 129) and from the unusual benzyne(trimethylphosphane)zirconocene by reaction with 3,3-disubstituted cyclopropenes (equation 130).

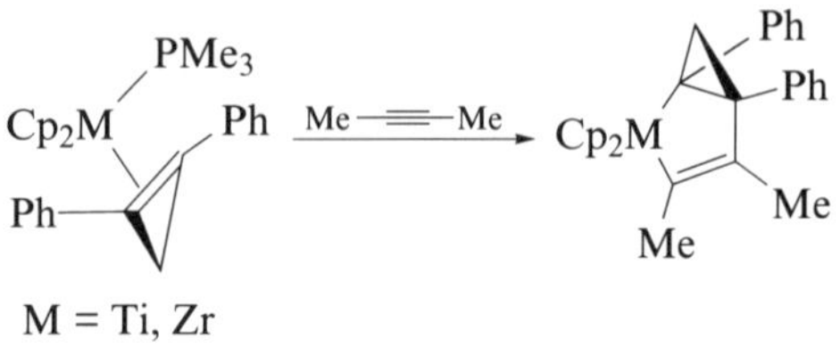

(129)

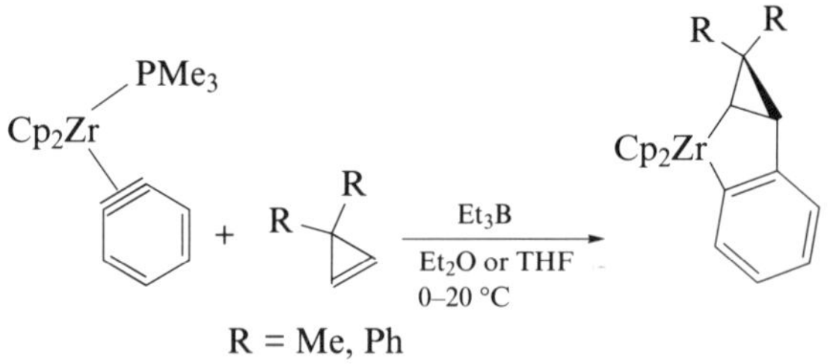

(130)

Some of the bicyclo[3.1.0]hexene zirconocene derivatives, especially those containing a phenyl group, rearrange quantitatively to the corresponding 4-vinyl-1-metalla-2-cyclobutene derivatives when heated in toluene to 60–80 °C for several hours (equation 131). The 6,6-dimethyl-2,3-diphenyl-2-zirconabicyclo[3.1.0]hex-3-ene derivative prefers ring-opening to the isomeric η^1,η^3-allylic complex upon thermolysis.

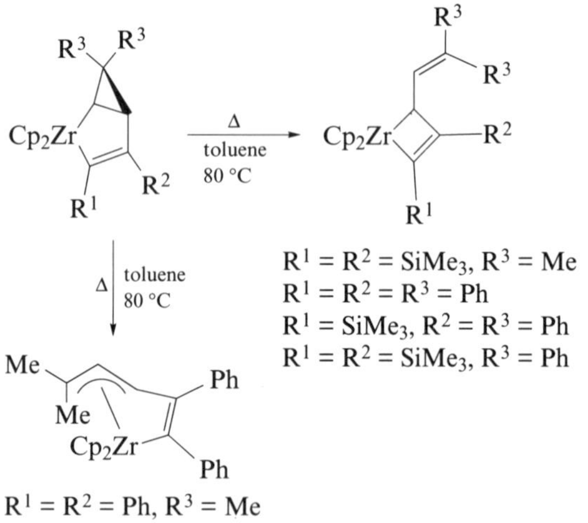

(131)

The coupling reaction[161] of (cyclopropylcarbinyl)WCp(CO)$_2$ and the carbonyl ligand, induced by trimethylphosphine (*see* Section II.D) gave initially an η^2-ketenyl complex, which under CO pressure (60 bar) adds a further carbonyl ligand to give a pentacoordinated η^1-cyclopropylketenyl tungsten complex (equation 132)[159]. This unusual cyclo-

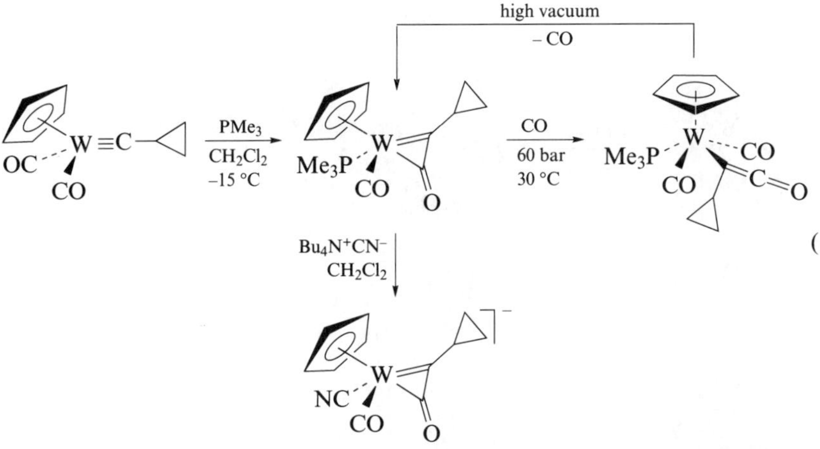

(132)

propylvinyl complex in which the metal is bonded to the internal vinylic carbon readily loses a carbonyl ligand under high vacuum at –30 °C, back to the η^2-ketene complex. The latter complex undergoes ligand substitution by cyanide to give the corresponding anionic η^2-cyclopropylketenyl complex.

b. π-Complexes of vinylcyclopropanes. The interaction of transition metal complexes with parent vinylcyclopropane (vcp) and unconjugated derivatives invariably leads to unstable η^2-complexes which usually rearrange by ring-opening reactions, typical of the metal in use. This is best illustrated by the reactions of vinylcyclopropanes with iron and palladium complexes. Irradiation of a 1% ethereal solution of the parent vinylcyclopropane and two equivalents of ironpentacarbonyl at –50 °C results in the formation of a 10:1 mixture of (η^2-vcp)$Fe(CO)_4$ and the η^1,η^3-allylic isomeric complex, obtained by allylic ring cleavage and carbonyl insertion, which can be separated by column chromatography at –20 °C (equation 133)[177]. Upon heating (η^2-vcp)$Fe(CO)_4$ rearranges with loss of a carbonyl ligand to (η^4-1,3-pentadiene)$Fe(CO)_3$ (3:1 *syn/anti* mixture) whereas the allylic isomer undergoes decarbonylation to the 1,3-σ,π-allylic complex.

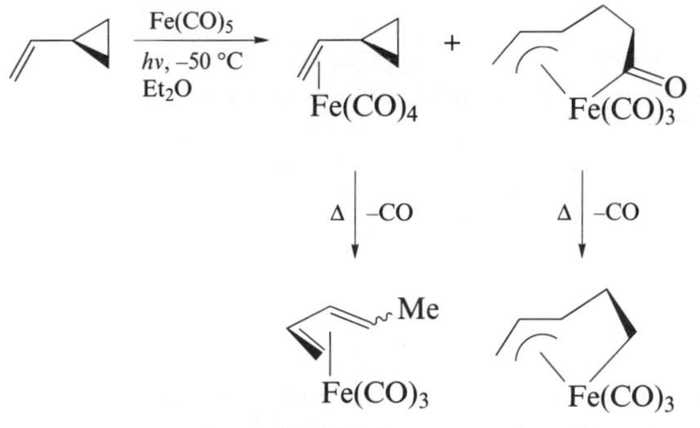

(133)

Numerous examples of vinylcyclopropane to diene rearrangement have been studied (e.g. equation 134)[178] and thoroughly reviewed in the aforementioned surveys and in this series[179], and hence will not be repeated here. The reader will also find in this series[180] ample examples of typical ring-opening reactions of cyclic vinylcyclopropanes (equation 135)[181] and divinylcyclopropanes (equation 136)[182] which lead to stable irontricarbonyl σ,π-allylic complexes.

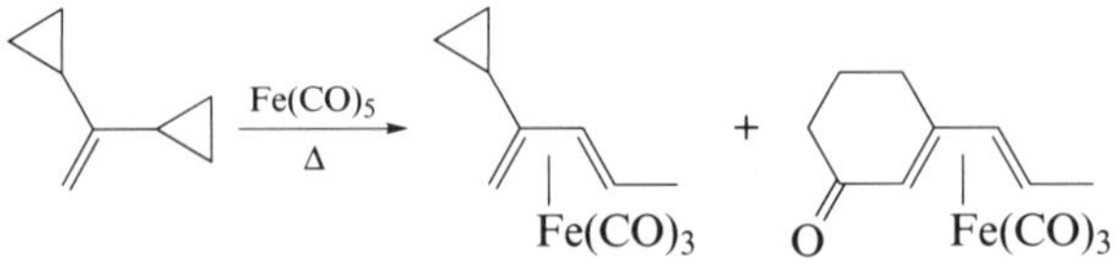

(134)

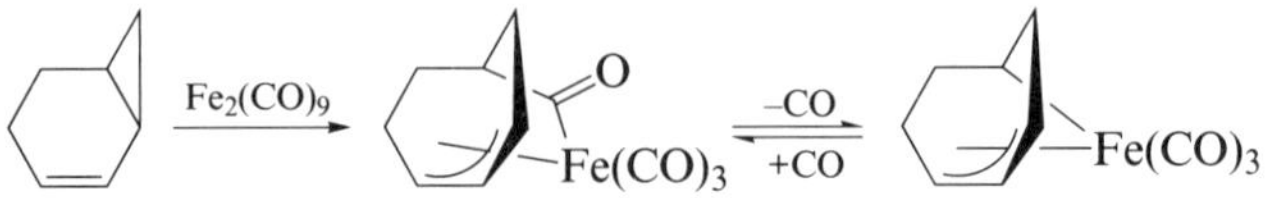

(135)

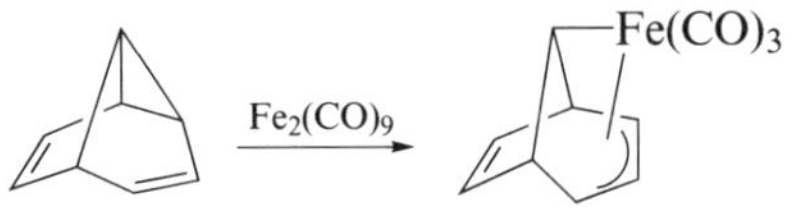

(136)

The analogous chloropalladation of vinylcyclopropane gave initially a π-complex that may involve (NMR analysis) additional Pd(II)–cyclopropane interaction (equation 137)[183,184] which first rearranges by ring cleavage and metal to organic ligand migration of a chloro ligand, to form a 1,4-η^1,η^2-complex, than to the final η^3-π-allylic complex. This mechanism has been further established by stereochemical studies of the chloro- and oxy-palladation of the chiral cyclic vinylcyclopropane (+)-2-carene[185].

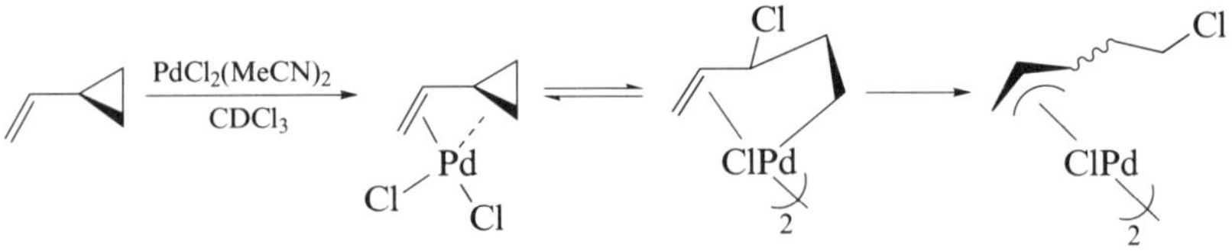

(137)

3. Conjugated vinylcyclopropane complexes

In sharp contrast to the generally labile η^2-vinylcyclopropane complexes, complexes of butadienylcyclopropane and higher vinylogue derivatives in which the metal is coordinated to more than one double bond are unequivocally stable. This allowed the application of standard synthetic methods to the preparation of polyenylcyclopropane complexes, such as cyclopropanation reactions of polyene complexes having at least one free double bond, as well as complexation of polyenylcyclopropanes by ligand exchange reactions. Consequently, a herd of acyclic and cyclic conjugated vinylcyclopropane complexes have been prepared over the past three decades and their structure and reactivity studied. These will be reviewed selectively here.

a. Acyclic η^4-dienyl complexes. Acyclic iron complexes of butadiene are important synthons in chiral synthesis, due to the *anti* (to metal) plane-selectivity imparted by the complex[186]. In this context, acyclic cyclopropyl-η^4-butadiene tricarbonyl iron complexes have been used to prepare chiral cyclopropanes. The complexes are readily prepared from the corresponding η^4-1,3,5-hexatriene iron complexes by one of three classical methods of cyclopropanation: (i) addition of dichlorocarbene under phase transfer conditions (PTC) (equation 138)[186] or Cu-catalyzed reaction with diazoacetate (equation 139)[187], (ii) 1,3-

cycloaddition of diazomethane followed by thermolysis (equation 140)[188,189] and (iii) addition of sulfur ylides to electrophilic double bonds (equation 141)[190,191].

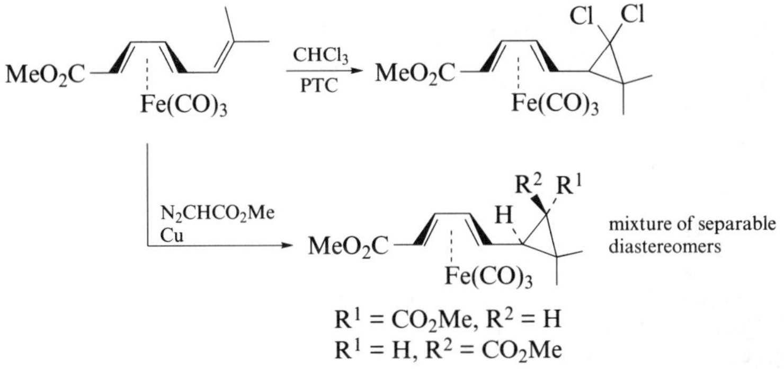

(138)

(139)

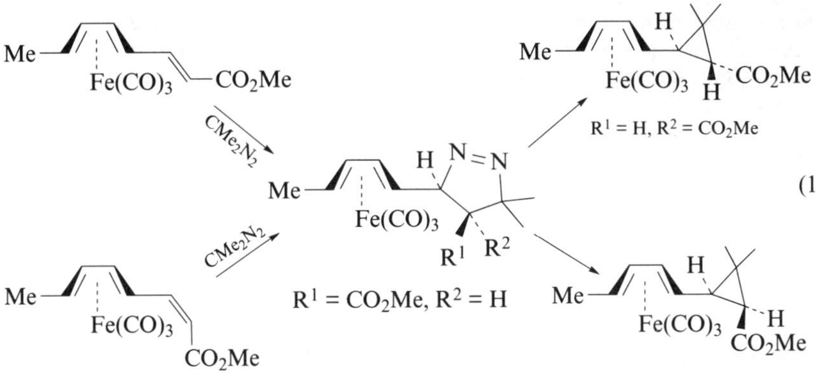

(140)

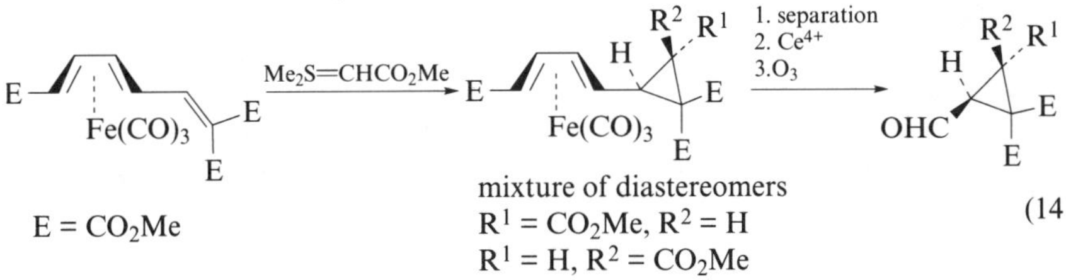

(141)

b. Cyclic complexes. (i) η^4-Bicyclo[4.1.0]hepta-2,4-diene (norcaradiene) complexes. Stabilization of the parent norcaradiene is achieved by η^4-coordination with tricarbonyliron. Two isomers of the complex are conceivable, the *anti*-isomer in which the metal fragment resides at the *exo*-face of the bicyclic system and the corresponding *syn*-isomer, where the metal is at the *endo*-position. Both complexes have been elegantly prepared by a formal disrotatory cyclobutene ring-opening of the corresponding *syn*- and *anti*-tricyclo[3.2.0.0^{2,4}]hept-6-ene iron tetracarbonyl complexes, induced by $Fe_2(CO)_9$ (equation 142)[192]. The *syn*- and *anti*-norcaradiene complexes display similar thermal stability, undergoing further disrotatory ring cleavage at 90.5 °C to η^4-cycloheptatrieneiron tricarbonyl with first-order rates of 2.51×10^{-5} and 3.16×10^{-5} s^{-1}, respectively.

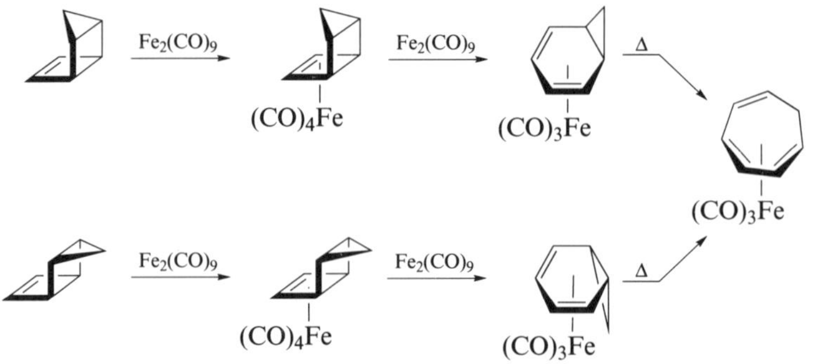

(142)

A related η^4-norcaradiene tricarbonyliron complex is obtained upon reaction of tricyclo[4.3.1.0^{1,6}]deca-2,4-diene with $Fe_3(CO)_{12}$ in boiling benzene (equation 143). However, the [4.3.1]propellane ring system is not retained in the analogous tricarbonylchromium complex. Instead, as suggested from solution NMR and solid state X-ray analyses, the complex assumes a homoaromatic structure, which is intermediate between a norcaradiene and a cycloheptatriene system (equation 144)[193,194]. It is noteworthy that the $Cr(CO)_3$ group prefers the same conformation as the $Fe(CO)_3$ group in the analogous norcaradiene iron complex.

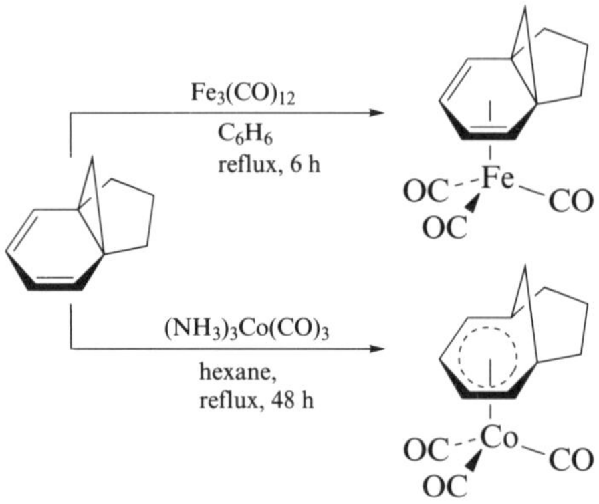

(143)

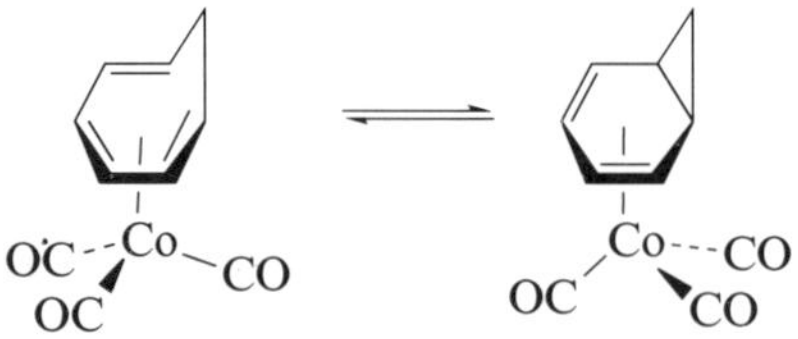

(144)

Further experimental and theoretical studies on the rotational barriers of the metal fragment in (cycloheptatriene)$Cr(CO)_3$ complexes[195] suggest that (cycloheptatriene)-$Cr(CO)_3$ complexes in general are in equilibrium with their norcaradiene valence isomers and their ground state conformation is controlled by the same electronic factors which effect the cycloheptatriene-norcaradiene equilibrium[195].

(ii) η^4-Spiro[2.5]octa-4,6-diene and η^5-spiro[2.5]octa-4,6-dienyl complexes. Spiro-[2.5]octa-4,6-diene tricarbonyliron complex (Scheme 6)[152] undergoes allylic hydrogen

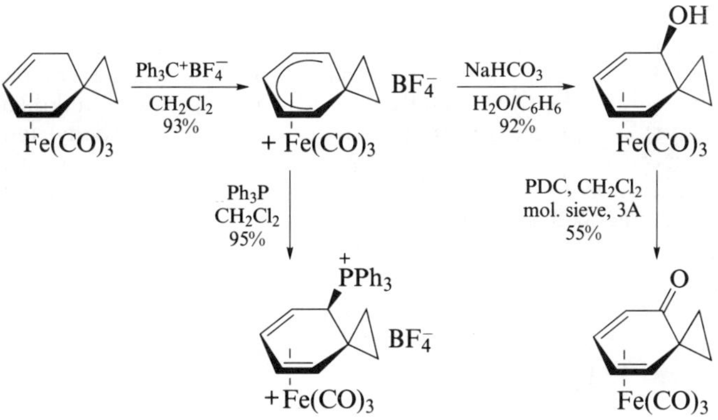

SCHEME 6

abstraction with trityl tetrafluoroborate to afford the interesting symmetrical η^5-spiro[2.5]-octa-4,6-dienylium cation complex, which is stabilized both by the metal and the conjugating cyclopropyl ring. Reaction with nucleophiles such as hydroxide and triphenylphosphine occurred exclusively at the cyclopropylcarbinyl position to give the corresponding allylic carbinol and phosphonium salt, respectively. The alcohol can be further oxidized with pyridinium dichromate in CH_2Cl_2 to the highly conjugated but stable cyclopropyl ketone complex. No ring opening products were observed in these reactions (*vide infra*).

(iii) η^4-Dispiro[2.0.2.4]deca-7,9-diene complexes. Stable (η^4-dispiro[2.0.2.4]deca-7,9-diene)$Fe(CO)_2L$ [L = CO, PPh_3, $P(OPh)_3$] iron complexes are readily prepared without ring cleavage by reaction of the hydrocarbon with $Fe(CO)_2L$ transfer agents (bda)$Fe(CO)_2L$ (bda = benzylideneacetone) (equation 145)[196]. The $Fe(CO)_3$ complex is also efficiently formed in the reaction with $Fe_2(CO)_9$. The corresponding tricarbonylruthenium complex was similarly obtained by the ligand exchange reaction with (COD)$Ru(CO)_3$. However, rearrangement of the dispiro hydrocarbon to *o*-ethylstyrene is observed when $Ru_3(CO)_{12}$ was employed.

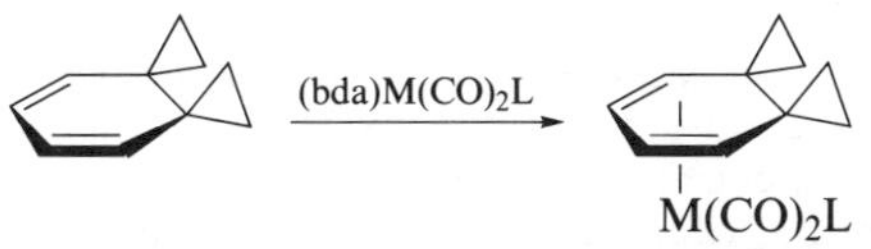

(145)

M = Fe, L = CO, PPh_3, $P(OPh)_3$
M = Ru, L = CO

Ring cleavage occurs when (η^4-dispiro[2.0.2.4]deca-7,9-diene)$Fe(CO)_3$ is treated with tetrafluoroboric acid, to give after exchange with PF_6^- a stable η^5-complex cation (equation 146) which, upon further reaction with $NaN(SiMe_3)_2$ in MeCN/THF, afforded the cross-conjugated vinylcyclopropane complex, specifically with a *Z*-configuration of the exocyclic ethylidene[152].

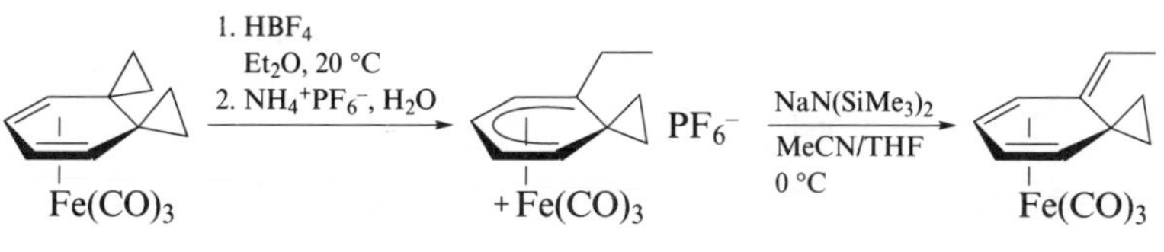

(146)

(iv) η^4-Bicyclo[5.1.0]octa-2,4-diene and η^5-bicyclo[5.1.0]octa-2,4-dienyl complexes. η^4-(Cycloheptatriene)$Fe(CO)_3$ undergoes cyclopropanation with CH_2I_2 under Simmons–Smith conditions, affording the parent homotropylidene complex (η^4-bicyclo[5.1.0]octa-2,4-diene)$Fe(CO)_3$[197] and with methyl diazoacetate[198] and dihalocarbenes CX_2 (X = Cl, Br)[199–202] to give the corresponding 8-*exo*-methoxycarbonyl and 8,8-dihalo complexes, respectively (equation 147). The parent homotropylidene complex undergoes a thermal degenerate isomerization at 75 °C, involving suprafacial migration of the cyclopropane ring together with a 1,5-H shift of the *syn*-allylic hydrogen[203].

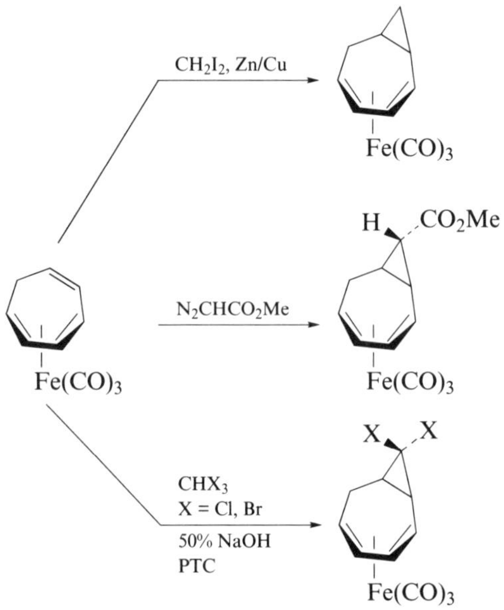

(147)

Much interest has been devoted to the related stable (η^5-bicyclo[5.1.0]octa-2,4-dienylium)$Fe(CO)_3$ cation complex, mainly in connection with the question of the possible homoaromaticity of these cations (equation 148)[204–206]. The complex was first prepared by the formal electrocyclic ring closure of protonated (cyclooctatetraene)$Fe(CO)_3$ at –60 °C[207,208] (Scheme 7) and has later been proved by NMR studies in magic acid solutions *not* to be the homotropyliumiron tricarbonyl ion[209,210]. The X-ray structure analysis of the related ruthenium cation phosphine complex [(η^5-bicyclo[5.1.0]octa-2,4-dienylium)-$Ru(PMe_2Ph)_3$]PF_6 confirmed the bicyclic structure of these ion complexes[211]. The analogous hetero azepine tricarbonyliron complexes (*anti* and *syn* isomers) have also been reported[212].

The iron complex interacts with nucleophiles such as $NaBH_4$ (and $NaBD_4$)[213,214] and NaOH[215] to give mainly the 2-*anti*-substitution products. This stereoselectivity was also observed (X-ray structure) using the bulky $Re(CO)_5^-$ as the nucleophile[216].

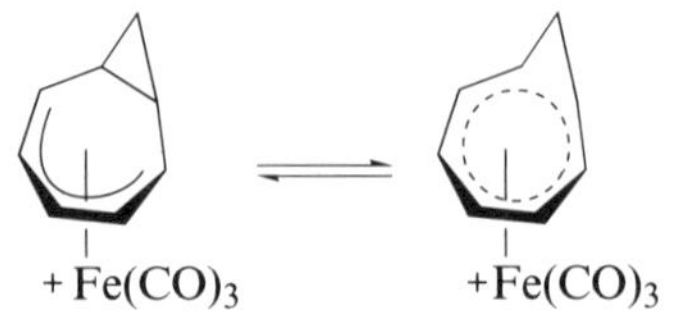

(148)

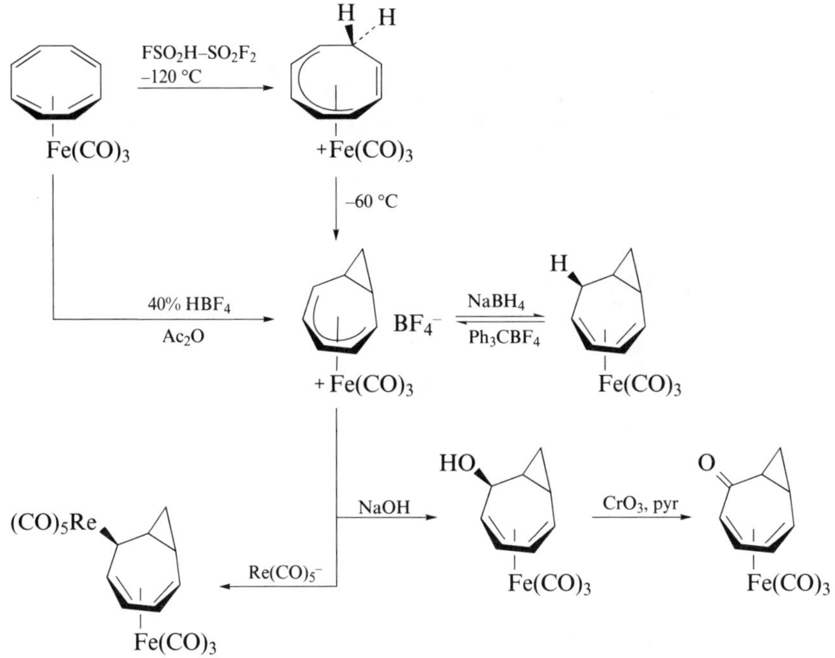

SCHEME 7

Access to alkyl substituted derivatives of the homotropylidene complexes is provided via the η^4-tropone iron complex by reaction with diazoalkanes, followed by mild thermolysis of the 3+2 pyrazoline adduct to give the corresponding homotropone complexes (equation 149)[217,218]. The 8,8-dimethyl derivative was used as starting material for the preparation of the fluxional (η^5-2,8,8-trimethylbicyclo[5.1.0]octa-2,4-dienylium)$Fe(CO)_3$ cation complex[219]. More recently (homotropone)$Fe(CO)_3$ was used for the synthesis of unique chiral 1,2-homoheptafulvene iron complexes[220,221].

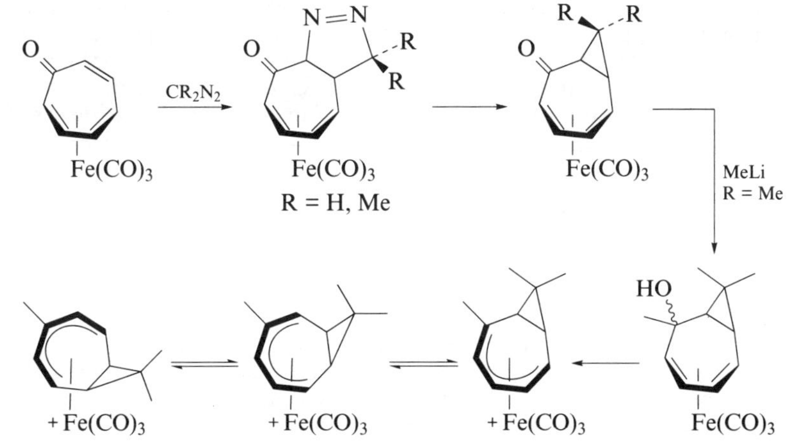

(149)

Interestingly, treating (η^4-cyclooctatetraene)$Fe(CO)_3$ with acetyl chloride under Friedel–Crafts reaction conditions yielded unexpectedly[222,223] the (η^2,η^3-8-*exo*-acetyl bicyclo[3.2.1]octadienylium)$Fe(CO)_3$ cation complex, presumably by rearrangement of the intermediate bicyclo[5.1.0]octadienylium isomer (Scheme 8). The structure of the rearranged cation was confirmed from the X-ray crystal structure and from the typical 1,3-σ,π-allylic products obtained upon nucleophilic reaction with $LiAlD_4$ and NaCN. The nucleophilic reaction of the more bulky iodide occurs, however, on the metal.

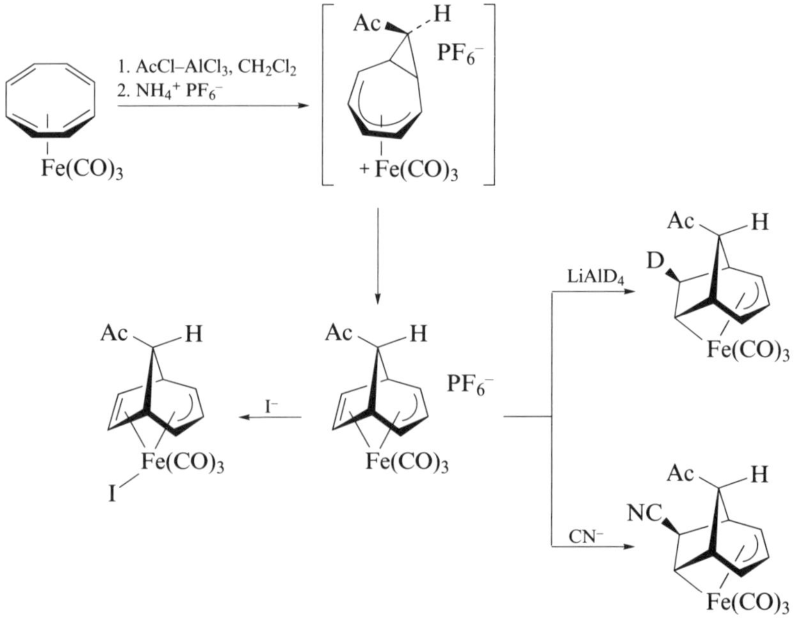

SCHEME 8

(v) η^6-Bicyclo[6.1.0]nona-2,4,6-triene complexes. Although the η^6-complexes of bicyclo[6.1.0]nona-2,4,6-triene with group 6 $M(CO)_3$ (M = Cr, Mo, W) have been synthesized quit awhile ago by ligand exchange reactions[224–226] of the triene with (dglm)$M(Co)_3$ (dglm = diethyleneglycoldimethyl ether), only recently has the structure of the molybdenum and tungsten complexes been correctly assigned by X-ray crystallography as the *syn*-isomers (equation 150)[227,228]. Related η^2,η^2-rhodium complexes of this system have also been briefly mentioned[229].

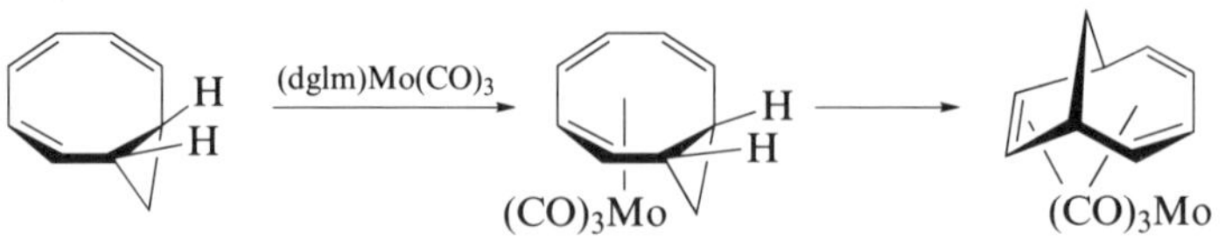

(150)

The η^6-complexes all undergo thermal rearrangement to η^2,η^4-bicyclo[4.2.1]nona-2,4,7-triene tricarbonyl complexes. The possible mechanism of the rearrangement of the molybdenum complex has been thoroughly investigated by deuterium labeling and kinetic studies[228].

F. Alkynylcyclopropane–Metal Derivatives

The organometallic chemistry of alkynylcyclopropanes involves primarily the formation and reactions of carbon-metal σ-bonds. Metals come essentially from the main group elements, with lithium playing a major role. The two metallation sites are the cyclopropyl and the acetylenic positions, which are expected to differ considerably in their acidity values (t-butylacetylene, pKa = 25[230], cyclopropane, pK_a = 46[183]) but less in the reactivity of their metal conjugated bases towards electrophiles.

Both main group and transition metal elements interact with the acetylenic triple bond in a variety of reactions, including hydrogenation, hydrometallation, hydration and cycloadditions. Notably, in most reactions the cyclopropane ring remains intact.

1. Cyclopropylethynyl-metal derivatives

Metallation of alkynylcyclopropanes at the acetylenic end is accomplished either by deprotonation or via metal–halide exchange reaction with strong bases. Metallation of ethynylcyclopropane may be affected by KOH in DMF, ethereal EtMgBr or preferably BuLi in THF (equation 151)[231]. All three metal acetylides react with methyl ketones to give the corresponding alcohols. However, the instability of cyclopropyl ketones towards bases, especially at the reaction conditions required by KOH (20 °C, 6h), and the sensitivity of cyclopropenyl double bonds in cyclopropenyl ketone derivatives towards addition reactions of alkylmagnesium compounds, make the alkyllithium (–78 °C, instant reaction) superior to the other reagents.

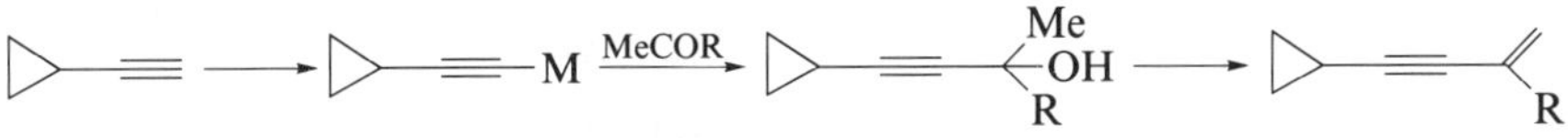

KOH-DMF	M = K	R = Me	(151)
EtMgBr-Et_2O	M = MgBr	R = Me, *c*-Pr	
BuLi-THF	M = Li	R = Me, *c*-Pr, 1,2-diisopropylcyclopropen-3-yl	

Alternatively, the reaction of cyclopropylethynylmagnesium bromide with cyclopropanone hemiacetal gives 1-(cyclopropylethynyl)cyclopropanol (equation 152)[232]. The reaction of cyclopropanone acetal with other alkynyl Grignard reagents serves as a general route to alkynylcyclopropanols. Similarly, alkynyllithium derivatives of vitamin D were coupled with cyclopropane carbonyl isoxazolidine to give the corresponding alkynyl–cyclopropyl ketones (equation 153)[233].

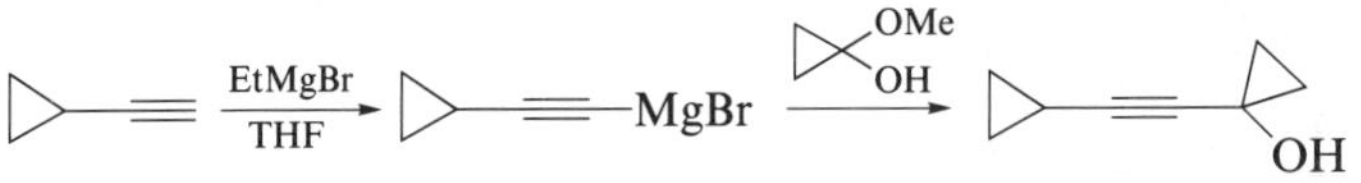

(152)

Cuprous cyclopropylacetylide, prepared from CuI and cyclopropylacetylene in ammoniacal solution couples with 4-iodonitrobenzene in pyridine, yielding 4-nitrophenyl cyclopropylacetylene (equation 154)[234]. Reaction of the acetylide with tropylium tetrafluoroborate in acetonitrile, in the presence of LiBr, affords 7-(cyclopropylethynyl)cycloheptatriene. The anion radicals obtained by reduction of these compounds were utilized for ESR spectroscopic analysis of the cyclopropyl β hyperfine splittings.

The discovery of an easy route to a series of 1-chloro-1-(trichlorovinyl)cyclopropanes from the thermal reaction of tetrachlorocyclopropene and olefins greatly promoted the availability and synthetic utility of alkynylcyclopropanes. Upon reductive elimination with two equivalents of *n*-BuLi in ether–hexane at –78 °C, a series of ring substituted

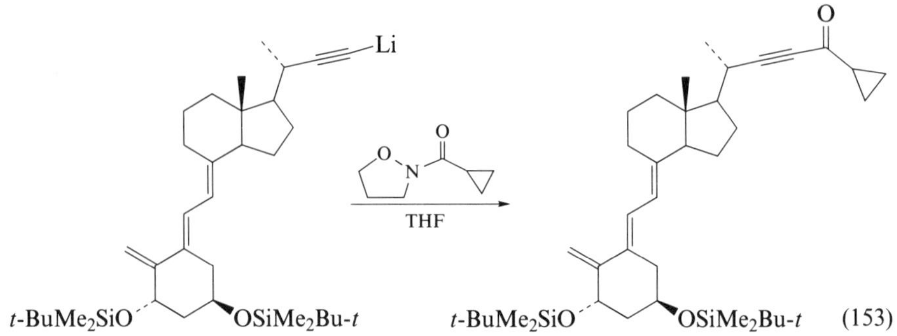

(153)

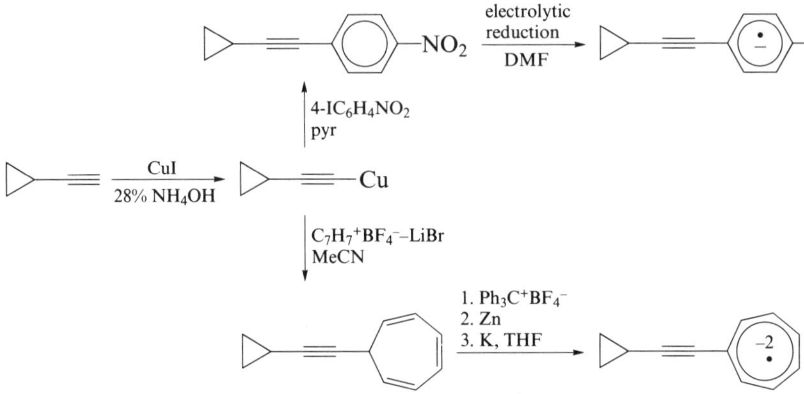

(154)

(1-chlorocyclopropyl)ethynyllithium compounds was readily obtained. These were quenched with water or methanol, giving the corresponding (1-cyclopropyl) acetylenes (equation 155)[235,236].

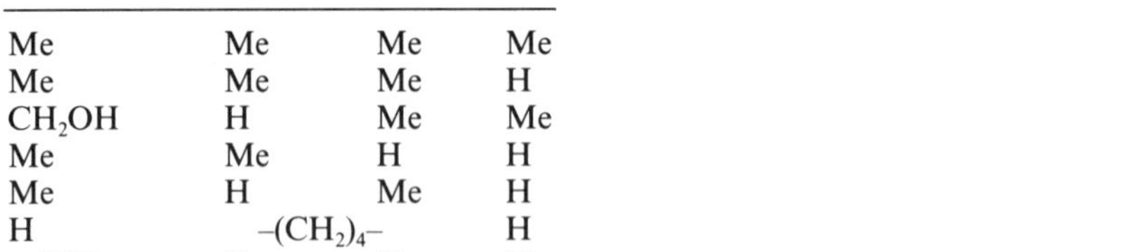

(155)

R^1	R^2	R^3	R^4
Me	Me	Me	Me
Me	Me	Me	H
CH_2OH	H	Me	Me
Me	Me	H	H
Me	H	Me	H
H	$-(CH_2)_4-$		H
n-C_3H_7	H	H	H
$SiMe_3$	H	H	H
H	H	H	H

Furthermore, the lithium acetylides could be trapped with a large variety of electrophiles to yield a plethora of 2-substituted (1-chlorocyclopropyl)acetylenes, as shown in equation 156 for 1-chloro-2,2,3,3-tetramethyl-1-ethynyl-1-cyclopropyllithium.

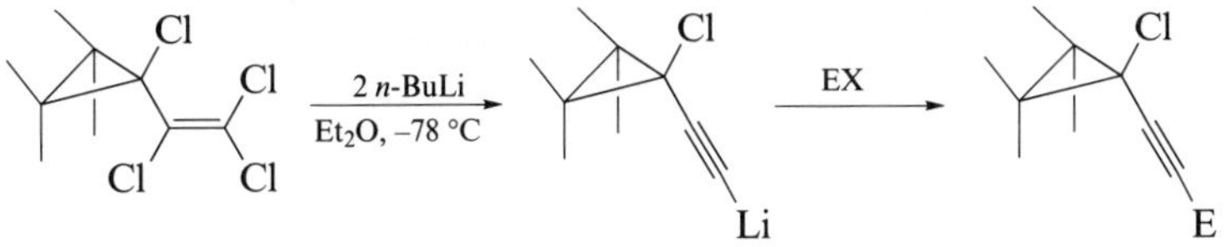

(156)

EX	E
CO_2	CO_2H
$ClCO_2Me$	CO_2Me
$ClSiMe_3$	$SiMe_3$
$(MeO)_2SO_2$	Me
MeI	Me
n-BuBr[a]	*n*-Bu
Me_2CO	Me_2COH
$(CH_2)_5CO$ (cyclic)	$(CH_2)_5COH$ (cyclic)
$(H_2CO)_n$	CH_2OH
NCS[b]	Cl
MeSSMe	SMe
$(MeCO)_2O$	COMe

[a] 30 mol% HMPT added
[b] *N*-Chlorosuccinimide

Use of chiral cyclopropylethynyllithium derivatives permits the elegant selective synthesis of labeled chiral vinylcyclopropanes, for stereochemical studies of the thermal vinyl cyclopropane–cyclopentene rearrangement[237]. Thus, reductive elimination of (1*S*,*trans*)-(2,2-dibromoethenyl)-1-methylcyclopropane with BuLi in pentane, followed by hydrolysis of the lithium acetylide, afforded (1*S*,*trans*)-2-ethynylmethylcyclopropane (equation 157).

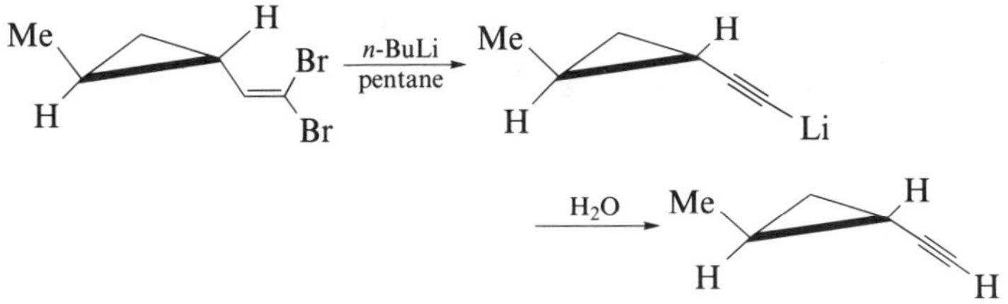

(157)

(1*S*, *trans*)-

Further regio- and stereospecific reductive addition of the acetylene with diisopropyl aluminum hydride (DIBAL) in CH_2Cl_2, followed by D_2O, gave (+)-(1*S*,*trans*)-*E*-2-(ethenyl-2-d_1)-methylcyclopropane (equation 158).

Similarly, DIBAL reduction, followed by hydrolysis, of the labeled (1*R*,*trans*)-2-(ethynyl-d_1)methylcyclopropane isomer resulted in the formation of (–)-(1*R*,*trans*)-*Z*-2-(ethenyl-2-d_1)-methylcyclopropane (equation 159).

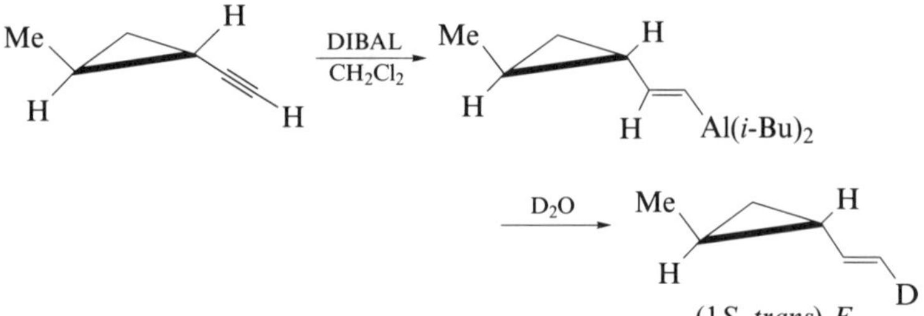

(158)

(1*S*, *trans*)-*E*

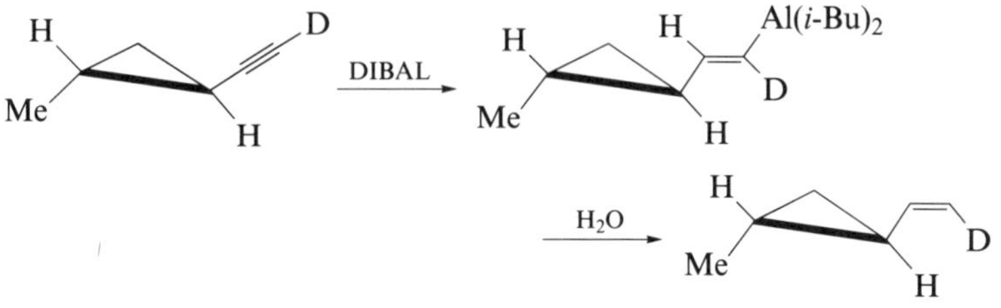

(159)

(1*R*, *trans*)-*Z*

2. 1-Alkynylcyclopropyllithium

When the acetylenic hydrogen of ethynylcyclopropanes is substituted by groups other than halogens, metallation at this end is blocked, leaving the cyclopropyl position as the next metallation site. 1-Chloro-1-alkynylcyclopropanes, e.g. 1-chloro-2,2,3,3-tetramethyl-1-(trimethylsilylethynyl)cyclopropane (equation 160), readily undergo a halogen–metal exchange reaction with *n*-BuLi in ether, affording 1-alkynyl cyclopropyllithium compounds. These react with electrophiles to give the corresponding disubstituted cyclopropylacetylenes[236,238].

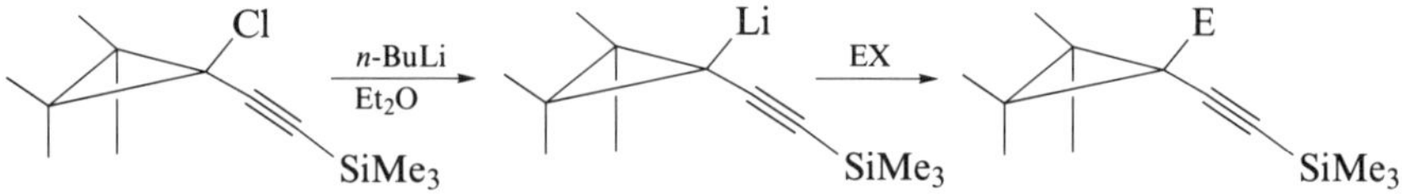

EX = H_2O, Me_3SiCl, CO_2, I_2, MeSSMe, 2,4,6-$Me_3C_6H_2SO_2ONMe_2$
E = H, $SiMe_3$, COOH, I, SMe, NMe_2 (160)

The organometallic chemistry of dicyclopropylacetylenes has been studied extensively. The parent compound was monometallated at the cyclopropyl position with *n*-BuLi in THF, yielding a separable organolithium compound. Upon quenching with a variety of electrophiles, this gave monosubstituted (cyclopropylethynyl)cyclopropanes (equation 161)[239]. Preparation of the analogous 1-(cyclopropylethynyl)cyclopropanol was described

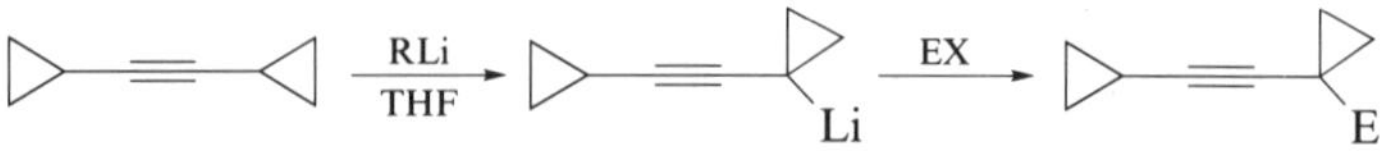

(161)

R = *n*-Bu, *t*-Bu, Ph
EX = CO_2, Me_3SiCl, *n*-BuCl, TsCl, TsBr, BrCN
E = COOH, Me_3Si, *n*-Bu, Cl, Br, CN

before (equation 152)[232] and the methyl analogue was obtained by reductive elimination of 1-(cyclopropylethynyl)-2,2-dibromo-1-methylcyclopropane with Bu_3SnH (equation 162)[231].

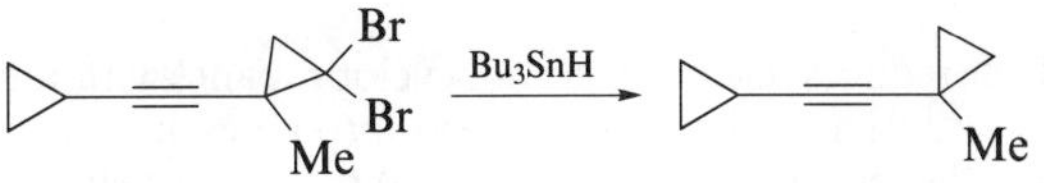

(162)

3. Dimetallation

Treating 1-chloro-1-(trichlorovinyl)cyclopropanes with three equivalents of BuLi afforded the the dilithiated acetylene This, upon reaction with excess trimethylsilyl chloride, yielded the bis(trimethylsilyl)ethynylcyclopropane (equation 163). The same results can be achieved using magnesium metal in THF as the metallating agent, instead of the preferable BuLi[236,238].

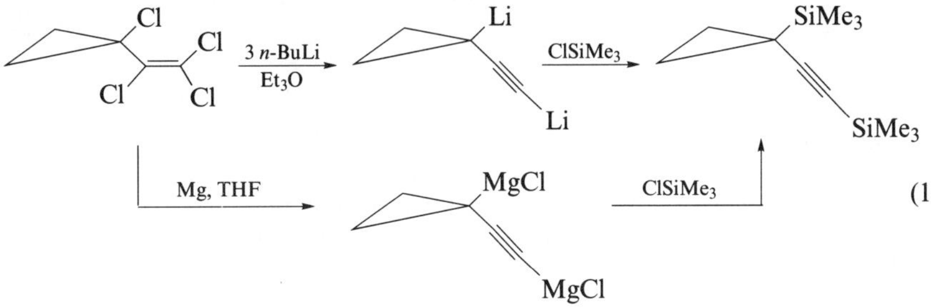

(163)

Dilithiation of dicyclopropylacetylene was best effected with 5-fold excess of *t*-BuLi-TMEDA complex. Quenching of the separable dilithio compound with electrophiles gave the corresponding disubstituted dicyclopropylacetylenes (equation 164)[239].

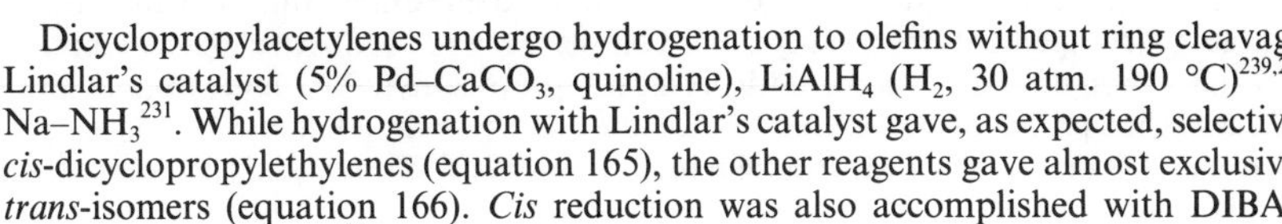

(164)

4. Reduction reactions

Dicyclopropylacetylenes undergo hydrogenation to olefins without ring cleavage with Lindlar's catalyst (5% Pd–$CaCO_3$, quinoline), $LiAlH_4$ (H_2, 30 atm. 190 °C)[239,240] and Na–NH_3[231]. While hydrogenation with Lindlar's catalyst gave, as expected, selectively the *cis*-dicyclopropylethylenes (equation 165), the other reagents gave almost exclusively the *trans*-isomers (equation 166). *Cis* reduction was also accomplished with DIBAL (*see above*) (equation 159)[237].

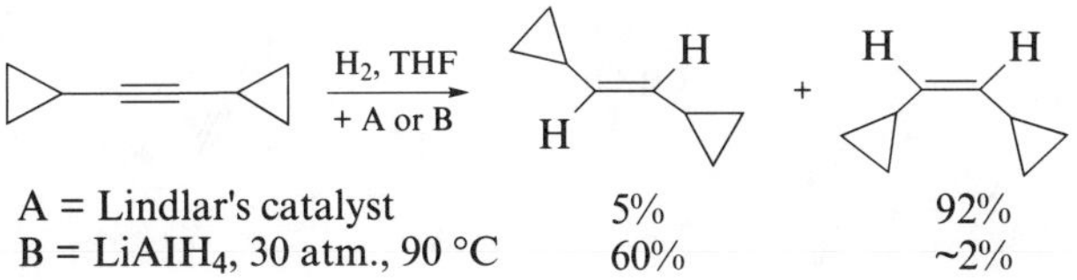

(165)

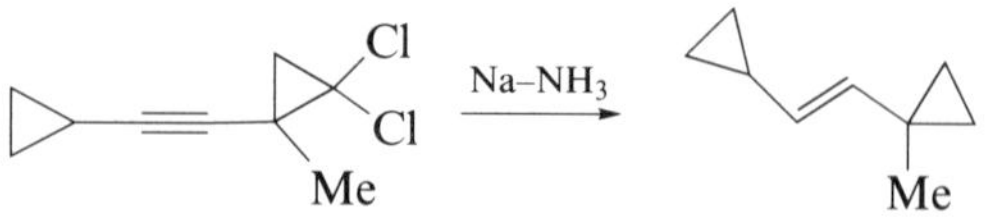

(166)

Tributyltin hydride reduces selectively bromocyclopropanes without affecting the triple bond (equation 167)[231]. The reduction of *gem*-dibromides is stepwise, which enables convenient access to the monobromoacetylenes and further, via elimination of HBr, to 3-(cyclopropylethynyl)cyclopropene.

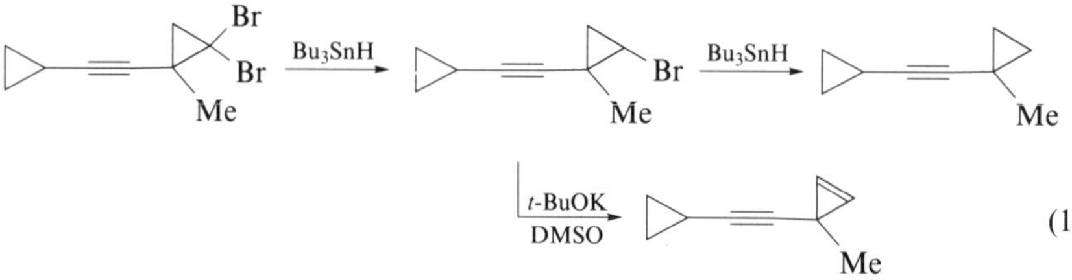

(167)

5. Hydration

Mercuric sulfate catalyzed hydration of cyclopropylacetylenes in aqueous sulfuric acid, like other monosubstituted alkynes, gave mainly the corresponding methyl ketone accompanied by small amounts of ring-opened prouducts (equation 168)[236]. Similar results were obtained using HgO in trichloroacetic acid, with catalytic amounts of BF_3–Et_2O and methanol.

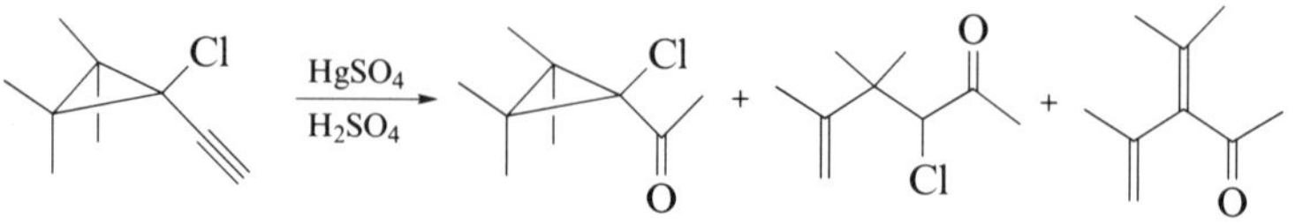

(168)

6. η^1-Transition metal complexes

Contrary to the rich organometallic chemistry of cyclopropylacetylenes with main group elements, σ-bonded transition metal complexes of the alkynes are rare. Nevertheless, the single example given in the literature is of a stable crystalline compound, which permits insight into the structure of cyclopropylethynyl complexes in the solid state.

η^1-Ketenyl complexes of the group 6 metals molybdenum and tungsten undergo a remarkable rearrangement to η^1-alkynyl σ-complexes with reductive deoxygenation by CO. Dicarbonyl(η^5-2,4-cyclopentadienyl)(cyclopropylethynyl)(trimethylphosphine) tungsten was thus prepared by heating (60 °C) the corresponding ketenyl W-complex in CH_2Cl_2 under CO pressure (70 bar) for 24 h (equation 169)[241]. The crystal structure of the

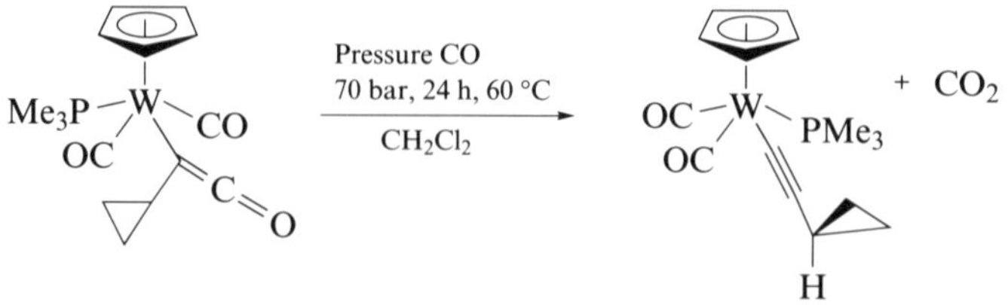

(169)

cyclopropylethynyl complex $Cp(CO)_2(PMe_3)W{-}C{\equiv}C(c\text{-}Pr)$ reveals a cyclopropyl group in the bisected conformation, with the ring pointing towards the phosphine group, away from the two *cis* CO ligands. The distal cyclopropane bond is significantly longer (153.8 pm) than that found in the free cyclopropane (150.9 pm), indicating that the metalalkynyl substituent acts as a donor group. This is in contrast to the π-acceptor nature of the metalalkynyl group (loc. cit.).

The ^{1}H and ^{13}C NMR spectra of the cyclopropyl group in CD_2Cl_2 show the expected high field chemical shifts at 0.4–0.8 and 1.37 ppm, and 3.1 and 8.3 ppm, respectively. An unusual long-range coupling of the phosphine ^{31}P and the cyclopropyl methylene ^{13}C was observed, $^{5}J\{^{31}P{-}^{13}C\} = 3.7$ Hz.

7. Reactions involving metal π-bonding

Ethynylcyclopropanes, like normal acetylenes, react with dicobalt octacarbonyl in ether to form stable dinuclear cluster-like hexacarbonyl complexes (equation 170)[236]. The complex with 1-chloro-2,2,3,3-tetramethylethynylcyclopropane reacts stereo- and regioselectively with norbornene in a typical Pauson-Khand reaction to give the *exo*-2-cyclopropyl substituted cyclopentenone (equation 171). Similarly, the reaction of 2-ethoxycyclopropylacetylene with cyclopentene in the presence of $Co_2(CO)_8$ under CO gave 3-(2-ethoxycyclopropyl)-*cis*-bicyclo[3.3.0]oct-3-en-2-one (equation 172)[242].

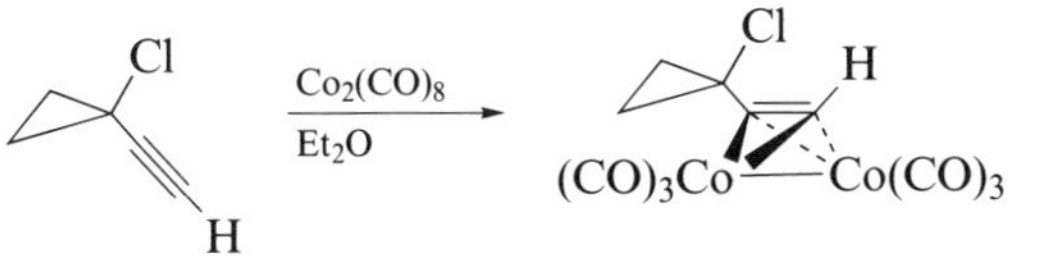

(170)

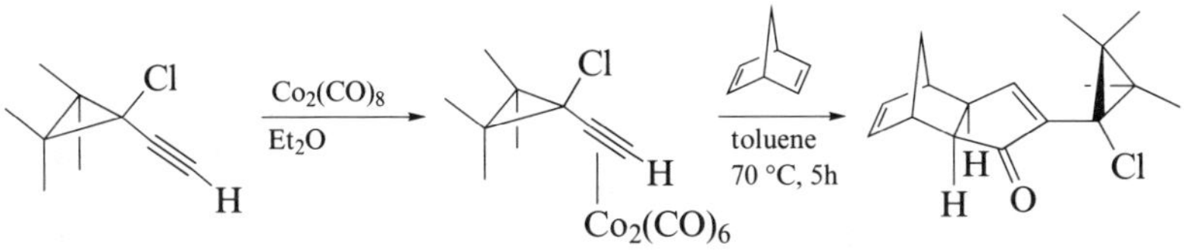

(171)

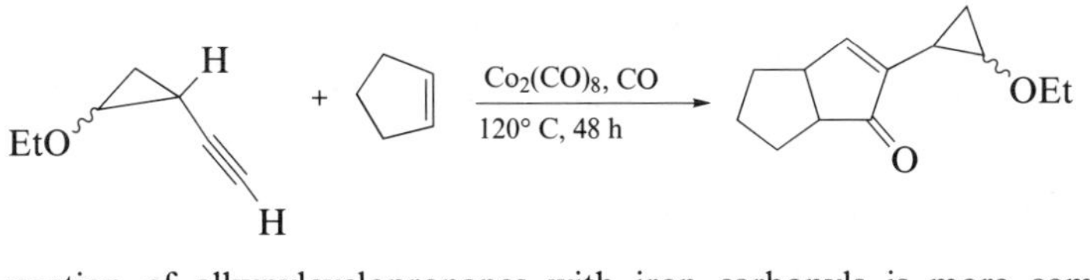

(172)

The reaction of alkynylcyclopropanes with iron carbonyls is more complicated. Irradiation of ethynylcyclopropane with iron pentacarbonyl in hexane (or CH_2Cl_2) solution gave an unidentified mixture of complexes, which upon oxidative degradation afforded the two isomeric 2,5- and 2,6-dicyclopropylbenzoquinones (equation 173)[243].

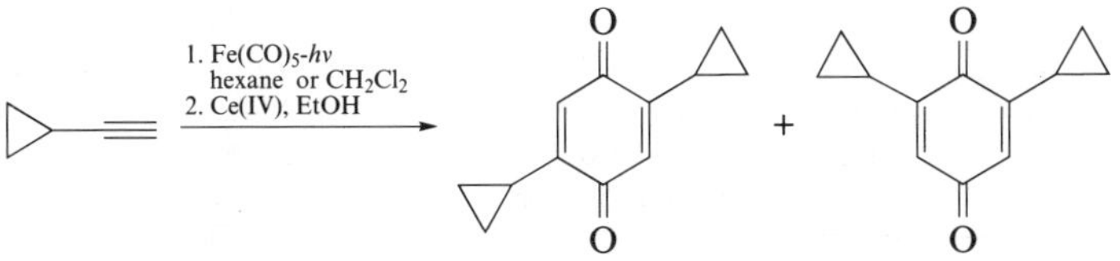

(173)

The analogous reaction with dicyclopropylacetylene gave a complex mixture of 8 dinuclear complexes[243]. The structure of the four main (>1%) cluster-like complexes is shown in equation 174. The product ratio could be changed by using $Fe_2(CO)_9$ or $Fe_3(CO)_{12}$ under thermal condition instead of $Fe(CO)_5$.

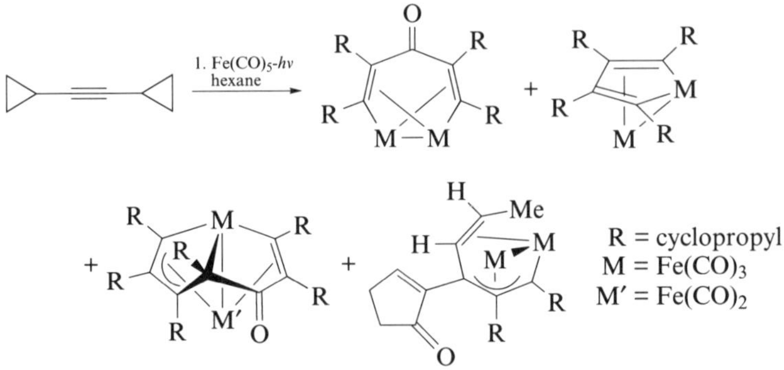

(174)

Finally, heating a neat 10:1 mixture of dicyclopropylacetylene with $Fe_3(CO)_{12}$ at high temperature (180 °C) for 2 h afforded [tetrakis(cyclopropyl)cyclopentadienone]$Fe(CO)_3$ together with the uncomplexed hexakis(cyclopropyl)benzene (equation 175)[244].

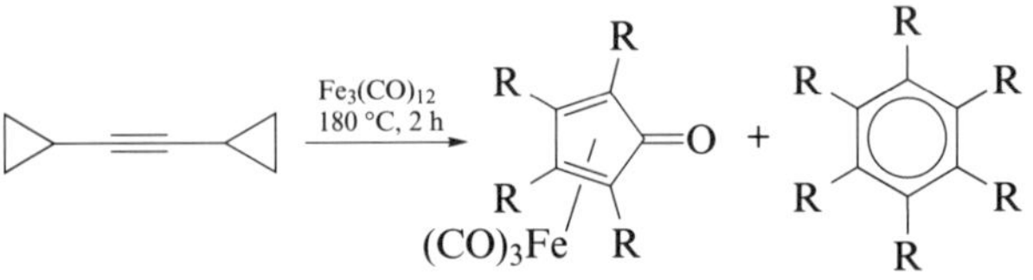

(175)

R = cyclopropyl

Noncarbonyl transition metal complexes catalyze dimerization and aromatic cyclotrimerization of ethynylcyclopropane. The product composition depends on the catalyst and the reaction conditions. Thus, $Co(acac)_2$ in the presence of phosphines and $AlEt_2Cl$ afforded either the dimer 1,3-dicyclopropyl-1-butyn-3-ene or a mixture of 1,2,4- and 1,3,5-tris(cyclopropyl)benzenes, whereas $Pd(OAc)_2$ gave the same dimer in the presence of PPh_3 but only a tris(cyclopropyl)fulvene in the absence of phosphines (equation 176)[245].

8. Cyclopropylalkynylcarbene complexes

Cyclopropylethynyllithium complexes react with $Cr(CO)_6$, followed by triethyloxonium tetrafluoroborate, in a typical Fischer metal carbene synthesis to give ethoxy cyclopropylpropynylidene chromium complexes (equation 177)[153,246,247].

The parent carbene complexes readily undergo the Michael addition with secondary amines and alcohols as nucleophiles (equation 178), giving selectively the corresponding *β*-substituted cyclopropyl-vinylcarbene complexes with the *E*-configuration. The reaction with ammonia gave a mixture of products consisting surprisingly of the Michael *Z*-diastereomeradduct together with the product of aminolysis (equation 178). The reversed diastereoselectivity is attributed to a favorable hydrogen bonding in the intermediate of the Michael reaction with ammonia. The Michael reaction with thiophenol proceeded with poor selectivity, yielding a diastereomeric mixture of $Z/E = 0.8$ (equation 178).

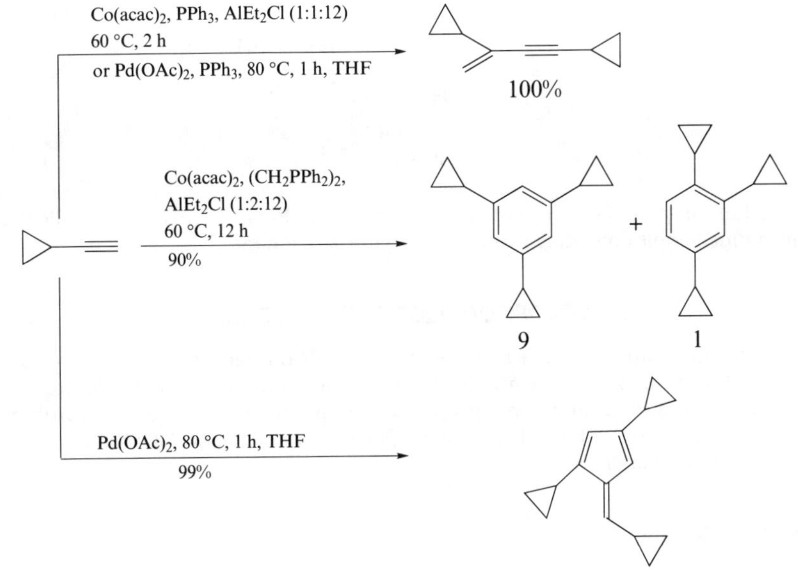

(176)

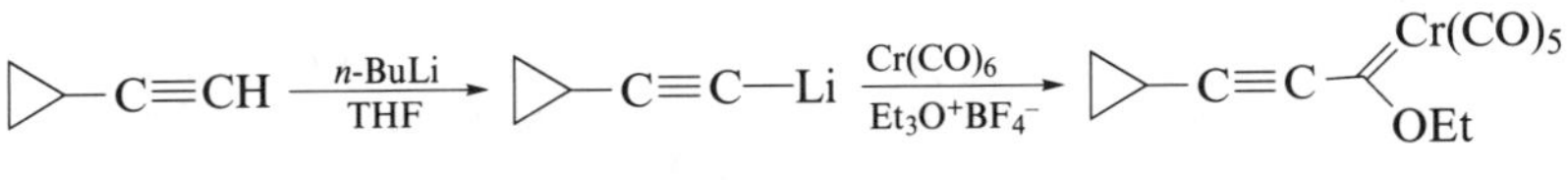

(177)

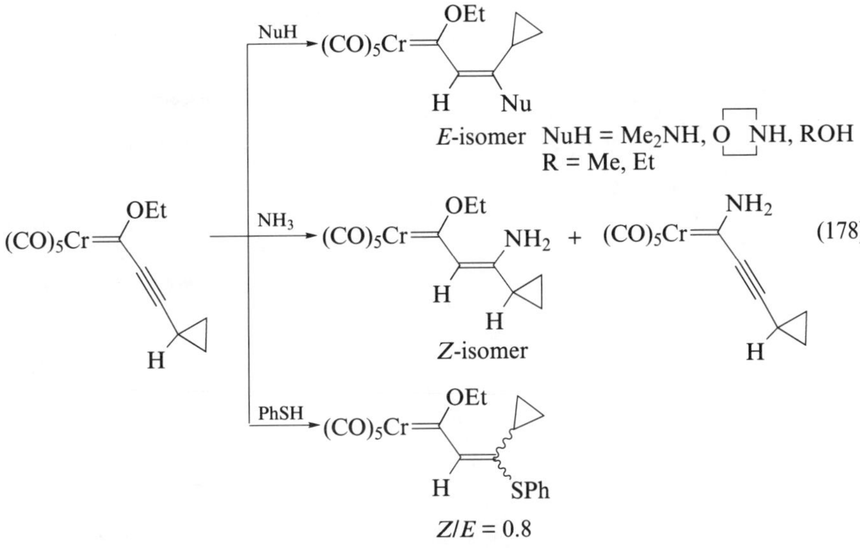

(178)

The more bulky 1-ethoxycyclopropyl group also induced a complete Z-diastereoselectivity to the Michael reaction of the ethynyl carbene complex with dimethylamine (equation 179)[246]. This is due to the sterically favored *anti*-position acquired by the bulky group in the intermediate. Unlike the parent cyclopropyl carbene complex, which gave only

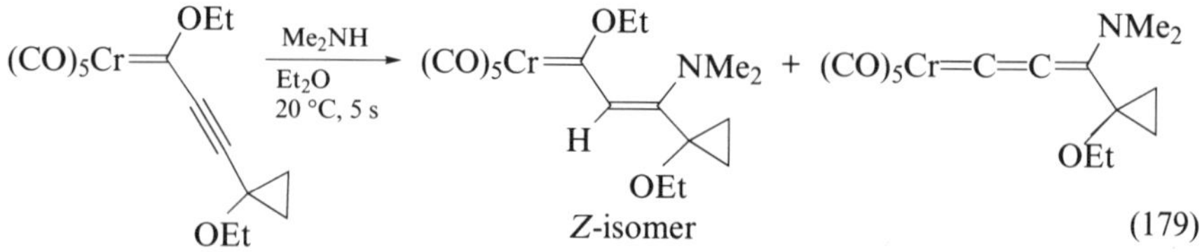

(179)

Michael addition products, the ethoxycyclopropyl analogue gave also the allenylidenechromium complex resulting from the elimination of ethanol.

III. CYCLOPROPYLIDENE COMPLEXES

Cyclopropylidene complexes are rare. In fact, there is only one report of isolated (μ-cyclopropylidene)diiron complexes which are stable in the solid state at room temperature (*vide infra*)[248]. More recently, a labile cyclopropylidene tungsten complex has been isolated which decomposes above –40 °C[42]. In all other reports[21,110] cyclopropylidene complexes are proposed as transient intermediates.

A. Tungsten Complexes

Irradiation of a benzene solution of (*endo*-7-chlorodibenzonorcaradienyl)WCp$(CO)_3$ through pyrex for 1 h at –15 °C, using a 450 W medium pressure Hg lamp, afforded the labile (dibenzonorcaradienylidene)WCp$(CO)_2$Cl in 37% isolated yield (equation 180)[42]. Continuous irradiation for 12 h results in rearrangement by cleavage of the cyclopropylidene ring, to give the corresponding stable η^2-dibenzocycloheptatetraene complex. The same cyclopropylidene-to-allene rearrangement occurs in low yields upon thermolysis of the carbene complex, and is accompanied by extensive decomposition.

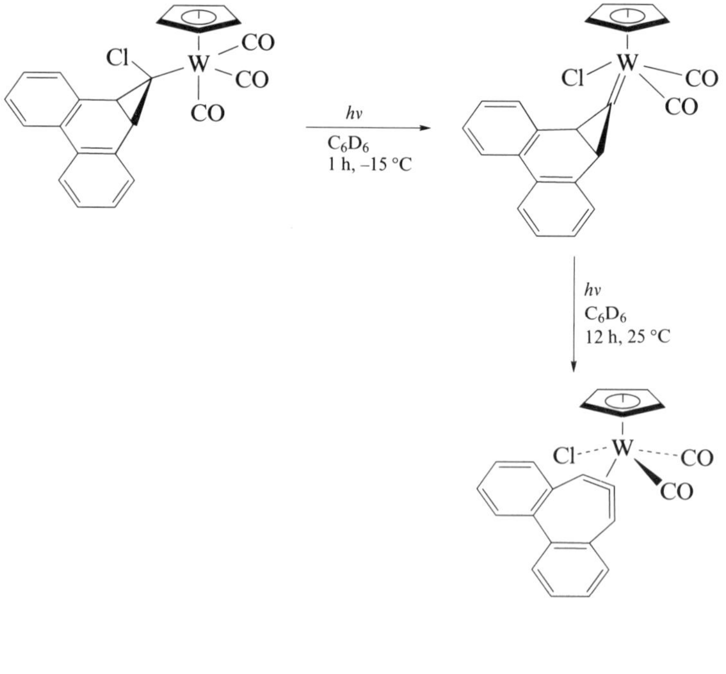

(180)

The cyclopropylidene W-complex appears to be indefinitly stable in the solid state only below –40 °C. It displays a characteristic downfield ^{13}C NMR (THF-d_8, –60 °C) resonance at δ 255.5 ($^1J_{cw}$ = 56.1 Hz), lower than the ligand carbonyl carbon resonances (δ 206.5 and 210.7).

A one-step intramolecular α-migratory elimination pathway is favored over a two-step dissociative mechanism since the reaction rate is relatively unaffected by the presence of exogenous CO. This also implies that if α-migration of chlorine is indeed the reaction mechanism, it must be substantially faster than exchange of CO from the solvent cage with exogenous CO. Analogous α-hydride eliminations in cyclopropyl metal complexes leading to allenes have been reported to pass through cyclopropylidene intermediate cations (Section II.B)[21].

B. Iron Complexes

1. μ-Cyclopropylidene complexes

Unusual bridging (μ-cyclopropylidene)diiron complexes having a tetrahedral carbene carbon have been studied as model intermediates in carbon–carbon bond formation in the Fischer-Tropsch synthesis[248]. The cyclopropylidene complexes *cis*- and *trans*-$[Cp(Co)Fe]_2(\mu\text{-}Co)(\mu\text{-}C_3H_4)$ were readily prepared by cyclopropanation in ether, of the corresponding *cis*- and *trans*-vinylidene complexes $[CpCoFe](\mu\text{-}CO)(\mu\text{-}CH_2)$ with diazomethane in the presence of CuCl (equation 181). Both isomers are air stable in the solid state. Solutions of the complexes are air stable for several hours, provided they are kept in the dark. The pure μ-cyclopropylidene isomers slowly interconvert in solution, like their parent μ-vinylidene and other alkylidene complexes. The final equilibrium ratio *cis*:*trans* = 4.8:1 is reached after two weeks.

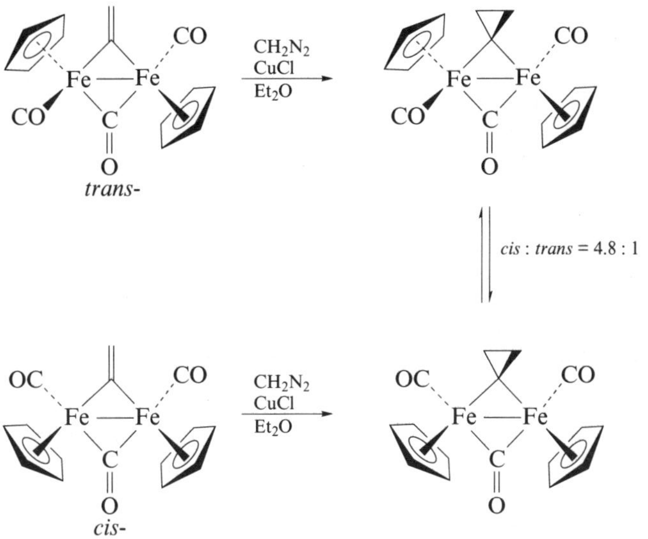

(181)

Single-crystal X-ray analysis of the *cis*-isomer reveals essentially a tetrahedral carbene with a spiro cyclopropane ring perpendicular to the C(carbene)—Fe—Fe plane. The Fe—C(carbene) bond distance of 194 pm is within the range of related diiron alkylidene complexes, and the average C—C bond length within the cyclopropylidene ligand of 151 pm is in good agreement with the value normally found in cyclopropane rings.

2. Cyclopropenylium complexes

Attempts to prepare (cyclopropenylium)Fp salt [Fp = $CpFe(CO)_2$] by α-hydride elimination with [(benzocyclobutenium)Fp]PF_6 did not afford the expected carbenium cation. Instead, ring opening occurs leading to [(η^2-allene)Fp]PF_6 complex (equation 182)[110]. Analogously, methoxy abstraction from either the racemic or optically active (*trans*-2,3-dimethyl-1-methoxycyclopropyl)Fp by trimethylsilyl trifluoromethanesulfonate (Me_3SiOTf) at –78 °C gave the corresponding (η^2-1,3-dimethylallene)Fp trifluoromethanesulfonate complex as a racemic mixture (equation 183)[21].

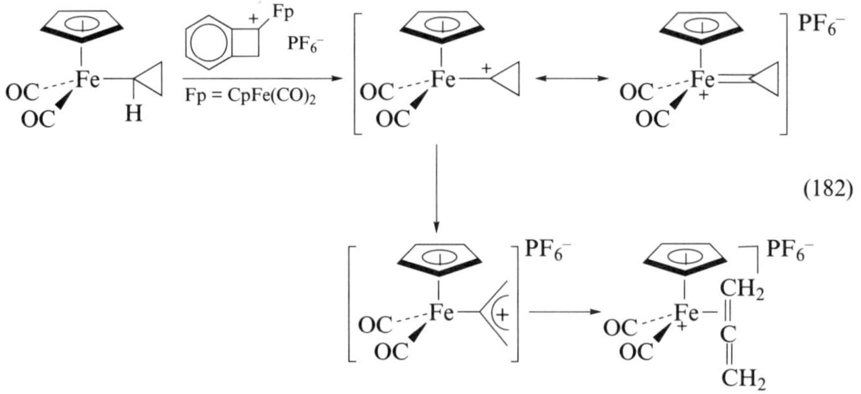

(182)

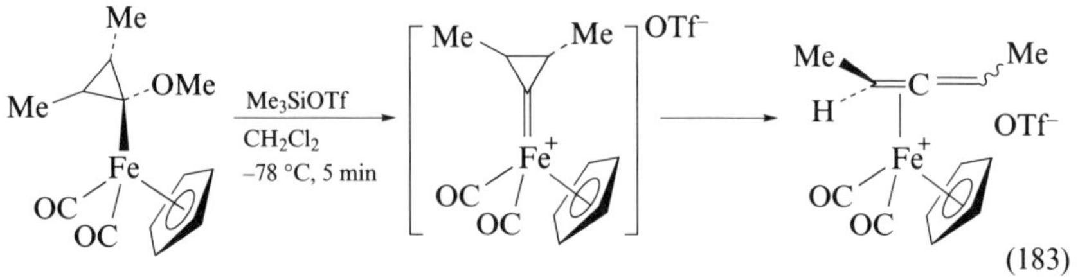

(183)

Both α-eliminations are presumably generating unstable metallocarbenium intermediates which spontaneouly ring-open to the corresponding allyl cation complexes, which subsequently collapse to the allenes. Evidence for a cyclopropenyl-to-allene rearrangement mechanism comes from the related α-migratory elimination observed in neutral tungsten complexes[42] where a cyclopropylidene intermediate was isolated (see Section II.A). The involvement of the allyl cation on the potential energy surface between the carbenium and allene complexes is consistent with the loss of chirality observed during the ring-opening process[21].

IV. METAL DERIVATIVES OF CYCLOPROPENE

Although organometallic derivatives of cyclopropenes have been previously investigated in considerable detail, this subject had received only a limited coverage in earlier reviews[249,250]. This is supplemented in the following section.

A. σ-Bonded Compounds

1. *1-Cyclopropenylmetal compounds*

a. Main group elements. When methylene chloride is added to a solution of 3-methyl-2-propenyllithium in THF at –35 °C, 2,3,3-trimethylcyclopropenyllithium is formed. This was established by hydrolysis to the parent cyclopropene, methylation to tetramethylcyclopropene or carboxylation to trimethylcyclopropene-1-carboxylic acid (equation 184)[251,252]. The same cyclopropenyllithium intermediate is obtained when *n*-BuLi is added to 1-chloro-2,3-dimethyl-2-butene[252,253] and when MeLi reacts with 1,1-dibromo-2,3-dimethyl-2-butene[252]. A mechanism involving the initial formation of vinylcarbene which cyclizes to cyclopropene, and subsequently metallated at the 1-position by excess of the alkyllithium reagent was suggsted[252] (equation 184).

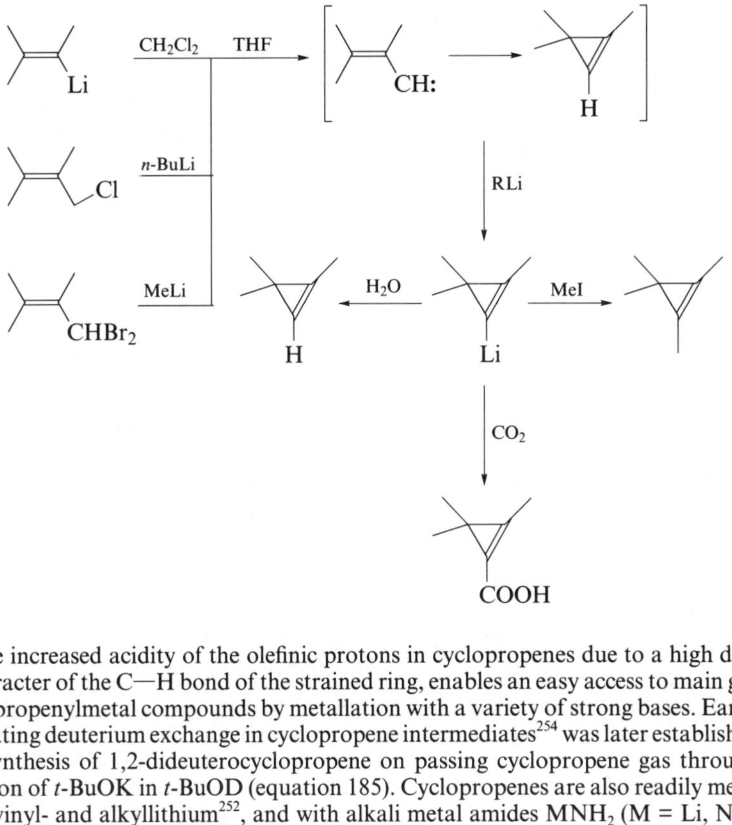

(184)

The increased acidity of the olefinic protons in cyclopropenes due to a high degree of s-character of the C—H bond of the strained ring, enables an easy access to main group 1-cyclopropenylmetal compounds by metallation with a variety of strong bases. Early work indicating deuterium exchange in cyclopropene intermediates[254] was later established[255] by the synthesis of 1,2-dideuterocyclopropene on passing cyclopropene gas through 10% solution of *t*-BuOK in *t*-BuOD (equation 185). Cyclopropenes are also readily metallated with vinyl- and alkyllithium[252], and with alkali metal amides MNH_2 (M = Li, Na, K) in liquid ammonia at –70 °C[256]. Subsequent alkylation of the 1-metallated cyclopropenes

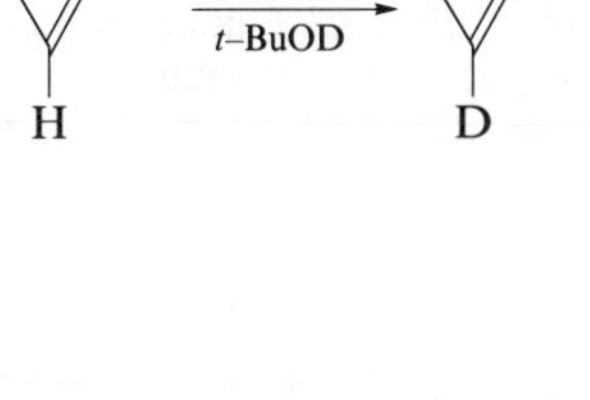

(185)

(preferebly 1-sodiocyclopropene) with a large variety of alkyl halides[257], epoxides and ketones[258] led to 1-alkylcyclopropenes, (1-cyclopropenyl)ethanols and (1-cyclopropenyl) carbinols, respectively (equation 186).

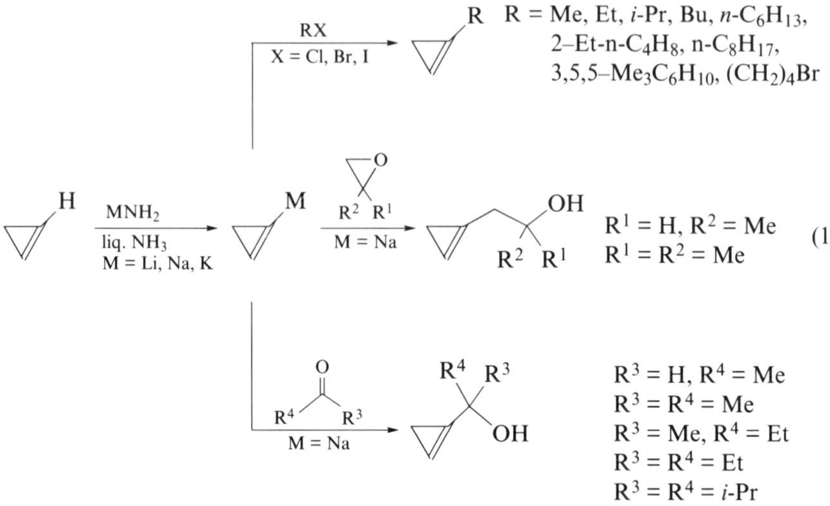

(186)

Use of excess of reagents and prolonged reaction times lead to dialkylation, dimerization, isomerizations and cross-coupling reactions (equation 187). In the presence of methoxy groups in the side chain, alkylation is followed by migration of the double bond to form the thermodynamically more stable methylenecyclopropanes (equation 188).

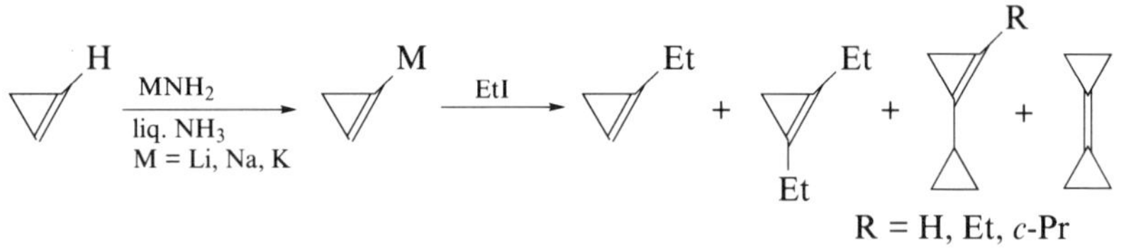

(187)

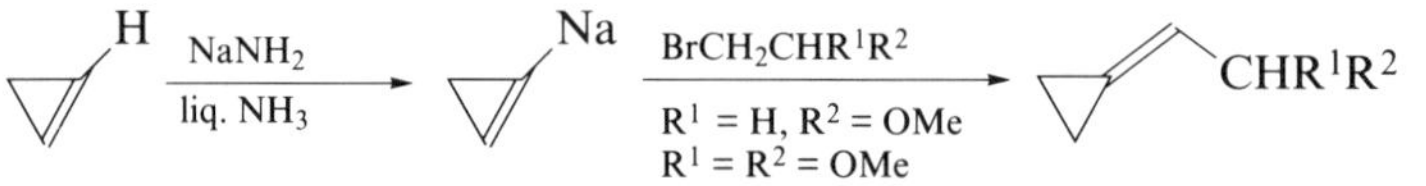

(188)

Cyclopropenone trimethylene acetals have been shown to give cyclopropenyl metal compounds of sodium, lithium, tin and zinc, which react with a variety of electrophiles[259]. 2,2-Bis(chloromethyl)-5,5-dimethyl-1,3-dioxane is readily prepared by didehydrochlorination with $NaNH_2$ in ammonia, and alkylated to give the alkyl-substituted acetal (equation 189). Hydrolysis in aqueous NH_4Cl provided the parent acetal which in turn is lithiated in THF with BuLi in the presence of HMPA or TMEDA. Alkylation and silylation of the lithiocyclopropenyl acetal is best achieved with TMEDA, whereas the reaction with carbonyl compounds is preferably done in the presence of HMPA. Metal exchange to the tributyltin compound is accomplished using Bu_3SnCl, and the bis(cyclopropenyl)zinc derivative is obtained by reaction with $ZnCl_2$ (equation 190). The cyclopropenylzinc acetal undergoes a palladium catalyzed Heck-type coupling reaction[260] with vinyl halides and aryl iodides to give vinyl and aryl cyclopropenyl acetals, respectively.

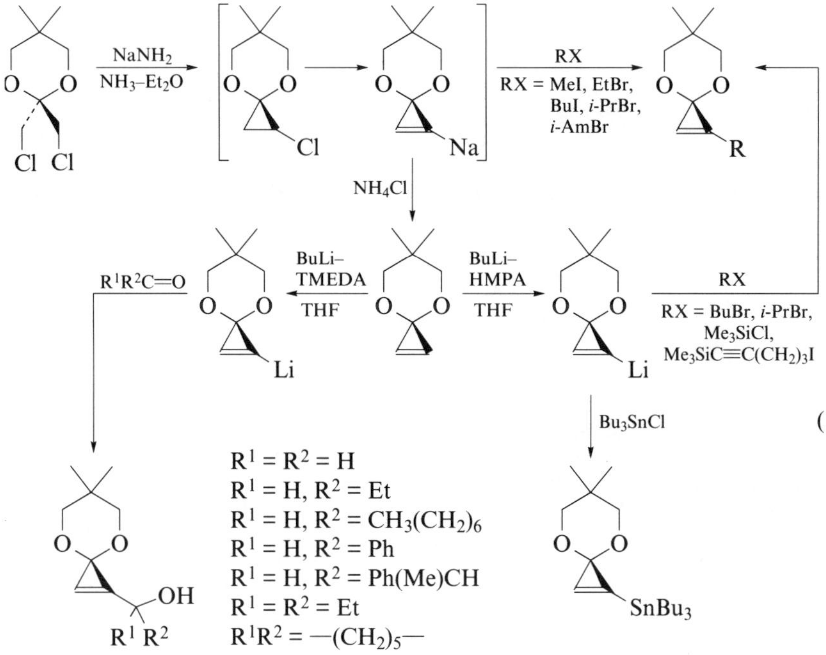

(189)

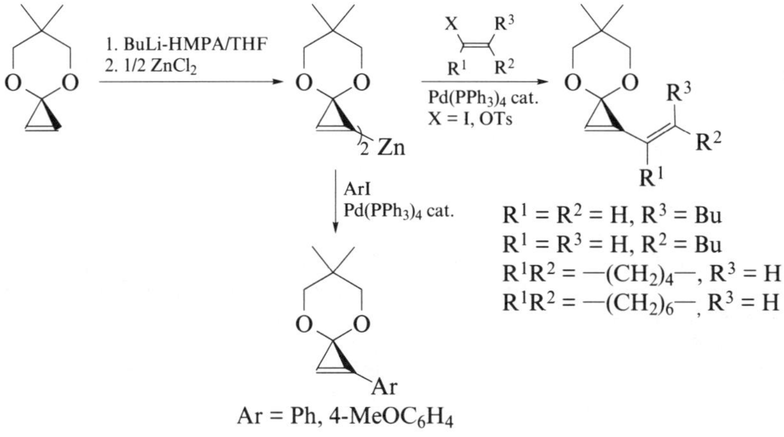

(190)

More recently, 3,3-dimethyl-1-trimethylsilylcyclopropene was metallated using lithium diisopropylamide (LDA) in THF[261]. Methylation, silylation and thiomethylation of the lithiocyclopropene intermediate afforded the corresponding 2-substituted cyclopropenes (equation 191).

b. Transition metal complexes. Transition metal complexes of cyclopropene, in which the metal is σ–bonded to an olefinic carbon are rare. Until recently, only one example was recorded in the literature[262] in which the cyclopropenylium–iron complex $[C_3Ph_2Fp]BF_4$ [Fp = $FeCp(CO)_2$] is treated with nucleophiles such as $NaBH_3CN$, NaOMe and KCN to

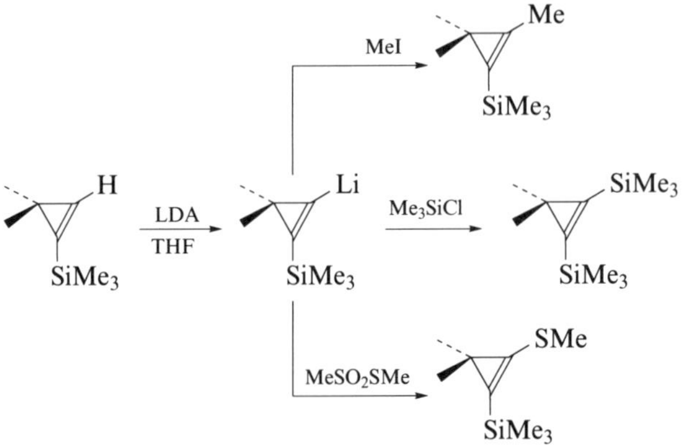

(191)

give the corresponding 3-substituted 1-Fp-cyclopropenes (equation 192). The 3-methoxy substituted complex is also obtained by spontaneous rearrangement of 1,2-diphenyl-3-[dicarbonyl(η^5-cyclopentadienyl)iron]-3-methoxycyclopropene obtained by the nucleophilic reaction of NaFp with the cyclopropenylium fluorosulfate $[C_3Ph_2(OMe)]SO_3F$ (equation 192)[262]. The presence of a methoxy group at the allylic 3-position of the cyclopropene makes the complex acid-sensitive. Indeed, the starting tetrafluoroborate complex $[C_3Ph_2Fp]BF_4$ is recovered upon treatment of the 3-methoxycyclopropene complex with ethereal tetrafluoroboric acid.

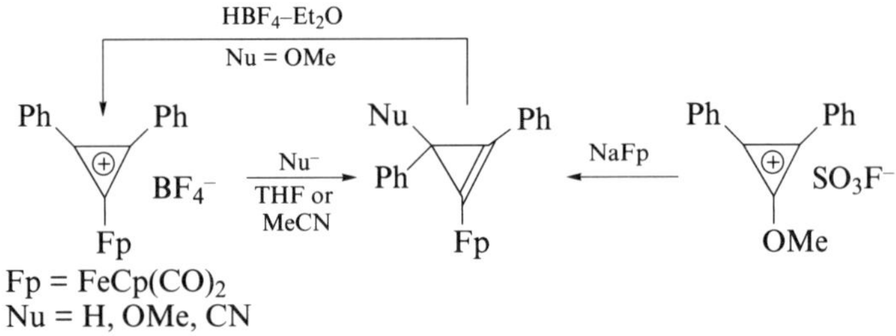

(192)

The first solid state structural confirmation of a 1-metallacyclopropene complex was recently provided by 1-[bis(triphenylphosphine)(η^5-cyclopentadienyl)ruthenium]-2-phenylcyclopropene-3-carbonitrile, prepared by a novel *fluoride* induced cyclization of γ-CN substituted cationic vinylidene complex $[Cp(PPh_3)_2Ru{=}C{=}C(Ph)CH_2CN]I$ in 85% yield (equation 193)[263]. The formation of the cyclopropene complex is reversible, and is effected by protonation with trifluoroacetic acid in MeCN. Ring cleavage is also effected by electrophilic attack of $Ph_3C^+PF_6^-$, affording a trityl cationic vinylidene complex which recloses with fluoride to the corresponding 3-trityl-substituted cyclopropenyl complex.

Interestingly, the structural features of the ruthenium complex are characteristic of regular cyclopropenes. The olefinic C=C bond length of 128.9 pm is close to the 129.6 ppm reported for cyclopropene, and so are the cyclopropene ring angles, 50.1°, 59.8°, and 70.1°, vs 51° and 64.5° in the noncoordinated compounds[250], The Ru–C distance of 203.4 is - typical for a Ru—C single bond. The Ru—C=C bond angle is 169.7 pm, close to a C(sp) hybridization bond angle, due to the ring strain.

Of special interest is the first recording of a cyclopropene *C*(metal) ^{13}C NMR signal which appeared at δ 126.2 ppm (t, J_{C-P} 23.0 Hz), considerably lower than the δ 108.7 ppm observed for cyclopropene[264].

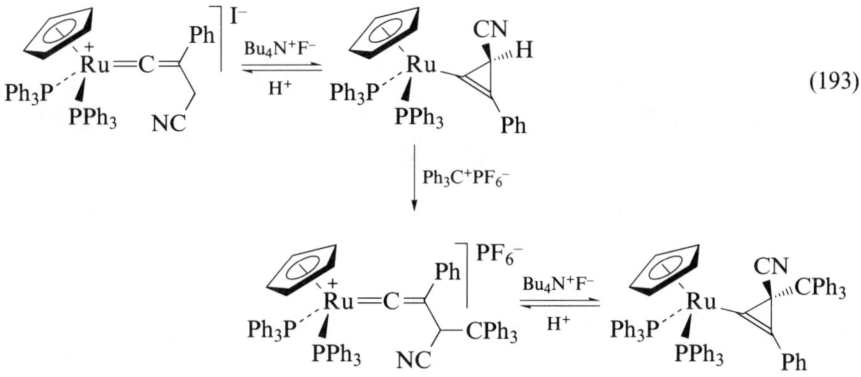

(193)

2. 3-Cyclopropenylmetal compounds

a. Preparation (i) From cyclopropenylium salts. Although it is expected that the reaction of cyclopropenylium cations with metal carbonyl anions will lead to 3-cyclopropenyl metal compounds, isolation of these complexes rarely occurs. Instead, formation of η^3-cyclopropenyl and η^3-oxocyclobutenyl complexes (Section V) is observed, presumably via a pathway involving 3-cyclopropenyl and 3-cyclopropenylcarbonyl metal compounds as intermediates (Scheme 9)[265].

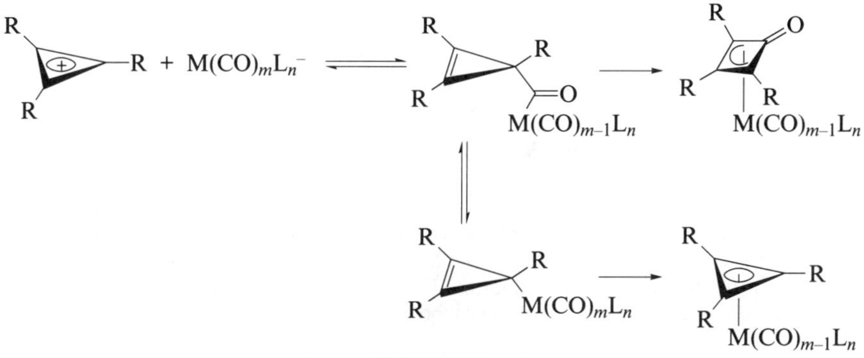

SCHEME 9

Isolation of 3-cyclopropenyl metal compounds by this method has been achieved so far for iron and rhenium metals only. Thus, the reaction of Na[CpFe(CO)$_2$] (NaFp) with cyclopropenylium salts at –70 °C, in THF, gave 3-Fp-cyclopropene complexes (equation 194)[266,267]. The X-ray crystal structure of the most stable iron complex 3-Fp-C_3Ph_3 exhibits a regular cyclopropene C—C single and double bond distances (151 and 129 pm), and a characteristic distance of 208 pm for the Fe—C σ-bond[267]. The ^{1}H NMR (CS_2) spectrum of the 3-Fp-C_3Ph_2H complex displays a singlet at $\delta = 2.63$ ppm, of the cyclopropenyl proton at the 3-position[266].

Analogously, the reaction of rhenate pentacarbonyl anion [Re(CO)$_5$]$^-$ as its sodium salt, with cyclopropenylium cations [C_3Ph_3]X (X = BF_4, PF_6) in THF, at –80 °C, afforded the octahedral σ–coordinated pentacarbonyl (η^1–1,2,3-triphenylcyclopropenyl)rhenium complex in 60–73% yield (equation 195)[268,269]. The ^{13}C NMR (acetone-d_6) spectrum dis-

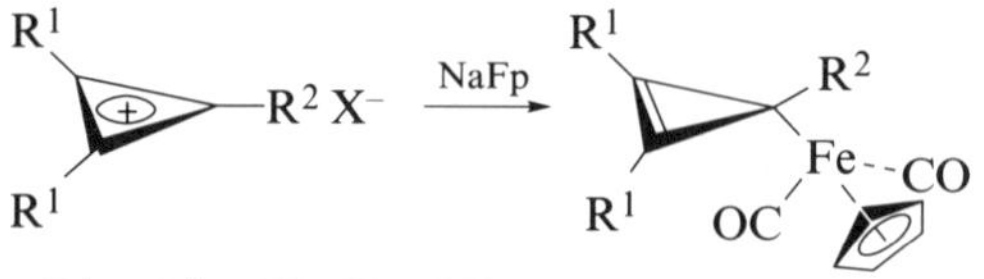

(194)

$R^1 = R^2 = Ph$; $X = BF_4$, Br
R^1 = t-Bu, R^2 = Me; $X = ClO_4$
R^1 = Ph, R^2 = H; $X = ClO_4$
Fp = $CpFe(CO)_2$

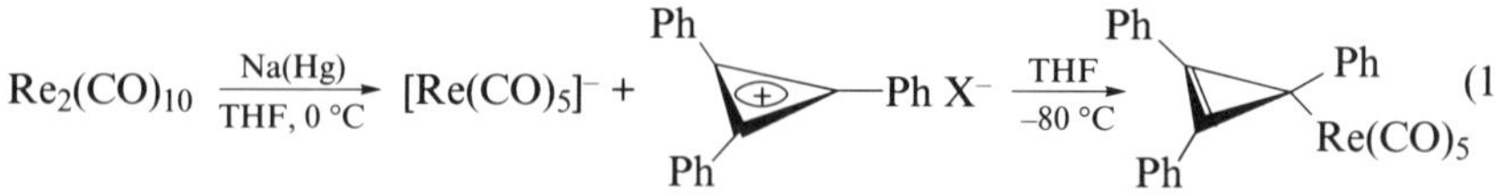

(195)

$X = BF_4$, PF_6

plays a characteristic high field resonance at $\delta = 12.76$ ppm of the ReC carbon, and a low field signal at $\delta = 160.95$ ppm of the uncoordinated olefinic carbons[268].

Notably, early attempts to similarly prepare cyclopropenyl complexes of group 6 molybdenum and tungsten, using $[CpM(CO)_3]^-$ anions (M = Mo, W) and $[C_3(Bu\text{-}t)_3]BF_4$, resulted in the electrophilic attack of the cyclopropenium cation on the peripheral cyclopentadienyl ligand, to give hydride complexes (equation 196)[270]. These air-sensitive hydride complexes readily react with CCl_4, to afford the corresponding air-stable chloro complexes.

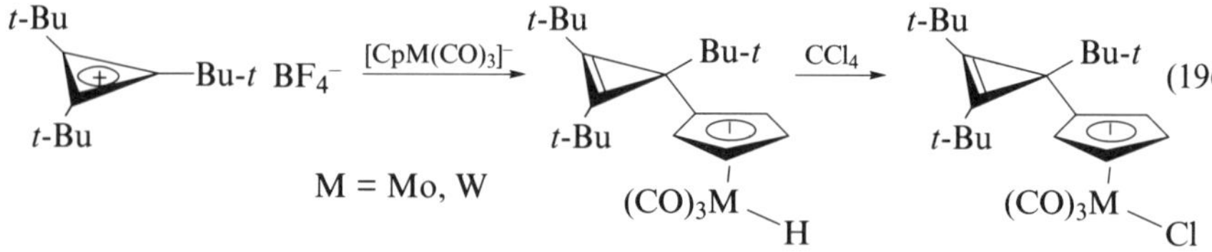

(196)

M = Mo, W

Ring enlargement to η^3-oxocyclobutenyl complexes, by carbonyl insertion into the three-membered ring, is generally observed in reactions of group 9 cobalt carbonyl anions with cyclopropenylium cations (equation 197)[271–275]. Formation of η^3-oxocyclobutenyl complexes also occurs with nitroso iron carbonyl anions[270,275]. These reactions are usually

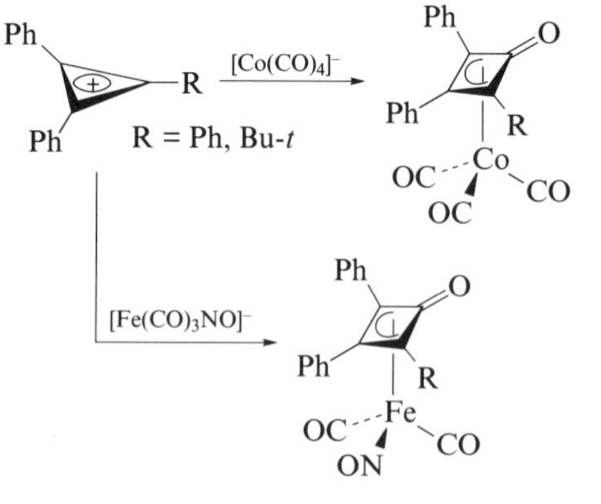

(197)

not selective and form in addition the corresponding η^3-cyclopropenyl complexes (Section V).

Since cobalt carbonyl anions react readily with a large variety of cyclopropenylcarbonyl chlorides to give selectively η^3-oxocyclobutenyl complexes (equation 198), it was suggested that reactions of cyclopropenylium cations and $[Co(CO)_4]^-$ involve direct electrophilic attack at the CO ligand rather than at cobalt[265,276]. However, observation of ring expansion products in the η^1-cyclopropen-3-yl rhenium complex (*vide infra*)[268,269] suggests that an alternative pathway involving initial electrophilic attack on the *metal*, followed by 1,2-migration of CO, and finally ring expansion, cannot be excluded (Scheme 9).

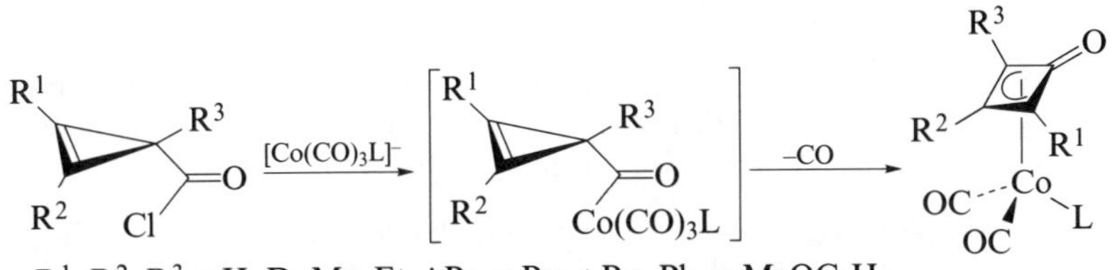

(198)

R^1, R^2, R^3 = H, D, Me, Et, *i*-Pr, *n*-Bu, *t*-Bu, Ph, *p*-$MeOC_6H_4$
L = CO, PEt_3, PPh_3, PPh_2Me, $PPhMe_2$

(ii) By decarbonylation of cyclopropenylcarbonyl metal complexes. Access to cyclopropenyl rhenium complexes was also achieved in high yields by decarbonylation of the corresponding labile η^1-cyclopropenylcarbonyl rhenium complexes in $CDCl_3$ at 20 °C (equation 199)[277,278]. Notably, the decarbonylation is accompanied by a formal 1,3-sigmatropic rearrangement to a *nonfluxional* η^1-cyclopropenyl rhenium complex, as was clearly demonstrated by using the deuterium-labeled acyl rhenium complex. The ^{1}H NMR of $(CO)_5Re(3\text{-}\eta^3\text{-}C_3(t\text{-Bu})H_2)$ displays the allylic proton resonance at δ =2.20 (d, J = 1.2 Hz) ppm, whereas in the ^{13}C NMR the Re*CH* carbon appears at 27.88 ppm. On attempting to decarbonylate the analogous manganese cyclopropenylcarbonyl complex in ether only $Mn_2(CO)_{10}$ and 1,2,3,4-tetraphenylbenzene are observed, but no evidence for (3-η^1-cyclopropenyl)manganese complex was detected (equation 200)[278].

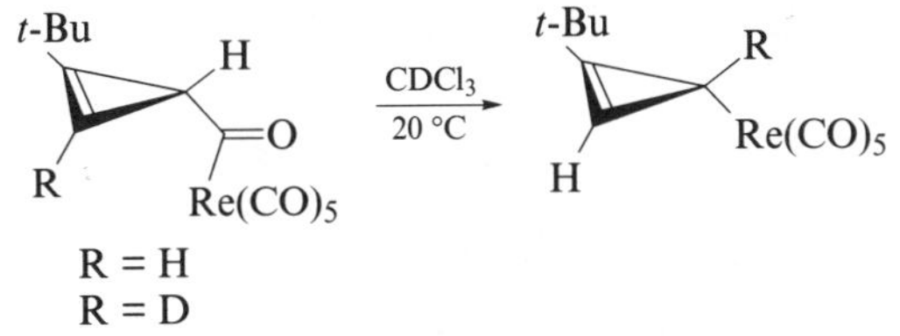

(199)

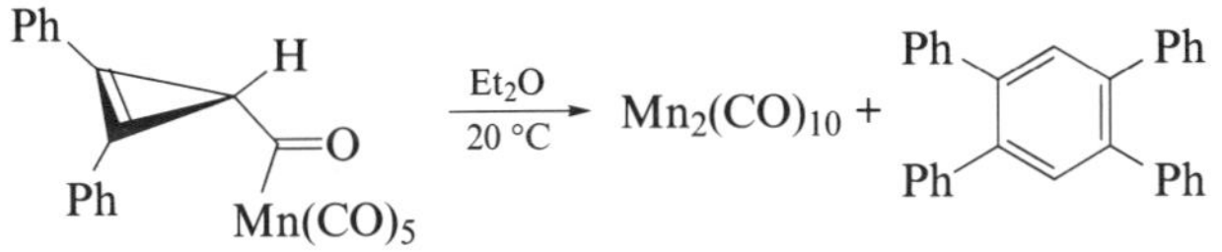

(200)

(iii) By oxidative addition reactions. Deprotonation of the platinum hydride d[16] planar complex *cis*-$(Ph_3P)_2PtH(SiMePh_2)$ with methyl- or butyl-lithium in ether, at 5 °C, followed by reaction with 3,3-dichlorodiphenylcyclopropene afforded the d[18] complex *cis*-$(Ph_3P)_2ClPt(C_3Ph_2Cl)$, in a formal oxidative addition of the carbon–halogen bond to the metal, together with *trans*-$(Ph_3P)_2PtCl(SiMePh_2)$ (equation 201)[279]. Alternatively, the same cyclopropenyl Pt-complex was also obtained as the sole product in 46% yield, by the

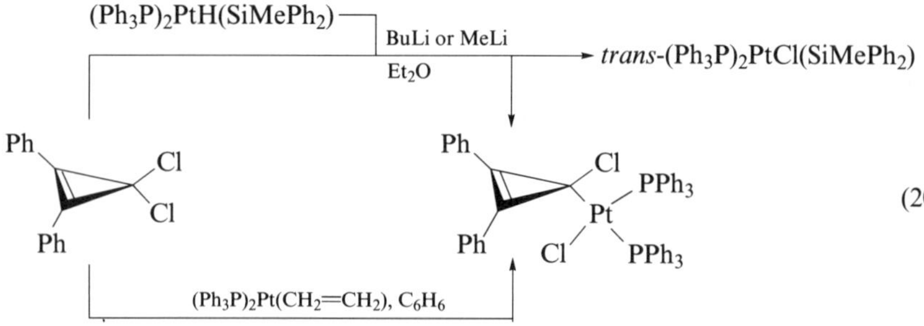

(201)

oxidative addition of the above dichlorodiphenylcyclopropene to η^2-ethylene complex $(Ph_3P)_2Pt(CH_2{=}CH_2)$. *cis*-$(Ph_3P)_2ClPt(C_3Ph_2Cl)$ is characterized by its ^{195}P NMR (benzene-d_6) spectrum which displays a signal at $\delta = -3353$ ppm (dd)[76].

b. Reactions. (i) With electrophiles. 3-Fp-triphenylcyclopropene reacts with Br_2, I_2 at –40 °C in CCl_4 and with SO_2 at –20 °C in CH_2Cl_2, and with HBF_4 in Et_2O, to give the corresponding 1,2,3-triphenylcyclopropenylium salts (equation 202)[267]. The reaction with $HgCl_2$ in benzene at 25 °C afforded the demetallated cyclopropenyl dimer in 92% yield, presumably via a cyclopropenyl radical coupling. Treatment of the cyclopropenyl-Fp complex with HCl in toluene gave a 1:1 mixture of 1,2-diphenylindene and FpCl (equation 203). It has been suggested that formation of the indene involves either removal of the metal (as FpCl) at an early stage, followed by acid catalyzed cyclization of uncoordinated 1,2,3-triphenylcyclopropene intermediate, or alternatively, initial ring opening to a Fp-allyl cation, cyclization and demetalation at the final stage[267]. Similarly, the Friedel-Crafts acylation of the complex gave 3-acyl-1, 2-diphenylindene (equation 203).

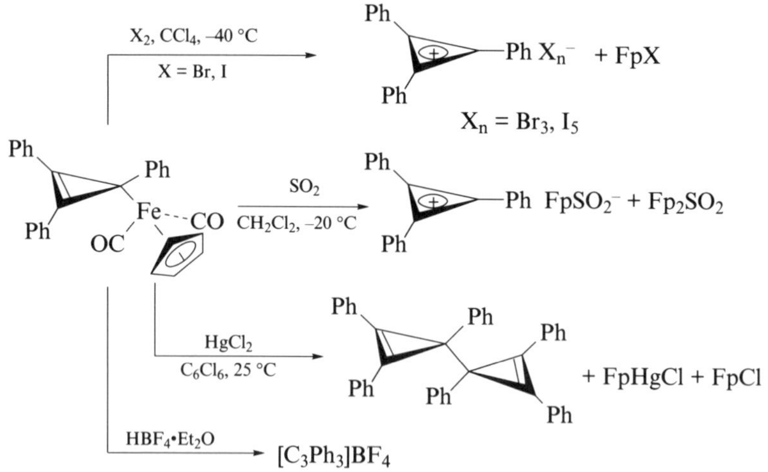

(202)

(ii) Photochemical and thermal reactions. Photolysis of 3-Fp-triphenylcyclopropene in ether at –40 °C, using a 254-nm low-pressure lamp (quartz), gave an unseparable mixture of the triphenyl η^3-oxocyclobutenyl complex and the free cyclobutenone ligand (equation 204)[267]. In contrast, irradiation of the octahedral rhenium carbonyl complex $(CO)_5Re(\eta^1\text{-}C_3Ph_3)$ with a high-pressure lamp in hexane at –20 °C gave the ring-expanded tetracarbonylrhenacyclobutadiene as the sole product in 25% yield (equation 205)[268].

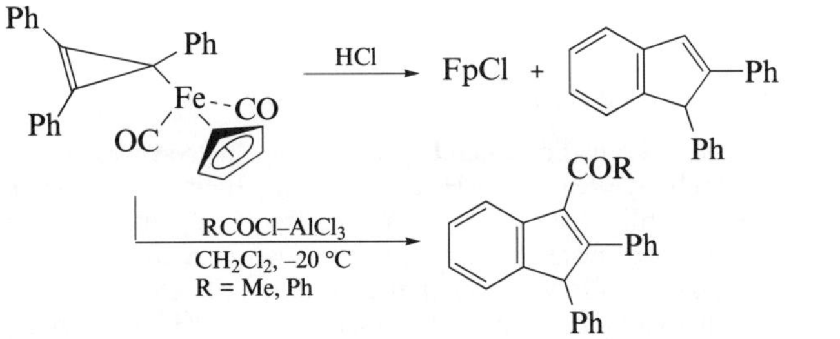

(203)

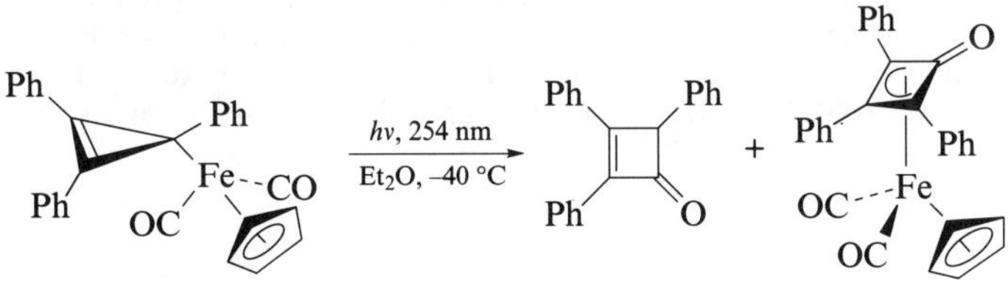

(204)

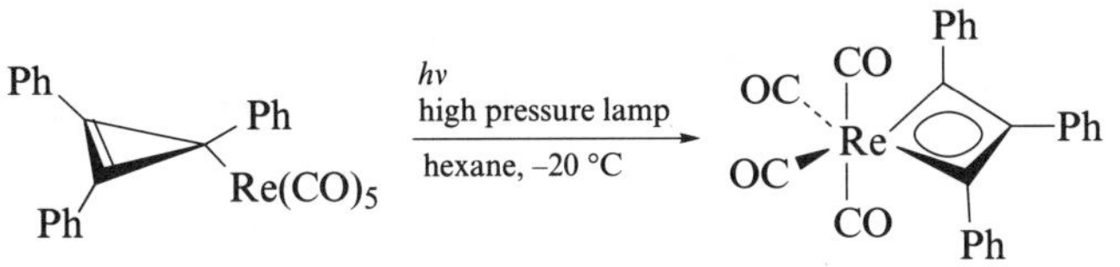

(205)

When the cyclopropenyl rhenium complex $(CO)_5Re(\eta^1\text{-}C_3Ph_3)$ was refluxed in hexane for 3–4 h, the isomeric η^3-oxocyclobutenyl and tricarbonyl(η^5-hexaphenylcyclohexadienyl)rhenium complexes are obtained, in addition to the tetracarbonyl rhenacyclobutadiene complex. In the presence of excess Me_3NO, a mixture of two interconvertible triphenyl rhenafuran complexes is obtained. Furthermore, if the thermal reaction is conducted in the presence of phosphorus ligands, addition and ring expansion occur, affording octahedral η^1-cyclobutenone complexes (equation 206)[268,269].

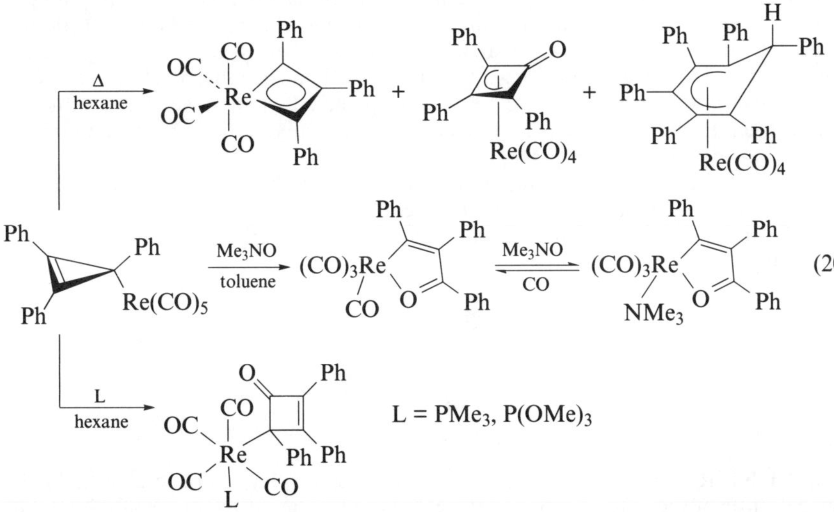

(206)

B. π-Complexes

1. Synthesis and characterization

a. Ligand exchange. Transition metal η^2-cyclopropene complexes are commonly prepared by ligand exchange reactions. However, unlike their stable nonstrained olefin counterparts, these π-coordinated cyclopropenes are frequently unstable under the reaction conditions, undergoing ring cleavage or rearrangement reactions (*vide infra*). The first group of isolable complexes of Pt was prepared by addition of excess of a cyclopropene to a solution of bis(triphenylphosphine)(ethylene)platinum in $CHCl_3$, benzene or THF, at ambient temperature, resulting in exchange of the ethylene ligand by cyclopropene (equation 207)[280]. An attempt to prepare analogous cyclopropene complexes by reaction of 3-methylcyclopropene with the monovalent and divalent complexes $IrCl(CO)(PPh_3)_2$, $[Rh(CO)_2Cl]_2$, $PdCl_2(PhCN)_2$ and $PtCl_2(C_2H_4)(Pyr)$ failed to retain the cyclic system[280].

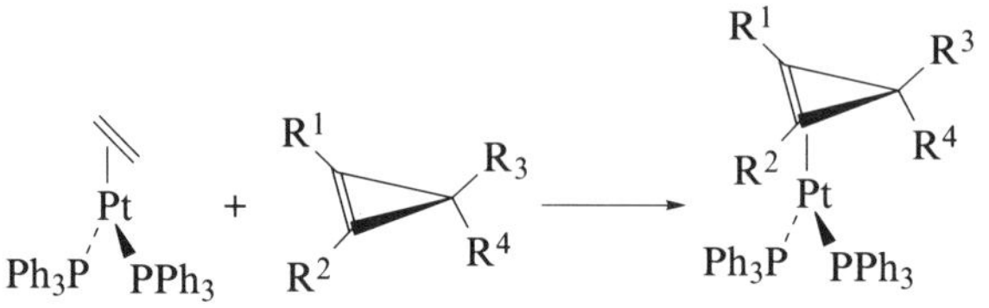

(207)

$R^1 = R^2 = R^3 = R^4 = H$
$R^1 = R^2 = R^3 = H, R^4 = Me$
$R^1 = R^2 = H, R^3 = R^4 = Me$
$R^1 = H, R^2 = R^3 = R^4 = Me$
$R^1 = R^2 = Me, R^3 = R^4 = H$
$R^1 = R^2 = R^3 = Me, R^4 = H$

The solid state structure of these platinum complexes was secured by X-ray analysis of the 1,2-dimethylcyclopropene derivative, which reveals a distorted square-planar Pt complex[281]. The distance between the Pt atom and the double bond carbons is 212 pm. In contrast, the distance between Pt and the allylic carbon (283 pm) is significantly longer, hence no interaction of the metal with this carbon is likely. The length of the C=C double bond is 150 pm, typical of Pt-coordinated double bonds. This is close to the allylic bond length (155 pm) but considerably longer than the C=C bond length in the free ligand (130 pm)[282]. The cyclopropene ring and the P—Pt—P plane are at an angle of 116°, and the vinylic methyls are tilted away from the metal, forming an angle of 112° between the cyclopropene ring and the C_1—C_2—C(Me) plane.

The structural and bonding features of this complex comply with the trigonal 'in-plane' conformational preference observed in d^{10} (olefin)ML_2 complexes. In molecular orbital terms, the dominant bonding interaction is between the b_2 HOMO of the ML_2 fragment and the ethylene π^* LUMO[283].

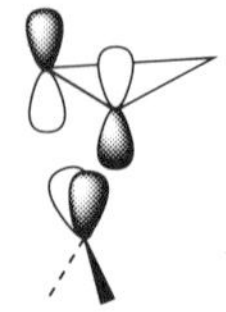

The ^1H NMR spectra of the platinum complexes display characteristic high-field chemical shifts of the vinyl protons of the complexed olefin, at the range of δ 2.56–3.02

ppm, approximately 5 ppm higher than the corresponding chemical shift in uncoordinated cyclopropene (7.65 ppm)[280]. The ^{195}Pt NMR spectrum of $(PMe_3)_2Pt(\eta^2$-3,3-$Me_2C_3H_2)$ in THF-d_8, δ -5102 ppm (relative to $[PtCl_6]^{2-}$ in D_2O as an external standard), is typical of an η^2-Pt-olefin complex[76].

Another unique series of η^2-cyclopropene-ML_2 complexes of group 10 are the nickel derivatives of cycloproparenes, obtained by treatment of difluorocyclopropabenzene (dfcb) with a variety of trigonal and tetragonal Ni complexes (equation 208)[284]. The analogous palladium complex $(PMe_3)_2Pd(\eta^2$-dfcb) was similarly prepared by reaction of dfcb with $CpPd(\eta^3$-allyl) and PMe_3, at –30 °C in pentane (equation 209)[285]. The presence of the electron-withdrawing fluorine substituents is crucial for the reaction success. The parent cyclopropabenzene and the bis(trimethylsilyl) derivative both give only ring-cleavage products under the same reaction conditions.

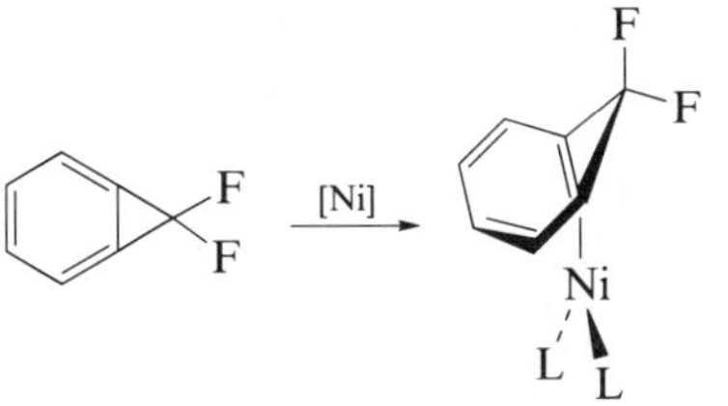

(208)

[Ni] = $L_2Ni(COD)$, L = PMe_3, PEt_3, dcpe [ethylenebis(cyclohexylphosphine)]
[Ni] = $(PPh_3)_4Ni$
[Ni] = $Ni(C_2H_4)_3$ + L_2, = TMEDA, bipy

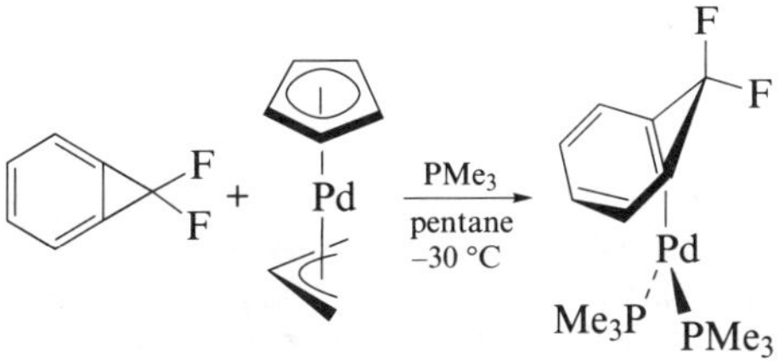

(209)

Interestingly, the structural features of these complexes, unveiled from the solid state structure of $(PEt_3)_2Ni(\eta^2$-dfcb) and their ^{13}C NMR spectra, are typical of the simple cyclopropene-$Pt(PPh_3)_2$ complex discussed before. Thus, the Ni–C distance of 194 pm, and the coordinated C=C bond length of 151 pm, are in agreement with the corresponding 212 pm and 150 pm distances found in $Pt(PPh_3)_2Pt(\eta^2$-1, 2-$C_3Me_2H_2)$[281]. Similarly, the chemical shifts of the coordinated vinylic carbon atoms experience an upfield displacement to the range of δ 33–52 ppm, more than 80 ppm higher than the corresponding signals in the free ligand.

The first reported example of a transition-metal complex containing both an unopened coordinated cyclopropene ring and CO ligands is a group 6 molybdenum complex, in which the cyclopropene double bond is part of a larger chelating ring (equation 210)[286]. The chelate complex was prepared by reaction of the 16-electron cation complex $[Cp^*(CO)_3Mo]BF_4$ with sodium 2,3-diphenyl-2-cyclopropene-1-carboxylate in dichloromethane. The anticipated tricarbonyl Mo–carboxylate intermediate undergoes a spontaneous intramolecular CO–olefin exchange and loss of a carbonyl group, to afford the chelating cyclopropenecarboxylate complex.

The molecular structure of the chelate complex was determined by a single-crystal X-ray diffraction study. The Mo distances to the cyclopropene ring carbon atoms, averaging 229

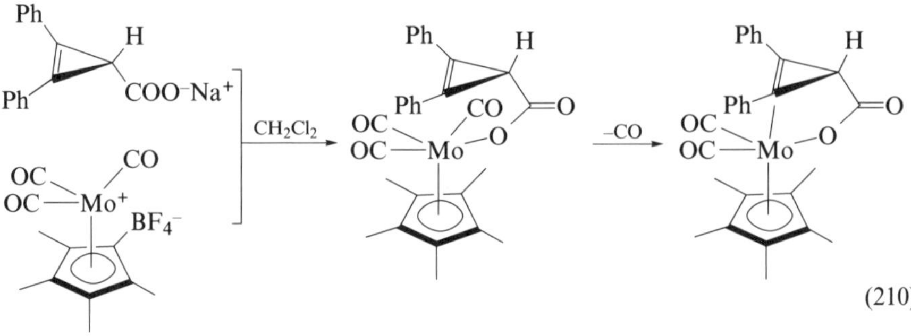

(210)

pm, the C=C bond length of 141 pm and the 38° *exo*-tilt of the Ph groups clearly reveal the η^2-olefin bonding. However, compared with the *shorter* M—C (212 pm) and *longer* C=C (150 pm) distances in the Pt complex $(PPh_3)_2Pt(2,3\text{-}C_3Me_2H_2)$ discussed above[281], a weaker M–olefin coordination interaction in molybdenum relative to platinum complexes is suggested.

The ^{13}C NMR spectrum in $CDCl_3$ at ambient temperature displays two Mo—CO (δ 222.2 and 228.79 ppm) and two characteristic high-field coordinated C=C (δ 58.64 and 71.91 ppm) resonances, indicating that the solid state structure of the Mo–chelate complex is maintained in solution. Conformational rigidity is generally observed in many η^2-complexes. Rotational barriers about the metal–olefin axes and conformational preferences in these complexes have been estimated using extended Hückel-type calculations[283].

More recently, a series of stable imido-tungsten complexes of 3,3-disubstituted cyclopropene $Cl_2(NAr)(PX_3)_2W(\eta^2\text{-}3,3\text{-}C_3R_2H_2)$ [Ar = Ph, 2,6-$C_6H_3Me_2$; PX_3 = $PMePh_2$, $P(OMe)_3$; R = Ph, RR = $(CH_2)_3O_2$] was prepared by substituting one of the phosphine ligands in $Cl_2(NAr)(PX_3)_3W$ by cyclopropene (equation 211)[287]. The complexes are relatively stable in the crystalline form but undergo ring-cleavage reactions in solution at room temperature (Section IV.B.2a). The stability of the η^2-cyclopropene complexes decreases as the steric bulk of the ancillary ligands increases, and for the same metal fragments, the diphenylcyclopropene derivatives are more stable than their acetal counterparts. Some of the precursor complexes with bulky ligands [e.g. PEt_2Ph, or N-2,6-$C_6H_3(Pr\text{-}i)_2$] do not form observable cyclopropene complexes.

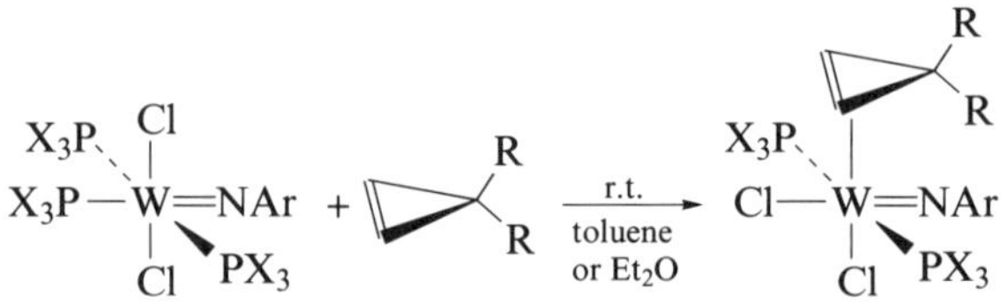

(211)

R = Ar = Ph; PX_3 = $PMePh_2$, $P(OMe)_3$
R = Ph, Ar = 2,6-$C_6H_3Me_2$; PX_3 = $P(OMe)_3$
RR = $O_2(CH_2)_3$; Ar = Ph; PX_3 = $PMePh_2$, $P(OMe)_3$
RR = $O_2(CH_2)_3$; Ar = 2,6-$C_6H_3Me_2$; PX_3 = $P(OMe)_3$

Characteristic of η^2-olefin and cyclopropene complex formation are the upfield shifts observed in the NMR resonances of the olefinic protons (δ =3.30–5.29 ppm) and carbons (δ = 59.9–72.4 ppm). For the phosphite complexes, these upfield shift resonances decrease as the steric bulk of the imido ligand increases, corresponding to weaker binding of the cyclopropene in the more sterically crowded molecule. Difference NOE spectroscopy indicates that the cyclopropene and the imido groups lie on the same side of the equatorial plane (*syn* rotamer). This was further confirmed by an X-ray diffraction study of the octahedral diphenyl phosphite complex $Cl_2(NPh)\{P(OMe)_3\}_2W(\eta^2\text{-}3,3\text{-}C_3Ph_2H_2)$, which also shows that the imino group and the two *trans*-located phosphine ligands occupy equatorial positions. The C=C (145 pm) and W—C (216 pm) bond lengths are typical of η^2-coordinated cyclopropenes (*vide supra*)[281,284].

A single group 5 niobocene–cyclopropene complex $Cp_2ClNb(C_3H_4)$ was synthesized by treating a toluene solution of dichloroniobocene with sodium amalgam in the presence of cyclopropene (equation 212)[288]. The isolated complex was characterized by IR and mass spectra, and by the ^{1}H NMR spectrum which displayed a sharp singlet of the cyclopentadienyl protons at δ 5.87 ppm and a broad multiplet of the cyclopropene protons at δ 1.43 ppm, in a ratio of 10:4.9. The presence of an intact cyclopropene ring in the complex was confirmed by reduction with HCl, which led to cyclopropane in >90% yield.

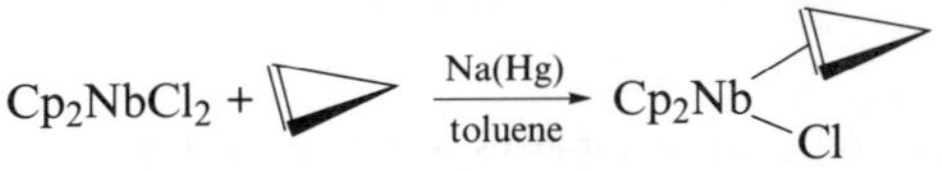

(212)

Group 4 elements, titanium and zirconium, give isolable crystalline η^2-cyclopropene complexes of titanocene and zirconocene. The zirconocene derivatives of 1,2- and 3,3-diphenylcyclopropene $Cp_2(PMe_3)Zr(\eta^2\text{-}Ph_2C_3H_2)$ are both obtained by olefin exchange reactions of cyclopropene with the corresponding η^2-butene zirconocene complex $Cp_2(PMe_3)Zr(\eta^2\text{-}CH_2{=}CHEt)$ (equation 213)[77,289]. However, whereas the reaction with 1,2-diphenylcyclopropene proceeds smoothly to give only the cyclopropene complex, a mixture of equal amounts of cyclopropene and the isomeric ring-opened diphenylpropenylidene zirconocene complexes is obtained in the reaction with 3,3-diphenylcyclopropene. Notably, these complexes do not equilibrate, nor do they interconvert upon heating. This suggests that the alkenylidene complex is a primary reaction product, and unlike the analogous tungsten complexes (*vide supra*)[287,290] is not formed from an intermediate cyclopropene complex.

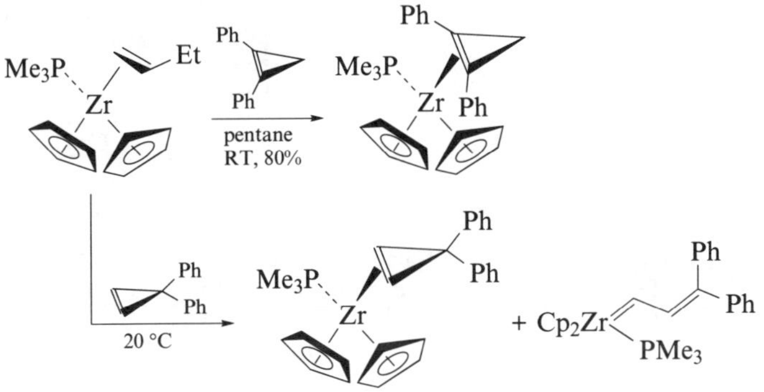

(213)

Exchange of one phosphine ligand by cyclopropene in the titanocene complex $Cp_2Ti(PMe_3)_2$ is the preferred method of preparing η^2-cyclopropene titanocene complexes $Cp_2(PMe_3)Ti(\eta^2\text{-}R_2C_3H_2)$ [R_2 = 1,2-Ph_2, 3,3-$(CH_2)_3O_2$] shown in equation 214[289]. The reactions in pentane take place spontaneously at ≤ 20 °C, producing stable, pentane insoluble complexes. In none of the reactions could any Ti-alkylidene complex be detected. However, prolonged standing of the diphenyl complex in THF at room temperature results in rearrangement to a titanacyclobutene (Section IV.B.2).

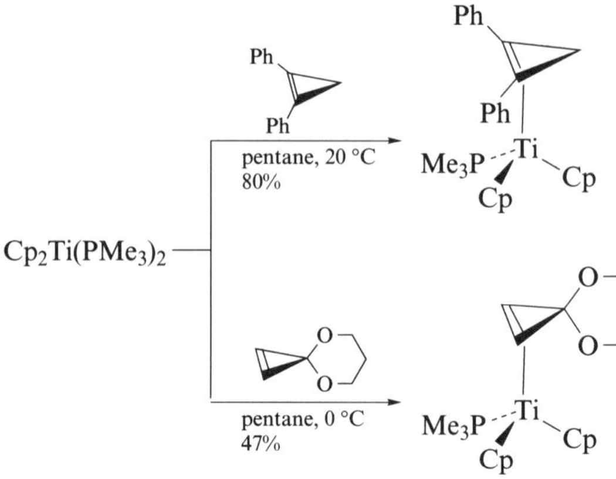

(214)

The stability of the metallocene complexes is strongly dependent on the nature of the cyclopropene substituents, and the reaction conditions. Thus, when equimolar amounts of 3,3-dimethylcyclopropene and $Cp_2Ti(PMe_3)_2$ react at 0 °C, a 2:1 mixture of alkylidene and cyclopropene complexes is formed. However, when excess of cyclopropene is used, a dicyclopropyl titanacycle is exclusively formed by oxidative coupling reaction of the intermediate cyclopropene complex (equation 215)[77]. The analogous zirconium oxidative-coupling product is obtained upon reaction of 3,3-dimethylcyclopropene with $Cp_2(PMe_3)Zr(\eta^2\text{-}CH_2{=}CHEt)$ (Section IV.B.2).

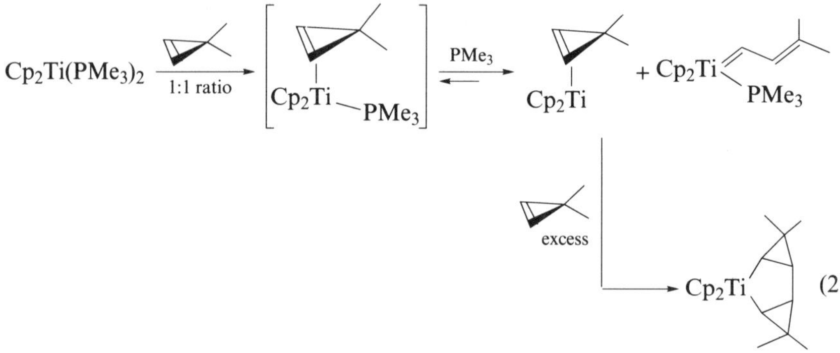

(215)

The structure of group 4 η^2-cyclopropene metallocenes was secured by their NMR spectra. The ^{13}C NMR spectra are of particular analytical significance, displaying high field signals of the coordinated olefin carbons at the range of δ 52–63 ppm[289]. Low-temperature studies indicate that these metallocenes are conformationally rigid at the organic ligand site, though at temperatures below –80 °C line broadening due to hindered rotation of the phosphorous ligands is observed.

b. Coupling of alkynes with alkylidene complexes. Coupling of alkynes with carbene ligands to form η^2-cyclopropene complexes is rare. The only reported example is the reaction of arylidene complexes of pentacarbonyltungsten with phenylacetylene at –80 °C, to give labile 1-phenyl-3-aryl-substituted η^2-cyclopropene W-complexes (equation 216)[290]. These complexes are stable only below –40 °C. Upon heating the reaction mixture to –30 °C, a rapid isomerization to the corresponding arylidenecarbene complexes occurs.

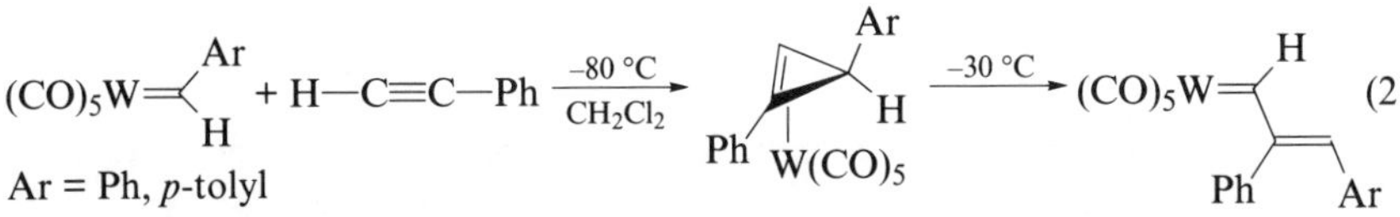

(216)

c. β-Hydride elimination. Although β-hydride elimination to give transition-metal π-complexes is a common reaction of σ-bonded alkyl-metal complexes[291,292], only few examples of η^2-cyclopropene complexes prepared in this way are reported in the literature. The successful cases are all zirconocene complexes of the type $Cp_2(PMe_3)Zr(\eta^2$-cyclopropene). The parent complex is prepared by thermal elimination of methane from methyl cyclopropyl zirconocene in the presence of Me_3P (equation 217)[293].

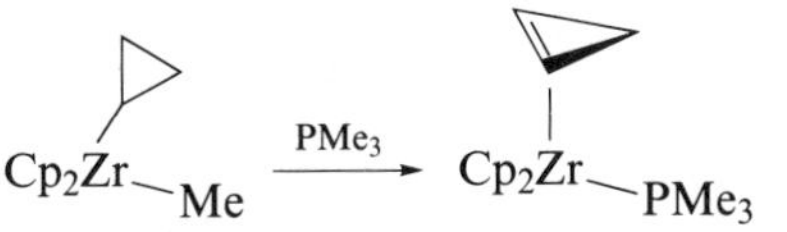

(217)

Zirconocene complexes of two strained cyclopropenes, bicyclo[3.1.0]hex-5-ene and bicyclo[4.1.0]hept-6-ene, are prepared similarly by warming benzene solutions of the corresponding mixtures of the stereoisomeric *exo–exo*, *exo–endo* and *endo–endo* bis-(bicycloalkyl)zirconocenes (equation 218)[41]. Since only the *exo–exo* and *exo–endo* isomers have at least one *cis* β-hydrogen required for the elimination, the product mixture contains two isomeric cyclopropene complexes and the unreactive *endo–endo* bis(bicycloalkyl)-zirconocenes.

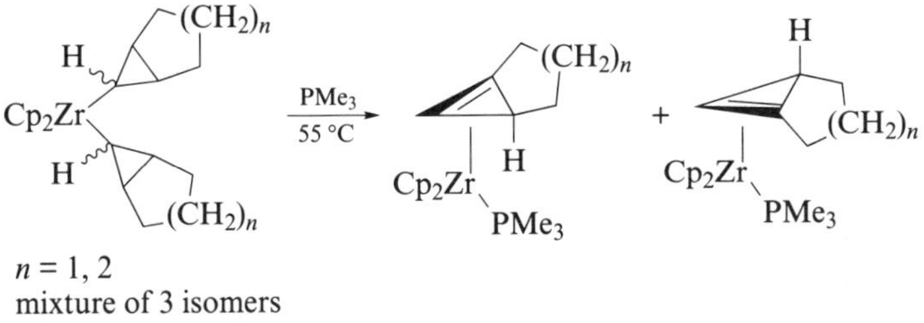

(218)

2. Reactions

a. Isomerization to vinyl carbene complexes. 1,3-Diaryl-substituted η^2-cyclopropene tungsten carbonyl complexes $(CO)_5W(\eta^2\text{-}1,3\text{-}C_3PhArH_2)$ (Ar = Ph, *p*-tolyl) are stable in solution only at –80 °C. They readily decompose in CH_2Cl_2, at >–40 °C, to the ring-opened vinyl alkylidene complexes $(CO)_5W{=}CHC(Ph){=}CHAr$ (equation 219; cf equation 216, Section IV.B.1b)[290].

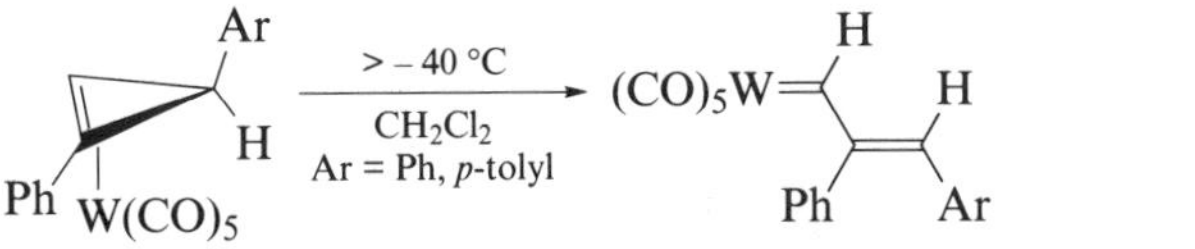

(219)

Similarly, the more stable imino-complexes $Cl_2(NAr)(PX_3)_2W(\eta^2\text{-}3,3\text{-}C_3R_2H_2)$ undergo facile ring cleavage reactions to the corresponding vinyl alkylidene complexes under thermal, photochemical and $HgCl_2$ catalyzed conditions[287]. The diphenylcyclopropene complexes give mixtures of the *syn* and *anti* isomers (equation 220) whereas the more labile cyclopropeneketal complexes give, in addition to the *syn* and *anti* isomers, s-*cis* vinyl alkylidene O-chelates (equation 221). The ratio between the products is dependent on the reaction conditions.

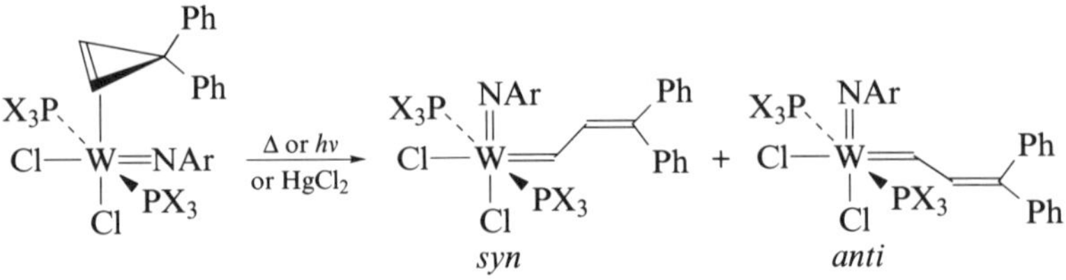

(220)

Ar = Ph, 2,6-$C_6H_3Me_2$; PX_3 = $PMePh_2$, $P(OMe)_3$

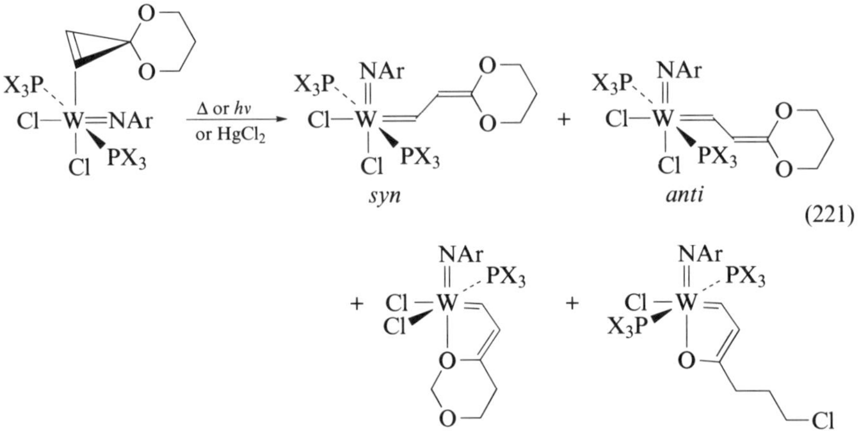

(221)

Since most of the η^2-cyclopropenyl complexes are obtained by ligand exchange reactions (cf Section IV.B.1.a) it is reasonable to believe that the reactions of cyclopropenes with octahedral complexes of tungsten (equation 222)[287] and rhuthenium (equation 223)[294], which lead directly to vinyl alkylidenes complexes, pass through an unstable η^2-cyclo-

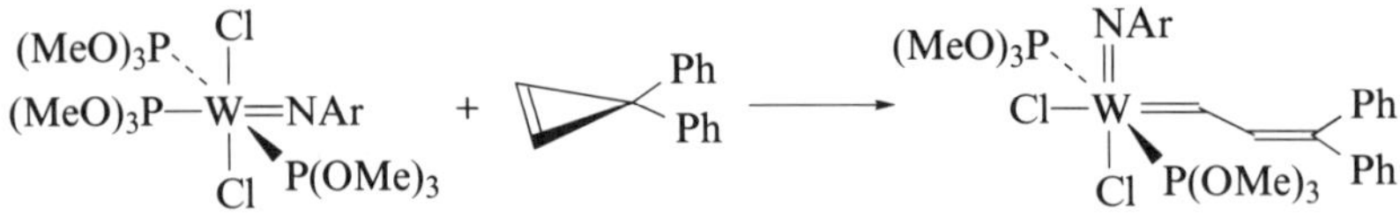

(222)

Ar = 2,6-$C_6H_3(Pr\text{-}i)_2$

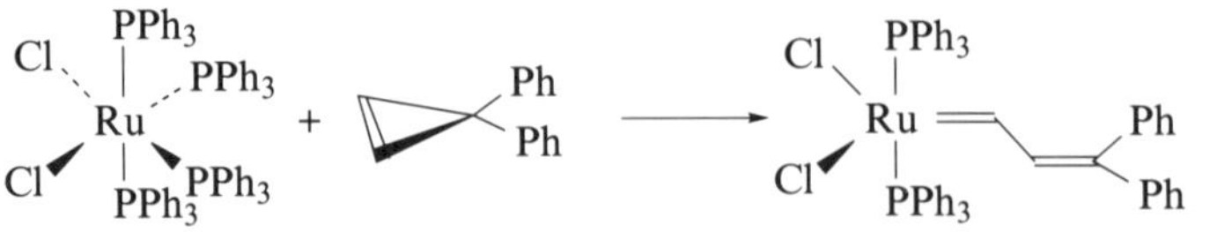

(223)

propene complex intermediate. However, one should bear in mind that this is not so in other cases, e.g. group 4 tetragonal metalocene complexes[77], where it was shown that the cyclopropene and vinyl alkylidene isomers (cf Section IV.B.1.a) *do not* interconvert (equation 224).

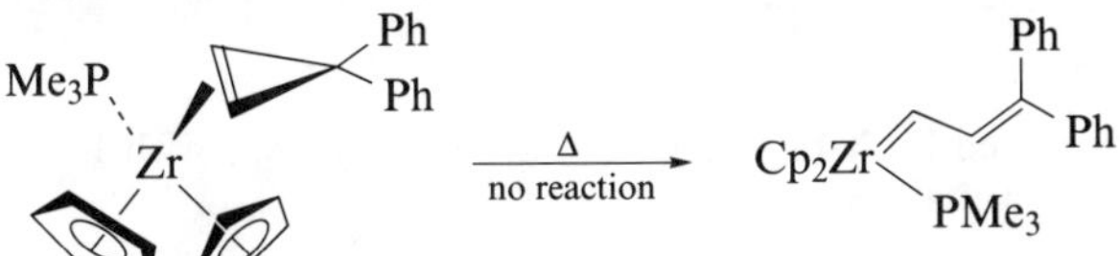

(224)

In light of the above results it is interesting to note that the reaction of diphenylcyclopropenone dimer spirolactone with ironenneacarbonyl yields a mixture of ring-opened vinyl carbene and η^4-vinylketene complexes, and these interconvert under addition (or removal) of CO (equation 225)[295a]. A possible pathway to vinylketene Fe-complexes, prepared earlier from cyclopropenes and ironcarbonyls[295b,296–298], may thus involve initial η^2-coordination, followed by ring cleavage to vinyl carbene and finally carbonylation to the ketene iron η^4-complexes. An analogous η^4-manganese complex is prepared similarly by the reaction of $CpMn(CO)_2THF$ with 3,3-dimethylcyclopropene complex (equation 226)[296].

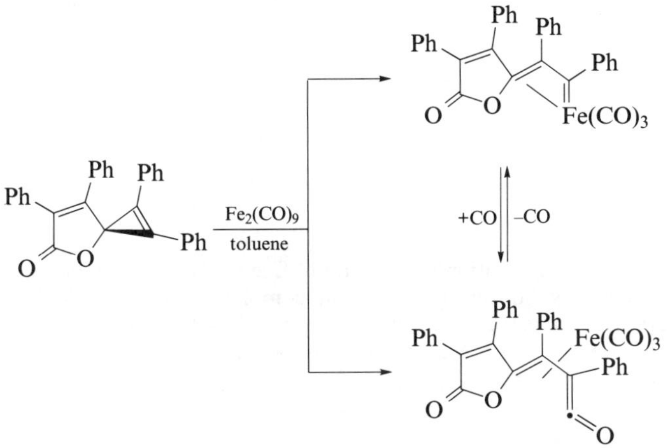

(225)

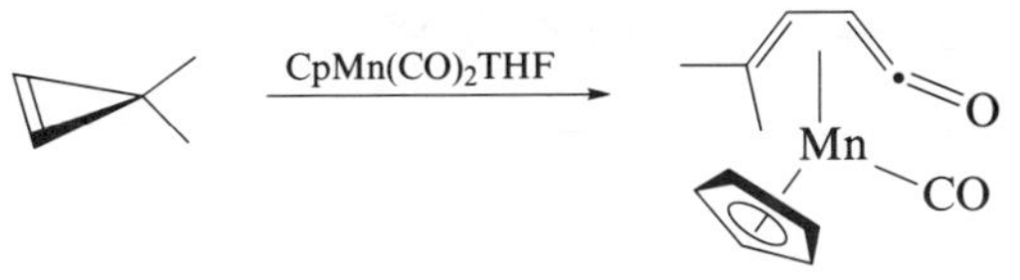

(226)

b. Rearrangement to metallacyclobutenes. Group 4 transition metals, titanocene and zirconocene η^2-cyclopropenyl derivatives undergo thermal ring enlargement to metallacyclobutenes[289]. With the titanocene derivative $Cp_2(PMe_3)Ti(\eta^2\text{-}1,2\text{-}Ph_2C_3H_2)$, this rearrangement occurs readily at 40 °C, and is accompanied by a spontaneous loss of the phosphine ligand to give the corresponding 16-electron titanacyclobutene complex (equation 227). The zirconocene analogue is, however, more stable, rearranging at 60 °C without the loss of PMe_3, to zirconacyclobutene $Cp_2(PMe_3)Zr(\eta^2\text{-}CPhCPhCH_2)$, whose

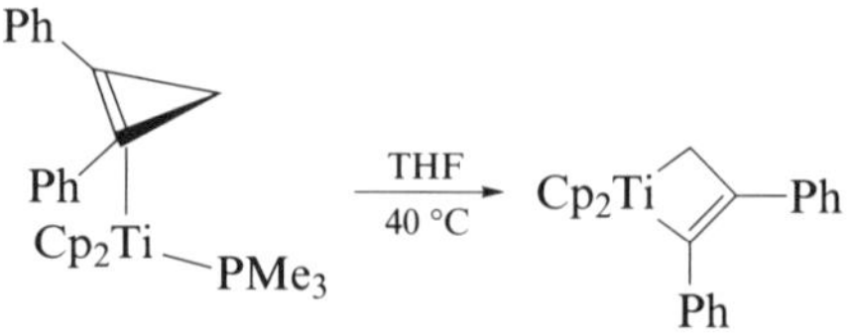

(227)

structure was determined by X-ray analysis. Further treatment with BEt_3 removes the phosphine ligand (equation 228).

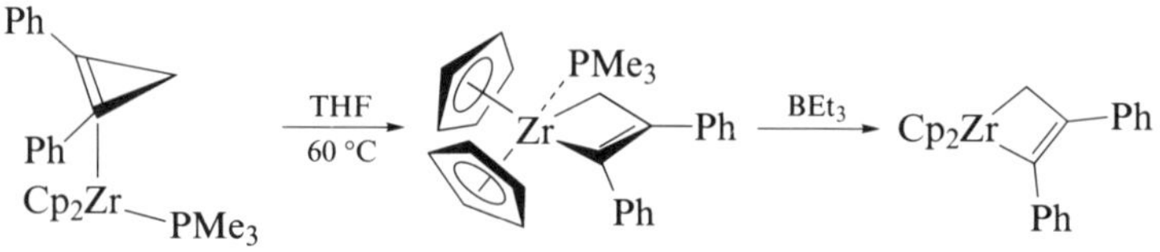

(228)

Notably, whereas the 1,1-difluorocyclopropabenzene gives an isolable η^2-cyclopropene complex upon reaction with CpPd(η^3-allyl) in the presence of PMe_3 (Section IV.B.1.a, equation 209), the analogous reaction with 1,1-bis(trimethylsilyl)cyclopropabenzene results in the ring-opened palladabenzocyclobutene complex (equation 229)[285]. It is thus likely that the reaction mechanism involves a cyclopropene complex intermediate, but a direct C—C activation pathway cannot be excluded.

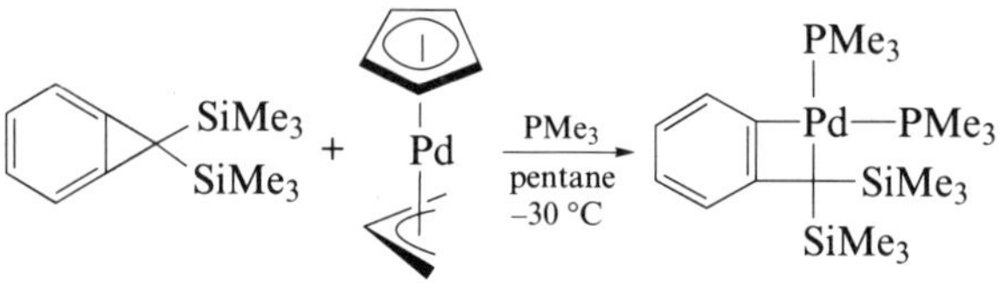

(229)

Direct metallacyclobutene formation by reaction of free tetrafluorocyclopropene and trigonal platinum complexes[299] or tetragonal iridium complexes (equation 230)[300] is also reported in the literature. These reactions may proceed via a labile η^2-cyclopropene complex intermediate. However, the stereoselectivity observed in the product complexes of

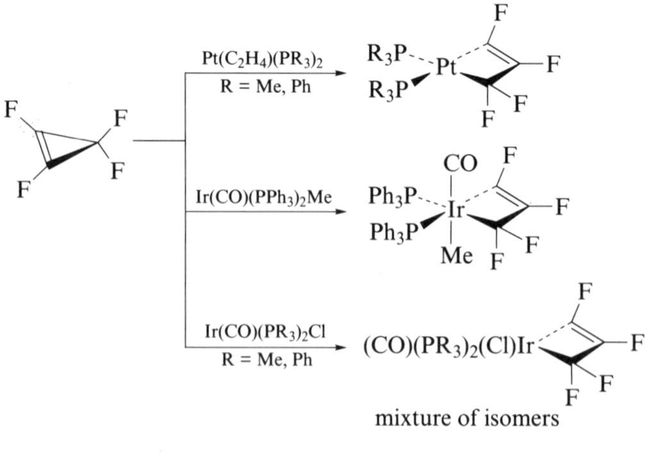

(230)

iridium suggests that the cyclopropene ring opening occurs by direct activation of the carbon–carbon bond[77].

The intermediacy of a metallacyclobutene is proposed upon reaction of the diphenylcyclopropenone dimer spirolactone with $CpCo(CO)_2$, ultimately yielding a η^4-vinylketene complex (equation 231)[295a]. Unlike the analogous iron complex (Section IV.B.2.a), no vinyl carbene complex was observed, and hence formation of the metallacyclobutene seems to be more likely.

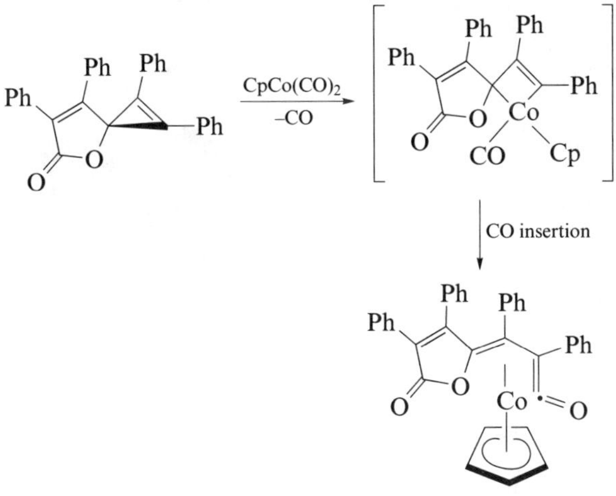

(231)

c. Reaction with alkynes and alkenes[250]. η^2-Zirconocene complexes of the strained cyclopropenes, bicyclo[3.1.0]hex-5-ene and bicyclo[4.1.0]hep-6-ene (or their precursors) (equation 218, Section IV.B.1.c), react with diphenylacetylene in benzene, to give the respective metallacyclopentene adducts as inseparable pairs of isomers. Connectivities in these adducts were confirmed by converting the complexes to their corresponding free hydrocarbons with HCl (equation 232)[41].

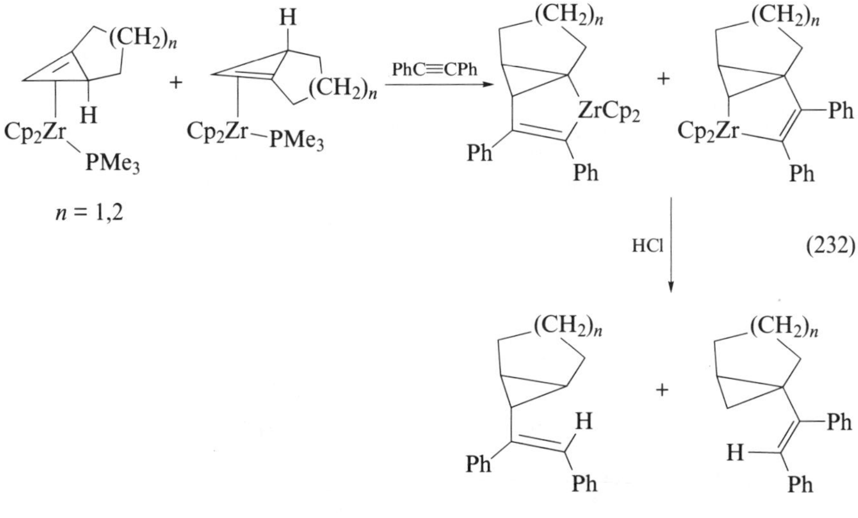

(232)

Similarly, paladium(0) complexes do not form observable η^2-cyclopropene complexes, but often catalyze cyclodimerization and cyclooligomerization reactions instead. The reaction mode is strongly dependent on the catalyst. This is exemplified by the reactions of 3,3-dimethoxycyclopropene, which in the presence of $Pd(dba)_2$ (dba = dibenzylideneacetone) gave only a tricyclic dimer, whereas a complex mixture of products is obtained when $Pd(dba)_2$ and PPh_3 are present (equation 233)[301]. When the rhodium carbonyl complex $(PPh_3)_2(CO)(Cl)Rh$ is used, insertion of a carbonyl group occurs together with the formation of a tetracyclic rhodacycloheptane oligomer (equation 234)[79].

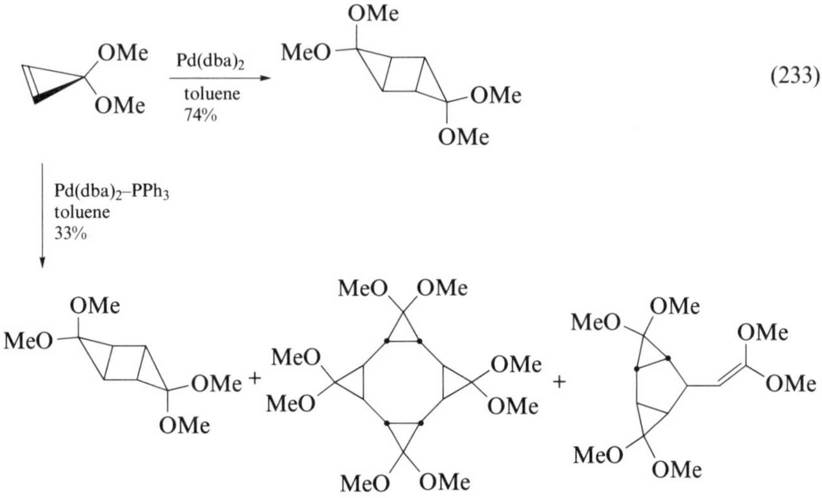

(233)

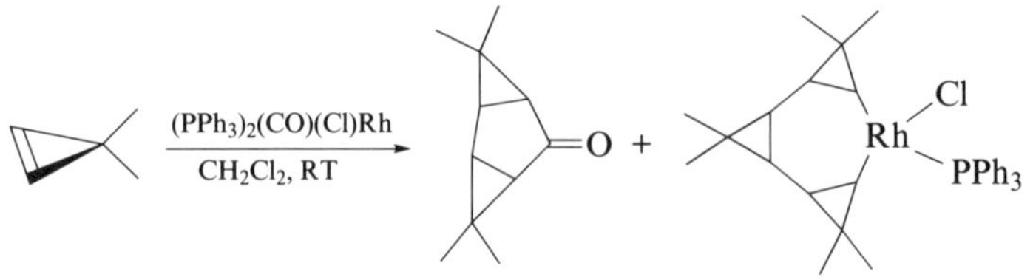

(234)

The intermediacy of η^2-cyclopropene complexes of nickel has been proposed in catalyzed 2+1 reactions of free cyclopropene with electron-poor olefins, to give vinylcyclopropanes. For example, the reaction of fumarate esters with 3,3-disubstituted cyclopropenes in the presence of $Ni(COD)_2$ catalyst gave vinyl-substituted *trans*-2,3-cyclopropane dicarboxylate esters (equation 235)[72,302]. However, when maleic esters were used instead, a mixture of both *cis* and *trans* vinylcyclopropane diesters is obtained.

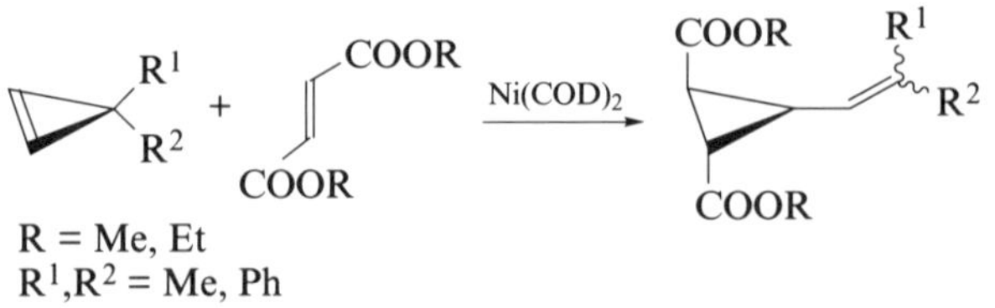

(235)

d. Oxidative coupling reactions. Although cyclodimerization of cyclopropenes to metallacyclopentanes is believed to involve the oxidative addition of cyclopropenes to intermediate η^2-cyclopropene complexes (Section II.A.3.c), the only direct indication for such a pathway is reported for the reaction of titanocene $Cp_2Ti(PMe_3)_2$ with 3,3-dimethylcyclopropene (equation 236, Section IV.B.1.a). The presence of titanocene $Cp_2Ti(\eta^2\text{-}3,3\text{-}Me_2C_3H_2)$ was observed by NMR when a 2:1 ratio of the two reagents are first reacted at –30 °C, to give a mixture containing the cyclopropene and the open-chain vinyl carbene complexes, which upon further reaction with excess cyclopropene at above 0 °C gives the titanocyclodimer.

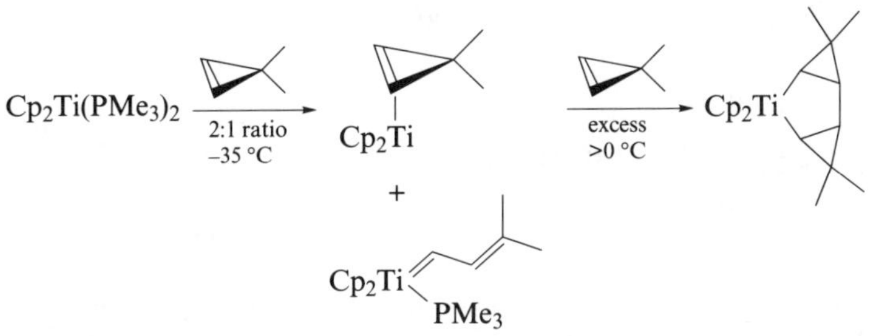

(236)

V. η^3-CYCLOPROPENYL COMPLEXES (METALLATETRAHEDRANES)

A. Preparation

1. Oxidative addition of cyclopropenylium salts

The major route to η^3-cyclopropenylium complexes $L_nM(C_3R_3)$ (metallatetrahedranes) is by oxidative addition reactions of cyclopropenylium salts to transition metal complexes of groups 5 (V), 6 (Mo, W), 8 (Fe, Ru), 9 (Co, Rh, Ir) and 10 (Ni, Pd, Pt). The addition is frequently accompanied by loss of one or more carbonyl, olefin or halogen auxiliary ligand. Concurrent formation of oxocyclobutenyl complexes by carbonyl insertion into the cyclopropenyl ring is often observed in reactions with group 9 cobalt triad and early transition metal complexes.

Metallatetrahedrane Oxocyclobutenyl complex

Chronologically, group 10 trigonal nickellatetrahedranes $L_2Ni(C_3R_3)$ (or dimers[303,304]) were first to be prepared, by oxidative addition reaction of $[C_3R_3]X$ (R = Me, *t*-Bu, Ph; X = Cl, Br, BF_4) to the plane tetragonal $Ni(CO)_4$, in MeOH (equation 237)[303,304]. When cyclopropenylium tetrafluoroborates are used, addition of an appropriate halide to the reaction mixture is required[304]. The analogous reaction of $[C_3Ph_3]ClO_4$ with the trigonal ethylene complex $(C_2H_4)Ni(PPh_3)_2$ in MeOH, followed by addition of Bu_4NPF_6, led to the hexafluorophosphate complex $[(PPh_3)_2Ni(C_3Ph_3]PF_6$ (equation 238)[305], whose structure was secured by X-ray analysis[306]. Similar reactions of cyclopropylium cations with plat-

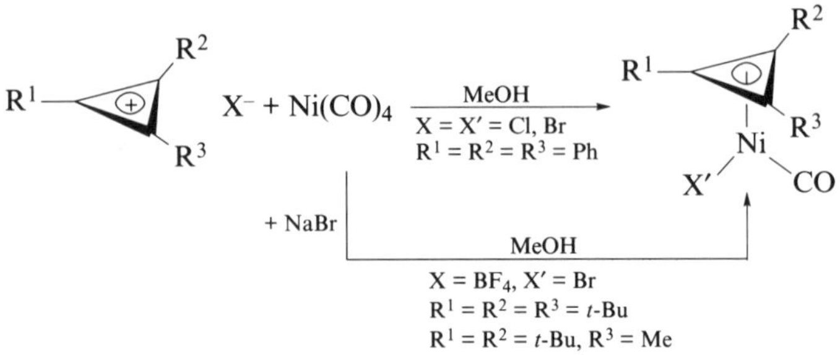

(237)

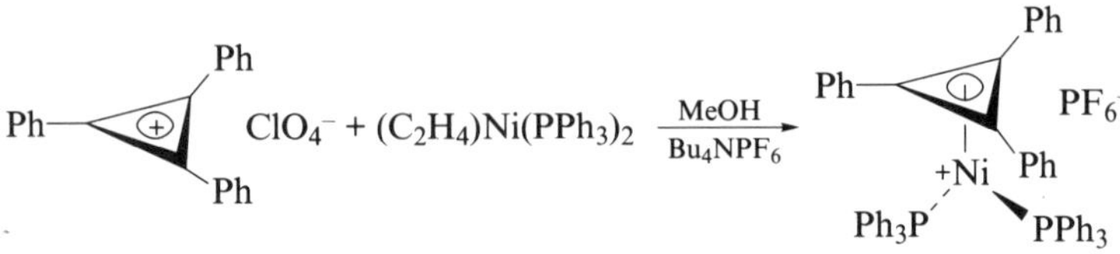

(238)

inum[307] and palladium[306] ethylene complexes in CH_2Cl_2-benzene gave the corresponding $[(PPh_3)_2M(C_3Ph_3)]X$ (M = Pt, Pd; X = ClO_4, PF_6) metallatetrahedranes; the hexafluorophosphates crystallize with one molecule of benzene. Notably, solid state analysis shows that in the nickel complex the P—Ni—P plane bisects the three-membered ring whereas in the platinum complex (equation 239)[307] the P—Pt—P plane parallels one of the ring C—C bonds. These conformational structures were studied theoretically by molecular orbital calculations of the extended Huckel type in connection with rotational barriers (ring-whizzing) in cyclic polyene-ML_2 complexes[306,308].

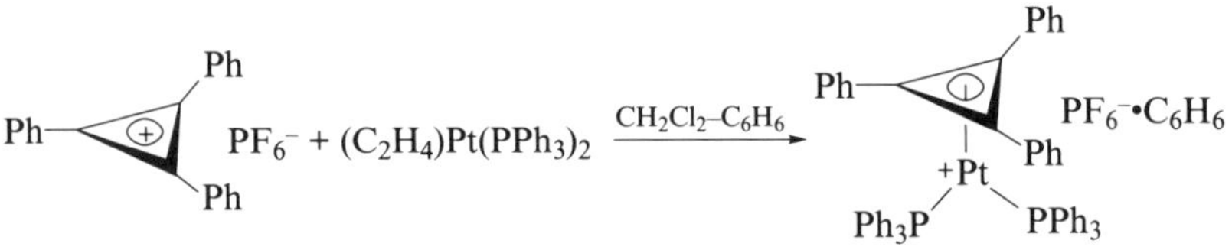

(239)

Group 8 (Fe)[275] and 9 (Co[275], Rh[309], Ir[310]) tetrahedral complexes of general structure $L_3M(C_3R_3)$ are best prepared by oxidative addition of $[C_3(t\text{-Bu})_3BF_4]$ to bis(triphenylphosphine) iminium (PPN) carbonyl complexes $[PPN][M(CO)_2L]$ (PPN^+ = $[PPh_3{=}N{=}PPh_3]^+$; M = Fe, L = NO; M = Co, Rh, Ir, L = CO) in THF or CH_2Cl_2 (equation 240). Interestingly, the less bulky phenyl (and naphthyl) substituted cyclopropenylium tetrafluoroborates react faster than their tri-*t*-Bu counterparts to give mixtures of nitroso ferratetrahedranes and ring carbonylation asymmetric oxocyclobutenyl complexes (equation 241)[275]. In the cobalt series, the use of metal carbonyl substrates other than $[PPN][Co(CO)_4]$, and changing to the more polar acetonitrile solvent, lead to mixtures of cobaltatetrahedranes and oxocyclobutenyl complexes (equation 242)[274, 275].

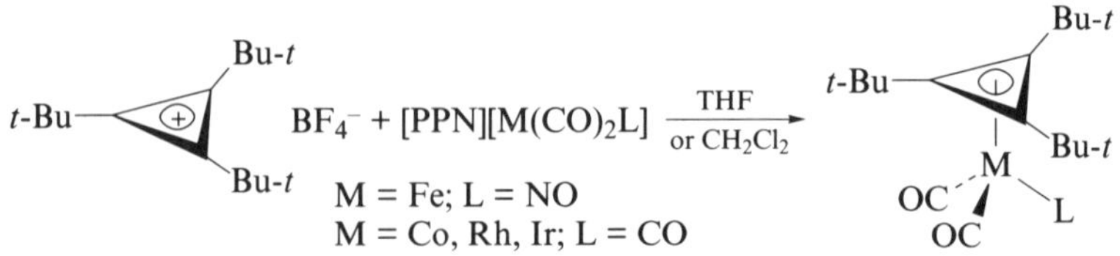

(240)

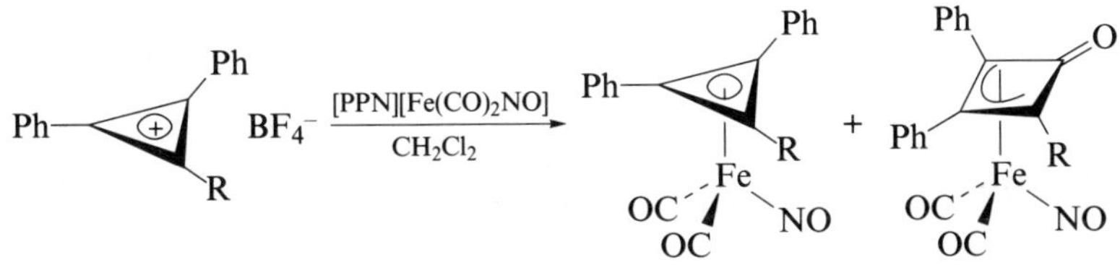

(241)

R = H,	6%	52%
R = Me,	65%	15%
R = *t*-Bu,	64%	22%
R = Ph,	36%	52%

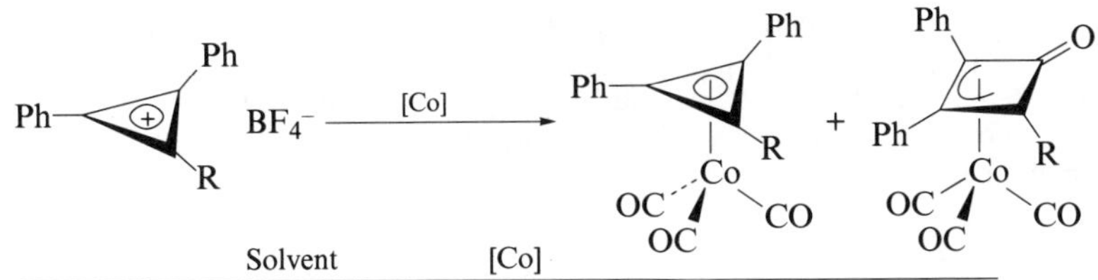

(242)

	Solvent	[Co]		
R = *t*-Bu,	$ClCH_2CH_2Cl$	$Co_2(CO)_8$	14%	41%
R = Ph,	$ClCH_2CH_2Cl$	$Co_2(CO)_8$	34%	33%
R = Ph,	MeCN	$Co_2(CO)_8$	—	87%
R = Ph,	MeCN	$Na[CO_3(CO)_{10}]$	7%	43%

Octahedral metallatetrahedranes $L_5M(C_3R_3)$ are prepared by oxidative addition reactions of cyclopropenylium salts to group 6 molybdenum and tungsten[311,312] and group 8 ruthenium complexes[313,314]. Reaction of $[C_3Ph_3]Br$ with $Mo(CO)_4L_2$ (L = bipy, phen, dpa = 2,2′-dipyridylamine, MeCN) in acetonitrile gave crystalline solvates $[(CO)_2L_2BrMo(C_3Ph_3)]MeCN$, in which the bromide ligand is *trans* to the three-membered ring as shown in equation 243 for the 2,2′-bipyridine complex. When the reaction was conducted in refluxing THF, the ring carbonylated oxocyclobutenyl crystalline solvates $[(CO)_2L_2BrMo(C_3Ph_3CO)]$ THF were obtained. More recently[312], oxidative addition of cyclopropenylium hexafluorophosphate to $Mo(CO)_3(MeCN)_3$ afforded $[(CO)_2(MeCN)_3Mo(C_3Ph_2R)]PF_6$, in 64% isolated yield (equation 244). Other ring-substituted molybdenum and tungsten complexes were likewise prepared and further reacted with cyclopentadienyl anion without isolation (*see below*). If, however, triphenylcyclopropenylium chloride or bromide is used in the reaction with $M(CO)_3(MeCN)_3$ (M = Mo, W) instead of hexafluorophosphate, ring-expanded oxocyclobutenyl complexes are ultimately formed.

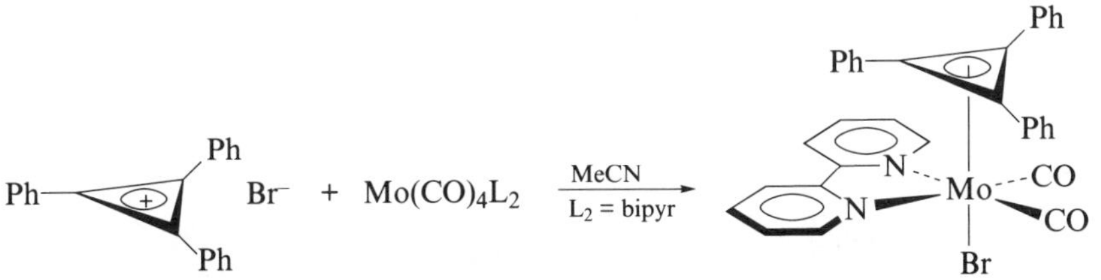

L_2 = 2,2'-bipyridine (bipy), 1,10-phenanthroline (phen), 2,2'-dipyridylamine (dpa) (243)

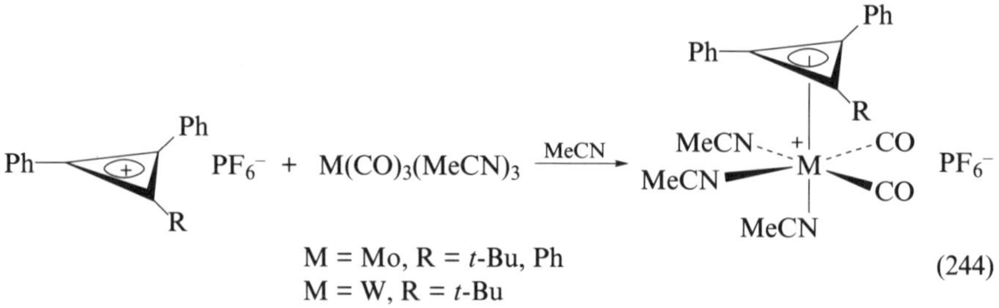

(244)

Oxidative addition reactions of cyclopropenylium salts $[C_3R_3]X$ (R = Me, Ph; X = Cl, Br, BF_4) with ruthenium cyclooctadienyl complex Cp(η^4-COD)ClRu or tetramers $(CpClRu)_4$ and $(Cp^*ClRu)_4$ (Cp* = η^5-pentamethylcyclopentadienyl) afforded $CpX_2Ru(C_3Ph_3)$ and $Cp^*X_2Ru(C_3Ph_3)$, respectively (equation 245)[313,314]. When the tetrafluoroborate salts are used in the reaction, addition of LiX (X = Cl, Br) is required. The mixed halo-complex $CpBrClRu(C_3Ph_3)$ was synthesized from $[C_3Ph_3]Br$ and CpBrRu(η^4-COD) in methanol[314].

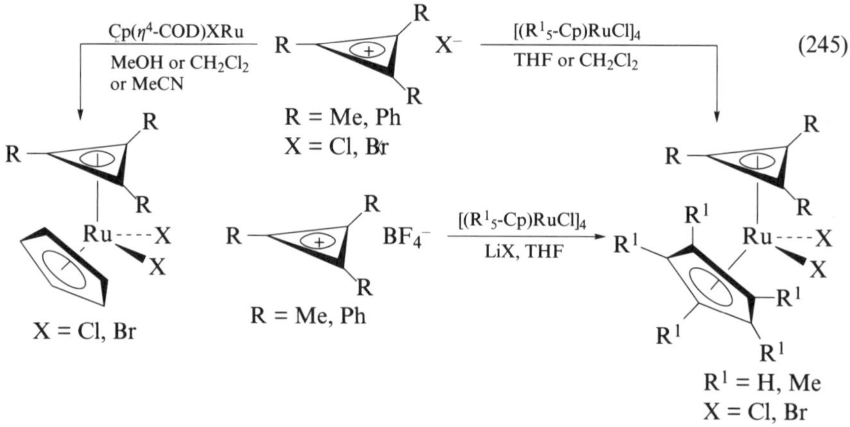

(245)

The only example of oxidative addition reaction effected by UV light is reported for the preparation of the unique group 5 vanadium complex $(CO)_4V(C_3(t\text{-Bu})_3)$[315]. Photolysis of a suspension of $[C_3(t\text{-Bu})_3]BF_4$ and $Na[V(CO)_6]$ in THF, using a medium-pressure mercury lamp, afforded the η^3-cyclopropenyl d^{16} V-complex in 81% yield (equation 246). While the solid state analysis reveals a distorted tetragonal-pyramidal structure, the metal–cyclopropenyl fragment appears to have a 3-fold symmetry on the ^1H NMR time scale in solution over a wide range of temperatures (+70 to –80 °C). This indicates a fast rotation about the metal-ring axis, presumably by the ring-whizzing mechanism suggested for group 10 trigonal complexes $(C_3R_3)ML_2$[306] (Section V.C.1).

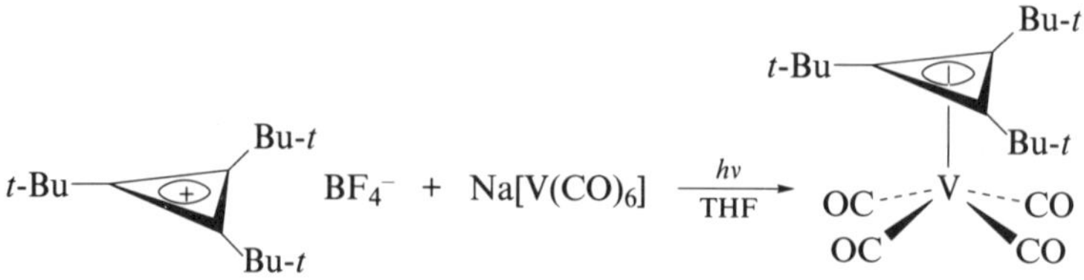

(246)

2. Metallacyclobutadiene rearrangements

Addition of trimethylphosphine to the tungstenacyclobutadinene complex $CpCl_2W[C(t\text{-}Bu)CMeC(t\text{-}Bu)]$ resulted in the rearrangement of the four-membered metallacycle into the tungstenatetrahedrane complex $Cp(PMe_3)Cl_2W(\eta^3\text{-}C_3Me(t\text{-}Bu)_2)$ (equation 247)[316]. X-ray analysis revealed a 'four-legged piano stool' structure, in which the C_3-ring and the PMe_3 ligand are *trans* to each other. This is the first metallatetrahedrane which shows conformational rigidity in solution. The ^{13}C NMR spectrum exhibits three cyclopropenyl ring carbon signals at 88.59, 64.30 and 56.65 ppm. Since metallacyclobutadienes are likely intermediates in metal catalyzed metathesis reactions of acetylenes, this experiment strongly supports the involvement of metallatetrahedranes in acetylene metathesis[317]. Similarly, addition of TMEDA to tungstenacyclobutadiene complex $Cl_3W[CMeCMeC(t\text{-}Bu)]$ in CH_2Cl_2 gave the octahedral 1:1 adduct $(TMEDA)Cl_3W[\eta^3\text{-}C_3Me_2(t\text{-}Bu)]$, whose metallatetrahedrane structure was confirmed by X-ray analysis (equation 248)[318]. The ^{13}C NMR spectrum (CD_2Cl_2) shows only two C_3-ring signals at 162.4 [$C(t$-Bu)] and 127.6 (2*C*Me) ppm, indicating fast rotation about the C_3–M axis. The pyridine complex $[(pyr)_2Cl_3W(\eta^3\text{-}C_3Me_2(t\text{-}Bu)]$ was similarly prepared by addition of two equivalents of pyridine to the trichloro tungstenacyclobutadiene complex[318].

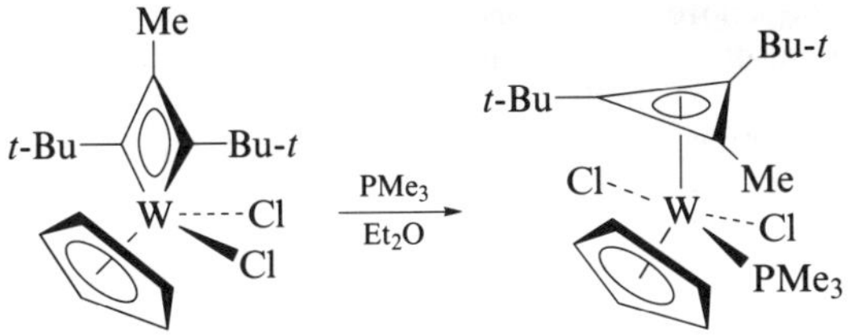

(247)

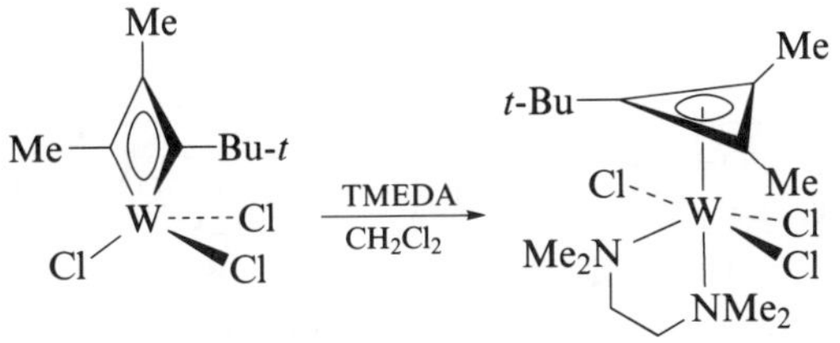

(248)

Rhenacyclobutadienes are more stable toward loss of an acetylene or further reaction with an internal acetylene than other metallacyclobutadienes[268,269,278,319]. However, whereas the low-temperature ^{13}C NMR spectra (–50 °C) of the rhenacyclobutadienes are consistent with their X-ray structure, coalescence of the ring signals is observed at room temperature, indicating a rapid exchange of the ReC_3-ring carbons. Since no loss of acetylene is observed, this was attributed to a rapid equilibrium of the rhenacyclobutadienes with their η^3-cyclopropenyl counterparts (equations 249 and 250)[268,269,319].

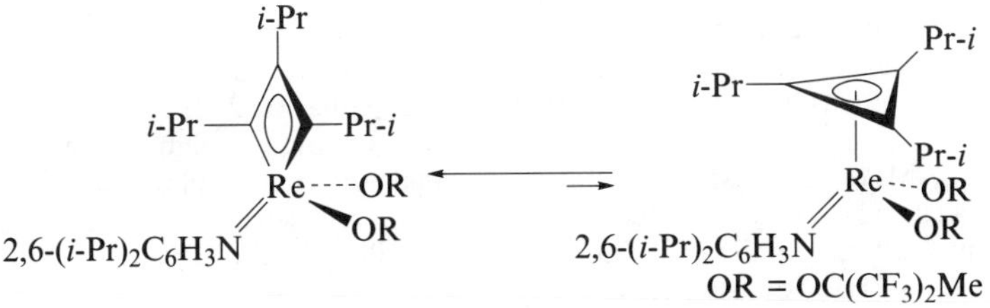

(249)

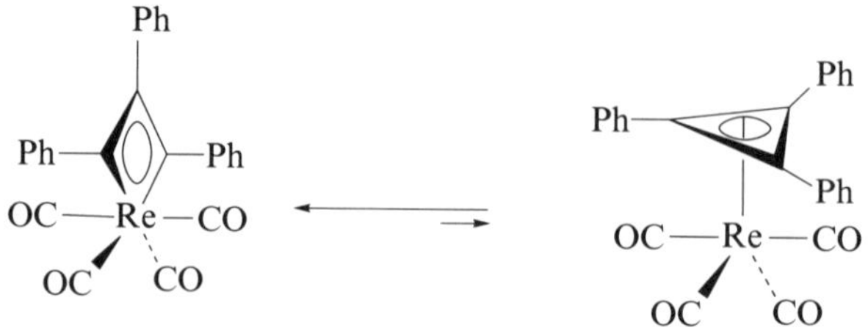

(250)

3. Reaction of carbyne complexes with alkynes

Molybdenum and tungsten carbyne (alkylidyne) complexes frequently undergo 2+2 cycloaddition reactions with alkynes to give the corresponding metallacyclobutadiene complexes[316–318]. However, carboxylic acid alkylidyne complexes of the type $(O_2CR)_3M{\equiv}CBu$-*t* (M = Mo, W; R = Me, *i*-Pr, *t*-Bu) react with internal alkynes $R^1C{\equiv}CR^1$ to form η^3-cyclopropenyl complexes (metallatetrahedranes) $(O_2CR)_3M(\eta^3$-$C_3R^1{}_2Bu$-*t*) (R^1 = Me, Et, Ph) (equation 251). The solid state analysis of $(O_2CMe)_3W(\eta^3$-C_3Et_2Bu-*t*) revealed the coordination geometry about the tungsten atom to be approximately pentagonal-bipyramidal with the three-membered ring system occupying one apical position, two acetate ligands occupying biequatorial sites and the third acetate ligand spanning an axial and equatorial position. The characteristic features in the ^{13}C NMR spectra of these complexes are two singlets at the relatively high field 70–80 ppm region of the three-membered ring carbons.

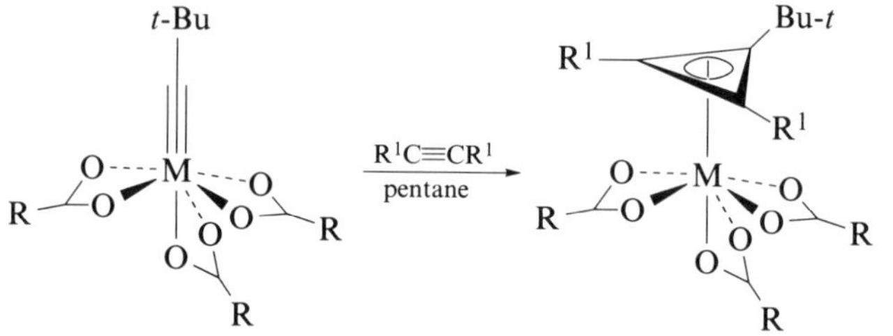

(251)

M = Mo, W
R = Me, *i*-Pr, *t*-Bu
R^1 = Me, Et, Ph

The reverse cleavage of the η^3-cyclopropenyl ring to acetylene and alkylidyne complexes was demonstrated by the reaction of $(pyr)_2Cl_3W(\eta^3$-C_3Me_2Bu-*t*) with three equivalents of LiOBu-*t* which afforded (pyr)(OBu-*t*)$_3W{\equiv}CMe$[318]. It is, however, not clear whether this degradation occurs directly from the η^3-cyclopropenyl cage or by first reverting to a tungstenacyclobutadiene ring and finally to an alkylidyne complex.

Cycloaddition can be induced by UV irradiation. Thus, photolysis of a solution of $Cp\{P(OMe)_3\}_2Mo{\equiv}CCH_2Bu$-*t* and diphenylacetylene in hexane affords the molybdenatetrahedrane $Cp\{P(OMe)_3\}_2Mo(\eta^3$-$C_3Ph_2CH_2Bu$-*t*) (equation 252)[321].

A closely related formation of η^3-cyclopropenyl molybdenum complex $Cp\{P(OMe)_3\}_2Mo(\eta^3$-$C_3Ph_2Me)$, possibly involving an interamolecular ligand cycloaddition within a carbyne–acetylene intermediate complex, has been accomplished by treating the acetylenic cation complex $[Cp\{P(OMe)_3\}_2Mo(\eta^2$-$PhC{\equiv}CPh)]BF_4$ with vinylmagnesium bromide in THF (equation 253)[321].

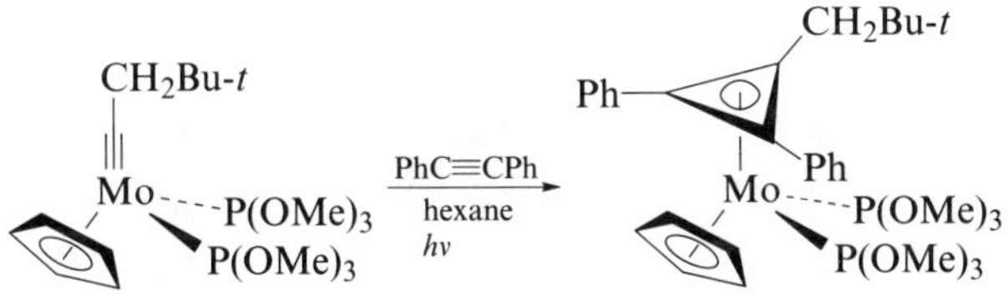

(252)

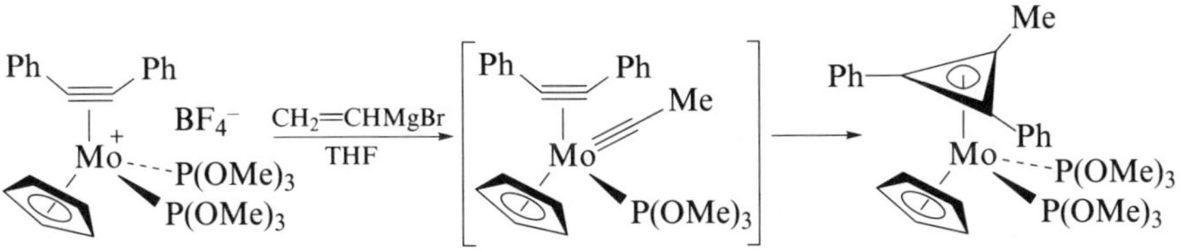

(253)

B. Reactions

1. Ligand exchange

a. Exchange of CO by phosphorus, nitrogen and oxygen ligands. Exchange of a carbonyl ligand by phosphine, phosphite and isonitrile groups has been mainly observed in tetragonal transition metal complexes $L_3M(\eta^3\text{-}C_3R_3)$ of group 8 (M = Fe) and 9 (M = Co, Rh, Ir). Monosubstitution occurs smoothly without ring cleavage when η^3-cyclopropenyl tricarbonyl complexes of group 9 elements and phosphorus ligands are refluxed in benzene (equation 254)[275,309]. Similar reaction of the isolobal nitroso dicarbonyl complexes of iron $(CO)_2(NO)Fe(\eta^3\text{-}C_3R_3)$ with *phosphite* ligands readily gave the corresponding mixed nitroso carbonyl phosphite complexes (equation 255)[322]. However, the reaction with *phosphines* yielded, in addition to the monosubstituted mixed η^3-cyclopropenyl complexes, ring-expanded mixed oxocyclobutenyl complexes.

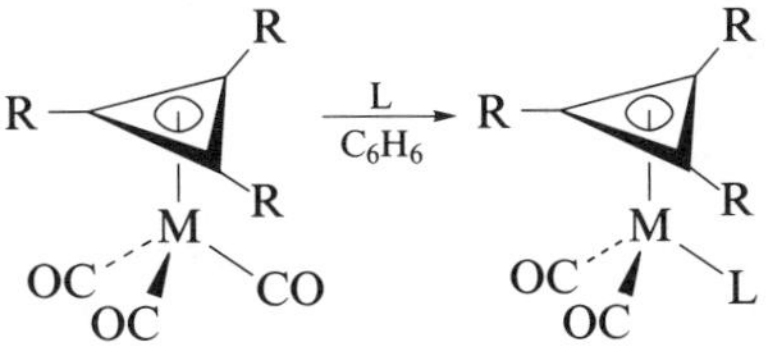

(254)

R = Ph, M = Co, L = PMe_2Ph, PPh_3, $P(OPh)_3$
R = *t*-Bu, M = Co, Rh, Ir, L = PMe_3
R = *t*-Bu, M = Ir, L = PPh_3, $P(OMe)_3$, *t*-BuNC

Kinetic studies of the substitution reaction of $(CO)_3M[\eta^3\text{-}C_3(t\text{-}Bu)_3]$ (M = Co, Rh, Ir) with $P(OEt)_3$ provide the first examples of dissociative CO substitution for carbonyl complexes of the cobalt triad (equation 256)[323].

The reaction rates are first order in the concentration of the complex and zero order in the concentration of the phosphite. High $\Delta H^{\#}$ and large positive $\Delta S^{\#}$ values are consistent with a dissociative mechanism, as is also the CO retardation of the reaction rates. This contrasts with the associative mechanism established for CO substitution reactions in related polyolefin and polyenyl metal carbonyls, in which ligand hapticity changes at the organic ligand occur along the substitution pathway[324,325].

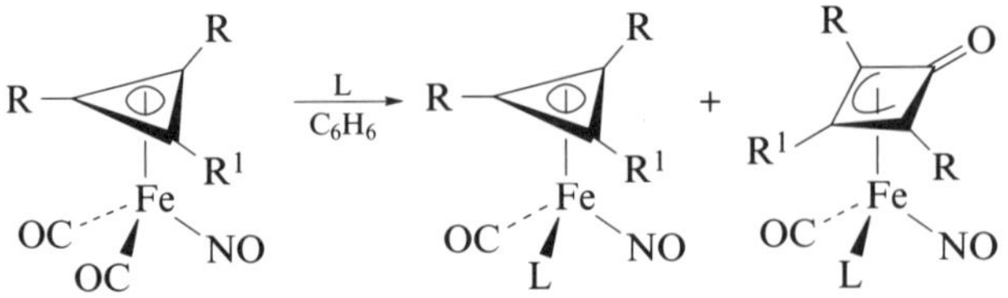

(255)

$R = R^1 = Ph, L = P(OMe)_3$	76%	
$R = R^1 = Ph, L = P(OPh)_3$	85%	
$R = R^1 = Ph, L = PPh_3$	38%	48%
$R = R^1 = Ph, L = PMe_2Ph$	11%	65%
$R = Ph, R^1 = Me, L = PPh_3$	12%	26%
$R = Ph, R^1 = H, L = PPh_3$	17%	25% (2:1 mixture with nonsymmetrical isomer)
$R = Ph, R^1 = t\text{-}Bu, L = PPh_3$	67%	
$R = R^1 = t\text{-}Bu, L = PMe_2Ph$	88%	
$R = R^1 = t\text{-}Bu, L = PMe_3$	72%	

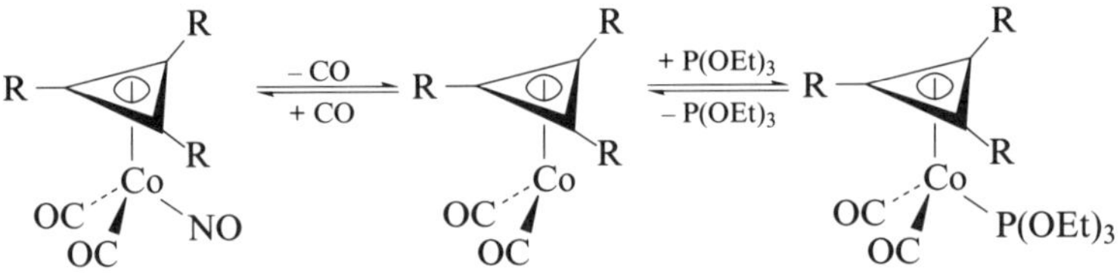

(256)

The Co complex is the most inert toward substitution. This is mainly because of the stronger Co—CO bond compared with the M—CO bonds of the other two complexes. The activation enthalpy for the dissociation reaction of CO from the Co complex (34.4 kcal mol^{-1}) is about 9 $kcal\,mol^{-1}$ higher than that for the Ir compound and about 12 $kcal\,mol^{-1}$ higher than that for the Rh compound. This correlates with the CO stretching frequencies which increase in the order Co<Ir<Rh, suggesting a decrease in M—CO π-backbonding in this order and a decrease in M—CO bond strength. This is consistent with other examples of the late transition metal triads in which the first-row elements form stronger M—CO bonds than their lower-row counterparts.

Measurements of the equilibrium constants of the reactions imply that the stabilities of the monosubstituted complexes are predominantly determined by steric effects of the ligand, reflecting a very crowded space around the metal in this system. The large ligands PPh_3 or $P(c\text{-}C_6H_{11})$ do not react with the cobalt complex.

The kinetics of the reaction of $(CO)_2(NO)Fe[\eta^3\text{-}C_3(t\text{-}Bu)_3]$ with PBu_3 obeys a two-term rate law, which suggests that the reaction occurs simultaneously by first-order dissociative (k_d) and second order associative (k_a) pathways (equation 257).

Kinetic studies at different phosphine concentrations indicated that the substitution reactions occur totally by a dissociative mechanism, while the ring expansion reaction is by an associative mechanism. The associative reaction could proceed by an 18-electron transition state involving either a bent NO or an η^3 to η^1 ring slippage mechanism. The bent NO mechanism seems more likely, because the ring slippage mechanism is known to result in the formation of oxocyclobutenyl product with ring expansion and because the isoelectronic cobalt complex does not react by a parallel associative pathway.

Reaction of $(CO)_2(NO)Fe(\eta^3\text{-}C_3Ph_3)$ with PBu_3 gives both CO substitution and ring-expansion products, whereas the reaction with PPh_3 follows only a dissociative pathway to yield monosubstituted product with the same rate constant as that obtained from the

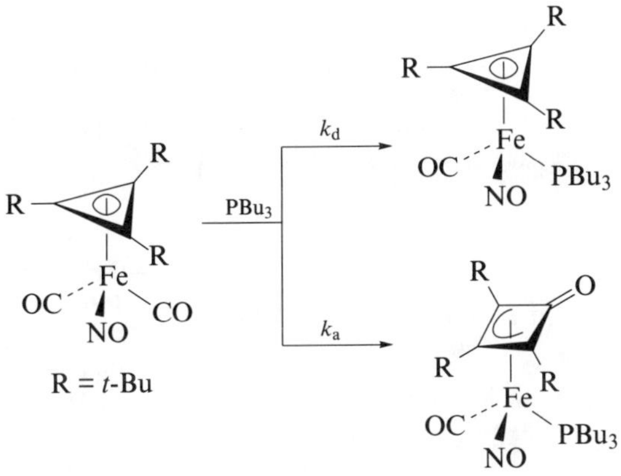

(257)

reaction with PBu_3. The rate for dissociation of CO from the metal is faster for the $(CO)_2(NO)Fe(\eta^3\text{-}C_3Ph_3)$ than for the analogous *t*-Bu compound. This is attributed to the electron-withdrawing phenyl substituents which decrease the electron density on the metal much more efficiently than does the electron-donating *t*-Bu group. This in turn is supported by the higher CO stretching frequencies of Ph compared with *t*-Bu complex.

The mechanism of oxycyclobutenyl $(CO)(NO)(PBu_3)Fe(C_3Ph_3CO)$ formation by the associative substitution reaction is suggested to involve $\eta^3\rightarrow\eta^1$ cyclopropenyl ring slippage followed by ligand migratory CO insertion, which in turn leads to the ring expansion product (equation 258)[265,269].

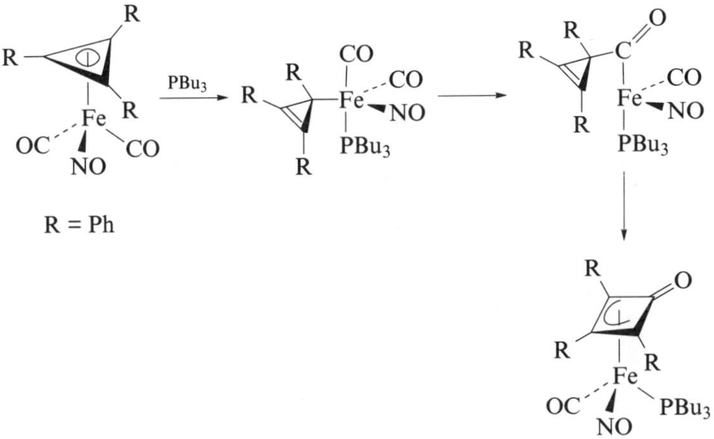

(258)

The trigonal nickel complex $(CO)ClNi(C_3Ph_3)$ undergoes CO exchange by phosphorus ligands but, unlike the tetragonal complexes of groups 8 and 9, monosubstitution takes place only with PPh_2Me (equation 259)[326]. With other phosphine and phosphites concomitant addition of another phosphorus ligand occurs, yielding tetragonal complexes of the type $(PL_3)_2ClNi(C_3Ph_3)$ (equation 260). Substitution and addition is also achieved upon reaction with THF, 2,2′-bipyridyl[326] and pyridine[327,328]. X-ray analysis of the complex with pyridine revealed the pyridine solvate molecular structure $[pyr_2ClNi(C_3Ph_3)]pyr$ in the solid state.

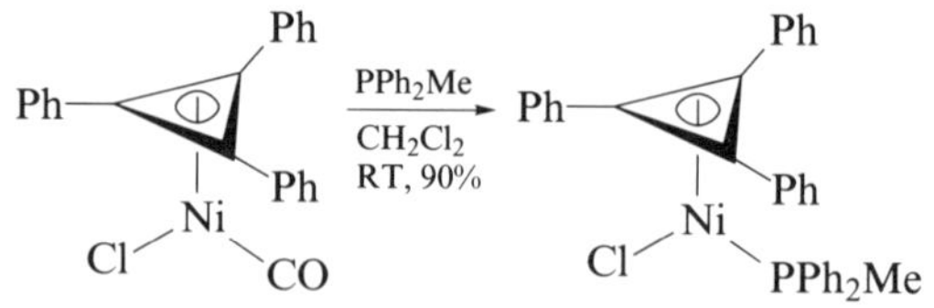

(259)

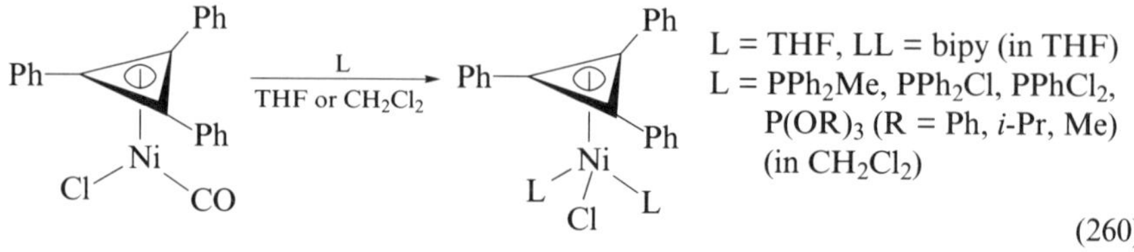

(260)

b. Exchange of CO by alkenes and alkynes. Reaction of $(CO)_3Ir[C_3(t\text{-}Bu)_3]$ with amine oxide (oxidative decarbonylation agent) followed by tetrafluoroethylene afforded the air-stable, trigonally coordinated monocarbonyl alkene iridium complex $(CO)(C_2F_4)Ir[C_3(t\text{-}Bu)_3]$ in its more hindered in-plane conformation (equation 261). Similarly, replacement of two carbonyl ligands in $(CO)_3Ir(C_3(t\text{-}Bu)_3]$ with 2-butyne gave the corresponding monocarbonyl alkyne complex $(CO)(MeC{\equiv}CMe)Ir[C_3(t\text{-}Bu)_3]$ (equation 262).

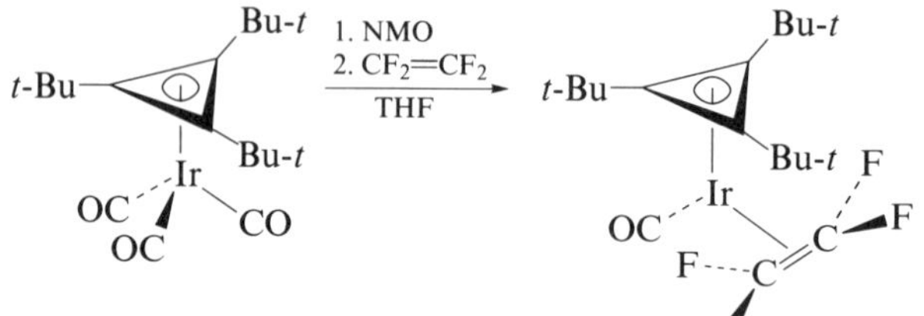

(261)

NMO = *N*-methylmorpholine *N*-oxide

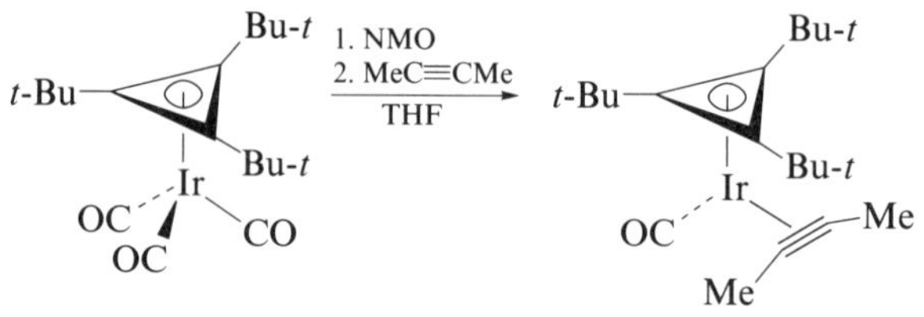

(262)

c. Exchange of ligands by cyclopentadienyl and related groups. The η^5-cyclopentadienyl (Cp) group is formally considered to occupy three of the coordination sites of a complex. Thus, replacement of the three auxiliary ligands of the tetragonal Ni complex $[Pyr_2BrNi(\eta^3\text{-}C_3Ph_3)]$pyr using cyclopentadienylthallium (CpTl) in benzene led to the first cyclopropenylium mixed sandwich complex $CpNi(\eta^3\text{-}C_3Ph_3)$ in 78% yield (equation 263)[330–332]. Similarly, $CpNi(\eta^3\text{-}C_3(t\text{-}Bu)_3]$ was prepared by reaction of the trigonal nickel complex $(Co)BrNi[\eta^3\text{-}C_3(t\text{-}Bu)_3]$ with CpNa in ethanol[304]. Interestingly, reaction of cyclopropenylium complex $(CO)BrNi(\eta^3\text{-}C_3Ph_3)$ with phosphacarborane $[7,9\text{-}B_9H_9CHPMe]Na$ in THF gave the unique nickelaphosphacarborane $(1,7\text{-}B_9H_9CHPMe)Ni(\eta^3\text{-}C_3Ph_3)$, isolobal to the cyclopentadienyl mixed sandwich complexes (equation 264)[333].

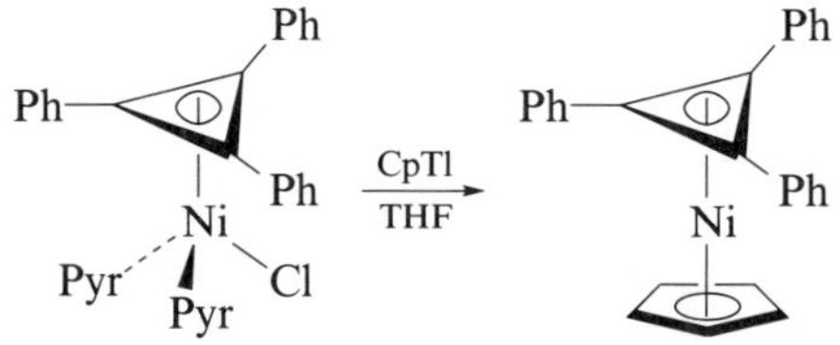

(263)

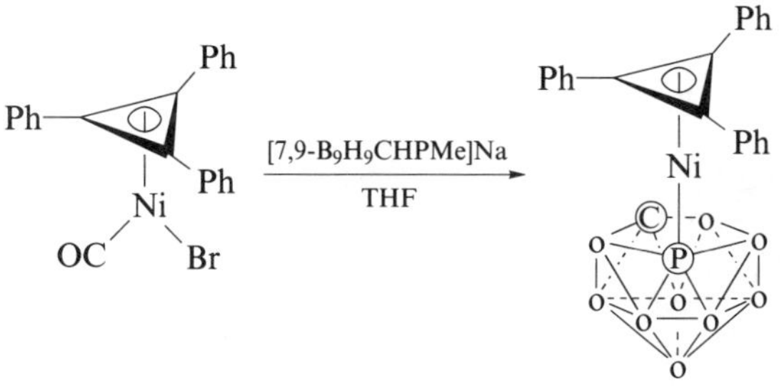

(264)

A similar substitution strategy is used for the preparation of group 6 transition metal complexes of type $Cp(CO)_2M(\eta^3\text{-}C_3Ph_3)$ (M = Mo, W). Early attempts to prepare the molybdenum complex by reaction of the octahedral molybdenum complex $(CO)_2(MeCN)_2ClMo(\eta^3\text{-}C_3Ph_3)$ with CpLi in THF resulted only in 10% yield[331]. However, modification of this method using the hexafluorophosphate salts $[(CO)_2(MeCN)_3M(\eta^3\text{-}C_3Ph_2R)]PF_6$ (M = Mo, W; R = *t*-Bu, Ph) and CpTl gave substantially higher yields (equation 265)[312].

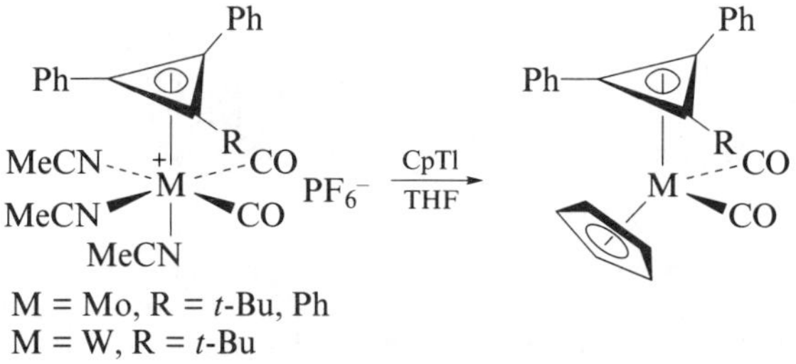

(265)

d. Metathetical exchanges of halogen and nitrogen ligands. Tetragonal dihalo ruthenium complexes $CpX_2Ru((\eta^3\text{-}C_3Ph_3)$, X = Cl, Br and $CpXX^1Ru(\eta^3\text{-}C_3Ph_3)$, X = Cl, X^1 = Br readily undergo halide metathesis using excess of KI in methanol, to give the diiodo complex $CpI_2Ru(\eta^3\text{-}C_3Ph_3)$[313,314]. Replacement of chloride and bromide in these and the related Cp* complexes can also be effected in refluxing CH_2Cl_2 containing aqueous HBr (48%) or HI (57%), affording the corresponding dibromo and diiodo complexes, respectively (equation 266)[314].

Similarly, metathetical exchange of the labile acetonitrile ligands in the octahedral molybdenum complex $(CO)_2(MeCN)_2BrMo(\eta^3\text{-}C_3Ph_3)$ with a variety of bidentate amines led to the corresponding amino complexes (equation 267)[311]. These form crystalline acetonitrile solvates suitable for X-ray determination, which confirm the octahedral structure of these complexes and the occurrence of the cyclopropenyl and bromide groups in a *trans* relationship.

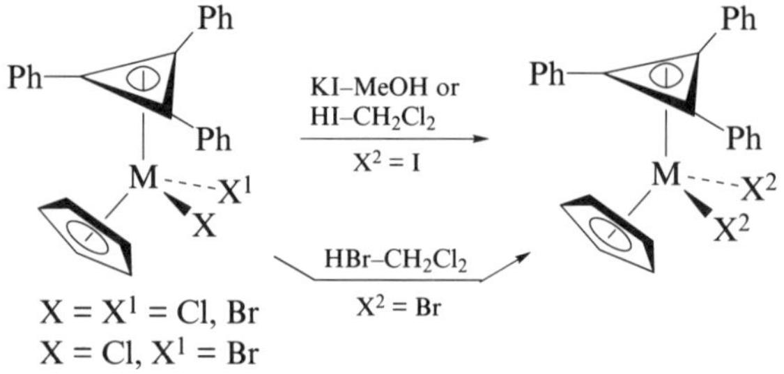

(266)

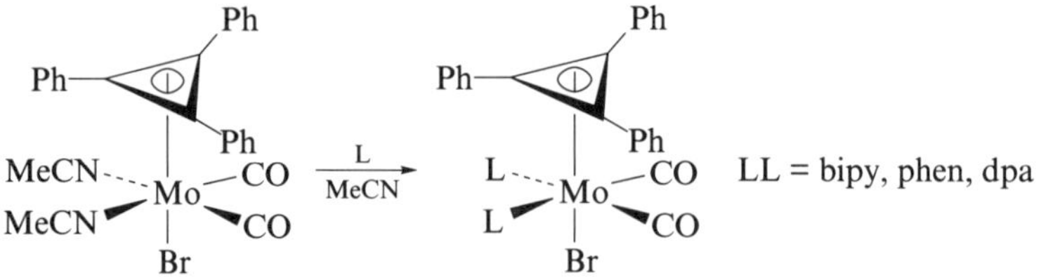

(267)

2. Ring enlargement reactions

Ring expansion of an η^3-cyclopropenyl ligand to an oxocyclobutenyl ligand can be induced by exogenous phosphorus ligands. This was demonstrated by the thermal reaction of $Cp(CO)_2Mo(C_3Ph_3)$ with tertiary phosphines and phosphites to yield oxocyclobutenyl complexes $Cp(CO)LMo(C_3Ph_3CO)$ [L = PMe_2Ph, $P(OPh)_3$] (equation 268)[312].

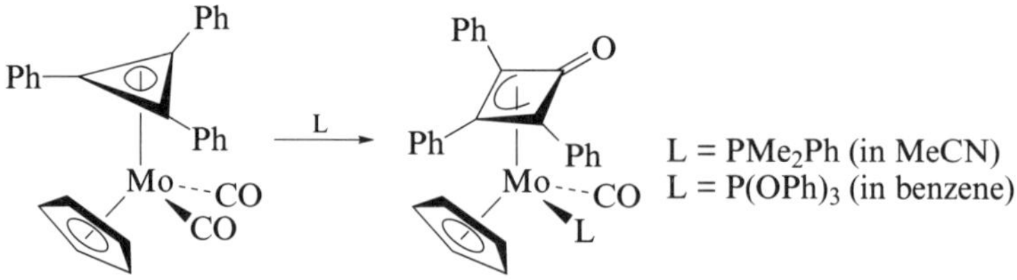

(268)

It should be noted, however, that many carbonyl metallatetrahedranes resist ring expansion. In fact, the analogous *t*-Bu complex $Cp(CO)_2Mo[C_3(t\text{-Bu})_3]$ did not react with PMe_2Ph under thermal (benzene, 80 °C) or photochemical (450 W Hanovia lamp, pyrex filter) conditions to give either ring expansion or ligand substitution chemistry. A similar lack of reactivity toward ring expansion has also been noted for group 9 triad of metallatetrahedranes (Co[275], Rh and Ir[309]), although products of CO substitution were obtained (see Section V.B.1.a). The isolobal nitroso ferratetrahedranes $(CO)_2(NO)Fe(C_3R_3)$ exhibit intermediate behavior which depends on both ring substitution and the phosphorus ligand reactant (see Section V.B.1.a)[275,322]. Kinetic studies of the ring expansion reactions of the iron complexes at different phosphine concentrations suggest an associative mechanism involving the linear-to-bent NO transformation rather than the $\eta^3 \rightarrow \eta^1$ cyclopropenyl ring slippage pathway (see Section V.B.1.a)[323].

3. Addition reactions

While oxidative addition and substitution reactions are widely used for the synthesis of metallatetrahedranes (Sections V.B.1 and V.B.2), addition reactions, in which the coordination number at the metal center increases without change of the oxidation state, are rare. One recent example is the trigonally coordinated monocarbonyl tetrafluoroethylene Ir-complex $(CO)(C_2F_4)Ir[\eta^3\text{-}C_3(t\text{-Bu})_3]$ which reacts with PMe_3 to give the stable 4-coordinate adduct $(CO)(PMe_3)(C_2F_4)Ir[\eta^3\text{-}C_3(t\text{-Bu})_3]$ in 97% yield (equation 269)[329]. Interestingly, the barrier to propeller rotation of the C_2F_4 ligand is low on the NMR time scale (14.3 kcal mol^{-1}) in contrast to the significantly higher barrier in the analogous allylic complex where the rotation is frozen on the NMR time scale. This supports the long suggested isolobal analogy between the η^3-cyclopropenyl and NO ligands, rather than with the allylic ligand.

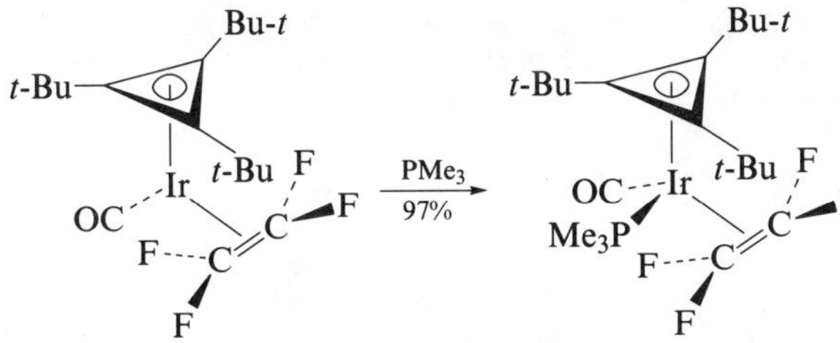

(269)

Finally, an interesting *reversible* elimination–addition reaction has been noted. When a sample of $Cp(PMe_3)Cl_2W[\eta^3\text{-}C_3Me(t\text{-Bu})_2]$ is heated to 70 °C, a broad NMR spectrum is observed that is characteristic of a rapidly equilibrating mixture of $Cp(PMe_3)Cl_2W[\eta^3\text{-}C_3Me(t\text{-Bu})_2]$, $CpCl_2W[\eta^3\text{-}C_3Me(t\text{-Bu})_2]$ and free PMe_3. However, $CpCl_2W[\eta^3\text{-}C_3Me(t\text{-Bu})_2]$ does not react with PEt_3, presumably for steric reasons[316].

C. Structural Aspects

1. Solid state structure

X-ray data on a large variety of metallatetrahedrans which differ in coordination number, local symmetry, transition metal atoms, ancillary ligands and C_3 ring substituents (Table 1) allow some generalizations to be made. Inspection of Table 1 reveals that, except for $L_2M(\eta^3\text{-}C_3R_3)$ and $L_4M(\eta^3\text{-}C_3R_3)$ complexes, the cyclopropenyl group is bound to the metal fragment in a symmetrical fashion, forming an elongated tetrahedron of atoms. The average M–C distances (MC_{AV}) in all complexes fall within the range of 199–223 pm, close to that of the Pt–C distance (212 pm) observed in η^2-cyclopropene complex[281] and Ni–C (194 pm) in cyclopropabenzene complexes[82], but significantly shorter than the W–C distance of 248 pm in octahedral complex[287]. Notably, the metal to C_3 ring-center distance (M–T) is usually shorter than that to the C_5 ring by *ca* 5 pm, except for the Mo complexes $Cp(CO)_2Mo(C_3R_3)$[312] where this is reversed.

Unlike most metallatetrahedrans, group 10 $L_2M(\eta^3\text{-}C_3R_3)$ cationic complexes and the group 5 vanadium complex $L_4M(\eta^3\text{-}C_3R_3)$ form distorted tetrahedrons in the solid state. In the Ni complex one M–C distance is significantly *shorter* than the other two, and the L—M—L plane bisects the cyclopropenyl ring. In contrast, the Pd, Pt and V complexes exhibit one M—C bond distance that is *longer* than the other two. In group 10 complexes the L—M—L plane approximately parallels a cyclopropenyl C—C bond[306,307]. Yet in solution these complexes exhibit fluxional behavior with fast rotation about the $M—C_3$ axis on the NMR time scale. These observations suggest that bonding in these complexes is essentially olefinic η^2 coordination. The different conformational structures in the solid state outline points on the reaction path between these geometries, the Pd and Pt complexes lying close to an η^2-geometry, whereas the Ni complex has the intermediate cross-over struc-

TABLE 1. X-ray data for η^3-cyclopropenyl complexes

Complex	M—C(pm)	MC_{AV} (M–T) (pm)	CC_{AV} (pm)	Reference
$[C_3Ph_3]ClO_4$			137	*a*
$Pt(\eta^2\text{-}C_3H_2Me_2)(PPh_3)_2$		212	150 (C=C) 154 (C—C)	281
$L_2M(\eta^3\text{-}C_3R_3)$				
$[(PPh_3)_2Ni(C_3Ph_3)]PF_6$	190, 201, 206	199	140	306
$[(PPh_3)_2Pd(C_3Ph_3)]PF_6.C_6H_6$	210, 211, 239	220	141	
$[(PPh_3)_2Pd(C_3Ph_3)]ClO_4$	210, 214, 233	219	141	
$[(PPh_3)_2Pt(C_3Ph_3)]PF_6.C_6H_6$	209, 209, 248	222	144	307
$L_3M(\eta^3\text{-}C_3R_3)$				
$(O_2CR)_3W[C_3(t\text{-Bu})Et_2]$	209, 211, 213	211	144	320
$(CO)_3Co(C_3Ph_3)$	200, 200, 202	201	142	274
$(CO)_3Ir[C_3(t\text{-Bu})_3]$	A, 218, 219 B, 213, 215	219 214, (202)	142	309
$[Pyr_2ClNi(C_3Ph_3)]Pyr$	190, 196, 197	194	142	328
$CpNi(C_3Ph_3)$	195, 196, 197	196	143	332
$L_4M(\eta^3\text{-}C_3R_3)$				
$(CO)_4V[C_3(t\text{-Bu})_3]$	203, 206, 242	217	150	315
$L_5M(\eta^3\text{-}C_3R_3)$				
$[(CO)_2(bipyr)_2BrMo(C_3Ph_3)]MeCN$	219, 220, 226	222, (206)	146	311
$Cp(CO)_2Mo(C_3Ph_3)$,	217, 222, 227	222, (205)	144	312
$Cp(CO)_2Mo[C_3Ph_2(t\text{-Bu})]$	219, 221, 228	222, (207)	144	
$Cl_3(TMEDA)W[C_3Me_2(t\text{-Bu})]$	204, 212, 213	210, (192)	145	318
$CpBr_2Ru(C_3Ph_3)$	218, 218, 223	220, (199)	143	313
$L_6M(\eta^3\text{-}C_3R_3)$				
$Cp(PMe_3)Cl_2W[C_3Me(t\text{-Bu})_2]$	214, 215, 220	216, (199)	145	316

[a]M. Sundaralingam and L.H. Jensen, *J. Am. Chem. Soc.*, **88**, 198 (1966).

ture[306]. A potential surface for this ring-whizzing motion was determined by extended Hückel molecular orbital calculations and was found to fit the experimental data[306,308,334,335].

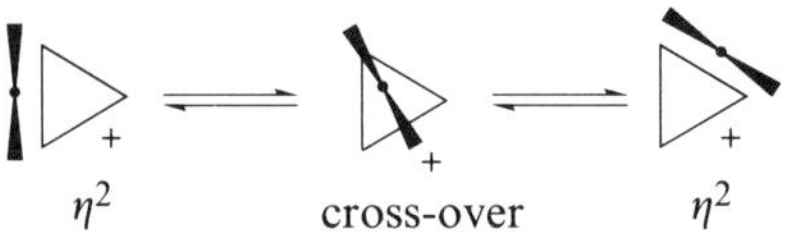

A particularly striking structural feature of the vanadium complex is the exceptional long distance (R) of 171 pm between the ring carbons closest to the metal compared to that in the Pt and Pd complexes (158 pm)[315]. This suggests that the vanadium complex approaches a metallacyclobutadiene structure, found in the analogous trigonal-bipyramidal rhenium complexes[315] (Section V.A.2).

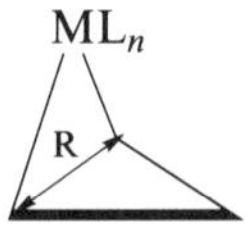

TABLE 2. Bent-back (tilt) angles in metallatetrahedranes

Complex	Bent-back (tilt) angles (deg)	Reference
$(CO)_3Ir[C_3(t\text{-}Bu)_3]$	19.3, 19.3, 15.0	309
$(CO)_3Co(C_3Ph_3)$	20.3, 22.2, 15.9	274
$Cp(PMe_3)Cl_2W[C_3Me(t\text{-}Bu)_2]$	36.7 (*t*-Bu), 29.5 (*t*-Bu), 25.3 (Me)	316

The average C—C distances (CC_{AV}) in the C_3 ring are all in the narrow range of 140–146 pm. They are longer than the C—C distance in the uncoordinated aromatic cyclopropenylium perchlorate (137 pm), but rather short compared with the coordinated C=C distance in η^2 cyclopropene complexes (150 pm). The ring substituents tilt back out of the plane of the three-membered ring, away from the metal. Typical back-bent (tilt) angles are given in Table 2.

2. Electronic structure

The nature of the bonding in (η^3-C_3R_3)ML_n (n = 2–5) metallatetrahedranes has been examined theoretically by qualitative bonding analysis based on extended Hückel calculations[305,306,308,334]. A typical example of the three dominant bonding molecular orbitals obtained from the orbital interaction diagram of the cyclopropenylium ring and group 10 (Ni, Pd, Pt) transition metal ML_2 unit is shown below.

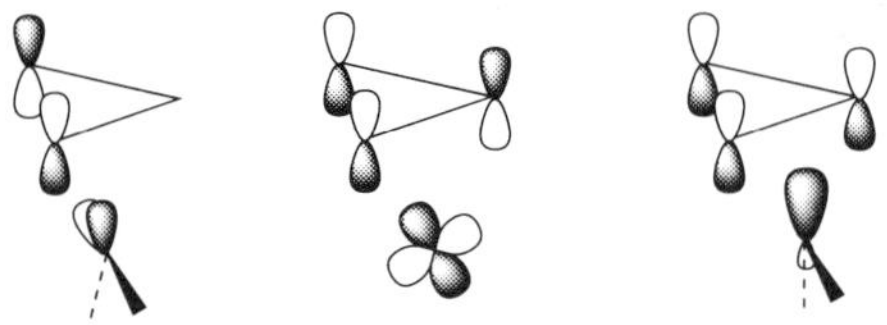

Electron distribution and bonding in group 9 $(CO)_3M[\eta^3\text{-}C_3(t\text{-}Bu)_3]$ (M = Co, Ir) and the isoelectronic $(CO)_2(NO)Fe[\eta^3\text{-}C_3(t\text{-}Bu)_3]$ complexes were determined from their He I and He II photoelectron spectra and the molecular orbital diagram for $(CO)_3Co(\eta^3\text{-}C_3H_3)$, generated by Fenske–Hall calculations (Figure 1)[310]. The ionization cross-sections indicate that the η^3-bound C_3R_3 ligand in these complexes is best described formally as a cation with a large amount of mixing and back bonding from the metal d_π orbitals to the $e_{\pi*}$ orbitals of the cyclopropenyl ring. A similar picture is obtained from the extended Hückel analysis of the bonding interactions between $CpBr_2Ru$ and $Cp(CO)_2Mo$ fragments (ML_5) and the C_3H_3 ligand[314].

More recently, a detailed study of the relative stabilities of isomeric metallatetrahedrane and metallacyclobutadiene complexes (possible intermediates in alkyne metathesis) has been carried out by *ab initio* molecular orbital theory[337]. Stability governing factors such as the ligand environment, number of valence electrons and type of transition metal were examined for a large number of $L_nM(C_3R_3)$ complexes (n = 2–6, M = transition metals of groups 7–10). Metallatetrahedrane complexes are found in general to be favored particularly among late transition metals of pseudo-octahedral $L_5M(C_3R_3)$ and tetrahedral $L_3M(C_3R_3)$ compounds.

3. Dynamic behavior

The activation barrier to rotation of the cyclopropenyl ligand about the M—C_3 axis has been shown to vary dramatically with the nature of the metal–ligand fragment[309,314,329].

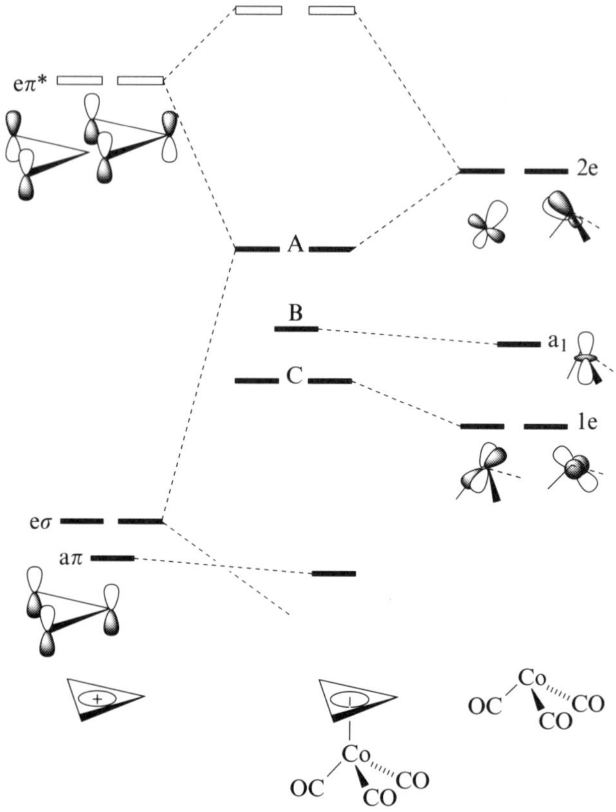

FIGURE 1. Molecular orbital diagram for $(CO)_3Co(\eta^3\text{-}C_3H_3)$ based on Fenske–Hall calculations[310]

Early attempts were made to correlate the presence or absence of this behavior with structural features pertaining to the M—C_3 interaction. For example, the conformational rigidity, on the NMR time scale, of $Cp(PMe_3)Cl_2W[C_3Me(t\text{-}Bu)_2]$ at ambient temperature, compared with the fast rotation of the cyclopropenyl ring at −80 °C in $Cp(CO)_2Mo(C_3Ph_3)$, was attributed to steric effects due to different coordination numbers at the metal, and apparently more tighter binding of the C_3 ring in the former complex[316]. However, it was later shown by ^{13}C NMR analysis that the cyclopropenyl ligands in $CpX_2Ru(C_3Ph_3)$ (X = Cl, Br, I) complexes, isoelectronic with $Cp(CO)_2Mo(C_3Ph_3)$, were also conformationally rigid[314]. Hence, there seems to be a lack of correlation between the barrier to cyclopropenyl rotation and structural or steric factors.

A more rational picture is based on the differences in the electronic structure between the isoelectronic ruthenium and molybdenum complexes. Indeed, analysis at the extended Hückel level of the bonding interactions between the C_3H_3 ligand and both $CpBr_2Ru$ and $Cp(CO)_2Mo$ fragments predicts a significantly higher barrier to cyclopropenyl rotation in the Ru system. This is due to the increase in the energies (destabilization) of all three highest occupied MOs of the Ru complex upon rotation, leading to a high rotational barrier.

TABLE 3. $\Delta G^{\#}$ values for rotation about the Ru—C_3 axis in $CpX_2Ru(\eta^3$-$C_3Me_3)$ complexes[314]

CpX_2 or Cp^*X_2	$\Delta G^{*\#}$ (298 K) (kcal mol^{-1})
$CpCl_2$	14.2
$CpBr_2$	14.9
CpI_2	15.4
Cp^*Cl_2	13.9
Cp^*Br_2	14.5
Cp^*I_2	14.7

Similar analysis for the Mo system showed an overall stabilization of the corresponding MOs, resulting in a relatively low rotational barrier for these complexes[314].

Eventually, substituting the bulky phenyl groups by the sterically less demanding methyl groups resulted in a series of ruthenium complexes $CpX_2Ru(C_3Me_3)$ and $Cp^*X_2Ru(C_3Me_3)$ (X= Cl, Br, I, Cp* = η^5-pentamethylcyclopentadienyl) which exhibit variable-temperature NMR behavior[314]. The rotational free energies of activation ($\Delta G^{\#}$) were obtained by line shape analysis of the ring methyl signals (Table 3). The energy barriers all fall within a narrow range of $\Delta G^{\#}$ = 14–15 kcal mol^{-1}, indicating lack of sensitivity to either the size of the halogen atom or the steric bulk of the C_5 ligand.

Group 9 tetrahedral complexes of the type $(CO)_2LM[\eta^3$-$C_3(t$-$Bu)_3]$ (M = Co, Rh, Ir; L = phosphorus ligand, CNBu-*t* and the iron nitroso analogue $(CO)(NO)(PMe_3)Fe[\eta^3$-$C_3(t$-$Bu)_3]$ also exhibit temperature-dependent NMR behavior[309]. Unlike the narrow energy range for cyclopropenyl rotation observed above in the ruthenium halo complexes, the activation barriers of this group span over a range of 11–18 kcal mol^{-1} (Table 4). Comparison of $\Delta G^{\neq}$ values within the Co triad complexes shows considerable increase in the rotational barrier on descending the group, in the order Co<Rh<Ir. Less clear are the steric and electronic effects of the ancillary ligands on the rotational barrier. The lower $\Delta G^{\#}$ observed for the more bulky PPh_3 iridium complex (15.8 kcal mol^{-1}) compared with that of the analogous PMe_3 complex (18.4 kcal mol^{-1}) suggests that activation energies are insensitive to steric hindrance such as ligand cone angles, but rather to electronic effects, e.g. ligand basicity of the phosphine groups[309].

TABLE 4. $\Delta G^{\neq}$ values for rotation about the M—C_3 axis in $L_3M(\eta^3$-$C_3(t$-$Bu)_3]$[309]

L_3M	$\Delta G^{\neq}$ (298 K) (kcal mol^{-1})
$(CO)_2(PMe_3)Co$	13.5
$(CO)_2(PMe_3)Rh$	16.9
$(CO)_2(PMe_3)Ir$	18.4
$(CO)_2(PPh_3)Ir$	15.8
$(CO)_2\{P(OMe)_3\}$	16.9
$(CO)_2(CNBu$-$t)Ir$	18.3
$(CO)(NO)(PMe_3)Fe$	11.2

VI. CYCLOPROPENYLIDENE COMPLEXES

The cyclopropenylidene ligand is predicted by theory to have a nucleophilic singlet ground state resulting from a large dipole moment (μ *ca* 3.4 D) and a large splitting of the σ and p orbitals on the divalent carbon, which amounts to 3.17 eV[338,339]. This wide energy gap is due to a combination of σ-inductive and π-resonative effects. The small bond angle at the constrained carbene center of cyclopropenylidene causes the σ-type orbital to be lowered in energy with an increased amount of s character, becoming the highest occupied molecular orbital (HOMO). On the other hand, the p-type orbital on the divalent carbon is raised in energy when mixing with the orbital of the ethylene moiety, becoming the lowest unoccupied molecular orbital (LUMO). Thus, stabilization of the singlet configuration is further achieved by the aromatic $4n+2$ ($n = 0$) π-system, and is expected to be further stabilized by π-conjugative electron-donating substituents, such as amino groups.

Cyclopropenylidene HOMO LUMO

Interaction of metals with cyclopropenylidene to form stable complexes has been widely studied[340] in the last two decades since the first reported synthesis of pentacarbonyl(2,3-diphenylcyclopropenylidene)chromium (*see below*)[341]. Two groups of cyclopropenylidene metal derivatives may be distinguished: neutral cyclopropenylidene complexes represented by two resonance forms, and the cationic cyclopropenylium σ-complexes (Scheme 10). The former are all transition metal complexes of groups 6 (Cr, Mo, W), 7 (Mn), 8 (Fe) and 10 (Pd, Pt), whereas the latter cationic σ-complexes are derived from both main group metals (Li, Mg) and group 10 (Pd, Pt) transition metals.

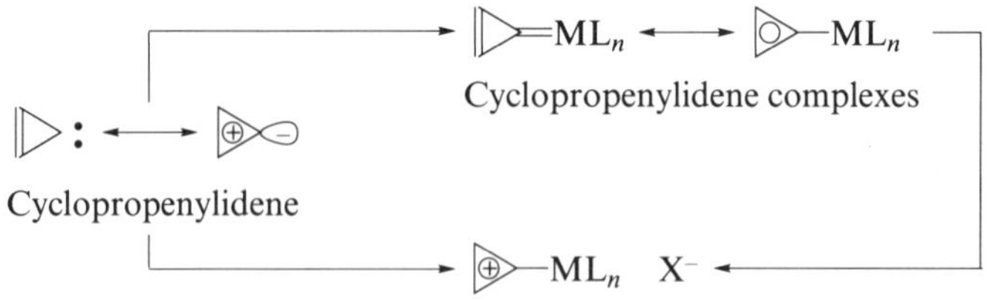

SCHEME 10

A. Cyclopropenylidene Complexes

1. Preparation

a. From 3,3-dichlorocyclopropenes. The reaction of 3,3-dichloro-1,2-diphenylcyclopropene with $Na_2Cr(CO)_5$ in THF gave the first stable cyclopropenylidene metal complex $(CO)_5Cr(C_3Ph_2)$ in 19.5% yield (equation 270)[341]. This is also the first metal complex in which the ligand carbene is not stabilized by hetero atoms[342]. X-ray structure analysis reveals an octahedrally coordinated chromium complex, with a Cr—$C_{carbene}$ distance of 205 pm, considerably shorter than the 221 pm bond length expected for a Cr—C_{sp^2} single

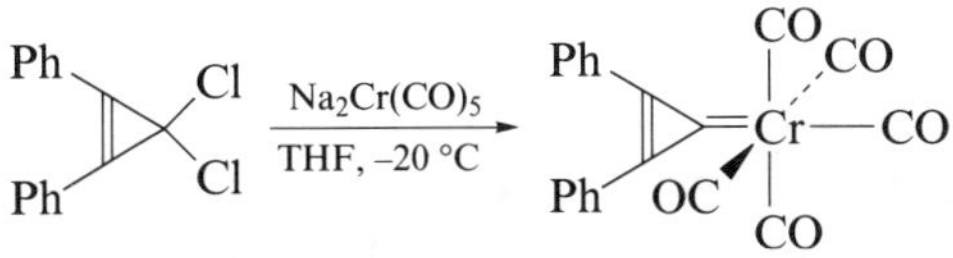

(270)

bond[343]. The mean C—C distance of 138 pm in the three-membered ring is not significantly different from that found for the triphenylcyclopropenylium cation (137 pm).

More recently, the analogous cyclopropenylidene molybdenum complexes $(MeCp)Cl_2Mn(R^1{}_2C_3Cl_2)$ were synthesized by reaction of dichlorocyclopropenes $R^1{}_2C_3Cl_2$(R^1 = *n*-Pr, *t*-Bu, Ph) with the anionic complex Na[(MeCp)$(CO)_2$Mn(ER_3)] (ER_3 = $SiMePh_2$, $GePh_3$, $SnPh_3$), prepared from the corresponding hydride complexes by reaction with NaH in THF (equation 271)[344,345]. X-ray structure analysis of the diphenyl complex (R^1 = Ph)[344] shows, as expected, that the three-membered ring plane coincides with the mirror plane of the metal fragment. Interestingly, the carbenoid Mn—C bond distance of 190 pm is significantly shorter than the corresponding Cr—C bond length (205 pm) in the corresponding pentacarbonyl chromium complex[346], but closer to the bond distance (196 pm) found in the analogous square-planar Pd complex (*see above*)[347]. The mean C—C bond length within the three-membered ring is 140 pm, typical of cyclopropenylidene complexes of Cr and Pd discussed above.

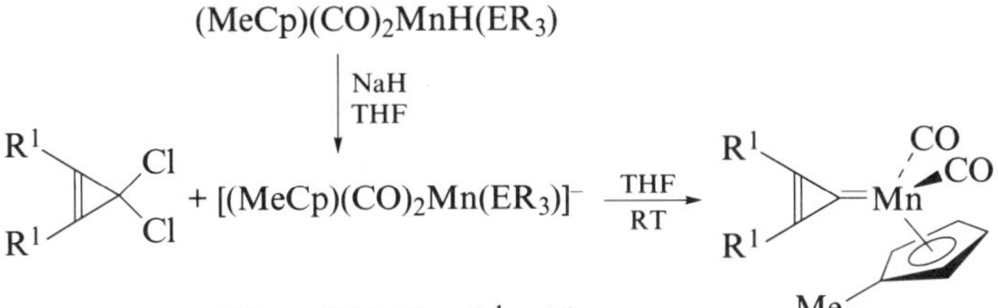

(271)

ER_3 = $SiMePh_2$; R^1 = Ph
ER_3 = $GePh_3$; R^1 = Pr
ER_3 = $SnPh_3$; R^1 = *t*-Bu

Similarly, the reaction of palladium black with 3,3-dichloro-1,2-diphenylcyclopropene in refluxing benzene afforded the dimeric μ-dichloropalladium cyclopropenylidene complex [Pd(C_3Ph_2)Cl_2], which upon further treatment with a variety of ligands gave the planar mononuclear complexes, assigned (from IR spectra) as *trans*-$Cl_2LPd(C_3Ph_2)$ [L = MeCN, THF, pyr, DMSO, $(C_6H_{11})_3P$] (equation 272)[348,349]. The same procedure is utlized to prepare the di-isopropyl- and di-*t*-butyl-cyclopropenylidene palladium dimer complexes, where π-conjugative interactions are small (equation 273)[350]. The dimers react with 2 equivalents of Bu_3P to give the *cis*-$Cl_2(Bu_3P)Pd(C_3R_2)$ (R = *i*-Pr, *t*-Bu) cyclopropenyl carbene complexes. The square-planar *cis* structure of these complexes is deduced from the X-ray structure analysis of the corresponding bis-dimethylamino complex *cis*-

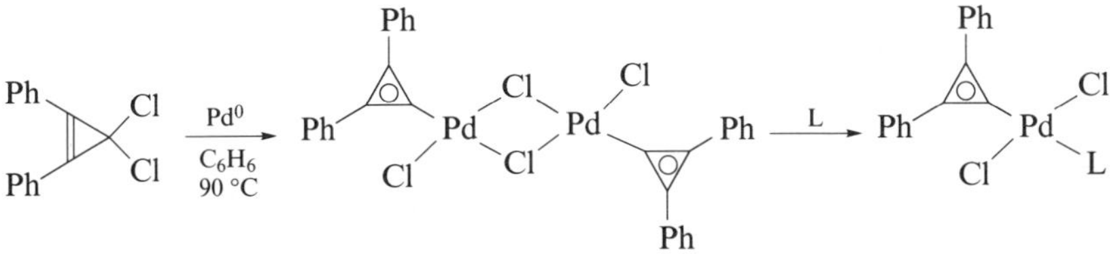

L = MeCN, THF, pyr, DMSO, $(C_6H_{11})_3P$ (272)

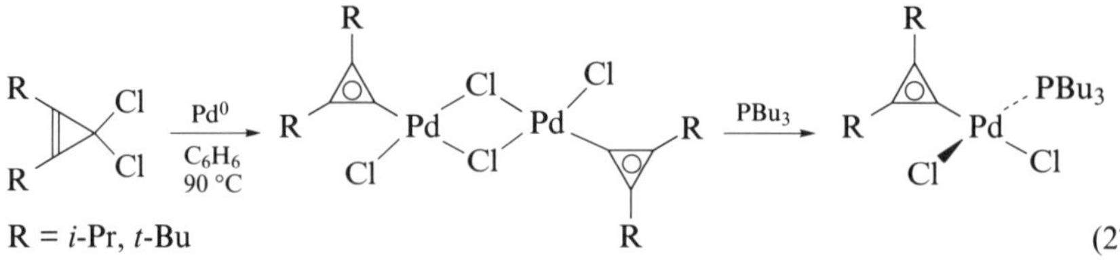

(273)

$Cl_2(Bu_3P)Pd(C_3\{Me_2N\}_2)$[347]. This plane is roughly a molecular mirror plane, bisecting the three-membered ring. Characteristic bond lengths are 196 pm for the carbenoid Pt—C bond, and the mean ring C—C bond length of 138 pm, the same as observed in the analogous chromium complex (*vide supra*). The ^{13}C NMR spectra of cyclopropenylidene palladium complexes is characterized by low-field resonances of both carbene and olefinic ring carbons, ranging between 125 to 198 ppm (see Section VI B. 3, Table 5).

The analogous reaction of dichlorocyclopropenes with platinum instead of palladium takes a different course. When 3,3-dichloro-1,2-di-*t*-butylcyclopropene is treated with Pt^0 (platinum black) in refluxing benzene, the stable octahedral bis(di-*t*-Bu-cyclopropenylidene)tetrachloroplatinum complex is obtained in 19% yield (equation 274). This Pt^{IV} complex is presumably obtained by two consecutive oxidative additions of dichlorocyclopropene to platinum (equation 274)[351]. Further reduction of the tetrachloroplatinum complex with $SnCl_2$ in CH_2Cl_2 afforded the Pt^{II}-complex bis(di-*t*-Bu-cyclopropenylidene) $PtCl_2$ ^{13}C and ^{195}Pt NMR studies indicate that the bonding interaction between the metal and the carbene ligand is mainly σ, with only negligible contribution of π-interaction, and the stereochemistry of these planar complexes is *cis*.

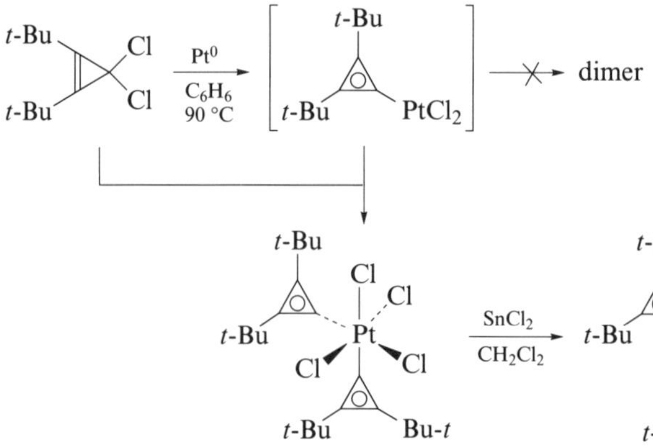

(274)

Nevertheless, synthesis of the monocyclopropenylidene platinum complex *cis*-$Cl_2(Bu_3P)Pt[C_3(t\text{-}Bu)_2]$ was achieved when 3,3-dichloro-1,2-di-*t*-butylcyclopropene was treated in refluxing benzene with the dimeric complex μ-$[(Bu_3P)Cl_2Pt]_2$ in a 2:1 ratio, instead of Pt^0 (equation 275)[351]. The reaction may involve a dimer intermediate analogous to that obtained in the palladium series above (equation 273).

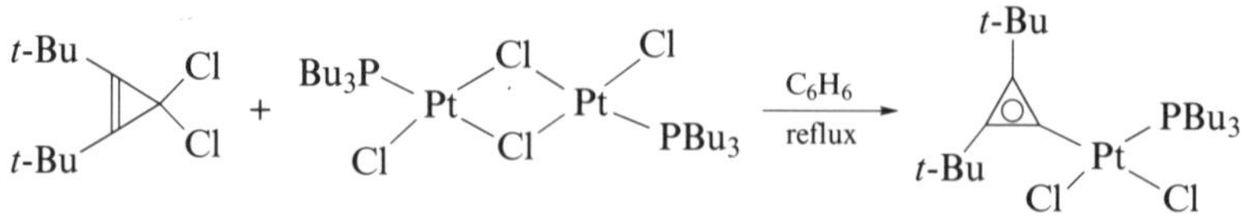

(275)

b. From halocyclopropenylium salts. When the cyclopropenylidene ligand is substituted by two dialkylamino groups, stable cyclopropenylium salts of the type $[(R_2N)_2ClC_3]X$ (R = Me, Et, *i*-Pr, X = Cl, I, BF_4, ClO_4) are readily prepared. These salts undergo oxidative addition reactions mainly with transition metal compounds of group 10, to give neutral cyclopropenylidene complexes[340]. The preparative procedures are quite similar to those employed in the previous section.

Thus, for group 10 palladium and platinum metals, a mixture of 1,2-bis(dialkylamino)-3-halocyclopropenylium halide and a slight excess of Pd or Pt black is refluxed in MeCN for 24h, affording the corresponding dimeric cyclopropenylidene μ-complexes $\{[(R_2N)_2C_3\}MX_2\}_2$ (R = Me, Et, *i*-Pr, M = Pd, Pt; X = Cl, I) (equation 276)[340,347,352]. The dimers further react with Bu_3P to give the planar mononuclear complex *cis*-$[(R_2N)_2C_3]MX_2(PBu_3)$.

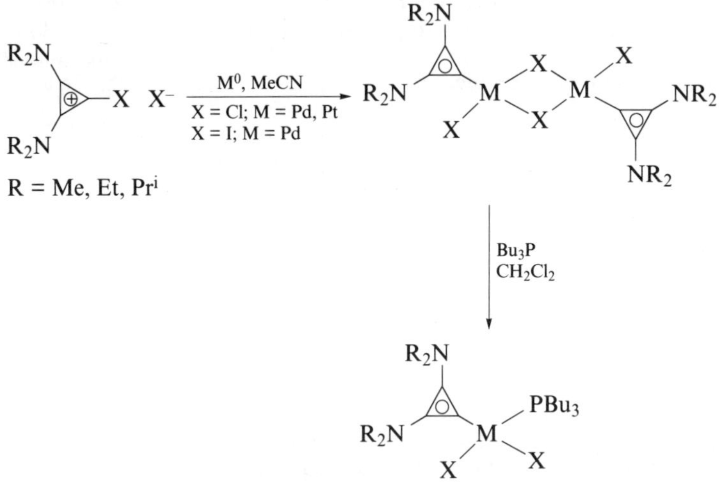

(276)

When the bis(isopropylamino)iodocyclopropenylium iodide is reacted with platinum black in acetonitrile, the reaction takes a different course, affording mainly the *trans*-bis[bis(diisopropylamino)cyclopropenylidene] diiodoplatinum complex (equation 277)[351]. A plausible pathway for this reaction involves two consecutive oxidative additions to platinum leading to the hexacoordinated intermediate Pt^{IV}-complex $[i\text{-}Pr_2N)_2C_3]PtI_4$, followed by reductive elimination of I_2 to form the product (cf Section VI. A. 1. a).

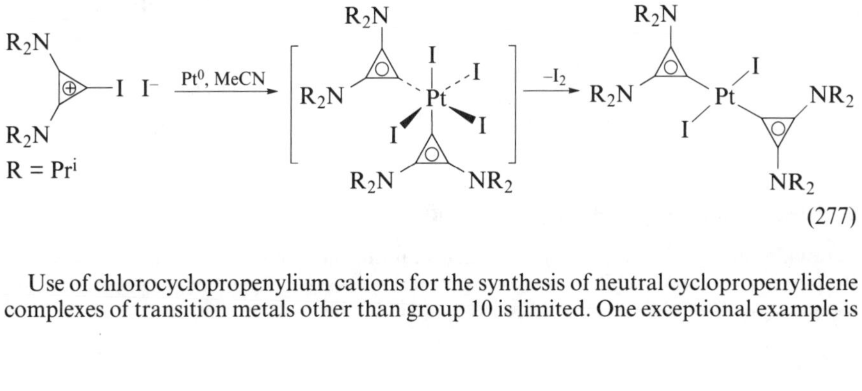

(277)

Use of chlorocyclopropenylium cations for the synthesis of neutral cyclopropenylidene complexes of transition metals other than group 10 is limited. One exceptional example is reported for iron. Reaction of chlorobis(diethylamino)cyclopropenylium perchlorate with dilithium tetracarbonyl ferrate in THF afforded bis(diethylamino)cyclopropenylidene iron tetracarbonyl in 45% yield (equation 278)[353].

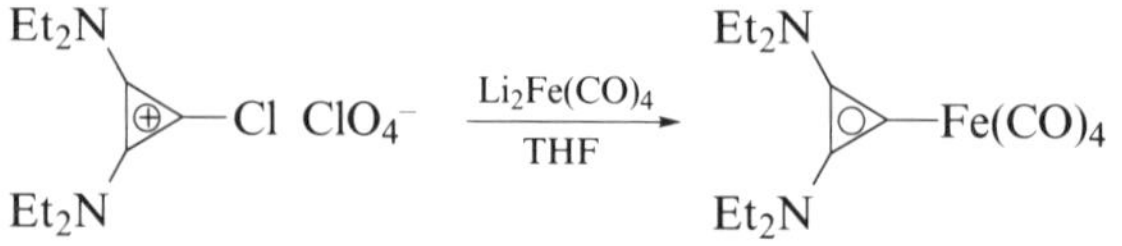

(278)

c. From alkoxycyclopropenylium salts. There are only two examples of *neutral* cyclopropenylidene complexes obtained from alkoxycyclopropenylium compounds, both of group 6 transition metals chromium and molybdenum. Reaction of 3-ethoxy-1,2-diphenylcyclopropenylium tetrafluoroborate with $Na_2M(CO)_5$. (M = Cr, Mo) in bis(2-dimethoxyethyl)ether (diglyme) gave (diphenylcyclopropylidene)$M(CO)_5$ in low (10%, M = Cr) to moderate (24%, M = Mo) yields (equation 279). The chromium complex was first prepared by an alternative route, via reaction of the pentacarbonylchromate with 3,3-diphenyl-1, 2-dichlorocyclopropene (see above, equation 270)[341].

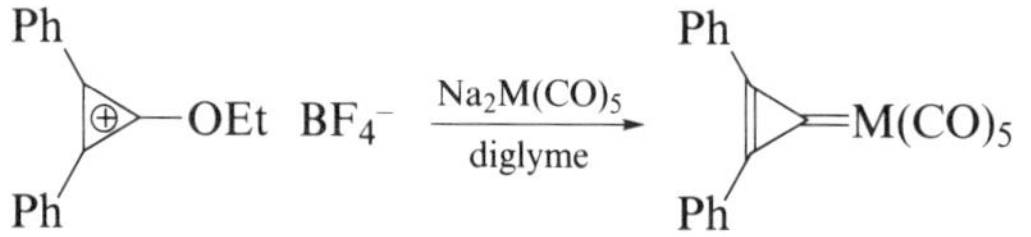

(279)

d. From lithiocyclopropenylium salts. Bis(diisopropylamino)lithiocyclopropenylium perchlorate $[i\text{-}Pr_2N)_2C_3Li]ClO_4$ undergoes double oxidative addition to $PtCl_2$ in ether solution, to form the plane-tetragonal *trans*-$[(i\text{-}Pr_2N)_2C_3]_2PtCl_2$ complex in poor (4%) yields (equation 280). Under similar conditions, reaction of the perchlorate with $PtI_2(COD)$ results in ligand exchange leading to a mixture of the *cis* and *trans* isomers of $[(i\text{-}Pr_2N)_2C_3]_2PtI_2$ (equation 281)[351]. The *cis*-isomer gradually rearranges to the *trans*-isomer. Replacement of one iodide ligand by chloride is achieved by successive treatment of the latter diiodide complex with CF_3SO_3Ag and Et_4NCl in CH_2Cl_2–MeCN.

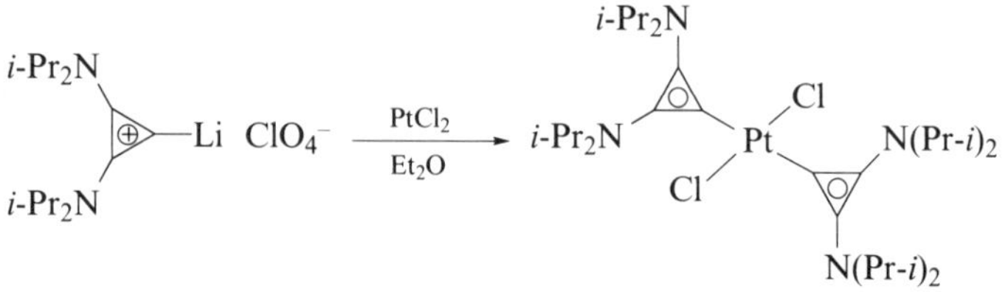

(280)

e. By desulfurization and deselenization. Group 6 octahedral cyclopropylidene complexes of type $(R_2C_3)M(M(CO)_5$ (R = Me_2N, i-Pr_2N, t-BuS, Ph; M = Cr, Mo, W) are synthesized by desulfurization and deselenation reactions of cyclopropenethiones and selones with $M(CO)_6$ (equation 282)[340].

B. Metallacyclopropenylium Compounds

A wide variety of cyclopropenylium metal cations has been prepared and studied over the past two decades, ranging from main group elements (Li, Mg) through transition metal σ-complexes (Fe, Mn, Pd, Pt). Extensive NMR studies suggest that the metal–carbene

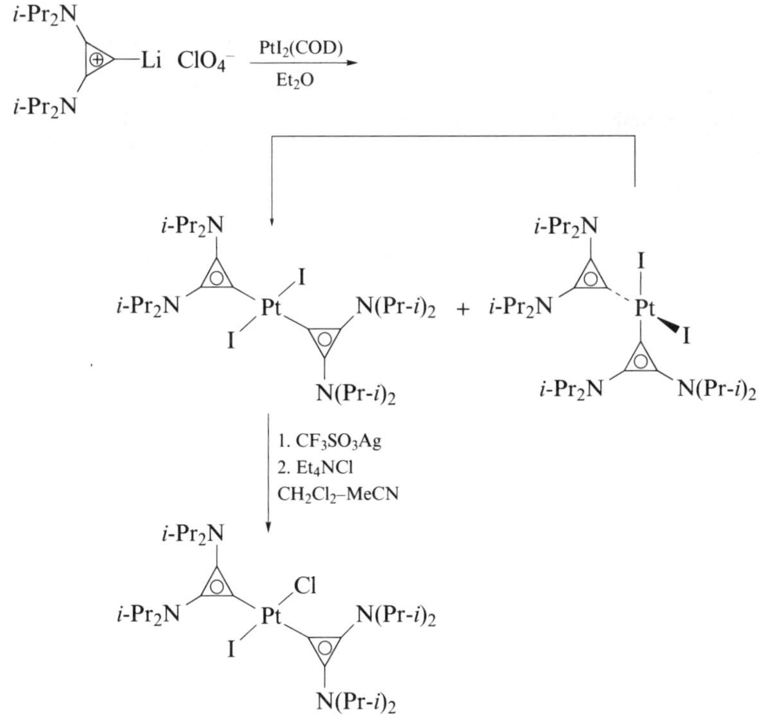

(281)

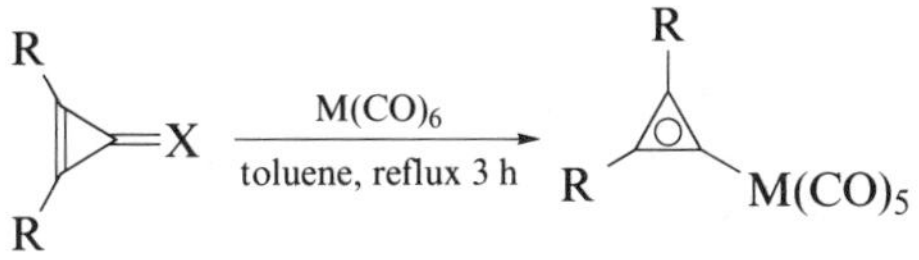

(282)

R = Me_2N, *i*-Pr_2N, *t*-BuS, Ph
X = S, Se
M = Cr, Mo, W

bond in all cyclopropenylium metal complexes is mainly σ in nature, with only minor contribution of $p\pi$–$d\pi$ backbonding interaction in the transition metal complexes[351].

cyclopropenylium σ-complexes

1. Preparation

a. Main group metal compounds. Stable cyclopropenylium metal compounds of this group are prepared either by direct metallation of bis(dialkylamino)cyclopropenylium perchlorates or metal halogen exchange reactions of bis(dialkylamino)halocyclopropenylium perchlorates, using alkyl- or aryl–metal compounds. Thus, bis(diisopropylamino)lithiocyclopropenylium perchlorate $[(i\text{-}Pr_2N)_2C_3Li]ClO_4$ was first prepared by metallation of bis(diisopropylamino)cyclopropenylium perchlorate $[(t\text{-}Pr_2N)_2C_3H]ClO_4$ with BuLi[351,356] (equation 283). Alternatively, the same lithio compound is prepared by metal halogen exchange reaction of the chlorocyclopropenylium derivative $[(i\text{-}Pr_2N)_2C_3Cl]ClO_4$ and BuLi (equation 283)[356]. The analogous magnesium derivatives $[(R^1R^2N)_2C_3(MgBr)]ClO_4$ ($R^1 = R^2 = i$-Pr; R^1 = Me, $R^2 = t$-Bu) are similarly prepared from the iodocyclopropenylium perchlorates $[(R^1R^2N)_2C_3I]ClO_4$ and PhMgBr (equation 284).

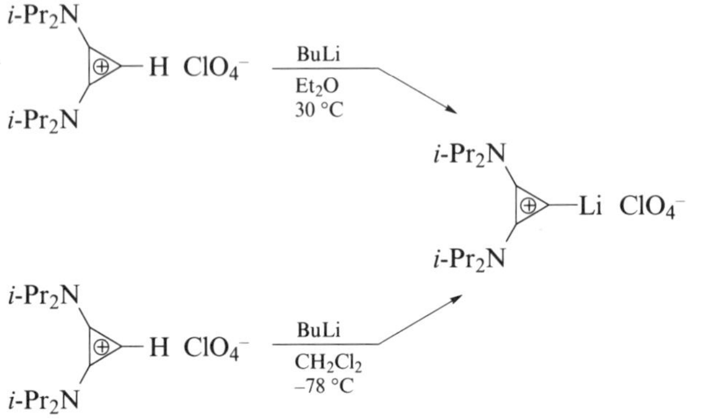

(283)

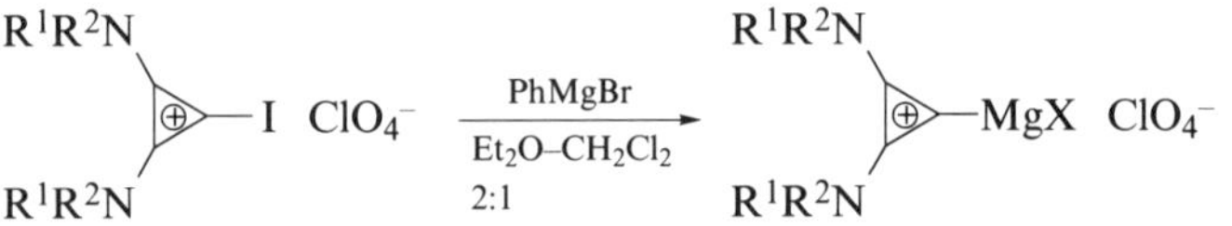

(284)

$R^1 = R^2 = i$-Pr
R^1 = Me, $R^2 = t$-Bu

b. Transition metal complexes. The main routes to the synthesis of cyclopropenylium complexes of the early transition metals of groups 8 and 9 are best exemplified by the various approaches to the cyclopentadienyldicarbonyliron (Fp) complexes, (equation 285)[266,353]. These include the three typical cyclopropenylium derivatives: the lithio-, chloro- and methoxy cyclopropenylium cations, which react with Fp derivatives. Thus, bis(diisopropylamino)lithiocyclopropenylium perchlorate and FpCl undergo metathesis in THF, forming $[(i\text{-}Pr_2N)_2C_3Fp]ClO_4$, and this and related cyclopropenylium Fp complexes are formed by oxidative addition reactions of the appropriate chloro and methoxy cyclopropenylium salts to NaFp and $FpSiMe_3$. Analogously, both chloro-and methoxy cyclopropenylium cations undergo oxidative addition reactions with sodium pentacarbonyl manganate affording $[(R_2\text{-cyclopropenylium})Mn(CO)_5]X$ salts (R = Ph; X = BF_4; R = Et_2N; X = ClO_4 (equation 286)[353].

Cyclopropenylium complexes of the late transition metals of group 9 (Rh) and 10 (Pd, Pt), like their early counterparts, can be prepared by metathesis reactions of bis(dialkylamino)lithiocyclopropenylium perchlorates $[(R_2N)_2C_3Li]ClO_4$ (R = Me, Et, *i*-Pr) and $M(Bu_3P)_2LX$ (M = Rh, L = CO, X = Cl; M = Pd, Pt; L = X = Cl, Br, I) (equation 287)[340,352]. But, unlike with early transition metals (*vide supra*), oxidative addition reactions of chlorocyclopropenylium salts with Pd and Pt give cyclopropylidene dimer complexes (Section IV. A.1.b). These dimers, however, react with 4 equivalents of phosphorus ligands, followed by $HClO_4$, to give cyclopropenylium complexes $[(R_2C_3)ML_2Cl]ClO_4$ (R = *i*-Pr, *t*-Bu, Ph, Me_2N, Et_2N, $i\text{-}Pr_2N$; M = Pt, Pd; L= Bu_3P) (equation 288)[340,350,351]. Alternatively, the dimers are first converted to cyclopropenylidene phosphine complexes $R_2C_3MLX_2$ (L = Bu_3P, pyr,

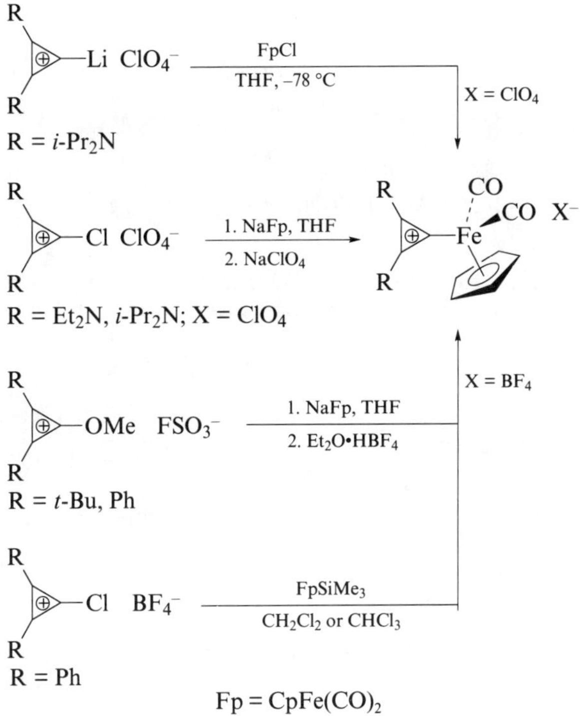
R
Li ClO4−
FpCl
THF, −78 °C
X = ClO4
R = i-Pr2N
Cl ClO4−
1. NaFp, THF
2. NaClO4
R = Et2N, i-Pr2N; X = ClO4
CO
CO X−
Fe
X = BF4
OMe FSO3−
1. NaFp, THF
2. Et2O•HBF4
R = t-Bu, Ph
Cl BF4−
FpSiMe3
CH2Cl2 or CHCl3
R = Ph
Fp = CpFe(CO)2

(285)

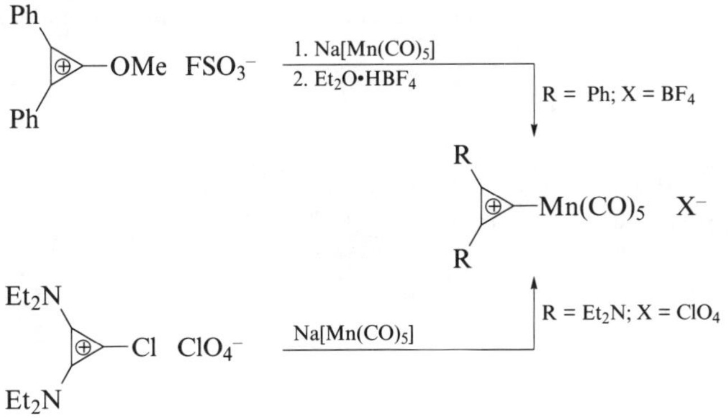
Ph
OMe FSO3−
1. Na[Mn(CO)5]
2. Et2O•HBF4
R = Ph; X = BF4
R
Mn(CO)5 X−
Et2N
Cl ClO4−
Na[Mn(CO)5]
R = Et2N; X = ClO4

(286)

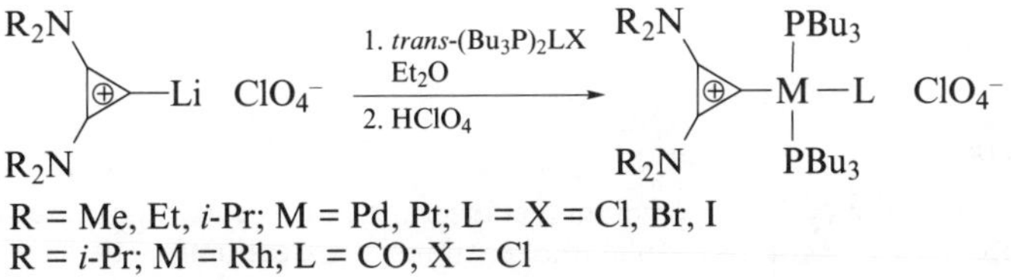
R2N
Li ClO4−
1. trans-(Bu3P)2LX
Et2O
2. HClO4
PBu3
M—L ClO4−
R = Me, Et, i-Pr; M = Pd, Pt; L = X = Cl, Br, I
R = i-Pr; M = Rh; L = CO; X = Cl

(287)

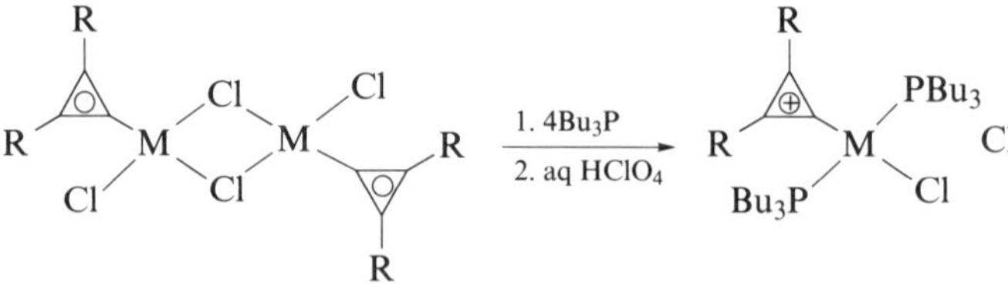

(288)

R = *i*-Pr, *t*-Bu, Ph, Me_2N, Et_2N, *i*-Pr_2N
M = Pd, Pt

MeCN, Me_2SO) using 2 equivalents of ligand (Section IV A.1.b), and subsequently by addition of phosphine and $HClO_4$ to the desired cyclopropenylium complexes (equation 289).

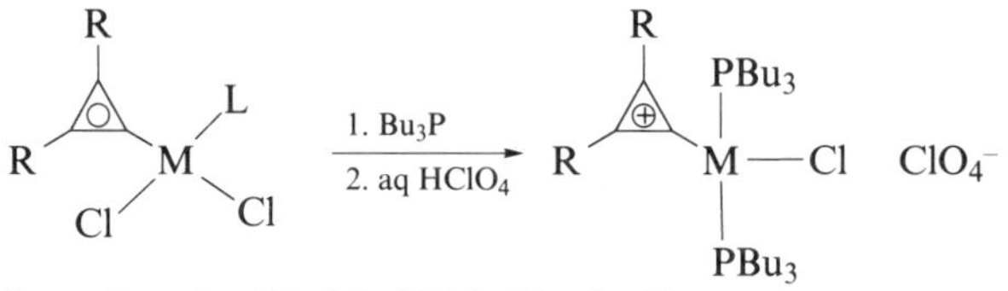

(289)

R = *i*-Pr, *t*-Bu, Ph, Me_2N, Et_2N, *i*-Pr_2N
M = Pd, Pt
L = Bu_3P, Pyr, MeCN, DMSO

Cyclopropenylidene palladium dimer complexes of type $[(R_2C_3)PdCl_2]_2$ undergo abstraction of chloride ligand by strong Lewis acids, to form cyclopropenylium dimer complexes $[(R_2C_3)PdCl]_2X_2$. The reaction is strongly dependent upon the substiuent R, and occurs only with π-resonance donor substituents. Thus, treatment of dimer complex $[((i$-$Pr_2N)_2C_3)PdCl_2]_2$ with $AgOSO_2CF_3$ or $SbCl_5$ in CH_2Cl_2 gave the corresponding trifluorosulfonate and hexachloroantimonate cyclopropenylium dimer complexes (equation 290)[358].

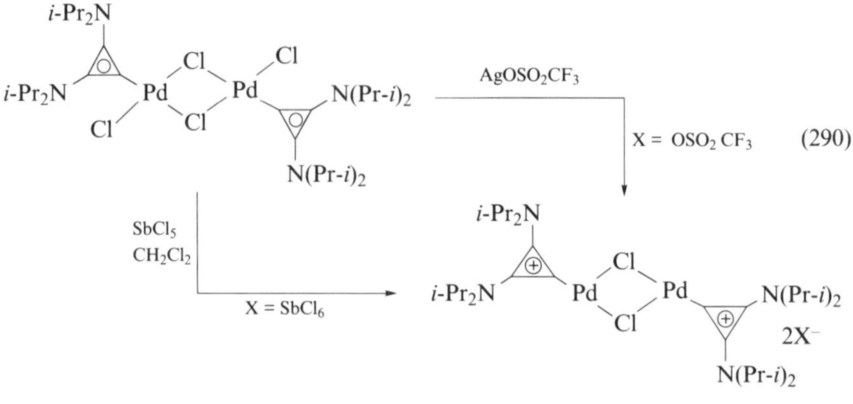

(290)

2. Reactions

All the reactions of cyclopropenylium metal compounds occur without cleavage of the three-membered ring. Moreover, in all the reactions, these cations participate as *σ-nucle-*

ophiles, leaving the stable aromatic cyclopropenylium π-system intact. Therefore, it is not surprising that the most studied metallacarbene cations are the bis(dialkylamino)lithiocyclopropenylium salts $[(R_2N)_2C_3Li]ClO_4$. These were shown by ^1H NMR and Li–Li exchange studies to have appreciable ionic character of the C—Li bond, which is larger than in PhC≡CLi[340].

Scheme 11 summarizes the reactions of bis(diisopropylamino)lithiocyclopropenylium perchlorate with a variety of electrophiles. These include acid hydrolysis[356], silylation[359], oxidative coupling with iodine to the symmetrical tetraisopropylamino triafulvalene dication[356], reaction leading to C—C bond formation, with electrophilic allenes[360] to give a zwitterionic dihydrotriafulvenes, with disubstituted cyclopropenylium perchlorate[356], and

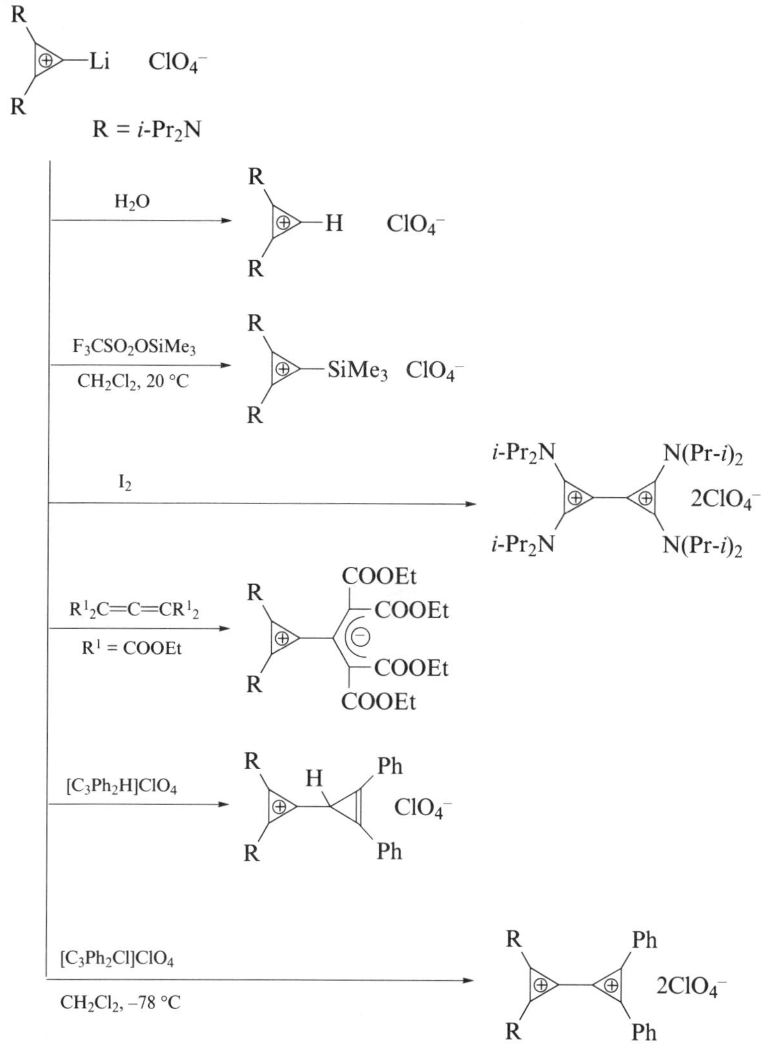

SCHEME 11

finally with diphenylchlorocyclopropenylium perchlorate[266,356,361] to give the nonsymmetrical triafulvalene dication.

Notably, the silylcyclopropenylium salt $[(i\text{-}Pr_2N)_2C_3SiMe_3]ClO_4$ reacts with fluoride ion in THF to give the highly reactive nucleophilic cyclopropenylidene intermediate $[(i\text{-}Pr_2N)_2C_3]$, which can be hydrolyzed to $[(i\text{-}Pr_2N)_2C_3H]ClO_4$ or trapped by tetrachlorocyclopropene, affording the highly symmetrical stable tetracationic salt $[\{(i\text{-}Pr_2N)_2C_3\}_3C_3]\cdot 4ClO_4$ in 40% yield (equation 291)[359]. The latter can also be obtained independently by direct coupling reaction of the lithiocyclopropenylium salt $[(i\text{-}Pr_2N)_2C_3Li]ClO_4$ and C_3Cl_4 in CH_2Cl_2.

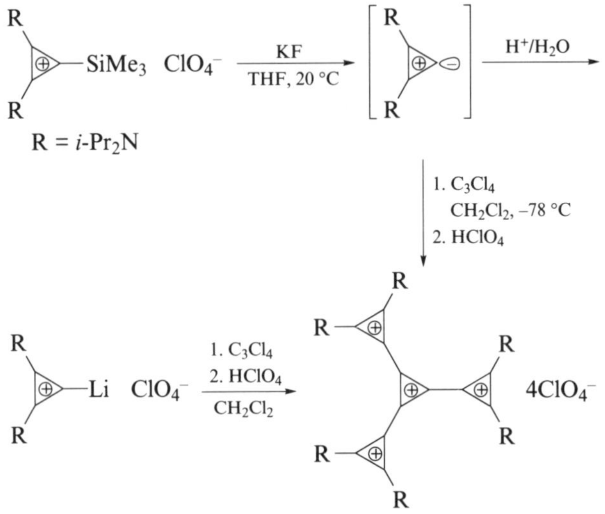

(291)

Bis(dialkylamino)cyclopropenylium magnesium halides, like their lithium counterparts, react as nucleophiles in various reactions, but their use is much less extensive. Like typical Grignard reagents they undergo hydrolysis, alkylation and reactions with carbonyl compounds at the carbenoid carbon (equation 292)[357].

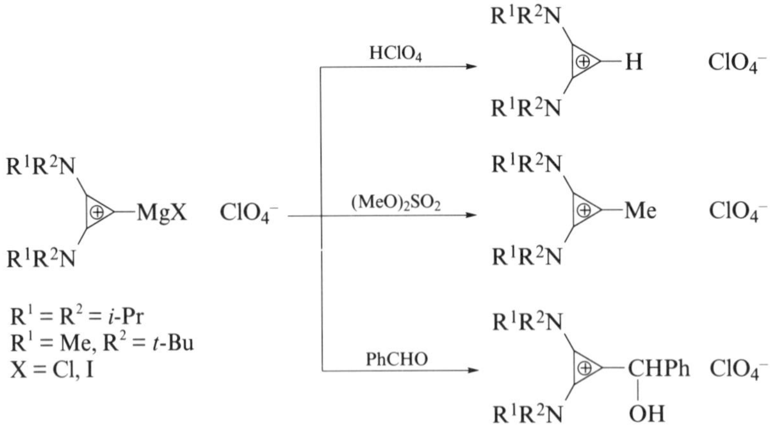

(292)

Studies of the reactivity of transition metal cyclopropenylium complexes are negligible. A single example on the acid hydrolysis of the hexachloroantimonate palladium dimer complex $[(i\text{-}Pr_2N)_2C_3(PdCl)]_2 \cdot 2SbCl_6$, and its demetallation using $SbCl_5$, is outlined in equation 293[358].

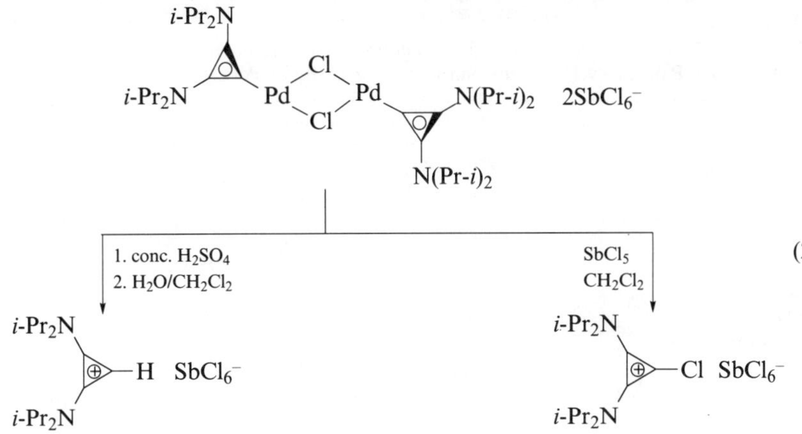

(293)

3. NMR spectra

^{13}C NMR spectra of cyclopropenylidene complexes are of much analytical use in studies of ligand and substituent electronic effects on the chemical shifts and coupling constants of the organic ligand. Both the carbene and olefin carbons resonate at low field (<100 ppm), with signals spanning over a wide range. The carbene carbon resonance may extend over 150 ppm in the range between δ 100–245 ppm. The olefinic carbones are less sensitive, resonating only within 60 ppm between δ 140–200 ppm. Table 5 for the carbonyl complexes of manganese[345] and rhodium[340] complexes illustrates the general effect of various ring substituents on the chemical shift of the three-membered ring carbons. In the absence of π-conjugating donor substituents the carbene resonances appear at a lower field than those of the adjacent olefin carbons, and are insensitive to the nature of simple alkyl and aryl substituents. Aryl but not alkyl groups cause a moderate upfield shift of the olefinic carbon resonance. However, the diisopropylamino group induces a striking upfield shift of both carbon signals, which is much more pronounced for the carbene, whose signal now appears at the highest field.

TABLE 5. ^{13}C NMR spectra of Mn and Rh cyclopropenylidene complexes

Substituent R	Ring carbons (ppm) C(carbene) (m, J_{CP} Hz)	C(olefin)	Reference
$(MeCp)Mn(C_3R_2)$			
Pr	240.4	191.7	345
t-Bu	238.9	196.1	345
Ph	243.1	178.9	345
$[trans\text{-}(CO)(Bu_3P)_2Rh(C_3R_2)]ClO_4$			
i-Pr_2N	147.8 (2t, 18.3)	149.9	340

A more thorough study on the effect of the metal, substituents, auxiliary ligands and the metal coordination sphere on the ^{13}C NMR spectra was carried out with group 10 palladium[235,347,350] (Table 6) and platinum[351] complexes (Table 7). For the Pd series it is striking to find only small differences in the chemical shifts of the carbons carrying the same R

TABLE 6. ^{13}C NMR spectra of palladium cyclopropenylidene complexes $[Cl_2Pd(C_3R_2)]_2$ and *cis*-$Cl_2(Bu_3P)Pd(C_3R_2)$, and cyclopropenylium salts [*trans*-$Cl(Bu_3P)_2Pd(C_3R_2)]ClO_4$

Substituent R	Ring carbons (ppm)		Reference
	C(carbene) (m, J_{CP} Hz)	C(olefin)	
$[Cl_2Pd(C_3R_2)]_2$			
i-Pr	181.2	195.7	350
t-Bu	183.1	198.0	350
Me_2N	99.9	150.9	347
cis-$Cl_2(Bu_3P)Pd(C_3R_2)$			
i-Pr	205.7 (d, <1.2)	195.1	350
t-Bu	205.2 (d, <1.2)	196.5	350
Ph	195.3 (d, 2.9)	175.2	350
Me_2N	125.0 (d, 6.0)	150.6	347
Et_2N	125.6 (d, 5.9)	149.5	340
i-Pr_2N	128.4 (d, 4.9)	147.4	340
[*trans*-$Cl(Bu_3P)_2Pd(C_3R_2)]ClO_4$			
i-Pr	205.4 (t, 9.8)	198.7	350
t-Bu	205.8 (t, 9.8)	199.7	350
Ph	195.2 (t, 10.7)	176.1	350
Me_2N	122.0 (t, 10.8)	151.5	350
i-Pr_2N	125.6 (t, 11.3)	147.8	340

TABLE 7. ^{13}C and ^{195}Pt NMR spectra of platinum cyclopropenylidene and cyclopropenylium complexes[351]

Compound	Ring carbons (δ, ppm)			
	C(carbene) (m, J_{CP} Hz)	C(olefin)	J_{PtC} Hz	^{195}Pt (δ, ppm)
cis-$Cl_4Pt[C_3(Bu\text{-}t)_2]_2$	166.6	195.3	1040	–715
cis-$Cl_2Pt[C_3(Bu\text{-}t)_2]_2$	178.2	192.4	1438	–3315
trans-$X_2Pt(C_3R_2)_2$				
R = *i*-Pr_2N, X = Cl	136.0	146.7	986	–3150
R = *i*-Pr_2N, X = I	131.8	146.6	935	–4718
R = *i*-Pr_2N, X_2 = ICl	134.6	146.9	956	–3928
cis-$Cl_2(Bu_3P)Pt[C_3(Bu\text{-}t)_2]$	184.8 (d, 9.0)	192.6	1400	
[*trans*-$Cl(Bu_3P)_2Pt(C_3R_2)]ClO_4$				
R = *t*-Bu	184.7 (d, 11.0)	195.0	1369	–4236
R = *i*-Pr_2N	110.3 (d, 10.8)	143.8	1374	–4324

substituent, whether it is a dimer or mononuclear cyclopropenylidene complex, or a cyclopropenylium carbenoid cation. Similarly, the chemical shifts are insensitive to the nature and relative position of the auxiliary groups. In contrast, Ph substituents cause an upfield shift of the olefinic carbon signal by *ca* 25 ppm, whereas a 40 and 80 ppm upfield shift of the olefinic and carbene carbons is observed, respectively, with dialkylamino substituents. Notably, ^{13}C–^{31}P coupling constants $^2J_{CP}$ are small in the *cis*- (1–6 Hz) compared to *trans*-complexes (10–11 Hz).

Platinum complexes add another dimension to the NMR analysis[351] of cyclopropenylidene complexes. ^{195}Pt NMR resonances for the Pt^{II}-complexes appear in the typical δ -3100 to -4700 ppm region, whereas the signal of the Pt^{IV}-complex *cis*-$Cl_4Pt(C_3R_2)_2$ appears upfield at –715 ppm. Coupling constants $^1J_{PtC}$ are found to be very sensitive to the nature of the ligand *trans* to the C(carbene) rather than to ring substituents. This is evident by comparing, for example, the small $^1J_{PtC}$ value in *trans*-$Cl_2Pt(C_3R_2)_2$ (986 Hz) to that of the *cis* counterpart (1438 Hz), and to the pair of [*trans*-$Cl(Bu_3P)_2Pt(C_3R_2)$]ClO_4 (R = *t*-Bu, *i*-Pr_2N) complexes (both *ca* 1370 Hz). Furthermore, a linear correlation between $^1J_{PtC}$ and the formal '*s*' % character of the C(carbene) ligating σ–hybrid orbital confirms the suggestion that the Pt—C bond is largely σ in nature, with only negligible π-backbonding contribution.

VII. METAL DERIVATIVES OF METHYLENECYCLOPROPANE

A. σ-Bonded Compounds

1. Metallation

Direct metallation of methylenecyclopropane with butyllithium in THF affords methylenecyclopropyllithium. This reacts with carbonyl electrophiles such as aldehydes, ketones and lactones by ring alkylation to give selectively 2-methylenecyclopropyl carbinols. No products of *exo* alkylation are isolated. Other bases such as *t*-BuOK and KH do not deprotonate methylenecyclopropane. Use of diethyl ether as the solvent, instead of THF, significantly reduced the rate of lithiation[362]. Similar reaction of the lithium reagent with ethylene oxide gave 2-(2-methylenecyclopropyl)ethanol (equation 294)[363]. In the reaction with ^{13}C-labeled ethylene oxide the addition of TMEDA to the reaction mixture is recommended[364].

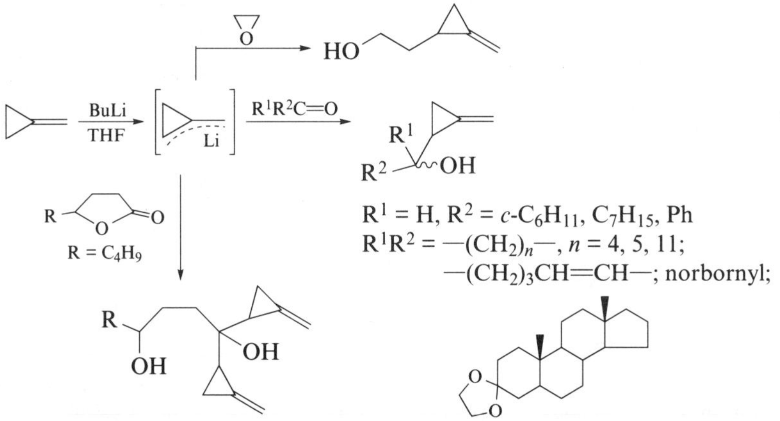

(294)

Deprotonation of (alkylcyclopropylidenemethyl)cyclopropanes (alkyl = methyl, cyclopropyl) with BuLi and subsequent reactions with various electrophiles afforded the corresponding ring-substituted methylenecyclopropanes (equation 295)[365]. When the lithiated compounds are treated with CO_2, carboxylic acids are obtained, together with isomeric lactones. These can be regarded as formal 3+2 adducts of the methylenecyclopropanes with CO_2 (equation 295)[366].

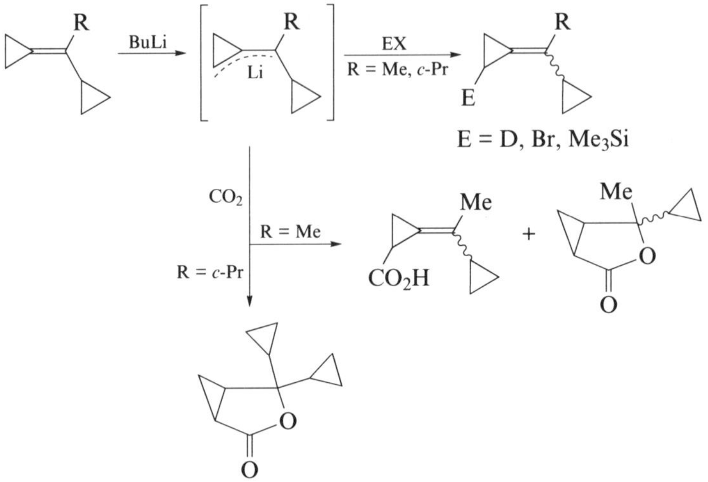

(295)

Silylation of lithiated 2-methyl-1-methylenecyclopropane is reported to give selectively *trans*-2-methyl-3-trimethylsilyl-1-methylenecyclopropane[367]. However, when 2-trimethylsilyl-1-methylenecyclopropane is again lithiated *in situ*, and subsequently reacted with Me_3SiCl, silylation occurs mainly at the already silylated carbon, giving 2,2-bis-(trimethylsilyl)-1-methylenecyclopropane, along with minor amounts of the *exo*-silylated isomeric cyclopropene and *syn* and *anti* trisilyl derivatives (equation 296)[368].

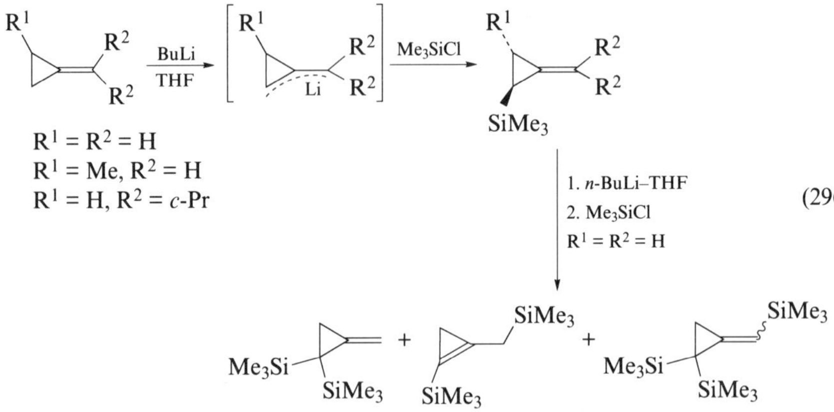

(296)

Under similar conditions reaction with alkyl bromides and benzyl chloride led exclusively to ring alkylation. With benzaldehyde, alkylation occurred solely at the *exo*-methylene position, while acetone gave approximately a 1:1 mixture of the isomeric ring- and *exo*-carbinols. These were further silylated and isolated as the corresponding siloxanes (equation 297).

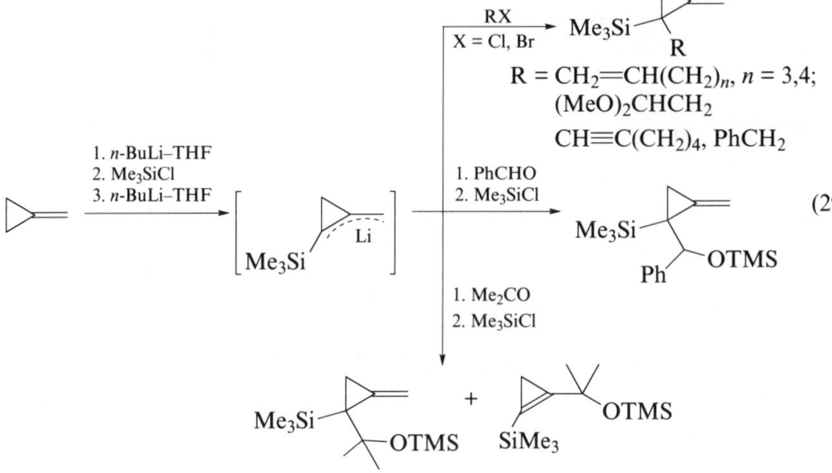

(297)

Exceptionally, the reaction of (diphenylmethylene)cyclopropane with BuLi did not result in the expected metallation. Instead, addition of the reagent to the double bond occurred, forming a cyclopropylcarbinyllithium compound (equation 298)[135].

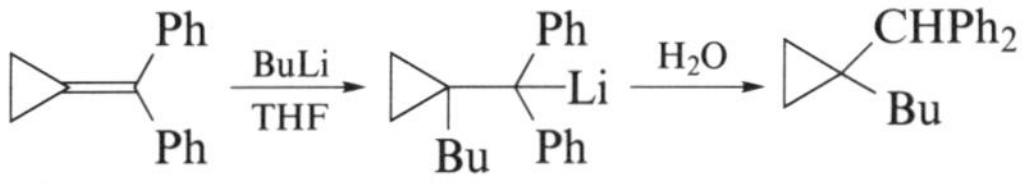

(298)

2. Metal–halogen exchange

Lithiation of 2-bromo-3,3-disubstituted-methylenecyclopropanes by metal halogen exchange reaction with EtLi in ether, followed by alkylation with epoxides, gave selectively ring-alkylated β-alcohols, derived from attack at the epoxide primary carbon (equation 299). When $R^1 \neq R^2$ a mixture of isomers is obtained[369].

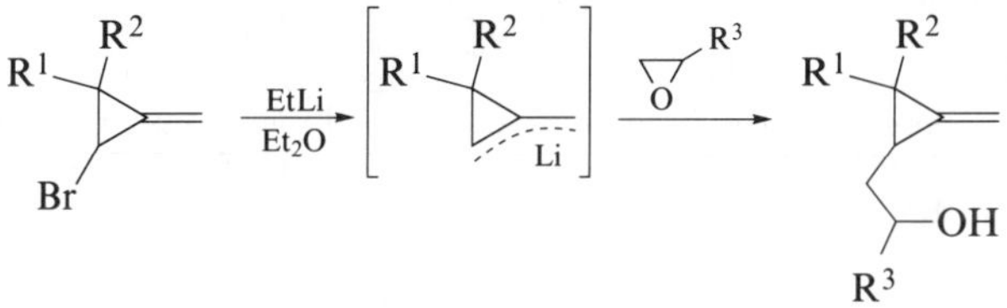

(299)

$R^1 = R^2 = Me$, $R^3 = H$, Me
$R^1 = Me$, $R^2 = Et$, $R^3 = H$, Me (isomeric mixture)
$R^1 = Me$, $R^2 = i\text{-}Pr$, $R^3 = H$, Me (isomeric mixture)
$R^1R^2 = —(CH_2)_n—$, $n = 3,4,5$, $R^3 = H$, Me

Lithiation of the exocyclic vinyl position of methylenecyclopropane is achieved by halogen exchange reaction of *t*-BuLi with (bromomethylene)cyclopropane in ether. The vinyl lithio compound readily reacts with electrophiles such as allyl bromide[51], epoxides[359], chlorosilanes[370] and hexacarbonylchromium[152], to give the corresponding alkylated, silylated and Fischer carbene complexes, respectively (equation 300).

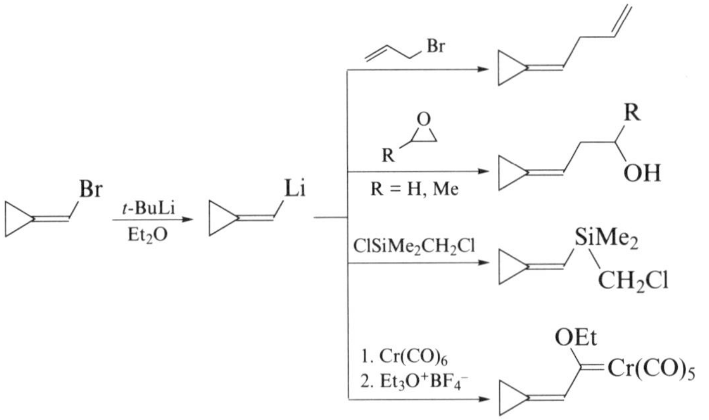

(300)

The metallation of cyclopropylidenephenylmethyl bromide with Mg in ether and THF yields an unstable vinylmagnesium compound, which undergoes ring cleavage of the proximal vinyl σ-bond, to give the acetylenic 4-phenylbutynylmagnesium bromide (equation 301)[371]. This species abstracts hydrogen from the solvent to give benzylidecyclopropane and 1-phenylbutyne, or disproportionate to form 1-phenylbut-3-en-1-yne and 1-phenylbutyne. The ring-cleavage mechanism is likely to involve a radical-like cyclopropylcarbinyl–homoallyl rearrangement, presumably via a four-centered cyclic transient.

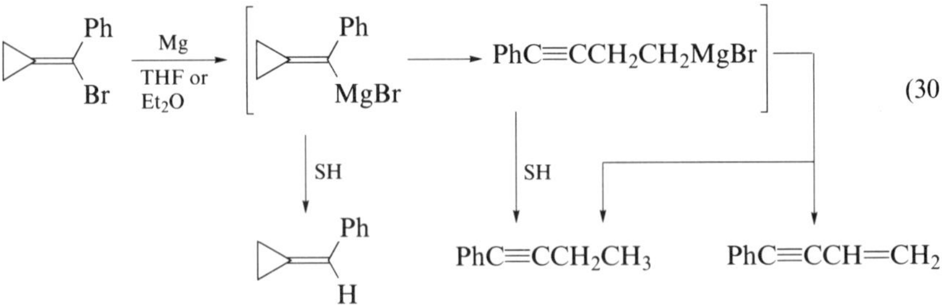

(301)

Notably, treatment of 2-(*p*-bromophenyl)-3,3-dimethyl-1-methylenecyclopropane with BuLi in THF followed by $HgCl_2$ led to mercuration of the aromatic ring (equation 302). This indicates that lithium-halogen exchange at the aromatic ring is preferable to direct lithiation of the three-membered ring[372].

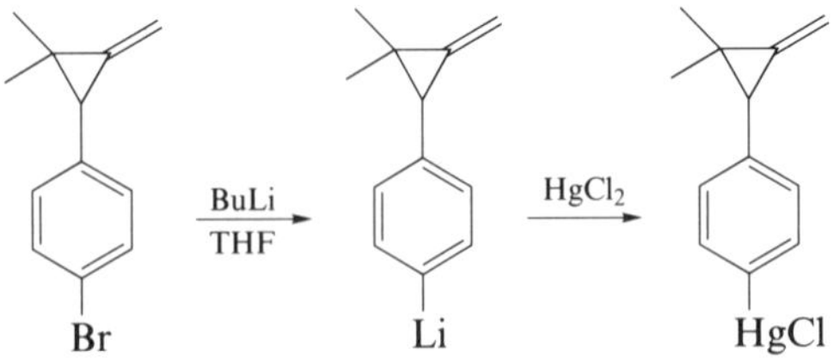

(302)

3. *Reactions with metals*

Methylenecyclopropane readily reacts with lithium powder in ether, by reductive cleavage of the proximal vinylic σ-bond, to give 2,4-dilithiobutene (equation 303)[373]. The reaction is believed to proceed via initial addition of two lithium atoms to the double bond, prior to ring cleavage. Similarly, the 'butterfly olefin' dicyclopropylidene undergoes reductive lithiation to the dilithio-propylidenecyclopropane, which upon standing at room temperature rearranges to the isomeric 1,6-dilithio-3-hexyne (equation 304)[374]. With larger alkyl substituents at the exocyclic methylene group lithiation is retarded, Pentylidenecyclopropane affords only 8% of 3-octane after prolonged (5 days) sonication and hydrolysis of the reaction mixture, whereas no reaction is observed with cyclohexylidenecyclopropane. Lithiation of ring-substituted methylenecyclopropanes proceeds normally, except for 2,2,3,3-tetramethyl-1-methylenecyclopropane which gives a monolithium diene as a result of LiH elimination (equation 305). Formation of the dilithio compounds is established by quenching the reaction mixture with a variety of electrophiles, to give the corresponding open-chain olefin derivatives. Since the two carbanion centers differ in reactivity, the reaction can be carried out stepwise with two different electrophiles, the first ending up at the homoallyl position and the second at the olefin internal position.

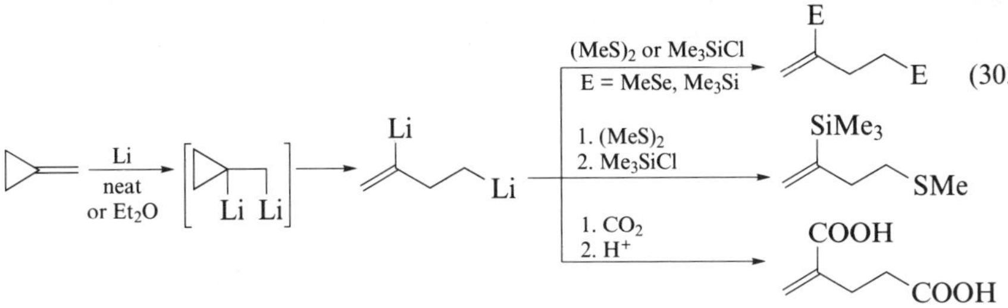

(303)

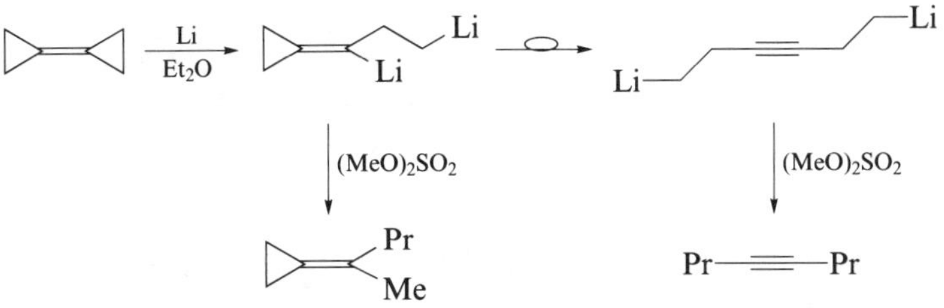

(304)

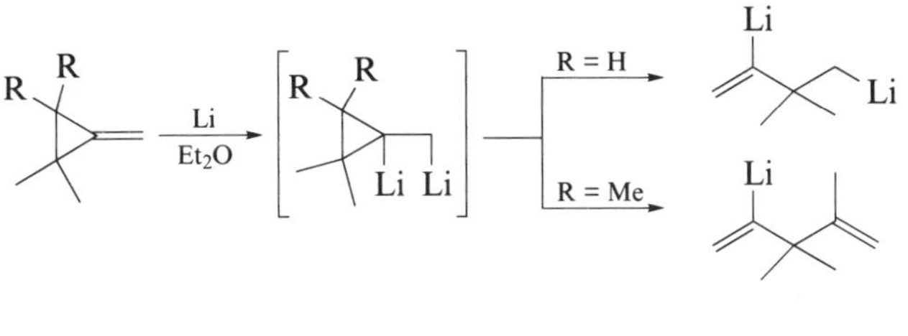

(305)

B. π-Complexes

The initial step in all interactions of methylenecyclopropanes with transition metal complexes, whether stoichiometric or catalytic, is believed to be the formation of an

η^2-complex. However, isolation of stable η^2-methylenecyclopropane complexes has only been achieved in a limited number of cases. In most reactions complexation is directly followed by intermediate complexes resulting from (a) addition to the double bond to give an intermediate cyclopropyl σ-complex, (b) ring-opening of the proximal vinyl bond to give a metallacyclic intermediate and (c) ring-opening of the distal allylic bond, forming a trimethylenemethane π-complex or the analogous σ-metallacyclic intermediate (Scheme 12). These pathways ultimately determine the reaction fate, and are depending on the metal, ring substituents and reaction conditions (*see below*)[73,367,375,376].

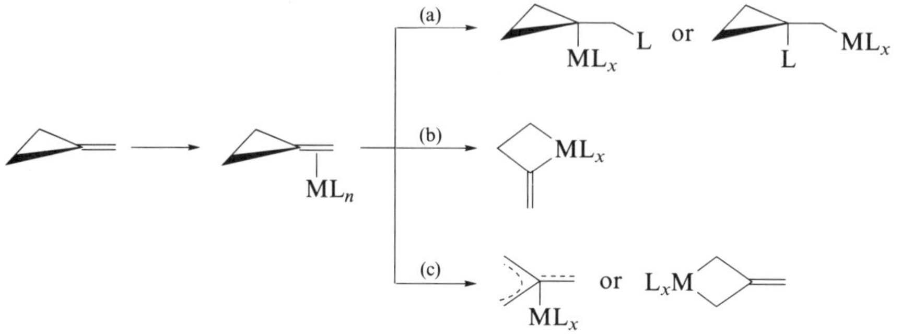

SCHEME 12

1. Preparation

a. Ligand exchange reactions. Feist's acid derivatives are the first methylenecyclopropanes to give isolable η^2-complexes. Reaction of *cis*-methylenecyclopropane-2,3-dicarboxylic anhydride (aF) and $Fe_2(CO)_9$ in refluxing ether or benzene gives a mixture of both *anti*- and *syn*-iron tetracarbonyl η^2-complexes (equation 306)[377]. Similarly, the reaction of Feist's *trans*- (tF) and *cis*-dimethyl esters (cF) with $Fe_2(CO)_9$ in refluxing hexane affords the *trans*- and *cis*-diester complexes, respectively (equations 307 and 308)[378]. Surprisingly, the crystal structure analysis of the *cis*-complex reveals that the more crowded *syn*-isomer

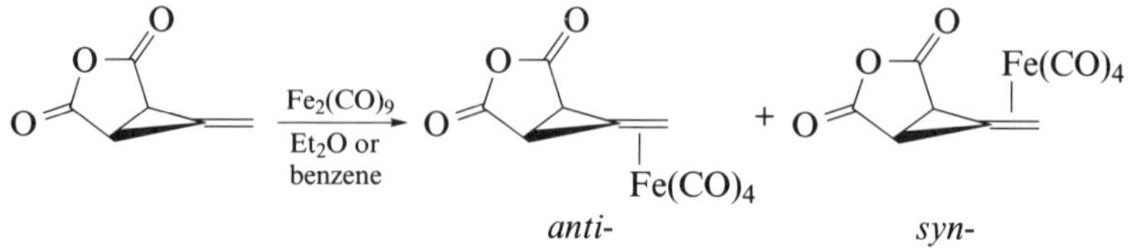

(306)

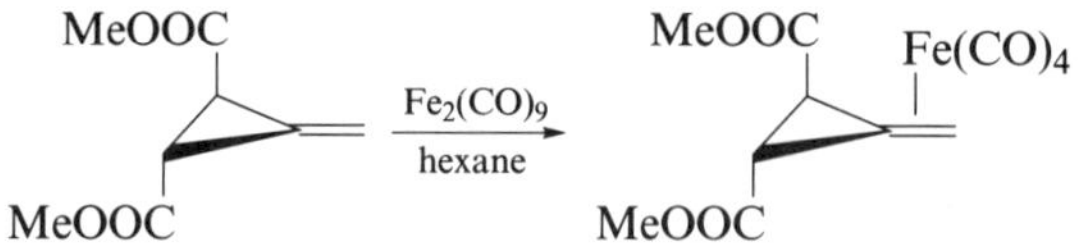

(307)

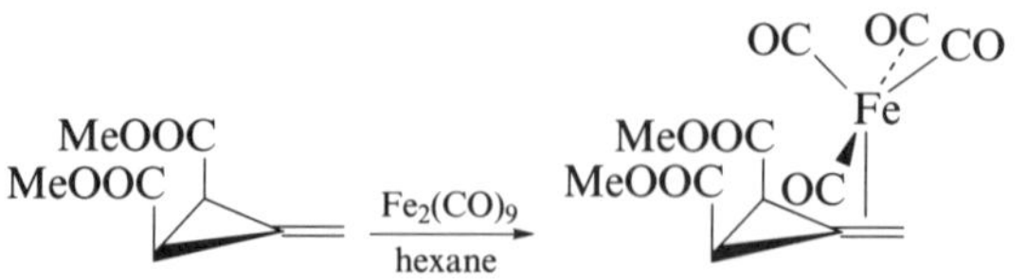

(308)

is exclusively formed[379]. This suggests the initial interaction of one of the ester carbonyl groups with the metal, and subsequent intramolecular transfer of the $Fe(CO)_4$ fragment to form the thermodynamically more stable olefin complex.

The solid state analysis of *cis*-Feist's ester iron tetracarbonyl (cF)$Fe(CO)_4$ further established the trigonal-bipyramidal structure of the complex, with the olefin occupying the more favorable equatorial position[380]. The coordinated olefinic C=C bond length of 141 pm has increased relative to that in the free Feist's acid (132 pm), and the methylene group is tilted out of the plane of the three-membered ring (away from the metal fragment) by 37.6°. The remainder of the cyclopropane ring shows no deviation from the regular geometry of uncomplexed counterparts. More recently, crystalline (2,2-diphenyl-1-methylenecyclopropane)$Fe(CO)_4$ was isolated from the analogous reaction of the free ligand with $Fe_2(CO)_9$ in benzene at 20 °C, whose trigonal bipyramidal structure conforms with that of the analogous Feist's ester complex[381].

Ligand exchange reactions are also applied to prepare the parent η^2-methylenecyclopropane (mcp) complexes (acac)Rh$(mcp)_2$ and $(Ph_3P)_2$Pt$(mcp)_2$, having two coordinated methylenecyclopropane ligands, and a large variety of mono- and di-Feist's dimethyl ester complexes of group 9 rhodium and iridium, and group 10 platinum[382]. Thus, the exchange of the ethylene ligands in complexes (acac)M$(ethylene)_2$ (M = Rh, Ir) and $(Ph_3P)_2$Pt-$(ethylene)_2$ with methylenecyclopropanes (tF and cF) readily gave the corresponding methylenecyclopropane complexes (equation 309). X-ray analysis of the parent rhodium complex (acac)Rh$(mcp)_2$ confirmed the η^2-square-planar structure of these complexes. Like in the analogous Feist's ester iron complex (*vide supra*), a significant elongation of the coordinated C=C bond to 142 pm compared with the free ligand (132 pm in Feist's acid), and an out-of-plane tilt of approximately 27° of the methylene group, are observed. Notably, the Rh—C bond distance to the ring carbon (207 pm) is shorter than that to the methylene carbon (213 pm).

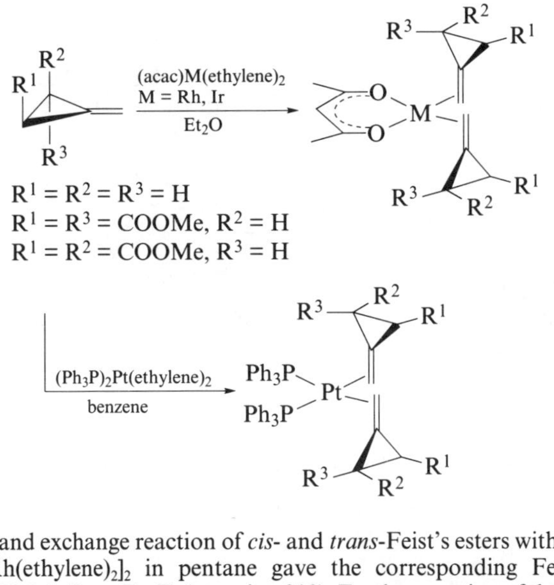

(309)

Analogous ligand exchange reaction of *cis*- and *trans*-Feist's esters with the dirhodium complex [μ-ClRh$(ethylene)_2]_2$ in pentane gave the corresponding Feist's esters η^2-complexes [μ-ClRh$L_2]_2$ (L = cF, tF) (equation 310). Further reaction of the latter complex [μ-ClRh$L_2]_2$ (L = tF) with cyclopentadienylthallium (CpTl) in CH_2Cl_2 afforded the monorhodium complex CpRh$(tF)_2$, and reaction with a mixture of both CpTl and dirhodium

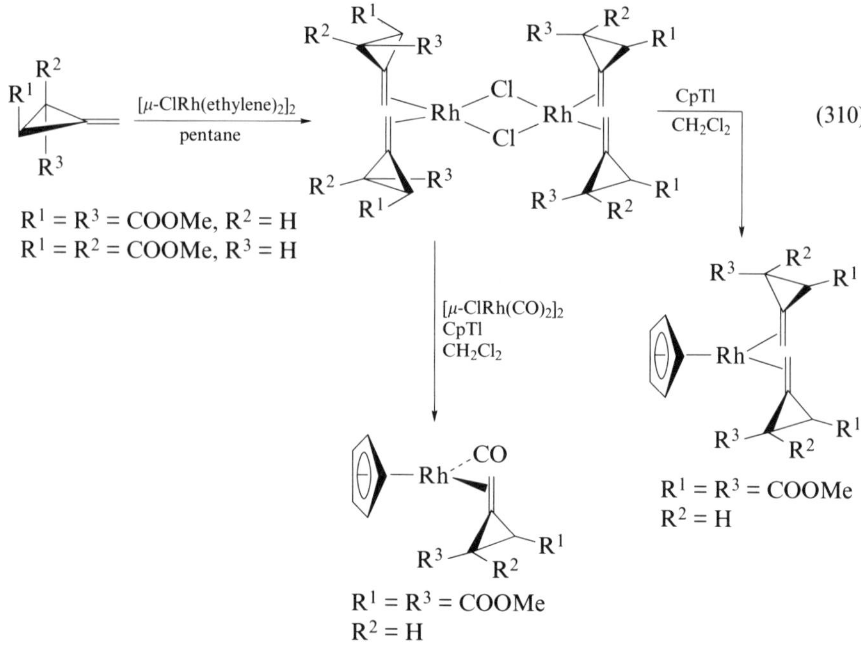

(310)

carbonyl complex $[\mu\text{-ClRh(CO)}_2]_2$ resulted in the formation of the mononuclear η^2-complex Cp(CO)Rh(tF).

Likewise, mononuclear complexes of rhodium and platinum containing only one methylenecyclopropane ligand are prepared by ligand exchange reactions of the Feist's esters with $(acac)Rh(CO)_2$ and *trans*-Cl_2(pyr)Pt(ethylene), giving complexes (acac)(CO)Rh(tF) and *trans*-Cl_2(pyr)PtL (L = cF, tF), respectively (equation 311).

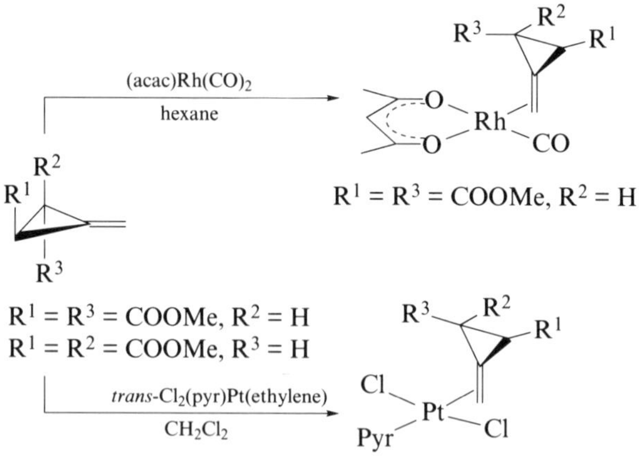

(311)

More recently, the parent mono- and di-methylenecyclopropane cyclopentadienylcobalt η^2-complexes were prepared by ligand exchange reactions[383]. Reaction of methylenecyclopropane (and its tetramethyl derivative) with $CpCo(ethylene)_2$ in pentane affords

$CpCo(mcp)_2$, which in turn can be further transformed to the mono-complex $CpCo(PPh_3)(mcp)$ by exchange of one methylenecyclopropane ligand with PPh_3 (equation 312). Although both complexes are isolable crystals, they are thermally less stable than the analogous Feist's ester complexes. $CpCo(mcp)_2$ readily undergoes thermal isomerization at 110 °C, to give cyclopropyl-substituted η^4-butadiene complexes (*see below*).

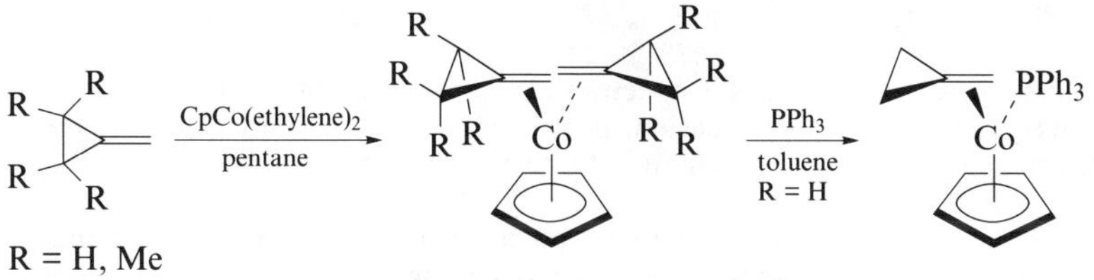

(312)

Nickel plays an extremely important role in the organometallic chemistry of methylenecyclopropane. Interaction of nickel complexes with a large variety of methylenecyclopropanes not only gives isolable η^2-complexes, but also prompts a herd of stoichiometric[367] and catalytic[73,384] ring-opening reactions, dimerizations, oligomerizations and cycloadditions (*see below*). The parent methylenecyclopropane and 2,2-dimethyl derivative form isolable bis(triphenylphosphine) η^2-complexes by a variety of ligand exchange reactions (equation 313)[385,386]. The analogous bipyridyl nickel complex of 2,2,3,3-tetramethyl-1-methylenecyclopropane, obtained upon reaction with (bipyr)Ni(COD), is also reported to be stable up to 40 °C, where it rearranges to the open-chain 2,3,3-trimethyl-1,4-pentadiene (equation 314)[73]. Interestingly, an equilibrium is attained between the nickel complexes of methylenecyclopropane and 1,2,2-trimethylcyclopropene, which could be shifted to either side by the proper use of excess of free olefin (equation 315)[386].

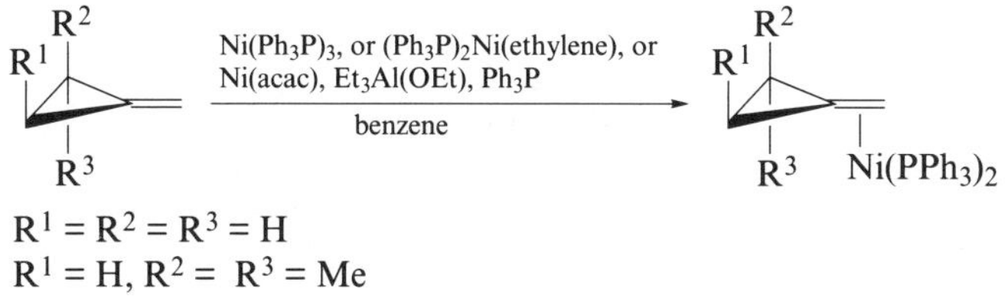

(313)

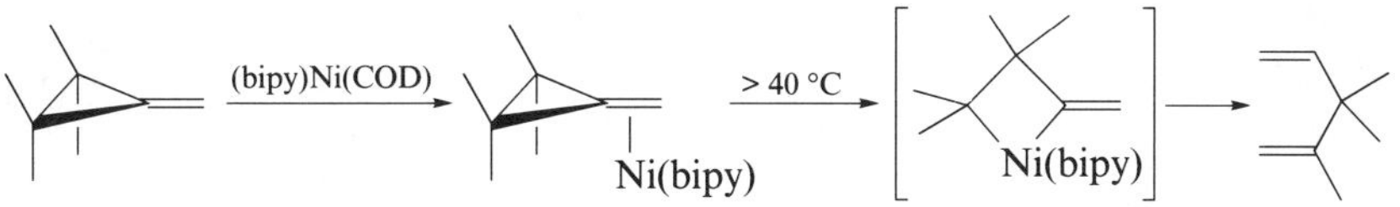

(314)

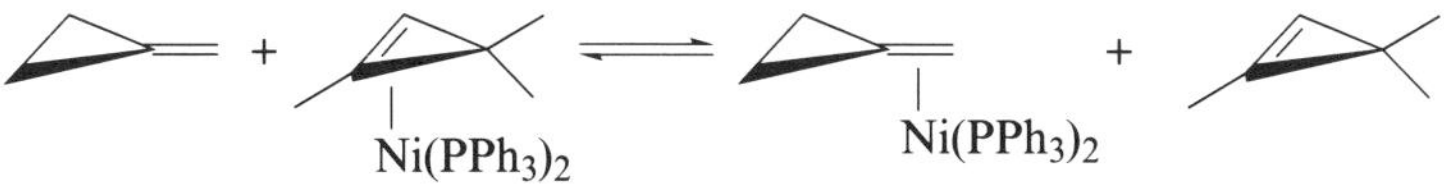

(315)

b. Cyclopropanation of allenes. Reaction of mono- and di-substituted allenes with benzylidenepentacarbonyl tungsten in pentane-dichloromethane at low temperatures results in regiospecific and stereoselective transfer of the carbene ligand to the allene, and η^2-coordination of the intermediate free methylenecyclopropanes, forming the corresponding methylenecyclopropane tungsten complexes (equation 316)[377]. Tetrasubstituted

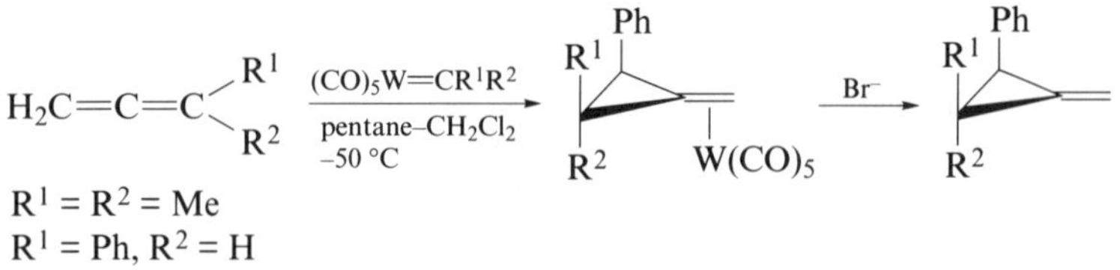

(316)

cumulenes, such as tetraphenylallene, tetraphenylbutatriene and tetraphenylhexapentaene do not undergo cyclopropanation, possibly for steric reasons. The free methylenecyclopropanes are obtained almost quantitatively upon decomplexation using Et_4NBr in dichloromethane.

X-ray structural analysis of 2,2-dimethyl-3-phenyl-1-methylenecyclopropane tungsten pentacarbonyl reveals an octahedral complex with characteristic W—C bond distance of 238 pm. The typical bond distances within the organic ligand are: 138 (complexed C=C), 148 (proximal C—C), 154 (distal C—C) pm, compared e.g. with 140, 148 and 154 pm, respectively, for the Feist's ester iron complex analogue (*see above*).

2. *NMR spectra*

Representative ^{1}H NMR chemical shifts of the methylene protons, and ^{13}C NMR chemical shifts of the methylene and cyclopropylidene carbons of a wide range of methylenecyclopropane complexes are gathered in Table 8. In general, the expected upfield shift of both methylene proton and coordinated carbon signals is observed. In the parent methylenecyclopropane and alkyl substituted complexes the methylene proton signals

TABLE 8. ^{1}H and ^{13}C NMR spectra of methylenecyclopropane complexes[382]

Complex	^{1}H NMR δ ppm ($^2J_{MH}$ Hz)	^{13}C NMR δ ppm ($^1J_{MC}$ Hz)	
		$C_{methylene}$	C_{ring}
(2,2-Me_2-3-Ph-mcp)$W(CO)_5$	3.52, 3.63[387]	44.5	92.0(15.2)
(tF)$Fe(CO)_4$	2.60[378]	23.0	54.9
(cF)$Fe(CO)_4$	2.45[378]	22.5	49.9
(aF)$Fe(CO)_4$	3.15 (*anti*), 3.23 (*syn*)[377]		
$(mcp)_2$Rh(acac)	1.94, 3.30	48.0 (12)	72.9 (20)
$(tF)_2$Rh(acac)	2.03, 3.79	45.4 (12)	73.3 (21)
$(cF)_2$Rh(acac)	2.30, 3.71	47.8 (12)	71.4 (30)
$[(tF)_2RhCl]_2$	1.82, 3.30		
$[(cF)_2RhCl]_2$	~2.70		
$(tF)_2$RhCp	1.24, 3.38		
(tF)(CO)Rh(acac)	1.97, 3.80		
(cF)(CO)RhCp	2.13, 2.97		
$(tF)_2$Ir(acac)	1.94, 3.71	30.4	49.8
$(cF)_2$Ir(acac)	2.16, 3.70		
(mcp)$Ni(PPh_3)_2$	2.52[83.3]		
(2,2-Me_2-mcp)$Ni(PPh_3)_2$	2.51, 2.35[386]		
$(mcp)_2Pt(PPh_3)_2$	~1.58		
$(tF)_2Pt(PPh_3)_2$	2.54, 2.91 (44)	28.2	54.9
$(cF)_2Pt(PPh_3)_2$	~2.17 (60)		
$(tF)_2PtCl_2$pyr	4.89, 4.94 (61, 62)	5.7 (136)	80.9 (354)
$(cF)_2PtCl_2$pyr	4.80 (60)		

appear upfield between δ 2.0–3.0 ppm, whereas in the Feist's ester derivatives resonances also appear at lower fields. The coordinated ring carbons appear upfield at the range of δ 20–50 ppm, considerably higher (δ 20–30 ppm) than their methylene carbon counterparts, which accordingly resonate at the lower field range of δ 50–80 ppm.

3. Reactions

a. Chloropalladation. In contrast to group 10 triad members nickel and platinum which gave isolable η^2-methylenecyclopropane complexes (*vide supra*), attempts to prepare analogous complexes of palladium, the central member of this triad, resulted unambiguously in addition reactions to the double bond and subsequent ring opening. In general, reaction of methylenecyclopropanes with $PdCl_2l_2$ led to distal ring cleavage by a formal 1,3-chloropalladation reaction, to form dimeric chloro-π-allylpalladium complexes (equation 317). The isomeric *exo*-methylene substituted derivatives do not undergo cyclopalladation and are stable under the reaction conditions. Treatment of the dimers with thallous acetylacetonate gave the monomeric acetylacetonato complexes. Parent methylenecyclopropane is exceptional in its reaction with $PdCl_2(PhCN)_2$ by giving di-μ-chloro-bis(2-chloro-1-methyl-π-allyl)dipalladium, which is formed apparently by *proximal* ring cleavage and hydride shift (equation 318)[388].

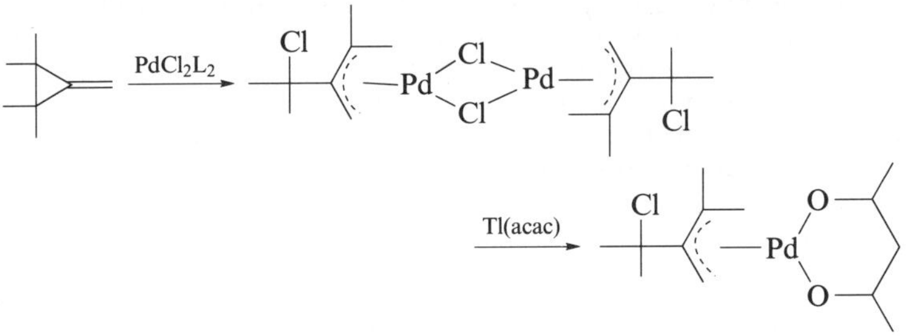

(317)

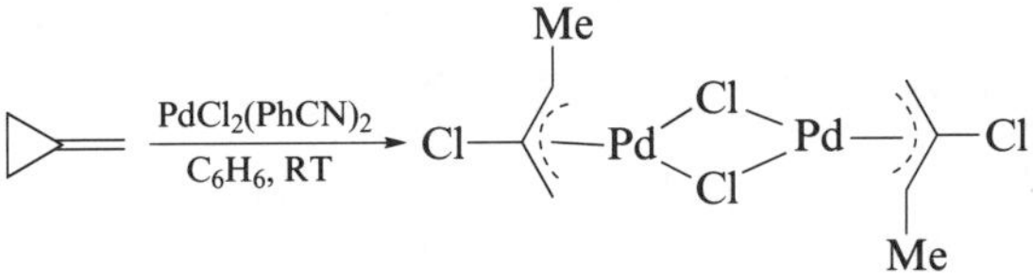

(318)

Chloropalladation of simple ring substituted alkyl- and aryl-methylenecyclopropanes usually gives a regioisomeric mixture of 2-(chloromethyl)-η^3-allylpalladium complexes. 2,2-Dimethyl-1-methylenecyclopropane readily reacts with $PdCl_2(PhCN)_2$ at room temperature in CH_2Cl_2, C_6H_6 or MeOH, to give a virtually quantiative yield of a 9:1 mixture of two noninterconvertible 2-(chloromethyl)allylic complexes (equation 319)[267,389,390]. Both *cis* and *trans*-2,3-dimethyl-1-methylenecyclopropane react similarly, giving an identical mixture of diastereomeric pairs of enantiomers, with one pair present in excess (equation 320). The ratio between the diastereomeric pairs is solvent-dependent, being

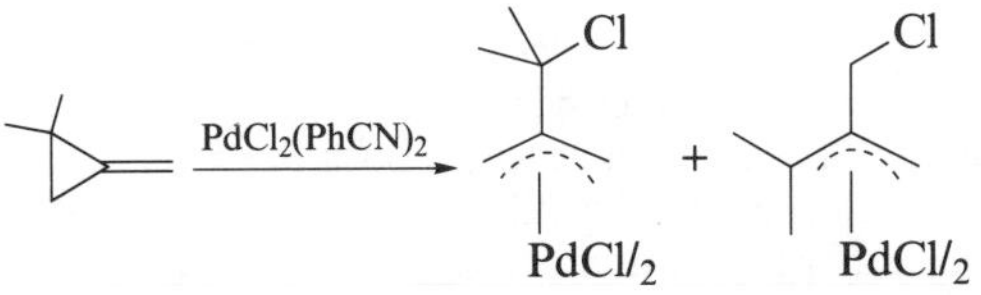

(319)

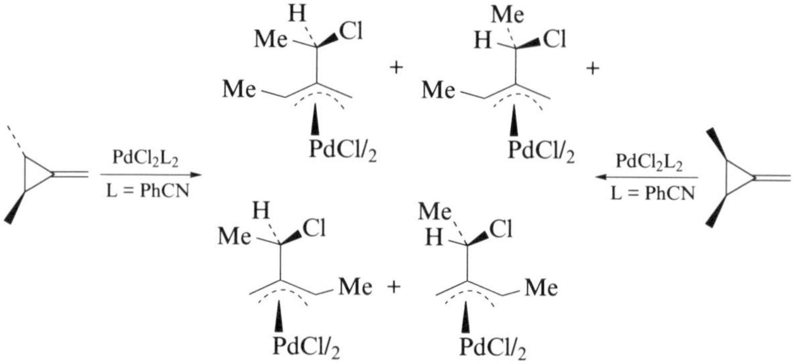

(320)

more selective (4:1 ratio) in polar $CDCl_3$ than in C_6D_6 (3:1 ratio). The lack of selectivity in the latter reaction is apparently due to rapid η^3 to η^1 to η^3 transformations of the organic ligand, which enables the transposition of the metal fragment from one face of the allyl ligand to the other. Since this process occurs after the chloropalladation step, no conclusions can be drawn from these examples as to the actual stereochemistry of the chloropalladation reaction.

2,2-Diphenyl-1-methylenecyclopropane undergoes similar chloropalladation reactions[391,392]. However, unlike the dimethyl counterpart, the reaction is non regioselective, giving a 1:1 ratio of the isomers, and the isomer having the tertiary chloride group slowly rearranges under the reaction conditions to the thermodynamically more stable primary chloromethyl isomer (equation 321)[388]. Isomerization is slower than solvolysis in methanol. Likewise, chloropalladation of 2-phenyl-1-methylenecyclopropane gives a mixture of three kinetic products, which interconvert under thermodynamic conditions to a mixture of the two primary chloride complexes, containing primarily the more stable Ph-*anti* (to the CH_2Cl group) isomer (equation 322).

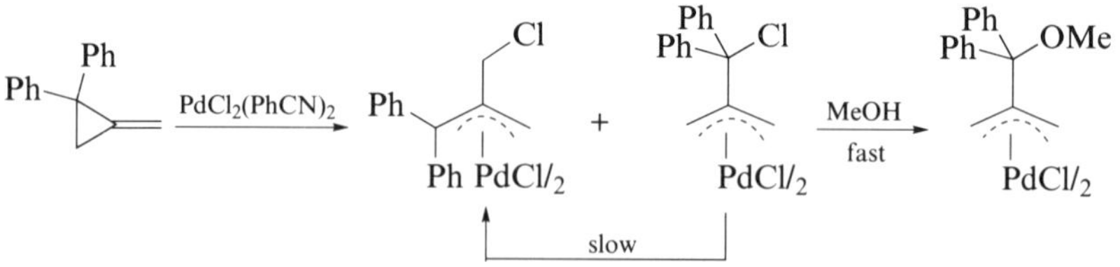

(321)

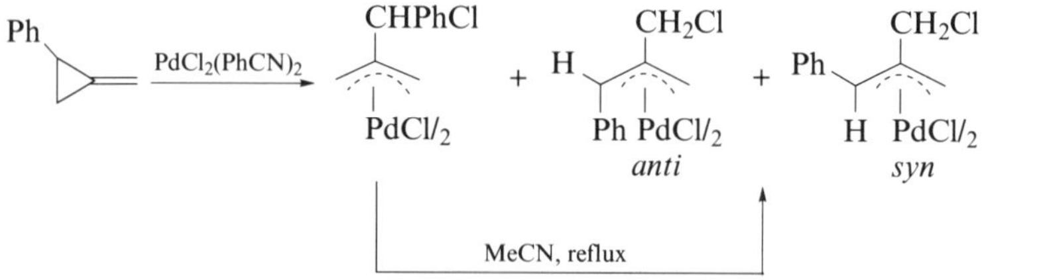

(322)

From a mechanistic point of view (*see below*) it is important to note that the chloropalladation of labeled 2,2-diphenyl-1-methylenecyclopropane-3,3-d_2 gave only *two* isomeric complexes, with absence of the isomer in which both the phenyl and deuterium reside on the allylic moiety (equation 323). This allows the exclusion of a symmetrically bound η^4-trimethylenemethane (TMM) intermediate or rapidly equilibrating η^3-TMM species,

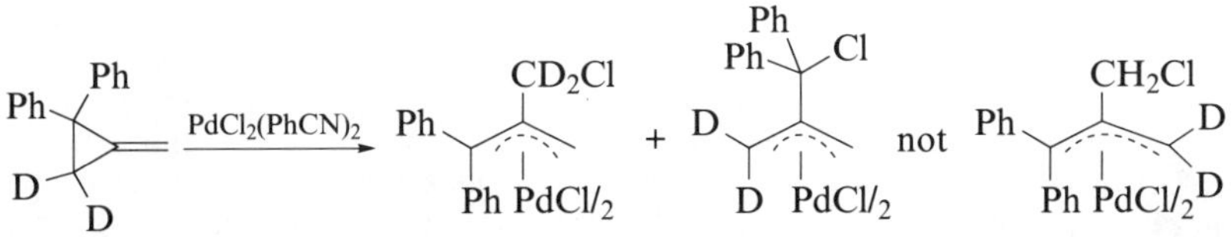

(323)

since both would result in the CH_2 and CD_2 termini becoming indistinguishable to the migrating chloride.

The true stereochemistry of 1, 3-chloropalladation is revealed in the reactions of *cis*-9-methylenebicyclo[6.1.0]nonane to give a single allylic complex (equation 324), and of *trans*-9-methylenebicyclo[6.1.0]nonane to give selectively a 4:1 mixture of *syn* and *anti* stereoisomeric allylic complexes (equation 325)[389,393]. In contrast to the noncyclic allylic complexes described above which interconvert under the reaction conditions, these monocyclic allylpalladium complexes are configurationally stable even under reflux in benzene for 8 h in the presence of 5 mol% PPh_3.

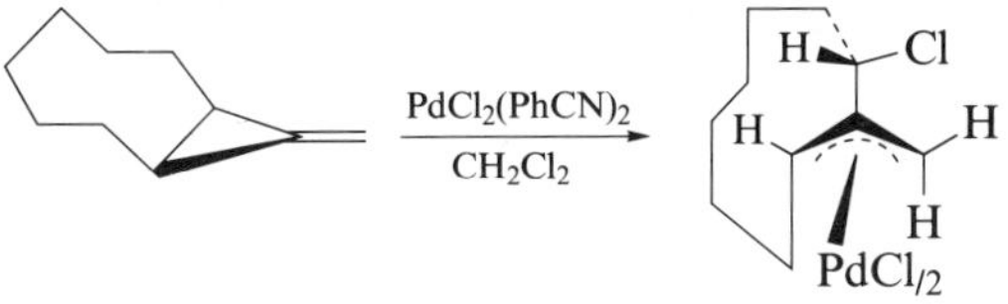

(324)

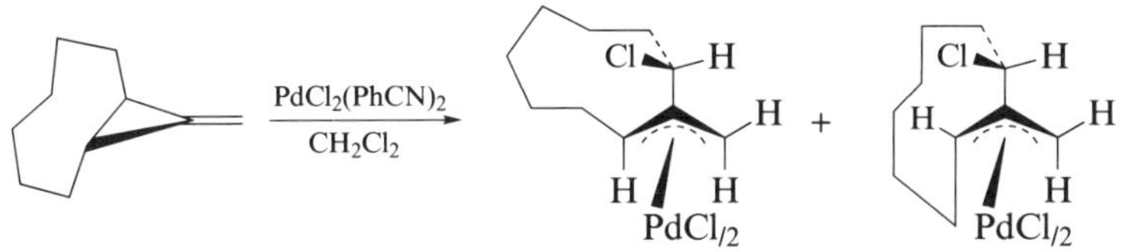

(325)

Formation of these products is rationalized in terms of suprafacial addition of the elements of Pd—Cl to a ring that is opening stereospecifically in a disrotatory mode, with the breaking bond bending away from the metal (dis-out) (equation 326). The absence of η^3 to η^1 to η^3 interconversions between the isomeric complexes is rationalized in terms of a severe steric blocking of the *anti* (to metal) allylic face of the complex by the nine-membered ring. This is evident from X-ray crystal structure analysis of the mononuclear acetylacetonate allylic complex obtained from chloropalladation of the *cis*-bicyclononane[394].

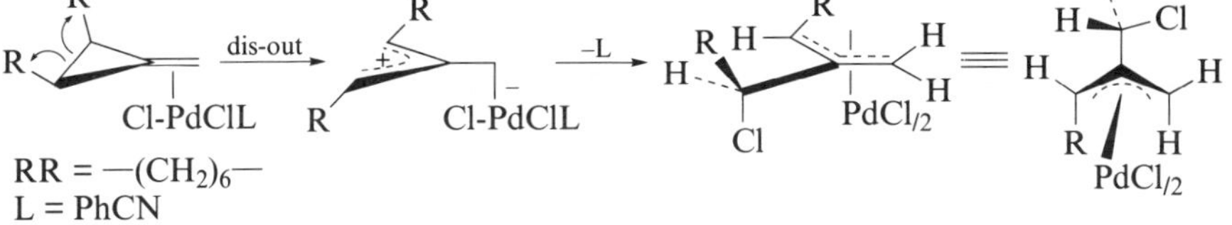

(326)

Interestingly, chloropalladation reaction of the more constrained *cis*-7-methylenebicyclo[4.1.0]heptane did not afford the expected dis-in kinetic product but rather the rearranged (η^3 to η^1 to η^3) thermodynamic isomer whose structure (as the acac mononuclear complex) was confirmed by X-ray crystallographic analysis (equation 327)[394]. More recently, 1-aryl-substituted derivatives of this bicyclic methylenecyclopropane (equation

328), the homologous 8-methylene-1-phenylbicyclo[5.1.0]octane (equation 329), and the tricyclic *exo*-3-methylene-2-phenyltricyclo[3.2.1.0^{2,4}]octane (equation 330) were chloropalladated to give mixtures of regioisomeric allyl complexes[220,395,396]. The ratio between these kinetic products appears to be dependent on ring strain, steric hindrance and the ability of the aryl group to stabilize positive charge. The kinetic products of chloropalladation of 7-methylene-1-phenylbicyclo[4.1.0] heptane, like its noncyclic phenyl-substituted counterparts (*vide supra*) interconvert under the reaction conditions and thus mask the true stereochemistry of the initial chloropalladation steps. Furthermore, the ability of the phenyl group to stabilize the positive charge of the zwitterionic η^1-intermediate could enable chloride attack from the *anti* (to metal) face, resulting in the low stereoselectivity observed in these reactions[395,396].

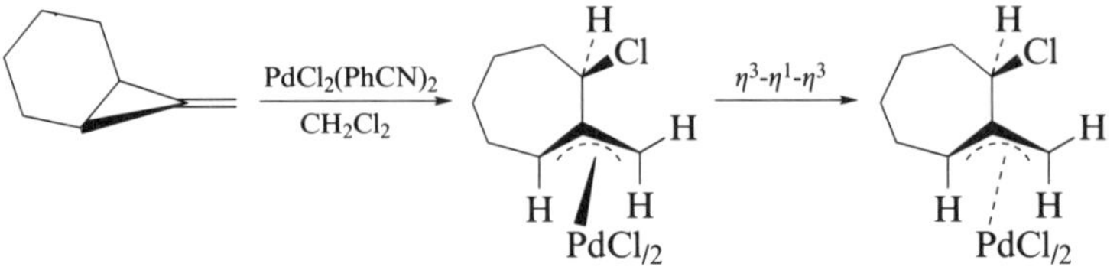

(327)

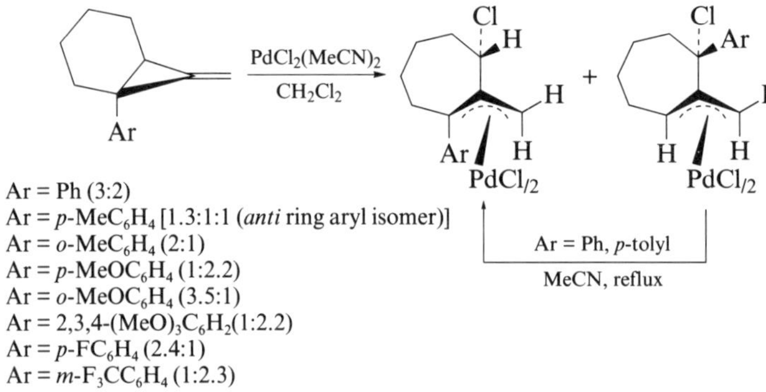

(328)

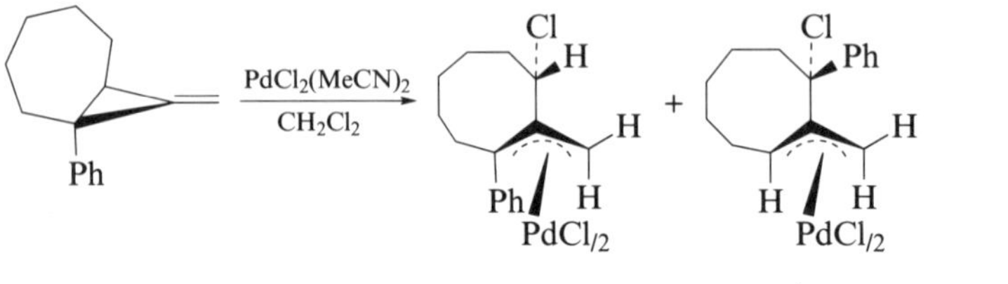

(329)

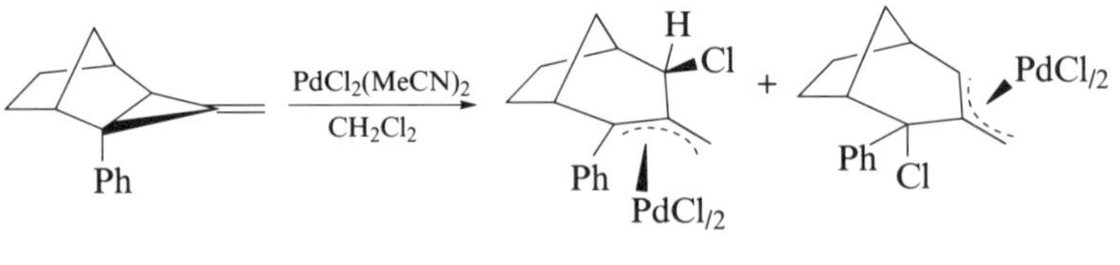

(330)

Molecular orbital calculations at the extended Hückel level indicate that the two disrotatory modes of ring cleavage (but not the symmetry-forbidden conrotatory mode) require similar activation energies for a model methylenecyclopropane–$PdCl_2(HCN)$ complex. The disrotatory motion of the carbon–carbon bond breaking away from the metal (dis-out) is very slightly favored over the dis-in mode of ring cleavage, on electronic grounds. This is in agreement with the experimental data, particularly for the bicyclic series, which show that the dis-in process is also feasible in *cis*-7-methylenebicyclo[4.1.0]heptane, where the dis-out process is geometrically made unfavorable[389].

Unlike the alkyl and aryl-substituted methylenecyclopropanes discussed above, both *cis*- and *trans*-Feist's esters undergo chloropalladation with *proximal* 1,2-ring opening, to give isomeric η^3-[3-chloro-1,2-bis(methoxycarbonyl)but-3-enyl]palladium complexes (equation 331)[397,398]. Formation of the but-3-enyl complexes is rationalized by sequential η^2-coordination, *syn*-addition to the double bond forming cyclopropylcarbinyl metal intermediate, and ring-cleavage. When the reaction is carried out in methanol, methoxypalladation occurs with the methoxy group attacking the terminal position of the coordinated double bond. This requires an inverse addition, followed by hydride shift, prior to the proximal ring opening (equation 332).

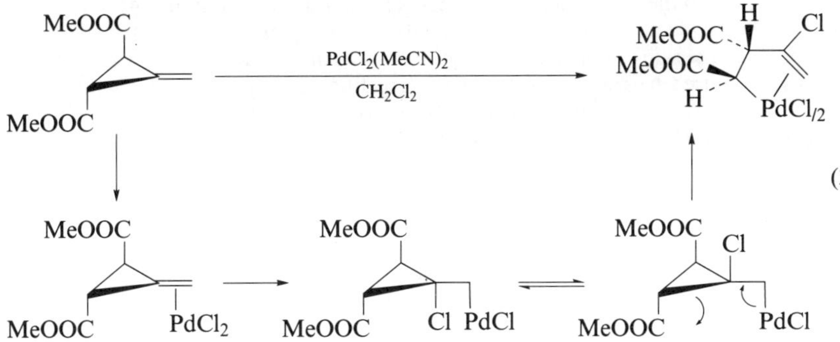

(331)

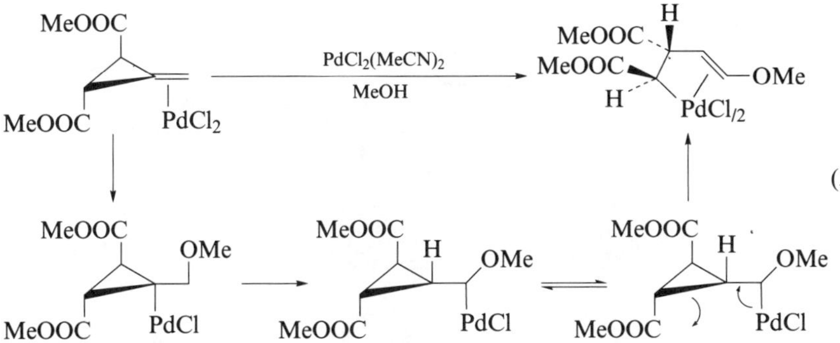

(332)

b. Hydroplatination. Methylenecyclopropane readily reacts with the platinum hydride complexes *trans*-$Pt(NO_3)(H)L_2$ (L = PEt_3, PMe_2Ph, PPh_3) in THF or benzene at room temperature, to give η^3-(1-methylallyl) platinum complexes which can be isolated as the hexafluorophosphate salts (equation 333)[399]. When the reaction was carried out with the deuterated complexes $Pt(NO_3)(D)L_2$ (L = PEt_3, PPh_3), the label appeared solely at the 2-position. This suggests that π-coordination of the metal is followed by Pt—H addition to the double bond to form cyclopropylcarbinyl–Pt complex, proximal ring opening to a but-3-enylplatinum complex, and finally rearrangement via β-H elimination and reinsertion, to give the η^3-(1-methylallyl)platinum complex.

Proximal ring opening also occurs when the Feist's esters undergo hydroplatination[400]. However, the reaction apparently stops at the early but-3-enyl stage (isolated as the

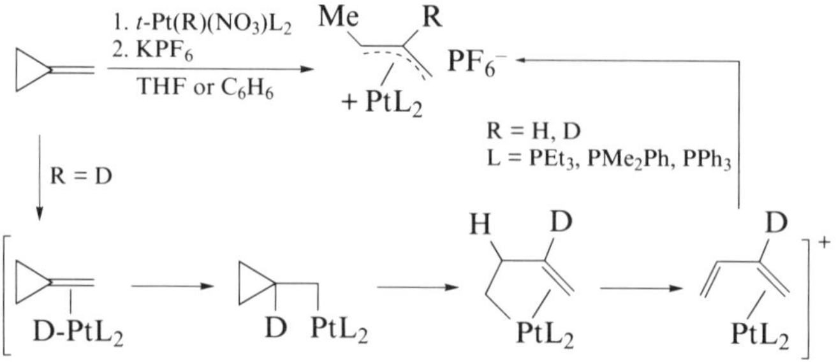

(333)

hexafluorophosphate or tetraphenylborate salt) as a result of coordination to the carbonyl group. Further double-bond isomerization takes place in the diastereomer obtained from *trans*-Feist's ester to the corresponding η^1-but-2-enyl complex (equation 344).

Interestingly, while no rearranged η^3-allylplatinum complex could be detected in solution for either the borate or phosphate salts, the crystals of the tetraphenylborate salt have the η^1-*allyl* structure whereas the crystals of the hexafluorophosphate salt have the η^3-*allyl* structure[401]. This reflects differences in stabilization of the complex structure due to crystal packing, and thus strongly suggests that the η^1-to-η^3 rearrangement is a low energy pathway.

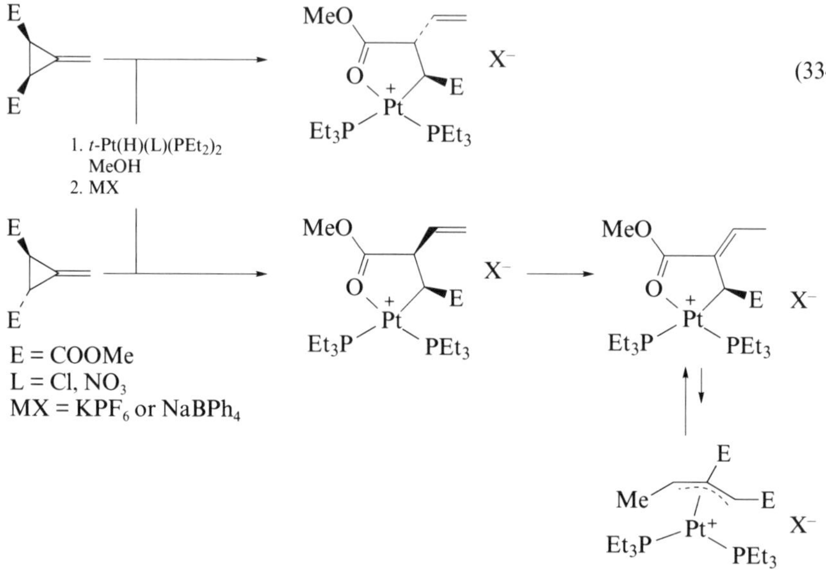

(334)

c. Reactions of η^2-methylenecyclopropane complexes. An early note on the thermal behavior of the bulky nickel complex (methylenecyclopropane)NiL_2 (L = tri-2-biphenylphosphite) reported that this complex rearranges to the corresponding η^4-butadiene complex (equation 335)[402], suggesting a proximal ring-opening pathway. Similarly, tetracarbonyl iron complexes of Feist's esters undergo stereoselective ring-opening reactions under thermal conditions (refluxing toluene) to butadiene complexes (equation 336). Interestingly, photolysis of *cis*-Feist's ester parallels the thermal reaction while irradiation of *trans*-Feist's ester leads to distal ring-opening products. Thermolysis in the presence of

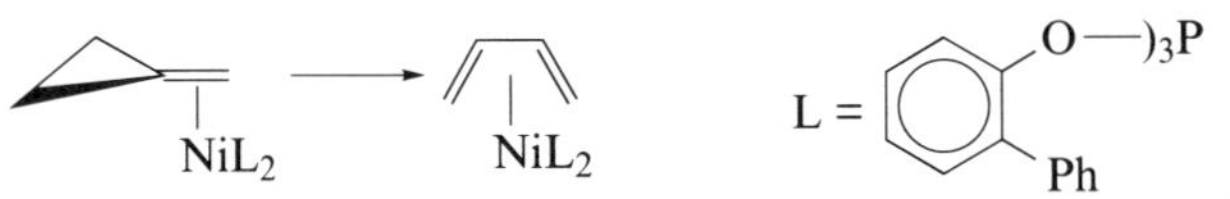

(335)

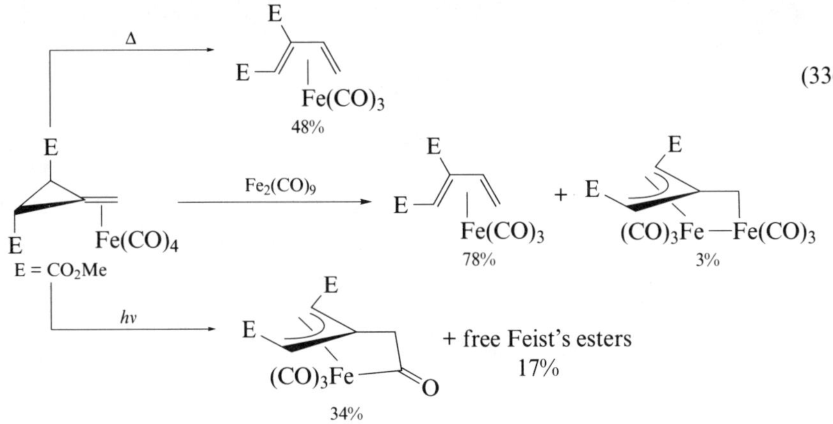

(336)

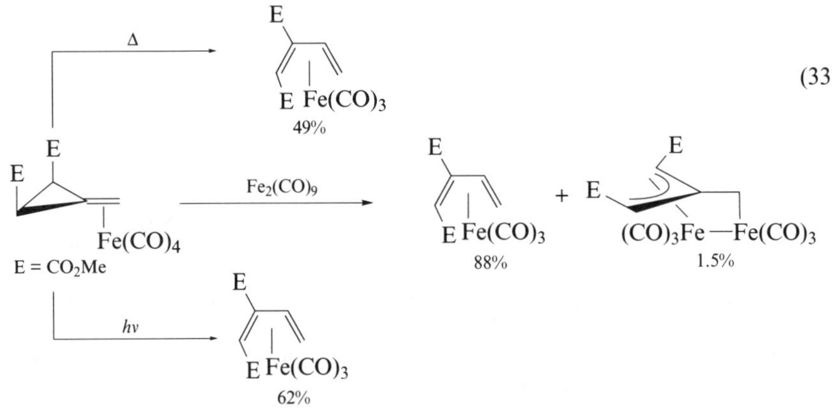

(337)

$Fe_2(CO)_9$ also gives small amounts (1.5–3%) of the distal ring-opened dinuclear σ,π-allylic complexes (equations 336 and 337)[378].

Thermolysis of bis(methylenecyclopropane)cyclopentadienylcobalt derivatives $CpCoL_2$ (L = mcp, 2,2,3,3-Me_4-mcp) led to isomeric butadiene complex dimers, in which one of the three-membered rings is retained (equation 338)[383]. Diphenylmethylenecyclopropane fails to give stable η^2-complexes of cobalt upon reaction with CpCo(ethylene)$_2$ but instead undergoes direct isomerization to a mixture of 1,1-diphenylbutadiene and fluxional dinuclear metallacyclopentadiene cobalt complexes (equation 339). Analogously, attempts to prepare η^2-complexes by the ligand exchange reaction (*see above*) of methylenecyclopropane and $CpCo(PPh_3)(CH_2{=}CHCN)$ gave only a codimer of the butadiene complex with acrylonitrile (equation 340).

A related proximal ring-opening reaction is observed when *trans*- or *cis*-2,3-bis(hydroxymethyl)methylenecyclopropanes react with $Fe_2(CO)_9$ in ether, affording a mixture of two noninterconvertible lactones (equation 341)[403,404]. It should, however, be noted

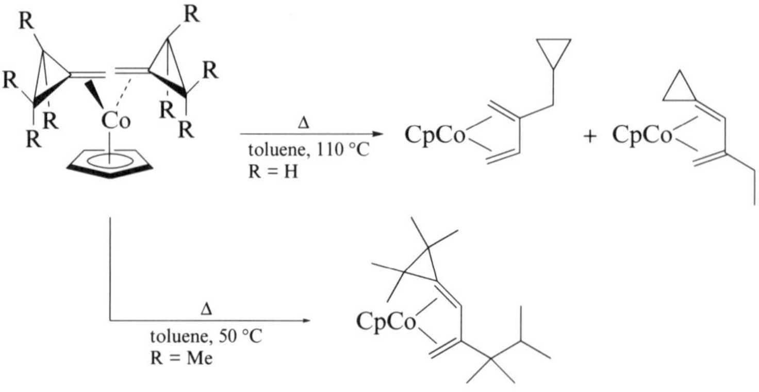

(338)

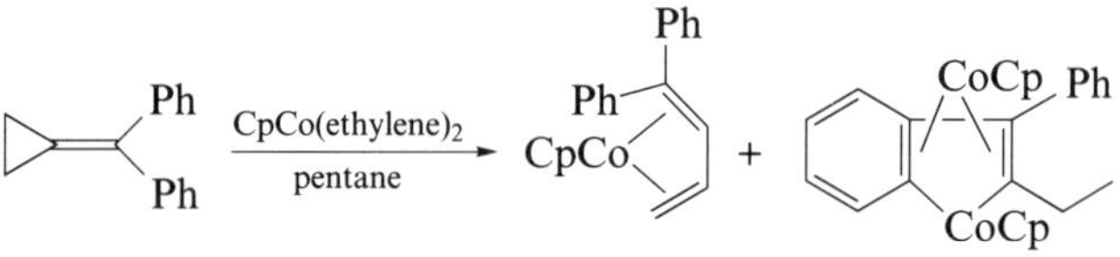

(339)

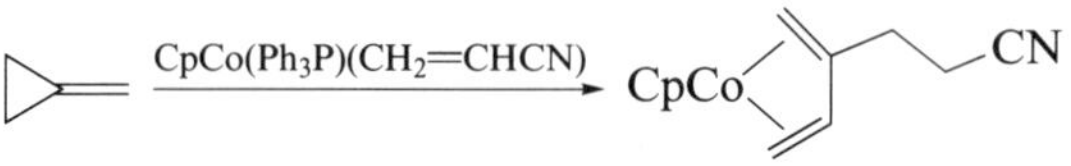

(340)

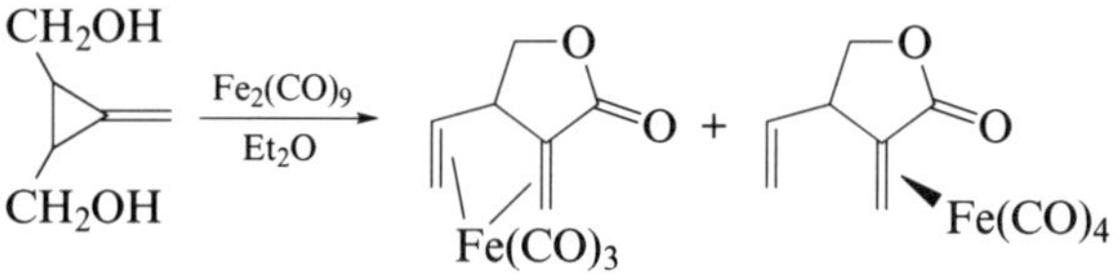

(341)

that reaction of the free Feist's esters[405], as well as other methylenecyclopropanes (*see below*), with excess of $Fe_2(CO)_9$ usually give trimethylenemethane complexes.

The reaction of 2-phenyl-1-methylenecyclopropane with $Fe_2(CO)_9$ gave, without isolation of the corresponding tetracarbonyl η^2-iron complex, a 3:2 mixture of both distal ring opening η^4-trimethylenemethane (TMM) complex (*see below*), and proximal ring-opening butadiene complex (equation 342)[381]. Both ring cleavage reactions are stereospecific, as shown by selective deuterium labeling of the 3-position. In contrast, the thermally stable (2,2-diphenyl-1-methylenecyclopropane)$Fe(CO)_4$ undergoes selective distal ring-opening upon reaction with $Fe_2(CO)_9$ (or Me_3NO) in benzene to give solely the η^4-TMM complex (equation 343).

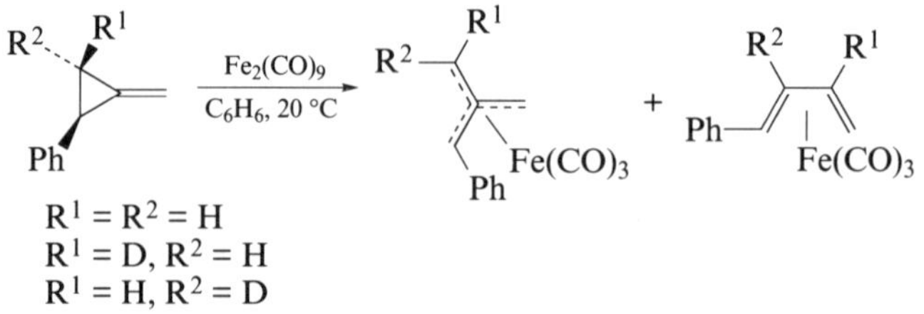

(342)

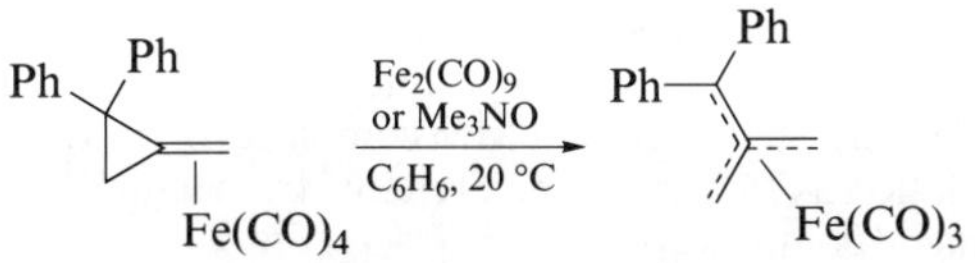

(343)

Based on the stereoselective ring opening reactions, it has been proposed that the butadiene complex formation involves initial decarbonylation, followed in turn by hydride abstraction to from (η^3-allyl)iron tricarbonyl hydride, disrotatory ring cleavage of the proximal σ-bond, and hydride insertion (Scheme 13)[381]. Analogously, a direct disrotatory distal ring-opening of the η^2-methylenecyclopropane iron tricarbonyl intermediate has been suggested for η^4-TMM complex formation (*see below*).

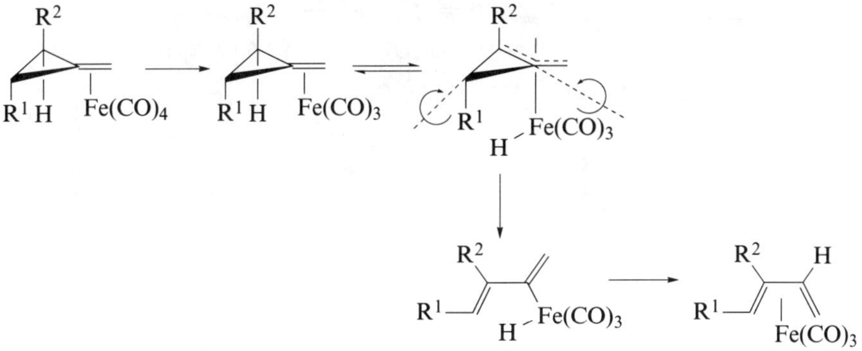

SCHEME 13

Finally, an unusual reaction between methylenecyclopropane and tricarbonyl(η^5-indenyl)methylmolybdenum gave, after prolonged standing at room temperature, the allylic dicarbonyl(η^5-indenyl)(η^3-1-methyl-2-acetylallyl)molybdenum complex (equation 344)[406]. It has been suggested that the reaction involves the initial formation of η^2-methylenecyclopropane acetylmolybdenum intermediate, followed by intramolecular coupling of the organic ligands. A proximal ring-opening rearrangement mode gives the final allylic complex.

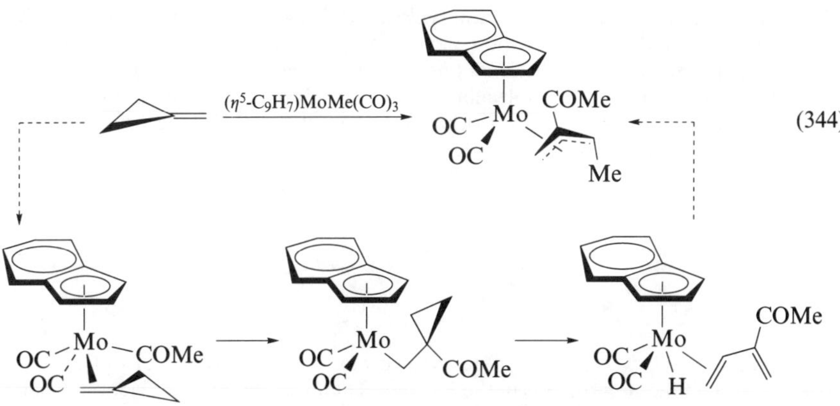

(344)

d. η⁴-Trimethylenemethane complexes. In principle, while the distal ring-opening of an η^2-methylenecyclopropane complex may lead to a η^4-trimethylenemethane (TMM) complex (equation 345), only a few isolated TMM complexes of iron (equation 346)[136,407–410] and molybdenum (equation 347)[411] have been prepared by reaction of methylenecyclopropane derivatives with metal complexes.

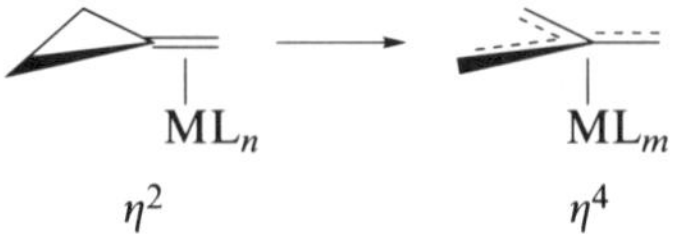

(345)

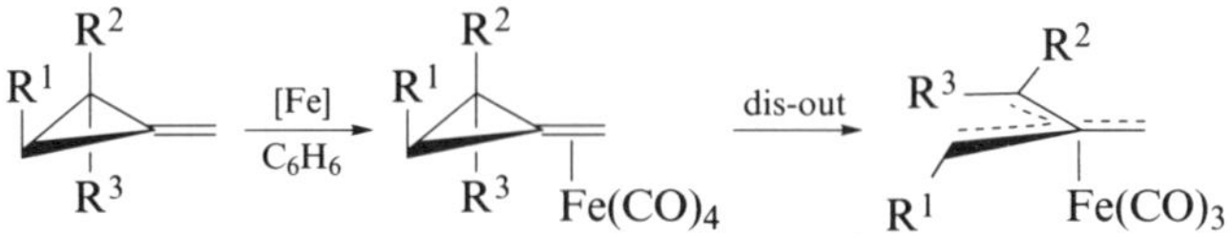

(346)

$R^1 = Ph, R^2 = R^3 = H$ $[Fe] = Fe_2(CO)_9$
$R^1 = Ph, R^2 = D, R^3 = H$ $Fe_2(CO)_9/Me_3NO$
$R^1 = Ph, R^2 = H, R^3 = D$ $Fe(CO)_5/Me_3NO$
$R^1 = H, R^2 = Me, R^3 = Ph$ (benzylideneacetone)$Fe(CO)_3$
$R^1 = H, R^2 = R^3 = Ph$
$R^1 = vinyl, R^2 = R^3 = H$
$R^1 = R^2 = CO_2Me, R^3 = H$
$R^1 = CO_2Me, R^2 = H, R^3 = CO_2Me$

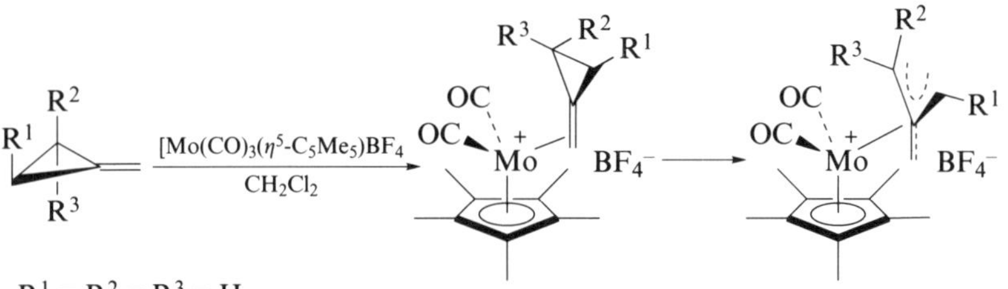

(347)

$R^1 = R^2 = R^3 = H$
$R^1 = R^2 = Me, R^3 = H$
$R^1 = R^3 = Me, R^2 = H$

As with the chloropalladation reaction (*vide supra*)[389] the rearrangement of η^2-methylenecyclopropane to η^4-TMM was shown experimentally to proceed stereoselectively by disrotatory ring cleavage of the distal σ-bond away from the metal (dis-out). This has been assessed by qualitative frontier molecular orbital considerations, which predict that the out-of-phase interaction between the σ-orbital and the metal orbital in the distal ring-opening of η^2-methylenecyclopropane complexes can be minimized by bending the bond up away from the metal (equation 348)[410,411].

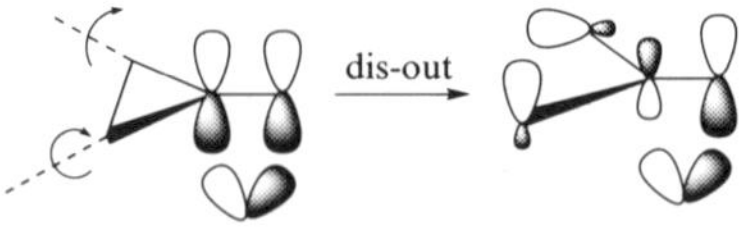

(348)

e. The Pauson–Khand reaction. Methylenecyclopropanes have been only recently utilized as the olefinic components in the Pauson–Khand cycloaddition reaction, where a cocycloaddition of alkynes, alkenes and carbon monooxide (from hexacarbonyldicobalt alkyne complex) generates cyclopentenones. Both the inter-[412] (equation 349) and intramolecular[413] (equation 350) reactions with methylenecyclopropanes proceed without ring cleavage, thus furnishing spirocyclopentenones. The reaction of the parent olefin with a variety of alkynes gives only poor yields (10–15%) of a 1:1 mixture of regioisomers when carried out in hexane. Better yields (25–81%) and improved selectivity are observed in a solvent-free system in which the acetylenic complex is applied to the surface of chromatography adsorbents such as SiO_2, Al_2O_3, $MgO{\cdot}SiO_2$ and Zeolite NaX. For steric reasons the intramolecular reaction of 1,6-enynes with a methylenecyclopropane terminator gives specifically the regioisomer with the carbonyl group adjacent to the spirocyclopropane ring. Diastereoselectivity is achieved by using chiral auxiliary groups adjacent to the triple bond, which ultimately leads to the enantioselective synthesis of spiro{cyclopropane-1,4′-bicyclo[3.3.0]oct-1-en-3-ones} (equation 351)[414].

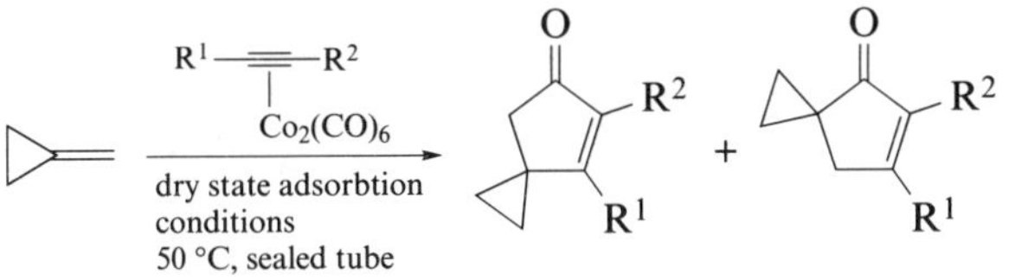

(349)

R^1 = H, R^2 = H, Me, Ph, *c*-Pr, vinyl, *i*-propenyl, $MeOCH_2$, $Me_2C(OH)$
$R^1 = R^2$ = Et
R^1 = Me, R^2 = Me_3Si

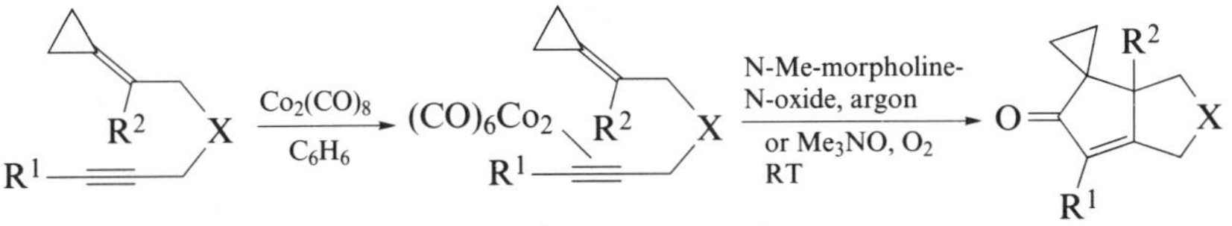

(350)

$R^1 = R^2$ = H, X = $CH(CO_2Me)$, $C(CO_2Me)_2$, $C(CO_2Me)(SO_2Ph)$, NTs
R^1 = Me, R^2 = H, X = $C(CO_2Et)_2$
R^1 = Me_3Si, R^2 = H, X = $C(CO_2Me)_2$
R^1 = H, R^2 = Me, X = $C(CO_2Me)_2$

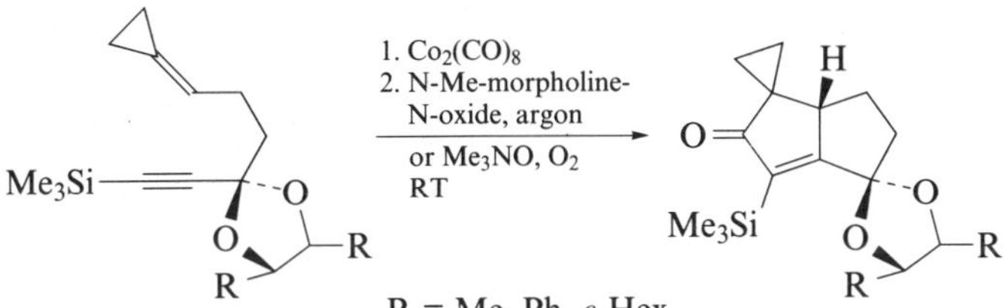

R = Me, Ph, *c*-Hex

(351)

4. *Catalyzed cycloaddition reactions*

A tremendous amount of work concerning metal-induced cycloadditions of methylenecyclopropane with olefins and alkynes has been done in recent years since the first reported nickel(0) catalyzed 3+2 cycloaddition of methylenecyclopropanes with electron-poor olefins (equation 352)[415] and the analogous palladium(0) codimerization (equation

353)[416,417]. Being among the most versatile and direct synthetic routes to functionalized cyclopentanes, these reactions have been thoroughly discussed in a number of excellent reviews[73,367,376,418,419] and thus only selected reactions will be illustrated here.

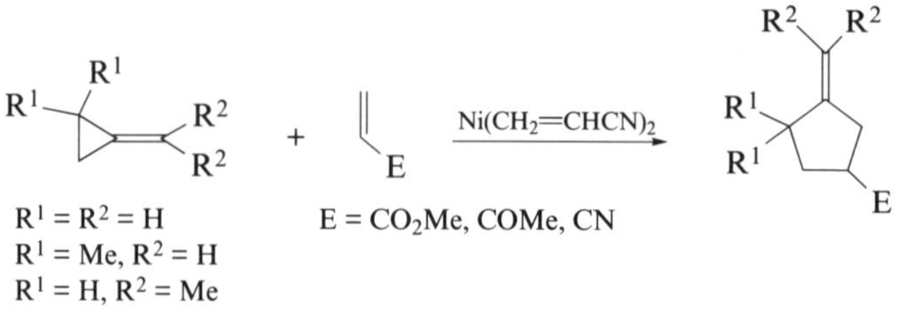

(352)

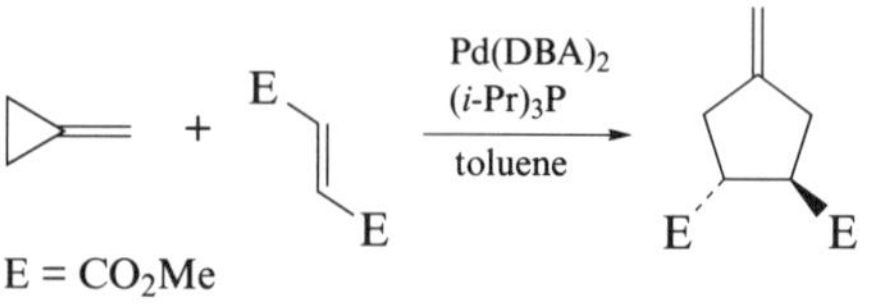

(353)

In general, two distinguished modes of ring cleavage reactions are conceivable: the 1,2-*proximal* ring-opening where a vinyl σ-bond is cleaved, and the 2,3-cleavage in which the *distal* σ-bond is ruptured. As a rule, alkyl ring-substituted methylenecyclopropanes undergo *nickel* catalyzed 3+2 cycloadditions with olefins selectively by the proximal ring-opening mode, whereas the analogous *palladium* catalyzed codimerizations proceed preferentially via the distal mode. (Scheme 14). However, codimerization of the *exo*-methylene- and phenyl-substituted methylenecyclopropanes with olefins occurs exclusively by distal ring-opening, irrespective of whether a Ni(0) or Pd(0) catalyst is employed.

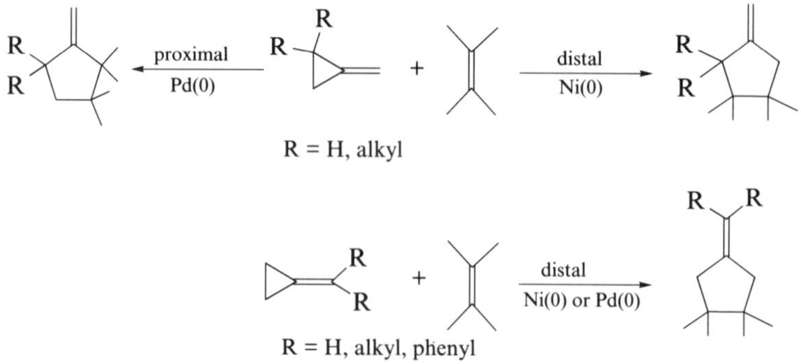

SCHEME 14

The mechanism of the codimerization reactions is depicted in Scheme 15[367]. It involves the initial coordination of the reactants (*i*) which is common to both proximal and distal reaction modes. For the proximal mode oxidative addition to the metal occurs next (*ii*) affording spirometallacyclopentane, which subsequently undergoes cyclopropylvinyl–homoallyl type rearrangement to metallacyclohexane (*iii*), and finally demetalation (*iv*) via reductive elimination. On the other hand, the distal pathway involves ring-opening to TMM intermediate complex (*v*), followed by successive oxidative addition to a σ,π-allyl complex (*vi*) and demetalation (*vii*).

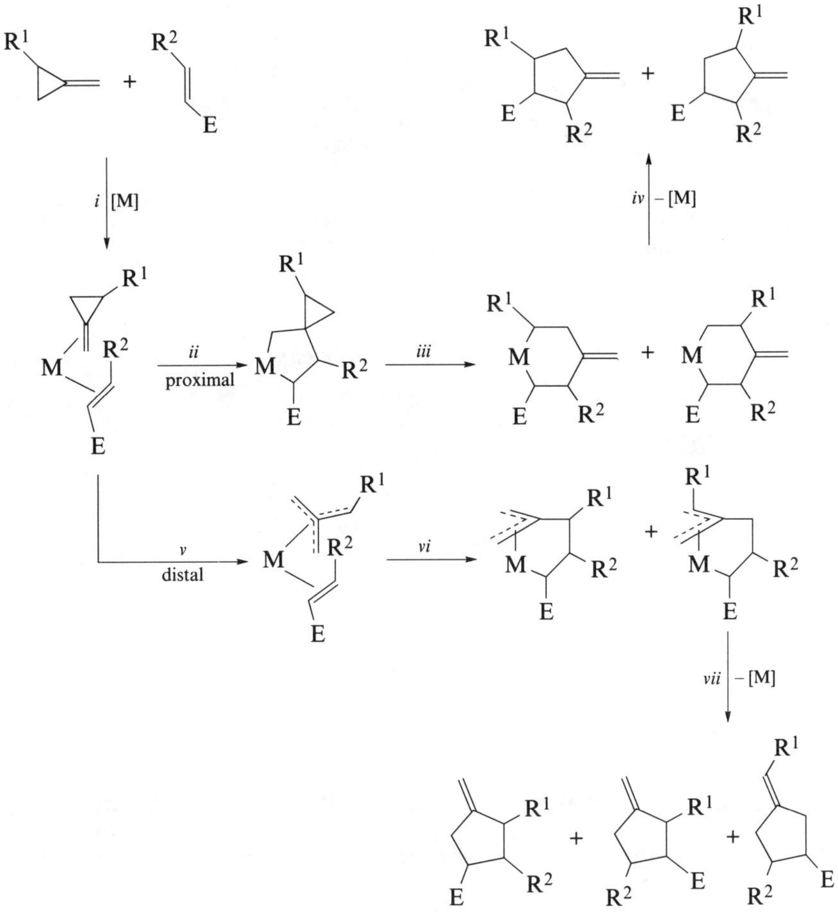

SCHEME 15

a. Intermolecular cycloadditions. While it has been established that the proximal versus distal chemoselectivity of the 3+2 codimerization reactions may eventually be controlled by adequate choice of the metal, these reactions nevertheless suffer from lack of regio- and stereoselectivity (Scheme 15). Further complications arise by competing reactions such as 2+2 dimerizations, 3+2 cyclodimerizations, trimerization[420] and ring cleavage to dienes. Selectivity between the various cycloadditions appears to be sensitive to steric rather than to electronic effects. Thus, progressive methyl ring substitution effects rather drastically the ratio of adducts in both Ni(0)-catalyzed dimerization (equation 354) and codimerization (equation 355) reactions[73]. However, the two double bonds in 2,3-dimethoxycarbonylnorbornadiene are almost equally active towards methylenecyclopropane, furnishing exclusively the *exo*-adducts (equation 356)[367].

Nickel(0)-catalyzed codimerization of methylenecyclopropanes with electron-deficient olefines are highly regiospecific, but show a rather poor stereoselectivity. Thus the asymmetric nickel(0)-catalyzed codimerization of methylenecyclopropanes with the chiral bornane derivatives of acrylic acid leads to the optically active 3-methylenecyclopen-

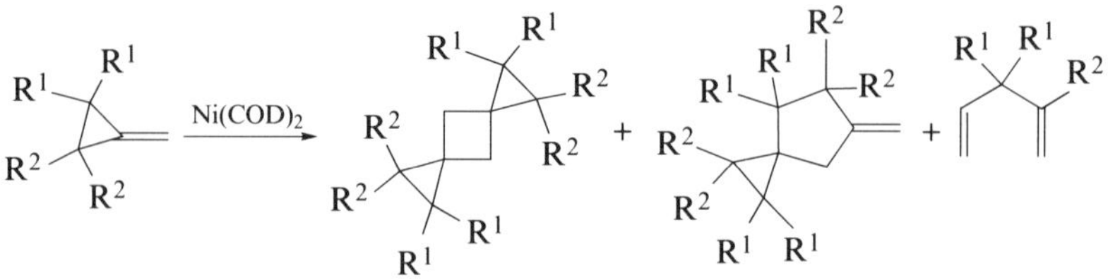

(354)

$R^1 = R^2 = H$	20%	80%	—
$R^1 = H, R^2 = Me$	87%	13%	0%
$R^1 = R^2 = Me$	70%	0%	30%

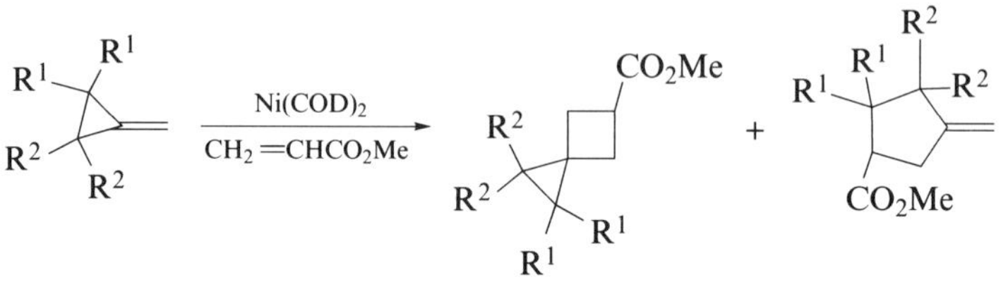

(355)

$R^1 = R^2 = H$	0%	100%
$R^1 = H, R^2 = Me$	19%	81%
$R^1 = R^2 = Me$	100%	0%

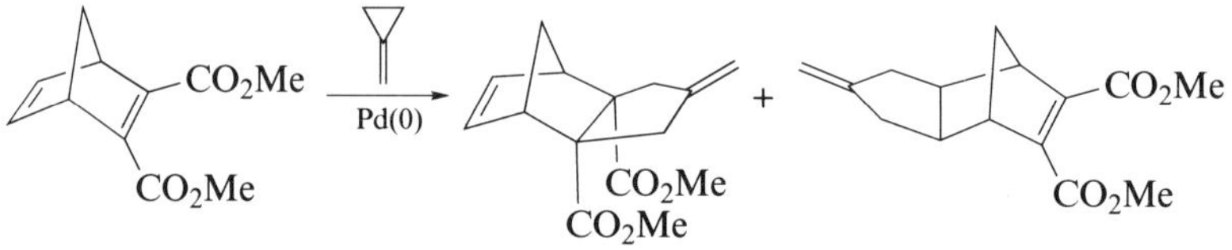

(356)

tanecarboxylic esters and amides in good yields (55–91%). The diastereomeric excess (de) accessible depends on the steric demand of the chiral auxiliaries. Values up to 98% de are achieved, particularly with the tricyclic bornane-sulfonamide derivatives (equation 357)[421,422].

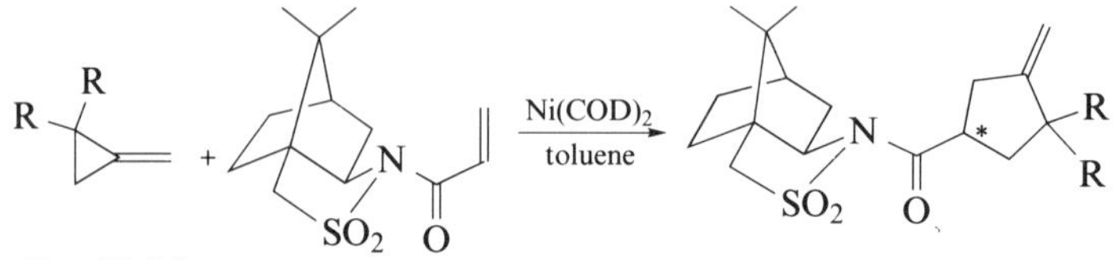

(357)

R = H, Me

However, when a series of electron-deficient *trans*-olefins is codimerized with methylenecyclopropane a mixture of stereoisomers is obtained, though only the regioisomers where the electron-withdrawing group is far from the *exo*-methylene group are formed (equation 358)[367].

b. Intramolecular cycloadditions. Metal-induced intramolecular 3+2 cycloadditions, unlike their intermolecular counterparts, are highly regioselective reactions, free of undesirable side reactions such as codimerization, ring cleavage to dienes and rearrangements. In constraint systems where the two addends are separated by a short chain, the intramol-

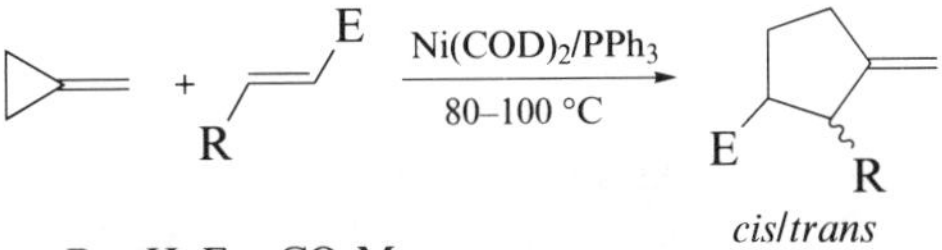

(358)

	cis/*trans*
R = H, E = CO_2Me	
R = Me, E = CO_2Me	22:78
R = Me, E = CHO	9:91
R = Pr, E = CO_2Me	27:73
R = $(CH_2)_2CO_2Me$, E = CO_2Me	28:72

ecular reactios are completely controlled by steric demands with no dependence on either the catalyst or the relative position of peripheral substituents. In more flexible systems the ring-cleavage regioselectivity is dictated by the catalyst. Hence both nickel [$Ni(COD)_2$–Ph_3P or $Ni(acac)_2$–DIBAL–Ph_3P] and palladium [$PdCl_2(Ph_3P)_2$–DIBAL] complexes catalyze the transannular 3+2 cycloaddition of methylenecyclopropane to olefin in 1-cyclopropylidene-5-alkylidenecyclooctane, giving [3.3.3]propellanes as the sole cycloaddition products (equation 359). The distal ring-opening mode of the methylenecyclopropane moiety, to give an intermediate complex in which the metal is coordinated to trimethylenemethane (TMM) and the olefin, has been implicated as the reaction pathway[423].

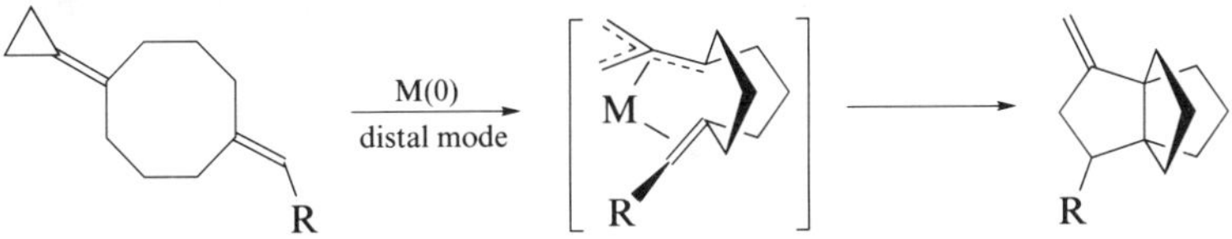

(359)

R = H, CO_2Et
M = $Ni(COD)_2/PPh_3$, $Ni(acac)_2$/DIBAL/PPh_3, $PdCl_2(PPh_3)_2$/DIBAL

Distal ring-opening is also observed in the intramolecular 3+2 cycloaddition reaction of a 3-pentenone diphenylmethylenecyclopropane derivative in the presence of bis(dibenzylideneacetone)palladium/tri-isopropyl phosphite catalyst, affording regioselectively a diphenylmethylene pentalenone system (equation 360)[424]. Similarly, thermal cyclization of

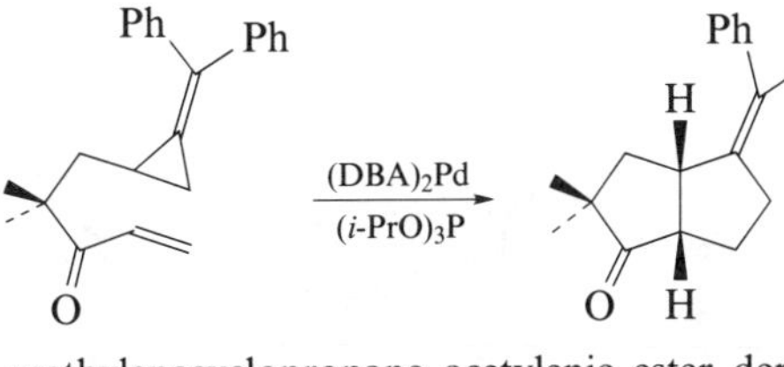

(360)

the corresponding methylenecyclopropane acetylenic ester derivative using the same $(DBA)_2Pd$-phosphite catalyst, or alternatively tetrakis(triphenylphosphine)palladium under mild sonication, gave regioselectively a 1:1 diastereomeric pentalene cycloadduct by the same distal mode, despite the reversed relative configuration of the peripheral electron-withdrawing substituent (equation 361).

Finally, when the tether between the two functional groups is elongated by one additional carbon, regioselectivity can be governed by the metal. Thus a Pd(0), which favors the distal ring-opening mode, induced the intramolecular cycloaddition of methylenecyclopropane propargylic ester derivative specifically to the corresponding hydrindanes, whereas the same substrate under Ni(0) catalysis gave solely the *anti*-Bredt diene cycloadduct of the proximal ring cleavage product (equation 362)[425].

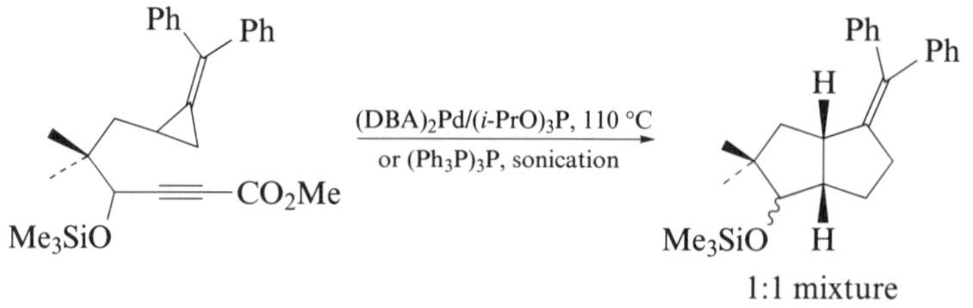

(361)

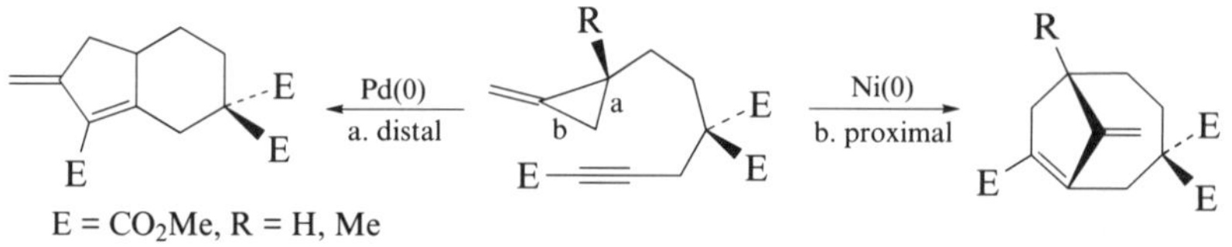

(362)

VIII. METHYLENECYCLOPROPENE COMPLEXES (TRIAFULVENES)

The only reported triafulvene in which the entire cross-conjugated π-system is coordinated to the metal is the unusual tetrasilylated η^4-irontricarbonyl complex, obtained by the intramolecular coupling of the acetylenic groups in a [14]macrocyclic oxasiladiyne, using $Fe_2(CO)_9$ in refluxing benzene (equation 363)[426]. The solid state X-ray structure reveals that the bond distances of both internal and external coordinated double bonds, 148 and 140 pm respectively, are greater than the corresponding bond distances (132 and 133 pm) in the parent triafulvene (determined by microwave spectra). In contrast, the σ-bond are only little affected by the complexation, becoming slightly shorter (142 pm) than the free hydrocarbon (144 pm). Unusual also is the out-of-plane bending of the exomethylene group towards the metal by 27.9°.

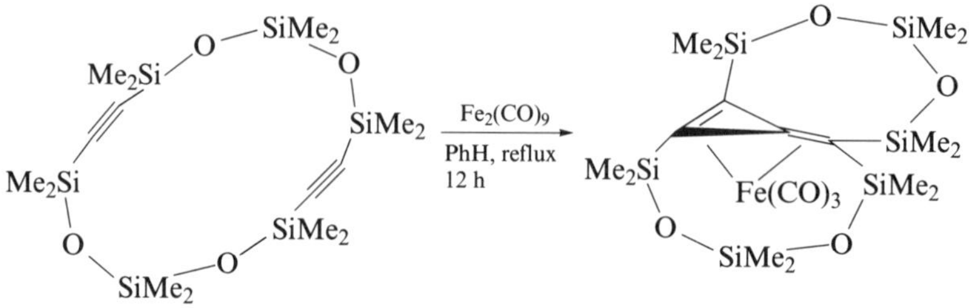

(363)

Attempt to prepare π-complexes of triafulvenes and related methylene cycloproparenes[285,427,428] directly by ligand exchange reaction with transition metal complexes resulted in metal insertion into the sigma bond, forming metallacyclic complexes. Thus reaction of the electron-poor triafulvene 1,2-diphenyl-3-dicyanomethylenecyclopropene with (ethylene)bis(triphenylphosphine)platinum in refluxing benzene gave two crystalline products whose platinacyclobutene structure was confirmed by X-ray structure analysis (equation 364)[429].

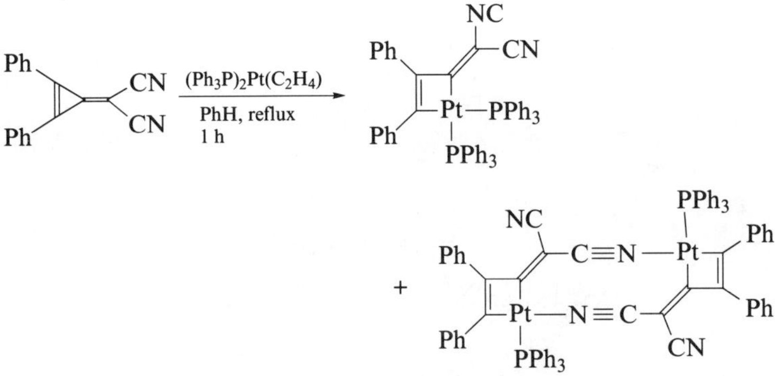

(364)

Bis(diisopropylamino)triafulvene, prepared by deprotonation of the corresponding methyl cyclopropenylium perchlorate (see Section VI.B.2) with BuLi, reacts with cyclopentadienyliron dicarbonyl iodide (FpI) to give the first triafulvene iron complex $[(\{i\text{-}Pr_2N\}_2C_3CH_2)FeCp(CO)_2]ClO_4$ (equation 365)[340]. The complex is stable in crystalline form, but is air-sensitive in solution. The ^{13}C NMR spectrum exhibits a high field methylene signal at $\delta = -18.8$ ppm, suggesting a major contribution of the cyclopropenylium σ-type canonic form rather than the η^2-triafulvene π-type complex form.

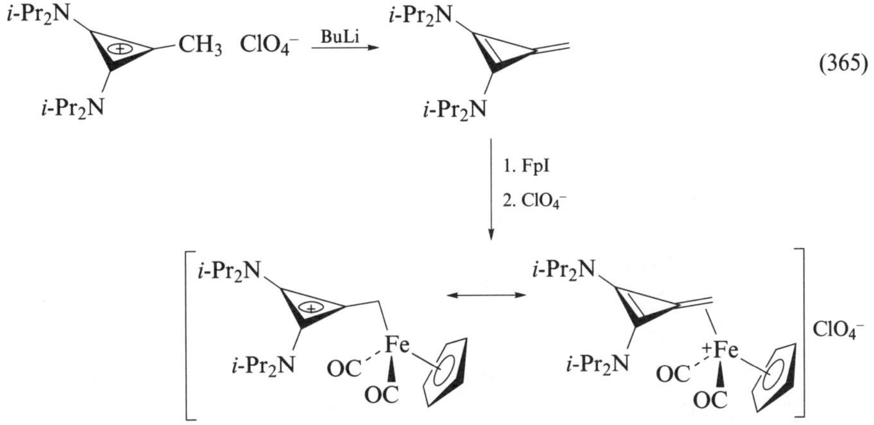

(365)

Similarly, a series of the related diaminocalicene carbonyl complexes (triapentafulvalenes) of group 6 was reported to be formed upon reaction of the parent bis(dialkylamino)calicene with $M(CO)_6$ (M = Cr, Mo, W). Likewise, a ferrocene analogue incorporating two diaminocyclopropenylium rings was synthesized by reaction of two equivalents of calicene with $FeCl_2$ (equation 366)[340].

More recently, a novel metal-substituted methylenecyclopropene (triafulvene) derivative was obtained when bis(propyne)zirconocene was treated with one equivalent of tris(pentafluorophenyl)borane, followed by excess of benzonitrile (equation 367)[430]. The first step involves alkynyl ligand coupling to give the isolable $Cp_2Zr(\mu\text{-}2,4\text{-hexadiyne})B(C_6F_5)_3$ betaine. This undergoes a formal intramolecular nitrile insertion into the Zr—C(sp^2) σ-bond of the adjacent alkenyl zirconocene unit, leading to the zirconium–boron triafulvene–betaine. X-ray analysis of the triafulvene confirmed the planar

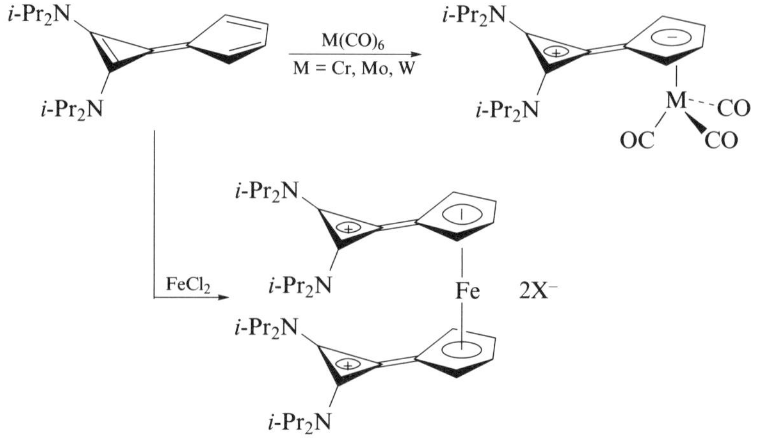

(366)

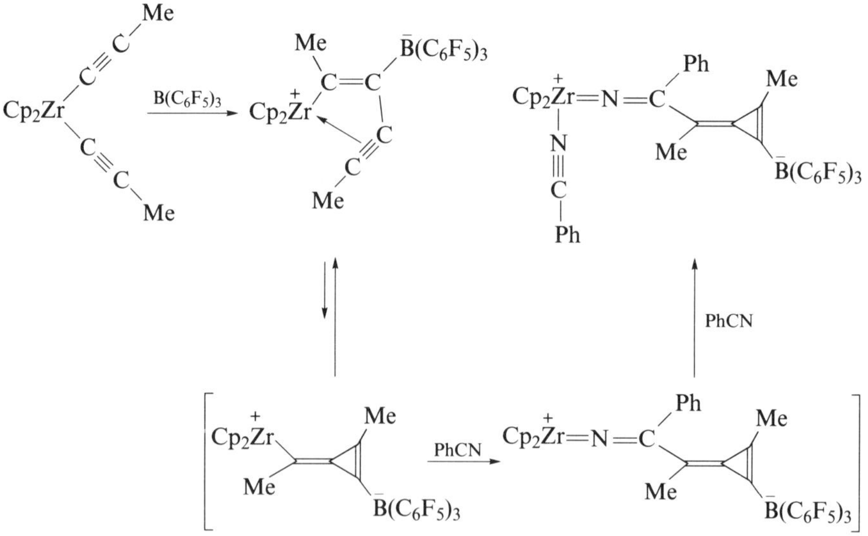

(367)

structure of the 4-ketimino methylenecyclopropane system. Characteristic bond lengths of the triafulvene system are: 134 (cyclopropene C=C), 137 (methylene C=C), 139 [cyclopropene C—C(B)] and 142 [cyclopropene C—C(Me)] pm.

Complexes of di- and tri-methylenecyclopropanes are unknown. However, the alkali metal salts of substituted trimethylenecyclopropane dianions and their corresponding radical anions are stable. The dianions are prepared by base-induced condensation of tetrachlorocyclopropene with three equivalents of malonic derivatives, or alternatively in two steps, via a zwitterionic aminotriafulvene (equation 368). Further oxidation with potassium persulfate gave the corresponding radical anions[431].

The potassium triafulvene thiolate $K[C_3(p\text{-}MeSC_6H_4)\{C(CN)_2\}S]$ is prepared similarly (equation 369)[432]. The synthesis of the lithium salt of an analogous cyclic hydrazidocalicene has also been reported (equation 370)[433].

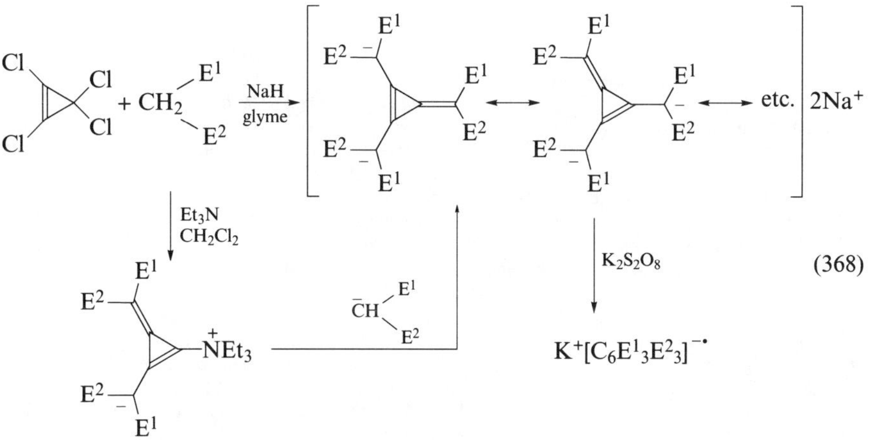

(368)

$E^1 = E^2 = COOMe, CN$
$E^1 = COOMe, E^2 = CN$

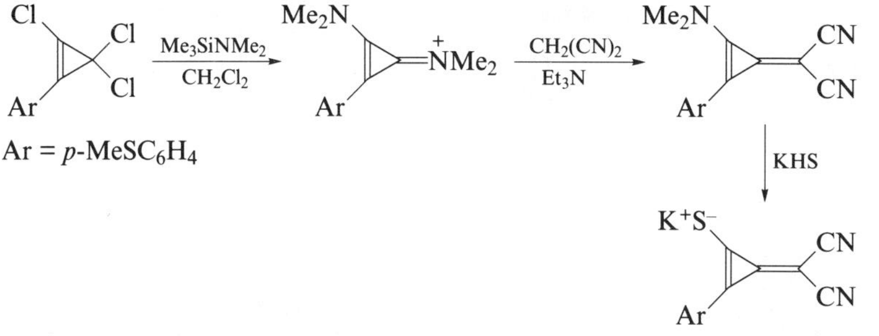

(369)

Ar = p-$MeSC_6H_4$

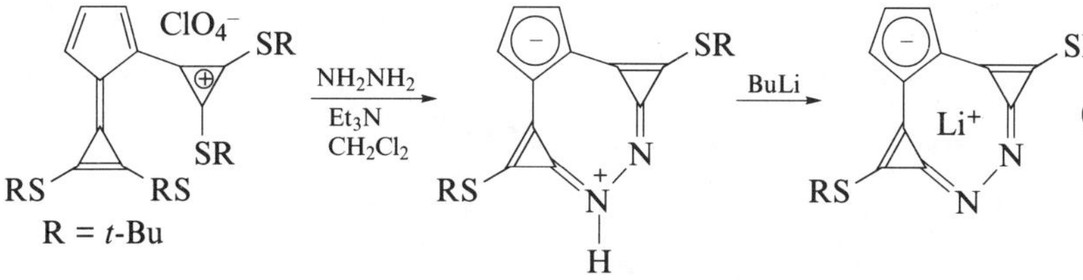

(370)

R = t-Bu

IX. ACKNOWLEDGMENTS

I thank Dr Joel. L. Wolk for reading the manuscript and for his useful comments, and my wife Riki for her encouragement and endless patience.

X. REFERENCES

1. Z. Rappoport (Ed.), *The Chemistry of the Cyclopropyl Group*, Vol. 1, parts 1 and 2, Wiley, Chichester, 1987.
2. D. Wendisch, in *Methoden Der Organischen Chemie (Houben-Weyl)*, Vol. 4/3 (Ed. E. Müller), Georg Thieme Verlag, Stuttgart, 1971.
3. Ch. Elschenbroich and A. Salzer, *Organometallics*, VCH Publ., Weinheim, 1989.

4. D. Seyferth and H. M. Cohen, *J. Organomet. Chem.*, **1**,15 (1963).
5. D. T. Longone and W. D. Wright, *Tetrahedron Lett.*, 2859 (1969).
6. J. D. Roberts and V. C. Chambers, *J. Am. Chem. Soc.*, **73**, 3176 (1951).
7. G. F. Reynolds, R. E. Dessy and H. H. Jaffe, *J. Org. Chem.*, **23**, 1217 (1958).
8. G. Boche and H. M. Walborsky, *Cyclopropane Derived Reactive Intermediates*, Wiley, Chichester, 1990.
9. K. Gawronska, J. Gawronski and H. M. Walborsky, *J. Org. Chem.*, **56**, 2193 (1991).
10. D. Seyferth and R. L. Lambert, Jr. *J. Organomet. Chem.*, **88**, 287 (1975).
11. C. P. Vlaar and G. W. Klumpp, *Angew. Chem., Int. Ed. Engl.*, **32**, 574 (1993).
12. A. Oku, Y. Ose, T. Kamada and T. Yoshida, *Chem. Lett.*, 573 (1993).
13. M. Rosenblum, *J. Organomet. Chem.*, **300**, 191 (1986).
14. A. Cutler, R. W. Fish, W. P. Giering and M. Rosenblum, *J. Am. Chem. Soc.*, **94**, 4354 (1972).
15. N. J. Conti and W. M. Jones, *Organometallics*, **7**, 1666 (1988).
16. Y. Omrčen, N. J. Conti and W. M. Jones, *Organometallics*, **10**, 913 (1991).
17. M. I. Bruce, M. N. Iqbal and F. G. A. Stone, *J. Organomet. Chem.*, **20**, 161 (1969).
18. D. J. Crowther, Z. Zhang, G. J. Palenik and W. M. Jones, *Organometallics*, **11**, 622 (1992).
19. A. P. Masters, M. Parvez, T. S. Sorensen and F. Sun, *Can. J. Chem.*, **71**, 230 (1993).
20. F. J. Manganiello, L. W. Christensen and W. M. Jones, *J. Organomet. Chem.*, **235**, 327 (1982).
21. J. R. Lisko and W. M. Jones, *Organometallics*, **4**, 612 (1985).
22. Y. Standstrøm and W. M. Jones, *Organometallics*, **5**, 178 (1986).
23. Y. Standstrøm, A. E. Kosiol, G. J. Palenik and W. M. Jones, *Organometallics*, **6**, 2079 (1987).
24. R. L. Trace and W. M. Jones, *J. Organomet. Chem.*, **376**, 103 (1989).
25. N. L. Jones and A. Abers, *Organometallics*, **2**, 490 (1983).
26. T. N. Mitchel and K. Kwetkat, *J. Organomet. Chem.*, **439**, 127 (1992).
27. D. Seyferth, R. L. Lambert, Jr. and M. Massol, *J. Organomet. Chem.*, **88**, 255 (1975).
28. T. Harada, T. Katsukhira, K. Hattori and A. Oku, *J. Org. Chem.*, **58**, 2958 (1993).
29. T. Butkowskyj-Walkiw and G. Szeimies, *Tetrahedron*, **42**, 1845 (1986).
30. M. Pohmakotr and S. Khosavanna, *Tetrahedron*, **49**, 6483 (1993).
31. H. Yamamoto, K. Kitatani, T. Hiyama and H. Nozaki, *J. Am. Chem. Soc.*, **99**, 5816 (1977).
32. K. I. Grandberg and V. P. Dyadchenko, *J. Organomet. Chem.*, **474**, 1 (1994).
33. H. M. Walborsky and M. P. Periasamy, *J. Organomet. Chem.*, **179**, 81 (1979).
34. E. Piers, C. K. Lau and I. Nagakura, *Tetrahedron Lett.*, 3233 (1976).
35. E. Piers, E. Banville, C. K. Lau and I. Nagakura, *Can. J. Chem.*, **60**, 2965 (1982).
36. E. Piers, C. K. Lau and I. Nagakura, *Can. J. Chem.*, **61**, 288 (1983).
37. A. de Meijere, F. Jackel, A. Simon, H. Borrmann, J. Kohler, D. Johnels and L. T. Scott, *J. Am. Chem. Soc.*, **113**, 3935 (1991).
38. (a) E. G. Perevalova, I. G. Bolesov, Y. T. Struchkov, I. F. Leschova, Y. S. Kalyuzhnaya, T. I. Voyevodskaya, Y. L. Slovokhotov and K. I. Grandberg, *J. Organomet. Chem.*, **286**, 129 (1985). (b) E. G. Perevalova, I. G. Bolesov, Y. S. Kalyuzhnaya, T. I. Voyevodskaya, L. G. Kuzmina, V. I. Korsunsky and K. I. Grandberg, *J. Organomet. Chem.*, **369**, 267 (1989).
39. M. Rosenblum, *Acc. Chem. Res.*, **7**, 122 (1974).
40. N. A. Petasis and E. I. Bzowej, *Tetrahedron Lett.*, **34**, 943 (1993).
41. J. Yin and W. M. Jones, *Organometallics*, **12**, 2013 (1993).
42. F. J. Feher, D. D. Gergens and J. W. Ziller, *Organometallics*, **12**, 2810 (1993).
43. H. Lehmkuhl, J. Grundke and R. Mynott, *Chem. Ber.*, **116**, 159 (1983).
44. H. Lehmkuhl, J. Grundke and R. Mynott, *Chem. Ber.*, **116**, 176 (1983).
45. J. Müller, C. Friedrich, P. E. Gaede, S. Sodemann and K. Qiao, *J. Organomet. Chem.*, **471**, 249 (1994).
46. J. M. Brown and K. Mertis, *J. Chem. Soc., Perkin Trans. 2*, 1993 (1973).
47. R. L. Phillips and R. J. Puddephatt, *J. Chem. Soc., Dalton Trans.*, 1732 (1978).
48. R. Häner, T. Maetzke and D. Seebach, *Helv. Chim. Acta*, **69**, 1655 (1986).
49. I. Reichelt and H.-U. Reissig, *Justus Liebigs Ann. Chem.*, 3895 (1983).
50. I. Reichelt and H.-U. Reissig, *Justus Liebigs Ann. Chem.*, 531 (1984).
51. W. Kirmse and K. Rode, *Tetrahedron*, **43**, 3187 (1987).
52. R. C. Petter, G. Kumaravel, D. G. Powers and C.-T. Chang, *Tetrahedron Lett.*, **32**, 449 (1991).
53. M. Pohmakotr and J. Ratchataphusit, *Tetrahedron*, **49**, 6473 (1993).
54. S. Hoz, in *The Chemistry of the Cyclopropyl Group*, Vol. 1, part 2 (Ed. Z. Rappoport), Chap. 19, Wiley, Chichester, 1987.

55. A.-D. Schlüter, H. Huber and G. Szeimies, *Angew. Chem., Int. Ed. Engl.*, **24**, 404 (1985).
56. E. Hahn, T. Maetzke, D. A. Plattner and D. Seebach, *Chem. Ber.*, **123**, 2059 (1990).
57. R. P. Zerger and G. D. Stucky, *J. Chem. Soc., Chem. Commun.*, 44 (1973).
58. D. A. Sanders and J. P. Oliver, *J. Am. Chem. Soc.*, **90**, 5910 (1968).
59. J. W. Moore, D. A. Sanders, P. A. Scherr, M. D. Glick and J. P. Oliver, *J. Am. Chem. Soc.*, **93**, 1035 (1971).
60. K. Tanaka, K. Minami, I. Funaki and H. Suzuki, *Tetrahedron Lett.*, **31**, 2727 (1990).
61. M. Lautens and P. H. M. Delanghe, *J. Org. Chem.*, **58**, 5037 (1993).
62. M. Lautens and P. H. M. Delanghe, *J. Org. Chem.*, **57**, 798 (1992).
63. D. Seyferth and F. M. Armbrecht, Jr., *J. Am. Chem. Soc.*, **91**, 2616 (1969).
64. H. Lekmkuhl and K. Mehler, *Justus Liebigs Ann. Chem.*, 1841 (1978).
65. H. Lehmkuhl, C. Naydowski, R. Benn and A. Rufinska, *J. Organomet. Chem.*, **216**, C41 (1981).
66. L. A. Paquette and R. Grée, *J. Organomet, Chem.*, **146**, 319 (1978).
67. K. Kubota, M. Nakamura, M. Isaka and E. Nakamura, *J. Am. Chem. Soc.*, **115**, 5867 (1993).
68. E. Nakamura, M. Isaka and S. Matsuzawa, *J Am. Chem. Soc.*, **110**, 1297 (1988).
69. M. Isaka and E. Nakamura, *J. Am. Chem. Soc.*, **112**, 7428 (1990).
70. L. R. Gilliom and R. H. Grubbs, *Organometallics*, **5**, 721 (1986).
71. L. R. Gilliom and R. H. Grubbs, *J. Am. Chem. Soc.*, **108**, 733 (1986).
72. P. Binger and H. M. Büch, *Top. Curr. Chem.*, **135**, 77 (1978).
73. P. Binger, B. Cetinkaya, M. J. Doyle, A. Germer and U. Schuchardt, *Fundam. Res. Homogeneous Catal.*, **3**, 271 (1979).
74. P. Binger, M. J. Doyle, J. McMeeking, C. Krüger and Y.-H. Tsay, *J. Organomet. Chem.*, **135**, 405 (1977).
75. T. A. Peganova, P. V. Petrovskii, L. S. Isaeva, D. N. Kravtsov, D. B. Furman, A. V. Kudryashev, A. O. Ivanov, S. V. Zotova and O. V. Bragin, *J. Organomet. Chem.*, **282**, 283 (1985).
76. R. Benn, R.-D. Reinhardt and A. Rufinska, *J. Organomet, Chem.*, **282**, 291 (1985).
77. P. Binger, P. Müller, R. Benn and R. Mynott, *Angew. Chem., Int. Ed. Engl.*, **28**, 610 (1989).
78. P. Binger, H. M. Buch, R. Benn and R. Mynnot, *Angew. Chem., Int. Ed. Engl.*, **21**, 62 (1982).
79. B. Cetinkaya, P. Binger and C. Krüger, *Chem. Ber.*, **115**, 3414 (1982).
80. E. Dunkelblum, *Isr. J. Chem.*, **11**, 557 (1973).
81. R. A. Periana and R. G. Bergmann, *J. Am. Chem. Soc.*, **108**, 7346 (1986).
82. K. B. Wiberg, J. V. McCluski and G. K. Schulte, *Tetrahedron Lett.*, **27**, 3083 (1986).
83. P. Knochel and J. F. Normant, *Tetrahedron Lett.*, **27**, 5727 (1986).
84. D. Beruben, I. Marek, J. F. Normant and N. Platzer, *Tetrahedron Lett.*, **34**, 7575 (1993).
85. S. Cohen and A. Yogev, *J. Am. Chem. Soc.*, **98**, 2013 (1976).
86. D. J. Patel, C. L. Hamilton and D. J. Roberts, *J. Am. Chem. Soc.*, **87**, 5144 (1965).
87. A. Maercker, P. Güthlein and H. Whittmayr, *Angew. Chem., Int. Ed. Engl.*, **12**, 774 (1973).
88. M. S. Silver, P. R. Schafer, J. E. Nordlander, C. Rüchardt and J. D. Roberts, *J. Am. Chem. Soc.*, **82**, 2646 (1960).
89. A. Maercker and J. D. Roberts, *J. Am. Chem. Soc.*, **88**, 1742 (1966).
90. E. Dunkelblum and S. Brenner, *Tetrahedron Lett.*, 669 (1973).
91. A. Maercker, *Justus Liebigs Ann. Chem.*, **732**, 151 (1970).
92. A. Maercker and U. Girreser, *Angew. Chem., Int. Ed. Engl.*, **29**, 667 (1990).
93. R. S. Brown, D. F. Eaton, A. Hosomi, T. G. Traylor and J. M. Wright, *J. Organomet. Chem.*, **66**, 249 (1974).
94. J. S. Philippo, J. Silbermann and P. J. Fagan, *J. Am. Chem. Soc.*, **100**, 4834 (1978).
95. J. S. Philippo and J. Silbermann, *J. Am. Chem. Soc.*, **104**, 2831 (1982).
96. M. S. Alnajjar, G. F. Smith and H. G. Kuivila, *J. Org. Chem.*, **49**, 1271 (1984).
97. A. J. Lucke and D. J. Young, *Tetrahedron Lett.*, **32**, 807 (1991).
98. A. J. Lucke and D. J. Young, *Tetrahedron Lett.*, **35**, 1609 (1994).
99. K. C. Pande and S. Winstein, *Tetrahedron Lett.*, 3393 (1964).
100. E. Vedejs and M. Salomon, *J. Org. Chem.*, **37**, 2075 (1972).
101. E. Vedejs and M. Salomon, *J. Chem. Soc., Chem. Commun.*, 1582 (1971).
102. A. Segnitz, E. Kelly, S. H. Taylor and P. M. Meitlis, *J. Organomet. Chem.*, **124**, 113 (1977).
103. K. C. Nicolaou, N. A. Petasis and D. A. Claremon, *Tetrahedron*, **41**, 4835 (1985).
104. K. C. Nicolaou, D. A. Claremon, W. E. Barnette and S. P. Seits, *J. Am. Chem. Soc.*, **101**, 3704 (1979).
105. S. Yamazaki, S. Katoh and S. Yamabe, *J. Org. Chem.*, **57**, 4 (1992).

106. S. Yamazaki, M. Tanaka, A. Yamaguchi and S. Yamabe, *J. Am. Chem. Soc.*, **116,** 2356 (1994).
107. L. J. Mérour, C. Charrier, J. Benaïm, J. L. Roustan and D. Commereuc, *J. Organomet. Chem.*, **39**, 321 (1972).
108. P. J. Krusic, P. J. Fagan and J. S. Philippo, *J. Am. Chem. Soc.*, **99**, 250 (1977).
109. W. P. Giering and M. Rosenblum, *J. Am. Chem. Soc.*, **93**, 5299 (1971).
110. L. Cohen, W. P. Giering, D. Kenedy, C. V. Magatti and A. Sanders, *J. Organomet. Chem.*, **65**, C57 (1974).
111. J. M. Brown, J. A. Conneely and K. Mertis, *J. Chem. Soc., Perkin Trans. 2*, 905 (1974).
112. G. E. Herberich, T. Carstensen, W. Klein and M. U. Schmidt, *Organometallics*, **12**, 1439 (1993).
113. J. K. Kochi and J. W. Powers, *J. Am. Chem. Soc.*, **92**, 137 (1970).
114. R. Poli, G. Wilkinson, M. Motevalli and M. B. Hursthouse, *J. Chem. Soc., Dalton Trans.*, 931 (1985).
115. H. Lehmkuhl and S. Fustero, *Justus Liebigs Ann. Chem.*, 1361 (1980).
116. H. Lehmkuhl, A. Rufinska, R. Benn, G. Schroth and R. Minott, *Justus Liebigs Ann. Chem.*, 317 (1981).
117. C. H. Jun and Y. G. Lim, *Bull. Korean Chem. Soc.*, **10**, 468 (1989).
118. M. Green and R. I. Hancock, *J. Chem. Soc. (A)*, 2054 (1967).
119. J. K. Stille and R. A. Morgan, *J. Am. Chem. Soc.*, **88**, 5135 (1966).
120. C. B. Anderson and B. J. Burreson, *J. Organomet. Chem.*, **7**, 181 (1967).
121. D. R. Coulson, *J. Am. Chem. Soc.*, **91**, 200 (1969).
122. E. Vedejs and M. F. Salomon, *J. Am. Chem. Soc.*, **92**, 6965 (1970).
123. A. Segnitz, P. M. Bailey and P. M. Meitlis, *J. Chem. Soc., Chem. Commun.*, 698 (1973).
124. E. Vedejs and P. D. Weeks, *J. Chem. Soc., Chem. Commun.*, 223 (1974).
125. R.C. Larock, S. S. Hershberger, K. Takai and M. A. Mitchel, *J. Org. Chem.*, **51**, 2450 (1986).
126. R. C. Larock, K. Takagi, J. P. Burkhart and S. S. Hershberger, *Tetrahedron,* **42**, 3759 (1986).
127. R. C. Larock, D. R. Leach and S. M. Bjorge, *J. Org. Chem.*, **51**, 5221 (1986).
128. R. Aumann, *J. Organomet. Chem.*, **76**, C32 (1974).
129. L. Cassar and J. Halpern, *J. Chem. Soc., Chem. Commun.*, 1082 (1970).
130. M. J. Doyle, J. McMeeking and P. Binger, *J. Chem. Soc., Chem. Commun.*, 376 (1976).
131. P. Binger, M. J. Doyle and R. Benn, *Chem. Ber.*, **116**, 1(1983).
132. W. D. Wulff, in *Comprehensive Organic Synthesis*, Vol. 5 (Ed. B. M. Trost), Pergamon Press, Oxford, 1991, p.1065.
133. M. Brookhart and W. B. Studabaker, *Chem. Rev.*, **87**, 411 (1987).
134. E. O. Fischer and A. Maasböl, *Angew. Chem., Int. Ed. Engl.*, **3**, 580 (1964).
135. J. A. Connor and E. M. Jones, *J. Organomet. Chem.*, **60**, 77 (1973).
136. J. A. Connor and E. M. Jones, *J. Chem. Soc., Dalton Trans.*, 2119 (1973).
137. J. W. Herndon, G. Chatterjee, P. P. Patel, J. J. Matasi, S. U. Tumer, J. J. Harp and M. D. Reid, *J. Am. Chem. Soc.*, **113**, 7808 (1991).
138. J. W. Herndon, M. Zora, G. Chatterjee, J. J. Matasi and S. U. Tumer, *Tetrahedron*, **49**, 5507 (1993).
139. J. W. Herndon and M. D. Reid, *J. Am. Chem. Soc.*, **116**, 383 (1994).
140. S. U. Tumer, J. W. Herndon and L. A. McMullen, *J. Am. Chem. Soc.*, **114**, 8394 (1992).
141. J. W. Herndon and M. Zora, *Synlett*, 363 (1993).
142. M. Brookhart, W. B. Studabaker and G. R. Husk, *Organometallics*, **4**, 943 (1985).
143. B. C. Söderberg, L. S. Hegedus and M. A. Sierra, *J. Am. Chem. Soc.*, **112** , 4364 (1990).
144. J. Montgomery, G. M. Wieber and L. S. Hegedus, *J. Am. Chem. Soc.*, **112**, 6255 (1990).
145. J. W. Herndon and S. U. Tumer, *Tetrahedron Lett.*, **30**, 4771 (1989).
146. J. W. Herndon and S. U. Tumer, *J. Org. Chem.*, **56**, 286 (1991).
147. H. U. Reissig, *Top. Curr. Chem.*, **144**, 73 (1988).
148. L. S. Hegedus, R. W. Bates and B. C. Söderberg, *J. Am. Chem. Soc.*, **113**, 923 (1991).
149. J. W. Herndon, S. U. Tumer and W.F.K. Schnatter, *J. Am. Chem. Soc.*, **110**, 3334 (1988).
150. J. W. Herndon and L. M. McMullen, *J. Am. Chem. Soc.*, **111**, 6854 (1989).
151. M. Zora and J. W. Herndon. *J. Org. Chem.*, **59**, 699 (1994).
152. A. de Meijere, A. Kaufman, R. Lackmann, H.-C. Militzer, O. Reiser, S. Schömenauer and A. Weier, in *Organometallics in Organic Synthesis 2* (Eds. H. Werner and G. Erker), Springer, Berlin, 1989, p.255.
153. M. Duetsch, R. Lackmann, F. Stein and A. de Meijere, *Synlett*, 324 (1991).
154. B. M. Trost and A. S. K. Hashemi, *J. Am. Chem. Soc.*, **116**, 2183 (1994).

155. H. Fischer, P. Hofmann, F. R. Kreissl, R. R. Schrock, U. Schubert and K. Weiss, *Carbyne Complexes*, VCH Publ., Weinheim, 1988.
156. E. O. Fischer, N. Hoa Tran-Huy and D. Neugebauer, *J. Organomet. Chem.*, **229**, 169 (1982).
157. J. D. Carter, K. B. Kingsbury, A. Wilde, T. K. Schoch, C. J. Leep, E. K. Pham and L. McElwee-White, *J. Am. Chem. Soc.*, **113**, 2947 (1991).
158. K. B. Kingsbury, J. D. Carter, A. Wilde, H. Park, F. Takusagawa and L. McElwee-White, *J. Am. Chem. Soc.*, **115**, 10056 (1993).
159. W. J. Sieber, M. Wolfgruber, N. Hoa Tran-Huy, H. R. Schmidt, H. Heiss, P. Hofmann and F. R. Kreissl, *J. Organomet. Chem.*, **340**, 341 (1988).
160. J. D. Carter, T. K. Schoch and L. McElwee-White, *Organometallics*, **11**, 3571 (1992).
161. A. Mayr and C. M. Bastos, *Prog. Inorg. Chem.*, **40**, 1 (1992).
162. F. R. Kreissl, W. J. Sieber, H. Keller, J. Riede and M. Wolfgruber, *J. Organomet. Chem.*, **320**, 83 (1987).
163. M. D. Mortimer, J. D. Carter and L. McElwee-White, *Organometallics*, **12**, 4493 (1993).
164. K. B. Kingsbury, J. D. Carter, L. McElwee-White, R. L. Ostrander and A. Rheingold, *Organometallics*, **13**, 1635 (1994).
165. S. Sarel, J. Yovell and M. Sarel-Imber, *Angew. Chem., Int. Ed. Engl.*, **7**, 5 (1968).
166. T. Hudlicky, T. M. Kuchan and S. M. Naqvi, *Org. React.*, **33**, 247 (1985).
167. Z. Goldschmidt and B. Crammer, *Chem. Soc. Rev.*, **17**, 229 (1988).
168. T. Hudlicky and J. W. Reed, in *Comprehensive Organic Synthesis*, Vol. 5 (Ed. B. M. Trost), Pergamon Press, Oxford, 1991, pp. 899–970.
169. E. Piers, in *Comprehensive Organic Synthesis*, Vol. 5 (Ed. B. M. Trost), Pergamon Press, Oxford, 1991, pp 971-1035.
170. R. I. Khusnutdinov and U. M. Dzhemilev, *J. Organomet. Chem.*, **471**, 1 (1994).
171. J. A. Landgrebe and L. W. Becker, *J. Org. Chem.*, **33**, 1173 (1968).
172. (a) J. P. Marino and L. J. Browne, *Tetrahedron Lett.*, 3241 (1976).
(b) J. P. Marino and L. J. Browne, *Tetrahedron Lett.*, 3245 (1976).
173. E. Piers and I. Nagakura, *Tetrahedron Lett.*, 3237 (1976).
174. S. Halazy and A. Krief, *Tetrahedron Lett.*, **22**, 4341 (1981).
175. S. Halazy and A. Krief, *Tetrahedron Lett.*, **22**, 2138 (1981).
176. P. Binger, P. Müller, F. Langhauser, F. Sandmeyer, P. Philipps, B. Gabor and R. Mynott, *Chem. Ber.*, **126**, 1541 (1993).
177. R. Aumann, *J. Am. Chem. Soc.*, **96**, 2631 (1974).
178. S. Sarel, R. Ben-Shoshan and B. Kirson, *Isr, J. Chem.*, **10**, 787 (1972).
179. G. Marr and B. W. Rockett, in *The Chemistry of the Metal-Carbon Bond*, Vol. 1 (Eds. F. R. Hartley and S. Patai), Chap.9, Wiley, Chichester, 1982.
180. J. K. Stille, in *The Chemistry of the Metal-Carbon Bond*, Vol. 2 (Eds. F. R. Hartley and S. Patai), Chap. 9, Wiley, Chichester, 1985.
181. R. Aumann, *J. Organomet. Chem.*, **47**, C29 (1973).
182. R. M. Moriarty, C.-L. Yeh, K.-N. Chen and K. C. Ramey, *J. Am. Chem. Soc.*, **93**, 6709 (1971).
183. M. A. Battiste and J. M. Coxon, in *The Chemistry of the Cyclopropyl Group*, Vol. 1, part 1 (Ed. Z. Rappoport), Chap. 6, Wiley, Chichester, 1987.
184. A. D. Kettley and J. A. Braatz, *J. Organomet. Chem.*, **9**, P5 (1967).
185. D. Wilhelm, J. E. Bäckvall, R. E. Nordberg and T. Norin, *Organometallics*, **4**, 1296 (1985).
186. R. Grée, *Synthesis*, 341 (1989).
187. A. Monpert, J. Martelli, R. Grée and R. Carrie, *Nouv. J. Chim.*, **7**, 345 (1983).
188. M. Franck-Neumann, *Pure Appl. Chem.*, **55**, 1715 (1983).
189. M. Franck-Neumann, D. Martina and M. P. Heitz, *J. Organomet. Chem.*, **301**, 61 (1986).
190. J. Martelli, R. Grée and R. Carrie, *Tetrahedron Lett.*, **21**, 1953 (1980).
191. A. Monpert, J. Martelli, R. Grée and R. Carrie, *Tetrahedron Lett.*, **22**, 1961 (1981).
192. W. Grimme and G. Köser, *J. Am. Chem. Soc.*, **103**, 5919 (1981).
193. W.-E. Bleck, W. Grimme, H. Günther and E. Vogel, *Angew. Chem., Int. Ed. Engl.*, **9**, 303 (1970).
194. R. L. Beddoes, P. F. Lindley and O. S. Mills, *Angew. Chem., Int. Ed. Engl.*, **9**, 304 (1970).
195. S. D. Reynolds and T. A. Albright, *Organometallics*, **4**, 980 (1985).
196. D. Wormsbächer, F. Edelmann, D. Kaufmann, U. Behrens and A. de Meijere, *Angew. Chem., Int. Ed. Engl.*, **20**, 696 (1981).
197. D. L. Reger and A. Gabrielli, *J. Organomet. Chem.*, **187**, 243 (1980).
198. Z. Goldschmidt and S. Antebi, *J. Organomet. Chem.*, **260**, 105 (1984).

199. G. A. Taylor, *J. Chem. Soc., Perkin Trans. 1*, 1716 (1979).
200. T. Ishizu, K. Harano, N. Hori, M. Yasuda and K. Kanematsu, *Tetrahedron*, **39**, 1281 (1983).
201. S. Antebi, Ph.D. Thesis, Bar-Ilan University, Ramat-Gan, 1983
202. P. T. Van Vuuren, R. J. Fletterick, J. Mainwald and R. E. Hughes, *J. Am. Chem. Soc.*, **93**, 4394 (1971).
203. R. Aumann, *Chem. Ber.*, **109**, 168 (1976).
204. R. F. Childs, A. Varadarajan, C. J. L. Lock, R. Faggiani, C. A. Fyfe and R. E. Wasylishen, *J. Am. Chem. Soc.*, **104**, 2452 (1982).
205. R. F. Childs, *Acc. Chem. Res.*, **17**, 347 (1984).
206. R. F. Childs, M. J. McGlinchey and A. Varadarajan, *J. Am. Chem. Soc.*, **106**, 5974 (1984).
207. A. Davison, W. McFarlane, L. Pratt and G. Wilkinson, *J. Chem. Soc.*, 4821 (1962).
208. M. Brookhart, E. R. Davis and D. L. Harris, *J. Am. Chem. Soc.*, **94**, 7853 (1972).
209. G. A. Olah, S. H. Yu and G. Liang, *J. Org. Chem.*, **41**, 2383 (1976).
210. G. A. Olah, G. Liang and S. Yu, *J. Org, Chem.*, **42**, 4262 (1977).
211. T. V. Ashworth, A. A. Chalmers, D. C. Liles, E. Meintjies, H. E. Oosthuizen and E. Singleton, *J. Organomet. Chem.*, **276**, C49 (1984).
212. R. Aumann and J. Knecht, *Chem. Ber.*, **111**, 3927 (1978).
213. R. Aumann, *Angew. Chem., Int. Ed. Engl.*, **12**, 574 (1973).
214. R. Aumann, *J. Organomet. Chem.*, **78**, C31 (1974).
215. J. D. Holmes and R. Pettit, *J. Am. Chem. Soc.*, **85**, 2531 (1963).
216. B. Niemer, J. Breimair, B. Wagner, K. Polborn and W. Beck, *Chem. Ber.*, **124**, 2227 (1991).
217. M. Franck-Neumann and D. Martina, *Tetrahedron Lett.*, 1759 (1975).
218. R. F. Childs and A. Varadarajan, *J. Organomet. Chem.*, **184**, C28 (1980).
219. R. F. Childs and A. Varadarajan, *Can. J. Chem.*, **63**, 418 (1985).
220. N. Morita, S. Ito, T. Asao, C. Kabuto, H. Sotokawa, M. Hatano and A. Tajiri, *Chem. Lett.*, 1527 (1990).
221. N. Morita, S. Ito, T. Asao, H. Sotokawa, M. Hatano and A. Tajiri, *Chem. Lett.*, 1639 (1990).
222. B. F. G. Johnson, J. Lewis and G. L. P. Randall, *J. Chem. Soc. (A)*, 422 (1971).
223. A. D. Charles, P. Diversi, B. F. G. Johnson, K. D. Karlin, J. Lewis, A. V. Rivera and G. M. Sheldrick, *J. Organomet. Chem.*, **128**, C31 (1977).
224. W. Grimme, *Chem. Ber.*, **100**, 113 (1967).
225. A. Salzer, *J. Organomet. Chem.*, **107**, 79 (1976).
226. A. Salzer, *J. Organomet. Chem.*, **117**, 245 (1976).
227. F. J. Liotta and B. K. Carpenter, *J. Am. Chem. Soc.*, **107**, 6426 (1985).
228. F. J. Liotta, G. Van Duyne and B. K. Carpenter, *Organometallics*, **6**, 1010 (1987).
229. R. Grigg and A. Sweeney, *J. Chem. Soc., Chem. Commun.*, 1248 (1971).
230. A. Streitwieser and D.M.E. Reuben, *J. Am. Chem. Soc.*, **93**, 1794 (1971).
231. O. M. Nefedov, I. E. Dolgii, L. B. Shvedova and R. A. Baidzhigitova, *Izv. Akad. Nauk SSSR, Ser. Khim.*, 1339 (1978); *Bull. Acad. Sci. USSR, Chem. Ser.*, 1164 (1978); *Chem. Abstr.*, **89**, 108235g (1978).
232. J. Salaün, *J. Org. Chem.*, **41**, 1237 (1976).
233. M. J. Calverley and C. A. S. Bretting, *Bio. Med. Chem. Lett.*, **3**, 1841 (1993).
234. C. E. Hudson and N. L. Bauld, *J. Am. Chem. Soc.*, **94**, 1158 (1972).
235. T. Liese and A. de Meijere, *Angew. Chem., Int. Ed. Engl.*, **21**, 65 (1982).
236. T. Liese and A. de Meijere, *Chem. Ber.*, **119**, 2995 (1986).
237. J. E. Baldwin and N. D. Ghatlia, *J. Am. Chem. Soc.*, **113**, 6273 (1991).
238. T. Liese, G. Splettstasser and A. de Meijere, *Tetrahedron Lett.*, **23**, 3341 (1982).
239. G. Köbrich and D. Merkel, *Justus Liebigs Ann. Chem.*, **761**, 50 (1972).
240. G. Köbrich, D. Merkel and K. W. Thiem, *Chem. Ber.*, **105**, 1683 (1972).
241. W. Sieber, M. Wolfgruber, D. Neugebauer, O. Orama and F. R. Kriessl, *Z. Naturforsch.*, **38 B**, 67 (1983).
242. S. Keyaniyan, M. Apel, J. P. Richmond and A. de Meijere, *Angew. Chem., Int. Ed. Engl.*, **24**, 770 (1985).
243. R. Victor, R. Ben-Shoshan and S. Sarel, *Tetrahedron Lett.*, 4211 (1973).
244. V. Usieli, R. Victor and S. Sarel, *Tetrahedron Lett.*, 2705 (1976).
245. U. M. Dzhemilev, R. I. Khusnutdinov, N. A. Shchadneva, O. M. Nefedov and G. A. Tolstikov, *Izv. Akad. Nauk SSSR, Ser. Khim.*, 2360 (1989); *Chem. Abstr.*, **112**, 197652z (1990).
246. M. Duetsch, F. Stein, R. Lackmann, E. Pohl, R. Herbst-Irmer and A. de Meijere, *Chem. Ber.*, **125**, 2051 (1992).

247. F. Stein, M. Duetsch, R. Lakmann, M. Noltemeyer and A. de Meijere, *Angew. Chem., Int. Ed. Engl.*, **30**, 1658 (1991).
248. E. L. Hoel, G. B. Ansell and S. Leta, *Organometallics*, **3**, 1633 (1984).
249. G. L. Closs, *Adv. Alicyclic Chem.*, **1**, 53 (1966).
250. B. Halton and M. G. Banwell, in *The Chemistry of the Cyclopropyl Group*, Vol. 1, part 2 (Ed. Z. Rappoport), Chap. 21, Wiley, Chichester, 1987.
251. G. L. Closs and L. E. Closs, *J. Am. Chem. Soc.*, **83**, 1003 (1961).
252. G. L. Closs and L. E. Closs, *J. Am. Chem. Soc.*, **85**, 99 (1963).
253. G. L. Closs and L. E. Closs, *J. Am. Chem. Soc.*, **83**, 2015 (1961).
254. K. B. Wiberg, R. K. Barnes and J. Albin, *J. Am. Chem. Soc.*, **79**, 4994 (1957).
255. E. A. Dorko and R. W. Mitchell, *Tetrahedron Lett.*, 341 (1968).
256. A. J. Schipperijn and P. Smael, *Recl. Trav. Chim. Pays-Bas*, **92**, 1121 (1973).
257. A. J. Schipperijn and P. Smael, *Recl. Trav. Chim. Pays-Bas*, **92**, 1159 (1973).
258. A. J. Schipperijn and P. Smael, *Recl. Trav, Chim. Pays-Bas*, **92**, 1298 (1973).
259. M. Isaka, S. Ejiri and E. Nakamura, *Tetrahedron*, **48**, 2045 (1992).
260. R. F. Heck, *Pure Appl. Chem.*, **50**, 691 (1978); B. M. Trost and T. R. Verhoeven, in *Comprehensive Organometallic Chemistry*, Vol. 8 (Ed. G. Wilkinson), Pergamon Press, Oxford 1982, pp. 779–938.
261. R. Walsh, S. Untiedt and A. de Meijere, *Chem. Ber.*, **127**, 237 (1994).
262. R. Gompper and E. Bartmann, *Angew. Chem., Int. Ed. Engl.*, **24**, 209 (1985).
263. P.-C. Ting, Y.-C. Lin, M.-C. Cheng and Y. Wang, *Organometallics*, **13**, 2150 (1994).
264. D. G. Morris, in *The Chemistry of the Cyclopropyl Group*, Vol.1, part 1 (Ed. Z. Rappoport), Chap.3, Wiley, Chichester 1987.
265. A. W. Donaldson and R. P. Hughes, *J. Am. Chem. Soc.*, **104**, 4846 (1982).
266. R. Gompper and E. Bartmann, *Angew. Chem., Int. Ed. Engl.*, **17**, 456 (1978).
267. R. Gompper, E. Bartmann and H. Nöth, *Chem. Ber.*, **112**, 218 (1979).
268. C. Löwe, V. Shklover and H. Berke, *Organometallics*, **10**, 3396 (1991).
269. C. Löwe, V. Shklover, H. W. Bosch and H. Berke, *Chem. Ber.*, **126**, 1769 (1993).
270. M. Green and R. P. Hughes, *J. Chem. Soc., Chem. Commun.*, 862 (1975).
271. C. E. Coffey, *J. Am. Chem. Soc.*, **84**, 118 (1962).
272. R. B. King and A. Efraty, *J. Organomet. Chem.*, **24**, 241 (1970).
273. J. Potenza, R. Johnson, D. Mastropaolo and A. Efraty, *J. Organomet. Chem.*, **64**, C13 (1974).
274. T. Chiang, R. C. Kerber, S. D. Kimball and J. W. Lauher, *Inorg. Chem.*, **18**, 1687 (1979).
275. R. P. Hughes, J. M. J. Lambert, D. W. Whitman, J. L. Hubbard, W. P. Henry and A. L. Rheingold, *Organometallics*, **5**, 789 (1986).
276. C. E. Chidsey, W. A. Donaldson, R. P. Hughes and P. F. Sherwin, *J. Am. Chem. Soc.*, **101**, 233 (1979).
277. P. J. Desrosiers and R. P. Hughes, *J. Am. Chem. Soc.*, **103**, 5593 (1981).
278. D. M. Desimone, P. J. Desrosiers and R. P. Hughes, *J. Am. Chem. Soc.*, **104**, 4842 (1982).
279. C. Müller and U. Schubert, *J. Organomet. Chem.*, **405**, C1 (1991).
280. J. P. Visser, A. J. Schipperijn and J. Lukas, *J. Organomet. Chem.*, **47**, 433 (1973).
281. J. J. de Boer and D. Bright, *J. Chem. Soc., Dalton Trans.*, 662 (1975).
282. P. H. Kasai, R. J. Myers, D. F. Eggers and K. B. Wiberg, *J. Chem. Phys.*, **30**, 512 (1959).
283. T. A. Albright, R. Hoffmann, J. C. Thibeault and D. L. Thorn, *J. Am. Chem. Soc.*, **101**, 3801 (1979).
284. H. Schwager, C. Krüger, R. Neidlein and G. Wilke, *Angew. Chem., Int. Ed. Engl.*, **26**, 65 (1987).
285. H. Schwager, R. Benn and G. Wilke, *Angew. Chem., Int. Ed. Engl.*, **26**, 67 (1987).
286. R. P. Hughes, J. W. Reisch and A. L. Rheingold, *Organometallics*, **4**, 241 (1985).
287. L. K. Johnson, R. H. Grubbs and J. W. Ziller, *J. Am. Chem. Soc.*, **115**, 8130 (1993).
288. S. Fredericks, and J. L. Thomas, *J. Am. Chem. Soc.*, **100**, 350 (1978).
289. P. Binger, P. Müller, A. T. Herrmann, P. Philipps, B. Gabor, F. Langhauser and C. Krüger, *Chem. Ber.*, **124**, 2165 (1991).
290. H. Fischer, J. Hofmann and E. Mauz, *Angew. Chem., Int. Ed. Engl.*, **30**, 998 (1991).
291. P. J. Davidson, M. F. Lapert and R. Pearce, *Chem. Rev.*, **76**, 219 (1976).
292. R. R. Schrock and G. W. Parshall, *Chem. Rev.*, **76**, 243 (1976).
293. S. L. Buchwald and R. B. Nielsen, *Chem. Rev.*, **88**, 1047 (1988).
294. S. T. Nguyen, L. K. Johnson and R. H. Grubbs, *J. Am. Chem. Soc.*, **114**, 3974 (1992).
295. (a) T. Valeri, F. Meier and E. Weiss, *Chem. Ber.*, **121**, 1093 (1988).
(b) G. Dettlaf, U. Behrens and E. Weiss, *Chem. Ber.*, **11**, 3019 (1978).

296. P. Binger, E. Cetinkaya and C. Krüger, *J. Organomet. Chem.*, **159**, 63 (1978).
297. G. Newton, N. S. Pantaleo, R. B. King and C.-K. Chu, *J. Chem. Soc., Chem. Commun.*, 10 (1979).
298. J. Klimes and E. Weiss, *Chem. Ber.*, **115**, 2606 (1982).
299. R. C. Hemond, R. P. Hughes, D. J. Robinson and A. L. Rheingold, *Organometallics*, **7**, 2239 (1988).
300. R. P. Hughes, M. E. King, D. J. Robinson and J. M. Spotts, *J. Am. Chem. Soc.*, **111**, 8919 (1989).
301. P. Binger and B. Biedenbach, *Chem. Ber.*, **120**, 601 (1987).
302. P. Binger, J. McMeeking and H. Schäfer, *Chem. Ber.*, **117**, 1551 (1984).
303. E. W. Gowling and S. F. A. Kettle, *Inorg. Chem.*, **3**, 604 (1964).
304. W. K. Olander and T. L. Brown. *J. Am. Chem. Soc.*, **94**, 2139 (1972).
305. C. Mealli, S. Midollini, S. Moneti and L. Sacconi, *Angew. Chem., Int. Ed. Engl.*, **19**, 931 (1980).
306. C. Mealli, S. Midollini, S. Moneti, L. Sacconi, J. Silvestre and T. A. Albright, *J. Am. Chem. Soc.*, **104**, 95 (1982).
307. M. D. McClure and D. L. Weaver, *J. Organomet. Chem.*, **54**, C59 (1973).
308. T. A. Albright, R. Hoffmann, Y. Tse and T. D'Ottavio, *J. Am. Chem. Soc.*, **101**, 3812 (1979).
309. R. P. Hughes, D. S. Tucker and A. L. Rheingold, *Organometallics*, **12**, 3069 (1993).
310. D. L. Lichtenberger, M. L. Hoppe, L. Subramanian, E. M. Kober, R. P. Hughes, J. L. Hubbard and D. S. Tucker, *Organometallics*, **12**, 2025 (1993).
311. M. G. Drew, B. J. Brisdon and A. Day, *J. Chem. Soc., Dalton Trans.*, 1310 (1981).
312. R. P. Hughes, J. W. Reisch and A. L. Rheingold, *Organometallics*, **4**, 1754 (1985).
313. R. P. Hughes, J. Robbins, D. J. Robinson and A. L. Rheingold, *Organometallics*, **7**, 2413 (1988).
314. R. Ditchfield, R. P. Hughes, D. S. Tucker, E. P. Bierwagen, J. Robbins, D. J. Robinson and J. A. Zakutansky, *Organometallics*, **12**, 2258 (1993).
315. R. B. Blunden, F. G. N. Cloke, P. B. Hitchcock and P. Scott, *Organometallics*, **13**, 2917 (1994).
316. M. R. Churchill, J. C. Fettinger, L. McCullough and R. R. Schrock, *J. Am. Chem. Soc.*, **106**, 3356 (1984).
317. R. R. Schrock, *Acc. Chem. Res.*, **19**, 342 (1986).
318. R. R. Schrock, S. F. Pedersen, M. R. Churchill and J. W. Ziller, *Organometallics*, **3**, 1574 (1984).
319. I. A. Weinstock, R. R. Schrock and W. M. Davis, *J. Am. Chem. Soc.*, **113**, 135 (1991).
320. R. R. Schrock, J. S. Murdzek, J. H. Freudenberger, M. R. Churchill and J. W. Ziller, *Organometallics*, **5**, 25 (1986).
321. M. Green, *J. Organomet. Chem.*, **300**, 93 (1986).
322. R. P. Hughes, J. M. J. Lambert and J. L. Hubbard, *Organometallics*, **5**, 797 (1986).
323. J. K. Shen, D.S. Tucker, F. Basolo and R. P. Hughes, *J. Am. Chem. Soc.*, **115**, 11312 (1993).
324. F. Basolo, *Polyhedron*, **9**, 1503 (1990).
325. J. M. O'Conner and C. P. Casey, *Chem. Rev.*, **87**, 307 (1987).
326. R. B. King and S. Ikai, *Inorg. Chem.*, **18**, 949 (1979).
327. D. L. Weaver and R. M. Tuggle, *J. Am. Chem. Soc.*, **91**, 6506 (1969).
328. R. M. Tuggle and D. L. Weaver, *Inorg. Chem.*, **10**, 2599 (1971).
329. R. P. Hughes and D. S. Tucker, *Organometallics*, **12**, 4736 (1993).
330. M. D. Rausch, R. M. Tuggle and D. l. Weaver, *J. Am. Chem. Soc.*, **92**, 4981 (1970).
331. R. G. Hayter, *J. Organomet. Chem.*, **13**, Pl (1968).
332. R. M. Tuggle and D. L. Weaver, *Inorg. Chem.*, **10**, 1504 (1971).
333. P. S. Welcker and L. J. Todd, *Inorg. Chem.*, **9**, 286 (1970).
334. E. D. Jemmis and R. Hoffmann, *J. Am. Chem. Soc.*, **102**, 2570 (1980).
335. J. Silvestre and T. A. Albright, *J. Am. Chem. Soc.*, **107**, 6829 (1985).
336. M. Sundaralingam and L. H. Jensen, *J. Am. Chem. Soc.*, **88**, 191 (1966).
337. Z. Lin and M. B. Hall, *Organometallics*, **13**, 2878 (1994).
338. R. Gleiter and R. Hoffmann, *J. Am.Chem. Soc.*, **90**, 5457 (1968).
339. T. J. Lee, A. Bunge and H. F. Schaefer III, *J. Am. Chem. Soc.*, **107**, 137 (1985).
340. Z. Yoshida, *Pure Appl. Chem.*, **54**, 1059 (1982).
341. K. Öfele, *Angew. Chem., Int. Ed. Engl.*, **7**, 950 (1968).
342. D. J. Cardin, B. Cetinkaya and M. F. Lappert, *Chem. Rev.*, **72**, 545 (1972).
343. T. Shono, T. Yoshimura, Y. Matsumura and R. Oda, *J. Org, Chem.*, **33**, 876 (1968).
344. U. Kirchgässner and U. Schubert, *Organometallics*, **7**, 784 (1988).
345. U. Kirchgässner, H. Piana and U. Schubert, *J. Am. Chem. Soc.*, **113**, 2228 (1991).

346. G. Huttner, S. Schelle and O. S. Mills, *Angew. Chem., Int. Ed. Engl.*, **8**, 515 (1969).
347. R. Wilson, Y. Kamitori, H. Ogoshi, Z. Yoshida and J. A. Ibers, *J. Organomet. Chem.*, **173**, 199 (1979).
348. K. Öfele, *Angew. Chem., Int. Ed. Engl.*, **8**, 916 (1969).
349. K. Öfele, *J. Organomet. Chem.*, **22**, C9 (1970).
350. Z. Yoshida and Y. Kamitori, *Chem. Lett.*, 1341 (1978).
351. S. Miki, T. Ohno, H. Iwasaki and Z. Yoshida, *J. Phys. Org, Chem.*, **1**, 333 (1988).
352. H. Konishi, S. Matsumoto, Y. Kamitori, H. Ogoshi and Z. Yoshida, *Chem. Lett.*, 241 (1978).
353. R. Gompper and E. Bartmann, *Justus Liebigs Ann. Chem.*, 229 (1979).
354. C. W. Rees and E. von Angerer, *J. Chem. Soc., Chem. Commun.*, 420 (1972).
355. T. Gilchrist, R. Livingston, C. W. Rees and E. von Angerer, *J. Chem. Soc., Perkin Trans. 1*, 2535 (1973).
356. R. Weiss, C. Priesner and H. Wolf, *Angew. Chem., Int. Ed. Engl.*, **17**, 446 (1978).
357. Z. Yoshida, H. Konishi, Y. Miura and H. Ogoshi, *Tetrahedron Lett.*, 4319 (1977).
358. R. Weiss and C. Priesner, *Angew. Chem., Int. Ed. Engl.*, **17**, 457 (1978).
359. M. Bertrand, G. Leandri and A. Meou, *Tetrahedron Lett.*, 1841 (1979).
360. R. Gompper and U. Wolf, *Justus Liebigs Ann. Chem.*, 1406 (1979).
361. Z. Yoshida, H. Konishi, S. Sawada and H. Ogoshi, *J. Chem. Soc., Chem. Commun.*, 850 (1977).
362. E. W. Thomas, *Tetrahedron Lett.*, **24**, 1467 (1983).
363. N. D. Lenn, Y. Shih, M. T. Stankovich and H. Liu, *J. Am. Chem. Soc.*, **111**, 3065 (1989).
364. J. E. Baldwin and W. C. Widdison, *J. Labelled Compd. Radiopharm.*, **28**, 175 (1990).
365. T. V. Akhachinskaya, M. Bachbuch, Y. K. Grishin, N. A. Donskaya and Y. A. Ustynyuk, *Zh. Org. Khim.*, **14**, 2317 (1978); Engl. Transl.; *J. Org. Chem. USSR*, **14**, 2139 (1978).
366. T. V. Akhachinskaya, N. A. Donskaya, Y. K. Grishin, V. A. Roznyatovskii and Y. S. Shabarov, *Zh. Org. Khim.*, **17**, 1429 (1981); Engl. Transl.: *J. Org. Chem. USSR*, **17**, 1271 (1981).
367. P. Binger and H. M. Büch, *Top. Curr. Chem*, **135**, 98 (1987).
368. E. Sternberg and P. Binger, *Tetrahedron Lett.*, **26**, 301 (1985).
369. G. Leandri, H. Monti and M. Bertrand, *Bull. Soc. Chim. Fr.*, 3015 (1974).
370. L. R. Robinson, G. T. Burns and T. J. Barton, *J. Am. Chem. Soc.*, **107**, 3935 (1985).
371. J.-L. Derocque and F.-B. Sunderman, *J. Org. Chem.*, **39**, 1411 (1974).
372. X. Creary, M. E. Mehrsheikh-Mohammadi and S. McDonald, *J. Org. Chem.*, **52**, 3254 (1987).
373. A. Maercker and K. D. Klein, *Angew. Chem., Int. Ed. Engl.*, **28**, 83 (1989).
374. A. Maercker and K.-D. Klein, *J. Organomet. Chem.*, **410**, C35 (1991).
375. P. Binger, A. Brinkmann and P. Wedemann, *Chem. Ber.*, **116**, 2920 (1983).
376. U. M. Dzhemilev, R. I. Khusnutdinov and G. A. Tolstikov, *J. Organomet. Chem.*, **409**, 15 (1991).
377. I. S. Krull, *J. Organomet. Chem.*, **57**, 373 (1973).
378. T. H. Whitesides and R. W. Slaven, *J. Organomet. Chem.*, **67**, 99 (1974).
379. T. H. Whitesides, R. W. Slaven and J. Calabrese, *Inorg. Chem.*, **13**, 1895 (1974).
380. T. A. Albright, J. K. Burdett and M.-H. Whangbo, *Orbital Interactions in Chemistry*, Wiley, New York, 1985.
381. A. R. Pinhas, A. G. Samuelson, R. Reisemberg, E. V. Arnold, J. Clardy and B. K. Carpenter, *J. Am. Chem. Soc.*, **103**, 1668 (1981).
382. M. Green, J. A. K. Howard, R. P. Hughes, S. C. Kellett and P. Woodward, *J. Chem. Soc., Dalton Trans.*, 2007 (1975).
383. P. Binger, T. R. Martin, R. Benn, A. Rufinska and G. Schroth, *Z. Naturforsch.*, **39b**, 993 (1984).
384. R. Noyori, T. Ishigami, N. Hayashi and H. Takaya. *J. Am. Chem. Soc.*, **95**, 1674 (1973).
385. L. S. Isaeva, T. A. Peganova, P. V. Petrovskii, D. B. Furman, A. V. Kudryashev, S. V. Zotova and O. V. Bragin, *J. Organomet. Chem.*, **258**, 367 (1983).
386. L. S. Isaeva, T. A. Peganova, P. V. Petrovskii and D. N. Kravtsov, *J. Organomet. Chem.*, **376**, 141 (1989).
387. H. Fischer, W. Bidell and J. Hofmann, *J. Chem. Soc., Chem. Commun.*, 858 (1990).
388. R. Noyori and H. Takaya, *J. Chem. Soc., Chem. Commun.*, 525 (1969).
389. T. A. Albright, P. R. Clemens, R. P. Hughes, D. E. Hunton and L. D. Margerum, *J. Am. Chem. Soc.*, **104**, 5369 (1982).
390. R. P. Hughes, D. E. Hunton and K. Schumann, *J. Organomet. Chem.*, **169**, C37 (1979).
391. B. K. Dallas and R. P. Hughes, *J. Organomet. Chem.*, **184**, C67 (1980).
392. B. K. Dallas R. P. Hughes and K. Schumann, *J. Am. Chem. Soc.*, **104**, 5380 (1982).

393. P. R. Clemens, R. P. Hughes and L. D. Margerum, *J. Am. Chem. Soc.*, **103**, 2428 (1981).
394. R. P. Hughes and C. S. Day, *Organometallics*, **1**, 1221 (1982).
395. W. Donaldson, *J. Organomet, Chem.*, **269**, C25 (1984).
396. (a) W. A. Donaldson, J. T. North, J. A. Gruetzmacher, M. Finley and D. J. Stepuszek, *Tetrahedron*, **46**, 2263 (1990).
(b) W. A. Donaldson, D. J. Stepuszek and J. A. Gruetzmacher, *Tetrahedron*, **46**, 2273 (1990).
397. M. Green and R. P. Hughes, *J. Chem. Soc., Chem. Commun.*, 686 (1974).
398. M. Green and R. P. Hughes, *J. Chem. Soc., Dalton Trans.*, 1880 (1976).
399. R. L. Phillips and R. J. Puddephatt, *J. Chem. Soc., Dalton Trans.*, 1736 (1978).
400. T. G. Attig, *Inorg. Chem.*, **17**, 3097 (1978).
401. C. P. Brock and T. G. Attig, *J. Am. Chem. Soc.*, **102**, 1319 (1980).
402. M. Englert, P. W. Jolly and G. Wilke, *Angew, Chem., Int. Ed. Engl.*, **10**, 77 (1971).
403. B. M. Chisnall, M. Green, R. P. Hughes and A. J. Welch, *J. Chem. Soc., Dalton Trans.*, 1899 (1976).
404. M. Green, R. P. Hughes and A. J. Welch, *J. Chem. Soc., Chem. Commun.*, 487 (1975).
405. I. S. Krull, *J. Organomet. Chem.*, **57**, 363 (1973).
406. M. Bottrill and M. Green, *J. Chem. Soc., Dalton Trans.*, 820 (1979).
407. R. Noyori, T. Nishimura and H. Takaya, *J. Chem. Soc., Chem. Commun.*, 89 (1969).
408. W. E. Billups, L. P. Lin and O. A. Gansow, *Angew. Chem., Int. Ed. Engl.*, **11**, 637 (1972).
409. W. E. Billups and L. P. Lin, *J. Organomet, Chem.*, **61**, C55 (1973).
410. A. R. Pinhas and B. K. Carpenter, *J. Chem. Soc., Chem. Commun.*, 17 (1980).
411. S. G. Barnes and M. Green, *J. Chem. Soc., Chem. Commun.*, 267 (1980).
412. W. A. Smit, S. L. Kireev, O. M. Nefedov and V. A. Tarasov, *Tetrahedron Lett.*, **30**, 4021 (1989).
413. A. Stolle, H. Becker, J. Salaün and A. de Meijere, *Tetrahedron Lett.*, **35**, 3517 (1994).
414. A. Stolle, H. Becker, J. Salaün and A. de Meijere, *Tetrahedron Lett.*, **35**, 3521 (1994).
415. R. Noyori, T. Odagi and H. Takaya, *J. Am. Chem. Soc.*, **92**, 5780 (1970).
416. P. Binger and U. Schuchard, *Angew. Chem., Int. Ed. Engl.*, **16**, 249 (1977).
417. P. Binger and U. Schuchard, *Chem. Ber.*, **114**, 3313 (1981).
418. D. M. T. Chan, in *Comprehensive Organic Synthesis*, Vol. 5 (Ed. B. M. Trost), Pergamon Press, Oxford, 1991, pp. 271–314.
419. T. Ohta and H. Takaya, in *Comprehensive Organic Synthesis*, Vol. 5 (Ed. B. M. Trost), Pergamon Press, Oxford, 1991, pp. 1185–1205.
420. P. Binger, A. Brinkmann and J. McMeeking, *Justus Liebigs Ann. Chem.*, 1065 (1977).
421. P. Binger and B. Schäfer, *Tetrahedron Lett.*, **29**, 529 (1988).
422. P. Binger, A. Brinkmann, P. Roefke and B. Schäfer, *Justus Liebigs Ann. Chem.*, 739 (1989).
423. S. Yamago and E. Nakamura, *Tetrahedron*, **45**, 3088 (1989); S. Yamago and E. Nakamura, *J. Chem. Soc., Chem. Commun.*, 1112 (1988).
424. R. T. Lewis, W. B. Motherwell and M. Shipman, *J. Chem. Soc., Chem. Commun.*, 948 (1988).
425. S. A. Bapuji, W. B. Motherwell and M. Shipman, *Tetrahedron Lett.*, **30**, 7107 (1989).
426. H. Sakurai, K. Hirama, Y. Nakadaira and C. Kabuto, *J. Am. Chem. Soc.*, **109**, 6880 (1987).
427. P. Müller, G. Bernardinelli and Y. Jacquier, *Helv. Chim. Acta*, **71**, 1328 (1988).
428. P. Müller, G. Bernardinelli and Y. Jacquier, *Helv. Chim. Acta*, **75**, 1995 (1992).
429. M. Lenardi, N. B. Pahor, M. Calligaris, M. Graziani and L. Randaccio, *Inorg. Chim. Acta*, **26**, L19 (1978).
430. B. Temme, G. Erker, R. Frohlich and M. Grehl, *J. Chem. Soc., Chem. Commun.*, 1713 (1994).
431. (a) T. Fukunaga, *J. Am. Chem. Soc.*, **98**, 610 (1976).
(b) T. Fukunaga, M. D. Gordon and P. J. Krusic, *J. Am. Chem. Soc.*, **98**, 611 (1976).
432. G. Arndt, T. Kämpchen, R. Schmiedel, G. Seitz and R. Sutrisno, *Justus Liebigs Ann. Chem.*, 1409 (1980).
433. Z. Yoshida, M. Shibata and T. Sugimoto, *Tetrahedron Lett.*, **25**, 4223 (1984).

CHAPTER **11**

Metal catalysed cyclopropanation

TAO YE and M. ANTHONY McKERVEY

School of Chemistry, The Queen's University, Belfast BT9 5AG, Northern Ireland, UK

I. INTRODUCTION 657
II. CYCLOPROPANATION USING DIAZOALKANES 658
III. CYCLOPROPANATION USING DIAZOCARBONYL COMPOUNDS 662
A. Cyclopropanation Using Alkyl Diazoesters/Diazoacetamides 662
B. Cyclopropanation Using Diazoketones 667
C. Cyclopropanation Using Dicarbonyl Diazomethanes 676
IV. CYCLOPROPANATION USING OTHER PRECURSORS 681
V. CHEMOSELECTIVITY IN METAL CATALYSED CYCLOPROPANATION 682
VI. REGIOSELECTIVITY IN METAL CATALYSED CYCLOPROPANATION 688
VII. STEREOCHEMICAL FEATURES OF ALKENE CYCLOPROPANATION 692
A. Stereoselectivity in Metal Catalysed Cyclopropanation 692
B. Diastereoselectivity in Metal Catalysed Cyclopropanation 695
C. Enantioselectivity in Metal Catalysed Cyclopropanation 697
VIII. CONCLUSION 702
IX. REFERENCES 702

I. INTRODUCTION

Although a metal catalysed decomposition of ethyl diazoacetate was originally described by Silberrad and Roy in 1906[1], it was to be many years before the value of this type of process for cyclopropanation of alkenes using transition metal catalysts was widely appreciated and reliable, efficient methods were developed. By the early 1960s, the reaction had become important in organic synthesis. Various transition metal compounds have been screened for catalytic cyclopropanation. Copper, rhodium and palladium compounds have

The chemistry of the cyclopropyl group, Vol. 2
Edited by Z. Rappoport

emerged as the highly efficient catalysts of choice. Today, the catalytic process is the most generally employed approach for the formation of cyclopropanes. This methodology has been extensively used in laboratory-scale synthesis as well as in the fine chemical industry. Catalytic processes enjoy the advantage over their non-catalytic counterparts of proceeding efficiently under milder conditions, thus leading to more energy-efficient processes. Furthermore, catalytic processes generally show high levels of selectivity that can be engineered into the addition process. Metal catalysed cyclopropanation has been a very active area of research in recent years. Of particular importance is the development of chiral rhodium and copper catalysts for use in asymmetric cyclopropanation where some very high ee values (>95%) have already been achieved. A numbers of recent reviews partially concerning the topics of metal catalysed cyclopropanation provide additional sources of information[2–12].

This chapter covers the use of metal catalysts for the addition of various diazo compounds to alkenes. Aspects of the regioselectivity, chemoselectivity and stereoselectivity, in particular the enantioselectivity, will be emphasized by means of recent examples.

II. CYCLOPROPANATION USING DIAZOALKANES

Diazoalkanes, in particular diazomethane, can efficiently transfer a methylene unit to olefinic double bonds via a metal catalysed process. The range of alkenes that may be used as substrates in this cyclopropanation is vast. The efficiency of the cyclopropanation of various types of alkenes can be very dependent upon the particular catalyst chosen for the reaction.

Both heterogeneous and homogeneous catalysts have been used in the cyclopropanation of simple alkenes; homogeneous catalysts are far more effective catalysts and are preferentially employed. Copper-based catalysts, for example copper(II) triflate, copper(II) acetylacetonate and copper(II) hexafluoroacetylacetonate, are suitable catalysts for the addition of diazoalkanes to simple alkenes to give cyclopropane adducts. Salomon and Kochi[13] implicated Cu(I) as the catalytically active oxidation state which can be derived from the reduction of copper(II) by diazo compounds. Some representative examples of copper catalysed cyclopropanations are shown in equations 1–6[13–16]. Cyclopropanation using diazomethane is a smooth and efficient reaction[13] (equations 1 and 2) whereas cyclopropanation using vinyl diazomethane proceeds in lower yields (equations 3 and 4)[14]. In the case of intramolecular cyclopropanation, CuI catalysed decomposition of diazo compound **1** resulted in a bicyclic adduct **2** which is the key intermediate in the synthesis of *d,l*-sirenin (equation 5)[15]. In another example, the intramolecular cyclopropanation of 2-acetamido-5-[*N*-(2-propenyl) methanesulphonamido]-4-diazocyclohexadien-1-one **3** gave the cycloprop[*c*]indolone ring system **4**, which is an analogue of the bioactive ring A of the antitumor antibiotic cc-1065. This reaction, which proceeded with essentially quantitative conversion, was achieved with a catalytic system consisting of trifluoropentanedionato copper(I) carbonyl and 1 equivalent of *N*-butylamine (equation 6)[16].

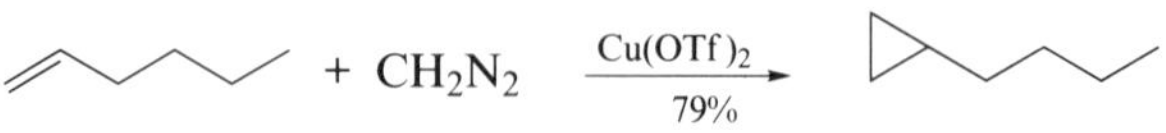

(1)

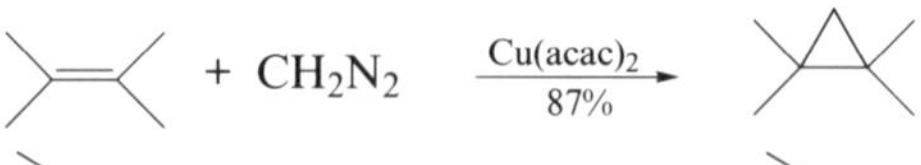

(2)

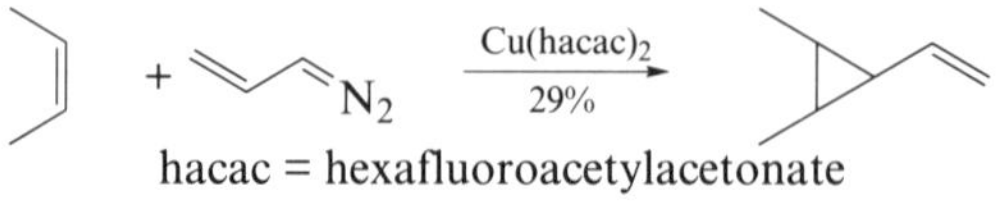

(3)

hacac = hexafluoroacetylacetonate

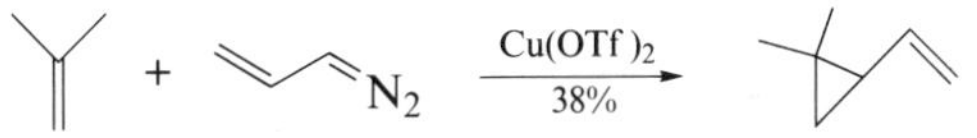

(4)

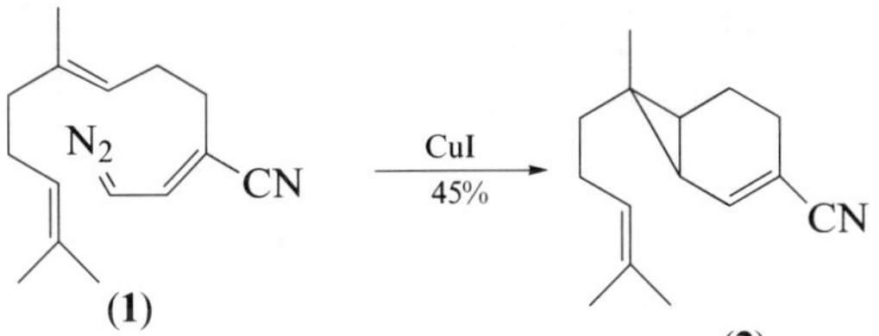

(1) (2) (5)

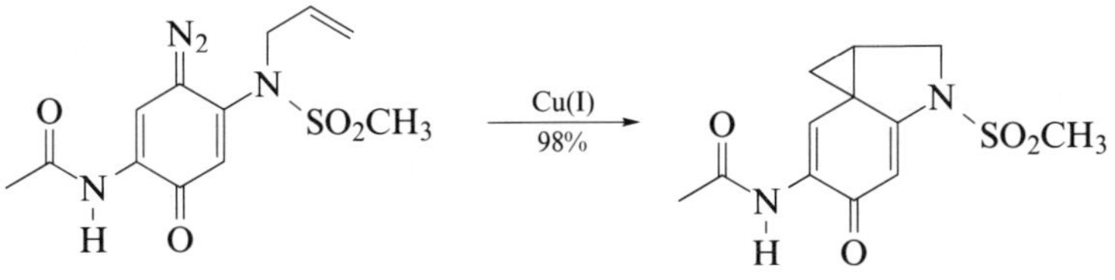

(3) (4) (6)

Palladium-based catalysts, especially palladium(II) acetate, $PdCl_2{\cdot}2PhCN$ and $[\eta^3\text{-}C_3H_5)PdCl]_2$, are very active catalysts for the cyclopropanation of aliphatic and strained cyclic olefins. Since palladium(II) readily coordinates with alkenes[17], the palladium(II) catalysed process may involve electrophilic addition of a palladium olefin complex onto the diazoalkane followed by loss of nitrogen to give a cyclopropane. Palladium(II) acetate was introduced by Teyssié and coworkers[18] in 1972 as an alternative to copper catalysts for diazo decomposition. Today, the palladium(II) derivatives are the most effective among transition metal catalysts for cyclopropanation reactions with diazoalkanes. The first example of palladium(II) catalysed cyclopropanation using diazomethane is shown in equation 7[18]. In general, palladium(II) catalysed cyclopropanation of simple or strained alkenes proceeds smoothly and with moderate to high yield (equations 7–16)[18–26]. In the case of cyclopropanation of allyl amines and allyl ethers, the cyclopropanations proceed smoothly and without evidence of ylide generation.[19,25,26].

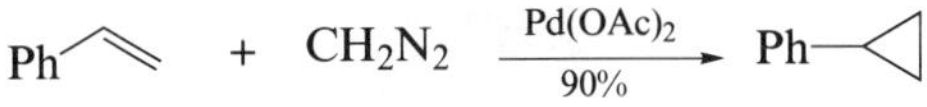

(7)

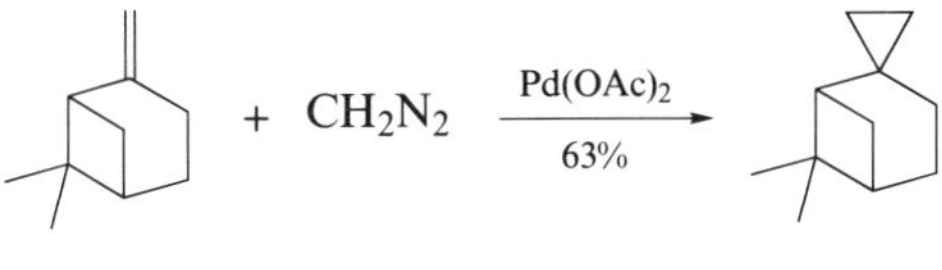

(8)

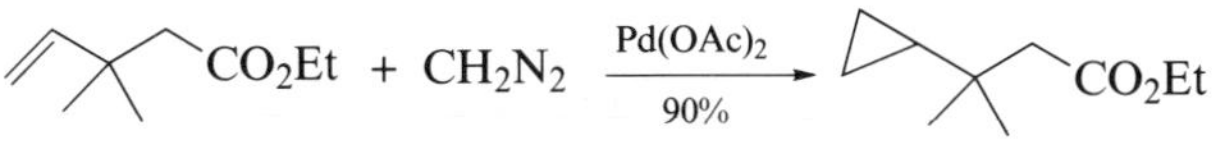

(9)

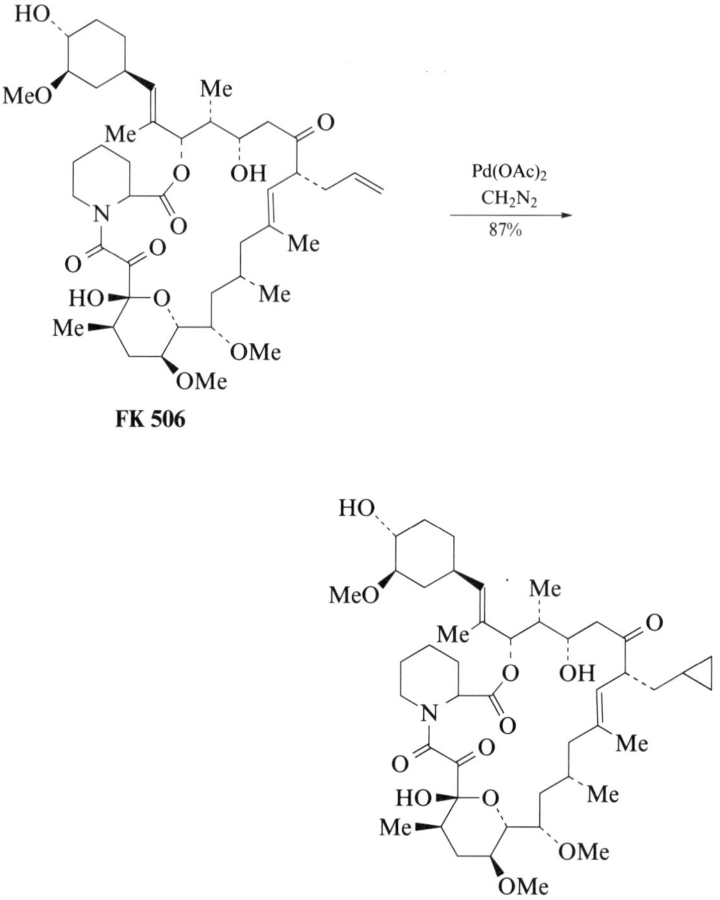

(10)

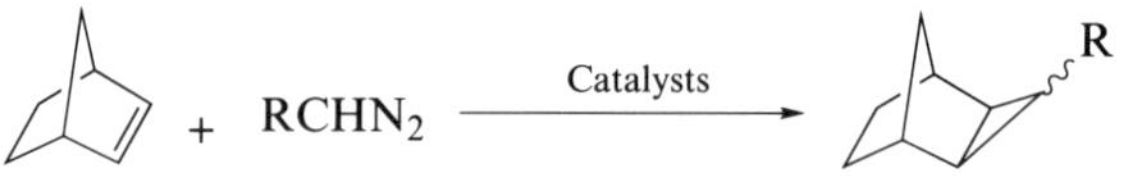

(11)

R=H, Catalyst=$Pd(OAc)_2$, yield=67%
R=CH_3, Catalyst=$(PhCN)_2PdCl_2$, yield=70%

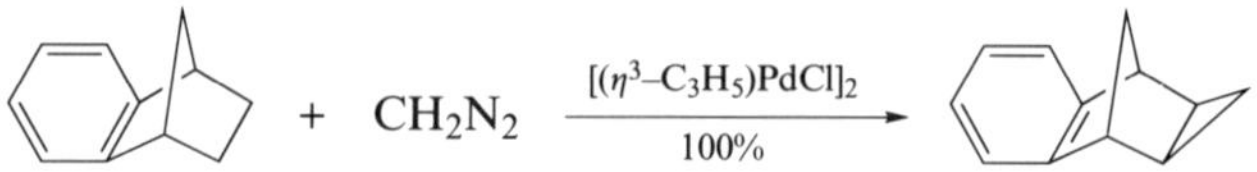

(12)

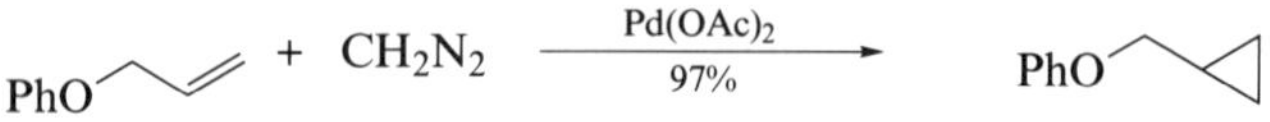

(13)

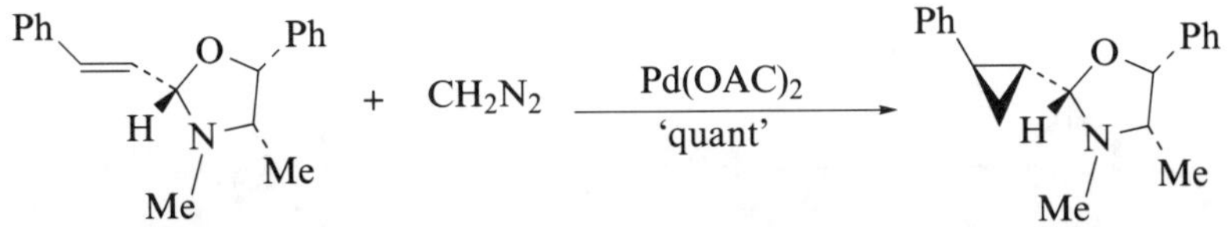

(14)

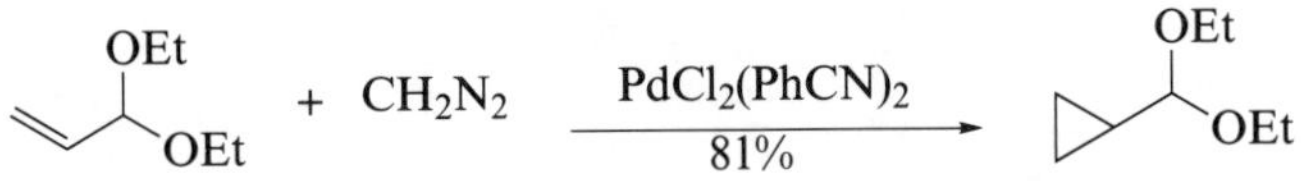

(15)

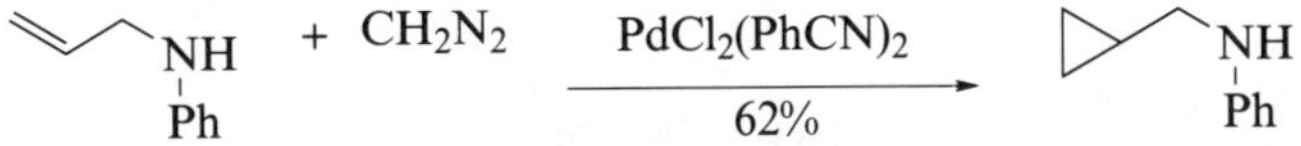

(16)

For cyclopropanation of very electron-rich alkenes such as vinyl ethers, copper(II) trifluoroacetate, copper(II) hexafluoroacetylacetonate or rhodium(II) acetate are the catalysts of choice. Copper trifluoroacetate catalysed cyclopropanation of vinyldiazomethane with dihydropyran gives the corresponding vinyl cyclopropane adduct in low yield (equation 17)[14]. In contrast, catalytic decomposition of phenyldiazomethane in the presence of various vinyl ethers results in high-yield phenylcyclopropane formation (equations 18 and 19)[27].

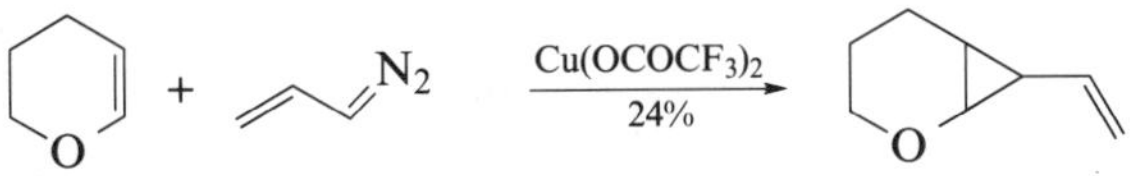

(17)

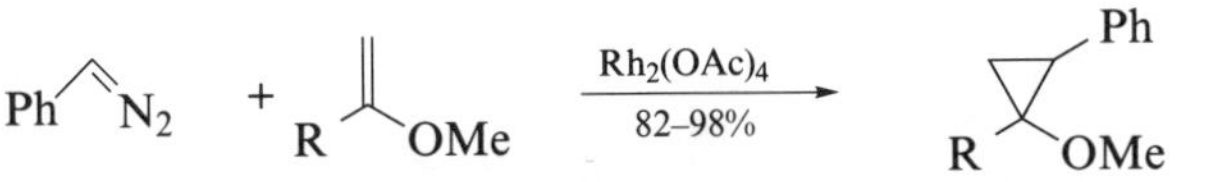

(18)

R=Ph, Me, *t*-Bu

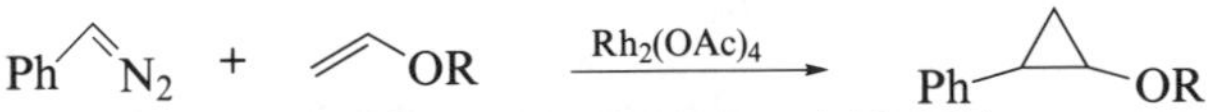

(19)

R = Et, Yield: 54%; R= *n*-Bu, yield: 92%

Palladium(II) compounds have unique characteristics suitable for efficient catalysed cyclopropanation of electron-deficient alkenes using diazoalkanes. Neither copper nor rhodium(II) catalysts have shown comparable reactivity with diazoalkanes, although these catalysts are superior to palladium(II) catalysts for cyclopropanation with diazocarbonyl compounds. A few examples of palladium(II) catalysed cyclopropanation of α,β-unsaturated carbonyl compounds with diazoalkanes are shown in equations 20–24[28–30].

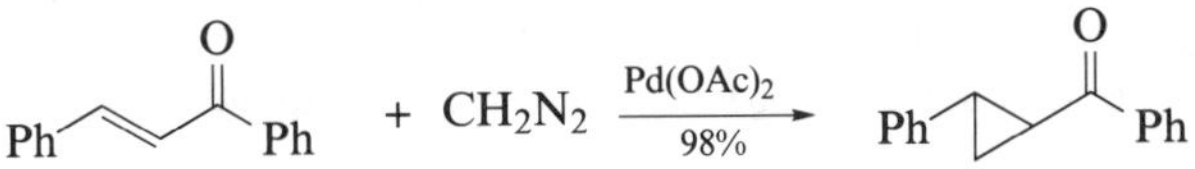

(20)

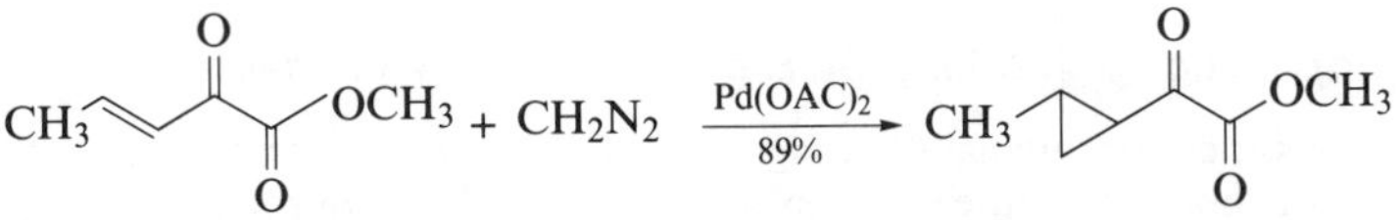

(21)

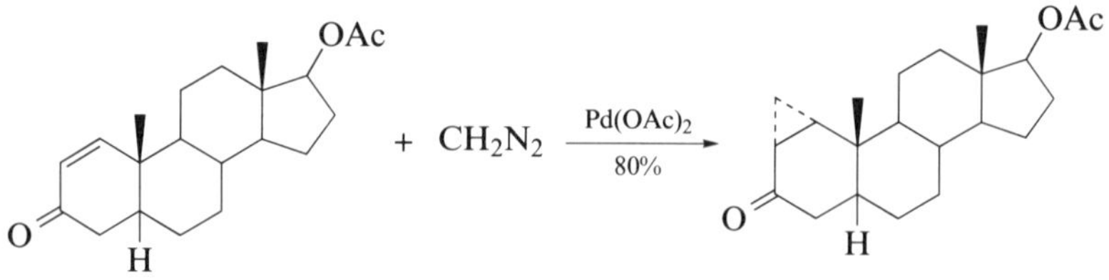

(22)

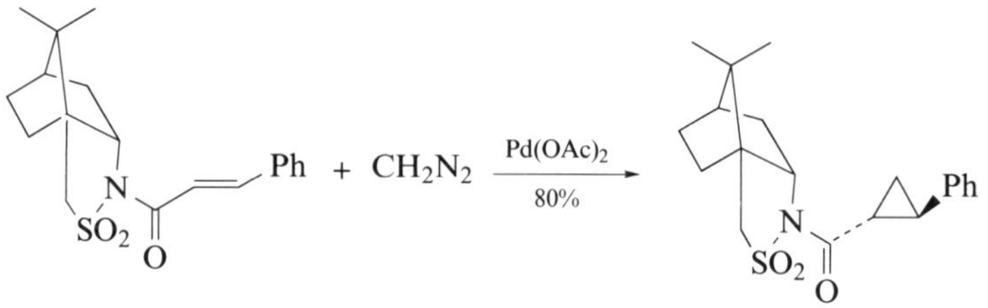

(23)

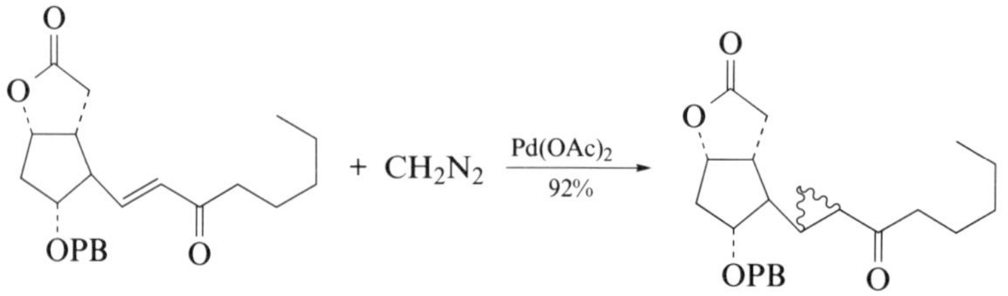

(24)

PB = *p*-Phenylbenzoyl

III. CYCLOPROPANATION USING DIAZOCARBONYL COMPOUNDS

Metal catalysed decomposition of diazocarbonyl compounds in the presence of alkenes provides a facile and powerful means of constructing electrophilic cyclopropanes. The cyclopropanation process can proceed intermolecularly or intramolecularly. Early work on the topic of intramolecular cyclopropanation (mainly using diazoketones as precursors) has been surveyed[31]. With the discovery of powerful group VIII metal catalysts, in particular the rhodium(II) derivatives, metal catalysed cyclopropanation of diazocarbonyls is currently the most fertile area in cyclopropyl chemistry. In this section, we will review the efficiency and versatility of the various catalysts employed in the cyclopropanation of diazocarbonyls. Cyclopropanations have been organized according to the types of diazocarbonyl precursors. Emphasis is placed on recent examples.

A. Cyclopropanation Using Alkyl Diazoesters/Diazoacetamides

Metal catalysed cyclopropanation using alkyl diazoesters has been confirmed as a useful synthetic method since an earlier review dealing with cyclopropanation chemistry appeared in 1970[32]. Catalysts containing various metals, e.g. ruthenium[33–35], rhodium[2–12,36], palladium[36–38], copper[4,31], cobalt[39–41], osmium[42,43], iron[44] and nickel[4] have been used in cyclopropanation with alkyl diazoesters. Among the transition metal based catalysts, rhodium(II) carboxylates appear the most generally effective in promoting the cyclopropanation of a wide range of alkenes with alkyl diazoacetates. Palladium(II) catalysts are suitable for the cyclopropanation of strained alkenes as well as styrenes. Homogeneous copper-based catalysts, although less effective than their rhodium counterparts, have the advantage of being inexpensive and are often used for cyclopropanation with alkyl diazoesters.

Intermolecular cyclopropanation using alkyl diazoesters has been extensively studied. The following representative examples illustrate the utility of this chemistry.

Intermolecular cyclopropanation reactions with ethyl diazoacetate have been employed for the construction of the cyclopropane-containing amino acid **7** (equation 25)[45]. Thus, rhodium(II) acetate catalysed decomposition of ethyl diazoacetate in the presence of D-cbz-vinylglycine methyl ester **5** afforded cyclopropyl ester **6** in 85% yield. Removal of the protecting group completed the synthesis of **7**. Another example illustrating intermolecular cyclopropanation can be found in Piers and Moss' synthesis of (±)-quadrone **8**[46] (equation 26). Intermolecular cyclopropanation of enamide or vinyl ether functions using ethyl diazoacetate has also been used in the synthesis of eburnamonine **9**[47], pentalenolactone E ester **10**[48] and (±)-dicranenone A **11**[49] (equations 27–29).

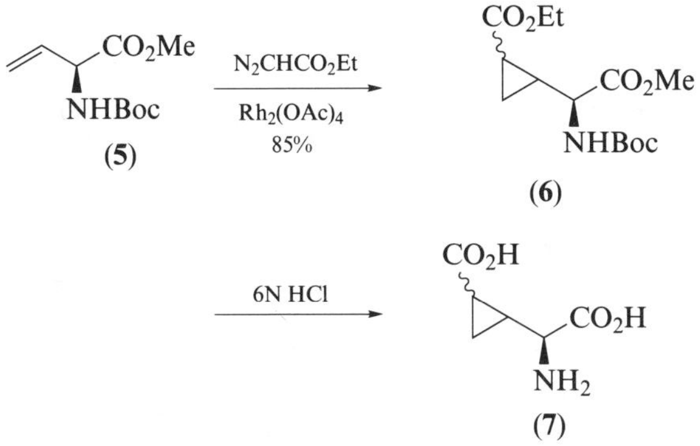

(25)

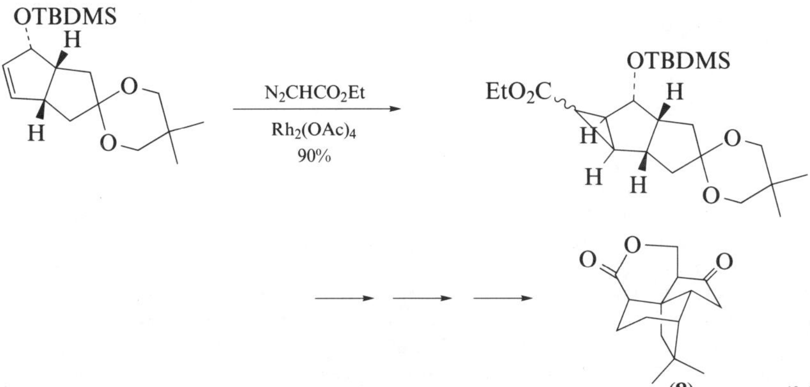

(8) (26)

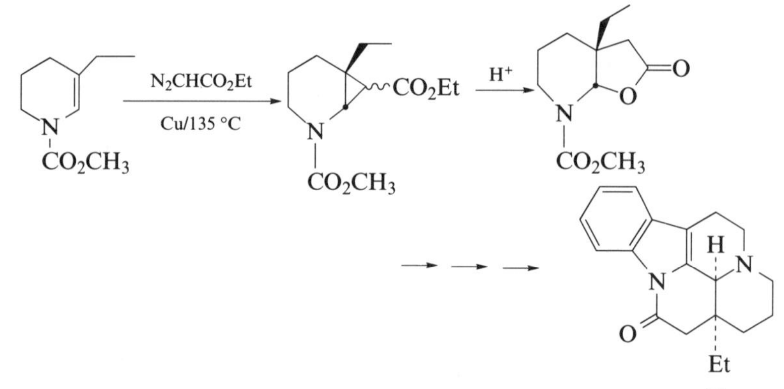

(9) (27)

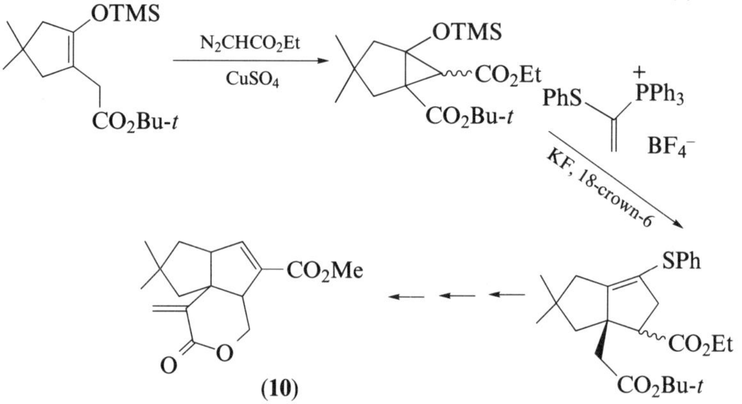

(28)

(11) (29)

Catalytic decomposition of vinyldiazomethane (**12**) containing an alkyl ester function alpha to the diazo group in the presence of vinyl ether **13** or diene **17** gave vinylcyclopropane (**14**) or divinylcyclopropane (**18**), respectively. Rhodium(II) carboxylates are the catalysts of choice for the decomposition of vinyldiazo compounds. In addition, the cleavage of the cyclopropyl ring of **14** resulted in a five-membered carbocyclic compound **15** or lactone **16**. The rearrangement chemistry of **14** is highly dependent on the presence of the ester functionality (equation 30)[50,51]. The *cis* isomers of the divinylcyclopropane **18** in general undergo facile Cope rearrangement to form seven-membered carbocycles **19** (equation 30)[52]. Two examples concerning the application of this chemistry are seen in the synthesis of (±)-ferruginine **20**[53] and (±)-anhydroecgonine methyl ester **21**[54] as shown in equations 31 and 32. In both cases, rhodium(II) octanoate ($Rh_2(OOct)_4$) is the catalyst of choice. Most work concerning this tandem cyclopropanation/Cope rearrangement approach leading to seven-membered ring formation has been reported by Davies' group and Davies has summarized his own work in a recent review[55].

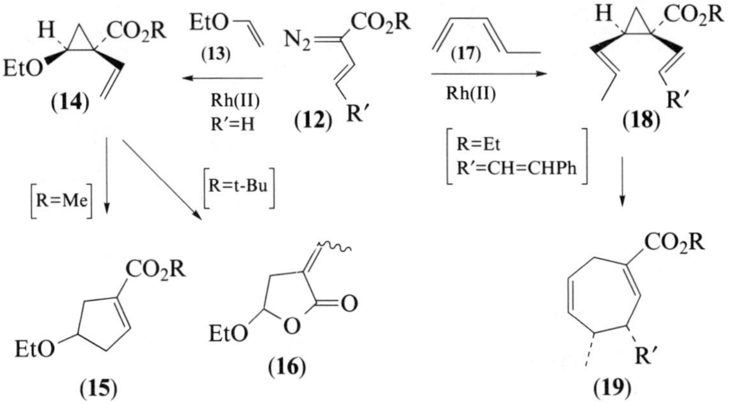

(30)

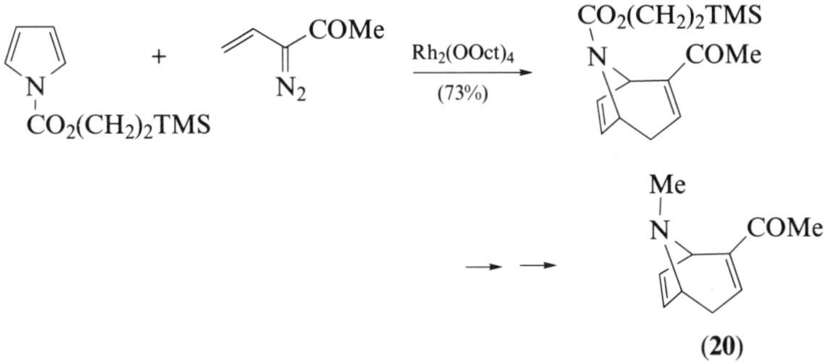

(31)

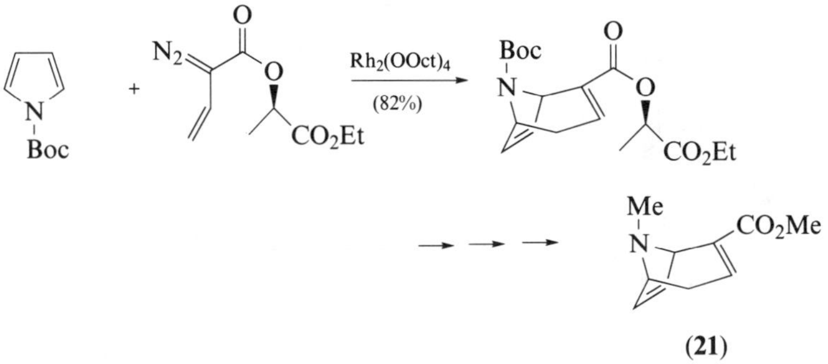

(32)

Intramolecular cyclopropanation using diazoesters is a powerful synthetic tool. Diazoesters are readily prepared from the corresponding alcohol via House's methods[56,57]. Numerous examples using the application of this transformation in synthesis have been reported. These include the potent synthetic pyrethroid NRDC 182 (**22**)[58], (1*R*)-(±)-*cis*-chrysanthemic acid (**23**)[59], the highly strained bicyclic system **24**[60], antheridic acid **25**[61,62] and cycloheptadiene **26**[63] (equations 33–37).

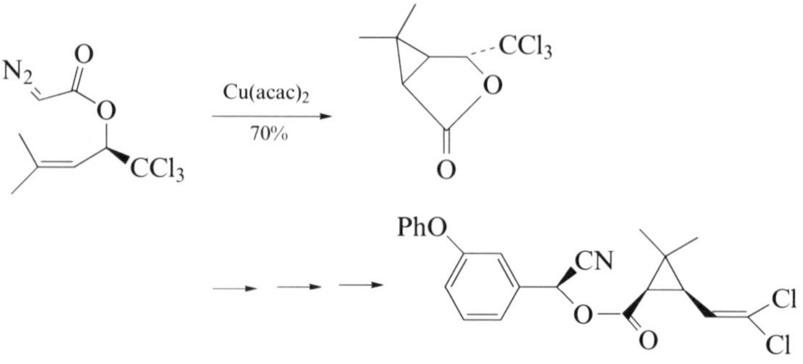

(22)

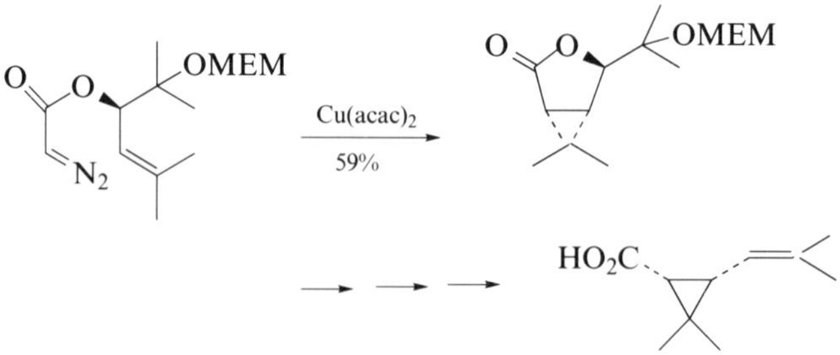

(23)

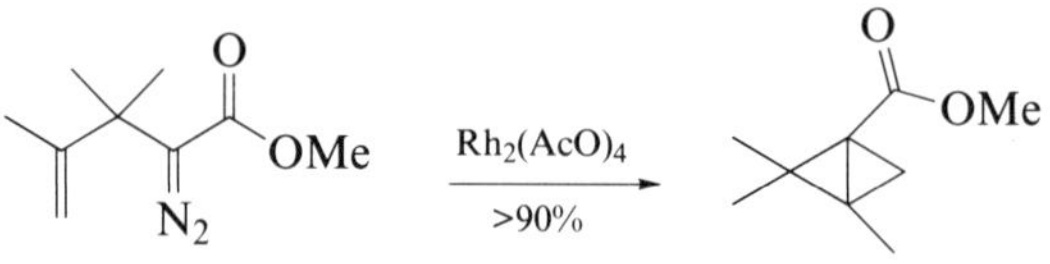

(24)

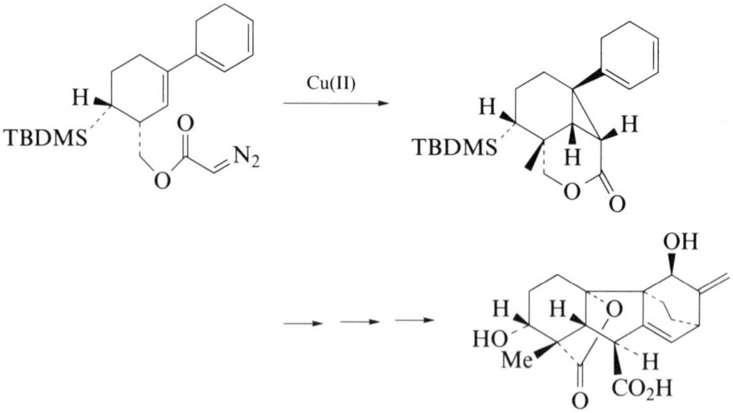

(25)

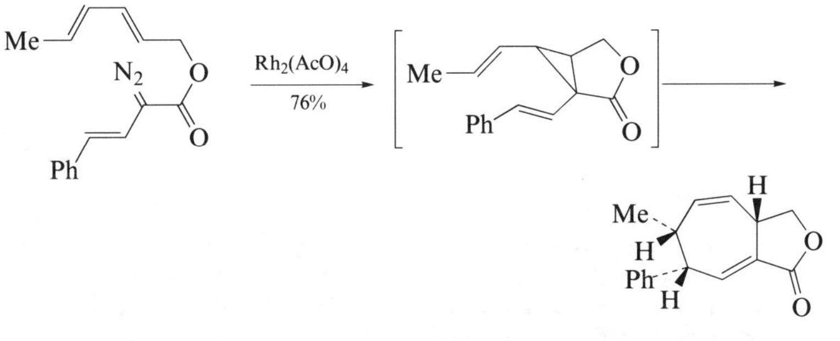

(37)

(26)

Catalysed cyclopropanation using diazoacetamides is rather less well developed. Doyle and coworkers have reported the intermolecular cyclopropanation of styrene with *N,N*-dimethyl diazoacetamide to give a cyclopropyl adduct in 74% yield (equation 38)[64]. A further example is found in a synthesis of α-(carboxycyclopropyl)glycine **27** where an intramolecular cyclopropanation of the diazoacetamide is the key step (equations 39)[65,66].

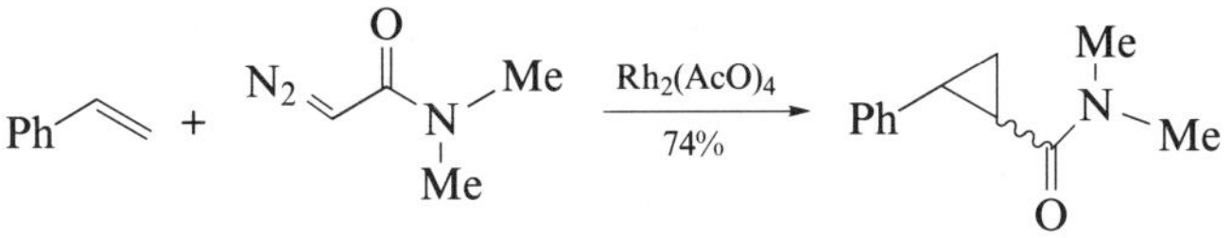

(38)

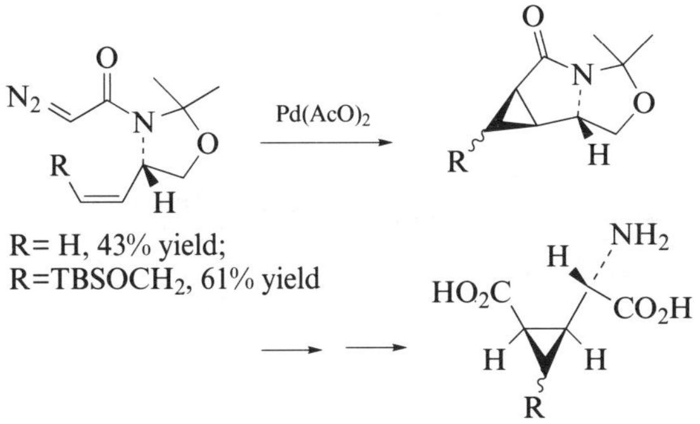

(39)

(27)

B. Cyclopropanation Using Diazoketones

The first copper catalysed intramolecular cyclopropanation of a diazoketone was reported by Stork and Ficini in 1961 (equation 40)[67]. During the following 16 years more than 160 examples of copper catalysed intramolecular cyclopropanation of diazoketones were reported[31]. Although the copper based catalysts have been used widely in cyclopropanation of diazoketones, in general copper catalysts require a relatively high reaction temperature and generally lack selectivity. Rhodium(II) carboxylates are the most active catalysts for cyclopropanation with diazoketones. The reaction can be carried out under ambient conditions to give cyclopropyl ketones with high yield. The efficiency of the cyclopropanation of diazoketones is also dependent upon the substituents present in the diazoketone or alkenes.

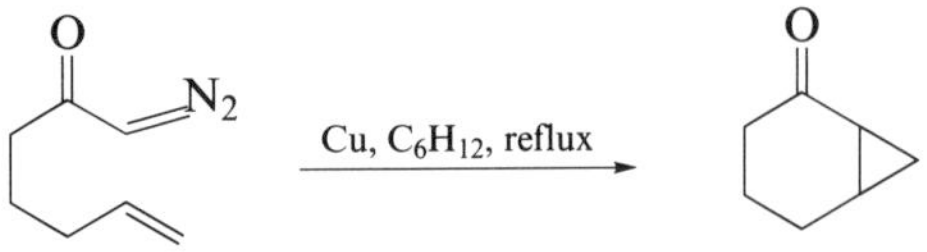

(40)

Intermolecular cyclopropanation of diazoketones is an effective method in organic synthesis. Wenkert and coworkers have applied this methodology to the synthesis of a substantial number of cyclopropane adducts **28**[68], **29**[69] and **30**[70] which are synthetic intermediates in the preparation of natural products (equations 41–43). Copper catalysts were chosen for these transformations. Another interesting application of intermolecular cyclopropanation is to be found in Daniewski's total synthesis of an aromatic steroid. Palladium(II) acetate catalysed decomposition of 4-bromo-1-diazo-2-butanone in the presence of *m*-methoxystyrene was used to give the cyclopropyl ketone **31** which was a key intermediate in the total synthesis (equation 44)[71].

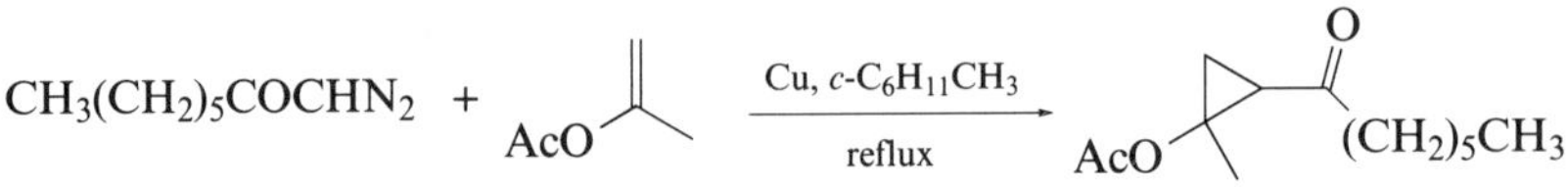

(41)

(28)

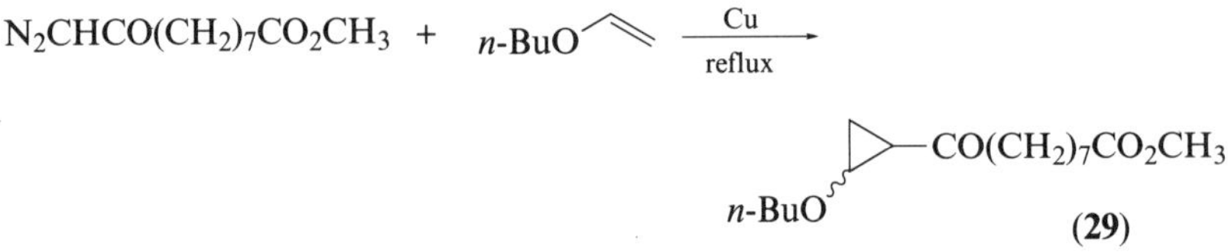

(42)

(29)

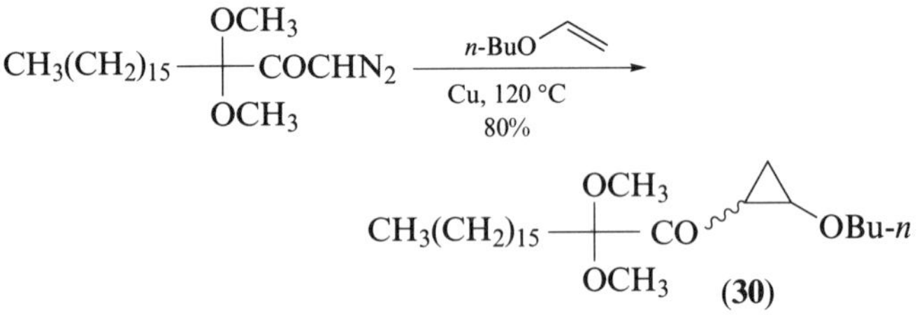

(43)

(30)

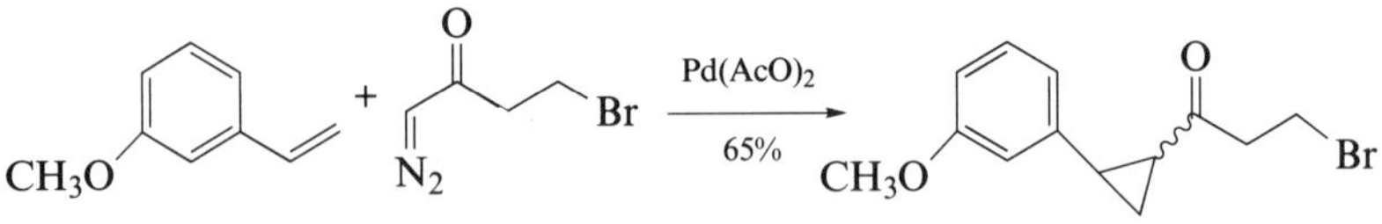

(44)

(31)

The intramolecular version of cyclopropanation using diazoketones has assumed strategic importance in organic synthesis. Metal catalysed decomposition of aliphatic diazoketones containing a remote carbon–carbon double bond results in bicyclic carbocycles. With β,γ-unsaturated α'-diazoketones, the resulting [2.1.0]bicyclic systems are unstable and undergo vinylogous Wolff rearrangement to give γ,δ-unsaturated carboxylic acid derivatives[72]. Two examples are shown in equations 45 and 46[73,74]. There is an exception to this type of transformation, which is shown in equation 47, where the copper or rhodium catalysed decomposition of diazoketones of type **32** leads to the highly strained tricyclic ketones **33** which are sufficiently stable to be isolated[75,76].

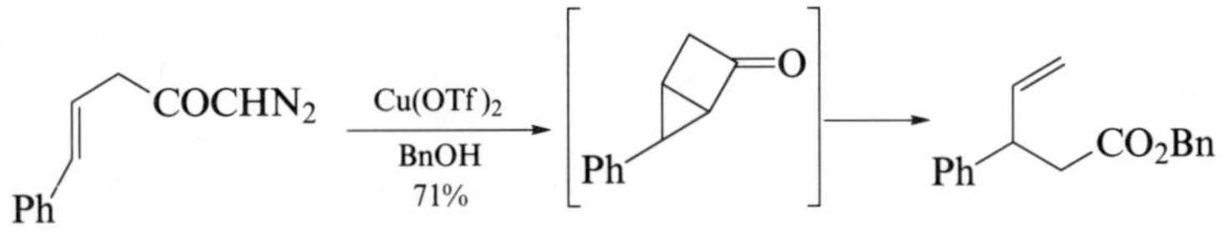

(45)

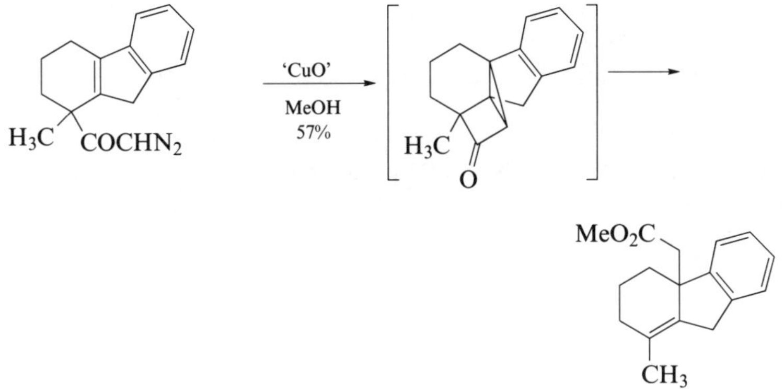

R=H, Catalyst=Cu:
<1% yield (46)

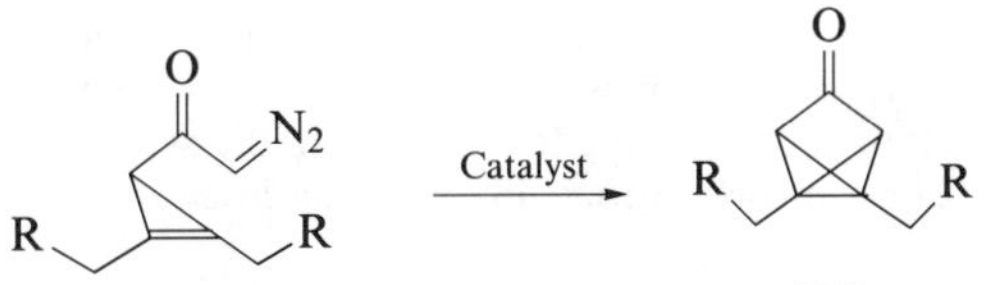

(32) **(33)**

R=OAc, Catalyst=
$Rh_2(AcO)_4$: 37% yield (47)

Intramolecular cyclopropanation of diazoketones to furnish [3.1.0] and [4.1.0] bicyclic systems are the most common and effective reactions in this category. Two recent examples are shown in equations 48 and 49. The bicyclic ketone **34** has been used in the synthesis of polycyclic cyclobutane derivatives[77], whereas ketone **35** is the key intermediate in the total synthesis of (±)-cyclolaurene[78]. When the olefinic double bond is attached to, or is part of, a ring system, the cyclopropanation process also works well. Copper oxide catalysed decomposition of diazoketone **36** produces the strained tricyclic ketone **37** in 86% yield (equation 50)[79]. In another case, in which the cyclopropanation of diazoketone **38** gave stereospecifically the cyclopropyl ketone **39**, copper sulphate catalysis was used. The cyclopropyl ketone **39** is the key intermediate in the total synthesis of (±)-albene **40** (equation 51)[80].

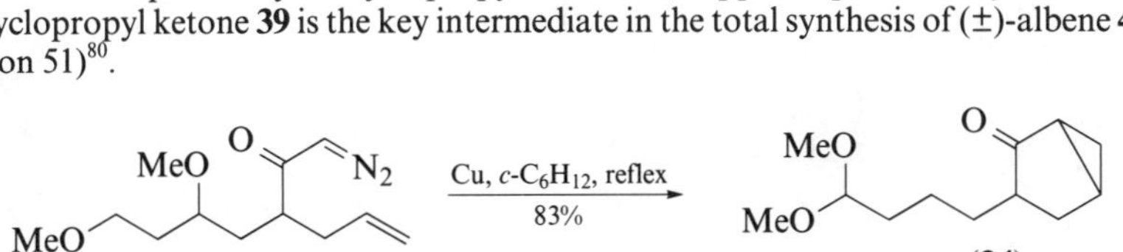

(48)

(34)

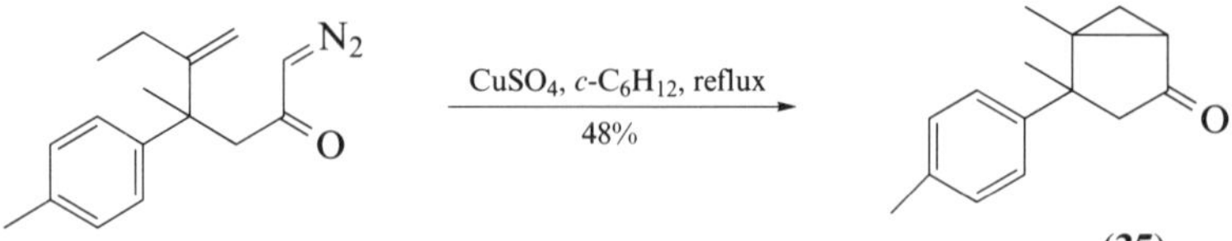

(35)

(49)

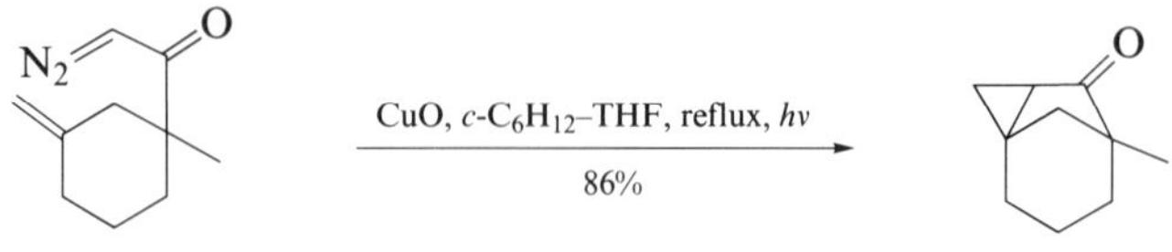

(36) (37)

(50)

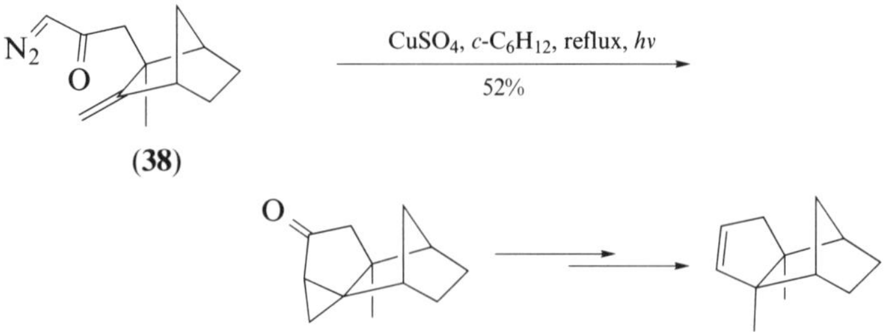

(38)

(39) (40)

(51)

Copper catalysed decomposition of diazoketone **41** furnishes tricyclic ketone **42** in 83% yield (equation 52) which provides the key building block in the synthesis of chysomelidial[81]. A few similar instances of cyclopropanation are to be found in the syntheses of (±)-hinesol **43**[82], (–)-acorenone B **44**[83], (±)-spirolaurenone **45**[84] and (±)-descarboxyquadrone **46**[85] (equations 53–56).

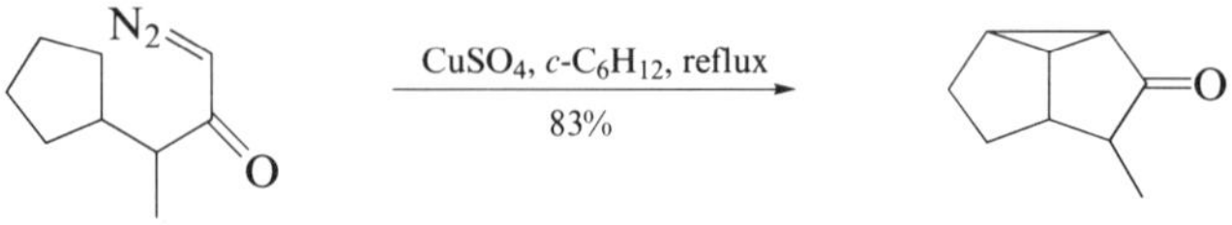

(41) (42)

(52)

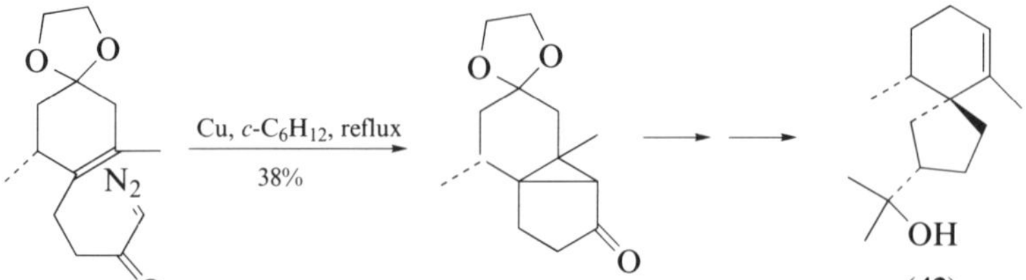

(43)

(53)

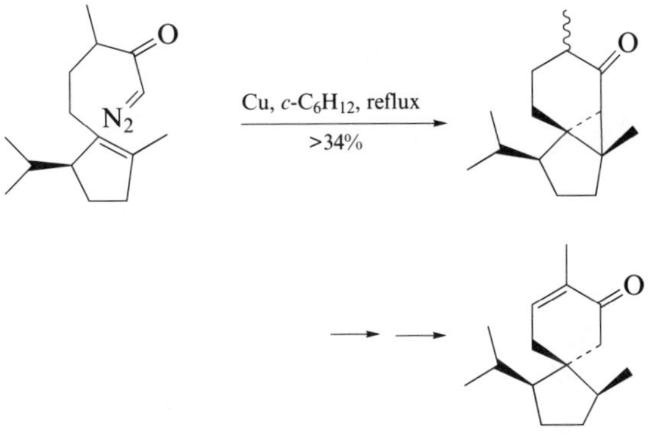

(44) (54)

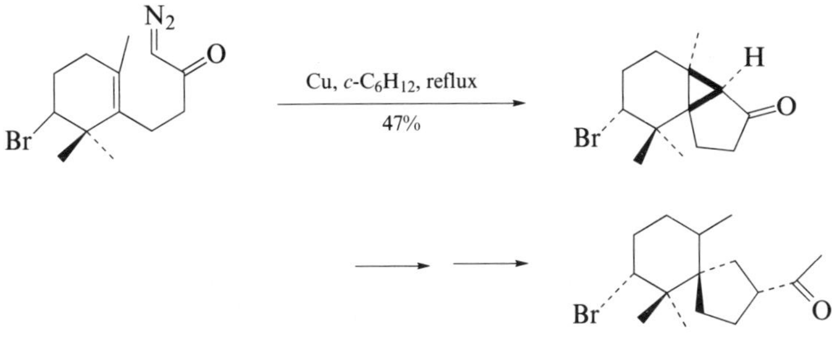

(45) (55)

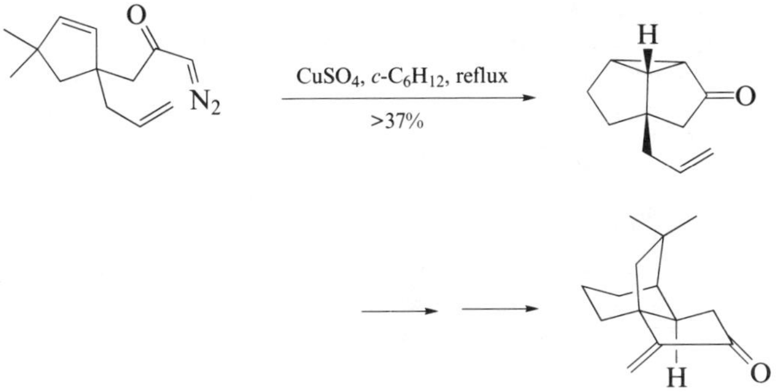

(46) (56)

Where there is an electron-rich olefinic double bond in the diazoketone as in the 2-substituted 3,4-dihydropyran derivative **47**, the intramolecular cyclopropanation produces oxatricyclic ketones, e.g. **48** (equation 57)[86,87]. Rhodium(II) acetate is the catalyst of choice for this transformation. An interesting application of this method is found in a stereoselective synthesis of (±)-*β*-chamigrene **49** (equation 58)[88].

Metal catalysed decomposition of dienoic diazoketones can give rise to vinyl cyclopropanes[89–92]. This cyclopropanation process is ring size dependent and only closures to

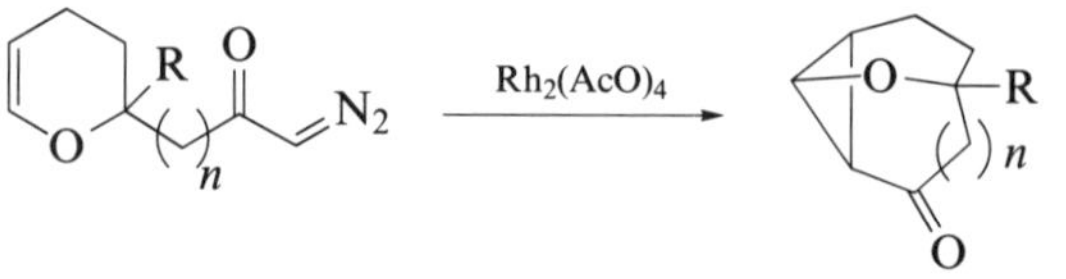

(**47**) (**48**)

R=H, n=0, yield=58%; n=1, yield=61%
R=CH_3, n=2, yield=95%; n=3, yield=80%

(57)

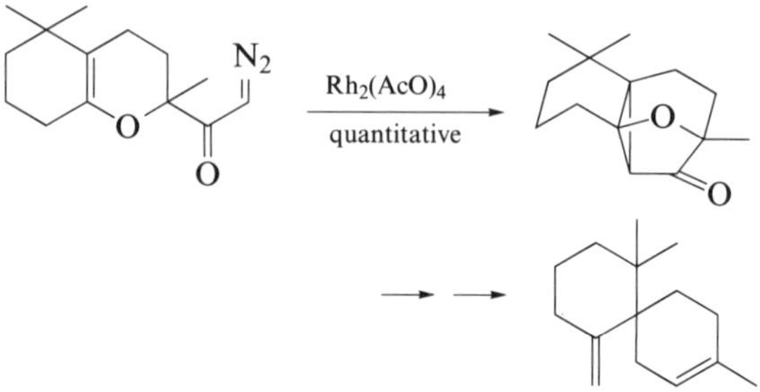

(**49**) (58)

five- and six-membered rings are synthetically useful. $Cu(acac)_2$ catalysed decomposition of dienoic diazoketone **50** furnishes vinyl cyclopropane **51** in high yield (equation 59). Pyrolysis of *cis* methylvinylcyclopropane **51b** with careful control of the pyrolysis temperature gives the 1,5-hydrogen shift product **53**. Selective ozonolysis of **53** leads to (±)-sarkomycin **54** (equation 60)[91,92]. A second rearrangement of vinyl cyclopropane **51** is possible at higher temperature, producing bicyclic ketone **52** (equation 59)[89,90]. The combination of intramolecular cyclopropanation with this rearrangement provides a facile approach to cyclopentene annellation. An interesting application of this approach is found as a step in the synthesis of (–)-verbenalol **55** (equation 61)[93]. A further example is found in a synthesis of (±)-sinularene **56** where the vinyl cyclopropane is again obtained via copper catalysis. In this case, the diazoketone bears a carbon–carbon triple bond as a masked *cis* double bond which meets the requirement for further stereocontrolled rearrangement (equation 62)[94].

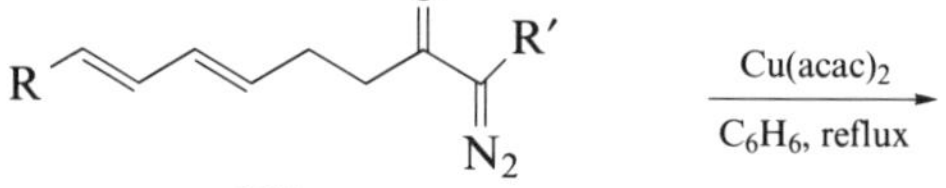

(**50**)

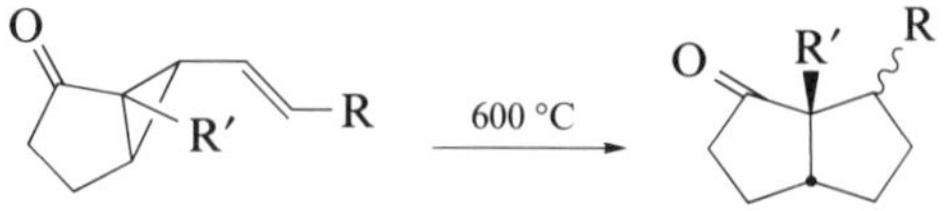

(59)

(**51**) (**52**)

(**a**) R=CH_3, R′=H;	94% yield,	70% yield
(**b**) R=H, R′=CH_3;	82% yield,	66% yield
(**c**) R=R′=CH_3;	75% yield,	63% yield

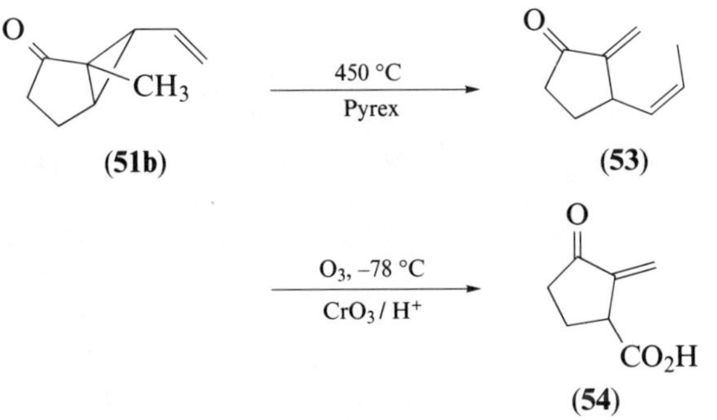

(51b) (53) (54)

(60)

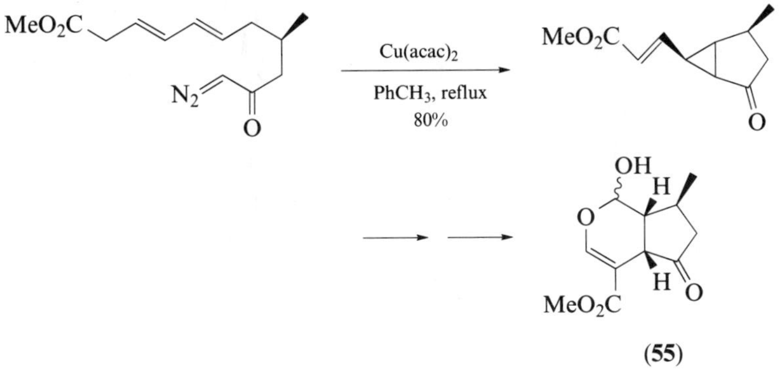

(55)

(61)

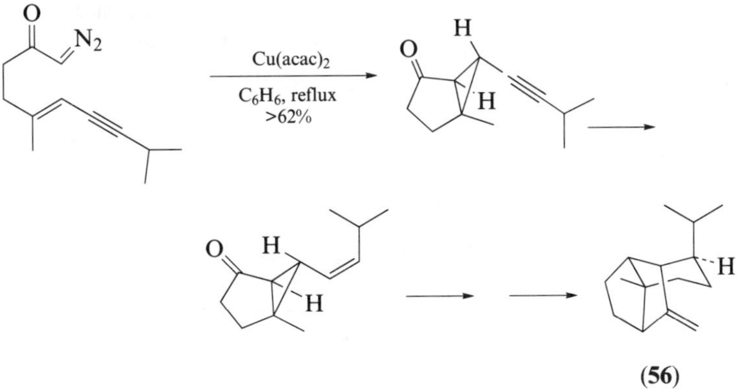

(56)

(62)

Where the diene function is attached to or is part of a carbocyclic system, the vinyl cyclopropane formation and subsequent rearrangement affords a reliable approach to the formation of tricyclic carbocycles. Some of the most elegant demonstrations of the use of this methodology in total synthesis of fused cyclopentanoid terpenes come from the work of Hudlicky and coworkers (equations 63–66)[90,95–100]. In these cases the diazoketones bear a carboxylate-substituted double bond of diene and the intramolecular cyclopropanation requires the combination of $CuSO_4$/$Cu(acac)_2$ as catalyst.

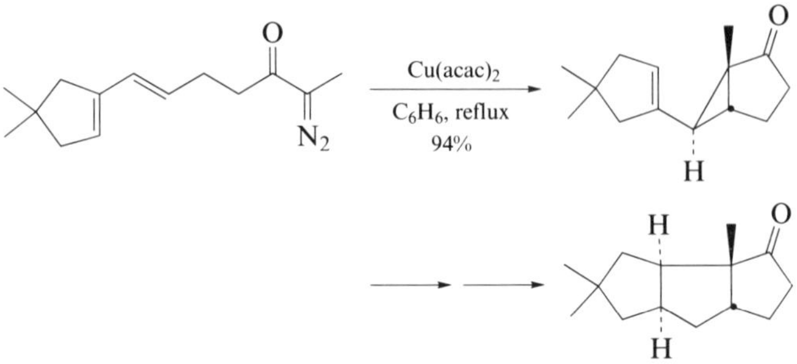

(±)-Hirsutene

(63)

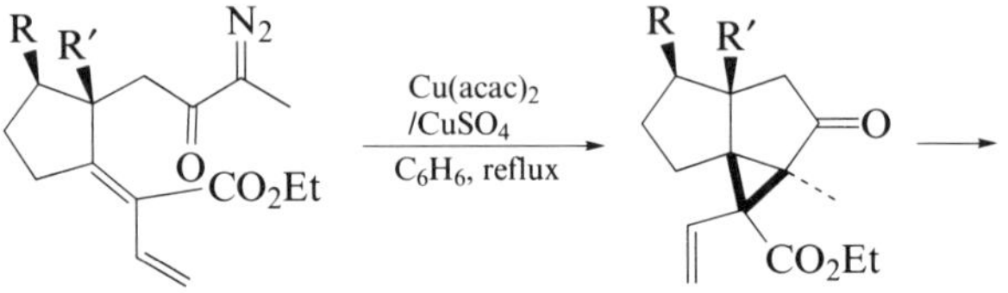

R = CH_3, R′ = H, 70–75% yield; R = H, R′ = CH_3, 61% yield

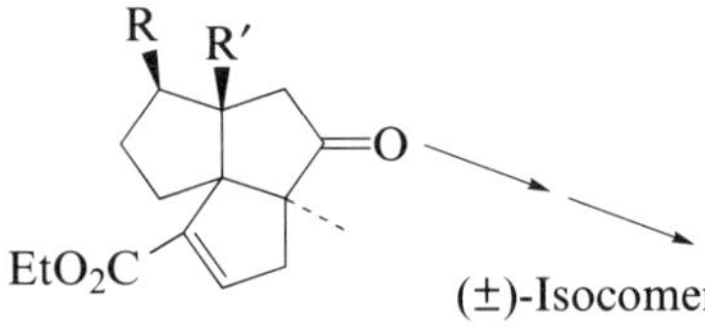

(±)-Isocomenic acid
(±)-Epiisocomenic acid
(±)-Retigeranic acid

(64)

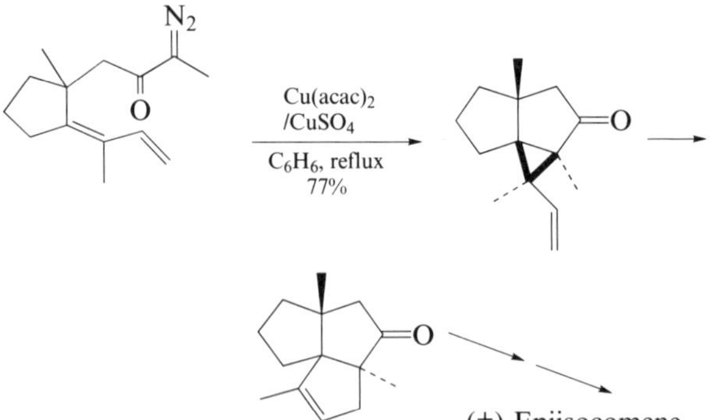

(±)-Epiisocomene

(65)

The intramolecular cyclopropanation of appropriate γ,δ-unsaturated α-diazoketones following a stereoselective catalytic reduction of the cyclopropyl ketone group provides a useful approach in diterpenoid synthesis. Some examples of the use of the cyclopropanation-reductive cleavage approach in synthesis are shown in equations 67 and 68[101-103].

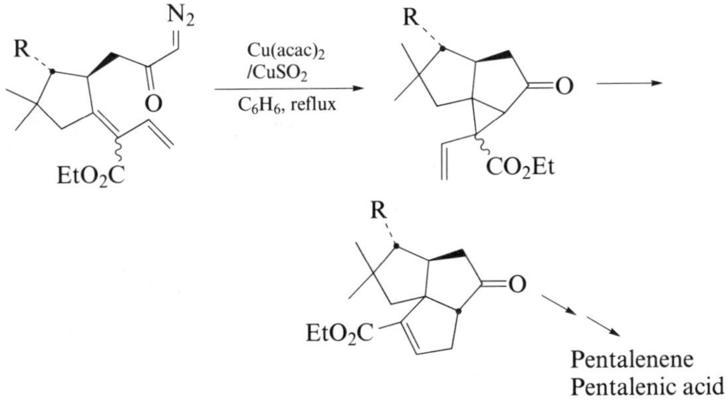

(66)

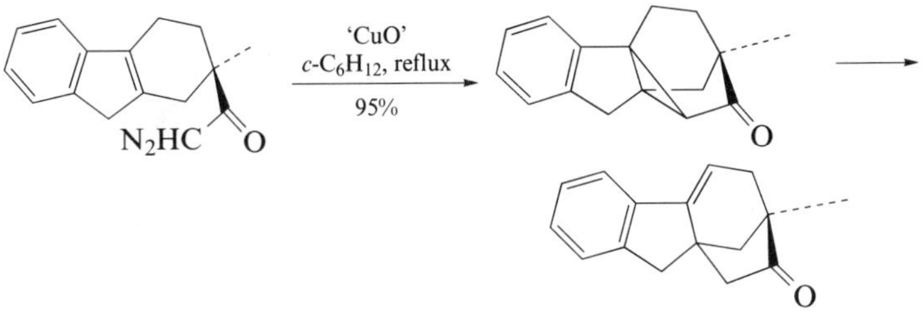

(67)

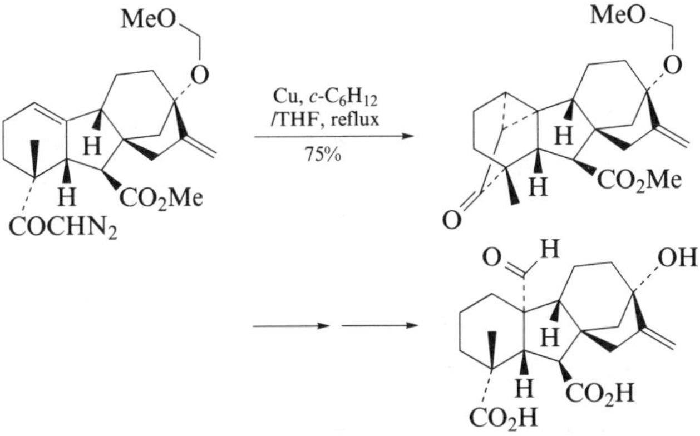

(68)

Where the carbon–carbon double bond is a part of an aromatic system, in general, cyclopropanation of diazoketones results in the formation of unstable cyclopropane adducts. For example, Saba[140] has shown that in the intramolecular cyclopropanation of diazoketone **57** the norcaradiene ketone **58** can be detected by low-temperature NMR and can be trapped in a Diels Alder reaction with 4-phenyl-1,2,4-triazoline-3,5-dione (equation 69). In addition, Wenkert and Liu have isolated the stable norcaradiene **60** from the rhodium catalysed decomposition of diazoketone **59** (equation 70)[105]. Cyclopropyl ketones derived from intramolecular cyclopropanation of hetereoaromatic diazoketones are also known and two representative examples are shown in equations 71 and 72[106]. Rhodium(II) compounds are the most suitable catalysts for the cyclopropanation of aromatic diazoketones.

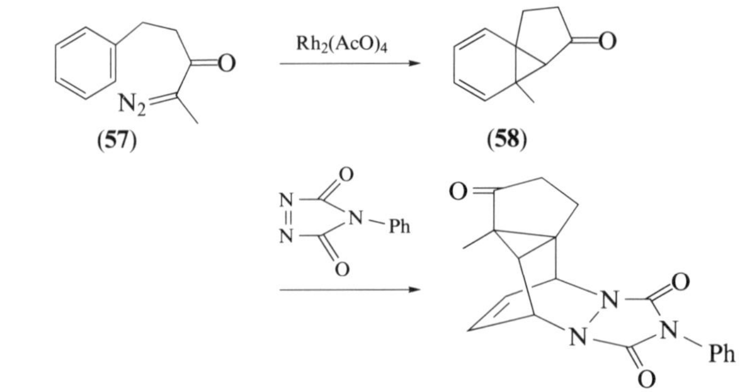

(69)

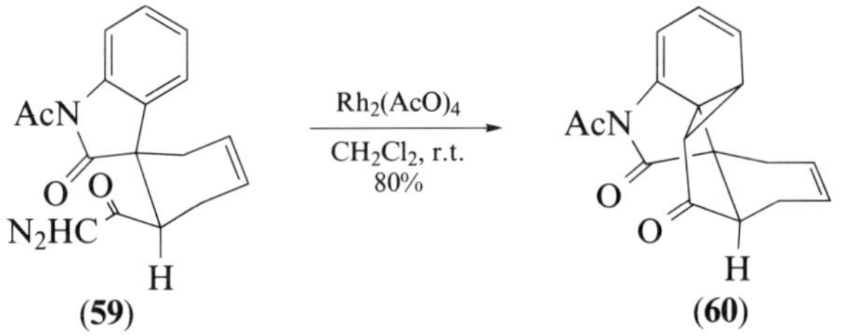

(70)

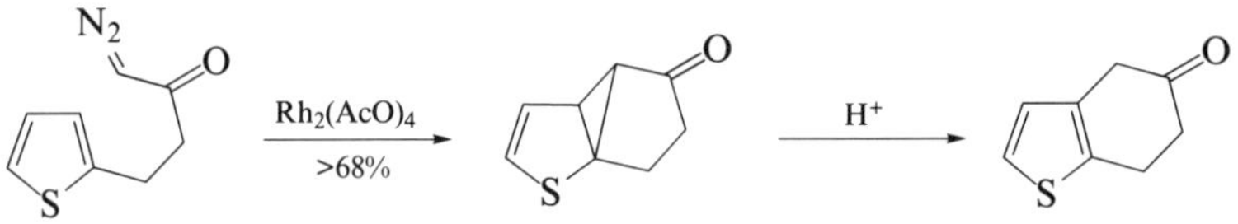

(71)

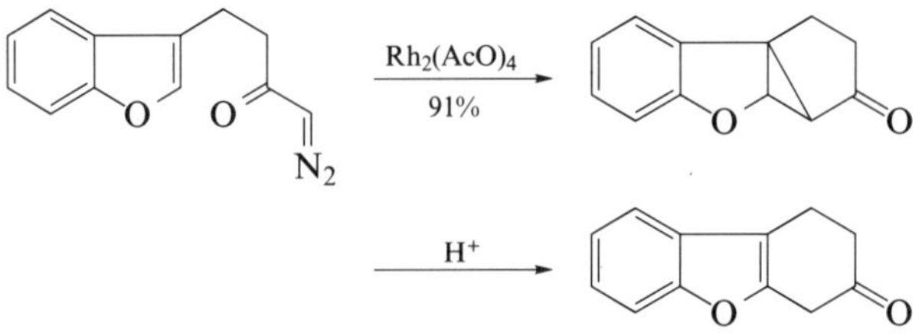

(72)

C. Cyclopropanation Using Dicarbonyl Diazomethanes

Electron-withdrawing substituents generally increase diazo compounds stability toward decomposition. Dicarbonyl diazomethane, which bears two carbonyl groups flanking the diazomethane carbon, are more stable than diazo compounds with only one carbonyl substituent. In general, metal catalysed decomposition of dicarbonyl diazomethane requires higher temperature than does monocarbonyl substituted diazomethane. As indicated before, rhodium(II) carboxylates are the most active catalysts for diazo decomposition. With dicarbonyl diazomethane, the rhodium(II) carboxylate-promoted cyclopropanation process can also be carried out under ambient conditions to afford a high yield of products.

With a range of methods available for the formation of 1,3-dicarbonyl compounds, the dicarbonyl diazomethanes can be readily prepared via a simple diazo transfer reaction with sulfonyl azide. This has made a vast array of dicarbonyl diazomethanes available, which enhances the versatility in organic synthesis. A selection of examples from recent literature to illustrate the versatility of the cyclopropanation using dicarbonyl diazomethane in the construction of natural products as well as other biologically active compounds is described below.

A few natural products which contain the cyclopropyl ring have been synthesized through metal catalysed cyclopropanation using dicarbonyl diazomethanes. (±)-Cycloeudesmol **63**, isolated from marine alga *Chondria oppositiclada*, was synthesized via a sequence involving a copper catalysed cyclopropanation of α-diazo-β-ketoester **61** to give the key intermediate **62** (equation 73)[107,108]. Similarly, the bicyclo[3.1.0]hexane derivative **65** was synthesized from the corresponding α-diazo-β-ketoester **64** via the catalytic method and was converted into (±)-trinoranastreptene **66** (equation 74)[109]. Intramolecular cyclopropanation of α-diazo-β-ketoesters **67** results in lactones **68** which are precursors to 1-aminocyclopropane-1-carboxylic acids **69** (equation 75)[110].

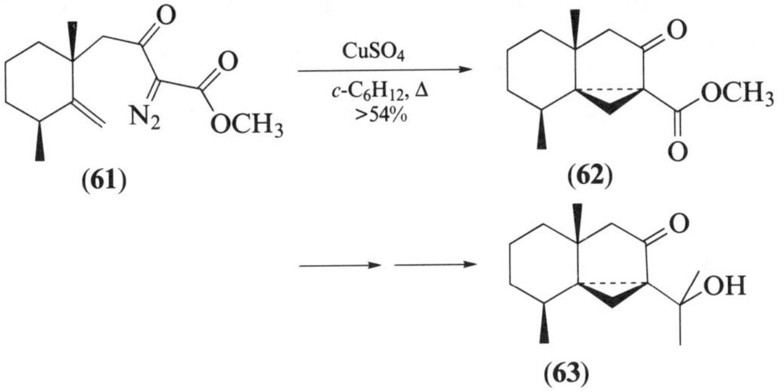

(73)

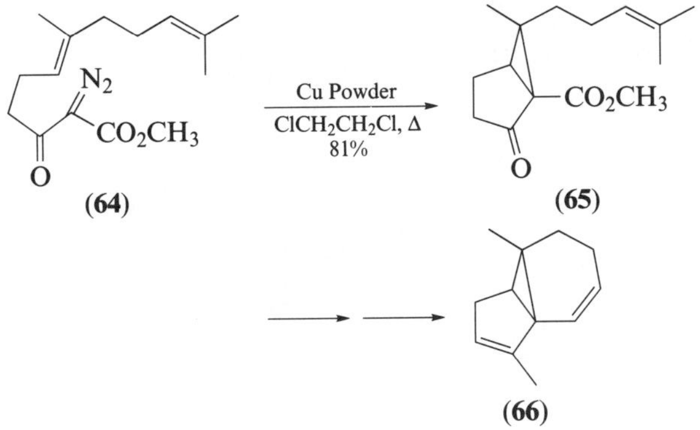

(74)

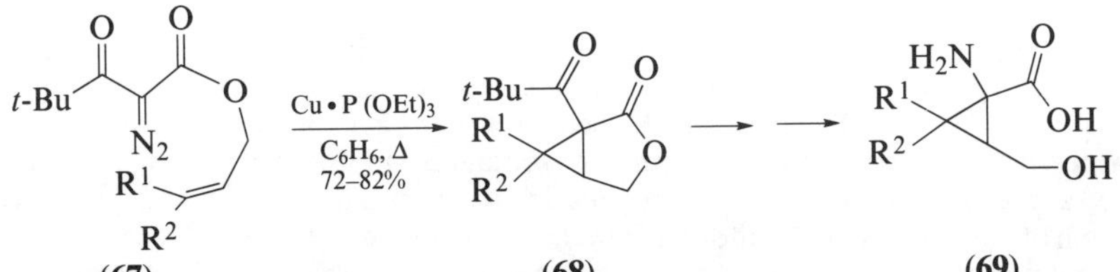

(75)

α-Diazo-β-ketoesters containing a freely rotating aliphatic chain with a carbon–carbon double bond in an appropriate position can undergo catalytic intramolecular cyclopropanation leading to five- or six-membered bicyclic systems. Fragmentation of the cyclopropyl ring can then lead to substituted cyclic products. Both Trost's[111] and Taber's groups[112,113] have used the above strategy to construct substituted cyclopentanones and Taber and coworkers have also applied this chemistry in the synthesis of (+)-isoneonepetalactone **70** (equation 76)[114]. Another application is to be found in the stereoselective synthesis of (±)-clavukerin A **72**; the cyclopropane adduct **71** obtained from the copper catalysed intramolecular cyclopropanation is the key intermediate for the synthesis (equation 77)[115,116].

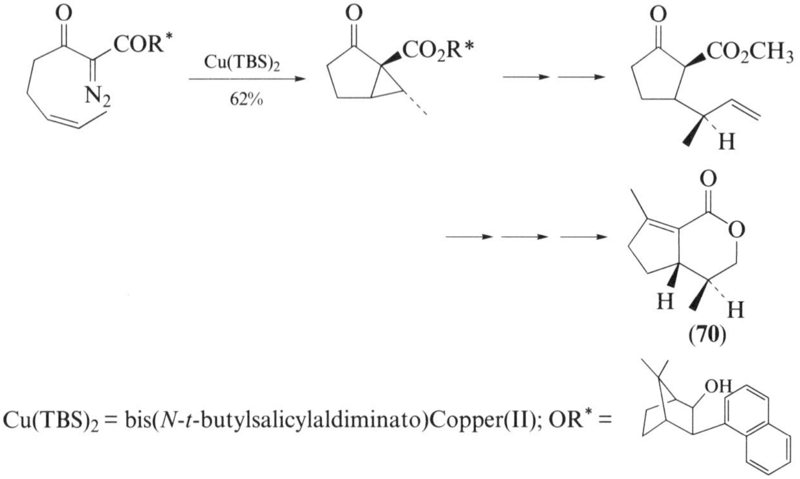

(76)

(70)

Cu(TBS)$_2$ = bis(*N*-*t*-butylsalicylaldiminato)Copper(II); OR* =

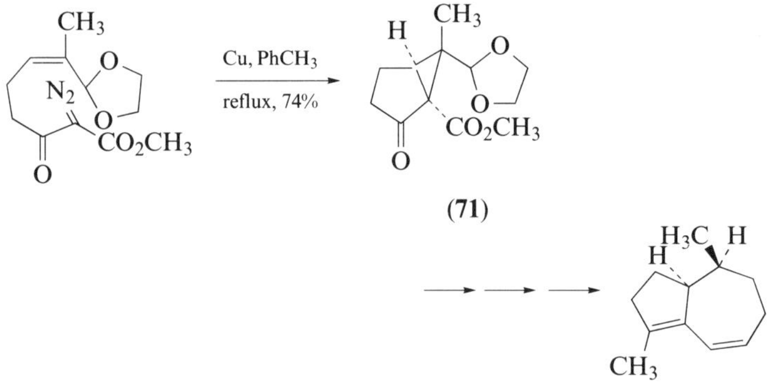

(71)

(77)

(72)

α-Diazo-β-ketoesters **73** containing a remote diene function under catalytic cyclopropanation afford vinylcyclopropane **74**. Further modification of **74** generated the divinylcyclopropane **75**, which on thermolysis furnished the Cope rearrangement product **76** (equation 78)[117–119]. The bicyclo[3.2.1]octane carbon skeleton (**76**) is a common structural feature on many naturally occurring substances. In another case, diazomalonate served as the precursor for the cyclopropanation reaction. Thus, copper catalysed decomposition of diazomalonate **77** afforded bicyclic lactone **78**, which was subjected to thermolysis and subsequent decarbomethoxylation to give guaiane synthons **79** (equation 79)[120].

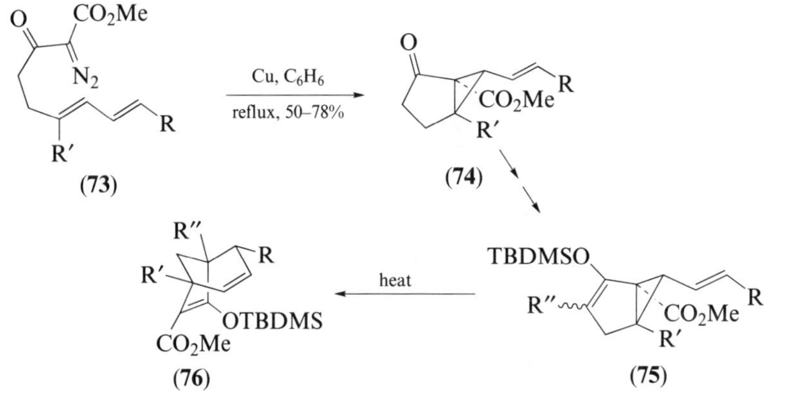

(73) (74) (75) (76) (78)

R=R′=R″=H
R=*i*-Pr, R′=R″=H
R=R″=H, R′=Me
R=R′=Me, R″=H
R=R′=H, R″=Me

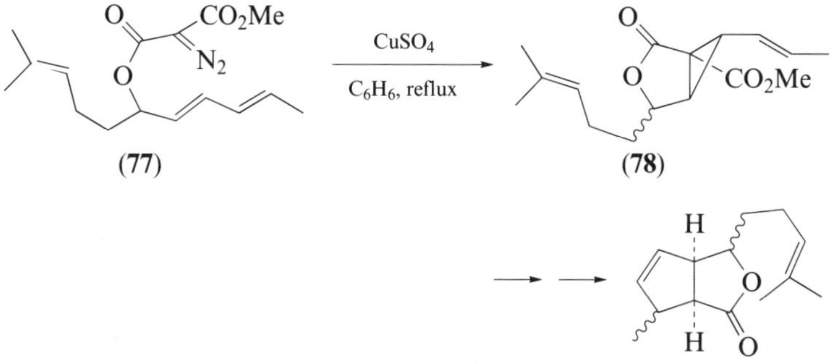

(77) (78) (79) (79)

Where the olefinic group is attached to or is part of a carbocyclic system, the catalytic intramolecular cyclopropanation of the α-diazo-β-ketoester with subsequent cleavage of the cyclopropane ring provides a useful strategy to bicyclic or polycyclic systems. (±)-Thapsane **80**, a sesquiterpene isolated from the root of a Mediterranean umbelliferous plant, *Thapsia villosa L*, has been synthesized using the above strategy as key step (equation 80)[121,122]. Similarly, coronafacic acid **81** was synthesized via a sequence involving intramolecular cyclopropanation as key step (equation 81)[123]. The combination of intramolecular cyclopropanation and cationic cyclization provides a facile and powerful approach to construction of complex natural products. A few elegant demonstrations of this chemistry in total synthesis have come from Corey's group and one representative example is shown in equation 82. Copper catalysed decomposition of α-diazo-β-ketoester **82** gave cyclopropyl ketoester **83** stereoselectively. The cyclopropyl ketoester **83** was converted to alcohol **84** in two further steps. Cyclization of **84** to pentacycle **85** leads to the required ring system directly and stereoselectively. The pentacycle **85** was then modified to (±)-cafestol **86** (equation 82)[124]. Variations of this approach have been used in the synthesis of (±)-atractyligenin[125]and (±)kahweol[126].

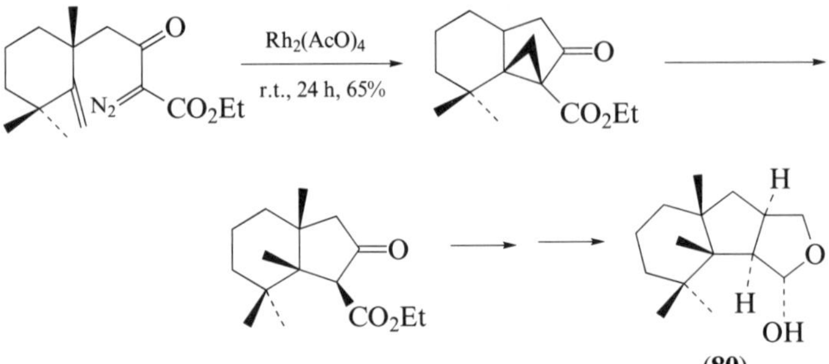

(80)

(80)

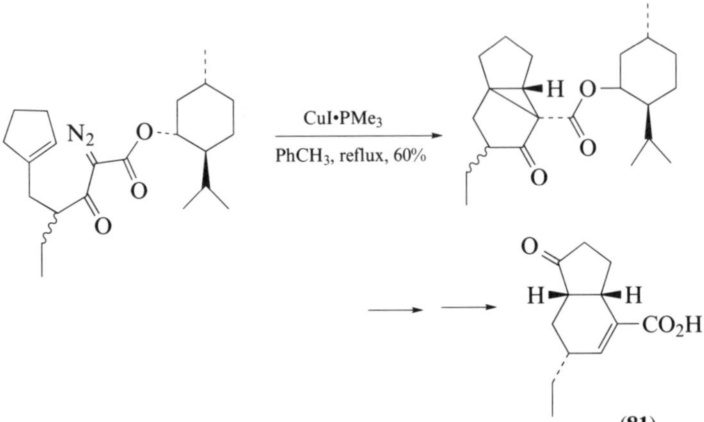

(81)

(81)

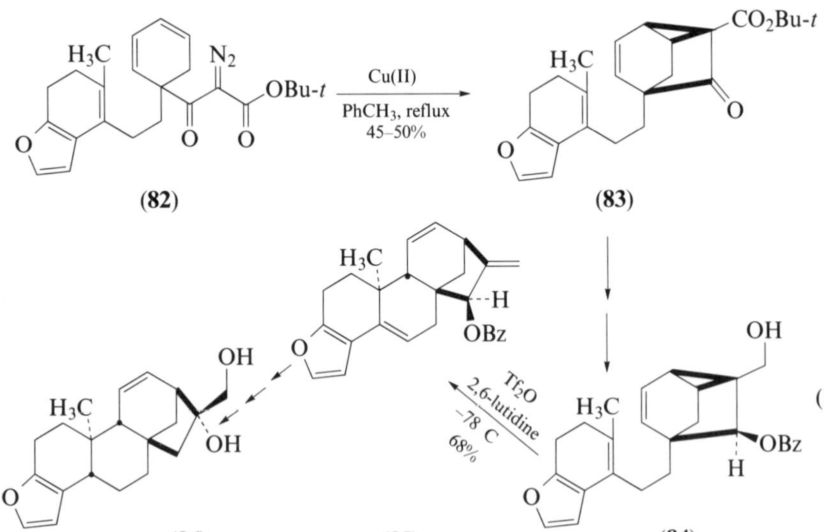

(82) **(83)** **(86)** **(85)** **(84)**

(82)

IV. CYCLOPROPANATION USING OTHER PRECURSORS

Carbenoid sources other than those derived from diazo precursors for catalytic cyclopropanation reactions are currently limited. Inter- and intramolecular catalytic cyclopropanation using iodonium ylide have been reported. Simple olefins react with iodonium ylides of the type shown in equations 83 and 84, catalysed by copper catalysts, to give cyclopropane adducts in moderate yield[127,128]. In contrast to the intermolecular cyclopropanation, intramolecular cyclopropanation using iodonium ylides affords high yields of products (equations 85 and 86)[129]. The key intermediate **88** for the 3,5-cyclovitamin D ring A synthon **89** was prepared in 80% yield as a diastereomeric mixture (70:30) via intramolecular cyclopropanation from iodonium ylide **87** (equation 87)[130].

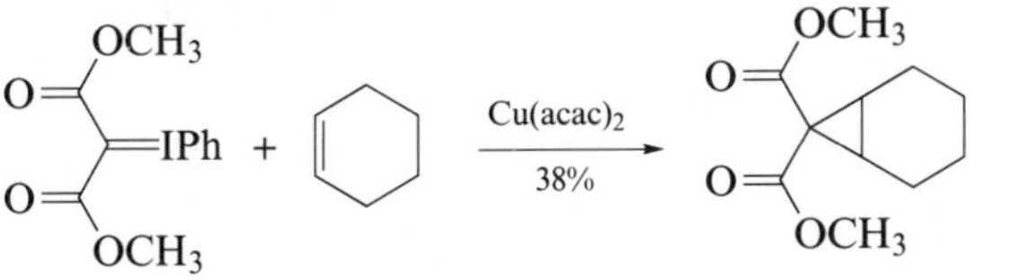

(83)

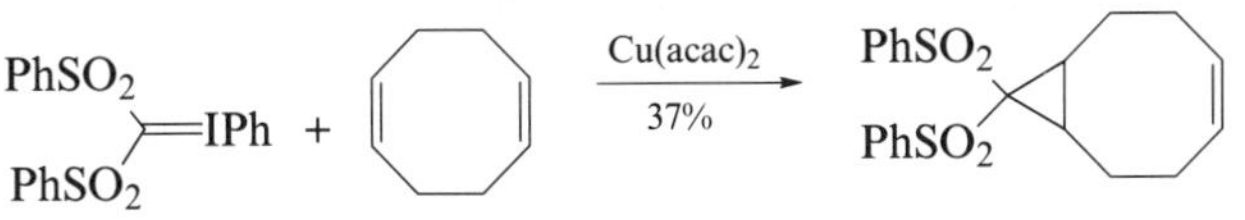

(84)

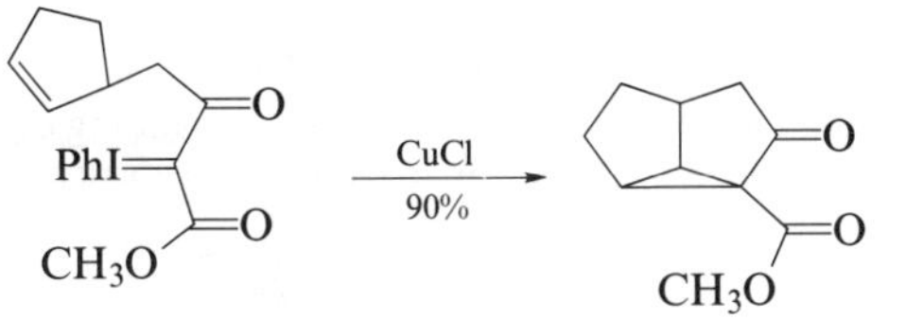

(85)

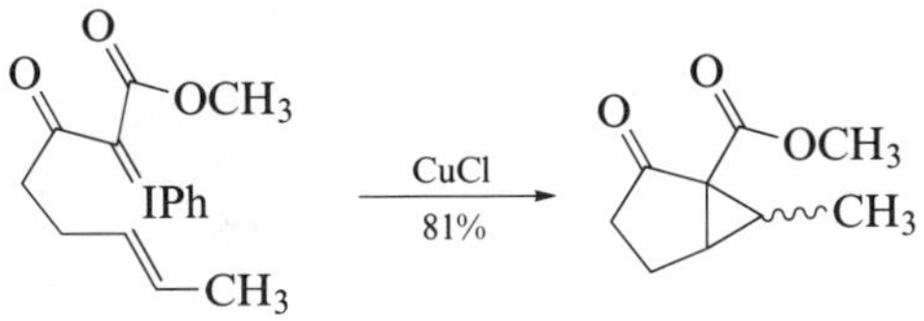

(86)

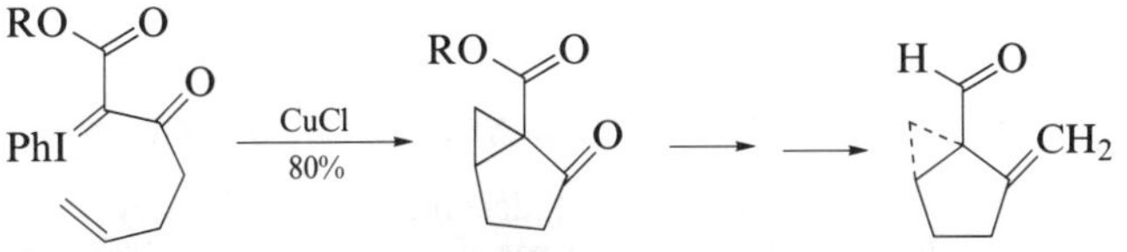

(87)

(87) **(88)** **(89)**

R=*l*-menthyl

Metal catalysed cyclopropanation using other types of intermediate is also possible. Lithiated *tert*-butyl alkyl sulphones bring about the cyclopropanation of various nonactivated alkenes under nickel(II) acetylacetonate catalysis (equation 88)[131,132]. Sulphonium ylides of type **90** react with simple alkenes under copper catalysis to give the corresponding cyclopropane adduct (equation 89)[113,134]. In this example the ylide (**90**) is the sulphonium equivalent of ethyl diazoacetate[134].

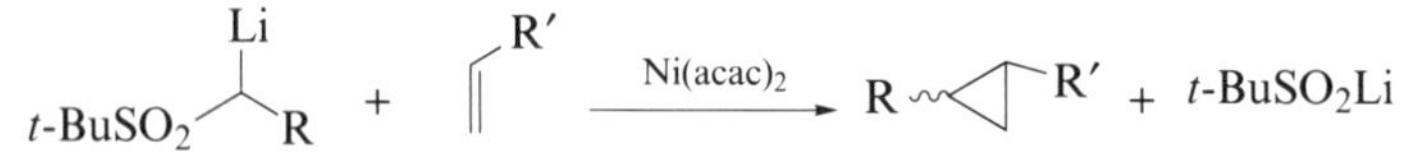

(88)

R=CH_3, R′=Ph, yield=60%; R=n-Bu, R′=$CH_3(CH_2)_3$, yield=92%;
R=CH_3, R′=$CH_3(CH_2)_9$, yield=83%; R=n-Bu, R′=$CH_3O(CH_2)_9$, yield=70%

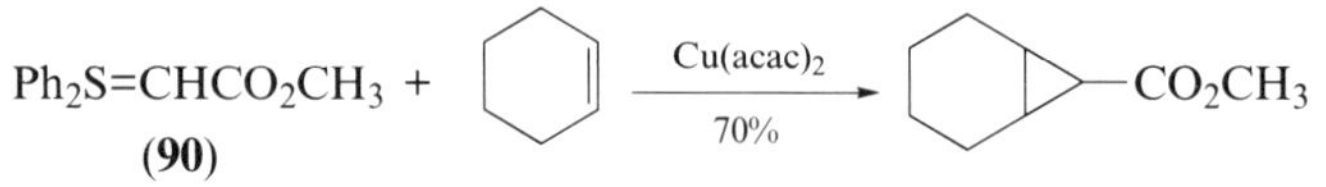

(**90**)

(89)

Cyclopropanation of olefins using bromomalonic ester and 1,8-diazabicyclo[5.4.0]-undec-7-ene (DUB) in the presence of a catalytic amount of copper(II) bromide gives 1,1-bis(alkoxycarbonyl)cyclopropane derivatives[135]. An example is shown in equation 90.

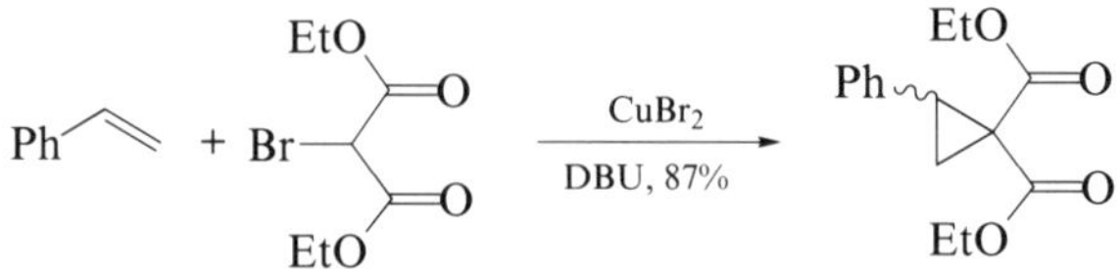

(90)

Palladium(0) catalysed cyclopropanation of electron-rich strained olefins using ketone α-carbonate or cyclic carbonates produce the corresponding cyclopropane adducts[136–139]. Two examples of this unusual cyclopropanation of norbornene are shown in equations 91 and 92.

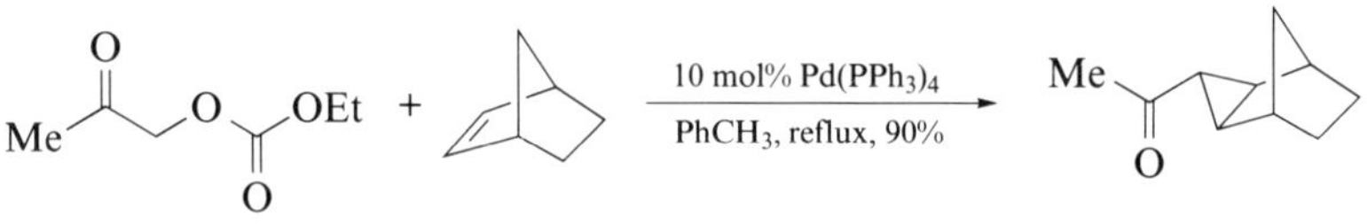

(91)

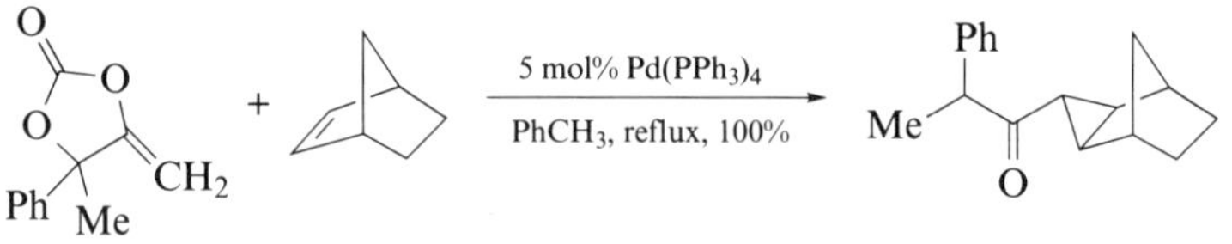

(92)

V. CHEMOSELECTIVITY IN METAL CATALYSED CYCLOPROPANATION

Carbenoids derived from the metal catalysed decomposition of diazo compounds undergo various chemical transformations. Control of chemoselectivity by choice of the appropriate catalyst has significantly increased the synthetic viability of catalytic cyclopropanation reactions. Intermolecular reaction of unsaturated alcohols with carbenoids derived from catalytic decomposition of alkyl diazoesters has been reported by Noels and coworkers[140]. In this case, the O—H insertion process is a competing side reaction. An example is shown in equation 93 where cyclopropanation is the favoured process[140]. In some cases, the O—H insertion reaction is the favoured process over the cyclopropanation. An important example in natural product synthesis where O—H insertion occurs exclusively is found in the reaction of diazomalonate **92** with the hydroxycyclohexene **91** (equation 94) in the synthesis of chorismic acid[141].

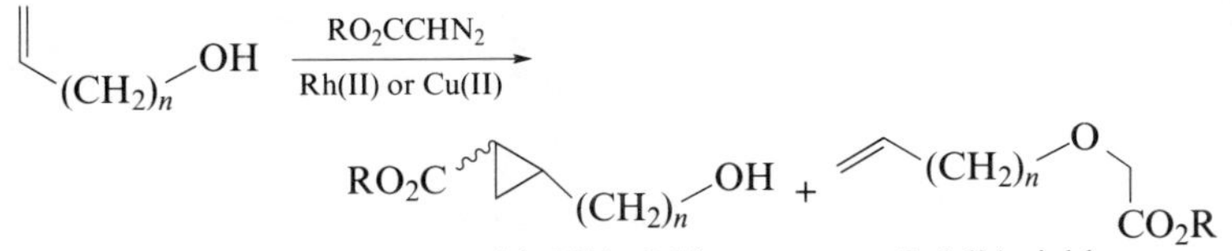

46–97% yield 3–36% yield

$R = CH_3, C_2H_5, n\text{-}C_4H_9, t\text{-}C_4H_9; n = 1,2,3$

(93)

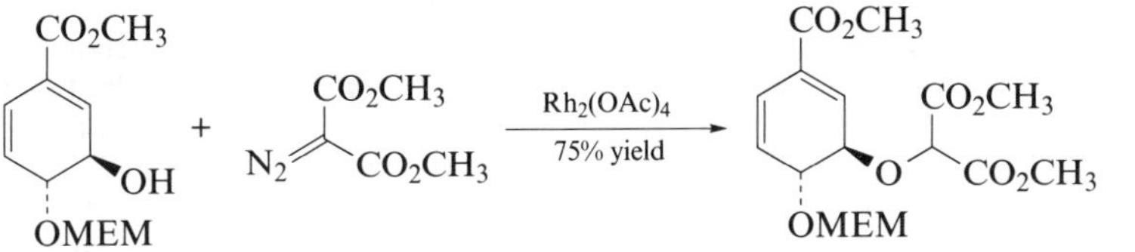

(91) **(92)**

$OMEM = OCH_2OCH_3$

(94)

Examples are known where intermolecular carbenoid transformations between diazomalonates or certain diazoketones and appropriate olefins result in competition between formation of cyclopropane and products derived from allylic C—H insertion[2,4]. For example, catalytic decomposition of ethyl diazopyruvate in the presence of cyclohexene gave the 7-*exo*-substituted norcarane **93** together with a small amount of the allylic C—H insertion product **94** (equation 95)[142,143]. In some cases, e.g. rhodium(II) decomposition of α-diazo-β-ketoester **95**, the major pathway afforded C—H insertion products **96** and **97** with only a small amount of the cyclopropane derivative **98**. In contrast, however, when a copper catalyst was employed for this carbenoid transformation, cyclopropane **98** was the dominant product (equation 96)[144]. The choice of the rhodium(II) catalyst's ligand can also markedly influence the chemoselectivity between cyclopropanation and C—H

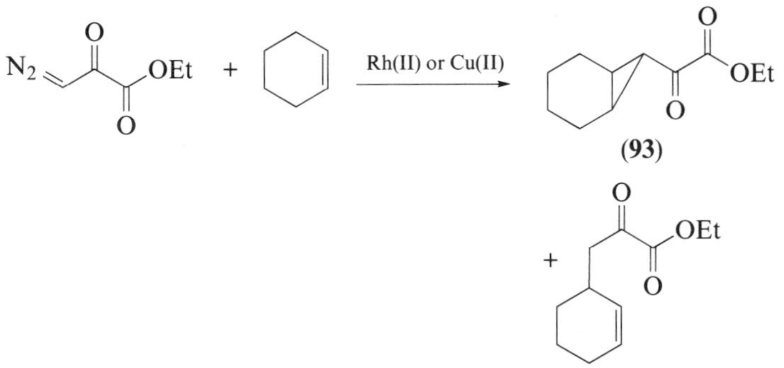

(93) **(94)**

(95)

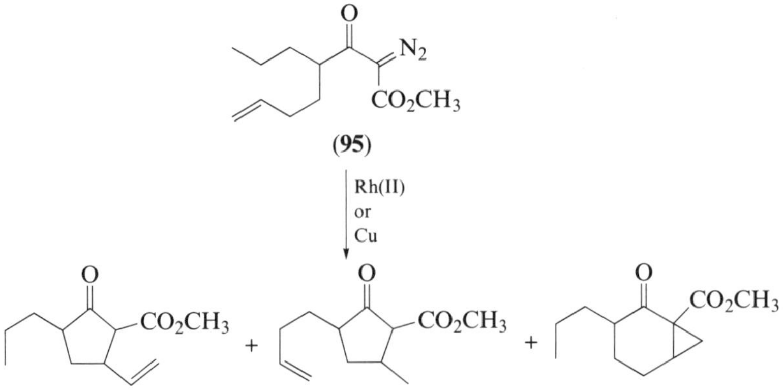

(95)

(96) (97) (98)

(96)

insertion. An example is shown in equation 97 where catalysis by rhodium(II) acetate produced both cyclopropanation product **99** and C—H insertion product **100**. However, only the cyclopropanation product **99** was formed with dirhodium(II) caprolactamate [$Rh_2(cap)_4$] whereas catalysis by rhodium(II) perfluorobutyrate [$Rh_2(pfb)_4$] gave C—H insertion product **100** exclusively[145]. Another interesting example in which the carbenoid reaction with a carbon–carbon double bond afforded cyclopropanation product **101** and α,β-unsaturated ketone **102** is shown in equation 98. Catalysis by dirhodium(II) caprolactamate [$Rh_2(cap)_4$] resulted in the exclusive formation of cyclopropanation product **101**[146]. Rhodium acetate catalysed decomposition of γ,δ-unsaturated diazoketone **103**, on the other hand, produces the cyclohexenone derivative **104** as the dominant product (equation 99)[147].

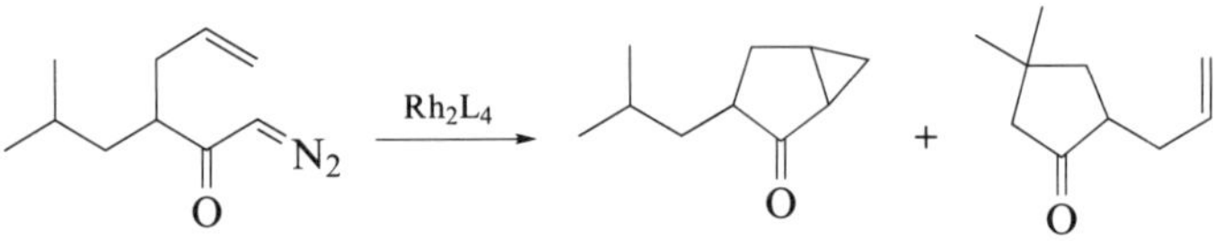

(97)

	Yield	(99)	(100)
$Rh_2(OAc)_4$	97%	44	56
$Rh_2(pfb)_4$	56%	0	100
$Rh_2(cap)_4$	76%	100	0

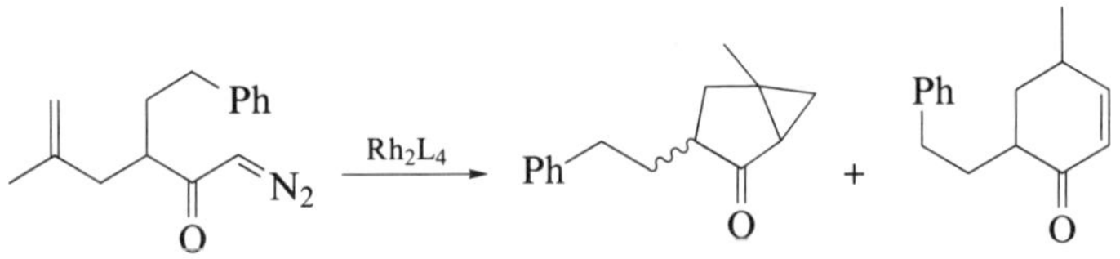

(98)

	Yield	(101)	(102)
$Rh_2(OAc)_4$	53%	87	13
$Rh_2(pfb)_4$	39%	72	28
$Rh_2(cap)_4$	61%	100	0

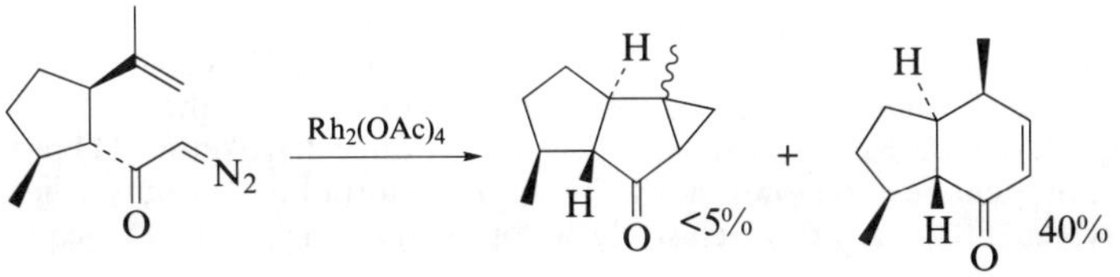

(99)

(103) **(104)**

Thus changing the ligands on dirhodium(II) can provide a switch which, in some cases, can turn competitive transformations 'on' or 'off'[146]. Other examples include the use of dirhodium(II) carboxamides to promote cyclopropanation and suppress aromatic cycloaddition[146]. For example, catalytic decomposition of diazoketone **105** with dirhodium(II) caprolactamate [$Rh_2(cap)_4$] provides only cyclopropanation product **106**. In contrast, dirhodium(II) perfluorobutyrate [$Rh_2(pfb)_4$] or dirhodium(II)triphenylacetate [$Rh_2(tpa)_4$] gave the aromatic cycloaddition product **107** exclusively (equation 100)[146,148]. Although we have already seen that rhodium(II) acetate catalysed decomposition of diazoketone **59**, which bears both aromatic and olefinic functionalities, afforded stable norcaradiene **60** (equation 70)[105], the rhodium(II) acetate catalysed carbenoid transformation within an acyclic system **(108)** showed no chemoselectivity (equation 101). However, when dirhodium(II) carboxamides were employed as catalysts for this type of transformation, only cyclopropanation product **109** was obtained (equation 101)[146].

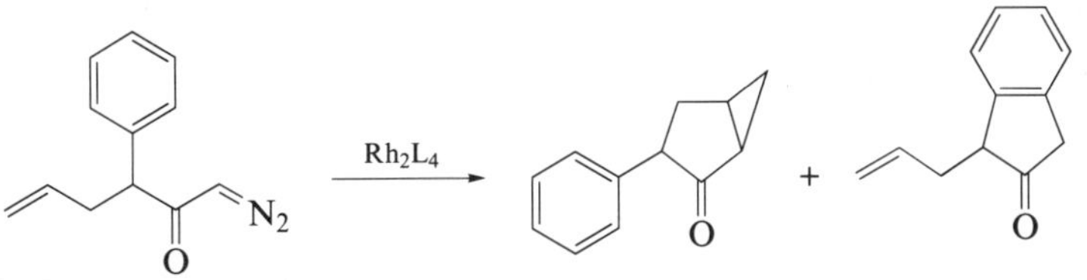

(100)

(105) **(106)** **(107)**

	Yield	(106)	(107)
$Rh_2(OAc)_4$	99%	67	33
$Rh_2(pfb)_4$	95%	0	100
$Rh_2(cap)_4$	72%	100	0
$Rh_2(tpa)_4$	83%	0	100

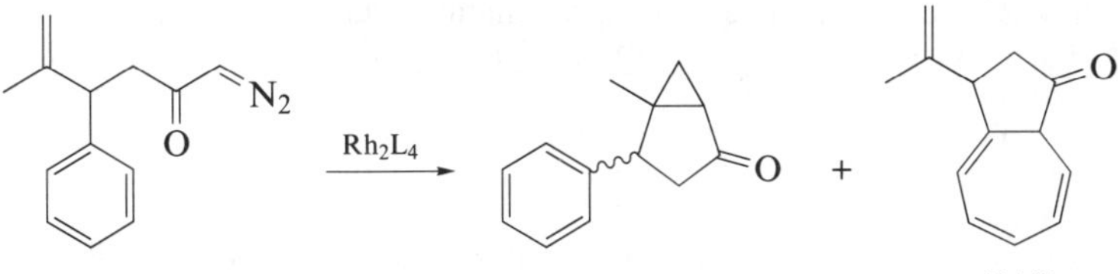

(101)

(108) **(109)** **(110)**

	Yield	(109)	(110)
$Rh_2(OAc)_4$	99%	53	47
$Rh_2(pfb)_4$	100%	35	65
$Rh_2(cap)_4$	100%	100	0

Carbenoid transformations involving competition between intramolecular cyclopropanation and β-hydride elimination have been investigated[149]. The chemoselectivity of these catalytic transformations can be effectively controlled by the choice of catalyst. Rhodium(II) trifluoroacetate catalysed decomposition of diazoketone **111** proceeds cleanly to give only enone **112**. However, rhodium(II) acetate or bis-(*N*-*t*-butylsalicyladiminato) copper(II) cu(TBs)$_2$ provides exclusively cyclopropanation product **113** (equation 102)[149].

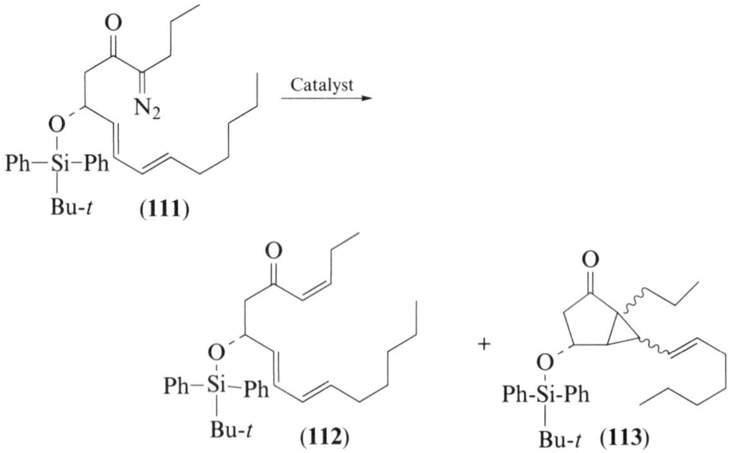

(102)

	Yield		
$Rh_2(OCOCF_3)_4$	90%	100	0
$Rh_2(OAc)_4$	55%	0	100
$Cu(TBS)_2$	78%	0	100

Metal catalysed decomposition of diazocarbonyl compounds in the presence of allylic sulfides, ethers, selenides, amines and halides may form allylic ylides which undergo [2,3]sigmatropic rearrangement. Ylide formation can often provide strong competition to the cyclopropanation reaction. The influence of the catalyst on the product distribution between intermolecular ylide formation and cyclopropanation has been investigated by Doyle and coworkers[150,151]. The chemoselectivity of intermolecular reactions of carbenoids derived from diazocarbonyl compounds with allylic halides is dramatically dependent on the nucleophilicity of the halide, the nature of the diazocarbonyl and the catalyst[150]. For example, Doyle has shown that allyl chloride and ethyl diazoacetate under rhodium acetate catalysis gave predominately cyclopropanation, whereas the corresponding reaction with allyl iodide gave entirely the product derived from ylide formation and sigmatropic rearrangement (equation 103)[150]. With substituted allyl ethers and diazocarbonyls, ylide generation occurs almost exclusively, and only small amounts of cyclopropanation products are formed, as shown, for example, in equation 104[152]. The intramolecular version of this reaction provided only products derived from the ylide rearrangement[153]. Catalytic decomposition of diazocarbonyls in the presence of allylic sulfides or allylic amines produces ylide derived products exclusively. In the case of diazocarbonyl **114**, cyclopropanation should give the favoured cyclopropane adduct **116**. However, instead, reaction with the sulfide occurred exclusively (equation 105)[154]. The cyclopropanation process is completely non-competitive with the sulfide ylide formation process. Intramolecular carbonyl ylide formation (followed by a subsequent cycloaddition) can also occur in competition with the cyclopropanation reaction. An illustrative example is rhodium(II) acetate catalysed decomposition of diazoketone **117** which leads to a 1:1 mixture of the internal dipolar cycloadduct **118** as well as the cyclopropanated product **119** (equation 106)[145].

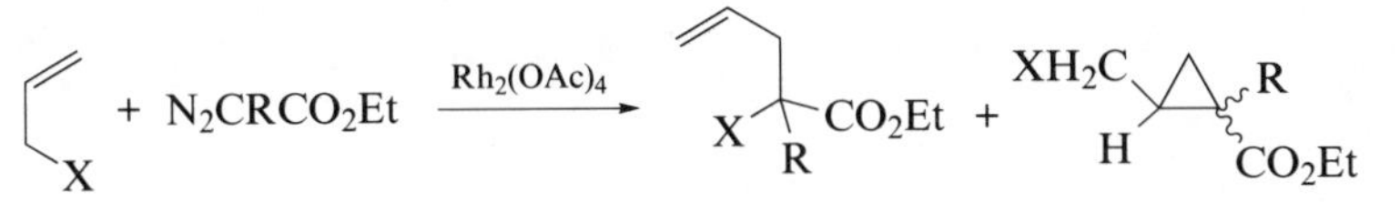

(103)

		Yield		
R=H,	X=I;	98%	100	0
R=H,	X=Br;	76%	28	72
R=H,	X=Cl;	95%	5	95
R=CO_2Et,	X=Br;	92%	93	7

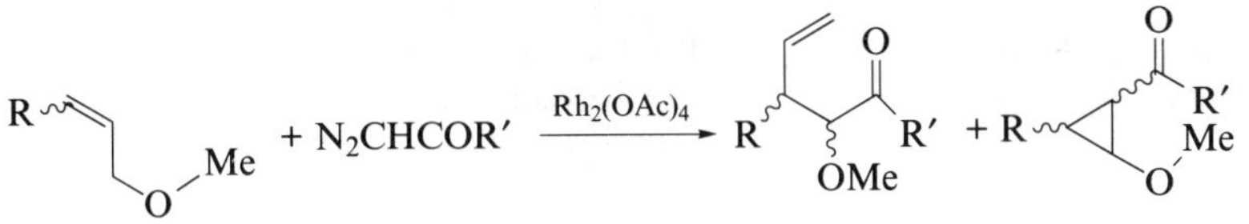

(104)

		Yield		
R=Ph,	R′=OEt;	95%	73	27
R=Ph,	R′=Ph;	86%	94	6
R=Me,	R′=OEt;	86%	92	8

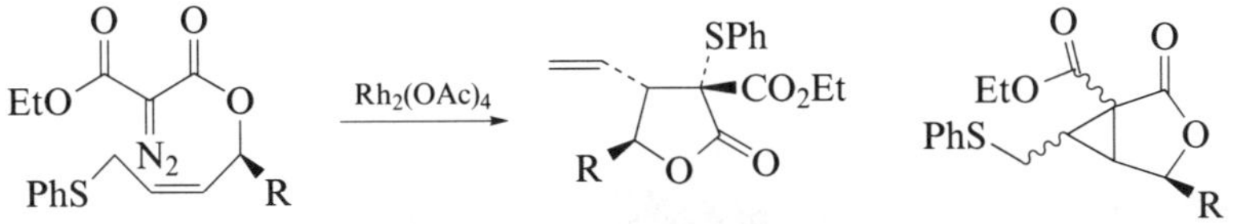

(105)

(114) **(115)** **(116)**

R = CH_3, C_3H_7, C_5H_{11}, $PhCH_2$; Yield = 53–70%

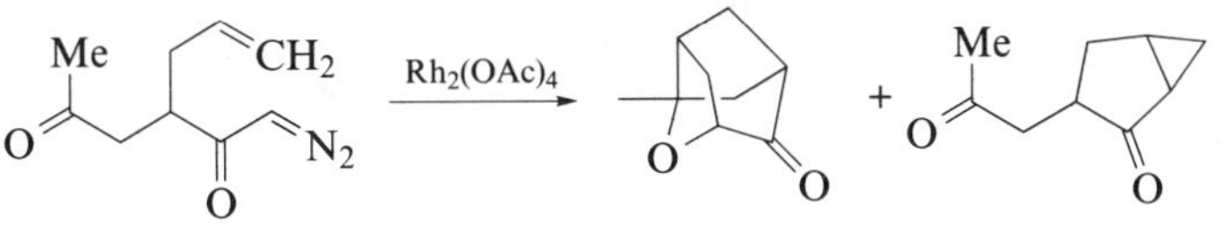

(106)

(117) **(118)** **(119)**

Certain diazoketones, for example diazopyruvate, alkyl 2-diazo-3-oxobutyrate or 3-diazo-2,4-pentanedione, react with vinyl ethers under metal catalysis to give dihydrofurans rather than cyclopropanes[2,4,12]. Most work on this type of transformation has been that of Wenkert[155–157] and Alonso[158,159] and their respective groups. A representative example is shown in equation 107. Finally, carbenoid dimerization is also a competitive reaction in metal catalysed intermolecular cyclopropanation. However, control of the chemoselectivity to favour the cyclopropanation is possible. In general, the dimeric product can be avoided by using excess of alkene or by very slow addition of the diazo compound to a mixture of alkene and catalyst[160].

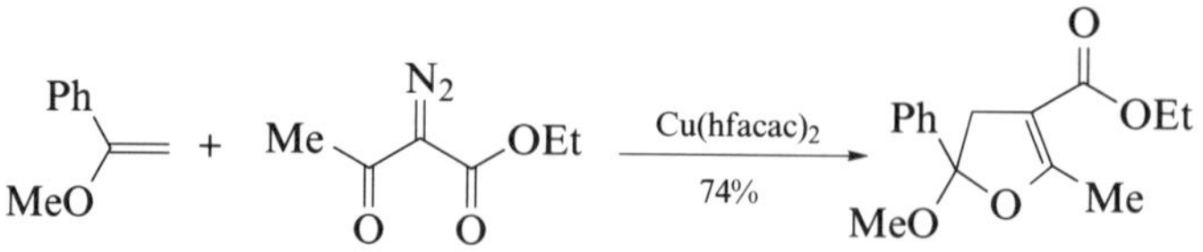

(107)

hfacac = heptafluoroacetylacetonate

VI. REGIOSELECTIVITY IN METAL CATALYSED CYCLOPROPANATION

In cases where more than one carbon–carbon double bond is present in the molecule, the possibility of selective cyclopropanation of one of them arises. Regiocontrol in intermolecular cyclopropanation of substituted dienes has been the subject of much investigation and considerable differences can occur, depending on the structure of the substrate, catalyst and the carbene substituents. With a 1-substituted terminal diene such as **120**, cyclopropanation, in general, occurs at the less-substituted double bond (equation 108)[22,26,37,161,162]. In this case, the nature of the catalyst and of the carbenoid precursors are less important in determining the regioselectivity.

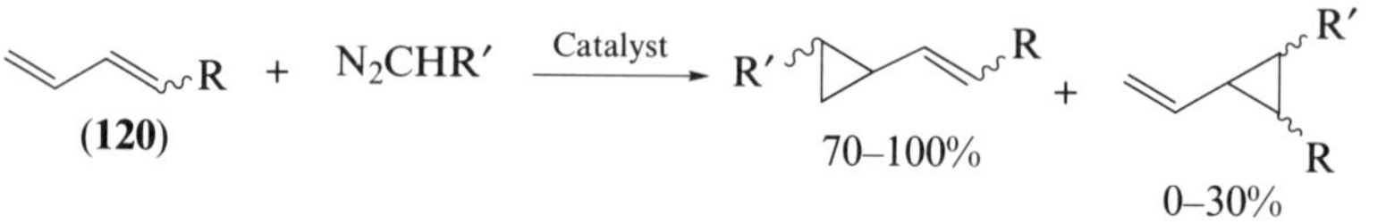

(108)

R=CH_3, Ph, OCH_3, OC_2H_5, Cl, $OSiMe_3$, OAc; R′=H, CH_3, CO_2Et
Catalyst=$Rh_2(OAc)_4$, $Rh_6(CO)_{16}$, $CuCl{\cdot}P(OPr\text{-}i)_3$, $Cu(OTf)_2$, $PdCl_2{\cdot}(PhCN)_2$

Intermolecular cyclopropanation of 2-substituted terminal diene **121** with rhodium or copper catalysts occurs preferentially at the more electron-rich double bond (equation 109)[37,162]. With a palladium catalyst, considerable differences in regiocontrol can occur, depending on the substituent of the diene. In general, palladium catalysed cyclopropanation occurs preferentially at the less substituted double bond (equation 110). However, with a stronger electron-donating substituent present in the diene, e.g. as in 2-methoxy-1,3-butadiene, the catalytic process results in exclusive cyclopropanation at the unsubstituted double bond (equation 110)[162].

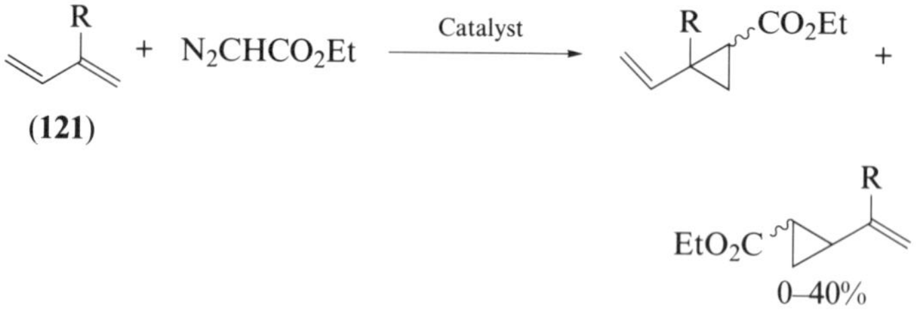

(109)

R= CH_3, Ph, OCH_3;
Catalyst= $Rh_2(OAc)_4$, $Rh_6(CO)_{16}$, $CuCl{\cdot}P(OPr\text{-}i)_3$, $Cu(OTf)_2$

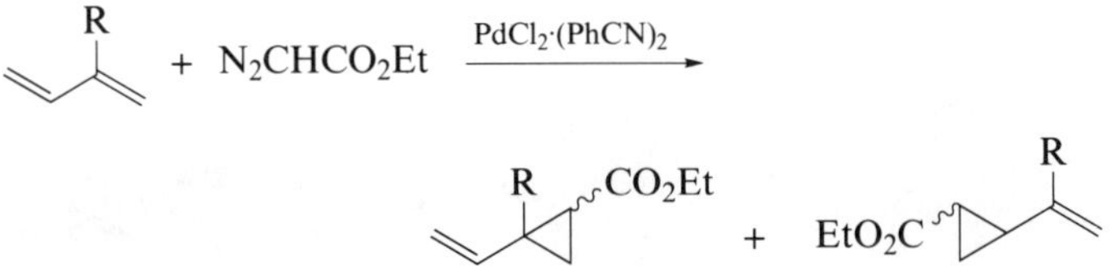

(110)

$R = CH_3$	33%	67%
$R = OCH_3$,	100%.	0%

Regiocontrol in the cyclopropanation of 1,4-disubstituted dienes depends both on the substitution and geometry of the diene as well as on the catalysts employed. Rhodium or copper catalysed reaction of ethyl diazoacetate with diene **122** provides vinylcyclopropane adducts **123** and **124**, regioselectively (76:24) with a preference for attack at the *Z* double bond (equation 111). In contrast, palladium catalysts show little selectivity between *Z* and *E* double bonds in the same reaction[161]. Other examples are shown in equations 112 and 113, in which the more electron-rich or less-hindered double bond is preferentially cyclopropanated[163,164]. With trienes containing two conjugated double bonds and one isolated double bond, cyclopropanation occurs preferentially at the conjugated diene functionality (equation 114)[161]. Regiocontrol in this type of cyclopropanation is in agreement with the results described above for simple substituted 1,3-dienes.

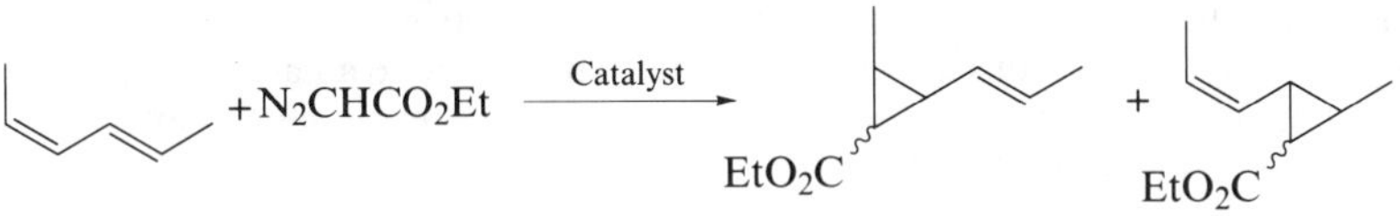

(111)

(122) (123) (124)

Catalyst	Yield	(123)	(124)
$Rh_2(OAc)_4$	87%	76	24
$Cu(OTf)_2$	87%	76	24
$Pd(OAc)_2$	60%	50	50

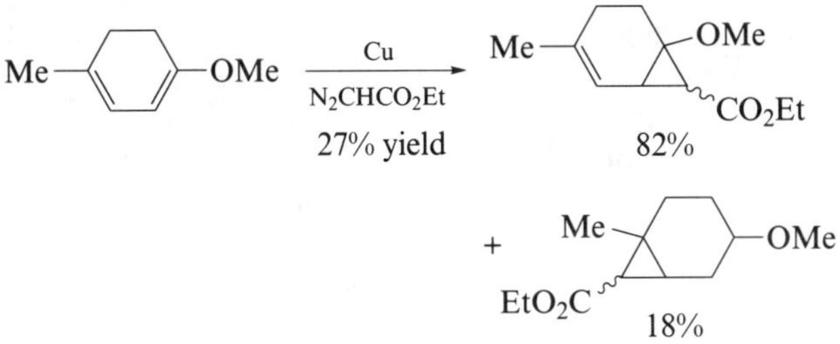

(112)

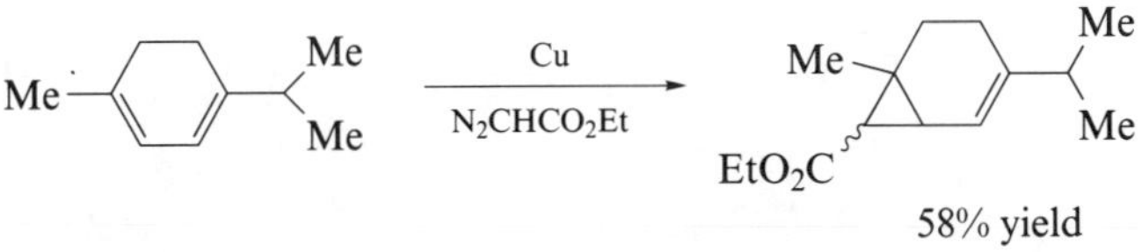

(113)

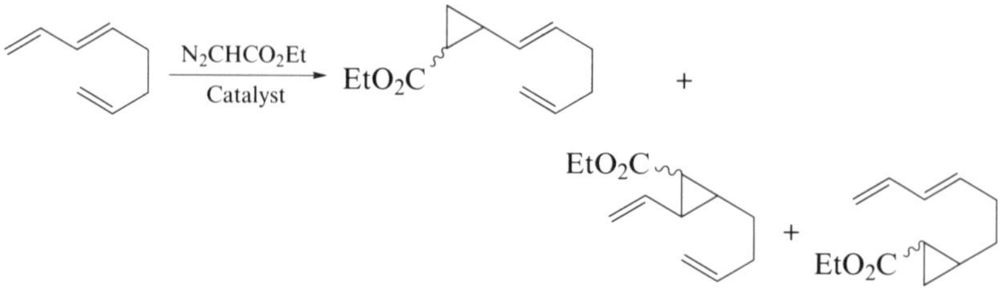

(114)

Catalyst	Yield			
$Rh_2(OAc)_4$	97%	87	10	traces
$(Cu(OTf)_2$	64%	56	8	traces

With non-conjugated dienes, site preference for cyclopropanation also depends on the electronic and steric factors of the individual double bonds as well as the nature of the carbenoid precursor and catalyst. Limonene **125**, containing a1,1-disubstituted and a trisubstituted double bond, under catalysed cyclopropanation with ethyl diazoacetat provides cyclopropanes **126** and **127** (equation 115). Cyclopropane **126**, resulting from cyclopropanation at the less substituted double bond, is the major regioisomer[165]. Palladium(II) acetate catalysed cyclopropanation of the same diene with diazomethane results in cyclopropane **126** as the only isolated product in 82% yield (equation 115)[19]. The use of 2,6-di-*tert*-butyl-4-methylphenyl diazoacetate **129** as carbenoid precursor provides significant enhancement in regiocontrol. Rhodium(II) acetamide ($Rh_2(acam)_2$) catalysed reaction of diene **128** with diazocarbonyl **129** provides high regioselective cyclopropanation to give **130** in 92% yield[166]. With molecule **131**, containing a double bond adjacent to an electron-withdrawing group, catalytic cyclopropanation occurs exclusively at the more reactive non-conjugated double bond (equation 117)[167].

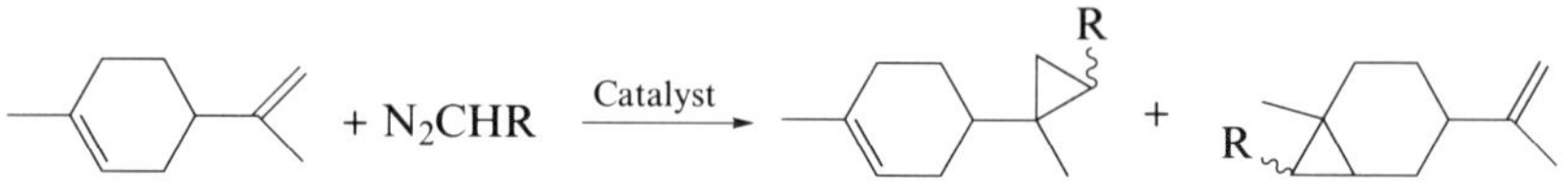

(125) **(126)** **(127)**

R	Catalyst	Yield		
CO_2Et	$Rh_2(OAc)_4$	98%	77	23
CO_2Et	$Cu(OTf)_2$	59%	80	20
CO_2Et	$PdCl_2{\cdot}(PhCN)_2$	32%	78	22
H	$Pd(OAc)_2$	82%	100	0

(115)

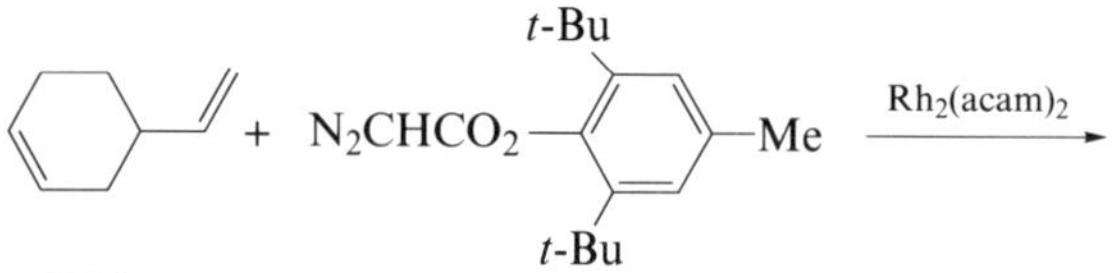

(128) **(129)**

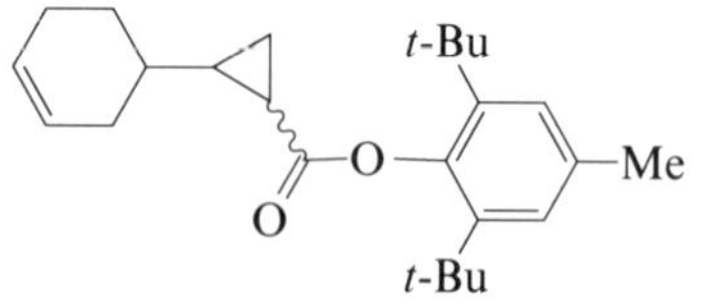

(116)

(130)

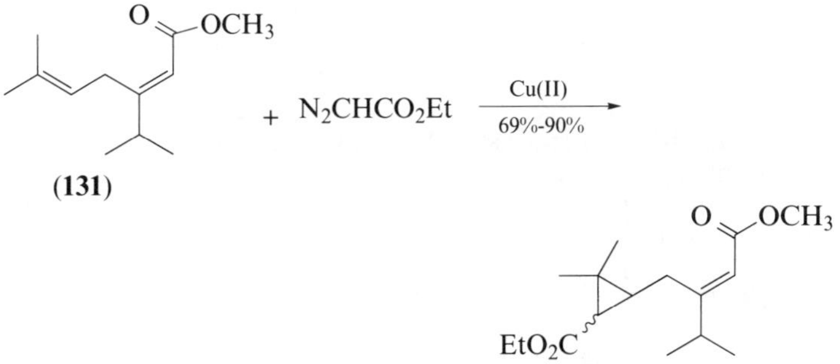

(131)

(117)

A few examples are available in which regiocontrol in the cyclopropanation of non-conjugated diene is catalyst-dependent. An early example is showed in equation 118. Copper(II) triflate catalysed cyclopropanation of diene **132** with diazomethane occurs preferentially at the less substituted double bond, whereas copper(II) acetylacetonate in contrast promotes cyclopropanation at the more substituted double bond (equation 118)[13]. Regiocontrol in the cyclopropanation of norbornene derivative **133** is strongly catalyst-dependent (equation 119). When diphenyldiazomethane is used as carbenoid precursor, the regioselectivity of this cyclopropanation is significantly enhanced[165].

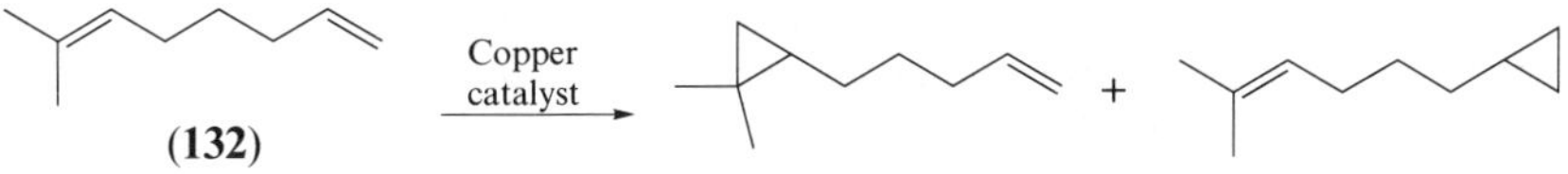

(132)

Catalyst=Cu(OTf)	29%	71%
Catalyst=$Cu(acac)_2$	91%	9%

(118)

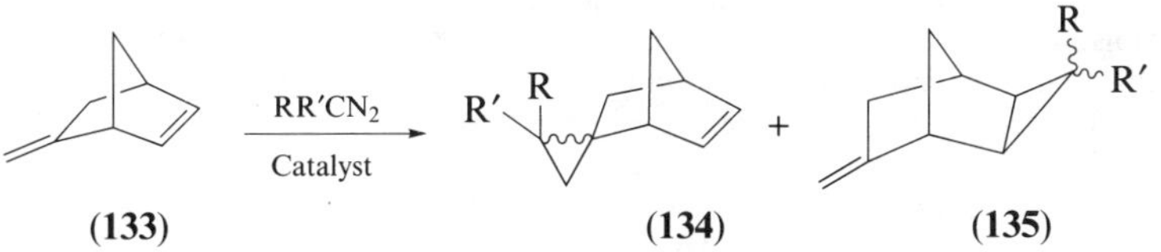

(133) (134) (135)

Diazo compound	Catalyst	(134)	(135)
R=R′=Ph	$Rh_2(OAc)_4$	99	1
R=R′=Ph	$PdCl_2\cdot(PhCN)_2$	2	98
R=CO_2Et, R′=H	$Rh_2(OAc)_4$	70	30
R=CO_2Et,R′=H	$PdCl_2\cdot(PhCN)_2$	14	86

(119)

Steric factors, in some cases, are the key to the regiocontrol in intermolecular cyclopropanation reactions. The inertness of trisubstituted double bonds to palladium catalysed cyclopropanation may explain the extremely regoselective cyclopropanation observed with FK 506 (equation 10)[20].

Regiocontrol in intramolecular cyclopropanation is mainly dependent on the site of the unsaturated centre related to the carbenoid centre. In other words, ring size of the bicyclic or polycyclic systems which derive from the cyclopropanation is the key to regiocontrol of this type of reaction. The regioselectivity is independent of the catalyst employed. Five-membered ring formation is favoured over the production of six- or seven-membered rings (equations 56, 59, 61, 63–66, 77 and 78). A particularly interesting example is shown in equation 64, in which the cyclopropanation occurs at the electron-deficient double bond resulting in the formation of a five-membered-ring product. Rhodium(II) catalysed cyclopropanation of diazocarbonyl precursor **136** provided a mixture of two bicyclic products **137** and **138** (equation 120). In contrast to the results described above for the intermolecular cyclopropanation, this intramolecular process shows almost no regioselectivity between the monosubstituted carbon–carbon double bond and the trisubstituted carbon–carbon double bond.

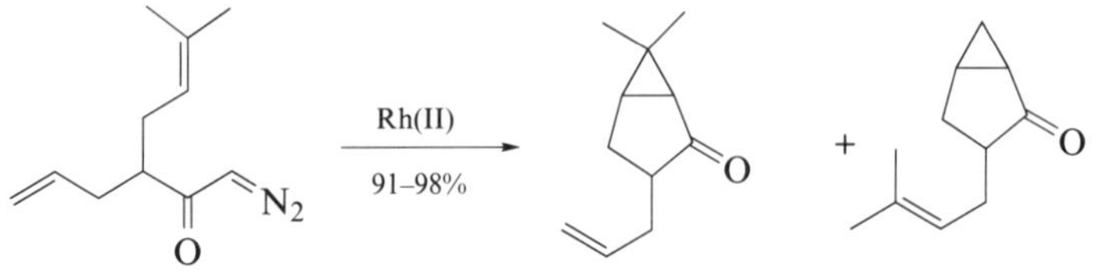

(120)

(136) **(137)** 45–54% **(138)** 46–55%

VII. STEREOCHEMICAL FEATURES OF ALKENE CYCLOPROPANATION

There are several important stereochemical features of alkene cyclopropanation with carbenoid precursors which bear directly on the efficacy of the reaction as a synthetic method. In the first place, cyclopropanation using diazoester precursors generally proceeds stereospecifically with respect to the geometry of the alkene. This has been established for the synthetically significant catalysts based on rhodium and copper. However, secondly, the stereoselectivity defining the spatial relationship between substituents on the carbenoid carbon and those at the double bond is not usually very pronounced and is governed not just by the catalyst used, but by the nature of the alkene and the diazo compound. For a detailed discussion and a few recent examples see Section VII.A. Diastereofacial differentiation occurs on cyclopropanation of substituted cycloalkenes. For example, only two stereoisomers are found in the product of cyclohexene **139** with methyl diazoacetate, the cyclopropane having been formed away from the methyl group (equation 121)[168]. These stereochemical features have been analysed in great detail by Maas for the literature up to 1985[4]. Yet another aspect of the stereoselectivity of catalysed cyclopropanation which is commanding rapidly growing attention is that of enantioselection through the use of chiral auxiliaries, the most attractive form of which is the development of chiral catalysts. Sections VII. B and VII. C will outline the recent development of the asymmetric synthesis in metal catalysed cyclopropanation.

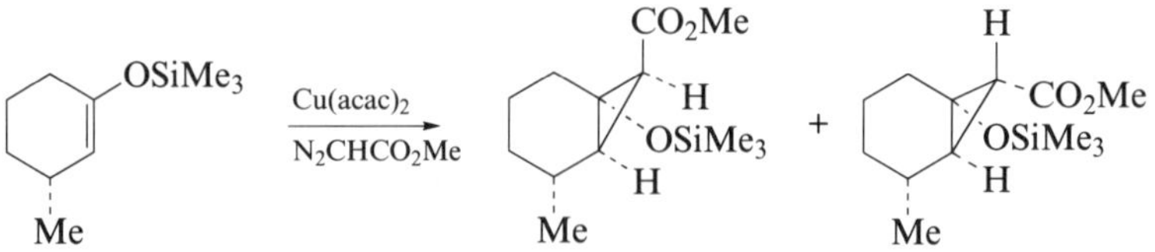

(121)

(139)

A. Stereoselectivity in Metal Catalysed Cyclopropanation

The stereoselectivity of metal catalysed intermolecular cyclopropanation has been examined in detail. With acyclic alkenes, *E*/*Z* ratios of substituted cyclopropanes in the range *ca* 1–3 in favour of the less congested *E* isomer are considered normal. With cyclic alkenes where the stereoselectivities among bicyclic products are usually expressed as *anti*/*syn* (or *exo*/*endo*) ratios, the preference is usually for the *anti* (*exo*) isomer, though here again the ratios are not large. Recent developments, however, show that considerable improvement in the stereoselectivity of the intermolecular cyclopropanation can be achieved, with values >90% recorded. A series of investigations with variously substituted diazo compounds shows that steric and electronic factors are very important in determining the stereoselectivity. The larger steric bulk of the diazocarbonyl favours the formation of *trans* (*anti*) isomer (equation 122)[166,169]. A comparison of stereocontrol for cyclopropanation of styrene with four types of diazo compound shows that high-level selectivity and variation is possible depending on the substituent of carbenoid precursor (equation 123)[27,162,170,171].

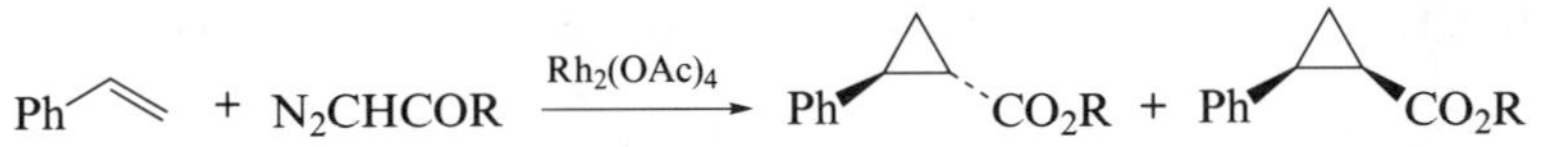

(122)

R=OEt	93% yield	62	38
R=OCMe(Pr-*i*)$_2$	94% yield	71	29
R=N(Pr-*i*)$_2$	53% yield	>98	<2

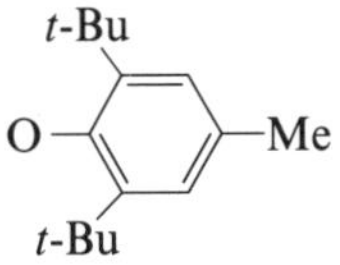

(aryloxy, above)	94% yield	84%	16%

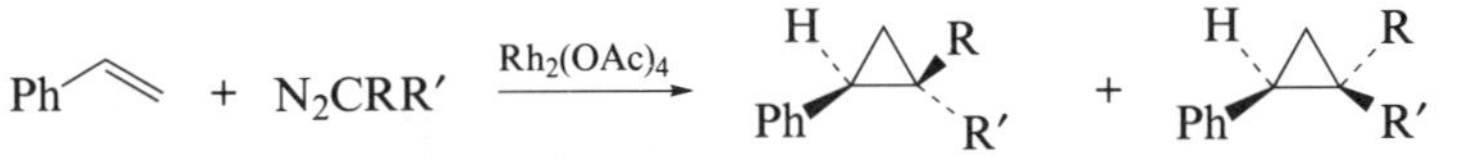

(123)

R	R′	Yield		
HC=CHPh	CO_2Et	94%	95	5
CO_2Et	NO_2	75%	89	11
COPh	NO_2	75%	14	86
H	Ph	38%	23	77

Stereocontrol in intermolecular cyclopropanation also depends on the structure of the unsaturated substrate. Early work concerning the influence of substrate on stereoselectivity has been summarized by Doyle[2]. In general, cyclopropanation of *cis*-disubstituted alkenes results in higher stereoselectivity than with monosubstituted alkenes and the steric bulk of the olefinic substituent enhances the stereoselectivity. However, the stereocontrol appears not simply to be caused by a steric factor. In comparable cases, the presence of halogen as an alkene substituent may cause a reversal of the normal stereoselectivity. A few examples which illustrate these effects are shown in equations 124[167,172–174].

Considerable variation in stereocontrol can also occur, depending on the catalyst employed (equation 125). In general, the various rhodium(II) carboxylates and palladium catalysts show little stereocontrol in intermolecular cyclopropanation[162,175]. Rhodium(II) acetamides and copper catalysts favour the formation of more stable *trans* (*anti*) cyclopropanes[162,166]. The ruthenium bis(oxazolinyl)pyridine catalyst [Ru(pybox-ip)] provides extremely high *trans* selectivity in the cyclopropanation of styrene with ethyl diazoacetate[43]. Furthermore, rhodium or osmium porphyrin complexes **140** are selective catalysts

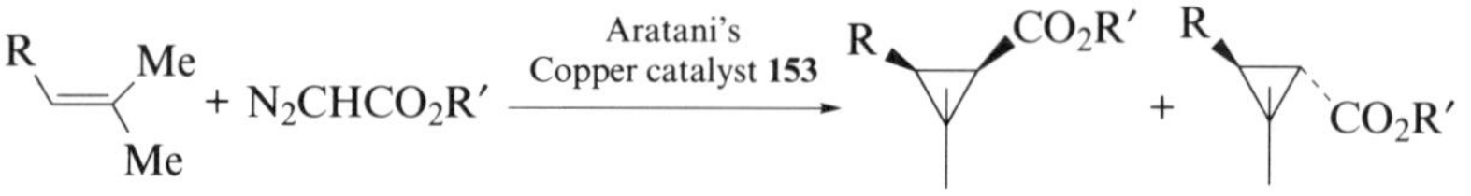

(124)

R = 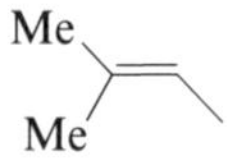 R′ = *l*-menthyl; 42% yield, 9 (*cis*) 91 (*trans*)

R = 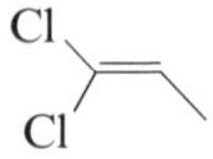 R′ = *l*-menthyl; 52% yield, 36 (*cis*) 64 (*trans*)

R = Cl_3CCH_2 R′ = *l*-menthyl; 54% yield, 85 (*cis*) 15 (*trans*)

R = Cl_2CHCH_2 R′ = *l*-menthyl; 57% yield, 86 (*cis*) 14 (*trans*)

R = 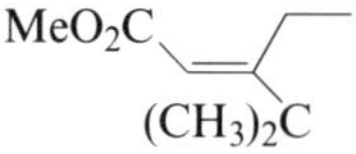 R′ = *l*-menthyl; 70% yield, 56 (*cis*) 44 (*trans*)

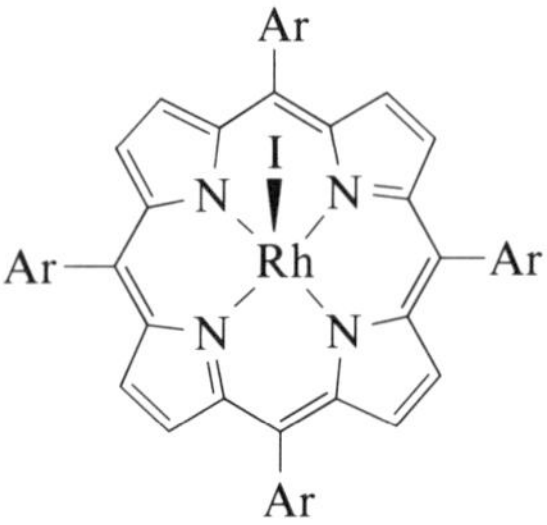

(140)

(**140**) M = Rh, Ar = Ph
(**140b**) M = Rh, Ar = (*R*)-1,1′-binaphth-2-yl
(**140c**) M = Rh, Ar = (*R*)-(1′-pyrenyl)-1-naphth-2-yl
(**140d**) M = Os, Ar = *p*-Tolyl

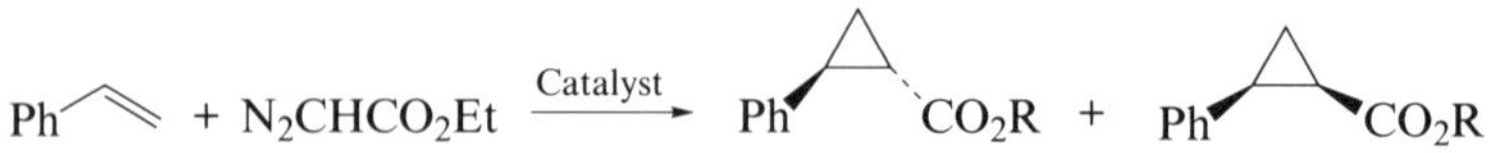

(125)

Catalyst	*trans*	*cis*
$Rh_2(OAc)_4$	62%	38%
$Cu(acac)_2$	72%	28%
$Rh_2(NHCOCH_3)_4$	68%	32%
Ru(pybox-ip)	91%	9%
140a	47%	53%
140b	30%	70%
140c	29%	71%
140d	9%	91%
$[CpFe(CO)_2(THF)^+BF_4^-$	15%	85%

which provide predominantly *cis*-stereoisomers with high-level selectivity in some cases[42,176–179]. The iron-based catalyst, $[CpFe(CO)_2(THF)]^+BF_4^-$, also provides high *cis*-selectivity[44].

The best stereoselectivity can be obtained by combining the appropriate choice of diazo precursor, alkene and catalyst. More recent work has provided some good examples where stereoselectivities up to 95% have been achieved (equations 116 and 126–128)[166,179–182].

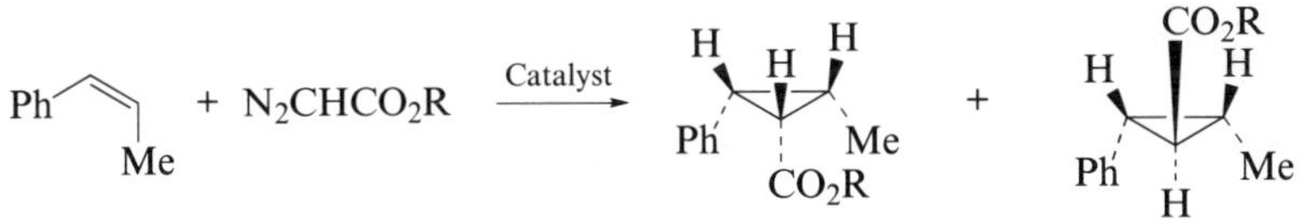

(126)

R = Et, Catalyst = **140c** — *cis* = 93% — *trans* = 7%

R = Bu-*t*, Catalyst = **141** — *cis* = 1% — *trans* = 99%

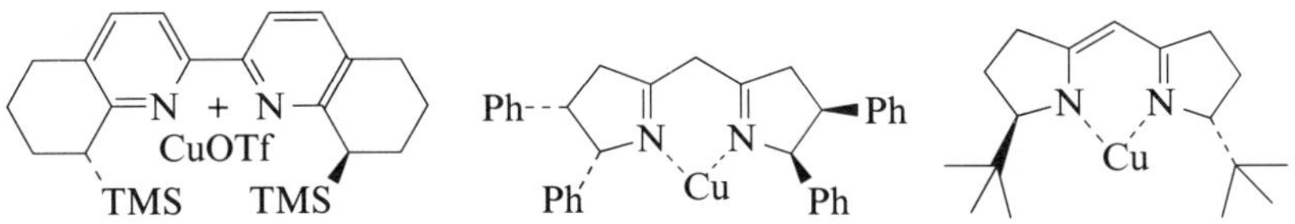

(**141**) (**142**) (**143**)

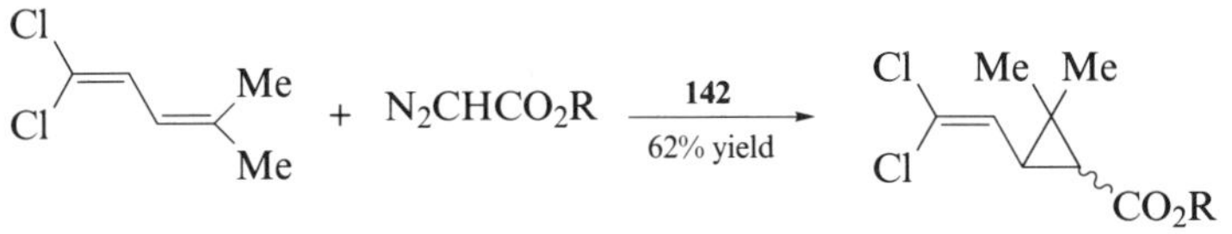

(127)

R = dicyclohexylmethyl — *trans*/*cis* = 99/1, 92% ee (*trans*)

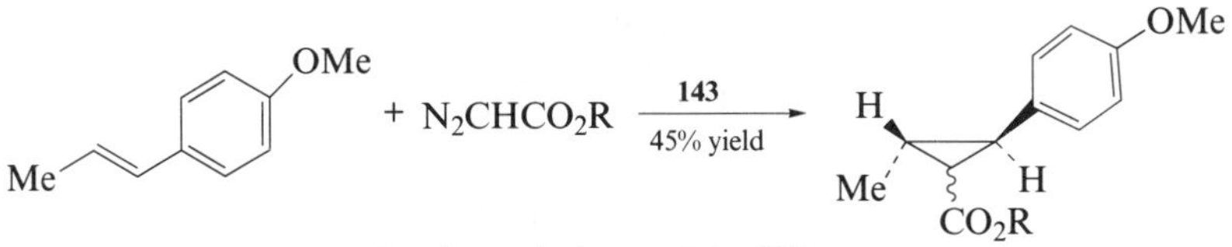

(128)

R= *l*-menthyl; *trans*/*cis*=98/2

B. Diastereoselectivity in Metal Catalysed Cyclopropanation

Stereosectivity is a broad term. The stereoselectivity in cyclopropanation which has been discussed in the above subsection, in fact, can also be referred to as diastereoselectivity. In this section, for convenience, the description of diastereoselectivity will be reserved for selectivity in cyclopropanation of diazo compounds or alkenes that are bound to a chiral auxiliary. Chiral diazoesters or chiral *N*-(diazoacetyl)oxazolidinone have been applied in metal catalysed cyclopropanation. However, these chiral diazo precursors and styrene yield cyclopropane products whose diastereomeric excess are less than 15% (equation 129)[183,184]. The use of several α-hydroxy esters as chiral auxiliaries for asymmetric intermolecular cyclopropanation with rhodium(II)-stabilized vinylcarbenoids have been reported by Davies and coworkers. With (*R*)-pantolactone as the chiral auxiliary, cyclopropanation of diazoester **144** with a range of alkenes provided cyclopropanes **145** in high yield with diastereomeric excess at levels of 90% (equation 130)[185,186]. Diastereocontrol in intramolecular cyclopropanation has also been investigated. Taber and coworkers have used naphthylborneol **149** as chiral auxiliary in the intramolecular cyclopropanation of α-diazo-β-ketoester **146** and moderately high levels of asymmetric induction have been achieved (equation 131)[187].

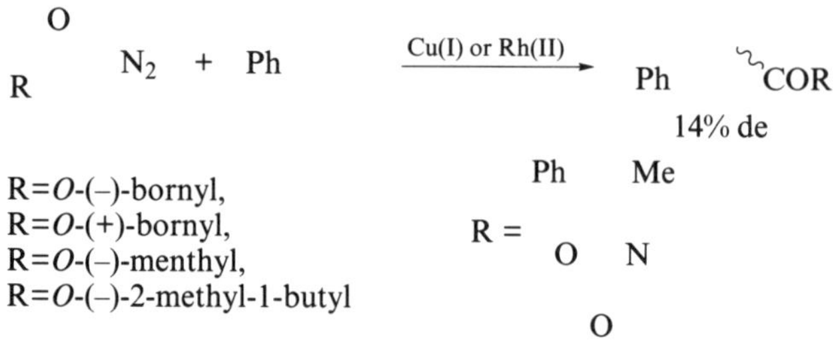

(129)

R=*O*-(–)-bornyl,
R=*O*-(+)-bornyl,
R=*O*-(–)-menthyl,
R=*O*-(–)-2-methyl-1-butyl

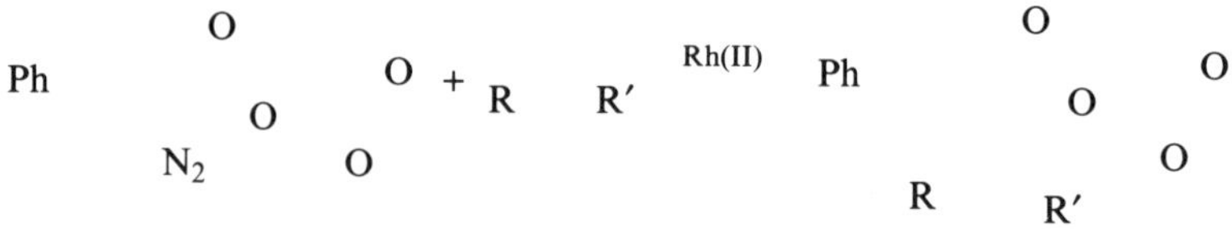

(130)

(**144**) (**145**)

R=Ph, R′=H	84% yield	97% de
R=4-ClC_6H_4, R′=H	92% yield	>95% de
R=4-$MeOC_6H_4$,R′=H	75% yield	>95% de
R=EtO, R′=H	71% yield	92% de
R=AcO, R=H	42% yield	90% de

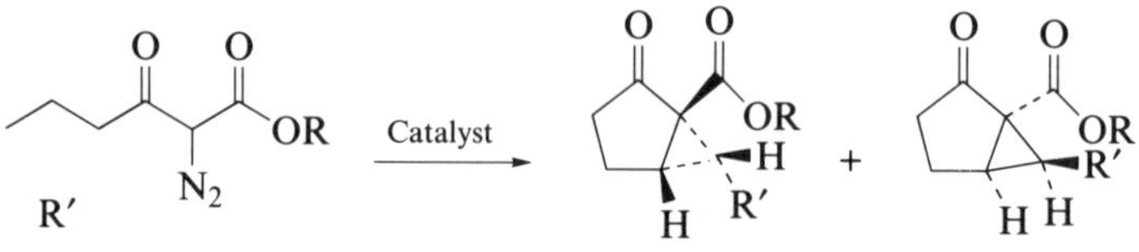

(131)

(**146**) (**147**) (**148**)

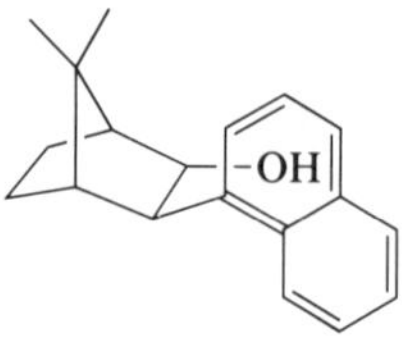

(**149**)

R′	R	Catalyst	Yield	**147**	**148**
H	**149**	$Cu(TBS)_2$	66%	28	72
H	**149**	$(TPP)RhCl_2$	64%	89	11
CH_3	**149**	$Cu(TBS)_2$	94%	66	34
CH_3	**149**	$(dppp)PdCl_2$	64%	80	20

$Cu(TBS)_2$ = bis(*N*-*t*-butylsalicylaldiminato)Copper (II)
TPP = mono (tetraphenylporphyrinato)
dppp = bis (1,3-diphenylphosphino)propane

Although the diastereocontrolled cyclopropanation generally uses a chiral diazo compound, there is one exception in which a chiral olefin was used to react with an achiral diazo compound. Thus, copper catalysed cyclopropanation of chiral butadiene–iron tricarbonyl complex **150** with methyl diazoacetate provided a 1:1 mixture of the *trans* (**151**) and *cis* (**152**) isomers (equation 132). The diastereomeric excess of both *trans* and *cis* are 90% and the decomplexation can be easily achieved by treating the adduct with trimethyl nitroxide in dichloromethane[188].

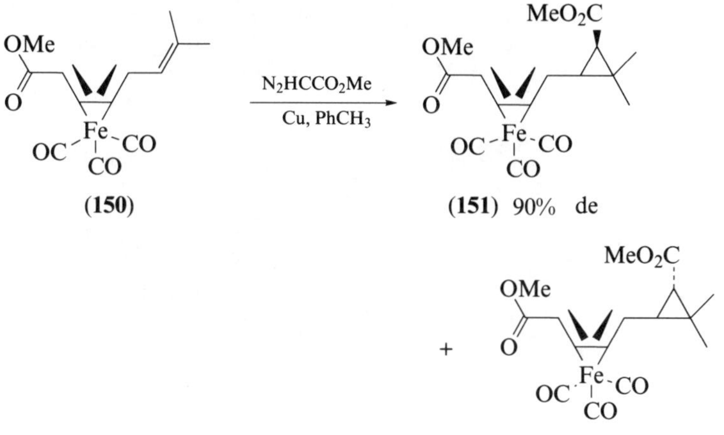

(**150**) (**151**) 90% de (**152**) 90% de (132)

C. Enantioselectivity in Metal Catalysed Cyclopropanation

There are several possibilities for asymmetric synthesis in catalysed cyclopropanation and very substantial progress has already been made especially with catalyst development. The option of covalently attaching chiral auxiliaries to diazo compounds or to substrates, e.g. alkenes for cyclopropanation, has been discussed above in the subsection on diastereoselectivity. The fact that many of the processes require metal catalysis makes the alternative option of using chiral catalysts particularly attractive and potentially more rewarding for commercial exploitation. The double option of combining the use of a chiral catalyst with a diazo compound carrying a chiral auxiliary is also available. For convenience, the double option is also included in this subsection.

Essentially all of the early studies were directed towards enantioselective cyclopropanation and Maas has reviewed the literature up to 1985[4]. The most successful of these early studies were those of Aratani and coworkers[172–174] who developed chiral copper(II) chelates of type **153** from salicylaldehyde and optically active amino alcohols with which to catalyse intermolecular cyclopropanation with diazoesters. Enantioselectivities exceeding 90% ee could be achieved in selected cases (equations 133 and 134) including the synthesis of permethrinic acid **154** and *trans*-chrysanthemic acid **155.**

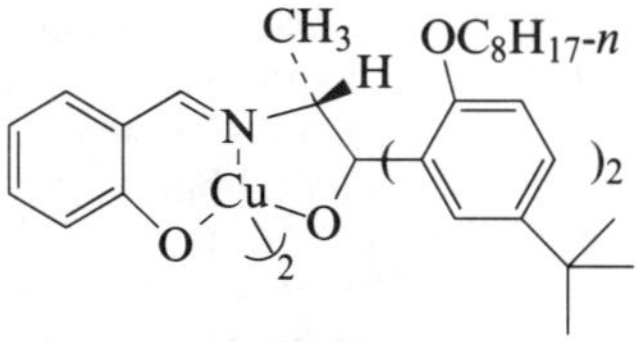

(**153**)

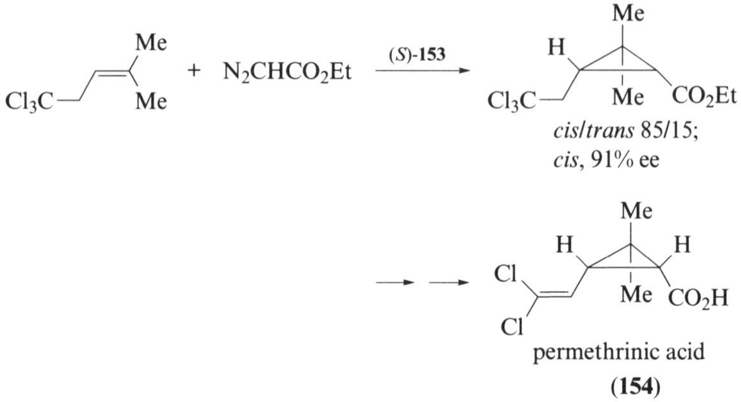

(133)

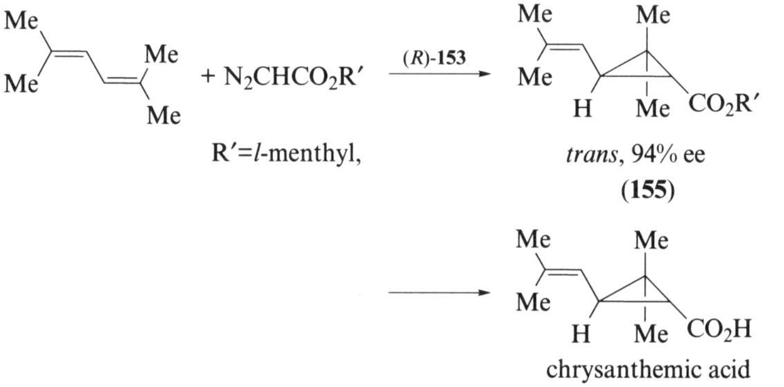

(134)

There are now several catalysts capable of achieving high levels (>90%) of enantioselectivity in cyclopropanation over a range of alkenes. The most successful of these are those with chiral ligands **156**, **157** and **158**. The catalysts developed by the groups of Pfaltz **156**[189–194], Masamune **157**[181,182] and Evans **158**[195,196] are all based on copper(II) or copper(I) with C_2-symmetric semicorrin or bis-oxazoline auxiliaries whereas Doyle's catalysts **159**[7,197–199] are rhodium(II) complexes bearing chiral carboxamide ligands. The active form of the former catalysts is copper(I) so that with the C_2-symmetric ligated catalyst, when prepared as a copper(II) complex, reduction prior to use is therefore necessary. Enantioselectivities greater than 90% have been realized with monosubstituted alkenes and dienes using the Pfaltz catalyst **156**, but di- and trisubstituted alkenes exhibited selectivities and yields inferior to those produced with the Aratani catalyst (**153**). Similar levels of enantiocontrol were obtained with the chiral bisoxazolone catalysts. A few representative examples are shown in equation 135. In addition, Masamune's catalysts also provide high levels of enantiocontrol in the cyclopropanation of di- and trisubstituted alkenes; an example has already been shown in equation 127 (catalyst **142** derived from ligand **157e**, *opposite below*). The Pfaltz catalyst **156** is also enantioselective in the cyclopropanation of 1,2-*trans*-disubstituted olefins with diazomethane (70–80% ee)[194]. Very recently, Ito and Katsuki[180] reported that the copper-based chiral bipyridine catalyst **141** gives high-level asymmetric induction in the cyclopropanation of styrene with *tert*-butyl diazoacetate (66–99% ee). Nishiyama and coworkers[43] have reported a new chiral ruthenium catalyst **160** which provides excellent stereo- and enantiocontrol in cyclopropanation of styrene with a range of alkyl diazoesters (>90% de; 86–96% ee). However, there is one report, by Matlin and coworkers[200], of the use of a copper(II) chelate possessing rigid trifluoroacetyl-(+)-camphor ligands for the cyclopropanation of styrene with 2-diazodimedone. Although an optical yield of 100% was reported for this intermolecular process, Dauben and coworkers[201] have found that the catalyst provided only a racemic product in an intramolecular cyclopropanation of an α-diazo-β-ketoester system. Doyle's Rh(II)-MEPY systems **159** are also very effective catalytically for cyclopropanation of monosubstituted alkenes with (+)- or (–)-menthyl diazoacetate or ethyl diazoacetate, although the levels of stereocontrol have not yet reached those produced with the chiral copper systems [197]. McKervey and coworkers[202,203] have developed prolinate derivatives of rhodium(II) **161** as homochiral catalyst for the intramolecular carbenoid transformation (*see below*). Davies and Hutcheson[204] have applied this type of catalyst to the reactions of the vinyldiazocarbonyl **163** with monosubstituted alkene **162** to give *E* vinylcyclopropane adduct **164** with high enantioselectivity (equation 136).

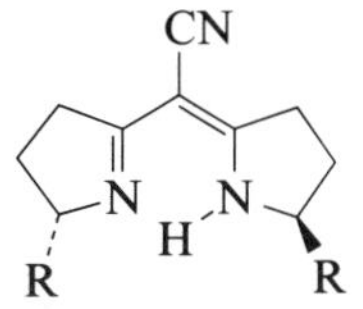

(**a**) R = CMe_2OH

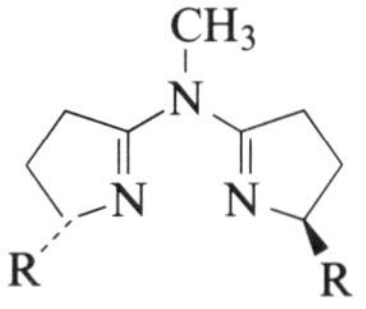

(**b**) R= CMe_2OSiMe_3
(**c**) R= CMe_2OSiMe_2Bu-*t*

Ligands of Pfaltz catalyst
(**156**)

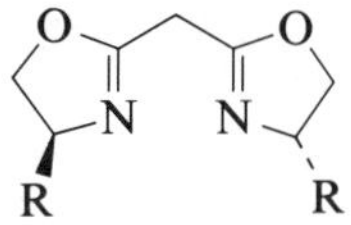

(**a**) R = Ph; (**b**) R = $PhCH_2$
(**c**) R = $CHMe_2$; (**d**) R = CMe_3

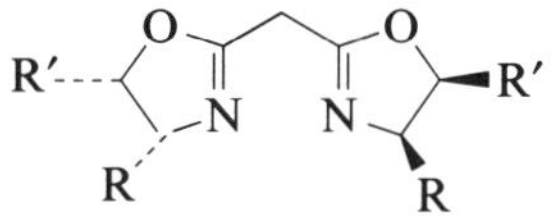

(**e**) R= R′=Ph
(**f**) R= Et, R′= Ph

Ligands of Masamune catalyst
(**157**)

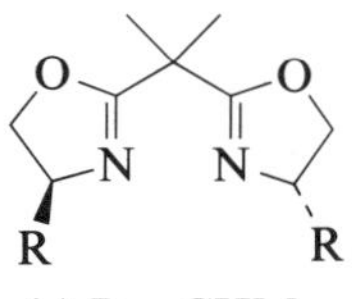

(**a**) R = $CHMe_2$
(**b**) R = CMe_3

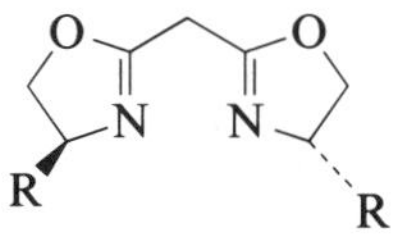

(**c**) R = $CHMe_2$
(**d**) R = CMe_3

Ligands of Evans catalyst
(**158**)

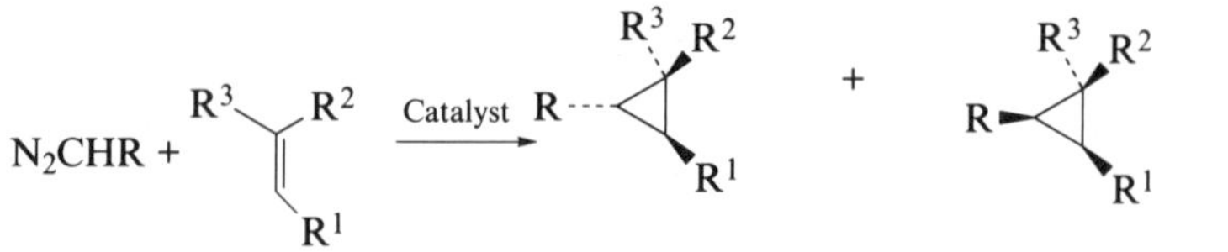

(135)

Catalyst	R	R^1	R^2	R^3	Yield	*trans*/*cis*	*trans* (ee)	*cis*(ee)
156a	CO_2-menthyl-*d*	Ph	H	H	70%	82/18	97%	95%
156b	CO_2-menthyl-*d*	Ph	H	H	89%	84/16	98%	99%
156c	CO_2Bu-*t*	Ph	H	H	75%	81/19	94%	95%
157d	CO_2-menthyl-*l*	Ph	H	H	72%	86/14	98%	96%
157e	CO_2-menthyl-*l*	Me	CMe_3	H	75%	88/12	95%	80%
158b	CO_2Et	Ph	H	H	77%	73/27	99%	97%
158b	CO_2Et	H	Me	Me	91%		99% ee	

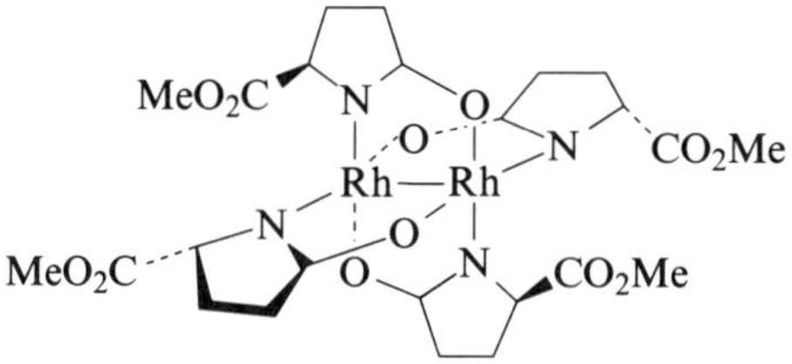

(159)

Doyle's catalyst: Rh(II)-MEPY

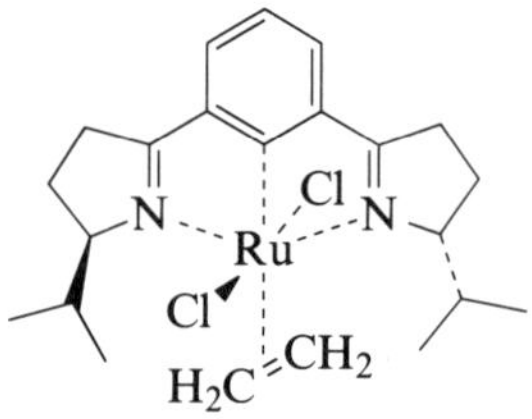

(160)

Ru(pybox-ip)

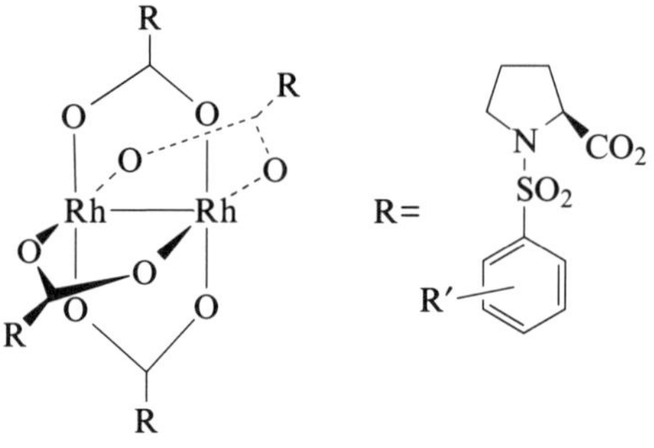

(161)

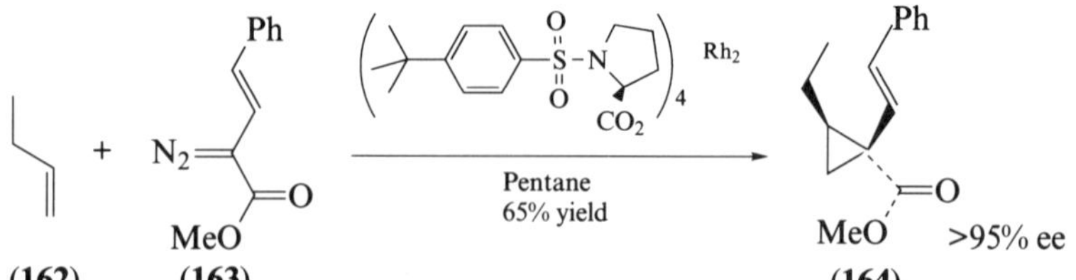

(162) **(163)** **(164)**

(136)

In intramolecular cyclopropanation, Doyle's catalysts (**159**) show outstanding capabilities for enantiocontrol in the cyclization of allyl and homoallyl diazoesters to bicyclic γ- and δ-lactones, respectively (equations 137 and 138)[198,205]. The data also reveal that intramolecular cyclopropanation of *Z*-alkenes is generally more enantioselective than that of *E*-alkenes in bicyclic γ-lactone formation[198]. Both Rh(II)-MEPY enantiomers are available and, through their use, enantiomeric products are accessible. In a few selected cases, the Pfaltz catalyst **156** also results in high-level enantioselectivity in intramolecular cyclopropanation (equation 139)[194]. On the other hand, the Aratani catalyst is less effective than the Doyle catalyst (**159**) or Pfaltz catalyst (**156**) in asymmetric intramolecular cyclopropanations[201]. In addition, the bis-oxazoline-derived copper catalyst **157b** shows lower enantioselectivity in the intramolecular cyclopropanation of allyl diazomalonate (equation 140)[206].

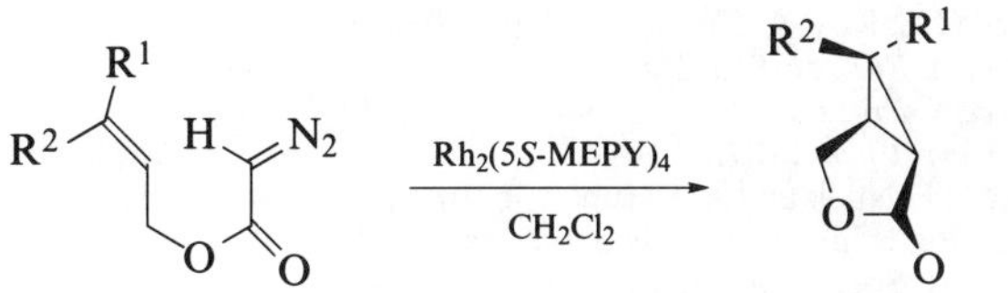

(137)

R^1=R^2=Me 98.6% ee
R^1=H; R^2=Ph, Et, $PhCH_2$ or (*n*-Bu)$_3$Sn > 98.6% ee
R^1=Ph; *n*-Pr; R^2=H 65–75% ee

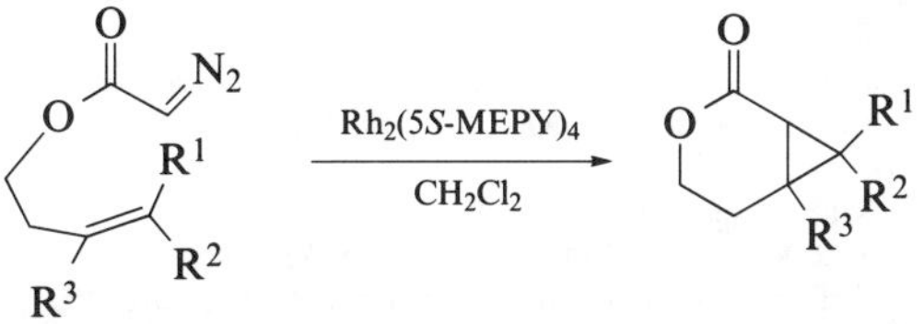

(138)

R^1=R^3=Me; R^3=H 77% ee
R^1=Et, Ph, Bn, TMS; R^2=R^3=H 80–90% ee
R^1=H; R^2=Et, Ph; R^3=H 73–82% ee
R^1=R^2=H; R^3 = Me 79% ee

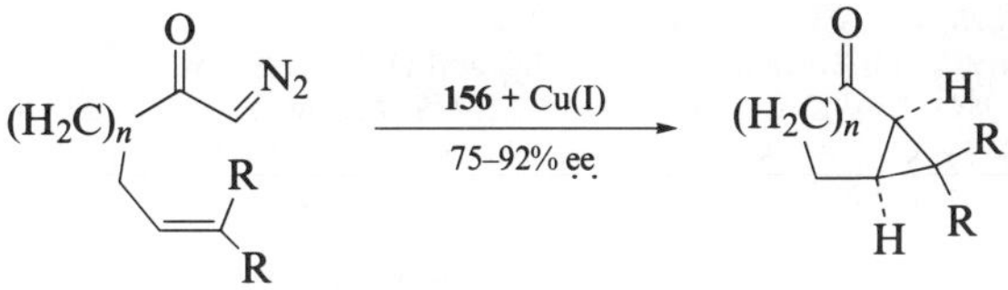

(139)

R=H, Me; *n*=1,2

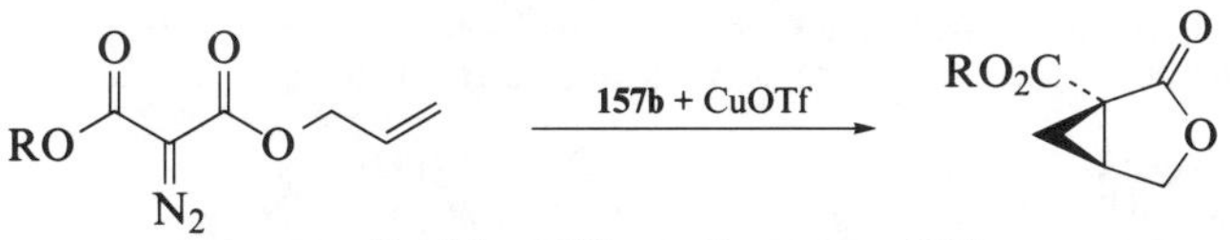

(140)

R= Me; 11% ee; R = *t*-Bu; 32% ee

VIII. CONCLUSION

The inclusion of a separate chapter on catalysed cyclopropanation in this latest volume of the series is indicative of the very high level of activity in the area of metal catalysed reactions of diazo compounds. Excellent, reproducible catalytic systems, based mainly on rhodium, copper or palladium, are now readily available for cyclopropanation of a wide variety of alkenes. Both intermolecular and intramolecular reactions have been explored extensively in the synthesis of novel cyclopropanes including natural products. Major advances have been made in both regiocontrol and stereocontrol, the latter leading to the growing use of chiral catalysts for producing enantiopure cyclopropane derivatives.

IX. REFERENCES

1. O. Silberrad and C. S. Roy, *J. Chem. Soc.*, 179 (1906).
2. M. P. Doyle, *Chem. Rev.*, **86**, 919 (1986).
3. M. P. Doyle, *Acc. Chem. Res.*, **19**, 348 (1986).
4. G. Maas, *Top. Curr. Chem.*, **137**, 75 (1987).
5. A. Demonceau, A. F. Noels and A. J. Hubert, in *Aspects of Homogeneous Catalysis* (Ed. R. Ugo), Vol. 6, D. Reidel Publishing Company, Lancaster, 1988, pp. 199–232.
6. J. Adams and D. M. Spero, *Tetrahedron*, **47**, 1765 (1991).
7. M. P. Doyle, *Recl. Trav. Chim. Pays-Bas*, **110**, 305 (1991).
8. A. Padwa and K. E. Krumpe, *Tetrahedron*, **48**, 5385 (1992).
9. M. P. Doyle, in *Catalytic Asymmetric Synthesis* (Ed. I. Ojima), VCH Publishers, New York, 1993, pp. 63–99.
10. P. Helquist, in *Comprehensive Organic Synthesis; Selectivity, Strategy and Efficiency in Modern Organic Chemistry* (Eds. B. M. Trost and I. Fleming), Vol. 4, Chapter 4.6, Pergamon, Oxford, 1991, pp. 951–997.
11. H. M. L. Davies, in *Comprehensive Organic Synthesis; Selectivity, Strategy and Efficiency in Modern Organic Chemistry* (Eds. B. M. Trost and I. Fleming), Vol. 4, Chapter 4.8, Pergamon, Oxford, 1991, pp. 1031–1067.
12. T. Ye and M. A. McKervey, *Chem. Rev.*, **94**, 1091 (1994).
13. R. G. Salomon and J. K. Kochi, *J. Am. Chem. Soc.*, **95**, 3300 (1973).
14. R. G. Salomon, M. F. Salomon, and T. R. Heyne, *J. Org. Chem.*, **40**, 756 (1975).
15. T. Mandai, K. Hara, M. Kawada and J. Nokami, *Tetrahedron Lett.*, **24**, 1517 (1983).
16. R. J. Sundberg and W.J. Pitts, *J. Org. Chem.*, **56**, 3048 (1991).
17. F. R. Hartley, *Chem. Rev.*, **73**, 163 (1973).
18. R. Paulissen, A. J. Hubert and P. Teyssié, *Tetrahedron Lett.*, 1465 (1972).
19. M. Suda, *Synthesis*, 714 (1981).
20. A. J. F. Edmunds, K. Baumann, M. Grassberger and G. Schulz, *Tetrahedron Lett.*, **32**, 7039 (1991).
21. J. K. H. Vorbruggen, *Synthesis*, 636 (1975).
22. Y. V. Tomilov, V. G. Bordakov, I. E. Dolgii and O. M. Nefedov, *Izv. Akad. Nauk SSSR. Ser. Khim.* 582 (1984); *Engl. Transl. Bull. Acad. Sci. USSR, Div Chem. Sci.*, 533 (1984).
23. Y. V. Tomilov, V. G. Bordakov, A. I. Lutsenko, S. O. Kozhinskii, I. E. Dolgii and O. M. Nefedov *Izv. Akad. Nauk. SSSR, Ser. khim.* 1338 (1987); *Engl. Transl., Bull. Acad. Sci. USSR, Div. Chem. Sci.*, 1234 (1987).
24. I. Dinulescu, L. N. Enescu, A. Ghenciulescu and M. Avram, *J. Chem. Res. (S)*, 456 (1978).
25. H. Abdallah, R. Grée and R. Carrié, *Tetrahedron Lett.*, **23**, 503 (1982).
26. Y. V. Tomilov, A. B. Kostitsyn, E. V. Shulishov and O. M. Nefedov, *Synthesis*, 246 (1990).
27. M. P. Doyle, J. H. Griffin, V. Bagheri and R. L. Dorow, *Organomet.*, **3**, 53 (1984).
28. U. Mende, B. Raduchel, W. Skuballa and H. Vorbruggen, *Tetrahedron Lett.*, 629 (1975).
29. B. Raduchel, U. Mende, G. Cleve, G.-A. Hoyer and H. Vorbruggen, *Tetrahedron Lett.*, 633 (1975).
30. J. Vallgarda and U. Hacksell, *Tetrahedron Lett.*, **32**, 5625 (1991).
31. S. D. Burke and P. A. Grieco, *Org. React.* (New York), **26**, 361 (1979).
32. V. Dave and E. Marnhoff, *Org. React.*, **18**, 217 (1970).

33. A. F. Noels, A. Demonceau, E. Carlier, A. J. Hubert, R.-L. Márquez-Silva and R. A. Sánchez-Delgado, *J. Chem. Soc., Chem. Commun.*, 783 (1988).
34. A. Demonceau, E. Saive, Y. de Froidmont, A. F. Noels, A. J. Hubert, I. T. Chizhevsky, I. A. Lobanova and V. I. Bregadze, *Tetrahedron Lett.*, **33**, 2009 (1992).
35. G. Maas, T. Werle, M. Alt and D. Mayer, *Tetrahedron*, **49**, 881 (1993).
36. A. J. Anciaux, A. J. Hubert, A. F. Noels, N. Petiniot and P. Teyssié, *J. Org. Chem.*, **45**, 695 (1980).
37. M. P. Doyle, R. L. Dorow, W. H. Tamblyn and W. E. Buhro, *Tetrahedron Lett.*, **23**, 2261 (1982).
38. M. W. Majchrzak, A. Kotelko and J. B. Lambert, *Synthesis*, 469 (1983).
39. A. Nakamura, A. Konishi, Y. Tatsuno and S. Otsuka, *J. Am. Chem. Soc.*, **100**, 3443 (1978).
40. A. Nakamura, A. Konishi, R. Tsujitani, M. Kudo and S. Otsuka, *J. Am. Chem. Soc.*, **100**, 3449 (1978).
41. G. Jommi, R. Pagliarin, G. Rizzi and M. Sisti, *Synlett*, 833 (1993).
42. D. A. Smith, D. N. Reynolds and L. K. Woo, *J. Am. Chem. Soc.*, **115**, 2511 (1993).
43. H. Nishiyama, Y. Itoh, H. Matsumoto, S.-B. Park and K. Itoh, *J. Am. Chem. Soc.*, **116**, 2223 (1994).
44. W. J. Seitz, A. K. Saha and M. M. Hossain, *Organomet.*, **12**, 2604 (1993).
45. R. Pellicciari, B. Natalini, M. Marinozzi, J. B. Monahan and J. P. Snyder, *Tetrahedron Lett.*, **31**, 139 (1990).
46. E. Piers and N. Moss, *Tetrahedron Lett.*, **26**, 2735 (1985).
47. E. Wenkert and T. Hudlicky, *J. Org. Chem.*, **53**, 1953 (1988).
48. J. P. Marino, C. Silveira, J. Comasseto and N. Petragnani, *J. Org. Chem.*, **52**, 4139 (1987).
49. J. Ollivier and J. Salaün, *J. Chem. Soc., Chem. Commun.*, 1269 (1985).
50. H. M. L. Davies and B. Hu, *J. Org. Chem.*, **57**, 3186 (1992).
51. H. M. L. Davies and B. Hu, *J. Org. Chem.*, **57**, 4309 (1992).
52. H. M. L. Davies, T. J. Clark and H. D. Smith, *J. Org. Chem.*, **56**, 3817 (1991).
53. H. M. L. Davies, E. Saikali and W. B. Young, *J. Org. Chem.*, **56**, 5696 (1991).
54. H. M. L. Davies and N. J. S. Huby, *Tetrahedron Lett.*, **33**, 6935 (1992).
55. H. M. L. Davies, *Tetrahedron*, **49**, 5203 (1993).
56. H. O. House and C. J. Blankley, *J. Org. Chem.*, **33**, 53 (1968).
57. C. J. Blankley, F. J. Scuter and H. O. House, in *Org. Synth.* Coll. Vol. 5 (Ed. H. E. Baumgarten), Wiley, New York, 1976, pp. 258–263.
58. C. E. Hatch III and J. S. Baum, *J. Org. Chem.*, **45**, 3281 (1980).
59. J. S. Yadav, S. V. Mysorkar and A. V. R. Rao, *Tetrahedron*, **45**, 7353 (1989).
60. M. S. Baird and H. H. Hussain, *Tetrahedron*, **43**, 215 (1987).
61. E. J. Corey and A. G. Myers, *J. Am. Chem. Soc.*, **107**, 5574 (1985).
62. E. J. Corey and H. Kigoshi, *Tetrahedron Lett.*, **32**, 5025 (1991).
63. H. M. L. Davies, M. J. McAfee and C. E. M. Oldenburg, *J. Org. Chem.*, **54**, 930 (1989).
64. M. P. Doyle, K.-L. Loh, K. M. DeVries and M. S. Chinn, *Tetrahedron Lett.*, **28**, 833 (1987).
65. K. Yamanoi and Y. Ohfune, *Tetrahedron Lett.*, **29**, 1181 (1988).
66. K. Shimamoto and Y. Ohfune, *Tetrahedron Lett.*, **31**, 4049 (1990).
67. G. Stork and J. Ficini, *J. Am. Chem. Soc.*, **83**, 4678 (1961).
68. E. Wenkert, R. A. Mueller, E. J. Reardon, Jr., S. S. Sathe, D. J. Scharf and G. Tosi, *J. Am. Chem. Soc.*, **92**, 7428 (1970).
69. E. Wenkert, B. L. Buckwalter, A. A. Craveiro, E. L. Sanchez and S. S. Sathe, *J. Am. Chem. Soc.*, **100**, 1267 (1978).
70. E. Wenkert, M. E. Alonso, B. L. Buckwalter and E. L. Sanchez, *J. Am. Chem. Soc.*, **105**, 2021 (1983).
71. A. R. Daniewski and T. Kowalczyk-Przewloka, *J. Org. Chem.*, **50**, 2976 (1985).
72. For details of this type of transformation, see Reference 12.
73. A. B. Smith, III, B. H. Toder and S. J. Branca, *J. Am. Chem. Soc.*, **106**, 3995 (1984).
74. B. Saha, G. Bhattacharjee and U. R. Ghatak, *J. Chem. Soc., Perkin Trans 1*, 939 (1988).
75. W. von E. Doering and M. Pomerantz, *Tetrahedron Lett.*, 961 (1964).
76. P. Dowd, P. Garner, R. Schappert, H. Irngartinger and A. Goldman, *J. Org. Chem.*, **47**, 4240 (1982).
77. M. Ihara, T. Taniguchi, K. Makita, M. Takano, M. Ohnishi, N. Taniguchi. K. Fukumoto and C. Kabuto, *J. Am. Chem. Soc.*, **115**, 8107 (1993).
78. A. Srikrishna and K. Krishnan, *Tetrahedron*, **48**, 3429 (1992).

79. N. Cagnoli, P. Ceccherelli, M. Curini, M. C. Marcotullio and E. Wenkert, *Synth. Commun.*, **17**, 1261 (1987).
80. A. Srikrishna and S. Nagaraju, *J. Chem. Soc., Perkin Trans. 1*, 657 (1991).
81. K. Kon and S. Isoe, *Tetrahedron Lett.*, **27**, 3399 (1986).
82. J. Lafontaine, M. Mongrain, M. Sergent-Guay, L. Ruest and P. Deslongchamps, *Can. J. Chem.*, **58**, 2460 (1980).
83. J. D. White, J. F. Ruppert, M. A. Avery, S. Torii and J. Nokami, *J. Am. Chem. Soc.*, **103**, 1813 (1981).
84. A. Murai, K. Kato and T. Masamune, *Tetrahedron Lett.*, **23**, 2887 (1982).
85. T. Imanishi, M. Matsui, M. Yamashita and C. Iwata, *Tetrahedron Lett.*, **27**, 3161 (1986).
86. J. Adams and M. Belley, *Tetrahedron Lett.*, **27**, 2075 (1986).
87. J. Adams, R. Frenette, M. Belley, F. Chibante and J. P. Springer, *J. Am. Chem. Soc.*, **109**, 5432 (1987).
88. J. Adams, C. Lepine-Frenette and D. M. Spero, *J. Org. Chem.*, **56**, 4494 (1991).
89. T. Hudlicky and J. P. Sheth, *Tetrahedron Lett.*, 2667 (1979).
90. T. Hudlicky, F.J. Koszyk, T. M. Kutchan and J. P. Sheth, *J. Org. Chem.*, **45**, 5020 (1980).
91. T. Hudlicky and F. J. Koszyk, *Tetrahedron Lett.*, **21**, 2487 (1980).
92. S. V. Govindan, T. Hudlicky and F. J. Koszyk, *J. Org. Chem.*, **48**, 3581 (1983).
93. M. Laabassi and R. Grée, *Tetrahedron Lett.*, **29**, 611 (1988).
94. E. Piers and G. L. Jung, *Can. J. Chem.*, **63**, 996 (1985).
95. T. Hudlicky, T. M. Kutchan, S. R. Wilson and D. T. Mao, *J. Am. Chem. Soc.*, **102**, 6351 (1980).
96. T. Hudlicky and R. P. Short, *J. Org. Chem.*, **47**, 1522 (1982).
97. R. P. Short, J.-M. Revol, B. C. Ranu and T. Hudlicky, *J. Org. Chem.*, **48**, 4453 (1983).
98. T. Hudlicky, L. D. Kwart, M. H. Tiedje, B. C. Ranu, R. P. Short, J. O. Frazier and H. L. Rigby, *Synthesis*, 716 (1986).
99. T. Hudlicky, M. G. Natchus and G. Sinai-Zingde, *J. Org. Chem.*, **52**, 4641 (1987).
100. T. Hudlicky, G. Sinai-Zingde, M. G. Natchus, B. C. Ranu and P. Papadopolous, *Tetrahedron*, **43**, 5685 (1987).
101. B. C. Ranu, M. Sarkar, P. C. Chakraborti and U. R. Ghatak, *J. Chem. Soc., Perkin Trans. 1*, 865 (1982).
102. R. D. Dawe, L. N. Mander and J. V. Turner, *Tetrahedron Lett.*, **26**, 363 (1985).
103. R. D. Dawe, L. N. Mander, J. V. Turner and X. Pan, *Tetrahedron Lett.*, **26**, 5725 (1985).
104. A. Saba, *Tetrahedron Lett.*, **26**, 4657 (1990).
105. E. Wenkert and S. Liu, *Synthesis*, 323 (1992).
106. A. Padwa, T. J. Wisnieff and E. J. Walsh, *J. Org. Chem.*, **54**, 299 (1989).
107. E. Y. Chen, *Tetrahedron Lett.*, **23**, 4769 (1982).
108. E. Y. Chen, *J. Org. Chem.*, **49**, 3245 (1984).
109. S. H. Kang, W. J. Kim and Y. B. Chae, *Tetrahedron Lett.*, **29**, 5169 (1988).
110. A. M. Koskinen and L. Muñoz, *J. Org. Chem.*, **58**, 879 (1993).
111. B. M. Trost and W. C. Vladuchick, *J. Org. Chem.*, **44**, 148 (1979).
112. D. F. Taber, S. A. Saleh and R. W. Korsmeyer, *J. Org. Chem.*, **45**, 4699 (1980).
113. D. F. Taber, K. R. Krewson, K. Raman and A. L. Rheingold, *Tetrahedron Lett.*, **25**, 5283 (1984).
114. D. F. Taber, J. C. Amedio, Jr. and K. Raman, *J. Org. Chem.*, **53**, 2984 (1988).
115. I. Shimizu and F. Aida, *Chem Lett.*, 601 (1988).
116. I. Shimizu and T. Ishikawa, *Tetrahedron Lett.*, **35**, 1905 (1994).
117. E. Piers and E. H. Ruediger, *J. Org. Chem.*, **45**, 1725 (1980).
118. E. Piers, G. L. Jung and N. Moss, *Tetrahedron Lett.*, **25**, 3959 (1984).
119. E. Piers, G. L. Jung and E. H. Ruediger, *Can. J. Chem.*, **65**, 670 (1987).
120. T. Hudlicky, S. V. Govindan and J.O. Frazier, *J. Org. Chem.*, **50**, 4166 (1985).
121. A. Srikrishna and K. Krishnan, *J. Chem. Soc., Chem.Commun.*, 1693 (1991).
122. A. Srikrishna and K. Krishnan, *J. Org. Chem.*, **58**, 7751 (1993).
123. S. Ohira, *Bull. Chem. Soc. Jpn.*, **57**, 1902 (1984).
124. E. J. Corey, G. Wess, Y. B. Xiang and A. K. Singh, *J. Am. Chem. Soc.*, **109**, 4717 (1987).
125. A. K. Singh, R. K. Bakshi and E. J. Corey, *J. Am. Chem. Soc.*, **109**, 6187 (1987).
126. E. J. Corey and Y. B. Xiang, *Tetrahedron Lett.*, **28**, 5403 (1987).
127. J. N. C. Hood, D. Lloyd, W. A. MacDonald and T. M. Shepherd, *Tetrahedron*, **38**, 3355 (1982).
128. L. Hatjiarapoglou and A. Varvoglis, N. W. Alcock and G. A. Pike, *J. Chem. Soc., Perkin Trans. 1*, 2839 (1988).

129. R. M. Moriarty, O. Prakash, R. K. Vaid and L. Zhao, *J Am. Chem. Soc.*, **111**, 6443 (1987).
130. R. M. Moriarty, J. Kim and L. Guo, *Tetrahedron Lett.*, **34**, 4129 (1993).
131. Y. H. Gai, M. Julia and J. N. Verpeaux, *Synlett*, 56 (1991).
132. Y. H. Gai, M. Julia and J. N. Verpeaux, *Synlett*, 269 (1991).
133. T. Cohen, G. Herman, T. M. Chapman and D. Kuhn, *J. Am. Chem. Soc.*, **96**, 5627 (1974).
134. B. Cimetière and M. Julia, *Synlett*, 271 (1991).
135. N. Kawabata, S. Yanao and J. Yoshida, *Bull. Chem. Soc. Jpn.*, **55**, 2687 (1982).
136. B. M. Trost and S. Schneider, *J. Am. Chem. Soc.*, **111**, 4430 (1989).
137. B. M. Trost and H. Urabe, *Tetrahedron Lett.*, **31**, 615 (1990).
138. S. Ogoshi, T. Morimoto, K. Nishio, K. Ohe and S. Murai, *J. Org. Chem.*, **58**, 9 (1993).
139. K. Ohe, H. Matsuda, T. Ishihara, S. Ogoshi, N. Chatani and S. Murai, *J. Org. Chem.*, **58**, 1173 (1993).
140. A. F. Noels, A. Demonceau, N. Petiniot, A. J. Hubert and P. Teyssié, *Tetrahedron*, **38**, 2733 (1982).
141. B. Ganem, N. Ikota, V. B. Muralidharan, W. S. Wade, S. D. Young and Y. Yukimoto, *J. Am. Chem. Soc.*, **104**, 6787 (1982).
142. S. Bien and Y. Segal, *J. Org. Chem.*, **42**, 1685 (1977).
143. E. Wenkert, M. E. Alonso, B. L. Buckwalter and E. L. Sanchez, *J. Am. Chem. Soc.*, **105**, 2021 (1983).
144. D. F. Table and R. E. Ruckle, Jr., *J. Am. Chem. Soc.*, **108**, 7686 (1986).
145. A. Padwa, D. J. Austin, S. F. Hornbuckle, M. A. Semones, M. P. Doyle and M. N. Protopopova, *J. Am. Chem. Soc.*, **114**, 1874 (1992).
146. A. Padwa, D. J. Austin, A.T. Price, M. A. Semones, M. P. Doyle and M. N. Protopopova, W. R. Winchester and A. Tran, *J. Am. Chem. Soc.*, **115**, 8669 (1993).
147. T. Honda, H. Ishige, M. Tsubki, K. Naito and Y. Suzuki, *J. Chem. Soc., Perkin Trans. 1*, 954 (1991).
148. S. Hashimoto, N. Watanabe and S. Ikegami, *J. Chem. Soc., Chem. Commun.*, 1508 (1992).
149. D. F. Taber and R. S. Hoerrner, *J. Org. Chem.*, **57**, 441 (1992).
150. M. P. Doyle, W. H. Tamblyn and V. Bagheri, *J. Org. Chem.*, **46**, 5094 (1981).
151. M. P. Doyle, J. H. Griffin, M. S. Chinn and D. van Leusen, *J. Org. Chem.*, **49**, 1917 (1984).
152. M. P. Doyle, V. Bagheri and N. K. Harn, *Tetrahedron Lett.*, **29**, 5119 (1988).
153. E. J. Roskamp and C. R. Johnson, *J. Am. Chem. Soc.*, **108**, 6062 (1986).
154. F. Kido, S. C. Sinha, T. Abiko and A. Yoshikoshi, *Tetrahedron Lett.*, **30**, 1575 (1989).
155. E. Wenkert, T. D. J. Halls, L. D. Kwart, G. Magnusson and H. D. H. Showalter, *Tetrahedron*, **37**, 4017 (1981).
156. E. Wenkert, M. E. Alonso, B. L. Buckwalter and E. L. Sanchez, *J. Am. Chem. Soc.*, **105**, 2021 (1983).
157. E. Wenkert, T. P. Ananthanarayan, V. F. Ferreira, M. G. Hoffmann and H. S. Kim, *J. Org. Chem.*, **55**, 4975 (1990).
158. M. E. Alonso, A. Morales and A. W. Chitty, *J. Org. Chem.*, **47**, 3747 (1982).
159. M. E. Alonso, P. Jano, M. I. Hernandez, R. S. Greemberg and E. Wenkert, *J. Org. Chem.*, **48**, 3047 (1983).
160. M. P. Doyle, D. van Leusen and W. H. Tamblyn, *Synthesis*, 787 (1981).
161. A. J. Anciaux, A. Demonceau, A. F. Noels, R. Warin, A. J. Hubert and P. Teyssié, *Tetrahedron*, **39**, 2169 (1983).
162. M. P. Doyle R. L. Dorrow, W. E. Buhro, J. H. Griffin, W. H. Tamblyn and M. L. Trudell, *Organomet.*, **3**, 44 (1984).
163. E. Wenkert, T. E. Goodwin and B. C. Ranu, *J. Org. Chem.*, **42**, 2137 (1977).
164. F. Bohlmann and W. Rotard, *Justus Liebigs Ann. Chem.*, 1211 (1982).
165. M. P. Doyle, L. C. Wang and K.-L. Loh, *Tetrahedron Lett.*, **25**, 4087 (1984).
166. M. P. Doyle, V. Bagheri, T. J. Wandless, N. K. Harn, D. A. Brinker, C. T. Eagle and K.-L. Loh, *J. Am. Chem. Soc.*, **112**, 1906 (1990).
167. A. Becalski, W. R. Cullen, M. D. Fryzuk, G. Herb, B. R. James, J. P. Kutney, K. Piotrowska and D. Tapiolas, *Can. J. Chem.*, **66**, 3108 (1988).
168. E. Kunkel, I. Reichelt and H. U. Reissig, *Justus Liebigs Ann. Chem.*, 512 (1984).
169. M. P. Doyle, K.-L. Loh, K. M. DeVries and M. S. Chinn, *Tetrahedron Lett.*, **28**, 833 (1987).
170. H. M. L. Davies, T. J. Clark and L. A. Church, *Tetrahedron Lett.*, **30**, 5057 (1989).
171. P. E. O'Bannon and W. P. Dailey, *Tetrahedron*, **46**, 7341 (1990).

172. T. Aratani, Y. Yoneyoshi and T. Nagase, *Tetrahedron Lett.*, 2599 (1977).
173. T. Aratani, Y. Yoneyoshi and T. Nagase, *Tetrahedron Lett.*, **23**, 685 (1982).
174. T. Aratani, *Pure Appl. Chem.*, **57**, 1839 (1985).
175. A. Demonceau, A. F. Noels and A. J. Hubert, *Tetrahedron*, **46**, 3889 (1990).
176. H. J. Callot and C. Piechocki, *Tetrahedron Lett.*, **21**, 3489 (1980).
177. H. J. Callot and C. Piechocki, *Tetrahedron*, **38**, 2365 (1982).
178. J. L. Maxwell, S. O'Malley, K. C. Brown and T. Kodadek, *Organomet.*, **11**, 645 (1992).
179. S. O'Malley and T. Kodadek, *Organomet.*, **11**, 2299 (1992).
180. K. Ito and T. Katsuki, *Tetrahedron Lett.*, **34**, 2661 (1993).
181. R. E. Lowenthal and S. Masamune, *Tetrahedron Lett.*, **32**, 7373 (1991).
182. R. E. Lowenthal, A. Abiko and S. Masamune, *Tetrahedron Lett.*, **31**, 6005 (1990).
183. P. E. Krieger and J. A. Landgrebe, *J. Org. Chem.*, **43**, 4447 (1978).
184. M. P. Doyle, R. L. Dorow, J. W. Terpstra and R. A. Rodenhouse, *J. Org. Chem.*, **50**,1663 (1985).
185. H. M. L. Davies and W. R. Cantrell, Jr. *Tetrahedron Lett.*, **32**, 6509 (1991).
186. H. M. L. Davies, N. J. S. Huby, W. R. Cantrell, Jr, and J. L. Olive, *J. Am. Chem. Soc.*, **115**, 9468 (1993).
187. D. F. Taber, J. C. Amedio, Jr. and K. Raman, *J. Org. Chem.*, **53**, 2984 (1988).
188. A. Monpert, J. Martelli, R. Grée and R. Carrié, *Tetrahedron Lett.*, **22**, 1961 (1981).
189. H. Fritschi, U. Leutenegger and A. Pfaltz, *Angew. Chem., Int. Ed. Engl.*, **25**, 1005 (1986).
190. H. Fritschi, U. Leutenegger, K. Siegmann, A. Pfaltz, W. Keller and C. Kratky, *Helv. Chim. Acta*, **71**, 1541 (1988).
191. H. Fritschi, U. Leutenegger and A. Pfaltz, *Helv. Chim. Acta*, **71**, 1553 (1988).
192. D. Müller, G. Umbricht, B. Weber and A. Pfaltz, *Helv. Chim. Acta*, **74**, 232 (1991).
193. H. Fritschi, U. Leutenegger, G. Umbricht, C. Fahrni, P. Vonmatt and A. Pfaltz, *Tetrahedron*, **48**, 2143 (1992).
194. A. Pfaltz, *Acc. Chem Res.*, **26**, 339 (1993).
195. D. A. Evans, K. A. Woerpel and M. M. Himman, *J. Am. Chem. Soc.*, **113**, 726 (1991).
196. D. A. Evans, K. A. Woerpel and M. J. Scott, *Angew, Chem., Int. Ed. Engl.*, **31**, 430 (1992).
197. M. P. Doyle, B. D. Brandes, A. P. Kazala, R. J. Pieters, M. B. Jarstfer, L. M. Watkins and C. T. Eagle, *Tetrahedron Lett.*, **31**, 6613 (1990).
198. M. P. Doyle, R. J. Pieters, S. F. Martin, R. E. Austin, C. J. Oalmann and P. J. Müller, *J. Am. Chem. Soc.*, **113**, 1423 (1993).
199. M. P. Doyle, A. V. Oeveren, L. J. Westrum, M. N. Protopopova and T. W. Clayton, Jr., *J. Am. Chem. Soc.*, **113**, 8982 (1991).
200. S. A. Matlin, W. J. Lough, L. Chan, D. M. H. Abram and Z. Zhou, *J. Chem. Soc., Chem. Commun.*, 1038 (1984).
201. W. G. Dauben, R. T. Hendricks, M. J. Luzzio and H. P. Ng, *Tetrahedron Lett.*, **31**, 6969 (1990).
202. M. A. McKervey and T. Ye, *J. Chem. Soc., Chem. Commun.*, 823 (1992).
203. M. Kennedy, M. A. McKervey, A. R. Maguire and G. H. P. Roos, *J. Chem.Soc., Chem. Commun.*, 361 (1990).
204. H. M. L. Davies and D. K. Hutcheson, *Tetrahedron Lett.*, **34**, 7243 (1993).
205. S. F. Martin, C. J. Oalmann and S. Liras, *Tetrahedron Lett.*, **33**, 6727 (1992).
206. A. M. P. Koskinen and H. Hassila, *J. Org. Chem.*, **58**, 4479 (1993).

CHAPTER **12**

Cycloproparenes

BRIAN HALTON

Department of Chemistry, Victoria University of Wellington, PO Box 600, Wellington, New Zealand

Fax: +64 4 495 5241; e-mail: halton@matai.vuw.ac.nz

I. INTRODUCTION . . . 708
II. SYNTHESIS OF CYCLOPROPARENES . . . 710
A. Historical Methods . . . 710
B. By Ring Contraction . . . 711
1. Of 3*H*-indazoles . . . 711
2. Of spiro-3*H*-pyrazoles . . . 713
C. By Ring Closure . . . 714
D. By Flash Vacuum Pyrolysis . . . 715
E. By Aromatization . . . 717
1. Of 7,7-dihalobicyclo[4.1.0]heptenes . . . 717
2. Of 1,6-dihalobicyclo[4.1.0]heptenes . . . 721
3. Of 1-halo and other 1,6-disubstituted bicyclo[4.1.0]heptenes . . . 724
4. Of oxygen-bridged 1,6-dihalobicyclo[4.1.0]heptenes . . . 727
F. Oxocycloproparenes (Benzocyclopropenones) . . . 727
G. Alkylidenecycloproparenes . . . 729
III. PHYSICAL AND THEORETICAL ASPECTS OF CYCLOPROPARENES . . . 733
IV. CYCLOPROPARENYL CATIONS, ANIONS AND RADICALS . . . 741
A. Cycloproparenyl Cations . . . 741
B. Cycloproparenyl Anions . . . 742
C. Cycloproparenyl Radicals . . . 743
V. REACTIONS OF CYCLOPROPARENES . . . 744
A. Reactions that Proceed without Ring Cleavage . . . 744
1. With electrophiles and nucleophiles . . . 744
2. With radicals . . . 745
3. In cycloadditions . . . 745
4. Upon metal complexation . . . 747
B. Reactions that Proceed with Ring Cleavage . . . 748
1. With electrophiles . . . 748
2. With radicals . . . 750
3. In cycloadditions . . . 751

The chemistry of the cyclopropyl group, Vol. 2
Edited by Z. Rappoport

4. With organometallic reagents. 753
5. Upon thermolysis and photolysis 756
VI. OXO- AND ALKYLIDENECYCLOPROPARENE CHEMISTRY 757
A. Physical and Theoretical Aspects 757
B. Chemical Reactivity 760
1. With electrophiles and nucleophiles 760
2. Upon oxidation 762
3. In cycloadditions 763
4. With organometallic reagents 764
5. Upon thermolysis and photolysis 765
VII. ACKNOWLEDGEMENTS 765
VIII. REFERENCES 765

I. INTRODUCTION

'In previous papers on the *Synthetical Formation of Closed Carbon-chains* I have dealt exclusively with the formation of such chains in the fatty series. In continuing these researches it appeared to me that very interesting results might be obtained if substances belonging to the aromatic series were also experimented with. In this way it might be possible to obtain derivatives of the following hydrocarbons:

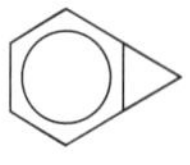
(**1**)

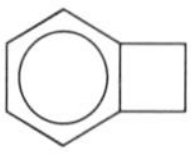
(**2**)

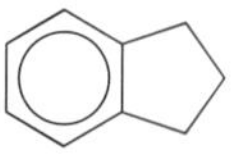
(**3**)

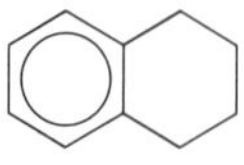
(**4**)

Although attempts to obtain the derivatives **1** and **2** have as yet been unsuccessful substances containing the benzene-ring and a 5-ring (**3**) and a benzene-ring and a 6-ring (**4**) may be synthesised by the following reactions.'

Recorded a little over 100 years ago[1], these words and the experiments of W. H. Perkin Jr. provided the foundation for the study of ring-fused aromatics, unquestionably an area of organic chemistry that has stood the test of time. This chapter addresses the chemistry of the cycloproparenes, the most highly strained class of compounds that can ensue from the fusion of a single ring to an aromatic framework and typified by the structure of cyclopropabenzene (**1**).

The existence of cyclopropabenzene as the most highly strained, isolable, member of the *ortho*-fused series of aromatic compounds was established almost thirty years ago[2]. The molecule, which has a decidely foul odour, is surprisingly stable and has a strain energy estimated[3] at 70 $kcal\,mol^{-1}$ and measured[4] as 68 $kcal\,mol^{-1}$. Dramatic improvements in organic synthesis from the time of the original report by Vogel and coworkers[2] have resulted in the preparation and characterization of various structural types of benzenoid cycloproparenes, e.g. cyclopropa[*l*]phenanthrene. Moreover, non-benzenoid systems containing 1,2-methylene fusion and cyclopropabenzenes constrained even further from the fusion of a second or even a third small carbocyclic ring have become prime targets for both synthetic and structural chemists. The principal reason for investigating this series of novel hydrocarbons has been to establish the limits to which stress and strain can be imposed upon a benzene ring and to examine the consequential effects that these force upon the bonding, structure and chemical reactivity of the molecules. Many studies have addressed these questions, but it is only in the very recent past that genuine insights have been forthcoming. Despite this, there still remains a need for further experimental and theoretical investigations.

Whereas strained ring systems are usually reactive and often unstable, molecules which satisfy the criteria for aromaticity exhibit enhanced stability. As is evident from the structural formula of **1**, the cycloproparenes set these features in juxtaposition as they are strained molecules in which a single carbon atom is fused across adjacent centres of an aromatic system. The interest of the experimentalist in strained molecules has been matched by the theoretician in the search for suitable models for developing the concepts of chemical bonding and aromaticity. The cycloproparenes have been particularly important in this regard as they meet the criterion for partial aromatic bond localization and consequent bond length alternation in the aromatic ring as proposed[5] by Mills and Nixon in 1930, viz. **1a** vs **1b**. The cation **5**, anion **6** and radical **7** derived from **1**, and also the ketone **8** and exocyclic methylene derivative **9**, are of interest in this respect.

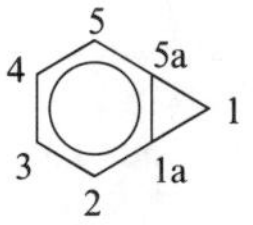

(1)

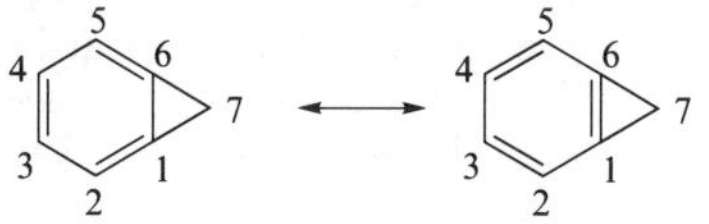

(1a) **(1b)**

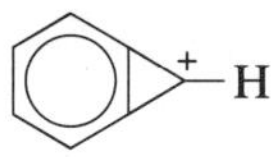

(5)

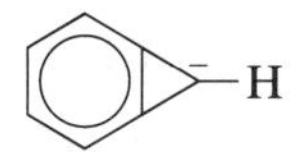

(6)

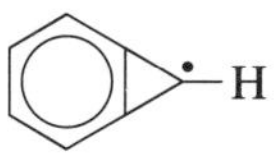

(7)

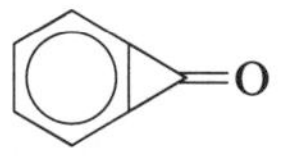

(8)

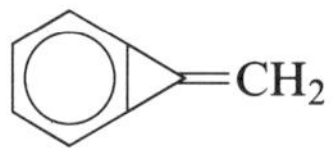

(9)

The systematic nomenclature for the cycloproparenes is confused because the 'fusion rule' (IUPAC Rule A 21.3) requires that 'at least two rings of five or more members' be present before the prefix *cyclopropa* may be used. Thus while 1*H*-cyclopropa[*a*]- and -[*b*]naphthalene are correct for **10** and **11**, respectively, 1*H*-cyclopropabenzene is incorrect for **1**. The Chemical Abstracts service and IUPAC are unanimous in naming **1** as bicyclo-[4.1.0]hepta-1,3,5-triene **1a**. Thus if the parent member is strictly named, not only does it differ from that of its higher homologues, but also it could be taken to imply a bond localized structure. Throughout this chapter parent **1** and its derivatives **5–9** are referred to as cyclopropabenzenes and numbered as shown for structure **1**.

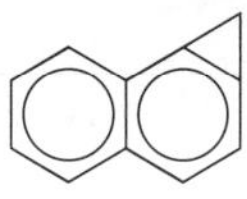

(10)

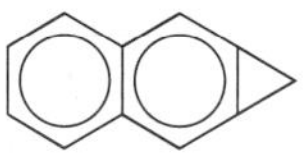

(11)

The cycloproparenes were reviewed in 1973[6], 1980[7] and 1989[8], and formed the subject of a research account[9] and report[10]. Detailed reviews of the cyclopropenes[11] and 1,3-bridged cyclopropenes[12] have strictly limited coverage of the cycloproparenes. The present contribution is designed to provide a thorough insight to the cycloproparenes and, while emphasis is placed upon the literature from the time of the last review, important early findings are included to provide understanding and continuity. *Chemical Abstracts* has been searched through Vol. 119 (1993).

II. SYNTHESIS OF CYCLOPROPARENES

The synthetic methods that have been employed for the preparation of the cycloproparenes fall into two distinct categories, namely those that commence with a preformed aromatic moiety to which is appended a three-membered ring, and those that build the molecular framework prior to aromatization. The sections that follow have been composed to provide a continuity of purpose rather than a strict division between these two axiomatic methodologies.

A. Historical Methods

Apart from the comment made by Perkin[1] that the synthesis of **1** had been unsuccessful, the appearance of cycloproparenes in the chemical literature did not occur until 1930 when De and Dutt[13] claimed that the decomposition of a series of aryliminosemicarbazones **12** resulted in the formation of the iminocyclopropa[*l*]phenanthrenes **13** (equation 1). A re-

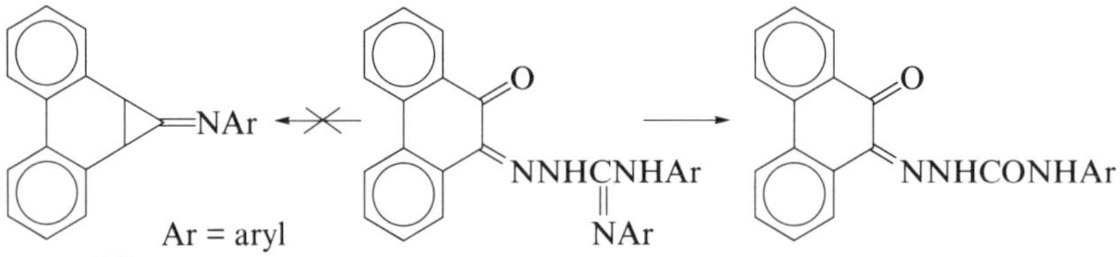

(1)

(13) **(12)** **(14)**

investigation of these decompositions[14] failed to reproduce the original findings; only the aryl semicarbazones **14** were isolated. The second claim to cycloproparene formation came in 1953 from Mustafa and Kamel who suggested[15] that secondary diazoalkanes added to quinone imides such as **15** to give the corresponding cycloproparene, e.g. **17**. When repeated, the products from these reactions matched those described, but they were shown[16,17] to be the products of addition to the quinone carbon–carbon double bond, namely **16** (equation 2). Base induced aromatization analogous to that proposed by the Egyptian workers[15] was deliberately employed by Ullman and Buncel[18] in 1963 in order to provide cycloproparenyl dianions, e.g. **19** (equation 3). Unfortunately no dianions were detected and the H/D exchange recorded was adequately explained from monoanion intervention. The aromatization of **16** and **18**, and analogous derivatives, by way of hydrogen migration has yet to be accomplished.

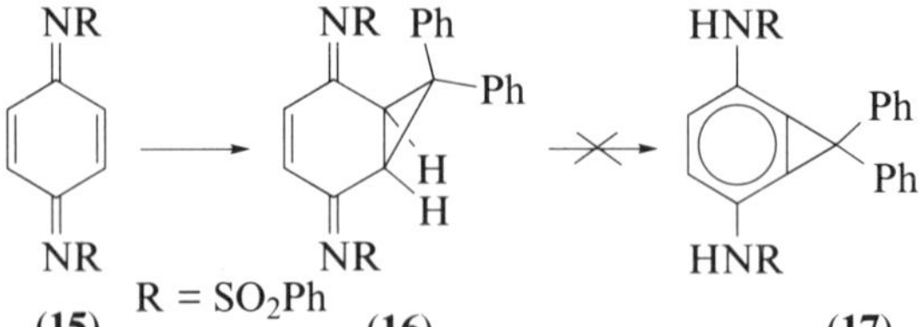

(2)

(15) **(16)** **(17)**

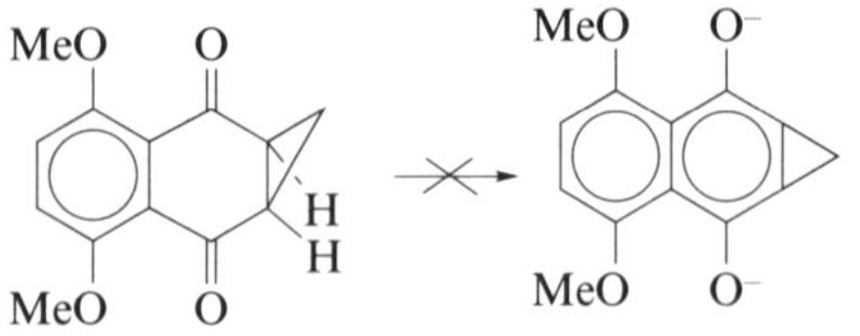

(3)

(18) **(19)**

B. By Ring Contraction

1. Of 3H-indazoles

The historical approaches discussed above are not without further significance. For example, the decomposition of the aryliminosemicarbazones **12** was thought to proceed by way of an unstable 3*H*-indazole[13] and it was by the specific use of such a compound that the first authenticated entry to the cycloproparene arena was attained. In 1964 Anet and Anet[19] found that irradiation of indazole **20** resulted in the loss of nitrogen and formation of the cyclopropabenzene ester **22** as the minor product of reaction (Scheme 1). Subsequent studies with a range of 3*H*-indazoles have shown that either singlet[20] or triplet[21] diradicals **21** may be involved in the photochemical deazetation, and that thermal decomposition[22–24] can also lead to cycloproparenes.

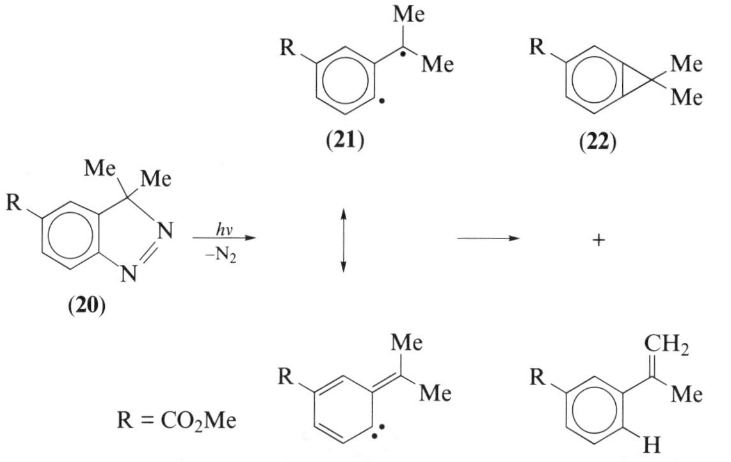

SCHEME 1

Upon thermolysis or photolysis a range of spiro-3*H*-indazoles lead to products by way of a spirocycloproparene. While Shechter and his group[22] were unable to establish that the spirocyclopropabenzene **24a** was definitely involved in the deazetation of **23a**, such intervention is a requirement[25] with the unsymmetrical derivatives **23b,c** (equation 4). In similar

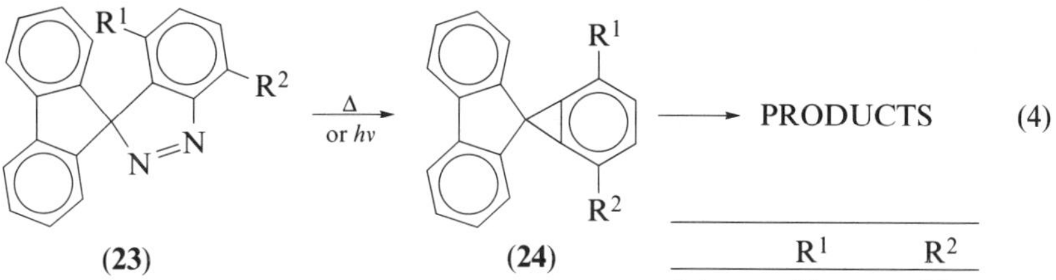

	R^1	R^2
(a)	H	H
(b)	H	Me
(c)	Me	H

manner, spirocycloproparenes are involved in the thermolysis and photolysis of a range of anthrone-10-spiroindazoles (Scheme 2)[23,24].

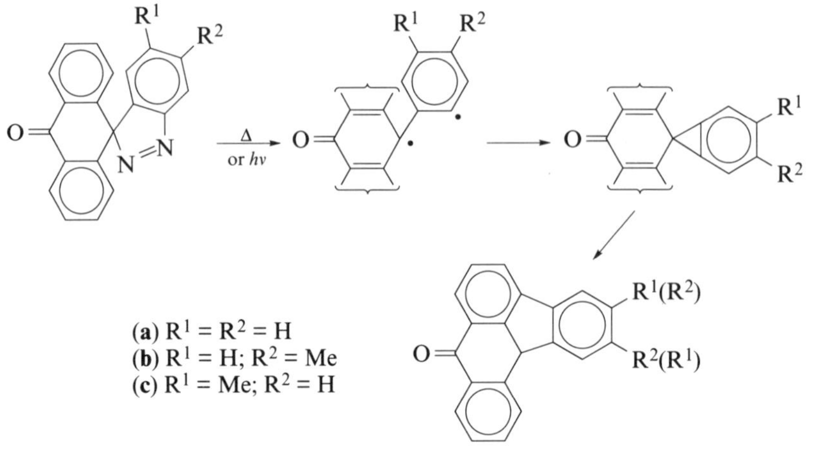

SCHEME 2

The 3*H*-indazole route is limited to 3,3-disubstituted derivatives because the monosubstituted analogues tautomerize to the aromatic 1*H*-isomer (equation 5). Moreover, 3,3-

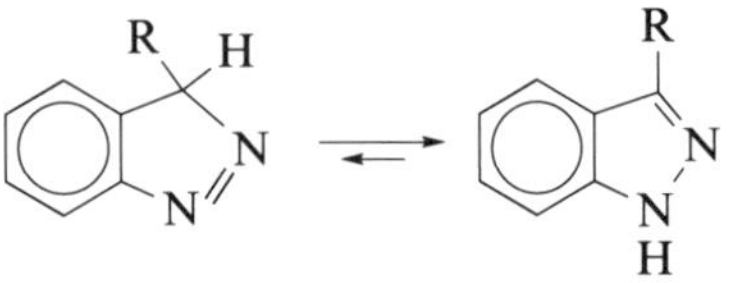

(5)

diaryl derivatives with an available *ortho* hydrogen do not provide cycloproparenes that are capable of isolation because of facile isomerization into a fluorene derivative[26] as shown for **25** (see also Section V.B.5). Thus the indazole route has strict limitations which, when

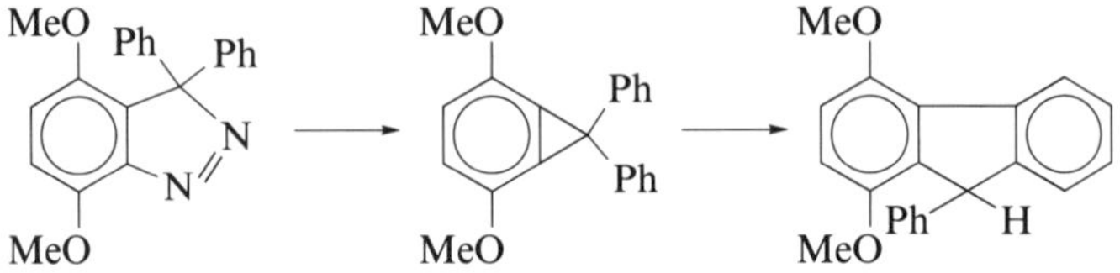

(25)

coupled with the requirements of substrate synthesis, have resulted in it receiving only limited attention. Nonetheless, an especially notable application lies in the synthesis of the only known cyclopropapyridine. Streith and his group[27] have found that photoextrusion of nitrogen from **26** provides the cyclopropa[*c*]pyridine **27** in 24% yield in direct analogy to the Anet and Anet synthesis[19] of the first cycloproparene. Heterocycle **27** is crystalline

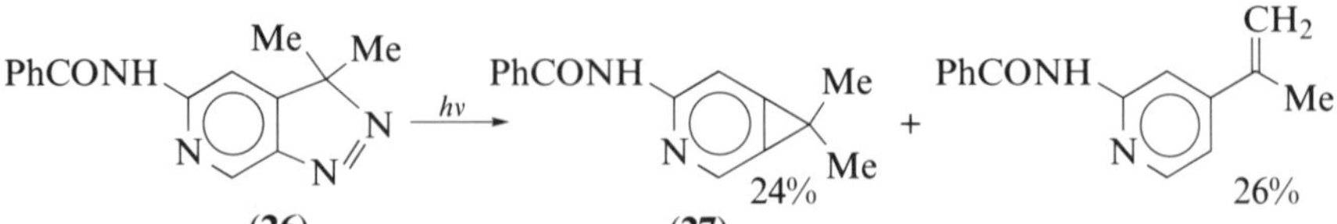

(26) **(27)**

and stable when stored in an inert atmosphere. In comparison, polyazatricycles **28** (equation 6) eject molecular nitrogen from the indazole moiety to give diradicals that can be intercepted with polar reagents or dienes. However, the intervention of cyclopropaheteroarenes was excluded[28]. It seems likely that the fused five-membered heterocycle exerts a pincing effect which forces the diradical centres too far apart for bonding.

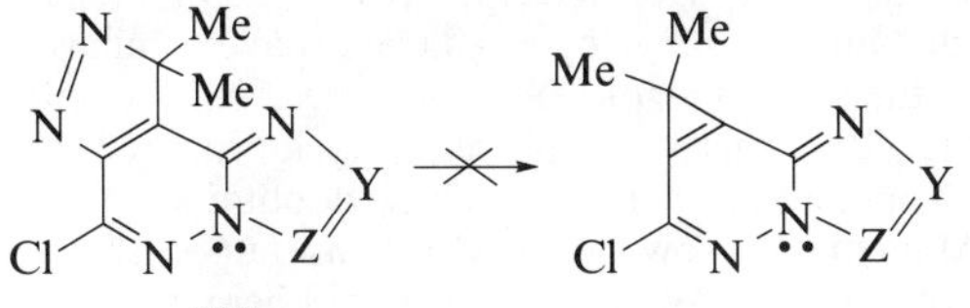

(28)

	Y	Z
(a)	CH	CH
(b)	N	CH
(c)	CH	N

(6)

2. *Of spiro-3*H*-pyrazoles*

The ring contraction of spiro-3*H*-pyrazoles to cycloproparenes was discovered and developed by Dürr and cowokers in 1969. The derivatives that this methodology has provided are more easily obtained than by other methods and they frequently appear in higher yields[29–32]. The general process involves the interaction of a diazocyclopentadiene with an alkyne to deliver what was thought to be a spiro-3*H*-pyrazole, e.g. **29**. Upon photolysis [1,5] carbon shifts lead to a 3*H*-indazole **30** which then ejects nitrogen to give the cycloproparene as discussed above (Scheme 3). In this way one of the few substituted cyclopropa[*a*]naphthalenes[29,30] **31c** and the only known spirocycloproparenes[32] **32a–c** were obtained.

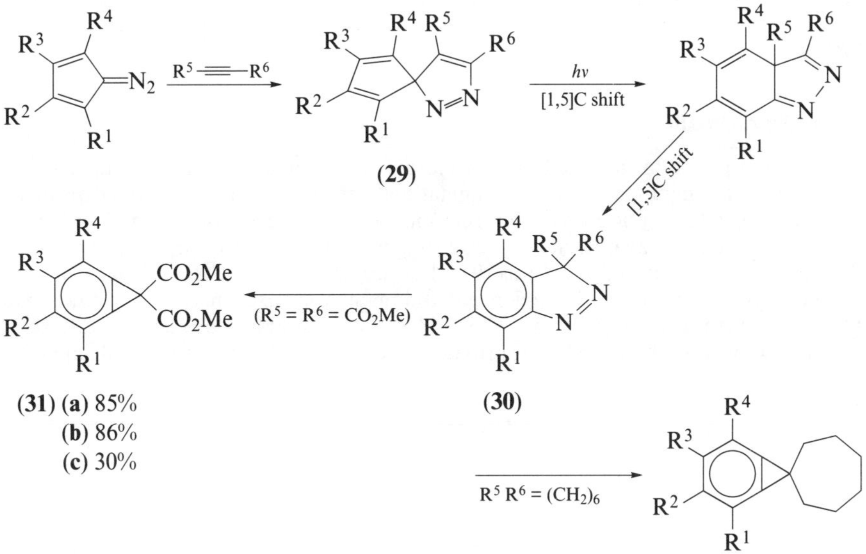

(a) $R^1 = R^2 = R^3 = R^4$ =Ph
(b) $R^1 = R^4$ = Ph; $R^2 = R^3$ = H
(c) R^1R^2 = benzo; $R^3 = R^4$ = H
(d) $R^1R^2 = R^3R^4$ = benzo

SCHEME 3

Much more recently, and with the advantage of sophisticated ^{13}C NMR techniques that have become available since the early 1970s, it has been shown[33] that the spiro-3*H*-pyrazole structure for **29a** is incorrect. It is known that the outcome of diazocyclopentadiene addition to dimethyl acetylenedicarboxylate is dependent upon the five-membered ring substituents[34,35]. In the case at hand, tetraphenyldiazocyclopentadiene adds to the alkyne to give **29a** as a labile product that rearranges under the reaction conditions to the 3*H*-indazole **30a** (Scheme 3); 1,3-di-*t*-butyldiazocyclopentadiene behaves similarly[33]. Thus in the formation of **31a** at least, the spiropyrazole **29a** is *not* the substrate and one must question the nature of the educt (**29** versus **30**) employed in cyclopropaplarene synthesis by the spiro-3*H*-pyrazole route. Nonetheless, there can be little doubt that spirocycle **29d** is the substrate employed by Mataka and coworkers[35,36] because, upon thermolysis, the corresponding indazole **30d** was isolated. What must be noted here is that the thermal reaction did not provide any of the cyclopropa[*l*]phenanthrene **31d**, but neither did independent photolysis of the isolable indazole **30d** in benzene solution; a 9,10-disubstituted phenanthrene is formed from diradical interaction with the solvent (equation 7).

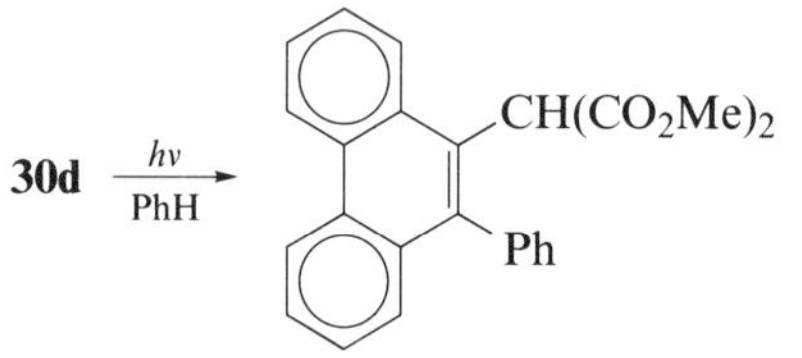

(7)

What has now become clear with this synthetic methodology is that the substrate may not be a spiro-3*H*-pyrazole but a 3*H*-indazole. However, this is of little consequence to the final outcome as the latter is a proposed intermediate in the photorearrangement of the former.

C. By Ring Closure

As an alternative to the diradical ring closure that occurs with the 3*H*-indazole ring contraction, the employment of a 1,3-elimination from an *ortho*, α-disubstituted aromatic has much appeal for cycloproparene formation, not least because of the simplicity of the process and the ready availability of the starting materials. It is not surprising, therefore, that such a report appeared as early as 1974. Radlick and Crawford[37] found that 1-bromo-2-(methoxymethyl)benzene (**33**) underwent lithium–halogen exchanged and cyclization to **1** upon treatment with butyllithium. The yield was a modest but acceptable 30%. In similar vein, the cyclization methodology provided 'rocketene' (**34**) albeit in 5% yield (equation 7a)[38].

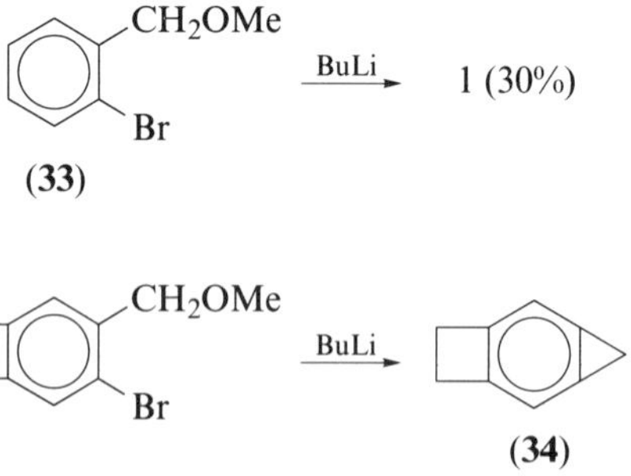

(7a)

Attempts to repeat[39] the synthesis of **1** by this methodology, and to extend the route to mono- and bisannelated derivatives, have been unsuccessful[40,41] even though the malodour of a cycloproparene is often detected. Clearly, the nature of the leaving group is of paramount importance. For the homologous cyclobutarene series, these intramolecular cyclizations provide a viable means of constructing the four-membered ring[42,43]. However, in the present case changing either the methoxy group of **33** for chloride, or the bromide for iodide[42] fails to provide for 1,3-cyclization. In comparison, with tosylate as leaving group[39,44] the yield of **1** improves but not to a viable level (*vide infra*). Attempts to prepare the angular cyclopropacyclobutabenzene **35** from an iodotosylate leads instead to a range of products[41] whose formation cannot exclude the intervention of **35** (equation 7b).

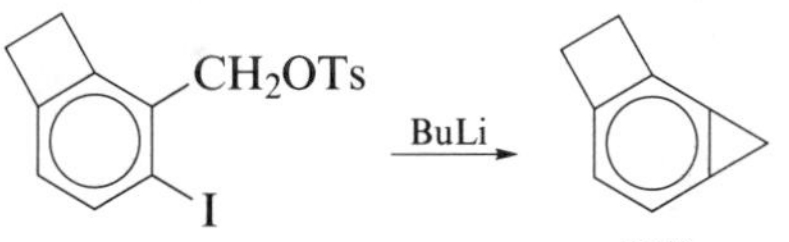

(35)

(7b)

The foregoing discussion makes clear that traditional functionalities are not easily persuaded to give an aromatic species fused to a three-membered ring. The quest to utilize somewhat more exotic substrates has only begun to be explored! McNichols and Stang[45] have recently discovered a two-step pathway that provides rocketene (**34**) in 32% overall yield. Cobalt-catalysed coupling[38] of hexa-1,5-diyne with the alkynylstannane depicted in equation 8 provides the cyclobutarenylstannane **36**. Upon reaction with butyllithium, **36**

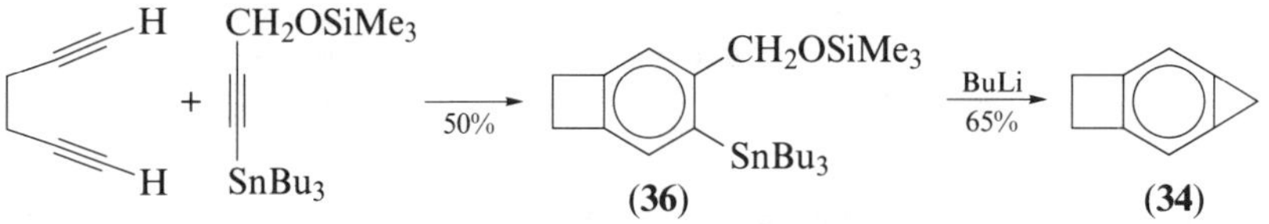

(36) **(34)**

(8)

undergoes 1,3-cyclization to give rocketene in 65% yield. The enhanced leaving ability of trimethylsilyloxy over methoxy is ably demonstrated and its efficacy in the preparation of other cycloproparenes is clearly worthy of examination.

To conclude this section two 'compound specific' reactions are worthy of mention. Firstly, the interaction of atomic carbon with benzene has been found to provide **1** albeit in 11% yield and as one of eleven hydrocarbon products[46]. Secondly, 4,5,7-tri-*t*-butylisobenzofuran has been reported to undergo furan-cyclopropenylaldehyde photorearrangement rather than 4π electrocyclization to the Dewar furan[47]. Steric factors undoubtedly dominate in yielding 2,3,5-tri-*t*-butylcyclopropabenzene-1-carbaldehyde; the product was obtained in solution and characterized by ^1H NMR only.

D. By Flash Vacuum Pyrolysis

The original synthesis of cyclopropabenzene (**1**) appeared[2] in 1965 and stemmed from the elegant methano[10]annulene work of Vogel and his group in Cologne. The synthetic strategy depends upon the facility by which methano[10]annulenes add acetylenic dienophiles to their norcaradiene valence bond isomeric forms **38** (Scheme 4) to give products that correspond to bridged cycloproparenes. Flash vacuum pyrolysis (fvp) effects Alder-Rickert retrodiene cleavage and liberation of the requisite cycloproparene hydrocarbon. Thus dimethyl acetylenedicarboxylate reacts with norcaradiene **37a** to give the *endo*-Diels-Alder product that fragments to **1** (45%) and dimethyl phthalate on flash vacuum pyrolysis[2] (Scheme 4). In like manner, the benzo-fused [10]annulene **37b** leads to

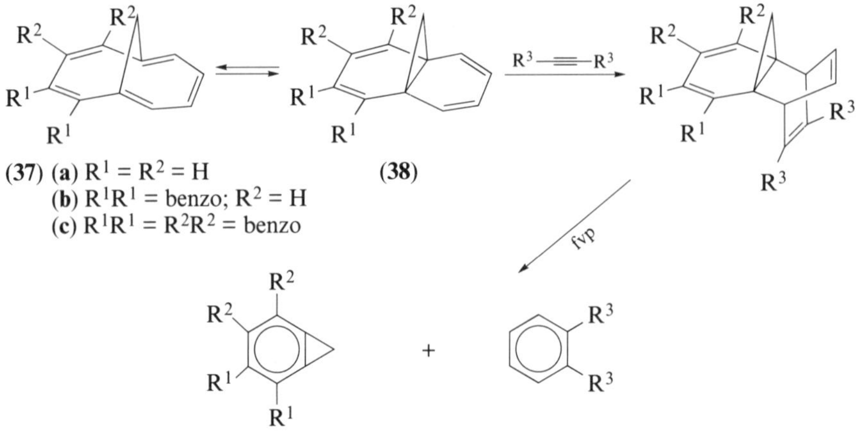

(37) (a) $R^1 = R^2 = H$
(b) R^1R^1 = benzo; $R^2 = H$
(c) $R^1R^1 = R^2R^2$ = benzo

(38)

(1) $R^1 = R^2 = H$
(10) R^1R^1 = benzo; $R^2 = H$
(39) $R^1R^1 = R^2R^2$ = benzo

$R^3 = CO_2Me$ or CN

SCHEME 4

cyclopropa[*a*]naphthalene (**10**). Here dicyanoacetylene as educt gives **10** from fvp in 68% yield, but with dimethyl acetylenedicarboxylate the yield is increased to an excellent 83%[48]. The compound is noticeably less stable than its linear isomer and decomposes on melting at 20 °C. The cycloproparene formed from fusion across the more olefin-like 9,10-double bond of phenanthrene (cyclopropa[*l*]phenanthrene, **39**) is also available[49] by fvp methods (Scheme 4) and is even less stable; it decomposes over a few days in the solid state at –78 °C.

While the synthesis of the requisite methano[10]annulenes is not always easy, the $[_\sigma 2 + _\sigma 2 + _\pi 2]$ cycloreversion has provided the essential structural variations of cyclopropa fusion into an aromatic moiety that complement other methodologies (Section II.E). Moreover, the sequence lends itself to further exploitation. Thus Vogel's group[50] has obtained the dicyanoacetylene adduct **40** which provides phenylacetylene on fvp (equation 9). It is likely that methylenecyclopropabenzene (**9**) is initially formed as a molecule too

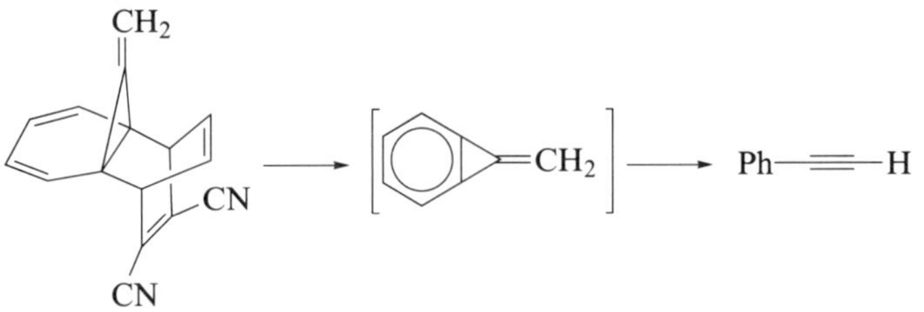

(40) (9) (9)

reactive for isolation. Application[51] of the flash vacuum pyrolysis method to dione **41** does not provide product capable of isolation. However, cyclopropabenzene-2,5-dione (**42**) is likely formed because an exchange Diels-Alder adduct is formed (12%) when the pyrolysis is performed in molten anthracene (equation 10). Finally, it is noteworthy that attempts[52] to provide the angular dicyclopropabenzene **44** have failed as the bismethanoannulene **43** adds dicyanoacetylene in the wrong sense (Scheme 5).

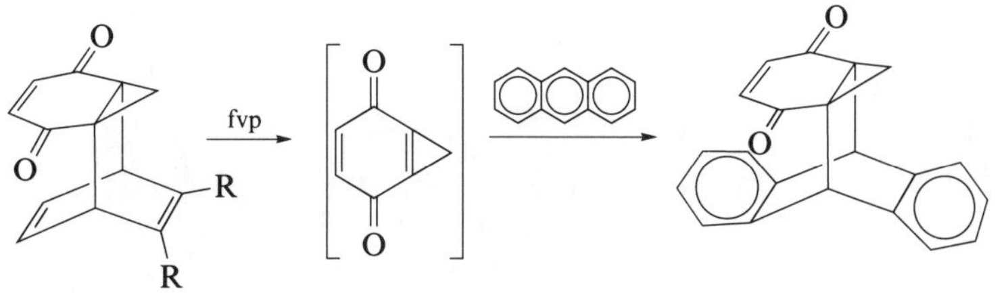

(**41**) R = CO_2Me (**42**) (10)

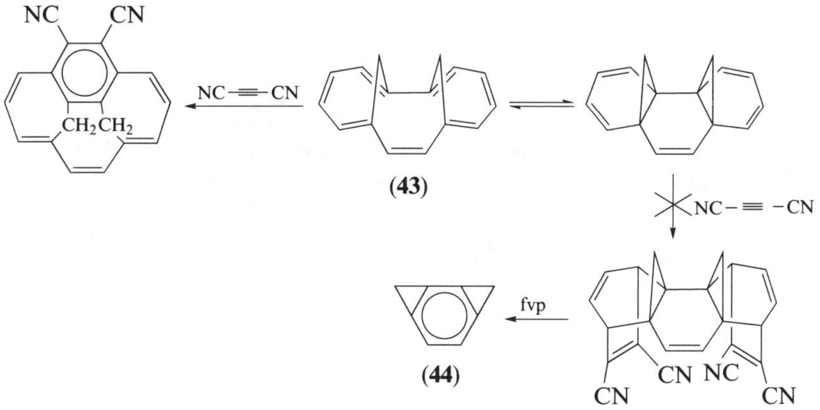

SCHEME 5

E. By Aromatization

The most efficacious routes to the cycloproparenes have proved to be those in which the molecular framework is constructed and aromaticity subsequently introduced. This last step may involve straightforward 1,2-eliminations or the more complex removal of an oxygen bridge. The former method was employed in many of the early studies while the latter is more recent in origin. However, there can be little doubt that the pathway to gain the greatest recognition is that devised by Billups and coworkers[53–55] which has provided easy and convenient laboratory syntheses of **1** and **11**.

1. Of 7,7-dihalobicyclo[4.1.0]heptenes

The double dehydrohalogenation of 7,7-dihalobicyclo[4.1.0]heptenes provides *the* method of choice for synthesis of simple linear cycloproparenes unsubstituted at the 1-position. The method was developed by Billups and coworkers[53–55] and it is now aptly referred to as the Billups route. The procedure involves aromatization of a 7,7-dihalobicycloheptene, e.g. **45**, via high-energy ring-fused intermediates as illustrated in Scheme 6 for the formation of **1**. The product from initial dehydrochlorination, bridged cyclopropene **46**, has been intercepted as a Diels-Alder cycloadduct[56], and $C_{(7)}$ of the synthon becomes $C_{(1)}$ of **1** as illustrated by a ^{12}C labelling study (Scheme 6)[57]; the method provides cyclopropa[*b*]naphthalene (**11**) very easily from naphthalene[54]. Much more recently Okazaki and coworkers[58] have found that a change in the phase transfer catalyst used in the preparation of **45** from cyclohexa-1,4-diene, coupled with changes in the procedure for

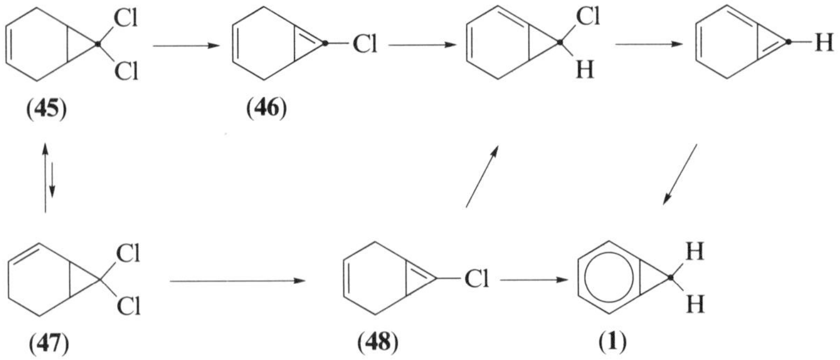

• = ^{12}C [from **45** only]

SCHEME 6

isolation of **1**, have provided product in approximately twice the yield (28–35%) and on a scale that can provide almost 0.5 mole of the highly odouriferous **1**. The procedure has allowed for the synthesis of 2- and 3-methylcyclopropabenzene (27 and 52% yield, respectively)[59], and some of the more novel cycloproparenes obtained this way are depicted in Table 1[54,55,58,60–68]. With the exception of the cyclopropacyclobutabenzene **35**[61,62] and the cyclopropannulenes **51**[67,68], (see below), the compounds depicted arise from a precursor in which the 7,7-dichlorobicyclo[4.1.0]hept-3-enyl moiety is in place. For the cyclopropannulenes **51** (Table 1) the immediate precursor has the 3,4-double bond replaced by a conjugated triene as depicted by **52** which successfully delivers **51a**; the elimination/

TABLE 1. Cycloproparenes from 7,7-dichlorobicyclo[4.1.0]heptenes

Compound	Yield(%)	Ref.	Compound	Yield(%)	Ref.
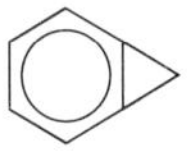 **(1)**	28–35	55,58	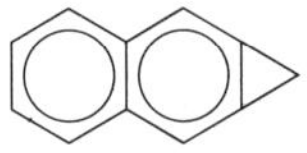 **(11)**	38	54,63
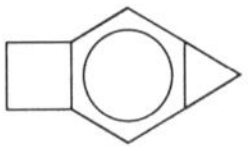 **(34)**	20–40	60,61	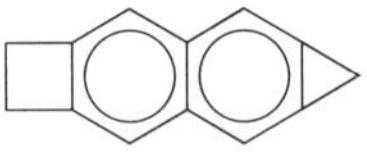 **(49)**	50	61,64
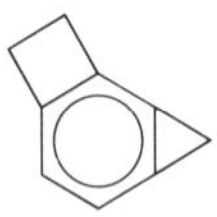 **(35)**	4–10	61,62	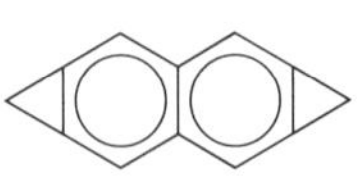 **(50)**	33	65,66
			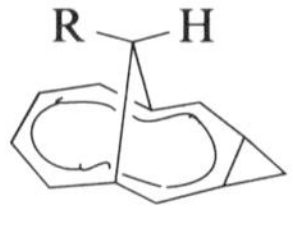 **(51)** (**a**) R = H	81	67
			(**b**) R = Cl	15	68

rearrangement sequences of Scheme 6 are presumed to apply. While **52** is the precise substrate employed by Vogel and Sombroek[67] it can be replaced by the more easily accessible tetracycle **53a** that undergoes base-induction triple dehydrochlorination to this same compound under the reaction conditions. In this way **53b** also affords **51b** albeit in 15% yield[68].

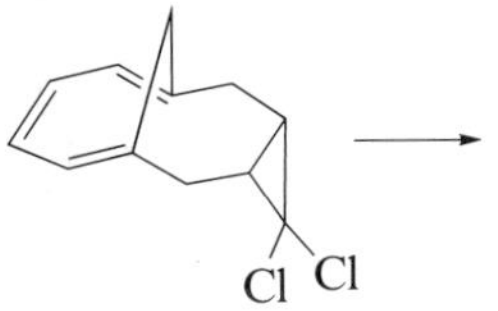

51a

(52)

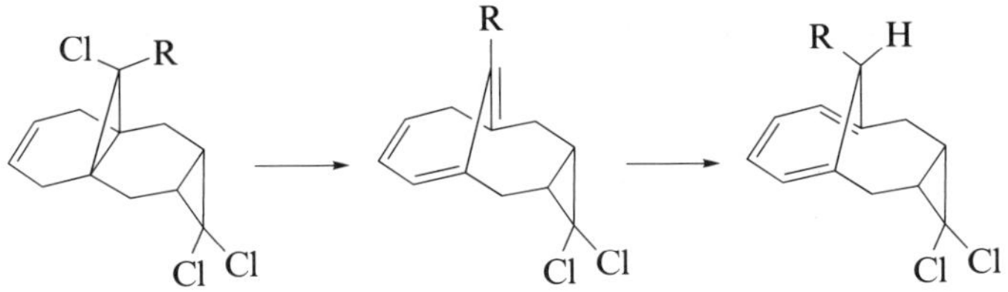

(53) (**a**) R = H
(**b**) R = Cl

It has been shown that neither the position of unsaturation nor its presence are prerequisites for the formation of **1** from a bicyclo[4.1.0]heptane ring system. Thus the 2,3-olefin **47** depicted in Scheme 6 affords **1** via **45** and/or **48** in comparable yield to that from **45** although the reaction is somewhat slower[61,69]. In like manner the presence of an appropriate number of leaving groups within the bicycloheptane allows for cycloproparene formation. Thus 2-bromo-7,7-dichlorobicyco[4.1.0]heptane loses three moles of hydrogen halide upon treatment with *t*-BuOK to give **1** in 33% yield[70]. In all probability the reaction proceeds via olefin **47**. With four halo substituents present **54** and **56** provide halogen-containing cyclopropabenzenes[71–75]. The 3,4,7,7-tetrahalides **54** give rise to the 3-halocyclopropabenzenes **55** in almost 50% yield and it is the $C_{(3)}$ [or $C_{(4)}$] substituent of **54** that is retained and $C_{(7)}$ becomes $C_{(1)}$ of **55** as shown by a labelling study[71]. Clearly this triple elimination proceeds without measureable skeletal rearrangement and a pathway analogous to that of Scheme 6 is likely. This is not the case with the 2,3,7,7-tetrahalides **56**. Whereas **56a,b** can provide only a single 2-halocyclopropabenzene **57** (48 and 28%, respectively)[71–74],

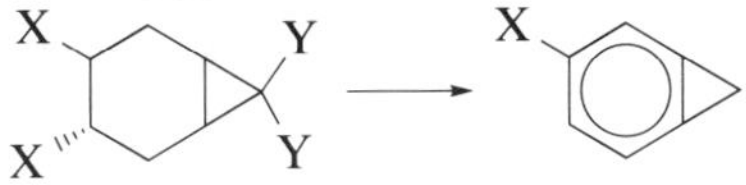

(54) **(55)**

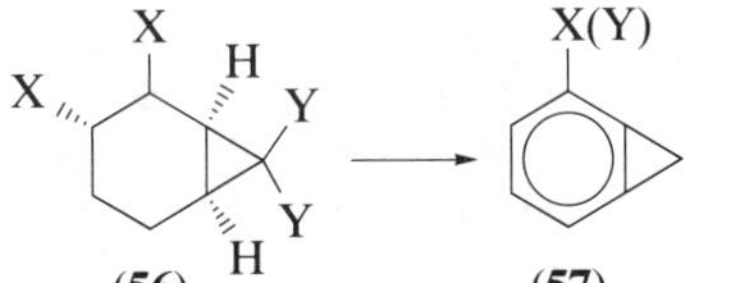

(56) **(57)**

	X	Y
(**a**)	Br	Br
(**b**)	Cl	Cl
(**c**)	Br	Cl

the 'mixed' halide **56c** gives rise[72] to a mixture of 2-bromo- and 2-chlorocyclopropabenzene (**57a** and **57b**), respectively, in which the skeletally rearranged isomer **57b** is the *major* product. Labelling studies[71] have shown that **56c** yields bromide **57a** without skeletal rearrangement whereas the $C_{(2)}$—Cl moiety of **57b** arises from migration of the $C_{(7)}$-labelled centre with one of its attendant chlorine substituents. While the pathway to **55** likely involves the mechanism of Scheme 6, the route for elimination/rearrangement from **56c** is less clear[71] and a discussion is beyond the scope of the present chapter.

While **1** is formed by loss of HCl from bicycloheptene **47**, the presence of such 'angular' rather than 'linear' three-membered ring fusion (with respect to the double bond) is not always compatible with cycloproparene formation. Thus tricycle **58**, with a tetrasubstituted double bond, fails to yield **35** while its disubstituted isomer does give the desired product, but only in low (4–10%) yield (equation 11)[61,62]. Similar problems have been

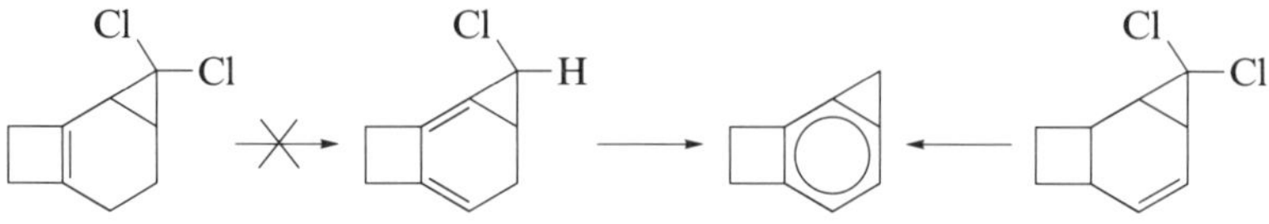

(58) **(35)** (11)

encountered in the attempted construction of other cycloproparenes (**10**[69], cyclopropa[*a*]anthracene and dicyclopropa[*a*,*g*]naphthalene[61]) when the endothermic migration of a tetrasubstituted olefinic linkage is required during the double dehydrohalogenation sequence. Moreover, the 'classical' methodology involving a 3,4-olefin as used for **1** and **11** fails for 1*H*-cyclopropa[*b*]anthracene[61] and 1*H*-cyclopropa[*b*]phenanthrene[76]. These results serve to show that in large measure the Billups route is limited to cyclopropabenzenes and -[*b*]naphthalenes.

The 7,7-dihalobicycloheptene route appears easily adaptable to the synthesis of 6π 5 atom cyclopropa[*c*]heteroarenes **61** as the three-membered ring is suitably located and the requisite precursors—the dichlorocarbene addition products of, for example, 2,5-dihydrofuran—are easily accessible. Unfortunately, treatment of halides **59** with *t*-BuOK fails to provide any evidence[56,77–79] for sought after **61**. Dehydrochlorination does occur but the strained 1,3-bridged cyclopropenes **60** ring expand to carbene or add a nucleophile faster than rearrangement and loss of a second molar equivalent of HCl. The precise outcome of these reactions is very dependent upon the nature of the ring system as the detailed studies show[56,78,79].

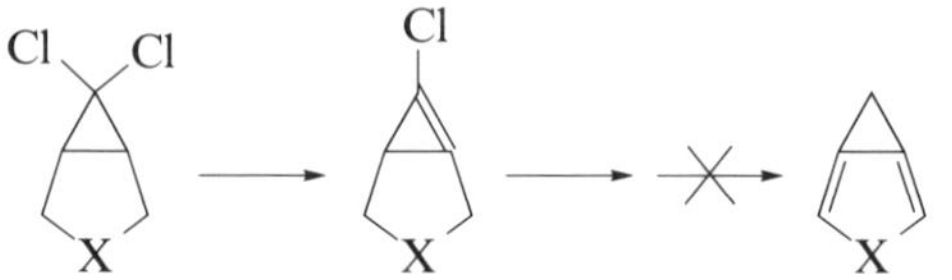

(59) **(60)** **(61)**

(**a**) X = S
(**b**) X = O
(**c**) X = CH_2

Both the 7,7-di-[80,81] or 7-monochlorocyclopropanes[82,83] **62** appeared to be plausible substrates for the synthesis of cyclopropa[*l*]phenanthrenes by way of monodehydrochlorination and rearrangement of the initially formed 1,3-bridged cyclopropene. However, the rearrangement does not occur and products arise from interception of the strained cyclopropene by nucleophile and from ring opening to a phenanthryl carbene (equation 12). In the context of cycloproparene synthesis it is now clear that bicyclo[4.1.0]hepta-1,3,6-trienes aromatize whereas their 2,4,6-isomers do not (Scheme 7)[8].

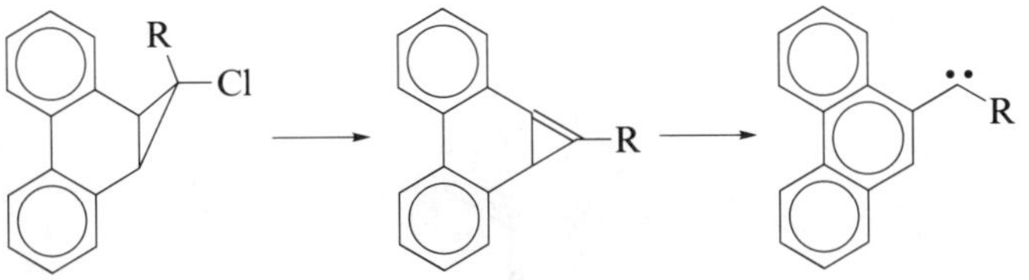

(12)

(62) (**a**) R = Cl
(**b**) R = H

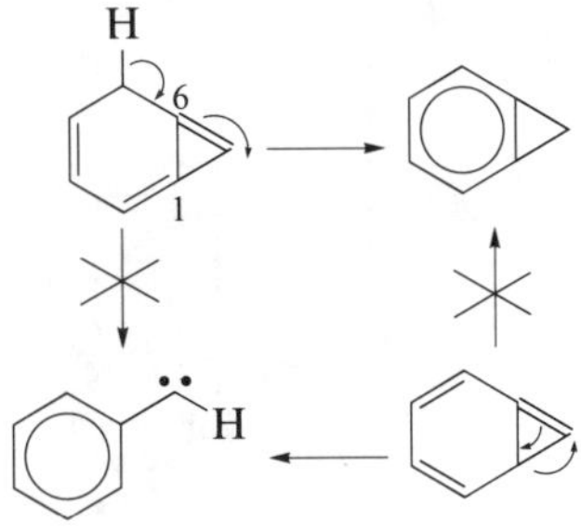

SCHEME 7

2. Of 1,6-dihalobicyclo[4.1.0]heptenes

Some of the earliest developments in cycloproparene chemistry employed the double dehydrochlorination of 1,6-dihalobicyclo[4.1.0]hept-3-enes and, because of the ready availability of perhalocyclopropenes, these were often targetted towards *gem*-dihalocycloproparenes. Thus 1,1-difluorocyclopropabenzene was obtained[84] from 1,6-dibromo-7,7-difluorobicyclo[4.1.0]hept-3-ene in 1968 and now[85] can be obtained in 50-gram quantities (60% yield) from optimization of the process. By enhancing the acidity of the protons that are removed in the eliminations, the yield of cycloproparene can became almost quantitative as occurs for 1,1-dichloro-2,5-diphenylcyclopropabenzene[86,87] (**63a**) and its naphthalene analogue[88,89] **63b** (equation 13). The requisite bicycloheptene substrates are easily available from Diels-Alder cycloaddition of an appropriate buta-1,3-diene to a (per)halocyclopropene. When the cyclopropene carries bulky flagpole substituents at $C_{(3)}$ the stereochemistry of cycloaddition to an (*E*)-1-substituted buta-1,3-diene is *exo* and the products have the configuration depicted in equation 13. The $C_{(3)}$ atom of the cyclopropene and the $C_{(1)}$ substituent of the diene become located on opposite faces

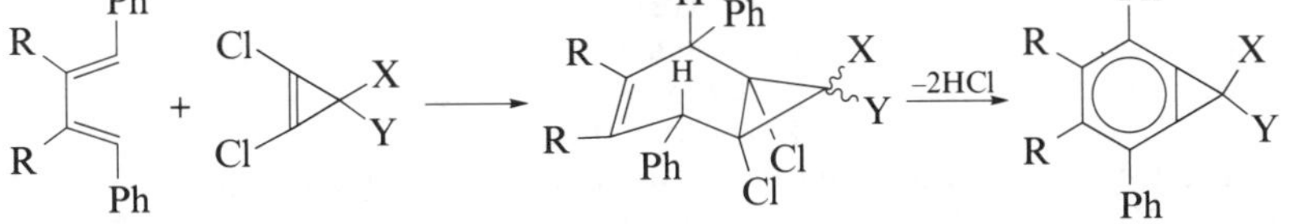

(13)

(63) (**a**) R = H; X = Y = Cl
(**b**) RR = benzo; X = Y = Cl
(**c**) R = H; X = Y = F
(**d**) R = H; X = Cl; Y = F

TABLE 2. Cycloproparenes from 1,6,7,7-tetrahalobicyclo[4.1.0]hept-3-enes

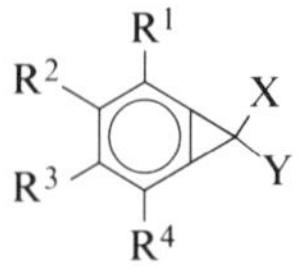

R^1	R^2	R^3	R^4	X	Y	Yield(%)	Ref.
H	H	H	H	F	F	80	84,85
H	H	H	H	Cl	F	37	92,93
H	Me	Me	H	F	F	78	94
H	Ph	Ph	H	F	F	58	94
H	$—(CH_2)_2—$		H	F	F	60	94
H	$—(CH_2)_3—$		H	F	F	86	94
H	$—(CH_2)_4—$		H	F	F	65	94
H	-benzo-[a]		H	F	F	70	95,96
H	-benzo-[a]		H	Cl	Cl	66	95,96
H	-2,3-naphtho-[b]		H	F	F	70	97,98
H	-2,3-naphtho-[b]		H	Cl	Cl	88	97,98
Ph	H	H	Ph	F	F	77	93
Ph	H	H	Ph	Cl	F	60	93
Ph	H	H	Ph	Cl	Cl	93	86,87
Ph	-benzo-[a]		Ph	Cl	Cl	83	88,89

[a]Benzo fusion implies a 1*H*-cyclopropa[*b*]naphthalene.
[b]2,3-Naptho fusion implies a 1*H*-cyclopropa[*b*]anthracene.

of the six-membered ring[90,91] and dehydrohalogenation proceeds in the usual antiperiplanar sense. The methodology has been employed and elegantly extended by Müller and coworkers[92–98] by using a variety of novel and effective 2,3-bridged buta-1,3-dienes, the outcomes of which are depicted in Table 2. The elimination sequence is not always successful[99] and antiperiplanar opening of the three-membered ring to give a dichloromethyl anion has been recorded[100].

The most significant advance in cycloproparene synthesis in the past decade has been provided by a convenient synthesis of 1-bromo-2-chlorocyclopropene as reported by Billups and coworkers[76]. This compound provides good access to an intriguing range of less common cycloproparenes because it is an effective dienophile in cycloadditions with orthoquinodimethanes and other 2,3-bridged buta-1,3-dienes. Thus Billups and coworkers[76,101–104] and Müller and his group[105,106] have been able to prepare most of the fascinating compounds depicted in Table 3 that include the cyclopropaquinoline[106].

By using 1,2- dibromocyclopropene as the dienophile and the removal of 2HBr, Anthony and Wege have prepared 5*H*-cyclopropa[*f*]isobenzofuran and -thiophene (Table 3) and demonstrated that the compounds have limited stability[78,107]. Moreover, these same authors have provided evidence for the existence of a cyclopropa[*c*]thiophene. The dibromothiabicycle **64** can lose HBr in only one direction and, with *t*-BuOK, elimination occurs but no discernable product ensues. However, in the presence of added isobenzofuran the bisadduct **65** (equation 14) is obtained; the authors have noted that sequential elimination–addition that would circumvent cyclopropathiophene formation cannot be excluded. Finally, it should be noted that whilst benzene has been strained by the fusion of a three-membered ring it has not yet been simultaneously subjected to such in-plane ring strain together with out-of-plane bending. Garratt and his group have prepared the cyclophane precursors **66** but preliminary experiments provided only intractable materials[108,109]; the study also included the 1,2-bis(trimethylsilyl)-3,3-dibromocyclopropane analogue of **66a**.

TABLE 3. Cycloproparenes from 1-bromo-6-chlorobicyclo[4.1.0]heptenes

Compound	Yield (%)	Ref.	Compound	Yield (%)	Ref.
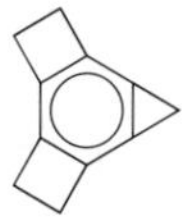	53	102	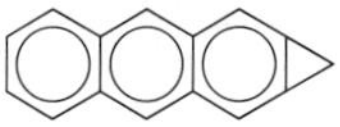	41	104
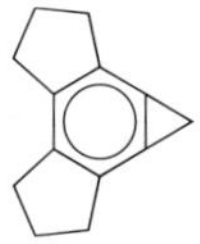	55	102	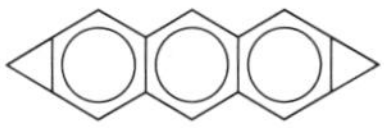	38	103
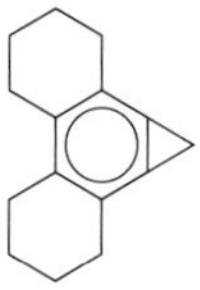	83	102	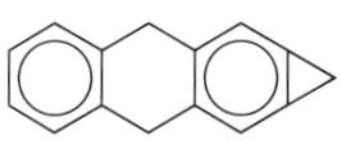	77	104
			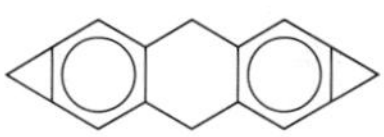	31	103
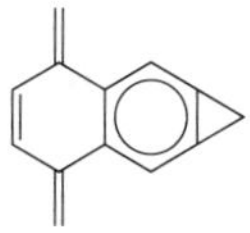	57	103	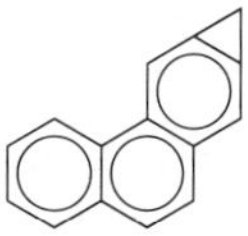	89	76
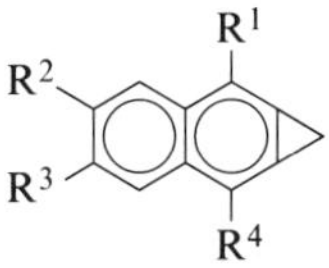			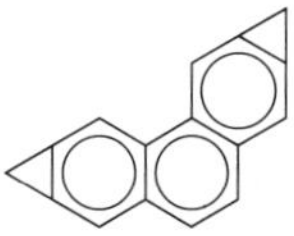	84	103
R^1 R^2 R^3 R^4			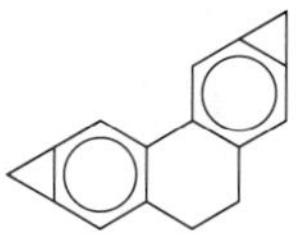	77	103
H H H H	98	105			
H Me H H	100	105			
Me Me H H	85	105			
Me H H Me	100	105			
H $-CH_2-$ H	52	103	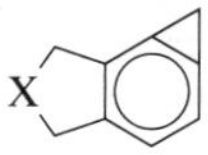		
	82	106	X = CH_2CH_2	80	101
			X = CH_2	44	101
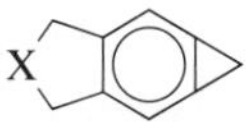			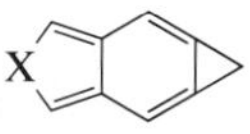		
X = CH_2CH_2	96	101	X = O	55[a]	107
X = CH_2	83	101	X = S	81[a]	107

[a]The substrate was a bridged 1,6-dibromocyclopropane.

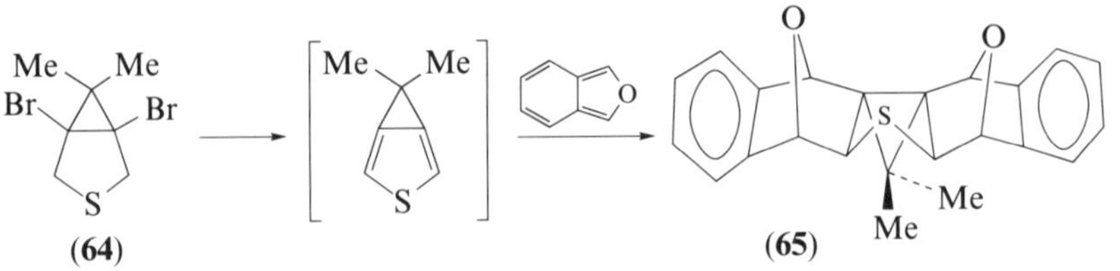

(64) (65) (14)

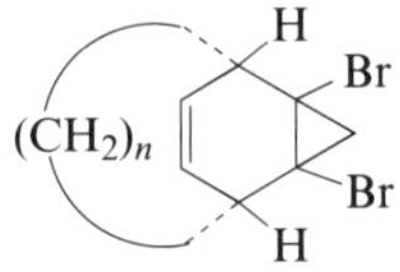

(66) (a) $n = 5$
(b) $n = 6$

3. Of 1-halo and other 1,6-disubstituted bicyclo[4.1.0]heptenes

As was noted in the foregoing discussion, modification of the Billups procedure to angular cycloproparenes was unsuccessful often because of the required migration of a benzenoid double bond in the elimination sequence[62,69,81]. The simple relocation of the two halogen substituents to the bridge positions of the ring fused cycloproparene progenitor is inappropriate, but the removal of hydrogen halide across the bridge bond can lead to a cycloproparene. This ingenious method devised by Müller, Thi and coworkers[110–112] has provided the *gem*-dihalocyclopropa[*a*]naphthalenes **68** (equation 15). Thus dihalocar-

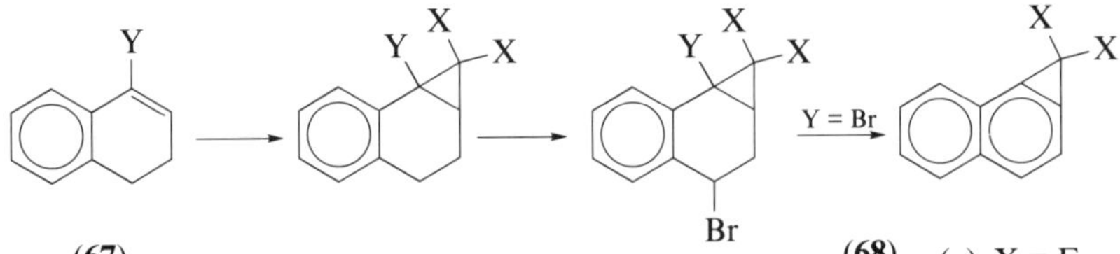

(67) (68) (15)

(a) X = F
(b) X = Cl
(10) X = H

benes add to a range of dihydronaphthalenes **67** to give products that can be brominated at the benzylic position. However, it is only with Y = Br that elimination with base is effective and yields **68**. The compounds can be obtained pure in solution but they are unstable above –30 °C; **68b** has been reductively dechlorinated *in situ* to **10**, which is also difficult to handle and obtain pure[112].

Another notable attempt to utilize dehydrobromination across the bridge bond of ring fused cyclopropanes was made by Brinker, Wüster and Maas[113] to generate dicycloproparene **71**. Although the compound was not obtained, the elegant chemistry of Scheme 8 emerged. The dibromotetrahydro derivative **69**, cunningly obtained by debrominative intramolecular coupling, readily ejects HBr with *t*-BuOK; however, **71** could not be detected between –70 and +20 °C. With one molar equivalent of the base and in the presence of diphenylisobenzofuran (dpibf) cyclopropene **70** is captured as the Diels-Alder adduct **72**. With excess base and dpibf, three bisadducts ensure and these same compounds also result from separate treatment of **72** under the reaction conditions (Scheme 8). It was concluded

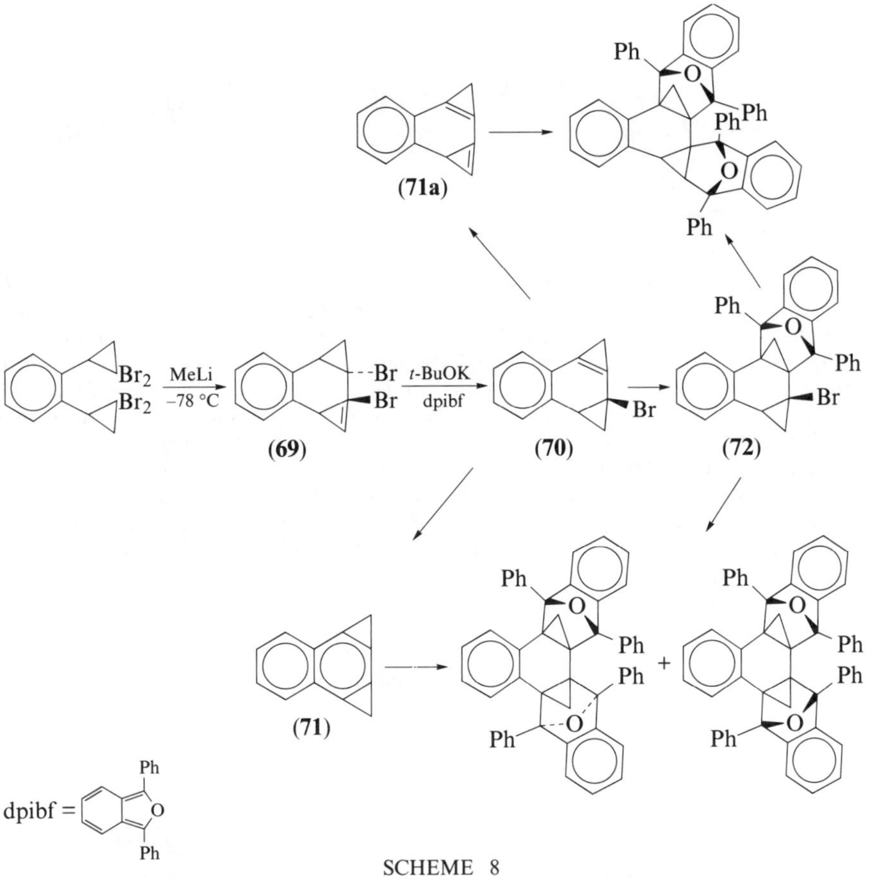

SCHEME 8

that in the dehydrobromination of **69** the constrained cyclopropene **70** was trapped as **72** more rapidly than it could lose HBr to give either **71** or its non-aromatised isomer (**71a**). Dicycloproparene **71** is unlikely to be involved and, at the time of writing, there is no known cycloproparene that contains two three-membered rings fused to a *single* aromatic ring.

The removal of substituents other than halogens from the 1- and 6-positions of a bicyclo[4.1.0]hept-3-ene has not been effective, but the methodology can be applied to specific non-linear homologues. Attempts to mirror the successful synthesis of cyclobutarenes from lead(IV) didecarboxylation of bicyclo[4.2.0]heptene-1,6-dicarboxylic acids to cycloproparenes is frustrated by the intermediate bridgehead bicycloheptene carbocation; the bridged cyclopropyl cation is captured by the remaining acid function and isobenzofuran-1(3*H*)-ones are formed[114]. In like manner, while utilization of **67** (Y = SPh or OAc) does provide carbene adducts as depicted by equation 15, neither these nor the derived sulphone (PhS→$PhSO_2$) aromatize to the corresponding cyclopropa[*a*]-naphthalenes[112]. In comparison, the only successful pathway to cyclopropa[*l*]-phenanthrenes is by way of a *syn*-elimination (see also Section II .D above).

The addition of dichlorocarbene to 9-phenylselenophenanthrene gives adduct (35%) that is easily oxidized to the selenoxide **73**. In solution at ambient temperature **73** loses

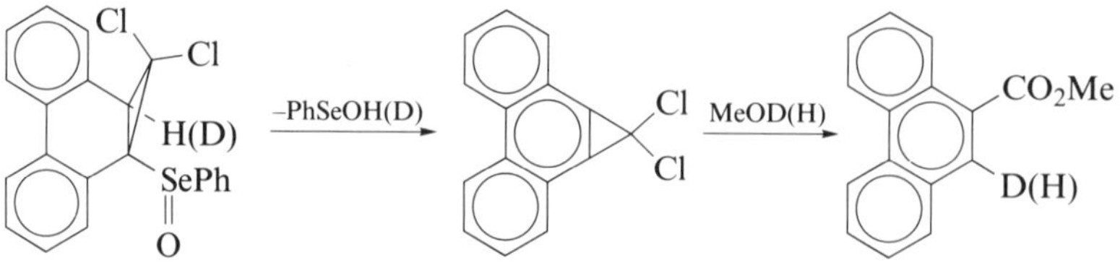

(73) **(74)**

benzeneselenenic acid (PhSeOH) and the dichlorocycloproparene **74** is formed[115,116]. As *gem*-dihalocycloproparenes are unstable and undergo facile opening of the three-membered ring (see Section V.B.1), **74** does not survive the reaction conditions but opens to provide an ester (equation 16). Complementary labelling studies confirm an intermediate with the symmetry of **74**. An extension to the $C_{(1)}$ unsubstituted analogue of **73**, viz. **75a**, has been frustrated by the difficulty of synthesis of the compound[49,117,118]. Nonetheless, Müller and coworkers have shown[118] that selenoxide loss does occur but that 1*H*-cyclopropa[*l*]phenanthrene (**39**) is not isolated; 9-hydroxymethylphenanthrene is formed instead, probably via opening of **39**. Definitive evidence for the formation of **39** has come from the ejection of dimethyl selenide[49,117] from ylide **75b** and of dimethyl sulphide[118] from **75c** (Scheme 9). The *syn*-eliminations proceed with proton abstraction from both $C_{(9b)}$ and $C_{(1)}$ to give a mixture of **39** and its $\Delta^{1(1a)}$-isomer **76** that are intercepted by furan as [4+2] cycloadducts **77–79** (40%) in a ratio of 20:13:7. It is reassuring to see that the favoured elimination is loss of the benzylic $H_{(9b)}$ with aromatization to **39**. In the absence of trapping agent no discernable products could be isolated. This is not too surprising when it is remembered that synthesis by fvp methods (Section II.D) gives **39** stable in the solid state for only a few days at –78 °C[49].

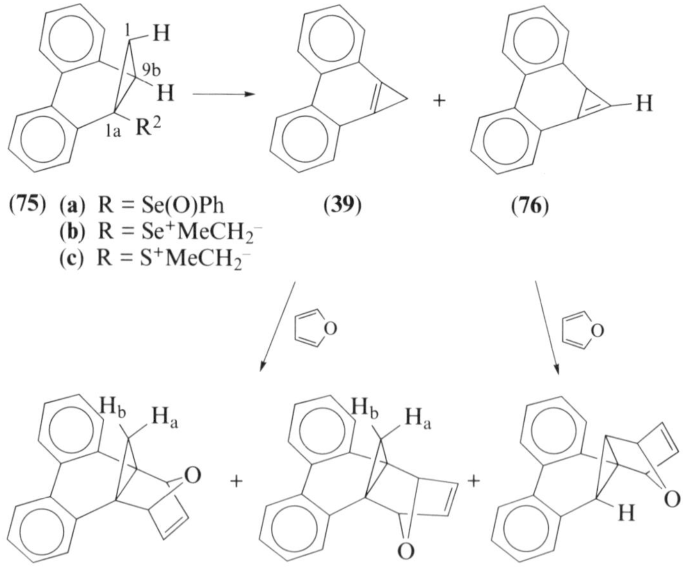

(75) **(a)** R = Se(O)Ph
(b) R = $Se^+MeCH_2^-$
(c) R = $S^+MeCH_2^-$

(39) **(76)**

(77) **(78)** **(79)**

SCHEME 9

4. Of oxygen-bridged 1,6-dihalobicyclo[4.1.0]heptenes

The concept of simultaneously removing the oxygen bridge of a 1,4-epoxybicycloheptene and the $C_{(1)}/C_{(6)}$ substituents was addressed in the early 1970s. Thus attempted dehydration of **80a** could not be brought about without cyclopropane ring opening[119,120]. More recently the 1-aza substituted derivative **80b** was addressed, but again with an unsuccessful outcome[121]. However, in 1986 Müller and Schaller[122] reported that low valent titanium[123] ($TiCl_3/LiAlH_4$) was effective in removing both the oxygen bridge and the halogen substituents of **80c** (equation 17) to give the desired cyclopropa[*b*]naphthalene in

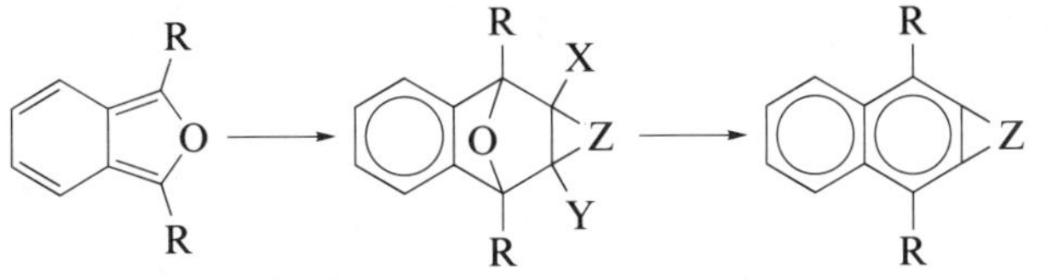

(17)

(**80**) (**a**) R = Ph; X = Y = H; Z = CH_2
(**b**) R = X = Y = H; Z = NR′
(**c**) R = Ph; X = Cl; Y = Br; Z = CH_2
(**d**) R = H; X = Cl; Y = Br; Z = CH_2

72% yield[122,124]. The details of the aromatization step are not well understood. It has been shown that treatment of **80c** with BuLi only effects lithiation at $C_{(1)}$ and thus some activation of the bridge oxygen atom by the titanium is likely. Although parent **11** can be obtained from **80d** the yield is noticeably lower (15%) and accompanied by 2- and 2,3-disubstituted naphthalenes that do not arise from ring opening of **11**. By employing $TiCl_3/MeLi$[124] the yield of **11** is increased to 60%. The low valent titanium sequence is also effective[124,125] for the cyclopropa[*g*]isoquinoline **81** but *not* for the 2,7-unsubstituted parent compound. It seems clear that the low valent titanium methodology requires that the centres to which the oxygen bridge is attached be activated and that the method of generation of the active species is rather important. Moreover, these latter syntheses illustrate further the utility of 1-bromo-2-chlorocyclopropene in cycloproparene synthesis.

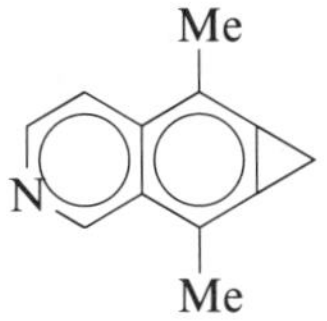

(**81**)

F. Oxocycloproparenes (Benzocyclopropenones)

The suggestion that oxocyclopropabenzene (benzocyclopropenone) (**8**), or more likely its open form, was involved as a reactive molecule in thermal decarboxylation of phthalic anhydride[126,127] and the didecarbonylation of indantrione[128] dates almost to the time of the first[19] cyclopropabenzene synthesis. It was only a few years later that the existence of **8** in solution was established[129–131]. Thus, upon photolysis in methanol, **82a,b** give[131] benzoate esters and, since methyl *p*-chlorobenzoate (42%) is the sole ester from **82b**, the rearrangement must proceed via an intermediate with the symmetry of **8** (Scheme 10). In like manner, lead(IV) oxidation of a range of aminotriazinones **83** provide rearranged and unrearranged esters[129,130]; the yields of the rearranged products (7–13%) provide a minimum estimate of the proportion of the reaction that proceeds through the oxocyclopropabenzene.

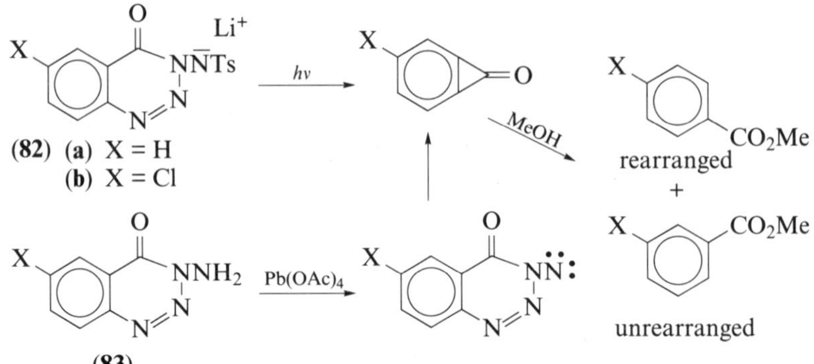

SCHEME 10

The photodecomposition of phthalic anhydride has received detailed attention[132] and **8** has been shown to be a definite but minor product with irradiation at 308 nm; the major path involves CO and CO_2 ejection to benzyne. Indeed, the use of low temperature photochemical methods to provide **8** has much appeal as equation 18 and the sequences of Scheme 11 attest. The deazetation[133] of **84** probably proceeds via carbene **85** which under-

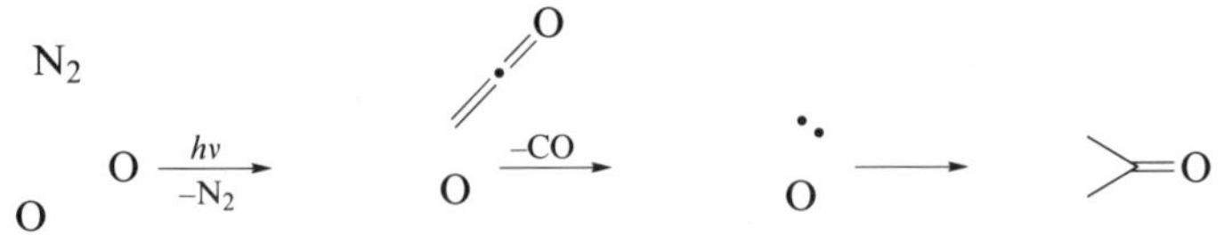

(18)

(84) **(85)** **(8)**

goes a Wolf-like rearrangement to **8**, a pathway that is not followed by the carbocyclic analogue with CH_2 instead of O in **85**[134,135]. Similar loss of carbon monoxide from **86** also provides[136] **8** (path *a*, Scheme 11) as a labile material that undergoes further photodecar-

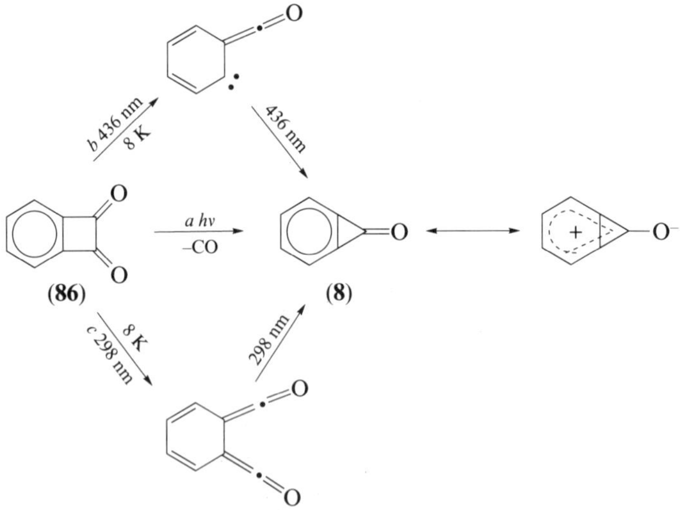

SCHEME 11

bonylation to benzyne even at 8 K. Nevertheless, IR data in argon at 8 K were obtained for each sequence[133,136] and the carbonyl stretching frequency recorded at 1838 cm^{-1} plausibly used to support the involvement of **8**. Very recently the photobehaviour of **86** (isolated in solid argon on a sapphire optical window) has been investigated in much detail using sophisticated narrow-band excitation and time-dependent spectroscopy[137,138]. With 2.84 eV (436 nm) irradiation, **86** is transformed through $^1(n_+\pi^*)$ absorption into **8** via a very reactive intermediate, likely the ketenecarbene of path *b*, Scheme 11. Ketone **8** can also be obtained by excitation of **86** to the $\pi\pi^*$ state with 4.17 eV (298 nm) excitation, in this case via the diketene of path *c*, Scheme 11. The carbonyl stretch of **8** was recorded[138] as about 1869 cm^{-1} and the earlier[133,136] 1838 cm^{-1} band could not be confirmed; Radziszewski and coworkers[132] recorded a value of 1852 cm^{-1} in 1992. ^{1}H and ^{13}C NMR data[139] have also been recorded (Section VI.A).

Ketone **8** and the naphthalene homologue can be generated in solution[140] from singlet oxygen addition to various alkylidenecycloproparenes (Section VI.B.2). However, attempts to construct **87** from 5,7-dibromodibenzo[*a,c*]cycloheptadienone were unsuccessful[141] and the only known condensed analogue is **88** that has been matrix isolated prior to photodecarbonylation[142].

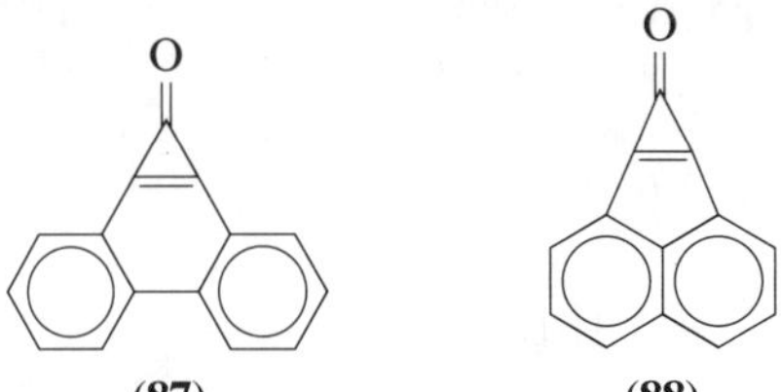

(87) **(88)**

The sole heterocycle derived cycloproparenone to be reported[143] is the thiophene derivative **90** from gas phase thermolysis of anhydride **89**; the product, like **8** and other oxocycloproparenes, was intercepted by hexa- and pentafluoroacetone (equation 19). More recent studies[144] using fvp and photolysis at 12 K have concluded that **90** is probably not involved in the fvp, and is even less likely in the photolysis, as it has comparable energy to its ring-opened isomeric precursor which is expected to interact with the trapping agent in exactly the same way.

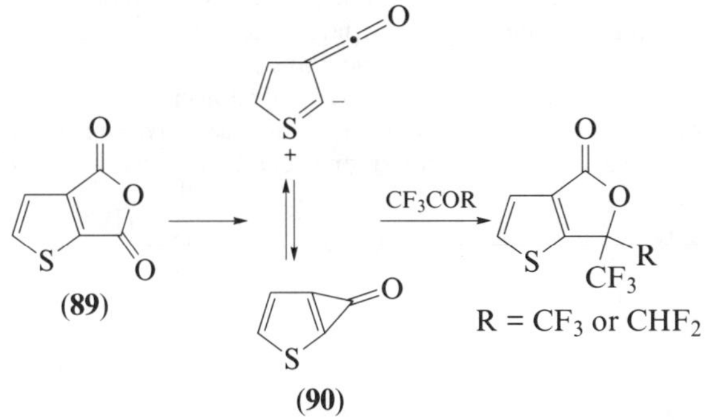

(19)

G. Alkylidenecycloproparenes

The expectation that methylenecyclopropabenzene (**9**) might be formed by ring contraction of methylenephthalide or methylenecoumaranone by fvp[145], or from an appropriate benzocyclobutarene by fvp or photolysis[134,135] in analogy to the obtention of **8** is not borne out (equation 20) as alternative lower-energy pathways are available. A more plausible approach employs a preformed ring system and the Vogel group[50] has found that **40**

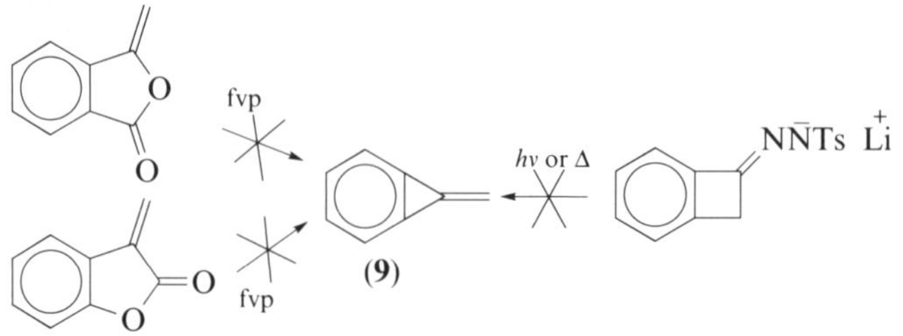

(20)

provides phenylacetylene upon fvp (equation 9, Section II.D). The logical product from Alder-Ricket cleavage is **9** which rearranges under the reaction conditions, but its presence has not been firmly established. There is no report yet available that substantiates the existence of this elusive molecule and it seems likely that **9**, like [3]radialene[146,147] and methylenecyclopropene[148–151], has limited stability. The higher homologue **92** also appears to be unknown even though triamine salt **91** was specifically prepared for subjection to Hofmann elimination (equation 21)[152]; to the best of our knowledge the outcome of this reaction has not been reported.

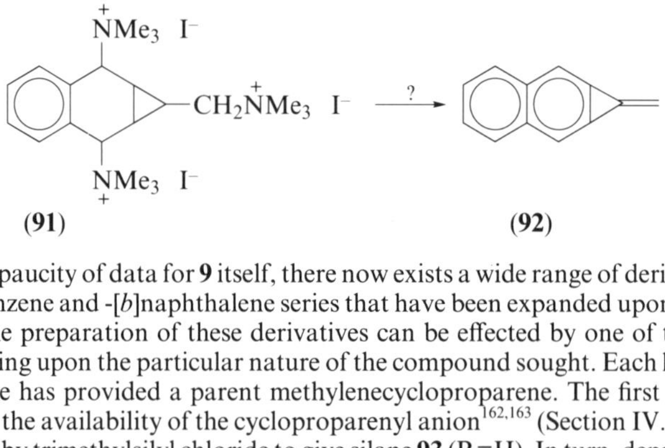

(21)

(91) **(92)**

Despite the paucity of data for **9** itself, there now exists a wide range of derivatives in the cyclopropabenzene and -[*b*]naphthalene series that have been expanded upon since a 1987 account[153]. The preparation of these derivatives can be effected by one of three distinct routes depending upon the particular nature of the compound sought. Each has its limitations and none has provided a parent methylenecycloproparene. The first method[154–161] depends upon the availability of the cycloproparenyl anion[162,163] (Section IV.B) which can be intercepted by trimethylsilyl chloride to give silane **93** (R=H). In turn, deprotonation of **93** at the benzylic position affords the stabilized α-silyl anion that gives alkylidene derivatives **94** (R=H) from interaction with an appropriate carbonyl compound in a Peterson olefination[153] (Scheme 12). The reaction sequence can be effected as a one-pot operation

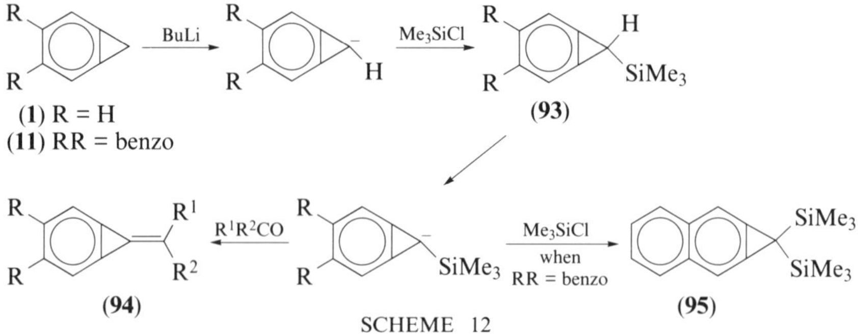

SCHEME 12

from **1** to **94** (R=H), but when applied to **11** modification is needed because monosilyl **93** (RR=benzo) defies isolation and reacts further to give disilyl **95**. The reaction stoichiometry can be easily adjusted to provide **95** in good (66%) yield[155] and, with *t*-BuOK, this is desilylated to the α-silyl anion that undergoes the Peterson process generally, in high yield. The synthetic sequence is limited essentially to non-enolizable aldehydes and ketones because of the strongly basic conditions that are required. The range of alkylidenecycloproparenes available is given in Table 4. To these must be added the first metal complex, 1,1-diphenylmethylene-1*H*-cyclopropa[*b*]naphthalene(η^6-tricarbonylchromium) prepared by Müller and coworkers[164] from the corresponding chromium complexed disilyl (Table 6, Sections III, IV.B and V.A.4) and characterized spectroscopically.

The most recent procedure to be adopted[165] employs the essentials of Scheme 12 but differs simply in the nature of the carbonyl compound employed; aldehydes and ketones are replaced by acid chlorides or acid cyanides. At the time of writing this procedure has been applied to the naphthalene series only to give **96** containing alkyl, electron-donating and electron-withdrawing groups at the exocyclic position (Scheme 13). The specific

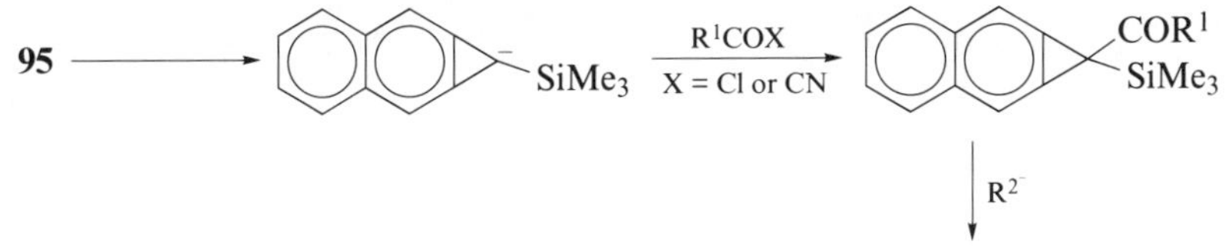

R^1 =	Me	Me	Et	OMe	NMe_2	OMe	NMe_2	NMe_2	NPh_2
R^2 =	H	Me	Et	OMe	OMe	CN	NMe_2	CN	NPh_2
(%)	41	54	55	52	58	62	60	36	65

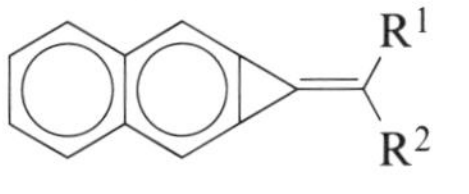

(96)

SCHEME 13

sequence involves classical nucleophilic addition–elimination at the modified carbonyl to give an acylcycloproparene that, in a separate step, adds a nucleophile (usually as its lithium salt) with subsequent Peterson olefination. Since each substituent is introduced separately, unsymmetrical products are easily prepared. However, good leaving groups, e.g. CN, need to introduced in the last step.

The third method for alkylidenecycloproparene synthesis provides triafulvene derivatives by way of cycloproparenylidene dimerization[166]. The reaction is limited in scope by the availability of appropriate carbene (carbenoid) precursors and thus far has provided **98a,b** only (equation 22). Attempts to extend the method into the naphthalene series[167] by using the known[89] naphthalene analogue of **97a** provides a very sensitive (air/light) deep red material that is fluorescent and, while uncharacterized, is likely the analogue of **98a**.

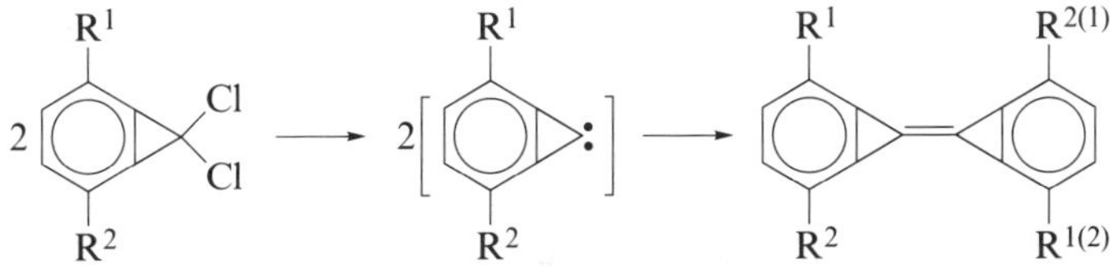

(22)

(97) **(a)**: $R^1 = R^2$ = Ph **(98)**

(b): R^1 = Ph; R^2 = 3-thienyl (structure)

TABLE 4. Alkylidenecycloproparenes from **1** and **11** via Peterson olefination employing aldehydes and ketones

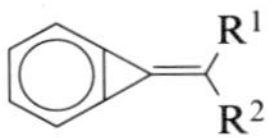

R^1	R^2	Yield (%)	μ	References
H	Ph	13		155
H	4-$MeOC_6H_4$	31		154
Ph	Ph	38	1.0	155,158
Ph	Me	10		155
4-$MeOC_6H_4$	4-$MeOC_6H_4$	34	1.9	155,158
4-$Me_2NC_6H_4$	4-$Me_2NC_6H_4$	24	2.2	157[c]
fluorenylidene		22	2.6	155,158
benzo[*c*]cycloheptatrienylidene[a]		9		155,158
dibenzo[*a,e*]cycloheptatrienylidene[b]		11	1.2	155,158

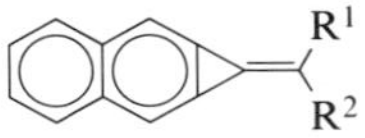

R^1	R^2	Yield (%)	μ	References
H	Ph	68		154,155
H	4-$MeOC_6H_4$	52	1.4	155,158
H	4-$Me_2NC_6H_4$	94	1.8	155,158
H	4-NCC_6H_4	17		159
H	4-$MeSC_6H_4$	61		160
H	4-ClC_6H_4	48		160
H	2,4-$(MeO)_2C_6H_3$	45		160
H	2,4,6-$(MeO)_3C_6H_2$	63		160
H	*t*-Bu	68		155
H	4-pyridyl	39	3.4	159
H	2-pyridyl	50		159
H	2-(1-methyl)pyrrolyl	87		159
Ph	Ph	95	0.4	154,155,158
Ph	CF_3	10		160
4-$O_2NC_6H_4$	4-$O_2NC_6H_4$	84		161
4-$MeOC_6H_4$	4-$MeOC_6H_4$	64	2.4	155,158
4-$Me_2NC_6H_4$	4-$Me_2NC_6H_4$	55	3.0	157[c]
fluorenylidene		96		156,158
benzo[*c*]cycloheptatrienylidene[a]		57		156,158
dibenzo[*a,e*]cycloheptatrienylidene[b]		51		156,158

[a] 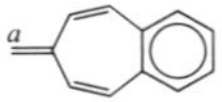 [b] 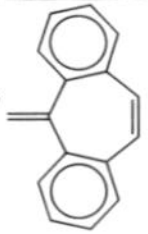

[c]The dipole moments recorded in this communication were inadvertently reversed.

A final inclusion here is quinone **100** that is available in 85% yield from cerium(IV) demethylation of diether **99**; it has yet to be transformed into alkylidene derivatives[168].

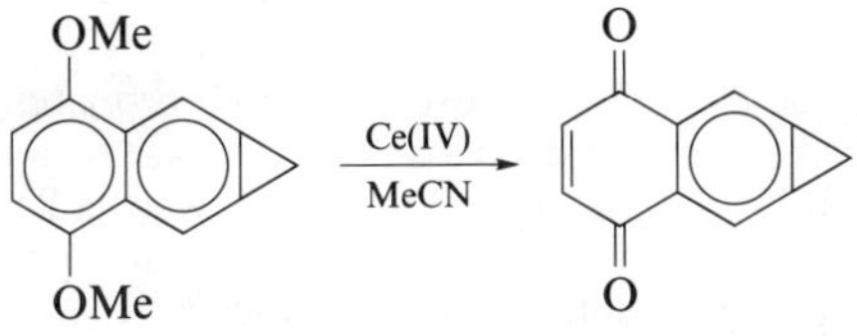

(99) **(100)**

III. PHYSICAL AND THEORETICAL ASPECTS OF CYCLOPROPARENES

Advances in spectroscopic and crystallographic techniques, and in computational methods, have allowed for detailed physical and theoretical analyses of the cycloproparenes. Central to a consideration of this interesting class of compounds is their role in debate over the Mills–Nixon effect—the concept of bond localization within the aromatic ring.

The Mills–Nixon hypothesis[5] had, as its foundation, certain differences in the chemical behaviour of indan (**3**) and tetralin (**4**) from which a localization of the aromatic π-bonds was predicted to occur in the direction depicted by **1a** rather **1b**. The original experimental evidence upon which the effect was based was shown to be erroneous, but calculations at various levels of theory[169–173] indicated that aromatic bond localization should exist and become more pronounced as the size of the annelated ring decreases[7,8]. In essence one can recognize that the D_{6h} structure of benzene has a symmetry such that both Kekulé structures must contribute equally. With C_s tetralin (**4**) (and the lower homologues) no such symmetry requirement exists and ring annelation could induce bond length alternation within the arene nucleus. As the strain imposed by the fused ring increases, the Mills–Nixon effect should increase. The hypothesis has been the subject of considerable discussion and the controversy is far from settled.

Quite recently there has been a series of publications advocating the existence of a significant Mills–Nixon effect in the ring-strained cycloalkarenes primarily due to rehybridization at the bridge carbon atoms[174–183]. These publications complement that of Hiberty and coworkers[184] at the VB SCF level. Moreover, it has been claimed[185] that electron paramagnetic resonance spectroscopic measurements on alkyl substituted cyclobutabenzenes lend support to a definite Mills–Nixon effect. In contrast, Apeloig and coworkers concluded[186] from 3-21G and 3-21G* level calculations on **1** that while there is significant distortion to the σ-molecular framework, **1** has only a small and generally insignificant π-distortion in the direction of **1a**. More recently, Baldridge and Siegel have concluded[187] that *ab initio* calculations must be performed at the level of 6-31G (D) or higher to avoid inferior results; using 6-31G (D), no Mills–Nixon effect was recorded for the simple annelated benzenes including tricyclopropabenzene (**101**). In an elegant study Stanger[188] (using 3-21G, MP2/3-21G and HF/6-31G* basis sets) concluded that the cycloproparenes and the cyclobutarenes will not exhibit measureable bond localization because the systems compensate for the imposed strain by forming 'banana' bonds. The conflict of the theoretical arguments[174–184, 186–189] now focuses on the levels of theory needed and

(101)

whether bond alternation is controlled by σ or π effects, as well as comparisons[190] with the experimentally determined geometries that provide information only about the σ frame. The observed chemistry of the cycloproparenes (Sections V–VII) is compatible with both models.

The structural analyses of various substituted cycloproparenes have continued to provide fascination for the past twenty years[191,192] not least because of the Mills–Nixon hypothesis. Crystallographic data from studies to 1989 were collected in tabular form in the last review[8] and they are not repeated here. Nonetheless, the past five years has seen the elegant work of Boese's group[193] continue to provide accurate structural data from crystal growth and crystal transfer to the diffractometer at low temperature. The range of fascinating compounds studied now includes *all* the known small ring fused cycloproparenes[194–196] and these are collected in Table 5 together with **11**[4] (that has been refined at low temperature[66]) and the dicyclopropanaphthalene **50**[66]. These experimental results show that cycloproparenes exhibit marked bond length and interbond angle distortions particularly about the sites of ring fusion. The three-membered ring is essentially coplanar with the arene nucleus as the tilt angle between the planes of the three- and six-membered rings is only 2–3°; the two such angles in the doubly fused **50** are in opposite senses. The structural parameters that have now been recorded can be accurately reproduced by calculations at the higher *ab initio* levels[196]. This contrasts with semiempirical studies which invariably overestimate grossly the length of the bridge bond[170–173, 197].

Examination of the data of Table 5 shows that in the cyclopropabenzenes it is the bridge bond that is the shortest (1.334–1.363 Å) and that the bonds adjacent to the fusion are next in order (1.363–1.385 Å)[194]. The interbond angle about the sites of fusion (angle α, Table 5) is widened slightly and falls in the range 121.9–126.3° while angle β is compressed to 109.2–117.7°. The obtuse angle for $C_{(1)}$—$C_{(1a)}$—$C_{(2)}$ is dramatically widened to 170.3–176.9°, emphasizing the distortion about the aromatic π-frame by the presence of the three-membered ring; this explains the shortening by 0.02 Å of the $C_{(1)}$—$C_{(1a)}$ bond of **1** compared to cyclopropene. The almost linear value of ε (176.9°) recorded for **35** represents the greatest distortion reported for a trigonal planar carbon atom. The data for **34**, **35** and **102** allow for an appraisal of multiple small-ring annelation. In comparison with both **1** and **2**, rocketene (**34**)[195] has a noticeably narrowed interbond angle at $C_{(2)}$ (109.2°) and, while the three- and four-membered ring σ-bonds are lengthened, it is bond **b** adjacent to the three-membered ring that has the largest increase (0.022 Å) in comparison to **1**. The cyclopropene ring pushes its electrons into the arene ring more strongly than does the cyclobutene ring and this is elegantly shown by the X–X deformation electron density map recorded by Boese and reproduced here (Figure 1). The 'bond path bond length' that follows the electron density contour is frequently close in dimension to that of a normal aromatic bond (1.395 Å) and a distinction between it and the internuclear separation is desirable. The impact of strain from the linear fusion of two small rings to benzene as in **34** appears to be complementary within the two halves of **34**. Angularly fused **35** has a cyclopropene bridge bond length (1.351 Å) that is only marginally longer than that of **34** (1.349 Å)[196]. Aromatic bond **b** that now connects the three- and four-membered rings is significantly shortened (1.363 Å) due to the non-linear fusion, and the added strain is exemplified by the external angle ε of 176.9°; the remaining bond lengths (Table 5) are consistent with expectations. For 'super rocketene' **102**, the three-membered ring bridge bond at 1.363 Å is the longest of the cyclopropabenzenes[196] (bond **a**, Table 5) and is attributed to the strain imposed by the two additional fused rings; the bonds of four-membered ring annelation are similarly lengthened. Once again the obtuse angles ε at $C_{(1a)}$ and $C_{(5a)}$ are unusually large (174.9°). The recorded geometries have been replicated by high-level *ab initio* computations. The structural details now in place allow three conclusions to be drawn: (i) annelation of a small ring to benzene results in the formation of banana bonds and thus internuclear distances can be misleading, (ii) any bond alternation is less than

TABLE 5. X-ray crystallographic geometries of selected cycloproparenes averaged on their respective highest molecular symmetry[a]

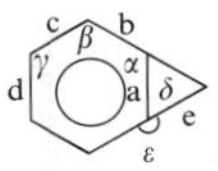

(1)

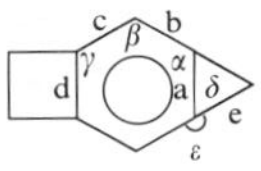

(34)

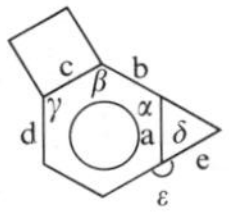

(35)

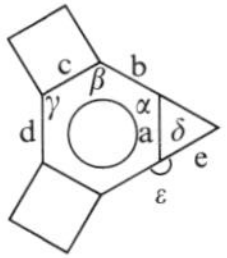

(102)

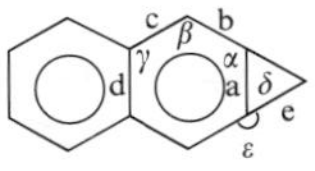

(11)

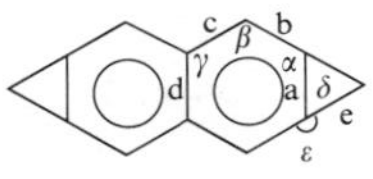

(50)

Bond	**1**[b]	**34**[c]	**35**[d,e]	**102**[d]	**11**[f]	**50**[f]	Angle	**1**[b]	**34**[c]	**35**[d,e]	**102**[d]	**11**[f]	**50**[f]
a	1.334(4)	1.349(1)	1.351(3)	1.363(4)	1.375(2)	1.360(3)	α	124.5(2)	126.3(1)	120.0(1)	121.9(3)	125.2(2)	124.9(2)
b	1.363(3)	1.385(1)	1.363(2)	1.372(4)	1.355(2)	1.351(3)	β	113.2(2)	109.2(1)	116.3(2)	117.7(2)	113.6(2)	114.8(2)
c	1.387(4)	1.405(1)	1.393(1)	1.411(3)	1.414(2)	1.437(3)	γ	121.4(2)	124.4(1)	123.7(2)	120.4(2)	121.1(2)	120.4(2)
d	1.390(5)	1.402(1)	1.384(2)	1.385(4)	1.448(2)	1.452(4)	δ	52.8(2)	53.2(1)	53.4(1)	53.8(2)	54.4(2)	53.7(1)
e	1.498(3)	1.508(1)	1.507(2)	1.506(4)	1.503(2)	1.499(3)	ε	171.7(2)	170.3(2)	176.9(1)	174.9(3)	171.6(2)	172.3(2)

[a]Bond distances (**a–e**) in Å and interbond angles (α–ε) in degrees; numbers in parentheses are the estimated standard deviations in the least significant digit.
[b]Data taken from Reference 194.
[c]Data taken from Reference 195.
[d]Data taken from Reference 196.
[e]Other bond lengths: $C_{(4)}$—$C_{(5)}$ 1.401(4), $C_{(5)}$—$C_{(5a)}$ 1.368(2), $C_{(5a)}$—$C_{(1)}$ 1.499(2) Å; other bond angles $C_{(3a)}$—$C_{(4)}$—$C_{(5)}$ 118.1, $C_{(5)}$—$C_{(5a)}$—$C_{(1a)}$ 125.7(2), $C_{(5)}$—$C_{(5a)}$—$C_{(1)}$ 170.7(2)°.
[f]Data taken from Reference 66.

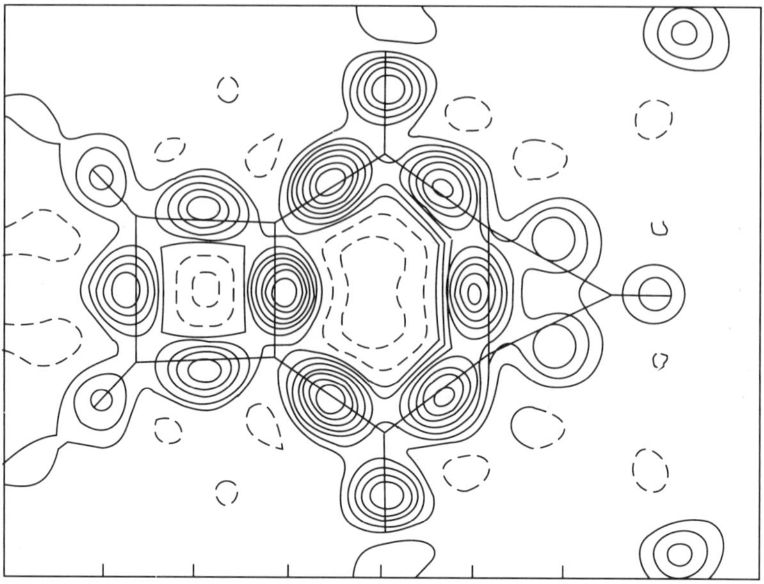

FIGURE 1. X–X difference electron density map of cyclopropa[*a*]cyclobuta[*d*]benzene (**34**) in the molecular plane. Reproduced with permission from *Advances in Strain in Organic Chemistry* (Ed. B. Halton), Vol. 2, JAI Press London, 1992, p 290

0.025 Å and is minor, and (iii) it is only the highest levels of *ab initio* calculations that are able to replicate the structural parameters, and of these the MP2/6-31G* basis set provides the best agreement.

In the context of structural analysis, cyclopropapyridine **27** has provided the data shown below[27]. The bridge bond length of 1.344 Å compares well with the value for **1** and the bonds adjacent to fusion are also short (1.355 Å) as expected. The $C_{(2)}$—N length (1.367 Å) is somewhat longer than the more normal N—$C_{(4)}$ length of 1.335 Å and interbond angles suffer distortions that match those of the hydrocarbon analogues. To our knowledge, the sole[198] example of a bond localized aromatic is the tri(benzocyclobutadieno)-benzene **103**.

bond lengths in Å

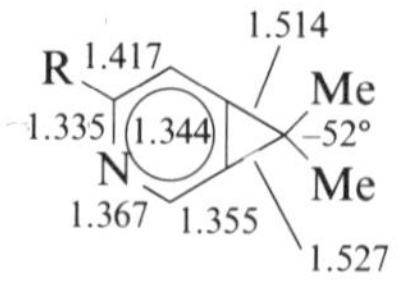

(27) R = PhCONH—

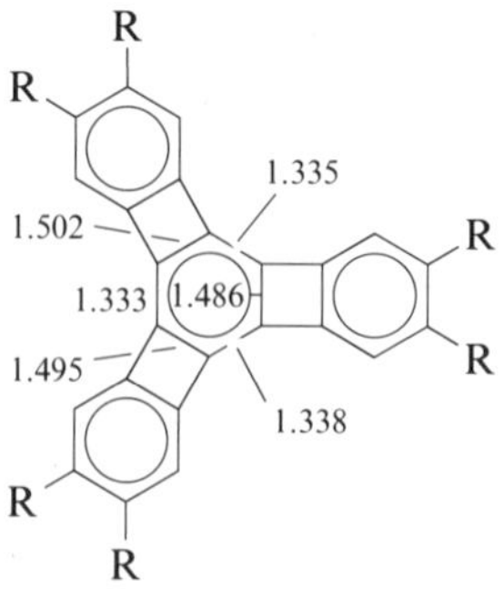

(103) R = $SiMe_3$

The structural parmeters[66] recorded for **11** and **50** show the cyclopropene σ-bonds to be very similar in length while the bridge bond of **50** (bond **a**, Table 5) falls between those of **1** and **11**. The reduction in length of bond **a** in **1** compared to benzene (0.062 Å), and in **11** compared to naphthalene (0.036 Å), illustrates that strain (68 kcal mol^{-1}) is more easily relieved in the naphthalene series. The double annelation in **50** does not allow for comparable strain relief and this is reflected in the larger deviation of bond **a** from naphthalene (0.051 Å). As in the benzene series, the cyclopropa[*b*]naphthalenes do not display any significant bond length alternation due to the fusion of the small ring.

Several cycloproparene complexes with chromium(0) have become available in recent years[164,199,200] and the available structural parameters are collected in Table 6 for the two cyclopropa[*b*]naphthalenes[199,200] **104** and **105** and for the cyclopropa[*b*]anthracene[164] **106**. In each case complexation is with the ring remote from the cycloproparenyl moiety and functional group modification at $C_{(1)}$ is possible with the chromium in place. Indeed, **105** is derived from **104** by desilylation (Section IV.B)[164,200]. The complexed rings of **104/105** match tricarbonylnaphthalenechromium reasonably well even for the central $C_{(2a)}$—$C_{(6a)}$ bond. Bond lengths for the uncomplexed ring are nicely comparable with **11** and **50**, with a bridge bond ($C_{(1a)}$—$C_{(7a)}$) of 1.384 Å in **104**. For **106**, the structure is essentially a superimposition of **11** on the corresponding anthracenetricarbonylchromium. Especially noteworthy here is that the structure of **106** represents the only X-ray analysis of a cyclopropanthracene currently available[164]. As the impact of the complexation in **104** and **105** has no marked effect on the geometry of the fused ring moiety, it is likely that the data from **106** about the sites of [4.1.0] ring fusion are a good approximation of those in an uncomplexed cyclopropa[*b*]anthracene.

Whereas semiempirical MO calculations do not provide reliable estimates of cycloproparene geometries, they do give reliable heats of formation and strain energies (SE) in comparison with both *ab initio* methods and experiment. The strain energies of **1** and **11**

TABLE 6. X-ray crystallographic geometries of tricarbonylchromium complexes of cycloproparenes[a]

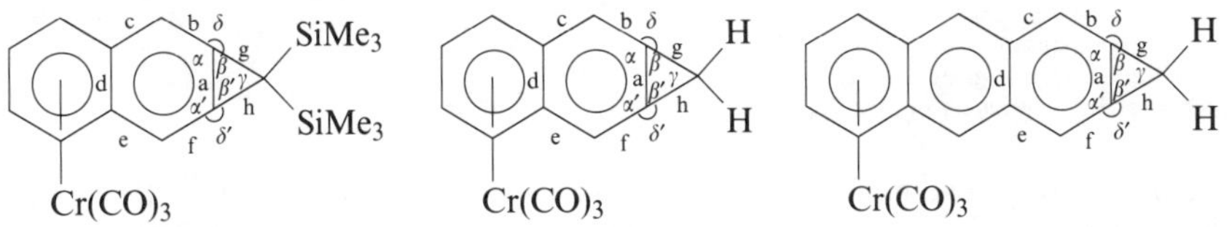

(104) **(105)** **(106)**

Bond	104[b]	105[b]	106[c]	Angle	104[b]	105[b]	106[c]
a	1.384(14)	1.368(7)	1.36(1)	α	124.5(11)	125.5(4)	126.1(7)
b	1.347(13)	1.321(7)	1.34(1)	α'	126.3(9)	124.3(4)	124.4(7)
c	1.452(16)	1.439(6)	1.43(1)	β	62.2(7)	63.0(3)	63.0(6)
d	1.441(13)	1.445(6)	1.464(9)	β'	63.3(7)	62.4(4)	62.5(6)
e	1.450(14)	1.433(6)	1.43(1)	γ	54.6(6)	54.6(3)	54.5(5)
f	1.335(17)	1.333(7)	1.34(1)	δ^d	~173.3	~171.5	~170.9
g	1.517(13)	1.487(8)	1.48(1)	δ'^d	~170.4	~173.3	~171.1
h	1.502(16)	1.494(8)	1.49(1)				

[a]Bond distances (**a–h**) in Å and interbond angles (α–δ') in degrees; numbers in parentheses are the estimated standard deviations in the last two digits.
[b]Data taken from Reference 200.
[c]Data taken from Reference 164.
[d]Estimated value using [360–(α+β)].
[e]Estimated value using [360–(α'+β')].

TABLE 7. Heats of formation (ΔH_f°) and strain energies (SE)[a] of selected cycloproparenes in kcal mol^{-1}

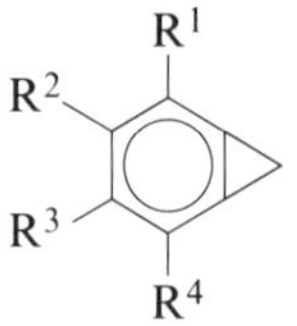

Entry No.	R^1	R^2	R^3	R^4	ΔH_f°	SE	Method	References
1	H	H	H	H	—	68	expt.	4
					90	70	3-21G*	3
					90	70	MNDO	3,201
2	H	—benzo—		H	104	67.8	expt.	4
					104	66	MNDO	202
					105	67	MNDO/3	201
3	H	H	—benzo—		109	71	MNDO	202
4	H	2,3—naphtho—[b]		H	128	69	MNDO	202
5	H	H	—2,3—naphtho—[c]		130	71	MNDO	202
6	—benzo—		—benzo—[d]		126	70	MNDO	202
7	H	$—CH_2—$		H	154	133	3-21G	203
					156	136	MNDO	201
8	H	H	$—CH_2—$		160	140	3-21G	203
					159	139	MNDO	201
9	$—CH_2—$		$—CH_2—$		237	217	3-21G	203
					227	207	MNDO RHF	204
10	H	$—(CH_2)_2—$		H	122	102	3-21G	203
11	H	H	$—(CH_2)_2—$		123	103	3-21G	203
12	H	—	—	H	190	170	3-21G*//3-21G	186
13	—	—	H	H	195	175	3-21G*//3-21G	186

[a] Relative to the parent arene.
[b] 2,3-Naphtho fusion gives cyclopropa[*b*]anthracene.
[c] 2,3-Naphtho fusion gives cyclopropa[*a*]anthracene.
[d] Dibenzo fusion gives cyclopropa[*l*]phenanthrene.

have been measured[4] as approximately 68 kcal mol^{-1} from the heats of Ag(I) catalysed methanolysis reactions. These values and those computed for a range of cycloproparenes[3,186,201–204] are collected in Table 7; heats of formation are also appended. It is assumed that stabilization due to aromaticity is the same in the cycloproparene as in the parent arene. The 15.6 kcal mol^{-1} higher strain in **1** over cyclopropene reflects the three-membered ring annelation to the benzene framework that results in added angle strain at the fusion sites. The excellent agreement between theory and experiment reinforces the value of the former as a predictive tool for molecules not yet synthesized. The strain energy caused by annelating a three-membered ring into an aromatic is essentially constant at *ca* 70 kcal mol^{-1}, but the small differences between structural isomers are important. For example, cyclopropa[*a*]naphthalene (**10**) is more strained by 4.8 kcal mol^{-1} than its linear isomer **11** (entries 3 and 2, respectively, Table 7); **10** is more difficult to handle than **11** and decomposes in solution at –30 °C. In this context is is noteworthy that cyclopropa[*l*]phenanthrene (**39**), which is less thermally stable and decomposes in solution at –60 °C, is expected to be slightly less strained than **10** (entries 6 and 3, respectively, Table 7). The difference of *ca* 1 kcal mol^{-1} likely reflects a pincing effect of the three-membered ring whereby the bay region protons have a little less steric interaction. The fusion of a second ring in an angular fashion is always expected to provide a more highly strained compound as exemplified by entries 4/5 and 7/8 of Table 7. The impact of a second ring to the molecular framework is approximately additive as illustrated by entries 1,7/8 and 9 (Table 7) and the values[201] of 68 and 120 kcal mol^{-1}, respectively, for **11** and **50**.

The data in Table 7 suggest that since the cyclopropabenzynes **107** and **108** (SE *ca* 190 kcal mol^{-1}) have been intercepted[186], the as yet unknown dicycloproparenes **109** and **44** (SE *ca* 135 kcal mol^{-1})[203] should also be detectable. Appropriate synthetic pathways have yet to be devised. The same comparison does not augur well for tricyclopropabenzene (**101**) as its projected[204] strain energy of *ca* 212 kcal mol^{-1} places it on the limit for existence; any aromatization of a preformed ring system would need to avoid sequential π-bond formation and be coupled with rapid interception as the activation barrier to conversion into cyclononatriyne is only 7.9 kcal mol^{-1}.

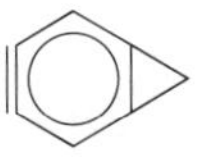

(107)

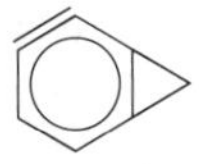

(108)

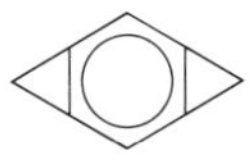

(109)

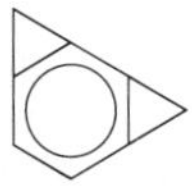

(44)

The photoelectron spectra of several cycloproparenes have now been recorded, but do not allow for conclusions on the presence or absence of bond alternation because any changes induced are below the level of detection. While the PE spectrum of **1** was recorded as early as 1973[205], the compound has been re-examined together with **11**, cyclopropa[*b*]anthracene and their *gem*-difluoro derivatives, rocketene (**34**) and its 1,1-difluoro derivative, and also difluoro **63c** and its derived hydrocarbon[195,206]. The unusual structure of **34** is reflected in its PE spectrum[195] as the first two ionization bands (π_2, π_3) are moved to lower energies in comparison to **1** and **2** and the splitting (0.84 eV) is larger (**1**: 0.66, **2**: 0.54 eV). More importantly, their average value (8.72 eV) is *ca* 0.1 eV lower than that expected assuming a simple additive effect of the two annelated rings (benzene: 9.25, **1** : 9.15, **2** : 8.93 eV) and this reflects the expansion of the aromatic ring in **34**; the results are in accord with the heat of formation[203] recorded in Table 7. From the more conventional derivatives[206] it is concluded that the inductive effect of the bridging CH_2 group is close to zero whereas that of the CF_2 group leads to positive ionization energy shifts (in agreement with previous experience). The hyperconjugative interaction of the local pseudo-π orbital of the CH_2 group provides a significant destabilization of the π orbitals of B_1 symmetry. The hyperconjugative abilities of the four-membered rings in various cyclobutarenes have been assessed[207] and an *ab initio* study of the outer valence ionization potentials and electron affinities of the cycloalkabenzenes including **1**, **34** and **109** has been reported[208].

The electronic absorption spectrum of a cycloproparene is very similar to that of the analogous dialkyl substituted aromatic and **1** [λ_{max} (cyclohexane) 252 (2.7), 258 (3.0), 264 (3.2), 270 (3.4), 277 nm (log ε 3.3)], **2** and *o*-xylene provide comparable spectra[8,10]. The presence of a second small fused ring has a noticeable impact. In the linear series the effect is a bathochromic shift that is more marked as the ring size becomes *smaller*; the largest bathochromic shift is for **34** [λ_{max} (cyclohexane) 284 ($\approx$3.0), 287.5 ($\approx$3.0), 294 nm (log $\varepsilon \approx 2.8$)]. In the angular series the shift is also to longer wavelength but the wavelength *increases* as the size of the fused ring increases[10,101]. This behaviour is consistent with the ability of the fused ring to participate in hyperconjugation and with changes in the configurational composition of the excited state[207]. Tabulations of UV data have appeared[10].

The infrared spectra of the cycloproparenes are characterized by a weak band that appears between 1660 and 1690 cm^{-1} due to a combination of the aromatic double bond stretch with the three-membered ring skeletal vibration. Angular fused cyclopropa[*a*]naphthalene (**10**) has the highest value (1687 cm^{-1}) while for linear **11** it is more usual at 1673 cm^{-1}, and cyclopropa[*b*]anthracene falls between these at 1678 cm^{-1}. The cyclopropabenzenes **1**, **34** and **102** show the stretch at 1662, 1664 and 1664 cm^{-1}, respectively.

NMR spectra recorded for the cycloproparenes and their alkylidene derivatives (Section VI.A) are fully compatible with molecules that display a diamagnetic ring current. Detailed arguments and chemical shift tabulations have been published[8,10] and consequently only a brief synopsis is provided here. The methylene protons of the cycloproparenes resonate in the range 3.05–3.60 ppm, and in the benzenoid series **1**, **34**, **35** and **102** show a singlet at 3.11, 3.08, 3.18 and 3.10 ppm, respectively. For the naphthalenes **10** and **11** it is at 3.40 and 3.42 ppm, respectively, and the downfield shift parallels that observed from toluene to 1- and 2-methylnaphthalene. The ring protons of **1** fall within the normal aromatic range with $H_{(2)}/H_{(5)}$ at 7.15 and $H_{(3)}/H_{(4)}$ at 7.19 ppm, respectively[209–211]. The most notable feature of the proton spectrum is that the magnitude of J_{meta} (0.3–0.7 Hz) and J_{para} (*ca* 1.9 Hz) is the *opposite* of a normal benzene derivative; this is likely a reflection of the bent bonds to $C_{(2)}/C_{(5)}$ from the three-membered ring fusion. In the carbon NMR spectrum[212] a similar feature is noted and $C_{(2)}/C_{(5)}$ resonate at high field, and generally in the range 105–115 ppm. The precise position is a function of the specific ring system and the substituents present and it can be outside of this range, cf. cyclopropa[*f*]isobenzofuran, 101.9 ppm[107]. The presence of the shielded carbon resonances is diagnostic for the cycloproparenes and again reflects the unusual bonding enforced by the ring fusion. The bridge carbons of **1** (125.4 ppm) and its homologues (119–125 ppm) are also shielded in comparison with cyclobutabenzene (145.6 ppm), again due to the presence of the three-membered ring. The carbons remote from the fusion sites ($C_{(3)}/C_{(4)}$) are largely unaffected by the presence of ring strain and resonate in the normal aromatic region. The impact of strain is clearly manifest in the magnitude of the $C_{(1)}$—H and $C_{(2)}$—H coupling constants. The former is *ca* 170 Hz, similar to the value found for cyclopropene (167 Hz)[213], while the latter represents a significant, but linear increase from **3** and **2**, and a value of *ca* 170 Hz is characteristic for this series of compounds. This increase in $^{1}J_{C_2\text{—}H}$ reflects the shortened $C_{(1a)}$—$C_{(2)}$ bond and the rehybridization that this causes. One-bond ^{13}C—^{13}C couplings have been recorded for **1** and, while the values for $C_{(1)}$—$C_{(1a)}$ and $C_{(1a)}$—$C_{(2)}$ are anomalous, they are explicable in terms of the Walsh model for cyclopropene[214]. These results corroborate rehybridization concepts advanced for the cycloalkabenzenes in 1968 whereby strain causes rehybridization of the bridgehead carbons, and orbitals with increased p character are used for bonding to the small ring[215]. Bonding to the adjacent *ortho* carbons of the aromatic ring must use an orbital with excess s character and inductive polarization is expected. This is fully consistent with the shift to higher field of $C_{(2/5)}$, the increase in $^{1}J_{C_2\text{—}H}$, $^{1}J_{C_1\text{—}H}$, and the ^{13}C—^{13}C NMR couplings discussed above.

Insofar as electron impact mass spectrometry is concerned, the cycloproparenes almost always display a molecular ion. The primary source of fragmentation is by loss of a $C_{(1)}$ substituent radical to provide a cycloproparenyl cation. However, labelling studies have shown that loss of $H^{\bullet}$ from **1** and **11** (at least) occurs only after complete scrambling of the carbon atoms[216,217]. The use of appearance energy measurements[218] for the loss of $Br^{\bullet}$ from ionized 2-bromocyclopropabenzene (**57a**), when coupled with thermochemical data, have led to the prediction that cation **110** is *more* stable than the phenyl cation by at least 27.6 kcal mol^{-1}; ΔH_f° of **110** is estimated at 311 kcal mol^{-1}.

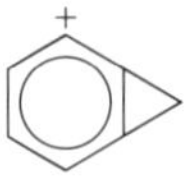

(110)

Negative-ion mass spectrometry has provided evidence for the $C_{(1)}$ 1-methylcyclopropabenzenyl anion and the radical anion of 1,1-dimethylcyclopropabenzene[219].

IV. CYCLOPROPARENYL CATIONS, ANIONS AND RADICALS

The formal removal of a $C_{(1)}$ substituent as anion, cation or radical from a cycloproparene provides the counter cycloproparenyl species **5–7** and suggestions on the existence of such entities predate the establishment of the cycloproparenes as a viable class of compounds[6,220].

A. Cycloproparenyl Cations

The cation **5** derived from **1** received theoretical consideration[221] as early as 1952 and was expected to be stabilized with positive charge distributed throughout the molecule[171,222]. Recent HF 6-31G* *ab initio* calculations by Maksic and coworkers[182] argue for a strong *reversed* Mills–Nixon deformation within the aromatic fragment due to strong π delocalization over the three-membered ring with consequential rehybridization of the bridgehead atoms.

The existence of cyclopropabenzenyl cations was suggested from exchange experiments[6] of various $C_{(1)}$ *gem*-dihalo derivatives with carbanions[89,223], fluoride[92,93,224] and hydride[225]; maintenance of the cycloproparenyl ring system in these reactions is illustrated by **111** (equation 23). However, the isolation and characterization of the salt **112** provided the first

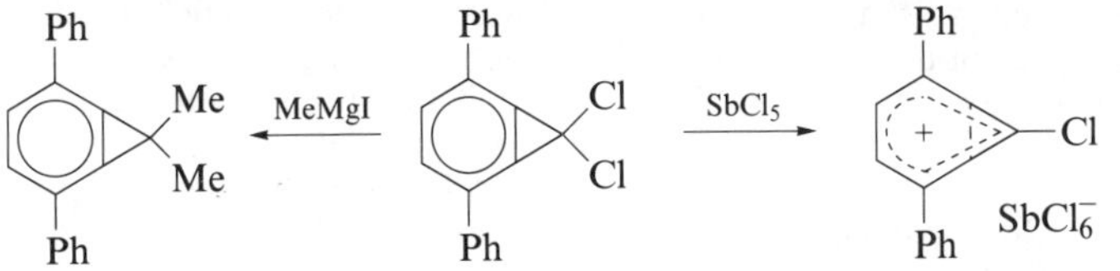

(111) **(63a)** **(112)** (23)

definitive evidence for the existence of cations[226,227]. Since that time the characterization of a large number of cycloproparenyl cations has been accomplished, especially 1-fluoro derivatives[94–96,111,228,229], and their spectroscopic data have been adequately discussed and tabulated elsewhere[8,10]. However, structural data are not available.

The parent ion **5** and its naphtho[*b*] analogue have been generated by treatment of **1** and **11**, respectively, with trityl tetrafluoroborate. The cations so formed are captured by water and the final product of reaction is the ring opened aryl aldehyde (equation 24)[230,231]. Kinetic measurements have shown[230] that abstraction of hydride ion from **1** is first order and displays a deuterium isotope effect of 6.5. To the best of the author's knowledge, the isolation and spectroscopic characterization of an unsubstituted cycloproparenyl cation has yet to be achieved.

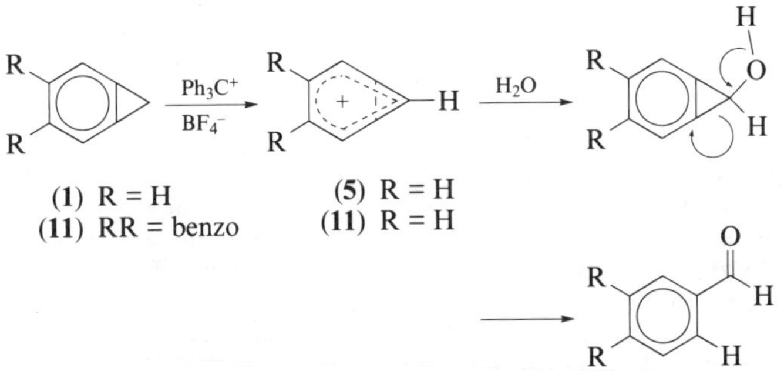

(1) R = H **(5)** R = H
(11) RR = benzo **(11)** R = H

(24)

The enchanced stability of the 2-cyclopropabenzenyl cation (**110**) over the phenyl cation has been discussed above (Section III).

B. Cycloproparenyl Anions

As the methylene protons of the cycloproparenes are formally benzylic, the compounds should be at least as acidic as the corresponding methyl arenes. Simple extended Hückel calculations[171] concur and predict some stabilization of **6** from charge delocalization; the cycloproparenyl anion (**6**) is expected to be accessible[222]. Indeed, treatment of **1** with butyllithium at –78 °C provides an organolithium that has been characterized by NMR and shown to be monomeric in solution[232]. Both the ^{1}H NMR ($H_{(1)}$ 1.91, $H_{(2/5)}$ 5.83, $H_{(3/4)}$ 6.13 ppm, respectively) and ^{13}C NMR ($C_{(1)}$ 31.3, $C_{(1a/5a)}$ 149.0, $C_{(2/5)}$ 99.1, $C_{(3/4)}$ 120.4 ppm, respectively) data recorded for **6** in THF are fully comparable with those of the tolyl anion[233,234]. The one-bond $^{1}J_{C_1-H}$ coupling constant for **6** is 127 Hz. The use of ^{6}Li labelling, carbon–lithium coupling, and carbon signal multiplicity—a triplet—leads to the conclusion that the species is monomeric in ether solution[232]. The comparability with the benzylic species supports anion **6** being stabilized by delocalization. Characterization of the more reactive[155,200] higher homologues has yet to be achieved.

The unpublished data referred to above[232] substantiate the earlier[162] presumption that **6** is present when **1** is treated with butyllithium. Indeed, reaction of the anion with trimethylsilyl chloride gives silyl **93** (R=H)[162] (Scheme 12, Section II.G). Subsequent cleavage of this with NaOH is some 64 times more rapid than for benzyltrimethylsilane and this gives a pK_a of *ca* 36 for **1**; STO-3G calculations[163] give a value of 33 and **1** is significantly more acidic than toluene. Silyls **93** (R=H) and **95** have been used in the synthesis of the alkylidenecycloproparenes **94** (Table 4, Section II.G)[154–158,161,164] and **96** (Scheme 13)[165] via Peterson olefination involving cyclopropabenzenyl and cyclopropa[*b*]naphthalenyl anion intervention, respectively. Particularly noteworthy is the fact that subjection of disilyl-containing chromium(0) complexes (**104**, or its anthracene analogue) to *t*-BuOK provides $C_{(1)}$ anion without the loss of chromium; in the case of **104** Peterson olefination has provided the corresponding complexed alkylidene derivative in high yield (equation 25)[164], but the compound is unstable and decomposes upon attempted purification.

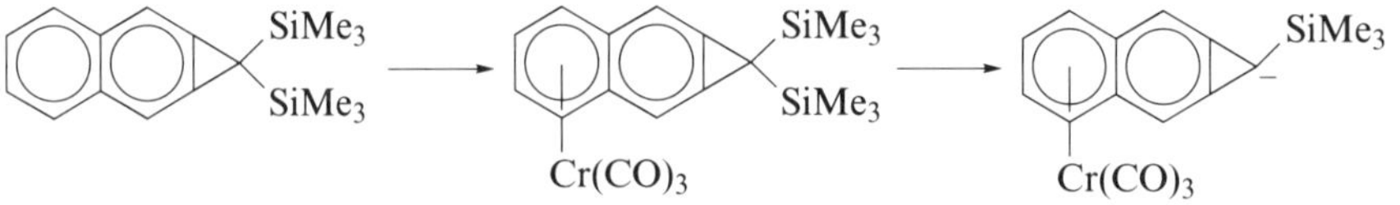

(95) **(104)**

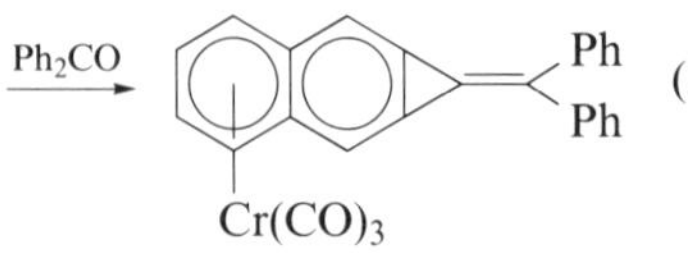

(25)

The lithiation of **1** has many synthetic applications beyond alkylidenecycloproparene synthesis. Thus Szeimies and Wimmer[232] have employed **6** for the synthesis of a range of derivatives in which the cyclopropabenzenyl framework is maintained (Scheme 14).

Until quite recently anion generation at a centre other than $C_{(1)}$ of **1** had not been contemplated. However, it has been found that when $C_{(1)}$ carries fluorine substituents, lithiation can be effected at $C_{(2)}$. Thus Neidlein and Kohl[235–238] have found that 1,1-difluorocyclopropabenzene can be lithiated at low temperature (–100 to –90 °C) with BuLi,

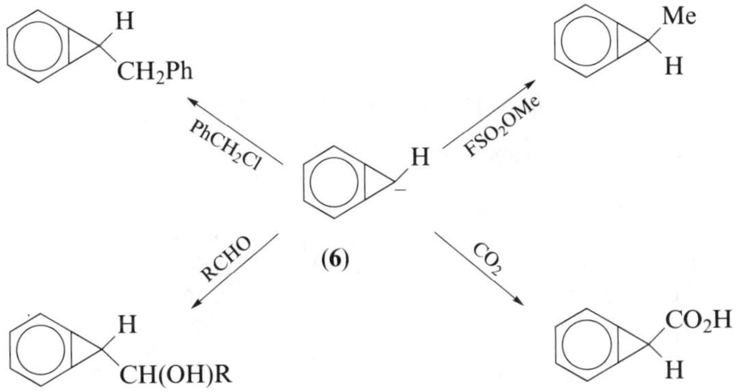

SCHEME 14

LDA or lithium 2,2,6,6-tetramethylpiperidide in the presence of TMEDA. The resultant anion (organolithium) is capable of interception with a wide range of electrophilic reagents (I_2, PhSSPh, MeOD, PhNCO, PhNCS, PhOCN, Ph_2CO, Me_2NCHO, $MgBr_2/CO_2$, TMSCl, etc.) thereby providing numerous 2-substituted *gem*-difluorocyclopropabenzenes as illustrated by Scheme 15. The impact of these reactions is electrophilic substitution at the 2-position of the cycloproparene. Furthermore, treatment of the cycloproparene with two equivalents of LDA interspersed with (excess) PhSSPh gives the 2- and 2,5-di-(thiophenyl) derivative, the latter in low yield[236]; the experimental procedure applied must result in sequential lithiation at $C_{(2)}$ and $C_{(5)}$ and *not* dianion formation.

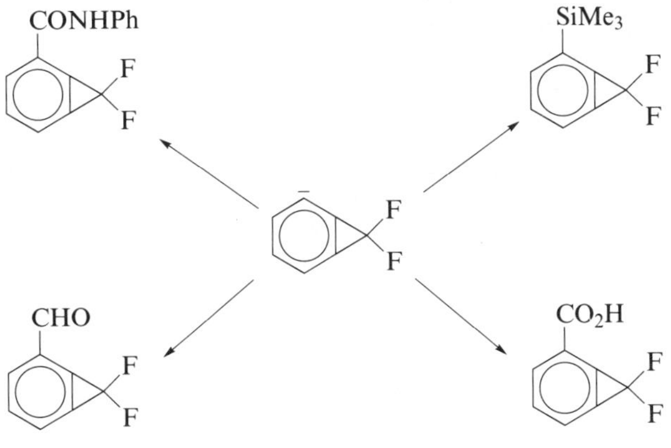

SCHEME 15

C. Cycloproparenyl Radicals

The earliest predictions[171] suggested that the cyclopropabenzenyl radical **7** should have a stability that lies between those of **5** and **6**. Attempts to generate the species by removal of a hydrogen atom from $C_{(1)}$ of **1** or a chlorine from **63a** have been unsuccessful[239]. The availability of the exocyclic olefins of Table 4 provides a potentially easier route to derivatives of **7** by radical addition to the exocyclic centre. Studies with such compounds have thus far failed[240] to provide any definitive evidence for the intervention of a cycloproparenyl radical despite conditions where parent **11** reacts to give product of ring cleavage (Section V.B.2). The generation of a cycloproparenyl radical thus remains a challenge.

Attempts to generate the radical anion of **11** by electron transfer resulted in fast oxidative dimerization such that $\mathbf{11}^{-\bullet}$ was not observed by ESR techniques[241].

V. REACTIONS OF CYCLOPROPARENES

The chemistry of the cycloproparenes is dominated by reactions in which cleavage of the three-membered ring occurs because of release of the high strain energy. This does not mean to say that reactions in which the ring system is retained cannot occur nor does it prevent additions to the bridge bond with subsequent cyclopropane (norcaradiene) ring opening. Calculation[3] and experiment[205] concur that the HOMO of **1**, b_1 is located between $C_{(1a)}$—$C_{(5a)}$ (the bridge bond) and $C_{(3)}$—$C_{(4)}$ and that it is higher in energy than the a_2 orbital. The various reacion pathways are systematically discussed in the following sections. For convenience, reactions that proceed remote from the strained ring fusion as well as those that give intermediates with the cycloproparene ring skeleton, e.g. addition to the strained bridge bond, are collected together as reactions without ring cleavage.

A. Reactions that Proceed without Ring Cleavage

1. With electrophiles and nucleophiles

Although a $C_{(1)}$ hydride can be removed from **1** and **11** by the trityl cation[230,231] and the ensuing cycloproparenyl cation captured with a suitable nucleophile (equation 24, Section IV.A), electrophilic aromatic substitution of the cycloproparenes is frustrated by reactions which result in opening of the three-membered ring (Section V.B.1). It is only by sterically blocking this part of the molecule that three-membered ring opening is prevented and reaction with the arene component observed.

Silylation of the cyclopropabenzenyl anion (**6**) with chlorotriisopropylsilane and a repetition of the sequence provides a sterically demanding disilyl which is nitrated[242] (67% HNO_3) at $C_{(3)}$ to give **113** in 58%; the use of ultrasound in the nitration improves the yield of product[243]. Attack of the electrophile at $C_{(3)}$ of **1** is clearly in accord with the location of the HOMO[3,205]. A range of transformations of the nitro functionality is possible as illustrated in Scheme 16[242]. Recent calculations[244] with HF/6-31G* and single-point

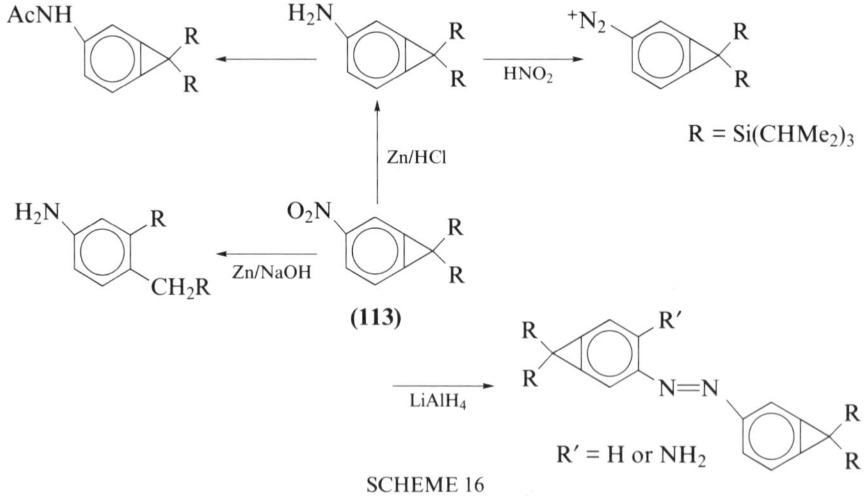

SCHEME 16

MP2(fc)//HF/6-31G* procedures have been used to examine theoretically the possible Wheland intermediates in the reaction of **1** with an electrophile (H^+) and these agree that the electrophile will attack at $C_{(3)}$; the results are used to support Mills–Nixon[5] bond localization in the direction originally proposed, viz. **1a**.

The interaction of *gem*-dihalocycloproparenes at $C_{(1)}$ with nucleophiles is consistent with ionization to a cycloproparenyl cation, cf. **5**, and subsequent capture of the nucleophile to give products of S_N1 substitution at the sp^3 centre as illustrated by equation 23 (Section IV.A). There are no known examples of nucleophilic aromatic substitution among the cycloproparenes because of an absence of appropriately functionalized compounds.

Treatment of 3- and 2-bromocyclopropabenzene **55a** and **57a**, respectively, with strong base (t-BuOK/NH_2^-) in THF/t-BuOH at ambient temperature effects dehydrobromination and formation of the cyclopropabenzynes **114** and **115** that are trapped as Diels-Alder adducts by furan (Scheme 17)[186,245]. Dehydrobromination to the 'linear' **114** from **55a** is highly regioselective in accord[186] with its lower energy (*ca* 5 kcal mol^{-1}). This difference reflects a matching of the distortions present in benzyne and **1** in the linear **114** but not in the angular **115**. The strain energies of **114** and **115** (*ca* 170 kcal mol^{-1}) (Table 7) are among the highest recorded and the interception of these reactive molecules suggests that dicyclopropabenzenes (SEs *ca* 135 kcal mol^{-1}) should also be capable of interception.

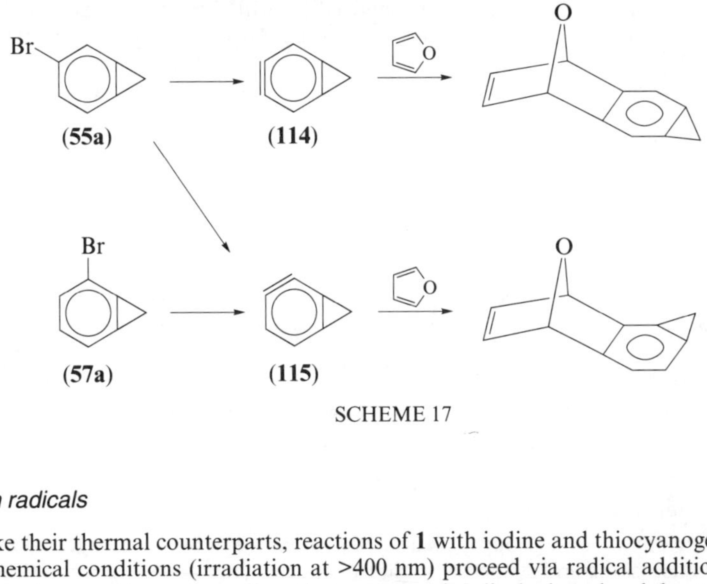

(55a) (114)

(57a) (115)

SCHEME 17

2. *With radicals*

Unlike their thermal counterparts, reactions of **1** with iodine and thiocyanogen under photochemical conditions (irradiation at >400 nm) proceed via radical addition to the bridge bond to provide a norcaradiene that opens to a 1,6-disubstituted cycloheptatriene[58]. The simple procedure has found use in synthesis[52,246] but it is by no means general. Radical reactions with NBS, halomethanes and thiols lead to products of three-membered ring cleavage only[247]. This is likewise observed with the homologous **11** where the norcaradiene intermediate is also a high-energy orthoquinodimethane[240] (Section V.B.2).

3. *In cycloadditions*

There is only one recorded case of cycloaddition to a cycloproparene that takes place other than across the bridge bond and that is with cyclopropa[*f*]isobenzofuran. In this case addition of dimethyl fumarate aromatizes the six-membered ring (equation 26) but the

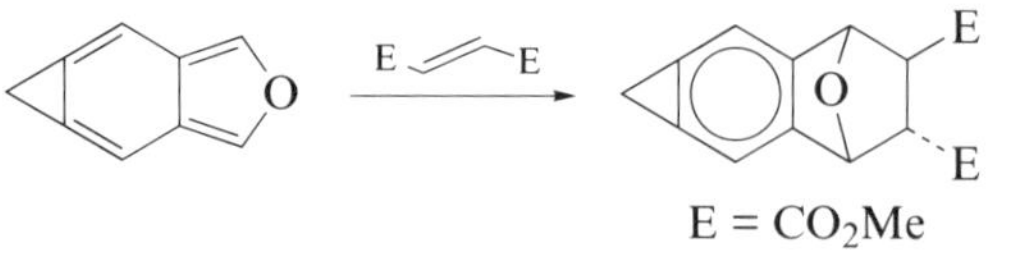

(26)

reaction is some four times *slower* than with isobenzofuran itself. This could be due to the increase in π character in the bridge bond caused by aromatization. Cycloadditions thought likely to occur remote from the cyclopropa fusion in annulene **51b**, and through the norcaradiene valence bond isomer, do not proceed at all[248]. Recent AM1 calculations show that **51a** (Table 1, Section II.E.1) is some 21 kcal mol^{-1} *more* stable than its ring closed form and this contrasts with the 7 kcal mol^{-1} that favours parent 1,6-methano[10]annulene over its norcaradiene isomer[249].

With its HOMO located at the bridge, **1** can formally participate as a 2π or 6π electron component in cycloaddition reactions. Products result from capture of the addend across the $C_{(1a)}$—$C_{(5a)}$ bond and the ensuing norcaradiene may or may not undergo electrocyclic ring opening. The earliest such report of Diels-Alder cycloaddition was made[250] in 1968. Thus reaction with buta-1,3-diene gives the norcaradiene **116** that remains ring closed. A closed norcaradiene is also obtained from the reaction of **1** with diphenylisobenzofuran[251,252], and from the more reactive **10**[112] and **39**[117,118] with furan (Schemes 9 and 18). The Diels-Alder addition of diphenylisobenzofuran to **1** gives product with the methylene and oxygen bridges *syn* and its formation is in competition with [2+2] addition to the three-membered ring σ-bond[253]. The product from **10** is also *syn*, but with **39** a *syn*/*anti* mixture of 3:2 is reported (Scheme 9, Section II.E.3)[117,118].

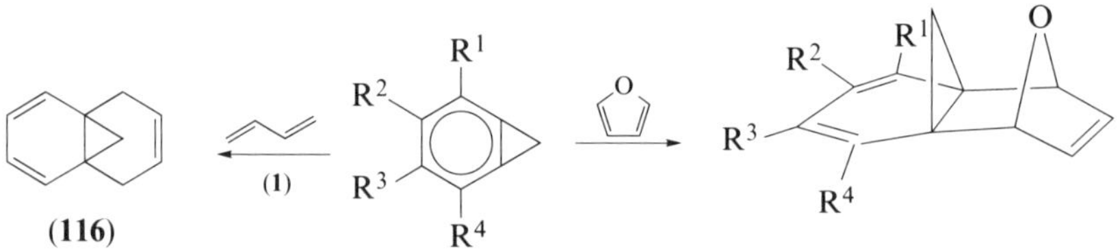

(116)

(1) $R^1 = R^2 = R^3 = R^4 = H$

(10) $R^1 = R^2 = H$; R^3R^4 = benzo

(39) $R^1R^2 = R^3R^4$ = benzo

SCHEME 18

The interception of cyclopropabenzene-2,5-dione (**42**) by [2+4] cycloaddition of anthracene across the bridge bond was alluded to earlier (equation 10, Section II.D) and a good number of cycloaddition reactions to **1** have been surveyed[8,10]. Thus reactions with the dienes dibromo-*o*-benzoquinone[254] and α-pyrone[250], various triazines[255,256] and tetrazines[257], as well as dipolar nitrile oxides[258], a mesoionic oxathiazolone and a 1,3-dithiolone[259,260] were discussed. In each of these cases initial addition to **1** is across the bridge bond to give a norcaradiene that may be stable[51,254,258], ring open to the corresponding heteroannulene[259,260] or eject a neutral molecule to provide the parent [10]-annulene ($-CO_2$)[250] or a heteroannulene ($-N_2$)[255–257]. More recently, additions of **1** to a range of cyclopentadienones, thiophene-1,1-dioxides and α-pyrones have been reported to yield methano[10]annulenes (Scheme 19)[261,262]. The reactions with cyclopentadienones proceed via **117** [*exo*-adduct **117** (R=Et; Ar=Ph) was isolated] but intermediates were not detected in the other cases.

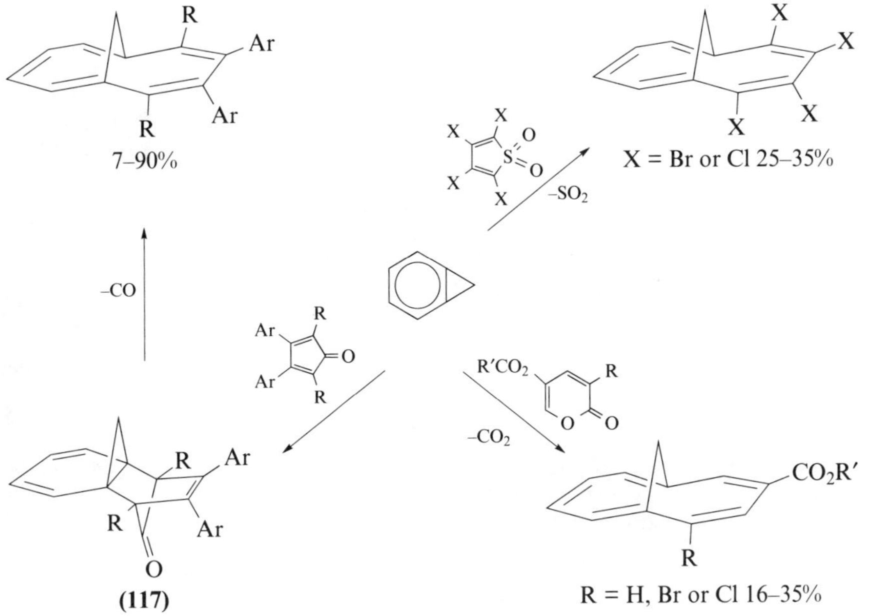

SCHEME 19

It has been found that dihalocarbenes add to **1**[263], **11**[264] and **50**[66]. The products, ring expanded cyclobutarenes, are plausibly accounted for by way of cheletropic addition to the strained bridge bond and subsequent rearrangement as illustrated for **1** in equation 27. However, the reaction mechanism has not been established.

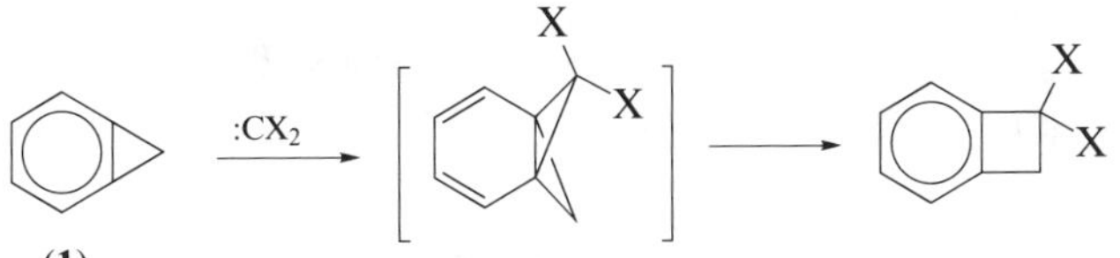

(27)

4. *Upon metal complexation*

It is only in recent years that the behaviour of the cycloproparenes towards organometallic reagents has been explored. Reactions occur either at the internal (bridge) or the latteral strained bond, the former to provide metallapropellanes and the latter to give ring expanded metallacyclobutarenes. The nature of the cycloproparene $C_{(1)}$ substituents, the metal and the coordinated ligands each play a role in determining the outcome of a given reaction.

With 1,1-difluorocyclopropabenzene and a range of nickel(0) complexes, nickelabicyclobutanes **118** (84–93%) are formed by loss of olefin or phosphane ligands and addition of the nickel atom across the bridge bond (Scheme 20)[256,266]. The products appear to be stable at ambient temperatures but are oxygen sensitive; the majority revert to cycloproparene in solution even below –20 °C. With (η^3-allyl)(η^5-cyclopentadienyl)palladium in the

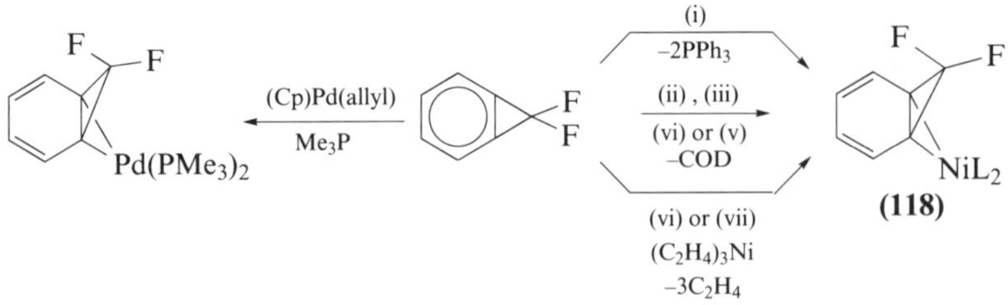

(118)

(i) $(Ph_3P)_4Ni$ (ii) $(Et_3P)_2Ni(COD)$ (iii) (dcpe)Ni(COD)
(iv) $(Me_3P)_2Ni(COD)$ (v) TMEDA (vi) bipy (vii) TEEDA
[COD = Cycloocta-1,5-diene; dcpe = ethenebis(dicyclohexyl)phosphine;
TMEDA = tetramethylethenediamine; bipy = 2,2′-bipyridyl;
TEEDA = tetraethylethenediamine.]

SCHEME 20

presence of trimethylphosphane, *gem*-difluorocyclopropabenzene also gives a bicyclobutane and this contrasts with parent **1** and its 1,1-bis(trimethylsilyl) analogue that yield ring opened products (Section V.B.4)[267].

Early attempts to utilize **1** as a donor for a chromium sandwich complex did not meet with success[268], but recent studies have now provided several η^6-chromium(0) complexes[164,199,200]. Whereas unsubstituted cycloproparenes undergo oxidative addition of the ring to the metal followed by carbon monoxide insertion (Section V.B.4), the 1,1-bis(trimethylsilyl) derivatives do not. Instead, reactivity is transferred to the arene and, with tris(acetonitrile)tricarbonylchromium, η^6-complexes are formed at the ring remote from the cycloproparene moiety (equation 28). However, the 1,1-disilyl derivative of **1** does not react and

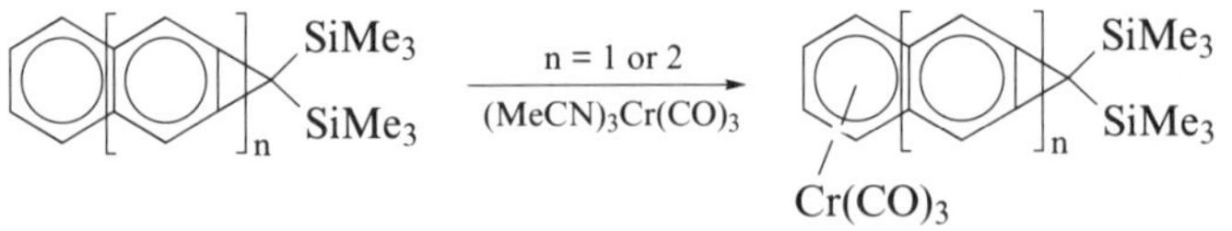

(28)

(95) n = 1 **(104)** n = 1

there is no known η^6-metal coordinated cyclopropabenzene. The two complexes obtained by this method have been desilylated with *t*-BuOK/*t*-BuOH to give the unsubstituted η^6-parents **105/106** via the α-silyl anion (Section IV.B) and crystallographic parameters have been recorded for **104–106** (Table 6, Section III)[164,199,200]. Under Peterson olefination conditions complex **104** is transformed into its 1-diphenylmethylene derivative[164] that could not be crystallized; attempted purification resulted in much decomposition. Alkylidenecycloproparenes have yet to be complexed directly with chromium reagents.

B. Reactions that Proceed with Ring Cleavage

1. *With electrophiles*

The three-membered ring of the cycloproparenes is opened easily by an electrophile and the attack can be at either the σ or the π framework. The reactions with halogens (non-photochemical)[58,59,61] and acids[250] give rise to benzyl derivatives that are best accounted for[3] by π capture of the electrophile (E^+) at the bridge, electrocyclic opening of the cyclopropyl cation to benzyl cation (path *a*, Scheme 21) and capture by the nucleophile.

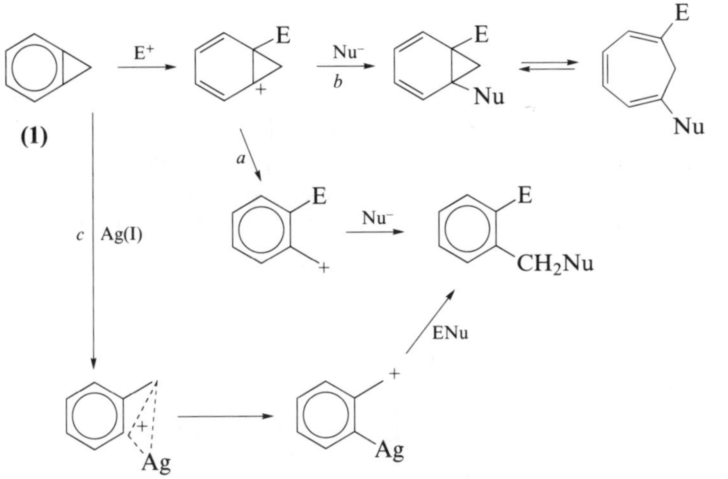

SCHEME 21

Regioselectivities observed with 2- and 3-methyl-[59] and 3-chlorocyclopropabenzene[73,74] are consistent with this pathway whereby π capture of the electrophile preferentially provides the more stable Wheland intermediate. Thus the 2-methyl derivative gives *m*-xylenes via ion **119** while the 3-methyl and the 3-chloro compounds also yield *m*-xylenes but via ion **120**. These are in accord with more recent experimental[101] and theoretical studies[244]. If the

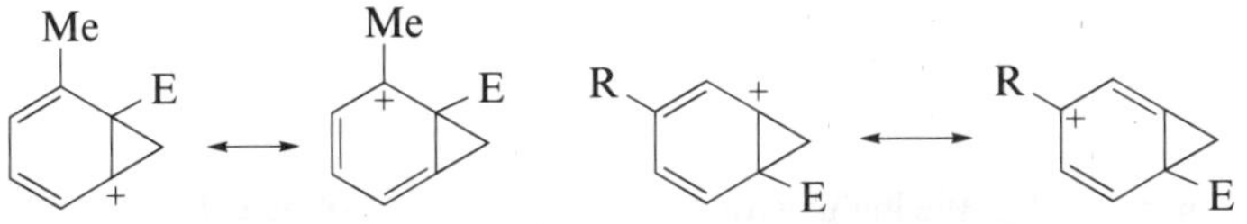

(119) **(120)** R = Me or Cl

norcaradienyl (cyclopropyl) cation is captured prior to opening, then cycloheptatrienes will ensue (path *b*, Scheme 21); low yields of 1,6-diiodocycloheptatriene (≤7%) have been observed[2,58] in the iodination of **1**. The formation of 1,6-disubstituted cyclohepta-1,3,5-trienes dominates under photochemical conditions with long-wavelength light (>400 nm). Thus the diiodo[52,58] and dithiocyanato[58] derivatives are formed from **1** in yields of 67 and 64%, respectivley (Section V.A.2); capture of a radical at $C_{(1a)}$ adequately accounts for the products.

Electrophilic cleavage of the cycloproparene three-membered ring is very effective in the presence of Ag(I) and this provides an excellent method of benzylation. Thus the silver(I) catalysed reactions of **1** with alcohols, amines and thiols proceed readily at 0 °C in aprotic media to give the corresponding benzyl derivatives in excellent yield[10,269]. Indeed, the Ag(I) catalysed methanolyses of **1** and **11** were used to determine the strain energies of the compounds[4]. These reactions most likely proceed by interaction of the metal ion with the strained σ-bond, ring cleavage, and nucleophilic capture of the benzyl cation thus formed (path *c*, Scheme 21). The regioselectivities recorded[59,61,73,74,101,186] are usually the *opposite* to those discussed above. Thus for 2-methylcyclopropabenzene the reaction yields substituted *o*-xylenes (equation 29) since the incipient *o*-substituted benzylic cation is the more stable. For the unsymmetrical small ring annelated system **35** and its higher homologue, the influence of additional strain is important[10,101] as the Ag(I) catalysed reactions are

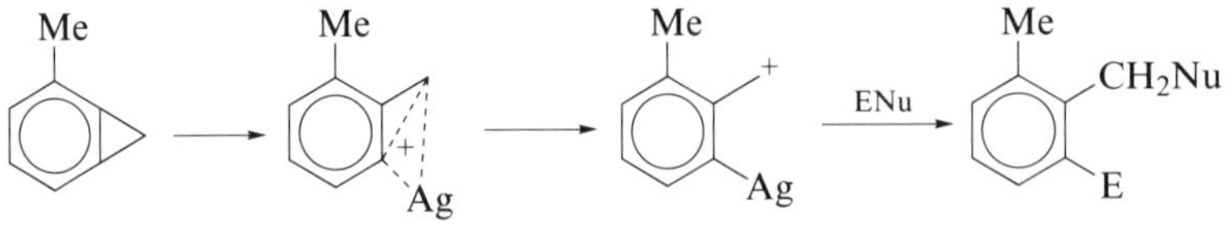

(29)

regiospecific and provide products via ion **121** (path *a*, Scheme 22). In contrast, path *b* of Scheme 22 becomes an effective competitor when the cycloproparene is less strained. Reactions with halogens also show regioselectivity but in the opposite direction[101]. It is

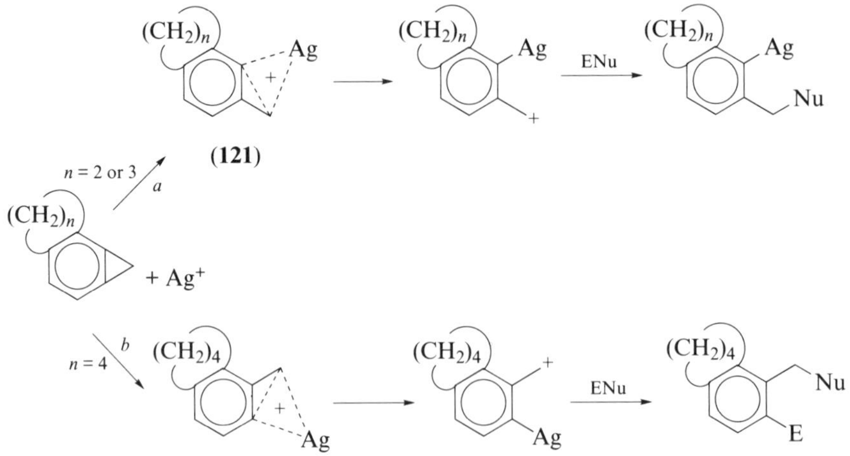

SCHEME 22

worthy of mention that the highly strained cycloproparene of Scheme 22 reacts with HCl to give products with the same regiochemistries as those from Ag(I)[101]. However, it is not known whether it is merely added strain that causes this effective protonation by the σ route (path *c*, Scheme 21).

Silver(I), Cu(II) and Hg(II) have been employed[269,270] in the dimerization of **1** and discussed elsewhere[8,10] as has the role of the first of these in the addition of alkenes, alkynes, allenes and conjugated dienes to cyclopropabenzene[269]. It is more than probable that product formation with Ag(I) is dictated by electrophilic addition of the metal and subsequent interaction of the organosilver benzylic cation with the hydrocarbon reagent to give ring-opened or ring-expanded products.

2. *With radicals*

Although some reactions of **1** with radicals proceed by ring expansion[58] (Section V.A.2), opening of the three-membered ring to provide benzyl derivatives is more common[240,247]. With **11** as substrate, benzocycloheptatriene formation is not observed[240] probably because of the need for a second high-energy orthoquinodimethane intermediate upon expanding the three-membered ring.

Christen[247] has found that **1** reacts with thiols and tri- and tetrahalomethanes under radical conditions to give products with the *opposite* orientation to that expected from reaction with an electrophile. Equation 30 depicts two such processes. In similar vein, **11** has been shown to interact with thiophenol and tributyltin hydride to give 3-substituted 2-methyl-

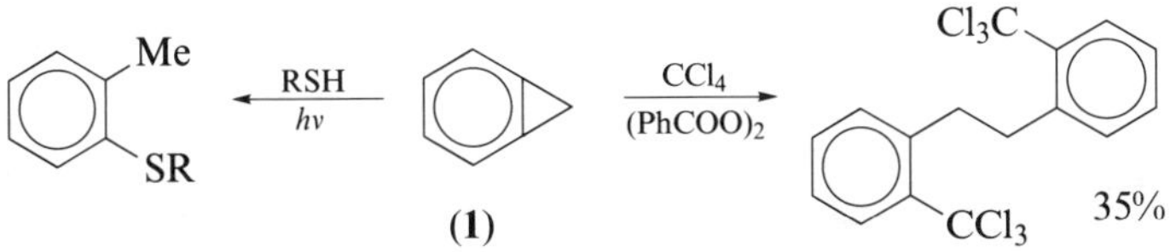

(30)

R = Et 61%
R = Ph 67%
R = 2-pyridyl 36%

naphthalenes[240]. In contrast, Ag(I) catalysed opening (in the dark) with thiophenol affords only 2-substituted product (equation 31). With both **1** and **11**, radical attack at $C_{(1a)}$ of the cycloproparene is followed by re-aromatization and opening to a benzylic (naphthalenyl) radical. Small quantities (5–10%) of the 'electrophilic' product are often isolated.

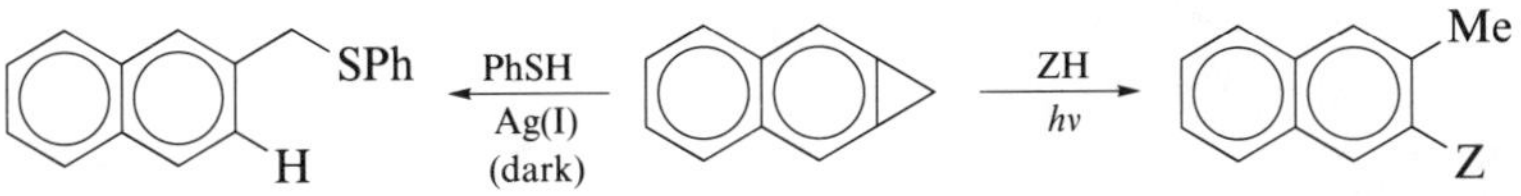

(31)

36% **(11)** Z = SPh 66%
Z = SnBu$_3$ 61%

3. *In cycloadditions*

Cycloadditions to the cycloproparenyl π framework with dienes and carbenes that likely proceed via norcaradiene intermediates have been discussed (Section V.A.3). Presented here are those processes that involve insertion into the strained three-membered ring σ bond.

The reaction of **1** with diphenylisobenzofuran (dpibf) gives norcaradiene product (Scheme 18, Section V.A.3) in competition with σ bond insertion[251–253]. When chloroform is the solvent there is no doubt that **1** acts as a 2σ donor and reacts with simple furans in a [2+2] reaction, and with dpibf in a [2+4] fashion, to give products of ring cleavage, e.g. **122** (equation 32)[253]. In THF, dpibf also behaves as a 4π component but adds to the bridge bond

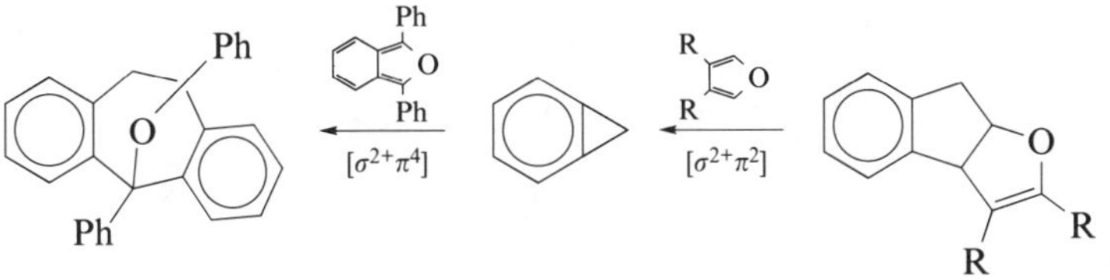

(32)

(122) (1) R = H or RR = benzo

of **1**[251] to give a product that does not transform to **122**[252]. Under the conditions for formation of **122**, **1** alone does not provide 9,10-dihydrophenanthrene, the known product of diradical ring opening[2]; a diradical intermediate is thus excluded. Steric and polar factors likely dominate and single electron transfer from dpibf is possible[252].

The interaction of **1** with dipolar reagents has led to bridged norcaradienes (Section V.A.3) of varying stability[258–260], but with electrophilic *C*,*N*-diphenylnitrone ring expanded **123** is formed (Scheme 23)[271]. Similarly[271,272], reaction with sulphonyl isocyanates provides isoindolinones **124**. Simple 1,3-dipoles such as methyl azidoformate and ethyl diazoacetate are without effect on **1**, indicating too large an energy gap between the

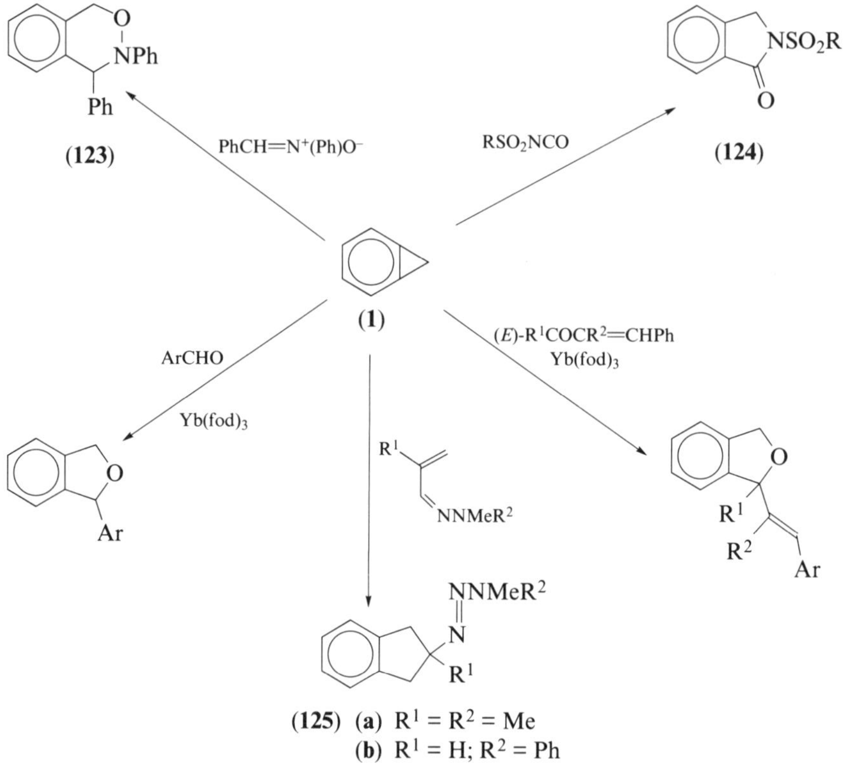

fod = 6, 6, 7, 7, 8, 8, 8,-heptafluoro-2, 2-dimethylocta-3,5-diene

SCHEME 23

electron-rich HOMO of **1** and the LUMO of the electrophilic dipole. Mechanistic information is lacking for these reactions, and distinction between attack at the π framework at $C_{(1a)}$ (path *a*, Scheme 21, Section V.B.1) and direct insertion into the σ bond cannot be made. Insertion into the three-membered ring of **11** to afford analogous $[\sigma^2+\pi^2]$ products has been effected with *N*-phenyltriazolinedione[252,264] and tetracyanoethene[264].

Perhaps of more interest is the recent account of $[\sigma^2+\pi^2]$ additions to **1** with a range of aromatic aldehydes and α,β-unsaturated carbonyl compounds. Neidlein and Krämer[273,274] have found that **1** reacts with electron-rich and electron-deficient aromatic aldehydes to give dihydroisobenzofurans by addition to the carbonyl bond (Scheme 23). Similar reactions take place with (*E*)-β-aryl-α,β-unsaturated aldehydes and ketones. At 47 °C, the reactions take a period of days and the yields are meagre. However, by employing $Yb(fod)_3$ as catalyst (1.5 mol%) yields are dramatically increased to *ca* 60–80%. It is likely that ytterbium is coordinated to the carbonyl oxygen atom thereby facilitating electrophilic attack at $C_{(1a)}$ of **1**; the reactions probably pass through a benzylic cation (path *a*, Scheme 21, Section V.B.1). Reactions with two hydrazones have provided the 2-substituted indans **125** in a reaction that is largely unaffected by added catalyst (Scheme 23). The reaction between diphenylcyclopropenone (and its thione equivalent) and **1** with ytterbium catalysis is less common[275]; an eight-membered heterocycle **126** results; the yield is good and the structure (X=O) is confirmed by X-ray analysis.

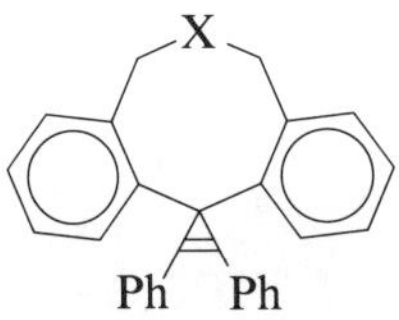

(**126**) X = O or S

4. *With organometallic reagents*

The oldest known[276] reaction of a cycloproparene with an organometallic reagent is the oxidative addition of **11** to $Fe_2(CO)_9$ followed by carbon monoxide insertion to give a metallaindanone. This same reaction (equation 33) occurs with **95** (where $C_{(1)}$ is substi-

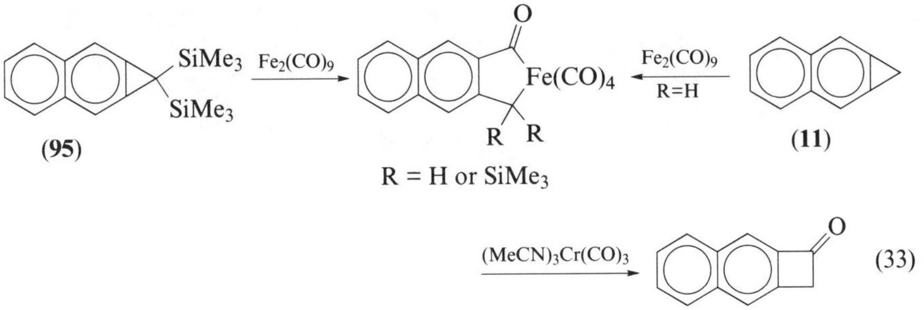

tuted with bulky trimethylsilyl groups)[164] and shows that the facility silicon has for diversion of chromium reagents away from the three-membered ring to yield η^6-complexes (Section V.A.4) is not paralleled with iron. *gem*-Difluorocyclopropabenzene gives the same ring system with $Fe_2(CO)_9$[238]. Reaction of **11** with tris(acetonitrile)tricarbonylchromium matches that with $Fe_2(CO)_9$, but proceeds further with reductive elimination of the metal to give a cyclobutanone (equation 33)[164,200]; cyclopropa[*b*]anthracene behaves analogously[164].

The formation of nickela- and palladabicyclobutanes from *gem*-difluorocyclopropabenzene has been recorded (Section V.A.4). Such bicyclobutane formation is diverted to a bridged nickelabicyclodecatriene with tris(ethene)nickel(0) when TEEDA is used in place of TMEDA (equation 34)[277]. However, the behaviour of **1**[278] and its disilyl

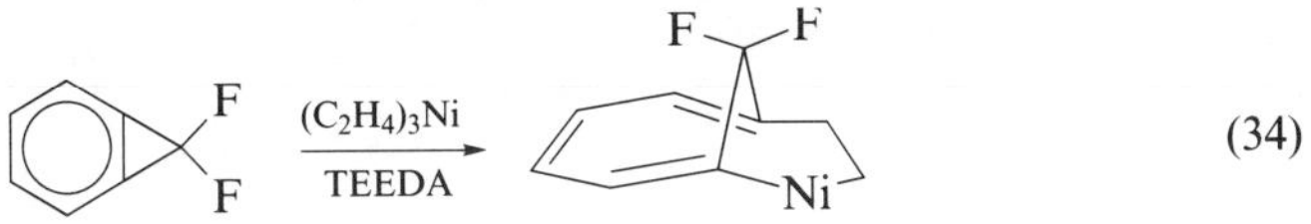

derivative[279] with the same reagent in the presence of various ligands leads to nickelacyclobutabenzenes, which survive ligand exchange (Scheme 24). Moreover, a choice of reagents and conditions pertains to the formation of products from **1**[278]. Palladacyclobutarenes have also been obtained from these same cyclopropabenzenes with (η^3-allyl)(η^5-cyclopentadienyl)palladium and trimethylphosphane (Scheme 25)[267]. For **1**, the initially formed palladacycle is unstable and the ring opens faster than ligand exchange can take place while the disilyl **127**, with two phosphane ligands, is isolated. The behaviour

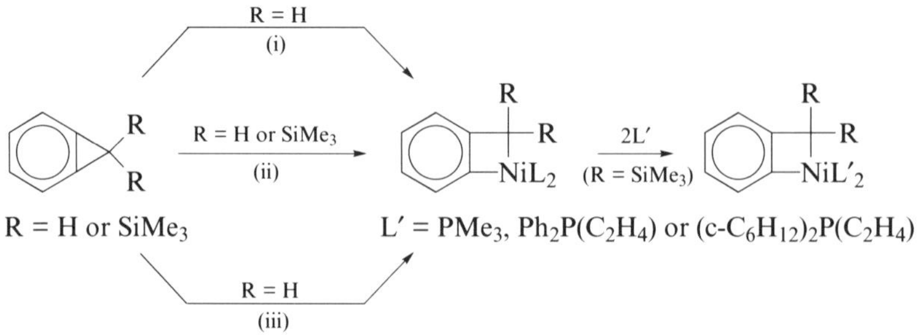

(i) $(R'_3P)_2Ni(COD)$ (R′ = Et or Bu), (ii) $(C_2H_4)_3Ni$ + L, (iii) $(Ph_3P)_4Ni$

SCHEME 24

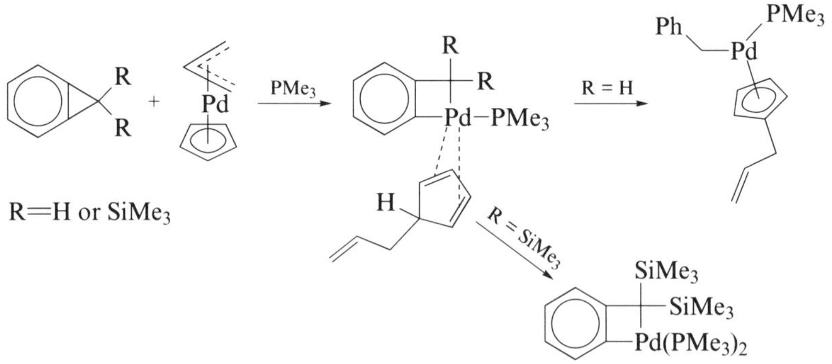

(127)

SCHEME 25

of homologue **11** has been assessed and the metallacyclobutarenes of Scheme 26 have been characterized[280]. The formation of metallacyclobutarenes by metal insertion into the strained cyclopropene σ bond has analogy to the interaction with Ag(I) discussed above (Section V.B.1).

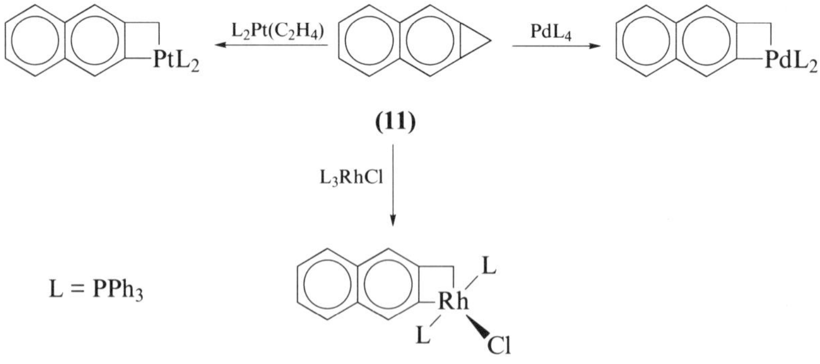

SCHEME 26

The foregoing examples have shown that a range of ligands can be employed in the synthesis of nickelacyclobutarenes. However, the fact that certain ligands react with a given substrate in one sense does not mean that they will react with another in the same way[8]; at this point in time little rationalization seems possible. Indeed, the nature of the ligand can have a marked impact upon the outcome of reaction. For example, **1** reacts with bis(trialkylphosphane)(cycloocta-1,5-diene)nickel(0) compounds. When the alkyl group is ethyl or butyl (propyl has not been examined), nickelacyclobutarenes result. However, when the bis(trimethylphosphane) analogue is used, two molecules of **1** are oxidatively added to the nickel complex with concomitant C—C bond formation and ejection of cyclooctadiene (Scheme 27)[281]. Complex **128a**, whose structure was confirmed by X-ray analysis, is obtained from **1** in 87% yield and is stable at ambient temperature under argon. It undergoes insertion reactions with the reductive elimination of the metal to give the range of compounds depicted in Scheme 27. Highly strained rocketene (**34**) behaves analogously and gives **128b** in 79% yield as a very air-sensitive metallacycle, but only at low temperature[195]. Like **128a, 128b** is carbonylated but gives the ring open bis(cycloheptatriene) **130** rather than the norcaradiene isomer **129b** and this is probably due to the presence of the strained four-membered rings. Hydrolysis of **128a,b** affords **131a,b**. A competition experiment between **1** and **34** for bis(trimethylphosphane)(cyclooctadiene)-nickel(0) (8:8:1) provided all three possible coupled products, namely **128a**, **128b** and the 'mixed' product in a ratio of 4:1:4. The result shows that the *less strained* **1** is incorporated to a *greater* extent than **34**. The outcome is inconsistent with a π bond localization analysis.

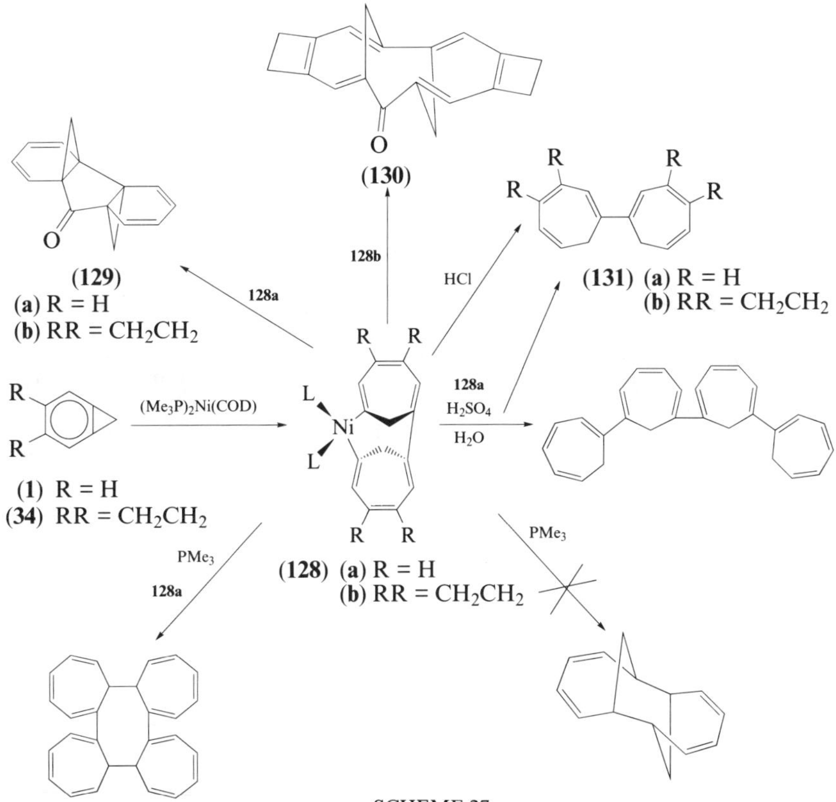

SCHEME 27

Reactions of the alkylidenecycloproparenes with rhodium(I) and platinum(0) reagents lead to metallacycles (Section VI.B.4)[282].

5. *Upon thermolysis and photolysis*

Heating cyclopropabenzene (**1**) to 80 °C in the vapour phase induces dimerization to 9,10-dihydrophenanthrene[2,250]; a diradical pathway is most likely (Scheme 28). While **11** behaves similarly, the coupling is in a head-to-tail sense to give dihydropentacene[252]. With radical initiators **1** is polymerized to poly(methylene-1,2-phenylene), a product that is also formed thermally but in much lower yield[283]. Flash vacuum pyrolysis (>500 °C) of **1** gives allene **133** together with a small amount of ethynylcyclopentadiene[284]. Product **133** is formed without randomization of the carbon skeleton and a Wolff-type rearrangement of carbene **132** is likely involved. The behaviour of **11** is strictly analogous, but the allene product is unstable and rearranges to 2-ethynylindene[285].

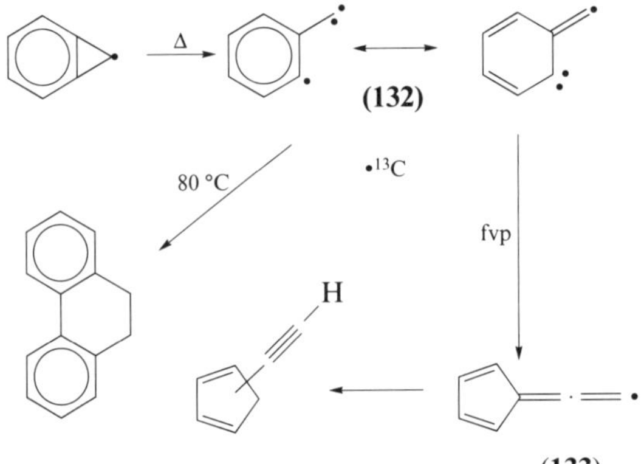

SCHEME 28

1,1-Dichloro-2,5-diphenylcyclopropabenzene (**63a**) dimerizes to give heptafulvalene derivatives[286–288] that have been adequately discussed elsewhere[8].

Carbene **132** is implicated in the photolysis of **1** since the observed[289] photodimerization to 9,10-dihydrophenanthrene and -anthracene is best explained by head-to-head and head-to-tail coupling of this species. Moreover, the fact that allene **134** is isolated[289, 290] as the major product from irradiation of diesters **31** (equation 35) is fully consistent with a photo-Wolff rearrangement of the carbene. The minor product here involves cyclization

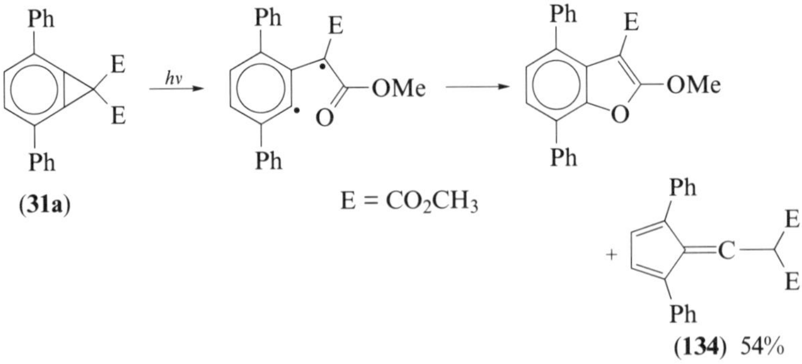

(35)

to benzofuran as illustrated by **31a**. Similar five-membered transition structures account for hydrogen atom transfer in the thermal and photochemical opening of $C_{(1)}$ substituted cycloproparenes that carry a hydrogen atom on the α carbon of the substituent chain; substituted styrenes are formed[89,223,290] and the reaction is well documented for arylcycloproparenes[6–8,26].

VI. OXO- AND ALKYLIDENECYCLOPROPARENE CHEMISTRY

A. Physical and Theoretical Aspects

Early reports[133,136] suggesting the transient existence of oxocyclopropabenzene (benzocyclopropenone) (**8**) were based on an infrared stretch at 1838 cm^{-1}. More recent studies[132,137,138] place the stretch in the range 1852–1869 cm^{-1} (Schweig and coworkers[138] 1869, Schweig and coworkers[137] 1857, Radziszewski and coworkers[132] 1852 cm^{-1}, respectively). However, the recent work of Schweig and colleagues has provided **8** in a matrix at 12 K from irradiation of cyclobutabenzene-1,2-dione (Section II.F)[138,139].

The recently available IR data[138] for **8** indicate a C—O stretch at 1869 cm^{-1} and a ring stretch at 1591 cm^{-1}. These compare with *ab initio* calculated values of 1869 and 1613 cm^{-1}. The electronic absorption spectrum was also measured, assigned and compared with that calculated. The $^1A_2$ ($n\pi^*$) symmetry forbidden band is centred at 344 nm, the $^1A_1$ ($\pi\pi^*$) band at 274 and the $^1B_2$ ($\pi\pi^*$) bands at 223 and 193 nm, respectively. In comparison to cyclobutabenzene-1,2-dione, the UV bands of **8** are shifted slightly to higher energies (*ca* 6 nm)[138].

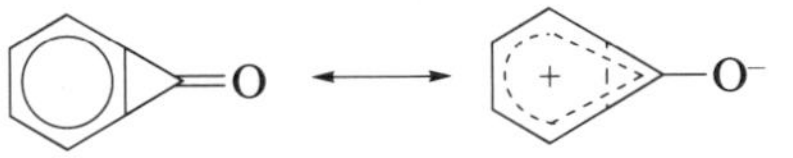

(8) **(8a)**

Both ^1H and ^{13}C NMR spectra have been recorded for **8** at 193 K in acetone-d_6 and at 213 K in a trichlorofluoromethane/1,2-dibromotetrafluoroethane/acetone-d_6 (5:5:1) mixture[139]. The fact that the compound has been handled under these conditions suggests that it will soon be the subject of chemical as well as physical study. The ^1H NMR spectrum (acetone-d_6) displays the AA′BB′ system expected for the aromatic protons at 7.82 and 8.28 ppm while the ^{13}C data show the quaternary carbons at 154.90 ppm (C=O) (cyclopropenone 155.1 ppm) and 139.33 ($C_{(1a/5a)}$). The non-equivalent methine signals resonate at 117.89 and 141.72 ppm and have been assigned to $C_{(3/4)}$ amd $C_{(2/5)}$, respectively[139]. These last assignments, made by comparison with cyclobutabenzene-1,2-dione and not verified[291], run counter to the expectation that $C_{(2)}/C_{(5)}$ of the cycloproparene are influenced markedly by the three-membered ring and appear at higher field (105–115 ppm) than a normal aromatic carbon resonance (Section III). For the alkylidenecyclopropabenzenes (see below) $C_{(1a/5a)}$ appear at *ca* 131 ppm, $C_{(2/5)}$ in the range 99–113 ppm and $C_{(3/4)}$ at *ca* 135 ppm. As the structure of **8** is expected to be stabilized by the polar form **8a**, and if the relationship that exists between cyclopropenone and methylenecyclopropene[151,292] has analogy with **8** and **9**, then **8** should derive more stabilization from its polar form than does **9**. On this basis the more logical assignment of the $C_{(2/5)}$ and $C_{(3/4)}$ resonances of **8** are 117.89 and 141.72 ppm, respectively.

Rotational constants have been calculated for **8** as *A* 5900, *B* 1700 and *C* 1300, but the geometry used for the ring system was not specified[293].

Unlike cyclopropabenzenones, alkylidenecycloproparenes are a well established class of compounds although parent **9**[50] and its -naphtho[*b*] analogue have yet to be obtained. The

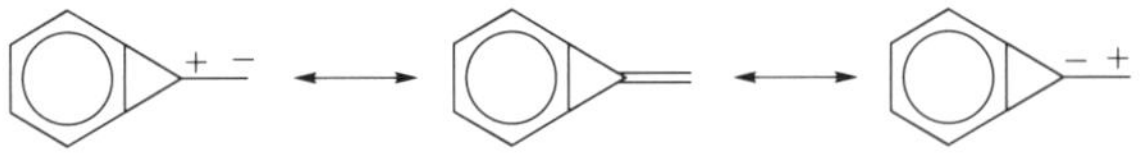

(9a) **(9)** **(9b)**

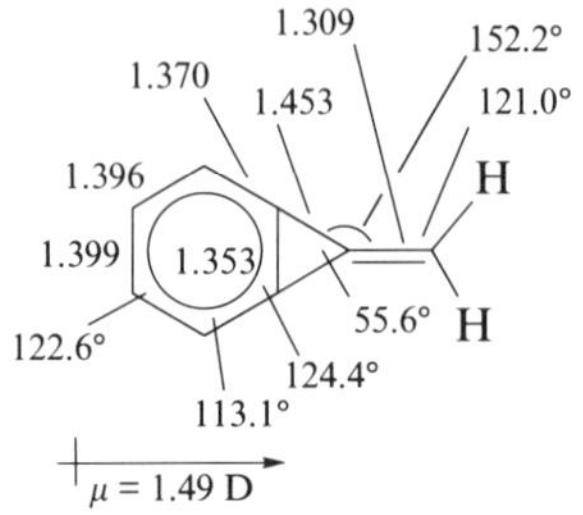

(9)

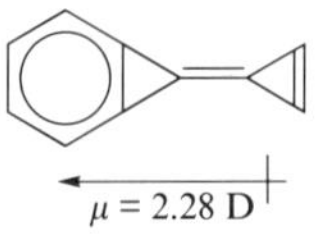

(135)

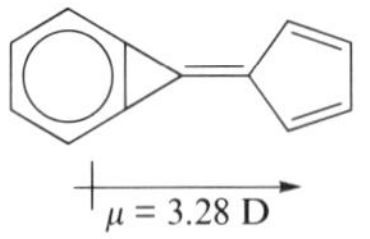

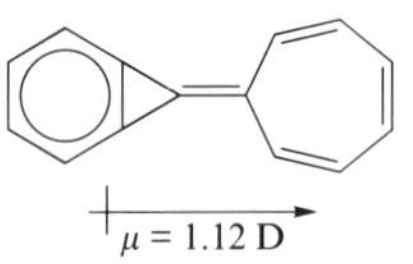

(136)

FIGURE 2. Calculated geometry of **9** (bond lengths in Å) and polarity of selected alkylidenecycloproparenes

known derivatives are stable, coloured, crystalline compounds that have a measureable polarity[153,294]. Early estimates of resonance energies (RE) and geometries of several benzo annelated derivatives (by semiempirical methods) appeared in 1965[295] and gave values close to expectation based on the number of benzenoid rings present, e.g. fluorenylidene **94a**, RE *ca* 100 kcal mol^{-1}. Not surprisingly, the geometries overestimate the bridge bond length. Recent calculations[202] predict the unknown parent **9** to have a strain energy *ca* 2 kcal mol^{-1} *less* than that of **1** because of stabilization by charge separation (μ 1.49 D) in the direction depicted by **9a**, but this does not prevent appropriate derivatives being polarized in the opposite direction and stabilized by a benzylic anion, cf. **9b** (Figure 2). Calculated bond lengths and angles for **9** and predicted dipole moments for several hydrocarbon derivatives are shown in Figure 2[202,203]. It is notable that cyclopropenylidenecyclopropabenzene (**135**) is the only hydrocarbon predicted to have its dipole oriented towards the six-membered ring and thus the cyclopropabenzenyl moiety appears to be a significant electron donor; cycloheptatrienylidene **136** is expected to have electron density directed towards the seven-membered ring (see below).

The only published structural investigation of a methylene derivative is that of 1-(diphenylmethylene)cyclopropabenzene **94c** (R=H; R^1=R^2=Ph)[155] and, while no new alkylidenecycloproparene.structures have been published since the 1989 review[8], the unpublished data reported therein have been discussed by Boese[193]. Thus the fluorenylidenecyclopropabenzene **94a** and the dibenzo[*a,e*]cycloheptatrien-7-ylidene **94b**[158], and the methylenecyclopropanaphthalenes **96a,b**[157,296] have been addressed. The effect of converting $C_{(1)}$ from sp^3 to sp^2 is to widen the $C_{(1a)}—C_{(1)}—C_{(5a)}$ angle by about 4° and further strain the remaining three-membered ring angles by *ca* 2°. The bridge bond is lengthened by

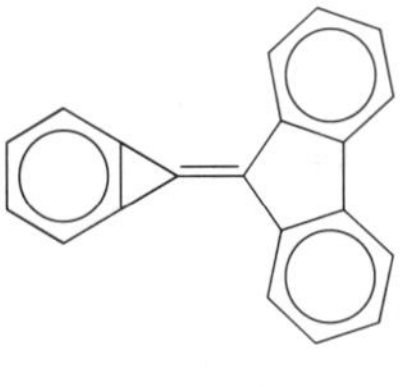

(**94a**)

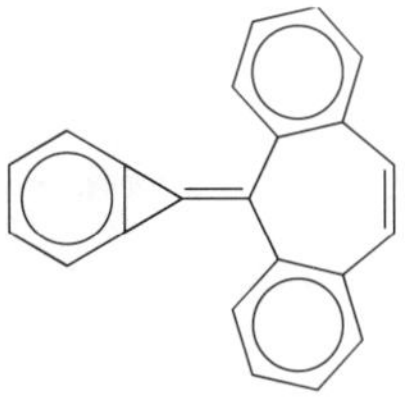

(**94b**)

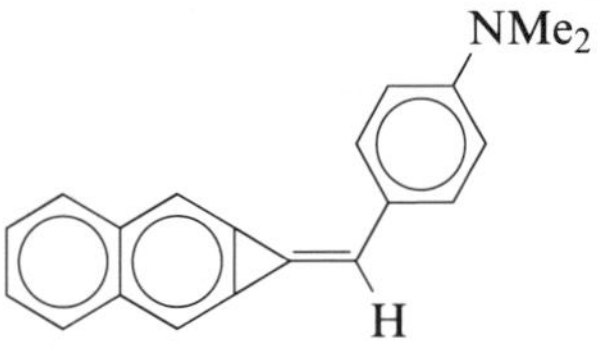

(**96a**)

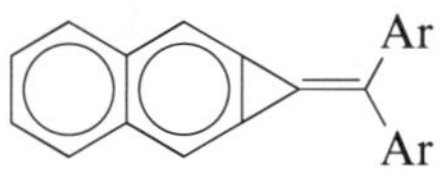

(**96b**) Ar = *p*-$Me_2NC_6H_4$

about 0.04 Å in the cyclopropabenzene derivatives and by *ca* 0.025 Å in the anilinonaphthalenes. The exocyclic double bond is in the range 1.329–1.347 Å and influenced by the known polarity of the compounds (Table 4, Section II. G)[153,157,158,294,297]. In fact, the precise parameters of the arene component are a function of whether the double bond substituents are electron withdrawing (**94a,b**) or electron donating (**96a,b**). The polarities of **94a,b** and **96a,b** lie in opposite directions and this shows that the cycloproparenyl moiety is ambiphilic. The dipole of **94b** (μ 1.1 D) should parallel its unknown parent **136** and be directed towards the seven-membered ring. The response to this electron donation and apparent enforced antiaromaticity is the formation of a *non-planar* seven-membered ring with the remote π bond tilted some 28° out of plane[193]. Equally significant is the twist angle of the anilino ring of **96a** from the plane containing the cycloproparene moiety[298]. For **96a** this is a mere 5°, and for the dianilino **96b** only 28°. These values are markedly lower than those for comparable heptafulvenes[299], e.g. 44.8° for **137**, and reflect the reduction in steric crowding about the exocyclic bond of **96** compared with **137**. Alkylidenecycloproparenes

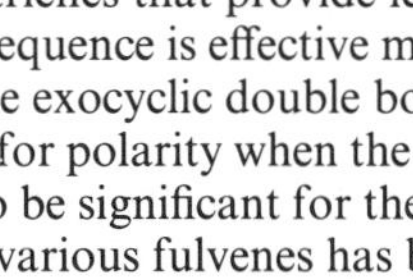

(**137**)

are also 1,6-didehydrocycloheptatrienes that provide less steric interference and notably smaller twist angles. A likely consequence is effective mesomerism because of plausible π orbital overlap; polarization of the exocyclic double bond by the *ipso*-carbon of the aryl substituent, a necessary corollary for polarity when the exocyclic group is severly rotated out of plane (cf **135**), is unlikely to be significant for the alkylidenecycloproparenes[300–302]. The impact of severe rotation on various fulvenes has been subject to MINDO/3 examination[303]. The experimental geometries and polarities of the alkylidene compounds are replicated by STO-3G and 6-31(d,p) calculations[304].

The UV spectra of the coloured aryl-substituted alkylidenecycloproparenes show long-wavelength absorption maxima, the positions of which are solvent dependent and in accord with the measured polarities (Table 4, Section II.G)[153,157,158,294,297]. Removal of the donating ability of an auxochrome such as *N,N*-dimethylamino in the aryl group by quaternization reverts the absorption maximum almost to that of the unsubstituted phenyl derivative[157]. Some of the compounds exhibit exceptional fluorescence that can be tuned with a nitrogen laser, e.g. dimethylanilino **96a** has an absolute quantum yield (Y_F) of 0.96[294,297].

In the infrared, the exocyclic methylene derivatives show stretches in the ranges 1750–1790 and 1510–1550 cm^{-1}, respectively, that compare well with the observed ranges of 1810–1880 and 1510–1550 cm^{-1} for substituted methylenecyclopropenes[11,305] and are due to coupling between the endo- and exocyclic π-bonds. The move to lower wavenumber of the higher-energy stretch is consistent with the polar nature of the compounds.

The NMR spectra of the alkylidenecycloproparenes have the aromatic proton resonances in the usual range. However, the precise chemical shift of the carbon atoms is a function of the exocyclic double bond substituents. For the *p*-substituted aryl derivatives depicted in Table 4 (Section II.G) an excellent linear correlation exists between δ and the Hammett constant σ_p^+ for *each* of the carbon atoms of the cycloproparenyl moiety and the exocyclic olefin centres[294,296]. This is consistent with mesomerism and the small twist angles of the substituents with respect to the cycloproparenyl plane[298] rather than the π polarization recorded for many other fulvenes[302]. In general, the exocyclic carbon $C_{(1')}$ falls in the range 104–112, $C_{(1)}$ 105–115, $C_{(1a/5a)}$ 129–132, $C_{(2/5)}$ 105–113 and $C_{(3/4)}$ 132–136 ppm, respectively. As occurs for the parent ring systems, the carbon atoms adjacent to the sites of ring fusion are most noticeably affected, giving rise to characteristic high-field resonances which are shifted even further upfield in the alkylidene derivatives.

Electron transfer to and from the exocyclic double bond of diphenylmethylenes **94c**[306] and **96c**[307] has been effected by electrochemical techniques. In each case reduction gives a stable anion radical while oxidation leads to a quasistable cation radical that can be further oxidized to a very short-lived dication. The spectroelectrochemical results[306,307] show that naphthalene **96c**$^{-\bullet}$ is more stable than the benzenoid **94**$^{-\bullet}$ as excitation requires

(94c)$^{-\bullet}$ **(96c)**$^{-\bullet}$

more energy (λ_{max} 519 versus 587 nm). In contrast, the cation radicals have similar stability (λ_{max} 480 versus 473 nm). However, the halfwave electrochemical oxidation potential of **96c** (0.81 eV) is higher than that of **94c** (0.68 eV) and this is ascribed to structure specific solvation. The anion radicals react with oxygen and regenerate **94c/96c**. The data argue convincingly for retention of the cycloproparenyl ring system and formulation as shown. The He(I) photoelectron spectra of these two compounds are similar[308] and their first vertical ionization potentials are essentially the same (7.15 eV); that for methylenecyclopropene is 8.41 eV[309].

B. Chemical Reactivity

1. With electrophiles and nucleophiles

Oxocycloproparenes are especially sensitive to electrophiles and nucleophiles and do not survive in solution but open to esters (Scheme 10, Section II.F)[130,131]. The chemistry of the alkylidenecycloproparenes is also dominated by reactions that involve ring opening and the compounds are far more sensitive to electrophiles[310] than to nucleophiles[140]. In general,

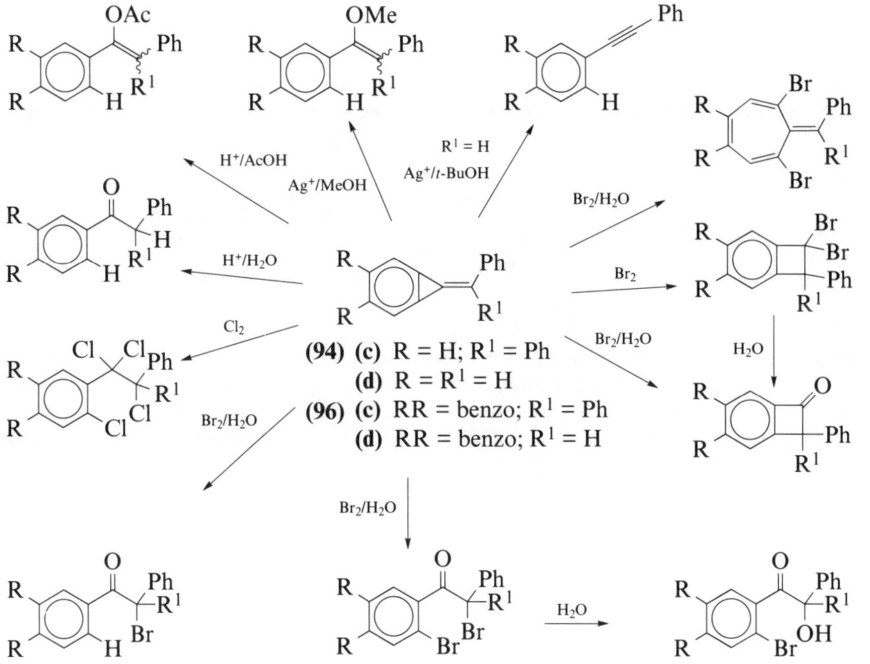

SCHEME 29

an electrophile is captured at the exocyclic centre and the resulting cycloproparenyl cation is then stabilized by one of a number of routes. However, examples of reactions with compounds carrying electron-donating substituents on the exocyclic double bond are lacking. Scheme 29 depicts a selection of the electrophilic processes recorded. It is notable that upon Ag(I) complexation, opening of the strained σ-bond proceeds as for the cycloproparenes. Addition of electrophilic bromine with concomitant ring expansion to a heptafulvene occurs for **94c** but not its homologue **96c**; reaction in the naphthalene series requires a high-energy orthoquinodimethane intermediate. Ring enlargement by cycloproparenylmethyl–cyclobutarenyl rearrangement is also recorded (Scheme 29). However, the number of compounds studied is few and does not reflect the range of compounds now known. Of particular note is the fact that cycloheptatrienylidene **94b** captures a proton at $C_{(1')}$, in accord with the dipole direction, and gives ring-cleaved and ring-expanded products[294].

Alkylidenecycloproparenes react slowly with nucleophiles and then only under forcing conditions[140]. Reactions with *t*-BuOK (refluxing THF, several days) provide heptafulvenes in *both* the benzene and naphthalene series by nucleophilic addition to the strained bridge bond and not to the exocyclic double bond (equation 36).

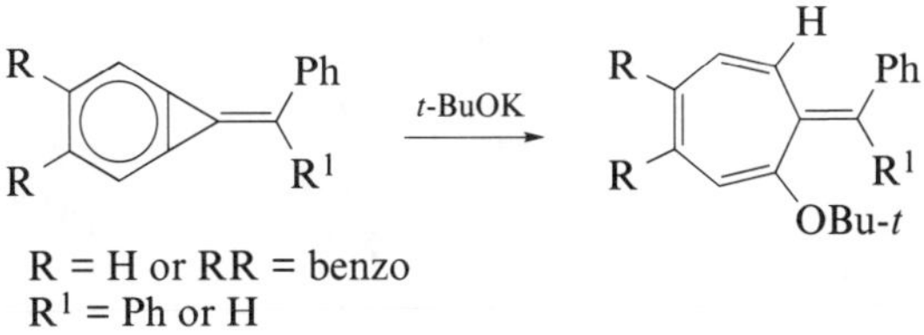

(36)

2. Upon oxidation

Oxidation of a methylenecycloproparene gives products of exocyclic double bond cleavage[140]. Epoxidation has been performed with *m*-chloroperoxybenzoic acid and 2-hydroxyethanones result (equation 37). The choice of reagent ensured that oxaspiropen-

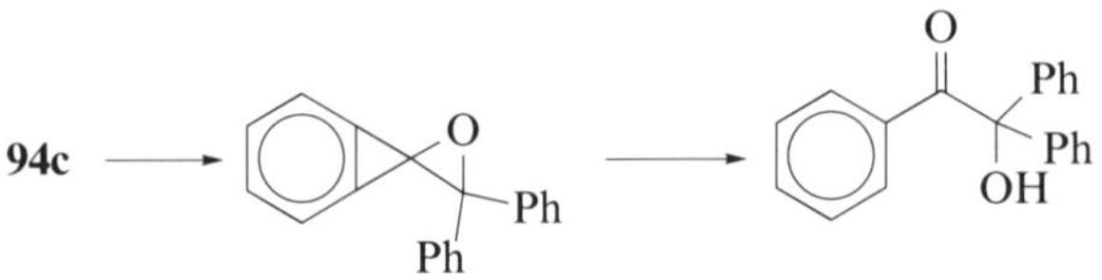

(138) (37)

tane **138** would not survive and the site of attack ($C_{(1)}$ versus $C_{(1')}$) is not known; an essential reaction with dimethyldioxirane has not been reported. With singlet oxygen, dioxetane intermediates account for product formation from the naphthalenes **96c,d** by electrocyclic opening (to oxocycloproparenes) and radical cleavage (to ethanones) (Scheme 30)[140]. The

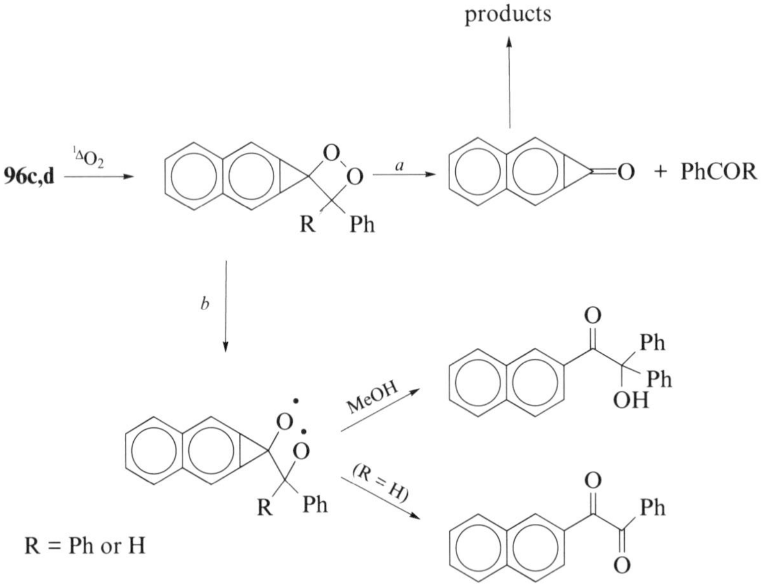

SCHEME 30

methylenecyclopropabenzene **94c** provides a complex product mixture from which only **139** has been characterized[311].

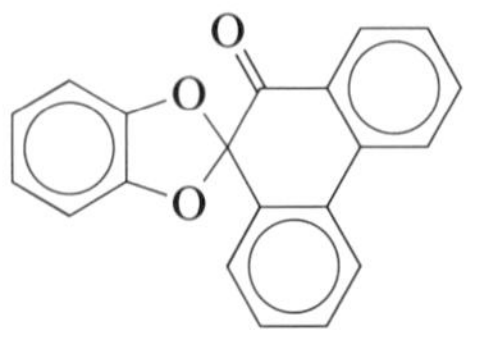

(139)

3. In cycloadditions

The alkylidenecycloproparenes provide two reactive sites for addition, namely the olefinic exocyclic double bond and the reactive bridge. In the naphthalene series all reactions occur at the exocyclic double bond and provide for strain relief (Scheme 31)[312]. These reflect a lower energy pathway than addition to the bridge with consequential orthoquinodimethane formation and loss of aromaticity in the remote ring; the outcome is supported by calculation[313]. The efficiency of diphenylisobenzofuran (dpibf) addition (Scheme 31) is markedly improved when ethylene glycol is used as solvent. For the benzenoid compounds, cycloadditions are difficult to effect but dpibf does add[313]. Crystallographic analysis shows that the product from **94c** results from addition to the bridge with retention of norcaradiene structure and corresponds to the Alder *endo*-product (equation 38). PM3-calculated[313] exothermicities concur and favour addition to the bridge bond by 1.5 kcal mol^{-1} with *endo*-product dominant by 0.4 kcal mol^{-1} from **94c**.

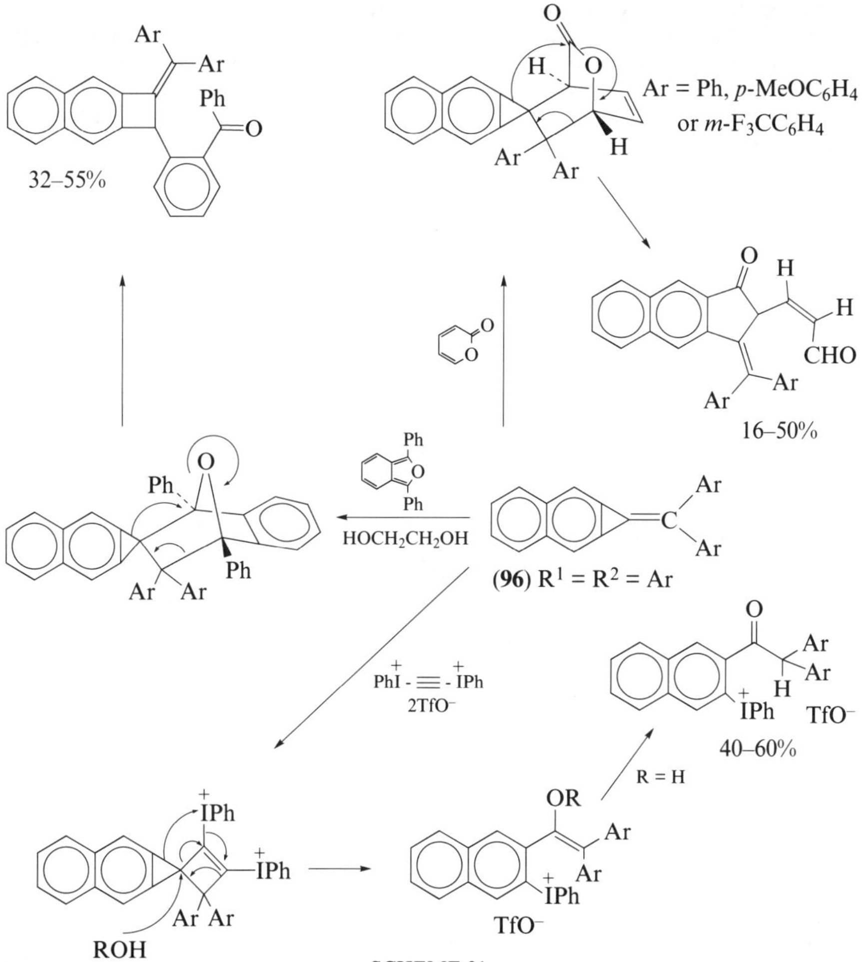

SCHEME 31

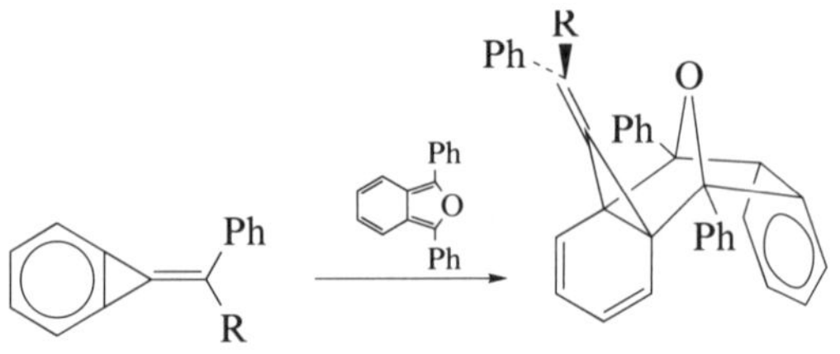

(94) R = Ph or p-MeOC$_6$H$_4$ (38)

4. *With organometallic reagents*

Although the chromium complex **104** has been transformed into its diphenylmethylene derivative[164], the known[282] reactions of alkylidenecycloproparenes with organometallic reagents provide products from interaction with the strained σ-bond. Tetrakis(triphenylphosphane)platinum(0) inserts metal to give platinacyclobutarenes (72–85%) that carry an exocyclic olefin; this is paralleled with 60–73% yields of **140** with chlorotris(triphenylphosphane)rhodium(I) (Scheme 32). In comparison, *trans*-chlorocarbonylbis(triphenylphosphane)rhodium(I) undergoes oxidative addition of the metal followed by migratory insertion of carbon monoxide into the σ bond to give rhodaalkylideneindan-2-ones **141** in 70–90% yield. Separate insertion of CO into the rhodacyclobutarenes has been examined only for **140** (RR=benzo, R^1=Ph) and this gives the isomeric indanone **142** (RR=benzo, R^1=Ph) as the major product in a reaction whose regiochemistry is temperature dependent; the preferred insertion of CO is to the weakest rhodium–carbon bond, viz. the arene–Rh bond.

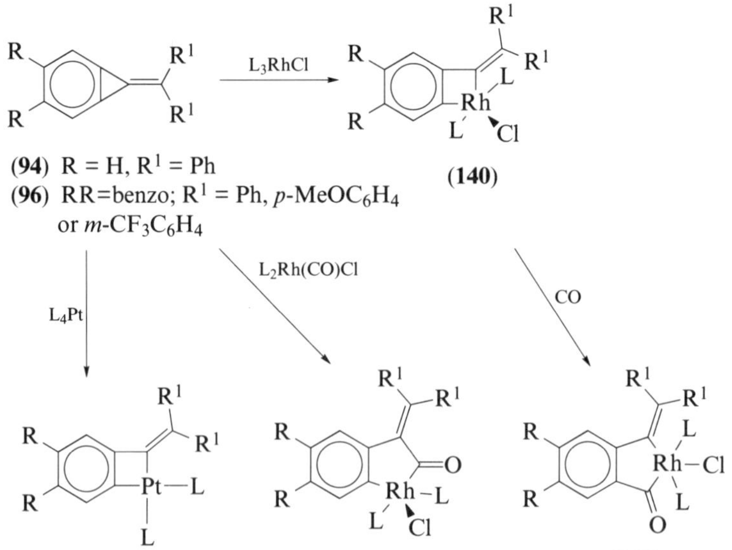

(94) R = H, R^1 = Ph
(96) RR=benzo; R^1 = Ph, p-MeOC$_6$H$_4$ or m-CF$_3$C$_6$H$_4$

(140) **(141)** **(142)**

SCHEME 32

5. Upon thermolysis and photolysis

Unlike their sp^3 progenitors, the alkylidenecycloproparenes show remarkable thermal stability and do not undergo easy ring cleavage; **96c** is recovered unchanged after several days in refluxing toluene[314]. However, under fvp conditions **96c** and its fluorenylidene analogue provide comparable, but complex, pyrolysates from which only dibenzoacephenanthrylenes have been isolated (Scheme 33)[315]. Limited as it is, this pyrolysis study shows a marked difference in behaviour to the simpler cycloproparenes (Section V.B.5).

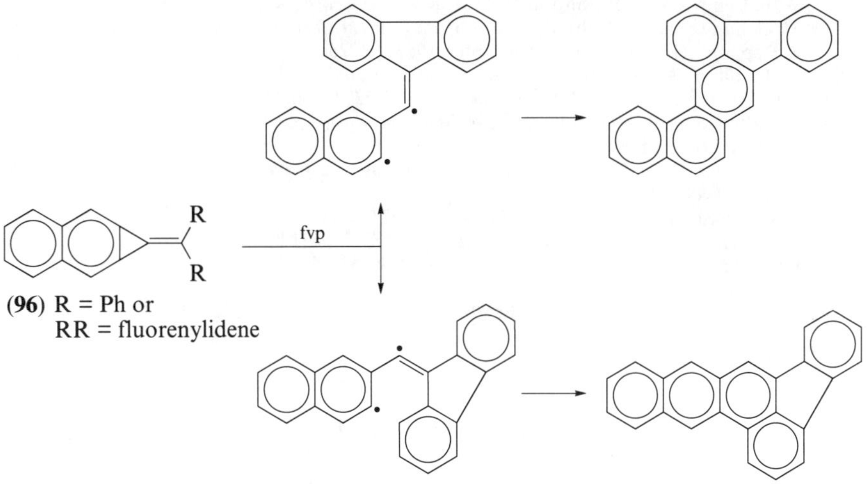

(**96**) R = Ph or
RR = fluorenylidene

SCHEME 33

There have been no accounts of alkylidenecycloproparene photolyses, but it has been suggested that the exocyclic propadienone **143** is stable to short-wavelength irradiation[316,317].

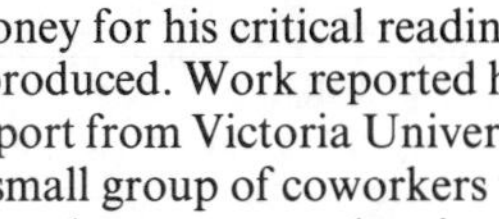

(143)

VII. ACKNOWLEDGEMENTS

The author thanks Dr M.J. Cooney for his critical reading of the manuscript and for the invaluable comments that this produced. Work reported herein from the author's laboratories has received financial support from Victoria University and has progressed because of the skill and dedication of a small group of coworkers who are named in the citations. Collaborations on aspects of cycloproparene chemistry with Professors Y. Apeloig (Technion), R. Boese (Essen) and P.J. Stang (Utah) have seen as stimulating as they have been productive.

VIII. REFERENCES

1. W. H. Perkin, *J. Chem. Soc.*, 1 (1888). The paragraph opening this chapter is taken verbatim from this source and is used by permission of the Royal Society of Chemistry.

2. E. Vogel, W. Grimme and S. Korte, *Tetrahedron Lett.*, 3625 (1965).
3. Y. Apeloig and D. Arad, *J. Am. Chem. Soc.*, **108**, 3241 (1986).
4. W. E. Billups, W. Y. Chow, K. H. Leavell, E. S. Lewis, J. L. Margrave, R. L. Sass, J. J. Shieh, P. G. Werness and J. L. Wood, *J. Am. Chem. Soc.*, **95**, 7878 (1973).
5. W. H. Mills and I. G. Nixon, *J. Chem. Soc.*, 2510 (1930).
6. B. Halton, *Chem. Rev.*, **73**, 113 (1973).
7. B. Halton, *Ind. Eng. Chem., Prod. Res. Dev.*, **19**, 349 (1980).
8. B. Halton, *Chem. Rev.*, **89**, 1161 (1989).
9. W. E. Billups, *Acc. Chem. Res.*, **11**, 245 (1978).
10. W. E. Billups, W. A. Rodin and M. M. Haley, *Tetrahedron*, **44**, 1305 (1988).
11. M. G. Banwell and B. Halton, 'Cyclopropenes', in *The Chemistry of the Cyclopropyl Group* (Ed. Z. Rappoport), Part 2, Wiley, Chichester, 1987, p.1223.
12. W. E. Billups, M. M. Haley and G.-A. Lee, *Chem. Rev.*, **89**, 1147 (1989).
13. S. C. De and D. N. Dutt, *J. Indian Chem. Soc.*, **7**, 537 (1930).
14. B. Halton, S. A. R. Harrison and C. W. Spangler, *Aust. J. Chem.*, **28**, 681 (1975).
15. A. Mustafa and M. Kamel, *J. Am. Chem. Soc.*, **75**, 2939 (1953).
16. G. W. Jones, D. R. Kerur, T. Yamazaki, H. Shechter, A. D. Woolhouse and B. Halton, *J. Org. Chem.*, **39**, 492 (1974).
17. A. G. Pinkus and J. Tsuji, *J. Org. Chem.*, **39**, 497 (1974).
18. E. F. Ullman and E. Buncel, *J. Am. Chem. Soc.*, **85**, 2106 (1963).
19. R. Anet and F. A. L. Anet, *J. Am. Chem. Soc.*, **86**, 525 (1964).
20. L. Schrader, *Tetrahedron*, **29**, 1833 (1973).
21. G. L. Closs, R. Kaplan and V. I. Bendall, *J. Am. Chem. Soc.*, **89**, 3376 (1967).
22. G. Baum, R. Bernard and H. Shechter, *J. Am. Chem. Soc.*, **89**, 5307 (1967).
23. K. Hirakawa, Y. Minami and S. Hayashi, *J. Chem. Soc., Perkin Trans. 1*, 577 (1982).
24. K. Hirakawa, T. Toki, K. Yamazaki and S. Nakazawa, *J. Chem. Soc., Perkin Trans. 1*, 1944 (1980).
25. W. Burgert, M. Grosse and D. Rewicki, *Chem. Ber.*, **115**, 309 (1982).
26. H. A. Staab and J. Ipaktschi, *Chem. Ber.*, **101**, 1457 (1968).
27. R. Bambal, H. Fritz, G. Rihs, T. Tschamber and J. Streith, *Angew. Chem., Int. Ed. Engl.*, **26**, 668 (1987).
28. S. J. Buckland, B. Halton and B. Stanovnik, *Aust. J. Chem.*, **40**, 2037 (1987).
29. H. Dürr and L. Schrader, *Angew. Chem., Int. Ed. Engl.*, **8**, 446 (1969).
30. H. Dürr and L. Schrader, *Chem. Ber.*, **103**, 1334 (1970).
31. H. Dürr, A. Ranade and I. Halberstadt, *Tetrahedron Lett.*, 3041 (1974).
32. H. Dürr and H. Schmitz, *Angew. Chem., Int. Ed. Engl.*, **14**, 647 (1975).
33. S. Braun, V. Sturm and K-O. Runzheimer, *Chem. Ber.*, **121**, 1017 (1988).
34. H. Dürr, L. Schrader and H. Seidl, *Chem. Ber.*, **104**, 391 (1971).
35. S. Mataka and M. Tashiro, *J. Org. Chem.*, **46**, 1929 (1981).
36. S. Mataka, T. Ohshima and M. Tashiro, *J. Org. Chem.*, **46**, 3960 (1981).
37. P. Radlick and H. T. Crawford, *J. Chem. Soc., Chem. Commun.*, 127 (1974).
38. C. J. Saward and K. P. C. Vollhardt, *Tetrahedron Lett.*, 4539 (1975).
39. B. Halton and D. L. Officer, unpublished observations (1979).
40. T. S. Chuah, J. T. Craig, B. Halton, S. A. R. Harrison and D. L. Officer, *Aust. J. Chem.*, **30**, 1769 (1977).
41. R. L. Hillard and K. P. C. Vollhardt, *Tetrahedron*, **36**, 2435 (1980).
42. W. E. Parham, L. D. Jones and Y. A. Sayed, *J. Org. Chem.*, **41**, 1184 (1976).
43. P. D. Brewer, J. Tagat, C. A. Hergruetner and P. Helquist, *Tetrahedron Lett.*, 4573 (1977).
44. K. Grohmann, G. Adler and T. Boccelari, as cited by R. L. Hillard and K. P. C. Vollhardt in *Tetrahedron*, **36**, 2435 (1980)—citation 11.
45. A. T. McNichols and P. J. Stang, *Synlett*, 971 (1992).
46. K. A. Biesiada, C. T. Koch and P. B. Shevlin, *J. Am. Chem. Soc.*, **102**, 2098 (1980).
47. S. Miki, M. Yoshida and Z.-i. Yoshida, *Tetrahedron Lett.*, **30**, 103 (1989).
48. S. Tanimoto, R. Schäfer, J. Ippen and E. Vogel, *Angew. Chem., Int. Ed. Engl.*, **15**, 613 (1976).
49. B. Halton, B. R. Dent, S. Bohm, D. L. Officer, H. Schmickler, F. Schophoff and E. Vogel, *J. Am. Chem. Soc.*, **107**, 7175 (1985).
50. F.-G. Klärner, B. M. J. Dogan, R. Weider, D. Ginsburg and E. Vogel, *Angew. Chem., Int. Ed. Engl.*, **25**, 346 (1986).
51. T. Watabe, K. Okada and M. Oda, *J. Org. Chem.*, **53**, 216 (1988).

52. E. Vogel, W. Püttmann, W. Duchatsch, T. Schieb, H. Schmickler and J. Lex, *Angew. Chem., Int. Ed. Engl.*, **25**, 720 (1986).
53. W. E. Billups, A. J. Blakeney and W. Y. Chow, *J. Chem. Soc., Chem. Commun.*, 1461 (1971).
54. W. E. Billups and W. Y. Chow, *J. Am. Chem. Soc.*, **95**, 4099 (1973).
55. W. E. Billups, A. J. Blakeney and W. Y. Chow, *Org. Synth.*, **55**, 12 (1976).
56. B. Halton, M. D. Diggins and A. J. Kay, *J. Org. Chem.*, **57**, 4080 (1992).
57. J. Prestien and H. Günther, *Angew. Chem., Int. Ed. Engl.*, **13**, 276 (1974).
58. R. Okazaki, M. O-oka, N. Tokitoh and N. Inamoto, *J. Org. Chem.*, **50**, 180 (1985).
59. L. K. Bee, P. J. Garratt and M. M. Mansuri, *J. Am. Chem. Soc.*, **102**, 7076 (1980).
60. D. Davalian and P. J. Garratt, *J. Am. Chem. Soc.*, **97**, 6883 (1975).
61. D. Davalian, P. J. Garratt, W. Koller and M. M. Mansuri, *J. Org. Chem.*, **45**, 4183 (1980).
62. D. Davalian, P. J. Garratt and M. M. Mansuri, *J. Am. Chem. Soc.*, **100**, 980 (1978).
63. A. R. Browne and B. Halton, *Tetrahedron*, **33**, 345 (1977).
64. D. Davalian and P. J. Garratt, *Tetrahedron Lett.*, 2815 (1976).
65. J. Ippen and E. Vogel, *Angew. Chem., Int. Ed. Engl.*, **13**, 736 (1974).
66. B. Halton, R. Boese, D. Blaser and Q. Lu, *Aust. J. Chem.*, **44**, 265 (1991).
67. E. Vogel and J. Sombroek, *Tetrahedron Lett.*, 1627 (1974).
68. M. G. Banwell, B. Halton, T. W. Hambly, N. K. Ireland, C. Papamihail, S. G. G. Russell and M. R. Snow, *J. Chem. Soc., Perkin Trans 1*, 715 (1992).
69. M. G. Banwell, R. Blattner, A. R. Browne, J. T. Craig and B. Halton, *J. Chem. Soc., Perkin Trans. 1*, 2165 (1977).
70. R. Neidlein and V. Poignée, *Chem. Ber.*, **121**, 1199 (1988).
71. B. Halton, C. J. Randall, G. J. Gainsford and W. T. Robinson, *Aust. J. Chem.*, **40**, 475 (1987).
72. B. Halton and C. J. Randall, *Tetrahedron Lett.*, **23**, 5591 (1982).
73. W. E. Billups, W. T. Chamberlain and M. Y. Asim, *Tetrahedron Lett.*, 571 (1977).
74. P. J. Garratt and W. Koller, *Tetrahedron Lett.*, 4177 (1976).
75. A. Kumar, S. R. Tayal and D. Devaprabhakara, *Tetrahedron Lett.*, 863 (1976).
76. W. E. Billups, L.-J. Lin, B. E. Arney, W. A. Rodin and E. W. Casserly, *Tetrahedron Lett.*, **25**, 3935 (1984).
77. B. Halton, J. H. Bridle and E. G. Lovett, *Tetrahedron Lett.*, **31**, 1313 (1990).
78. I. J. Anthony, Y. B. Kang and D. Wege, *Tetrahedron Lett.*, **31**, 1315 (1990).
79. B. Halton and E. G. Lovett, *Struct. Chem.*, **2**, 147 (1991).
80. W. E. Billups, L. P. Lin and W. Y. Chow, *J. Am. Chem. Soc.*, **96**, 4026 (1974).
81. W. E. Billups, L. E. Reed, E. W. Casserly and L. P. Lin, *J. Org. Chem.*, **46**, 1326 (1981).
82. B. Halton and D. L. Officer, *Tetrahedron Lett.*, **22**, 3687 (1981).
83. B. Halton and D. L. Officer, *Aust. J. Chem.*, **36**, 1167 (1983).
84. E. Vogel, S. Korte, W. Grimme and H. Günther, *Angew. Chem., Int. Ed. Engl.*, **7**, 289 (1968).
85. C. Glück, V. Poignée and H. Schwager, *Synthesis*, 260 (1987).
86. B. Halton and P. J. Milsom, *J. Chem. Soc., Chem. Commun.*, 814 (1971).
87. B. Halton, P. J. Milsom and A. D. Woolhouse, *J. Chem. Soc., Perkin Trans. 1*, 731 (1977).
88. A. R. Browne and B. Halton, *J. Chem. Soc., Chem. Commun.*, 1341 (1972).
89. A. R. Browne and B. Halton, *J. Chem. Soc., Perkin Trans. 1*, 1177 (1977).
90. Y. Apeloig, D. Arad, M. Kapon and M. Wallerstein, *Tetrahedron Lett.*, **28**, 5917 (1987).
91. P. Müller, G. Bernardinelli, D. Rodriguez, J. Pfyffer and J.-P. Schaller, *Helv. Chim. Acta*, **71**, 544 (1988)
92. P. Müller, R. Etienne, J. Pfyffer, N. Pineda and M. Schipoff, *Tetrahedron Lett.*, 3151 (1978).
93. P. Müller, R. Etienne, J. Pfyffer, N. Pineda and M. Schipoff, *Helv. Chim. Acta*, **61**, 2482 (1978).
94. P. Müller and D. Rodriguez, *Helv. Chim. Acta*, **69**, 1546 (1986).
95. P. Müller and H.-C. Nguyen-Thi, *Tetrahedron Lett.*, **21**, 2145 (1980).
96. P. Müller and H. C. N. Thi, *Isr. J. Chem.*, **21**, 135 (1981).
97. P. Müller and M. Rey, *Helv. Chim. Acta*, **64**, 354 (1981).
98. P. Müller and D. Rodriguez, *Helv. Chim. Acta*, **66**, 2540 (1983).
99. B. Halton and D. L. Officer, *Aust. J. Chem.*, **36**, 1291 (1983).
100. K. Mackenzie, A. S. Miller, K. W. Muir and L. Manojlovic, *Tetrahedron Lett.*, **24**, 4747 (1983).
101. W. E. Billups and W. A. Rodin, *J. Org. Chem.*, **53**, 1312 (1988).
102. W. E. Billups, B. E. Arney and L.-J. Lin, *J. Org. Chem.*, **49**, 3436 (1984).
103. W. E. Billups, M. M. Haley, R. C. Claussen and W. A. Rodin, *J. Am. Chem. Soc.*, **113**, 4331 (1991).
104. W. E. Billups, E. W. Casserly and B. E. Arney, *J. Am. Chem. Soc.*, **106**, 440 (1984).

105. P. Müller and D. Rodriguez, *Helv. Chim. Acta*, **68**, 975 (1985).
106. P. Müller and J.-P. Schaller, *Helv. Chim. Acta*, **73**, 1228 (1990).
107. I. J. Anthony and D. Wege, *Tetrahedron Lett.*, **28**, 4217 (1987).
108. P. J. Garratt, D. Payne and A. Tsotinis, *Pure Appl. Chem.*, **62**, 525 (1990).
109. P. J. Garratt and A. Tsotinis, *J. Org. Chem.*, **55**, 84 (1990).
110. P. Müller and H.-C. Nguyen-Thi, *Helv. Chim. Acta*, **67**, 467 (1984).
111. P. Müller and H. C. N. Thi, *Chimia*, **39**, 362 (1985).
112. P. Müller, G. Bernardinelli and H.-C. G. Nguyen-Thi, *Helv. Chim. Acta*, **72**, 1627 (1989).
113. U. H. Brinker, H. Wüster and G. Maas, *Angew. Chem., Int. Ed. Engl.*, **26**, 577 (1987).
114. N. Galloway and B. Halton, *Aust. J. Chem.*, **32**, 1743 (1979).
115. B. R. Dent and B. Halton, *Tetrahedron Lett.*, **25**, 4279 (1984).
116. B. R. Dent and B. Halton, *Aust. J. Chem.*, **39**, 1789 (1986).
117. B. R. Dent and B. Halton, *Aust. J. Chem.*, **40**, 925 (1987).
118. P. Müller, H. C. N. Thi and J. Pfyffer, *Helv. Chim. Acta*, **69**, 855 (1986).
119. M. P. Cava and K. Narasimhan, *J. Org. Chem.*, **36**, 1419 (1971).
120. K. Giebel and J. Heindl, *Tetrahedron Lett.*, 2133 (1970).
121. K. Narasimhan and P. R. Kumar, *J. Org. Chem.*, **48**, 1122 (1983).
122. P. Müller and J.-P. Schaller, *Chimia*, **40**, 430 (1986).
123. H. N. C. Wong, *Acc. Chem. Res.*, **22**, 145 (1989).
124. P. Müller and J.-P. Schaller, *Helv. Chim. Acta*, **72**, 1608 (1989).
125. P. Müller and J.-P. Schaller, *Tetrahedron Lett.*, **30**, 1507 (1989).
126. E. K. Fields and S. Meyerson, *J. Chem. Soc., Chem. Commun.*, 474 (1965).
127. R. F. C. Brown, D. V. Gardner, J. F. W. McOmie and R. K. Solly, *Aust. J. Chem.*, **20**, 139 (1967).
128. R. F. C. Brown and R. K. Solly, *Aust. J. Chem.*, **19**, 1045 (1966).
129. J. Adamson, D. L. Forster, T. L. Gilchrist and C. W. Rees, *J. Chem. Soc., Chem. Commun.*, 221 (1969).
130. J. Adamson, D. L. Forster, T. L. Gilchrist and C. W. Rees, *J. Chem. Soc. (C)*, 981 (1971).
131. M. S. Ao, E. M. Burgess, A. Schauer and E. A. Taylor, *J. Chem. Soc., Chem. Commun.*, 220 (1969).
132. J. G. Radziszewski, B. A. Hess and R. Zahradnik, *J. Am. Chem. Soc.*, **114**, 52 (1992).
133. O. L. Chapman, C.-C. Chang, J. Kole, N. R. Rosenquist and H. Tomioka, *J. Am. Chem. Soc.*, **97**, 6586 (1975).
134. H. Dürr, H. Nickels and W. Philippi, *Tetrahedron Lett.*, 4387 (1978).
135. M. A. O'Leary and D. Wege, *Tetrahedron Lett.*, 2811 (1978).
136. O. L. Chapman, K. Mattes, C. L. McIntosh, J. Pacansky, G. V. Calder and G. Orr, *J. Am. Chem. Soc.*, **95**, 6134 (1973).
137. J. G. G. Simon, N. Münzel and A. Schweig, *Chem. Phys. Lett.*, **170**, 187 (1990).
138. J. G. G. Simon, A. Schweig, Y. Xie and H. F. Schaefer, *Chem. Phys. Lett.*, **200**, 631 (1992).
139. J. G. G. Simon and A. Schweig, *Chem. Phys. Lett.*, **201**, 377 (1993).
140. S. J. Buckland, B. Halton and P. J. Stang, *Aust. J. Chem.*, **41**, 845 (1988).
141. T. L. Gilchrist and C. W. Rees, *J. Chem. Soc. (C)*, 1763 (1969).
142. O. L. Chapman, J. Gano, P. R. West, M. Regitz and G. Maas, *J. Am. Chem. Soc.*, **103**, 7033 (1981).
143. M. G. Reinecke, L.-J. Chen and A. Almqvist, *J. Chem. Soc., Chem. Commun.*, 585 (1980).
144. J. H. Teles, B. A. Hess and L. J. Schaad, *Chem. Ber.*, **125**, 423 (1992).
145. R. Bloch and P. Orvane, *Tetrahedron Lett.*, **22**, 3597 (1981).
146. E. A. Dorko, *J. Am. Chem. Soc.*, **87**, 5518 (1965).
147. H. Hopf and G. Mass, *Angew. Chem., Int. Ed. Engl.*, **31**, 931 (1992).
148. W. E. Billups, L.-J. Lin and E. W. Casserly, *J. Am. Chem. Soc.*, **106**, 3698 (1984).
149. S. W. Staley and T. D. Norden, *J. Am. Chem. Soc.*, **106**, 3699 (1984).
150. T. D. Norden, S. W. Staley, W. H. Taylor and M. D. Harmony, *J. Am. Chem. Soc.*, **108**, 7912 (1986).
151. S. W. Staley, T. D. Norden, W. H. Taylor and M. D. Harmony, *J. Am. Chem. Soc.*, **109**, 7641 (1987).
152. I. A. D'yakonov, T. V. Mandel'shtam and É. M. Kharicheva, *J. Org. Chem. USSR*, **6**, 2260 (1970); *Chem. Abstr.*, **74**, 42213 (1971).
153. B. Halton and P. J. Stang, *Acc. Chem. Res.*, **20**, 443 (1987).
154. B. Halton, C. J. Randall and P. J. Stang, *J. Am. Chem. Soc.*, **106**, 6108 (1984).

155. B. Halton, C. J. Randall, G. J. Gainsford and P. J. Stang, *J. Am. Chem. Soc.*, **108**, 5949 (1986).
156. B. Halton, S. J. Buckland, Q. Mei and P. J. Stang, *Tetrahedron Lett.*, **27**, 5159 (1986).
157. B. Halton, Q. Lu and P. J. Stang, *J. Chem. Soc., Chem. Commun.*, 879 (1988).
158. B. Halton, S. J. Buckland, Q. Lu, Q. Mei and P. J. Stang, *J. Org. Chem.*, **53**, 2418 (1988).
159. B. Halton and T. W. Davey, unpublished observations (1993).
160. Q. Lu, Ph. D. Thesis, Victoria University of Wellington, 1989.
161. B. Halton, Q. Lu and P. J. Stang, *Aust. J. Chem.*, **43**, 1277 (1990).
162. C. Eaborn, R. Eidenschink, S. J. Harris and D. M. R. Walton, *J. Organomet. Chem.*, **124**, C27 (1977).
163. C. Eaborn and J. G. Stamper, *J. Organomet. Chem.*, **192**, 155 (1980).
164. P. Müller, G. Bernardinelli and Y. Jacquier, *Helv. Chim. Acta*, **75**, 1995 (1992).
165. A. T. McNichols, P. J. Stang, D. M. Addington and B. Halton, *Tetrahedron Lett.*, **35**, 437 (1994).
166. R. Neidlein, V. Poignée, W. Kramer and C. Glück, *Angew. Chem., Int. Ed. Engl.*, **25**, 731 (1986).
167. M. J. Cooney and B. Halton, unpublished observations (1994).
168. B. Halton, A. J. Kay and Z. M. Zha, *J. Chem. Soc., Perkin Trans. 1*, 2239 (1993).
169. H. C. Longuet-Higgins and C. A. Coulson, *Trans. Faraday Soc.*, **42**, 756 (1946).
170. C. S. Cheung, M. A. Cooper and S. L. Manatt, *Tetrahedron*, **27**, 701 (1971).
171. B. Halton and M. P. Halton, *Tetrahedron*, **29**, 1717 (1973).
172. M. J. S. Dewar and H. S. Rzepa, *J. Am. Chem. Soc.*, **100**, 58 (1978).
173. M. K. Mahanti, *Indian J. Chem.*, **19**, 149 (1980).
174. M. Eckert-Maksic, Z. B. Maksic, M. Hodoscek and K. Poljanec, *Int. J. Quantum Chem.*, **42**, 869 (1992).
175. Z. B. Maksic, M. Eckert-Maksic, D. Kovancek and D. Margetic, *J. Mol. Struct. (Theochem)*, **260**, 241 (1992).
176. J. E. Bloor, M. Eckert-Maksic, M. Hodoscek, Z. B. Maksic and K. Poljanec, *New J. Chem.*, **17**, 157 (1993).
177. M. Hodoscek, D. Kovacek and Z. B. Maksic, *Theochem*, **100**, 213 (1993).
178. W. Koch, M. Eckert-Maksic and Z. B. Maksic, *J. Chem. Soc., Perkin Trans. 2*, 2195 (1993).
179. W. Koch, M. Eckert-Maksic and Z. B. Maksic, *Int. J. Quantum Chem.*, **48**, 319 (1993).
180. M. Eckert-Maksic, Z. B. Maksic, M. Hodoscek and K. Poljanec, *J. Mol. Struct. (Theochem)*, **285**, 187 (1993).
181. M. Hodoscek, D. Kovacek and Z. B. Maksic, *Theor. Chim. Acta*, **86**, 343 (1993).
182. Z. B. Maksic, M. Eckert-Maksic and K. H. Pfeifer, *J. Mol. Struct.*, **300**, 445 (1993).
183. D. Kovacek, D. Margetic and Z. B. Maksic, *J. Mol. Struct.*, **104**, 195 (1993).
184. P. C. Hiberty, G. Ohanessian and F. Delbecq, *J. Am. Chem. Soc.*, **107**, 3095 (1985).
185. A. G. Davies and K. M. Ng, *J. Chem. Soc., Perkin Trans. 2*, 1857 (1992); D. V. Avila, A. G. Davies, E. R. Li and M. K. Ng, *J. Chem. Soc. Perkin Trans. 2*, 355 (1993).
186. Y. Apeloig, D. Arad, B. Halton and C. J. Randall, *J. Am. Chem. Soc.*, **108**, 4932 (1986).
187. K. Baldridge and J. S. Siegel, *J. Am. Chem. Soc.*, **114**, 9583 (1992).
188. A. Stanger, *J. Am. Chem. Soc.*, **113**, 8277 (1991).
189. K. Jug and A. M. Köster, *J. Am. Chem. Soc.*, **112**, 6772 (1990).
190. R. Benassi, S. Ianelli, M. Nardelli and F. Taddei, *J. Chem. Soc., Perkin Trans. 2*, 1381 (1991).
191. E. Carstensen-Oeser, B. Müller and H. Dürr, *Angew. Chem., Int. Ed. Engl.*, **11**, 422 (1972).
192. B. Halton, T. J. McLellan and W. T. Robinson, *Acta Crystallogr., Sect. B*, **B32**, 1889 (1976).
193. R. Boese, 'Structural Studies of Strained Molecules', in *Advances in Strain in Organic Chemistry* (Ed. B. Halton), Vol. 2, JAI Press, London, (1992), p. 191.
194. R. Neidlein, D. Christen, V. Poignée, R. Boese, D. Bläser, A. Gieren, C. Ruiz-Pérez and T. Hüber, *Angew. Chem., Int. Ed. Engl.*, **27**, 294 (1988).
195. D. Bläser, R. Boese, W. A. Brett, P. Rademacher, H. Schwager, A. Stanger and K. P. C. Vollhardt, *Angew. Chem., Int. Ed. Engl.*, **28**, 206 (1989).
196. R. Boese, D. Bläser, W. E. Billups, M. M. Haley, A. H. Maulitz, D. L. Mohler and K. P. C. Vollhardt, *Angew. Chem., Int. Ed. Engl.*, **33**, 313 (1994).
197. A. Toyata, *J. Chem. Soc., Perkin Trans. 2*, 85 (1984).
198. R. Diercks and K. P. C. Vollhardt, *J. Am. Chem. Soc.*, **108**, 3150 (1986); A. Stanger and K. P. C. Vollhardt, *J. Org. Chem.*, **53**, 4889 (1988).
199. P. Müller, G. Bernardinelli and Y. Jacquier, *Helv. Chim, Acta*, **71**, 1328 (1988).
200. P. Müller, G. Bernardinelli, Y. Jacquier and A. Ricca, *Helv. Chim. Acta*, **72**, 1618 (1989).

201. H. Dürr, personal communication (1988). I thank Professor Dürr for making these data available.
202. Y. Apeloig, personal communication (1989). I thank Professor Apeloig for making these results available.
203. Y. Apeloig, M. Karni and D. Arad, 'Cyclopropabenzenes and Alkylidenecyclopropabenzenes. A Synergistic Relationship Between Theory and Experiment', in *Strain and Its Implications* (Eds. A. de Meijere and S. Blechert), Reidel, Dordrecht, 1989, p. 457.
204. M. J. S. Dewar and M. K. Holloway, *J. Chem. Soc., Chem. Commun.*, 1188 (1984).
205. F. Brogli, E. Giovanni, E. Heilbronner and R. Schurter, *Chem. Ber.*, **106**, 961 (1973).
206. J. Lecoultre, E. Heilbronner, P. Müller and D. Rodriguez, *Collect. Czech. Chem. Commun.*, **53**, 2385 (1988).
207. C. Santiago, R. W. Gandour, K. N. Houk, W. Nutakul, W. E. Cravey and R. P. Thummel, *J. Am. Chem. Soc.*, **100**, 3730 (1978).
208. V. Galasso, *Chem. Phys. Lett.*, **166**, 415 (1990).
209. M. A. Cooper and S. L. Manatt, *J. Am. Chem. Soc.*, **92**, 1605 (1970).
210. H. Günther and G. Jikeli, *Chem. Rev.*, **77**, 599 (1977).
211. H. Günther, A. Shyoukh, D. Cremer and K.-H. Frisch, *Justus Liebigs Ann. Chem.*, 150 (1978).
212. H. Günther, G. Jikeli, H. Schmickler and J. Prestien, *Angew. Chem., Int. Ed. Engl.*, **12**, 762 (1973).
213. H. Günther and H. Seel, *Org. Magn. Reson.*, **8**, 299 (1976).
214. H. Günther and W. Herrig, *J. Am. Chem. Soc.*, **97**, 5594 (1975).
215. A. Streitwieser, G. Ziegler, P. Mowery, A. Lewis and R. Lawler, *J. Am. Chem. Soc.*, **90**, 1333 (1968).
216. G. A. Singy, J. Pfyffer, P. Müller and A. Buchs, *Org. Mass Spectrom.*, **11**, 499 (1976).
217. E. Wentrup-Byrne, F. O. Gülaçar, P. Müller and A. Buchs, *Org. Mass Spectrom.*, **12**, 636 (1977).
218. E. Uggerud, D. Arad, Y. Apeloig and H. Schwatz, *J. Chem. Soc., Chem. Commun.*, 1015 (1989).
219. J. H. Bowie, T. Blumenthal, M. H. Laffer, S. Janposri and G. E. Gream, *Aust. J. Chem.*, **37**, 1447 (1984).
220. M. A. Battiste and B. Halton, *J. Chem. Soc., Chem. Commun.*, 1368 (1968).
221. J. D. Roberts, A. Streitwieser and C. M. Regan, *J. Am. Chem. Soc.*, **74**, 4579 (1952).
222. O. Sinanoglu, *Tetrahedron Lett.*, **29**, 889 (1988).
223. B. Halton, A. D. Woolhouse and P. J. Milsom, *J. Chem. Soc., Perkin Trans. 1*, 735 (1977).
224. P. Müller, *J. Chem. Soc., Chem. Commun.*, 895 (1973).
225. P. Müller, *Helv. Chim. Acta*, **57**, 704 (1974).
226. B. Halton, A. D. Woolhouse, H. M. Hugel and D. P. Kelly, *J. Chem. Soc., Chem. Commun.*, 247 (1974).
227. B. Halton, H. M. Hugel, D. P. Kelly, P. Müller and U. Burger, *J. Chem. Soc., Perkin Trans. 2*, 258 (1976).
228. U. Burger, P. Müller and L. Zuidema, *Helv. Chim. Acta*, **57**, 1881 (1974).
229. P. Müller and D. Rodriguez, *Helv. Chim. Acta*, **69**, 1546 (1986).
230. P. Müller, *Helv. Chim. Acta*, **56**, 500 (1973).
231. W. E. Billups and W. Y. Chow, unpublished observations (1978)—citation 39 in Reference 9.
232. G. Szeimies and P. Wimmer, unpublished observations; P. Wimmer, Diplomarbeit, University of Munich (1985). I thank Professor Szeimies for making these data available.
233. V. R. Sandel and H. H. Freedman, *J. Am. Chem. Soc.*, **85**, 2328 (1963).
234. R. Breslow and J. Schwarz, *J. Am. Chem. Soc.*, **105**, 6795 (1983).
235. R. Neidlein and M. Kohl, *Helv. Chim. Acta*, **73**, 1497 (1990).
236. R. Neidlein, T. Constantinescu and M. Kohl, *Phosphorus, Sulfur and Silicon*, **59**, 165 (1991).
237. R. Neidlein and M. Kohl, personal communication (1991). I thank Professor Neidlein for providing these results.
238. M. Kohl, Ph. D. Thesis, Universität Heidelberg, 1990.
239. K. U. Ingold, personal communication (1985). I thank Professor Ingold for this information.
240. C. C. L. Chai, B. Halton and M. A. E. Starr, unpublished observations (1994). M. A. E. Starr, M.Sc. Thesis, Victoria University of Wellington, 1994.
241. M. Baumgarten, B. Halton and K. Müllen, unpublished observations (1993).
242. R. Neidlein and D. Christen, *Helv. Chim. Acta*, **69**, 1623 (1986).
243. R. Neidlein, personal communication (1991). I thank Professor Neidlein for providing these results.

244. M. Eckert-Maksic, Z. B. Maksic and M. Klessinger, *J. Chem. Soc., Perkin Trans. 2*, 285 (1994).
245. B. Halton and C. J. Randall, *J. Am. Chem. Soc.*, **105**, 6310 (1983).
246. R. Okazaki, T. Hasegawa and Y. Shishido, *J. Am. Chem. Soc.*, **106**, 5271 (1984).
247. D. Christen, Ph.D. Thesis, Universität Heidelberg, 1986.
248. B. Halton and S. G. G. Russell, *Aust. J. Chem.*, **45**, 911 (1992).
249. P. Warner, personal communication (1994). I thank Professor Warner for performing these calculations.
250. S. Korte, Ph.D. Thesis, Universität Köln, 1968.
251. U. H. Brinker and H. Wüster, *Tetrahedron Lett.*, **32**, 593 (1991).
252. B. Halton and S. G. G. Russell, *Aust. J. Chem.*, **43**, 2099 (1990).
253. K. Saito, H. Ishihara and S. Kagabu, *Bull. Soc. Chem. Jpn.*, **60**, 4141 (1987).
254. E. Vogel, J. Ippen and V. Buch, *Angew. Chem., Int. Ed. Engl.*, **14**, 566 (1975).
255. M. L. Maddox, J. C. Martin and J. M. Muchowski, *Tetrahedron Lett.*, **21**, 7 (1980).
256. J. C. Martin and J. M. Muchowski, *J. Org. Chem.*, **49**, 1040 (1984).
257. R. Neidlein and L. Tadesse, *Helv. Chim. Acta*, **71**, 249 (1988).
258. M. Nitta, S. Sogo and T. Nakayama, *Chem. Lett.*, 1431 (1979).
259. H. Kato, S. Toda, Y. Arikawa, M. Masuzawa, M. Hashimoto, K. Ikoma, S.-Z. Wang and A. Miyasaka, *J. Chem. Soc., Perkin Trans. 1*, 2035 (1990).
260. H. Kato and S. Toda, *J. Chem. Soc., Chem. Commun.*, 510 (1982).
261. R. Neidlein, M. Kohl and W. Kramer, *Helv. Chim. Acta*, **72**, 1311 (1989).
262. R. Neidlein, personal communication (1990). I thank Professor Neidlein for providing these results.
263. S. Kagabu and K. Saito, *Tetrahedron Lett.*, **29**, 675 (1988).
264. I. Durucasu, N. Saraçoglu and M. Balci, *Tetrahedron Lett.*, **32**, 7097 (1991).
265. H. Schwager, Ph.D. Thesis, Ruhr-Universität Bochum, 1986.
266. H. Schwager, C. Krüger, R. Neidlein and G. Wilke, *Angew. Chem., Int. Ed. Engl.*, **26**, 65 (1987).
267. H. Schwager, R. Benn and G. Wilke, *Angew. Chem., Int. Ed. Engl.*, **26**, 67 (1987).
268. A. J. Lee, R. C. Ugolick, J. G. Fulcher, S. Togashi, A. B. Bocarsly and J. A. Gladysz, *Inorg. Chem.*, **19**, 1543 (1980).
269. W. E. Billups, W. Y. Chow and C. V. Smith, *J. Am. Chem. Soc.*, **96**, 1979 (1974).
270. T. Shirafuji and H. Nozaki, *Tetrahedron*, **29**, 77 (1973).
271. S. Kagabu, K. Saito, H. Watanabe, K. Takahashi and K. Wada, *Bull. Chem. Soc. Jpn.*, **64**, 106 (1991).
272. S. Kagabu and T. Inoue, *Chem. Lett.*, 2181 (1989).
273. R. Neidlein and B. Krämer, *Chem. Ber.*, **124**, 353 (1991).
274. R. Neidlein and B. Krämer, personal communication (1991). I thank Professor Neidlein for providing these results.
275. R. Neidlein, B. Krämer and C. Krüger, *Z. Naturforsch.*, **45b**, 1577 (1990).
276. F. A. Cotton, J. M. Troup, W. E. Billups, L. P. Lin and C. V. Smith, *J. Organomet. Chem.*, **102**, 345 (1975).
277. R. Benn, H. Schwager and G. Wilke, *J. Organomet. Chem.*, **316**, 229 (1986).
278. R. Neidlein, A. Rufinska, H. Schwager and G. Wilke, *Angew. Chem., Int. Ed. Engl.*, **25**, 640 (1986).
279. C. Krüger, K. Laakmann, G. Schroth, H. Schwager and G. Wilke, *Chem. Ber.*, **120**, 471 (1987).
280. P. J. Stang, L. Song and B. Halton, *J. Organomet. Chem.*, **388**, 215 (1990).
281. R. Mynott, R. Neidlein, H. Schwager and G. Wilke, *Angew. Chem., Int. Ed, Engl.*, **25**, 367 (1986).
282. P. J. Stang, L. Song, Q. Lu and B. Halton, *Organometallics*, **9**, 2149 (1990).
283. K. T. Lim and S. K. Choi, *J. Polym. Sci., Part C: Polym. Lett.*, **24**, 645 (1986).
284. C. Wentrup, E. Wentrup-Byrne and P. Müller, *J. Chem. Soc., Chem. Commun.*, 210 (1977).
285. C. Wentrup, E. Wentrup-Byrne, P. Müller and J. Becker, *Tetrahedron Lett.*, 4249 (1979).
286. H. M. Hugel, D. P. Kelly, A. R. Browne, B. Halton, P. J. Milsom and A. D. Woolhouse, *J. Chem. Soc., Perkin Trans. 1*, 2340 (1977).
287. J. A. Fahey, H. M. Hugel, D. P. Kelly, B. Halton and J. B. Williams, *J. Org. Chem.*, **45**, 2862 (1980).
288. W. H. Robinson, E. J. Ditzel, H. M. Hugel, D. P. Kelly and B. Halton, *J. Org. Chem.*, **46**, 5003 (1981).
289. H. Dürr and H.-J. Ahr, *Tetrahedron Lett.*, 1991 (1977).

290. E. Lüddecke, H. Rau, H. Dürr and H. Schmitz, *Tetrahedron*, **33**, 2677 (1977).
291. A. Schweig, personal communication (1993). I thank Professor Schweig for his correspondence on this matter.
292. S. M. Bachrach, *J. Org. Chem.*, **55**, 4961 (1990).
293. R. D. Brown, P. D. Godfrey and M. Rodler, *J. Am. Chem. Soc.*, **108**, 1296 (1986).
294. B. Halton, *Pure Appl. Chem.*, **62**, 541 (1990).
295. M. J. S. Dewar and G. J. Gleicher, *Tetrahedron*, **21**, 3243 (1965).
296. B. Halton, Q. Lu and P. J. Stang, *J. Org. Chem.*, **55**, 3056 (1990).
297. B. Halton, Q. Lu and W. H. Melhuish, *J. Photochem. Photobiol., A: Chem.*, **52**, 205 (1990).
298. Y. Apeloig, R. Boese, D. Blaser, B. Halton and A. H. Maulitz, in preparation (1994).
299. M. Hanack, K. Ritter, I. Stein and W. Hiller, *Tetrahedron Lett.*, **27**, 3357 (1986).
300. W. Bauer, T. Laube and D. Seebach, *Chem. Ber.*, **118**, 764 (1985).
301. M. Neuenschwander, 'Fulvenes', *Supplement A: in The chemistry of Double Bonded Functional Groups* (Ed. S. Patai), Part 2, Wiley, Chichester, 1989, p. 1131.
302. D. J. Sardella, C. M. Keane and P. Lemonias, *J. Am. Chem. Soc.*, **106**, 4962 (1984).
303. O. Takahashi and O. Kikuchi, *Tetrahedron*, **46**, 3803 (1990).
304. R. Boese and A. H. Maulitz, personal communication (1993). I thank Professor Boese for providing these results.
305. T. Eicher and J. L. Weber, *Top. Curr. Chem.*, **57**, 1 (1975).
306. K. Ashley, F. Sarfarazi, S. J. Buckland, J. K. Foley, Q. Mei, B. Halton, P. J. Stang and S. Pons, *Can. J. Chem.*, **65**, 2062 (1987).
307. K. Ashley, K. J. Foley, Q. Mei, J. Ghoroghchian, F. Sarfarazi, J. Cassidy, B. Halton, P. J. Stang and S. Pons, *J. Org. Chem.*, **51**, 2089 (1986).
308. T. Koenig, T. Curtiss, R. Winter, K. Ashley, Q. Mei, P. J. Stang, S. Pons, S. J. Buckland and B. Halton, *J. Org. Chem.*, **53**, 3735 (1988).
309. S. W. Staley and T. D. Norden, *J. Am. Chem. Soc.*, **111**, 445 (1989).
310. S. J. Buckland, B. Halton and P. J. Stang, *Aust. J. Chem.*, **40**, 1375 (1987).
311. G. J. Gainsford, D. L. Officer and B. Halton, *Acta Crystallogr., Sect. C*, **47C**, 2397 (1991).
312. A. T. McNichols, P. J. Stang, B. Halton and A. J. Kay, *Tetrahedron Lett.*, **34**, 3131 (1993).
313. B. Halton, A. J. Kay, A. T. McNichols, P. J. Stang, R. Boese, T. Haumann, Y. Apeloig and A. H. Maulitz, *Tetrahedron Lett.*, **34**, 6151 (1993).
314. B. Halton, unpublished observations (1989).
315. R. F. C. Brown and B. Halton, unpublished observations (1989).
316. S. Murata, T. Yamamoto and H. Tomioka, *J. Am. Chem. Soc.*, **115**, 4013 (1993).
317. S. Murata, T. Yamamoto, H. Tomioka, H.-k. Lee, H.-R. Kim and A. Yabe, *J. Chem. Soc., Chem. Commun.*, 1258 (1990).
318. W. E. Billups, D. J. McCord and B. R. Maughon, *J. Am. Chem. Soc.*, **116**, 8831 (1994).
319. W. E. Billups, D. J. McCord and B. R. Maughon, *Tetrahedron Lett.*, **35**, 4493 (1994).
320. M. Saunders, A. Jiménez-Vázquez, J. R. Cross, W. E. Billups, C. Gesenberg and D. J. McCord, *Tetrahedron Lett.*, **35**, 3869 (1994).
321. B. Halton, M. J. Cooney and H. Wong, *J. Am. Chem. Soc.*, **116** (1994), in press.

Note Added in Proof

Billups *et al.* have reported the synthesis of the first two triscycloproparenes using bromochlorocyclopropene (Section II.E.2)[318], Ag(I) catalysed cycloproparene dimerizations (Section V.B.1)[319], and the addition of **11** to C_{60} via σ-bond opening (Section V.B.3)[320]. Dimethyldioxirane epoxidation of **96c** has provided the first oxaspiropentene derivative—the naphtho analogue of **138** (Section VI.B.2)[321].

CHAPTER **13**

[1.1.1]Propellanes

PIOTR KASZYNSKI

Department of Chemistry, Vanderbilt University, Nashville, Tennessee 37235, USA

and

JOSEF MICHL

Department of Chemistry and Biochemistry, University of Colorado, Boulder, Colorado 80309-0215, USA
Fax: +1(303)492-0799; e-mail: michl@eefus.colorado.edu

I. INTRODUCTION . . . 773
II. STRUCTURE . . . 775
A. Molecular Structure . . . 775
B. Electronic Structure . . . 777
C. Stability . . . 779
III. SPECTROSCOPY . . . 783
A. Nuclear Magnetic Resonance . . . 783
B. Vibrational Spectra . . . 784
C. Photoelectron and Electron Transmission Spectra . . . 785
D. Electronic Spectra . . . 786
IV. SYNTHESIS . . . 787
A. Intramolecular Carbene Insertion . . . 787
B. Anionic Cyclization . . . 787
C. Elimination . . . 790
V. REACTIONS . . . 792
A. Introduction . . . 792
B. Nucleophilic Attack . . . 794
C. Radical Attack . . . 795
D. Electrophilic Attack . . . 806
E. Reduction . . . 809
VI. CONCLUSIONS . . . 809
VII. ACKNOWLEDGMENT . . . 809
VIII. REFERENCES . . . 809

I. INTRODUCTION

[1.1.1]Propellane (**1a**), the smallest member of the [*p.q.r*]propellane family, had long been the subject of speculation and computations[1,2] when its first successful synthesis was

The chemistry of the cyclopropyl group, Vol. 2
Edited by Z. Rappoport

reported by Wiberg and Walker in 1982[3]. This improbable-looking hydrocarbon was found to be surprisingly stable in solution up to temperatures of about 100 °C. Initially, the existence and stability of **1a** were primarily a curiosity since the laborious 12-step low-yield synthesis from diethyl phenylmalonate precluded practical applications. The seminal discovery in the laboratory of Szeimies[4] of an efficient two-step synthesis of **1a** from commercially available materials and the analogous synthesis of several of its derivatives opened an exciting period of rapidly expanding chemistry of [1.1.1]propellanes, bicyclo[1.1.1]pentanes and their oligomers, [*n*]staffanes.

Previous reviews of [1.1.1]propellane chemistry were written by Ginsburg[5], Wiberg[6] and Tobe[7]. The present review is intended to provide a comprehensive and updated survey of the structure, properties, synthesis and reactions of [1.1.1]propellanes. Structures of all known compounds of this class are shown in Chart 1.

Tricyclo[1.1.1.0^{1,3}]pentanes

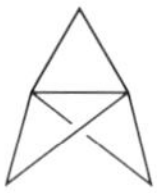

(1a)

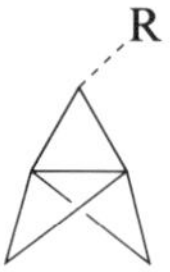

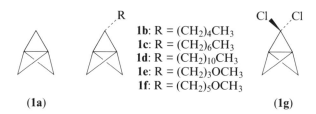

1b: R = $(CH_2)_4CH_3$
1c: R = $(CH_2)_6CH_3$
1d: R = $(CH_2)_{10}CH_3$
1e: R = $(CH_2)_3OCH_3$
1f: R = $(CH_2)_5OCH_3$

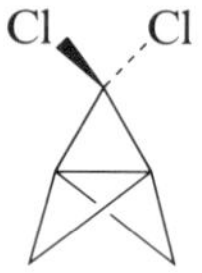

(1g)

Tetracyclo[4.1.0.0^{1,5}.0^{2,6}]heptanes

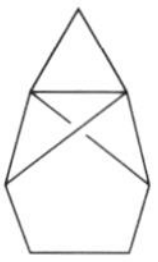

(1h)

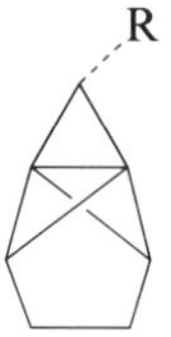

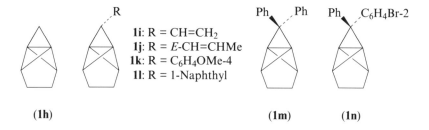

1i: R = CH=CH$_2$
1j: R = *E*-CH=CHMe
1k: R = C_6H_4OMe-4
1l: R = 1-Naphthyl

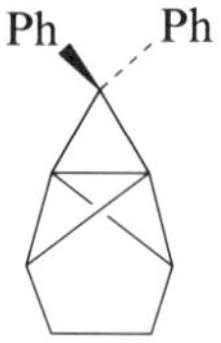

(1m)

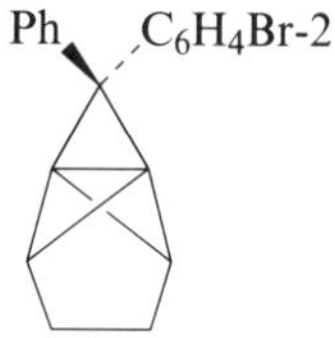

(1n)

Tetracyclo[5.1.0.0^{1,6}.0^{2,7}]octanes

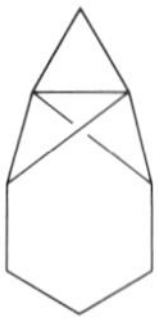

(1o)

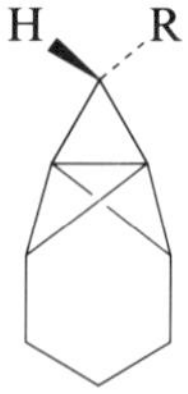

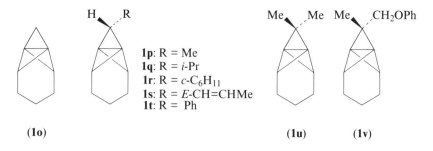

1p: R = Me
1q: R = *i*-Pr
1r: R = *c*-C_6H_{11}
1s: R = *E*-CH=CHMe
1t: R = Ph

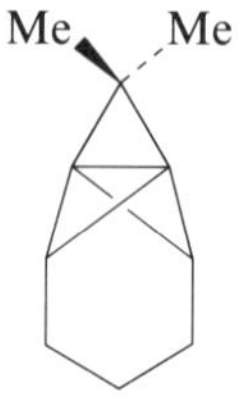

(1u)

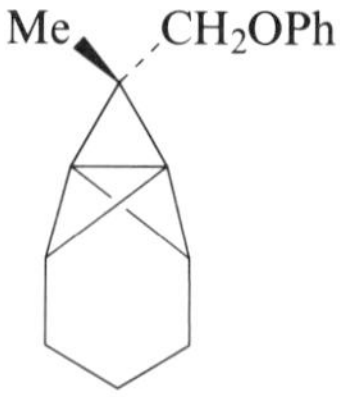

(1v)

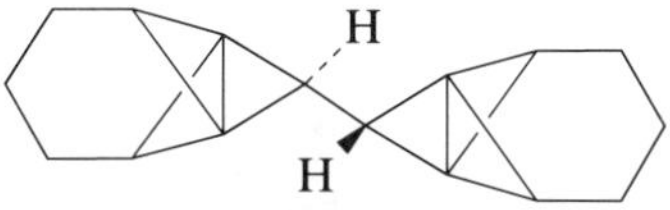

(1w)

Spiro[cyclopentane-1,8′-tetracyclo [5.1.0.$0^{1,6}$.$0^{2,7}$]octane]

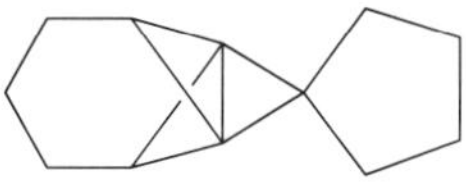

(1x)

Spiro[cyclohexane-1,8′-tetracyclo [5.1.0.$0^{1,6}$.$0^{2,7}$]octane]

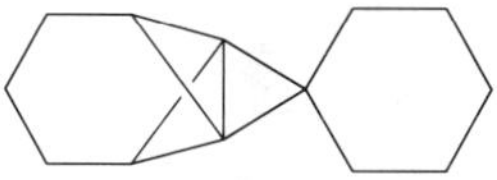

(1y)

Dispiro[tetracyclo[5.1.0.$0^{1,6}$.$0^{2,7}$]octane-8,1′-cyclohexane-4′,8″-tetracyclo[5.1.0.$0^{1,6}$.$0^{2,7}$]octane]

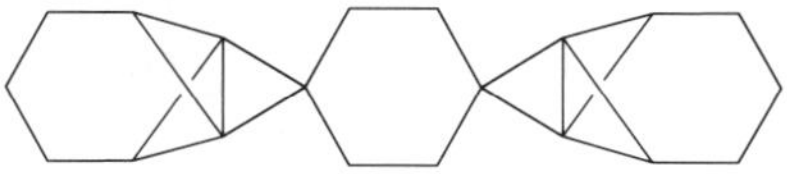

(1z)

Benzo[*c*]tetracyclo[5.1.0.$0^{1,6}$.$0^{2,7}$]oct-3-ene

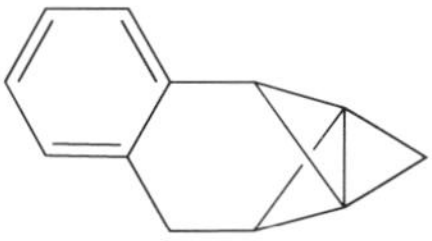

(1aa)

Pentacyclo[6.1.0.$0^{1,7}$.$0^{2,8}$.$0^{3,5}$]nonane

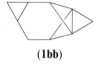

(1bb)

CHART 1

II. STRUCTURE

A. Molecular Structure

The molecular structures of [1.1.1]propellane (**1a**) and its derivatives have been determined by a variety of methods (Table 1) and clearly reveal the two main characteristics that dominate any discussion of their properties: (i) the long central bond, and (ii) the unusual valence angles at all the carbon atoms, but particularly the bridgeheads. The central bond length ranges from 1.577 Å (**1w**) to 1.605 Å (**1a**), whereas the bridgehead-methylene distances lie between 1.499 Å and 1.599 Å (both in **1z**) but mostly correspond well to the bond length in cyclopropane (1.514 Å)[8,9]. The geometry at the two bridgehead carbons is 'inverted', i.e. all four bonds of each carbon are directed to the same side of a plane. At these carbons, the valence angles of the C2C1C4 type are about 95°, similar to those in bicyclo[1.1.1]pentane (**2**)[10], and those of the C1C3C2 type are about 58°, close to those in cyclopropane (**2**)[8,9]. The CCC valence angles at the bridge CH_2 carbons are about 63°, again very similar to those in cyclopropane.

TABLE 1. Selected interatomic distances and angles in [1.1.1]propellane and its derivatives

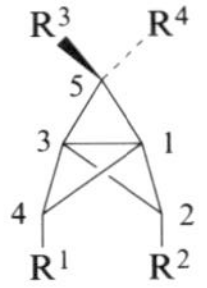

Compound					Distances (Å)			Angles (deg)			
1	R^1	R^2	R^3	R^4	C1—C3	C1—C2	C1—C5	C1—C2—C3	C1—C5—C3	C1—C3—C2	C2—C1—C5
a	H	H	H	H	1.60[a]	1.522[a]					
					1.596[b]	1.525[b]		63.1[b]			95.1[b]
					1.605[c,d]	1.525–1.555[c,d]		63.0–63.7[c,d]		57.9–58.8[c,d]	92.4–99.0[c,d]
					1.593[c,e]	1.512–1.539[c,e]		62.4–62.8[c,e]		57.9–59.7[c,e]	94.4–96.9[c,e]
h	$(CH_2)_2$		H	H	1.587[f]	1.529[f]	1.517[f]	62.48,62.50[f]	62.95[f]	58.69–58.84[f]	98.04,98.03[f]
m	$(CH_2)_2$		Ph	Ph	1.592[d,g]	1.517–1.534[d,g]	1.521,1.529[d,g]	62.7,62.9[d,g]	62.9[d,g]	58.7,59.1[d,g]	97.2–98.4[d,g]
					1.586[e,g]	1.524–1.526[e,g]	1.517,1.524[e,g]	62.7[e,g]	62.9[e,g]	58.7[e,g]	97.8–98.2[e,g]
o	$(CH_2)_3$		H	H	1.585[f]	1.524–1.538[f]	1.519,1.522[f]	62.31,62.36[f]	62.81[f]	58.43–59.22[f]	95.42,96.08[f]
w	$(CH_2)_3$		H	*h*	1.577[i]	1.513–1.529[i]	1.515[i]	62.46[i]	62.72[i]	58.30[i]	96.20[i]
z	$(CH_2)_3$		*j*		1.587[g,k]	1.499–1.599[g,k]	1.531,1.532[g,k]	60.7,63.7[g,k]	62.4[g,k]	57.8,58.7[g,k]	94.7–97.1[g,k]
					1.601[g,l]	1.508–1.560[g,l]	1.534,1.552[g,l]	62.5,63.4[g,l]	*m*	57.4,59.8[g,l]	*m*

[a] IR/Raman spectrum: Reference 15.
[b] Electron diffraction: Reference 25.
[c] X-ray diffraction: Reference 26.
[d] Molecule **A** in the crystal.
[e] Molecule **B** in the crystal.
[f] X-ray diffraction: Reference 20.
[g] X-ray diffraction: Reference 45.
[h] Two propellane cages connected through this bond.
[i] From X-ray diffraction: Reference 61.
[j] Two propellane cages spiro connected through two $—CH_2CH_2—$ bridges.
[k] One propellane cage in the molecule.
[l] Second propellane cage in the molecule.
[m] Not listed.

The distance between the two propellane subunits in **1w** is 1.494 Å, somewhat shorter than the corresponding distance in bicyclopropyl (1.517 Å)[11]. The HCH angle in the methylene group is 116° in **1a**. The PhCPh angle is somewhat less in the diphenyl derivative **1m**, 113.1° and 114.3° in the two inequivalent molecules in the crystal.

A theoretical study of the geometries of several even more highly strained dehydro derivatives of **1a** has appeared[12].

B. Electronic Structure

Hybridization. The simplest way to think about bonding in [1.1.1]propellane is to characterize the state of hybridization at the two types of carbon atoms and then combine the hybrids pairwise into localized bond orbitals. In ordinary molecules, these have electron occupancy close to two, and the corresponding antibonds are nearly unoccupied. Bond delocalization can then be introduced by considering interactions of the occupied localized bond orbitals with the unoccupied localized antibond orbitals.

There are two ways in which the state of hybridization at the carbon atoms of [1.1.1]-propellane has been deduced. The results disagree dramatically (Figure 1), and at least one is clearly wrong. The hybridization at the carbon atoms can be derived from: (i) the empirical correlations between the s orbital character of a carbon hybrid orbital and the ^{13}C–^{1}H and ^{13}C–^{13}C NMR coupling constants across the bonds it forms, (ii) the analysis of computed wavefunctions, using one of several possible schemes.

Hybridization from NMR coupling constants[13]. Using the standard correlation formula[14], the ^{1}H–^{13}C coupling constant (J_{CH} = 163.7 Hz)[3,13,15] corresponds to $sp^{2.1}$ hybridization for the carbon hybrids used in the CH bonds. This leaves $sp^{4.7}$ hybrids on the bridge carbon for use in the C1C2 bond. The ^{13}C–^{13}C coupling constant for this bond is $J_{CC} = 9.9 \pm 0.1$ Hz. Another empirical correlation formula[14] yields $sp^{8.6}$ hybridization for the hybrid used by the bridgehead carbon. This is a nearly pure p orbital, with only 10% s character. The remaining bridgehead orbital available for the formation of the central bond therefore is $sp^{0.5}$ hybridized. This is a nearly pure s orbital, with 70% s character. The conclusion is that the central bond is mostly s in character.

The use of the newer empirical equations relating percent s character to ^{13}C–^{1}H and ^{13}C–^{13}C NMR coupling constants,[16] $J_{CH} = 5.70(\%s) - 18.4$ and $J_{AB} = 0.0621(\%s_A)(\%s_B) - 10.2$, yields $sp^{2.1}$ for the hybrid used by the bridge carbon in the CH bond, $sp^{4.5}$ for the hybrid used by it in the peripheral CC bond, $sp^{4.6}$ for the hybrid used by the bridgehead carbon in the peripheral CC bond and $sp^{1.2}$ for the hybrid used by the bridgehead carbon in the central CC bond. This corresponds to 46% s character in the hybrid used in the central bond. Although this is a less extreme value, it is still heavily tilted towards s character for this orbital.

The picture that emerges from this analysis is easily visualized. Since the three peripheral CC bonds formed by the bridgehead carbon are approximately at right angles to each

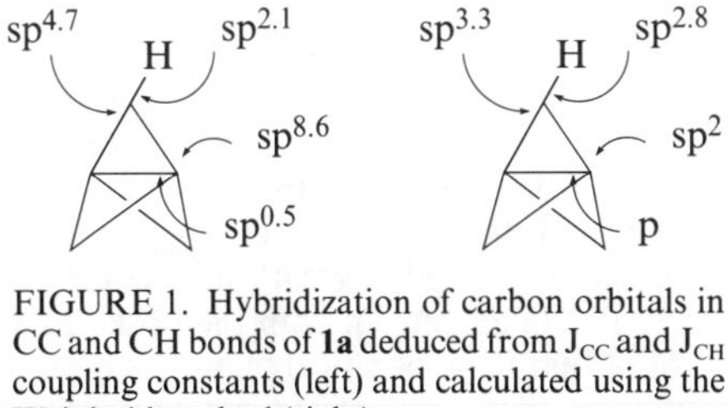

FIGURE 1. Hybridization of carbon orbitals in CC and CH bonds of **1a** deduced from J_{CC} and J_{CH} coupling constants (left) and calculated using the Weinhold method (right)

other, their formation calls for the use of unhybridized $2p_x$, $2p_y$, and $2p_z$ atomic orbitals. The remaining 2s orbitals of the bridgehead carbons have spherical symmetry and no directional preference in bond formation. Their proximity permits them to combine into the bonding orbital of the central bond. When the bent nature of bonds in three-membered rings is recognized, the picture changes somewhat, but the high p character of the hybrid orbitals used by the bridgehead carbons for the peripheral bonds and the high s character of the hybrid orbitals used by them for the central bond, deduced from the NMR coupling constants, are clearly accounted for.

Computed hybridization. The above picture of hybridization of the carbon atoms in [1.1.1]propellane is in complete disagreement with numerical quantum chemical calculations. The pioneering Hartree–Fock calculations[1], performed at the minimum basis set and the 4-31G levels, were analyzed in terms of (nonorthogonal) hybrids and the authors concluded that the bridge carbon used an $sp^{2.7}$ hybrid in the CH bond and an $sp^{4.3}$ hybrid in the C1C2 bond. These values are in mediocre agreement with the values deduced above from NMR. For the bridgehead carbon, the calculation gave an $sp^{1.3}$ hybrid for the peripheral C1C2 bonds, and an $sp^{4.1}$ hybrid for the central CC bond. The use of nonorthogonal hybrids makes these numbers somewhat difficult to interpret, but clearly, the discrepancy with the NMR values is spectacular, as the conclusion from the theoretical calculation is that the central bond is mostly p in character. The basic conclusion from the theoretical approach does not appear to be sensitive to the method of calculation, and a more recent GVB result[17] was $sp^{4.5}$ for the hybrids used in the central bond.

We have performed a HF/6-311G** calculation for the purposes of this review, and analyzed the results in terms of Weinhold's natural hybrid orbitals[18]. We consider the natural hybrid orbitals to be optimal for this type of analysis. We find an even more extreme disagreement: the orthogonal hybrids used by the bridgehead carbon for the peripheral C1C2 bonds are nearly exactly sp^2, leaving an almost pure (99.6%) p orbital for the formation of the central bond. The hybridization calculated for the bridge carbon is $sp^{2.8}$ for the hybrid used in the CH bond and $sp^{3.3}$ for that used in the C1C2 bond. For this carbon, the disagreement with the NMR analysis is less striking.

The theoretical result is again easy to visualize in simple terms[19]. The molecule can be thought of as a combination of two coaxial staggered planar methyl radicals bound by their p orbitals, in which three CH_2 bridges have replaced three pairs of hydrogen atoms, forming two strongly bent CC bonds each. The deviation of the orbital axis from the line joining the carbon nuclei is 33.5° at the bridgehead and 24.9° at the bridge carbon.

Note that either picture of hybridization suggests that an orbital of an attacking electrophile, radical or nucleophile will have equally easy access to the orbital used by a bridgehead carbon to make the central bond as does the orbital of the other bridgehead carbon. In either case, an attack by a reagent will sterically resemble reactions on the 1s orbital of a terminal hydrogen atom in a bond such as H—H or C—H. It has indeed been pointed out that the large electron density located outside the cage next to the bridgehead carbons is responsible for the reactivity of [1.1.1]propellane towards electrophiles[20,21].

Failure of the empirical NMR method. Which picture of hybridization is right? If the central CC bond is composed primarily of 2s orbitals, as suggested by the NMR analysis, it will be quite low in energy. If it is composed primarily of 2p orbitals, it will be high in energy. There is evidence from a combination of photoelectron[22] and electron energy loss spectroscopy[23] that the central bond is heavily represented in the highest occupied molecular orbital of [1.1.1]propellane, since its length is greatly affected by excitation into the lowest triplet state. This notion is also in agreement with the observed chemical reactivity. We believe that there is no doubt that the theoretical picture is basically correct and that the empirical NMR correlations fail in this case. The central bond is primarily p in character.

The failure of the empirical NMR coupling constant correlation is actually not very surprising. These correlations have a reputation of breaking down in highly strained hydrocarbons[24], of which [1.1.1]propellane surely is a prime example, and do not do particularly well even in bicyclo[1.1.0]butane.

Biradicaloid character and bond delocalization. The Weinhold type of analysis provides two additional insights. First, while the electron occupancy of the bond orbitals in the molecule is generally close to two, as expected, for the central bond orbital it is only 1.83. The occupancy of the antibonding orbitals is generally very close to zero, but for the central antibond it is 0.15. This shows that the central bond has a mildly biradicaloid character, in keeping with its increased length. It is in keeping with the fact that the accumulation of charge between the bridgehead nuclei is calculated to be only about 80% of that in a normal C—C bond at the bond critical point[21]. Still, it corresponds to expectations for a typical cyclopropane bond, worth about 60 kcal mol^{-1} [3,19], even though the bond in [1.1.1]propellane is longer[15,25,26]. A measurement of electron density in the central bonds of several propellanes[20] indeed indicated a value lower than usual, and an accumulation of density in the region on the outside of this bond.

Second, the ordinary Lewis structure describing the [1.1.1]propellane molecule is less adequate than is usual for hydrocarbons. This is demonstrated by the fairly large interaction matrix elements between the occupied localized orbital of the central bond and the vacant localized peripheral antibonding orbitals, and between the unoccupied localized central antibond orbital and the occupied localized peripheral bond orbitals. Bonds in [1.1.1]propellane are quite a bit more delocalized than is usual in saturated hydrocarbons, and this is perhaps the reason why the empirical correlations for NMR coupling constants fail.

Detailed analysis. A fair number of detailed theoretical discussions of the nature of bonding in [1.1.1]propellane have appeared over the years in addition to those already mentioned. We make no attempt to summarize them here and refer the interested reader to a few of the original articles[17,27–34].

C. Stability

[1.1.1]Propellane and many of its derivatives are liquids at ambient temperature. Some derivatives such as **1m**, **1w** and **1z** melt at higher temperatures, 60–62 °C, 70–73 ° and 171 °C (dec), respectively. The parent propellane itself melts at about –11°C,[26] and propellanes **1h** and **1o** are solids at –78 °C[35].

Considering their high expected strain energy, [1.1.1]propellane and its derivatives possess substantial thermal stability. The neat parent hydrocarbon **1a** is moderately stable at room temperature, particularly in the absence of air, but after a few hours, a considerable amount of polymer builds up[36,37]. A 3–5% solution can be stored in a refrigerator for weeks without noticeable losses.

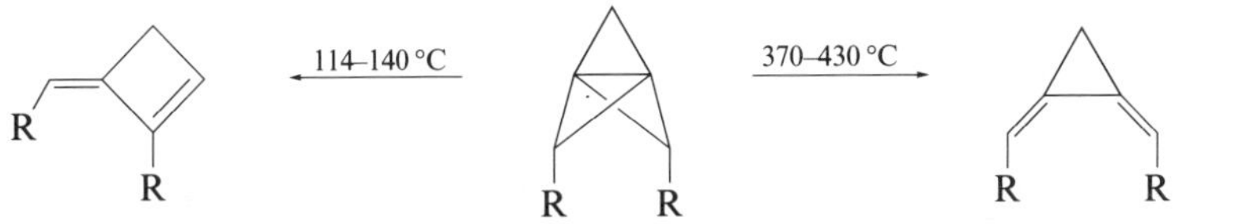

(1)

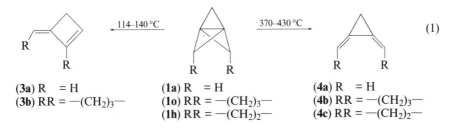

(**3a**) R = H
(**3b**) RR = $-(CH_2)_3-$

(**1a**) R = H
(**1o**) RR = $-(CH_2)_3-$
(**1h**) RR = $-(CH_2)_2-$

(**4a**) R = H
(**4b**) RR = $-(CH_2)_3-$
(**4c**) RR = $-(CH_2)_2-$

At elevated temperatures, **1a** rearranges to methylenecyclobutene (**3a**) with a half-life of 5 min in the gas phase at 114 °C[3]. The derivative **1o** decomposes completely within 30 min in C_6D_6 solution at 140 °C[38] and yields **3b** along with some polymeric products. It has been suggested that the formation of **3** is due to an acid-catalyzed rearrangement (equation 1). In pyridine solvent, the NMR signals of **1o** disappeared after 9 h at 140 °C, and only polymeric materials were produced[38]. Pyrolysis of **1a**, **1o** and **1h** in a flow system at 430° C leads to the formation of the primary products **4a**, **4b** and **4c**, respectively, with retention of the central bond[35,38] (equation 1). Methylenecyclopropane derivatives were also found to be major thermal decomposition products of propellanes **1m** and **1p**[35]. The propellane **1g** is the least thermally stable of those known, and survives for less than 0.5 h at ambient temperature in solution[39].

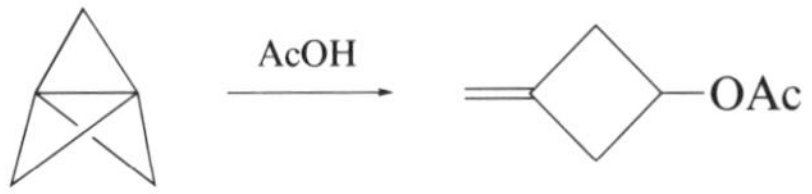

(2)

(1a) **(5)**

The enthalpy of reaction of **1a** with acetic acid to give methylenecyclobutyl acetate (**5**) (equation 2) has been measured[15] and is -35.2 ± 0.4 kcal mol^{-1}. This provided the enthalpy of formation of **1a** in liquid phase, 77.9 ± 1 kcal mol^{-1}. Assuming the enthalpy of vaporization to be 6.4 ± 0.5 kcal mol^{-1}, the enthalpy of formation of **1a** in gas phase was estimated to be 84 ± 1 kcal mol^{-1} [15]. This corresponds well to the value of 89 kcal mol^{-1} calculated at the HF 6-31G* level[40]. Using a set of group equivalents[40] the strain energy of **1a** was estimated to be about 100 kcal mol^{-1} [16,41], only about 20 kcal mol^{-1} more than three times the value for cyclopropane. It is interesting to note that relatively little strain energy is released when the central bond in **1a** is broken. Thus, the hydrogenolysis of **1a** to form bicyclo[1.1.1]pentane (**2**, equation 3) is only moderately exothermic. Using the experimental heat of formation of gasous **2** (46.1 ± 1.1 kcal mol^{-1})[42] the enthalpy of the reaction is -37.9 kcal mol^{-1}, close to the calculated values[15] of -39.7 kcal mol^{-1} (MP2 6-31G*) and -46.0 kcal mol^{-1} (MP3 6-31G*). Using the experimental bond strength of H_2 (104.1 kcal mol^{-1}), this yields 67 kcal mol^{-1} for the strength of the central bond in **1a**[3,19,43].

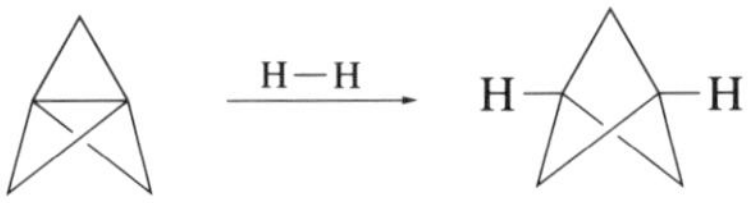

(3)

(1a) **(2)**

The addition of a hydrogen atom to **1a** to yield the bicyclo[1.1.1]pent-1-yl radical (**6**) has been calculated (UHF/MP2 6-31G*) to be exothermic by only 47 kcal mol^{-1} [19]. The addition of the second hydrogen atom to the radical to form bicyclo[1.1.1]pentane was calculated (MP3 6-311G*) to be exothermic by 104.4 kcal mol^{-1} [43] (equation 4). By this argument, the strength of the central bond in **1a** is about 60 kcal mol^{-1} [3,19] and is comparable to that in cyclopropane.

(4)

(1a) **(6)** **(2)**

TABLE 2. Selected NMR chemical shifts in [1.1.1]propellanes (C_6D_6)

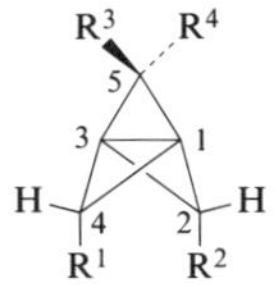

1	R^1	R^2	R^3	R^4	^{1}H NMR H^2	H^4	$R^1 = H$	$R^2 = H$	$R^4 = H$	^{13}C NMR C^1	C^2	C^3	C^4	C^5	Ref.
a	H	H	H	H	1.77	1.77	1.77	1.77	1.77	1.23	73.91	1.23	73.91	73.91	37
a[a]	H	H	H	H	2.06	2.06	2.06	2.06	2.06	1.0	74.1	1.0	74.1	74.1	3,15
b–f[a,b]	H	H	alkyl	H	1.74	2.69	2.16	1.74	2.60	5.3	71.9	5.3	69.2	91.0	46
g[c]	H	H	Cl	Cl	3.06	3.06	2.26	2.26	—	*d*					39
h	$(CH_2)_2$		H	H	2.73	2.73	—	—	2.32	11.78	84.11	11.78	84.11	70.73	4
i	$(CH_2)_2$		CH=CH$_2$	H	2.59	3.44	—	—	3.47	17.42	83.05	17.42	79.33	89.23	45
j	$(CH_2)_2$		*E*-CH=CHMe	H	2.59	3.45	—	—	3.53	17.33	82.93	17.33	78.78	88.87	45
k	$(CH_2)_2$		4-C_6H_4OMe	H	2.74	3.37	—	—	4.22	17.75	84.50	17.75	79.36	90.75	45
l	$(CH_2)_2$		1-Naphthyl	H	2.83	3.18	—	—	4.38	*d*					45
m	$(CH_2)_2$		Ph	Ph	3.05	3.05	—	—	—	22.75	82.27	22.75	82.27	110.71	45
n	$(CH_2)_2$		2-C_6H_4Br	Ph	2.89	3.19	—	—	—	*d*					45
o	$(CH_2)_3$		H	H	2.75	2.75	—	—	1.55	9.42	86.51	9.42	86.51	66.55	45
p	$(CH_2)_3$		Me	H	2.52	3.56	—	—	2.10	15.05	84.66	15.05	81.90	76.23	45
q	$(CH_2)_3$		*i*-Pr	H	2.49	3.48	—	—	1.90	14.21	85.05	14.21	81.78	91.96	45
r	$(CH_2)_3$		*c*-C_6H_{11}	H	2.50	3.50	—	—	1.98	13.75	85.17	13.75	81.84	90.56	45
s	$(CH_2)_3$		*E*-CH=CHMe	H	2.62	3.65	—	—	2.73	15.14	83.42	15.14	81.99	85.51	45
t	$(CH_2)_3$		Ph	H	2.72	3.41	—	—	3.41	14.96	86.11	14.96	82.46	87.26	45
u	$(CH_2)_3$		Me	Me	3.48	3.48	—	—	—	19.26	79.87	19.26	79.87	83.86	61
v	$(CH_2)_3$		CH_2OPh	Me	3.48	3.48	—	—	—	19.48	81.45	19.48	80.39	86.57	61
w	$(CH_2)_3$		*e*	H	2.50[f]	4.10[f]	—	—	2.13	16.02	87.14[f]	16.02	82.02[f]	75.15	61
x	$(CH_2)_3$		$(CH_2)_4$		3.25	3.25	—	—	—	19.55	82.13	19.55	82.13	94.38	45
y	$(CH_2)_3$		$(CH_2)_5$		3.55	3.55	—	—	—	17.90	79.33	17.90	79.33	93.35	45

(continued)

TABLE 2. (*continued*)

1	R^1	R^2	R^3	R^4	^{1}H NMR: H^2	H^4	$R^1=H$	$R^2=H$	$R^4=H$	^{13}C NMR: C^1	C^2	C^3	C^4	C^5	Ref.
z	$(CH_2)_3$		*g*		3.48	3.48	—	—	—	18.54	79.75	18.54	79.75	92.88	45
aa	$-2\text{-}C_6H_4CH_2-$		H	H	2.73	3.28	—	—	1.63, 1.76	7.57	86.23	7.57	84.29	67.42	45
bb	$-c\text{-}C_3H_4CH_2-$		H	H	2.49	3.18	—	—	1.40	9.45	83.45	9.45	88.72	65.30	45

[a] Recorded in $CDCl_3$.
[b] NMR shifts for all alkyl-substituted derivatives **1b**–**1f** are constant within ± 0.1 ppm for ^{1}H spectra and ± 0.5 ppm for ^{13}C spectra.
[c] Recorded in THF-d_8.
[d] Not reported.
[e] Two [1.1.1]propellane cages connected through this bond.
[f] The shifts may be reversed.
[g] Two [1.1.1]propellane cages spiro connected through two $-CH_2CH_2-$ bridges.

TABLE 3. Selected $^4J_{H-H}$ coupling constants in [1.1.1]propellanes[a]

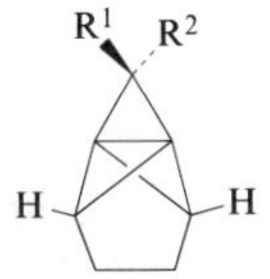

1	R^1	R^2	$^4J_{HH}$ (Hz)
i	H	CH=CH_2	4
j	H	*E*-CH=CHMe	4
k	H	C_6H_4OMe-4	4.5
l	H	1-Naphthyl	4
n	Ph	C_6H_4Br-2	4.5

[a]Reference 45. H—H coupling to a proton in position R^1 was not detected

III. SPECTROSCOPY

A. Nuclear Magnetic Resonance

Proton spectra. In ^{1}H NMR spectra, protons in the unsubstituted bridge appear between δ 1.40 and 2.69 (Table 2), to be compared with δ 0.22 in cyclopropane[44]. The usual values of geminal coupling constants $^2J_{HH}$ are 2 Hz, and the W-type $^4J_{HH}$ coupling constants typically are about 4 Hz (Figure 2, Table 3)[45,46]. These values are somewhat smaller than those found in derivatives of bicyclo[1.1.1]pentane[47]. The $^1J_{CH}$ coupling constant is about 160 Hz. It was measured for **1a** (163.7 Hz)[3,13,15] and **1o** (CH, 159 Hz and CH_2, 162 Hz)[45].

The chemical shift of [1.1.1]propellane protons is somewhat sensitive to the solvent used for the measurement. For instance, the ^{1}H NMR shift in **1a** changes from δ 2.06[3,15] in $CDCl_3$ to δ 1.77[37] in C_6D_6 and δ 1.97[48] in a 1:4 diethyl ether–$CDCl_3$ mixture.

Carbon spectra. The ^{13}C chemical shifts of the unsubstituted bridge carbons usually are about δ 65–74 (cf cyclopropane[49] δ 3.5). The bridgehead carbons resonate at about δ 1–23. Some insight into the origin of these values was obtained from a measurement of the individual components of the shielding tensor[50]. For the bridge carbon, these are δ 43 along the bisectrix of the CH_2 angle, δ 57 perpendicular to the HCH plane and δ 138 perpendicular to the CCC plane. For the bridgehead carbon, they are δ 35 in the direction of the threefold axis and δ -11 perpendicular to it. The bridge carbon assignments were based on IGLO calculations, which accounted quite well for the value of all the tensor components. The isotropic shifts were recently recalculated using IGLO/II, with similar results[43].

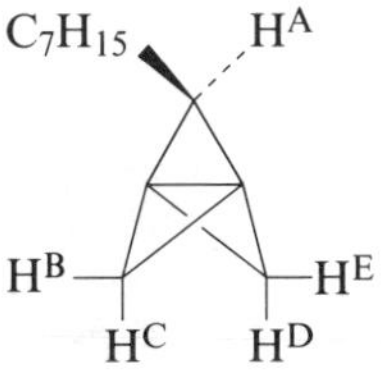

FIGURE 2. H—H coupling constants in **1c**: J_{AC} = 4.5, J_{BE} = 7.6, J_{BC} = 2.3, J_{DE} = 1.7 (Reference 46)

^{13}C NMR chemical shifts are also solvent-dependent but to a smaller degree than the proton shifts (Table 2).

The C1C2 *J* coupling constant is 9.9 Hz[13].

B. Vibrational Spectra

Small-ring propellanes exibit a characteristic intense IR band at low frequencies: 515 cm^{-1} in [3.2.1]propellane, 530 cm^{-1} in [2.2.1]propellane, 574 cm^{-1} in [2.1.1]propellane and 603 cm^{-1} in [1.1.1]propellane. (The first three values were determined in argon matrix and the last one in CS_2 solution)[51]. In an initial analysis based on MNDO calculations,[51] this was attributed to an antisymmetric combination of several peripheral CC stretching motions. It was referred to as a 'bobbing' mode in which the two bridgehead atoms move in unison perpendicular to the plane of the three peripheral carbon atoms in [1.1.1]-propellane and an analogous plane in the larger propellanes, while the hydrogen atoms move as a counterpoise. Table 4 shows that the frequency of this characteristic vibrational mode is somewhat structure sensitive.

Several other vibrational modes are characteristic of the propellane structures, such as the C—H stretching vibrations of the CH_2 groups at the nearly constant frequencies of 3000 cm^{-1} (symmetric) and 3060 cm^{-1} (antisymmetric), almost the same as in cyclopropane itself (see Table 4).

A subsequent detailed analysis[15] of infrared and Raman spectra of [1.1.1]propellane and [1.1.1]propellane-d_6 was based on comparison with HF/6-31G* calculations. Rotational fine structure in the IR spectrum was analyzed and yielded structural information (Section II.A). IR intensities were discussed in terms of molecular charge distribution and suggested a large electron population at the bridgehead carbons. The analysis provided a full force field, which was subsequently compared with that of the related molecules, bicyclo[1.1.0]butane[52] and bicyclo[1.1.1]pentane[53]. The authors concluded that with regard to structural data, force constants and dipole moment derivatives [1.1.1]propellane resembles cyclopropane, while bicyclo[1.1.1]pentane resembles cyclobutane. Despite its unusual geometry, [1.1.1]propellane is best thought of as three fused cyclopropane rings.

TABLE 4. IR asymmetric C1C2C3 stretching frequencies in [1.1.1]propellanes

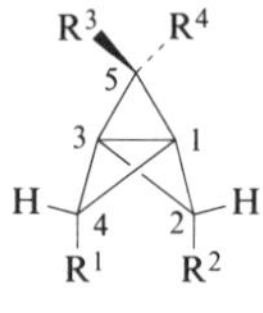

1	R^1	R^2	R^3	R^4	Medium	$\tilde{\nu}$ (cm^{-1})	Ref.
a	H	H	H	H	gas phase	612	15
g	H	H	Cl	Cl	Ar matrix	642	39
h	$(CH_2)_2$		H	H	hexane	609	45
m	$(CH_2)_2$		Ph	Ph	KBr	635	45
o	$(CH_2)_3$		H	H	pentane	595	45
z	$(CH_2)_3$		*a*		KBr	573	45
aa	2-$C_6H_4CH_2$—		H	H	hexane	596	45
bb	—*c*-$C_3H_4CH_2$—		H	H	hexane	615	45

[a] Two [1.1.1]propellane cages spiro connected through two —CH_2CH_2— bridges.

The strikingly large intensity of the antisymmetric C—C stretching ('bobbing') mode of [1.1.1]propellane has been traced to changes in atomic dipoles at the bridgehead carbons during motion along this normal mode[53]. Qualitatively, one might imagine that the central C—C bond is quite polarizable and responds to the development of a difference in electronegativity of the two bridgehead carbons as their valence angles change in opposite ways during the normal mode motion.

A calculation of the IR and Raman spectra of [1.1.1]propellane at the MP2/6-31G* level has been reported recently[54] and is in very good agreement with experiment.

Additional information on the vibrations of [1.1.1]propellane was obtained from the analysis of its electron energy loss spectrum[23].

C. Photoelectron and Electron Transmission Spectra

These spectroscopic techniques tie the states of [1.1.1]propellane to those of its radical cation and radical anion and are usually interpreted as semiquantitative indicators of the nature of occupied and unoccupied molecular orbitals of the neutral species, respectively, through the use of Koopmans' theorem.

The photoelectron spectrum of [1.1.1]propellane[22] shows a fairly low first ionization potential of 9.74 eV and a series of additional bands in fair agreement with the calculated molecular orbital energies. The lowest-energy peak in the spectrum can be quite safely assigned to ionization out of a totally symmetrical molecular orbital that is predominantly localized in the central CC bond. The shape of the band is Franck–Condon allowed and strikingly narrow, suggesting that the equilibrium geometries of the neutral and the ionized species are very similar. This agrees with the molecular orbital calculations reported by the authors,[22] and also with subsequent ones[43], according to which the central bond is actually slightly shorter in the radical cation than in the neutral molecule. The authors point out that much of the density that is removed upon ionization originates in the regions outside the central bond.

A value of the vertical ionization potential calculated[43] using the energy difference method at the MP3/6-311G* level, 9.5 eV, is in excellent agreement with the observed value.

An unresolved puzzle remains[22]: the first two vibrational peaks in the first band in the photoelectron spectrum are separated only by 360 ± 20 cm^{-1}, less than the lowest frequency vibration observed in neutral propellane (529 cm^{-1}), and very much less than its lowest totally symmetric vibration (908 cm^{-1}). Yet, the authors' calculations[22] suggest that the lowest frequency totally symmetrical vibration of the radical cation will be at higher and not lower frequencies. The authors suggested that the vibrational structure may be due to vibronic mixing with the lowest excited state of the radical cation.

The electron transmission spectrum of [1.1.1]propellane[23] shows a high-energy resonance similar to that observed in cyclopropane, but the first resonance lies already at 2.04 eV, an unusually low electron attachment energy for a saturated hydrocarbon (a more usual value is about 6 eV). This corresponds to a very low energy of the lowest unoccupied molecular orbital, in agreement with theoretical expectations (the LUMO is computed to be the antibonding combination of the 2p orbitals of the two bridgehead carbons aimed at each other, with some contributions from peripheral CC bonds[19]). The width of the first resonance and its vibrational structure, combined with the vibrational electron energy loss spectra, suggest that in the radical anion state the length of the central CC bond is significantly increased, and the peripheral bonds are lengthened as well. This is in agreement with a UHF/6-31G** calculation, which yielded 1.88 Å for the central CC bond in the radical anion.

The difference between the lengths of the central CC bond in the equilibrium geometries of the radical cation of [1.1.1]propellane (a one-electron bond, nearly the same length as in the neutral molecule) and of its radical anion (a three-electron bond, much longer than in the neutral molecule), well reproduced by *ab initio* calculations, emphasizes the inadequacy of the usual simplistic concepts of bonding based on Hückel theory and neglect of overlap. An explicit introduction of overlap, however, permits a qualitative rationalization of the difference.

D. Electronic Spectra

No UV absorption studies of [1.1.1]propellane seem to have been reported, but a recent measurement of the electron energy loss spectrum provided a wealth of information about the electronically excited states of this molecule[23].

The lowest excited singlet state occurs at 7.26 eV and is partly Rydberg in character. The spectral peak is narrow, suggesting little change in equilibrium geometry upon excitation. The excitation appears to correspond to a HOMO to LUMO promotion and is thus primarily concentrated in the central CC bond. This excited state is analogous to other σ–σ^* excitations in that the excited state can be viewed as a resonance state of two structures represented by contact ion pairs, with a positive charge on one bridgehead and a negative charge at the other[55]. Unlike the σ–σ^* triplet, this state is not repulsive but bound, due to the electrostatic attraction of the charges and the absence of a large geometrical distortion is not surprising. At higher energies, Rydberg states and Feshbach resonances are observed.

The lowest triplet state is observed at 4.1 eV (first observed peak) to 4.70 eV (vertical excitation). The calculated adiabatic excitation energies are lower (3.2 eV at the MP3 6-31G* level)[15]. The spectral band shows a long progression in the stretching vibration of the central CC bond (980 cm^{-1} in the triplet), and it appears that this bond is greatly stretched in the excited state. This corresponds to simple expectations for a triplet σ–σ^* excitation, with a strongly antibonding repulsive interaction in a biradical-type excited state wavefunction, and to the calculated[15] bond length of 1.83 Å in the triplet state (UHF/6-31G*), nearly the value typical of bicyclo[1.1.1]pentanes. It is possible that the band origin lies below 4.1 eV and is too weak to be observed. This would be compatible with the calculated values and with the quite high quenching constants of various triplet sensitizers by [1.1.1]propellane (Table 5).

A theoretical study of inelastic electron scattering by [1.1.1]propellane showed good agreement with observations[56].

TABLE 5. Excited state quenching rate constants for **1a** in benzene[a]

	Donor	E (kcal mol^{-1})	$k_q \times 10^6$ ($M^{-1}s^{-1}$)
T:	Triphenylene	67	6.8
	Benzophenone	69	9.9[b]
	Phenanthrene	62	0.92
	Fluorenone	53	0.063
S:	Pyrene	77	140

[a] Reference 96.
[b] $k_q = 7.6 \times 10^6$ $M^{-1}s^{-1}$ in MeCN.

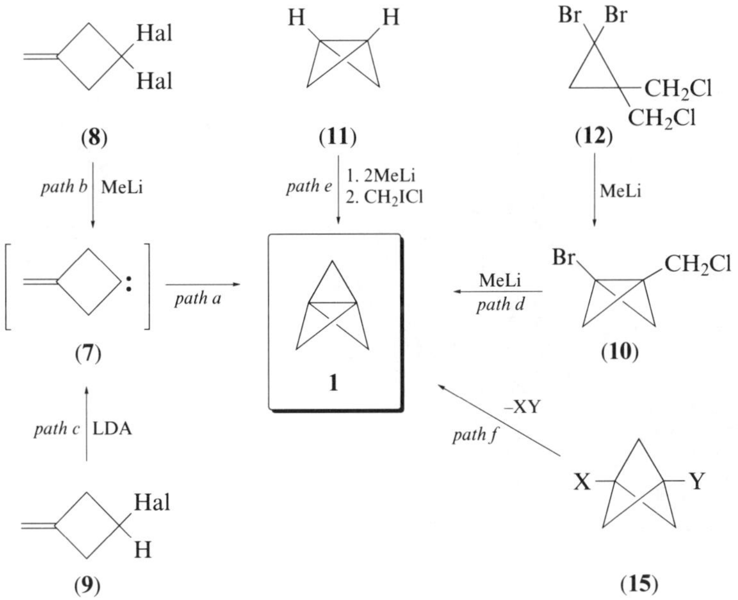

SCHEME 1

IV. SYNTHESIS

The known synthetic approaches to [1.1.1]propellanes can be divided into three general categories (Scheme 1): (i) intramolecular carbene insertion in a 3-methylenecyclobut-1-ylidene, (ii) intramolecular anionic ring closure in a bridgehead-substituted bicyclo-[1.1.0]butane, and (iii) 1,3-elimination in a bicyclo[1.1.1]pentane.

The parent [1.1.1]propellane (**1a**) is readily accessible by route (ii) in two steps from commercial materials. It can be prepared neat[37,57] or, more conveniently, as a 2–3 % solution in diethyl ether[4,58] or pentane[48,59]. Stable propellanes substituted with alkyl[35,46,60,61] or aryl[45] groups, as well as the thermally unstable 2,2-dichloro[1.1.1]propellane[39] (**1g**), have also been reported (Chart 1).

A. Intramolecular Carbene Insertion

As was suggested earlier based on theoretical considerations[3], 3-alkylidenecyclobutylidenes (**7**) efficiently cyclize intramolecularly to [1.1.1]propellanes[35,45,61,62] (path a). The carbenes were generated either by metal–halogen exchange[35,45,61,62] in the corresponding 1, 1-dihalides **8** (path *b*) or by deprotonation[45,62] of 1-halides **9** (path *c*) followed by α-elimination. The reaction appears to be general and several tetracyclic and spirocyclic [1.1.1]propellanes were prepared in good yields (Table 6).

B. Anionic Cyclization

Cyclization of derivatives of bicyclo[1.1.0]butane (**10**) represents a second important method for the synthesis of [1.1.1]propellanes[35,45] (path *d*). Metal–halogen exchange at the

TABLE 6. Generation of [1.1.1]propellanes from cyclobutanes

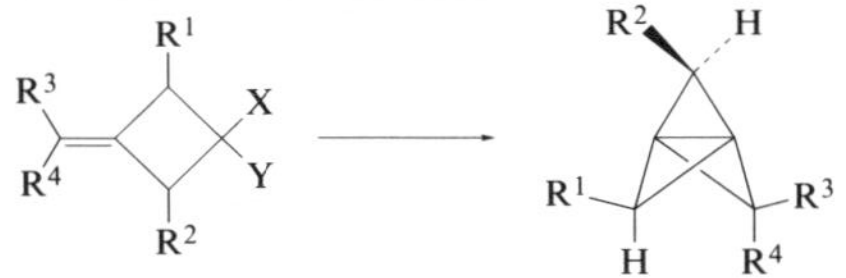

(**8**) X,Y = Halogen
(**9**) X = H,Y = Halogen
(**1**)

	X	Y	R^1	R^2	R^3	R^4	**1**	Yield (%)	References
From **8** and MeLi									
a	Br	Br	H	H	H	H	**a**	76	35,61,62
b	Br	Br	$—(CH_2)_2—$		H	$CH{=}CH_2$	**i**	41	45
c	Br	Cl	$—(CH_2)_2—$		H	*E*-CH=CHMe	**j**	56	45
d	Br	Cl	$—(CH_2)_2—$		H	C_6H_4OMe-4	**k**	85	45
e	Br	Cl	$—(CH_2)_2—$		H	1-naphthyl	**l**	35	45
f	Br	Cl	$—(CH_2)_2—$		Ph	Ph	**m**	88	45
g	Br	Cl	$—(CH_2)_2—$		C_6H_4Br-2	Ph	**n**	77	45
h	I	I	$—(CH_2)_3—$		H	H	**o**	43	61,62
i	Br	Cl	$—(CH_2)_3—$		H	*E*-CH=CHMe	**s**	31	45
j	Br	Br	$—(CH_2)_3—$		H	Ph	**t**	35	45
k	Br	Cl	$—(CH_2)_3—$		H	Ph	**t**	48	62
l	Br	I	$—(CH_2)_3—$		Me	Me	**u**	57	61
m	Br	I	$—(CH_2)_3—$		CH_2OPh	Me	**v**	95	61
n	Br	I	$—(CH_2)_3—$		H	*a*	**w**	94	35,61
From **9** and LDA									
a	Cl	H	$—(CH_2)_3—$		H	H	**o**	62	45,61,62
b	Br	H	$—(CH_2)_3—$		H	*i*-Pr	**q**	77	45
c	Cl	H	$—(CH_2)_3—$		$—(CH_2)_4—$		**x**	44	45
d	Cl	H	$—(CH_2)_3—$		$—(CH_2)_5—$		**y**	50	45
e	Cl	H	$—(CH_2)_3—$		*b*		**z**	18	61

[a] Two [1.1.1]propellane cages connected through this bond.
[b] Two [1.1.1]propellane cages spiro connected through two $—CH_2CH_2—$ bridges

TABLE 7. Generation of [1.1.1]propellanes from bicyclo[1.1.0]butanes

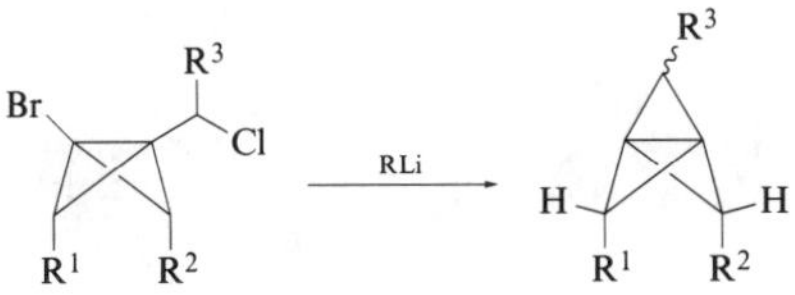

(10) **(1)**

10	R^1	R^2	R^3	**1**	Yield (%)	References
a	H	H	H	**a**	30	65
b	$—(CH_2)_2—$		H	**h**	47	35,45
c	$—(CH_2)_3—$		H	**o**	71	4,35,45
d	$—(CH_2)_3—$		Me	**p**	66	35,45
e	$—(CH_2)_3—$		*i*-Pr	**q**	60	35,45
f	$—(CH_2)_3—$		*c*-C_6H_{11}	**r**	57	35,45
g	$—2\text{-}CH_2C_6H_4—$		H	**aa**	10	35,45
h	$—CH_2C_3H_4—$		H	**bb**	67	35,45

relatively acidic bridgehead position of **10**, followed by elimination (intramolecular nucleophilic substitution), provides high yields of **1** (Table 7). The elimination can proceed by intramolecular analogs of the S_N2 but also the S_N2' substitution mechanism[45] (equation 5). Cyclopropanation of the doubly lithiated bicyclo[1.1.0]butane derivatives of **11** (path *e*) gives the corresponding propellanes in yields comparable to those obtained in the reaction of **10** (Table 8)[45].

The tetrahalides **12** treated with an alkyllithium[4,48,57,58,60,63–65] or with sodium-doped lithium metal[57] undergo two sequential metal–halogen exchange and nucleophilic

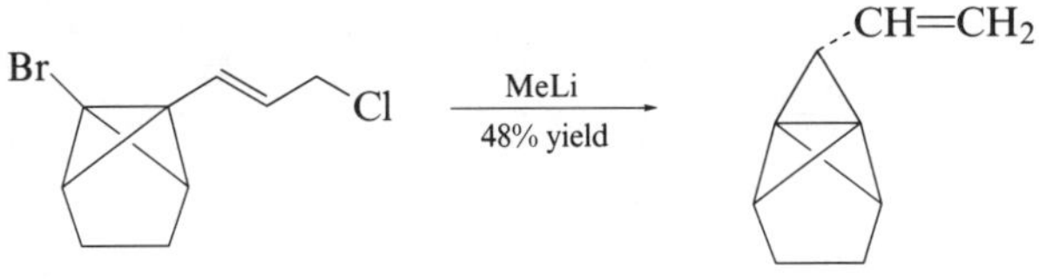

(5)

(1i)

TABLE 8. Generation of [1.1.1]propellanes from bicyclo[1.1.0]butanes

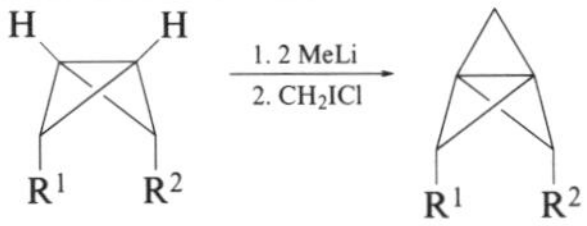

(11) **(1)**

11	R^1	R^2	**1**	Yield (%)	References
a	$—(CH_2)_2—$		**h**	50	35,45
b	$—2\text{-}CH_2C_6H_4—$		**aa**	13	35,45

TABLE 9. Generation of propellanes from cyclopropanes

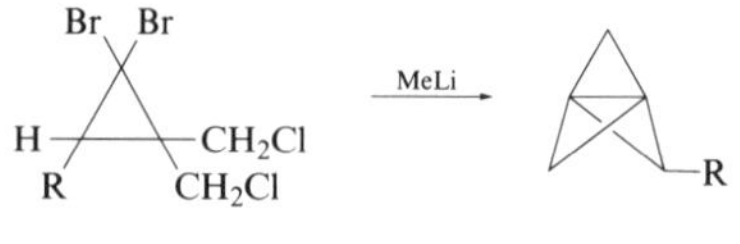

(12) (1)

12	R	1	Yield (%)	References
a	H	a	30–70	48,57,58,63–65
			25–38[a]	57
b	n-C_5H_{11}	b	34–60	46,60
c	n-C_7H_{15}	c	54	46
d	n-$C_{11}H_{23}$	d	21	46
e	$(CH_2)_3OCH_3$	e	57	46
f	$(CH_2)_5OCH_3$	f	40	46

[a] Lithium metal was used.

substitution processes via the bicyclo[1.1.0]butane derivative **10**[65], leading to the [1.1.1] propellanes **1** (Table 9). This reaction is currently the best method for the preparation of the parent [1.1.1]propellane (**1a**).

C. Elimination

Bicyclo[1.1.1]pent-1-yl anions (**13**) and radicals (**14**) carrying a suitable leaving group in position 3 undergo a 1,3-elimination reaction (path *f*, an intramolecular analog of S_N2 and S_H2 substitution, respectively) and yield [1.1.1]propellane (equations 6 and 7; Table 10).

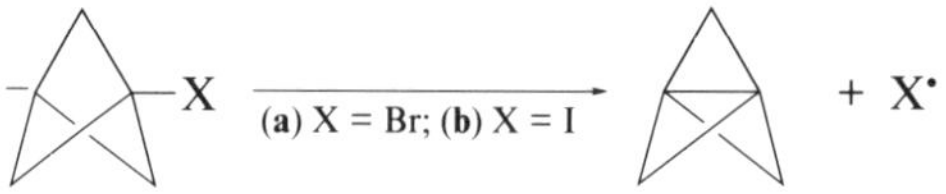

(13) (1a) (6)

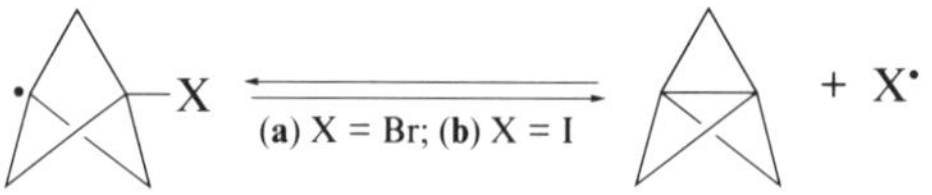

(14) (1a) (7)

Thus, dehalogenation of 1,3-dibromobicyclo[1.1.1]pentane (**15a**) with n-BuLi was used in the original synthesis[3,15] of **1a**, and the treatment of 1,3-diiodobicyclo[1.1.1]pentane (**15b**) with NaCN in DMSO is a convenient and efficient source of neat **1a**[37]. Deiodination of **15b** with hydroxide anion in ethanol is a second-order process with a rate constant of $2.72 \pm 0.08 \times 10^{-3}\ \mathrm{L\,mol^{-1}s^{-1}}$ at 25 °C[66]. Other iodides such as **15**, X = I; Y = MeO, Et_3N^+, $C_5H_5N^+$ are more inert than **15b**, and they react only with stronger nucleophiles such as organomagnesium halides or organolithiums with liberation of **1a**[67]. The deiodination of **15b** is apparently a common process and other bases such as alkyl and aryl organomagnesium or

TABLE 10. Formation of [1.1.1]propellanes from bicyclo[1.1.1]pentanes

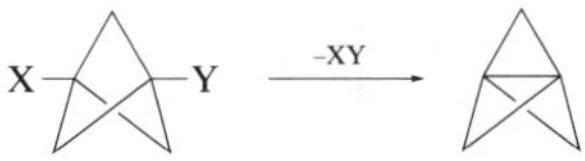

(15) (1a)

15	X	Y	Reagent	Yield (%)	References
a	Br	Br	*t*-BuLi	46	3,15
b	I	I	NaCN	88	37
			OH^-	100	66
			PhMgBr	58	67
			R_3P	*a*	67
			i-Pr_2NH	*a*	67
			Bu_3SnH	*b*	113
c	Br	H	*t*-BuLi	40[c]	68
d	PhS	PhS	Li[d]	*e*	71
e	Br	Br[f]	MeLi	18[g]	39
f	$C_5H_5N^+$	I	PhLi	*a*	67

[a] Formation of **1a** is observed.
[b] Formation of **1a** is postulated.
[c] Isolated as 3-*t*-butylbicyclo[1.1.1]pentane-1-carboxylic acid.
[d] Reduction with lithium 4,4′-di-*t*-butylbiphenyl.
[e] Not isolated.
[f] 2,2-Dichloro derivative.
[g] Propellane **1g** was trapped with I_2 and isolated as 2,2-dichloro-1,3-diiodobicyclo[1.1.1]pentane.

lithium reagents, hindered amines, e.g. *i*-Pr_2NH, collidine, and phosphines (Bu_3P, Ph_3P), produce various amounts of **1a**. In these reactions, the base apparently performs an S_N2 substitution on a halogen to yield a bridgehead anion **13a** or **13b** that carries another halogen in position 3, and this then undergoes a 1,3-elimination (intramolecular S_N2 substitution).

The treatment of 1-bromobicyclo[1.1.1]pentane (**15c**) with *t*-butyllithium presumably yields transient [1.1.1]propellane via the bridgehead anion **13a**, since 3-*t*-butylbicyclo[1.1.1]pentane-1-carboxylic acid is obtained after carboxylation[68]. However, the corresponding chloride is stable under these conditions (equation 8) and the iodide gives the bicyclo[1.1.1]pentyl anion **16** by metal–halogen exchange (equation 9)[69]. An *ab initio* study of the deprotonation reaction of 1-chlorobicyclo[1.1.1]pentane showed a 10% elongation of the C—Cl bond and a high negative charge on chlorine (–0.479) in the anion **13c**[70].

(13c) (1a) (8)

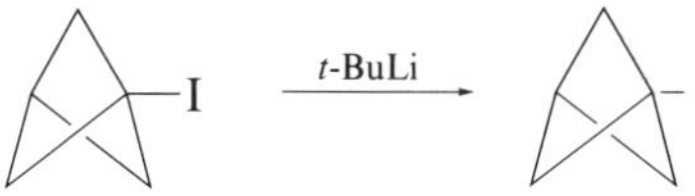

(16) (9)

The treatment of the double bridgehead sulfide, 1,3-bis(phenylthio)bicyclo[1.1.1]-pentane (**15d**), with lithium 4,4′-di-*t*-butylbiphenyl also induces reductive elimination and leads to [1.1.1]propellane[71]. However, as the required bicyclo[1.1.1]pentanes **15** are most conveniently prepared from the corresponding propellanes, they do not really offer an alternative synthetic route to [1.1.1]propellanes. They are best considered as convenient propellane storage materials[37].

V. REACTIONS

A. Introduction

The central bond of [1.1.1]propellanes (**1**) is the center of their reactivity, and in many ways, it is useful to think of it as somewhat akin to the π bond in an alkene. The strengths of the two are comparable, and both are susceptible to electrophilic and radical attack. The main difference is that the central bond in **1a** is apparently somewhat susceptible to nucleophilic attack as well, whereas the π bond in unsubstituted ethylene is not. In both cases, introduction of electron-withdrawing groups enhances reactivity towards nucleophiles.

Reactions of **1** with nucleophiles, radicals or electrophiles normally lead to a cleavage of the central CC bond and to the formation of the bridgehead anions, radicals and cations, respectively. The anions and radicals react with electrophiles and radicalophiles, respectively, giving access to a variety of functionalized bicyclo[1.1.1]pentanes and their oligomers, [*n*]staffanes. The bridgehead cations usually rearrange to the corresponding methylenecyclobutyl cations before reacting with a nucleophile. Only those that carry an electronegative substituent in position 3 are sufficiently long-lived to be trapped unrearranged (Scheme 2).

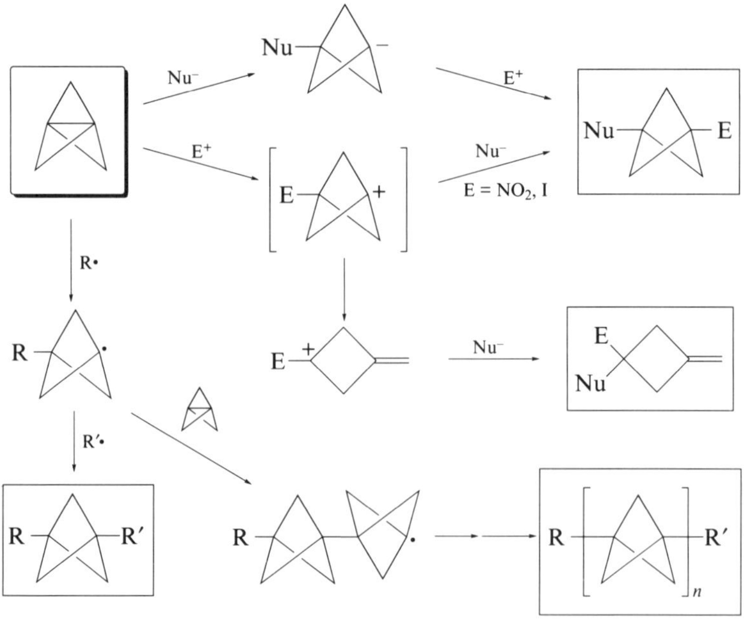

SCHEME 2

Reactions of [1.1.1]propellane that preserve the tricylic structure are not known. One such reaction was investigated computationally at the MINDO/3 level: the generation of the α-halo anions **17** and subsequently the carbene **18** from chloro- and fluoro[1.1.1] propellane (equation 10). The authors concluded that it should be possible to generate the anion **17** under relatively mild phase transfer conditions[72].

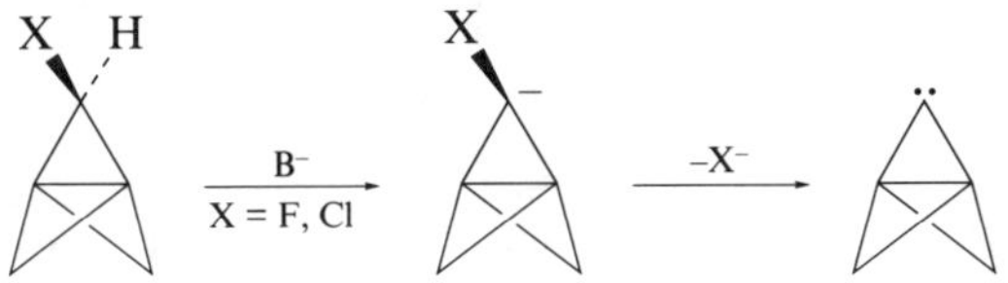

(10)

(17) **(18)**

Before proceeding to a discussion of the reactions of the central bond of [1.1.1] propellane, it will be useful to describe the properties of the first-formed intermediates, the bicyclo[1.1.1]pent-1-yl anion **16**, radical **6**, and cation **19**.

Bicyclo[1.1.1]pent-1-yl anion ***16***. The basicity and electrophilicity of **16** relative to ordinary tertiary alkyl anions are not easy to predict without a computation. On the one hand, the high percent of s character in the hybrid holding the lone pair on the bridgehead carbon atom suggests that this anion will have quite low basicity. On the other hand, the transannular interaction between the two bridgehead carbons can be expected to increase the basicity, due to a relief in transannular closed-shell repulsions when a lone pair is replaced by a bond at the bridgehead. Indeed, the computed (MP2/6-31G*) inter-bridgehead distance decreases from 1.974 Å in the anion to 1.872 Å in bicyclo[1.1.1] pentane[43].

It appears that the combined effects of high s character and transannular interaction produce a basicity somewhat higher than that of the vinyl anion. The gas-phase proton affinity of the parent bridgehead anion (**16**) is 411 ± 3.5 kcal mol^{-1} [73]. This value is close to those of the cyclopropyl anion (408 ± 5 kcal mol^{-1}) and the vinyl anion (407 ± 3 kcal mol^{-1}), in agreement with MP3/6-31 + G* calculations[43].

Bicyclo[1.1.1]pent-1-yl radical ***6***. The high s character of the bridgehead orbital carrying the radical center suggests that **6** will be highly reactive relative to ordinary tertiary alkyl radicals. The effect of transannular interaction is not easy to predict in an *a priori* fashion, since it corresponds to an interaction of a half-filled orbital with a closed shell. The interbridgehead distance in **6** has been calculated at the UMP2/6-31G* level[43] to be 1.797 Å and at the UHF/6-31G* level[74] to be 1.814 Å, compared to 1.872 Å at the MP2/6-31G* level[43] and 1.870 Å at the HF/6-31G* level[74] calculated for bicyclo[1.1.1]pentane. This suggests that the destabilization by transannular interaction is smaller in the radical, reducing its reactivity.

In principle, the bridgehead bicyclo[1.1.1]pent-1-yl radical **6** can release more of its strain energy by opening its strained cage in a unimolecular rearrangement to the 3-methylenecyclobut-1-yl radical **20**. The rearrangement, however, is remarkably slow, nowhere near as fast as that of the analogous cations. MINDO/3 calculations[75] suggested that the transformation of **6** to yield **20** (equation 11) is indeed strongly exothermic

(11)

(6) R = H **(20)** R = H

(41.8 kcal mol^{-1}), but requires a large activation energy of 25.5 kcal mol^{-1}. More recent calculations at the UHF/6-31G* level predicted the activation energy for the rearrangement of **6** to be 25.8 kcal mol^{-1} [76]. Experimental efforts[76,77] to detect a rearrangement of the bridgehead radical **6** to **20** have given negative results, implying a slow rate for the rearrangement. The authors estimated initially that the activation energy for the transformation exceeds 14.3[77], and later 26[76] kcal mol^{-1}. A recent experimental study of the rearrangement of the probably more rearrangement-prone 3-phenylbicyclo[1.1.1]pent-1-yl radical found the rate constant to be 5.5×10^4 s^{-1} at 150 °C and the activation barrier to be 21 ± 3 kcal mol^{-1} [76].

*Bicyclo[1.1.1]pent-1-yl cation **19**.* The high s character of the exocyclic carbon orbitals suggests a very high energy and nucleophilicity for the bridgehead cation. However, the effect of transannular interaction should now be particularly stabilizing, since it corresponds to the interaction of a closed shell with an empty orbital. The solvolysis of 1-chlorobicyclo[1.1.1]pentane is actually 3 times faster than that of *t*-butyl chloride[78]! Solvolysis of the bridgehead bromide **15c** is also significantly accelerated relative to *t*-butyl bromide[79,80]. The extraordinary ease of formation of the bridgehead bicyclo[1.1.1]pentyl cation (**19**) is revealed even by calculations at the simplest levels. According to an early CNDO calculation, the stability of **19** lies between that of the isopropyl cation and the planar *t*-butyl cation[78]. A recent *ab initio* study at several levels of correlated theory concluded that hydride abstraction at the tertiary position is about 2 kcal mol^{-1} easier from bicyclo[1.1.1]pentane (**2**) than from isobutane[43].

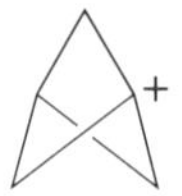

(**19**)

The importance of transannular interaction for the stabilization of the bridgehead cation **19** was revealed in numerous calculations[43,81–86] and is obvious from the inter-bridgehead distance in this species, calculated to be 1.525 Å at the MP2/6-31G* level[43]. This can be compared with bicyclo[1.1.1]pentane (**2**), for which this level of calculation yields 1.872 Å (recall that the reported experimental values are 1.845[87] and 1.874[88] Å) and with [1.1.1]propellane (**1a**), for which 1.592 Å is calculated at this level of theory[43] and 1.60 Å is observed[15,25,26]. The bridgehead cation **19** clearly contains a central CC bond and it is probably best to view it as a protonated [1.1.1]propellane[43,86,89]. The calculated (HF/4-31G; footnote 24 in Reference 82) proton affinity of [1.1.1]propellane is 229 kcal mol^{-1}, very close to that of ammonia. The cation can also be viewed as a 'CH^+-trimethylmethylene complex'[83].

Calculations at all levels suggest that the parent bridgehead cation **19** with C_{3v} symmetry is at best only a shallow local minimum on the $C_5H_7^+$ potential energy surface, and most likely only a transition state[43,82,83,86]. Unsubstituted **19** has indeed not been observed or trapped experimentally and appears to rearrange immediately upon formation. Thus, reactions of **1** with an electrophile generally lead to the formation of the trapping products expected from the 3-methylenecyclobutyl cation[78,79]. Only those rare derivatives of **19** that carry an electronegative substituent such as I or NO_2 on the other bridgehead have yielded unrearranged trapped products[64,66,67,90].

B. Nucleophilic Attack

Aliphatic[35] and some aromatic[91] Grignard reagents add to [1.1.1]propellanes to give the bicyclo[1.1.1]pent-1-ylmagnesium reagents (**21a**). These reactions are rather slow and are performed in refluxing ether solution. In contrast, organolithium reagents react with

TABLE 11. Formation and protonation of bridgehead organometallic reagents

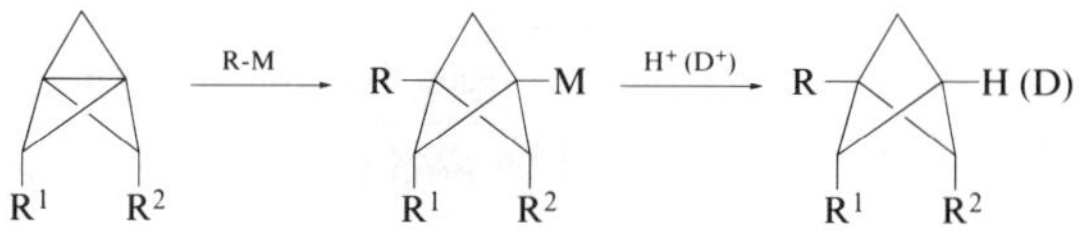

(21) (a) M = MgBr
(b) M = Li

R	M	Isotopic purity	Yield (%)[a]	Reference
1a $R^1,R^2 = H$				
t-Bu[b]	Li	—	78	68
cubyl	Li	—	—	116
1,4-cubanediyl	Li	—	—	116
H-C_8H_{12}-1-yl[c]	Li	—	—	116
C_8H_{12}-1,4-diyl[c]	Li	—	—	116
1b $R^1 = H, R^2 = (CH_2)_4Me$				
t-Bu	Li[d]	—	—	60
1o $R^1,R^2 = —(CH_2)_3—$				
Et	MgBr	—	21	35
i-Pr	MgBr	—	46	35
t-Bu	MgBr	—	32	35
t-Bu	Li[d]	>96	43–96	92,93
$PhCH_2$	MgBr	—	68	35
allyl	MgBr	—	67	35
Ph	MgBr	—	67	35
Ph	Li[d]	>96	34	92,93

[a] Yields based on starting [1.1.1]propellanes.
[b] Formed *in situ* from **15c**.
[c] C_8H_{12} stands for bicyclo[2.2.2]octane-1,4-diyl.
[d] Oligomers **[*n*]21**.

[1.1.1]propellane readily and yield substituted [*n*]staffane oligomers **[*n*]21b** or polymers as products[4,59,60,92–94]. The degree of oligomerization strongly depends on the RLi/propellane ratio[93,94]. These reactions are fast in hydrocarbon solutions at temperatures above –30 °C, and are sensitive to the nature of the alkyllithium[4,93]. Subsequent transformations of the carbanions **21** include a typical range of reactions with electrophiles[35,64,68,69,71,95]. Upon quenching of the RLi-induced oligomerization reaction mixture with a source of D^+, incorporation of deuterium (>95%) at the bridgehead position was observed[60,92,93] (Table 11), but quenching with other electrophiles was unsuccessful[94]. The course of the anion-induced oligomerization reaction of **1** is sensitive to the reaction conditions, and it is now suspected that one-electron transfer radical processes are involved[94]. Propellane **1a** generated *in situ* from 1-bromobicyclo[1.1.1]pentane (**15c**) reacts with excess *t*-BuLi and forms the anion **[1]21b** (R = *t*-Bu)[68,69].

C. Radical Attack

Radical attack on the central bond in [1.1.1]propellane **1** occurs 2–3 times faster than attack on styrene[96] and yields bridgehead bicyclo[1.1.1]pent-1-yl radicals[6]. Laser flash photolysis techniques were used to measure the rate constants for the reactions of **1a** with five different radicals (Table 12)[96–99]. The addition of the phenylthiyl radical to **1a** is

TABLE 12. Rate constants for radical attack on [1.1.1]propellane and styrene[a]

Radical	Rate constant $[M^{-1} s^{-1}]$ at RT	
	[1.1.1]Propellane	Styrene
t-BuO·	2.8×10^6	0.91×10^6
PhS·	6.2×10^7	2.2×10^7
Et_3Si·	6.0×10^8	2.0×10^8
Cl·$\{C_6H_6\}$[b]	3.1×10^9	*c*
p-$MeOC_6H_4COO$·	1.0×10^7	5.5×10^6

[a] Reference 96.
[b] Complex.
[c] Not measured.

reversible at room temperature[97]. The radical character of reactions of **1a** with I_2, PhSH and $PhICl_2$ was confirmed by using tribromonitrosobenzene as a spin trap and monitoring the reaction by ESR[100].

Carbenes react with **1a** at comparable rates, and rate constants of $6.7 \pm 0.8 \times 10^6\ M^{-1}s^{-1}$ [97] and $6.7 \times 10^7\ M^{-1}\ s^{-1}$ [99] have been obtained for the reactions of diphenylcarbene and chlorophenylcarbene, respectively. Both series of experiments show a trend of more facile reaction of **1a** with more electropositive centers. The triplet 1,4-diradical **22a** (X = Ph : λ_{max} = 330 nm)[97], formed by the addition of diphenylcarbene to [1.1.1]propellane, rearranges to the diene **23a** with an unusually long lifetime of 9.7 μs[97] (equation 12). It has been suggested that the mutual orthogonality of the singly occupied orbitals is responsible for slow intersystem crossing and thus the long lifetime for **22a**. The 1,4-diradical **22b**, however, has a much shorter lifetime, less than 30 ns[99]. Dichloro- and dibromocarbene similarly react with **1o** yielding a mixture of dimethylenecyclobutanes **24** and **25**[62].

(12)

(22) (**a**) X = Ph (**23**)
(**b**) X = Cl

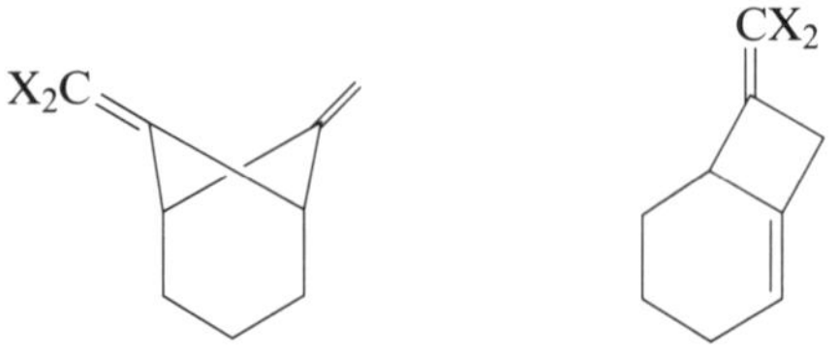

(24) X = Cl,Br **(25)**

Radical addition reactions provide synthetic access to 1-substituted and 1,3-disubstituted bicyclo[1.1.1]pentanes. They have been reviewed recently[6], but a large number of new examples has accumulated in the literature since. Under suitable conditions, these processes produce significant yields of terminally substituted oligomers of bicyclo[1.1.1]pentane, [*n*]staffanes (Table 13). The course of the reaction depends on the

TABLE 13. Formation of bicyclo[1.1.1]pentanes and [*n*]staffanes by reaction of [1.1.1]propellanes with radicals

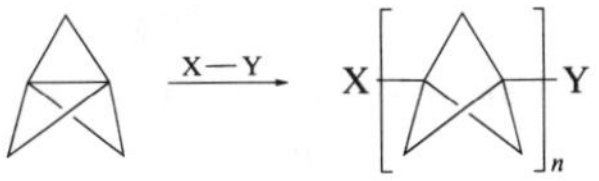

Reagent	Conditions[a]	Yield (%)[b]	References
	Radical Source: **R—H**		
Carbon radicals			
R: $CH(COOEt)_2$	E	*c,d*	59,104
$C(COOEt)_3$	E	28,11,1,0.2	59,104
$CPh(COOEt)_2$	E	*c,d*	59,104
CH(COMe)COOMe	E,N	45[d]	59,64,104
CH(CN)COOMe	E,N	45[d]	59,64,104
CH(Me)OEt	E	>22,16,6,4,2[e]	48
	E	13,12,5[f]	62
CCl_3	N	43	64,110
	P	70,20	65,101
	P	35,26,9	*g*
CCl_2Me	E	>21,14,2	109
CH_2COMe	N	33[d]	50
CH(Me)COMe	N	72[d]	64
CH(Me)COOMe	N	45,18	64
CH(Cl)COOMe	N	75	64
CCl(Me)COOMe	N	65	64
COOMe	N	40	64
	P	21,18,9,3,1	48,59,103
2-Tetrahydrofuranyl	N	44,24[d]	64
2- and 4-Dioxolanyl	N	50[d,h]	64
$CH(Me)NEt_2$	N	22	64
Other radicals			
R: H	P	*c*,13,9,6,2,0.3	105
$PO(OR')_2$	E	45,23,5	59,104,107
	N	58	64
PPh_2	O	*c*	64
SPh	E	98	4,64,107
SCOMe	E	*c*	107
$SiEt_3$	N	40[h]	64
$SnMe_3$	E	50–57	112
$SnBu_3$	E	28	65
	E	72[f]	62
$Sn(cyclohexyl)_3$	E	56	112
$SnPh_3$	E	58	112
	Radical source: **R—X**		
X = Cl			
R: CCl_3	N	79	64
	P	70,20	65,101
		54,4[f]	62
CCl_2Me	E	>18,18	109
SO_2Me	E	*c*,50	48

(*continued*)

TABLE 13 (*continued*)

Reagent	Conditions[a]	Yield (%)[b]	References
	O	15,2	120
SO_2Ph	E	72	48
$SO_2C_6H_4CH_3$-*p*	O	19,32	120
SO_2Cl	O	20,10	120
OBu-*t*	N	41	64,110
X = *Br*			
R: *t*-Bu	E	16	48
	E	30	115
	N	36[f]	62
CH_2Ph	E	*c*	48,59
	N	50[f]	62
$CH_2CH{=}CH_2$	E	11[i]	113
CH_2Cl	N	31[f]	62
CCl_3	O	100	64,110
$CH(COOEt)_2$	E	>60[e]	113
	E	68	65
CH_2COOMe	E	>60[e]	48,113
	P	>6,11,4	48
CN	P	*c*,45,11	64,110
$CHBr_2$	P	88	65
CBr_3	P	83	65
CF_2Br	P	85,10	118
$—CF_2CF_2—$	P	40,7	118
X = *I*			
R: Me	N	25[f]	62
	N	42	109
	P	65	65
CF_3	E	75	67
n-Bu	E	46	48
	P	14,16,8[e]	48
s-Bu	E	47	48
alkyl	E	>60[e]	113,114
Ph	P	>20[e]	48,114
$CH_2CH{=}CH_2$	E	11[j]	113
$—C_8H_6—$[k]	O	55[l,m], 78[m,n]	116
$—C_8H_{12}—$[o]	O	74[l,m], 81[m,n]	116
$—C_5H_6—$[p]	P	95[l,m]	48
	P	60	119
$3\text{-BuC}_5H_6—$	P	>27,12,4.7,1.4[e]	48
$3\text{-PhC}_5H_6—$	P	*c*,*d*	48
	Radical source: **X—X**		
X: SPh	E	45	64,110
	E	63,27	35,63
SMe	E	40,12,0.4,0.1	35,48,63,102
SEt	E	18,4	63
$S(CH_2)_2COOEt$	E	5	63
SCOMe	E	2.7,14.4,2.6,0.3,0.05	48,59,104
SePh	E	38	64,110
NO_2	E	25	64
$P(OEt)_2$	E	27	48,104

(*continued*)

TABLE 13 (*continued*)

Reagent	Conditions[a]	Yield (%)[b]	References
Br	E	13	48
	E	37	65
q	E	31,21	115,117
I	E	88	64,110,48,65
	E	61	37
I	E	18[r]	39
PhCO	E	1.7–4.5, 2.3–3.8, 0.5–0.7, 0.1[e]	48
MeCO	E	58	58

[a] Solvent: E, diethyl ether; P, pentane; N, none; O, others.
[b] Yields of [*n*]staffanes based on [1.1.1]propellane are given in the sequence $n = 1, 2, \ldots$.
[c] Yield not reported.
[d] Oligomers formed.
[e] Yields of the isolated subsequent products.
[f] 2,4-Trimethylene[1.1.1]propellane (**1o**).
[g] G. S. Murthy, K. Hassenrück and J. Michl, unpublished results.
[h] Mixture of two products.
[i] 3,3′-Dibromo[2]staffane only.
[j] 1,3-Diiodobicyclo[1.1.1]pentane only.
[k] Cubane-1,4-diyl from 1,4-diiodocubane.
[l] 1:1 Adduct.
[m] Yield based on the starting iodide.
[n] 1:2 Adduct.
[o] Bicyclo[2.2.2]octane-1,4-diyl from 1,4-diiodobicyclo[2.2.2]octane.
[p] Bicyclo[1.1.1]pentane-1,3-diyl from 1,3-diiodobicyclo[1.1.1]pentane.
[q] 1,2-Dibromotetrachloroethane and PPh_3.
[r] 2,2-Dichloro[1.1.1]propellane (**1g**).

nature of the reactant, on its concentration and relative molar ratio to the propellane [48,59,63,64,101], and also on the solvent used[48]. The rate constant for the reaction of the bridgehead 1-bicyclo[1.1.1]pentyl radical **6** with propellane **1** is expected to be rather low[64], but when the rate of termination of the radical chain, e.g. by hydrogen abstraction, is lower still, the oligomerization of **1** to [*n*]staffanes may dominate. For instance, in the reaction of **1a** with THF, the ratio of the rate constant of hydrogen abstraction to the rate constant of propagation was estimated to be 1.8×10^{-3}, despite the relatively high bridgehead C—H bond dissociation energy in the bicyclo[1.1.1]pentane product[64].

The oligomerization reactions require a separation step for the isolation of pure individual products. The separation is relatively easy for the lower members of the series, since the increment is five carbon atoms, but becomes increasingly difficult for the higher molecular weight products. Pure individual substituted derivatives have usually been isolated only up to $n = 4$ or 5.

Sometimes the isolation of individual members of the series is not an issue, as in polymer synthesis. Pure neat propellane polymerizes spontaneously in a matter of hours at room temperature[37,46]. The process can be suppressed by dilution with a solvent or addition of a small amount of a radical inhibitor. Although a possible catalytic role of impurities and Teflon-coated container walls has not been ruled out completely rigorously, it appears likely that this may be a genuine example of a process in which two closed-shell molecules react to produce a biradical which then triggers oligomerization and polymerization. A SINDO1 computational study has led to the proposal that the reaction proceeds through a [2]staffane-3,3′-diyl triplet formed by the interaction of two monomers followed by intersystem crossing[102].

The first propellane to be polymerized by reaction with a base was **1o**[92,93] and **1a** followed soon after[59]. Subsequent investigations[36,103] confirmed the initial findings, but all the studies were limited by the intractability of the polymers, which could not be dissolved nor melted despite the only low average polymerization degree obtained in radical-induced[59,104,105], anion-induced[59,104] and spontaneous[36,37] polymerization of parent **1a**. The atactic polymers of 2-pentyl[1.1.1]propellane (**1b**)[46,60] and 2-methoxypropyl[1.1.1] propellane (**1e**)[46], prepared more recently, have the great advantage of being soluble. Relatively high molecular weight samples (10^5 dalton) have been prepared and studied[103,106].

Addition across the H—H bond. A mixture of the parent [*n*]staffane hydrocarbons (**[*n*]2**) has been prepared in a good total yield by a radical oligomerization of [1.1.1]-propellane in *n*-pentane solution at –110 °C induced by hydrogen atoms generated in a microwave discharge[105]. There is no indication that *n*-pentane itself is attacked by the bicyclo[1.1.1]pentyl radicals **6** (no deuterium incorporation from pentane-d_{12}). The lower members of the oligomeric hydrocarbon series have now been individually isolated pure up to $n = 7$, whereas the higher members have only been obtained from this reaction as a mixture with an average polymerization degree of about ten to fifteen. Higher average polymerization degrees, up to about 100 as estimated from end group analysis by solid state NMR, were obtained when the hydrogen-atom induced oligomerization was performed inside a zeolite, avoiding precipitation[94].

Additions across the C—H Bonds. Many compounds with an activated CH bond, such as cyanoacetic esters or chloroform, oligomerize with propellanes in ethereal solution[48,59,101,104,107–109]. However, since diethyl ether itself can add across its activated α C—H bonds[48,62,64], mixtures can result, and some of the oligomerization reactions, such as the reaction with formate esters[48,59], are best performed in a hydrocarbon solution[48,59,104,105], or without any solvent[62,64,110].

There are strong indications that bond strengths are not the only controlling factor determining the ease of hydrogen abstraction by the bridgehead radical and the rate of radical addition to [1.1.1]propellane, and that electronegativity effects play an important role. Some of the reported observations are puzzling. Thus, several seemingly good hydrogen donors (dimethyl acetal, methyl isobutyrate) do not undergo radical addition to [1.1.1]propellane (**1a**)[64], and others, like methoxyacetonitrile and methylthioacetonitrile, react in only marginal yields[107]. Other observations are easier to rationalize. Thus, chloroform, $D(\mathrm{CH}) = 96\ \mathrm{kcal\,mol^{-1}}$, reacts readily with **1a**, and the reaction with triethylsilane, $D(\mathrm{SiH}) = {\sim}90\ \mathrm{kcal\,mol^{-1}}$, yields mostly the CH rather than the originally expected SiH insertion product[64]. This suggests strongly that the transition state for hydrogen abstraction is stabilized by contributions from resonance structures that place a positive charge on the bridgehead carbon in the bridgehead radical **6**, presumably due to the stabilization of the charge by transannular interaction. Abstraction of acidic hydrogens would then be facilitated. Although such speculation may be correct, kinetic data on abstraction rates are needed before any reliable understanding emerges.

Additions across P—H, S—H, Si—H and Sn—H bonds. Compounds with P—H bonds such as dialkyl phosphites[59,64,104,107], diphenylphosphine[64] and tetrasubstituted phosphoranes[111] react with [1.1.1]propellane, giving series of oligomers,[59,107], or only monomeric products[64,111]. Reagents with S—H bonds (thiophenol[4,64,107] and thiolacetic acid[107]) yield only monomeric products. Silicon[64] and tin[62,65,112] hydrides also react with neat propellanes to give monomeric products. The latter react much faster than the former.

Addition across carbon—halogen bonds. Alkyl iodides[48,62,65,109,113,114] and also activated bromides[48,65,107,113,115] (e.g. methyl bromoacetate) react with **1a** thermally[62] or under UV irradiation to give 1,3-disubstituted bicyclo[1.1.1]pentanes when the reaction is performed in diethyl ether, and the insertion of a single bicyclo[1.1.1]pentane cage can be viewed as the standard reaction pattern.

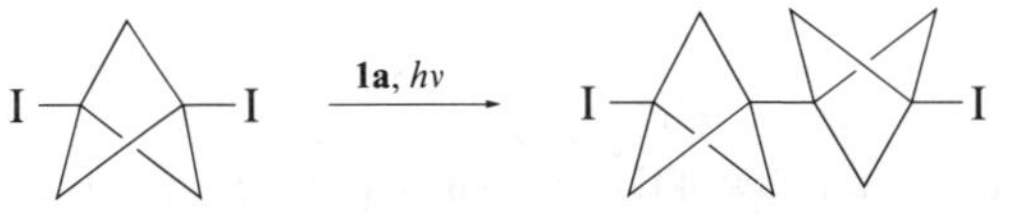

([1]15b) **([2]15b)** (13)

Oligomeric $3,3^{(n-1)}$-disubstituted [*n*]staffanes were obtained[48] in special cases with concentrated solutions of **1a** in pentane. 1,3-Diiodobicyclo[1.1.1]pentane (**[1]15b**, obtained from **1a** and iodine) reacts with excess [1.1.1]propellane in pentane under UV irradiation to give a high yield of the poorly soluble **[2]15b** only[48,65] (equation 13) and higher [*n*]staffanes are not formed. Under similar conditions, *n*-butyl iodide, 3-butyl-1-iodobicyclo[1.1.1]pentane, iodobenzene, and 3-phenyl-1-iodobicyclo[1.1.1]pentane give oligomeric series[48,114]. Several other aromatic iodides reacted with **1a** under the same conditions only to a small degree if at all[48].

The product **[2]15b** represents the best entry to doubly bridgehead functionalized [2]staffanes at the moment.

The bridgehead iodides, 1,4-diiodobicyclo[2.2.2]octane and 1,4-diiodocubane, reacted with [1.1.1]propellane in a dilute benzene solution only to give the products **26** and **27** in a ratio that can be controlled by reaction conditions, and no oligomers were observed[116] (equation 14).

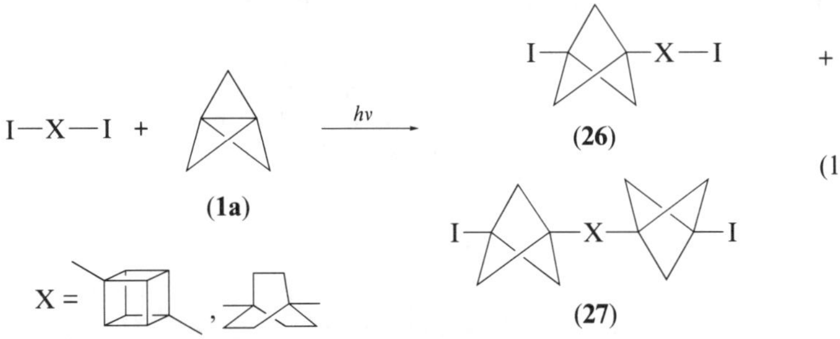

(14)

Allyl iodide and bromide react with **1a** to yield the dihalides **[1]15b** and **[2]15a**, respectively, in about 15% yield. The usual adducts were not observed[113] (equation 15).

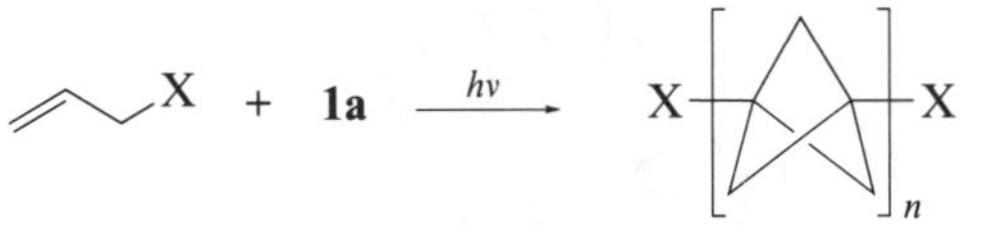

X = Br,I

([n]15a) X = Br, $n = 2$
([n]15b) X = I, $n = 1$

(15)

1,2-Dibromotetrachloroethane reacts with **1a** in the presence of PPh_3 and yields the dibromides **[1]15a** and **[2]15a** in 31% and 21% yields, respectively[115], or only **[2]15a** in 14% yield[117]. In contrast, dibromotetrafluoroethane yields a mixture of the two oligomers **[*n*]28** and a double insertion product **29**[118] (equation 16). The bromide **[2]15a** has also been

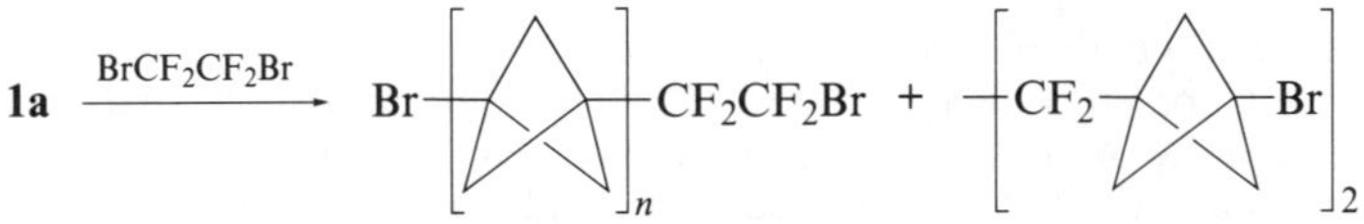

([*n*]28) **(29)** (16)

reported to form in 26% yield during irradiation of 1-bromo-3-(tribromomethyl)-bicyclo[1.1.1]pentane with **1a**[65,119]. Benzyl bromide[48,59,62], cyanogen bromide[64,110] and diethyl bromomalonate[65] react with **1a** giving mostly the expected monomeric adducts, but the formation of small quantities of the dibromide **[2]15a** (typically less than 10%) is also observed in the reaction.

Most alkyl chlorides are not active enough to react with propellanes. However, carbon tetrachloride[64,65,101], *gem*-trichlorides[109] and sulfonyl chlorides[48,104,120] react easily to give the first few oligomeric derivatives of [*n*]staffanes. The radical reaction of **1a** with sulfuryl chloride yields the expected oligomers **[*n*]30** but also the compounds **31** and **32** are formed in 12% and 8% yield, respectively[120].

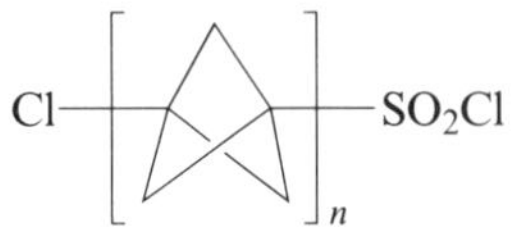

([*n*]30)
$n = 1,2$

(31) X = Cl
(32) X = SO_2Cl

Addition across the halogen–halogen bonds. Iodine[37,48,64,65,110,121] and bromine[48,65,121] themselves react with [1.1.1]propellane to give 1,3-dihalobicyclo[1.1.1]pentanes **[1]15b** and **[1]15a**, respectively. Iodine reacts with **1a** cleanly, but bromine itself or as a complex with dioxane gives a significant amount of **33a**[65,121] (equation 17). Reaction of chlorine with **1a** gives a complex mixture of products, but reaction of **1a** with $PhICl_2$ yields the dichloride **15g** and the tetrachloride **33b** in 3:2 ratio[121]. In the presence of chloroform, this reaction yielded 23% of 1-iodo-3-(trichloromethyl)bicyclo[1.1.1]pentane[65,122]. Reaction of 2,4-trimethylene[1.1.1]propellane (**1o**) with iodine yielded only a ring-opened product **34**[62].

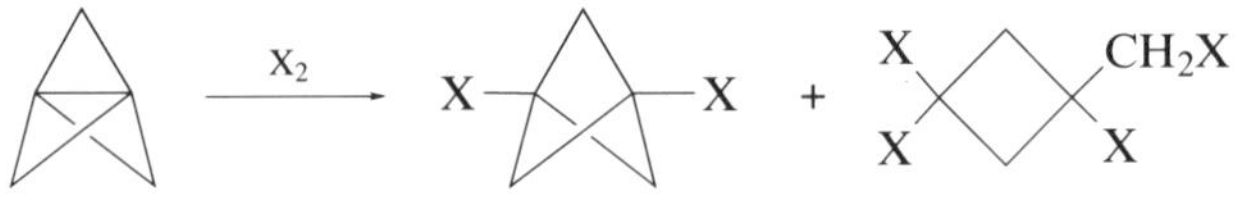

(17)

(1a) **(15)** (**a**) X = Br (**b**) X = I (**g**) X = Cl **(33)** (**a**) X = Br (**b**) X = Cl

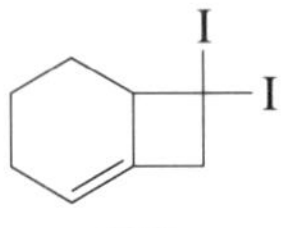

(34)

Addition across S—S and Se—Se bonds. Organic disulfides[35,48,59,63,64,104,110,123] readily react with propellanes, and the degree of oligomerization depends on the ratio of the reagents[35,63]. The addition of diacetyl disulfide gives the oligomeric series and provides particularly convenient access to doubly terminal dithiols[48,124]. The hindered di-*t*-butyl and dipivaloyl disulfides do not react with **1a** in diethyl ether[123]. Diphenyl diselenide reacts with **1a** and yields only the first adduct[64,110].

Addition across the P—P and P—C bonds. Tetraethyl hypophosphite was successfuly added to **1a**. Upon subsequent air oxidation the diphosphonate **35**, and a significant amount of the more complex structure **36** were isolated[48,104] (equation 18).

Three phosphonites $(RO)_2PR$ have also been found to add to **1a**. The products contained the bicyclo[1.1.1]pentane cage inserted into the P—C bond and were isolated after oxidation to stable pentavalent phosphorus derivatives[125]. In a related process involving

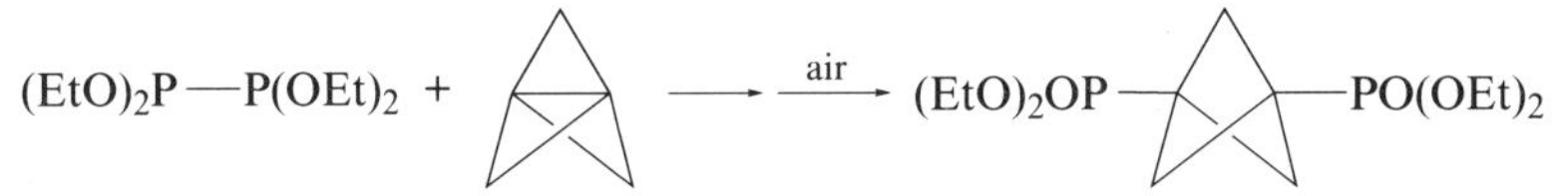

(18)

(1a) (35)

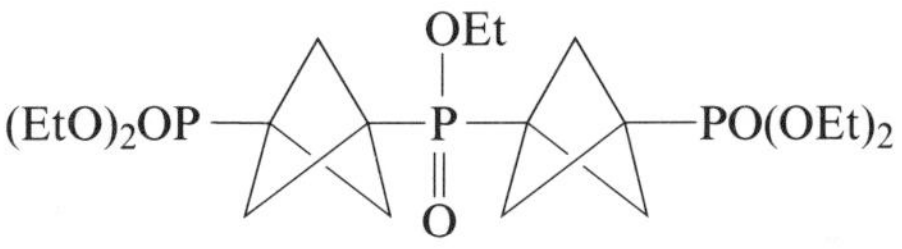

(36)

β-scission, **1a** was found to react with dimethyl benzyl phosphite to yield dimethyl 3-benzylbicyclo[1.1.1]pentane-1-phosphonate[125].

Reactions with carbonyl compounds. The reaction of acetaldehyde with [1.1.1]-propellane turned out differently than anticipated, and yielded the formal 1:2 adduct **37a**[58,64,110] (Scheme 3). It has been proposed[64,110] that the radical **38** (R = MeCO) adds to the carbonyl group instead of abstracting the aldehydic hydrogen, and the resulting oxy

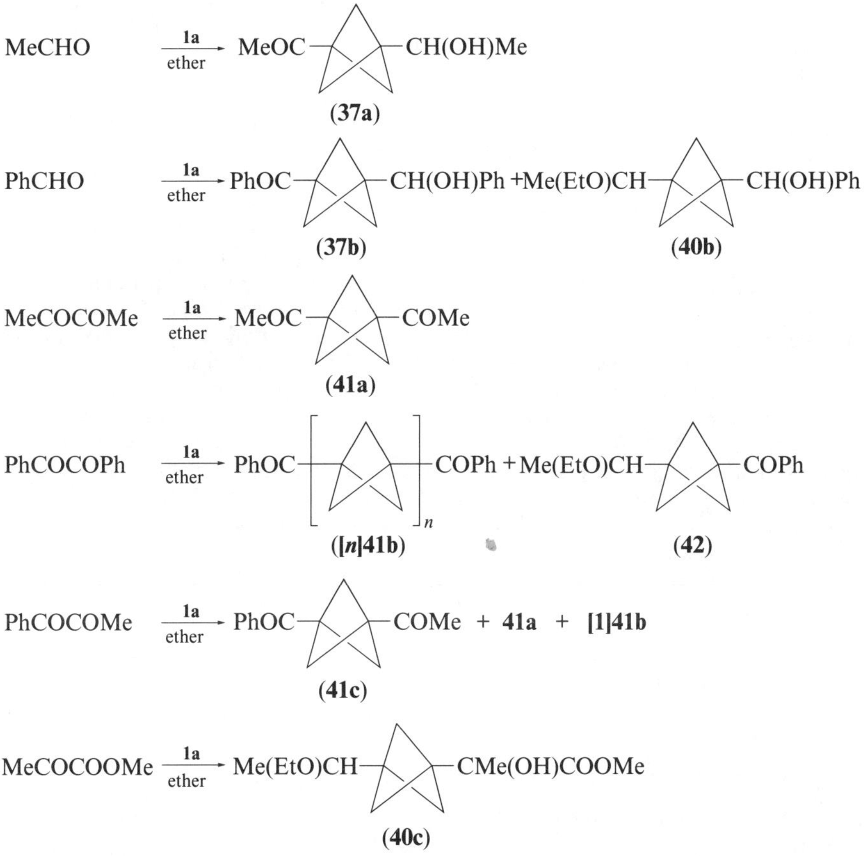

SCHEME 3

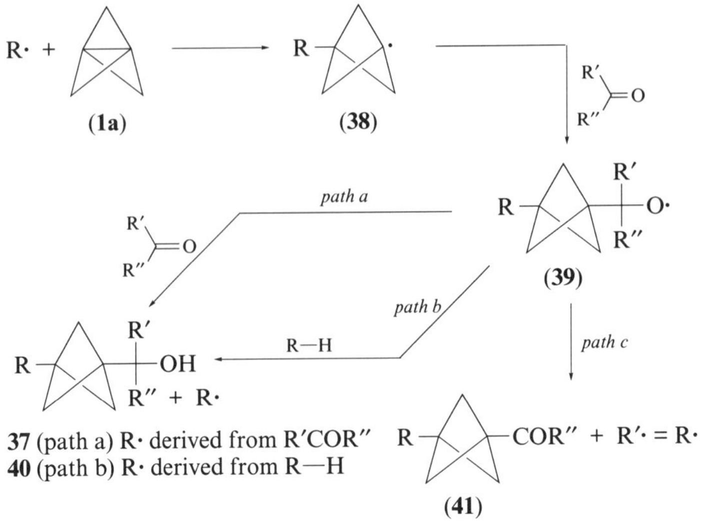

SCHEME 4

radical **39** (R = MeCO, R′ = H, R″ = Me) abstracts the hydrogen, forming **37a** (Schemes 3 and 4). Other, less electrophilic aliphatic aldehydes react with neat propellane (but not its diethyl ether solution[48]), giving mixtures of products[64].

The high propensity of the brigehead bicyclo[1.1.1]pentyl radicals for nucleophilic addition to the carbonyl group has been further exemplified in reactions of **1a** with activated ketones in diethyl ether, which readily form the products **40** or **41**[64,110]. The accumulated data (Scheme 3) led to a general mechanistic scheme to explain the results (Scheme 4)[48].

The initially formed oxy radical **39** may react via one of three routes depending on (i) the hydrogen-atom donating ability of the carbonyl compound (path a) and other reaction partners present, e.g. the solvent (path b), and (ii) the fragmentation propensity of the oxy radical (path c). Both substituents in the products **37**, which are formed along path a, are derived from the same carbonyl molecule (e.g., acetaldehyde). In the products **40** obtained via path b the substituents originate in two different molecules (e.g. methyl pyruvate and diethyl ether in **40c**)[48]. Benzaldehyde forms both products **37b** and **40b** in a ratio that depends on the relative concentrations of benzaldehyde and propellane[48,64]. Products **41** obtained via path c formally result from insertion of **1a** between two carbonyl groups (e.g. biacetyl)[48,58]. Benzil gives the oligomeric products **[*n*]41b** along with some by-products, e.g. **42** (cf. Scheme 3)[48].

Insufficiently activated carbonyl compounds such as esters (including oxalate), thioesters, carbonates and ketones do not yield carbonyl adducts[48,64].

Other reactions. [1.1.1]Propellane reacts with tribromonitrosobenzene to yield a paramagnetic compound for which a bis-nitroxyl structure **43** was proposed. When the

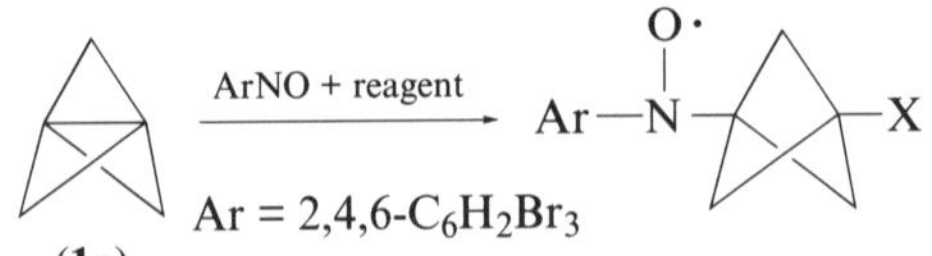

Ar = 2,4,6-$C_6H_2Br_3$

(1a)

Reagent	X	Compound
none	ArNO·	**43**
I_2	I	**44**
PhSH	PhS	**45**
$PhICl_2$	Cl	**46**

(19)

reaction was carried out in the presence of I_2, PhSH or $PhICl_2$, it produced other paramagnetic species, for which the nitroxyl structures **44**–**46** were suggested[100] (equation 19).

The reaction of nitric oxide with **1a** in CS_2 yields 90% of product **47**[64] (equation 20). It is believed that the species attacking the propellane is $NO_2\cdot$. The resulting bridgehead radical adds to CS_2 and **47** is formed as an end result of a series of oxygen transfer reactions. The same reaction run without CS_2 gives mostly an ill-identified olefin, presumably a methylenecyclobutane derivative. When **1a** reacted with N_2O_4 in CH_2Cl_2 or CS_2, olefins were obtained, but in diethyl ether 1,3-dinitrobicyclo[1.1.1]pentane was isolated in 25% yield. It is possible that the attacking species is NO_2^+[64].

The reaction of the weakly electrophilic 2-nitrophenylsulfenyl chloride with [1.1.1]-propellane produces a complex mixture of products, and the cyclobutanone **48** has been isolated in 10% yield (equation 21). The formation of the product has been attributed to the presence of oxygen in the reaction mixture[65,126].

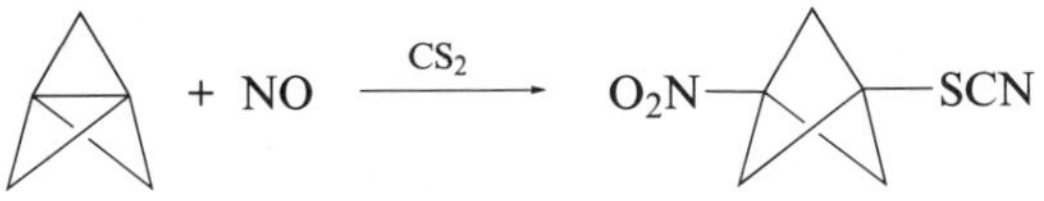

(20)

(1a) **(47)**

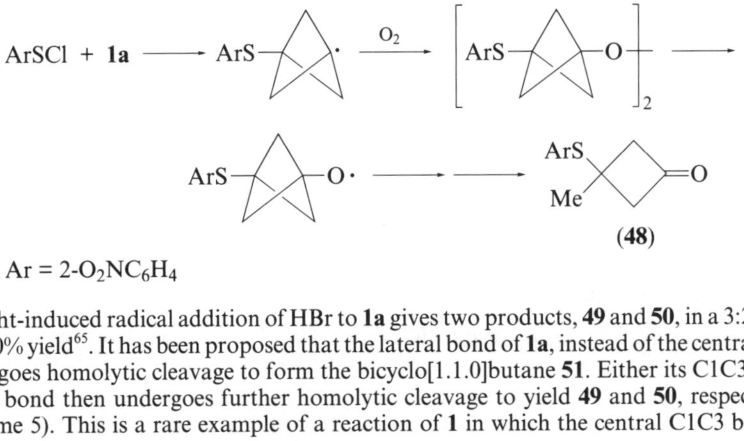

(21)

(48)

Ar = 2-$O_2NC_6H_4$

Light-induced radical addition of HBr to **1a** gives two products, **49** and **50**, in a 3:2 ratio and 80% yield[65]. It has been proposed that the lateral bond of **1a**, instead of the central one, undergoes homolytic cleavage to form the bicyclo[1.1.0]butane **51**. Either its C1C3 or its C1C2 bond then undergoes further homolytic cleavage to yield **49** and **50**, respectively (Scheme 5). This is a rare example of a reaction of **1** in which the central C1C3 bond is preserved.

[1.1.1]Propellane reacts spontaneously with electron-deficient olefins to yield high molecular weight polymers (Table 14). Electronic structures of both reaction partners are set up for the formation of alternating copolymers[60,127–129], and in the case of maleic anhydride, the reaction is also face-specific[127]. In the propagation step the electron-deficient radical in position α to the electron-withdrawing group attacks **1**. The resulting nucleophilic bridgehead radical readily adds to the double bond of the olefin, propagating the polymerization chain. As a result, almost perfectly alternating polymers are formed even in the presence of a large excess of the olefin. NMR and elemental analysis showed that the ratio of bicyclo[1.1.1]pentane unit to the aliphatic part in the polymer is essentially 1:1 with 1:1 to 1:5 ratio of **1** and the olefin, but it drops to 2:3 when the olefin is in 123-fold excess. The polymerization works better in hydrocarbon solutions than in ether and is suppressed by radical inhibitors[129].

In contrast to the efficient copolymerization reaction described above, reactions of **1a** with strongly electron-deficient olefins and acetylenes (dicyanoacetylene, dimethyl acetylenedicarboxylate, tetracyanoethylene and dichlorodicyanobenzoquinone) produce

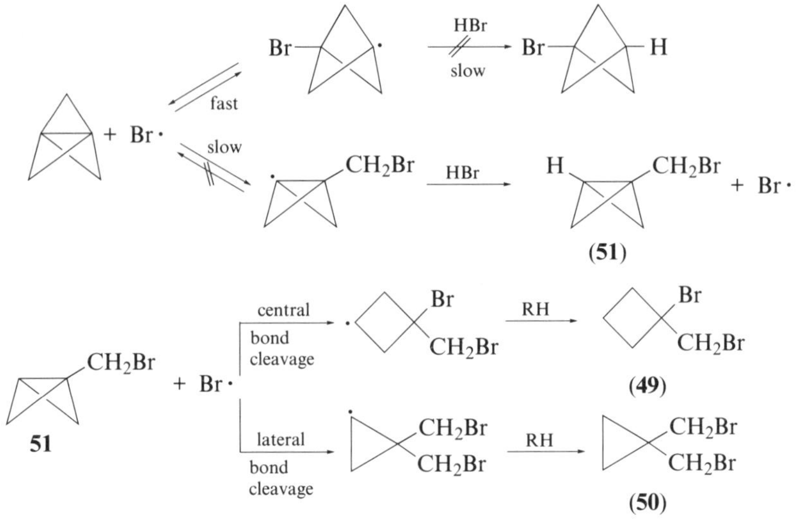

SCHEME 5

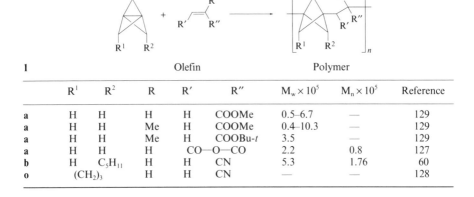

only ring-opened products, e.g. **52**[64,130]. It has been proposed that the initially formed 1,5-biradical **53** opens to the 1,3-biradical **54**, which collapses to the product (Scheme 6). The degree of zwitterionic character in **53** is unclear. It would promote the step **53** → **54**, which is expected to be slow in a nonpolar biradical (cf the discussion in Section V.A)[64,130], unless the closure of the cyclopropane ring is simultaneous with the opening of the bicyclo[1.1.1]pentane cage.

D. Electrophilic Attack

[1.1.1]Propellanes are sensitive to acids. Protonation of [1.1.1]propellane with acetic acid yielded 3-methylenecyclobutyl acetate (**5**)[3,15]. The formation of 3-methylenecyclobutyl

TABLE 14. Copolymerization of [1.1.1]propellane and electron-deficient olefins

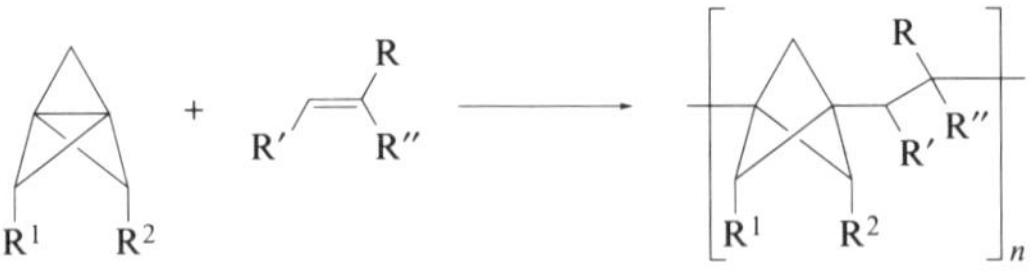

1			Olefin			Polymer		
	R^1	R^2	R	R′	R″	$M_w \times 10^5$	$M_n \times 10^5$	Reference
a	H	H	H	H	COOMe	0.5–6.7	—	129
a	H	H	Me	H	COOMe	0.4–10.3	—	129
a	H	H	Me	H	COOBu-*t*	3.5	—	129
a	H	H	H	CO—O—CO		2.2	0.8	127
b	H	C_5H_{11}	H	H	CN	5.3	1.76	60
o	$(CH_2)_3$		H	H	CN	—	—	128

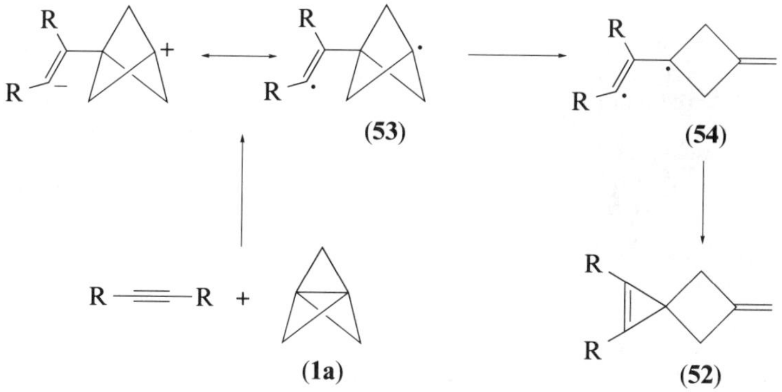

SCHEME 6

fluoride in reaction of **1a** with XeF_2 in ether was attributed to fast protonation reaction with traces of the HF present (footnote 6 in Reference 67). The initial proton addition step may lead either to the cation **19** or perhaps directly to the bicyclo[1.1.0]butyl-1-carbinyl cation (**55**). The former cation would easily rearrange to the latter, which further rearranges to **56** and eliminates H^+ to give **57** or is trapped with the nucleophile to give **5**. The formation of product **58** was attributed to a proton-catalyzed rearrangement of **1**, presumably followed by electrophilic addition to another molecule of propellane[46,64] (equation 22). Similar results were obtained when **1a** was treated with $AgBF_4$, $[Rh(CO)_2Cl]_2$, or with complexes of Pt(0), Pt(II), Pd(II) and Ir(I). Typically, more complex olefins were formed along with **57** and **58**[64].

Reaction of [1.1.1]propellane with $Hg(OAc)_2$ also gives the methylenecyclobutyl cation which was trapped with the acetate anion to give **59**[65].

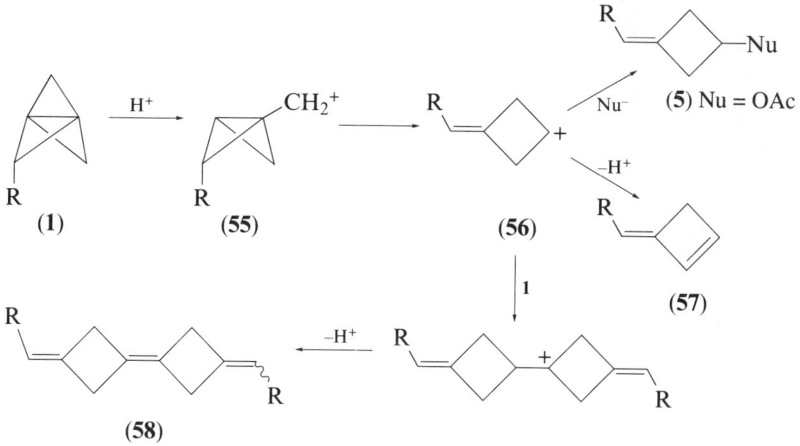

(22)

(59)

Trialkylboranes add to propellane to give a zwitterion that can rearrange to give **60** or react with another molecule of **1a** to give **61**. Both organoboron compounds were oxidized with hydrogen peroxide and isolated as the corresponding alcohols in 65% and 21% yield, respectively[65,131] (equation 23).

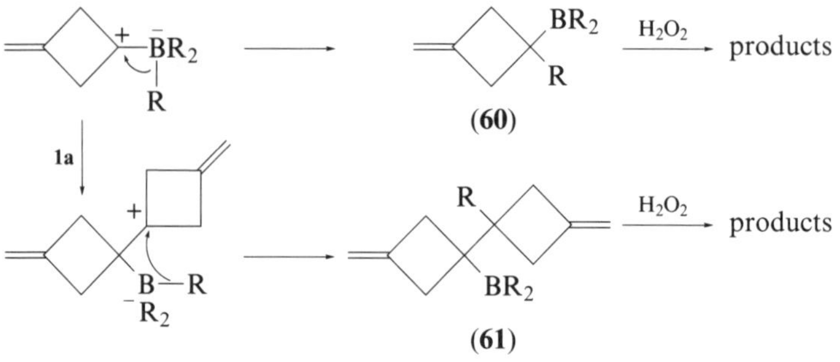

(23)

In some electrophilic reactions of **1** the bicyclic ring system is preserved. It has been postulated that in the reaction of 1,3-diiodobicyclo[1.1.1]pentane (**15b**) with sodium methoxide, [1.1.1]propellane generated *in situ* reacts with iodine in the presence of pyridine[67,132] or with MeOI,[66] an I^+ source also generated *in situ*, to give the 3-iodobicyclo[1.1.1]pent-1-yl cation **62**[67,90]. This cation is subsequently trapped with a nucleophile such as MeOH, N_3^-, pyridine or triethylamine to give **63** (Table 15)[66,67,132].

TABLE 15. Nucleophilic substitution via the bicyclo[1.1.1]pentyl cation

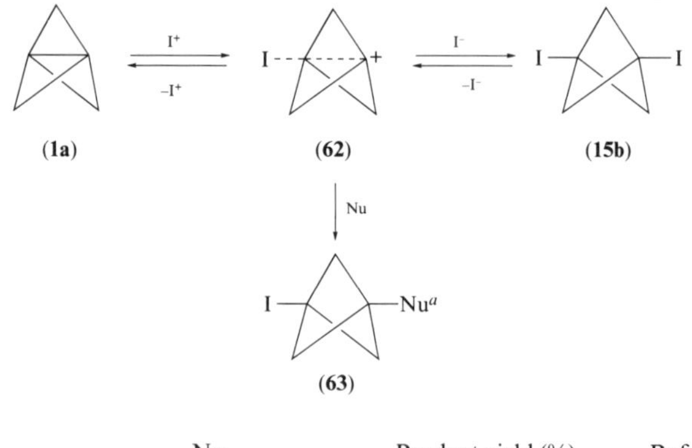

From	Nu	Product yield (%)	Reference
1a	Pyridine	37	67
15b	Pyridine	67	67
	MeO^-	68	66,67
	Et_3N	63	67
	DABCO	31	67
	N-Methylmorpholine	76	67
	4,4′-bipyridine	73	67
	N_3^-	*b*	66

[a] The Nu moiety in (**63**) is positively charged when the nucleophile is an amine.
[b] Major product.

E. Reduction

The central bond in **1** undergoes reductive cleavage with dissolving metals. For instance, reaction of **1a** and **1o** with lithium 4,4′-di-*t*-butylbiphenyl in dimethyl ether yields poorly soluble and incompletely characterized organometallic derivatives in moderate yields, presumably the corresponding 1,3-dilithiobicyclo[1.1.1]pentanes (**64**)[95] (equation 24). The

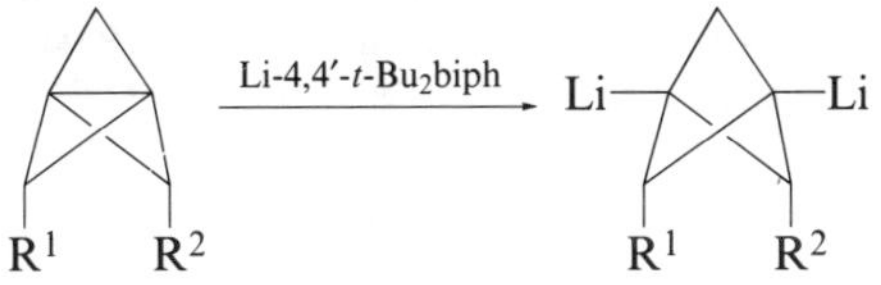

(24)

(**1**) (**64**)

reaction performed under other conditions (lithium biphenyl, THF) gave much worse results.

Reaction **1o** with lithium metal in ethylamine yields the corresponding hydrocarbon in 63% yield[62].

VI. CONCLUSIONS

In a dozen years since the initial discovery, [1.1.1]propellane has been transformed from a laboratory curiosity to a readily available starting material for organic synthesis. The original preparation involved tedious assembly of a bicyclo[1.1.1]pentane derivative as a precursor, the simple approach developed more subsequently turned the tables and bicyclo[1.1.1]pentanes are now best synthesized from [1.1.1]propellane! This simple five-carbon structure has provided a fertile playground and intriguing challenges for chemists of many kinds, from theoreticians and spectroscopists to kineticists and polymer scientists. The combination of unique structure, reasonable ease of handling and high reactivity that [1.1.1]propellane offers promises a continued growth of applications to various areas of chemistry. Personally, we are most intrigued by its rod-like oligomers, the [*n*]staffanes, and we have reviewed their chemistry recently[133].

VII. ACKNOWLEDGMENT

We are grateful to the National Science Foundation for support (CHE-9318469).

VIII. REFERENCES

1. M. D. Newton and J. M. Schulman, *J. Am. Chem. Soc.*, **94**, 773 (1972).
2. W.-D. Stohrer and R. Hoffmann, *J. Am. Chem. Soc.*, **94**, 779 (1972).
3. K. B. Wiberg and F. H. Walker, *J. Am. Chem. Soc.*, **104**, 5239 (1982).
4. K. Semmler, G. Szeimies and J. Belzner, *J. Am. Chem., Soc.*, **107**, 6410 (1985).
5. D. Ginsburg, in *The Chemistry of the Cyclopropyl Group* (Ed. Z. Rappoport), Wiley, New York, 1987, p. 1193.
6. K. B. Wiberg, *Chem. Rev.*, **89**, 975 (1989).
7. Y. Tobe, in *Carbocyclic Cage Compounds: Chemistry and Applications* (Eds. E. Osawa and O. Yonemitsu), VCH, New York, 1992, p. 125.
8. R. J. Butcher and W. J. Jones, *J. Mol. Spectrosc.*, **47**, 64 (1964).
9. W. J. Jones and B. P. Stoicheff, *Can. J. Phys.*, **42**, 2259 (1964).
10. A. C. Friedli, V. M. Lynch, P. Kaszynski and J. Michl, *Acta Crystallogr., Sect. B*, **B46**, 377 (1990).

11. O. Bastiansen and A. de Meijere, *Angew. Chem. Int. Ed. Engl.*, **5**, 124 (1966).
12. V. Balaji and J. Michl, *Pure Appl. Chem.*, **60**, 189 (1988).
13. R. M. Jarret and L. Cusumano, *Tetrahedron Lett.*, **31**, 171 (1990).
14. K. Frei and H. J. Bernstein, *J. Chem. Phys.*, **38**, 1216 (1963).
15. K. B. Wiberg, W. P. Dailey, F. H. Walker, S. T. Waddell, L. S. Crocker and M. Newton, *J. Am. Chem. Soc.*, **107**, 7247 (1985).
16. M. D. Newton, J. M. Schulman and M. M. Manus, *J. Am. Chem. Soc.*, **96**, 17 (1974).
17. R. P. Messmer and P. A. Schultz, *J. Am. Chem. Soc.*, **108**, 7407 (1986).
18. A. E. Reed, L. A. Curtiss and F. Weinhold, *Chem. Rev.*, **88**, 899 (1988).
19. D. Feller and E. R. Davidson, *J. Am. Chem. Soc.*, **109**, 4133 (1987).
20. P. Seiler, J. Belzner, U.Bunz and G. Szeimies, *Helv. Chim. Acta*, **71**, 2100 (1988).
21. K. B. Wiberg, R. F. Bader and C. D. H. Lau, *J. Am. Chem. Soc.*. **109**, 985 (1987).
22. E. Honegger, H. Huber, E. Heilbronner, W. P. Dailey and K. B. Wiberg, *J. Am. Chem. Soc.*, **107**, 7172 (1985).
23. O. Schafer, M. Allan, G. Szeimies and M. Sanktjohanser, *J. Am. Chem. Soc.*, **114**, 8180 (1992).
24. L. B. Krivdin and G. A. Kalabin, *Progress in NMR Spectroscopy*, **21**, 293 (1989).
25. L. Hedberg and K. Hedberg, *J. Am. Chem. Soc.*, **107**, 7257 (1985).
26. P. Seiler, *Helv. Chim. Acta*, **73**, 1574 (1990).
27. K. B. Wiberg, *J. Am. Chem. Soc.*, **105**, 1227 (1983).
28. K. Jug and S. Buss, *J. Comput. Chem.*, **6**, 507 (1985).
29. A. B. Pierini, H. F. Reale and J. A. Medrano, *J. Mol. Struct.* (*Theochem*), **148**, 109 (1986).
30. P. Politzer and K. Jayasuriya, *J. Mol. Struct.* (*Theochem*), **135**, 245. (1986).
31. K. B. Wiberg, R. F. W. Bader and C. D. H. Lau, *J. Am. Chem. Soc.*, **109**, 1001 (1987).
32. T. S. Slee, in *Modern Models of Bonding and Delocalization*; (Eds. J. F. Liebman and A. Greenberg), VCH, New York, 1988, p. 63.
33. M. S. Gordon, K. A. Nguyen and M. T. Carroll, *Polyhedron*, **10**, 1247 (1991).
34. J. M. Cullen, *J. Comput. Chem.*, **13**, 901 (1992).
35. G. Szeimies, in *Strain and Its Implications in Organic Chemistry* (Eds. A. de Meijere and S. Blechert), NATO ASI Series Vol. 273, Kluwer, Dordrecht, 1989, p. 361.
36. A.-D. Schlüter, *Polym. Commun.*, **30**, 34 (1989).
37. F. Alber and G. Szeimies, *Chem. Ber.*, **125**, 757 (1992).
38. J. Belzner and G. Szeimies, *Tetrahedron Lett.*, **27**, 5839 (1986).
39. S. J. Hamrock and J. Michl, *J. Org. Chem.*, **57**, 5027 (1992).
40. K. B. Wiberg, *J. Comput. Chem.*, **5**, 197 (1984).
41. K. B. Wiberg, *Angew. Chem., Int. Ed. Engl.*, **25**, 312 (1986).
42. M. P. Kozina, S. M. Pimenova, V. A. Lukyanova and L. S. Surmina, *Dokl. Akad. Nauk SSSR*, **283**, 661 (1985); *Chem. Abstr*, **103**, 184723n (1985).
43. K. B. Wiberg, C. M. Hadad, S. Sieber and P. v. R. Schleyer, *J. Am. Chem. Soc.*, **114**, 5820 (1992).
44. K. B. Wiberg and B. J. Nist, *J. Am. Chem. Soc.*, **83**, 1226 (1961).
45. J. Belzner, B. Gareiss, K. Polborn, W. Schmid, K. Semmler and G. Szeimies, *Chem. Ber.*, **122**, 1509 (1989).
46. H. Bothe and A.-D. Schlüter, *Chem. Ber.*, **124**, 587 (1991).
47. K. B. Wiberg and V. Z. Williams, Jr., *J. Org. Chem.*, **35**, 369 (1970).
48. P. Kaszynski, A. C. Friedli and J. Michl, *J. Am. Chem. Soc.*, **114**, 601 (1992).
49. J. J. Burke and P. C. Lauterbur, *J. Am. Chem. Soc.*, **86**, 1870 (1964).
50. A. M. Orendt, J. C. Facelli, D. M. Grant, J. Michl, F. H. Walker, W. P. Dailey, S. T. Waddell, K. B. Wiberg, M. Schindler and W. Kutzelnigg, *Theor. Chim. Acta*, **68**, 421 (1985).
51. J. Michl, G. J. Radziszewski, J. W. Downing, K. B. Wiberg, F. H. Walker, R. D. Miller, P. Kovacic, M. Jawdosiuk and V. Bonacic-Koutecky, *Pure Appl. Chem.*, **55**, 315 (1983).
52. K. B. Wiberg, S. T. Waddell and R. E. Rosenberg, *J. Am. Chem. Soc.*, **112**, 2184 (1990).
53. K. B. Wiberg, R. E. Rosenberg and S. T. Waddell, *J. Phys. Chem.*, **96**, 8293 (1992).
54. N. V. Riggs, U. Zoller, M. T. Nguyen and L. Radom, *J. Am. Chem. Soc.*, **114**, 4354 (1992).
55. J. Michl, *Acc. Chem. Res.*, **23**, 127 (1990).
56. C. Winstead, Q. Sun and V. McKoy, *J. Chem. Phys.*, **97**, 9483 (1992).
57. J. Belzner, U. Bunz, K. Semmler, G. Szeimies, K. Opitz and A.-D. Schlüter, *Chem. Ber.*, **122**, 397 (1989).
58. P. Kaszynski and J. Michl, *J. Org. Chem.*, **53**, 4593 (1988).
59. P. Kaszynski and J. Michl, *J. Am. Chem. Soc.*, **110**, 5225 (1988).

60. K. Opitz and A.-D. Schlüter, *Angew. Chem., Int. Ed. Engl.*, **28**, 456 (1989).
61. G. Kottirsch, K. Polborn and G. Szeimies, *J. Am. Chem. Soc.*, **110**, 5588 (1988).
62. J. Belzner and G. Szeimies, *Tetrahedron Lett.*, **28**, 3099 (1987).
63. U. Bunz, K. Polborn, H.-U. Wagner and G. Szeimies, *Chem. Ber.*, **121**, 1785 (1988).
64. K. B. Wiberg and S. T. Waddell, *J. Am. Chem. Soc.*, **112**, 2194 (1990).
65. N. S. Zefirov, L. S. Surmina, N. K. Sadovaya, A. V. Blokhin, M. A. Tyurekhodzhaeva, Y. N. Bubnov, L. I. Lavrinovich, A. V. Ignatenko, Y. K. Grishin, O. A. Zelenkina, N. G. Kolotyrkina, S. V. Kudrevich and A. S. Koz'min, *J. Org. Chem. USSR.*, **26**, 2002 (1990).
66. K. B. Wiberg and N. D. McMurdie, *J. Am. Chem. Soc.*, **113**, 8995 (1991).
67. J. L. Adcock and A. A. Gakh, *J. Org. Chem.*, **57**, 6206 (1992).
68. E. W. Della, D. K. Taylor and J. Tsanaktsidis, *Tetrahedron Lett.*, **31**, 5219 (1990).
69. E. W. Della and D. K. Taylor, *Aust. J. Chem.*, **44**, 881 (1991).
70. J. M. Lehn and G. Wipff, *Tetrahedron Lett.*, **21**, 159 (1980).
71. K. B. Wiberg and S. T. Waddell, *Tetrahedron Lett.*, **29**, 289 (1988).
72. A. I. Ioffe, and O. M. Nefedov, *Izv. Akad. Nauk SSSR, Ser. Khim.*, 2313 (1988); *Bull. Acad. Sci. USSR Div. Chem. Sci*. (*Engl. Transt.*), **37**, 2081 (1988).
73. S. T. Graul and R. R. Squires, *J. Am. Chem. Soc.*, **112**, 2517 (1990).
74. A. J. McKinley, P. N. Ibrahim, V. Balaji and J. Michl, *J. Am. Chem. Soc.*, **114**, 10631 (1992).
75. J. R. Bews, C. Glidewell and J. C. Walton, *J. Chem. Soc., Perkin Trans. 2*, 1447 (1982).
76. E. W. Della, P. E. Pigou, C. H. Schiesser and D. K. Taylor, *J. Org. Chem.*, **56**, 4659 (1991).
77. B. Maillard and J. C. Walton, *J. Chem. Soc., Chem. Commun.*, 900 (1983).
78. K. B. Wiberg and V. Z. Williams, Jr., *J. Am. Chem. Soc.*, **89**, 3373 (1967).
79. E. W. Della and D. K. Taylor, *Aust. J. Chem.*, **43**, 945 (1990).
80. C. A. Grob, C. X. Yang, E. W. Della and D. K. Taylor, *Tetrahedron Lett.*, **32**, 5945 (1991).
81. J. M. Lehn and G. Wipff, *Chem. Phys. Lett.*, **15**, 450 (1972).
82. J. E. Jackson and L. C. Allen, *J. Am. Chem. Soc.*, **106**, 591 (1984).
83. J. Chandrasekhar, P. v. R. Schleyer and H. B. Schlegel, *Tetrahedron Lett.*, 3393 (1978).
84. E. W. Della and C. H. Schiesser, *J. Chem. Res.* (*S*), 172 (1989).
85. E. W. Della and C. H. Schiesser, *Tetrahedron Lett.*, **28**, 3869 (1987).
86. E. W. Della, P. M. W. Gill, and C. H. Schiesser, *J. Org. Chem.*, **53**, 4354 (1988).
87. J. F. Chiang and S. H. Bauer, *J. Am. Chem. Soc.*, **92**, 1614 (1970).
88. A. Almenningen, B. Andersen and B. A. Nyhus, *Acta Chem. Scand.*, **25**, 1217 (1971).
89. K. B. Wiberg, *Tetrahedron Lett.*, **26**, 599 (1985).
90. J. L. Adcock, A. A. Gakh, J. L. Pollitte and C. Woods, *J. Am. Chem. Soc.*, **114**, 3980 (1992).
91. A. C. Friedli, Ph.D. Dissertation, University of Texas at Austin, 1992.
92. A.-D. Schlüter, *Angew. Chem., Int. Ed. Engl.*, **27**, 296 (1988).
93. A.-D. Schlüter, *Macromolecules*, **21**, 1208 (1988).
94. R. W. Morrison and J. Michl, unpublished results.
95. U. Bunz and G. Szeimies, *Tetrahedron Lett.*, **31**, 651 (1990).
96. P. F. McGarry and J. C. Scaiano, private communication.
97. P. F. McGarry, L. J. Johnston and J. C. Scaiano, *J. Am. Chem. Soc.*, **111**, 3750 (1989).
98. P. F. McGarry, L. J. Johnston and J. C. Scaiano, *J. Org. Chem.*, **54**, 6133 (1989).
99. J. C. Scaiano and P. F. McGarry, *Tetrahedron Lett.*, **34**, 1243 (1993).
100. V. A. Vasin, I. Y. Bolusheva, E. P. Sanaeva, L. S. Surmina, N. K. Sadovaya, A. S. Koz'min, and N. S. Zefirov, *Dokl. Akad. Nauk SSSR*, **305**, 621 (1989); *Proc. Acad. Sci. USSR*, *Engl. Trans*, **305**, 94 (1989).
101. N. S. Zefirov, N. K. Sadovaya, L. S. Surmina, I. A. Godunov, A. S. Koz'min, K. A. Potekhin, A. V. Maleev and Y. T. Struchkov, *Izv. Akad. Nauk SSSR, Ser. Khim.*, 2648 (1988); *Bull. Acad. Sc*; USSR *Div. Chem. Sci.* (*Engl. Transl.*) **37**, 2388 (1988).
102. K. Jug and A. Poredda, *J. Am. Chem. Soc.*, **113**, 761 (1991).
103. A.-D. Schlüter, H. Bothe and J. M. Gosau, *Makromol. Chem.*, **192**, 2497 (1991).
104. J. Michl, P. Kaszynski, A. C. Friedli, G. S. Murthy, H.-C. Yang, R. E. Robinson, N. D. McMurdie and T. Kim, in *Strain and Its Implications in Organic Chemistry*, (Eds. A. de Meijere and S. Blechert), NATO ASI Series Vol. 273, Kluwer, Dordrecht, 1989, p. 463.
105. G. S. Murthy, K. Hassenrück, V. M. Lynch and J. Michl, *J. Am. Chem. Soc.*, **111**, 7262 (1989).
106. H. Bothe and A.-D. Schlüter, *Adv. Mater.*, **3**, 440 (1991).
107. P. Kaszynski, Ph.D. Dissertation, University of Texas at Austin, 1991.
108. H. K. Chang and J. Michl, unpublished results.

109. U. Bunz and G. Szeimies, *Tetrahedron Lett.*, **30**, 2087 (1989).
110. K. B. Wiberg, S. T. Waddell and K. C. Laidig, *Tetrahedron Lett.*, **27**, 1553 (1986).
111. K. P. Dockery and W. G. Bentrude, 204th ACS National Meeting, Washington, DC, August 23–28, 1992; Book of Abstracts: ORGN 375.
112. D. S. Toops and M. R. Barbachyn, 203rd ACS National Meeting, San Francisco, CA, April 5–10, 1992; Book of Abstracts: ORGN 050.
113. P. Kaszynski, N. D. McMurdie and J. Michl, *J. Org. Chem.*, **56**, 307 (1991).
114. P. Kaszynski, A. C. Friedli, N. D. McMurdie and J. Michl, *Mol. Cryst. Liq. Cryst.* **191**, 193 (1990).
115. R. Gleiter, K.-H. Pfeifer, G. Szeimies and U. Bunz, *Angew. Chem., Int. Ed. Engl.*, **29**, 413 (1990).
116. K. Hassenrück, G. S. Murthy, V. M. Lynch and J. Michl, *J. Org. Chem.*, **55**, 1013 (1990).
117. O. Schafer, M. Allan, G. Szeimies and M. Sanktjohanser, *Chem. Phys. Lett.*, **195**, 293 (1992).
118. M. A. Tyurekhodzhaeva, A. A. Bratkova, A. V. Blokhin, V. K. Brel, A. S. Koz'min and N. S. Zefirov, *J. Fluorine Chem.*, **55**, 237 (1991).
119. A. V. Blokhin, M. A. Tyurekhodzhaeva, N. K., Sadovaya and N. S. Zefirov, *Izv. Akad. Nauk SSSR, Ser. Khim.*, 1933 (1989); *Bull. Acad. Sci. USSR, Div. Chem. Sci.*, (*Engl. Transl.*) **38**, 1779 (1989).
120. N. K. Sadovaya, A. V. Blokhin, L. S. Surmina, M. A. Tyurekhodzhaeva, A. S. Koz'min, and N. S. Zefirov, *Izv. Akad. Nauk SSSR, Ser. Khim.* 2451 (1990); *Bull. Acad. Sci. USSR, Div. Chem. Sci.* (*Engl. Transl.*) **39**, 2224 (1990).
121. N. S. Zefirov, L. S. Surmina, N. K. Sadovaya and A. S. Koz'min, *Izv. Akad. Nauk SSSR, Ser. Khim.*, 2871 (1987); *Bull. Acad. Sci. USSR, Div. Chem. Sci.* (*Engl. Transl.*) **36**, 2670 (1987).
122. K. A. Potekhin, A. V. Maleev, A. Y. Kosnikov, E. N. Kurkutova, Y. T. Struchkov, L. S. Surmina, N. K. Sadovaya, A. S. Koz'min and N. S. Zefirov, *Dokl. Akad. Nauk SSSR*, **304**, 367 (1989); *Chem. Abstr*; **110**, 163902z (1989).
123. A. C. Friedli, P. Kaszynski and J. Michl, *Tetrahedron Lett.*, **30**, 455 (1989).
124. Y. S. Obeng, M. E. Laing, A. C. Friedli, H. C. Yang, D. Wang, E. W. Thulstrup, A. J. Bard and J. Michl, *J. Am. Chem. Soc.*, **114**, 9943 (1992).
125. K. P. Dockery and W. G. Bentrude, *J. Am. Chem. Soc.*, **116**, 10332 (1994).
126. N. S. Zefirov, N. K. Sadovaya, L. S. Surmina, K. A. Potekhin, A. V. Maleev, Y. T. Struchkov, V. V. Zhdankin and A. S. Koz'min, *Sulfur Lett.*, **8**, 21 (1988).
127. J.-M. Gosau, A.-D. Schlüter, *Chem. Ber.*, **123**, 2449 (1990).
128. H. Bothe and A.-D. Schlüter, *Makromol. Chem., Rapid Commun.*, **9**, 529 (1988).
129. V. S. Reddy, C. Ramireddy, A. Qin and P. Munk, *Macromolecules*, **24**, 3973 (1991).
130. K. B. Wiberg and S. T. Waddell, *Tetrahedron Lett.*, **28**, 151 (1987).
131. Y. N. Bubnov, L. I. Lavrinovich, A. V. Ignatenko, N. K. Sadovaya, L. S. Surmina, A. S. Koz'min, and N. S. Zefirov, *Izv. Akad. Nauk SSSR, Ser. Khim.*, 210 (1989); *Bull. Acad. Sci. USSR, Div. Chem. Sci.* (*Engl. Transl.*) **38**, 198 (1989).
132. J. L. Adcock and A. A. Gakh, *Tetrahedron Lett.*, **33**, 4875 (1992).
133. P. Kaszynski and J. Michl, in *Advances in Strain in Organic Chemistry* (Ed. B. Halton), Vol. 4, 1994 (in press).

CHAPTER **14**

Long-lived cyclopropylcarbinyl cations

GEORGE A. OLAH, PRAKASH V. REDDY and G. K. SURYA PRAKASH

Loker Hydrocarbon Research Institute and Department of Chemistry, University of Southern California, Los Angeles, CA 90089-1661, USA
Fax: 213-740-6679; e-mail: OLAH @METHYL·USC·EDU

I. INTRODUCTION 814
II. GENERAL METHODS OF PREPARATION OF LONG-LIVED CYCLOPROPYLCARBINYL CATIONS 816
A. Ionization of Alcohols or Halides 816
B. Hydride Ion Abstraction from Hydrocarbons 816
C. Neighboring Group Participation 817
D. Protonation of Carbonyl Compounds 817
E. Addition of Carbenes to Unsaturated Compounds 817
F. From Allyl Alcohols 817
III. PRIMARY CYCLOPROPYLCARBINYL CATIONS 818
A. Equilibrating (Degenerate) Cations 818
1. NMR studies of $C_4H_7^+$ cation 818
2. Theoretical studies of $C_4H_7^+$ ion 820
3. IR studies of $C_4H_7^+$ cation 820
4. 1-Methylcyclopropylcarbinyl (1-methylcyclobutyl) cation 821
5. Solvolytic generation of substituted $C_4H_7^+$ cations 822
B. Static Cations 824
1. Bicyclo[1.1.0]butyl-1-carbinyl cation 824
2. Nortricyclylcarbinyl cation 824
IV. SECONDARY AND TERTIARY CYCLOPROPYLCARBINYL CATIONS 825
A. Static Cations 825
1. Tricyclopropylcarbinyl cation 825
2. α-Methylcyclopropylcarbinyl cations 825
3. α,α-Dimethylcyclopropylcarbinyl cations and their derivatives . . . 826
4. 1-Cyclopropylcycloalkyl cations 827
5. Tertiary 8,9-dehydro-2-adamantyl and 2,4-dehydro-5-homoadamantyl cations 827

The chemistry of the cyclopropyl group, Vol. 2
Edited by Z. Rappoport

6. Arylcyclopropylcarbinyl cations 828
7. 3-Nortricyclyl cations 828
8. 3-Tetracyclo[3.3.1.0^{2,8}.0^{4,6}]nonyl cation 829
9. Nortricyclylcarbinyl cations 829
10. Tertiary barbaralyl cations 830
11. 9-Cyclopropyl-9-xanthyl cation 831
12. Bridgehead barrelyl and bullvalyl cations 831
13. α-Cyclopropylvinyl cations 832
14. Substituted bicyclo[3.1.0]hexenyl cations 833
15. Allyl substituted cyclopropylcarbinyl cations 834
B. Equilibrating (Degenerate) Cations 834
1. 1-(1-Methylcyclopropyl)ethyl cation and its β-methyl derivatives . . 835
2. 2-Bicyclo[*n*.1.0]alkyl cations 836
3. 3-Homonortricyclyl cation 837
4. 8,9-Dehydro-2-adamantyl cation 838
5. 2,8-Dimethyl-8,9-dehydro-2-adamantyl cation 838
6. 4-Phenyl-2,5-dehydro-4-protoadamantyl cation 839
7. 4-Methyl-2,5-dehydro-4-protoadamantyl cation 839
8. 2,4-Dehydro-5-homoadamantyl cation 840
9. 9-Barbaralyl cation 840
10. Bicyclo[3.1.0]hexenyl cation 843
V. SPIROCYCLOPROPYLCARBINYL CATIONS 843
A. General Aspects 843
B. 3-Spirocyclopropylbicyclo[2.2.1] and Related Cations 844
C. 2-Spirocyclopropyl Cyclohexyl Cation 846
D. Phenonium Ions (Spirocyclopropylbenzonium Ions) 846
E. Benzo-2-nortricyclyl Cations 850
VI. CYCLOPROPYLCARBINYL DICATIONS 851
VII. EXTENT OF POSITIVE CHARGE DELOCALIZATION INTO THE CYCLOPROPYL GROUP 852
VIII. X-RAY CRYSTALLOGRAPHIC STUDIES 855
IX. CONCLUSIONS AND OUTLOOK 856
X. REFERENCES 856

I. INTRODUCTION

The nature of the $C_4H_7^+$ cation, ever since J. D. Roberts first carried out his pioneering solvolytic studies on the cyclobutyl and cyclopropylcarbinyl derivatives[1,2], has been under intense investigation by many groups of researchers using various techniques[3]. Roberts initially proposed a pentacoordinated tricyclobutonium structure for the cationic intermediate and named it as a 'nonclassical' ion (for the first time the 'nonclassical' name was used). Bartlett's view that 'among the nonclassical ions, the ratio of conceptual difficulty to the molecular weight reaches a maximum with the cyclopropylcarbinyl–cyclobutyl system' explains the enigma behind the structure of the $C_4H_7^+$ cation[3].

Solvolytic studies provided the first structural indication for almost every carbocationic intermediate and the $C_4H_7^+$ ion is no exception. Roberts observed that the solvolysis of cyclopropylcarbinyl or cyclobutyl systems and the diazotative deamination reactions of cyclopropylcarbinyl amine or cyclobutyl amine gave similar product mixtures consisting of cyclopropylcarbinyl, cyclobutyl and allylcarbinyl derivatives in essentially the same ratio[1,2]. A common cationic intermediate of C_{3v} structure, the tricyclobutonium ion **1**, was proposed to explain the solvolytic behavior. The tricyclobutonium ion structure was, however, soon replaced by the equilibrating bridged bicyclobutonium ions **2** as further experimental results were not consistent with it[4]. The deamination of cyclopropylcarbinyl-1-^{14}C-amine led to cyclopropylcarbinol and cyclobutanol in which the label is scrambled to all the methylene carbons. However, the scrambling of the label was not uniform as expected for the tricyclobutonium ion **1**. The deamination gave cyclopropylcarbinol, cyclobutanol and allylcarbinol in a ratio of 48%, 47% and 5%, respectively. The ^{14}C label was found to be 53.2% at C1, and 48% at C2 and C3 positions in the cyclopropylcarbinol. In the cyclobutanol products, the label was 35.8% at C2 and C4, and 28% at C3. Thus the methylene transposition, although significant, is not complete. These results in conjunction with the abnormally large solvolytic reactivity of cyclopropylcarbinyl and cyclobutyl substrates were accounted for by suggesting three rapidly equilibrating nonclassical unsymmetrical 'bicyclobutonium ions' **2** as intermediates (equation 1). Instantaneous equilibration of these structures would be expected to give complete scrambling of the label just as would have also resulted from the tricyclobutonium ion **1**.

(1) (2) (1)

Brown, on the other hand, favored[5] equilibrating classical cyclopropylcarbinyl cations **3** as intermediates, with puckered cyclobutyl cation **2a** involved in the interconversion of these classical cations (equation 2). He alternatively ascribed the unusual solvolytic rates of cyclopropylcarbinyl systems to σ-conjugation with the cyclopropyl group, and argued against the σ-participation (by using Gassman–Fentiman type of linear free energy relations) in the solvolysis of 1-arylcyclopropylcarbinyl substrates.

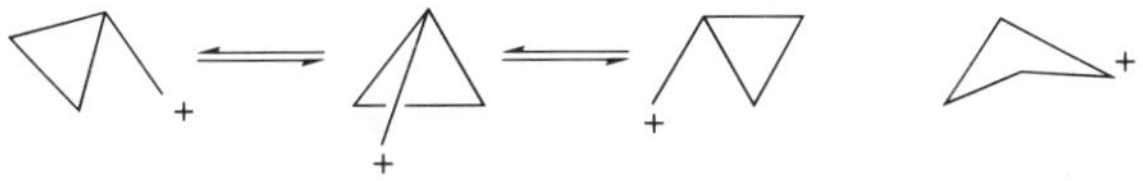

(3) (2a) (2)

The observation of carbocationic intermediates, directly obtained from suitable precursors at low temperatures in superacidic, low nucleophilic media, was achieved by the pioneering efforts of Olah and coworkers[6]. Consideration of ^{1}H and ^{13}C NMR chemical shifts and coupling constants (both J_{H-H} and J_{C-H}) provided means of distinguishing between classical, nonclassical and partially bridged species. In the nonclassical structures, the charge density on the cationic center is relatively low, and hence substantially shielded cationic chemical shifts are expected. Because of the large chemical shift range, ^{13}C NMR is particularly well suited for the studies. Using combined ^{1}H and ^{13}C NMR studies, it was shown very clearly that the parent cyclopropylcarbinyl cation is nonclassical (i.e. its structure cannot be adequately described using conventional Lewis valence bond structures). Olah, Schleyer and coworkers introduced a further criterion[7] for distinguishing nonclassical from classical ions, which involves comparing the sum of all the ^{13}C NMR chemical shifts of a given carbocation with that of its neutral hydrocarbon progenitor. For nonclassical cations the difference in the chemical shifts ($\Sigma\delta R^{+}-\Sigma\delta RH$) is typically substantially less than 200, whereas it is in excess of 350 ppm for classical cations. This approach is very useful in the overall assignment of classical vs nonclassical structures.

Saunders, Telkowski and Vogel developed[8a] another powerful method for distinguishing nonclassical from classical cations using the deuterium isotopic perturbation method[8b].

By introducing deuterium into ions undergoing rapid degenerate equilibria, the equilibrium constant is shifted away from unity, i.e. the degeneracy is lost. As a result, pairs of nuclei, that otherwise give a single absorption for the degenerate structures, now give two separate signals. The degree of separation of these signals in the deuteriated ions is significantly large (80–100 ppm or larger) for equilibrating classical cations. For nonclassical ions (having a single minimum), on the other hand, very small splitting (usually less than two ppm) are observed. On this basis the 9-pentacyclo[4.3.0.0^{2,4}.0^{3,8}.0^{5,7}]nonyl cation (Coates' cation), $\Delta\delta = 0.1$[9], and 2-norbornyl cation[10], $\Delta\delta = 2$ were shown to be static nonclassical ions, and cyclopropylcarbinyl and 1-methylcyclobutyl cations were shown to be equilibrating nonclassical ions (*vide infra*).

The field of long-lived cyclopropylcarbinyl cations has been previously reviewed[11], emphasizing NMR spectroscopic study of the cations, and their equilibration via degenerate and nondegenerate structures. Hence it was not considered necessary here to report a comprehensive review of the NMR spectral data for all cyclopropylcarbinyl cations, unless it was necessary for the present discussion which centers on recent developments in the field while providing sufficient background for the reader. The reader is referred to previous articles and reviews for further information.

II. GENERAL METHODS OF PREPARATION OF LONG-LIVED CYCLOPROPYLCARBINYL CATIONS

A. Ionization of Alcohols or Halides

The usual methods of preparation of the carbocations under long-lived ion conditions are applicable to the preparation of the cyclopropylcarbinyl cations. They can be prepared, for example, by ionizing the corresponding alcohols in protic superacids such as FSO_3H, SbF_5-FSO_3H or CF_3SO_3H or by the ionization of the corresponding halides in SbF_5 or other superacidic Lewis acids in low nucleophilicity solvents such as SO_2ClF (equation 3). Certain tertiary cations, such as the tricyclopropylcarbenium ion, can be prepared even in weakly sulfuric acid[12]. Alternatively, cation salts such as tetrafluoroborates can be prepared by the reaction of selected tertiary cyclopropylmethyl alcohols with *in situ* generated anhydrous HBF_4. The simplest of these cations, namely the parent cyclopropylcarbinyl cation, could be prepared from the cyclopropylmethanol with antimony pentafluoride in sulfuryl chloride fluoride.

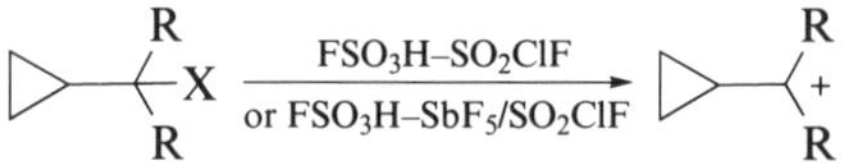

(3)

X = OH, F, Cl or Br
R = Alkyl, aryl, cyclopropyl or H

B. Hydride Ion Abstraction from Hydrocarbons

The tertiary hydrogen alpha to the cyclopropyl ring, in suitable cases, can be abstracted as hydride ion by strong superacids such as FSO_3H–SbF_5 in SO_2ClF (equation 4)[13]. The

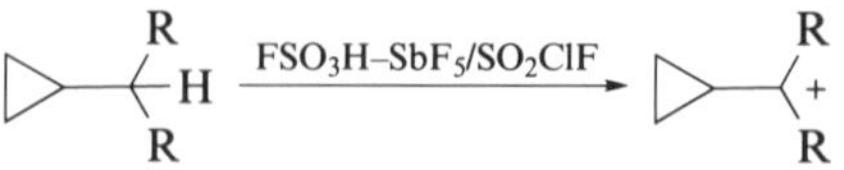

(4)

high energy of activation required for this reaction together with competing protolytic ring cleavage limits, however, the practical preparation in many cases.

C. Neighboring Group Participation

In solvolytic reactions it was observed that cyclopropylcarbinyl cations are formed as transient intermediates from the homoallylic alcohols[14] (equation 5).

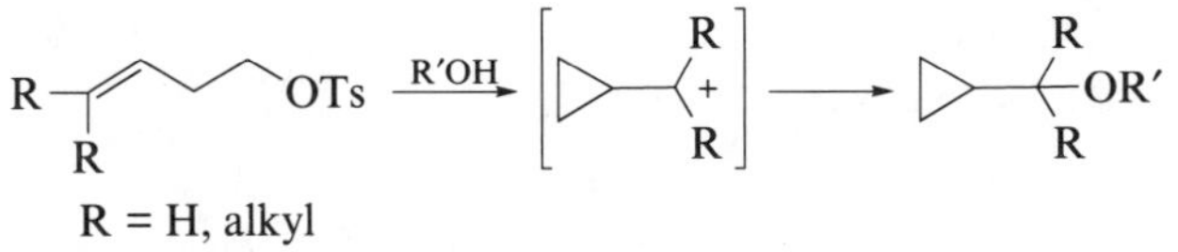

(5)

β-Arylethyl halides, upon ionization with SbF_5 in SO_2ClF, give the ethylenebenzenium ions, having characteristics of cyclopropylcarbinyl cations (equation 6). In these ions, the positive charge is significantly delocalized into the cyclopropyl ring[15].

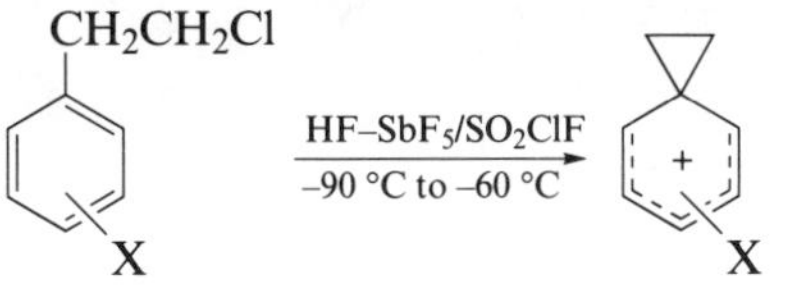

(6)

D. Protonation of Carbonyl Compounds

Alkyl cyclopropyl ketones or dicyclopropyl ketones in fluorosulfuric acid/sulfuryl chloride fluoride at about –90 °C undergo O-protonation to α-hydroxycyclopropylcarbinyl cations (equation 7). These cations are essentially protonated ketones, as the positive charge heavily resides on the oxygen atom[16]. X-ray structural studies have also been carried out on hydroxycyclopropylcarbinyl cation salts, since they are easily obtained as crystalline materials[17].

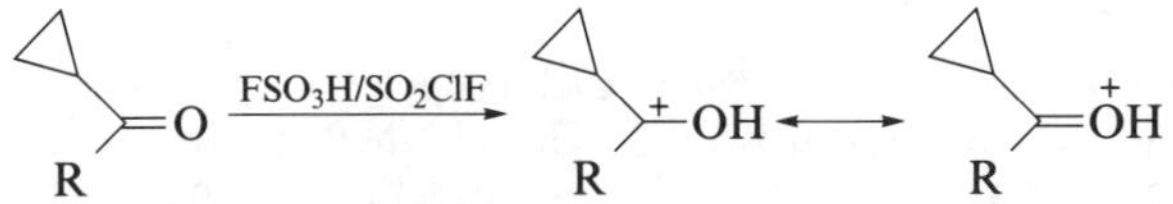

(7)

E. Addition of Carbenes to Unsaturated Compounds

Some cyclopropylcarbinyl cations, such as aromatic cyclopropyl cyclopropenium ions (**4a** and **4b**), have been prepared by the addition of the corresponding carbenes to cyclopropyl acetylenes[18] (equation 8).

F. From Allyl Alcohols

In suitable cases, allylic alcohols can be converted to the cyclopropylcarbinyl cations by reaction with superacids. The reaction involves the rearrangement of the initially formed allyl cation to the homoallyl cation by a 1,2-hydride transfer followed by its cyclization to the cyclopropylcarbinyl cation[19] (equation 9).

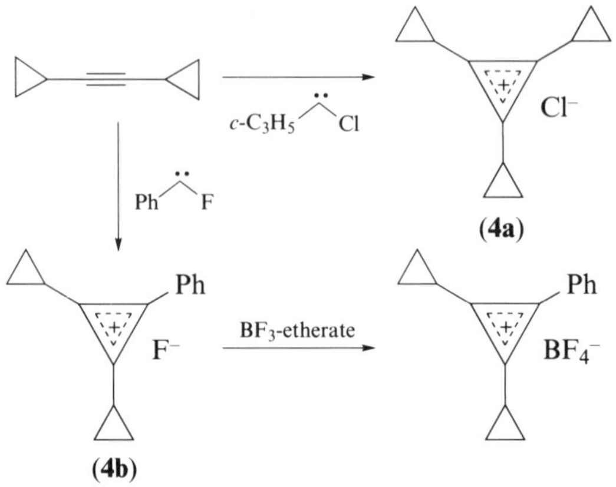

(8)

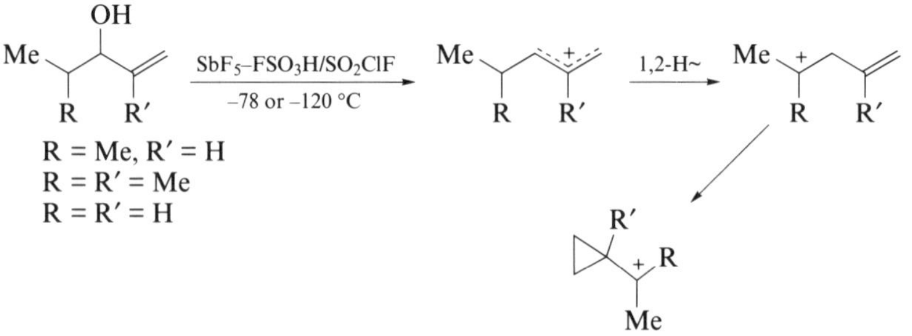

(9)

III. PRIMARY CYCLOPROPYLCARBINYL CATIONS

A. Equilibrating (Degenerate) Cations

1. NMR studies of $C_4H_7^+$ cation

Olah and coworkers obtained the parent cyclopropylcarbinyl cation and characterized it by both ^{1}H and ^{13}C NMR spectroscopy[20]. The ^{1}H NMR spectrum of the cation shows two overlapping quartets ($J = 8$ and 6.5 Hz) for the methine protons, and two sets of doublets for the methylene protons, δ^1H 4.64 and 4.21. Thus the methylene hydrogens are stereochemically non equivalent, which is unexpected for classical cyclopropylcarbinyl or cyclobutyl cations. The ^{13}C NMR spectrum shows only two signals: δ^{13}C 108.4 (CH), and 55.2 (CH_2, $J_{CH} = 180$ Hz). Thus all three methylene carbons are identical, and the protons attached to each of them are nonequivalent. Equilibrating classical cyclopropylcarbinyl cations can be ruled out not only based on substantially shielded averaged methylene carbon ^{13}C NMR shift, but also due to the presence of nonequivalent geminal protons on each of the methylene carbons. A pentacoordinated nonclassical bicyclobutonium ion can account for the observed results. The proton NMR absorptions at δ 4.64 and 4.21 are assigned to the *endo*- and *exo*-methylene hydrogens, respectively, based on the NMR spectra of the stereospecifically deuteriated cyclopropylcarbinyl cations[21]. The *endo*-deuteriated cyclopropylcarbinyl cation was prepared from alcohol **5a**, whereas a 1:1 mixture of *endo*- and *exo*-deuteriated cations were prepared from alcohol **5b** (equations 10 and 11).

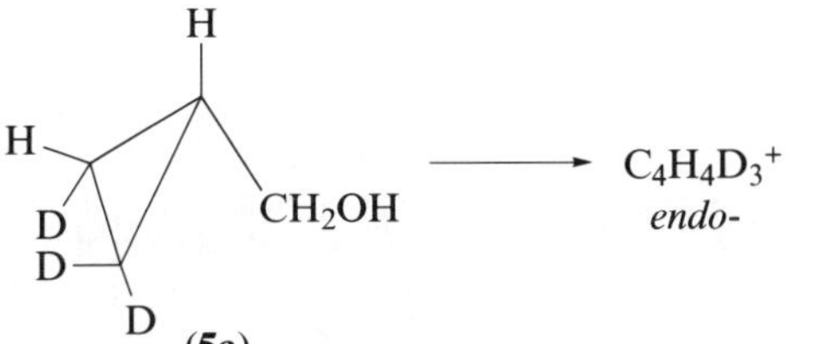

(5a) (10)

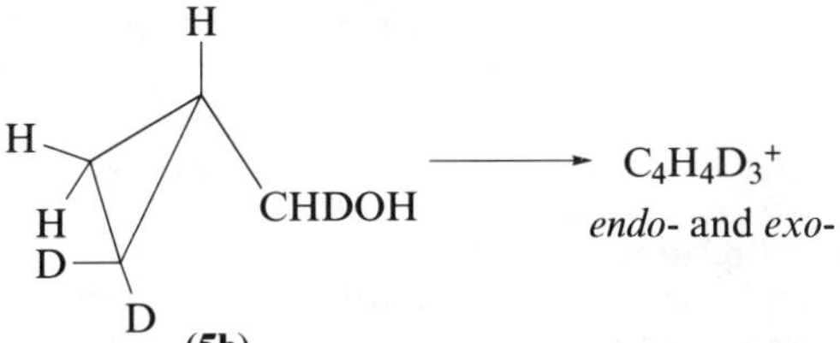

(5b) (11)

The *endo* and *exo* deuterium atoms show isotope effects of different magnitudes on the ^{13}C NMR signals. In the *endo* deuteriated cation the CHD carbon signal ($\delta^{13}C$ 56.69) is deshielded by 1.5 ppm, while in the *exo* deuteriated cation the CHD signal ($\delta^{13}C$ 50.51) is shielded by 4.7 ppm, with respect to the unlabeled cation. Such stereochemically distinctive deuterium isotopic perturbation of equilibria were earlier found in the work of Saunders and Siehl[22]. They observed stereochemically distinctive deuterium isotopic perturbation on the ^{13}C NMR signals of the monodeuterio cyclopropylcarbinyl cation (a mixture of *endo* and *exo* isomers, obtained from α-monodeuteriated cyclopropylcarbinol). The nonequivalent isotope shifts shown by methylene protons excludes classical equilibrating $C_4H_6D^+$ ions. The deuterium equilibrium isotope effects were also calculated using *ab initio* force constants, which reproduced the NMR results[23]. Saunders and Siehl have also measured the isotope induced shifts on ^{13}C NMR signals for the cation derived from α,α-dideuteriated cyclopropylcarbinol. The unlabeled methylene carbon was shielded with respect to that of unlabeled cation by 1.77 ppm at –135 °C and 1.24 ppm at –107 °C.

Olah, Roberts and coworkers observed[24] temperature-dependent chemical shifts for the $C_4H_7^+$ ion, prepared from cyclopropylcarbinol-1-^{13}C. They suggested an equilibration involving nonclassical bicyclobutonium ion **2** and the bisected cyclopropylcarbinyl cation **3** (equation 12).

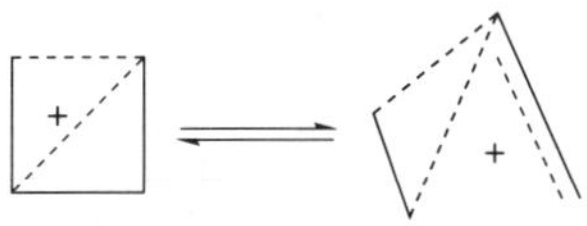

(2) **(3)** (12)

Distinct evidence for the equilibration of bicyclobutonium with a minor isomer, bisected cyclopropylcarbinyl cation, comes from the ultra-low temperature CPMAS studies of Myhre, Webb and Yannoni[25]. They have observed a major isomer, the bicyclobutonium ion, with a ^{13}C chemical shift of 15 ppm for the pentacoordinated carbon, and a minor bisected cyclopropylcarbinyl cation, whose cationic center's chemical shift was found to be at 235 ppm. The NMR chemical shifts of the cation are also comparable with those calculated by the IGLO method at that temperature[26,27]. The energies of these cations were shown to be nearly the same ($\Delta\Delta H^0 = 0.05$ kcal mol^{-1}).

2. Theoretical studies of $C_4H_7^+$ ion

Theoretical calculations at various levels have been carried out by several groups[28–35] for the characterization of the potential energy surface of the $C_4H_7^+$ ion, and these were reviewed in detail earlier[11]. Cyclopropylcarbinyl cation can exist in two conformers which are stationary points on the potential energy surface: the bisected and the perpendicular conformers **6** and **7**. In the bisected conformation, the vacant p-orbital is parallel to the cyclopropane ring and is well aligned with the C2—C3 and C2—C4 bonding orbitals. Maximum orbital overlap is possible in this conformation. In the perpendicular conformer the vacant p-orbital is perpendicular to the plane of the ring and its overlap with C2—C3 or C2—C4 orbitals is effectively cancelled. Therefore the cyclopropylcarbinyl cations exist in the lower energy bisected conformation whenever sterically permissible.

The planar cyclobutyl cation and the perpendicular cyclopropylcarbinyl cation are of nearly equal energy and much less stable than the bisected cyclopropylcarbinyl or bicyclobutonium ion (*ca* 36 kcal mol^{-1}) while the latter two structures have very similar energies. Inclusion of correlation at the MP4SDQ/6-31G*//MP2-6-31G* levels showed that bicyclobutonium ion is more stable than the bisected cyclopropylcarbinyl cation only by 0.7 kcal mol^{-1}. Selected structural parameters for the bisected and perpendicular conformers (**6** and **7**) of the cyclopropylcarbinyl cation as well as the bicyclobutonium ion **2**, as calculated at the MP2/6-31G* level, are as shown[26].

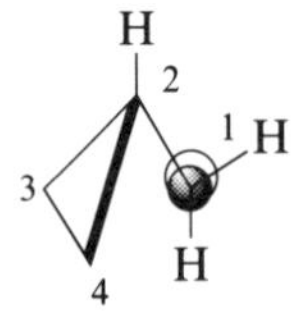

(6)
Bisected conformer

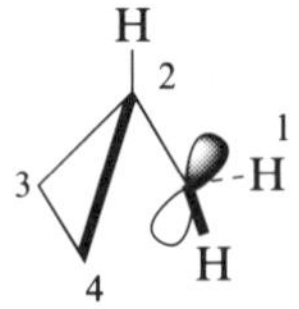

(7)
Perpendicular conformer

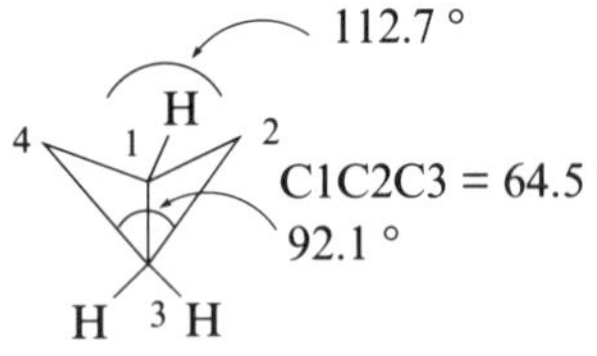

(2)
Bicyclobutonium ion

Bond lengths (Å):

C1—C2 = 1.355	C1—C2 = 1.410	C1—C2 = C1—C4 = 1.424
C2—C4 = C2—C3 = 1.644	C2—C4 = C2—C3 = 1.530	C2—C3 = C3—C4 = 1.647
C3—C4 = 1.414	C3—C4 = 1.499	C1—C3 = 1.649

3. IR studies of $C_4H_7^+$ cation

Based on Roberts' suggestion[36], Vancik, Gabelica, Sunko and Schleyer generated the $C_4H_7^+$ ion from cyclopropylcarbinyl and cyclobutyl chloride in SbF_5 matrices at 77 to 200 K, and the IR spectra were recorded[37]. The spectra were matched with the vibrational frequences of the bicyclobutonium ion and cyclopropylcarbinyl cations calculated at the MP2/6-31G* level, scaled by a factor of 0.94. The experimental frequencies matched the calculated ones, and the spectrum corresponded to a 1:1 mixture of the bicyclobutonium ion and the bisected cyclopropylcarbinyl cation. The ion irreversibly rearranges to the 1-methylallyl cation, the global minimum on the potential energy surface of the $C_4H_7^+$ cation, at 230 K. The IR results confirmed the ultra-low temperature cross-polarization magic-angle spinning NMR results of Myhre, Webb and Yannoni[25] and show that the $C_4H_7^+$ ion is a rapidly equilibrating mixture of two nonclassical structures of nearly equal energy, namely the bicyclobutonium ion and the bisected cyclopropylcarbinyl cation.

4. *1-Methylcyclopropylcarbinyl (1-methylcyclobutyl) cation*

The 1-methylcyclobutyl cation **8**, which can also be considered as equilibrating 1-methylcyclopropylcarbinyl cations, was shown to have a nonclassical structure by ^{13}C NMR spectroscopic studies in conjunction with Saunders' isotopic perturbation of equilibria techniques.

The ion could be easily prepared from either 1-methylcyclobutanol or 1-methylcyclopropylmethyl alcohol as well as from their halide derivatives (equation 13). At above –25 °C the cation rearranges to cyclopropylmethylcarbinyl cation (*vide infra*) with an activation barrier of about 20 kcal mol^{-1}. At equilibrium, the 1-methylcyclobutyl cation is found to be only about 2% (K_{eq} = 50) showing that the latter cation is about 2 kcal mol^{-1} more stable than the 1-methylcyclobutyl cation[38] (equation 14).

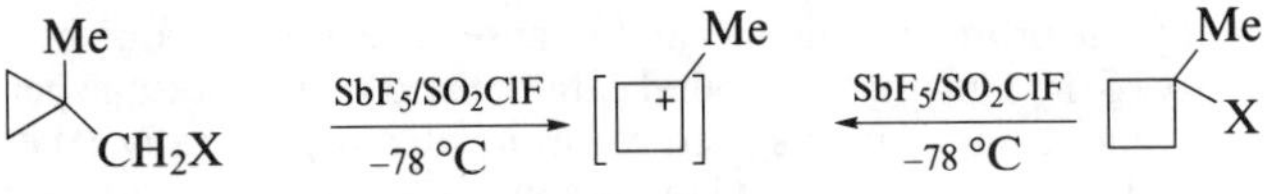

X = OH, Cl (**8**) (13)

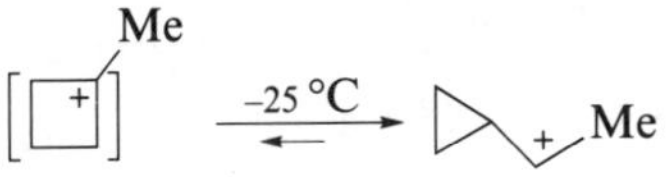

(14)

(**8**) $K_{eq} = 50$
ΔG^0 (25 °C) = 2 kcal mol^{-1}

The 1-methylcyclobutyl cation at –80 °C shows three absorptions in its ^{13}C NMR spectrum at δ^{13}C 163.1 (C$^+$), 48.7 (CH$_2$) and 25.4 (CH$_3$), suggesting that it undergoes degenerate equilibration of σ-delocalized 1-methylcyclopropylcarbinyl cations[20b]. The ^{13}C NMR at –156 °C, however, shows the splitting of the methylene carbons into two distinct absorptions at δ^{13}C 72.2 and –2.83[39]. The high field absorption is typical of nonclassical cations, although Sorensen and Kirchen once proposed[39] an sp^3-hybridized carbon to account for the results. Deuterium isotope effects studies on the ^{13}C NMR chemical shifts of the cation as well as studies on related cations by Prakash, Olah and coworkers showed the involvement of equilibrating σ-delocalized 1-methylbicyclobutonium ions **9**[40,41]. It was found that the unlabeled methylene carbons of the dideuteriated cation (equation 15) are

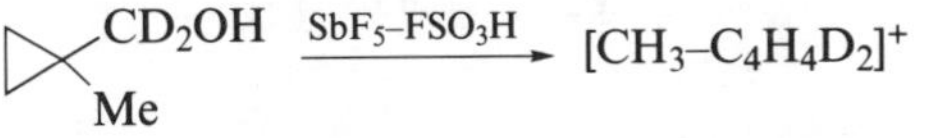

(15)

shielded compared to that of unlabeled cation by 1.27 ppm at –50 °C, and by 1.41 ppm at –90 °C, showing a definite equilibrium isotope effect. The deuteriated methylene carbon was also deshielded from the other methylenes by 2.91 ppm at –50 °C. These results and similar equilibrium isotope effect studies of Siehl[42] on the α-monodeuteriated and α,α-dideuteriated 1-methylcyclobutyl cation show the equilibration among σ-delocalized 1-methylcyclobutyl cations **9a** (equation 16). The relative deshielding of the deuteriated

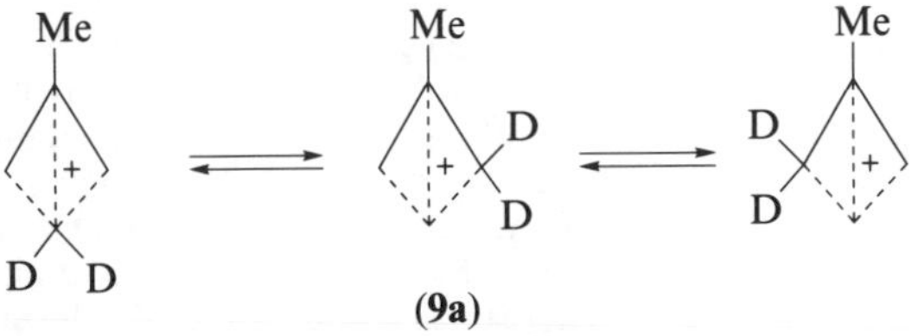

(16)

(**9a**)

carbons indicates that in the nonclassical structures the deuterium is less favored on the pentacoordinated carbon.

The nonclassical σ-delocalized structures were also indicated by the shielding of the cationic center of trideuteriomethylcyclobutyl cation **10** (δ^{13}C 161.3), compared to that of the unlabeled cation (δ^{13}C 162.4)[43].

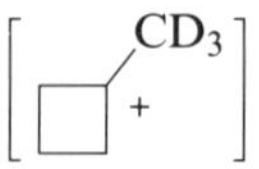

(10)

The σ-delocalized structures are nonplanar, although ^{1}H NMR shows only one absorption for all the methylene protons. Saunders and Krause have given unequivocal evidence for the nonplanarity of the 1-methylcyclobutyl cation based on the isotopic perturbation of the equilibrium among nonplanar structures using stereospecifically labeled α,β,α'-trideuteriated 1-methylcyclobutyl cation **11** (equation 17)[44]. The cation showed two peaks

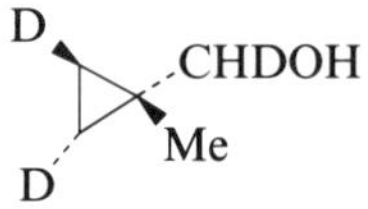

$\longrightarrow$ $[CH_3\text{–}C_4H_3D_3]^+$ (17)

(11)

in a ratio of 2:1 for the methylene protons (unlike only one for the unlabeled cation), which are also slightly deshielded with respect to those of the unlabeled ion. The observation of the splitting of the methylene protons clearly proves that the cation is nonplanar. The isotope-included deshielding of the methylene protons was interpreted as involving an equilibration with an unspecified minor species. At the MP2(FC)/6-31G* level, it was found that the planar 1-methylcyclobutyl cation is less stable than the bent 1-methylcyclobutyl cation (i.e. the bicyclobutonium ion **9**) by 9.3 kcal mol^{-1}. The MP2/6-31G* optimized bond lengths (in Å) for the bent 1-methylcyclobutyl cation are as shown (*vide supra*)[45].

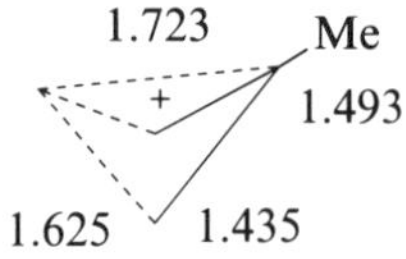

(9)

All the obtained evidence gathered thus far clearly shows the nonclassical nature of the 1-methylcyclobutyl cation.

5. Solvolytic generation of substituted $C_4H_7^+$ cations

Although under solvolytic conditions carbocations are formed only as short-lived (transient) species, it is of interest to discuss here some recent work involving substituted $C_4H_7^+$ cations which are relevant to the general discussions in the review. Wiberg and coworkers have carried out extensive studies on solvolysis reactions involving 3-substituted cyclobutyl tosylates **12** (equation 18), and rationalized the observed products by *ab initio* molecular orbital calculations at the MP2/6-31G* level[46,47].

They have postulated that the solvolysis of 3-substituted cyclobutyl tosylates **12** (X = Alkyl, Ar, Cl, OEt, SiR'_3) proceeds through the initial formation of bicyclobutonium ion in the rate determining step, which rearranges stereospecifically to the cyclopropyl-

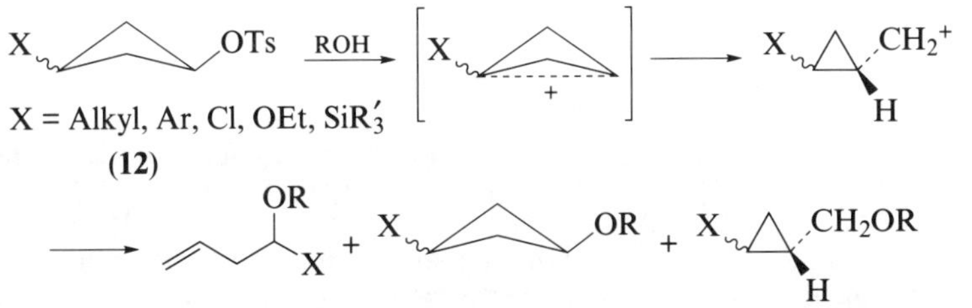

(18)

carbinyl/homoallyl cation. The calculations showed that the 3-substituents lowered the energy barrier for the rearrangement to the cyclopropylcarbinyl cation. The *trans*-2-methylcyclopropylcarbinyl cation is 2.4 kcal mol^{-1} more stable than the *cis*-2-methyl analogue, which is reflected in the nature of the products of the reaction; in general the *cis*-substituted cyclopropylcarbinyl cations give mainly the homoallylic products, whereas the *trans*-2-alkyl substituted cations give the cyclopropylcarbinyl products. Since the interconversion of the cyclobutyl and cyclopropylcarbinyl cation is stereospecific, the *cis*-3-alkyl cyclobutyl tosylate gives the *cis*-2-alkyl cyclopropylcarbinyl cation, which invariably gives the homoallylic acetates in acetolysis reactions (equation 19). The *trans*-3-alkylcyclobutyl tosylate, on the other hand, gives the *trans*-2-alkylcyclopropylcarbinyl cation as the intermediate, which gives the cyclopropylcarbinyl acetates as the major products (equation 20).

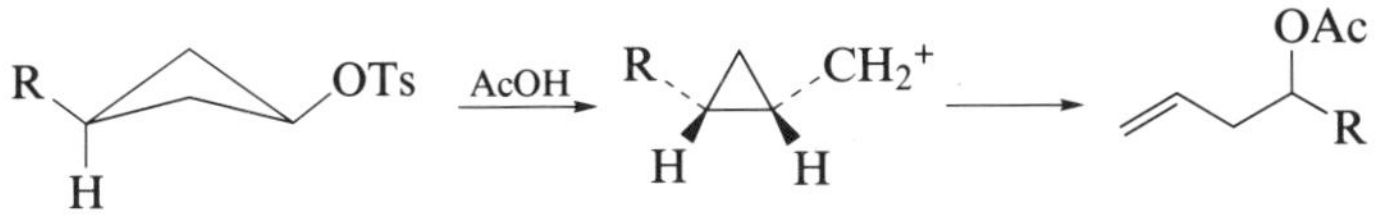

(19)

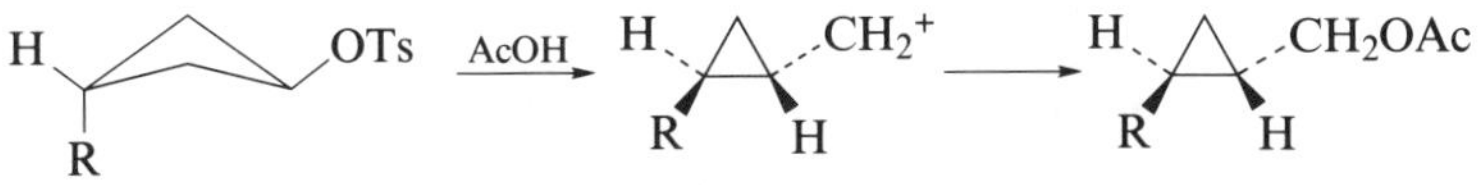

(20)

The 2-aryl substituted cyclopropylcarbinyl cations have partial homoallylic character, whose contribution to the resonance hybrid increases when strong electron-withdrawing substituents (e.g. phenyl) are attached at the C2. Thus, 3-arylcyclobutyl tosylates on acetolysis give the homoallylic acetates predominantly, through the intermediate formation of the 2-arylcyclopropylcarbinyl cations (equation 21).

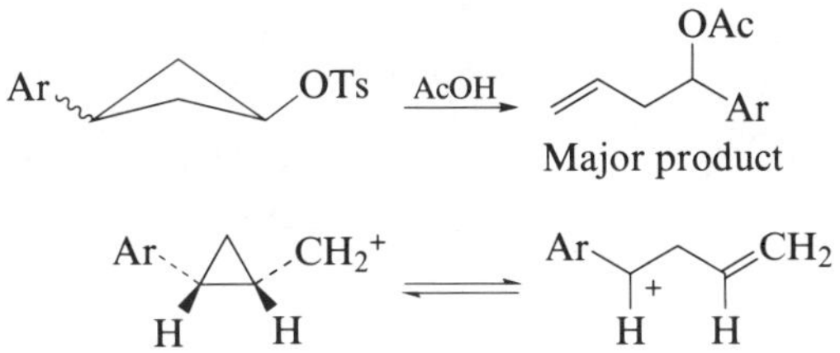

(21)

The *cis*- or *trans*-2-silyl group virtually has no effect on the energy of the cation, and especially for the *cis*-silyl cation a nearly symmetrical structure was obtained at the MP2/6-31G* level of theory. 3-Halo and 3-alkoxycyclobutyl tosylates on solvolysis give only the cyclobutyl products, presumably involving the cyclobutyl cation as the tight ion pair. The MP2/6-31G* calculated structures indicated considerable cross-ring interactions in these cyclobutyl cations also. Presumably because of the tight ion pair character, these ions do not undergo isomerization to the more stable bisected cyclopropylcarbinyl cations.

B. Static Cations

1. Bicyclo[1.1.0]butyl-1-carbinyl cation

Wiberg and McMurdie have prepared bicyclo[1.1.0]butyl-1-carbinyl cation **13** by the ionization of bicyclo[1.1.1]pentyl-1-bromide with SbF_5/SO_2ClF at –120 °C[47]. The cation showed the following ^{13}C NMR absorptions: $\delta^{13}C$ 224.0 (C^+), 134.8 (C1), 65.0 (C3), 46.4 (C2). It is interesting that solvolysis of the bicyclo[1.1.1]pentyl-1-bromide or other derivatives give products derived from the intermediacy of 3-methylenecyclobutyl cation **14**, rather than the bicyclobutylcarbinyl cation (equation 22). In nucleophilic solvents, solvent

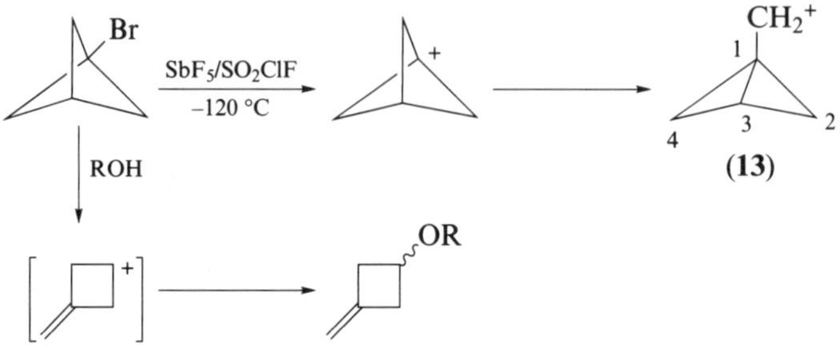

(13)

(14)

(22)

attack presumably occurs on the homoallylic carbon concurrent with the formation of the contact ion pair of the bicyclobutylcarbinyl cation. As shown by the calculations of Wiberg and coworkers[47], the ion on the other hand may possess some homoallylic character, the predominant contributing structure in the nucleophilic solvents. The structure of the bicyclobutylcarbinyl cation is confirmed by matching the observed NMR chemical shifts with those calculated by the IGLO method at the basis set II//MP2/6-31G* [IGLO $\delta^{13}C$: 229.4 (C^+), 129.9 (C1), 67.9 (C3), 57.3 (C2)].

2. Nortricyclylcarbinyl cation

Schmitz and Sorensen have prepared the primary, secondary and tertiary nortricyclylcarbinyl cations (*vide infra*) from the respective alcohols under the standard conditions of ionization in SO_2ClF[48]. They found that the primary cation **15** is a static species. Thus the stabilization provided by the nortricyclyl group is comparable in magnitude with that of the ferrocenyl group. The cation was stable up to –20 °C. One significant point about the cation is that it has a high barrier (not yet measured because of the instability of the cation at higher temperatures) of rotation across the C1—C1′ bond. The two hydrogens on the cationic center show distinct ^1H NMR absorptions at 7.83 (*anti*) and 7.61 (*syn*) ppm. The sterically enforced bisected nature of the cyclopropyl group in this system makes it possible to observe the primary cation. Theoretical calculations at MNDO and STO-3G suggested that the primary cation **15** has enhanced vinyl-bridging character compared to the parent cyclopropylcarbinyl system[49] (equation 23).

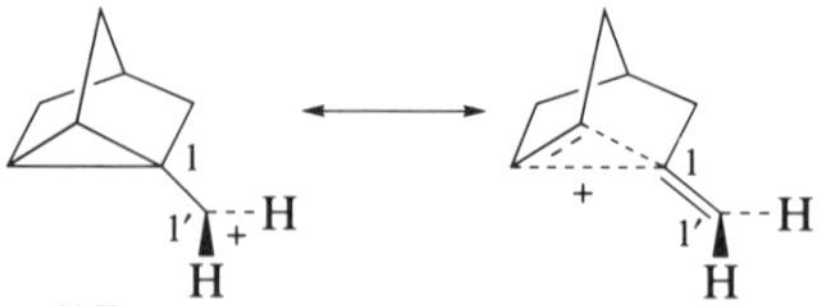

(15)

(23)

IV. SECONDARY AND TERTIARY CYCLOPROPYLCARBINYL CATIONS

A. Static Cations

Several secondary and tertiary substituted cyclopropylcarbinyl cations such as dicyclopropylmethyl cation, (α-methylcyclopropyl)carbinyl cation and 1,α,α-trimethylcyclopropylcarbinyl cation were discussed in the early review[11].

1. Tricyclopropylcarbinyl cation

The tricyclopropylcarbinyl cation **16** is among the earliest stable carbocations to have been prepared[12,50a]. It is stable even in neat sulfuric acid. Earlier studies reported the UV absorptions for this cation at λ_{max} 270 nm with extinction coefficient exceeding 20,000. The proton NMR at 300 MHz showed well resolved absorptions for the methine and methylene protons at 2.6 (m) and 3.0 (s) ppm, respectively. ^{13}C NMR showed at δ^{13}C 280.5 (C$^+$), 32.5 (CH) and 30.8 (CH$_2$). The X-ray structure of cation **16** has not yet been obtained although its salts can be isolated[50b].

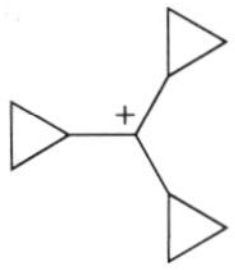

(16)

2. α-Methylcyclopropylcarbinyl cations

From solvolytic studies of isotopically labeled substrates it was shown that cyclopropylcarbinyl–cyclobutyl interconversion is stereospecific[51,52]. The stereospecific interconversion of cyclobutyl cations to the corresponding cyclopropylcarbinyl cation was also cleanly observed in superacid medium, and was used to prepare otherwise unstable *cis*-(α-methylcyclopropyl)carbinyl cation **17**[53]. Thus ionization of *cis*-2-chloro- or *cis*-3-chloro-1-methylcyclobutane in SbF_5–SO_2ClF at –135 °C yielded the *cis*-isomer which rapidly rearranged irreversibly into the *trans*-isomer **18** at about –100 °C. The *trans*-isomer **18** is the only cation obtained when the preparation was carried out at –80 °C, or when prepared from the cyclopropylmethyl carbinol[20b,38,50a,c] (equation 24).

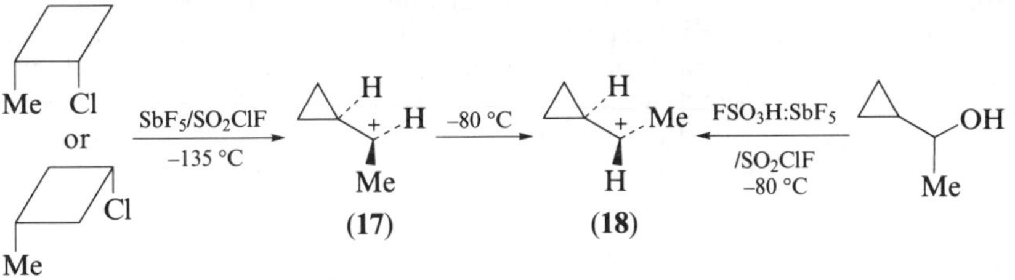

(24)

Gas-phase protonation of spiropentane **19** using $^3HeT^+$ or D_3^+ as protonating agents provided initially the corner protonated spiropentane, which rearranged into 1-methylcyclobutyl cation (i.e. bicyclobutonium ion **9**)[45]. Under mild protonating conditions, *cis*- and *trans*-2-methylcyclopropylcarbinyl cations **17** and **18** were observed. These cations were presumably formed through the isomerization of the 1-methylcyclobutyl cation **9** into the *cis*-isomer **17**, followed by its interconversion to the *trans*-isomer **18** through the intermediacy of the 1-ethylallyl cation (equation 25).

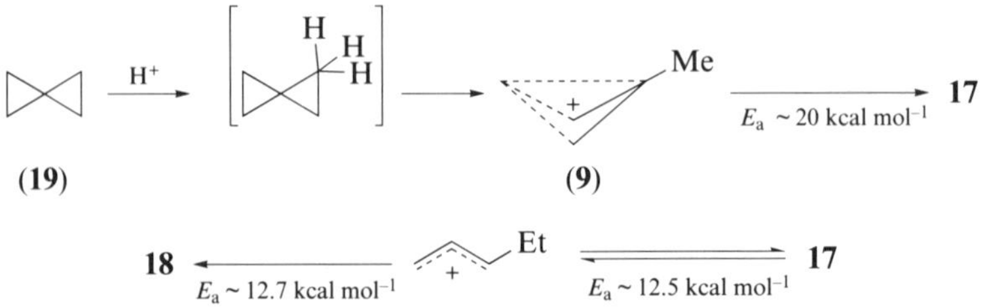

(25)

Ab initio MO calculations at the MP2(FC)/6-31G* level of theory gave estimates of the activation barrier of 1-methylcyclobutyl cation to the α-methylcyclopropylcarbinyl cation of about 20.0 kcal mol^{-1}. At MP2/6-31G* + ZPE, the *trans*-isomer is 2 kcal mol^{-1} more stable than the *cis*-isomer, whereas the bent 1-methylcyclobutyl cation **9** is less stable than the *trans*-1-methylcyclopropylcarbinyl cation **18** by 0.8 kcal mol^{-1} (*vide supra*). The MP2/6-31G* optimized bond lengths (in Å) for the *cis* and *trans* isomeric cations are as shown:

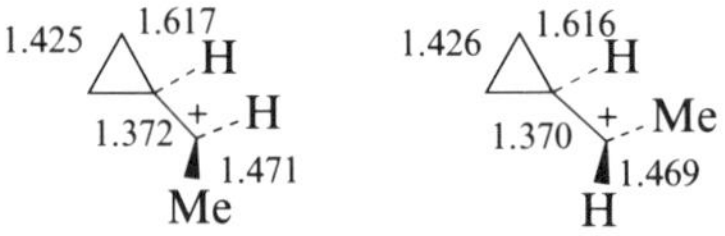

(17) **(18)**

3. α,α-Dimethylcyclopropylcarbinyl cations and their derivatives

Olah and coworkers[50a,c] obtained the α,α-dimethylcyclopropylcarbinyl cation **20** from the corresponding alcohol by ionization with SbF_5/SO_2ClF. The cation shows distinct absorptions for the two methyl groups in the proton and ^{13}C NMR. The *cis*-methyl group is shielded (δ^1H 3.57) compared to the *trans*-methyl (3.68 ppm). The ^{13}C NMR shows the following absorptions: $\delta^{13}C$ 279.9 (C^+), 59.1 (CH), 53.1 (CH_2), 39.1 (*exo* CH_3), 30.1 (*endo* CH_3). The nonequivalence of the absorptions up to –35 °C for the methyl groups suggests a rotational barrier of 8–10 kcal mol^{-1} around the cyclopropyl CH—C_α bond. Kabakoff and Namanworth[13] measured this rotational barrier as 13.7 kcal mol^{-1} with an Arrhenius pre-exponential factor A of 12.2. Theoretical calculations at the 3-21G level gave a value of 18 kcal mol^{-1} for this barrier[54]. The effect of the 1-methyl substitution (as in **20a**) in the 1,α,α-trimethylcyclopropylcarbinyl cation is to shield the cationic center's chemical shift by about 3.4 ppm. The 1-methyl substituent does not affect the rotational barrier to any significant extent[54]. The 1-methyl substituent in other related cations facilitates the rearrangement into the ring-opened cations followed by other reactions leading to polymerization[54].

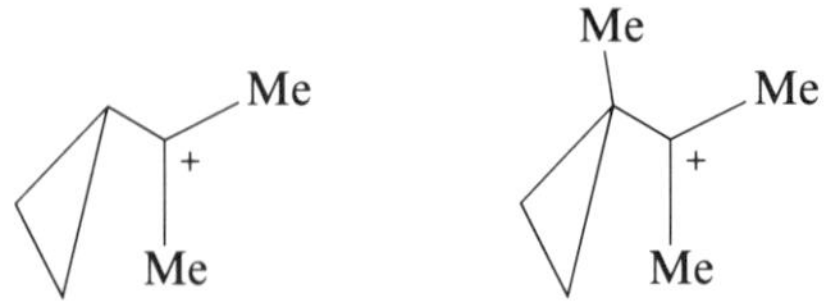

(20) **(20a)**

4. *1-Cyclopropylcycloalkyl cations*

Similar to the α,α-dimethylcyclopropylcarbinyl cation **20**, the 1-cyclopropylcyclopentyl and 1-cyclopropylcyclohexyl cations **21** and **22** show restricted rotation of the cyclopropyl

(21) **(22)**

group, with the result that the α-methylene carbons show distinct NMR chemical shifts[39,55]. The α-methylene carbons in the latter cation appear at 51 and 40.1 ppm, whereas in the former cation they are at 49.5 and 44.2 ppm, respectively. These cations exist in the bisected conformation[39,55]. The 1-cyclopropylcyclobutyl cation **23** similarly shows the α-methylenes at 49.9 and 42.4 ppm. **23** is, however, unstable and rearranges to the cyclopentylidenemethyl cation **24** even at –100 °C[39] (equation 26).

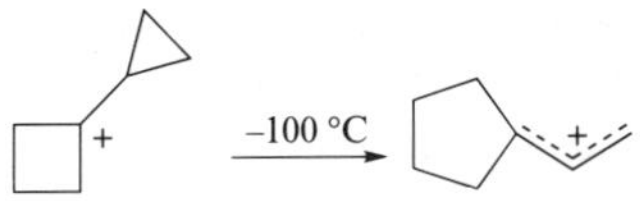

(23) **(24)** (26)

5. *Tertiary 8,9-dehydro-2-adamantyl and 2,4-dehydro-5-homoadamantyl cations*

The 2-methyl, 2-cyclopropyl and 2-phenyl substituted 8,9-dehydro-2-adamantyl cations **25** were prepared from their respective alcohols using fluorosulfuric acid in sulfuryl chloride fluoride at low temperatures (equation 27). The relative extent of charge delocalization in these cations was estimated by comparing their NMR spectra. The ions are nonequilibrating static cations, as shown by their proton NMR spectra[56].

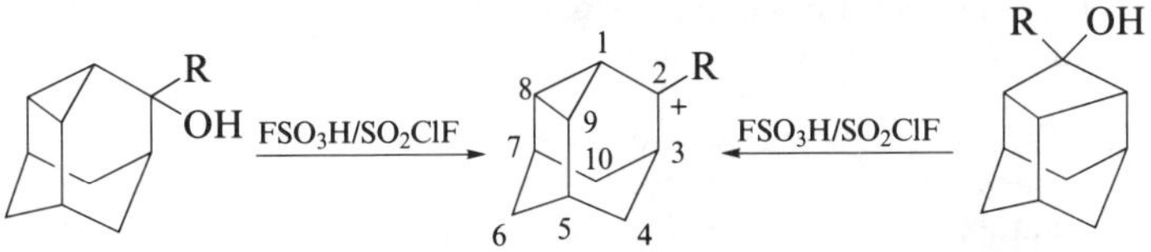

(25) (27)

R = Me, cyclopropyl, Ph

The tertiary 2-methyl-8,9-dehydro-2-adamantyl cation **25**, R = Me is stable up to 10 °C, and shows a much deshielded absorption for the cationic center (δ^{13}C 274.4), which is relatively shielded compared with the 2-methyl-2-adamantyl cation (δ^{13}C 323) showing some charge delocalization into the cyclopropane ring. The C8 and C9 carbons are also deshielded (δ^{13}C 100.7), comparable to other static cyclopropylcarbinyl cations.

The 1,2-dimethyl-8,9-dehydro-2-adamantyl cation **26**[56], as expected, is also a static cation and stable up to –10 °C. Its NMR behavior (δ^{13}C of C^+ = 266.3) can be compared with the 1,α,α-trimethylcyclopropylcarbinyl cation **20a** (δ^{13}C of C^+ = 276.6)[54]. The introduction of the 1-methyl group causes shielding of the cationic centers in both cases (the cationic center is shielded with respect to the demethylated analogue by 8 ppm in the former case, and it is shielded by 3.4 ppm in the latter case).

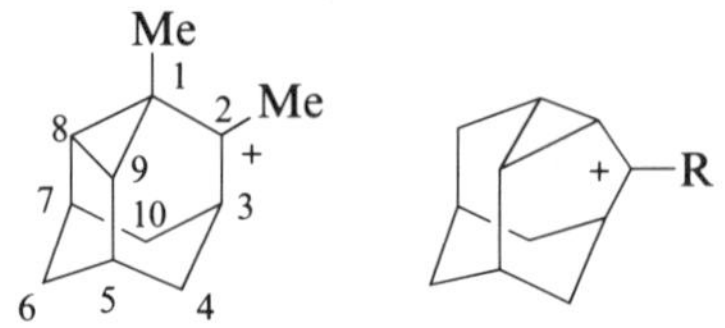

(26) **(26a)**

R = Me, Ph

The tertiary 5-methyl and 5-phenyl-2,4-dehydro-5-homoadamantyl cations, **26a**, were prepared[57] by two different routes: either from the corresponding alcohols or the corresponding 2-*endo*-hydroxyhomoadamant-4-enes. The two cations were shown to be classical carbocations with varying degrees of charge delocalization.

6. Arylcyclopropylcarbinyl cations

Tertiary 1-aryl-1-cyclopropylethyl cations **27** show the expected deshielding of the cationic center chemical shift with increasing electron demand, as measured by σ^+, or $\sigma_\alpha C^+$ of the substituents on the aryl rings[58]. Thus, $\delta^{13}C(C^+)$ values of 227.8, 247.6 and 251.1 were observed for the *p*-OMe, *p*-H and *p*-CF_3 substituents, respectively.

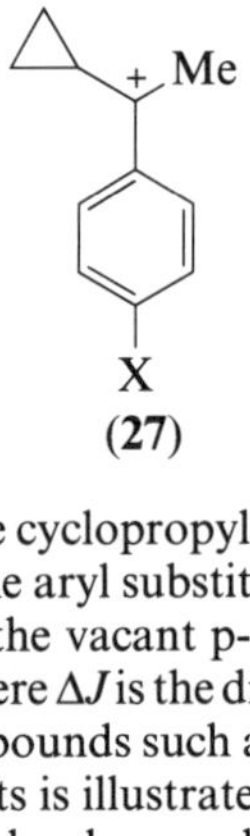

(27)

The C—H coupling constants of the cyclopropyl methine carbons are also linearly correlated with the electron demand of the aryl substituents and, in addition, to the dihedral angle between the C—H orbital and the vacant p-orbital, as expressed by the equation: $\Delta J = (1 + 0.6\sigma^+)(10.9 - 14.3\cos^2\theta)$, where ΔJ is the difference in coupling constant between the cation and the neutral model compounds such as carbonyls. The dependence of the J values on the nature of the substituents is illustrated by the J values of 174, 177, 179 and 183 Hz for the C1′—H bond of 1-arylcyclopropylethyl cation for the *para* substituents OMe, Me, H and CF_3, respectively[59].

7. 3-Nortricyclyl cations

The participation of the cyclopropyl group is at its maximum in its bisected conformation. Hence, 3-nortricyclyl cations **28** would be expected to be highly stabilized, as the

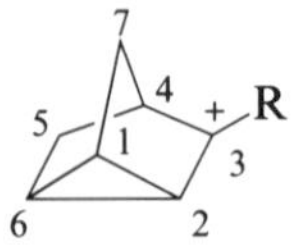

(28)

R = H, Me, Et, Ph, Cl, F

cyclopropyl group in these cations is in the ideal bisected conformation with respect to the vacant p-orbital of the cationic center[60].

Comparison of the ^{13}C absorptions for the cationic center of the 3-nortricyclyl cation **28**, R = H (258.5 ppm) with that of the α-methylcyclopropylcarbinyl cation **18** (250.8 ppm), however, shows that the cationic center is not preferentially stabilized in the former cation. Comparison of the ^{13}C absorption for the tertiary 3-methyl-3-nortricyclyl cation **28**, R = Me (293.2 ppm) with that of 2-methyl-8,9-dehydro-2-adamantyl cation **25**, R = Me (274.4 ppm) and dimethylcyclopropylcarbinyl cation **20** (281.9 ppm) also reveals that no exceptional stabilization is provided by the cyclopropane ring in the 3-nortricyclyl skeleton. Based on these results Olah and Liang have proposed that, due to steric inhibition of hyperconjugation in this system, limiting nonclassical cation structure is prevalent[61]. However, considerable charge delocalization into the cyclopropane ring is indicated by the much deshielded absorptions for the C1 and C6 carbons of the cyclopropane ring ($\Delta\delta^{13}C$ C1—C2 = 25). A number of other 3-substituted-3-nortricyclyl cations (R = Et, Ph, F and Cl) were prepared by ionizing the corresponding alcohols with Magic Acid. It was shown that the 3-fluoro-3-nortricyclyl cation **28**, R = F is the most stable cation among these, and is stable up to –20 °C in SO_2ClF. The strong back-donation of the fluorine is shown by its large coupling constant to the attached carbon (J_{C-F} = 420 Hz), and the relatively less deshielded absorptions for the cyclopropane ring carbons. All of these ions are thoroughly characterized by ^1H and ^{13}C NMR spectroscopy[60,61].

8. *3-Tetracyclo[3.3.1.0^{2,8}.0^{4,6}]nonyl cation*

3-Tetracyclo[3.3.1.0^{2,8}.0^{4,6}]nonyl cation **29** (3-triasteranyl cation) has been prepared by the ionization of 3-triasteranol in $FSO_3H/SO_2ClF/CD_2Cl_2$ at low temperature (equation 28)[62]. The ^1H NMR spectrum shows the proton at the secondary cationic center at δ^1H 10.25. The corresponding shift of the dicyclopropylmethyl cation is at δ^1H 8.14[50c]. The α and β methine protons are observed at δ^1H 3.89 and 2.97, respectively, indicating substantial positive charge delocalization into the annulated cyclopropane rings. The ion **29** has also been obtained by the ionization of tricyclo[3.3.2.0^{2,8}]nonan-3-ene-6-ol in low yield[62].

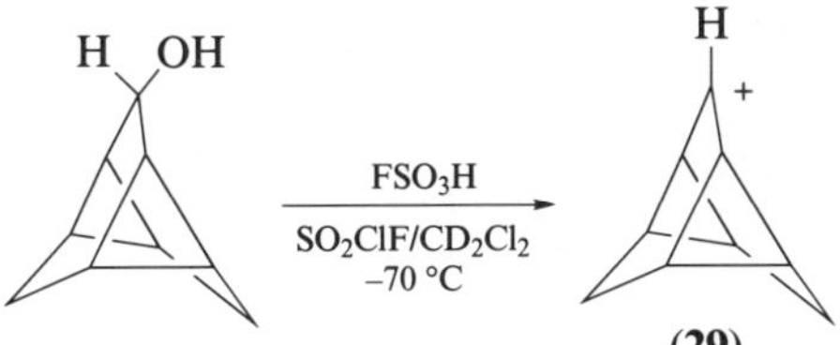

(29) (28)

9. *Nortricyclylcarbinyl cations*

Schmitz and Sorensen have prepared the secondary and tertiary nortricyclylcarbinyl cations **30** and **31** from the respective alcohols[48,49].

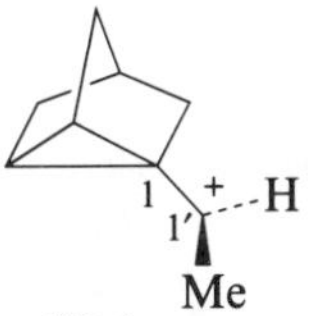

(30a)

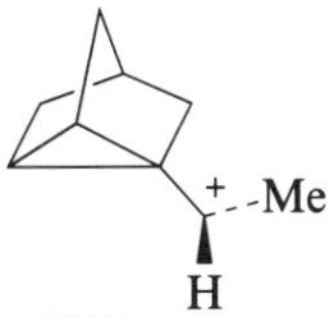

(30b)

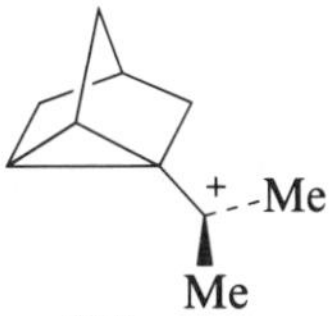

(31)

The rotational barrier for the secondary 1′-methyl cation **30** could not be measured experimentally, but the barrier for the tertiary 1′,1′-dimethyl cation **31** could be estimated from the observed line broadening at + 80 °C as 18 kcal mol^{-1}. Theoretical calculations at the STO-3G level gave a value of 17.7 kcal mol^{-1} for the latter cation. The secondary cationic system is a mixture of two isomers (*cis*-**30a** and *trans*-**30b**) due to the restricted rotation across the C1—C1′ bond. In the tertiary dimethyl cation **31**, the methyls expectedly show distinct NMR absorptions: δ^1H: 3.14 (*cis*-Me), 2.60 (*trans*-Me); δ^{13}C: 30.2 (*cis*-Me), 33.7 (*trans*-Me). The cationic center's ^{13}C absorptions are also interesting in that they are highly shielded compared with other related cations, including cyclopropylcarbinyl cations. The tertiary cationic center has δ^{13}C of 249.6, the secondary cationic center δ^{13}C of 220.8 and the primary cationic center δ^{13}C of 191.4. Calculations at MNDO and STO-3G suggested that these cations have enhanced vinyl-bridging character compared to the simple cyclopropylcarbinyl model[49].

10. Tertiary barbaralyl cations

The 9-methylbarbaralyl cation **32a** was prepared from the ionization of 9-methyl-9-barbaralol, 2- or 3-methylbicyclo[3.2.2]non-3,6,8-trien-2-ols with FSO_3H in SO_2ClF–SO_2F_2, and characterized by ^{1}H and ^{13}C NMR spectra[63a]. The ^{13}C NMR spectrum at –129 °C showed absorptions typical of cyclopropylcarbinyl cation, and no significant

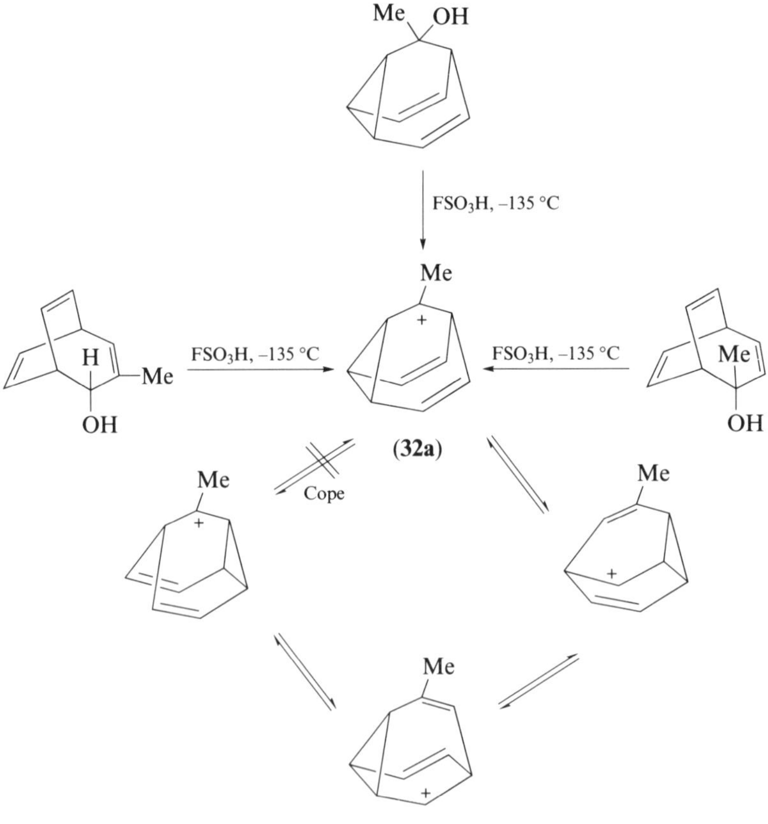

(29)

charge delocalization into the double bonds. δ^{13}C: 72.5 (C1), 86.2 (C2, C8), 116.2 (C3, C7), 130.2 (C4, C6), 260.0 (C9), 33.2 (CH_3). The ion undergoes partially degenerate Cope-like rearrangements at temperatures of −135 °C to −115 °C. By using the transfer of spin saturation method, the rearrangement barrier for this cation was obtained as 7.6 kcal mol^{-1} at −129 °C, which is lower than that expected for the Cope rearrangement. The unexpectedly low activation barrier suggested that the ion undergoes divinylcyclopropylcarbinyl–divinylcyclopropylcarbinyl rearrangement rather than a Cope rearrangement (equation 29).

The 1,9-dimethyl cation **32b** was also prepared by a similar method from the 1,9-dimethyl-9-barbaralol (equation 30) and was found to have a low barrier of degenerate rearrangement of about 7.6 kcal mol^{-1} [63a].

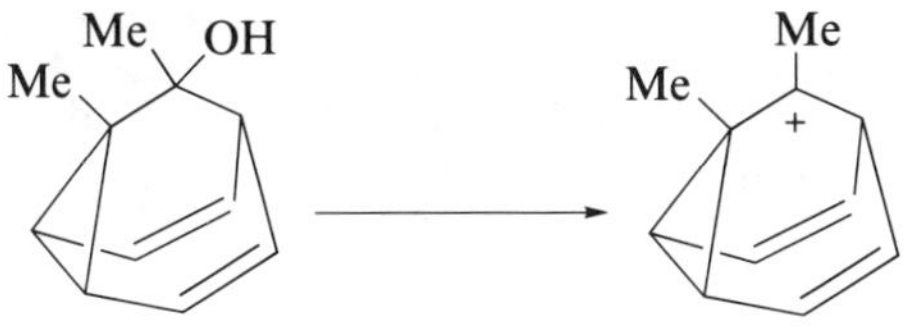

(30)

(32b)

The 9-ferrocenyl-9-barbaralyl cation **32c** was found to be stable even in trifluroacetic acid[63b]. The ^{13}C NMR chemical shifts of the cation are similar to that of the protonated barbaralone, reflecting the high stabilizing ability of the ferrocenyl group.

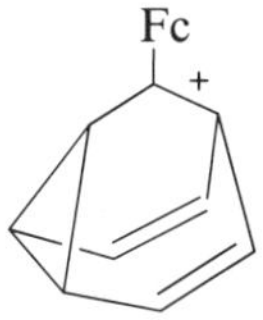

(32c)

11. 9-Cyclopropyl-9-xanthyl cation

The 9-Cyclopropyl-9-xanthyl cation **33** is so stable that the positive charge is hardly delocalized into the cyclopropyl group. The alpha and beta hydrogens of the cyclopropyl group show absorptions at δ^1H 2.96, 1.38 and 1.96, comparable to that of cyclopropylammonium ion (δ^1H 2.93 and 0.93), rather than a typical cyclopropylcarbinyl cation, such as dimethylcyclopropylcarbinyl cation (δ^1H 3.4, 4.0)[64].

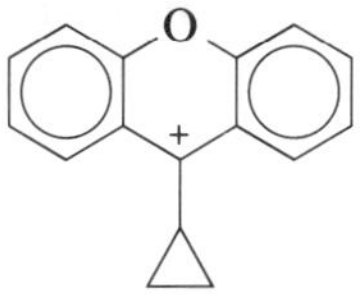

(33)

12. Bridgehead barrelyl and bullvalyl cations

All attempts at preparing the bridgehead 1,5-trishomobarrelenediyl dication **34** were unsuccessful[65]. However, bridgehead monocations such as 1-trishomobarrelyl and 1-

trishomobullvalyl cations **35** and **36** were prepared from the respective chlorides. The charge in these cations is delocalized symmetrically into all the adjacent cyclopropyl groups[65,66].

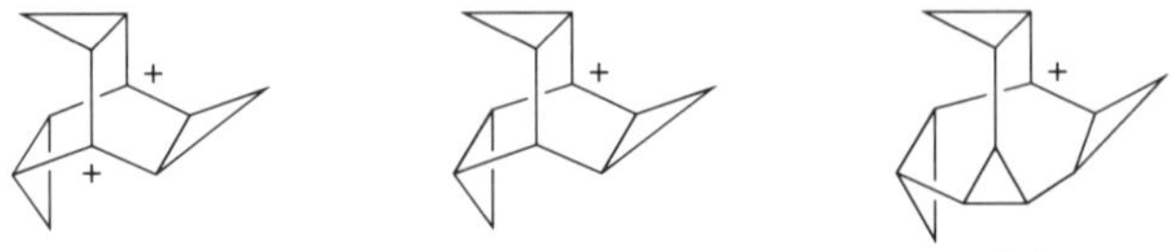

(34) **(35)** **(36)**

13. α-Cyclopropylvinyl cations

Siehl and coworkers have prepared cyclopropylcyclopropylidenemethyl cation **37**, by protonation of bis(cyclopropylidene)methane with FSO_3H–SbF_5 at liquid nitrogen temperature (equation 31)[67]. The vinyl cationic center is stabilized by the α-cyclopropyl group.

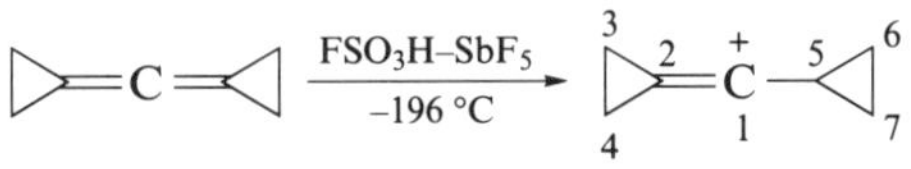

(37) (31)

It shows ^{13}C absorptions (–135 °C) at $\delta^{13}C$ 234.18 (C^+), 21.21 (C2), 38.15 (C3, C4), 51.67 (C5), 43.9 (C6, C7). Compared to dicyclopropylcarbinyl cation, the ^{13}C signal for the cationic center is shielded by 20 ppm, indicating cyclopropyl group participation in the stabilization of the cation. The chemical shifts were compared with the values obtained from GIAO-SCF and GIAO-MP2 methods.

An α-cyclopropylvinyl cation, 1-cyclopropyl-3-methylbuta-1,2-dienyl cation **38**, has been prepared by Siehl by ionizing the corresponding acetylenic alcohol with SbF_5 (equation 32)[68].

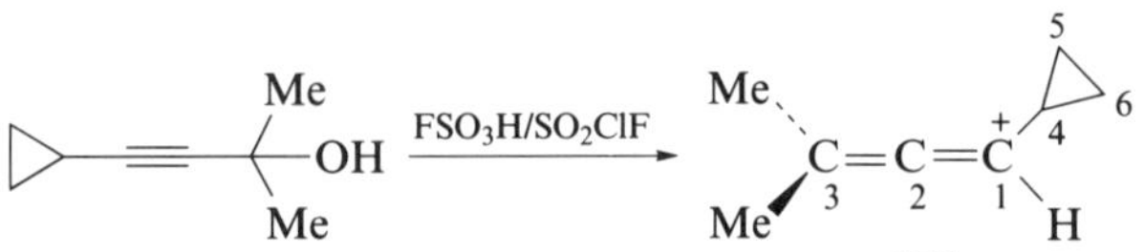

(38) (32)

The cation **38** at –99 °C shows six ^{13}C absorptions at $\delta^{13}C$ 124.45 (C3), 238.89 (C1), 127.31 (C2), 43.69 (CH_3), 41.64 (C5, C6), 25.74 (C4). At –136 °C, the cation showed two distinct absorptions for the methyl carbons, indicating the preferred bisected geometry for this cation. Charge delocalization into the cyclopropyl ring is shown by the deshielded absorptions for the cyclopropyl methylene absorptions as compared to the precursor alcohol ($\Delta\delta$ 33), and in analogy with other cyclopropylcarbinyl cations. The absorptions for the methyl carbons coalesced into a single peak at around –120 °C. From line shape analysis a rotational barrier of 7.2 kcal mol^{-1} was estimated. When the bisected conformation is not readily available for the intermediate α-cyclopropylvinyl cations, the cations are usually not formed in solvolytic reactions; e.g., on attempted solvolysis of 6-spirocyclopropyl-1-cyclohexen-1-yl triflate it undergoes sulfur–oxygen cleavage to give the alcohol in trifluoroethanol[69].

Other cyclopropylidene-substituted vinyl cations, **39** and **40**, have also been prepared[70].

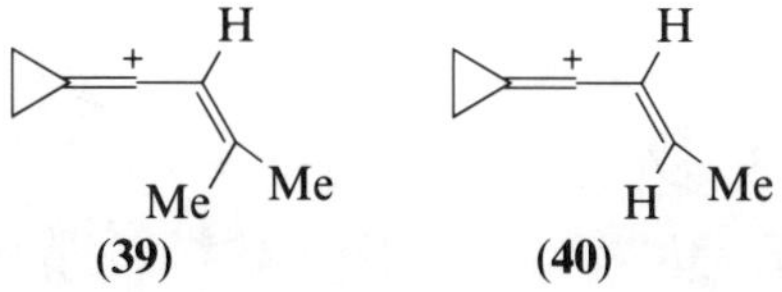

The reactive cyclopropylidene cyclopropylcarbinyl cation **37** was generated from the corresponding bromide and silver hexafluoroantimonate, and was reacted *in situ* with olefins such as cyclohexene to form the addition products (equation 33). On the other hand, under the same conditions, the isopropylidene cyclopropylcarbinyl cation **41** rearranged spontaneously by ring expansion (equation 34)[71].

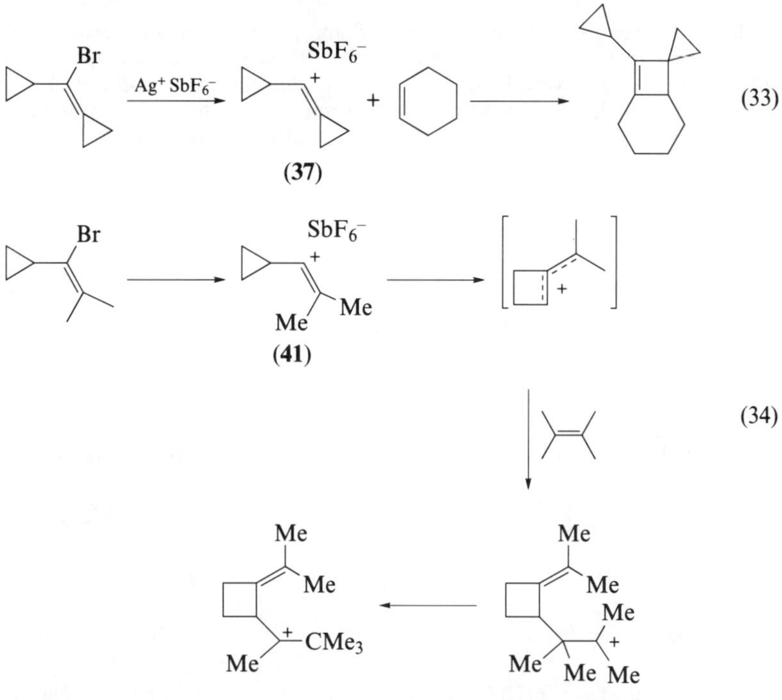

The stabilization of the vinyl cation by the α-cyclopropyl group was calculated to be significantly less than that by the phenyl group. The theoretical rotational barrier of the α-cyclopropylvinyl cation is less than that of the cyclopropylethyl cation, presumably due to the stabilization of the intermediate perpendicular conformation by the overlap of the σ-bonds with the π-electrons of the C=C bond[72].

14. Substituted bicyclo[3.1.0]hexenyl cations

Several substituted bicyclo[3.1.0]hexenyl cations **42** and **43** were prepared by Olah and coworkers[73], and they all showed extensive charge delocalization into the cyclopropyl group as shown by the significantly deshielded ^{13}C absorptions for the methylene carbons (89–124 ppm).

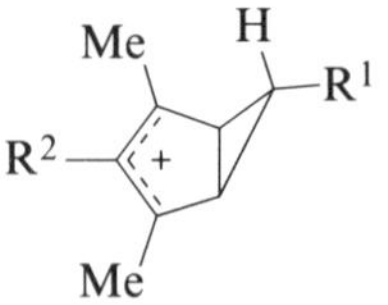

(42)

$R^1 = R^2 = H$
$R^1 = H, R^2 = Me$
$R^1 = R^2 = Me$

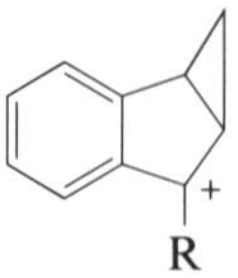

(43)

R = H, Me, Et, *i*-Pr

15. Allyl substituted cyclopropylcarbinyl cations

The dienylic cations **44** with cyclopropyl and phenyl groups were also prepared and characterized by the protonation of respective fulvenes[74]. Other cyclopropyl substituted allyl cations include acyclic 1,3- and 1,4-disubstituted allyl cations **45** and **46**[75]. The charge in these cations is localized mainly on the carbon adjacent to the cyclopropyl group. The rearrangement of these cations at higher temperatures was also studied[76].

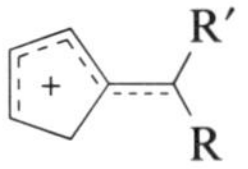

(44)

R = cyclopropyl, R′ = Ph
R = R′ = cyclopropyl

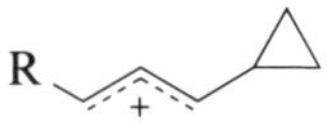

(45)

R = Me, cyclopropyl

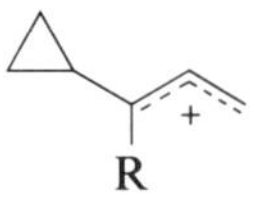

(46)

R = Me, Ph, cyclopropyl

Recently, in attempts to prepare hindered allyl cations, 2-adamantylidene-1,1-dicyclopropylethyl cation **47** has been prepared by the ionization of the corresponding 1,3-diol with FSO_3H or SbF_5–FSO_3H (equation 35)[77]. The ^{13}C spectrum of the cation at –80 °C

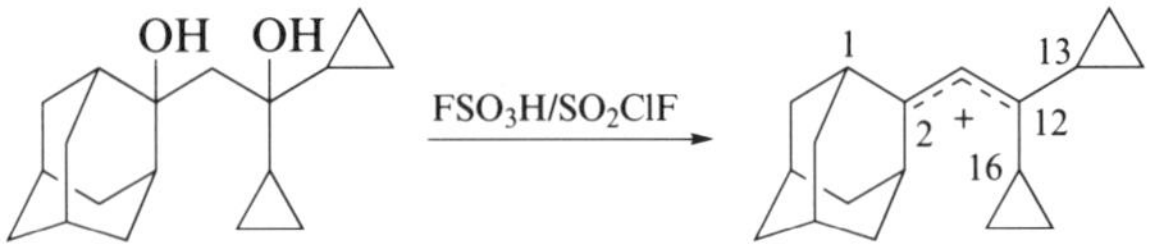

(47) (35)

shows absorptions at $\delta^{13}C$ 207.9 and 244.3, respectively, for C2 and C12. The cyclopropyl methine carbons also show a single absorption at $\delta^{13}C$ 31.6, and so are the absorptions for the methylene carbons $\delta^{13}C$ 29.6. Upon cooling to –90 °C, the methine ^{13}C absorptions at 31.6 ppm collapse into the base line, based on which a low barrier to rotation across the C11—C13 bond (<5 kcal mol^{-1}) was estimated. The low barrier could be explained as due to the weakening of the C11—C12 bond by the cyclopropyl group participation. The allyl cation thus exists as an unsymmetrically delocalized species.

B. Equilibrating (Degenerate) Cations

There are several examples of secondary and tertiary cyclopropylcarbinyl cations where equilibria are degenerate; i.e., the structures that are equilibrating are identical in the absence of deuterium or any other labels. The degenerate equilibria can be broken at lower temperatures, where the ^{13}C NMR spectra show typical unsymmetrical structures. Discussed below are some of the examples of the degenerate equilbria.

1. 1-(1-Methylcyclopropyl)ethyl cation and its β-methyl derivatives

The 1-(1-methylcyclopropyl)ethyl cation **48** undergoes degenerate rearrangement at –80 °C, whereas it is a static classical ion at –95 °C[78]. At –80 °C, the ^{13}C NMR absorptions for the two methyl groups (23.8 and 16.2 ppm at –95 °C) merge into a single peak at 20.8 ppm, from which a barrier for degenerate rearrangement of about 9 kcal mol^{-1} was estimated. It was suggested that the equilibration may proceed through the 1,2-dimethylcyclobutyl cation **49** involving a 1,2-hydride transfer. Chandrashekhar and Schleyer's calculations at the STO-3G level gave a barrier of 7.9 kcal mol^{-1} for the latter hydride migration[79]. The cation **48** irreversibly rearranges to the 1,1,3-trimethylallyl cation **50** above –35 °C (equation 36).

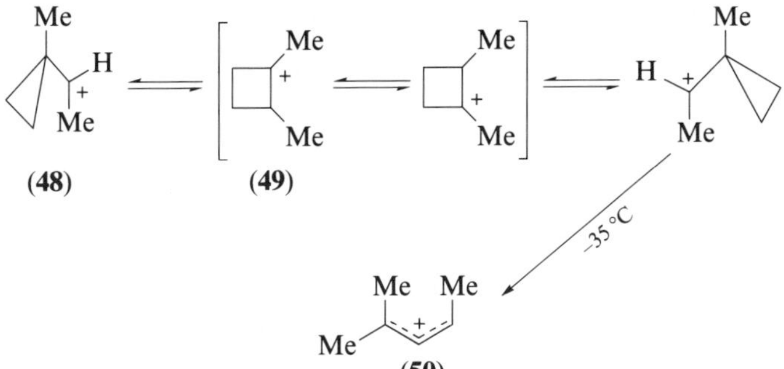

(48) (49) (50)

(36)

1-(*cis*-1,2-Dimethylcyclopropyl)ethyl and 1-(*cis*-2,3-dimethylcyclopropyl)ethyl cations, **51** and **53**, prepared from their corresponding alcohols with antimony pentafluoride (equations 37 and 38), undergo degenerate equilibria even at temperatures as low as –135 °C, indicating that the barriers for their rearrangements are probably less than 5 kcal mol^{-1} [80].

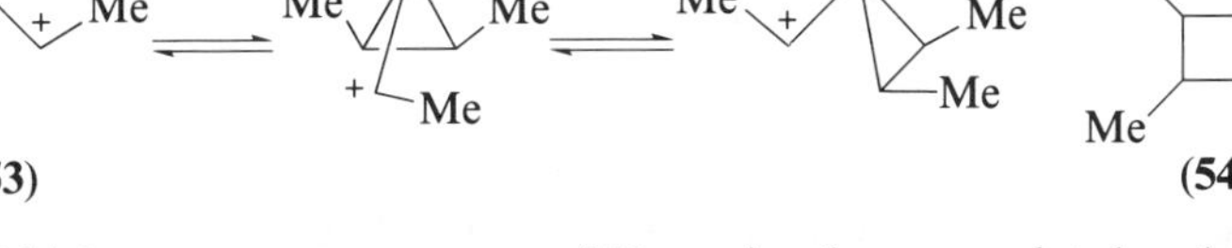

(51) (52)

(37)

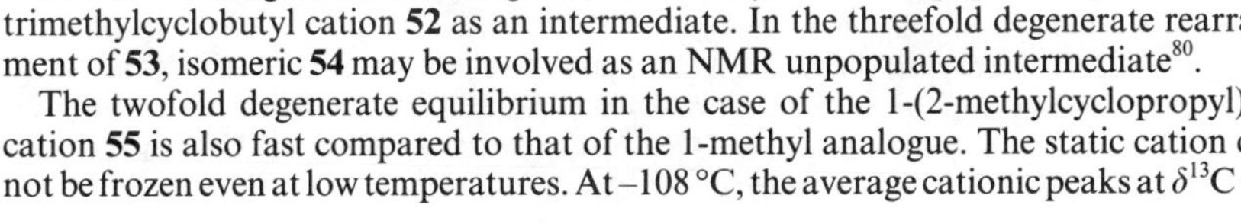

(53) (54)

(38)

The twofold degenerate rearrangement of **51** may involve unpopulated puckered 1,2,4-trimethylcyclobutyl cation **52** as an intermediate. In the threefold degenerate rearrangement of **53**, isomeric **54** may be involved as an NMR unpopulated intermediate[80].

The twofold degenerate equilibrium in the case of the 1-(2-methylcyclopropyl)ethyl cation **55** is also fast compared to that of the 1-methyl analogue. The static cation could not be frozen even at low temperatures. At –108 °C, the average cationic peaks at δ^{13}C 167.4

merged into the base line, again giving indication of the barrier to the rearrangement to be less than 5 kcal mol^{-1} involving unpopulated **55a**[80] (equation 39).

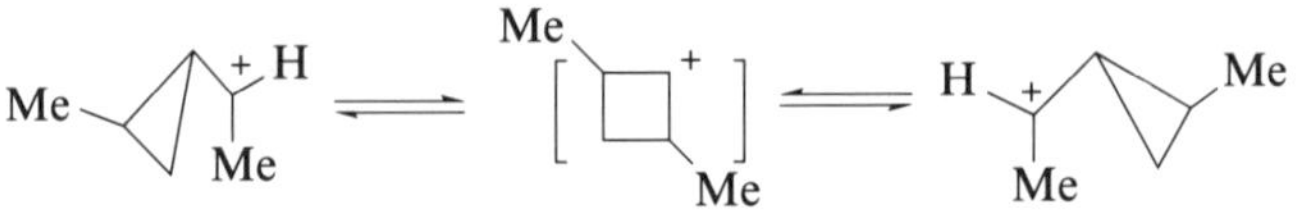

(39)

(55) **(55a)**

Timberlake and coworkers have studied the degenerate rearrangement of pentacyclopropylethyl cation **56** (involving 1,2-cyclopropyl shifts) under long-lived stable ion conditions[81,82] (equation 39a). The rearrangement could not be frozen even at –80 °C. However, additivity of ^{13}C NMR chemical shift analysis[7] indicates the classical trivalent nature of the carbocation.

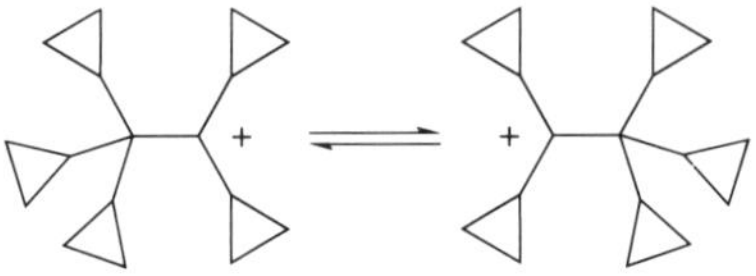

(39a)

(56)

2. 2-Bicyclo[n.1.0]alkyl cations

Attempts have been made to generate a number of 2-bicyclo[*n*.1.0]alkyl cations under stable ion conditions[55]. Ionization of bicyclo[3.1.0]hexan-2-ol **57** in SbF_5/SO_2ClF even at –140 °C gave the rearranged cyclohexenyl cation **59**, presumably through the bicyclo[3.1.0]hex-2-yl cation **60** (equation 40). However, ionization of bicyclo[4.1.0]heptan-2-

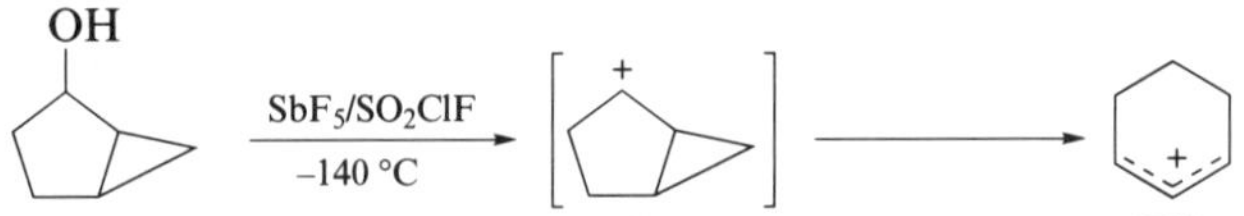

(40)

(57) **(60)** **(59)**

ol **58** in SbF_5/SO_2ClF at –100 °C gave the bicyclo[4.1.0]hept-2-yl cation **61**, which showed seven signals in the ^{13}C NMR at δ^{13}C 238.9, 109.6, 76.5, 55.4, 38.1, 25.8 and 23.5, indicating it to be a static secondary cyclopropylcarbinyl cation. As the solution temperature was raised to –50 °C, signals due to C2, C6 and C3, C5 averaged out at 179.8 and 32.3 ppm, respectively, indicating that the cation is undergoing degenerate equilibration at higher temperatures presumably through the unpopulated bicyclo[3.1.1]hept-6-yl cation **62**. Above –50 °C, the cation irreversibly rearranged to the cycloheptenyl cation **63**[57] (equation 41).

The 1-methylbicyclo[4.1.0]hept-2-yl cation **64** obtained from the corresponding alcohol is degenerate even at –140 °C. The ion is stable only below –80 °C and above it rearranges to the 3-methylcycloheptenyl cation **65**. An activation energy barrier of less than 6.0 kcal mol^{-1} is predicted for the degenerate cyclopropylcarbinyl rearrangement which may proceed through 6-methylbicyclo[3.1.1]hept-6-yl cation **66**. Attempted preparation of bicyclo[5.1.0]oct-2-yl and bicyclo[6.1.0]non-2-yl cations **67** and **68** led to degenerate bicyclo[3.3.0]oct-1-yl and 1-cyclopropyl-1-cyclohexyl cations **69** and **22**, respectively, through extensive rearrangement processes[55] (equation 42).

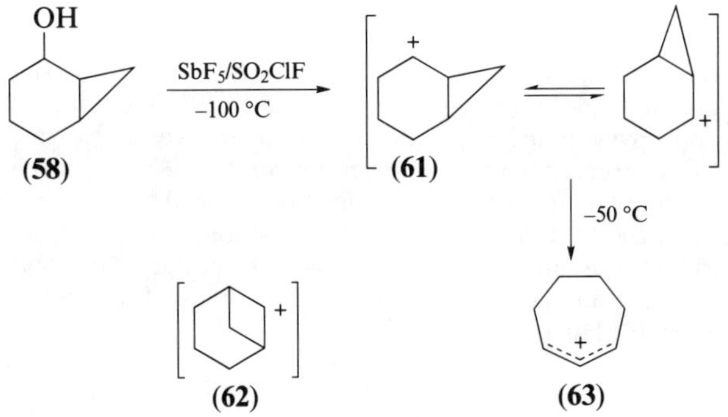

(41)

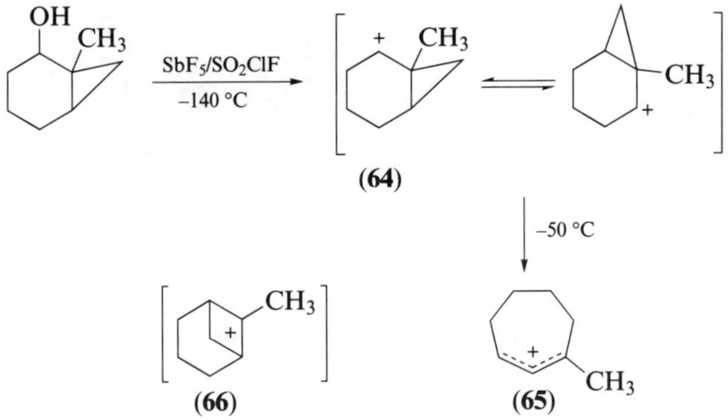

(42)

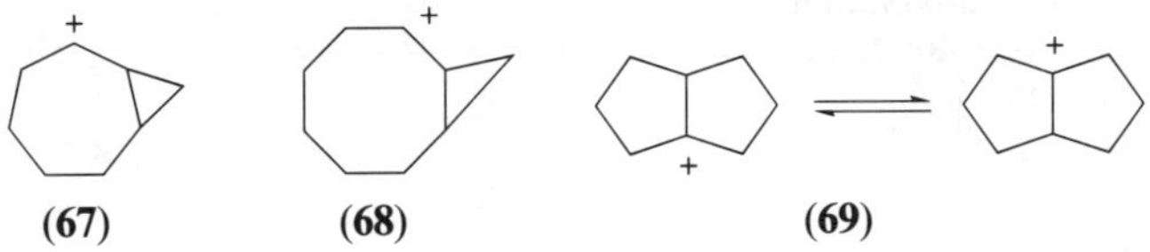

3. 3-Homonortricyclyl cation

The 3-homonortricyclyl cation **70** was prepared by the isomerization of bicyclo[3.2.1]oct-3-en-2-yl cation **71** at 20 °C in SbF_5/SO_2ClF solution[83]. The ion shows a threefold degenerate rearrangement between –85 °C to 20 °C. At 20 °C the C4, C6, C8 and C1, C3, C7 carbons become equivalent with an average of 36.19 ppm and 135.8 ppm, respectively (equation 43). Below –80 °C the cation is a static secondary cyclopropylcarbinyl cation with the cationic center chemical shift at $\delta^{13}C$ 234.1.

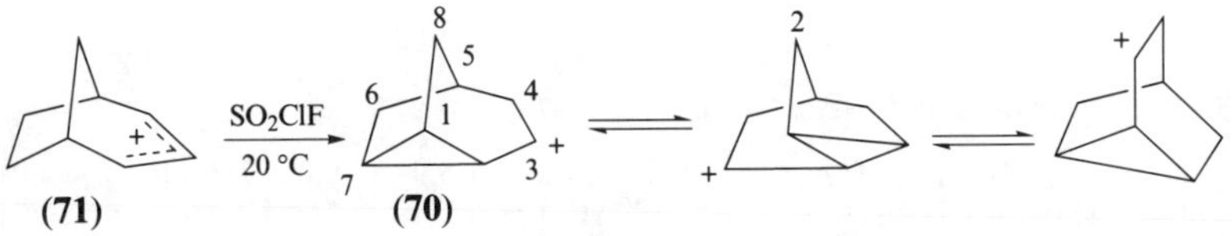

(43)

4. 8,9-Dehydro-2-adamantyl cation

Baldwin and Foglesong observed unusually fast rates for the solvolysis reactions of 8,9-dehydro-2-adamantyl 3,5-dinitrobenzoate. It is nearly 10^6 times more reactive than the analogous cyclopropyl systems, and about 10^8 times more reactive than the 2-adamantyl esters[84]. It was also observed that the 2-deuteriated and the 2-tritiated analogues undergo complete label scrambling to the C8 and C9 positions. The observations were rationalized by proposing a threefold degenerate equilibrium among partially charged delocalized cyclopropylcarbinyl type cations **72** (equation 44). The bridged bicyclobutonium ion-like structure **73** was thought to be an intermediate for these equilibrations, which was further confirmed by extended Huckel MO calculations[85].

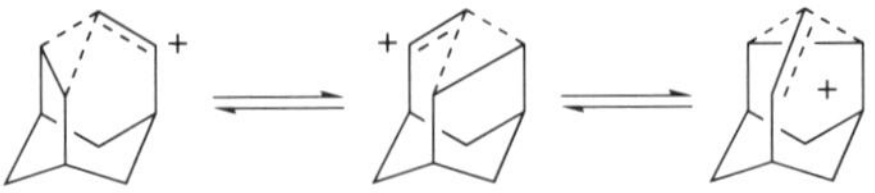

(44)

(72)

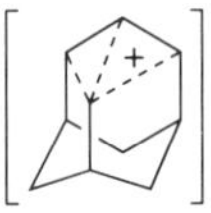

(73)

Olah and coworkers' ^{13}C NMR study[56,86] of the 8,9-dehydro-2-adamantyl cation, obtained from the corresponding alcohol in superacidic medium, showed the equivalence of C2, C8 and C9 carbons (δ^{13}C 157.0) in agreement with the conclusions of solvolytic studies[84,85]. Even at –120 °C the structure of the cation could not be frozen out to a static cation, showing the extremely fast equilibration of the threefold degenerate cyclopropylcarbinyl cation **74**. An identical NMR spectrum was obtained from the ionization of the 2,5-dehydro-4-protoadamantanol, which prompted the suggestion of the intermediacy of the 2,5-dehydro-4-protoadamantyl cation **75**. The ion rearranges to an allylic cation **76** at –78 °C[56,86] (equation 45).

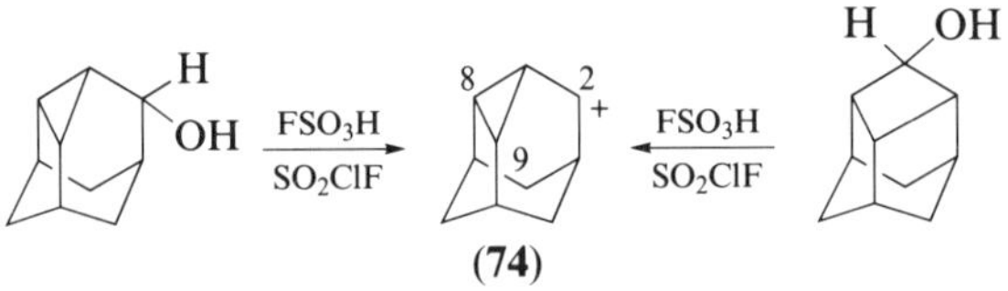

(74)

(45)

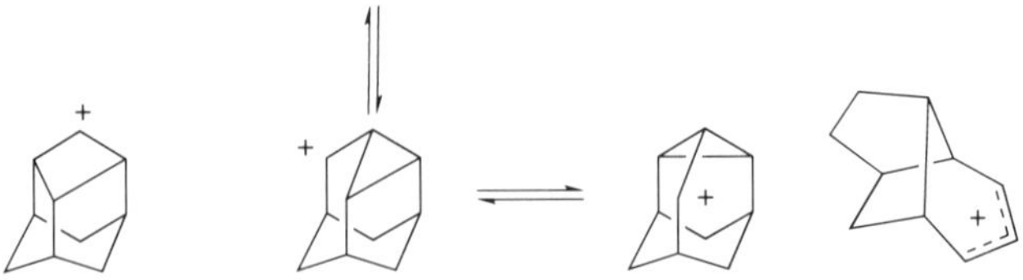

(75) **(76)**

5. 2,8-Dimethyl-8,9-dehydro-2-adamantyl cation

Degenerate equilibration of the tertiary 2,8-dimethyl-8,9-dehydro-2-adamantyl cation **77** was observed between –26 °C and –128 °C. At –128 °C, the ^{13}C NMR spectrum showed a static cation (10 ^{13}C peaks) (equation 46). Raising the temperature of the solution resulted in the appearance of degenerate equilibrium (8 ^{13}C peaks). The methyl carbons, for example, show a single absorption, δ^{13}C 29.7, at –26 °C, whereas at –126 °C they appear as two distinct absorptions (δ^{13}C 32.8 and 26.5). Similarly, the C2 and C8 carbons also show a single peak, δ^{13}C 202.7 ppm, whereas at –126 °C two peaks were observed: δ^{13}C 253.9 and 151.0. From the dynamic NMR studies it was shown that the activation barrier for the degenerate equilibrium is 7.4 kcal mol^{-1} at –112 °C[87]. This activation barrier is similar to those observed for other 1-cyclopropylethyl cations. The 1-(*cis*-1,2-dimethylcyclopropyl)ethyl and 1-(2-methylcyclopropyl)ethyl cations **51** and **55** have an activation barrier for the degenerate rearrangement of less than 5 kcal mol^{-1}. The degenerate equilibrium in all these cases may proceed through the substituted cations as intermediates (*vide supra*)[80].

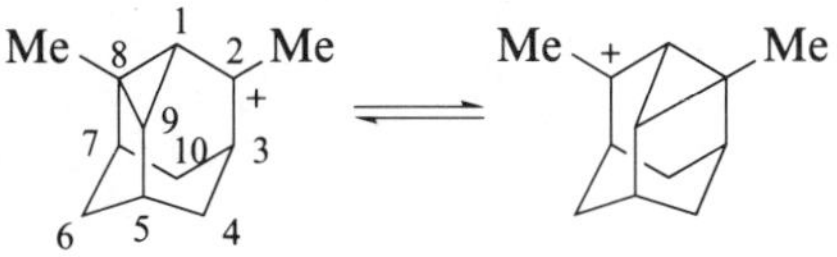

(77)

(46)

6. 4-Phenyl-2,5-dehydro-4-protoadamantyl cation

An interesting degenerate rearrangement is provided by the 4-phenyl-2,5-dehydro-4-protoadamantyl cation **78**[87]. The cation is a static classical cation at –130 °C (δ^{13}C 222.0). At –40 °C, however, it shows time-averaged NMR chemical shifts for (C2, C3, C5), (C1, C6, C8) and (C7, C9, C10) carbons. That is, the distinct absorptions for C2 and (C3, C5) disappear at –40 °C, and a new single absorption emerges. A rearrangement barrier of 6.9 kcal mol^{-1} was estimated from the coalescence of C2, C3 and C5 peaks at –110 °C. The rearrangement was postulated to proceed through the intermediacy of the 1-phenyl-8,9-dehydro-2-adamantyl cation **79** (equation 47).

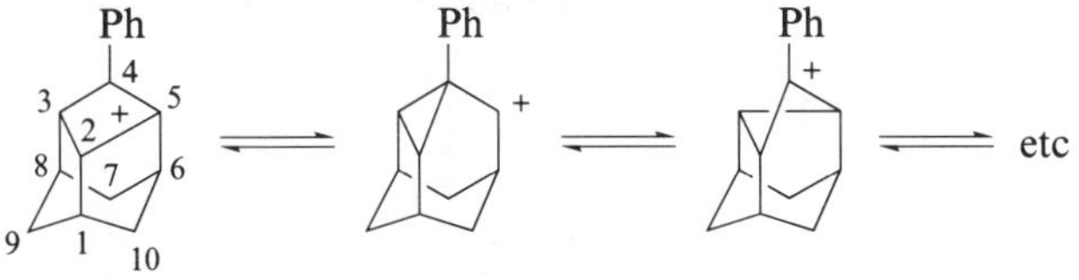

(78) **(79)**

(47)

7. 4-Methyl-2,5-dehydro-4-protoadamantyl cation

Carbocation **80** is obtained by the ionization of either 2-hydroxy-1-methyl-8,9-dehydroadamantane or 4-*endo*-hydroxy-4-*exo*-methyl-2,5-dehydroprotoadamantane in SbF_5/SO_2ClF solution at –120 °C[58]. The same ion was also obtained upon ionization of 4-*endo*-hydroxy-4-methyl-2,5-dehydroprotoadamant-4-ene. The isomeric 1-methyl-8,9-dehydro-2-adamantyl cation **81** was not formed under the reaction conditions. Ion **80** is stable up to –45 °C. The ^{13}C NMR spectrum of the ion **80** shows only five peaks at δ^{13}C 187 (C4), 83.3 (C2, C3, C5), 50.5 (C1, C6, C8), 42.1 (C7, C9, C10) and 26.2 (Me), indicating that the ion undergoes threefold degenerate rearrangement, fast on the NMR time scale even at –120 °C involving most probably the unpopulated ion **81** (equation 48).

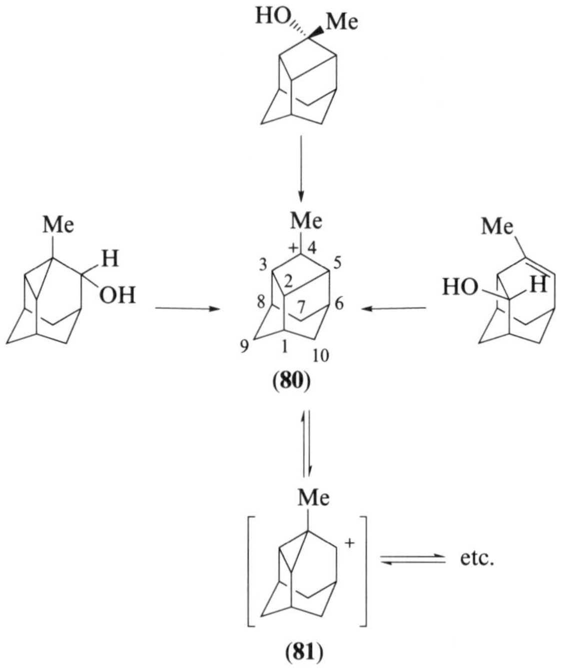

(48)

8. 2,4-Dehydro-5-homoadamantyl cation

The secondary ion **82** shows degenerate equilibration at –45 °C (8-line ^{13}C NMR spectrum), but is static at –110 °C. It was shown that the rearrangement takes place through the participation of the C3—C4 bond rather than the C2—C3 bond, and thus through a cyclopropylcarbinyl–cyclobutyl interconversion[57] (equation 49).

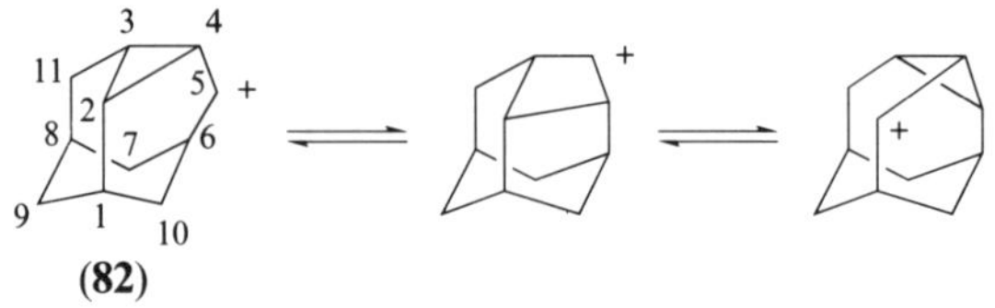

(49)

9. 9-Barbaralyl cation

Schleyer and coworkers investigated the nature of the 9-barbaralyl cation by solvolysis studies of the corresponding tosylate[88]. The acetolysis of 9-deuterio-9-barbaralyl tosylate gave 9-barbaralyl acetate with complete deuterium scrambling, whereas solvolysis in more nucleophilic aqueous acetone gave the product alcohols with the label exchanged to only positions C3, C7 and C9 (equation 50). Solvolysis of 4-deuteriobicyclo[3.2.2]nona-2,6,8-trien-4-yl 3,5-dinitrobenzoate also resulted in the formation of the 9-barbaralols with deuterium scrambled to positions 1, 2, 8, 4, 5 and 6, with only trace amounts at C3 and C7[89] (equation 51).

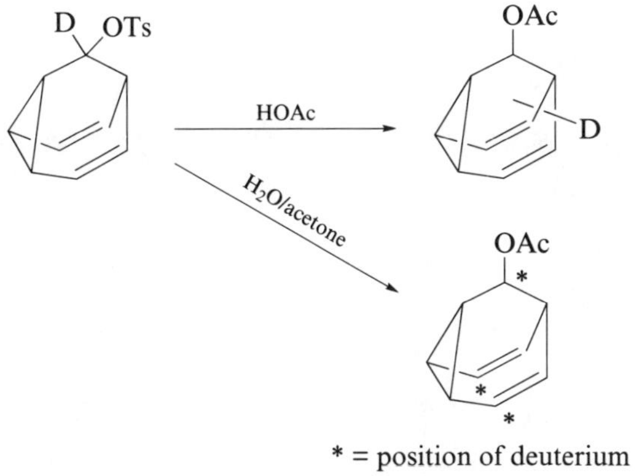

(50)

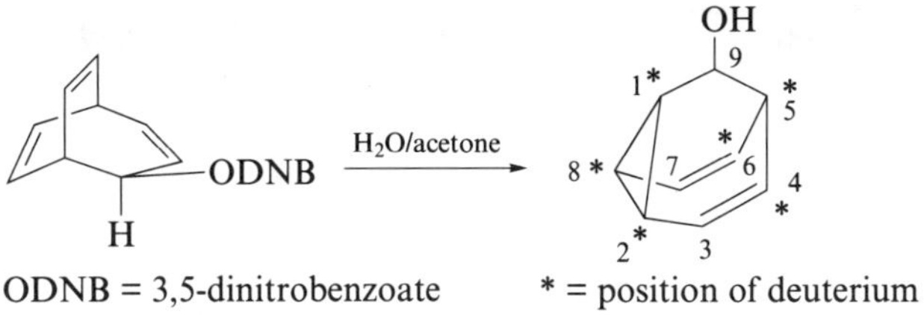

(51)

Ahlberg and coworkers[90] have prepared the ion **83** ($C_9H_9^+$) from bicyclo[3.2.2]nona-3,6,8-trien-2-ol and fluorosulfuric acid at –135 °C (equation 52). The cation showed only

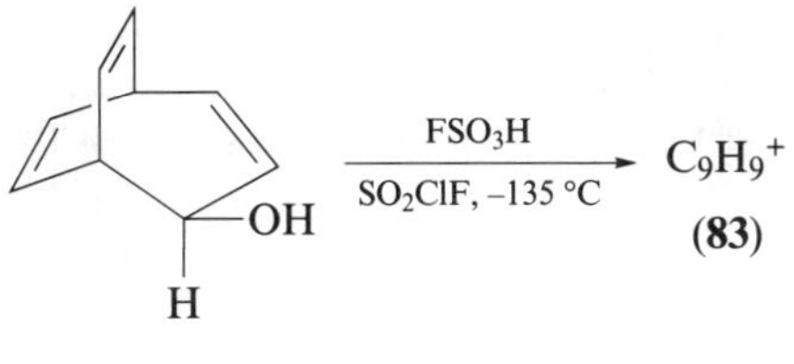

(52)

one signal as a sharp singlet in the proton NMR at δ^1H 6.59 and there were no detectable signals in the ^{13}C NMR spectrum. A mono ^{13}C labeled cation, prepared from the C3—^{13}C labeled bicyclo[3.2.2]nona-3,6,8-trien-2-ol, however, showed a single bond signal at δ^{13}C 118.5. At –150 °C the signal was split into two peaks at δ^{13}C 101 and 152 with relative intensities of 6:3. The equivalence of all carbons and protons in the cation at a temperature as low as –130 °C suggested that the cation undergoes a sixfold partially degenerate equilibration involving a divinylcyclopropyl carbinyl–divinylcyclopropylcarbinyl cation rearrangement, with an activation barrier of 5 kcal mol^{-1}. On the basis of deuterium isotopic perturbation studies of the ^{13}C and ^{2}H labeled cation, asymmetrically charge delocalized species **83b** of D_{3h} symmetry was excluded for the observed cation[91] (equation 53).

Extensive theoretical studies have been carried out in order to rationalize the rearrangement pathways in the cationic system[92]. The characterization of the potential surface of $C_9H_9^+$ cations[93] at MP2-, MP3- and MP4(SDQ)/6-31G* levels showed that the open 9-barbaralyl cation **83a** is more stable than the completely charge delocalized (nonclassical) structure of D_{3h} symmetry, and the bicyclo[3.2.2]nona-3,6,8-trien-2-yl cation **83b** by 6.9

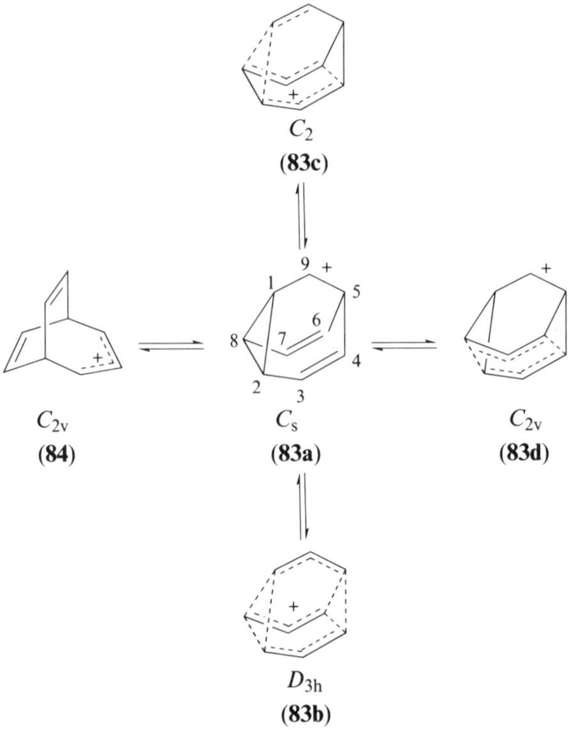

(53)

and 4.6 kcal mol^{-1}, respectively. The transition state for the sixfold degenerate rearrangement (equation 54) was found to be of C_2 symmetry **83c** with an activation energy of 3.6 kcal mol^{-1}. Other rearrangements involving higher energy structures were also located on the potential energy surface.

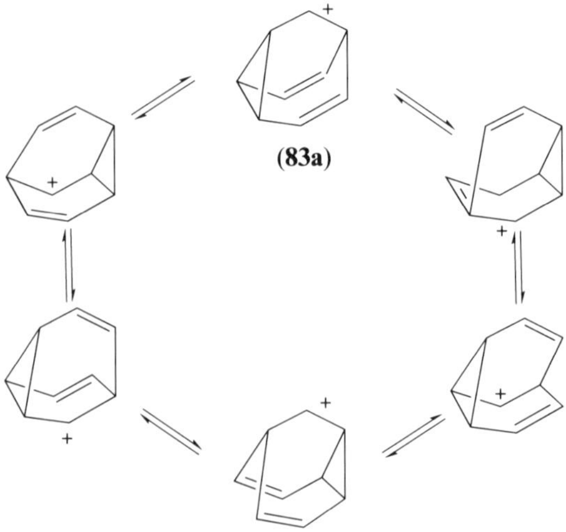

(54)

The potential energy surface in the vicinity of the $C_9H_9^+$ cation is rather flat and characterized by reaction paths connecting 181,440 different forms of **83a**, 90,720 different forms of **84** and 30,240 forms of **83b**!

The NMR chemical shifts were also calculated using the IGLO method on the MP2/6-31G* optimized geometries of **83a**, **83b** and **84**. Unfortunately no resolved ^{13}C NMR data for the cation are yet known experimentally at the lowest accessible temperatures for solution NMR (–150 °C). The IGLO calculated chemical shifts for 9-barbaralyl cation **83a** (C_s) are: δ^{13}C 82.7 (C1), 91.4 (C2, C8), 125.2 (C3, C7), 140.3 (C4, C6), 51.1 (C5) and 249.2 (C9). The bond orders at MP2/6-31G* are also interesting in that they are only 0.54 Å for C1—C8 (C1—C2) and 0.84 Å for C4—C5 (C5—C6), which further reflects on the extremely low barriers for the degenerate rearrangements of the cations.

The 9-barbaralyl cation **83a** is stable only at or below –135 °C, and on warming to –125 °C it is irreversibly converted into 1,4-dihomotropylium cation **85** (equation 54). The

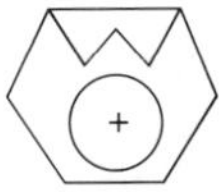

(85)

latter cation is calculated to be 8.3 kcal mol^{-1} more stable than the 9-barbaralyl cation at MP2/6-31G* [94]. The homoaromatic stabilization of 1,4-dihomotropylium cation was calculated to be rather small, only about 3 kcal mol^{-1}. The IGLO calculated NMR chemical shifts and the calculated bond orders are typical of a 6 π-electron aromatic system.

10. Bicyclo[3.1.0]hexenyl cation

Berson and coworkers[95] observed that the parent bicyclo[3.1.0]hexenyl cation **86** undergoes circumambulatory migration with an activation barrier of 15.1 kcal mol^{-1} (equation 55).

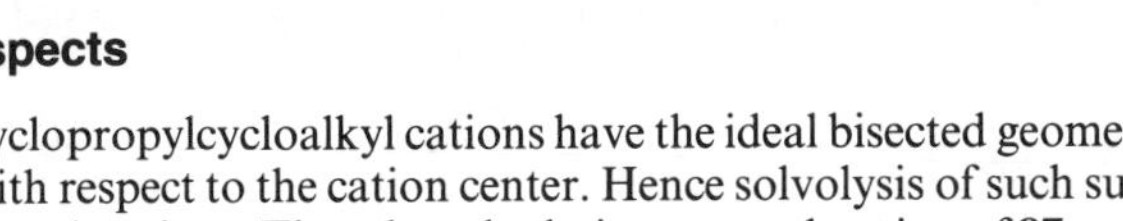

(86) (55)

V. SPIROCYCLOPROPYLCARBINYL CATIONS

A. General Aspects

The α-spirocyclopropylcycloalkyl cations have the ideal bisected geometry of the cyclopropyl group with respect to the cation center. Hence solvolysis of such substrates reveals enormous rate accelerations. Thus the solvolytic rate acceleration of **87**, as compared with the analogous α,α-dimethyl system **88**, is in the range of 10^5 [96]. Even the tertiary substrate **89** undergoes enhanced solvolytic rates as compared with its α,α-dimethyl analogue **90**.

(87) **(88)**

Relative Rate: 1.9×10^5 : 1

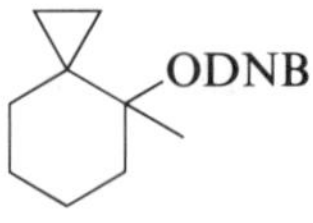

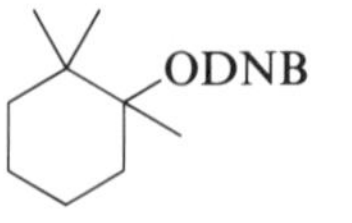

(89) (90)

Relative Rate: 2.3×10^4 : 1

The cyclopropyl group, in addition to the conjugative effect, also shows an electron-withdrawing inductive effect (*–I*), although much smaller in magnitude than the methylene group. The spiro[cyclopropane-1,2′-adamantyl]tosylate **91**, in which case the cyclopropyl group is enforced in the perpendicular conformation with respect to the developing cationic center, undergoes solvolysis 1000 times slower than the 1-adamantyl tosylate **92**[97].

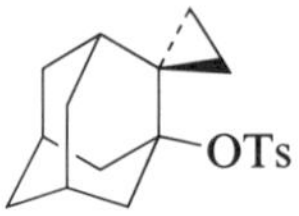

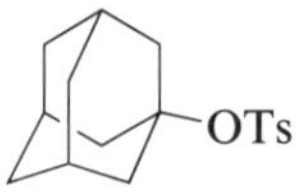

(91) (92)

Relative Rate: 10^3 : 1

B. 3-Spirocyclopropylbicyclo[2.2.1] and Related Cations

The participation of the α-spirocyclopropyl group results in significant rate enhancements even in *endo*-2-norbornyl systems. Whereas the solvolysis of *endo*-2-norbornyl *p*-nitrobenzoate is about 350 times slower than that of the *exo*-derivatives, the relative reactivities of *exo*- and *endo*-3-spirocyclopropyl-2-norbornyl derivatives are virtually identical[98]. Thus, the usual σ-participation in the 2-norbornyl cation is overwhelmed by the spirocyclopropyl group participation. This further indicates that factors other than steric effects are responsible for the solvolysis rates of secondary 2-norbornyl derivatives. The 3-spirocyclopropyl group causes a rate enhancement of about 10^4 even in the tertiary 2-methyl-2-norbornyl substrates **93**. Schleyer and coworkers[99] similarly found an *exo*/*endo* rate ratio of 12 for the solvolysis of spiro[cyclopropane-3′-benzonorbornen]-2′-yl *p*-nitrobenzoates **94**. The corresponding tertiary system **95**, however, retained high *exo*/*endo* rate ratios, probably because of increased steric effects in this tertiary system[100]. The [3-spirocyclopropyl]-2-norbornyl cation **96** spontaneously rearranges to the allylic, 2-methylbicyclo[3.2.1]oct-3-en-2-yl cation **96a** under long-lived stable ion conditions (equation 56).

(93)

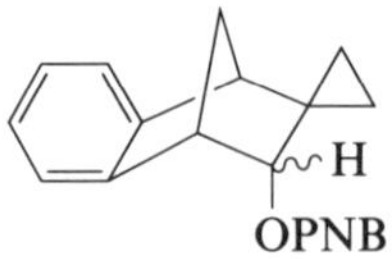

(94)

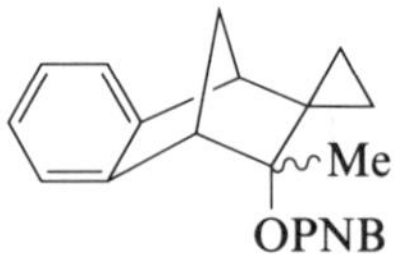

(95)

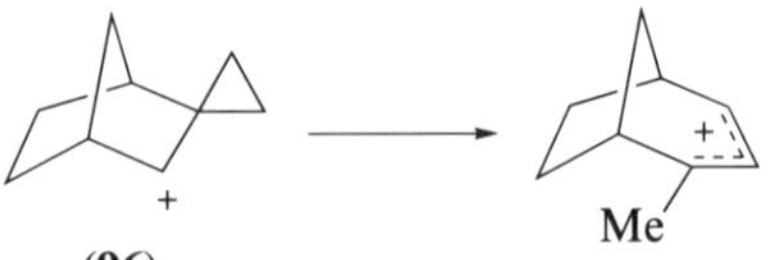

(96) (96a) (56)

It was, therefore, not possible to observe the secondary 3-spirocyclopropyl-2-norbornyl cation **96**[101].

However, the corresponding tertiary analogues, 2-methyl-, 2-phenyl- and 2-cyclopropyl substituted cations **97**, have been prepared at low temperatures, and characterized by ^{13}C NMR spectroscopy (equation 57). The cations were obtained by ionization of their corresponding alcohols with SbF_5/FSO_3H in SO_2ClF[102].

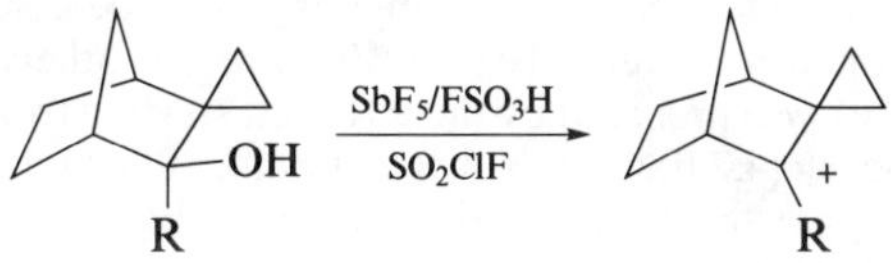

R = Me, Ph, cyclopropyl **(97)** (57)

The chemical shifts for these cations were also calculated by IGLO at DZ//STO-3G or DZ//3-21G levels, and are in good accord with the observed chemical shifts.

The 3-spirocyclopropyl-2-cyclopropyl-2-norbornyl cation **97**, R = $c\text{-}C_3H_5$ is stable even up to −20 °C, whereas the phenyl and methyl analogues rearranged to the allylic cations **98** at −70 °C and −90 °C, respectively (equation 58).

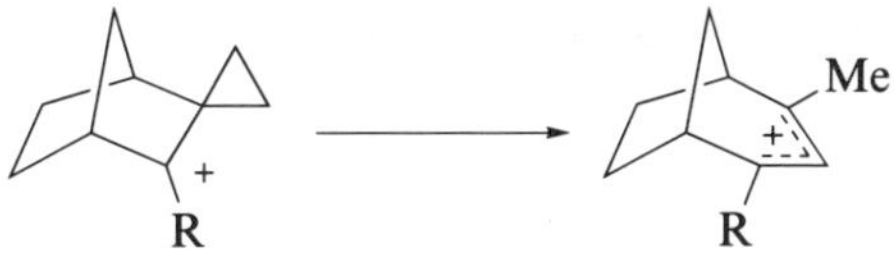

(97) **(98)** (58)

R = Me, Ph

The quaternary carbons of the tertiary cations **97** have chemical shifts of $\delta^{13}C$ 69.2, 61.1 and 58.8 for the Me, Ph and cyclopropyl substituted ions, respectively, which shows that progressively less charge is delocalized into the spirocyclopropyl ring going from methyl to phenyl to cyclopropyl. Consequently, the order of the stabilizing effects of the substituents on the cationic center is: cyclopropyl > Ph > Me.

Comparison of C1 and C6 ^{13}C chemical shifts showed that the σ-participation from the 2-norbornyl ring is significantly reduced in the 2-methyl analogue, whereas in the cyclopropyl and phenyl analogues it has essentially vanished. The STO-3G calculated structures show that the spirocyclopropyl participation is mainly from the *exo*-C—C bond. The ^{13}C NMR studies of these cations adequately accounted for the vanishingly low values of solvolytic k_{exo}/k_{endo} rate constants, and show that 3-spirocyclopropyl groups effectively compete with the C1—C6 σ-bond participation in the 2-norbornyl cation framework.

The next higher homologue, 1,3′-spirocyclopropylbicyclo[2.2.2]oct-2-yl cation **99**, could not be directly observed. Ionization of 3-spirocyclopropylbicylo[2.2.2]octan-2-ol in FSO_3H/SO_2ClF gave the rearranged allylic cation **100** derived from the transiently formed secondary cation **99** (equation 59). Just as in the secondary 2-norbornyl analogues, the

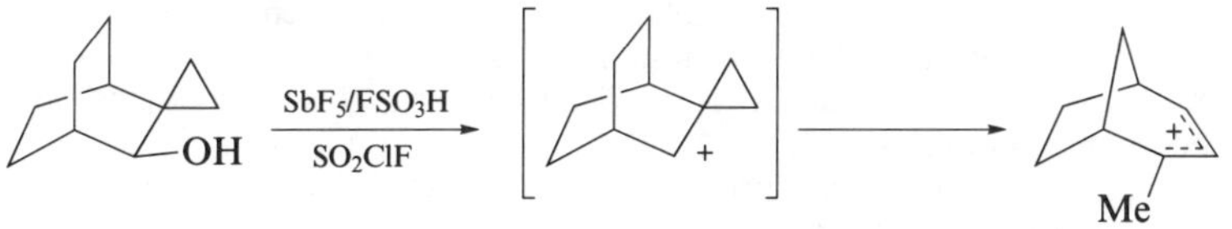

(99) **(100)** (59)

spirocyclopropyl group in this cation is in the favored bisected conformation resulting in its spontaneous rearrangement. The effect of the adjacent spirocyclopropyl group on the cationic centers can only be observed by NMR for the tertiary systems[103].

C. 2-Spirocyclopropyl Cyclohexyl Cation

Stabilization by the bisected spirocyclopropyl group was employed in the successful preparation of the secondary cyclohexyl cation **101**, which is otherwise unobtainable as a long-lived ion[101]. Ion **101** was prepared by three routes, starting from 2-spirocyclopropylcyclohexanol, *trans*-bicyclo[4.2.0]octan-1-ol or bicyclo[4.1.0]hept-1-yl methanol and SbF_5/SOClF at –78 °C. The bisected nature of the cyclopropyl group is indicated by a single ^{13}C absorption for the cyclopropyl methylene groups. The ion is stable up to –10 °C, where it rearranges to the bicyclo[3.3.0]oct-1-yl cation **69** (equation 60).

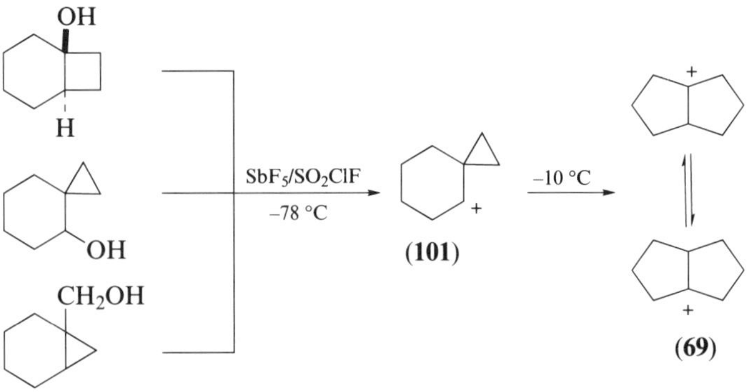

(60)

D. Phenonium Ions (Spirocyclopropylbenzonium Ions)

Phenonium ions, the symmetrically aryl bridged carbocations, were postulated as reaction intermediates as early as in 1949 by Cram[104]. His pioneering studies on the solvolytic study of the stereochemistry of *erythro*- and *threo*-3-phenyl-2-butyl tosylates provided unambiguous evidence for the intermediacy of phenonium ions. Acetolysis of *threo* tosylate gave an enantiomeric mixture of the *threo* acetates, whereas the acetolysis of *erythro* tosylate gave the diastereomeric *erythro* acetates. This retention of diastereomeric configuration can only be explained by the intermediacy of the phenonium ions. The phenonium ion from the *threo* tosylate is *meso* and hence is expected to give the observed racemic mixture, whereas the phenonium ion from the *erythro* tosylate is dissymetric and thus diastereomeric products are formed (see Scheme 1). The kinetic studies of the 2-phenylalkyl tosylates are much complicated by the involvement of both solvent assisted (k_s) and anchimerically assisted (k_Δ) pathways, whose magnitudes are solvent dependent. The k_Δ pathway predominates in low nucleophilicity solvents such as trifluoroacetic acid[105], while in the relatively more nucleophilic acetic acid the neighboring group participation is marginal. Brown's earlier postulates of equilibrating open carbocations based on the lack of significant rate enhancement in acetolysis studies turned out to be no longer valid[106]. The review by Lancelot, Cram and Schleyer covers this topic exhaustively and is not discussed here any further[104].

Phenonium ions, more appropriately called spirocyclopropylbenzenium or ethylenebenzenium ions, have been directly observed[107] in low nucleophilicity solvents in superacidic media, and well characterized as having symmetrical bridging. The ^{13}C NMR

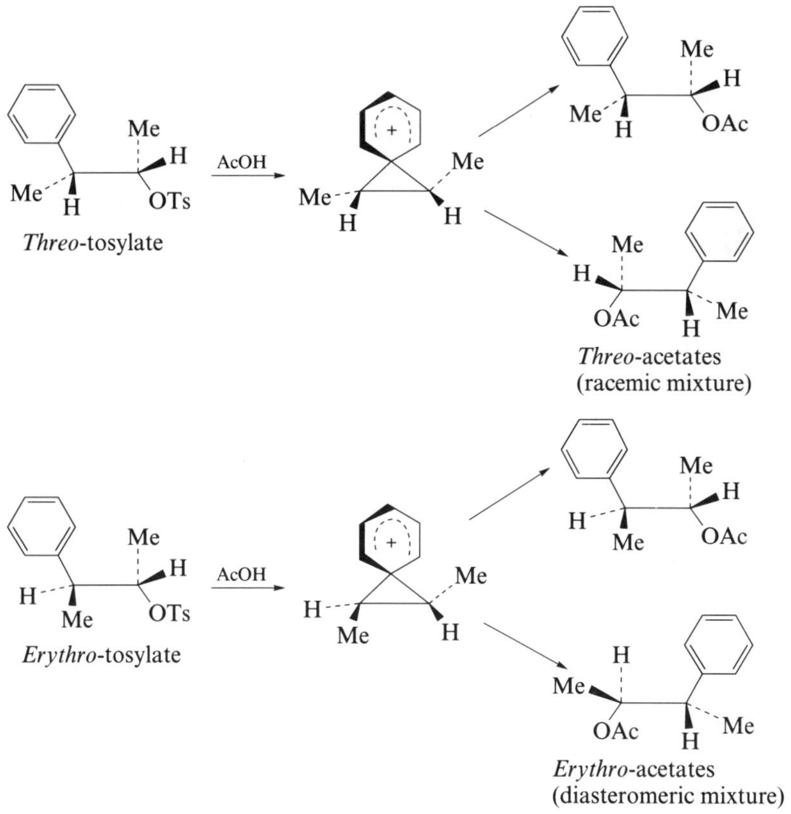

SCHEME 1

spectra of the cations show deshielded absorptions for the methylene carbons of the cyclopropyl ring, showing the cyclopropylcarbinyl cation character of these ions. The ^{13}C absorptions for the formal aromatic rings are also significantly deshielded, typical of the arenium ions.

The parent spirocyclopropylbenzenium ion **102** can be prepared by the ionization of β-phenylethyl chloride in HF–SbF_5–SO_2ClF at –90 °C followed by warming to –60 °C. At higher temperatures (–27 to –5 °C), the ion isomerizes to the α-methylbenzyl cation **103** with an activation energy of 13 kcal mol^{-1}, as shown by the ^1H NMR kinetic study (equation 61). The rearrangement probably proceeds through a partially delocalized primary 2-phenylethyl cation **104**[107b,c]. The ^{13}C NMR spectrum of the ethylenebenzenium ion **102** showed $\delta^{13}C$ 68.8 (C1), 171.8 (C2, C6), 133.4 (C3, C5), 155.4 (C4), 60.7 (CH_2). The equivalence of the C3, C5 and C2, C6 carbons as well as the methylene carbons show that the ion has C_{2v} symmetry. The deshielded absorptions for the CH_2 carbons are similar to those in other cyclopropylcarbinyl cations.

Collisional activation mass spectroscopic studies (CAMS) showed that, in the gas phase, the ethylenebenzenium ion is formed from the ionization of β-phenylethyl bromide or iodide (equation 62). However, ionization of the corresponding chloride gave instead protonated benzocyclobutene **105** (equation 63). The lifetime of ethylenebenzenium ion was estimated as about 10^{-5} s[108].

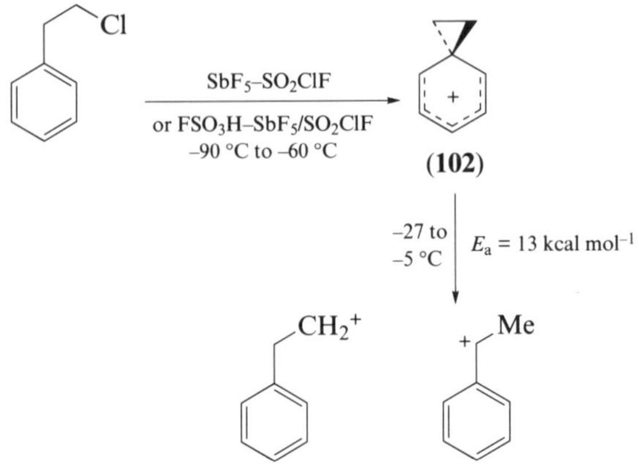

(104) (103)

(61)

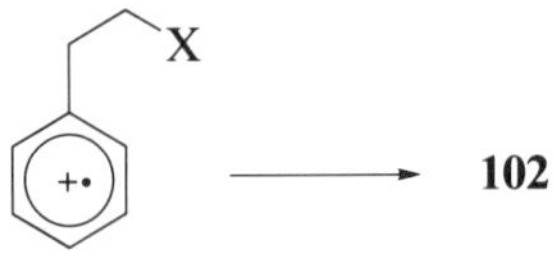

X = Br, I

(62)

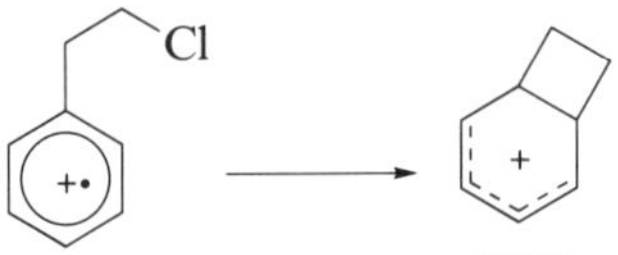

(105)

(63)

Hehre's STO-3G calculations[109a] showed, as expected, that the spirocyclopropylbenzenium ion **102** is more stable than the open β-phenylethyl cation. Recently, Sieber, Schleyer and Gauss calculated the structure of the spirocyclopropylbenzenium ion, and other alternate structures at the correlated *ab initio* levels (MP2/6-31G*)[109b]. The geometry of the cation at MP2/6-31G* shows the following bond lengths: C1—C2 (1.419 Å), C2—C3 (1.384 Å), C3—C4 (1.401 Å), C1—CH_2 (1.625 Å). The NMR chemical shift calculations by the IGLO method could not reproduce the experimental values, but GIAO-MP2 calculations, which involve correlation effects, gave well agreeable ^{13}C NMR chemical shifts: δ^{13}C 71.9 (C1), 168.1 (C2, C6), 140.8 (C3, C5), 153.6 (C4), 61.5 (CH_2). They have favored the pentacoordinated nonclassical phenonium ion **106** rather than the conventionally written ethylenebenzenium structure for this cation.

(106)

A reinvestigation of all available results by Olah, Prakash and coworkers[110] has shown that the postulated nonclassical phenonium structure is not supported by the available evidence. Experimental data, including the difference in the additivity of chemical shifts of the cation compared with the parent hydrocarbon precursor, showed the ion to be indeed a spirocyclopropylbenzenium ion **102** with a 4π-cyclohexadienyl system with simultaneous cyclopropylcarbinyl delocalization[110]. The ion has also been generated by the new route of protonation of benzocyclobutene in HF–SbF_5 (equation 64).

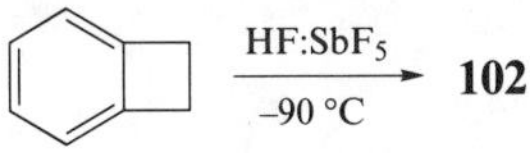

(64)

Olah and coworkers have prepared and characterized[111] several other ethylenebenzenium ions such as ethylene-2,4,6-trimethylbenzenium ion **107**, ethylene-4-methylbenzenium ion **108** and ethylene-4-methoxybenzenium ion **109** from the corresponding chlorides by ionization with SbF_5–SO_2ClF at −78 °C. The deshielding of the cyclopropyl methylene carbons gradually decreases in the order: *p*-OMe < *p*-Me < *p*-H, reflecting the decreased demand on the cyclopropyl ring for the stabilization of the cation on increasing the electron-donating ability of groups on the aromatic ring.

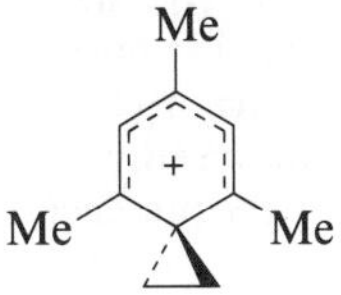

(107)

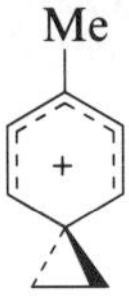

(108)

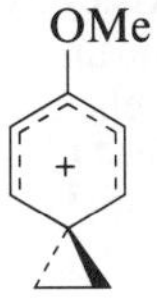

(109)

The existence of other substituted ethylenebenzenium ion such as (dimethylethylene)benzenium ions (**110** and **111**), (methylphenylethylene)benzenium ion **112** and (tetramethylethylene)benzenium ion **113** were initially based on ^{1}H NMR studies[112,113]. However, subsequent ^{13}C NMR studies indicated very little support for such bridged structures[107b].

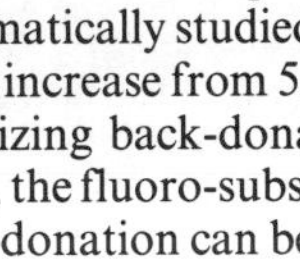

(110)

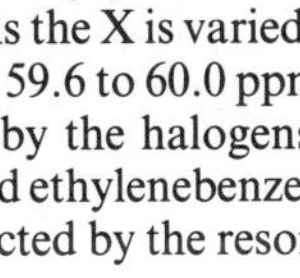

(111)

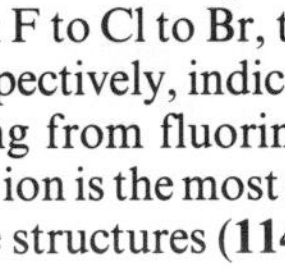

(112)

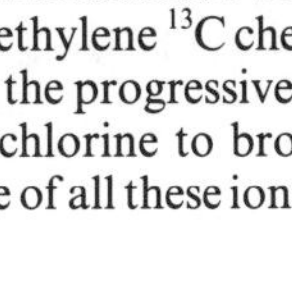

(113)

The effect of the *p*-halogens (X) on the stabilities of the ethylenebenzenium ions was also systematically studied[114]. As the X is varied from F to Cl to Br, the methylene ^{13}C chemical shifts increase from 55.8 to 59.6 to 60.0 ppm, respectively, indicating the progressively less stabilizing back-donation by the halogens going from fluorine to chlorine to bromine. Thus, the fluoro-substituted ethylenebenzenium ion is the most stable of all these ions. The back-donation can be depicted by the resonance structures (**114-F**).

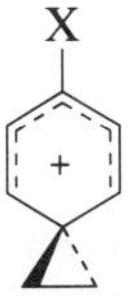

(114)
X = F, Cl, Br

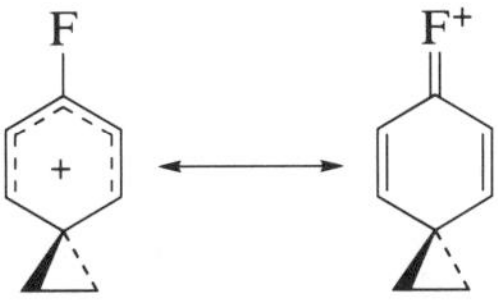

(114-F)

It was also possible to prepare stable ethylene-(4-X)-naphthalenium ions **115** (X= Me, Ph, OMe) and 9,9-ethylene-(10-X)-anthracenium ions **116** (X = H, Br) under long-lived stable ion conditions[114].

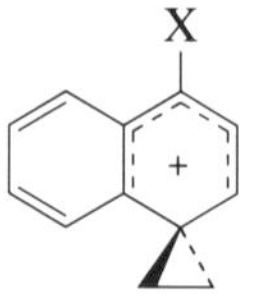

X = Me, Ph, OMe

(115)

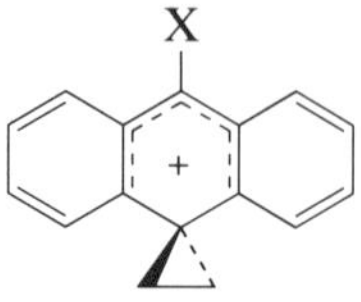

X = H, Br

(116)

E. Benzo-2-nortricyclyl Cations

The benzo-2-nortricyclyl cations **117** and **118**, prepared from the respective alcohols or benzo-2-norbornene under stable ion conditions, show characteristics of both nortricyclyl cations and ethylenebenzenium ions[115]. The spirocyclic carbon (C11) of the parent benzo-2-nortricyclyl cation **117** has δ^{13}C 84.1, similar to that of the ethylenebenzenium ion (68.8 ppm). The symmetrical nature of the ion is indicated by the presence of only 9 peaks in its ^{13}C NMR. Compared to the ethylenebenzenium ion the charge in the benzo-2-nortricyclyl cation is significantly delocalized into the aromatic ring. Substitution at C6 by methoxy, methyl and chloro groups (as in **118**) results in the progressively less deshielding of the cyclopropyl carbons in that order. That is, the methoxy group is the most stabilizing group, and hence little positive charge is delocalized into the cyclopropyl ring, resulting in its least deshielded ^{13}C absorptions. Compared to the unsubstituted cation **117**, the 6-chloro cation **118**, R = Cl shows relatively shielded ^{13}C absorptions for the cyclopropyl carbons (C1 and C2) indicating that the chloro group can stabilize the cationic center better than hydrogen through back-donation.

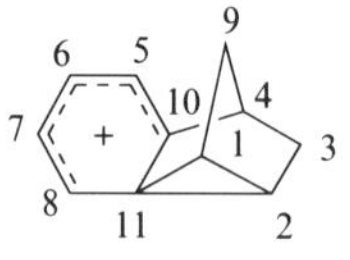

(117)

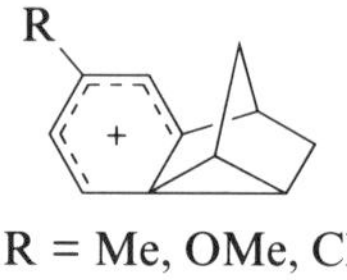

R = Me, OMe, Cl

(118)

C9-Halo benzonortricyclyl cations **119** were also prepared from *exo*-2-*anti*-7-dihalo-benzonorbornanes at –78 ° C (equation 65). The preferential ionization of the halogen at

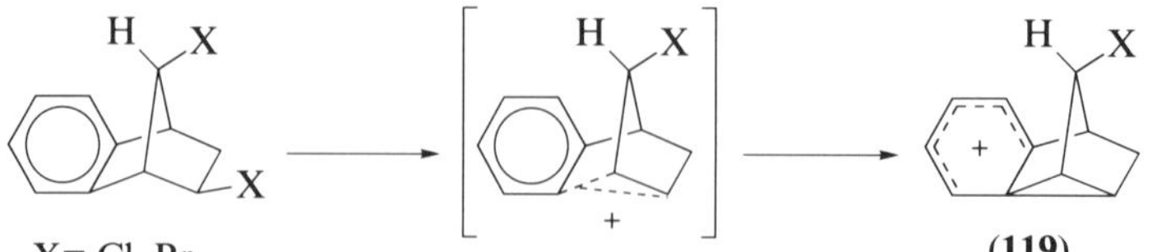

X= Cl, Br **(119)** (65)

the C2 rather than C9 carbons suggests that the initially formed ion is a nonclassical pentacoordinate ion, which rearranges to the observed benzonortricyclyl cation. The formation of these benzonortricyclyl ions, in spite of their inherent strain, shows significant pπ-pπ interaction between the aromatic ring and the electron-deficient center.

Olah, Prakash and coworkers[110] have prepared the tetrafluorobenzonortricyclyl cation **120** by the ionization the 2-fluoro(tetrafluoro)benzonorbornene in superacidic medium (equation 66).The highly deactivated aromatic ring still participates in the arenium ion formation.

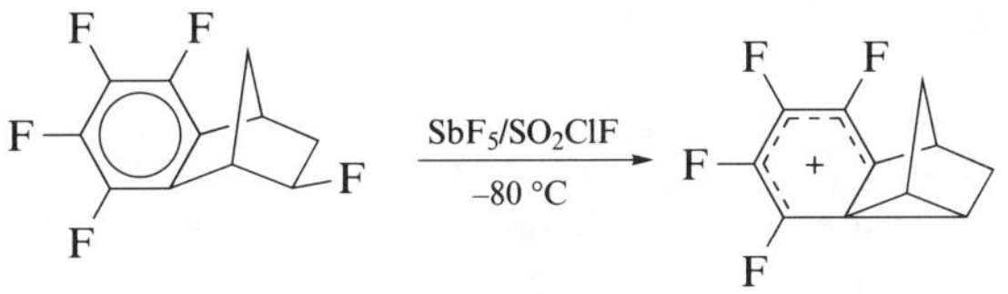

(120)

(66)

VI. CYCLOPROPYLCARBINYL DICATIONS

Carbodications, i.e. doubly positively charged carbocations, are considered relatively unstable owing to the charge–charge repulsion. If the cationic centers, however, are stabilized by phenyl or cyclopropyl groups or the two charges are separated by at least two carbon atoms, such carbodications could be prepared and studied. The charge is extensively delocalized into the substituents in these cations, with the result that in most cases the cationic center's ^{13}C NMR chemical shifts are even somewhat shielded compared to that of the similar monocations. There are only a few carbodications that are stabilized by cyclopropyl groups. The nature of 2,6-dicyclopropyl-2,6-adamantanediyl dication **121**[116], *trans*-cyclopropane-1,2-bis(diphenylmethylium)dication **122**[117] and 2,6-disubstituted *anti*-tricyclo[5.1.0.$0^{3,5}$]octan-2,6-diyl dications **123**[118] were discussed in detail in the earlier review and will be briefly mentioned here[11].

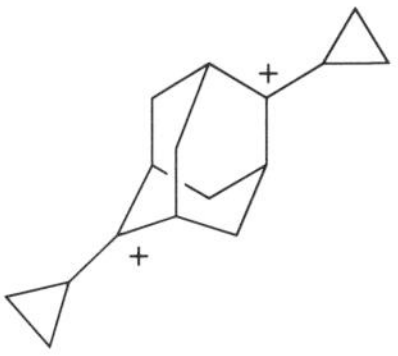

(121)

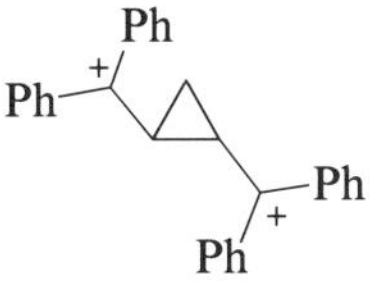

(122)

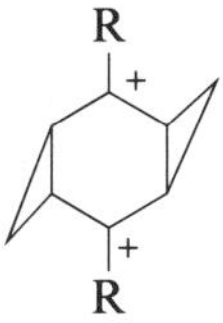

(123)
R = Me, Ph, cyclopropyl

2,6-Dicyclopropyl-2,6-adamantanediyl cation **121** has a δ^{13}C of 277.1 ppm, for the cationic center which appears to be much deshielded compared to the analogous 2,6-diphenyl dication, **123**, R = Ph (δ^{13}C 252.3. However, such deshielding is due to factors other than the relative charge stabilization, and is quite general for all cyclopropylcarbinyl cations. The dication's cationic centers are actually shielded by 17.2 ppm, compared to that of the monocarbocation, 2-cyclopropyl-2-adamantyl cation (δ^{13}C 294.2), showing that the positive charge is delocalized into the cyclopropane rings. The charge delocalization into the cyclopropane ring is reflected by the fact that the cyclopropyl methylenes in the dications (δ^{13}C 60.4) are relatively more deshielded than those of the monocation (δ^{13}C 48.9). The same trend is also observed in the 2,6-dicyclopropyl-*anti*-tricyclo[5.1.0.$0^{3,5}$]octan-2,6-diyl dication, whose cationic center (δ^{13}C 260.8) is deshielded compared to that of the corresponding 2,6-diphenyl dication (δ^{13}C 235.4) but shielded with respect to the related monocation, tricyclopropyl carbenium ion **16** (δ^{13}C 280.5). The secondary *anti*-tricyclo[5.1.0.$0^{3.5}$]octan-2,6-diyl dication **123**, R = H spontaneously rearranges to the

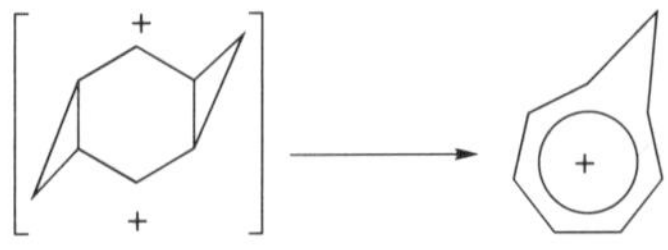

(123), R = H (124) (67)

homotropylium cation **124** (equation 67). The tertiary methyl, phenyl and cyclopropyl analogues **123** are stable dicationic species, which are static (nondegenerate).

Attempted preparation of *trans*-α,α′-tetramethylcyclopropane-1,2-dimethylium dication **125** resulted in the formation of ring-opened 2,6-dimethylhepta-2-dienyl cation **126** (equation 68)[117]. The tetraphenyl substituted dication **122**, however, is a stable species, and the charge is significantly delocalized into the aromatic rings (δ^{13}C 220.9) as compared to its monocation analogue, diphenylcyclopropylmethyl cation (δ^{13}C 234.4)[50a].

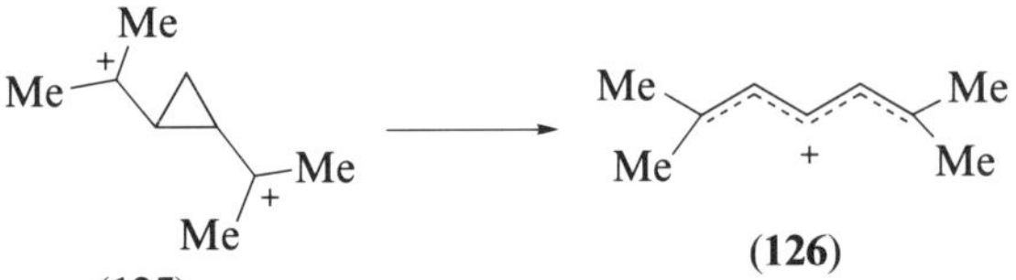

(125) (126) (68)

The tetracyclopropyl analogue, the *trans*-cyclopropane-1,2-diylbis(dicyclopropylmethylium)dication **127**, has also been prepared, and showed the following ^{13}C NMR

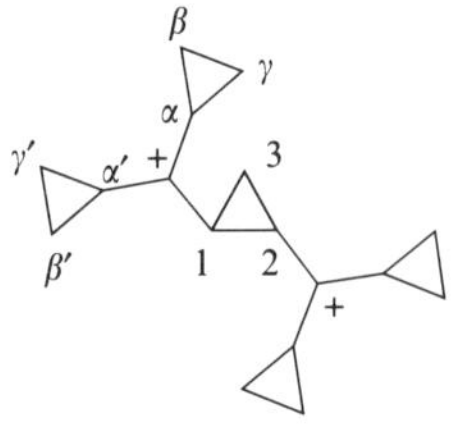

(127)

absorptions at δ^{13}C (−73 °C): 39.3 (C_α, $C_{\alpha'}$), 37.6, 37.3, 35.9, 34.9, 33.7, 32.5 (C3) and 264.1 (C^+)[119]. The ion shows reversible temperature-dependent behavior in its ^{13}C NMR spectrum. The absorptions from 37.6 to 33.7 ppm, presumably for the four distinct β-methylene carbons of the peripheral cyclopropanes and C1, C2 carbons, coalesced to give a single broad peak at δ^{13}C 37.3 at −50 °C, from which a barrier to rotation around the C^+—C_α or C^+—$C_{\alpha'}$ was estimated as 10–12 kcal mol^{-1}. The barrier to rotation around C^+—C1 or C^+—C2 is even smaller as both C_α and $C_{\alpha'}$ carbons show identical ^{13}C chemical shifts (δ^{13}C 39.3) at −73 °C. Olah and Schleyer's chemical shift additivity criterion showed a classical dicationic nature and the difference of the summation of the chemical shifts of all carbons in the cationic species is greater than in the related hydrocarbon by at least 893 ppm, which is about twice the value for a typical classical monocarbocation[119].

VII. EXTENT OF POSITIVE CHARGE DELOCALIZATION INTO THE CYCLOPROPYL GROUP

Solvolytic rate constants of tertiary cyclopropylcarbinyl *p*-nitrobenzoates **129** and **130** revealed that each cyclopropyl group causes a rate enhancement of the order of 10^2–10^3

over the isopropyl group (of **128** and **129**), which has a similar steric requirement as the cyclopropyl group[120].

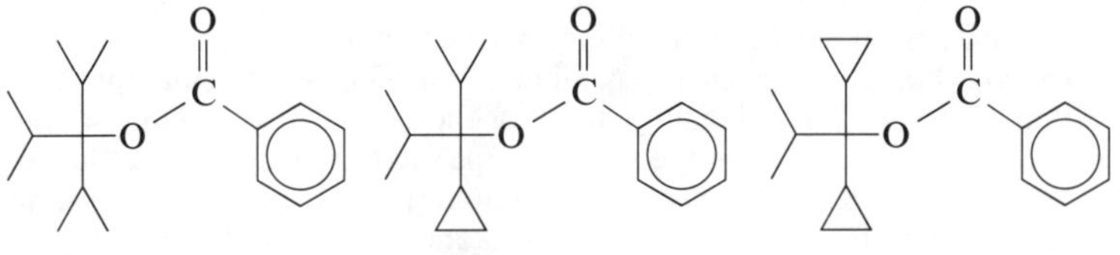

(**128**) (**129**) (**130**)

Relative Rates: 1.0 246 23,500

The cyclopropyl group, even far removed from the reaction center, as in *p*-cyclopropylphenyldimethylcarbinyl chloride **131**, can cause significant rate enhancement over the unsubstituted compound **132**[121].

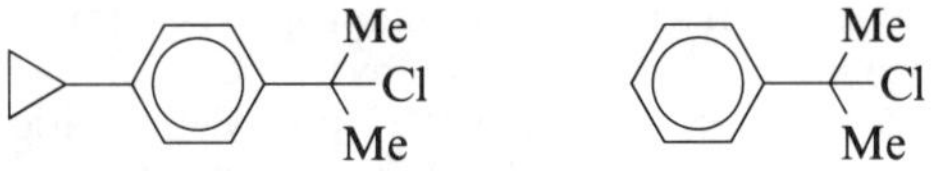

(**131**) (**132**)

Relative Rate: 154 1

These rate constant ratios provide indirect evidence for the large relative stabilizing ability of the cyclopropyl group compared to the other alkyl groups or to hydrogen (note, however, that they only indicate the relative energy differences of the carbocationic transition states vs the precursors). The cyclopropyl group is a better cation stabilizing group than phenyl as shown by product studies[50a] of the electrophilic addition reaction to the 1-cyclopropyl-2-phenyl ethylenes. Oxymercuration and hydrochlorination of isomeric olefins **133** and **134** gave the alcoholic and chlorinated products with OH and Cl groups on the carbons attached to the cyclopropyl groups (equation 69).

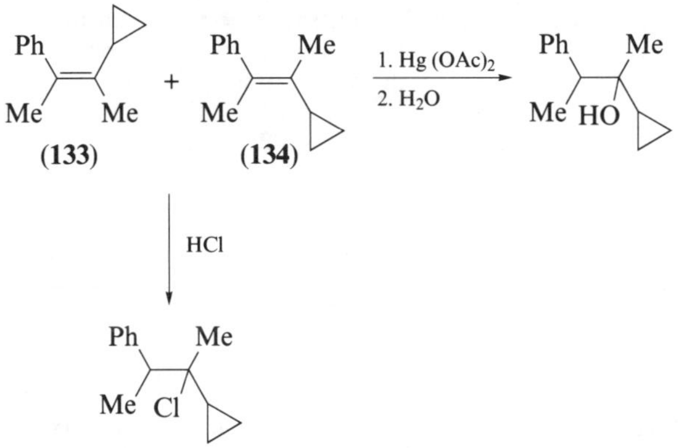

(69)

The product studies complement kinetic studies which show that the cyclopropyl group is a better electron-releasing group than the phenyl group when attached to electron-deficient cationic centers, although the relative magnitude of participation is not clear. A detailed study of the correlation of σ^+ with $\delta^{13}C$ (C^+) showed that phenyl and cyclopropyl groups are comparable in their stabilizing abilities[50,59,122]. The thermodynamics of carbocation formation from the precursor alcohols throws also much light on the relative stabilizing effects of the cyclopropyl and phenyl groups. Deno and coworkers[123] found that the pK_{R^+} of tricyclopropylcarbinyl cation (–2.31) is much higher than that of the triphenylmethyl cation (–6.63), i.e. the equilibrium constant for the formation of the tricyclopropylcarbinyl cation **16** in a given acid medium is much higher than that for the triphenylmethyl cation. The superior stabilizing ability of the cyclopropyl group over phenyl in the gas phase is also revealed from the relative appearance potentials of cyclopropyl and phenyl substituted cations. The appearance potentials (in kcal mol^{-1}) for the formation of a series of carbocations from the corresponding alkanes (equation 70) is as follows: R = H (0), R = Me (36), R = Ph (55), R = cyclopropyl (58)[124].

$$RCH_3 + e^- \longrightarrow RCH_2^+ + 2e^- + H^\bullet \qquad (70)$$

The higher the appearance potential, the greater the stability of the cation, and hence the relative order of the stabilizing abilities in the gas phase is: cyclopropyl > phenyl >> methyl.

The gas-phase heats of formation obtained from pulsed ion cyclotron resonance (ICR) spectroscopy showed that the tertiary 1-cyclopropyl-1-methylethyl cation (**20**) is more stable than the 1-phenyl-1-methylethyl cation by 0.8 kcal mol^{-1}, while the secondary 1-cyclopropylethyl cation (**18**) is less stable than the 1-phenylethyl cation by 4.8 kcal mol^{-1}[25]. Thus a substantial reversal of the stabilization of the phenyl over cyclopropyl groups is observed. The results were also rationalized by STO-3G calculations for the isodesmic reaction involving proton transfer (equation 71).

(71)

ΔG^0(STO-3G) = –2.2 kcal mol^{-1} (R = Me); 3.6. kcal mol^{-1} (R = H).

Quantitative estimates of the thermodynamic stabilities of various phenyl and cyclopropyl substituted cyclopropenium ions were carried out by their pK_{R^+} measurements[18]. The pK_{R^+} values for 1,2,3-tricyclopropylcyclopropenium, 1,2-dicyclopropyl-3-phenylcyclopropenium ion and 1-cyclopropyl-2,3-diphenylcyclopropenium ion were determined to be 10.0, 7.09 and 5.04, respectively. Thus, replacement of each phenyl group by a cyclopropyl group enhances the stability of the carbocation by two pK_{R^+} units (2.74 kcal mol^{-1}). These results were also supported by the isodesmic reaction of equation 72 for which the energies were optimized at the HF/3-21G*//HF/3-21G* level[18].

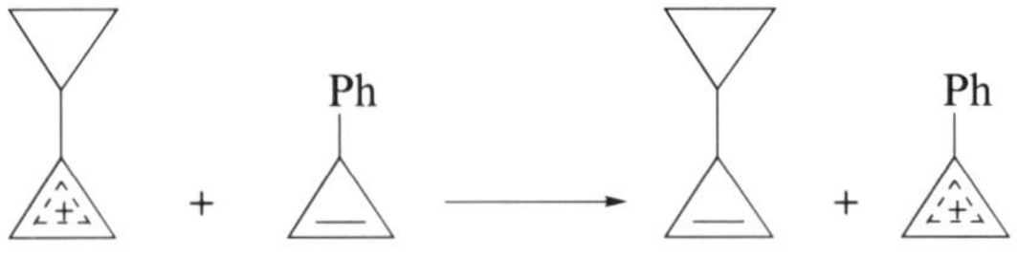

(72)

ΔH = 1.3 kcal mol^{-1} at HF/3-21G*//HF-3-21G*

Arnett and Hofelich measured heats of reaction of a variety of alcohols with SbF_5/FSO_3H in sulfuryl chloride fluoride to form their respective carbocations at constant temperature (–40 °C). In this superacid medium there were no ion-pair complications[126] and hence reliable calorimetric data were obtained for various cyclopropyl and phenyl substituted cations. The heats of reaction for the formation of tricyclopropylcarbinyl cation (–59.2 kcal mol^{-1}), trityl cation (–49.0 kcal mol^{-1}) and *tert*-butyl cation (–35.5 kcal mol^{-1}) show that the relative order of the stabilization of the cationic center is: cyclopropyl >

TABLE 1. Comparison of ^{13}C NMR chemical shifts of the cationic centers of cyclopropylcarbinyl and related cations

Carbocation	$\delta^{13}C\ (C^+)$
$(CH_3)_3C^+$	328.5
$(CH_3)_2CH^+$	318.1
$(HO)_3C^+$	165.1
$(HO)_2CH^+$	176.1
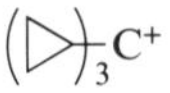	270.9
	253.0
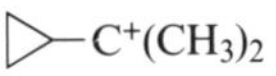	279.9
	252.2
$(Ph)_3C^+$	211.2
$(Ph)_2CH^+$	198.7
$PhC^+(CH_3)_2$	254.2
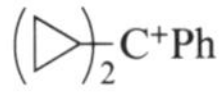	260.4
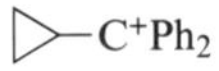	234.3

phenyl > Me. However, the relative order is changed to: Ph > cyclopropyl > Me when the data for the heats of formation of Ph_2XC^+ were compared [ΔH_r (kcal mol^{-1}): X = Ph (–49), cyclopropyl (–44.3), Me (–37.5][127].

^{13}C NMR chemical shifts can also be used in estimating the relative charge density on the cationic centers. In substituted cations, the less positive charge present on the cationic center, the more upfield is its ^{13}C chemical shift. The data in Table 1 show this trend. The trihydroxymethyl cation has a $\delta^{13}C$ of 165.1 ppm which is highly shielded compared to that of the *tert*-butyl cation ($\delta^{13}C$ 328.5 ppm, $\Delta\delta$ = 163.4 ppm). The cationic center of dicyclopropylmethyl cation (253.0 ppm) is similarly much shielded with respect to that of isopropyl cation (318.1 ppm), showing enhanced charge delocalizing ability of the cyclopropyl group over the methyl[122,128].

The relationship between charge density and the NMR chemical shifts is, however, only qualitative and should be used with caution. Other factors, such as neighboring anisotropic effects of the substituents, should also be considered. The cationic center of triphenylmethyl cation ($\delta^{13}C$ 211.2 ppm), for example, is much shielded from that of tricyclopropylmethyl cation (270.9 ppm), which may erroneously lead to the conclusion that a phenyl is more stabilizing than the cyclopropyl group.

VIII. X-RAY CRYSTALLOGRAPHIC STUDIES

X-ray crystallographic studies of the cyclopropylcarbinyl cations have so far been mainly confined to the hydroxycyclopropylcarbinyl cations, generated from the corresponding carbonyl compounds. Typically, the corresponding hexafluoroantimonate salts were prepared by treating the carbonyl compounds with HF–SbF_5, and single-crystal X-ray diffractions were carried out. Childs and coworkers have obtained several X-ray structures on such cations as **135–140**[17].

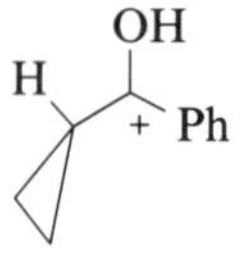

(135)

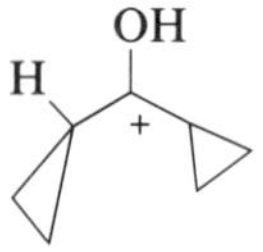

(136)

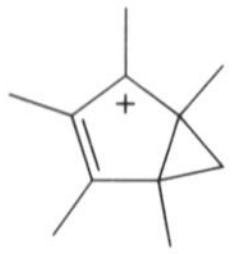

(137)

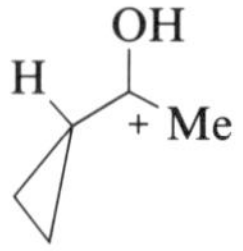

(138)

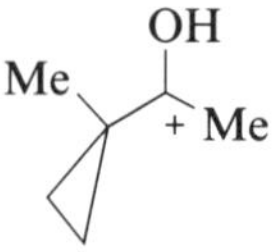

(139)

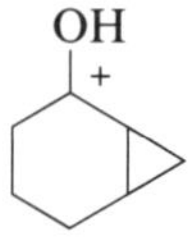

(140)

Moss and coworkers have determined[18] X-ray structures for the tricyclopropylcyclopropenium and 1,2-dicyclopropyl-3-phenylcyclopropenium ions **4a** and **4b** with SbF_6^- and BF_4^- counterions, respectively. In these structures it was found that all the cyclopropyl groups are orthogonal to the cyclopropenium residues. The phenyl group was in the same plane as the cyclopropenium moiety.

IX. CONCLUSIONS AND OUTLOOK

The successful preparation and study of an ever-increasing number of cyclopropylcarbinyl cations as long-lived (stable) carbocations greatly extended our knowledge of these significant intermediates involved in electrophilic reactions of cyclopropylcarbinyl systems. Together with previous reviews (References 3 and 11) the present chapter discusses mainly advances in recent years with relevant background and should be of use to all those interested in this fascinating and significant field. There is no doubt that future work in the area will add much more to our understanding of cyclopropylcarbinyl cations including their reactivity and structural aspects.

X. REFERENCES

1. J. D. Roberts and R. H. Mazur, *J. Am. Chem. Soc.*, **73**, 2509 (1951).
2. J. D. Roberts and V. C. Chambers, *J. Am. Chem. Soc.*, **73**, 5034 (1951).
3. For reviews, see: P. D. Bartlett, *Nonclassical Ions*, W. A. Benjamin, New York, 1965, p. 272; P. Vogel, *Carbocation Chemistry*, Elsevier, Amsterdam, 1985, p. 350; P. Ahlberg, G. Jonsall and C. Engdahl, *Adv. Phys. Org. Chem.*, **19**, 223 (1983); H. G. Richey, Jr. in *Carbonium Ions* (Eds. G. A. Olah and P. v. R. Schleyer), Vol. III, Wiley, New York, 1972, p. 1201. Also see: T. T. Tidwell, in *The Chemistry of the Cyclopropyl Group* (Ed. Z. Rappoport), Part 1, Chap. 10, Wiley, Chichester, 1987, p. 565.
4. R. H. Mazur, W. N. White, D. A. Semenov, C. C. Lee, M. S. Silver and J. D. Roberts, *J. Am. Chem. Soc.*, **81**, 4390 (1959).
5. H. C. Brown, *The Nonclassical Ion Problem*, Plenum Press, New York, 1977.
6. G. A. Olah, G. K. S. Prakash and J. Sommer, *Superacids*, Wiley, New York, 1985.
7. P. v. R. Schleyer, P. Lenoir, P. Mison, G. Liang, G. K. S. Prakash and G. A. Olah, *J. Am. Chem. Soc.*, **102**, 683 (1980).
8. (a) M. Saunders, L. Telkowiski and M. R. Kates, *J. Am. Chem. Soc.*, **99**, 8070 (1977).
   (b) M. Saunders, J. Chandrashekhar and P. v. R. Schleyer, in *Rearrangements in Ground and Excited States* (Ed. P. D. Mayo), Vol. 1, Academic Press, New York, p. 8, 1980.
9. R. M. Coates and E. R. Fretz, *J. Am. Chem. Soc.*, **97**, 2538 (1975).
10. M. Saunders and M. R. Kates, *J. Am. Chem. Soc.*, **102**, 6868 (1980).
11. G. A. Olah, V. P. Reddy and G. K. S. Prakash, *Chem. Rev.*, **92**, 69 (1992).

12. N. C. Deno, H. G. Richey, Jr., J. S. Liu, J. D. Hedge, J. J. Houser and M. J. Wisotsky, *J. Am. Chem. Soc.*, **84**, 2016 (1962).
13. D. S. Kabakoff and E. Namanworth, *J. Am. Chem. Soc.*, **92**, 3234 (1970).
14. K. L. Servis and J. D. Roberts, *J. Am. Chem. Soc.*, **86**, 3773 (1964).
15. C. J. Lancelot, D. J. Cram and P. v. R. Schleyer, in *Carbonium Ions* (Eds. G. A. Olah and P. v. R. Schleyer), Vol. III, Wiley, New York, 1972, p. 1347.
16. G. A. Olah, R. J. Spear, P. C. Hiberty and W. J. Hehre, *J. Am. Chem. Soc.*, **98**, 7470 (1976).
17. R. F. Childs, M. D. Kostyk, C. J. Lock and M. Mahendran, *J. Am. Chem. Soc.*, **112**, 8912 (1990); R. F. Childs, R. Faggiani, C. J. L. Lock., M. Mahendran and S. D. Zweep, *J. Am. Chem. Soc.*, **108**, 1692 (1986); S. K. Chadda, R. F. Childs, R. F. Faggiani and C. J. L. Lock, *J. Am. Chem. Soc.*, **108**, 1694 (1986); R. F. Childs, R. Faggiani, C. J. Lock and M. Mahendran, *J. Am. Chem. Soc.*, **108**, 3613 (1986); R. F. Childs, R. Faggiani, C. J. Lock and A. Varadarajan, *Acta Crystallogr.*, **C40**, 1291 (1984).
18. R. A. Moss, S. Shen, K. K. Jespersen, J. A. Potenza, H.-J. Schugar and R. C. Munjal, *J. Am. Chem. Soc.*, **108**, 134 (1986).
19. H. Mayr and G. A. Olah, *J. Am. Chem. Soc.*, **99**, 510 (1977).
20. (a) G. A. Olah, D. P. Kelly, C. L.Jeuell and R. D. Porter, *J. Am. Chem. Soc.*, **92**, 2544 (1970).
(b) G. A. Olah, C. L. Jeuell, D.P. Kelly and R. D. Porter, *J. Am. Chem. Soc.*, **94**, 146 (1972).
21. W. J. Brittain, M. E. Squillacote and J. D. Roberts, *J. Am. Chem. Soc.*, **106**, 7280 (1984).
22. M. Saunders and H.-U. Siehl, *J. Am. Chem. Soc.*, **102**, 6868 (1980)
23. M. Saunders, K. E. Laidig and M. Wolfsberg, *J. Am. Chem. Soc.*, **111**, 8989 (1989).
24. S. J. Staral, I. Yavari, J. D. Roberts, G. K. S. Prakash, D. J. Donovan and G. A. Olah, *J. Am. Chem. Soc.*, **100**, 8016 (1978).
25. P. C. Myhre, G. G. Webb and C. S. Yannoni, *J. Am. Chem. Soc.*, **112**, 8992 (1990).
26. M. Saunders, K. E. Laidig, K. B. Wiberg and P. v. R. Schleyer, *J. Am. Chem. Soc.*, **110**, 7652 (1988).
27. M. Schindler, *J. Am. Chem. Soc.*, **109**, 1020 (1987).
28. T. S. Sorensen and R. P. Kirchen, *J. Am. Chem. Soc.*, **99**, 6687 (1977).
29. M. J. S. Dewar and C. Reynolds, *J. Am. Chem. Soc.*, **106**, 6388 (1984).
30. W. J. Hehre and P. Hiberty, *J. Am. Chem. Soc.*, **97**, 396 (1975).
31. W. J. Hehre, *Acc. Chem. Res.*, **8**, 369 (1975).
32. W. J. Hehre and P. C. Hiberty, *J. Am. Chem. Soc.*, **96**, 302 (1974).
33. B. A. Levi, E. S. Blurock and W. J. Hehre, *J. Am. Chem. Soc.*, **101**, 5537 (1979).
34. W. Koch, B. Liu and D. DeFrees, *J. Am. Chem.Soc.*, **110**, 7325 (1988).
35. M. L. McKee, *J. Phys. Chem.*, **90**, 4908 (1986).
36. J. D. Roberts, *Proc. Robert A. Welch Found. Conf. Chem. Res.*, **34**, 312 (1990).
37. H. Vancik, V. Gabelica, D. E. Sunko and P. v. R. Schleyer, *J. Phys. Org. Chem.*, **6**, 427 (1993).
38. M. Saunders and J. Rosenfeld, *J. Am. Chem. Soc.*, **92**, 2548 (1970).
39. T. S. Sorensen and R. P. Kirchen, *J. Am. Chem. Soc.*, **99**, 6687 (1977).
40. G. K. S. Prakash, M. Arvanaghi and G. A. Olah, *J. Am. Chem. Soc.*, **107**, 6017 (1985).
41. G. A. Olah, G. K. S. Prakash, D. Donovan and I. Yavari, *J. Am. Chem. Soc.*, **107**, 6017 (1985).
42. H.-U. Siehl, *J. Am. Chem. Soc.*, **107**, 3390 (1985).
43. K. L. Servis and F.-F. Shue, *J. Am. Chem. Soc.*, **102**, 7233 (1980).
44. M. Saunders and N. Krause, *J. Am. Chem. Soc.*, **110**, 8050 (1988).
45. P. Cecchi, A. Pizzabiocca, G. Renzi, F. Grandinetti, C. Sparapani, P. Buzek, P. v. R. Schleyer and M. Speranza, *J. Am. Chem. Soc.*, **115**, 10338 (1993).
46. K. B. Wiberg, D. Shobe and G. L. Nelson, *J. Am. Chem. Soc.*, **115**, 10645 (1993).
47. (a) K. B. Wiberg and N. McMurdie, *J. Org. Chem.*, **58**, 5603 (1993).
(b) K. B. Wiberg and N. McMurdie, *J. Am. Chem. Soc.*, **116**, 11990 (1994).
48. L. R. Schmitz and T. S. Sorensen, *J. Am. Chem. Soc.*, **104**, 2600 (1982).
49. L. R. Schmitz and T. S. Sorensen, *J. Am. Chem. Soc.*, **104**, 2605 (1982).
50. (a) G. A. Olah, P. W. Westerman and J. Nishimura, *J. Am. Chem. Soc.*, **96**, 3548 (1974).
(b) G. A. Olah, J. J. Svoboda and A. T. Ku, *Synthesis*, 492 (1973).
(c) C. U. Pittman, Jr. and G. A. Olah, *J. Am. Chem. Soc.*, **87**, 5123 (1965); G. A. Olah and P. W. Westerman, *J. Am. Chem. Soc.*, **95**, 7530 (1973).
51. K. B. Wiberg and G. Szeimies, *J. Am. Chem. Soc.*, **92**, 571 (1970).
52. Z. Majerski and P. v. R. Schleyer, *J. Am. Chem. Soc.*, **93**, 665 (1971).

53. A. C. Falkenberg, K. Ranganayakulu, C. R. Schmitz and T. S. Sorensen, *J. Am. Chem. Soc.*, **106**, 178 (1984).
54. G. A. Olah, R. J. Spear, P. C. Hiberty and W. J. Hehre, *J. Am. Chem. Soc.*, **98**, 7470 (1976).
55. G. A. Olah, G. K. S. Prakash and T. N. Rawda, *J. Org. Chem.*, **45**, 965 (1980).
56. G. A. Olah, G. Liang, K. A. Babiak, T. K. Morgan, Jr. and R. K. Murray, Jr., *J. Am. Chem. Soc.*, **100**, 1494 (1978).
57. G. A. Olah, G. Liang, K. A. Babiak, T. K. Morgan, Jr. and R. K. Murray, Jr., *J. Am. Chem. Soc.*, **98**, 576 (1976).
58. For a review see: G. K. S. Prakash and P. S. Iyer, *Rev. Chem. Int.*, **9**, 65 (1988).
59. D. P. Kelly, G. J. Farquharson, J. J. Giansiracusa, W. A. Jensen, H. M. Hugel, A. P. Porter, I. J. Rainbow and P. H. Timewell, *J. Am. Chem. Soc.*, **103**, 3539 (1981).
60. G. A. Olah and G. Liang, *J. Am. Chem. Soc.*, **95**, 3792 (1973).
61. G. A. Olah and G. Liang, *J. Am. Chem. Soc.*, **97**, 1920 (1975).
62. U. Biethan, W. Fauth and H. Musso, *Chem. Ber.*, **110**, 3636 (1977).
63. (a) C. Engdahl and P. Ahlberg, *J. Am. Chem. Soc.*, **101**, 3940 (1979).
(b) T. S. Abram and W. E. Watts, *J. Organomet. Chem.*, **105**, C16 (1976).
64. N. Deno, J. J. Jaruzelski and A. Schriesheim, *J. Am. Chem. Soc.*, **77**, 3044 (1955).
65. G. K. S. Prakash, T. N. Rawdah and G. A. Olah, *Angew. Chem., Int. Ed. Engl.*, **22**, 390 (1983); A. De Meijere, O. Schallner, C. Weitemeyer and W. Spielmann, *Chem. Ber.*, **112**, 908 (1979).
66. A. De Meijere and O. Schallner, *Angew. Chem., Int. Ed. Engl.*, **12**, 399 (1973).
67. H.-U. Siehl, T. Mulleer, J. Gauss, P. Buzek and P. v. R. Schleyer, *J. Am. Chem. Soc.*, **116**, 6384 (1994).
68. H.-U. Siehl, *J. Chem. Soc., Chem. Commun.*, 635 (1984).
69. C. J. Collins, A. G. Martinez, R. M. Alvarez and J. A. Aguirre, *Chem. Ber.*, **117**, 2815 (1984).
70. H.-U. Siehl and E. W. Koch, *J. Org. Chem.*, **49**, 575 (1984).
71. W. Brennenstuhl and M. Hanack, *Tetrahedron Lett.*, **25**, 3437 (1984).
72. Y. Apeloig, P. v. R. Schleyer and J. A. Pople, *J. Org. Chem.*, **42**, 3004 (1977).
73. G. A. Olah, G. Liang and S. P. Jindal, *J. Org. Chem.*, **40**, 3259 (1975).
74. G. A. Olah, G. K. S. Prakash and G. Liang, *J. Org. Chem.*, **42**, 661 (1977).
75. G. A. Olah and R. Spear, *J. Am. Chem. Soc.*, **97**, 1539 (1975).
76. K. Rajeswari and T. S. Sorensen, *J. Am. Chem. Soc.*, **95**, 1239 (1973); N. Ikazawa and T. S. Sorensen, *Can. J. Chem.*, **95**, 2355 (1978).
77. G. A. Olah, V. P. Reddy, J. Casanova and G. K. S. Prakash, *J. Org. Chem.*, **57**, 6431 (1992).
78. G. A. Olah, D. Donovan and G. K. S. Prakash, *Tetrahedron Lett.*, 4779 (1978).
79. J. Chandrashekhar and P. v. R. Schleyer, *Tetrahedron Lett.*, 4057 (1979).
80. G. A. Olah, G. K. S. Prakash and T. Nakajima, *J. Am. Chem. Soc.*, **104**, 1031 (1982).
81. Y. M. Jun and J. W. Timberlake, *Tetrahedron Lett.*, **23**, 1761 (1982).
82. M. Burch, J. Y. Moo, A. E. Ludtke, M. Schneider and J. W. Timberlake, *J. Org. Chem.*, **51**, 2969 (1986).
83. G. A. Olah and G. Liang, *J. Am. Chem. Soc.*, **98**, 7026 (1976).
84. J. E. Baldwin and W. D. Foglesong, *J. Am. Chem. Soc.*, **90**, 4303 (1968).
85. J. E. Baldwin and W. D. Foglesong, *J. Am. Chem. Soc.*, **90**, 3410 (1968).
86. G. A. Olah, G. Liang, K. A. Babiak and R. K. Murray, Jr., *J. Am. Chem. Soc.*, **96**, 6794 (1974).
87. R. K. Murray, Jr., T. M. Ford, G. K. S. Prakash and G. A. Olah, *J. Am. Chem. Soc.*, **102**, 1865 (1980).
88. J. C. Barborak, J. Daub, D. M. Follweiler and P. v. R. Schleyer, *J. Am. Chem. Soc.*, **91**, 7760 (1969).
89. C. Barborak and P. v. R. Schleyer, *J. Am. Chem. Soc.*, **92**, 3184 (1970); see also: J. B. Grutzner and S. Winstein, *J. Am. Chem. Soc.*, **92**, 3186 (1970); **94**, 2200 (1972).
90. C. Engdahl, G. Jonsall and P. Ahlberg, *J. Am. Chem. Soc.*, **105**, 891 (1983).
91. C. Ahlberg, C. Engdahl and G. Jonsall, *J. Am. Chem. Soc.*, **103**, 1583 (1981); C. Engdahl, G. Jonsall and P. Ahlberg, *J. Chem. Soc., Chem. Commun.*, 626 (1979).
92. R. Hoffmann, W.-D. Stohrer and M. Goldstein, *J. Bull. Chem. Soc. Jpn.*, **45**, 2513 (1972); M. J. Goldstein and R. Hoffmann, *J. Am. Chem. Soc.*, **93**, 6193 (1971); M. B. Huang and G. Jonsall, *Tetrahedron*, **41**, 6055 (1985); M. B. Huang, O. Goscinski, G. Jonsall and P. Ahlberg, *J. Chem. Soc., Perkin Trans. 2*, 305 (1983).
93. D. Cremer, P. Svensson, E. Kraka and P. Ahlberg, *J. Am. Chem. Soc.*, **115**, 7445 (1993).

94. D. Cremer, P. Svensson, E. Kraka, Z. Konkoli and P. Ahlberg, *J. Am. Chem. Soc.*, **115**, 7457 (1993).
95. P. Vogel, M. Saunders, N. M. Hasty and J. A. Berson, *J. Am. Chem. Soc.*, **93**, 1551 (1971).
96. T. Tsuji, I. Moritani, S. Nishida and G. Tadokoro, *Bull. Chem. Soc. Jpn.*, **40**, 2344 (1967).
97. B. R. Ree and J. C. Martin, *J. Am. Chem. Soc.*, **92**, 1660 (1970).
98. C. F. Wilcox, Jr. and R. G. Jesaitis, *Tetrahedron Lett.*, 5065 (1970); *Chem. Commun.*, 1046 (1967).
99. D. Lenoir, P. v. R. Schleyer and J. Ipaktschi, *Justus Liebigs Ann. Chem.*, **750**, 28 (1971).
100. D. Lenoir, W. Roll and J. Ipaktschi, *Tetrahedron Lett.*, 3073 (1976).
101. G. K. S. Prakash, A. P. Fung, G. A. Olah and T. N. Rawdah, *Proc. Natl. Acad. Sci. U.S.A.*, **84**, 5092 (1982).
102. G. A. Olah, V. P. Reddy, G. Rasul and G. K. S. Prakash, *J. Org. Chem.*, **57**, 1114 (1992).
103. V. P. Reddy, G. A. Olah and G. K. S. Prakash, *J. Org. Chem.*, **58**, 7622 (1993).
104. D. J. Cram, *J. Am. Chem. Soc.*, **71**, 3863, 3871, 3875 (1949).
105. J. E. Nordlander and W. G. Deadman, *J. Am. Chem. Soc.*, **90**, 1590 (1968); J. E. Nordlander and W. J. Kelly, *J. Am. Chem. Soc.*, **91**, 9956 (1969); P. v. R. Schleyer and C. J. Lancelot, *J. Am. Chem. Soc.*, **91**, 4297 (1969).
106. H. C. Brown and C. J. Kim, *J. Am. Chem. Soc.*, **93**, 5765 (1971).
107. (a) G. A. Olah and R. D. Porter, *J. Am. Chem. Soc.*, **93**, 6877 (1971)
(b) G. A. Olah, R. J. Spear and D. A. Forsyth, *J. Am. Chem. Soc.*, **98**, 6284 (1976)
(c) G. A. Olah, R. J. Spear and D. A. Forsyth, *J. Am. Chem. Soc.*, **99**, 2615 (1977).
108. C. Koppel and F. W. McLafferty, *J. Am. Chem. Soc.*, **98**, 8293 (1976); C. Koppel, C. C. Van de Sande, N. M. M. Nibbering, T. Nishishita and F. W. McLafferty, *J. Am. Chem. Soc.*, **99**, 2883 (1977).
109. (a) W. J. Hehre, *J. Am. Chem. Soc.*, **94**, 5919 (1972).
(b) S. Sieber, P. v. R. Schleyer and J. Gauss, *J. Am. Chem. Soc.*, **115**, 6987 (1993).
110. G. A. Olah, N. J. Head, G. Rasul and G. K. S. Prakash, unpublished results.
111. G. A. Olah and R. D. Porter, *J. Am. Chem. Soc.*, **92**, 7627 (1970); G. A. Olah and A. M. White, *J. Am. Chem. Soc.*, **91**, 5801 (1969).
112. G. A. Olah and C. U. Pittman, *J. Am. Chem. Soc.*, **87**, 3509 (1965); G. A. Olah, M. B. Comisarow, E. Namanworth and B. Ramsey, *J. Am. Chem. Soc.*, **89**, 5259 (1967).
113. G. A. Olah, M. B. Comisarow and C. J. Kim, *J. Am. Chem.Soc.*, **91**, 1458 (1969); G. A. Olah and R. D. Porter, *J. Am. Chem. Soc.*, **93**, 6877 (1971).
114. G. A. Olah and B. P. Singh, *J. Am. Chem. Soc.*, **104**, 5618 (1982); G. A. Olah, B. P. Singh and G. Liang, *J. Org. Chem.*, **49**, 2922 (1984); S. Winstein and L. Eberson, *J. Am. Chem. Soc.*, **87**, 3506 (1965).
115. G. A. Olah and G. Liang, *J. Am. Chem. Soc.*, **97**, 1920 (1975).
116. G. K. S. Prakash, V. V. Krishnamurthy, M. Arvanaghi and G. A. Olah, *J. Org. Chem.*, **50**, 3985 (1985).
117. G. A. Olah, J. L. Grant, R. J. Spear, J. M. Bollinger, A. Seroamz and G. Sipos, *J. Am. Chem. Soc.*, **98**, 2501 (1976).
118. G. K. S. Prakash, A. P. Fung, T. N. Rawdah and G. A. Olah, *J. Am. Chem. Soc.*, **107**, 2920 (1985).
119. G. A. Olah, V. P. Reddy, G. Lee, J. Casanova and G. K. S. Prakash, *J. Org. Chem.*, **58**, 1639 (1993).
120. H. Hart and J. M. Sandri, *J. Am. Chem. Soc.*, **81**, 320 (1959).
121. H. C. Brown and J. D. Cleveland, *J. Org. Chem.*, **41**, 1792 (1976).
122. G. A. Olah, G. K. S. Prakash and G. Liang, *J. Org. Chem.*, **42**, 2666 (1977).
123. N. C. Deno, H. G. Richey Jr., J. S. Liu, D. N. Lincoln and J. D. Turner, *J. Am. Chem. Soc.*, **87**, 4533 (1965).
124. R. W. Taft, R. H. Martin and F. W. Lampe, *J. Am. Chem. Soc.*, **87**, 2490 (1965).
125. J. F. Wolf, P. G. Harch, R. W. Taft and W. J. Hehre, *J. Am. Chem. Soc.*, **97**, 2902 (1975).
126. E. M. Arnett and T. C. Hofelich, *J. Am. Chem. Soc.*, **105**, 2889 (1983).
127. E. M. Arnett and T. C. Hofelich, *J. Am. Chem. Soc.*, **104**, 3522 (1982).
128. G. A. Olah and A. M. White, *J. Am. Chem. Soc.*, **92**, 7627 (1970).

CHAPTER **15**

Spiroannulated cyclopropanes

KIRILL A. LUKIN

Department of Chemistry, The University of Chicago, 5735 South Ellis Avenue, Chicago, Illinois 60637, USA
Fax: (+1)312-702-0805; e-mail: luki@midway.uchicago.edu

and

NIKOLAI S. ZEFIROV

Department of Chemistry, Moscow State University, Moscow, Russia
Fax: 7-095-939-02-90; e-mail: zefirov@synth.chem.msu.su

I. INTRODUCTION 862
II. NOMENCLATURE OF SPIROANNULATED CYCLOPROPANES . . 862
III. STRAIN IN SPIROANNULATED CYCLOPROPANES 863
IV. SYNTHESIS OF SPIROANNULATED CYCLOPROPANES 864
A. Reductive Dehalogenation of *gem*-(Dihalomethyl)cycloalkanes 864
B. Cyclopropanation 865
C. Cyclization of *gem*-Disubstituted Cyclopropanes 866
D. Alkylation of Compounds Having Acidic CH_2 Groups with 1,2-Dibromoethane 867
E. Synthesis of SPC via Cycloaddition Reactions 867
F. Miscellaneous Reactions 868
V. NONACTIVATED SPIROANNULATED CYCLOPROPANES AND SOME OF THEIR USEFUL APPLICATIONS 869
VI. CONJUGATION BETWEEN CYCLOPROPYL FRAGMENTS AND ITS INFLUENCE ON CHEMICAL PROPERTIES OF SPIROANNULATED CYCLOPROPANES 870
VII. THE ROTANES 871
A. [3]Rotane 872
B. [4]Rotane 872
C. [5]Rotane 873
D. [6]Rotane 873
VIII. OLIGOMERS OF SPIROANNULATED CYCLOPROPANES: THE TRIANGULANES 874
A. Synthesis of Triangulanes 875
B. Geometry of Triangulanes 878
C. Spectroscopic Identification of Triangulanes 880

The chemistry of the cyclopropyl group, Vol.2
Edited by Z. Rappoport

D. Chemistry of Triangulanes 880
1. Thermal stability 880
2. Ring opening in triangulanes 880
3. Stereospecific functionalization of triangulanes 881
IX. CONCLUDING REMARKS 882
X. REFERENCES 883

I. INTRODUCTION

The first question that one might ask after reading the title of this chapter is whether the separation of spiroannulated cyclopropanes from other substituted cyclopropanes is justified. Does the spiroannulation really change the chemistry of these compounds? Analysis of the literature conducted by the authors with this question in mind suggested that only few types of spiroannulated cyclopropanes display unusual behavior or can be considered as targets to search for that. Namely, the 'activation' was observed in (1) compounds where the formation of the spirocenter is associated with an additional strain, and (2) compounds possessing conjugated cyclopropyl fragments. However, the importance of these types of spiroannulated cyclopropanes is growing since the imagination of chemists created the whole classes of compounds constructed from these fragments. For example, cyclic oligomers of cyclopropylidene ([*n*]rotanes) are constructed only from conjugated spiroannulated polycyclopropanes. Both types of activation are present in oligomers of spiroannulated cyclopropanes (triangulanes). The chemistry of these unusual compounds was the primary focus of the authors of this chapter. The description of nonactivated spiroannulated cyclopropanes with similar chemistry to those of simple substituted cyclopropanes will be limited to general methods of their preparation and some useful synthetic applications.

II. NOMENCLATURE OF SPIROANNULATED CYCLOPROPANES[1]

The prefix 'spiro' before the name of a compound indicates the presence of spiroannulated fragments (i.e. cycles annulated via one common atom, the spirocenter). In spiroannulated cyclopropanes (SPC) the spirocenter is a quaternary carbon atom. In the names of compounds possessing *monocyclic* spiroannulated fragments, the number of atoms in each ring separated by the spirocenter is indicated in increasing order in brackets. The numbering of atoms starts from the smallest ring. For example, compound **1** is spiro[2.3]hexanone-4.

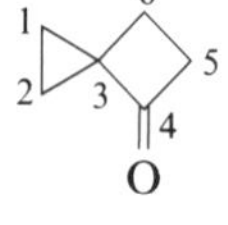

(1)

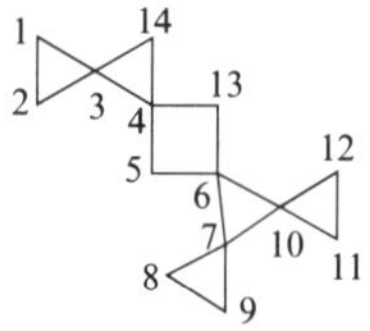

(2)

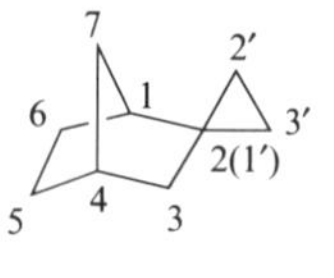

(3)

Polyspiro compounds are indicated by prefixes dispiro, trispiro, etc. which correspond to the number of spirocenters. The systematic numbering of atoms which are separated by spirocenters starts from the smallest terminal ring and proceeds by the *shortest* path to the second spirocenter, etc. When the last spirocenter is reached, one counts backward to the starting point. The numbers for the atoms are assigned according to the same principle. For example, the name of compound **2** is pentaspiro[2.0.1.0.2.0.2.0.1.1]tetradecane.

In spiroannulated *polycyclic* compounds, each fragment is described as an independent molecule. The names of these fragments are given in alphabetical order and are separated

TABLE 1. Calculated (MM2) and experimental heats of formation and strain energy in some spiroannulated cyclopropanes

Compound	ΔH_f(kcal mol^{-1}) MM2	ΔH_f(kcal mol^{-1}) exptl	Strain energy (kcal mol^{-1})	Strain energy of spiroannulation (kcal mol^{-1})
Spiro[2.2]pentane	43.96	44.26	63.6	8.6
Spiro[2.3]hexane	29.71 (32.3)	—	53.4	0.4
Spiro[2.4]heptane	1.97	—	33.13	−0.5
Spiro[2.5]octane	−10.71	—	26.2	−1.3
Dispiro[2.0.2.1]heptane **4**	76.0	72.9	100.4	18.0
Trispiro[2.0.0.2.1.1]octane **5**	108.0	102.9	137.15	27.1

by the numbers which are assigned to the spirocenter in each fragment. Thus, compound **3** is named spiro(bicyclo[2.2.1]heptane-2,1′-cyclopropane).

III. STRAIN IN SPIROANNULATED CYCLOPROPANES

Evaluation of the strain energy in SPC was a subject of numerous calculations, which provided substantially different results. Unfortunately, very limited experimental data on the heats of formation of these compounds[2,3] preclude the evaluation of the validity of the calculations. The results of nonempirical (4-31G//STO-3G) and molecular mechanics (MM2) calculations were in reasonable agreement with experimental data, while the semiempirical methods resulted in substantial deviations[4]. Even better results were obtained by Aped and Allinger using the MM3 method[5]. Available experimental data on the heats of formation of SPC and matching calculated values are collected in Table 1.

It is particularly interesting to evaluate the extra strain energy caused by spiroannulation. The last column of Table 1 shows that spiroannulation of cyclopropane to five-, six- and higher-membered rings is not associated with any extra strain (i.e. the strain energy of the molecule is equal to the sum of the strain energy of cyclopropane and the strain energy of the larger ring). On the other hand, the formation of spiro[2.3]hexane and spiro[2.2]pentane is associated with an extra strain of 0.4 and 8.6 kcal mol^{-1}, respectively. In accord with these findings, spiroannulation of cyclopropane to five- and higher-membered carbocycles did not result in any changes in the geometry of either fragment. However, in spiro[2.2]pentane and spiro[2.3]hexane substantial elongation of distal bonds and shortening of the bonds connected to the spirocenter was observed (Figure 1)[6–8]. These changes in the geometry result in an increase in the angle between carbon–carbon bonds at the spirocenter in each fragment and, consequently, in a partial release of strain.

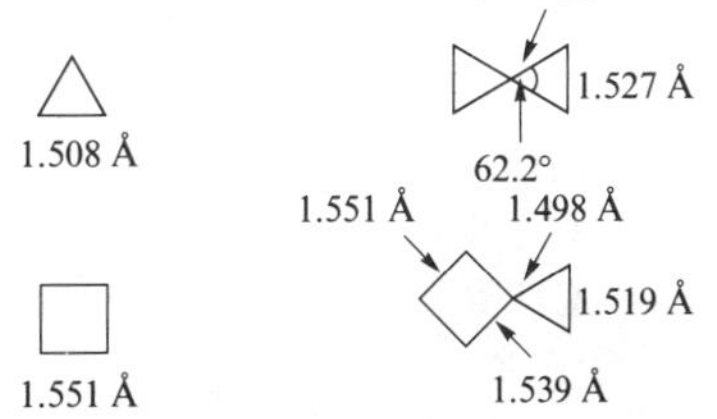

FIGURE 1. Comparison of the geometries of cyclopropane, cyclobutane, spiropentane and spirohexane

The considerations above suggest that spiropentane and, to a certain extent, also spirohexane derivatives are the objects of choice in the search for new chemistry uncommon to 'normal' (nonactivated) cyclopropanes.

Even more exciting results might be expected from polyspiranes constructed from the fragments of spirohexane or spiropentane. For example, in oligomers of spiroannulated cyclopropanes (triangulanes, **6**)[9] the addition of each cyclopropyl fragment results in the formation of a new spiro center which is associated with extra strain of 9 kcal mol^{-1}. This means that in trispirononane, **5**, the strain energy arising from spiroannulation is already higher than the strain in one of cyclopropyl fragments. Although adjustments in the geometry of these molecules (see Section VIII), resulting in a partial release of strain, has been observed for compounds **4** and **5** (see Table 1), the accumulation of the strain energy in triangulanes is nevertheless impressive[9].

(4) **(5)** **(6)**

IV. SYNTHESIS OF SPIROANNULATED CYCLOPROPANES

Existing synthetic methodologies allow convenient preparation of different types of SPC. A selection of the most general methods, as well as some unusual synthetic transformations, are given below.

A. Reductive Dehalogenation of *gem*-(Dihalomethyl)cycloalkanes

Reductive dehalogenation is an efficient method of synthesis of cyclopropanes spiroannulated to five- and higher-membered carbocycles (i.e. compounds in which spiroannulation does not result in accumulation of extra strain)[10]. The required *gem*-(dihalomethyl)cycloalkanes are usually prepared by halogenation of the precursor diols (equation 1). The cyclization is most efficiently accomplished in the Zn–alcohol–water system[10,11]. For example, spiro[2.5]octane **7** was prepared in 91% yield using this procedure[11]. This method is useful even for a one-step preparation of bis-spirocyclopropyl compounds as exemplified in equation 2[12]. However, the application of the reductive dehalogenation method to the synthesis of more strained SPC (i.e. spirohexane or spiropentane) often leads to rearranged products. For example, methylenecyclopentane was the only product obtained from bis(bromomethyl)cyclobutane (equation 3)[13].

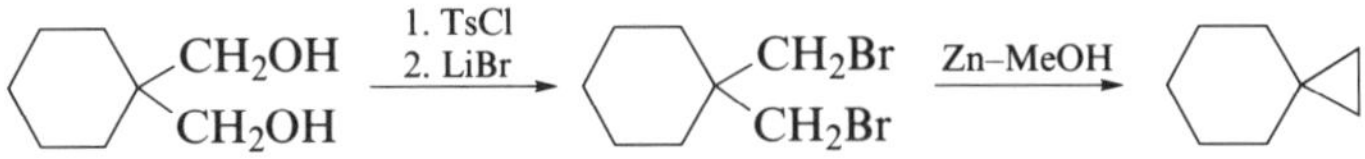

(1)

(7)

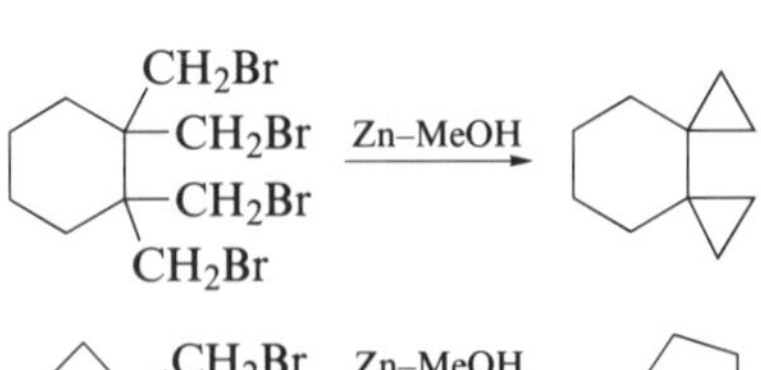

(2)

(3)

Despite the demonstration of reductive cyclization of tetrabromopentaerythritol **8** to spiropentane almost a century ago[14], the formation of methylenecyclobutane and other side products (equation 4) decreases the preparative importance of the method.

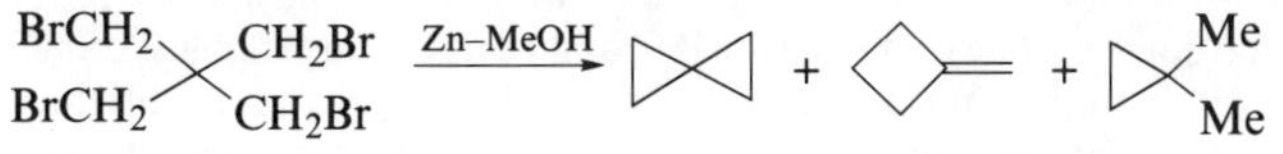

(4)

(8)

It was believed that Lewis acid properties of the Zn^{2+} or of the intermediate organozinc compound **9** were responsible for the rearrangement[15]. Indeed, it was later demonstrated that electrochemical reductive cyclization[16] or complexation of zinc species with the tetrasodium salt of EDTA resulted in the formation of unrearranged products, such as spiropentane and spirohexane[15,17]. However, it was found that the use of other reducing agents, such as alkali metals or alkyllithiums (with gegen cations that do not have Lewis acid properties), still resulted in the formation of rearranged products, especially in nonpolar solvents[18]. This result indicates that several side reactions might be responsible for the failure of cyclization of possible reaction intermediates.

It is not surprising that the synthesis of even more strained SPC (e.g. triangulanes, see Section VIII) has never been accomplished by using reductive cyclization, even if electrochemical methods were applied[19]. The only exception is the preparation of 7-methylenedispiro[2.0.2.1]heptane **11** from dibromide **10** (equation 5)[20]. In this case, a participation of the double bond in the stabilization of an intermediate may account for the success of the cyclization.

(9)

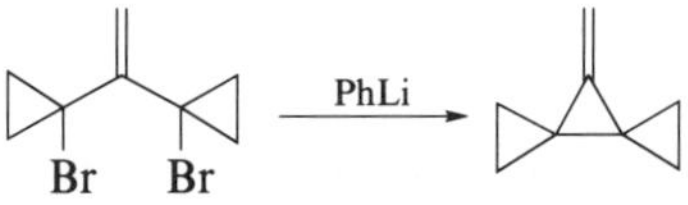

(5)

(10) **(11)**

B. Cyclopropanation

The introduction of carbenes and carbenoids into synthetic organic chemistry revolutionized the synthesis of cyclopropane derivatives[21]. In particular, cyclopropanation of methylenecycloalkanes became a very useful method for the preparation of SPC. Moreover, since cycloaddition of carbenes to olefins involves a very fast *concerted process* (i.e. it eliminates any intermediates during the formation of the three-membered ring)[21], the method is equally efficient for the preparation of both unstrained and highly strained compounds.

The Simmons–Smith reaction, which involves cyclopropanation by the Zn–CH_2I_2 system (equation 6), is a very popular variant of this method[22]. However, an application of the Simmons–Smith reaction to methylenecycloalkanes often requires the use of high excess (up to ten equivalents) of the reagent, which complicates the workup and lowers the yields.

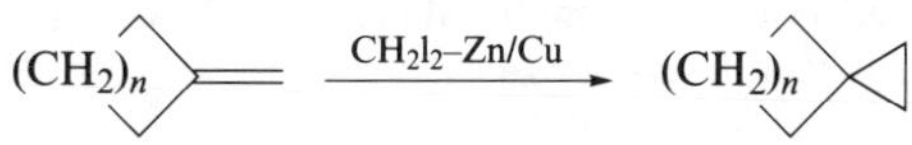

(6)

Although many recent improvements in the preparation of the Simmons–Smith reagent might be helpful[23,24], the authors of this chapter would recommend one to consider an alternative two-step cyclopropanation procedure, which includes cycloaddition of dichloro- or dibromocarbene to methylenecycloalkane[25] followed by reductive dehalogenation (equation 7)[26]. The first reaction is usually carried under phase transfer conditions and presents a very simple and efficient procedure. Reduction of *gem*-dihalocyclopropanes with lithium in *tert*-butanol or with sodium in liquid ammonia usually proceeds without complications and with high yield.

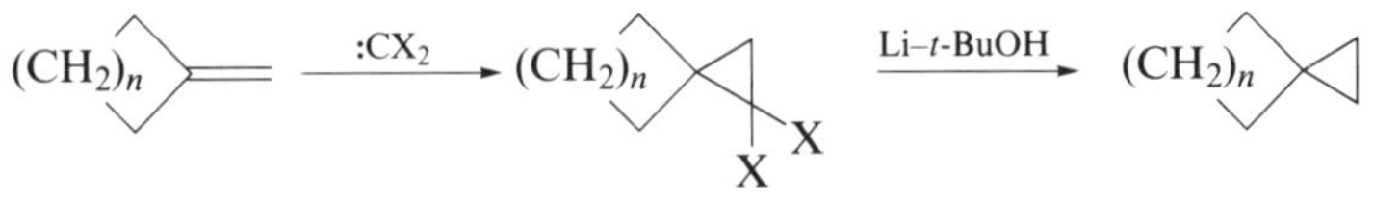

(7)

X = Cl, Br

It was demonstrated that palladium acetate catalyzed cyclopropanation of methylenecyclopropanes is an effective method of preparation of spiropentane derivatives (equation 8; see also equation 36 below)[9].

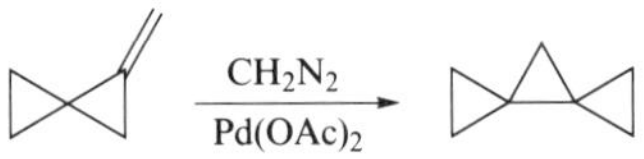

(8)

Rhodium acetate catalyzed cyclopropanation of methylenecycloalkanes with diazocarbonyl compounds (equation 9) provides a direct method for preparation of functionalized SPC[27].

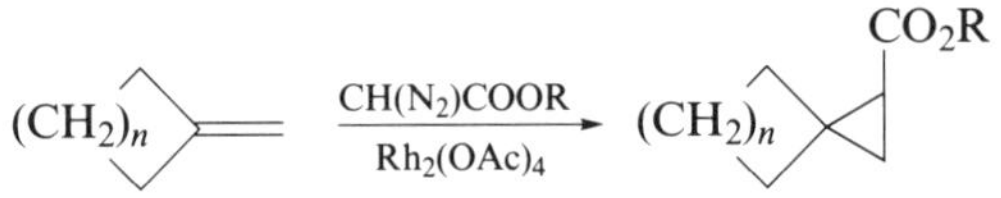

(9)

gem-Dibromocyclopropanes are intermediates in the synthesis of cyclopropanes described above, and are useful starting materials for various transformations[28]. For example, their low-temperature lithiation yields bromolithiocarbenoid **12**, which can be trapped with electrophiles, thus providing access to functionalized SPC (equation 10).

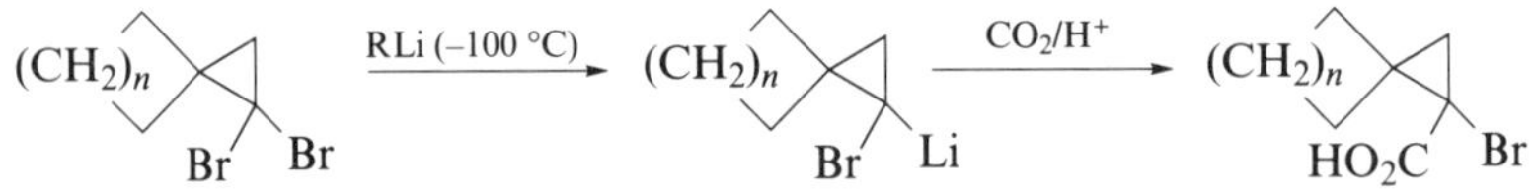

(10)

(12)

C. Cyclization of *gem*-Disubstituted Cyclopropanes

This obvious approach is not widely used, because the required starting materials are generally less available than the precursors described above for the three-membered ring formation. However, the method might be very useful for the synthesis of bis-spirocyclopropyl compounds. For example, dispiroketone **13** was prepared according to equation 11 and was used as a starting material in the synthesis of [5]rotane[29]. Even highly strained dispiro[2.0.2.2]octene derivative **14** was synthesized using this approach (equation 11)[30].

(13)

(11)

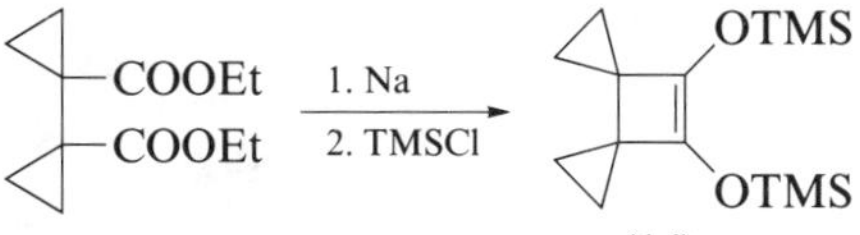

(14)

D. Alkylation of Compounds Having Acidic CH_2 Groups with 1,2-Dibromoethane

Preparation of spirohepta[2.4]diene-4,6 (**15**) from cyclopentadiene and 1,2-dibromoethane (equation 12) is a nice demonstration of the possibilities of this method[31].

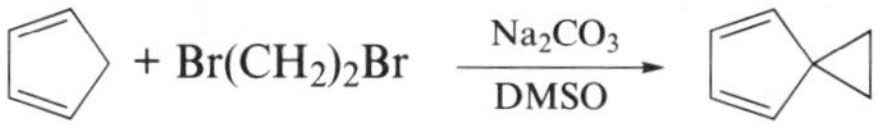

(12)

(**15**)

E. Synthesis of SPC via Cycloaddition Reactions

The highly strained double bond in methylenecyclopropane displays enhanced reactivity in cycloaddition reactions. In addition to 'normal' [4+2] cycloaddition to 1,3-dienes (e.g. equation 13)[32], methylenecyclopropane and its derivatives have a pronounced tendency to undergo thermal [2+2] cycloaddition reactions. For example, thermal dimerization of methylenecyclopropane in the gas phase results in formation of isomeric dispirooctanes **16** and **17** (equation 14)[33]. This unusual cyclization is considered to proceed via a stepwise radical mechanism involving the intermediacy of biradical **18** (equation 15)[34]. Equation 15 demonstrates that methylenecyclopropanes possessing substituents capable of stabilizing intermediate radicals undergo efficient [2+2] dimerization even

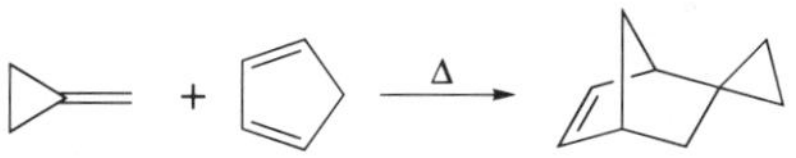

(13)

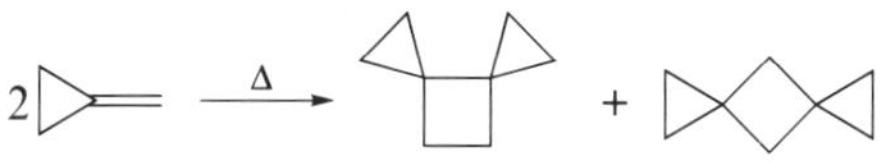

(14)

(**16**) (**17**)

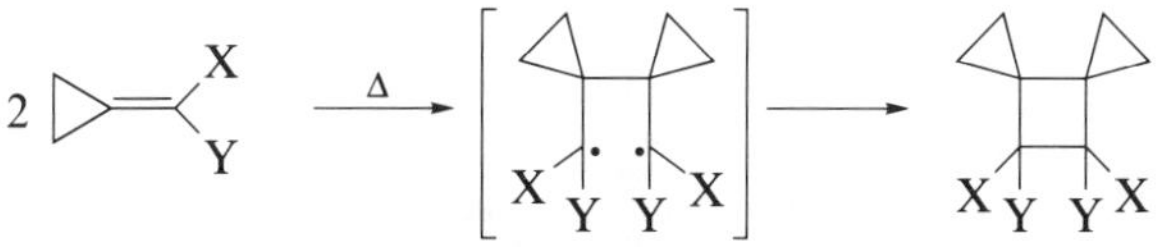

(15)

X = CH=CHOEt, Y = H (**18**)
X = Y = Cl

under mild conditions[34–36]. These reactions proceed in a highly regioselective manner and present a useful synthetic route to dispiro[2.0.2.2]octane derivatives.

The more reactive bicyclopropylidene (**19**) displays even greater tendency to undergo [2+2] cycloadditions[37]. For example, only 1,3-cyclopentadiene reacts with this olefin in the [4+2] mode, while 1,3-butadiene and 1,3-cyclohexadiene give products resulting mainly from [2+2] cycloaddition (equation 16)[37].

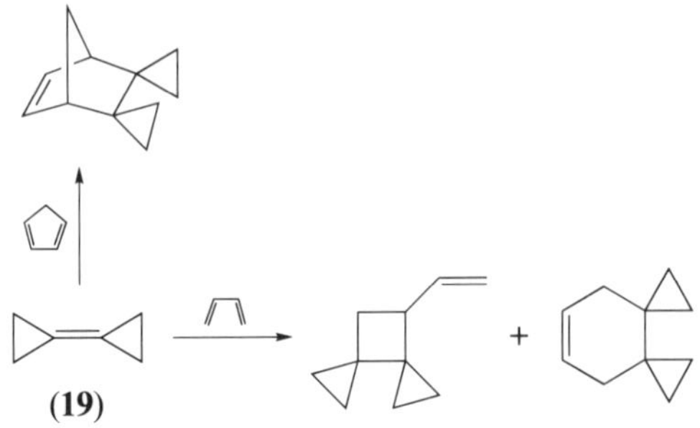

(16)

F. Miscellaneous Reactions

Diels–Alder cycloadditions of the readily available spiro[2.4]heptadiene (**15**) are a useful entry to spiro(cyclopropane-1,7′-norbornene) and related polycyclic compounds (equation 17)[38]. 1,1-Diacetylcyclopropane was found to undergo unusual base catalyzed cyclizations resulting in spiroannulated cyclopropanes (equation 18)[39,40].

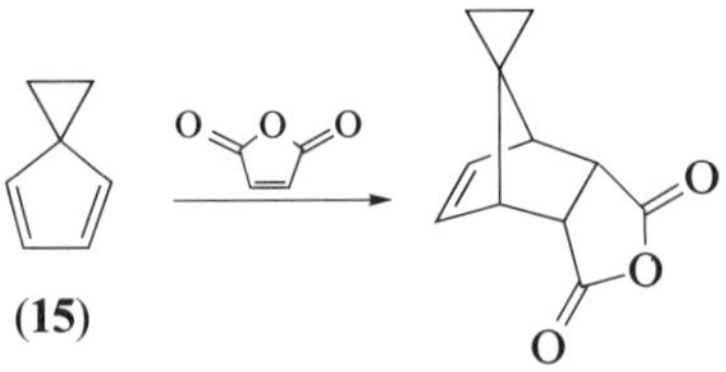

(17)

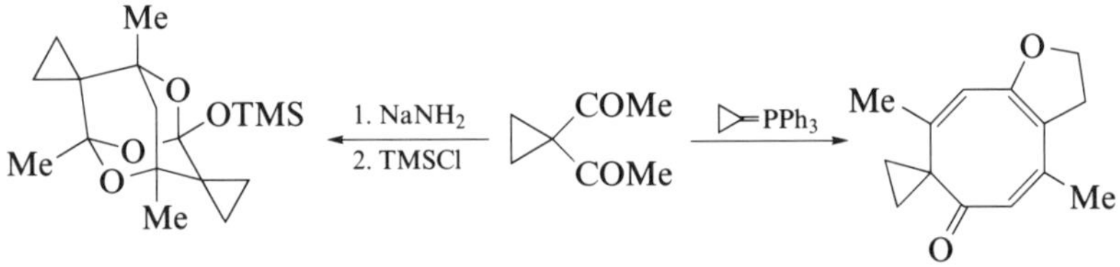

(18)

Methylenecyclopropane undergoes an interesting reaction with cobalt carbonyl complexed acetylenes yielding spiro[2.4]heptenone derivatives (equation 19)[41].

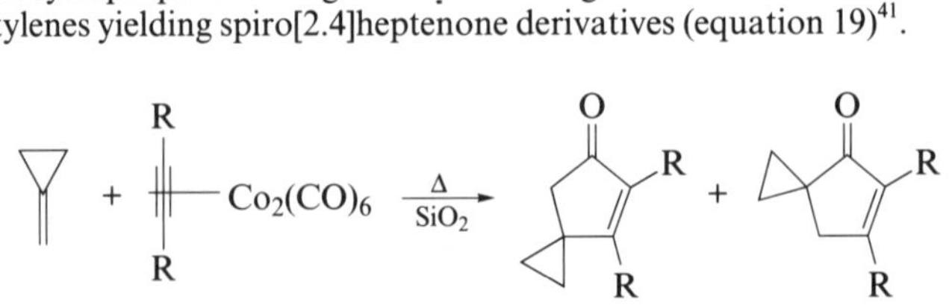

(19)

V. NONACTIVATED SPIROANNULATED CYCLOPROPANES AND SOME OF THEIR USEFUL APPLICATIONS

It was demonstrated in Section 3 that spiroannulation of cyclopropane to four- and higher-membered rings does not cause an increase in the strain energy of the molecule. It is not surprising, then, that the chemistry of these compounds does not differ from that of simple alkyl substituted cyclopropanes. Since exciting reviews on the chemistry of cyclopropanes are available[42], the authors of this chapter will limit the description of properties of SPC of these types to some of their interesting applications.

The important application of SPC in organic synthesis is their conversion into the *gem*-dimethyl group, which is a common fragment in many natural products as well as in synthetically made, physiologically active compounds. This transformation utilizes hydrogenation of the cyclopropyl fragment over a noble metal catalyst. Equation 20 presents a typical example of the introduction of the *gem*-dimethyl group into a complex molecule via a SPC intermediate[43].

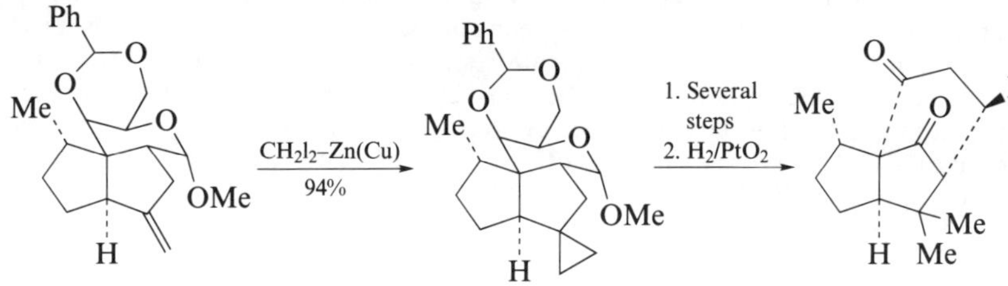

(20)

Some natural compounds containing SPC fragments display high biological activity. For example, ptaquiloside **20** was found to cleave DNA. The proposed mechanism of its action is outlined in equation 21 and includes the formation of a conjugated dieneone system followed by nucleophilic opening of the cyclopropyl fragment with DNA[44].

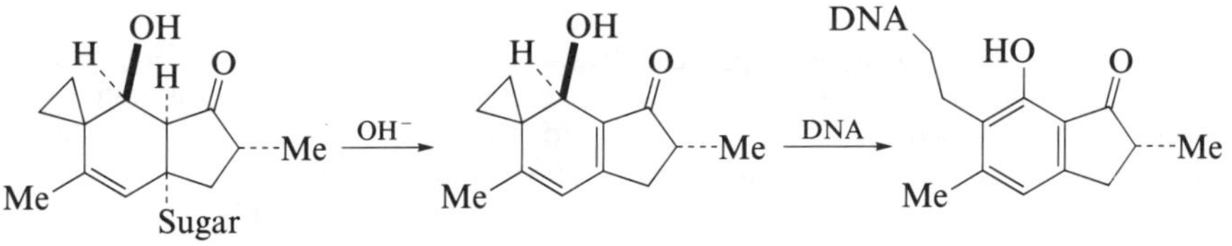

(20) (21)

Unusual compounds containing SPC fragments (e.g. **21**) were recently isolated from plants[45].

'Activated' SPCs can be expected to demonstrate more interesting chemical behavior. The additional strain which is associated with the formation of the spirocenter in spiropen-

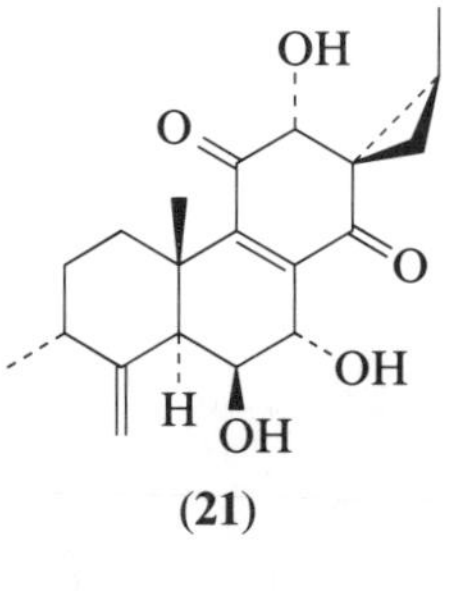

(21)

tane and its triangulane derivatives can be considered as one of the activating factors. The chemistry of these compounds is discussed in Section VIII. Below, we describe some interesting chemical reactions which were observed in the second type of activated SPC, namely spiranes containing *conjugated* cyclopropyl fragments.

VI. CONJUGATION BETWEEN CYCLOPROPYL FRAGMENTS AND ITS INFLUENCE ON CHEMICAL PROPERTIES OF SPIROANNULATED CYCLOPROPANES

The phenomenon of conjugation between two cyclopropyl fragments is the subject of extensive studies[46]. Figure 2 represents the orbitals resulting from this interaction. The conjugation between cyclopropyl fragments does not affect the UV spectra of the corresponding molecules; however, it is manifested in their photoelectron (PE) spectra. Rotation of cyclopropyl fragments in nonrigid molecules like bicyclopropyl makes their PE spectra complicated, due to contribution from different conformers[46]. Consequently, the fixation of a conformation in SPC possessing adjacent three-membered rings makes these compounds perfect models for conjugation studies. The interaction between cyclopropyl fragments could be quantitatively described by the resonance integral for the neighboring 2p atomic orbital of the conjugated cyclopropyl groups. The experimental values for [4]rotane (**22**) (–1.50 eV) and [3]rotane (**23**) (–2.05 eV) are comparable to the resonance integral found in 1,1′-bicyclopropyl (**24**) (–1.76 eV)[47,48]. Not surprisingly, separation of cyclopropyl fragments by one or more methylene groups cancels the conjugation.

(22) **(23)** **(24)**

Interactions between three-membered rings in SPC which are observed in their PE spectra could be used to account for some unusual properties of these compounds. For example, tetracyanoethylene (TCNE) was found to react with the cyclopropyl fragment of hydrocarbon **25**, giving a formal insertion product **26** (equation 22)[12]. For another

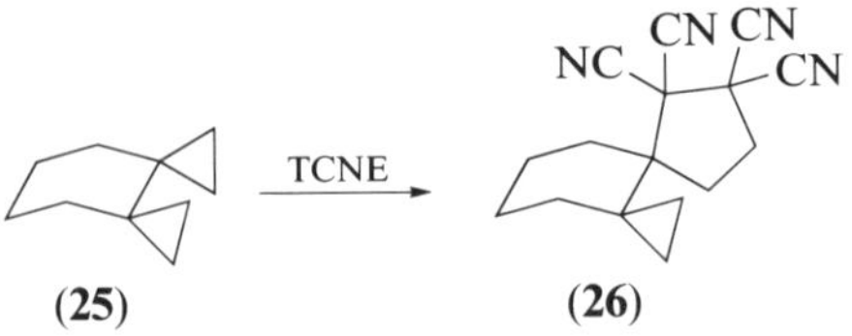

(22)

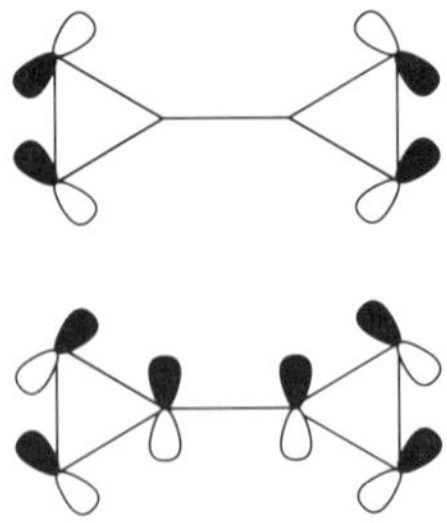

FIGURE 2. Bonding interactions between two conjugated cyclopropyl fragments

example of this type of reaction, see Reference 49. However, bicyclopropyl (**24**), also possessing conjugated three-membered rings, which are freely rotating, is inert to TCNE.

Oxidation with ozone is another interesting reaction characteristic of SPC with conjugated cyclopropyl fragments. It was demonstrated that ozone oxidized an α-methylene group in alkyl substituted cyclopropanes, yielding the corresponding ketones (equation 23)[50]. However, in SPC with conjugated cyclopropyl fragments, competitive attack on the three-membered ring was observed (equation 24)[51]. This direction of the ozone attack becomes exclusive in the absence of activated methylene groups (equation 25)[51]. Cycloaddition of ozone to the carbon–carbon bond of the cyclopropyl fragment, yielding intermediate **27**, was suggested to account for the mechanism of the reaction[51]. Fitjer and Conia demonstrated that conjugated cyclopropyl fragments in SPC could undergo cascade nucleophilic rearrangements, providing access to cyclic polyspirocyclobutanes (equation 26)[52].

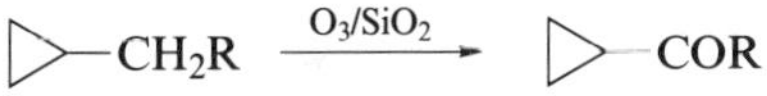

(23)

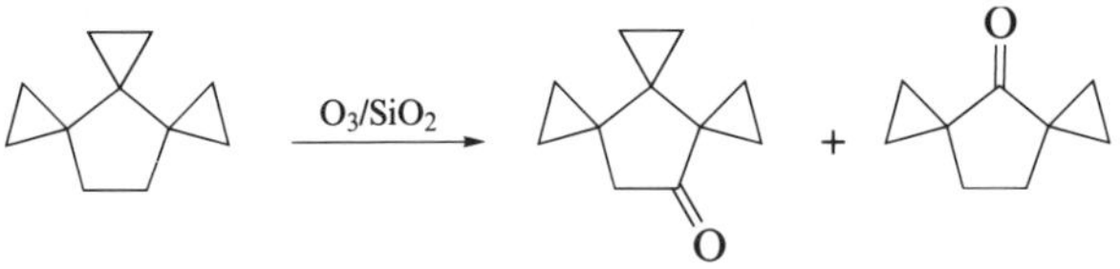

(24)

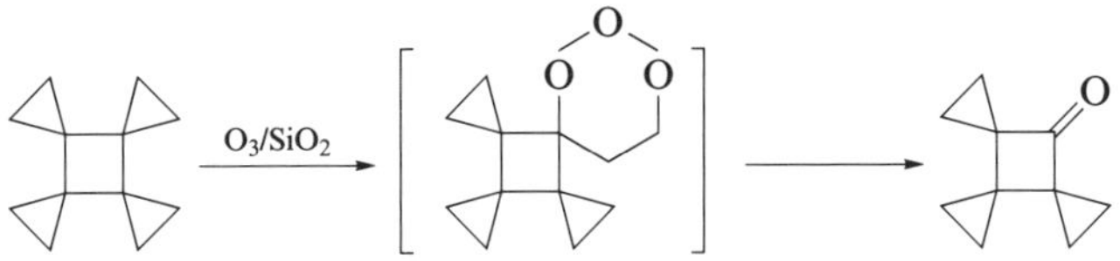

(25)

(27)

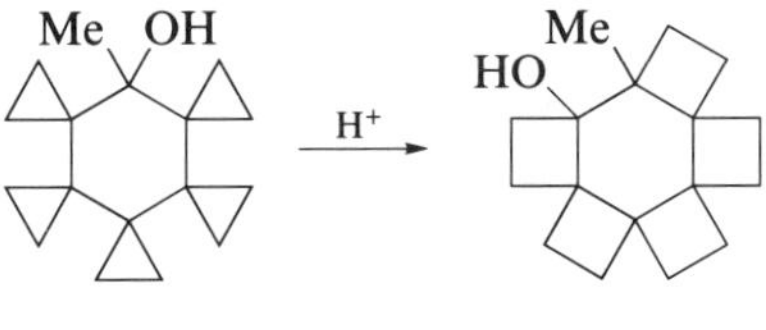

(26)

VII. THE ROTANES

The *rotanes* are an unusual type of hydrocarbon constructed from conjugated spiroannulated cyclopropyl fragments. Structurally, *rotanes* **28** can be considered as cyclic oligomers of cyclopropylidene.

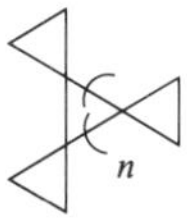

(28)

As demonstrated in Section IV, compounds possessing two spiroannulated cyclopropyl fragments could be successfully prepared by using double cyclopropanation or double cyclization of the corresponding bismethylene or bis(*gem*-dihalomethyl) precursors.

However, the introduction of a third, fourth, etc. cyclopropyl fragment requires the development of more elaborate synthetic procedures based on sequential introduction of cyclopropyl fragments into the appropriate precursors. The syntheses of rotanes accomplished in the 1970s demonstrated interesting solutions to this problem.

A. [3]Rotane (trispiro[2.0.2.0.2.0]nonane 23)[20,53]

An improved method of preparation of this compound is based on cyclopropanation of 7-methylenedispiro[2.2.1]heptane (**11**)[19,53,54]. This olefin was originally prepared by reductive cyclization of dibromide **10** (equation 5)[20]. A more convenient method is based on cyclopropanation of bicyclopropylidene **19** with chloromethylcarbene, followed by dehydrochlorination (equation 27)[19,54]. Additional information about [3]rotane, a member of the triangulane family, appears in Section VIII.

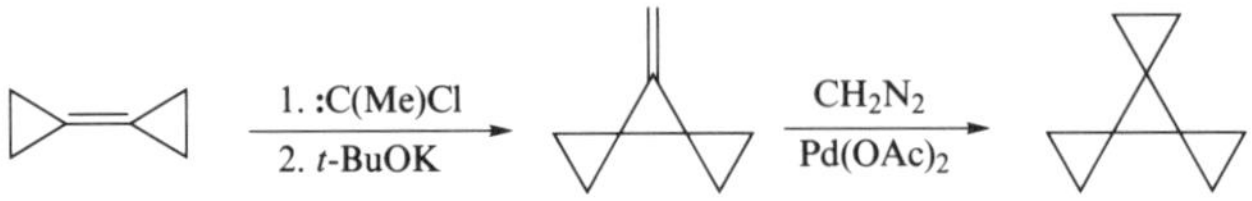

(27)

(19) **(11)** **(23)**

B. [4]Rotane (tetraspiro[2.0.2.0.2.0.2.0]dodecane, 22)

The first multistep synthesis of this compound was described by Conia and coworkers[30,55]. Its key step was condensation of bicyclopropyl diester **29** to the bis-TMS-enediol **14** (equation 28). Stepwise desilylation followed by Wittig methylenation and cyclopropanation gave the desired rotane **22**. It was found later that bicyclopropylidene **19** can undergo thermal [2+2] cyclodimerization to yield [4]rotane **22** (equation 29)[56]. Although the yield of [4]rotane prepared by this method was only 20% (due to the isomerization of bicyclopropylidene into methylenespiropentane **30**), the availability of the starting material[57] makes this method reasonably attractive.

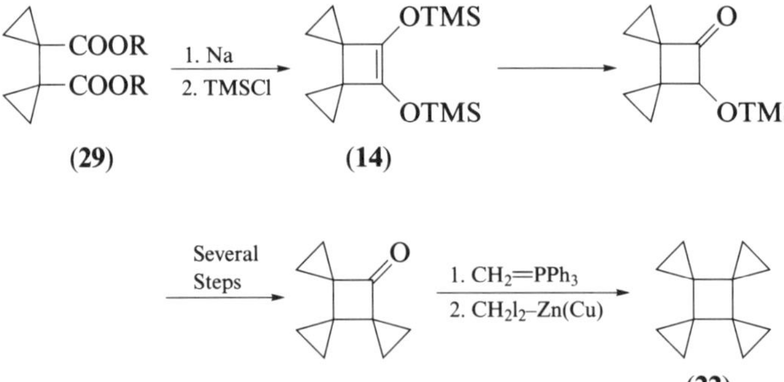

(29) **(14)** **(22)**

(28)

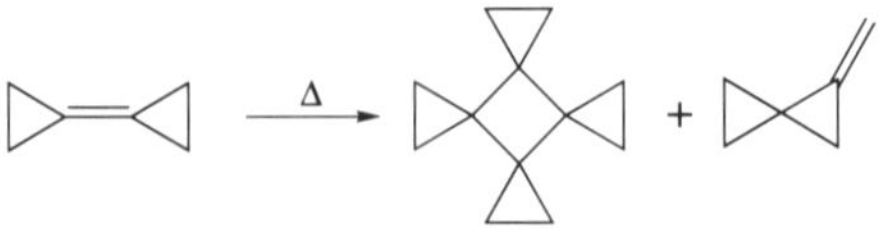

(29)

(19) **(22)** 20% **(30)**

C. [5]Rotane (pentaspiro[2.0.2.0.2.0.2.0.2.0]pentadecane, 31)

[5]Rotane was first synthesized by Rippol and Conia[29]. The key steps included the synthesis of dispirononanone **13** (equation 11). Condensation of **13** with formaldehyde provided the tetrahydroxymethyl derivative **32**, which was converted into the tetraspiro compound **33** by tosylation-bromination followed by reductive cyclization. Subsequent Wittig methylenation and cyclopropanation of the ketone **33** completed the synthesis (equation 30).

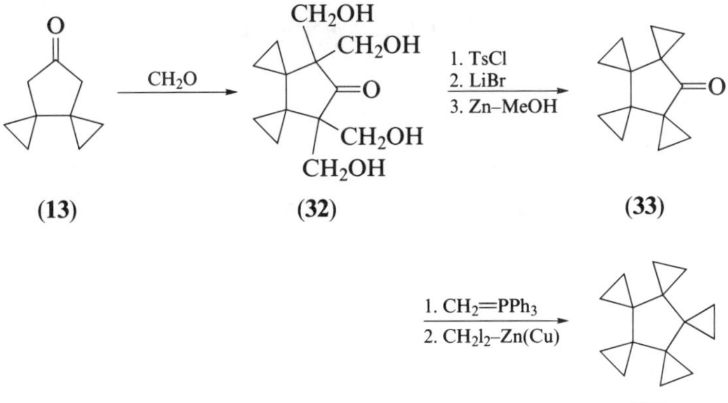

(13) **(32)** **(33)**

(31)

(30)

D. [6]Rotane (hexaspiro[2.0.2.0.2.0.2.0.2.0.2.0]octadecane, 34)

The synthesis of [6]rotane was reported by Proksh and de Meijere[58]. The precursor dispiro[2.2.2.2]decane (**35**) was oxidized with ozone on silica to give diketone **36** as a major product. Wittig methylenation and cyclopropanation gave tetraspiro compound **37**. This sequence (oxidation, methylenation and cyclopropanation) was repeated to give the desired rotane **34** (equation 31). At the same time, Fitjer developed a nice method of successive 'homologization' of rotanes[59]. The key steps include regiospecific cycloaddition of *p*-nitrophenyl azide to the double bond of the rotane precursor **38** followed by nitrogen extrusion from resulting triazoline **39** to yield imide **40**. Hydrolysis of **40** gave ketone **41**, which was converted into [4]rotane **22** via Wittig methylenation and cyclopropanation, or into the precursor for the [5]rotane **42** via reaction with cyclopropylidenephosphorane

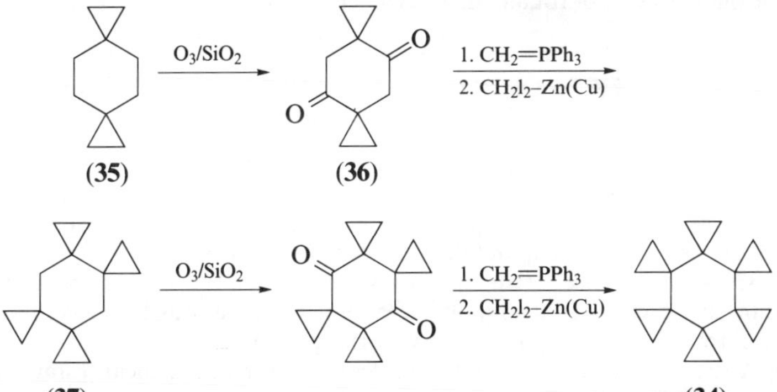

(35) **(36)**

(37) **(34)**

(31)

(equation 32). The synthetic utility of the method was demonstrated by preparation of [4]-, [5]- and [6]rotanes from olefin **38**[59].

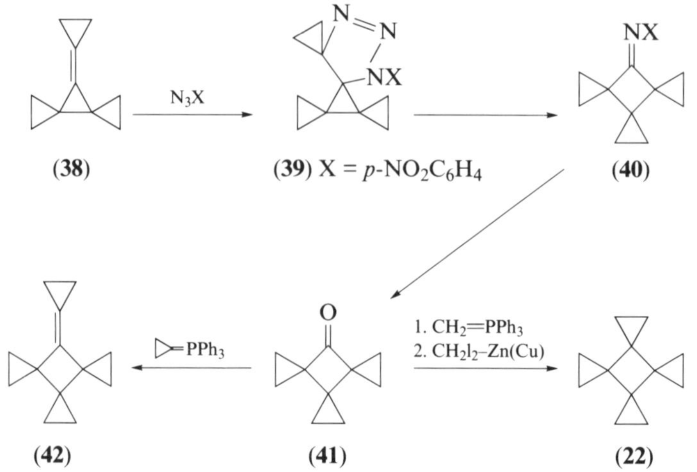

(32)

X-ray structural studies of [5]- and [6]rotanes (**31** and **34**) demonstrated that the geometry of the cyclopropyl fragments, as well as of the central ring, was not disturbed as a result of the spiroannulation[60]. (However, structural data for [3]- and [4]rotanes reported in Reference 60 were found to be incorrect[61,62].) In contrast, substantial changes in the geometry were observed in spiroannulated fragments in [3]- and [4]rotane[61,62], compared to cyclopropane and cyclobutane. Elongation of the distal bonds and shortening of the bonds connected to spirocenters were found in the peripheral cyclopropyl fragments of both compounds **22** and **23**. Bond lengths in the central fragments were significantly smaller than in cyclopropane or cyclobutane, respectively[61,62]. These observations are in agreement with the strain-induced changes in the geometry of SPC which were discussed in Section III and indicated that even in rotanes **31** and **34** the formation of the spirocenter was not associated with an additional strain.

[6]Rotane (**34**) can be considered as one of the most rigid derivatives of cyclohexane. The barrier for the inversion of the six-membered ring in this compound was found to be 22 kcal mol^{-1} [59,63].

The chemistry of rotanes remains unexplored. Ozonolysis of [4]rotane (see equation 25) may be the only reaction of rotanes described in the literature[51].

VIII. OLIGOMERS OF SPIROANNULATED CYCLOPROPANES: THE TRIANGULANES

Triangulanes are a unique class of polycyclic hydrocarbons constructed from spiroannulated cyclopropanes. Because of a stereochemical diversity, the number of isomeric triangulanes sharply increases with increasing the number of cyclopropyl fragments. Consequently, one may speak about the 'land of triangulanes'.

For classification purposes, triangulanes have been divided into three types—linear (**43**), branched (**4**) and cyclic (**44**)[9]. Compounds **45** and **46** can be considered as precursors to unknown, extremely strained [4]- and [5]cyclotriangulanes[64,65].

An application of graph theory to the problem of isomerism in linear triangulanes provided a program which allowed one to follow the development of a generation tree for each

(43)

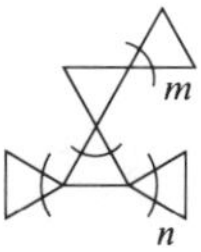

(4)

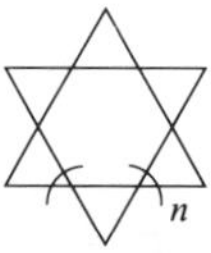

(44)

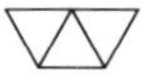

(45)

(46)

isomer[9]. Using this method one can find that [4]triangulane **5** exists as a pair of enantiomers, [5]triangulane as a pair of geometrical isomers **47** and **48** and [9]triangulane as a mixture of twenty isomers of which sixteen are chiral[9].

(5)

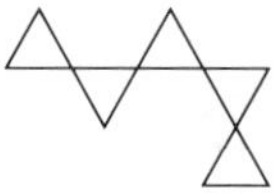

(47)

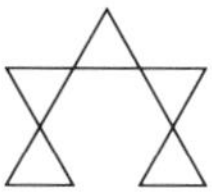

(48)

Despite the isomerism curiosity, an unusual combination of properties (high strain energy, unique geometry, conjugation between three-membered rings) makes these compounds a valuable synthetic target. While the cyclic triangulanes continue to intrigue organic chemists[66], recent synthetic efforts provided useful methodologies for the synthesis and study of linear and branched triangulanes. These accomplishments are summarized in the following paragraphs.

A. Synthesis of Triangulanes

The smallest triangulane derivative, i.e. spiropentane, was synthesized by Gustavson in 1896[14a] and by Zelinski and Kraevich[14b] by the reductive cyclization of tetrakis-(bromomethyl)methane (**8**) with Zn–NaOH (equation 4). However, it was shown in Section IV that the reductive cyclization methodology cannot be applied to the synthesis of higher homologs of triangulanes. Only eighty years later did developments in carbene cycloadditions to olefins facilitate syntheses of [3]- and [4]triangulanes **6**, **5** and **23** as outlined in equations 33–35[53,67,68]. However, a general efficient methodology was still required for the synthesis of higher homologs. Since triangulanes can be considered as spiroannulated *oligomers* of cyclopropanes, it is appropriate to apply terminology and, hopefully, methodology of polymer chemistry to the synthesis of these compounds. Zefirov's group developed a general method for the synthesis of *linear triangulanes* (**43**) utilizing procedures for the propagation and termination of the chain of spiroannulated three-membered rings, based on cycloadditions of carbenes to derivatives of methylenecyclopropane[9,19]. The two-step transformation, namely the addition of chloro(methyl)carbene[69] to an olefin, followed

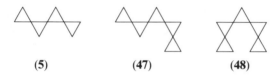

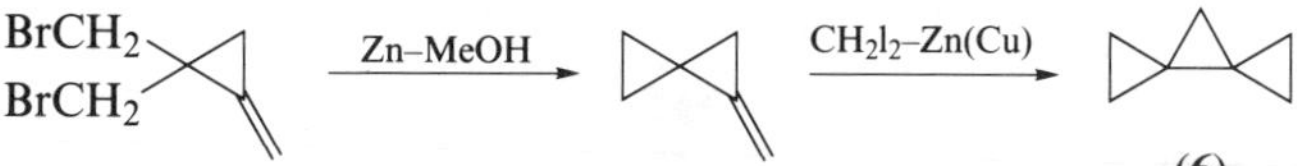

(6) (33)

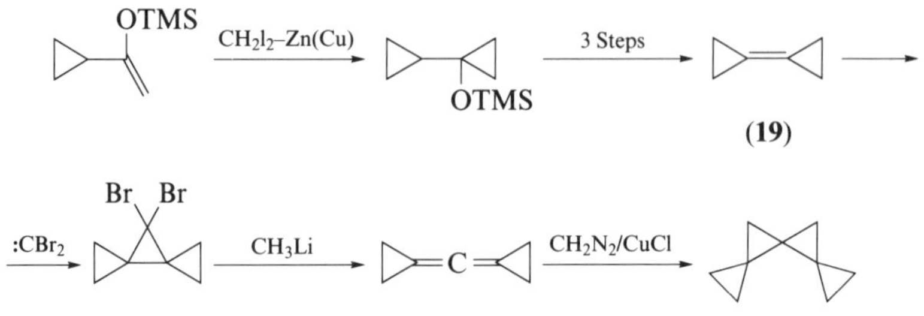

(19)

(5)

(34)

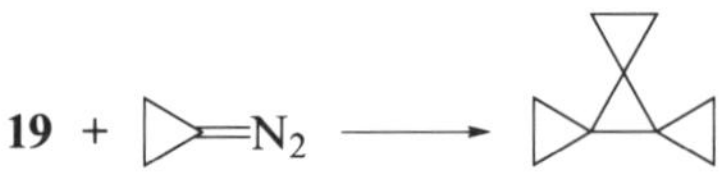

(23)

(35)

by dehydrochlorination (equation 36), was found very useful for the chain propagation. The best results in the termination, i.e. cyclopropanation of the last methylenecyclopropyl fragment in the molecule, were achieved using the diazomethane–palladium acetate system.

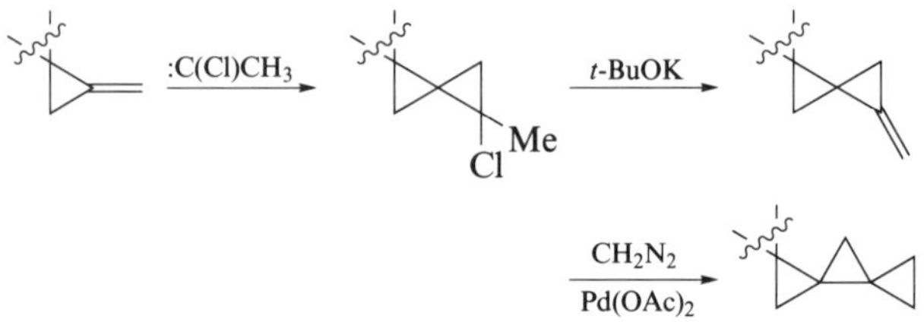

(36)

The efficiency of these procedures was demonstrated by the synthesis of linear triangulanes containing up to six cyclopropyl fragments. Steric interactions were found to strongly influence the direction of chain development and, consequently, the isomeric content of the resulting triangulanes[9]. For example, a 3:1 ratio of the *trans* to *cis* isomer (**47** to **48**) was determined in [5]triangulane. None of the *cis-cis*-isomer **49** was found in [6]triangulane prepared by using this methodology[9].

The application of the above procedures to bicyclopropylidene **19** or triangulane derivatives, containing the bicyclopropylidene fragment, allows the preparation of *branched triangulanes*. Thus [3]rotane **23**, as well as [5]- and [6]triangulanes (**50** and **51**), were prepared starting from bicyclopropylidene[54,70]. However, the generality of this approach is limited by the availability of the required bicyclopropylidene derivatives[20,67,71]. A new procedure for the conversion of triangulane esters into (cyclopropylidene)triangulanes (equation 37), based on the Kulinkovich reaction[72], is helpful in expanding the scope of this method[57].

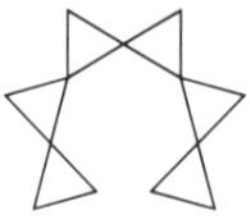

(49)

(50)

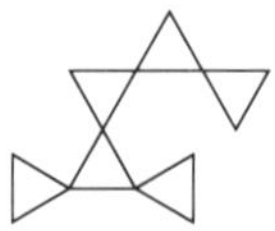

(51)

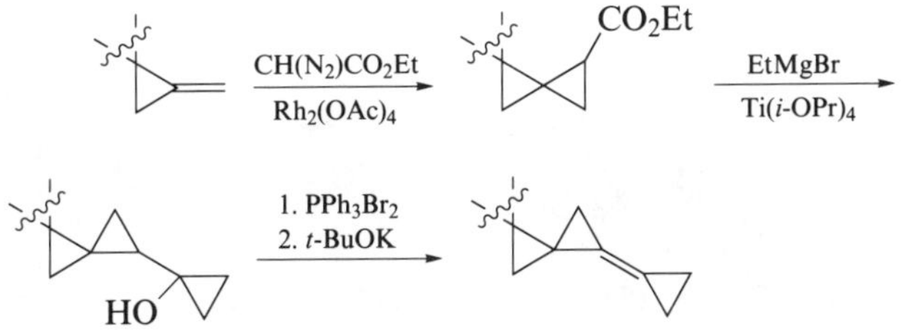

(37)

Zefirov's group suggested a general method for the synthesis of *branched triangulanes* **4** by using two separate procedures for the chain branching[70,73]. The first one is based on the transformation of a methylenecyclopropyl fragment into a vinylidenecyclopropyl moiety. Practically, this was achieved by the cycloaddition of dibromocarbene, followed by treatment of the resulting dibromocyclopropyl fragment with methyllithium (equation 38)[70].

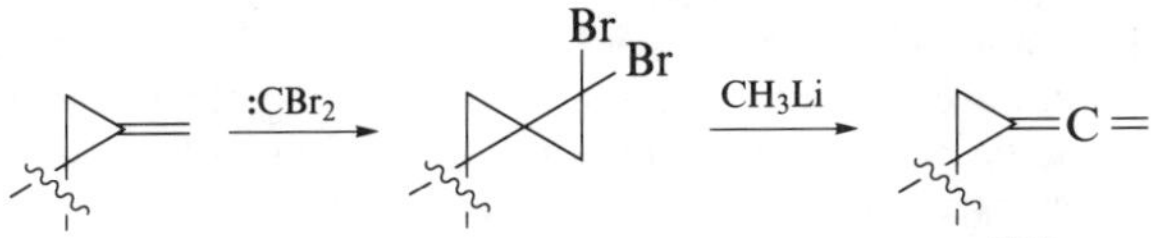

(38)

(52)

The simultaneous propagation of the synthetic chain in two directions using double addition of (chloro)methylcarbene to compounds of the type **52** was found inefficient[74]. Instead, the terminal double bond was converted into a (tetrahydropyranyloxymethyl)-cyclopropyl fragment (cf **53**) via rhodium acetate catalyzed cyclopropanation with ethyl diazoacetate, followed by reduction and protection of the resulting alcohol (equation 39)[70]. While the remaining double bond in compound **52** can be used for the propagation of the synthetic chain in one direction, the (tetrahydropyranyloxymethyl)cyclopropyl fragment can be later transformed into a new methylenecyclopropane fragment which is used to develop the chain in the second direction.

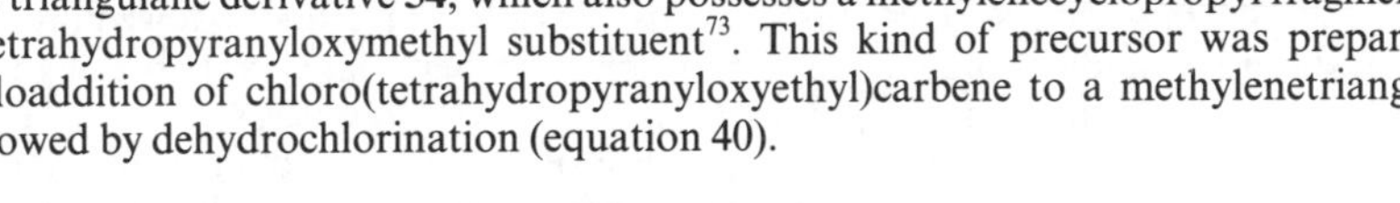

(39)

(53)

The second method for the chain branching is based on the preparation of the alternative triangulane derivative **54**, which also possesses a methylenecyclopropyl fragment and a tetrahydropyranyloxymethyl substituent[73]. This kind of precursor was prepared by cycloaddition of chloro(tetrahydropyranyloxyethyl)carbene to a methylenetriangulane followed by dehydrochlorination (equation 40).

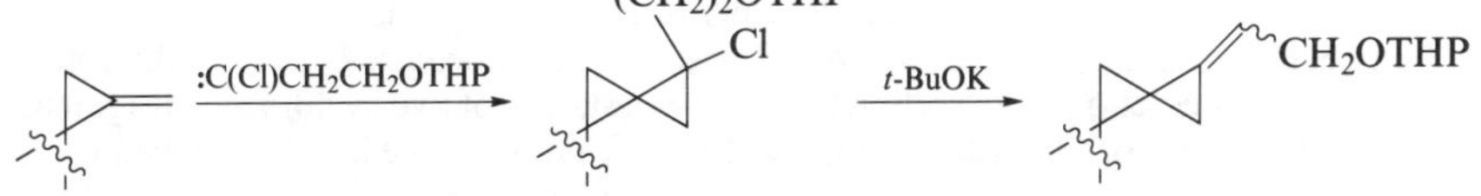

(54) (40)

Both procedures were successfully used for the preparation of branched triangulanes **50** and **51** as well as for the previously unavailable [6]triangulane **55**[70,73,75].

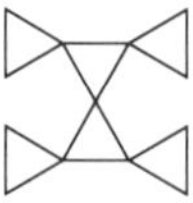

(55)

(56) $n = 0, 1, 2$

The synthetic efforts described above provided general procedures which were used for the preparation of various isomers of triangulanes, containing up to six three-membered rings, and it may be applicable for the synthesis of triangulanes, containing up to eight cyclopropyl fragments. However, a decrease of the total yield with growing number of steps, as well as low stereoselectivity of carbene cycloadditions, diminish the efficiency of the synthesis of higher homologs of triangulanes.

Zefirov's group reported the synthesis of bis-methylene derivatives of triangulanes **56** which are prospective starting materials for the synthesis of high molecular weight oligomers of spiroannulated cyclopropanes[76]. However, the efficient synthesis of the latter compounds requires development of a method for coupling of various triangulane fragments. The only known procedure for this kind of coupling utilizes the cycloaddition of cyclopropylidene or corresponding ylidene derivatives of triangulanes to unsaturated triangulanes, according to equation 35[55]. Although this method gives only modest yields of the desired products even if an excess of the olefinic component is used, it remains the only way to synthesize triangulanes possessing more then eight three-membered rings. For example, the synthesis of highly symmetric [10]triangulane **57** was achieved by de Meijere group using this method (equation 41)[77].

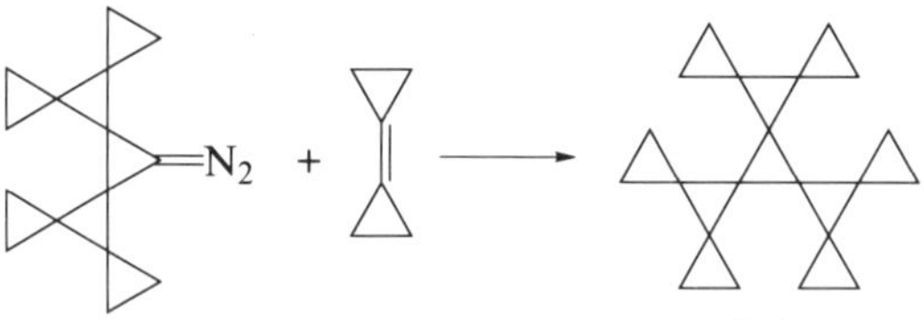

(41)

(57)

B. Geometry of Triangulanes

The studies of the strain distribution in spiroannulated systems (Section III) and its influence on the geometry of molecules required determination of the structure of triangulanes. Boese and coworkers published the X-ray structure of spiropentane[7] and revised structure of [3]rotane **29**[61]. Recent improvements of synthetic methods allowed preparation and structural study of other highly symmetric branched triangulanes bearing up to ten cyclopropyl fragments (e.g. compounds **55** and **57**)[71,73,77]. However, structural data on linear triangulanes remained unavailable because of difficulties obtaining these compounds in crystalline form. These problems were overcome by X-ray studies of derivatives of branched triangulanes bearing linear fragments (e.g. compounds **58–60**)[78,79].

The available data on the geometry of triangulanes have been reviewed[79,80]. In accord with theoretical predictions (see Section III) spiroannulation of two cyclopropyl fragments causes the elongation of the distal bonds and the shortening of the bonds attached to the spirocenter. This results in the increase of the angle at the spirocenter from 60° to 62.2° and a partial release of strain.

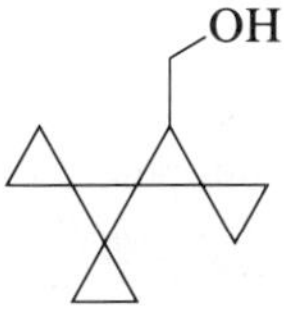

(58)

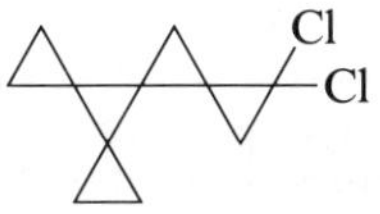

(59)

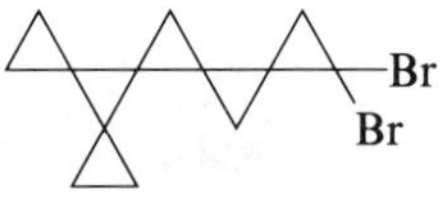

(60)

Superposition of these effects in higher triangulanes accounts for substantial differences in bond lengths in cyclopropyl fragments as outlined, for example, in Figure 3[79]. Fortunately, it was demonstrated that a simple empirical additive scheme can nicely describe the geometry of all known triangulanes, containing up to six cyclopropyl fragments[79].

The chain of three-membered rings in linear triangulanes has a unique helix-like geometry (e.g. in all *trans*-[*n*]triangulane **61**). This allows cyclopropyl fragments which are separated by several three-membered units to potentially reside very close in space. Thus, triangulanes are very useful models for studies of proximity-induced reactions.

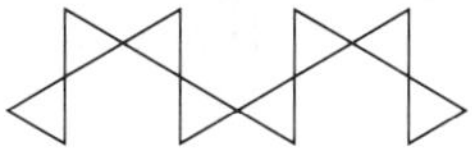

(61)

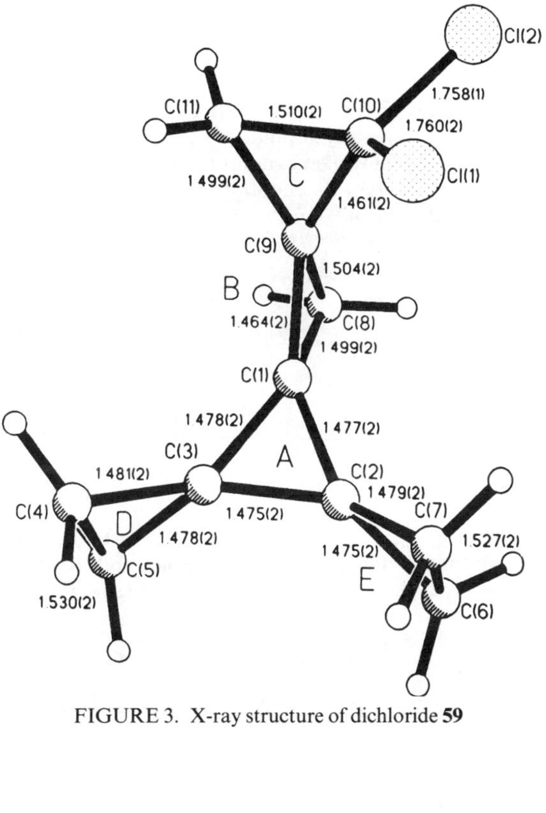

FIGURE 3. X-ray structure of dichloride **59**

C. Spectroscopic Identification of Triangulanes

NMR is a standard method for identification of hydrocarbons; triangulanes are no exception. While the ^{1}H NMR spectra of triangulanes become more and more complicated with the growing number of cyclopropane fragments, ^{13}C NMR spectroscopy was found very useful for the assignment of the configuration of triangulanes[81]. For example, it was demonstrated that chemical shifts of spiroatoms were influenced by the number of neighboring spirocenters. The dependencies of carbon chemical shifts of terminal and internal methylene groups on the number and the configuration of the other three-membered rings were determined and provided an additivity scheme for the calculation of chemical shifts[81].

D. Chemistry of Triangulanes

1. Thermal stability

Despite the high strain energy which increases with the number of three-membered rings in the molecule, triangulanes demonstrate high thermal stability. Thus spiropentane isomerizes to methylenecyclobutane only at temeratures higher than 360 °C (equation 42)[82]. [10]Triangulane **57** melts at 200 °C without decomposition[77].

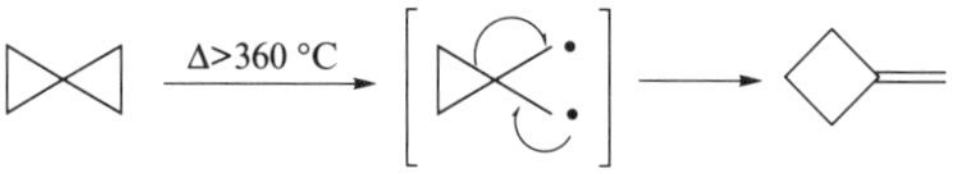

(42)

(62)

The studies of the thermal stability of substituted spiropentanes demonstrated the possibility of their isomerization via biradical mechanism at temperatures higher than 300 °C. It was estimated that biradical **62** collapses back to spiropentane ten times faster than it undergoes ring opening to methylenecyclobutane[83]. This makes thermal isomerization of substituted triangulanes under controlled conditions synthetically important. For example, *cis*- [3]triangulane derivative **63** was smoothly converted into the *cis-trans*-isomer **64** in the gas phase at 370 °C (equation 43)[84].

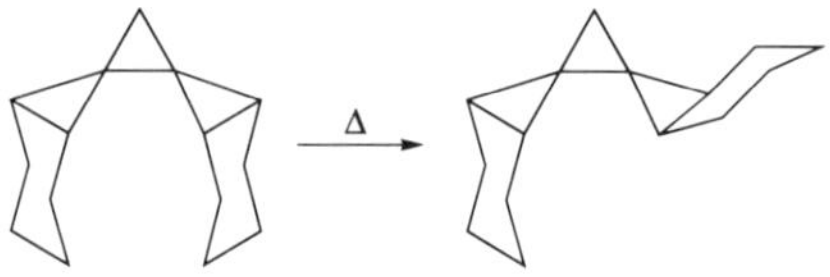

(43)

(63) **(64)**

2. Ring opening in triangulanes

Surprisingly, spiropentane was shown to be more resistant to the ring opening by transition metal compounds than cyclopropane. For example, while Rh(I) and Pd(II) destroy the skeleton of spiropentane, yielding π-allylic complex **65** (equation 44), in the presence

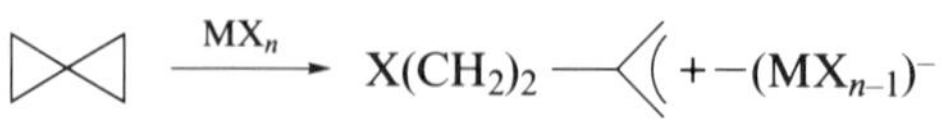

(44)

M = Rh(I), Pd(II) **(65)**

of Pt(II) it is intact[85,86]. However, this decreased reactivity can be accounted for by considering steric restrictions. Indeed, 1,1-dimethylcyclopropane was also inert to Zeise salt[87].

Recently, it was found that the triangular skeleton can be easily destroyed in reactions which most likely include an electron transfer step. For example, the reduction of *gem*-dichloro[3]triangulane **66** with lithium in *t*-butanol (which is the standard method for the synthesis of cyclopropanes) unexpectedly gave 50% of unsaturated hydrocarbon (equation 45)[88]. An even more complex mixture of hydrogenolysis products was obtained in the reduction of dichloride **67**[88].

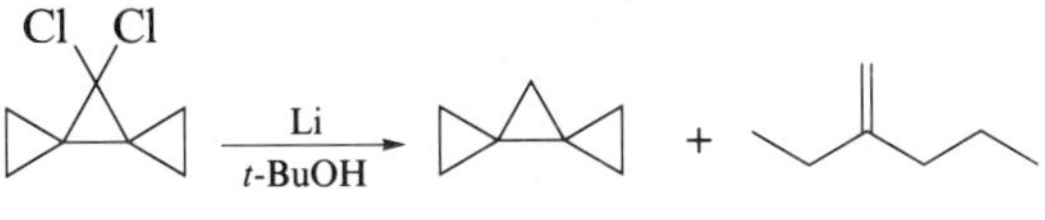

(45)

(66)

(67)

Attempted conversion of *gem*-dibromotriangulanes **68** and **71** to the corresponding vinylidene derivatives, e.g. **69** was also accompanied by an unusual rearrangement, yielding cyclobutene derivatives (equations 46 and 47)[89]. On the other hand, a similar reaction of dibromide **71** gave the desired allene **72** in 85% yield (equation 48)[71].

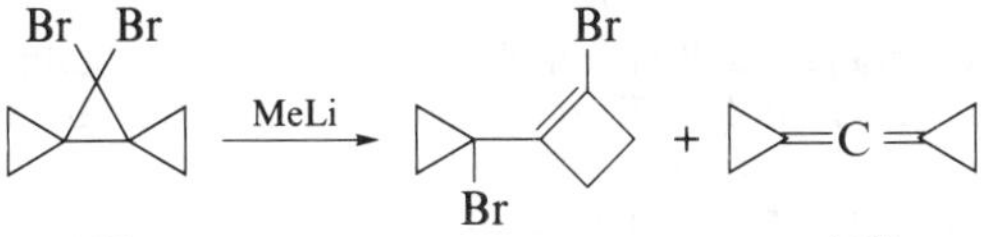

(46)

(68) **(69)**

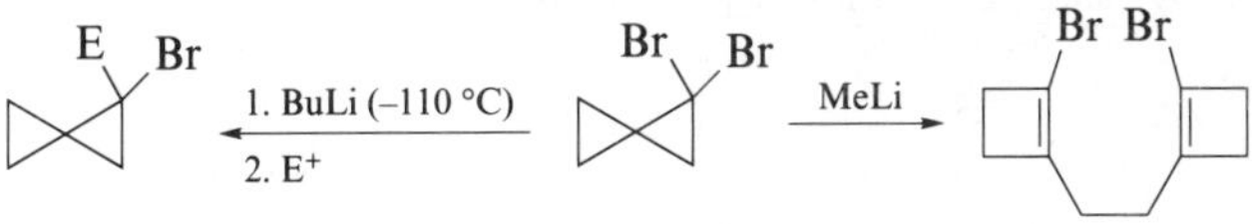

(47)

(70) Major Product

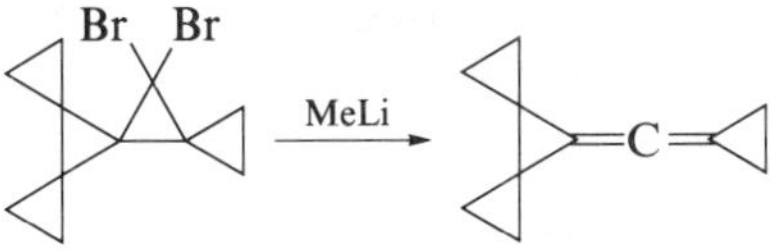

(48)

(71) **(72)**

Treatment of dibromospiropentane **70** with butyllithium at –110 °C also results in formation of expected bromolithio derivative, which was subsequently trapped by electrophiles (equation 47)[90].

3. Stereospecific functionalization of triangulanes

Synthesis of more complicated triangulanes necessitates the development of procedures for their stereospecific functionalization. Enhanced acidity of hydrogens in cyclopropyl fragments and the unique geometry of triangulanes makes the directed metalation methodology[91,92] particularly promising to achieve this goal. Eaton and Lukin applied the carboxamido group-directed metalation of triangulanes with magnesium amides for their stereospecific functionalization[93]. They demonstrated that metalation of isomeric [3]triangulane amides **73** and **74** gave only one product, namely *β-cis*-anion **76**, which was quenched with a variety of electrophiles, thus uncovering a method for the stereospecific functionalization of the activated ring (equation 49).

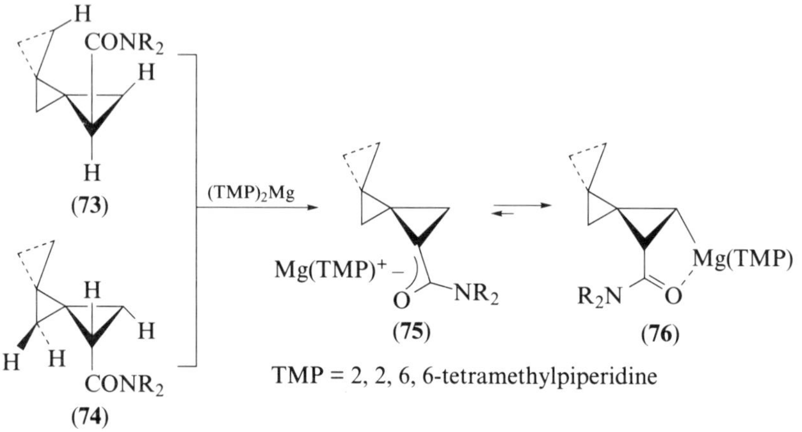

(49)

This result can be rationalized by the higher thermodynamic stability of the chelated *β*-enolate **76** over 'normal' *α*-enolate **75** in triangulanes[93]. In the absence of the *α*- or *β-cis*-hydrogens, the directed metalation was extended to adjacent and even remote cyclopropyl fragments, thus demonstrating the unique through-space activation of C—H bonds in triangulanes. For example, a carboxylic group was specifically introduced into position 5 in compound **77**, by using this method (equation 50)[93].

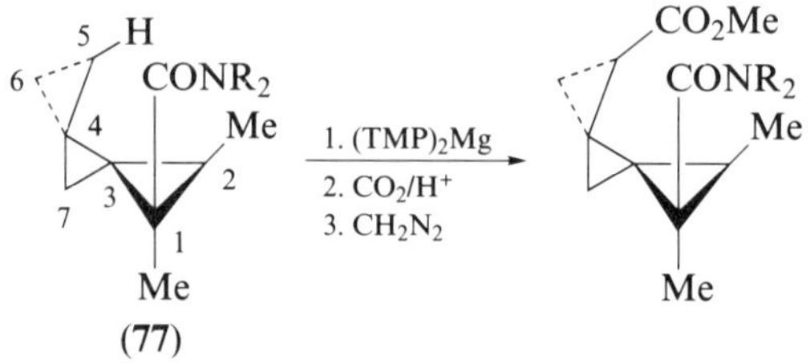

(50)

IX. CONCLUDING REMARKS

In this chapter we tried to demonstrate that existing synthetic methods allow convenient preparation of various types of SPC. Analysis of the literature suggests that the spiroannulation in most cases does not result in a change of the chemistry of the cyclopropyl fragment, and SPC can be treated as normal derivatives of cyclopropane.

This review was primarily focused on the SPC with unusual chemical behavior and the search for the factors responsible for this 'activation'. The chemistry of such compounds mainly remains unexplored, so that the authors tried to mark interesting direction for its development.

In addition to the importance of the 'activated' SPC for understanding the relationship between strain and chemical properties, these compounds have many potential practical applications. For example, a combination of a high energy content and remarkable thermal stability makes triangulanes valuable additives for special fuels. Unusual geometry and rigid skeleta of the molecules of SPC open the possibility for use of these compounds as frames for the controlled arrangement of functional groups in space for the design of future materials.

Finally, the authors hope that this review will encourage more chemists to creativley consider the use of SPC in their own work.

X. REFERENCES

1. *IUPAC Nomenclature of Organic Chemistry*, 3rd ed., Butterworth, London. 1971.
2. M. P. Kozina, V. S. Mastryukov and E. M. Milvitskaya, *Russian Chem. Reviews*, **51**, 765 (1982).
3. K. B. Wiberg, *Angew. Chem., Int. Ed. Engl.*, **25**, 312 (1986).
4. (a) J. Kao and L. Radom, *J. Am. Chem. Soc.*, **100**, 760 (1978).
   (b) A. I. Ioffe, V. A. Svyatkin and O. M. Nefedov, *Izv. Akad. Nauk SSSR, Ser. Khim.*, 801 (1987); *Chem. Abstr.*, **108**, 111418b (1988).
   (c) S. P. Zilberg, A. I. Ioffe and O. M. Nefedov, *Bull. Acad. Sci. USSR, Div. Chem. Sci.*, **32**, 228 (1983).
   (d) H. Dodziuk, *Bull. Polon. Acad. Sci.*, **34**, 49 (1986).
   (e) K. Rusmussen and C. Tosi, *J. Mol. Struct.*, **121**, 233 (1985).
5. P. Aped and N. L. Allinger, *J. Am. Chem. Soc.*, **114**, 1 (1992).
6. G. Dallinga, R. K. Van Der Draai and L. H. Tonemann, *Recl. Trav. Chim. Pays-Bas*, **87**, 897 (1968).
7. R. Boese, D. Blaser, K. Gomann and U. H. Brinker, *J. Am. Chem. Soc.*, **111**, 1501 (1989).
8. K. Kovacevic, Z. B. Naksic and A. Mogus, *Croatica Chem. Acta*, **52**, 249 (1979).
9. N. S. Zefirov, S. I. Kozhushkov, T. S. Kuznetzova, O. V. Kokoreva, K. A. Lukin, B. I. Ugrak and S. S. Tratch, *J. Am. Chem. Soc.*, **112**, 7702 (1990).
10. Ya. M. Slobodin and M. V. Blinova, *J. Gen. Chem. USSR*, **24**, 633 (1954).
11. R. W. Shortridge, R. A. Craig, K. W. Greulee, J. M. Derfer and C. A. Boord, *J. Am. Chem. Soc.*, **70**, 976 (1948); P. Leriverend and J. M. Conia, *Bull. Soc. Chim. France*, 116 (1966).
12. D. S. Magrill, J. Altmann and D. Ginsburg, *Isr. J. Chem.*, **7**, 479 (1969).
13. Ya. M. Slobodin, M. V. Blinova, *J. Gen. Chem. USSR*, **23**, 2109 (1953).
14. (a) G. Gustavson, *J. Pract. Chem.*, **54**, 97 (1896).
    (b) N. D. Zelinski and V. P. Kraevich, *J. Russ. Phys. Chem. Soc.*, **44**, 1870 (1912).
15. D. E. Applequist, G. F. Fanta and B. W. Henrickson, *J. Org. Chem.*, **23**, 1715 (1958).
16. M. Rifi, *Tetrahedron Lett.*, 1043 (1969).
17. D. E. McGreer, *Can. J. Chem.*, **38**, 1638 (1960).
18. A. I. D'yachenko, E. L. Protasova and O. M. Nefedov, *Bull. Acad. Sci. USSR, Div. Chem. Sci.*, **28**, 1093 (1979); A. I. D'yachenko, A. I. Ioffe, E. L. Protasova and O. M. Nefedov, *Bull. Acad. Sci. USSR, Div. Chem. Sci.*, **28**, 1331(1979).
19. N. S. Zefirov, K. A. Lukin, S. I. Kozhushkov, T. S. Kuznetzova, A. M. Domarev and I. M. Sosonkin, *Zh. Org. Khim.*, **25**, 312 (1989); *Chem. Abstr.*, **111**, 194165m (1989).
20. L. Fitjer, *Angew. Chem., Int. Ed. Engl.*, **15**, 762 (1976).
21. W. Kirmse, *Carbene Chemistry*, Academic Press, New York, 1964.
22. H. E. Simmons, T. L. Cairns, S. A. Vladuchick and C. M. Hoiness, *Org. React.*, **20**, 1 (1973).
23. K. Maruoka, Y. Fukutani and H. Yamamoto, *J. Org. Chem.*, **50**, 4412 (1985); E. C. Friederich, J. M. Domek and R. Y. Pong, *J. Org. Chem.*, **50**, 4640 (1985).
24. O. Repic and S. Vögt, *Tetrahedron Lett.*, **23**, 2729 (1982).
25. W. P. Weber and G. W. Gokel, in *Reactivity and Structure*, Vol. 11 (Ed. K. Hafner), Springer, Berlin, 1977; C. M. Starks and C. Liotta, *Phase Transfer Catalysis: Principles and Techniques*, Academic Press, New York, 1978.
26. A. R. Pinder, *Synthesis*, 425 (1980).
27. M. P. Doyle, *Chem. Rev.*, **86**, 919 (1986).

28. M. G. Bawell and M. E. Reum, in *Advances in Strain in Organic Chemistry* (Ed. B. Halton), Vol. 1, JAI Press, Greenwich, CT, 1991.
29. J. L. Rippol and J. M. Conia, *Tetrahedron Lett.*, 979 (1969)
30. J. M. Conia and J. M. Denis, *Tetrahedron Lett.*, 3545 (1969).
31. N. S. Zefirov, T. S. Kuznetziva, S. I. Kozhushkov, L. S. Surmina and Z. A. Rashchupkina, *J. Org. Chem. USSR*, **19**, 474 (1983).
32. W. Adam, M. Dorr, K. Hill, E. M. Peters and H. G. von Schnering, *J. Org. Chem.*, **50**, 587 (1985).
33. P. Binger, *Angew. Chem., Int. Ed. Engl.*, **11**, 433 (1972).
34. W. R. Dolbier, D. Lomas and P. Tarant, *J. Am. Chem. Soc.*, **90**, 3594 (1968).
35. J. L. Rippol, *Tetrahedron*, **33**, 389 (1977).
36. F. Kienzle and J. Stadlwieser, *Tetrahedron Lett.*, **32**, 551 (1991).
37. D. Kaufmann and A. de Meijere, *Angew. Chem., Int. Ed. Engl.*, **12**, 159 (1973).
38. R. Y. Levina, N. N. Mezentsova and O. V. Lebedev, *J. Gen. Chem. USSR*, **25**, 1055 (1955).
39. S. I. Kozhushkov, T. S. Kuznetzova, D. S. Yufit, Yu. T. Struchkov, O. V. Kokoreva and N. S. Zefirov, *Dokl. Akad. Nauk SSSR*, **312**, 118 (1990); *Chem. Abstr.*, **113**, 171216f (1990).
40. N. S. Zefirov, T. S. Kuznetzova, S. I. Kozhushkov and O. V. Kokoreva, *Zh. Org. Khim.*, **26**, 756 (1990); *Chem. Abstr.*, **113**, 131610y (1990).
41. W. A. Smit, S. L. Kireev, O. M. Nefedov and V. A. Tarasov, *Tetrahedron Lett.*, **30**, 4021 (1989).
42. Z. Rappoport (Ed.), *The Chemistry of the Cyclopropyl Group*, Wiley, New York, 1987.
43. H. Park, I. I. Canalda and B. Fraser–Reid, *J. Org. Chem.*, **55**, 3009 (1990).
44. H. Kigoshi, Y. Imamura, K. Mizuta, H. Niwa and K. Yamada, *J. Am. Chem. Soc.*, **115**, 3056 (1993).
45. M. Moir, P. Ruedi and C. H. Euguster, *Helv. Chim. Acta*, **56**, 261 (1973).
46. M. Klessinger and P. Rademacher, *Angew. Chem., Int. Ed. Engl.*, **18**, 826 (1979).
47. R. Gleiter, R. Haider, J.-P. Barnier, J.-M. Conia, A. de Meijere and W. Weber, *J. Chem. Soc., Chem. Commun.*, 130 (1979).
48. P. Hemmerrsbach and M. Klessinger, *Tetrahedron*, **36**, 1337 (1980).
49. H. Wenck, A. de Meijere, F. Gerson and R. Gleiter, *Angew. Chem., Int. Ed. Engl.*, **25**, 335 (1986).
50. E. Proksh and A. de Meijere, *Angew. Chem., Int. Ed. Engl.*, **15**, 761 (1976).
51. A. de Meijere, *Angew. Chem., Int. Ed. Engl.*, **18**, 809 (1979).
52. L. Fitjer and J. M. Conia, *Angew. Chem., Int. Ed. Engl.*, **18**, 868 (1979).
53. L. Fitjer and D. Wehle, *Angew. Chem., Int. Ed. Engl.*, **12**, 332 (1973).
54. I. Erden, *Synth. Commun.*, **16**, 117 (1986).
55. J. M. Denis, P. Le Perchec and J. M. Conia, *Tetrahedron*, **33**, 399 (1977).
56. P. Le Perchec and J. M. Conia, *Tetrahedron Lett.*, 1587 (1970).
57. A. de Meijere, S. I. Kozhushkov, T. Spaeth and N. S. Zefirov, *J. Org. Chem.*, **58**, 502 (1993).
58. E. Proksh and A. de Meijere, *Tetrahedron Lett.*, 4851 (1976).
59. L. Fitjer, *Angew. Chem., Int. Ed. Engl.*, **15**, 763 (1976).
60. T. Prange, C. Pascard, A. de Meijere, U. Behrens, J.-P. Barnier and J.-M. Conia, *Nouv. J. Chim.*, **4**, 321 (1980).
61. R. Boese, T. Miebach and A. de Meijere, *J. Am. Chem. Soc.*, **113**, 1743 (1991).
62. A. Almenningen, O. Bastiansen, B. N. Cyvin, S. Cyvin, L. Fernholt and C. Romming, *Acta Chem. Scand.*, **A 38**, 31 (1984).
63. L. Fitjer, K. Justus, P. Puder, M. Dittmer, C. Hassler and M. Noltemeyer, *Angew. Chem., Int. Ed. Engl.*, **30**, 436 (1991).
64. K. B. Wiberg and J. V. McClusky, *Tetrahedron Lett.*, **28**, 5411 (1987).
65. L. Skattebol, *J. Org. Chem.*, **31**, 2789 (1966).
66. N. S. Zefirov, T. S. Kuznetzova, O. V. Eremenko and O. V. Kokoreva, *Mendeleev Commun.*, 91 (1993).
67. W. R. Dolbier, D. Lomas and P. Tarrant, *J. Am. Chem. Soc.*, **90**, 3594 (1968).
68. L. Fitjer and J. M. Conia, *Angew. Chem., Int. Ed. Engl.*, **12**, 761 (1973).
69. S. Arora and P. Binger, *Synthesis*, 801 (1974).
70. K. A. Lukin, S. I. Kozhushkov, A. A. Andrievski, B. I. Ugrak and N. S. Zefirov, *J. Org. Chem.*, **56**, 6176 (1991).
71. S. Zollner, H. Buchholtz, R. Boese, R. Gleiter and A. de Meijere, *Angew. Chem., Int. Ed. Engl.*, **30**, 1518 (1991).
72. O. G. Kulinkovich, S. V. Sviridov and D. A. Vasilevski, *Synthesis*, 234 (1991).

73. N. S. Zefirov, S. I. Kozhushkov, B. I. Ugrak, K. A. Lukin, O. V. Kokoreva, D. S. Yufit, Yu. T. Struchkov, S. Zoelner, R. Boese and A. de Meijere, *J. Org. Chem.*, **57**, 701 (1992).
74. K. A. Lukin, A. Yu. Masunova, S. I. Kozhushkov, T. S. Kuznetsova, B. I. Ugrak, V. A. Piven and N. S. Zefirov, *J. Org. Chem. USSR*, **27**, 422 (1991).
75. K. A. Lukin, S. I. Kozhushkov, A. A. Andrievski, B. I. Ugrak and N. S. Zefirov, *Mendeleev Commun.*, 51 (1992).
76. K. A. Lukin, A. Yu. Masunova, B. I. Ugrak and N. S. Zefirov, *Tetrahedron*, **47**, 5769 (1991).
77. S. I. Kozhushkov, T. Haumann, R. Boese and A. de Meijere, *Angew. Chem., Int. Ed. Engl.*, **32**, 401 (1993).
78. D. S. Yufit, S. I. Kozhushkov, K. A. Lukin, N. S. Zefirov, Yu. T. Struchkov and A. de Meijere, *Doklady Chemistry*, **320**, 258 (1991).
79. K. A. Lukin, S. I. Kozhushkov, N. S. Zefirov, D. S. Yufit and Yu. T. Struchkov, *Acta Crystallogr. Sect. B*, **49**, 704 (1993).
80. R. Boese, in *Advances in Strain in Organic Chemistry* (Ed. B. Halton), Vol. 2, JAI Press, Greenwich, CT, 1992.
81. K. A. Lukin, S. I. Kozhushkov and N. S. Zefirov, *Magn. Reson. Chem.*, **29**, 774 (1991).
82. M. C. Flowers and H. M. Frey, *J. Chem. Soc.*, 5550 (1961).
83. J. J. Gajewski and L. T. Burka, *J. Am. Chem. Soc.*, **94**, 8857 (1972).
84. Y. Fukuda, Y. Yamamoto, K. Kimura and Y. Odaira, *Tetrahedron Lett.*, 877 (1979).
85. A. D. Ketley, J. A. Bratz and J. Craig, *J. Chem. Soc., Chem. Commun.*, 1117 (1970).
86. R. Rossi, P. Diversi and L. Porri, *J. Organometal. Chem.*, **47**, C 21 (1973).
87. B. C. Menon and R. E. Pincock, *Can. J. Chem.*, **47**, 3327 (1968).
88. K. A. Lukin, A. A. Andrievski and N. S. Zefirov, *Dokl. Akad. Nauk SSSR*, **321**, 521 (1991); *Chem. Abstr.*, **116**, 128207 (1992).
89. K. A. Lukin, N. S. Zefirov, D. S. Yufit and Yu. T. Struchkov, *Tetrahedron*, **48**, 9977 (1992).
90. K. A. Lukin and N. S. Zefirov, unpublished results.
91. V. Snieckus, *Chem. Rev.*, **90**, 928 (1990).
92. D. G. Gallagher, C. G. Garrett, R. P. Lemieux and P. Beak, *J. Org. Chem.*, **56**, 853 (1991).
93. P. E. Eaton and K. A. Lukin, *J. Am. Chem. Soc.*, **115**, 11370 (1993).

Author index

This author index is designed to enable the reader to locate an author's name and work with the aid of the reference numbers appearing in the text. The page numbers are printed in normal type in ascending numerical order, followed by the reference numbers in parentheses. The numbers in *italics* refer to the pages on which the references are actually listed.

Abboud, J.-L.M. 239 (72), 240 (76), *256*
Abdullah, H. 659 (25), *702*
Abelt, C.J. 354 (36), *407*
Abers, A. 500, 510 (25), *648*
Abiko, A. 292 (210), *315*, 695, 698 (182), *706*
Abiko, T. 686 (154), *705*
Abrahams, S.C. 197 (284a), *219*
Abram, D.M.H. 699 (200), *706*
Abram, T.S. 831 (63b), *858*
Achiba, Y. 60 (60), *131*
Achmatowicz, B. 303 (306), *317*
Adam, R. 271 (47a), *312*
Adam, W. 200 (299), *220*, 247 (114b), 248 (117), *259*, 867 (32), *884*
Adamczak, O. 229 (35, 37), 231 (43), 234 (51), 245 (99, 101, 103), 246 (105), *252, 253, 258, 259*, 357–359, 361, 378, 387, 389, 390 (41), *407*, 450–453, 457 (225), *467*
Adamowicz, L. 109 (185b), *134*
Adams, J. 291 (196), *315*, 658, 663 (6), 671 (86–88), *702, 704*
Adams, R. 412 (5), *460*
Adams, W.J. 145 (36), *213*
Adamsky, F. 200 (299), *220*
Adamson, J. 727 (129, 130), 760 (130), *768*
Adcock, J.L. 790, 791 (67), 794 (67, 90), 798, 807 (67), 808 (67, 90, 132), *811, 812*
Addington, D.M. 731, 742 (165), *769*
Adiwidjaja, G. 186 (236), *218*, 294 (236), *316*
Adler, G. 715 (44), *766*
Adlington, R.M. 170 (134), *216*, 302 (297), *317*
Agopovich, J.W. 176–178 (163), *216*
Aguirre, J.A. 832 (69), *858*
Ahlberg, P. 111 (208, 210), *136*, 364, 375, 381, 391, 392, 394, 396, 397 (50, 57), 402 (50), 403 (50, 57), 405 (57), *407, 408*, 414 (28), 440 (167–174), 441 (28, 171–174, 177, 180), 442 (28, 169, 180), 459 (282), *461, 465, 468*, 814 (3), 830, 831 (63a), 841 (90–93), 843 (94), 856 (3), *856, 858, 859*
Ahlquist, B. 210, 211 (340), *221*
Ahlrichs, R. 69 (75), *131*
Ahmar, M. 268 (39), *311*
Ahmed, J. 185 (213), *217*
Ahmed, M.S. 116 (228), *136*
Ahr, H.-J. 756 (289), *771*
Aida, F. 678 (115), *704*
Akhachinskaya, T.V. 620 (365, 366), *655*
Alber, F. 779, 781, 787, 791, 792, 799, 800, 802 (37), *810*
Albin, J. 569 (254), *653*
Albrecht, F.X. 324 (31b), *338*
Albright, T.A. 350, 370, 373 (28), *407*, 552 (195), 578, 580 (283), 589 (306), 590 (306, 308), 592, 601 (306), 602 (306, 335), 603 (306, 308), 625 (380), 629, 631, 633, 638 (389), *651, 653–655*
Alcock, N.W. 681 (128), *704*
Aldrich, P.D. 146 (46, 47), *214*
Al-Dulayymi, J. 289 (166), *314*
Alex, A. 128 (279), *137*
Ali, M.B. 287 (140), *313*
Allan, M. 778, 785, 786 (23), 799, 801 (117), *810, 812*

Allen, F.H. 83, 86, 87 (99a–d), *132*, 140 (8–12), 141 (18), 144 (9, 11), 146 (8, 9), 147 (8), 148 (9), 155 (8, 9), 156 (8), 159 (95), 160, 161, 175, 177 (8), 180, 185 (95), 190 (11), 198 (12), 201 (10), 202 (12), *213, 215*, 416 (45–47), 417 (45), 418 (45, 46), 422 (45), 456 (45–47), *462*
Allen, L.C. 48, 49, 51, 74 (32a–c), 83, 86 (32a–c, 105b, 105c), 95 (105b, 105c), 96–98, 124 (32a–c), *130, 132*, 147, 177 (53), 187 (238), *214, 218*, 418 (65), *462*, 794 (82), *811*
Allinger, N.L. 77 (92a, 92c), *132*, 149 (62), *214*, 228 (24b), 235 (54, 58), 239 (73), *251, 253, 254, 256*, 389 (98), *409*, 863 (5), *883*
Allred, E.L. 110 (195), *135*
Almenningen, A. 85, 86 (140), *133*, 153, 154 (77, 78), 155 (78), 204 (317), 210, 211 (340), *214, 220, 221*, 452 (244), *467*, 794 (88), *811*, 874 (62), *884*
Almqvist, A. 729 (143), *768*
Alnajjar, M.S. 513 (96), *649*
Alonso, M.E. 668 (70), 683 (143), 687 (156, 158, 159), *703, 705*
Alsenoy, C.van 209 (333), *221*
Alt, M. 663 (35), *703*
Alti, G.de 60 (59), *131*
Altmann, J. 864, 870 (12), *883*
Altmeier, P. 201 (303), *220*, 271 (47b), *312*
Alvarez, R.M. 832 (69), *858*
Ambler, P.W. 284 (126, 127), *313*
Amedio, J.C.Jr. 292 (224), *315*, 678 (114), 696 (187), *704, 706*
Amezua, M.G. 322, 333 (21), *337*
Amos, R.D. 84, 85, 96, 98, 101, 102 (130), 104 (167d, 168b), 106 (174), 107, 108 (178), 109 (167d), *133, 134*
Ananthanarayan, T.P. 687 (157), *705*
Anciaux, A.J. 663 (36), 688, 689 (161), *703, 705*
Andersen, B. 204 (317), *220*, 794 (88), *811*
Anderson, C.B. 519 (120), *650*
Ando, M. 266 (25), *311*
Andose, J.D. 77 (92e), *132*
André, J.-M. 83, 86 (109a), *132*
André, M.-C. 83, 86 (109a), *132*
Andrews, A.M. 146 (49, 50), *214*
Andrews, G.D. 470 (44, 59), 477, 479 (59), *489*
Andrews, U.H. 470 (52, 54), *489*
Andrievski, A.A. 876, 877 (70), 878 (70, 75), 881 (88), *884, 885*
Anet, F.A.L. 451 (229), 455 (257), *467*, 711, 712 (19), *766*
Anet, R. 711, 712 (19), *766*
Angerer, E.von (354, 355), *655*
Angus, P.M. 264 (19), *311*
Anisimov, V.M. 176 (155), *216*
Annamalai, A. 476 (172), *491*
Annunziata, R. 225 (13), 241 (84), *250, 257*
Ansell, G.B. 566, 567 (248), *653*
Antebi, S. 554 (198, 201), *651, 652*
Anthony, I.J. 720 (78), 722 (78, 107), 723, 740 (107), *767, 768*
Antipin, M.Y. 176 (155), *216*
Ao, M.S. 727, 760 (131), *768*
Aoki, K. 170 (132), *216*
Aoyama, H. 278 (71), *312*
Aped, P. 863 (5), *883*
Apel, M. 289 (169), *314*, 563 (242), *652*
Apeloig, Y. 47 (21a), 102 (160), *130, 133*, 708 (3), 722 (90), 733 (186), 738 (3, 186, 202, 203), 739 (203), 740 (218), 744 (3), 745, 749 (186), 758 (202, 203), 759, 760 (298), 763 (313), *766, 767, 769–772*, 833 (72), *858*
Apoita, M. 323, 331 (28a), *338*
Applequist, D.E. 412, 427 (10), *461*, 865 (15), *883*
Arad, D. 708 (3), 722 (90), 733 (186), 738 (3, 186, 203), 739 (203), 740 (218), 744 (3), 745, 749 (186), 758 (203), *766, 767, 769, 770*
Arai, H. 270 (43), *311*
Arai, I. 283 (109, 110), *313*
Arai, M. 473, 477 (145), *491*
Arai, Y. 283 (110), *313*
Araki, S. 301, 302 (293), *317*
Aratani, T. 292 (202), *315*, 693, 697 (172–174), *706*
Arbuzov, B.A. 199 (298), *220*
Arenal, I. 305 (315), *317*
Arias, J. 111 (206), *136*
Arif, A.M. 456 (265), *468*
Arikawa, Y. 746, 751 (259), *771*
Armbrecht, F.M.Jr. 506 (63), *649*
Armesto, D. 322 (21), 323, 331 (28a, 28b), 333 (21), *337, 338*
Armstrong, R. 412 (8), *460*
Arndt, G. 646 (432), *656*
Arnett, E.M. 854 (126), 855 (127), *859*
Arney, B.E. 720 (76), 722, 723 (76, 102, 104), *767*
Arney, J.S. 485 (279), *493*
Arnold, D.R. 470 (32), 477 (187), *489, 491*
Arnold, E.V. 625, 636, 637 (381), *655*
Arnold, W.T. 116 (226), *136*
Arora, S. 875 (69), *884*
Arvanaghi, M. 447 (208), *466*, 821 (40), 851 (116), *857, 859*
Asao, T. 555 (220, 221), 632 (220), *652*
Asaoka, M. 280 (83), *312*
Åsbrink, L. 60 (58), *131*

Ascensio, G. 457 (270), *468*
Ashley, K. 760 (306–308), *772*
Ashworth, T.V. 554 (211), *652*
Asim, M.Y. 719, 749 (73), *767*
Aso, Y. 301, 302 (292), *317*
Asuncion, L.A. 473, 477, 479 (144), *491*
Atavin, E.G. 191 (260), *219*
Atovmyan, L.O. 176 (157), *216*
Attig, T.G. 633 (400), 634 (401), *656*
Atwater, M.A.M. 423 (101), *463*
Augart, N. 197 (282), *219*
Augelli-Szafran, C.E. 292 (223), *315*
Aumann, R. 420 (84), *463*, 521 (128), 549 (177, 181), 554 (203, 212–214), *650–652*
Austin, D.J. 291 (199, 200a, 200c), *315*, 684 (145, 146), 685 (146), 687 (145), *705*
Austin, R.E. 292 (215), *315*, 698, 701 (198), *706*
Averina, N.V. 296 (253), *316*
Avery, M.A. 670 (83), *704*
Avila, D.V. 733 (185), *769*
Avram, M. 659 (24), *702*

Babiak, K.A. 827 (56), 828, 836 (57), 838 (56, 86), 840 (57), *858*
Babler, J.H. 301 (285), *317*
Bachbuch, M. 620 (365), *655*
Bachrach, S.M. 757 (292), *771*
Bäckvall, J.-E. 128 (280, 281), *137*, 267 (30), 268 (34), 282 (108), *311, 313*, 550 (185), *651*
Bader, R.F. 778 (21), *810*
Bader, R.F.W. 64, 65 (67a–d), 68 (67a–d, 72a, 72b), 74, 75 (78, 79), 76 (88a–c), 107 (67a–d, 79), 123, 124 (79), *131, 132*, 144, 204 (29), *213*, 361, 365 (44), 375 (44, 80, 81), 376 (80, 81), 377 (80), 378 (44, 80), 379 (44), 380 (86), 393 (80, 81), 394 (44), *407–409*, 414, 415, 434, 435 (30), *461*, 779 (31), *810*
Badger, G.M. 342, 382, 390 (4), *406*
Badoui, E. 263 (11), *311*
Baeckstrom, P. 325 (35), *338*
Baenziger, N.C. 434 (149), *465*
Baer, T. 116 (229), *136*
Baert, F. 174 (143–145), *216*
Baeyer, A.von 77 (90a), *132*
Baghal-Vayjooee, M.H. 77, 120 (89b), *132*
Bagheri, V. 661 (27), 686 (150, 152), 690 (166), 693 (27, 166), 695 (166), *702, 705*
Bagryanskaya, I.Yu. 434 (148), *464*
Baidzhigitova, E.A. 557, 561, 562 (231), *652*
Bailey, P.L. 181 (190), *217*
Bailey, P.M. 519 (123), *650*
Bailey, S.M. 224 (9), 237 (9, 69), *249, 255*
Baird, M.S. 289 (166), *314*, 665 (60), *703*
Baird, N.C. 384 (88), *409*
Baird, P.D. 170 (134), *216*
Bak, B. 145 (38), *213*
Baker, C. 60 (53), *130*
Bakken, P. 153, 154 (77), 210, 211 (340), *214, 221*
Bakshi, R.K. 679 (125), *704*
Balaban, A.T. 344, 345 (22), *406*, 413, 415, 416, 419, 427, 450, 459 (20), *461*
Balaji, V. 777 (12), 793 (74), *810, 811*
Balcerzak, P. 288 (149, 153), *314*
Balci, M. 747, 752 (264), *771*
Baldridge, K. 733 (187), *769*
Baldwin, J.E. 84, 85 (132), 98 (159), 101 (132), 102 (159), 126 (132), *133*, 170 (134), 187 (240), 188 (250, 251), *216, 218*, 302 (297), *317*, 470 (39, 44, 55, 59, 62, 64), 473 (149, 155–158), 474 (161, 162), 475 (162–164, 166–169), 476 (166–169, 174, 175, 177), 477 (59, 62, 64, 149, 155–158, 162, 163, 166), 478 (200, 201), 479 (59, 62, 64), 481, 482 (268, 270), 484 (166, 270), 485 (162–164), 487 (166, 270, 282), *489, 491–493*, 559, 561 (237), 619 (364), *652, 655*, 838 (84, 85), *858*
Ballard, R.E. 44, 48, 60 (17), *129*
Bally, T. 231 (44), *253*
Balme, G. 268 (40), *311*
Balog, J. 459 (283), *468*
Balquist, J.M. 470, 480 (30), *488*
Bambal, R. 712, 736 (27), *766*
Banciu, M. 344, 345 (22), *406*, 413, 415, 416, 419, 427, 450, 459 (20), *461*
Bandara, B.M.R. 300 (276), *316*
Banville, E. 501 (35), *648*
Banwell, M.G. 568, 572, 587 (250), *653*, 709 (11), 718 (68), 719 (68, 69), 720, 724 (69), 760 (11), *766, 767*
Bapuji, S.A. 643 (425), *656*
Baranović, G. 191 (261), *219*
Barbachyn, M.R. 797, 800 (112), *812*
Barbeaux, P. 301 (281), 308 (333), *316, 318*
Barborak, J.C. 441 (178), *465*, 840 (88, 89), *858*
Bard, A.J. 802 (124), *812*
Barden, T.C. 474 (162), 475 (162–164), 477 (162, 163), 485 (162–164), *491*
Barinelli, L.S. 268 (41), *311*
Barkhash, V.A. 431, 434, 444 (133), *464*
Barluenga, J. 308 (336), *318*
Barnard, J.A. 471 (126), *490*
Barnardinelli, G. 737 (199, 200), 742 (200), 748 (199, 200), 753 (200), *769*
Barnes, R.K. 569 (254), *653*
Barnes, S.G. 638 (411), *656*
Barnette, W.E. 514 (104), *649*
Barnier, J.-P. 870 (47), 874 (60), *884*

Barone, V. 121, 122 (250), *137*
Barrett, A.G.M. 273 (53), *312*
Barsa, E.A. 473, 477 (152–154), 478 (152, 153), 479 (152–154), *491*
Bartell, L.S. 145 (36), 149 (61), 155 (85), 160 (101, 106), 178 (173b), 181 (197), 187, 190 (85), *213–217*
Bartlett, P.D. 814, 856 (3), *856*
Bartlett, R.J. 104 (168c), *134*, 392 (104, 105, 107), *409*
Bartmann, E. 571, 572 (262), 573 (266, 267), 576 (267), 610 (353), 612 (266, 353), 616 (266), 629 (267), *653, 655*
Bartmess, J.E. 249 (122), *260*
Barton, B.D. 471 (109), *490*
Barton, D.H.R. 320 (1), *337*
Barton, T.J. 419 (73), *462*, 622 (370), *655*
Barzaghi, M. 384–386 (93), 394 (93, 119), *409, 410*, 425 (109), *463*
Basch, H. 60 (53), *130*
Basolo, F. 595 (323, 324), 600 (323), *654*
Bastiansen, O. 85, 86 (140), *133*, 143 (19, 20), 144 (20), *213*, 452 (244), *467*, 777 (11), *810*, 874 (62), *884*
Bastos, C.M. 536, 548 (161), *651*
Bates, R.W. 527 (148), *650*
Battiste, M. 431 (136), 448 (212), *464, 466*
Battiste, M.A. 44, 48, 126 (18), *129*, 453 (247), *467*, 550, 557 (183), *651*, 741 (220), *770*
Bau, R. 185 (211), *217*
Bauder, A. 209 (334), *221*, 471 (87), *490*
Bauer, F. 470 (46), 477 (195, 196), *489, 492*
Bauer, I. 193, 194 (273), 209 (332), *219, 220*
Bauer, S.H. 194 (274), 204 (317), 208 (329), *219, 220*, 456 (264), *468*, 794 (87), *811*
Bauer, W. 759 (300), *772*
Bäuerle, P. 434 (150), *465*
Bauld, N.L. 77, 78 (93), *132*, 290 (185), *314*, 384 (91), *409*, 557 (234), *652*
Baum, G. 198 (287), *219*, 429, 430 (128), *464*, 711 (22), *766*
Baum, J.S. 665 (58), *703*
Baum, R. 478 (204), *492*
Baumann, K. 659, 691 (20), *702*
Baumgartel, H. 231 (44), *253*
Baumgarten, M. 744 (241), *770*
Bawell, M.G. 866 (28), *884*
Baxter, S.G. 456 (265), *468*
Bays, J.P. 282 (106), *313*
Beagley, B. 170 (137), *216*
Beak, P. 882 (92), *885*
Beasley, G.H. 455 (255), *467*, 477 (185), *491*
Beauchamp, J.L. 114 (220), *136*
Beauchamp, R.N. 176 (163), 177 (163–165), 178 (163, 164), *216*
Beaudet, R.A. 83, 86, 88 (116), *133*, 145 (40), *213*
Becalski, A. 690, 693 (167), *705*
Beck, W. 554 (216), *652*
Becker, H. 639 (413, 414), *656*
Becker, J. 756 (285), *771*
Becker, L.W. 544 (171), *651*
Becker, P. 61 (63a), *131*
Beckhaus, H.-D. 147–149 (57), *214*, 247 (111, 112), *259*
Beddoes, R.L. 552 (194), *651*
Bee, L.K. 718, 748, 749 (59), *767*
Behrens, U. 553 (196), 585 (295b), *651, 653*, 874 (60), *884*
Beitat, A. 477 (195), *492*
Belec, C.A. 445 (193), *466*
Belfield, K.D. 477 (199), *492*
Bell, H.M. 445 (197), *466*
Bella, J. 440 (170), *465*
Beller, A.J. 110 (195), *135*
Belley, M. 671 (86, 87), *704*
Bellott, E.M.Jr. 162 (119), *215*
Belzner, J. 204, 205 (319), *220*, 262 (3–5), *311*, 776 (20, 45), 778, 779 (20), 780 (38), 781–784 (45), 787 (45, 57, 62), 788 (45, 62), 789 (45, 57), 790 (57), 796–798, 800, 802, 809 (62), *810, 811*
Benaïm, J. 515 (107), *650*
Benard, R. 711 (22), *766*
Benassi, R. 734 (190), *769*
Bendall, V.I. 711 (21), *766*
Bender, C.F. 487 (289), *494*
Bender, C.O. 326 (40), *338*
Benedetti, E. 170–172 (129), *216*
Benedetti, F. 267 (27), *311*
Benesi, A. 456 (263), *468*
Bengtson, G. 189 (255), *218*
Benjamin, W.A. 814, 856 (3), *856*
Benn, R. 506 (65), 508 (76–78), 519 (116), 521 (131), 546 (65), 576 (76), 579 (76, 285), 581, 582, 585 (77), 586 (285), 587 (77), 626, 635 (383), 644 (285), *649, 650, 653, 655*, 748 (267), 753 (267, 277), *771*
Bennett, W. 412 (8), *460*
Beno, M.A. 456 (264), *468*
Ben-Shoshan, R. 549 (178), 563, 564 (243), *651, 652*
Benson, J.E. 211 (343), *221*
Benson, R.C. 110 (199d), *135*
Benson, S. 77, 120 (89b), *132*
Benson, S.W. 104 (166b), 126 (277a–e), *134, 137*, 145 (31), *213*, 479 (216–218), 480 (218), *492*
Benterud, B. 210, 211 (340), *221*
Bentley, T.W. 444 (189), *466*
Bentrude, W.G. 800 (111), 802, 803 (125), *812*

Beran, K. 190, 195, 196 (258), *218*
Bergbreiter, D.E. 290 (174), *314*
Berger, R.J. 444 (191), *466*
Bergeron, G. 481 (235), *492*
Bergholz, R. 56, 57, 75 (46a), *130*
Bergman, R.G. 469 (5), 473, 477, 480 (146–148), *488, 491*
Bergmann, E.D. 342, 344, 382, 390 (8), *406*
Bergmann, R.G. 509 (81), *649*
Berke, H. 573 (268, 269), 574 (268), 575 (268, 269), 576 (269), 577, 593 (268, 269), 597 (269), *653*
Berks, A.H. 263 (7), *311*
Bernabé, M. 283 (118), 289 (118, 161), 305 (314, 315, 321), *313, 314, 317*
Bernal, I. 157 (91), 192 (267), *215, 219*
Bernardi, A. 297 (259), *316*
Bernardinelli, G. 644 (427, 428), *656*, 722 (91), 724, 725 (112), 731, 737, 742 (164), 746 (112), 748, 753, 764 (164), *767–769*
Berndt, A. 429 (128, 131), 430 (128, 129, 131), 431 (131), *464*
Bernett, W.A. 177 (166), *216*
Bernlöhr, W. 147–149 (57), *214*
Bernstein, D. 191 (263), *219*
Bernstein, H.J. 777 (14), *810*
Berry, R.J. 182 (199–201), 183 (199, 201), 190, 195 (256), *217, 218*
Berson, J.A. 104 (166a), *134*, 419 (76), 423 (99), 432, 433, 435 (142, 143), 436 (142), 437 (157), 453 (247), *463–465, 467*, 469 (7, 8), 470, 480 (30), 484 (276–278), 485 (277–279), 486 (276, 278), *488, 493*, 843 (95), *859*
Berthelot, M. 470 (72), *489*
Berthelot, P.E.M. 470 (69), *489*
Berti, F. 267 (27), *311*
Bertram, A.K. 289 (164), *314*
Bertrán, J. 125 (266), *137*
Bertrand, M. 615, 616 (359), 621 (369), 622 (359), *655*
Bertsch, A. 402 (125), *410*, 453 (248), *467*
Beruben, D. 276 (64), 277 (65), *312*, 509 (84), *649*
Bessard, Y. 287 (145–147), *314*
Bessmertnykh, A.G. 149 (63), *214*
Bestmann, G. 186 (225), *218*
Bestul, A.B. 242 (87), *257*
Betz, P. 197 (280), *219*
Bews, J.R. 793 (75), *811*
Bhattacharjee, G. 668 (74), *703*
Bhaumik, A. 151 (75), *214*
Bhupathy, M. 300 (275), *316*
Bidell, W. 628 (387), *655*
Biedenbach, B. 588 (301), *654*
Biegler-König, F.W. 76 (88a, 88c), *131, 132*
Bielfeldt, T. 285 (132), *313*
Biellmann, J.-F. 266 (26), *311*
Bien, S. 683 (142), *705*
Bierbaum, V.M. 113, 114 (217), *136*
Bieri, G. 60 (58), *131*
Bierwagen, E.P. 591, 592, 599, 603–605 (314), *654*
Biesiada, K.A. 715 (46), *766*
Biethan, U. 829 (62), *858*
Billups, W.E. 188 (244), 193, 194 (272), *218, 219*, 248 (119), *260*, 262 (2f), *310*, 353 (30), *407*, 638 (408, 409), *656*, 708 (4), 709 (9, 10, 12), 717 (53–55), 718 (54, 55), 719 (73), 720 (76, 80, 81), 722, 723 (76, 101–104), 724 (81), 730 (148), 734 (4, 196), 735 (196), 738 (4), 739 (10, 101), 740 (10), 741 (10, 231), 744 (231), 746 (10), 749 (4, 10, 73, 101, 269), 750 (101, 269), 753 (276), *766–771*
Bingel, W.A. 56, 57, 75 (46b), *130*
Binger, P. 507 (72, 73), 508 (74, 77–79), 521 (130, 131), 547 (176), 581, 582 (77, 289), 585 (77, 289, 296), 587 (77), 588 (72, 79, 301, 302), 620 (367, 368), 624 (73, 367, 375), 626 (383), 627 (73), 635 (383), 640 (73, 367, 416, 417), 641 (73, 367, 420), 642 (367, 421, 422), *649–651, 653–656*, 867 (33), 875 (69), *884*
Bingham, R.C. 430 (125), *464*
Binkley, J.S. 85 (136b), 104 (168a), *133, 134*, 391 (100), 392 (103), *409*
Binkley, R.S. 323 (26), *337*
Binkley, R.W. 320 (4), 335 (51), *337, 338*
Binsch, G. 280 (81), *312*, 364 (59), *408*
Birch, A.J. 457 (270), *468*
Birney, D.M. 453 (247), *467*
Bitter, I. 302 (298), *317*
Bjorge, S.M. 519 (127), *650*
Björkman, E.E. 128 (281), *137*
Black, K.A. 474 (161), *491*
Blaeser, D. 471 (89), *490*
Blain, D.A. 401 (123), *410*, 454 (251), *467*
Blakeney, A.J. 717 (53, 55), 718 (55), *767*
Bland, J. 170–172 (128), *215*
Blankley, C.J. 665 (56, 57), *703*
Blasco, J. 305 (320), *317*
Bläser, D. 193, 194 (272), 201 (306), 202 (307), *219, 220*, 353 (29, 30), *407*, 734, 735 (194–196), 739, 755 (195), *769*, 718, 734, 735, 737, 747 (66), 759, 760 (298), *767, 771*, 863 (7), *883*
Blattner, R. 719, 720, 724 (69), *767*
Bleck, W.-E. 552 (193), *651*
Bley, W.R. 110 (199c), *135*
Blinova, M.V. 864 (10, 13), *883*

Bloch, R. 729 (145), *768*
Blokhin, A.V. 789, 790, 797 (65), 798 (65, 118–120), 799, 800 (65), 801 (65, 118), 802 (65, 119, 120), 805, 807, 808 (65), *811, 812*
Blomberg, M.R.A. 128 (280, 283), *137*
Bloor, J.E. 733 (176), *769*
Bloy, V. 306 (323), *317*
Blumenthal, T. 740 (219), *770*
Blunden, R.B. 592, 602 (315), *654*
Blurock, E.S. 820 (33), *857*
Boates, T.L. 149 (61), *214*
Boatz, J.A. 83, 96–98 (124), 102 (161), 128 (282), *133, 137*
Bocarsly, A.B. 748 (268), *771*
Boccelari, T. 715 (44), *766*
Boche, G. 224 (2), *249*, 419, 420, 423 (72), *462*, 469 (16), *488*, 499 (8), *648*
Bock, C.W. 74 (84), *131*, 226, 237 (19b), *250*, 377 (84), 387 (95, 96), 389 (96), 393, 404 (84), *408, 409*
Bock, H. 365 (60), *408*
Boenke, M. 477 (196), *492*
Boer, J.J.de 578–581, 601, 602 (281), *653*
Boer, J.S.A.M.de 156, 157 (89), 175 (149), 181 (192, 195, 196), 182 (192, 196), *215–217*
Boerth, D.W. 417 (49), *462*
Boese, R. 140, 187 (5), 188 (243, 244, 246), 189 (254), 190 (5), 193, 194 (272, 273), 197 (281, 282), 201 (5, 306), 202 (307), 211 (342), *212, 218–221*, 229 (35, 37), 231 (43), 234 (51), 245 (99, 101, 103), 246 (105), *252, 253, 258, 259*, 264 (18), *311*, 353 (29, 30), 357–359, 361, 378, 387, 389, 390 (41), *407*, 450–453, 457 (225), *467*, 471 (89), *490*, 718 (66), 734 (66, 193–196), 735 (66, 194–196), 737 (66), 739 (195), 747 (66), 755 (195), 758 (193), 759 (193, 298, 304), 760 (298), 763 (313), *767, 769, 771, 772*, 863 (7), 874 (61), 876 (71), 877 (73), 878 (61, 71, 73, 77, 80), 880 (77), 881 (71), *883–885*
Boettcher, R.J. 322 (19), *337*
Boeykens, M. 270 (42b), *311*
Bofill, J.M. 121, 123 (249), *136*, 430 (125), *464*
Bogaard, M.P. 107, 108 (179), *134*
Bogey, M. 195 (276), *219*
Boggs, J.E. 83 (100, 102a, 105d, 107, 108b, 110), 84, 85 (131), 86 (100, 102a, 105d, 107, 108b, 110), 88 (102a), 93 (102a, 145), 95 (102a, 105d, 131), *132, 133*, 144 (24), 160 (105), 176 (24), 177 (168), 183 (207), *213, 215–217*
Bohlmann, F. 689 (164), *705*
Böhm, M. 357, 359, 361, 378, 389, 390 (42), *407*, 450–453, 457 (226), *467*
Böhm, M.C. 366 (61), *408*
Bohm, S. 716, 726 (49), *766*
Boikess, R.S. 110 (199b), *135*
Bois, C. 164 (121), *215*
Bojilova, A. 300 (274), *316*
Boldrini, G.P. 273 (54), *312*
Bolesov, I.G. 160 (107), 191 (260), *215, 219*, 477 (178), *491*, 501 (38a, 38b), *648*
Bollinger, J.M. 439 (166), *465*, 851, 852 (117), *859*
Bolte, O. 303 (303b), *317*
Bolton, J.R. 122 (253), *137*
Bolusheva, I.Y. 796, 805 (100), *811*
Bonacic-Koutecky, V. 481 (246), *493*, 784 (51), *810*
Bonacorsi, S. 470 (64), 477, 479 (64, 181), *489, 491*
Bonnell, D.W. 246 (109), *259*
Bonser, S.M. 280 (85), *312*
Boord, C.A. 864 (11), *883*
Boord, C.E. 471 (83), *489*
Booze, J.A. 116 (229), *136*
Borchert, A.E. 471 (91), *490*
Bordakov, V.G. 659 (22, 23), 688 (22), *702*
Borden, W.T. 116, 117 (227), 126, 127 (275), *136, 137*, 370 (67), *408*, 469 (9), 477 (198), 481, 482, 484 (269), *488, 492, 493*
Borgias, B.A. 197 (283), *219*
Borrmann, H. 501 (37), *648*
Bosch, H.W. 573, 575–577, 593, 597 (269), *653*
Bostrom, R.E. 83, 86, 89, 95 (106b), *132*
Bothe, H. 781, 783, 787, 790 (46), 797 (103), 799 (46, 103), 800 (46, 103, 106), 805, 806 (128), 807 (46), *810–812*
Botskor, I. 186 (233), *218*
Bottrill, M. 637 (406), *656*
Boucher, D. 175 (151, 152), *216*
Bouma, W.J. 96, 98 (150), 116 (225), *133, 136*
Bouman, T.D. 83, 86 (108), 110 (189, 193a), *135*, 393, 396 (112), *409*
Boutonnet, F. 273 (57), *312*
Bovce, H.H. 470 (58), *489*
Bovey, F.A. 420 (89), *463*
Bowers, M.T. 145 (40), *213*
Bowie, J.H. 740 (219), *770*
Bowman, N.S. 440, 441 (176), *465*
Bowman, W.R. 271 (48), *312*
Boyd, R.H. 77 (92b), *132*, 229 (33), *252*
Boyd, R.J. 116 (226), *136*
Boykin, D.W.Jr. 418 (64), *462*
Boys, S.F. 56 (48a, 48b), *130*
Braatz, J.A. 550 (184), *651*
Bradley, C.H. 419, 425 (80), *463*
Bradley, J.N. 471 (119), *490*

Bragin, O.V. 508 (75), 627 (385), *649, 655*
Brahms, J.C. 190, 195, 196 (258), *218*
Braish, T.F. 275 (61), *312*
Brammer, L. 159, 180, 185 (95), *215*
Branca, S.J. 668 (73), *703*
Brandes, B.D. 292 (213), *315*, 698 (197), *706*
Brandt, S. 282 (107), *313*
Branton, G.R. 479 (210), *492*
Brard, L. 212 (344), *221*
Bratkova, A.A. 798, 801 (118), *812*
Bratz, J.A. 881 (85), *885*
Brauman, J.I. 110 (199b), *135*, 419, 420, 423, 428 (70), *462*, 471 (133), *490*
Braun, M. 176 (158), *216*
Braun, S. 714 (33), *766*
Bregadze, V.I. 663 (34), *703*
Bréhin, P. 164 (121), *215*
Breimair, J. 554 (216), *652*
Breit, B. 197 (280, 281), *219*
Brel, V.K. 798, 801 (118), *812*
Bremer, M. 110, 111 (202g), *135*, 393, 396, 397 (116), *410*, 446 (203, 204), 449 (216), *466*
Brennan, M.E. 453 (247), *467*
Brennenstuhl, W. 833 (71), *858*
Brenner, S. 512 (90), *649*
Breslow, R. 742 (234), *770*
Brett, A.M. 74 (84), *131*, 387 (95, 96), 389 (96), *409*
Brett, W.A. 202 (307), *220*, 734, 735, 739, 755 (195), *769*
Bretting, C.A.S. 557 (233), *652*
Breuckmann, R. 229 (35, 37), 231 (43), 234 (51), 245 (99, 101, 103), 246 (105), *252, 253, 258, 259*, 357–359, 361, 378, 387, 389, 390 (41), *407*, 450–453, 457 (225), *467*, 470 (46), 477 (190, 196), *489, 492*
Brewer, P.D. 715 (43), *766*
Bridle, J.H. 720 (77), *767*
Bright, D. 578–581, 601, 602 (281), *653*
Brinker, D.A. 690, 693, 695 (166), *705*
Brinker, U.H. 470 (43), 471 (89), *489, 490*, 724 (113), 746, 751 (251), *768, 771*, 863 (7), *883*
Brinkmann, A. 624 (375), 641 (420), 642 (422), *655, 656*
Brisdon, B.J. 591, 599, 602 (311), *654*
Britt, C.O. 93 (145), *133*, 183 (207), *217*
Brittain, W.J. 818 (21), *857*
Broadhurst, M.J. 394 (120), *410*, 419 (83), 442 (182), *463, 465*
Brock, C.P. 634 (401), *656*
Brodkin, G.I. 434 (148), *464*
Brogli, F. 739, 744 (205), *770*
Brookhart, M. 262 (2c), 285 (2c, 134), 294 (237), *310, 313, 316*, 420 (85), 445 (198, 200), 446 (198), 449 (217), *463, 466*, 522 (133), 524, 525 (142), 554 (208), *650, 652*
Brookhart, M.S. 423 (101), *463*
Brooks, W.V.F. 151 (75), *214*
Brouwer, D.M. 431 (134), *464*
Brown, D. 111 (206), *136*
Brown, D.S. 271 (48), *312*
Brown, H.C. 417 (54), 445 (197), *462, 466*, 815 (5), 846 (106), 853 (121), *856, 859*
Brown, J.M. 458 (272, 274), *468*, 503 (46), 516 (111), *648, 650*
Brown, K.C. 290 (179), *314*, 694 (178), *706*
Brown, M.F. 286 (137, 138), 294, 295 (244), *313, 316*
Brown, M.S. 471 (128), *490*
Brown, R.D. 188 (247, 248), *218*, 757 (293), *771*
Brown, R.F.C. 727 (127, 128), 765 (315), *768, 772*
Brown, R.S. 418 (68), *462*, 513 (93), *649*
Brown, T.L. 589, 598 (304), *654*
Browne, A.R. 718 (63), 719, 720 (69), 721, 722 (88, 89), 724 (69), 731, 741 (89), 756 (286), 757 (89), *767, 771*
Browne, L.J. 545 (172a, 172b), *651*
Bruce, M.I. 500 (17), *648*
Brückner, D. 458 (275), *468*
Brun, J.le 301, 302 (294), *317*
Brunn, E. 176 (158), *216*
Bruns, R.E. 471 (108), *490*
Brunsvold, W.R. 437 (159), *465*
Brupbacher, T. 471 (87), *490*
Bryan, R.F. 429 (118), *464*
Bryant, T. 299 (271), *316*
Bublak, W. 156, 157 (90), *215*
Bubnov, Y.N. 789, 790, 797–802, 805, 807 (65), 808 (65, 131), *811, 812*
Büch, H.M. 507, 588 (72), 620, 624, 640–642 (367), *649, 655*
Buch, V. 746 (254), *771*
Buchert, M. 294, 295 (240), *316*
Buchholtz, H. 876, 878, 881 (71), *885*
Buchholz, H. 188 (243), *218*
Buchler, U. 231 (44), *253*
Buchs, A. 740 (216, 217), *770*
Buchwald, S.L. 583 (293), *653*
Buchwalter, S.L. 470 (38), *489*
Buck, H.M. 428 (117), *464*
Buckert, U. 77 (92a), *132*
Buckingham, A.D. 107, 108 (179), *134*
Buckland, S.J. 713 (28), 729 (140), 730, 732, 742 (156, 158), 758, 759 (158), 760 (140, 306, 308, 310), 761, 762 (140), *766, 768, 769, 772*
Buckles, R.E. 412 (3), *460*
Buckwalter, B.L. 668 (69, 70), 683 (143), 687 (156), *703, 705*

Budzelaar, P.H.M. 123 (259), *137*, 377, 393, 404 (84), *408*, 430 (130), *464*
Buehl, M. 110, 111 (202a–e), *135*
Buehler, N.E. 322 (19), *337*
Buenker, R.J. 112 (213), *136*, 481 (232), *492*
Bühl, B. 446 (203), *466*
Bühl, M. 96, 98 (152), 110, 111 (201b, 201c), *133, 135*, 393, 396, 397 (116), *410*, 429–431 (131), *464*
Buhro, W.E. 663 (37), 688 (37, 162), 693 (162), *703, 705*
Bulai, A.Kh. 176 (157), *216*
Buldzelaar, P.H.M. 96, 98 (151), *133*
Buncel, E. 710 (18), *766*
Bunge, A. 606 (339), *654*
Bunker, P.R. 482 (271), *493*
Bunz, U. 204, 205 (319), *220*, 262 (4), *311*, 776, 778, 779 (20), 787 (57), 789, 790 (57, 63), 795 (95), 797 (109), 798 (63, 109, 115), 799 (63, 115), 800 (109, 115), 801 (115), 802 (63, 109), 809 (95), *810–812*
Buraev, V.I. 432, 433, 435 (145), *464*
Burch, M. 836 (82), *858*
Burdett, J.K. 350, 370, 373 (28), *407*, 625 (380), *655*
Burger, U. 291 (193), *314*, 741 (227, 228), *770*
Burgert, W. 711 (25), *766*
Burgess, E.M. 727, 760 (131), *768*
Burgess, K. 303 (307), *317*
Bürgi, H.-B. 140 (1), *212*, 364 (48), *407*, 451 (233), *467*
Burie, J. 175 (151, 152), *216*
Burka, L.T. 880 (83), *885*
Burke, J.J. 783 (49), *810*
Burke, N.A.D. 344, 345 (21), *406*, 413, 415, 416, 419, 445, 450 (22), *461*
Burke, S.D. 662, 667 (31), *702*
Burkert, U. 389 (98), *409*
Burkhalter, J.F. 471 (112), *490*
Burkhart, J.P. 519 (126), *650*
Burkholder, C.R. 287 (144), 288 (151, 152), *314*
Burns, E.G. 446 (205), *466*
Burns, G.R. 98 (156), *133*
Burns, G.T. 622 (370), *655*
Burreson, B.J. 519 (120), *650*
Burzlaff, H. 197 (285a–c), *219*
Busch, T. 197 (280), *219*
Busetti, V. 170–172 (130), *216*
Bush, S.F. 93 (146), *133*
Buss, S. 779 (28), *810*
Buss, V. 418 (63), *462*
Butcher, R.J. 775 (8), *809*
Butcher, S.S. 451 (228), *467*
Butkowskyj-Walkiw, T. 500, 502, 505 (29), *648*
Butsugan, Y. 301, 302 (293), *317*
Butz, V. 199 (295–297), *219*
Buxton, L.W. 146 (47), *214*
Buzek, P. 110, 111 (201a), *135*, 394, 396 (118), *410*, 417 (58), 425 (110), *462, 463*, 822, 825 (45), 832 (67), *857, 858*
Bzowej, E.I. 502 (40), *648*

Cagnoli, N. 291 (188), *314*, 669 (79), *704*
Cairns, T.L. 865 (22), *883*
Calabrese, J. 625 (379), *655*
Calabrese, J.C. 197 (284b), *219*, 446 (205), *466*
Calder, G.V. 728, 729, 757 (136), *768*
Calligaris, M. 644 (429), *656*
Callot, H.J. 694 (176, 177), *706*
Calverley, M.J. 557 (233), *652*
Cameron, D.M. 470 (35), *489*
Cameron, T.S. 156, 159 (87), *215*
Campos, P. 333 (48), *338*
Canalda, I.I. 869 (43), *884*
Cantrell, W.R.Jr. 292 (219, 220), *315*, 695 (185, 186), *706*
Capon, B. 452 (243), *467*
Caponecchi, A.J. 471 (120), *490*
Carboni, B. 282, 290 (101, 102), *313*
Carboni, R.A. 445 (196), *466*
Cardin, D.J. 606 (342), *654*
Cargill, R.L. 469 (4), *488*
Cargle, V.H. 471 (93–96), *490*
Carlacci, L. 126 (272), *137*, 481 (266), *493*
Carlier, E. 663 (33), *703*
Carlsen, J. 206 (323), *220*
Caroll, M.A. 459 (284), *468*
Carpenter, B.K. 104 (166a), *134*, 469 (13, 14, 24), 477, 479 (180), 484, 485 (277, 278), 486 (278), 487 (283, 285–287), *488, 491, 493, 494*, 556 (227, 228), 625, 636, 637 (381), 638 (410), *652, 655, 656*
Carr, R.W. 471 (111), *490*
Carrié, R. 282, 290 (101), *313*, 550 (187), 551 (190, 191), *651*, 659 (25), 697 (188), *702, 706*
Carroll, M.T. 779 (33), *810*
Carroll, P.J. 183 (208, 209), 184 (208), 191 (264), 201 (302), *217, 219, 220*, 277 (68), *312*
Carroll, S. 246 (109), *259*
Carstensen, T. 517 (112), *650*
Carstensen-Oeser, E. 734 (191), *769*
Carter, C.G. 473, 477 (155–158), *491*
Carter, J.D. 535 (157, 158), 536 (160), 537 (157, 158), 538 (160, 163), 539 (158, 160, 164), 540 (164), 541 (160, 164), 542 (158), 543 (158, 160), *651*
Carter, W.L. 473, 477, 480 (146, 148), *491*
Casanova, J. 834 (77), 852 (119), *858, 859*

Case, M.G. 418 (63), *462*
Caserio, M.J. 427 (113), *463*
Casey, C.P. 285 (133), *313*, 487 (281), *493*, 595 (325), *654*
Casper, D. 290 (178), *314*
Cassada, D.A. 83, 86, 89, 93 (122), *133*, 181, 182 (191), 183 (202), *217*
Cassar, L. 521 (129), *650*
Casserly, E.W. 720 (76, 81), 722, 723 (76), 724 (81), 730 (148), *767, 768*
Casserly, W.E. 722, 723 (104), *767*
Cassidy, J. 760 (307), *772*
Castells, J. 430 (125), *464*
Castiglioni, C. 105–107 (169a–c), *134*
Cativiela, C. 305 (320), *317*
Caton, P.C. 882 (93), *885*
Caufield, C.E. 323 (23, 24), *337*
Cava, M.P. 727 (119), *768*
Cazes, B. 268 (39), *311*
Cazes, D. 481 (235), *492*
Ceccherelli, P. 291 (187, 188), *314*, 669 (79), *704*
Cecchi, P. 822, 825 (45), *857*
Cederbaum, L.S. 60 (55), *131*
Cekovic, Z. 273 (58), *312*
Celerier, J.P. 304 (313), *317*
Cencek, W. 113 (214), *136*
Cervan, P.B. 469 (12), *488*
Cesak, J. 77, 78 (93), *132*
Cetinkaya, B. 507 (73), 508, 588 (79), 606 (342), 624, 627, 640, 641 (73), *649, 654*
Cetinkaya, E. 585 (296), *654*
Cha, J.K. 268 (33), *311*
Chaboteaux, G. 301 (284, 287), 302 (287), *317*
Chadda, S.K. 344, 345 (21), *406*, 413, 415, 416 (22), 417 (60), 419 (22), 434 (60), 445, 450 (22), *461, 462*, 817, 855 (17), *857*
Chae, Y.B. 291 (198), *315*, 677 (109), *704*
Chai, C.C.L. 743, 745, 750, 751 (240), *770*
Chakrabarti, P. 124 (262), *137*, 296 (251), *316*
Chakraborti, P.C. 674 (101), *704*
Chalmers, A.A. 554 (211), *652*
Chamberlain, N.F. 413 (11), *461*
Chamberlain, W.T. 719, 749 (73), *767*
Chambers, T.S. 470 (75), *489*
Chambers, V.C. 499 (6), *648*, 814 (2), *856*
Champion, R. 188 (247), *218*
Chan, D.M.T. 640 (418), *656*
Chan, L. 699 (200), *706*
Chandrasekhar, J. 47 (22), 113, 114 (216), *130, 136*, 345 (26), 353 (33), *407*, 413 (24), 449 (221), 458 (24, 273, 281), 459 (281), *461, 466, 468*, 794 (83), *811*
Chandrashekhar, J. 815 (8b), 835 (79), *856, 858*
Chang, C.-C. 728, 729, 757 (133), *768*
Chang, C.-T. 504, 505 (52), *648*
Chang, H.K. 800 (108), *811*
Chang, M.C. 84, 85 (127), *133*, 143, 144 (22), *213*, 471 (86), *490*
Chang, S. 229 (33), *252*
Chang, S.-J. 77 (92b), *132*
Chang, S.S. 242 (87), *257*
Chang, W. 470 (45), *489*
Chapman, O.L. 419, 420 (78), *463*, 728 (133, 136), 729 (133, 136, 142), 757 (133, 136), *768*
Chapman, T.M. 682 (133), *705*
Chapuisat, X. 481 (235, 248–251, 253–256), *492, 493*
Charbonnier, F. 267 (31), *311*
Charles, A.D. 556 (223), *652*
Charrier, C. 515 (107), *650*
Charton, M. 44, 48 (1), *129*, 140 (2), *212*, 226, 238 (20b), *250*, 416 (40), *461*
Chatani, N. 309 (350, 352), *318*, 682 (139), *705*
Chatt, J. 48, 49, 71 (38), *130*
Chatterjee, G. 523 (137, 138), 525 (138), 530 (137, 138), 531, 533 (138), *650*
Cheer, C.J. 191 (263), *219*
Chen, B. 290 (174), *314*
Chen, C. 301, 302 (290, 291), *317*
Chen, E.Y. 677 (107, 108), *704*
Chen, J. 277 (67), *312*
Chen, J.D. 471 (114), *490*
Chen, J.P. 470 (57), *489*
Chen, K. 149 (62), *214*
Chen, K.-N. 550 (182), *651*
Chen, K.S. 122 (251c), *137*
Chen, L.-J. 729 (143), *768*
Cheng, A.K. 455 (257), *467*
Cheng, J.C. 105 (172b), *134*
Cheng, M.-C. 572 (263), *653*
Chesick, J.P. 470 (37), *489*
Cheung, C.S. 733, 734 (170), *769*
Chiacchio, U. 291 (200d), *315*
Chiang, J.F. 177 (166), 204 (317), *216, 220*, 794 (87), *811*
Chiang, T. 574, 590, 602, 603 (274), *653*
Chibante, F. 671 (87), *704*
Chickos, J.S. 225 (10, 11, 13), 235 (58), 241 (84), 242 (87), *249, 250, 254, 257*, 476 (172), *491*
Chidsey, C.E. 575 (276), *653*
Childs, R.F. 110 (200), *135*, 160 (108), *215*, 341 (3), 344, 345 (20, 21), 346, 349 (3), 361 (45), 394, 402 (121, 122), 405 (3), *406, 407, 410*, 413 (19, 22), 414 (25), 415, 416 (19, 22), 417 (59–61), 419 (19, 22, 77), 420 (77, 87, 90, 91), 421 (19, 91–93), 422 (91–93, 97, 98), 423 (77,

Childs, R.F. (*cont.*)
92, 93, 100), 424 (77, 102), 425 (25, 77, 104), 426 (25), 427 (19, 116), 430 (25), 431 (137), 432 (138–141), 433 (25, 137–141, 146), 434 (60), 435 (25, 61, 137–141, 152), 436 (140, 155), 437 (158), 438 (164), 439 (87), 441, 442 (25), 443 (184), 445 (22), 448, 449 (25), 450 (19, 22, 25), 451 (25, 236), 452 (25, 237), 453 (246), 454, 456 (25), 459 (19, 25), *461–465*, *467*, 554 (204–206), 555 (218, 219), *652*, 817, 855 (17), *857*
Chinchilla, R. 297 (262), *316*
Chinn, M.S. 290 (177), *314*, 667 (64), 686 (151), 693 (169), *703, 705*
Chisnall, B.M. 635 (403), *656*
Chitty, A.W. 687 (158), *705*
Chizhevsky, I.T. 663 (34), *703*
Choe, J.-I. 83, 86, 88, 93 (102c), *132*, 175 (148), *216*
Choi, S.K. 756 (283), *771*
Chon, S.-L. 114 (219), *136*
Choplin, A. 83, 86, 88, 93 (101), *132*
Chou, P.K. 237 (67), *255*
Chou, T.C. 110 (195), *135*
Chow, W.Y. 708 (4), 717 (53–55), 718 (54, 55), 720 (80), 734, 738 (4), 741, 744 (231), 749 (4, 269), 750 (269), *766, 767, 770, 771*
Christen, D. 201 (306), *220*, 353 (29), *407*, 452 (239), *467*, 734, 735 (194), 744 (242), 745, 750 (247), *769–771*
Christensen, L.W. 500 (20), *648*
Christl, M. 176 (158), 202, 203 (309), *216, 220*, 308 (332), *318*, 458 (275), 459 (285, 287), *468*
Christoph, G.G. 366 (61), *408*, 456 (264), *468*
Chu, K.S. 289 (163), *314*
Chuah, T.S. 715 (40), *766*
Chui, C.-K. 585 (297), *654*
Chu-Moyer, M.Y. 275 (62), *312*
Chung, A.L.H. 356, 357, 382 (39), *407*
Chung, L.H. 247 (114b), 248 (117), *259*
Church, L.A. 290 (180), *314*, 693 (170), *705*
Churchill, M.R. 593 (316, 318), 594 (316, 318, 320), 601 (316), 602 (316, 320), 603, 604 (316), *654*
Churney, K.L. 224 (9), 237 (9, 69), 246 (110), 247 (112), *249, 255, 259*
Cianciosi, S.J. 475 (165–169), 476 (165–169, 174, 175), 477, 484 (166), 487 (166, 282), *491, 493*
Ciganek, E. 451 (230), *467*
Cimetière, B. 682 (134), *705*
Ciorba, V. 344, 345 (22), *406*, 413, 415, 416, 419, 427, 450, 459 (20), *461*
Citterio, A. 274 (59), *312*
Clardy, J. 366 (61), 401 (123), *408, 410*, 454 (251), *467*, 625, 636, 637 (381), *655*
Clardy, J.C. 353 (34), *407*
Claremon, D.A. 514 (103, 104), *649*
Clark, B.C.Jr. 344, 345 (16), *406*, 413, 415, 416, 419, 427, 444, 450, 459 (17), *461*
Clark, R.A. 470 (50), *489*
Clark, T. 48, 49, 83, 85–90, 92–95 (33), 96 (33, 151, 152), 98 (151, 152), 128 (279), *130, 133, 137*, 156, 159, 178, 179, 181 (88), 199 (295), *215, 219*, 417 (52), *462*
Clark, T.J. 290 (180, 182), *314*, 664 (52), 693 (170), *703, 705*
Clarke, F.W. 418 (63), *462*
Clarke, S.C. 453 (247), *467*
Claussen, R.C. 722, 723 (103), *767*
Clayton, T.W. 698 (199), *706*
Clegg, W. 181 (190), *217*
Clemens, P.R. 629 (389), 631 (389, 393), 633, 638 (389), *655*
Cleve, G. 661 (29), *702*
Cleveland, J.D. 853 (121), *859*
Cloke, F.G.N. 592, 602 (315), *654*
Close, R. 417 (52), *462*
Closs, G.L. 470 (38), *489*, 568 (249), 569 (251–253), *653*, 711 (21), *766*
Closs, L.E. 569 (251–253), *653*
Coates, R.M. 448 (213), *466*, 816 (9), *856*
Coburn, J.F.Jr. 235 (60), *254*
Cocks, A.T. 471 (126), *490*
Coffey, C.E. 574 (271), *653*
Cohen, H.M. 499, 500 (4), *648*
Cohen, L. 516, 566, 568 (110), *650*
Cohen, N. 284 (128), *313*
Cohen, S. 510 (85), *649*
Cohen, T. 300 (275, 277), 301 (280), *316*, 682 (133), *705*
Collins, C.J. 832 (69), *858*
Collins, J.B. 47 (21a), 102 (160), *130, 133*
Collins, J.R. 60 (57), *131*, 377, 393, 404 (84), *408*
Colobert, F. 268 (36), *311*
Comasseto, J. 663 (48), *703*
Cometta-Morini, C. 121, 122 (246), 123 (246, 254b), *136, 137*
Comisarow, M.B. 849 (112, 113), *859*
Commereuc, D. 515 (107), *650*
Coms, F.D. 470 (40), *489*
Concellon, J.M. 308 (336), *318*
Condé-Petiniot, N. 290 (172), *314*
Conia, J.M. 864 (11), 866 (29, 30), 870 (47), 871 (52), 872 (30, 55, 56), 873 (29), 874 (60), 875 (68), 878 (55), *883, 884*
Conneely, J.A. 516 (111), *650*
Connor, J.A. 522 (135, 136), 523–525 (136), 621 (135), 638 (136), *650*
Constantinescu, T. 742, 743 (236), *770*

Conti, N.J. 499 (15, 16), *648*
Cook, D. 440 (172), 441 (172, 178), *465*
Cooke, D.J. 470 (51), *489*
Cooke, M.P.Jr. 271 (45), *311*
Cooke, R.S. 470 (52, 54), *489*
Coolidge, M.B. 477 (198), *492*
Cooney, M.J. 457 (269), *468*, 731 (167), *769*
Cooper, D.L. 57, 58, 66, 69, 71, 73, 76, 77, 82, 83 (51), *130*
Cooper, M.A. 733, 734 (170), 740 (209), *769, 770*
Coops, J. 243 (95), *258*
Coppens, P. 61 (63b), 62 (65), *131*
Corey, E.J. 470 (49), *489*, 665 (61, 62), 679 (124–126), *703, 704*
Cortes, D. 485 (279), *493*
Corver, H.A. 443 (184), *465*
Cotter, B.R. 325 (38), *338*
Cotton, F.A. 753 (276), *771*
Coulson, C.A. 48, 49, 55–57, 68, 75, 122 (23), *130*, 733 (169), *769*
Coulson, D.R. 519 (121), *650*
Counotte-Potman, A. 454 (252), *467*
Cowley, A.H. 456 (265), *468*
Cowley, B.R. 321 (13), *337*
Cox, J.D. 229 (31), 243 (89), *251, 258*, 387, 388 (97), *409*
Cox, K.W. 85, 86 (142), *133*, 199 (289), *219*
Coxon, J.M. 44, 48, 126 (18), *129*, 550, 557 (183), *651*
Cradock, S. 182, 183 (201), *217*
Craig, J. 881 (85), *885*
Craig, J.T. 715 (40), 719, 720, 724 (69), *766, 767*
Craig, N.C. 83, 86, 95 (105b), *132*, 147 (53), 176 (160), 177 (53, 160, 161, 164), 178 (160, 161, 164), *214, 216*, 418 (65), *462*
Craig, R.A. 864 (11), *883*
Cram, D.J. 817 (15), 846 (104), *857, 859*
Crammer, B. 469 (25), *488*, 544 (167), *651*
Crane, P.M. 470 (53), *489*
Craveiro, A.A. 668 (69), *703*
Cravey, W.E. 739 (207), *770*
Crawford, H.T. 714 (37), *766*
Crawford, K.S.K. 243 (89), *258*
Crawford, R.J. 470 (34, 35), 472 (143), 473 (145), 477 (143, 145), 484 (143), *489, 491*
Creary, X. 622 (372), *655*
Cremer, D. 44 (9–13), 47 (22), 48 (9–13), 49 (9, 11–13), 50 (11, 12), 61 (11), 65 (13, 68, 69), 66 (9–12), 67 (9–13, 71), 68 (9–13, 68, 69), 70 (12), 71 (9, 13), 73 (9, 12, 13, 71), 74 (9–13), 75 (10), 76 (9–11), 77, 78 (10), 79 (9, 10), 83 (10, 12, 71), 84 (10, 133), 85 (10, 133, 135, 138a, 138b), 86 (10), 93 (135), 96 (9, 71), 97 (9, 13, 71), 98 (9, 71), 102, 103 (162, 163), 104 (135, 162, 163), 105 (163), 109 (135), 110 (203), 111 (203, 205, 208–211), 118 (240), 123 (259), *129–131, 133, 135–137*, 144, 178 (25), *213*, 227 (21c), 242 (88b), 249 (122), *250, 258, 260*, 340 (2), 341 (3), 342 (2, 13), 343 (13), 345 (27), 346 (2, 3), 347 (13), 349 (3), 350, 351 (2), 354 (2, 35), 355, 356 (13), 357 (27), 361 (27, 44, 46), 364 (27, 49–58), 365 (44), 374 (77), 375 (27, 44, 49–58, 80), 376 (27, 80, 82, 83), 377 (27, 49, 56, 80, 82, 84), 378 (2, 27, 44, 49, 54, 56, 80), 379 (27, 44, 85), 380 (27, 49, 54, 82, 83, 85), 381 (27, 49–58), 386 (94), 387, 388 (54, 56), 390 (27), 391 (49–58, 101, 102), 392 (49–58, 101, 102, 106, 108), 393 (80, 82–84), 394 (44, 49–58), 395 (49), 396, 397 (49–58), 398 (49), 399 (27, 49, 82, 83), 401 (27, 82, 83), 402 (35, 49–55), 403 (49–58), 404 (27, 82–84), 405 (3, 13, 57), *406–409*, 414 (25–30), 415 (30), 416 (26, 43, 44), 417 (43, 44, 50), 421 (27), 422 (44), 424 (103), 425 (25, 27, 103, 111), 426 (25, 27, 103), 427, 429 (111), 430 (25, 111, 130), 433 (25), 434 (30), 435 (25, 26, 30), 440 (169), 441 (25, 28, 180), 442 (25, 28, 169, 180), 443 (183), 444 (190), 448 (25, 190), 449 (25, 214), 450, 451 (25), 452 (25, 238–240), 453 (238), 454 (25, 26), 455 (258), 456 (25, 258), 459 (25), *461–467*, 740 (211), *770*, 841 (93), 843 (94), *858, 859*
Cremer, S. 83, 86, 89, 90, 94, 95 (96), *132*
Cremer, S.E. 185 (213), *217*
Criegee, R. 470 (36), *489*
Crisma, M. 170–172 (128–130), *215, 216*
Crocker, L.S. 204 (315), *220*, 237 (68), *255*, 776, 777, 780, 781, 783, 784, 786, 790, 791, 794, 806 (15), *810*
Cronauer, R. 271 (47a), *312*
Crooks, E. 228 (24a, 25, 26), *251*
Cross, P.C. 102, 104 (164a), *134*
Crossland, I. 179, 180 (182), 193 (271), *217, 219*
Crossley, R.W. 471 (120, 122), *490*
Crowther, D.J. 500 (18), *648*
Crumrine, D.S. 320 (6, 7), *337*
Csizmadia, I.G. 452 (243), *467*
Csöregh, I. 161 (114–116), 167 (114–116, 127), 168 (116), 169 (114, 115, 127), *215*
Cui, Y.-P. 452 (243), *467*
Cullen, J.M. 779 (34), *810*
Cullen, W.R. 690, 693 (167), *705*

Cunico, R.F. 280 (79), 289 (163), *312, 314*
Curini, M. 291 (187, 188), *314*, 669 (79), *704*
Curtiss, L.A. 374, 384, 393 (78), *408*, 778 (18), *810*
Curtiss, T. 760 (308), *772*
Cusumano, L. 777, 783, 784 (13), *810*
Cutler, A. 499, 502 (14), *648*
Cutler, T.P. 329 (47b), *338*
Cyvin, B.N. 874 (62), *884*
Cyvin, S. 874 (62), *884*
Czugler, M. 161 (114), 167, 169 (114, 127), *215*

Dagdagan, O.A. 228 (24b), 235 (54, 58), 239 (73), *251, 253, 254, 256*
Dailey, B.P. 83, 86, 89 (121), *133*
Dailey, W.P. 183 (208, 209), 184 (208), 190 (258), 191 (264), 195, 196 (258), 204 (315), *217–220*, 237 (68), *255*, 277 (68), 294 (225–227), *312, 315*, 693 (171), *705*, 776, 777 (15), 778 (22), 780, 781 (15), 783 (15, 50), 784 (15), 785 (22), 786, 790, 791, 794 (15), 797 (50), 806 (15), *810*
Daily, W.P. 110 (197), *135*
Dakkouri, A. 186 (235), *218*
Dakkouri, M. 83, 86, 89, 94, 95 (123), *133*, 150 (64, 65), 185 (215), 186 (215, 217–220, 232, 234, 235), *214, 217, 218*
Dallas, B.K. 630 (391, 392), *655*
Dallinga, G. 471 (84), *489*, 863 (6), *883*
Damrauer, M. 113, 114 (217), *136*
Dancsó, A. 304 (312), *317*
Danheiser, R.L. 297, 299 (258), *316*
Daniewski, A.R. 668 (71), *703*
Danishefsky, S.J. 275 (62), *312*
Dannacher, J. 231 (44), *253*
Dannenberg, J.J. 455 (262), *467*
Darborn, G.T.. 110 (193d), *135*
Dass, S.C. 151 (75), *214*
Daub, J. 840 (88), *858*
Dauben, H.J. 394 (120), *410*, 420 (88), *463*
Dauben, H.J.Jr. 451 (235), *467*
Dauben, W.G. 292 (203), *315*, 323 (27c), *337*, 699, 701 (201), *706*
Davalian, D. 718 (60–62, 64), 719 (61), 720 (61, 62), 724 (62), 748, 749 (61), *767*
Dave, P.R. 209 (335), *221*
Dave, V. 663 (32), *702*
Davey, T.W. 730, 732 (159), *769*
Davidson, E.R. 93 (144a), 126, 127 (275), *133, 137*, 391 (99), *409*, 481 (269), 482 (269, 272, 273), 484 (269), *493*, 778, 780, 785 (19), *810*
Davidson, P.J. 583 (291), *653*
Davidson, R.B. 48, 83, 86, 90 (26b), *130*
Davies, A.G. 733 (185), *769*
Davies, H.M.L. 290 (180–183), 292 (218–220), *314, 315*, 658, 663 (11), 664 (50–54), 665 (55, 63), 693 (170), 695 (185, 186), 699 (204), *702, 703, 705, 706*
Davies, S.G. 284 (126, 127), *313*
Davis, B.R. 471 (117), *490*
Davis, E.R. 554 (208), *652*
Davis, W.M. 593 (319), *654*
Davison, A. 420 (84), *463*, 554 (207), *652*
Dawe, R.D. 674 (102, 103), *704*
Day, A. 591, 599, 602 (311), *654*
Day, B.W. 158, 159 (93, 94), *215*
Day, C.S. 631 (394), *656*
De, S.C. 710 (13), *766*
Deadman, W.G. 846 (105), *859*
Deakyne, C.A. 83, 86, 95 (105b, 105c), *132*, 147, 177 (53), 187 (238), *214, 218*, 418 (65), *462*
De Barros Neto, B. 471 (108), *490*
deBoer, Th.J. 120 (241–244), *136*
Decius, J.C. 102, 104 (164a), *134*
Declercq, J.-P. 164 (120), 176 (156), 179 (178, 183–188), 180 (183, 185–188), 200 (300), *215–217, 220*
Decleva, P. 60 (59), *131*
Deejroongraung, K. 228 (24a, 25, 26), *251*
Deezer, A.E. 233 (49), *253*
DeFrees, D. 820 (34), *857*
DeFrees, D.J. 123 (255), *137*, 417 (57), *462*
Dehmlow, E.V. 288 (148, 155, 157, 159), *314*
Dehnicke, K. 429 (119), *464*
Dejroongruang, K. 457 (269), *468*
Dekaprilevich, M.O. 310 (357), *318*
Delanghe, P.H.M. 281 (97), 282 (98), *313*, 505 (61), 506 (62), *649*
Delbecq, F. 733 (184), *769*
Delker, G.L. 48, 49, 71 (42), *130*
Della, E.W. 791 (68, 69), 794 (76, 79, 80, 84–86), 795 (68, 69), *811*
Demaison, J. 175 (151, 152), *216*
De Maré, G.R. 83, 86, 88, 93 (102b), *132*
DeMayo, P. 320 (1), *337*
De Meijere, A. 44, 48 (3), *129*, 233 (49), *253*, 831 (65), 832 (65, 66), *858*
De Mesmaeker, A. 179, 180 (183, 186), *217*
Demjanov, N.J. 226, 238 (20a), *250*
Demjanow, N.J. 470 (79), *489*
Demonceau, A. 290 (176), *314*, 663 (33, 34), 683 (140), 688, 689 (161), 693 (175), *703, 705, 706*
Demondeau, A. 658, 663 (5), *702*
Demoulliens, A. 440 (170), *465*
Demuynck, C. 195 (276), *219*
Denis, A. 267 (31), 268 (34), *311*
Denis, J.M. 866 (30), 872 (30, 55), 878 (55), *884*
Denis, R.C. 273 (57), *312*

Denmark, S.E. 280 (91), *312*
Deno, N. 831 (64), *858*
Deno, N.C. 344 (24), *406*, 413, 419 (23), *461*, 816, 825 (12), 854 (123), *857, 859*
Dent, B.R. 716 (49), 726 (49, 115–117), 746 (117), *766, 768*
DePuy, C.H. 113, 114 (217), *136*, 472 (142), *491*
Derfer, J.M. 471 (83), *489*, 864 (11), *883*
Derocque, J.-L. 622 (371), *655*
Descotes, G. 294 (229), *315*
Desimone, D.M. 575, 593 (278), *653*
Deslongchamps, P. 670 (82), *704*
Desrosiers, P.J. 575 (277, 278), 593 (278), *653*
Dessy, R.E. 499, 500 (7), *648*
Destombes, J.L. 195 (276), *219*
Destro, R. 421 (94), *463*
Dettlaf, G. 585 (295b), *653*
Deuring, L.A. 197 (284a), *219*
Deuter, J. 155, 156 (83), *214*
Devaprabhakara, D. 719 (75), *767*
Devaquet, A.J.P. 370 (70), *408*, 415 (37), 430 (124), 437 (160), *461, 464, 465*
Devos, M.J. 301, 302 (288), *317*
DeVries, K.M. 290 (177), *314*, 667 (64), 693 (169), *703, 705*
deVries, L. 413 (12), *461*
Dewar, M.J.S. 44 (8a–c), 48, 49, 71 (34a, 34b, 36, 37, 43), 74, 82 (8a–c), 96–98 (43), 110 (199a), 114, 115 (222), *129, 130, 135, 136*, 227 (21a, 21b), 235 (62), 242 (88c), *250, 255, 258*, 342, 348 (12), 353 (32), 355 (38), 356, 357 (39), 363 (38, 47), 364 (47), 370, 373 (65), 374 (76), 382 (39, 65), *406–408*, 415 (35), 430 (125), 455 (256, 260), 456 (260), 457 (268), *461, 464, 467, 468*, 470 (70), *489*, 733, 734 (172), 738, 739 (204), *769, 770*, 820 (29), *857*
Dewarr, M.J.S. 758 (295), *771*
Deycars, S. 122 (251a), *137*
Deyrup, C.L. 448 (212), 453 (247), *466, 467*
Diaz, A. 440 (172), 441 (172, 178), 445, 446 (198), 449 (220), *465, 466*
Diaz, M. 111 (206), *136*
Di Blasio, B. 170–172 (129), *216*
Dick, B. 452 (239, 240), *467*
Diercks, R. 736 (198), *769*
Dietrich, H. 194 (275), *219*
Diggins, M.D. 717, 720 (56), *767*
Dill, B. 471 (103), *490*
Dill, J.D. 47 (21a), 83, 86 (95), 102 (160), *130, 132, 133*
DiMagno, S.G. 192 (266), *219*
Dinsmore, C.J. 291 (201a, 201b), *315*
Dinulescu, I. 659 (24), *702*
Dirlam, J.P. 440 (172), 441 (172, 178), *465*
Disch, R.L. 235 (62), *255*, 446 (203), 452 (241), 457 (268), *466–468*
Ditchfield, R. 109 (184a–d), *134*, 384 (92), *409*, 591, 592, 599, 603–605 (314), *654*
Dittmer, M. 874 (63), *884*
Ditzel, E.J. 756 (288), *771*
Diversi, P. 556 (223), *652*, 881 (86), *885*
Dockery, K.P. 800 (111), 802, 803 (125), *812*
Dodson, R.M. 412 (4), *460*
Dodziuk, H. 863 (4d), *883*
Doering, W. 126 (268d), *137*
Doering, W.v. 455 (255), *467*
Doering, W.v.E. 188 (249), *218*
Doering, W.von E. 235 (60), *254*, 413 (11), *461*, 469 (17, 17), 470 (48, 33, 48), 473 (150, 151, 153, 154, 150, 151, 153, 154), 474 (160, 160), 477 (150, 151, 153, 154, 185, 190, 196, 199, 150, 151, 153, 154, 185, 190, 196, 199), 478 (153, 203, 153, 203), 479 (150, 151, 153, 154, 212, 150, 151, 153, 154, 212), 480 (220, 221, 220, 221), 481 (150, 151, 150, 151), *488, 489, 491, 492*, 669 (75), *703*
Dogan, B.M.J. 716, 730, 757 (50), *766*
Doi, T. 296 (250), *316*
Dojarenko, M. 470 (79), *489*
Dolbier, W.R. 867, 868 (34), 875, 876 (67), *884*
Dolbier, W.R.Jr. 83, 86 (104), *132*, 287 (144), 288 (151, 152), *314*
Dolgii, I.E. 557, 561, 562 (231), *652*, 659 (22, 23), 688 (22), *702*
Domalski, E.S. 246 (110), 247 (112), *259*
Domarev, A.M. 865, 872, 875 (19), *883*
Domek, J.M. 280 (87), *312*, 866 (23), *883*
Domenicano, A. 141 (13), 159 (96, 97, 99), *213, 215*
Domnin, I.N. 191 (262, 265), 192 (268), *219*, 235 (61), *254*
Doms, L. 209 (333), *221*
Donaldson, A.W. 573, 575, 597 (265), *653*
Donaldson, W. 632 (395, 396a, 396b), *656*
Donaldson, W.A. 575 (276), *653*
Donovan, D. 821 (41), 835 (78), *857, 858*
Donovan, D.J. 819 (24), *857*
Donskaya, N.A. 149 (63), *214*, 620 (365, 366), *655*
Doorn, R.van 120 (245), *136*
Döpp, D. 320 (5, 6), *337*
Dorko, E.A. 194 (274), *219*, 471 (120, 122), *490*, 569 (255), *653*, 730 (146), *768*
Dorow, R.L. 292 (221), *315*, 661 (27), 663 (37), 688 (37, 162), 693 (27, 162), 695 (184), *702, 703, 705, 706*
Dorr, M. 867 (32), *884*
Dorrity, M.J. 309 (344), *318*

Dorsch, D. 268 (32), *311*
D'Ottavio, T. 590, 603 (308), *654*
Doubleday, C. 126 (269, 272, 273), *137*, 481 (261, 266, 267), *493*
Dougerthy, R.C. 356, 357, 382 (39), *407*
Dougherty, D.A. 469 (12), 470 (40), 481 (262), *488, 489, 493*
Dougherty, R. 370, 373, 382 (65), *408*
Dowd, P. 202 (310, 311), *220*, 273 (55), 289 (162), *312, 314*, 470 (45), *489*, 669 (76), *703*
Downing, J.W. 784 (51), *810*
Doyle, M.J. 507 (73), 508 (74), 521 (130, 131), 624, 627, 640, 641 (73), *649, 650*
Doyle, M.P. 262 (2a, 2i), 290 (2a, 177), 292 (2i, 213–215, 221), 294 (238), *310, 311, 314–316*, 658 (2, 3, 7, 9), 661 (27), 663 (2, 3, 7, 9, 37), 667 (64), 684 (145, 146), 685 (146), 686 (150–152), 687 (2, 145, 160), 688 (37, 162), 690 (165, 166), 691 (165), 693 (2, 27, 162, 166, 169), 695 (166, 184), 698 (7, 197–199), 701 (198), *702, 703, 705, 706*, 866 (27), *884*
Draai, R.K.van der 471 (84), *489*
Drabløs, F. 150, 151 (68), *214*
Drake, J. 291 (191), *314*
Draux, M. 157 (91), *215*
Dreizler, H. 182 (203), 186 (225), *217, 218*
Drew, M.G. 591, 599, 602 (311), *654*
Driessen, P.B.J. 438 (162), *465*
Drumright, R.E. 157 (92), *215*
Druzhinina, A.I. 229 (33), *252*
Du, P. 116, 117 (227), *136*
Dubourg, A. 164 (120), *215*
Dubrulle, A. 175 (151, 152), *216*
Dubus, H. 195 (276), *219*
Duchatsch, W. 716, 745, 749 (52), *767*
Duetsch, M. 530 (153), 564 (153, 246, 247), 565 (246), *650, 652, 653*
Dumont, W. 263 (11), 298 (265–267), *311, 316*
Dumright, R.E. 417 (48), *462*
Duncan, A.B.F. 111, 112 (212a), *136*
Duncan, J.L. 98 (154–156), *133*, 186 (222), *218*
Duncanson, L.A. 48, 49, 71 (38), *130*
Dung, N.H. 150 (66), *214*
Dunitz, J.D. 124 (262), *137*, 140 (1), 183 (206), *212, 217*, 364 (48), *407*, 451 (233), *467*
Dunkelblum, E. 508 (80), 512 (90), *649*
Dunlap, L.H.Jr. 247 (115), *259*
Dunlap, S.E. 303 (308), *317*
Dunogués, J. 280, 285 (80), *312*
Dupuis, M. 121, 123 (247), *136*
Durandetti, S. 280 (90), *312*
Durig, J.R. 150 (64), 160 (102, 104), 181 (193), 182 (193, 199–201), 183 (199, 201), 186 (226, 227, 230, 235), 191 (261), *214, 215, 217–219*
Durig, J.R.Jr. 83, 86, 94, 105 (115), *132*
Durmaz, S. 83, 86, 89, 90, 94, 95 (96), *132*, 481 (231), *492*
Dürr, H. 713 (29–32), 714 (34), 728 (134), 734 (191), 738, 739 (201), 756 (289, 290), 757 (290), *766, 768–771*
Durucasu, I. 747, 752 (264), *771*
Durup, J. 481 (235), *492*
Dusseau, Ch.H.V. 120 (241), *136*
Dutt, D.N. 710 (13), *766*
Du Vernet, R.B. 444 (185), *465*
Duyne, G.van 353 (34), 401 (123), *407, 410*
Dvořák, D. 167 (126), *215*
D'yachenko, A.I. 865 (18), *883*
Dyadchenko, V.P. 501 (32), *648*
D'yakonov, I.A. 730 (152), *768*
Dyas, C. 418 (65), *462*
Dyczmons, V. 153 (80), *214*
Dzhemilev, U.M. 544 (170), 564 (245), 624, 640 (376), *651, 652, 655*
Dzikliñsha, A. 288 (150), *314*
Dziwok, K. 185 (212), *217*

Eaborn, C. 730, 742 (162, 163), *769*
Eagle, C.T. 292 (213), *315*, 690, 693, 695 (166), 698 (197), *705, 706*
Eastman, R.H. 418 (64), *462*
Eaton, D.F. 418 (63), *462*, 513 (93), *649*
Eaton, R. 459 (283), *468*
Ebbrecht, T. 477 (195), *492*
Ebel, M. 266 (24), *311*
Eberson, L. 849, 850 (114), *859*
Eckert-Maksić, M. 83, 86 (109b), *132*, 188 (246), 191 (261), 201 (305), *218–220*, 401 (123), *410*, 454 (251), *467*, 733 (174–176, 178–180), 744, 749 (244), *769, 771*
Edelmann, F. 553 (196), *651*
Edge, D.G. 122 (251c), *137*
Edmiston, C. 65 (70c), *131*
Edmunds, A.J.F. 659, 691 (20), *702*
Edwards, J.P. 280 (91), *312*
Effenberger, F. 434 (150), *465*
Efraty, A. 574 (272, 273), *653*
Egawa, T. 145, 199, 208 (35), *213*
Eggers, D.F. 578 (282), *653*
Egorova, T.G. 432, 433, 435 (145), *464*
Ehrhardt, C. 69 (75), *131*
Eicher, T. 760 (305), *772*
Eidenschink, R. 730, 742 (162), *769*
Eigen, P. 294 (235), *316*
Eijck, B.P.van 162 (117), *215*
Ejiri, S. 570 (259), *653*

Elahmad, S. 275 (63), *312*
Elia, G. 341, 346, 349, 405 (3), *406*
Elinson, M.N. 172 (139), *216*, 310 (357), *318*
Ellinger, Y. 98, 101, 102, 105 (158), *133*
Elliott, C.S. 471 (140), *491*
Elliott, R.J. 458 (274), *468*
Ellis, D. 98 (155), *133*
Ellis, R.J. 470 (47), 471 (92), *489, 490*
Elmes, P.S. 188 (247), *218*
Elrod, L.F. 305 (318), *317*
Elschenbroich, Ch. 498 (3), *647*
Elser, W.R. 322 (16), *337*
Emrich, R. 205 (321), 206 (321, 322), 207 (322), *220*, 429 (121), *464*
Emsley, J.W. 110 (194), *135*
Enden, L.van den 209 (333), *221*
Endo, Y. 84, 85 (127), *133*, 143, 144 (22), 145 (44), 186 (222), *213, 218*, 471 (86), *490*
Ene, D. 292 (214), *315*
Enescu, L.N. 659 (24), *702*
Engdahl, C. 440 (168, 173, 174), 441 (173, 174), *465*, 814 (3), 830, 831 (63a), 841 (90, 91), 856 (3), *856, 858*
Engelman, C. 418 (65), *462*
Engler, A. 189 (255), *218*
Engler, E.M. 77 (92e), *132*
Englert, M. 634 (402), *656*
Epling, G.A. 328 (46), *338*
Epple, K.J. 185, 186 (221), *218*
Erden, I. 872, 876 (54), *884*
Eremenko, O.V. 875 (66), *884*
Ergle, W. 470 (43), *489*
Eriksson, L.A. 116 (232), *136*
Erker, G. 645 (430), *656*
Es-Sayed, M. 289 (170), *314*
Essén, H. 68 (72a), *131*, 380 (86), *409*
Etienne, R. 722, 741 (92, 93), *767*
Etter, J.B. 281, 285 (95), *313*
Etter, M.C. 167 (126), *215*
Euguster, C.H. 869 (45), *884*
Evans, D.A. 292 (209), *315*, 698 (195, 196), *706*
Evans, W.H. 224 (9), 237 (9, 69), *249, 255*
Evanseck, J.D. 487 (288), *494*
Evrard, G. 263 (11), *311*
Ewbank, J.D. 145 (37), *213*
Ewing, E.E. 470 (33), *489*
Ezaki, A. 306 (325), *317*

Fabregue, E. 164 (120), *215*
Facelli, J.C. 110 (197), *135*, 783, 797 (50), *810*
Factor, R.E. 322 (22), 325 (37), 329 (47c), 333 (37), *337, 338*
Fagan, P.J. 197 (284b), *219*, 446 (205), *466*, 513 (94), 515 (94, 108), *649, 650*
Faggiani, R. 344, 345 (21), 394, 402 (121, 122), *406, 410*, 413, 415, 416 (22), 417 (59, 60), 419 (22), 420 (91), 421, 422 (91–93), 423 (92, 93), 434 (60), 445, 450 (22), *461–463*, 554 (204), *652*, 817, 855 (17), *857*
Fahey, J.A. 756 (287), *771*
Fahrni, C. 292 (208), *315*, 698 (193), *706*
Fairfax, D.J. 291 (200d), *315*
Falcetta, M.F. 450 (224), *466*
Falck, J.R. 267 (28), *311*
Falconer, W.E. 471 (115), *490*
Falkenberg, A.C. 825 (53), *858*
Fan, K. 84, 85, 95 (131), *133*, 144, 176 (24), *213*
Fang, W. 230 (38), 231 (43), 232 (38), 247 (113), *252, 253, 259*
Fanta, G.F. 865 (15), *883*
Fariña, F. 303 (310), *317*
Farley, B. 374 (74), *408*, 452, 455, 456 (242), *467*
Farona, M.F. 208 (327), *220*
Farquharson, G.J. 828, 853 (59), *858*
Fastabend, U. 288 (157), *314*
Fath, J. 199 (293, 294), *219*
Fatima, A. 303 (309), *317*
Faucher, H. 121, 122 (250), *137*
Faul, M.M. 292 (209), *315*
Faulkner, T.R. 105 (172a), *134*
Fauth, W. 829 (62), *858*
Fedoryñski, M. 288 (158), *314*
Fedorynsky, M. 288 (150), *314*
Fedukovich, S.K. 172 (139), *216*
Fee, G.-A. 709 (12), *766*
Feeney, J. 110 (194), *135*
Fegley, G.J. 297 (260, 261), *316*
Feher, F.J. 503, 566, 568 (42), *648*
Feller, D. 391 (99), *409*, 778, 780, 785 (19), *810*
Feng, A.S. 192 (266), *219*
Ferguson, L.N. 226, 238 (20c), *250*, 340, 350 (1), *406*
Fernández, D. 289 (161), *314*
Fernández, M.D. 283, 289 (118), 305 (314, 321), *313, 317*
Fernández-Alvarez, E. 283 (118), 289 (118, 161), 305 (315, 321), *313, 314, 317*
Fernandez-Simon, J.L. 308 (336), *318*
Fernholt, L. 874 (62), *884*
Ferreira, V.F. 687 (157), *705*
Feshbach, H. 68 (73), *131*
Fessenden, R.W. 122 (251b), *137*
Fessner, W.-D. 477 (190), *492*
Fettinger, J.C. 593, 594, 601–604 (316), *654*
Ficini, J. 667 (67), *703*
Field, L.D. 444 (188), *465*
Fields, E.K. 727 (126), *768*

Fierens, P.J.C. 232 (45), *253*
Figge, L. 477 (190), *492*
Fink, M.J. 192 (269, 270), *219*
Finley, M. 632 (396a), *656*
Finnerty, M.A. 235 (62), *255*, 451, 457 (234), *467*
Finzel, R.B. 445 (193), *466*
Fischer, E.O. 522 (134), 534 (156), *650, 651*
Fischer, H. 374 (75), *408*, 534 (155), 581–583 (290), 628 (387), *651, 653, 655*
Fish, R.W. 499, 502 (14), *648*
Fisher, A.M. 279 (77), *312*
Fisher, S.A. 302 (299), *317*
Fitjer, L. 280 (81), *312*, 865 (20), 871 (52), 872 (20, 53), 873 (59), 874 (59, 63), 875 (53, 68), 876 (20), *883, 884*
Fitzgerald, G.B. 104 (168c), *134*
Flechtner, T.W. 334, 335 (50c), *338*
Fleischer, U. 110 (188h, 192a, 201c, 202e–g), 111 (201c, 202e–g), *135*, 393, 396 (111, 115, 116), 397 (116), *409, 410*, 446 (204), *466*
Fleming, A. 300 (273), *316*
Fleming, I. 263 (9a, 9b), 280 (94), *311, 313*, 415, 451 (36), *461*
Fleming, S.A. 297 (263), *316*
Fletterick, R.J. 554 (202), *652*
Flitcroft, T.L. 239 (73), *256*
Flood, E. 83, 86, 95 (105d), *132*
Flowers, M.C. 471 (139), 479 (207, 215), *491, 492*, 880 (82), *885*
Flygare, W.H. 110 (199d), *135*, 146 (46, 47), 187 (240), *214, 218*
Foglesong, W.D. 838 (84, 85), *858*
Föhlisch, B. 179 (180), *217*
Foley, J.K. 760 (306, 307), *772*
Follweiler, D.M. 840 (88), *858*
Fontani, P. 282, 290 (101, 102), *313*
Fontanille, M. 295 (249), *316*
Ford, G.P. 48, 49, 71, 96–98 (43), *130*
Ford, P.W. 478 (200), *492*
Ford, R.G. 83, 86, 88 (116), *133*
Ford, T.M. 418 (63), *462*, 839 (87), *858*
Formicheva, M.B. 418 (65), *462*
Forst, W. 471 (110), *490*
Forster, D.L. 727 (129, 130), 760 (130), *768*
Förster, T. 48, 49, 55 (24), *130*
Forsyth, D.A. 846, 847 (107b, 107c), 849 (107b), *859*
Foster, J.M. 56 (48a), *130*
Foster, J.P. 374, 384, 393 (78), *408*
Foucaud, A. 301 (282), *316*
Fouchet, B. 271 (46), 301, 302 (294), *312, 317*
Fournet, G. 268 (40), *311*
Franck-Neumann, M. 306 (323), *317*, 551 (188, 189), 555 (217), *651, 652*
Franke, R. 110 (192a), *135*, 393, 396 (115), *410*
Franklin, J.L. 44, 74, 83 (5b), 114 (219), *129, 136*, 246 (109), *259*
Fraser-Reid, B. 869 (43), *884*
Fray, G.I. 419 (74), *463*
Frazier, J.O. 673 (98), 678 (120), *704*
Fredericks, S. 581 (288), *653*
Freedman, H.H. 742 (233), *770*
Freedman, T.B. 475 (166, 167), 476 (166, 167, 174–176), 477, 484 (166), 487 (166, 280, 282), *491, 493*
Frei, B. 280 (82), *312*
Frei, K. 777 (14), *810*
Freiser, B.S. 113, 114 (215), *136*
Frend, M.A. 471 (119), *490*
Frenette, R. 671 (87), *704*
Frenking, G. 208 (330), *220*, 377, 393, 404 (84), *408*
Fretz, E.R. 448 (213), *466*, 816 (9), *856*
Freudenberger, J.H. 594, 602 (320), *654*
Freudenberger, R. 262 (6), *311*
Freund, A. 470 (65), *489*
Frey, H.M. 469 (2, 3), 470 (47), 471 (92, 139, 140), 479 (207, 209, 210, 215), *488–492*, 880 (82), *885*
Fridh, C. 60 (56), *131*
Friederich, E.C. 866 (23), *883*
Friedli, A.C. 775 (10), 783, 787, 789, 790 (48), 794 (91), 797 (48, 104), 798 (48, 104, 114), 799 (48), 800 (48, 104, 114), 801 (48, 114), 802 (48, 104, 123, 124), 804 (48), *809–812*
Friedmann, N. 470 (58), *489*
Friedrich, C. 503 (45), *648*
Friedrich, E.C. 280 (87–89), *312*, 394 (120), *410*, 417 (54), 419 (81), *462, 463*
Friese, C. 186 (236), *218*, 294 (236), 301 (279), *316*
Frisch, K.-H. 740 (211), *770*
Frisch, M.J. 85 (136d), *133*, 392 (104), *409*
Fritchie, C.J.Jr. 64 (66b), *131*
Fritsch, F.N. 143, 144 (20), *213*
Fritschi, H. 292 (204–206), *315*, 698 (189–191, 193), *706*
Fritz, H. 712, 736 (27), *766*
Froelicher, S.W. 113, 114 (215), *136*
Frohlich, R. 645 (430), *656*
Froidmont, Y.de 290 (176), *314*, 663 (34), *703*
Frutos, M.P.de 283, 289 (118), 305 (321), *313, 317*
Frutos, P.de 289 (161), *314*
Fryling, J.A. 283 (114), *313*
Fryxell, G.E. 291 (192), *314*
Fryzuk, M.D. 690, 693 (167), *705*
Fuchs, E. 197 (280), *219*
Fuchs, P.L. 275 (61), *312*

Fuchs, R. 157 (91), 192 (267), *215, 219*, 239 (74), 243 (95), *256, 258*
Fueno, T. 481 (259), *493*
Fugami, K. 279 (76), *312*
Fugiel, R.A. 419, 420 (78), *463*
Fujimoto, H. 48 (26d, 40), 49, 71 (40), 83, 86, 90 (26d), *130*
Fujisawa, T. 284 (124), *313*
Fujita, H. 305 (319), *317*
Fukayama, T. 84 (126), *133*
Fukuda, Y. 880 (84), *885*
Fukumoto, K. 669 (77), *703*
Fukunaga, T. 646 (431a, 431b), *656*
Fukutani, Y. 280 (92), *312*, 866 (23), *883*
Fukuyama, K. 303 (311), *317*
Fukuyama, T. 143, 144 (21), 145 (21, 35), 155 (81), 199, 208 (35), 209 (333), *213, 214, 221*
Fukuzawa, S.-I. 279 (75), *312*
Fulcher, J.G. 748 (268), *771*
Funaki, I. 265 (22, 23), *311*, 505 (60), *649*
Fung, A.P. 845, 846 (101), 851 (118), *859*
Furlani, T.R. 126 (271, 272), *137*, 481 (265, 266), *493*
Furman, D.B. 508 (75), 627 (385), *649, 655*
Furst, G.T. 212 (344), *221*
Furuta, K. 306 (324), *317*
Fusheng Feng 160 (104), *215*
Fustero, S. 519 (115), *650*
Futagawa, T. 283 (119–122), *313*
Fyfe, , C.A. 420–422 (91), *463*
Fyfe, C.A. 394, 402 (121), *410*, 554 (204), *652*

Gabelica, V. 417 (58), *462*, 820 (37), *857*
Gabor, B. 547 (176), 581, 582, 585 (289), *651, 653*
Gabrielli, A. 554 (197), *651*
Gaede, P.E. 503 (45), *648*
Gai, Y. 301, 302 (289), *317*
Gai, Y.H. 682 (131, 132), *705*
Gainsford, G.J. 719, 720 (71), 730–732, 742, 758 (155), 762 (311), *767, 769, 772*
Gajewski, J.J. 469 (10, 11, 21), 470 (60, 61, 63), 478 (204), *488, 489, 492*, 880 (83), *885*
Gakh, A.A. 790, 791 (67), 794 (67, 90), 798, 807 (67), 808 (67, 90, 132), *811, 812*
Galasso, V. 109 (185a), *134*, 739 (208), *770*
Gallagher, D.G. 882 (92), *885*
Gallardo, O. 161, 167 (115, 116), 168 (116), 169 (115), *215*
Gallego, M.G. 323, 331 (28a, 28b), *338*
Galloway, N. 725 (114), *768*
Gallup, G.A. 60 (57), *131*
Gammon, B.E. 235 (58), *254*
Gandour, R.W. 450 (223), *466*, 739 (207), *770*
Ganem, B. 683 (141), *705*
Ganesh, S. 296 (251), *316*
Gannett, T.P. 322 (21), 325 (34), 333 (21), *337, 338*
Gano, D.R. 128 (282), *137*
Gano, J. 729 (142), *768*
Gansow, O.A. 638 (408), *656*
Gao, J.-N. 69 (74), *131*
Gao, Y. 302 (301), *317*
Garcia-Grandas, S. 297 (262), *316*
Gardner, D.V. 727 (127), *768*
Gareau, Y. 291 (200a), *315*
Gareiss, B. 262 (5), *311*, 776, 781–784, 787–789 (45), *810*
Garin, D.L. 470 (51, 56), *489*
Garner, P. 202 (311), *220*, 669 (76), *703*
Garratt, P.J. 342, 344, 382, 390 (7), *406*, 415, 424, 427, 431 (39), *461*, 718 (59–62, 64), 719 (61, 74), 720 (61, 62), 722 (108, 109), 724 (62), 748 (59, 61), 749 (59, 61, 74), *767, 768*
Garrett, C.G. 882 (92), *885*
Gassman, P.G. 280 (85), *312*
Gassmann, P.G. 274 (60), *312*
Gasteiger, J. 419 (71), *462*
Gatilov, Yu.V. 434 (148), *464*
Gatti, C. 384–386 (93), 394 (93, 119), *409, 410*, 425 (109), *463*
Gaudin, J.M. 268 (37), *311*
Gauss, J. 67, 73, 83 (71), 85, 93 (135), 96–98 (71), 104, 109 (135), 110 (190a, 190b, 191), 111 (211), *131, 133, 135, 136*, 249 (122), *260*, 364, 375 (56), 377 (56, 84), 378, 381 (56), 386 (94), 387, 388 (56), 391, 392 (56, 101, 102), 393 (84, 114), 394 (56), 396 (56, 114), 397, 403 (56), 404 (84), *408, 409*, 416, 417 (43), 425, 427, 429 (111), 430 (111, 130), *462–464*, 832 (67), 848 (109b), *858, 859*
Gavezzotti, A. 481 (227), *492*
Gaw, J.F. 104 (168b), *134*
Gawronska, K. 499 (9), *648*
Gawronski, J. 499 (9), *648*
Geertsen, J. 110 (193b, 193c), *135*
Gehrlach, E. 179 (180), *217*
Geise, H.J. 145 (36), 209 (333), *213, 221*
Geisler, S.E. 471 (128), *490*
Geißler, E. 207 (325), *220*
Genêt, J.-P. 267 (30, 31), 268 (34, 36, 37), *311*
George, B.E. 344, 345 (21), *406*, 413, 415, 416, 419, 445, 450 (22), *461*
George, P. 74 (84), *131*, 226, 237 (19b), *250*, 387 (95, 96), 389 (96), *409*
Gergens, D.D. 503, 566, 568 (42), *648*
Germain, G. 174 (144), *216*
Germer, A. 507, 624, 627, 640, 641 (73), *649*

Gerratt, J. 57, 58, 66, 69, 71, 73, 76, 77, 82, 83 (51), *130*
Gerry, M.C.L. 178 (175), 182 (205), *217*
Gerson, F. 871 (49), *884*
Getty, S.J. 126, 127 (275), *137*, 175 (148), *216*, 481, 482, 484 (269), *493*
Ghatak, U.R. 668 (74), 674 (101), *703, 704*
Ghatlia, N.D. 470, 477, 479 (62), *489*, 559, 561 (237), *652*
Ghenciulescu, A. 659 (24), *702*
Ghiro, E. 273 (57), *312*
Ghoroghchian, J. 760 (307), *772*
Giam, C.S. 116 (228, 233), 117 (233), *136*
Giansiracusa, J.J. 447 (209), *466*, 828, 853 (59), *858*
Giebel, K. 727 (120), *768*
Gieren, A. 201 (306), *220*, 353 (29), *407*, 734, 735 (194), *769*
Giering, W.P. 499, 502 (14), 515 (109), 516 (109, 110), 566, 568 (110), *648, 650*
Gilbert, K.E. 470 (55), *489*
Gilchrist, T. (355), *655*
Gilchrist, T.L. 415, 451 (36), *461*, 727 (129, 130), 729 (141), 760 (130), *768*
Gill, G.B. 415, 451 (36), *461*
Gill, P.M.W. 794 (86), *811*
Gillespie, R.J. 159 (98), *215*
Gillette, G.R. 48, 49, 71 (44), *130*
Gillies, C.W. 176 (160, 163), 177 (160–165), 178 (160, 161, 163, 164, 174), *216, 217*
Gillies, J.Z. 177 (165), *216*
Gilliom, L.R. 507 (70, 71), *649*
Ginsburg, D. 236 (66), 237 (69), *255*, 716, 730, 757 (50), *766*, 774 (5), *809*, 864, 870 (12), *883*
Giovanni, E. 739, 744 (205), *770*
Girreser, U. 512 (92), *649*
Givens, R.S. 323 (27b), *337*
Givens, R.W. 323 (26), *337*
Gladysz, J.A. 748 (268), *771*
Glanzmann, M. 444 (185), *465*
Gleicher, G.J. 356, 357, 382 (39), *407*, 758 (295), *771*
Gleiter, R. 44, 48 (4), *129*, 188 (243), 193, 194 (272), 208, 209 (328), *218–220*, 264 (17), *311*, 353 (34), 366 (61), 401 (123), *407, 408, 410*, 418 (63), 450 (224), 454 (251), *462, 466, 467*, 606 (338), *654*, 798–801 (115), *812*, 870 (47), 871 (49), 876, 878, 881 (71), *884, 885*
Glenar, D.A. 475 (169), 476 (169, 170), *491*
Glick, M.D. 505 (59), *649*
Glidewell, C. 178 (173a), 179 (177a, 177b), *216, 217*, 793 (75), *811*
Glinka, T. 300 (278), *316*
Glück, C. 709, 720 (8), 731 (166), 733, 734, 739–741, 746, 755–758 (8), *766, 769*
Glukhovtsev, M.N. 44, 48 (20), *129*, 342, 344 (10), 353 (33), 382, 390 (10), *406, 407*
Goddard, W.A. 481 (243), *493*
Godfrey, P.D. 188 (247), *218*, 757 (293), *771*
Godunov, I.A. 797, 799, 800, 802 (101), *811*
Goebel, P. 235 (60), *254*
Gokel, G.W. 866 (25), *883*
Gold, E.H. 427 (114), 430 (123), 433 (114), *464*
Goldberg, A.H. 481 (262), *493*
Golden, D.M. 77, 120 (89a), *132*
Golding, B.T. 264 (19), *311*
Goldman, A. 669 (76), *703*
Goldmann, A. 202 (311), 205, 206 (321), *220*
Goldschmidt, Z. 469 (25), *488*, 544 (167), 554 (198), *651*
Goldstein, E. 60, 112, 113 (52), *130*
Goldstein, M. 841 (92), *858*
Goldstein, M.J. 415, 425 (34), *461*
Goldwhite, H. 145 (40), *213*
Golić, M. 191 (261), *219*
Gomann, K. 471 (89), *490*, 863 (7), *883*
Gompper, R. 571, 572 (262), 573 (266, 267), 576 (267), 610 (353), 612 (266, 353), 615 (360), 616 (266), 629 (267), *653, 655*
Goodman, , A.L. 418 (64), *462*
Goodman, J.L. 470 (41), *489*
Goodman, M. 305 (316), *317*
Goodwin, T.E. 689 (163), *705*
Gordon, A.J. 226, 238 (20e), *250*
Gordon, A.S. 479 (214), *492*
Gordon, M.D. 646 (431b), *656*
Gordon, M.S. 83, 96–98 (124), 102 (161), 128 (282), *133, 137*, 779 (33), *810*
Gordymova, T.A. 432, 433, 435 (145), *464*
Gore, J. 268 (39, 40), *311*
Gosau, J.-M. 797, 799, 800 (103), 805 (127), *811, 812*
Goscinski, O. 440 (170), *465*, 841 (92), *858*
Göthling, W. 188 (246), *218*, 289 (165), *314*
Goto, N. 74, 79 (80), *131*
Gottlieb, C.A. 195 (277), *219*
Gottstein, W. 179 (180), *217*
Gough, K.M. 108, 109 (180), *134*
Goumzili, M.E. 271 (46), *312*
Govindan, S.V. 671, 672 (92), 678 (120), *704*
Gowling, E.W. 589 (303), *654*
Grand, A. 121, 122 (250), *137*
Grandberg, K.I. 160 (107), *215*, 501 (32, 38a, 38b), *648*
Grandinetti, F. 822, 825 (45), *857*
Grant, D.M. 110 (195), *135*, 783, 797 (50), *810*
Grant, J.L. 851, 852 (117), *859*
Grant, M.D. 110 (197), *135*
Grassberger, M. 659, 691 (20), *702*

Gratkowski, C. 289 (170), *314*
Graul, S.T. 793 (73), *811*
Graupner, F. 285 (132), *313*
Gravel, D. 273 (57), *312*
Graziani, M. 644 (429), *656*
Gream, G.E. 740 (219), *770*
Grée, R. 506 (66), 550 (186, 187), 551 (190, 191), *649, 651*, 659 (25), 672 (93), 697 (188), *702, 704, 706*
Greemberg, R.S. 687 (159), *705*
Green, M. 519 (118), 574 (270), 594 (321), 625, 628 (382), 633 (397, 398), 635 (403, 404), 637 (406), 638 (411), *650, 653–656*
Greenberg, A. 44, 74 (5a), 83 (5a, 95, 104), 86 (95, 104), *129, 132*, 140 (6, 7), 191 (263), *213, 219*, 224 (3–6), 225 (3, 18), 226 (18, 20d), 227 (18, 23), 229 (33), 232 (45), 235 (58, 59), 238 (20d), 243 (94), 246 (3), 247 (3, 116), 248 (120), *249–254, 258–260*, 451 (232), *467*
Greenlee, K.W. 471 (83), *489*
Grehl, M. 645 (430), *656*
Greig, G. 123 (256a, 256b), *137*
Greulee, K.W. 864 (11), *883*
Grev, R.S. 83 (94), 94 (148), 96, 98 (94, 148), *132, 133*
Grieco, P.A. 662, 667 (31), *702*
Griffin, G.W. 336 (54), *338*
Griffin, J.H. 661 (27), 686 (151), 688 (162), 693 (27, 162), *702, 705*
Griffin, M.T. 158, 159 (93), *215*
Grigg, R. 309 (344, 345), *318*, 556 (229), *652*
Grignon-Dubois, M. 280, 285 (80), *312*
Grimm, U.W. 471 (122), *490*
Grimme, W. 402 (125), *410*, 453 (248), *467*, 551 (192), 552 (193), 556 (224), *651, 652*, 708, 715 (2), 721, 722 (84), 749, 751, 756 (2), *766, 767*
Grimmeis, A.M.H. 197 (285a–c), *219*
Grishin, Y.K. 620 (365, 366), *655*, 789, 790, 797–802, 805, 807, 808 (65), *811*
Grob, C.A. 794 (80), *811*
Grohmann, K. 455 (262), *467*, 715 (44), *766*
Groner, P. 186 (226, 227), *218*
Gross, M.L. 117 (237), *136*
Grosse, M. 711 (25), *766*
Groth, P. 197 (286), *219*
Grubbs, R.H. 507 (70, 71), 580, 581 (287), 584 (287, 294), 601 (287), *649, 653*
Gruenbaum, W.T. 325 (39), *338*
Gruetzmacher, J.A. 632 (396a, 396b), *656*
Grundke, J. 503 (43, 44), *648*
Grunewald, G.L. 323 (25a, 25b), 327, 328 (45), *337, 338*
Grushina, O.E. 149 (63), *214*
Grützmacher, H. 83, 96 (125), *133*
Grutzner, J.B. 345 (25), *407*, 413 (24), 440, 441 (171), 458 (24), *461, 465*, 840 (89), *858*
Guelzim, A. 174 (143, 144), *216*
Guillory, J.P. 160 (101, 106), 181 (197), *215, 217*
Guillot, N. 200 (300), *220*
Guittet, E. 263 (11), 301 (281), *311, 316*
Gülaçar, F.O. 740 (217), *770*
Günther, H. 48, 83, 86, 90, 96 (27), *130*, 552 (193), *651*, 717 (57), 721, 722 (84), 740 (210–214), *767, 770*
Guo, L. 681 (130), *705*
Gussoni, M. 105–107 (169a–c), *134*
Gustavson, G. 470 (78), *489*, 864, 875 (14a), *883*
Güthlein, P. 511 (87), *649*
Gutman, D. 477 (191, 192, 197), *492*
Gutner, N.M. 232 (45), 241 (82), *253, 257*
Gutsche, C.D. 469 (18), *488*
Gwinn, W.D. 102, 104 (164b), *134*

Ha, T.-K. 113 (214), 121–123 (246), *136*
Haas, C.K. 48, 49, 71 (42), *130*
Haase, H.L. 384 (90), *409*
Hackett, P. 329 (47a), *338*
Hacksell, U. 661 (30), *702*
Hadad, C.M. 60, 61, 84–86 (62), *131*, 263 (7b), *311*, 780, 783, 785, 793, 794 (43), *810*
Haddon, R. 422 (95, 96), *463*
Haddon, R.C. 366 (63), 368 (63, 64), 370, 373 (68, 71), 394 (64, 68, 117), 395 (64), *408, 410*, 425 (106–108), 426 (106), 429 (107, 122), 430 (107, 126), *463, 464*
Hadjiarapoglou, L. 310 (353, 354), *318*
Hafner, K. 478 (206), *492*
Hagen, E.L. 116 (224), *136*
Hagen, G. 147, 149 (58), *214*
Hagen, K. 147, 149 (58), *214*
Häger, J. 471 (114), *490*
Hahn, E. 505 (56), *649*
Haider, R. 870 (47), *884*
Haire, J.J. 324 (31b), *338*
Halazy, S. 545 (174), 546 (174, 175), *651*
Halberstadt, I. 713 (31), *766*
Haley, G.J. 353 (34), *407*
Haley, M.M. 188 (244), 193, 194 (272), *218, 219*, 262 (2f), *310*, 353 (30), *407*, 709 (10, 12), 722, 723 (103), 734, 735 (196), 739–741, 746, 749 (10), *766, 767, 769*
Hall, H.R.Jr. 310 (358), *318*
Hall, M.B. 61 (63b), *131*, 603 (337), *654*
Hallman, J.H. 239 (74), 243 (95), *256, 258*
Halls, T.D.J. 687 (155), *705*
Halow, I. 224 (9), 237 (9, 69), *249, 255*
Halpern, J. 521 (129), *650*

Halton, B. 201 (304), *220*, 246 (106), *259*, 453 (247), *467*, 568, 572, 587 (250), *653*, 709 (6–8, 11), 710 (14, 16), 713 (28), 715 (39, 40), 716 (49), 717 (56), 718 (63, 66, 68), 719 (68, 69, 71, 72), 720 (8, 56, 69, 71, 72, 77, 79, 82, 83), 721 (86–89), 722 (86–89, 99), 724 (69), 725 (114), 726 (49, 115–117), 729 (140), 730 (153–159, 161), 731 (89, 155, 165, 167), 732 (154–159, 161), 733 (7, 8, 168, 171, 186), 734 (8, 66, 171, 192), 735, 737 (66), 738 (186), 739, 740 (8), 741 (6, 8, 89, 171, 220, 223, 226, 227), 742 (154–158, 161, 165), 743 (171, 240), 744 (241), 745 (186, 240, 245), 746 (8, 117, 248, 252), 747 (66), 749 (186), 750 (240), 751 (240, 252), 752 (252), 754 (280), 755 (8), 756 (8, 252, 282, 286–288), 757 (6–8, 89, 223), 758 (8, 153, 155, 157, 158, 294, 296), 759 (153, 157, 158, 294, 297, 298), 760 (11, 140, 157, 294, 296–298, 306–308, 310), 761 (140, 294), 762 (140, 311), 763 (312, 313), 764 (282), 765 (314, 315), *766–772*
Halton, M.P. 733, 734, 741, 743 (171), *769*
Ham, S.W. 470 (45), *489*
Hamamoto, I. 301 (283), *317*
Hamberger, H. 440, 441 (172, 175), *465*
Hambly, T.W. 718, 719 (68), *767*
Hamdouchi, C. 299 (269), *316*
Hameka, H. 109 (181a–c), *134*
Hamelin, J. 301, 302 (294), *317*
Hamilton, C.L. 511 (86), *649*
Hamilton, J.G. 57, 59, 69, 75, 112 (50), *130*, 144, 145 (30), *213*
Hamony, M.D. 190, 195 (256), *218*
Hamrock, S.J. 780, 781, 784, 787, 791, 799 (39), *810*
Hamzaoui, F. 174 (145), *216*
Hanack, M. 297 (256, 257), *316*, 759 (299), *771*, 833 (71), *858*
Hancock, K.G. 321 (14), 334 (49), *337, 338*
Hancock, R.I. 519 (118), *650*
Handwerker, B.M. 302 (299), *317*
Handy, N.C. 84, 85, 96, 98, 101, 102 (130), 104 (168b), 106 (174), 110 (193d), *133–135*
Häner, R. 504 (48), *648*
Hann, J.W. 428 (117), *464*
Hansen, A.E. 110 (189, 193a, 198), *135*, 393, 396 (112), *409*
Hansen, H.-J. 473 (159), *491*
Hanzawa, Y. 268 (35), *311*
Hara, K. 306 (325), *317*, 658 (15), *702*
Harada, T. 500 (28), *648*
Harano, K. 554 (200), *652*
Harch, P.G. 854 (125), *859*
Hargittai, I. 141 (13), 145 (34), 159 (98–100), *213, 215*
Hariharan, P.C. 44, 96 (6), *129*, 430 (125), *464*
Harmata, M. 275 (63), *312*
Harmony, M.D. 83 (102c, 106a–c, 122), 84 (134), 85 (139a, 142), 86 (102c, 106a–c, 122, 139a, 142), 88 (102c), 89 (106a–c, 122), 93 (102c, 122), 95 (106a–c), *132, 133*, 144 (27), 147 (55), 175 (148), 181 (191, 194), 182 (191), 183 (202), 187 (27, 241), 190 (258, 259), 194 (27), 195 (27, 241, 258), 196 (241, 258), 199 (289, 290), *213, 214, 216–219*, 730 (150, 151), 757 (151), *768*
Harn, N.K. 686 (152), 690, 693, 695 (166), *705*
Harp, J.J. 523, 530 (137), *650*
Harring, L.S. 281, 282 (96), *313*
Harris, D.L. 419, 425 (80), 440 (167, 171, 172), 441 (171, 172, 178), *463, 465*, 554 (208), *652*
Harris, J.M. 418 (63), *462*
Harris, S.J. 730, 742 (162), *769*
Harris, W.C. 93 (146), *133*
Harrison, R.J. 104 (168c), *134*
Harrison, S.A.R. 710 (14), 715 (40), *766*
Harsányi, K. 302 (298), *317*
Hart, H. 435 (153), 449 (218), *465, 466*, 853 (120), *859*
Hartley, F.R. 659 (17), *702*
Hartman, A. 64 (66a), *131*, 176 (154), *216*
Harusawa, S. 299 (272), *316*
Harvey, D.F. 286 (137, 138), 294, 295 (244, 245), *313, 316*
Hase, H.L. 60 (54c), *130*
Hasegawa, A. 186 (229), *218*
Hasegawa, M. 280 (78), *312*
Hasegawa, T. 745 (246), *771*
Haselbach, E. 231 (44), *253*
Hashemi, A.S.K. 534 (154), *650*
Hashemi, M.M. 420, 425 (86), 426 (112), *463*
Hashimoto, M. 746, 751 (259), *771*
Hashimoto, S. 685 (148), *705*
Hashmi, A.S.K. 309 (348), *318*
Hassel, O. 143 (19), *213*
Hassenrück, K. 795 (116), 797 (105), 798 (116), 800 (105), 801 (116), *811, 812*
Hassila, H. 701 (206), *706*
Hassler, C. 874 (63), *884*
Hasty, N.M. 843 (95), *859*
Hasty, N.M.Jr. 432, 433, 435 (142, 143), 436 (142), *464*
Hatajima, T. 282 (100), 308 (335), *313, 318*
Hatano, M. 555 (220, 221), 632 (220), *652*
Hatch, C.E.III 665 (58), *703*
Hatjiarapoglou, L. 681 (128), *704*
Hattori, K. 500 (28), *648*

Hauck, J. 211 (341), *221*
Haumann, T. 211 (342), *221*, 264 (18), *311*, 763 (313), *772*, 878, 880 (77), *885*
Hay, P.J. 481 (243), *493*
Hayashi, M. 186 (229), *218*
Hayashi, N. 627 (384), *655*
Hayashi, S. 711 (23), *766*
Hayashi, T. 268 (38), *311*
Hayes, E.F. 481 (233), *492*
Hayter, R.G. 598, 599 (331), *654*
Haywood-Farmer, J. 344, 345 (17), *406*
Haywood-Farmer, J.S. 448 (212), *466*
He, J. 477 (199), *492*
He, Z. 85 (138a, 138b), *133*, 392 (106, 108), *409*
Head, N.J. 848, 849, 851 (110), *859*
Head-Gordon, M. 85 (137), *133*, 392 (105, 110), *409*
Healy, E.F. 114, 115 (222), *136*, 430 (125), *464*
Hechtl, W. 419 (71), *462*
Heck, R.F. 570 (260), *653*
Hecker, M. 161 (114–116), 167 (114–116, 127), 168 (116), 169 (114, 115, 127), *215*
Hedberg, K. 83 (108b), 85 (143), 86 (108b, 143), *132, 133*, 143, 144 (20), 177 (168, 171), 178 (171), 204 (316), *213, 216, 220*, 776, 779, 794 (25), *810*
Hedberg, L. 83 (108b), 85 (143), 86 (108b, 143), *132, 133*, 177 (168, 171), 178 (171), 204 (316), *216, 220*, 776, 779, 794 (25), *810*
Hegedus, L.S. 295 (248), *316*, 525 (143, 144), 526 (143), 527 (148), *650*
Hehre, W.J. 44 (6), 47 (21b), 48–50 (30), 83, 86, 88, 93 (102d), 96 (6), 102 (160), 117–119 (238c), 123 (255), *129, 130, 132, 133, 136, 137*, 370 (66, 69, 70), 371, 379 (66, 69), 384 (92), 394 (69), *408, 409*, 414 (31), 415 (31, 37), 425 (31, 105), 430 (31, 124), 435 (105, 151), 437 (105, 151, 160), 451 (31), *461, 463–465*, 817 (16), 820 (30–33), 826, 827 (54), 848 (109a), 854 (125), *857–859*
Heijdenrijk, D. 175 (149), *216*
Heilbronner, E. 48–50, 56, 57 (31a, 31b), *130*, 356 (40), 365 (60), *407, 408*, 414, 415 (32), *461*, 739 (205, 206), 744 (205), *770*, 778, 785 (22), *810*
Heindl, J. 727 (120), *768*
Heiss, H. 535, 536, 548 (159), *651*
Heiszman, J. 302 (298), *317*
Heitz, M.P. 551 (189), *651*
Heldmann, C. 182 (203), *217*
Hell, Z. 291 (186), *314*
Helm, D.van der 158, 159 (93, 94), *215*
Helmchen, J.G. 268 (32), *311*
Helquist, P. 282 (106, 107), 286 (136), *313*, 658, 663 (10), *702*, 715 (43), *766*
Hemmerrsbach, P. 870 (48), *884*
Hemond, R.C. 586 (299), *654*
Hencher, J.L. 194 (274), *219*
Hendricks, R.T. 292 (203), *315*, 699, 701 (201), *706*
Hendricksen, D.K. 83, 86, 89, 95 (106a, 106b), *132*
Hendrickson, T. 459 (283), *468*
Henniges, H. 309 (347), *318*
Henrickson, B.W. 865 (15), *883*
Henry, W.P. 574, 590, 595, 600 (275), *653*
Herb, G. 690, 693 (167), *705*
Herberich, G.E. 517 (112), *650*
Herbst-Irmer, R. 564, 565 (246), *652*
Hergruetner, C.A. 715 (43), *766*
Herkert, T. 207 (325), *220*
Herman, G. 682 (133), *705*
Herman, M.S. 470 (41), *489*
Hernandez, M.I. 687 (159), *705*
Herndon, J.W. 294, 295 (242), *316*, 523 (137–141), 525 (138–140, 145, 146), 527 (139), 528 (139, 140, 149, 150), 529 (140, 151), 530 (137, 138, 141, 149), 531 (138, 140), 533 (138, 150), *650*
Herr, R. 185 (211), *217*
Herrera, S. 111 (206), *136*
Herrig, W. 740 (214), *770*
Herrmann, A.T. 581, 582, 585 (289), *653*
Hershberger, S.S. 519 (125, 126), *650*
Hertkorn, N. 458 (276, 280), 459 (286), *468*
Herzog, C. 176 (158), *216*
Hess, B.A. 729 (144), *768*
Hess, B.A.J. 728, 729, 757 (132), *768*
Hess, B.A.Jr. 417 (54), 457 (269), *462, 468*
Hesse, D.G. 225 (11), 241 (84), 242 (87), *250, 257*
Hevesi, L. 301 (284), *317*
Hewkin, C.T. 181 (190), *217*
Hey, Von E. 429 (119), *464*
Heydinger, J.A. 336 (53), *338*
Heydt, H. 197 (280), *219*
Heydtmann, H. 83, 86, 89, 94, 105 (112), *132*, 471 (103), *490*
Heyne, T.R. 658, 661 (14), *702*
Hiberty, P. 820 (30, 32), *857*
Hiberty, P.C. 733 (184), *769*, 817 (16), 826, 827 (54), *857, 858*
Hickey, M.J. 77 (92b), *132*, 229 (33), *252*
Hidaka, Y. 471 (134), *490*
Hilderbrandt, R.L. 175 (147), 199 (291), *216, 219*
Hill, K. 867 (32), *884*
Hillard, R.L. 715 (41, 44), *766*
Hiller, W. 759 (299), *771*

Hillig, K.W.II 146 (49, 50), *214*
Himman, M.M. 698 (195), *706*
Hinde, A.L. 457 (270), *468*
Hinman, M.M. 292 (209), *315*
Hinton, J.F. 110 (186), *134*, 393, 396 (113), *409*
Hintze, F. 267 (29), *311*
Hippler, H. 477 (193), *492*
Hirai, K. 294 (234), *316*
Hirakawa, K. 711 (23, 24), *766*
Hirama, K. 644 (426), *656*
Hiroi, K. 469 (15), *488*
Hirose, C. 145 (39, 41), *213*
Hirota, E. 84, 85 (127), *133*, 143, 144 (22), 145 (44), 186 (222, 228), *213, 218*, 471 (86), *490*
Hirshfeld, F.L. 64 (66a), *131*, 176 (154), *216*
Hisamichi, H. 305 (322), *317*
Hitchcock, P.B. 592, 602 (315), *654*
Hixson, S.S. 334, 335 (50a), *338*
Hiyama, T. 501 (31), *648*
Hnyk, D. 110, 111 (202a), *135*, 393, 396, 397 (116), *410*
Ho, K.-K. 303 (307), *317*
Hoa Tran-Huy, N. 534 (156), 535, 536, 548 (159), *651*
Hobe, M. 263 (11, 12), 264 (13), *311*
Hodge, J.D. 816, 825 (12), *857*
Hodoscek, M. 733 (174, 177, 176, 177, 180, 181), *769*
Hoel, E.L. 566, 567 (248), *653*
Hoerrner, R.S. 686 (149), *705*
Hofelich, T.C. 854 (126), 855 (127), *859*
Hoffman, G.J. 475, 476 (169), *491*
Hoffmann, H.M.R. 272 (50, 51), *312*
Hoffmann, M.G. 687 (157), *705*
Hoffmann, R. 48 (26a–d, 28, 40), 49, 71 (40), 83, 86, 90 (26a–d), 125, 126 (263), *130, 137*, 355 (37), 363 (37, 47), 364 (47), *407*, 415 (33, 34), 418 (67), 425 (34), 438 (163), 451 (33, 231), 455 (231), *461, 462, 465, 467*, 480 (222–226), *492*, 578, 580 (283), 590 (308), 602 (334), 603 (308, 334), 606 (338), *653, 654*, 773 (2), *809*, 841 (92), *858*
Hoffmann, W. 232 (45), *253*
Hofmann, J. 581–583 (290), 628 (387), *653, 655*
Hofmann, P. 534 (155), 535, 536, 548 (159), *651*
Hogeveen, H. 438 (162), 439 (165), 444 (186), 449 (219), *465, 466*
Hohenberg, P. 61, 64 (64), *131*
Hoiness, C.M. 865 (22), *883*
Holder, A.J. 235 (62), *255*, 457 (268), *468*
Holloway, M.K. 353 (32), *407*, 738, 739 (204), *770*
Holloway, R.L. 77, 78 (93), *132*
Hollowell, C.D. 155, 187, 190 (85), *215*
Holm, T. 243 (91), *258*
Holm, U. 471 (135), *490*
Holmes, J.D. 419 (75), *463*, 554 (215), *652*
Holmes, J.L. 226, 237 (19c), 249 (122), *250, 260*
Holmes, T. 305 (317), *317*
Holt, E.M. 170–172 (128), *215*, 305 (318), *317*
Holtzclaw, J.R. 93 (146), *133*
Holtzhauer, K. 123 (254a, 254b), *137*
Holzawrth, G. 105 (172a), *134*
Hömberger, G. 287 (142), *313*
Hommes, N.J.R.v.E. 110, 111 (201c), *135*, 393, 396, 397 (116), *410*
Hon, M.-Y. 469 (26), *488*
Honda, T. 684 (147), *705*
Honegger, E. 48–50, 56, 57 (31a, 31b), *130*, 778, 785 (22), *810*
Hongu, A. 280 (78), *312*
Hood, J.N.C. 681 (127), *704*
Hopf, H. 151 (72), 187, 194 (237), 210 (339), *214, 218, 221*, 730 (147), *768*
Hopkinson, A.C. 117–119 (239), *136*
Hopkinson, A.C.J. 121 (248), *136*
Hoppe, D. 267 (29), *311*
Hoppe, M.L. 590, 603, 604 (310), *654*
Hori, N. 554 (200), *652*
Horn, E. 167 (125), *215*
Hornbuckle, S.F. 291 (192, 199), *314, 315*, 684, 687 (145), *705*
Horne, W. 185 (213), *217*
Horner, D.A. 94, 96, 98 (148), *133*
Horowitz, G. 478 (203), *492*
Horsley, J.A. 481 (241, 242), *493*
Horspool, W.M. 323, 331 (28a, 28b), *338*
Hosoda, A. 297 (255), *316*
Hosomi, A. 278 (71), *312*, 513 (93), *649*
Hossain, M.B. 158, 159 (93, 94), *215*
Hossain, M.M. 285, 286 (135), 290 (178), *313, 314*, 663, 694 (44), *703*
Hosseini, S. 242 (87), *257*
Houk, K.N. 125 (265), *137*, 189 (252), *218*, 450 (223), 457 (268), *466, 468*, 487 (284, 288), *494*, 739 (207), *770*
Houle, F.A. 114 (220), *136*
House, H.O. 665 (56, 57), *703*
Houser, J.J. 816, 825 (12), *857*
Hout, R.F.Jr. 48–50 (30), *130*
Howard, A.E. 83, 86, 88, 93 (102c), *132*
Howard, J.A.K. 625, 628 (382), *655*
Hoye, T.R. 291 (201a, 201b), 308 (339–342), *315, 318*
Hoyer, G.-A. 661 (29), *702*
Hoyer, H. 242 (87), *257*
Hoz, S. 505 (54), *648*
Hrib, N.J. 200 (301), *220*

Hrovat, D.A. 116, 117 (227), *136*, 477 (198), 481, 482, 484 (269), *492, 493*
Hsu, E.C. 105 (172a), *134*
Hu, B. 290 (183), *314*, 664 (50, 51), *703*
Hu, C.-M. 277 (67), *312*
Hu, N.X. 301, 302 (292), *317*
Hu, Y. 208 (327), *220*
Huang, J.-L. 296, 297 (254), *316*
Huang, M.-B. 69 (74), 116, 117 (230, 235), *131, 136*, 440 (170), 441 (179), *465*, 841 (92), *858*
Huang, Y.-Z. 296, 297 (254), 301, 302 (290, 291), *316, 317*
Hubbard, J.L. 574 (275), 590 (275, 310), 595, 600 (275, 322), 603, 604 (310), *653, 654*
Huber, H. 419 (71), *462*, 505 (55), *649*, 778, 785 (22), *810*
Hüber, T. 734, 735 (194), *769*
Huber-Patz, U. 202 (311), 209 (332), *220*
Hubert, A.J. 290 (172, 176), *314*, 658 (5), 659 (18), 663 (5, 33, 34, 36), 683 (140), 688, 689 (161), 693 (175), *702, 703, 705, 706*
Hübner, T. 201 (306), *220*, 353 (29), *407*
Huby, N.J.S. 292 (220), *315*, 664 (54), 695 (186), *703, 706*
Hudlicky, T. 291 (195, 197), 299 (271), 300 (273), *315, 316*, 469 (22, 26, 28, 29), *488*, 544 (166, 168), *651*, 663 (47), 671, 672 (89–92), 673 (90, 95–100), 678 (120), *703, 704*
Hudson, C.E. 116 (228, 233), 117 (233), *136*, 557 (234), *652*
Hugel, H.M. 741 (226, 227), 756 (286–288), *770, 771*, 828, 853 (59), *858*
Hughes, L. 122 (251a), *137*
Hughes, R.E. 554 (202), *652*
Hughes, R.P. 573 (265), 574 (270, 275), 575 (265, 276–278), 579 (286), 586 (299, 300), 590 (275, 309, 310), 591 (312–314), 592 (313, 314), 593 (278), 595 (275, 309, 322, 323), 597 (265), 599 (312–314), 600 (275, 309, 312, 322, 323), 601 (312, 329), 602 (309, 312, 313), 603 (309, 310, 314, 329), 604 (310, 314), 605 (309, 314), 625, 628 (382), 629 (389, 390), 630 (391, 392), 631 (389, 393, 394), 633 (389, 397, 398), 635 (403, 404), 638 (389), *653–656*
Huisgen, R. 77 (90b), *132*, 419 (71), 452 (245), *462, 467*
Hunt, W.J. 481 (243), *493*
Hunter, T.H. 471 (115), *490*
Hunton, D.E. 629 (389, 390), 631, 633, 638 (389), *655*
Huntsman, W.D. 470 (57), *489*
Hursthouse, M.B. 517, 518 (114), *650*
Husi, R. 294 (230), *315*
Husk, G.R. 285 (134), *313*, 524, 525 (142), *650*
Hussain, H.H. 289 (166), *314*, 665 (60), *703*
Hutcheson, D.K. 292 (218), *315*, 699 (204), *706*
Huttner, G. 607 (346), *654*
Huyffer, P.S. 320 (5, 6), *337*
Hyman, A.S. 225 (10, 13), 241 (84), *249, 250, 257*
Hymen, B. 478 (204), *492*

Ianelli, S. 734 (190), *769*
Ibata, T. 310 (356), *318*
Ibers, J.A. 96 (147), *133*, 141 (15), 156 (15, 86), 159 (15), *213, 215*, 607–609, 618 (347), *655*
Ibrahim, M.A. 96, 98 (152), *133*
Ibrahim, M.R. 74 (86), *131*
Ibrahim, P.N. 793 (74), *811*
Icheln, D. 303 (303b), *317*
Ignatenko, A.V. 789, 790, 797–802, 805, 807 (65), 808 (65, 131), *811, 812*
Ihara, M. 669 (77), *703*
Ii, A. 279 (76), *312*
Iijima, K. 170 (136, 137), *216*
Iijima, T. 145 (42, 43), 146 (42), 181 (198), 210 (42), *213, 217*
Ikai, S. 597 (326), *654*
Ikazawa, N. 834 (76), *858*
Ikeda, M. 307 (329), *317*
Ikegami, S. 685 (148), *705*
Ikoma, K. 746, 751 (259), *771*
Ikota, N. 683 (141), *705*
Imai, T. 283 (123), *313*
Imamoto, T. 282 (99, 100), 308 (335), *313, 318*
Imamura, K. 869 (44), *884*
Imanishi, T. 271 (49), 291 (194), *312, 314*, 670 (85), *704*
Inagaki, S. 56, 57, 69 (47), 74 (80), 75 (47), 79 (47, 80), 80, 81, 96–98 (47), 128 (278), *130, 131, 137*
Inamoto, N. 717, 718, 745, 748–750 (58), *767*
Ingold, C.K. 412 (2), *460*, 471 (80), *489*
Ingold, K.U. 122 (251a, 252), *137*, 743 (239), *770*
Inomata, K. 280 (93), *313*
Inoue, T. 751 (272), *771*
Ioffe, A.I. 140 (4), *212*, 793 (72), *811*, 863 (4b, 4c), 865 (18), *883*
Ipaktschi, J. 712, 757 (26), *766*, 844 (99, 100), *859*
Ippen, J. 716 (48), 718 (65), 746 (254), *766, 767, 771*
Iqbal, M.N. 500 (17), *648*
Ireland, N.K. 718, 719 (68), *767*

Irie, T. 448 (212), 453 (247), *466, 467*
Irngartinger, H. 155, 156 (83), 198 (288), 202 (288, 309–311, 313), 203 (309), 205 (321), 206 (321, 322), 207 (322), 208 (328), 209 (328, 331, 332), 210 (336, 337), 211 (341), *214, 219–221*, 264 (17), *311*, 429 (121), *464*, 669 (76), *703*
Isaev, I.S. 432, 433, 435 (145), 438 (161), *464, 465*
Isaeva, L.S. 508 (75), 627 (385, 386), 628 (386), *649, 655*
Isagawa, K. 287 (143), *314*
Isaka, M. 506 (67–69), 570 (259), *649, 653*
Ishibashi, H. 307 (329), *317*
Ishifune, M. 308 (338), *318*
Ishigami, T. 627 (384), *655*
Ishige, H. 684 (147), *705*
Ishihara, H. 746, 751 (253), *771*
Ishihara, T. 309 (350, 352), *318*, 682 (139), *705*
Ishikawa, S. 195 (278), *219*
Ishikawa, T. 678 (116), *704*
Ishitani, Y. 56, 57, 69, 75, 79–81, 96–98 (47), *130*
Ishizawa, S. 268 (35), *311*
Ishizu, T. 554 (200), *652*
Isoe, S. 670 (81), *704*
Israel, R.J. 264 (16), *311*
Ito, K. 695, 698 (180), *706*
Ito, S. 555 (220, 221), 632 (220), *652*
Ito, T. 64 (66c), *131*
Itoh, K. 270 (43), 292 (212a, 212b), *311, 315*, 663, 693, 698 (43), *703*
Itoh, Y. 268 (38), *311*, 663, 693, 698 (43), *703*
Ivanov, A.O. 508 (75), *649*
Ivanov, C. 300 (274), *316*
Iwanaga, K. 306 (324), *317*
Iwasaki, H. 608–612, 618, 619 (351), *655*
Iwasaki, M. 116, 117 (234a), *136*
Iwata, C. 271 (49), 291 (194), *312, 314*, 670 (85), *704*
Iwata, S. 60 (60), *131*
Iyer, P.S. 828 (58), *858*
Iyer, R.S. 286 (136), *313*

Jackel, F. 501 (37), *648*
Jackman, L.M. 456 (263), *468*
Jackson, J.E. 48, 49, 51, 74, 83, 86, 96–98, 124 (32a, 32b), *130*, 794 (82), *811*
Jackson, R.F.W. 181 (190), *217*
Jacobs, C.A. 190, 195, 196 (258), *218*
Jacobs, G.D. 83, 86, 89 (120), *133*, 177 (167), *216*
Jacobson, R.A. 211 (343), *221*
Jacoby, D. 304 (313), *317*
Jacquier, Y. 644 (427, 428), *656*, 731 (164), 737 (164, 199, 200), 742 (164, 200), 748 (164, 199, 200), 753 (164, 200), 764 (164), *769*
Jaffe, H.H. 499, 500 (7), *648*
Jager, J. 294 (235), *316*
Jahn, R. 202 (310, 313), 205 (321), 206 (321, 322), 207 (322), 209 (332), 210 (336, 337), *220, 221*
Jähne, G. 264 (17), *311*
Jalkanen, K.J. 106 (174), *134*
James, B.R. 690, 693 (167), *705*
James, M.N.G. 422 (98), *463*
Janiak, R. 206 (323), *220*, 363, 364 (47), *407*, 455 (261), *467*
Jankowski, P. 303 (305), *317*
Jano, P. 687 (159), *705*
Janoschek, R. 65, 74, 77, 82, 96 (76b), *131*
Janposri, S. 740 (219), *770*
Jarret, R.M. 777, 783, 784 (13), *810*
Jarstfer, M.B. 292 (213), *315*, 698 (197), *706*
Jaruzelski, J.J. 831 (64), *858*
Jason, M.E. 141, 156, 159 (15), *213*, 353 (34), *407*
Jaszunski, M. 109 (185b), 110 (193e), *134, 135*
Jaw, J.Y. 271 (45), *311*
Jawdosiuk, M. 784 (51), *810*
Jayasuriya, K. 779 (30), *810*
Jean, Y. 481 (240–242, 248–251, 256, 257), *493*
Jeffers, P. 471 (125), *490*
Jeffers, P.M. 471 (129), *490*
Jeffery, S.M. 264 (15), *311*
Jemmis, E.D. 47 (21a), 102 (160), *130, 133*, 602, 603 (334), *654*
Jenkins, J.A. 419 (76), 423 (99), 437 (157), *463, 465*
Jennison, C.P.R. 402 (124), *410*, 454 (253), *467*
Jensen, F. 477 (194), *492*
Jensen, F.R. 451 (229), *467*
Jensen, L.H. 603 (336), *654*
Jensen, M.S. 277 (66), *312*
Jensen, W.A. 828, 853 (59), *858*
Jesaitis, R.G. 844 (98), *859*
Jespersen, K.K. 817, 854, 856 (18), *857*
Jessen, S.M. 161, 164, 165 (113), 166 (113, 122), 170 (122), *215*
Jeuell, C.L. 818 (20a, 20b), 821, 825 (20b), *857*
Jian-Qi, W. 48–50, 56, 57 (31a), *130*
Jiao, H. 353 (33), *407*
Jie, C. 363, 364 (47), *407*, 455, 456 (260), *467*
Jikeli, G. 740 (210, 212), *770*
Jimbo, H. 181 (198), *217*
Jiménez, P. 239 (72), 240 (76), *256*
Jindal, S.P. 432, 433 (144), *464*, 833 (73), *858*

Jochem, K. 156, 159 (87), *215*
Jogun, K.H. 434 (150), *465*
Johnels, D. 501 (37), *648*
Johnson, B.F.G. 556 (222, 223), *652*
Johnson, B.L. 453 (247), *467*
Johnson, C.E.Jr. 420 (89), *463*
Johnson, C.R. 686 (153), *705*
Johnson, D.C. 197 (284b), *219*
Johnson, D.W. 471 (121, 124), *490*
Johnson, L.K. 580, 581 (287), 584 (287, 294), 601 (287), *653*
Johnson, M.P. 192 (269, 270), *219*
Johnson, R. 574 (273), *653*
Johnson, R.P. 322, 333 (21), *337*
Johnson, T. 325 (35), *338*
Johnston, L.J. 122 (252), *137*, 795 (97, 98), 796 (97), *811*
Jolly, P.W. 634 (402), *656*
Jommi, G. 663 (41), *703*
Jonas, V. 208 (330), *220*
Jonczyk, A. 288 (149, 150, 153, 158), *314*
Jones, A.V.K. 447, 448 (207), *466*
Jones, E.M. 522 (135, 136), 523–525 (136), 621 (135), 638 (136), *650*
Jones, G.W. 710 (16), *766*
Jones, L.D. 715 (42), *766*
Jones, M.Jr. 455 (254), *467*
Jones, N.L. 500, 510 (25), *648*
Jones, P.G. 147, 149 (60), 161 (110, 112), 162, 163 (110), 164, 165 (112), 166 (123), 172 (138), 173 (140), 174 (141, 142, 146), 176 (153), 177 (169, 170, 172), 178 (169), 179 (172), *214–216*
Jones, S. 459 (283), *468*
Jones, W.J. 775 (8, 9), *809*
Jones, W.M. 499 (15, 16), 500 (18, 20–24), 502 (41), 510 (21, 22, 24), 566–568 (21), 583, 587 (41), *648*
Jongejan, E. 120 (241–243), *136*
Jonsäll, G. 440 (168, 170, 173), 441 (173, 177, 179), 459 (282), *465, 468*, 814 (3), 841 (90–92), 856 (3), *856, 858*
Jordan, K.D. 356 (40), *407*
Jordan, K.I. 450 (224), *466*
Jorgensen, C.W. 48–50 (29), *130*
Jørgensen, P. 104, 109 (167a), 110 (193b), *134, 135*
Jørgensen, W.L. 235 (62), 245 (100), *255, 258*, 345 (25), *407*, 370 (67, 72, 73), 371 (72, 73), 372 (73), 373 (72, 73), 378 (73), 379 (72, 73), 386, 394 (73), *408*, 413 (24), 415, 425, 430 (38), 458 (24), 454 (249), *461, 467*
Jorgenson, W.L. 448 (213), *466*
Joucla, M. 271 (46), 301, 302 (294), *312, 317*
Juers, D.F. 329 (47a), *338*
Jug, K. 481 (228), *492*, 733 (189), *769*, 779 (28), 798, 799 (102), *810, 811*
Julia, M. 301, 302 (289), *317*, 682 (131, 132, 134), *705*
Jun, C.H. 519 (117), *650*
Jun, Y.M. 836 (81), *858*
Jung, G.L. 672 (94), 678 (118, 119), *704*
Jung, L. 266 (24), *311*
Juntunen, S.K. 282 (108), *313*
Justnes, H. 176 (160), 177, 178 (160, 161), *216*
Justus, K. 874 (63), *884*

Kaarsemaker, S. 243 (95), *258*
Kabakoff, D.S. 816, 826 (13), *857*
Kabat, M. 284 (128), *313*
Kabat, M.M. 303 (306), *317*
Kabuto, C. 555, 632 (220), 644 (426), *652, 656*, 669 (77), *703*
Kaesz, H.D. 394 (120), *410*, 419 (81), 420 (84), *463*
Kafarski, P.K. 185 (213), *217*
Kagabu, S. 746 (253), 747 (263), 751 (253, 271, 272), *771*
Kagan, H. 481 (235), *492*
Kai, Y. 183 (206), *217*
Kaifu, N. 195 (278), *219*
Kaiser, F.-J. 198 (287), *219*
Kaji, A. 265 (21, 23), 301 (283), *311, 317*
Kajtar-Peredy, M. 304 (312), *317*
Kakefu, T. 56, 57, 69, 75, 79–81, 96–98 (47), *130*
Kalabin, G.A. 779 (24), *810*
Kalasinsky, V.F. 181, 182 (193), *217*
Kalaus, G. 304 (312), *317*
Kalcher, J. 65, 74, 77, 82, 96 (76b), *131*
Kalinowski, H.-O. 183 (206), *217*
Kallfaß, D. 208 (328), 209 (328, 331, 332), *220*
Kalyuzhnaya, E.S. 160 (107), *215*
Kalyuzhnaya, Y.S. 501 (38a, 38b), *648*
Kamada, T. 499 (12), *648*
Kamaratos, E. 471 (112), *490*
Kamath, A.P. 336 (52), *338*
Kambara, H. 145, 199, 208 (35), *213*
Kamel, M. 710 (15), *766*
Kamimura, A. 301 (283), *317*
Kamitori, Y. 607 (347, 350), 608 (347), 609 (347, 352), 612 (350, 352), 618 (347, 350), *655*
Kamiya, Y. 282 (100), 308 (335), *313, 318*
Kamm, K.S. 324 (31a), *338*
Kämpchen, T. 646 (432), *656*
Kanai, Y. 270 (43), *311*
Kanaoka, Y. 278 (70), *312*
Kaneko, C. 305 (322), *317*
Kanemasa, S. 290 (184), *314*

Kanematsu, K. 554 (200), *652*
Kang, D. 74, 83 (83), *131*
Kang, S.H. 291 (198), *315*, 677 (109), *704*
Kang, Y.B. 720, 722 (78), *767*
Kao, J. 60 (54b), *130*, 417 (49), *462*, 863 (4a), *883*
Kaplan, R. 711 (21), *766*
Kapon, M. 722 (90), *767*
Karadakov, P.B. 57, 58, 66, 69, 71, 73, 76, 77, 82, 83 (51), *130*
Karl, R.R.Jr. 208 (329), *220*
Karlin, K.D. 556 (223), *652*
Karni, M. 738, 739, 758 (203), *770*
Karpenko, N.A. 241 (82), *257*
Karplus, M. 481 (247), *493*
Kasai, P.H. 578 (282), *653*
Kashimura, S. 306 (325), 308 (338), *317, 318*
Kashiuchi, M. 310 (356), *318*
Kasprzycka-Guttman, T. 235 (52), *253*
Kass, S.R. 237 (67), *255*
Kassir, J.M. 291 (200a, 200d, 200e), *315*
Kaszynski, P. 775 (10), 783 (48), 787 (48, 58, 59), 789, 790 (48, 58), 791 (113), 795 (59), 797 (48, 59, 104, 107), 798 (48, 59, 104, 113, 114), 799 (48, 58, 59), 800 (48, 59, 104, 107, 113, 114), 801 (48, 113, 114), 802 (48, 59, 104, 123), 804 (48, 58), 809 (133), *809–812*
Kataeva, O.N. 199 (298), *220*
Katagiri, T. 287 (139), *313*
Kates, M.R. 448 (213), *466*, 815 (8a), 816 (10), *856*
Kato, H. 746, 751 (259, 260), *771*
Kato, K. 670 (84), *704*
Kato, S. 481 (260), *493*
Katoh, S. 306 (326), *317*, 514 (105), *649*
Katsuhira, T. 500 (28), *648*
Katsuki, T. 292 (212a, 212b), *315*, 695, 698 (180), *706*
Katsumata, S. 60 (60), *131*
Kattija-Ari, M. 83, 86, 88, 93 (102c), *132*
Katz, S.A. 302 (299), *317*
Katz, T.J. 427 (114), 430 (123), 433 (114), *464*
Kaufman, A. 530, 552, 553, 622 (152), *650*
Kaufman, C. 289 (162), *314*
Kaufman, P. 289 (162), *314*
Kaufmann, D. 553 (196), *651*, 868 (37), *884*
Kaufmann, E. 345 (26), *407*, 413 (24), 449 (221), 458 (24, 273, 281), 459 (281), *461, 466, 468*
Kaupert, C. 83, 86, 89, 94, 105 (112), *132*
Kausch, M. 449 (216), *466*
Kawabata, N. 682 (135), *705*
Kawada, M. 658 (15), *702*
Kawashima, H. 270 (43), *311*
Kay, A.J. 717, 720 (56), 733 (168), 763 (312, 313), *767, 769, 772*
Kazala, A.P. 292 (213), *315*, 698 (197), *706*
Kazimirchik, I.V. 229 (34), *252*, 282 (105), *313*
Keane, C.M. 759, 760 (302), *772*
Keck, G.E. 321 (10, 11), 322 (19), 325 (34), *337, 338*
Keese, R. 327 (44), *338*
Keiderling, T.A. 105 (172c), *134*, 476 (172, 173), *491*
Keith, T.A. 64, 65, 68, 107 (67d), *131*, 375, 376, 393 (81), *408*
Keller, C.E. 419 (71, 82), *462, 463*
Keller, H. 536 (162), *651*
Keller, W. 292 (206), *315*, 698 (190), *706*
Kellett, S.C. 625, 628 (382), *655*
Kelley, D.F. 471 (109), *490*
Kellogg, M.S. 323 (27c), *337*
Kelly, D.P. 434 (147), 447 (209), *464, 466*, 741 (226, 227), 756 (286–288), *770, 771*, 818 (20a, 20b), 821, 825 (20b), 828, 853 (59), *857, 858*
Kelly, E. 514, 519 (102), *649*
Kelly, M.B. 186 (234), *218*
Kelly, W.J. 846 (105), *859*
Kelsey, D.R. 430 (125), *464*
Kemmer, P. 308 (332), *318*
Kempe, T. 366 (61), *408*
Kenedy, D. 516, 566, 568 (110), *650*
Kennard, O. 83, 86, 87 (99d), *132*, 159, 180, 185 (95), *215*, 416, 456 (47), *462*
Kennedy, M. 292 (217), *315*, 699 (203), *706*
Kerber, R.C. 574, 590, 602, 603 (274), *653*
Kerl, K. 471 (135), *490*
Kerr, J.A. 123 (256c), *137*
Kerur, D.R. 710 (16), *766*
Ketley, A.D. 881 (85), *885*
Kettle, S.F.A. 589 (303), *654*
Kettley, A.D. 550 (184), *651*
Keyaniyan, S. 191 (265), *219*, 289 (165, 169), *314*, 563 (242), *652*
Kharasch, M.S. 225 (16), 227 (22), 241 (82), 243 (90), *250, 257, 258*
Kharicheva, É.M. 730 (152), *768*
Kharraf, Z.el 294 (229), *315*
Khosavanna, S. 501, 504 (30), *648*
Khusnutdinov, R.I. 544 (170), 564 (245), 624, 640 (376), *651, 652, 655*
Kido, F. 686 (154), *705*
Kiefer, E.F. 427 (113), *463*
Kiegiel, J. 284 (128), *313*
Kiel, D.G. 471 (112), *490*
Kienzle, F. 868 (36), *884*
Kiers, C.T. 175 (149), 210 (336), *216, 221*
Kigoshi, H. 665 (62), *703*, 869 (44), *884*
Kikuchi, O. 759 (303), *772*
Kikuchi, T. 308 (337), *318*
Kilpatrik, J.E. 366, 370 (62), *408*

Kim, C.J. 846 (106), 849 (113), *859*
Kim, H.-R. 765 (317), *772*
Kim, H.S. 687 (157), *705*
Kim, J. 280 (86), *312*, 681 (130), *705*
Kim, T. 797, 798, 800, 802 (104), *811*
Kim, W.J. 291 (198), *315*, 677 (109), *704*
Kimball, S.D. 574, 590, 602, 603 (274), *653*
Kimpe, N.de 179 (178), *217*, 270 (42a, 42b), *311*
Kimura, K. 60 (60), *131*, 880 (84), *885*
Kimura, M. 145 (43), *213*
Kincaid, S. 459 (283), *468*
King, H.F. 126 (271, 272), *137*, 481 (265, 266), *493*
King, M.E. 586 (300), *654*
King, R.B. 574 (272), 585 (297), 597 (326), *653, 654*
King, R.C. 212 (344), *221*
King, R.K. 323 (23, 24), *337*
Kingsbury, K.B. 535, 537 (157, 158), 539 (158, 164), 540, 541 (164), 542, 543 (158), *651*
Kinugasa, H. 308 (338), *318*
Kirby, A.J. 174 (146), *216*
Kirby, S.P. 225 (15), 229 (33), 232 (45), *250, 252, 253*
Kirchen, R.P. 820 (28), 821, 827 (39), *857*
Kirchgässner, U. 607 (344, 345), 617 (345), *654*
Kireev, S.L. 639 (412), *656*, 868 (41), *884*
Kirin, V.N. 234 (51), *253*
Kirklinl, D.R. 246 (110), 247 (112), *259*
Kirkpatrick, J.L. 448 (213), *466*
Kirmse, W. 232 (45), *253*, 287 (142), *313*, 504, 505, 622 (51), *648*, 865 (21), *883*
Kirsch, G. 145 (37), *213*
Kirschning, A. 303 (303a), *317*
Kirson, B. 549 (178), *651*
Kiseleva, N.N. 235 (56), 241 (82), *254, 257*
Kishnan, R. 104 (168a), *134*
Kispert, L.D. 418 (65), *462*
Kistiakowsky, G.B. 470 (75), *489*
Kitatani, K. 501 (31), *648*
Kitchen, D.B. 48, 49, 51, 74, 83, 86, 96–98, 124 (32b), *130*
Kizer, K.L. 186 (230), *218*
Klages, U. 280 (81), *312*
Klahn, B. 153 (80), *214*
Klärner, F.-G. 200 (299), *220*, 236 (65, 66), 245 (100), *255, 258*, 357, 359, 361, 378, 389, 390 (42), *407*, 450–453, 457 (226), *467*, 716, 730, 757 (50), *766*
Kleibömer, B. 188 (247), *218*
Klein, A.W. 83, 86, 88 (117), *133*, 147 (54), 175 (150), 186 (54), *214, 216*
Klein, I.E. 126 (268a), *137*
Klein, K.-D. 623 (373, 374), *655*
Klein, W. 517 (112), *650*
Kleschick, W.A. 264 (20), *311*
Klessinger, M. 744, 749 (244), *771*, 870 (46, 48), *884*
Klever, C. 241 (82), *257*
Klimes, J. 585 (298), *654*
Klose, R. 48, 49, 83, 85–90, 92–96 (33), *130*, 156, 159, 178, 179, 181 (88), *215*
Klumpp, G.W. 499 (11), *648*
Klusik, H. 430 (129), *464*
Kmiecik-Laweynowicz, G. 294 (232), *315*
Knecht, J. 554 (212), *652*
Knochel, P. 183 (206), *217*, 307 (327, 328), *317*, 509 (83), *649*
Knoll, K. 207 (324, 325), *220*
Knutson, D. 427 (113), *463*
Kobayashi, S. 284 (125), *313*
Kobayashi, Y. 268 (35), 297 (255), *311, 316*
Kober, E.M. 590, 603, 604 (310), *654*
Köbrich, G. 560 (239), 561 (239, 240), *652*
Koch, C.T. 715 (46), *766*
Koch, E.W. 832 (70), *858*
Koch, W. 110, 111 (202e–g), 114–116, 126, 128 (223), *135, 136*, 201 (305), *220*, 377 (84), 393 (84, 116), 396, 397 (116), 404 (84), *408, 410*, 417 (57), 444 (189), 446 (203, 204), *462, 466*, 733 (178, 179), *769*, 820 (34), *857*
Kochanski, E. 48, 49, 60, 71 (39), *130*
Kocharian, A.K. 150 (67), *214*
Kochi, G.K. 122 (251c), *137*
Kochi, J.K. 517 (113), *650*, 658, 691 (13), *702*
Koda, G. 306 (325), *317*
Kodadek, T. 290 (179), *314*, 694 (178, 179), 695 (179), *706*
Koenig, T. 760 (308), *772*
Kohl, M. 742 (235–238), 743 (236), 746 (261), 753 (238), *770, 771*
Köhler, F.H. 458 (276, 280), 459 (286), *468*
Kohler, J. 501 (37), *648*
Kohn, W. 61, 64 (64), *131*
Kokoreva, O.V. 230, 231 (41), *252*, 863, 864, 866 (9), 868 (39, 40), 875 (9, 66), 876 (9), 877, 878 (73), *883–885*
Kole, J. 728, 729, 757 (133), *768*
Kolesov, V.P. 225 (14), 227 (23), 229 (33), 230, 231 (41), 232 (45, 47), 233 (48, 49), 235 (53, 55, 58), 238 (14, 70, 71), 241 (82), *250–254, 256, 257*
Kollat, P. 179 (180), *217*
Koller, W. 718 (61), 719 (61, 74), 720, 748 (61), 749 (61, 74), *767*
Kollmar, H. 83, 86, 89, 90, 94, 95 (96), 126 (276b), *132, 137*, 374 (75), 382, 383 (87), *408, 409*, 481 (231), *492*
Kolonits, M. 159 (99), *215*

Kolotyrkina, N.G. 789, 790, 797–802, 805, 807, 808 (65), *811*
Komornicki, A. 98, 101, 102, 105 (158), *133*
Kon, K. 670 (81), *704*
Kondakov, D.Y. 307 (330), *318*
Kondo, S. 288 (156), *314*
König, J. 479 (211), *492*
König, W.A. 303 (303b), *317*
Konings, M.S. 192 (266), *219*
Konishi, A. 663 (39, 40), *703*
Konishi, H. 609, 612 (352), 616 (357, 361), *655*
Konkoli, Z. 102–104 (162, 163), 105 (163), 111 (208), *133, 136*, 364, 375, 381, 391, 392, 394, 396, 397, 402, 403 (50), *407*, 414, 441, 442 (28), *461*, 843 (94), *859*
Koopmans, T. 60 (61), *131*
Kopf, J. 191 (265), 192 (268), *219*
Koppel, C. 847 (108), *859*
Koptyug, V.A. 431 (132), 432, 433 (145), 434 (132), 435 (145), 438 (161), 444 (132), *464, 465*
Korkor, O. 290 (181), *314*
Korkowski, P.F. 308 (339), *318*
Korp, J.D. 192 (267), *219*
Korsmeyer, R.W. 678 (112), *704*
Korsunsky, V.I. 501 (38b), *648*
Korte, S. 708, 715 (2), 721, 722 (84), 746, 748 (250), 749, 751 (2), 756 (2, 250), *766, 767, 771*
Kos, A.J. 96, 98 (151), *133*, 444 (189), 458, 459 (281), *466, 468*
Köser, G. 551 (192), *651*
Kosiol, A.E. 500 (23), *648*
Koskinen, A.M. 677 (110), *704*
Koskinen, A.M.P. 291 (189, 190), *314*, 701 (206), *706*
Kosnikov, A.Y. 802 (122), *812*
Köster, A.M. 733 (189), *769*
Koster, J.B. 429 (120), *464*
Kostikov, R.R. 285 (132), *313*
Kostitsyn, A.B. 282 (103), *313*, 659, 688 (26), *702*
Kostyk, M.D. 160 (108), *215*, 417, 435 (61), *462*, 817, 855 (17), *857*
Koszyk, F.J. 671, 672 (90–92), 673 (90), *704*
Kotelko, A. 663 (38), *703*
Kotel'nikova, T.A. 229 (34), *252*
Kottirsch, G. 776, 781, 787, 788 (61), *811*
Koutecky, J. 481 (246), *493*
Kovacevic, K. 863 (8), *883*
Kovacic, P. 784 (51), *810*
Kovancek, D. 733 (175), *769*
Kowalczyk-Przewloka, T. 668 (71), *703*
Kozelka, J. 164 (121), *215*
Kozhinskii, S.O. 659 (23), *702*
Kozhushkov, S.I. 230 (40, 41), 231 (41), *252*, 282 (105), 308 (334c), *313, 318*, 863, 864 (9), 865 (19), 866 (9), 867 (31), 868 (39, 40), 872 (19, 57), 875 (9, 19), 876 (9, 70), 877 (70, 73, 74), 878 (70, 73, 75, 77–79), 879 (79), 880 (77, 81), *883–885*
Kozina, M.P. 225 (14, 14), 227 (23), 229 (34), 232 (45, 47), 233 (48, 49), 234 (51), 235 (53, 55, 58), 238 (14, 14, 70, 71), 241 (82), *250–254, 256, 257*, 780 (42), *810*, 863 (2), *883*
Koz'min, A.S. 208 (326), *220*, 477 (178), *491*, 789, 790 (65), 796 (100), 797 (65, 101), 798 (65, 118, 120), 799, 800 (65, 101), 801 (65, 118), 802 (65, 101, 120–122), 805 (65, 100, 126), 807 (65), 808 (65, 131), *811, 812*
Kraemer, W.P. 60 (55), *131*
Kraevich, V.P. 864, 875 (14b), *883*
Krajewski, J. 303 (306), *317*
Kraka, E. 44, 48, 49 (9, 11, 13), 50, 61 (11), 65 (13, 68, 69), 66 (9, 11), 67 (9, 11, 13, 71), 68 (9, 11, 13, 68, 69), 71 (9, 13), 73 (9, 13, 71), 74 (9, 11, 13), 76 (9, 11), 79 (9), 83 (71, 97), 86–88, 90–95 (97), 96 (9, 71, 97), 97 (9, 13, 71), 98 (9, 71), 110 (203), 111 (203, 205, 208, 210), 123 (259), *129, 131, 132, 135–137*, 144, 178 (25), *213*, 227 (21c), 242 (88b), *250, 258*, 340, 342 (2), 345 (27), 346, 350, 351 (2), 353 (33), 354 (2), 357 (27), 361 (27, 44), 364 (27, 49, 50, 52, 54, 57), 365 (44), 375 (27, 44, 49, 50, 52, 54, 57, 80), 376 (27, 80, 82, 83), 377 (27, 49, 80, 82, 84), 378 (2, 27, 44, 49, 54, 80), 379 (27, 44, 85), 380 (27, 49, 54, 82, 83, 85), 381 (27, 49, 50, 52, 54, 57), 386 (94), 387, 388 (54), 390 (27), 391, 392 (49, 50, 52, 54, 57, 101), 393 (80, 82–84), 394 (44, 49, 50, 52, 54, 57), 395 (49), 396, 397 (49, 50, 52, 54, 57), 398 (49), 399 (27, 49, 82, 83), 401 (27, 82, 83), 402 (49, 50, 52, 54), 403 (49, 50, 52, 54, 57), 404 (27, 82–84), 405 (57), *406–409*, 414 (25–30), 415 (30), 416 (26, 43, 44), 417 (43, 44, 50), 421 (27), 422 (44), 425, 426 (25, 27), 430, 433 (25), 434 (30), 435 (25, 26, 30), 440 (169), 441 (25, 28), 442 (25, 28, 169), 443 (183), 444 (190), 448 (25, 190), 449–451 (25), 452 (25, 238), 453 (238), 454 (25, 26), 456, 459 (25), *461, 462, 465–467*, 841 (93), 843 (94), *858, 859*
Krall, R.E. 248, 249 (121), *260*
Krämer, B. 752 (273–275), *771*
Kramer, W. 731 (166), 746 (261), *769, 771*
Krasnoperov, L.N. 477 (197), *492*

Krass, N. 289 (170), *314*
Kratky, C. 292 (206), *315*, 698 (190), *706*
Krause, N. 822 (44), *857*
Kravtsov, D.N. 508 (75), 627, 628 (386), *649, 655*
Kreissl, F.R. 534 (155), 535 (159), 536 (159, 162), 548 (159), *651*
Kreiter, C.G. 394 (120), *410*, 419 (70, 81), 420 (70, 84), 423, 428 (70), *462, 463*
Kremer, P.W. 185 (213), *217*
Krewson, K.R. 678 (113), *704*
Krief, A. 263 (11, 12), 264 (13), 298 (265–268), 301 (281, 284, 287, 288), 302 (287, 288), 308 (333), *311, 316–318*, 545 (174), 546 (174, 175), *651*
Krieger, P.E. 695 (183), *706*
Krieger, W. 471 (114), *490*
Kriessl, F.R. 562 (241), *652*
Krigas, T.M. 83, 86, 89 (120), *133*, 177 (167), *216*
Krishnamurthy, V.V. 851 (116), *859*
Krishnan, K. 669 (78), 679 (121, 122), *703, 704*
Krishnan, R. 85 (136c, 136d), *133*, 392 (104), *409*
Kristensen, H. 336 (54), *338*
Krivdin, L.B. 779 (24), *810*
Krogh-Jespersen, K. 47 (22), 116, 117 (231), *130, 136*, 280 (84), *312*
Kron, J. 179 (181), *217*
Krough-Jespersen, K. 294 (232, 233), *315, 316*
Krow, G.R. 439 (166), *465*
Kruchten, E.M.G.A.van 449 (219), *466*
Krüger, C. 197 (280), *219*, 429 (120), *464*, 508 (74, 79), 579 (284), 581 (284, 289), 582 (289), 585 (289, 296), 588 (79), *649, 653, 654*, 747 (266), 752 (275), 753 (279), *771*
Krull, I.S. 624, 627, 628 (377), 636 (405), *655, 656*
Krumpe, K.E. 291 (200b), *315*, 658, 663 (8), *702*
Krusic, P.J. 515 (108), 646 (431b), *650, 656*
Ku, A.T. 825, 853 (50b), *857*
Kuan, C.-P. 280 (79), *312*
Kubodera, H. 116, 117 (234b), *136*
Kubota, K. 506 (67), *649*
Kuchan, T.M. 544 (166), *651*
Kucharski, S. 392 (105), *409*
Kuchitsu, K. 84 (126), 85, 86 (139b), *133*, 143, 144 (21), 145 (21, 32, 35), 155 (81, 85), 181 (197), 187 (85), 188 (245), 190 (85), 199, 208 (35), 209 (333), *213–215, 217, 218, 221*
Kuczkowski, R.L. 85, 86 (139a), *133*, 146 (49–51), *214*
Kudo, M. 663 (40), *703*
Kudrevich, S.V. 789, 790, 797–802, 805, 807, 808 (65), *811*
Kudryashev, A.V. 508 (75), 627 (385), *649, 655*
Kuebler, N.A. 60 (53), *130*
Kuhn, D. 682 (133), *705*
Kuivila, H.G. 513 (96), *649*
Kukolich, S.G. 146 (48), *214*
Kulinkovich, O.G. 308 (334a, 334b), *318*, 876 (72), *885*
Kumar, A. 719 (75), *767*
Kumar, P.R. 727 (121), *768*
Kumaravel, G. 504, 505 (52), *648*
Kunkel, E. 692 (168), *705*
Kunz, E. 268 (32), *311*
Kunz, T. 292 (207), *315*
Kuo, G.-H. 286 (136), *313*
Küppers, H. 166, 170 (122), *215*
Kurihara, T. 299 (272), *316*
Kurkutova, E.N. 802 (122), *812*
Kurokawa, N. 284, 292 (129), *313*
Kurtz, D.W. 325 (35), *338*
Kurtz, H.A. 344, 345 (23), 354 (36), 374 (74), *406–408*, 413, 415, 416, 419, 427, 440, 444, 450 (21), 452 (242), 455 (242, 259), 456 (21, 242, 259), 457, 459 (21), *461, 467*
Kutateladze, A.G. 326 (41), *338*
Kutchan, T.M. 469 (22), *488*, 671, 672 (90), 673 (90, 95), *704*
Kutney, J.P. 690, 693 (167), *705*
Kutzelnigg, W. 110 (187a, 187b, 188a–c, 188h, 192a, 192b, 197, 202g), 111 (202g), 125 (264), *135, 137*, 393, 396 (111, 115, 116, 116), 397 (116, 116), *409, 410*, 446 (204), *466*, 783, 797 (50), *810*
Kutzelnigg, W.J. 110, 111 (201c), *135*
Kutznetsova, T.S. 230 (40, 41), 231 (41), *252*
Kuzmina, L.G. 501 (38b), *648*
Kuznetsova, T.S. 282 (105), *313*, 877 (74), *885*
Kuznetziva, T.S. 867 (31), *884*
Kuznetzova, T.S. 863, 864 (9), 865 (19), 866 (9), 868 (39, 40), 872 (19), 875 (9, 19, 66), 876 (9), *883, 884*
Kuzubova, L.I. 432, 433, 435 (145), *464*
Kuzuya, M. 449 (218), *466*
Kwant, P.W. 444 (186), *465*
Kwart, L.D. 673 (98), 687 (155), *704, 705*
Kwart, L.R. 299 (271), *316*
Kwetkat, K. 500 (26), *648*
Kwiatkowski, S. 183 (206), *217*
Kybart, C. 429–431 (131), *464*
Kybett, B.D. 246 (109), *259*

Laabassi, M. 672 (93), *704*
Laakmann, K. 753 (279), *771*

Laane, J. 186 (231, 232, 234), *218*
Labaudiniere, L. 276 (64), 277 (65), *312*
Labeots, L.A. 459 (284), *468*
Laber, G. 413 (11), *461*
Lachmann, J. 185 (212), *217*
Lackmann, R. 530 (152, 153), 552, 553 (152), 564 (153, 246, 247), 565 (246), 622 (152), *650, 652, 653*
Ladon, L.H. 225 (10, 13), 241 (84), *249, 250, 257*
Laffer, M.H. 740 (219), *770*
Lafferty, W.A. 85, 86 (139a), *133*
Lafontaine, J. 670 (82), *704*
Laidig, K.C. 797–800, 802–804 (110), *812*
Laidig, K.E. 83, 86, 88 (98, 111), 89 (98), 93 (98, 111), 94, 95 (98), 111 (207), *132, 136*, 144, 149, 181, 182, 186 (26), *213*, 417 (51, 57), *462*, 819 (23, 26), 820 (26), *857*
Laidig, W.D. 104 (168c), *134*
Laidler, K.J. 471 (136), *490*
Laing, M.E. 802 (124), *812*
Laity, J.L. 394 (120), *410*, 420 (88), 451 (235), *463, 467*
Lajoie, G.A. 170 (134), *216*
Lakshin, A.M. 235 (61), *254*
Lalowski, W. 186 (225), *218*
Lambert, J.B. 445 (193), *466*, 470 (48), *489*, 663 (38), *703*
Lambert, J.M.J. 574, 590 (275), 595, 600 (275, 322), *653, 654*
Lambert, L.R.Jr. 48, 49, 71 (42), *130*
Lambert, R.L.Jr. 499 (10), 500 (10, 27), 505 (10), *648*
Lambs, L. 266 (26), *311*
Lamer, W. 262 (6), *311*
Lamiot, J. 174 (145), *216*
Lamm, M.L. 83, 86, 89 (121), *133*
Lampe, F.W. 854 (124), *859*
Lancelot, C.J. 817 (15), 846 (105), *857, 859*
Landgrebe, J.A. 544 (171), *651*, 695 (183), *706*
Lane, P. 96 (153), *133*
Lane, S. 459 (283), *468*
Lang, R. 176 (158), 202, 203 (309), *216, 220*
Langhauser, F. 547 (176), 581, 582, 585 (289), *651, 653*
Langhoff, S.R. 482 (273), *493*
Lapert, M.F. 583 (291), *653*
Lapiccirella, A. 106 (175), *134*
Lappert, M.F. 606 (342), *654*
Larock, R.C. 519 (125–127), *650*
Larsson, A. 102–104 (162, 163), 105 (163), *133*
Lathan, W.A. 44, 96 (6), *129*
Lau, C.D.H. 68 (72b), 74, 75 (78, 79), 107, 123, 124 (79), *131*, 144, 204 (29), *213*, 361, 365, 375, 378, 379 (44), 380 (86), 394 (44), *407, 409*, 414, 415, 434, 435 (30), *461*, 778 (21), 779 (31), *810*
Lau, C.K. 501 (34–36), *648*
Laube, T. 446 (202), *466*, 759 (300), *772*
Lauher, J.W. 156 (86), *215*, 574, 590, 602, 603 (274), *653*
Laurie, V.W. 83 (101, 105a, 105c), 85 (139a), 86 (101, 105a, 105c, 139a), 88, 93 (101), 95 (105a, 105c), *132, 133*, 177 (159), 186 (224), 187 (238, 239), 190 (257), *216, 218*
Lautens, M. 281 (97), 282 (98), *313*, 505 (61), 506 (62), *649*
Lauterbur, P.C. 783 (49), *810*
Lavrikand, P.B. 444 (185), *465*
Lavrinovich, L.I. 789, 790, 797–802, 805, 807 (65), 808 (65, 131), *811, 812*
Lawler, R. 740 (215), *770*
Layton, E.M.Jr. 65 (70d), *131*
Lazzeretti, P. 106 (175), 109 (185c), *134*
Leach, D.R. 519 (127), *650*
Leandri, G. 615, 616 (359), 621 (369), 622 (359), *655*
Leavell, K.H. 708, 734, 738, 749 (4), *766*
Lebedev, O.V. 868 (38), *884*
Lebedeva, N.D. 232 (45), 235 (56), *253, 254*
Lechoux, L. 266 (24), *311*
Lecomte, P. 298 (267, 268), *316*
Lecoultre, J. 739 (206), *770*
Lee, A.J. 748 (268), *771*
Lee, C. 274 (60), *312*
Lee, C.C. 417 (55), *462*, 815 (4), *856*
Lee, C.-H. 211 (342), *221*, 264 (18), *311*
Lee, G. 852 (119), *859*
Lee, H.-k. 765 (317), *772*
Lee, J.-Y. 310 (358), *318*
Lee, R.E. 458 (279), *468*
Lee, R.K.Y. 471 (126), *490*
Lee, T.J. 84, 85, 96, 98, 101, 102 (130), *133*, 606 (339), *654*
Leep, C.J. 535, 537 (157), *651*
Leffers, W. 288 (148), *314*
Legon, A.C. 146 (46, 47, 52), *214*
Legrand, E. 176 (156), *216*
Lehd, M. 477 (194), *492*
Lehmann, M.S. 176 (156), *216*
Lehmkuhl, H. 503 (43, 44), 506 (64, 65), 519 (115, 116), 546 (64, 65), *648–650*
Lehn, J.M. 48, 49, 60, 71 (39), *130*, 791 (70), 794 (81), *811*
Leininger, H. 458 (275), *468*
Leiserowitz, L. 162 (118), *215*
Lemal, D.M. 247 (115), *259*
Lemieux, R.P. 882 (92), *885*
Lemonias, P. 759, 760 (302), *772*
Lenardi, M. 644 (429), *656*

Lenartz, H.W. 232 (45), *253*
Lenn, N.D. 619 (363), *655*
Lennartz, H.-W. 189 (254), *218*, 229 (35, 37), 231 (43), 234 (51), 236 (65, 66), 245 (99–101, 103), 246 (105), *252, 253, 255, 258, 259*, 357 (41, 42), 358 (41), 359, 361, 378 (41, 42), 387 (41), 389, 390 (41, 42), *407*, 450–453, 457 (225, 226), *467*, 477 (190), *492*
Lenoir, D. 444 (187), 447 (210), *465, 466*, 844 (99, 100), *859*
Lenoir, P. 815 (7), *856*
Leone, S.R. 471 (113), *490*
Le Perchec, P. 872 (55, 56), 878 (55), *884*
Lepine-Frenette, C. 291 (196), *315*, 671 (88), *704*
Leriverend, P. 864 (11), *883*
Leroy, G. 83, 86 (109a), *132*, 477 (182–184), *491*
Leschova, I.F. 160 (107), *215*, 501 (38a), *648*
Leslie, D.H. 471 (85), *490*
Leslie, D.R. 447 (209), *466*
Less, R. 459 (287), *468*
Leta, S. 566, 567 (248), *653*
Leung, C.T.W. 271 (48), *312*
Leusen, D.van 686 (151), 687 (160), *705*
Leutenegger, U. 292 (204–206, 208), *315*, 698 (189–191, 193), *706*
Levi, B.A. 820 (33), *857*
Levin, I.W. 98, 101, 105 (157), *133*
Levin, R.D. 114, 116 (218), *136*, 249 (122), *260*
Levina, R.Y. 868 (38), *884*
Levina, R.Ya. 477 (178), *491*
Levine, R.A. 105 (170c), *134*
Levy, G.C. 110 (196), *135*
Lew, C.S.Q. 452 (243), *467*
Lewis, A. 740 (215), *770*
Lewis, D. 342, 382, 390 (5), *406*, 415, 424, 427, 431 (39), *461*, 471 (125), *490*
Lewis, D.K. 471 (128), 475 (169), 476 (169, 170), *490, 491*
Lewis, E.J. 280 (88, 89), *312*
Lewis, E.S. 708, 734, 738, 749 (4), *766*
Lewis, J. 556 (222, 223), *652*
Lewis, R.J. 174 (146), *216*
Lewis, R.T. 294 (231), *315*, 643 (424), *656*
Lewis-Bevan, W. 178 (175), *217*
Lex, J. 716, 745, 749 (52), *767*
Lhommet, G. 304 (313), *317*
Li, E.R. 733 (185), *769*
Li, H. 178 (175), *217*
Li, J.C. 190 (257), *218*
Li, L.-Q. 299 (271), *316*
Li, Y. 487 (288), *494*
Li, Y.S. 182 (200), 186 (230), *217, 218*
Li, Z. 268 (41), *311*
Liang, C. 48, 49, 51, 74, 83, 86, 96–98, 124 (32c), *130*
Liang, G. 419 (83), 427, 428 (115), 430 (115, 127), 432, 433 (144), 442 (182), 445 (201), 447 (210), *463–466*, 554 (209, 210), *652*, 815 (7), 827 (56), 828 (57), 829 (60, 61), 833 (73), 834 (74), 836 (57), 837 (83), 838 (56, 86), 840 (57), 849 (114), 850 (114, 115), 853, 855 (122), *856, 858, 859*
Liang, G.A. 394 (120), *410*
Liang, S. 211 (342), *221*, 264 (18), *311*
Liao, Q. 296 (252a, 252b), *316*
Liao, Y. 301, 302 (291), *317*
Lias, S.G. 114, 116 (218), *136*, 249 (122), *260*
Lichtenberger, D.L. 590, 603, 604 (310), *654*
Lichter, R.L. 110 (196), *135*
Licke, G. 334 (49), *338*
Li Du 158, 159 (94), *215*
Liebman, J.F. 44, 74 (5a, 5c), 75 (5c), 83 (5a, 5c, 95, 104), 86 (95, 104), *129, 132*, 140 (7), *213*, 224 (3–6), 225 (3, 10, 11, 13), 226 (19a, 20d), 227 (23), 229 (33), 231 (44), 232 (45, 46), 235 (58, 59, 62), 237 (19a), 238 (20d), 239 (72), 241 (84, 86), 242 (87), 243 (89, 90), 245 (104), 246 (3, 107), 247 (3, 104, 116), 248 (120), 249 (122), *249–260*, 451 (232, 234), 457 (234, 267), *467, 468*, 470 (71), *489*
Liebman, J.L. 114, 116 (218), *136*
Lien, M.H. 117–119 (239), *136*
Lien, M.M.H. 121 (248), *136*
Liese, T. 289 (167), *314*, 558 (235, 236), 560, 561 (236, 238), 562, 563 (236), 618 (235), *652*
Liles, D.C. 554 (211), *652*
Lim, K.T. 756 (283), *771*
Lim, Y, G. 519 (117), *650*
Lin, C. 478 (205), *492*
Lin, F.-T. 459 (283), *468*
Lin, L.-J. 720 (76), 722, 723 (76, 102), 730 (148), *767, 768*
Lin, L.P. 638 (408, 409), *656*, 720 (80, 81), 724 (81), 753 (276), *767, 771*
Lin, M.C. 471 (136), *490*
Lin, Y.-C. 572 (263), *653*
Lin, Z. 603 (337), *654*
Lincoln, D.N. 854 (123), *859*
Lind, G.J. 83, 86, 88, 93 (102c), *132*
Lindeman, S.V. 149 (63), 172 (139), *214, 216*
Linden, A. 156, 159 (87), *215*
Lindh, R. 459 (282), *468*
Lindley, P.F. 552 (194), *651*
Lindquist, R.H. 479 (213), *492*
Liotta, C. 866 (25), *883*
Liotta, F.J. 556 (227, 228), *652*
Liras, S. 292 (216), *315*, 701 (205), *706*

Lisini, A. 60 (59), *131*
Lisko, J.R. 500, 510, 566–568 (21), *648*
Lister, D.G. 145 (41), *213*
Litrico, A. 291 (200d), *315*
Litterst, E. 208, 209 (328), *220*
Little, R.D. 322 (18a, 18b), *337*
Little, T.S. 83, 86, 94, 105 (115), *132*, 150 (65), 160 (102, 104), 186 (235), *214, 215, 218*
Litvinov, I.A. 199 (298), *220*
Liu, B. 74, 83 (83), 110, 111 (202f), 114–116, 126, 128 (223), *131, 135, 136*, 393, 396, 397 (116), *410*, 417 (57), *462*, 820 (34), *857*
Liu, H. 619 (363), *655*
Liu, H.-L. 69 (74), *131*
Liu, H.-W. 170 (135), *216*
Liu, J.S. 816, 825 (12), 854 (123), *857, 859*
Liu, M.T.H. 289 (164), *314*
Liu, S. 675, 685 (105), *704*
Liverton, N.J. 200 (301), 201 (302), *220*
Livingston, R. (355), *655*
Lizunova, T.L. 310 (357), *318*
Llano, C.de 356, 357, 382 (39), *407*
Lloyd, D. 342, 344, 382, 390 (6), *406*, 681 (127), *704*
Lluch, J.M. 125 (266), *137*
Lo, D.H. 374 (76), *408*, 430 (125), 455 (256), *464, 467*
Lobanova, I.A. 663 (34), *703*
Lock, C.J. 817, 855 (17), *857*
Lock, C.J.L. 160 (108), *215*, 344, 345 (21), 394, 402 (121, 122), *406, 410*, 413, 415, 416 (22), 417 (59–61), 419 (22), 420 (91), 421 (91–93), 422 (91–93, 97), 423 (92, 93), 434 (60), 435 (61), 445, 450 (22), *461–463*, 554 (204), *652*
Lodder, A.E. 428 (117), *464*
Loehlin, J.H. 167 (126), *215*
Loerzer, T. 155, 156 (83), *214*
Loew, L.M. 418 (67), *462*
Löfström, C. 282 (108), *313*
Loggins, S.A. 235 (62), *255*, 451, 457 (234), *467*
Loh, K.-L. 290 (177), *314*, 667 (64), 690 (165, 166), 691 (165), 693 (166, 169), 695 (166), *703, 705*
Lohr, W. 231 (44), *253*
Lok, S. 305 (316), *317*
Lomas, D. 867, 868 (34), 875, 876 (67), *884*
London, F. 109 (182a, 182b), *134*
Long, F. 487 (280), *493*
Longone, D.T. 499 (5), *648*
Longuet-Higgins, H.C. 733 (169), *769*
Loopstra, B.O. 156, 157 (89), *215*
Lopata, A.D. 186 (226, 227), *218*
Lopez-Domingo, C. 239 (72), *256*
Lorenz, K.T. 290 (185), *314*
Lough, W.J. 699 (200), *706*
Louguinine, W. 241 (82), *257*
Lovas, F.J. 85, 86 (139a), *133*
Lovelace, T.C. 469 (28), *488*
Lovett, E.G. 720 (77, 79), *767*
Löwe, C. 573 (268, 269), 574 (268), 575 (268, 269), 576 (269), 577, 593 (268, 269), 597 (269), *653*
Lowe, C. 170 (134), *216*
Lowe, M.A. 476 (171), *491*
Lowenthal, R.E. 292 (210, 211), *315*, 695, 698 (181, 182), *706*
Lu, Q. 718 (66), 730, 732 (157, 158, 160, 161), 734, 735, 737 (66), 742 (157, 158, 161), 747 (66), 756 (282), 758 (157, 158, 296), 759 (157, 158, 297), 760 (157, 296, 297), 764 (282), *767, 769, 771*
Lu, S.-H. 452 (245), *467*
Lübbe, F. 437 (156), *465*
Lucke, A.J. 513 (97, 98), *649*
Lüddecke, E. 756, 757 (290), *771*
Ludtke, A.E. 836 (82), *858*
Luettke, W. 56, 57, 75 (46a, 46b), *130*
Luh, T.-Y. 273 (56), *312*
Lukas, J. 578, 579 (280), *653*
Lukas, K.L. 198, 202 (288), *219*
Lukin, K.A. 230 (40), *252*, 282 (104, 105), *313*, 863, 864 (9), 865 (19), 866 (9), 872 (19), 875 (9, 19), 876 (9, 70), 877 (70, 73, 74), 878 (70, 73, 75, 76, 78, 79), 879 (79), 880 (81), 881 (88–90), 882 (93), *883–885*
Lukin, P.M. 176 (155, 157), *216*
Lukina, M.Y. 340, 350 (1), *406*
Luk'yanova, V.A. 229 (34), 230, 231 (41), 234 (51), 235 (53, 55, 58), 241 (82), *252–254, 257*, 780 (42), *810*
Lund, A. 116, 117 (235), *136*
Lund, K.P. 294, 295 (245), *316*
Lunell, S. 116 (230, 232, 235), 117 (230, 235), *136*
Lunetta, S.E. 280 (88), *312*
Luo, Y.-R. 226, 237 (19c), *250*
Lustgarten, R.K. 445 (198, 200), 446 (198), 449 (217), *466*
Lustzyk, J. 122 (251a), *137*
Lutsenko, A.I. 659 (23), *702*
Lüttke, W. 83, 86, 88 (118), *133*, 147, 148 (56), 153 (77, 79), 154 (77), 155, 156 (83), 210, 211 (340), *214, 221*, 233 (49), *253*
Lutz, R.E. 418 (64), *462*
Luzzio, M.J. 292 (203), *315*, 699, 701 (201), *706*
Lynch, D.C. 321 (12), *337*

Lynch, T.R. 472, 477, 484 (143), *491*
Lynch, V.M. 775 (10), 795 (116), 797 (105), 798 (116), 800 (105), 801 (116), *809, 811, 812*
Lyshchikov, A.N. 176 (157), *216*
Lyu, P.-C. 191 (263), *219*

Maas, G. 187, 194 (237), 199 (293–297), *218, 219*, 282, 290 (102), *313*, 470 (31), *489*, 658 (4), 663 (4, 35), 687, 692, 697 (4), *702, 703*, 724 (113), 729 (142), 730 (147), *768*
Maasböl, A. 522 (134), *650*
Macdonald, J.N. 179 (179), *217*
MacDonald, W.A. 681 (127), *704*
Mac Dougall, L. 380 (86), *409*
MacDougall, P.L. 68 (72b), *131*
MacGillavry, C.H. 161, 164 (111), *215*
Machida, K. 145 (44), *213*
Machida, M. 278 (70), *312*
Machinek, R. 56, 57, 75 (46a), *130*, 188 (246), 211 (341), *218, 221*
Mack, H. 297 (257), *316*
Mackay, D. 402 (124), *410*, 454 (253), *467*
Mackenzie, K. 722 (100), *767*
Mackor, E.L. 431 (134), *464*
MacLean, C. 431 (134), *464*
Maddox, M.L. 746 (255), *771*
Maekawa, Y. 326 (41), *338*
Maercker, A. 511 (87), 512 (89, 91, 92), 623 (373, 374), *649, 655*
Maestro, M.C. 303 (310), *317*
Maetzke, T. 504 (48), 505 (56), *648, 649*
Magarian, R.A. 158, 159 (93, 94), *215*
Magatti, C.V. 516, 566, 568 (110), *650*
Magdesieva, N.N. 296 (253), *316*
Magnusson, G. 687 (155), *705*
Magrill, D.S. 864, 870 (12), *883*
Maguire, A.R. 292 (217), *315*, 699 (203), *706*
Mahanti, M.K. 733, 734 (173), *769*
Mahendran, M. 160 (108), *215*, 344, 345 (21), 394, 402 (122), *406, 410*, 413, 415, 416 (22), 417 (59, 61), 419 (22), 421 (92), 422 (92, 97, 98), 423 (92, 100), 435 (61), 445, 450 (22), *461–463*, 817, 855 (17), *857*
Mahidol, C. 300 (276), *316*
Mahinya, E.F. 241 (82), *257*
Mahler, J.E. 419 (79), *463*
Maier, G. 193, 194 (273), 205 (320, 321), 206 (321, 322), 207 (322), 209 (332), *219, 220*, 290 (173), *314*, 429 (121), 451 (227), *464, 467*
Maier, J.P. 231 (44), *253*
Maillard, B. 794 (77), *811*
Mainwald, J. 554 (202), *652*
Majchrzak, M.W. 663 (38), *703*
Majerski, Z. 825 (52), *857*
Maki, A.G. 85, 86 (139a), *133*
Makita, K. 669 (77), *703*
Makosza, M. 300 (278), *316*
Maksic, Z. 366, 370 (62), *408*
Maksic, Z.B. 56, 57, 71, 75, 76 (45), 83, 86 (109b), *130, 132*, 201 (305), *220*, 733 (174–183), 739, 741 (182), 744, 749 (244), *769, 771*
Maksimović, L. 188 (246), *218*
Maleev, A.V. 797, 799, 800 (101), 802 (101, 122), 805 (126), *811, 812*
Malins, R.J. 471 (106, 107, 130), *490*
Mallard, W.G. 249 (122), *260*
Malon, B.J. 235 (60), *254*
Malone, J.F. 309 (344), *318*
Malpass, J.R. 419 (73), *462*
Malsch, C.-D. 429 (121), *464*
Malsch, K.-D. 205, 206 (321), *220*
Mamatyuk, V.I. 432, 433, 435 (145), *464*
Manatt, S.L. 427 (113), *463*, 733, 734 (170), 740 (209), *769, 770*
Mancini, F. 273 (54), *312*
Mandai, T. 658 (15), *702*
Mandel'shtam, T.V. 730 (152), *768*
Mander, L.N. 674 (102, 103), *704*
Manganiello, F.J. 500 (20), *648*
Mangette, J.E. 326 (41), *338*
Manhart, S. 185 (210), *217*
Manion, M.L. 478 (204), *492*
Manojlovic, L. 722 (100), *767*
Mansuri, M.M. 718 (59, 61, 62), 719 (61), 720 (61, 62), 724 (62), 748, 749 (59, 61), *767*
Manus, M.M. 777, 780 (16), *810*
Mao, D.T. 673 (95), *704*
Mapelli, C. 170 (131), *216*, 305 (318), *317*
Marchand, A.P. 48, 49, 71 (37), *130*, 209 (335), *221*
Marco, J.L. 305 (314, 321), *317*
Marcotullio, M.C. 291 (187, 188), *314*, 669 (79), *704*
Marcus, A. 459 (283), *468*
Marcus, R.A. 471 (101), *490*
Marek, I. 276 (64), 277 (65), *312*, 509 (84), *649*
Margerum, L.D. 629 (389), 631 (389, 393), 633, 638 (389), *655*
Margetic, D. 733 (175, 183), *769*
Margrave, J.L. 246 (109), *259*, 708, 734, 738, 749 (4), *766*
Mariano, P.S. 323 (29), *338*
Marino, J.P. 545 (172a, 172b), *651*, 663 (48), *703*
Marinozzi, M. 663 (45), *703*
Marnhoff, E. 663 (32), *702*
Márquez-Silva, R.-L. 663 (33), *703*
Marr, G. 549 (179), *651*

Marsden, C.J. 177, 178 (171), *216*
Marsh, P. 197 (284a), *219*
Marshall, D.C. 479 (209), *492*
Marstokk, K.-M. 105 (171), *134*, 151 (70–74), 152 (76), 161 (109), *214, 215*
Martelli, J. 550 (187), 551 (190, 191), *651*, 697 (188), *706*
Martin, E. 179 (180), *217*
Martin, H.-D. 469 (6), *488*
Martin, J.C. 746 (255, 256), *771*, 844 (97), *859*
Martin, M.V. 303 (310), *317*
Martin, R.H. 854 (124), *859*
Martin, S. 153–155 (78), *214*
Martin, S.F. 292 (215, 216), *315*, 698 (198), 701 (198, 205), *706*
Martin, T.R. 626, 635 (383), *655*
Martina, D. 551 (189), 555 (217), *651, 652*
Martinez, A.G. 832 (69), *858*
Maruoka, K. 280 (92), *312*, 866 (23), *883*
Marvell, E.N. 478 (205), *492*
Mas, R.H. 157 (92), *215*, 417 (48), *462*
Masamba, W. 179, 180 (183, 186), *217*
Masamune, S. 292 (210, 211), *315*, 447, 448 (207), *466*, 695, 698 (181, 182), *706*
Masamune, T. 670 (84), *704*
Mash, E.A. 283 (111–117), *313*
Massa, W. 198 (287), *219*, 429, 430 (128, 131), 431 (131), *464*
Massol, M. 500 (27), *648*
Masters, A.P. 500 (19), *648*
Mastropaolo, D. 574 (273), *653*
Mastryukov, V.S. 145 (33), 146 (45), 149 (62), 191 (260), 199 (291, 292), *213, 214, 219*, 863 (2), *883*
Masunova, A.Yu. 877 (74), 878 (76), *885*
Masuzawa, M. 746, 751 (259), *771*
Mataka, S. 714 (35, 36), *766*
Matasi, J.J. 523 (137, 138), 525 (138), 530 (137, 138), 531, 533 (138), *650*
Mateescu, Gh.D. 434 (147), *464*
Mathey, F. 384 (90), *409*
Mathur, S.N. 83, 86 (102c, 106c), 88 (102c), 89 (106c), 93 (102c), 95 (106c), *132*, 147 (55), 181 (194), *214, 217*
Mathy, P. 301 (284), *317*
Matlin, S.A. 699 (200), *706*
Matro, A. 294 (233), *316*
Matsuda, H. 309 (352), *318*, 682 (139), *705*
Matsuda, K. 290 (184), *314*
Matsui, M. 291 (194), *314*, 670 (85), *704*
Matsuki, T. 301, 302 (292), *317*
Matsumoto, H. 663, 693, 698 (43), *703*
Matsumoto, S. 609, 612 (352), *655*
Matsumura, Y. 607 (343), *654*
Matsuura, H. 265 (22), *311*
Matsuzaki, K. 303 (311), *317*
Matsuzawa, S. 506 (68), *649*
Matt, P.von 292 (208), *315*
Mattauch, B. 308 (332), *318*
Mattes, K. 728, 729, 757 (136), *768*
Matthews, D.A. 64 (66d), *131*
Mattson, M. 282 (108), *313*
Mattson, M.N. 282 (106), *313*
Matturro, M.G. 353 (34), *407*
Matz, J.R. 353 (34), *407*
Maulitz, A.H. 353 (30), *407*, 734, 735 (196), 759 (298, 304), 760 (298), 763 (313), *769, 771, 772*
Maury, L. 164 (120), *215*
Mauz, E. 581–583 (290), *653*
Maxwell, J.L. 694 (178), *706*
Mayants, L.S. 471 (102), *490*
Mayer, A. 456 (263), *468*
Mayer, D. 663 (35), *703*
Mayr, A. 536, 548 (161), *651*
Mayr, H. 345 (26), *407*, 413 (24), 449 (221), 457 (270), 458 (24, 273, 281), 459 (281), *461, 466, 468*, 817 (19), *857*
Mazhar-Ul-Haque 185 (213), *217*
Mazur, R.H. 417 (55), *462*, 814 (1), 815 (4), *856*
McAdoo, D.J. 116 (228, 233), 117 (233), *136*
McAfee, M.J. 665 (63), *703*
McBride, E.F. 334, 335 (50a), *338*
McCall, J.M. 329 (47a), *338*
McClure, M.D. 590, 601, 602 (307), *654*
McCluskey, R.J. 471 (111), *490*
McCluski, J.V. 509, 601 (82), *649*
McClusky, J.V. 263 (7a, 7b), *311*, 874 (64), *884*
McCullough, J.J. 320 (4), *337*
McCullough, L. 593, 594, 601–604 (316), *654*
McDonald, S. 622 (372), *655*
McElwee-White, L. 535 (157, 158), 536 (160), 537 (157, 158), 538 (160, 163), 539 (158, 160, 164), 540 (164), 541 (160, 164), 542 (158), 543 (158, 160), *651*
McEwan, A.B. 456 (266), *468*
McEwen, A.B. 342 (11), *406*
McFarlane, N.S. 418 (64), *462*
McFarlane, W. 420 (84), *463*, 554 (207), *652*
McGarry, P.F. 786 (96), 795 (96–99), 796 (96, 97, 99), *811*
McGeehan, G.M. 302 (296), *317*
McGhee, D.B. 471 (120), *490*
McGlinchey, M.J. 110 (200), *135*, 420 (90), *463*, 554 (206), *652*
McGreer, D.E. 865 (17), *883*
McIntosh, C.L. 728, 729, 757 (136), *768*
McIver, J. 481 (266, 267), *493*
McIver, J.W. 126 (269, 272, 273), *137*, 481 (261), *493*
McIver, R.T. 123 (255), *137*

McKean, D.C. 98 (154), 104, 120 (165), *133, 134*
McKee, M.L. 44, 74, 82 (8c), 110, 111 (202c), *129, 135*, 342, 348 (12), 393, 396, 397 (116), *406, 410*, 820 (35), *857*
McKelvey, R.D. 324, 325 (33), *338*
McKern, I.D. 447 (209), *466*
McKervey, M.A. 292 (217), *315*, 658, 663 (12), 668 (72), 687 (12), 699 (202, 203), *702, 703, 706*
McKinley, A.J. 793 (74), *811*
McKoy, V. 786 (56), *810*
McLafferty, F.J. 239 (73), 247 (113), *256, 259*
McLafferty, F.W. 847 (108), *859*
McLellan, T.J. 734 (192), *769*
McMeeking, J. 508 (74), 521 (130), 588 (302), 641 (420), *649, 650, 654, 656*
McMillen, D.F. 77, 120 (89a), *132*
McMullen, L.A. 523, 525, 528, 529, 531 (140), *650*
McMullen, L.M. 528, 533 (150), *650*
McMurdie, N. 822, 824 (47a, 47b), *857*
McMurdie, N.D. 790 (66), 791 (66, 113), 794 (66), 797 (104), 798, 800 (104, 113, 114), 801 (113, 114), 802 (104), 808 (66), *811, 812*
McMurry, J.E. 353 (34), *407*, 450 (224), *466*
McNally, D. 77 (92b), *132*, 229 (33), *252*
McNesby, J.R. 479 (214), *492*
McNichols, A.T. 715 (45), 731, 742 (165), 763 (312, 313), *766, 769, 772*
McOmie, J.F.W. 727 (127), *768*
Mealli, C. 589 (305, 306), 590, 592, 601, 602 (306), 603 (305, 306), *654*
Medinger, K.S. 83, 86 (104), *132*
Medrano, J.A. 779 (29), *810*
Meerts, W.L. 186 (223), *218*
Meester, M.A.M. 161, 164 (111), *215*
Mehler, K. 506, 546 (64), *649*
Mehrsheikh-Mohammadi, M.E. 622 (372), *655*
Mei, Q. 730, 732, 742 (156, 158), 758, 759 (158), 760 (306–308), *769, 772*
Meier, F. 585, 587 (295a), *653*
Meijere, A.de 83, 86, 88 (118), *133*, 147, 149 (59), 151 (72), 153 (79), 155 (82), 188 (242, 243, 246), 189 (255), 191 (262, 265), 211 (341, 342), *214, 218, 219, 221*, 264 (18), 285 (132), 289 (165, 167–170), 308 (334c), 309 (346, 347), *311, 313, 314, 318*, 416 (41), *461*, 501 (37), 530 (152, 153), 552 (152), 553 (152, 196), 558 (235, 236), 560, 561 (236, 238), 562 (236), 563 (236, 242), 564 (153, 246, 247), 565 (246), 571 (261), 618 (235), 622 (152), 639 (413, 414), *648, 650–653, 656*, 777 (11), *810*, 868 (37), 870 (47), 871 (49–51), 872 (57), 873 (58), 874 (51, 60, 61), 876 (71), 877 (73), 878 (61, 71, 73, 77, 78), 880 (77), 881 (71), *884, 885*
Meintjies, E. 554 (211), *652*
Meinwald, J. 455 (257), *467*
Meitlis, P.M. 514 (102), 519 (102, 123), *649, 650*
Melaga, W.P. 430 (127), *464*
Melhuish, W.H. 759, 760 (297), *771*
Melius, C.F. 236 (64), *255*
Mélot, J.-M. 301 (282), *316*
Memmesheimer, H. 197 (281), *219*
Mende, U. 661 (28, 29), *702*
Mendelson, S.A. 301 (280), *316*
Menéndez-Velázquez, A. 297 (262), *316*
Menon, B.C. 881 (87), *885*
Menzinger, M. 471 (138), *491*
Meou, A. 615, 616, 622 (359), *655*
Merényi, R. 179, 180 (186), *217*, 455 (254), *467*
Merkel, D. 560 (239), 561 (239, 240), *652*
Merola, J.S. 157 (92), *215*, 417 (48), *462*
Mérour, L.J. 515 (107), *650*
Mertis, K. 503 (46), 516 (111), *648, 650*
Messsmer, R.P. 778, 779 (17), *810*
Meyer, F.E. 309 (346, 347), *318*
Meyers, A.I. 289 (170), 297 (263, 264), *314, 316*
Meyerson, S. 727 (126), *768*
Mezentsova, N.N. 868 (38), *884*
Michelsen, K. 211 (341), *221*
Michl, J. 110 (195, 197), *135*, 481 (246), *493*, 775 (10), 777 (12), 780, 781 (39), 783 (48, 50), 784 (39, 51), 786 (55), 787 (39, 48, 58, 59), 789, 790 (48, 58), 791 (39, 113), 793 (74), 795 (59, 116), 797 (48, 50, 59, 104, 105), 798 (48, 59, 104, 113, 114, 116), 799 (39, 48, 58, 59), 800 (48, 59, 104, 105, 113, 114), 801 (48, 113, 114, 116), 802 (48, 59, 104, 123, 124), 804 (48, 58), 809 (133), *809–812*
Midollini, S. 589 (305, 306), 590, 592, 601, 602 (306), 603 (305, 306), *654*
Miebach, T. 874, 878 (61), *884*
Miki, S. 608–612, 618, 619 (351), *655*, 715 (47), *766*
Miles, W.H. 285 (133), *313*
Milewski-Mahrla, B. 185 (214), *217*
Militzer, H.-C. 530, 552, 553, 622 (152), *650*
Millen, D.J. 146 (52), *214*
Miller, A.S. 722 (100), *767*
Miller, D.L. 117 (237), *136*
Miller, J.S. 197 (284c), *219*
Miller, L.S. 455 (262), *467*
Miller, M.A. 235 (62), *255*, 457 (268), *468*
Miller, R.D. 784 (51), *810*
Miller, S.L. 470 (58), *489*

Millerand, M.A. 77 (92c), *132*
Millevolte, A.J. 48, 49, 71 (44), *130*
Mills, O.S. 552 (194), 607 (346), *651, 654*
Mills, W.H. 709, 733 (5), *766*
Milsom, P.J. 721, 722 (86, 87), 741 (223), 756 (286), 757 (223), *767, 770, 771*
Mil'vitskaya, E.M. 469 (20), *488*, 863 (2), *883*
Minami, K. 265 (21, 23), *311*, 505 (60), *649*
Minami, Y. 711 (23), *766*
Minaskanian, G. 200 (301), *220*
Minato, T. 128 (278), *137*
Mineta, H. 283 (123), *313*
Minichino, C. 121, 122 (250), *137*
Minkin, V.I. 44, 48 (20), *129*, 342, 344, 382, 390 (10), *406*
Minott, R. 519 (116), *650*
Mioduski, J. 455 (257), *467*
Mishra, A. 470 (34), *489*
Mison, P. 447 (210), *466*, 815 (7), *856*
Misumi, A. 306 (324), *317*
Mitchel, M.A. 519 (125), *650*
Mitchel, T.N. 500 (26), *648*
Mitchell, J. 353 (34), *407*
Mitchell, R.W. 569 (255), *653*
Mitra, R.B. 301, 302 (286), *317*
Miura, Y. 616 (357), *655*
Miyama, H. 471 (118), *490*
Miyasaka, A. 746, 751 (259), *771*
Mizuno, K. 209 (333), *221*, 287 (143), *314*
Mizuta, K. 869 (44), *884*
Mlinaric-Majerski, K. 280 (85), *312*
Mlynek, C. 151 (72), *214*
Mo, Y.K. 434 (147), *464*
Mochel, A.R. 93 (145), *133*, 183 (207), *217*
Moffatt, J.R. 418 (63), *462*
Moffitt, W.E. 48, 49, 55–57, 68, 75, 122 (23), *130*
Mogus, A. 863 (8), *883*
Mohamadi, F. 289 (160), *314*
Mohler, D.L. 353 (30), *407*, 734, 735 (196), *769*
Moir, M. 869 (45), *884*
Molander, G.A. 281 (95, 96), 282 (96), 285 (95), *313*
Møllendal, H. 105 (171), *134*, 151 (69–74), 152 (69, 76), 161 (109), *214, 215*
Møller, C. 85 (136a), *133*
Monahan, J.B. 663 (45), *703*
Moneti, S. 589 (305, 306), 590, 592, 601, 602 (306), 603 (305, 306), *654*
Mongrain, M. 670 (82), *704*
Monpert, A. 550 (187), 551 (191), *651*, 697 (188), *706*
Montaudo, G. 418 (64), *462*
Montgomery, D. 278 (73), *312*
Montgomery, J. 525 (144), *650*
Monti, H. 621 (369), *655*
Moo, J.Y. 836 (82), *858*
Moore, C.B. 471 (105), *490*
Moore, C.M. 334, 335 (50b), *338*
Moore, J.A. 476 (176), *491*
Moore, J.W. 505 (59), *649*
Moorehead, A.W. 248 (119), *260*
Morales, A. 687 (158), *705*
Moreno, M. 125 (266), *137*
Morgan, R.A. 519 (119), *650*
Morgan, T.K.Jr. 418 (63), *462*, 827 (56), 828, 836 (57), 838 (56), 840 (57), *858*
Mori, A. 283 (109, 110), *313*
Moriarty, R.M. 308 (331), *318*, 550 (182), *651*, 681 (129, 130), *705*
Morimoto, T. 309 (351), *318*, 682 (138), *705*
Morita, N. 555 (220, 221), 632 (220), *652*
Moritani, I. 843 (96), *859*
Morokuma, K. 454 (250), *467*, 481 (260), *493*
Morris, D.G. 44, 48 (16), *129*, 572 (264), *653*
Morrison, R.W. 795, 800 (94), *811*
Morse, P.M. 68 (73), *131*
Morse, R.L. 321 (15), *337*
Mortimer, C.T. 233 (49), *253*
Mortimer, M.D. 538 (163), *651*
Morvant, M. 290 (174), *314*
Moscowitz, A. 105 (172a), *134*
Moser, C. 481 (241, 242), *493*
Mosher, H.S. 105 (172a), *134*
Moss, N. 663 (46), 678 (118), *703, 704*
Moss, R.A. 125 (265), *137*, 280 (84), 294 (232, 233), *312, 315, 316*, 817, 854, 856 (18), *857*
Motevalli, M. 517, 518 (114), *650*
Motherwell, W.B. 285 (131), 294 (231), *313, 315*, 643 (424, 425). *656*
Mowery, P. 740 (215), *770*
Moylan, C.R. 471 (133), *490*
Muchowski, J.M. 746 (255, 256), *771*
Mueller, G.W. 471 (122), *490*
Mueller, N. 56 (49), *130*
Mueller, P.H. 291 (200e), *315*
Mueller, R.A. 668 (68), *703*
Mühler, C. 60 (54c), *130*
Muir, K.W. 722 (100), *767*
Mulholland, D.L. 420 (87), 424 (102), 433 (146), 439 (87), *463, 464*
Mulleer, T. 832 (67), *858*
Müllen, K. 744 (241), *770*
Müller, B. 734 (191), *769*
Müller, C. 384 (89), *409*, 575 (279), *653*
Müller, D. 698 (192), *706*
Müller, G. 156, 157 (90), 185 (211, 212), *215, 217*, 264 (17), *311*, 458 (280), *468*
Müller, H. 459 (285, 287), *468*
Müller, J. 503 (45), *648*

Müller, P. 292 (214, 215), *315*, 508 (77), 547 (176), 581, 582, 585 (77, 289), 587 (77), 644 (427, 428), *649, 651, 653, 656*, 722 (91–98, 105, 106), 723 (105, 106), 724 (110–112), 725 (112), 726 (118), 727 (122, 124, 125), 731 (164), 737 (164, 199, 200), 739 (206), 740 (216, 217), 741 (92–96, 111, 224, 225, 227–230), 742 (164, 200), 744 (230), 746 (112, 118), 748 (164, 199, 200), 753 (164, 200), 756 (284, 285), 764 (164), *767–771*
Müller, P.J. 698, 701 (198), *706*
Müller, U. 287 (145), *314*
Munjal, R.C. 817, 854, 856 (18), *857*
Munk, P. 805, 806 (129), *812*
Muñoz, L. 291 (189, 190), *314*, 677 (110), *704*
Münzel, N. 729, 757 (137), *768*
Murai, A. 670 (84), *704*
Murai, S. 309 (350–352), *318*, 682 (138, 139), *705*
Murakami, A. 195 (278), *219*
Muralidharan, V.B. 683 (141), *705*
Murata, S. 765 (316, 317), *772*
Murayama, E. 263 (10), 279 (74a, 74b), 308 (337), *311, 312, 318*
Murdie, N. 263 (7b), *311*
Murdzek, J.S. 594, 602 (320), *654*
Murr, T. 285 (132), *313*
Murray, C.K. 294 (243, 246, 247), 295 (246, 247), *316*
Murray, J.-S. 74 (81), 96 (153), *131, 133*
Murray, M.J. 471 (82), *489*
Murray, R.K.Jr. 264 (16), *311*, 418 (63), *462*, 827 (56), 828, 836 (57), 838 (56, 86), 839 (87), 840 (57), *858*
Murthy, G.S. 795 (116), 797 (104, 105), 798 (104, 116), 800 (104, 105), 801 (116), 802 (104), *811, 812*
Musso, H. 829 (62), *858*
Mustafa, A. 710 (15), *766*
Muthard, J.L. 366 (61), *408*
Muthusubramanian, L. 301, 302 (286), *317*
Myers, A.G. 665 (61), *703*
Myers, M. 300 (277), 301 (280), *316*
Myers, R.J. 578 (282), *653*
Myhre, P.C. 417 (58), *462*, 819, 820, 854 (25), *857*
Mynott, R. 503 (43, 44), 508 (77, 78), 547 (176), 581, 582, 585, 587 (77), *648, 649, 651*, 755 (281), *771*
Mysorekar, S.V. 665 (59), *703*

Nafie, L.A. 105 (172b, 172c), *134*, 475 (166, 167), 476 (166, 167, 174–176), 477, 484 (166), 487 (166, 282), *491, 493*
Nagakura, I. 501 (34–36), 545 (173), *648, 651*
Nagaraju, S. 669 (80), *704*
Nagase, T. 693, 697 (172, 173), *706*
Nägele, U.M. 297 (256), *316*
Nagi, Sh.M. 434 (148), *464*
Naito, K. 684 (147), *705*
Nájera, C. 297 (262), *316*
Nakadaira, Y. 644 (426), *656*
Nakai, H. 283 (110), *313*
Nakajima, T. 835, 836, 839 (80), *858*
Nakamura, A. 663 (39, 40), *703*
Nakamura, E. 506 (67–69), 570 (259), 643 (423), *649, 653, 656*
Nakamura, M. 506 (67), *649*
Nakashima, T. 447, 448 (207), *466*
Nakata, M. 84 (126), *133*, 143–145 (21), 188 (245), *213, 218*
Nakatani, H. 307 (329), *317*
Nakatsuji, H. 116, 117 (234b), *136*
Nakayama, T. 746, 751 (258), *771*
Nakazawa, S. 711 (24), *766*
Naksic, Z.B. 863 (8), *883*
Namanworth, E. 816, 826 (13), 849 (112), *857, 859*
Nandi, M. 296 (251), *316*
Nandi, R.N. 175 (148), *216*
Nangia, P.S. 126 (277b), *137*, 479 (217), *492*
Naqvi, S.M. 291 (195), *315*, 469 (22), *488*, 544 (166), *651*
Narasimhan, K. 727 (119, 121), *768*
Nardelli, M. 734 (190), *769*
Narjes, F. 303 (303a, 303b, 304), *317*
Nasakin, O.E. 176 (155, 157), *216*
Nasielski, J. 232 (45), *253*, 327 (44), *338*
Natalini, B. 663 (45), *703*
Natalis, P. 246 (109), *259*
Natchus, M.G. 291 (197), *315*, 673 (99, 100), *704*
Naumov, V.A. 199 (298), *220*
Naydowski, C. 506, 546 (65), *649*
Naylor, R.D. 225 (15), 229 (33), 232 (45), *250, 252, 253*
Nazarova, L.F. 232 (45), 235 (56), *253, 254*
Nease, A.B. 182 (200), *217*
Nebzydosky, J.W. 453 (247), *467*
Nefedov, O.M. 140 (4), *212*, 282 (103), *313*, 557, 561, 562 (231), 564 (245), 639 (412), *652, 656*, 659 (22, 23, 26), 688 (22, 26), *702*, 793 (72), *811*, 863 (4b, 4c), 865 (18), 868 (41), *883, 884*
Negishi, E.-I. 263 (8), 309 (343), *311, 318*
Neidlein, R. 201 (306), *220*, 353 (29), *407*, 579, 581 (284), *653*, 719 (70), 731 (166), 734, 735 (194), 742 (235–237), 743 (236), 744 (242, 243), 746 (257; 261, 262), 747 (266), 752 (273–275), 753 (278), 755 (281), *767, 769–771*
Nelson, A.D. 434 (149), *465*
Nelson, G. 85, 86 (142), *133*, 199 (289), *219*

Nelson, G.L. 110 (196), *135*, 417, 418 (56), *462*, 822 (46), *857*
Nelson, K.A. 283 (111–113, 116), *313*
Nemba, R.M. 477 (184), *491*
Nes, G.J.H.van 155 (84), *214*
Nesbitt, D.J. 471 (113), *490*
Neuenschwander, M. 759 (301), *772*
Neugebauer, D. 534 (156), 562 (241), *651, 652*
Newman-Evans, R.H. 477, 479 (180), *491*
Newton, G. 585 (297), *654*
Newton, J.H. 105 (170a–c), *134*
Newton, M. 204 (315), *220*, 237 (68), *255*, 776, 777, 780, 781, 783, 784, 786, 790, 791, 794, 806 (15), *810*
Newton, M.D. 44, 56, 57, 75 (7), 123 (260), 124 (261), *129, 137*, 773 (1), 777 (16), 778 (1), 780 (16), *809, 810*
Newton, M.G. 170 (131), *216*
Ng, H.P. 292 (203), *315*, 699, 701 (201), *706*
Ng, K.M. 733 (185), *769*
Nguyen, K.A. 779 (33), *810*
Nguyen, M.T. 144, 204 (28), *213*, 785 (54), *810*
Nguyen, P. 423 (100), *463*
Nguyen, S.T. 584 (294), *653*
Nguyen-Dang, T.T. 64, 65, 68, 107 (67a, 67b), *131*, 361, 365 (44), 375 (44, 81), 376 (81), 378, 379 (44), 393 (81), 394 (44), *407, 408*, 414, 415, 434, 435 (30), *461*
Nguyen-Thi, H.-C. 722 (95, 96), 724 (110–112), 725 (112), 726 (118), 741 (95, 96, 111), 746 (112, 118), *767, 768*
Nibbering, N.M.M. 120 (245), *136*, 847 (108), *859*
Nicholas, A.M.de P. 477 (187), *491*
Nicholas, K.M. 268 (41), *311*
Nicholson, J.M. 458 (272), *468*
Nickels, H. 728 (134), *768*
Nicolaou, K.C. 514 (103, 104), *649*
Nicolini, M. 274 (59), *312*
Niedoba, S. 264 (15), *311*
Nielsen, R.B. 583 (293), *653*
Niemer, B. 554 (216), *652*
Niessen, W.von 60 (55, 58), *131*
Niimoto, Y. 279 (75), *312*
Niiranen, J.T. 477 (197), *492*
Nijveldt, D. 144 (23a–c), 145 (23b), 147 (23b, 23c), 149 (23a–c), 153 (23c), 154 (23a–c), *213*, 471 (88), *490*
Nikishin, G.I. 172 (139), *216*, 310 (357), *318*
Nilssen, A.V. 150 (67), *214*
Ninbari, F. 291 (194), *314*
Ninghai Hu 170 (132), *216*
Nishida, S. 262 (1), 283 (123), *310, 313*, 843 (96), *859*
Nishimura, J. 825, 826, 852, 853 (50a), *857*
Nishimura, M. 284 (124), *313*
Nishimura, T. 638 (407), *656*
Nishio, A. 481 (259), *493*
Nishio, K. 682 (138), *705*
Nishio, K.-I. 309 (351), *318*
Nishishita, T. 847 (108), *859*
Nishiyama, H. 270 (43), *311*, 663, 693, 698 (43), *703*
Nist, B.J. 783 (44), *810*
Nitta, M. 746, 751 (258), *771*
Niwa, H. 869 (44), *884*
Nixdorf, M. 205, 206 (321), *220*, 264 (17), *311*, 429 (121), *464*
Nixon, A. 420, 439 (87), *463*
Nixon, I.G. 709, 733 (5), *766*
Nobes, R.H. 96, 98 (150), *133*
Noble, W.J.le 342, 344, 382, 390 (9), *406*
Noels, A.F. 290 (172, 176), *314*, 658 (5), 663 (5, 33, 34, 36), 683 (140), 688, 689 (161), 693 (175), *702, 703, 705, 706*
Noga, J. 392 (105), *409*
Nokami, J. 658 (15), 670 (83), *702, 704*
Noltemeyer, M. 874 (63), *884*
Norbury, D. 179 (179), *217*
Nordberg, R.E. 550 (185), *651*
Norden, T.D. 144 (27), 187 (27, 241), 194 (27), 195 (27, 241), 196 (241), *213, 218*, 730 (149–151), 757 (151), 760 (309), *768, 772*
Norden, T.N. 190 (259), *218*
Nordlander, J.E. 511 (88), *649*, 846 (105), *859*
Norin, T. 550 (185), *651*
Normant, J.F. 276 (64), 277 (65), 307 (327), *312, 317*, 509 (83, 84), *649*
North, J.T. 632 (396a), *656*
Northing, J. 471 (129), *490*
Norton, C. 445 (194), *466*
Nöth, H. 573, 576, 629 (267), *653*
Nour, E.M. 186 (232), *218*
Nowick, J.S. 297, 299 (258), *316*
Noyori, R. 627 (384), 629, 630 (388), 638 (407), 639 (415), *655, 656*
Nozaki, H. 501 (31), *648*, 750 (270), *771*
Nugent, W.A. 290 (175), *314*
Nunome, K. 116, 117 (234a), *136*
Nussbaumer, M. 193, 194 (273), *219*
Nutakul, W. 739 (207), *770*
Nuttall, R.L. 224 (9), 237 (9, 69), *249, 255*
Nyhus, B.A. 204 (317), *220*, 794 (88), *811*

Oaks, F.L. 333 (48), *338*
Oalmann, C.J. 292 (215, 216), *315*, 698 (198), 701 (198, 205), *706*
O'Bannon, P.E. 183, 184 (208), 191 (264), *217, 219*, 294 (225–227), *315*, 693 (171), *705*
Obeng, Y.S. 802 (124), *812*

Oberhammer, H. 83, 86 (110), *132*
Occolowitz, J.L. 458 (272), *468*
O'Connor, E.J. 282 (107), *313*
O'Connor, J.M. 595 (325), *654*
Oda, K. 278 (70), *312*
Oda, M. 287 (141), *313*, 420, 425 (86), *463*, 716 (51), *766*
Oda, R. 607 (343), *654*
Oda, Y. 305 (319), *317*
Odagi, T. 639 (415), *656*
Odaira, Y. 880 (84), *885*
Oddershedde, J. 110 (193b), *135*
Oettle, W.F. 323 (27b), *337*
Oeveren, A.V. 698 (199), *706*
Öfele, K. 606 (341), 607 (348, 349), 610 (341), *654, 655*
Officer, D.L. 715 (39, 40), 716 (49), 720 (82, 83), 722 (99), 726 (49), 762 (311), *766, 767, 772*
Ogliaruso, M. 420 (85), 458 (272), *463, 468*
Ogoshi, H. 607, 608 (347), 609 (347, 352), 612 (352), 616 (357, 361), 618 (347), *655*
Ogoshi, S. 309 (351, 352), *318*, 682 (138, 139), *705*
O'Grady, B.V. 471 (131), *490*
Ogura, F. 301, 302 (292), *317*
Ohanessian, G. 733 (184), *769*
Ohe, K. 309 (350–352), *318*, 682 (138, 139), *705*
Ohfune, Y. 284 (129, 130), 292 (129, 222), *313, 315*, 667 (65, 66), *703*
Ohira, S. 679 (123), *704*
Ohishi, M. 195 (278), *219*
Ohkata, K. 444 (185), *465*
Ohnishi, M. 669 (77), *703*
Ohno, M. 284 (125), *313*
Ohno, T. 608–612, 618, 619 (351), *655*
Ohra, T. 271 (49), *312*
Ohshima, T. 714 (36), *766*
Ohshima, Y. 188 (245), *218*
Ohta, H. 306 (325), *317*
Ohta, K. 116, 117 (234b), *136*, 303 (311), *317*
Ohta, T. 640 (419), *656*
Ojha, , N.D. 418 (66), *462*
Okada, K. 287 (141), *313*, 716 (51), *766*
Okada, M. 307 (329), *317*
Okarma, P.J. 353 (34), *407*
Okazaki, R. 717, 718 (58), 745 (58, 246), 748–750 (58), *767, 771*
Oki, T. 471 (134), *490*
Okiye, K. 145 (41), *213*
Oku, A. 499 (12), 500 (28), *648*
Okuma, K. 306 (325), *317*
Olah, G. 445 (201), *466*
Olah, G.A. 202 (308), *220*, 394 (120), *410*, 417 (54), 419 (83), 427, 428 (115), 430 (115, 127), 432, 433 (144), 434 (147), 439 (166), 442 (182), 444 (188, 192), 447 (192, 208, 210), 457 (270), *462–466, 468*, 554 (209, 210), *652*, 815 (6, 7), 816 (11), 817 (16, 19), 818 (20a, 20b), 819 (24), 820 (11), 821 (20b, 40, 41), 825 (11, 20b, 50a–c), 826 (50a, 50c, 54), 827 (54–56), 828 (57), 829 (50c, 60, 61), 831, 832 (65), 833 (73), 834 (74, 75, 77), 835 (78, 80), 836 (55, 57, 80), 837 (83), 838 (56, 86), 839 (80, 87), 840 (57), 845 (101, 102), 846 (101, 103, 107a–c), 847 (107b, 107c), 848 (110), 849 (107b, 110–114), 850 (114, 115), 851 (11, 110, 116–118), 852 (50a, 117, 119), 853 (50a–c, 122), 855 (122, 128), 856 (11), *856–859*
Olander, W.K. 589, 598 (304), *654*
Oldenburg, C.E.M. 665 (63), *703*
O'Leary, M.A. 728 (135), *768*
Oliva, A. 125 (266), *137*
Olive, J.L. 292 (220), *315*, 695 (186), *706*
Olivella, S. 121, 123 (249), 126 (274), *136, 137*, 430 (125), 437 (156), *464, 465*
Oliver, J.P. 505 (58, 59), *649*
Ollerenshaw, J. 470 (39), *489*
Ollivier, J. 663 (49), *703*
Olson, L.P. 470 (63), *489*
Olsson, L. 111 (205), *135*, 364, 375, 377, 378, 380, 381, 391, 392, 394–399, 402, 403 (49), *407*
O'Malley, S. 694 (178, 179), 695 (179), *706*
Omrčen, Y. 499 (16), *648*
Onak, T. 111 (206), *136*
O'Neal, H.E. 479, 480 (218), *492*
O'Neil, H.E. 126 (277c, 277d), *137*
Ono, N. 301 (283), *317*
Onoda, Y. 279 (74b), *312*
Onuma, S. 170 (136), *216*
O-oka, M. 717, 718, 745, 748–750 (58), *767*
Oosthuizen, H.E. 554 (211), *652*
Opitz, K. 262 (4), *311*, 787, 789, 790 (57, 60), 795, 800, 805, 806 (60), *810, 811*
Orama, O. 562 (241), *652*
Ordronneau, C. 445 (195), *466*
Orendt, A.M. 110 (197), *135*, 783, 797 (50), *810*
Orgias, R.M. 422 (97), *463*
Orpen, A.G. 159, 180, 185 (95), *215*
Orr, G. 728, 729, 757 (136), *768*
Orvane, P. 729 (145), *768*
Osamura, Y. 84, 85, 98, 101, 102, 126 (129), *133*, 481 (263), *493*
Osanai, K. 301, 302 (295), *317*
Ose, Y. 499 (12), *648*
Osina, E.L. 145 (33), 146 (45), 199 (291), *213, 214, 219*
Ostlund, N.S. 125 (267), *137*

Ostrander, R.L. 539–541 (164), *651*
Ostrowski, S. 300 (278), *316*
Oth, J.F.M. 121, 122 (246), 123 (246, 254b), *136, 137*, 183 (206), *217*, 247 (114a), 248 (117), *259*, 419 (83), 440, 441 (176), 455 (254), *463, 465, 467*
Otsubo, T. 301, 302 (292), *317*
Otsuji, Y. 287 (143), *314*
Otsuka, S. 663 (39, 40), *703*
Ott, K.H. 471 (90), *490*
Otte, A.R. 272 (50, 51), *312*
Otto, A.H. 111 (211), *136*, 249 (122), *260*, 364, 375, 377, 378, 381, 387, 388, 391, 392, 394, 396, 397, 403 (56), *408*, 425, 427, 429, 430 (111), *463*
Overberger, C.G. 418 (64), *462*, 471 (91), *490*
Overend, J. 105 (170d), *134*
Owens, K.G. 212 (344), *221*
Ozier, I. 186 (223), *218*, 471 (87), *490*

Pacansky, J. 121, 123 (247), *136*, 728, 729, 757 (136), *768*
Paddon-Row, M.N. 356 (40), *407*, 450 (224), *466*
Padwa, A. 291 (192, 199, 200a–e), *314, 315*, 658, 663 (8), 675 (106), 684 (145, 146), 685 (146), 687 (145), *702, 704, 705*
Paetow, M. 267 (29), *311*
Page, M. 126 (269, 273), *137*, 481 (261, 267), *493*
Pagenkopf, G.K. 418 (63), *462*
Pagliarin, R. 663 (41), *703*
Pahor, N.B. 644 (429), *656*
Paik, Y.H. 202 (310, 311), *220*, 289 (162), *314*
Painter, C.E. 471 (120), *490*
Palamareva, M. 179 (178), *217*
Palenik, G.J. 500 (18, 23), *648*
Palke, W.E. 57, 59, 69, 75, 112 (50), *130*, 144, 145 (30), *213*
Palm, R. 471 (90), *490*
Pan, D.K. 69 (74), *131*
Pande, K.C. 513 (99), *649*
Panshin, S.Y. 225 (11), *250*
Pantaleo, N.S. 585 (297), *654*
Panyachotipun, C. 300 (276), *316*
Papadopolous, P. 291 (197), *315*, 673 (100), *704*
Papamihail, C. 718, 719 (68), *767*
Paquette, L.A. 235 (62, 62), *255*, 262 (2b), 277 (69), *310, 312*, 344, 345 (19), 366 (61), 394 (120), *406, 408, 410*, 413, 415, 416 (18), 419 (18, 73, 83), 427 (18), 430 (127), 439 (166), 440 (18), 442 (18, 182), 444 (185), 450 (18, 223), 456 (264), 457 (267), 459 (18), *461–466, 468*, 506 (66), *649*
Parfonry, A. 176 (156), *216*
Parham, W.E. 715 (42), *766*
Park, H. 535, 537, 539, 542, 543 (158), *651*, 869 (43), *884*
Park, S.-B. 663, 693, 698 (43), *703*
Parker, D.W. 188 (250, 251), *218*
Parker, V.B. 224 (9), 237 (9, 69), *249, 255*
Parks, A.T. 160 (106), *215*
Parlier, A. 295 (249), *316*
Parrington, B.D. 361 (45), *407*, 432, 433, 435 (138, 141), *464*
Parshall, G.W. 583 (292), *653*
Parsons, P.J. 309 (346), *318*
Parvez, M. 500 (19), *648*
Pasau, P. 298 (266, 267), *316*
Pascard, C. 874 (60), *884*
Pasteris, R.L. 320 (8, 9), *337*
Patai, S. 224 (8), *249*
Patapoff, T.W. 474, 475, 477, 485 (162), *491*
Patel, D.J. 511 (86), *649*
Patel, P.P. 523, 530 (137), *650*
Pauling, L. 178, 179 (176), *217*, 366, 370 (62), *408*
Paulini, K. 290 (171), *314*
Paulissen, R. 659 (18), *702*
Pauzat, F. 98, 101, 102, 105 (158), *133*
Pavone, V. 170–172 (129), *216*
Pawlak, J.M. 459 (284), *468*
Payne, D. 722 (108), *768*
Pearce, R. 583 (291), *653*
Pearce, R.A.R. 98, 101, 105 (157), *133*
Pearson, R.Jr. 83, 86, 88, 93 (101), *132*
Pedersen, L.D. 104 (166a), *134*, 484 (275–278), 485 (275, 277, 278), 486 (275, 276, 278), 487 (275), *493*
Pedersen, S.F. 593, 594 (318), *654*
Pedley, J.B. 225 (15), 229 (33), 232 (45), *250, 252, 253*
Pedone, C. 170–172 (129), *216*
Peganova, T.A. 508 (75), 627 (385, 386), 628 (386), *649, 655*
Pellicciari, R. 663 (45), *703*
Penadés, S. 305 (315), *317*
Penn, R.E. 83, 86 (100), *132*, 160 (105), *215*
Penner, A.P. 471 (110), *490*
Peperle, W. 242 (87), *257*
Peretta, A.T. 83, 86, 95 (105a), *132*
Perevalova, E.G. 160 (107), *215*, 501 (38a, 38b), *648*
Periana, R.A. 509 (81), *649*
Periasamy, M.P. 501 (33), *648*
Perichon, J. 280 (90), *312*
Perkin, W.H. 708 (1), *765*
Perretta, A.T. 177 (159), *216*
Person, W.B. 105 (170a–d), *134*
Petasis, N.A. 502 (40), 514 (103), *648, 649*
Pete, J.-P. 417 (53), *462*
Peters, C.W. 471 (85), *490*

Peters, D. 342, 382, 390 (5), *406*, 415, 424, 427, 431 (39), *461*
Peters, E.-M. 176 (158), 188 (242), 189 (255), 200 (299), 206 (323), 207 (324, 325), 210 (338), *216, 218, 220, 221*, 363, 364 (47), *407*, 455 (261), 456 (263), *467, 468*
Peters, E.M. 867 (32), *884*
Peters, K. 147–149 (57), 176 (158), 189 (255), 196 (279), 200 (299), 206 (323), 207 (324, 325), 210 (338), *214, 216, 218–221*, 363, 364 (47), *407*, 455 (261), 456 (263), *467, 468*
Peterson, J.R. 235 (62), *255*, 457 (267), *468*
Peterson, M.R. 83, 86, 88, 93 (102b), *132*
Petiniot, N. 663 (36), 683 (140), *703, 705*
Petit, H. 304 (313), *317*
Petragnani, N. 663 (48), *703*
Petrov, V.I. 149 (63), *214*
Petrovskii, P.V. 508 (75), 627 (385, 386), 628 (386), *649, 655*
Petter, R.C. 504, 505 (52), *648*
Pettersen, A. 150 (67, 68), 151 (68), *214*
Pettersson, L. 128 (281), *137*
Pettit, R. 419 (71, 75, 79, 82), *462, 463*, 554 (215), *652*
Pews, R.G. 418 (66), *462*
Peyerimhoff, S.D. 112 (213), *136*, 481 (232), *492*
Pfaltz, A. 292 (204–206, 208), *315*, 698 (189–194), 701 (194), *706*
Pfeifer, K.-H. 193, 194 (272), *219*, 798–801 (115), *812*
Pfeifer, K.H. 733, 739, 741 (182), *769*
Pflederer, J.L. 321 (11), *337*
Pfyffer, J. 722 (91–93), 726 (118), 740 (216), 741 (92, 93), 746 (118), *767, 768, 770*
Pham, E.K. 535, 537 (157), *651*
Philippi, W. 728 (134), *768*
Philippo, J.S. 513 (94, 95), 515 (94, 108), *649, 650*
Philipps, P. 547 (176), 581, 582, 585 (289), *651, 653*
Phillips, R.L. 504 (47), 633 (399), *648, 656*
Piana, H. 607, 617 (345), *654*
Piechocki, C. 694 (176, 177), *706*
Pierens, R.K. 107, 108 (179), *134*
Pierini, A.B. 779 (29), *810*
Piers, E. 501 (34–36), 544 (169), 545 (173), *648, 651*, 663 (46), 672 (94), 678 (117–119), *703, 704*
Pieters, R.J. 292 (213, 215), *315*, 698 (197, 198), 701 (198), *706*
Pietro, W.J. 48–50 (30), *130*
Pigou, P.E. 794 (76), *811*
Pike, G.A. 681 (128), *704*
Pikulik, I. 451 (236), 452 (237), 453 (246), *467*
Pilcher, G. 229 (31), 243 (89), *251, 258*, 387, 388 (97), *409*
Pilling, M.J. 477 (197), *492*
Pimenova, S.M. 230, 231 (41), 232 (47), 233 (48, 49), 235 (53, 55, 58, 61), 241 (82), *252–254, 257*, 780 (42), *810*
Pincock, R.E. 448 (212), 449 (215), *466*
Pinder, A.R. 866 (26), *884*
Pineda, N. 722, 741 (92, 93), *767*
Pineock, R.E. 881 (87), *885*
Pinhas, A.R. 625, 636, 637 (381), 638 (410), *655, 656*
Pinkus, A.G. 710 (17), *766*
Piotrowska, K. 690, 693 (167), *705*
Pipkin, O.A. 471 (121), *490*
Pirrung, M. 302 (296), *317*
Pirrung, M.C. 170, 171 (133), *216*, 303 (308), *317*
Pittman, C.U. 849 (112), *859*
Pittman, C.U.Jr. 418 (65), *462*, 825, 826, 829, 853 (50c), *857*
Pitts, W.J. 658 (16), *702*
Piven, V.A. 877 (74), *885*
Pizzabiocca, A. 822, 825 (45), *857*
Plas, H.C.van der 454 (252), *467*
Plate, A.F. 469 (20), *488*
Plattner, D.A. 505 (56), *649*
Platzer, N. 277 (65), 295 (249), *312, 316*, 509 (84), *649*
Plesset, M.S. 85 (136a), *133*
Plotkin, V. 192 (268), *219*
Poblet, J.M. 440 (170), *465*
Pochan, J.M. 187 (240), *218*
Podder, R.K. 302 (300), *317*
Pododensin, C. 239 (72), *256*
Podosenin, A. 451, 457 (234), *467*
Pohl, E. 564, 565 (246), *652*
Pohmakotr, M. 180 (189), *217*, 501 (30), 504 (30, 53), *648*
Poignée, V. 201 (306), *220*, 353 (29), *407*, 719 (70), 721, 722 (85), 731 (166), 734, 735 (194), *767, 769*
Polborn, K. 262 (5), *311*, 554 (216), *652*, 776, 781 (45, 61), 782–784 (45), 787, 788 (45, 61), 789 (45, 63), 790, 798, 799, 802 (63), *810, 811*
Poli, R. 517, 518 (114), *650*
Politzer, P. 74 (81), 96 (153), *131, 133*, 779 (30), *810*
Poljanec, K. 733 (174, 176, 180), *769*
Polley, J.S. 418 (63), *462*
Pollitte, J.L. 794, 808 (90), *811*
Pollok, T. 185 (211), *217*
Pomerantz, M. 235 (60), *254*, 669 (75), *703*
Pong, R.Y. 280 (87), *312*, 866 (23), *883*
Pons, S. 760 (306–308), *772*

Popelier, P.L.A. 64, 65, 68, 107 (67d), *131*, 375, 376, 393 (81), *408*
Pople, J.A. 44 (6), 47 (21a, 21b, 22), 85 (136b–d, 137), 96 (6), 102 (160), 104 (168a), 109 (183a, 183b), 114 (221), 117–119 (238a–c), *129, 130, 133, 134, 136*, 384 (92), 391 (100), 392 (103–105, 109, 110), *409*, 430 (125), *464*, 481 (234), 483 (274), *492, 493*, 833 (72), *858*
Poppinger, D. 47 (22), 116 (225), *130, 136*
Poredda, A. 798, 799 (102), *811*
Porri, L. 881 (86), *885*
Porter, A.P. 828, 853 (59), *858*
Porter, R.D. 434 (147), *464*, 818 (20a, 20b), 821, 825 (20b), 846 (107a), 849 (111, 113), *857, 859*
Potekhin, K.A. 208 (326), *220*, 797, 799, 800 (101), 802 (101, 122), 805 (126), *811, 812*
Potenza, J. 574 (273), *653*
Potenza, J.A. 817, 854, 856 (18), *857*
Poulter, C.D. 110 (199b), *135*
Poupart, M.-A. 277 (69), *312*
Powers, D.G. 504, 505 (52), *648*
Powers, J.W. 517 (113), *650*
Prakash, G.K. 417 (54), *462*
Prakash, G.K.S. 444 (188, 192), 447 (192, 208, 210), *465, 466*, 815 (6, 7), 816 (11), 819 (24), 820 (11), 821 (40, 41), 825 (11), 827 (55), 828 (58), 831, 832 (65), 834 (74, 77), 835 (78, 80), 836 (55, 80), 839 (80, 87), 845 (101, 102), 846 (101, 103), 848, 849 (110), 851 (11, 110, 116, 118), 852 (119), 853, 855 (122), 856 (11), *856–859*
Prakash, O. 308 (331), *318*, 681 (129), *705*
Prakash-Reddy, V. 417 (54), *462*
Praly, J.-P. 294 (229), *315*
Prange, T. 874 (60), *884*
Prange, U. 419 (83), 440, 441 (176), *463, 465*
Prapansiri, V. 300 (276), *316*
Pratt, A.C. 324, 325 (30, 32), 336 (32), *338*
Pratt, G.L. 471 (116), *490*
Pratt, L. 420 (84), *463*, 554 (207), *652*
Prestien, J. 717 (57), 740 (212), *767, 770*
Price, A.T. 684, 685 (146), *705*
Prichard, H.O. 471 (104), *490*
Priesner, C. 612 (356), 614 (358), 615, 616 (356), 617 (358), *655*
Prinzbach, H. 202 (313), *220*, 477 (190), *492*
Pritchard, D.E. 56 (49), *130*
Pritytskaya, T.S. 308 (334a), *318*
Pritzkow, H. 83, 96 (125), *133*
Prokhorov, A.I. 176 (157), *216*
Proksh, E. 871 (50), 873 (58), *884*
Prosperi, T. 106 (175), *134*
Protasova, E.L. 865 (18), *883*
Protopopova, M.N. 292 (214), *315*, 684 (145, 146), 685 (146), 687 (145), 698 (199), *705, 706*
Prout, K. 170 (134), *216*
Puddephatt, R.J. 504 (47), 633 (399), *648, 656*
Puder, P. 874 (63), *884*
Pues, C. 429, 430 (128), *464*
Pulay, P. 104, 109 (167b), 110 (186, 202g), 111 (202g), *134, 135*, 393, 396 (113, 116), 397 (116), *409, 410*, 446 (204), *466*
Pullman, B. 342, 344, 382, 390 (8), *406*
Puranik, D.B. 192 (269, 270), *219*
Purvis, G.D. 392 (104), *409*
Püttmann, W. 716, 745, 749 (52), *767*
Putz, B. 440, 441 (176), *465*
Pyo, S. 268 (33), *311*

Qi, Y. 247 (113), *259*
Qiao, K. 503 (45), *648*
Qin, A. 805, 806 (129), *812*
Qin, X.-Z. 117 (236a, 236b), *136*
Qtaitat, M. 186 (235), *218*
Quast, H. 206 (323), 207 (324, 325), *220*, 299 (270), *316*, 363, 364 (47), *407*, 455 (261), 456 (263), *467, 468*

Raajca, A. 457 (271), *468*
Rabinovitch, B.S. 126 (268a–c), *137*, 470 (76, 77), 471 (76, 77, 100, 109, 112, 132, 137, 141), 472 (141), 479 (76), 480 (141), *489–491*
Rademacher, P. 202 (307), *220*, 734, 735, 739, 755 (195), *769*, 870 (46), *884*
Radesca, L. 300 (273), *316*
Radlick, P. 714 (37), *766*
Radom, L. 44 (6), 47 (21b), 60 (54b), 96 (6, 150), 98 (150), 102 (160), 116 (225), 117–119 (238b, 238c), *129, 130, 133, 136*, 144, 204 (28), *213*, 384 (92), *409*, 457 (270), *468*, 785 (54), *810*, 863 (4a), *883*
Raduchel, B. 661 (28, 29), *702*
Radziszewski, G.J. 784 (51), *810*
Radziszewski, J.G. 728, 729, 757 (132), *768*
Raghavachari, K. 85 (137), 114 (221), 117–119 (238a), *133, 136*, 392 (105, 109, 110), *409*
Raghavachari, R. 430 (126), *464*
Ragunathan, N. 475 (166, 167), 476 (166, 167, 176), 477, 484 (166), 487 (166, 282), *491, 493*
Raimondi, M. 57, 58, 66, 69, 71, 73, 76, 77, 82, 83 (51), *130*
Rainbow, I.J. 828, 853 (59), *858*
Rajeswari, K. 834 (76), *858*
Rall, M. 181, 182 (191), 190 (259), *217, 218*

Raman, K. 292 (224), *315*, 678 (113, 114), 696 (187), *704, 706*
Ramasubbu, N. 189 (253), *218*
Ramey, K.C. 550 (182), *651*
Ramig, K. 300 (275), *316*
Ramireddy, C. 805, 806 (129), *812*
Ramos, A. 323, 331 (28a), *338*
Ramos, M.N. 105–107 (169c), *134*
Ramsay, D.A. 85, 86 (139a), *133*
Ramsden, C.A. 430 (125), *464*
Ramsey, B. 849 (112), *859*
Ranade, A. 713 (31), *766*
Rancourt, J. 273 (57), *312*
Randaccio, L. 644 (429), *656*
Randall, C.J. 719, 720 (71, 72), 730 (154, 155), 731 (155), 732 (154, 155), 733, 738 (186), 742 (154, 155), 745 (186, 245), 749 (186), 758 (155), *767–769, 771*
Randall, G.L.P. 556 (222), *652*
Randic, M. 56, 57, 71, 75, 76 (45), *130*, 366, 370 (62), *408*
Ranganayakulu, K. 825 (53), *858*
Ranu, B.C. 291 (195, 197), *315*, 673 (97, 98, 100), 674 (101), 689 (163), *704, 705*
Rao, A.V.R. 665 (59), *703*
Rappoport, Z. 224 (1), *249*, 498, 499 (1), *647*, 869 (42), *884*
Rashchupkina, Z.A. 867 (31), *884*
Rasul, G. 845 (102), 848, 849, 851 (110), *859*
Ratchataphusit, J. 504 (53), *648*
Rathmann, F.H. 470 (74), *489*
Raths, H.-C. 288 (155), *314*
Ratner, V.V. 199 (298), *220*
Rau, H. 756, 757 (290), *771*
Rausch, M.D. 598 (330), *654*
Raw, T.T. 178 (174), *217*
Rawda, T.N. 827, 836 (55), *858*
Rawdah, T.N. 444, 447 (192), *466*, 831, 832 (65), 845, 846 (101), 851 (118), *858, 859*
Rawlings, B.J. 302 (297), *317*
Ray, S.C. 302 (300), *317*
Reale, H.F. 779 (29), *810*
Reardon, E.J.Jr. 668 (68), *703*
Reber, G. 156, 157 (90), *215*
Rebollo, H. 200 (299), *220*
Reddy, V.P. 816, 820, 825 (11), 834 (77), 845 (102), 846 (103), 851 (11), 852 (119), 856 (11), *856, 858, 859*
Reddy, V.S. 805, 806 (129), *812*
Redmore, D. 469 (18), *488*
Ree, B.R. 844 (97), *859*
Reed, A.E. 374, 384, 393 (78), *408*, 778 (18), *810*
Reed, J.W. 469 (28, 29), *488*, 544 (168), *651*
Reed, L.E. 720, 724 (81), *767*
Rees, C.W. (354, 355), *655*, 727 (129, 130), 729 (141), 760 (130), *768*
Rees, J.C. 444, 447 (192), *466*
Regan, C.M. 741 (221), *770*
Reger, D.L. 554 (197), *651*
Regitz, M. 197 (280, 281), 209 (331), *219, 220*, 729 (142), *768*
Rego, C.A. 146 (52), *214*
Rehberg, G.M. 308 (340–342), *318*
Rehberg, R. 453 (247), *467*
Reichel, F. 110 (203, 204), 111 (203, 205), *135*, 249 (122), *260*, 364, 375 (49, 50, 53, 56), 377 (49, 56, 84), 378 (49, 56), 380 (49), 381 (49, 50, 53, 56), 387, 388 (56), 391, 392 (49, 50, 53, 56), 393 (84), 394 (49, 50, 53, 56), 395 (49), 396, 397 (49, 50, 53, 56), 398, 399 (49), 402 (49, 50, 53), 403 (49, 50, 53, 56), 404 (84), *407, 408*, 414 (27, 28), 421 (27), 424 (103), 425 (27, 103, 111), 426 (27, 103), 427, 429, 430 (111), 441, 442 (28, 180), *461, 463, 465*
Reichell, F. 111 (211), *136*
Reichelt, I. 504, 505 (49, 50), *648*, 692 (168), *705*
Reid, M.D. 523 (137, 139), 525, 527, 528 (139), 530 (137), *650*
Reiff, W.M. 197 (284c), *219*
Reimann, W. 202, 203 (309), *220*
Reinecke, M.G. 729 (143), *768*
Reingold, I.D. 291 (191), *314*
Reinhardt, G. 402 (125), *410*, 453 (248), *467*
Reinhardt, R. 299 (270), *316*
Reinhardt, R.-D. 508, 576, 579 (76), *649*
Reisch, J.W. 579 (286), 591, 599–602 (312), *653, 654*
Reisemberg, R. 625, 636, 637 (381), *655*
Reiser, O. 530, 552, 553, 622 (152), *650*
Reisinger, F. 434 (150), *465*
Reissig, H.-H. 290 (171), *314*
Reissig, H.-U. 262 (2e), 292 (207), 294 (239–241), 295 (240), *310, 315, 316*, 504, 505 (49, 50), *648*
Reissig, H.U. 526 (147), *650*, 692 (168), *705*
Remington, .B. 84, 85, 96, 98, 101, 102 (130), *133*
Renzi, G. 822, 825 (45), *857*
Repic, O. 866 (24), *883*
Replogle, E.as. 483 (274), *493*
Replogle, E.S. 392 (105, 109), *409*
Reuben, D.M.E. 557 (230), *652*
Reum, M.E. 866 (28), *884*
Reutrakul, V. 300 (276), *316*
Revol, J.-M. 673 (97), *704*
Rewicki, D. 711 (25), *766*
Rey, M. 722 (97), *767*
Reynders, P. 83, 86, 89, 93 (114), *132*
Reynolds, C. 820 (29), *857*
Reynolds, D.N. 663, 694 (42), *703*

Reynolds, G.F. 499, 500 (7), *648*
Reynolds, K. 278 (73), *312*
Reynolds, S.D. 552 (195), *651*
Rezvukhin, A.I. 432, 433, 435 (145), *464*
Rheingold, A. 539–541 (164), *651*
Rheingold, A.L. 574 (275), 579 (286), 586 (299), 590 (275, 309), 591 (312, 313), 592 (313), 595 (275, 309), 599 (312, 313), 600 (275, 309, 312), 601 (312), 602 (309, 312, 313), 603, 605 (309), *653, 654*, 678 (113), *704*
Ricca, A. 737, 742, 748, 753 (200), *769*
Rice, J.E. 84, 85, 96, 98, 101, 102 (130), 104 (168b), *133, 134*
Richards, W.G. 458 (274), *468*
Richey, H.G. 445, 446 (198), *466*, 478 (202), *492*
Richey, H.G.Jr. 417 (54), *462*, 814 (3), 816, 825 (12), 854 (123), 856 (3), *856, 857, 859*
Richmond, J.P. 289 (169), *314*, 563 (242), *652*
Riede, J. 185 (211), *217*, 536 (162), *651*
Riegel, B. 412 (4), *460*
Rieke, R.D. 322 (17), *337*
Rifi, M. 865 (16), *883*
Rigby, H.L. 673 (98), *704*
Riggs, N.V. 144, 204 (28), *213*, 785 (54), *810*
Rihs, G. 712, 736 (27), *766*
Rimmelin, A. 470 (36), *489*
Rimoldi, J. 459 (283), *468*
Ringold, C.E. 170 (131), *216*
Rippol, J.L. 866 (29), 868 (35), 873 (29), *884*
Risaliti, A. 267 (27), *311*
Ritter, K. 759 (299), *771*
Ritter von Onciul, A. 199 (295), *219*
Rivera, A.V. 556 (223), *652*
Rizzi, G. 663 (41), *703*
Robaugh, D.A. 477 (188), *492*
Robba, M. 150 (66), *214*
Robbins, J. 591, 592, 599 (313, 314), 602 (313), 603–605 (314), *654*
Robbins, J.D. 324, 325 (33), *338*
Roberts, D.J. 511 (86), *649*
Roberts, J.D. 248, 249 (121), *260*, 412 (8, 10), 417 (55), 427 (10, 113), 445 (196), *460–463, 466*, 499 (6), 511 (88), 512 (89), *648, 649*, 741 (221), *770*, 814 (1, 2), 815 (4), 817 (14), 818 (21), 819 (24), 820 (36), *856, 857*
Roberts, L.R. 285 (131), *313*
Roberts, M. 305 (317), *317*, 440, 441 (172, 175), *465*
Roberts, P.J. 429 (120), *464*
Roberts, S.W. 264 (14), *311*
Robertson, L.R. 470 (33), *489*
Robin, M.B. 60 (53), 111, 112 (212b), *130, 136*
Robinson, D.J. 586 (299, 300), 591, 592, 599 (313, 314), 602 (313), 603–605 (314), *654*
Robinson, L.R. 622 (370), *655*
Robinson, R.E. 797, 798, 800, 802 (104), *811*
Robinson, W.H. 756 (288), *771*
Robinson, W.T. 719, 720 (71), 734 (192), *767, 769*
Roby, K.R. 93 (144b), *133*
Rockett, B.W. 549 (179), *651*
Rode, K. 504, 505, 622 (51), *648*
Rodenhouse, R.A. 292 (221), *315*, 695 (184), *706*
Rodewald, H. 155, 156 (83), 202 (310, 313), 205, 206 (321), 209 (332), 210 (336), *214, 220, 221*
Rodewald, L.B. 472 (142), *491*
Rodin, W.A. 262 (2f), *310*, 709 (10), 720 (76), 722, 723 (76, 101, 103), 739 (10, 101), 740, 741, 746 (10), 749 (10, 101), 750 (101), *766, 767*
Rodios, N.A. 300 (274), *316*
Rodler, M. 757 (293), *771*
Rodriguez, D. 722 (91, 94, 98, 105), 723 (105), 739 (206), 741 (94, 229), *767, 768, 770*
Rodwell, W.R. 96, 98 (150), *133*
Roefke, P. 642 (422), *656*
Rogers, D.W. 228 (24a, 24b, 25, 26), 230 (38), 231 (43), 232 (38), 235 (54, 58, 60, 62, 62), 239 (72, 73), 247 (113), *251–256, 259*, 451 (234), 457 (234, 267, 269), *467, 468*
Rogerson, C.V. 419, 420, 423–425 (77), 438 (164), *463, 465*
Roginskii, S.Z. 470 (74), *489*
Rohmer, M.-M. 48, 49, 71 (41), *130*
Roll, W. 844 (100), *859*
Rollefson, G.K. 479 (213), *492*
Romanow, W.J. 212 (344), *221*
Romine, J.L. 297 (263), *316*
Rømming, C. 150 (67, 68), 151 (68), 167 (124), *214, 215*, 874 (62), *884*
Romo, D. 297 (264), *316*
Rondan, N.G. 125 (265), *137*, 450 (223), *466*, 487 (284), *494*
Roos, B. 48, 49, 71 (41), *130*
Roos, B.O. 459 (282), *468*
Roos, G.H.P. 292 (217), *315*, 699 (203), *706*
Rose, H. 402 (125), *410*, 453 (248), *467*
Rose, T.L. 470 (53), *489*
Rosenberg, R.E. 784 (52, 53), 785 (53), *810*
Rosenblum, M. 499 (13, 14), 502 (13, 14, 39), 515, 516 (109), *648, 650*
Rosenburg, J.L.Jr. 419 (79), *463*

Rosenfeld, J. 116 (224), *136*, 821, 825 (38), *857*
Rosenquist, N.R. 728, 729, 757 (133), *768*
Rosenstock, H.M. 231 (44), *253*
Roskamp, E.J. 686 (153), *705*
Rossi, R. 881 (86), *885*
Rotard, W. 689 (164), *705*
Roth, E.A. 155, 187, 190 (85), *215*
Roth, H.D. 116, 117 (231), *136*, 188 (249), *218*, 354 (36), *407*
Roth, W.R. 189 (254), *218*, 229 (35, 37), 231 (43), 232 (45), 234 (51), 236 (65, 66), 245 (99–101, 103), 246 (105), *252, 253, 255, 258, 259*, 357 (41, 42), 358 (41), 359, 361, 378 (41, 42), 387 (41), 389, 390 (41, 42), *407*, 450–453, 457 (225, 226), *467*, 469 (17), 470 (46), 477 (190, 195, 196), 479 (211, 212), *488, 489, 492*
Rouchy, P. 200 (300), *220*
Roustan, J.L. 515 (107), *650*
Roux, M.V. 239 (72), 240 (76), *256*
Rowland, C. 481 (244), *493*
Rowley, M. 263 (9b), *311*
Roy, C.S. 657 (1), *702*
Roznyatovskii, V.A. 620 (366), *655*
Rozsondai, B. 159 (100), *215*
Rüchardt, C. 147–149 (57), *214*, 247 (111, 112), *259*, 511 (88), *649*
Rücker, C. 202 (313), *220*
Ruckle, R.E.Jr. 683 (144), *705*
Rudler, H. 295 (249), *316*
Rue, R.R. 65 (70e), *131*
Ruedenberg, K. 65 (70a–e), *131*
Ruedi, P. 869 (45), *884*
Ruediger, E.H. 678 (117, 119), *704*
Ruest, L. 670 (82), *704*
Rufinska, A. 506 (65), 508 (76), 519 (116), 546 (65), 576, 579 (76), 626, 635 (383), *649, 650, 655*, 753 (278), *771*
Ruhkamp, J. 477 (196), *492*
Rui, M. 105–107 (169c), *134*
Ruiter, B.de 211 (343), *221*
Ruiz, J.M. 114, 115 (222), *136*
Ruiz-Pérez, C. 201 (306), *220*, 353 (29), *407*, 734, 735 (194), *769*
Rulin, F. 469 (28), *488*
Runge, W. 44, 48 (15), *129*
Rüngeler, W. 200 (299), *220*
Runzheimer, K.-O. 714 (33), *766*
Ruppert, J.F. 670 (83), *704*
Rusmussen, K. 863 (4e), *883*
Russell, J.J. 477 (191), *492*
Russell, S.G.G. 718, 719 (68), 746 (248, 252), 751, 752, 756 (252), *767, 771*
Ruthven, D.M. 471 (123), *490*
Ryadnenko, V.L. 235 (56), 241 (82), *254, 257*
Rychnovsky, S.D. 280 (86), *312*
Rydberg, D.B. 308 (339), *318*
Rykowski, A. 300 (278), *316*
Rzepa, H.S. 733, 734 (172), *769*

Saal, W.von der 299 (270), *316*
Saba, A. 310 (355), *318*, 675 (104), *704*
Sabio, M.L. 452 (241), *467*
Sacconi, L. 589 (305, 306), 590, 592, 601, 602 (306), 603 (305, 306), *654*
Sachdev, K. 473, 477 (150, 151), 478 (203), 479, 481 (150, 151), *491, 492*
Sack, T.M. 117 (237), *136*
Sadlej, A. 110 (193e), *135*
Sadovaya, N.K. 789, 790 (65), 796 (100), 797 (65, 101), 798 (65, 119, 120), 799, 800 (65, 101), 801 (65), 802 (65, 101, 119–122), 805 (65, 100, 126), 807 (65), 808 (65, 131), *811, 812*
Sael-Imber, M. 544 (165), *651*
Sagae, H. 278 (71), *312*
Saha, A.K. 290 (178), *314*, 663, 694 (44), *703*
Saha, B. 668 (74), *703*
Saicic, R. 273 (58), *312*
Saidi, M.R. 279 (76), *312*
Saigo, K. 280 (78), *312*
Saikali, E. 664 (53), *703*
Saito, K. 746 (253), 747 (263), 751 (253, 271), *771*
Saito, S. 145 (44), 186 (222), 195 (278), *213, 218, 219*
Saive, E. 290 (176), *314*, 663 (34), *703*
Sakai, M. 361 (45), *407*, 431 (137), 432 (138), 433, 435 (137, 138), 440 (172), 441 (172, 178), 447, 448 (207), 449 (220), 458 (272), *464–466, 468*
Sakai, S. 279 (75), *312*
Sakai, T. 303 (311), *317*
Sakurai, H. 644 (426), *656*
Sakurai, T. 64 (66c), *131*
Salaün, J. 469 (23, 27), *488*, 557, 561 (232), 639 (413, 414), *652, 656*, 663 (49), *703*
Salaün, J.R.Y. 262 (2d, 2h), *310*
Saleh, S.A. 678 (112), *704*
Salem, L. 48–50 (29), *130*, 365 (60), *408*, 481 (235–242, 244, 245, 248, 250, 252, 258), *492, 493*
Salomon, M. 513 (100), 514, 519 (101), *649*
Salomon, M.F. 519 (122), *650*, 658, 661 (14), *702*
Salomon, R.G. 658 (13, 14), 661 (14), 691 (13), *702*
Salvadori, J. 126 (274), *137*
Salzer, A. 498 (3), 556 (225, 226), *647, 652*
Samdal, S. 161 (109), *215*
Samizio, F. 287 (141), *313*
Samuel, C.J. 324, 325 (33), *338*

Samuel, S.D. 235 (62), *255*, 451, 457 (234), *467*
Samuelson, A.G. 625, 636, 637 (381), *655*
Sana, M. 477 (183, 184), *491*
Sanaeva, E.P. 796, 805 (100), *811*
Sánchez, B. 305 (314), *317*
Sanchez, E.L. 668 (69, 70), 683 (143), 687 (156), *703, 705*
Sánchez-Delgado, R.A. 663 (33), *703*
Sandel, V.R. 742 (233), *770*
Sanders, A. 516, 566, 568 (110), *650*
Sanders, D.A. 505 (58, 59), *649*
Sandmeyer, F. 547 (176), *651*
Sandri, J.M. 853 (120), *859*
Sanktjohanser, M. 778, 785, 786 (23), 799, 801 (117), *810, 812*
Santarsiero, B.D. 422 (98), *463*
Santiago, C. 739 (207), *770*
Santini, A. 170–172 (129), *216*
Santo, K. 299 (272), *316*
Santone, P. 200 (301), *220*
Saraçoglu, N. 747, 752 (264), *771*
Sardella, D.J. 759, 760 (302), *772*
Sarel, S. 469 (19), *488*, 544 (165), 549 (178), 563 (243), 564 (243, 244), *651, 652*
Sarel-Imber, M. 469 (19), *488*
Sarfarazi, F. 760 (306, 307), *772*
Sargeson, A.M. 264 (19), *311*
Sarkar, A. 296 (251), *316*
Sarkar, A.K. 280 (94), *313*
Sarkar, M. 674 (101), *704*
Sarkar, R.K. 302 (300), *317*
Sarr, M. 471 (125), *490*
Sasaki, S. 301, 302 (295), *317*
Sass, R.L. 708, 734, 738, 749 (4), *766*
Sastry, C.K. 151 (75), *214*
Sastry, K.V.L.N. 151 (75), *214*
Sathe, K.M. 296 (251), *316*
Sathe, S.S. 668 (68, 69), *703*
Sato, M. 305 (322), *317*
Sato, T. 263 (10), 279 (74a, 74b), 308 (337), *311, 312, 318*
Saunders, M. 83, 86, 88, 93 (111), 111 (207), 116 (224), *132, 136*, 417 (57), 432, 433, 435, 436 (142), 444 (191), 445 (199), 448 (213), *462, 464, 466*, 815 (8a, 8b), 816 (10), 819 (22, 23, 26), 820 (26), 821 (38), 822 (44), 825 (38), 843 (95), *856, 857, 859*
Sawada, M. 265 (23), *311*
Sawada, S. 616 (361), *655*
Saward, C.J. 714, 715 (38), *766*
Sawaryn, A. 377, 393, 404 (84), *408*
Sax, A.F. 65, 74, 77, 82, 96 (76b), *131*
Saxton, R.G. 419 (74), *463*
Sayed, Y.A. 715 (42), *766*
Scaiano, J.C. 795 (97–99), 796 (97, 99), *811*
Scarrow, R.C. 197 (283), *219*
Schaad, L.J. 457 (269), *468*, 729 (144), *768*
Schachtschneider, J.M. 77, 78 (91a), *132*
Schaefer, H.F. 84, 85 (129, 130, 132), 96 (130, 149), 97 (149), 98 (129, 130, 149), 101 (129, 130, 132), 102 (129, 130), 126 (129, 132), *133*, 470 (42), 481 (263, 264, 268, 270), 482 (268, 270), 484 (270), 487 (270, 289, 290, 292, 293), *489, 493, 494*, 729, 757 (138), *768*
Schaefer, H.F.III 94, 96 (148), 98 (148, 159), 102 (159), 126 (270), *133, 137*, 606 (339), *654*
Schaefer, H.III 83, 96, 98 (94), *132*
Schäfer, B. 642 (421, 422), *656*
Schäfer, H. 588 (302), *654*
Schäfer, H.N. 197 (285a–c), *219*
Schäfer, L. 145 (37), 209 (333), *213, 221*
Schafer, O. 778, 785, 786 (23), 799, 801 (117), *810, 812*
Schafer, P.R. 511 (88), *649*
Schäfer, R. 716 (48), *766*
Schäfer, W. 353 (34), 384 (90), *407, 409*
Schafer, W. 450 (224), *466*
Schaller, J.-P. 722 (91, 106), 723 (106), 727 (122, 124, 125), *767, 768*
Schallner, O. 831 (65), 832 (65, 66), *858*
Schamp, N. 179 (178), *217*, 270 (42a), *311*
Schank, K. 294 (235), *316*
Schapin, I.Yu. 146 (45), *214*
Schappert, R. 669 (76), *703*
Scharf, D.J. 668 (68), *703*
Scharfenberg, P. 159 (100), *215*
Schauer, A. 727, 760 (131), *768*
Schaumann, E. 186 (236), *218*, 294 (236), 301 (279), 303 (303a, 303b, 304), *316, 317*
Scheffer, J.R. 322 (17), *337*
Schei, S.H. 155 (82), *214*
Schelle, S. 607 (346), *654*
Scheller, K. 471 (122), *490*
Scheller, M.E. 280 (82), *312*
Schenk, H. 161, 164 (111), 175 (149), 181, 182 (192), 210 (336), *215–217, 221*
Scherbine, J.P. 301 (280), *316*
Scherr, P.A. 505 (59), *649*
Scheuermann, H.-J. 280 (81), *312*
Schick, A. 193, 194 (273), *219*
Schieb, T. 716, 745, 749 (52), *767*
Schier, A. 156, 157 (90), 185 (210, 214), *215, 217*
Schiesser, C.H. 794 (76, 84–86), *811*
Schindler, M. 110 (187b, 188a–h, 197, 202g), 111 (202g), *135*, 393, 396 (111, 116), 397 (116), *409, 410*, 430 (125), 446 (204), 449 (216), *464, 466*, 783, 797 (50), *810*, 819 (27), *857*
Schipoff, M. 722, 741 (92, 93), *767*

Schipperijn, A.J. 569 (256), 570 (257, 258), 578, 579 (280), *653*
Schlag, E.W. 470, 471 (76, 77), 479 (76), *489*
Schlegel, H.B. 104 (167c, 168a), 109 (167c), *134*, 794 (83), *811*
Schlesinger, A.H. 412 (6), *460*
Schleyer, P.v.R. 47 (21a, 21b, 22), 48, 49 (33), 60, 61 (62), 65 (76a, 76b), 74 (76a, 76b, 86), 77 (76a, 76b, 92e), 82 (76a, 76b), 83 (33, 111), 84 (62), 85 (33, 62), 86 (33, 62, 111), 87 (33), 88 (33, 111), 89, 90, 92 (33), 93 (33, 111), 94, 95 (33), 96 (33, 76a, 76b, 151, 152), 98 (151, 152), 102 (160), 110 (201a–c, 202a–g), 111 (201a–c, 202a–g, 207, 211), 113 (216), 114 (216, 221, 223), 115, 116 (223), 117–119 (238a–c), 120 (76a), 123 (259), 126, 128 (223), *130–133, 135–137*, 156, 159, 178, 179, 181 (88), *215*, 249 (122), *260*, 342 (11), 345 (26), 353 (33), 364 (56), 375 (56, 79), 377 (56, 84), 378, 381 (56), 384 (92), 387, 388, 391, 392 (56), 393 (84, 116), 394 (56, 79, 118), 396 (56, 116, 118), 397 (56, 116), 403 (56), 404 (84), *406–410*, 413 (24), 417 (52, 57, 58), 418 (62, 63), 425 (110, 111), 427 (111), 429 (111, 131), 430 (111, 130, 131), 431 (131), 441 (178), 444 (189), 446 (203, 204), 447 (210), 449 (216, 221), 456 (266), 457 (270), 458 (24, 273, 281), 459 (281), *461–466, 468*, 780, 783, 785, 793 (43), 794 (43, 83), *810, 811*, 815 (7, 8b), 817 (15), 819 (26), 820 (26, 37), 822 (45), 825 (45, 52), 832 (67), 833 (72), 835 (79), 840 (88, 89), 844 (99), 846 (105), 848 (109b), *856–859*
Schlosberg, R.H. 434 (147), *464*
Schlosser, M. 287 (145–147), *314*
Schlötz, K. 446 (204), *466*
Schlüter, A.-D. 124 (262), *137*, 262 (4, 6), *311*, 505 (55), *649*, 779 (36), 781, 783 (46), 787 (46, 57, 60), 789 (57, 60), 790 (46, 57, 60), 795 (60, 92, 93), 797 (103), 799 (36, 46, 92, 93, 103), 800 (36, 46, 60, 103, 106), 805 (60, 127, 128), 806 (60, 128), 807 (46), *810–812*
Schmelzer, A. 48–50, 56, 57 (31a, 31b), *130*
Schmickler, H. 716 (49, 52), 726 (49), 740 (212), 745, 749 (52), *766, 767, 770*
Schmid, W. 262 (5), *311*, 776, 781–784, 787–789 (45), *810*
Schmidbaur, H. 156, 157 (90), 185 (210–212, 214), *215, 217*
Schmidt, H.R. 535, 536, 548 (159), *651*
Schmidt, M.U. 517 (112), *650*
Schmidt, T. 361 (46), *407*, 477 (179), *491*
Schmiedel, R. 646 (432), *656*
Schmitz, C.R. 825 (53), *858*
Schmitz, H. 713 (32), 756, 757 (290), *766, 771*
Schmitz, L.R. 824, 829 (48, 49), 830 (49), *857*
Schnatter, W.F.K. 528, 530 (149), *650*
Schneider, F.W. 471 (100), *490*
Schneider, K.-A. 429 (121), *464*
Schneider, M. 836 (82), *858*
Schneider, S. 309 (349), *318*, 682 (136), *705*
Schnering, H.G.v. 196 (279), *219*, 363, 364 (47), *407*, 455 (261), 456 (263), *467, 468*
Schnering, H.G.von 147–149 (57), 176 (158), 200 (299), 206 (323), 207 (324, 325), 210 (338), *214, 216, 220, 221*, 867 (32), *884*
Schoch, T.K. 535 (157), 536 (160), 537 (157), 538, 539, 541, 543 (160), *651*
Schoeller, W. 197 (280), *219*
Schoeller, W.W. 481 (229, 230), *492*
Schoetz, K. 110, 111 (202g), *135*, 393, 396, 397 (116), *410*
Scholl, B. 473 (159), *491*
Schöllkopf, K. 434 (150), *465*
Schömenauer, S. 530, 552, 553, 622 (152), *650*
Schönwalder, K.H. 434 (150), *465*
Schophoff, F. 716, 726 (49), *766*
Schötz, K. 449 (216), *466*
Schrader, L. 711 (20), 713 (29, 30), 714 (34), *766*
Schrauzer, G.N. 420 (84), *463*
Schreiber, K. 412 (7), *460*
Schriesheim, A. 831 (64), *858*
Schrock, R.R. 534 (155), 583 (292), 593 (316–319), 594 (316–318, 320), 601 (316), 602 (316, 320), 603, 604 (316), *651, 653, 654*
Schröder, B. 329 (47a), *338*
Schröder, G. 419 (83), 440, 441 (176), 444 (185), 455 (254), *463, 465, 467*
Schroth, G. 519 (116), 626, 635 (383), *650, 655*, 753 (279), *771*
Schrumpf, G. 83, 86 (114, 117), 88 (117), 89, 93 (114), *132, 133*, 147 (54, 60), 149 (60), 153–155 (78), 161 (110, 112), 162, 163 (110), 164, 165 (112), 166 (123), 172 (138), 173 (140), 174 (141, 142), 175 (150), 176 (153), 177 (169, 170, 172), 178 (169), 179 (172), 186 (54), *214–216*
Schubert, U. 179 (181), 185 (214), *217*, 534 (155), 575 (279), 607 (344, 345), 617 (345), *651, 653, 654*
Schuchard, U. 640 (416, 417), *656*
Schuchardt, U. 507, 624, 627, 640, 641 (73), *649*
Schugar, H.-J. 817, 854, 856 (18), *857*

Schuler, R.H. 122 (251b), *137*
Schulman, J.M. 123 (260), 124 (261), *137*, 235 (62), *255*, 446 (203), 452 (241), 457 (268), *466–468*, 773 (1), 777 (16), 778 (1), 780 (16), *809, 810*
Schulte, G.K. 509, 601 (82), *649*
Schultz, A.J. 188 (251), *218*
Schultz, G. 159 (99), *215*
Schultz, P.A. 778, 779 (17), *810*
Schulz, G. 659, 691 (20), *702*
Schulz, J.C. 114 (220), *136*
Schulz, T.-J. 285 (132), *313*
Schumann, K. 629 (390), 630 (392), *655*
Schumm, R.H. 224 (9), 237 (9, 69), *249, 255*
Schurter, R. 739, 744 (205), *770*
Schuster, D.I. 320 (2), 327 (43), *337, 338*, 435 (154), *465*
Schwaager, H. 747 (265, 266), 748, 753 (267), *771*
Schwab, J.M. 476 (176), *491*
Schwager, H. 202 (307), *220*, 579 (284, 285), 581 (284), 586, 644 (285), *653*, 721, 722 (85), 734, 735, 739 (195), 753 (277–279), 755 (195, 281), *767, 769, 771*
Schwartz, H. 740 (218), *770*
Schwarz, H. 444 (188, 189), *465, 466*
Schwarz, J. 742 (234), *770*
Schwarz, W.H.E. 69 (74), *131*
Schweig, A. 60 (54c), *130*, 384 (89, 90), *409*, 729 (137–139), 757 (137–139, 291), *768, 771*
Schwendeman, R.H. 85, 86 (139a), *133*, 160 (103), 177 (167), *215, 216*
Schwendman, R.H. 83, 86, 89 (120), *133*
Scolastico, C. 297 (259), *316*
Scott, D.S. 471 (117), *490*
Scott, L. 422 (96), *463*
Scott, L.T. 366, 368 (63), *408*, 420, 425 (86), 426 (112), 437 (159), 450 (224), 455 (254), 457 (269), *463, 465–468*, 501 (37), *648*
Scott, M.J. 698 (196), *706*
Scott, P. 592, 602 (315), *654*
Scott, W.R. 449 (215), *466*
Scuter, F.J. 665 (57), *703*
Seakins, P.W. 477 (197), *492*
Sears, A.B. 469 (4), *488*
Sebastiano, R. 274 (59), *312*
Sedrati, M. 306 (323), *317*
Seebach, D. 183 (206), *217*, 504 (48), 505 (56), *648, 649*, 759 (300), *772*
Seefeld, M.A. 273 (53), *312*
Seeger, R. 47 (21a), 85 (136b), 102 (160), *130, 133*, 392 (103), *409*
Seel, H. 740 (213), *770*
Seeman, J.I. 323 (27c), *337*
Seetula, J.A. 477 (191), *492*
Segal, G.A. 60, 112, 113 (52), *130*, 476 (171), *491*
Segal, Y. 683 (142), *705*
Segnitz, A. 514 (102), 519 (102, 123), *649, 650*
Seidenspinner, H.-M. 299 (270), *316*
Seidl, E.T. 470 (42), *489*
Seidl, H. 714 (34), *766*
Seidl, P. 440, 441 (172), *465*
Seidler, M.D. 197 (283), *219*
Seiler, P. 124 (262), *137*, 204 (318, 319), 205 (319), *220*, 776 (20, 26), 778 (20), 779 (20, 26), 794 (26), *810*
Sein, U.I. 477 (178), *491*
Seits, S.P. 514 (104), *649*
Seitz, G. 198 (287), *219*, 646 (432), *656*
Seitz, W.J. 290 (178), *314*, 663, 694 (44), *703*
Seki, M. 128 (278), *137*
Sekino, H. 392 (105), *409*
Selden, C.B. 473, 477 (149), *491*
Semenov, D.A. 815 (4), *856*
Semenow, D.A. 417 (55), *462*
Seminario, J.M. 96 (153), *133*
Semmingsen, D. 197 (286), *219*
Semmler, K. 262 (3–5), *311*, 774 (4), 776 (45), 781 (4, 45), 782–784 (45), 787 (4, 45, 57), 788 (45), 789 (4, 45, 57), 790 (57), 795, 797, 800 (4), *809, 810*
Semones, M.A. 291 (199, 200d, 200e), *315*, 684 (145, 146), 685 (146), 687 (145), *705*
Sengupta, S.K. 176 (160), 177, 178 (160, 161), *216*
Sergeeva, T.A. 296 (253), *316*
Sergent-Guay, M. 670 (82), *704*
Seroamz, A. 851, 852 (117), *859*
Servis, K.L. 817 (14), 822 (43), *857*
Setser, D.W. 471 (100, 141), 472, 480 (141), *490, 491*
Sévricourt, M.C.de 150 (66), *214*
Sevrin, M. 301 (284, 288), 302 (288), *317*
Seyfarth-Jacob, A. 186 (219), *218*
Seyferth, D. 48, 49, 71 (42), *130*, 499 (4, 10), 500 (4, 10, 27), 505 (10), 506 (63), *648, 649*
Seyse, R.J. 470 (53), *489*
Shabarov, Y.S. 620 (366), *655*
Shabarov, Yu.S. 149 (63), *214*
Shafiq, M. 320 (1), *337*
Sharpless, K.B. 302 (301), *317*
Shary-Tehrany, S. 77 (92b), *132*, 229 (33), *252*
Shatavsky, M. 445 (194), *466*
Shavitt, I. 392 (104), *409*
Shaw, G.S. 344, 345 (21), *406*, 413, 415, 416, 419, 445, 450 (22), *461*
Shchadneva, N.A. 564 (245), *652*
Shea, J.A. 146 (47), *214*

Shechter, H. 710 (16), 711 (22), *766*
Shefter, E. 364 (48), *407*, 451 (233), *467*
Sheldrick, G.M. 173 (140), *216*, 556 (223), *652*
Shen, J.K. 595, 600 (323), *654*
Shen, S. 817, 854, 856 (18), *857*
Shen, Y. 296 (252a, 252b), 301, 302 (290), *316, 317*
Shepherd, T.M. 681 (127), *704*
Sheridan, J. 145 (41), 179 (179), *213, 217*
Sherrill, C.D. 470 (42), *489*
Sherwin, M.A. 323 (26), *337*
Sherwin, P.F. 575 (276), *653*
Sheth, J.P. 671, 672 (89, 90), 673 (90), *704*
Shevlin, P.B. 715 (46), *766*
Shiba, T. 305 (319), *317*
Shibata, M. 646 (433), *656*
Shida, T. 116, 117 (234b), *136*
Shieh, C. 229 (33), *252*
Shieh, J.J. 708, 734, 738, 749 (4), *766*
Shih, Y. 619 (363), *655*
Shiki, Y. 186 (229), *218*
Shimada, K. 301, 302 (295), *317*
Shimamoto, K. 284 (130), *313*, 667 (66), *703*
Shimazaki, A. 297 (255), *316*
Shimizu, I. 678 (115, 116), *704*
Shin, J.-H. 449 (222), *466*
Shiner, C.S. 279 (77), *312*
Shipman, M. 643 (424, 425), *656*
Shirafuji, T. 750 (270), *771*
Shishido, Y. 745 (246), *771*
Shklover, V. 573 (268, 269), 574 (268), 575 (268, 269), 576 (269), 577, 593 (268, 269), 597 (269), *653*
Shleider, I.A. 438 (161), *465*
Shobe, D. 417, 418 (56), *462*, 822 (46), *857*
Shono, T. 308 (338), *318*, 607 (343), *654*
Shoppee, C.W. 412 (1), *460*
Short, R.P. 673 (96–98), *704*
Shortridge, R.W. 864 (11), *883*
Showalter, H.D.H. 687 (155), *705*
Shue, F.-F. 822 (43), *857*
Shulishov, E.V. 282 (103), *313*, 659, 688 (26), *702*
Shvedova, L.B. 557, 561, 562 (231), *652*
Shyoukh, A. 740 (211), *770*
Sibille, S. 280 (90), *312*
Sieber, S. 60, 61, 84–86 (62), 110, 111 (201a), *131, 135*, 249 (122), *260*, 364 (56), 375 (56, 79), 377, 378, 381, 387, 388, 391, 392 (56), 394 (56, 79, 118), 396 (56, 118), 397, 403 (56), *408, 410*, 425 (110, 111), 427, 429, 430 (111), *463*, 780, 783, 785, 793, 794 (43), *810*, 848 (109b), *859*
Sieber, W. 562 (241), *652*
Sieber, W.J. 535 (159), 536 (159, 162), 548 (159), *651*
Siegbahn, P.E.M. 128 (280, 281, 283), *137*
Siegel, J.S. 353 (31), *407*, 733 (187), *769*
Siegmann, K. 292 (206), *315*, 698 (190), *706*
Siehl, H.-U. 444 (187), 447 (211), *465, 466*, 819 (22), 821 (42), 832 (67, 68, 70), *857, 858*
Siemion, I.Z. 470 (66), *489*
Siepert, G. 357, 359, 361, 378, 389, 390 (42), *407*, 450–453, 457 (226), *467*
Sierra, M.A. 525, 526 (143), *650*
Silbermann, J. 513 (94, 95), 515 (94), *649*
Silberrad, O. 657 (1), *702*
Silveira, C. 663 (48), *703*
Silver, M.S. 417 (55), *462*, 511 (88), *649*, 815 (4), *856*
Silvestre, J. 589, 590, 592, 601 (306), 602 (306, 335), 603 (306), *654*
Simandiras, E.D. 84, 85, 96, 98, 101, 102 (130), 104 (168b), *133, 134*
Simkin, B.Y. 342, 344, 382, 390 (10), *406*
Simkin, B.Ya. 44, 48 (20), *129*
Simmons, H.E. 865 (22), *883*
Simon, A. 188 (242), 189 (255), *218*, 501 (37), *648*
Simon, J.G.G. 729, 757 (137–139), *768*
Simon, R.J. 477, 479 (180), *491*
Simonetta, M. 412 (9), 421 (94), *460, 463*, 481 (227), *492*
Simons, J. 104, 109 (167a), *134*
Simpson, J.M. 478 (202), *492*
Sinai-Zingde, G. 291 (197), *315*, 673 (99, 100), *704*
Sinanoglu, O. 741, 742 (222), *770*
Sinclair, G.C. 447 (209), *466*
Singh, A.K. 679 (124, 125), *704*
Singh, B.P. 849, 850 (114), *859*
Singh, N.P. 266 (26), *311*
Singleton, E. 554 (211), *652*
Singy, G.A. 740 (216), *770*
Sinha, S.C. 686 (154), *705*
Sinke, G.C. 225 (17), 241 (82), 243 (90), *250, 257, 258*
Sipos, G. 851, 852 (117), *859*
Sisti, M. 663 (41), *703*
Siu, A.K.Q. 481 (233), *492*
Sivapalan, M. 423 (100), *463*
Siveramakrishnan, H. 200 (301), *220*
Skaarup, S. 145 (38), *213*
Skancke, A. 60 (54a), 83, 86 (102a, 103, 104, 105d, 107, 108a, 113), 88 (102a), 89 (113), 93 (102a, 113), 94 (113), 95 (102a, 105d), *130, 132*, 470 (71), *489*
Skancke, P.N. 85, 86 (140), *133*, 454 (250), *467*
Skattebol, L. 874 (65), *884*
Skelton, B. 180 (189), *217*
Skinner, G.B. 471 (127), *490*

Skinner, H.A. 239 (73), *256*
Skowron, J.F.III 268 (33), *311*
Skuballa, W. 661 (28), *702*
Slater, N.B. 471 (97–99), *490*
Slaven, R.W. 624 (378), 625 (379), 628, 635 (378), *655*
Slayden, S.W. 243 (89), *258*
Slee, T.S. 74 (82), *131*, 361, 365 (44), 375 (44, 80), 376, 377 (80), 378 (44, 80), 379 (44), 393 (80), 394 (44), *407, 408*, 414, 415, 434, 435 (30), *461*, 779 (32), *810*
Sliepcevich, C.M. 471 (121, 124), *490*
Slobodin, Ya.M. 864 (10, 13), *883*
Slovokhotov, Y.L. 501 (38a), *648*
Slovokhotov, Yu.L. 160 (107), *215*
Sluis, P.van der 202 (312), *220*
Smael, P. 569 (256), 570 (257, 258), *653*
Smart, B.E. 241 (85), *257*
Smiser, M. 247 (111, 112), *259*
Smit, W.A. 639 (412), *656*, 868 (41), *884*
Smith, A. 123 (256c), *137*
Smith, A.B.III 200 (301), 201 (302), 212 (344), *220, 221*, 668 (73), *703*
Smith, C.V. 749, 750 (269), 753 (276), *771*
Smith, D.A. 663, 694 (42), *703*
Smith, D.M. 419 (83), *463*
Smith, F.T. 126 (276a), *137*, 480 (219), *492*
Smith, G.F. 513 (96), *649*
Smith, H.D. 290 (181, 182), *314*, 664 (52), *703*
Smith, L.A. 451 (229), *467*
Smith, N.K. 235 (58), *254*
Smith-Vosejpka, L.J. 487 (281), *493*
Smutny, E.J. 427 (113), *463*
Snieckus, V. 882 (91), *885*
Snow, M.R. 167 (125), *215*, 718, 719 (68), *767*
Snyder, J.P. 663 (45), *703*
Snyder, R.G. 77, 78 (91a, 91b), *132*
Sodemann, S. 503 (45), *648*
Söderberg, B.C. 295 (248), *316*, 525, 526 (143), 527 (148), *650*
Soepert, G. 236 (65, 66), 245 (100), *255, 258*
Sogo, S. 746, 751 (258), *771*
Solé, A. 121, 123 (249), *136*, 430 (125), *464*, 437 (156), *465*
Solly, R.K. 727 (127, 128), *768*
Somayajula, K. 459 (283), *468*
Sombroek, J. 718, 719 (67), *767*
Sommer, J. 815 (6), *856*
Song, L. 754 (280), 756, 764 (282), *771*
Sonnenberg, J. 413 (12), 447 (206), *461, 466*
Sootome, N. 308 (337), *318*
Sorensen, T.S. 500 (19), *648*, 820 (28), 821 (39), 824 (48, 49), 825 (53), 827 (39), 829 (48, 49), 830 (49), 834 (76), *857, 858*
Soria, M.L. 303 (310), *317*
Sorokin, V. 208 (326), *220*
Sosonkin, I.M. 865, 872, 875 (19), *883*
Søtofte, I. 179, 180 (182), 193 (271), *217, 219*
Sotokawa, H. 555 (220, 221), 632 (220), *652*
Soufi, J. 288 (155), *314*
Soum, A. 295 (249), *316*
Sousa, L.R. 324, 325 (33), *338*
Spaeth, T. 308 (334c), *318*, 872 (57), *884*
Spangler, C.W. 710 (14), *766*
Spanka, C. 301 (279), *316*
Sparapani, C. 822, 825 (45), *857*
Spear, R. 834 (75), *858*
Spear, R.J. 427, 428, 430 (115), *464*, 817 (16), 826, 827 (54), 846, 847 (107b, 107c), 849 (107b), 851, 852 (117), *857–859*
Speer, D.V. 192 (266), *219*
Spek, A.L. 202 (312), *220*
Spencer, K.M. 476 (174–176), *491*
Speranza, M. 822, 825 (45), *857*
Spero, D.M. 291 (196), *315*, 658, 663 (6), 671 (88), *702, 704*
Spielmann, W. 831, 832 (65), *858*
Spina, K.P. 301 (285), *317*
Spitzer, R. 366, 370 (62), *408*
Spitzer, W.A. 323 (27c), *337*
Spitznagel, G.W. 48, 49, 83, 85–90, 92–96 (33), 113, 114 (216), *130, 136*, 156, 159, 178, 179, 181 (88), *215*, 417 (52), *462*
Spitzner, D. 189 (255), *218*
Splettstasser, G. 560, 561 (238), *652*
Spotts, J.M. 586 (300), *654*
Springer, J.P. 366 (61), *408*, 671 (87), *704*
Spyroudis, S. 310 (353), *318*
Squicciarini, M.P. 470 (61), *489*
Squillacote, M.E. 818 (21), *857*
Squires, R.R. 113, 114 (215), *136*, 458 (279), 459 (288), *468*, 793 (73), *811*
Sridharan, V. 309 (345), *318*
Srikrishna, A. 669 (78, 80), 679 (121, 122), *703, 704*
Srinivasan, R. 208 (327), *220*
Srivastava, V.P. 305 (317), *317*
Srnak, A. 291 (195), *315*
Staab, H.A. 712, 757 (26), *766*
Stadlwieser, J. 868 (36), *884*
Staley, S.W. 83, 86 (102c, 122), 88 (102c), 89 (122), 93 (102c, 122), *132, 133*, 144 (27), 175 (148), 181, 182 (191), 183 (202), 187 (27, 241), 190 (259), 194 (27), 195 (27, 241), 196 (241), 199 (290), *213, 216–219*, 358 (43), *407*, 730 (149–151), 757 (151), 760 (309), *768, 772*
Stalick, B.K. 96 (147), *133*
Stam, C.H. 156, 157 (89), 175 (149), 181 (192, 195, 196), 182 (192, 196), *215–217*, 454 (252), *467*
Stamatis, N. 429–431 (131), *464*

Stambach, J.F. 266 (24), *311*
Stammer, C.H. 170 (128, 131), 171, 172 (128), *215, 216*, 305 (317, 318), *317*
Stamper, J.G. 730, 742 (163), *769*
Standstrøm, Y. 500 (22, 23), 510 (22), *648*
Stang, P.J. 201 (304), *220*, 715 (45), 729 (140), 730 (153–158, 161), 731 (155, 165), 732 (154–158, 161), 742 (154–158, 161, 165), 754 (280), 756 (282), 758 (153, 155, 157, 158, 296), 759 (153, 157, 158), 760 (140, 157, 296, 306–308, 310), 761, 762 (140), 763 (312, 313), 764 (282), *766, 768, 769, 771, 772*
Stanger, A. 202 (307), *220*, 733 (188), 734, 735 (195), 736 (198), 739, 755 (195), *769*
Stangl, R. 176 (158), *216*
Stankovich, M.T. 619 (363), *655*
Stanovnik, B. 713 (28), *766*
Staral, J.S. 427, 428, 430 (115), *464*
Staral, S.J. 819 (24), *857*
Starks, C.M. 866 (25), *883*
Starova, G. 192 (268), *219*
Starr, M.A.E. 743, 745, 750, 751 (240), *770*
Stawitz, J. 299 (270), *316*
Steele, W.V. 235 (58), *254*
Stein, F. 530 (153), 564 (153, 246, 247), 565 (246), *650, 652, 653*
Stein, I. 759 (299), *771*
Stein, S.E. 477 (188), *492*
Steinberg, H. 120 (241–244), *136*
Steinmetz, M.G. 322 (19), *337*
Stenstrøm, Y. 151 (72), *214*
Stephens, P.J. 105 (172b, 172c, 173), 106 (174), *134*, 476 (171), *491*
Stephenson, D.S. 280 (81), *312*
Stepuszek, D.J. 632 (396a, 396b), *656*
Sternberg, E. 620 (368), *655*
Stevens, E.D. 62 (65), *131*
Stevens, R.M. 481 (241, 242), *493*
Stevenson, , E.H. 471 (82), *489*
Stevenson, G.R. 235 (62), *255*
Stevenson, P. 278 (73), *312*
Stevenson, T.A. 140 (6), *213*, 225–227 (18), 243 (94), *250, 258*
Stewart, J.J.P. 430 (125), *464*
Stewart, J.M. 418 (63), *462*
Stezowski, J.J. 179 (180), *217*, 434 (150), *465*
Stigliani, W.M. 187 (239), 190 (257), *218*
Still, W.C. 289 (160), *314*
Stille, J.K. 519 (119), 549 (180), *650, 651*
Stirling, C.J.M. 264 (14, 15), *311*
St.John, W.M. 481 (233), *492*
Stohmann, F. 241 (82), *257*
Stohrer, W.-D. 48 (28), *130*, 363, 364 (47), *407*, 434 (150), 451, 455 (231), *465, 467*, 841 (92), *858*
Stohrer, W.-W. 773 (2), *809*
Stoicheff, B.P. 775 (9), *809*
Stolarsky, V. 282 (106), *313*
Stoll, A.T. 263 (8), *311*
Stolle, A. 639 (413, 414), *656*
Stone, A.J. 107 (176), *134*
Stone, F.G.A. 500 (17), *648*
Storer, J.W. 457 (268), *468*
Stork, G. 667 (67), *703*
Storr, R.C. 415, 451 (36), *461*
Story, P.R. 344, 345 (16), *406*, 413, 415, 416, 419, 427, 444 (17), 445 (199), 450, 459 (17), *461, 466*
Streith, J. 712, 736 (27), *766*
Streitwieser, A. 192 (266), *219*, 557 (230), *652*, 740 (215), 741 (221), *770*
Strich, A. 128 (281), *137*
Strongin, R.M. 212 (344), *221*
Strozier, R.W. 450 (223), *466*
Struchkov, Y.T. 310 (357), *318*, 501 (38a), *648*, 797, 799, 800 (101), 802 (101, 122), 805 (126), *811, 812*
Struchkov, Yu.T. 149 (63), 160 (107), 172 (139), 176 (155), 208 (326), *214–216, 220*, 868 (39), 877 (73), 878 (73, 78, 79), 879 (79), 881 (89), *884, 885*
Stryker, J.M. 272 (52), *312*
Stucky, G.D. 48, 49 (42), 64 (66d), 71 (42), *130, 131*, 505 (57), *649*
Studabaker, W.B. 262 (2c), 285 (2c, 134), 294 (237), *310, 313, 316*, 522 (133), 524, 525 (142), *650*
Stufflebeme, G. 290 (185), *314*
Stull, D.R. 225 (17), 241 (82), 243 (90), *250, 257, 258*
Stull, P.D. 291 (192), *314*
Sturm, V. 714 (33), *766*
Stutchbury, A.P. 271 (48), *312*
Styger, C. 471 (87), *490*
Su, C.-F. 190 (259), *218*
Subra, R. 121, 122 (250), *137*
Subramanian, L. 590, 603, 604 (310), *654*
Subramanian, P. 302 (302), *317*
Suda, M. 659, 690 (19), *702*
Suga, H. 242 (87), *257*
Sugimoto, T. 646 (433), *656*
Sugimura, T. 283 (119–122), 287 (139), *313*
Sugita, H. 287 (143), *314*
Sugiyama, K. 271 (49), *312*
Sukirthalingam, S. 309 (345), *318*
Sullivan, J.F. 182 (200, 201), 183 (201), *217*
Sulmon, P. 270 (42a, 42b), *311*
Sulzbach, H.M. 324, 326 (42), *338*
Summerville, R.H. 444 (185), *465*
Sun, F. 500 (19), *648*
Sun, Q. 786 (56), *810*
Sundaralingam, M. 603 (336), *654*

Sundberg, R.J. 658 (16), *702*
Sunderman, F.-B. 622 (371), *655*
Sunko, D.E. 417 (58), *462*, 820 (37), *857*
Surmina, L.S. 780 (42), 789, 790 (65), 796 (100), 797 (65, 101), 798 (65, 120), 799, 800 (65, 101), 801 (65), 802 (65, 101, 120–122), 805 (65, 100, 126), 807 (65), 808 (65, 131), *810–812*, 867 (31), *884*
Sustmann, R. 437 (156), *465*
Sutcliffe, L.H. 110 (194), *135*
Sutrisno, R. 646 (432), *656*
Suzuki, H. 195 (278), *219*, 265 (22), *311*, 505 (60), *649*
Suzuki, N. 307 (330), *318*
Suzuki, T. 290 (184), *314*
Suzuki, Y. 684 (147), *705*
Sveiczer, A. 151 (72), *214*
Svensson, P. 111 (208–210), *136*, 364, 375, 381, 391, 392, 394, 396, 397 (50, 51, 53, 57), 402 (50, 51, 53), 403 (50, 51, 53, 57), 405 (57), *407, 408*, 414 (28), 424–426 (103), 440 (169), 441 (28, 180), 442 (28, 169, 180), *461, 463, 465*, 841 (93), 843 (94), *858, 859*
Sviridov, S.V. 308 (334a, 334b), *318*, 876 (72), *885*
Svoboda, J.J. 825, 853 (50b), *857*
Svyatkin, V.A. 140 (4), *212*, 863 (4b), *883*
Swatton, D.W. 435 (153), *465*
Sweeney, A. 556 (229), *652*
Swenson, J.R. 48 (26d, 40), 49, 71 (40), 83, 86, 90 (26d), *130*
Swenton, J.S. 320 (3a, 3b), 327 (44), *337, 338*
Swenton, J.W. 323 (27a), *337*
Sydnes, L.K. 150 (67, 68), 151 (68), 167 (124), *214, 215*
Szabo, A. 125 (267), *137*
Szabó, G.T. 291 (186), *314*
Szabo, K.J. 84, 85 (133), 111 (209), 118 (240), *133, 136*, 340, 342, 346, 350, 351 (2), 354 (2, 35), 364, 375 (51–53, 55, 58), 378 (2), 381, 391, 392, 394, 396, 397 (51–53, 55, 58), 402 (35, 51–53, 55), 403 (51–53, 55, 58), *406–408*, 416, 417, 422 (44), 424–426 (103), 443 (183), 444, 448 (190), 449 (214), 455, 456 (258), *462, 463, 465–467*
Szalanski, L.B. 182 (205), *217*
Szántay, C. 304 (312), *317*
Szeimies, G. 204, 205 (319), *220*, 262 (3–5), *311*, 500, 502 (29), 505 (29, 55), *648, 649*, 742 (232), *770*, 774 (4), 776 (20, 45, 61), 778 (20, 23), 779 (20, 35, 37), 780 (35, 38), 781 (4, 37, 45, 61), 782–784 (45), 785, 786 (23), 787 (4, 35, 37, 45, 57, 61, 62), 788 (35, 45, 61, 62), 789 (4, 35, 45, 57, 63), 790 (57, 63), 791, 792 (37), 794 (35), 795 (4, 35, 95), 796 (62), 797 (4, 62, 109), 798 (35, 62, 63, 109, 115), 799 (37, 63, 115, 117), 800 (4, 37, 62, 109, 115), 801 (115, 117), 802 (35, 37, 62, 63, 109), 809 (62, 95), *809–812*, 825 (51), *857*
Szeimies, G.J. 124 (262), *137*

Taber, D.F. 292 (224), *315*, 678 (112–114), 686 (149), 696 (187), *704–706*
Table, D.F. 683 (144), *705*
Tabuchi, S. 292 (212a), *315*
Taddei, F. 734 (190), *769*
Tadesse, L. 746 (257), *771*
Tadokoro, G. 843 (96), *859*
Taft, R.W. 854 (124, 125), *859*
Tagat, J. 715 (43), *766*
Tagliavini, E. 273 (54), *312*
Taguchi, M. 181 (198), *217*
Taguchi, T. 297 (255), *316*
Tai, A. 283 (119–122), 287 (139), *313*
Tajiri, A. 555 (220, 221), 632 (220), *652*
Takabayashi, F. 145, 199, 208 (35), *213*
Takagi, K. 519 (126), *650*
Takahashi, H. 284 (125), 308 (335), *313, 318*
Takahashi, K. 751 (271), *771*
Takahashi, O. 759 (303), *772*
Takahashi, T. 296 (250), 307 (330), *316, 318*
Takai, K. 519 (125), *650*
Takano, M. 669 (77), *703*
Takase, K. 266 (25), *311*
Takaya, H. 627 (384), 629, 630 (388), 638 (407), 639 (415), 640 (419), *655, 656*
Takeda, Y. 288 (156), *314*
Takei, H. 280 (83), *312*
Takemoto, Y. 271 (49), *312*
Takenouchi, K. 280 (83), *312*
Takeyama, T. 282 (100), *313*, 471 (118), *490*
Takhistov, V.V. 235 (61), *254*
Takikawa, Y. 301, 302 (295), *317*
Takiyama, N. 282 (99, 100), *313*
Takusagawa, F. 535, 537, 539, 542, 543 (158), *651*
Tal, T. 76 (88c), *132*
Tal, Y. 64, 65, 68 (67b), 76 (88a), 107 (67b), *131*, 375, 376, 393 (81), *408*
Taleb-Bendiab, A. 146 (51), *214*
Tamagawa, K. 145 (43), 175 (147), *213, 216*
Tamaru, Y. 279 (76), *312*
Tamblyn, W.H. 663 (37), 686 (150), 687 (160), 688 (37, 162), 693 (162), *703, 705*
Tamura, M. 170–172 (128), *215*
Tamura, Y. 307 (329), *317*
Tanaka, K. 170 (136), *216*, 265 (21–23), *311*, 505 (60), *649*
Tanaka, M. 514 (106), *650*
Tanaka, S. 279 (76), *312*

Tanaka, T. 265 (23), 291 (194), *311, 314*
Tanaka, Y. 306 (325), *317*
Tanatar, S. 470 (67), *489*
Tang, R. 145 (40), *213*
Tang, T. 76 (88a), *131*
Tang, T.-H. 452 (243), *467*
Tang, Y. 296, 297 (254), *316*
Tanida, H. 448 (212), 452 (245), 453 (247), *466, 467*
Taniguchi, N. 669 (77), *703*
Taniguchi, T. 669 (77), *703*
Tanimoto, S. 716 (48), *766*
Tanko, J. 469 (26), *488*
Tanko, J.M. 157 (92), *215*, 417 (48), *462*
Tao, F. 278 (72), 288 (154), *312, 314*
Tapiolas, D. 690, 693 (167), *705*
Tarakanova, A.V. 234 (51), 235 (53, 55, 58), 241 (82), *253, 254, 257*, 469 (20), *488*
Tarant, P. 867, 868 (34), *884*
Tarasov, V.A. 639 (412), *656*, 868 (41), *884*
Tardy, D.C. 471 (106, 107, 130, 132), *490*
Tarrant, P. 875, 876 (67), *884*
Tashiro, M. 714 (35, 36), *766*
Tatsuno, Y. 663 (39), *703*
Tavanaiepour, I. 209 (335), *221*
Tayal, S.R. 719 (75), *767*
Taylor, D.K. 791 (68, 69), 794 (76, 79, 80), 795 (68, 69), *811*
Taylor, E.A. 727, 760 (131), *768*
Taylor, G.A. 554 (199), *652*
Taylor, M.D. 200 (301), *220*
Taylor, N.J. 402 (124), *410*, 454 (253), *467*
Taylor, R. 83, 86, 87 (99d), *132*, 159, 180, 185 (95), *215*, 416, 456 (47), *462*
Taylor, S.H. 514, 519 (102), *649*
Taylor, W.H. 83, 86, 89, 93 (122), *133*, 144 (27), 183 (202), 187 (27, 241), 194 (27), 195 (27, 241), 196 (241), 199 (290), *213, 217–219*, 730 (150, 151), 757 (151), *768*
Teles, J.H. 729 (144), *768*
Telkowiski, L. 815 (8a), *856*
Temme, B. 645 (430), *656*
Teretake, S. 452 (245), *467*
Terhune, R.W. 471 (85), *490*
Terpstra, J.W. 292 (221), *315*, 695 (184), *706*
Tetzlaff, C. 199 (293, 295, 297), *219*, 271 (47a), *312*
Texier-Boullet, F. 301 (282), *316*
Teyssié, P. 290 (172), *314*, 659 (18), 663 (36), 683 (140), 688, 689 (161), *702, 703, 705*
Thaddeus, P. 195 (277), *219*
Theys, R.D. 285, 286 (135), *313*
Thibeault, J.C. 578, 580 (283), *653*
Thiel, W. 83, 86, 89, 94, 105 (112), *132*, 430 (125), *464*
Thiele, G.F. 478 (206), *492*
Thiem, K.W. 561 (240), *652*
Thinapong, P. 180 (189), *217*
Thio, J. 440, 441 (176), *465*
Thomas, A.P. 280 (94), *313*
Thomas, B.E.IV 189 (252), *218*
Thomas, E.W. 619 (362), *655*
Thomas, J.L. 581 (288), *653*
Thomsen, M.W. 302 (299), *317*
Thorn, D.L. 578, 580 (283), *653*
Thulstrup, E.W. 802 (124), *812*
Thummel, R.P. 739 (207), *770*
Thynne, J.C.C. 123 (256a, 256b), *137*
Tidwell, T.T. 44, 48 (19), *129*, 234 (50), *253*, 416, 417 (42), *462*, 814, 856 (3), *856*
Tiedje, M.H. 673 (98), *704*
Tiekink, E.R.T. 167 (125), *215*
Tietz, J.V. 175 (148), *216*
Tilborg, W.J.M.van 120 (241, 245), *136*
Timberlake, J.W. 477 (189), *492*, 836 (81, 82), *858*
Timewell, P.H. 828, 853 (59), *858*
Timofeeva, A.Y. 282 (104), *313*
Timofeeva, L.P. 229 (34), 230 (40), 234 (51), *252, 253*
Timovfeeva, L.P. 235 (53, 55, 58), 241 (82), *253, 254, 257*
Tinant, B. 176 (156), 179 (183–188), 180 (183, 185–188), 200 (300), *216, 217, 220*
Ting, P.-C. 572 (263), *653*
Tjaden, E.B. 272 (52), *312*
Tobe, Y. 774 (7), *809*
Tochtermann, W. 210 (338), *221*
Toda, S. 746, 751 (259, 260), *771*
Todd, L.J. 598, 602 (333), *654*
Toder, B.H. 668 (73), *703*
Togashi, S. 748 (268), *771*
Töke, L. 291 (186), 302 (298), *314, 317*
Toki, T. 711 (24), *766*
Tokitoh, N. 717, 718, 745, 748–750 (58), *767*
Tokue, I. 155 (81), *214*
Tolbert, L.M. 287 (140), *313*, 457 (271), *468*
Tollefson, M.B. 324, 326 (42), *338*
Tolstikov, G.A. 564 (245), 624, 640 (376), *652, 655*
Tomilov, Y.V. 282 (103), *313*, 659 (22, 23, 26), 688 (22, 26), *702*
Tomioka, H. 294 (234), *316*, 728, 729, 757 (133), 765 (316, 317), *768, 772*
Toneman, L.H. 471 (84), *489*
Tonemann, L.H. 863 (6), *883*
Toniolo, C. 170–172 (128–130), *215, 216*
Toops, D.S. 797, 800 (112), *812*
Topolski, M. 271 (44), *311*
Torii, S. 670 (83), *704*
Torisawa, Y. 297 (255), *316*
Toriyama, K. 116, 117 (234a), *136*
Torok, D.S. 283 (117), *313*

Toscano, V.G. 455 (255), *467*
Tosi, C. 863 (4e), *883*
Tosi, G. 668 (68), *703*
Toth, G. 291 (186), *314*
Toth, K. 284 (128), *313*
Toyata, A. 734 (197), *769*
Trace, R.L. 500, 510 (24), *648*
Trachtman, M. 74 (84), *131*, 226, 237 (19b), *250*, 387 (95, 96), 389 (96), *409*
Traeger, J.C. 116 (228), *136*, 243 (92), *258*
Trætteberg, M. 147 (56, 58, 59), 148 (56), 149 (58, 59), 151 (72), 153, 154 (77, 78), 155 (78), 188 (242), 191 (260, 262), 210, 211 (340), *214, 218, 219, 221*, 451 (228), 452 (244), *467*
Tran, A. 684, 685 (146), *705*
Tran, D. 111 (206), *136*
Tratch, S.S. 863, 864, 866, 875, 876 (9), *883*
Trautz, M. 470 (73), *489*
Traylor, T.G. 418 (63, 68), *462*, 513 (93), *649*
Trendafilova, A. 300 (274), *316*
Tribble, M.T. 77 (92c), *132*
Trimitsis, G. 459 (283), *468*
Trimitsis, G.B. 458 (277), *468*
Trimitsis, M. 459 (283), *468*
Trinks, U.P. 303 (308), *317*
Troe, J. 477 (193), *492*
Trombini, C. 273 (54), *312*
Trost, B.M. 309 (348, 349), *318*, 534 (154), 570 (260), *650, 653*, 678 (111), 682 (136, 137), *704, 705*
Trotman-Dickenson, A.F. 123 (256c), *137*, 471 (115), *490*
Troup, J.M. 753 (276), *771*
Trucks, G.W. 85 (137), *133*, 392 (109), *409*
Trudell, M.L. 688, 693 (162), *705*
Tsanaktsidis, J. 791, 795 (68), *811*
Tsang, J.W. 305 (316), *317*
Tsang, W. 77, 120 (89c), *132*, 477 (186), *491*
Tsay, Y.-H. 429 (120), *464*, 508 (74), *649*
Tschamber, T. 712, 736 (27), *766*
Tse, C.-W. 469 (26), *488*
Tse, Y. 590, 603 (308), *654*
Tseng, C.-Y. 321 (13), *337*
Tseng, J. 111 (206), *136*
Tsotinis, A. 722 (108, 109), *768*
Tsubuki, M. 684 (147), *705*
Tsuda, K. 288 (156), *314*
Tsuge, O. 290 (184), *314*
Tsuji, J. 296 (250), *316*, 710 (17), *766*
Tsuji, T. 448 (212), 452 (245), 453 (247), *466, 467*, 843 (96), *859*
Tsujitani, R. 663 (40), *703*
Tsushima, K. 297 (255), *316*
Tucker, D.S. 590 (309, 310), 591, 592 (314), 595 (309, 323), 599 (314), 600 (309, 323), 601 (329), 602 (309), 603 (309, 310, 314, 329), 604 (310, 314), 605 (309, 314), *654*
Tuggle, R.M. 597 (327, 328), 598 (330, 332), 602 (332), *654*
Tukada, H. 285 (133), *313*
Tumer, S.U. 294, 295 (242), *316*, 523 (137, 138, 140), 525 (138, 140, 145, 146), 528 (140, 149), 529 (140), 530 (137, 138, 149), 531 (138, 140), 533 (138), *650*
Tuncay, A. 459 (284), *468*
Turner, A.B. 418 (64), *462*
Turner, D.W. 60 (53), *130*
Turner, J.D. 854 (123), *859*
Turner, J.V. 674 (102, 103), *704*
Turner, R.B. 235 (60), *254*
Turongsomboon, G. 300 (276), *316*
Turrión, C. 239 (72), *256*
Typke, V. 83, 86, 89, 93, 95 (119), *133*, 175 (152), 186 (216–218, 233), *216, 218*
Tyurekhodzhaeva, M.A. 789, 790, 797 (65), 798 (65, 118–120), 799, 800 (65), 801 (65, 118), 802 (65, 119, 120), 805, 807, 808 (65), *811, 812*

Uda, H. 470 (49), *489*
Ueda, T. 145, 199, 208 (35), *213*
Ueda, Y. 271 (49), *312*
Uggerud, E. 740 (218), *770*
Ugolick, R.C. 748 (268), *771*
Ugrak, B.I. 310 (357), *318*, 863, 864, 866, 875 (9), 876 (9, 70), 877 (70, 73, 74), 878 (70, 73, 75, 76), *883–885*
Ukaji, Y. 280 (93), 284 (124), *313*
Ullenius, C. 478 (200, 201), *492*
Ullman, .F. 710 (18), *766*
Umani-Ronchi, A. 273 (54), *312*
Umbricht, G. 292 (208), *315*, 698 (192, 193), *706*
Untiedt, S. 571 (261), *653*
Urabe, H. 682 (137), *705*
Urch, C.J. 263 (9a), *311*
Usieli, V. 564 (244), *652*
Uskokovic, M.R. 284 (128), *313*
Ustynyuk, Y.A. 620 (365), *655*

Vågberg, J.O. 267 (30), 268 (34), *311*
Vaid, R.K. 308 (331), *318*, 681 (129), *705*
Vajda, E. 110, 111 (202a), *135*, 393, 396, 397 (116), *410*
Valeri, T. 585, 587 (295a), *653*
Valle, G. 170–172 (128, 129), *215, 216*
Vallgarda, J. 661 (30), *702*
Van-Catledge, F.A. 417 (49), *462*
Vancik, H. 417 (58), *462*, 820 (37), *857*
Van Der Draai, R.K. 863 (6), *883*
Van Derveer, D.G. 188 (250, 251), *218*

Van de Sande, C.C. 847 (108), *859*
Van Dine, G.W. 418 (62), *462*
Van Duyne, G. 454 (251), *467*, 556 (228), *652*
Van Meerssche, M. 176 (156), 179 (183–188), 180 (183, 185–188), *216, 217*
Van Vechten, D. 44, 74, 75, 83 (5c), *129*, 243 (89), 246 (107), 247 (116), *258, 259*
Van Vuuren, P.T. 554 (202), *652*
Varadarajan, A. 110 (200), *135*, 394, 402 (121), *410*, 418 (69), 420 (69, 90, 91), 421 (69, 91, 93), 422 (91, 93), 423 (69, 93), 424 (69, 102), 425 (69, 104), 437 (158), *462, 463, 465*, 554 (204, 206), 555 (218, 219), *652*, 817, 855 (17), *857*
Vargas, R.M. 285, 286 (135), *313*
Varushchenko, R.M. 229 (33), *252*
Varvoglis, A. 310 (353, 354), *318*, 681 (128), *704*
Vasella, A. 294 (228, 230), *315*
Vasilevski, D.A. 876 (72), *885*
Vasilevskii, D.A. 308 (334a, 334b), *318*
Vasin, V.A. 796, 805 (100), *811*
Vaultier, M. 282, 290 (101, 102), *313*
Vecht, J.R.van der 120 (244), *136*
Vedejs, E. 513 (100), 514 (101), 519 (101, 122, 124), *649, 650*
Veldhuizen, B.van 454 (252), *467*
Ven, L.J.M. 428 (117), *464*
Venkatesan, A.M. 292 (223), *315*
Venkatesan, K. 189 (253), *218*
Verhoeven, T.R. 570 (260), *653*
Verkade, J.G. 211 (343), *221*
Vermeer, H. 384 (89), *409*
Verpeaux, J.-N. 301, 302 (289), *317*, 682 (131, 132), *705*
Victor, R. 563 (243), 564 (243, 244), *652*
Viehe, H.G. 179, 180 (183, 186), *217*
Vigneron, J.-P. 306 (323), *317*
Vijaya, S. 60, 112, 113 (52), *130*
Vilkov, L.V. 145 (33), 199 (291), *213, 219*
Villa, R. 297 (259), *316*
Villarica, K.A. 487 (280), *493*
Villarreal, J.R. 186 (231), *218*
Villegas, M.D.D.de 305 (320), *317*
Vilsmaier, E. 199 (293–297), 201 (303), *219, 220*, 271 (47a, 47b), *312*
Viossat, B. 150 (66), *214*
Visser, J.P. 578, 579 (280), *653*
Vlaar, C.P. 499 (11), *648*
Vladuchick, S.A. 865 (22), *883*
Vladuchick, W.C. 678 (111), *704*
Voelter, W. 303 (309), *317*
Vogel, E. 357, 359, 361, 378, 389, 390 (42), *407*, 450 (226), 451 (226, 227), 452, 453, 457 (226), *467*, 469 (1), 471 (90), *488, 490*, 552 (193), *651*, 708, 715 (2), 716 (48–50, 52), 718 (65, 67), 719 (67), 721, 722 (84), 726 (49), 730 (50), 745 (52), 746 (254), 749 (2, 52), 751, 756 (2), 757 (50), *766, 767, 771*
Vogel, M. 427 (113), *463*
Vogel, P. 116 (224), *136*, 432, 433, 435, 436 (142), *464*, 814 (3), 843 (95), 856 (3), *856, 859*
Vogelsanger, B. 209 (334), *221*, 471 (87), *490*
Vogt, C. 210 (338), *221*
Vogt, J. 141 (17), *213*, 231 (44), *253*
Vogt, S. 866 (24), *883*
Volatron, F. 440 (170), *465*
Volger, H.C. 439 (165), *465*
Volkenburgh, R.van 471 (83), *489*
Vollhardt, K.P.C. 202 (307), *220*, 353 (30), *407*, 714 (38), 715 (38, 41, 44), 734, 735 (195, 196), 736 (198), 739, 755 (195), *766, 769*
Volltrauer, H.N. 160 (103), *215*
Volz, W.E. 456 (264), *468*
Vonderwahl, R. 413 (11), *461*
Vonmatt, P. 698 (193), *706*
Vorbruggen, H. 661 (28, 29), *702*
Vorbruggen, J.K.H. 659 (21), *702*
Vos, A. 144 (23a–c), 145 (23b), 147 (23b, 23c), 149 (23a–c), 153 (23c), 154 (23a–c), 155 (84), *213, 214*, 471 (88), *490*
Vos, M.J.de 301 (284), *317*
Voyevodskaya, T.I. 160 (107), *215*, 501 (38a, 38b), *648*
Vries, L.de 431 (135), 447 (206), *464, 466*
Vrtilek, J.M. 195 (277), *219*

Waage, E.V. 126 (268c), *137*, 471 (137), *491*
Wada, K. 266 (25), *311*, 751 (271), *771*
Waddell, S.T. 110 (197), *135*, 204 (315), *220*, 237 (68), *255*, 776, 777, 780, 781 (15), 783 (15, 50), 784 (15, 52, 53), 785 (53), 786 (15), 789 (64), 790 (15, 64), 791 (15, 71), 792 (71), 794 (15, 64), 795 (64, 71), 797 (50, 64, 110), 798–800, 802–804 (64, 110), 805 (64), 806 (15, 64, 130), 807 (64), *810–812*
Wade, K. 444 (188), *465*
Wade, P.A. 183 (209), *217*, 277 (68), *312*
Wade, W.S. 683 (141), *705*
Wagemann, R. 199 (295), *219*
Wagman, D.D. 224 (9), 237 (9, 69), *249, 255*
Wagner, B. 554 (216), *652*
Wagner, F.E. 185 (211), *217*
Wagner, H.-U. 789, 790, 798, 799, 802 (63), *811*
Wagner, P. 111, 112 (212a), *136*, 189 (255), *218*
Wahab, A.-M.A.A. 294 (235), *316*
Wakamiya, T. 305 (319), *317*

Walborsky, H.M. 224 (2), *249*, 271 (44), *311*, 412 (7), *460*, 469 (16), *488*, 499 (8, 9), 501 (33), *648*
Waldraff, C.A.A. 294 (228), *315*
Walker, F.H. 110 (197), *135*, 204 (315), *220*, 237 (68), *255*, 774 (3), 776 (15), 777, 780, 781 (3, 15), 783 (3, 15, 50), 784 (15, 51), 786 (15), 790, 791 (3, 15), 794 (15), 797 (50), 806 (3, 15), *809, 810*
Wallasch, M. 377, 393, 404 (84), *408*
Waller, F.J. 290 (175), *314*
Wallerstein, M. 722 (90), *767*
Wallis, T.G. 366 (61), *408*
Wallwork, A.L. 146 (52), *214*
Walsh, A.D. 48, 49 (25, 35), 71 (35), *130*
Walsh, C.T. 170 (135), *216*
Walsh, E.J. 675 (106), *704*
Walsh, R. 123 (257), *137*, 469 (3), *488*, 571 (261), *653*
Walson, K.N. 402 (124), *410*
Walton, D.M.R. 730, 742 (162), *769*
Walton, J.C. 793 (75), 794 (77), *811*
Wan, C.-C. 48 (26d, 40), 49, 71 (40), 83, 86, 90 (26d), *130*
Wandless, T.J. 690, 693, 695 (166), *705*
Wang, A.-Y. 83, 86, 94, 105 (115), *132*, 160 (104), *215*
Wang, D. 802 (124), *812*
Wang, I.S.Y. 481 (247), *493*
Wang, J.L. 158, 159 (93), *215*
Wang, L.C. 690, 691 (165), *705*
Wang, S.-Z. 746, 751 (259), *771*
Wang, Y. 48, 49, 71 (42), *130*, 572 (263), *653*
Wang, Y.C. 208 (329), *220*, 456 (264), *468*
Ward, M.D. 197 (284b, 284c), *219*
Wardeiner, J. 56, 57, 75 (46a), *130*
Warin, R. 290 (172), *314*, 688, 689 (161), *705*
Warner, H.E. 146 (52), *214*
Warner, J.M. 470 (60), *489*
Warner, P. 394 (120), *410*, 419 (80, 83), 425 (80), 440, 441 (172), 442 (181, 182), *463, 465*, 746 (249), *771*
Warner, P.M. 344, 345 (18), 402 (125), *406, 410*, 413, 415, 416, 419, 427, 450 (15), 452 (245), 453 (248), 459 (15), *461, 467*
Washburn, W.N. 458 (278), *468*
Wasserman, E. 487 (291, 293), *494*
Wasylishen, R.E. 394, 402 (121), *410*, 420–422 (91), *463*, 554 (204), *652*
Watabe, T. 716 (51), *766*
Watanabe, H. 751 (271), *771*
Watanabe, M. 263 (10), 279 (74a, 74b), *311, 312*
Watanabe, N. 685 (148), *705*
Watanabe, T. 279 (74b), *312*
Watkins, L.M. 292 (213), *315*, 698 (197), *706*
Watson, D.G. 159, 180, 185 (95), *215*
Watson, K.N. 454 (253), *467*
Watson, W.H. 209 (335), *221*
Watts, W.E. 831 (63b), *858*
Wayner, D.D.M. 116 (226), *136*, 470 (32), *489*
Weaver, D.L. 590 (307), 597 (327, 328), 598 (330, 332), 601 (307), 602 (307, 332), *654*
Webb, G.G. 417 (58), *462*, 819, 820, 854 (25), *857*
Weber, B. 698 (192), *706*
Weber, E. 161 (114–116), 167 (114–116, 127), 168 (116), 169 (114, 115, 127), *215*
Weber, J.L. 760 (305), *772*
Weber, R.J. 478 (204), *492*
Weber, W. 289 (168), *314*, 870 (47), *884*
Weber, W.H. 471 (85), *490*
Weber, W.P. 866 (25), *883*
Wedemann, P. 624 (375), *655*
Weeks, , P.D. 519 (124), *650*
Wege, D. 720 (78), 722 (78, 107), 723 (107), 728 (135), 740 (107), *767, 768*
Wehle, D. 280 (81), *312*, 872, 875 (53), *884*
Wehrmann, R. 430 (129), *464*
Weider, R. 716, 730, 757 (50), *766*
Weier, A. 530, 552, 553, 622 (152), *650*
Weiner, W.P. 197 (283), *219*
Weingarten, M.D. 291 (200e), *315*
Weinhold, F. 374, 384, 393 (78), *408*, 778 (18), *810*
Weinstock, I.A. 593 (319), *654*
Weinstock, R.B. 374, 384, 393 (78), *408*
Weisman, S.A. 456 (265), *468*
Weiss, E. 585 (295a, 295b, 298), 587 (295a), *653, 654*
Weiss, K. 534 (155), *651*
Weiss, R. 197 (285a–c), *219*, 612 (356), 614 (358), 615, 616 (356), 617 (358), *655*
Weitemeyer, C. 831, 832 (65), *858*
Welch, A.J. 635 (403, 404), *656*
Welcker, P.S. 598, 602 (333), *654*
Weller, F. 429 (119), *464*
Wellington, C.A. 479 (208), *492*
Wenck, H. 871 (49), *884*
Wendeloski, J.J. 84–86 (128), *133*
Wendisch, D. 44, 48 (2), *129*, 340, 350 (1), *406*, 498 (2), *647*
Weng, W.-W. 273 (56), *312*
Wenkert, E. 291 (187, 188), *314*, 663 (47), 668 (68–70), 669 (79), 675 (105), 683 (143), 685 (105), 687 (155–157, 159), 689 (163), *703–705*
Wentrup, C. 756 (284, 285), *771*
Wentrup-Byrne, E. 740 (217), 756 (284, 285), *770, 771*
Werle, T. 663 (35), *703*
Werness, P.G. 708, 734, 738, 749 (4), *766*
Werthemann, D.P. 324 (31a), *338*

Wertz, D.H. 77 (92c), *132*
Wertz, J.E. 122 (253), *137*
Wess, G. 679 (124), *704*
West, P.R. 729 (142), *768*
West, R. 48, 49, 71 (44), *130*
Westbook, J.D. 280 (84), *312*
Westerman, P.W. 825, 826 (50a, 50c), 829 (50c), 852 (50a), 853 (50a, 50c), *857*
Westheimer, F.H. 77 (92d), *132*
Westrum, E.F. 241 (82), *257*
Westrum, E.F.Jr. 225 (17), 243 (90), *250, 258*
Westrum, L.J. 698 (199), *706*
Whangbo, M.H. 350, 370, 373 (28), *407*, 625 (380), *655*
White, A.H. 107, 108 (179), *134*, 180 (189), *217*
White, A.M. 849 (111), 855 (128), *859*
White, D.H. 353 (34), *407*
White, J.D. 277 (66), *312*, 670 (83), *704*
White, W.N. 417 (55), *462*, 815 (4), *856*
Whiteside, R.A. 114 (221), 117–119 (238a), *136*
Whitesides, T.H. 624 (378), 625 (379), 628, 635 (378), *655*
Whitman, D.W. 574, 590, 595, 600 (275), *653*
Whitmore, F.C. 471 (81), *489*
Whittmayr, H. 511 (87), *649*
Wiberg, K. 863 (3), *883*
Wiberg, K.B. 60, 61 (62), 74 (77–79, 85), 75 (78, 79), 83 (98, 111), 84 (62, 128), 85 (62, 128, 141, 142), 86 (62, 98, 111, 128, 141, 142), 88 (98, 111), 89 (98), 93 (98, 111), 94, 95 (98), 107 (79), 110 (197), 111 (207), 123, 124 (79, 258), 125 (258), *131–133, 135–137*, 140 (3), 144 (26, 29), 149, 181, 182, 186 (26), 199 (289), 204 (29, 314a, 314b, 315), *212, 213, 219, 220*, 237 (68, 69), 242 (88a), *255, 258*, 262 (2g), 263 (7a, 7b), *310, 311*, 353 (34), *407*, 417 (51, 54, 56, 57), 418 (56), *462*, 470 (68, 76), 471, 479 (76), *489*, 509 (82), 569 (254), 578 (282), 601 (82), *649, 653*, 774 (3, 6), 776 (15), 777 (3, 15), 778 (21, 22), 779 (27, 31), 780 (3, 15, 40, 41, 43), 781 (3, 15), 783 (3, 15, 43, 44, 47, 50), 784 (15, 51–53), 785 (22, 43, 53), 786 (15), 789 (64), 790 (3, 15, 64, 66), 791 (3, 15, 66, 71), 792 (71), 793 (43), 794 (15, 43, 64, 66, 78, 89), 795 (6, 64, 71), 796 (6), 797 (50, 64, 110), 798–800, 802–804 (64, 110), 805 (64), 806 (3, 15, 64, 130), 807 (64), 808 (66), *809–812*, 819, 820 (26), 822 (46, 47a, 47b), 824 (47a, 47b), 825 (51), *857*, 874 (64), *884*
Wiberg, W. 44, 48, 74, 110, 111, 124 (14), *129*
Wicha, J. 303 (305, 306), *317*
Widdison, W.C. 619 (364), *655*
Wieber, G.M. 525 (144), *650*
Wiedenmann, K.-H. 186 (233), *218*
Wieder, G.M. 471 (101), *490*
Wienand, A. 294 (239, 241), *316*
Wierig, A. 161, 167 (115, 116), 168 (116), 169 (115), *215*
Wilante, C. 477 (184), *491*
Wilcox, C.F. 418 (67), *462*
Wilcox, C.F.Jr. 401 (123), *410*, 454 (251), *467*, 844 (98), *859*
Wilde, A. 272 (50, 51), *312*, 535, 537 (157, 158), 539, 542, 543 (158), *651*
Wilhelm, D. 550 (185), *651*
Wilk, B. 280 (84), *312*
Wilke, G. 579 (284, 285), 581 (284), 586 (285), 634 (402), 644 (285), *653, 656*, 747 (266), 748 (267), 753 (267, 277–279), 755 (281), *771*
Wilkenloh, J. 288 (159), *314*
Wilkinson, D.L. 185 (212), *217*
Wilkinson, G. 420 (84), *463*, 517, 518 (114), 554 (207), *650, 652*
Willcott, M.R. 469 (4), 471 (93, 94, 96), *488, 490*
Willerhausen, , P. 429–431 (131), *464*
Williams, F. 117 (236a, 236b), *136*
Williams, J.B. 756 (287), *771*
Williams, J.H. 107, 108 (177, 178), *134*
Williams, R.B. 413 (11), *461*
Williams, R.E. 444 (188), *465*
Williams, R.M. 297 (260, 261), *316*
Williams, R.V. 344, 345 (23), 354 (36), 374 (74), *406–408*, 413, 415, 416, 419, 427, 440, 444, 450 (21), 452 (242), 455 (242, 259), 456 (21, 242, 259), 457, 459 (21), *461, 467*
Williams, V.Z.Jr. 783 (47), 794 (78), *810, 811*
Willis, M.R. 415, 451 (36), *461*
Wilson, A.R. 673 (95), *704*
Wilson, E.B. 162 (119), *215*
Wilson, E.B.Jr. 102, 104 (164a), *134*
Wilson, J.D. 394 (120), *410*, 420 (88), 451 (235), *463, 467*
Wilson, J.W. 321 (13), *337*
Wilson, R. 607–609, 618 (347), *655*
Wilson, S.R. 292 (223), *315*
Wimmer, P. 742 (232), *770*
Winchester, W.R. 684, 685 (146), *705*
Wingert, H. 209 (331), *220*
Winkler, K. 470 (73), *489*
Winnewisser, G. 182 (205), *217*
Winnewisser, M. 182 (205), *217*
Winstead, C. 786 (56), *810*

Winstein, S. 110 (199b), *135*, 344, 345 (14, 15), 352, 353 (14), 361 (45), 365, 366 (14, 15), 374 (14), 394 (120), *406, 407, 410*, 412 (3, 5–7, 9), 413 (12–14, 16), 415 (13, 14, 16), 416 (14, 16), 419 (14, 16, 70, 80, 81), 420 (14, 70, 84, 85), 423 (70), 425 (80), 427 (14, 16), 428 (70), 431 (136, 137), 432 (138–140), 433, 435 (137–140), 436 (140), 440 (167, 171, 172, 175), 441 (171, 172, 175, 178, 179), 442 (181), 444 (14, 16), 445 (194, 195, 198, 200), 446 (198), 447 (206), 449 (217, 220), 450 (14, 16), 458 (272), 459 (14, 16), *460–466, 468*, 513 (99), *649*, 840 (89), 849, 850 (114), *858, 859*
Winter, R. 760 (308), *772*
Winzenberg, K.N. 200 (301), *220*
Winzer, M. 189 (254), *218*
Wipff, G. 791 (70), 794 (81), *811*
Wirth, D. 384 (91), *409*
Wisnieff, T.J. 675 (106), *704*
Wisotsky, M.J. 816, 825 (12), *857*
Wittaker, D. 444, 447 (192), *466*
Witzig, C. 294 (230), *315*
Wlostowski, M. 294 (233), *316*
Woerpel, K.A. 292 (209), *315*, 698 (195, 196), *706*
Wojtowicg, H. 287 (144), *314*
Wolf, B. 290 (173), *314*
Wolf, H. 612, 615, 616 (356), *655*
Wolf, J.F. 854 (125), *859*
Wolf, U. 615 (360), *655*
Wolfgang, R. 471 (138), *491*
Wolfgruber, M. 535 (159), 536 (159, 162), 548 (159), 562 (241), *651, 652*
Wolfsberg, M. 819 (23), *857*
Wolinski, K. 110 (186), *134*, 393, 396 (113), *409*
Wolmershäuser, G. 201 (303), *220*, 271 (47b), *312*
Wong, H.N.C. 469 (26), *488*, 727 (123), *768*
Wong, M. 186 (223), *218*
Woo, L.K. 663, 694 (42), *703*
Wood, J.L. 708, 734, 738, 749 (4), *766*
Wood, J.T. 485 (279), *493*
Woodard, R.W. 302 (302), *317*
Woods, C. 794, 808 (90), *811*
Woods, W.G. 445 (196), *466*
Woodward, P. 625, 628 (382), *655*
Woodward, R.B. 355, 363 (37), *407*, 415 (33), 438 (163), 445 (194), 451 (33), *461, 465, 466*
Woolhouse, A.D. 710 (16), 721, 722 (87), 741 (223, 226), 756 (286), 757 (223), *766, 767, 770, 771*
Wormsbächer, D. 553 (196), *651*
Wortmann, O. 477 (196), *492*
Wovkulich, P.M. 284 (128), *313*
Wright, J.M. 513 (93), *649*
Wright, J.S. 481 (241, 242), *493*
Wright, W.D. 499 (5), *648*
Wu, I.-W. 475, 476 (169), *491*
Wu, S. 179, 180 (183, 186), *217*
Wulff, W.D. 294 (243, 246, 247), 295 (246, 247), *316*, 522 (132), *650*
Wüllen, C.van 110 (192a, 192b), *135*, 393, 396 (115), *410*
Wurrey, C.J. 182 (199, 200), 183 (199), *217*
Würthwein, E.-U. 210 (338), *221*, 377, 393, 404 (84), *408*
Wüstefeld, M. 477 (195), *492*
Wüster, H. 724 (113), 746, 751 (251), *768, 771*

Xiang, Y.B. 679 (124, 126), *704*
Xiang Zhu 150 (65), *214*
Xiao, C. 307 (328), *317*
Xie, Y. 96–98 (149), *133*, 729, 757 (138), *768*
Xifan, P. 674 (103), *704*
Xinhua Ji 158, 159 (94), *215*
Xu, J.-H. 323 (23), *337*
Xu, L. 278 (72), 288 (154), *312, 314*
Xu, S.L. 291 (200a, 200c, 200d), *315*

Yabe, A. 765 (317), *772*
Yabushita, S. 481 (259), *493*
Yadav, J.S. 665 (59), *703*
Yaeger, D.L. 110 (193c), *135*
Yamabe, S. 128 (278), *137*, 306 (326), *317*, 514 (105, 106), *649, 650*
Yamada, K. 182 (204, 205), *217*, 869 (44), *884*
Yamago, S. 643 (423), *656*
Yamaguchi, A. 514 (106), *650*
Yamaguchi, I. 145 (44), *213*
Yamaguchi, K. 481 (259), *493*
Yamaguchi, Y. 84, 85 (129, 132), 98 (129, 159), 101 (129, 132), 102 (129, 159), 126 (129, 132, 270), *133, 137*, 481 (263, 264, 268, 270), 482 (268, 270), 484, 487 (270), *493*
Yamamoto, A. 268 (38), *311*
Yamamoto, H. 280 (92), 283 (109, 110), 306 (324), *312, 313, 317*, 501 (31), *648*
Yamamoto, I. 303 (311), *317*
Yamamoto, S. 84 (126), *133*, 143, 144 (21), 145 (21, 35), 188 (245), 195 (278), 199, 208 (35), *213, 218, 219*
Yamamoto, T. 765 (316, 317), *772*
Yamamoto, Y. 880 (84), *885*
Yamanoi, K. 292 (222), *315*, 667 (65), *703*
Yamashita, K. 454 (250), *467*
Yamashita, M. 291 (194), *314*, 670 (85), *704*
Yamashita, T. 280 (78), *312*
Yamashita, Y. 296 (250), *316*, 474 (160), *491*

Yamazaki, H. 170 (132), *216*
Yamazaki, K. 711 (24), *766*
Yamazaki, S. 306 (326), *317*, 514 (105, 106), *649, 650*
Yamazaki, T. 60 (60), *131*, 305 (316), *317*
Yamazaki, Y.T. 710 (16), *766*
Yamin, A. 292 (223), *315*
Yamomoto, H. 866 (23), *883*
Yanai, T. 301 (283), *317*
Yanao, S. 682 (135), *705*
Yaneda, S. 441 (179), *465*
Yang, C.X. 794 (80), *811*
Yang, D.C: 294 (243, 246, 247), 295 (246, 247), *316*
Yang, H.-C. 797, 798, 800, 802 (104, 124), *811, 812*
Yang, L.R. 149 (62), *214*
Yannoni, C.S. 417 (58), *462*, 819, 820, 854 (25), *857*
Yanovskii, A.I. 208 (326), *220*
Yasuda, M. 554 (200), *652*
Yasui, K. 279 (76), *312*
Yasui, S.C. 476 (173), *491*
Yau, A.W. 471 (104), *490*
Yavari, I. 819 (24), 821 (41), *857*
Ye, T. 658, 663 (12), 668 (72), 687 (12), 699 (202), *702, 703, 706*
Yeh, C.-L. 550 (182), *651*
Yeh, M.C.P. 307 (328), *317*
Yelekci, K. 470 (57), *489*
Yeroushalmi, S. 424 (102), *463*
Yin, J. 502, 583, 587 (41), *648*
Yin, L. 116, 117 (230), *136*
Yin, T.-K. 470 (57), *489*
Yip, Y.-C. 469 (26), *488*
Yogev, A. 510 (85), *649*
Yokelson, H.B. 48, 49, 71 (44), *130*
Yoneda, R. 299 (272), *316*
Yoneyoshi, Y. 693, 697 (172, 173), *706*
Yoshida, E. 278 (70), *312*
Yoshida, J. 682 (135), *705*
Yoshida, K. 145 (44), *213*
Yoshida, M. 470 (32), *489*, 715 (47), *766*
Yoshida, T. 499 (12), *648*
Yoshida, Z. 279 (76), *312*, 441 (179), *465*, 606 (340), 607 (347, 350), 608 (347, 351), 609 (340, 347, 351, 352), 610 (340, 351), 611 (351), 612 (340, 350–352), 615 (340), 616 (357, 361), 617 (340), 618 (340, 347, 350, 351), 619 (351), 645 (340), 646 (433), *654–656*
Yoshida, Z.-i. 715 (47), *766*
Yoshihara, K. 306 (325), *317*
Yoshikawa, K. 74, 79 (80), *131*
Yoshikawa, M. 283 (120), *313*
Yoshikoshi, A. 686 (154), *705*
Yoshimura, T. 607 (343), *654*
Yoshioka, M. 284 (125), *313*
Yoshizawa, T. 282 (100), *313*
Young, D.J. 513 (97, 98), *649*
Young, J.E.Jr. 155, 187, 190 (85), *215*
Young, S.D. 683 (141), *705*
Young, W.B. 664 (53), *703*
Youngs, W.J. 208 (327), *220*
Yovell, J. 469 (19), *488*, 544 (165), *651*
Yu, J. 267 (28), *311*
Yu, S.H. 554 (209, 210), *652*
Yu, T. 278 (72), *312*
Yufit, D.S. 868 (39), 877 (73), 878 (73, 78, 79), 879 (79), 881 (89), *884, 885*
Yukimoto, Y. 683 (141), *705*
Yus, M. 308 (336), *318*

Zahradnik, R. 728, 729, 757 (132), *768*
Zakutansky, J. 282 (106), *313*
Zakutansky, J.A. 591, 592, 599, 603–605 (314), *654*
Zaman, F. 303 (309), *317*
Zanasi, R. 106 (175), 109 (185c), *134*
Zarate, E.A. 208 (327), *220*
Závada, J. 167 (126), *215*
Zdrojewski, T. 294 (233), *316*
Zeeb, S. 150 (64), *214*
Zefirov, N.S. 208 (326), *220*, 230, 231 (41), *252*, 282 (104, 105), 308 (334c), *313, 318*, 789, 790 (65), 796 (100), 797 (65, 101), 798 (65, 118–120), 799, 800 (65, 101), 801 (65, 118), 802 (65, 101, 119–122), 805 (65, 100, 126), 807 (65), 808 (65, 131), *811, 812*, 863, 864 (9), 865 (19), 866 (9), 867 (31), 868 (39, 40), 872 (19, 57), 875 (9, 19, 66), 876 (9, 70), 877 (70, 73, 74), 878 (70, 73, 75, 76, 78, 79), 879 (79), 880 (81), 881 (88–90), *883–885*
Zelenkina, O.A. 789, 790, 797–802, 805, 807, 808 (65), *811*
Zelinski, N.D. 864, 875 (14b), *883*
Zellweger, D. 291 (193), *314*
Zerbi, G. 77, 78 (91b), 105–107 (169a–c), *132, 134*
Zercher, C. 268 (34), *311*
Zerger, R.P. 505 (57), *649*
Zeroka, D. 109 (181c), *134*
Zeya, M. 436 (155), *465*
Zha, Z.M. 733 (168), *769*
Zhang, J.H. 197 (284c), *219*
Zhang, L.J. 470 (57), *489*
Zhang, W. 273 (55), *312*
Zhang, Y. 309 (343), *318*
Zhang, Z. 500 (18), *648*
Zhao, L. 308 (331), *318*, 681 (129), *705*
Zhdankin, V.V. 208 (326), *220*, 805 (126), *812*
Zhi, L. 291 (200b), *315*
Zhixing, C. 76 (87), *131*

Zhou, Z. 699 (200), *706*
Zhou, Z.-L. 296, 297 (254), *316*
Zhu, Y.-F. 305 (316), *317*
Ziegler, G. 740 (215), *770*
Zilberg, S.P. 863 (4c), *883*
Ziller, J.W. 503, 566, 568 (42), 580, 581, 584 (287), 593 (318), 594 (318, 320), 601 (287), 602 (320), *648, 653, 654*
Zilm, K.W. 110 (195), *135*
Zimmerman, G.A. 320 (4), *337*
Zimmerman, H.E. 320 (2, 3a, 3b, 4–9), 321 (10–15), 322 (16, 17, 18a, 18b, 19–22), 323 (23, 24, 25a, 25b, 26, 29), 324 (30, 31a, 31b, 32, 33, 42), 325 (30, 32–35, 36a, 36b, 37–39), 326 (40–42), 327 (43–45), 328 (45, 46), 329 (47a–d), 333 (21, 37, 48), 334 (49, 50a–c), 335 (50a–c, 51), 336 (32, 52, 53), *337, 338*, 355, 363 (38), *407*, 414, 415 (32), 435 (154), *461, 465*
Zimmermann, P. 458 (277), *468*
Ziólkowska, W. 288 (158), *314*
Zoebisch, E.G. 430 (125), *464*
Zoelner, S. 877, 878 (73), *885*
Zoeren, E.van 162 (117), *215*
Zoller, U. 144, 204 (28), *213*, 785 (54), *810*
Zöllner, S. 188 (243, 246), *218*, 876, 878, 881 (71), *885*
Zolotoi, A.B. 176 (155, 157), *216*
Zora, M. 523 (138, 141), 525 (138), 529 (151), 530 (138, 141), 531, 533 (138), *650*
Zotova, S.V. 508 (75), 627 (385), *649, 655*
Zozom, J. 176–178 (160), *216*
Zubkov, V.A. 418 (65), *462*
Zuidema, L. 741 (228), *770*
Zurawski, B. 125 (264), *137*
Zweep, S.D. 344, 345 (21), *406*, 413, 415, 416 (22), 417 (59), 419, 445, 450 (22), *461, 462*, 817, 855 (17), *857*

Index compiled by K. Raven

Subject index

Ab initio calculations,
  for cyclopropanation 515
  for cycloproparenes 733, 734, 757
  for cycloproparenyl cations 741
  for homoconjugation 390–403
  for metallatetrahedranes 603
  σ-Acceptor ability 96
Acetylenic dienophiles, reactions of 715
Acylcarbonylcyclopropanes, synthesis of 290
Acylcycloproparenes, reactions of 731
2-Adamantylidene-1,1-dicyclopropylethyl cation 834
Adiabatic frequencies 102
Alder–Rickert cleavage 715, 730
Alkenes — *see also* Cycloalkenes
  cyclopropanation of, metal-catalysed 657–702
Alkenylcyclopropanes, synthesis of 267–269, 285
Alkoxycarbonylcyclopropanes, synthesis of 290
Alkoxycyclopentenones, synthesis of 531
Alkoxycyclopropanes, structure of 179
Alkoxycyclopropenylium salts, reactions of 610
Alkoxycyclopropylcarbene complexes — *see also* Cyclopropyl(methoxy)carbene complexes
  NMR spectra of 522
Alkylcyclopropanes — *see also* Methylcyclopropanes, Trialkylcyclopropanes
  structure of 147–152
Alkyl dihalides, reductive 1,3-elimination reactions of 262, 263
Alkylidenecyclopropabenzenes — *see also* Methylenecyclopropabenzene
  geometry of 758
Alkylidenecyclopropanaphthalenes, geometry of 758
2-Alkylidenecyclopropanemethanols, synthesis of 282
Alkylidenecycloproparenes 757 — *see also* Methylenecycloproparenes
  geometry of 758, 759
  in cycloadditions 763, 764
  ion radicals of 760
  IR spectra of 760
  NMR spectra of 760
  oxidation of 762
  photolysis of 765
  reactions of 729
    with nucleophiles 761
    with organometallics 764
  synthesis of 729–733
  thermolysis of 765
  UV spectra of 759, 760
Alkylidyne–carbonyl coupling reactions 536
Alkylidyne complexes, reactions with alkynes 594, 595
1,2-Alkyl migration 336
Alkylthiocyclopropanes, structure of 179
Alkyne–metallocene complexes 547
Alkynes, metathesis of 593
Alkynylcyclopropanes — *see also* Ethynylcyclopropanes
  metal derivatives of 557–566
  synthesis of 268
Alkynylstannanes, cobalt-catalysed coupling of 715
Allenes,
  cyclopropanation of 627, 628
  synthesis of 756
Allenylidenechromium complexes 566
Allenylidenecyclopropane, structure of 188
Allylcyclopropylcarbinyl cations 834
  σ, π-Allylic complexes 549, 550
Allylic derivatives,
  cyclization of 270–273
  cyclopropanation of 686
Allylic ylides, [2,3]sigmatropic rearrangement of 686
AM1 calculations 746

Aminocyclopropanecarboxylic acids,
derivatives, synthesis of 264, 265, 271
structure of 170–172
Aminocyclopropanes, structure of 181, 182
Aminotriafulvenes, zwitterions of 646, 647
Aminotriazinones, oxidation of 727, 728
Angle strain, in cycloproparenes 738
Anisyl migration 322
Annulenes — *see also* Cyclopropannulenes, Heteroannulenes
homoaromaticity in 450
synthesis of 746
Anthrone-10-spiroindazoles, photolysis of 711, 712
Antiaromaticity 342
Antibonding odd-electron density 334, 335
Antihomoaromaticity 342
Aromatic bond localization 709, 733
Aromaticity,
in-plane 342
loss of 322
radial 342–344
spherical 342, 343
three-dimensional 342–344
π-Aromaticity 342, 343
σ-Aromaticity 82, 83, 342, 343
4-Arylcyclohexenones, photorearrangement of 321–323
4-Arylcyclopentenones, photorearrangement of 322
Arylcyclopropanes — *see also* Cyanoarylcyclopropanes, Phenylcyclopropanes
structure of 156–160
synthesis of 263, 285
Arylcyclopropylcarbinyl cations 823, 828
1-Arylcyclopropylcarbinyl compounds, solvolysis of 815
Arylthiocyclopropanes, structure of 179
Associative substitution 595
Asterenes, structure of 210
Asymmetry parameters 147, 155, 159, 179
Atomic basins 65, 68, 76
Atomic charge 74
Atomic energies 74, 75
Atomic subspaces 65
Atomic volume 74
Atomization energies 76
Aza-di-π-methane rearrangement 323
Azidoformates, cycloadditions of 751
Azocyclopropanes, synthesis of 270

Back-bent (tilt) angles, in metallatetrahedranes 603
Baeyer strain energy 77, 79
Banana bonds 733
Barbaralanes, structure of 207, 208
Barbaralenes, homoaromaticity in 450, 455
Barbaralyl cations 830, 831, 840–843
Barrelene, photorearrangement of 323, 324
Barrelyl cations 831, 832
Bending force constant 77
Bent bond character, of CC bonds in cyclopropane 64
Bent bond orbitals 56
Bent bonds 144, 146, 150, 154, 204, 206
convex 97
Benzoannelation 245, 246
Benzocyclopropenones — *see* Oxocycloproparenes
Benzo-2-nortricyclyl cations 850, 851
Benzoquinones, cycloaddition to cycloproparenes 746
Bicycle rearrangement 329–334
Bicyclic[3.1.0] ketones, synthesis of 321
2-Bicyclo[*n*.1.0]alkyl cations 836, 837
Bicyclo[1.1.0]butanes 123, 124
structure of 198, 199, 202
$\delta^{1,3}$-Bicyclo[1.1.0]butene, enthalpy of formation of 236, 237
Bicyclobutonium cation 417, 815, 818, 820, 822, 823
Bicyclo[1.1.0]butyl-1-carbinyl cation 824
Bicyclobutyl cations 427
Bicyclo[6.2.0]decapentaene, potential homoantiaromaticity in 361
Bicyclo[4.1.0]hepta-2,4-diene, homoaromatic stabilization of 236
Bicyclo[4.1.0]hepta-1,3,5-trienes — *see* Cyclopropabenzenes
Bicyclo[4.1.0]hepta-1,3,6-trienes, aromatization of 720, 721
Bicyclo[4.2.0]heptene-1,6-dicarboxylic acid, didecarboxylation of 725
Bicyclo[4.1.0]heptenes — *see also* Halobicyclo[4.1.0]heptenes
enthalpies of formation of 234
Bicyclo[3.2.0]hept-3-yl cations 449
Bicyclo[3.1.0]hexanes,
structure of 199
synthesis of 308
Bicyclo[3.1.0]hex-2-enes 470
Bicyclo[3.1.0]hex-3-en-2-ones, photorearrangement of 327
Bicyclo[3.1.0]hexenyl cations 431–439, 833, 834, 843
electron delocalization in 433
homoconjugation in 371
potential homoantiaromaticity in 378
rearrangement of 435–439
structure of 434, 435
Bicyclo[3.1.0]hex-2-yl cation 448
Bicyclo[3.2.2]nona-3,6,8-trien-2-yl cation 841, 842

Bicyclo[4.2.0]octadiene, bond homoconjugation in 352, 353
Bicyclo[3.2.1]octadienyl cations, equilibrium with homotropenylium cations 425
Bicyclo[5.1.0]octadienyl cations 419
Bicyclo[5.1.0]octanes, structure of 200, 201
Bicyclo[3.2.1]oct-3-en-2-yl cations 844
  isomerization of 837
Bicyclo[3.3.0]oct-1-yl cation 846
Bicyclo[1.1.1]pentanes, structure of 204
Bicyclo[2.1.0]pentanes 470 — *see also* 2-Methylenebicyclo[2.1.0]pentanes
Bicyclo[2.1.0]pentenes,
  electron delocalization in 435
  homoantiaromaticity in 236, 379, 450, 454
  homoconjugation in 371
Bicyclo[1.1.1]pent-1-yl anion 793
Bicyclo[1.1.1]pent-1-yl cation 794
Bicyclo[1.1.1]pent-1-yl radical 793, 794
Bicyclopropenyls, structure of 193
Bicyclopropyl, PE spectrum of 870
Bicyclopropyl derivatives, structure of 149, 150
Bicyclopropylidene,
  cyclopropanation of 872
  structure of 187
Billups method 717, 720, 724
Bis(cycloheptatriene)s, synthesis of 755
Bis(cyclopropyl)titanocene, synthesis of 502
Bishomoantiaromatic cations 449, 450
Bishomocyclopropenium cations 444–447
Bishomotropenylium cations 440–444
Bisnorcaradiene, valence tautomerism in 380, 381
Bond critical point 66
Bond delocalization 355–357
Bond energies 74–77
1,3-Bond formation 262–279
Bonding,
  super-σ 70
  three-centre 69, 70, 73
  π 70, 73
Bond order 66
Bond path 144
Bornane-sulphonamide derivatives, cycloadditions of 642
Bridged species, protonated 327
Bullvalenes, homoaromaticity in 455
Bullvalyl cations 831, 832
3-Butenyl derivatives, cyclization of 273–276

Cage molecules, structure of 202–212
Calicenes — *see also* Hydrazidocalicenes
  reactions of 645
Carbene insertion–semipinacol rearrangement 529
Carbocation formation, thermodynamics of 853, 854
Carbocyclization 513, 514, 519, 520
Carbometallation 506, 507, 546
Carbon,
  atomic, reactions with benzene 715
  divalent 329
Carbon atoms, planar 47
Carbonylcyclopropanes — *see also* Acylcarbonylcyclopropanes, Alkoxycarbonylcyclopropanes
  structure of 160–174
  synthesis of 264
Carbonyl insertion 574
Carboxamide substituent, effect of 188, 201
Carbyne complexes, reactions with alkynes 594, 595
Caronic acid, structure of 164
Catemer motif 162, 164, 165, 167
Charge-transfer excited states 537
Chelotropic addition 747
Chemoselectivity, in metal-catalysed cyclopropanation 658, 682–688
Chiral bipyridine catalysts, for cyclopropanation 698
Chiral *N*-(diazoacetyl)oxazolidinone, in cyclopropanation 695
Chiral diazoesters, in cyclopropanation 695–697
Chiral metal catalysts, for cyclopropanation 697–701
Chloropalladation, of methylenecyclopropane complexes 629–633
Cholesterol/*i*-cholesterol system 412, 413
Chromium complexes, with cycloproparenes 737, 748
CI calculations 332
Circumambulatory rearrangement 435–439, 843
Clathrates 167–169
Coates' cation 816
Cobalt catalysts, for cyclopropanation 663
Codimerization 521
Combustion calorimetry 247
  π-Complexes 68–73
Configuration analysis 74, 79
Conformation, definition of 141, 142
Conformational rigidity, of metallatetrahedranes 604
Conjugation,
  cyclopropyl 346
  definition of 347, 348
  π-Conjugation 347, 348
  σ-Conjugation 347, 348
  with cyclopropyl group 815
Cope rearrangement 533
  degenerate 323

Copper catalysts, for cyclopropanation 657–659, 661, 663, 664, 666, 668–675, 677–683, 686, 688–694, 696, 697
Coupled cluster (CC) theory 392
Coupling constants, in determination of hybridization 777, 778
Covalent bonding,
conditions for 65
indications of 62
Cremer–Kraker criteria, for covalent bonding 377
Cumulenes, enthalpies of formation for 225
Cyanoarylcyclopropanes, structure of 157
Cyanocyclopropanecarboxylic acids, structure of 166, 167
Cyanocyclopropanes,
structure of 175, 176, 190
substituent effects in 93
Cyanocyclopropenes, structure of 190
cis-1-Cyano-2-isobutenylcyclopropane, epimerization in 474
1-Cyano-2-methylcyclopropanes, isomerism in 473
trans-1-Cyano-2-(phenylethynyl)cyclopropane, correlation of rate constants for 478
Cyanophenyl migration 322, 327
Cyclization reactions 509, 510, 866, 867
Cycloaddition reactions 867, 868 — *see also* Photocycloaddition reactions
of cyclopropyl–iron complexes 502
of methylenecyclopropanes 639–644
Cycloalkenes — *see also* Cyclopropenes
cyclopropanation of 229
Cyclobutabenzene-1,2-dione, irradiation of 757
Cyclobutane,
Gaussian functions for 58
relationship to cyclopropane 47, 48
structure of 145, 194, 199, 208
Cyclobutanones, synthesis of 753
Cyclobutarenes — *see also* Palladacyclobutarenes
synthesis of 715, 747
Cyclobutarenylstannanes, 1,3-cyclization of 715
Cyclobutenium cation, energy surface of 418
Δ^1-Cyclobutenone complexes 577
Cyclobutenyl cations 412, 427
homoconjugation in 371
Cyclobutyl tosylates, solvolysis of 822, 823
Cyclodimerization 521, 588
1,5,9-Cyclododecatriyne, no-bond homoconjugation in 352, 353
Cycloethane 227
1,4-Cycloheptadienes 451
Cycloheptadienones, synthesis of 530, 531
Cycloheptatrienecarboxylic acids, homoconjugation in 412, 413
Cycloheptatrienes — *see also* 1, 6-Didehydrocycloheptatrienes, 1,1-Dihalocycloheptatrienes
homoaromaticity in 374, 450–453
homoconjugation in 352, 353, 359
Cycloheptenyl cation 836, 837
2,5-Cyclohexadienones, photorearrangement of 319–321
Cyclohexadienyl cations 431–439
photoisomerization of 432
structure of 434, 435
Cyclohexane, CH_2 group of, as reference group 74, 75
cis,cis,cis-1,4,7-Cyclononatriene,
as possible homoaromatic molecule 456, 457
no-bond homoconjugation in 353
Cyclooctatetraene, reactions of 419
Cyclooctatrienyl cation 423
Cyclooligomerization 588
Cyclopentadienones, cycloaddition to cycloproparenes 746
Cyclopentadienylcarbinyl cation, circumambulation in 437
Cyclopentadienylthallium 625
Cyclopentenones — *see* Alkoxycyclopentenones, 4-Arylcyclopentenones, Cyclopropylcyclopentenones
Cyclopentenyl cation 444
Cyclopentylidenemethyl cation 827
Cyclophanes, precursors of 722, 724
Cyclopropaanthracenes,
synthesis of 720
X-ray geometry of 737
Cyclopropa[*b*]anthracenes,
IR spectra of 739
PE spectra of 739
reactions with organometallics 753
Cyclopropabenzene-2,5-dione 746
formation of 716, 717
Cyclopropabenzene ester 711
Cyclopropabenzenes 708, 709 — *see also* Alkylidenecyclopropabenzenes, Halocyclopropabenzenes, Methylcyclopropabenzenes, Tricyclopropabenzenes
structure of 201, 202
Cyclopropabenzenyl anions 740, 742
silylation of 744
Cyclopropabenzynes,
formation of 745
strain energies in 739
Cyclopropacyclobutabenzene, synthesis of 715, 718
Cyclopropaheteroarenes 713
synthesis of 720

Cyclopropaisobenzofurans,
cycloadditions of 745, 746
NMR spectra of 740
synthesis of 722, 723
Cyclopropa[*g*]isoquinolines, synthesis of 727
Cyclopropa[*a*]naphthalenes — *see also gem*-Dihalocyclopropa[*a*]naphthalenes
strain energies in 738
synthesis of 713, 716, 725
Cyclopropa[*b*]naphthalenes,
geometry of 737
synthesis of 727
Cyclopropa[*b*]naphthalenyl anion 742
Cyclopropanation 514, 515
asymmetric 283, 284, 292, 297, 658
diastereoselective 506
enthalpies of reaction of 228, 229
intramolecular 658
metal-catalysed 657, 658
chemoselectivity in 682–688
diastereoselectivity in 695–697
enantioselectivity in 658, 697–701
regioselectivity in 658, 688–692
stereoselectivity in 658, 692–695
using diazoacetamides 662–667
using diazoalkanes 658–662
using diazoesters 662–667
using diazoketones 667–676
using dicarbonyl diazomethanes 676–680
using iodonium ylides 681
of acetylenes 235
of active methylene compounds 302–304
of alkenes 657–702
of allenes 627, 628
of bicyclopropylidene 872
of carbon–carbon multiple bonds 279–295
of cycloalkenes 229
of methylenecycloalkanes 865, 866
of Michael acceptors 296–302
with diazo compounds 304–306
Cyclopropane,
chemical bonding of 61
complexes of 146
epimerization in 475, 476
formation of 125, 126
geometry of 84–86
protonated 114–116
ring opening of 126–129
ring strain in 145
stereomutation in 126, 487
structure of 45–48, 143–146
Cyclopropanecarboxaldehyde, conformation of 160
Cyclopropanecarboxamides — *see also* Methylenecyclopropanecarboxamides
structure of 174
Cyclopropanecarboxylic acids — *see also* Aminocyclopropanecarboxylic acids, Cyanocyclopropanecarboxylic acids
complexes of 162
cyclic dimers of 162, 163
structure of 161–170
Cyclopropanecarboxylic anhydrides, structure of 172
Cyclopropanecarboxylic esters, synthesis of 304
Cyclopropane-1,1-dicarboxylic esters, synthesis of 302
Cyclopropane-1,2-dicarboxylic esters, synthesis of 306
Cyclopropanehexacarboxylates, structure of 172, 173
Cyclopropane–propene isomerization 470, 471
Cyclopropane ring, geometry of 416
Cyclopropanes
— *see also* Alkenylcyclopropanes, Alkoxycyclopropanes, Alkylcyclopropanes, Alkynylcyclopropanes, Aminocyclopropanes, Arylcyclopropanes, Azocyclopropanes, Carbonylcyclopropanes, Cyanocyclopropanes, Germylcyclopropanes, Halocyclopropanes, Heterocyclopropanes, Isocyanocyclopropanes, Lithiocyclopropanes, Nitrocyclopropanes, Siloxycyclopropanes, Silylcyclopropanes, Sulphinylcyclopropanes, Sulphonylcyclopropanes, Thiocyclopropanes, Vinylcyclopropanes, Vinylidenecyclopropanes
1,2-dioxygenated, synthesis of 294, 295
gem-disubstituted, cyclization of 866, 867
fullerene — *see* Fullerene cyclopropanes
fused 123–125
lithiation of 504, 505
spiroannulated — *see* Spiroannulated cyclopropanes
stereomutations in 470, 471, 479–484, 486, 487
thermochemistry of,
compared to benzenes 237–242
compared to cyclobutanes 242–249
compared to olefins 227–237
1,1,2-trisubstituted with electron-withdrawing groups 301
trans-1-R^1-2-R^2-Cyclopropanes, correlation of rate constants for 476–479
Cyclopropanethiol, substituent effects in 95
Cyclopropannulenes, synthesis of 718
Cyclopropanols,
structure of 179

Cyclopropanols (*cont.*)
substituent effects in 95
synthesis of 264, 278, 279, 282–284, 308
Cyclopropanone, structure of 187
Cyclopropa[*b*]phenanthrenes, synthesis of 720
Cyclopropa[*l*]phenanthrenes 714 — *see also* Iminocyclopropa[*l*]phenanthrenes
strain energies in 738
synthesis of 716, 720, 725
Cyclopropapyridines,
geometry of 736
synthesis of 712
Cyclopropaquinolines, synthesis of 722, 723
Cycloproparenes 708, 709 — *see also* Acylcycloproparenes, Alkylidenecycloproparenes, Dicycloproparenes, *gem*-Dihalocycloproparenes, Oxocycloproparenes, Spirocycloproparenes
dimerization of 750
in cycloadditions 745–747, 751–753
IR spectra of 739
isomerization of 712
mass spectra of 740
NMR spectra of 740
PE spectra of 739
photolysis of 756, 757
physical aspects of 733–740
reactions of,
with electrophiles 744, 745, 748–750
with nucleophiles 745
with organometallics 747, 748, 753–756
without ring cleavage 744–748
with radicals 745, 750, 751
with ring cleavage 748–757
structure of 201, 202, 734
synthesis of 710
by aromatization 717–727
by flash vacuum pyrolysis 715–717
by ring closure 714, 715
by ring contraction 711–714
theoretical aspects of 733–740
thermolysis of 756, 757
tricarbonylchromium complexes of 737
UV spectra of 739
Cycloproparenyl anions 730, 742, 743
Cycloproparenyl cations 741, 742, 761
as reaction intermediates 745
Cycloproparenyl dianions 710
Cycloproparenylidene dimerization 731
Cycloproparenylmethyl–cyclobutarenyl rearrangement 761
Cycloproparenylquinones, synthesis of 733
Cycloproparenyl radicals 743, 744
Cyclopropathiophenes, synthesis of 722, 723
Λ^2-Cyclopropene complexes,
reactions of 583–589
synthesis of 578–583
Cyclopropenes — *see also* Cyanocyclopropenes, Halocyclopropenes, Methylenecyclopropenes, Nitrocyclopropenes, Silylcyclopropenes
enthalpies of formation of 235
structure of 190–194
synthesis of 263
Cyclopropeneselones, deselenization of 610
Cyclopropenethiones, desulphurization of 610
Cyclopropenium cations 412 — *see also* Cyclopropylcyclopropenium cations, Homocyclopropenium cations
Cyclopropenium complexes 568
Cyclopropenones — *see also* Benzocyclopropenones, Dihalocyclopropenones, Dihydroxycyclopropenones
aromaticity of 248, 249
cycloaddition to cycloproparenes 752
structure of 194–197
Cyclopropenyl cation 197
Λ^3-Cyclopropenyl complexes,
dynamic behaviour of 603–605
reactions of 595–601
structure of 601–603
synthesis of 589–595
Cyclopropenylidene complexes,
NMR spectra of 617–619
synthesis of 606–610
Cyclopropenylidene radical 195
Cyclopropenylidenes, structure of 194–198
Cyclopropenylium salts 197 — *see also* Alkoxycyclopropenylium salts, Halocyclopropenylium salts, Metallacyclopropenylium salts, Silylcyclopropenylium salts
oxidative addition reactions of 589–592
1-Cyclopropenyl–metal compounds 569–573
3-Cyclopropenyl–metal compounds,
reactions of 576, 577
synthesis of 573–576
Cyclopropenyl ring slippage 597
Cyclopropylacyl compounds, synthesis of 271
Cyclopropyl acylsilanes, synthesis of 297, 299, 300
Cyclopropyl alcohols, synthesis of 263
Cyclopropylalkynylcarbene complexes 564–566
Cyclopropylalkynyl–metal compounds 557–560
Cyclopropylamines,
substituent effects in 95
synthesis of 264, 270
Cyclopropyl anion 113, 114
Cyclopropylbis(Λ^4-diene) complexes, synthesis of 503

Cyclopropyl bonds, π-character of 340
Cyclopropylborane, substituent effects in 93
Cyclopropylcarbene complexes — *see also* Alkoxycyclopropylcarbene complexes
  reactions of 524–534
  synthesis of 522–524
Cyclopropylcarbinyl cations 814–816 — *see also* Allylcyclopropylcarbinyl cations, Arylcyclopropylcarbinyl cations, Cyclopropylcarbinyl dications, Cyclopropylidenecyclopropylcarbinyl cations, Dicyclopropylcarbinyl cations, Hydroxycyclopropylcarbinyl cations, Methylcyclopropylcarbinyl cations, Spirocyclopropylcarbinyl cations, Tricyclopropylcarbinyl cations
  bisected 819, 820
  conformation of 160
  energy surface of 418
  equilibrating 818–823, 834–843
  IR spectra of 820
  NMR spectra of 818, 819
  primary 818–824
  secondary 825–843
  solvolytic generation of 822, 823
  static 824–834
  structure of 417
  substituent effects in 93
  synthesis of,
    by addition of carbenes to cyclopropylalkynes 817, 818
    by hydride ion abstraction from hydrocarbons 816, 817
    by ionization of alcohols or halides 816
    by protonation of carbonyls 817
    from allyl alcohols 817, 818
    with neighbouring group participation 817
  tertiary 825–843
  X-ray studies of 855, 856
Cyclopropylcarbinyl–cyclobutyl interconversion 814, 825, 840
Cyclopropylcarbinyl dications 851, 852
Cyclopropylcarbinyl–homoallyl rearrangement 622
Cyclopropylcarbinyl–metal compounds, synthesis of,
  by carbocyclization 513, 514, 519, 520
  by cyclopropanation 514, 515
  by hydrometallation 519
  by insertion reactions 521
  by metal–halogen exchange 512, 513, 515–519
  by reactions with metals 511, 512
  in catalytic reactions 521, 522
Cyclopropylcarbinyl radical 513
Cyclopropylcarbonitriles, synthesis of 264
Cyclopropylcarbonyl complexes 500
Cyclopropylcarbonyl halides, conformation of 160
Cyclopropylcarbyne complexes,
  reactions of,
    without ring cleavage 535–538
    with ring cleavage 538–544
  synthesis of 534, 535
Cyclopropyl cation 117–120
1-Cyclopropylcycloalkyl cations 827
Cyclopropylcyclopentenones, synthesis of 530
Cyclopropylcyclopropenium cations 817, 818
  p*K* values for 854
Cyclopropyldicarbinyl diradical, opening of 324
Cyclopropylethanols, conformation of 151
Cyclopropyl ethers, synthesis of 294
1-Cyclopropylethyl cations 839
Cyclopropyl fragments, conjugation between 870, 871
Cyclopropyl group,
  extent of positive charge delocalization into 852–855
  substituent effects on geometry of 86–96
Cyclopropyl β hyperfine splittings 557
Cyclopropylidene–allene rearrangement 566
Cyclopropylidene complexes 566–568
Cyclopropylidenecyclopropylcarbinyl cations 832, 833
Cyclopropylidenemethanone, structure of 188
Cyclopropylidenes, structure of 187–189
Cyclopropyl isocyanates, structure of 182, 183
Λ^2-Cyclopropylketenyl complexes 536
Cyclopropyl ketones,
  conformation of 160
  protonation of 817
  synthesis of 306
Cyclopropyl–metal compounds, synthesis of 498
  by addition reactions 506–508
  by cyclization reactions 509, 510
  by decarbonylation and metal migration 510, 511
  by exchange reactions 499–505
  by insertion reactions 508, 509
  direct 499
Cyclopropylmethanethiol, conformation of 151
Cyclopropylmethanols,
  reactions of 816
  synthesis of 264, 266
Cyclopropyl(methoxy)carbene complexes,
  cycloadditions of 526–533
  reactions with alkenes 525, 526
  rearrangement of 533
Cyclopropylmethoxycarbenium complexes, synthesis of 524

(Cyclopropylmethyl)amine, conformation of 151
Cyclopropylmethylcarbinol, reactions of 825
Cyclopropylmethylene anion, substituent effects in 95
Cyclopropylmethyl radicals 273
Cyclopropylmethylselenobenzene, synthesis of 514
Cyclopropyl oxide anion, substituent effects in 95
Cyclopropyl phenyl sulphides, synthesis of 264, 265
Cyclopropylphosphanes, structure of 185
Cyclopropylphosphine, substituent effects in 95
Cyclopropyl phosphonates, synthesis of 294
Cyclopropyl radical 47, 48, 120–123
Cyclopropyl radical cation 47, 48, 60, 61, 116, 117
Cyclopropyl ring, pinching effect of 738
1,2-Cyclopropyl shifts 836
Cyclopropyl-to-metal migration 510
Cyclopropyl trimethylsilyl ethers, synthesis of 279
α-Cyclopropylvinyl cations 832, 833
9-Cyclopropyl-9-xanthyl cation 831
Cyclopropylzinc reagents 276, 277
Cyclopropyne 236, 237
Cyhalothric acid 167

Decarbonylation 575
 oxidative 598
Dehalogenation, reductive 864, 865
8,9-Dehydro-2-adamantyl cations 827–829, 838, 839
Dehydrochlorination, of 1,6-dihalobicyclo[4.1.0]hept-3-enes 721
2,4-Dehydro-5-homoadamantyl cations 827, 828, 840
3,5-Dehydrophenyl cation, no-bond homoconjugation in 352, 353
2,5-Dehydro-4-protoadamantyl cations 838–840
Delocalization 79–82
 σ-Delocalization, in 1-methylcyclobutyl cations 821, 822
Deltic acid, structure of 197
Deuterium isotopic perturbation method 815, 816, 819
Dewar furans 715
Dewar resonance energies 382
Diamagnetic susceptibility 420
 in cycloheptatriene 451
Diastereoselectivity, in metal-catalysed cyclopropanation 695–697
Diazoacetates, cycloadditions of 751
Diazoalkanes, in cyclopropanation 658–662
Diazocarbonyl compounds, in cyclopropanation 662–680
Diazocyclopentadienes, reactions of 713, 714
Diazotative deamination 814
Dibenzoacephenanthrylenes, synthesis of 765
Δ^2-Dibenzocycloheptatetraene complexes 566
Dicyanoacetylene, reactions of 716
Dicyclopropanaphthalenes,
 synthesis of 720
 X-ray studies of 734, 735
Dicycloproparenes,
 strain energies in 739
 synthesis of 724, 725
Dicyclopropylacetylenes 560, 561
 dilithiated 561
Dicyclopropylbenzoquinones, synthesis of 563
Dicyclopropylcarbinyl cations 832
Dicyclopropylmercury, synthesis of 500
Dicyclopropylmethyl cation 825
1,6-Didehydrocycloheptatrienes 759
Didehydrophenyl cation 342
Diels–Alder reaction 721
Difference electron density distribution 62–64
Dihalobicyclo[4.1.0]heptenes, aromatization of 717–724
Dihalocarbenes, cycloaddition to cycloproparenes 746
1,6-Dihalocycloheptatrienes, synthesis of 749
gem-Dihalocyclopropa[*a*]naphthalenes, synthesis of 724
Dihalocyclopropanes, synthesis of 287, 288
gem-Dihalocycloproparenes,
 formation of 721, 722, 726
 reactions with organometallics 753
Dihalocyclopropenes,
 cycloaddition to orthoquinodimethanes 722
 synthesis of 287, 288
Dihalocyclopropenones, structure of 196
gem-(Dihalomethyl)cycloalkanes, reductive dehalogenation of 864, 865
1,4-Dihomotropylium cation 843
9,10-Dihydroanthracene, synthesis of 756
Dihydroisobenzofurans, synthesis of 752
Dihydropentacene, synthesis of 756
9,10-Dihydrophenanthrene 751
 synthesis of 756
Dihydrotetrazines, as possible homoaromatic molecules 454
Dihydrotriafulvenes 615
Dihydroxycyclopropenones, structure of 197
Di-π-methane rearrangement 323–327 — *see also* Aza-di-π-methane rearrangement, Oxa-di-π-methane rearrangement
 regiochemistry of 324, 325
 reverse 334
 stereochemistry of 325

1,1-Dimethylcyclopropabenzene radical anion 740
1,2-Dimethylcyclopropanes, stereomutations in 471
α, α-Dimethylcyclopropylcarbinyl cations 826, 829
Dioxetanes,
as intermediates 762
thermolysis of 321
1,2-Diphenylcyclopropanes, isomers of 472, 473
C, N-Diphenylnitrone, cycloaddition to cycloproparenes 751
1,3-Diradicals, rearrangement of 336
1,4-Diradicals, in photorearrangements 322
Dissociative substitution 595
π-Distortion, to σ-molecular framework 733
Divinylcyclopropanes,
isomerism in 473
synthesis of 509, 510, 545
Divinylcyclopropylcarbinyl–divinylcyclopropylcarbinyl rearrangement 831, 841
σ-Donor ability 96
Double bonds, π-character of 377
Double-bond twisting 324
Dunitz–Schomaker strain energy 77

Elassovalene, potential homoaromaticity in 374
Electron delocalization 67, 355–357
Electron-density analysis 404
Electron-density description, of substituted cyclopropanes 90–93
Electron-density distribution 48, 61–68, 144, 149, 154, 202–205, 375–381
Laplace concentration of 68
of homotropenylium cation 426
Electronegativity, of carbon 75
Electronic spectroscopy, of [1.1.1]propellanes 786, 787
Electron transmission spectroscopy, of [1.1.1]propellanes 785, 786
syn-Elimination 726
1,3-Elimination,
from *ortho*-α-disubstituted aromatics 714, 715
of HX 264–270
of two heteroatoms 262–264
Enantioselectivity, in metal-catalysed cyclopropanation 658, 697–701
Ene reaction 506
Energy densities 65
Enolate delocalization 327
Enthalpies of combustion 246
Enthalpies of condensation 243
Enthalpies of dissociation 77
Enthalpies of formation 224–226
of bicyclo[4.1.0]heptenes 234
of CH group 246, 247
of cyclopropane/cyclobutane 243
of cycloproparenes 738
of cyclopropenes 235
of methylcyclopropane 227
of polyenes 232
of quadricyclane 247
of spiro-joined cyclopropanes 229–231
of tercyclopropyls 233
Enthalpies of hydrogenation 229, 236
Enthalpies of olefination 244
Enthalpies of protonation, of ketones 424
Enthalpies of rearrangement 247
Enthalpies of sublimation 224, 225
Enthalpies of vaporization 224, 225
of cyclopropane/cyclobutane 243
of spiro-joined cyclopropanes 230, 231
1,4-Epoxybicycloheptenes, aromatization of 727
Ethene–carbene cycloaddition reaction 47, 48
1-Ethylallyl cation 826
Ethylene-(4-*X*)-anthracenium ions 850
Ethylenebenzenium cations 817, 846–850
Ethylene-(4-*X*)-naphthalenium ions 850
1-Ethyl-2-methylcyclopropanes, isomerism in 471, 473, 480
Ethynylcyclopentadiene, synthesis of 756
Ethynylcyclopropanes,
structure of 175
synthesis of 303
Exchange reactions, metathetical 599, 600

Favorskii-like mechanism 320
Feist's acid, structure of 189
Feist's esters 624–627
isomerism in 188
Fenske–Hall calculations 603
Fischer carbene complexes 286, 522
Fischer–Tropsch synthesis 567
Force constants 102
Force field calculations 389, 390
Formylcyclopropylcarbonitriles, synthesis of 304
Förster–Coulson–Mofitt orbitals 48, 49, 55–60
Free-rotor effect 324
Friedel–Crafts reaction 556
Fullerene cyclopropanes, structure of 212
Fulvalenes — *see* Triafulvalenes, Triapentafulvalenes
Fulvenes — *see also* Heptafulvenes, Homofulvene, Triafulvenes, Tris(cyclopropyl)fulvenes, Pentafulvenes
protonation of 834
Furan–cyclopropenylaldehyde photorearrangement 715

Furanones, synthesis of 530, 531, 533

Gassman–Fentiman linear free energy relations 815
Gaussian functions 58
Geminal overlap 58
Geometrical angle 66
Germylcyclopropanes,
structure of 185, 186
substituent effects in 95
3-21G level calculations 733
Gradient vector field 64, 65
Grob fragmentation 329, 331

Halobicyclo[4.1.0]heptenes — *see also* Dihalobicyclo[4.1.0]heptenes, Tetrahalobicyclo[4.1.0]heptenes
aromatization of 724–726
Halocyclopropabenzenes,
reactions of 579
synthesis of 719
Halocyclopropanes — *see also* Dihalocyclopropanes
structure of 176–179
substituent effects in 94, 95
Halocyclopropenes — *see* Dihalocyclopropenes, Polyhalocyclopropenes, Tetrahalocyclopropenes
Halocyclopropenylium salts, reactions of 609, 610
Halocyclopropylidenacetates, structure of 189
1-Halo-2-(methoxymethyl)benzene, reactions of 714
Hartree–Fock (HF) method 391
Heck coupling reaction 570
Heptafulvenes — *see also* Homoheptafulvenes
synthesis of 756, 761
Heteroannulenes, synthesis of 746
Heteroaromaticity 342, 343
Heterocyclic molecules, structure of 145
Heterocyclopropanes, geometry of 96–98
Hexacyanotrimethylenecyclopropanide anion 197
Hexacyclopropylethane, structure of 148, 149
Hexa-1,5-diyne, cobalt-catalysed coupling of 715
Hexakis(cyclopropyl)benzene, synthesis of 564
Hexaquinacene, as possible homoaromatic molecule 456
C_{16}-Hexaquinanacene, no-bond homoconjugation in 352, 353
HF/6-31G* calculations 733, 744
Hohenberg–Kohn theorem 64
Homoallyl cations 417, 823
energy surfaces of 418
Homoallylic tosylates 823
Homoantiaromaticity 235, 236, 245, 342, 360–362
Homoaromatic anions 457–459
Homoaromaticity 235, 236, 245, 342, 360–362
bond 399–401
criteria for 413, 414
definition of,
energy-based 381–390
general 399–403
Winstein's 365, 366
electron-density distribution description of 378–381
environmental effects on 404
example of 368–370
no-bond 401, 402
of (bicyclooctadienylium)$Fe(CO)_3$ cations 554
PMO description of 370–373
role of 415
three-dimensional 449
Homo-π-aromaticity 342, 343
Homo-σ-aromaticity 342, 343
Homoaromatic stabilization energies 382–389
Homoaromatic systems, neutral 450–457
Homoconjugation,
ab initio investigations of 390–403
bond 341, 350, 352, 353, 414
criteria for 413, 414
cyclobutyl 449
cyclopropyl 340, 341, 346
definition of,
chemical 355–360, 364
topological 348–355, 360
in acyclic systems 416–418
no-bond 341, 350, 352, 353, 414
potential energy surface and 362–364
role of 415
Homoconjugative bonds 350
definition of, electron density based 375–378
description by bond orders 373–375
Homoconjugative interactions 350
Homocyclopropenium cations 427–431 — *see also* Trishomocyclopropenium cations
boron analogues of 430, 431
circumambulation in 437
structure of 428, 429
theoretical calculations for 430
Homodesmotic reactions 74, 75, 387–389
Homofulvene 431
Homoheptafulvenes, complexes of 555
Homoheteroaromaticity 343
HOMO–LUMO crossing 332
3-Homonortricyclyl cation 837
Homopentafulvenes 478

Homotropenylium cations 342, 418, 419 — *see also* Hydroxyhomotropenylium cations, Trishomotropenylium cations
  ab initio calculations for 394–399
  as example of homoaromaticity 368–371, 426
  circumambulation in 437
  metal carbonyl complexes of 419, 420
  photorearrangement of 438
  ring inversion in 423
  structure of 421–423
  theoretical calculations for 425–427
  thermochemistry of 423–425
Homotropone complexes 555
Homotropylidene,
  complexes of 554, 555
  homoaromaticity in 455
Homotropylium cations 851, 852
Hückel calculations,
  for cycloproparenyl anions 742
  for metallatetrahedranes 603
Hückel orbital systems 49, 361, 370, 371
Hybridization 144, 148
  in cyclopropenes 190
  in [1.1.1]propellanes 777, 778
Hybrid orbitals 49, 50, 55–57
  non-orthogonal 55, 57, 58
  orthogonal 57, 58
  s-character of 79
Hydrazidocalicenes 646, 647
Hydride complexes, synthesis of 574
  α-Hydride elimination 568
Hydrindanes, synthesis of 643, 644
Hydrogen bonding,
  in alkylcyclopropanes 150
  in cyclopropanecarboxylic acids 162–170
  in cyclopropane complexes 146
Hydrometallation 519
  of methylenecyclopropane complexes 633, 634
Hydroxycyclopropylcarbinyl cations, synthesis of 817
Hydroxyhomotropenylium cations 420, 424
9-Hydroxymethylphenanthrene, synthesis of 726
Hyperconjugation 412

Imido–tungsten complexes, of cyclopropenes 580
Imino complexes 583, 584
Iminocyclopropa[*l*]phenanthrenes, formation of 710
Inclusion crystals 167, 168, 205
Indan 709, 733
Indantrione, didecarbonylation of 727
3*H*-Indazoles, ring contraction of 711–713

Infrared spectroscopy 98–106
  of cycloproparenes 739, 757, 760
  of cyclopropylcarbinyl cations 820
Insertion reactions,
  migratory 597
  of cyclopropane 126–129
  O—H 683
Interactions, non-bonded 77
Interbond populations 79
Internal modes 102
Interorbital angles 59
Interpath angles 66, 77
Intrinsic frequencies 102
Iodonium ylide, in cyclopropanation 681
Ionization potentials 60
Iron carbenoids 285, 286
Iron catalysts, for cyclopropanation 663
Isobenzofuranones, synthesis of 725
Isobenzofurans — *see also* Dihydroisobenzofurans, Tri-*t*-butylisobenzofurans
  cycloaddition to cycloproparenes 746, 751
Isocyanocyclopropanes, substituent effects in 93
Isodesmic reactions 96, 384–387
Isoindolinones, synthesis of 751
Isolobal analogy 601

Jatropholones, structure of 200

Kekulé structures 733
Ketenecarbenes 728, 729
Λ^1-Ketenyl complexes 562
α-Ketocyclopropylsulphones, synthesis of 501
Koopmans theorem 60

Laplace concentration 380
Ligand substitution reactions 535, 536
Lindlar's catalyst 561
Lithiocyclopropanes, substituent effects in 95
Lithiomethylenecyclopropanes,
  alkylation of 620
  mercuration of 622
  silylation of 620
  synthesis of 619–623
Lithium(phenylthio)cyclopropylcuprate, reactions of 501
Lumisantonin, as photoproduct 320

Magnetic properties, calculation of 404
Magnetic susceptibility 398
  of homotropenylium cation 426
Mass spectrometry, of cycloproparenes 740
Maximum electron density (MED) path 65, 375–378
Mercury–aluminium exchange reactions 505
Metal–halogen exchange reactions 499, 500,

512, 513, 515–519
Metallacyclobutadienes, rearrangement of 593, 594
Metallacyclobutarenes, synthesis of 747, 753–755, 764
Metallacyclobutenes, synthesis of 585–587
Metallacyclopentene complexes 541
Metallacyclopentenes, synthesis of 587
Metallacyclopropenylium salts,
reactions of 610, 614–617
synthesis of 611–614
Metallaindanones, synthesis of 753
Metallapropellanes, synthesis of 500, 501, 747
Metallatetrahedranes,
dynamic behaviour of 603–605
reactions of 595–601
structure of 601–603
synthesis of 589–595
Metalloalkylideneindan-2-ones, synthesis of 764
Metallocene complexes 585
Methano[10]annulenes,
potential homoaromaticity in 374
reactions of 715
synthesis of 746
valence tautomerism in 380, 381
1,2-Methanobuckminsterfullerene, structure of 212
1-Methylcyclobutyl cations 816, 821, 822, 825, 826
Methylcyclopropabenzenes, synthesis of 718
Methylcyclopropanes — *see also* 1,2-Dimethylcyclopropanes, 1-Ethyl-2-methylcyclopropanes, 1,1,2,2-Tetramethylcyclopropanes
deuteriated, isomers of 471, 472
enthalpies of formation of 227
substituent effects in 95
Methylcyclopropylcarbinyl cations 821–826, 829 — *see also* α,α-Dimethylcyclopropylcarbinyl cations, 1,α,α-Trimethylcyclopropylcarbinyl cations
1-(1-Methylcyclopropyl)ethyl cation 835, 836
2-Methylenebicyclo[2.1.0]pentanes 470
Methylenecoumaranone, ring contraction of 729, 730
Methylenecycloalkanes, cyclopropanation of 865, 866
Methylenecyclopropabenzene, formation of 716
Methylenecyclopropanecarboxamides, structure of 188
Methylenecyclopropane rearrangement 188
Methylenecyclopropanes — *see also* Lithiomethylenecyclopropanes
π-complexes of 623, 624
NMR spectra of 628, 629
reactions of 629–644
synthesis of 624–628
cycloadditions of 639–644
dimerization of 641, 642
metal derivatives of 619–623
structure of 187, 189
synthesis of 303
Methylenecycloproparenes, complexes of 644, 645
Methylenecyclopropenes,
complexes of 644–647
structure of 194–197
Methylenecyclopropylmethanols, synthesis of 306
Methylenephthalide, ring contraction of 729, 730
Michael addition 514, 564, 565
α-Migratory elimination 567
Mills–Nixon effect 733, 734, 741, 745
MINDO/3 calculations 759
Mobius array 325
Mobius orbital systems 49, 361, 370, 371
Molecular orbital crossing 332
Molecular orbital description, of substituted cyclopropanes 87–90
Molecular orbitals 48
Møller–Plesset (MP2) perturbation theory 391, 392
Most-substituted bond hypothesis 480
MP2(fc)//HF/6-31G* calculations 745
MP2/3-21G basis sets 733

Nickelabicyclobutanes, synthesis of 747, 753
Nickelabicyclodecatriene, synthesis of 753
Nickelacyclobutabenzenes, synthesis of 753, 754
Nickelacyclobutarenes, synthesis of 754, 755
Nickel catalysts, for cyclopropanation 663, 682
Nitrile oxides, cycloaddition to cycloproparenes 746
Nitrocyclopropanes,
structure of 183, 184
substituent effects in 93, 94
synthesis of 264, 267, 271
Nitrocyclopropenes, structure of 191
NMR chemical shifts, and homoaromaticity 361, 393, 396–399
Norbornadiene, through-space interactions in 356, 357
7-Norbornadienyl cations 445, 446
Norbornene, polymerization of 507
Norbornenyl cations 412, 445, 446
Norbornenyl–palladium complexes 519
Norcaradialene, bond homoconjugation in 352, 353
Norcaradiene epoxides, structure of 200

Norcaradienes,
as isomers of methano[10]annulenes 715
complexes of 551, 552
geometry of 452, 453
homoaromaticity in 450–454
homoconjugation in 359, 371
synthesis of 746, 751
Norcaradienyl cation 749
Norcarane derivatives, conformation of 199
Nortricyclenes, structure of 208
Nortricyclylcarbinyl cations 824, 829, 830
3-Nortricyclyl cations 828, 829
Nuclear magnetic resonance spectroscopy 109–111 — *see also* NMR chemical shifts
of alkoxycyclopropylcarbene complexes 522
of cycloproparenes 740, 757, 760
of cycloproparenyl anions 742
of cyclopropenylidene complexes 617–619
of cyclopropylcarbinyl cations 818, 819
of metallatetrahedranes 605
of methylenecyclopropane complexes 628, 629
of [1.1.1]propellanes 783, 784
of triangulanes 880

Octabisvalenes, structure of 202, 203
Octahedranes, structure of 210, 211
Oligomerization/dehydrogenation reactions 232–235, 241, 242
One-electron properties 106–109
Orbital energies 60
Orbital interactions, nature of 349–351
Orbital overlap 366–370
Orbitals 48–61
π, bridged 51, 52, 97
Orbital symmetry 415
Osmium catalysts, for cyclopropanation 663, 693, 694
Overshoot-backup pathway 329
Oxa-di-π-methane rearrangement 323
Oxasiladiynes, [14]macrocyclic 644
Oxidative addition reactions 575, 576
Oxidative coupling 507, 589
Oxiranes,
complexes of 146
structure of 152, 178
Λ^3-Oxocyclobutenyl complexes 573
Oxocycloproparenes,
IR spectra of 757
NMR spectra of 757
reactions of 760, 761
synthesis of 727–729
UV spectra of 757
Oxyallyl zwitterions 320
Oxymercuration 508
Oxymetallacyclic complexes 539, 541

Palladabenzocyclobutene complexes 586
Palladabicyclobutanes, synthesis of 747, 748, 753
Palladacyclobutarenes, synthesis of 753, 754
Palladium catalysts, for cyclopropanation 657, 659–663, 667, 668, 682, 689–691, 693
Pauson–Khand reaction 563, 638, 639
Pentacyclononyl cations 816
Pentacyclopropylethyl cation, rearrangement of 836
Pentafulvenes — *see* Homopentafulvenes
Pentalenes, synthesis of 643, 644
Pentamethylbenzene 431
Pentylidenecyclopropane, sonication of 623
Peptide analogues 170
[3]Pericyclyne, no-bond homoconjugation in 352, 353
[*n*]Pericyclynes, synthesis of 501
Perpendicular orbitals 322
Perturbational MO (PMO) theory 370–373
Peterson olefination 730–732, 742
Phenanthrenes
— *see* Cyclopropa[*b*]phenanthrenes, Cyclopropa[*l*]phenanthrenes, 9-Hydroxymethylphenanthrenes, 9-Phenylselenophenanthrene
Phenolic photoproducts 327
Phenonium ions 846–850
Phenylacetylene, synthesis of 716
Phenylcyclopropanes — *see also* 1,2-Diphenylcyclopropanes
stereomutations in 474, 475, 484, 485, 487
9-Phenylselenophenanthrene, reactions of 725, 726
Phosphatriafulvenes, structure of 197
Phosphonium cyclopropylides, structure of 185
Photochemical electron transfer 543
Photocycloaddition reactions, of cyclopropyl(methoxy)chromium carbene complexes 526, 527
Photodeazetation 711
Photoelectron spectroscopy,
of cycloproparenes 739
of [1.1.1]propellanes 785
of spiroannulated cyclopropanes 870
Photooxidation, of cyclopropylcarbyne complexes 537, 538, 542–544
Photo-Wolff rearrangement 756
Phthalic anhydride,
photodecomposition of 728
thermal decarboxylation of 727
Pi orbital axis vector (POAV) analysis 366–370
Pitzer strain 79
Pivot process 331
Polyacetylenes, cyclic, as possible homoaromatic molecules 456, 457
Polycyclic systems, structure of 198–212

Polyenes, enthalpies of formation of 232
Polyhalocyclopropenes, reactions of 289
Poly(methylene-1,2-phenylene), synthesis of 756
Potential energy surface 362–364
of homotropenylium cation 426
Prismanes, structure of 208, 209
Promolecular density 69
Propadienones 765
Propane,
CH_2 group of, as reference group 75
Gaussian functions for 58
[1.1.1]Propellanes 124, 125, 773–775
electronic spectra of 786, 787
electronic structure of 777–779
electron transmission spectra of 785, 786
enthalpies of formation of 236, 237
molecular structure of 775–777
NMR spectra of 783, 784
PE spectra of 785
polymerization of 799, 800, 805
reactions of 792–794
with electrophiles 806–808
with nucleophiles 794, 795
with radicals 795–806
reduction of 809
stability of 779–783
structure of 203–205
synthesis of 262
by anionic cyclization 787–790
by carbene insertion 787
by elimination reactions 790–792
vibrational spectra of 784, 785
Propellasilanes, dilithiated, reactions of 500, 501
Propene, relationship to cyclopropane 47, 48
Pyrazolines, nitrogen extrusion from 304, 305
Pyrethroid insecticides 167, 174
Pyrones, cycloaddition to cycloproparenes 746

Quadricyclanes,
enthalpies of formation of 247
structure of 209, 210
Quenching, selective 321
Quinone imides, reactions of 710

[3]Radialene,
structure of 194
thermochemistry of 231
Radical scavengers 517
Radical stabilization energies 476–478
Rearrangements,
Type-A 319–321
Type-B 327–329
Reference states, diagonal 75
Regioselectivity, in metal-catalysed cyclopropanation 658, 688–692
Resonance energies 358, 381–390
calculation of 404
Reversibility, microscopic 334
Rhenacyclobutadienes 576
Rhenafuran complexes 577
Rhodacyclobutane complexes 509
Rhodium catalysts, for cyclopropanation 657, 658, 663–667, 672, 676, 680, 683–696
Ribbon delocalization 342, 343
Ribbon orbitals 49
Ring critical point 67
Ring current, calculation of 420
Ring enlargement reactions 600
Ring inversion, of homocyclopropenium cation 427, 428
Ring opening,
of cyclopropane 126–129
of triangulanes 880, 881
Ring strain,
analysis of,
in terms of atomic energies 74, 75
in terms of bond energies 75–77
in cyclopropane 145
Ring-whizzing motion 602
Rocketene,
reactions with organometallics 755
synthesis of 714, 715
Rotanes,
synthesis of 871–874
X-ray studies of 874
Rotational barriers,
in metallatetrahedranes 604, 605
of silyl group 185, 186
Ruthenium catalysts, for cyclopropanation 663, 693
Rydberg orbitals 50, 55

Santonin, photorearrangement of 319, 320
Saunders' isotopic perturbation of equilibria technique 821
SCF-CI calculations 332
SCF orbitals 55, 60
Schrock cyclopropylcarbene complexes 522
1-Seleno-1-vinylcyclopropanes, rearrangement of 546, 547
Semibullvalenes,
homoaromaticity in 450, 455, 456
photorearrangement of 323
structure of 206
Siloxycyclopropanes, synthesis of 279
α-Silyl anion 730
Silylcyclopropanes,
structure of 185, 186
substituent effects in 95
Silylcyclopropenes, structure of 192
Silylcyclopropenylium salts, reactions of 616
Simmons–Smith reaction 280–285

Singlet–triplet switches 322
Spin–orbit coupling 322
Spiroannulated cyclopropanes,
  enthalpies of formation of 229–231
  enthalpies of vaporization of 230, 231
  nomenclature of 862
  nonactivated, applications of 869, 870
  oligomers of 874–882
  PE spectra of 870
  strain in 862–864
  synthesis of,
    by alkylation 867
    by cyclization 866, 867
    by cycloaddition 867, 868
    by cyclopropanation 865, 866
    by reductive dehalogenation 864, 865
Spiroannulation, strain energy of 863
Spirocycloproparenes, formation of 711, 713
Spirocyclopropylbenzenium ions 846–850
3-Spirocyclopropylbicyclo[2.2.1] cations 844–846
1,3′-Spirocyclopropylbicyclo[2.2.2]oct-2-yl cation 845
Spirocyclopropylcarbinyl cations 843–851
Spirocyclopropylcyclohexenyl triflates, solvolysis of 832
2-Spirocyclopropylcyclohexyl cation 846
Spirocyclopropyllactams, synthesis of 277, 278
3-Spirocyclopropyl-2-norbornyl cations 844, 845
Spiro-3*H*-indazoles, ring contraction of 711
Spiro-metallacyclobutanes 509
Spiropentanes 470, 471
  protonation of 825, 826
Spiro-3*H*-pyrazoles, ring contraction of 713, 714
Stannanes — *see* Alkynylstannanes, Cyclobutarenylstannanes, Tetracyclopropyltin
Stereoisomerization 324
  photochemical 334–336
Stereoselectivity, in metal-catalysed cyclopropanation 658, 692–695
STO-3G calculations 759
  for cycloproparenes 742
Strain energies 242–249
  conventional 73, 74
  of cyclopropabenzene 708
  of cycloproparenes 737–739
Stretching strain 79
Structural determination, experimental techniques for 141
Substituent effects,
  additivity of 147
  in fluorocyclopropanes 176–178
  on benzene ring 159
Substituent–ring interactions, electron density model of 88–96
Sulphinylcyclopropanes, structure of 180
Sulphonylcyclopropanes,
  structure of 180
  synthesis of 264
Sulphonyl isocyanates, cycloaddition to cycloproparenes 751
Super-rocketene, geometry of 734, 735
Surface delocalization 68–73, 82, 83, 342, 343, 380
Surface orbitals 69

Tamoxifen, conformation of 159
Tebe reagent 507
Tercyclopropyls, enthalpies of formation of 233
Tetracyclo[3.2.0.0.0]heptanes, structure of 209, 210
Tetracyclo[3.3.1.0.0]nonanes, structure of 210, 211
3-Tetracyclononyl cations 829
Tetracyclo[8.2.2.$2^{2,5}$.$2^{6,9}$]-1,5,9-octadecatriene, no-bond homoconjugation in 353
Tetracyclopropyltin, synthesis of 500
Tetrahalobicyclo[4.1.0]heptenes, aromatization of 722
Tetrahalocyclopropenes, reactions of 586
Tetrahedranes — *see also* Netallatetrahedranes
  structure of 205–207
Tetralin 709, 733
1,1,2,2-Tetramethylcyclopropane, *cis*/*trans* isomerism in 480
Tetrazines — *see also* Dihydrotetrazines
  cycloaddition to cycloproparenes 746
Thermal equilibration 322
Thiocyclopropanes
  — *see* Alkylthiocyclopropanes, Arylthiocyclopropanes
Thiophene-1,1-dioxides, cycloaddition to cycloproparenes 746
Through-space interactions 341, 350
  in norbornadiene 356, 357
Tin–lithium exchange reactions 505
Topology, of cyclopropane 45–48
Transition states, frozen 363
Triafulvalenes, dications of 615
Triafulvenes 644–647 — *see also* Aminotriafulvenes, Dihydrotriafulvenes, Phosphatriafulvenes
  structure of 194
Trialkylcyclopropanes, synthesis of 276, 277
Triangulanes 874, 875
  branched 876
  geometry of 878, 879
  linear 875
  NMR spectra of 880

Triangulanes (*cont.*)
ring opening in 880, 881
stereospecific functionalization of 881, 882
synthesis of 875–878
thermal stability of 880
X-ray studies of 878, 879
Triapentafulvalenes 645
3-Triasteranyl cation 829
Triasterenes, structure of 210, 211
Triazines, cycloaddition to cycloproparenes 746
Triazinones — *see* Aminotriazinones
Tri(benzocyclobutadieno)benzene, geometry of 736
Tri-*t*-butylcyclopropabenzene-1-carbaldehydes, synthesis of 715
Tri-*t*-butylisobenzofurans, photorearrangement of 715
Tricyclobutabenzene, bond homoconjugation in 352, 353
Tricyclobutonium cation 814, 815
Tricyclo[2.2.1.0]heptanes, structure of 208
Tricyclo[2.1.0.0^{1,3}]pentane, synthesis of 263
Tricyclo[1.1.1.0]pentanes, structure of 203, 204
Tricyclo[2.1.0.0]pentanes, structure of 202
Tricyclopropabenzene,
bond homoconjugation in 352, 353
bond localization in 733
Tricyclopropylcarbenium cations 816, 851
Tricyclopropylcarbinyl cations 825
1,1,3-Trimethylallyl cation 835
1,α,α-Trimethylcyclopropylcarbinyl cation 825, 826
Trimethylenecyclopropane, structure of 194
Trimethylene diradicals 47, 48, 477, 479–484, 487
(Trimethylsilylvinyl)cyclopropanes, synthesis of 297
Triplet diradicals 320, 327
Triplet methylene 487
Triquinacene,
as possible homoaromatic molecule 456, 457
no-bond homoconjugation in 353
Tris(cyclopropyl)benzenes, synthesis of 564, 565
Tris(cyclopropyl)fulvenes, synthesis of 564, 565
Trishomocyclopropenium cations 418, 447–449
Trishomotropenylium cations 444
Trithiodeltate dianion, structure of 198
Tunable diode laser spectroscopy 476

Ultraviolet spectroscopy 111–113
of cycloproparenes 739, 757, 759, 760

Valence bond calculations, spin-coupled 58, 73
Valence bond orbitals 48, 55, 57–59
Valence tautomerism 363, 380
Van der Waals repulsion, of aryl groups 331
VB SCF level 733
Vibrational spectroscopy 98–106, 476
of [1.1.1]propellanes 784, 785
Vinylalkylidene complexes 583
Vinylcyclopropane,
as reference compound for definition of homoconjugation 358, 359
chloropalladation of 550
resonance energy of 383
substituent effects in 95
synthesis of 470, 471
Vinylcyclopropane–cyclopentene rearrangement 470, 559
correlation of rate constants for 478, 479
Vinylcyclopropane–metal compounds 544–556
Vinylcyclopropanes — *see also* Divinylcyclopropanes, 1-Seleno-1-vinylcyclopropanes, (Trimethylsilylvinyl)cyclopropanes
cis/*trans* isomerism in 471
photolysis of 336
structure of 144, 149, 153–156
synthesis of 300, 302, 303, 324
Vinylcyclopropenes, structure of 190
Vinylidenecyclopropanes, synthesis of 287
Λ^4-Vinylketene complexes 585
Virial partitioning method 62, 64, 65, 74
Virtual orbitals 50
Volume delocalization 343, 344

Walsh orbitals 48–55
refined 51–54
Wheland intermediates 745, 749
Wolff-type rearrangement 728, 756

X-ray studies,
of cycloproparenes 734–737
of cyclopropylcarbinyl cations 855, 856
of homotropenylium cations 421
of metallatetrahedranes 601–603
of rotanes 874
of triangulanes 878, 879
X–X difference electron density maps 735, 736

Zeise's dimer 509
Zero-flux surfaces 65, 76
Zimmerman–Schuster oxyallyl zwitterion 320
Zwitterions, dark generation of 320, 328

Index compiled by P. Raven